C0-CFI-036

Encyclopedia *of* REPRODUCTION

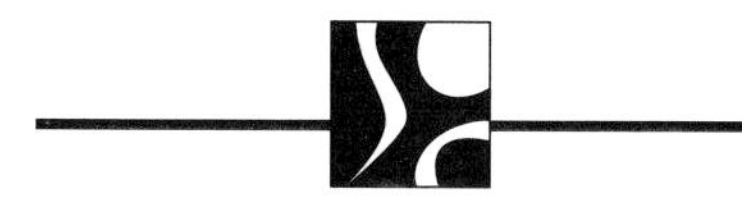

Volume 4 Pro–Z

Editorial Board

Encyclopedia *of* REPRODUCTION

Volume 4 Pro–Z

Editors-in-Chief

Ernst Knobil

H. Wayne Hightower Professor in the Medical Sciences
and
Ashbel Smith Professor,
University of Texas Health Sciences Center
Houston, Texas

Jimmy D. Neill

Distinguished Professor
University of Alabama at Birmingham

ACADEMIC PRESS
San Diego London Boston New York Sydney Tokyo Toronto

This book is printed on acid-free paper. ♾

Academic Press
a division of Harcourt Brace & Company
525 B Street, Suite 1900, San Diego, California 92101-4495, USA
http://www.apnet.com

Academic Press Limited
24-28 Oval Road, London NW1 7DX, UK
http://www.hbuk.co.uk/ap/

Library of Congress Catalog Card Number: 98-84463

International Standard Book Number: 0-12-227020-7 (set)
International Standard Book Number: 0-12-227021-5 (vol. 1)
International Standard Book Number: 0-12-227022-3 (vol. 2)
International Standard Book Number: 0-12-227023-1 (vol. 3)
International Standard Book Number: 0-12-227024-X (vol. 4)

PRINTED IN THE UNITED STATES OF AMERICA
98 99 00 01 02 03 MM 9 8 7 6 5 4 3 2 1

Contents

S

T

Contents of Other Volumes

VOLUME 1

VOLUME 2

VOLUME 3

P

Contents by Subject Area

GAMETES, FERTILIZATION, AND EARLY EMBRYOGENESIS

REPRODUCTION IN HUMANS AND EXPERIMENTAL PRIMATES

DOMESTIC ANIMALS

MAMMALIAN REPRODUCTION

AVIAN REPRODUCTION

REPTILES AND AMPHIBIA

FISH, ELASMOBRANCHII, AND CYCLOSTOMES

INVERTEBRATES

REPRODUCTIVE BEHAVIOR

FEMALE REPRODUCTIVE SYSTEMS

MALE REPRODUCTIVE SYSTEMS

PREGNANCY

LACTATION

Preface

The publication of the *Encyclopedia of Reproduction* comes at a most opportune time. Hardly a day goes by when the news media do not report some new dimension in the treatment of infertility or, conversely, controversies associated with the control of fertility and the ethical issues raised by both. Organismal cloning is a matter of constant debate, and the pharmacological correction of erectile dysfunction has become a preoccupation of international dimensions. Procreation remains a subject of universal interest to every segment of society, from scientists to students, from science reporters to the proverbial person on the street.

The present work should serve as a convenient and comprehensive source of information encompassing all aspects of the subject of reproduction as it relates to the entire animal kingdom. It should be as useful to the expert exploring reproductive phenomena outside his or her own field as it is to students and to the educated public at large. Topics for inclusion were initially generated by forming a matrix of systems (gametes, fertilization, and early embryogenesis; reproductive behavior; female reproductive systems; male reproductive systems; pregnancy; and lactation) and of groups of animals (humans and experimental primates; domestic animals; mammals; birds; reptiles and amphibia; fish, elasmobranchii, and cyclostomes; and invertebrates).

A group of outstanding Associate Editors having expertise in one of more of these areas was then recruited. The preliminary list of entries prepared by the Editors was refined and expanded at a meeting with these Associate Editors, who then identified the appropriate authors. Manuscripts were critically reviewed by the Associate Editors and finally scrutinized by us and the editorial staff at Academic Press.

In a work of this kind, errors of omission and of commission are inevitable and we should appreciate having them called to our attention for correction in possible future editions. The 542 entries constituting the work each contain a glossary of terms, a summary introduction, cross-references to related articles, and a reading list. A standard subject index and an index of reproductive systems and zoological groupings are provided.

Each entry was written to be self-contained, inevitably leading to some overlap of content. We do not view this as a weakness, but instead, believe that it will facilitate a reader's search for information by reducing the number of entries that have to be consulted.

The completion of this project demanded the best efforts of a large number of participants. Chief among them are the 700 authors, especially those who wrote articles on short notice so that the publication deadline could be met. The stellar group of 15 Associate Editors, each of whom possesses great breadth of knowledge and who, as a group, span the spectrum of expertise from zoology to animal husbandry to obstetrics and gynecology, also rendered exceptional service.

Finally, we acknowledge the indispensable contributions of the staff at Academic Press: Jasna Markovac, Editor-in-Chief for Biomedical Science, who originally conceived of the Encyclopedia; and Chris Morris, Gail Rice, and Erika Conner, Major Reference Works editors who provided ongoing management of the project.

Ernst Knobil
Jimmy D. Neill

Guide to the Encyclopedia

ORGANIZATION

The *Encyclopedia of Reproduction* is organized to provide the maximum ease of use for its readers. All of the articles are arranged in a single alphabetical sequence by title. Articles whose titles begin with the letters A to En are in Volume 1, articles with titles from Ep through L are in Volume 2, then M through Pri in Volume 3, and Pro to Z in Volume 4.

Volume 4 also includes a complete subject index for the entire work, an alphabetical list of the contributors to the encyclopedia, and a glossary of key terms used in the articles.

Article titles generally begin with the key noun or noun phrase indicating the topic, with any descriptive terms following. For example, "Uterus, Human" is the article title rather than "Human Uterus," and "Migration, Birds" is the title rather than "Bird Migration." This is done so that the same phenomenon or feature can be studied across various groups. For example, all the articles on female reproductive systems in humans, other mammals, birds, etc., appear in one sequence in the Fe- section of the encyclopedia.

INDEX

The Subject Index in Volume 4 contains more than 20,000 entries. The subjects are listed alphabetically and indicate the volume and page number where information on this topic can be found. In addition, the Table of Contents by Subject Area also functions as an index, since it lists all the topics covered in a given area; e.g., the encyclopedia includes 90 different articles dealing with reproduction in invertebrates.

OUTLINE

Each entry in the Encyclopedia begins with a topical outline that indicates the general content of the article. This outline serves two functions. First, it provides a brief preview of the article, so that the reader can get a sense of what is contained there without having to leaf through the pages. Second, it highlights important subtopics that will be discussed within the article. For example, the article "Fallopian Tube" includes subtopics such as "Tubal Disorders," "Tubal Sterilization," and "Assisted Reproductive Techniques Involving the Tube."

The outline is intended as an overview and thus it lists only the major headings of the article. In addition, extensive second-level and third-level headings will be found within the article.

GLOSSARY

The Glossary section contains terms that are important to an understanding of the article and that may be unfamiliar to the reader. Each term is defined in the context of the article in which it is used. Thus the same term may appear as a glossary entry in two or more articles, with the details of the definition varying slightly from one article to another. The encyclopedia has approximately 4,250 glossary entries.

In addition, Volume 4 provides a comprehensive glossary that collects all the core vocabulary of reproductive biology in one A-Z list. This section can be consulted for definitions of terms not found in the individual glossary for a given article.

DEFINING STATEMENT

The text of each article in the encyclopedia begins with an introductory paragraph that defines the topic under discussion and summarizes the content of the article. For example, the article "Energetics in Reproduction" begins with the following statement:

> Energetics of reproduction is defined as the amount of energy that an animal expends to reproduce. The energetics of reproduction can include the costs of gamete manufacture, synthesis of secondary sexual characteristics and sex-attractant chemicals (pheromones), and reproductive behavior including territorial defense, nest building, courtship rituals, and parental care. . . .

CROSS-REFERENCES

Almost all articles in the Encyclopedia have cross-references to other articles. These cross-references appear at the conclusion of the article text. They indicate articles that can be consulted for further information on the same topic or for other information on a related topic. For example, the article "Osteoporosis" contains references to the articles "Estrogen Replacement Therapy" and "Menopause."

BIBLIOGRAPHY

The Bibliography section appears as the last element in an article. The reference sources listed there are the authors' recommendations of the most appropriate materials for further research on the given topic. The bibliography entries are for the benefit of the reader and thus they do not represent a complete listing of all the materials consulted by the author in preparing the article.

COMPANION WORKS

The *Encyclopedia of Reproduction* is one of a series of multivolume reference works in the life sciences published by Academic Press. Other such titles include the *Encyclopedia of Human Biology*, *Encyclopedia of Cancer*, *Encyclopedia of Toxicology*, *Encyclopedia of Immunology*, and *Encyclopedia of Microbiology*.

Progesterone Actions on Behavior

Anne M. Etgen

Albert Einstein College of Medicine

GLOSSARY

estradiol An ovarian steroid hormone whose secretion during the follicular phase of the estrous cycle, prior to progesterone, is often required to prepare the brain to respond to the behavioral actions of progesterone.

lordosis The principal receptive component of reproductive behavior in female rodents produced in response to copulatory stimulation (e.g., mounting), consisting of immobilization, concave flexion of back muscles, extension of the limbs, lateral deviation of the tail, and elevation of the head and rump.

proceptive behaviors Behaviors performed by estrous females to solicit mounting attempts from males; examples include hopping/darting locomotion and ear wiggling in female rats.

receptive behaviors Behaviors performed by estrous females that permit or facilitate mounts and intromissions by males, e.g., lordosis.

receptors Cellular proteins that recognize and bind to specific chemical messengers (e.g., hormones and neurotransmitters) and transduce the chemical signal into a change in cellular function.

ventral tegmental area An area of the ventral midbrain which is an important mediator of progesterone facilitation of lordosis behavior in female hamsters and possibly other species.

ventromedial nucleus of the hypothalamus A region of the basal forebrain which contains high levels of estradiol-induced progesterone receptors and which mediates many of the effects of progesterone on reproductive behaviors in female rodents.

In most mammals, the period during which females will mate with males is often limited to the hours or days surrounding ovulation, thereby maximizing the probability that a female will be inseminated by a male. In many species, the sequential actions of estradiol secreted during the follicular phase of the estrous cycle followed by a surge of progesterone in the periovulatory period are responsible for this temporal coordination of female reproductive physiology and behavior. The behavioral actions of progesterone are mediated by actions on specific populations of brain cells. In most cases, the capacity of the brain to respond to progesterone with changes in reproductive behavior requires prior exposure to estradiol (i.e., estradiol priming). In addition to behavioral actions related to the establishment of pregnancy, progesterone also modulates feeding, maternal, and aggressive behaviors in female rats. In birds, progesterone participates in the facilitation of nesting and incubation behaviors and in the inhibition of aggression. This article focuses on the neural sites and mechanisms of progesterone regulation of reproductive behaviors in female mammals, particularly rodents.

I. PROGESTERONE REGULATION OF REPRODUCTIVE BEHAVIOR

A. Species Differences

In female rodents, such as rats, mice, hamsters, and guinea pigs, estradiol and progesterone act synergistically to control the onset, quality, and duration of behavioral sexual receptivity. Female mating behaviors are abolished after ovariectomy and can be restored by sequential replacement therapy with estradiol followed 24–48 hr later by progesterone. The frequency and intensity of lordosis behavior in response to male mounting attempts peak several hours after systemic administration of progesterone, and high levels of behavioral receptivity are maintained for 6–12 hr. The prolonged period of estradiol "priming" is a prerequisite for the facilitatory actions of progesterone on reproductive behaviors both in gonadally intact and in ovariectomized female rodents. That is, in the absence of estradiol priming, progesterone has no measurable effects on female mating responses. In ovariectomized female rats, high levels of estradiol can promote behavioral receptivity without the addition of progesterone. Nevertheless, ovarian progesterone is needed for the full display of female reproductive behaviors in gonadally intact female rats, and the time course of behavioral responsiveness correlates with the rise and subsequent decline in circulating progesterone. Moreover, even though estradiol alone is sufficient to enhance lordosis (receptive) behavior in female rats, the display of proceptive behaviors is much more dependent on the addition of progesterone.

Although the participation of progesterone in the expression of female rodent reproductive behaviors is clearly documented, progesterone is not required for receptive and proceptive sexual behaviors in all mammals. For example, in reflex ovulators such as rabbits and cats, estradiol alone appears to determine the quality and duration of sexual responsiveness. In other species, the temporal requirements for estradiol and progesterone facilitation of reproductive behavior differ from those in rodents. In sheep, sustained elevations in circulating progesterone during the luteal phase of the estrous cycle, followed by a precipitous decline, appear to prepare the brain to respond behaviorally to the preovulatory increase in plasma estradiol. Whether this pattern is true for all ungulates has yet to be determined. In dogs as in rodents, both estradiol and progesterone contribute to the facilitation of female reproductive behaviors. However, behavioral sexual receptivity persists for a week or more after ovulation, during which time estradiol levels decline while progesterone secretion from the corpus luteum increases. The role of progesterone action in the brain in the expression of female reproductive behaviors in nonhuman primates has not been thoroughly explored. Some data suggest that increased levels of circulating progesterone derived from the corpus luteum after ovulation reduce copulation because progesterone action on the vaginal epithelium reduces female attractiveness to males.

B. Facilitatory and Inhibitory Effects of Progesterone in Rodents

In rodents, exogenous administration of progesterone to ovariectomized, estradiol-primed females has biphasic effects on reproductive behavior. When first administered to estrogen-primed rats, hamsters, or guinea pigs, progesterone synergizes with estradiol to produce maximal behavioral responsiveness. In addition, this exposure to progesterone renders animals refractory to a subsequent injection of progesterone given 24–36 hr later. This "sequential inhibitory" effect of progesterone is believed to limit the duration of behavioral estrus to the periovulatory phase of the estrous cycle. Therefore, progesterone may be important for both the initiation and the termination of estrous responsiveness in female rodents, especially those with a long luteal phase (e.g., guinea pigs).

It is interesting that hamsters and guinea pigs are much more sensitive than rats to the sequential inhibitory effects of progesterone on lordosis. Indeed, studies in which hormone was administered to ovary-intact females at different points of the estrous cycle suggest that under physiological conditions, progesterone may play little or no active role in the termination of estrous behavior in female rats. Rather, behavioral responsiveness declines as a function of decreasing levels of plasma progesterone. However, progesterone acting in combination with prolactin does inhibit sexual receptivity during late pregnancy and lactation in female rats.

II. NEURAL SITES OF PROGESTERONE ACTIONS ON BEHAVIOR

That the absence of reproductive behavior in ovariectomized rodents results from withdrawal of ovarian steroids rather than changes in pituitary hormone secretion was demonstrated by experiments showing that estradiol and progesterone restore female reproductive behavior in hypophysectomized female rats. Considerable effort has gone into identifying the specific brain sites at which progesterone acts to regulate female reproductive behavior in rodents. These localization studies have provided fundamental information to investigators who are attempting to determine the cellular and molecular mechanisms of hormonal regulation of reproductive behavior.

A. Progesterone Action in the Hypothalamus

The results of several experimental approaches are consistent with the interpretation that the hypothalamus is a major site of progesterone facilitation of both receptive and proceptive sexual behaviors in female rodents. Autoradiographic examination of rat brain sections following systemic administration of radiolabeled steroid hormones first showed that several regions of the hypothalamus and preoptic area accumulate significant amounts of radiolabeled estradiol. Radioligand binding studies then demonstrated (i) that the hypothalamus and preoptic area contain binding sites for estradiol with characteristics expected of estrogen receptors (stereospecificity, high affinity, and limited capacity) and (ii) that administration of behaviorally effective priming doses of estradiol increases the density of progesterone receptors in the hypothalamus and preoptic area. Lesions of several hypothalamic sites also interfere with hormonal facilitation of reproductive behavior. Finally, implants of small amounts of ovarian hormones and antihormones directly into the brain confirmed that progesterone facilitation and sequential inhibition of estrous behavior involves direct actions of the hormone in the hypothalamus. Within the hypothalamus, the most sensitive site of progesterone action in rats, hamsters, and guinea pigs appears to be the lateral aspect of the ventromedial nucleus or its anatomical equivalent. In rats, application of progesterone directly into the ventromedial nucleus of estradiol-primed females facilitates the expression of the full complement of both receptive (lordosis) and proceptive behaviors.

B. Progesterone Action in the Midbrain

There is evidence that the behavioral actions of progesterone are not limited to the hypothalamus, especially in hamsters. In this species, implants of progesterone into the ventromedial hypothalamus of estrogen-primed females produce relatively low levels of lordosis responsiveness. Moreover, lesions of both the ventromedial hypothalamus and the ventral tegmental area inhibit behavioral receptivity in hamsters. The latter observation led investigators to apply progesterone simultaneously to the ventromedial hypothalamus and to the contralateral ventral tegmental area of estrogen-primed female hamsters. This combination of progesterone stimulation of both the midbrain and hypothalamus produced robust lordosis behavior in a high percentage of animals. Hormone implant studies in rats also implicate the midbrain reticular formation and the habenula as sites of progesterone facilitation of receptive and proceptive behaviors in rats.

Electrophysiological studies in freely behaving animals also suggest that the midbrain is a particularly important site of progesterone modulation of lordosis behavior in hamsters. Within minutes of administration of progesterone to estradiol-primed hamsters, there is a change in the pattern of activity of neurons recorded in the midbrain. For example, some neurons that had previously been inactive begin to fire spontaneously, whereas firing of other neurons abruptly ceases. Changes in neuronal firing become progressively more apparent during the next several hours until high levels of lordosis responsiveness are seen. At that time, the functional properties of the majority of midbrain neurons are altered. It has been suggested that the midbrain is a more important site of progesterone action in hamsters than in other rodents because lordosis behavior in this species is accompanied by a much more sustained period of immobility than in other species (i.e., many minutes vs seconds in rats and guinea pigs). Therefore, widespread changes in the functional activity of midbrain neurons may be required to suppress locomotion during behavioral sexual receptivity in hamsters. Nonetheless, experimental evidence indicates that progesterone action in the midbrain contributes to the facilitation of lordosis behavior in other rodents as well.

III. MECHANISMS OF PROGESTERONE ACTIONS ON BEHAVIOR

There are still many unanswered questions regarding the cellular and molecular actions of hormones in the brain. In particular, the question of how hormone action in specific neuronal populations alters the probability that females will exhibit appropriate behavioral responses when approached or mounted by males is fundamental to understanding hormonal coordination of female reproductive physiology and behavior. However, knowledge of the neural circuitry underlying hormone-dependent mating behaviors in female rodents has led to significant progress in answering this question. Current concepts of the mechanisms by which progesterone acts in the hypothalamus and midbrain to regulate female reproductive behaviors are summarized in the following sections.

A. Progesterone Action in the Hypothalamus

The brains of all vertebrate species examined have significant numbers of gonadal steroid hormone receptor-containing neurons in brain regions such as the hypothalamus, preoptic area, and limbic forebrain. Smaller numbers of hormone-sensitive cells are also found in the midbrain. These receptors are ligand (i.e., hormone)-activated transcription factors which alter cell function by enhancing and suppressing the expression of specific genes. Because the actions of progesterone on female reproductive behavior are estrogen dependent, the initial discovery of neural progesterone receptors was followed by experiments to determine whether the expression of these receptors is regulated by estradiol. In the hypothalamus of ovariectomized females, progesterone receptor concentrations increase significantly in response to administration of behaviorally effective estradiol doses. Within the hypothalamus, the ventromedial nucleus is one of the most sensitive areas for estradiol induction of progesterone receptors. There is a close temporal correlation between estradiol elevation of hypothalamic progesterone receptors and the ability of progesterone to facilitate female reproductive behaviors. Similarly, the time course of the decline in estrogen-induced hypothalamic progesterone receptors parallels the decline in behavioral receptivity. Brain cell nuclear accumulation of progesterone receptors also precedes the display of progesterone-facilitated lordosis behavior in estrogen-primed rats and guinea pigs. Moreover, progesterone receptor antagonists such as RU 38486 inhibit progesterone-stimulated behavioral responses when administered systemically or when applied directly into the ventromedial hypothalamus. Thus, a large body of evidence is consistent with the hypothesis that estradiol-regulated hypothalamic progesterone receptors, especially those in the ventromedial hypothalamus, mediate the facilitatory effects of progesterone on female rodent reproductive behaviors. However, the cellular and molecular events occurring subsequent to progesterone-induced accumulation of receptors in brain cell nuclei are poorly understood.

The disappearance of progesterone receptors from hypothalamic cell nuclei following administration of the hormone (i.e., agonist-induced downregulation) has been proposed to underlie the sequential inhibitory effects of progesterone on sexual receptivity in female rodents. When animals are given a second injection of hormone a day after a lordosis-facilitating progesterone treatment, there is a very low level of cell nuclear accumulation of progesterone receptors in the hypothalamus. A role for progesterone receptor downregulation in determining the duration of behavioral responsiveness is supported by the finding that administration of the progesterone receptor antagonist RU 38486 accelerates the loss of progesterone receptors from hypothalamic cell nuclei and results in early termination of lordosis behavior in rats and guinea pigs. However, certain observations in guinea pigs suggest that receptor downregulation may be too simple a mechanism to explain progesterone sequential inhibition. In female guinea pigs, intracranial administration of a protein synthesis inhibitor can block the inhibitory effects of progesterone on lordosis behavior without interfering with the hormone's initial facilitatory effects. Therefore, the inhibitory actions of progesterone on female reproductive behavior may require hormone-dependent synthesis of some new protein(s) that interferes with the cellular processes that mediate progesterone facilitation. This interpretation is compatible with the relatively long time course required for the development of progesterone inhibition of mating behavior.

B. Progesterone Action in the Midbrain

Estradiol-regulated progesterone receptors are unlikely to mediate the behavioral actions of progesterone in the midbrain. The concentration of progesterone receptors in the midbrain is much lower than that in the hypothalamus, and administration of priming doses of estradiol does not increase midbrain progesterone receptor levels. Application of minute amounts of estradiol directly into the ventromedial hypothalamus, a procedure which precludes exposure of the midbrain to estrogen, is also sufficient to prime female rodents to respond to the facilitatory actions of progesterone on both receptive and proceptive sexual behaviors. Furthermore, midbrain implantation of progesterone conjugated to bovine serum albumin, a large hydrophilic protein that is impermeable to cell membranes (thereby blocking access of progesterone to intracellular receptors), facilitates lordosis behavior in estrogen-primed female rats and hamsters. In contrast, conjugated progesterone does not facilitate estrous behavior when implanted into the ventromedial hypothalamus.

The mechanism of progesterone's behavioral actions in the midbrain may involve interactions of the hormone or its reduced metabolites with a membrane receptor that normally mediates the actions of the neurotransmitter γ-aminobutyric acid (GABA). Activation of receptors of the $GABA_A$ subtype by GABA normally inhibits neuronal firing. Although progesterone interacts rather weakly with $GABA_A$ receptors, certain progesterone metabolites which can be formed locally in the brain are potent modulators of $GABA_A$ receptors. Several behavioral studies have utilized implants of progesterone metabolites with differential efficacy to modulate $GABA_A$ receptor function or nonsteroidal $GABA_A$ receptor agonists and anatagonists into the ventral midbrain of estrogen-primed female rodents. The results of such studies are consistent with the conclusion that the lordosis-facilitating actions of progesterone exerted at the midbrain could result from metabolism and subsequent interaction of the metabolites with $GABA_A$ receptors.

See Also the Following Articles

Estrous Cycle; Lordosis; Mating Behaviors, Mammals

Bibliography

Barfield, R. J., Glaser, J. H., Rubin, B. S., and Etgen, A. M. (1984). Behavioral effects of progestin in the brain. *Psychoneuroendocrinology* **9**, 217–231.

Baum, M. J., Everitt, B. J., Herbert, J., and Keverne, E. B. (1977). Hormonal basis of proceptivity and receptivity in female primates. *Arch. Sexual Behav.* **6**, 173–192.

Blaustein, J. D., and Olster, D. O. (1989). Gonadal steroid hormone receptors and social behavior. In *Advances in Comparative and Environmental Physiology* (J. Balthazart, Ed.), Vol. 3, pp. 31–104. Springer-Verlag, Berlin.

DeBold, J. F., and Frye, C. A. (1994). Progesterone and the neural mechanisms of hamster sexual behavior. *Psychoneuroendocrinology* **19**, 563–579.

Etgen, A. M. (1984). Progestin receptors and the activation of female reproductive behavior: A critical review. *Horm. Behav.* **18**, 411–430.

Feder, H. H. (1981). Estrous cyclicity in mammals. In *Neuroendocrinology of Reproduction, Physiology and Behavior* (N. T. Adler, Ed.), pp. 279–348. Plenum, New York.

Pfaff, D. W., Schwartz-Giblin, S., McCarthy, M. M., and Kow, L. M. (1994). Cellular and molecular mechanisms of female reproductive behaviors. In *The Physiology of Reproduction* (E. Knobil and J. D. Neill, Eds.), 2nd ed., pp. 107–220. Raven Press, New York.

Rose, J. D. (1990). Brainstem influences on sexual behavior. In *Brainstem Mechanisms of Behavior* (W. R. Klemm and R. P. Vertes, Eds.), pp. 407–463. Wiley, New York.

Sodersten, P. (1985). Estradiol–progesterone interactions in the reproductive behavior of female rats. In *Current Topics in Neuroendocrinology: Actions of Progesterone on the Brain* (D. Ganten and D. Pfaff, Eds.), Vol. 5, pp. 141–174. Springer-Verlag, Berlin.

Progesterone Actions on Reproductive Tract

Cindee R. Funk and Francesco J. DeMayo

Baylor College of Medicine

GLOSSARY

angiogenesis The development of blood vessels in newly formed tissue.

blastocyst A postmorula fertilized egg consisting of an outer layer of cytotrophoblast cells, a fluid-filled cavity (the blastocoele), and an inner mass of embryonic cells connected at the embryonic pole of the blastocyst.

cytokine A cell proliferation agent isolated from and stimulates growth of blood lymphocytes.

endometrium The inner uterine cellular mass consisting of the luminal and glandular epithelium and the underlying stroma which nurtures the developing embryo until a placenta is formed.

endothelial cell The layer of epithelial cells which lines the cavities of the heart as well as the linings of blood and lymph vessels.

epithelium The cellular covering of internal and external body surfaces; in the uterus, the luminal epithelium lines the inner most cavity, whereas the glandular epithelium, although connected to the lumen, is interspersed throughout the underlying stroma and becomes secretory when under the influence of progesterone.

myometrium The outer cellular compartment of the uterus consisting of a layer of longitudinal and a layer of circular smooth muscle cells.

ovariectomize To surgically remove the ovaries in order to eliminate production of the female steroids, estrogen and progesterone.

parturition The later stage of pregnancy, also known as labor, which starts with the onset of regular uterine contractions and ends with expulsion of both the fetus and the placenta.

placenta The mammalian pregnancy organ which develops upon fertilization of the ovum and which links the mother and the fetus for endocrine secretion and exchange of blood-soluble nutrients and waste products.

stroma The supporting tissue or matrix of an organ; in the uterus, it is the central cell mass separating the luminal epithelium and myometrium. Composed of many different cell types, it responds to attachment of an embryo by differentiating or "decidualizing" and facilitates formation of a placenta.

trophoblast The peripheral cells of the blastocyst, which mediates attachment of the embryo to the endometrium, provides protection for the developing fetus, and eventually develops into the placenta.

For many years it has been known that female steroids play important roles in determining the fate of a fertilized mammalian egg, and many studies have been undertaken to understand the roles of these hormones in the uterus. The role of progesterone is paramount with respect to transforming the nonpregnant uterus into an enriched environment specifically suited for the developing embryo and for maintenance of the pregnant state, yet little is known about the actual signaling mechanisms involved. In order to understand the mechanisms that govern progesterone-mediated responses in human pregnancy, the rodent has been used extensively as a convenient model system. Although evolutionarily very different, the two species have a similar hemochorial or an "invasive" type of placenta and also have many similarities in terms of hormone-responsive gene expression in the uterus. This article summarizes the work that has been done at the molecular level to understand the role of progesterone during pregnancy.

I. OVERVIEW AND IMPORTANCE OF PROGESTERONE

Like many developmental processes, the factors involved in determining a successful mammalian pregnancy are intricately balanced, depending on the precise timing of two totally independent events, namely, egg fertilization and uterine sensitization. Upon development of the fertilized egg to the stage of a blastocyst, and upon induction of the uterus to become "receptive" to implantation, these two independent processes converge, whereupon the blastocyst adheres to the receptive epithelium and establishes a unique unit which eventually develops into the placenta. Initially, the growing embryo is completely dependent on the immediate uterine environment for both protection and nourishment until the placenta is formed. Later, the placenta makes a direct link between the fetal unit and the maternal blood supply which is maintained until birth and, in most species, takes on an endocrine role in which it assumes the role of steroid biosynthesis and ensures the maintenance of pregnancy through birth.

The uterine environment is therefore crucial in determining the success rate of a given pregnancy, and this in turn is critically dependent on the intricate interplay between the two female steroid hormones, estrogen and progesterone. Although estrogen initiates the uterine transformation associated with the very early stages of pregnancy, progesterone has been traditionally regarded as the "pregnancy" hormone because of its roles during both implantation and placental maintenance, which is continued throughout parturition. Once the egg is fertilized, progesterone-induced responses ultimately determine the success of a given pregnancy. The importance of progesterone is clearly noted in clinical situations in which progesterone antagonists, such as RU 486, are used to both prevent implantation of a fertilized ovum and to prematurely terminate an already established pregnancy. The aging uterus also exemplifies the importance of progesterone with respect to embryo implantation. The natural decline of circulating hormone in aging women negatively influences the success rates associated with *in vitro* fertilizations. Clinical trials utilizing a twofold increase in the dose of progesterone in older women were drastically improved and slightly exceeded the success rates observed in the younger control group. Improvement in the success rate clearly underscores the importance of progesterone in properly sensitizing the endometrium during implantation.

In the past several years, numerous reports have identified gene targets influenced by exposure to progesterone and have implicated these gene products as contributors to various aspects of pregnancy. However, a clear picture that unifies these postulated biochemical pathways does not yet exist. In an attempt to understand progesterone regulation throughout pregnancy, we have tried to summarize the critical aspects of the published literature and to explain it in a way that will help organize it in a stepwise progression through the process of conception and birth.

II. MECHANISMS OF REGULATING PROGESTERONE ACTIVITY

There are several levels involved in mediating the response to a particular hormone. In the uterus, progesterone mediates its effects by binding to and activating the progesterone receptor. When activated, the receptor migrates to the nucleus and initiates transcription of progesterone target genes. The regulation of progesterone activity occurs both through the mediation of progesterone synthesis and through the expression, activation, and coactivation of the hormone receptor.

A. Progesterone Synthesis

Hormone synthesis and secretion is controlled by a series of positive and negative feedback loops involving several different circulating hormones. Briefly, the corpus luteum forms upon release of a mature follicle from the ovary. Maintenance of the corpus luteum during early pregnancy is primarily under the control of prolactin in the rodent and trophoblast-derived human chorionic gonadotrophin in the human. Progesterone is primarily synthesized in the corpus luteum of both species until a functional placenta is formed. Once formed, the placenta then assumes the primary role of steroid production during the mid- and late stages of pregnancy as the corpus luteum gradually regresses. The triggering of parturition is coincident with a rapid decline of circulating progesterone in both rodent and human pregnancies.

B. Receptor Expression

The progesterone receptor (PR) is activated by binding of ligand and acts as a nuclear transcription factor in its respective target tissues. The gene encoding progesterone receptor is thought to be in single copy, producing as many as nine different transcripts through differential splicing. Although as many as four different protein isoforms of the receptor have been reported in various species, the two largest forms, PR-A (81–83 kDa) and PR-B (116–120 kDa), are the most abundant forms found in human, chick, and rodent tissue and are thought to regulate gene transcription either by transcriptional activation through PR-B or by repression through the dominant repression of PR-B by the PR-A isoform. Consequently, the relative proportion of the PR isoforms in any given target cell will ultimately determine if a specific gene will be expressed in the cell upon hormonal stimulation. A progesterone receptor knockout mouse line with a PGK-neo insert just downstream of the second ATG site of the gene (ablating PR-A and PR-B expression) is phenotypically compromised in all aspects of female reproduction, showing no courtship behavior, being unable to ovulate or induce a uterine decidual response, and having diminished mammary ductal branching. The overall phenotype of the knockout mouse clearly emphasizes the importance of PR-mediated events in all aspects of female reproduction.

In general, the level of available progesterone receptor is induced in most reproductive tissue by estrogens, growth factors, and cAMP, but the level of available receptor decreases in response to progesterone and may explain the need for maintenance of estrogen in order to maintain continued induction of the progesterone receptor throughout pregnancy.

C. Receptor Activation

In the absence of hormone, the human and avian progesterone receptors exist as inactive multiprotein complexes. The inactive avian receptor complex is primarily located in the cytosol and is composed of the heat shock proteins, hsp90 and hsp70 (in some cases); a combination of at least two of the three immunophilins, FKBP52, FKBP54, or cyclophilin-40; and a critically important receptor complex association protein, p23, which has been shown to be essential for receptor complex association and activation by ligand. Upon binding of hormone, however, these complexes completely dissociate and undergo a series of conformational changes and modifications which ultimately result in activation of the receptor complex, translocation to the nucleus, and binding of the receptor as an active dimer to target DNA sequences.

Progesterone responsive genes contain specific sequences known as progesterone response elements (PREs) upstream of their promoters which mediate binding of the activated receptor dimer to the DNA target. However, a given progesterone response may be altered if other steroid response elements (SREs) are located near the PRE. Many hormone-responsive genes can be regulated by several different hormones in a time- or tissue-specific manner and often have more than one SRE upstream of their promoters. Binding of different transcriptional units to these SREs can induce an interaction between different hormonal regulators resulting in either a synergistic or an antagonistic response. This type of coordinate induction is common in estrogen- and progesterone-responsive genes. Consequently, steroid hormones can act either independently by binding of activated receptor to only its specific SRE or coordinately by cooperative binding of two different activated complexes to adjacent SREs. In some cases, antiprogestins have been shown to induce transrepressed genes that are regulated by interactive transcriptional complexes involving PR.

Additional transcriptional "coactivators" and "corepressors" are also involved which help mediate conformational changes of the transcriptional complex. Even though the molecular mechanisms governing cooperative interaction of dual receptor complexes have not yet been clearly explained, both estrogen and progesterone play critical roles during pregnancy and can mediate their regulatory effects independently, synergistically, or antagonistically. Thus, progesterone-mediated events are complex, involving one or more levels of regulation including hormone synthesis, receptor synthesis, receptor activation, and receptor–receptor interaction.

III. STEROID ACTION IN UTERINE PHYSIOLOGY

The uterus is a complex tissue consisting of three major cell compartments: a central luminal epithelium which branches into and forms the glandular epithelium; the underlying stroma cell compartment which is composed of many different cell types, including fibroblasts and endothelial and blood cells; and a surrounding layer of longitudinal and circular muscle known as the myometrium. All the uterine cell compartments have receptors for both estrogen and progesterone, but each particular cell type has a unique response to these steroids. In a coordinate manner, these individual cellular responses determine the overall responsive state of the uterus.

In the rodent uterus, estrogen primarily induces the epithelium to proliferate. This estrogenic response occurs in immature, mature, postmenopausal, and hyperplastic uterine epithelium. Estrogen can also stimulate stroma to proliferate; however, this induction is limited to the stroma of immature animals. In the adult, the stroma only responds to progesterone following induction by estrogen. In the mature animal, progesterone administration alone does not induce proliferation of either the epithelium or the stroma, but in conjunction with estrogen, it actually antagonizes estrogen-induced proliferation of the epithelium and simultaneously induces both epithelial differentiation and stroma cell proliferation. This sequential triggering of uterine cell proliferation and differentiation is required for the induction of the secretory phase of pregnancy and the induction of receptivity.

In the rodent, receptivity requires that the endometrium be initially primed with estrogen to stimulate epithelial cell proliferation and induce expression of the progesterone receptor, and then it is subsequently sensitized by progesterone for a minimum of 48 hr and exposed to a pulsatile release of nidatory estrogen to actually open the window of receptivity. Within 12–24 hr of the nidatory surge, the progesterone-sensitized endometrium becomes receptive to blastocyst attachment but goes on to become refractory (nonreceptive) to implantation about 30 hr after exposure to estrogen. Some of the earlier literature attributed the importance of estrogen priming to the induction of the mitotic index, induction of PR, and to the augmentation and modulation of progesterone-induced responses. It is now known that both estrogen and progesterone induce unique sets of genes which are intricately involved in mediating and maintaining the receptive and pregnant uterus.

At the molecular level, numerous studies have been undertaken to try to identify and elucidate the critical signaling pathways that mediate hormonal response during pregnancy. To date, an extensive amount of work has been forthcoming in terms of estrogen-specific targets which is probably attributable to the direct effect of estrogen on cellular proliferation. On the other hand, progesterone-mediated responses are in large part dependent on estrogen for induction of PR; this complicates the ability to decipher the role of progesterone-mediated actions independently of estrogen. In the past several years, major advances have been made with the construction of mutant (knockout) mice which ablate both the estrogen receptor and the progesterone receptor. Although viable, both mouse lines are impaired with respect to reproductive function and should prove to be critical tools in deciphering the independent signaling pathways involved in reproductive physiology.

IV. PROGESTERONE-DEPENDENT GENE EXPRESSION DURING PREGNANCY

A. Preimplantation

1. *Steroid Receptors*

At the time of ovulation, the uterine milieu is dominated by estrogen which prepares the endometrial environment for subsequent sensitization by progesterone. Two of the most important genes induced directly by estrogen encode the estrogen and progesterone receptors. Estrogen-responsive cells maintain a basal level of estrogen receptor so that they are able to detect and respond rapidly to the hormone, and many of these cells are stimulated to synthesize the receptor in response to the hormone. Estrogen induces expression of progesterone receptor in all compartments of the uterus as well. Thus, one of the primary functions of estrogen is to set the stage for subsequent hormonal stimulation.

2. *Growth Factors*

Growth factors play a large role in mediating the early stages of pregnancy. Although elucidation of their function(s) remains to be fully determined, their spatial distribution suggests that they may regulate critical signaling pathways in growth and differentiation in the uterus. One of the most extensively studied uterine growth factors is epidermal growth factor (EGF). Both EGF activity and synthesis of its receptor (EGF-R) are induced by estrogen in the mature uterus. EGF induces not only cell proliferation but also expression of the progesterone receptor. However, other members of the EGF family, which bind to and activate the EGF receptor, are regulated by progesterone. It is theorized that active EGF (or some of its related family members) is synthesized in the uterine epithelium and interacts with the EGF receptor on the preimplanted blastocyst to mediate attachment. The receptor is also found in the underlying stroma and may play a role in proliferation. Other members of the EGF family are amphiregulin, heparin-binding epidermal growth factor-like growth factor (HB-EGF), and transforming growth factor-α (TGF-α). These family members bind to the EGF receptor with variable affinities; both TGF-α and HB-EGF bind the receptor with approximately a 100-fold stronger affinity than EGF. The importance of different family members binding to the same receptor probably involves mediating specific effects in response to changing hormonal influences since they are clearly subject to different regulation *in vivo*.

Another important uterine growth factor is insulin-like growth factor-1 (IGF-1), which is an important mediator of growth hormone action and a mitogen for fibroblastic and epithelial cells in culture. It is induced in response to estrogen in rat uterine epithelium and is thought to mediate estrogen-induced proliferation. Progesterone, on the other hand, induces IGF-1 in the stroma and slowly stimulates the production of IGF-binding protein-1 (IGFBP-1), which specifically binds to and negatively modulates the activity of IGF-I and -II. IGFBP-1 is the major secretory protein in human decidua and thus antagonizes the estrogenic induction of IGF-1 in the progesterone-dominated uterus. IGF is only a weak mitogen but potentiates the action of other growth factors including EGF. The actual importance of these growth factors *in vivo* is not known, but they are likely to play critical roles during pregnancy.

3. *Immediate Early Genes*

Predictably, some of the immediate early genes (c-*fos*, c-*jun*, and *jun-B*) are differentially expressed in the epithelium and myometrium of the rat uterus in response to hormone. Estrogen specifically induces expression of c-*myc*, c-*fos*, and *jun-B*, and it represses c-*jun* expression in the epithelium but apparently induces c-*jun* expression in the myometrium. Although ineffective when administered alone, progesterone attenuates the estrogenic induction of c-*fos* and completely blocks the repressive effect of estrogen on epithelial c-*jun* expression. Whereas estrogen induces c-*myc* expression in the epithelium, progesterone stimulates expression in the stromal cell compartment. During pregnancy, Myc protein is seen in epithelial cells on Days 1 or 2, in stromal cells on Days 3 or 4, and in the decidualizing stroma at the site of implantation on Days 5 or 6, suggesting a potential role for Myc in regulating DNA synthesis in these proliferating uterine cells. Taken together, these findings suggest that specific induction of individual immediate early genes may be involved in determining the cell-specific and progesterone-dependent proliferative responses of estrogen and progesterone.

4. *Related Genes*

There are multiple examples of other estrogen-induced genes expressed during the estrogen-dominated phase of pregnancy which are downregulated or repressed as the uterine milieu becomes dominated by progesterone. One of these is the secreted iron-binding glycoprotein, lactoferrin, which has been implicated as a mitogen for epithelial cells and is involved in hematopoiesis. In the uterus, it is induced by estrogen, but progesterone specifically inhibits this induction and decreases the level of protein in the uterus. Likewise, the gene encoding mucin episialin (*muc-1*) is abundantly expressed in the epithelium until just prior to attachment of the blastocyst. On Day 4 of pregnancy, the level of mucin drastically declines in response to progesterone. Since expression of *muc-1* can prohibit attachment of the embryo, progesterone-mediated "repression" is also an important regulatory mechanism involved in facilitating uterine receptivity. Estrogen has also been implicated in stimulating phosphatidylinositol metabolism in the uterine epithelium, but progesterone clearly antagonizes this effect as well. Because of the role of phosphatidylinositol in stimulating cell growth through the induction of protein kinase C by diacylglyceride, and also in its role in prostaglandin synthesis, this pathway could play a key role in determining the hormonal control of epithelial proliferation in the uterus.

B. Attachment and Implantation

1. *Growth Factors*

During implantation, the uterus is dominated by progesterone. As mentioned earlier, one mechanism thought to be involved in blastocyst attachment is mediated through the binding of the EGF family of growth factors to EGF receptor on the blastocyst. Amphiregulin is an implantation-specific, progesterone-regulated member of this family. Its expression surges in the uterine epithelium on Day 4 of pregnancy, accumulating specifically at the site of attachment, and this induction is inhibited with antiprogestins. This is an example of a growth factor which is specifically regulated by progesterone. HB-EGF is another member of the EGF family which is regulated by progesterone in stromal cells but also by estrogen in the epithelium. Its expression pattern is consistent with the induction of proliferation of these two cell compartments and it is therefore a primary candidate for mediating the mitogenic signals in the endometrium. Like amphiregulin, HB-EGF is expressed in the luminal epithelium solely at the site of implantation 6 or 7 hr prior to blastocyst attachment; thus, these are candidates for direct signaling to the attaching blastocyst. TGF-α is less well characterized but also binds to and activates the EGF receptor. It is produced by preimplantation embryos and trophoblast cells, significantly increasing trophoblast outgrowth *in vitro*, and probably plays a similar role during the postimplantation stages of pregnancy.

Platelet-derived growth factor (PDGF), which is produced by macrophages as well as the blastocyst, also induces proliferation of trophoblastic cells *in vitro*. PDGF receptor is selectively expressed on the early embryo and appears to be involved in modulating implantation *in vivo*. This growth factor cannot stimulate proliferation by itself, and it is often found to work in concert with EGF and IGF. Thus, it may help mediate attachment as well as estrogen-induced proliferation of uterine epithelial cells. It may also play a role later in pregnancy since it has growth-promoting activity for smooth muscle cells *in vitro* and is expressed in higher concentrations in the myometrium during mid- to late stages of pregnancy. It is therefore likely that it plays an important role in the growth and repair of myometrial tissue during the later stages of both menstruation and pregnancy.

2. *Cytokines*

Leukemia-inhibitory factor (LIF) is a pleiotrophic cytokine with multiple activities that parallel those of IL-6, but it also plays a crucial role in embryo attachment. Although structurally unrelated to IL-6, the functional similarity of IL-6 and LIF may be due to the similarities of the two receptor complexes. Uterine expression of LIF is essential for embryo implantation and is induced just prior to the onset of implantation. Its expression is under maternal control and is predominant in the progesterone-dominated uterus. In the mouse, LIF is clearly a requirement for embryo implantation since healthy blastocysts are unable to successfully attach to the endometrium of a LIF-deficient mouse. Although there is a correlation between the expression of LIF in the rodent and human endometrium, it has not yet been proven that LIF plays an essential role in human pregnancy. Additional studies on LIF-binding protein show a general induction by progesterone of the binding protein (which is a soluble receptor) that inhibits biological activity of LIF through possible sequestering of LIF, thus preventing binding to cell-associated receptors. As a consequence of progesterone induction of the soluble receptor, LIF activity is probably indirectly downregulated during the later stages of pregnancy by progesterone.

3. *Related Genes*

In addition to IL-6 and colony-stimulating factor (CSF), there are several other genes which have been associated with mediating attachment and/or implantation due to their differential expression patterns, but a direct effect has not been demonstrated. Uteroglobin, for example, is induced by progesterone in the uterine epithelium, and although it is a major secretory protein in the rabbit, it is very weakly expressed in other species. Other genes, such as calcitonin, ferritin heavy chain polypeptide, and P-glycoprotein (mdr-1b), are also induced in the luminal epithelium in response to progesterone and may play important roles during implantation, but these also have not been well characterized.

C. Invasion

Restructuring of the extracellular matrix (ECM) in the stromal cell compartment is crucial in both stroma cell decidualization and hemochorial (invasive) types of implantations. Both the human and rodent have hemochorial types of pregnancies in which mediation of trophoblast invasion is clearly dependent on the activities of matrix-associated proteinases and growth factors. Progesterone indeed plays a major role in this remodeling and coordinately regulates multiple aspects involved in this remodeling.

1. *Growth Factors*

Keratinocyte growth factor (KGF) is a member of the fibroblast growth factor (FGF) family and is present in primate endometrial tissue and its secretions. It is induced specifically by progesterone and is postulated to play a role in remodeling of the uterus through binding to heparan sulfate proteoglycans (HSPGs) in the matrix. Other members of the FGF family include acidic- and basic-FGF (b-FGF). Both are expressed in the pregnant uterus and have potent mitogenic and chemotactic activity for endothelial cells, making them potent angiogenic factors. Although both are present in the uterine matrix as inactive complexes with HSPG, it is b-FGF expression which is induced by progesterone and is essential for mediating endothelial cell proliferation and differentiation in cultured ECM. Once FGF is secreted into the matrix, it is bound to HSPG as an inactive complex which both stabilizes and protects FGF from proteolytic degradation. Heparin and heparinases, which are then released from infiltrating mast cells, platelets, and macrophages during remodeling of the matrix, release b-FGF from the inactive complex and facilitate binding of the activated growth factor to its receptors. It initially binds to low-affinity receptors (i.e., syndecan, a cell surface HSPG on uterine epithelial cells) and then to high-affinity receptors located on both epithelial and endothelial cells *in vivo*.

2. *Matrix-Associated Substrates*

A number of matrix components are produced in the stromal cell compartment during pregnancy to help guide and anchor the implanting embryo into the endometrium, including tenacin, fibronetin, laminin, heparin, entactin, and chondroitin sulfate. Tenasin is an extracellular matrix protein which is specifically deposited in the stroma immediately subjacent to the site of embryo attachment just prior to embryo attachment. Normally at low levels during the normal estrous cycle of the mouse, it is induced on Day 4.5 of pregnancy and declines precipitously once decidualization has been completed in the cell and noticeably declining by Day 5.5 in the primary decidual zone, the initial site of decidualization. In the mouse uterus, tenacin expression is stimulated in the progesterone-dominated uterus, serving as an early marker for uterine receptivity, and may mediate embryo attachment or facilitate invasion of the developing trophoblast.

Fibronectin is formed in the decidual basement membrane in response to progesterone stimulation of stromal cells and forms a dense fibrillar network with collagens I and IV, fibrin, heparin, and DNA. This fibronectin-based network is prevalent during the early stages of implantation and decreases as the trophoblast invades, becoming minimally apparent in the rodent uterus by Day 10 of pregnancy. It is probably involved in initiating invasion of the developing trophoblast. Conversely, laminin networks which include laminin and collagenase IV are less apparent at the early stages of attachment and become more prevalent as the decidual response progresses. Laminin and desmin show similar expression patterns which reflect the need for a trophoblast-induced factor in addition to progesterone. Although a direct association between these components has not yet been demonstrated, it is possible that matrix-associated networks containing one or both of these are involved in halting invasion of the trophoblast.

HSPG is produced by uterine epithelial cells along with keratin sulfate and is thought to mediate implantation (through interactions with amphiregulin and HB-EGF) and/or invasion [through interaction with FGFs, KGF, PDGF, and vascular endothelial growth factor (VEGF)] of the developing embryo. HSPG has also been associated with releasing angiogenic factors (such as acidic and basic FGFs) as it is degraded through the process of matrix remodeling. Hyaluronic acid is another matrix-associated glycosaminoglycan which is hydrophilic and attracts large amounts of water into the matrix, causing expansion of the extracellular space for penetration of migrating cells and invasive processes that lead to both embryo implantation and vascularization. This is a uterine-mediated response and probably involves regulation by progesterone. Apolipoprotein J is also a secretory glycoprotein which binds lipids and membrane-associated proteins and is abundantly expressed in the epithelial glands on Day 2.5, but it is not expressed at the time of implantation. Rather, it is expressed in the stroma throughout Day 8 of pregnancy, suggesting a potential role in restructuring of the uterine matrix.

3. *Metalloproteinases*

Although decidualizing stromal cells secrete urokinase-type plasminogen activator into the matrix in response to induction by prostaglandin, interaction of the trophoblast with fibronetin and laminin in the ECM can also enhance the trophoblastic production of some of the plasminogen activators. The proteinases, stromelysin-1 (MMP-3) and matrix-associated plasminogen activators, are both specifically inhibited by progesterone in cultured decidualizing stroma cells. MMP-3 degrades proteoglycan core protein, collagens types II, IV, and V, fibronectin, and laminin, and it can also activate other matrix metalloproteinases such as interstitial collagenase (MMP-1) and the 92-kDa gelatinase type IV collagenase (MMP-9). Plasminogen activators interact with the MMPs in order to optimize proteolysis of the ECM. Conversely but in a predictable manner, progesterone elevates expression of one of the more potent plasminogen activator inhibitors (PAI-1) and also induces one of the metalloproteinase inhibitors (TIMP-3). Thus, progesterone apparently plays a key role in limiting the degree of trophoblast invasion into the uterine stroma.

4. *Vascularization*

i. Induction of Vascularization Vascularization is in part mediated by prostaglandin production which, when blocked, inhibits implantation and decreases the vascular permeability normally associated with implantation. Prostaglandin, PGE-2, is induced in both decidualizing tissue and placental macrophages, and in the rodent, receptors for PGE-2 are induced with progesterone in the uterine stroma. Angiogenic growth factors associated with the uterus are frequently associated with heparin in the endometrial matrix and include VEGF, IGFs, EGF, and TGFs, but only some of these are hormonally regulated. Progesterone induces expression of at least one (b-FGF) of two potent matrix-associated angiogenic factors (a-FGF and b-FGF) and may also be involved in increasing vasodialatory responses.

VEGF is another angiogenic mitogen for endothelial cells and is a potent stimulator of microvascular permeability which is synthesized by the decidualizing stroma. Its transcription is rapidly induced in the rat uterus by both estrogen and progesterone, with responses as early as 1 hr for estrogen and 6 hr for progesterone. VEGF may therefore play a critical role in both the estrogen- and the progesterone-dominated uterus.

ii. Repression of Vascularization Although there are several examples of progesterone-related inducers of angiogenesis, uncontrolled induction could prove fatal for both the mother and the fetus. One study implicates thrombospondin-1 (TSP-1) as a mediator of this induction in the human endometrium during the normal cycle. TSP-1 is expressed at high levels during the secretory phase and was shown to be induced with progesterone in isolated stromal cells and also suppressed by RU 486. This study therefore implicates progesterone as a potential repressor of angiogenesis. Considering the dual role of progesterone in this process, it seems evident that the hormone must mediate its specific regulatory effects through the use of additional cell-specific factors.

5. *Placental Development*

Little is known about the actual signals involved in the formation of a self-contained placenta. Homozygous embryos deficient in progesterone receptor are born in a predictable Mendelian ratio, indicating that progesterone has little effect on the embryonic determinants of placental formation; however, this does not eliminate the need for maternal progesterone in placental development. In the rodent, the best characterized candidate for mediating placental formation is the colony-stimulating factor, CSF-1. It is exclusively expressed in the epithelium and dramatically increases starting on Day 3 of pregnancy—the start of the progestational uterus. Although CSF-1 mRNA levels can exceed 100-fold, peaking on Days 14–16 of pregnancy, protein levels continue to climb throughout pregnancy, exceeding a 1000-fold induction at term, implying additional roles for the cytokine. Its role in placental development is supported by the accumulation of CSF-1 receptor mRNA in the decidua starting on Day 6 which declines once the mature placenta is formed, and also by its ability to initiate differentiation of human trophoblasts in culture. Its additional roles may involve mediating uterine immunological events since it is chemotaxic for and regulates proliferation of mononuclear phagocytes and induces macrophages to synthesize additional cytokines. There are a variety of other cytokines which are produced by the placenta and are postulated to be involved in mediating local immunosuppression during the early stages of pregnancy.

There are other less well-characterized growth factors implicated in placental formation. Tumor necrosis factor TGF-β inhibits placental cell proliferation in culture, thus suggesting an important role in placental development. In addition to growth factors, other genes such as the porcine uteroferrin gene, which transports iron from the maternal uterus to the developing conceptus, are also stimulated by progesterone. Uteroferrin plays a unique role in communicating between the mother and the fetus and may be involved in placental formation.

D. Parturition

1. *Steroid Levels*

The onset of parturition is coincident with a rapid decline of circulating progesterone in both human and rodent pregnancies. In the rodent, progesterone declines to its lowest level on Day 21 (parturition) and is coincident with a small increase in estrogen on Day 21 of pregnancy. In humans, estrogen is produced simultaneously with progesterone throughout pregnancy where a drop in progesterone could relieve progesterone-mediated repression of estrogen-induced genes. It is not clear if this decline in hormone is the cause or the consequence of parturition, and the decline may vary in different species. In humans, the administration of either prostaglandins or oxytocin can induce the onset of labor; however, blocking of the progesterone receptor can also induce premature labor. As such, this issue is not clearly understood. As a consequence of declining progesterone levels, estrogen-dependent gene expression could also play an important role in the production of labor-specific factors.

2. *Oxytocin*

In humans, parturition is regulated by an increase in uterine levels of oxytocin, a potent uterotonic agent which is used to induce labor in women. However, in rodents, although the levels of uterine oxytocin are induced 150-fold at term, oxytocin is not required for delivery in oxytocin-deficient mice. It binds and activates two sets of receptors: one on myometrial cells which, when activated, stimulates uterine contractions; and the other on decidua cells which initiates the generation of $PGF_{2\alpha}$. $PGF_{2\alpha}$ diffuses into the adjacent myometrium and augments the oxytocin-induced contractions while simultaneously softening the cervix. In sheep, induction of $PGF_{2\alpha}$ by oxytocin can be further increased in the presence of the antiprogestin, RU 486, implicating a role of progesterone in repressing oxytocin activity. Subsequent studies have shown that not only is estrogen a strong inducer of both uterine oxytocin (OT) and oxytocin receptor (OTR) gene expression in the rat uterine epithelium but also progesterone unexpectedly "amplifies" the induction of OT and minimally attenuates the induction of OTR; however, progesterone does in fact completely reverse the estrogen-induced binding of OT to its receptor, suggesting a posttranscriptional regulatory mechanism.

3. *Gap Junctions*

Disruption of labor in rats using RU 486 rapidly reduces the levels of β_1 and β_2 connexins in gap junctions of luminal epithelium while simultaneously inducing the level of α_1 connexin (Cx43) in the gap junctions of the myometrium. In ovariectomized rats, progesterone negatively regulates Cx43, whereas estradiol induces its expression. These findings strongly implicate the need for progesterone in maintaining cell–cell communication in the epithelium throughout the onset of labor, whereupon a decline in progesterone at parturition is involved in establishing more important cell–cell communication in the contractile myometrium which is necessary for successful delivery of the fetus.

V. SUMMARY OF PROGESTERONE-RESPONSIVE GENES

To date, there is only limited knowledge available regarding the biochemical pathways involved in progesterone regulation of gene expression in the uterus during pregnancy. Since the first characterization of uteroglobin in the rabbit uterus as a progesterone-induced gene, several more genes have been identified as being targets of progesterone induction. They are diverse in nature, including growth factors, cytokines, receptors, binding proteins, proteinases, and proteinase inhibitors, but they all interrelate in a time-dependent manner to ensure a successful pregnancy. For example, growth factors are specifically induced in uterine tissue in response to progesterone and are thought to help mediate embryo implantation and invasion. This steroid can also induce factors which direct and control remodeling of the extracellular matrix. The pleiotrophic nature of progesterone regulation becomes apparent during angiogenesis in which it can both stimulate production of angiogenic factors and limit the amount of vascularization by inducing angiogenic inhibitors, clearly emphasizing the importance of other cellular factors in determining the overall hormonal effect *in vivo*.

There is a growing literature which identifies specific targets of progesterone regulation in the uterus, but the information is very diverse and defined signaling pathways need to be elucidated to unify the current knowledge of the field. Future research will certainly benefit from utilizing the receptor mutant (knockout) mouse lines to analyze differentially expressed genes *in vivo*, to develop receptor-deficient cell lines to study step-specific signaling pathways, and to investigate communication pathways between estrogen- and progesterone-mediated responses (crosses between the PR and ER mutant mouse lines). Regardless, progesterone is crucial in determining the fate of a healthy developing embryo, and understanding the intricate interactions involved in mediating its overall effect will certainly aid in clinical approaches to improving female reproduction and related diseases.

See Also the Following Articles

Estrogens, Overview; Oxytocin; Placenta: Implantation and Development; Progesterone Actions on Behavior; Progesterone Effects and Receptors, Subavian Species; Progestins

Bibliography

Baulieu, E. E. (1989). Contraception and other clinical applications of RU 486, an antiprogesterone at the receptor. *Science* **245**, 1351–1357.

Beato, M., Herrlich, P., and Schutz, G. (1995). Steroid hormone receptors: Many actors in search of a plot. *Cell* **83**, 851–857.

Edwards, R. G. (1995). Physiological and molecular aspects of human implantation. *Hum. Reprod.* **10**(Suppl. 2), 1–13.

Haimovici, F., and Anderson, D. J. (1993). Cytokines and growth factors in implantation. *Microsc. Res. Technique* **25**, 201–207.

Kraus, W. L., and Kazenellenbogen, B. S. (1993). Regulation of progesterone receptor gene expression and growth in the rat uterus: Modulation of estrogen actions by progesterone and sex steroid hormone antagonists. *Endocrinology* **132**, 2371–2379.

Lefebvre, D. L., Farookhi, R., Giaid, A., Neculcea, J., and Zingg, H. H. (1994). Uterine oxytocin gene expression. II. Induction by exogenous steroid administration. *Endocrinology 134*(6), 2562–2566.

Lubahn, D. B., Moyer, J. S., Golding, T. S., Couse, J. F., Korach, K. S., and Smithies, O. (1993). Alteration of reproductive function but not prenatal sexual development after insertional disruption of the mouse estrogen receptor gene. *Proc. Natl. Acad. Sci. USA* **90**, 11162–11166.

Lydon, J. P., DeMayo, F. J., Funk, C. R., *et al.* (1995). Mice lacking progesterone receptor exhibit pleiotropic reproductive abnormalities. *Genes Dev.* **9**, 2266–2278.

Pollard, J. W. (1990). Regulation of polypeptide growth factor synthesis and growth factor-related gene expression in the rat and mouse uterus before and after implantation. *J. Reprod. Fertil.* **88**, 721–731.

Psychoyos, A. (1973). Hormonal control of ovoimplantation. *Vitamins Horm.* **31**, 201–256.

Psychoyos, A., Nikas, G., and Gravanis, A. (1995). The role of prostaglandins in blastocyst implantation. *Hum. Reprod.* **10**(Suppl. 2), 30–41.

Reynolds, L. P., Killilea, S. D., and Redmer, D. A. (1992). Angiogenesis in the female reproductive system. *FASEB J.* **6**, 886–892.

Stancel, G. M., Baker, V. V., Hyder, S. M., Kirkland, J. L., and Loose-Mitchell, D. S. (1993). Oncogenes and uterine function. In *Oxford Reviews of Reproductive Biology* (S. R. Milligan, Ed.), pp. 1–42. Oxford Univ. Press, London.

Weitlauf, H. M. (1988). Biology of implantation. In *The Physiology of Reproduction* (E. Knobil and J. D. Neill, Eds.), pp. 231–262. Raven Press, New York.

Progesterone Effects and Receptors, Subavian Species

Marina Paolucci
Università degli Studi di Napoli

Noemi Custodia and Ian P. Callard
Boston University

GLOSSARY

dimerization A term used to signify the coupling of two molecules (monomers) to form dimers; can be composed of two identical molecules (homodimers) or two different molecules (heterodimers).

diplotene A stage in the first meiotic prophase in which DNA is duplicated, after which the oocyte enters a resting phase.

domain A term used with reference to functional subdivisions of biological molecules.

germinal vesical The nucleus of the oocyte after activation of maturation by progestins.

immunocytochemistry A method whereby antibodies to specific proteins of interest are used in combination with specific tags so as to identify the cellular source of the protein in question.

meiosis The process of division of germ cells in which the DNA content is first duplicated and then the chromosome number is halved prior to union of sperm and egg.

oogenesis The mitotic division of early germ cells (oogonia) in the ovary which give rise to oocytes.

Encyclopedia of Reproduction VOLUME 4

polar body The extruded second nucleus formed after the completion of the first meiotic division of the oocyte.

progestins 21-carbon steroids which have a characteristic effect on the nuclei in cells of the uterus; these include progesterone, the primary hormone of pregnancy maintenance in mammals; also involved in control of oviduct function in nonmammals.

transcription The process whereby genetic (DNA) information is copied into ribonucleic acid prior to protein synthesis.

zinc fingers The region of steroid receptors and other transcription factors which complexes with zinc to form a finger-like process which interacts directly with the DNA to induce transciption.

Progesterone and other cholesterol-derived 17 and 20 hydroxylated steroids (progestins) are mainly secreted by the ovary. Throughout subavian vertebrates progestins demonstrate a variety of activities. Progestins are important hormones for oocyte maturation in fishes, amphibians, and reptiles; in combination with estradiol they are responsible for the growth and development of the oviduct, the maintenance of embryos, as well as modulation of contractile activity. They are also involved in the negative feedback control of reproductive cycles through actions in the hypothalamus, for behavioral phenomena and the inhibition of vitellogenesis. The biological actions of progestins are mediated by specific receptors within the target organs which bind the hormone with high affinity. The hormone–receptor complex regulates gene activation and transcription in the nucleus by binding to specific sites on the DNA.

I. STRUCTURE AND BIOSYNTHESIS

The basic structure of progesterone consists of four rings, yielding the cyclopentanoperhydrophenanthrene molecule. Cholesterol is the precursor and its conversion to progesterone is common to the synthesis of all five classes of bioactive steroids (androgens, estrogens, progestins, glucocorticoids, and mineralocorticoids). Although this is the principal progestin in nonmammals, such as reptiles and elasmobranchs, in teleost fish and amphibia it is recognized that 17, 20, and 21 hydroxylated progestins are important.

In all subavian vertebrates, the ovary is the main source of progesterone. It is synthesized and secreted by follicles and follicle-derived luteal tissues as in birds and mammals. Although during the preovulatory phase follicles produce mainly estrogens, progesterone is also a quantitatively important product, particularly in egg-laying species (reptiles and elasmobranchs). After ovulation the follicle undergoes remodeling and transformation to the corpus luteum, which survives for variable times according to species. In this process, the granulosa layer becomes hypertrophic and hyperplastic and is invaded by thecal elements carrying blood vessels. The transformation of granulosa cells that accompanies corpus luteum formation is defined as luteinization. The corpora lutea are active in steroidogenesis, producing mainly progestins, which are important for subsequent modification of the oviduct during pregnancy and egg transport.

II. PROGESTERONE EFFECTS

In lower vertebrates, progesterone is a versatile hormone, exerting a variety of effects on diversified target organs generally associated with reproduction. The following is a description of progesterone effects on the gonads, the secondary sexual characters, the liver, and the central nervous system in subavian vertebrates. Typically, progesterone and estrogen interact to bring about proper reproductive function, and in many instances the action of progesterone is dependent on prior exposure to estrogen, which induces progesterone receptors.

A. The Gonads

An important direct effect of progesterone on the female gonads, first observed in amphibians, is the induction of oocyte maturation. During oogenesis, oogonia undergo mitotic divisions yielding primary oocytes. In some groups oogonia are a reservoir for

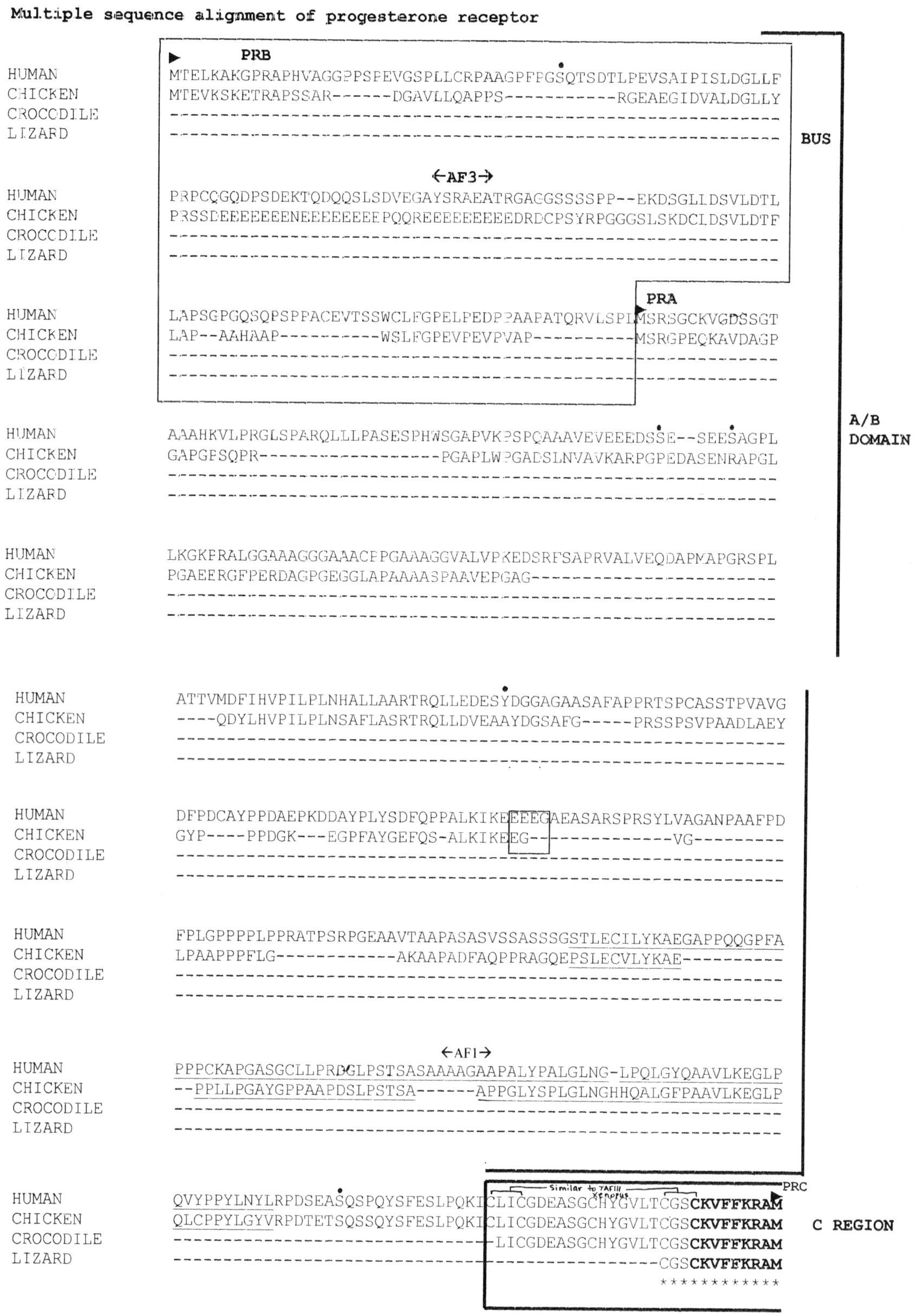

Multiple sequence alignment of progesterone receptor

```
                   PRB
HUMAN      MTELKAKGPRAPHVAGGPPSPEVGSPLLCRPAAGPFPGSQTSDTLPEVSAIPISLDGLLF
CHICKEN    MTEVKSKETRAPSSAR------DGAVLLQAPPS-----------RGEAEGIDVALDGLLY
CROCODILE  ------------------------------------------------------------
LIZARD     ------------------------------------------------------------   BUS

                              ←AF3→
HUMAN      PRPCQGQDPSDEKTQDQQSLSDVEGAYSRAEATRGAGGSSSSPP--EKDSGLLDSVLDTL
CHICKEN    PRSSDEEEEEEENEEEEEEEEPQQREEEEEEEEEDRDCPSYRPGGGSLSKDCLDSVLDTF
CROCODILE  ------------------------------------------------------------
LIZARD     ------------------------------------------------------------

                                                        PRA
HUMAN      LAPSGPGQSQPSPPACEVTSSWCLFGPELPEDPPAAPATQRVLSPLMSRSGCKVGDSSGT
CHICKEN    LAP--AAHAAP----------WSLFGPEVPEVPVAP----------MSRGPEQKAVDAGP
CROCODILE  ------------------------------------------------------------
LIZARD     ------------------------------------------------------------

HUMAN      AAAHKVLPRGLSPARQLLLPASESPHWSGAPVKPSPQAAAVEVEEEDSSE--SEESAGPL   A/B
CHICKEN    GAPGFSQPR-----------------PGAPLWPGADSLNVAVKARPGPEDASENRAPGL    DOMAIN
CROCODILE  ------------------------------------------------------------
LIZARD     ------------------------------------------------------------

HUMAN      LKGKPRALGGAAAGGGAAACPPGAAAGGVALVPKEDSRFSAPRVALVEQDAPMAPGRSPL
CHICKEN    PGAEERGFPERDAGPGEGGLAPAAAASPAAVEPGAG------------------------
CROCODILE  ------------------------------------------------------------
LIZARD     ------------------------------------------------------------

HUMAN      ATTVMDFIHVPILPLNHALLAARTRQLLEDESYDGGAGAASAFAPPRTSPCASSTPVAVG
CHICKEN    ----QDYLHVPILPLNSAFLASRTRQLLDVEAAYDGSAFG-----PRSSPSVPAADLAEY
CROCODILE  ------------------------------------------------------------
LIZARD     ------------------------------------------------------------

HUMAN      DFPDCAYPPDAEPKDDAYPLYSDFQPPALKIKEEEEGAEASARSPRSYLVAGANPAAFPD
CHICKEN    GYP----PPDGK---EGPFAYGEFQS-ALKIKEEG--------------VG---------
CROCODILE  ------------------------------------------------------------
LIZARD     ------------------------------------------------------------

HUMAN      FPLGPPPPLPPRATPSRPGEAAVTAAPASASVSSASSSGSTLECILYKAEGAPPQQGPFA
CHICKEN    LPAAPPPFLG------------AKAAPADFAQPPRAGQEPSLECVLYKAE----------
CROCODILE  ------------------------------------------------------------
LIZARD     ------------------------------------------------------------

                              ←AF1→
HUMAN      PPPCKAPGASGCLLPRDGLPSTSASAAAAGAAPALYPALGLNG-LPQLGYQAAVLKEGLP
CHICKEN    --PPLLPGAYGPPAAPDSLPSTSA------APPGLYSPLGLNGHHQALGFPAAVLKEGLP
CROCODILE  ------------------------------------------------------------
LIZARD     ------------------------------------------------------------

                                              Similar to TAFIII Xenopus   PRC
HUMAN      QVYPPYLNYLRPDSEASQSPQYSFESLPQKICLICGDEASGCHYGVLTCGSCKVFFKRAM
CHICKEN    QLCPPYLGYVRPDTETSQSSQYSFESLPQKICLICGDEASGCHYGVLTCGSCKVFFKRAM   C REGION
CROCODILE  ------------------------------LICGDEASGCHYGVLTCGSCKVFFKRAM
LIZARD     ---------------------------------------------CGSCKVFFKRAM
                                                          ***********
```

FIGURE 1 Multiple sequence alignment of progesterone receptor. The progesterone receptor mRNA contains the sequence of PR-B, PR-A, and a proposed PR-C. The first codon (arrowheads) corresponds to the B isoform. The A isoform starts at Met165

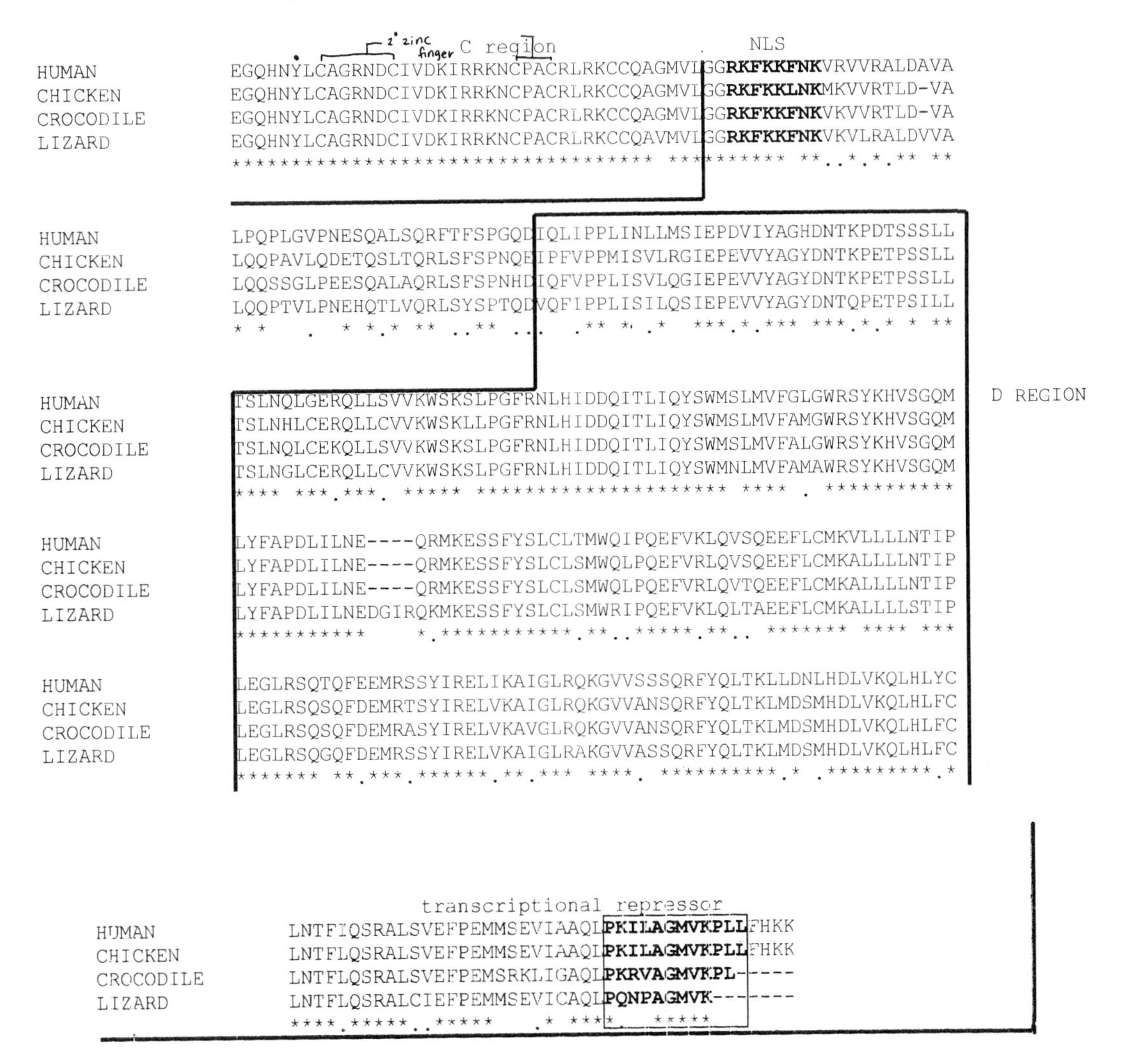

and a proposed C isoform starts at position Met595. The first characteristic region of the PR is the A/B region. This region is very rich in proline residues (~15%) and the difference in length is characteristic of each species. This segment also contains the BUS region (boxed), a β-upstream segment that is rich in Ser and Pro residues and is responsible for the differences between isoforms B and A. In the BUS region there is a transactivation domain (AF3); even though its boundaries have not been established, there is evidence that it modulates the activity of AF1 and AF2. In the A/B region there is also another transactivation domain, AF1 (underlined), that is contained within both PR-Z and PR-B (amino acids 456–546). It is also rich in proline residues (~16%). The next important region is the C domain (DNA-binding domain), the hinge between the DBD and the hormone-binding domain. The E region (hormone-binding domain) is 26–72% homologous among species. It contains two dimerization sites as well as the DBD. It also contains a transactivation domain (AF2) and a transcriptional repressor sequence that is responsible for repression functions. There are six basal phosphorylation sites in the entire PR sequence (●).

oocytes and oogenesis occurs throughout reproductively active life (many teleosts and reptiles and all amphibia). In other vertebrates (elasmobranchs, a few teleosts and reptiles, and birds and mammals), oogonia are transformed into primary oocytes during embryogenesis, leaving no stem germ cells (oogonia) behind for the repetition of the seasonal multiplication of oogonia by mitosis. In these species, at the time of birth the ovaries contain a limited number of primary oocytes halted at diplotene. Oocytes enter meiosis but further progression through the cell cycle beyond prophase is temporarily blocked at the diplotene stage. Oocytes may remain in this stage for many years or be induced to mature further by a combination of gonadotropins and steroids. At ovulation, induced by luteinizing hormone, the mature oocyte resumes the first meiotic division. This involves migration of the germinal vesicle to the periphery of the animal pole, and germinal vesicle breakdown (GVBD), chromosome condensation, and elimination of the first polar body occur. Meiosis is again arrested, now at the metaphase of the second meiotic division. Progesterone in amphibians and related 17α, 20β-dihydroxyprogesterone and 20β-S (17α,20β, 21 trihydroxy-4-pregnen-3-one) in teleosts have been shown to be potent initiators of GVBD and oocyte maturation (maturation-inducing steroids). Possibly, progesterone is also responsible for oocyte maturation in reptiles.

Initiation of meiosis by progestins is perhaps the best example of the nongenomic effects of steroid hormones, in contrast to their better known direct intracellular role as regulators of transcription. The observations are compatible with an action of progesterone at the level of the oocyte plasma membrane, probably mediated by a progesterone-binding molecule, and confirmed by the presence of a progestin (progesterone, 17α,20β progesterone, and 20β S) binding receptor residing in the oocyte plasma membranes of teleosts and amphibians. Progesterone also may play a role in the male gonad and sperm maturation, as suggested by the presence of a specific receptor in the testis of the spiny dogfish *Squalus acanthias*. Progesterone receptors (PRs) have been found to be characteristically associated with spermatids, suggesting that progesterone actions are primarily related to spermiogenesis and spermiation.

In summary, available evidence suggests that progesterone exerts both direct and indirect effects on the gonads. In the female progesterone and/or its metabolites are important oocyte maturation factors; in the male a role for progesterone may be suggested but with a lower level of confidence. The presence of a PR associated with the cytoplasmic membrane of target cells is indicative of nongenomic effects of progesterone .

B. Secondary Sexual Characters

The female reproductive tract is the primary target of progesterone in both oviparous and viviparous subavian species. Although estradiol and progesterone synergize in the regulation of cellular differentiation in the bird oviduct, there is a paucity of information to support a similar role for progesterone in other nonmammalian vertebrates. However, in the toad *Bufo bufo*, injections of progesterone cause the secretion of "albumen" into the oviducts. This effect of progesterone seems to be a consequence of estrogen priming of the oviducal glands and induction of protein synthesis. The jelly-like albuminous material envelops the eggs during their passage through the oviduct, protecting the eggs from dehydration and fixing the eggs to each other and the substratum.

In reptiles, progesterone has been reported to stimulate the oviduct of some snakes but not others. In the turtle *Chrysemys picta*, progesterone decreases oviducal contractility as much as in mammals. Progesterone is involved in the events of ovulation, capsule formation, and oviposition in elasmobranchs. In general, in viviparous species during pregnancy, progesterone maintains the quiescent state of the oviduct to prevent premature expulsion of the embryo. In oviparous species, such an action allows time for oviduct synthesis of the eggshell and addition of other proteins to the egg. The duration, temporal pattern, and quantity of progesterone produced by the postovulatory follicles or corpora lutea varies with the species and is correlated with their mode of reproduction (oviparous vs viviparous); this suggests a causal relationship between egg retention and the longevity of the corpus luteum.

C. Central Nervous System

Progesterone, like estradiol, appears to be involved in the regulation of follicular development, ovulation, and reproductive behavior in subavian vertebrates. Thus, implants of progesterone in the hypothalamus of some lizards prevent follicular development and ovulation presumably by negative feedback at steroid-sensitive sites. In support of this, progesterone binding sites have been found in the hypothalamus of lizards, and these, along with estrogen binding sites, are presumably involved in the regulation of reproductive behavior by these steroids.

D. Progesterone and the Control of Vitellogenesis

Vitellogenin, the precursor of yolk protein, is normally synthesized by the liver of reproductively mature females under the control of estradiol and transported to the ovary.

It has been suggested that progesterone is physiologically important as a coregulator (with estradiol) of vitellogenin synthesis during the ovarian cycle. In addition to a potential role in the timing of follicular development, progesterone may prevent further vitellogenesis during pregnancy in most viviparous species. There is evidence that progesterone inhibits vitellogenesis in several reptiles (lizards and turtles) as well as elasmobranchs (skate and spiny dogfish). However, the mode of action (direct or indirect) is not known. The presence of both estrogen receptor and PR in the liver suggests an interaction in the control of vitellogenin gene transcription. It has been suggested that the disappearance of vitellogenin in the ancestors of mammals is associated with the role of progesterone as a primary regulator of pregnancy and the evolution of an alternative mode of embryo–fetal nutrition in the form of the placenta. Certain elasmobranchs and reptiles have evolved well-developed placental structures and these species have very much reduced egg yolk content. Thus, progesterone may play a key role in the evolution of viviparity by (i) regulating oviduct growth, function, and contractility and (ii) suppressing vitellogenin production while favoring placentation.

III. MECHANISM OF PROGESTERONE ACTION

From recent studies in elasmobranchs and reptiles, PR occurs as two distinct isoforms: PR-A and PR-B. PR-A is a truncated form of PR-B and therefore has a lower molecular weight. The ratio of the PR isoforms varies in different target tissues, suggesting that their differential expression may be critical for appropriate cellular responses to progesterone. As for other steroid hormones, progesterone is considered to enter the cell passively and binds to its receptor complexed with heat shock proteins. The binding of ligand causes release of the heat shock proteins, receptor dimerization, nuclear translocation, and interaction with specific DNA response elements. Ultimately, this cascade of events results in the synthesis of specific proteins. The exact localization of PR within the cell is still a matter of debate, but ligand (steroid)-bound PR is localized within the nucleus. As discussed previously, there is evidence, particularly from amphibian and teleost species, that progestins act through a membrane receptor in male and female gametes (sperm and eggs) to facilitate the process of final maturation.

A. Evidence for PR in Subavian Species

A PR can be demonstrated in both cytosol and nuclear extracts prepared from the oviduct of oviparous and viviparous reptiles and from one oviparous elasmobranch, as well as the elasmobranch testis. In these groups PR resembles the PR-A and PR-B of mammals and birds. Studies in the oviparous turtle *C. picta* suggest that two isoforms of PR exist and these are referred to as isoform A (115 kDa) and isoform B (88 kDa) based on homology with avian and mammalian PR isoforms. As in birds, in the turtle, each isoform has both low- and high-affinity binding sites for progesterone. In birds, the high-affinity sites represent the active receptor, but the role of the lower affinity sites is unclear. Using immunocytochemical detection, PR was seen in the turtle oviduct at all times of the seasonal cycle. Immunostaining of the epithelium was most intense in the preovulatory and early postovulatory stages. The presence of PR in the oviduct correlates well with the functional role of the epithelium, which is responsible for the secretion of a complex proteinaceous gel-like material (albumen) and mucus, essential for lubrication of the passage of the eggs down the oviduct. PR heterogeneity has also been observed in an oviparous lizard (*Podarcis s. sicula*) and the water snake *Nerodia sipedon*, a viviparous species; the presence of PR in the oviduct of this species during pregnancy suggests it is of physiological importance. PRs have also been identified in the oviduct and liver of the oviparous elasmobranch, *Raja erinacea*, and the available evidence suggests that PR-A is located in the liver and PR-B in the oviduct, suggesting tissue-specific expression patterns.

In all vertebrate species studied to date, PR fluctuate throughout the reproductive cycle and show a correlation with the plasma steroids, estradiol and progesterone, suggesting that PR is hormonally regulated.

B. Structure of the Progesterone Receptor in Subavian Species

The intracellular PR is a member of a superfamily of related proteins mediating the cellular effects of steroid hormones, thyroid hormones, vitamin D, and retinoic acid. As determined for mammals and birds, the PR of subavian species is organized into three distinct structural domains: the N-terminal domain, the DNA-binding domain, and the hormone-binding domain.

Multiple sequence alignment of PR from the crocodile and lizard indicates that PR mRNA contains sequences for PR-A, PR-B, and a proposed PR-C. The first start codon corresponds to the B isoform. The A isoform starts at the Met165 and proposed C isoform at Met595. The A/B region (domain) is rich in proline residues (~15%) and its length is species specific. This domain also contains the "BUS" region (Fig. 1, boxed area), an upstream region rich in the amino acids serine and proline; this region is absent in the A isoform. The BUS region contains a transactivation domain (AF3), which may modulate the activities of AF1 (amino acids 456–546, present in PR-A and PR-B) and AF2. The AF1 region (amino acids 456–546) is present in the A/B domain of both PR-A and PR-B. This is rich in the amino acid proline (~16%). The DNA-binding domain (DBD), or zinc finger domain, contains the typical two zinc finger motifs that are 99% conserved among species and are similar to the TAF III in the South African frog, *Xenopus*. The D domain is the hinge region between the DBD and the E or ligand-binding domain, which is 26–72% homologous between the vertebrate species. In addition to the ligand-binding domain, it contains two dimerization sites, a transactivation region (AF2) and a transcriptional repressor sequence. There are six basal phosphorylation sites in the entire sequence (Fig. 1, solid circles).

IV. CONCLUSIONS

Progesterone, produced by the ovarian follicles and corpora lutea, facilitates oviduct secretion, egg retention, and suppression of oviducal contractions. In addition, it and related hydroxylated derivatives (e.g., 20β S) are important oocyte maturation-inducing substances. In oviparous species, involution of the corpus luteum around the time of oviposition leads to a drop in the levels of circulating progesterone, which permits or induces oviducal contractions, leading to oviposition. In contrast, corpora lutea persist until near parturition in live-bearing species. Progesterone acts via a specific intracellular receptor (PR), which resembles the PR of mammals and birds. The two forms of PR of lower vertebrates fluctuate throughout the reproductive cycle and may be differentially regulated by steroid hormones. Thus, it can be said that the basic functions of progesterone with regard to diverse reproductive processes and expression of "reproductive" genes in vertebrates were established at the beginning of vertebrate evolution ~400 million years before present.

Acknowledgments

This work was supported by Italian MURST 40–60% to MP, an NICHD Training Grant Fellowship to NC, and grants from the NSF to IPC.

See Also the Following Articles

Androgens, Subavian Species; Estrogen Effects and Receptors, Subavian Species; Vitellogenins and Vitellogenesis

Bibliography

Callard, I. P., and Callard, G. V. (1987). Sex steroid receptors and non-receptor binding proteins. In *Hormones and Reproduction in Fishes, Amphibians and Reptiles*, pp. 355–384. Plenum, New York.

Callard, I. P., and Klosterman, L. (part A); Callard, G. V. (part B) (1988). Reproductive physiology. In *Physiology of Elasmobranch Fishes*, pp. 277–312. Springer-Verlag, Berlin.

Callard, I. P, Fileti, L. A., Perez, L. E., Sorbera, L. A., Giannoukos, G., Klosterman, L. L., Tsang, P., and McCraken, J. A. (1992). Role of the corpus luteum and progesterone in the evolution of vertebrate viviparity. *Am. Zool.* 32, 264–275.

Chester-Jones, I., Ingleton, P. M., and Phillips, J. G. (1987). *Fundamentals of Comparative Vertebrate Endocrinology.* Plenum, New York.

McDonnell, D. P. (1995). Unraveling the human progesterone receptor signal transduction pathway. *Trends Endocrinol. Metab.* 6, 133–138.

Progestins

Thomas P. Burris

The R. W. Johnson Pharmaceutical Research Institute

GLOSSARY

conceptus The product of conception, or embryo.

corpus luteum A progesterone-secreting endocrine structure formed within the ovary at the site of the ruptured follicle immediately following ovulation.

decidual cells Lipid- and glycogen-rich secretory cells formed within the endometrium from stromal cells in response to implantation of the conceptus and the presence of progesterone.

decidualization The process of differentiation of the stromal cells of the endometrium into decidual cells.

endometrium The inner portion of the uterus composed of two cellular layers: a surface layer of epithelial cells overlaying a layer of stromal cells.

follicular phase The phase of the ovarian cycle during which the follicle grows and develops leading to ovulation. This estrogen-dominated phase of the cycle corresponds to the proliferative phase of the uterus.

luteal phase The phase of the ovarian cycle following ovulation corresponding to the life span of the corpus luteum. This progesterone-dominated phase of the cycle corresponds to the secretory phase of the uterus.

luteotrophic Having a stimulatory action on the development and function of the corpus luteum.

myometrium The muscular wall of the uterus.

parturition Childbirth.

progestational Referring to the ability of an agent to stimulate alterations in the uterus necessary for implantation and development of a fertilized ovum.

progesterone response element A specific DNA sequence that is recognized and bound by the progesterone receptor.

steroidogenesis The biological synthesis of steroids.

Progestins are substances that are either naturally occurring in the organism or synthetic in nature that are able to induce the modifications in the uterus essential for implantation and development of a fertilized ovum. Various progestins are intermediates in the biosynthetic pathways of many biologically active steroids, including androgens, estrogens, glucocorticoids, and mineralocorticoids; thus, they are found in all steroidogenic tissues. However, natural progestins are primarily secreted by the ovaries and placenta. Synthetic progestins are used clinically to treat dysfunctional uterine bleeding and endometriosis. Their most common use is in combination with an estrogen for contraception and hormone replacement therapy.

I. PROGESTIN STRUCTURE

Progestins, like all steroid hormones, are composed of a ring complex composed of three cyclohexane rings (A, B, and C) and a cyclopentane ring (D) (Fig. 1A). Various substitutions on this ring complex, termed the steroid nucleus, give rise to numerous steroid hormones with diverse biological functions. The steroid hormones are classified based on their chemical structure or their biological activity. Based on their biological activity, the sex steroids are composed of three classes: progestins (also known as gestagens and progestagens), androgens, and estrogens. Generally, these three classes conform to structural parent compounds as shown in Fig. 1B. Progestins have a parent compound composed of the pregnane (C_{21}) nucleus, whereas androgens and estrogens have parent androstane (C_{19}) and estrane (C_{18}) nuclei, respectively. However, this is not a strict rule since many steroids with androstane and estrane parent rings have progestational activity. This is best illustrated by the number of synthetic steroids with nonpregnane parent rings that are potent progestins (Fig. 2).

A. Natural Progestins

The biosynthetic pathway for progestins, as well as other steroids, begins with cholesterol (Fig. 3). Steroidogenic cells internalize cholesterol (complexed with low-density or high-density lipoproteins; LDL or HDL) from the blood via a specific receptor-mediated event. The origin of this cholesterol is ei-

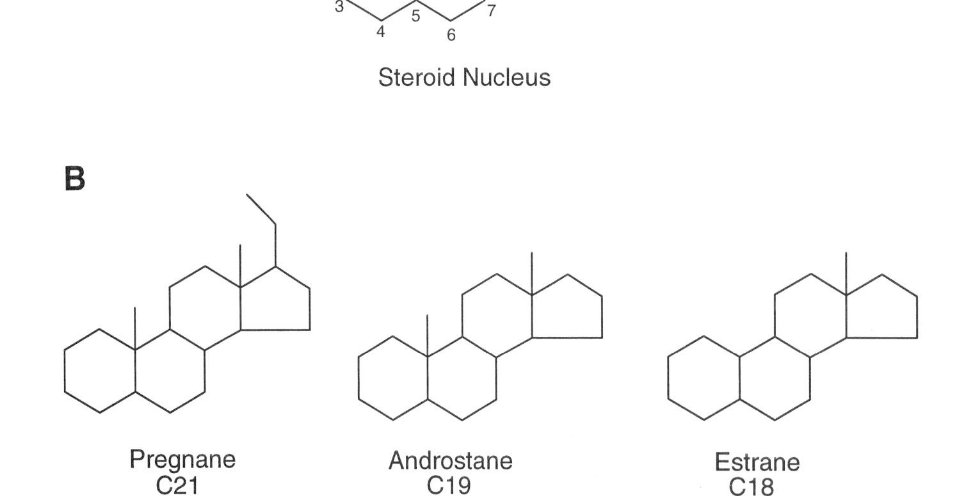

FIGURE 1 General chemical structure and classification of sex steroids. (A) Structure of the "steroid nucleus" illustrating the schemes for classification of the specific cyclohexane and cyclopentane rings along with the numbering system for the carbon atoms. (B) Parent steroid structures for the three sex steroid hormones: pregnane (progestins), androstane (androgens), and estrane (estrogens). The number beneath the ring name refers to the number of carbon atoms within the parent structure.

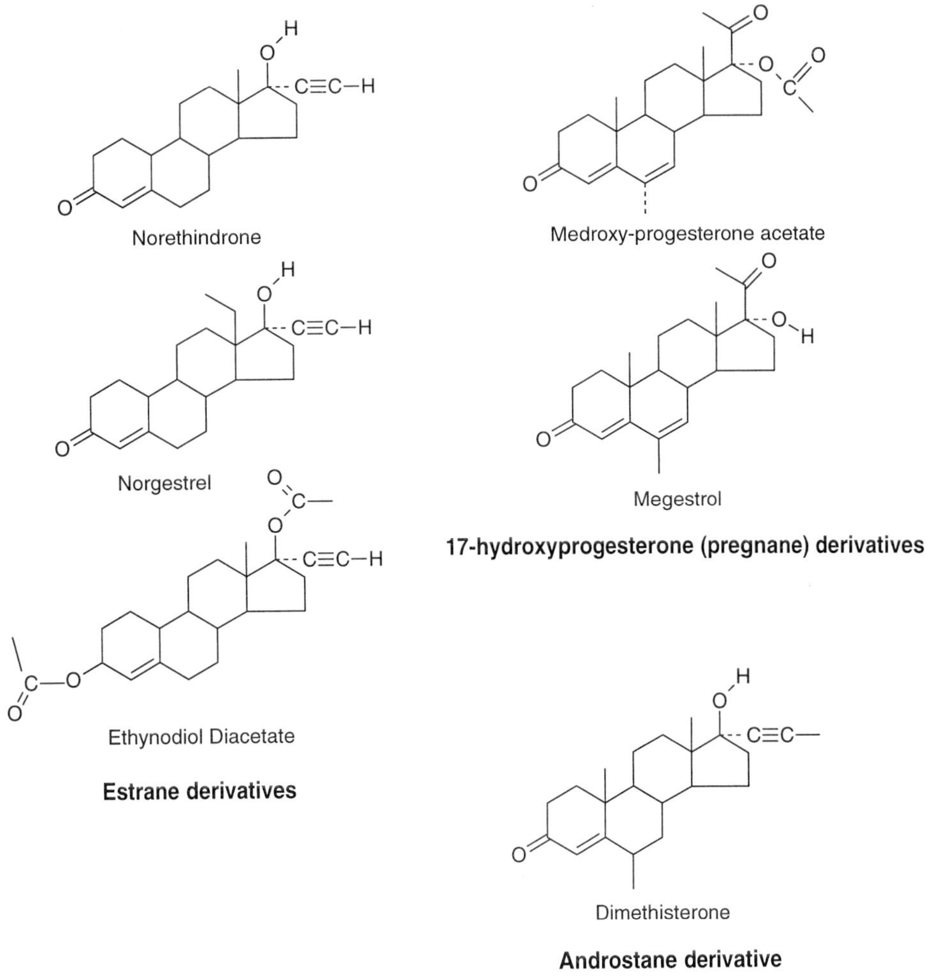

Norethindrone

Norgestrel

Ethynodiol Diacetate

Estrane derivatives

Medroxy-progesterone acetate

Megestrol

17-hydroxyprogesterone (pregnane) derivatives

Dimethisterone

Androstane derivative

FIGURE 2 Common synthetic progestins. Various classes of synthetic progestins are illustrated, falling into the classifications of estranes, pregnanes, and androstanes based on their parent steroid structure.

ther from intestinal absorption (diet) or from hepatic biosynthesis. A very minor fraction of the cholesterol is produced *de novo* by the steroidogenic cells from acetate.

Within the mitochondria of the steroidogenic cell, the side chain of cholesterol is cleaved by cholesterol side chain cleavage enzyme (P450ssc) to produce pregnenolone (Δ_5-P) (Fig. 3, reaction A). This enzymatic reaction is the rate-limiting step for the production of progestins and all other steroid hormones and is the step that is specifically regulated in the ovary by gonadotropins secreted by the anterior pituitary gland. At this step, Δ_5-P can act as a substrate for either steroid 17α-hydroxylase (P45017α) or 3β-hydroxysteroid dehydrogenase (3β-HSD) to yield 17-hydroxypregnenolone (17 OH-Δ_5-P) or progesterone (P), respectively (Fig. 3, reactions B and C). Progesterone, along with estradiol, are the primary sex steroids secreted by the ovaries. Progesterone is the most active natural progestin and is the principal progestational steroid in most species. 17-Hydroxyprogesterone (17 OH-P) is produced from the substrates 17-hydroxypregnenolone or progesterone catalyzed by either P45017α or 3β-HSD (Fig. 3, reactions B and C). 20-Hydroxyprogesterone (20 OH-P) is produced by 20-hydroxysteroid dehydrogenase using progesterone as a substrate (Fig. 3, reaction D). In rabbits, the secretion of both progesterone

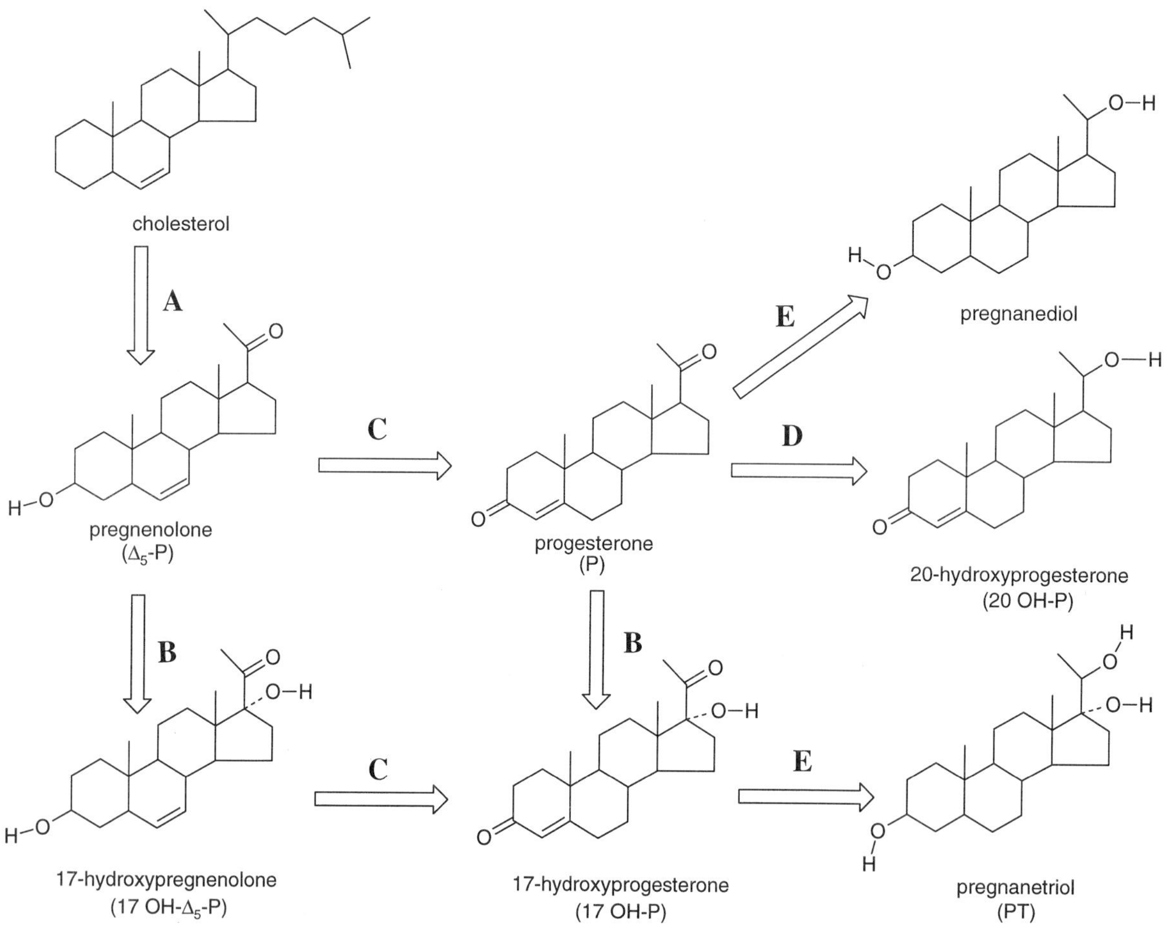

FIGURE 3 Biosynthetic pathway for the production of progestins from cholesterol. The letters refer to specific enzymes that catalyze the corresponding reactions: A, cholesterol side chain cleavage enzyme (P450ssc); B, steroid 17α-hydroxylase (P45017α); C, 3β-hydroxysteroid dehydrogenase (3β-HSD); D, 20-hydroxysteroid dehydrogenase; and E, 3-hydroxysteroid oxidoreductase and 3-oxoΔ_4 steroid reductase. Beneath the names of the specific steroids are commonly used abbreviations for the steroids.

and 20-hydroxyprogesterone by the corpus luteum is significant, and although 20-hydroxyprogesterone is much less potent a progestin than progesterone, 20-hydroxyprogesterone is produced in quantities more than 10-fold more than progesterone. Thus, 20-hydroxyprogesterone is considered the principal progestin in this species. Pregnanediol and pregnanetriol are two other naturally occurring progestins resulting from reductive metabolism of progesterone and 17-OH progesterone, respectively (Fig. 3, reaction E). These two progestins have much weaker progestational activity than progesterone and their biological significance is unknown; however, they are major metabolites of progesterone and 17-hydroxyprogesterone and are both eliminated in their glucuronide forms.

B. Synthetic Progestins

The recognition of the utility of progestins in the clinic, along with the low bioavailability of orally administered progesterone (and other natural progestins), has encouraged the pursuit of synthetically prepared progestins. Based on chemical structure, synthetic progestins can be divided into four groups: (i) isomers of progesterone (e.g., retroprogesterone), (ii) derivatives containing an androstane parent ring nucleus, (iii) derivatives containing an estrane parent ring nucleus, and (iv) derivatives of 17-hydroxyprogesterone (pregnane parent ring). Examples of the latter three classifications are illustrated in Fig. 2. The most common clinically used synthetic progestins are derivatives containing an estrane ring and derivatives of 17-hydroxyprogesterone. Although 17-hydroxyprogesterone itself is essentially inactive in progestational assays, various derivatives of this steroid have been found to be orally active.

II. PHYSIOLOGY OF PROGESTINS

As described previously, there are many steroids with progestational activity; however, since progesterone is the principal natural progestin in most species, the discussion of progestin physiology will focus on this particular steroid.

A. Regulation of Progesterone Production

Progesterone is formed in all steroidogenic tissues, such as the ovary, testis, adrenal gland, and placenta, as an intermediate in the production of all other steroid hormones. The specific steroid hormone produced by a steroidogenic cell is determined by the enzymatic milieu within that cell.

Progesterone is secreted by the corpus luteum within the ovary during the second half (luteal phase) of the female reproductive cycle, which occurs immediately following ovulation. The function of progesterone secreted during the luteal phase is to prepare the uterus for implantation of the fertilized ovum. At this stage, the most important factor in the stimulation of progesterone synthesis and secretion from the corpus luteum is luteinizing hormone. After implantation, progesterone is also required for maintenance of pregnancy. In some animals, the length of the pregnancy is no longer than the normal length of the luteal phase of the cycle; thus, there is no need for prolongation of corpus luteum function. However, most mammals have evolved a mechanism to shorten the life span of the corpus luteum in the event of a nonfertile cycle. In these mammals, the length of pregnancy lasts longer than the normal luteal phase, thus there is a need for prolongation of the life span of the corpus luteum in the event of pregnancy. In order for prolongation of the life span of the corpus luteum during pregnancy maternal recognition is required, and without this recognition the corpus luteum will regress and another ovarian cycle will ensue.

In humans, as well as in many other mammals, this "rescue" of the corpus luteum is mediated by a luteotropic hormone, chorionic gonadotropin, secreted by the developing trophoblast immediately following implantation of the fertilized ovum. In rats and mice, the initial rescue of the corpus luteum is mediated by prolactin secreted by the pituitary gland in response to a neural reflex to the mating stimulus. In these animals, during the second half of gestation, the placenta produces two luteotropic hormones, placental lactogen (with prolactin-like activity) and chorionic gonadotropin, that are responsible for maintenance of progesterone secretion from the corpus luteum.

In some mammals, including primates, the corpus luteum does not remain the major source of progesterone during pregnancy. In humans, as early as the second or third month of gestation the placenta commences production and secretion of progesterone. During this "luteal–placental" shift, the ability of the corpus luteum to secrete progesterone wanes while the secretory capacity of the placenta grows until the corpus luteum is no longer essential to sustain gestation. The placenta continues to secrete progesterone until birth.

B. Actions of Progestins on Target Organs

The principal physiological action of progestational steroids is to prepare the female reproductive tract for pregnancy and to provide continuing nutritive support for the conceptus during gestation. For progesterone to affect its target organs in the reproductive tract, these tissues must first have been exposed to an estrogen in order to induce progesterone receptor expression. In the uterus, following the proliferative actions of estradiol during the follicular phase of the reproductive cycle, progesterone induces endometrial differentiation. The mitotic activity of endometrial cells is reduced and the uterine glands become more coiled and begin to accumulate glycogen in vacuoles. This development of the "secretory endometrium" continues under the influence of progesterone, causing increased secretion by the uterine glands and increased vascularization and coiling of the spiral arteries within the stroma. These modifications of the endometrium are necessary to provide support and sustenance for the conceptus until implantation and formation of the placenta.

Late in the secretory phase, stromal cells near the uterine vasculature differentiate into decidual cells by enlarging and accumulating lipids and glycogen. In the event of pregnancy, the continuing exposure to progesterone stimulates the decidual reaction, transforming the entire stroma into a layer of decidual cells. However, if the cycle is nonfertile and the corpus luteum is not maintained, several changes are noted. The decidual cells rapidly disappear and the endometrium regresses due to the decreasing levels of progesterone. This regression is most notable in primates, in which the abrupt decline in progesterone causes endometrial cell degeneration and sloughing and thus menstruation.

Progesterone also acts on the oviduct regulating its contractions and delaying the transport of the ova to the uterus. This delay is necessary to allow the uterus time to develop the required environment for the conceptus to implant. Progesterone acts on the uterine myometrium by decreasing its contractility and sensitivity to oxytocin, thus preventing organized contractions that could lead to expulsion of the fetus. In some species, such as the sheep, progesterone plays a fundamental role in parturition. In sheep, a precipitous decrease in progesterone precedes parturition, facilitating the onset of labor by increasing uterine contractility. Another target organ of progesterone is the mammary gland. Progesterone acts in concert with estrogen to stimulate alveolar development and increase the eventual capacity of the glands to secrete milk. A thermogenic effect of progesterone has been well characterized and has been utilized as a marker of ovulation. In humans, if the body temperature is measured each day of the menstrual cycle an increase in body temperature of 0.5C at midcycle can be detected. This increase in temperature correlates to ovulation and is caused by the increase in progesterone secretion due to entry into the luteal phase.

C. Clinical Use of Progestins

Contraception represents the major clinical use of progestins. Used in combination with an estrogen, oral contraceptives are more than 99% effective at preventing pregnancy. The estrogen/progestin combination oral contraceptives have a myriad of effects on the reproductive tract, many of which are capable of interfering with fertility. The dominant effect is inhibition of ovulation; however, other actions include alterations of the endometrium, cervical mucus, and tubular mobility. Although addition of progestins to the contraceptive formulation is primarily to prevent uncontrolled estrogen-mediated uterine proliferation, progestins themselves have effects that can be exploited for their potential contraceptive actions. These actions include blocking sperm passage into the uterus by thickening of the cervical mucus and decreasing the movement of the ovum through the fallopian tubes. Since the endometrium must be in the proper condition for implantation to occur, any disturbance of this state by exogenously administered progestin may be sufficient to prevent pregnancy. The contraceptive effects of progestins are adequate, considering the effectiveness (97 or 98%) of progestin-only contraceptives (known as the "minipill").

Progestins are also used in combination with estrogens in hormone replacement therapy. Again, the addition of progestins is to prevent the uncontrolled proliferative actions of estrogen and decrease the possibility of endometrial hyperplasia and carcinoma. Other disorders of the reproductive system in which progestins are utilized as therapeutic agents include dysfunctional uterine bleeding, dysmenorrhea, endometriosis, and cancers of the breast and endometrium.

III. MECHANISM OF ACTION OF PROGESTINS

Progestins, like all steroids, exert their biological effects by altering the pattern of gene expression within hormone-responsive cells. Steroids are lipophilic and easily diffuse across the plasma membrane of their target cells. Once within the hormone-responsive cell, steroids interact with specific intracellular receptors that are ligand-inducible transcription factors belonging to the nuclear receptor superfamily. Steroids induce a conformational change in their cognate receptors, allowing dimerization and interaction of the receptor dimer with specific DNA sequences within the genome of the cell. These specific DNA sequences or hormone response elements (HRE) are found in the regulatory or promoter region of target genes and are responsible for directing the active receptor to the appropriate steroid-responsive genes. Once bound to the HRE, the hormone-activated receptor is able to interact with the cellular machinery responsible for transcription of genes, thereby affecting the rate of mRNA production and thus protein production (Fig. 4).

Progestins exert their actions by binding to the progesterone receptor, the domain organization of which is illustrated in Fig. 5. The progesterone receptor belongs to a large family of transcription factors known as the nuclear receptor superfamily that includes not only the steroid receptors but also the receptors for thyroid hormone, retinoids, and vitamin D. Members of this superfamily have a very well-conserved domain structure (Fig. 5). The amino-

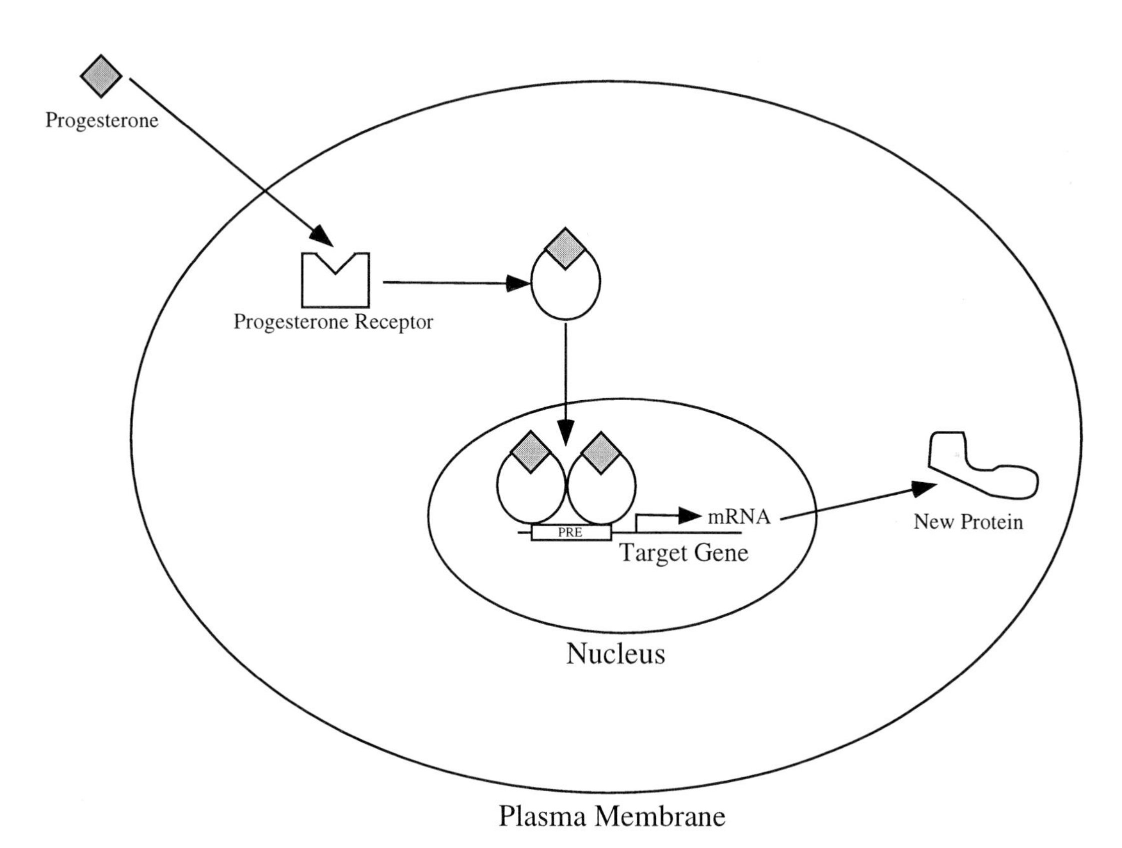

FIGURE 4 Molecular mechanism of action of progestins. Progestins, like all steroids, exert their biological effects by altering the pattern of gene expression within hormone-responsive cells. Progestins are hydrophobic and diffuse across the plasma membrane of their target cells. Once within the progestin-responsive cell, progestins interact specifically with the progesterone receptor and induce a conformational change allowing dimerization and interaction of the receptor dimer with specific DNA sequences within the genome of the cell (progesterone-response elements; PRE's). These PRE's are found in the regulatory or promoter region of target genes and are responsible for directing the active receptor to the appropriate progestin-responsive genes. Once bound to the PRE, the liganded receptor is able to interact with the cellular machinery responsible for transcription of genes, thereby affecting the rate of mRNA production and thus protein production.

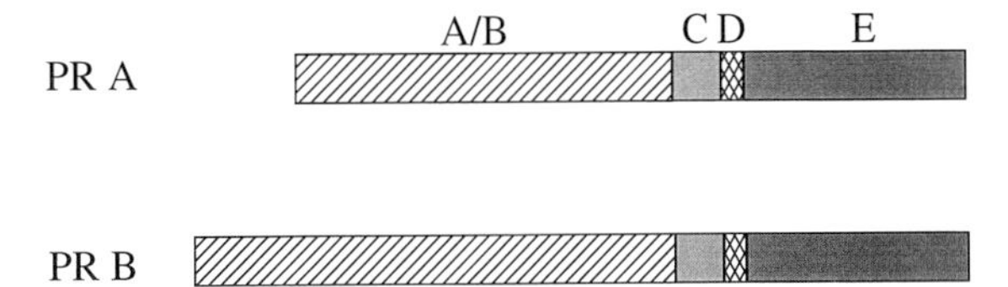

FIGURE 5 Schematic illustrating the domain structure of the progesterone receptors and the differences between the two molecular forms (A and B). Letters indicate the functional domains of the receptor. In the human, the B form is 164 amino acids longer than the A form; however, these receptors are otherwise structurally identical because they arise from alternative translational initiation within the same gene. A/B, variable amino terminal; C, DNA binding; D, hinge; E, ligand binding.

terminal domain (A/B region) is variable in length and is the least conserved domain between superfamily members. Centrally located in the receptor is a well-conserved DNA-binding domain (C region) that contains two zinc finger motifs. Amino acid sequences within this domain are responsible for specifying the DNA sequence to which the receptor binds and thus which genes the receptor regulates. The hinge domain (D region) follows the DNA-binding domain and serves as a link to the ligand-binding domain (E region). The ligand-binding domain is responsible for the interaction of the receptor with the steroid and mediates the conformational changes required for receptor action. Specific subdomains within both the A/B and E regions of the progesterone receptor mediate the interaction of the receptor with the transcriptional machinery required for modulation of gene expression. The progesterone receptor exists in two molecular forms, A and B, that differ in the length of their amino-terminal A/B domains (Fig. 5). In the human, the B form is 164 amino acids longer than the A form; however, these receptors are otherwise structurally identical because they arise from alternative translational initiation within the same gene.

See Also the Following Articles

Corpus Luteum; Hormonal Contraception; Steroidogenesis

Bibliography

Baniahmad, A., Burris, T. P., and Tsai, M. J. (1994). The nuclear hormone receptor superfamily. In *Mechanism of Steroid Hormone Regulation of Gene Transcription* (M. J. Tsai and B. W. O'Malley, Eds.), pp. 1–24. R. G. Landes, Austin, TX.

Gore-Langton, R. E., and Armstrong, D. T. (1994). Follicular steroidogenesis and its control. In *Physiology of Reproduction* (E. Knobil and J. D. Neill, Eds.), 2nd ed., Vol. I, pp. 571–627. Raven Press, New York.

Gunnet, J. W., and Dixon, L. (1995). Sex hormones. In *Kirk–Othmer Encyclopedia of Chemical Technology*, 4th ed., Vol. 13, pp. 433–480. Wiley. New York.

Milgrom, E. (1990). Steroid hormones. In *Hormones* (E. E. Baulieu and P. A. Kelly, Eds.), pp. 386–437. Hermann, New York.

Murad, F., and Kuret, J. A. (1990). Estrogens and progestins. In *Goodman's and Gilman's Pharmacological Basis of Therapeutics*, 8th ed., pp. 1384–1412. Pergamon, New York.

Niswender, G. D., and Nett, T. M. (1994). Corpus luteum and its control in infraprimate species. In *Physiology of Reproduction* (E. Knobil and J. D. Neill, Eds.), 2nd ed., Vol. I, pp. 781–816. Raven Press, New York.

Yen, S. S. C. (1990). Clinical endocrinology of reproduction. In *Hormones* (E. E. Baulieu and P. A. Kelly, Eds.), pp. 444–481. Hermann, New York.

Zeleznik, A. J., and Benyo, D. F. (1994). Control of follicular development, corpus luteum function, and the recognition of pregnancy in higher primates. In *Physiology of Reproduction* (E. Knobil and J. D. Neill, Eds.), 2nd ed., Vol. II, pp. 751–782. Raven Press, New York.

Prolactin, Overview

Nadine Binart, Vincent Goffin, Christopher J. Ormandy, and Paul A. Kelly

INSERM

GLOSSARY

null mutation Targeted disruption (knockout) of a single gene in mice.

signal transduction The biochemical pathways by which proteins send signals from the cell surface to the nucleus.

Prolactin affects more physiological processes than all other pituitary hormones combined. Among these are the regulation of mammary gland development, initiation and maintenance of lactation, immune modulation, osmoregulation, and behavioral modification.

I. INTRODUCTION

At the cellular level, prolactin (PRL) exerts mitogenic, morphogenic, or secretory activities. This raises the question of the mechanisms by which a single hormone can modulate so many unrelated functions. The diversity of PRL actions is directed by the combination of at least four parameters: (i) structural polymorphism of PRL, (ii) local PRL production, (iii) wide distribution of PRL receptor (PRLR), and (iv) existence of different PRLR isoforms signaling through divergent intracellular pathways and activating different target genes.

In this article, we first discuss PRL-producing cells, gene regulation, structural organization of the PRL receptor as well as its tissue distribution, and we then focus on recent progress in deciphering the molecular mechanisms by which PRL acts on target cells. Finally, the phenotypes associated with the knockout of the PRL receptor will be reviewed as a means of demonstrating which of the multiple functions of this hormone are only mediated by the PRL receptor and which can be taken over by another hormone or cytokine.

II. PROLACTIN

A. Pituitary Prolactin

1. *Lactotrophic Cells*

Early in the twentieth century, changes were observed in the histology of the anterior pituitary gland of women during pregnancy. A new cell type was identified, which during late pregnancy and postpartum periods constitutes the most common pituitary cell type, later identified as lactotrophic cells exhibiting secretory granules of PRL. Lactotrophic cells can be identified by either erythrosin or carmosin stains. They comprise 20–50% of total anterior pituitary cells and usually develop peripherally to somatotrophic cells.

2. *Regulation of Secretion*

i. Inhibitors Pituitary PRL secretion is under general negative control by the hypothalamus. Dopamine is the major factor responsible for PRL inhibition. It binds to specific D_2 receptors on lactotropic cells, inhibiting cyclic adenosine monophosphate levels. Therefore, dopamine inhibition of the adenylate cyclase system would offer a direct means of inhibiting PRL production.

Gonadotropin-releasing hormone-associated peptide has been proposed as a prolactin-inhibiting factor. This peptide inhibits prolactin secretion presumably by interacting with specific receptors on lactotrophic cells.

A number of other hypothalamic and peripheral factors have specific effects on prolactin secretion. γ-Aminobutyric acid and, in some cases, somatostatin also specifically inhibit PRL secretion.

ii. Activators Estrogens and thyrotropin-releasing hormone (TRH) stimulate PRL secretion. In addition to inducing hypertrophy of the lactotrophic cells, estradiol increases PRL production by directly stimulating PRL gene transcription. As early as 20 min after treatment with the steroid, increased synthesis of PRL mRNA and subsequently of the protein is observed, which suggests a direct regulation of PRL transcription by the estradiol–receptor complex.

TRH specifically increases the release of PRL *in vivo* as well as in pituitary cells in primary culture. It binds to specific receptors identified on lactotrophic cells of the anterior pituitary. Following binding of TRH to its receptors, hydrolysis of inositol phospholipids occurs, resulting in the formation of intracellular inositol phosphate and diacylglycerol. An increase in intracellular Ca^{2+} occurs as a result of the mobilization of intracellular pools and the stimulation of voltage-regulated channels. PRL secretion (exocytosis) may be stimulated directly by Ca^{2+} or by subsequent phosphorylation of proteins, through a calmodulin-dependent protein kinase.

In experimental animals, neurotensin, vasoactive intestinal peptide, epidermal growth factor, and bombesin are potent stimulators of secretion. Opiate peptides and morphine also stimulate secretion; they act by blocking the inhibition of prolactin secretion without affecting spontaneous PRL release.

B. Extrapituitary Prolactin

A recent review drew attention to extrapituitary PRL. After hypophysectomy of female rats, 10–20% of lactogenic activity could be detected in their serum compared to the controls; this lactogenic activity gradually increased to 50% within 2 months. This observation elucidated several vital functions served by PRL and raised the possibility that extrapituitary PRL compensates, at least in part, for a deficiency in pituitary PRL. The most established extrapituitary sites of PRL secretion are the decidua and myometrium, some immune cells, and brain, with emerging evidence for PRL synthesis by the skin and exocrine glands, including mammary, sweat, and lacrimal. The overall concept of PRL as an endocrine/paracrine/autocrine factor is schematized in Fig. 1.

C. Structure and Isoforms of PRL

The rapid development of cloning technology in the 1970s allowed the determination of the nucleotidic sequence of PRL cDNAs from several species. In humans, the amino acid sequence of PRL (also referred to as lactogenic hormone or luteotropic hormone based on its biological properties) revealed a protein of 199 amino acids. As anticipated from earlier structural studies, the primary structure of PRL appeared closely related to that of two other hormones: growth hormone, also of pituitary origin, and placental lactogen, secreted by mammalian placenta. Today, genetic, structural, binding, and functional studies of these three hormones have clearly demonstrated their belonging to a unique family of proteins.

PRL has been reported to exist in different forms within the pituitary. For example, a 16-kDa variant, generated by an enzymatic cleavage of native 23-kDa prolactin, has been identified, and this variant displays antiangiogenic properties mediated by a receptor different from the classical lactogen receptor. PRL also exists in a glycosylated form; in the circulation, PRL exists primarily in monomeric form, although dimeric and even oligomeric forms are routinely observed. These are important because they are frequently measured (e.g., by radioimmunoassay), but their biological activity is lower than that of the monomeric form of the hormone.

III. PROLACTIN RECEPTORS

A. Isoforms

In the 1970s the PRLR was identified as a specific, high-affinity, membrane-anchored protein; in 1988, the cDNA encoding rat PRLR was first isolated in

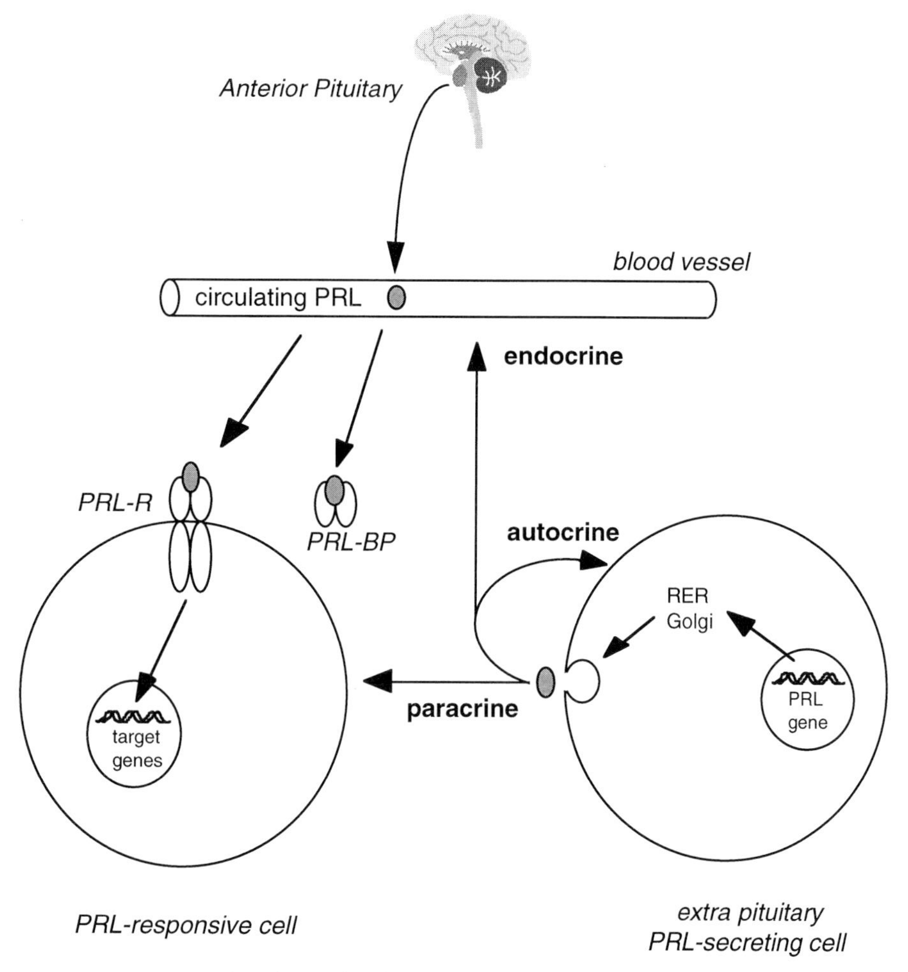

FIGURE 1 Schematic representation of prolactin (PRL)-secreting cells and PRL target cells (modified from Ben-Jonathan *et al.*, 1996). Circulating PRL and variants are derived from the anterior pituitary (lactotrophic cells) and from extrapituitary cells (endocrine). Locally produced PRL can affect either adjacent cells (paracrine) or the PRL-secreting cell itself (autocrine). Processing of PRL can occur at both the production site and the target tissues. The soluble PRL-binding protein (PRL-BP) corresponds to the extracellular domain of the membrane receptor for PRL.

our laboratory. The gene encoding PRLR is located within a cluster of cytokine receptor loci (on chromosomes 5 and 15p12–13 in human and mouse, respectively). It contains at least 10 exons for an overall length larger than 160 kb. Contrary to PRL, for which a unique transcript encodes a unique mature protein, several isoforms of membrane-bound PRLR resulting from alternative splicing of the primary PRLR transcript have been identified. These different PRLR isoforms differ in the length and composition of their cytoplasmic tail and are referred to as short, intermediate, or long PRLR with respect to their size. In addition to the membrane-anchored PRLR, soluble isoforms have also been identified (PRL-binding protein).

In the early 1990s, sequence comparison with newly identified membrane receptors led to the identification of a new family of receptors termed class 1 cytokine receptors, including receptors for PRL, several interleukins, granulocyte-colony stimulating factor, granulocyte macrophage-colony stimulating factor, leukemia inhibitory factor, oncostatin M, erythropoietin, thrombopoietin, and the obesity factor, leptin. All these receptors are single-pass membrane chains that contain stretches of highly conserved amino acids, both in the extracellular and in the intracellular domains. Typically, a cytokine receptor extracellular domain (ECD) is composed of a domain of ~200 amino acids, referred to as the cytokine receptor homology region which can be divided into two subdomains (D_1 and D_2), each showing analogies with the fibronectin type III module. It seems that ligand interactions are primarily driven by the conserved fibronectin-like domains. Two highly conserved features within cytokine receptor ECDs are found in the PRLR: the first is two pairs of disulfide-linked cysteines in the N-terminal subdomain D_1 and the second is a pentapeptide termed "WS motif" (Trp-Ser-any amino acid-Trp-Ser) found in the membrane proximal region of the C-terminal subdomain D_2. These features are required for proper folding and cell trafficking of the receptor as well as for ligand binding. The 3D structure of genetically engineered hPRLR ECD has been determined by crystallographic analysis and is exclusively composed of loops and β sheets.

The cytoplasmic domain of PRLR contains two regions, called Box-1 and Box-2, relatively conserved within cytokine receptors. Box-1 is a membrane-proximal region composed of 8 amino acids highly enriched in proline and hydrophobic residues. Due to the particular structural properties of proline residues, the conserved P-x-P (x = any amino acid) motif within Box-1 is assumed to adopt the consensus folding specifically recognized by transducing molecules containing a SH3 domain. The second consensus region, Box 2, is much less conserved than Box-1 and consists of the succession of hydrophobic, negatively charged then positively charged residues. It is noteworthy that Box-2 is not found in the short PRLR isoform.

B. Activation by Homodimerization

Two regions of hPRL are involved in the binding of the hormone to the PRLR (binding sites 1 and 2). Activation of the PRLR involves ligand-induced, sequential receptor dimerization. In the first step, the interaction of PRL binding site 1 with one receptor molecule occurs and leads to the formation of an inactive H_1:R_1 (1 hormone:1 receptor) complex. Formation of this complex appears to be a prerequisite for PRL binding site 2 to interact with another molecule of receptor, which leads to an active trimeric complex (H_1:R_2) composed of one molecule of hormone and one receptor homodimer.

C. Signal Transduction

Signaling cascades activated by cytokine receptors have been and continue to be extensively studied and deciphering the connections between tyrosine kinases, serine/threonine kinases, phosphatases, Stat proteins, and other transducers has become a major challenge of modern molecular biology. A schematic representation of the current knowledge of PRLR-activated signaling pathways is provided in Fig. 2.

1. *The Receptor-Associated JAK2 Tyrosine Kinase*

Although cytoplasmic domain of the PRLR is devoid of any intrinsic enzymatic activity, hormonal stimulation leads to tyrosine phosphorylation of several cellular proteins, including the receptors themselves. The first major step in understanding PRLR signaling was the identification of JAK2, a member of the Janus tyrosine kinase family, as the tyrosine kinase responsible for the observed phosphorylations. We have shown that JAK2 is constitutively associated with the PRLR. There is strong evidence that Box-1 is the site of interaction with JAK2. Although the mechanism by which the kinase is activated remains poorly understood, it is usually assumed that receptor dimerization upon ligand binding brings two JAK2 molecules close to each other, which allows transphosphorylation and subsequent activation of the enzyme. JAK2 is thought to be the initial element upstream of several, if not all, signaling pathways of PRLR. Presumably, there are several other substrates than the receptors, including Stat proteins and receptor-associated phosphatases.

2. *Stat Proteins*

Stat (signal transducer and activators of transcription) proteins are a family of latent cytoplasmic factors recently described as involved in cytokine receptor signaling. Stats are assumed to associate through their SH2 domains with phosphotyrosines of acti-

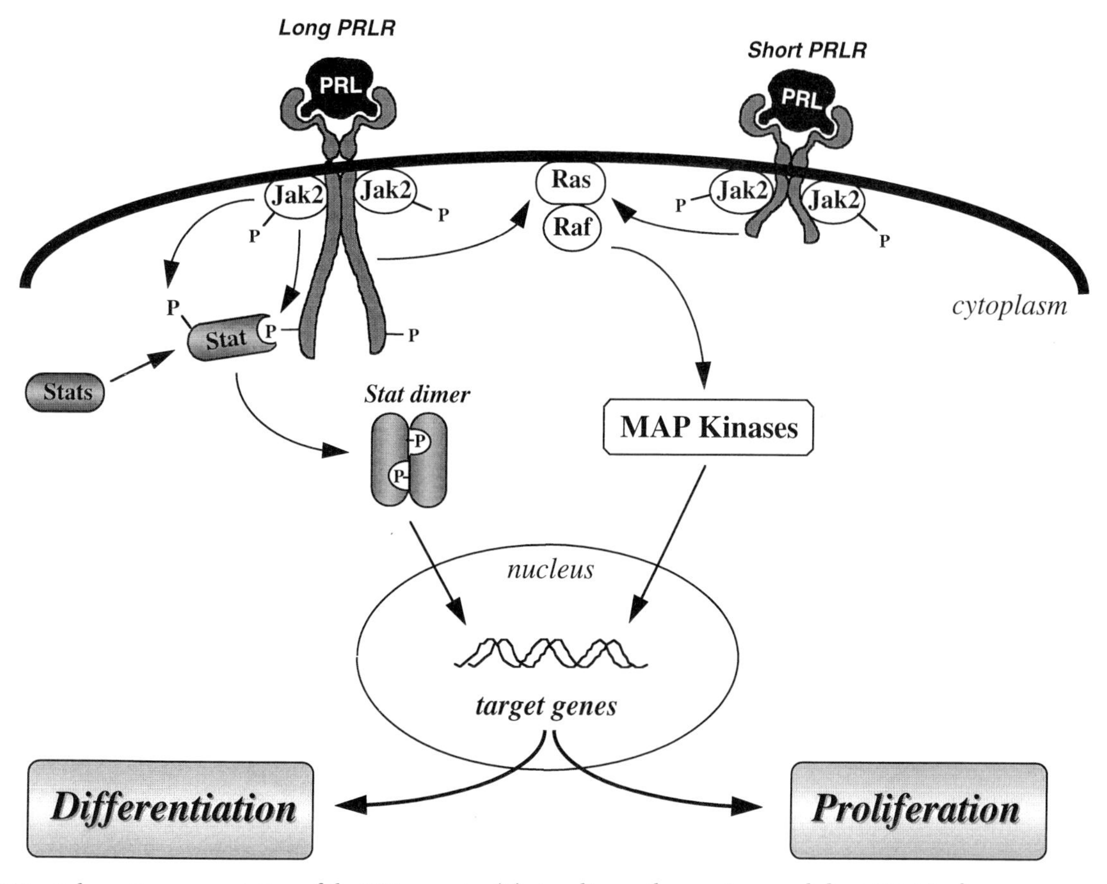

FIGURE 2 Schematic representation of the PRL receptor (R) signaling pathways. Long and short PRLR isoforms are represented. PRLR activates Stat1, Stat3, and, mainly, Stat5. Whether the short PRLR isoform activates the Stat pathway is currently unknown. The MAP kinase (MAPK) pathway involves the Ras–Raf–MAPK cascade and is presumably activated by both PRLR isoforms. Connections between the JAK–Stat and MAPK pathways have been suggested.

vated receptors, leading to their phosphorylation by JAK2. Stats then dissociate from the receptor through a mechanism that remains unknown, homo- or heterodimerize, and translocate to the nucleus where they interact with and activate specific DNA elements found in the promoters of target genes. Three members of the Stat family have been clearly identified as transducer molecules of PRL receptors. Stat5, previously referred to as mammary gland factor, is the major Stat activated by PRLR. It has been demonstrated that the DNA-binding activity of Stat5 requires phosphorylation of a single tyrosine residue that is mediated by JAK2 but not by Src kinases, confirming Stat proteins as Janus kinase substrates. Stat3 and Stat1 are also activated, although to lesser extents. The downstream effects mediated by Stat1 and Stat3 remain unknown.

3. *Src Kinases*

Fyn, a member of the Src family of tyrosine kinases, is associated with the PRLR and activated by PRL stimulation in the rat T lymphoma Nb2 cell line. Recently, association of the PRLR with Src has also been reported after PRL stimulation in lactating rat hepatocytes. The targets of these kinases remain to be identified.

4. *Tyrosine Phosphatases*

Involvement of tyrosine phosphatase in PRLR signaling has been suggested, although such associations remain poorly documented. Since many intracellular signaling steps require tyrosine phosphorylation (e.g., receptors, Janus kinases, and Stats), one would anticipate that tyrosine phosphatases would be negative regulators of signal transduction. In fact, it has recently been reported that the phosphatase PTP-1D (renamed SHP-2), identified as a JAK2 substrate, acts as a positive regulator of PRLR-dependent induction of β-casein gene transcription.

5. *Ras/Raf/MAP Kinase Pathway*

Signaling through MAP kinases (MAPK) involves the Shc/Sos/Grb2/Ras/Raf/MAPK cascade. Activation of the MAPK pathway has been reported in different biological systems under PRL stimulation. Whether activation of the MAP cascade requires JAK2, Fyn (or any Src kinase), or any other pathway is currently unknown.

D. Tissue Distribution

PRLRs are expressed at variable levels by a wide variety of adult tissues, including many in which the activity of prolactin is unknown. The expression of short and long forms has been shown to vary as a function of the stage of the estrous cycle, pregnancy, and lactation. However, little information is available during fetal development. We have recently determined the cellular distribution and developmental expression of the PRLR in the late gestational fetal rat by *in situ* hybridization, immunocytochemistry, and radioligand binding. These studies showed that the mRNA encoding both the short and long isoforms was widely expressed in tissues from all three germ layers: In addition to the classical target organs of PRL, tissues not known previously to contain PRL receptors, such as olfactory neuronal epithelium and bulb, trigeminal and dorsal root ganglia, cochlear duct, brown adipose tissue, submandibular glands, whisker follicles, tooth primordia, and proliferative and maturing chondrocytes of developing bone, also expressed PRLR. There was also a high level of expression of receptor mRNA in the fetal adrenal cortex, gastrointestinal and bronchial mucosae, renal tubular epithelia, choroid plexus, thymus, liver pancreas, and epidermis. The level of PRL receptor mRNA and protein actually increased between Days 17.5 and 20.5 of pregnancy in a number of tissues, suggesting that lactogenic hormones such as prolactin and placental lactogens may play important roles in fetal and neonatal development. Among others, these studies suggest novel roles for lactogenic hormones in olfactory differentiation and development and may provide insight into new mechanisms whereby lactogenic hormones regulate neonatal behavior and maternal–infant interactions.

IV. BIOLOGICAL ACTIONS OF PROLACTIN

A. Seven Broad Categories

The multiple actions of prolactin correlate well with the widespread distribution of its receptors. Prolactin has been identified in all mammals studied thus far as well as in birds, reptiles, amphibians, and fishes.

The full spectrum of PRL functions in mammals is not completely understood. PRL was originally isolated by its ability to stimulate mammary development and lactation, and it was shown also to be luteotropic by promoting the formation of the corpus luteum. In the now classical reviews by Nicoll and Bern (1972) and Nicoll (1980), over 85 biological functions have been described for PRL which fall into seven broad categories: reproduction and lactation, water and salt balance, growth and morphogenesis, metabolism, behavior, immunoregulation, and effects on the ectoderm and skin. Since the publication of these reviews, numerous other biological functions of prolactin have been identified. The investigation of the effects of PRL and of the PRL-regulated genes has centered on the mammary gland and recently on the immune system, and it is in these tissues that the developmental, regulatory, and pathological effects of prolactin are best known. The recent availability of mice in which the PRLR has been eliminated (knockout) emphasizes the requirement of PRL in several biological functions.

B. Knockout Mice

The coding region of the mouse PRLR gene was isolated. Exons 4 and 5 were each found to contain a pair of extracellular cysteine residues: Loss of just one of these cysteines results in complete lack of hormone-binding activity (see Section III,A). A targeting vector was prepared in which exon 5 was replaced by a neomycine cassette, resulting in the creation of an in-frame stop codon. The expression

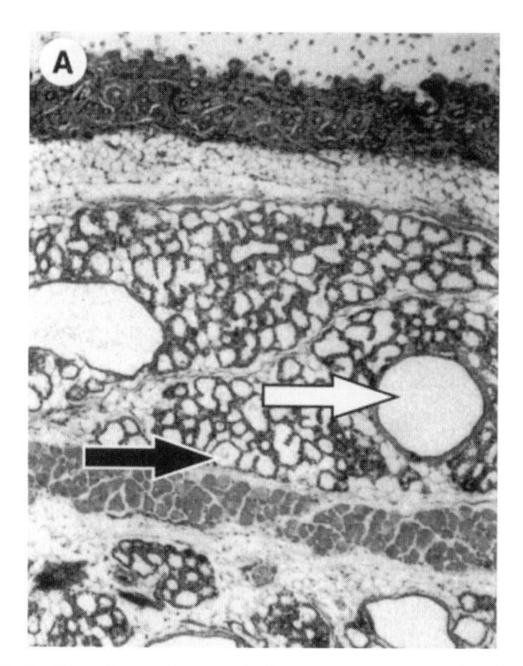

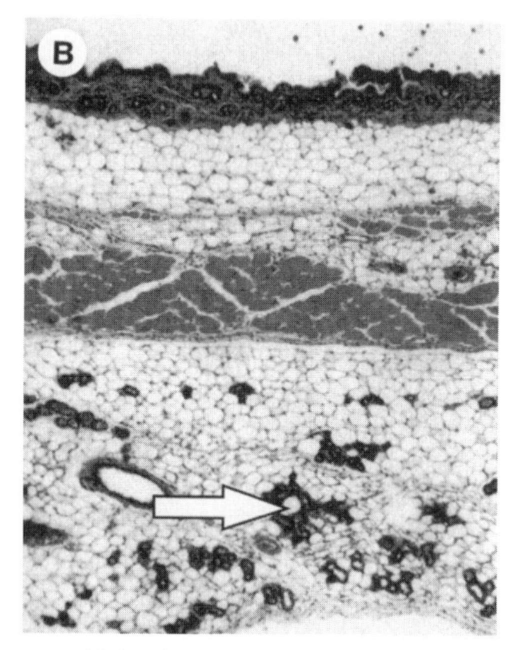

FIGURE 3 Histology of second thoracic mammary glands. (A) F1 PRLR$^{+/+}$ female 48 hr postpartum. (B) F1 PRLR$^{+/-}$ female unable to lactate 48 hr postpartum (magnification, ×100). The tissue is oriented with the skin on the top, followed to the bottom by adipose tissue, mammary epithelial cells, alveoli (black arrow) and ducts (white arrows), and muscle.

of the mutated PRLR gene was checked using reverse transcription-polymerase chain reaction (RT-PCR) and Northern blotting, and the absence of the protein was demonstrated by Western blot and ligand binding.

When 6- to 8-week-old PRLR heterozygote (PRLR$^{+/-}$) F1 females were mated with males, most of their first litter died within 24 hr, and the entire litter perished by 48 hr. All pups were observed to attach to the nipple and suckle; however, dead pups were dehydrated, with loose skin and loss of weight. The examination of their stomach contents showed air bubbles but no milk present, indicating that PRLR$^{+/-}$ females were unable to lactate. When assessed by pup survival, this phenotype was not apparent following the second pregnancy, in which all F1 PRLR$^{+/-}$ females produced surviving pups. Histological examination of the mammary glands from the F1 animals after 48 hr of their first lactation showed that lactational performance was correlated with the degree of mammary gland development. The glands from nonlactating mothers showed very little development when compared to the highly developed state of the PRLR$^{+/+}$ mammary glands (Fig. 3). These results demonstrate that two functional PRLR alleles are required for efficient lactation and that this phenotype in heterozygotes is primarily due to a deficit in the degree of mammary gland development.

PRLR$^{-/-}$ females are infertile. This is indicative of the presence of a number of reproductive deficiencies: Fewer eggs are fertilized, oocytes at the germinal vesicle stage are released from the ovary, and fragmented embryos are found. The number of eggs ovulated was reduced in the PRLR$^{-/-}$ females compared to the controls. Most important, fertilized eggs develop poorly to the blastocyst stage. Single cell fertilized eggs were recovered at all stages studied, suggesting for most oocytes that an arrest of development occurred immediately after fertilization (Fig. 4). In order to test if the absence of development of the fertilized eggs in the PRLR$^{-/-}$ females was due to the mother (oviduct environment) or the egg, transplantation experiments were performed. PRLR$^{+/+}$ and PRLR$^{-/-}$ females were mated to PRLR$^{+/+}$ males and fertilized single or two-cell stage embryos were flushed from their oviducts and reimplanted into the oviducts of pseudopregnant foster PRLR$^{+/+}$ mothers. Most of the eggs (PRLR$^{+/+}$ and PRLR$^{+/-}$) produced normal embryos. When this experiment was repeated using PRLR$^{-/-}$ males (to exclude the possible mater-

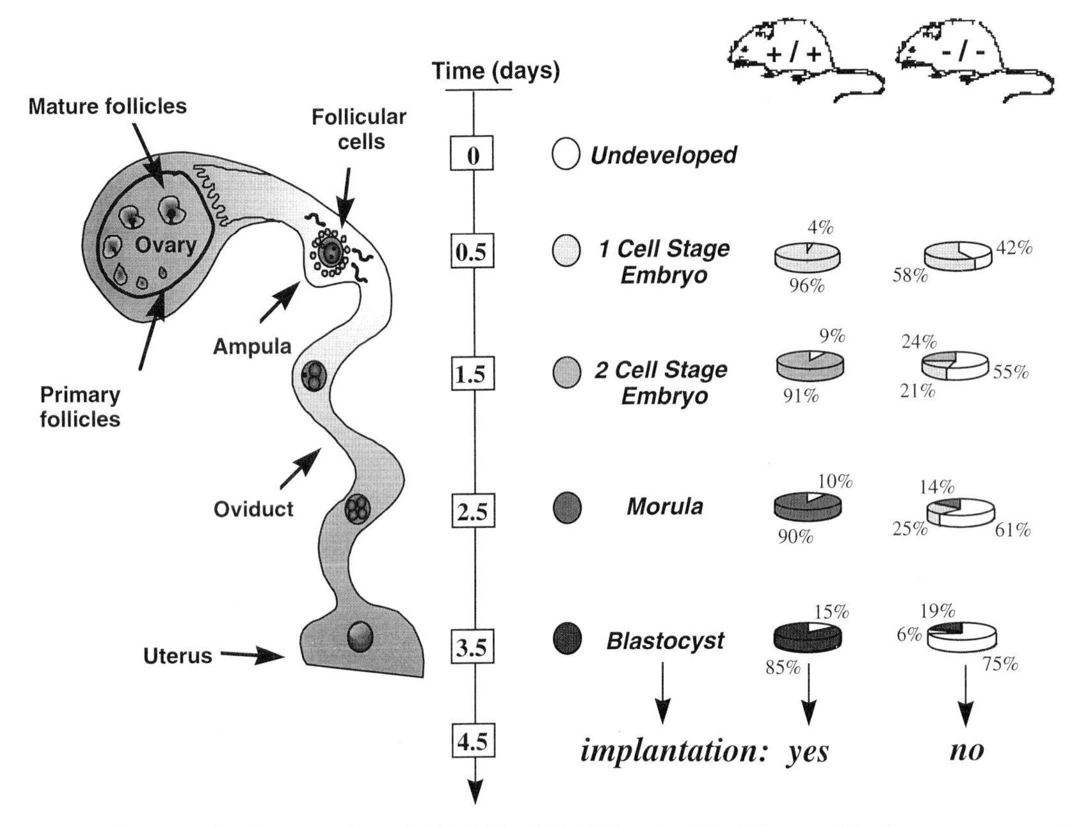

FIGURE 4 Preimplantation development of eggs in PRLR$^{+/+}$and PRLR$^{-/-}$ mice. The left part of the figure represents the normal development of a fertilized egg from ovary through oviduct before implantation in uterus. The vertical arrow indicates the time scale (in days) from ovulation (Day 0) to implantation (Day 4.5). On the right, pie charts represent the ratio (in percentage) of eggs found at each development stage in PRLR$^{+/+}$ versus PRLR$^{-/-}$ mice: Undeveloped oocytes are depicted in white (including oocytes at the germinal vesicle stage, oocytes that have excluded the first polar body, and degenerated embryos), and one-cell and two-cell stage embryos, morula, and blastocysts are symbolized by increasing shades of gray.

nal contribution of PRLR to the embryo) normal embryos were also recovered, demonstrating that the eggs (even PRLR$^{-/-}$) are viable and that the environment of the embryo in the oviduct is deficient. The absence of PRLR in female mice thus results in reduced ovulation, reduced fertilization, and almost complete arrest of preimplantation development. The small number of embryos that progress to the blastocyst stage are released into an environment refractory to implantation. The outcome is complete sterility.

Almost every aspect of reproduction is altered in these animals, including lactation, maternal behavior, and delayed male fertility, unambiguously demonstrating that the PRLR is a key regulator of reproduction. The ability of this new model to provide novel insights into the function of lactogenic hormones and their receptor illustrates the power of the knockout approach to discover unknown roles for well-investigated molecules. In view of the putative immunomodulatory role of PRL, we are currently analyzing the immune phenotype of these knockout animals.

Acknowledgments

We thank A. Pezet and P. Clement-Lacroix for help with the figures.

See Also the Following Articles

Estrogen Action on the Female Reproductive Tract; Estrogens, Overview; Hypothalamic-Hypophysial Complex; Gn-RH (Gonadotropin-Releasing Hormone); Prolactin Inhibitory Factors

Bibliography

Ben-Jonathan, N. (1994). Regulation of prolactin secretion. In *The Pituitary Gland* (H. Imura, Ed.), pp. 261–283. Raven Press, New York.

Ben-Jonathan, N., Mershon, J. L., Allen, D. L., and Steinmetz, R. W. (1996). Extrapituitary prolactin: Distribution, regulation, functions, and clinical aspects. *Endocr. Rev.* **17**, 639–669.

Bole-Feysot, C., Goffin, V., Edery, M., Binart, N., and Kelly, P. A. (1998). Prolactin and its receptor: Actions, signal transduction pathways and phenotypes observed in prolactin receptor knockout mice. *Endocr. Rev.,* **19**, 225–268.

Boutin, J. M., Jolicoeur, C., Okamura, H., Gagnon, J., Edery, M., Shirota, M., Banville, D., Dusanter-Fourt, I., Djiane, J., and Kelly, P. A. (1988). Cloning and expression of the rat prolactin receptor, a member of the growth hormone/prolactin receptor gene family. *Cell* **53**, 69–77.

Goffin, V., Shiverick, K. T., Kelly, P. A., and Martial, J. A. (1996). Sequence–function relationships within the expanding family of prolactin, growth hormone, placental lactogen and related proteins in mammals. *Endocr. Rev.* 17, 385–410.

Ihle, J. N. (1996). STATs: Signal transducers and activators of transcription. *Cell* **84**, 331–334.

Kelly, P. A., Djiane, J., Postel-Vinay, M. C., and Edery, M. (1991). The prolactin/growth hormone receptor family. *Endocr. Rev.* **12**, 235–251.

Nicoll, C. S. (1980). Ontogeny and evolution of prolactin's functions. *Fed. Proc.* **39**, 2563–2566.

Nicoll, C. S., and Bern, H. (1972). On the actions of PRL among the vertebrates: Is there a common denominator? In *Lactogenic Hormones* (G. E. W. Wolstenholme and J. Knight, Eds.), pp. 299–337. Churchill Livingstone, London.

Ormandy, C. J., Camus, A., Barra, J., Damotte, D., Lucas, B. K., Buteau, H., Edery, M., Brousse, N., Babinet, C., Binart, N., and Kelly, P. A. (1997). Null mutation of the prolactin receptor gene produces multiple reproductive defects in the mouse. *Genes Dev.* **11**, 167–178.

Sinha, Y. N. (1995). Structural variants of prolactin: Occurrence and physiological significance. *Endocr. Rev.* **16**, 354–369.

Somers, W., Ultsch, M., De Vos, A. M., and Kossiakoff, A. A. (1994). The X-ray structure of the growth hormone–prolactin receptor complex. *Nature* **372**, 478–481.

Prolactin, Actions of

James A. Rillema

Wayne State University School of Medicine

GLOSSARY

alveolus A small anatomical sac-like structure that serves as the functional unit of many tissues, including the mammary gland, prostate gland, and salivary gland.

amenorrhea The absence of menses during the menstrual cycle in women.

immune system The cells (white blood cells and other progenitor cells) and antibodies that serve to protect an organism from undesirable foreign cells, viruses, and other molecules.

luteolytic Describing a stimulus that causes regression of the corpus luteum in the ovary.

phosphatidylinositol pathway A pathway that involves the catalytic degradation of specific phospholipids in the plasma membranes of cells. The products of this degradation are diglycerides (DG) and inositol triphosphate (IP_3). IP_3 releases calcium ions from the endoplasmic reticulum within cells. The calcium ions along with DG activate an intracellular enzyme called protein kinase C.

pituitary gland An epithelial body at the base of the brain that secretes several hormones including prolactin. It is often referred to as the "master gland" because of its diverse functions.

polyamines Substances that facilitate the formation of proteins and RNA in cells. A variety of growth-promoting factors increase the synthesis of these substances.

prostaglandins 20-carbon compounds that are formed from polyunsaturated free fatty acids. They exert a number of effects in cells and are associated with fever in illness. Most of the therapeutic effects of aspirin can be attributed to the inhibition of prostaglandin synthesis.

protein kinase C An intracellular serine–threonine kinase that catalyzes phosphorylation of intracellular proteins and alters their activities.

Ras–MAP kinase pathway A signaling pathway involving the activation of several molecules arranged in a series cascade. The final product in this pathway is MAP (mitogen-activated protein) kinase, which is translocated from the cytosol to the nucleus. In the nucleus, activated MAP kinase phosphorylates proteins that in turn regulate production of RNAs from genes.

Stat proteins Cytoplasmic proteins that are phosphorylated by tyrosine kinases and then alter RNA production from genes in the nucleus.

Prolactin is a pituitary hormone that has a wealth of diverse biological functions in vertebrate animals. Prolactin is a large peptide hormone with a molecular weight of about 23,000, with small differences among various species; minor forms of prolactin include those that are phosphorylated, glycosylated, or truncated fragments.

I. INTRODUCTION

Of the more than 100 physiological effects of prolactin that have been identified, most concern the regulation of processes associated with reproduction, immune functions, salt and water balance, growth, and maternal behavior. In mammals the best recognized effects of prolactin are the stimulation of lactation (milk formation) and growth of the mammary gland. In submammalian species prolactin regulates a plethora of diverse processes, including salt and water balance in fish, amphibian growth and metamorphosis, and crop "milk" secretion in pigeons and doves. Prolactin initiates its biological responses by binding to receptor glycoproteins embedded in the plasma membranes of its target cells. Recent experimental work has uncovered several molecular signaling pathways by which prolactin expresses its effects on biological processes.

II. SPECIFIC BIOLOGICAL EFFECTS OF PROLACTIN

A. Effects Associated with Reproduction

1. Mammary Gland

The mammary gland in various species is composed of several to many thousand milk-producing units called alveoli. These units consist of a unicellular layer of epithelial cells arranged in a spheroid structure (Fig. 1). The alveolar epithelial cells take up nutrients from the blood and secrete milk products into the alveolar lumen. Each alveolus is connected to a duct through which milk flows, eventually being externalized through a lactiferous pore. During pregnancy, prolactin, along with several other hormones (estrogen, progesterone, growth hormone, glucocorticoid, and insulin), stimulates mammary development, including the alveoli. In the periparturient period, increased secretion of prolactin coupled with glucocorticoids, and the relative absence of progesterone and estrogen, initiates lactation (lactogenesis). During lactation, prolactin, in concert with insulin, glucocorticoid, and thyroid hormone, stimulates and maintains milk production.

2. Crop Milk

In both male and female pigeons and doves, prolactin stimulates a thickening of the mucosal epithelium of the crop sac. Desquamated cells are eventually sloughed off into the lumen of the crop sac, forming crop milk. Crop milk, which has the consistency of cottage cheese, is regurgitated by males and females

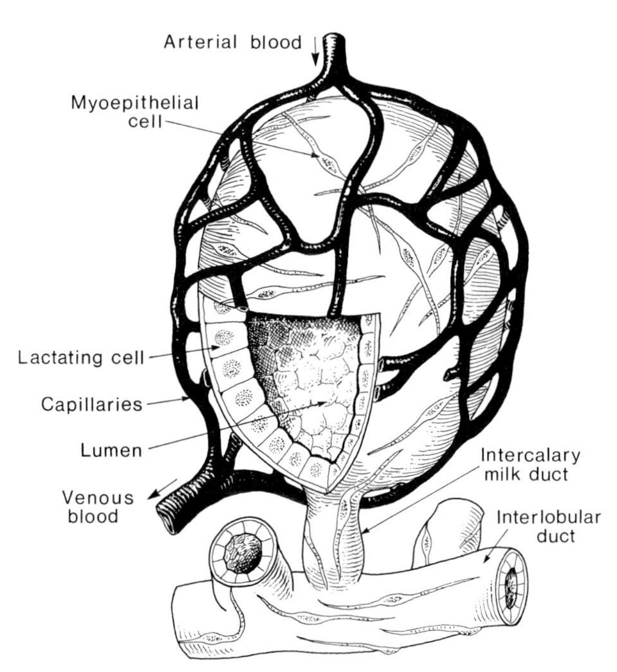

FIGURE 1 Structure of the mammary gland.

alike into the mouths of the offspring and is used to nurture the young.

3. *Female Reproductive Functions*

In the ovaries of rats, mice, and hamsters, prolactin functions in synergy with the gonadotropins under certain physiological conditions to maintain the corpus luteum and increase progesterone secretion; under other conditions prolactin may have a luteolytic function, which reduces progesterone secretion. In other species, including the human, prolactin does not increase corpora luteal function or progesterone secretion. In fact, chronic hyperprolactinemia in humans is associated with amenorrhea in women and reduced androgen levels in males. Prolactin reduces the size of male and female reproductive organs in a variety of avian species. Prolactin thus has diverse effects on female reproductive processes, many of which are species and situation specific.

4. *Male Reproductive Functions*

In several rodent species, prolactin synergizes with the androgens in males to stimulate spermatogenesis as well as growth and function of the secondary sex organs. In avian species, however, prolactin causes the regression of the male sex organs. In humans and other species, the physiological role of prolactin remains to be established.

5. *Behavioral Effects*

In a variety of avian species, prolactin causes brooding behavior and a drive to incubate (sit on) eggs; in some species prolactin also stimulates brood patch formation on the bird's breast. Evidence also suggests that prolactin enhances mothering behavior in rodents and stimulates nest building behavior by rabbits. In fish, prolactin regulates a variety of parental care behaviors. In salamanders, prolactin regulates their migration from land to water; the migratory patterns of specific bird species are also regulated by prolactin.

B. Regulation of the Immune System

A number of experimental studies have shown that the immune system is compromised in animals in which the pituitary gland has been removed; the administration of prolactin or growth hormone restores immunocompetence in these animals. Prolactin may thus be an important hormone for maintaining normal function of the immune system.

C. Growth-Promoting Effects

Prolactin stimulates growth processes in many species, including fish, amphibians, reptiles, birds, and mammals. Prolactin stimulates mammary gland development and growth of the prostate gland and seminal vesicles. In amphibians and reptiles, prolactin stimulates tail growth and limb regeneration. In birds, prolactin stimulates growth of the crop sac in pigeons and doves, brood patch development in the white-throated sparrow, and feather growth in selected species.

D. Salt and Water Balance

Prolactin has osmoregulatory actions in all vertebrates. In teleost fish, prolactin regulates sodium and water transport across the gills, kidneys, urinary bladder, and gut. In the amphibians, ion transport across the skin and urinary bladder may be regulated by prolactin. Prolactin also has diverse effects on salt and water balance in mammals.

III. PROLACTIN RECEPTOR ACTIVATION MECHANISM

Prolactin initiates its effects on target cells by binding to its "receptor" (Fig. 2). The prolactin receptor is a glycoprotein molecule that is embedded in the plasma membrane of target cells. The extracellular portion of the receptor has a high-affinity binding site for prolactin. After prolactin binds to its receptor, an enzyme, Janus kinase-2 (JAK-2), constitutively bound to the intracellular tail of the receptor, becomes activated. JAK-2 is a tyrosine kinase that inserts phosphate groups on intracellular proteins and thus modifies their activities. Prolactin receptors have been identified on a variety of tissues including the mammary gland, prostate gland, seminal vesicles, kidney, adrenal gland, ovary, liver, testis, uterus, brain, choroid plexus, adipose tissue, pancreas, and epididymus. The physiological functions of prolactin for all these tissues are not known.

IV. MOLECULAR SIGNALING MECHANISMS FOR PROLACTIN

Subsequent to the activation of JAK-2 kinase by prolactin, several molecular signaling pathways are activated which culminate in the biological effects of prolactin (Fig. 2). Most of these signaling pathways cause a change in the production of messenger RNA molecules from the genes in the nucleus of the cells (transcription); the messenger RNA molecules are coded for the synthesis (translation) of specific proteins including the milk proteins.

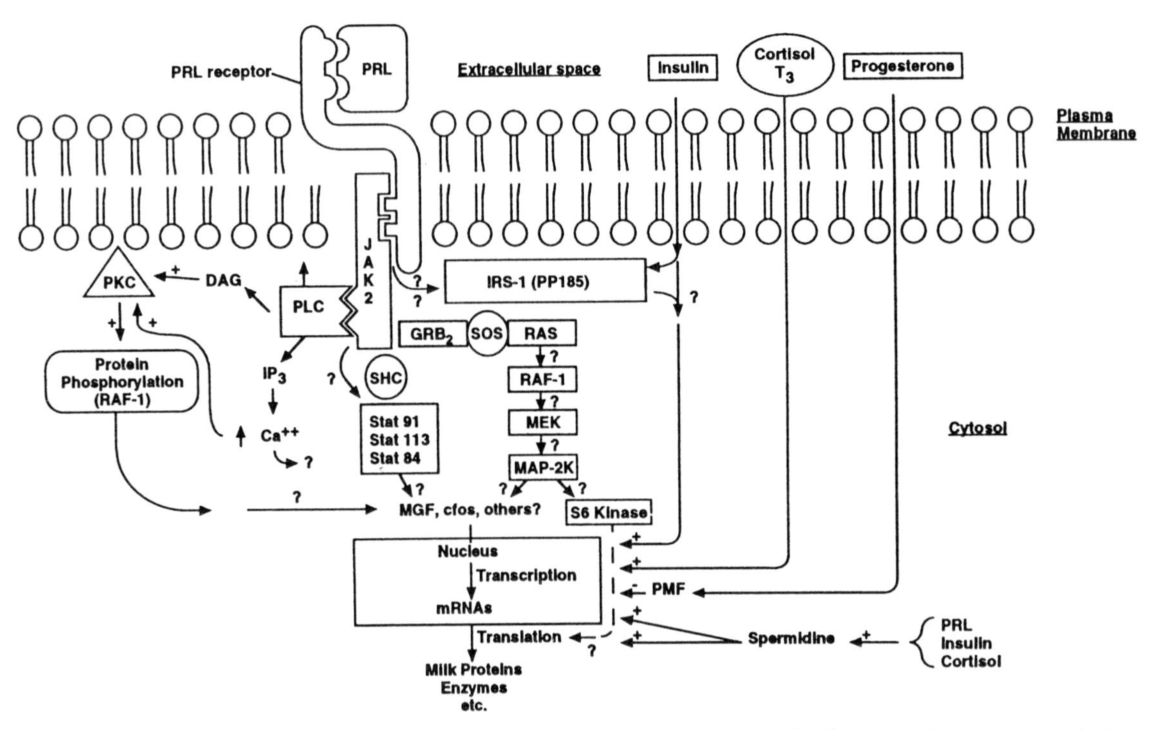

FIGURE 2 Signaling mechanisms for prolactin and other hormones in the mammary gland. PRL, prolactin; T_3, triiodothyronine; JAK-2, janus kinase 2; PLC, phospholipase C; DAG, diacylglycerol; PKC, protein kinase C; IP_3, inositol triphosphate; Stat, signal transducers and activators of transcription; MAP, mitogen-activated protein; mRNA, messenger ribonucleic acid; IRS, insulin receptor substrate.

Signaling pathways that are known to be activated under specific physiological conditions are as follows:

1. An elevation of intracellular calcium ion levels, which in turn alters many cellular metabolic processes
2. An elevation of the intracellular level of the polyamine spermidine, which increases the rates of messenger RNA production in the nucleus and protein synthesis
3. An increased rate of prostaglandin synthesis (not shown in Fig. 2), which alters many cellular metabolic processes
4. A stimulation of the phosphatidyl inositol pathway which involves a stimulation of phospholipase C, the generation of diacylglycerol and inositol triphosphate, the activation of protein kinase C(PKC), and the phosphorylation of specific intracellular proteins including RAF-1; the phosphorylated proteins then alter intracellular processes
5. An enhanced tyrosyl phosphorylation of Stat (signal transducers and activators of transcription) proteins which go directly to the nucleus and alter the rate of formation of several messenger RNAs from the genes
6. A stimulation of the MAP kinase pathway which involves a tandem of signaling molecules associated in the following sequence: GRB_2-SOS-RAS-[RAF-1]-MEK-[MAP-2K]. The MAP-2K is transported to the nucleus where it alters the production of specific messenger RNAs from genes.

A combination of these molecular pathways is likely involved in the prolactin regulation of most of its biological responses, including the mammary gland. Much ongoing research focuses on precisely how these pathways integrate with the signaling pathways for other hormones that work in synergy with prolactin to elicit biological responses.

See Also the Following Articles

Lactogenesis; Pituitary Gland; Prolactin Inhibitory Factors; Prolactin, in Nonmammals; Prolactin, Overview

Bibliography

Ben-Jonathan, N., Mershou, J. L., Allen, D. L., and Steinmetz, R. W. (1996). Extrapituitary prolactin: Distribution, regulation, functions, and clinical aspects. *Endocr. Rev.* **17**, 639–669.

Groner, B., and Gouilleux, F. (1995). Prolactin-mediated gene activation in mammary epithelial cells. *Curr. Opin. Genet. Dev.* **5**, 587–594.

Horseman, N. D., and Yu-Lee, L. (1994). Transcriptional regulation by the helix bundle peptide hormones: Growth hormone, prolactin, and hematopoietic cytokines. *Endocr. Rev.* **15**, 627–649.

Kelly, P. A., Djiane, J., Postel-Vinay, M. C., and Edery, M. (1991). The prolactin/growth hormone receptor family. *Endocr. Rev.* **12**, 235–251.

Kooijman, R., Hooghe-Peters, E. L., and Hooghe, R. (1996). Prolactin, growth hormone, and insulin-like growth factor-1 in the immune system. *Adv. Immunol.* **63**, 377–454.

Nicoll, C. S. (1974). Physiological actions of prolactin. In *Handbook of Physiology*, Section 7, Vol. 4, pp. 253–292. American Physiological Society, Washington, DC.

Rillema, J. A. (1986). *Actions of Prolactin on Molecular Processes.* CRC Press, Boca Raton, FL.

Rillema, J. A. (1994). Development of the mammary gland and lactation. *Trends Endocrinol. Metab.* **5**, 149–154.

Tucker, H. A. (1994). Lactation and its hormonal control. In *The Physiology of Reproduction,* pp. 1065–1098. Raven Press, New York.

Yu-Lee, L. (1997). Molecular actions of prolactin in the immune system. *Proc. Soc. Exp. Biol. Med.*, in press.

Prolactin Inhibitory Factors

Michael Selmanoff
University of Maryland

GLOSSARY

dopamine The catecholamine neurotransmitter synthesized from l-DOPA by the action of tyrosine hydroxylase.

gonadotropin-releasing hormone A hormone that stimulates luteinizing hormone and follicle-stimulating hormone secretion from the anterior pituitary gland.

lactotroph The anterior pituitary cell type which synthesizes and secretes prolactin

median eminence The portion of the medial basal hypothalamus from which the portal hypophysial vessels arise.

perikarya Neuronal cell bodies.

prolactin The prolactational trophic hormone secreted from the anterior pituitary gland.

somatostatin The hypothalamic growth hormone-inhibiting hormone which inhibits growth hormone secretion from the anterior pituitary gland.

thyrotropin-releasing hormone A hormone that stimulates thyroid-stimulating hormone secretion from the anterior pituitary gland.

I. HYPOTHALAMIC CONTROL IS INHIBITORY

It was recognized early that the tonic regulation of prolactin (PRL) by the hypothalamus is inhibitory. In 1954 Everett observed that transplantation of the pituitary to the kidney capsule resulted in markedly increased PRL secretion, in distinct contrast to other trophic hormones of the anterior pituitary gland. Isolation of the pituitary from the hypothalamus by pituitary stalk section similarly was shown to result in increased PRL secretion in both experimental animals and humans. Indeed, incubation of pituitaries *in vitro* was found to result in marked secretion of PRL into the medium, an amount which in a single day far exceeded that initially present in the gland. Hypothalamic extracts were soon shown to inhibit the profound PRL hypersecretion observed from pituitary tissue incubated *in vitro*, and this was used as a bioassay system by workers trying to isolate the postulated hypothalamic prolactin inhibitory factor (PIF). The inhibitory influence of the hypothalamus could be removed by electrolytic lesions of the medial basal hypothalamus (MBH) which resulted in dramatic increases in serum PRL levels and pituitary PRL synthesis.

Since the 1950s it had been noted that administration of major tranquilizers (e.g., barbiturates), antihypertensives (e.g., reserpine and α-methyl-dopa), and antipsychotics (e.g., chlorpromazine and haloperidol) to some patients could cause breast enlargement and galactorrhea. Later, other more specific drugs affecting the metabolism of the biogenic amines norepinephrine, epinephrine, dopamine, and serotonin were found to produce profound changes in PRL secretion which were associated with changes in hypothalamic PIF activity, which was at that time measured by *in vitro* bioassay. Hence, attention was focused on hypothalamic bioaminergic neurons as ones which must be closely associated with the PIF neurosecretory neurons or which themselves might be the PIF neurosecretory neurons. In 1968, van Maanen and Smelik advanced the hypothesis that the tuberoinfundibular dopamine (TIDA) neurons were in fact PIF neurosecretory neurons, and that the neurotransmitter dopamine (DA) was PIF. At about the same time, electron microscopic (EM) studies of the median eminence (ME) persuaded Kobayashi and Matsui (1969, p. 26) to conclude that "it seems likely that dopamine in the palisade layer is transferred to the adenohypophysis and affects the cells directly." However, the "dopamine is PIF hypothesis" was widely ignored for several years due principally to the fact that the other hypothalamic-releasing and -inhibiting factors discovered at that time all were peptides [i.e., gonadotropin-releasing hormone (GnRH), thyrotropin-releasing hormone(TRH), and somatostatin], and to several reports of peptidic hypothalamic compounds with PIF activity. By the late 1970s, several additional compelling lines of evidence were developed, consistent with the DA hypothesis. Taken together, this body of evidence convinced Weiner and colleagues, and later Ben-Jonathan, to promote DA from the status of a putative PIF to that of the most physiologically significant prolactin inhibitory hormone (PIH) known.

II. DOPAMINE AS A PHYSIOLOGICALLY SIGNIFICANT PIH

A. Neuroanatomy of the TIDA Neurons

Perhaps the most fundamental evidence supporting the hypothesis of Weiner *et al.* is the neuroanatomy of the tuberoinfundibular dopaminergic neurons themselves (Fig. 1). The TIDA neuronal cell bodies, which number about 2800 in the rat, are situated in the arcuate and anterior periventricular nuclei (A_{12} cell group of Dahlström and Fuxe) with axonal projections passing ventrally to the region of the median eminence. Here the terminal boutons are observed to be in immediate contact with the perivascular space of the primary portal capillaries in the external layer of the median eminence (Fig. 2).

This dopaminergic system constitutes one of the most dense projections known in mammalian brain,

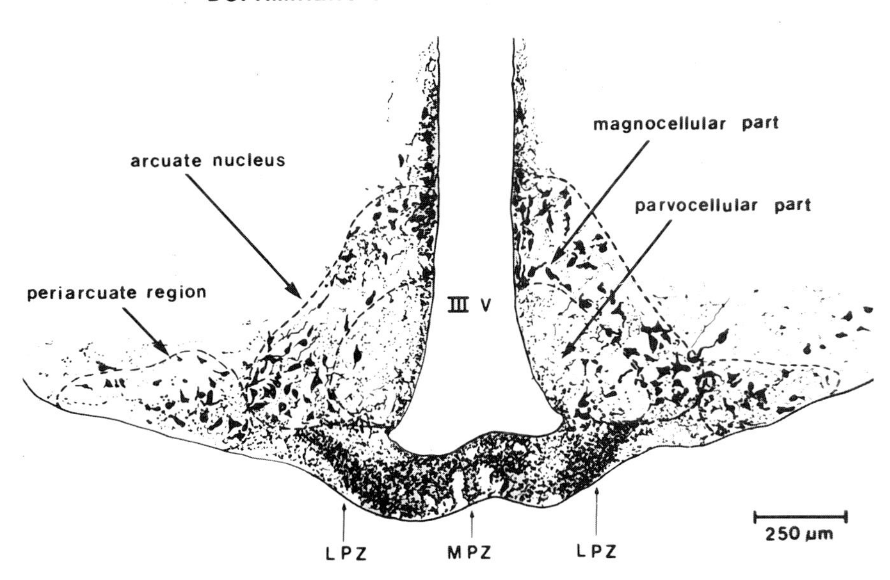

FIGURE 1 Coronal section of the rat arcuate nucleus and median eminence region of the medial basal hypothalamus illustrating the location of DA neuronal perikarya (large profiles) and nerve terminals (small profiles). The drawing is based on tyrosine hydroxylase immunoreactivity. Note that the DA cell bodies are predominantly in the magnocellular portion of the arcuate nucleus and the periarcuate region, while dense collections of DA nerve terminals are present in the lateral (LPZ) and medial (MPZ) pallisade zones of the median eminence (reproduced with permission from Fuxe *et al.*, 1985a. © Springer-Verlag).

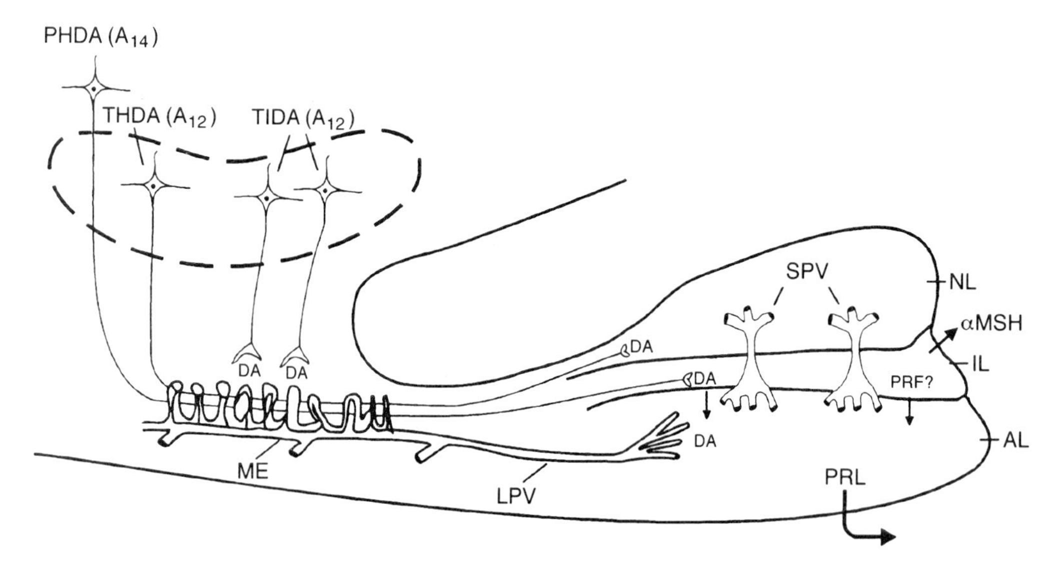

FIGURE 2 Parasagittal section through the medial basal hypothalamus and pituitary of the rat schematically illustrating the location of axonal projections of the A_{12} and A_{14} DA neurons. THDA (rostral A_{12}), tuberohypophysial dopaminergic; TIDA (A_{12}), tuberoinfundibular dopaminergic; PHDA (A_{14}), periventricular–hypophysial dopaminergic; DA, dopamine; ME, median eminence; LPV, long portal vessels; SPV, short portal vessels; PRF, prolactin-releasing factor; PRL, prolactin; αMSH, α-melanocyte-stimulating hormone; NL, neural lobe; IL, intermediate lobe; AL, anterior lobe (modified with permission from Neill and Nagy, 1994).

as morphometric EM analysis indicates that fully one-third of the terminal boutons in significant portions of the external layer of the ME are dopaminergic. The concentration of DA in the midportion of the ME exceeds by 1.3- to 58-fold that of each of the other dopamine terminal projection fields in brain, including the corpus striatum. Degeneration studies of the external layer of the ME in completely deafferented animals indicate that all of the perikarya of the TIDA neurons lie within the MBH. Some of the rostral TIDA cells (A_{12} cell group) also are thought to project to the posterior pituitary lobe (THDA, tuberohypophysial dopaminergic neurons), whereas other dopamine-containing cell bodies situated in the periventricular nucleus (A_{14} cell body group) are reported to innervate the intermediate pituitary lobe (PHDA, periventricular–hypophysial dopaminergic neurons) (Fig. 2).

B. Pituitary Tissue Cultured *in Vitro*

Pituitary organ cultures provided the first strong evidence that low concentrations of catecholamines, including DA at concentrations of 10^{-9} M, could directly inhibit PRL secretion, presumably by acting at the level of the pituitary lactotrophs. Later, high-affinity (K_D {{approxequal}} 10^{-8} M), low-capacity DA receptor binding was demonstrated in dispersed anterior pituitary cell cultures, and the binding affinity of DA receptor agonists was directly correlated with *in vitro* PRL secretion. This receptor K_D is in the range to be physiologically relevant considering the concentrations of DA detected in hypophysial portal blood (1–100 n*M*). Similar to DA receptors in the corpus striatum, anterior pituitary DA receptors were demonstrated to become supersensitive following MBH lesion, and DA receptors were visualized on pituitary lactotrophs at the EM level. Subsequently, these lactotroph DA receptors were shown to be of the D_2 receptor subtype.

C. Hypothalamic Deafferentation

Complete surgical deafferentation of the MBH results in a marked reduction in norepinephrine within the island, while the DA concentration remains relatively unaffected. This indicates that the TIDA neurons remain intact in this neurosurgical preparation, while the noradrenergic afferents to the MBH are severed. Importantly, serum PRL titers do not increase following deafferentation, indicating that the hypothalamic inhibitory mechanism resides in the MBH and does not involve noradrenergic, adrenergic, or serotonergic neurons, all of which are MBH afferents. Subsequent lesion of the ME, or treatment with the tyrosine hydroxylase inhibitor α-methyl-para-tyrosine in such deafferented animals, results in a rapid and marked increase in circulating PRL concentrations. Hence, it was concluded that the TIDA neurons are likely involved in the hypothalamic inhibitory mechanism.

D. Hypothalamic Extracts and DA Infusions

Shaar and Clemens substantially strengthened the hypothesis by showing that the PIF activity of hypothalamic extracts could be accounted for by their catecholamine content as determined by direct assay. Moreover, alumina absorption (of catecholamines) or monoamine oxidase digestion of the catecholamine-containing extracts rendered them inactive, while proteolytic digestion had no effect on their PIF activity. Infusion of DA into the third cerebral ventricle, or directly into the long portal vessels, was found to decrease serum PRL levels. At this time it was observed that systemic infusion of DA or its immediate precursor, l-DOPA, rapidly depresses PRL secretion in normal human subjects and in hyperprolactinemic patients.

E. Portal Blood Dopamine Measurements

The groups of Porter and Neill both carried out studies in which DA was measured in samples of portal blood collected from rats. Hypophysial portal blood concentrations were found to be in the range of 1–100 n*M*, approximately 10- to 20-fold greater than those in the systemic circulation and sufficient to substantially inhibit PRL secretion *in vivo*. Moreover, portal blood DA concentrations were found to vary with the estrous cycle and during pregnancy. Portal blood concentrations were reported to be greater in females than in males and greater in medially than in laterally situated portal vessels. These portal blood findings corroborated reports that median eminence DA turnover was greater in females than in males, and that turnover in the medial ME exceeded that in the lateral ME by 1.8-fold. One would expect that the ability of the TIDA neurons to release physiologically relevant concentrations of DA into the portal circulation would be impacted by the efficacy of the terminal bouton DA reuptake mechanism. While the TIDA neurons express the mRNA encoding the DA reuptake transport protein (DAT), and the DA catabolite DOPAC (thought to be formed principally from released and recaptured DA) is present in the ME, the K_m for DA uptake into ME DA terminals is 26- to 43-fold less than that observed for DA terminals in the corpus striatum. This decreased efficacy of the DA reuptake process in the ME is consistent with the TIDA neurons being neurosecretory. Nonetheless, that the DAT knockout mouse exhibits chronically increased DA release from the TIDA neurons indicates that DA reuptake in these neurons is indeed a physiologically significant process.

F. Pharmacological Manipulations

A tremendous literature exists demonstrating that antidopaminergic drugs cause increased PRL secretion and vice versa. This literature is strikingly consistent and includes work carried out both *in vivo* and *in vitro*, in both male and female mammals, and in numerous species, including rats and several other experimental animals, nonhuman primates, and humans. Drugs which are DA receptor antagonists (e.g., haloperidol, pimozide, perphenazine, clozapine, butaclamol, and spiroperidol), which deplete DA storage granules (reserpine), or which interfere with DA biosynthesis (e.g., α-methyl-para-tyrosine and α-methyl-dopa) antagonize the inhibitory action of DA and result in increased PRL synthesis and secretion. Conversely, drugs which are DA receptor agonists (e.g., dihydroergocornine, apomorphine, bromocryptine, and piribedil) or which increase DA concentrations in brain and/or pituitary by precursor loading (l-DOPA), monoamine oxidase inhibition (e.g., pargyline, iproniazid, and nialamide), or reuptake blockade (e.g., benztropine and imipramine) enhance the inhibitory action of DA and decrease PRL secretion. Moreover, some of these compounds can block the PRL-stimulating effects of the aforementioned antidopaminergic drugs. The DA receptor agonists and antagonists act directly at the pituitary lactotroph DA receptor, while compounds affecting DA metabolism (i.e., synthesis, transport, storage, release, or reuptake) act at the hypothalamus.

III. DOPAMINE AS THE PIH CONTROLLING BASAL PRL SECRETION

Critical to the acceptance of the DA is PIF hypothesis was work performed by Neill and colleagues demonstrating that the portal blood DA concentration they measured directly could account for about 70–85% of the inhibition of PRL secretion under conditions of basal secretion in α-methyl-para-tyrosine-treated animals. Their inability to account for 15–30% of the normal inhibition is unclear. It may be accounted for by one or more of the following possibilities: (i) the measured portal blood DA concentration in control animals was underestimated because of hemodynamic and or anesthetic effects attendant to portal blood collection, (ii) the sustained DA infusion signal employed in these studies was less potent in inhibiting PRL secretion than would be a pulsatile DA signal (pulsatility likely characterizes the endogenous DA signal to the lactotroph), (iii) the inferred DA concentration in the interstitial fluid surrounding the pituitary lactotrophs in control animals was underestimated because any contribution of posterior lobe or intermediate lobe DA (Fig. 2) would not be represented in the portal blood DA determinations, (iv) the portal blood DA concentration achieved by peripheral DA infusion was less than that measured in the systemic circulation, and/or (v) existence of another PIF(s).

Basal PRL secretion is regulated by circulating levels of PRL. This short-loop autofeedback effect (Fig. 3) was first demonstrated by Clemens and Meites in 1968. It appears to be a homeostatic mechanism whereby basal PRL secretion is maintained within its normal range by subtle changes in the activity of the TIDA neurons. Hypoprolactinemia, induced by hypophysectomy, administration of DA agonists such as bromocryptine, or passive immunization

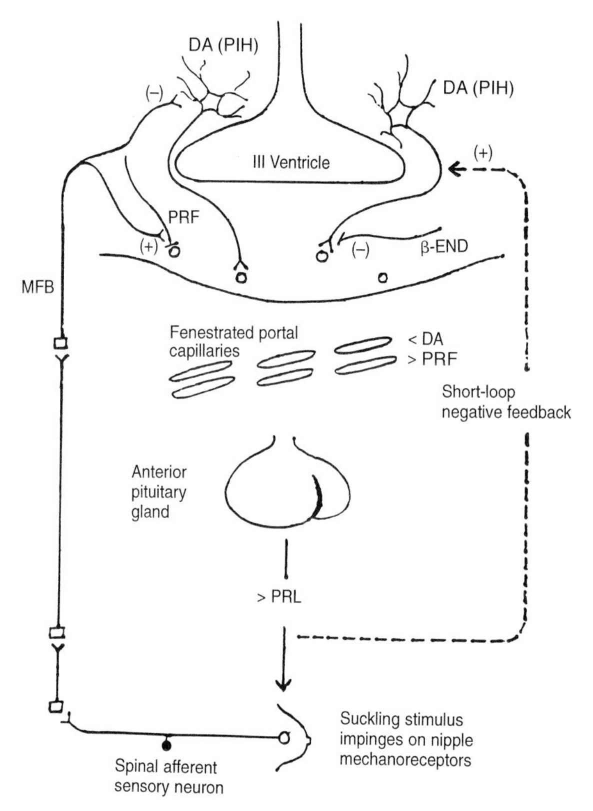

FIGURE 3 Schematic representation of the PRL short-loop autofeedback mechanism (right) and the hypothalamic control of suckling-induced PRL release (left). DA, dopamine; PIH, prolactin inhibitory hormone; PRF, prolactin releasing factor; MFB, medial forebrain bundle; β-END, β-endorphin.

with PRL antibody, reduces TIDA neuronal activity. Conversely, PRL stimulates DA synthesis, turnover and release, and steady-state tyrosine hydroxylase mRNA levels in the TIDA neurons. That this stimulatory action of PRL on TIDA neuronal activity is effective in suppressing basal PRL secretion from the anterior pituitary gland has been demonstrated in both male and female rats. The PRL autofeedback effect is temporally correlated with demonstrated increases in TIDA neuronal turnover, is reversible, and is dose dependent. Circulating PRL is thought to access the arcuate–median eminence region by retrograde flow in the pituitary portal vessels or via the cerebrospinal fluid. PRL binding sites have been visualized by autoradiography in the arcuate–median eminence region and PRL receptor mRNA expression has been reported in arcuate neurons.

IV. PIH AND PROLACTIN-RELEASING FACTOR MEDIATION OF PRL SECRETORY EPISODES

Hypothalamic control of PRL secretory episodes, such as those associated with stress, proestrus, pregnancy, and suckling, may logically result from decreased PIH release, increased prolactin-releasing factor (PRF) release, or perhaps both from the ME. Numerous reports demonstrate that decreases in TIDA neuronal activity accompany stress-induced, proestrous surge and the pregnancy diurnal and nocturnal PRL surges. Consider the evidence consistent with the PIH DA mediating suckling-induced PRL release (Fig. 3). Suckling-induced PRL release is one of the most dramatic and well-characterized neuroendocrine reflex responses known. The response is elicited by a precise, controllable stimulus, has a rapid onset, is of tremendous magnitude, and is stimulus limited. In rats, the PRL response to a litter of hungry pups is detectable within 3–5 min of the onset of suckling and attains peak values some 30- to 40-fold greater than baseline at 20–30 min of suckling. Selmanoff and Wise first reported that ME DA turnover decreased during an acute 30-min suckling episode at 10 but not 20 days postpartum, when the response is markedly blunted or absent. This finding was verified in both medial and lateral aspects of the ME and has been corroborated by three other groups. Electrical stimulation of a mammary nerve trunk in a lactating rat produces a modest 5- to 10-fold PRL increase, but despite the diminished PRL response in this anesthetized preparation, this simulated suckling stimulus produces 50–70% decreases in portal blood DA concentrations collected over 10- to 20-min intervals. Finally, Rondeel and coworkers reported a 55% decrease in suckling-induced DA released into push–pull perfusates of the arcuate–median eminence region of awake, freely behaving lactating rats.

That DA withdrawal provokes rapid and marked increases in PRL secretion from lactotrophs has been demonstrated unequivocally in dispersed pituitary cells continuously perfused on columns or coverslips. DA removal from the perfusate provokes rapid 8- to 10-fold increases in PRL release within 30 sec.

Finally, the quantities of PRL secreted during suckling episodes are comparable only to the quantities secreted by the pituitary gland when removed from dopaminergic inhibition. To date, no putative PRF causes PRL release anywhere near the magnitude of the suckling-induced PRL response, whereas antidopaminergics do. Serum PRL data presented by Selmanoff and Wise indicate that blockade of the dopaminergic mechanism in lactating rats with α-methyl-para-tyrosine fully accounts for the PRL levels observed with suckling. Taken together, these findings are consistent with the hypothesis that the dopaminergic mechanism alone mediates suckling-induced PRL release.

V. OTHER PROLACTIN-INHIBITING FACTORS

Perhaps the greatest amount of evidence for a hypothalamic PIF other than DA is for the inhibitory neurotransmitter γ-aminobutyric acid (GABA). GABA was first proposed to be a PIF by Schally and coworkers and a tuberoinfundibular GABAergic (TIGA) system of neurons exists which could be neurosecretory in nature. Four groups have demonstrated direct GABA inhibition of PRL secretion at the pituitary and GABA receptor binding has been observed in the rat and human anterior pituitary gland. However, portal blood GABA concentrations have not consistently been found to be greater than peripheral blood levels, and GABA is about 100 times less potent than DA in inhibiting PRL secretion. Hence, its physiologic role remains enigmatic. The gonadotropin-associated peptide (GAP) was initially reported to exhibit PIF activity; however, significant theoretical problems exist in accepting GAP as a PIF, and other workers have not been able to replicate the original findings. Some evidence exists for the existence of other PIFs, including somatostatin and endothelin.

See Also the Following Articles

GnRH (Gonadotropin-Releasing Hormone); Prolactin, Actions of; Prolactin Secretion, Regulation of

Bibliography

Ben-Jonathan, N. (1985). Dopamine: A prolactin-inhibiting hormone. *Endocr. Rev.* **6**, 564–589.

Ben-Jonathan, N., Oliver, C., Weiner, H. J., Mical, R. S., and Porter, J. C. (1977). Dopamine in hypophysial portal plasma of the rat during the estrous cycle and throughout pregnancy. *Endocrinology* **100**, 452–458.

Ben-Jonathan, N., Laudon, M., and Garris, P. A. (1991). Novel aspects of posterior pituitary function: Regulation of prolactin secretion. *Front. Neuroendocrinol.* **12**, 231–277.

Bossá, R., Fumagalli, F., Jaber, M., Giros, B., Gainetdinov, R. R., Wetsel, W. C., Missale, C., and Caron, M. C. (1997). Anterior pituitary hypoplasia and dwarfism in mice lacking the dopamine transporter. *Neuron* **19**, 127–138.

Caron, M., Beaulieu, M., Raymond, V., Gagne, B., Drouin, J., Lefkowitz, R. J., and Labrie, F. (1978). Dopaminergic receptors in the anterior pituitary gland. *J. Biol. Chem.* **253**, 2244–2253.

Chiu, S., and Wise, P. M. (1994). Prolactin receptor mRNA localization in the hypothalamus by *in situ* hybridization. *J. Neuroendocrinol.* **6**, 191–199.

Crowley, W. R., Shyr, S.-W., Kacsoh, B., and Grosvenor, C. E. (1987). Evidence for stimulatory noradrenergic and inhibitory dopaminergic regulation of oxytocin release in the lactating rat. *Endocrinology* **121**, 14–20.

Demarest, K. T., McKay, D. W., Riegle, G. D., and Moore, K. E. (1981). Sexual differences in tuberoinfundibular dopamine nerve activity induced by neonatal androgen exposure. *Neuroendocrinology* **32**, 108–113.

Demarest, K. T., McKay, D. W., Riegle, G. D., and Moore, K. E. (1984). Biochemical indices of tuberoinfundibular dopaminergic neuronal activity during lactation: A lack of response to prolactin. *Neuroendocrinology* **36**, 130–137.

Frantz, A. G. (1973). The regulation of prolactin secretion in humans. In *Frontiers in Neuroendocrinology* (W. F. Ganong and L. Martini, Eds.), pp. 337–374. Oxford Univ. Press, New York.

Fuxe, K., Agnati, L. F., Kalia, M., Goldstein, M., Andersson, K., and Härfstrand, A. (1985a). Dopaminergic systems in the brain and pituitary. In *Basic and Clinical Aspects of Neuroscience: The Dopaminergic Systems* (E. Flückiger, E. E. Müller, and M. O. Thorner, Eds.), pp. 11–25. Springer-Verlag, Berlin.

Fuxe, K., Agnati, L. F., Calza, L., Andersson, K., Giardino, L., Benfenati, F., Camurri, M., and Goldstein, M. (1985b). Quantitative chemical neuroanatomy gives new insights into the catecholamine regulation of the peptidergic neurons projecting into the median eminence. In *Catecholamines: Neuropharmacology and Central Nervous System—Theoretical Aspects, Neurology and Neurobiology* (E. Usdin, A. Carlsson, A. Dahlström, and J. Engel, Eds.), Vol. 8B, pp. 441–449. A. R. Liss, New York.

Gudelsky, G. A., and Porter, J. C. (1981). Sex-related difference in the release of dopamine into hypophysial portal blood. *Endocrinology* **109**, 1394–1398.

Jaffe, R. B. (1981). Physiologic and pathophysiologic aspects of prolactin production in humans. In *Prolactin* (R. B. Jaffe, Ed.), pp. 181–217. Elsevier, Amsterdam.

Kobayashi, H., and Matsui, T. (1969). Fine structure of the median eminence and its functional significance. In *Frontiers in Neuroendocrinology* (W. R. Ganong and L. Martini, Eds.), pp. 3–46. Oxford Univ. Press, New York.

Kordon, C., Drouva, S. V., Martínez de la Escalera, G., and Weiner, R. I. (1994). Role of classic and peptide neuromediators in the neuroendocrine regulation of luteinizing hormone and prolactin. In *The Physiology of Reproduction* (E. Knobil and J. Neill, Eds.), Vol. 1, pp. 1621–1682. Raven Press, New York.

MacLeod, R. M. (1976). Regulation of prolactin secretion. *Front. Neuroendocrinol.* **4**, 169–194.

Martínez de la Escalera, G., and Weiner, R. I. (1992). Dissociation of dopamine from its receptor as a signal in the pleiotropic hypothalamic regulation of prolactin secretion. *Endocr. Rev.* **13**, 241–255.

Meites, J., Lu, K.-H., Wuttke, W., Welsch, C. W., Nagasawa, H., and Quadri, S. K. (1972). Recent studies on functions and control of prolactin secretion in rats. *Recent Prog. Horm. Res.* **38**, 471–526.

Moore, K. E., and Lookingland, K. J. (1995). Dopaminergic neuronal systems in the hypothalamus. In *Psychopharmacology: The Fourth Generation of Progress* (F. E. Bloom and D. J. Kupfer, Eds.), pp. 245–256. Raven Press, New York.

Neill, J. D. (1980). Neuroendocrine regulation of prolactin secretion. *Front. Neuroendocrinol.* **6**, 129–155.

Neill, J. D., and Nagy, G. M. (1994). Prolactin secretion and its control. In *The Physiology of Reproduction* (E. Knobil and J. Neill, Eds.), Vol. 1, pp. 1833–1860. Raven Press, New York.

Park, S.-K., and Selmanoff, M. (1991). Dose-dependent suppression of postcastration luteinizing hormone secretion exerted by exogenous prolactin administration in male rats: A model for studying hyperprolactinemic hypogonadism. *Neuroendocrinology* **53**, 404–410.

Reymond, M. J., Speciale, S. G., and Porter, J. C. (1983). Dopamine in plasma of lateral and medial hypophysial portal vessels: Evidence for regional variation in the release of hypothalamic dopamine into hypophysial portal blood. *Endocrinology* **112**, 1958–1963.

Rondeel, J. M. M., de Greef, W. J., Visser, T. J., and Voogt, J. L. (1988). Effect of suckling on the *in vivo* release of thyrotropin-releasing hormone, dopamine and adrenaline in the lactating rat. *Neuroendocrinology* **118**, 287–294.

Schally, A. V., Redding, T. W., Arimura, A., Dupont, A., and Linthicum, G. L. (1977). Isolation of gamma-amino butyric acid from pig hypothalami and demonstration of its prolactin release-inhibiting (PIF) activity *in vivo* and *in vitro*. *Endocrinology* **100**, 681–691.

Seeburg, P. H., Mason, A. J., Stewart, T. A., and Nikolics, K. (1987). The mammalian GnRH gene and its pivotal role in reproduction. *Recent Prog. Horm. Res.* **43**, 69–98.

Selmanoff, M., (1981). The lateral and medial median eminence: Distribution of dopamine, norepinephrine, and luteinizing hormone-releasing hormone and the effect of prolactin on catecholamine turnover. *Endocrinology* **108**, 1716–1722.

Selmanoff, M. (1985). Rapid effects of hyperprolactinemia on basal prolactin secretion and dopamine turnover in the medial and lateral median eminence. *Endocrinology* **116**, 1943–1952.

Selmanoff, M., and Gregerson, K. A. (1985). Suckling decreases dopamine turnover in both medial and lateral aspects of the median eminence of the rat. *Neurosci. Lett.* **57**, 25–30.

Selmanoff, M., and Wise, P. M. (1981). Decreased dopamine turnover in the median eminence in response to suckling in the lactating rat. *Brain Res.* **212**, 101–115.

Shaar, C. J., and Clemens, J. A. (1974). The role of catecholamines in the release of anterior pituitary prolactin *in vitro*. *Endocrinology* **95**, 1202–1212.

van Maanen, J. H., and Smelik, P. G. (1968). Induction of pseudopregnancy in rats following local depletion of monoamines in the median eminence of the hypothalamus. *Neuroendocrinology* **3**, 177–186.

Weiner, R. I., and Ganong, W. F. (1978). Role of brain monoamines and histamine in regulation of anterior pituitary secretion. *Physiol. Rev.* **58**, 905–976.

Weiner, R. I., Cronin, M. J., Cheung, C. Y., Annunziato, L., Faure, N., and Goldsmith, P. C. (1979). Dopamine: A prolactin inhibitory hormone. In *Neuroendocrine Correlates in Neurology and Psychiatry* (E. E. Müller and A. Agnoli, Eds.), pp. 41–55. Elsevier, Amsterdam.

Prolactin in Nonmammals

Gregory M. Weber
North Carolina State University

E. Gordon Grau
University of Hawaii at Manoa and Hawaii Institute of Marine Biology

GLOSSARY

brood pouch Section of the posterior abdominal skin of some birds which becomes defeathered, highly vascularized, and endematous around egg laying.

crop milk Substance derived from cells of the mucosal epithelial lining of the crop sac that accumulate lipid and then degenerate and are eventually sloughed off in a cheesy mass.

crop sac A chamber located at the end of the esophogus connecting to the stomach that is used for the storage of food in some birds.

vitellogenesis Processes of vitellogenin synthesis by the liver and uptake and cleavage into yolk proteins by the oocyte.

vitellogenin A lipoglycophosphoprotein precursor of yolk that is synthesized by the liver and released into the bloodstream.

In nonmammalian vertebrates, prolactin (PRL) has many reported reproductive actions, including effects on gonads, vitellogenin synthesis, the development of sexual and parental accessory organs and structures, and sexual and parental behaviors. Actions of PRL in reproduction are diverse and they display a high degree of variability even down to the species level. Even the regulation of changes in PRL cell activity varies from one species to another. Our knowledge about PRL in reproduction in nonmammalian vertebrates is not evenly distributed among these vertebrate groups. There is extensive knowledge about birds, with more studies on turkeys alone than on reptiles. There is also growing information on PRL's involvement in amphibian courting behavior and the development of related sexual accessories and production of pheromones. While studies on fishes are many, few are in-depth.

I. INTRODUCTION

Versatility of action, a prolactin (PRL) hallmark, dissuades one from making generalizations about the role of PRL in reproduction in nonmammals. Since there are over 20,000 species of teleosts alone, it is not surprising that there is a wide diversity of actions of PRL in reproduction among nonmammals, and even within each vertebrate class. Tied to the variety of actions is the diversity in PRL cell regulation and in requirements for target response. Prolactin cell activity is known to be regulated by a myriad of external and internal stimuli. This activity is affected by photoperiod and temperature in many species. Prolactin cell activity in an individual is often influenced by other members of its species; these influences include physical and visual contact with eggs, offspring, and mates. Previous sexual or parental experience has also been reported to be a factor determining the response to exogenous PRL treatment. Hormonal regulators of PRL cells include hypothalamic, hypophysial, and gonadal hormones. Prolactin has also evolved many synergistic, supportive, and antagonistic relationships with other hormones, and thus changes in the hormone milieu can dramatically affect PRL's response. Prolactin has sites of action in contact with the systemic circulation and in the brain. Diurnal changes in PRL levels and responsiveness of tissues have also been reported. Furthermore, many putative PRL-influenced events occur simultaneously, making it difficult to specify the basis for changes in PRL levels over these time periods.

II. SEXUAL PHASE

Prolactin has been reported to have actions in nonmammalian vertebrates pertaining to many components of the sexual phase of reproduction. The sexual phase of reproduction comprises all activities leading to fertilization in oviparous animals or parturition in ovoviviparous and viviparous animals. These actions include effects on development of gonads, sexual accessory organs, and secondary sexual characteristics and effects on behavior including migration to a breeding ground, courtship, and nest building.

A. Gonads

1. Birds

In birds, PRL generally has antigonadal actions, shutting down gonadal function as part of the transition from egg laying to parental care, which includes incubation and brooding. Prolactin treatment induces gonadal regression in most species of birds studied. Often this can be overcome by the addition of follicle-stimulating hormone (FSH), suggesting that PRL acts in part by decreasing gonadotropin (GtH) release. Luteinizing hormone (LH) can be increased by injections of PRL antiserum or antiserum to the PRL-releasing factor, vasoactive intestinal peptide (VIP). Prolactin's inhibition of GtH release appears to be controlled in part through the hypothalamus. Prolactin injections have been shown to decrease gonadotropin-releasing hormone (GnRH) levels in the hypothalamus of laying turkeys but do not appear to affect the ability of GnRH to stimulate the *in vitro* release of GtH from pituitary glands obtained from incubating turkeys. Prolactin may therefore act through the central nervous system to inhibit GtH release. This notion is supported by the finding that in ring doves, intracerebroventricular administration of PRL is effective at doses below those required for peripheral administration to inhibit GtH secretion and gonadal activity. Receptor-binding studies have identified saturable, high-affinity binding sites for PRL in the brains of several avian species. There is evidence that blood-borne PRL gains access to the cerebrospinal fluid via receptor-mediated transport processes in the choriod plexus. Whether PRL-like molecules may also be synthesized in the central nervous system of nonmammalian vertebrates has not been determined.

There is evidence that PRL also acts at the gonad to block GtH-induced steroid synthesis. In the Gifujidori hen, PRL blocks *in vitro* estradiol-17β (E_2) synthesis by the numerous smaller ovarian follicles and not by the larger follicles. Prolactin injections in the turkey were found to increase the number of atretic follicles and to inhibit cytochrome P450 aromatase gene expression in small white follicles, suggesting that PRL reduces E_2 production by inhibiting this gene. It has been reported that small white follicles of the turkey have greater levels of PRL receptor mRNA than do larger follicles.

Some species of birds appear to lack gonadal response to PRL, and others display stimulatory or inhibitory effects, depending on the time of day of administration. Injections of PRL failed to alter ovarian growth in response to GtH injections or photoperiod stimulation in white-crowned sparrow, *Zonotrichia leucophrys gambelii*. In the photorefractory white-throated sparrow, *Zonotrichia albicollis*, PRL injections have been reported to inhibit ovarian responses to exogenous FSH and LH when given early in the day but to enhance the gonadal response if the PRL is administered at midday.

2. *Reptiles*

Prolactin has been shown to inhibit the actions of GtH on gonadal tissues in reptiles. Prolactin inhibits ovarian growth stimulated by GtH administration in the lizards *Diposaurus dorsalis*, *Phrynosoma cornutum*, and *Anolis carolinensis*. Attenuation by PRL is restricted to vitellogenic females in *A. carolinensis*. These reports are consistent with evidence that PRL inhibits production of the egg yolk precursor, vitellogenin, by the liver. Administration of mammalian and avian PRL failed to reduce blood progesterone levels in the turtle, *Chrysemys picta*, despite their ability to decrease LH-stimulated progesterone synthesis from cultured ovaries. Prolactin had no effect on the testis of *A. carolinensis*.

3. *Amphibians*

There is little direct evidence that PRL has a role in gonadal development in anurans. Nevertheless, seasonal changes in plasma levels of PRL appear to coincide with changes in ovarian growth in the green frog, *Rana esculenta*. Annual patterns of change in serum PRL differ between the sexes in *R. esculenta*, with PRL levels in males positively correlated with changes in androgen levels. Prolactin may stimulate gonadal development in part by increasing the sensitivity of GtH cells to GnRH via paracrine action. Prolactin cells are often in contact with GtH cells in the anuran pituitary, and *in vitro* studies have revealed that PRL added to the culture medium with enzymatically dispersed anterior pituitary cells of the bullfrog, *Rana catesbeiana*, increase LH release into the medium in response to GnRH. There is evidence that PRL may be an important hormone for production of vitellogenin in anurans.

As in many birds, PRL treatment inhibits gonadal development in the crested newt, *Triturus cristatus carnifex* Lauranti, and this action is reversible by subsequent FSH administration or blocked by simultaneous FSH administration. Reduction in testosterone (T) has been observed in response to PRL injections in this species. On the other hand, T, LH, and GnRH have been reported to increase PRL blood levels. As in the green frog, annual patterns of change in plasma PRL differ between the sexes of the crested newt. In males, PRL and T reach maximal levels during the reproductive period, possibly acting together to induce courtship behavior and development of pheromone-producing structures. In females, PRL peaks just prior to increases in blood vitellogenin levels. Prolactin actions on gonadal development have received little attention in other species of urodeles.

4. *Fishes*

Gonadotropic and antigonadal effects of PRL treatment have been reported in teleosts. Mammalian PRL has been reported to have antigonadal actions on ovary and testis in golden shiner, *Notemigonus cryso-*

leucas, and ovary of the Indian catfish, *Heteropneustes fossilis* (Bloch). There were temperature-dependent, diurnal changes in PRL effectiveness on gonadal maturation in the golden shiner. Although in fish maintained at 24°C, PRL injections given early or late in the light phase repressed gonadal maturation, they had no effect in fish maintained at 15°C when administered early in the light phase. In most other studies PRL has shown gonadotropic effects and stimulated steroid production. One of the earliest reports of steroidogenic actions in fishes was the stimulation of 3β-hydroxy-Δ^5-steroid in the ovaries of the cichlid *Aequidens pulcher*. Since these early studies, purified fish PRLs have become available. Injections of purified salmon PRL into hypophysectomized male killifish, *Fundulus heteroclitus*, increased plasma concentrations of T but had no effect on *in vitro* steroid production. Furthermore, the injections prevented the decline in gonadal weight in both males and females following hypophysectomy. Salmon PRL was also found to potentiate GtH-induced E_2 and 17α-hydroxy,20β-dihydroprogesterone release from rainbow trout ovarian follicles *in vitro*. Conversely, injections of the progestin reduced PRL levels in immature rainbow trout, suggesting a negative feedback mechanism. Estradiol-17β and T injections were ineffective. In addition, an inverse relationship between blood 17α-hydroxy,20β-dihydroprogesterone and PRL following ovulation was observed in female rainbow trout and chum salmon. One of two PRL variants purified from the tilapia, *Oreochromis mossambicus*, stimulated E_2 synthesis by vitellogenic follicles of the guppy, *Poecilia reticulata*. Nevertheless, no changes in PRL cell activity were observed over the reproductive cycle of a closely related species, *Poecilia latipinna*. In a study using homologous hormones, both PRL variants were found to affect T production by testes of tilapia *in vitro*. The tilapia PRLs stimulated T production by testes from courting males but not from noncourting males. *In vitro* pretreatment with PRL, followed by LH exposure, enhanced T production by testes from courting males but attenuated production from minced testis from noncourting males. The effect of the PRLs appears to be direct. High levels of specific PRL binding sites have been found in the ovaries and testes of tilapia, and PRL–receptor mRNA was found in tilapia testes using a probe based on homologous PRL–receptor DNA.

B. Sexual Accessory Organs and Secondary Sexual Characteristics

1. *Birds*

In birds, PRL's actions on the oviduct generally parallel its effects on gonads, suggesting action mediated through sex steroids or sensitivity to sex steroids. Prolactin decreases oviduct weight in domestic hens and the turkey. In the turkey, oviduct regression was observed in response to the infusion of VIP into the third ventricle of the brain, which simultaneously elevated blood PRL levels. Prolactin counteracted GtH effects on the ovary and oviduct in photorefractory white-throated sparrows but only on oviduct growth in white-crowned sparrows. However, ovine PRL injections did not affect increases in ovary or oviduct weight in response to photostimulation or exogenous GtH administration in photostimulated white-crowned sparrows, suggesting a possible seasonal change in response to PRL. On the other hand, PRL administration has been shown to promote oviduct development in several avian species including canaries. We know of no report of PRL's effects on sexual accessory organs in male birds. In cockerels, PRL injections decrease comb size, a secondary sexual characteristic.

2. *Reptiles*

Reports on effects of PRL on sexual accessories in reptiles are scarce. Prolactin was found to act synergistically with steroid hormones to stimulate oviduct development in the Texas horned lizard, *P. cornutum*, but in *A. carolinensis* it depressed oviduct weight, and it countered increases in oviduct weight induced by exogenous GtH treatment in the lizard *D. dorsalis*. In male lizards, PRL or FSH stimulated development of the epididymis in *Lacerta sicula campestris*, whereas PRL was without effect on accessory organs of *A. carolinensis*.

3. *Amphibians*

One of the earliest reports of a reproductive effect of PRL in amphibians was that of PRL stimulating oviducal jelly in the toad, *Bufo arenarum*. This effect was not observed in two other anurans, *R. pipiens* and *R. catesbeiana*. Recently, PRL, measured by homologous radioimmunoassay, has been found to be elevated in some but not all female toads, *Bufo japonicus,* near breeding ponds during the breeding season, suggesting that PRL does not induce migration. Some of these females ovulate before they reach the pond. It was suggested that elevated PRL may have stimulated oviducal jelly secretion, although no direct studies were conducted. Oviduct development and oviducal jelly secretion also appear to require PRL in the red-bellied newt, *Cynops pyrrhogaster*. Estradiol-17β and T were not able to induce complete development of the oviduct without PRL. Oviducal jelly in the oviduct was stimulated much more by a combination of E_2 and PRL than by either hormone alone. It is interesting that PRL was reported to increase oviducal jelly synthesis in a toad and a newt, both of which change their normal terrestrial-adapted physiology for a temporary stay in fresh water at breeding time, but PRL has no effect on *Rana* species, which normally live in an aquatic environment. Perhaps making oviducal jelly requires osmotic regulation by PRL in terrestrial amphibians.

Prolactin has pronounced effects on physical features related to courtship in urodeles. The development of these structures is also influenced by steroid sex hormones. Nuptial pad development in male red-spotted newts, *Notophthalmus viridescens*, appears to be regulated by PRL, androgen, and thyroid hormones. Nuptial color in the tail of male red-bellied newts is regulated by PRL and androgen. An increase in male-characteristic tail height can be induced with PRL treatment and inhibited with E_2 in the red-bellied newt and the sword-tailed newt, *Cynops ensicauda*. Prolactin and androgens act synergistically in the development of the abdominal gland and its production and secretion of a female-attracting pheromone in male red-bellied and red-spotted newts. The pheromone from the red-bellied newt is a decapeptide, sodefrin, and is the first peptide pheromone reported in a vertebrate.

4. Fishes

Prolactin synergizes with T or testosterone propionate (TP) to stimulate seminal vesicles of hypophysectomized, castrated, and sexually inactive Indian catfish (*H. fossilis* Bloch) and hypophysectomized and castrated gobies (*Gillichthys mirabilis*). In hypophysectomized gobies, the seminal vesicles were partially maintained by PRL or TP replacement therapy separately, whereas their combined treatment resulted in complete maintenance. Luteinizing hormone, human chorionic gonadotropin (hCG), and C-21 steroids, but not PRL or FSH, caused ovipositor elongation in the rose bitterling, *Rhodeus ocellatus ocellatus*.

C. Vitellogenin Synthesis

1. Birds

Prolactin may promote ovarian growth through actions on vitellogenin synthesis. The egg yolk precursor, vitellogenin (VTG), is produced by the liver of females in response to E_2 in all vertebrates that make yolk. Evidence suggests prolactin and/or growth hormone (GH) may be modulators or synergists for this process in some species. In birds, E_2 could not induce VTG synthesis by hepatocytes in a primary monolayer culture system that employed cells obtained from a young male domestic fowl. However, ovine PRL or bovine GH alone or in combination with E_2 induced the synthesis and secretion of VTG into the culture medium. It is interesting that in the chicken, PRL, a hormone associated with gonadal regression, is also important for VTG synthesis. One possible explanation for this apparent inconsistency is that PRL is working through GH receptors and it is GH that normally supports VTG synthesis.

2. Reptiles

In reptiles, GH appears to be a necessary synergist for VTG synthesis, whereas PRL appears to inhibit VTG synthesis, at least in lizards and turtles. Estradiol-17β injections are able to induce VTG synthesis in male turtles. Vitellogenin can be induced *in vitro* from liver explants from female turtles, and from male turtles only if they are first treated with E_2 *in vivo*. Even the addition of GH to the culture medium with E_2 could not induce the synthesis of VTG from liver explants of intact male turtles. However, even though E_2 or GH separately could not induce VTG synthesis in liver explants from hypophysectomized males, whether they had intact gonads or were castrated, the combination of E_2 and GH was effective. Thus, GH is necessary for E_2-induced VTG synthesis, and another pituitary factor or pituitary-induced nongonadal factor appears to override this stimulation. This factor is believed to be PRL since PRL inhibits E_2-induced vitellogenesis in the turtle and normal seasonal ovarian development and vitellogenesis in lizards.

3. *Amphibians*

Bullfrog (*R. catesbeiana*) PRL and bullfrog GH stimulate VTG synthesis *in vitro* from liver explants from the green frog, *R. esculenta*. Estradiol-17β stimulates VTG synthesis by liver explants of both males and females during winter stasis but not during the summer refractory period. Estradiol-17β levels are highest in the plasma of females during the summer, and thus the livers are probably refractory to E_2 induction of VTG at this time. Prolactin induced VTG synthesis in males during both winter and summer phases but only during the winter phase in females. In contrast, GH was active in both sexes during both seasonal phases. Prolactin induced liver explants to release VTG into the medium, and the amounts released were higher for winter males than for summer males and higher for winter males than for winter females. Thus, there are seasonal and sex differences in PRL action on vitellogenesis in *R. esculenta*. In females, pituitary homogenate takes longer to stimulate VTG production than does E_2, and E_2 does not synergize with pituitary homogenate. Therefore, PRL and GH appear to stimulate VTG production via a different mechanism than E_2 in *R. esculenta*. Actions of PRL on vitellogenesis have not been investigated in urodeles, but plasma PRL levels appear to increase in concert with plasma VTG levels in the crested newt when examined by radioimmunoassay based on PRL purified from the red-bellied newt.

4. *Fishes*

Prolactin and GH stimulation of VTG synthesis *in vitro* has been detected in teleost fishes. Vitellogenin synthesis was stimulated from livers of female and male European sea bass, *Dicentrarchus labrax*, by E_2, ovine PRL, or seabream GH, except when the livers came from spawning fish of either sex. Prolactin and GH also had a potentiating effect on VTG synthesis in primary hepatocyte cultures from the Japanese eel, *Anguilla japonica*. Growth hormone was a much more potent synergist than PRL in the induction of VTG synthesis in a primary culture of female silver eel (*A. anguilla* L.) hepatocytes. Several teleost PRLs were tested in this species, and although purified salmon PRL showed some activity, trout and tilapia PRL were without effect. Prolactin-specific receptors in the eel liver have been reported, affirming that the liver is responsive to PRL. The lack of a VTG response to PRL in the silver eel may be due to the use of heterologous hormones, or the PRL receptors in the liver may serve to mediate one of the many other functions of PRL. There are reports of PRL having no action or even inhibitory actions on vitellogenesis in other fish species. For example, PRL has been reported to depress vitellogenesis in an elasmobranch, the dogfish shark.

D. Sexual Behavior

1. *Birds*

Courtship and nest-building behavior appear to be primarily regulated by GtHs and steroids in birds. Prolactin stimulates migration behavior in some species of birds, and elicits other physiological changes associated with migration including increased feeding behavior. Nocturnal restlessness, which is evidence of the onset of migration behavior in species that normally migrate at night, increased feeding behavior, and fat deposition are induced by ovine PRL injections in white-crowned sparrows. Pituitary levels of PRL were elevated in white-crowned sparrows not only prior to migration to the breeding grounds in the spring but also before migration in the fall, well after the breeding season had ended and gonads had regressed. Physiological changes induced by PRL in preparation for migration to a breeding site, such as changes in feeding habits and deposition of fat, or away from a breeding site, such as shutting down reproductive function, might suggest that these changes are reproductive, but they also may be viewed as changes that meet metabolic needs.

2. *Amphibians*

There is strong evidence that PRL induces migration in some urodeles. The need for some salamandrian species to return to water (water drive) to reproduce is one of the classic examples of PRL's effects on behavior. This phenomenon, often referred to as a second metamorphosis, has been studied because the animal must also undergo many physiological changes to survive in the aquatic habitat. Injection of antiserum to newt PRL into aquatic-living newts changed their preference to terrestrial habitats. This effect was reversed by injections of PRL, suggesting the involvement of endogenous PRL in water drive. The toad, *B. japonicus*, also returns to water to reproduce, and PRL serves osmoregulatory functions in this species, but PRL does not appear to initiate a behavior analogous to water drive.

Prolactin appears to be intimately associated with courtship behavior in urodeles. As has already been discussed, PRL is involved with migration to water and the development of accessory sex organs, including nuptial pads and the abdominal gland that produces the female-attracting pheromone. Prolactin is also involved in the development of secondary sexual characteristics, including the appearance of nuptial coloration and an increase in tail fin height in males. It is also worth noting that PRL in combination with E_2 appears to elicit the production and release of a male-attracting pheromone produced within or released through the oviduct of the female newt, even following ovariectomy. Similarly, the female attractant is produced by castrated males in response to a combination of PRL and TP. Sexually inactive animals of both sexes do not normally respond to the attractant of the opposite sex but can be induced to do so by injections of PRL combined with hCG. These studies were conducted using the red-bellied newt, *C. pyrrhogaster*, in which the action of PRL in the regulation of courtship has been most fully described.

Prolactin in conjunction with androgens acts to elicit tail vibrating, a courtship display, in male newts. The induction of this courtship display in castrated, hypophysectomized males requires a combination of androgen and PRL. Furthermore, injection of courting males with antiserum against newt PRL decreased both frequency and incidence of display compared with injection of courting males with preimmune serum. This decrease in courtship was accompanied by a preference for terrestrial habitat. Prolactin in combination with GtH also increases the size of Mauthner neurons, which are involved in tail movement in the red-bellied newt. As with males, treatment of the ovariectomized females with PRL and TP increased male-type courtship display.

3. *Fishes*

Migration from seawater to fresh water for reproduction is accompanied by an increase in PRL cell activity or blood PRL levels in several teleost fishes. Prolactin is a central hormone for osmoregulation of teleost fish in fresh water. It has been observed to increase in sockeye and chum salmon as they approach and enter rivers on their way to spawning grounds. Thus, this increase may be involved in migration behavior as well as osmoregulation. Changes in PRL have also been inversely correlated with changes in progestin levels associated with ovulation in chum salmon, suggesting that changes in PRL during migration may serve a gonadal function or be influenced by gonadal steroids. Similar changes in PRL cell activity have been observed with seawater-to-freshwater spawning migration in the three-spine stickleback, *Gasterosteus aculeatus*. Prolactin cell activity appears to increase in the spring as the animal migrates to fresh water. Injections of PRL induce physiological changes that allow seawater-adapted sticklebacks to live in fresh water. As in salmonids, evidence is not yet sufficient to determine whether the increase in PRL cell activity initiates migration or is tied solely to the animal's freshwater osmoregulatory requirements.

Prolactin facilitates nest building in the paradise fish, *Macropodus opercularis*. The male paradise fish constructs a bubble nest by gulping air and then making bubbles out of air and mucus. Nest-building behavior is increased in the paradise fish by treatment with a combination of PRL with hCG or androgen. The nest-building behavior itself is elicited by treatment with GtH or androgen alone, but PRL increases mucus cells and mucus production, which aids in nest building. While mucus production can be induced by PRL administration in both males and females, the effect is more pronounced in the males, which normally build the nests.

III. PARENTAL PHASE

Prolactin has many reported actions in parenting. Among these are the development of accessory organs and tissue changes, including brood patch and crop sac development in male and female birds, brood pouch development in male seahorses, and the production of integumentary mucus that is used in some fish to feed young. Prolactin has also been found to play a role in the onset or maintenance of incubation and brooding behavior in birds and fishes. We are not aware of evidence that PRL plays a role in the parental phase of reproduction in reptiles or amphibians.

A. Parental Care Accessory Structures and Parental-Derived Nutrition

1. *Birds*

The brood patch is formed at the time of egg laying; it aids in transfer of heat to the eggs. The brood patch is a section of the posterior abdominal skin which becomes defeathered, highly vascularized, and edematous, thus increasing heat production and transfer. Prolactin and sex steroids are required for full development of the brood patch. Both parents in the California quail and laughing gull develop a brood patch, incubate the eggs, and respond to PRL together with either T or E_2 for brood patch formation. Although both starling parents incubate eggs, only E_2 synergizes with PRL to promote brood patch formation in either sex. Testosterone, but not E_2, acts synergistically with PRL in two species of phalaropes in which only the males form a brood patch and incubate eggs. Brood patch formation can be induced in the sex of several species in which that sex does not normally form a brood patch or incubate eggs. The cowbird lays its eggs in the nest of other species for incubation and thus has no need for a brood patch; studies have found that a brood patch is not induced by combined E_2 and PRL treatment in this species.

The crop sac is a large chamber used for food production and storage, located at the end of the esophagus and connecting to the proventriculus, the avian stomach. At about the time of hatching in some passerine birds of the family Columbiformes (pigeons and doves), the two lateral lobes of the crop wall are thickened and begin to produce a substance called crop "milk." Crop milk is derived from cells of the mucosal epithelial lining that accumulate lipid and then begin to degenerate and are eventually sloughed off in a cheesy mass. The crop milk is regurgitated to feed the hatchlings. This ability of PRL to cause hyperplasia of the crop sac mucosa is the best known action of PRL in birds and has served to provide the most widely used bioassay of PRL activity. Specific binding of PRL to the crop sac, localized in the epithelium, has been reported. There is evidence that cyclic AMP may mediate PRL's actions on the crop sac. Recently, it has been found that PRL acts on the liver to cause the release of a factor called synlactin. Synlactin synergizes with PRL to bring about effects in target tissues. Among these effects is hyperplasia of the crop sac mucosal epithelium.

2. *Fishes*

The male seahorse, *Hippocampus guttulatus*, incubates its eggs in a ventral pouch. Development of the pouch is impaired by castration or hypophysectomy. Prolactin works in conjunction with adrenal cortical-stimulating hormone and corticoids to maintain the pouch. Adrenal cortical-stimulating hormone and corticoids maintain the structure of the pouch, and PRL acts to stimulate proliferation of its epithelial lining. The action of PRL may be in part mediated through corticosterones since PRL can stimulate the adrenal equivalent, the interrenal gland, in this species.

Integumental mucus secretion is stimulated in response to PRL in many teleost fishes. In some fishes, mucus is a source of nourishment for the young. In these species, mucus increases during brooding, and the young feed upon the mucus on the parent's body soon after hatching. This effect of PRL was first described in the discus fish, *Symphysodon discus*. Prolactin has even been detected in the mucus of the Midas cichlid, *Cichlasoma citrinellum*, and studies suggest that it promotes growth in the young. Prolactin's ability to stimulate production of integumentary mucus has been tested only in intact animals and so it is not known if PRL synergizes with other hormones in this action.

B. Parental Care Behavior

1. Birds

Evidence that PRL is involved in parental care in birds derives from many lines of study. Exogenous PRL effects on incubation and brooding have been demonstrated in numerous species employing a variety of procedures. Much controversy surrounds the role of PRL in the onset of incubation in birds, although there is strong evidence that PRL is at least involved in the maintenance of this behavior. Systemic injection as well as infusions of PRL into the brain induce incubation in laying birds and extend the length of incubation. The hypothesis that PRL may act in the brain to affect parental behavior is supported by reports that PRL-binding activity is concentrated in the preoptic area of the hypothalamus of the ring dove and several song birds; this area has been implicated in the regulation of parental behavior in birds. Moreover, PRL binding is lower in the preoptic area of the brown-headed cowbird (*Molothrus ater*), which does not exhibit parental care, than in that of two nesting song birds, the red-winged blackbird (*Agelaius phoeniceus*) and the European starling (*Sturnus vulgaris*).

There are many examples in which PRL is elevated in the pituitary or in the blood during periods of incubation and brooding. Unfortunately, it is difficult to determine whether these changes in PRL are associated with the behavior per se or with other PRL-influenced processes which may be occurring at the same time, such as the development of the brood patch, production of crop milk, or gonadal regression. Elevated PRL levels can be extended in canaries and Wilson's phalaropes by extending egg incubation time, suggesting a positive feedback by egg contact or nesting behavior on PRL release. This feedback is not shared by the gray-headed albatross or pied flycatcher. In both of these species PRL release is not extended by delayed hatching, suggesting an endogenous control over the duration of PRL release. In pied flycatchers, PRL levels drop in response to a delay in hatching but rebound once the females are exposed to newly hatched young. The length of time PRL levels remain elevated after hatching in this species can be extended by repeatedly replacing older hatchlings with younger hatchlings. Similarly, PRL levels are reduced early if brooding is terminated early. Thus, there is flexibility in the response of the pied flycatcher to the exogenous stimuli of hatchlings. Interestingly, the length of time that elevated PRL levels were maintained in the males, which do not brood but do feed the young, was decreased by early termination of brooding by the female but was not affected by prolonged exposure of the female to hatchlings. This effect on the males is similar to the effect that the sight of incubating female pigeons and doves has on maintaining elevated PRL in the males of these species. Feeding of young by nonbrooding parents is not accompanied by elevated PRL levels in many avian species. It is worth noting that although replacement of hatchlings can extend elevated PRL levels for up to 12 days in female pied flycatchers, it can extend parental care even longer—up to 23 days. In many avian species, PRL levels drop after hatching. Thus, sustained elevation in PRL levels is not necessary for the continuation of parental care.

The regulation of PRL cell activity associated with increasing photoperiod and the onset of incubation has been investigated in several avian species. Regulation is best characterized for domestic fowl, particularly the turkey hen. Prolactin rises slowly in the hen with increasing photoperiod, then abruptly increases at the onset of incubation. Photostimulation initiates recruitment and hypertrophy of lactotrophs, increases in hypothalamic VIP content, the number of hypothalamic VIP immunoreactive neurons, and VIP binding sites on the anterior pituitary. The transition from laying to incubation in the turkey appears to occur because serotonin causes an increase in VIP release, which then increases PRL. The increase in PRL observed with increasing photoperiod appears to be mediated through photoperiod effects on serotonin's ability to affect VIP release. This regulation results in substantial increases in pituitary and blood PRL as well as PRL mRNA in incubating hens compared with laying hens.

2. Fishes

Parental care behavior has been attributed to PRL in some teleosts. Fanning of eggs, which increases water circulation, appears to be regulated by PRL in sticklebacks, bluegill sunfish, a wrasse, and some cichlid species. In cichlids, "calling movements" to fry and reduced feeding (suggested to prevent the cannibalism of young during brooding) have been attributed to PRL. Most of the support for PRL's involvement in these actions comes from studies in which mammalian PRLs have been injected into the fish and the specific behavior has then been observed to be induced or increased. Tilapia PRL was found to be effective in stimulating fanning behavior in other cichlids, and implantation of a second pituitary lobe containing a nearly homogeneous population of PRL cells increased fanning behavior in sticklebacks. Prolactin cell activity was shown to be increased in male three-spined sticklebacks displaying fanning behavior, based on ultrastructure morphometry and the [^{3}H]-lysine incorporation rate of PRL cells, providing further evidence that PRL influences fanning behavior. Studies on PRL-induced parental behavior have not been conducted using hypophysectomized fish. Thus, any of several hormones may mediate or synergize these actions. For example, induction of fanning requires cotreatment with PRL and gonadal steroids in several species.

See Also the Following Articles

Amphibian Reproduction, Overview; Avian Reproduction, Overview; Fish, Modes of Reproduction in; Prolactin, Overview; Reptilian Reproduction, Overview

Bibliography

Buntin, J. D. (1993). Prolactin–brain interactions and reproductive function. *Am. Zool.* **33**, 229–243.

El Halawani, M. E., Mauro, L. J., Phillips, R. E., and Youngren, O. M. (1990). Neuroendocrine control of prolactin and incubation behavior in gallinaceous birds. In *Progress in Comparative Endocrinology* (A. Epple, C. G. Scanes, and M. H. Stetson, Eds.), pp. 678–684. Wiley-Liss, New York.

Goldsmith, A. R. (1983). Prolactin in avian reproductive cycles. In *Hormones and Behaviour in Higher Vertebrates* (J. Balthazart, R. Prove, and R. Gilles, Eds.), pp. 375–387. Springer-Verlag, Berlin.

Mazzi, V., and Vellano, C. (1987). Prolactin and reproduction. In *Hormones and Reproduction in Fishes, Amphibians, and Reptiles* (D. O. Norris and R. E. Jones, Eds.), pp. 87–115. Plenum, New York.

Nicoll, C. S. (1974). Physiological actions of prolactin. In *Handbook of Physiology, Endocrinology, Vol. 4, Part 2: The Pituitary Gland and Its Neuroendocrine Control* (E. Knobil and W. H. Sawyer, Eds.), pp. 253–292. Williams & Wilkins, Baltimore.

Polzonetti-Magni, A., Carnevali, O., Yamamoto, K., and Kikuyama, S. (1995). Growth hormone and prolactin in amphibian reproduction. *Zool. Sci.* **12**, 683–694.

Prolactin Secretion, Regulation of

György M. Nagy, Pál Gööz, Katalin M. Horváth, and Béla E. Tóth
Semmelweis University Medical School

GLOSSARY

dopamine A catecholamine with both neurohormone and neurotransmitter properties which is produced by the hypothalamic–arcuate nucleus and is an established prolactin-inhibiting factor.

paracrine/autocrine regulation A mechanism by which cells can send a signal to a neighboring target cell (paracrine signaling) or to itself, in which case the signal molecule and its receptor are produced by the same cell (autocrine signaling).

prolactin A single polypeptide chain composed of 197–199 amino acids with three disulfide bonds. There is only one prolactin gene in the haploid genome. In the human it is located on chromosome 6.

signal transduction The recognition and binding of a signaling molecule to a complementary receptor protein which can initiate a cascade of specific membrane-bound and intracellular events; thus, each cell will respond in a programmed and characteristic way.

Prolactin is one of the most versatile pituitary hormones in both number and diversity of the physiological properties it regulates. Prolactin plays a decisive role in preparation and maintenance of mammary gland milk production, maternal behavior during lactation, and normal immune function. Prolactin also has a definitive role in response to stress, osmoregulation, and angiogenesis. Prolactin is synthesized in the anterior lobe of the pituitary gland. However, several prolactin-related proteins are also produced in many mammalian tissues, including the brain, uterus, placenta, and cells of the immune system. The physiological functions of extrahypophysial prolactin-related proteins are not entirely known in most cases; thus, discussion will be limited to the regulation of pituitary prolactin secretion.

I. HYPOTHALAMIC REGULATION OF PROLACTIN SECRETION

There are two major mechanisms by which the hypothalamus can influence prolactin secretion from the pituitary gland: inhibition by a prolactin-inhibiting factor (PIF) and stimulation by a prolactin-releasing factor (PRF).

A. Background

It was first postulated in 1954 by Everett that a factor produced by the hypothalamus released into hypophysial portal blood inhibits pituitary prolactin secretion. Indeed, the existence of a PIF was later demonstrated in hypothalamic extract. Observations, that (i) drugs interfering with catecholamine (including dopamine) synthesis, metabolism, or receptor action altered prolactin secretion and (ii) the presence of high concentrations of dopamine in both the median eminence (ME) and portal blood led investigators, such as R. M. MacLeod, to conclude that dopamine is the major hypothalamic PIF. To date, research has failed to identify peptidic or other non-peptidic candidates as PIFs.

B. Inhibition

Dopamine, the best established hypothalamic hypophysiotrophic inhibitor of prolactin secretion, is produced by neurons in the arcuate–periventricular nucleus of the hypothalamus (Fig. 1). This hypothalamic dopaminergic system can functionally be divided to two different parts. Tuberoinfundibular dopaminergic (TIDA) neurons originate from the medial and dorsal subdivision of the arcuate nucleus in the rostrocaudal direction and terminate near the capillary loops of the ME. After dopamine is released from these terminals, it travels down via the bloodstream of the long portal vessels (LPV) to the anterior lobe (AL). Their activity can be affected by ovarian steroids and by prolactin feedback. Tuberohypophysial dopaminergic (THDA) neurons originate from the most rostral portion of the arcuate nucleus and the periventricular region of the hypothalamus and terminate at the neurointermediate lobe (NIL) of the pituitary gland. Dopamine from these terminals is transported by the short portal vessels (SPV) to the region of the AL adjacent to the NIL and to the transitional territory between AL and stalk-ME (Fig. 1). THDA neurons are extremely sensitive to small changes in osmolarity, i.e., dehydration.

The relative contribution of these two dopaminergic systems in the control of prolactin secretion is still a relevant issue. Involvement of the TIDA system in prolactin regulation is well established; however, the physiological significance of THDA neurons has recently been demonstrated by NIL lobectomy, denervation, or dehydration which have interfered with

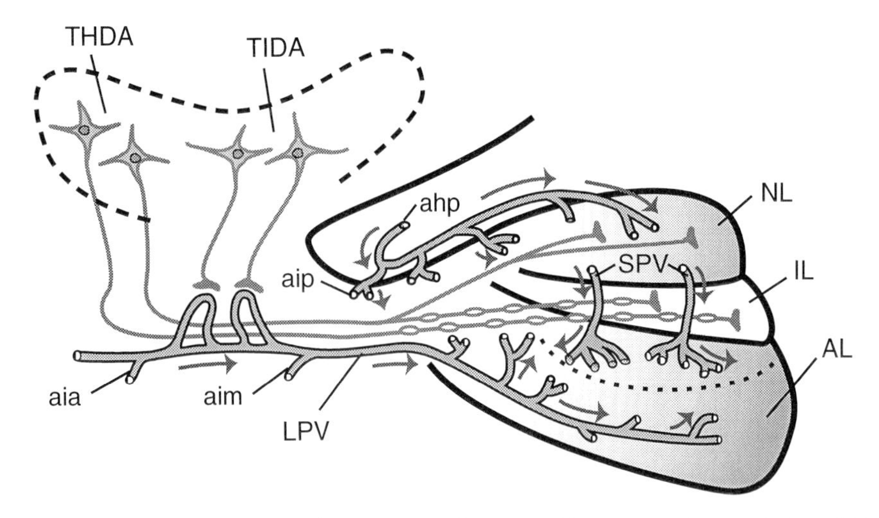

FIGURE 1 Recent model for the hypothalamic- and NIL-neuroendocrine and/or IL-paracrine regulation of AL prolactin secretion. THDA, tuberohypophysial dopaminergic neurons; TIDA, tuberoinfundibular dopaminergic neurons; LPV, long portal vessel; SPV, short portal vessels; NL, neural lobe; IL, intermediate lobe; AL, anterior lobe; aia, anterior infundibular artery; aim, medial infundibular artery; aip, posterior infundibular artery; ahp, posterior hypophysial artery.

neurogenic- (i.e., suckling) and hormonal (i.e., estrogen)-induced prolactin secretion. Thus, dopamine of hypothalamic origin which is delivered to the adenohypophysis by way of LPV and SPV (from TIDA or THDA, respectively) seems quantitatively sufficient to account for the inhibition of prolactin secretion (Fig. 1).

C. Stimulation

One can conclude that by having an established PIF that tonically and permanently suppresses the secretory function of mammotropes, the simplest way to increase prolactin release would be to reduce hypothalamic inhibition. There are conflicting reports about the changes in dopaminergic neuronal activity during surges of prolactin, such as suckling-, stress-, and estrogen-induced responses, suggesting that this is not the case. Comparison of currently available data is almost impossible and has led to the conclusion that disinhibition of dopamine itself cannot account for stimulus-induced prolactin releases.

Recent evidence, obtained by M. E. Freeman and colleagues, strongly suggests that dopamine might also be regarded as a prolactin-releasing agent under certain conditions. Dopamine enhancement of prolactin secretion raises the question of the relevance of this phenomenon. A physiological role is supported by those observations when a brief suckling stimulus in lactating or estradiol treatment in combination with progesterone in ovariectomized (OVX) rats rendered mammotropes to respond with an increase of prolactin release.

In most cases, dopamine concentration in stalk blood is in the range that is mostly ineffective or has only a weak inhibitory effect *in vitro*. One possible explanation for these contradictions is that "responsiveness" agents or mechanisms exist for potentiating dopamine to inhibit as well as stimulate prolactin release. Ascorbic acid, as shown by S. Shin, is a major candidate for the enhancement of inhibition by dopamine. Conversely, S. F. Frawley and colleagues have provided evidence that α-melanocyte-stimulating hormone (α-MSH) from the intermediate lobe (IL) can function as a stimulatory-responsiveness factor *in vitro*. An obvious possibility is that dopamine itself can play this responsiveness regulatory role. Small changes in the local concentration of dopamine may be able to desensitize or sensitize mammotropes at the level of the receptor or somewhere in the signal transduction cascade (*vida infra*). This role of dopamine needs to be investigated.

In addition to dopamine, two major compounds, thyrotropin-releasing hormone (TRH) and vasoactive intestinal peptide (VIP), have been considered as physiologically relevant PRF's of hypothalamic origin. TRH- and VIP-producing nerve cells are localized in the parvicellular portion of the hypothalamic paraventricular nucleus which terminates at the external zone of the ME. They are secreted into the portal blood and have high-affinity membrane receptors on mammotropes. The physiological role of TRH and VIP is fairly puzzling. TRH can stimulate prolactin release both *in vivo* and *in vitro* only when dopaminergic input has been previously reduced or absent. On the other hand, recent findings suggest that VIP is also produced by mammotropes themselves and is established as an autocrine stimulatory factor of prolactin release (*vida infra*). In summary, the physiological significance of these mechanisms and factors will require further exploration.

II. INTRAHYPOPHYSIAL OR LOCAL REGULATION OF PROLACTIN SECRETION

It is generally accepted that the other two parts of the pituitary gland, neural lobe (NL) and IL, have a significant influence on AL prolactin secretion. An interesting feature of the AL is the regionalization of its blood flow. Blood drains from the NL via the SPV to the central zone of the AL (Fig. 1). It is comparable to the LPV found between the ME and the AL. This is the anatomical basis of a neuroendocrine communication between NL and the central zone of the AL. Since most of the SPV only traverse the IL, a paracrine regulatory mechanism can also be in operation between IL and AL. Moreover, many neuropeptides, growth factors, and enzymes have been localized in different subpopulations of AL cells. They can be released into the extracellular fluid and must have some autocrine or paracrine regulatory role affecting different functions of the same or neighboring cells, respectively.

A. Neuroendocrine and Paracrine Regulation by the Neuro-Intermediate Lobe

During the past decade, several experimental findings have emphasized the role of the NIL in the regulation of prolactin secretion. Surgical removal or denervation of the NIL results in an elevation of basal level of plasma prolactin in cycling or lactating female as well as in male rats. It is supposedly caused by the missing inervation of THDA, which tonically inhibits prolactin secretion (*vida supra*). Moreover, NIL removal and denervation blocks or attenuates several well-known stimulus-induced secretory responses of mammotropes (i.e., suckling- and mating-induced nocturnal surges and the peak phase of the proestrous surge), suggesting that a PRF in the NIL exists. These observations led the group of N. Ben-Jonathan to investigate the PRF activity of the NIL. Indeed, perchloric acid extracts of the NIL can release prolactin *in vitro* from cultured AL cells. However, the chemical nature and the site of production (i.e., NL, IL, or both) of the NIL's PRF remain to be elucidated.

The previously mentioned PRF differs from another factor, found only in the IL, that mediates the acute estradiol-induced recruitment of mammotropes (from the non-secretory pool to the secretory one) in OVX rats. It has been proposed that αMSH may be the mammotrope-recruitment factor (MRF). Most likely, MRF diffuses to the AL rather than being released into the SPV. Therefore, it is a paracrine factor. In this respect, it is worth mentioning that MRF can induce mammotropes to release prolactin exclusively from the inner zone of the AL, which is in close vicinity to the IL. Taking into consideration that in humans IL cells do not form a distinct lobe, this paracrine regulatory role may be more obvious.

B. Paracrine Regulation within the Anterior Lobe

Paracrine interactions between different AL cell types have been studied extensively by C. Denef and colleagues. Paracrine interactions with mammotropes have been demonstrated most convincingly

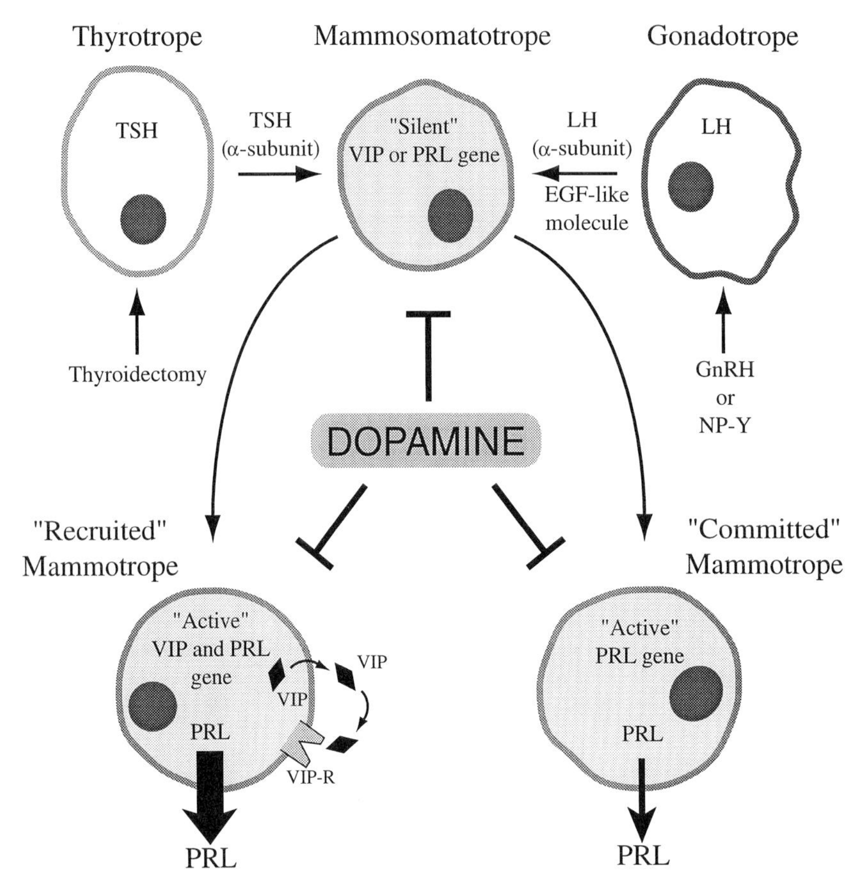

FIGURE 2 Examples for the paracrine communication between thyrotrope vs mammotrope (mediated by α subunit of LH) and thyroidectomy or gonadotrope vs mammotrope (mediated by the α subunit of LH or an EGF-like molecule) during the cytodifferentiation of early postnatal life. The autocrine mode of action of VIP on mammotropes and dominant inhibitory function of dopamine are also indicated. See text for abbreviations and further explanations.

for three cell types: gonadotropes, folliculostellate cells, and corticotropes. Evidence regarding mammotrope and gonadotrope interactions has been collected on reaggregated pituitary cell cultures obtained from animals during the early postnatal life. This paracrine interaction, however, is probably more common and may operate *in* vivo in monkeys and also in women during the menstrual cycle. It has been postulated that angiotensin-II (AII) might be the agent mediating this effect. Despite the fact that the AL of rat and human expresses all components of the renin–angiotensin system, its role has not been positively identified.

On the other hand, it seems likely that gonadotropes also play a role in the postnatal development of mammotropes (Fig. 2). Signals from gonadotropes, controlled by gonadotropin-releasing hormone or neuropeptide-Y, may affect "progenitor" cells (i.e., mammosomatotropes) favoring differentiation into mammotropes and inhibiting differentiation into somatotropes. Recent experiments suggest that the substance released by gonadotropes that stimulates mammotrope development is an epidermal growth factor (EGF)-like molecule. Another possibility is the α-subunit of luteinizing hormone (LH) as the mediator for inducing mammotrope development. Both of these challenging hypotheses must be further clarified.

Folliculostallate cells inhibit secretion of prolactin when they are cocultured with mammotropes. Surprisingly, folliculostallete cells also antagonize the inhibition of prolactin release by dopamine. The identity of this paracrine inhibitory responsiveness factor is also unknown. Because it was observed in the paracrine communication between both gonadotropes or folliculostallate cells and mammotropes, intimate contact is not required for these influences.

One of the best characterized paracrine communications in the rat anterior pituitary is a cholinergic system located in a subpopulation of corticotropes. They probably exert a tonic inhibitory effect on prolactin release, but only when they have been cultured in the presence of glucocorticoids. This dependence on glucocorticoids suggests that the cholinergic–paracrine influence mainly operates *in vivo* during conditions in which levels of plasma glucocorticoids are high for a prolonged period of time (i.e., in environmental or immune stresses).

It seems unlikely that such paracrine interactions are important in an acute release of prolactin, but rather they may be involved in gradual or chronic changes in prolactin regulation. It is also possible that these local mediators can function in certain restricted regions of the AL at given physiological situations.

C. Autocrine Regulation at the Level of Mammotropes

Autocrine regulation of pituitary mammotropes is the latest mode of the local regulation (Fig. 2). VIP, a known hypothalamic–hypophysiotropic stimulator of prolactin release, has been demonstrated to be synthesized by the AL and is present in mammotropes, as assessed with the combination of the reverse hemolytic plaque assay and immunocytochemical procedure by G. M. Nagy and co-workers. It has been concluded that the peculiar ability of mammotropes to release prolactin spontaneously at a high rate may be due to the tonic positive feedback of VIP that is produced, released, and accepted by mammotropes. The action of dopamine to suppress prolactin secretion, therefore, would occur by way of antagonizing the stimulatory effect of VIP or by inhibiting VIP secretion or both.

IV. PERIPHERAL ORGANS REGULATING PROLACTIN SECRETION

Several peripheral organs provide physiologically important signals to the hypothalamohypophysial regulatory mechanism of prolactin secretion. Among them, the regulatory roles of ovarian steroids and glucocortocoids of the adrenal glands are well established. A definitive role of uterine, placental, and lymphocytic regulation of pituitary prolactin secretion, though ubiquitous, has not been entirely characterized. OVX reduces and estrogen, alone or in combination with progesterone, increases prolactin secretion. These effects are mediated by influencing hypothalamic dopaminergic neurons (probably TIDA), by directly affecting mammotropes, and/or by indirectly releasing an unidentified MRF (*vida supra*) from the IL. Plasma levels of prolactin increase significantly after adrenalectomy (ADX), whereas the effect of ADX can be reversed by administration of corticosteroids. Similarly, the synthetic glucocorticoid dexamethasone (DEX) decreases prolactin release manifested in acute stimulus-like suckling, stress, or TRH treatment.

V. CELLULAR AND MOLECULAR MECHANISMS IN THE REGULATION OF PROLACTIN SECRETION

Any of the known physiologically relevant secretagogues that affect prolactin secretion bind to specific cell-surface receptors that act as signal tranducers. Receptors, for example, for TRH, AII, VIP, and dopamine, have been molecularly cloned, sequenced, and shown to be members of the 7-transmembrane spanning, GTP-binding protein (G-protein)-linked receptor family.

As a general statement, it can be said that VIP stimulates adenylate cyclase (AC) and consequently cyclic AMP (cAMP) and protein kinase A. Involvement of a stimulatory G-protein (G_s) is well established in this signal transduction cascade. In the case of dopamine, inhibitory coupling of D_2 receptors to AC mediated by inhibitory G-protein (G_i) has been demonstrated. Furthermore, dopamine also lowers cytoplasmic Ca^{2+} levels in mammotropes by decreasing Ca^{2+} influx. The third possibility is that G_i, parallel with its effect on cAMP, may directly open K^+ channels in the plasma membrane which makes it harder to depolarize the cell and therefore enhances the inhibitory effect of dopamine. In contrast, TRH activates phospholipase C. This signal is transmitted by two recently identified G-proteins, G_q and G_{11}. Activation of TRH receptor results in a simultaneous activation of protein kinase C and increased entry of Ca^{2+} through voltage-dependent Ca^{2+} channels.

Considering the multiplicity of these receptors and the signal transduction cascade present on prolactin cells, it is important to understand how they integrate the multiple stimuli that may differ both temporally and chemically. It is obvious that more precise knowledge is required about signal transduction mechanisms of the pituitary D_2 dopamine receptor to understand the physiological significance of the dominant and tonic inhibitory regulation of mammotropes. A rarely discussed but well-known feature of D_2 receptors, as established in other G_i-coupled inhibitory receptors, is the dependence of cells on the inhibitory tone of DA. This is evidenced by an immediate withdrawal reaction (elevation of prolactin release) by mammotropes after disrupting dopaminergic tone (*vida supra*). Studies of the cellular and molecular regulation of D_2 dopaminergic responsiveness (also termed desensitization/sensitization) have been difficult to perform due to the lack of cell lines expressing D_2 receptors. Transfection of different mammalian cell lines with D_2 receptors has been used in most studies. The greatest disadvantage, however, is in extrapolating these results to the regulatory mechanisms that exists in normal cells or tissues. However, recent studies have indicated that the same mechanisms which have been found on transfected cell lines may operate in an animal *in vivo*. For example, 10-min suckling stimulus results in a dramatic decrease in prolactin responsiveness to dopamine in addition to a marked increase in prolactin responsiveness to TRH or AII and a low concentration of dopamine. Nevertheless, these data suggest that the regulation of responsiveness has a significant influence on the secretory activity of mammotropes. At the same time, this raises important questions: (i) How is this change in responsiveness regulated? (at the level of the receptor or functionally uncoupling of G-protein from the receptor); and (ii) is the factor regulating responsiveness dopamine? Because of data about a new class of prolactin regulators (*vida supra*) and also about an active role of mammotropes in the mechanism of responsiveness, we may have to view the regulation of prolactin secretion differently.

See Also the Following Articles

Hypothalamic–Hypophysial Complex; Neuroendocrine Systems; Prolactin Inhibitory Factors

Bibliography

Ben-Jonathan, N. (1994). Regulation of prolactin secretion. In *The Pituitary Gland* (H. Imura, Ed.), 2nd ed, pp. 261–283. Raven Press, New York.

Denef, C. (1994). Paracrine mechanisms in the pituitary. In *The Pituitary Gland* (H. Imura, Ed.), 2nd ed, pp. 351–378. Raven Press, New York.

Freeman, M. E. (1994). The neuroendocrine control of the ovarian cycle of the rat. In *Physiology of Reproduction* (E. Knobil and J. D. Neill, Eds.), 2nd ed., pp. 613–658. Raven Press, New York.

Houslay, M. D. (1995). Compartmentalization of cyclicAMP phosphodiesterases, signaling "crosstalk," desensitization and the phosphorylation of G_i-2 add cell specific personalization to the control of the levels of the second messenger cyclicAMP. *Adv. Enzyme Regul.* **35**, 303–338.

Lamberts, S. W. J., and MacLeod, R. M. (1990). Regulation of prolactin secretion at the level of the lactotroph. *Physiol. Rev.* **70**, 279–318.

Neill, J. D., and Nagy, G. M. (1994). Prolactin secretion and its control. In *Physiology of Reproduction* (E. Knobil and J. D. Neill, Eds.), 2nd ed., pp. 1833–1860. Raven Press, New York.

Prostaglandins

see Eicosanoids

Prostate Cancer

James M. Kozlowski
Northwestern University Medical School

GLOSSARY

carcinoma of prostate (CaP) Adenocarcinoma involving the prostatic acinus.

central zone The region of the prostate thought to be derived from the mesonephric duct which accounts for 20% of prostate volume.

cytokeratin A family of intermediate filaments uniquely expressed by epithelial cells.

digital rectal examination In prostate cancer, this evaluation typically reveals nodularity or induration.

human kallikrein 2 A trypsin-like protease produced by prostate epithelial cells.

peripheral zone The posterior aspect of the prostate which is derived from the urogenital sinus and accounts for 75% of CaP.

prostate-specific antigen A cytoplasmic serine protease which is produced by prostate epithelial cells.

prostate-specific membrane antigen A transmembrane glycoprotein frequently expressed by prostate cancer cells.

prostatic acid phosphatase A lysomal protein whose serum levels are often elevated in locally advanced and metastatic CaP.

reverse-transcriptase polymerase chain reaction A molecular technique which permits the identification of mRNA derived from circulating tumor cells.

transition zone An area of the prostate that is located adjacent to the urethra and is the site of origin of benign prostatic hyperplasia and 25% of CaP.

transrectal ultrasound An imaging study that permits visualization of the prostate zones and precise tissue sampling.

The prostate is a compound tubuloalveolar gland whose base abuts the bladder neck and whose apex merges with the membranous urethra to rest on the urogenital diaphragm. Adenocarcinoma of the prostate is the most common cancer in American men. It constitutes a major factor in the health of the male population in the United States and many other countries in the Western Hemisphere.

I. INTRODUCTION

Projected 1998 estimates from the American Cancer Society suggest that carcinoma of prostate (CaP) will account for 29% (184,500 cases) of newly diagnosed cancers in males. The disease is also projected to account for 13% (39,200 events) of all male cancer deaths. CaP is the most prevalent cancer in men. Occult CaP has been identified in over 30% of autopsy specimens in males older than 50 years of age. The rate of detection of occult CaP increases progressively with age, tripling by the ninth decade. Most of these histologic or occult carcinomas exhibit a slow growth rate and will not adversely effect the quality or duration of life. In contrast, those prostate cancers which become clinically manifest often acquire features which facilitate tumor invasion and metastasis. About 25% of men with clinically apparent CaP die of the disease.

II. ETIOLOGY

Advancing age is the strongest risk factor for the development of CaP. By age 90, this neoplasm is identified in 67% of male cadavers that undergo autopsy. The postpubertal presence of an intact hypothalamic–pituitary–testicular axis is critical since castration prior to the onset of puberty markedly diminishes the risk of developing CaP.

Being of the black race is another strong etiologic factor. African American men have the highest incidence and mortality from CaP in the world. In contrast, the Japanese and mainland Chinese have the lowest rates of CaP. Many mechanisms have been postulated to account for this marked predilection of African American men, including (i) higher fetal exposure to circulating androgens; (ii) increased expression of prostatic 5α reductase which converts testosterone to the more active androgen, dihydrotestosterone; and (iii) increased dietary fat consumption.

About 9% of CaP cases have been estimated to result from the inheritance of mutated CaP susceptibility genes. Risk of developing CaP increases two- or threefold in the presence of a first-degree relative affected with this cancer. The risk increases to ninefold for an individual with two first-degree relatives with CaP. Hereditary CaP tends to develop prior to the age of 55. Recent linkage studies have identified the hereditary prostate cancer 1 locus on chromosome 1q24–25. In contrast, most cases of sporadic prostate cancer are associated with hypermethylation of the regulating sequences of the glutathione S-transferase gene locus.

A Western diet rich in saturated fatty acids may predispose to the development of CaP and facilitate its subsequent growth. Sexual activity, sexually transmitted infection, and antecedent vasectomy have a very tenuous association with the development of CaP.

III. EPIDEMIOLOGY

The identical prevalence of histologic (or unsuspected) CaP at autopsy in various populations throughout the world is not reflected in an identical prevalence of clinical CaP. For example, the incidence of CaP is highest in Scandinavian countries and lowest in Asia. Although the incidence of histologic carcinoma in an oriental male living in Japan and a male of Japanese descent living in Hawaii is essentially the same, the age-adjusted prevalence of clinical CaP for the latter cohort is about 10-fold greater. However, these Japanese Hawaiians have a prevalence rate for CaP which approaches only 50% of that observed in the indigenous local white population. When black and white males in the United States are compared with American Indians and Hispanics, the latter groups are at relatively low risk of dying of CaP. All these observations suggest that genetic and epigenetic influences play a role in the development of these disease. Some adverse environmental factors which have been proposed include decreased exposure to ultraviolet light, high-fat diet, elevated levels of cadmium, and decreased levels of zinc.

IV. PATHOLOGY

Primary adenocarcinomas of the prostate accounts for over 95% of all prostatic malignancies. These tumors are thought to originate from the adluminal (or secretory) cells of the prostatic acinus. About 70% of CaP originate from the peripheral zone (PZ). The transition zone and central zone account for

20 and 10% of these tumors, respectively. Multiple tumor foci of variable volume and grade are commonly encountered and support of the concept of a "field change" effect for this neoplasm. Most prostate cancers arise in close proximity to the capsule. Tumor cells may exit the gland by following a pathway of least resistance along the perineural spaces, by direct invasion of the prostatic capsule, and via intraglandular vascular/lymphatic channels. Histologically, the characteristic double layer of cells lining the acinus may be replaced by a single layer or a piling of cells. The basal cell layer is lost. Another histologic hallmark of CaP is the loss of intervening or enveloping stroma which results in a "back-to-back" configuration of the acini. Invasion of the intraprostatic perineural spaces has been seen in 85% of incidental and early carcinomas. Most CaP stain positively for cytokeratin (CK) 18. CKs 5/15 are generally not identified because these CKs are typically associated with the basal cell layer, which is no longer present in CaP. Immunohistochemical studies have also demonstrated that CaP commonly expresses prostatic-specific antigen (PSA), prostatic acid phosphatase (PAP), prostate-specific membrane antigen (PSMA), and nuclear androgen receptor.

Grading is defined as an effort to use histologic characteristics of a tumor to predict its biologic activity. The Gleason classification scheme is the most popular and employs low-power magnification (×40–100) in order to evaluate the glandular pattern of the tumor and its relationship to the stromal compartment (Table 1). Five tumor grades progressing from the most (1) to the least (5) differentiated are recognized. The Gleason's score consists of the sum of the most and next to most prevalent grade. Possible pattern scores range from 2 to 10, with the former being indicative of a very well-differentiated tumor and the latter of the highly undifferentiated variety. The Gleason tumor score correlates with the risk of developing metastatic disease. Gleason's scores of

TABLE 1
Histologic Patterns of Adenocarcinoma of the Prostate[a]

Pattern	*Margins of tumor areas*	*Gland pattern*	*Gland size*	*Gland distribution*	*Stromal invasion*
1	Well-defined	Single, separate, round	Medium	Closely packed	Minimal, expansile
2	Less defined	Single, separate, rounded, but more variable	Medium	Spaced up to one gland diameter, average	Mild, in larger stromal planes
3	Poorly defined	Single, separate, more irregular	Small, medium, or large	Spaced more than one gland diameter, rarely packed	Moderate, in larger or smaller stromal planes
		Rounded masses of cribriform or papillary epithelium	Medium or large	Rounded masses with smooth sharp edges	Expansile masses
4	Ragged, infiltrating	Fused glandular masses or "hypernephroid"	Small	Fused in ragged masses	Marked, through smaller planes
5	Ragged, infiltrating	Almost absent; few tiny glands or signet ring cells	Small	Ragged analplastic masses of epithelium	Severe, between stromal fibers or destructive
		Few small lumina in rounded masses of solid epithelium; central necrosis?	Small	Rounded masses and cords with smooth, sharp edges	Expansile masses

[a] Reproduced with permission from D. F. Gleason, Veterans Administration Cooperative Urological Research Group: Histologic grading in clinical staging of prostatic carcinoma, In *Urologic Pathology: The Prostate* (M. Tannenbaum, Ed.), pp. 171–197, Lea & Febiger, Philadelphia, 1977.

2–4, 5–7, and 8–10 are associated with metastases in 20, 40, and 75% of cases, respectively.

V. SIGNS/SYMPTOMS

A traditional precept is that early stage (or organ-confined) CaP does not cause symptoms. This is logical given the PZ origin of most of these tumors. Locally advanced CaP is commonly associated with urethral and bladder neck obstruction. Most of these tumors also exhibit evidence of extraprostatic disease with involvement of the periprostatic tissues, seminal vesicles, and bladder base. These patients will complain of irritative and obstructive voiding symptoms, including increased frequency, dysuria, nocturia, hesitancy, intermittency, urgency, and a perception of incomplete bladder emptying. Urinary retention, hematuria, and urinary tract infections may also occur. Patients with metastatic disease will often complain of constitutional symptoms such as being easily fatigued and weight loss. Bulky lymph node metastases may be associated with lower extremity edema. Patients with skeletal metastases will often complain of multifocal bone pain and may exhibit signs of lower extremity weakness due to spinal cord compression.

VI. SCREENING/DIAGNOSIS

Traditionally, the digital rectal examination (DRE) has been the principal method of obtaining objective evidence to support consideration of a diagnosis of CaP. The classical features of CaP on DRE include a firm palpable nodule with indistinct margins; irregular, stony areas of induration; obliteration of the lateral sulci or median furrow; and induration with loss of capsular integrity in the region of the prostatic base, implying seminal vesicle infiltration. The limitations of DRE as an indicator of the presence and extent of CaP are well documented. For example, CaP will be documented in only 39% of patients who undergo biopsy on the basis of an abnormal DRE alone. Alternate causes of a nodular induration include a prostatic calculus, prostatitis, tuberculosis, focal infarction, a postbiopsy tissue reaction, and a spheroid of benign prostatic hyperplasia. DRE will miss many prostate cancers and will often underestimate the volume/extent of existing disease.

None of the current physical imaging modalities are sufficiently sensitive/specific to provide great assistance in the diagnosis of early stage prostate cancer. Currently, the anatomic information provided by transrectal ultrasound (TRUS) is superior to that obtained by conventional computed tomographic (CT) scanning or magnetic resonance imaging (MRI). Classically, CaP presents as a hypoechoic area on TRUS. Unfortunately, many prostate cancers do not manifest this typical feature. The relatively recent advent of color Doppler TRUS and transrectal MRI may increase the accuracy of these imaging techniques.

The advent of PSA testing has dramatically enhanced our ability to scan at-risk populations. PSA is a single-chain glycoprotein with a molecular weight of about 34 kDa. The human PSA gene is located on chromosome 19. PSA is a cytoplasmic serine protease which shares homology with the human kallikreins. PSA is translated as an inactive pre–pro molecule. Following passage through the intracellular secretory pathway, the signal peptide is cleaved, yielding the pro form of the protein. Conversion of pro-PSA to the mature, enzymatically active PSA requires the action of human kallikrein 2. The latter, like PSA, is produced by the prostate epithelium, is androgen-regulated, and shares a 78% amino acid homology with PSA. Mature PSA can exist in a "free" form or bound to other proteins, particularly α1-antichymotrypsin. Most of the PSA which enters the systemic circulation is derived from prostatic epithelial cells which have undergone programmed cell death. These dying cells release PSA into the interstitial compartment; from there it gains access to the systemic circulation. The interstitial barrier is thought to be more fragile in cancer, thus accounting for the increased likelihood to detect PSA elevations in that condition.

PSA has numerous biological functions. It is primarily responsible for the lysis of the seminal coagulam. PSA is a regulator of the insulin-like growth factors (IGFs). In fact, PSA cleaves IGF from its binding protein, releasing the free moiety and thus facilitating the mitogenic impact of the IGFs on prostate cells. PSA possesses properties of a weak matrix degrading protease and its release may facilitate the matrix degradation seen in CaP. PSA is also a mitogen for osteoblasts and may play a role (along with many other factors) in the osteoblastic proliferation seen in metastatic CaP. Of interest, PSA has been identified in a number of nonprostatic tissues, including periurethral glands, perianal glands, salivary glands, benign and malignant breast tissue, the uterus, and a number of solid tumors capable of the constitutive expression of the steroid receptor super family.

The normal PSA value should be ≤4 ng/ml. Recently, adjustments in the standard reference range have been suggested as a function of advancing age or ethnic background. Most serum PSA is complexed with other proteins, particularly α1-antichymotrypsin and α2-macroglobulin. A small fraction of PSA occurs in the free uncomplexed molecular form. The percentage of free PSA decreases in CaP. It is thought that posttranslational modifications in cancer increase the affinity of PSA for these binding proteins. It has also been proposed that there may be an increased production of these binding proteins by CaP cells. Assessment of the percentage of free PSA is commonly performed in men with a total PSA elevation between 4 and 10 ng/ml. If the percentage of free PSA is >24%, only 10% of such patients have CaP. Utilizing this approach, about 25% of false-negative prostate biopsies can be avoided. A PSA density (PSA ÷ prostate volume) of >0.15 is thought to define an at-risk population. Similarly, the PSA velocity (rate of change over time) is another useful barometer. A PSA velocity of >0.75 documented over a 12-month interval reflects a worrisome degree of proliferative activity. Both of these abnormalities would prompt ultrasound-guided prostate needle biopsies.

False-positive PSA elevations have been associated with prostatitis, DRE, ejaculation, lower tract instrumentation, needle biopsy, and prostatic infarction and have been shown following cardiac bypass. About 20% of patients with CaP have a normal PSA.

About 20% of men with CaP are currently diagnosed by DRE alone; 45% have a normal DRE but an elevated PSA; and about 37% demonstrate abnormalities on both tests. Annual screening (DRE + PSA) should commence at age 50 for white men and for men with a negative family history. Screening should probably commence at age 40 for African Americans and for any individual with one or more first-degree relatives with CaP.

Patients with an abnormal DRE and/or an abnormal PSA profile are generally advised to undergo TRUS biopsies. This involves the use of a spring-loaded biopsy gun that permits the acquisition of a 17-mm tissue core. Most patients will undergo systematic "sextant" biopsies of the base, midzone, and apex of both prostatic lobes. In addition, any suspicious hypoechoic area (or area of palpable abnormality) will also be sampled. Serious postbiopsy complications are unusual. Most men will experience a small amount of blood in the urine and stool for several days and blood in ejaculate for several weeks. Risks of infection are surprisingly low if proper precautions are taken. Negative biopsies do not exclude a diagnosis of prostate cancer. A tissue sampling error is inherit in the procedure and about 20% of patients who are subjected to a second biopsy scenario will have CaP confirmed.

This approach to screening has dramatically increased the detection of organ-confined prostate cancer. For example, nearly 60% of men were found to have evidence of advanced-stage disease when screening relied on DRE only. With the advent of PSA, only 37% of patients are identified with evidence of advanced-stage disease. This modern approach to screening may also account for the recent observation that the mortality rate from CaP is on the decline.

VII. STAGING

Staging is a clinical effort to identify the phase of the natural history of prostate cancer that exists in a given patient by documenting the site (s) and degree of tumor involvement. The most frequently used staging systems are those of Whitmore and TNM classification that is advocated by the American Joint Committee on Cancer (Table 2).

Stage T_1 (A) is a designation employed for a CaP that is clinically unsuspected and discovered on tissue removed to relieve bladder neck obstruction (stages T_{1a}, T_{1b}, A_1, or A_2) or because of an elevated serum PSA (stage T_{1c}). Patients with clinically suspected, organ-confined, histologically confirmed prostate cancer are classified as having stage T_2 (B)

TABLE 2
Staging Designations for Carcinoma of Prostate[a]

Description	Clinical Stage: UICC (AJCC, 1992)	Hopkins (modified Jewett)	Memorial (modified Whitmore, 1990)
Disease localized to prostate			
Clinically unsuspected, incidental histologic finding	T_1	A_1	A
Focal, low-grade	T_{1a}[b]	A_1[b]	A_1[c]
Intragland lump diffuse or high grade	T_{1b}[d]	A_2	A_2
Tumor identified, needle biopsy (e.g., PSA elevated)	T_{1c}		
Risk recognized clinically (confined to prostate)	T_2	B	B
Tumor confined to one lobe surrounded by normal tissue		B_1	
<2 cm (Whitmore)		B_{1N}	B_1
>2 cm			B_2
Half a lobe or less	T_{2a}		
More than half a lobe, but not both	T_{2b}		
Tumor in both lobes	T_{2c}	B_2	B_3
Disseminated disease			
Periprostatic, extends through capsule	T_3	C	C
Lateral sulcus			C_1
Unilateral	T_{3a}		
Bilateral	T_{3b}		
Base of seminal vesicle and/or lateral sulcus	T_{3c}		C_2
>Base of seminal vesicle and/or other structure	T_{3c}		C_3
Tumor fixed: invades adjacent structure other than seminal vesicle	T_4		
Bladder neck, extend sphincter and retcum	T_{4a}		
Levator or pelvic wall	T_{4b}		
Distant		D	D
Pelvic lymph nodes	N_{1-3}[e]	D_1	D_1
Bones, lung, etc.	M_{1a-1c}[f]	D_2	D_2
Elevated acid phosphatase only		D_0	D_0

Note. Abbreviations used: UICC, International Union Against Cancer; AJCC, American Joint Committee on Cancer. Note that T_0 category is now listed as no evidence of primary tumor (Schroeder, 1992).

[a] Reproduced with permission from Kozlowski and Grayhack (1996).

[b] Tumor present in 5% or less of tissue.

[c] Tumor present in more than three microscopic foci.

[d] Tumor present in more than 5% of tissue.

[e] N_1, one regional node $\leq$2 cm; N_2, one regional node >2 cm but <5 cm or multiple regional nodes, none 5 cm: N_3, regional node >5 cm.

[f] M_{1a}, nonregional nodes; M_{1b}, bone; M_{1c}, other site.

disease. Tumors that have demonstrated periprostatic extension on clinical examination but have no evidence of distant metastasis are classified as stages T_3 and T_4 (C). Tumors that have metastasized to lymph nodes, bone, or other organs are classified as stages T_{1-4}, N_{1-3}, and M_{1a-c}.

A variety of procedures are available to facilitate the ultimate staging assignment. The DRE itself, if properly performed, may permit the identification of capsular perforation and/or probable seminal vesicle involvement. The TRUS may also demonstrate evidence of obvious extraprostatic extension and facilitate acquisition of tissue samples from such areas, particularly the seminal vesicles. Additional physical imaging studies (CT scans or MRI) are generally reserved for patients with PSA levels $\geq$10 ng/ml or for those individuals with obvious evidence of bulky locally advanced cancers. Additional serum markers, such as PAP and bone isoenzyme of alkaline phosphatase, are occasionally helpful. Reverse-transcriptase polymerase chain reaction for PSA and PSMA mRNA has recently been employed to detect the presence of circulating prostate tumor cells and may facilitate the detection of subclinical metastatic disease. The ProstaScint scan is the most recent addition to our diagnostic/staging armamentarium. ProstaScint is a murine monoclonal antibody directed against PSMA. It can detect disease not readily demonstrated by traditional imaging modalities.

Surgery does play a role in the staging process. The evaluation of the pelvic lymph nodes can yield important information which may dramatically alter treatment planning. The lymph nodes at greatest risk constitute the medial group of the external iliac lymph node chain. These lymph nodes can be removed by a standard open surgical approach, a minilap variation of that approach, or laproscopically.

VIII. NATURAL HISTORY/ PROGNOSIS

Many tumor cell features (or phenotypes) have been associated with CaP endowed with the capacity for invasion and systemic dissemination (Table 3). Tumors endowed with these features are capable of rapid intraglandular growth and extraprostatic spread following natural anatomic planes and conduits. The pelvic lymph nodes and the skeletal system are the most prominently identifiable sites of initial dissemination. Involvement of lung, liver, and other anatomic areas occurs infrequently and such involvement is usually a premorbid event.

CaP are biologically heterogeneous and contain androgen-dependent, androgen-sensitive, and androgen-independent subpopulations. A majority of CaPs maintain a relative or absolute dependence on androgen for much of their natural life. Most of the patients that succumb to metastatic CaP die as a result of the uncontrolled proliferation of hormone-refractory cells.

The time frame for development and progression of carcinoma of the prostate is highly variable. Less than 10% of patients with a clinically well-differentiated focal carcinoma (stages T1a or A1) exhibit progressive disease in follow-up intervals approaching 10 years. Approximately 50% of patients with clinical stage T2 disease will exhibit progression to multiple extraorgan disease in 5 years. The progression rate is relatively constant at 10% per year. Patients with untreated stage C (T3) carcinoma demonstrate a 10–20% yearly and 60% 5-year progression rate. Among men who are untreated, the average survival is about 8–10 years for patients with stage A/B disease, 3–6 years for those with stage C cancers, and 2 or 3 years for patients with metastatic (stage D disease).

IX. TREATMENT OPTIONS

A. Organ-Confined Disease

The treatment of organ-confined CaP must be individualized. Important variables which impact decision making include age, associated medical comorbidities, anticipated life expectancy, the assigned clinical stage, tumor grade, the pretreatment PSA

TABLE 3
Prostate Cancer: Adverse Tumor Phenotypes

Gleason grade 4 or 5	Tumor volume > 1 cc	Tetraploid/aneuploid DNA content
Abnormal cell shape/motility	Negative cell surface charge	Nuclear androgen receptor expression
↑ Neovascularity	Gains of chromosome 7 and 8	Deletion of metastasis suppressor gene on short arm of chromosome 11
↑ IGFs	↑ Expression of u-PA and its receptors	↑ TGF-α and EGFR
Upregulation of bcl-2	↑ $p185^{erbB-2}$ and $p160^{erbB-3}$	Androgen receptor gene mutations
↑ Thymosin β-15	↓ E-cadherin	↓ α-Catenin
	↑ Endothelin	TGF-β1 type I and II receptor mutations

level, and the patient's preference after a discussion of the risks/benefits of available options. The patient has essentially five treatment options: "watchful waiting," androgen ablation, radiation therapy (external beam versus interstitial), radical prostatectomy, and cryotherapy.

The watchful waiting option may be attractive for men with low-grade/low-volume CaP and for those who do not have a 10-year projected life expectancy. If the tumor is organ-confined, evidence of extracapsular extension will occur in 40% of patients at 5 years and in virtually all patients at 10 years. Metastasis to lymph nodes and bone will occur with an anticipated frequency of 8% at 5 years and about 30% at 10 years. CaP is an anticipated cause of death in 1% of these patients at 5 years and 25% at 10 years.

Androgen ablation is effective palliative treatment which is usually reserved for men with locally advanced or metastatic CaP. It is occasionally used for men with organ-confined disease who are too frail to undergo more meaningful treatment. About 80% of patients will objectively respond to this therapy which most commonly consists of a luteinizing hormone-releasing hormone agonist (± antiandrogen). As an alternative, scrotal orchidectomy can be performed. The average duration of response is approximately 2 or 3 years within this patient group. Ultimately, hormone-refractory tumor populations arise. Side effects include loss of libido, erectile dysfunction, and hot flashes.

Radiation therapy is potentially curative treatment. Most commonly, external beam therapy involves the administration of high-energy electrons or photons. Using CAT scans and computers, very precise dosimetry planning is possible. Some patients with bulky/necrotic tumors might be best served with heavy particle therapy (neutrons or photons). Recent studies have suggested that a course of neoadjuvant androgen ablation (for 2–6 months) enhances the efficacy of external beam radiation. An alternative to external beam radiation is interstitial or brachytherapy. Most commonly, this involves the use of transrectal ultrasonography to precisely guide the placement of ^{125}I or ^{103}Pd seeds into the prostate. Tumors of high grade/high volume may be treated with a combination regimen including androgen ablation, an abbreviated course of external beam radiation, and brachytherapy. Currently, there is no established consensus regarding which of these radiation therapy options is most effective. The majority of patients undergo neoadjuvant hormonal therapy followed by conformal external beam radiation. Curative therapy is generally associated with a decrease in the serum PSA <0.5 ng/ml and negative posttreatment prostate biopsies, which are usually performed 12–18 months following completion of therapy. About 20–25% of patients achieve definitive cure. Another 25% achieve excellent local control and are not troubled by tumor progression. Fifteen-year survival rates of about 50% have been reported. Radiation failure is generally treated with watchful waiting or androgen ablation. Salvage prostatectomy is infrequently used because of a high complication rate and lack of efficacy in the majority of patients. Potential side effects of radiation therapy include cystitis, proctitis with rectal bleeding, altered bowel function, and impotence.

Another potentially curative treatment is radical prostatectomy, which implies the removal of the entire prostate gland, the seminal vesicles, and the ampulla of the deferens. Reconstruction of the urinary tract involves modification of the bladder neck and its anastomosis to the membranous urethra. Preservation of the pelvic splanchnic nerves (nerves for erection) is technically possible but may be imprudent if tumor is located in close proximity to the nerve bundles. The operation can be performed through a retropubic or perineal approach. The former permits concomitant lymph node sampling and represents the most facile approach for nerve sparing. Following surgery, an indwelling urethral catheter is left in place for several weeks until the anastomosis heals. The anticipated operative mortality is 0.8% in healthy patients. Incisional pain (which is generally modest) and blood loss (possibly requiring transfusion) are common generic complications. Impotence can occur following the procedure. About 50–60% of patients will return to baseline sexual activity if both neurovascular bundles can be preserved. It may take 12–18 months for this recovery to take place, however. Patients who are rendered impotent can have sexual function restored using a wide variety of techniques, including a vacuum erection device, intraurethral PgE (MUSE), the intercavernosal injection of PgE or other vasodilators, insertion of a penile prosthesis, or the use of a new oral agent (sildenafil) which inhibits type 5 phosphodiesterase activity. Another potential complication of radical prostatectomy is urinary incontinence. About 95–97% of patients achieve excellent social control. They should be pad-free at night and during the day with normal activity. Some leakage will occur when patients are very active. The remaining patients are totally incontinent and generally require placement of an artificial sphincter or the performance of a sling procedure. Radical prostatectomy is curative in about 70% of patients with organ-confined disease. Postoperatively, the serum PSA should revert to undetectable. If this does not happen or if the surgical margins are significantly compromised, adjuvant radiation therapy can be provided with the anticipation of excellent local control.

Another treatment option that may be applicable to patients with organ-confined disease is cryotherapy. This entails the use of either liquid nitrogen or gas systems to induce freezing, which is dynamically monitored on TRUS. The long-term efficacy of this approach has not been established. The results to date would not rank it in parity with other potentially curative treatment options. Numerous complications have been reported, including urethral obstruction, impotence, and rectal injury with subsequent fistula formation.

B. Locally Advanced Disease

Most patients with clinically stage T3/4 (C) disease are not perceived to be good candidates for radical prostatectomy. Many of these patients will undergo a diagnostic lymph node dissection. If the pelvic lymph nodes are free of tumor, most will be advised to undergo external beam radiation therapy (± neoadjuvant androgen ablation). If the pelvic lymph nodes are positive, most patients will undergo pharmacologic androgen ablation or scrotal orchidectomy. Occasionally, transurethral surgery will be required for the relief of obstructive voiding symptoms.

C. Metastatic Disease

Treatment-naive patients who present with lymph nodes or skeletal metastases are considered optimal candidates for androgen-ablative therapy. About 80% of these patients will demonstrate evidence of both objective and subjective improvement. The duration of response varies from a few months to several years, with an average duration of benefit of about 18–24 months. At that point, hormone-refractory disease becomes evident. Fifty percent of these patients will die within the first year following relapse. The majority of the remainder will succumb to the disease process or related causes within 2 years. In an attempt to extend the duration of benefit to androgen ablation, the concept of "intermittent" hormonal therapy has emerged. Optimal candidates are those patients who achieve near undetectable serum PSA levels. Conceptually, these patients are "cycled" on a 9- to 12-month basis. While they are "off therapy" the patient is given a respite from the side effects of treatment and the tumor is allowed to slowly repopulate in the presence of testosterone. It is hoped that such an approach might prolong the duration of androgen responsiveness.

Patients with hormone-refractory metastatic CaP have a poor prognosis, with a mean survival of <1 year. For some of these patients, an attempt at second-line hormonal therapy is reasonable. For example, if the patient has been maintained on an antiandrogen, such therapy should be stopped because a paradoxical enhancement of tumor growth has been observed in about 30–40% of such patients. If the patient has not been exposed to an antiandrogen, the use of such therapy has been associated with modest and transient benefit in about 30% of patients. Alternative approaches to second-line hormonal therapy include ketoconazole plus hydrocortisone, aminoglutethimide plus hydrocortisone, or hydrocortisone alone. Once again, about 30% of patients will transiently benefit from such strategies.

Systemic chemotherapy is generally reserved for patients who are overtly symptomatic and have failed first- and/or second-line hormonal therapy. For these patients, chemotherapy is administered for its palliative value. Cure is not currently obtainable. Mitoxantrone and prednisone is a currently popular combination regimen. Other efficaous two-drug regimens involve the use of estramustine with one of the following agents: vinblastine, vinorelbine, paclitaxel, or etopside. In general, these combinations are associated with a favorable PSA response in about 50% of patients. Relief of pain and decrease in measurable disease have been reported in 30% of patients. Unfortunately, the duration of benefit is relatively transient. Many other experimental therapies have been or are being evaluated, including the use of antigrowth factor agents (suramin), differentiation-inducing therapy (phenylbutyrate), the use of antiangiogenesis agents (TNP-470 and linomide), the use of agents which interfere with tumor cell adhesion (citrus pectin), the induction of programmed cell death (Thapsigargin), and the use of gene therapy.

The proper care of these unfortunate patients requires that great attention be paid to various palliative care issues. The somatic and visceral pain associated with advanced-stage malignancy should be managed utilizing the World Health Organization analgesic "step ladder." External beam radiation therapy can also be quite effective for the treatment of localized skeletal pain and can also be used for the relief of ureteral and bladder neck obstruction in some patients. Multifocal skeletal discomfort can be effectively managed utilizing the systematic calcium analog strontium-89. Anemia, constipation, persistent neuropathic pain, and depression represent some additional problems that require prompt and appropriate management.

X. CONCLUSION

Despite recent advances, the etiology of prostate cancer remains obscure. Curative treatment is available for many patients with organ-confined disease. Unfortunately, curative therapy is not available for patients with locally advanced CaP. Sophisticated scientific approaches are beginning to yield significant insights into the genetic and biochemical aberrations associated with aggressive prostate cancer. In all probability, this enhanced insight will translate into the development of novel therapies whose impact will be well tolerated, effective, and durable. Until then, it is imperative that we provide these patients with compassionate and insightful care that will translate into an improvement in the quality of their life.

See Also the Following Articles

Prostate Gland; Prostate-Specific Antigen

Bibliography

Carter, R. E., Feldman, A. R., and Coyle, J. T. (1996). Prostate-specific membrane antigen is a hydrolase with substrate and pharmacologic characteristics of a neuropeptidase. *Proc. Natl. Acad. Sci. USA* **93**, 749–753.

Esper, P. S., and Pienta, K. J. (1997). Supportive care in the patient with hormone refractory prostate cancer. *Sem. Urol. Oncol.* **15**(1), 56–64.

Gronberg, H., Isaacs, S. D., Smith, J. R., Carpten, J. D., Bova, G. S., Freije, D., Xu, J., Meyers, D. A., Collins, F. S., Trent, J. M., Walsh, P. C., and Isaacs, W. B. (1997). Characteristics of prostate cancer in families potentially linked to the hereditary prostate cancer 1 (HPC1) locus. *J. Am. Med. Assoc.* **278**, 1251–1255.

Keetch, D. W., McMurtry, J. M., Smith, D. S., Andriole, G. L., and Catalona, W. J. (1996). Prostate specific antigen density versus prostate specific antigen slope as predictors of prostate cancer in men with initially negative prostatic biopsies. *J. Urol.* **156**, 428–431.

Kozlowski, J. M., and Grayhack, J. T. (1996). Carcinoma of the prostate. In *Adult and Pediatric Urology* (Gillenwater *et al.*, Eds.), 3rd ed., Vol. 2, pp. 1575–1713. Mosby, St. Louis, MO.

Landis, S. H., Murray, T., Bolden, S., and Wingo, P. A. (1998). Cancer statistics, 1998. *CA Cancer J. Clin.* **48**, 6–29.

Millikan, R., and Logothetis, C. (1997). Update of the NCCN Guidelines for Treatment of Prostate Cancer. *Oncology* **11A**, 180–193.

Oesterling, J. E. (1996, April). Molecular PSA: The next frontier in PCa screening. *Contemp. Urol.*, 76–92.

Pienta, K. J. (1997). *Advances in the Treatment of Metastatic Prostate Cancer.* Biomedical Communications, Univ. of Michigan, Ann Arbor.

Young, C. Y. F., Seay, T., Hogen, K., Charlesworth, M. C., Roche, P. C., Klee, G. G., and Tindall, D. J. (1996). Prostate-specific human kallikrein (hK2) as a novel marker for prostate cancer. *Prostate Suppl.* **7**, 17–24.

Zhang, Y., Zippe, C. D., Van Lente, F., Klein, E. A., and Gupta, M. K. (1997). Combined nested reserve transcription-PCR assay for prostate-specific antigen and prostate-specific membrane antigen in detecting circulating prostatic cells. *Clin. Cancer Res.* **3**, 1215–1220.

Prostate Gland

Lynn Janulis and Chung Lee
Northwestern University Medical School

GLOSSARY

androgen A male hormone which possesses masculinizing activities. Testosterone produced from the testis is one such hormone.

estrogen A female sex hormone which is usually produced by the ovary. However, in males, estrogen can be produced by a metabolic conversion of androgen in the body.

growth factors Soluble polypeptides that travel across small and discrete distances to regulate cellular activities through interactions with respective receptors.

prostatic ductal system A single functional unit of the prostate. All glandular structures within a ductal system share a single ductal orifice which opens into the urethra.

prostatic urethra The portion of the urethra that penetrates the prostate.

urogenital sinus A stromal structure surrounding the embryonic urethra which later develops into the male accessory organ, such as the prostate and seminal vesicle.

verumontanum An elevation of the mucosa by the entrance of the ejaculatory ducts and the utricle. This structure is an anatomical landmark of the prostatic urethra.

The function of the prostate is to secrete prostatic fluid that facilitates the capacity of sperm to fertilize ova. The prostate is of great clinical importance because it is a frequent site of infection and benign and malignant growth. Each of these conditions will have profound consequences in health in men. In order to implement means of prevention, diagnosis, and management for these conditions, it is important that we have an understanding of the biology of the normal prostate.

I. INTRODUCTION

The prostate is a male sex accessory gland. It is also an androgen-sensitive organ in that its growth and regression depends on the presence or absence of circulating androgen. The manner in which androgen acts in this organ plays a central role in its growth. However, other factors work in concert with androgen to provide a homeostatic environment for the prostate. This article will cover topics of developmental biology, anatomy, embryology, physiology, endocrinology, and cellular mechanisms of normal prostatic growth. We shall discuss mainly the human prostate. Whenever pertinent, discussion will also include studies of animal models.

II. GROSS ANATOMY AND HISTOLOGY

A. Structural Biology

The adult prostate is a firm and elastic gland resembling a horseshoe-shaped chestnut that lies between the urinary bladder and the pelvic floor. It is transpierced by the urethra and ejaculatory ducts. The prostate is made up of periurethral glands and is not lobulated. It has a base and an apex and anterior, posterior, and inferolateral surfaces. The base is the upper surface below the bladder neck and the apex is the lowest part. The posterior surface lies in front of the lower rectum and is separated by the rectovesical fascia. The ejaculatory ducts pierce the posterior surface just below the bladder and pass obliquely through the gland with separate openings into the prostatic urethra.

The prostate consists of several glandular and nonglandular components that are fused tightly together within a common capsule (Figs. 1–3). The nonglandular tissues are concentrated anteromedially and consist of the preprostatic sphincter, striated sphincter, and anterior fibromuscular stromal. The three major glandular regions are the peripheral zone, the central zone, and the transition zone. The peripheral zone represents about 70% of the glandular part of the prostate. This zone forms the lateral and posterior aspect of the organ. The ducts of the peripheral zone open into the distal prostatic urethra and extend laterally in the coronal plane. The central zone occupies 25% of the glandular component of the prostate.

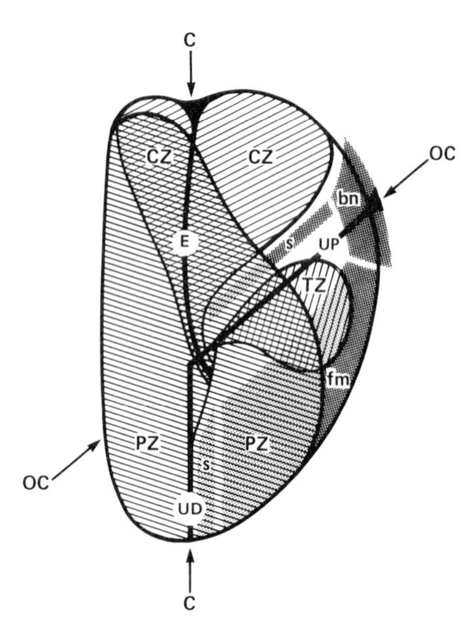

FIGURE 1 Sagittal diagram of distal prostatic urethral segment (UD), proximal urethral segment (UP), and ejaculatory ducts (E) showing their relationships to a sagittal section of the anteromedial nonglandular tissues. bn, bladder neck; fm, anterior fibromuscular stroma; s, preprostatic and distal striated sphincters. These structures are shown in relation to a three-dimensional representation of the glandular prostate. CZ, central zone; PZ, peripheral zone; TZ, transitional zone. Coronal plane (C) of Fig. 2 and oblique coronal plane (OC) of Fig. 3 are indicated by arrows (from McNeal, 1988; reproduced with permission of Lippincott-Raven).

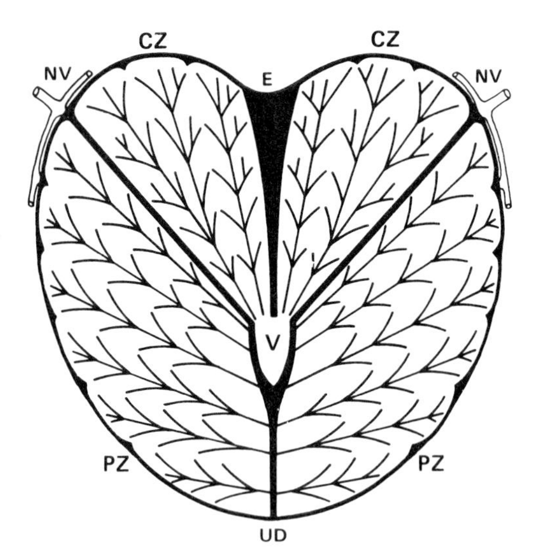

FIGURE 2 Coronal section diagram of prostate showing location of central zone (CZ) and peripheral zone (PZ) in relation to distal urethral segment (UD), verumontanum (V), and ejaculatory ducts (E). Branching pattern of prostatic ducts is indicated; subsidiary ducts provide uniform density of acini along entire main duct course. Neurovascular bundle (NV) is located at the junction between the central zone and the peripheral zone (from McNeal, 1988; reproduced with permission of Lippincott-Raven).

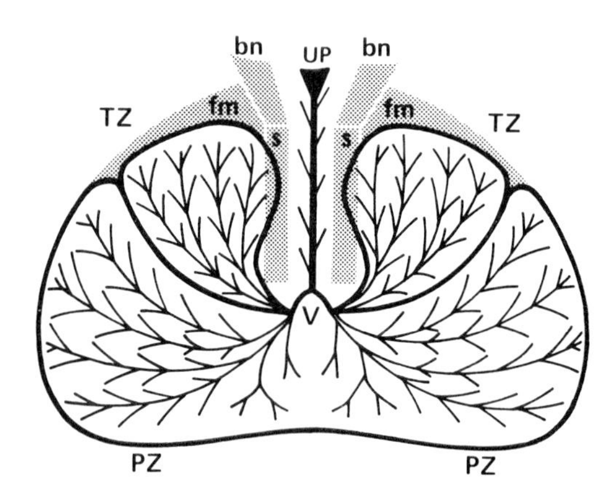

FIGURE 3 Oblique coronal section diagram of prostate showing location of peripheral zone (PZ) and transition zone (TZ) in relation to proximal urethral segment (UP), verumontanum (V), preprostatic sphincter (s), bladder neck (bn), and periurethral region with periurethral glands. Branching pattern of prostatic ducts is indicated; medial transition zone ducts penetrate into sphincter. fm, anterior fibromuscular stroma (from McNeal, 1988; reproduced with permission of Lippincott-Raven).

This zone is wedge shaped and surrounds the ejaculatory ducts with its apex at the verumontanum and its base by the bladder neck. Ducts in this zone extend radially and drain into the urethra near the ejaculatory ducts. The transition zone is the smallest glandular part of the prostate, comprising 5–10% of the prostate. This zone consists of two independent small lobes with ducts draining into the posteriolateral aspects of the urethral wall. For a detailed description of the human prostate, see the original paper by McNeal.

B. The Prostatic Urethra

The portion of the urethra that penetrates the prostate is considered the prostatic urethra. It is about 3 cm long and is the widest part of the male urethra. It is divided into proximal and distal segments of approximately equal length, with an abrupt anterior angulation of its posterior wall at the midpoint. In cross section, the prostatic urethra appears crescentic in outline, which is formed on the posterior wall by an elevation of the mucous membrane and subjacent tissue, called urethral crest. Verumontanum is an elevation in the middle of the length of the urethral crest showing the slit-like orifice of the prostatic utricle. Openings of the two ejaculatory ducts are situated on each side of the orifice. The prostatic utricle is a remnant of the female reproductive tract. It is a blind diverticulum about 6 mm in length that extends upward and backward into the prostatic substance.

The proximal urethral segment is surrounded by a layer of smooth muscle fibers, forming the preprostatic sphincter. Tiny ducts and abortive acinar systems are scattered along the length of this segment. The distal prostatic urethra is the segment below the openings of the ejaculatory ducts. The distal segment possesses a thin layer of smooth muscle. It is also surrounded by a sphincter with small-diameter striated muscle fibers separated by connective tissue.

C. Blood and Lymphatic Supply

The primary artery of the prostate is a branch of the inferior vesical artery that arises from the hypergastric artery. The veins from the prostate drain into a series of large channels in the lateral prostatic ligaments, forming the prostatic venous plexus. The veins in the puboprostatic space join the veins from the penis forming the plexus of Santorini. The lymphatics surrounding each prostatic acinus unite to form channels that course to the surface of the prostate forming the periprostatic plexus. They eventually join the lymph channels from the urethra and drain into the internal iliac nodes.

D. Innervation

The human prostate receives an abundant innervation from both sympathetic (noradrenergic) and parasympathetic (cholinergic) nerves via the prostatic nervous plexus. The parasympathetic fibers are derived input from the sacral segment of the spinal cord (S2–S4), whereas the sympathetic fibers are derived input from the hypergastric nerves (T10–L2). Intrinsic adrenergic innervation of the prostate appears to course along the capsule and trabecular smooth muscle, and this innervation is prominent around the ducts of the prostate. There is evidence indicating that expulsion of the contents of the prostatic ducts during emission is under adrenergic control. Intrinsic cholinergic innervation appears to be much less dense than adrenergic innervation. Cholinergic nerves are related closely to the glandular epithelium, suggesting a role for them in prostatic secretion. Neural peptides have been identified in the nerve fibers of the prostate, including vasoactive intestinal polypeptide, neuralpeptide Y, substance P, enkephalins, calcium gene-related peptide, and somatostatin. The role of nitric oxide as a regular prostatic secretion and excretion is now being considered. While the secretory and contractile role of the autonomic nerve supply has wide acceptance, a prostatic trophic role for these nerves or neural transmitters has not been disseminated, despite strong evidence to support their existence.

III. EMBRYOLOGY AND DEVELOPMENTAL BIOLOGY

The prostate gland develops from the pelvic portion of the urogenital sinus at about 10–12 weeks of gestation. The ductal network within the prostate originates from the prostatic buds. These buds arise from the endodermal urogenital sinus immediately below the urinary bladder. The buds penetrate into the Müllerian mesoderm, which develops into the utricle, and the mesonephric mesoderm, which develops into the ejaculatory ducts. Furthermore, these buds proliferate rapidly, lengthen, arborize, and canalize. Ultimately, the most extensive areas of proliferation of the prostate occur adjacent to the areas of the ejaculatory ducts and the verumontanum. The endodermal buds invade the abundant surrounding mesenchyme of the urogenital sinus, which is responsible for the development of the definitive prostate. By 13 weeks of gestation, 70 primary ducts are present and exhibit secretory cytodifferentiation. Normal development of the Wolffian and Müllerian ducts and urogenital sinus-derived structures is dependent on the support of androgen and Müllerian-inhibiting substance of the fetal testis that begins at about the eighth week of gestation. Unlike the development of the Wolffian duct derivatives, which are dependent on testosterone, the growth and differentiation of the prostate is dependent on 5α-dihydrotestosterone.

IV. PHYSIOLOGY

The prostate is a male sex accessory organ. Its growth and development depend on endocrine control. The prostate is an exocrine gland, and its secretions compose 15% by volume of the seminal fluid. Prostate secretions play a role in the dissolution of coagulated semen upon ejaculation. Prostatic fluid is a highly complex and heterogeneous mixture of organic and inorganic compounds. Compounds such as zinc, magnesium, calcium, and citrate in the seminal fluid are mainly derived from the prostate. Nitrogenous compounds, phosphorylcholine and polyamines, are found in abundance in prostatic fluids. A host of proteolytic enzymes are found in prostatic secretions. Prostatic-specific antigen, prostatic acid phosphatase, and human kallikrein-2 are the most abundant proteases in prostatic secretions. Their presence in the circulation has been used as a marker for the diagnosis and management of prostatic malignancy.

Prostatic secretory activity is under both endocrine and neural controls. Androgen is required for this activity. The degree of androgen stimulation determines the amount and the composition of prostatic secretions. Stimulation of appropriate pelvic nerves results in production and/or expulsion of prostatic fluid. Administration of sympathetic and parasympathetic agonists leads to increased secretions of prostatic fluid by different mechanisms. α-Adrenergic receptor agonists cause expulsion of fluid by smooth muscle contraction. Parasympathetic agents cause increased secretion of fluid that has a different composition from that induced by androgen.

V. ENDOCRINOLOGY

The prostate is the target tissue of many endocrine systems. Although androgen plays a significant part in prostatic development, growth, and maintenance, other endocrine factors participate to provide a homeostatic environment.

A. Androgen

The prostate is dependent on androgen for growth, differentiation, and maintenance of structural and functional integrity. The testis is the major source of androgen. Removal of androgenic support can be accomplished by bilateral orchiectomy that will lead to rapid prostatic regression. The subsequent replacement of androgen to a castrated host will reactivate prostatic growth. Therefore, androgen is by far the most potent mitogen to the prostate, and testosterone is the major circulating androgen. However, the proximal mitogen for the prostate is 5α-dihydrotestosterone, which is converted from testosterone by 5α-reductase This enzyme has been localized in the human prostate. Thus, this is the rationale for using 5α-reductase inhibitors to decrease the level of 5α-dihydrotestosterone in the prostate in the clinical treatment of benign prostatic hyperplasia.

B. Nonandrogenic Role of the Testis

A widely held view is that the role of the testis in prostatic growth is through its ability to produce androgen. Recent experimental results support strongly the concept that the testis also produces a nonandrogenic substance(s), most likely protein in nature, which works in concert with androgen to promote prostatic growth. This nonandrogenic testicular factor can either stimulate prostatic growth directly or act indirectly to enhance the sensitivity of the prostate to androgen. This concept offers a possible explanation for the paradox of continued prostatic growth in aging men in the face of a declining level of circulating androgen.

C. Estrogen

A role for estrogen in prostatic growth has been proposed. Estrogen receptors have been detected in prostatic stromal cells. Therefore, it was theorized that the stroma may be the target of estrogen in prostatic growth. Recently, a new member of the estrogen receptor family (estrogen receptor-β) has been recognized. Its role in prostatic growth remains to be determined. In addition, estrogen action in prostatic growth can also be mediated through an interaction with sex hormone-binding globulin.

D. Other Endocrine Factors

Prolactin has a positive effect on the prostate. It can synergize with androgen to promote prostatic growth. It can also delay the rate of castration-induced regression in the prostate.

Oxytocin has been implicated in prostatic growth. Oxytocin has been shown to increase 5α-reductase activity. An increase in levels of oxytocin has been associated with benign growth of the prostate in men and in dogs.

Both receptors for luteinizing hormone (LH) and gonadotropin-releasing hormone (GnRH) have been detected in prostate cancer cells. Therefore, in addition to their effects on circulating levels of androgen, these hormones of pituitary origin may have a direct impact on the prostate, especially the malignant prostate.

Other hormones, such as progesterone, relaxin, activin, and inhibin, have been implicated for having a role in prostatic growth. These compounds have been detected in the prostate. However, their functional role in prostatic growth is still to be determined.

VI. THE PROSTATIC DUCTAL SYSTEM

A prostatic ductal system is defined as a single prostatic functional unit in which all glandular structures share a single drainage duct into the urethra. Each prostatic ductal system can be traced from its opening on the urethra wall. These orifices lead to individual tubular structures from which branches and subbranches are formed in a manner like the branching pattern of a tree. The human prostate consists of more than 30 such ductal systems. In the rat ventral prostate, each lobe consists of 8 such ductal systems. The rat ductal system is simpler in structural organization than that of the human counterpart. In the following sections, the prostatic ductal system of the rat will be discussed.

A. Local Variation in Epithelial Cells

Studies in the rat prostate revealed that the axis of the prostatic ductal system may be divided into three regions. Owing to the distance from the urethral orifice of the duct, the entire axis of the ductal system can be designated as proximal, intermediate, and distal regions (Figs. 4 and 5). Morphological features of epithelial cells are different in these re-

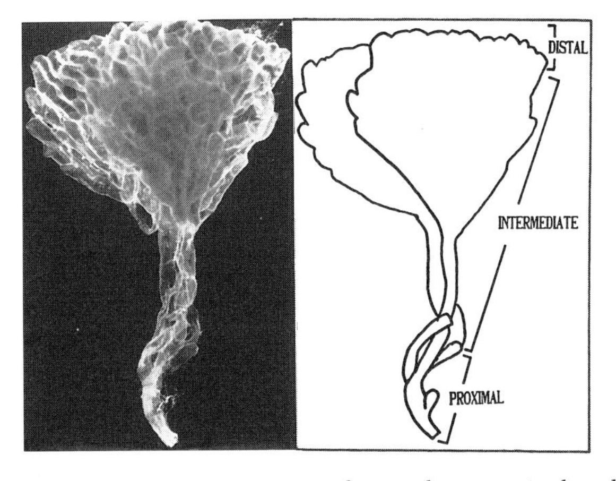

FIGURE 4 Gross appearance of a single prostatic ductal system of the ventral prostate from an adult rat. (Right) A diagrammatic presentation of the ductal system on the left. Note the branches and subbranches extending from the proximal end to distal tips of the ductal system. Actual size = 3 mm (reproduced with permission from C. Lee, J. A. Sensibar, S. M. Dudek, *et al., Biol. Reprod.* 43, 1079, 1990).

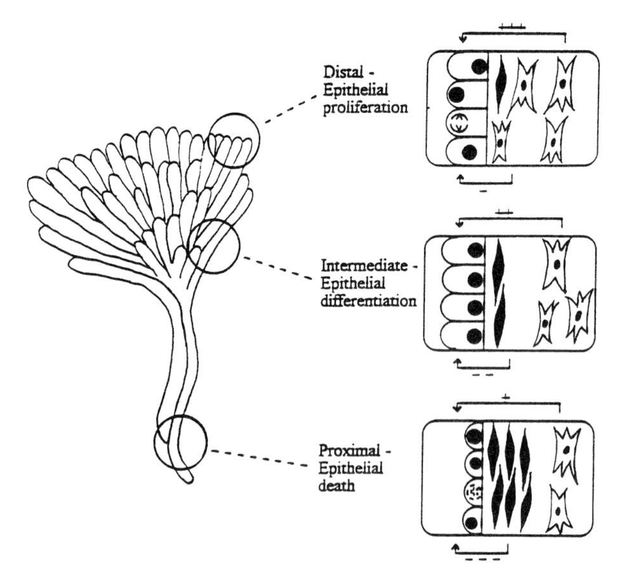

FIGURE 5 Proposed effects of stromal heterogeneity within the prostatic ductal system. In the distal region, abundant fibroblasts may elaborate positive growth factors, inducing epithelial proliferation. Increased smooth muscle cells in the intermediate region may block the positive factors or produce inhibitory factors for epithelium, yielding mitotic quiescence. In the proximal region, high levels of negative growth factors derived from the smooth muscle cells may induce epithelial apoptosis (from J. A. Nemeth and C. Lee, *Prostate* **28**, 124, 1996; reproduced with permission of Wiley-Liss, Inc., a subsidiary of John Wiley & Sons, Inc.).

gions. In the distal region (the tips of the ductal system, equivalent to the top of the tree) the epithelial cells are tall and columnar in shape. Nuclei are frequently located in the apical portion of these cells. Mitotic figures can be identified occasionally in cells of this region. These cells do not contain specific differentiation proteins. These observations indicate that these cells are actively undergoing cellular proliferation and they do not produce prostate-specific differentiation products.

Epithelial cells in the intermediate region (equivalent to the majority of the body of the tree) are also tall and columnar. These cells have basally located nuclei and are mitotically quiescent. They have the ability to produce differentiation products. Cell death is not evident in either of these two regions.

Cells in the proximal region, a region immediately adjacent to the urethra (equivalent to the tree trunk), are low, cuboidal in shape, or flat. These cells do not produce differentiation proteins. Interestingly, they are actively undergoing cell death or apoptosis. These observations have permitted us to recognize that epithelial cells lining the prostatic ductal system are not the same. Rather, their shape and functional activities are different in different regions of the ductal system. The human prostate exhibits a similar but more complicated structure in the ductal system.

B. Local Variation in Stromal Cells

The prostatic stromal cells are a group of pleiomorphic mesenchymal cell types with the potential of changing cellular phenotype in response to the changing environment within the prostate. The exact cell type and nomenclature of prostatic stromal cells have been poorly characterized. Basically, two major types of stromal cells can be recognized (in addition to blood-borne cells, nerve cells, and endothelial cells) based on immunohistochemical studies. One type is smooth muscle α actin-positive and vimentin-negative cells, and the other type is smooth muscle α actin-negative and vimentin-positive cells. For the purpose of the present discussion, the former are designated as smooth muscle cells and the latter as fibroblasts.

One of the most prominent features of the prostatic stroma is the multiple layers of smooth muscle cells surrounding the proximal region. The number of smooth muscle cells declines as we trace upward along the axis of the ductal system. A continuous single layer of smooth muscle cells covers the ducts of the intermediate region. Smooth muscle cells surrounding the distal region become sparse and discontinuous. Therefore, there is a gradient in smooth muscle cells along the ductal system, being most abundant in the proximal region and least abundant in the distal region.

Areas in the prostatic stroma that are not occupied by smooth muscle cells are filled with fibroblasts. Fibroblasts in the prostate are often separated from epithelial cells by smooth muscle cells which are located immediately adjacent to epithelial cells. An exception is in the distal region, where smooth muscle cells are sparse and form a discontinuous layer surrounding the ducts. Therefore, in the distal region, fibroblasts have an opportunity for direct contact with epithelial cells (Fig. 5).

C. Stromal–Epithelial Interaction

The previous findings showing regional variation in stromal and epithelial cells have provided new insights into the fundamental biology of the prostate. A comparison between stromal and epithelial cells in different regions of the ductal system has revealed an interesting correlation. In the proximal region, there is extensive apoptosis in the epithelial population that is associated with multiple layers of smooth muscle cells in the stroma. On the other hand, in the distal region, the extensive proliferative activity in epithelial cells is associated with a sparse and discontinuous layer of smooth muscle cells. The distal region is also the portion of the ductal system in which fibroblasts come in contact with epithelial cells. The concept of prostatic stromal–epithelial interaction states that the fate of prostatic epithelial cells is dictated by the cellular activity of the adjacent stromal cells. With this concept, we postulate that smooth muscle cells are able to produce a substance(s) that is inhibitory to epithelial cells, and fibroblasts are able to produce a stimulatory substance(s). This hypothesis explains epithelial cell apoptosis in the proximal region and proliferation in the distal region. We also postulate that in the intermediate region, the stimulatory action of fibroblasts and the inhibitory action of smooth muscle cells are able to work in concert. The net result of this interaction is the mitotically quiescent and differentiated epithelial cells in this region. The following section provides partial support for this hypothesis.

D. Role of Growth Factors

Growth factors are a group of polypeptides that travel across small, discrete, and very localized distances to regulate cellular activities through interaction with respective receptors. They are potent mediators of cellular proliferation, differentiation, and apoptosis. Under normal conditions, growth factors mediate their action in the prostate through a paracrine mechanism, in which a growth factor is produced by one cell type (e.g., stromal cell) and impacts locally on another cell type (epithelial cell). The growth factors of interest in the prostate include epidermal growth factor, transforming growth factor-α (TGF-α), insulin-like growth factor, TGF-β, basic fibroblast growth factor, and keratinocyte growth factor (KGF). The present discussion will focus on two growth factors, KGF and TGF-β.

KGF is a stromally derived mitogenic growth factor. Epithelial cells are its targets because they contain KGF receptors. KGF is predominantly localized in fibroblasts of the prostate. Therefore, fibroblasts may be stimulatory to prostatic epithelial cells through, at least, KGF production. TGF-β is a potent inhibitor to prostatic epithelial cells and it induces epithelial apoptosis. TGF-β receptors are expressed predominantly in prostatic epithelial cells. TGF-β1 is localized in smooth muscle cells of the rat prostate. Therefore, smooth muscle cells may play a negative growth regulatory role in epithelial cells through the production of TGF-β1.

E. Action of Androgen

The recognition of a regional heterogeneity in cellular activity within the prostatic ductal system has inspired us to question the mechanism of androgen action in prostatic cells. The conventional concept of the action of androgen has been that of a stimulatory one. This is because prostatic growth requires the support of an adequate level of circulating androgen. However, current observations indicate that androgen can also stimulate epithelial cells to undergo cell death. We now realize that prostatic epithelial cells do not respond to androgen in a uniform manner, even though they contain the same level of androgen receptors. Some cells proliferate while others die in response to androgen in the same ductal system. These interesting observations have prompted us to search for a mechanism that can explain this seemingly paradoxical phenomenon. The recognition that prostatic smooth muscle cells are concentrated in the proximal region has allowed us to reevaluate the mode of androgen action in the prostate. Because smooth muscle cells in the prostate produce the inhibitory growth factor, TGF-β1, in response to androgen, the androgenic effect on prostatic epithelial cells in the proximal region can be considered inhibitory. On the other hand, fibroblasts are able to come in contact with epithelial cells in the distal region. Because fibroblasts produce KGF in response to androgen, the androgenic effect on prostatic epithelial cells in the distal region can be considered stimulatory. Therefore, the effect of androgen on prostatic epithelial cells is not necessarily stimulatory nor inhibitory. Rather, the outcome of an androgenic effect on epithelial cells depends on the composition of adjacent stromal cells.

F. Effect of Androgen Deprivation

Androgen deprivation, as by bilateral orchiectomy in a host, results in rapid prostatic atrophy that is associated with massive epithelial apoptosis. Changes in the stroma take place at a very slow rate. This event is associated with an inactivation of a set of androgen-dependent genes coupled with an activation of a series of other genes. An examination of cellular morphology in the prostatic ductal system during castration-induced atrophy has revealed an interesting phenomenon. Epithelial apoptosis is initiated at the distal region, where cell proliferation is prominent prior to castration. The apoptotic process spreads to the intermediate region during the peak of prostatic atrophy (Day 4 postcastration in rats). Surprisingly, at this stage, epithelial cells in the proximal region do not undergo apoptosis and remain viable for as long as androgen deprivation lasts. Therefore, a reverse gradient of epithelial cell viability exists along the axis of the ductal system during castration-induced prostatic atrophy. Again, the effect of androgen deprivation on prostatic epithelial cells is not exclusively cytotoxic. The outcome of the effect of androgen deprivation on prostatic epithelial cells varies according to the specific region of the ductal system that these cells occupy.

An important growth factor involved in castration-induced prostatic atrophy is TGF-β1. The expression of TGF-β1 in the prostate increased following castration and reached a peak by 8 days after castration. The expression TGF-β receptors also increased after castration. This increase in TGF-β signaling is consistent with an increase in epithelial apoptosis. A population of newly formed smooth muscle cells surrounds the ducts of distal and intermediate regions. These smooth muscle cells express TGF-β1. At the same time, stromal cells surrounding ducts of the proximal region show a reduction in the expression of both smooth muscle actin and TGF-β1. The prostatic ductal systems of long-term castrated rats displayed a similar reduction in smooth muscle actin and TGF-β1.

VII. IMPLICATIONS FOR PROSTATIC DISEASES

The previous discussion represents a brief account of the basic biology of the prostate. The prostate is a male sex accessory gland which is the site of frequent benign and malignant overgrowth, constituting a major health concern. An understanding of the prostate's anatomy, embryology, physiology, endocrinology, and ductal system will allow us to develop the means for prevention, diagnosis, and management of these disorders. The use of α-adrenergic blockers and 5α-reductase inhibitors for the treatment of benign prostatic hyperplasia are two examples. The recognition of the existence of a regional cellular heterogeneity within the prostatic ductal system has provided new insights into cellular mechanisms of androgen action. We now understand that the effect of androgen on prostatic epithelia is not a direct one. Rather, the outcome of the effect of androgen on proliferation or inhibition of epithelial growth depends on the type of adjacent stromal cells. This knowledge has not yet been applied to the treatment of prostatic cancer using the antiandrogenic approach.

There is a dynamic coexistence between epithelial and stromal compartments of the prostatic ductal system that is characterized by an active dialogue of cell-to-cell signaling, which influences proliferation, differentiation, and apoptosis. Deviation from the normal programming of the cell-to-cell interaction in either compartment may contribute to the abnormal control of prostatic growth, as is the case in benign prostatic hyperplasia and in prostatic cancer.

The regional cellular heterogeneity in the prostatic ductal system has important correlates with regard to the normal choreography of prostatic growth, spatial distribution of malignant transformation, and hyperplastic morphology. If epithelial homeostasis in the ductal system is maintained by a balance between cell proliferation in the distal tip and apoptosis at the proximal end, epithelial cells must migrate from the distal end toward the proximal end along the ductal system. Although analogous examples of cell migration have been observed in other tissues, evidence for this model has not been available in the prostate. A provocative corollary of this model is that the predominance of cellular proliferation at the distal tip could convey an association of an increased risk of malignant transformation. Autopsies of men who have died of prostatic cancer have shown that malignancy originates more commonly at the periphery of the gland, which corresponds to the distal region of the ductal system.

It has been speculated that benign prostatic hyperplasia develops from an imbalance of cell proliferation and cell death. The prostatic ductal system maintains its homeostasis through the mechanism of a highly orchestrated program of cell proliferation, cell migration, and cell death. Prostatic hyperplasia could result as a consequence of an inequality in the rate of cell proliferation, migration, and death in the ductal system.

VIII. CONCLUSION

This brief review of the biology of the prostate emphasizes the limitations of current knowledge in this field. The subjects of anatomy, embryology, physiology, endocrinology, and the ductal system are crucial to our understanding of diseases of the prostate. Many areas of the biology of the prostate merit continued and intensive investigation. These areas include the role of basal cells, the regionalization of stromal production of growth factors, the mechanism of androgen action, the impact of epithelial cells on the stroma, and the etiology of prostatic cancer and benign prostatic hyperplasia. It is anticipated that additional knowledge will facilitate the development of means for prevention, diagnosis, and management of prostatic diseases.

See Also the Following Articles

Growth Factors; Prostate Cancer; Prostate-Specific Antigen

Bibliography

Kuiper, G. G., Enmark, E., Pelto-Huikko, M., *et al.* (1996). Cloning of a novel receptor expressed in rat prostate and ovary. *Proc. Natl. Acad. Sci. USA* **93**, 5925–5930.

Lee, C. (1981). Physiology of castration-induced regression in rat prostate. In *The Prostatic Cell: Structure and Function* (G. Murphy, A. A. Sandberg, and J. P. Karr, Eds.), pp. 145–159. A. R. Liss, New York.

Lee, C. (1997). Biology of the prostatic ductal system. In *Prostate: Clinical and Basic Aspects* (R. Nez, Ed.), pp. 53–71. CRC Press, Boca Raton, FL.

Lee, C., and Grayhack, J. T. (1989). Overall perspectives on the role of prolactin in the prostate. In *Prolactin and Lesions in Breast, Uterus, and Prostate* (H. Nagasawa, Ed.), p. 153. CRC Press, Boca Raton, FL.

Lee, C., Kozlowski, J. M., and Grayhack, J. T. (1997). Intrinsic and extrinsic factors controlling benign prostatic growth. *Prostate* **31**, 131–138.

Mann, T., and Lutwak-Mann, C. (1981). *Male Reproductive Function and Semen*. Springer-Verlag, New York.

McNeal, J. E. (1988). Normal histology of the prostate. *Am. J. Surg. Pathol.* **12**, 619–633.

Nemeth, J. A., Sensibar, J. A., White, R. R., *et al.* (1997). Prostatic ductal system in rats: Tissue-specific expression and regional variation in stromal distribution of transforming growth factor-β1. *Prostate* **33**, 64–71.

Nemeth, J. A., Zelner, D. J., Lang, S., *et al.* (1998). Keratinocyte growth factor in rat ventral prostate: Androgen-independent expression. *J. Endocrinol.* **156**, 115–125.

Nicholson, H. D., and Jenkin, L. (1995). Oxytocin and prostatic function. *Adv. Exp. Med. Biol.* **395**, 529–538.

Rosner, W. (1994). Estradiol causes the rapid accumulation of c-AMP in human prostate. *Proc. Natl. Acad. Sci. USA* **91**, 5402–5405.

Prostate-Specific Antigen

Ricardo Beduschi
The University of Michigan

Joseph E. Oesterling
Ann Arbor, Michigan

GLOSSARY

free prostate-specific antigen (PSA) Proportion of total serum PSA which is not complexed with serine protease inhibitors present in the serum.

prostate cancer The most frequently detected malignancy in men and the second leading cause of cancer death in the male population.

prostate-specific antigen Glycoprotein present in the human serum recognized as the best tumor marker available for prostate cancer.

radical prostatectomy Surgical ablation of the prostate; the most used definitive treatment for prostate cancer.

tumor marker A specific and easily accessible protein that can be used for the evaluation and management of a certain malignancy.

Prostate-specific antigen (PSA) was identified in the seminal plasma almost a quarter of a century ago, and since then it has been intensively investigated by those devoted to the cause of understanding prostate cancer. Years of clinical and basic research has made PSA a superior clinical tool for the diagnosis and management of this malignancy. Several studies have demonstrated that PSA alone detects considerably more malignancies than digital rectal examination or any other isolated parameter. Of importance, most (85–90%) of these tumors have histopathological features of clinically important, life-threatening cancer.

I. INTRODUCTION

Although widely accepted as the clinical tool of choice for diagnosing and monitoring prostate malignancy, prostate-specific antigen (PSA) is by no means perfect. It is neither cancer nor tissue specific, and alone it is not sensitive or specific enough to be considered the "perfect" tumor marker. This lack of sensitivity and specificity has led researchers to investigate various methods with the purpose of optimizing the clinical use of PSA by improving its diagnostic accuracy. Concepts such as PSA density, PSA velocity, and age-/race-specific PSA reference ranges have been proposed and tested in clinical practice. Although useful, they do not address all the concerns. Moreover, investigations have shown that PSA is present in the human serum in several molecular forms and that the concentration of free and complexed PSA (PSA bound to α1-antichymotripsin) may vary according to specific pathologic states present in the prostate gland.

Despite all that is yet to be learned about PSA, it is clear that PSA brought the diagnosis of prostate cancer into a new era, and apparently it will continue to play a major role in the management of this disease in upcoming years. The objective of this article is to review what is known about PSA, the most used clinical tumor marker currently available in medicine.

II. HISTORICAL FACTS ABOUT PSA

Several independent groups in the early 1970s that were searching for more specific markers than acid phosphatase, which is used in forensic medicine (investigation of rape), discovered prostate antigens which later would be identified as the same molecule, known today as PSA. PSA was first identified in seminal plasma by Hara *et al.* in 1971, who initially called it "γ-seminoprotein." Independently, Li and Beling, using chromatographic techniques, also identified a protein from human seminal plasma, which they called E_1 because of its electrophoretic mobility. The E_1 antigen was characterized as a major antigenic component of human seminal plasma. In 1978, Sensabaugh determined an antigen in the human semen with a molecular weight of approximately 30 kDa that he named p30, and further studies revealed the prostate gland as the most likely site for the synthesis of this protein. In 1979, Graves and associates developed an immunoassay for p30 and applied it to the identification of rape victims. In the same year, Wang and associates isolated what was initially thought to be a tissue-specific antigen and called it "prostate-specific antigen." This antigen was found in both benign and malignant prostatic tissue. Two years later, the same authors concluded that this tissue protein was immunologically identical to that originally discovered in seminal plasma. Further comparison between sera and prostatic tissue obtained from patients with prostatic malignancy also revealed immunologic identity. Papsidero, extending investigations on PSA to clinical settings, examined serum PSA from patients with prostate cancer and in normal patients using immunoelectrophoresis and noticed activity in patients with advanced prostate cancer as opposed to normal patients. These initial findings led to the development of a more sensitive immunoenzymatic assay capable of detecting 0.1 ng/ml of PSA. Kuriyama *et al.* also demonstrated the potential of PSA as a prostatic tumor marker. Since then, there have been many applications of PSA to detect prostate cancer , and today the scientific literature contains over 3000 articles on this topic.

III. BIOCHEMICAL CHARACTERISTICS OF PSA

PSA is a single-chain serine protease of 237 amino acids and very similar in primary structure to other proteases of the human kallikrein family. It is encoded by a gene designated *hKLK3*, which is located on the long arm of chromosome 19. The molecular weight of PSA has recently been determined to be 28.43 kDa. The molecular structure of the molecule is composed of five disulfide bonds due to the presence of 10 cysteine residues, and the active site of the enzyme is composed of 3 amino acids: histidine 41, aspartate 96, and serine 189. The expression of the PSA gene is androgen dependent and mainly, but not only, restricted to the prostatic epithelium. After being synthesized in the ductal epithelium and acini of the prostate, PSA is secreted into the lumina of the prostatic ducts to become a component of the seminal plasma. In the seminal plasma, it acts as serine protease, exhibiting proteolitic activity similar to chymotrypsin. Its physiological function is to liquefy the seminal coagulum that forms at ejaculation, releasing the spermatozoa. A recent investigation suggests that PSA may also modulate cell growth and prostate cancer activity.

Despite original beliefs, PSA is neither a tissue-specific antigen nor a gender-specific antigen. Through immunohistochemical and immunoassay methods, this protein has also been detected in female and male periurethral glands, anal glands, normal and malignant breast tissue, ovary, endometrium, liver, colon, kidney, and virtually any tissue expressing hormone receptor activity.

IV. MOLECULAR FORMS OF PSA AND THEIR CONCENTRATION IN SERUM AND SEMINAL FLUID

Subsequent investigation into the molecular and biochemical properties of PSA has revealed that PSA exists in several forms and in different concentrations in serum and seminal fluid. The concentration of PSA in serum is approximately 106-fold lower than that in seminal fluid and normally maintained below 3.0 ng/ml. PSA is capable of forming covalently linked complexes with certain serine protease inhibitors found in the serum, and as a result it exists in the serum in several molecular forms (Table 1). The main forms include a noncomplexed or free form and complexes of PSA with the serine protease inhibitors, α1-antichymotrypsin (PSA-ACT) and α2-macroglobulin (PSA-MG). Other minor forms include complexes with protein C inhibitor (PSA-PCI), α1-antitrypsin (PSA-AT), and inter-α-trypsin (PSA-IT). Total PSA refers to the sum of all immunodetectable species of PSA, primarily free PSA and PSA-ACT. The predominant bound form of PSA is the PSA-ACT, which accounts for approximately 90% of all complexed forms of PSA. Until recently, the PSA-MG form was undetectable with available immunoassays. An assay capable of measuring the PSA-MG form has recently been developed, and further information regarding this molecular form of PSA should be available soon. PSA bound to PCI has not been found in plasma from patients with prostate cancer, whereas PSA-IT complexes have not yet been studied in detail. Free PSA possesses at least one antigenic epitope that becomes unavailable once binding with ACT occurs. Antibodies targeted against this free epitope represent the basis for the development of immunohistochemical assays that distinguish free from total PSA.

TABLE 1
Molecular Forms of PSA[a]

Formal name	*Common name*	*Description*
Total PSA	t-PSA	All immunodetectable forms in serum; primarily f-PSA and complexed PSA
Free PSA	f-PSA	Noncomplexed PSA; may be proteolytically active or inactive in seminal fluid; only inactive in serum
PSA complexes	PSA-ACT	PSA covalently bound to α1-antichymotrypsin inhibitor; major immunodetectable form in serum
	PSA-AMG	PSA covalently linked and encapsulated by α2-macroglobulin; not detected by any available immunoassay
	PSA-PCI	PSA covalently bound to protein C inhibitor; not detected in serum
	PSA-AT	PSA covalently bound to α1-antitrypsin; trace amounts in serum
	PSA-AT	PSA covalently bound to α1-antitrypsin; trace amounts antitrypsin; trace component in serum
	PSA-IT	PSA covalently bound to inter-α trypsin; trace amounts in serum

[a] Adapted from R. T. McCormack, H. Lilja, J. E. Oesterling *et al.*, Molecular forms of prostate-specific antigen and the human kallikrein gene family: A new era. *Urology* 45, 729–744, 1995.

In the seminal fluid, PSA has been found in glycosylated forms and in a nicked or inactive form (in which the bonds at specific residues have been cleaved). Most PSA isolated from seminal fluid is enzymatically active, whereas 15–30% is found in the nicked (inactive) form. Due to a very high concentration of PSA, the extracellular protein inhibitors do not play a significant role in seminal plasma. The main serine protease present in seminal fluid is PCI, which complexes 5% of PSA.

V. PSA ASSAYS

The presence of antigenic epitopes with unique binding characteristics distributed over the PSA molecule provides the necessary features for the development of a series of very specific and sensitive assays to be used in determining the presence and concentration of the various PSA components in body fluids. Most of the immunologic agents used in PSA assays are generated against free PSA. The polyclonal antibody conjugates used in some commercial assays usually include a subpopulation of antibodies that bind to the region of the PSA molecule to which ACT binds. Consequently, ACT will block the epitopes and preclude binding of this polyclonal subpopulation to the complex. As a result, fewer polyclonal antibodies will bind to each molecule of PSA-ACT than to free PSA, attenuating the response to PSA-ACT. This explanation represents the basis for the development of assays that can distinguish free from total PSA.

Currently, the following five PSA assays have FDA approval for clinical use: Tandem-R assay (Hybritech, San Diego, CA), Tandem-E assay (Hybritech, San Diego, CA), IMx assay (Abbott Laboratories, Chicago, IL), Tosoh assay (Tosoh), and Immunolite assay (Diagnostic Products Corporation). Despite some individual intrinsic characteristics, the clinical results provided are very similar although not identical.

VI. THE ROLE OF PSA IN EARLY PROSTATE CANCER DIAGNOSIS

Although discovered 25 years ago, PSA did not receive significant attention as a clinical tool for the early detection of prostate cancer until 1986. Classically utilized as a clinical marker for follow-up of patients who had undergone treatment for prostate cancer, only in recent years has PSA been recommended and widely used as a screening test for this malignancy. Today, PSA is considered the most valuable clinical tool available for the diagnosis of prostate cancer. Serum PSA detects twice as many prostate cancers as digital rectal examination (DRE), and nearly two-thirds of these tumors are early stage, potentially curable cancers. Several recent studies have shown that serum PSA, when used as a screening test, increases the detection rate of prostate cancer above that achieved with DRE.

Although unique, PSA has important limitations that prevent it from being the ideal tumor marker. PSA is not cancer specific, which limits the marker's ability to differentiate malignancy from benign conditions that can frequently produce elevations in the serum PSA concentration, such as benign prostatic hyperplasia (BPH), acute prostatitis, acute urinary retention, and prostatic ischemia. Despite several studies demonstrating that the median PSA concentration is significantly and consistently higher for men with organ-confined prostate cancer than for men with BPH, it is also true that serum PSA elevation is not always observed in men with prostate cancer. In order to reliably detect prostate malignancy at an early stage, a low serum PSA cutoff level is used in screening programs (4.0 ng/ml). Use of this cutoff level is associated with a considerable number of false-positive findings and false-negative findings (65% false-positive and 20% false-negative rate), demonstrating PSA's lack of power in discriminating malignant from benign conditions of the prostate. On the other hand, at least 20% of men with biopsy-proven cancer at the time of diagnosis have a serum PSA concentration within the reference range of 0.0–4.0 ng/ml. If the total serum PSA concentration had been solely used, these diagnoses would have been missed. This lack of sensitivity and specificity has led researchers to propose various methods as a means of improving the ability of total serum PSA in detecting prostate cancer.

VII. IMPROVING BOTH THE SENSITIVITY AND SPECIFICITY OF PSA TESTING

To improve the ability of this tumor marker to detect clinically significant prostate cancer the capability of PSA to distinguish BPH from prostate cancer and to identify clinically significant, nonpalpable cancer in a reliable manner must be enhanced. With the purpose of optimizing the clinical use of PSA, several concepts have been developed and tested in clinical practice.

A. PSA Density

PSA density (PSAD) is mathematically defined as the quotient of the serum PSA concentration divided by the volume of the prostate gland as determined by transrectal ultrasonography. Thus, it correlates the serum PSA concentration with the size of the prostate gland and it is based on the premise that a mildly elevated serum PSA level associated with a small prostate may indicate prostate cancer, whereas the same PSA level in a patient who has a large gland may indicate BPH only. Despite the initial enthusiasm for this new idea, some later reports have shown that PSAD fails to demonstrate improved cancer detection when compared to total PSA.

Currently, the use of PSAD is recommended for patients who have a PSA value in the high-normal or mildly elevated range (4.0–10.0 ng/ml). In this range, PSAD may be useful in avoiding unnecessary negative biopsies. Meanwhile, when age-specific reference ranges are used in a screening population, PSAD does not provide additional relevant information when compared to PSA alone. Currently, PSAD may be considered a second-line screening parameter to safely reduce the number of biopsies performed in patients with negative DRE and TRUS results and a serum PSA level below 10.0 ng/ml but above the age-specific limit of normal. Patients with a PSAD >0.15 are at a higher risk for harboring prostate cancer.

B. PSA Velocity

PSA velocity (PSAV) is defined as the change in serum PSA over time. This concept was introduced as an attempt to identify cancers that could not be detected by total PSA or DRE. The rationale for developing such a concept was to use serial serum PSA measurements to evaluate men at risk for harboring prostate cancer in order to avoid the interrelated variables of PSA concentration, BPH, cancer volume, and differentiation that can affect a single PSA value. It is known that the serum PSA concentration is directly proportional to the amount of benign epithelium and cancer volume. However, a given volume of poorly differentiated cancers produces less PSA than an equal volume of a well-differentiated prostate tumor. PSA velocity is extremely helpful for monitoring men with rising PSA levels and a total PSA in the normal range. In this range, it helps to differentiate between the expected age-related PSA rise and the increase in PSA that suggests the presence of prostate cancer and the need for prostate biopsy. If PSAV is determined at 1.5- to -2.0-year interval, <5% of men without prostate cancer will have a PSAV of 0.75% ng/ml per year or greater and more than 70% of men with prostate cancer have a PSAV of at least 0.75% ng/ml per year. Thus, every patient with a normal PSA value and PSAV >0.75 ng/ml per year should have a prostate biopsy to exclude prostate malignancy.

C. Age-/Race-Specific Reference Ranges

This concept was developed and introduced by Oesterling and colleagues in 1993. Assessing the relationship between age, PSA, and prostate volume, they demonstrated that the serum PSA concentration increased over time in a population with no evidence of cancer at a ratio of 0.04 ng/ml/year. They also noticed a closer relationship of serum PSA with prostate volume (r = 0.55) rather than with age (r = 0.43). Statistical analysis of these variables showed that 30% of the serum PSA variance was due to prostate volume and 5% was accounted for by age. From these data and using the 95th percentile, age-specific reference ranges were calculated and recommended for use in clinical practice (Table 2). The use of these ranges potentially increases the sensitivity of PSA by detecting more cases of organ-confined disease in younger men while increasing specificity of PSA by eliminating prostate biopsies in older men. Subsequent reports by other authors corroborated these findings, indicating that serum PSA was related to age even after adjusting for volume of the gland.

Recently, the normal age distribution of serum PSA was reevaluated by Anderson *et al.* They concluded that the age-specific reference range proposed by Oesterling had upper limits that were too high for men younger than 60 years of age and potentially too low for 70- to 79-year-old men. Therefore, they

TABLE 2
Age-Specific Reference Range for Serum PSA[a]

Age range	*Reference range (ng/ml)*		
	Asians	*Blacks*	*Whites*
40–49	0–2	0–2	0–2.5
50–59	0–3	0–4	0–3.5
60–69	0–4	0–4.5	0–4.5
70–79	0–5	0–5.5	0–6.5

[a] From T. Richardson and J. E. Oesterling, Age-specific reference ranges for serum PSA. *Urol. Clin. North Am.* **24**(2), May 1997.

proposed the following upper limits of normality: 1.5 ng/ml for men 40–49 years of age, 2.5 ng/ml for men 50–59 years, 4.5 ng/ml for men 60–69 years, and 7.5 ng/ml for men 70–79 years. These limits still maintain 95% specificity across all ages. Recent investigations have supported the use of age-specific reference ranges in men <60 years of age, although some investigators believe that this concept requires further investigation before it should be widely utilized for older men.

Although it demonstrated the effectiveness of the new concept in improving the clinical utility of the PSA test, this study had one important limitation. The data were generated from a homogeneous group of white men, which could represent a bias since the disease may have different behavior among different populations. To address this matter, Oesterling *et al.* studied a similar cohort of healthy Japanese men and found lower age-specific reference ranges. Morgan and coworkers, investigating the use of this concept in a large group of black men, concluded that the previous ranges determined for white men would have missed 41% of the cancers in this population. Therefore, a different set of age-specific ranges was established for black men. The recommended age-/race-specific reference range is shown in Table 2.

Age-/race-specific reference range should be the starting point for enhancing PSA testing performance. If a patient presents with an abnormal PSA (Table 2), he should undergo further prostatic evaluation either by another concept or by prostate biopsy. If the PSA level is normal according to the age-/race-specific reference range and the DRE is normal, the patient should be followed periodically with prostatic evaluation and PSA velocity.

D. Percentage Free PSA (Proportion of Free PSA to Total PSA)

Since the discovery that PSA exists in the serum in several molecular forms, free PSA (noncomplexed form of PSA) and its proportion to total PSA (percentage free PSA) have been under intense investigation as a mean of increasing the clinical utility of PSA in the early detection of prostate cancer. Preliminary investigations showed that men with prostate cancer have a higher proportion of PSA complexed to ACT than men with BPH. Thus, measuring this proportion could be a more reliable method than total PSA alone to differentiate benign from malignant prostate disease. Lilja and coworkers determined the serum concentration of the molecular forms of PSA in men with prostate cancer and BPH. The proportion of free PSA to total PSA (percentage free PSA) was shown to be significantly lower in patients with prostate cancer (mean value, 0.18) than in those with BPH (mean value, 0.28). This difference was statistically significant when total PSA was <10 ng/ml. Subsequently, Christensson and coworkers demonstrated a substantial increase in specificity from 55 to 73% and a minimal decrease in sensitivity when using a percentage free PSA cutoff of 0.18. With these promising initial findings, further studies were conducted to define the true role of the percentage free PSA in prostate cancer diagnosis. The results pointed toward the use of percentage free PSA as a means of increasing cancer detection and eliminating negative prostate biopsies in patients with total PSA values in the upper aspect of normal or mildly elevated. Later, an optimal range of the total PSA for the use of percentage free PSA was defined. Percentage free PSA offers the greatest advantage to the PSA test when the total PSA value is between 3.0 and 10.0 ng/ml. The ability of percentage free PSA to distinguish early, curable cancer from BPH over this range is superior to total PSA by more than 20%. Thus, the range of 3.0–10.0 ng/ml for total PSA appears appropriate for the clinical use of percentage free PSA. This range is called the "reflex range," indicating that for these total PSA concentrations, the laboratory instrument would go into a reflex mode, measuring the amount of free PSA in the sample and calculating the percentage free PSA value for the specific sample. There is also a clinical rationale for accepting this reflex range of 3.0–10.0 ng/ml. It is based on the fact that the risk of prostate cancer is extremely low if total PSA value is <3.0 ng/ml and extremely high (over 50%) when PSA level is >10.0 ng/ml, independently of the percentage free PSA value.

Several studies from different institutions, evaluating a considerable number of different assays, have confirmed the ability of percentage free PSA to improve specificity in detecting prostate cancer during screening programs with no loss of sensitivity. This fact is especially noticeable for the subset of patients whose total PSA values are mildly elevated (4.1–10.0 ng/ml).

In addition, some investigators have suggested a role for percentage free PSA in reducing the number of prostate biopsies in men at high risk of developing prostate cancer. Today, most urologists recommend prostate biopsy only for men whose PSA levels are higher than 4.0 ng/ml. However, 13–20% of men with PSA levels between 2.6 and 4 ng/ml are likely to be diagnosed with prostate cancer within 3–5 years.

Addressing the issue of whether or not percentage free PSA can eliminate negative biopsies, Vashi and colleagues concluded that for total PSA of 3.0 or 4.0 ng/ml, using a 19% cutoff for percentage free PSA would result in detection of 90% of all cancers; the biopsy rate would be 73%, and the cancer detection rate would be 44%. Thus, for every 1.7 biopsies performed, one cancer is detected. For the 4.1–10.0 ng/ml range of total PSA, the most appropriate cutoff for percentage free PSA is 24% if the objective is to decrease the number of unnecessary biopsies (increase specificity). At this value, 95% of the cancers would be detected and 13% of the negative biopsies could be avoided. Thus, for seven negative biopsies avoided, one cancer goes undetected. However, with annual prostatic evaluations, the truly significant prostate cancer would still be detected at a curable stage considering that the doubling time of these tumors is approximately 3 or 4 years.

Despite some controversies regarding the issue of whether free PSA varies with age, the use of percentage free PSA is recommended as a way to discriminate patients with early prostate cancer from men with BPH in the total PSA "reflex range" of 2.5–3.0 to 10.0 ng/ml. In this range, percentage free PSA improves the clinical performance of the PSA test by 20–25%. With the total PSA level of 4.0 ng/ml or less, percentage free PSA enhances cancer detection with an acceptable increase in the number of prostate biopsies performed. When the total PSA value is slightly elevated (4.1–10.0 ng/ml), percentage free PSA decreases the number of unnecessary biopsies. Both of these outcomes are highly desirable when attempting to diagnose organ-confined prostate cancer in a diligent manner.

VIII. THE ROLE OF PSA IN DETERMINING THE PRESENCE OF PROSTATE CANCER

The advent of PSA has definitively made a difference in the way that prostate cancer is currently managed. Currently, there is a consensus that serum PSA is the best clinical tool not only for the diagnosis but also for monitoring treatment outcome for clinically localized prostate cancer, regardless of the adopted therapeutic option.

Pretreatment PSA level has been shown to be a powerful predictor of subsequent rises in PSA despite the elected treatment. It is well established that therapeutic success is highly related to preoperative PSA levels. Men diagnosed with prostate cancer who have serum PSA levels of 20 ng/ml or higher have less of a chance of remaining free of recurrence than men who have serum PSA levels below 20 ng/ml, and men with serum PSA levels of 4.0 ng/ml or less tend to have a favorable outcome despite the therapeutic option they choose.

PSA has also been established as the most important intermediate end point predictor of therapeutic success or failure because it is generally assumed that a rising PSA is indicative of active, progressive disease. On the other hand, the assumption that a rising PSA level correlates with survival has not been validated. The available data are insufficient to predict when this biochemical failure (PSA rising) will be translated in clinical symptoms and if these will cause the death of individuals with PSA failure.

Another controversial issue regarding the use of PSA in the follow-up evaluation of men who have undergone definitive treatment for prostate malignancy is the definition of what would be the ideal PSA range for each therapeutic option available. An undetectable PSA level is the expected and the accepted definition of cure following definitive prostate cancer therapy. Nevertheless, the so-called "undetectable" ranges vary according to the elected treatment, and a consensual value has not been defined.

Furthermore, the assays used to measure the serum PSA level, although similar, are not identical, and physicians should be aware of this before assuming any rise in PSA as a biochemical failure. Addressing this matter, specific undetectable ranges or residual cancer detection limits (RCDLs) have been defined for the emerging assays used in serum PSA determination. This has made it possible for the practicing urologist to easily assess the presence of an elevated serum PSA concentration following prostate cancer treatment. The Tandem-R PSA assay had a RCDL estimated value of 0.2 ng/ml because all treated patients presenting this serum PSA concentration (or a rising PSA pattern) developed recurrent, progressive disease with time. The IMx PSA assay has a lower RCDL value of 0.1 ng/ml. Recently, a third generation of ultrasensitive PSA assays was introduced. With analytical threshold levels <0.05 ng/ml, they have increased our ability to detect biochemical relapse in patients who have undergone radical prostatectomy at a much earlier phase.

After surgical removal of the prostate, which is considered the most effective therapy for localized prostate cancer, an undetectable serum PSA level serves as the gold standard for disease-free status. The majority of patients presenting a rising PSA pattern following radical prostatectomy will eventually develop clinical evidence of persistent disease. A PSA level higher than the predicted value immediately after or within 2 years of the surgery usually indicates residual disease, whereas a PSA level rising after 2 years of surgery usually reflects local recurrence. Assessing PSA kinetics in the postoperative period may allow some distinction regarding the anatomical site of the persistent tumor (local or distant). The median PSA doubling time for those patients that progressed to distant metastases was 4.3 months compared to 11.7 months for the group that had clinically detected local failure. PSA velocity is another useful concept for monitoring posttreatment of prostate cancer. A PSA velocity of 0.75 ng/ml/year is observed in more than 90% of patients with local recurrence, and 50% of patients with distant lesions had a PSA velocity >0.75 ng/ml/year. The precise identification of the site of recurrence is essential in determining the most adequate adjuvant therapy for the patient displaying an elevation in the serum PSA concentration.

Following external radiation therapy (RxT), an elevated pretreatment total PSA value has been shown to correlate with treatment success rate in several postradiation studies. Patients with a serum PSA >15.0 ng/ml tend to have a low disease-free progression rate despite the clinical stage of the disease. Conversely, a PSA level <15.0 ng/ml was predictive of a different treatment outcome within the same stage groups. Noteworthy is the fact that the best RxT results are achieved with patients whose serum PSA level at diagnosis is <4.0 ng/ml. Studies of the alterations occurring in the serum PSA during and following RxT revealed a precipitous increase in the serum levels of this antigen, presumably resulting from cellular damage and subsequent spillage of high amounts of the PSA molecule into the blood circulation. Following this acute phase, serum PSA levels dropped rapidly for a period of 3 months. After this period, a continuous, slower fall in the PSA levels may persist for 12 or more months. The ideal PSA level at which a patient can be considered free of disease following RxT has not been established. Authors have reported several different PSA nadir values (4.0, 2.0, 1.5, 1.0, and 0.5 ng/ml), but no standard value has been determined. The time required to reach this level also does not seem to be relevant. Based on recent investigations, a serum PSA level <0.5 ng/ml seems appropriate for defining a successful outcome, and treatment failure should be acknowledged after three consecutive increases in the serum PSA level.

For patients who undergo antiandrogen therapy for prostate cancer, serum PSA monitoring allows a more adequate and thorough therapeutic approach and a more precise evaluation of the disease progression/relapse. For these reasons, serum PSA monitoring during hormonal therapy has assumed a fundamental role in the care of patients with advanced prostate cancer and an is an objective factor indicating candidates for nonhormonal salvage therapy. The concept of using PSA for monitoring hormone therapy for prostate cancer is based on the premise that PSA synthesis is under hormonal regulation and that antiandrogen therapy may produce a direct effect on the serum PSA concentration. Clinical and experimental evidence has shown that androgen deprivation usually results in a lower serum PSA level. This effect is due to a reduction in the population of viable prostatic epithelium (benign and malignant) and a decreased expression of PSA mRNA in the remaining cells. Hormonal withdrawal does not reduce PSA production to zero, and once reintroduced there is usually a rapid response by the cells resulting in synthesis of PSA. Androgen cell receptors regulate PSA synthesis and so mutations in these receptors may alter their specificity and affinity to the several available steroids. Consequently, PSA production in the presence or absence of androgens may vary according to the receptors abnormalities, making it possible for tumors to progress despite low serum PSA levels. Although widely accepted as an excellent marker in measuring response to treatment in PCa patients being hormonally managed, there is no consensus as to which PSA parameter to follow. Commonly, reported parameters include pretreatment PSA, the nadir posttreatment value, the rate of PSA decrease, and the percentage of serum PSA decrease. Some studies indicate that timing for the normalization of serum PSA concentration following therapy is the most precocious and most significant predictor of treatment outcome among the PSA parameters commonly employed. Responders usually display a steady fall in serum PSA values during the early phases of treatment and nadir values are commonly attained after 2 months of treatment and almost always at 6 months. The posttreatment nadir PSA value also correlates with patient outcome. Initial studies indicated that a PSA nadir <10.0 ng/ml is associated with longer time to progression and survival. Recent investigations support the use of 4.0 ng/ml as the most adequate nadir value in the follow-up of patients under antiandrogen therapy. In summary, a normal PSA values (<4.0 ng/ml) at 6 months usually means a longer time to progression and improved survival. Normalization of serum PSA at 3 months is the earliest and most significant predictor of treatment outcome. Prostate cancer that escapes hormone therapy usually presents a rising PSA 6–12 months prior to any clinical or radiological sign of disease recurrence. Serial serum PSA measurements are indicated for all patients undergoing hormonal therapy; the first evaluation should be made after 3 months of treatment and evaluations should be repeated at 6- to 12-month intervals. In the absence of elevations, other staging tests are not indicated (e.g., bone scan) unless a clinical sign of progression ensues. A rising serum PSA concentration reliably predicts disease progression and may be an indication for alternative forms of therapy.

IX. SUMMARY

It is clear that PSA is a superior tumor marker, whether the issue is diagnosis, staging, or monitoring response to treatment. Nevertheless, it is not the ideal tumor marker. As a result, basic science and clinical investigation continue to look for new ways to enhance the clinical utility of the PSA test.

Because PSA is not specific for prostate cancer and this malignancy develops in men at an age when other benign conditions are very prevalent, several concepts have been developed and are being investigated to improve the sensitivity and specificity of the PSA test. These include molecular forms of PSA (percentage free PSA), PSA density, age-specific reference ranges, and PSA velocity. All four parameters have their proponents and their critics and are being used to varying degrees by clinicians to improve their prostate cancer detection rate. Regardless of the approach used, PSA testing is clearly detecting more organ-confined, clinically significant, and curable prostate cancer. Regarding monitoring treatment outcome for prostate cancer, the availability of posttreatment serum PSA levels provides a more accurate identification of persistent disease regardless of the adopted therapeutic option. Thus, given the proper time and attention to the clinical use of this tumor marker, it is reasonable to expect that the ultimate goal of any cancer detection and treatment program will be achieved—a decline of the mortality rate.

See Also the Following Articles

Male Reproductive Disorders, Overview; Prostate Cancer; Prostate Gland; Tumors of the Reproductive System, Overview

Bibliography

Benson, M. C., Whang, I. S., Pontuk, A., *et al.* (1992). Prostate specific antigen density: A means of distinguishing benign prostatic hypertrophy and prostate cancer. J. *Urol.* **147**, 815.

Brawer, M. K., Chetner, M. P., Beatie, J., *et al.* (1992). Screening for prostatic carcinoma with prostate-specific antigen. *J. Urol.* **147**, 841.

Carter, H. B., Pearson, J. D., Metter, E. J., *et al.* (1992). Longitudinal evaluation of prostate specific antigen levels in men with and without prostate disease. J. *Am. Med. Assoc.* **267**, 2215.

Catalona, W. J., Smith, D. S., Ratliff, T. L., *et al.* (1991). Measurements of prostate-specific antigen in serum as a screening test for prostate cancer. *N. Engl. J. Med.* **324**, 1156.

Catalona, W. J., Hudson, M. A., Scardino, P. T., *et al.* (1994). Selection of optimal prostate-specific antigen cutoffs for early detection of prostate cancer: Receiver operating characteristics curves. *J. Urol.* **152**, 2037.

Glenski, W. J., Malek, R. S., Myrtle, J. F., *et al.* (1992). Sustained substantially increased concentration of prostate specific-antigen in the absence of prostatic malignant disease: An unusual clinical scenario. *Mayo Clinic Proc.* **67**, 249.

Graves, H. C. B. (1995). Nonprostatic sources of prostate-specific antigen: A steroid hormone-dependent phenomenon? *Clin. Chem.* **41**, 7.

Lilja, H., Christensson, A., Dahlän, U., *et al.* (1991). Prostate-specific antigen in human serum occurs predominantly in complex with α-1-antichymotrypsin. *Clin. Chem.* **37**, 1618.

Morgan, T. O., Jacobsen, S. J., McCarthy, W. F., *et al.* (1996). Age-specific reference ranges for serum prostate-specific antigen in black men. *N. Engl. J. Med.* **335**, 304.

Oesterling, J. E. (1991). Prostate specific antigen: A critical assessment of the most useful tumor marker for adenocarcinoma of the prostate. *J. Urol.* **145**, 907.

Oesterling, J. E. (1993, November). Prostatic tumor markers. *Urol. Clin. North Am.* **20**(4).

Oesterling, J. E. (1997, May). Prostate-specific antigen: The best prostatic tumor marker. *Urol. Clin. North Am.* **24**(2).

Oesterling, J. E., Jacobsen, S. J., Chute, C. G., *et al.* (1993). Serum prostate-specific antigen in a community-based population of healthy men: Establishment of age-specific reference ranges. *J. Am. Med. Assoc.* **270**, 860.

Oesterling, J. E., Kumamoto, Y., Tsukamoto, T., Girman, C. J., Guess, H. A., Masumori, N., *et al.* (1995). Serum prostate-specific antigen in a community-based population of healthy Japanese men: Lower values than for similarly aged white men. *Br. J. Urol.* **75**, 347.

Olson, C. A., and Benson, M. C. (1996, November). Localized prostate cancer. *Urol. Clin. North Am.* **23**(4).

Partin, A. W., Criley, S. R., Subong, E. N., *et al.* (1996). Standard versus age-specific prostate specific antigen reference ranges among men with clinically localized prostate cancer: A pathological analysis. *J. Urol.* **155**, 1336.

Sershon, P. D., Barry, M. J., and Oesterling, J. E. (1993). Serum PSA values in men with histologically confirmed BPH versus patients with organ-confined prostate cancer. *J. Urol.* **149**, 421A. [Abstract]

Stenman, U. H., Leinonen, J., Alfthan, H., *et al.* (1991). A complex between prostate-specific antigen and alpha-1-antichymotrypsin is the major form of prostate-specific antigen in serum of patients with prostatic cancer: Assay of the complex improves clinical sensitivity for cancer. *Cancer Res.* **51**, 222–226.

Protein Hormones of Primate Pregnancy

Gerald J. Pepe and Eugene D. Albrecht
Eastern Virginia Medical School and The University of Maryland School of Medicine

GLOSSARY

autocrine action Regulation of cell function by a chemical messenger produced in the same cell.

chemical messengers Substances produced by endocrine or nonendocrine cells and which bind to a receptor on or in a target cell to elicit control and/or function of the target cell.

cytotrophoblast Cells derived from the outer wall of the blastocyst which act as a stem cell population that undergoes rapid division and fusion to form the syncytiotrophoblast layer of the placenta.

decidua Differentiated cells of the uterine endometrium which comprise the maternal portion of placenta.

diabetogenic hormones Chemical messengers that antagonize the actions of insulin on peripheral tissues and indirectly promote an increase in pancreatic insulin secretion.

endocrine action Activation of a cell or tissue by a chemical messenger produced in cells of another tissue and which is secreted into the blood.

paracrine action Regulation of cell function by a chemical messenger produced in a neighboring cell.

syncytiotrophoblast The multinucleated syncytium of the placenta which functions as the site of hormone synthesis and exchange of chemicals and nutrients between the maternal and fetal blood.

During primate pregnancy, the placenta produces significant quantities of steroid and protein hormones that act as chemical messengers to ensure the maintenance of pregnancy, maternal homeostasis, and development of the fetus. The placental hormones modulate the maternal–fetal dialogue that exists during pregnancy and which regulates maternal metabolic processes (e.g., carbohydrate and calcium metabolism) upon which fetal growth and maturation are critically dependent.

I. INTRODUCTION

The establishment and maintenance of pregnancy culminating in the birth of a healthy newborn are dependent on the coordinated secretion of chemical messengers within and between the mother, fetus, and placenta. Thus, the placenta produces steroid and protein hormones, neuropeptides, and growth factors which modulate metabolic processes in maternal and fetal organ systems and also regulate hormone production by endocrine glands of the mother and fetus. Several of the placental hormones also act in an autocrine and/or paracrine manner to regulate the growth and differentiation of the placental cytotrophoblast and syncytiotrophoblast. Clearly, placental hormone secretion is one of the most important determinants of successful maternal adaptation to pregnancy in virtually all mammalian species. This article summarizes the key peptide hormones produced by the human placenta, describes their primary effects on maternal metabolic function, and outlines their major actions on other hormonal systems integral to pregnancy maintenance, maternal homeostasis, and fetal development.

II. SECRETION AND ROLE OF THE PROTEIN HORMONES OF THE PLACENTA

A. Chorionic Gonadotrophin

Human chorionic gonadotropin (hCG) is a glycosylated protein heterodimer composed of noncovalently bound α and β subunits. βhCG is structurally similar to the β subunit of pituitary luteinizing hormone (LH), with the exception of the last 28 amino acids. While α subunit genes are expressed in the pituitary and placenta, the β subunit gene is only expressed in the placenta. Following their differentiation from cytotrophoblast, the syncytiotrophoblast becomes the primary site of CG production. In human pregnancy, maternal CG levels rise markedly following implantation, peak at approximately 8–12 weeks of gestation, and decline thereafter (Fig. 1). Although secreted predominantly into the maternal circulation, CG has also been detected in fetal serum. The production of CG appears to be restricted to human and nonhuman primates. Although the qualitative and quantitative pattern of CG secretion in the great apes resembles that in the human, in nonhuman primates, such as the baboon and the rhesus and bonnet monkey, the secretion of CG appears to be restricted to the first one-third of gestation. Despite this difference, the primary biologic action of CG, namely, to "rescue" the corpus luteum and maintain the production of progesterone and consequently pregnancy, appears to be conserved. Thus, administration of antibodies against CG reduces luteal progesterone synthesis and causes pregnancy termination. In addition to this role, placental CG stimulates the fetal testis to produce testosterone, which is necessary for masculinization of a male fetus. Several factors (e.g., gonadotropin-releasing hormone, opioids, interleukins, and growth factors) of placental origin acting in an autocrine/paracrine manner have the capacity to modulate CG production and secretion.

B. Chorionic Somatomammotropin and Growth Hormone Variant

The chorionic somatomammotropin–growth hormone (CS–GH) gene family is composed of five linked genes, of which the CS-A and CS-B genes

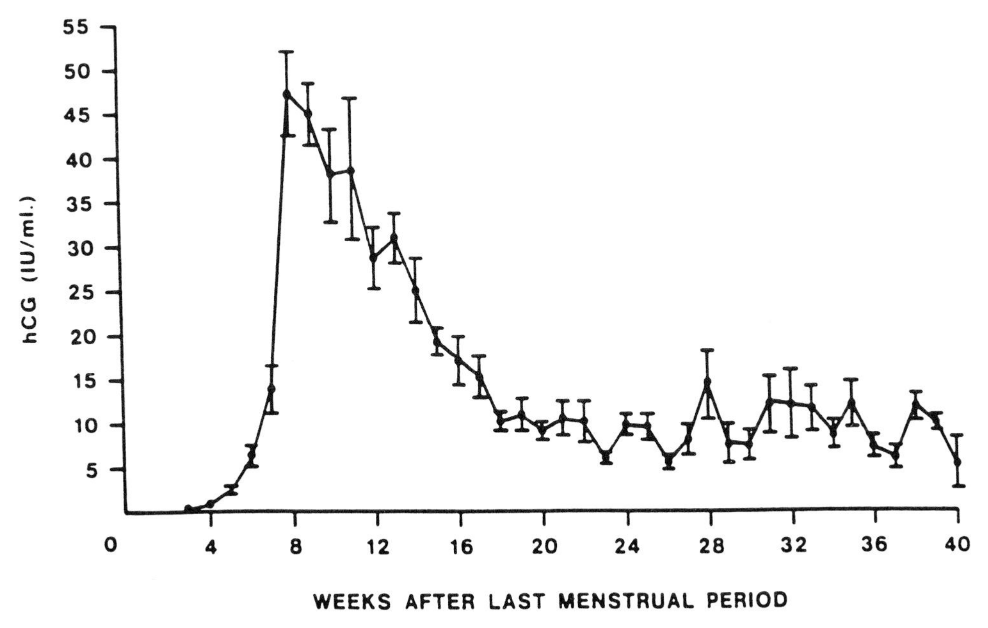

FIGURE 1 Mean serum hCG levels throughout normal pregnancy. Arithmetic scale used on ordinate. Bars represent SEM (reproduced with permission from G. D. Braunstein, J. Rasor, H. Danzer, D. Adler, and M. E. Wade, Serum human chorionic gonadotropin levels throughout normal pregnancy, *Am. J. Obstet. Gynecol.* **126**, 678–681, 1976).

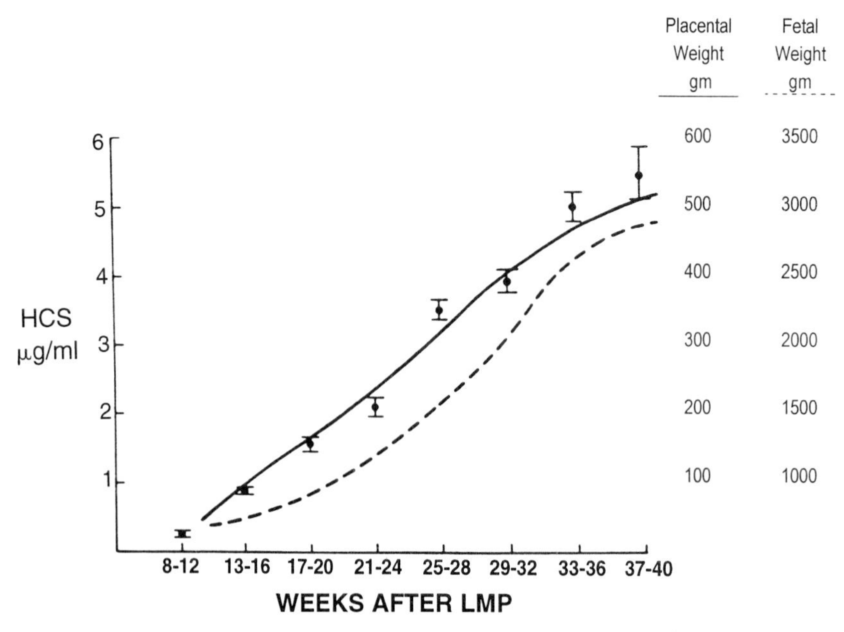

FIGURE 2 Mean (± SE) concentration of serum human somatomammatropin (HCS) (●) at 4-week intervals during gestation in relation to placental (solid line) and fetal (broken line) weights. LMP, last menstrual period [reproduced with permission from M. M. Grumbach *et al*:, *HCS-Physiology: Hormonal Effects of Peptide Hormones* (S. A. Berson and R. S. Yalow, Eds.), pp. 797–802, North Holland, Amsterdam, 1973].

encode CS, also known as human placental lactogen. CS, a 192-amino acid protein which is 96% homologous to GH of pituitary origin (GH-N), is synthesized by the placental syncytiotrophoblasts once the latter are formed from cytotrophoblast. In contrast to CG, the levels of CS increase steadily during human (Fig. 2) and nonhuman primate pregnancy and parallel the increase in placental syncytiotrophoblast mass. Levels of CS in maternal serum are approximately 1000-fold greater than those in fetal blood.

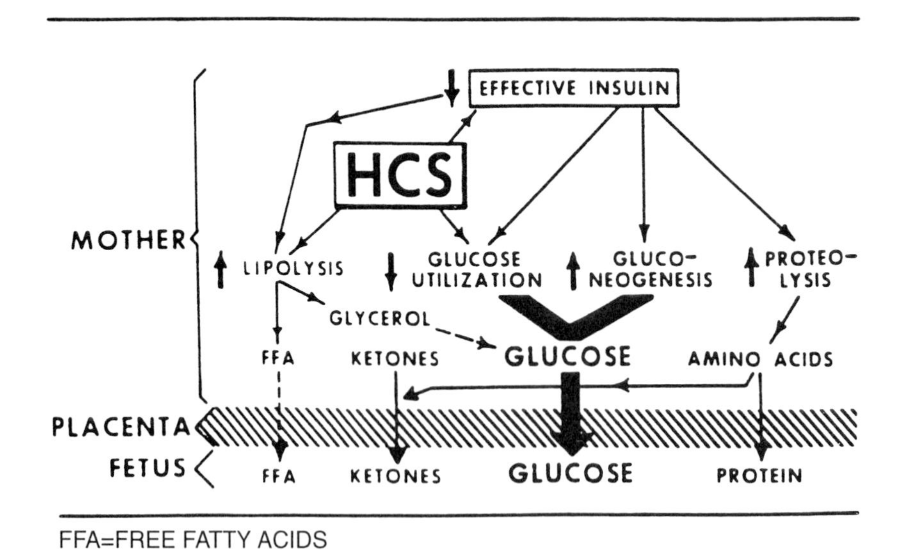

FIGURE 3 The tonic regulatory effect of HCS on maternofetal substrate supply (reproduced with permission from M. M. Grumbach, *Methods in Investigative and Diagnostic Endocrinology*, North Holland, Amsterdam, 1973).

It has been suggested that the major role of CS is to regulate intermediary metabolism to ensure delivery of glucose to the fetus. Thus, as shown in Fig. 3, CS promotes peripheral insulin resistance in virtually all maternal tissues except adipocytes, thereby increasing lipolysis and proteolysis while decreasing maternal glucose utilization and increasing gluconeogenesis and thus maternal glucose levels. As a consequence (Fig. 3), maternal tissues selectively use free fatty acids for energy while diverting glucose to the fetus to promote growth. Although CS has been shown to elicit a direct effect on metabolism by fetal tissues in sheep, this role remains to be established in the human. CS has also been shown to stimulate growth of the mammary glands during gestation but does not increase milk production, i.e., lactogenesis. There is controversy as to whether CS levels are increased in women who have pregnancy-induced diabetes and low levels have been found in some, but not all, women with placental insufficiency or fetal distress.

Placental growth hormone variant (GH-V), which differs from GH-N in 13 amino acid substitutes dispersed throughout the protein, is synthesized by the syncytiotrophoblast and maternal levels steadily increase throughout the course of human pregnancy (Fig. 4). GH-V is the predominant GH in the maternal blood by the second trimester as the production of GH-N, which is increased early in pregnancy, declines steadily thereafter (Fig. 4). Like GH-N, GH-V binds to a circulatory GH-binding protein, which is a secreted form of the GH receptor. GH-V is not secreted into the fetal circulation and thus it would appear that this product of the placenta does not elicit a direct effect on fetal growth. It has been proposed, however, that GH-V is a major regulator of hepatic as well as placental insulin-like growth factor (IGF)-I production during primate pregnancy. Thus, in human and nonhuman primates, the maternal levels of IGF-I increase steadily during pregnancy (Fig. 4) and parallel the increase in GH-V production. By regulating IGF-I production, placental GH-V also appears to regulate maternal pituitary GH-N production. Thus, the decrease in GH-N levels after Week 10 of gestation in humans is most likely due to feedback inhibition of pituitary GH-N production by the increased levels of IGF-I.

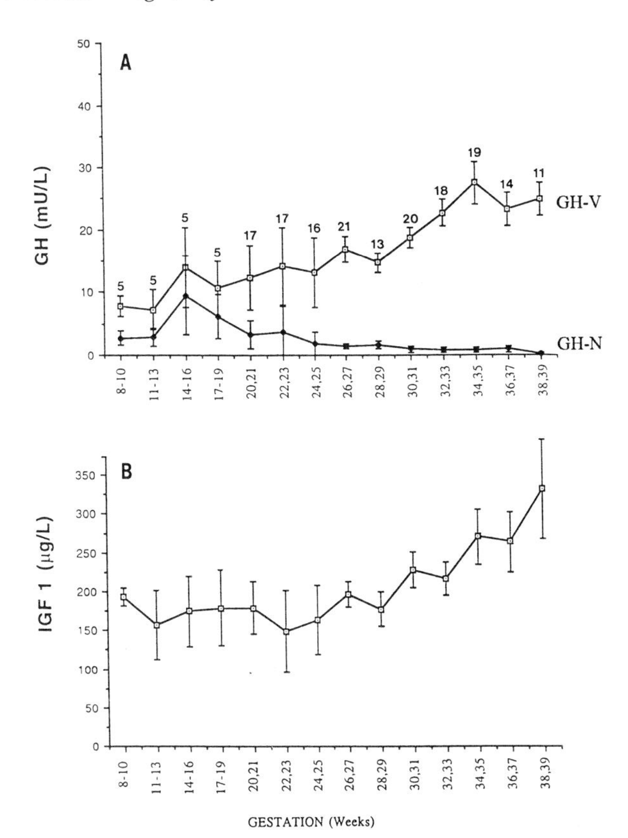

FIGURE 4 Transverse study of maternal plasma GH (A) and IGF-I (B) levels during pregnancy ($n = 186$). Each point represents the mean ± SEM of values from individual samples obtained in pregnant women at the indicated periods of pregnancy expressed. Number of individual assays in GH and IGF-I for each gestational stage is indicated in A on top of the vertical bars (modified and reproduced with permission from V. Mirlesse, F. Frankenne, E. Alsat, M. Poncelet, G. Hennen, and D. Evain-Brion, Placental growth hormone levels in normal pregnancy and in pregnancies with intrauterine growth retardation, *Pediatr. Res.* **34**, 439–442, 1993).

C. Growth Factors

Growth factors are polypeptides which are secreted into the circulation or the extracellular fluid attached to large proteins termed binding proteins and which subsequently bind to specific cell surface membrane receptors to initiate a rapid chain of events culminating in DNA duplication and cell division. The human placenta produces several growth factors which act in an autocrine/paracrine manner to regulate trophoblast function or are secreted into the maternal and/

or fetal circulations to elicit endocrine effects on peripheral tissues.

1. Endocrine Actions of Placental Growth Factors

i. IGF-I and IGF-II IGF-I and -II are produced by the syncytiotrophoblast and cytotrophoblast and secreted into the maternal circulation. While IGF-I concentrations increase with advancing gestation (Fig. 4), those of IGF-II remain relatively unchanged. The ability of the IGFs to elicit their potential biologic effects is influenced by binding to high-affinity IGF-binding proteins (IGFBPs), six of which have been identified. In humans, the majority (>80%) of the IGFs in maternal circulation are bound to IGFBP-3, the levels of which also increase slightly during late gestation. While IGF-I elicits effects on intermediary metabolism, the precise role of IGF-II in maternal tissue remains to be determined. IGF-I, IGF-II, and IGFBP-3 are also detected in fetal blood at levels which are approximately 50% of those measured in the mother. However, because fetal tissues express the genes for the IGFs, the extent to which fetal IGF levels reflect production by the placenta remains to be determined. Recent studies have shown that there is a significant correlation between neonatal weight and the fetal levels of IGF-I, but not IGF-II, or the levels of IGF-I and -II in the mother. It has been suggested, therefore, that IGF-I has a role in the control of intrauterine fetal growth.

ii. Inhibins Inhibins are glycoproteins that belong to the transforming growth factor-β (TGF-β) superfamily and are characterized by their ability to suppress the secretion of follicle-stimulating hormone (FSH) by the pituitary gland. The inhibins are composed of an α subunit and one of two β subunits (β_A or β_B) giving rise to two functional glycoproteins, inhibin A ($\alpha\beta_A$) and inhibin B ($\alpha\beta_B$). Using an assay which detected the α subunit, it has been shown that inhibin concentrations rise during early gestation, peak at 9 or 10 weeks, fall to a plateau at approximately 15 weeks, and subsequently rise to reach another peak prior to term (Fig. 5). Recent evidence indicates that the inhibin measured in maternal serum is inhibin A. In the baboon, in which maternal serum inhibin levels also increase throughout the second half of pregnancy, placental inhibin production appears to be regulated primarily by the mass of functioning trophoblast. In the amniotic fluid compartment in which the fetus is contained, both inhibin A and B are measured. Because only inhibin A is of placental origin, inhibin B must be produced by the membranes of the placenta and/or by fetal tissues such as the adrenal gland or testis. Although several functions, including inhibition of placental CG and pituitary FSH production, have been ascribed to inhibin, its precise role in human pregnancy remains to be determined.

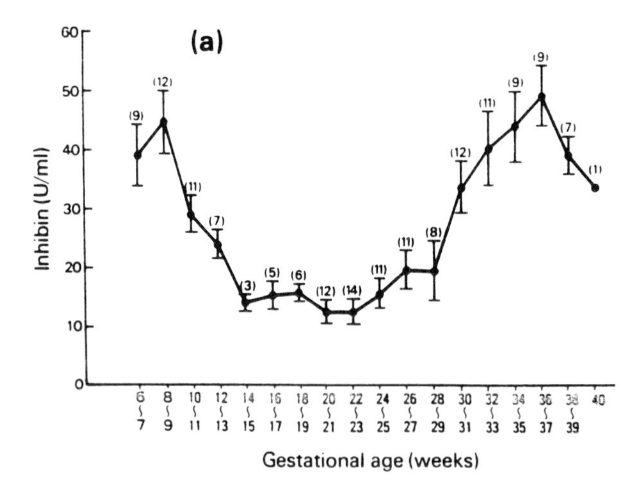

FIGURE 5 Serum levels (mean ± SEM) of inhibin in pregnant women throughout gestation (6–40 weeks). Values in parentheses refer to numbers of serum samples for measurement (modified and reproduced with permission from T. Tabei, K. Ochiai, Y. Terashima, and N. Takanashi, Serum levels of inhibin in maternal and umbilical blood during pregnancy, *Am. J. Obstet. Gynecol.* **164**, 896–900, 1991).

iii. Activin Activin, a homodimer ($\beta_A\beta_A$ = activin A; $\beta_B\beta_B$ = activin B) or heterodimer ($\beta_A\beta_B$ = activin AB) of the same two β subunits used to form inhibin, elicits biologic effects which are opposite to those of inhibin. Compared to the very low levels produced during the menstrual cycle, activin A is markedly elevated early in gestation, remains consistent through Week 24, and subsequently increases to a peak at term. There are no studies of the regulation of this potent mitogen.

2. *Autocrine/Paracrine Effects of Placental Growth Factors*

In addition to eliciting endocrine effects on peripheral tissues, several of the placental growth factors act in an autocrine and/or paracrine manner to regulate trophoblast growth and differentiation as well as hormone production. Thus, receptors for these various growth factors have been identified in the cytotrophoblast and/or syncytiotrophoblast as well as in nontrophoblast cells of the placenta. Current evidence is consistent with a major role of IGF-II and epidermal growth factor (EGF). Thus, under *in vitro* culture conditions, IGF-II and EGF stimulated mitosis and/or differentiation of human placental villous cytotrophoblasts into syncytiotrophoblasts. A physiologic role for IGF-II in placental development is substantiated by the observation that mice lacking a functional IGF-II gene as a result of gene targeting had smaller placentas and fetuses. Because IGFBP-3 appears to be expressed by syncytiotrophoblast, this binding protein may also play a role in regulating the action of IGF-II locally within the placenta. The ontogeny and cell-specific regulation of placental IGFBP-3 remains to be examined. The EGF receptor also binds TGF-α and thus this growth factor, which is produced primarily by cytotrophoblast, may exert actions on syncytiotrophoblast. It is apparent that placental peptide signaling is complex. Moreover, it remains to be shown whether these and/or other factors are operative *in vivo* and by what mechanism(s) they are regulated. Finally, growth factors including IGF-II and IGFBP-1 and IGFBP-2 are also produced by the maternal uterus and its differentiated endometrial cells known as the decidua. The impact of decidual secretions on placental growth and function *in vivo* during the time of implantation and early stages of pregnancy are currently under investigation.

D. Hypothalamic-like Neuropeptides

Several of the neuropeptides produced within the hypothalamus and/or pituitary are also elaborated by the human and nonhuman primate placenta. Although this includes corticotrophin-releasing hormone (CRH), gonadotropin-releasing hormone (GnRH), proopiomelanocortin (POMC), oxytocin, neuropeptide Y, thyrotropin-releasing hormone (TRH), and somatostatin, we will focus on CRH, GnRH, and POMC.

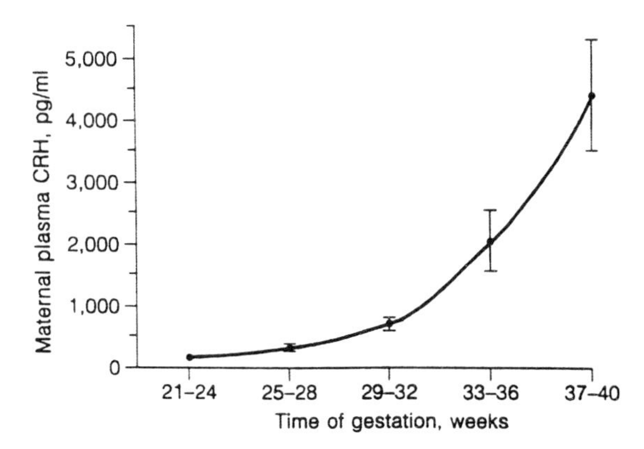

FIGURE 6 Mean plasma CRH concentrations in eight women followed sequentially during the second half of pregnancy (reproduced with permission from R. S. Goland, I. M. Conwell, W. B. Warren, and S. L. Wardlaw, Placental corticotropin-releasing hormone and pituitary–adrenal function during pregnancy, *Neuroendocrinology* 56, 742–749, 1992).

1. *CRH*

Placental CRH is identical in immunoreactivity and bioactivity to hypothalamic CRH and in human pregnancy maternal plasma CRH concentrations progressively increase in the second and third trimesters and fall precipitously after delivery (Fig. 6). In baboons, plasma CRH concentrations peak in the first half of pregnancy, then decline to lower levels for the remainder of gestation. In both species, placental CRH is also secreted into the fetus but at concentrations 100-fold lower than those in the mother. Consistent with plasma levels, human placental CRH levels also increase markedly in the final 5 weeks of pregnancy. CRH has been localized within both the cytotrophoblast and syncytiotrophoblast layers of the placenta and is also released from the fetal membranes *in vitro*. In twin pregnancies, maternal CRH levels increase earlier in gestation and rise to higher levels, reflecting in part the greater trophoblast mass and earlier time of delivery. In gestations complicated by pregnancy-induced hypertension, intrauterine fetal growth retardation, fetal asphyxia, premature rupture of membranes, and premature labor, maternal plasma CRH levels are also increased.

Several physiological roles for placental CRH have been proposed, including the regulation of the onset of parturition, uterine blood flow, and maternal pituitary adrenocorticotropin (ACTH) production. Indeed, the progressive increase in maternal plasma CRH concentrations in human pregnancy is accompanied by a corresponding rise in the concentrations of maternal ACTH and cortisol (Fig. 7). However, up to 80% of the CRH in maternal plasma is complexed to a binding protein produced in the placenta and liver, and although the maternal levels of this binding protein decline during the second half of pregnancy, concentrations still exceed that of CRH by more than 10-fold. Because CRH bound to this protein is biologically inactive and the unbound and thus physiologically active levels of CRH are not necessarily elevated, it remains to be determined whether the maternal pituitary gland is subject to stimulation by placental CRH. It is also possible, however, that the changes in maternal ACTH and thus cortisol reflect increased responsivity of the maternal pituitary gland to other hormones. For example, arginine vasopressin (AVP), the production of which is increased during pregnancy, has been

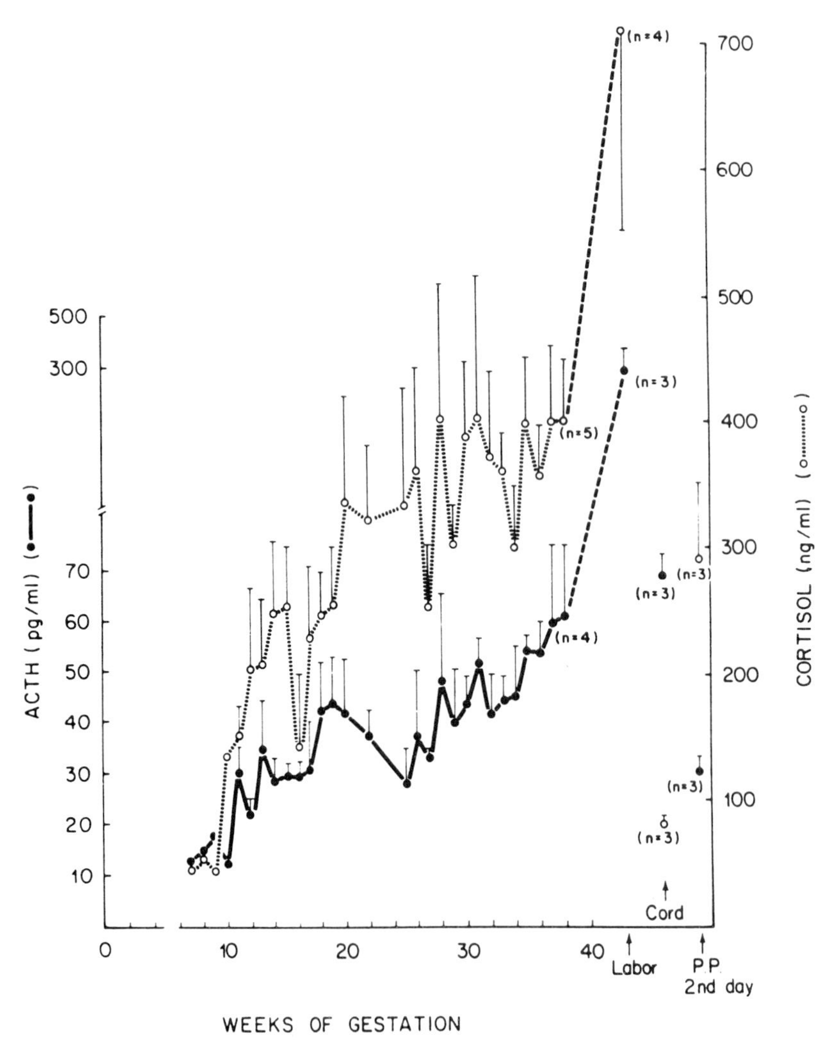

FIGURE 7 Circulating ACTH throughout gestation (reproduced with permission from B. R. Carr, C. R. Parker, Jr., J. D. Madden, P. C. MacDonald, and J. C. Porter, Maternal plasma adrenocorticotropin and cortisol relationships throughout human pregnancy, *Am. J. Obstet. Gynecol.* **139**, 416–422, 1981).

shown to enhance ACTH production both *in vitro* and *in vivo*.

In contrast to the negative feedback effects of cortisol on hypothalamic CRH production, cortisol apparently increases CRH output by cultures of human placental trophoblasts. However, the effects of cortisol *in vivo* remain to be established since maternal peripheral plasma CRH concentrations are not increased by the administration of synthetic glucocorticoids and the placenta which takes up maternal cortisol actively catabolizes this cortisol to its inactive metabolite cortisone. In addition, progesterone, which is produced by the placenta and therefore present in very high levels, suppresses the release of CRH by cultured trophoblast cells.

2. *GnRH*

Placental GnRH is produced by the cytotrophoblast and has the capacity to stimulate CG production by syncytiotrophoblasts which express the GnRH receptor. Moreover, GnRH receptor levels appear to parallel the time course of CG secretion and thus are at their highest at Week 9 of gestation, then decline thereafter to nondetectable values by term. Several factors produced in the placenta can enhance (e.g., activin and estradiol) or inhibit (e.g., inhibin, opioids, and progesterone) trophoblast release of GnRH.

3. *POMC*

POMC, the common glycoprotein precursor that gives rise to the opioid β-endorphin (β-END) in the brain and the hormones ACTH and α-melanocyte-stimulating hormone in the pituitary, is also expressed in the placenta in which all three products are produced. In contrast to the pituitary, however, the placenta actively converts β-END to an acetylated product, Acβ-END. *In vitro*, β-END decreases trophoblast secretion of GnRH, CG, and progesterone. The placental neuropeptide CRH has the capacity to stimulate release of ACTH by trophoblast cells in culture, a response which is not inhibited by glucocorticoid or enhanced by AVP. Thus, the regulation of placental POMC differs from that in the pituitary in which glucocorticoids inhibit and AVP enhances the action of CRH. The role of placental POMC may be restricted to specific trophoblast functions since *in vivo* administration of a bolus of CRH into the umbilical artery (i.e., into the placenta) of the baboon does not enhance maternal or fetal ACTH levels. Moreover, although β-END increases throughout gestation and rises markedly during labor, these changes reflect contributions from the pituitary and not the placenta since levels of Acβ-END are not increased.

E. Placental-Specific Proteins

The human placenta synthesizes and secretes four distinct pregnancy-associated plasma proteins (PAPP) termed PAPP-A, PAPP-B, PAPP-C, and PAPP-D. PAPP-C has been shown to be identical to the pregnancy-specific glycoprotein SP1, whereas PAPP-D is immunologically identical to CS. Although PAPP-B has not been fully characterized, human PAPP-A is a homotetrameric glycoprotein with a molecular weight of approximately 820,000. The placentas of nonhuman primates, including the baboon, also produce PAPP-A, which exhibits physiochemical and immunologic characteristics very similar to those of the human glycoprotein.

The precise function(s) of SP-1 and PAPP-A during pregnancy remains to be determined. However, levels of both proteins progressively increase with advancing gestation and appear to be regulated by the mass of functioning syncytiotrophoblasts. It has been suggested that PAPP-A is a reliable marker of placental development and viability since maternal peripheral concentrations of PAPP-A increase with advancing gestation, whereas low levels are associated with early pregnancy failure in humans and baboons.

III. SECRETION AND ROLES OF THE PROTEIN HORMONES OF THE MATERNAL ENDOCRINE GLANDS

A. Pancreatic Insulin, Glucagon, and Somatostatin

Pregnancy is associated with an increase in the production and serum levels of CS, IGF-I, GH-V, and cortisol, hormones that act to realign maternal metabolism to help provide for the metabolic demands of pregnancy and ensure that an adequate supply of glucose is made available to the fetus. However, in addition to increasing blood glucose levels, these hormones of pregnancy are diabetogenic and thus also increase the resistance of peripheral tissues of the mother to the action of insulin. Because insulin is required for glucose uptake and utilization as well as storage as glycogen, the formation and storage of lipids, and for protein synthesis, it is imperative that the maternal pancreatic β cells compensate and produce increasing levels of insulin to ensure that intermediary metabolism is balanced and blood glucose levels do not become excessive. Thus, not only are insulin levels increased in pregnant women but also responsivity of the pancreas to changes in nutrients such as glucose is greatly exaggerated. Moreover, it appears that the pancreatic β cells actually increase in size (hypertrophy) during pregnancy. In the absence of these changes, insulin resistance would be increased and fat catabolism enhanced, thereby resulting in metabolic derangements including hyperglycemia, which can disrupt normal intrauterine development.

The α cells of the pancreas are also active during pregnancy and produce the polypeptide hormone glucagon, the basal levels of which increase in late gestation. Glucagon acts to break down glycogen in liver but not muscle, enhance glucose formation from noncarbohydrate sources (i.e., gluconeogenesis), and increase breakdown of lipids. In pregnancy, the pancreatic α cells exhibit an exaggerated response to physiologic stimulators such as amino acids and inhibitors such as glucose. However, there does not appear to be any peripheral resistance to glucagon in human pregnancy.

The pancreatic D cells produce the polypeptide hormone somatostatin, which is also synthesized in the hypothalamus and which inhibits production of pituitary GH. In the pancreas, somatostatin release is increased by the same factors which induce (i.e., glucose) or inhibit (i.e., the hormone epinephrine) insulin secretion. Although maternal levels of somatostatin are similar in pregnant and nonpregnant women, the increase in pancreatic somatostatin induced following ingestion of a meal in nonpregnant women is abolished in late gestation.

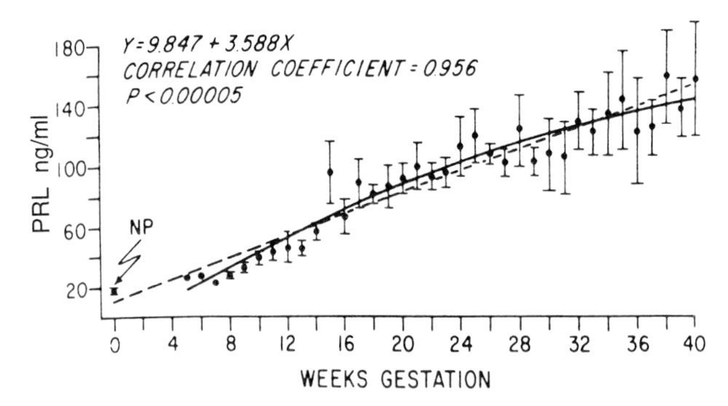

FIGURE 8 Circulating prolactin throughout gestation (reproduced with permission from L. A. Rigg, A. Lein, and S. S. Yen, Pattern of increase in circulating prolactin levels during human gestation, *Am. J. Obstet. Gynecol.* **129**, 454–456, 1977).

B. Pituitary Prolactin

As a consequence of the modification in the hormonal milieu induced in the mother by pregnancy, the maternal pituitary gland undergoes marked changes. Thus, corticotrophe function (i.e., production of ACTH) is increased, whereas the activities of gonadotrophs, which produce LH and FSH, and somatotrophs, which produce GH-N, decrease due to negative feedback inhibition by the increased production of the steroid hormones estrogen and progesterone and IGF-I, respectively. Consequently, LH, FSH and GH-N levels in maternal plasma remain very low during human and nonhuman primate pregnancy. In contrast, pituitary lactotrophes, which are the site of production of the polypeptide hormone prolactin, undergo hyperplasia and hypertrophy resulting in an increased production of prolactin (Fig. 8). These changes are presumed to be due to increased production of estrogen, which acts directly on the pituitary to increase prolactin production or responsivity to stimulatory agents. The increase in maternal prolactin is considered essential to preparation of the mammary gland for subsequent postpartum lactation. Although the maternal decidua is also a source of prolactin, decidual prolactin is secreted into the fetus and not the mother. Prolactin secreted into the fetus may modulate fetal adrenal androgen production at a time when the fetal pituitary does not synthesize ACTH.

The biochemical nature of circulating prolactin is different in pregnant and nonpregnant women. In the nonpregnant state, prolactin circulates as an N-linked glycosylated glycoprotein termed G-PRL. During pregnancy, increasing amounts of nonglycosylated prolactin appear in maternal serum, and by late gestation, the ratio of G-PRL to nonglycosylated prolactin is significantly reduced. The precise role(s) of these prolactin forms remains to be determined, although it has been suggested that nonglycosylated prolactin exhibits more biological activity than G-PRL.

C. Thyroid Hormone Function

The ability of the thyroid to produce the hormones thyroxine (T_4) and triiodothyronine (T_3) requires the glycoprotein TSH, which acts on the thyroid to increase iodine uptake and T_4 and T_3 formation and secretion. Regulation of thyroid hormone production is elicited by feedback inhibition by the thyroid hormones on pituitary TSH. Once released, the majority (>99%) of the thyroid hormones are bound to serum-binding proteins, the most significant of which is thyroxine-binding globulin (TBG) which is synthesized in the liver. During pregnancy, the maternal hypothalamic–pituitary–thyroid axis must compensate for and is influenced by the new endocrine milieu. Thus, during early gestation the high levels of placental CG secreted into the mother are sufficient to bind to the TSH receptor and increase thyroid hormone production, which in turn results in a decrease in serum TSH. With the decrease in CG thereafter, maternal pituitary TSH levels return to normal. As a consequence of the marked increase in placental estrogen production, there is a concomitant increase in maternal TBG levels which transiently reduces the unbound levels of T_4 and T_3 and results in activation and enhanced function of the pituitary-thyroid axis (Fig. 9). Although the levels of total (i.e., bound plus biologically active unbound) T_4 are presumed to be increased in pregnancy, the unbound levels of T_4 and T_3 remain normal or decrease slightly and thus pregnant women are euthyroid. Thyroid function in pregnancy is also modulated by the availability of iodine. Because the trophoblast has the catabolic deiodinase type III enzyme that converts T_4 and T_3 to inactive reverse-T_3

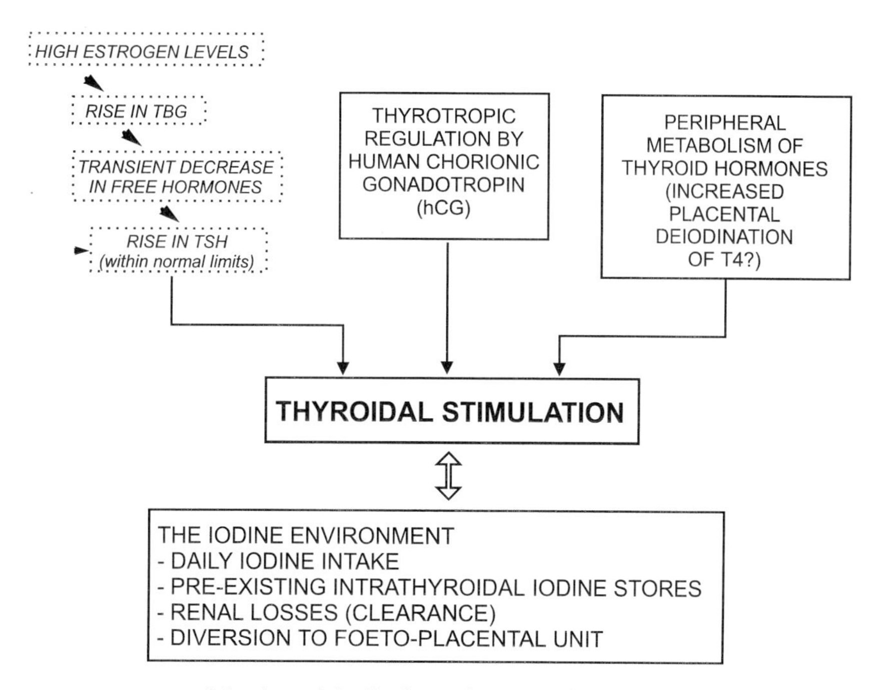

FIGURE 9 Schematic representation of the thyroid feedback regulatory mechanisms operative during pregnancy (adapted with permission from D. Glinoer, The regulation of thyroid function in pregnancy: Pathways of endocrine adaptation from physiology to pathology, *Endocr. Rev.* **18**, 404–433, 1997).

and diiodotyrosine, respectively, the placenta is to some extent a barrier between the maternal and fetal hypothalamic–pituitary–thyroid axes. Thus, maternal TSH and T_4/T_3 do not cross the placenta very efficiently and in cases in which the fetus exhibits congenital thyroid deficiency, the amount of maternal T_4 that crosses the placenta at term is only sufficient to maintain fetal T_4 levels at 50% of normal. Finally, it has also been shown that the human placenta expresses the deiodinase type II enzyme, which converts inactive T_4 to biologically active T_3, indicating that the placenta is also a thyroid hormone target tissue.

D. Parathyroid Hormone and Vitamin D_3

Calcium homeostasis is controlled by the integrated actions of parathyroid hormone (PTH), vitamin D_3 and its active metabolite 1,25-dihydroxy vitamin D_3. Thus, a decrease in serum ionized calcium induces the secretion of PTH, which acts on bone to rapidly increase uptake of calcium into osteocytes and subsequent secretion into blood, while concomitantly acting to enhance demineralization of bone by activating osteoclast activity. PTH also enhances calcium reabsorption by the kidney and the activity of renal 1α-hydroxylase enzyme-catalyzing formation of 1,25-dihydroxy vitamin D_3, which then acts to enhance calcium uptake from the gastrointestinal tract. During pregnancy, metabolic demands for calcium are markedly increased, particularly since normal fetal skeletal mineralization requires that more than 30 g of calcium be supplied to the fetus by the mother over the course of gestation. The movement of calcium from the mother to the fetus is an active process which is regulated in part by a calcium-ATPase in the placenta which resides primarily in the basal membranes of the syncytiotrophoblast. The regulation of placental calcium transport is under intense investigation. Although total calcium levels in the mother steadily decline with advancing gestation due to the pregnancy-associated decrease in serum albumin, serum ionized calcium concentrations remain relatively stable (Table 1). Although maternal serum PTH concentrations decline during gestation, the positive calcium balance must be achieved by increasing absorption of calcium by the small intestine. Indeed, the levels of 1,25-dihydroxy vitamin D_3 in maternal serum are increased in gestation. The increase in 1,25-dihydroxy vitamin D_3 reflects enhanced production by the placenta and decidua, both of which express the 1α-hydroxylase enzyme. Although the primary regulator of 1,25-dihydroxy vitamin D_3 is PTH, which acts to increase the 1α-hydroxylase enzyme in the kidney and possi-

TABLE 1
Serum and Urinary Values during Pregnancy and Postpartum

	Second trimester	*Third trimester*	*Postpartum*	*Significance*
Ionized Ca^{2+} (mmol/liter)	5.04 ± 0.04	5.01 ± 0.05	5.04 ± 0.05	NS
PTH (ng/dl)	12.0 ± 1.4	15.3 ± 1.8	23.1 ± 2.6	$p < 0.001$
$1,25(OH)_2D$ (pg/ml)	113 ± 8	101 ± 6	50 ± 3	$p < 0.01$
25(OH)D (ng/ml)	45 ± 3	47 ± 5	37 ± 5	NS
UCa (mg)	287 ± 30	291 ± 31	109 ± 13	$p < 0.00001$
UNa (mEq)	98 ± 8	96 ± 6	91 ± 6	NS
UCr (mg)	1104 ± 45	1126 ± 48	1038 ± 49	NS
DBP (mm Hg)	63 ± 1	62 ± 1	63 ± 3	NS

Note. Abbreviations used: NS, not significant; PTH, parathyroid hormone; $1,25(OH)_2D$, 1,25-hydroxyvitamin D; 25(OH)D, 25-hydroxyvitamin D; UCa, urinary calcium; UNa, urinary sodium; UCr, urinary creatinine; DBP, diastolic blood pressure (reproduced with permission from E. W. Seely, E. M. Brown, D. M. DeMaggio, D. K. Weldon, and S. W. Graves, A prospective study of calciotropic hormones in pregnancy and post partum: Reciprocal changes in serum intact parathyroid hormone and 1,25-dihydroxyvitamin D, *Am. J. Obstet. Gynecol.* **176**, 214–217, 1997).

bly the placenta, estrogen, prolactin, and GH-V have also been shown to enhance the activity of this enzyme.

E. Relaxin

Relaxin is a polypeptide hormone composed of dissimilar A and B chains linked by disulfide bonds. Although relaxin shares extensive structural similarity to insulin, relaxin has no insulin-like activity. The corpus luteum expresses one of the two nonallelic genes for relaxin and is thus the major source of this hormone during pregnancy. Indeed, relaxin levels are higher in the ovarian vein draining the corpus luteum-bearing ovary than in the ovary without a corpus luteum. It has been proposed that luteal relaxin production is regulated by pituitary gonadotropin immediately after ovulation and by placental CG during Weeks 3–10 of pregnancy. Recent studies have confirmed that maternal relaxin levels are highest early in gestation, fall slightly thereafter, and subsequently plateau. In contrast to studies in rodents, there is no prelabor surge in maternal relaxin production in humans, although levels may be increased during labor. Although the precise function(s) of relaxin in human pregnancy remains to be established, relaxin appears to have a role in facilitating parturition by promoting cervical ripening and initiating the release of collagenases important to relaxation of the pelvic symphysis.

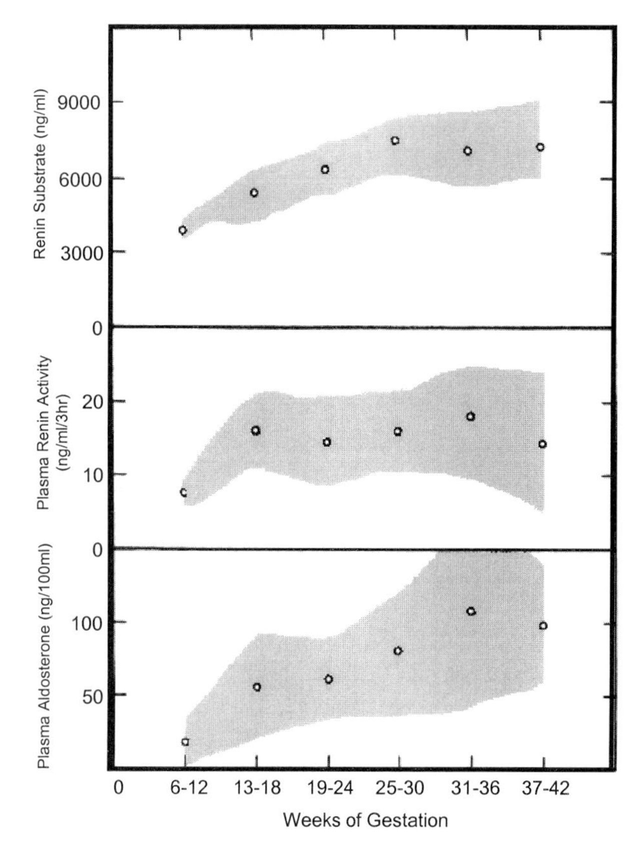

FIGURE 10 Changes in plasma renin activity, renin substrate, and aldosterone levels during normal pregnancy. The mean level is represented by the open circles and 1 ± SD is represented by the hatched area (reproduced with permission from M. H. Weinberger, N. J. Kramer, L. P. Peterson, *et al.*, *Hypertension in Pregnancy*, p. 263, Wiley, New York, 1976).

F. Renin–Angiotensin II–Aldosterone and AVP

Pregnancy is associated with striking changes in the maternal cardiovascular system, including a 30–50% increase in plasma volume, blood volume, and cardiac output. The changes in volume are due to the estrogen-induced increase in plasma renin activity which results in enhanced production of angiotensin II and consequently increased adrenal production of the salt-retaining steroid hormone aldosterone (Fig. 10). The increase in renin–angiotensin may also reflect the diuretic effects of the increasing concentrations of progesterone and its competitive antagonism of aldosterone action leading to an inhibition of renal sodium reabsorption. Although the kidney and liver are typically the principal sites of the biosynthesis of renin and angiotensin substrate, the latter are also expressed in the decidua and placenta. Renin has been localized to the cytotrophoblast and is present in greater amounts in the first trimester than in term human placentas. Moreover, renin secretion by placental cell cultures was inhibited by angiotensin II in a manner characteristic of regulation within the kidney. In the nonpregnant state, angiotensin II enhances vasoconstriction which in part increases blood pressure. This action of angiotensin II is reduced in pregnancy presumably by estrogen, which has been shown in various animal models to suppress vascular responsivity to angioten-

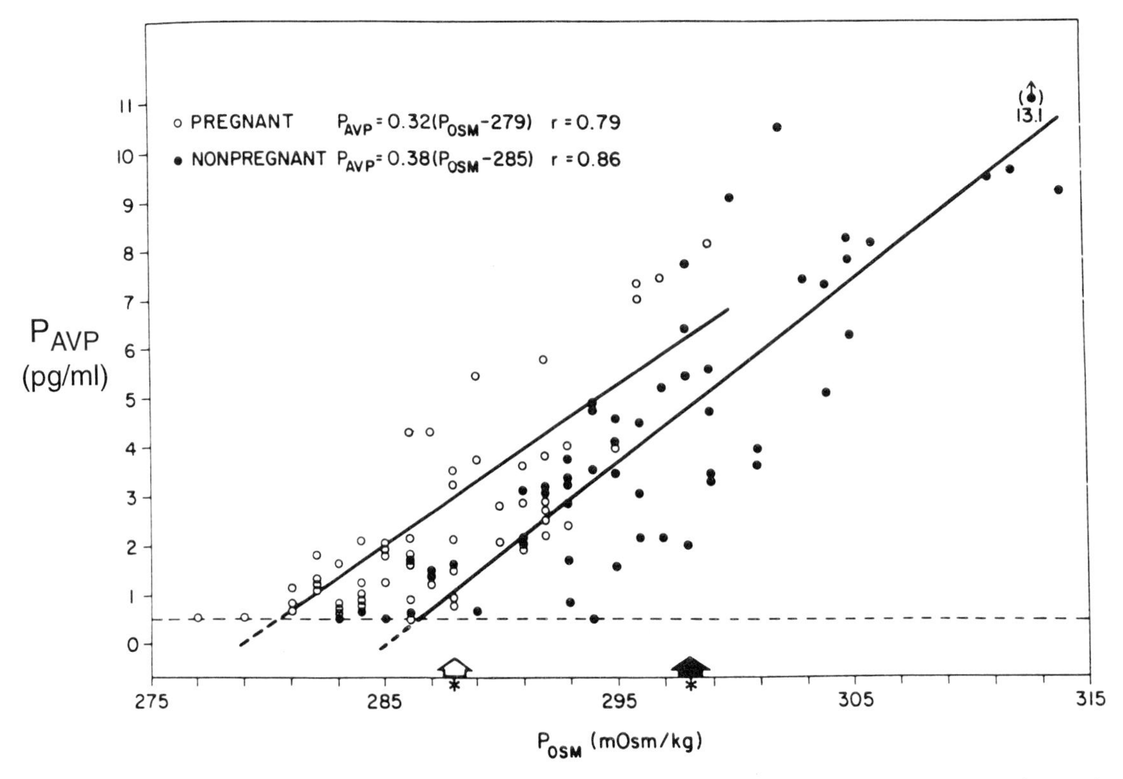

FIGURE 11 The relationship between plasma osmolality (Posm) and vasopressin (Pavp) during infusion of 5% saline over 2 hr in pregnant women. Open symbols represent third-trimester values and closed symbols represent values 6–8 weeks postpartum. The arrows represent thirst thresholds (reproduced with permission from J. M. Davison, E. A. Gilmore, J. Durr, G. L. Roberts, and M. D. Lindheimer, Altered osmotic thresholds for vasopressin secretion and thirst in human pregnancy, *Am. J. Physiol.* 246, F105–F109, 1984).

sin II. Consequently, despite large increases in volume, maternal blood pressure is not increased during pregnancy.

Plasma osmolality and thus volume are also regulated by the hypothalamic octapeptide AVP, which acts on the kidney to enhance water reabsorption. Throughout the course of gestation, plasma osmolality (275 mOsmol/kg) is significantly lower than that in nonpregnant women (285 mOsmol/kg). Moreover, in contrast to the nonpregnant state in which AVP levels are reduced by water loading, the latter does not occur in pregnancy, indicating that the osmolality set point has been reduced (Fig. 11). The factors regulating this change remain to be elucidated, although it has been shown that increased levels of angiotensin II stimulate hypothalamic AVP production and also induce thirst.

IV. SUMMARY

It is apparent that the placenta, via the production of peptide and steroid hormones, plays a primary role in regulating maternal organ systems as well as in development of the trophoblast itself. Thus, during the periimplantation period the newly formed placental syncytiotrophoblast produce CG which "rescues" the corpus luteum, resulting in the continued secretion of progesterone which acts to prevent uterine contractility and thus maintain the pregnancy. The continued secretion of CG, which appears to be regulated by neuropeptides and/or growth factors also produced in discrete cells of the placenta, stimulates the maternal pituitary–thyroid axis as well as relaxin secretion by the corpus luteum and activates testicular function essential for normal sexual development of the male fetus. As the cytotrophoblast proliferate to increase syncytiotrophoblast mass, CS production is increased, which by its glucogenic and lipolytic actions ensures that glucose is made available to the fetus and not utilized by maternal tissues which are in turn provided an adequate supply of free fatty acids. To ensure that maternal glucose levels do not become excessive, the maternal pancreas secretes exaggerated amounts of insulin to maintain homeostasis. The cytotrophoblast and syncytiotrophoblast continue to produce growth factors and various neuropeptides which act in an autocrine/paracrine manner to facilitate continued growth of the placenta and a marked increase in the production of estrogen and progesterone. Estrogen increases the production of renin–angiotensin II–aldosterone essential for maintenance of appropriate sodium balance to support a requisite increase in plasma volume and concomitantly acts to diminish the vasoconstrictive effects of angiotensin II to maintain blood pressure within normal limits. The increase in estrogen production also stimulates the pituitary to produce prolactin, which acts with estrogen to enhance renal formation of the active calcitropic hormone 1,25-dihydroxy vitamin D_3, also produced by the placenta. The increased production of 1,25-dihydroxy vitamin D_3 ensures that maternal calcium balance is maintained and that the fetus is provided 100–200 mg of calcium per day to support its development. Prolactin and CS also act to enhance mammary development essential to support lactation in the postpartum period.

See Also the Following Articles

Activin and Activin Receptors; Chorionic Gonadotropin, Human; Decidua; Fetal Adrenals; Growth Factors; Human Placental Lactogen (Human Chorionic Somatomammotropin); IGF (Insulin-like Growth Factors); Inhibins; Pregnancy, Maintenance of; Prolactin, Overview

Bibliography

Albrecht, E. D., and Pepe, G. J. (1988). Endocrinology of pregnancy. In *Non-Human Primates in Perinatal Research* (Y. W. Brans and T. J. Kuehl, Eds.), pp. 13–78. Wiley, New York.

Glinoer, D. (1997). The regulation of thyroid function in pregnancy: Pathways of endocrine adaptation from physiology to pathology. *Endocr. Rev.* **18**, 404–433.

Han, V. K. M. (1993) Growth factors in placental growth and development. In *Molecular Aspects of Placental and Fetal Membrane Autocoids* (G. E. Rice and S. P. Brennecke, Eds.), pp. 395–445. CRC Press, Boca Raton, FL.

Hild-Petito, S., Donnelly, K. M., Miller, J. B., Verhage, H. G., and Fazleabas, A. T. (1995). A baboon (*Papio anubis*) simulated-pregnant model: Cell specific expression of insulin-like growth factor binding protein-1 (IGFBP-1), type I IGF receptor (IGF-1R) and retinol binding protein (RBP) in the uterus. *Endocr. J.* **3**, 639–651.

Matzuk, M. M., Kumar, T. R., Shou, W., Coerver, K. A., Lau, A. L., Behringer, R. R., and Finegold, M. J. (1996). Transgenic models to study the roles of inhibins and activins in reproduction, oncogenesis, and development. *Rec. Prog. Horm. Res.* **51**, 123–154.

Pepe, G. J., and Albrecht, E. D. (1990). Regulation of the primate fetal adrenal cortex. *Endocr. Rev.* **11**, 151–176.

Pepe, G. J., and Albrecht, E. D. (1995). Actions of placental and fetal adrenal steroid hormones in primate pregnancy. *Endocr. Rev.* **16**, 608–648.

Petraglia, F., Florio, P., Nappi, C., and Genazzani, A. R. (1996). Peptide signaling in human placenta and membranes: Autocrine, paracrine, and endocrine mechanisms. *Endocr. Rev.* **17**, 156–186.

Stock, M. K., and Metcalfe, J. M. (1994). Maternal physiology during gestation. In *The Physiology of Reproduction* (E. Knobil and J. D. Neill, Eds.), pp. 947–983. Raven Press, New York.

Strauss, J. F. I., Gafvels, M., and King, B. F. (1995). Placental hormones. In *Endocrinology* (L. J. DeGroot, Ed.), Vol. 3, pp. 2171–2206.

Tulchinsky, D., and Little, A. B. (1994). *Maternal–Fetal Endocrinology*, 2nd ed. Saunders, Philadelphia.

Protozoa

O. Roger Anderson
Columbia University

GLOSSARY

autogamy Sexual reproduction by fusion of gametes from a single-parent organism resulting in self-fertilization.

binary fission Asexual reproduction yielding two daughter cells by a nearly equal division of the parent cell.

budding Asexual production of daughter cells by cellular fission producing one or more usually smaller progeny, either simultaneously or in succession, by protrusion of the cytoplasm forming a bud that is released by constriction and separation at the point of attachment.

gametogamy Sexual reproduction by fusion of gametes (syngamy) to produce a zygote.

gamontogamy Sexual reproduction by pairing of parent cells and exchange of gametes usually resulting in cross-fertilization.

multiple fission Asexual reproduction by successive or simultaneous divisions of the parent cell resulting in numerous genetically identical daughter cells.

Protozoa are highly diverse and widely distributed, single-celled organisms dwelling in aquatic and terrestrial habitats. Some are parasitic. Free-living (nonparasitic) species range in abundance from hundreds to millions per liter in aquatic habitats and thousands to hundreds of thousands per gram of soil in terrestrial habitats. Most members have animal nutritional characteristics and many exhibit some form of locomotion during part of their life cycle. However, some are also photosynthetic; hence, they are not included in the animal kingdom.

I. INTRODUCTION

Some taxonomists place Protozoa in the kingdom Protista, encompassing algae and protozoa. Others designate a separate kingdom for "Protozoa," excluding some of the photosynthetic forms that are placed in the plant kingdom. In other treatments, Protozoa are considered a phylum. The category Protozoa is used here as a convenient label for single-celled eukaryotic organisms with representatives that have heterotrophic nutrition, though some may be photosynthetic. Although protozoa are among the earliest evolutionary forms of unicellular organisms, they have evolved highly varied modes of reproduction—far too diverse to be fully described in a brief summary. Therefore, only some major representative patterns will be presented, with special attention to sexual reproduction.

A. Major Protozoan Groups

The free-living protozoa are classified in three main groups: flagellates—moving by action of undulating, whip-like appendages called flagella; sarcodines—ameboid, floating, or creeping organisms extending finger-shaped or elongate pseudopodia; and ciliates—moving by the flickering motion of fine, hair-like undulating appendages, often covering part or all of the cell surface. These three groups also include some parasitic species (invading the cells and tissues of animal and plant hosts). Other major parasitic protozoa are classified in other groups such as Apicomplexa (e.g., malaria), Microspora, and Myxozoa.

B. Reproductive Patterns

Reproductive patterns will be exemplified by some free-living species. Protozoa exhibit various reproductive strategies in response to environmental demands, including asexual reproduction to rapidly increase a population when favorable conditions occur and sexual reproduction involving the exchange of genetic material. Sexual reproduction is comparable to that of animals, but it occurs in remarkably diverse ways among major groups of protozoa. Sexual reproduction patterns span a range from widespread release of numerous motile gametes that optimizes dispersal and likelihood of interbreeding to highly conservative mechanisms in which gametes or only reproductive nuclei are exchanged between closely joined pairs of individuals. In the latter case, gametes are not released into the environment. This increases protection of the gametes or gamete nuclei while also maximizing successful completion of fertilization.

II. ASEXUAL REPRODUCTION

A. Fission and Budding

Protozoa reproduce asexually, typically by fission or budding. Fission is binary or multiple. In binary fission a parent cell undergoes mitotic division (nuclear replication followed by splitting of the cell) to yield two daughter cells of nearly equal size and with identical genetic composition to the parent. Flagellates typically divide longitudinally beginning at the flagellar end of the cell and eventually separating by rupture of a very thin cytoplasmic strand at the posterior end connecting the two daughter cells. Some amebas contract into a somewhat discoidal shape and divide by constriction approximately along the diameter of the cell. Binary fission in ciliates is transverse, producing two daughter cells that at first are connected end to end and then separate. In multiple fission, the parent cell divides repeatedly, yielding more than two genetically identical daughter cells. This mechanism increases numbers rapidly in response to favorable environments. Some species reproduce by forming bud-like daughter cells, each with a nucleus produced by mitotic division. The daughter nuclei, surrounded by a thin layer of cytoplasm, are "pinched off" from the surface of the parent cell. The daughter cells may be smaller than the parent cell initially and subsequently grow to a mature size.

B. Budding and Metamorphosis

The sessile, tentacle-bearing stage of suctoria (a ciliate) produces a ciliated swarmer that is released through a pore in the cell surface (Fig. 1) and undergoes a primitive form of metamorphosis. The adult stage is attached to the substratum by a stalk and contains tentacles that are used to snare prey. During

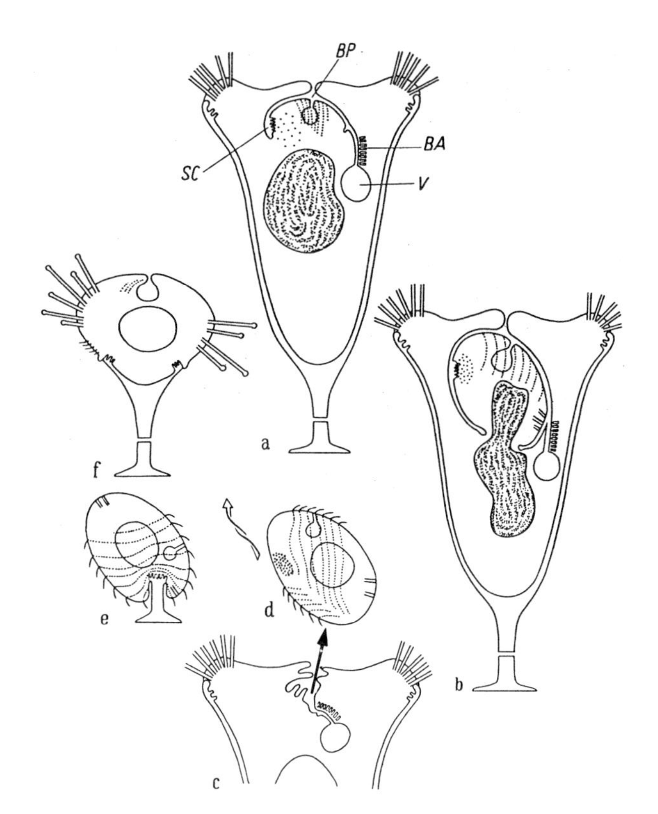

FIGURE 1 Asexual reproduction by release of a ciliated swarmer in the suctorian *Acineta tuberosa*. Diagram of budding and metamorphosis with release of swarmer from brood pouch (BP) and its transformation into a mature stalked cell (reproduced with permission from Grell, 1973).

swarmer production, a brood pouch situated above the nucleus invaginates to form a bud of cytoplasm. The parent nucleus elongates and protrudes into the developing cytoplasmic bud. The nuclear bud, containing a portion of the multiple chromosomes in the mother nucleus, pinches off to form the nucleus of the ciliated swarmer. After budding from the cytoplasm within the brood pouch, the swarmer is released through a birth pore. It swims away and eventually settles down and undergoes metamorphosis. A short stalk is produced on the ventral surface and gradually elongates. Concurrently, feeding tentacles are produced. The brood pouch and associated structures differentiate from the surrounding cytoplasm forming the adult stage, thus completing the life cycle. Other protozoa also undergo systematic changes in cell shape and function during phases of the life cycle. However, these are not directly comparable to metamorphosis in higher forms of life in which tissues and organs undergo transformations to yield different life stages. Hence, the term metamorphosis is not widely applied in protozoology.

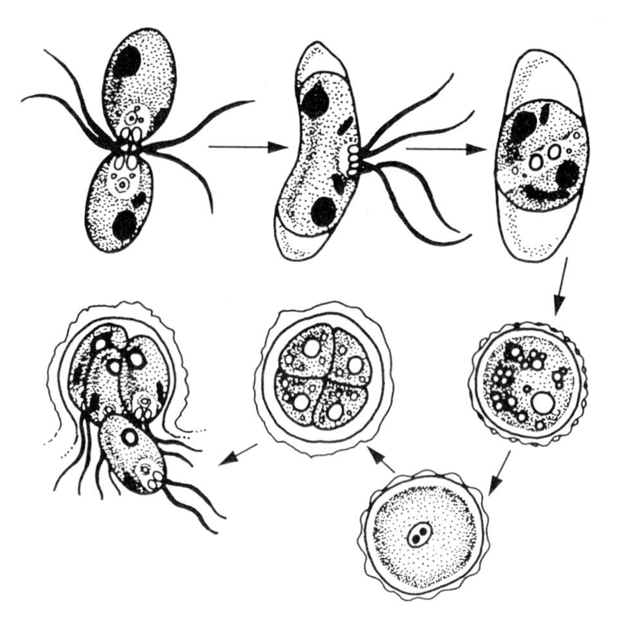

FIGURE 2 Sexual reproduction (gametogamy) in the flagellate *Chlamydomonas*. Fusion of two haploid gametes yields a zygote that eventually divides meiotically to form four flagellated daughter cells (gametes) released into the environment (reproduced with permission from Anderson and Druger, 1997, © 1997 by the National Science Teachers Association).

III. SEXUAL REPRODUCTION

Sexual reproduction occurs when meiotic (haploid) products of nuclear division fuse to form diploid offspring. Since the fusing nuclei are contributed by different parents, the offspring may contain a combined set of genetic traits different from those of either parent. Consequently, sexual reproduction offers an advantage of enhancing genetic diversity. This can produce offspring that are better adapted to the environment. Sexual reproduction in protozoa is conveniently classified into three groups: gametogamy, in which gametes from two parent individuals are released into the environment and subsequently fuse to form the offspring; autogamy, in which the gametes are derived from a single parent cell and may fuse within the body of the parent cell; and gamontogamy, in which compatible reproductive individuals unite (gametes are not released directly into the environment) before the exchange of gametes or gamete nuclei.

A. Gametogamy

In the simplest form, two cells of the same size and appearance, but of different genetic composition (haploid gametes), are attracted to one another perhaps by a chemical stimulus and fuse during fertilization. Typically, at least one of the pair is motile. The product is a zygote (diploid cell) that gives rise to the next generation. This is exemplified by some flagellates (e.g., *Chlamydomonas*; Fig. 2). The motile gametes of the same size are called isogametes and the process is isogamety. Upon attraction, the cells unite at the anterior flagellated end, the membranes of the two cells merge, the cytoplasm becomes joined, and the nuclei fuse, completing the fertilization step. The resulting zygote with a diploid nucleus undergoes a period of cytoplasmic reorganization, which may include a dormant stage. The nucleus undergoes repeated division by meiosis to produce four haploid flagellated daughter cells that are released into the environment. They may divide repeatedly by mitosis to form a clone, and they eventually serve as gametes. They pair and fuse to form a zygote, thus completing the life cycle.

The advantages of gametogamy include wide dissemination of the gametes and increased likelihood of cross-fertilization. The gametes of some species are self-sustaining, being photosynthetic or otherwise capable of obtaining nourishment, and are motile, thus increasing chances that they will disperse and fuse with gametes from other individuals of different genetic makeup. Disadvantages include the possible loss of gametes through trauma or predation during their dispersal in the environment or failure to encounter a compatible gamete of the proper mating type.

Anisogamety and oogamety are forms of gametogamy in which the gametes are of markedly different size and morphology. In oogamety, one gamete corresponds to an egg (usually larger and nonmotile) and the other a sperm (smaller and motile). The sperm cells released by one parent individual swim to the nonmotile oogametes of another where fusion and fertilization occur. The resulting zygote gives rise to progeny by developmental stages that are varied among different protozoan taxa. Oogamety occurs in *Volvox*, a large gelatinous spherical colony of numerous flagellated cells. The sperm-producing colony releases a packet of haploid flagellated sperm that swim to the oogamete-producing colony. The motile sperm are released from the packet at the surface of the receptive colony and penetrate its gelatinous wall. The sperm swim to the nonmotile gamete ("egg") in which fusion and fertilization occur, producing a zygote that eventually develops into a new colony.

Among the foraminifera, a group of marine sarcodines living within chambered calcitic shells, the parent produces numerous flagellated gametes that are released into the environment. These fuse in pairs, producing a zygote that matures to form a small calcitic shell, initially with a few chambers. Additional chambers of increasing size are added, culminating in the adult stage capable of releasing gametes, thus completing the life cycle.

B. Autogamy

Autogamy is illustrated in the well-documented case of *Actinophrys sol* (Fig. 3). *Actinophrys* is a floating sarcodine with a central nucleus surrounded by a layer of denser cytoplasm and a frothy external cytoplasmic envelope. Fine ray-like pseudopodia radiate out from the frothy envelope and are used to capture prey. During autogamy, the diploid nucleus undergoes division to form two daughter cells contained within a cyst wall. Each of the daughter cells undergoes two stages of meiotic division to produce a haploid gamete. These fuse within the protective space of the cyst wall to produce a zygote. During the first meiotic division, only one nucleus of the pair survives; the other degenerates. The surviving nucleus divides to produce two haploid nuclei; however, again only one survives. Thus, at the end of this process, two haploid daughter cells, each with a single haploid nucleus, coexist within the cyst wall (Fig. 3g). Each of the cells differentiates into a mature gamete. One may produce pseudopodia and the other withdraws somewhat at the point contacted by the pseudopodia of its partner. The gametes unite, followed by nuclear fusion to produce a zygote. The zygote may enter a resting phase (Fig. 3i) but eventually differentiates to form mature progeny with the characteristic radiating rays of pseudopodia (Fig. 3a).

Autogamy as evidenced in *A. sol* has the advantage of protecting the gametes within the surrounding cyst wall, thus enhancing chances of survival and increasing the probability of fertilization. However, this provides less opportunity for mixing of genetic material since the gametes arise from the same parent cell. This is fundamentally a process of self-fertilization as also occurs in animals and can help rid the genome of deleterious mutations by pairing of recessive genes that are fully expressed in the offspring, but this also leads to homozygosity.

C. Gamontogamy

Gamontogamy can occur with gamete formation (reproductive cells) or by production of gamete nuclei that are exchanged. In both cases, however, the reproductive organisms pair with one another. When

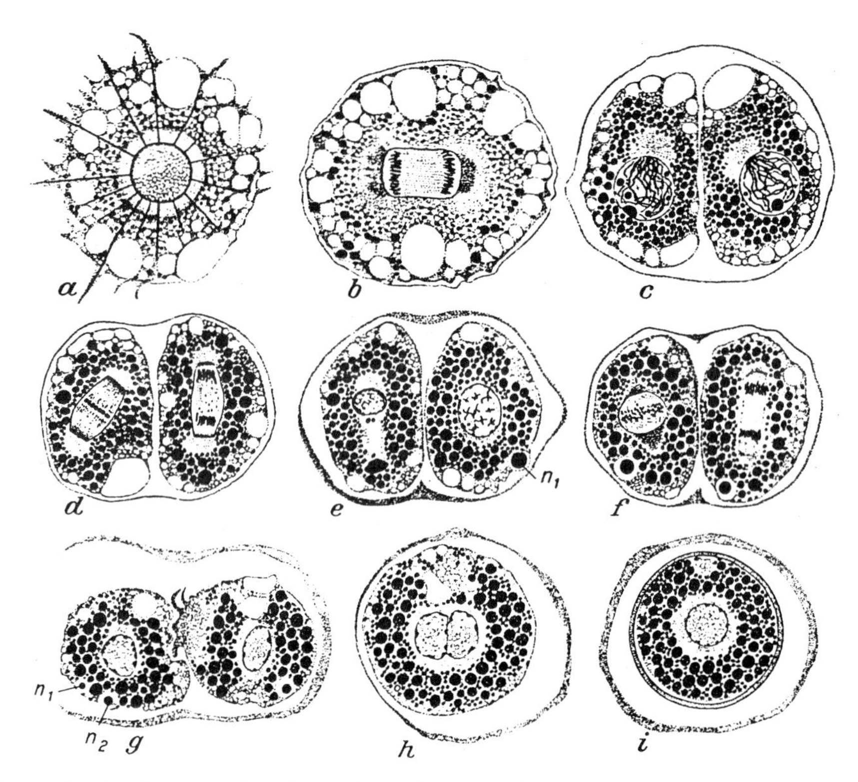

FIGURE 3 Sexual reproduction (autogamy) in *Actinophrys sol*. Repeated meiotic divisions of the parent cell (c–f) are followed by gamete pair formation, one with pseudopodia (g). Fusion of the gametes (h) within the wall of the parent cell produces a diploid zygote (i) (reproduced with permission from Grell, 1973).

gametes are produced, each gamete, consisting of a haploid nucleus and surrounding cytoplasm, fuses with a compatible gamete from the other partner in the pair. The gametes may be flagellated or ameboid depending on the species. Gamontogamy without gametes occurs typically in ciliates in which compatible individuals fuse by contact of the cells at their anterior surfaces, forming a cytoplasmic connection, and haploid nuclei are exchanged. The motile nuclei act as sperm nuclei. Nonmotile egg nuclei remain within each pair and are fertilized by the migratory sperm nuclei.

D. Gamontogamy with Gamete Production

Although there are variations in detail, gamontogamy with gamete production in foraminifera is illustrated by *Discorbis mediteranensis* (Fig. 4). Two mature, compatible organisms are attracted and migrate toward one another. Upon contact of their shells, the organisms use their contractile rhizopodia (branched filamentous pseudopodia) to reorient the shells, bringing the flattened ventral surfaces into contact and forming a united pair. They are joined by a cement substance secreted by the pair. The ventral surface of each shell contains an aperture permitting exchange of gametes. During reproduction, the internal walls of the shells dissolve, producing a common space in which gametes can mingle and compatible pairs fuse. The entire cytoplasm of the reproducing organisms is consumed to produce the flagellated gametes (typically several hundred) that swim about within the two united shells and fuse to form zygotes. The zygotes give rise to offspring that disperse and mature into adults with a typical multichambered calcitic shell. In other species (e.g., *Patellina corrugata*), more than two gamonts may unite, forming a cluster of organisms with their ventral surfaces oriented toward the center of the cluster. They se-

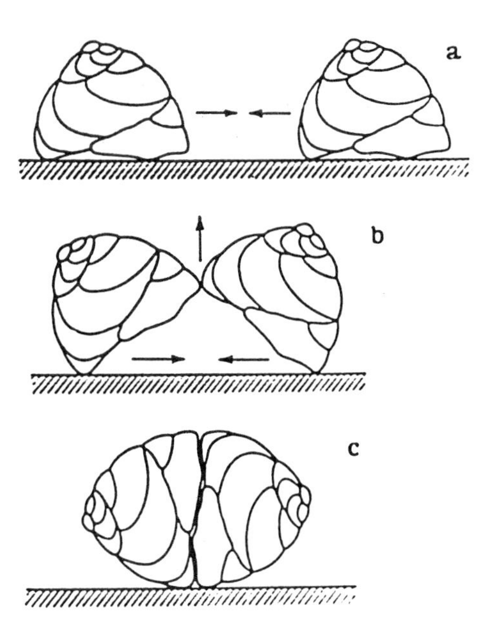

FIGURE 4 Sexual reproduction (gamontogamy) in the benthic foraminifera *Discorbis mediterranensis*. Mating pairs migrate toward each other (a), reorient their shells (b), and pair along the ventral surfaces (c). Gametes are exchanged, followed by fertilization within the spaces inside the shells (reproduced with permission from Grell, 1973).

crete a membrane that covers them and anchors them to the substrate. Gamete dispersal and fusion take place within this protective covering enclosing the cluster. Mating studies have confirmed that individuals are of different mating types or sexes. Opposite types either pair, as in the case of *D. mediteranensis*, or form a cluster containing a mixture in varying proportions of the two mating types as with *P. corrugata*. Gamontogamy, as does autogamy, provides protection for the gametes during formation and fusion, thus increasing the probability of successful fertilization and initiation of the next generation. However, unlike autogamy, gametes come from at least two parent organisms, thus increasing the possibility of genetic diversity among the offspring.

E. Hermaphroditism (Monoecy)

Protozoa lack tissue level organization. Hence, they do not exhibit true hermaphroditism as is found in higher organisms in which male and female reproductive tissue occur within the same individual. Autogamy, however, can be considered a form of hermaphroditism since gametes of two different mating types ("male" and "female") are produced by the same individual. The term "monoecy" is preferred in protozoology. Monoecy occurs when two mating types of gametes are produced within the same clone of cells or within an individual. In contrast, dioecy is used when male and female reproductive cells are produced in different individuals. Colonial flagellates (e.g., *Eudorina* and *Volvox*) include species that are monoecious and produce internal sperm-producing and egg-producing colonies within the same parent organism. Conjugation in ciliates includes the production of sperm and egg nuclei within each of the mating individuals and is conceptually a form of hermaphroditism.

IV. COMPLEX LIFE CYCLES

Complex life cycles of protozoa involve several different stages of asexual and sexual reproduction as illustrated for the large ameboid slime mold *Physarum polycephalum* (Fig. 5). The plasmodium is diploid and multinucleate. It lives in moist terrestrial environments and feeds by engulfing microbiota and organic matter on its ventral surface. During adverse environmental conditions such as drying, the plasmodium contracts and forms a condensed resting stage known as a sclerotium. The sclerotium is composed of numerous cysts cemented together. Hence, fragmentation of the sclerotium is not fatal and may aid dispersal. The sclerotium can revert to a plasmodium when the environment becomes sufficiently moist. At some point, the plasmodium reorganizes to form a mass of stalked sporangia containing haploid spores. These are released into the air and, after settling on a suitably moist surface, germinate to form small, haploid migratory amebas. These transform into flagellated gametes. They swim for a while and fuse to form a diploid zygote. The ameboid zygote feeds upon bacteria and other microbiota and grows in size to form the large multinucleated plasmodium. Asexual reproduction occurs by fragmentation of the plasmodium and by spore production. The sexual reproductive phase is limited to the production of flagellated swarmers and their fusion.

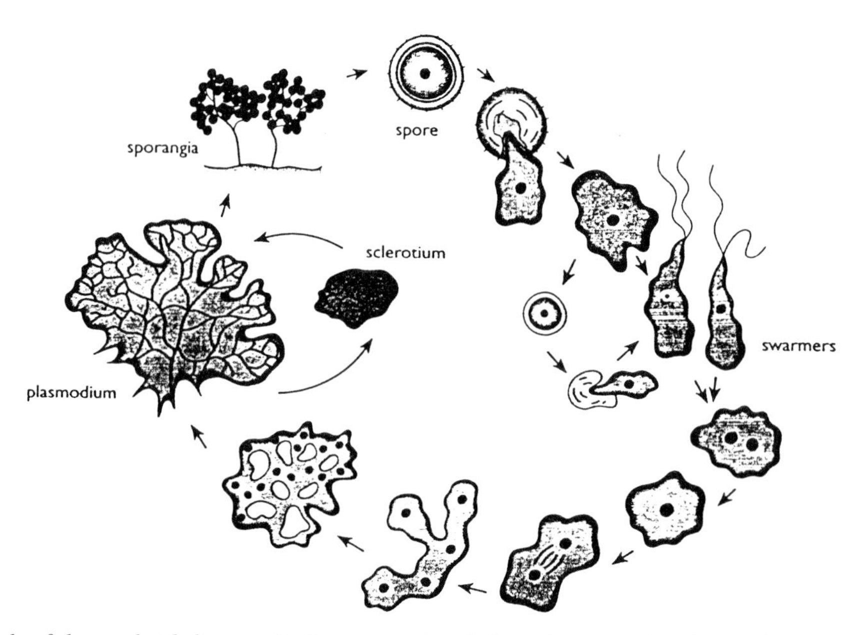

FIGURE 5 Life cycle of the ameboid slime mold *Physarum polycephalum* with asexual and sexual reproductive stages. Asexual reproduction by fragmentation of the giant plasmodium and by haploid spore formation is followed by release of flagellated swarmers (gametes) that pair and fuse to form a zygote. This matures into a new large multinucleated plasmodium (reproduced with permission from Anderson and Druger, 1997, © 1997 by the National Science Teachers Association).

Complex life cycles enhance survival by permitting wide dispersal of spores or other fission products during asexual reproduction and increased probability of enhanced genetic diversity by sexual reproduction.

Bibliography

Anderson, O. R. (1988). *Comparative Protozoology: Ecology, Physiology, Life History*, pp. 375–423. Springer-Verlag, Berlin.

Anderson, O. R., and Druger, M. (Eds.) (1997). *Explore the World Using Protozoa*. National Science Teachers Association and Society of Protozoologists, Arlington, VA.

Cavalier-Smith, T. (1993). Kingdom Protozoa and its 18 phyla. *Microbiol. Rev.* 57, 953–994.

Corliss, J. O. (1994). An interim utilitarian ("user-friendly") hierarchical classification and characterization of the protists. *Acta Protozool.* 33, 1–51.

Grell, C. (1973). *Protozoology*. Springer-Verlag, Berlin.

Lee, J. J., Hutner, S. H., and Bovee, E. C. (Eds.) (1985). *An Illustrated Guide to the Protozoa*. Society of Protozoologists, Lawrence, KS.

Patterson, D. J. (1993). Protoza: Evolution and systematics. In *Progress in Protozoology* (K. Hausmann and N. Hÿlsmann, Eds.), pp. 1–14. Fisher, Stuttgart.

Sleigh, M. (1989). *Protozoa and Other Protists*, pp. 70–96. Arnold, London.

Steidinger, K. A., and Walker, L. M. (Eds.) (1984). *Marine Plankton Life Cycle Strategies*, pp. 19–66. CRC Press, Boca Raton, FL.

Pseudocyesis

Joseph F. Mortola
Cook County Hospital

GLOSSARY

conversion disorder A syndrome characterized by the presence of one or more symptoms that are suggestive of a physical disorder which occurs in the absence of that physical disorder and which the individual does not consciously produce.

delusion A false idea which is believed without rational basis and which is contrary to commonly held societal beliefs.

hypothalamic amenorrhea The absence of menstrual bleeding based on failure of the hypothalamus to produce appropriate amounts of gonadotropin-releasing hormone. The majority of cases are behaviorally induced.

neuroleptic Any of several classes of compounds that are used to treat psychosis. The most common mechanism of action is through dopamine antagonism.

phenothiazine The largest class of antipsychotic drugs; it includes thorazine and stelazine.

pseudocyesis A syndrome seen in humans and several other species characterized by symptoms of pregnancy in the absence of such a pregnancy.

Pseudocyesis is defined as the false idea that one is pregnant despite clear medical evidence to the contrary. The symptoms of pseudocyesis are impressive and remarkably similar to pregnancy itself. They include menstrual irregularities, abdominal enlargement, and the sensation of fetal movement. Galactorrhea has been reported in almost half of all cases.

I. PREVALENCE

In the Western world, pseudocyesis is now exceedingly rare. Since the availability of accurate pregnancy tests, the prevalence of the condition has dropped dramatically. In the 1940s, it was estimated to occur in 1 of 250 pregnancies. By the 1970s, the prevalence had dropped to 6 of 22,000. The prevalence of the disorder remains high, however, in some geographic areas, such as those in which there is exceedingly high societal value placed on pregnancy or where modern pregnancy testing is less readily available. An example of such a culture is in parts of South Africa, where "lobola" is practiced. Lobola is the custom whereby the bride's family receives a monetary award from the groom's family if the bride is of proven fertility.

II. HISTORICAL ASPECTS

The historical aspects of pseudocyesis are of importance in appreciating the differences in the prevalence rates of associated psychopathology which have accompanied the disorder in recent years. The first account of pseudocyesis dates back to Hippocrates, who reported 12 cases of the disorder. Since that time, cases have been reported throughout history, including the case of the Queen of England, Mary Tudor (1516–1558), who had two episodes of pseudocyesis. Perhaps the most carefully described case of pseudocyesis is that of Anna O, a patient of Sigmond Freud; Freud wrote an extensive monograph describing her symptoms and treatment. From the perspective of psychiatric nosology, the case of Anna O clearly fell within what at the time was classified as a neurotic disorder and as such was considered highly amenable to psychoanalysis. In current psychiatric nomenclature, the case of Anna O, and the majority of the cases before the availability of rapid, inexpensive, widely available pregnancy tests, would be classified as a conversion disorder (or hysterical neurosis, conversion type). In a review of pseudocyesis written in 1989, O'Grady and Rosenthal characterized such individuals with neurotic level functioning as having "pseudocyesis vera."

III. CONVERSION DISORDERS

The diagnostic criteria for conversion disorder include (i) a change in physical functioning, suggesting a physical disorder, (ii) a temporal association with a psychosocial stressor which is apparently related to a psychological conflict and the onset of the symptoms, (iii) the person is not consciously producing the symptom, and (iv) the symptom cannot be explained by a physical disorder. Although pregnancy is not a physical "disorder," substitution of the word "condition" for disorder makes this set of criteria applicable to the case of pseudocyesis with a neurotic basis. Individuals with conversion disorders often, but not invariably, manifest hysterical personality traits. These include excessive need for approval, vanity, exaggerated expressions of emotions, self-centeredness, and the need for immediate gratification. Often such individuals exhibit "la belle indifference" which is an apparent marked disconcern for the physical complaint.

IV. DELUSIONAL DISORDERS

The advent of modern pregnancy tests and ultrasonography has resulted in a marked decrease in the prevalence of pseudocyesis as a conversion symptom. Currently, women with access to medical care who fiercely cling to the notion that they are pregnant, despite incontrovertible evidence to the contrary, are most likely to be delusional. Delusions are false, fixed ideas which are not consistent with societal beliefs. They occur most commonly with schizophrenia and are often associated with mania and psychotic depressions. As a result, it is important to consider these diagnoses in the differential diagnosis when presented with a patient with pseudocyesis. If these disorders are excluded, the most likely diagnosis is delusional disorder, somatic type. The diagnostic criteria for the disorder are given in Table 1. Delusional disorders in aggregate are rare, with an incidence of 0.3%. Only a small percentage of these individuals have somatic delusions (with paranoid delusions being much more common), and only a small percentage of those with somatic delusions have the delusion of pseudocyesis.

TABLE 1
Diagnostic Criteria for Delusional Disorder

Nonbizzarre delusions (i.e., those involving events that actually occur in life) of at least 1 month duration
Auditory or visual hallucinations if present at all are not prominent
Apart from the delusion(s), behavior is not obviously odd or bizarre
The delusions are present in the absence of acute manic or major depressive episodes
The patient is not schizophrenic.

V. EVALUATION

Evaluation of the patient with pseudocyesis requires attention to both the gynecologic and the psychiatric aspects of the condition. Since the majority of patients present with amenorrhea, gonadotropin measurement in concert with a careful history as to the patients age and the presence of hot flashes will be useful in distinguishing patients with menopause, chronic anovulation, or hypothalamic amenorrhea. A follicle-stimulating hormone (FSH) level higher than 30 mIU/ml implies a perimenopausal or menopausal state. If both the luteinizing hormone (LH) and FSH levels are low or low-normal, hypothalamic amenorrhea is diagnosed. The latter patients may have elevated prolactin levels. The presence of pseudocyesis and hypothalamic amenorrhea is consistent with the simultaneous development of a delusional system and hypothalamic dysfunction in response to stress. Cases have been reported in which pseudocyesis occurs in the setting of anorexia nervosa, although this is extremely rare in modern times since women with anorexia function in the neurotic spectrum.

VI. PATHOPHYSIOLOGY

Both Zurate *et al.* and Tulandi *et al.* have reported an increased response of LH to gonadotropin-releasing hormone challenge in women with pseudocyesis. This response is reminiscent of polycystic ovary syndrome. Indeed, some women with pseudocyesis manifest the full triad of polycystic ovary syndrome, including amenorrhea, hyperandrogenism, and obesity. Hyperprolactinemia is observed more often than would be expected, although it is does not appear integral to the pathophysiology of the disorder. Animal studies of pseudocyesis clearly demonstrate that the disorder is not limited to humans. However, little insight into the human condition has been gained by study of these models. In the majority of these animals, pseudocyesis is associated with elevated prolactin. In these species, unlike the human, prolactin is luteotropic. In the classical animal form of pseudocyesis, persistent corpus luteum function is at the basis of the pathophysiology. In the human, this is much more rarely seen. In humans, any condition which may result in amenorrhea can be the basis of pseudocyesis.

Although the false conviction that one is pregnant is usually triggered by an episode of amenorrhea, the converse may also occur. Amenorrhea may result from the stress attendant upon the psychological conflict which the individual woman experiences. Thus, hypothalamic amenorrhea of a psychogenic etiology may be the result of the psychological upset and further contribute to the false notion that the woman is pregnant. This mechanism was more commonly seen when conversion-type pseudocyesis was the predominant form of the disorder.

VII. TREATMENT

Patients with pseudocyesis and fixed delusional systems should be considered candidates for neuroleptic therapy. Appropriate neuroleptic therapy includes phenothiazines such as thorazine or stelazine and haloperidol. Careful assessment should be conducted for the presence of schizophrenia or major affective disorder.

Women with prolonged episodes of amenorrhea are susceptible to osteoporosis and a theoretically increased risk of cardiovascular disease. The appropriate treatment is dependent on evaluation of the prolactin and thyroid-stimulating hormone levels. Patients with elevated prolactin levels are candidates for bromocriptine or cabergoline. Those with hypothyroidism should be treated with thyroid replacement. Individuals with hyperthyroidism should be treated with antithyroid medications. Patients who are hypoestrogenic in the absence of thyroid disorder or hyperprolactinemia are candidates for replacement therapy using either a menopausal replacement regimen or oral contraceptive pills if they are nonsmokers under the age of 50 or cigarette smokers under the age of 35. Patients with pseudocyesis in the presence of a polycystic ovary-like syndrome may also be treated with cyclic medroxyprogesterone acetate, although their desire to become pregnant makes this a less attractive alternative than oral contraceptive pills.

See Also the Following Articles

Amenorrhea; Galactorrhea

Bibliography

American Psychiatric Association (1987). *Diagnostic and Statistical Manual of Mental Disorders*, 4th ed., pp. 199–203, 257–259. American Psychiatric Association Press, Washington, DC.

Bray, M. A., Muneyyiric-Delale, O., Kofinas, G. D., and Reyes, F. I. (1991). Circadian, ultradian and episodic gonadotropin and prolactin secretion in human pseudocyesis. *Acta Endocrinol.* **124**, 501–509.

Brenner, B. N. (1976). Pseudocyesis in blacks. *S. Afr. Med. J.* **50**, 1757–1759.

Brown, E., and Barglow, P. (1964). Pseudocyesis: A paradigm of psychophysiological interactions. *Arch. Gen. Psychiatr.* **24**, 221–229.

Fried, P. H., Rakoff, A. E., Schopbach, R. R., and Kaplan, A. J. (1951). Pseudocyesis: A psychosomatic study in gynecology. *J. Am. Med. Assoc.* **145**, 1329–1335.

O'Grady, J. P., and Rosenthal, M. (1989). Pseudocyesis: A modern perspective on an old disorder. *Obstet. Gynecol. Surv.* 44, 500–511.

Tulandi, T., McInnes, R. A., and Tolis, G. (1982). Pseudocyesis: Pituitary function before and after resolution of symptoms. *Obstet. Gynecol.* 59, 119–121.

Yen, S. S. C., Rebar, R. W., and Quesenberry, W. (1976). Pituitary function in pseudocyesis. *J. Clin. Endocrinol. Metab.* 43, 132–236.

Zurate, A., Canales, E. S., Soria, J., *et al.* (1974). Gonadotropin and prolactin secretion in pseudocyesis. *Ann. Endocrinol. (Paris)* 35, 445–452.

Pseudopregnancy

Mary S. Erskine
Boston University

GLOSSARY

corpus luteum A structure formed from ovarian granulosa and thecal cells after ovulation which secretes the progesterone necessary for endometrial growth, implantation, and maintenance of the blastocyst in early pregnancy.

decidualization The differentiation of placental-like tissue (deciduomas) within the uterine endometrium in response to trauma.

delayed pseudopregnancy A phenomenon in which the initiation of pseudopregnancy by cervical stimulation is delayed until the following ovulation when luteinization can occur.

prolactin A hormone secreted by the anterior pituitary which is responsible for maintaining corpus luteum function after mating in the rat.

tuberhypophyseal dopamine neurons Dopamine-releasing neurons whose cell bodies are located within the arcuate and periventricular nuclei of the hypothalamus and whose terminals are located in the posterior and intermediate lobes of the pituitary.

tuberinfundibular dopamine neurons Neurons whose cell bodies are located within the arcuate and periventricular nuclei of the hypothalamus and whose terminal areas are within the median eminence. Dopamine released by these neurons reaches the anterior pituitary through the pituitary–portal vessels.

Pseudopregnancy is a term which refers to a postmating or postovulatory period during which ovarian follicular development ceases and functional activity of the corpus luteum is maintained but in which pregnancy does not occur. In most of the Eutherian mammals for which information is available, pseudopregnancy is induced by an infertile mating, and the neurogenic stimulus provided by coital stimulation results in a prolongation of corpora luteal activity beyond that which occurs without mating stimulation. Mating can induce pseudopregnancy both in spontaneously ovulating species such as the rat and in reflexively ovulating species such as the rabbit. In the dog, a spontaneous period of corpora luteal activation may occur at the end of estrus even without mating; because it may be associated with mammary gland development and lactation, deposition of fat, and ovarian acyclicity, this period is also termed pseudopregnancy. In either case and despite the absence of developing fetuses, the endocrine changes associated with pseudopregnancy are similar to those seen during early pregnancy, the most prominent of which is an increased production of progesterone by the corpus luteum. This increase in progesterone results in development of the uterine endometrium, a prolonged diestrum which is characterized by continuous leucocytic vaginal smears, a lack of sexual or estrous behavior, a cessation of ovarian follicular development, and a potential for decidualization of the endometrium in response to traumatic insult.

The duration of pseudopregnancy is variable across species. In rodent species, such as the rat, mouse, and hamster, the duration of pseudopregnancy is approximately half the length of gestation, a duration which coincides with the period during pregnancy when the corpus luteum is sustained by the pituitary. When placental tissue fails to develop and the luteotrophic support for the corpus luteum is withdrawn, pseudopregnancy ends. Presumably, the abbreviated life span of the corpus luteum during pseudopregnancy serves to reduce the reproductive and metabolic effort expended on a pregnancy when a mating was infertile. In other species, such as the ferret, the duration of pseudopregnancy is the same as the length of gestation.

I. INITIATION OF PSEUDOPREGNANCY BY MATING OR VAGINOCERVICAL STIMULATION

Like other neuroendocrine reflexes such as suckling-induced prolactin secretion, pseudopregnancy is induced by neurogenic stimuli. Stimulation of the uterine cervix by penile intromissions received from males or by artificial mechanical or electrical vaginocervical stimulation (VCS) initiates the period of pseudopregnancy, and this occurs despite the absence of fertilization and pregnancy. Afferent neural input required for pseudopregnancy in the rat is carried by sensory afferents within the pelvic nerve which innervate caudal portions of the reproductive tract including the rostral vaginal wall and the cervix. Transection of the pelvic nerve bilaterally completely prevents mating-induced pseudopregnancy. The occurrence of pseudopregnancy is dependent on the female receiving sufficient numbers of intromissions from males, and too few intromissions will fail to prolong corpus luteum function. On the other hand, repeated matings in which the female receives very high numbers of intromissions and ejaculations neither prolong pseudopregnancy nor enhance progesterone secretion beyond normal limits. Therefore, the occurrence of pseudopregnancy requires that a threshold amount of VCS be received by the female. Other elements of mating stimulation which influence the attainment of this threshold are the ejaculatory stimulus and the timing of the interval between intromissions.

II. PATTERNS OF PROLACTIN SECRETION DURING PSEUDOPREGNANCY

Prolongation of corpora luteal function and pseudopregnancy is a consequence of the luteotrophic actions of one or more anterior pituitary hormones. Luteinizing hormone (LH) initiates the biochemical and morphological changes which occur in the postovulatory follicle as the corpus luteum develops, and LH is the primary luteotrophin throughout pseudopregnancy in many species. In the rat, sensitivity of the corpus luteum to LH is dependent on secretion of prolactin, and prolactin is secreted in a characteristic manner throughout pseudopregnancy as well as for the first 10 days of pregnancy. Prolactin via binding to ovarian prolactin receptors maintains the number of luteal LH receptors and thereby increases progesterone secretion. As shown in Fig. 1 (top), prolactin is secreted normally during the rat estrous cycle in a single surge on the afternoon of the day of proestrus. Following mating or VCS (Fig. 1, bottom), this pattern is altered substantially and prolactin is secreted in a uniquely episodic pattern for the 10- to 12-day period of pseudopregnancy. Mating initiates two daily surges of prolactin secretion which begin approximately 36 hr after VCS and persist undiminished for the majority of pseudopregnancy. These surges, which in the rat are approximately 10 times higher than basal circulating prolactin concentrations, occur late in the afternoon just before lights off (diurnal surge) and early in the morning just before lights on (nocturnal surge).

The expression of the surges is linked to a circadian

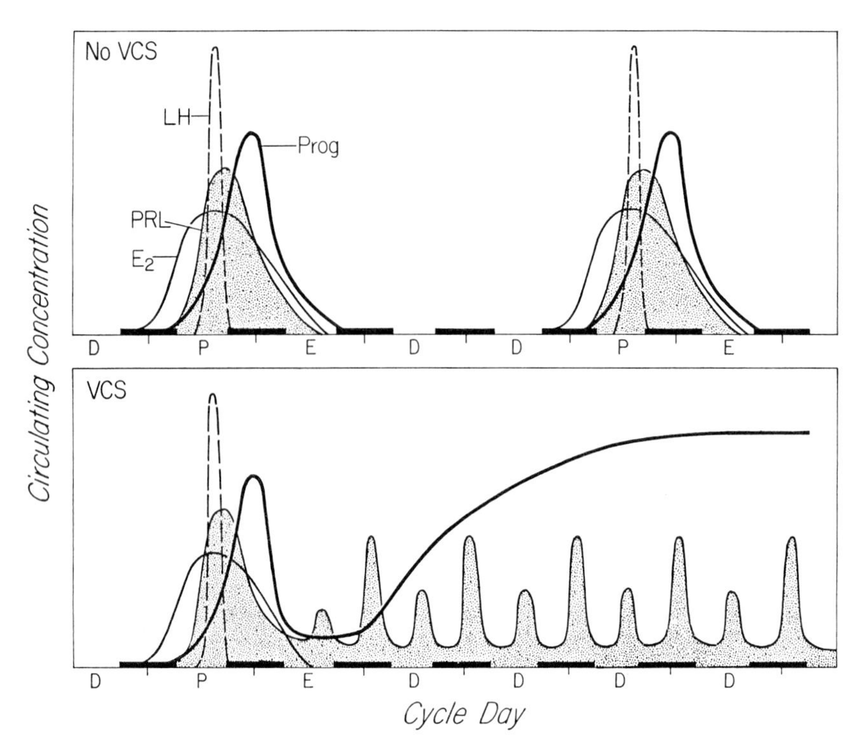

FIGURE 1 Diagrammatic representation of the patterns of circulating levels of luteinizing hormone (LH), prolactin (PRL), estradiol (E_2), and progesterone (Prog) on two separate days of proestrus in females not receiving VCS (top) and in females receiving VCS and becoming pseudopregnant (bottom). Shaded areas indicate circulating prolactin (reproduced with permission from M. S. Erskine, Prolactin release after mating and genitosensory stimulation in females, *Endocr. Rev.* **16**, 508–528, 1995. © The Endocrine Society).

mechanism since both surges can be entrained and phase-shifted to the light/dark cycle, can free-run in the absence of light/dark cues, and are eliminated by lesion of the suprachiasmatic nuclei of the hypothalamus. In addition, the timing of the surges is independent of when mating stimulation is received by the female; regardless of the time of day when mating occurs or VCS is given, the diurnal and nocturnal prolactin surges occur at the standard nocturnal and diurnal surge times.

The mating-induced prolactin surges are expressed in the ovariectomized/adrenalectomized female given VCS, showing that circulating steroids are not required for induction of prolactin surges. However, the likelihood that the surges will be initiated by mating, the magnitude of the induced surges, and the persistence of the surge pattern after mating are modulated by circulating ovarian steroids. In general, estrogen enhances the magnitude of the diurnal surge and prolongs its expression, whereas progesterone stimulates and prolongs the nocturnal surge. Exogenous administration of estrogen to an unmated or unstimulated female will induce repeated diurnal surges, whereas treatment with progesterone will induce pseudopregnancy. Both steroids can prolong the duration of an established pseudopregnancy by as much as 4 days.

III. TERMINATION OF PSEUDOPREGNANCY

Termination of pseudopregnancy and reinitiation of ovarian cyclicity result from a waning of progesterone secretion as the corpus luteum regresses. The process of luteal regression is a consequence of several factors. Evidence suggests that prostaglandin $F_{2\alpha}$ is produced by the uterus and has a local luteolytic effect on the corpus luteum. Hysterectomy has been shown to prolong the period of pseudopregnancy by several days. In addition, the endothelial lining of the uterus produces a factor which has an inhibitory action on prolactin secretion by direct action on pituitary lactotrophs. This inhibition of prolactin secretion produces an effective withdrawal of luteotrophic support. Lastly, the initial decreases in production of progesterone by the corpus luteum on Days 10 and 11 combined with increases in estradiol secretion on Days 11 and 12 appear to inhibit the secretion of the diurnal and nocturnal prolactin surge, respectively. The presence of deciduomas within the uterus prolongs pseudopregnancy, and this is thought to be the consequence of the production of one or more luteotrophic factors produced by decidual tissue.

IV. PSEUDOPREGNANCY AND THE NEURAL MNEMONIC

A single mating experience during estrus initiates neuroendocrine changes which persist for many days; therefore, such stimulation is said to initiate a mnemonic or memory of the mating stimulation received. The processes by which the memory is established and maintained are unknown. However, it has been suggested that there are two separate components to the mnemonic: a short-term memory, which involves assimilation and quantification of the genitosensory stimulation received to attain the sensory induction threshold, and a longer term memory, which is responsible for persistence of the twice-daily prolactin surges throughout the 10- to 12-day postmating period of pseudopregnancy. Attainment of the induction threshold requires that the number of intromissions received by the female be recorded and stored sequentially until sufficient numbers have occurred for initiation of pseudopregnancy. The existence of a short-term mnemonic involved in the recording and storage of VCS was most clearly demonstrated in a study by Edmonds *et al.* (1972) showing that experimentally prolonging the interval between individual intromissions by up to 1 hr did not prevent the occurrence of pseudopregnancy. Therefore, information about the numbers of intromissions received can be retained for relatively prolonged periods until the total received exceeds the sensory threshold for induction of pseudopregnancy. The persistence of prolactin surges throughout pseudopregnancy and without subsequent reinforcing of mating stimulation indicates that a longer term memory of mating exists which is responsible for the repeated daily expression of prolactin surges. The longer term mnemonic can be observed under experimental conditions by administering artificial VCS to cycling rats at times in the estrous cycle when mating would not normally occur. In contrast to females mated on proestrus, animals given VCS on diestrus continue to show estrous cyclicity for several days after mating, and the onset of pseudopregnancy does not occur until approximately 36–72 hr later (delayed pseudopregnancy). The onset of pseudopregnancy is delayed until after the next ovulation when a new set of corpora lutea can be stimulated by prolactin and their function prolonged. Prolactin surges are not required during the 36- to 72-hr postmating period for delayed pseudopregnancy to occur. Therefore, the genitosensory information is retained for several days before the luteotrophic actions of prolactin are initiated. The neural processes underlying the establishment of either of these mnemonics are unknown, but it appears that one site that may be involved is the medial amygdala since interruption of neural activity in this area around the time of mating prevents establishment of pseudopregnancy.

V. NEUROENDOCRINE CONTROL OF PSEUDOPREGNANCY

Very little information is available on the mechanisms by which the genitosensory stimulation received during mating is transduced into the prolactin surges of pseudopregnancy. The ascending fibers within the ventral noradrenergic bundle carry afferent information important for this process since transection of these fibers or lesion of this area prevent the occurrence of pseudopregnancy. As indicated, some evidence suggests that the medial amygdala is important for establishment of the mnemonic(s) required for pseudopregnancy, and several hypothalamic nuclei and the preoptic area are involved in regulation of one or more of the prolactin surges induced by mating. Studies involving lesion of several specific nuclei have demonstrated that (i) the medial preoptic area appears to tonically inhibit expression of the nocturnal prolactin surge and stimulate the diurnal surge, (ii) the dorsomedial–ventromedial nuclei of the hypothalamus facilitate the expression of both nocturnal and diurnal surges, and (iii) the suprachiasmatic nuclei are important for the expression of both surges (Gunnet and Freeman, 1983). Neuromodulatory peptides produced in the paraventricular nucleus have also been implicated in the stimulation of mating-induced prolactin surges.

It is known that both inhibitory and stimulatory factors are involved in regulation of the prolactin secretion during pseudopregnancy. Prolactin secretion is under the tonic influence of hypothalamic prolactin inhibitory factor(s), the most potent and the best studied being dopamine. Dopamine released from tuberoinfundibular and tuberohypophyseal dopamine neurons inhibits prolactin secretion from pituitary lactotrophs via its binding to D_2 dopamine receptors. During pseudopregnancy, a rhythm of dopaminergic activity occurs within the hypothalamus which is inverse to that of the prolactin surges, suggesting that prolactin surges may result, at least in part, from disinhibition of prolactin secretion at the times of the diurnal and nocturnal surges. Although VCS has been demonstrated to decrease concentrations of dopamine in the hypophyseal portal system at the time of the diurnal surge on the day after mating, the acute and prolonged effects of VCS on dopaminergic activity in tuberoinfundibular dopamine neurons have not been examined systematically.

Several neuroendocrine factors are known to exert stimulatory actions on prolactin secretion. Among those that have been shown to influence the diurnal and/or nocturnal surges of pseudopregnancy are opioid peptides, serotonin, VIP, and oxytocin. However, since each of these factors can stimulate prolactin at other times in the reproductive cycle, it is not clear which, if any, of these prolactin-releasing factors are selectively involved in pseudopregnancy. Further studies are required to elucidate those neuroendocrine factors which are stimulated by VCS, are involved in the establishment of the mnemonics of pseudopregnancy, and are necessary for initiation and maintenance of the luteotrophic endocrine conditions associated with pseudopregnancy.

See Also the Following Articles

Corpus Luteum; LH (Luteinizing Hormone); Prolactin, Actions of

Bibliography

Edmonds, S., Zoloth, S. R., and Adler, N. T. (1972). Storage of copulatory stimulation in the female rat. *Physiol. Behav.* 8, 161–164.

Erskine, M. S. (1995). Prolactin release after mating and genitosensory stimulation in females. *Endocr. Rev.* **16**, 508–528.

Gunnet, J. W., and Freeman, M. E. (1983). The mating-induced release of prolactin: A unique neuroendocrine response. *Endocr. Rev.* 4, 44–61.

Niswender, G. D., and Nett, T. M. (1994). Corpus luteum and its control in infraprimate species. In *The Physiology of Reproduction* (E. Knobil and J. D. Neill, Eds.), 2nd ed., pp. 781–816. Raven Press, New York.

Puberty Acceleration

John G. Vandenbergh
North Carolina State University

GLOSSARY

estrus A period of sexual receptivity associated with ovulation in spontaneously ovulating female mammals, often associated with characteristic vaginal cellular changes.

pheromone A substance or blend of substances produced by one organism, transmitted through the environment, inducing a behavioral, developmental, or physiological change in the recipient of the same species.

priming pheromone A chemical signal produced by one or more individuals resulting in a developmental or physiological change, usually over the long term, in the recipient animals of the same species.

puberty acceleration The acceleration of puberty in females of many mammalian species through direct contact with the adult male or by exposure to a male urinary pheromone; also known as the Vandenbergh effect.

puberty inhibition The delay of puberty among female mammals housed in groups due to direct physical interactions or to a female urinary pheromone.

In mammals, the timing of puberty is a finely tuned event that is partly dependent on genetic heritage, and it is influenced by a series of environmental factors spanning the period from early embryonic development to the onset of puberty. Pheromones produced by adult males are one of the most important factors known to accelerate the onset of female puberty in several mammals. Pheromonal acceleration of puberty in the laboratory mouse was first described by Vandenbergh and has been termed the Vandenbergh effect.

I. PUBERTY ACCELERATION OF FEMALE MICE

The female laboratory mouse is usually expected to attain puberty at 45–60 days of age. However, puberty can occur as early as 28 days of age if the female is housed singly and exposed to an adult male or to urine from the male. Puberty in females housed in groups is significantly delayed, apparently due to female-to-female inhibitory urinary pheromones.

A. Androgen Dependency

Adult male house mice produce a chemical signal, probably a peptide, in their urine that induces earlier onset of puberty among females. This compound, or blend of compounds, is under androgen control. If the mouse is castrated, his urine loses its ability within 1 week to accelerate the onset of puberty. If the castrated male is injected with testosterone, the puberty-accelerating effect of his urine is restored. Similarly, female mice do not normally produce puberty-accelerating pheromones in their urine, but if injected with testosterone for several days, their urine then accelerates puberty in juvenile females.

Stimulated by the finding that the presence of puberty-accelerating pheromone in male urine is dependent on androgen levels, follow-up experiments were conducted. First, urine from trained fighter mice that had been dominant in every encounter was applied to the nares of juvenile females in comparison to applications of urine from the subordinate males. Only the urine from the dominant mice accelerated the onset of puberty. Second, when urine was collected from individual male mice living in a social group, the puberty-accelerating effect of the urine was strongest for the dominant male and reduced in a dose-dependent fashion for animals of lower social rank. Extensive attempts to isolate the specific chemical or chemicals serving as a puberty-accelerating pheromone in male urine have met with limited success. Using the uterine weight of juvenile females as a bioassay, the protein fraction of urine was shown to contain the active component. Further column chromatography resulted in the isolation of an active fraction from male urine at a peak molecular weight of 860. Tests of this fraction were positive for peptides.

B. Natural Populations

The puberty acceleration effect of the male occurs among house mice living in seminatural conditions as well as in the laboratory, as shown by Drickamer and students. Wild house mice were placed in large outdoor enclosures and exposed to male urine, female urine, or water. Frequent trapping to monitor population density revealed that the populations exposed to male urine reached the highest density, and those exposed to group female urine attained the lowest density. Water-exposed populations were intermediate. These results suggest that puberty-accelerating pheromones can result in higher density populations, whereas puberty-inhibiting pheromones, from the group females, can result in lower population densities.

II. OTHER SPECIES

The presence of an adult male or a pheromone from the male can accelerate the onset of puberty in a number of species in addition to the house mouse. In three species of *Microtus*, reflex ovulators, breeding can be advanced by as much as 2 weeks. The possibility that puberty could be advanced in domestic farm animals has also been explored. In the cow, sheep, pig, and goat, puberty has been advanced between 1 and 4 weeks as a result of exposure to an adult male or to presumptive priming pheromones from the male. One primate, the tamarin (*Saguinas fuscicollis*), has shown an advance in the first age of conception in the presence of male stimulation.

III. RELATIONSHIP TO PUBERTY INHIBITION

It is interesting to note that puberty inhibition due to exposure to other females or to their urine also occurs in mice and in several other species of mammals. This inhibiting effect balances out the male acceleratory effect; in fact, it can override the male's effect. Female mice produce the urinary inhibitory pheromone when crowded together for several days. This suggests that the crowding due to high population density in the wild could feedback on the population and slow the rate of puberty attainment. The puberty inhibition effect has been demonstrated under field conditions. Crowded females in a seminatural population produce the puberty-inhibiting pheromone, whereas those in a sparse population do not. Thus, some support is available for the theory that there could be a density-dependent factor operating through a pheromonal mechanism to regulate mouse populations.

The puberty-acceleration effect is just one of many reproductively related pheromonal effects that have been described.

See Also the Following Articles

Bruce Effect; Estrus; Pheromones; Puberty, in Nonprimates; Whitten Effect

Bibliography

Drickamer, L. C. (1986). Puberty-influencing chemosignals in house mice: Ecological and evolutionary considerations. In *Chemical Signals in Vertebrates* (D. Duvall, D. Muller-Schwarze, and R. M. Silverstein, Eds.), pp. 441–455. Plenum, New York.

Drickamer, L. C., and Mikesic, D. G. (1990). Urinary chemosignals, reproduction, and population size of house mice (*Mus domesticus*) living in field enclosures. *J. Chem. Ecol.* **16**, 2955–2968.

Hadley, M. E. (1992). *Endocrinology*, 3rd ed. Prentice Hall, Englewood Cliffs, NJ.

Nelson, R. J. (1995). *An Introduction to Behavioral Endocrinology*. Sinauer, Sunderland, MA.

Ojeda, S. R., Aguado, L. I., and Smith, S. (1983). Neuroendocrine mechanisms controlling the onset of female puberty: The rat as a model. *Neuroendocrinology* 37, 306–313.

Vandenbergh, J. G. (1969). Male odor accelerates female sexual maturation in mice. *Endocrinology* **84**, 658–660.
Vandenbergh, J. G. (1987). Regulation of puberty and its consequences on population dynamics of mice. *Am. Zool.* **27**, 891–898.
Vandenbergh, J. G. (1994). Pheromones and mammalian reproduction. In *Physiology of Reproduction* (E. Knobil and J. D. Neill, Eds.), Vol. 2, pp. 343–362. Raven.
Vandenbergh, J. G., Finlayson, F. S., Dobrogosz, W. J., Dills, S. S., and Kost, T. A. (1976). Chromatographic separation of puberty accelerating pheromone from male mouse urine. *Biol. Reprod.* **15**, 260–265.

Puberty, in Humans

Thomas A. Klein
Thomas Jefferson University

GLOSSARY

adrenarche Onset of increased adrenal androgen secretion, usually without clinical manifestation. Often used synonymously with pubarche.

gonadarche Onset of clinical evidence of increased gonadal sex steroid secretion or (in males) increased testicular size.

gonadostat Hypothalamic–pituitary mechanism which regulates gonadotropin secretion; modulated by various hormonal and neural influences.

gonadotropin-releasing hormone pulse generator Network of hypothalamic neurons which secrete gonadotropin-releasing hormone in a synchronized pulsatile manner into the pituitary portal circulation.

juvenile pause Period of slow growth without sexual maturation, between age 2 years and the onset of puberty, accompanied by low gonadotropin levels; also called the "prepubertal hiatus."

menarche Onset of vaginal bleeding, usually anovulatory.

pubarche Onset of pubic and axillary hair growth.

thelarche Onset of female breast development.

Puberty may be defined as the period during which the individual achieves reproductive competence, along with its associated endocrine and physical developmental changes. It is the transitional state between childhood and adulthood, during which both somatic and psychological maturation take place, and it is frequently tumultuous. This article will be confined to the timing, endocrinology, and somatic changes of puberty and the growth which accompanies it. Psychologic issues are not discussed, nor are abnormalities such as precocious or delayed puberty.

I. TIMING OF PUBERTY

The entire process of pubertal maturation normally takes place over a period of 4 or 5 years, but in the female menarche is a discrete event which can be timed accurately. Menarche therefore often serves as a marker for the onset of puberty, even though it is a rather late component and of course is not applicable to males. In industrialized Western countries for which data are available, the average age of menarche has decreased by 2 or 3 months per 10

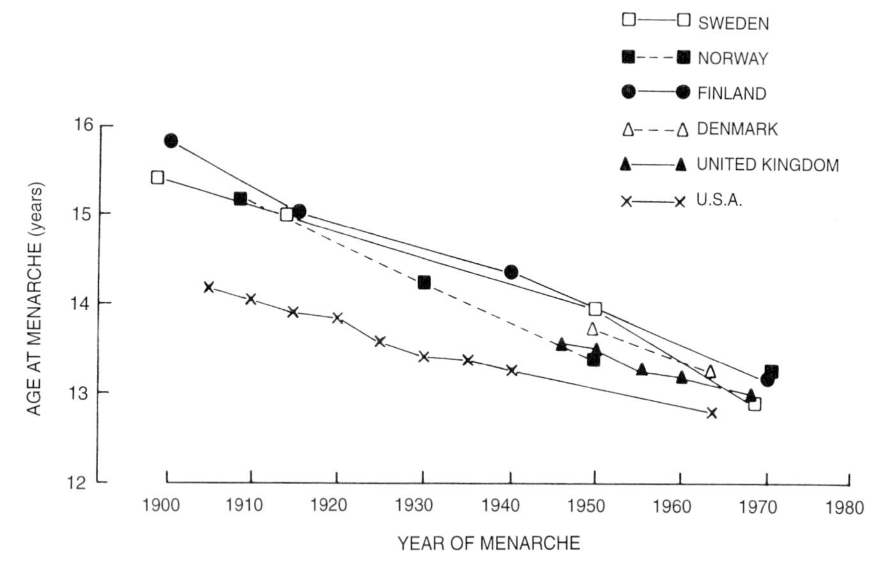

FIGURE 1 Secular decline in age of menarche in six countries (from Dewhurst, 1984, p. 27).

years for 150 years up to about 1970 (Fig. 1). Since then, there appears to have been little change. As of 1979, the median age of menarche for the United States was reported to be 12.8 years.

The age of menarche is influenced by socioeconomic conditions, geographic location, altitude, exposure to light, nutrition, general health, and genetic factors. In the United States, menarche occurs earlier in black females than in whites, and socioeconomic factors do not appear to be responsible for this discrepancy. The onset of puberty may be delayed by malnutrition, chronic disease, severe obesity, and strenuous physical activity. Early menarche has been associated with mild obesity, lower altitudes, lower latitudes, blindness, urban milieu, and a family history of early menarche.

There is evidence that a critical body mass (47.8 kg) must be reached in order for menarche to occur. However, this concept is controversial since many exceptions have been reported. It has also been suggested that a shift in body fat proportion from 16 to 23% is the critical requirement, but this also is not always true. As explained more fully below, current evidence indicates that gonadal steroids initiate and control pubertal growth through the secretion of growth hormone (GH) and insulin-like growth factor-I (IGF-I). This endocrine mechanism is a major influence on both weight and body fat composition.

II. NEUROENDOCRINOLOGY

The dramatic physical changes of human pubertal development are accompanied by equally dramatic functional alterations of the hypothalamic–pituitary–ovarian axis (Table 1). However, these alterations are part of a continuum of sexual maturation which begins with gonadal differentiation in the first trimester of fetal life and progresses into advanced age.

TABLE 1
Hypothalamic–Pituitary Changes Accompanying Puberty

Diminished central inhibition of pulsatile GnRH secretion
Decreased sensitivity of pituitary and hypothalamus to negative gonadal steroid feedback
Increased GnRH pulse amplitude and frequency
Increased gonadotropin pulse amplitude and frequency, initially sleep-related
Increase in ratio of circulating LH to FSH
Increased bioactivity of gonadotropins, secondary to modified glycosylation
Increased pituitary sensitivity to GnRH stimulation
Development in girls of positive feedback of estradiol on gonadotropin secretion
Increased growth hormone pulse frequency and amplitude

A. Fetal Life and Infancy

Fetal concentrations of the gonadotropins follicle-stimulating hormone (FSH) and luteinizing hormone (LH) reach adult levels in the second trimester of gestation as a result of pulsatile secretion of gonadotropin-releasing hormone (GnRH) by the hypothalamus. The ontogeny of the hypothalamic pulse generator is not well understood, but hypothalamic neuronal cells grown in culture have been found to have an intrinsic ability to form networks and secrete bursts of GnRH. Gonadotropin levels fall somewhat in the last trimester, presumably under negative feedback regulation by maternal and placental sex steroids, and then rise again after birth. During infancy, gender-related differences have been observed in gonadotropin secretion. In girls, serum FSH levels may reach adult or on occasion castrate levels quite rapidly, whereas LH levels rise slowly to low adult levels. In both sexes, immunoassayable FSH concentrations exceed those of LH. The FSH response to exogenously administered GnRH is initially much greater than that of LH, a relationship which is reversed during later puberty. In boys, both FSH and LH rise slowly during childhood and early puberty. Corresponding gender-related differences in GnRH pulsatility have been found in primates. These differences have been attributed to effects of gonadal androgens on the developing fetal brain in males.

The fetal and neonatal gonads are able to respond to stimulation by gonadotropins. Thus, ovarian follicle maturation and atresia occur in the female fetus and infant, and adult levels of serum testosterone or estradiol may be found briefly during the first few months of infancy.

B. Early Childhood

From about 2 years of age until the onset of puberty (a period which has been called the "juvenile pause" or "prepubertal hiatus"), the hypothalamic–pituitary system (the gonadostat) maintains FSH and LH secretion at low levels, at or below the limits of detectability by conventional immunoassays. Newer highly sensitive assays have demonstrated episodic nocturnal gonadotropin secretion during this period, however. Prepubertal girls may experience episodes of nonovulatory follicular activity, occasionally manifested by breast stimulation, but serum estradiol concentrations are generally not above 10 pg/ml. In boys, Leydig cells and seminiferous tubules are relatively inactive and serum testosterone levels usually do not exceed 20 ng/dl. This juvenile pause appears to be partially a result of negative feedback by gonadal steroids, functioning at an extreme degree of sensitivity at both the hypothalamic and pituitary levels. The gonadostat is 6–15 times more sensitive to negative feedback in childhood than after puberty.

In addition to heightened negative feedback control, there is indirect evidence for a central mechanism of GnRH inhibition during childhood. In individuals with gonadal dysgenesis (and therefore without gonadal steroids to induce negative feedback), serum gonadotropin levels during juvenile pause are very low, similar to those in normal children at comparable ages (Fig. 2). This suggests the existence of intrinsic nonsteroidal suppression of endogenous GnRH release. In normal prepubertal children, as well as in other primates, the pituitary and gonads have been shown to respond to appropriate endocrine stimulation (Fig. 3). Clearly, timing of the onset of puberty is not dependent on maturation of these organs—they are capable of response at all ages. Rather, the events of puberty originate with the ability of the hypothalamus to generate GnRH in a pulsatile rhythm with specific frequency and amplitude. The pulsatile characteristics of the GnRH release are critical to its effect. If the pulsatility is disrupted by disease or long-acting GnRH analogs, gonadotropin synthesis and release are diminished.

C. The GnRH Pulse Generator

Exactly how the hypothalamic pulse generator acquires its capabilities remains the subject of active investigation. Broadly, two processes must be involved: desensitization of the gonadostat to negative feedback by gonadal steroids and decreased intrinsic central inhibition of GnRH secretion. The mechanism of this latter inhibition is unknown, but observations of certain human disorders may be suggestive. Hydrocephalus or subarachnoid cysts, which cause increased intracranial pressure but do not specifically affect the hypothalamus, may be associated

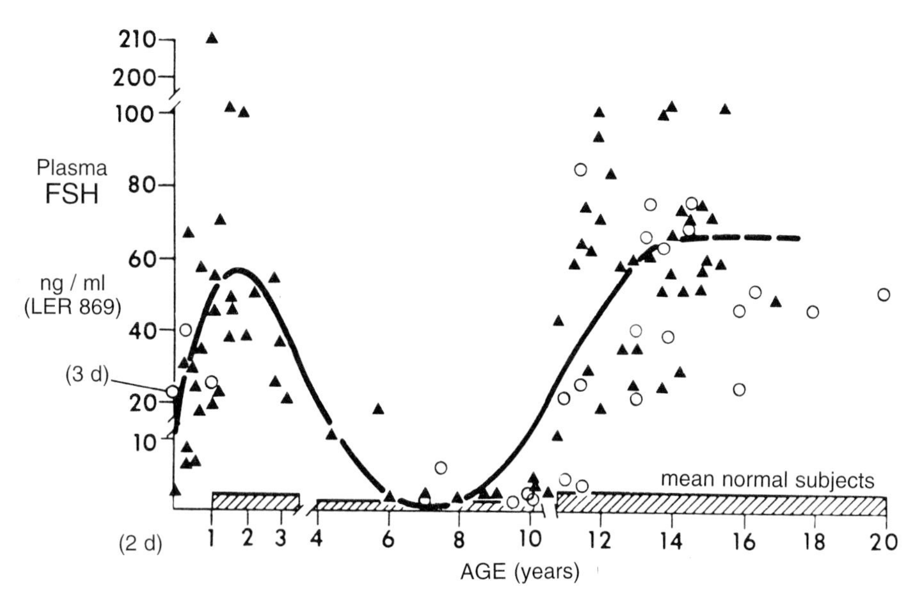

FIGURE 2 Variation of plasma FSH levels with age in 58 patients with gonadal dysgenesis. ▲, XO karyotype; ○, structural abnormalities of the X chromosome and mosaics. The hatched area indicates mean range for FSH levels in normal females (from F. A. Conte, M. M. Grumbach, and S. L. Kaplan, A diphasic pattern of gonadotropin secretion in patients with the syndrome of gonadal dysgenesis, *J. Clin. Endocrinol. Metab.* 40, 670, 1975).

with precocious puberty, which is reversed when the pressure is relieved. Also, the fact that posterior, but not anterior, hypothalamic lesions often cause precocious puberty indicates an anatomic area for further investigation. Another such site is the pineal gland and its endocrine product, melatonin. Observations in children with pineal tumors, plus studies in animals with light-dependent reproductive behavior, suggest that melatonin could be an important suppressor of GnRH pulse generation. However, controlled studies have failed to support such a role for melatonin in normal human pubertal development.

Opioid peptides have also been proposed as mediators of an inhibitory signal to the prepubertal hypothalamic pulse generator because of their inhibitory effects on gonadotropin secretion in adults. However, administration of opioid antagonists to prepubertal children has not resulted in increased gonadotropin secretion.

It has been suggested that adrenal androgens might contribute to the origins of puberty because adrenarche, the increased secretion of the adrenal androgens dehydroepiandrosterone, its sulfated conjugate, and androstenedione, usually occurs at age 8 years, about 2 years before gonadarche, the beginning of adult gonadal function. Adrenarche is eventually manifested by the growth of axillary and pubic hair and an increase in size of the zona reticularis of the adrenal cortex. Adrenarche is not accompanied by significant changes in pituitary adrenocorticotropin secretion or adrenal secretion of cortisol.

Despite the fact that adrenarche normally precedes gonadarche, however, observations in children with premature adrenarche, precocious puberty, hypogonadism, and hypoadrenalism (Addison's disease) lead to the conclusion that adrenarche and gonadarche are independent phenomena. The initiating cause of adrenarche remains obscure. Several pituitary peptides have been proposed as adrenal androgen stimulators, but convincing evidence is currently lacking that any one of these is responsible for adrenarche in normal children.

D. The Pubertal Transition

Recent studies have increased our knowledge of some of the details of pubertal pituitary dynamics. Sensitive immunofluorescence assays and bioassays, together with computer-assisted analysis of gonadotropin pulsatility, have shown that increased LH pulse amplitude (burst mass), rather than increased pulse frequency, is predominant during puberty in

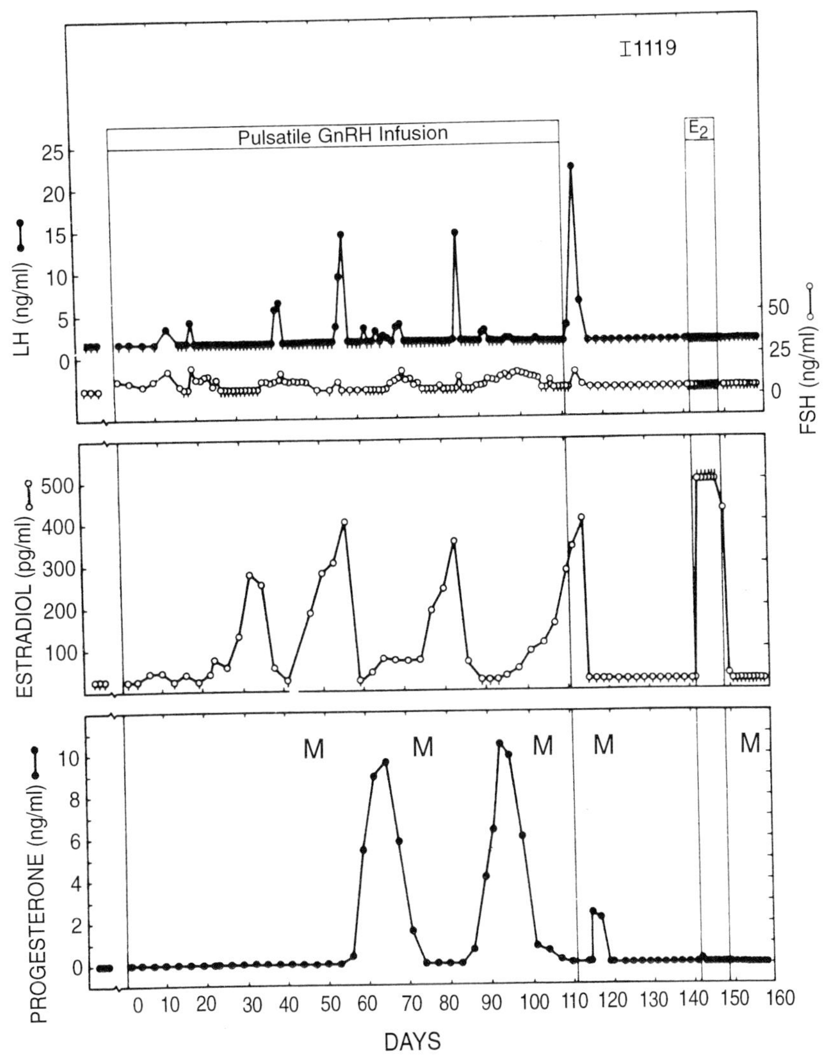

FIGURE 3 Induction of ovulatory menstrual cycles in an immature rhesus monkey by the administration of a pulsatile GnRH regimen (1 mg/min for 6 min once per hour). (Top) ○, FSH; ●, LH. (Middle) Estradiol. (Bottom) Progesterone. Menstruation indicated by M (from L. Wildt, G. Marshall, and E. Knobil, Experimental induction of puberty in the infantile female rhesus monkey, *Science* **207**, 1373, 1980).

boys. The burst mass of LH increases 30-fold by the end of puberty, compared with childhood levels, whereas the burst frequency doubles. Gonadotropin bioassays, e.g., estimating FSH by measuring aromatase activity in rat Sertoli cells and LH by testosterone production in rat Leydig cells, show changing relationships with immunoassay results during the pubertal transition. Levels of biologically active LH rise several times higher than immunoassayable LH, and the same is true to a lesser extent for FSH. Glycosylation of the gonadotropin molecules is essential for their biological activity, and changes in the glycosyl moieties take place during the years of puberty.

Increasing biological activity of gonadotropins is but one of several endocrine changes which culminate in adult pituitary–gonadal relationships. An early change in both sexes is increased REM sleep-related pulsatile release of LH, and to a lesser degree FSH, presumably in response to increased GnRH activity during sleep. This circadian variation in gonadotropin release can be observed with sensitive assays before puberty as well, but during early puberty there is a striking increase in the nocturnal frequency and especially the amplitude of the pulses. As puberty progresses, the diurnal variation becomes less pronounced as the duration of increased gonadotropin pulse amplitude extends during the night, and eventually throughout the 24-hr period with lower amplitude (the adult state). At full maturation in girls, LH and FSH pulse amplitude and frequency vary with the phases of the menstrual cycle, under modulation by ovarian steroids.

Another endocrine change in early puberty is the increasing sensitivity of the pituitary to GnRH stimulation. This is mediated partially by a self-priming effect of GnRH on the gonadotrope cells, inducing greater numbers of its own cell surface receptors and augmenting the other effects of increased GnRH pulse frequency and amplitude.

The increased gonadotropin activity has progressive effects on the gonads, leading to higher levels of steroid secretion and to gamete maturation. The final endocrine step in puberty is the development in girls of a positive estradiol feedback effect on LH release, making ovulatory cycles possible. Estradiol enhances the pituitary's midcycle release of LH selectively, while acting in concert with inhibin to blunt the FSH response. Inhibin levels rise throughout puberty in both boys and girls, though to higher levels in boys. It appears that for positive feedback to occur the pituitary must be exposed to an adequate estradiol concentration (200–300 pg/ml) for an adequate period of time (>36 hr). This in turn requires FSH priming of the dominant ovarian follicle so that adequate estradiol is secreted and GnRH priming of the pituitary so that an adequate pool of releasable gonadotropins is available. The entire process is somewhat slow to develop, probably explaining why ovulation usually is delayed for several months following menarche.

III. GONADAL AND SOMATIC CHANGES

A. Gonadal Changes

The size of the ovaries increases in a linear fashion from birth until completion of puberty; ovarian diameter rises from 0.7–0.9 cm during childhood to 2–10 cm in adults. All components of the ovary are involved, including greater number and size of antral follicles as well as increased volume of medullary stroma. Although growth and atresia of follicles take place, preovulatory (Graafian) follicles are not found in prepubertal ovaries. Serum estradiol levels begin to rise between ages 8 and 10 years, resulting in the onset of secondary sexual changes as described below.

In males the pattern of gonadal growth is different. After growing very slowly during childhood, testicular volume increases from a prepubertal size of 2–4 ml to the adult volume of 12–25 ml, beginning at 9.5–13.5 years of age. Most of this growth is due to increased diameter and tortuosity of the seminiferous tubules, associated with the appearance of Sertoli cells and spermatocytes. Spermatogenesis is generally thought to be present beginning at ages 12–16 years, but release of mature spermatozoa may occur as early as Tanner stage 2 of development. Few Leydig cells are found before puberty, and their development is thought to precede the rise in testicular volume. Daytime levels of testosterone begin to rise at approximately 10 years of age, with nocturnal elevations occurring a few months earlier. The rise to adult levels, an increase of 20-fold or more, may be quite rapid once the process has begun.

B. Somatic Changes

In girls, rising estradiol levels induce the uterus to increase in length from 2 or 3 cm in childhood to its adult size of 5–8 cm. Its shape changes from tubular to globular and the corpus increases in size relative to the cervix. As estradiol stimulation increases, the endometrium proliferates, resulting in vaginal bleeding. The vaginal mucosa thickens and becomes rugated. Its pH decreases and a white or clear discharge may appear. The labia majora, labia minora, and hymen also thicken.

Progressive changes take place in the development of the breasts (in girls) and pubic hair. These changes occur in a predictable sequence, as described by Tanner (Table 2), and are related in a fairly linear fashion to increases in circulating gonadotropin and estradiol levels (Fig. 4) and to bone age. The age range at which the various pubertal events occur has also been reported by Tanner (Fig. 5), but it should be noted that his studies involved European children,

TABLE 2
Tanner Stages of Pubertal Maturation

Boys: Genital development
Stage 1: Prepubertal; testes, scrotum, and penis about the same size and proportion as in early childhood
Stage 2: Enlargement of scrotum and testes; reddening of scrotal skin
Stage 3: Enlargement of penis, primarily in length; further growth of testes and scrotum
Stage 4: Increased size of penis with growth in breadth and development of glans; further growth of testes and scrotum; darkening and rugation of scrotal skin
Stage 5: Adult size and proportion of genitalia
Girls: Breast development
Stage 1: Prepubertal; elevation of papilla only
Stage 2: Breast bud stage; elevation of breast and papilla as small mound; increased areolar diameter
Stage 3: Further enlargement and elevation of breast and areola, which have a single continuous contour
Stage 4: Projection of areola and papilla to form a secondary mound above the level of the breast
Stage 5: Mature stage; projection of papilla only, because of recession of the areola to the breast level
Both sexes: Pubic hair
Stage 1: Prepubertal; pubic vellus same as on abdomen
Stage 2: Sparse growth of long, slightly pigmented downy hair, straight or slightly curled, primarily at base of penis or along labia majora
Stage 3: Hair becomes darker, coarser, and more curled, spreading sparsely over the junction of the pubes
Stage 4: Hair is adult in type, covering a smaller area than in adults, without spread to thighs
Stage 5: Adult hair in quantity and type, with spread to medial thighs

Note. Modified from Tanner (1962).

whereas pubertal maturation occurs at least 6 months earlier in the United States. Recent data describing the onset of pubertal changes in American girls are shown in Table 3. The entire process of pubertal maturation requires about 4.5 years (range, 1.5–6.0 years). The onset of puberty is usually signaled by an acceleration of growth followed by breast budding in girls (thelarche), growth of axillary and pubic hair (pubarche), and finally menarche. In some children, pubarche may precede breast budding and in about 20% of girls the first manifestation of puberty is pubic hair growth. Considerable variation occurs in all aspects of somatic pubertal changes, including age of onset, age of completion, and magnitude and velocity of development. This variation often causes needless anxiety for all concerned.

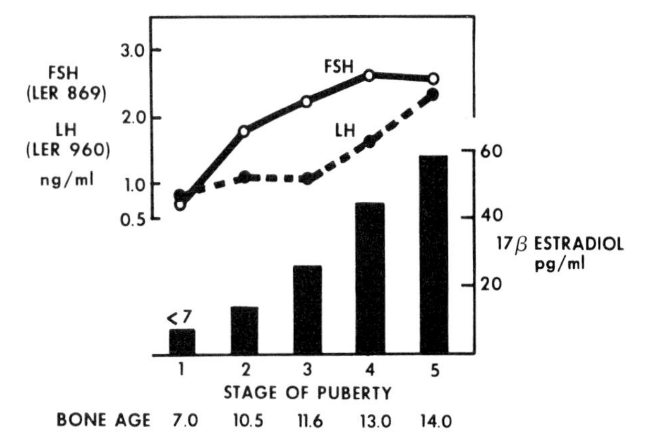

FIGURE 4 Mean plasma estradiol, FSH, and LH concentrations in prepubertal and pubertal females, by stage of maturation. Similar relationships are seen in males in relation to testosterone and gonadotropin levels [from M. M. Grumbach, Onset of puberty, In *Puberty: Biologic and Psychosocial Components* (S. R. Berenberg, Ed.), p. 3, H. E. Stenfert Kroese, Leiden, 1975].

In males, the first visible evidence of puberty is testicular enlargement, followed by progressive growth of the penis as well as growth, reddening, and rugation of the scrotum and proliferation of pu-

TABLE 3
Age of Onset of Pubertal Changes in American Girls

Pubertal change	*Mean age (years)*	*Standard deviation*
Breast, Tanner stage 2		
White	9.96	1.82
African-American	8.87	1.93
Pubic hair, Tanner stage 2		
White	10.51	1.67
African-American	8.78	2.00
Menarche		
White	12.66	1.20
African-American	12.16	1.21

Note. Modified from M. E. Herman-Giddens, E. J. Slora, R. C. Wasserman, C. J. Bourdony, M. V. Bhapkar, G. G. Koch, and C. M. Hasemeier, Secondary sexual characteristics and menses in young girls seen in office practice: A study from the pediatric research in office settings network, *Pediatrics* 99, 505–512, 1997.

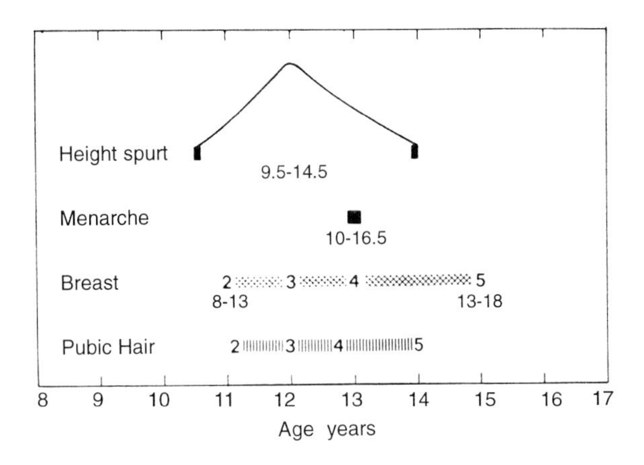

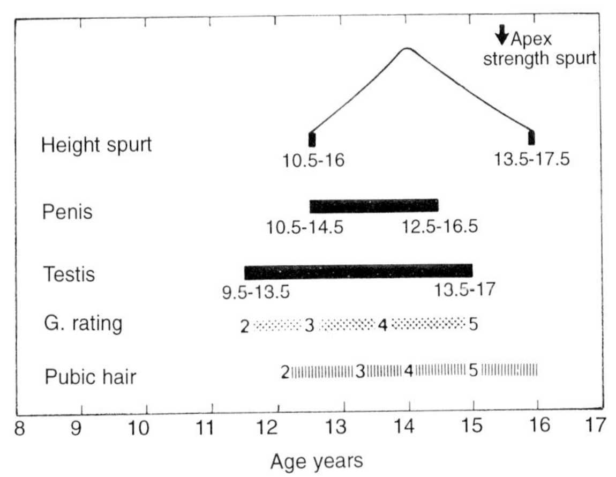

FIGURE 5 Ages of occurrence of pubertal events in females (top) and males (bottom). Horizontal bars represent average ages of occurrence; numbers within bars indicate maturational stage, and numbers below bars indicate range of ages. Most changes occur about 6 months earlier in American children (from Marshall and Tanner, 1969, 1970).

bic hair (Table 2). The prostate and seminal vesicles enlarge, and there is growth of the larynx with deepening of the voice. In both sexes, serum alkaline phosphatase concentrations increase.

IV. GROWTH

A. Clinical Course

During childhood, from age 2 years until puberty, both boys and girls grow at a fairly constant rate of 5 or 6 cm per year in height and about 2.5 kg per year in weight. This growth appears to depend on adequate nutrition and normal levels of GH and thyroid hormones. With the onset of puberty, rapid and sexually dimorphic changes occur in body size and proportion.

The pubertal growth spurt in girls occurs at age 11 or 12 years, roughly at the time of breast budding and about 2 years earlier than in boys. Peak height velocity (PHV) of 8 cm per year is reached about 2 years after the onset of the growth spurt and 6–12 months prior to menarche. The PHV is sustained for only a few months, after which the rate of height increase decelerates over the next 2 years and then ceases with estrogen-induced epiphyseal fusion. During this period, fat deposition is twice as rapid as in boys. In boys, the PHV is over 9 cm/year (range, 7–12 cm/year). This fact, plus the additional prepubertal growth permitted by the later onset of the growth spurt, probably explains why males tend to become taller than females. Muscle mass also increases, and the android pattern of fat deposition is established.

B. Endocrinology

Hormonal requirements for the pubertal growth spurt include GH, gonadal steroids, IGF-I, thyroid hormone, and glucocorticoids. The interactions among these hormones are complex and are reviewed elsewhere. Briefly, the direct mediator of growth is IGF-I, which is produced in greatest quantity by the liver but also by cartilage and various other tissues. Growth hormone, secreted by the anterior pituitary in pulsatile bursts under hypothalamic regulation, stimulates the synthesis of IGF-I. Gonadal steroids in turn stimulate the secretion of GH, but they can also increase bone growth without IGF-I, though to a lesser degree. GH bioavailability and action are also modulated by changes in the GH receptor which occur during adolescence.

The sensitivity of GH secretion to estrogens is extraordinarily high because manifestations of growth are seen before any of the other signs of sexual development. Also, in agonadal children minuscule doses of ethinyl estradiol (100 ng per kilogram body weight) produce increased height and GH pulse amplitude, without the other expected estrogenic effects. In boys, testosterone similarly stimulates GH secretion, though this effect is transient since circulating GH and IGF-I levels fall in late puberty despite sustained high testosterone production. In fact, high doses of both estradiol and testosterone inhibit hepatic IGF-I secretion. Recent studies suggest that testosterone may exert its GH-stimulatory effect through aromatization to estrogens.

See Also the Following Articles

Adrenarche; GnRH; GnRH Pulse Generator; Menarche; Neuroendocrine Systems; Puberty, Precocious

Bibliography

Clark, P. A., and Rogol, A. D. (1996). Growth hormones and sex steroid interactions at puberty. *Endocrinol. Metab. Clin. North Am.* **25**, 665–681.

Dewhurst, J. (1984). *Female Puberty and Its Abnormalities*. Churchill Livingstone, New York.

Grumbach, M. M., and Styne, D. M. (1991). Puberty: Ontogeny, neuroendocrinology, physiology and disorders. In *Williams' Textbook of Endocrinology* (J. D. Wilson and D. W. Foster, Eds.), 8th ed. Saunders, Philadelphia.

Grumbach, M. M., Sizonenko, P. C., and Aubert, M. L. (Eds.) (1990). *Control of the Onset of Puberty*. Williams & Wilkins, Baltimore.

Marshall, W. A., and Tanner, J. M. (1969). Variations in the pattern of pubertal changes in girls. *Arch. Dis. Child* **44**, 291–303.

Marshall, W. A., and Tanner, J. M. (1970). Variations in the pattern of pubertal changes in boys. *Arch. Dis. Child* **45**, 13–23.

Plant, T. M. (1988). Puberty in primates. In *The Physiology of Reproduction* (E. Knobil and J. D. Neill, Eds.). Raven Press, New York.

Tanner, J. M. (1962). *Growth at Adolescence*, 2nd ed. Blackwell Scientific, Oxford, UK.

Veldhuis, J. D. (1996). Neuroendocrine mechanisms mediating awakening of the human gonadotropic axis in puberty. *Pediatr. Nephrol.* **10**, 304–317.

Puberty, in Nonhuman Primates

Tony M. Plant

University of Pittsburgh School of Medicine

GLOSSARY

hypogonadotropism A hormonal state characterized by low circulating concentrations of the pituitary gonadotropins, luteinizing hormone and follicle-stimulating hormone.

hypothalamic gonadotropin-releasing hormone pulse generator A neuroendocrine timing mechanism resident within the hypothalamus (anteroventral forebrain) that governs the pulsatile release of GnRH (a decapeptide that provides the stimulus for gonadotropin secretion from the anterior pituitary) into the hypophysial portal circulation.

insulin-like growth factors Peptides secreted by the liver in response to growth hormone that mediate some of the growth-promoting actions of growth hormone.

melatonin A hormone secreted by the pineal gland.

nonhuman primates Nonhuman primates comprise the three suborders Lemuroidea, Tarsioidea, and Anthropoidea. Anthropoidea is further subdivided into the infraorders Platyrrhini (New World monkeys and marmosets) and Catarrhini (higher primates), the latter of which contains the Old World monkeys, great apes, and humans. The understanding of the control of the onset of puberty in nonhuman primates is based largely on studies of *Macaca*, a genus of the family of Old World monkeys, and the terms "monkey" and "primate" have been used loosely in the text to describe this family.

prepubertal developmental stages Infancy, birth until approximately 4 months of age; the juvenile period, approximately 4 months of age until the onset of puberty.

Puberty is the stage of development during which an animal first becomes capable of reproducing sexually and is characterized by maturation of the genital organs and the development of secondary sexual characteristics. The study of puberty may be approached from two major perspectives. The first addresses the mechanisms that time the onset of this developmental event and that dictate the tempo at which it proceeds. The second is concerned with the impact of gonadal maturation during puberty on growth and behavior.

I. INTRODUCTION

In nonhuman primates, the onset of puberty can occur during the first year of life, as in the marmoset, but more typically it is delayed for at least 2 or 3 years, as in the Old World monkeys, or even longer, as in the great apes. Although maturation of the gonads at puberty is initiated by increased secretion of the gonadotropic hormones, the pituitary response is activated, in turn, by the upregulation of a hypothalamic neuronal network responsible for the pulsatile discharge of gonadotropin-releasing hormone (GnRH) into the hypophysial portal circulation. Interestingly, this neuroendocrine system, which is known as the hypothalamic GnRH pulse generator, is operational during infancy and, as a result, pituitary gonadotropin secretion is elevated at this stage of primate development. Later in infancy, however, the GnRH pulse generator is brought into check and a hypogonadotropic state ensues guaranteeing that the gonads of the prepubertal monkey are held in a characteristic state of quiescence. Since pulsatile stimulation with exogenous GnRH readily elicits menstrual cyclicity and spermatogenesis in the premenarcheal female or juvenile male, respectively, the pituitary–gonadal axis during postnatal development may be viewed as a "slave" to the brain. Accordingly, the key determinant of the onset and tempo of primate puberty is the neurobiological mechanism responsible for the increased release of GnRH at this stage of development and the physiological control system that sets the hypothalamus in motion at an appropriate developmental and chronological age.

II. ONTOGENY OF THE GnRH PULSE GENERATOR

GnRH neurons are exceptional in that they originate in the olfactory placode early in embryonic life and soon migrate to the hypothalamus. By the middle of the 5 months of gestation in the monkey, the hypothalamus of the fetus probably has its full complement of approximately 1000–2000 diffusely distributed GnRH neurons. Moreover, the elevated levels of circulating gonadotropins in the monkey fetus at midgestation strongly suggest that the fundamental properties of the GnRH pulse generator are extant at this early stage of development. Later in gestation, the activity of the GnRH pulse generator is diminished, presumably as a consequence of an inhibitory action of fetal placental steroids. With loss of this inhibitory signal at the time of parturition, GnRH pulse generator activity is robustly expressed during the first few months of infancy. The gonadotrophs of the pituitary respond to this hypophysiotropic drive and, in the male, luteinizing hormone (LH) discharges drive the Leydig cells of the neonatal testis to secrete testosterone in a mode comparable to that produced by the fully mature testis. By 6 months of age, however, the GnRH pulse generator has been brought into check, leading to the hypogonadotropic state that imposes gonadal quiescence until the prepubertal phase of development is terminated by a reaugmentation of GnRH pulse generator activity.

The prepubertal hiatus in GnRH pulse generator activity, which is a unique hallmark of development in higher primates, appears to be fully preserved in the absence of testicular inputs. Indeed, agonadal

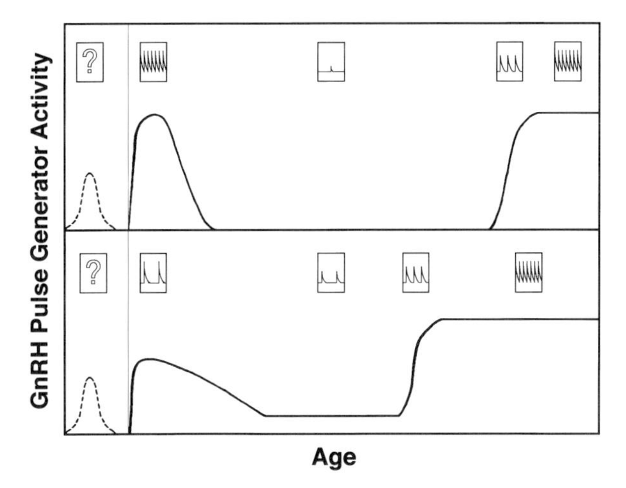

FIGURE 1 A schematic representation of the open loop ceiling for GnRH pulse generator activity from embryonic development until adulthood in a representative male (top) and female (bottom) monkey. The overall hypophysiotropic drive to the gonadotroph, which is dependent on both the frequency and the amplitude of pulsatile GnRH secretion, is represented by the curved lines. Broken curved lines indicate phases of development in which data on pulse generator activity are scant. The inset boxes provide schematic patterns of frequency and amplitude of GnRH discharges at critical stages of development. The fine vertical line indicates birth.

male primates, in which GnRH pulse generator activity is unimpeded by testicular feedback, have provided useful paradigms with which to examine the developmental modulation of GnRH pulse generator activity during this and other critical stages of development (Fig. 1). By infancy, the GnRH pulse generator of the male hypothalamus has acquired the capacity to operate at a circhoral frequency typical of that of the postpubertal animal. The intensity of the break on the GnRH pulse generator during prepubertal development in the male is such that unequivocal evidence of GnRH pulsatility has not been obtained at this stage of development. The lifting of the prepubertal brake on the male GnRH pulse generator, which triggers the initiation of puberty, occurs surprisingly rapidly in the agonadal situation, with pulse frequency accelerating explosively over a period of 30–40 days at the end of the prepubertal phase of development. From the foregoing considerations, it may be concluded that any developmental changes there may be in the sensitivity of the GnRH pulse generator to inhibition by testicular steroids do not play pivotal roles in imposing the fundamental pattern in ontogenetic activity of this neuroendocrine system. Thus, the notion that the transition into puberty in male primates is triggered by a gonadostat has long been rejected.

The postnatal development of GnRH pulse generator activity in the female differs in several respects from that in the male, and these sex differences are most notable in the agonadal situation (Fig. 1). In the infantile female, the GnRH pulse generator does not operate at the circhoral adult frequency. Additionally, the prepubertal brake on the female GnRH pulse generator is applied less forcibly and for a shorter duration, and during this phase of development unequivocal slow frequency pulsatile GnRH release is manifest in the absence of the ovaries. Because the hypophysiotropic drive to the gonadotroph is not completely removed during prepubertal development in the female, the hypogonadotropism of this stage of development is relative. Consequently, the prepubertal ovary is driven to produce hormones which feedback on the central component (hypothalamus and pituitary) of this neuroendocrine axis and, together with the prepubertal brake on the GnRH pulse generator, limit gonadotropin secretion during this phase of development. It is important to reiterate, however, that the primary determinant that times both the placing and the lifting of the prepubertal brake on the GnRH pulse generator in the female, as in the male, is nongonadal. The tempo at which female puberty unfolds, once it has been initiated by a lifting of the nongonadal brake upon the GnRH pulse generator, is dictated, in part, by ovarian steroid feedback mechanisms. Interestingly, these steroid influences on the unfolding of GnRH pulse generator activity during puberty may be related to developmental changes in the activity of the neuroendocrine axis regulating growth hormone secretion. The role of the pubertal testis in regulating the tempo of puberty in male monkeys has not been systematically studied.

Although not established empirically, it is likely that sexual differentiation of the ontogeny of pulsatile GnRH release in monkeys results primarily from prenatal exposure of the fetal hypothalamus to testicular testosterone secretion. Elevated circulating testosterone concentrations during infancy, however, may also program the tempo of the reawakening of the GnRH pulse generator during subsequent pubertal development.

In both male and female, the GnRH pulse generator is subjected to diurnal modulation during infancy, with increased activity of this neuroendocrine system being observed at night. During prepubertal development, a similar diurnal modulation of the GnRH pulse generator is observed in the female in which the prepubertal brake on GnRH release is less marked than in the male. Although the lifting of the prepubertal restraint on the GnRH pulse generator is initially manifest at night in both sexes, nighttime augmentation of pulsatile GnRH release persists into adulthood only in the male.

Before the neurobiology of the pubertal reaugmentation of pulsatile GnRH release is discussed, it is first necessary to reiterate that if the primate GnRH pulse generator was not brought into check during infancy, gonadal activation and development of secondary sexual characteristics would be initiated during the first year of life in species such as the Old World monkeys and great apes, in which the onset of puberty normally occurs at 3 and 8 years of age, respectively.

III. NEUROBIOLOGY OF THE PUBERTAL REAUGMENTATION OF THE GnRH PULSE GENERATOR

The most parsimonious speculation that may currently be proposed to account for the characteristic pattern of GnRH pulse generator activity from birth until puberty in primates, i.e., periods of robust pulsatility during early infancy and puberty that are separated by a prolonged period of relative quiescence during the greater part of prepubertal development, is as follows: The attenuation of pulsatile GnRH release in infancy and the maintenance of this reduced hypophysiotropic drive to the gonadotroph during the remainder of prepubertal development is produced either by the loss of a stimulatory or the addition of an inhibitory input to or within the GnRH pulse generator. According to this view, the pubertal reaugmentation of pulsatile GnRH release would be occasioned by the reversal of the restraining input initially imposed during infancy.

The finding that repetitive stimulation of the GnRH neuronal network of the prepubertal monkey with a glutamate receptor agonist results in the immediate activation of an "adult-like" pattern of pulsatile GnRH release that leads to precocious testicular function (Fig. 2) suggests that the prepubertal restraint on pulsatile GnRH release is determined by developmental changes in an upstream input to the neurons responsible for the secretion of this peptide. Curiously, this restraining input to the GnRH pulse generator of the prepubertal hypothalamus does not appear to be associated with marked impairment in the biosynthesis of the decapeptide. That the prepubertal restraint on the GnRH pulse generator is imposed by an inhibitory neuronal input is supported by ultrastructural studies indicating a decline in axosomatic input to GnRH neurons at the onset of puberty. The most logical candidate for an inhibitory prepubertal signal is γ-aminobutyric acid (GABA), the major inhibitory neurotransmitter in the brain. Indeed, Terasawa and associates at the Wisconsin Regional Primate Research Center have shown that hypothalamic release of GABA declines with the onset of puberty, and that interruption of GABA synthesis or action by local administration of antisense ologonucleotides for the mRNA of the GABA synthesizing enzyme, glutamic acid decarboxylase, or an antagonist to the $GABA_A$ receptor, respectively, elicits GnRH release during prepubertal development. It is therefore tempting to speculate that a developmentally regulated remodeling of axosomatic GABA synapses may be a key event underlying the initiation and termination of the prepubertal brake on the GnRH pulse generator and therefore the timing of the onset of puberty. In addition to the GABA hypothesis, a prepubertal diminution in the tone of excitatory neurotransmitters or neuromodulators, such as glutamate, norepinephrine, and neuropeptide Y, has also been implicated in the pubertal reaugmentation of GnRH pulse generator activity.

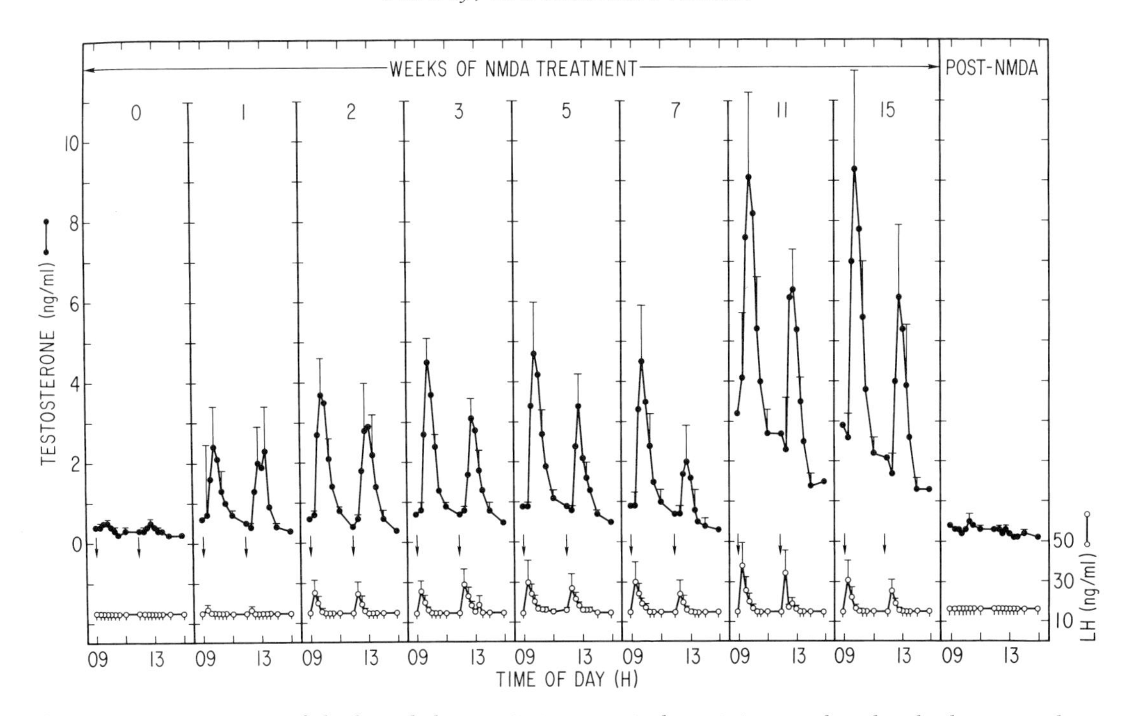

FIGURE 2 Premature activation of the hypothalamic–pituitary–testicular axis in prepubertal male rhesus monkeys induced by repetitive stimulation of hypothalamic GnRH release with brief iv infusions of a glutamate receptor agonist administered once every 3 hr (frequency of spontaneous GnRH release in intact adult males). The monkeys were 15 or 16 months of age at the beginning of treatment. During treatment, testicular volume increased progressively from 0.75 to 5.5 cm^3, spermatogenesis was initiated, and somatic growth rate exceeded that observed during spontaneous puberty, which occurs at approximately 30 months of age (reproduced with permission from T. M. Plant, V. L. Gay, G. R. Marshall, and M. Arslan, Puberty in monkeys is triggered by chemical stimulation of the hypothalamus, *Proc. Natl. Acad. Sci. USA* **86**, 2506–2510, 1989).

In addition to potential neuronal signals that might mediate the prepubertal brake to the GnRH neuron, the possibility that the hiatus in GnRH release at this stage of development also involves glial inputs must be considered. In contrast to the scant innervation of GnRH neurons, glial ensheathment of both GnRH perikarya and terminals is substantial. One major candidate for a glial factor as a developmental control of GnRH release is transforming growth factor-α (TGF-α). This growth factor, which is produced by glial cells of the hypothalamus, has been shown to stimulate GnRH release in the rat, and during postnatal development in the monkey TGF-α gene expression is inversely related to the activity of the GnRH pulse generator.

The foregoing notions regarding the neurobiological basis of the prepubertal hiatus in GnRH pulse generator activity, however, remain to be integrated into a unifying model to account for the onset of puberty in primates.

IV. TIMING OF THE PUBERTAL REAUGMENTATION OF THE GnRH PULSE GENERATOR

Whatever neurotransmitter or neuromodulator or combination thereof is involved in dictating the postnatal ontogeny of pulsatile GnRH release, the neurobiological event that leads to removal of the prepubertal brake on the GnRH pulse generator must, presumably, be timed by a specific developmental cue. Such a cue might be generated by a "pubertal clock" or by the attainment of a particular state of somatic maturation. With regard to the latter possibility, the attainment of a particular proportion of

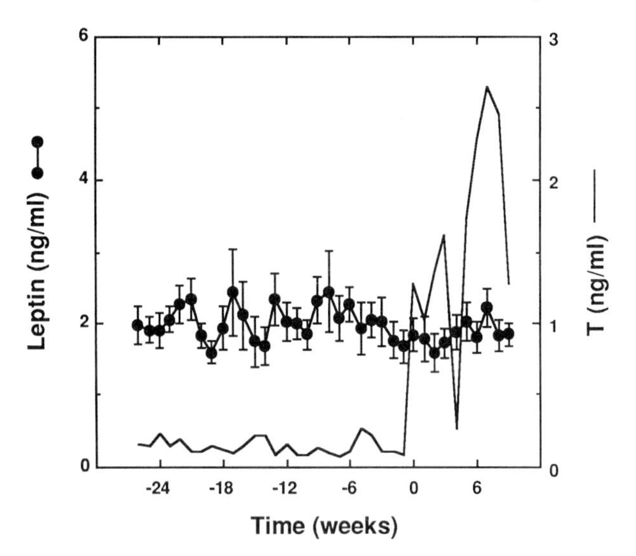

FIGURE 3 The absence of a relationship between the circulating plasma leptin concentrations (●) and the onset, at Week 0, of nocturnal testosterone secretion (—) in the male rhesus monkey. The dramatic increase in testosterone secretion indicates that the pubertal reaugmentation of pulsatile GnRH release has occurred. At Week 0 the monkeys ranged in age from 27 to 31 months of age (reproduced with permission from T. M. Plant and A. R. Durrant, Circulating leptin does not appear to provide a signal for triggering the initiation of puberty in the male rhesus monkey (*Macaca mulatta*), *Endocrinology* **138**, 4505–4508, 1997).

fat has long been argued by Frisch and colleagues to be requisite for the onset of puberty in girls. Interest in this notion has recently been rekindled because of the recent discovery of leptin, a protein derived from adipocytes, which provides the hypothalamus with a somatic signal that relays information on fat mass to central neural control systems regulating feeding behavior. Moreover, infertility in leptin-deficient mice may be reversed by treatment with this adipocyte protein, and circulating leptin concentrations rise is association with the onset of puberty in both boys and girls. In the monkey, however, a rise in circulating leptin concentrations does not precede the pubertal reaugmentation of GnRH pulse generator activity, as reflected by initiation of nocturnal LH and testicular testosterone secretion (Fig. 3). The reason for the difference in the peripubertal pattern in circulating leptin in male monkeys and boys is most probably related to differences in the proportional increase of fat to the overall enlargement in body mass during this stage of development. In any event, it must be concluded that, in the monkey, this fat protein is unlikely to provide the signal for triggering the initiation of puberty. Moreover, since the fundamental qualitative hallmarks of the ontogeny of the hypothalamic–pituitary–gonadal axis in monkey and man appear to be identical, it would seem unlikely that the developmental cue that times the lifting of the prepubertal brake on the GnRH pulse generator of higher primates, and therefore the onset of puberty, is species specific.

V. MODULATION OF THE TIMING OF THE ONSET OF PUBERTY

As in other species, the timing of puberty in primates may be modulated to varying degrees by several factors, including those related to season, nutrition, and social status. In an outdoor environment, most monkeys exhibit seasonal breeding with mating and birth seasons in the autumn and spring, respectively. Under these conditions, season imposes a quantum effect on the initiation of puberty and, presumably, therefore on the lifting of the prepubertal brake upon the GnRH pulse generator. In the female monkey, for example, the onset of puberty, as reflected by first ovulation, is usually manifest during the autumn of the fourth year of life. However, when first ovulation does not occur at this age it is seen to be either advanced or delayed by a period of 1 year. The principal factor responsible for seasonal modulation of the onset of puberty appears to be photoperiod, with the prolonged nighttime melatonin secretion of short days providing a permissive signal for robust GnRH pulse generator activity. Season probably does not program the time at which the prepubertal brake on the GnRH pulse generator is lifted, but rather the long days of spring and summer impair GnRH pulse generator activity if the developmental brake is lifted at this time of year. Analogous control systems probably also underlie perturbations in the timing of the onset of primate puberty imposed by other, albeit less well-documented, environmental factors such as nutrition and social status.

VI. CONSEQUENCES OF THE PUBERTAL REAUGMENTATION OF THE GnRH PULSE GENERATOR

The immediate consequence of increased pulsatile GnRH release at the initiation of puberty is to upregulate the gonadotrophs of the anterior pituitary. During the prepubertal hiatus in GnRH pulse generator activity, the gonadotrophs are relatively unresponsive to GnRH stimulation and contain diminished amounts of LH and follicle-stimulating hormone (FSH). This hypogonadotropic state is presumably associated with decreased expression of the genes encoding the GnRH receptor, the gonadotropin subunits, and paracrine factors, such as activin and follistatin, that are involved in the regulation of LH and FSH release from the pituitary of the mature animal.

Termination of the hypogonadotropic state at the end of the prepubertal phase of development results in a dramatic acceleration in the growth of the testis. Between 3 and 4.5 years of age in the rhesus monkey, testicular volume increases from approximately 2.5 to 25 ml. During this time, Sertoli cell number, which sets the spermatogenic ceiling of the adult testis, proliferates under the enhanced pubertal gonadotropin drive. Similarly, mature Leydig cells appear and testicular testosterone secretion is initiated. Combined FSH and androgen stimulation of the Sertoli cell activates the germinal epithelium for the first time and spermatogenesis is initiated.

In contrast to the testis, the ovary grows in a rectilinear fashion from infancy to adulthood. Moreover, growth and atresia of ovarian follicles occur throughout infantile and juvenile development, although Graafian follicles are not observed during prepubertal development. In the monkey, circulating estradiol concentrations rise between 2.5 and 3 years of age and menarche is observed. As in the human female, the postmenarcheal phase of development in the monkey is also characterized by a variable phase of adolescent infertility that is the result of a high incidence of anovulatory and short luteal phase cycles.

Monkey puberty is associated with an activation of the neuroendocrine axis governing growth that, as in man, seems to be triggered by enhanced gonadal steroid secretion and leads to peaks in circulating concentrations of growth hormone, insulin-like growth factors (IGFs) and IGF-binding proteins. Under the combined influence of sex steroids and IGFs, the pubertal growth spurt and skeletal maturation are initiated. In the female monkey, the peak in the pubertal growth spurt precedes menarche. Similarly, the changes in hormonal status brought about by maturation of gonadal function lead to changes in sexual, agonistic, and social behaviors.

VII. SUMMARY

The activation of the pituitary–gonadal axis, which underlies the transition into puberty, is triggered in nonhuman higher primates by a reaugmentation of pulsatile GnRH release. This hypothalamic event takes place after a prolonged diminution in GnRH pulse generator activity that occupies the greater part of prepubertal development. The neurobiological trigger for the pubertal increase in GnRH release in both male and females is independent of the gonad and probably involves structural and functional remodeling of afferent inputs to the GnRH neuronal network. The precise nature and relative importance of these inputs are not fully understood. The tempo at which puberty progresses once initiated is determined by gonadal steroid feedback actions. Puberty is also characterized by a peak in the activity of the neuroendocrine activity governing growth, which underlies the growth spurt at this stage of development, and by steroid-induced changes in affective behaviors. Puberty and the control of this critical developmental phase in nonhuman primates are strikingly similar to those in man.

See Also the Following Articles

GnRH (Gonadotropin-Releasing Hormone); GnRH Pulse Generator; Primates, Nonhuman; Puberty Acceleration; Puberty, in Humans; Puberty, in Nonprimates

Bibliography

Mann, D. R., Akinbami, M. A., Gould. K. G., Paul, K. and Wallen, K. (1998). Sexual maturation in male rhesus monkeys: Importance of neonatal testosterone exposure and social rank. *J. Endocrinol.*, in press.

Ojeda, S. R., Ma, Y. J., and Rage, F. (1995). A role for TGFα in the neuroendocrine control of female puberty. In *The Neurobiology of Puberty* (T. M. Plant and P. A. Lee, Eds.), pp. 103–117. Journal of Endocrinology Limited, Bristol, UK.

Perera, A. D., and Plant, T. M. (1997). Ultrastructural studies of neuronal correlates of the pubertal reaugmentation of hypothalamic gonadotropin-releasing hormone (GnRH) release in the rhesus monkey (*Macaca mulatta*). *J. Comp. Neurol.* **385**, 71–82.

Plant, T. M. (1994). Puberty in primates. In *The Physiology of Reproduction* (E. Knobil and J. D. Neill, Eds.), Vol. 2, 2nd ed., pp. 453–485. Raven Press, New York.

Plant, T. M. (1996). Environmental factors and puberty in non-human primates. Acta *Paediatr.* **85**(Suppl. 417), 89–91.

Schwanzel-Fukuda, M., Jorgenson, K. L., Bergen, H. T., Weesner, G. D., and Pfaff, D. W. (1992). Biology of normal luteinizing hormone-releasing hormone neurons during and after their migration from the olfactory placode. *Endocrinol. Rev.* **13**, 623–634.

Suter, K. S., Pohl, C. R., and Plant, T. M. (1998). The pattern and tempo of the pubertal reaugmentation of oxen-loop pulsatile gonadotropin-releasing hormone release assessed indirectly in the male rhesus monkey (*Macaca mulatta*). *Endocrinology* **139**, 2774–2783.

Tanner, J. M., Wilson, M. E., and Rudman, C. G. (1990). Pubertal growth spurt in the female rhesus monkey: Relation to menarche and skeletal maturation. *Am. J. Human Biol.* **2**, 101–106.

Terasawa, E. (1995). Mechanisms controlling the onset of puberty in primates: The role of GABAergic neurons. In *The Neurobiology of Puberty* (T. M. Plant and P. A. Lee, Eds.), pp. 139–151. Journal of Endocrinology Limited, Bristol, UK.

Wilson, M. E. (1995). IGF-I administration advances the decrease in hypersensitivity to oestradiol negative feedback inhibition of serum LH in adolescent female rhesus monkeys. *J. Endocrinol.* **145**, 121–130.

Puberty, in Nonprimate Mammals

Douglas L. Foster
University of Michigan

Francis J. P. Ebling
University of Cambridge

GLOSSARY

gonadostat theory The theory that the increase in gonadotropin-releasing hormone secretion which underlies puberty results from a decrease in the ability of circulating sex steroids to inhibit gonadotropin-releasing hormone release.

gonadotropin-releasing hormone A decapeptide produced by neurons in the basal forebrain which project to the median eminence; causes synthesis and secretion of gonadotropins from the anterior pituitary gland.

Encyclopedia of Reproduction VOLUME 4

gonadotropins A collective term for luteinizing hormone and follicle-stimulating hormone, glycoproteins secreted by the anterior pituitary gland which exert tropic effects on the gonads.

puberty The attainment of reproductive function, culminating in production of mature gametes and reproductive behavior.

sex steroids Steroidal products of the gonads, chiefly testosterone and estradiol before puberty.

Puberty is the transition into adulthood: the time in life when mature gametes are first produced and reproductive activity is initiated. In most species, external indications of this transition include gradual changes in behavior and body appearance. Thus, puberty can be considered to be a process. In the female, some regard puberty as an event because of some abrupt overt change, for example, first menstruation in primates, vaginal opening in rodents, or first estrus which occurs in many species. Regardless of whether they are gradual or abrupt, these changes are due to the increasing production of sex steroids by the gonads, the ovaries or testes, in response to the increasing secretion of gonadotropins from the anterior pituitary gland at the base of the brain. This increased activity of the pituitary gland is, in turn, regulated by the increased secretion of gonadotropin-releasing hormone (GnRH) by the hypothalamus. It is clear that the GnRH secretory system develops prenatally and plays a role in the initial development of the gonads before birth. During puberty, the secretion of GnRH increases markedly. The release of this key neuropeptide into the microcirculation supplying the pituitary gland is regulated by both stimulatory and inhibitory neurons. Exactly when this cascade of hormones begins to increase is determined by a variety of signals routed to the brain, both internal signals relating to growth and external signals relating to the individual's environment.

I. INTRODUCTION

The benefits of understanding the pubertal process in children are clear with respect to the treatment of disorders of sexual maturation. It is also of considerable importance in that this understanding would allow manipulation of the onset of reproduction in other animal species. For example, decreasing the generation interval by advancing the time of puberty may be of advantage in many domesticated animal species used for production of food and fiber. In endangered species, ensuring that puberty is attained before death in a greater proportion of the population is advantageous. By contrast, delaying or even preventing sexual maturity would be of great benefit in certain populations of prolific household pets (dogs and cats) and species of pests (rats and mice).

Mechanisms timing the transition into adulthood (i.e., puberty) have only been studied in any detail in but a handful of the more than 4000 known species of mammals. The information available indicates that fundamental mechanisms underlying the pubertal process are highly conserved. Differences that have been noted provide fascinating examples of adaptations of a species to its own particular environment. Many involve the use of external cues to time puberty to a certain season or alter its progression in relation to population density and composition.

We commonly associate puberty with the very earliest stage of the life cycle, largely because our focus is often on long-lived species, such as the human being in which puberty occurs about one-sixth or one-seventh (age 12 or 13 years) the way through life (70+ years) (Fig. 1, bottom right). This is not the case with the vast majority of mammals, which are small, have a relatively short life expectancy, and live outdoors. For example, deer mice living in the

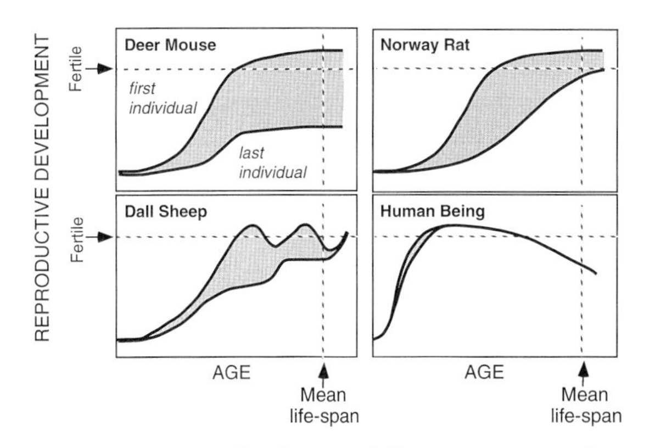

FIGURE 1 Timing of puberty in different species in relation to life span (redrawn from Bronson, 1989).

seasonally harsh environment of Michigan have an average life span of only 18 weeks, yet puberty does not occur until 5 weeks at the earliest (about one-third of life span) (Fig. 1, top left). Indeed, small rodents born toward the end of their breeding season may not even attain puberty during the first year. Such mammals have evolved a strategy whereby they adopt a winter survival physiology, thereby partitioning energy into fat reserves, increased thermogenesis, molting to a winter pelage, but arresting sexual maturation, and only initiate puberty the following spring. Many individuals will not ever become sexually mature because they will not survive this overwintering process.

II. PUBERTY AS AN INCREASE IN RELEASE OF GONADOTROPIN-RELEASING HORMONE

The study of how puberty occurs is fundamentally an investigation of the neuroendocrine systems controlling the increase in secretion of GnRH. Generally, the research concentrates on (i) mechanisms whereby the GnRH neurons increase their activity and (ii) mechanisms timing when the increase in GnRH activity occurs during the life cycle to initiate reproduction at the appropriate growth stage and in optimal social and physical surroundings.

A. GnRH Neurons

Embryologically, the GnRH neurons arise within the olfactory placode and migrate to the forebrain early in life. In many species, their final destination is a loose continuum from the rostral forebrain (medial septum and diagonal band of Broca) through the preoptic area and anterior hypothalamus to the mediobasal hypothalamus, though in mustelids and primates a large proportion are clustered in the mediobasal hypothalamus. GnRH is released into the pituitary portal system as a discrete pulse, but how the approximately 1000 or so GnRH neurons typically present in a mammalian brain communicate to produce this coordinated release of the peptide is unresolved. The importance of GnRH neurons to the pubertal process is evident when they fail to migrate properly or, in an experimental mutant (hypogonadal mouse strain), when adults cannot produce mature GnRH peptide. Puberty does not occur because the cascade of hormones to stimulate the ovaries or testes is not initiated. Treatment with exogenous GnRH or transplantation of GnRH-containing neurons into the appropriate area of the brain can initiate the pubertal process in such individuals, demonstrating the key role of this neuropeptide in the pubertal process. After the initial migration of GnRH neurons, whether their number, distribution, and morphology remains the same throughout life is controversial. What is evident is that this network is capable of secreting GnRH long before puberty, beginning in mid- to late gestation (fetal sheep) or in the neonatal period (rodents). This secretion early in life is greater in males than in females and may be important for initial testicular development (see Section V). Thus, puberty results from the reactivation of GnRH secretion.

The release of GnRH subsequently becomes diminished in most species, resulting in a period of juvenile reproductive quiescence which varies greatly in length in relation to the ecology and life span of each species. Why the GnRH secretory system becomes inactive is as an important and poorly understood question as why it subsequently reactivates to drive puberty. It is clear that some ability to synthesize and release GnRH is maintained during the juvenile period: GnRH mRNA is high in the mouse and rat throughout this period, and no further increases in mRNA levels are detected as puberty is completed. The ability to release GnRH can be demonstrated experimentally in the juvenile period, for example, after stimulation with agonists mimicking the endogenous neurotransmitter glutamate.

It therefore seems that after the initial development of the GnRH secretory system, these neurons are a relatively passive component of the neuroendocrine cascade which drives puberty, and that the key regulation occurs by upstream mechanisms. The greatly increased GnRH secretion during puberty may reflect changes in the activity of existing inputs impinging on their soma or terminals in the median eminence or may result from neuronal plasticity; that is, a change in density or type of synapses or neuropeptide receptors impinging upon GnRH neurons. Furthermore, it is not clear if the activation of the GnRH neurosecretory system results from an increase in activity of stimulatory neurons, a decrease in activity of inhibitory neurons, or a more complex combination of both types of input. Pharmacological studies reveal that a huge range of neurotransmitters (dopamine, noradrenaline, and GABA) and neuropeptides (e.g., β-endorphin and neuropeptide Y) can influence GnRH secretion, but neuroanatomical studies indicate that GnRH neurons possess very few receptors for these compounds. Thus, it is controversial as to which neurochemical systems directly innervate GnRH neurons. The role of the excitatory amino acid neurotransmitter glutamate has attracted much attention because of the substantial effects of pharmacological manipulations on reproductive development, but glutamate is likely to be just one of many neurochemical pathways which ultimately regulate GnRH release. Finally, some of these interneurons must be sensitive to the inhibitory feedback action of sex steroids as indicated in the following section.

B. Gonadostat Hypothesis

The innate ability to produce high levels of GnRH is reestablished at the end of the juvenile period, but the secretion of GnRH remains low. This is because the pacing of the final stages of puberty is governed by the sensitivity to the inhibitory feedback action of gonadal steroids. Evidence for this type of inhibitory system exists in many species in which the potential to produce high-frequency GnRH pulses can be revealed experimentally in the immature individual. The "gonadostat" hypothesis offers a conceptual explanation for why they are not normally occurring earlier at this young age. This hypothesis, the origins of which may be traced back nearly half a century, has been refined over the years so that it now includes the regulation of pulsatile GnRH secretion. The hypothesis, as it relates to the female, states that during the late juvenile stage the system governing the secretion of GnRH is capable of producing high-frequency GnRH pulses, but it does not do so because it is very sensitive to the inhibitory feedback action of ovarian steroids such as estradiol. As the time of puberty approaches, the sensitivity then decreases, allowing GnRH secretion to increase and, as a consequence, the ovary is stimulated by gonadotropins to function in an adult-like manner.

Interestingly, although the original hypothesis was based on work conducted in the female rat, controversy has arisen concerning whether it is applicable to this rodent. Moreover, in the rhesus monkey, another well-studied species, the hypothesis may only apply to the final stages of sexual maturation. In the sheep, the hypothesis provides a relatively complete explanation for the timing of puberty (Fig. 2).

Although the hypothesis is useful in explaining

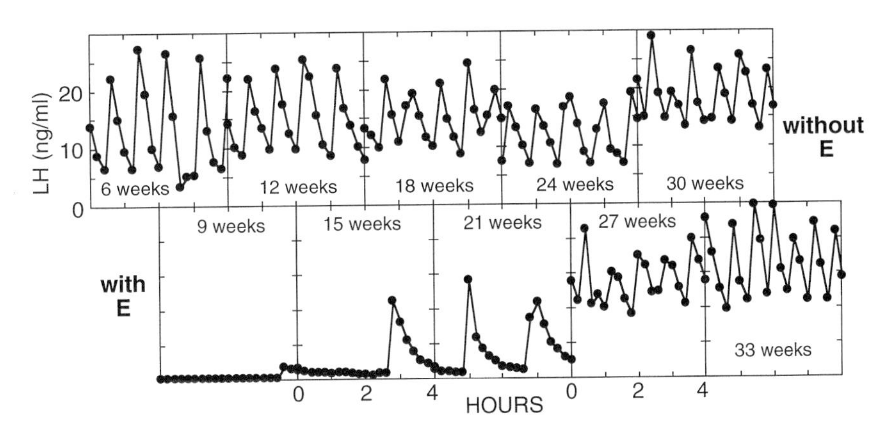

FIGURE 2 Decrease in sensitivity to estradiol inhibition of luteinizing hormone (LH) secretion during the pubertal transition in the female sheep. Endogenous ovarian steroids were removed at ovariectomy at 3 weeks of age, and a Silastic implant releasing physiological levels of estradiol was alternatively replaced and removed every 3 weeks. (Top) LH 3 weeks after removal of the steroid feedback signal. (Bottom) Patterns of LH secretion 3 weeks after insertion of estradiol implant. Note that this individual is capable of producing high-frequency LH pulses in the absence of the steroid feedback signal at all ages, but in the presence of estradiol the lamb only expresses that capacity to increase LH secretion at around 27 weeks of age, an age when intact lambs would begin their first estrous cycles (redrawn from Ebling, F. J. P., Kushler, R. H., and Foster, D. L. (1990). *Neuroendocrinology* 52, 229–237).

why low circulating concentrations of estradiol are effective in suppressing GnRH pulse frequency before puberty, the limitation of extending this hypothesis is that there is currently no viable neurobiological explanation for how sensitivity to estrogen feedback inhibition is determined mechanistically. This explanation must include how steroids communicate with GnRH neurons because these neurons do not themselves contain nuclear receptors for steroids. The regulation must reside in the control of their presynaptic neurons. A second limitation of the gonadostat hypothesis is that it may be describing a change of homeostatic feedback loops at puberty. Clearly, a decrease in the effectiveness of gonadal steroid negative feedback on gonadotropin release is necessary to allow the final stages of sexual maturation, but this could also be explained simply by a greater drive to GnRH secretion and, because of this, greater amounts of steroid are required for its inhibition. Proponents of this alternative, "direct-drive" hypothesis argue that the resetting of the "gonadostat" does not really occur or that at best it does and is a secondary event which produces the pubertal increase in GnRH release.

III. SIGNALS, SENSORS, AND PATHWAYS RELATED TO GROWTH CUES

Reproductive scientists have long recognized that the initiation of sexual maturation is more closely associated with body growth than with chronological age. In studies across a wide range of species, a restricted diet results in retarded growth and delayed puberty, demonstrating the critical need to achieve species-specific growth or metabolic targets for reproduction to be initiated. What are the signals to the GnRH secretory system that these somatic requirements have been met during normal growth? Over 30 years ago, Kennedy and Mitra provided an hypothesis based on energetics. They proposed that the brain might somehow detect the decrease in basal metabolic rate which occurs during growth. This decrease is necessary to maintain a stable core temperature because the rate of increase of the body mass (increase in heat production) exceeds the rate of the increase of surface area (heat dissipation). They believed that the first transition between an infertile and fertile state (puberty) is tightly coupled to changes in somatic metabolism. This prophetic statement may also apply to the adult and to the transitions between fertility and infertility which depend on changes in energetics as the level of nutrition varies. Clearly, the understanding of what peripheral substances are used to provide information to the brain about energy metabolism is one of the fundamental pieces of information needed to solve the puzzle of why puberty occurs at a relatively predictable stage of growth rather than at a fixed age.

Peripheral signals such as IGF-1, insulin, and glucose availability have been implicated in the control of GnRH release. At this stage of our understanding, we might consider these and other signals as providing a metabolic "gate" for puberty to proceed. Recent attention has focused on a hormone secreted from white fat cells, leptin, to provide a biochemical explanation for how the brain is actually able to detect the level of fat in the body. That fatness is an important determinant for initiation of reproductive activity has long been championed by Frisch through her demographic studies. Leptin is thought to provide a feedback signal from calorie deposits (fat) regulating central control of appetite and metabolism. Thus, mice with a genetic deficiency in leptin production (ob/ob) or in leptin receptors (db/db) are obese. Importantly, such mice lacking leptin signaling pathways do not enter puberty, and fertility can be induced in ob/ob mice by treatment with exogenous leptin. Leptin treatment advances puberty in rats raised on a restricted diet which would otherwise have delayed puberty as a correlate of low body fat content, suggesting that leptin can partially override the negative reproductive consequences of reduced caloric intake. Certainly, leptin satisfies a number of criteria that one might expect a peripheral signal to meet if it were to influence reproductive function. However, whether its pattern in the peripheral circulation meets the criterion of a temporal signal for puberty remains to be determined. It is likely that no one substance alone provides a unique signal to time puberty. Perhaps leptin will contribute to a constellation of such signals reflecting the general availability of metabolic fuel to allow puberty to proceed. If a number of substances are used to assess metabolic state, then it will be of interest to determine if species differences exist regarding which substances are used as primary signals.

IV. SIGNALS, SENSORS, AND PATHWAYS RELATED TO EXTERNAL CUES

Achievement of the appropriate metabolic state is all that seems to be required to initiate puberty in the human being. In other species, including subhuman primates, the timing of puberty is more complex. In addition to attaining this growth-related state, they must rely on additional cues to time puberty largely because they live outdoors in conditions which they cannot control. Thus, if one considers that internal metabolic signals provide a permissive gate which allows puberty to proceed when energy balance is appropriate, then species-specific external factors precisely time the initiation of reproductive activity. These factors may be both seasonal and social.

A. Seasonal Factors

Typically, spring and summer are considered the "seasons of plenty" because food supply and temperature are optimal, allowing sufficient energy intake and reduced energy expenditure, facilitating the lactational support of the offspring. Thus, most species living in a natural environment use seasonal cues to time puberty, such that the birth of their first offspring occurs in the optimal season. Although seasonal timing of puberty is the rule, this has been lost in the human being; we have the ability to control our environment, the result of which has been the loss of selection pressure for a single birth season. Other exceptions are some of the domesticated species (cattle and pigs) which man has selectively bred to generate year-round production of food.

1. *Food Supply*

Of all the environmental factors timing the rate of the pubertal progression, the amount of available food is the most important. There is no exception to this. Nutritional modulation of GnRH secretion has been demonstrated in all species, including man, although as discussed previously the exact metabolic cues which are detected by the brain remain unknown. During periods of reduced food availability, small, short-lived species with high metabolism and reduced growth do not even attain puberty before they die. In longer lived species, puberty is delayed until more food becomes available. In times of famine or in human hunter–gather societies living in an unpredictable environment, puberty occurs at older ages. Interestingly, the age at menarche has gradually decreased by 3 or 4 years over the past 150 years in developed countries. However, body weight, a gross indicator of energy balance and energy reserves, has remained the same, suggesting that young women are attaining the critical state of energy metabolism earlier in life because of both a quantitative and qualitative improvement in diet.

2. *Nonnutritional Plant Cues*

Environmental factors not directly related to calories can be used to predict the appropriate time of the year for first mating. A direct plant cue, a phenolic compound derived from the new shoots of grass, induces puberty in the montane vole which lives in the high Rocky Mountains. This predictive cue, since it has no intrinsic nutritive value, allows such a short-lived species to attain adulthood and reproduce at the appropriate time of year. As we learn more about the life cycles of more wild animal species, surely we will discover many additional interesting environmental cues which are used to time puberty in a highly variable environment. Only when these signals are identified can we learn more about the variety of sensors and pathways which are used to regulate GnRH secretion.

3. *Photoperiod*

Many species use a less direct approach to monitor season to predict the optimal time for breeding. Information about length of day (photoperiod) provides a stable seasonal cue and is the most common environmental signal timing puberty in species which have evolved in temperate and arctic latitudes. The use of photoperiod as a cue becomes more advantageous as body size and life span increase because the day-to-day variation in immediate environment is less critical to survival and well-being. In species which have a life span >1 year, it is common to find that the photoperiodic mechanism which first times puberty is reused later in life to time subsequent breeding seasons. Because in temperate regions spring and summer are the optimal seasons of birth, the time of year when puberty and thus mating occurs depends on length of gestation, providing that there is no secondary delay of embryonic development (i.e., delayed implantation). Those species with very short pregnancies (a few days or weeks) mate and deliver offspring during the increasing and long day lengths of the spring and summer as do those with a pregnancy of nearly a year (horse). By contrast, those with intermediate-length gestation (~5 months; sheep, goat, and deer) must conceive during the decreasing and short days of the autumn to deliver their young in the spring.

i. Transduction of Photoperiod Information: Melatonin How does the developing individual monitor the seasonal changes in photoperiod? Postnatally, information about length of day (hours of light vs hours of darkness) is relayed from the eye to the circadian clock in the hypothalamus, the suprachiasmatic nucleus. This structure ultimately modifies the secretion of the hormone melatonin from the pineal gland via a multisynaptic pathway. The consequence of this pathway in which photoperiod regulates the circadian rhythm of pineal gland function is that melatonin is only secreted for the duration of the dark phase of the photoperiod. Long days produce a short duration of high melatonin secretion, whereas short days induce a long duration. These patterns of melatonin form a neurochemical record of photoperiod experience which modulates the seasonal timing of puberty. In some species, this photoperiod history begins prenatally through transfer of the melatonin signal from the mother across the placenta to the fetus. The pattern of circulating melatonin provides a neurochemical code for night length, and this information ultimately regulates the GnRH neurosecretory system. It is clear that the circadian melatonin signal provides a temporal cue and is not directly progonadal or antigonadal. Thus, the effect of a repeated long-duration melatonin signal is to accelerate reproductive development in a species that undergoes puberty in short days, such as the rhesus monkey or sheep. However, this delays reproductive maturation in a species which undergoes puberty in long days where a short nightly melatonin signal is used as the positive signal (e.g., the hamster).

In addition to puberty, melatonin provides temporal information which regulates many seasonal physiological responses, including fat metabolism, hibernation, thermogenesis, and molting. Given this range of neuroendocrine actions, it is no surprise that there are multiple sites of action within the brain and pituitary gland; however, the precise site at which GnRH secretion is regulated is controversial. It is unlikely that melatonin acts directly on GnRH perikarya. Studies in sheep and rodents using receptor autoradiography and *in situ* hybridization have not found melatonin receptors in many of the forebrain regions which contain GnRH perikarya. Lesion studies in hamsters and microimplant studies in sheep indicate a site of action in the mediobasal hypothalamus, but receptor autoradiography has not revealed melatonin binding at this site in all the species which are known to respond reproductively to melatonin. Interestingly, one region which contains a high density of melatonin binding sites in all subhuman species is the pars tuberalis of the pituitary gland; studies in sheep in which the hypothalamus and pituitary gland are surgically disconnected indicate that the melatonin receptors in the pars tuberalis have a role in seasonal regulation of prolactin secretion.

We have little understanding of the cellular and molecular action of melatonin; this is of particular interest because of the long lag time (several weeks) between changes in the pattern of melatonin secretion and the appropriate GnRH response. It has been speculated that this time lag reflects an action of melatonin in inducing changes in morphology by remodeling unknown elements within the brain which thereby modulate GnRH secretion. Whatever the site and mechanism of melatonin action, it is evident that many classes of rhythms become involved in information transfer about season to the developing individual. The annual cycle of photoperiod changes the circadian melatonin rhythm to influence the ultradian brain rhythms that modulate the circhoral production of GnRH necessary for puberty (Fig. 3).

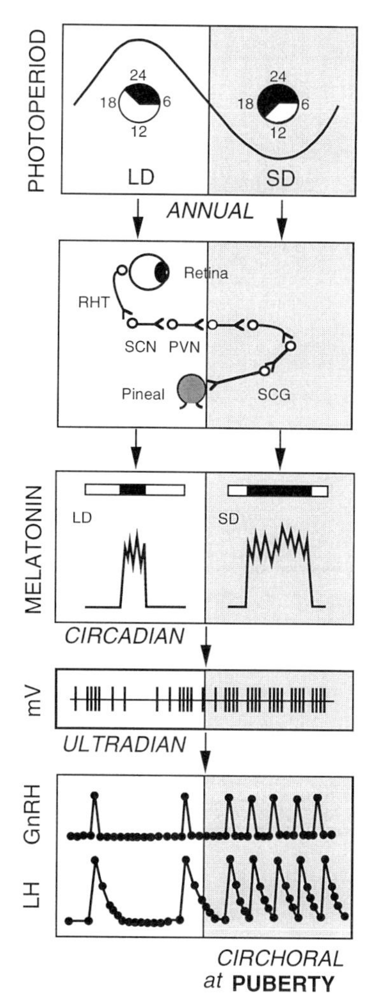

FIGURE 3 Interplay among rhythms to time puberty that involve day length, the pineal gland, and electrical activity of the brain to culminate in high-frequency GnRH/LH secretion. LD, long day; SD, short day; RHT, retinohypothalamic tract, SCN, suprachiasmatic nucleus; PVN, paraventricular nucleus; SCG, superior cervical ganglion. (Redrawn from Foster, 1994.)

ii. Photoperiod and Nutrition as Codeterminants Timing Puberty The profound importance of photoperiod, as well as its interaction with growth in timing the initiation of reproductive cycles, is illustrated in Fig. 4. In this example, groups of young female sheep were all born at the same time of year (spring), and the growth rates were altered by changing the availability of food. Females of this species require a decreasing day length to attain puberty. In this experiment, females (group A) which were well fed from birth grew rapidly, and first ovulations (puberty) occurred at the usual age (~30 weeks) during the decreasing day lengths of autumn. Other lambs (groups B–D) were placed on a restricted diet for varying periods after weaning. The onset of reproductive cycles is delayed in such females despite their having experienced the appropriate photoperiod for initiation of ovulations. Puberty did not occur when they were growth retarded because metabolic cues did not provide the appropriate message to the brain to reduce the high sensitivity to steroid negative feedback to permit high GnRH secretion. This underscores the importance of information about metabolism and somatic growth. When such females were provided additional food to induce growth during the autumn and winter breeding season (groups B and C), cycles began at an earlier stage of growth (i.e., a smaller body size—35 kg) than for controls (i.e., 45 kg). In retrospect, this also raises the possibility that the normally growing females (group A) had achieved the appropriate metabolic state and stage of growth for puberty much earlier in the year (August), but day lengths were too long at those younger ages (i.e., 25 weeks) to reduce hypersensitivity to estradiol feedback inhibition. Later in the year (October), when day length became much shorter, sensitivity to estradiol negative feedback could be reduced to initiate reproductive cycles (i.e., 30 weeks of age). Finally, when the phase of rapid growth was induced during the spring anestrous season (group D), the lambs remained anovulatory despite their growth well beyond the normal size required to initiate reproductive cycles. These females continued to be hypersensitive to estradiol negative feedback because of the long days of summer, and it was not until the decreasing day lengths of autumn that reproductive cycles began. Thus, in this example in the sheep, both day length and growth serve as codeterminants timing the initiation of reproductive cycles through photoperiod and as growth-related cues timing the decrease in sensitivity to steroid feedback and expression of the high-frequency GnRH pulses.

B. Social Factors

Pheromonal and tactile cues from other members of the same species can accelerate or decelerate the

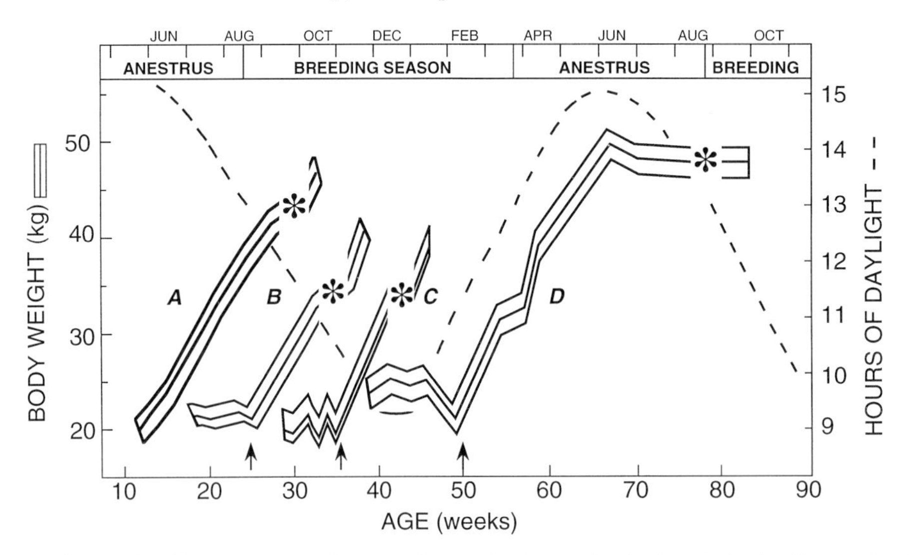

FIGURE 4 Season and growth influence timing of onset of reproductive cycles in female lambs (*mean body weight and age at first luteal phase). They were born in spring (March) and were raised outdoors (broken line, photoperiod). Some were fed *ad libitum* after weaning at 10 weeks of age (group A). Others were placed on a restricted diet of similar composition (groups B–D); at various ages (arrows), feeding *ad libitum* was begun in food-restricted lambs. Note that puberty can only occur during the breeding season and when growth is sufficient [redrawn from D. L. Foster and K. D. Ryan, Puberty in the lamb: Sexual maturation of a seasonal breeder in a changing environment, In *Control of the Onset of Puberty II* (P. C. Sizonenko, M. L. Aubert, and M. M. Grumbach, Eds.), pp. 108–142, Williams & Wilkins, Baltimore, 1990].

timing of puberty; such mechanisms are particularly common in mice, rats, hamsters, and voles. The urine of male mice contains a substance that accelerates sexual maturation in females; the identity of the pheromone remains uncertain, although it may be an amine. A potent puberty-inhibiting pheromone from the urine of grouped female mice is also under investigation. The substance is of interest because its production is dependent on population density, and therefore it provides a mechanism modulating population growth by increasing the generation interval (delaying puberty). Such mechanisms are not confined to small animal species because the timing of puberty in both pigs and cattle can be altered by male and female stimuli. The steroidal pheromones androstenol and androstenone produced by mature boars are potent inducers of estrus in virgin pigs. There is ample evidence that nonpheromonal cues, those involving aversive emotional stimuli, can profoundly influence the transition into adulthood; thus, in social species dominant individuals will suppress puberty in conspecifics. Finally, the females of many mammalian species are not spontaneous ovulators, that is, ovulation will only occur if cervical stimulation occurs during mating. This phenomenon is common in the mustelid (stoats, badgers, mink, etc.) and cat families. If one considers puberty as the final production of mature gametes, then in such induced ovulators one might consider that mating itself is the final trigger to puberty.

V. SEX DIFFERENCES IN TIMING OF PUBERTY

The reproductive strategies of males and females can differ as much as those between two species. Most of the reproductive effort of the typical female mammal is invested after mating, and she expends considerable energy in the growth and early development of her offspring. The typical polygynous male uses an entirely different strategy to time sexual activity. Most of the energy for his reproductive effort is expended before mating, primarily in competition for females. He produces large numbers of offspring, but does little to promote their survival. Thus, males and females have evolved under different selective pressure, and the timing of puberty is often different in the two sexes. This is especially apparent in the sheep, in which the male begins puberty at about 10 weeks of age and the female at about 30 weeks. This does not simply reflect the difference between the time course of activation of the testes and ovaries. Rather, this difference resides within brain mechanisms controlling GnRH secretion. Administration of testosterone to the female lamb before birth advances the pubertal increase in GnRH secretion from 30 weeks of age to 10 weeks of age. This prenatal exposure to testicular steroids reduces the reliance on photoperiod. Whereas the female sheep must experience a decreasing day length, the male does not use this or any other known photoperiod requirement to activate its reproductive system. In females exposed experimentally to testosterone before birth, a change in length of day is no longer required to initiate puberty. A corollary is that early androgen exposure allows the neuroendocrine events that underlie puberty to become activated at a much earlier stage of somatic development than occurs in the female. This sex difference provides an example of a strategy used to allow the male, with its longer spermatogenic cycle (months), to begin its maturational process before the female, which only requires a few days to develop a follicle and ovulate once it has received the appropriate internal (growth cues) and external (photoperiod) cues that a pregnancy could be established.

VI. RECURRENT PUBERTY: THE CASE OF THE SEASONAL BREEDER

After puberty, most long-lived species living in natural conditions become reproductively inactive for several months to prevent mating at times of year which would lead to births during unfavorable seasons (winter). To do this, the seasonal breeder has retained a key mechanism used during sexual maturation—the ability to change sensitivity to steroid feedback inhibition of GnRH secretion. However, for an adult attempting to become infertile, the mechanism changes its direction. Sensitivity to steroid negative feedback becomes increased and hypogonadotropism is produced for several months. Then, after this extended period of reproductive inactivity to prevent unwanted pregnancies, breeding begins again when the young adult reverses the mechanism and reduces its sensitivity to steroid feedback inhibition to allow high GnRH secretion to begin. This repeated seasonal reactivation of the reproductive system in the adult has been considered by some to be an "annual puberty." Indeed, it is clear that the annual period of seasonal quiescence in the adult shares many fundamental similarities with the prepubertal period, at least at the level of GnRH neurosecretion and below (pituitary and gonads). Both are characterized by hypogonadotropism due to reduced GnRH secretion because of high sensitivity to estradiol negative feedback of gonadal steroids. Thus, the annual reversal of this mechanism to cause seasonal reactivation of the pituitary–gonadal axis in the adult could be thought of as puberty. However, this view must be tempered by the consideration that many stimuli can activate the reproductive system through GnRH: photoperiod cues, plant factors, seasonal cues, and increased nutrition. Until we understand the different neural inputs to the GnRH neurosecretory system and the differences between factors which uniquely activate the reproductive system the first time in relation to other times, this discussion remains largely one of semantics.

See Also the Following Articles

Energetics of Reproduction; Photoperiodism, in Vertebrates; Puberty, Precocious; Puberty Acceleration; Puberty, in Humans; Puberty, in Nonhuman Primates; Seasonal Reproduction, Mammals

Bibliography

Aubert, M. L., Guraz, N. M., D'Alleves, V., Pierroz, D. D., Arsenijevic, Y., and Sizonenko, P. C. (1995). Body weight, growth hormone, insulin-like growth factors, and neuropeptide Y as potential factors for the control of the initial sexual maturation. In *Frontiers in Endocrinology, Puberty: Basic and Clinical Aspects* (C. Bergada and J. A. Moguilevsky, Eds.), pp. 343–356. Ares-serono Symposia, Rome.

Bronson, F. H. (Ed.) (1989). *Mammalian Reproductive Biology*. Univ. of Chicago Press, Chicago.

Brooks, A. N., McNeilly, A. S., and Thomas, G. B. (1995). Role of GnRH in the ontogeny and regulation of the fetal hypothalamo–pituitary–gonadal axis in sheep. *J. Reprod. Fertil. Suppl.* 49, 163–175.

Cameron, J. L., Hansen, P. P., McNeil, T. H., Koerker, D. J., Clifton, D. K., Rogers, K. V., Bremner, W. J., and Steiner, R. A. (1985). In *Adolescence in Females* (C. Flagmini, S. Ventruroli, and J. R. Givens, Eds.), pp. 59–78. Yearbook Med, Pub., Chicago.
Cheung, C. C., Thornton, J. E., Kuijper, J. L., Weigle, D. S., Clifton, D. K., and Steiner, R. A (1997). Leptin is a metabolic gate for the onset of puberty in the female rat. *Endocrinology* **138**(2), 855–858.
Ebling, F. J. P., and Foster, D. L. (1989). Seasonal breeding—A model for puberty? In *Control of the Onset of Puberty III* (H. A. Delemarre-van de Waal, T. M. Plant, G. P. van Rees, and J. Schoemaker, Eds.), pp. 253–264. Elsevier, Amsterdam.
Foster, D. L. (1980). Comparative development of mammalian females: Proposed analogies among patterns of LH secretion in various species. In *Problems in Pediatric Endocrinology* (C. La Cauza and A. W. Root, Eds.), pp. 193–210. Academic Press, New York.
Foster, D. L. (1994). Puberty in the sheep. In *The Physiology of Reproduction* (E. Knobil and J. D. Neill, Eds.), 2nd ed., pp. 411–451. Raven Press, New York.
Foster, D. L., Ebling, F. J. P., Ryan, K. D., and Yellon, S. M. (1989). Mechanisms timing puberty: A comparative approach. In *Control of the Onset of Puberty III* (H. A. Delemarre-van de Waal, T. M. Plant, G. P. van Rees, and J. Schoemaker, Eds.), pp. 227–249. Elsevier, Amsterdam.
Foster, D. L., Nagatani, S., Bucholtz, D. B., Tsukamura, H., Tanaka, T., and Maeda, K.-I. (1997). Links between nutrition and reproduction: Signals, sensors and pathways controlling GnRH secretion. In *Nutrition and Reproduction* (W. Hansel and S. McCann, Eds.). LSU Press, Baton Rouge.
Frisch, R. E. (1990). Body fat, menarche, fitness and fertility. In *Adipose Tissue and Reproduction* (R. E. Frisch, Ed.), pp. 1–26. Karger, Basel.
Kennedy, G. C., and Mitra, J. (1963). Body weight and food intake as initiating factors for puberty in the rat. *J. Physiol.* **166**, 408–418.
Ojeda, S. R., and Urbanski, H. F. (1994). Puberty in the rat. In *The Physiology of Reproduction* (E. Knobil and J. D. Neill, Eds.), 2nd ed., pp. 363–409. Raven Press, New York.
Plant, T. M. (1994). Puberty in primates. In *The Physiology of Reproduction* (E. Knobil and J. D. Neill, Eds.), 2nd ed., pp. 453–485. Raven Press, New York.
Plant, T. M., and Lee, P. A. (Eds.) (1995). *The Neurobiology of Puberty*. Journal of Endocrinology, Ltd., Bristol, UK.

Puberty, Precocious

Leo Plouffe, Jr.
Eli Lilly and Company

GLOSSARY

adrenarche Pubic hair development.
menarche Onset of menses.
Tanner staging System of staging physical development of specific areas (e.g., breast and pubic hair) for the assessment of pubertal development.
thelarche Breast development.

Normal pubertal development in women is regulated by a complex series of interactions and proceeds in an orderly fashion, delineated by clear physical changes. Precocious puberty generally refers to the appearance of puberty before age 8, though this has recently come under scrutiny. Physical examination and history are essential elements in the evaluation of patients presenting with this problem.

Encyclopedia of Reproduction VOLUME 4

I. BASIC CONSIDERATIONS IN FEMALE PUBERTAL DEVELOPMENT

A wide variety of factors contribute to the onset of puberty. The exact mechanisms are still under investigation. Recent proposals include a complex interplay of neurotransmitter elements as well as growth factors.

On clinical grounds, female puberty is characterized by the classic triad of thelarche (breast development), adrenarche (pubic hair development), and menarche (onset of menses) specific to females. These progress in an orderly fashion. Thelarche is usually seen first, followed by adrenarche and menarche. In up to 15% of Caucasians and 50% of blacks, adrenarche may precede thelarche. In all instances of normal pubertal development, menarche is the last event to occur in the progression. Puberty in both sexes is also associated with dramatic changes in height and weight which are a reflection of the intricate mechanisms that regulate growth.

It is usually stated that 95% of females will have the onset of puberty between ages 8 and 13. Thelarche is usually the first sign followed by adrenarche. In 15% of individuals, however, the reverse is true. Axillary hair usually is seen about 1 year following the development of pubic hair. Menarche occurs on average at age 12.5 years, with a normal range between 10 and 16 years. Menarche usually occurs at the same time as the weight spurt, which occurs about 6 months after the height spurt.

By the most commonly accepted definition, precocious puberty in a female is diagnosed whenever there is evidence of pubertal development prior to age 8. Such precocity may manifest in all axes of pubertal development or a single of its components, such as premature adrenarche. The appropriateness of age 8 as a cutoff for the diagnosis of precocious puberty will be discussed.

II. BASIC ELEMENTS IN THE EVALUATION OF FEMALE PUBERTAL DEVELOPMENT

In order to address specific disorders of female puberty, it is essential to have a good understanding of the basic clinical skills that are needed to assess, characterize, and track pubertal development. The history and physical examination are key elements in the evaluation of pubertal disorders. They cannot be overemphasized.

In a simplistic way, three concurrent events are necessary for normal sexual development: (i) normal adrenal androgen input, (ii) normal estrogenic input, and (iii) normal growth factors. Most of these three events can be assessed largely on clinical grounds. Seldom are further tests indicated.

The history should proceed systematically (Table 1). It is important to show great sensitivity in obtaining this information. The patient and family are often extremely anxious about the condition and its long-term implications. In many instances, it becomes necessary to obtain prior medical records and these should be sought with great diligence.

The most commonly accepted method to assess and describe thelarche and pubarche is the Tanner method of staging (Table 2). While this classification is basic, its proper interpretation continues to be a challenge for many. A recent multicenter study demonstrated that 6% of experienced pediatricians were unable to reliably assess the Tanner stage among peripubertal subjects. It should also be noted that the

TABLE 1
Elements to Elicit in a Female with Precocious Puberty

Timing of appearance of signs of precocious puberty
Age
Order of appearance (height/thelarche/adrenarche/menarche)
Progression of development (rapid vs slow)
Development history
Medical history
Chronic ongoing medical illness
Severe medical illness requiring specific therapy
Medication cytotoxic agents, and radiotherapy
Socioeconomic factors: nutrition/physical and sexual abuse
Exposure to exogenous hormones
Iatrogenic: medications
Environmental
Illicit drug use (marijuana)
Family history of disordered puberty

TABLE 2
Tanner Staging

Stage	*Breast*	*Pubic hair*
1	Preadolescent, with elevation of papilla only	None
2	Elevation of breasts and papilla to form a small mound; areola enlarges	Sparse, lightly pigmented, long and straight downy hair along the labis
3	Further enlargement of breasts and areola with no separation of their contours	Hair is considerably darker, coarser, and more curled; the hair spreads sparsely over the labis
4	Projection of the areola and papilla to form a secondary mound above the level of the breast	Hair, adult in type, covers a smaller area than in the adult and does not extend onto the thighs
5	Resemble a mature female; areola has recessed to the general contour of the breast	Hair is adult in consistency and type, with extension onto the thigh

Tanner staging method was developed for a primarily white European cohort and that its full applicability across ethnic and racial groups has yet to be confirmed.

The development of pubic and axillary hair is the manifestation of androgen input. This is usually of adrenal origin but may occasionally reflect the production of androgens from the ovaries and very rarely from ectopic sites or exogenous sources. By contrast, thelarche and menarche reflect estrogen input, usually of ovarian origin.

Menarche is reported as the onset of vaginal bleeding secondary to uterine endometrial shedding. It is always wise for the clinician to remember that there are many etiologies for vaginal bleeding. In particular, the appearance of vaginal bleeding in the absence of other manifestations of sexual development should prompt due consideration of causes other than shedding of a stimulated endometrium.

One element in the evaluation of pubertal disorders which often receives too little attention by generalist physicians is the assessment of growth. This is best expressed as growth velocity and represents the amount of growth that occurs during a given period. It is expressed in units/year. Linear growth velocity is therefore generally expressed in cm/year. It represents a key element in the evaluation of pubertal development. Obviously, the period of maximum increase in length/height is during the first 6 months of life, approximating 16 or 17 cm. This rapidly declines to 10 cm during the second year of life, 8 cm in the third year of life, and 7 cm in the fourth year. Beyond this stage, the linear growth velocity stabilizes at 4–6 cm/year. Linear growth velocity increases again with the onset of puberty and ultimately peaks around 9 or 10 cm/year. This latter figure is often referred to as the peak (linear) growth velocity.

Much care is needed to assess the height of individuals at each visit. Basic guidelines are presented in Table 3. Each measurement should be plotted on

TABLE 3
Guidelines for Height Measurement

Use a reliable measurement tool
- Stadiometer

Positioning
- Fully erect
- Back of head, thoracic spine, buttocks, and heels all in line
- Shoes removed

Use same observer as much as possible
Take three measurements at each sampling
Try to measure always at same time of day
Measure every 3 months at the earliest
Plot measurements on growth chart and grow velocity chart

special graphs to ensure appropriate monitoring and follow-up. It is important to realize the difference between charts recording absolute height, weight vs height, and those designed to monitor linear growth velocity.

Several laboratory investigations complement the basic physical evaluation of the patient with pubertal precocity. The initial steps in the evaluation of these disorders continue to be a thoughtful history, physical examination, and formulation of a differential diagnosis prior to proceeding with advanced diagnostic algorithms.

In summary, precocious puberty should be suspected whenever there is evidence of pubertal development prior to the age of 8. In order to establish the presence of this condition, the clinician must master key elements in history taking and physical examination as outlined previously.

TABLE 4
Partial Listing of GnRH-Dependent Causes of Precocious Puberty

Idiopathic/familial
CNS lesions
  Tumors of hypothalamus, pineal, or cortex including hamatoma, craniopharyngioma, glioma
  Infection, including toxoplasmosis, encephalitis, meningitis
  Neurocutaneous syndromes, neurofibromatosis
  Developmental defects, including microcephaly, tuberous sclerosis, aqueduct stenosis, craniosynostosis
  Trauma
  Miscellaneous; Sturge–Weber syndrome, diffuse encephalopathy, idiopathic epilepsy
Other
  Juvenile primary hypothyroidism
  Silber syndrome

III. ETIOLOGIC CONDITIONS OF PRECOCIOUS PUBERTY: CLASSIFICATION

Numerous classification systems have been proposed for the etiologies of precocious puberty, but one system has gained broad acceptance. It divides precocious puberty in two major classes: gonadotropin-releasing hormone dependent (GnRH dependent) and gonadotropin-releasing hormone independent (GnRH independent).

GnRH-dependent conditions imply activation of the hypothalamic–pituitary axis and a normal progression of pubertal events, though they occur prematurely and often at an accelerated pace. By contrast, GnRH-independent conditions present with either a disorderly progression of puberty or the appearance of a single component of pubertal development. Some clinicians prefer to reserve the term GnRH-independent causes strictly for cases with a disorderly progression of pubertal landmarks, such as the appearance of menarche followed by adrenarche and thelarche. The term isolated precocious puberty is then used for those cases which manifest a single element of puberty, such as isolated adrenarche.

It should be obvious that a simple history and physical examination is usually sufficient to determine in which of these two categories a patient belongs. Having achieved this basic understanding, one must contemplate the different diagnoses which make up each category, as outlined in Tables 4 and 5.

TABLE 5
Partial Listing of GnRH-Independent Causes of Precocious Puberty

Idiopathic premature pubarche
Idiopathic premature thelarche
Idiopathic premature menarche
Ovarian
  Ovarian tumors: malignancy, lutemas
  McCune–Albright syndrome
Adrenal lesions
  Congenital adrenal hyperplasia, Cushing's syndrome, tumors
Iatrogenic
  Androgen or estrogen administration, vitamins, oral contraceptives

IV. WORK-UP OF PRECOCIOUS PUBERTY

The decision of when to initiate a work-up for precocious puberty frequently proves to be a major challenge. Certain cases are clearly obvious, particularly prior to age 5 or 6. Beyond this, one must weigh the risks of passing off the symptoms as normal without performing any investigation against those of generating undue anxiety by suggesting a potentially fatal condition exists and investing an inordinate amount of time in a complex work-up. Clearly, the history and physical examination are the first building blocks in the decision algorithm.

At this point, it is appropriate to discuss a diagnosis of constitutional precocious puberty. Obviously, a family history of a mother or siblings having presented early puberty is reassuring. However, in no circumstances should such a history be viewed as establishing the diagnosis. In all cases, constitutional precocious puberty is a diagnosis of exclusion.

The appropriateness of age 8 as a cutoff for precocious puberty has recently come under scrutiny. In a large multicenter study conducted in the offices of American pediatricians in a community setting, the high prevalence of early pubarche and to some extent thelarche was demonstrated. This was most obvious in the black population. These findings may relate to the documented decline in the age of onset of menarche and support the concept that all elements of pubertal development may be appearing at an earlier age. The same study shows that there may be important interracial differences in the chronology of pubertal development. Thelarche and adrenarche appear to display the greatest variability, whereas the age of onset of menarche seems to be more concordant between all groups.

Once it has been decided that it is appropriate to initiate a work-up for precocious puberty, a few simple principles should guide the work-up. First, one should derive from the history and physical exam whether the presentation is one of GnRH-dependent or GnRH-independent precocious puberty. The rest of the investigations will be driven based on this decision. A complete physical examination is indicated in any case but should be focused more on certain elements depending on the suspected etiology (Table 6). It should be emphasized that examination of the external genitalia is clearly in order in the evaluation of these patients. On the other hand, a full pelvic examination is seldom indicated except in those cases in which a pelvic tumor is strongly suspected. While many clinicians claim they can feel "a pubertal uterus" on examination, I feel strongly that this is a dubious affirmation and that there are a number of less invasive tests with higher accuracy which are now available to the clinician. The pelvic examination is sometimes best delayed to a subsequent visit. Rarely is examination under anesthesia required in this age group.

TABLE 6
Key Elements to Be Considered in the Physical Examination in a Female with a Pubertal Development Aberrancy

Height, weight, vital signs
Evidence of masculinization: deepening of voice, hirsutism
Skin: café-au-lait, hirsutism, cutis laxa, red/purple striae
Hair: Tanner staging, axillary hair, coarse or sparse hair
CNS exam: gross CNS examination, peripheral vision and olfaction
Eyes: fundus
Facies: dysmorphic features
Chest and breast development: Tanner staging
Abdomen: masses
Genitals: external inspection, clitoromegaly, pelvic examination

In addition to the history and physical examination, radiographic bone age is helpful in establishing the presence of precocious puberty. This is accomplished through a simple radiography of the wrist. Interpretation is performed through one of several bone age atlases. An experienced radiologist is needed for proper interpretation of these tests and the results unfortunately demonstrate a fairly large degree of variability. Current efforts are under way to establish a reliable computerized method to determine bone age.

One major challenge in the evaluation of precocious puberty is the assessment of the patient who presents with adrenarche or thelarche only. This could represent isolated precocious puberty or be the first manifestation of GnRH-dependent precocious puberty. Several strategies have been proposed to help distinguish between these two conditions. These are still under investigation but include pelvic ultrasonography, gonadotropin testing, and adrenocorticotropin hormone-stimulation testing. Currently, this continues to be a dilemma for the clinician, who must rely on his or her judgment to decide how to proceed in the diagnostic algorithm.

When GnRH-dependent precocious puberty is suspected, the work-up focuses on the central nervous system, as the site for the GnRH pulse generator, as well as other endocrine organs which may affect the hypothalamic–pituitary axis. Basic components of this line of investigation are listed in Table 7. Beyond the initial group of tests, several investigations require great expertise in their performance and interpretation. These are best reserved to subspecialists. A discussion of the investigations is beyond the scope of this article. Two points are emphasized, however.

Measurement of serum follicle-stimulating hormone (FSH) and luteinizing hormone (LH) may be helpful in establishing the activation of the GnRH-dependent axis. The older assay systems, however, were frequently unable to distinguish between prepubertal and early pubertal levels. Recent highly sensitive assays may have overcome this problem. It is important to request that such a specific assay system be used. There has been much controversy as to what imaging modality is best indicated for the evaluation of the central nervous system in these patients. The evolving consensus is that magnetic resonance imaging (MRI) is the preferred approach.

TABLE 7
Investigations in the Work-Up of Suspected GnRH-Dependent Precocious Puberty

Hormonal studies
- FSH and LH
- Estradiol
- Prolactin
- Thyroid function tests

X-ray studies
- Bone age
- CNS imaging (MRI)

Advanced tests (usually in coordination with consultant)
- Assessment of the growth hormone axis
- Assessment of the adrenal axis
- Dynamic hormonal stimulation
  - GnRH, GnRH pump + retest, GnRH + clomiphene
- Visual fields/Goldman perimetry
- Pelvic ultrasound
- Cytogenetic analysis
- DNA diagnostic tests

TABLE 8
Investigations in the Work-Up of Suspected GnRH-Independent Precocious Puberty

Hormonal studies
- With evidence of estrogenization: estradiol, prolactin, thyroid function tests, human chorionic gonadotropin

With evidence of androgen excess: DHEAS, testosterone, 17α-hydroxyprogesterone, other

X-ray studies
- Bone age
- Abdominal imaging
- Pelvic imaging

Advanced tests (usually in coordination with consultant)
- Assessment of the adrenal axis
- DNA diagnostic tests
- Other

The work-up of a GnRH-independent presentation centers on target organs, such as the ovary and the adrenal, and primary pathology at the site of presentation (Table 8). Again, certain cautions are in order. Pelvic ultrasonography may be of benefit in certain patients. The findings focus on the evaluation of the ovaries, for evidence of activity, as well as the uterus, which may illustrate estrogen activity. However, it must be remembered that the ovary in the peripubertal stage is often multicystic; this is a normal finding and does not reflect pathology. Thyroid disease can also stimulate the ovary and produce multiple ovarian cysts. It is absolutely critical for the clinician to recognize these findings and avoid being lured into performing any type of surgery on these ovaries.

V. MANAGEMENT OF PRECOCIOUS PUBERTY

The management of precocious puberty represents a complex interplay of the science and the art of medicine. Once a specific condition has been identified, the most appropriate management must be selected. This field of medicine has been evolving very rapidly. Many new therapeutic options are available,

TABLE 9
Management of GnRH-Dependent Precocious Puberty

Correct underlying cause (e.g., neurosurgery for a tumor or thyroid therapy)
Gonadotropin-suppressive therapy
Consider addition of peripheral hormone receptor blocker (e.g., cyproterone acetate)
Supportive therapy

many of which are still in the process of being evaluated. When and how to use the evolving therapies lies more in the realm of the art of medicine than the science at this point. In addition, it is critical for the clinician to be sensitive to the special needs of the teenager who presents with precocious puberty. Above all, one should make sure that one's clinical skills are appropriate and up-to-date to manage this condition. In many instances, a referral to a subspecialist in the field is most appropriate.

Certain management steps are shared between all etiologies of precocious puberty, including due attention for emotional and psychological support of the patient and her family. In addition, close assessment of height and weight is needed. Close follow-up is needed to further confirm the diagnosis and monitor any therapeutic intervention.

The general management principles for GnRH-dependent conditions are listed in Table 9. The major breakthrough in this area has been the introduction of long-acting gonadotropin-releasing hormone agonist (LA-GnRHa). Within a few months of initiation of this form of therapy, the signs of pubertal development regress. More important, this treatment modality yields an improved final adult height over untreated patients. The approach to GnRH-independent conditions is presented in Table 10. This approach is primarily focused on the underlying etiologic process, even though there is now increasing acceptance of the addition to LA-GnRHa in the treatment of these conditions. In addition, agents which block the effects of hormones on target tissues can be of great assistance.

TABLE 10
Management of GnRH-Independent Precocious Puberty

Correct underlying cause (e.g., surgery for an adrenal tumor or steroid therapy for CAH)
Gonadal suppression
High dose progestin
Testolactone will prevent steroidogenesis from the ovary
Long-acting GnRH agonist
Hormone receptor blockade
High dose progestin
Cyproterone acetate
Other
Supportive therapy

VI. CONCLUSION

Precocious puberty occurs when there is evidence of pubertal development before the age of 8. This definition may soon be revised in light of recent evidence. The etiologies of precocious puberty are varied but can be divided into GnRH-dependent and GnRH-independent causes. The diagnosis and management of these patients is complex but follows a well-defined algorithm. The long-term prognosis is generally excellent, particularly in light of recently introduced therapeutic modalities.

See Also the Following Articles

Adrenarche; GnRH (Gonadotropin-Releasing Hormone); Menarche; Puberty, in Humans

Bibliography

Albanese, A., Hall, C., and Stanhope, R. (1995). The use of a computerized method of bone age assessment in clinical practice. *Horm. Res.* 44(Suppl. 3), 2–7.

Aritaki, S., Takagi, A., Someya, H., and Jun, L. (1997). A comparison of patients with premature thelarche and idiopathic true precocious puberty in the initial stage of illness. *Acta Paediatr. Japon.* 39, 21–27.

Balducci, R., Boscherini, B., Mangiantini, A., Morellini, M., and Toscano, V. (1994). Isolated precocious pubarche: An approach. *J. Clin. Endocrinol. Metab.* 79, 582–589.

Bourguignon, J. P., Jaeken, J., Gerard, A., and de Zegher, F. (1997). Amino acid neurotransmission and initiation of puberty: Evidence from nonketotic hyperglycinemia in a female infant and gonadotropin-releasing hormone secretion by rat hypothalamic explants. *J. Clin. Endocrinol. Metab.* **82**, 1899–1903.

Bridges, N. A., Cooke, A., Healy, M. J., Hindmarsh, P. C., and Brook, C. G. (1995). Ovaries in sexual precocity. *Clin. Endocrinol.* **42**, 135–140.

Ehrhardt, A. A., and Meyer-Bahlburg, H. F. (1994). Psychosocial aspects of precocious puberty. *Horm. Res.* **41**(Suppl. 2), 30–35.

Griffin, I. J., Cole, T. J., Duncan, K. A., Hollman, A. S., and Donaldson, M. D. (1995). Pelvic ultrasound findings in different forms of sexual precocity. *Acta Paediatr.* **84**, 544–549.

Haber, H. P., Wollmann, H. A., and Ranke, M. B. (1995). Pelvic ultrasonography: Early differentiation between isolated premature thelarche and central precocious puberty. *Eur. J. Pediatr.* **154**, 182–186.

Hayward, C., Killen, J. D., Wilson, D. M., *et al.* (1997). Psychiatric risk associated with early puberty in adolescent girls. *J. Am. Acad. Child Adol. Psych.* **36**, 255–262.

Hedlund, G. L., Royal, S. A., and Parker, K. L. (1994). Disorders of puberty: A practical imaging approach. *Sem. Ultrasound CT MR* **15**, 49–77.

Herman-Giddens, M. E., Slora, E. J., Wasserman, R. C., *et al.* (1997). Secondary sexual characteristics and menses in young girls seen in office practice: A study from the Pediatric Research in Office Settings network. *Pediatrics* **99**, 505–512.

Ibanez, L., Bonnin, M. R., Zampolli, M., *et al.* (1995). Usefulness of an ACTH test in the diagnosis of nonclassical 21-hydroxylase deficiency among children presenting with premature pubarche. *Horm. Res.* **44**, 51–56.

Kaplan, S. A. (1990). Growth and growth hormone: Disorders of the anterior pituitary. In *Clinical Pediatric Endocrinology* (S. A. Kaplan, Ed.), pp. 1–62. Saunders, Philadelphia.

Kauli, R., Galatzer, A., Kornreich, L., *et al.* (1997). Final height of girls with central precocious puberty, untreated versus treated with cyproterone acetate or GnRH analogue. A comparative study with re-evaluation of predictions by the Bayley–Pinneau method. *Horm. Res.* **47**, 54–61.

Kornreich, L., Horev, G., Blaser, S., *et al.* (1995). Central precocious puberty: Evaluation by neuroimaging. *Pediatr. Radiol.* **25**, 7–11.

Oostdijk, W., Rikken, B., Schreuder, S., *et al.* (1996). Final height in central precocious puberty after long term treatment with a slow release GnRH agonist. *Arch. Dis. Childhood* **75**, 292–297.

Paul, D., Conte, F. A., Grumbach, M. M., and Kaplan, S. L. (1995). Long-term effect of gonadotropin-releasing hormone agonist therapy on final and near-final height in 26 children with true precocious puberty treated at a median age of less than 5 years. *J. Clin. Endocrinol. Metab.* **80**, 546–551.

Rage, F., Hill, D. F., Sena-Esteves, M., *et al.* (1997). Targeting transforming growth factor alpha expression to discrete loci of the neuroendocrine brain induces female sexual precocity. *Proc. Natl. Acad. Sci. USA* **94**, 2735–2740.

Rosenfield, R. L. (1996). The ovary and female sexual maturation. In *Pediatric Endocrinology* (M. A. Sperling, Ed.), pp. 329–385. Saunders, Philadelphia.

Shankar, R. R., and Pescovitz, O. H. (1995). Precocious puberty. *Adv. Endocrinol. Metab.* **6**, 55–89.

Soliman, A. T., Al Lamki, M., Al Salmi, I., and Asfour, M. (1997). Congenital adrenal hyperplasia complicated by central precocious puberty: Linear growth during infancy and treatment with gonadotropin-releasing hormone analog. *Metab. Clin. Exp.* **46**, 513–517.

Styne, D. M. (1997). New aspects in the diagnosis and treatment of pubertal disorders. *Pediatr. Clin. North Am.* **44**, 505–529.

Verrotti, A., Ferrari, M., Morgese, G., and Chiarelli, F. (1996). Premature thelarche: A long-term follow-up. *Gynecol. Endocrinol.* **10**, 241–247.

Puerperal Infections

John W. Riggs and Jorge D. Blanco
University of Texas Medical School, Houston

GLOSSARY

asymptomatic bacituria The presence of 100,000 or more colonies of a bacterial pathogen per milliliter of urine on two consecutive clean-catch midstream voided specimens in the absence of signs or symptoms of urinary tract infection.

childbed fever Puerperal fever; usually refers to the epidemic puerperal infections as were studied by Semmelweis in the 1800s.

cystitis An inflammation or infection of the urinary bladder.

endometritis A clinically apparent, postpartum infection of the upper reproductive tract.

mastitis Inflammation of the breast usually due to bacterial infection. This infection is subclassified as epidemic, occurring in hospitals in association with epidemics in nurseries, or endemic, occurring sporadically among nonhospitalized women.

polymicrobial infection An infection characterized by the growth of multiple pathogenic organisms.

pyelonephritis Inflammation of the kidney particularly due to local bacterial infection.

wound infection An infection of a surgical site, usually referring to the skin and subcutaneous tissue.

The most common infections in the puerperium include endometritis, wound infection, urinary tract infections, and mastitis with endometritis following cesarean section being the most common. These infections continue to be a major source of morbidity and mortality following birth despite many advances in microbiology and antibiotic development.

I. ENDOMETRITIS

Endometritis is an infection of the upper reproductive tract particularly the endometrium, myometrium, and possibly the adnexa. Several terms are used to describe this spectrum of infections: endometritis, endomyometritis, endoparametritis, and even "phlegmon." The term endometritis will be used to describe a clinically apparent, postpartum infection of the upper reproductive tract.

A. History

It is difficult for most modern clinicians to imagine a time when the possibility of transmission of disease between individuals was not widely accepted but postpartum pelvic infections provided a crucial link in our understanding of the importance of aseptic technique. Infection following childbirth, known as childbed fever, was unfortunately common throughout history and was a common cause of mortality. While Oliver Wendell Holmes had theorized in 1843 that obstetricians were capable of carrying infection between women, little credibility was attributed to his conjecture by his colleagues.

In 1846 Ignaz Semmelweis confronted a 10% mortality rate among women and newborns due to sepsis in Vienna's general hospital. Several important observations were made by him that supported the conclusion that women were being infected by their doctors. Patients on the obstetrical wards became infected much more frequently than those on the midwives' wards. Women who became infected were usually those women on the row of beds which was examined that day by the physicians. Immediately before making their morning rounds on the obstetrical service, the physicians assigned to that ward performed autopsies on women who had just died of childbed fever. No antiseptic technique was utilized by the physicians after leaving the autopsy suite. Perhaps the greatest insight by Semmelweis occurred after his friend died from sepsis following a knife injury which he incurred during one of the puerperal autopsies. The institution of antiseptic technique in his unit produced dramatic declines in the maternal mortality rate, but his 1861 publication was ridiculed and he died in an asylum in 1865 after many years of mockery by his colleagues.

B. Risk Factors

The single greatest risk factor for endometritis is cesarean section with a 1 or 2% rate of endometritis following vaginal delivery compared to 38.5% after a cesarean. The theoretical reasons for the higher infection rate following cesarean section are numerous. Cesarean section involves greater manipulation of the uterus and endometrial cavity; for example, manual extraction of the placenta can increase this risk. Spillage of intrauterine and probably vaginal fluid into the intraperitoneal cavity, suture placement through the uterine wall which probably produces some tissue necrosis, and hematoma or seroma formation may support bacterial growth. Any of these factors could contribute to infection. Besides the method of delivery, the next two most important risk factors for endometritis are the length of labor and the length of rupture of membranes. The longer the length of either labor or rupture of membranes or both, the higher the rate of infection. Also, women who undergo cesarean section when not in labor develop endometritis much less frequently than women who have been in labor. Therefore, some patients may already be infected at the time of cesarean section. In labor, their infection is subclinical, but shortly after surgery the infection may progress and hence become symptomatic. Other risk factors help predict endometritis. Lower socioeconomic status appears to be an important predictor of infection, possibly because there may be a difference in flora, hygiene, or nutrition that may lead to a higher rate of infection. Young age is also associated with infection. Factors such as intrauterine catheters, the number of vaginal exams, obesity, and anemia have not consistently been shown to increase infection rates.

C. Microbiology

Like many genital tract infections, puerperal endometritis is a polymicrobial infection. Table 1 lists the microbiologic isolates from patients with endometritis after a cesarean section. A majority of patients grew anaerobic organisms (80%) with the most common isolate being the *Bacteroides* species, which is notorious for abscess formation and drug resistance. The next most common isolates are the aerobic gram-negative bacilli such as *Escherichia coli* and gram-positive organisms such as streptococci. *Mycoplasma hominis*, *Ureaplasma urealyticum*, and *Chlamydia trachomatis* play a limited role, if any, in the development of puerperal infections. Group A streptococci caused lethal epidemics of childbed fever in the last century and occasionally reappear today. However, this organism is much less frequently lethal today.

TABLE 1
Bacteriologic Isolates from the Endometrium of Patients with Endometritis

Microorganism	*Patients with isolate (%)*
Anaerobes	
Bacteroides bivius	45
Bacteroides fragilis	15
Other *Bacteroides* sp.	25
Fusobacterium sp.	21
Peptococcus sp.	18
Peptostreptococcus sp.	20
Clostridium sp.	14
Aerobes	
Escherichia coli	16
Other gram-negative bacilli	14
Enterococcus	22
Group B streptococcus	10
Other gram-positive cocci	7

D. Diagnosis

Fevers are common following delivery, occurring in 6% of women in the first 24 hr. Of vaginally delivered women, this fever resolves spontaneously in 80% but in only 30% following cesarean delivery. Other sources need to be excluded when fever occurs, including atelectasis, viral diseases, and pyelonephritis. Endometritis is often a diagnosis of exclusion. The diagnosis requires exclusion of other infections such as cystitis, wound infection, or pneumonia. The diagnostic criteria for endometritis include fever and uterine tenderness with the exclusion of another infected site, so a thorough physical exam is suggested. Abdominal and pelvic examination, evaluation for costovertebral angle or leg tenderness, as well as lung auscultation should usually be included. Speculum examination is necessary to exclude foreign bodies, but endometrial cultures are not typically performed for diagnosis. "Pure" endometrial cultures are difficult to obtain because of contamination from the cervix. Therefore, some investigators question their need in most routine patients, except when rare isolates and resistant or contagious organisms are suspected. Cultures can be obtained to identify resistant organisms and to assist in antibiotic selection in those patients who do not respond to initial treatment by using a multiluminal, protected catheter systems. Commonly, patients with endometritis will have an elevated white blood cell count. Blood cultures are important to obtain, but they are positive in only 10–15% of patients. As stated previously, endometritis is a polymicrobial infection, and isolation of a single organism from the blood does not imply that other organisms are not involved in the pelvic infection. These results are helpful when they reveal an organism which is not being covered by the treatment regimen.

E. Management

With the administration of a single-agent antibiotic such as cephazolin at the time the umbilical cord is clamped, prevention of endometritis may be achieved in 50% of women undergoing cesarean section who are in labor or have ruptured membranes. More broad-spectrum antibiotics are no more effective at prevention. However, prophylaxis does not guarantee prevention.

TABLE 2
Commonly Used Antibiotics in Postpartum Endometritis

Combinations
Clindamycin–gentamicin
Clindamycin–aztreonam
Ticarcillin–claunlanic acid
Ampicillin–sulbactam
Cephalosporins
Cefoxitin
Cefotaxime
Cefoperazone
Ceftizoxime
Cefotetan
Extended-spectrum penicillins
Mezlocillin
Piperacillin

Two stages of intraperitoneal infection have been recognized which have guided the choice of antibiotics: (i) septicemia with gram-negative aerobes and (ii) an anaerobic (usually *Bacteroides* species)-dominated infection which may result in abscess formation. Initial treatment with antibiotics effective against both aerobes and anaerobes decreases the rate of failure to respond and, more important, markedly reduces the rate of abscess formation. Since the initial studies utilizing clindamycin–gentamicin for endometritis, many single agents (second- and third-generation cephalosporins, extended spectrum penicillins, and other combinations) have been developed with the appropriate coverage (Table 2). Due to the small difference in clinical outcomes, it is unlikely that any one antibiotic will ever be proven to be superior to the others.

During treatment for endometritis, the patient ideally should be reassessed daily. After 24 hr with no fever, 95% of patients will remain afebrile without any additional antibiotics. Approximately 95% of patients with endometritis will respond to one of these initial antibiotic regimen, but if the patient fails to respond within 48–72 hr, a very thorough reassessment is critical and should include a pelvic examination. Evaluation should strive to exclude such diagnoses as drug fever, superficial phlebitis, resistant bacteria, wound infection, hematoma, pelvic abscesses, or septic pelvic thrombophlebitis. It is not uncommon for abdominal wound infections to be initially diagnosed as endometritis because wound tenderness is very difficult to distinguish from uterine tenderness. Also, after 48 hr without improvement, a change in antibiotics is commonly indicated if there are no other findings at the time of reassessment. If the patient fails to respond after an additional 48 hr with the new antibiotics, the possibility of retained placental fragments, abscess, or septic pelvic thrombophlebitis should be highly considered. An ultrasound or CT scan of the pelvis can be very helpful to demonstrate a hematoma, abscess, septic pelvic thrombophlebitis, or even retained products. If retained placental products are recognized, curettage may be necessary. Rarely, drainage of an abscess or hematoma may be required either via needle aspiration in selected cases or, more commonly, transabdominally.

Septic pelvic thrombophlebitis is a rare complication of patients with endometritis. This condition is characterized by protracted, spiking fevers that persist despite appropriate antibiotic therapy. Interestingly, these patients may have little, if any, other physical findings and feel well other than the high-fever spikes. A CT scan may show clots within the ovarian venous system. Clots may also be present in the uterine venous system; however, these are extremely difficult to see even with the CT scan. In these patients, a therapeutic trial of intravenous heparin is indicated and defervescence confirms the diagnosis of septic pelvic thrombophlebitis. This treatment is usually continued for 7–10 days.

II. URINARY TRACT INFECTION

Due to the high rate of asymptomatic bacituria in women and the increased risk of urinary retention following delivery, urinary tract infections are common postpartum. The rate of postpartum urinary tract infections is also increased when the urinary tract has been catheterized in labor, at delivery, or in the postpartum state. Although catheterization increases the risk of the development of urinary tract infection postpartum, it may be necessary to decrease the risk of urinary retention. The most common organisms isolated in the postpartum period from patients who develop a urinary tract infection are *E. coli*, *Citrobacter* species, and *Enterobacter*, of which *E. coli* is the most common isolate. Common symptoms of a urinary tract infection are dysuria, frequency, and urgency and, if the infection ascends into the upper urinary tract, pyelonephritis with fever, flank pain, costovertebral angle tenderness, and even septic shock may occur. Pyuria, bactiuria, hematuria, and leukocytosis variably accompany these symptoms and cultures of the urine should be monitored for the presence of antibiotic-resistant organisms.

A variety of antibiotics are effective in the treatment of the puerperal patient with urinary tract infection. Oral macrodantin or a cephalosporin are useful for cystitis, but for pyelonephritis the first-generation parenteral cephalosporins are ideal. Severely ill patients may require amnioglycosides or antibiotic combination. Because there is a high prevalence of *E. coli* resistant to ampicillin and penicillin, it is best not to utilize these drugs as a first line of therapy. Women with pyelonephritis are also at risk for the development of septic shock and adult respiratory distress syndrome and so should be monitored for these complications.

III. MASTITIS

The development of mastitis may occur in any puerperal patient, especially those that are nursing their infants. Seven to 11% of nursing mothers develop inflammation and/or acute mastitis. Mastitis can be divided into two types: epidemic and endemic. The epidemic type occurs usually in conjunction with *Staphylococcus aureus* nursery epidemics. This type is characterized by infection of the lobular apparatus of the breast. Endemic (also termed nonepidemic) mastitis is an acute puerperal cellulitis that extends along the periglandular connective tissue and is often associated with a nipple fissure. Breast engorgement is a common, noninfectious complication distinguished from mastitis by bilateral, fullness, and tenderness. Mastitis is usually unilateral. Milk-stasis, characterized by engorgement, pain, incomplete emptying of the breast, edema, and erythema, may progress to acute mastitis with increased pain, chills, and fever. If severe mastitis develops, malaise and chills may also be noted. Common organisms are the staphylococci and the streptococci. One key to management is effective emptying of the breast. Other measures of benefit include moist heat and the use of antiinflammatory agents. In patients thought to have mastitis, antibiotics should be utilized promptly. Penicillin, ampicillin, oxacillin, dycloxicillin, erythromycin, and several of the cephalosporins have all been utilized. The important concern for most of these infections is to provide appropriate coverage for staphylococcus and streptococci and to encourage breast drainage.

About 5–10% of patients who develop mastitis will develop a breast abscess if untreated. If this occurs, antibiotic coverage for anaerobic organisms may be necessary. If the patient does not respond promptly, or if there is a fluctuant mass, drainage of the abscess should be considered.

See Also the Following Articles

Cesarean Section; Endometrium; Infections in Pregnancy; Puerperium

Bibliography

Creasy, R. K. (Ed.) (1997). *Management of Labor and Delivery*. Blackwell Science, Malden, MA.

Duff, P. (1993). Antibiotic selection for infections in obstetric patients. *Sem. Perinatol.* 17, 367–378.

Gibbs, R. S., and Sweet, R. L. (1995). *Infectious Diseases of the Female Genital Tract*. Williams & Wilkins, Baltimore.

Monga, M., and Oshiro, B. T. (1993). Puerperal infections. *Sem. Perinatol.* 17, 426–431.

Soper, D. E. (1993). Infections following cesarean section. *Curr. Opin. Obstet. Gynecol.* 5, 517–520.

Puerperium

Harish M. Sehdev

Cooper Hospital/University Medical Center

GLOSSARY

endometritis Postpartum uterine infection that involves the decidua and myometrium and can also involve the parametrium.

lochia Vaginal discharge in the initial postpartum period resulting from the sloughing of decidual tissue.

postpartum blues Transient state of irritation, anxiety, restlessness, and tearfulness that occurs in the first few days of the puerperium.

puerperal febrile morbidity Temperature >100.4°F (38.0°C) on two or more occasions (>6 hr apart) during the first 10 postpartum days, excluding the first 24 hr.

The first 6 postpartum weeks are considered as the puerperal period. It is that time during which the reproductive tract returns to its nonpregnant state and ovulation, in those women who are not breast-feeding, is reestablished.

Encyclopedia of Reproduction VOLUME 4

I. UTERINE INVOLUTION AND ENDOMETRIAL REGENERATION

A. Uterine Involution

After delivery of the fetus and placenta, the uterus undergoes tremendous change. Upon delivery, the contracted uterus weighs approximately 1 kg and is positioned just below the level of the umbilicus (the approximate size of a 20-week gestation). This myometrial contracture compresses the intrauterine vessels and facilitates hemostasis, and on histologic sectioning, the postpartum uterus will appear ischemic. By 2 or 3 weeks postpartum, the uterus has contracted into the true pelvis (the approximate size of a 12-week gestation) and weighs about 300 g. By the completion of the puerperal period, the uterus will return to its previous nonpregnant size. During this period, the lower uterine segment, which in late pregnancy is a well-defined structure, shrinks back into an almost unidentifiable portion of tissue that separates the uterine corpus from the cervix. This involution of the uterus is marked by a decrease in the connective tissue framework and by primarily a decrease in the size of the individual muscle cells with only a minimal decrease in the number of muscle cells.

B. Regeneration of the Endometrium

The basal decidua remains after separation and delivery of the placenta and fetal membranes. After delivery, this layer appears irregular and is infiltrated with blood. The basal decidua differentiates into two layers within 72 hr postpartum, and the superficial layer is extruded in the lochia as it becomes necrotic. Regeneration of the endometrium occurs from the basal layer which stays intact, and the regenerative process is complete within 3 weeks. The reparative process involves an infiltrate of inflammatory cells that begins immediately after delivery and persists for approximately 10 days, and it is proposed that this inflammatory response acts as an antimicrobial barrier. Following this initial response, plasma cells appear as the leukocyte response abates. This process may be histologically visible up to several months later, and, again, it is representative of the normal process of regeneration.

Regeneration of the placental site takes twice as long as it does for the rest of the uterine cavity. After delivery of the placenta, the placental site is about 9 cm across, approximately half its size prior to delivery. During the next 2 weeks, this area decreases by another 50%. Shortly after delivery, the vessels of the placental bed become thrombosed and hyalinized. Through extension of the endometrium from the margins of the placental site, and eventual downgrowth of tissue, the thrombosed and necrotic tissue marking the placental insertion site is exfoliated and is associated with a transient increase in uterine bleeding between Postpartum Days 7 and 14. This restorative process aids in preventing scar tissue formation.

II. PHYSIOLOGIC CHANGES

A. Urinary Tract

Due mostly to compression by adjacent vessels as well as the enlarging uterus, there is a progressive dilation of the maternal renal pelvises and ureters above the pelvic brim (right greater than left) throughout the course of pregnancy. These changes resolve, without damage, during the puerperium. In the initial postpartum period, cystometric evaluation reveals an increased bladder capacity, which decreases over the 6 weeks of the puerperium. As is the case during pregnancy, this continued increase in bladder capacity and the dilated collecting system predisposes women to an increased risk of urinary tract infections.

Less than 10% of women will develop stress incontinence after delivery. Development of stress incontinence is associated with obstetrical factors such as birth weight, infant head circumference, and length of second stage, whereas delivery by cesarean section seems to protect against the development of stress incontinence. The etiology for the incontinence seems to be impaired function of those muscles in and around the urethra, and most women with postpartum incontinence will have normalized function by 12 weeks postpartum. A small percentage of women will also develop transient bladder hypotonia. Factors related to the development of bladder hypotonia seem to be use of an epidural for anesthesia and prolonged labors.

During pregnancy, renal function changes dramatically. Both glomerular filtration and endogenous creatinine clearance increase by 50% early in pregnancy and remain elevated until delivery. Both of these parameters normalize by the eighth week postpartum. Renal plasma flow, which increases by 25% in the first half of pregnancy and then decreases in the last trimester, continues to decrease, for reasons unknown, to levels below that of prepregnancy until about 6 months postpartum, after which it finally normalizes by 1 year.

B. Cardiovascular and Hematologic

During pregnancy, plasma volume increases by approximately 50%. During delivery, plasma volume decreases by about 1 liter due to blood loss. During pregnancy, extravascular fluid volume has also increased dramatically, and during the initial 2 postpartum days, the decreased plasma volume due to blood loss is compensated for by shifting of the extracellular fluid into the vascular space. By about 7 days postpartum, the plasma volume closely approximates prepregnancy levels.

During pregnancy, cardiac output, heart rate, and stroke volume all increase. Immediately after delivery, and these parameters remain elevated. Cardiac output will remain elevated for over 2 days postpartum secondary to increased stroke volume from increased venous return. By 10–14 days postpartum, these parameters will return to prepregnant levels. In the first postpartum week, both systolic and diastolic blood pressure will show a temporary rise.

After delivery, there can be a significant leukocytosis that is predominantly granulocytic. During the first few days after delivery, there is increased fibrinolytic activity and increased levels of fibrinogen. These levels return to normal 7–14 days after delivery.

C. Endocrine

1. *Ovulation and Menstruation*

The return of menses postpartum is greatly influenced by breast-feeding. Regardless of lactation status, estrogen levels fall immediately postpartum. Estrogen levels begin to rise by approximately 14 days after delivery in women who are not lactating, whereas they remain low in those who are lactating. While estrogen levels are affected by lactation, follicle-stimulating hormone levels are identical in both sets of women. Therefore, it is believed that in the immediate puerperal period, ovulation is suppressed by elevated prolactin levels. It is not until the third postpartum week that prolactin levels fall into the normal range for women who are not lactating; in lactating women, prolactin levels do not fall until the sixth postpartum week.

In nonlactating women, ovulation has been demonstrated as early as 4 weeks postpartum with the mean time to ovulation being approximately 70 days. In 70% of nonlactating women, menses will resume by 12 weeks, and prior to discharge plans for contraception should be discussed and implemented. In lactating women, onset of ovulation is delayed as explained by the prolonged increase in prolactin levels. Delay of ovulation depends on the frequency and length of time that the mother lactates. In women who exclusively breast-feed, the risk of ovulation is as low as 3% at 1 month and increases to 60% at 1 year. Even if a women plans to breast-feed, contraception should be discussed and instituted. Use of low-dose combination oral contraceptive pills (begun 1 month postpartum) does lower the fat, protein, phosphorus, and calcium concentration of breast milk and causes a small decrease in lactation performance and infant weight gain verses controls. In lactating women the use of progestin-only pills and levonorgestrel subdermal implants (after the fourth week) have no effect on infant growth and lactation.

2. *Thyroid Function*

During pregnancy, total thyroid hormone (thyroxin and triiodothyronine) levels increase. However, there is also an increase in thyroid-binding globulin that is induced by the increased estrogen levels of pregnancy; therefore, free T_4 levels remain unchanged and the pregnant woman remains euthyroid. Also, during pregnancy, thyroid gland volume increases by about one-third. By 4 weeks postpartum, thyroid hormone levels will return to prepregnancy levels, whereas it will take up to 3 months for the thyroid gland itself to return to its prepregnant size.

In the postpartum period, it has been shown that women are at increased risk for developing autoimmune thyroiditis; thus, those with prolonged complaints of lethargy and fatigue should be evaluated for thyroid dysfunction. During pregnancy, those with subclinical autoimmune thyroid dysfunction may improve secondary to the suppressed immune function (both cell mediated and humoral) associated with pregnancy. Autoimmune thyroid dysfunction may flare postpartum when immune suppression is relieved and it is usually associated with an initial period of hyperthyroidism. This process is mediated primarily by antibodies that cause destruction of thyroid tissue (as proven by histologic evaluation) and not by stimulation as in Graves' disease. Patients usually present with symptoms of palpitations and fatigue, and up to one-third of these patients will ultimately develop hypothyroidism, whereas the majority become euthyroid over the ensuing months. Of those that do become hypothyroid, the majority will eventually return to a euthyroid state, and postpartum thyroiditis may recur in subsequent pregnancies.

III. MATERNAL CARE

Immediately postpartum, the mother is at risk for significant bleeding, and care should be directed toward observation for excessive bleeding and uterine atony. After delivery of the placenta, contraction of the uterine musculature prevents excessive bleeding. Upon external palpation, the uterus should be firm and palpable at about the level of the umbilicus. If the uterus fails to contract well, excessive amounts of blood can accumulate within the uterus without initial evidence of external bleeding. In this situation, the uterus will be soft and enlarged on external palpation. Contraction of the uterus can be affected with external massage, intravenous oxytocin, and, if needed, intramuscular prostaglandin and methergine.

Early ambulation should be encouraged after the effects of regional anesthesia have worn off. Early ambulation has been shown to decrease the risk of postpartum venous thrombosis and pulmonary embolism. Early ambulation and feeding can also help reduce the risk of postpartum constipation.

During the initial postpartum period, special attention must be directed toward bladder function. Care must be directed toward avoiding overdistension of the maternal bladder, and several factors are involved in the risk of overdistension. It is common to give intravenous fluids postpartum, and, when coupled with the release of the antidiuretic effect of oxytocin when it is discontinued, the bladder can rapidly fill. Regional anesthesia, lacerations, and unevacuated vaginal hematomas can all impair emptying of the bladder and bladder sensation. Sensation and voiding are further impaired with overdistension of the bladder, and this overdistension increases the risk for urinary tract infections. If the new mother has not voided within 6 hr and cannot void on her own, the bladder should be drained by catheterization.

Episiotomies and vaginal lacerations are common in women undergoing vaginal deliveries. For first- or second-degree lacerations, pain control can be obtained with ibuprofen (which will also aid in decreasing pain from uterine cramping) and perineal care can be accomplished with simple cleansing during a shower or bath. Patients with third- and fourth-degree lacerations or mediolateral episiotomies usually experience more pain and are at greater risk for hematomas and perineal infections. Pain relief and cleansing can be aided by use of cold sitz baths. Cold sitz baths can decrease muscle spasm, reduce edema, decrease hematoma formation (through vasoconstriction), and can help in pain relief through a decrease in nerve conduction. Prolapsed hemorrhoids can also be the cause of significant pain postpartum, and excision can provide immediate pain relief.

Other considerations of immediate maternal care include diet and immunizations. If a women is not lactating, then her dietary requirements are those of nonpregnant women. If a woman is breast-feeding, her diet should include increased calories and protein. It is recommended to continue iron supplementation for several months postpartum. Prior to discharge, women who are not immune to rubella should be immunized, and women who are D-negative and not isoimmunized should be given anti-D globulin if they deliver a D-positive baby.

IV. CLINICAL ASPECTS OF MATERNAL RECOVERY

Postpartum, women will experience lochia for 3–8 weeks. Lochia consists of decidual tissue, epithelial cells, red blood cells, and microorganisms. During the initial 3 days postpartum, the lochia is primarily red in color as a result of blood (and should not be more than a normal period in quantity). For the next week, the lochia will appear paler in color, with the exception of when the placental eschar sloughs and transient heavy vaginal bleeding occurs (on approximately Days 7–14). The fluid content of lochia decreases after 1.5 weeks postpartum and contains an increased concentration of leukocytes. At this time, lochia will be white or yellowish in color. Tampons can be used if there are no significant lacerations and if they are changed frequently. Evaluation for infection should be sought if the lochia becomes foul smelling.

During the first week postpartum, women will experience significant physiological diuresis. During pregnancy, extracellular water is significantly increased along with plasma volume and serum osmolality decreases. The elevated estrogen levels of pregnancy promote fluid retention, and as estrogen levels fall, the stimuli to retain fluid is lessened. The gravid uterus also increases venous pressure in the lower extremities, and as the "obstruction" is relieved with delivery and uterine contraction, the increased venous pressure of the lower extremities is relieved allowing hydrostatic forces to cause extracellular water to shift into the intravascular compartment.

Upon delivery of the amniotic fluid, baby, placenta, and blood, there is an immediate loss of approximately 10–12 pounds. During the physiologic diuresis of the first 7 days, another 4–6 pounds are lost. By the end of the puerperium, about 30% of women will return to their prepregnancy weight, with the majority of women attaining their prepregnancy weight by 6 months postpartum. Breast-feeding and age do not affect weight loss, whereas early return to work, smoking, and primiparity have been found to be associated with increased weight loss during the puerperium. Those women who gain over 20 pounds during their pregnancy are more likely to retain weight postpartum.

If the delivery has been uncomplicated, physical activity beyond ambulation can be initiated immediately and increased as tolerated. Sexual intercourse should be refrained from for the initial 2 or 3 weeks postpartum secondary to bleeding and risk of infection. After this time, intercourse can be resumed depending on perineal discomfort and desire. Approximately one-third of women resume intercourse by the end of the puerperium. Women who breastfeed may experience longer periods of discomfort with intercourse due to vaginal dryness and atrophy associated with estrogen suppression.

The enlarging gravid uterus places constant pressure on the abdominal wall, and this causes changes in the elastic fibers of the skin. Postpartum, the abdominal wall will be soft and "loose." Return of the abdominal wall to its prepregnant appearance is usual and takes several weeks. Exercise decreases the "recovery" time for the abdominal wall to regain its prepregnant appearance.

V. POSTPARTUM DEPRESSION

In the early puerperal period, up to 70% of mothers may experience a mild and transient depression that is termed the "postpartum blues." This psychological state includes a wide range of symptoms, e.g., mood lability, irritability, restlessness, lethargy, confusion, forgetfulness, insomnia, weeping, and even negative feelings toward the newborn. While this state may last up to 10 days, it usually resolves within 3 days, and, therefore, therapy other than support is not indicated. This mild depression shows no correlation with hormonal changes, social class, economic status, or personality types. Evidence suggests that neurotransmitters (lower levels of tryptophan) and changes in the hypothalamic–pituitary adrenal and thyroid axises may play a role in the development of postpartum blues, postpartum depression, and nonpuerperal major depression.

If signs and symptoms of depression persist beyond 10 days, then evaluation should be sought, including evaluation of thyroid function because puerperal hypothyroidism can present with similar symptoms. Approximately 8–15% of women will have true postpartum depression, and its signs and symptoms are similar to those in nonpregnant women with depression. Several studies have shown that a significant percentage of patients developing postpartum depression had previous episodes of depression or other trying life experiences and situations unrelated to their pregnancies that played a role in developing depression. Those women with spousal conflicts or undesired pregnancies seem to be at greater risk for developing more severe depression. If emotional support and attention are not helpful in alleviating symptoms, then tricyclic antidepressants or serotonin uptake inhibitors (which have less side effects than the tricyclic antidepressants) have been shown to be effective. Symptoms of postpartum depression may last for 1 year, and prognosis for recovery is usually good. Those women who have experienced postpartum depression have a 50% chance of recurrence of postpartum depression in subsequent pregnancies.

The incidence of puerperal psychosis is approximately 0.14–0.26%. The diagnosis can be made if the patient expresses suicidal thoughts or actions, and generally the signs and symptoms are similar to those of people who experience nonpuerperal psychosis. Patients with puerperal psychosis present with signs of manic depression along with disorientation and confusion. Patients with puerperal psychosis usually have a more favorable prognosis than those with nonpuerperal psychosis, with a usual duration of illness lasting up to 12 weeks. Treatment involves hospitalization, and therapeutic interventions (including lithium, neuroleptics, tricyclic antidepressants, and electroconvulsive therapy) have not been shown to enhance recovery.

VI. PUERPERAL INFECTION

In the past, puerperal febrile morbidity accounted for a significant number of maternal deaths. As aseptic techniques and practice became standard and antibiotics have become available, the number of maternal deaths attributable to infectious etiologies has significantly decreased. Infection now accounts for approximately 5% of all maternal deaths and <1 death per 100,000 live borns.

A. Endometritis

Endometritis is a major cause of febrile morbidity. It complicates 1–3% of vaginal deliveries and 5–15% of deliveries by cesarean section. Risk factors associated with endometritis include young age, prolonged rupture of membranes, multiple internal exams and the use of internal monitors, chorioamnionitis, lower socioeconomic status, cesarean section, and lower genital tract infection or colonization by group B streptococci, gonorrhea, and bacterial vaginosis. Endometritis is an ascending, polymicrobial infection caused by organisms that reside in the bowel and that also colonize the lower genital tract. Group B streptococci, anaerobic streptococci, and both aerobic gram-negative bacilli (*Escherichia coli, Klebsiella,* and *Proteus* species) and anaerobic gram-negative bacilli (namely, *Bacteroides— fragilis* and *disiens*) are the most common pathogens of endometritis. *Chlamydia trachomatis* is implicated in late endometrial infections and is not commonly associated with early onset endometritis. Prophylactic antibiotic use with first-generation cephalosporins in women undergoing cesarean sections (especially in those with prolonged labor courses or extended time with rupture of membranes) has been clearly shown to reduce the frequency of endometritis.

Findings in patients with endometritis include temperature elevation (>100.4°F), abdominal pain, uterine tenderness, tachycardia, chills, and a foul-smelling discharge (either scant or profuse). White blood cell count will range between 15 and 30,000 cells/ml. Prior to initiating antibiotic therapy, it is advocated by many to obtain aerobic and anaerobic blood cultures, but the incidence of positive blood cultures is only 10–25%.

Treatment of endometritis following vaginal delivery can be accomplished with single agent therapy utilizing penicillins or cephalosporins. Those with moderate to severe infections (including those delivered by cesarean section) require combination therapy that is broad spectrum and includes two or three agents (gentamicin and clindamycin, clindamycin and aztreonam, or a penicillin with gentamicin and clindamycin or metronidazole). Upon initiation of antibiotic therapy, 90% of patients will improve within 72 hr. When a patient has been afebrile for more than 24 hr, antibiotic therapy can be withdrawn (unless the patient has a staphylococcal bacteremia requiring an extended course of intravenous and oral antibiotics).

B. Complications of Endometritis

Failure to respond appropriately to antibiotic therapy includes resistant organisms and several more serious complications of endometritis. In patients with endometritis, about 3–5% will develop a wound infection, and if diagnosed clinically (incisional erythema, tenderness, or induration), the wound should be opened. The opened wound should be drained and inspected to ensure integrity of the fascial closure. The wound should be irrigated up to three times a day, and antibiotic therapy should include coverage against staphylococci until the patient responds and evidence of cellulitis has resolved. Rarely, a wound infection is complicated by necrotizing fasciitis. If necrotizing fasciitis is diagnosed, complete wound debridement (which can be extensive) and treatment with broad-spectrum antibiotics is mandated.

Up to 1% of patients with endometritis may develop a pelvic abscess. Abscesses can develop in the anterior or posterior cul de sac and in the broad ligament, and they usually involve *Bacteroides*. Despite antibiotic therapy, patients continue to experience temperature elevations, lower abdominal tenderness, and may have a palpable mass adjacent to the uterus on bimanual examination. Diagnostic radiographic studies include ultrasound and computerized tomography (CT scan) or magnetic resonance imaging (MRI). While ultrasound examination is cheaper, the other two modalities have a higher diagnostic sensitivity. Treatment involves drainage and continued antibiotic coverage against anaerobic and coliform pathogens.

Another rare complication of endometritis (<1%) is septic pelvic vein thrombosis (SPVT). One or both ovarian veins (usually the right vein) are involved as a result of pathogenic drainage of the placental veins (most commonly into the veins of the upper portion of the uterus) which also causes injury to the ovarian vascular endothelium and initiates thrombosis. In most cases, patients with SPVT look well but continue to have high temperatures and may have flank pain and gastrointestinal symptoms. CT scan and MRI examination are the diagnostic tests with greatest sensitivity, and treatment involves therapeutic heparinization for 10 days with broad-spectrum antibiotics. Patients should respond to therapy within 72 hr, and pulmonary embolism is extremely rare.

C. Other Causes of Puerpera Febrile Morbidity

While endometritis is a common cause of puerperal febrile morbidity, other causes of infection and temperature elevation should be considered in the postpartum patient. Within the first 24 hr postpartum, and especially if delivered by cesarean section, puerperal fevers and infection could be due to respiratory conditions that include atelectasis and pneumonia (aspiration or bacterial). Pyelonephritis and urinary tract infections are also common and can be difficult to diagnose in the early puerperal period, especially after abdominal delivery. Postpartum fevers can arise from breast engorgement (up to 15% of all postpartum patients) and bacterial mastitis. Also, febrile morbidity in the postpartum patient can be the result of thrombophlebitis of the deep or superficial veins.

See Also the Following Articles

Decidua; Endometrium; Postpartum Depression; Puerperal Infections

Bibliography

Bonnar, J., Franklin, M., Nott P. N., *et al.* (1975). Effect of breast-feeding on pituitary–ovarian function after childbirth. *Br. Med. J.* 4, 82.

Dinsmoor, M. J., Newton, E. R., and Gibbs, R. S. (1991). A randomized, double-blind, placebo-controlled trial of oral antibiotic therapy following intravenous antibiotic therapy for postpartum endometritis. *Obstet. Gynecol.* 77, 60.

Duff, P. (1985). Pathophysiology and management of postcesaren endometritis. *Obstet. Gynecol.* 67, 26.

Duff, P. (1987). Prophylactic antibiotics for cesarean delivery: A simple cost-effective strategy for prevention of postoperative morbidity. *Am. J. Obstet. Gynecol.* 157, 794.

Duff, P. (1993). Septic pelvic vein thrombophlebitis. In *Obstetric and Perinatal Infections* (D. Charles, Ed.), pp. 104–108. Mosby/Year Book, St. Louis, MO.

Jausson, R., Dahlberg, P. A., Winsa, B., *et al.* (1987). The postpartum period constitutes an important risk for the development of clinical Graves' disease in young women. *Acta Endocinol.* **116**, 321.

Lazarus, J. H., and Othman, S. (1991). Review: Thyroid disease in relation to pregnancy. *Clin. Endocrinol.* **34**, 91.

Monheit, A. G., Cousins, L., and Resnik, R. (1980). The puerperium: Anatomic and physiologic readjustments. *Clin. Obstet. Gynecol.* **23**, 973.

Scholl, T. O., Hediger, M. L., Schall, J. I., *et al.* (1995). Gestational weight gain, pregnancy outcome, and postpartum weight retention. *Obstet. Gynecol.* **86**, 425.

Stein, G. (1982). The maternity blues. In *Motherhood and Mental Illness* (I. F. Brockington and R. Kumar, Eds.). Academic Press, London.

Stowe, Z. N., and Nemeroff, C. B. (1995). Women at risk for postpartum-onset major depression. *Am. J. Obstet. Gynecol.* **173**, 639.

Viktrup, L., Lose, G., Rolff, M., *et al.* (1992). The symptoms of stress incontinence caused by pregnancy or delivery in the primipara. *Obstet. Gynecol.* **79**, 945.

Rabbits

Josephine B. Miller
University of Illinois at Chicago

GLOSSARY

estradiol The "ultimate" luteotropic hormone in rabbits which supports progesterone synthesis by the corpus luteum.

induced ovulation Ovulation of follicles in response to a coital-induced surge of luteinizing hormone or by the injection of luteinizing hormone, human chorionic gonadotropin, or gonadotropin-releasing hormone.

pseudopregnancy A reproductive state characterized by prolonged maintenance of the corpora lutea and secretion of progesterone in the absence of a conceptus. Hormone profiles parallel those seen in the first half of pregnancy. Pseudopregnancy occurs in rabbits in response to an infertile mating or from the injection of luteinizing hormone/human chorionic gonadotropin.

uteroglobin The major protein in the uterine fluid at the time of implantation.

Rabbits are an important species for reproductive research because they are prolific and adapt to a variety of breeding environments. They do not exhibit an estrous cycle; rather, the mature doe has follicles within the ovary at all times which will ovulate in response to the stimulus of mating or an injection of human chorionic gonadotropin, luteinizing hormone (LH), or gonadotropin-releasing hormone. Thus, as an induced ovulator, this species has been widely used to study mechanisms involved in the process of LH surge generation, ovulation, implantation, and pregnancy maintenance and parturition since these events can be precisely timed with respect to the initiating signal, i.e., a surge of LH or a coital stimulus.

I. GENERAL CHARACTERISTICS

A. Taxonomy

The order Lagomorpha consist of two families, the Ochotonidae (pikas) and Leporidae (rabbits and hares). Both families have many native genera and species throughout the world since rabbits were bred by the Romans beginning in the first century BCE and carried in and released from ships along trading routes to provide a meat source. There are three common genera: Lepus, which includes hares and jackrabbits; Sylvilagus, which includes the cotton-tailed rabbit common in the United States; and Oryctolagus, which includes the European rabbit (*Oryctolagus cuniculus*) common to North America in both the domestic and the feral form (a revision of the domestic form to the wild form). Lagomorpha differ from rodents in that they have four rather than two upper incisors. The additional pair of incisors, called peg teeth, are small, with rounded edges, and can be seen behind the large incisors in the upper jaw.

B. Physiology

The relative ease in breeding has produced over 50 well-established breeds providing the basis for studying the genetics of hair color, morphology, size inheritance, blood typing and components, and various pathologies. While over 20 inbred strains are described in the literature, only a limited number are currently being maintained in active breeding colonies. The Watanabe rabbit, which has a point mutation in the low-density lipoprotein receptor resulting in hyperlipemia, has been used for the study of atherogenesis and various transgenic rabbits have been produced. However, the New Zealand White and the Dutch Belted are the two breeds most commonly used in research. The lack of pigment and large ears in the New Zealand White breed makes it especially easy to obtain blood samples.

Adult rabbits weigh 2–6 kg and have a life expectancy of 5 or 6 years. There are 44 pairs of chromosomes with a standardized karotype. Their skeleton is frail, comprising only 8% of their weight, and the thoracic cavity is relatively small compared to its large abdominal cavity. The teeth grow continuously (10–12 cm/year); thus, malocclusion can lead to overgrowth. Rabbits are herbivores and can be extremely sensitive to changes in diet and environmental factors. Their very large cecum comprises 10–15% of the total body weight. In addition, they are coprophagic and will ingest large amounts of protein and vitamin-rich feces ("night stool") directly from their anus. As a result, the stomach of a healthy rabbit is never empty. They produce bile at about six times the rate of dogs and cats and many have an atropinase enzyme in their blood which renders atropine inactive.

C. Reproductive Behavior

The rabbit is both prolific and adaptable to a variety of breeding environments; however, temperature can influence reproductive behavior and fecundity and should range between 55 and 70°C. Light conditions have little effect on reproductive efficiency.

Courtship behavior in caged male rabbits involves tail flagging, urination, and licking the genitalia. For breeding, a doe should be taken to a buck's cage since bucks mark their cage with secretions from the chin gland. If she is receptive, she will quickly raise her hindquarters to allow copulation. Intromission is usually accomplished after several rapid copulatory movements and ejaculation follows on the first intromission. Following ejaculation, the buck falls off backwards or sideways since, at the time of ejaculation, both hindfeet are off the ground and the buck may also emit a characteristic cry. A gelatinous vaginal plug may form, but this is usually lost shortly after copulation. If copulation does not take place within a few minutes the doe should be removed from the cage.

Female rabbits can be bred at 5 or 6 months of age with a breeding life of about 2 years. Although domestic rabbits do not show a true estrous cycle, they do show some variation in behavioral receptivity probably due to waves of follicular development. Vaginal cytology does not change reliably with receptivity; however, the vulva, which is pale pink or white and dry during nonreceptive times, becomes a darker pink or red and swollen and moist when receptive. Toward the end of the receptive period, the vulva is dark red to purplish. If the doe is bred during the receptive period, the vulva once again becomes light pink throughout gestation.

Ovulation occurs spontaneously approximately 10 or 11 hr after copulation or after an injection of luteinizing hormone (LH) or human chorionic gonadotropin (hCG). Pregnancy can be detected by palpitation after about 12 days of pregnancy. The gestation period is 31 or 32 days, with larger litter size (10–12) having a slightly shorter gestation than smaller litters (<4). Parturition usually occurs during the morning and kindling usually takes <30 min. Several days before parturition, the doe collects and builds a nest with materials such as hay, straw, excelsior, and plucked body hair. A nest is built under conditions of pregnancy, pseudopregnancy, or hormonal manipulations. As the young are born, the doe eats the placental membranes and severs the umbilical cord. Does do not usually retrieve pups which have strayed from the nest. Lactation is associated with lack of sexual receptivity. Nursing lasts only a few minutes and occurs only once a day, generally during the early morning. Milk production is maximal at 3 weeks, then gradually falls off and contains 12.3% protein, 13.1% fat, 1.9% lactose, and 2.3% minerals. The best milk substitute is synthetic kitten milk. Animals weaned at 8 weeks weigh 1600–1800 g compared to birth weights of about 50–60 g. Does are usually bred when the pups have been weaned (4–7 weeks); however, if the litter is removed immediately following parturition, the does will breed for up to 36 days following delivery.

Rabbits may also be bred using artificial insemination. Semen can be collected using an artificial vagina in conjunction with a teaser doe or with a rabbit skin worn over the arm. Insemination can be accomplished with semen (20–50 $\times$ 10^6 spermatozoa/0.3–0.7 ml) deposited near the cervix using an insemination pipette, directly into the uterus or oviduct, or injected intraperitoneally through a 20-gauge needle 12–16 hr before ovulation. If artificial insemination is used, does must be stimulated to ovulate either by mating to a vasectomized male or with hormone injection. Antibodies can be produced against the ovulation-inducing hormones, preventing subsequent inductions.

II. FEMALE REPRODUCTION

A. Regulation of Gonadotropin-Releasing Hormone Secretion

Estrus rabbits exhibit low pulses of gonadotropin-releasing hormone (GnRH), LH, and estrogen. After coitus, GnRH levels rise 40-fold within 30–60 min which is paralleled by increases in LH. Increases in follicle-stimulating hormone (FSH) and prolactin follow. Thus, the initial signal for the GnRH/LH surge is increasing electrical activity in the arcuate, premammillary, and posterolateral hypothalamic neurons. GnRH neurons have extensive interneuronal contacts which are probably crucial in initiating the LH surge. Several neural peptides and transmitters have been implicated in triggering the LH surge, including adrenergic (norepinephine, NE) and cholinergic transmitters as well as neuropeptide Y (NPY) and endorphins. Like NE, the effect of NPY depends on estrogen. NPY stimulates GnRH release in the presence of estrogen but inhibits GnRH release in ovariectomized animals. This effect of NPY can be blocked by 1-adrenergic receptor antagonist, suggesting that NE is likely the upstream stimulator of GnRH release. This increase in NE is associated with an increase in tyrosine hydroxylase activity within 15 min after coitus. NE is probably also involved in regulating ultradian GnRH. The events triggering the LH surge are rapid since catecholamine-blocking drugs can prevent ovulation if given within 1 min of coitus and apparently are widely distributed since a 360° deafferentation is required to prevent ovulation. GnRH synthesis and release is the final signal from the median eminence to the pituitary; however, steroids regulate the pituitary response since estrogen upregulates and progesterone downregulates pituitary GnRH receptors.

B. Ovarian Function

1. *Follicular Development and Ovulation*

The ovary is approximately 1.5 $\times$ 0.5 cm and is surrounded by a bursa which communicates freely with the peritoneal cavity. Rabbits do not exhibit a true estrous cycle and rabbit ovaries contain various sizes of follicles throughout their reproductive life. Variations in receptivity suggest that there are waves of follicular development approximately every 12–14 days, resulting in waves in estrogen production. Granulosa cells of the preovulatory follicle are one to four layers thick and tend to be columnar in the periphery and cuboidal within the follicle. The peripheral cell nuclei are toward the basal region and contain irregular cytoplasmic projections which extend into the antrum. Gap junctions form in the follicle during the antral phase of development. Preovulatory follicles synthesize predominately estradiol and testosterone. Follicular development and estrogen production appear to be delayed in pregnancy and pseudopregnancy and in follicles adjacent to corpora lutea, suggesting that progesterone itself may regulate estrogen production. This is consistent with the presence of progesterone receptors in granulosa cells of preovulatory follicles. Ovulation, occurring 10 or 11 hr after an LH surge, is normally associated not only with follicular rupture but also with ovum maturation and the secretion of estradiol, testosterone, and 17-OH progesterone. Although the precise mechanisms involved in this ovulatory event have not been worked out, essential components appear to be (i) an increase in cAMP and the activation of protein kinase A; (ii) the synthesis and activation of collagenase; and (iii) the synthesis, after approximately 4 or 5 hr, of both prostaglandins E_2 and $F_{2\alpha}$. Steroids, especially progesterone, may also be involved in the process of follicular rupture since steroid synthesis inhibitors can prevent the induction of ovulation by hCG. Similar results have been found with agents perturbing protein kinase C (PKC) generation. Other agents can be shown to influence ovulation using the *in vitro* perfused ovary as a model. These include interleukin-1, angiotensin II (possibly by stimulating prostaglandin production), glandular kalikreins, plasminogen activator (possibly activating procollagenase), bradykinin, and histamine. While these agents will cause ovulation *in vitro*, they are not able to induce ovum maturation or steroid production.

2. *The Corpus Luteum*

i. Luteinization and Corpus Luteum Formation After ovulation, luteininzing granulosa and thecal cells increase in diameter and accumulate both lipid droplets and aggregations of intracellular membranes. Luteal cells are linked by gap junctions which form as septate-like contact zones between adjacent cells and between villi and plasma membrane of the same cell. Between Days 2 and 9, luteinizing cells increase in diameter from approximately 15 to 32 mm. On Day 9, smaller cells are also evident (15 mm); however, it is not clear whether these small and large cells represent fundamentally distinct cell types. During the luteinization process, steroid production shifts from that of estradiol and testosterone to the singular production of progesterone. Estradiol is not synthesized by rabbit corpora lutea, whereas the interstitial portion of the ovary secretes large quantities of 20α-dihydroprogesterone in response to LH. Luteinization is accompanied by rapid angiogenesis and, in general, serum progesterone levels parallel luteal blood flow which in turn reflects mean arterial pressure. There does not appear to be hormonal regulation of luteal blood flow nor is there significant variation in the metabolic clearance rate of progesterone throughout pregnancy or pseudopregnancy.

ii. Hormonal Regulation of Progesterone Secretion by the Corpus Luteum Rabbit corpora lutea contain receptors for prolactin, estradiol, and LH. Prolactin increases lipid accumulation in the luteal cell; however, it is not required for progesterone secretion. Estradiol, rather than LH, is the well-established luteotropic hormone in the rabbit since (i) progesterone secretion is maintained in hypophysectomized rabbits supplied with exogenous estradiol, (ii) progesterone synthesis is inhibited by estrogen antagonists, and (iii) LH cannot sustain progesterone secretion if ovarian follicles have been removed by irradiation. Thus, estradiol has been called the ultimate luteotropin in the rabbit. This estrogen does not need to act locally since progesterone secretion by corpora lutea transplanted beneath the kidney capsule can be supported by serum levels of estradiol supplied exogenously or by estrogen secreted from the follicular portion of the ovary. The rabbit corpus luteum also has a well-characterized adenylyl cyclase which is responsive to both LH and catecholamines. Catecholamines probably are not important physiologically since denervated ovaries and corpora lutea removed to ectopic sites produce the usual patterns of progesterone. *In vivo*, high doses of hCG/LH from an injection or from a mating stimulus result in luteolysis due to the ovulation of estrogen-producing follicles. Treatment with exogenous estradiol prevents this decrease in progesterone, suggesting that corpora lutea are not regulated by LH in the presence of estradiol. *In vitro*, LH can increase progesterone output from dissociated luteal tissue; however, corpora lutea which are not exposed to dissociation show less response. It has been proposed that estrogen maintains the viability of luteal cells which have the intrinsic capacity to secrete progesterone autonomously without stimulation by LH. Thus, the functional role of the adenyl cyclase system in the rabbit remains elusive.

iii. Mechanism of Estrogen Action on the Corpus Luteum The mechanism by which estrogens regulate steroidogenesis in the corpus luteum has been the subject of investigation for many years. It seems clear that estrogen acts via the classical estrogen–receptor complex and the presence and characteristics of the estrogen receptor have been well documented by autoradiography, immunocytochemistry, and biochemical analysis. Biochemical and morphometric evidence indicate that estradiol has little effect on the activity or synthesis of key steroidogenic enzymes; rather, estradiol appears to regulate the mobilization of cholesterol from cellular storage sites and its conversion to pregnenolone. Thus, estrogen treatment and progesterone secretion are not correlated with lipoprotein utilization, 3-hydroxy-3-methylglutaryl-CoA activity, the levels of cytochrome P450 cholesterol side chain cleavage enzyme (P450scc), acyl CoA:cholesterol acyltransferase, or cholesterol ester hydrolase activity. Recent studies suggest that the mitochondrial protein, steroidogenic acute regulatory (StAR) protein, may mediate the luteotropic effects of estrogen to promote the synthesis of progesterone. Estrogen also synergizes with insulin-like growth factor-I to stimulate progesterone secretion and activates the δ; isoform of PKC.

iv. Luteal Regression Progesterone secretion peaks around Days 10–12 following ovulation and if does are not pregnant, corpora lutea regress by about Day 16. Functional changes (i.e., loss of progesterone secretion) precede visible changes in morphology, although macrophages can be seen to invade the corpus luteum as regression proceeds. Using an experimental model in which pseudopregnant rabbits are treated with estradiol-containing silastic implants, the withdrawal of such implants results in a dramatic drop in progesterone output, an invasion of macrophages, an increase in the surface area of lipid droplets containing cholesterol ester, and a concomitant reduction in the surface area of the inner mitochondrial membrane. These changes can be reversed with estrogen replacement which results in restimulation of progesterone secretion and synthesis of the StAR protein, despite the presence of macrophages. This suggests that the increase in luteal macrophages is a consequence of rather than an initiator of luteal regression and is consistent with the hypothesis that estrogen regulates progesterone synthesis by controlling the transport/translocation of cholesterol from lipid droplets or from the cytosolic compartments to cytochrome P450scc located on the inner mitochondrial membrane. The precise signal through which luteal regression is initiated in the rabbit remains unknown. Receptors for estradiol parallel the pattern of progesterone secretion and there is no apparent loss of estrogen receptors preceding the fall in progesterone.

Luteolytic substances appear to be produced by the rabbit endometrium and hysterectomy will prolong progesterone secretion during pseudopregnancy for a few days. In addition, estrogen will prolong progesterone secretion in hysterectomized but not uterine-intact rabbits. This suggest that the uterine factor alters luteal responsiveness to estradiol. The production of a uterine luteolysin may be regulated by progesterone since luteal regression occurs rapidly and permanently in animals treated with specific progesterone antagonists on Days 5–7 (the time of implantation) which also causes endometrial regression. This luteolytic effect is not seen in rabbits hysterectomized on Day 1 of pseudopregnancy and can be prevented by coadministration of exogenous estradiol. This suggests that progesterone inhibits the production of some uterine factor which may interfere with the action of estradiol and demonstrates the tight coupling of the uterus and the ovary in regulating progesterone secretion. As the corpus luteum begins to regress, such a putative luteolytic substance may be produced. Prostaglandin $F_{2\alpha}$, known to be the uterine luteolysin in rodents, ruminants, horse, pig, and most other species, is secreted by the uterus and corpus luteum of the rabbit. In addition, progesterone secretion is reduced through Day 12 of pseudopregnancy by an injection of $PGF_{2\alpha}$. This effect of $PGF_{2\alpha}$ can be prevented with exogenous estradiol or the presence of a conceptus. After Day 15 of pregnancy, the corpus luteum becomes insensitive to the luteolytic effects of exogenous $PGF_{2\alpha}$. During pseudopregnancy, the increase in $PGF_{2\alpha}$ synthesis and secretion by both the uterus and the corpus luteum occurs only after progesterone secretion has begun to fall (i.e., after Day 12). This suggests that while $PGF_{2\alpha}$ is a potent uterine luteolytic agent, it is probably not responsible for the initiation of luteal regression, but may play a role in the final stages of luteal regression.

v. Maternal Recognition of Pregnancy The mechanism by which progesterone is maintained during pregnancy is not completely established; however, there is evidence that the conceptus exerts both a luteotropic role to support luteal progesterone secretion and an "antiluteolytic" role to prevent or counteract inhibitory influences ($PGF_{2\alpha}$) secreted by the uterine endometrium. The luteotropic substance secreted by the placenta has not yet been isolated. Partial purification studies suggest that it is an acidic protein, sensitive to heat and trypsin, with a molecular weight ≧12,000–14,000 kDa. It is probably not similar to hCG but may have GnRH-like activity in that partially purified fractions stimulates LH secretion from perfused pituitaries. Results suggest that the placental luteotropin does not maintain progesterone secretion independently; instead, estradiol, along with the placental factor, is required to support pregnancy. In addition, the placental luteotropin appears to act directly on the ovary since (i) removal of the conceptus on Days 12 or 18 results in rapid luteal regression without a change in serum LH levels, (ii) the placenta is required in both intact and hypophysectomized estrogen-treated rabbits to sustain progesterone production throughout pregnancy, and (iii) both placental extracts and estradiol are required to stimulate progesterone production by luteal cells *in vitro*.

C. Uterine Function

1. *Uterine Modifications after Ovulation*

During the first 4 days following ovulation, the endometrium proliferates and differentiates in preparation for implantation which occurs on Day 7. Differentiation results in both structural changes, including changes in intermediate filaments, cell–cell junctions, and secretory products, and stage-specific changes in the uterine epithelial surface proteins/glycoproteins. Two such epithelial glycoproteins are a 42-kDa protein (related to haptoglobin) and CD44. Both are expressed just prior to implantation although their potential roles in the implantation process are not known. Ovarian steroids also enhance the secretion of various proteins into both the oviductal and the uterine fluids. As the embryo passes through the isthmus of the oviduct, a mucin layer is added to the embryo. The thickness of this layer correlates with the ability of transplanted embryos to implant in the uterus. During the peri-implantation period and just prior to trophoblast attachment, the uterine fluid contains large quantities of a carbohydrate-rich protein fraction termed β-glycoprotein, which is secreted by the ciliated epithelial cells. The major protein found in the uterine fluid at this time, however, is uteroglobin.

2. *Uteroglobin*

Uteroglobin (UG), also called blastokinin, is not specific to the uterus but has been identified in the oviduct, the male genital tract, the tracheobronchial tree, and the digestive tract of both males and females. It is not a glycoprotein and binds progesterone and related compounds with an affinity constant of 4.1×10^{-7} M for progesterone (P) and a binding stoichiometry of 1 P:2 UG. UG also binds retinol and retinoic acid at a site distinct from that of progesterone, but it does not bind estrogens, androgens, or corticosteroids. In the lung, UG (identical to the Clara cell 10-kDa protein) is present in clusters of bronchial and bronchiolar epithelial cells but not in goblet cells.

i. Hormonal Regulation of UG Synthesis In the uterus, UG is synthesized and secreted in response to progesterone beginning on Day 3 of pregnancy, increases in absolute amount until Day 6, and then decreases so that after Day 10 UG is almost undetectable in uterine fluid. There is no evidence that the conceptus itself can synthesize UG *de novo* or degrade it, although the conceptus accumulates UG within protein vacuoles and crystalloid bodies. The secretion of UG is preceded first by an increase in uterine progesterone receptor mRNA which is at a maximum on Day 3 and then by an increase in the UG mRNA. Estradiol can also stimulate UG synthesis but is less effective than progesterone. This pattern can be reproduced by the administration of progesterone to stimulate UG synthesis and secretion followed by estradiol to induce its disappearance. Prolactin can augment the secretion of UG in response to estrogen and progesterone presumably by regulating one or more uterine proteins that bind to the UG promoter. In the lung, UG synthesis is regulated by glucocorticoids rather than by ovarian hormones, which is consistent with the presence of progesterone, glucocorticoid, and estradiol response elements in the UG gene 5′ promoter region. In addition, activation of the UG gene may be mediated by both general and tissue-specific nuclear factors. Nuclear factors of the

Sp transcription family require at least two other endometrial-derived proteins. It has been suggested that estradiol inhibits UG synthesis and secretion by regulating the proliferation and differentiation of endometrial cells in response to progesterone. Thus, ovarian steroids may regulate production and secretion of UG at three levels: (i) by affecting the cell cycle program of endometrial cells, (ii) by regulating the total cellular transcriptional activity and/or mRNA processing activity of the uterine epithelium, and (iii) by specifically modifying the rate of transcription of the UG gene.

ii. Action of Uteroglobin Because of its high amino acid sequence homology to calmodulin, it has been suggested that UG is related to a family of calcium-binding proteins. UG inhibits phospholipase A_2 activity (but not trypsin, papain, elastase, chymotrypsin, or subtilisin) and, as a result, decreases arachidonic acid release and prostaglandin formation. This is similar to the actions of lipocortins (the glucocorticoid-induced antiphospholipase A_2 proteins) which are structurally similar to UG. This suggest that at least one function of UG is to protect the embryo from maternal immune and inflammatory responses during implantation. Like lipocortin, UG may bind to the surface of blastomeres (and spermatozoa) to protect these cells against maternal phagocytes and to mask cell surface antigens. This is consistent with results showing that UG can inhibit the invasiveness of prostatic tumor cell lines. Thus, the progesterone-dependent secretion of an anti-inflammatory and immunosuppressive phospholipase A_2 inhibitory protein into the uterine lumen at the time of implantation could have three main functions: (i) to inhibit prostaglandin production in the areas of the endometrium which are not directly involved in implantation, (ii) to regulate endometrial and possibly blastocyst prostaglandin synthesis at the implantation site, and (iii) to prevent maternal inflammatory and/or immune response from damaging the implanting embryo. A parallel role for UG has been proposed within the lung, in which UG is thought to protect surfactant from phospholipase A_2 hydrolysis and prevent the development of chronic inflammatory lung disease.

3. *Uterine Changes during Implantation*

The rabbit endometrial epithelium differentiates prior to the time of blastocyst implantation, which is characterized by a loss of surface negativity, a change in glycocalyx morphology, and an increase in the protein and polysaccharide composition of the apical epithelial surface. At the time of attachment, i.e., 7 days postovulation, the blastocyst is considerably expanded and is oriented with the embryonic disk facing the mesometrial endometrium, whereas the abembryonic trophoblast faces the antimesometrial endometrium. Contact between trophoblast and uterine epithelium is preceded by an increased stickiness of blastocyst coverings and the uterine epithelial surface. This is followed by dissolution of the barrier formed by the blastocyst coverings through enzymatic activity at the implantation site. Other changes associated with the implantation site are increases in the synthesis of prostaglandins, phospholipase, and platelet-activating factor (PAF). Both prostaglandins (which can be sequestered by the blastocyst) and PAF are thought to increase local endometrial vascular permeability in response to the implanting blastocysts. The rabbit placenta does not synthesize progesterone; thus, the maintenance of pregnancy depends on continued progesterone secretion from the corpus luteum. Ultrastructural studies show that relaxin is synthesized and secreted from the syncytiotrophoblast of the rabbit placenta and stored in membrane-bound granules; however, the role of relaxin in the rabbit is unclear. Parturition is the result of withdrawal of progesterone which initiates uterine contractions by (i) decreasing the resting membrane potential of the myometrial cell causing an increased amplitude and synchrony of myometrial contractions, (ii) releasing inhibition of prostaglandin-mediated contractile activity, and (iii) allowing oxytocin secretion to increase.

III. MALE REPRODUCTION

A. The Testis

The testes (approximately 3 × 0.8 cm) are within scrotal sacs which communicate freely with the peritoneal cavity. The glandular walls of the vas deferens enlarge to form the ampullary dilation before entering the seminal vesicle. The seminal vesicle has two anterior lobes and empties through a single ejaculatory duct into the urethra. During maturation, there is a steady increase in testis volume and in tubular length. At around Day 50 of age, there is an increase in serum FSH coincident with the onset of spermatogenesis (which takes approximately 48 days to complete). LH receptors and plasma testosterone increase between Days 65 and 100 of age. The volume of the ejaculate ranges from 0.05 to 1.5 ml, with a sperm concentration of 0.5–3.5 $\times$ 10^8/ml. Fructose and occasionally glucose and glycerylphosphorylcholine are found in seminal fluid. Sperm counts show some cyclicity with maximum counts occurring approximately every 3 days.

B. Fertilization

Following ejaculation, motile sperm are continuously transported from the anterior vagina into the uterus at a rate that increases as ovulation approaches. Capacitation of sperm takes 4–8 hr in the female reproductive tract. Attachment of rabbit sperm to the egg is mediated by sperm antigens, which bind to carbohydrate moities of the zona pellucida (ZP) proteins. The rabbit ZP proteins, identified as R55, R75, and R45, correspond to the ZP1, -2, and -3 class of proteins of other species. The sperm proteins thought to bind to the ZP proteins are produced in the epidydimis and seminal vesicles and adhere to the sperm plasma membrane during its passage before ejaculation. One such sperm protein, Sp17, has an amino acid sequence which is similar to that of the N-terminal of human testis cAMP-dependent protein kinase and lies on the apical surface of sperm. Antibodies against rabbit Sp17 inhibit fertilization both *in vitro* and *in vivo*. Other proteins, termed "spermadhesins," may also contribute to sperm binding to the ZP. This initial attachment initiates the acrosome reaction and exposes and/or activates the hydrolytic enzymes within the acrosome, including proacrosin/acrosine, which aid in sperm passage through the zona pellucida. Although the specific signaling pathways have not been confirmed in the rabbit, data from other species suggest that ZP-receptor activation involves induction of tyrosine kinase activity, G-protein activation, a rise in intracellular DAG levels, protein kinase C activation, an increase in calcium levels, and fusion of the outer acrosomal and plasma membranes. This results in exocytosis of the acrosomal contents. At the time of the acrosome reaction, the sperm is still attached to the ZP through the acrosomal cap. Proacrosin/acrosin remains on the acrosomal cap even when spermatozoa are no longer associated with it and appears to function as a secondary sperm ZP-binding protein and to aid in sperm penetration through the ZP. Monoclonal antibodies to acrosin can inhibit fertilization, demonstrating the importance of acrosin in this process. In the rabbit, the proacrosin/acrosin system remains on the surface of the inner acrosomal membrane for at least several hours after the acrosome reaction. This system is still present on about 25% of sperm isolated from the perivitiline space of the ovum and is correlated with the ability of these perivitiline sperm to fertilize freshly ovulated oocytes.

See Also the Following Articles

Cricetidae; Luteinization; Microtinae; Pregnancy, Maternal Recognition of; Pseudopregnancy; Rodentia; Transgenic Animals

Bibliography

Barros, C., Crosby, J. A., and Moreno, R. D. (1996). Early steps of sperm–egg interactions during mammalian fertilization. *Cell Biol. Int.* **20**, 33–39.

Gadsby, J. E., and Lancaster, M. E. (1989). Rabbit placental-conditioned medium stimulates progesterone accumulation by granulosa–lutein cells in culture: Preliminary characterization of a placental luteotropic hormone. *Biol. Reprod.* **40**, 239–249.

Holt, J. A. (1989). Regulation of progesterone production in the rabbit corpus luteum. *Biol. Reprod.* **40**, 201–208.

Keyes, P. L., Kostyo, J. L., and Towns, R. (1994). The autonomy of the rabbit corpus luteum. *J. Endocrinol.* **143**, 423–431.

Kundu, G. C., Mantile, G., Miele, L., Cordella-Miele, E., and Mukherjee, A. B. (1996). Recombinant human uteroglobin suppresses cellular invasiveness via a novel class of high-affinity cell surface binding site. *Proc. Natl. Acad. Sci. USA* **93**, 2915–2919.

Manning, P. J., Ringler, D. H., and Newcomer, C. E. (Eds.) (1994). *The Biology of the Laboratory Rabbit,* 2nd ed. Academic Press, New York.

Marcinkiewicz, J. L., and Bahr, J. M. (1993). Identification and preliminary characterization of luteotropic activity in the rabbit placenta. *Biol. Reprod.* 48, 403–408.

Marcinkiewicz, J. L., Moy, E. S., and Bahr, J. M. (1992). Change in responsiveness of rabbit corpus luteum to prostaglandin F-2 alpha during pregnancy and pseudopregnancy. *J. Reprod. Fertil.* 94, 305–310.

Miele, L., Cordella-Miele, E., and Mukherjee, A. B. (1987). Uteroglobin: Structure, molecular biology, and new perspectives on its function as a phospholipase A_2 inhibitor. *Endocr. Rev.* 8, 474–490.

Miller, J. B., and Pawlak, C. M. (1994). Characterization and physiological variation in prostaglandin, prostacyclin and thromboxane synthesis by corpora lutea, nonluteal and uterine tissues during pseudopregnancy in the rabbit. *Life Sci.* 54, 341–353.

Ramirez, V. D., and Soufi, W. L. (1994). The neuroendocrine control of the rabbit ovarian cycle. In *The Physiology of Reproduction* (E. Knobil and J. D. Neil, Eds.), pp. 585–611. Raven Press, New York.

Richardson, R. T., Yamasaki, N., and O'Rand, M. G. (1994). Sequence of a rabbit sperm zona pellucida binding protein and localization during the acrosome reaction. *Dev. Biol.* 165, 688–701.

Thie, M., Bochskani, R., and Kirchner, C. (1986). Glycoproteins in rabbit uterus during implantation. *Histochemistry* 84, 73–79.

Townson, D. H., Wang, X. J., Keyes, P. L., Kostyo, J. L., and Stocco, D. M. Expression of the steroidogenic acute regulatory protein in the corpus luteum of the rabbit: Dependence upon the luteotropic hormone, estradiol-17β. *Biol. Reprod.* 55, 868–874.

Wiltbank, M. C.. Dysko, R. C., Gallagher, K. P., and Keyes, P. L. (1988). Relationship between blood flow and steroidogenesis in the rabbit corpus luteum. *J. Reprod. Fertil.* 84, 513–520.

Radioimmunoassay

Terry M. Nett and Jennifer M. Malvey

Colorado State University

GLOSSARY

accuracy The extent to which the mean of an infinite number of measurements of a ligand agrees with the known amount of ligand that is present.

acromegaly A condition caused by excessive secretion of growth hormone in adults that leads to hyperplasia of the nose, jaws, fingers, and toes.

adjuvant Any substance that when mixed with an antigen enhances antigenicity and produces a superior immune response.

affinity The degree of force with which an antibody binds a ligand.

antibody A protein produced by immune cells in response to the introduction of an antigen and that has the ability to combine specifically with the antigen that stimulated its production.

antigen A high-molecular-weight substance, usually protein or protein–carbohydrate complex, which when foreign to an animal stimulates production of specific antibodies.

antigenic site That portion of a molecule that stimulates production of an antibody or to which an antibody binds.

antigenicity Potency as an antigen.

antiserum Serum containing antibodies; usually from an animal that has been immunized.

gamma globulin A fraction of blood plasma rich in antibodies.

hapten A small molecule that by itself is not antigenic, but when attached to a large protein it becomes highly antigenic.

immunization Injection of a specific antigen foreign to the host to promote formation of antibodies.

immunoglobulin An antibody.

immunologic response The reaction of an animal to injection of an antigen.

ligand A substance which an antibody binds with high affinity; in the context of this article, it is a substance to be measured by radioimmunoassay.

logit A mathematical conversion used to convert a sigmoidal curve into a straight line.

microcurie A commonly used measure of radioactivity that refers to mass of a radioactive substance in which the number of nuclear disintegrations are 2.2×10^6 per minute.

monoclonal antibody Identical antibody molecules produced from a single immune cell; antibodies produced by one cell recognize only a single antigenic site.

parallelism The ability of a radioimmunoassay to quantify the same concentration of ligand in a biological sample independent of the volume of sample measured.

polyclonal antibodies Antibodies produced by multiple immune cells. Each immune cell produces an antibody that recognizes a unique antigenic sites on the same molecule.

precision The degree to which a given set of measurements of a particular ligand agrees with the mean of those measurements; an assessment of variability in an assay.

primary antibody The antibody used to measure the concentration of ligand in biological samples.

secondary antibody The antibody used to precipitate the primary antibody.

sensitivity The smallest amount of ligand that can be measured.

Sepharose A small polysaccharide bead to which large proteins (such as antibodies) may be attached.

specific activity The amount of radioactivity contained in a specified mass of ligand.

specificity Lack of interference from substances other than the ligand intended to be measured.

standard curve An inhibition curve for a radioimmunoassay established by measuring the degree to which differing amounts of a purified ligand inhibit the binding of a radiolabeled ligand to antibody. The concentration of ligand in biological samples is obtained by comparing the inhibition produced by the biological sample to the inhibition produced by the standards.

substrate A substance that when acted on by an enzyme is converted to a product with different chemical properties.

titer The concentration of antibody molecules in serum; usually expressed as the dilution of serum that results in binding of approximately 50% of the radiolabeled ligand.

Radioimmunoassay is a technique used to quantify substances present in minute amounts in biological fluids and tissues. It is based on the interaction between an antibody specific for the substance of interest and a radioactive form of the substance. Quantification is achieved by measuring the ability of the substance of interest (referred to hereafter as ligand) to compete with a radioactive form of the ligand for binding sites on an antibody. The technique was first developed in 1960 by Dr. Rosalyn Yalow who was awarded a Nobel Prize for this work. Radioimmunoassay is widely used as a research tool and has many applications such as clinical diagnosis of endocrine abnormalities, monitoring drug therapies, detecting the presence of illegal drugs in blood or urine, and measuring levels of pesticides in water and food products. Three components are needed for development of a radioimmunoassay: an antibody specific for the ligand to be quantified, a radioactive form of the ligand, and a method for separating the ligand bound to antibody from that remaining free in solution. Development and importance of each of these components will be discussed. Procedures for evaluating the reliability of a radioimmunoassay also will be considered.

I. GENERATION OF ANTISERA

Central to the development of a radioimmunoassay is generation of an antiserum specific for a particular ligand of interest. Antibodies for radioimmunoassay have been generated in a variety of species including rats, horses, sheep, guinea pigs, goats, turkeys, mice, and rabbits. When choosing a species for immunization, there are several criteria one should consider. These include the quantity of ligand available for immunization, the amount of antiserum needed, and the foreignness of the ligand to the species being immunized. The rabbit is a species that is relatively small and therefore does not require large amounts of ligand for immunization, but it is large enough to provide several milliliters of antiserum from a single bleeding. Therefore, rabbits are probably the most widely utilized species for generation of antiserum for radioimmunoassay. For the remainder of this discussion, it will be assumed that the antiserum for radioimmunoassay is generated in rabbits.

A. Proteins or Other Large Molecules

Foreign proteins or other large molecules generate an immunologic response when injected into rabbits. The antigenicity of a molecule is a function of its size, chemical composition, and degree of foreignness to the animal being immunized. To enhance the immunological response and to prolong the period over which the protein is absorbed into the bloodstream, proteins are normally mixed with an adjuvant prior to injection into a rabbit. The protein–adjuvant mixture is injected at multiple sites, either intradermally or subcutaneously. This stimulates the immune system to produce antibodies (immunoglobulins) against the foreign protein over a period of several weeks. After the initial immunization, rabbits are administered additional injections of the protein–adjuvant mixture (booster injections) at 30- to 60-day intervals to maximize production of antibodies. Beginning 1 week after the first booster injection, blood may be collected at weekly intervals. The concentration of antibodies in the blood of rabbits (titer) will vary with time after immunization. Following the initial immunization and after each booster immunization there should be an increase in concentration of antibodies in the blood of the animal that persists for several weeks. After this, the concentration of antibodies in blood will gradually decrease over a period of several weeks. If a booster immunization is administered when the concentration of antibody in blood is decreasing, there is often an exaggerated response to the antigen resulting in extremely high concentrations of antibody. Antiserum with a high titer is very useful for radioimmunoassay. A typical profile of antibody concentration in blood of an immunized rabbit is depicted in Fig. 1. Since dramatic changes in the titer of the antiserum may occur between bleedings, particularly following a booster, each bleeding should be evaluated for the concentration of antibodies present.

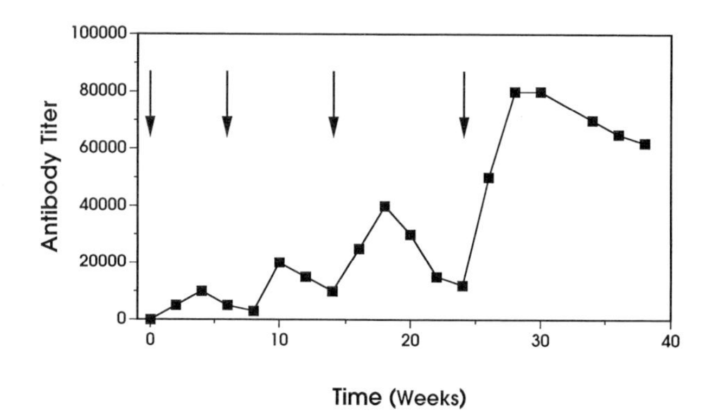

FIGURE 1 Antibody titer in a rabbit immunized against gonadotropin-releasing hormone conjugated to bovine serum albumin. The first arrow on the left indicates the time of the initial immunization and subsequent arrows indicate booster immunizations. Titer is expressed as the dilution of serum that binds approximately 5×10^{-15} mol of radioiodinated GnRH.

Since an antigenic site consists of only 4–6 amino acids and since proteins are composed of many amino acids (usually more than 35), it is possible that a single rabbit will produce multiple antibodies to a single protein, each directed to a different antigenic site of the protein (Fig. 2). Serum from a rabbit containing multiple antibodies to a single protein is referred to as polyclonal antiserum. In addition to the polyclonal antibodies that bind the ligand of interest, serum from an immunized rabbit also will contain antibodies to virtually every foreign protein to which the rabbit has ever been exposed (i.e., bacteria, viruses, toxins, etc.).

B. Production of Antisera against Small Molecules That Are Not Normally Antigenic

Small molecules (i.e., those with a molecular weight of <3000) normally do not elicit an immunological response when injected into rabbits. However, if these small molecules are chemically linked to a large molecule they may stimulate formation of antibodies specific for the small molecules by functioning as a hapten. Immunologists have taken advantage of this characteristic to generate antibodies to a myriad of small molecules including virtually all steroid hormones, neurotransmitters, peptide hormones, thyroid hormones, vitamins, antibiotics,

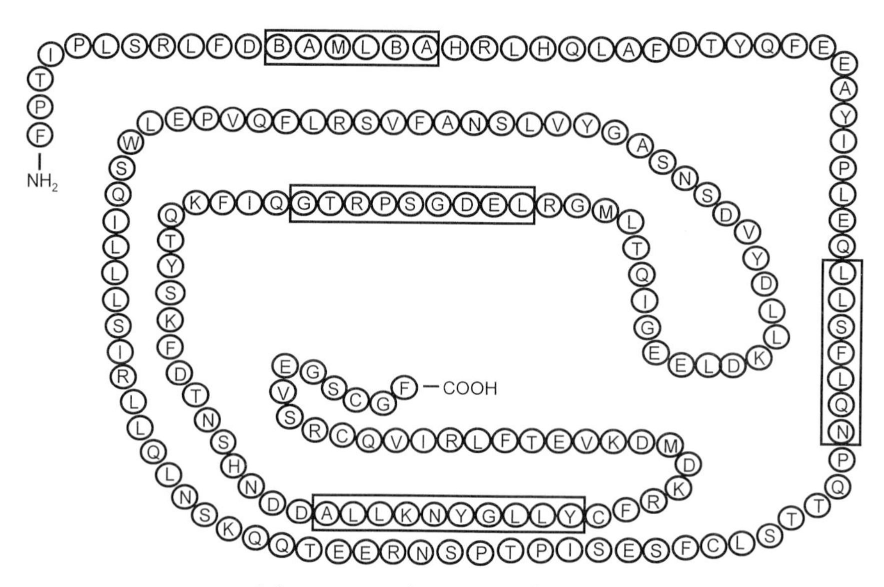

FIGURE 2 Diagrammatic representation of the amino acid sequence of a protein (GH). Boxes around groups of amino acids indicate potential antigenic sites. Since there are numerous antigenic sites in GH, a rabbit immunized against GH will produce antibodies against several of the antigenic sites. Although each of these antibodies will be specific for GH, they will each bind to a different antigenic site on GH. This is referred to as a polyclonal antiserum.

drugs (legal and illegal), pesticides, toxins, and food metabolites.

Bovine serum albumin, ovalbumin, and keyhole limpet hemocyanin are large molecules commonly used as carriers to which haptens are chemically conjugated. Each of these protein molecules possesses multiple sites to which haptens may be attached. To maximize generation of antibodies to a hapten, it is desirable to have a minimum of 10 hapten molecules linked to each molecule of the carrier protein.

Most haptens do not have a reactive group that permits them to be linked directly to carrier molecules. Therefore, they must be chemically altered before they can be attached to the carrier molecule. In most instances, the chemical alteration involves preparation of a derivative of the hapten that includes a free carboxylic acid. The carboxylic acid can then be linked directly to lysine residues in the carrier molecule. An example of how a steroid hormone is altered so that it may be conjugated to a carrier molecule is shown in Fig. 3.

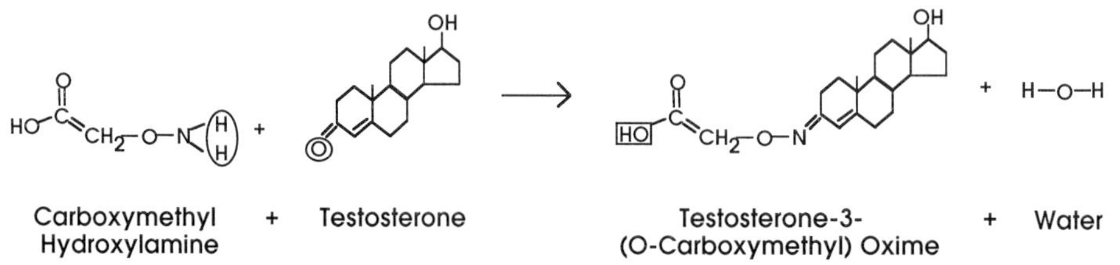

FIGURE 3 Modification of the structure of a steroid hormone (testosterone) so that it can be attached to a large protein. When carboxymethyl hydroxylamine is reacted with testosterone, the amine group of the carboxymethyl hydroxylamine attaches to the ketone group on testosterone. The two hydrogen atoms on the carboxymethyl hydroxylamine and the oxygen atom on testosterone that are removed to form water are circled to demonstrate where the reaction occurs. The box enclosing the hydroxyl group (HO) on testosterone-3-(*O*-carboxymethyl) oxime indicates the active site on the steroid derivative that reacts with amino groups on the carrier protein to form a steroid–protein conjugate.

II. PREPARATION OF A RADIOACTIVE LIGAND

In radioimmunoassay, quantification is achieved by measuring the amount of ligand bound to antibody. To measure the amount of a ligand bound to an antibody, it is necessary to use a form of the ligand that can be easily detected in minute quantities. In radioimmunoassay, this is accomplished by utilizing a radioactive form of the ligand of interest. Specificity and sensitivity of the radioimmunoassay are directly related to the nature and properties of the radioactive ligand. As indicated previously, antisera obtained from rabbits contain many different antibodies. Some of these are directed to the compound of interest, but each antiserum also contains antibodies to a variety of other proteins to which the rabbit had been exposed previously. Therefore, it is imperative that the radioactive ligand is pure. If the radioactive ligand is contaminated with other proteins to which the rabbit has formed antibodies, those antibodies will bind the contaminating proteins and may interfere with quantification of the compound of interest.

A. Proteins

The radioisotope most commonly used to label proteins is ^{125}iodine (^{125}I). It can be conveniently attached to tyrosine residues in the protein by several different means. Radioactive iodine may be purchased from several manufacturers as a solution of the sodium salt (Na-^{125}I). To attach radioactive iodine to proteins (radioiodination), it is necessary to convert the radioiodine into a gaseous state. This is accomplished by oxidation. Several different oxidizing agents may be used to generate gaseous iodine. A detailed discussion of the procedures for radioiodination is beyond the scope of this article. The reactions in Fig. 4 indicate how radioiodine may be incorporated into tyrosine residues of proteins.

The oxidant used to catalyze this reaction has the potential to damage the protein into which the radioiodine is being incorporated. Therefore, it is desirable to minimize the amount of oxidant used or the duration of the reaction. Either of these precautions will reduce the oxidative damage to the protein being radioiodinated.

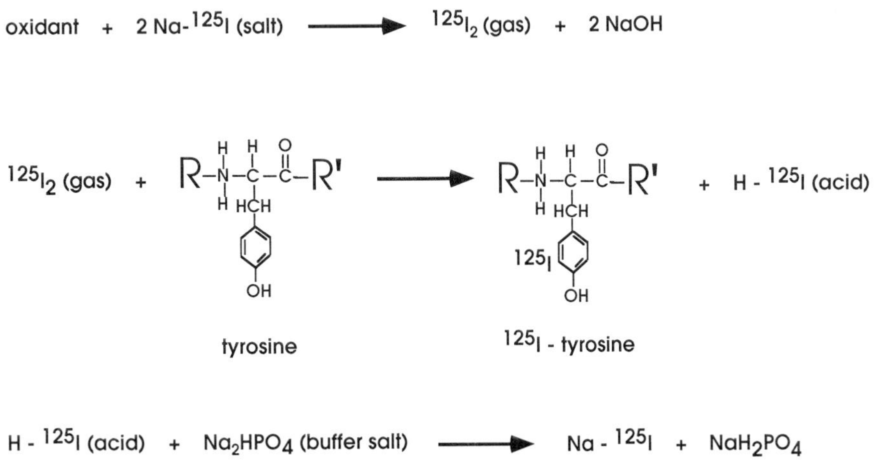

oxidant + 2 Na-^{125}I (salt) → $^{125}I_2$ (gas) + 2 NaOH

$^{125}I_2$ (gas) + tyrosine → ^{125}I - tyrosine + H - ^{125}I (acid)

H - ^{125}I (acid) + Na_2HPO_4 (buffer salt) → Na - ^{125}I + NaH_2PO_4

FIGURE 4 Reactions that result in the incorporation of iodine into protein. The radioactive iodine (^{125}I) in the form of a solution of sodium salt (Na-^{125}I) reacts with an oxidizing agent (oxidant) to produce a gaseous form of the radioactive iodine ($^{125}I_2$) and also generates sodium hydroxide (NaOH). The radioactive iodine gas reacts with tyrosine residues in proteins to form a radioiodinated tyrosine plus hyrdoiodic acid (H-^{125}I). Buffer salts in the reaction mixture then convert H-^{125}I back to Na-^{125}I so that it can be recycled in the reaction. The R and R' groups on the ends of the tyrosine residue are indicative of continued peptide chains in a protein.

To be certain that a molecule of radioiodine is incorporated into each molecule of protein, normally a slightly greater amount of Na-^{125}I than protein is included in the reaction mixture. Therefore, it is unlikely that all the radioiodine in any given reaction is incorporated into protein. This necessitates separation of the ^{125}I bound to protein from the remaining Na-^{125}I. Separation of protein-bound ^{125}I and free Na-^{125}I is normally accomplished by column chromatography. Contents of the radioiodination mixture are applied to a column containing a resin that separates compounds based on size and/or charge. Since the protein containing the radioiodine is much larger than Na-^{125}I, the radioiodinated protein passes through the column much more rapidly than Na-^{125}I (Fig. 5). The radioiodinated protein is saved and the Na-^{125}I is discarded. By knowing the amount of protein that was used for the reaction and calculating the quantity of ^{125}I incorporated into tyrosine residues in the protein, it is possible to determine the specific activity of the radioiodinated protein. This is expressed as amount of radioiodine (usually in microcuries, μCi) per unit of protein (usually in micrograms, μg); for example, the specific activity of a protein might be 100 μCi/μg of protein. Specific activity is one of the primary factors that determines sensitivity of a radioimmunoassay.

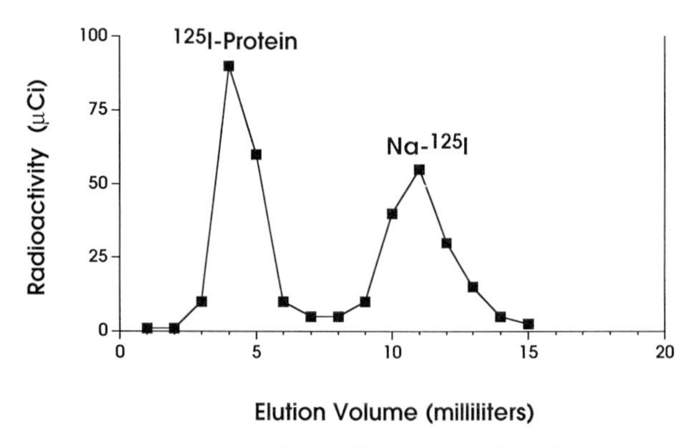

FIGURE 5 Separation by gel filtration of radioiodinated protein from Na-^{125}I that was not incorporated into protein. Large molecules such as proteins pass through the gel filtration column rapidly, whereas small molecules such as salts pass through more slowly. Those fractions containing the protein (fractions 4 and 5) are saved and used for radioiodinated ligand in the radioimmunoassay, whereas the remaining fractions containing Na-^{125}I are discarded.

B. Haptens

Many haptens to which antisera have been generated do not contain a tyrosine residue, so they cannot be radioiodinated directly. In some cases, radioactive forms of haptens containing either radioactive hydrogen [tritium (^{3}H)] or carbon (^{14}C) are available commercially. However, in many cases, it is not feasible to obtain ^{3}H or ^{14}C forms of the hapten. Furthermore, it is more expensive, both in time and materials, to use ^{3}H or ^{14}C forms of the hapten for radioimmunoassay than it is to use a radioiodinated ligand. Therefore, many laboratories have opted to develop forms of the hapten that may be radioiodinated. Haptens can be linked to a derivative of tyrosine [tyrosine methyl ester (TME)] using procedures similar to those described previously for linking a hapten to protein (Fig. 6). The hapten–TME conjugate can then be radioiodinated using the same procedures described previously for proteins.

As eluded to in the previous paragraph, there are several advantages to using radioiodinated haptens for radioimmunoassay compared to ^{3}H or ^{14}C forms of the hapten. Perhaps the most compelling of these is to produce a radioactive ligand with a much higher specific activity. Radioactivity is released 75 times more rapidly from ^{125}I than from ^{3}H and radioactivity can be detected only at the time it is released from the substance of interest. Therefore, it takes approximately 75 times as many molecules of hapten containing ^{3}H as the radioisotope as it does of a hapten containing ^{125}I as the radioisotope to generate the same amount of radioactivity. Using ^{14}C as the radioisotope is far worse; ^{125}I releases its radioactivity over 34,000 times more rapidly than ^{14}C. Thus, radioiodinated haptens require considerably less mass to release a defined amount of radioactivity than haptens containing ^{3}H or ^{14}C as the radioisotope. Since the sensitivity of a radioimmunoassay is dependent on the amount of radioactivity released from radiolabeled ligand bound to antibody, use of haptens labeled with ^{125}I results in assays that are two or three orders of magnitude more sensitive than assays utilizing haptens labeled with ^{3}H or ^{14}C. In addition, ^{125}I can be counted directly, without the addition of special fluid to detect the radioactivity. Both ^{3}H and ^{14}C require addition of a special scintillation fluid (which is very expensive and requires an extra manipulation

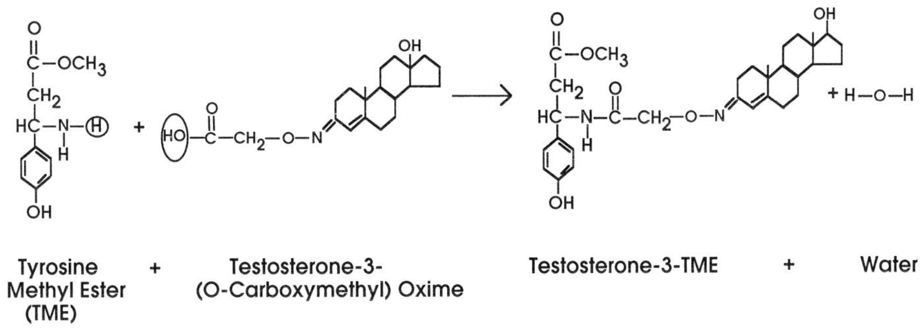

Tyrosine Methyl Ester (TME) + Testosterone-3-(O-Carboxymethyl) Oxime → Testosterone-3-TME + Water

FIGURE 6 Preparation of a steroid derivative that can be radioiodinated. In this case tyrosine methyl ester (TME) is covalently linked to testosterone-3-(*O*-carboxymethyl) oxime, the same derivative of testosterone that was used to prepare the steroid–protein conjugate for generating antibodies to testosterone (see Fig. 3).

during an assay) before the radioactivity can be detected. Thus, both reagent and labor costs associated with radioimmunoassay are greatly increased when ^{3}H or ^{14}C are used as the radioisotope.

III. METHODS FOR SEPARATING LIGAND–ANTIBODY COMPLEXES FROM UNBOUND LIGAND

Another important component of a radioimmunoassay is an efficient means for separating ligand bound to antibody from ligand that remains free in solution. This may be accomplished by several different means. The particular method of choice depends on the ligand being quantified and the biological medium in which the ligand is being measured. The ideal method should provide a complete separation of antibody-bound and free ligand and yet be unaffected by nonspecific substances present in the reaction mixture. The method should also be rapid, simple, and inexpensive. Methods of separation most commonly used include solid-phase primary antibodies, solid-phase adsorption of ligands, chemical precipitation of antibody–ligand complexes, and immunoprecipitation of antibody–ligand complexes.

A. Solid-Phase Primary Antibody

An important property of antibodies is their ability to bind to plastic surfaces. This characteristic is used to advantage in the solid-phase procedure to separate antibody–ligand complexes from free ligand. For this procedure, a solution of primary antibody is added to polystyrene tubes and the antibody nonspecifically sticks to the surface of the tube within a short time. The remaining solution is decanted and the tubes can be stored dry at room temperature for several months. For the radioimmunoassay, sample and radioactive ligand are added to the tubes and they are allowed to react with the antibody adhered to the tube. At the end of the reaction, separation is achieved by decanting the solution from the tube. The antibody–ligand complexes remain in the tube, whereas free ligand is removed in the solution that was decanted. This procedure is most reliable when used to quantify concentrations of ligand in samples containing little or no other protein. This method of separation of antibody-bound from free ligand is simple and does not require expensive equipment. The major limitation of this procedure arises when it is used in an assay system designed to quantify ligand in samples of serum or plasma (or other media that has large amounts of soluble protein). In this case, proteins in the sample (that are present in much higher concentrations than the antibody) may adhere nonspecifically and displace some of the antibody bound to the plastic tube. This reduces the amount of primary antibody attached to the plastic tube that is available for binding ligand. This can result in an inaccurate estimate of the amount of ligand in the sample.

B. Solid-Phase Adsorption of Ligands

A second method for separating antibody-bound from free ligand is to adsorb the free ligand to insoluble particles such as charcoal, talc, or similar compounds at the end of the reaction. This procedure is also simple and results in rapid separation of antibody-bound and free ligand. Unfortunately, charcoal and talc only bind small molecules so use of this procedure is limited to quantification of relatively small ligands (i.e., those having a molecular weight of <2 kDa). There are also other disadvantages to this technique. These adsorbents have a high affinity for ligand which results in a competition between the adsorbent and the antibody for the ligand. Unless the conditions employed for the separation are stringent, it is possible for the adsorbent to dislodge (strip) some of the ligand bound to antibody. Since the basis for quantification in a radioimmunoassay is the measurement of radioactive ligand bound to antibody, this may lead to inaccurate quantification of ligand.

C. Chemical Precipitation of Antibody–Ligand Complexes

Chemical means to precipitate antibody–ligand complexes have also been used to separate ligand bound to antibody from free ligand. Perhaps the most widely practiced of the chemical methods is the use of polyethylene glycol. Polyethylene glycol at a final concentration of 12% will cause antibodies (and ligands bound to them) to precipitate. This procedure is rapid and inexpensive and does not suffer from the disadvantages described for antibody-coated tubes or adsorption of free ligand to insoluble particles. However, this procedure cannot be used for all radioimmunoassays. The characteristic of the polyethylene glycol that causes antibodies to precipitate also induces precipitation of other large proteins. Thus, if one is using radioimmunoassay to quantify a large protein (e.g., pituitary hormones), the free ligand (pituitary hormone) will also be precipitated. Since quantification is achieved by measuring the amount of ligand bound to antibody, if both antibody-bound and free ligand are precipitated, quantification cannot be achieved.

D. Immunoprecipitation of Antibody–Ligand Complexes

Perhaps the most adaptable method for separating free ligand from that bound to antibody is the "double-antibody" procedure. This procedure has gained widespread acceptance because it can be applied to virtually any radioimmunoassay system regardless of the size and chemical properties of the ligand being quantified. The double-antibody method does not have any of the limitations listed for the procedures described previously. Immunoprecipitation of antibody–ligand complexes requires a "second" (precipitating) antibody generated against immunoglobulins from the species in which the primary antibody was produced. For example, if the primary antibody was produced in rabbits, then immunoglobulins from rabbits might be used to immunize sheep. The antiserum to rabbit immunoglobulin (second antibody) produced in sheep is then used to precipitate the primary antibody produced in a rabbit during the radioimmunoassay. The second antibody is added to the assay mixture after the reaction of primary antibody and ligand is complete. Reaction of the secondary antibody with the primary antibody–ligand complex yields an insoluble immune complex leaving the "free" ligand in solution which is decanted to effect separation. To ensure formation of a visible precipitate, nonspecific immunoglobulin (usually in the form of serum) from the species in which the primary antibody was produced is added to the reaction mixture at a final dilution of approximately 1:2000.

At this point, one might question why an immune precipitate does not occur when the primary antibody binds ligand. For immune complexes to form a conglomerate large enough to precipitate, there must be a precise ratio of ligand to antibody. This ratio never occurs during the primary antibody–ligand reaction because the ligand is always in excess. A large excess of either ligand or antibody prevents formation of immune conglomerates large enough to precipitate. Therefore, there must always be a precise ratio between the primary and secondary antibodies for an immune complex to precipitate.

The primary disadvantage of the second antibody procedure is that it is relatively slow compared to other methods used for separating antibody-bound from free ligand (from 12 to 72 hr may be required to effect separation). However, it is possible to attach the second antibody to an insoluble matrix such as Sepharose or iron particles. In this case, the ratio of primary to secondary antibody is unimportant as long as the secondary antibody is in excess. Thus, a large excess of second antibody can be added to speed the rate of separation. This modification can shorten the reaction time to as little as 30 min. This latter procedure has all the advantages of both the double-antibody and other separation methods described previously.

IV. THEORY OF RADIOIMMUNOASSAY

In the preceding sections, each of the individual components required for a radioimmunoassay has been considered. In this section, use of those components to develop a quantitative assay will be discussed. The first step in developing a quantitative assay is to determine the dilution of primary antibody to be used in the assay. In general, a dilution of primary antibody that binds between 30 and 50% of the radioactive ligand in the absence of nonradioactive ligand is chosen. At this dilution, sensitivity and precision of the assay are optimized (see next section). Quantitation is achieved by measuring the ability of the unlabeled ligand to compete with radioactive ligand for binding sites on the antibody (Fig. 7). If the quantity of primary antibody and radioactive ligand are held constant, then the amount of radioactive ligand bound to antibody is a function of the quantity of nonradioactive ligand present in the assay mixture. That is, as the amount of nonradioactive ligand in the reaction mixture is increased, the amount of radioactive ligand bound to antibody is reduced. Evaluation of the amount of radioactive ligand bound to antibody in the presence of varying amounts of nonradioactive ligand is used to generate a standard curve (Fig. 8). By comparing the inhibition of binding of radioactive ligand in an unknown sample with that caused by a standard of known concentration, it is possible to determine the quantity of ligand present in the test sample.

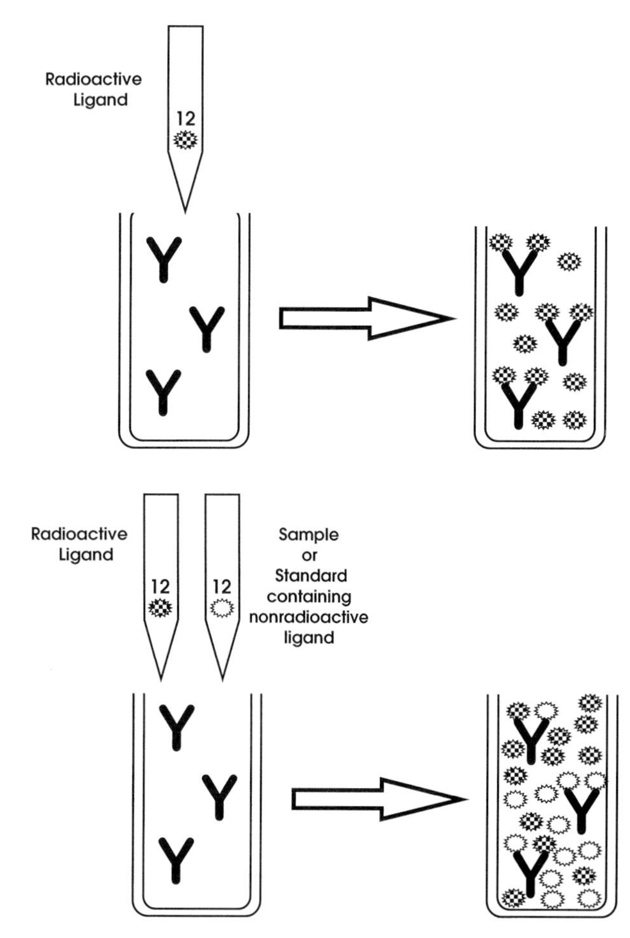

FIGURE 7 For radioimmunoassay, antibody (denoted by Y) and radioactive ligand (checkered starburst) are combined in concentrations such that ~50% of the radioactive ligand is bound to antibody (upper panel). This is defined as B_0 and forms the basis for calculating the data in a radioimmunoassay. If unlabeled ligand (empty starburst) is added to the mixture of antibody and radioactive ligand, it will compete with the radioactive ligand for binding sites on the antibody, thereby reducing the amount of radioactive ligand bound to antibody (lower panel). It should be noted that each antibody molecule is capable of binding two molecules or ligand.

Three different formats have been used to depict the ability of unlabeled ligand to inhibit binding of radioactive ligand to antibody (Fig. 8). The format used in Fig. 8 (top) has largely been abandoned because it does not permit easy assessment of parallelism (see next section). Both of the other methods are still widely used; however, the logit-log depiction (Fig. 8, bottom) permits easiest visualization of lin-

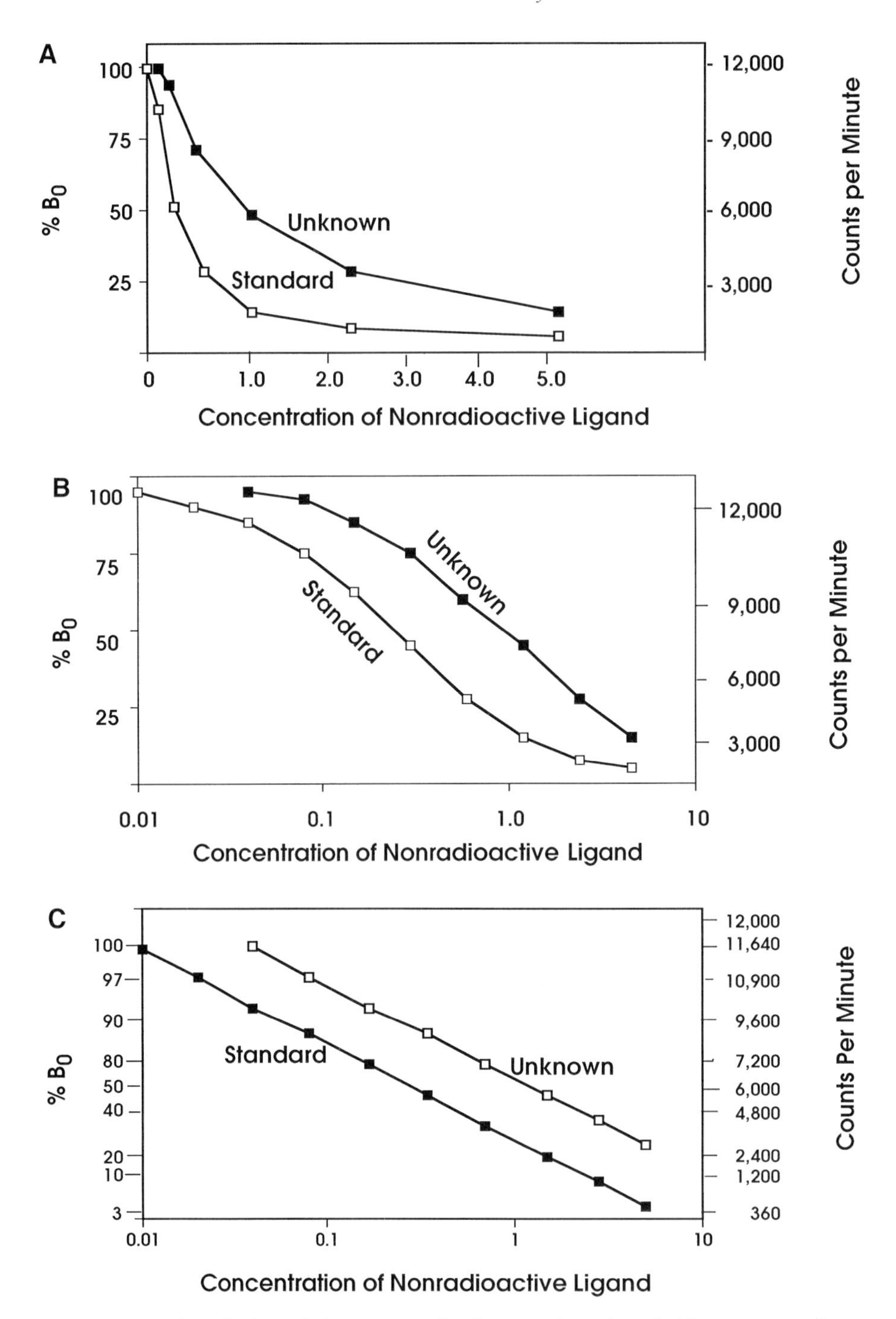

FIGURE 8 When the amount of antibody and the amount of radioactive ligand are held constant in the presence of varying quantities of nonradioactive ligand, the amount of radioactive ligand bound to antibody decreases as the quantity of nonradioactive ligand in the assay mixture increases. By adding increasing concentrations of nonradioactive ligand to the assay mixture, it is possible to generate a standard curve. Data from the standard curve can be plotted using many different formats. Three commonly used formats are depicted here. (A) Data are plotted using linear scales for both the *y*-axis and the *x*-axis. (B) Data are plotted using a linear scale for the *y*-axis and a logarithmic scale for the *x*-axis. (C) Data are plotted using a logit scale for the *y*-axis and a logarithmic scale for the *x*-axis. Presentation of data using either the linear-log plot or the logit-log plot permits easy visualization of parallelism between standard and unknown. When data are plotted using the linear–linear format, parallelism is not easily visualized.

earity of the standard curve and parallelism of unknowns.

V. PARAMETERS FOR ASSESSING RELIABILITY OF A RADIOIMMUNOASSAY

For an assay to be considered valid, it must accurately quantify the compound of interest in biological samples. The four criteria most often used to assess the reliability of a radioimmunoassy are specificity, sensitivity, accuracy, and precision. Each of these will be considered separately and their importance to development of a valid radioimmunoassay will be discussed.

A. Specificity

The most important parameter for determining reliability of a radioimmunoassay is to demonstrate that it quantifies only the ligand which it is intended to measure. This characteristic is known as specificity. Specificity is defined as the lack of interference from substances other than the one to be measured. The primary determinant of specificity of a radioimmunoassay is the purity of the radioactive ligand. If the radioactive ligand is pure, then it is very likely that any compound that inhibits binding of the radioactive ligand to antiserum is very similar, if not identical, to the radioactive ligand. It is important to remember that when using polyclonal antibodies for a radioimmunoassay, the antiserum contains immunoglobulins to virtually every foreign substance that has gained entry into the bloodstream of the rabbit. Therefore, unless the radioactive ligand is pure, the potential for nonspecificity of the assay system exists.

If the radioimmunoassay is to be used to measure concentrations of a ligand in serum or plasma, then it is necessary to demonstrate that other components in blood do not interfere with measurement of the ligand of interest. For example, if one is interested in measuring concentrations of growth hormone (GH) in serum, it is necessary to demonstrate that substances other than GH do not inhibit binding of radioactive GH to antibody. First, it should be demonstrated that serum samples that lack GH do not inhibit binding of radioactive GH to antibody. Such samples may be obtained from individuals that have had their pituitary gland removed. Second, concentrations of GH should be higher than normal in blood samples from individuals with acromegaly. In addition, if different quantities of serum obtained from an acromegalic person are measured in the assay system, the inhibition curve obtained using different quantities of serum should be parallel to that obtained with the GH standard (Fig. 9). Likewise, if GH used for standard is added to sample that contains little or no GH, then the inhibition curve obtained from the sample with the added GH should be parallel to that obtained with standard. Parallelism indicates that the kinetics by which the standard (in this case GH) and the substance in serum inhibit binding of radioactive GH to antibody are the same.

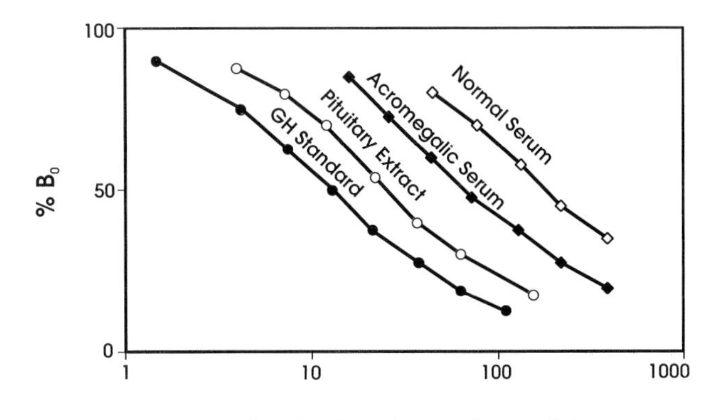

FIGURE 9 Concentrations of growth hormone (GH) in pituitary extract (an extract of the endocrine gland that synthesizes GH in the body), serum from an acromegalic patient (an adult that secretes excessive amounts of GH), and serum from a normal individual. Several different volumes of pituitary extract and serum were evaluated in the radioimmunoassay so that parallelism of biological samples with the standard preparation could be assessed. Note that curves produced by the standard and biological samples inhibiting binding of radioactive GH to the antibody are all parallel. This indicates that factors other than GH do not appear to interfere in the radioimmunoassay. If the curves were nonparallel, it would indicate that the substance inhibiting binding of radioactive GH to antibody was binding to the antibody with a different affinity than GH and thus was probably not GH. This would indicate the assay was invalid and could not be used to quantitatively assess concentrations of GH in biological samples.

If a substance in serum other than GH were inhibiting binding of radioactive GH to antibody, it is unlikely that kinetics of its binding to antibody would be identical to those of GH. In this case, an inhibition curve that was nonparallel to the standard would be obtained. This would suggest that the assay system was nonspecific and invalid for quantification of GH.

B. Sensitivity

Perhaps the second most important criterion for a radioimmunoassay is its sensitivity. If it does not have sufficient sensitivity to quantify the ligand of interest in biological samples, then it is not very useful.

Historically, radioimmunoassays have been among the most sensitive methods for quantifying ligands. Sensitivity is defined as the smallest amount of ligand which can be measured by a radioimmunoassy and is a function of the affinity of antibody for the ligand and the specific activity of the radioactive ligand. It is determined by measuring the variation associated with binding of radioactive ligand in the absence of unlabeled ligand (Fig. 10). There is little one can do to influence the affinity of the antibody. However, in some cases it is possible to increase specific activity of the radioactive ligand by increasing the number of radioactive atoms incorporated into the ligand. By increasing the specific activity of the ligand a smaller mass of radioactive ligand will generate the same number of radioactive particles. Therefore, less mass of radioactive ligand is needed in the radioimmunoassay and the amount of nonradioactive ligand required to inhibit binding of radioactive ligand to antibody is reduced, i.e., the assay is more sensitive. Another approach some investigators have used to increase the sensitivity of a radioimmunoassay is to incubate standards or samples with the antiserum for a period of time (usually 24 hr) prior to the addition of radioactive ligand. Under the incubation condition used for a radioimmunoassay, the reaction of antibody with ligand is virtually irreversible. Therefore, if the standard or sample is allowed to react with antibody prior to the addition of radioactive ligand, the nonradioactive ligand in the standard or sample will have a better opportunity to bind antibody since it is not competing with radioactive ligand for the limited number of binding sites on the antibody. After this reaction has neared completion, the radioactive ligand is added. Now, it can only bind to sites on the antibody that are not already taken by the nonradioactive ligand in the standard or sample. This technique may be used to increase the sensitivity of a radioimmunoassay by 4- to 10-fold (Fig. 11) and is useful when attempting to quantify ligands that are present at extremely low concentrations.

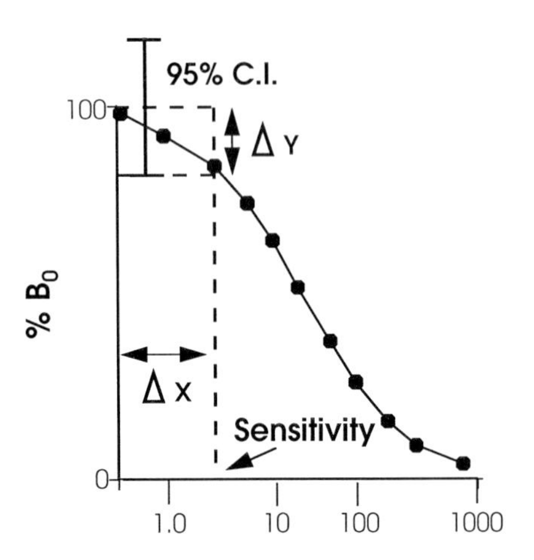

FIGURE 10 To determine the sensitivity of a radioimmunoassay, it is necessary to determine the variation associated with binding of radioactive ligand to antibody in the absence of nonradioactive ligand (B_0). A 95% confidence interval is determined for B_0. The lower limit of the 95% confidence interval is then used to determine the minimum amount of unlabeled ligand that can be accurately differentiated from no ligand simply by reading that value from the *x*-axis. That point represents the sensitivity of the radioimmunoassay.

C. Precision

Precision is defined as the extent to which a given set of measurements for the same sample agrees with the mean of those measurements and is determined by taking repeated measurements of the same sample. Precision is normally expressed as the coefficient of variation (standard deviation divided by the mean) of a group of measurements of the same sample. Precision will vary at different points along the stan-

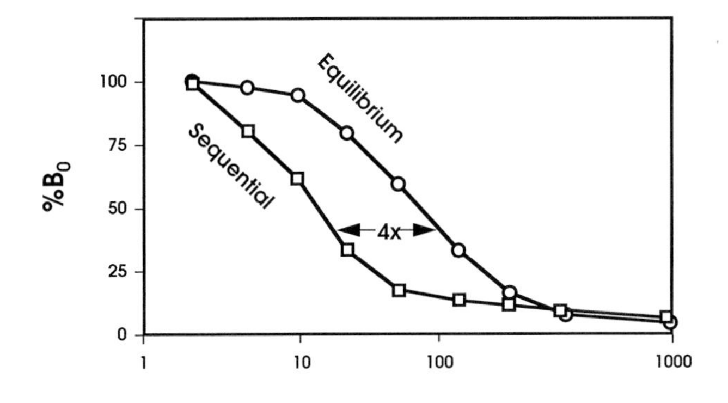

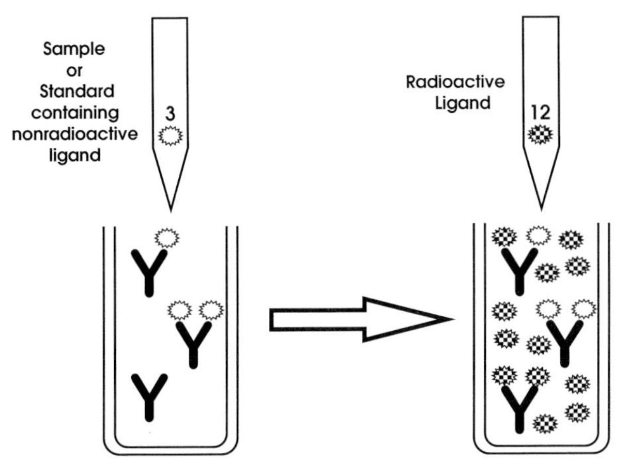

FIGURE 11 Inhibition of binding of radioactive ligand to antibody by unlabelled ligand added either at the same time as radioactive ligand (equilibrium) or several hours prior to addition of radioactive ligand (sequential). Allowing unlabeled ligand to react with antibody for several hours prior to addition of radioactive ligand increases the sensitivity of the radioimmunoassay. This occurs because the reaction of ligand with antibody is nearly irreversible under the conditions employed for radioimmunoassay. Therefore, by permitting unlabeled ligand to bind to the antibody first, radioactive ligand can only bind to antibody that does not have unlabeled ligand bound to it. For example, by examining Fig. 7, you will note that when 12 units of radioactive ligand and 12 units of unlabeled ligand were added simultaneously to the assay tube containing antibody, the binding of radioactive ligand was reduced by 50%. In contrast, the lower panel of this figure shows that under the same conditions, if the unlabeled ligand is allowed to compete for antibody prior to addition of labeled ligand, then only 3 units of unlabeled ligand reduce the binding of labeled ligand by 50%. Thus, an increase in sensitivity of fourfold is achieved.

dard curve. Therefore, at least three estimates of precision should be made for a given radioimmunoassay in samples containing low, medium, and high concentrations of ligand. This will provide estimates of error associated with measurements at three different points on the standard curve.

Precision of an assay can be influenced by variation in pipetting of each of the reagents added to a radioimmunoassay (normally there is 1 or 2% variation associated with pipetting of each reagent) and by the accuracy with which the radioactivity is quantified at the end of the assay. For example, the error associated with counting of radioactivity is equal to the square root of the total radioactive events (counts) detected divided by the total counts (i.e., if 100 counts are accumulated, the error associated with counting is 10%, whereas if 10,000 counts are accumulated, the error associated with counting is 1%). Usually, a minimum of 2500 counts of radioactivity should be accumulated to maintain the error associated with quantifying the radioactivity below 2%.

Precision should be determined each time a radioimmunoassay is performed and estimates should be obtained for variation within an assay (within-assay coefficient of variation) and between different replications of the same assay (between-assay coefficient of variation). In general, the coefficients of variation for most assays should be <20%, but the degree of variation that is acceptable is left to the judgment of individual investigators.

D. Accuracy

Accuracy is the extent of agreement between the mean of an infinite number of measurements of a particular ligand and the exact known amount of ligand present. Assessment of accuracy requires that the ligand of interest be measured using a procedure in addition to radioimmunoassay and that the results of the two methods of measurements be compared. If the ligand of interest is available in a pure state, probably the most common method used to assess accuracy is to compare the results obtained by radioimmunoassay with those obtained by weighing the ligand (Fig. 12). If there is good agreement between the two techniques, comparison of the data should

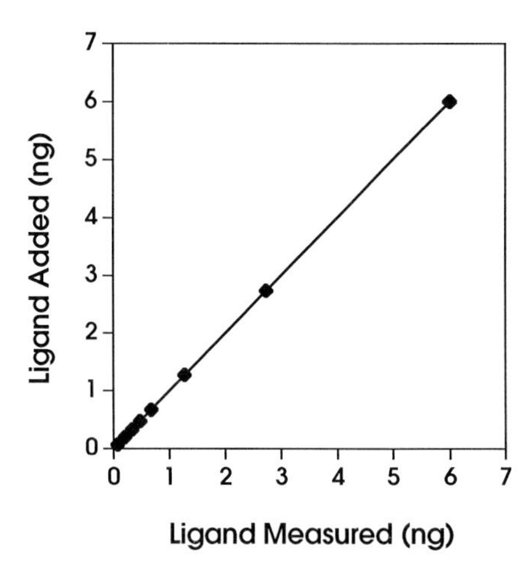

FIGURE 12 To determine if a radioimmunoassay accurately assesses the concentration of ligand in biological samples, it is necessary to evaluate the amount of ligand in the sample by another means. For the purpose of assay validation, this is often achieved by weighing a known amount of ligand (in this case, GH) and adding varying amounts to a biological fluid that contains little or no GH (i.e., serum from a dwarf). The concentrations of GH in the serum samples is then determined by radioimmunoassay, and a correlation between the amount added (y-axis) and the amount measured (x-axis) is calculated. Under ideal conditions, the amount assayed will exactly mimic the amount added so that the slope of the line obtained will be 1.0.

result in a line with a slope of 1.0 and a very high degree of correlation between results generated with each method.

VI. FINAL THOUGHTS

Radioimmunoassay has been one of the most widely utilized techniques for measuring concentrations of substances in biological media for the past four decades. Its utility, sensitivity, and specificity have been unparalleled. In recent years, there have been many modifications to the classic radioimmunoassay that several laboratories find attractive. Perhaps the most popular of these is the use of something other than a radioisotope for quantifying the amount of ligand bound to antibody. Enzymes that convert an uncolored substrate to a product with color have been used by many investigators to replace radioactivity in immunoassays. In this case, the radioactive ligand used in a radioimmunoassay is replaced by a ligand linked to an enzyme. At the end of the immunoassay, quantitation is achieved by incubating an uncolored substrate with the antibody–ligand complexes. The amount of color produced in the reaction mixture is proportional to the amount of ligand–enzyme bound to the antibody. The amount of color is used to calculate results in exactly the same way as the amount of radioactivity is used to quantify ligand in a radioimmunoassay.

Although enzymes attached to ligand have been used to replace radioactive ligand in immunoassays, this alteration has not resulted in enhanced performance of the assays. However, it does positively impact safety considerations for technicians exposed to radioactivity and eliminates the need to dispose of radioactive waste. Even with these potential advantages, it is likely to be years before most laboratories switch from radioimmunoassay to enzyme immunoassay systems. The equipment needed to perform enzyme immunoassays and to quantify the color changes at the end of the assay is different than the equipment needed for radioimmunoassay. Since the equipment for either type of assay is quite expensive, most laboratories are reluctant to switch once a particular type of assay is established.

A second significant development in the field of radioimmunoassay that has been utilized for analysis of some substances has been the use of monoclonal antibodies. Unlike polyclonal antiserum, the monoclonal antibodies recognize only a single antigenic site on the ligand. Therefore, the assays are extremely specific and there is a reduced likelihood of interference from other compounds present in the biological medium being analyzed. This characteristic has prompted the development of monoclonal antibodies for numerous substances in recent years. However, production of monoclonal antibodies is much more difficult and expensive than production of polyclonal antisera. Furthermore, many monoclonal antibodies bind ligand with much lower affinity than polyclonal antisera. Since the sensitivity of a radioimmunoassay is, in part, dependent on the affinity of the antibody for ligand, many monoclonal antibodies are not useful for radioimmunoassay even though they bind ligand very specifically. Therefore, even with the advent of assay systems not utilizing radioactive ligands and the availability of monoclonal antibodies for many substances, the classical radioimmunoassay is still widely used and will continue to be for the foreseeable future.

Bibliography

Diczfalusy, E. (1969). *Immunoassay of Gonadotropins*. Bogtrykeriet Forum, Copenhagen.

Jaffe, B. M., and Behrman, H. R. (1979). *Methods of Hormone Radioimmunoassay*. Academic Press, New York.

Niswender, G. D., and Nett, T. M. (1977). Biological and immunological assay of gonadotropic and gonadal hormones. In *Reproduction in Domestic Animals* (H. H. Cole and P. T. Cupps, Eds.), pp. 119–142. Academic Press, New York.

Niswender, G. D., Akbar, A. M., and Nett, T. M. (1975). Use of specific antibodies for quantification of steroid hormones. *Methods Enzymol.* **36**, 16–34.

Yalow, R. S. (1990). Radioimmunoassay of peptide hormones: Problems and pitfalls. In *Hormones: From Molecules to Disease* (E. Baulieu and P. A. Kelly, Eds.), pp. 486–488. Chapman & Hall, New York.

Rats

see Rodentia

Receptors for Hormones, Overview

Kevin J. Catt

National Institutes of Health

GLOSSARY

endocytosis The uptake of proteins and other molecules into the cell. Many ligand–receptor complexes become clustered in plasma membrane structures called clathrin-coated pits, which invaginate into the cytoplasm and form endocytic vesicles containing the internalized proteins.

glycoprotein Many receptors and other cell surface proteins are glycosylated and contain from one to several carbohydrate chains on their extracellular regions.

G protein Guanine nucleotide regulatory (or binding) proteins serve as molecular switches that are activated when binding guanosine triphosphate (GTP) and become deactivated when their intrinsic GTPase activity hydrolyzes the bound GTP to guanosine diphosphate. G proteins are involved in many aspects of cell regulation and are of major importance in signal transduction at the plasma membrane level.

plasma membrane The phospholipid bilayer membrane that surrounds the cytoplasm of all eukaryotic cells. It contains numerous intrinsic proteins that span the lipid bilayer, including a panoply of receptors, ion channels, and transporters.

protein phosphorylation The covalent attachment of phosphate groups to serine, threonine, and tyrosine groups of proteins by specific kinases. Protein phosphorylation often leads to activation or inhibition of enzymatic and other regulatory functions and to interactions between signal transduction proteins.

signal transduction The translation of information reaching the cell surface into chemically coded signals that regulate intracellular effector systems in the cytoplasm and nucleus.

All cells express a wide variety of cell surface proteins through which they interact with the extracellular environment. Many of these proteins possess high-affinity binding sites that act as sensors for physiological regulatory substances in the interstitial fluid. They include numerous receptors for hormones and other ligands that regulate cellular responses. Hormonal ligands include peptides, proteins, steroids, and lipid derivatives that are formed locally or carried in the circulation from their sites of origin in other tissues. The traditional peptide and steroid hormones originate in the various endocrine organs, including the pituitary, adrenal, gonads, thyroid, and parathyroid glands, and constitute a relatively small proportion of the plethora of stimulatory and inhibitory factors that influence cellular function. However, they are of major importance in the control of cell growth, metabolism, and secretion in endocrine target tissues throughout the body. The specific binding of hormones to their cognate receptors in the plasma membrane is followed by the generation of biochemical signals that are released into the cell interior and influence the numerous cytoplasmic and nuclear processes that control cellular function.

I. INTRODUCTION

The receptor concept was introduced in pharmacology to account for the selective manner in which many drugs exert their actions on well-defined cellular responses. The term "receptive substance" was later modified to "receptor" to indicate the identity of a specific molecule that mediates the actions of drugs and hormones on their target cells. Many of the hormones are present in the circulation at picomolar (10^{-12} M) to nanomolar (10^{-9} M) concentrations, and their receptors are characterized by correspondingly high affinity (sensitivity) as well as high selectivity (specificity) for their ligands. The peptide, protein, and glycoprotein hormones, including gonadotropin-releasing hormone, prolactin, and pituitary gonadotropins, are specifically bound at the surface of their target cells by receptors that are embedded in the plasma membrane. The steroid hormones, including estradiol, progesterone, and testosterone, diffuse into all cells but are bound selectively by high-affinity receptor proteins that are present within their target cells. The active hormone–receptor complexes then act in the nucleus to regulate gene transcription and the expression of specific proteins.

II. PEPTIDE HORMONE RECEPTORS

The binding of peptide and protein hormones to their receptors depends on a rapid and strong interaction between the ligand and the extracellular binding domain of its receptor. The exposed portion of the receptor molecule on the cell surface is hydrophilic since it usually contains a high proportion of polar and charged amino acids and up to three or four carbohydrate chains. The size of the extracellular region varies from a few dozen amino acids up to several hundred residues that form a large ligand-binding domain on the cell surface. The placement of the hormone binding site varies according to the nature and size of the receptor. In many receptors, the hormone binding site is located in the extracellular region. However, in some (such as adrenergic receptors) it involves amino acids in the transmembrane region of the receptor, and in others (such as the parathyroid hormone receptor) it includes both extracellular and transmembrane residues. Whatever its location, the binding of hormone to the receptor site leads to a major conformational transition in the receptor protein. This includes changes in the cytoplasmic regions of the receptor that enhance its interaction with signal transduction proteins at the inner surface of the plasma membrane.

The location of receptors at the cell surface results from their anchoring in the plasma membrane by one or more hydrophobic regions, each containing about 23 amino acids arranged as a helical structure that traverses the lipid bilayer. The receptors for growth factors and cytokines have a single membrane-spanning domain, whereas many of those for neurotransmitters and peptide hormones contain seven such domains that are grouped as a bundle of helices within the lipid bilayer. Another major group is the family of ligand-gated receptor channels, which are composed of subunits that contain four transmembrane domains. While one major function of the transmembrane helices is to anchor the receptors in the plasma membrane, they are also of importance in the operation of the receptor during its activation by the agonist. Thus, in the seven-transmembrane domain family of receptors, the helices undergo movements that transmit the agonist-induced conformational change to the cytoplasmic aspect of the receptor, which interacts with one or more guanyl nucleotide regulatory (G) proteins that initiate signal transduction. In growth factor receptors, the transmembrane domains participate in interactions during ligand-induced dimerization, which in turn leads to the activation of intrinsic tyrosine kinase domains in their cytoplasmic regions. Receptors for cytokines, such as prolactin and growth hormone, also undergo dimerization after binding their agonists but do not possess intrinsic tyrosine kinase activity and are instead phosphorylated by cytoplasmic tyrosine kinases. In the receptor channels, the second transmembrane domains of each subunit participate in the formation of the ion channels and contain hydrophilic uncharged amino acids that favor the rapid influx of hydrated ions while the channel is open.

A. G Protein-Coupled Receptors

The G protein-coupled receptor (GPCR) family is an extremely large and diverse group of more than 1000 seven-transmembrane domain proteins that mediate the actions of numerous extracellular ligands. These include peptide and glycoprotein hormones; biogenic amines, purines, and other transmitters; alkaloids; prostanoids; numerous tastants and odorants; and early developmental signals. The archetypal receptor of this family is visual rhodopsin, a seven-transmembrane domain protein that interacts with the G protein, transducin, during photoreception in the retinal rod cells. The molecular cloning of the β-adrenergic receptor in 1986, and the recognition of its similarity to rhodopsin, was followed by the identification of numerous seven-transmembrane domain receptors which are coupled to a variety of heterotrimeric G proteins that resemble transducin (Fig. 1). Agonist binding to ligand recognition sites in GPCRs leads to changes in receptor conformation that alter the orientations of the transmembrane helices and expose previously masked regions of the intracellular loops. These regions, and sometimes others in the carboxy-terminal cytoplasmic tail of the receptor, then interact with a heterotrimeric G protein at the inner aspect of the cell membrane. When activated, these proteins interact with specific effector systems that include plasma membrane enzymes and ion channels (Table 1). Each of the several G proteins is composed of α, β, and γ subunits and can in turn activate a specific subset of signaling pathways and phenotypic cell responses. Thus, G_q and G_{11} activate phospholipase C and promote calcium mobilization; G_s activates adenyl cyclase and increases cyclic AMP production; and G_i inhibits adenyl cyclase and activates growth signaling pathways. Although most receptors are selectively coupled to G_s, G_i, or G_q, several have been found to interact with more than one G protein. When receptors have multiple subtypes, these are often coupled to distinct G proteins. Examples of this include the muscarinic and serotonin receptors, whose several subtypes are individually coupled to $G_{q/11}$, G_s, or G_i.

The localization of the heterotrimeric G proteins at the inner surface of the cell membrane depends on the presence of covalently attached lipids in the α and γ subunits. These increase the hydrophobicity of the subunits and favor the association of α and $\beta\gamma$ subunits with each other and with the plasma membrane. Each of the G protein α, β, and γ subunits is encoded by several genes. About 20 α subunits have been identified and are categorized into four main classes: α_s and α_i, which respectively activate and inhibit adenylate cyclase; $\alpha_{q/11}$, which activates phospholipase C; and others (α_{12}, α_{13}, α_z) of unknown function. There are at least 6 β and 12 γ subunits, so the potential number of $\alpha\beta\gamma$ trimers would exceed 1000. The β and γ subunits are released from the $\alpha\beta\gamma$ heterotrimer as tightly associ-

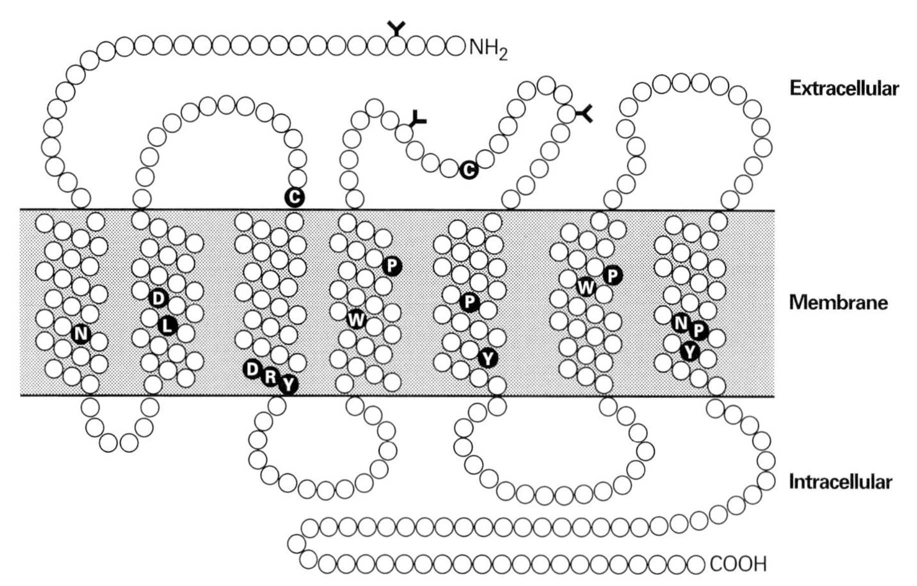

FIGURE 1 Secondary structure of the angiotensin AT_1 receptor, a typical G protein-coupled receptor. The seven hydrophobic helices are shown within the plasma membrane. The amino acids shown in black are conserved among many GPCRs and are involved in the structure of the receptor and its transduction mechanism. Three carbohydrate chains (Y) are shown attached to the extracellular region of the receptor. Several residues (not shown) in the second and third intracellular loops, and the intracellular C-terminal tail of the receptor, are involved in G protein coupling, desensitization, and endocytosis of the receptor.

TABLE 1
Plasma Membrane Receptors and Effector Systems

Receptor	*Primary effector system*
ACTH, β-adrenergic, calcitonin, CRF, dopamine D_1, FSH, GHRH, glucagon, LH, MSH, PACAP, PTH, serotonin $5HT_3$, TSH, vasopressin V_2, VIP	Adenylyl cyclase: activation (increased cyclic AMP)
Adenosine A_1, α_2-adrenergic, angiotensin AT_1, muscarinic M_1 and M_3, opiates, serotonin $5HT_1$, somatostatin	Adenylyl cyclase: inhibition (decreased cyclic AMP)
ANP	Guanyl cyclase: activation (increased cyclic GMP)
α_2-Adrenergic, angiotensin AT_1, GnRH, histamine H_1, muscarinic M_1, M_3, and M_5, PDGF, serotonin $5HT_2$, thrombin, vasopressin V_1	Phospholipase C: activation (increased inositol trisphosphate)
EGF, FGF, IGF-1, insulin, PDGF	Receptor tyrosine kinase: activation (tyrosine phosphorylation)
Activin, inhibin, MIS, TGF-β	Intrinsic receptor serine kinase: activation (serine, threonine phosphorylation)
Nicotinic acetylcholine	Sodium channel: activation (Na^+ influx)
AMPA, kainate, NMDA	Sodium/calcium channel: activation (Na^+, Ca^{2+} influx)
GAGA, glycine	Chloride channel: activation (Cl^- influx)
Muscarinic M_2	Potassium channel: activation (increased K^+ efflux)

Note. Abbreviations used: ACTH, adrenocorticotropin; AMPA, α-amino-3-hydroxy-5-methyl-isoxazole 4-propionic acid; ANP, atrial natriuretic peptide; CRF, corticotropin-releasing factor; EGF, epidermal growth factor; FGF, fibroblast growth factor; FSH, follicle-stimulating hormone; GABA, γ-aminobutyric acid; GnRH, gonadotropin-releasing hormone; GHRH, growth hormone-releasing hormone; 5-HT, 5-hydroxytryptamine; IGF, insulin-like growth factor; LH, luteinizing hormone; MIS, Müllerian-inhibiting substance; MSH, melanocyte-stimulating hormone; NMDA, *N*-methyl-D-aspartic acid; PACAP, pituitary adenylate cyclase-activating peptide; PDGF, platelet-derived growth factor; PTH, parathyroid hormone; TRH, thyrotropin-releasing hormone; TGF-β, transforming growth factor-β; TSH, thyroid-stimulating hormone, thyrotropin; VIP, vasoactive intestinal peptide.

ated $\beta\gamma$ complexes that act on a wide range of effector proteins, including certain isoforms of adenyl cyclase and phospholipase C, ion channels, and protein and lipid kinases. $\beta\gamma$ subunits are known to be important mediators of the growth and proliferative responses that are elicited by ligand activation of many GPCRs. For example, the $\beta\gamma$ subunits released after activation of many GPCRs cause activation of the MAP kinase cascade, a major pathway to growth and mitogenic responses in many hormone-stimulated cells. The guanosine diphosphate (GDP)-bound heterotrimeric G proteins are inactive until their interaction with the cytoplasmic surface of the agonist-activated receptor catalyzes the exchange of guanosine triphosphate (GTP) for GDP. Binding of GTP causes immediate dissociation of the G protein into its α-GTP and $\beta\gamma$ subunits, both of which can activate a variety of effector systems within the cell. The Gα subunit is a weak GTPase and hydrolyzes its bound GTP to GDP at a rate that determines the duration of the signaling process and the temporal characteristics of the cellular response to hormonal stimulation.

As noted previously, the seven-transmembrane helices of GPCRs are grouped into a bundle that surrounds a central cavity which corresponds to the space into which retinal is bound in the rhodopsin molecule. The polar side chains of the amino acids in each helix are thought to be oriented into the central cavity, whereas the hydrophobic side chains face the membrane lipids. All GPCRs contain a disulfide bond that connects the first and second extracellular loops and is necessary to maintain the integrity of the receptor. Although there is relatively little amino acid sequence similarity among most GPCRs, many contain a group of conserved residues that are mainly located in the transmembrane and cytoplasmic regions of the receptor (Fig. 1). Some of these are required for the structural arrangement of the receptor, and others are essential for the conserved mechanism of activation of the receptor during occupancy by agonist ligands. Consistent with the extremely high specificity of the 1000 or more GPCRs for their individual ligands, there is no generalized conservation of amino acid residues in the extracellular regions of these receptors. However, common features are present in these regions of receptor families (e.g., for glycoprotein hormones) and within those of receptor subtypes.

In addition to their signal transduction properties, most G protein-coupled receptors exhibit two other characteristic features: desensitization and endocytosis. During desensitization, the initial phase of agonist-stimulated receptor signaling is followed by a diminution or loss of activity that serves to limit the cellular response to receptor activation. Desensitization is caused by phosphorylation of serine or threonine residues located in the inner loops or cytoplasmic tail of the receptor. This is catalyzed by protein kinases that are activated during G protein-mediated intracellular signaling or by a family of specific receptor kinases that phosphorylate the ligand-activated receptor. During endocytosis, the hormone–receptor complex is internalized after clustering in specialized regions of the cell membrane termed coated pits. The coated pits invaginate into the cytoplasm and form vesicular structures that contain the hormone–receptor complexes. Agonist-induced endocytosis is followed by recycling of receptors to the cell membrane and variable degrees of ligand and receptor degradation. It is probable that receptor endocytosis is accompanied by dephosphorylation and resensitization of the receptors that are recycled back to the plasma membrane.

Several GPCRs have been found to exist in mutant forms that are associated with hormone resistance or with constitutive overactivity in the absence of agonist stimulation. Inactivating mutations are present in the adrenocorticotropin hormone receptor of patients with familial glucocorticoid deficiency, the growth hormone-releasing hormone receptor of a strain of dwarf mice, the V2 receptor of patients with X-linked diabetes insipidus, the luteinizing hormone (LH) receptor of familial gonadal resistance, and the follicle-stimulating hormone receptor of women with primary ovarian failure. Activating mutations in GPCRs have been found in the thyroid-stimulating hormone receptor in hyperfunctioning thyroid adenomas and autosomal-dominant hyperthyroidism, the LH receptor in familial male precocious puberty, the parathyroid hormone (PTH) receptor in familial hypoparathyroidism, the PTH/PTHrP receptor in metaphyseal dysplasia, and the rhodopsin receptor in congenital night blindness. These mutations are believe to act by disrupting intramolecular constraints that maintain the receptor in its inactive conformation in the absence of the agonist. An additional form of activating mutation, which mimics agonist binding to the melanocortin-1 receptor, is responsible for dark coat color in the mouse and fox.

B. Growth Factor Receptors

The cellular receptors for growth factors, such as insulin, insulin-like growth factor-I (IGF-I), epidermal growth factor (EGF), and platelet-derived growth factor, contain only a single transmembrane domain. These receptors possess a large N-terminal extracellular domain that contains the ligand binding site and a C-terminal intracellular domain that initiates signal generation. Growth factor receptors are activated by ligand-induced dimerization that leads to phosphorylation of tyrosine residues in their cytoplasmic domain. This is mediated by intrinsic tyrosine kinase domains in the cytoplasmic regions that initially phosphorylate tyrosine residues in the paired receptors and subsequently in other intracellular proteins. Such receptors are also termed "receptor tyrosine kinases." Important functions of tyrosine phosphorylation are to form binding sites for the attachment of adaptor proteins that link the activated receptors to other effector proteins (often enzymes) in the cytoplasm that transmit signals into the nucleus, and to activate gene expression and growth responses. Many growth factors undergo endocytosis after agonist binding, as described for GPCRs.

Signal transduction from the plasma membrane to intracellular effector systems is mediated by a series of highly specific protein–protein interactions that depend on modular binding domains in the complementary proteins. Such interactions often require the presence of an intermediate adaptor protein that contains the binding modules and serves solely as a link between the two proteins. Adaptors such as Grb2 and Shc are small proteins that contain two distinct types of binding modules (termed SH2 and SH3) that respectively interact with phosphotyrosine motifs (e.g., on growth factor receptors) and proline-rich motifs in GTP exchange proteins and several protein kinases.

C. Receptor Serine Kinases

This family of receptors also possesses a single transmembrane domain and is characterized by the presence of a serine kinase domain in the intracellular region of the receptors. These receptors are activated by inhibins and activins as well as by transforming growth factor-β (TGF-β) and Müllerian-inhibiting substance. Other receptors in this family are involved in the control of early embryonic development and bone formation. Their protein ligands are disulfide-linked dimers that are rich in cysteine residues and have an extended conformation that includes a "knot" formed by the several cysteines and other conserved residues. The serine kinase receptors have a relatively small extracellular domain and exist as type I and type II isoforms that undergo heterodimerization during agonist binding. The specificity of ligand binding is determined by the type II receptor, and the formation of a heterodimeric complex with the type I receptor is necessary for the generation of agonist-induced signaling responses.

D. Cytokine Receptors

Cytokines comprise a heterogeneous group of small proteins that are released during activation of immune and hematopoietic cells and exert local regulatory actions on numerous tissues and cell types. Cytokines exert pleiotropic actions as regulators of immune responses and inflammatory reactions as well as cell development, differentiation, and proliferation. The cytokines are also notable for their redundancy in that several distinct cytokines may elicit similar cellular responses. Structurally, cytokines are characterized as "helix bundle" proteins due to their content of four long α-helices that are arranged in an antiparallel manner. Cytokines include several interleukins and hematopoietic factors as well as hormones such as prolactin and growth hormone. Other cytokines are the interferons and the tumor necrosis factor α and leptin. The cytokine receptors have a single transmembrane domain and a relatively large extracellular domain that contains the ligand binding site and sequences required for dimerization. The cytoplasmic domains are of variable length and contain proline-rich regions that participate in binding to adaptor proteins and downstream signaling systems. Several of the cytokine receptors are released from the cell surface as soluble binding proteins that enter the circulation and retain high affinity for their cognate ligands. One such soluble binding domain is released from the GH receptor and has been utilized to determine the nature of the interaction between growth hormone and its receptor site. The binding of GH and prolactin involves the sequential association of one hormone molecule with two receptor molecules—the first with high affinity and the second with low affinity—to produce the dimeric form of the receptor.

Many cytokine receptors exist as homo- or heterodimeric complexes or associate with other protein subunits. However, higher ligand:receptor ratios occur between some of the cytokines and their receptors. In addition, many cytokine receptors tend to form multimeric complexes that contain a common signaling subunit. This feature can account for the ability of several cytokines to elicit similar cellular responses and is the structural basis of the functional redundancy of cytokines. Such common chains appear to be necessary for high-affinity ligand binding as well as for mediating signal transduction. The cytokine receptors do not possess intrinsic protein kinase activity, but their dimerization or oligomerization leads to recruitment of cytoplasmic tyrosine kinases (termed Janus kinases or Jaks) that phosphorylate specific tyrosine residues in their cytoplasmic domains. The Jaks also phosphorylate adaptor proteins and signaling proteins termed STATS, which form dimers and move into the nucleus to initiate gene transcription. A small but growing group of cytokines includes the chemoattractant factors or chemokines, which differ from the majority of this family by acting on G protein-coupled receptors to initiate their intracellular signaling pathways and elicit cellular responses. Some of the chemokine receptors have been found to act as essential coreceptors during the entry of the HIV virus into macrophages and T cells.

III. STEROID HORMONE RECEPTORS

The gonads and the adrenal gland secrete a variety of steroid hormones, including estrogen, progesterone, androgens, glucocorticoids, and mineralocorticoids. These lipophilic molecules diffuse into cells throughout the body and are selectively bound by highly specific receptors located in the cytoplasm or nucleus of their target cells. The hormone receptors are latent transcription factors that are associated with heat shock proteins in the absence of their steroid ligands. After hormone binding, the receptors are freed from the heat shock proteins and form dimers that bind with high affinity to short DNA sequences [hormone response elements (HREs)] located in the promoter regions of hormone-regulated genes. Thyroid hormone, vitamin D, retinoic acid, and 9-*cis* retinoic acid also bind to and activate specific intracellular receptors that act as nuclear transcription factors in their respective target cells. Most of the members of this superfamily of intracellular receptors are associated with the nucleus in the absence of hormone and become more tightly bound to DNA and initiate transcription once activated by the ligand. Steroid hormones are also present in insects (ecdysone) and plants (brassinosteroids) and are important regulators of growth and development. In addition to their actions at the nuclear level, several steroid hormones exert rapid, nongenomic effects at the plasma membrane that result in early signaling responses and other actions on cellular function.

A. Functional Domains

The steroid hormone receptors contain several molecular domains that subserve specific functions in the quiescent and activated receptors. These include ligand binding, interaction with heat shock proteins, receptor dimerization, binding to DNA, and transcriptional activation of the target gene (Fig. 2). The functional domains vary in size, and some functions involve more than a single domain of the receptor molecule. There are two dimerization domains, and up to three activation domains have been identified

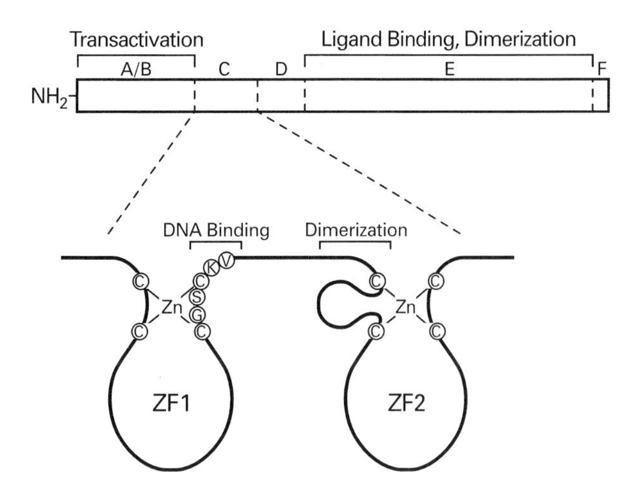

Receptors	DNA Binding Domains	Response Elements
GR, MR, PR, AR	-Cys-**Gly-Ser**-Cys-Lys-**Val**-	AGAACAnnnTATXCX
ER	-Cys-**Asn-Gly**-Cys-Lys-**Ala**-	AGGTCAnnnTGACCT
TR, VDR, RAR, RXR	-Cys-**Asn-Gly**-Cys-Lys-**Gly**-	AGTTCA(n_x)AGTTCA (n_x)

FIGURE 2 Structural features of a typical steroid hormone receptor. (Top) The modular arrangement of the several functional domains of the receptor. A/B, transcriptional activation; C, DNA binding; D, hinge region; E, ligand binding, dimerization, heat shock protein binding, and transactivation; F, repressor function. (Middle) The formation of zinc fingers and the adjacent dimerization region by the amino acids of the receptor's binding domain. The zinc fingers of the receptor dimer bind within the major groove of the DNA to nucleotides that comprise the hormone response elements in the promoter regions of hormone-regulated genes. (Bottom) The major DNA-binding domains of the steroid hormone receptors and the sequences and orientations of their DNA response elements.

in different regions of the receptor. In general, the N-terminal region of the receptor is concerned largely with transactivation and the C-terminal region with ligand binding. The ligand-binding domain occupies most of the C-terminal half of the molecule, which also participates in the association with heat shock proteins, receptor dimerization, nuclear localization, and transactivation. The extreme C-terminal region also contains a transcriptional inhibitory function that constitutively represses the activation domains and becomes sequestered into the more compact conformation of the agonist–receptor complex. This and some of the other functions involve only small regions of the receptor molecule, but others require much of its C-terminal portion. The dimerization domain includes leucine-rich sequences that favor interactions between receptors and is located toward the C-terminal end of the ligand-binding domain. An interaction between the ligand-binding domain and heat shock proteins maintains the unliganded receptor in its inactive state.

Adjacent to the ligand-binding domain are the hinge region, which participates in receptor folding and localization to the nucleus, and the DNA-binding domain. The latter region contains two specialized amino acid sequences, termed zinc fingers, that engage in DNA binding and dimerization. The remainder of the N-terminal portion of the receptor is highly variable in size and amino acid sequence and is essential for the transcriptional activation of individual hormone-regulated genes. This region of the receptor recruits coactivator and coinhibitor proteins and interacts with the transcription factor complex that binds to the TATA box or other initiator elements to promote gene transcription.

B. Receptor Phosphorylation

All steroid receptors are phosphoproteins, and many undergo a rapid increase in phosphorylation after binding their respective hormones. This covalent modification appears to enhance the receptor's capacity for gene activation and may facilitate the transfer of receptors from the cytoplasm to the nucleus. In some cases, hormone-induced phosphorylation occurs after binding of the activated receptor to its HRE and is catalyzed by a DNA-dependent protein kinase. In addition to such ligand-dependent phosphorylation, certain hormone receptors are phosphorylated during cell stimulation by agonists that increase intracellular cyclic AMP levels. For example, dopamine causes phosphorylation and activation of estrogen and progesterone receptors, effects that could account for its central actions on sexual behavior in female rats. Some of the synergistic actions of adenyl cyclase activation may result from cooperative interactions between steroid receptors and cAMP-dependent transcription factors at the DNA-binding level. Several growth factors (EGF, IGF-I, and TGF-α) have also been found to activate estrogen receptors and stimulate gene transcription, presumably by promoting tyrosine phosphorylation. It is clear that more than one mechanism can induce the conformational changes in steroid hormone receptors that are required for DNA binding and gene transcription. Such "cross talk" between peptide and steroid hormone receptors increases the range of signaling pathways through which plasma membrane receptors can influence gene transcription and cell growth. Furthermore, steroid hormones can increase the expression of growth factors and can enhance the signaling responses generated by plasma membrane receptors. The ability of the major hormonal stimuli to interact in this manner could provide positive feedback between their signaling pathways and enhance the biological responses of their target cells.

C. Receptor Activation and DNA Binding

In the absence of their activating ligands, the receptors for estrogen (ER), progesterone (PR), androgen (AR), glucocorticoids (GR), and mineralocorticoids (MR) exist as inactive complexes in association with heat shock proteins. After binding their respective hormones, the steroid–receptor complexes undergo conformational changes and assume a more compact structure. This leads to exposure of domains that favor their association into dimers and removes the inhibitory action of the C-terminal region. The dimeric receptors attach to HREs in the promoter region of steroid-responsive genes by their DNA-binding domains and induce gene transcription through their transactivation domains. In contrast to the traditional steroid hormone receptors, the receptors for thyroid hormone, vitamin D, retinoic acid, and 9-*cis* retinoic acid (TR, VDR, RAR, and RXR) do not form complexes with heat shock proteins. These receptors have relatively short N-terminal transactivation domains and bind to DNA in the absence of hormone, either as monomers or dimers. In the presence of hormone, these receptors also form heterodimers, especially with RXR, that can regulate target genes by activation or repression of their transcriptional activity. In addition to promoting receptor dimerization and binding to the HRE, the hormone seems to be necessary for optimal transactivation as well as for the rapid dissociation of the receptor from its DNA-binding element. When activated by ligand binding, these receptors form homodimers (or heterodimers with RXR) that bind tightly to their HREs and stimulate gene transcription. Some of the steroid receptors (e.g., for androgens) bind to DNA half-sites and exert inhibitory effects on gene transcription. The two subgroups of "steroid-type" receptors differ in their initial locations and modes of activation by hormones, but they act in a similar fashion to initiate gene transcription in their target cells.

The DNA-binding domains of steroid hormone receptors are highly conserved and possess two adjacent zinc finger motifs that contain four cysteine residues coordinately linked to a zinc atom. The binding specificity of each steroid receptor is encoded by three amino acids at the base of the first zinc finger. Some of the amino acids in the second zinc finger form a dimerization domain and others interact with the DNA helix adjacent to the HRE. The DNA sequences of the HREs, and their spacing and orientation, are responsible for their selectivity in binding the individual hormone receptors.

D. Transactivation of Target Genes

The mechanism by which steroid hormone receptors activate target gene expression involves several steps that are associated with binding to their cognate DNA response elements. These include the recruitment of multiple coregulatory proteins and interaction with the general transcription factors that mediate receptor regulation of the RNA polymerase transcriptional machinery. Two important coactivators are the cyclic AMP response element-binding protein and the general steroid coactivator, SRC-1. Several receptor coactivators promote histone acetylation and thereby contribute to the remodeling of repressed chromatin that is necessary for assembly of the preinitiation complex that ultimately stimulates transcription. Conversely, coinhibitors may recruit histone deacetylases to the promoter and increase repression at the chromatin template. The importance of coregulatory proteins in steroid hormone action is indicated by the recent finding that deletion of the *SRC-1* gene causes partial hormone resistance in mice.

E. Steroid Hormone Antagonists

Steroid hormone antagonists exert their inhibitory actions not only by competing for agonist binding to the receptor but also by inducing partial conformational changes that expose the dimerization domains and promote the formation of receptor dimers. Due to their incomplete change in conformation, the dimeric antagonist–receptor complexes bind with high affinity to the HRE but cannot initiate transcription due to the inhibitory action of the C-terminal repressor function on their activation domain(s). Many antagonists promote receptor phosphorylation as well as DNA binding, and some can act as partial agonists. This could result from antagonist-induced exposure of activation functions in the receptor that are not inhibited by the C-terminal suppressor sequence. The tight binding and slow dissociation of the inactive antagonist–receptor complexes at the HRE favor their competition with agonist-activated receptors for DNA binding and the consequent inhibition of steroid hormone action. It is likely that certain antagonists increase the ability of the receptor to interact with corepressor proteins that inhibit the initiation of transcription at the DNA level.

F. Reproductive Hormone Receptors

All members of the steroid hormone receptor family share most of the common characteristics described previously. The receptors for gonadal steroid hormones are of major importance in the development and function of the reproductive system and will be described in more detail to indicate their specific features as major regulators of hormone target tissues.

1. *Estrogen Receptors*

Estrogen receptors are abundantly expressed in all female reproductive tissues and also in the brain, pituitary gland, myocardium, and vascular smooth muscle cells. They are also present in many breast cancers and serve as therapeutic targets for receptor antagonists such as tamoxifen. Upon activation by estradiol, the ER binds to a relatively specific response element, the ERE, and initiates transcription of a wide variety of estrogen-regulated genes. In addition to the originally defined estrogen receptor (ERα), a second form (ERβ) has been identified and is present in several tissues, including the ovary, breast, prostate, and decidua, as well as brain, bone, heart, and vascular smooth muscle. The α and β receptors can form heterodimers, raising the possibility that there are at least two signaling pathways in tissues that express both subtypes.

Some estrogen antagonists, such as tamoxifen, behave as partial agonists and exert stimulatory actions in the uterus and other tissues. Recently developed compounds, such as raloxifene, inhibit estrogen action in the breast and uterus but exert estrogen-like actions on bone and lipid metabolism. There is currently great interest in the clinical applications of these and other compounds with the ability to act as selective estrogen receptor modulators (SERMs). The recent demonstration that prolonged treatment with tamoxifen significantly reduces the incidence of breast cancer has emphasized the potential importance of SERMs for the tissue-selective therapy of conditions arising from inappropriate estrogen action in a variety of hormone target cells.

2. *Progesterone Receptors*

Progesterone receptors are present not only in the female reproductive tract but also in blood vessels, the urinary tract, prostate stromal cells, and many meningiomas. The human PR exists as two forms, termed PR-A and PR-B, both of which are derived from a single gene. The relative abundance of the two forms differs during the menstrual cycle, and uterine fibroids often contain an excess of PR-A. Both PR-A and PR-B exist as inactive forms in association with heat shock proteins within the nucleus, in which hormone binding causes their dissociation from the heat shock proteins, dimerization, and phosphorylation. Highly effective progesterone antagonists such as RU 486 (mifepristone) and related compounds have been derived from 19-nor-testosterone. These agents act as pseudoagonists by binding to and partially activating the PR, leading to phosphorylation and dimerization. However, the activation domains remain under the inhibitory control of the C-terminal repression domain and cannot initiate transcription. The PR-B acts as a positive regulator of many progesterone-responsive genes, whereas the PR-A has a more limited range of positive actions. In fact, the PR-A can inhibit PR-B function as well as the activities of other steroid hormone receptors, including the ER and the AR. The inhibitory action of RU 486 includes an effect on PR-A which impairs not only progesterone's actions but also those of estrogen. These effects probably account for the ability of RU 486 to act as an antiestrogen in several tissues and also as a potent antiglucocorticoid. The RU 486-bound PR-B interacts with a corepressor that inhibits its transcriptional activity, probably by masking its transactivation domains.

3. *Androgen Receptors*

The AR belongs to a small subgroup of closely related steroid receptors that includes PR, GR, and MR. These four receptors share considerable sequence homology and interact with a common HRE to activate gene transcription. The AR is expressed in the testis, prostate, epidermis, and genital skin fibroblasts as well as in the pituitary, kidney, larynx, and liver. The receptor has high affinity for dihydrotestosterone and testosterone and lower affinity for weaker androgens and progesterone. Agonist binding to the cytoplasmic receptor causes its release from the inactive HSP complex followed by translocation to the nucleus, where it positively or negatively influences gene transcription. In addition to binding to the same HRE as the PR, GR, and MR, the dimeric AR interacts with androgen-specific elements that include binding sequences for other transcription proteins. Like other steroid hormone receptors, the AR also interacts with coactivator proteins that modulate its biological actions and contribute to the specificity of androgen action. In addition to its nuclear actions on gene transcription, the AR has several cytoplasmic actions that include tyrosine phosphorylation and activation of the MAP kinase pathway, with consequent effects on cell proliferation or differentiation. Androgens play a major role in the growth of the prostate gland and act in conjunction with peptide growth factors. The progression of prostate cancer is reduced by androgen ablation, but hormone resistance often develops and is sometimes associated with mutations in the androgen receptor. The androgen receptor also exhibits a number of mutations and deletions that result in androgen insensitivity and are associated with clinical manifestations that range in severity from male pseudohermaphroditism to micropenis, hypospadia, and cryptorchidism.

See Also the Following Articles

ANDROGENS; ESTROGENS, OVERVIEW; GROWTH FACTORS; PROGESTERONE ACTIONS ON REPRODUCTIVE TRACT; STEROID HORMONES, OVERVIEW

Bibliography

Beato, M., and Sanchez-Pacheco, A. (1996). Interaction of steroid hormone receptors with the transcription initiation complex. *Endocr. Rev.* **17**, 587–609.

Donnelly, D., Findlay J. B. C., and Blundell, T. M. (1994). The evolution and structure of aminergic G protein-coupled receptors. *Receptors Channels* **2**, 61–78.

Drewett, J., and Garbers, D. L. (1994). The family of guanylyl cyclase receptors and their ligands. *Endocr. Rev.* **15**, 133–162.

Dufau, M. L. (1998). The luteinizing hormone receptor. *Annu. Rev. Physiol.* **60**, 461–496.

Gutkind, J. S. (1998). The pathways connecting G protein-coupled receptors to the nucleus through divergent mitogen-activated protein kinase cascades. *J. Biol. Chem.* **273**, 1839–1842.

Hu, Z.-Z., Zhuang, L., and Dufau, M. L. (1998). Prolactin receptor gene diversity: Structure and regulation. *Trends Endocrinol. Metab.*, in press.

Matthews, L. S. (1994). Activin receptors and cellular signaling by the receptor serine kinase family. *Endocr. Rev.* **15**, 310–325.

O'Malley, B. W., Schrader, W. T., Mani, S., Smith, C., Weigel, N. L., Conneely, O. M., and Clark, J. H. (1995). An alternative ligand-independent pathway for activation of steroid receptors. *Recent Prog. Horm. Res.* **50**, 333–347.

Pratt, W. B., and Toft, D. O. (1997). Steroid receptor interactions with heat shock proteins and immunophilins. *Endocr. Rev.* **18**, 306–360.

Premont, R. T., Inglese, J., and Lefkowitz, R. J. (1995). Protein kinases that phosphorylate activated G protein-coupled receptors. *FASEB J.* **9**, 175–182.

Quigley, C. A., DeBellis, A., Marschke, K. B., El-Awady, M. K., Wilson, E. M., and French, F. S. (1995). Androgen receptor defects: Historical, clinical, and molecular perspectives. *Endocr. Rev.* **16**, 271–321.

Reveli, A., Massobrio, M., and Tesarik, J. (1998). Nongenomic actions of steroid hormones in reproductive tissues. *Endocr. Rev.* **19**, 3–17.

Shibata, H., Spencer, T. E., Onate, S. A., Jenster, G., Tsai, S. Y., Tsai, M. J., and O'Malley, B. W. (1997). Role of co-activators and co-repressors in the mechanism of steroid/thyroid receptor action. *Recent Prog. Horm. Res.* **52**, 141–164.

Smith, C. L. (1998). Cross-talk between peptide growth factor and estrogen receptor signaling pathways. *Biol. Reprod.* **58**, 627–632.

Strader, C. C., Fong, T. M., Graziano, M. P., and Tota, M. R. (1995). The family of G-protein-coupled receptors. *FASEB J.* **9**, 745–754.

Van Biesen, T., Luttrell, L. M., Hawes, B. E., and Lefkowitz, R. J. (1996). Mitogenic signaling via G protein-coupled receptors. *Endocr. Rev.* **17**, 698–714.

Van der Geer, P., Hunter, T., and Lindberg, R. A. (1994). Receptor protein tyrosine kinases and their signal transduction pathways. *Res. Cell Biol.* **10**, 251–337.

Wells, J. A. (1995). Binding in the growth hormone receptor complex. *Proc. Natl. Acad. Sci. USA* **93**, 1–6.

Reflex (Induced) Ovulation

Arnold L. Goodman

Boca Raton, Florida

GLOSSARY

exteroceptors Sensory (afferent) neural receptors that are activated by external stimuli or environmental cues; in contrast to interoceptive (afferent) neural receptors that are activated by internal stimuli.

hypophysiotropic Describing hormones or other biochemical agents that regulate (stimulate or inhibit) the secretion of anterior pituitary hormones.

neuroendocrine reflex A physiological mechanism by which an external or internal stimulus activates afferent (exteroceptive or interoceptive) and then central neurons to alter the secretory rate and blood levels of a specific hormone, i.e., a neural stimulus triggers a stereotypical endocrine response.

ovulation The final stage of oogenesis in which the oocyte is released from the ruptured ovarian follicle and (typically) completes meiosis to become the haploid ovum in preparation for fertilization.

surge mode of gonadotropin secretion A brief (hours long) episode of accelerated hormone secretion that rapidly produces a transient, monophasic pulse or "surge" in blood hormone levels many-fold above the tonic levels that ordinarily prevail.

syngamy The fusion of gametes (ovum and sperm) initiating fertilization.

tonic mode of gonadotropin secretion The classical homeostatic (or negative feedback) set-point relationship between the minute-to-minute secretion of pituitary gonadotropic and gonadal hormones characterized by (i) persistently "low" blood levels of gonadotropins in intact individuals producing adequate amounts of gonadal hormones to inhibit the overproduction of gonadotropins and (ii) persistently "high" blood levels of gonadotropins in individuals with impaired gonadal hormonal production (e.g., after castration or menopause).

Among well-established physiological criteria that distinguish mammals by reproductive pattern is the dichotomous relationship between copulation and ovulation: On one side of this dichotomy, ovulation occurs only after—and as a coordinated neuroendocrine response to—copulation, thus the designation "reflex ovulation" "(coitus-)induced ovulation," or even more formally, "obligate reflex ovulation." On the other side, ovulation occurs without dependency on coition or even on the proximity of the male and is considered "spontaneous." The dichotomy is not always strict. There are instances of "facultative" reflex ovulation: In some species, copulation can trigger ovulation in otherwise spontaneous ovulators made anovulatory by experimental intervention, whereas in others copulation reportedly hastens spontaneous ovulation. Reflex ovulators include species as taxonomically diverse, and as geographically remote, as the rabbit, ferret, house cat, beaver, vole, raccoon, mink, llama, camel, and bottlenose dolphin. Spontaneous ovulators include the higher primates (macaques, apes, and humans) as well as more common livestock (cattle, sheep, horses, pigs, and goats) and laboratory animals (rats, mice, and guinea pigs).

I. NEUROENDOCRINE COMPONENTS

A. Regulation of the Ovulatory Surge of Luteinizing Hormone: The Crux of the Disparity between Reflex and Spontaneous Ovulators

Although not all species have been studied in comparable detail, a vast body of evidence can be adduced to support the reasonable inference that the physiological trigger for ovulation in virtually all (eutherian) mammalian species is hormonal and indeed is the same signal, namely, a blood-borne surge of luteinizing hormone (LH). Even in ovaries transplanted to remote sites or otherwise denervated, the (pre-)ovulatory LH surge (or injection of exogenous LH) initiates a complex cascade of ovarian events that comprise the process of ovulation in virtually all mammals: final ripening and swelling of the Graafian follicle, focal dissolution and rupture of the follicular wall, release of the enclosed oocyte, and (in most species) resumption of meiosis and extrusion of polar bodies to produce the haploid ovum that awaits fertilization.

In both reflex and spontaneous ovulators, the preovulatory surge of LH levels in blood is the direct result of a large and rapid increase in LH secretion from gonadotrope cells in the anterior pituitary gland (i.e., a surge of LH secretion). Moreover, in both reflex and spontaneous ovulators, both the tonic and surge modes of LH secretion are absolutely dependent on a hypophysiotropic-releasing hormone [LHRH or gonadotropin-releasing hormone (GnRH)] produced by hypothalamic neurosecretory cells.

Despite this common dependence on GnRH, there is a divergence in the regulation of tonic and surge modes of LH secretion in spontaneous versus reflex ovulators: In spontaneous ovulators in which fertility of the female is linked or limited to certain seasons of the year (e.g., sheep), the tonic mode of LH secretion and its regulation by GnRH is unequivocally sensitive to and altered by external environmental cues (e.g., the amount of daylight or the daily light:dark ratio). In contrast, the onset of the preovulatory LH surge mode in all spontaneous ovulators is presumably regulated solely by neuroendocrine signals arising internally from components of the hypothalamic–pituitary–ovarian (HPO) axis, which may not necessarily include an antecedent "surge" in GnRH secretion (i.e., GnRH secretion is necessary but only permissive).

In reflex ovulators, whether or not tonic LH secretion is seasonally dependent, initiation of the surge mode of LH secretion is ipso facto a neuroendocrine response triggered by external signals arising from copulation that traverse the HPO axis. This neuroendocrine reflex necessarily includes activation of a "surge mode" of GnRH secretion.

Thus, the crux of the physiological distinction between reflex and spontaneous ovulation traces to the neuroendocrine activation of the surge mode of LH secretion by the hypothalamic hypophysiotropic hormone, LHRH (GnRH). Clearly, the adjective, "reflex," in the term, "reflex ovulation," must be understood not as a modifier of "ovulation," per se, which occurs at the ovary, but rather as a qualifier describing the release of hypothalamic LHRH (GnRH) and the ovulation-causing LH surge from the anterior pituitary.

B. Copulation-Induced Ovulation as a Neuroendocrine Reflex

Analogous with a spinal reflex arc, a neuroendocrine reflex arc can be viewed as composed of a sensory receptor(s) within a receptive field, an afferent neural limb, central "switching" (interneuronal) and transducing (neurosecretory) elements, an efferent endocrine limb, and, finally, the effector cells, tissue, or organ responsive to the effector hormone. A straightforward example of a neuroendocrine reflex is suckling-induced milk ejection: Stimulation of mammary proprioreceptors by sucking at the nipple is conveyed by mammary nerves (afferent limb) to the central nervous system; other central interneurons convey the signal to hypothalamic magnocellular neurons (transducing elements) to stimulate the neurosecretion of oxytocin into capillaries in the posterior pituitary. The secreted oxytocin hormone (efferent endocrine limb) circulates to the mammary gland wherein it stimulates the contraction of myoepithelial cells (effector cells) to propel milk through ducts from secretory cells in the gland through pores in the nipple.

Unfortunately, copulation-induced ovulation has proven to be a more complex and arcane neuroendocrine reflex. Despite intensive and often ingenious investigation over several decades, key aspects of the neuroendocrine mechanisms subserving copulation-induced ovulation remain only poorly understood.

1. *Exteroceptive Stimuli and the Afferent Limb*

A fundamental obstacle to mapping the afferent limb of this neuroendocrine reflex arc is the difficulty encountered in identifying precisely the exteroceptive field(s) and receptor(s) activated by external stimuli accompanying copulation. Careful and thorough experiments in rabbits, the best studied example of all reflex ovulators, by Walter Heape as early as 1905 and by later workers convincingly eliminated individual likely candidates, including stimulation of the vulva, vagina, or cervix and deposition of semen or sperm. Instead, rather than containing a single, obligatory sensory modality, the female's exteroceptive field for copulatory stimuli (which Ramirez and Soufi term the "copuloceptive" field) extends well beyond the genitalia and likely includes receptors for tactile, olfactory, visual, and auditory stimuli. (Indeed, it is well-known to rabbit breeders that ovulation and pseudopregnancy occur in does housed together and apart from a buck and can be avoided by preventing does from mounting each other, or what is quaintly euphemized as "jumping.") Copulation presumably needs to activate several modalities in parallel since any one alone is neither necessary nor sufficient to activate the reflex arc (at least in rabbits).

2. *Central Pathways and Transducing Elements*

In light of the difficulty of identifying precise exteroceptive stimuli provided by copulation, tracing this neuroendocrine reflex through the central nervous system to the anterior pituitary gland has posed enormous experimental challenges. Although knowledge of the central aspects of the reflex arc remains fragmentary, findings from a variety of experimental approaches have provided important insights into key intracranial components.

i. Anatomical Studies First and foremost, careful anatomical studies by G. W. Harris and others implicated a vascular link, the hypophysial stalk portal vessels, rather than neural tracts as the important pathway from brain to anterior pituitary. These portal vessels connect capillaries in the median eminence of the hypothalamus with capillaries in the anterior pituitary and serve as the major physiological conduit for all the hypothalamic hypophysiotropic hormones, including GnRH (LHRH), to reach anterior pituitary secretory cells. Other studies involving precisely placed stereotaxic lesions demonstrated the essential participation of the hypothalamus in coitus-induced ovulation, whereas extrahypothalamic lesions were generally ineffective in blocking the reflex. Still other lesion experiments (in rabbits) demonstrated anatomical separation of brain areas involved in coitus-induced ovulation and in estrous behavior (see Section I,B,3,iii).

ii. Electrophysiological Studies Two kinds of electrophysiological studies have been used to investigate reflex ovulation in the rabbit: electrical stimulation of specific brain areas to provoke ovulation and single-unit and multiunit recordings to monitor neural activity before and after mating. Results from the former implicated the hypothalamus as an important participant in the reflex arc, whereas results from the latter of a diffuse distribution of neural activity reinforced the notion that multiple and perhaps independent afferent neural limbs and interneurons are activated by coitus.

iii. Pharmacological Studies C. H. Sawyer and others, using carefully timed administration of pharmacological agents known to interfere with neural transmission, were able to block postcoital ovulation in estrous rabbits. These early studies with antagonists implicated both cholinergic- and catecholaminergic-mediated pathways in the neural regulation of postcoital LH release. Later studies, in which norepinephrine injected into the brain's third ventricle evoked LH surges indistinguishable from postcoital surges, corroborated the participation of a catecholamine-mediated pathway as an activator of the hypothalamic–pituitary components of the neuroendocrine reflex. Recent studies have linked the neurotransmitter neuropeptide Y (NPY) to GnRH secretion from the hypothalamus and to tonic LH secretion, but its precise role, if any, in mediating the postcoital LH surge is unclear.

iv. Push–Pull Perfusion Studies Recent studies using precisely placed push–pull perfusion cannulae in rabbits have done much to confirm and extend current knowledge on the relationship between candidate brain neurotransmitters, especially norepinephrine, female steroid hormones, GnRH release, and LH secretion in this reflex ovulator.

3. Endocrine Effector Limb and Ovulation

i. An Effector Limb Cascade Because of the vascular connection between hypothalamus and hypophysis, the hypophysiotropic LHRH released from hypothalamic neurons can be construed as both a neuroendocrine/neurosecretory transduction element and an intermediate (or proximate) endocrine effector carried by the hypophysial portal blood to pituitary gonadotropes, which represent intermediate (or proximate) effector cells that respond, in turn, by releasing the LH surge. As explained in Section I,A, the LH surge is considered the ultimate endocrine efferent limb that stimulates ovulation at the ovary, the ultimate effector organ.

ii. Comparative Temporal Aspects of the LH Surge and Ovulation In the rabbit, copulation is quite brief, lasting only a few seconds, but a single congress is sufficient to trigger the release of the LH surge, which lasts at least 4 hr, with a peak about 2 hr after mating. Ovulation typically occurs 9–12 hr postcoitum. In other species (ferret, cat, and mink), the interval from mating to ovulation is longer (24–72 hr), but the comparatively prolonged interval seems attributable to ovarian rather than hypothalamic–pituitary factors.

iii. Ovarian Regulation of the Coitus-Induced LH Surge Ovarian hormones, presumably chiefly estradiol-17β, are required not only to induce the behavioral estrus that permits copulation but also to maintain the responsiveness of key components of the neuroendocrine reflex to the copulatory stimulus. This inference originally derived from the observation that the pattern of postcoital LH release in estradiol-treated rabbits mated soon after ovariectomy mimicked the postcoital pattern in intact does, whereas postcoital LH release in estradiol-treated rabbits ovariectomized more than 30 days earlier was impaired. Recent push–pull perfusion studies (see Section I,B,2,iv) have confirmed that GnRH release in response to infusions of norepinephrine or NPY depends on adequate preexposure to estradiol. Despite earlier expectations to the contrary, ovarian hormones are not required after copulation to sustain or amplify the LH surge (at least in the rabbit) since bilateral ovariectomy within minutes after mating does not alter the typical postcoital pattern of LH in blood.

II. PHYSIOLOGICAL IMPLICATIONS

A. Evolutionary or Adaptive (In-)Significance

Reflex (vs spontaneous) ovulation, like other differences in reproductive patterns, inevitably provokes speculation of potential evolutionary significance or adaptive advantage. Just as the ovulatory LH surge seems virtually universal in mammals, so too is the evanescent life span of mature mammalian gametes once released from their gonads. In light of mature gametes' fleeting existence, copulation-induced ovulation, along with the implicit copulation-induced ejaculation of sperm into the vagina, seems an especially efficient means of ensuring the well-timed fusion of viable sperm and ovum. In nonprimates, however, copulation occurs only during brief periods of female sexual receptivity (estrus). Since estrus is synchronized with the presence of mature follicles and the preovulatory LH surge in (nonprimate) spontaneous ovulators, estrus alone is a sufficient strategy to restrict copulation, and the vaginal deposition of sperm, to the periovulatory interval. Thus, reflex ovulation and estrus can be seen as redundant mechanisms ensuring timely syngamy; that this kind of redundancy is especially advantageous, more primitive, or more specialized is moot. Clearly, since many species of both kinds of ovulator are extant, neither type of ovulation has impaired the fecundity or compromised the survival of members of the respective groups.

B. Regulation of Tonic Gonadotropin Secretion

Although the regulation of the preovulatory LH surge differs between reflex and spontaneous ovulators, the negative feedback regulation of the tonic secretion of LH and FSH by ovarian hormones appears largely indistinguishable in the two groups. In addition to the inhibitory feedback of ovarian steroids, there is evidence that rabbit granulosa cells also produce an inhibin-like substance that can preferentially inhibit FSH (but not LH) secretion from rabbit pituitary cells.

C. Reproductive Cycles

Since obligate reflex ovulators require copulation to induce the secretion of the ovulatory LH surge, an obvious physiological consequence of such a requirement is the absence of a luteal phase in unmated females, i.e., corpora lutea only arise from ruptured follicles after ovulation. While perhaps an unnecessarily pedantic point, reflex ovulators such as the rabbit do not, in fact, have ovarian cycles. Reflex ovulators do, however, exhibit seasonal reproductive cycles and pass between anestrus and estrus.

Mating in reflex ovulators is followed by pregnancy or pseudopregnancy, depending on the fertility of the male (or jumping female). Until mating occurs, the cohort of mature follicles is presumably continuously replenished by new follicles as older follicles become atretic; thus, the ovaries do not cycle between follicular and luteal phases but remain instead in a kind of steady state of readiness while the unmated female is in estrus.

III. HISTORICAL IMPORTANCE

Studies earlier in this century of reflex ovulators such as the rabbit became cornerstones in the foundation not merely of our current knowledge of the LH surge as the physiological trigger of ovulation but also, more importantly, of our understanding of the hypothalamic regulation of all anterior pituitary hormones in all mammals, and probably in all vertebrates as well. As indicated in Section I,B,2,i, careful anatomical studies in the rabbit by G. W. Harris, a framer and fervent advocate of the neurovascular hypothesis, along with the work of other neuroendocrine pioneers demonstrated the physiological importance of the vascular link, the hypophysial portal vessels, between the hypothalamus and adenohypophysis. Based in large part on his many experiments in rabbits, Harris was among the first to propose that hormones secreted by hypothalamic neurons into the primary capillary plexus circumscribing the median eminence and carried by the stalk portal vessels to the anterior lobe are key regulators of the secretion of LH and of other pituitary hormones.

IV. CLINICAL CONNECTIONS

Even though ovulation in women, as in other primates, occurs spontaneously during the ovarian/menstrual cycle, there is a long-standing and an often misunderstood connection, now historical, between pregnancy diagnosis in women and reflex ovulation in rabbits. Soon after it was learned that the urine of pregnant women, but not of nonpregnant women, contained a substance with LH activity [now known to be human chorionic gonadotropin (hCG)], a bioassay commonly called a "rabbit test" was devised to diagnose or confirm pregnancy. In this bioassay, extracts of urine samples were injected intravenously into estrous female rabbits, which were killed 2 days later to inspect the ovaries for freshly ruptured follicles, a hallmark of LH/hCG action and thus a positive indicator of pregnancy. Made obsolete by over-the-counter immunochemical-based pregnancy test kits, the rabbit test was widely used for many years, even though many patients confused a positive result with "the rabbit died."

Finally, "ovulation induction" or an "ovulation-induction protocol," produced pharmacologically by exogenous gonadotropins in the female partner of infertile couples as a prelude to *in vitro* fertilization among other assisted reproductive technologies, should be recognized as wholly distinct from the natural physiological pattern of induced ovulation in reflex ovulators.

See Also the Following Articles

GnRH (Gonadotropin-Releasing Hormone); LH (Luteinizing Hormone); Neuroendocrine Systems; Ovulation

Bibliography

Karsh, F. J., Bowen, J. M., Caraty, A., Evans, N. P., and Moenter, S. M. (1997). Gonadotropin-releasing hormone requirements for ovulation. *Biol. Reprod.* **56**, 303–309.

Ramirez, V. D., and Soufi, W. L. (1994). The neuroendocrine control of the rabbit ovarian cycle. In *The Physiology of Reproduction* (E. Knobil and J. D. Neill, Eds.), 2nd ed., pp. 585–611. Raven Press, New York.

Rowlands, I. W., and Weir, B. J. (1984). Mammals: Non-primate Eutherians. In *Marshall's Physiology of Reproduction* (G. E. Lamming, Ed.), Vol. 1, 4th ed., pp. 455–658. Churchill Livingstone/Longman, Edinburgh, UK.

Sowers, J. R. (Ed.) (1980). *Hypothalamic Hormones, Benchmark Papers in Human Physiology,* Vol. 14. Dowden, Hutchinson & Ross, Stroudsburg, PA.

Spies, H. G., Pau, K.-Y., and Yang, S.-P. (1997). Coital and estrogen signals: A contrast in the preovulatory neuroendocrine networks of rabbits and rhesus monkeys. *Biol. Reprod.* **56**, 310–319.

van Tienhoven, A. (1983). *Reproductive Physiology of Vertebrates.* Cornell Univ. Press, Ithaca, NY.

Regulation of Sertoli Cells

Michael D. Griswold
Washington State University

GLOSSARY

cycle of the seminiferous epithelium In the mammalian testis, germ cells that have entered spermatogenesis are organized in the seminiferous tubule in specific repeating cellular associations that encompass all stages of spermatogenesis and result in asynchronous sperm production.

paracrine interactions A hormonal type of cell-to-cell regulation that generally occurs over short distances between cooperating cell types.

pituitary–testicular axis Refers to the control of testicular function via pituitary hormones and the feedback regulation of the pitutitary gland by products of the testis.

I. INTRODUCTION

Spermatogenesis is controlled via direct stimulation of the gonad by pituitary hormones and the secretion of pituitary hormones is in turn controlled via feedback from products of the gonad (Fig. 1). The components of this pituitary–testicular axis include luteinizing hormone (LH) and follicle-stimulating hormone (FSH) from the pituitary and testosterone and inhibin from the testis. Testosterone and FSH are the primary regulators of spermatogenesis and both act via receptors located in Sertoli cells or on the plasma membrane of Sertoli cells. The sole function of LH in the male is the stimulation of testosterone production from Leydig cells. Inhibin, a feedback inhibitor of FSH secretion by the pituitary, is also made by the Sertoli cells. In addition, paracrine and autocrine factors from both Sertoli and germ cells are important in the functioning of both cell types.

In spermatogenesis, well-defined groups of germ cells interact with Sertoli cells in a cyclic pattern. These recurring groups of germ cells define the cycle of the seminiferous epithelium during which sperm are produced along the length of the tubule in an asynchronous fashion. While it is important to know what Sertoli cells and germ cells make, it is equally important to know when they make it. The temporal control of this cycle and ultimately of spermatogenesis is a result of well-timed signaling interactions between Sertoli cells, germ cells, and the pituitary gland.

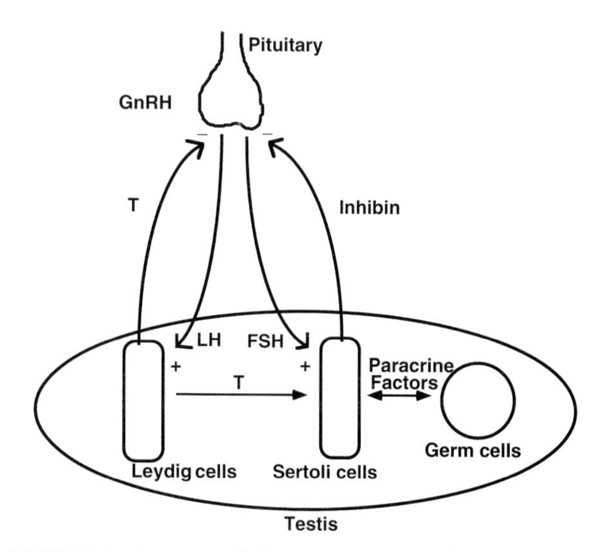

FIGURE 1 Diagram of the major aspects of the pituitary–testicular axis. Secretions from the pituitary regulate testis function by ultimately affecting Sertoli cells. Sertoli cells make inhibin that feeds-back on the release of FSH, whereas Leydig cells make androgens that inhibit LH secretion. LH, luteinizing hormone; FSH, follicle-stimulating hormone; T, testosterone; GnRH, gonadotropin-releasing hormone.

II. ROLE OF FSH

The target cell for the action of FSH in the male is the Sertoli cell. Both the cDNA and the gene for the FSH receptor have been cloned and characterized. The FSH receptor is expressed and is active in the Sertoli cells early in testis development and throughout the reproductive life of the animal. Many of the studies examining the actions of FSH have been done on cultured Sertoli cells from prepubertal rodents. The Sertoli cells from rodents of this age are easily placed in culture and respond to FSH with increased levels of cAMP, increased protein synthesis, and increased estradiol production. As the rat enters puberty, the response of Sertoli cells both in culture and *in vivo* changes. There is a large increase in the phosphodiesterase activity in the cells and the accumulation of cAMP and the subsequent stimulation of specific protein synthesis is curtailed. At least in adult rodents, the response of Sertoli cells to FSH is attenuated in the adult and many of the biochemical activities of FSH in the prepubertal animal appear to be assumed by testosterone in the adult. These findings have led to considerable controversy concerning the role of FSH in male reproduction.

Recently, the overall biological role of FSH in the mammalian testes has been clarified. Testosterone administration to mice deficient in gonadotropin-releasing hormone (GnRH) in the absence of FSH was found to be sufficient to achieve testicular maturation and fertility. GnRH-deficient mutant mice treated with testosterone implants had normal spermatogenesis but reduced testis size and germ cell numbers. FSH treatment, in addition to testosterone treatment, resulted in GnRH-deficient mice with quantitatively normal spermatogenesis and testes of normal size. These studies showed that FSH treatment during the first 2 weeks of life increased Sertoli cell numbers and total sperm production by the mouse testis. Treatment of normal rats during their prepubertal stage with additional exogenous FSH produced larger than normal testes and higher total germ cell numbers. The latest clarification occurred when the FSH β gene was knocked out. This resulted in a line of mice lacking FSH in which the males are fertile but have smaller than normal testes and reduced germ cell numbers. In humans, a FSH receptor null mutation has been reported in five male humans. The men showed variable degrees of spermatogenic failure but were not infertile. None of the men had normal sperm parameters and testicular size was reduced, but two of the men had fathered two children each. These results utilizing mice and men with null mutations in the GnRH, FSH, or FSH receptor genes have shown clearly that FSH is not required for fertility in mice or man.

The most important function of FSH in the male is probably the stimulation of Sertoli cell mitosis during testis development. The actions of FSH in the prenatal and newborn rat as a Sertoli cell mitogen are critical in the ultimate spermatogenic capability of the testis. The size of the Sertoli cell population is limiting to overall sperm production and FSH influences the spermatogenic capability by influencing the size of the Sertoli cell population. Thus, FSH may not be necessary for fertility but clearly plays an important role in the number of spermatozoa an organism can produce. This role can be significant in the reproductive competition within a species.

III. REGULATION BY ANDROGENS

The important regulatory role of androgens in spermatogenesis is well recognized and accepted. Animal models, primarily rodents, deprived of the consequences of androgen action by hypophysectomy or by genetic mutations [such as the testicular feminized mutation (*Tfm*)] exhibit arrested spermatogenesis. In mice and humans bearing *Tfm*, in which functional androgen receptors are absent, only limited germ cell development into primary spermatocytes is observed. Androgen receptors have been demonstrated to be present in peritubular cells and in the Sertoli cells of the mammalian testis. There is still some controversy regarding the possible presence of androgen receptors in germ cells but available biological and genetic evidence suggests that Sertoli cells are the focus of androgen action in the regulation of spermatogenesis. Despite the wealth of biological information pertaining to the importance of androgen action in spermatogenesis and detailed molecular information on the androgen receptor, there is little reliable information about the molecular events stimulated by testosterone in Sertoli cells. Results from experiments utilizing cultured Sertoli cells have shown little or no effect of testosterone on the synthesis of specific macromolecules. In addition to regulating the synthesis of key genes in the Sertoli cells, it is possible that testosterone plays a general trophic role in Sertoli cells in the adult testis. The mode of action of testosterone in spermatogenesis remains one of the major enigmas in male reproduction.

IV. PARACRINE FACTORS AND GERM CELLS

An important question is whether the germ cells are actively involved in the regulation of Sertoli cells or whether they are passively acted upon by Sertoli cells. Clues have come from experiments designed to examine the dynamics of the cycle of the seminiferous epithelium. The commonly used diagram of the cycle of the seminiferous epithelium in the rat consists of a series of 14 columns that describe a different group of cellular associations found in the testis. The cycle is similar in different species although the length of the cycle may vary between species. In rodents the stages or cellular associations are arranged in a consecutive and linear fashion along the seminiferous tubule, whereas in humans or other primates the cycle may be organized along the tubule in a spiral or patchy manner. The net result of the cycle is the same in all species, i.e., asynchronous sperm production along the length of the tubule so that a constant source of sperm are available during the breeding period.

In recent years a number of techniques have been developed that allow for the investigation of molecular events taking place in Sertoli cells and germ cells during this cycle. All of these techniques have shown that the secretions by Sertoli cells and thus the immediate environment in the seminiferous epithelium vary as a function of the stage of the cycle. Sertoli cells associated with germ cells at one stage of the cycle express different genes than those in other parts of the cycle. Analyses of a number of the products of Sertoli cells have been done and the data are suggestive of a dual mode of Sertoli cell function. One mode includes the stages of meiotic division in which the secretion of a group of Sertoli cell products is maximal and a second mode includes the spermiation stages in which the secretion of a different group of products is maximal. In other words, the actions of Sertoli cells are important in the meiotic divisions and in spermatid differentiation. In addition, Sertoli cells appear to influence stem cell and spermatogonial division in several parts of the cycle.

This cyclic activity of Sertoli cells could derive from an inherent internal cycle in Sertoli cells or from signals from the developing germ cells. Culture studies suggest that in the absence of germ cells the cyclic activity of the Sertoli cells ceases; therefore, it appears that signals, most likely in the form of paracrine factors coming from germ cells, are important in regulating the cycle. Germ cells have been shown to synthesize a number of paracrine factors, including a form of basic fibroblast growth factor.

Most discussions of Sertoli cell–germ cell communication regard the germ cells as a single entity. The communication network is perhaps more complex than has been perceived, with different types of germ cells providing different signaling molecules. In addition to interactions between germ cells in the basal compartment and the pituitary and Leydig cells, interactions between different germ cell subtypes may also be important. Our information with regard to signaling from germ cell products is very limited. Significant problems in the study of germ cells are the lack of viability in culture and difficulties in determining cell purity.

See Also the Following Articles

FSH (Follicle-Stimulating Hormone); Inhibins; Leydig Cells; LH (Luteinizing Hormone); Sertoli Cells, Function; Sertoli Cells, Overview; Spermatogenesis, Hormonal Control of

Bibliography

Clermont, Y. (1975). Kinetics of spermatogenesis in mammals. *Physiol. Rev.* 52, 198–204.

Fritz, I. (1978). Sites of actions of androgens and follicle stimulating hormone on cells of the seminiferous tubule. In *Biochemical Actions of Hormones* (G. Litwack, Ed.), Vol. V, pp. 249–278. Academic Press, New York.

Griswold, M. D. (1993). Action of FSH on mammalian Sertoli cells. In *The Sertoli Cell* (M. D. Griswold and L. D. Russell, Eds.), pp. 493–508 Cache River Press, Clearwater, FL.

Griswold, M. D. (1995). Interactions between germ cells and Sertoli cells in the testis. *Biol. Reprod.* **52**(2), 211–216.

Jegou, B. (1993). The Sertoli–germ cell communication network in mammals. *Int. Rev. Cytol.* **147**, 25–96.

Kumar, T. R., Wang, Y., Lu, N., and Matzuk, M. (1997). Follicle stimulating hormone is required for ovarian follicle maturation but not male fertility. *Nature Genet.* **15**, 201–204.

Meachem, S. J., McLachlan, R., de Kretser, D., Robertson, D. M., and Wreford, N. G. (1996). Neonatal exposure of rats to recombinant follicle stimulating hormone increases adult Sertoli cell and spermatogenic cell numbers. *Biol. Reprod.* **54**, 36–44.

Singh, J., and Handelsman, D. J. (1996). Neonatal administration of FSH increases Sertoli cell numbers and spermatogenesis in gonadotropin-deficient (*hpg*) mice. *J. Endocrinol.* **151**, 37–48.

Singh, J., O'Neill, C., and Handelsman, D. J. (1995). Induction of spermatogenesis by androgens in gonadotropin-deficient (*hpg*) mice. *Endocrinology* **136**(12), 5311–5321.

Tapanainen, J. S., Aittomaki, K., Vaskivuo, T., and Huhtaniemi, I. T. (1997). Men homozygous for an activating mutation of the follicle-stimulating hormone (FSH) receptor gene present variable suppression of spermatogenesis and fertility. *Nature Genet.* **15**, 205–206.

Zirkin, B. R., Awoniyi, C., Griswold, M. D., Russell, L. D., and Sharpe, R. (1994). Is FSH required for adult spermatogenesis? *J. Androl.* **15**(4), 273–276.

Relaxin, Mammalian

Russell V. Anthony
Colorado State University

GLOSSARY

cervix The neck-like structure of the uterus that separates the body of the uterus from the vagina.

corpora lutea The structures formed on the ovary following ovulation, at the site of the ruptured follicle, which is the major site of progesterone production and is a site of relaxin production.

endometrium The mucosal lining of the uterus, consisting of an epithelial lining and a lamina propria containing tubular glands.

mammogenesis The growth and development of the mammary gland during pregnancy.

myometrium The muscular wall of the uterus, consisting of an inner, circular layer of smooth muscle and an outer, longitudinal layer of smooth muscle.

pubic symphysis The cartilaginous joint between the two pubic bones.

relaxin A small two-chain polypeptide hormone that stimulates softening of the cervix and preparation of the birth canal for parturition.

Relaxin is a polypeptide hormone that is structurally related to insulin but exerts its main actions on the reproductive tract. It is produced by a variety of tissues in the mammals that synthesize it, and it has a variety of biological actions, including preparation of the birth canal for parturition, development of the mammary gland, and growth of the uterus.

I. INTRODUCTION

Relaxin, or more correctly its actions, was first reported in 1926 by Professor Hisaw when he treated virgin guinea pigs with sera derived from pregnant guinea pigs or rabbits and discovered a pronounced relaxation of the pubic symphysis. This activity was soon to be extracted from pregnant sow corpora lutea, but interest in relaxin waned until the 1950s, when considerable insight into its biological functions was obtained with partially purified preparations of porcine relaxin. Many of these observations were again confirmed during the late 1970s and early 1980s when highly purified relaxin became available, and a more thorough understanding of relaxin's structure, site of synthesis, and function evolved. However, the past decade has provided exciting insight into relaxin's action within reproductive tissues as well as potential roles in nonreproductive tissues (e.g., cardiovascular system). The focus of this article will be on the secretion by, and action on, the mammalian reproductive tract.

II. STRUCTURE OF RELAXIN

Relaxin is a two-chain polypeptide that is approximately 6 kDa in mass and structurally related to insulin and the insulin-like growth factors. Since the initial report on the purification of pig relaxin in 1974, relaxin has been purified to homogeneity from a variety of mammalian and nonmammalian species. The isolation of highly purified relaxin allowed structural characterization of the hormone as well as determination of the amino acid sequences for the α and β chains. In other species, the amino acid sequence has been inferred from nucleotide sequencing of complementary DNAs (cDNAs). Relaxin is synthesized as preprorelaxin, with the "pre" segment composed of a 22- to 25-amino acid leader peptide. The leader peptide is located at the NH2 terminus of prorelaxin and is removed cotranslationally by signal peptidase in the endoplasmic reticulum. Prorelaxin consists of three distinct domains (Fig. 1A): the β chain, C peptide, and α chain, translated in that order. The α and β chains are retained in mature relaxin, but the C or connecting peptide (100 amino acids in length) is removed before secretion (Fig. 1A). The enzymes responsible for this final processing step have not been positively identified but are likely members of the family of prohormone convertases. This arrangement of functional domains in prorelaxin and the processing steps that it undergoes are very similar to those of insulin, an ancestoral relative of relaxin.

Mature relaxin is composed of the α and β chains connected by two interchain disulfide bridges (Fig. 1A). There is an additional disulfide bridge that connects two cystine residues in the α chain. These disulfide bridges are critical to relaxin's secondary structure and contribute to the tertiary structure of human relaxin depicted in Fig. 1B. Crystallography data demonstrated that the α chain contains two short α-helices, whereas the β chain possesses a single α-helix. Although not evident in Fig. 1B, if one examines the central axis of the β chain α-helix, relaxin appears triangular in shape with the back of the triangle formed by the α chain lying across the β chain helix at nearly right angles. The overall wedge-like topography of relaxin is likely important for interaction with its receptor. Unfortunately, the relaxin receptor has not been purified and structurally characterized, such that crytallographic studies of the relaxin–relaxin receptor complex are unavailable. This type of characterization immensely enhanced our understanding of the mechanism of action of growth hormone and prolactin, and similar studies with relaxin would provide needed insight into relaxin's actions.

As noted earlier, the amino acid sequence of several species of relaxin was inferred from cDNA sequences. This also led to structural characterization of the relaxin gene. Relaxin is encoded by a fairly simple gene that is composed of only two exons (Fig. 2). Exon 1 encodes the 5′ untranslated region, the leader peptide, the β chain, and the amino-terminal portion of the C peptide, whereas exon 2 encodes the carboxy-terminal portion of the C peptide, the α chain, and the 3′ untranslated region. The two exons are separated by an ~3700-base pair intron. In most species examined relaxin is encoded by a single-copy gene, but in the human and great apes there are at least two relaxin genes. Human relaxin gene 2 (human relaxin-2) appears to be the predominant gene

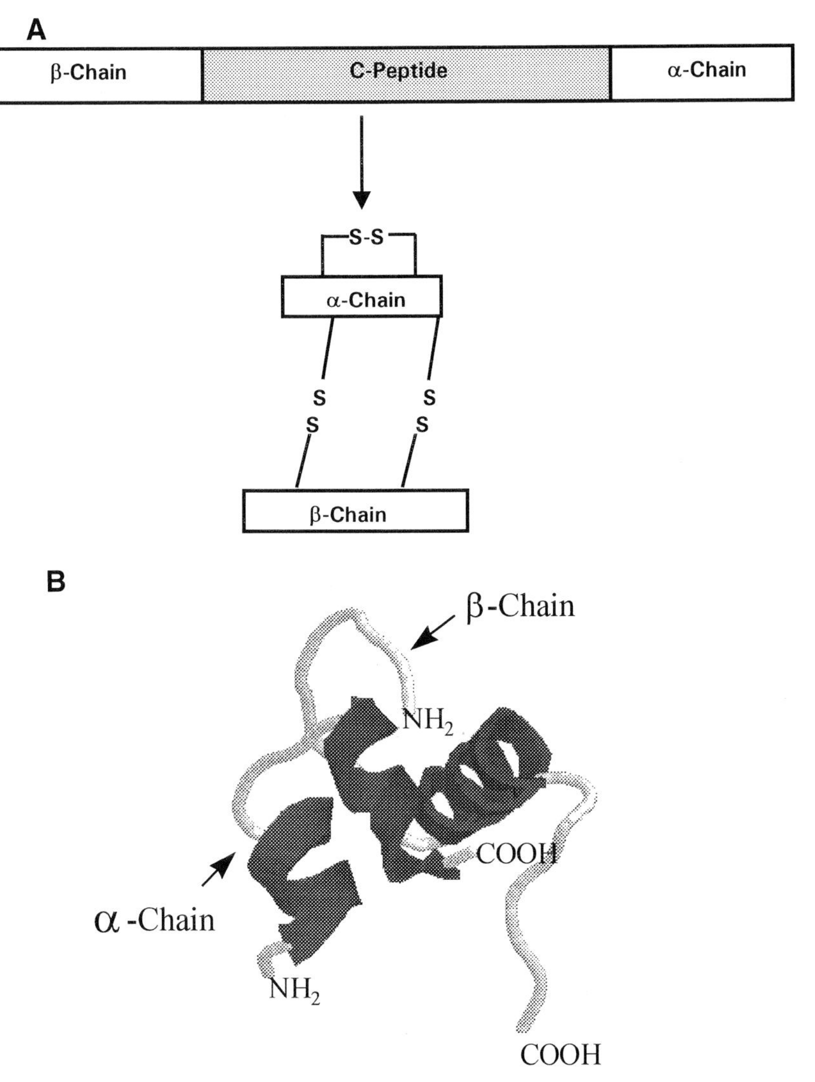

FIGURE 1 Prorelaxin (A) is a single-chain polypeptide composed of three distinct peptide domains that is processed into mature relaxin composed of two polypeptide chains. Relaxin contains two interchain disulfide bridges and one intrachain disulfide loop which aid in forming the tertiary structure depicted in B.

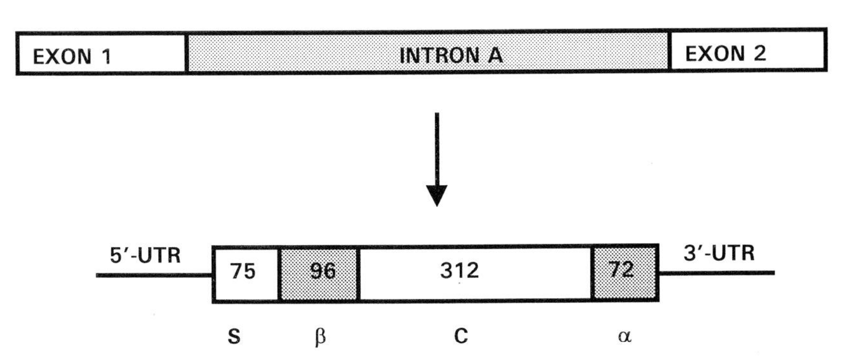

FIGURE 2 Structure of the human relaxin gene 1 and its mRNA. The gene is composed of two exons and one intron. The mRNA encompasses the 5′ untranslated region (5′-UTR); an open reading frame encoding the leader peptide (S), β chain (β), C peptide (C), and α chain (α); and the 3′ untranslated region (3′-UTR). The numbers inside the boxes indicate the number of nucleotides encoding each domain.

expressed in reproductive tissues, but human relaxin-1 is expressed by the placenta and decidua. Nothing is known about the transcriptional regulation of the relaxin gene other than what can be inferred by nucleotide sequence analysis. There is a consensus TATA box in the pig gene but not in the two human genes. The nonconsensus TATA-box sequences in the human genes are located just downstream of GC boxes, and the latter may possibly take over the role of directing assembly of the preinitiation complex. However, until the functionality of the putative *cis*-acting elements are examined experimentally, any discussion of transcriptional regulation of relaxin is speculative at best.

Although relaxin or relaxin-like activity has been identified and purified from a variety of species, in some species a "true" relaxin has not been found. The best example of this is in sheep and cattle. In both these species, relaxin-like activity has been identified in reproductive tissues, but relaxin has never been purified from domestic ruminant tissues, at least to the point of being able to obtain an amino acid sequence verifying its existence. In sheep, a genomic clone was identified that contained what appeared to be the exon 2 sequence, but no exon 1 sequence could be found. It was suggested from these studies that a functional relaxin gene may have been lost from these species due to evolutionary divergence. However, this conclusion did not explain the identification of "relaxin" bioactivity in sheep and cattle tissues and the ability to purify this activity from cattle corpora lutea. The recent identification and characterization of a relaxin-like factor in these species and others may help clarify the situation. The relaxin-like factor was originally identified as an "insulin-like" peptide predominantly expressed by testicular Leydig cells. This peptide has been identified in pigs, humans, and mice and is expressed by both male and female gonads in cattle. The sequence of the relaxin-like factor retains many of same hallmarks believed to be important for relaxin binding to its receptor. Accordingly, it is possible that ruminant relaxin-like factor may serve as a replacement for a classical relaxin in these species. More important, these data indicate that there are additional members of this ancestoral gene family that may or may not have overlapping functions.

III. SOURCES AND SECRETION OF RELAXIN

Relaxin was originally isolated from pregnant corpora lutea and for a long time it was considered a luteal hormone of pregnancy. However, it is now known that relaxin can be expressed by a variety of reproductive tissues in most species that have been examined. Besides the human, the pig and the rat have been the most extensively studied in regard to source, secretion profiles, sites of action, and biological activities. The predominant source of relaxin in both of these species is the corpora lutea. In the pig, relaxin mRNA and protein can be detected in luteal cells during the estrous cycle and early pregnancy, but the levels of relaxin in peripheral circulation are marginal at best and remain relatively low throughout most of gestation (Fig. 3). Most of the relaxin synthesized by the pregnant corpora lutea accumulates within dense secretory granules, with little being secreted until the last 2 days before parturition. At this time, relaxin is released as a large secretory surge (Fig. 3), reaching concentrations well over 100 ng/ml about 14 hr before birth of the first piglet. In the rat, a similar profile is observed. Relaxin becomes detectable by Day 10 of pregnancy, ranges from 40 to 80 ng/ml at Day 14, but then remains relatively constant until Day 20 of gestation, at which time the peripartum surge of relaxin secretion begins, resulting in serum concentrations exceeding 150 ng/ml just prior to pupping.

Like the pig and rat, the major source of relaxin during pregnancy in the human is the corpus luteum. However, the profile of relaxin concentration in maternal plasma is quite different (Fig. 3). Plasma levels of relaxin rise early in pregnancy, peak toward the end of the first trimester (Fig. 3), and decline and remain relatively low through the remainder of gestation. In contrast to the pig and rat, there does not appear to be an antepartum peak of relaxin release. The concentration profile of human relaxin during pregnancy is quite similar to that of human chorionic gonadotropin (hCG), and there is good evidence that hCG will stimulate luteal secretion of relaxin. The human does not require the corpus luteum for maintenance of pregnancy during the latter half of gestation, and it is possible that other tissues secrete suffi-

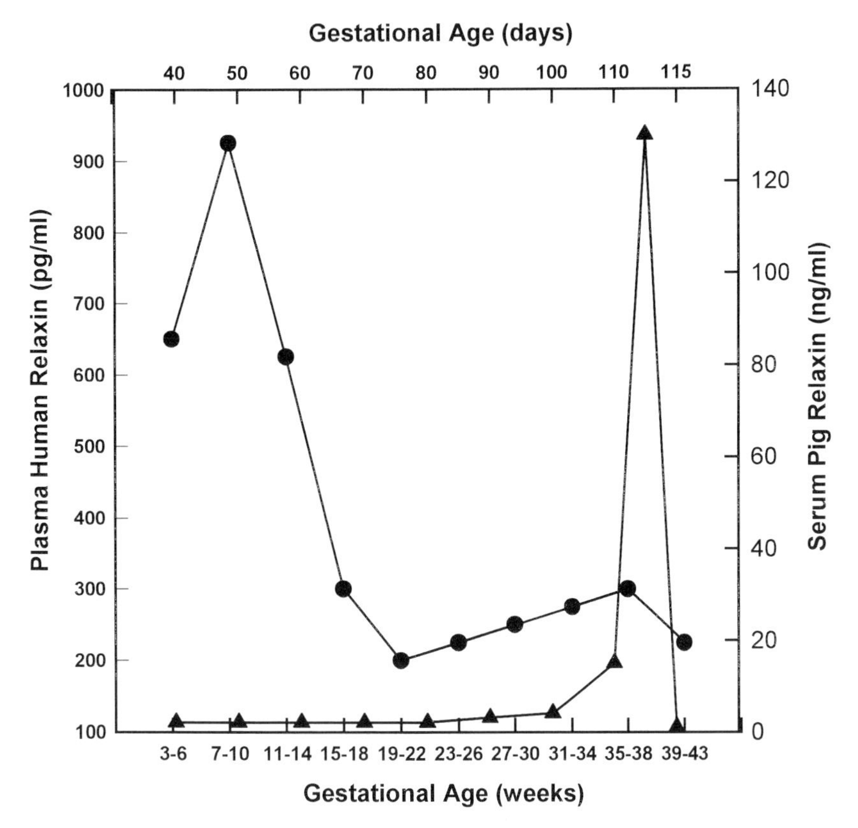

FIGURE 3 The gestational profiles of human relaxin (●) and pig relaxin (▲). Note the considerable difference in the concentration profiles in maternal sera during the course of gestation between these two species.

cient quantities, acting in a local fashion, to meet whatever requirement for relaxin these tissues have. Alternatively, the profile of relaxin secretion during pregnancy may infer that it does not play the same role during the antepartum period as that of pig and rat relaxin.

In all three species, there is good evidence that relaxin is synthesized by the uteroplacental unit. In women, expression of both the H1 and H2 genes has been detected in decidual and placental tissues, with expression of the H2 gene predominating. Furthermore, monoclonal antibodies for human relaxin-2 have identified the decidua parietalis, cytotrophopblast, and syncytiotrophoblast as sites of relaxin production. Although it is questionable if the amount of relaxin produced by the uteroplacental unit has systemic effects in women, it is likely that it has important local actions. There is strong evidence for relaxin production by the uterine epithelium in pigs and rats, again providing the opportunity for local actions within the uteroplacental unit. Furthermore, among these three species, relaxin and/or relaxin mRNA has been identified within ovarian follicular cells (theca interna), mammary epithelial cells, and hypothalamic and neurohypophyseal tissues. All these nonovarian sites of relaxin production raise the possibility for autocrine and paracrine actions of relaxin.

The primary source of relaxin varies among other species. Placental tissues are the primary source of relaxin in the cat, dog, horse, golden hamster, and rabbit, whereas the uterus is the primary source of relaxin in the guinea pig. Furthermore, in several of these species, secondary sources of relaxin have been identified. Although relaxin production has been examined primarily in the female, it is also produced in the male. The prostate appears to be the site of relaxin production in the human, dog, and armadillo, whereas the seminal vesicle epithelium is the site of relaxin production in boars. Differences in the site of production between boars and other species may be related to the proportion of seminal plasma produced by the prostate and seminal vesicles. The boar has a voluminous ejaculate, and the primary source of this volume is the seminal vesicles. Interestingly, examination of the male reproductive tract for relaxin production in rats and mice provided no evidence for its synthesis in these species. As noted earlier, relaxin-like factor is produced by Leydig cells in humans, pigs, and mice, and it is possible that this polypeptide may be more important in the male than is relaxin. Furthermore, if relaxin-like factor has similar properties and activities with relaxin, it may be that at least some of the reports of relaxin activity in the male could have been detecting relaxin-like factor.

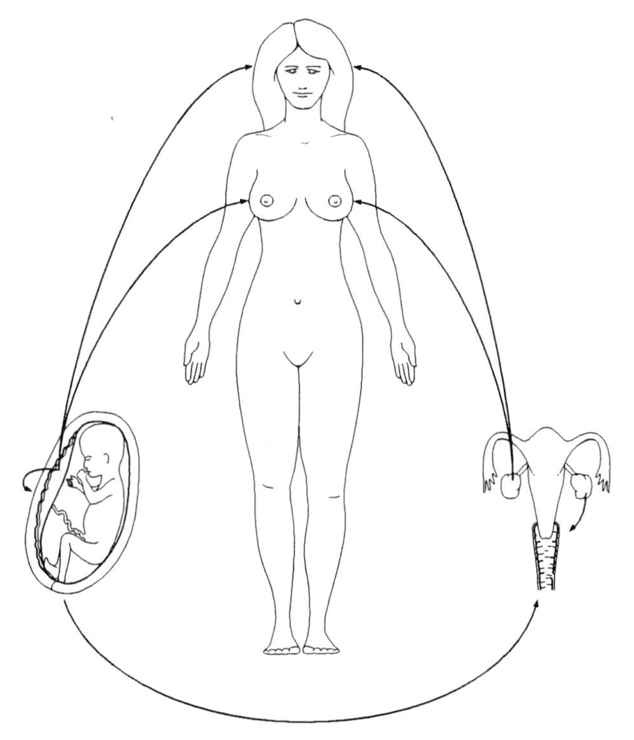

FIGURE 4 The major sources (ovarian and uteroplacental) and sites of action of relaxin. The major source of relaxin in women is the corpus luteum, but it is also produced by the placenta and decidua. Relaxin has effects on the neuroendocrine, mammary, and reproductive systems, including actions on myometrial contractility and cervical ripening. However, not all these effects have been demonstrated in women, and it is possible that some of these actions may be more or less species specific.

IV. BIOLOGICAL ACTIONS OF RELAXIN

The late Professor David Porter once gave a seminar titled, "Relaxin: A Protean Hormone." This title truly reflected what we know of this hormone's biological actions. The actions of relaxin are appropriately grouped into its classical endocrine effects on the reproductive tract and its newer paracrine/autocrine effects, some of which are not in the reproductive tract. Our understanding of the former dates back to the earliest investigations of relaxin and the latter primarily within the past decade. Figure 4 depicts the source and potential sites of relaxin action.

Professor Hisaw originally described relaxin activity from the standpoint that it caused relaxation of the pubic symphysis ligament in guinea pigs, and relaxin bioactivity is still often assessed in "guinea pig units," although it is often bioassayed in mice. Transformation of the interpublic ligament into an elastic structure does not occur within all species late in pregnancy, but it does in mice and guinea pigs in response to exogenous relaxin and there is a pronounced transformation of this joint during the later half of pregnancy in these species. Some relaxation of the pelvic ligaments occurs in women during the first half of pregnancy, when relaxin concentrations in plasma are maximal. However, it remains to be demonstrated in women that endogenous or exogenous relaxin will recapitulate this phenomenon.

Cervical softening and growth is a well-characterized effect of relaxin in rats and pigs, and the effects of exogenous relaxin appear to simulate the cervical changes that take place in preparation for parturition. Relaxin-induced growth of the cervix is attributed to increased water content and increased production of collagen, proteoglycans, and hyaluronic acid. These effects, coupled with remodeling of the extracellular matrix components within the cervical stroma, result in softening or ripening of the cervix. Many of the prepartum changes in the cervix do not occur following ovariectomy of pregnant pigs and rats, and the changes can be restored by relaxin replacement therapy. Furthermore, monoclonal antibody immunoneutralization of rat relaxin results in impaired antepartum growth and softening of the cervix, thereby impeding the progression of parturition. Both the rat and the pig exhibit an antepartum surge release of luteal-derived relaxin that can induce these cervical changes; however, as noted earlier, the human does not exhibit a similar profile. Cervical softening has been induced in women with pig relaxin, but these results cannot be mirrored with human relaxin. This raises the question as to whether relaxin plays a role in cervical ripening in women, and it appears that the answer may be "no." However, this does not mean that relaxin does not play an important role during the antepartum period in women. Recent evidence indicates that decidual relaxin is acting in a paracrine/autocrine fashion (Fig. 4) to stimulate metalloproteinase activity in fetal membranes during the periparturient period. By stimulating metalloproteinase activity, such as collagenase, relaxin may be responsible for the partial degradation of the amniotic and chorionic connective tissues necessary for proper fetal membrane rupture.

Another bioassay of relaxin is based on its ability to inhibit myometrial contractions in isolated uterine strips. Experimental evidence in the rat and pig demonstrated the ability of relaxin to reduce uterine contractions *in vivo*. This has been most eloquently demonstrated in the rat, in which relaxin is able to inhibit the frequency and amplitude of uterine contractions, and there is strong evidence in the rat that relaxin during the latter half of pregnancy is responsible for the reduction in uterine contractile activity. It was suggested that the antepartum release of relaxin in pigs and rats may prevent the onset of uterine contractions during the immediate prepartum period, more or less synchronizing the onset of labor when the effects of relaxin could be overridden by oxytocin or prostaglandin. However, immunoneutralization experiments in the rat do not support this hypothesis since the timing of labor onset was not altered if rat relaxin was neutralized during the last half of pregnancy.

For a long time it was thought that relaxin may promote mammogenesis during pregnancy, but only recently has there been strong experimental evidence to support this hypothesis. Nipple development during the latter half of pregnancy was dramatically impaired when rat relaxin was immunoneutralized, and relaxin replacement therapy reestablished the mammary parenchymal growth that occurs in pigs during pregnancy. Furthermore, specific-relaxin binding sites have been identified in rat and pig mammary tissues. Relaxin is produced by human and guinea pig mammary tissue, and it is possible that relaxin acts in an autocrine/paracrine role to promote development of the mammary parenchyma in these species. Further evidence for mammary actions in the human is provided by the ability of relaxin to inhibit growth and promote differentiation of human MCF-7 cells. Although not exhaustive, the available evidence supports a mammogenic role in the species studied to date.

The ability of relaxin to stimulate uterine growth in rats was demonstrated in the 1950s and was recently documented in pigs. The growth of uterine tissues in response to exogenous relaxin results from increased water content and increased DNA and protein synthesis. These effects can be seen in the absence of ovarian steroids, but the response is augmented in the presence of estrogen when physiological treatment paradigms are used. These actions may be mediated by relaxin's ability to stimulate the production of insulin-like growth factors and their binding proteins within uterine endometrial tissues of the pig and human. Recent evidence demonstrated the production of relaxin by uterine epithelial tissue of pigs and rats, and it had already been documented that relaxin is produced within the uteroplacental unit of humans. Relaxin is capable of exerting its uterotropic actions during early pregnancy in the pig as well as in ovariectomized gilts given replacement therapy to mimic the steroid environment during the estrous cycle and early pregnancy. It is during the luteal phase of the estrous cycle and during early pregnancy that the uterus produces relaxin. Relaxin is likely acting locally to promote uterine growth since the amount of relaxin produced by the uterus is sufficiently small so that it probably does not impact concentrations in peripheral vasculature. However, only small quantities are needed to produce significant uterine growth because an intramuscular injection of 25 μg was able to stimulate significant uterine growth in 80-kg pigs. The local uterine production of relaxin may be important for promoting uterine function during early pregnancy, but the uterotropic effects demonstrated late in pregnancy are likely mediated by luteal relaxin.

Another location where the local production of relaxin may be important is within the ovarian follicle. Relaxin is produced by follicular cells in the rat and pig, and in the pig this production has been localized to the theca interna. Two lines of evidence point to its local actions within the follicle. First, relaxin stimulates *in vitro* DNA synthesis and cell proliferation of pig granulosa and theca cells. Again, this action of relaxin appears to be mediated by stimulating the production of insulin-like growth factor-I by granulosa cells. The second action of relaxin in the preovulatory follicle is the stimulation of plasminogen activator, collagenase, and proteoglycanase activity. The actions of these enzymes to alter the connective tissue structure of the follicular wall are thought to be involved in the growth of the follicle as well as the ovulatory process. Indeed, there is *in vitro* evidence that human relaxin can induce ovulation in perfused gonadotropin-primed rat ovaries. The role of intraovarian relaxin needs further investigation, but assessment of its necessity in the growth and rupture of ovarian follicles awaits an appropriate relaxin deficient model.

As indicated earlier, relaxin production has been demonstrated in the male reproductive tract, particularly in the prostate and seminal vesicles. Since relaxin concentrations are low to undetectable in peripheral vasculature samples, it is unlikely that relaxin has endocrine actions in the male. There is evidence that endogenous relaxin in seminal plasma contributes to normal sperm motility in humans and pigs. However, the lack of experimental evidence makes it difficult to assign a specific role for relaxin in the male.

Although not necessarily germane to the actions of relaxin on the mammalian reproductive tract, considerable data are currently being accumulated demonstrating both the production and the action of relaxin in a variety of nonreproductive tissues, including the central nervous system, cardiac atria, gastrointestinal tract, kidney, and lung. Some of the reported effects of relaxin within the central nervous system include increased blood pressure via induction of vasopressin release and increased drinking. Considerable evidence indicates that relaxin is a potent vasoactive agent, and it may mediate its effects both centrally and systemically. Some of these actions do not seem related to reproduction, but it is possible that relaxin may be involved in some of the adaptive changes that occur during pregnancy in blood hemodynamics and cardiovascular regulation. Unfortunately, it is not feasible at this time to draw firm conclusions or generate any overall models regarding the role of relaxin in nonreproductive tissues.

One aspect of relaxin action that needs considerable emphasis in the future is its receptor. The relaxin receptor has not been purified or structurally characterized in any species. This has impeded our ability to define this hormones mechanism of action and, consequently, to refine our measures of relaxin response. The adenylate cyclase–cyclic AMP second messenger pathway has been implicated in mediating relaxin's actions, especially within uterine myometrial tissue. However, the protein kinase C pathway and tyrosine kinase activity have been implicated as being responsive to relaxin. Whether or not the activation of these second-messenger pathways is in direct association with the relaxin receptor remains to be determined. Based on the fact that relaxin is structurally and evolutionarily related to insulin and the insulin-like growth factors, it is easy to hypothesize that the relaxin receptor may be a tyrosine kinase-type receptor. Regardless of the end result, considerable emphasis needs to be placed on defining relaxin's mechanism of action.

See Also the Following Articles

Cervix; Chorionic Gonadotropin, Human; Corpus Luteum; IGF (Insulin-like Growth Factors); Relaxin, Nonmammalian

Bibliography

Hisaw, F. L. (1926). Experimental relaxation of the pubic ligament of the guinea pig. *Proc. Soc. Exp. Biol. Med.* **23**, 661–663.

MacLennan, A. H., Bryant-Greenwood, G., and Tregear, G. (1995). *Progress in Relaxin Research*. World Scientific, Singapore.

Sherwood, O. D. (1994). Relaxin. In *Physiology of Reproduction* (E. Knobil and J. D. Neill Eds.), pp. 861–1009. Raven Press, New York.

Sherwood, O. D., and O'Byrne, E. M. (1974). Purification and characterization of porcine relaxin. *Arch. Biochem. Biophys.* **160**, 185–196.

Steinetz, B. G., Schwabe, C., and Weiss, G. (1982). Relaxin: Structure, function and evolution. *Ann. N. Y. Acad. Sci.* **380**.

Weiss, G., and Goldsmith, L. T. (1996). Relaxin. In *Reproductive Endocrinology, Surgery, and Technology* (E. Y. Adashi, J. A. Rock, and Z. Rosenwaks, Eds.), pp. 827–839. Lippincott-Raven, New York.

Relaxin, Nonmammalian

Thomas J. Koob

Shriners Hospital for Children and Mount Desert Island Biological Laboratory

GLOSSARY

cervical softening A process which transforms the cervix from an inextensible to a highly compliant tissue in order to allow passage of the term fetus.

elasmobranchs The class of fish which includes the sharks, skates, and rays.

lecithotrophic viviparity A reproductive mode in which the ovulated eggs contain all the organic nutrients for embryonic development but incubation to term occurs in the uterus.

oviparity A reproductive mode in which shelled or encapsulated eggs are laid soon after ovulation and embryonic development occurs outside of the mother.

pubic symphysis The juncture between the two pubic bones in the pelvis which in some mammals is necessarily modified to facilitate delivery of relatively large fetuses.

relaxin A disulfide-bonded, dimeric peptide hormone similar in structure to insulin produced by the corpora lutea, uterus, and placenta in mammals and responsible for mediating connective tissue alterations for cervical softening and pubic symphysis relaxation at parturition and for inhibiting uterine myometrial contractions during late pregnancy.

tunicates A subphylum of solitary or colonial chordates with sac-like bodies in which the notochord, the embryonic precursor of the spinal column in vertebrates, appears early in development but disappears in the adult form.

viviparity A reproductive mode in which embryonic development occurs in the uterus and fully formed progeny are delivered at parturition.

Relaxin-like peptides have been isolated from ovaries of both oviparous and viviparous nonmammalian vertebrates. Based on molecular size and composition, cross-reactivity with anti-porcine relaxin antibodies, and the ability to induce relaxin-mediated events in mammalian bioassays, these peptides appear to be members of the relaxin family. Experimental studies suggest that relaxin regulates physiological events during reproductive cycles in birds and elasmobranchs that are analogous to relaxin-mediated processes associated with mammalian pregnancy. These processes include inhibition of myometrial contractions and induction of cervical softening. The origin and actions of relaxin in nonmammalian species suggest that this reproductive hormone evolved early in vertebrate history to mediate morphological and physiological changes connected with delivery of eggs in oviparous species and embryos in live-bearing species.

Encyclopedia of Reproduction VOLUME 4

I. INTRODUCTION

Are there reasons to suppose that relaxin may be a critical endocrine regulator in nonmammalian species given that its principal biological action in mammals is associated with pregnancy and parturition? Relaxin governs connective tissue remodeling in the cervix and pubic symphysis in many mammalian species, allowing a well-timed and unimpeded delivery of the term fetus. Do similar structures exist in nonmammalian species which require remodeling in preparation for passage of eggs or embryos? Relaxin regulates the uterine myometrium in some mammals. Does it exhibit similar actions in egg-laying or live-bearing fish, reptiles, or birds? Relaxin has been implicated in regulating lactation. Could it be involved in other glandular processes peculiar to the reproductive processes of specific taxa? Major sources of relaxin in mammals include corpora lutea, placenta, and uterus. Are analogous tissues present in nonmammalian species? Estrogen priming is required for relaxin action in mammals. Is estrogen available in nonmammalian species at the right time? This article examines these questions and provides available evidence to conclude that relaxin is a phylogenetically widespread endocrine factor which has homologous, analogous, and novel functions in both oviparous and viviparous nonmammalian species.

II. PHYLOGENETIC SURVEY

A. Aves

In 1950, Hisaw and Zarrow reported finding in extracts of chicken ovaries a relaxin-like bioactivity as detected by the guinea pig pubic symphysis assay. Relaxin activity was later found in crude extracts of rooster testes. Only recently has the ovarian factor responsible for this cross-species bioactivity been identified as a true relaxin. A relaxin-like peptide was isolated from deyolked follicles of laying hens by the standard procedures for purification of mammalian relaxins. Material immunoreactive with homologous porcine relaxin antibodies eluted from size-exclusion and ion-exchange chromatography was exactly like authentic porcine relaxin. The purified peptide exhibited a molecular weight on SDS–PAGE comparable to porcine relaxin at approximately 6 kDa. The dose dependency of this peptide paralleled that of homologous porcine relaxin in a specific radioimmunoassay with porcine relaxin.

Verification that this peptide has relaxin-like activity was demonstrated in the estrogen-primed mouse uterine contractility bioassay. Inhibition of spontaneous uterine contractions *in vitro* resulted from exposure to the isolated chicken peptide comparable to that caused by the porcine relaxin standard. While the molecular structure of this molecule has yet to be determined, this observation established fairly convincingly that relaxin is an ovarian hormone in the hen and likely in other birds as well. As such, it supports earlier observations suggesting that relaxin may be an important endocrine regulator in oviparous species in widely divergent taxa.

B. Reptilia

Although relaxin-like molecules have not been unequivocally identified in any member of the class Reptilia, several lines of evidence suggest that certain reproductive events may require regulation by a relaxin-like hormone in this group. In addition, ovarian and follicular dynamics and the endocrinology associated with ovulatory cycles and gestation allow the possibility that relaxin could be a biosynthetic product of the reptilian ovary.

Oviparity is the predominant reproductive mode within the class Reptilia. Entirely oviparous groups, such as turtles, alligators, and crocodiles, produce relatively large eggs contained in substantial, calcareous eggshells. Following ovulation, eggs are held in the oviduct while the shell is deposited around the egg. Further egg retention can last from several days to weeks before oviposition. Within the squamates, i.e., the lizards and snakes, reproduction is either oviparous or viviparous. In oviparous squamates egg production is similar to that in other oviparous reptiles except that the eggshell has less calcium and less albumen is deposited around the egg. Some species of lizards and many snake species reproduce via viviparous mechanisms. Chorioallantoic placenta develop in these species for respiratory demands, waste management, and, in some species, to augment nutrition supplied in the ovulated egg. Eggs and embryos are retained *in utero* until development is completed—in some cases for months.

Some degree of egg retention occurs in all reptiles. The uterus must accommodate the eggs in oviparous species and the developing embryos to term in viviparous species (Figs. 1 and 2). Uterine activity must adapt to the various degrees of egg retention but must assist oviposition or parturition at the appropriate time. The distal end of the uterus is bounded by a constriction, analogous to the mammalian cervix, which confines the egg/embryos during retention but must allow passage of the eggs at oviposition or embryos at parturition. The participation of a relaxin-like factor in regulating these activities appears plausible, especially given the established endocrine mechanisms associated with these events.

Corpora lutea form from follicles following ovulation in all reptiles examined to date. The life of corpora lutea corresponds to the duration of egg retention. Progesterone is the principal steroid produced by corpora lutea and progesterone levels remain elevated while they remain active. Progesterone has been shown to inhibit uterine myometrial contractions in oviparous species, as it does in mammals. Given the parallel between endocrine patterns and uterine/cervical activity in reptiles and these same physiological mechanisms in mammals, the possibility that relaxin is involved in regulating reproductive tract function in reptiles seems likely. We must await future studies for unequivocal evidence for reptilian relaxin.

C. Osteichthyes

There is no evidence suggesting that relaxin is produced or may be involved in reproductive events in the teleosts. Bony fishes lack Müllerian duct derivatives, including oviduct, uterus, and cervix. The gonoduct leading from the ovary to the genital pore is an extension of the ovarian stroma. There are no spatially limiting skeletal structures impeding delivery of eggs or embryos. Therefore, the absence of structures homologous with Müllerian duct derivatives in other vertebrates may have obviated the necessity to evolve relaxin-like peptides.

However, the ovary in bony fish functions in place of Müllerian ducts and viviparity is prevalent in certain taxa. If relaxin is a hormone which has evolved concomitantly with the live-bearing strategy, as has been postulated, it might have emerged in these fish. Embryonic development in viviparous teleosts occurs either in the follicles or in the lumen of the ovary. Regulation of egg/embryo retention and delivery of term embryos may require physiological mechanisms similar to those governed by relaxin in other vertebrates.

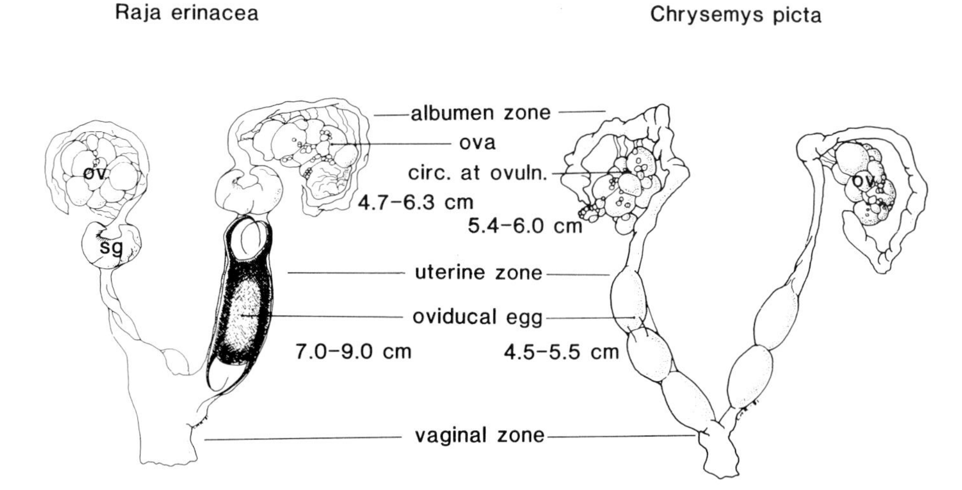

FIGURE 1 Schematic diagram of the morphologically distinct regions within reproductive tracts of representative oviparous species. ov, ovary; sg, shell gland (reproduced with permission from Koob and Callard, 1982).

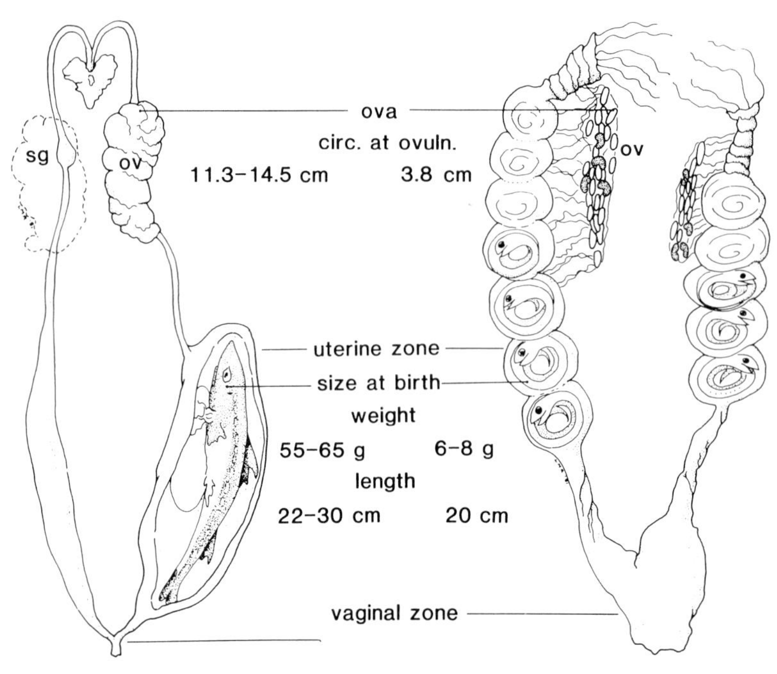

FIGURE 2 Schematic diagram of the morphologically distinct regions of the reproductive tract in representative viviparous species. ov, ovary; sg, shell gland (reproduced with permission from Koob and Callard, 1982).

D. Chondrichthyes

The elasmobranchs offer exceptional opportunities for examining the presence, functional role, and evolution of relaxin in nonmammalian species. Both oviparity and disparate forms of viviparity are well represented within the class. The reproductive endocrinology in female elasmobranchs resembles that in mammals to such an extent that it has been postulated to be the archetype for terrestrial vertebrates. Moreover, all elasmobranchs produce at term relatively large eggs or embryos. In oviparous species, large egg capsules are transported through the uterine portion of the reproductive tract and exit through a sphincter which is entirely analogous to the cervix (Fig. 1). In viviparous species, large, precocial young develop *in utero* for protracted periods and are finally delivered through a cervix requiring structural modification (Fig. 2). It might be predicted that relaxin-like molecules are ubiquitous throughout the taxon. The identification and characterization of relaxin-like molecules in three elasmobranch species which employ distinct reproductive modes supports this prediction.

The first, unequivocally identified, elasmobranch relaxin was obtained from ovaries of pregnant sand tiger sharks, *Carcharias taurus*, thereby confirming a report demonstrating relaxin-like activity in ovarian extracts from this same species. Using methodologies identical to those for the purification of porcine relaxin, including differential solubility in organic solvents, size exclusion, and ion-exchange chromatography coupled with relaxin-specific bioassays, a peptide was purified which, like all mammalian relaxins, is composed of two disulfide-bonded chains and is similar to these relaxins in overall molecular size (~6 kDa). Amino acid analysis of the purified peptide showed a relaxin-like primary structure, but additional amino acids were found in this relaxin compared to porcine relaxin. Structural similarity of the purified peptide with porcine relaxin was also demonstrated by radioimmunoassay using antibodies to porcine relaxin.

The purified sand tiger shark peptide was tested for *in vivo* bioactivity in the estradiol-primed mouse pubic symphysis assay and the estradiol-primed guinea pig symphysis assay. The shark peptide failed to induce pubic symphysis relaxation in the mouse, but it was effective at increasing the mobility of the guinea pig pubic symphysis. *In vitro* tests with uterine strips from estrogen-primed mice and guinea pigs demonstrated that the shark peptide was able to inhibit spontaneous contractions in uterine strips from guinea pigs but not those from mice. Thus, the sand tiger shark relaxin exhibited typical relaxin activities in these heterologous assays, but only in guinea pigs and not mice.

A second shark relaxin was subsequently isolated from the lecithotrophic viviparous spiny dogfish, *Squalus acanthias*. Ovaries from pregnant females were again the source for extraction and the peptide was isolated by the standard means as noted previously. The peptide exhibited a similar molecular size at 5.88 kDa and was demonstrated to be composed of two disulfide-bonded chains. Circular dichroism of the dogfish relaxin was identical to that of the sand tiger peptide, the spectrum of both of which varied only slightly from that of porcine relaxin. In contrast to the sand tiger peptide, the dogfish relaxin was effective at relaxing the interpubic ligament in mice, but like the sand tiger peptide it was even more effective in the guinea pig pubic symphysis assay.

The presence of relaxin-like peptides in elasmobranchs is not restricted to viviparous species. A disulfide-bonded, relaxin-like peptide was extracted from ovaries of the reproductively active little skate, *Raja erinacea*, which, like all skates, is oviparous. While certain structural features of this peptide differed from porcine relaxin, it nonetheless was effective in the mouse interpubic ligament assay and at inhibiting spontaneous contractions in the uterine strip bioassay. This was the first identification of relaxin in an oviparous species and, together with the experimental studies on the little skate reproductive tract, raises several intriguing questions about the origin and evolution of relaxin function.

E. Tunicates

Relaxin-like peptides have been identified recently from ascidian ovaries using the conventional methods for the isolation of mammalian relaxins described previously. In radioimmunoassays for porcine relaxin, the purified peptide exhibited parallel reactivity, suggesting at least partial amino acid sequence homology. Relaxin-like bioactivity was demonstrated in the rat uterine contractility assay. The fraction containing the relaxin-like peptides lowered both the amplitude and the frequency of spontaneous uterine contractions *in vitro*, as did comparable porcine relaxin preparations. Taken together, these observations suggest that relaxin-like hormones appeared early in chordate evolution, and that the appearance of relaxin in the evolutionary history is not necessarily restricted to viviparous species.

III. SOURCE OF RELAXIN IN NONMAMMALIAN SPECIES

In mammals, the principal sources of relaxin are the corpora lutea, placenta, and uterus. In the nonmammalian species from which relaxin-like peptides have been isolated, all were extracted from ovaries. Only in the hen and the ascidians have specific compartments which contain relaxin been identified. Using antibodies to the porcine hormone, it was discovered that immunoreactive relaxin could be found only in the granulosa cells of postovulatory follicles in chicken ovaries. Relaxin was not found in cells of the largest preovulatory follicles nor in the developing follicles, suggesting that relaxin biosynthesis does not commence until after ovulation.

The immunoreactive relaxin-like peptides from ascidians were extracted from entire ovaries of spawning specimens of *Herdmania momus*. Immunohistochemical tests using anti-porcine antibodies localized immunoreactive material in the follicle cells surrounding oocytes at all stages of maturation. The same reactions showing antirelaxin reactivity were found in another species, *Ascidia nigra*. In the seasonally spawning *Ciona intestinalis*, immunoreactive relaxin was found in the follicle cells of mature oocytes only. Given the observation that immunoreactive relaxin was not detected in other ovarian compartments nor in nonovarian tissues, the specific localization of relaxin indicates that most likely it is produced in the follicle cells.

The relaxin peptides identified from the three elasmobranch species were extracted from whole ovaries. The exact ovarian compartment was not determined in these studies and has yet to be examined. However, correlative information is available on the ovarian cycles in these species and thus provides some indications about possible sources.

In the sand tiger shark, relaxin was extracted from ovaries of pregnant females. Sand tiger sharks are viviparous, but embryonic development is oophagous. Following ovulation of the first group of ova from which the two surviving embryos emerge, ova are continually ovulated during early pregnancy as supplemental nourishment. While the exact dynamics of follicle development, corpora lutea formation, and ovulation remain to be described, it seems likely that relaxin originates in an ovarian compartment associated with folliculogenesis.

Spiny dogfish are viviparous, with pregnancies lasting nearly 2 years. Corpora lutea form after ovulation and remain active for the first half of pregnancy. Elevated progesterone levels parallel the life of the corpora lutea, the source of circulating progesterone titers. Development of follicles for the subsequent pregnancy proceeds slowly during this period but increases during the second half of pregnancy, when circulating estradiol levels are elevated. The relaxin extracted from ovaries of pregnant dogfish could have originated in any of these compartments.

In reproductively active little skates, *R. erinacea*, the species from which the skate relaxin described previously was extracted, ovaries contain developing follicles of graded sizes, preovulatory follicles, corpora lutea, and atretic follicles of various sizes. This pattern of follicular compartments is typical of oviparous elasmobranchs because pairs of eggs are produced every few days for periods lasting months. Developing and preovulatory follicles are responsible for the circulating titers of estradiol and testosterone. Corpora lutea produce primarily progesterone. Since all these follicular compartments generate reproductive hormones, any one could be the source of the isolated relaxin peptide.

Other potential sources of relaxin in nonmammalian species have not been explored. Both a metabolically active endometrium and placenta, two sources of mammalian relaxin, are present in many nonmammalian species. The uterus in most nonmammalian viviparous vertebrates is biosynthetically active, especially in matrotrophic forms. Whether the uteri in these species produce peptide endocrine factors is currently unknown. Chorioallantoic placenta develop in many squamate reptiles and yolk sac placenta form in some viviparous sharks. Nonmammalian placenta could be a potential source of peptide hormones including relaxin.

Relaxin has also been identified in extracts of testes from nonmammalian species. Rooster testes yielded relaxin bioactivity. The dogfish testis has been shown to be a source of relaxin. The functional significance of relaxin production in males remains obscure.

IV. FUNCTION OF RELAXIN IN NONMAMMALIAN SPECIES

For mammalian species, a critical understanding of the function of relaxin has relied on the measurement of circulating titers during pregnancy. For example, significant circulating levels of relaxin are typical of mid- to late pregnancy in most mammals, with marked elevations occurring just before parturition. Elevated levels of relaxin correlate with morphogenetic and biomechanical changes in the uterine cervix and pubic symphysis. Unfortunately, circulating relaxin levels have not been measured during reproductive cycles in any nonmammalian species. Insight into putative relaxin functions in nonmammalian species can only be gained through experimental studies. The following sections summarize the experimental efforts to determine the function of relaxin in nonmammalian species.

The principal actions of relaxin in mammals are associated with late pregnancy and parturition, namely, softening of the uterine cervix and relaxation of the interpubic ligament. In addition, relaxin is thought to inhibit uterine contractions late in pregnancy when progesterone levels have declined in preparation for parturition. Relaxin has also mammotrophic actions, promoting growth and differentiation of mammary gland tissue in some species. Experimental studies have demonstrated relaxin actions in nonmammalian species that are similar to these actions in mammals.

A. Pigeon Crop Sac

Experimental studies on mice, pigs, and rats have implicated relaxin in promoting mammary gland growth and development, affecting particularly the epithelium and myoepithelium of the ducts and the fat pads. The absence of mammary glands would seem to preclude a "mammotrophic" action for relaxin in nonmammalian species. However, there are tissues, such as the crop sac in columbid birds, that are functionally analogous to the mammary gland. The glandular lobes of the crop sacs in pigeons and doves produce a lipid-rich nutritive material which is regurgitated to feed the offspring. The production of this material is mediated in part by prolactin, as it is in mammals. Recent experimental studies have clearly demonstrated that exogenously administered mammalian relaxin has effects on the pigeon crop sac that are similar to its actions on mammary glands.

Local subdermal injection of purified guinea pig relaxin into the crop of pigeons caused a marked growth and differentiation of the crop mucosal epithelium, including an increase in surface area and thickness and an increase in the number of epithelial cells. These changes appeared similar to those occurring during the natural crop sac activation during incubation and hatching. Relaxin treatment also induced significant changes in the vascularity of the crop sac. Capillaries were dilated after relaxin treatment, suggesting one action of the hormone could be to increase blood flow in conjunction with its trophic action on the mucosa. While the native role for relaxin in birds and the need for interaction with other hormones like prolactin in controlling crop sac morphogenesis and differentiation are still unknown, the results of the experimental studies clearly suggest that relaxin may have an important role in regulating crop sac function in columbid birds.

B. Uterine Myometrium

A standard method for assessing the bioactivity of relaxin preparations is the inhibition of spontaneous contractions of uterine strips *in vitro*. All the nonmammalian relaxins previously described exhibit the capacity to inhibit these contractions in mammalian uterine strips. Do these relaxins act *in vivo* in the species of origin to regulate uterine myometrial activity? While only a few experiments have been performed regarding this question, the experimental data that do exist indicate that relaxin may regulate uterine myometrial activity in nonmammalian vertebrates, as it does in mammals.

The best evidence derives from experiments on the uterus of pregnant spiny dogfish, *Squalus acanthias*. Purified dogfish relaxin was used in these experiments both *in vitro* and *in vivo*, thereby obviating problems of interpretation when heterologous preparations are used.

Treatment of pregnant dogfish with dogfish relaxin reduced the frequency of *in vivo* myometrial contractions. This effect also occurred in isolated uterine strips. When animals were primed with estradiol, relaxin was effective at inhibiting the frequency of uterine contractions. However, pretreatment with progesterone eliminated the relaxin effect. Since progesterone levels are low and estradiol levels are high in late pregnant dogfish, the native relaxin may act during this time to reduce the frequency of uterine contractions and thereby contribute to embryo retention.

C. Cervix

The cervix in nonmammalian species functions in an analogous manner to the uterine cervix in mammals, both in oviparous and viviparous species. It is the final barrier through which eggs or embryos emerge at oviposition and parturition. The possibility that relaxin mediates alterations in the connective tissue thereby increasing compliance of the cervix has been studied primarily in elasmobranchs. Although circulating levels of relaxin have not been determined, the accumulated experimental evidence suggests that at least one action of relaxin is to regulate structural and biomechanical properties of the reproductive tract, an action similar to that in mammals in late pregnancy.

Administration of porcine relaxin in pregnant spiny dogfish resulted in a significant increase in the extensibility of the cervix. Both the natural and maximal cross-sectional areas were significantly greater in relaxin-treated females and estradiol priming enhanced the magnitude of relaxin effect. Relaxin was effective only in late-pregnant females and was tissue specific in that it did not affect the extensibility of the proximal uterine sphincter. Bovine insulin caused an equivalent response. The overall result was premature loss of near-term fetuses. These observations suggest that (i) the cervix in this viviparous species must remain noncompliant during the latter phases of gestation in order to retain fetuses *in utero*, (ii) a well-timed increase in cervical compliance is essential for normal parturition, and (iii) dogfish relaxin mediates cervical softening in this species in much the same way as it does in mammals.

The reproductive tract in the oviparous little skate also responds to exogenous treatment with a relaxin analog by changing its biomechanical properties. Treatment of reproductively active female skates with porcine insulin increased the extensibility of three morphologically distinct regions of the reproductive tract: the isthmus just below the shell gland, the uterus, and cervix. The most dramatic effect was observed in cervical compliance. Estrogen priming appeared unnecessary for this action of insulin, probably because circulating estradiol is cyclically elevated in egg-laying females. Porcine relaxin had no effect on the tissue compliance. This structural specificity might be expected since circular dichroism studies on purified little skate relaxin have established that the endogenous relaxin in this species is structurally more similar to porcine insulin than porcine relaxin. Given the identification of a relaxin-like peptide from ovaries in this species and the effects of porcine insulin on cervical compliance, these observations indicate the relaxin plays an important role in egg retention and oviposition in oviparous elasmobranchs.

V. CONCLUSIONS

The reproductive tracts in birds, reptiles, and elasmobranchs accommodate the passage of relatively large eggs and embryos. The uterus adapted morphogenetic, structural, and physiological mechanisms to retain eggs in oviparous species and embryos in viviparous species. Both the connective tissue stroma and the smooth muscle participate in accommodating and retaining eggs or embryos in the uterine lumen. Thus, the uterus in nonmammalian vertebrates is a dynamic structure that is capable of remarkable modifications during reproductive cycles to allow accommodation and passage of eggs and embryos. The cervix in nonmammalian vertebrates operates to confine the conceptus during egg/embryo retention *in utero* but to allow unimpeded passage during oviposition or parturition. A well-timed increase in tissue compliance is necessary for normal delivery of progeny, be they eggs or embryos. That relaxin is involved in regulating these reproductive events in at least some nonmammalian vertebrates seems clear from the evidence accumulated to date.

Corpora lutea are formed from postovulatory follicles in all vertebrate taxa except birds. All the vertebrate relaxin-like peptides were extracted from ovaries that contained corpora lutea. The only case in which relaxin was purified from vertebrate ovaries not containing corpora lutea was that of the chicken. However, even in this species, relaxin was found only in postovulatory follicles.

Relaxin evolved in both oviparous and viviparous nonmammalian species, perhaps in response to similar selective pressures. Species producing relatively large offspring either by oviparous or viviparous means faced identical problems with respect to delivering progeny to the external environment. Morphological constraints on reproductive tract tissues and associated structures and regulatory challenges in the proper timing for delivery are common elements in animals producing large young. The appearance of relaxin-like peptides early in vertebrate history may have contributed to the evolution of these successful reproductive modes.

See Also the Following Articles

Avian Reproduction, Overview; Elasmobranch Reproduction; Hormonal Control of the Reproductive Tract, Subavian Vertebrates; Relaxin, Mammalian; Reptilian Reproduction, Overview

Bibliography

Brackett, K. H., Fields, P. E., Dubois, W., Chang, S.-M. T., Mather, F. B., and Fields, M. J. (1997). Relaxin: An ovarian hormone in an avian species (*Gallus domesticus*). *Gen. Comp. Endocrinol.* **105**, 155–163.

Bullesbach, E. E., Gowan, L. K., Schwabe, C., Steinetz, B. G., O'Byrne, E., and Callard, I. P. (1986). Isolation, purification, and the sequence of relaxin from spiny dogfish (*Squalus acanthias*). *Eur. J. Biochem.* **161**, 335–341.

Bullesbach, E. E., Swchwabe, C., and Callard, I. P. (1987). Relaxin from an oviparous species, the skate (*Raja erinacea*). *Biochem. Biophy. Res. Commun.* **143**, 273–280.

Callard, I. P., and Koob, T. J. (1993). Endocrine regulation of the elasmobarnch reproductive tract. *J. Exp. Biol.* **266**, 368–377.

Callard, I. P., Klosterman, L. L., Sorbera, L. A., Fileti, L. A., and Reese, J. C. (1989). Endocrine regulation in elasmobranchs: Archetype for terrestrial vertebrates. *J. Exp. Zool. Suppl.* **2**, 23–34.

Georges, D., Tashima, L., Yamamoto, S., and Bryant-Greenwood, G. D. (1990a). Relaxin-like peptide in ascidians. I. Identification of the peptide and its mRNA in ovary of *Herdmania momus*. *Gen. Comp. Endocr.* **79**, 423–428.

Georges, D., Viguier-Martinez, M. C., and Poirier, J. C. (1990b). Relaxin-like peptide in ascidians. II. Bioassay and immunolocalization with anti-porcine relaxin in three species. *Gen. Comp. Endocr.* **79**, 429–438.

Koob, T. J., and Callard, I. P. (1982). Relaxin: Speculations on its physiological importance in some non-mammalian species. *Ann. N. Y. Acad. Sci.* **380**, 163–173.

Reinig, J. W., Daniel, L. N., Schwabe, C., Gowan, L. K., Steinetz, B. G., and O'Byrne, E. M. (1981). Isolation and characterization of relaxin from the sand tiger shark (*Odontaspis taurus*). *Endocrinology* **109**, 537–543.

Steinetz, B. G., Beach, V. L., and Kroc, R. L. (1959). The physiology of relaxin in laboratory animals. In *Recent Progress in the Endocrinology of Reproduction* (C. W. Loyd, Ed.), pp. 389–423. Academic Press, New York.

Reproductive Senescence, Human

Charles V. Mobbs
Mt. Sinai School of Medicine

GLOSSARY

atresia The largely autonomous process of programmed cell death of ovarian follicles, unrelated to ovulation, which leads to an approximately exponential decay of follicles during aging.

corpora lutea The residual of the follicle, after ovulation (release of oocyte), which continues to function by secreting progesterone.

gonadotropin A collective term for follicle-stimulating hormone and luteinizing hormone, which regulate the development and maturation of the follicle.

luteinization A process during which the ovum is released from the mature follicle (ovulation), followed by the development of the corpora lutea from the residual of the follicle.

menopause The cessation of normal menstrual cycles in women. Clinically defined by the World Health Organization as the final menstrual period following which no menses occur for 12 consecutive months.

oophorectomy Removal of the ovaries.

ovarian follicle A structure in the ovary consisting of an oogonium or oocyte (both precursors of the ovum or egg) and its surrounding support cells.

primordial follicle Precursor of the mature ovarian follicle.

Menopause, the cessation of normal ovarian cycles in women, is one of the earliest, most robust, and most profound physiological changes associated with aging. At 20 years of age, over 90% of women ovulate; by age 55, essentially no women ovulate.

The clinical significance of menopause is based on two main effects: loss of fertility and loss of female sex hormones, especially estradiol. Age-related loss of fertility has historically been accepted as natural and perhaps even desirable. Such acceptance was logical when, as until this century, most women died before they underwent menopause, and indeed a leading cause of death was childbirth. However, in the modern era, with a greatly extended life span, the phenomenon of menopause, the female "biological clock," may significantly reduce women's choices in balancing career and family (limitations with which men need not contend, even though women live longer than men). An even more important clinical aspect of menopause is the loss of circulating estradiol. It has become increasingly clear that estrogen can play an important role in ameliorating age-related pathologies, including osteoporosis, cardiovascular disease, and possibly cognitive impairments. The significance of these salubrious effects of estrogen is demonstrated by the observation that estrogen replacement therapy decreases all-cause mortality rate in postmenopausal women, an effect which does not even account for the improvement in quality of life. These considerations raise the possibility that an intervention which could postpone the age of menopause, though no such intervention currently exists, might be of considerable clinical value.

The proximal cause of the postmenopausal loss of ovarian steroids appears to the almost complete depletion of ovarian follicles at menopause. Nevertheless, because aging is associated with numerous biological changes, the biological state produced by natural menopause in older women is not identical to the state produced by surgical menopause in younger women or related to the normal nadir of estrogen during menstrual cycles. For example, as discussed later, although even in young women estrogen plays a role in maintaining bone mass, the loss of bone mass during aging, independent of estrogen (observed in both sexes), makes older women particularly vulnerable to the loss of estrogen after menopause, and thus makes hormone replacement therapy particularly important in older postmenopausal women. One implication of the age-dependency of menopause is that even though ovarian depletion is the proximal cause of menopause, changes in reproductive neuroendocrine function also occur during aging which may be largely independent of ovarian function but may interact with age-related ovarian impairments and contribute to cessation of cycles. An even more important implication is that clinical effects of estrogen may be more important in older postmenopausal women than in younger women.

I. MENOPAUSE AND POSTMENOPAUSE

Natural menopause is retrospectively defined by the World Health Organization as the final menstrual period following which no menses occur for 12 consecutive months unrelated to obvious causes such as pregnancy, lactation, or oophorectomy, especially after the age of 45; in such circumstances essentially no women will ovulate again. However, it must still be understood that while this definition is useful for analysis and discussion, the precise menopausal event is still not entirely captured by such a definition. For example, the ovaries may still be functional at some level up to 1 year after the final menstrual period, and for all practical purposes women will generally be sterile, often with very low levels of ovarian steroids, up to a decade before the last menstrual period. Nevertheless, by this definition several studies have estimated that the mean age of natural menopause (at least in Caucasian women) is between 49 and 52 years.

The most salient feature of human menopause is that plasma estradiol and estrone fall to minimal levels (equivalent to levels observed in oophorectomized women, although some estrogen continues to be produced by peripheral fat) within a year after the final menstrual period; after this period, estrogens do not decrease further with increasing age. Concomitant with, and perhaps earlier than, the loss of estrogens is a decrease in the ovarian hormone inhibin to minimal (postoophorectomy) levels. In contrast to estrogens and inhibin, progesterone and testosterone are not as clearly decreased after menopause. At the time of menopause, the ovary continues to be the major source of testosterone, possibly due to the persistence of hilus or luteinized stromal cells well after the final menstrual period. Thus, after menopause, testosterone and androsterone are still significantly elevated above levels observed in oophorectomized women.

The loss of ovarian estradiol and inhibin is associated with an almost complete depletion of ovarian follicles at the time of menopause (<1000 follicles at 51 years of age compared to about 1 million primordial follicles at birth and 25,000 follicles at age 37). While 1000 follicles may thus be considered to be a threshold for maintaining menstrual cycles, it has been reported that some women have continued to exhibit menstrual cycles with as few as 200 follicles. The primary cause of follicular depletion with age is not ovulation (which occurs at most only about 500 times during the reproductive life span) but rather the phenomenon of atresia, a largely autonomous process of follicular apoptosis which leads to an approximately exponential decay of follicles during aging. However, as discussed later, the rate of atresia may accelerate just before menopause.

Consistent with the loss of ovarian estrogens and inhibin, gonadotropin levels increase dramatically after menopause, reaching maximal levels 2 or 3 years after the final menstrual period. However, in contrast to the stable levels in estrogens after menopause, beginning 3 years after menopause gonadotropin secretion steadily decreases with age. The available evidence indicates that if postmenopausal women are given steroid replacement, gonadotropins respond [in both inhibitory ("negative feedback") and stimulatory ("positive feedback") modes] appropriately compared to the responses of young women. Thus, at least in the early postmenopausal period, gonadotropin responses to steroids are relatively normal, and the menopausal changes in gonadotropins can thus be attributed largely, if not entirely, to ovarian impairments (e.g., follicular depletion). However, more than 10 years after menopause, there appear to be primary neuroendocrine impairments which lead to decreased gonadotropin secretion. Nevertheless, at menopause or even prior to menopause there may be more subtle impairments in the neuroendocrine regulation of gonadotropins which have not been properly assessed to date.

II. THE TRANSITION TO MENOPAUSE

Fertility in women is maximum at around age 25 years, with a marked decrease beginning around age 35 years, and the mean age at the birth of the last child is about 40 years. The causes of the decline in fertility have not been ascertained. After the age of 26 until about 3 years before menopause, menstrual cycles are regular and entail normal ovulation but become gradually shorter due to a shortening of the follicular phase. Menstrual cycles do not generally become irregular until about age 45–48, about 4 years before menopause. The earliest detectable neuroendocrine marker of reproductive senescence in women is a gradual rise in serum follicle-stimulating hormone (FSH), detectable at about 30 years of age and becoming more marked by age 40. The data on luteinizing hormone (LH) during the perimenopausal period are more complex, but recent studies with large numbers of women indicate that levels of LH also rise gradually during the perimenopausal period, although the rise first becomes detectable at about 35 years of age and becomes more marked after age 40. The elevation in gonadotropins reflects decreases in inhibin and estradiol, which are most important for the inhibition of FSH and LH, respectively. Thus, it seems clear that inhibin begins to decrease even during regular (though shortening) cycles, whereas estradiol levels are not obviously decreased until the perimenopausal period of irregular cycles. Although inhibin and estradiol are both produced from the granulosa cells, inhibin decreases before estradiol probably because production of inhibin presumably reflects the mass of antral follicles as well as the dominant follicle, whereas estradiol is produced primarily by the dominant follicle under stimulation by FSH, which is elevated with age.

Although the number of follicles at all stages of development decrease exponentially with age, there appears to be an acceleration (doubling) in the rate at which follicles are lost beginning around age 38, at which time an apparently critical number (25,000) of follicles is reached. The cause of this accelerated loss of follicles is not known, although the elevations of gonadotropins which occur secondary to prior follicular depletion (especially decreased inhibin) are obvious candidates. Interestingly, if this acceleration of the loss of follicles did not occur, the otherwise simple exponential loss of follicles with age would predict that the approximate threshold for menopause (1000 follicles) would not occur until age 70, suggesting menopause (in the absence of neuroendocrine impairments) would be delayed until that age. Thus, the accelerated perimenopausal atresia, possibly involving a neuroendocrine mechanism, may constitute a promising target for intervention to delay menopause.

III. MECHANISMS OF MENOPAUSE

Because the virtual depletion of follicles coincides so closely with the final menstrual period, it has long been generally assumed that menopause is caused by follicular depletion. There is little doubt that the decline of plasma estradiol to minimal levels within a year after the final menstrual cycle is due to follicular depletion. On the other hand, changes in neuroendocrine function are also associated with menopause. Some of these changes, especially the rise in FSH and later LH, are probably secondary to follicular depletion. However, some neuroendocrine impairments occur which appear to be independent of follicular depletion. The most robust of these changes is that following the gradual rise of gonadotropins which occur in the perimenopausal period and accelerate immediately after the final menstrual period, gonadotropin levels begin to decrease beginning about 10 years after menopause. Interestingly, the decline of FSH is first clear about 10 years after menopause, whereas the decline of LH is first clear about 20 years after menopause, just as the rise of FSH precedes the rise in LH in the perimenopausal period. The decline in gonadotropin secretion, several years after the last ovulation, appears to be due to a hypothalamic defect since pituitary responses to gonadotropin-releasing hormone appear to be intact at this age. Although gonadotropin secretion in the presence of minimal estradiol appears to decline with age, both the stimulatory and the inhibitory effects of estradiol on gonadotropin levels appear to be intact with age. Nevertheless, it is possible that subtle impairments in these responses may be demonstrable by more refined protocols.

The observation that there are at least some apparently ovary-independent age-related impairments in the regulation of gonadotropins raises the question of whether the exact timing of the final menstrual period is determined entirely by follicle depletion or whether age-related changes in neuroendocrine function, in combination with the profound ovarian impairments characteristic of the perimenopausal period, may play a role in the timing of menopause. This question is given particular piquancy by the observation that some women continue to exhibit menstrual cycles with as few as 200 follicles, whereas others stop exhibiting cycles with over 1000 follicles.

Studies in rodents clearly indicate that in these species neuroendocrine impairment plays an important role in determining the precise age at which estrous cycles cease. In broad terms, female rodents exhibit age-related reproductive impairments similar to the impairments exhibited by humans: Fertility begins to decline shortly after puberty, ovulatory cycles become irregular sometime before cessation, and ovulatory cycles cease somewhat later than half the average life span. Furthermore, in rodents ovariectomized well past the cessation of ovulatory cycles, the secretion of gonadotropins is significantly attenuated compared to the secretion exhibited in younger rats after ovariectomy.

In rats, even before estrous cycles become irregular, the preovulatory surge of LH decreases with age. Furthermore, the experimental production of an LH surge after ovariectomy and steroid replacement is profoundly impaired before, and even more so after, estrous cycles have ceased. In addition, direct counts of follicles in rats suggest that there are enough mature follicles and developed corpora lutea to sustain estrous cycles, if there are no neuroendocrine impairments. A simple interpretation of these data in rats is that although follicles continue to develop before and after loss of cycles, the preovulatory surge in LH gradually decreases and eventually becomes insufficient to produce ovulation. The anovulatory state is self-sustaining since the constant secretion of steroids prevents the production of an LH surge. These data suggest that neuroendocrine impairments play a major role in the cessation of estrous cycles in rats. Mice exhibit a similar profile of neuroendocrine impairments associated with the loss of estrous cycles, although direct follicular counts suggest that in this species loss of cycles occurs in association with profound follicular depletion.

However, there are important differences in the endocrine profile of premenopausal humans and in rodents shortly before and after cessation of reproductive cycles. First, whereas in humans there is clear evidence of functional follicular depletion, especially as evidenced by elevated gonadotropin levels, well before menopause, in rats there is little similar functional evidence. Thus, whereas in humans FSH and LH begin to rise before loss of menstrual cycles and inhibin and estradiol levels decrease, the opposite appears to be the case in rats, in which both baseline and preovulatory LH levels decrease as rats get older, even before cycles become irregular. Furthermore, it appears that estradiol levels actually increase in the transition to acyclicity, and well after cessation of cycles the ovaries continue to produce substantial levels of estradiol. Early after cessation of cycles (around 13 months in rats, about halfway through the average life span), rats predominantly exhibit a constant vaginal pattern of constant estrous (indicative of relative dominance of estradiol over progesterone). At this stage, estradiol levels are relatively constant, intermediate between the highest (proestrous) and lowest (diestrous) levels exhibited in young cycling rats. In addition, constant estrous rats exhibit a physiological state concomitant with constant estrogenic stimulation, including constant cornified vaginal epithelia, constantly elevated uterine weight, and constant sexual receptivity. The ovaries of such rats usually exhibit numerous developed follicles, but few or no corpora lutea, consistent with continuing ovarian activity without ovulation. Of particular pertinence in this regard, numerous manipulations which impact on neuroendocrine function, including agents which stimulate catecholamine levels, will cause estrous cycles to begin again in rats exhibiting constant estrous, and after such treatments the ovaries will once again contain corpora lutea. Nevertheless, much later (around 18–20 months of age, approximately 80% through the average life span), rats begin to exhibit a vaginal cell pattern referred to as anestrous state. The anestrous vaginal cell profile is indicative of complete follicular depletion, and rats exhibiting this pattern are characterized by low plasma estradiol (essentially at levels characteristic of the ovariectomized rat), no cornified and few of any other kind of cells from vaginal smears, atrophic uteri of weight similar to that exhibited by ovariectomized rats, and loss of sexual receptivity. The ovaries of rats exhibiting anestrous vagial smears are characterized by no obvious follicle or luteal elements. Nevertheless, in this late stage of ovarian failure LH levels do not rise, though there is perhaps a trend toward elevated FSH. Thus, this late stage of rat reproductive senescence, observed in rats near the mean life span, is similar in most respects to the endocrine profile of women two decades after menopause.

Female mice generally show a profile of reproductive senescence which is intermediate between rats and humans. Thus, in mice there is evidence of functional follicular depletion (lower estradiol) even before cycles cease. Furthermore, at the time cycles cease in mice, the ovaries are almost completely depleted of follicles, and at 14 months of age, the ovaries of mice which continue to cycle exhibit twice as many follicles as the ovaries of mice which have ceased cycling. As in women, after cycles cease in mice some ovarian activity continues, as indicated by some cornified vaginal epithelial, uterine hypertrophy, and levels of estradiol somewhat higher than levels observed after ovariectomy. However, also as in women, shortly after the loss of ovarian cycles it becomes clear that follicles are virtually depleted and both vaginal cells and the uterus become atrophied. Furthermore, in contrast to rats but as in women, plasma LH increases concomitant with the loss of ovarian function. Nevertheless, LH in anestrous mice, while higher than baseline levels in the presence of a normal intact ovary, never reaches the levels observed in young ovariectomized mice. Thus, this age-related impairment in LH regulation, occurring well after follicular depletion, is similar to the impairment observed in older women about 20 years after the final menstrual period.

Since both ovarian and neuroendocrine impairments occur during reproductive senescence, what may be said about the functional importance of these different components in limiting the duration of fertility? That is, is it possible to ascertain the tissue whose failure is the proximal cause of menopause? Given the interdependence of neuroendocrine and ovarian function, and given that both functions are progressively impaired with age, it may be impossible to resolve the question of whether the ovarian or the neuroendocrine component is more important to menopause. However, the question may be recast in a more pragmatic context: Would manipulation of ovarian or neuroendocrine function be more likely to delay menopause? This question is of more than academic interest, given that the loss of fertility and estradiol not only may have social and economic impact on some women but also may increase morbidity and mortality in postmenopausal women.

As previously mentioned, rats show clear neuroendocrine impairments before cycles cease; the most dramatic being the decrease in magnitude of the preovulatory LH surge. Shortly after loss of cycles, cycles can be reinitiated by a variety of interventions which act through neurotransmitters which appear to stimulate a preovulatory surge of LH. Therefore, it seems likely that interventions which act to elevate gonadotropin secretion in older rats to youthful levels, especially the preovulatory surge in LH, could delay the cessation of ovulation in rats. Nevertheless, even in rats it is possible that estrous cycles could be restimulated by enhancement of ovarian function.

In humans, there is little evidence of decreased gonadotropin secretion before loss of menstrual cycles; indeed, the neuroendocrine changes which are observed before menopause (elevated basal gonadotropins as opposed to the lower LH surge observed in rats) appear to indicate appropriate responses to ovarian follicular depletion. Furthermore, follicular depletion appears complete within a year after the final menstrual cycle. Nevertheless, a rigorous analysis of the preovulatory LH surge in perimenopausal women has not been reported, so it is conceivable that some impairment in this function may yet be demonstrated before menopause, similar to such impairments in rats. However, even if reduced secretion of LH were the immediate cause of menopause, because the decreased LH surge appears to be a primary cause in the cessation of estrous cycles in rats, restoring completely normal neuroendocrine function after the onset of menopause would presumably only restore menstrual cycles for a maximum of approximately 1 year (at which point follicular depletion would result in complete loss of ovarian production of estradiol). On the other hand, as described previously, the rate of follicular loss accelerates in the last decade before menopause; if this accelerated atresia did not occur, the ovary might not be a limiting factor in the continuation of menstrual cycles, and indeed it has been estimated that without this accelerated atresia the age of menopause might be approximately age 70. A plausible, though speculative, hypothesis is that this accelerated atresia is a result of the elevated FSH which characterizes perimenopausal cycles. Conceivably, inhibition of the elevation of gonadotropins which occurs in the perimenopausal period might decelerate the accelerated atresia and thus postpone menopause. Thus, a neuroendocrine intervention might, in humans as in rats, also postpone loss of cycles. However, the nature of such intervention would most likely be the opposite in humans compared to rats. In women the transition to menopause entails elevated gonadotropin secretion, possibly leading to accelerated atresia; this atresia-related follicular depletion, leading to the disappearance of mature follicles, is the most likely proximal cause of cessation of menstrual cycles and certainly the cause of the loss of ovarian production of estradiol. In contrast, in rats the transition to acyclicity entails lower gonadotropin secretion (especially a decreased preovulatory LH surge), presumably the proximal cause of failure to ovulate in the presence of mature follicles; this failure to ovulate in the presence of mature follicles is the most likely proximal cause of cessation of cycles in rats.

The complex interaction of ovarian and neuroendocrine impairments has been addressed directly in heterochronic ovarian graft studies, especially in mice. As mentioned, at the time estrous cycles cease, mice exhibit profound impairments in the steroid-induced LH surge and also exhibit almost complete follicular depletion. When young ovaries are grafted under the kidney capsules of young ovariectomized mice, the host young mice will begin to exhibit estrous cycles which continue almost as long as those in intact mice of the same age. However, when ovaries from older cycling mice are grafted into ovariectomized young mice, few if any cycles are observed. In contrast, when young ovaries are grafted into previously acyclic and ovariectomized older mice, hosts will usually exhibit a reinitiation of estrous cycles, although these cycles will not continue as long as the cycles supported by young hosts given young ovarian grafts (correlated with impairments in the LH surge observed in mice at this age). Such studies demonstrate a clear interaction between age-related impairments in ovarian function and age-related impairments in neuroendocrine function. Thus, impairments in both ovarian and neuroendocrine function contribute to loss of cycles, and the coexistence of these impairments produces functional loss greater than would be observed with impairments in only one locus. These studies also suggest that at least in mice, restoring neuroendocrine function to youthful levels without restoring ovarian function in older cycling mice would be unlikely to extend reproductive life span substantially, whereas restoring ovarian function to youthful levels (if such were possible) without restoring neuroendocrine function would extend reproductive life span substantially. Nevertheless, these studies also indicate that to obtain maximum reproductive life span, it would be necessary to restore both neuroendocrine and ovarian function to youthful levels.

An additional level of interaction between the ovaries and neuroendocrine function was indicated by variations of these studies. When mice were ovariectomized when young and, after reaching middle age, given young ovarian grafts, they supported estrous cycles almost as long as young mice, in contrast to the middle-aged mice that were ovariectomized at the time of grafting, which supported cycles for a much shorter time. Furthermore, when old acyclic mice were ovariectomized 2 months before grafting, these mice exhibited significantly more cycles than mice ovariectomized at the time of grafting. Similar results were obtained when a steroid-induced LH surge, rather than cycles after grafting, was used to assess neuroendocrine function. These data suggest that neuroendocrine impairments are to some extent caused by ovarian secretions, possibly estradiol. However, neuroendocrine impairments occur even in the absence of the ovaries, and by the mean life span, these ovary-independent neuroendocrine impairments significantly reduce the ability to support estrous cycles even in long-term ovariectomized mice given young ovarian grafts.

The impaired ability to produce an estrogen-dependent LH surge in middle-aged rodents, along with the reduced estrogen receptors observed in aging rodents, has led to the suggestion that aging is associated with a decreased responsiveness to estrogen. However, the regulation of LH by estrogen is very complex and biphasic, whereas the data reported using simpler responses to estrogen do not uniformly indicate an impaired sensitivity to estrogen in middle-aged rodents; indeed, some studies suggest an increased sensitivity to estrogen. A more likely cause of the impaired LH surge is a perturbation of the circadian rhythm system, upon which the LH surge is also exquisitely sensitive.

IV. CONSEQUENCES OF MENOPAUSE

Although until recently menopause was viewed as a benign and natural transition, it is increasingly clear that loss of estrogen after menopause predisposes to diseases and mortality. However, because early studies suggested that estrogen replacement therapy could be a risk factor for other diseases, in particular reproductive cancers, the use of estrogen replacement therapy continues to be somewhat controversial. Nevertheless, more refined estrogen replacement regimes have demonstrated that estrogen replacement therapy dramatically reduces osteoporosis, cardiovascular disease, and all-cause mortality in postmenopausal women. The effect of estrogen replacement therapy on the risk of breast and uterine cancer continues to be somewhat unclear, but because cardiovascular disease causes far more deaths than cancer in women (as well as men), the net effect of estrogen replacement therapy is to decrease mortality rate. In addition, there is increasing evidence that estrogen replacement therapy may protect against age-related impairments in cognitive function. Therefore, there is an increasing consensus among clinicians that for most women, estrogen replacement therapy after menopause is recommended. Nevertheless, since estrogen replacement therapy does not perfectly mimic the youthful pattern of steroid secretion, there is considerable room for improvement in this treatment.

Although the health risks of menopause have only recently been appreciated, women have always recognized two major nuisance effects of menopause which, while not generally life threatening, could greatly impact quality of life. First, hot flushes constitute the classic sign of menopause and occur in a majority of postmenopausal women. These vasomotor disturbances entail a sudden sensation of heat which radiates across the body accompanied by flushing and perspiration, lasting for a few minutes, and sometimes followed by a chill, palpitations, and/

or a sense of anxiety. These episodes can disrupt sleep and are disturbing enough to motivate many postmenopausal women to seek medical care. The second major nuisance effect is a general urogenital atrophy, in particular atrophy of the vagina, vulva, and urethra. Urogenital atrophy involves a thinning of the walls of these structures and a dramatic decrease in sebaceous gland secretion. Genital dryness can be irritating, can make sexual intercourse painful, and can predispose toward certain kinds of infections. Urinary tract atrophy can contribute to incontinence. Estrogen replacement is the most effective means to attenuate these effects of menopause.

Less immediately evident to women undergoing menopause are two potentially life-threatening complications: osteoporosis and cardiovascular disease. Osteoporosis affects more than 20 million women in the United States and is major risk factor for hip fractures, which in turn are a major source of loss of mobility in older women. Bone density is maximal around 30 years of age and then slowly decreases; in women, the decrease in bone density sharply accelerates immediately after menopause. Although calcium supplementation and exercise are undoubtedly helpful, most practitioners agree that estrogen replacement therapy is an essential component in the regimes intended to prevent or treat osteoporosis.

Although the use of estrogen replacement therapy for the prevention and treatment of osteoporosis has long been standard, recent data suggest that a possibly even more important complication of menopause is an increased risk of death due to cardiovascular disease. Several large studies over the past decade have convincingly demonstrated that estrogen replacement therapy can reduce the risk of cardiovascular disease by as much as 50% or even more. Because cardiovascular disease is the major cause of death in women (as in men), such a large effect of estrogen translates into decreased all-cause mortality rate in postmenopausal women on estrogen replacement therapy vs women who are not on estrogen replacement therapy. Although there has historically been concern about the use of estrogen replacement therapy as a risk factor for cancers of the reproductive tract, recent studies suggest that the magnitude of such risk is much smaller with current steroid formulations than was the case with early formulations. Also, although this risk is still probably real, in most women the beneficial effects of estrogen replacement therapy far outweigh the risks.

Very recent data suggest that the decreased estrogen associated with menopause may also be a risk factor for age-related impairments in cognitive function and possibly even Alzheimer's disease. Currently, there are no definitive large studies, and these observations must be considered preliminary. Nevertheless, a substantial body of data is quickly accumulating suggesting that estrogen, through unknown mechanisms, appears to be neuroprotective under a variety of stresses. As more data accumulate, it is likely that these studies will provide even further evidence of the value of estrogen replacement therapy for most postmenopausal women. Nevertheless, since the mechanisms underlying most beneficial effects of estrogen are not known, and since estrogen replacement neither completely prevents the diseases described previously nor is completely without risk, much more work is needed to develop improved treatments for the postmenopausal condition.

See Also the Following Articles

Estrogen Replacement Therapy; Estrogens, Overview; Follicular Atresia; Menstrual Cycle; Osteoporosis

Bibliography

Ahmed Ebbiary, N. A., Lenton, E. A., and Cooke, I. D. (1994). Hypothalamic–pituitary ageing: Progressive increase in FSH and LH concentrations throughout the reproductive life in regularly menstruating women. *Clin. Endocrinol.* **41**, 199–206.

Andrews, W. C. (1996). Menopause and hormone replacement: Introduction. *Obstet. Gynecol.* **87**(2, Suppl.), 1S–15S.

Gosden, R. G., and Faddy, M. J. (1994). Ovarian aging, follicular depletion, and steroidogenesis. *Exp. Gerontol.* **29**(3/4), 265–274.

McKinlay, S. M. (1996). The normal menopause transition: An overview. *Maturitas* **23**, 137–145.

Rossmanith, W. G. (1995), Gonadotropin secretion during aging in women: Review article. *Exp. Gerontol.* **30**(3/4), 369–381.

Villablanca, A. C. (1996). Coronary heart disease in women. *Postgrad. Med.* **100**(3), 191–202.

Wise, P. M., Scarbough, K., Larson, G. H., Lloyd, J. M., Weiland, N. G., and Chiu, S. (1991). Neuroendocrine influences on aging of the female reproductive system. *Frontiers Neuroendocrinol.* **12**, 323–356.

Reproductive Senescence, Nonhuman Mammals

Anne N. Hirshfield and Jodi A. Flaws
University of Maryland School of Medicine

GLOSSARY

fertility The ability to produce offspring.

menopause The period of cessation of menstruation.

oocyte A female germ cell that has entered the final stages of maturation and is incapable of undergoing mitosis.

oogonia Immature female germ cells which can multiply by mitotic division.

senescence The process of growing old; aging.

Reproductive senescence is the decline, and ultimately the cessation, of reproductive function as a consequence of increasing age. This phenomenon occurs in both males and females. However, in most cases, the symptoms are more dramatic and have more far-reaching consequences in the female because she usually bears the larger part of the reproductive burden. The particular manifestations and the severity of reproductive decline vary widely among species. At one extreme are human females, who spend nearly one-third of their lives in a postreproductive state. At the other extreme are large tortoises and turtles, which experience little, if any, diminished fertility through their long life spans.

I. INTRODUCTION

Reproductive senescence is often the earliest and clearest manifestation of aging in an organism. Senescence manifests as a decline in function of a number of organ systems. In aging animals, diminished function can be measured in virtually every parameter—elasticity of connective tissues diminishes, respiratory volume declines, and ability to fight infections is reduced. It is hypothesized that these detrimental changes reflect a natural, metabolically dictated exhaustion process involving the gradual accumulation of metabolic-derived insults. Some damage is quite harmful and is repaired by a variety of mechanisms, but aging damage is hypothesized to be more benign and/or more difficult to repair. When these types of damage accumulate and become very numerous, they collectively become significant.

Most species experience a gradual decline in reproductive capacity with age, mirroring the gradual decline in all parameters as animals approach the end of their life span. However, nearly all wild animals are capable of reproducing as long as they live. Only female humans and some old world monkeys, some cetaceans, and some domestic species are known to experience a significant postreproductive period.

These diverse patterns reflect the differences in each species' life history. A survey of the animal kingdom reveals a dazzling variety of reproductive strategies. Some animals, such as mayflies and Pacific Coast salmon, "put all their eggs in a single basket," expending all their energy in producing one enormous clutch of eggs. Others, such as the medaka fish or the domestic chicken, extend their egg laying over several weeks or months. This strategy requires that some energy must be diverted away from egg production toward support of continued survival. The longer the period of survival, the more energy that must be expended on nonreproductive activities. Still other animals, such as humans and polar bears, bear singlet offspring over periods of several years and provide extensive nurture after birth. This strategy requires division of energy expended on reproduction in both gamete-producing activity, pregnancy, and postpartum nurturance and activities which promote extended survival. Postpartum nurturance of offspring puts a premium on long-term survival: If parents die before their young are independent, this substantially reduces the likelihood that their offspring will live to maturity . Despite the great variety in these reproductive strategies, success is measured by a single outcome: the total number of offspring that survive to reproduce. The pressures of natural selection favor those adaptations that maximize this outcome.

When viewed from the evolutionary perspective, it is clear that reproduction and senescence are inextricably intertwined. It appears that longevity can only be achieved by diverting energy away from the mechanics of reproduction. This makes evolutionary "sense" only if extended survival increases the likelihood of reproductive success in offspring. There would be no selective advantage in diverting energy away from reproductive processes if longevity did not confer any benefits upon the offspring. However, there would be a selective advantage in extending the reproductive period over a longer time span if the climate, food supply, or predator population were variable. Some offspring would therefore be more likely, by chance, to be born under more favorable conditions and be more likely to survive. In animals, such as humans, longevity has the selective advantage of enabling learning to be passed from parent to child, increasing adaptation and flexibility in response to a challenging and changing environment.

The average individual life span of a species is thus a component of the organism's reproductive strategy, a consequence of the selective pressures that molded the reproductive plan. The number of offspring, and their chances of survival to maturity, depends on the amount of parental investment (e.g., on the number and size of eggs and on the amount of care, if any, given to offspring), and this in turn influences the parent's survival probability. The trade-off between reproduction and "personal growth" made by an adult in any one year will affect its probability of surviving that year and thereby influence the possible trade-offs in all future years.

II. WILD ANIMALS

A. Fish

Some of the most fascinating life histories with respect to reproductive senescence can be found among fishes. Fishes show three types of senescence: sudden death at spawning (e.g., pacific salmon), gradual senescence (guppies, medaka, and many other teleosts), and negligible senescence with long life and indeterminate growth (as in sturgeon). Fish with gradual senescence are known to experience a decrease in reproductive capacity with age; egg quality decreases with age and it is a common acquacultural practice to replace older female broodstock. However, reproductive capability appears to persist throughout life. Unlike mammals, fish continue to produce new oocytes (by proliferation of oogonia) through their adult life spans, so depletion of oocytes with age is not an issue. In one study of male platyfish, some laboratory animals with almost twice the average life span had testes that contained all stages of spermatogenesis and pituitaries that were immunocytochemically similar to those of young animals. Unfortunately, information concerning reproductive capacity in long-lived fishes is lacking.

B. Birds

Songbirds and other small birds have short lives in the wild, with few surviving more than 1 or 2 years. However, some birds are very long lived. Large seabirds such as albatrosses may survive nearly a century. In captivity, parrots can live 70 years or more and cockatoos have been known to live 80 years. Although some captive birds lose the capacity to reproduce in extreme old age, a pair of wild-caught sarus cranes in the National Zoo produced two chicks annually for nearly 30 years until the female died. The male is now paired with another mate and continues to reproduce.

C. Reptiles

Tortoises and turtles are the longest lived creatures on earth. Although they routinely survive over 100 years, in many species their reproductive capacity does not diminish with age. Turtles, tortoises, and crocodiles exhibit indeterminate growth: The longer they live, the larger they get. This phenomenon appears to be the key to their longevity. Most mammals and birds have a rapid growth phase which is followed by a prolonged period of existence at a constant adult body size. Constant adult body size (determinate growth) is an important adaptation for warm-blooded creatures living on land because it permits the evolution of appropriate bodily dimension for the species' particular environmental niche. However, determinate growth has a long-term cost: At maturation, these animals lose their ability to increase or replace functional units such as nephrons, alveoli, and neurons. Aging is the loss of these now irreplaceable functional units. In contrast, some reptiles, amphibians, and fishes can replace worn out or damaged parts throughout life. Like fish, many reptiles and fish continue to produce new ova throughout life, through mitotic division of germ cells (oogonia). In fact, for some reptiles and fish, the relationship between age and reproduction is quite the reverse of that seen in mammals. Clutch size increases with age in king snakes, some turtles, and some lizards. In other turtles and snakes, however, there is evidence of diminished reproductive output with advancing age. One report documents a marked decline in egg production in alligators after 30 years of age.

D. Mammals

Short-lived, fast-breeding mammals (small rodents, lagomorphs, etc.) seldom, if ever, attain sufficient longevity in the wild to experience age-related effects on fertility. Although they may be able to survive for many years in captivity, in the wild they have a very high rate of mortality that is almost completely independent of age. A study of wild brown rats on a Maryland farm revealed that only 5% of the animals survived a full year, and a study of white-footed mice in Michigan indicated that mean life expectancy from birth was from 17.4 to 31.5 weeks, depending on the severity of the winter (whereas white-footed mice can survive 6–8 years in captivity). Only 9–38% of cotton-tailed rabbits in California lived a full year and most common shrews in England die before their first birthday. Thus, most, if not all, individuals die before they are old enough to experience a significant decline in fertility. In a Mediterranean population of rabbits, the oldest animals (>24 months) represented only 5% of the females in the population, yet they had the greatest number of embryos per litter.

In populations in which mortality is not as high, the effects of increasing age of the female become apparent. An investigation of the European varying hare (*Lepus timidus*) in Sweden showed that fertility remained fairly constant from 2 to 7 years of age, but the number of litters per year and the number of young per litter dropped precipitously in Year 9, with virtual cessation of reproduction by Year 10.

Survival curves of large, slow-breeding mammals show age-specific mortality rates, with mortality highest in the young and old cohorts. Unfortunately, information about fertility in the longest lived wild animals is scarce or lacking. Elephants can occasionally live 60–70 years and it has been suggested that some very old females are postreproductive. However, a detailed study of the African elephant (*Loxodonta africana*) demonstrated that 33% of the 50- to 60-year-old cohorts were pregnant and only one of the five very old elephants examined appeared to have no macroscopic follicles remaining in its ovaries. The incidence of pregnancy in American bison was highest in the 2- to 12-years of age cohorts (87% were pregnant) and then declined gradually, reaching 21% pregnant in the oldest cohorts (25–35 years of age). Female giraffes in the wild can reproduce at least until they are 20 years old; the longevity record for giraffes in captivity is 28 years. Fin whales have survived as long as 96 years and sei whales live as long as humans, but their reproductive life spans are unknown.

Despite the lack of detailed information, it is probably safe to conclude that few animals in the wild experience an extended postreproductive period comparable to the menopause of the human female. A notable exception is the short-finned pilot whale (*G. macrorhynchus*). A study of 483 of these creatures revealed that the estimated age of females ranged up to 63 years, yet none of the 92 females over 36 years old were pregnant. In fact, the rate of pregnancy appeared to decline beyond age 20. All females over

40 years old had apparently ceased to ovulate and ovarian follicles >1 mm were found in only 6 of the 49 females. Interestingly, the decline in fertility is accompanied by an increase in lactational activity. Although young pilot whales begin to take solid food within the first year postpartum, offspring of older mothers nurse for much longer periods of time than those of young mothers (the lactational period was estimated to be 2.8 years for mothers in the 15- to 20-year-old age group versus 6.4 years for whales over age 30). Surprisingly, lactating whales as old as 51 years were found, presumably 11 years after ovulation had ceased; apparently, the last calf may be suckled until puberty (8 years in females and 11 years in males). It is significant that the postreproductive period of these whales is close to the maximum time taken for the young to reach puberty, suggesting that selection favored females that phased out the investment in bearing young as the probability of living long enough to rear them declined.

The only other whale known to have a natural, long postreproductive period is the sperm whale. Both pilot whales and sperm whales are deep-diving mammals, suggesting that the postreproductive period is an adaptation that provides "grandmotherly baby-sitters" to tend the young (who are incapable of deep dives) while their mothers feed at great depth. Alternatively, postreproductive females could serve as repositories of essential information for their family group, leading them to the best feeding grounds and eliminating the need for energetically costly, random 2000-ft dives. A similar function may explain the selective advantage conferred by postreproductive wild horses, cattle, and elephants on their species; these grandmothers might be a critical resource for remembering the best migratory routes and the location of water holes

III. DOMESTIC AND CAPTIVE ANIMALS

Animals in captivity greatly exceed the expected life spans of their wild relatives. From these animals, we can learn about the length of the potential life span, and we can accurately measure changes in productivity with advancing age. However, the broader significance of these reproductive changes is not clear. Reproductive parameters in young captive individuals are often very different from those of their wild counterparts. Some species breed poorly, if at all, in captivity even when young. Individual caging, controlled lighting, *ad libitum* food supply, and constant temperatures of the laboratory alter the impact of environmental cues that profoundly influence reproductive parameters in wild populations. Thus, baseline fertility in captive populations is different from that of wild populations. Furthermore, the older individuals in these captive populations usually represent a subpopulation that is never encountered in the wild. Therefore, decline in fertility in this group is a somewhat artificial phenomenon, restricted to laboratory conditions and not characteristic of the species' natural life history. Finally, it should be noted that domestic animals and laboratory animals have been bred for specific characteristics seldom, if ever, including longevity. Dairy cows, for example, have been selected for high milk production or high amounts of butterfat in the milk. Only a few individuals are retained for breeding purposes; most are killed for other reasons before they have reached the upper limits of their species' potential life span. Reproductive success in cows follows the pattern seen in most long-lived mammals. Reproductive rates increase as cows mature (2–5 years), peak at 5–9 years, and then decrease from 9 to 12 years. Most are culled from the herds long before the end of the first decade. Breed type is also a major source of variation in the reproductive life span of beef cattle.

A. Primates

Data on reproductive senescence in female monkeys are equivocal. Tamarin monkeys over 17 years of age were anovulatory; they had altered ovarian hormone profiles suggesting continued production of steroids by the ovarian interstitial tissue. Rhesus monkeys are considered to be "extremely old" by the time they are 25 years old; however, females in provisioned populations are fecund well into their third decade of life: One monkey at the University of Maryland is 35 years old and still cycling regularly. In Japanese monkeys, there is a measurable decline in ovarian activity by 21–25 years of age, with cessation of cyclicity at 27 years, near the end of the life span.

B. Rodents

Although the average life span of rodents in the wild is only a few months, rats can survive under laboratory conditions for longer than 2 years. However, signs of reproductive aging can be detected well within the first year. The first symptoms are subtle changes in patterns of secretion of the reproductive hormones. Beginning at about 6 months of age, litter sizes decrease compared with younger rats, a phenomenon that appears to be due to an increase in early embryonic abnormalities. By 8–12 months, estrous cycles become irregular and ultimately cease. Few irregularly cycling rats achieve pregnancy when mated, and even those rats that were still cycling regularly at 8–12 months were less likely to reproduce successfully.

As in humans, the total number of oocytes in the ovaries declines with age, but unlike humans there is no obvious acceleration in the rate of loss over time, and very aged rats were found to have some follicles remaining in their ovaries (at 1000 days postpartum, approximately 500 remained). The rate of loss of oocytes in mice is strain specific, indicating a strong genetic determinant.

The difficulty of correlating results from studies of laboratory animals with the human condition is evident from observations of dietary influences on the reproductive life span of rodents. Lifelong moderate caloric restriction substantially extends the overall life span of rats and significantly extends their reproductive life span (as measured by maintenance of regular estrous cyclicity) significantly. Severe caloric restriction greatly extends both reproductive life span and overall longevity in laboratory mice. Thus, caloric restriction may serve as an antidote to aging. Alternatively, it may simply redress a pathological situation: The continual availability of food may lead to overeating in caged rodents which could be hazardous to health when coupled with their diminished level of physical activity in confined quarters. It is significant in this regard that in human females higher weight may be associated with later menopause, the inverse relationship from that seen in rodents. Similar lack of correlation between rodents and humans pertains to the influence of pregnancy on the duration of the fertile life span. Frequent pregnancy, or chronic progesterone administration to mimic pregnancy, results in extended reproductive life span in rats. However, no such correlation appears to exist for humans.

IV. CONCLUSIONS

The variations in the relationship between fecundity and senescence can only make sense when placed within the context of such factors as demographics, duration of the infantile period, number of young, and the species' ecological niche—the organism's overall life history strategy. Unfortunately, "almost everything we know about the physiological mechanisms underlying reproductive development has been learned from experiments done with a very few highly domesticated stocks, mostly rodents and livestock" (Bronson, 1989). Also, "even for the species that we know best, like our laboratory and domesticated animals, we still have only a hazy idea of the environment in which their reproductive systems evolved. Until we understand that we cannot begin to make sense out of this seemingly endless reproductive diversity" (Short, 1984). It is tragic that many species are likely to vanish from the face of the earth before the details of their reproductive life histories are known.

See Also the Following Articles

Fertility and Fecundity; Menopause; Reproductive Senescence, in Humans

Bibliography

Adams, C. E. (1984). Reproductive senescence. In *Reproduction in Mammals, Book 4: Reproductive Fitness* (C. R. Austin and R. V. Short, Eds.), pp. 210–233. Cambridge Univ. Press, Cambridge, UK.

Bronson, F. H. (1989). *Mammalian Reproductive Biology*. Univ. of Chicago Press, Chicago.

Norris, K. S., and Pryor, K. (1991). Some thoughts on grandmothers. In *Dolphin Societies: Discoveries and Puzzles* (K.

Pryor and K. S. Norris, Eds.). Univ. of California Press, Berkeley.

Short, R. V. (1984). Species differences in reproductive mechanisms. In *Reproduction in Mammals, Book 4: Reproductive Fitness* (C. R. Austin and R. V. Short, Eds.). Cambridge Univ. Press, Cambridge, UK.

Vom Saal, F. S., and Finch, C. E. (1988). Reproductive senescence: Phenomena and mechanisms in mammals and selected vertebrates. In *The Physiology of Reproduction* (E. Knobil and J. D. Neill, Eds.), pp. 2351–2403. Raven Press, New York.

Wise, P. M. (1989). Aging of the female reproductive system: A neuroendocrine perspective. In *Neuroendocrine Perspectives* (E. E. Muller and R. M. MacLeod, Eds.), Vol. 7, pp. 117–168. Springer-Verlag, New York.

Wise, P. M., Krajnak, K. M., and Kashon, M. L. (1996). Menopause: The aging of multiple pacemakers. *Science* 273, 67–70.

Reproductive Technologies, Overview

Alan Trounson

Monash University

GLOSSARY

blastocyst A stage of embryo development with a nest of inside cells or inner cell mass cells that are surrounded by a fluid-filled cavity and an outer ring of trophectoderm cells.

cloning The insertion of nuclei from an embryo, embryo cell line, fetal cell line, or adult somatic cell line into the cytoplasm of a mature egg that has had its own chromatin removed. These embryos may develop to term in some animal species.

cryopreservation The freezing in ice-forming solutions or vitrification in glass-forming solutions of cells or tissues and their storage at low temperature.

ICSI Intracytoplasmic sperm injection, involving the direct injection of a sperm into the center of an egg or oocyte.

IVF *In vitro* fertilization or fertilization in the laboratory, outside the body.

IVM *In vitro* maturation; refers to the completion of cellular events in the oocyte that allows for fertilization and full developmental competence.

preimplantation embryo The fertilized egg and developing embryo up to the blastocyst stage, at which time the embryo normally implants in the wall of the uterus.

superovulation The induction of multiple follicular growth with fertility drugs which are usually purified preparations of follicle-stimulating hormone.

zona pellucida The glycoprotein shell surrounding the oocyte; composed of three major glycoproteins—ZP1, ZP2, and ZP3.

zygote A single-celled fertilized egg with a male and a female pronucleus.

Reproductive technologies have been developed for animal breeding, in which the primary interest is accelerated genetic gain in productivity, and in human reproductive medicine for the treatment of human infertility and inherited genetic disease. In animals, the techniques are referred to as egg (or ovum) transfer, embryo transfer, *in vitro* maturation, and embryo transfer, or embryo multiplication and transfer (EMT) when cloning or nuclear transplants are involved. In human medicine, the techniques are referred to as assisted conception, *in vitro* fertilization (IVF), or IVF and embryo transfer. Essentially, many of the procedures are the same in animal breeding and human medicine and, where applicable and appropriate, can be interchanged.

I. INTRODUCTION

Like many animal species, human sperm are capable of fertilizing the mature human egg (oocyte) in culture *in vitro*. This permits the option of manipulating maturation, fertilization, and developmental events *in vitro* and enables the consequent therapies for human infertility known as reproductive technologies. Importantly, increased access to the oocyte made possible by clear visualization of Graafian follicles and their aspiration under vaginally guided ultrasound has allowed *in vitro* fertilization (IVF) techniques to flourish. In 1994 in Australia, births resulting from the use of reproductive technologies accounted for 1.0% of all births. Data from the same year indicate that approximately one-third of the patients were classified as male infertility, one-quarter had blocked Fallopian tubes, and one-quarter had both tubal disease and male infertility. The remainder had endometriosis, were idiopathic, used donor oocytes or embryos, or used IVF techniques for preimplantation embryo genetic diagnosis. It is now very clear that the reproductive technologies have been adopted by reproductive medicine as a preferred therapy for human infertility, reducing the demand for surgical repair procedures, artificial insemination, and a number of endocrine treatments.

II. EMBRYO PRODUCTION

The preference for reproductive technologies is to induce superovulation for the recovery of numerous oocytes (usually 8–14) by the treatment of the female partner with gonadotropin-releasing hormone (GnRH) analog for as long as necessary (14–40 days) to downregulate pituitary gonadotropin secretion and hence ovarian steroid production. Multiple follicular growth is then achieved by daily administration of gonadotropin preparations, which are now usually purified human follicle-stimulating hormone (FSH) or recombinant human FSH, for as long as is necessary to generate a cohort of growing follicles. When the leading follicle is 1.7–2.0 cm in diameter, as measured by ultrasound, the patients are given an ovulating dose of human chorionic gonadotropin, the treatment with GnRH analog is terminated, and oocytes are recovered by follicular aspiration under vaginally guided ultrasound 35–38 hr later.

Oocytes surrounded by their cumulus cloud are readily identified in follicular aspirates under dissecting microscopy using transmitted light. The oocytes are placed into culture medium for 4–6 hr to enable the spontaneous completion of oocyte maturation, recognized by extrusion of the first polar body into the perivitelline space between the oocyte cell and the glycoprotein shell that surrounds the oocyte, known as the zona pellucida.

Sperm obtained by masturbation is separated from seminal plasma components by buoyant gradient density centrifugation and motile sperm added to the oocytes at a concentration of 1.0 to 20×10^4 sperm/ml. Sperm may be left with oocytes for 12 to ~16 hr but it is becoming more common to remove the oocytes from the sperm solution after 1 hr and to begin culture in appropriate embryological media. This brief exposure to sperm is sufficient to select a fertilizing spermatozoon and further incubation with sperm does not increase fertilization rates. Pronuclei are visible in the cytoplasm of human oocytes from 12 to 20 hr after insemination. Two centrally located pronuclei and the presence of a second polar body in the perivitelline space confirms fertilization. Oocytes with single pronuclei may also be normally diploid because of the occasional lack of synchronization in the timing of formation of male and female pronuclei.

The first cleavage division will occur 22–34 hr after insemination and further cleavage divisions will occur at 14- to 18-hr intervals. Cleavage may not be synchronous and individual cells may become blocked in their cell cycle or fragment. The pronuclear zygote (one-celled fertilized oocyte) and early cleavage stage embryo uses pyruvate and lactate in preference to glucose, is sensitive to phosphate ions, and changes in pH. There is also a preference to avoid the use of serum in culture media for early embryos. Consequently, embryo culture media need to be carefully designed and made to quality assurance standards for the production of viable embryos. In the absence of serum, human serum albumin and amino acids, together with vitamins, are necessary for normal embryo development.

It is usual to transfer embryos on the second or third day of culture to the uterus of the female partner. At this time, embryos are usually 2–6 cells (Day 2) and 8–12 cells (Day 3). However, recent research has shown a benefit for increased viability, and hence implantation rates, for the culture of embryos for 5 or 6 days by which time the embryo is termed a blastocyst and has differentiated into inner cell mass cells and trophectoderm and has a fluid-filled cavity. Using optimized culture conditions, approximately 60% of pronuclear zygotes will develop to blastocysts in 6 days of culture. Pregnancy rates for the transfer of one to three blastocysts are usually in excess of 40% and implantation rates for embryos in excess of 22%. While it is not essential, embryos can be grown to blastocysts in coculture systems with a range of feeder-cell types.

It is possible to create a tear in the zona pellucida using micromanipulation needles or to digest a hole in the zona with acidified solutions (pH ~3.0). This is known as "assisted hatching" and may improve the chance of implantation for some embryos transferred to the uterine cavity on Day 3. This is particularly important for embryos with thickened zonae. Blastocysts usually hatch out of their zonae by Day 6 and it is usual to completely remove the zonae of blastocysts enzymatically with Pronase solutions.

Embryos are transferred to the uterus, because there is no survival advantage of placing early cleavage-stage embryos into the Fallopian tubes. Pregnancy rates increase with increasing numbers of embryos transferred but so does the multiple implantation rate. For highly viable embryos, the numbers transferred should be restricted to one or two and the remainder cryopreserved (frozen) for subsequent use by the patient. Multiple implantations can result in the need for fetal reduction and an increased risk of complete loss of pregnancy.

III. MICROFERTILIZATION TECHNIQUES FOR MALE INFERTILITY

Decreasing semen quality results in decreasing fertilization rates by insemination. Decreasing sperm motility and increasing rates of abnormal sperm morphology adversely affect the probability of fertilization. There are also sperm that do not acrosome react in the presence of oocytes and, more particularly, the zona. When few sperm are produced, there may be insufficient motile sperm for fertilization by insemination.

Micromanipulation techniques have been developed for the direct insertion of sperm into the cytoplasm of oocytes [intracytoplasmic sperm injection(ICSI)]. This requires the isolation of a few sperm and their immobilization, which is usually achieved by crushing the tail or midpiece with the injection pipette. The immobilized spermatozoon is drawn into a fine sharpened glass pipette (7–10 m outer diameter) and the pipette pushed through the zona and into the oocyte. The oocyte is held stationary on a suction pipette with the first polar body and adjacent chromosomes at 90° to the injection pipette to avoid disruption of the oocyte's chromosomes on the metaphase plate. Oocyte cytoplasm is aspirated up the injection pipette to ensure penetration of the oocyte's plasma membrane and the sperm injected into the cytoplasm. The plasma membrane will seal when the pipette is withdrawn.

No specific treatment is required for sperm to decondense and form a male pronucleus after ICSI. The oocyte cytoplasmic proteases are able to remodel the sperm nucleus and the microinjected sperm is capable of activating the oocyte by release of the sperm-derived calcium "oscillogen." It is not essential to remove human sperm acrosomal membranes or to induce the acrosome reaction. The latter, however, is essential for sperm penetration of the zona pellucida and gamete fusion in the normal fertilization process.

Fertilization rates after ICSI are often better than those obtained by insemination, probably because any immature oocytes are seen after removal of the follicular cells and are discarded. Some reduction in fertilization is observed with injection of grossly abnormal sperm and those which are completely immotile when recovered. Sperm recovered from the epididymis and from the testis can be used successfully for ICSI without any marked reduction in fertilization rate. Embryo development and fetal development to term is not different to that observed for mature ejaculated sperm. There have been some reports of increased sex chromosomal abnormalities in concepti after ICSI and this is believed to be due to the occurrence of mosaicism in sex chromosomes of severely infertile men.

There are increasing concerns about gene deletions in the Y chromosome of infertile men [e.g., the deleted in azoospermia (DAZ) gene] and mutations in the androgen receptor on the X chromosome. As a consequence, infertility may be inherited by some children conceived by ICSI. Counseling of patients should include this possibility and some attempt made to diagnose such genetic variants for infertile men prior to ICSI.

IV. CRYOPRESERVATION OF OOCYTES, EMBRYOS, AND OVARIAN TISSUE

Currently, human oocytes cannot be cryopreserved and it is calculated that $<1\%$ of mature oocytes will survive, fertilize, and develop to term. Once the oocyte is fertilized, it may be successfully frozen at the pronucleate stage (zygote) and during early cleavage stages. Embryos can also be frozen at the morula (16 to 32 cells) or blastocyst stages (>50 cells). In the human, zygotes and embryos are cryopreserved by conventional slow cooling techniques in solutions that contain 1,2-propanediol or dimethyl sulfoxide. This enables equilibrium to be maintained between the intracellular and extracellular compartments which effectively dehydrates the embryonic blastomeres during cooling and the phase of ice formation. When sufficient dehydration is achieved, intracellular cryoprotectant levels prevent the formation and growth of lethal intracellular ice and the embryos may be rapidly cooled and stored in liquid nitrogen ($-196°C$). Approximately 60–85% of embryos survive thawing and removal of the cryoprotectant and these implant at about the same rate as unfrozen embryos. By cryopreserving embryos for patients, pregnancy expectations can be substantially increased from the one cycle of IVF treatment. Embryos can be stored frozen for an unlimited period of time, but in some jurisdictions, legislative or regulation imposes maximum storage periods of 5 or 10 years.

It is now also possible to cryopreserve ovarian tissue for patients. Small pieces of ovarian cortex containing the primordial follicles will survive cryopreservation and, when thawed and transplanted to severe immune incompetent mice, will begin functioning apparently normally. Concerns have been raised about the possible transfer of disease or infection with an autotransplant of ovarian tissue and this could prevent the use of ovarian tissue freezing by patients with cancer, particularly those involving migratory or blood-borne cancer cells. Currently, there is no record of a successful ovarian autotransplantation of cryopreserved cortical tissue in the human.

Rapid cooling techniques have been developed for cell and tissue cryopreservation. By using relatively high solute concentrations in the cryopreservation medium and rapid cooling and warming techniques, glass rather than ice forms. Glass formation is termed vitrification and this process avoids the formation of ice, which is dangerous to cell survival. As yet, these techniques have not been adopted for human embryo cryopreservation but it is likely they will be in the near future.

V. PREIMPLANTATION GENETIC DIAGNOSIS

Embryonic blastomeres may be removed from the developing embryo by partial digestion of the zona and aspiration of one or several nucleated blastomeres. The single cells may be stained with chromosome-specific fluorescent-labeled probes for *in situ* hybridization (FISH). The majority of embryos have been assessed for age-related aneuploidies involving chromosomes X, Y, 13, 18, and 21. The embryos of some patients are also examined for chromosomal translocations. The first polar body may also be analyzed by FISH to enable interpretation of probable aneuploidies in the resulting embryos although abnormalities may be overestimated from missing or extra chromatids by polar body analysis. These procedures are usually applied to women over the age of 35 years and can be used to increase the number of chromosomally normal diploid embryos for transfer and hence the birth of chromosomally normal babies. However, some problems still exist in mosaicism of chromosomal numbers in cells of a significant proportion of embryos ($\sim 12\%$). While this may be a normal situation in early development, it is usually recommended that chromosomally mosaic embryos not be transferred to patients.

Determination of embryo sex by FISH for patients with sex-linked genetic disease is also widely available. These procedures maintain the carriers in the population by selecting only female embryos for transfer but prevent affected male children from being born. Preimplantation diagnosis can be strongly preferred to prenatal diagnosis and therapeutic termination of pregnancy.

Single gene disorders can be specifically identified using the polymerase chain reaction (PCR). The accuracy of diagnosis from PCR of one or several blastomeres is in excess of 90% although allele dropout may be a problem in heterozygous states. Multiplex PCR for at least two separate regions of affected genes usually allows detection of allele dropout and prevents misdiagnosis of affected embryos. The genetic diseases being diagnosed in embryos include different thalassemia mutations, cystic fibrosis, familial adenomatous polyposis coli, sickle cell anemia, and myotonic dystrophy. As the methods improve and the range of PCR probes expands for specific disease states, there will be a gradual increase in the demand for preimplantation diagnosis for genetic disorders.

VI. EMBRYONIC MANIPULATION AND NUCLEAR TRANSFER

Embryo manipulations are performed in livestock breeding and include the mechanical splitting of embryos to produce demigenetically identical embryos. This can effectively increase pregnancy outcome in cattle by having two embryos instead of one. As a result, the production of calves can be increased to ≈105% per embryo instead of ≈65–70% for single embryos. It is also possible to clone embryos by nuclear transfer and this has been reported for sheep and cattle. Embryonic blastomeres can be fused into enucleated oocytes and the resultant embryo develops in culture and to term when replaced in the uterus of a recipient ewe or cow. Cultured embryonic cells, fetal cells, and even adult somatic cells of sheep have been used for cloning by nuclear transfer. There is little interest in using these techniques in the human although identical twins occur naturally. It would be considered unethical to apply cloning techniques in human reproductive medicine.

See Also the Following Articles

Blastocyst; Cloning Mammals by Nuclear Transfer; Cryopreservation of Embryos; Cryopreservation of Sperm; Implantation; Infertility; In Vitro Fertilization; Sterility

Bibliography

Lacham-Kaplan, O., and Trounson, A. (1997). Fertilization and embryonic developmental capacity of epididymal and testicular sperm and immature spermatids and spermatocytes. *Reprod. Med. Rev.* **6**, 55–68.

Lancaster, P., Shafir, E., Hurst, T., and Huang, J. (1997). *Assisted Conception Australia and New Zealand 1994 and 1995*, Assisted Conception Series No. 2. AIHW National Perinatal Statistics Unit, Sydney, Australia. (ISSN 1038-7234)

Ng, S. C., Bongso, A., and Trounson, A. O. (1996). Update on micromanipulation techniques for assisted conception. *Curr. Opin. Obstet. Gynaecol.* **8**, 171–177.

Trounson, A. O., and Bongso, A. (1996). Fertilization and development in humans. In *Current Topics in Developmental Biology* (R. A. Pedersen and G. P. Schatten, Eds.), Vol. 32, pp. 59–101. Academic Press, San Diego.

Verlinsky, Y., Munné, S., Simpson, J. L., *et al.* (1997). Current status of preimplantation diagnosis. *J. Assisted Reprod. Genet.* **14**, 72–75.

Wilmut, I., Schnieke, A. E., McWhir, J., Kind, A. J., and Campbell, K. H. S. (1997). Viable offspring derived from fetal and adult mammalian cells. *Nature* **385**, 810–813.

Wood, C., Shaw, J., and Trounson, A. (1997). Cryopreservation of ovarian tissue: Potential "reproductive insurance" for women at risk of early ovarian failure. *Med. J. Aust.* **166**, 366–369.

Reproductive Toxicology

Robert E. Chapin
National Institute of Environmental Health Sciences

GLOSSARY

Computer-Assisted Semen Analysis (CASA) The automated analysis of sperm swimming patterns; used to help identify treatment effects on sperm. Poor CASA values may correlate with reduced fertility, although the degree of this correlation is not yet known.

estromimetics Compounds that mimic the effect of estrogen.

selective reproductive toxicant A compound that causes reproductive toxicity at doses that do not detectably affect other systems in the body.

sperm granuloma A circumferential wall of phagocytic monocytes and neutrophils surrounding a core of sperm that have escaped their normal tubular environment. When this occurs in the excretory ducts, it may block sperm transit through those ducts and reduce fertility, even though spermatogenesis is normal.

toxicant A poison of man-made origin.

toxin A poison of natural origin (bee venom, urushiol from poison ivy, etc.).

window of vulnerability A specific stage of development in which an organism is especially vulnerable to one type of toxicity. For example, gonocyte migration occurs during an ≈2-day window prenatally: Blocking cell migration during this stage would prevent gonad formation, whereas the same exposure occurring in the following week would have no effect on gonadal development.

Reproductive toxicology is the discipline that determines which compounds are toxic to reproduction and determines the mechanisms by which those toxicants work. Historically, workers in this field have defined this area as studying induced defects in producing and releasing functional gametes, whereas nonspecialists have also included compounds that adversely affect *in utero* development. These conceptual lines between reproduction and development have become blurred with the realization, in recent years, that there are *in utero* exposures that have their primary manifestations on the reproductive system in the subsequent adults. Overall, the field encompasses the areas of male reproductive function, female reproductive function, and *in utero* growth and development. Some would also include postnatal growth and development to the point of being a reproductively competent adult.

I. INTRODUCTION

Reproduction is like other aspects of health in that occupational or unintentional anthropogenic exposures that degrade the process of producing a healthy child are broadly considered unacceptable. The process of determining which exposures are toxic to reproduction, at which doses and biological target, and by which mechanisms, is reproductive toxicology.

For ease of discussion, and to mimic the way that research fields are often divided, the discussion is divided into male, female, and developmental toxicities.

The most compelling way to identify a compound as being toxic is to show an effect on the function of the reproductive system. A compound that inhibits the midcycle luteinizing hormone rise, for example, would inhibit ovulation and produce sterility in that female; such a compound is clearly a female reproductive toxicant. Similarly, if all we know about a compound is that it reduces sperm count, we can be reasonably sure that it will be a functional reproductive toxicant. Less convincing are those data that, in the absence of a functional change, show some measurable effect on a component of the system. This leads to considerable debate about whether small and reversible changes (such as a temporary alteration in, for example, mitochondrial function in one cell type) constitute toxicity and warrant labeling certain compounds as reproductive toxicants. This pit can be dodged by terming some changes "adverse" and labeling others as simply "effects."

Once a compound is shown to produce a change in function, the common approach is to identify the probable target tissue, and then if one suspects a direct effect on that tissue (as opposed to an indirect effect mediated by reduced hormonal stimulation, for example), take the target tissue and study the effects in isolation. Culture studies require that the investigator know how much of the compound (or active metabolite) is in the tissue or in the circulation so that relevant amounts of the compound will be used *in vitro*. The *in vitro* use of m*M* levels of a toxicant when the tissue is exposed to m*M* levels *in vivo* is poor toxicology; such data are hard to relate back to the whole animal and ultimately do not lead to a better understanding of how a compound works and whether such a compound poses a risk for humans.

II. MALE REPRODUCTIVE TOXICITY

A compound is termed a male reproductive toxicant when it adversely affects the male reproductive system. The most gross manifestations of such toxicity are sterility and azoospermia.

If the compound directly targets a cell type of the seminiferous epithelium, there will be some initial change in the structure of the epithelium (inhibited sperm release, Sertoli cell vacuoles, increased cell death, and/or the more difficult to spot reduced cell division) which, with continued dosing, progresses to encompass widespread cell death and eventual atrophy of the epithelium. The atrophy produced by most exposures leaves Sertoli cells and residual spermatogonia in the tubules; some chemotherapeutics also kill these spermatogonia.

Toxicity in the epididymis can take the form of broken tubules, followed by the formation of a sperm granuloma, which impedes sperm passage through the duct. Alternatively, recent work has found a correlation between a secreted protein and the resulting motility of the sperm and fertility of the animal (Klinefelter *et al.*, 1997). Thus, occult alterations in the secretory capabilities of the epithelium can impair the function of the epididymis and of the animal.

Effects in the accessory glands (prostate, seminal vesicles, coagulating gland, etc.) are often limited to reductions in the amount of androgen-dependent secretory product formed by the gland. Often these reductions are secondary to other changes in the androgen status of the animal. The lack of variety of responses for these glands after adult exposures is probably due to the fact that (i) few people have looked for more subtle effects and (ii) significant treatment-induced changes in structure (inflammation, cancer, sloughing of the epithelium, etc.) are exceedingly rare events.

Some compounds (trimethyl phosphate and α-chlorhydrin) can selectively reduce sperm motility through mechanisms that are still relatively poorly known.

The adult male rodent appears relatively robust to compounds that target the hormonal status of the reproductive system. When the exposure is sufficient, reduced serum and tissue levels of testosterone can dampen spermatogenesis (although the effect is often mild and appears to require large reductions in plasma T to be effective) and will reduce libido and perhaps epididymal function. There are greater consequences of exposure to such compounds when the subject is exposed during development (*vide infra*).

III. FEMALE REPRODUCTIVE TOXICITY

The clearest separation between female reproductive toxicity and developmental toxicity can be made before conception: Compounds that reduce the ability of the adult female to become pregnant can be confidently labeled as female toxicants, whereas something that impairs her ability to deliver healthy offspring may target either the mother or the fetus, and much scientific digging is required to identify the real target of toxicity.

Several compounds may work by acting at the level of the hypothalamus and/or pituitary. At least two formamidine pesticides, a dithiocarbamate, δ-9-tetrahydrocannabinol, and several estromimetics have all been shown to impact female fertility. An additional effect of such compounds could be behavioral receptivity alteration.

Toxicants can also interfere with the process of follicular development. Several exposures have been shown capable of inducing death of primordial follicles: Polycyclic hydrocarbons (as in cigarette smoke), chemotherapeutics, ionizing radiation, and 4-vinylcyclohexane have been shown to reduce the pool of primordial follicles from which develop the eggs that will eventually be released at ovulation. This reduced pool of oocytes leads to premature menopause.

Few compounds have been described as targeting the process of follicular maturation; mono-(2-ethylhexyl) phthalate has been shown to reduce numbers of granulosa cells and also to reduce estradiol production and block ovulation.

Spindle poisons will inhibit oocyte maturation and fertilization in the female tract. Compounds that block spindle action (e.g., the antifungal compound benomyl or its active metabolite carbendazim) produce such effects.

Since implantation and the initial stages of placental development are dependent on steroid hormones and interactions between the uterus and embryo, it is reasonable to expect that compounds that mimic steroid signals will modulate such events. This remains an understudied area, but chlordecone (Kepone), methoxychlor, and estrogenic congeners of DDT have been shown to interfere with the process of normal implantation.

The adult female appears more vulnerable to hormonal perturbations than the adult male rodent. Not only is the adult affected but also perinatal and developmental exposures can have long-lasting effects (*vide infra*).

IV. DEVELOPMENTAL TOXICITY

Developmental toxicity refers to adverse effects that result from toxicant exposure during a subadult stage of development. This broad definition includes the traditional field of teratology (the study of structural deformations resulting from *in utero* exposure) and encompasses reduced growth and behavioral disturbances as well.

The archetypal teratogen, and the compound that gave the modern field of developmental toxicology its raison d'etre, is thalidomide. Sadly, despite an enormous number of studies examining possible mechanisms of thalidomide toxicity, we still are ignorant of the true mechanism that can explain the effects in humans.

The developing organism has cells in organizational and maturational stages that often will not occur again in that animal's life. Thus, one of the basic tenets in developmental toxicity is that a developmental toxicant has that activity because it interacts with cells at specific points in the cell cycle, or "hits" cells at particular points in their development, when they express molecules and have signaling requirements that they will not express again. The uniqueness of each stage in development is demonstrated by the fact that many developmental toxicants produce their effects only in a narrow time window during development. Thus, toxicant exposure during all of development might produce the same effect as exposure only on Gestational Day (GD) 10, for example (in a rat), suggesting that the key target process occurs at GD 10. Such targets can include signaling molecules the slow onset of expression of toxicant catabolizing enzymes which would allow a toxic compound to accumulate to effective levels near the target molecule, or the sequential expression of cell–cell contact signaling systems that convey important information about cell placement and migration. These are just some of the unique features that make the developing organism more vulnerable to toxicants.

This concept of specific periods of vulnerability has found new importance in the area of endocrine disruptors. The developing reproductive organs have been shown to pass through periods in which they are exquisitely sensitive to changes in the endocrine melieu. Too much estrogen-like signal during development has been shown, in at least a few labs, to produce changes in the functioning of the adult prostate, for example. Too much or too little thyroid hormone during the early nursing period of a rodent can permanently alter the size of the adult testis by changing the day at which Sertoli cells stop dividing. Because such windows of special vulnerability are not found in adults, the concept that the fetus/neonate is much more vulnerable to the effects of these "endocrine disruptors" than is the adult animal is currently accepted. Indeed, this has been shown clearly for numerous compounds.

Prior to ≈5 years ago, a major challenge in the field of developmental toxicity was to separate toxic effects on the dam from alterations in the offspring. That is, what proportion of a developmental defect can be attributed to effects occurring primarily in the mother? This is mostly a problem when dealing with developmental toxicity testing studies, in which some maternal toxicity must be evidenced at the highest dose level to ensure that sufficiently high doses have been tested. Clearly, when terata are selectively produced (i.e., when there are fetal effects in the absence of maternal effects), this problem vanishes.

The advent of molecular biological techniques has led to the exploration of gene expression during development and the finding of genes which control the expression of other genes. By altering the expression of a single homeobox gene, the expression or timing of expression of dozens of other genes could be affected. Thus, considerable effort is currently spent identifying the cascades of gene expression that underlie normal development. From this, it is hoped that new pathways in the process of abnormal development will be identified.

V. REGULATORY TOXICOLOGY

Developmental and reproductive toxicity studies meet the "real world" in the regulatory arena, where the laboratory findings are translated into acceptable exposure limits for the general population or for specific subsets of the population.

The key points in the process of determining the degree of risk posed by a given exposure are (i) the activity of a compound and the dose at which the effects are produced (this can be any effect, in any species); (ii) the amount of compound to which humans are exposed; and (iii) the mechanism of toxicity (if known; knowledge of mechanism is rare). The process of risk assessment combines all this information and an exposure level is derived that is predicted to produce minimal toxicity in the human population. The most useful data are from studies of human subjects or populations, but animal studies play a key role as well, providing data that cannot be obtained from humans (tissue distributions, development of toxicity, etc.). Animal studies most useful are those that determine functional effects and find and evaluate markers of exposure, markers of effect, and measures of internal dose in the same report (or series of reports). These elements form the kind of data that are most useful to regulators.

The process of risk assessment is the process of translating uncertainty into acceptable human exposure levels. The uncertainty is magnified with fewer data and is reduced when more data are available. Specifically, uncertainty is reduced by the presence of data showing similar effects in multiple species (ideally, including humans), having biomarkers of effect (e.g., cholinesterase inhibition after carbamate exposure), or providing knowledge of mechanism. The uncertainties of extrapolating across species, and accounting for interindividual variation, are an occupational hazard for the risk assessor, who uses a standard body of data to support what have become the default assumptions of a factor of 10 to account for interspecies differences and a factor of 10 to account for interindividual differences.

Finally, the specific process of assessing risks and the amount of data that a registrant company needs to supply to the regulatory agency differ for chemicals and foods/drugs. Regulatory agencies dealing with exposures to pesticides, metals, or industrial chemicals have had much less information to utilize. Regulators dealing with drugs have (eventually) a great deal of human data to compare against the animal data. To date, the regulations require the testing of a greater number of generations of compounds being added to foods than those required for external exposures. This is changing as the world's regulatory agencies attempt the huge task of harmonizing their minimal data requirements for product registration.

VI. ENDOCRINE DISRUPTORS

In recent years, there has been increased recognition of the importance of a normal hormonal melieu during development and the concept that transiently changing this melieu during development can lead to irreversible changes in the adult animal.

Although much of the attention is focused on environmental estrogens, it is clear from both laboratory and field studies that altering levels of thyroid hormones, vitamin A status, testosterone, or gestational exposure to dioxin or TCDD-like compounds can have different and profound effects on the endocrine system, including the reproductive system. Endocrine disruption involves many more hormone systems than just estrogen.

The unknowns in this area are many and include (i) the doses at which such changes occur (are the changes seen at environmental levels or are much higher exposures required?); (ii) the possible interactions of unrelated toxicants (synergy); and (iii) the mechanisms through which these toxicants might act (receptor mediated or are there nonreceptor-mediated signals that are important?).

Relatively new to this area are phytoestrogens, compounds derived from plants that share some signals with estrogen. For example, some phytoestrogens have been shown to accelerate puberty in rodents, make the size of the sexually dimorphic nucleus in the hypothalamus more female, and alter sex-related behaviors. These appear to work through both receptor-mediated and receptor-independent mechanisms. Phytoestrogens will likely play a role in treating the symptoms and problems of menopause (bone loss and cardiovascular disease) and may be able to do so without the troublesome cancer-causing side effects of pure steroids. These same compounds whose effects are adverse when given in large amounts prenatally may be of significant benefit to the body after the ovary's output of estrogens ceases.

Bibliography

Chapin, R. E., Stevens, J. T., Hughes, C. L., Kelce, W. R., Hess, R. A., and Daston, G. P. (1996). Endocrine modulation of reproduction. Symposium proceedings. *Fundam. Appl. Toxicol.* **29**(1), 1–17.

Daston, G. P., Gooch, J. W., Breslin, W. J., Shuey, D. L., Nikiforov, A. I., Fico, T. A., and Gorsuch, J. W. (1997). Environmental estrogens and reproductive health: A discussion of the human and environmental data. *Reprod. Toxicol.* **11**, 465–481.

Hood, R. (Ed.) (1989). *Developmental Toxicology: Risk Assessment and the Future.* van Nostrand Reinhold, New York.

Klinefelter, G. R., Laskey, J. W., Ferrel, J., Suarez, J. D., and Roberts, N. L. (1997). Discriminant analysis indicates a single sperm protein is predictive of fertility following exposure to epididymal toxicants. *J. Androl.* **18**, 139–150.

Lamb, J. C., IV, and Foster, P. M. D. (Eds.) (1988). *Physiology and Toxicology of Male Reproduction.* Academic Press, San Diego.

Sipes, I. G., McQueen, C. A., and Gandolfi, A. J. (Eds.) (1997). *Comprehensive Toxicology, Vol. 10: Reproductive and Endocrine Toxicology.* Elsevier, London.

Witorsch, R. J. (Ed.) (1995). *Reproductive Toxicology*, 2nd ed. Raven Press, New York.

Reptilian Reproduction, Overview

David Crews
University of Texas at Austin

GLOSSARY

associated reproductive pattern A pattern of reproduction in which the secretion of gonadal sex steroid hormones, the production of gametes, and the display of sexual behavior occur synchronously. Species exhibiting this pattern of reproduction tend to be found in predictable environments with prolonged conditions beneficial to breeding.

constant reproductive pattern A pattern of reproduction in which the secretion of gonadal sex steroid hormones, the production of gametes, and the display of sexual behavior do not occur synchronously. Species exhibiting this pattern of reproduction tend to be found in unpredictable environments with brief periods suitable for breeding.

dissociated reproductive pattern A pattern of reproduction in which the secretion of gonadal sex steroid hormones, the production of gametes, and the display of sexual behavior do not occur synchronously. Species exhibiting this pattern of reproduction tend to be found in predictable environments with restricted breeding opportunities.

parthenogenesis Reproduction by cloning. This form of reproduction is found in some lizards and fish. Parthenogenetic vertebrate species consist entirely of females.

temperature-dependent sex determination A mechanism of sex determination found in many reptiles in which the temperature of the incubating egg during the midtrimester of embryogenesis establishes the gonadal sex of the individual.

Reptiles tend to be restricted to temperate and tropical regions and exhibit a seasonal pattern of reproduction in which birth of offspring occurs during restricted periods of the year. Perhaps most interesting is the variety of ways in which this pattern of reproduction is achieved among reptiles. For example, many species of snakes and freshwater turtles exhibit reproductive patterns in which cycles of gonadal activity and sexual behavior are dissociated and peak at different times of year.

I. COSTS OF REPRODUCTION

Reproduction is the single most costly event in an animal's life and hence patterns of reproduction have evolved to maximize an individual's contribution of genetic information to the next generation. Timing of reproduction in a population is determined by when the most offspring survive and when parents, most often females, are capable of energetically supporting the production of viable young at the least cost to themselves. Furthermore, timing of reproduction not only reflects current conditions but also is influenced by factors regulating the total lifetime production of offspring. Thus, the pattern of reproduction exhibited will be influenced by a variety of factors, including generation time, age to reproduction, life expectancy, and age-specific mortality, as well as by the predictability of the environment, ecological niche, and body size.

The energetic demands of reproduction in reptiles can be considerable. For example, in tropical anolis lizards as much as one-fourth of the female's net metabolizable energy is directed to egg production in a breeding season and some North American iguanid lizard species expend this much energy in the production of a single clutch; thus, the energetic costs of reproduction increase tremendously when more than one clutch is produced per season. In addition to energetic demands there are associated risks of mortality from predation during courtship or mating, complications of pregnancy (in viviparous animals) or gravidity (in oviparous animals), and subsequent parturition and oviposition. Costs to males may also be high, particularly in terms of the aggressive and courtship behaviors associated with reproduction.

II. TIMING AND PATTERNS OF REPRODUCTION

Reproduction is considered seasonal when individuals in a population breed at a specific time of year. Usually, all individuals are synchronized in their reproductive activity, but this is not necessarily the case in all species. For example, in sea turtles, males breed each year, but females may breed only every third year such that each year different cohorts of females lay eggs. Thus, seasonal reproductive cycles are not necessarily synonymous with annual reproductive cycles; for example, an individual that breeds in the spring every 2 or 3 years has a seasonal cycle but not an annual one. In aseasonal (or acyclic) reproduction the individuals in a population are not synchronized and breeding may occur continuously or year-round. Aseasonal reproduction may also include reproductive patterns in which individuals of species breed in response to specific but erratic environmental cues. For example, reptiles living in harsh environments such as the deserts of midwestern Australia may breed only when occasional and unpredictable rains inundate an area.

Tropical species have traditionally been considered reproductively acyclic. However, the categorization of a species as acyclic implies that each individual of that species is constantly reproductively active. In reality, only a fraction of the population is breeding or in reproductive condition at any one time, and individuals are breeding on a seasonal, cyclic basis but not synchronously.

Not all aspects of reproduction, including gonadal growth, sex steroid hormone secretion, and sexual behavior, are expressed at the same time. In most species that have been studied with respect to the environmental cues and physiological mechanisms influencing reproduction, these three events are functionally associated (Fig. 1). Such species tend to inhabit environments with prolonged and predictable periods suitable for reproduction. However, in environments having brief but predictable benign periods, gonadal activity and sexual behavior are expressed at different times or are dissociated. In particularly harsh environments species may exhibit a constant reproductive pattern in which gonads are maintained in a near ready state and sexual behavior is expressed in response to particular environmental cues. Such differing patterns of reproductive activity indicate that a wide variety of physiological mechanisms have evolved to meet the demands of reproduction in different environments.

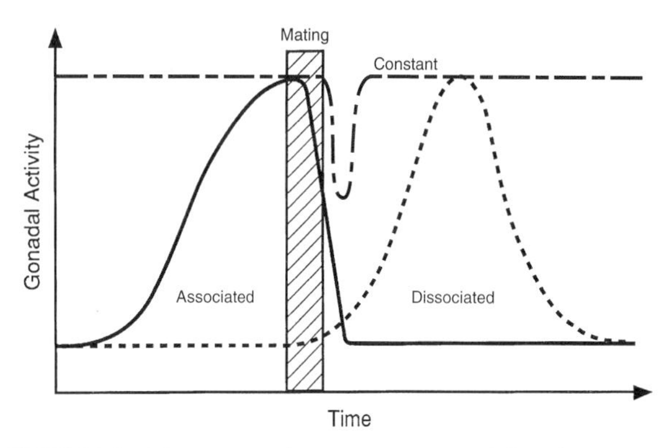

FIGURE 1 Three reproductive patterns exhibited in vertebrates, including reptiles. Gonadal activity is defined in terms of the maturation and shedding of gametes and/or the secretion of sex steroid hormones. In individuals exhibiting the associated reproductive pattern, gonadal activity increases immediately prior to mating. In species-dissociated reproductive pattern, gonadal activity is minimal at mating. In species exhibiting the constant reproductive pattern, the gonads are maintained at or near maximum activity throughout the year.

III. ENDOGENOUS AND ENVIRONMENTAL REGULATION OF REPRODUCTION

Mechanisms that control seasonal reproduction may be viewed as lying on a continuum between two extremes. At one extreme, in preprogrammed or closed control mechanisms, seasonal reproduction is entirely determined by endogenous cycles and is not influenced by external cues. At the other extreme of this continuum lies labile control, in which seasonal reproduction occurs as a result of responses to exogenous cues. In predictable environments, individuals of a population that rely on preprogrammed control mechanisms may have an advantage because they are ready to reproduce and need not rely on external influences to initiate reproduction. In less predictable environments, preprogrammed control may be disadvantageous if reproductive responses occur under inappropriate conditions. Labile control mechanisms, although they may retard a reproductive response until conditions are favorable, result in fewer risks in unpredictable environments.

Many reptiles exhibit intermediate control mechanisms between these two extremes. For example, preparatory events leading to reproduction, including gonadal growth, can be influenced by endogenous factors, but final stages of maturation as well as reproductive behavior can be controlled by integration of specific cues. Another intermediate stage of control is represented by the "hourglass" mechanism, in which external events initiate but have no further influence on an endogenously controlled cycle.

Different environmental situations provide unique cues that can be exploited to predict the time when reproduction is most likely to succeed. In both benign BS predictably harsh environments, such as found at extreme latitudes and altitudes, photoperiod and temperature are reliable cues associated with the prediction of spring or summer-like conditions. While tropical environments do not undergo seasonal extremes in photoperiod and temperature, often there are regularly recurring wet and dry periods. Some desert environments similarly may have seasonal rainfall. Other environments appear to lack regularly occurring external cues. These would include areas that have favorable conditions year-round, as in some tropical environments. Finally, there are environments in which favorable conditions for reproduction are unpredictable, including some deserts and certain high mountain areas.

The environmental factors affecting reproduction in reptiles have received much study. In general, day length and temperature both serve as proximate cues for gonadal maturation. For example, in the male green anole lizard, temperature is of primary importance in the induction of gonadal recrudescence, whereas photoperiod influences testicular regression. Females are also sensitive to these cues, but there is an additional requirement of sufficient relative humidity if ovarian growth is to occur.

Social cues are also important regulators of reptilian reproductive cycles. Male courtship behavior (in particular, extension of the dewlap) facilitates ovarian development in the green anole lizard. In the Canadian red-sided garter snake the stimulus of mating initiates the neuroendocrine changes that stimulate ovarian growth.

Nutrition, usually in the form of stored fat, has a profound influence on reproduction in reptiles. Many species of northern temperate snakes reproduce facultatively, often every 2 or 3 years rather than on a strictly annual basis. Such facultative reproduction may occur in part as a result of an influence of nutritional reserves on reproduction. For example, in the Canadian red-sided garter snake, females of less than average body weights at a given snout–vent length are less likely to reproduce than heavier individuals. Although low body weight significantly influences ovarian recrudescence in this species, body weight (within normal ranges) does not influence sexual attractivity or receptivity of females. Since female Canadian red-sided garter snakes can store viable sperm for years, as do a number of reptiles, it is not surprising that ovarian responses and sexual behavior can be independently regulated by factors such as body weight. In a related species, the Mexican garter snake, body condition more directly influences the reproductive cycle; nutritional condition of females regulates female attractivity and the incidence of male sexual behavior and testicular cycles.

In addition to environmental cues, some species utilize endogenous cues to regulate timing of reproduction. Endogenous circannual cycles are defined by the occurrence of complete reproductive cycles on a near annual basis under constant conditions (e.g., constant darkness or a nonvarying 10-hr light : 14-hr dark photoregimen and constant temperature). Interactions between endogenous cycles in sensitivity to environmental stimuli and the regular progressions of changes in the environment regulate the period (length) of the cycles. Few long-term studies have investigated endogenous circannual cycles in reptiles, although evidence with parthenogenetic whiptail lizards indicates such cycles may occur. Several species of reptiles do exhibit seasonal variation in sensitivity to exogenous cues. This is primarily manifested by a refractory period following the breeding season and extending to midwinter. This has been documented most extensively in the green anole lizard. Females of this species become unresponsive to long photoperiods and warm temperatures after the breeding season. The lack of response may be due, at least in part, to the presence of large atretic follicles. Male anoles also exhibit refractoriness to exogenous cues during late summer. Again, circadian rhythmicity to exogenous cues has been implicated as mediating the response.

Finally, there often are interactions between endogenous events and seasonal cues. Thus, different environmental factors can alter reproduction during particular gonadal phases. Such considerations complicate experimental approaches to environmental influences on reproduction.

IV. NEUROENDOCRINE CONTROL OF REPRODUCTION

The specific cues utilized for regulating reproduction, be they internal or external, must be integrated in such a way so that gonadal products are produced and sexual behavior is expressed at the appropriate time(s). The neuroendocrine system has evolved to perform this function in a precise, and often complex, way. Redundancy often is found in neuroendocrinological controlling mechanisms so that removal or addition of particular substances may not always result in a single predictable result. In other words, the notion that one mechanism regulates all reproductive events or even a single reproductive event is overly simplistic.

Reptiles possess similar neuroendocrinological regulating mechanisms as other vertebrates that have been studied. Sensory systems and specific brain areas serve to integrate external and internal events to bring about changes in the reproductive system. The eyes and the pineal appear to be involved in the influence of photoperiod on the reproductive system, and in snakes and turtles the pineal appears to be involved in the transduction of temperature. Chemical signals are detected by the vomeronasal system in lizards and snakes and probably other reptiles as well. The integration of these cues is not well understood, but, as in other vertebrates, specific nuclei in the limbic system receive input from these sensory systems. Many of these brain areas also concentrate steroid hormones, and areas such as the preoptic area, anterior hypothalamus, and ventromedial hypothalamus are involved in the regulation of social and sexual behaviors. Specific areas of the hypothalamus secrete gonadotropin-releasing hormone (GnRH) that influences release of gonadotropin from the anterior pituitary. Gonadotropins in turn stimulate gonadal maturation and steroid hormone production. Steroid hormones feed back on the hypothalamus and the pituitary to modulate subsequent function of these structures. Administration of exogenous GnRH stimulates release of gonadotropin and repeated pulses of GnRH administered to females increase the frequency of aggressive and reproductive behaviors of conspecific males and induce receptive behavior in females. Finally, both steroid hormones and behavioral context regulate the expression of genes coding for sex hormone receptors in limbic nuclei.

In contrast to species with an associated reproductive pattern, in the Canadian red-sided garter snake, which exhibits a dissociated pattern of reproductive cycle (Fig. 1), neither hypophysectomy nor administration of exogenous GnRH or a number of other neuroendocrine substances affect male sexual behavior. In fact, in this species a prolonged period of cold temperature followed by warming, thereby simulating the natural course of events of hibernation and spring emergence, initiates courtship behavior of males. Pituitary and gonadal hormones are not involved in the activation of mating behavior, although melatonin from the pineal and the eyes appear to be essential.

Although there is a large amount of information available on neuroendocrinological control of reproduction in reptiles, little is known of the interactions responsible for normal seasonal changes. For example, female green anole lizards undergo a seasonal cycle in which their ovaries recrudesce in the spring, produce eggs for approximately 12 weeks, and then regress in late summer. Sexual behavior is linked to the ovarian cycle and can be stimulated by the administration of the ovarian hormones estrogen and progesterone. There is seasonal change in sensitivity of the brain of ovariectomized female green anoles to the stimulatory effects of estrogen and progesterone on sexual receptivity. Paralleling this is the seasonal refractoriness of ovarian responses to exogenous follicle-stimulating hormone. Thus, in the green anole lizard, although there is a suggestion of seasonal changes at the brain and gonad, potential interactions between these events over an entire reproductive season are not understood.

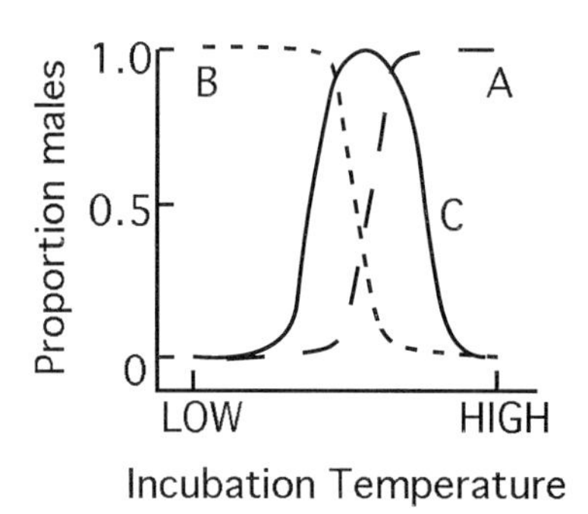

FIGURE 2 Response of hatchling sex ratio to incubation temperature in various egg-laying reptiles. These graphs represent only the approximate pattern of the response and are not drawn according to any single species. The following are the three patterns recognized: (A) only females produced from low incubation temperatures, with males at high temperatures; (B) only males produced from low incubation temperatures, with females at high temperatures; and (C) only females produced at the temperature extremes, with male production at the intermediate incubation temperatures. Genotypic sex determination also occurs in reptiles with the result that the hatchling sex ratio is fixed at 1:1 despite incubation conditions.

V. TEMPERATURE-DEPENDENT SEX DETERMINATION

Reptiles exhibit two basic forms of sex determination. All snakes and apparently all viviparous lizards exhibit genotypic sex determination, having either male or female heterogamety. However, many, but by no means all, egg-laying reptiles lack sex chromosomes. Instead, all crocodilians, most turtles, and many lizards, including the tuatara, depend on the temperature experienced by the embryo during the middle third of incubation to determine the sex of the offspring. This process is termed temperature-dependent sex determination (TSD). Interestingly, no snakes and no viviparous lizards appear to exhibit TSD, relying instead on genotypic sex determination.

In some species low temperatures produce males and high temperatures produce females, whereas in other species the opposite pattern occurs; in still other species extreme incubation temperatures produce females and intermediate temperatures produce males (Fig. 2). For example, in the red-eared slider turtle the spectrum of temperature is slight, extending over no more than 5–7°C, with the transitional temperature range (those temperatures yielding a mixed sex ratio rather than all males or all females) <1°C. Furthermore, the effect of temperature during this window of sensitivity is quantitative, having a cumulative effect on sex determination.

Incubation temperature is transduced in the urogenital system and perhaps the brain of the developing embryo and acts by modifying the activity as well as the temporal and spatial sequence of enzymes and sex hormone receptors. The temperature-typical hormone milieus that are created in the urogenital system in turn determine the type of gonad formed. Because the genetic machinery necessary for the formation of both ovaries and testes are present in the embryo, this means that temperature must activate one gonad-determining cascade while inhibiting the complementary gonad-determining cascade.

The temperature experienced by the embryo not only determines gonadal sex but also influences sexual differentiation. Indeed, it appears that much of the variation observed within the sexes of TSD reptiles can be traced back to the incubation temperature of the eggs, much like intrasexual variation in polytocous mammals can be traced to the position of the fetus relative to the sex of its neighbors *in utero*.

This is demonstrated particularly well in the leopard gecko, in which adult variations in morphology, endocrine physiology, behavior, and brain chemistry both between and within a sex are affected profoundly by the temperature experienced as an egg. For example, in the leopard gecko the male is the larger sex. However, females from a male-biased incubation temperature grow faster and larger than do females from a female-biased incubation temperature, growing as rapidly and as large as males from lower, female-biased incubation temperatures. When the effects of gonadal sex and incubation temperature are experimentally dissociated using estrogen to sex-reverse eggs incubating at a male-biased temperature, estrogen-determined females grow according to the incubation temperature rather than their gonadal sex.

VI. PARTHENOGENESIS

Reptiles are the only amniote vertebrates (mammals, birds, and reptiles) that exhibit obligate parthenogenesis. Such species consist only of female individuals that do not need to mate (and hence do not require sperm to activate embryogenesis) in order to reproduce. In such cloning species only female offspring are produced that are genetically identical to the mother. In obligate parthenogenesis there is no reduction in ploidy (indeed, usually there is an increase) prior to meiosis. Parthenoforms can arise by mutation through hybridization between sexual species; indeed, in many instances we actually know which species hybridized to create the parthenogen. Obligate parthenogenesis appears to be restricted to the lizards; the Brahminy blind snake may also be unisexual, although this has not been confirmed. There are about 40 species of lizards, including the Australian geckos, eurasian rock lizards, and southwestern whiptail lizards, that reproduce in this manner.

The evolution of a cloning reproductive pattern does not mean that the genes for maleness were lost. Administration of aromatase inhibitor during a specific time in embryonic development will result in the complete suppression of the ovary-determining cascade and the activation of the testis-determining cascade, resulting in individuals having testes, producing motile sperm as adults, and mating as males. This effect is not simply the result of sex steroid hormones determining gonad type directly, because administration of steroid hormones or their antagonists does not alter the normal (ovary) development or primary sex structure. This suggests that the key to parthenogenetic reproduction may lie in the dual process of activation of the ovary-determining cascade and the concomitant suppression of the testis-determining cascade.

See Also the Following Articles

Amphibian Reproduction, Overview; Female Reproductive System, Reptiles; Fish, Modes of Reproduction in; Male Reproductive System, Reptiles

Bibliography

Crews, D. (1980). Interrelationships among ecological, behavioral and neuroendocrine processes in the reproductive cycle of *Anolis carolinensis* and other reptiles. In *Advances in the Study of Behavior, Volume 11* (J. S. Rosenblatt, R. A. Hinde, C. G. Beer, and M. C. Busnel, Eds.), pp. 1–74. Academic Press, New York.

Crews, D. (1992). Diversity of hormone–behavior relations. In *Introduction to Behavioral Endocrinology* (J. Becker, M. Breedlove, and D. Crews, Eds.), pp. 143–186. MIT Press/Bradford, Cambridge, MA..

Crews, D. (1996). Temperature-dependent sex determination: The interplay of steroid hormones and temperature. *Zool. Sci.* **13**, 1–13.

Crews, D., and Moore, M. C. (1986). Evolution of mechanisms controlling mating behavior. *Science* **231**, 121–125.

Crews, D., and Silver, R. (1985). Reproductive physiology and behavior interactions in nonmammalian vertebrates. In *Handbook of Behavioral Neurobiology, Vol. 7: Reproduction* (N. T. Adler, D. W. Pfaff, and R. W. Goy, Eds.), pp. 101–182. Plenum, New York.

Crews, D., Bergeron, J. M., Flores, D., Bull, J. J., Skipper, J. K., Tousignant, A., and Wibbels, T. (1994). Temperature-dependent sex determination in reptiles: Proximate mechanisms, ultimate outcomes, and practical applications. *Dev. Genet.* **15**, 297–312.

Reptilian Reproductive Cycles

Valentine A. Lance
Center for Reproduction of Endangered Species, Zoological Society of San Diego

GLOSSARY

fat body Abdominal fat vesicles found in all temperate zone squamates. Fat bodies build up in the nonbreeding season and decline as the gonads develop.

ovoviviparity The retention of eggs in the oviduct of female reptiles until the young hatch and are thus born live.

parthenogenesis Reproduction in females without fertilization by males. Parthenogenetic species are generally polyploid and there are no known males.

recrudescence The reinitiation of gonadal development in seasonally breeding reptiles.

refractory period A period of several weeks or months following the breeding season during which the gonads fail to respond to environmental stimuli.

sexual segment of the kidney A portion of the kidney found only in male lizards and snakes that hypertrophies in response to testosterone during the spermatogenetic cycle. It is believed to contribute lipid secretions to the semen.

vitellogenesis The process of yolk formation in which yolk precursor protein is synthesized in the liver in response to estrogen and transported via the bloodstream to the developing oocyte.

There are approximately 6500 species of extant reptiles divided into four orders: Squamata (lizards and snakes), more than 6000 species; Chelonia (turtles and tortoises), 220 species; Crocodilia (crocodiles, alligators, and caimans), 23 species; and Rhynchocephalia (the sphenodon or tuatara), 2 species. There is information on the reproductive cycle of a large number of lizards and snakes, tortoises and turtles, the American alligator and the Nile crocodile, and the tuatara. Detailed hormonal data are available for only a very few species.

I. INTRODUCTION

The major differences between mammalian reproduction and reptilian reproduction are (i) the vitellogenic phase of follicular development; (ii) the occurrence of sperm storage in the female reproductive tract in some species, and hence the fertilization of eggs without the presence of a male and, conversely, the storage of sperm in the epididymis of many species of turtle for many months after spermiogenesis and well before mating; (iii) the occurrence of parthenogenesis in a number of lizard species, and (iv) the importance of the environment and, in particular, temperature as a regulator of reproduction.

All Chelonia, rhynchocephalians, and crocodilians lay eggs. There is an enormous variability in the number of eggs produced among the reptiles, from as high as 10 clutches of up to 90 eggs per clutch in a single nesting season in leatherback sea turtles (*Dermochelys coriacia*) to clutches of only 1 egg in some geckos. Most squamates are egg layers, but viviparity has evolved many times in both lizards and snakes.

Because reptiles are poikilotherms or exotherms (i.e., body temperature is the same temperature as ambient) the most important regulator of reproduction is environmental temperature. Photoperiod plays a role in some species, but only if temperature is adequate. Another important environmental trigger for reproduction in many species is seasonal rainfall.

II. MALE REPRODUCTIVE CYCLE

There is no typical male reproductive cycle. There are tropical species that appear to be sexually active throughout the year and species living in extreme environments that may mate only once in 4 years if conditions are too harsh. In these temperate zone species at the extremes of their range it may take 2 years or more for a single spermatogenetic cycle to be completed, thus emphasizing the importance of temperature in reptilian reproductive physiology. There are species in which the testicular and mating activity cycles correspond, species in which the two occur at different times of the year, and species intermediate between these two types. Accordingly, some authors have attempted to classify male reptile reproductive cycles by type of annual spermatogenic and mating activity. Terms such as prenuptial and postnuptial are also used to describe the type of spermatogenic cycle in which spermatogenesis takes place before mating or after mating. The development of a mature spermatozoon from a spermatogonium is similar to that seen in mammals. Eight stages of the spermatogenetic cycle have been described for reptiles according to the appearance of the seminiferous tubules. Only two types of cycle will be presented here, but readers should be aware that many variations on these two extremes occur.

In the type I cycle, hormonal, testicular, and mating activity correspond. In these species (Fig. 1), there is a single mating season during which circulating testosterone is high and spermatogenesis and copulation are completed. In snakes and lizards exhibiting this type of cycle the abdominal fat bodies start to build up in late summer after mating has been completed. The fat bodies remain enlarged until the onset of the following breeding season and decline as the gonads increase in mass. The renal sex segment enlarges in response to the increase in testosterone secretion at the onset of spermatogenesis (only in lizards and snakes) and regresses after mating is complete and testosterone levels have returned to baseline. The lipid secretions of the sexual segment are added to the semen, and in some species these secretions contribute to the vaginal plug found in females after mating. Many lizards, snakes, crocodiles, and alligators exhibit this type of cycle. Often in this type of cycle there is a secondary increase in testosterone in the fall with some mating activity. Figures 1–4 indicate winter and summer in the

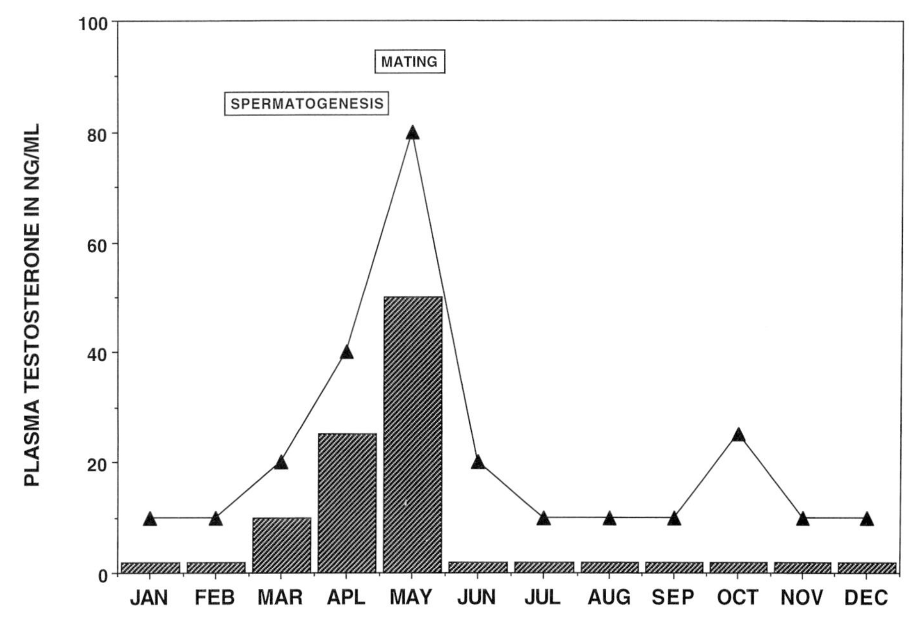

FIGURE 1 Simplified diagram of a type I or prenuptual reproductive cycle of male reptile. ▲, circulating plasma testosterone levels in ng/ml; bars, testis mass. The fat body cycle (not shown) is the reciprocal of the testis cycle, i.e., is greatest in September and October and declines in March–May. In many species, a secondary rise in testosterone is seen in late summer to early fall often associated with mating activity.

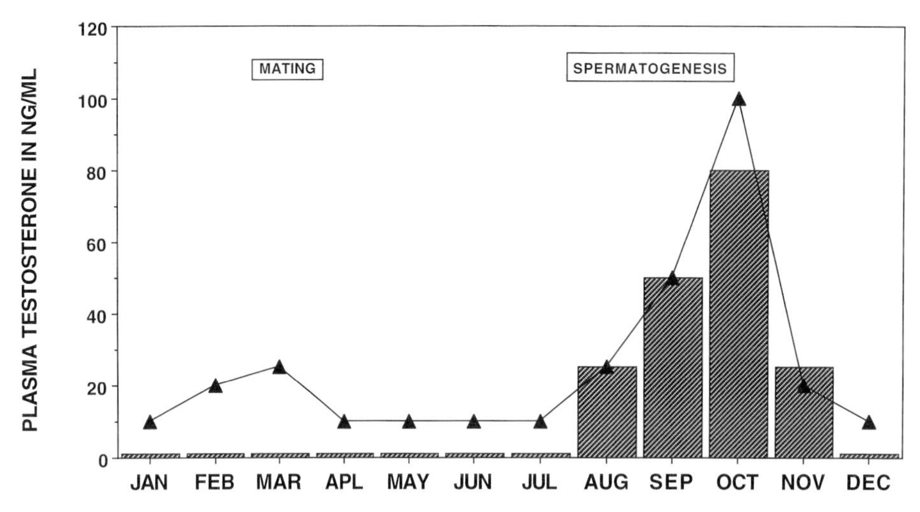

FIGURE 2 Simplified diagram of a type II or postnuptial male reproductive cycle. ▲, plasma testosterone levels in ng/ml; bars, testis mass. Mating occurs when the testes are atrophied and plasma testosterone is relatively low. Semen from the epididymides stored over winter is used to inseminate the female. Spermatogenesis and highest testosterone levels occur in the late summer, well after spring mating.

Northern Hemisphere. Obviously, temperate zone reptiles from the Southern Hemisphere will show a cycle in which breeding occurs in the austral summer, i.e., December and January.

Most temperate zone turtles and a number of squamates display a type II cycle in which spermatogenesis is initiated in midsummer well after the spring mating period (Fig. 2). Spermatogenesis is completed by late summer and the spermatozoa are expelled from the seminiferous tubules and stored in the epididymis. During the winter months the epididymides are packed full of spermatozoa, whereas the testes remain completely regressed. On emergence in spring the males undergo courtship and copulation using the spermatozoa stored in the epididymis to fertilize the females. At this time the testes are still fully regressed. After mating is complete there is a short refractory period and then testicular recrudescence begins and the spermatogenetic cycle resumes. Plasma testosterone shows two peaks: a small peak associated with mating in the spring when the testes are regressed and a larger peak in late summer at spermiogenesis.

A. Hormones

The steroid hormones secreted by the reptilian testis are similar to those of birds and mammals. Testosterone is the principal steroid secreted by the Leydig cells, but smaller amounts of androstenedione and other androgens are also found. In those species in which both the histology of the testis and plasma testosterone have been studied, highest hormone levels correlate with stage 6 when spermatogenesis is at its peak. Low levels of estradiol have been found in the plasma of male snakes, but estradiol is below detectable levels in male turtles and alligators, and aromatase has not been located in the Leydig cells of these species. There is no information on inhibin in reptiles. Published information on plasma testosterone in reptiles presents an enormous range of values—species in which maximum levels are as high as 1500 ng/ml (Bolson tortoise) to species in which maximum levels are only 1 or 2 ng/ml. Some of this disparity may be due to species differences in plasma steroid-binding protein and some due to the method of sampling. Testosterone secretion is inhibited by stress such that blood samples that are taken many hours after an animal was captured in the field do not reflect the actual levels in nature. The pituitary hormones, follicle-stimulating hormone (FSH) and luteinizing hormone (LH), have been purified from turtle and alligator and appear to have similar roles to their mammalian counterparts, but there are very few studies in which these hormones have been measured throughout the breeding cycle. No pituitary hormones from squamates have been isolated and sequenced and there is still uncertainty as to the nature of the gonadotropin(s) in this group. It has been suggested that snakes and lizards have only a single gonadotropin that has both FSH and LH properties.

In some species, steroids produced by the adrenocortical tissue also show a seasonal cycle in concert with the hormones of the testis. In the desert tortoise plasma corticosterone in males is significantly higher than that in females throughout the year, and the annual corticosterone cycle is almost identical to that of testosterone. The significance of this cycle and the sex differences remain obscure.

III. FEMALE REPRODUCTIVE CYCLE

All female reptiles, whether egg laying or viviparous, undergo a vitellogenic phase of ovarian development. During the vitellogenic phase the ovarian follicles secrete estradiol, which stimulates the liver to synthesize and secrete the yolk precursor protein, vitellogenin. Vitellogenin, a complex phospholipoprotein that binds calcium and other divalent cations, is taken up at specific receptor sites by the developing oocytes where it is processed into yolk. In most species examined there is a close correlation between plasma estradiol and vitellogenin (or plasma calcium, which rises in concert with vitellogenin). The vitellogenic phase can be as short as 3 weeks or as long as 3 months. In turtles and tortoises it is generally a very long and gradual process during which there is slow follicular growth and a relatively low and constant production of vitellogenin. In some snakes and lizards plasma calcium, a marker for vitellogenin, can increase more than fourfold during the short vitellogenic phase. Temperate zone female squamates also show abdominal fat body cycles that are inversely correlated with the gonadal cycle: The fat body regresses as the ovarian follicles develop. Again, as in the case of the males, there is enormous variability in the reproductive cycle from species to species and even within a species depending on the latitude. Turtle species at the northern limit of their range will produce a single clutch, whereas the same species is capable of producing up to four clutches in the southern portion of its range. Only two types of ovarian cycle will be presented, but again many variations on these two themes occur.

Type I is typical of many temperate zone squamates and the American alligator in which vitellogenesis, mating, and ovulation all occur within 1 or 2 months in early summer, often referred to as a prenuptial follicular phase (Fig. 3). Estradiol increases in early spring and reaches a peak just prior to ovulation. A surge of progesterone occurs at the point of ovulation, and in many squamates that develop a corpus luteum, progesterone remains high during the time the eggs are in the oviduct. In alligators and turtles the surge of progesterone is relatively short-lived. After oviposition there is a refractory period and the ovary remains relatively inactive until the following year. In many species there is an initiation of ovarian development and a second period of courtship just before the winter.

The type II ovarian cycle is typical of many temperate zone turtles. There is a long, slow vitellogenic and follicular growth phase beginning in midsummer that is halted during the winter months and resumes on emergence in spring. Plasma estradiol and calcium (indicative of vitellogenin) are both elevated during this phase, as is plasma testosterone. Mating occurs prior to the onset of the follicular growth phase and thus the cycle can be referred to as postnuptial (Fig. 4).

A. Hormones

Circulating estradiol levels during the follicular phase in reptiles are generally higher than those seen in female mammals. Peak levels of estradiol as high as 5 ng/ml have been reported. Female reptiles differ from mammals in that the ovaries secrete a substantial amount of testosterone during the follicular phase. The amount of testosterone can be as high as 30 ng/ml in some female varanid lizards, but in most species it is <5 ng/ml. Testosterone levels do not appear to correlate with estradiol but are related to the number of preovulatory follicles. In multiclutched species such as sea turtles, the levels of testosterone are highest at the beginning of the nesting season and decline with each successive clutch as the nesting season progresses. Plasma progesterone is low during the follicular phase, peaks at ovulation, continues to be secreted while the eggs are in the

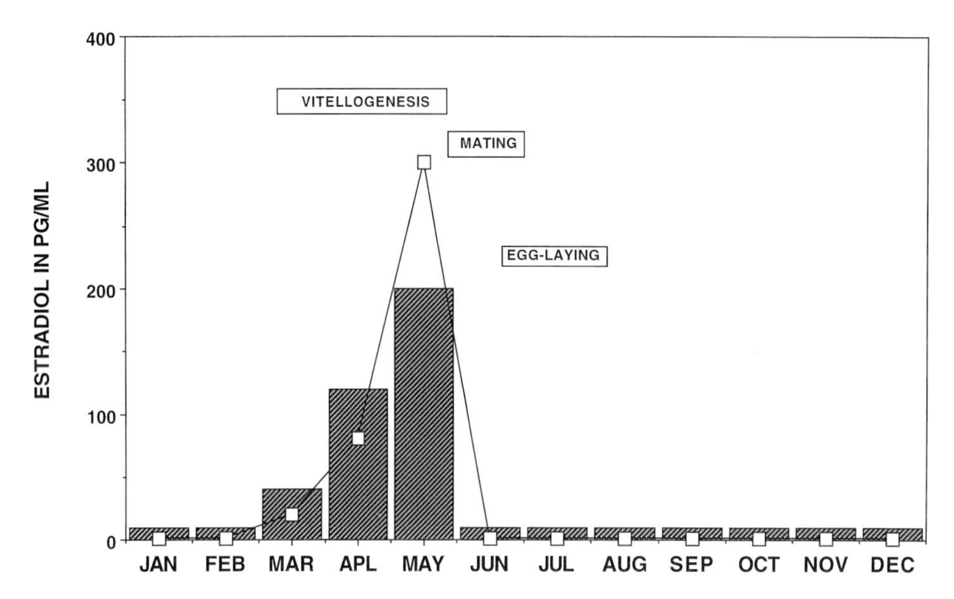

FIGURE 3 Simplified diagram of a type I or prenuptial female reproductive cycle. □, plasma estradiol in pg/ml; bars, ovarian mass. Abdominal fat body mass (not shown) is the reciprocal of ovarian mass, i.e., is greatest in September–February and declines rapidly in April and May.

oviduct, and generally returns to baseline just prior to oviposition. Viviparous snakes and lizards have high circulating progesterone during most of the gestation period.

As is the case with male reptiles, there is no clear picture regarding circulating gonadotropins and reproduction. Alligator brain contains two gonadotropin-releasing hormone (GnRH) molecules identical to the two in chicken brain. Other reptiles also have two GnRH-like molecules, one of which is identical to the chicken-II and a second molecule that may be similar to the mammalian form or the first chicken form or unknown. The mammalian form of GnRH causes an increase in circulating steroids when in-

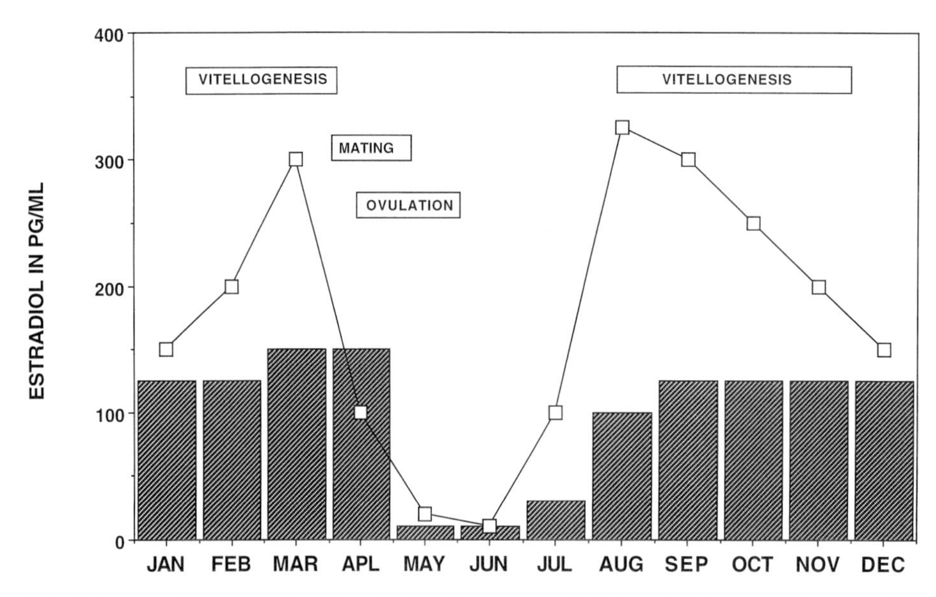

FIGURE 4 Simplified diagram of a type II female cycle. □, plasma estradiol in pg/ml; bars, ovarian mass. This cycle is typical of many temperate zone turtles. Follicular growth and vitellogenesis begins in late summer when peak circulating estradiol levels are seen. Follicular growth stops during the winter months and then resumes just prior to mating in spring.

jected into alligators but is inactive in snakes and most turtles. The posterior pituitary hormone arginine vasotocin is the reptilian homolog of oxytocin that is believed to be released at oviposition along with oviductal prostaglandins. Oxytocin will induce oviposition if the eggs are in the oviduct in most reptiles.

See Also the Following Articles

VITELLOGENINS AND VITELLOGENESIS; VIVIPARITY AND OVIPARITY

Bibliography

Callard, I. P., and Ho, S. M. (1980). Seasonal reproductive cycles of reptiles. *Prog. Reprod. Biol.* **5**, 5–38.

Fitch, H. S. (1970). Reproductive cycles of lizards and snakes, Publ. No. 52. Univ. of Kansas Museum of Natural History, Lawrence.

Fitch, H. S. (1982). Reproductive cycles in tropical reptiles, Publ. No. 96. Univ. of Kansas Museum of Natural History, Lawrence.

Ho, S. M., Kleis, S., McPherson, R., Heisermann, G. J., and Callard, I. P. (1982). Regulation of vitellogenesis in reptiles. *Herpetologica* **38**, 40–50.

Lance, V. (1984). Endocrinology of reproduction in male reptiles. *Symp. Zool. Soc. London* **52**, 357–383.

Licht, P. (1984). Reptiles. In *Marshall's Physiology of Reproduction, Vol. 1. Reproductive Cycles of Vertebrates* (G. E. Lamming, Ed.), pp. 206–282. Churchill Livingstone, Edinburgh, UK.

Norris, D. O. (1997). *Vertebrate Endocrinology*, 3rd ed. Academic Press, San Diego.

Respiratory Distress Syndrome

Rebecca A. Simmons
University of Pennsylvania

GLOSSARY

alveolus An air sac in the lung that exchanges gas.

anatomic dead space The volume of the conducting airways.

atelectasis Collapse of airspaces.

compliance The volume change per unit pressure change in the lung.

cyanosis Bluish discoloration of the skin resulting from an inadequate amount of oxygen in the blood.

glucocorticoids Steroid hormones synthesized by the adrenal cortex; essential for the utilization of carbohydrates, fats, and protein by the body and for a normal response to stress.

respiratory distress syndrome The condition of a newborn infant in which the lungs are imperfectly expanded.

surfactant A substance in the lung that reduces surface tension.

Respiratory distress syndrome, or hyaline membrane disease (HMD), is the most common cause of respiratory failure in the newborn, occurring in approximately 1 or 2% of newborns. Prior to 1970, over half of infants with HMD died. Recently, improved methods of treatment, especially surfactant replacement, have dramatically reduced the mortality rate.

I. INTRODUCTION

Hyaline membrane disease occurs primarily in premature infants and is more common in whites than nonwhites and in male than in female infants. When corrected for gestational age, the incidence of hyaline membrane disease (HMD) is significantly increased in gestational diabetes and insulin-dependent mothers without vascular disease. Cesarean section, particularly if labor has not begun, can predispose the infant to HMD. There is some evidence that prolonged rupture of membranes protects against HMD. When gestational age is controlled, some studies have shown that there is a significant reduction in the incidence of HMD after prolonged rupture of membranes for more than 24 hr and a greater reduction after 48 hr.

Hyaline membrane disease develops when an infant attempts to ventilate an immature lung with small respiratory units that inflate with difficulty and do not remain gas filled between respiratory efforts. Surfactant deficiency is the primary underlying process contributing to the development of HMD. Surfactant, synthesized by alveolar type II cells, is a unique complex of lipids and proteins. The major surface-active lipid is dipalmitoylphosphatidylcholine. At least four proteins (A–D) are associated with surfactant and participate in its metabolism and function. These proteins play a role in stabilization of the surface film, recycling of surfactant back into type II cells, and immunologic functions. The control of synthesis and secretion of surfactant is a complex process. Several hormones play an important role in this process, including cortisol, thyroid hormone, prolactin, testosterone, and β-adrenergic agents.

In the absence of an adequate amount of surfactant, surface tension at the interface between alveolar gas and the alveolar wall is increased, and the lung becomes progressively atelectatic. The infant's work of breathing is increased, and cyanosis ensues as the volume of the lung decreases. In more severely affected infants, respiratory failure rapidly develops if the infant is not mechanically ventilated. In some infants, HMD is self-limiting. The surge of glucocorticoids and β-adrenergic agents that occurs at birth stimulates maturation of the lung, and recovery occurs within 48–72 hr.

II. PATHOPHYSIOLOGY

Lungs from infants who have died of HMD show marked atelectasis. Diffuse atelectasis and alveolar ducts that are lined with a homogenous hyaline-staining material characterize this disease. The hyaline membranes consist of plasma clots containing fibrin, other plasma constituents, and cellular debris. Pulmonary capillaries and veins are congested and interstitial edema is evident.

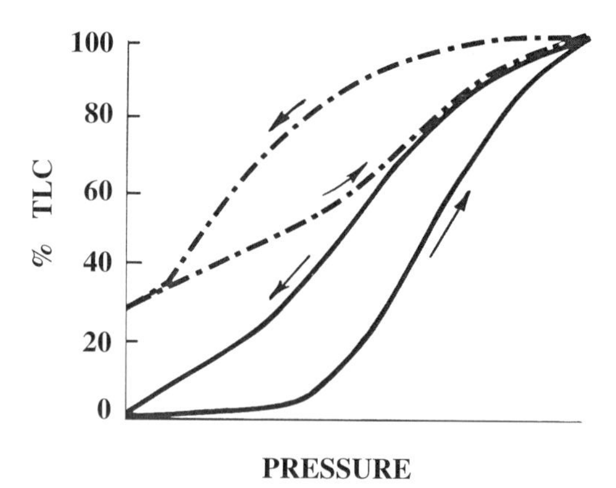

FIGURE 1 Pressure–volume studies of immature (solid lines) and mature (dotted lines) fetal rabbits. Inflation volumes (up arrows) are consistently lower than deflation volumes (down arrows) at identical pressures in the same lung.

Because surfactant content and function are low in the infant with HMD, surface tension is increased and the lung is less compliant. Tidal volumes are low and retractive forces are high. As a result, atelectasis is a key component of HMD, producing respiratory failure in the newborn. Pressure volume loops illustrating the effects of increased surface tension are shown in Fig. 1.

Surfactant distribution in the lung is uneven and normal compliant lung units coexist with abnormal compliant lung units. Very little gas flow goes to the poorly ventilated lung units and alveolar PO_2 is quite low. A low alveolar PO_2 causes vasoconstriction of the blood vessels supplying the lung unit. Unoxygenated blood is diverted to more normal parts of the lung via right-to-left shunts. The degree of hypoxemia in the infant with HMD is directly correlated with the size of the poorly ventilated compartment of the lung and the oxygen tension in the alveoli of that compartment. The more compliant parts of the lung in the infant with HMD receive a disproportionate share of the ventilation and are ventilated out or proportional to their perfusion resulting in a large physiologic dead space.

III. CLINICAL FEATURES

The onset of HMD is variable. Very premature infants (<1000 g) usually fail to expand their lungs at birth and they have severe respiratory failure from delivery. Others have progressive atelectasis over several hours. In the unventilated infant, clinical signs include cyanosis, tachypnea, intercostal and substernal retractions, and grunting. Grunting respiration is caused by exhaling against a partially closed glottis. Grunting maintains an intrathoracic pressure that is higher than atmospheric pressure and may help to prevent atelectasis. The infant breathes at a rapid rate in an attempt to increase minute volume; however, dead space volume is also increased so that alveolar ventilation is low. Airway resistance is often increased because the small airways are compressed by collections of interstitial fluid. In addition, small airways are damaged by the large transpulmonary pressures that are needed to inflate the lung. Decreased compliance and increased resistance result in a marked increase in the work of breathing. The very small infant soon tires and arterial carbon dioxide tension rises.

Infants who have mild disease usually rapidly improve after 72 hr. Improvement in respiratory function is often associated with diuresis. Very low birth weight infants usually require mechanical ventilation and will have a more prolonged course.

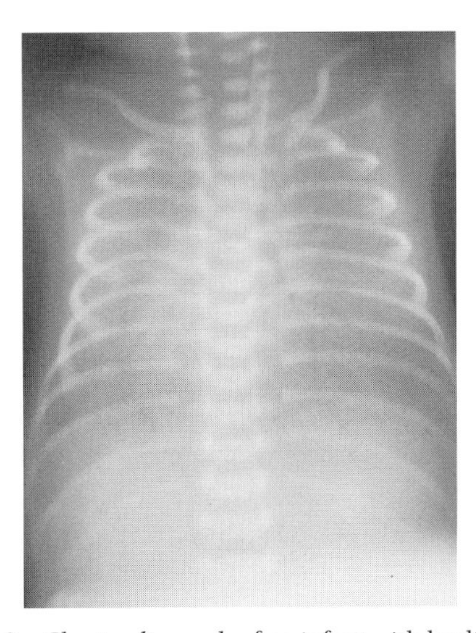

FIGURE 2 Chest radiograph of an infant with hyaline membrane disease. The lung volume is reduced, the parenchyma has a diffuse reticular granular pattern, and air bronchograms are present.

IV. RADIOLOGIC FEATURES

The chest X ray of the infant with HMD has a very characteristic appearance (Fig. 2). There is a diffuse reticulogranular pattern of increased density, which is usually uniform in distribution. The densities result from diffuse atelectasis and interstitial edema. The bronchial tree is clearly outlined by air against the poorly aerated lung (air bronchogram). Once positive pressure ventilation is established, these findings can resolve in infants with mild disease.

V. PRENATAL PREDICTION

Surfactant is secreted into fetal lung fluid which enters the amniotic cavity. The concentration of surfactant in amniotic fluid is related to the maturity of the lung. Studies have shown that the concentrations of the phospholipid lecithin and sphingomyelin are equal in amniotic fluid at midgestation. After 34–36 weeks there is twice as much lecithin as sphingomyelin. In normal pregnancy the L/S ratio increases from 1 at 32 weeks and to 2 by 35 weeks. The incidence of HMD is only 0.5% for an L/S ratio of 2 or more but 100% for an L/S ratio of <1. Phosphatidylinositol in amniotic fluid progressively increases until 36 weeks and then decreases. At 36 weeks phosphatidylglycerol (PG) appears and increases until term. The presence of PG is also a strong indicator of lung maturity. The presence of PG combined with an L/S ratio $\geqq 2$ is associated with a very low risk of HMD.

VI. PREVENTION

In 1972 Liggins and Howie reported that the administration of β-methasone to women in premature labor at least 2 days before delivery significantly reduced the incidence of HMD in infants who were born at <34 weeks' gestation. Despite numerous studies supporting these earlier findings, until 1994 only 15% of eligible infants in the United States received antenatal glucocorticoids. In 1994, a panel of scientific advisors sponsored by the National Institutes of Health developed the following recommendations for the use of antenatal corticosteroids: All fetuses between 24 and 34 weeks gestation who are at risk of preterm delivery should be considered candidates for antenatal treatment with corticosteroids. Patients who are eligible for therapy with tocolytics also should be eligible for treatment with antenatal corticosteroids. Treatment consists of two doses of 12 mg of β-methasone given intramuscularly 24 hr apart or four doses of 6 mg of dexamethasone given 12 hr apart. Optimal benefit begins 24 hr after initiation of therapy and lasts 7 days. Because treatment with corticosteroids for less than 24 hr is still associated with significant reduction in neonatal morbidity and mortality rates, antenatal corticosteroids should be given unless immediate delivery is anticipated. In premature rupture of membranes at less than 30–32 weeks of gestation in the absence of clinical chorioamnionitis, use of antenatal corticosteroid is recommended because of the high risk of intraventricular hemorrhage at these early gestatonal ages. In complicated pregnancies in which delivery before 34 weeks' gestation is likely, use of antenatal corticosteroids is recommended unless there is evidence that corticosteroids will have an adverse effect on the mother or delivery is imminent.

VII. TREATMENT

Adequate resuscitation is extremely important and can reduce the morbidity and mortality of infants who are at risk for HMD. Resuscitation measures should include expansion of the lungs with positive pressure and assisted ventilation if the infant cannot maintain an adequate respiratory effort. Secretion of surfactant is impaired by inadequate expansion of the lungs and therefore many believe that all infants under 1000 g should be intubated at delivery. Oxygen should be given to maintain the PaO_2 between 50 and 70 mm Hg.

Results of several large clinical trials have shown that administration of exogenous surfactant substantially reduces the mortality from HMD. Currently, there are two preparations of surfactant that are commercially available: a synthetic surfactant that primarily consists of desaturated phosphatidylcholine plus the alcohol of palmitic acid (Exosurf), and a preparation that is derived from minced calf lung with added synthetic desaturated phosphatidylcholine (Survanta). Exosurf lacks the surfactant-associated proteins. Survanta only contains SP-B and SP-A. Depending on the size of the infant and the severity of the disease, babies with HMD receive one dose of surfactant at the time of delivery and between one and three doses during the next 24 hr.

It was recently shown that a combination of antenatal glucocorticoids and surfactant therapy significantly reduced morbidity and mortality rates to very low levels in a group of low birth weight babies. This finding underscores the importance of antenatal steroids in addition to surfactant.

Survival rates of infants with HMD have declined dramatically in the past 10 years and infants who weigh ≧1000 g have an 90% survival rate.

Bibliography

Avery, M. E., and Mead, J. (1959). Surface properties in relation to atelectasis and hyaline membrane disease. *Am. J. Dis. Child* **97**, 517–521.

Bose, C., Corbet, A., and Bose, G. (1990). Improved outcome at 28 days of age for very low birth weight infants treated with a single dose of a synthetic surfactant. *J. Pediatr.* **117**, 947–952.

Boughton, K., Gandy, G., and Gairdner, D. (1970). Hyaline membrane disease. II: Lung lecithin. *Arch. Dis. Child* **45**, 311–316.

Clements, J. A. (1962). Surface phenomena in relation to pulmonary function (sixth Bowditch lecture). *Physiologist* **5**, 11–14.

Collaborative Group on Antenatal Steroid Treatment (1981). Effect of antenatal dexamethasone on the prevention of respiratory distress syndrome. *Am. J. Obstet. Gynecol.* **141**, 276–285.

Doyle, L. W., Kitchen, W. H., and Ford, G. W. (1986). Effects of antenatal steroid therapy on mortality and morbidity in very low birthweight infants. *J. Pediatr.* **108**, 287–295.

Farrell, P. M., and Avery, M. E. (1975). Hyaline membrane disease. *Am. Rev. Respir. Dis.* **111**, 657–662.

Fujiwara, T., Chila, S., and Watobe, Y. (1980). Artificial surfactant therapy in hyaline membrane disease. *Lancet* **1**, 55–60.

Gilstrap, L. C., Christensen, R., and Clewell, W. H. (1994). Effect of corticosteroids for fetal maturation on perinatal outcomes: NIH consensus statement. *J. Am. Med. Assoc.* **12**, 1–9.

Gregory, G. A., Kitterman, J. A., and Phibbs, R. H. (1971). Treatment of idiopathic respiratory distress syndrome with continuous airway pressure. *N. Engl. J. Med.* **284**, 1333–1340.

Hack, M., Fanaroff, A., and Klaus, M. (1976). Neonatal respiratory distress following elective deliver: A preventable disease? *Am. J. Obstet. Gynecol.* **126**, 43–50.

Kendig, J. W., Notter, R. H., and Cox, C. (1988). Surfactant replacement therapy at birth: Final analysis of a clinical trial and a comparison with previous trials. *Pediatrics* **82**, 756–762.

Liggins, G. C., and Howie, R. N. (1972). A controlled trial of antepartum glucocorticoid treatment for prevention of the respiratory distress syndrome in premature infants. *Pediatrics* **50**, 515–522.

Merritt, T. A., Hallman, M., and Bloom, B. T. (1986). Prophylactic treatment of very premature infants with human surfactant. *N. Engl. J. Med.* **315**, 785–791.

Shapiro, D. L., Notter, R. H., and Morin, F. C. (1985). Double-blind randomized trial of a calf lung surfactant extract administered at birth to very premature infants for prevention of respiratory distress syndrome. *Pediatrics* **76**, 593–599.

Rete Testis

see Testis, Overview

Rh Factor

see Erythroblastosis Fetalis

Rhodnius prolixus

K. G. Davey
York University

GLOSSARY

aedeagus The extensible portion of the intromittent organ or penis of most insects.

allatostatin A neuropeptide which acts to suppress the synthesis of juvenile hormone by the corpus allatum.

allatotropin A neuropeptide which stimulates the synthesis of juvenile hormone by the corpus allatum.

bursa copulatrix The portion of the female tract of the insect which receives the male genitalia, analogous to the vagina of mammals.

chorion The egg shell surrounding the insect egg.

corpus allatum A nonnervous endocrine gland of insects which secretes juvenile hormone.

corpus cardiacum An endocrine gland of insects consisting of the secretory terminals of neurosecretory cells in the brain together with intrinsic neurosecretory cells.

ecdysteroid A family of steroids related to ecdysone, the hormone controlling molting in insects.

hemolymph The blood of insects, contained in an open system (the hemocoel).

juvenile hormone The hormone secreted by the corpus allatum of insects controlling metamorphosis and many aspects of reproduction.

mesenteron A capacious nondigestive anterior portion of the midgut of some insects where the undigested food is stored and from which food is released gradually to the intestine for digestion.

spermatheca The organ in female insects which stores spermatozoa transferred by the male.

Rhodnius prolixus is a member of the family Reduviidae of the order Hemiptera. Like many of the other members of this family, it is an obligate blood feeder. It inhabits South and Central America and is occasionally found in the southern United States. Primarily associated with large birds, it also infests a wide variety of hosts as diverse as rodents and aramadillos. It domesticates easily and is often found living in close association with humans (see box). It is also easily reared in the laboratory.

I. INTRODUCTION

In the late 1920s, Sir Patrick Buxton, the head of the London School of Hygiene and Tropical Medicine, brought some living specimens of *Rhodnius prolixus* back from South America. Their usefulness as experimental models was immediately recognized by V. B. Wigglesworth, the founder of insect physiology, who had recently joined the London School. The model has been exploited by Wigglesworth and others for more than 60 years in describing the functioning of the principal organ systems of insects, in the analysis of basic questions in development, and, above all, in establishing the basic facts of insect endocrinology.

The advantages of *R. prolixus* as an experimental animal are several. It is easily reared in the laboratory. It can be maintained in standard incubators and can be fed on a variety of hosts, most commonly laboratory rabbits. It is remarkably resistant to starvation, and the later larval stages will survive many months without food. Development and reproduction depend on the taking of a blood meal. The imbibition of a meal of adequate size sets in train the chain of endocrine events leading to molting to

RHODNIUS AND HUMANS

Rhodnius prolixus easily forms a close association with humans. Hiding in cracks and crevasses in the dwelling during the day, it emerges at night in search of food. It locates its host by a combination of infrared and CO_2 sensors, and since the sleeping victim may normally have only the head exposed, the insect is frequently drawn to the lips, which are often the site of penetration of the proboscis. This has earned *Rhodnius* the common name "kissing bug." The proboscis is exceedingly fine, not much greater in internal diameter than a red blood cell, and as the proboscis is inserted, the insect injects saliva containing anesthetic and anticoagulant proteins, and the penetration is practically undetectable. The flexible proboscis probes beneath the skin until it encounters a small vessel. Stimulated by the detection of ATP by sense organs in the proboscis, *Rhodnius* then takes in a very large meal of blood (several times the unfed weight of the insect; a 65-mg adult female may imbibe as much as 250 mg of blood) aided by a powerful pump in the pharynx. Although the initial penetration of the proboscis is virtually undetectable, humans may suffer a subsequent profound local reaction to the bite as the result of the injection of the saliva during the feeding process. Some humans may become hypersensitive and develop anaphylaxis with successive bites. Researchers should be alert to this possibility.

In order to reduce the volume and weight of the meal, a brisk diuresis begins almost as soon as the insect has begun to feed, and some drops of urine are passed during feeding. The causative agent of Chagas' disease, the protozoan *Trypanosoma cruzi*, is a trypanosome which is transmitted by the so-called "posterior station" route. The infective stages of the trypanosome (after a period of development in the insect gut) take up a position in the rectum and are passed in the urine onto the surface of the host, from which the parasite penetrates the skin, often aided by the scratching of the bite by the host. While the trypanosome initially penetrates only tissue at the site of entry, the infection may eventually become more general and involve a variety of tissues, including the heart.

the next developmental stage: Absent the meal, no growth or molting occurs. The availability of a specific signal of this sort has been enormously useful in establishing the sequence of endocrine events controlling development. *Rhodnius* is tolerant of many surgical procedures, including the removal of entire endocrine organs, parts of the brain, decapitation at various levels, or being joined together in parabiosis. Finally, the principal endocrine organs of the head are laid out in almost diagrammatic simplicity so that decapitation at various levels can remove different organs.

II. THE CORPUS ALLATUM AND EGG PRODUCTION

The *Rhodnius* model has been particularly useful in establishing the network of hormones that control egg production (Fig. 1). The central role of the corpus allatum in governing egg production was first demonstrated in *Rhodnius* by Wigglesworth in 1936, who showed that this endocrine gland was essential for vitellogenesis. Two factors govern egg production: feeding and mating. While an unfed female will make very few eggs, fed females make a number of eggs depending on the amount of blood stored in the mesenteron. Mated females make more eggs than virgin females. They also lay the completed eggs quickly, whereas virgin females retain the completed eggs for longer periods against the possibility that mating might ensue (in insects, eggs are fertilized after the shell, or chorion, is laid down).

The number of eggs produced is governed by juvenile hormone (JH) secreted by the corpus allatum. Females which have been allatectomized (surgical removal of the corpus allatum) make very few eggs very slowly. Virgin females which receive supplements of JH will make the mated number of eggs, and mated and virgin females will make the same, larger than the normal mated, number when the corpus allatum is freed from its inhibitory nervous

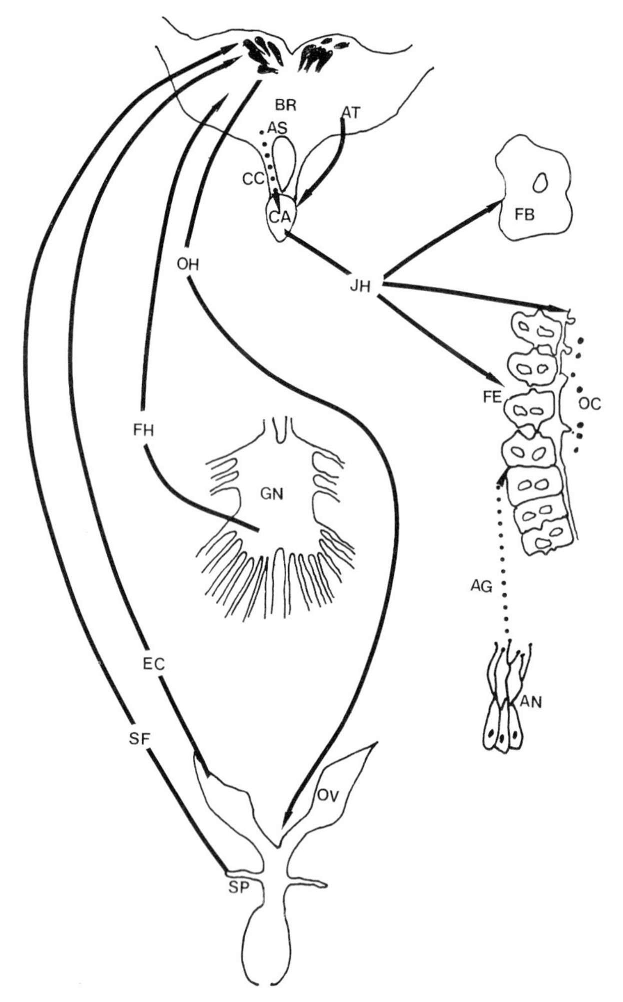

FIGURE 1 Diagram illustrating the hormonal interrelationships controlling egg production in *Rhodnius prolixus*. Solid lines indicate stimulatory influences and dotted lines inhibitory. When the female feeds, a neuropeptide "feeding hormone" (FH) is released from the fused ganglia (GN) in the thorax into the abdomen and carried forward to the brain (BR) by the circulation. This sets in train the other endocrine events by relieving the inhibition of the corpus allatum (CA) by allatostatin(s) carried to the CA by nerves. Allatotropin(s) from the brain and/or corpus cardiacum (CC) stimulate the CA, which produces juvenile (JH). JH stimulates the fat body (FB) to produce vitellogenin which is carried by the hemolymph to the ovary. JH also acts on the follicular epithelium (FE), causing the cells to shrink and large lateral spaces to appear through which the vitellogenin gains access to the oocyte surface. JH acts on the oocyte membrane to increase the number of vitellogenin binding sites. The action of JH of the follicular epithelium is antagonized by antigonadotropin (AG), a neuropeptide from abdominal neurosecretory organs (AN). Ovulation and oviposition are controlled by an ovulation hormone (OH), mytropic neuropeptide from cerebral neurosecretory cells. A spermathecal factor (SF) from the spermathecae, signaling matedness, and an ecdysteroid (EC) from the ovary, signaling the presence of mature eggs, are both required for the release of ovulation hormone.

connection to the brain. Mated and virgin females make eggs at the same rate: The difference in the numbers of eggs produced is a result of the timing of the cessation of egg production. Egg production ceases in a mated female when the amount of blood remaining in the mesenteron reaches a specific weight. In virgins, the cessation of egg production is similarly related to the weight of the meal retained in the mesenteron, although the value is somewhat higher.

III. INPUTS CONTROLLING EGG PRODUCTION

A. Meal Size

As indicated previously, it is the weight of the blood meal remaining in the mesenteron which determines when the corpus allatum will cease to secrete JH. The weight of the mesenteron appears to be monitored by a series of four paired sense organs in the ventral body wall beneath the mesenteron. These sense organs are pressure sensitive and nonadapting: They continue to respond at a constant rate to a particular pressure until the pressure is changed. While the sense organs are connected to the central nervous system via the segmental nerves, information from them does not reach the brain by a strictly nervous route. On the one hand, severing the ventral nerve cord between the brain and the first thoracic ganglion does not prevent egg production. On the other hand, reducing the flow of the hemolymph to the brain by severing the dorsal aorta inhibits egg production, and this inhibition is relieved by JH treatment or by severing the nervous connections between the brain and the corpus allatum. By directly monitoring the electrical activity in 10 large neurosecretory cells in the brain which are known to release the ovulation hormone in response to feeding, it has been possible to show that feeding results in the appearance in the hemolymph of a neuropeptide originating in the fused mesothoracic–metathoracic–abdominal ganglia which acts on the brain to activate the neuroendocrine axis governing reproduction. While there is no direct evidence that the corpus allatum is relieved of its inhibition by the brain by this route, it is the best explanation of the experimental results. In addition to the inhibition imposed by the nerves from the brain to the corpus allatum, presumably due to allatostatins, there is also evidence for stimulation of the corpus allatum by an allatotropic factor(s) from the brain and corpus cardiacum. Nothing is known about the factors controlling the release of these allatotropins.

B. Mating

As noted previously, mated females of *R. prolixus* not only make more eggs but also lay their eggs as soon as they are made: Egg laying begins on the third or fourth day after feeding and is complete by about the eighth day. On the other hand, virgin females, which make their reduced number of eggs at the same rate and the same time as mated females, seldom begin to oviposit before the tenth or eleventh day, and the eggs are trickled out slowly over the next 2 weeks. Matedness is signaled to the endocrine system by a hormone from mated spermathecae. Implanting mated spermathecae into virgins induces egg production and oviposition characteristic of mated females, whereas removing the spermathecae prevents the increase in egg production and oviposition normally attendant upon mating. Ovulation, the release of eggs from the ovary to the oviducts, and their subsequent oviposition are under the control of an ovulation hormone originating in 10 large neurosecretory cells in the pars intercerebralis of the brain. This hormone, a relatively large (about 8.5 kDa) peptide of the FMRFamide family, is a myotropin which acts on the muscles of the ovary to expel the mature eggs.

An examination of the titers of myotropic activity in the hemolymph of females reveals that there is a brief release of the hormone immediately after feeding in both mated and virgin females. In mated females, but not in virgins, there is a second release beginning about 4 or 5 days after feeding and coinciding with the principal bout of oviposition in mated females. This release cannot be dependent solely on the factor from the spermatheca because that factor is presumably present throughout the gonotrophic cycle. A second factor, emanating from the ovaries, signaling that mature eggs are present, is also essential. That factor is an ecdysteroid: The titer of ecdysteroids rises in both virgin and mated females on the fourth or fifth day after feeding.

Because the myotropin cells in the brain are large and easily accessible, it has been possible to relate electrophysiological events to release: Release of the hormone is signaled by an increase in tonic bursts. These electrophysiological signs of release are increased simply by applying 20-H ecdysone directly to the brain of a mated female deprived of ecdysteroid by earlier ovariectomy. The ecdysteroid appears to act on the myotropin cells via a pathway involving biogenic amines.

IV. JH AND THE FOLLICLE CELLS

As in many other insects, JH acts in *R. prolixus* at a number of sites, such as the fat body and the ovary. While other model insects have been used to work out the mode of action of JH on the fat body, where JH directs the synthesis of the principal yolk protein, vitellogenin, *R. prolixus* has been particularly useful in studying the control by JH of the entry of vitellogenin into the follicle. In insects each oocyte is surrounded by a single-celled layer of follicular epithelium, and the vitellogenin in the hemolymph must pass through this layer in order to gain access to the oocyte surface from which it is taken up into the developing oocyte by receptor-mediated endocytosis. The vitellogenin enters the follicle through large lateral spaces which appear between the cells of the follicular epithelium at the beginning of vitellogenin uptake. The appearance of these spaces, termed patency, is under the control of JH. In *R. prolixus*, patency involves a reduction in volume of the cells and a rearrangement of the cell junctions. Using the reduction in volume as an assay for JH action, K. G. Davey and colleagues have shown that JH acts on the cell membrane via a cascade involving protein kinase C to activate a specific JH-sensitive Na^+/K^+-ATPase, thereby bringing about a rapid and reversible reduction in cell volume to about 60% of the original volume. These facts suggest that JH may act in this case via a membrane receptor, and a JH binding site has been identified in follicle cell membranes from

R. prolixus with a dissociation constant in the low nanomolar range.

In females of *R. prolixus* which have been deprived of JH by allatectomy (surgical removal of the corpus allatum), a small number of follicles are still produced; these grow very slowly. Significantly, when the follicle cells from such ovaries are exposed to JH *in vitro*, the cells do not shrink. Evidently cells which have developed in the absence of JH lack the capacity to respond to JH later in their development. The evidence suggests that the lack of response is traceable to a lack of both the JH-binding protein and the special JH-sensitive Na^+/K^+-ATPase. Thus, JH has two actions on the follicle cells: a priming action which directs the development of the cellular machinery necessary for the second action; and a regulatory action by which the cells respond to the hormone by shrinking. These two types of action of JH have since been shown to be more generally applicable (Wyatt and Davey, 1996).

The regulatory action of JH on the follicle cells is antagonized by an antigonadotropin, a neuropeptide released from paired abdominal neurosecretory organs in each of the principal abdominal segments. The neuropeptide is small, but its sequence has not yet been determined. It acts at very low concentrations (low nanomolar or femtomolar range) and is not an antagonist of JH binding.

V. THE MALE SYSTEM

Rhodnius prolixus has also been useful in exploring the control of the male system, partly because the development of the system is dependent on imbibing a large blood meal, as described for the female, but also because the various components of the system are laid out with diagrammatic simplicity. Each of the paired testes consists of six seminiferous tubules each connected by a vas efferens to the vas deferens, an expansion of which serves as a seminal vesicle. The accessory glands of the male consist on each side of four finger-like lobes. Each lobe is a single layer of secretory cells surrounding a reservoir. Three of the lobes are transparent and produce the proteins which form the spermatophore, whereas the fourth lobe is full of a milky, opaque secretion which causes contractions in the female ducts. The lobes of the accessory glands on each side lead to a separate duct, which eventually joins with the vas deferens in the paired bulbous ejaculatorius. The two bulbi unite to form the ductus ejaculatorius leading to the intromittent organ.

A. Control of Spermatogenesis

Rhodnius prolixus has been useful in determining the role of the morphogenetic hormones, ecdysone and JH, in spermatogenesis. As in other insects, the spermatogonia are organized into cysts, and the divisions of the cells within a cyst are synchronous. There is a low basal rate of mitotic divisions among the spermatogonia. Classical extirpation experiments combined with hormone replacement therapy have demonstrated that ecdysone accelerates this basal rate and that JH inhibits the ecdysone stimulated increase but not the basal rate. Thus, at each of the first four larval molts, the stimulatory effect of the secretion of ecdysone, initiated by the blood meal, is antagonized by the nearly simultaneous secretion of JH. However, when the fifth-stage larva feeds, the action of ecdysone is not antagonized by JH, the secretion of which is suppressed in order to permit metamorphosis to occur, and mitotic divisions are particularly exuberant. Many cysts enter meiosis and spermiogenesis follows so that spermatozoa are formed by the time the adult male ecloses.

However, given the basal rate of mitosis, those insects which do not have an opportunity to feed may produce spermatozoa inappropriately early in somatic development. That possibility is accommodated by the selective autolysis of the most differentiated cysts in unfed insects. Feeding suppresses the autolysis. This suppression of autolysis was previously associated with the secretion of ecdysone. However, unlike many other insects, *R. prolixus* continues to produce spermatozoa in the adult male, and autolysis can be observed in unfed adults. Ecdysteroids are absent from adult males of *R. prolixus*, and the suppression of autolysis by feeding appears to depend on a factor secreted from the head.

B. The Control of the Accessory Glands

Rhodnius prolixus was the first insect in which JH was demonstrated to be essential for the production of the accessory gland secretion. Since that original demonstration by Wigglesworth in 1936, it is now clear that JH is released by feeding as in the female. As the glands fill, however, JH-directed synthesis of the proteins in the transparent accessory glands slows. If the male mates, the glands become depleted, and JH-directed synthesis resumes. These facts suggest that the insect is able to sense the amount of secretion in the reservoirs of the glands. The means by which the distension of the glands is detected, and the mode of transmission of this information to the corpus allatum, is not known, but the brain does not appear to be involved in this aspect of the control of the corpus allatum. The situation is rendered more complex by two considerations. First, there appears to be a parallel influence on protein synthesis in the gland because a neuropeptide from the brain and corpus cardiacum has been shown to increase protein synthesis *in vitro*. The relation of this control to that imposed by JH is not known. Second, one of the proteins synthesized in the transparent accessory gland is also secreted into the hemolymph of the male in considerable quantities and may function in the male as a storage protein.

VI. COPULATION AND SPERM TRANSPORT

During copulation, which lasts about 30 min, the eversible aedeagus acts as a mold for the formation of the spermatophore. The viscous transparent secretion flows into the aedeagus while it is inserted into the bursa copulatrix of the female. It forms a gelatinous mass, clotting in response to a lower pH produced by secretions of the ejaculatory bulb. The spermatozoa are inserted into a slit in the neck of the pear-shaped spermatophore, and the opaque accessory secretion is placed just behind this slit. When the spermatophore is complete, it is ejected from the intromittent organ so that the slit containing the spermatozoa encloses the opening of the common oviduct in the anterodorsal wall of the bursa copulatrix.

Over the next 5 or 6 hr, the spermatozoa are transported up the oviduct to the spermathecae, the storage organs for the spermatozoa. The spermathecae in *R. prolixus* are not homologous to the spermathecae of other insects (the homologous spermatheca functions as a cement gland, the secretion of which is used to fasten the eggs to the substrate on which they are laid), but they are paired outpouchings of the common oviduct. *Rhodnius prolixus* was the first animal in which it was shown unequivocally that the semen was transported within the female only by contractions of the female ducts. These are stimulated by the opaque accessory secretion. In matings in which the male has been surgically deprived of the opaque gland, a normal spermatophore containing sperms is produced, but the sperms do not leave the bursa. Conversely, if a spermicide is injected into a normal spermatophore so as to kill the spermatozoa, the dead spermatozoa are transported to the spermathecae. Finally, if the muscles of a female are paralyzed by exposure to pure nitrogen (this treatment does not alter the motility of the spermatozoa) immediately after mating, the sperms are not transported to the spermathecae until the supply of oxygen is restored. Application of the opaque secretion to the bursa induces powerful peristaltic contractions of the common oviduct, but the identity of the active principle has never been determined.

See Also the Following Articles

Accessory Glands, Insects; Allatostatins; Corpus Allatum; Corpus Cardiacum; Ecdysteroids; Juvenile Hormone

Bibliography

Davey, K. G. (1996). Hormonal control of the follicular epithelium during vitellogenin uptake. *Invert. Reprod. Dev.* **30**, 249–254.

Davey, K. G. (1997). Hormonal controls on reproduction in female Heteroptera. *Arch. Insect Biochem. Physiol.* **35**, 443–453.

Huebner, E., Harrison, R., and Yeow, K. (1994). A new feeding technique for experimental and routine culturing of the insect *Rhodnius prolixus*. *Can. J. Zool.* **72**, 2244–2247.

Wigglesworth, V. B. (1936). The function of the corpus allatum in the growth and reproduction of *Rhodnius prolixus* (Hemiptera). *Q. J. Microsc. Sci.* **79**, 91–121.

Wyatt, G. R., and Davey, K. G. (1996). Cellular and molecular actions of juvenile hormone. II. Roles of juvenile hormone in adult insects. *Adv. Insect Physiol.* **26**, 1–155.

Rhombozoa

John S. Pearse
University of California, Santa Cruz

GLOSSARY

axial cell A central cell running the full length of the fully developed worm; there are three in a row in the earliest infective stage of the worm.

axoblasts Cells within the axial cell that divide and develop into new worms or into the infusorigen that produces gametes; sometimes called *agametes* or *germ cells*.

Dicyemida The major class of Rhombozoa; sometimes used as the phylum name instead of Rhombozoa.

infusoriform larvae Free-swimming, nonfeeding larvae that develop from zygotes within the axial cell of the rhombogen and escape the host excretory organ, presumably to infect new hosts.

infusorigen A stage within the axial cell of the rhombogen with an outer layer of oocytes surrounding an axial cell containing nonflagellate sperms; sometimes considered to be a hermaphroditic gonad of the rhombogen.

jacket cells Covering cells surrounding the axial cell of the nematogen and rhombogen; ciliated in Dicyemida and nonciliated in Heterodicyemida; also called *trunk* or *somatic cells*.

nematogen A fully developed worm attached to the epithelium of the host excretory organ and consisting of a single elongate axial cell surrounded by a layer of jacket cells. Axoblasts within the axial cell produce more nematogens or rhombogens.

polar cells Cells that attach to the epithelium of the host excretory organ; found at the apical end of nematogens and rhombogens.

rhombogen The final stage of the life cycle; similar to a nematogen except it produces infusorigens within the axial cell and, after zygotes are formed, infusoriform larvae.

stem nematogen The initial infective worm in the host excretory organ; like the nematogen except smaller, unattached, and with three axial cells in a row rather than just one.

Rhombozoans are marine, worm-like symbionts found in the excretory organs of octopuses, cuttlefishes, and bottom-dwelling squids. The body organization is unique: A central cell is surrounded by a single layer of outer cells. Cells within the central cell can develop into more worms that leave the central cell to reinfect the host or they can develop into a sexual form that produces both eggs and sperms. Self-fertilization or parthenogenesis occurs within the central cell and the resulting larvae are released into the host excretory organ; from there, they leave in the host urine, presumably in search of new hosts.

I. INTRODUCTION

Rhombozoa includes about 65 species known exclusively from the excretory organs of cephalopods. They are ubiquitous in some host species, and when present they can occur in enormous numbers, seemingly replacing the excretory organs themselves. However, there is no evidence that they are detrimental to their hosts. Their organization—with an elongate central cell, the axial cell, surrounded by a layer of jacket cells and a distinctive set of polar cells at the attachment end—is unique (Fig. 1). However, their simple, vermiform shape is superficially similar to the free-swimming stage of the Orthonectida, and the two groups have been placed together in the phylum Mesozoa. The two groups differ in several ways: Orthonectids are not vermiform when in the host but are single or small groups of cells within hypertrophied host cells, and the free-living, worm-like, sexual form of orthonectids does not possess an axial cell but instead is full of gametes. Moreover, recent molecular analyses using 18S rDNA indicate a deep separation between the two groups.

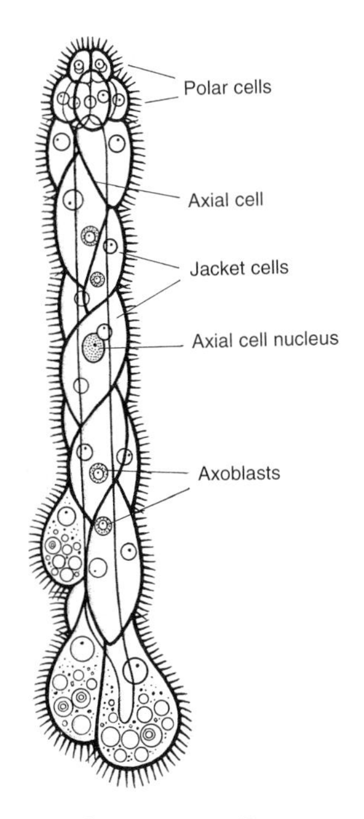

FIGURE 1 Dicyemid nematogen (from Pearse *et al.*, *Living Invertebrates*, 1987; after B. H. McConnaughey, 1951).

The phylum Rhombozoa comprises two groups: the Dicyemida, which includes most species, and the Heterodicyemida, with only two poorly known species that are distinguished from dicyemids mainly by the lack of cilia on the thin, possibly syncytial, jacket cells. Although rhombozoans appear simple morphologically, they have a complex life cycle (Fig. 2). Moreover, molecular data (18S rDNA) suggest that they are not primitive but rather aligned with triploblastic animals.

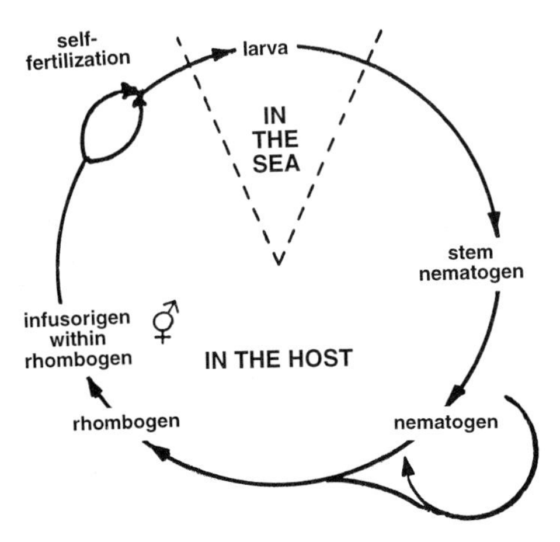

FIGURE 2 Life cycle of rhombozoan (from Pearse *et al.*, *Living Invertebrates*, 1987).

II. INFECTION AND GROWTH IN HOST

The only free-living phase in the rhombozoan life cycle is the infusorigen larva, which consists of 37 or 39 cells, including 2 apical cells, a layer of ciliated cells, a ring of capsule cells, and 4 central cells. Where these larvae reside, how long they live, or how they infect their host remain unknown. However, because only bottom-dwelling cephalopods are infected, the infective larvae probably occur on or in bottom sediments. The earliest stage in the cephalopod hosts, found swimming freely in the excretory organs, is the stem nematogen, which has 3 elongate axial cells aligned in a row, and each contain numerous tiny cells called axoblasts, agametes, or germ cells. Stem nematogens attach to the walls of the excretory organs with their polar cells and transform into vermiform nematogens that grow by cell enlargement to reach lengths of 0.5–10 mm. The final number of cells in the nematogen is species specific, with 10 or 11 polar cells, 12–17 jacket cells, and a single axial cell.

Axoblasts propagate more axoblasts and also produce more worms within the axial cell of the nematogen. New nematogens are produced when an axoblast divides and one cell becomes the new axial cell, whereas the other continues to divide until it engulfs the new axial cell, forming the jacket and polar cells. When fully developed, a new nematogen leaves the axial cell of the parent nematogen and attaches to the wall of the host excretory organ, where it can grow to full size by cell enlargement. Asexual replication can continue in this manner until enormous numbers of nematogens are attached to the epithelia of the excretory organ.

III. SEXUAL REPRODUCTION

Under still unknown conditions, nematogens transform into rhombogens, in which the axoblasts within the axial cells develop into infusorigens that can be considered to be another form of the worm or simply as hermaphroditic gonads. Infusorigens consist of an inner axial cell full of sperms surrounded by outer cells. The outer cells undergo meiosis and become ova. The sperms fertilize the ova while still within the axial cell of the rhombogen. There is no evidence for cross-fertilization, but pseudogamy may occur in which the sperms only activate the ova and diploidy is attained by fusion of one of the polar bodies with the ovum.

Zygotes develop into embryos within the axial cell of the rhombogen. Early cleavage is spiral, and blastomeres from the animal pole form the outer ciliated cells, which overgrow the blastomeres of the vegetal pole by epiboly. Development, which is the simplest known for a metazoan, is highly programmed, with four to eight cleavage cycles producing 37–39 cells; in some species 2 cells normally undergo cell death. The resulting infusoriform larvae leave the axial cell, become free in the host's excretory organ, and leave the host with its urine, presumably to eventually infect new hosts.

See Also the Following Articles

Marine Invertebrate Larvae; Orthonectida; Parasites and Reproduction, Effects on Hosts

Bibliography

Furuya, H., Tsuneki, K., and Koshida, Y. (1992). Development of the infusorifom embryo of *Dicyema japonicum* (Mesozoa: Dicyemidae). *Biol. Bull.* **183**, 248–257.

Hochberg, F. G. (1983). The parasites of cephalopods: A review. *Mem. Natl. Museum Victoria 44*, 109–145.

Katayama, T., Wada, H., Furuya, H., Satoh, N., and Yamamoto, M. (1995). Phylogenetic position of the dicyemid Mesozoa inferred from 18S rDNA sequences. *Biol. Bull.* **189**, 81–90.

Lapan, E. A., and Morowitz, H. (1972). The Mesozoa. *Sci. Am.* **227**(6), 94–101.

McConnaughey, B. H. (1951). The life cycle of the dicyemid Mesozoa. *Univ. California (Berkeley) Publ. Zool.* **55**(4), 295–336.

Pawlowsky, J., Montoya-Burgos, J.-I., Fahrni, J. F., West, J., and Zaninetti, L. (1996). Origin of the Mesozoa inferred from 18S rDNA gene sequences. *Mol. Biol. Evol.* **13**, 1128–1132.

Rhythms, Lunar and Tidal

Peter P. Fong
Gettysburg College

GLOSSARY

circalunar rhythm A series of repeated responses which occur over a period of time approximating 34 days.

The activity patterns of aquatic animals are often cyclic in nature. These cycles can be diurnal (≈24 hr), semimonthly, monthly, or annual. In the ocean, tides are a particularly conspicuous and predictable component of daily water movement. Thus, it is not surprising that the activities of a large number of marine animals living in shallow water are regulated by tides. Since tides are controlled to a great extent by the gravitational force of the moon, these two environmental cues (moonlight and tides) can be used by animals to regulate their reproductive activities. The use of moonlight or tides to regulate the timing of reproductive events can function for the organism in two ways: (i) For species that broadcast their gametes into the sea or that aggregate in large numbers before releasing gametes, synchronized spawning increases the likelihood of fertilization, and (ii) for species that release larval stages, movement of larvae during favorable tides increases the chances of being transported to locations where either predation is decreased or food is plentiful.

I. INVERTEBRATES

Some of the most famous examples of lunar-controlled reproductive activity are the mass spawnings of marine invertebrates such as palolo worms and tropical reef corals. In the south Pacific, palolo worms swarm at the sea surface to release large numbers of eggs and sperm. This occurs regularly during the third-quarter moons in October and November. Native peoples have known of the tight synchrony of swarming for generations and gather together in shallow water to collect and feast on the swimming worms. Such lunar-regulated swarming in marine worms is not limited to tropical seas.

The polychaete worm *Platynereis dumerilii* swarms and spawns in the Bay of Naples with a circalunar rhythm of approximately 34 days. Worms maintained in the laboratory under natural light conditions spawn at the same time as animals in the field, with a maximum between the last and first-quarter moons. If instead the worms experience constant illumination, the lunar rhythm is lost and animals spawn without any synchrony. Such a populationwide reproductive periodicity is based on an endogenous reproductive rhythm stimulated to enter a final phase of gamete maturation during certain phases of the moon.

A similar reproductive rhythm in a congeneric species, *Platynereis bicanaliculata,* occurs on the Pacific coast of North America in Monterey Bay, California. Here, worms swarm at night when extremely high tides occur at midnight mainly around the time of spring new moon (Fig. 1). This lunar synchrony can be mimicked in the laboratory by employing artificial moonlight at night. The worms remain in their tubes during the "moonlight" period but swarm on dark nights immediately after the moonlight has been turned off (Figs. 2A and 2B).

A large number of tropical Pacific reef corals spawn synchronously. Mass spawnings of gametes of over 100 coral species off of the Great Barrier Reef on the same nights between full moon and last-quarter moon during the late spring have been reported by numerous authors. These were widespread events occurring on reefs separated by up to 500 km. Corals not only release gametes during certain phases of the moon but also release their planula larvae. The release of these larvae around the time of last-quarter moon on the Great Barrier Reef has been reported

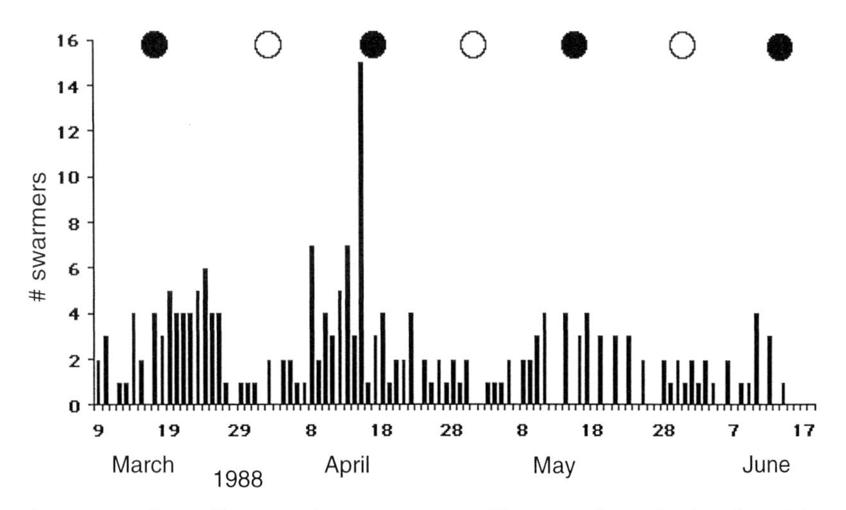

FIGURE 1 Number of swarming (sexually mature) marine worms, *Platynereis bicanaliculata* from Monterey Bay, California. Worms were collected from tide pools, placed in aquaria with running seawater, and exposed to ambient photoperiod. Peaks of swarming occurred around the time of the ambient new moon. ●, new moon; ○, full moon.

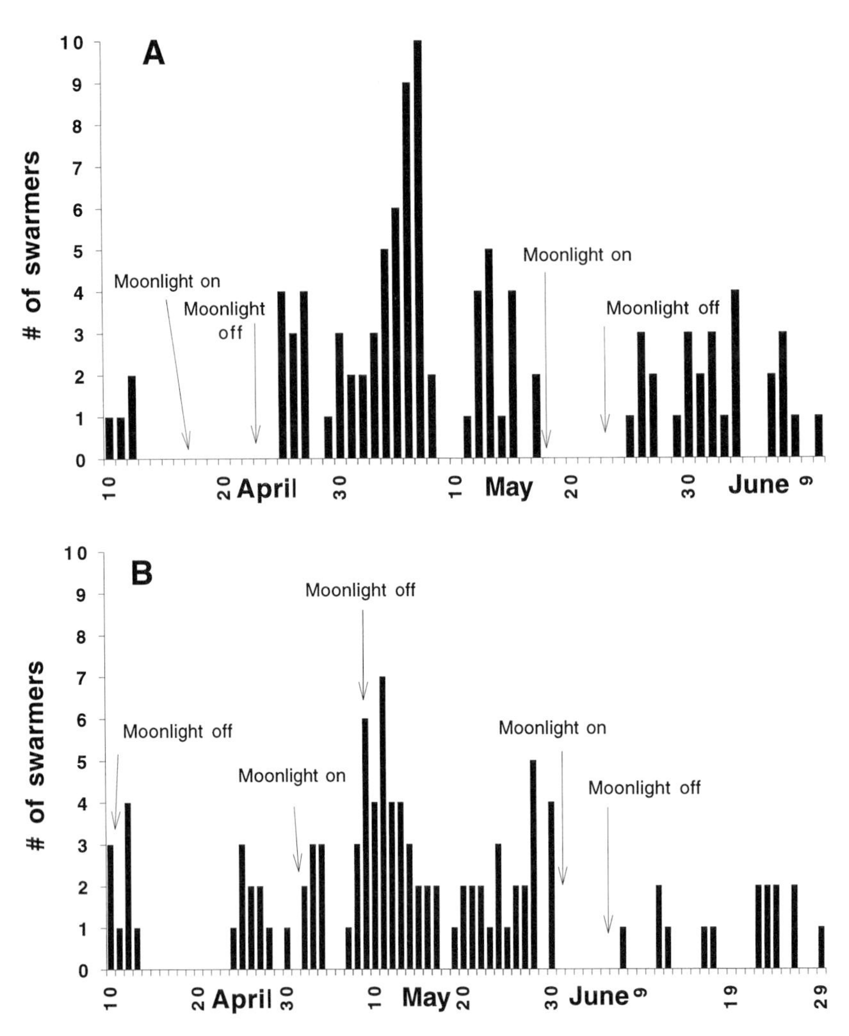

FIGURE 2 Number of swarming *Platynereis bicanaliculata* under conditions of artificial moonlight. "Moonlight on" was a period of six nights during which a dim light (artificial moonlight) was turned on at night for 12 hr. (A) Artificial moonlight during the period of the ambient full moon; (B) artificial moonlight during the period of ambient new moon.

in the coral *Pocillopora damicornis*. Thus, the release of both gametes and larvae occurs on or near the same moon phase in this area of Australia.

Sea urchins have been shown to have lunar periodicity regulating their gonadal growth and the timing of spawning. The southern California sea urchin *Centrostephanus coronatus* has a lunar-regulated reproductive rhythm, which is not synchronized by tidal forces. Ripe gonads were found mainly after the full moons of July and August and spawning occurred during the third-quarter moon.

Crabs have internal fertilization and release larvae directly into the water. Once released, the larvae hatch and begin to feed in the plankton. The hatching periodicities of crabs can be measured by sampling the plankton at various times over a lunar month. Along the east coast of the United States, the hatching rhythms of two intertidal crabs (*Uca* sp. and *Panopeus herbstii*) show two monthly peaks coinciding with early evening high spring tides. By contrast, several subtidal species in the same area release larvae mainly during all outgoing tides. The differences in larval release may have evolved to minimize predation and stranding. The larvae of the intertidal species are subject to intense predation by visual predators such as fishes. Thus, a nocturnal release will reduce predation. Moreover, release of larvae just after the high tide means swifter transport away from the area, again minimizing predation but also reducing the chances of getting stranded at low tide. By contrast, the subtidal species, by virtue of the fact that they are not subjected to exposure, can release their larvae at any ebbing tide.

II. VERTEBRATES

Some of the best examples of lunar and tidally regulated reproductive rhythms among vertebrates are marine fishes. The mummichog (*Fundulus heteroclitus*) from Florida spawns with a lunar pattern, a semilunar pattern, and a tidal pattern. From February to April, spawning peaks occurred at full moon during spring tides. From April to September, spawning peaks were smaller and occurred at both full and new moon spring tides. During both of these lunar patterns, mummichogs spawned mainly during the highest spring tides.

Perhaps the most famous case of a lunar-controlled reproductive rhythm in a fish is that of the grunion, *Leuresthes tenuis*. This small southern California fish swims atop the crashing waves onto sandy beaches where it releases eggs and sperm. This occurs at night during spring high tides after full and new moons. The eggs are buried in sand but hatch and release 2 weeks later when extremely high tides flood them. As in the case of palolo worms, humans know when the fish will spawn and gather to net the wriggling animals.

In summary, the reproductive rhythms of aquatic animals are coordinated by both lunar and tidal forces. The extent to which such synchrony exists in different species and the functional significance of this synchrony is still a topic of scientific debate and inquiry. Moreover, our knowledge of the underlying molecular mechanisms regulating such synchrony in aquatic animals is extremely limited. There is, however, current intense interest in the area of molecular regulation of circadian rhythms in fruit flies and fungi. Clock genes are essentially internal oscillators that generate circadian rhythmicity, and they do so by synthesizing proteins at specific times during a 24-hr cycle. The oscillator must respond to environmental stimuli via photoreceptors, thereby allowing the organism to gauge time. Such clocks have been identified and in some cases cloned. The genes *period* and *timeless* in *Drosophila* and *frequency* in the mold *Neurospora* are true oscillators. Mutant genotypes lacking such genes are arhythmic.

Unfortunately, none of the model organisms whose clock genes are currently being studied are marine. However, the molecular mechanisms which link the perception of real time from photoperiodic information with the circadian synthesis of proteins would also be viable for gauging monthly time using lunar cues. Future studies employing intertidal organisms which have a lunar or monthly reproductive rhythm should elucidate just how widespread clock genes are as a mechanism of regulating various types of photoperiodic timing.

See Also the Following Articles

Circadian Rhythms; Circannual Rhythms; Photoperiodism, Vertebrates

Bibliography

Caspers, H. (1984). Spawning periodicity and habitat of the palolo worm *Eunice viridis* (Polychaeta: Eunicidae) in the Samoan Islands. *Mar. Biol.* **79**, 229–236.

Fong, P. P. (1993). Lunar control of epitokal swarming in the polychaete *Platynereis bicanaliculata* from California. *Bull. Mar. Sci.* **52**, 911–924.

Hauenschild, C. (1960). Lunar periodicity. *Cold Spring Harbor Symp. Quant. Biol.* **25**, 491–497.

Hsiao, S.-M., Greeley, M. S., and Wallace, R. A. (1994). Reproductive cycling in female *Fundulus heteroclitus*. *Biol. Bull.* **186**, 271–284.

Kay, S. A., and Millar, A. J. (1995). New models in vogue for circadian clocks. *Cell* **83**, 361–364.

Kennedy, B., and Pearse, J. S. (1975). Lunar synchronization of the monthly reproductive rhythm in the sea urchin *Centrostephanus coronatus* Verrill. *J. Exp. Mar. Biol. Ecol.* **17**, 323–331.

Palmer, J. D. (1974). *Biological Clocks in Marine Organisms: The Control of Physiological and Behavioral Tidal Rhythms*. Wiley, New York.

Palmer, J. D. (1995). *The Biological Rhythms and Clocks of Intertidal Animals.* Oxford Univ. Press, New York.
Pearse, J. S. (1990). Lunar reproductive rhythms in marine invertebrates: Maximizing fertilization? In *Advances in Invertebrate Reproduction 5* (M. Hoshi and O. Yamashita, Eds.). Elsevier, Amsterdam.
Robertson, D. R., Peterson, C. W., and Brawn, J. D. (1990). Lunar reproductive cycles of benthic-brooding reef fishes: Reflections of larval biology or adult biology? *Ecol. Monogr.* 60(3), 311–329.
Salmon, M., Seiple, W. H., and Morgan, S. G. (1986). Hatching rhythms of fiddler crabs and associated species at Beaufort, North Carolina. *J. Crustacean Biol.* 6, 24–36.
Tanner, J. E. (1996). Seasonality and lunar periodicity in the reproduction of Pocilloporid corals. *Coral Reefs* 15, 59–66.

Rodentia

F. H. Bronson
University of Texas at Austin

GLOSSARY

biome A region of the earth having a distinct form of vegetation, e.g., a tropical forest.

fossorial Having the behavior pattern of digging or burrowing; living in holes or burrows.

monestrous Having a single fertile cycle during a given mating season.

murid One of the family Muridae, the Old World mice and rats.

opportunistic breeding The characteristic in some species of trying to reproduce whenever warranted by immediate conditions (e.g., brood availability) as opposed to rigid seasonality enforced by a predictive cue like day length.

photoperiod The period of time within a day in which an organism is exposed to light, varying with the season of the year.

polyestrus Having two or more fertile cycles during a given mating season.

seasonal breeding The characteristic in some species of having reproductive behavior that varies according to season, such as not reproducing in winter.

Rodents account for about 40% of the approximately 4000 species of mammals. In terms of sheer numbers they account for perhaps as much as 60% of all mammals. Rodents are almost worldwide in distribution, and they are found in every biome, including tundra and the harshest deserts, as well as the more benign biomes. Included in the order Rodentia are species well adapted for terrestrial, arboreal/climbing, arboreal/gliding, fossorial and semiaquatic existences, and some that have opted for a communal existence with humans. Given such a diversity of life history strategies and habitats, generalities about the reproduction of rodents are often meaningless.

I. INTRODUCTION

Rodent classification is controversial, particularly in regard to the murids. This article presents a list of the 29 families cited by Corbet and Hill and a brief look at the reproductive characteristics of each. Fourteen of the 29 families contain <10 species and 7 contain only a single species.

II. APLODONTIDAE (MOUNTAIN BEAVER)

The single species in this family, the mountain beaver, is not really a beaver (nor does it normally live in mountains). In appearance it resembles a moderately sized (up to 1.8 kg) muskrat without a tail. It lives in coastal North America from British Columbia to northern California. Ovulation apparently is spontaneous. The mountain beaver is monestrous and exhibits a well-defined breeding season of 1 or 2 months, timed so that its two to four young are born in the early summer. The environmental factor controlling the breeding season is unknown, but it is probably photoperiod. Gestation is about 1 month. Young mountain beavers mature during their second year of life.

III. SCIURIDAE (SQUIRRELS)

This is the second largest family of rodents, with over 250 species, including, among others, the tree squirrels, palm squirrels, ground squirrels, chipmunks, marmots, and some but not all of the flying squirrels. Sciurids of one kind or another can be found worldwide except in the general region of Australia, Madagascar, southern South America, and the deserts of the Arabian peninsula. In size, the sciurids range from small (several species weigh <50 g) to relatively large (the marmots, up to 8 kg). As might be expected given the diversity within this family, the reproductive characteristics of the sciurids vary immensely. Ovulation is reflexive in some species and spontaneous in others. Some species, such as the marmots and some of the ground squirrels, are monestrous with a short, well-defined breeding season that is coupled tightly to emergence from hibernation. Photoperiodic regulation has been established in several cases. Other sciurids, such as the tree squirrels of the midlatitudes, breed seasonally but sometimes more than once a season, again probably regulated by photoperiod. Still other sciurids, such as the palm squirrels of tropical India, are polyestrus and routinely breed year-round. The sciurids inhabiting semidesert environments probably reproduce opportunistically in relation to rainfall and may use secondary plant compounds associated with newly emerging green vegetation as predictive cues. All sciurids produce litters of several young.

IV. GEOMYIDAE (POCKET GOPHERS)

There are 35 species in this family, all adapted morphologically for a fossorial existence. They can be found wherever the soil is suitable for digging from southern Canada to Panama. In size, they range up to 800 g. They are routinely polyestrus and estrous cycles are short (7–10 days). At the higher latitudes there is a short breeding season, after which one litter of two to eight young is produced. Continuous breeding is seen at the lower latitudes and, correlatively, there is a smaller litter size. Reports of the length of gestation vary widely even for the same species of geomyids—from 18 to 51 days in one case. Pocket gophers mature in 4–6 months.

V. HETEROMYIDAE (POCKET MICE)

The name refers to the occurrence of large external cheek pouches in which seeds are stored. There are 60 species in this family, some of which (the kangaroo rats and kangaroo mice) are strongly modified morphologically for jumping; the rest are mouse-like in appearance. Heteromyids of one species or another can be found continuously from western Canada to northwestern South America, usually in dry environments. They range in size from 10 to 200 g. All heteromyids are polyestrus, producing litters of two to eight young. The length of the estrous cycle varies from 4 or 5 days in some species to several weeks in others. At least some species have a postpartum estrus. Gestation for most heteromyids is on the order of 3 or 4 weeks. Maturation to puberty can range from 3 to 9 months, depending among other things on latitude of residence. Breeding is strictly seasonal at the higher latitudes and is probably cued by photoperiod. At the lower latitudes, breeding can be continuous, throughout the year, with or without seasonal tendencies depending on rainfall patterns. Some heteromyids are known to use secondary plant compounds in newly emerging variation as predictive cues to promote successful opportunism in deserts, in which rainfall and hence vegetative growth is episodic and unpredictable.

VI. CASTORIDAE (BEAVERS)

There are two species of beaver, one in North America and the other in Europe and Asia. These are among the largest rodents, with some individuals weighing up to 45 kg. Both are monestrus with estrous cycles of 2 weeks duration. Beavers are strictly seasonal breeders. Mating takes place in midwinter and births of two to four young follow in 3 or 4 months. Sexual maturation may require up to 2 years. Monogamy apparently is common, at least in the North American species. Female beaver display a unique morphological characteristic—the urogenital canal and GI tract empty into a common cloaca.

VII. ANOMALURIDAE (SCALY-TAILED SQUIRRELS)

There are seven species of scaly-tailed squirrels, most having a membrane between the front and back legs that allows them to glide like the flying squirrels of the family Sciuridae. All anomalurids are found in the tropical forests of west and central Africa. They vary in size from small to moderately large (15 g to 1 kg). Little is known about the reproduction of scaly-tailed squirrels. Some are reported to produce only one young annually in a relatively restricted season; others are reported to breed year-round, producing two to four young per litter. Undoubtedly, they reproduce in relation to rainfall and hence their food supply, but little else is known.

VIII. PEDETIDAE (SPRING HARE)

The single species in this family resembles a small (3 or 4 kg) kangaroo with a long bushy tail. It is found in arid environments from Kenya to South Africa. The spring hare is polyestrus and it reproduces continuously when allowed by rainfall and hence food availability. Females produce one young at a time, and usually average 3 or 4 offspring per year. Gestation has been reported to require 60–80 days. There is a postpartum estrus.

IX. MURIDAE (RATS, MICE, VOLES, LEMMINGS, HAMSTERS, GERBILS, JIRDS, ETC.)

With 1160 species, this is the largest family in the class Mammalia. There are over 130 species of the genus *Rattus* alone, one of which is the common laboratory rat. Murids of various kinds are found almost worldwide and they include species that are adapted for terrestrial, arboreal, semiaquatic, and fossorial existences, the latter resembling moles more than rats and mice. Most of the mammals that have opted for a communal existence with humans are found in this order, including the house mouse. Murids range in adult size from 5 g to well over 1 kg. In general, members of this family are herbivores or omnivores that are heavily preyed upon by other vertebrates. Their short life expectancy is balanced by the potential for an assembly-line production of litter after litter of large numbers of offspring that mature rapidly. Some voles mature in less than 20 days. A postpartum estrus is common and, energetic conditions permitting, females that are both pregnant and suckling a litter are seen often in the wild. Estrous cycles tend to be short and ovulation can be either spontaneous or reflexive. Photoperiodic regulation and strict seasonal breeding is common at the higher latitudes, but even on the arctic tundra some individual lemmings in some populations are insensitive to variation in day length and thus can reproduce opportunistically during the winter under the snow if energetic conditions permit. Opportunism without regulation by seasonal predictors is the hallmark of this family in the lower part of the temperate zone and the tropics. Two subgroups of murids—the cricetids and microtines—are discussed in detail elsewhere in this encyclopedia.

X. GLIRIDAE (DORMICE)

There are 20 species of dormice spread widely throughout Europe, Asia, and Africa, both above and below the Sahara desert. Most dormice look like small squirrels, but without a distinct neck, and most are more or less arboreal in their habits. They range in size from 15 to ~200 g. At the higher latitudes the glirids hibernate during the winter, producing one or two litters of 2–10 young during their short breeding season. Photoperiodic regulation of seasonality is probable. The estrous cycle lasts 10 days and gestation lasts 20–30 days. Dormice at these latitudes typically breed only after their first winter. Little is known about the dormice that live south of the Sahara, but it is probable that those in the tropical regions do not hibernate. It is also probable that the tropical dormice have the potential to reproduce continuously with seasonal peaks depending on rainfall and hence food availability.

XI. SELEVINIIDAE (DESERT DORMOUSE)

The single species in this family lives in the deserts of Kazakstan. It is small (20 g) and looks like the other dormice. It is rare and little is known about its reproduction except that it reproduces seasonally and produces its young in litters of six to eight.

XII. ZAPODIDAE (JUMPING MICE AND BIRCH MICE)

Representatives of the 17 species in this family reside from eastern Europe through Asia to China (all birch mice and one species of jumping mouse) and throughout much of North America south to the Mexican border (the rest of the jumping mice). Jumping mice are strongly modified for jumping; they can jump up to 2 m when startled, using their long tails as balance organs. Birch mice are only slightly modified for jumping. All zapodids hibernate. At high altitude and high latitude jumping mice hibernate for as much as 8 or 9 months, allowing them only 3 or 4 months to produce one litter of 3–10 young which must grow and achieve sufficient fatness to withstand hibernation. This is an amazing overwintering strategy for a mammal of such small size. As latitude and/or altitude decreases, two or three litters can be produced each year. Photoperiodic regulation is probable at all latitudes. Gestation varies from less than 3 to 5 weeks for the various species of zapodids.

XIII. DIPODIDAE (JERBOAS)

These small to medium-sized rodents (<40 to >400 g depending on species and season) are strongly modified for jumping; their back legs are at least four times as long as their front legs and their tails are exceptionally long and used for balance. Many jerboas have long rabbit-like ears. There are 30 species of jerboas, all inhabiting arid areas from the Sahara to the Gobi desert. In Asia the dipodids are polyestrus hibernators, producing up to three litters of 1–10 young each breeding season. This suggests photoperiodic control, but since these are desert animals it is probable that their reproduction is also heavily dependent on rainfall and hence food availability. It is not known if they use secondary plant compounds as predictive cues in the harsher desert habitats. Ovulation apparently is reflexive. Gestation periods are reported to range from 3 to 7 weeks. In Africa, some jerboas become dormant for a short time; others show no indication of hibernation and breed continuously year-round, rainfall and hence food availability permitting. The available evidence suggests that at least some jerboas are reflex ovulators.

XIV. HYSTRICIDAE (OLD WORLD PORCUPINES)

These are large (1.5–30 kg), long-lived rodents, of which there are 11 species. The hystricids inhabit a large variety of environments in Africa, southern Europe, southern Asia, and the East Indies. They are generally polyestrus, breeding more or less continuously, albeit slowly, even at the higher latitudes, e.g., in South Africa. In some locales they reproduce only one litter a year; in other locales they produce as many as three litters of one to four young. Estrous cycles last 4 or 5 weeks and gestation has been reported to last from as little as 7 weeks to as much as 4 months. Maturation is routinely slow, taking 9–18 months to reach sexual maturity.

XV. ERETHIZONTIDAE (NEW WORLD PORCUPINES)

This family of 11 species is found from the Arctic coast of North America to Equador and northern Argentina. All are moderate to large in size (1.5–18 kg) and relatively long-lived. The single species in North America is polyestrus and it reproduces seasonally, probably regulated by photoperiod. The estrous cycle lasts 25–30 days. Ovulation usually is restricted to the right ovary and is probably reflexive. Mating occurs in fall or early winter. Usually, a single young is produced about 7 months later. In the tropics, at least some of the porcupines reproduce continuously and have incorporated a postpartum estrus in their reproductive cycle. Typical litter size is one.

XVI. CAVIDAE (CAVIES)

This family of 16 species includes a variety of cavies, which can weigh up to 1 kg. Also included in this family is the rabbit-like mara, which can weigh up to 16 kg. One of the cavies is the guinea pig, whose domestication can be traced back the Incas. Cavies can be found throughout South America except in the Amazon basin. They are polyestrus. Estrous cycles are on the order of 2 or 3 weeks. Ovulation is reflexive in at least some species. Gestation periods range from <3 weeks to over 7 weeks, depending on the species. Usually there is a postpartum estrus. Reproduction at very high altitude is seasonal with up to four litters of several young produced in a season. At lower altitudes, they are potentially continuous breeders with seasonality imposed in some areas by rainfall cycles and possibly photoperiod at the highest latitudes. One of the cavies, a cui, holds the mammalian record for early maturation—it conceived when it was 9 days old.

XVII. HYDROCHAERIDAE (CAPYBARA)

The capybara, the only species in this family, has been said to look like a cross between a guinea pig and a hippopotamus. It is the largest extant rodent; females can weight up to 60 kg. Capybaras are semiaquatic and can be found from Panama to Uruguay and Argentina on the east side of the Andes. They typically produce one litter each year but occasionally two are produced. Births can occur throughout the year, but they peak just prior to the rainy season. Gestation has been reported to last from 110 to 150 days. Litters can contain up to eight young. A year is required to attain puberty.

XVIII. DINOMYDAE (PACARANA)

The single species in this family, the pacarana, looks like a large (up to 15 kg) guinea pig and is found only in Venezuela, Columbia, and eastern Bolivia. It is very rare and once was feared to be extinct. Little is known about its reproduction except that it produces one or two young with a gestation period on the order of 8 months.

XIX. DASYPROCTIDAE (AGOUTIS, ACOUCHIS, AND PACAS)

The 14 species in this family range from 1 to 4 kg (agoutis) or from 6 to 12 kg (pacas). They are found from east central Mexico to southern Brazil. All dasyproctids are polyestrus, with estrous cycles usually reported to last about 40 days. Gestations range from 100 to 120 days. The paca produces one offspring at a time; the agoutis and acouchis produce one to three. Reproduction ranges from seasonal, with births timed to occur in the dry season, to continuous year-round, but at a slow rate of about two litters per year. Whether other factors besides rainfall patterns and hence food availability influences this variation is not known.

XX. CHINCHILLIDAE (CHINCHILLAS AND VISCACHAS)

The six species in this family are found in western and southern South America, most often in the Andes and always on relatively barren ground. The chinchillas actually may now exist only in the domesticated form. In size, chinchillas generally weigh <1 kg; the plains viscachas can weigh up to 8 kg, and the mountain viscachas are intermediate in size. The latter look like rabbits with long, bushy tails. The chinchillids are polyestrus, with estrous cycles lasting from 35 to 60 days. Gestation periods range from 110 to 160 days. Captive chinchillas usually produce two litters of two or three young each during the winter. Viscachas also reproduce continuously in captivity but usually seasonally in the wild, possibly because of photoperiodic regulation or negative energy balance caused by low temperature and food shortage or both. One to four young are produced usually once or twice during the breeding season. Viscachas are colonial with colonies of several hundred individuals having been reported. Two interesting oddities can be seen in the viscachas. The plains viscacha sheds 200–400 ova per ovulation, 10% of which are fertilized and 1% of which implant. Only the right ovary is functional in the adult mountain viscacha.

XXI. CAPROMYIDAE (HUTIAS)

There are 14 species of hutias, ranging in weight from 500 g to 2 kg. They are arboreal and now found only in the West Indies. Most are approaching extinction due to human disturbance. In appearance, hutia run like squirrels but look like fat rats. They are polyestrus and estrous cycles have been reported to last 10 days to 1 month. They breed year-round in captivity and probably also do so in the wild. Gestation periods vary from 110 to 140 days and at least one species does not mature until almost a year of age. Litter sizes vary from one to six.

XXII. MYOCASTORIDAE (COYPU)

The coypu or nutria (the name of its fur) is the only species in this family. It is semiaquatic and weighs 5–10 kg. It looks like a large robust rat. It is native to southern South America, where it reproduces year-round. It is polyestrus and a reflex ovulator. The coypu's estrous cycle averages 25 days in length, and its gestation averages 130 days. It exhibits a postpartum estrus. Average litter size is five. Huge colonies of these animals have been established—often accidentally by escapees from fur farms—in many parts of Europe, Asia, and North America. In these areas the species is viewed sometimes as a valuable fur animal and sometimes as a pest.

XXIII. OCTODONTIDAE (DEGUS, ETC.)

There are nine species of octodonts, including the degus, viscacha rats, rock rats, the chozchoz, and the coruro. The representatives of this family are limited to Peru, Bolivia, Argentina, and Chile. They range in weight from 50 to 300 g. Most octodonts, like the rat-like degus, are communal. Degus live in complex burrow systems in the Andes and exhibit group territoriality. The gopher-like coruro is heavily modified for a fossorial existence; they are also communal but they live at lower altitudes. Only the reproduction of the degus has been studied to any degree. Degus are monestrus, continuous breeders in the laboratory but seasonal in the wild, possibly because of energetic constraints. The timing of the breeding season depends on altitude and latitude and is restricted largely to the summer at high altitude and latitude. The gestation period is 90 days and 1–10 young are produced per litter. Sexual maturity has been reported to vary from a matter of weeks to a year, possibly due to pheromonal regulation. One of the degus has abdominal testes even during the breeding season.

XXIV. CTENOMYIDAE (TUCO-TUCOS)

There are 38 species of these small (200 g to 1 kg) burrowing rodents, all modified morphologically for a fossorial existence. In appearance they resemble the pocket gophers of North America. They are found from the lowlands to the altiplano and from Bolivia to Tierra del Fuego. They are monestrous, producing one litter a year, the timing of which is influenced by the pattern of rainfall at the lower latitudes and temperature and food availability at the higher latitudes. Photoperiodic regulation at the higher latitudes seems probable. Ovulation is reflexive. Gestation is 100–120 days, with litter sizes from one to five. There is a postpartum estrus but apparently mating seldom occurs at that time.

XXV. ABROCOMIDAE (CHINCHILLA RATS)

Little is known about the two species of chinchillones or chinchilla rats. Both are relatively small, on the order of 200–300 g, and, as the name implies, they look like a cross between a chinchilla and a rat. Chinchilla rats live on the altiplano from Boliva south through Chile and at lower altitudes in northwestern Argentina. They are colonial. They produce litters of several young and their gestation has been reported to be 115–120 days.

XXVI. ECHIMYIDAE (SPINY RATS)

This is the third largest family of rodents, with 63 species. They are small to moderate-sized (20–700 g) forest-dwellers, rat-like in appearance, and found from Honduras south to southeastern Brazil. They have the potential to reproduce throughout the year but their reproductive success probably varies with the rainfall cycle. The duration of the estrous cycle of one species is 3 weeks. Ovulation is reflexive in at least some species. Gestation length varies from 60 to 120 days in this family and litter sizes vary from one to seven.

XXVII. THRYONOMYIDAE (CANE RATS)

The two species of cane rats are widespread in Africa below the Sahara desert. They look like large (up to 9 kg) rats with extra large heads They have the capacity to reproduce continuously, but seasonality is imposed where the rainfall pattern is seasonal. They have a 6-day estrous cycle. Ovulation apparently is reflexive. Gestation period is reported to be about 150 days, litter size is one to eight, and puberty occurs at about 1 year of age.

XXVIII. PETROMURIDAE (AFRICAN ROCK RAT)

Very little is known about the single species in this family, which is also known as the dassie rat. It is relatively small (100–300 g), squirrel-like in appearance except for a less brushy tail, and found in arid areas of southwestern Africa. It is reported to produce only one litter a year. It probably produces that litter opportunistically in relation to rainfall and hence food availability.

XXIX. BATHYERGIDAE (AFRICAN MOLE RATS)

The 11 species in this family are found throughout Africa south of the Sahara desert. They are relatively small (up to 250 g), fossorial, and gopher-like in appearance. They are often colonial. They apparently show a wide variety of seasonal reproductive patterns throughout the broad range they inhabit. Bathyergids are monestrous and strictly seasonal at the higher latitudes (e.g., South Africa). At the lower latitudes they are either seasonal or aseasonal depending on rainfall patterns. Gestation periods range from 40 to 90 days, with litter sizes ranging from 3 to 10. Testes are internal in at least one species. The behavior of one species of bathyergid, the naked mole rat, has been intensively studied because its social organization resembles that of the altruistic social insects: one queen, a few kings, and many infertile workers that spend their lives serving the queen.

XXX. CTENODACTYLIDAE (GUNDIS)

The five species in this family live in the deserts and semideserts of the northern half of Africa. In appearance they resemble small (150–350 g) guinea pigs. Gundis are polyestrus, and the estrous cycle lasts 3 or 4 weeks. They apparently reproduce once a year depending on rainfall. Gestation period is about 2 months, and litter size ranges from one to three. Gundis are slow to mature; puberty occurs at 9–12 months of age.

See Also the Following Articles

Cricetidae (Hamsters and Lemmings); Estrus; Global Zones and Reproduction; Microtinae (Voles); Naked Mole Rats; Photoperiodism, Vertebrates; Rabbits; Seasonal Breeding, Mammals

Bibliography

Corbet, G. B., and Hill, J. E. (1991). *A World List of Mammalian Species*, 3rd ed. Oxford Univ. Press, Oxford, UK.

Hayssen, V., van Tienhoven, A., and van Tienhoven, A. (1993). *Asdell's Patterns of Mammalian Reproduction*. Cornell Univ. Press, Ithaca, NY.

Nowak, R. M. (1991). *Walker's Mammals of the World*, Vols. I and II, 5th ed. Johns Hopkins Univ. Press, Baltimore.

Rowland, I. W., and Weir, B. J. (1984). Mammals: Non-primate eutherians. In *Marshall's Physiology of Reproduction*, 4th ed. (G. E. Lamming, Ed.). Churchill Livingstone, London.

Rotifera

Robert Lee Wallace
Ripon College

GLOSSARY

amictic A phase in the life cycle of monogonont rotifers in which reproduction is accomplished by diploid, ameiotic parthenogenesis.

amphoteric A type of monogonont female capable of producing both amictic and mictic eggs.

diapause A dormant period of arrested growth and development characterized by an encysted stage with greatly slowed metabolism.

eutely (*eutelic*) A condition found in metazoans in which there is a predetermined number of cells.

heterogony The alternation of generations between asexual and sexual phases.

mictic A phase in the life cycle of monogonont rotifers in which reproduction is sexual (mixis), resulting in a diapausing, diploid embryo called a resting egg.

mixis Sexual reproduction.

morphotype A morphologically distinct body form present within rotifer populations whose polymorphic feature(s) undergo a generational, nongenetic change in phenotype.

orthoclone A clone established after many generations of subculturing the ith offspring of every generation; usually specified as young or old orthoclones.

oviparous The condition in which females lay eggs that hatch outside the body.

ovoviviparous The condition in which females lay eggs that hatch within the body with subsequent release of live offspring.

resting egg A resistant (encysted), thick-walled, diapausing, diploid embryo with highly sculptured outer surfaces usually produced by sexual reproduction in monogonont rotifers; these eggs always hatch into amictic females.

subitaneous egg A thin-walled, rapidly developing egg produced by an amictic female that hatches into a young, diploid female that is either amictic or mictic.

tocopherol (vitamin E) An organic compound capable of inducing the mictic cycle in a genus of monogonont rotifers.

vitellarium A large yolk gland situated between the ovary and oviduct in bdelloids and monogononts; together the ovary and the vitellarium comprise the germovitellarium.

I. DESCRIPTION

A. Introduction

Phylum Rotifera includes about 2000 species of minute (usually 50–500 μm), short-lived, eutelic, freshwater and marine micrometazoans that are closely related to a group of obligatorily parasitic worms, the Acanthocephala. Two main features separate rotifers from other micrometazoans (Fig. 1). First, the apical end usually bears a ciliated region, the corona, that may be embellished by various folds, lips, or ear-like extensions. Second, the ventral region of the pharynx, the mastax, is modified to house a complex series of articulating hard jaws (trophi) composed of chitin and scleroprotein.

Although a small phylum, rotifers can be very important in certain freshwater environments because they feed on a variety of bacteria and algae and then, in turn, are fed upon by higher trophic levels. Rotifers are widely distributed and possess a high reproductive potential, occasionally attaining population densities of >5000 individuals per liter. In some unusual habitats and in cultures much higher population densities have been recorded (>10^5 individuals per liter).

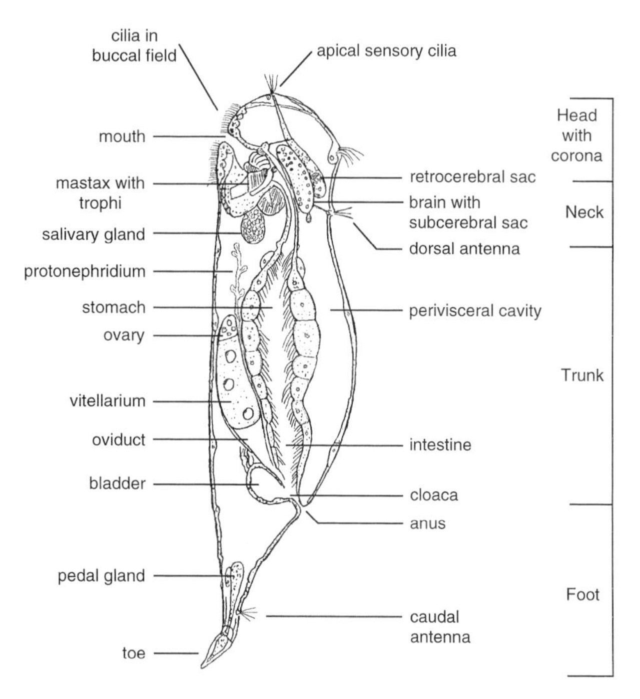

FIGURE 1 Lateral view of a generalized rotifer (modified with permission from Wallace and Snell, 1991).

Aquaculturalists have exploited the enormous reproductive potential of rotifers to develop intensive aquaculture systems that use mass-culture techniques to produce extremely large quantities (kilograms) of rotifer biomass each day. This biomass is then used to feed the young of commercially important species of crustaceans and fish.

Because of their ease of culture, rotifers are useful model organisms for aquatic toxicity testing, studies of population dynamics, and investigations of aging. Unfortunately, even after nearly 300 years of study, our understanding of rotifers is incomplete and based largely on detailed studies of only a small number of species.

B. Literature of the Discipline

There is no journal in which students of rotifers routinely publish their research. However, besides the general texts of invertebrate zoology and freshwater biology, and the advanced works listed in the Bibliography, one may consult the published volumes of the International Rotifer Symposia (to date eight symposia have been held).

C. Taxonomy

Most classification schemes recognize three classes of rotifers (Seisonidea, Bdelloidea, and Monogononta). The first two classes comprise rotifers with paired gonads. On that basis, Seisonidea and Bdelloidea are sometimes placed within class Digononta, leaving class Monogononta separate. However, recent cladistical studies reject Digononta as a paraphyletic taxon. The unique anatomy and obligatory sexual reproduction of Seisonidea are used to separate them from the other two classes which are grouped into superclass Eurotatoria. The mode of reproduction varies dramatically among the classes (see Section II,A). Relationships of the Rotifera to the Acanthocephala have not yet been fully resolved. Traditional, morphologically based classification schemes place Acanthocephala as a sister group to Rotifera. However, recent molecular analyses using mitochondrial 16S rRNA and nuclear 18S rRNA genes place Acanthocephala within the Rorifera as a sister group to the Bdelloidea.

1. *Class Seisonidea*

This class is composed of a monogeneric taxon (*Seison*) of only two large ($\leq$3,000 μm) dioecious, marine species that live on gills of the leptostracan crustacean, *Nebalia* (Fig. 2). Sexes are of similar size and morphology. Females have ovaries without vitellaria. Males lack copulatory organs.

2. *Class Bdelloidea*

This small class ($\approx$350 species) possesses a relatively uniform morphology characterized by a wormlike body and paired ovaries with vitellaria (Fig. 3). Males have never been reported in the class, so these rotifers apparently reproduce exclusively by parthenogenesis. Bdelloids usually are found in sediments, among plant debris, or crawling on the surfaces of aquatic plants. Some species inhabit the capillary water films formed in soils or covering mosses. Many species are capable of tolerating extreme environmental conditions, including freezing and desiccation (anhydrobiosis).

3. *Class Monogononta*

Comprising the largest class of rotifers ($\approx$1600 species), monogononts are all dioecious, but in many

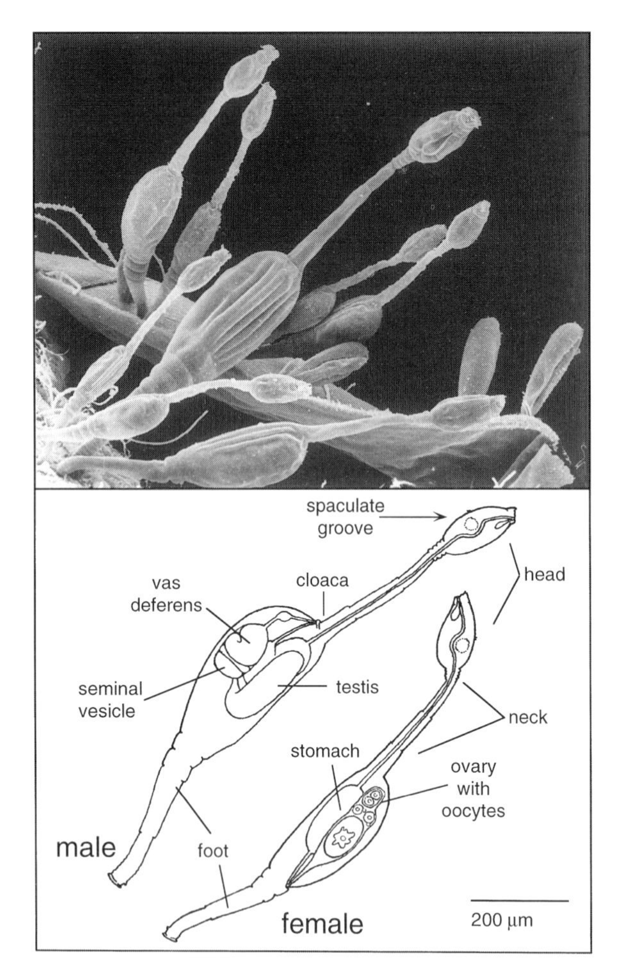

FIGURE 2 *Seison nebaliae*. (Top) Scanning electron photomicrograph of *Seison* on the crustacean *Nebalia*: male, center; female, lower center; juveniles, smallest individuals; eggs, ovoid structures to the right (photomicrograph courtesy of Giulio Melone). (Bottom) Internal schematic views of a female and male emphasizing the digestive and reproductive tracts (modified from Ricci *et al.*, 1993, with kind permission from Kluwer Academic Publishers and the authors).

species males have never been observed. The archetypal life cycle is characterized by relatively long periods of female parthenogenesis that occasionally are broken by brief outbreaks of sexual reproduction (see Section II,A,3). Males usually are structurally reduced, often with a vestigial gut that functions in energy storage (Fig. 4). Adult females are benthic, free swimming, or sessile.

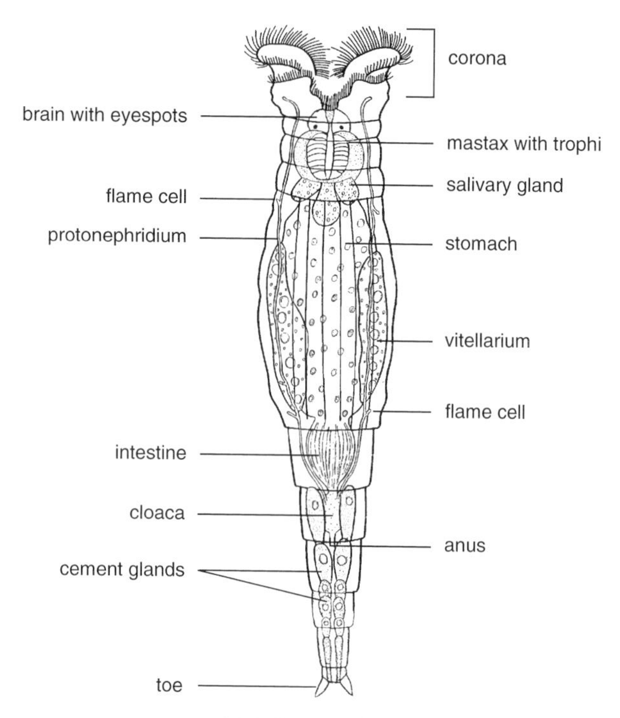

FIGURE 3 A typical bdelloid rotifer (≈200–300 μm long) (modified with permission from Wallace and Snell, 1991).

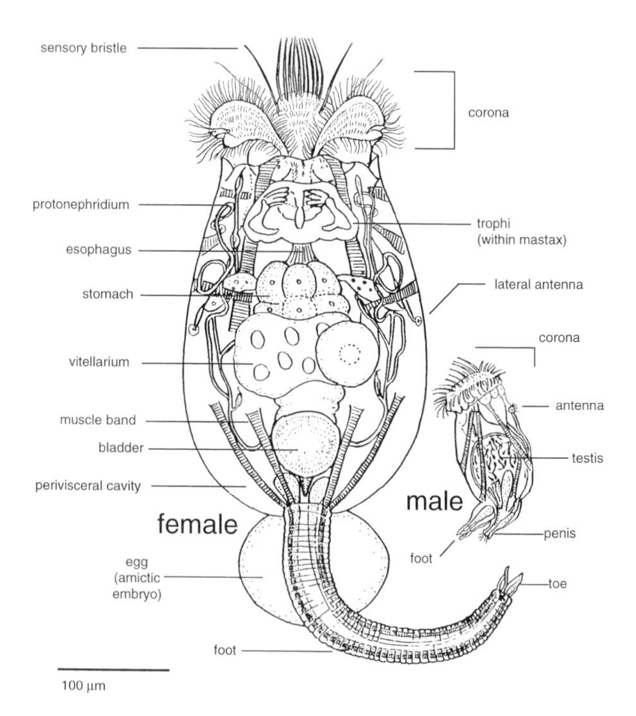

FIGURE 4 Female and male *Brachionus plicatilis*, a monogonont rotifer (modified with permission from Wallace and Snell, 1991).

II. MODES OF REPRODUCTION

A. Life Cycles

1. *Seisonidea*

In seisonids, reproduction is entirely bisexual. Although no studies have been published on the chromosomes of this genus, a sex ratio of approximately 1:1 suggests that sex determination is based on a XY/XX or XO/XX mechanism. Gametogenesis takes place by ordinary meiosis with production of two polar bodies. There is little sexual dimorphism in seisonids (see Section IV).

2. *Bdelloidea*

Bdelloid reproduction is exclusively by asexual parthenogenesis; no males have been observed in this class. Two equational divisions produce two polar bodies and give rise to a diploid embryo.

3. *Monogononta*

Reproduction is more complex in monogononts than in the other two classes; here both asexual and sexual phases are found (Fig. 5). When natural populations of monogonont rotifers are sampled, most often they are found to be reproducing asexually by diploid (2n), ameiotic parthenogenesis with the production of a single polar body. The parthenogenetic egg (subitaneous egg) is thin walled and develops rapidly into a female (Figs. 4–6). This asexual reproductive cycle is called the amictic cycle and it may go on indefinitely. In some species, it is the only type of reproduction known because males have never been observed. However, not all amictic reproduction results in a rapidly developing embryo; a few clones of a *Synchaeta* species can produce amictic eggs that enter a short diapause.

Many species periodically switch from asexual to sexual reproduction; thus, a type of heterogony is present in the class. While sexual reproduction (mixis) has not been seen in all monogononts, it is assumed that all species are capable of sexual reproduction or that sex was an ancestral feature that has been lost. In some localities mixis has never been recorded in the natural population despite being ob-

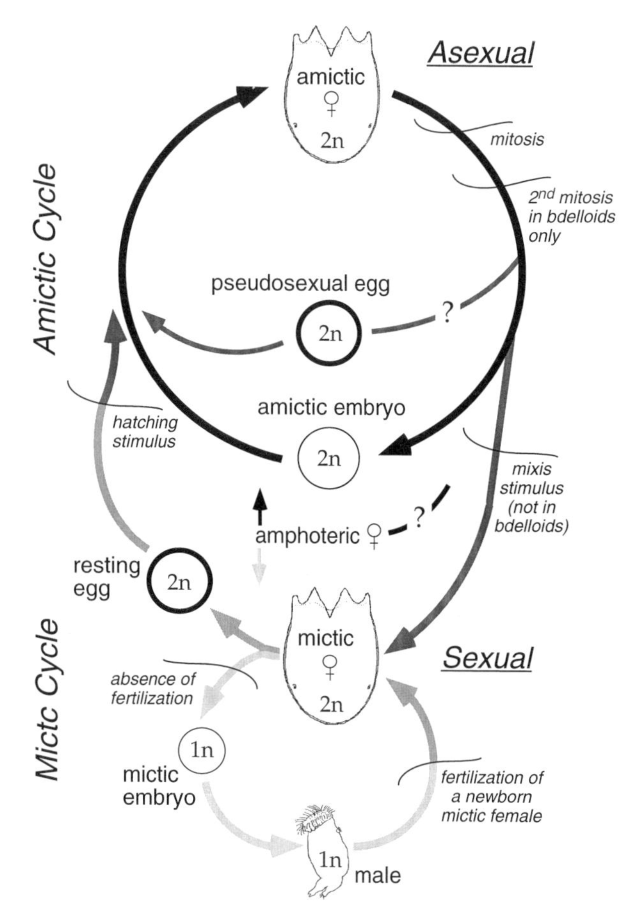

FIGURE 5 Generalized life cycle of class Monogononta. The life cycle of class Bdelloidea is comparable to only the parthenogenetic portion (upper cycle).

served in laboratory cultures begun from individuals isolated from the same habitat.

The intermittent switch from repeated asexual generations to sexual reproduction occurs when amictic females begin to lay subitaneous eggs that develop into diploid, mictic females. Mictic females are usually indistinguishable from amictic females except for the type of egg they produce. If they have not mated soon after birth, mictic females produce haploid (1n) male eggs with two polar bodies (Fig. 6). However, if impregnated, they will produce a different kind of fertilized diploid egg, again with two polar bodies. These eggs are termed resting eggs because they require a diapause of varied length (Fig. 6). Resting eggs possess a thick wall and a surface

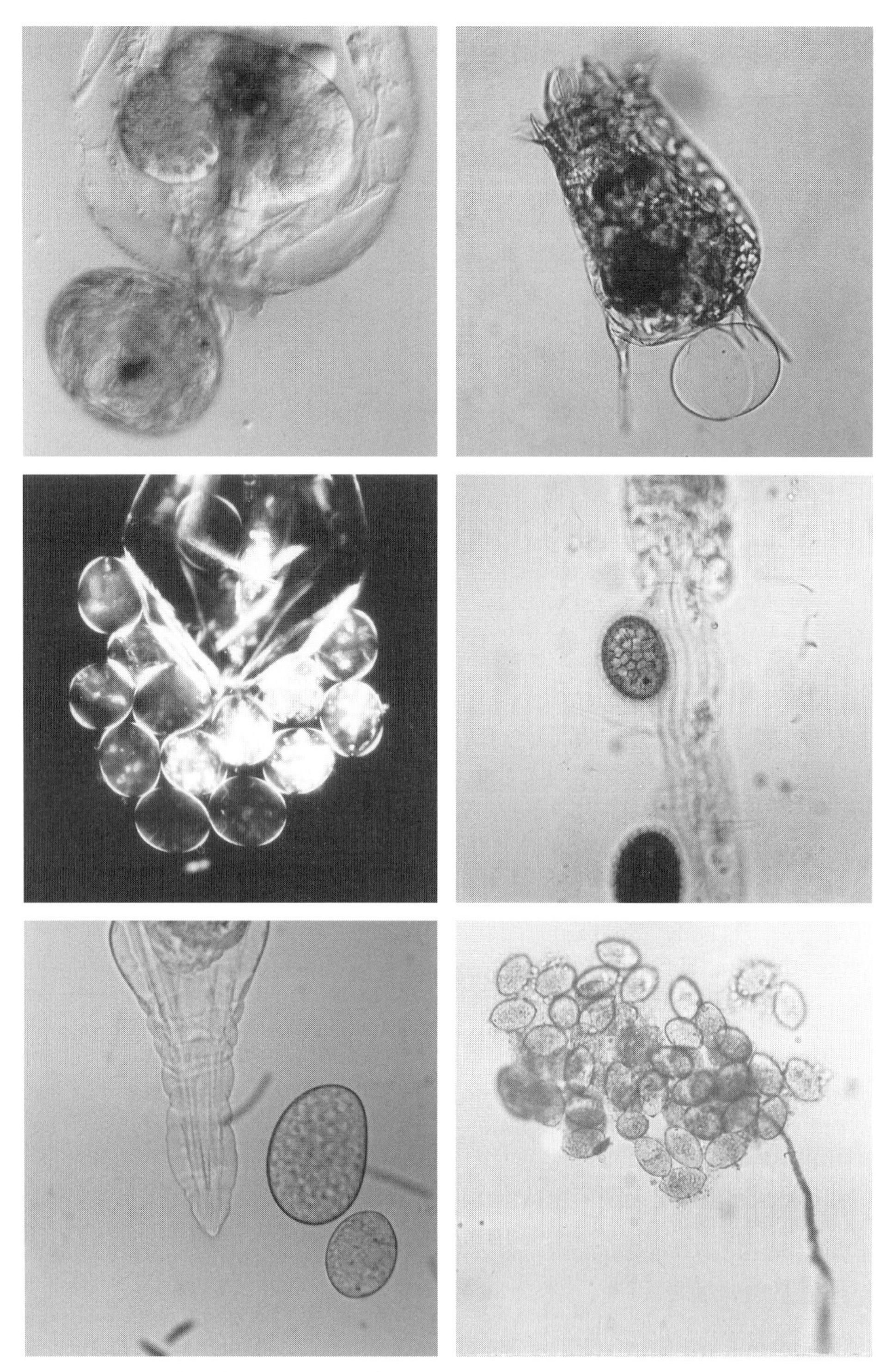

FIGURE 6 Photomicrographs of rotifer eggs. (Top, left) Subitaneous egg attached to an amictic female. (Top, right) Shell of a hatched subitaneous egg attached to an amictic female. (Middle, left) Male (unfertilized mictic) eggs (dark field illumination). (Middle, right) Resting (fertilized mictic) egg held within the jelly tube of a sessile rotifer. (Lower, left) Amphoteric (subitaneous and male) eggs. (Lower, right) Clutch of bdelloid eggs produced under culture conditions. (All from class Monogononta except that in the lower left, which is from class Bdelloidea.) Typical egg size range: size = 25–60 × 40–140 μm; egg volume = 10^5 to 10^6 μm^3 (lower left photomicrograph courtesy of Aydin Orstan; others by R. L. Wallace).

coat that is often highly sculptured and ornamented. A result of the monogonont life cycle is that males and amictic females developing from subitaneous eggs are impaternate, whereas amictic females developing from resting eggs are paternate. Consequently, the oocytes of monogonont females are bipotent, at least in their early stages of development. Mictic females produce oocytes that, if unfertilized, develop parthenogenetically into males; if fertilized, the oocytes develop into diapausing eggs that will hatch as amictic females. On the other hand, amictic females produce eggs that may develop into additional amictic females or into mictic females. Therefore, the future reproductive status of monogonont females is determined during the development of the oocyte.

In a few species a third type of female is recognized. Amphoteric females are capable of producing both amictic and mictic eggs (Fig. 6). These forms are common to only a few genera (*Conochiloides* and *Sinantherina*), but they are rare in others (*Asplanchna*) and undescribed (absent?) in most. The biology of amphoteric production and its significance to population dynamics remain unclear.

Initiation of mixis may not ensure that 100% of the population becomes mictic. In fact, continuation of some portion of the population in the amictic cycle will guarantee persistence of the population past the sexual period. Otherwise, the population will crash after resting egg production. The sex ratio of a population with mictic female production will vary greatly depending on the length of time the mictic phase has been under way and on the population in question.

B. Genetics

Our knowledge of rotifer genetics is very incomplete, but the reasons for this paucity of information are easily understood: The chromosomes of rotifers are very small ($\approx$0.5–5 μm) and difficult to study using traditional karyotyping techniques. However, we do know from published observations that chromosome number in monogononts and bdelloids ranges from 10 to 14. Additionally, one case of hypotriploidy has been reported in monogononts. In both bdelloids and monogononts, parthenogenesis is believed to be ameiotic and to lack crossing over. Thus, barring mutational events, parthenogenesis in rotifers produces genotypes that are identical; variation in isozymic signatures among individuals isolated from a single habitat is therefore attributed to the expansion of individual clones that were each established from a single founding adult or resting egg. While information on the genomic size of rotifers is very limited, recent studies of three bdelloids and one monogonont species indicate that the total nuclear DNA content ranges from 0.7 to 1.0 pg (i.e., approximately 10^9 base pairs).

Sex determination in monogononts is genetic, with males being haploid and females diploid, but the mechanism of this control is unknown. The genetics of seisonids has yet to be explored.

C. Individual Reproductive Effort

Rotifers are eutelic and so the total number of oocytes available for reproduction is fixed at birth. Thus, oocyte number determines the maximum potential fecundity (MPF) of the species. These potentials vary according to species but are in the range of 10–30 offspring per female. The realized fecundity is usually well below that of the MPF. Realized fecundity also varies depending on the type of egg produced. Amictic eggs are much larger than unfertilized mictic eggs (male eggs), but given similar dietary regimens the total egg volume produced by each type of female (amictic vs mictic) will be about the same. Thus, the female egg ratio (eggs per female) will be higher for mictic females producing male eggs than for amictic females producing subitaneous eggs. In mictic females production of unfertilized eggs is about two to seven times that of fertilized (resting) eggs. This difference is no doubt due to the large difference in relative investment in each type of egg. In general, fecundity of all females varies widely with food quantity and quality and temperature having the most significant effect. Parental age also is a factor in determining fecundity in both monogonont and bdelloid rotifers. Orthoclones established from young females have higher reproductive potentials than those established from older females. Obviously observations of this phenomenon are restricted to laboratory cultures because these age effects cannot be tracked in natural populations.

D. Population Reproductive Potentials

The parameters of egg development time, birth and death rates, and population growth rates have been extensively examined for a variety of bdelloid and monogonont species. Using both laboratory cultures and natural populations, these studies attempt to identify and understand the relative importance of all factors regulating population dynamics. As might be expected the most important of these regulating factors are food quantity and quality and temperature.

1. *Laboratory Cultures*

Laboratory cultures traditionally have used the technique of life table analysis. This rigorous approach gives an age-specific record of all births and deaths within a cohort. These data are then used to calculate important population statistics, e.g., maximum life span and age-specific survivorship curves, net reproduction rate and age-specific fecundity, intrinsic rate of natural increase, generation time, and egg development time. Obviously, such efforts are very labor-intensive and yield population growth parameters based on only a few animals (usually fewer than 50 individuals are used). On the other hand, sophisticated, computer-controlled, turbidostat and single- and dual-stage chemostat culture systems are able to provide similar information, but based on substantially larger populations. Large-scale aquaculture has provided an abundant literature, albeit directed toward the practical efforts of growing massive quanities of rotifers. While these enterprises usually are based on simple batch-culture techniques, they are of an enormous scale. In some systems scuba divers are used to clean the sides of the tanks.

2. *Natural Populations*

Tracking the dynamics of natural populations has been done for decades. While not as rigorous as laboratory methods, these studies have provided good estimates of the population statistics for a variety of species. As expected, food and temperature are dominant factors controlling growth in natural populations. However, competition (both interference and exploitive), predation, and parasitism also play crucial roles in the dynamics of most species.

III. MODES OF SEXUALITY

All sexual reproduction in the phylum is obligatorily bisexual. In seisonids sex results in embryos that undergo a relatively quick development and hatch into juveniles. In monogononts sex results in a diapausing, diploid embryo called a resting egg which is resistant to harsh conditions. Production of resting eggs via parthenogenesis (i.e., amictic or pseudosexual egg production without males) has been reported (Fig. 5) but remains an inadequately explored phenomenon.

The critical feature leading from the asexual to the sexual cycle in monogononts is the production of a female embryo that develops into a mictic female (see Section II,A,3). This switch is mediated by a specific inducing factor along with other factors that modify the rate of mictic production. So far, most of the detailed research on mictic female induction has focused on only three genera (*Asplanchna*, *Brachionus*, and *Notommata*), basically because these species are easily induced to produce mictic females in culture. In these genera, the inducing factors fall into two categories: exogenous (population density, photoperiod, and dietary tocopherol) and endogenous (parental-age effects). The environmental factors that induce mixis are triggered before oviposition in oviparous species, whereas in ovoviviparous species the embryo may be influenced throughout its developmental period. Initiation of mixis has often been described as a response to a deteriorating environment. An alternative view, however, argues that mixis is optimized when the habitat is capable of supporting a large healthy population and provides good dietary conditions for the production of energetically costly resting eggs.

IV. ANATOMY OF THE REPRODUCTIVE SYSTEMS

Rotiferan classes are separated based on their reproductive biology. Paired gonads are present in Bdelloidea and Seisonidea, but males are unknown in bdelloids. Class Monogononta has only one gonad; however, this organ begins its embryological development as a bilateral structure. Reproductive organs are ventral to the digestive tract in all classes.

A. Females

The female reproductive system consists of ovary, vitellarium, and follicular layer, but in seisonids the vitellarium is absent. The follicular layer surrounds both ovary and vitellarium and forms the thin-walled oviduct in some species. The oviduct opens into a cloaca and is of ectodermal origin, at least in monogononts. The ovary is a small syncytial mass of closely packed nuclei placed like a cap at the anterior end of the vitellarium. The vitellarium is usually spherical or elongated into a tubular structure; it is easily recognized by its large nuclei. Amictic and mictic females of most monogonont species cannot be distinguished other than by observing the kinds of eggs they produce: amictic (subitaneous eggs), mictic (male or resting eggs), or amphoterics (subitaneous and mictic eggs).

B. Males

1. *Seisonidea*

In seisonids, males and females are about the same size, although, depending on the species being considered, different authors report one sex larger than the other. The U-shaped, male genital apparatus consists of a pair of large, elongate testes that unite caudally and extend anteriorly to form a large organ concealing part of the digestive system. Sperms leaving the testes enter a large seminal vesicle where they receive additional processing. First, each sperm is compressed into a ciliated bulb. Then two sperms are attached together, tightly rolled, and encysted with a secretion. This processing is not analogous to spermatophore production in other invertebrates; in the spermatophores of other taxa, numerous sperms are packed inside a capsule. Therefore, use of the term spermatophore for the encysted sperm pair (ESP) may not be appropriate. From the large seminal vesicle the ESP ($\approx 1.5 \times 8\ \mu m$) is moved anteriorly to the vas deferens. From there a ciliated duct communicates with the cloaca. Although male seisonids lack a copulatory organ, a spatulate groove on the head is believed to be involved in transfer of the ESP to the female (Fig. 2) (see Section VI,A).

2. *Monogononta*

In monogononts, males are usually smaller than females (0.25–0.75 times) and are often structurally reduced, but various grades of this sexual dimorphism may be found in this class. Extremes are represented by fully developed males in *Rhinoglena* and dwarf males in *Conochilus*. In many monogononts the digestive system consists of rudiments. These males do not feed, nor can they assimilate dissolved organic compounds in appreciable levels.

The testis in monogononts is a rather large, round or oblong sac containing fewer than 50 sperms. It connects to a ciliated vas deferens that may lead to an eversible cirrus or a protrusive penis. Usually one (rarely two) pair of accessory (prostate) glands is present.

V. REPRODUCTIVE PHYSIOLOGY AND CONTROL

A. Asexual Reproduction

There is no asexual reproduction in seisonids. In monogononts (and probably bdelloids) oocytes separate sequentially from the ovary and attach to the vitellarium. After joining with the vitellarium the egg is penetrated by a tube-like structure which then supplies it with yolk. After a period of nutritional uptake the egg detaches and it begins development. Cleavage begins after oviposition, which appears to occur shortly after completion of the oocyte growth phase.

Disposition of the egg after development varies among the classes. Both species of seisonids are oviparous, laying their eggs exclusively on their crustacean host. Oviparous bdelloids usually lay their eggs on a substratum, either loosely or firmly attached with a cement. It is not known whether bdelloids show a preference for particular substrata, but some species lay their eggs in collective clutches (Fig. 6). Oviparous monogononts commonly carry their eggs attached to the mother by a fine thread (*Brachionus*), release them into the plankton (*Notholca*), or deposit them onto a substratum (*Euchlanis*). A few embed their eggs in a massive gelatinous secretion that surrounds most of the animal (*Lacinularia*). In some species oviposition involves a series of elaborate behaviors. In *Euchlanis* these behaviors result in the female depositing her eggs on preferred substrata. Ovoviviparous bdelloids and monogononts retain their embryos within the body of the mother until hatching (e.g., *Rotaria* and *Asplanchna*, respectively).

Little is known about the periodicity of oviposition, either on a diel basis or over the course of the fecund period of the female. Some species oviposit at regular intervals, whereas others lay their eggs in pulses. Temperature, light, food levels, and maternal age seem to be important variables in oviposition intervals. Egg hatching is generally considered to be solely a function of embryonic development time, with ambient temperature being the most important factor influencing development. Thus, eggs should hatch at intervals conmensurate with time of oviposition and water temperature. However, there are a few cases in which egg hatching within a population is clustered. For example, in one species of *Sinantherina* egg hatching is somehow modified so that all the eggs that will hatch on a given day do so within a few hours of one another. Clustering of egg hatching may be due to simultaneous oviposition, a synchronization of development times after oviposition, or a delay in hatching.

B. Sexual Reproduction

Sexuality remains incompletely studied. A few examples will suffice to illustrate our lack of understanding. Nothing is known about reproductive physiology in seisonids. The conditions inducing mixis have been examined in only a few monogonont genera. Males have not been described for every monogonont species, and it is only an assumption that all species are capable of producing males or that male production is a plesiomorphic character in the class. Except for a few species, most male monogonts have degenerate guts and do not feed; thus, they are dependent on energy provided by their mother during oocyte development. While apparently no one has assessed male fertility as a function of the age of their mothers, maternal age may have an important effect on male offspring, should vitellarial nutrition diminish significantly as the female ages.

C. Adult Reproductive Life History

The life stages of several species of bdelloid and monogonont females are known. From that research we recognize that temperature and diet are important factors influencing duration of the prereproductive, reproductive, and postreproductive periods. Furthermore, in monogonont females the relative length of these periods may differ in amictic and unfertilized mictic females. Other than a much shorter life span, little is known about the life stages of male rotifers.

Many bdelloids are capable of anhydrobiosis, which involves a slow loss of metabolic water and other changes at the cellular level. Survivorship of anhydrobiotic animals (tuns) is not always 100%, but these rotifers have been revived even after 20 years of desiccation. This phenomenon is thought to represent only a pause (lag) in the reproductive life of the bdelloids that has little influence on fecundity. Unfortunately, only a few studies addressing these questions have been performed.

D. Sexual Behavior

Information on sexual behavior in rotifers is limited to a few genera of monogononts. These studies indicate that only the males play an active role in mate recognition and mating behaviors. Males apparently swim randomly and encounter females by chance alone. Their rapid swimming speed, when compared to females, is probably adapted to increase probability of encounters with potential mates. Recognition of a mate is made only when the male makes a head-on encounter. This is due to the presence of a species-specific, glycoprotein signal on the surface of the female's integument. Chemoreceptors on the male's corona respond to this mate recognition pheromone (MRP) by initiating a series of specific behaviors. In some species males apparently do not discriminate between amictic and mictic females, so the MRP must be a glycoprotein common to both. Recent research indicates that the carbohydrate moiety of the MRP is critical for mate recognition, but its biochemistry has yet to be elucidated.

After mate recognition, the male initiates precopulatory behaviors. These include circling of the female and localization of her corona. Copulation begins when the male's penis (cirrus) becomes attached (fixed) to the female's integument. During this time his corona remains in contact with the female. Penial fixation, which is thought to be aided by secretions from the prostate gland, may be to almost any part of the female. However, the penis is usually attached where the female's integument is not thickened (e.g., the corona). When copulation begins, the male disengages his corona from the female and a few (two or three) sperms are transferred hypodermically. When sperm transfer is complete (1 or 2 min) the male dissociates himself and swims away.

Both male and female age play an important role in rotifer reproductive physiology. In some species a male's responsiveness to females first increases and then later decreases as he ages. Likewise, newborn females apparently provide more stimulus for males than older females. As a result, probability of mating after a male–female encounter decreases as a function of female age, with the strongest response being with newborn females.

VI. MODES OF FERTILIZATION

A. Seisonidea

After insemination sperms become embedded in a syncytium (mantle) surrounding the oocytes. This tissue acts as a kind of seminal receptacle. Fertilization occurs between the first and second meiotic divisions, by which time the oocyte is in the oviduct. The nucleus then enters a resting stage while oocyte growth is completed. After growth is completed, the second meiotic division occurs. Information on spermatogenesis in seisonids is lacking.

Mating in seisonids has never been observed, but workers suggest that the ESP is transferred to the male's head region (ventral spatulate organ) and by appropriate movements the male attaches it to the female's genital region.

B. Monogononts

In monogononts most mating is accomplished by a firm attachment of the male penis to the female body (fixation). Insemination usually is accomplished by hypodermic impregnation, but in a few species sperms are deposited directly into the cloaca. Although examined in only a few species, maternal age appears to be important to both male mating behavior (see Section V,D) and the male's ability to fertilize mictic females (i.e., as determined by successful production of resting eggs). Likewise, fertilization success also is a function of the female's age. Production of resting eggs declines with age of first mating of the mictic female. Little is known about male sterility. The only published study indicates that a single recessive mutation can produce sterility in an *Asplanchna* species.

Observations indicate that sperms enter the oocyte before the first meiotic division. Penetration is thought to initiate transferal of nutritive material from the vitellarium, but the underlying mechanism that controls this event is unknown. Fusion of the haploid sperm and egg nuclei does not occur until after the maturation divisions have been completed. In fact, each pronucleus continues its independent divisions through the first meiotic division. Separate divisions continue through the beginning cleavage stages; thereafter, fusion of the male and female nuclei occurs. This independent existence of pronuclei is found in other invertebrates.

VII. MODES OF DEVELOPMENT

A. Vitellarial Function

From an ultrastructural and light microscopical perspective the vitellaria of monogononts varies depending on the sexual status of the female. Vitellaria of fertilized mictic females are more opaque than those of amictic females. No doubt this is due to the presence of the numerous shell-secreting materials and lipids. Both gastric glands and vitellarium of an *Asplanchna* species have been found to undergo postmitotic polyploidization of their nuclei. While the oocytes are diploid, gastric glands and vitellarium were shown to possess amounts of DNA equal to as many as eight duplications of the diploid genome. Presumably this duplication enables a speedy synthesis of mRNA to ensure rapid protein synthesis for the developing oocytes.

B. Egg Development in Monogononts

1. *Amictic Embryos*

A holoblastic, unequal, modified spiral cleavage of determinate type has been reported in a few monogonont species. For a detailed examination of early embryogenesis, gastrulation, and organogenesis, consult the reviews of Thane (1974) and Gilbert (1989).

Development of amictic oocytes is accompanied by a dramatic increase in oocyte volume (>1500×). This is due to the flow of materials through a cytoplasmic bridge from the vitellarium. Where examined, it has been noticed that oocytes of oviparous monogononts have yolk reserves, but that those of ovoviviparous species do not. Thus, eggs developing outside of the mother may have all the energy needed for embryogenesis, whereas ovoviviparous embryos may assimilate maternal nutrients *in utero*.

Parthenogenetic eggs of oviparous, amictic monogononts and bdelloids possess two relatively thin, chitinous envelopes or shells, whereas the eggshells of those ovoviviparous species examined are thinner and contain no chitin. A thinner shell in ovoviviparous eggs may be related to transfer of maternal nutrients to the embryo while *in utero*.

2. *Male Embryos*

Male embryogenesis apparently parallels that of the female embryo to a certain point of organogenesis (e.g., in *Asplanchna*, formation of the cerebral ganglion). At that point certain organs either degenerate (stomach) or fail to develop altogether (mastax).

Spermatogenesis begins late in the embryogenesis of the unfertilized mictic egg and includes, as its major features, cell elongation, flagellar differentiation, and development of the nuclear membranes. Developing along with the sperms are atypical germs cells which eventually degenerate. In recently matured males, sperms outnumber the atypical cell type by a ratio of 2:1. As they mature, the atypical cells extrude immobile, rod-like structures approximately $2 \times 10^{-15}\,\mu m$ in size. These rods appear to facilitate passage of sperms through the female integument.

3. *Diapausing Embryos (Resting Eggs)*

Three complex shells develop to surround the resting egg. The outer layer is secreted by the growing oocyte, whereas the dense middle and very thin inner layers are products of the embryonic ectoderm. The latter two are chitinous. Resting egg embryogenesis, as well as postdormancy development in the diapaused embryo, is poorly known and in need of considerable research. While resting eggs have been reported to remain viable in sediments for more than 30 years, they are not as resistant to freezing or drying as are anhydrobiotic bdelloids. The timing of resting egg hatching is highly variable, and it seems that for both natural populations and cultures these eggs hatch at irregular intervals. Genotypic variability of these bisexual eggs may account for this variability. Experimental work indicates that resting egg hatching is influenced by a variety of factors, including temperature, light, salinity, and oxygen concentration. Hatching also is stimulated when resting eggs are exposed to ultraviolet (UV) light, dilute solutions of hydrogen peroxide, and certain prostaglandins. It has been hypothesized that active oxygen, either generated by UV photolysis or from the peroxide itself, oxidizes fatty acids within the dormant embryo. These fatty acids induce production of prostaglandins, which then initiate hatching.

D. Phenotypic Variation

Phenotypic plasticity is very common in rotifers. The environment may influence the phenotype of offspring in rather important ways, including changes in spine length (*Brachionus* and *Keratella*), body size (*Asplanchna*), and disproportionate growth of other regions of the body (*Asplanchna*). Each phenotypic variation is termed a morphotype. Environmental factors such as temperature, chemicals released by predators, and maternal diet are responsible for inducing these variations.

VIII. LARVAE AND METAMORPHOSIS

With few exceptions, newborn rotifers possess all the adult features and, except for their smaller size, closely resemble their mothers. Thus, there is no true larval stage. However, there are cases in which the newborns look different from their mothers. Individuals of the planktonic species *Polyarthra* that have recently hatched from resting eggs lack paddle-like appendages (aptera generation) normally present in the adult. However, these paddles, which are used to avoid predators by permitting rapid skipping movements through the water, are regained in the next generation (typica generation). The juvenile motile stages of sessile rotifers have a very different body form from their mothers, so these families do appear to possess a larval stage (Monogononta: Collothecidae and Flosculariidae). While these young are not larvae in the classical meaning of the term, they undergo rather dramatic changes in behavior upon encountering a potential substratum and undergo a type of metamorphosis after settlement. Therefore, because of the conceptual parallel to marine sessile invertebrates, workers usually refer to these stages as larvae. Metamorphosis of the larva involves degenerative changes (loss of eyespots and cilia), extensive morphological changes (changes in body:foot ratios and expansion of the corona), and production of a gelatinous matrix (tube).

See Also the Following Article

ACANTHOCEPHALA

Bibliography

Clément, P., and Wurdak, E. (1991). Rotifera. In *Microscopic Anatomy of Invertebrates, Volume 4: Aschelminthes* (F. W. Harrison and E. E. Ruppert, Ed.), pp. 219–297. Wiley-Liss, New York.

Edmondson, W. T. (1959). Rotifera. In *Fresh-Water Biology* (W. T. Edmondson, Ed.), 2nd ed., pp. 420–494. Wiley, New York.

Gilbert, J. J. (1974). Dormancy in rotifers. *Trans. Am. Microsc. Soc.* **93**, 490–513.

Gilbert, J. J. (1983a). Rotifera. In *Reproductive Biology of Invertebrates, Volume I: Oogenesis, Oviposition, and Oosorption* (K. G. Adiyodi and R. G. Adiyodi, Eds.), pp. 181–209. Wiley, New York.

Gilbert, J. J. (1983b). Rotifera. In *Reproductive Biology of Invertebrates, Volume II: Spermatogenesis and Sperm Function* (K. G. Adiyodi and R. G. Adiyodi, Eds.), pp. 181–193. Wiley, New York.

Gilbert, J. J. (1988). Rotifera. In *Reproductive Biology of Invertebrates, Volume III: Accessory Sex Glands* (K. G. Adiyodi and R. G. Adiyodi, Eds.), pp. 73–80. Oxford & IBH, New Delhi.

Gilbert, J. J. (1989). Rotifera. In *Reproductive Biology of Invertebrates, Volume IV, Part A: Fertilization, Development, and Parental Care* (K. G. Adiyodi and R. G. Adiyodi, Eds.), pp. 179–199. Oxford & IBH, New Delhi.

Gilbert, J. J. (1992). Rotifera. In *Reproductive Biology of Invertebrates, Volume V: Sexual Differentiation and Behaviour* (K. G. Adiyodi and R. G. Adiyodi, Eds.), pp. 115–136. Oxford & IBH, New Delhi.

Gilbert, J. J. (1993). Rotifera. In *Reproductive Biology of Invertebrates, Volume VI, Part A: Asexual Propagation and Reproductive Strategies* (K. G. Adiyodi and R. G. Adiyodi, Eds.), pp. 231–263. Oxford & IBH, New Delhi.

Gilbert, J. J. (1995). Structure, development and induction of a new diapause stage in rotifers. *Freshwater Biol.* **34**, 263–270.

Gilbert, J. J., and Williamson, C. E. (1983). Sexual dimorphism in zooplankton (Copepoda, Cladocera, and Rotifera). *Annu. Rev. Ecol. Syst.* **14**, 1–33.

Nogrady, T., Wallace, R. L., and Snell, T. W. (1993). *Rotifera, Volume 1: Biology, Ecology and Systematics. Guides to the Identification of the Microinvertebrates of the Continental Waters of the World* (H. J. Dumont, Ed.). SPB Academic, The Hague. [This volume is an introduction to a multivolume work that upon completion will provide a complete taxonomic treatise of the Rotifera]

Pennak, R. W. (1989). *Fresh-Water Invertebrates of the United States*, 3rd ed. Wiley, New York.

Pourriot, R. (1979). Rotifères du sol. *Rev. d'Ecol. Biol. Sol.* **16**, 279–312.

Ricci, C. (1992). Rotifera: Parthenogenesis and heterogony. In *Selected Symposia and Monographs, Volume 6: Sex Origin and Evolution* (R. Dallai, Ed.), pp. 329–341. Unione Zoologica Italiana, Mucchi, Modena, Italy.

Ricci, C., Melone, G., and Sotgia, C. (1993). Old and new data on Seisonidea (Rotifera). *Hydrobiologia* **255/256**, 495–511.

Starkweather, P. L. (1987). Rotifera. In *Animal Energetics, Volume 1: Protozoa through Insects* (T. J. Pandian and F. J. Vernberg, Eds.), pp. 159–183. Academic Press, New York.

Thane, A. (1974). Rotifera. In *Reproduction of Marine Invertebrates, Volume I: Acoelomate and Pseudocoelomate Metazoans* (A. C. Giese and J. S. Pearse, Eds.), pp. 471–484. Academic Press, New York.

Wallace, R. L., and Snell, T. W. (1991). Rotifera. In *Ecology and Classification of North American Freshwater Invertebrates* (J. Thorp and A. Covich, Eds.), pp. 187–248. Academic Press, New York.

RU-486

see Antiprogestins

Ruminants

William W. Thatcher
University of Florida

GLOSSARY

bipartite uterus A type of uterus having a large uterine body and relatively short uterine horns found in ruminant species.

buck An intact caprine male.

bull An intact bovine male.

caruncle A convex connective tissue (stromal) structure of the endometrium that is covered by a single layer of luminal epithelium and characteristic of the uterus of ruminants.

cotyledon A differentiated region of the chorioallantoic placenta in ruminants that interdigitates with a caruncle to form a specialized structure for transport of selected nutrients and gases between maternal and fetal circulations.

cow A bovine female that has produced a calf.

doe A caprine female of any age and reproductive experience.

ewe An ovine female of any age or reproductive experience.

heifer A bovine female that has not produced a calf.

placentome The functional caruncle–cotyledon of ruminant placentae.

ram An intact ovine male.

steer A castrate bovine male.

wether A castrate ovine male.

Ruminants are vital to the human race for the conversion of plants to food (e.g., meat and milk) and also as a source of work power in many cultures. Indeed, the dairy cow has been referred to as the foster mother of the human race because of its production of milk for human consumption. As developing countries increase their gross national income, consumer preference and demand for milk and meat products increases. Ruminants have developed a diversity of strategies to reproduce efficiently under a variety of environmental conditions. Sheep are seasonal breeders that are photosensitive and are referred to as short-day breeders. With their normal gestation length of 147 days, birth of the newborn occurs in the spring when the probability of survival is greatest. Although cattle have the potential to cycle throughout the year, the reproductive system is sensitive to environmental inputs associated with nutrition and heat stress that influence reproductive rates. With limited nutrient quality and intake, long periods of acyclicity (anestrus) and enhanced rates of early embryonic death (heat stress) will occur.

I. GENDER DEVELOPMENT AND PUBERTY

The ovary and testis are the primary reproductive organs because they produce the female (ovum) and male (sperm) gametes and sex hormones critical for sexual behavior, provide for induction of gamete development, and are responsible for production of hormones regulating reoccurring estrous cycles and maintenance of pregnancy. Primordial germ cells of the urogenital ridge, derived from endoderm of the yolk sac, are present in gonads as early as 26 (cattle) and 29 (sheep) days after conception. The embryonic origin of the testes is the primary sex cords, which develop into the medulla of the urogenital ridge to form the germinal epithelium of the seminiferous tubules. Differentiation of the gonads to testes becomes clear at 34 or 35 days with the appearance of the tunica albuginea. Ovaries develop in the cortex from the secondary sex cords of the genital ridge and the primitive germ cells become oogonia, which undergo mitotic divisions to form oocytes in the embryonic ovary. The primary oocytes begin meiosis in the embryo, and at birth most have reached the diplotene stage within prophase I of meiosis which is the first of four meiotic stages. At this stage, they are identified as oocytes. Initially, the germinal ridges are undifferentiated or bisexual in that differentiation can be into the male or the female reproductive system. At fertilization sex is determined as XX or XY chromosomal constitutions and is maintained by mitotic division in the primary germ cells and all cell types of the embryo. If the sex chromosome genotype is XX, then the ovary will develop, but if a Y chromosome is present the gonad will develop into a testis. Clearly, the Y chromosome contains genes involved in initiating testicular differentiation. Once the undifferentiated gonad develops into a testis, differentiation of both internal and external reproductive tissues follows. The testis produces Müllerian-inhibiting factor (MIF) and testosterone which regulate the formation of male structures. The Sertoli cells produce MIF that causes regression of the Müllerian duct system. Testosterone, produced by the Leydig cells, stimulates the Wolffian duct system that becomes the epididymis and vas deferens. The conversion of testosterone to dihydrotestosterone (DHT) allows DHT to induce formation of the prostate, penis, prepuce, and scrotum. Descent of the testes into the scrotum finalizes development of the external genitalia. In the absence of a testis and these regulatory agents, the Müllerian duct system persists, the Wolffian duct system regresses, and the genitalia retain their female characteristics leading to development of oviducts, uterus, vagina, vulva, and clitoris. Primary ovarian follicles are formed during the prenatal period of the female (e.g., 110–130 days of gestation for bovine fetuses) in which oocytes become surrounded by a squamous follicular epithelium of somatic cells (granulosa cells) and a basement membrane. The number of primordial follicles in the bovine ovary at puberty averages 133,000 and declines with age in association with subsequent follicular developmental processes, atresia, and ovulation. Wide variations in number of follicles within the primordial pool exist among animals.

Puberty is defined in the female as the age at first expressed estrus with an ovulation. In the male, definition of puberty is less specific and defined as the age at which the sexual organs are developed functionally, sexual behavior is evident, and reproduction is possible. A more functional definition has been used for bulls, in which puberty is evident upon collection of the first ejaculate with a minimum of 50 million sperm per milliliter that have at least 10% progressive motility. Age of puberty is 10–12 months for bulls, 3–5 months for bucks, and 4–6 months for rams. However, numerous genetic (sex and breed) and environmental (nutritional status, social interactions, temperature, and photoperiod) factors influence onset of puberty. For both male and female, puberty occurs when gonadotrophins [follicle-stimulating hormone (FSH) and luteinizing hormone (LH)] are produced in sufficient amounts to initiate gamete production. The control of gonadotropin secretion is markedly different between the sexes. In the bull dynamic and continuous endocrine changes occur early after birth. Frequent discharges of LH occur at approximately 2 months of age. Induced LH surges do not stimulate increased serum testosterone secretion until about 6 months of age, which indicates that Leydig cells may become more sensitive to LH as puberty approaches. Concentrations of FSH do not change appreciably with age. The concurrent increase in LH and testosterone as puberty approaches in males (bulls, rams, and bucks) supports the concept that sexual maturation may occur through a desensitization of the hypothalamic–hypophyseal complex to gonadal steroids. In heifers a similar desensitization occurs but with a different control system. Ovariectomy of prepubertal heifers results in increased frequency of LH pulses which can be blocked by administration of estradiol. Frequency of LH pulses increases progressively during the 50 days approaching puberty. Associated with this increase was a decrease in estradiol receptors in the medial basal hypothalamus. This suggests that increased LH secretion is due to a desensitization of the hypothalamic–hypophyseal complex to negative feedback of estradiol. Such a system has been well characterized in sheep. In cattle, reoccurring patterns of dominant follicle development and atresia occur prior to puberty. However, production of estradiol by these follicles under a low-progesterone environment will not elicit a preovulatory surge of LH until desensitization to negative estradiol feedback diminishes. When that happens, estradiol from a developing dominant follicle exerts a positive feedback effect that induces a LH surge and ovulation approximately 28–30 hr later. Exposure to progesterone may accelerate the process of desensitization to negative estradiol feed back. Average age of puberty for females of various species is 11–13 months for dairy cattle, 10–15 months for *Bos taurus* and 17–27 months for *Bos indicus* beef cattle, 6–9 months for ewes, and 5–7 months for does. Onset of puberty is not indicative of sexual maturation. For example, animals conceiving to first estrus may have considerable problems with parturition. Puberty in sheep occurs when they reach 40–50% of their mature weight but breeding is not recommended until they reach 65% of their mature weight. A similar analogy is recommended for dairy cattle, which reach puberty at 35–45% of their mature weight but breeding is not suggested until they reach 55% of their mature weight. Feeding animals below a recommended plane of nutrition will delay puberty. Under various conditions that differentially regulate time of puberty (e.g., diminished or enhanced planes of nutrition) body weight at puberty is not changed markedly, although age to obtain this weight is altered considerably. Introduction of a mature male during the prepubertal period will induce an earlier occurrence of puberty in ewes, does, and cattle.

II. ESTROUS CYCLE

Sheep, cattle, and goats have estrous cycles of about 17, 21, and 20 days, respectively. Variation in length of estrous cycles occurs in each of these species. For example, a normal range in estrous cycle length for cattle is 17–24 days. Periods of the estrous cycle are estrus, metestrus, diestrus, and proestrus. These occur sequentially and in a cyclic manner. Estrus is considered to be the period of time when the female is receptive to the male and will stand for mating (12–18 hr in cattle, 24–36 hr in sheep, and 30–40 hr in does). An ovulatory surge of LH occurs during estrus (Day 0) and initiates events which culminate in ovulation about 30 hr later. In cattle estrus terminates before ovulation, which is conducive to the use of artificial insemination in which semen can be deposited into the anterior os of the cervix before ovulation to ensure good conception rates. The period of metestrus begins with the end of estrus and lasts for approximately 3 days. It is the period when the corpus luteum (or corpora lutea with multiple ovulations) forms. Ovulation actually occurs during metestrus in ewes and cattle. With ovulation of the estrogenic preovulatory follicle and withdrawal from estradiol, there is increased uterine vascularity and metestrus bleeding which is much more prominent in heifers than in mature cows. Metestrus bleeding is due to capillary breakage and occurs at approximately 1.5–2 days after end of estrus. Diestrus is the period during the estrous cycle when the corpus luteum is fully functional beginning on Days 5, 4, and 4 for cows (lasting 10–14 days), ewes (lasting 10–12 days), and does (lasting 13–15 days), respectively. The period of proestrus begins with regression of the corpus luteum and continues to the onset of estrus. During the proestrus period, continued growth and development of the preovulatory follicle occurs with production of estradiol that will induce estrous behavior. Proestrus lasts 3 or 4 days in cattle and 2 or 3 days in sheep and goats. Concentrations of progesterone in milk or plasma, produced by the corpus luteum, are maximum in middiestrus. Utilizing transrectal ultrasonography, follicular dynamics (continual process of growth and regression of ovarian follicles) during the estrous cycle of cattle have been described with measurement of follicle size (>2 mm) and numbers on a within-animal basis. There are two to four follicular waves during each estrous cycle in cattle. Each wave of follicle development is characterized by recruitment of a cohort of follicles that are 6–9 mm in diameter. One follicle within the cohort is selected and becomes the largest follicle of the cohort. When this follicle is 2 mm larger than the second largest cohort, the selected follicle is considered dominant and continues to grow in a linear manner before entering a plateau phase. Dominance is extended for a period until recruitment of a new cohort is inhibited. With atresia of the dominant follicle, a second follicle wave is initiated that undergoes the sequential processes of recruitment, selection, and dominance. Preceding each follicle wave there is a temporal rise in plasma FSH that induces follicular recruitment from the pool of follicles that are 2–5 mm in size. The process of dominant follicle selection is not totally understood but likely is associated with the follicle that has the greatest sensitivity to FSH (e.g., greater number of FSH receptors) and can survive in an environment of decreasing FSH induced by inhibin production from the dominant follicle. The reduction in FSH prevents recruitment of a new follicle wave. The selected dominant follicle continues to grow but becomes atretic because LH secretion is not sufficient to maintain continued growth in the presence of a corpus luteum and a high-progesterone environment. With regression of the corpus luteum, a dominant follicle of either the second or third follicle wave will continue its growth due to enhanced basal secretion of LH and will secrete estradiol. High concentrations of estradiol and low progesterone induce a preovulatory surge of LH which induces resumption of meiosis in the oocyte and subsequent ovulation. These dynamic changes in follicle development occur in both *B. taurus* and *B. indicus* species of cattle. Follicle dynamics have been characterized in sheep, but they do not have the distinct wave-like patterns found in cattle. This may be related to the occurrence of a higher ovulation rate in sheep in which strong follicular dominance mechanism would not be conducive to the development of multiple ovulatory follicles.

Following ovulation, thecal and granulosa cells differentiate into small and large steroidogenic cells, respectively. Furthermore, some small cells appear to differentiate into large luteal cells during the estrous cycle. LH appears to be the luteotrophic hormone in sheep and cattle, with the small luteal cells containing LH receptors and increasing progesterone secretion in response to LH. Luteolysis is induced by pulsatile release of $PGF_{2\alpha}$ from endometrial epithelium during late diestrus. Uterine secretion of luteolytic pulses of $PGF_{2\alpha}$ in sheep is dependent on effects of progesterone, estrogen, and oxytocin on the uterine luminal epithelium. Progesterone acts to increase phospholipid stores (arachidonic acid source) and cyclooxygenase enzymatic activity necessary for conversion of arachidonic acid to $PGF_{2\alpha}$. Oxytocin secreted by the corpus luteum and posterior pituitary acts through oxytocin receptors present on endometrial epithelium at specific stages of the estrous cycle to stimulate release of luteolytic pulses of $PGF_{2\alpha}$. Progesterone ensures the potential for uterine release of luteolytic $PGF_{2\alpha}$. However, progesterone acting through the progesterone receptor suppresses the luteolytic mechanism by inhibiting induction of estradiol and oxytocin receptors. Eventually, progesterone downregulates the progesterone receptor which permits induction of estradiol and oxytocin receptors. Follicular estradiol secretion upregulates oxytocin receptors. Indeed, cauterization of ovarian follicles delays corpus luteum regression. During the luteolytic period, many of the luteolytic $PGF_{2\alpha}$ pulses are coincident with pulses of oxytocin. However, oxytocin pulses can occur without an associated pulse of $PGF_{2\alpha}$, and conversely $PGF_{2\alpha}$ pulses can occur without an oxytocin pulse. Nevertheless, $PGF_{2\alpha}$ secreted by the uterus passes via a local countercurrent exchange system between the uteroovarian vein and ovarian artery to induce regression of the corpus luteum. This system of luteolytic control appears to be present in ewes, does, and cows.

III. PREGNANCY

A. Development of the Embryo

After fertilization, embryos remain near or at the ampullary–isthmic junction of the oviduct before entering the uterus at 72 hr in ewes or 72–96 hr in cows. Sheep embryos fail to develop beyond the early blastocyst stage if confined to the oviduct due to the absence of factors essential for further development or the presence of embryotoxic factors. Transuterine migration of blastocysts between the uterine horns is rare in monovulatory ewes and cows. However, transuterine migration occurs in sheep, but not cows, following multiple ovulations from the same ovary. Elongation of sheep, goat, and cow conceptuses represents the first step toward implantation. Conceptus development in water buffalo is similar to that for the cow but slightly faster to Day 8. Details of conceptus development in the family Camelidae, which includes camels, llama, and alpaca, are not available. Females in those species are induced ovulators and ovulation can be induced in bactrian camels by intravaginal deposition of seminal plasma. Ovulation occurs about 24–26 hr postcoitum in llama and alpaca and 30–48 hr postcoitum in camels. In camels, ovulation from the right and left ovaries occurs with equal frequency; however, 98–100% of the conceptuses develop in the left uterine horn. The right uterine horn seems incompetent to support conceptus development beyond about 50 days. Embryonic death losses are also very high in alpaca and llama and the left uterine horn contains the conceptus in almost 100% of the cases.

Sheep blastocysts are spherical on Days 4 (0.14 mm) and 10 (0.4 mm) and elongate to the filamentous form by Days 12 (1.0 × 33 mm), 14 (1 × 68 mm), and 15 (1 × 150–190 mm). Elongation continues until the trophoblast extends through the uterine body and into the contralateral uterine horn by Days 16 or 17 of pregnancy. Cow blastocysts are spherical on Days 8 or 9 (0.17 mm diameter), oblong or tubular on Days 12 or 13 (1.5 × 3.0 mm diameter), and oblong to filamentous on Days 13 or 14 (1.4 × 10 mm), 14 or 15 (2 × 18 mm), 16 or 17 (1.8 × 50 mm), and 17 or 18 (1.5 × 160 mm). By Days 17 or 18 the bovine conceptus occupies about two-thirds of the gravid uterine horn, the entire gravid uterine horn by Days 18–20, and extends into the contralateral uterine horn by Day 24.

Conceptus development during the preimplantation period for goats, buffalo, and Camelidae is similar to that for sheep and cows since all these species have a central type of implantation with placentae of the diffuse and cotyledonary type. Elongation of bovine and ovine conceptuses is a prerequisite for the central type of implantation which involves trophoblast primarily; the embryonic disc is not required in sheep. Implantation in sheep, cow, buffalo, goat, and Camelidae is superficial and is of the noninvasive central type with increasing trophectoderm–uterine epithelial cell apposition and adhesion with no permanent erosion of uterine epithelium. Ruminant endometrium is composed of both caruncular and intercaruncular tissue and the process of implantation involves both areas. Trophoblast cells of sheep remain flat until Day 12 before assuming a columnar shape by Day 14 and becoming more secretory. Microvilli are abundant over the surface of the trophoblast prior to Day 14 but become shorter and less dense after Day 14 and finally disappear as apposition with the uterine epithelium ensues. Microvilli on the surface of endometrial epithelium are in contact with the smooth plasma membranes of trophoblast by Day 15 and penetrate into folds in trophectoderm which leads to close adhesion between these tissues. Trophectoderm cells develop villous projections by Day 13 which project into the lumina of uterine glands during the third week of gestation. These microvilli immobilize the conceptus and facilitate uptake of secretions from uterine epithelium. Similar modifications of the bovine trophoblast occur but slightly later. Adhesion between trophectoderm and caruncular epithelium occurs by Day 16, but tenuous attachment between trophectoderm and uterine epithelium does not occur until Days 18 or 19.

B. Maintenance of Pregnancy

Maintenance of the corpus luteum and pregnancy is induced by an antiluteolytic signal in ruminants that has been identified as interferon-τ (IFN-τ) produced by trophectoderm. IFN-τ exerts a paracrine, antiluteolytic effect on the endometrium to inhibit endometrial production of luteolytic $PGF_{2\alpha}$ pulses. Other conceptus and/or uterine products secreted during pregnancy [e.g., PGE_2 and platelet-activating factor (PAF)] may exert secondary luteal protective effects contributing to maintain the functional corpus luteum. Mechanisms for conceptus-induced maintenance of the corpus luteum are similar for sheep, cattle, and goats. The presence of the conceptus in the uterus prevents luteolysis by secreting IFN-τ from the trophectoderm that attenuates the uterine secretion of luteolytic pulses of $PGF_{2\alpha}$. The IFN-τ acts through its receptor to (i) directly inhibit expression of estradiol receptors which reduces the induction of oxytocin receptors and thereby attenuates the mechanisms for oxytocin to induce the episodic release of luteolytic $PGF_{2\alpha}$, (ii) directly inhibit synthesis of endometrial OTR, and/or (iii) inhibit biosynthetic components of the cascade leading to the mobilization and metabolism of arachidonic acid to $PGF_{2\alpha}$. Available results indicate that luminal epithelium of the endometrium of pregnant ewes, cows, and goats have few or no receptors for estradiol and oxytocin. The degree to which $PGF_{2\alpha}$ secretion is attentuated by the conceptus and/or IFN-τ may vary between sheep, cattle, and goats, but the pulsatile release of $PGF_{2\alpha}$ required for luteolysis is absent. It seems logical that variation in conceptus production of IFN-τ would account for some of the embryonic losses associated with an inability to maintain the corpus luteum. However, conclusive studies to improve embryo survival with IFN-τ supplementation in early pregnancy have not been performed. Cytokines produced by the uterus in response to IFN-τ also may regulate pregnancy recognition signaling and conceptus development. Binucleate cells appear in trophoblast of sheep and cow conceptuses by Days 16 and 19, respectively, and migrate into the uterine epithelium soon thereafter. Multinucleate giant cells are also present and form a syncytium as they become stretched into a sheet of tissue. Binucleate cell migration occurs throughout pregnancy, but multinucleate cell formation ceases by about Day 28 in sheep. Binucleate cell formation, the presence of multinucleate giant cells, and migration of binucleate cells are associated with degeneration of endometrial surface epithelium. This is followed by removal of endometrial epithelium by phagocytosis and restoration of endometrial surface epithelium. Functions of binucleate cells include (i) removal and replacement of the endometrial surface epithelium, (ii) establishment of a syncytium between maternal and conceptus compartments to provide immunological protection of the conceptus, (iii) production of chorionic hormones, and (iv) production and transport of macromolecules (e.g., pregnancy-specific protein B).

C. Placentation

The extraembryonic mesoderm originates from the embryonic disc and migrates between the trophectoderm and endoderm. This mesodermal layer divides and combines with trophectoderm to form the chorion and with endoderm to form the yolk sac. The mesoderm also contributes to formation of the amnion and allantois. These membranes form the placenta to allow for histotrophic and hematotrophic nutrition of the embryo/fetus. The cotyledonary placenta is found in most ruminants. Chorionic villi are restricted to very distinct oval areas of the chorion (cotyledons) which overlie caurncles of the uterus. The fetal cotyledons and uterine caruncles form a unit known as the placentome. The number of caruncles and therefore potential placentomes per uterus range from 70 to 142 in cattle, 60 to 150 in sheep, 160 to 180 in goats, and 75 to 170 in buffalo. Ruminants have an epitheliochorial placenta. In the cow, there is a clear interdigitation of chorionic and maternal epithelial villi, and on the fetal side there is a distinct endothelium, basement membranes of endothelium and chorion, and cells of the chorion. On the maternal side there is the maternal epithelium, basement membrane of the epithelium, collagen processes, basement membrane, and endothelium of the maternal capillaries. Variation in placental architecture occurs such that in sheep, in which fetal binucleate cells are evident, the uterine epithelium forms a partial syncytium and appears to lack a distinct basement membrane. Development of the placenta brings embryonic and maternal capillaries into close proximity to facilitate exchange of nutrients and blood gases (O_2) from mother to conceptus and waste products in the reverse direction. Intimate contact also occurs between the allantochorion and uterine epithelium between the caruncles. Only in this area are openings of uterine glands found which provide various protein secretions that are taken up by areolae on the surface of the allantochorionic membranes. In the first half of pregnancy there is a rapid growth of placental membranes and accumulation of allantoic and amniotic fluids. The presence of the fluids allows for pregnancy diagnosis by such techniques as rectal palpation or ultrasound scanning. During the second half of pregnancy, growth of the fetus dominates with the greatest increase in fetal body weight occurring during the last month. Progesterone is produced throughout pregnancy primarily by the corpus luteum (e.g., cattle and goats) and/or the corpus luteum and placenta (e.g., sheep). Placental steroidogenic activity, as determined by measurements of conjugated estrogens (e.g., estrone sulfate), increases throughout pregnancy in conjunction with fetoplacental growth, development, and well-being. Other placental hormones (e.g., placental lactogen) are secreted and are indicative of placental function. Collectively, these hormones are important for maintenance of pregnancy, partitioning of nutrients from the mother to support fetal growth, and coordinated development of the maternal mammary gland to support continued nutrition and health of the newborn.

D. Parturition

The process of parturition marks the termination of pregnancy, at which time the fetus is capable of being maintained outside of the uterus. Initiation of parturition centers around activation of the fetal hypothalamic–pituitary–adrenal axis. With approaching parturition, the fetal adrenal cortex becomes increasingly sensitive to ACTH. The rise in fetal corticoids activates appropriate enzymes of the placentome to convert progestins to androgens and subsequently estrogens. A gradual decline in maternal plasma concentrations of progesterone is associated with a concurrent increase in free estradiol. Associated with these alterations in endocrine state are increases in uterine $PGF_{2\alpha}$ and PGE_2 secretion that lead to regression of the corpus luteum. As a result of alterations in steroid secretion, increased uterine sensitivity to oxytocin, and production of $PGF_{2\alpha}$, myometrial contractility becomes more coordinated and intensifies as parturition approaches. Uterine contractions in combination with cervical relaxation advance the fetus into the cervix and anterior vagina, which also stimulates the reflex release of oxytocin. This increases myometrial contractions and leads to expulsion of the fetus. The parturition process involves more than just uterine contractions. Softening of collagen in the cervix, uterus, vagina, and pelvic ligaments must occur to minimize resistance during expulsive force elicited by coordinated uterine and abdominal contractions. The hormone relaxin, produced by the ovary, is critical to this process. Signs of impending parturition are increased udder development with secretions taking on a yellowish-opaque appearance indicative of colostrum. Edema of the udder is also evident. Marked relaxation of the ligaments associated with the pelvis are also symptomatic of impending parturition. Signs of discomfort and restlessness do not appear until noticeable dilation of the cervix has occurred. Passage of the chorioallantois membranes and fluids to the vulva causes further dilation of the cervix. Rupture of the membranes and release of fetal fluids provides lubrication of the birth canal. Generally, the three stages of parturition or labor encompass (i) onset of uterine contractions until the cervix is dilated fully (12 hr for sheep and considerably less than 12 hr in cattle) , (ii) expulsion of the fetus (0.5 hr for singles and longer for twins in sheep; <1 hr in cattle), and (iii) expulsion of the placenta (2–4 hr in sheep and <16 hr in cattle). Retention of the placentae beyond 6 hr in cattle is considered abnormal and is an important disorder that can be influenced by diets affecting calcium mobilization in the prepartum period and certain deficiencies such as selenium.

IV. POSTPARTUM PERIOD

In the general sense, the postpartum period begins with parturition and ends with complete involution of the uterus. Uterine regression has been characterized most extensively in cattle. In the cow at parturition, when the cotyledonary villi pull out of the caruncular mass, they leave a mass of tissue that is to large to be contracted back into the endometrium but is not massive enough to tear off at parturition.

The large blood vessels in the endometrium constrict at the base of the caruncle and close off the circulation to the septa. This results in cellular necrosis and sluffing of all of the maternal cells that have been involved actively in the placentome. The septal caruncular mass is lost at 5–10 days after parturition, whereas in the true deciduous-type placentae the maternal tissue involved in the placenta is lost acutely after passage of the fetus. This is why the bovine placenta is classified generally as cotyledonary, epitheliochorial, and delayed deciduate. Immediately following delivery of the fetoplacental unit, the large postpartum uterus undergoes dramatic morphological and histological changes that lead to reestablishment of a uterine environment conducive to initiating a pregnancy. Uterine involution is composed of three overlapping processes: reduction in size, loss of tissue, and repair of tissue. Early reduction of uterine size results from vasoconstriction and peristaltic contractions that persist for several days. Uterine lochial fluid is composed of both normal and degenerated cells from the tunica mucosa and caruncles and from bleeding occurring during shedding of the fetal membranes. The maximum amount of uterine lochial fluid is present during the first 48 hr and then decreases between Days 10 and 20 and is absent by approximately 24 days postpartum. Tissue loss and repair occur simultaneously in the early postpartum period, but gradually the process of tissue repair becomes predominant and the uterine mucosa is restored. Regeneration of epithelial cells on the caruncular surface begins as early as Day 12 postpartum and the caruncular surface is covered with an epithelial layer on Day 30. At the end of the repair process (e.g., 40 days), caruncles regressed to smooth, oblong, epithelial-covered avascular knobs (4–8 mm in diameter) and appear as rows of white discs in a pink and glandular endometrium. The myometrium also undergoes its most dramatic changes during the first 20–30 days postpartum. The muscle layers (circular and longitudinal) decreased in mass by reduction in both cell size and number. Muscle fibers appear normal and the entire myometrium is greatly reduced in size by Day 30. Suckling appears to enhance the rate of involution of the three major uterine tissue layers (endometrium and both circular and longitudinal myometrial muscle fibers).

Since pituitary content of FSH is not suppressed after parturition, growth of ovarian follicles begins as early as 7 days postpartum. LH secretion is not restored until after 14 days postpartum, and first ovulation will occur in many cows between 11 and 15 days after parturition. Over 90% of these first ovulations occur on the ovary opposite the previous pregnant uterine horn, and the local inhibitory influence of the previous pregnant uterine horn is no longer evident by 30 days postpartum. If ovulation and corpus luteum development occur early in the postpartum period, life span of the first corpus luteum is usually reduced resulting in a short cycle. Cattle that have an array of parturition problems and metabolic disturbances in the early postpartum period (dystocia, prolapsed uterus, retained fetal membranes, hypocalcemia, ketosis, and metritis) will tend to have a delay in uterine involution and be less fertile. Lactating dairy cows that are in marked negative energy balance (their energy output associated with maintenance and milk production is greater than their energy intake) may have a prolonged period before resumption of ovarian cycles. Postpartum beef cows that are suckling calves have a long postpartum anestrous period (no ovarian activity). Resumption of ovarian cycles can be restored partially with temporary weaning of calves. Indeed, if appropriate estrous synchronization schemes involving the use of progesterone or progestins (synthetic progesterone-like molecules) are coupled with temporary weaning, then estrous cycles that are fertile can be induced with some degree of success. A better body condition at calving with minimal losses in body weight and body condition of cows during lactation in dairy and beef cows is associated with an earlier return to cyclic activity and better fertility.

V. CONTROL OF REPRODUCTIVE CYCLES

Synchronization of estrus in ruminants will aid in reproductive management. Large numbers of animals in estrus at the same time will reduce labor associated with detection of estrus. The breeding season can be shortened because more animals become pregnant during the early part of the breeding season. Animals that have synchronized estrus and are inseminated will have grouped patterns of parturition that will allow better management and health care at parturition. Furthermore, synchronization of estrus and inseminations will better optimize timing of births for the most favorable conditions associated with nutrient availability for survival of mother and newborn or conditions associated with optimal markets. In cyclic ruminants, the occurrence of estrus is controlled by secretion of progesterone from the corpus luteum. Progesterone inhibits the preovulatory surge of LH such that final development, maturation, and ovulation of the preovulatory follicle will not occur until progesterone declines in association with regression of the corpus luteum. There are two strategies to control life span of the corpus luteum that will permit synchronization of estrus. The first strategy utilizes long-term administration of a progestogen so that the corpus luteum will regress spontaneously during progestogen exposure and estrus will occur in a synchronized manner following withdraw of the progestogen. This strategy has been very effective in sheep in which a progestogen is administered orally, injected, implanted subcutaneously, or released from an intravaginal device that is impregnated with progestogen. Usually at the time of progestogen withdrawal, equine chorionic gonadotropin (eCG) is given to stimulate follicle development so that this system of estrous synchronization will be effective in cycling and anestrous ewes. In goats, a similar scheme is effective with the insertion of intravaginal progestogen sponges that are placed in the vagina for 17–22 days with eCG given on removal of the sponges. Long-term progestogen treatment (e.g., 14 days) in cattle results in good synchronization of estrus but reduced fertility at the synchronized estrus. This is attributed to induction of a persistent large follicle that undergoes premature maturation.

A second alternative for controlling life span of the corpus luteum is to administer $PGF_{2\alpha}$, the natural luteolytic agent, to induce regression of the corpus luteum. Following injection of $PGF_{2\alpha}$, corpus luteum regression occurs within 24–48 hr, followed by estrus and ovulation within 2 or 3 days. A limitation of using injections of $PGF_{2\alpha}$ is that the corpus luteum is responsive to $PGF_{2\alpha}$ only during certain developmental stages. For example, in cattle, $PGF_{2\alpha}$ is not effective during the first 5 days of the cycle. This limitation can be avoided by inseminating all cows detected in heat during 5 or 6 days and injecting the remaining animals with $PGF_{2\alpha}$ on Days 6 or 7 and inseminating them at detected estrus. A second alternative is to inject $PGF_{2\alpha}$ twice, 11 days apart for heifers or 14 days apart for cows, and inseminate all animals detected in estrus following the second injection of $PGF_{2\alpha}$. This system hypothetically results in all animals having a corpus luteum that is responsive to $PGF_{2\alpha}$ at the second injection.

An additional refinement in synchronization schemes has been the combination of a progestogen for a short period of time (e.g., 7 days) combined with an injection of $PGF_{2\alpha}$ 1 day prior to removal of the progestogen. Such a scheme results in a synchronized estrus at approximately 64–72 hr after progestogen withdrawal. A commercial scheme approved for beef cattle involves a combination of estrogen and short-term progestogen treatment. On the first day of treatment, an injection of estradiol valerate and the progestogen (Norgestomet) is given, and a Norgestomet implant is inserted and left in place for a 9-day period. Estrus is detected on the second or third day after progestogen withdrawal.

Further refinements in systems for estrous synchronization have been developed with the realization that follicle waves normally occur during the estrous cycle of cows. Thus, it is critical to synchronize follicle wave development with corpus luteum regression. This has been achieved in two ways: (i) injection of gonadotropin-releasing hormone (GnRH) followed 7 days later with injection of $PGF_{2\alpha}$ and inseminations are made at detected estrus, and (ii) injection of estradiol benzoate at the time of progestogen insertion and removal of the progestogen 9 days later. Both these systems induce turnover of a dominant follicle and recruitment of a new follicle that will induce estrus following either induction of corpus luteum regression with $PGF_{2\alpha}$ or removal of the progestogen. Currently, major efforts are under way to further refine the systems to precisely control the time of ovulation by inducing a preovulatory surge of LH. This can be done by injection of GnRH (e.g., at 48 hr after an injection of $PGF_{2\alpha}$) or injection of estradiol (e.g., at 24 hr) after removal of a progestogen. Such systems allow for a timed insemination without the need for estrus detection.

VI. SUMMARY

Ruminants will continue to be a valuable source of food (milk and meat) and work power in the world. Our current understanding of reproductive physiology and endocrinology in ruminants has advanced to the point that major strides have been made in maximizing reproductive performance utilizing systems of reproductive management. As we approach the twenty-first century, we can look forward to the implementation of programs to induce an earlier onset of puberty, induction of fertile estrus in anestrous animals during the postpartum period, increases in embryo survival, and advancements in reproductive technology that will permit transfer of customized embryos that are produced *in vitro*.

See Also the Following Articles

Estrous Cycle; Parturition, Nonhuman Mammals; Pigs

Bibliography

Anderson, L. H., McDowell, C. M., and Day, M. L. (1996). Progestin-induced puberty and secretion of luteinizing hormone in heifers. *Biol. Reprod.* **54**, 1025–1031.

Bearden H. J., and Fuquay, J. W. (1997). *Applied Animal Reproduction*, 4th ed. Prentice Hall, Upper Saddle River, NJ.

Hafez. E. S. E. (1993). Folliculogenesis, egg maturation and ovulation. In *Reproduction in Farm Animals* (E. S. E. Hafez, Ed.), 6th ed., pp. 114–143. Williams & Wilkins, Media, PA.

King, G. J., and Thatcher, W. W. (1993). *Pregnancy. Reproduction in Domesticated Animals*, World Animal Science Vol. B9, pp. 229–270. Elsevier, Amsterdam.

Knickerbocker, J. J., Drost, M., and Thatcher, W. W. (1986). Endocrine patterns during the initiation of puberty, the estrous cycle, pregnancy and parturition in cattle. *Curr. Ther. Theriogenol.* **2**, 117–125.

Seals

Daniel P. Costa and Daniel E. Crocker
University of California, Santa Cruz

GLOSSARY

embryonic diapause Delayed implantation of the embryo in the uterine wall. Reactivation of the blastocyst is followed by a placental gestation.

metabolic overhead The proportion of a lactating female's energy expenditure that is used to meet her own metabolic needs as opposed to milk production.

otariid Eared seals (sea lions and fur seals). Otariids are characterized by the use of foreflipper propulsion.

phocid Earless or "true" seals. Phocids are characterized by the use hindflipper propulsion.

polygyny Mating system in which males are able to monopolize access to multiple females.

The Pinnipedia are a suborder of aquatic carnivores that includes the Odobenidae or walrus, the Otariidae or eared seals (sea lions and fur seals), and the Phocidae or earless seals ("true" seals). Reproduction in pinnipeds is characterized by delayed puberty, embryonic diapause, and, with a few exceptions, synchronous annual reproductive cycles. All pinniped species appear to be polygynous but the degree varies considerably from very slight to extreme. Females give birth to a single precocial offspring, and all parental care is provided solely by females. Pinnipeds are unique among mammals because they feed in the marine environment and reproduce on land or ice, requiring a spatial and temporal separation of feeding from lactation.

I. REPRODUCTIVE CYCLES AND PHYSIOLOGY

Reproduction in pinnipeds is generally characterized by embryonic diapause and synchronous seasonal cycles. The key adaptation, embryonic diapause, allows pinnipeds to utilize the oceans for food resources and exhibit terrestrial parturition. By synchronizing reproductive cycles, embryonic diapause also allows mating and parturition to occur at the same time, which reduces the cost of these energetically expensive, terrestrial behaviors. Most pinnipeds reproduce annually, although walrus show a 2-year cycle and the Australian sea lion exhibits a supra-annual (17.6-month) nonseasonal breeding cycle.

Studies of pinniped reproductive anatomy indicate that their estrous cycle is controlled in the same fashion as most mammals. The time between birth and estrus is associated with active growth of ovarian follicles and increases in progesterone levels. Estrus behavior coincides with a peak in estrogen secretion, when one large follicle is ovulated. Twinning is highly unusual. Ovulation results in the formation of a corpus luteum, with a continued rise in progesterone levels. Changes in hormone levels may cause females to initiate weaning. In males, testosterone levels, testicular mass, and spermatogenesis increase prior to and peak during the breeding season. These seasonal alterations in male endocrine physiology are probably important triggers in stimulating reproductive behaviors such as territory or harem defense.

While the precise length of gestation for most pinniped species has not been determined, active gestation is generally thought to last 8 to 9 months. Some experimental evidence has suggested that the timing of implantation is associated with the activation of reproductive hormones and is dependent on females achieving a certain body composition. A number of studies have suggested that photoperiod plays an important role in activating reproduction and synchronizing pinniped reproductive cycles, although other local factors can also influence reproduction. Such is the case for the Australian sea lion, a species that utilizes a seasonally stable, although nutrient-poor, marine environment. Because there is no advantage to a temporally fixed lactation period, lactation periods have been extended as has the placental phase of pregnancy. In this way, the daily costs of gestation and lactation are reduced. Other species may use environmental cues such as sea surface temperature to fine-tune their reproductive timing to variations in local conditions.

II. TYPES OF MATING SYSTEMS

Pinnipeds exhibit diverse mating systems. Many species of pinnipeds are highly polygynous and sexually dimorphic, whereas others show little dimorphism and are essentially monogamous. It appears that the distribution of females and environmental factors such as prey availability and distribution are important determining factors in the evolution of mating systems. The type of mating system is also strongly influenced by breeding habitat. Pinnipeds can be divided into three groups according to their breeding habitats: (i) species that breed on land, (ii) species that breed on fast ice, and (iii) species that breed on pack ice. Species inhabiting each of these habitats show different breeding behaviors.

A. Land-Breeding Pinnipeds

Extreme polygyny is characteristic of most land-breeding pinnipeds. Of the 21 species of land-breeding pinnipeds, 18 are highly polygynous and sexually dimorphic (Fig. 1). This includes all of the sea lions and fur seals, both species of elephant seals, *Mirounga*, and the gray seal, *Halichoerus grypus*. The remaining 3 species of land-breeding seals, 2 species of monk seals, *Monachus*, and harbor seals *Phoca vitulina*, mate in the water near land and exhibit lesser degrees of polygyny. In general, land-breeding pinnipeds aggregate on beaches, rocks, or flat areas on islands. These island aggregations provide several potential advantages, including parturition sites for rearing pups, lack of terrestrial predators, and, for the lactating otariids, proximity to food resources. This clumping of females, together with estrus synchrony, provides the opportunity for males to control access to and mate with a great many females (Fig. 2). This polygyny takes two main forms: (i) female or harem defense, in which males fight for social

FIGURE 1 A harem master mates with a female northern elephant seal. Notice how much larger the male is compared to the female.

FIGURE 2 A colony of Antarctic fur seals at the beginning of the breeding season. The pups are small and all black, females are intermediate size and have a blond ventral surface. The males are much larger and more sparsely distributed. A male is in the lower right.

status that confers access to clumped females (e.g., elephant seals) (Fig. 3a), and (ii) resource defense, in which males compete for territories that include breeding substrate needed by females (e.g., northern fur seals, *Callorhinus ursinus*) (Fig. 3b). Within the context of resource defense, there may be an element of female choice among territories that resembles a modified form of Lek (e.g., Stellar's sea lion, *Eumetopias jubatus,* and California sea lion, *Zalophus californianus*). Those species that breed in the water show a lesser degree of polygyny, in part because the potential for males to defend territories or females is more limited.

B. Ice-Breeding Pinnipeds

Of the 13 species of pinnipeds that breed on ice, 11 exhibit slight polygyny or facultative monogamy. Behavioral observations suggest that two ice-breeding species, the Weddell seal, *Leptonychotes weddellii*, and the walrus, *Odobenus rosmarus*, are moderately polygynous. However, because all ice-breeding species mate in the water, copulations and hence the degree of polygyny are difficult to observe.

Females that breed on pack ice have access to vast areas of breeding substrate, and while females of many species are widely dispersed throughout the pack and are generally observed alone with their pup or in a triad with a male, other species do aggregate loosely during the breeding season. For all pack ice breeders, the approaching breakup of the unstable pack ice poses a threat to the new offspring. Most females give birth during the short period of time when the ice is most stable and come into estrus synchronously upon weaning. This synchrony, together with the degree of spatial separation, limits the ability of males to mate with many females. It is likely that males of these species desert females after copulation and attempt to pair with another preestrus female.

In contrast, seals that breed on fast ice (expansive areas of ice attached to land) are generally found in small, well-spaced colonies within which females are widely separated from each other. Colony formation is facilitated by the fact that seals can only breed along the shore or near cracks and holes in the fast ice. In the Arctic, predation pressure may also serve to limit clumping of females. Females which breed in the fast ice also show reproductive synchrony, which, in combination with loose aggregations, makes it difficult for males to monopolize either females or reproductive habitat. As a result the pre-

FIGURE 3 (a) Male elephant seals fight to determine who will become the dominant male on the breeding beach. (b) Male Antarctic fur seals fight to establish territorial boundaries.

dominant mating strategy is facultative monogamy or slight polygyny.

III. PARENTAL INVESTMENT

The initial utilization of the marine environment by pinnipeds occurred at a time when coastal upwelling was at a cyclic high and thus presented an abundant, diverse, and essentially untapped food resource. However, the need to return to shore to provision the young necessitated a spatial and/or temporal separation of feeding from lactation. The ideal solution required a balance between the conflicting demands of optimal provisioning of the pup onshore and the need to maximize energy acquisition at sea. Within the pinnipedia, the phocids, otariids, and odobenids have evolved different strategies to cope with the separation of feeding and lactation. All conceive young during the previous reproductive season and exhibit a period of delayed implantation that usually lasts 2 or 3 months. Actual fetal development occurs over a 9-month period, during which mothers feed at sea either almost continuously, as is the case for phocids, or intermittently, as is seen in the otariids and odobenids. Because the nursing period can be extremely long in these species, females are often lactating and gestating at the same time.

A. Patterns of Parental Care

Fasting during lactation is a unique component of the life history pattern of marine mammals. With the exception of bears and whales, no other mammals are capable of producing milk without feeding. In pinnipeds, two main lactation strategies are employed. Most phocid mothers remain on or near the rookery from the time their pup is born until it is weaned. Body reserves stored prior to parturition fuel the female's metabolic needs and are utilized for milk production. Although some phocids, most notably harbor, ringed (*Phoca hispida*), and Weddell seals, feed during lactation, most of the maternal investment is derived from body stores. Weaning is abrupt and occurs after a minimum of 4 days of nursing (hooded seal, *Cystophora cristata*) to a maximum of 6 weeks (Weddell seal) (Fig. 4).

In contrast, otariid mothers only stay with their pups the first week or so after parturition and then periodically go to sea to feed, returning to suckle their pup on the rookery. Feeding trips vary from 1 to 14 days depending on the species, and shore visits to the pup, which has been fasting, last 1–3 days. The pups are weaned from a minimum of 4 months in the subpolar fur seals (northern and Antarctic, *Arctocephalus gazella*) to up to 3 years in the equatorial Galapagos fur seal (*A. galapagoensis*). In the otariids that inhabit temperate latitudes, pups are usually weaned within a year of birth, although weaning age can vary both within and between species as a function of seasonal and site-specific variations in environmental conditions. Walruses can feed their offspring for up to 3 years, both while on shore and in the water (Fig. 5).

These differences between the reproductive pattern of phocid and otariid seals in part relate to the amount of stored energy at parturition. Most phocids studied are capable of storing the energy required for the entire lactation interval, whereas otariids must feed during the lactation interval. The differences in the reproductive strategies of phocids and otariids are reflected in the amount of time spent nursing the pup versus the time spent feeding at sea and are dependent on a number of variables, including maternal energy/nutrient investment, maternal mass, metabolism and energy stores, duration of the lactation interval, milk composition, trip duration, and distance to the foraging rounds.

B. Cost of Reproduction

In pinniped males the cost of reproduction is limited to the cost of finding and maintaining access to

FIGURE 4 A Weddell seal mother and pup on fast ice in McMurdo Sound, Antarctia. Weddell seals are in the family Phocidae, or true seals and lack external ears.

FIGURE 5 An Antarctic fur seal mother and pup resting on a tussock grass mound at South Georgia Island in the South Atlantic. Fur seals are members of the family otariidae and have external ears and long hind flippers.

estrous females. In terrestrially breeding pinnipeds larger male body size is adaptive since it confers both an advantage in fighting and allows the male to maintain terrestrial territories longer. In addition, larger animals can fast longer because they have a lower mass-specific metabolic rate than smaller animals. In species that compete for females in the water, males are approximately the same size as females. For the aquatic breeders, underwater agility is more important than large size when competing for mates, and large energy reserves may not be as necessary because males are able to make short foraging trips and then return to the breeding area without losing territory or females. However, even among aquatic breeders, large males may have an advantage over small males because they can reduce the number of foraging trips and thus spend a greater proportion of time in the water competing for females.

The cost of reproduction for pinniped females can be broken down into the energetic requirements for gestation (the cost of producing a fetus) and lactation (the cost of nursing young until weaning). Although no direct measurements on the cost of gestation are available for marine mammals, investigations of terrestrial eutherians suggest that the cost of producing a fetus is insignificant relative to the costs associated with lactation. If fetal birth mass relative to maternal mass is used as a relative index of the cost of gestation, there are no differences among pinnipeds in gestational costs. However, lactation strategies and maternal investment vary markedly between phocids and otariids, which suggests that lactation costs may differ among species. For their size, phocids have significantly shorter lactation duration and a much higher rate of pup growth than do other marine mammals. Otariid and odobenid (walruses) young have slower growth rates and much longer lactation periods.

Comparisons of maternal investment patterns must consider differences in the behavior and metabolic rates of the mother and her young. Rapid growth of pup adipose tissue stores and proportionately little growth in lean tissue characterize short lactation periods, such as those seen in phocids. As a result, although phocid pups are weaned at an early age, they are not truly nutritionally independent at that time; they rely on maternally derived energy, stored as blubber, for days to months after weaning. In contrast, other marine mammals, such as otariids, odobeniids, sea otters, and odontocete cetaceans, wean their offspring much later. Longer lactation periods allow the young to acquire a greater proportion of lean tissue growth but also require that more energy be supplied to the young in support of its maintenance metabolism. At weaning, the young are not as reliant on stored energy reserves and may even have experience in foraging independently. However, longer lactation periods do require greater total maternal investment.

In addition to physiological factors, the duration of lactation may be dictated by the habitat of the marine mammals. This is especially apparent for phocids and is an important factor that permits this group to breed on ice. The shortest lactation interval and fastest growth rates for pups occur in pack ice breeding seals such as hooded and harp seals, whereas the longest lactation intervals are seen in species that breed on fast ice such as Weddell and ringed seals (*Phoca hispida*). Lactation intervals can be long in these species because fast ice is quite stable. Conversely, pack ice is a very unstable breeding substrate and can disappear at any time. The shortened lactation interval of pack ice seals ensures that the pup is weaned prior to the breakup of the pack. Island-breeding phocids that feed far offshore, such as elephant and gray seals, show an intermediate pattern. These animals may not be able to feed during the lactation period because the food resource is too distant. As a result, they concentrate the investment interval, which allows a greater proportion of stored maternal resources to go into milk production rather than maternal maintenance metabolism and thus reduces maternal overhead. The comparatively long lactation interval of Weddell, ringed, and harbor seals probably reflects the ability of these females to augment their maternal reserves by feeding during the lactation period. With the proximity of food, short feeding trips, by supplying additional energy, may decrease the pressure on the females to shorten the investment interval and reduce metabolic overhead. It is unlikely that short-duration feeding trips can supply sufficient energy to support the rapid growth rates of phocid pups, and studies indicate that most of the energy and materials supplied to the pup are still derived from maternal body reserves even in those species that feed during lactation.

C. Body Size and Maternal Resources: The Role of Maternal Overhead

The ability of a pinniped female to fast while providing milk to her offspring is related to the size of her energy and nutrients reserves and the rate at which she utilizes them. When food resources are far from the breeding grounds, as may occur for some phocids, the optimal solution is to maximize the amount of energy and nutrients provided to the young and to minimize the amount expended by the mother. The term "metabolic overhead" has been used to define the amount of energy a female pinniped expends on herself while ashore suckling her pup. Either increasing body mass or reducing the duration of lactation can minimize metabolic overhead. The strategy for small phocid seals (i.e., ringed and harbor seals) is to maintain short lactation periods or to feed during lactation. Larger phocids such as elephant seals are able to maintain longer lactation intervals without feeding since metabolic overhead is reduced due to their lower (mass-specific) metabolic rate and disproportionately larger blubber energy stores.

D. Maternal Metabolism

Variation in maternal metabolism, at least within the otariids, appears to be linked to ambient temperature. Galapagos fur seal females that live in the warm equatorial environment exhibit fasting metabolic rates on land that are only 1.1 times the predicted basal metabolic rate. In contrast, northern and Antarctic fur seals which inhabit the cold subpolar environment exhibit metabolic rates 3.4 times predicted levels. Decreases in an animal's onshore metabolism may be achieved by a reduction in activity as observed in most otariids or by periodic breathing as occurs in phocids.

The ability to store energy also differs between otariids and phocids. The available data indicate that phocid mothers store significantly more fat than otariids. Body composition of lactating phocid females ranges from 24.5% fat for harbor seals to 47% for harp seals. In comparison, Galapagos, northern, and Antarctic fur seals and California and Australian sea lions (*Neophoca cinerea*) show 26, 22, 19, 13, and 8.3% body fat, respectively. With the exception of Galapagos fur seals, these otariids appear to maintain onshore fasting metabolic rates similar to those of phocids once corrected for body mass. Overall, the ability of phocid seals to store most, if not all, of the maternal energy and nutrients prior to arriving onshore may be attributed to large body size, low metabolic overhead, and greater lipid reserves.

E. Energy Investment and Trip Duration

While most phocids fast throughout the lactation interval, otariid females spend from 0.5 to 14 days at-sea foraging between onshore suckling periods. However, differences in the pattern of energy investment and trip duration seen in the pinnipeds are consistent with those predicted by central place foraging theory. Otariid mothers making short feeding trips provide their pups with less milk energy than mothers that make long trips, and this is correlated with the distance to the offshore feeding grounds.

Otariids such as the Steller sea lion (*Eumetopias jubatus*) make trips of relatively short duration (approximately 36 hr), feed near shore, and travel short distances to the feeding grounds. Northern fur seals feed up to 100 km offshore and make trips of 7 days duration. As predicted by the model, Steller sea lions deliver considerably smaller amounts of milk energy (0.8 MJ $\times$ kg$^{-.75}$) per visit to their pup than northern fur seals (4.6 MJ $\times$ kg$^{-0.75}$). A similar pattern is observed for fur seals species. Inshore feeding species (i.e., Galapagos fur seals) forage for less than 24 hr between shore visits to suckle their pups, whereas species which feed offshore, such as Antarctic, northern, and Juan Fernandez fur seals, may spend anywhere from 4 to 12 days at sea foraging.

Optimization of foraging behavior is also observed for phocids. Island-breeding species, such as elephant and gray seals, represent extreme examples of offshore feeders that utilize highly dispersed or distant prey resources and make as few trips as possible per reproductive event. Because the reproductive pattern of these phocids is less constrained by the time it takes to travel and exploit distant prey, they are able to utilize more dispersed and patchy food resources. In fact, by spreading the acquisition of prey energy over many months at sea, northern elephant seal females only need to increase their daily food intake by approximately 12% to cover the entire cost of milk production and maternal metabolism. As a result, phocids may have a reproductive pattern that is better suited for dealing with dispersed, unpredictable, or distant prey resources than that of otariids.

While lactation enables pinnipeds to supply their pups with the appropriate amount of energy, regardless of foraging strategy, the length of time that females are able to fast while lactating places a limit both on the duration of investment and on the total amount of energy that can be invested in the pup. To compensate for limits on lactation duration, all pinniped mothers supply their pups with milk of significantly greater energy density than the prey consumed. However, increasing the energy content of milk by increasing the lipid content has a disadvantage. Because the high-energy density of pinniped milk is achieved by decreasing the water content and without changing the protein content, the protein to energy ratio in the milk is extremely low. As a result, for those species with the most energy-dense milk, offspring may be provided with sufficient energy to fuel metabolism but be limited in their ability to deposit lean tissue mass. This is especially important in species that have shortened lactation intervals such as hooded seals. In these species, pups receive a similar amount of total energy but in much smaller total quantities of milk. Because milk protein content is independent of lactation duration, these pups receive less total protein than pups which nurse over longer periods. This constraint is reflected by the fact that phocid postnatal growth is characterized by the accumulation of large adipose tissue stores with little lean tissue growth. For example, northern elephant and harp seal pups are born almost without fat but are composed of approximately 50% lipid at weaning.

F. Variation in Milk Composition

The variation in milk fat among marine mammals is correlated with the duration of time that the mother and pup are together and with maternal attendance patterns. For example, the highest milk fat contents are observed in those species that spend only a few days to weeks with their pups, and the lowest contents are observed in species that wean their pups after many months. Hooded seals are the most impressive of the phocids: They have a 4-day lactation period, and their milk is 65% lipid. Not surprisingly, pups nursing on such lipid-rich milk show the highest rates of postnatal growth. In otariids, the lipid, and therefore energy content of the milk, increases with trip duration. Females that spend short times on shore with their pups between foraging trips have milk with higher lipid levels. However, in otariids, the correlation between milk fat content and trip duration is complicated by the fact that otariid females with long trip durations often inhabit higher latitudes where the highly seasonal environment appears to favor shorter lactation periods. Thus, the higher milk fat in these species may reflect both the shorter feeding trips and the shorter total lactation period. Overall, for both otariids and phocids, the use of a fat, energy-rich milk can be seen as a strategy which allows mothers to transfer high levels of energy in a very short period of time.

Bibliography

Atkinison, S. (1997). Reproductive biology of seals. *Rev. Reprod.* **2**, 175–194.

Bonner, W. N. (1984). Lactation strategies in pinnipeds: Problems for a marine mammalian group. In *Physiological Strategies in Lactation* (M. Peaker, R. G. Vernon, and C. H. Knight, Eds.), Symposium of the Zoological Society of London No. 51, pp. 253–272. Academic Press, London.

Bonness, D. J., and Bowen, W. D. (1996). The evolution of maternal care in pinnipeds. *Bioscience* **46**, 645–654.

Bowen, W. D. (1991). Behavioral ecology of pinniped neonates. In *The Behaviour of Pinnipeds* (D. Renouf, Ed.), pp. 66–127. Chapman & Hall, London.

Costa, D. P. (1991a). Reproductive and foraging energetics of pinnipeds: Implications for life history patterns. In *The Behaviour of Pinnipeds* (D. Renouf, Ed.), pp. 300–344. Chapman & Hall, London.

Costa, D. P. (1991b). Reproductive and foraging energetics of high latitude penguins, albatrosses and pinnipeds: Implications for life history patterns. *Am. Zool.* **31**, 111–130.

Costa, D. P. (1993). The relationship between reproductive and foraging energetics and the evolution of the Pinnipedia. In *Recent Advances in Marine Mammal Science* (I. Boyd, Ed.), Symposium of the Zoological Society of London No. 66, pp. 293–314. Oxford Univ. Press, Oxford, UK.

Kovacs, K. M., and Lavigne, D. M. (1986). Maternal investment and neonatal growth in phocid seals. *J. Anim. Ecol.* **55**, 1035–1051.

Kovacs, K. M., and Lavigne, D. M. (1992). Maternal investment in otariid seals and walruses. *Can. J. Zool.* **70**, 1953–1964.

Oftedal. O. T., Boness, D. J., and Tedman, R. A. (1987). The behavior, physiology, and anatomy of lactation in the Pinnipedia. *Curr. Mammal.* **1**, 175–245.

Trillmich, F. (1990). The behavioral ecology of maternal effort in fur seal and sea lions. *Behaviour* **114**, 3–20.

Seasonal Reproduction, Birds

Alistair Dawson
Institute of Terrestrial Ecology

GLOSSARY

circadian rhythms Cycles of biochemical, physiological, or behavioral functions which oscillate with a periodicity of about 24 hr in the absence of external stimuli but which are normally entrained to exactly 24 hr by external stimuli (e.g., dawn).

gonadotropin-releasing hormone A decapeptide synthesized by neurons with cell bodies in the preoptic region of the hypothalamus and axons which terminate in the median eminence, adjacent to the pituitary gland. In birds two GnRHs have been identified, designated GnRH-I and -II. Only GnRH-I (Gln^8-GnRH) is thought to play a direct role in the control of seasonal breeding. Secretion of GnRH-I stimulates synthesis and secretion of the pituitary gonadotropins.

gonadotropins Hormones which stimulate gonadal maturation. There are two gonadotropins synthesized by the pituitary: luteinizing hormone and follicle-stimulating hormone.

prolactin A pituitary hormone which stimulates production of milk in mammals and crop milk in pigeons and which is thought to have a variety of other roles, including antigonadal or antigonadotropic effects, throughout vertebrates.

vasoactive-intestinal polypeptide (VIP) A 28-amino acid peptide present in neurons of the basal hypothalamus with projections to the median eminence. VIP acts as a neurotransmitter and, in birds, is the releasing hormone for prolactin.

Birds inhabiting seasonally changing environments restrict their breeding attempts to the period of the year when survival of young is greatest. Accurate timing of breeding requires that information from environmental cues is relayed to the gonadotropin-releasing hormone (GnRH) neurons to modulate secretion of GnRH. This leads to the appropriate physiological, morphological, and behavioral changes. The major cue is often photoperiod, which defines a window within which birds are physiologically able to breed. Fine-tuning within this window is achieved through nonphotoperiodic stimuli.

I. ULTIMATE AND PROXIMATE FACTORS

Natural selection favors those genotypes raising the greatest number of young which themselves survive to breed. In seasonally changing environments, it therefore follows that natural selection will favor genotypes whose progeny are reared at the most auspicious time of year. Attempts to breed earlier or later than the optimal breeding season will be nullified by disproportionate mortality of young and possibly also of adults. The ultimate factors are those which directly or indirectly affect survival and which therefore ultimately lead to the evolution of particular breeding seasons.

Ultimate factors are the concern of ecology: They determine why birds breed when they do. However, they may have little or no value as predictive information enabling individuals to time their reproductive effort correctly. For example, in great tits, the dominant ultimate factor, as with most species, is the availability of food on which the young are fed—in this case caterpillars. Breeding is timed so that nestlings are as near as possible 9 days old (when their demand for food is greatest) when the availability of caterpillars is maximal. The appearance of caterpillars cannot itself be used by the parent as a cue to time egg laying. For young to hatch just prior to maximum food availability requires that a cascade of physiological and behavioral processes (acquisition of territory, mate selection, gonadal maturation, nest building, copulation, ovulation, egg laying, and incubation) is set in train many weeks beforehand. Rather, timing of this sequence is brought about by proximate factors. These are reliable environmental cues that the individual can perceive and use as predictive information. They may themselves have little or no effect on survival. The most important proximate factor outside the tropics is the seasonal change in day length.

In summary, ultimate factors dictate why species breed when they do, whereas proximate factors are the means by which individuals time their breeding appropriately.

A. Ultimate Factors Which Determine Breeding Seasons

Reproduction is energetically demanding. Consequently, food supply tends to be the major ultimate factor. Food supply is dependent directly or indirectly on the primary production of an ecosystem and this varies seasonally. Outside the tropics, temperature and day length affect primary production, whereas within the tropics rainfall is more important. In the vast majority of species, therefore, breeding is seasonal. In most nontropical species, breeding occurs during spring or summer, whereas for tropical species, breeding seasons are frequently related to the patterns of rainfall.

The importance of food availability as an ultimate factor is possibly more acute in birds than in mammals because they do not have the flexibility that feeding young on milk permits. In mammals, adequate milk production only requires a sufficient supply of the food on which adults feed. In birds, the supply of food specifically suitable for the young has to be abundant. In altricial species (those which raise dependent young in a nest), in addition to satisfying their own needs, parents have to provide young with adequate amounts of high-quality food to ensure rapid growth rates. In precocial species (those whose young hatch in a sufficiently advanced state of development to feed independently), food has to be abundant enough for the young themselves to find ample amounts despite their inexperience.

Since different species utilize different food resources, breeding seasons vary accordingly. In general, for species in mid- and high latitudes, breeding begins some time during spring, but the length of the breeding season varies. At one extreme are species which lay eggs only during a short period early in the year (e.g., rooks, herons, and owls—March and April), and at the other extreme are species which lay later and for a longer period (e.g., partridges and quail—April–August). In addition, breeding seasons tend to be shorter at higher latitudes. However, there are species that do not breed only in spring or summer. Eleanora's falcons, for example, feed their offspring on young migrating passerines and so they breed during the autumn when the number of migrants is highest. Crossbills feed on pine cones, which become available unpredictably throughout the year. Crossbills are therefore opportunistic breeders and have been recorded breeding in most months of the year.

For most terrestrial species within the tropics, food availability is related to seasonal rainfall rather than to day length or temperature and breeding is timed accordingly. However, for some tropical species, particularly seabirds, food supply may remain fairly constant throughout the year, and so such species are not constrained to breed only at one particular time of year. Several such species breed at regular intervals of less than 12 months.

Breeding is not the only energetically demanding aspect of a bird's annual cycle. Renewal of the feathers (molt) is also energetically demanding and is vital for survival. Birds normally molt immediately after breeding: There is little or no overlap between the two activities. Both require comparatively abundant food supply. Therefore, breeding seasons have not evolved simply to occupy the whole period of food abundance; there has to be a compromise between the requirements of breeding and the need to molt. Birds that breed late will suffer the penalty of less time to molt, resulting in poorer quality feathers. This will increase the risk of mortality and reduce the chances of breeding successfully in the following year. In some large species (e.g., albatrosses), the combined period required for breeding and molt exceeds 12 months, so they only attempt to breed every other year.

Migration may impose an additional constraint. Migratory species need a period of hyperphagia to lay down sufficient energy reserves for their migration. This too must coincide with a period of comparative food abundance. All these energetic demands act as ultimate factors. Other aspects, such as weather, predation, and the availability of nest sites or nesting material, may also act as ultimate factors. Breeding seasons evolve as the optimal compromise between all of the factors that ultimately affect survival.

B. Control of Breeding Seasons by Proximate Factors

As already explained, ultimate factors may have little utility in conveying information to the individual as to when to begin the physiological changes that culminate in egg laying. This, instead, is the role of proximate factors. An ideal proximate factor is one which changes predictably during the year and reliably between years so that it can act as a calendar. Outside the tropics, the seasonal change in day length fulfills these criteria. In comparison, other environmental variables, such as temperature, rainfall, and food supply, are uncertain and unreliable.

It is probably true to say that all temperate zone birds are highly photoperiodic. Change in day length is used to time gonadal maturation and gonadal regression and the onset and rate of molt. Even some tropical species have been shown to be photoperiodic: Experimentally imposed lighting regimes can alter their reproductive responses even though such birds would naturally experience no significant changes in day length.

Day length is the major proximate cue outside the tropics. By controlling the time of gonadal maturation and the time of gonadal regression, it defines the window within which birds are physiologically able to breed. This may leave a certain amount of flexibility as to exactly when individuals actually do breed. The degree of flexibility varies between species. Those with longer potential breeding seasons have a greater capacity for flexibility, and this will be greatest in opportunistic breeders such as crossbills. Exactly when individuals breed is the result of fine-tuning involving a variety of extrinsic and intrinsic factors. Since the extrinsic factors need be less predictive than day length, they are often the same as ultimate factors, such as weather conditions, food availability, and access to a nest site. Within the tropics, where there is no reliable photoperiodic cue, these act as the major proximate factors. Intrinsic factors include variables such as nutritional status and social status of the individual. In many species, subdominant individuals breed later, and, in some species, many individuals do not even attempt to breed.

II. PHOTOPERIODIC CONTROL OF BREEDING SEASONS

A. Differences in Breeding Seasons

The breeding seasons of nontropical species vary considerably and so too must the way in which they are controlled by day length. Figure 1 shows the breeding seasons (at 52°N) of three extensively studied species, European starlings, house sparrows, and Japanese quail, to illustrate this variety. Gonadal maturation occurs at approximately the same time in spring in all three species, and this is brought about by the increase in day length (photostimulation). However, the timing of gonadal regression differs.

In starlings, gonadal regression occurs during May, before the summer solstice. This is due to birds becoming photorefractory—refractory to the stimulatory effect of long days. Long days cause photorefractoriness. Termination of photorefractoriness (recovery of photosensitivity) occurs in October due to the effect of short days. Therefore, starlings are photorefractory for 6 months of the year and photosensitive for the other 6 months. In experimental situations, starlings continuously exposed to long days remain photorefractory indefinitely.

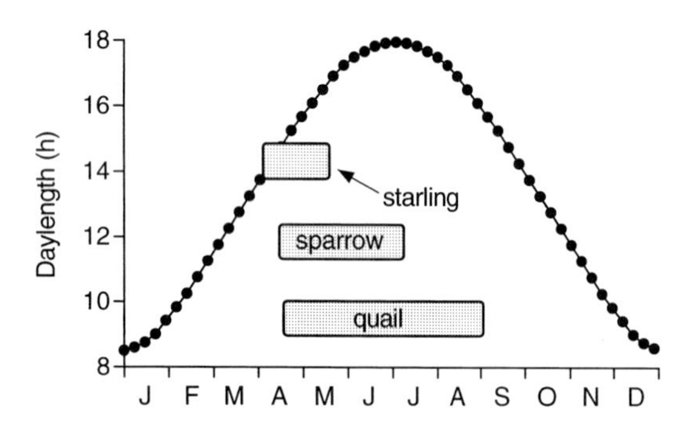

FIGURE 1 The periods of gonadal maturity (potential breeding season) in European starlings, house sparrows, and Japanese quail under natural changes in day length at 52°N. The timing of gonadal maturation, during spring, is similar in all three species, but the timing of gonadal regression varies considerably.

In quail, gonadal regression does not occur until the short days (and possibly decreasing temperature) of autumn. However, the day length when this occurs still exceeds that which caused gonadal maturation during spring. Quail are therefore said to become "relatively" photorefractory. This differs from the "absolute" photorefractoriness of starlings. Quail exposed continuously to long days never show gonadal regression, and following short day-induced gonadal regression, quail remain photosensitive: An increase in day length always stimulates renewed gonadal maturation.

Sparrows are intermediate. Although gonadal regression does not occur until soon after the solstice, as day length begins to decrease, sparrows become absolutely photorefractory, like starlings. It is possible that a brief period of relative photorefractoriness precedes the development of absolute photorefractoriness. Under natural conditions, sparrows become photosensitive again during short days of autumn, but, unlike starlings, under experimentally imposed continuous long days, spontaneous gonadal regrowth does eventually occur.

In all three species, breeding seasons are asymmetrical with respect to change in day length and this asymmetry is brought about by either absolute or relative photorefractoriness. Molting in all three species begins during gonadal regression.

These three examples encompass most of the breeding patterns found in nontropical species. One exception is the opportunistically breeding crossbills. These species are photoperiodic: A decrease in day length causes gonadal regression and molt follows. However, after molt there follows a prolonged period during which birds are capable of breeding. Food availability acts as the proximate factor dictating the exact timing of breeding.

If nontropical species are deprived of photoperiodic information, by keeping them under constant 12-hr days, they undergo repeated cycles of gonadal maturation which occur at approximately 12-month intervals. Whether these "circannual rhythms" are evidence of a circannual clock, which is normally synchronized by seasonal changes in day length, is unclear. Certainly, the pattern of gonadal maturation in such conditions is somewhat different from that under natural cycles. There is a prolonged period during which the gonads remain fairly mature. Then gonadal regression occurs, a molt is completed, and renewed gonadal maturation begins soon afterwards. This pattern is similar to the natural cycles of species within the tropics. In such species, nonphotoperiodic proximate cues determine the exact time of breeding. Breeding is followed by gonadal regression, molt, and renewed gonadal maturation. The gonads remain in a state of "readiness to breed" for a considerable portion of the year.

B. Photostimulation

In experimental situations, in which birds are transferred from short days to various longer days, the rate of gonadal maturation increases as day length increases. The exact nature of this relationship differs between species. In some species a "critical day length," below which gonadal maturation is very slow but above which it is rapid, can clearly be defined, e.g., white-crowned sparrows and Japanese quail. This critical day length marks the beginning of the breeding season. In other species this asymptote is less clear. There can be modest gonadal maturation under short days, and the increase in the rate of maturation may be more linearly related to the increase in day length, e.g., house sparrows. During this period, environmental information other than day length may act as additional proximate cues, affecting the later stages of gonadal maturation, particularly in females, and the exact timing of breeding.

C. Photorefractoriness

The majority of species from mid- and high latitudes probably become absolutely photorefractory in a manner similar to starlings or sparrows. In such species, prolonged exposure to day lengths in excess of a particular species-specific day length—another critical day length—eventually leads to photorefractoriness. This critical day length is comparatively easy to define. In experimental situations, birds exposed to a constant day length less than this critical day length will remain photosensitive indefinitely, whereas those exposed to a longer day length will eventually become photorefractory. The longer the day length to which birds are exposed (beyond the critical day length) the sooner they become photorefractory. Thus, at higher latitudes, the breeding season is briefer.

The critical day length for the induction of photorefractoriness need be no longer than that for photostimulation. Photostimulation simply occurs more rapidly than photorefractoriness. In fact, in most temperate zone species, breeding occurs during day lengths which exceed the critical day length for the induction of photorefractoriness. Therefore, breeding occurs at the end of the photosensitive phase during the period when photorefractoriness is being induced.

Most nontropical species become absolutely photorefractory. The termination of breeding by relative photorefractoriness is rare and has only been confirmed in species of quail (although it appears to be similar to mechanisms commonly found in mammals). The physiological relationship between relative and absolute photorefractoriness, i.e., whether they are extremes of the same underlying mechanism or totally unrelated mechanisms, remains unclear. Although relative photorefractoriness requires a decrease in day length to become manifest, it too is caused by long days. In absolutely photorefractory species, the longer the day length, the sooner birds become photorefractory. In relatively photorefractory species, the longer the day length, the longer the subsequent "short" day need be to cause regression. If absolutely photorefractory species are transferred from long days to short days before they become photorefractory, the gonads regress. This regression is slower than that caused by photorefractoriness, and, unlike short-day-induced regression in quail, it is not followed by molt.

III. PHYSIOLOGICAL CONTROL OF SEASONAL GONADAL CYCLES

A. Neural Mechanisms Involved in Photoperiodism

The environmental stimuli that act as proximate factors need to be transformed into the appropriate behavioral and reproductive outputs at the optimal times. This is achieved via a relay of neural and endocrine signals which pivots around the GnRH

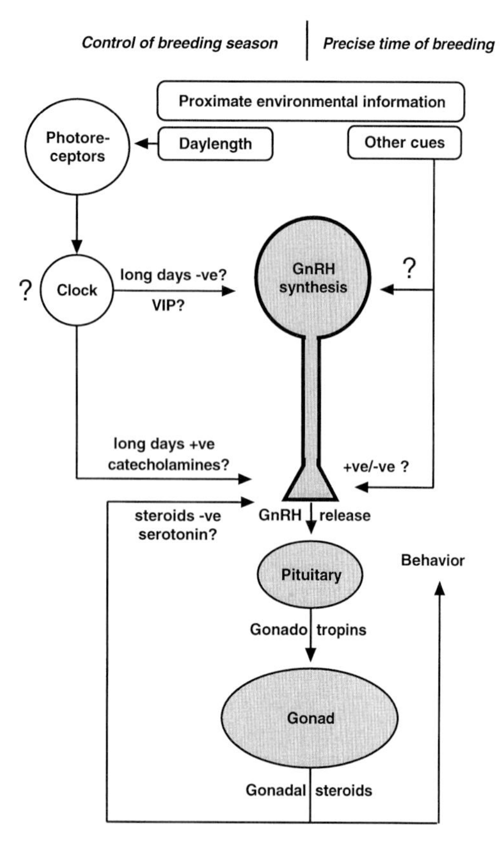

FIGURE 2 A schematic representation of the hypothalamo–pituitary–gonad axis showing how environmental cues may regulate gonadal maturation and the timing of breeding. The duration of the breeding season is largely controlled by seasonal change in day length (left). Increasing day length during spring stimulates release of GnRH (photoperiodic drive) from the median eminence sufficiently to override gonadal steroid negative feedback. This results in an increase in gonadotropin secretion and gonadal maturation, thus defining the beginning of the breeding season. Prolonged exposure to long days leads to an inhibition of GnRH synthesis and gonadal regression at the end of the breeding season. Exactly when within the potential breeding season individuals actually breed may be influenced by other environmental cues, e.g., food availability, weather, or a suitable nest site. Nutritional status and behavioral interactions may also be important, particularly in females.

neurons (Fig. 2). Gonadal maturity ultimately depends on the rate of release of GnRH from the hypothalamus.

While the endocrine system downstream of the hypothalamus is broadly similar in birds to that in mammals, the mechanism by which photoperiodic information is relayed to the GnRH neurons is different. In mammals, information on the presence or absence of light is conveyed along a retinohypothalamic tract to the suprachiasmatic nuclei (SCN), where it is integrated with information from a circadian clock. This photoperiodic information passes via the superior cervical ganglion to the pineal gland, which produces melatonin during darkness but not during periods of light. The resultant pattern of melatonin is used as the signal to control gonadal maturation. In birds, light is detected by photoreceptors which are neither retinal nor pineal. They are thought to lie within the hypothalamus, possibly in the lateral septum and tuberal hypothalamus. Although measurement of photoperiod in birds also involves a circadian clock, they have no structure analogous to the mammalian SCN. Furthermore, although birds have a well-defined daily pattern of melatonin, this is not used as a signal to regulate gonadal maturation. Rather, birds appear to use an entirely neural link between photoreceptors and the GnRH neurons. There is evidence that the neurons involved contain vasoactive-intestinal polypeptide (VIP).

B. Photoperiodic Control of GnRH Neurons

Seasonality centers on changes in activity of the GnRH neurons. The exact nature of these changes, and how they are controlled, remains far from clear. The description provided here, although based on currently available data, should be regarded only as a potential model.

Breeding seasons are not symmetrical with respect to the annual cycle in day length. Consequently, GnRH secretion cannot simply be directly proportional to day length. This means that day length must have two separate effects on GnRH neurons. One concerns photostimulation and the other photorefractoriness. In birds which become absolutely photorefractory, the annual cycle can be divided into distinct phases.

1. *Photorefractory Phase*

During this phase, hypothalamic GnRH levels are low and little if any GnRH is secreted from the median eminence since the gonads remain fully regressed and plasma gonadotropin levels are low. The lack of GnRH secretion is not due to gonadal steroid feedback since gonadectomy has no effect. We can conclude that little GnRH is being synthesized.

2. *Recovery of Photosensitivity*

During short days, stored GnRH increases. Since little or no GnRH was released during the preceding photorefractory period, this increase cannot be due to a decrease in release. Presumably short days stimulate renewed GnRH synthesis. Release of GnRH must remain low since gonadotropin levels show little or no apparent increase. However, unlike the situation during photorefractoriness, the inhibition of GnRH release is at least in part due to gonadal steroid feedback. Gonadectomy of photosensitive birds results in an increase in circulating gonadotropins, although the magnitude of this increase varies between species. In photosensitive birds, therefore, the hypothalamo–pituitary–gonad axis is "switched on" but operating at a low level of activity. It is ready to respond to an increase in day length.

3. *Photostimulation*

When photosensitive birds experience an increase in day length, there is an increase in circulating gonadotropins, leading to gonadal maturation. Presumably this is due to an increase in GnRH release. This "photoperiodic drive" on GnRH release must be sufficient to overcome the negative feedback of gonadal steroids. Two recent pieces of evidence tentatively suggest that photostimulation acts primarily by controlling secretion of GnRH from the neuron terminals rather than synthesis in the cell bodies:(i) Lesions within the posterior infundibular nucleus block photostimulation without affecting stored GnRH, and (ii) photostimulation results in the appearance of *fos*-like protein within neurons located in the same region. This increased release of GnRH is not associated with a decrease in stored GnRH: in fact, it is possible that the reverse is true. Therefore, GnRH synthesis must increase at least sufficiently to compensate for the increased release.

4. *Onset of Photorefractoriness*

After a period of exposure to day lengths in excess of the critical day length, circulating gonadotropins decrease, presumably reflecting a decrease in GnRH secretion. Stored GnRH also decreases, so synthesis must decrease. However, the exact timing of these events, i.e., whether the decrease in synthesis precedes the decrease in release, remains unclear. One possible explanation is that photorefractoriness is directly due to long days downregulating GnRH synthesis. The other effect of long days, that on photoperiodic drive specifically affecting release rate, may persist. However, if no GnRH is being synthesized, none can be released.

In this model, day length has two distinct effects: One whereby long days and short days switch off and switch on GnRH synthesis, respectively, leading to the transition between photosensitivity and photorefractoriness and vice versa; the other by which GnRH release rate (photoperiodic drive) increases as day length increases. This dual effect of day length gains support from the observation that photorefractoriness is associated with increased synaptic input to the GnRH cell bodies in the preooptic area, whereas photostimulation appears to involve changes in the basal hypothalamus. However, an alternative explanation of photorefractoriness, in which long days directly inhibit the release of GnRH and this later feeds back to inhibit synthesis, cannot be ruled out. Gonadal regression in relative photorefractory quail is not associated with a decrease in stored GnRH. In fact, stored GnRH may increase, suggesting that gonadal regression is due to decreased release of GnRH rather than decreased synthesis.

This description tacitly assumes that both photoperiodic processes, photostimulation and photorefractoriness, are entirely neural with regard to the GnRH neurons. However, it has been suggested that peripheral hormone systems may play a role in the development of photorefractoriness. One of these is the thyroid. Certainly, in the absence of thyroid hormones, birds do not become photorefractory under long days. However, this may be one comparatively obvious manifestation of a more general requirement for thyroid hormones in neuronal mechanisms. Prolactin has also been implicated. There is a photoperiodically driven seasonal cycle in prolactin with peak levels which coincide with the onset of gonadal regression. While there is some evidence that prolactin can be antigonadal and antigonadotropic, there is no evidence that it induces photorefractoriness. One reason to doubt a causal role for prolactin is that prolactin levels are very high during incubation, and multibrooded species clearly do not become photorefractory while incubating early broods. In the hypothalamus there is an inverse correlation between VIP and GnRH, but again no causal role for VIP in the development of photorefractoriness has yet been demonstrated.

There may be a physiological relationship between photorefractoriness and prebasic molt. The prebasic molt does not occur during periods of gonadal maturity for sound ecological reasons. In most species molt always begins at the start of the photorefractory period and never occurs at any other time. Gonadal steroids inhibit molt, which may partly account for the relationship. However, this pattern persists in gonadectomized birds, implying a more fundamental physiological relationship, the nature of which remains obscure.

Nonphotoperiodic proximate factors must interact with the system at some level. They are unlikely to have an effect during the photorefractory period. However, there is some evidence that nonphotoperiodic proximate cues, particularly behavioral interactions, can modify the timing of the beginning and end of photorefractoriness, which suggests a possible effect on GnRH synthesis. There is clear evidence that nonphotoperiodic cues act to fine-tune the later stages of gonadal maturation and the exact timing of breeding. These effects may be brought about by altering the rate of GnRH release. Investigation of these relationships has only recently begun.

See Also the Following Articles

Avian Reproduction, Overview; Circadian Rhythms; Circannual Rhythms; Global Zones and Reproduction; GnRH (Gonadotropin-Releasing Hormone); Nutritional Factors and Reproduction; Photoperiodism, Vertebrates; Seasonal Reproduction, Mammals; Seasonal Reproduction, Fish

Bibliography

Ball, G. F. (1993). The neural integration of environmental information by seasonally breeding birds. *Am. Zool.* **33**, 185–199.

Cockrem, J. F. (1995). Timing of seasonal breeding in birds, with particular reference to New Zealand birds. *Reprod. Fertil. Dev.* 7, 1–19.

Harvey, S. (1997). *Perspectives in Avian Endocrinology*. Society for Endocrinology, Bristol, CT.

Murton, R. K., and Westwood, N. J. (1977), *Avian Breeding Cycles*. Clarendon, Oxford, UK.

Nicholls, T. J., Goldsmith, A. R., and Dawson, A. (1988). Photorefractoriness in birds and comparison with mammals. *Physiol. Rev.* **68**, 133–176.

Perrins, C. M., and Birkhead, T. (1983). *Avian Ecology*. Blackie, London.

Saldahna, C. J., Leak, R. K., and Silver, R. (1994). Detection and transduction of daylength in birds. *Psychoneuroendocrinology* **19**, 641–656.

Sharp, P. J. (1996). Strategies in avian breeding cycles. *Anim. Reprod. Sci.* **42**, 505–513.

Stetson, M. H. (1988). *Processing of Environmental Information in Vertebrates*. Springer-Verlag, Berlin.

Wingfield, J. C., and Farner, D. S. (1993). Endocrinology of reproduction in wild species. In *Avian Biology* (D. S. Farner, J. R. King, and K. C. Parkes, Eds.), Vol. IX, pp. 163–372. Academic Press, London.

Seasonal Reproduction, Fish

Jon P. Nash
The University of Sheffield

GLOSSARY

biological clock Term implying an underlying physiological mechanism that times a measurable rhythm or other type of biological timekeeping.

circadian rhythm An endogenous and persistent cyclical variation in the intensity of a physiological or behavioral process with a period of about 24 hr when under constant conditions.

circannual rhythm Same as circadian rhythm, but with a period of about 1 year.

endogenous Of rhythms or biological timekeeping ultimately controlled from within the organism with some kind of physiological biological clock.

exogenous Of rhythms or biological timekeeping that arise solely, or mainly, as a direct response to environmental signals.

period The duration of one complete cycle of a rhythmic variation.

phase A reference point in the cycle of a rhythm.

photoperiod Usually the light phase of a light–dark cycle, although can also refer to the whole light–dark cycle.

reproductive strategy Unique assembly of coadapted reproductive traits characteristic of an individual from a common gene pool which enable the organism to optimize reproductive success in a particular environment.

ultimate factors Abiotic and biotic environmental factors that affect survival and through the process of evolution induce changes in gene frequencies.

Almost all fish, particularly those living in temperate and polar waters, have a seasonal reproductive cycle. Reproductive seasonality is the synchronization of reproductive processes to specific and optimal times of the year so as to maximize reproductive success. Reproductive seasonality is controlled by physiological mechanisms that give the organism a sense of time and/or a measure of seasonal environmental variations or cues, allowing specific temporal control of various components of the reproductive process.

I. INTRODUCTION

The daily rotation of earth on its axis and its annual orbit around the sun cause an annual cycle of variation in the angle of incidence of solar radiation. As latitude increases, the changing altitude of the sun causes an annual cycle of variation in day length (Fig. 1) and solar intensity. Solar radiation is the ultimate driving force in nature, providing energy for photosynthesis, thereby, causing a strong seasonal cycle of primary production. Changes in solar radiation also drive complex seasonal variations in the world's climate. Depending on the geographical location, air and water temperatures, precipitation (resulting in changes in river flows, flooding, etc.), air movements, and water currents all show cycles of variation with an average periodicity of 1 year. Through the process of natural selection, fish in temperate and polar regions have responded to these seasonal variations by synchronizing their reproduction to specific and predictable times of the year in order to maximize reproductive success. Even near the equator and in the ocean depths, where there is little or no seasonal variation in day length or temperature, the environment can still have a predictable annual periodicity (such as rainy seasons or nutrient cycles) and, hence, many fish inhabiting

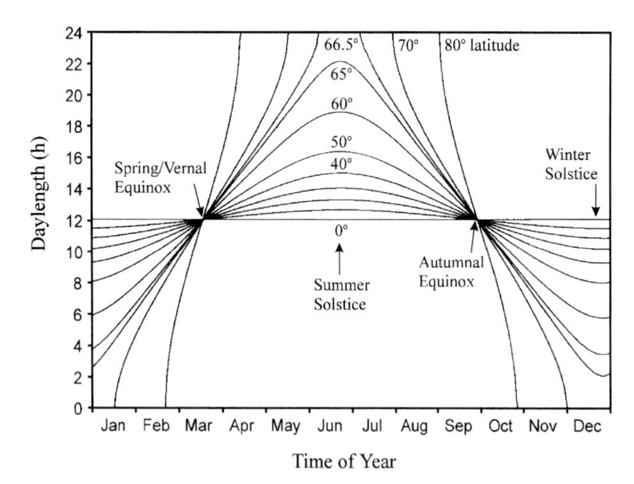

FIGURE 1 Day length as a function of time during the year for various northern latitudes.

these environments still show adaptations for seasonal reproduction.

There are two general approaches to the study of reproductive seasonality in fishes: (i) understanding the evolutionary causes of seasonal reproduction and (ii) determining the physiological mechanisms that mediate temporal control of reproduction. They are not, however, mutually exclusive but rather complementary.

II. EVOLUTIONARY CAUSES OF REPRODUCTIVE SEASONALITY

A. The Ultimate Causes of Reproductive Seasonality

The aim of seasonal reproduction is to optimize reproductive or Darwinian fitness (i.e., maximal survival of progeny and postbreeding adult survival in iteroparous species). Ultimately, it is seasonal variations in abiotic (i.e., physical factors such as light, temperature, oxygen levels, turbidity, and water flow) and biotic (i.e., food, plant cover, and competition and predation) environmental variables, termed "ultimate factors" (Fig. 2), that cause a specific time or times of the year to be optimal for certain reproductive processes. Furthermore, certain reproductive processes may also occur at specific times of the year (and day) so that there is temporal (and spatial) synchrony with other conspecifics and the opposite sex (see Section II,E).

B. Reproductive Strategies and Their Influence on the Timing of Reproduction

Conventionally, it is usually stated that the sole cause of reproductive seasonality is to synchronize the production of gametes to a time when environmental conditions (i.e., ultimate factors) are conducive to the survival and growth of the offspring. This is, however, an oversimplification of the complex evolutionary interactions involved in reproductive seasonality. While reproductive seasonality is an adaptation that ultimately optimizes for the survival of the progeny, natural selection does not act solely at the gamete level, i.e., a single component of fitness, but instead acts to select for the fittest average phenotype, i.e., the total reproductive value. The fittest average phenotype is dependent on the temporal optimization of the whole reproductive process, consisting of a number of different reproductive components or traits. This unique assembly of adaptive reproductive traits characteristic of a particular species is known as its reproductive strategy (Fig. 2). What constitutes the optimal time of the year for each component in the reproductive process is strictly dependent on the organism's specific reproductive strategy. Reproductive strategies involve many complex aspects of life history organization, such as trade-offs between somatic growth and maintenance versus reproduction, decisions on the numbers and size of the offspring, and investment in current offspring versus future reproductive attempts (i.e., semelparity versus interoparity), as well as the timing of reproduction. Moreover, the reproductive process is itself made up of a complex series of integrated components, such as gonadal growth and maturation, migration, mating, spawning, fertilization, embryogenesis, larval development, and parental care. Every component will be affected by a different array of ultimate factors and will be under a different intensity of seasonally dynamic selection pressure (particularly if forced to occur at a different time of the year). Each component will therefore be under a different level (if any) of pressure to synchronize

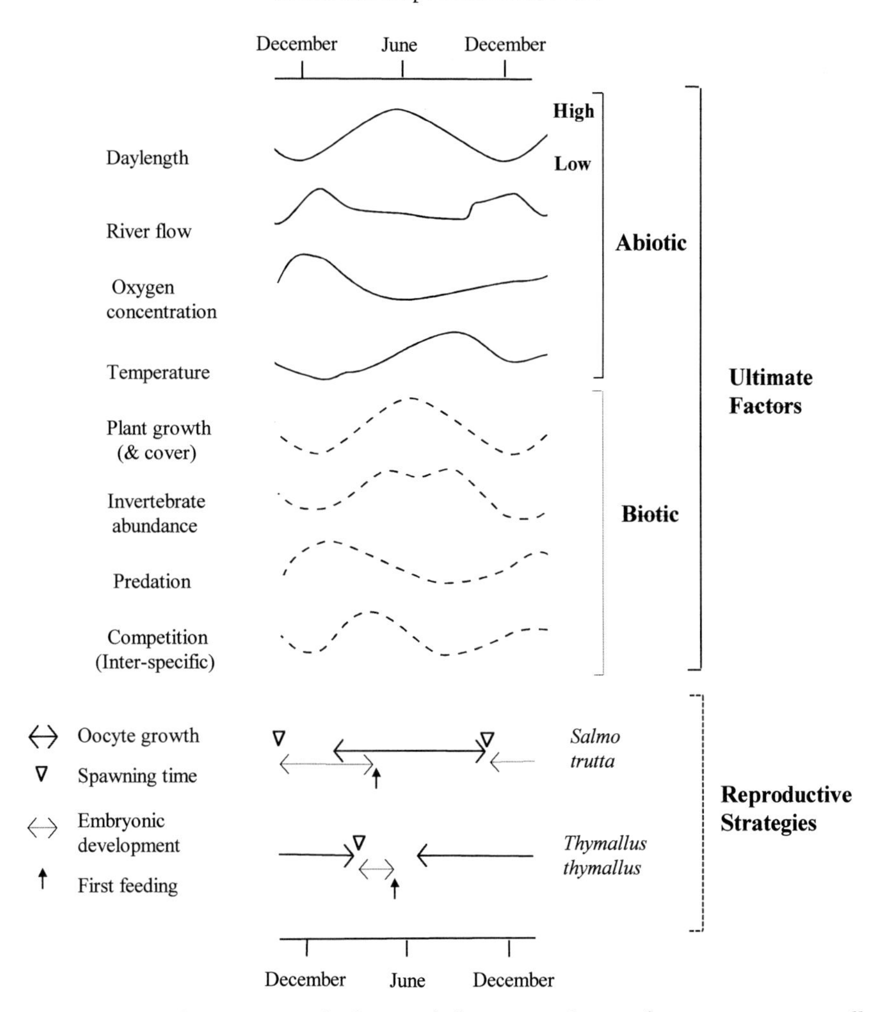

FIGURE 2 Hypothetical example of a stream in which annual changes in ultimate factors cause temporally dynamic selection pressures and two contrasting reproductive strategies of fish that have synchronized four components of their reproductive cycle to an optimal time of the year.

to a specific and maybe separate time of the year; determined by interactions between the ultimate factors and the various trade-offs imposed by the particular reproductive strategy of the fish. Furthermore, many components in the reproductive cycle require a finite period of time for completion, which forces different reproductive components to occur at separate times of the year and in some fish can cause the reproductive cycle to be spread over the entire year. This constraint has important consequences on the subsequent pattern of reproduction and is discussed in greater detail in Sections II,C and II,D.

It is true, however, that the strongest and most overriding selection forces are those that act directly on survival and growth of the fry. In temperate latitudes, the optimal time for production of fry, at least at the exogenous feeding stage, generally coincides with the spring when several important ultimate factors, such as food availability, warmth, and plant cover, become favorable. Moreover, the offspring have the greatest chance of assimilating sufficient resources during the summer to survive the following winter (or breed before the onset of winter for semelparous fishes). In general, the overriding aim of the many reproductive strategies is to produce fry that are capable of exogenous feeding at some time during the spring or early summer.

The actual optimal time for the production of offspring for a particular species, however, depends more specifically on the particular reproductive strategy of that species. Some fish actually produce their gametes at times of the year that are not obviously favorable in terms of the ultimate factors (i.e., very early or late in the spring and even in the autumn for many salmonids) but, in doing so, gain advantages specific to their reproductive strategy. The brown trout (*Salmo trutta*), for example, actually lays its eggs in midwinter, the most inhospitable time of the year (Fig. 2). The female, however, lays very large eggs, which have sufficient resources to fuel an extended period of larval development. Since they do not need to start feeding until the early spring, the large, well-developed fry gain an early competitive advantage. Another advantage of this reproductive strategy is that the adult trout can devote the entire summer (when conditions are optimal) to feeding and converting resources directly into gonadal growth. In contrast, the grayling (*Thymallus thymallus*), which lives sympatrically with *S. trutta*, spawns in early April and lays a larger number of eggs that are considerably smaller than those of *S. trutta* (Fig. 2). The small eggs, however, hatch quickly, resulting in fry that start first feeding in early May. This reproductive strategy has the advantage of releasing a greater number of gametes at a less harsh time, although it has none of the advantages gained by early spawning. Both these reproductive cycles have a very different sequence of seasonal reproduction, which is a function of their reproductive strategy. On average, however, neither is less successful.

When attempting to understand seasonal reproduction in fish it is vital to take into account their reproductive strategies, particularly since fish are one of the most diverse vertebrates in terms of their modes of reproduction.

C. Reproductive Anticipation

Many components in the series of sequential events that make up the reproductive process require a finite period of time for completion. Processes such as gonadal recrudescence and embryonic and larval development are constrained by energetic and developmental time constraints and thus can each take weeks or even months to complete. In many of the larger species of salmonid, gadoid, and perciforms with synchronous spawning and a high gonadosomatic index, gonadal recrudescence may, for example, occur over a period of 6 months (Fig. 2 shows two specific examples). Furthermore, processes such as migration and development of secondary sexual characteristics can also take time to complete, although they generally occur concurrently with the former processes. This forces many reproductive stages to be spread over a considerable period of time and can cause the reproductive cycle to occupy almost the entire year (see Fig. 2). This places an important constraint on the optimal timing of various reproductive events since the entire reproductive process cannot simply occur when conditions become optimal. Since the temporal selection pressures tend to be greatest nearer to the end of this sequence of reproductive events, many of the events are to a certain extent anticipatory i.e., the preceding (preparatory) stages of reproduction, in which the ultimate factors are exerting a weaker effect, start in anticipation of optimal times which are near the end of the process.

There still exists temporal dynamic selection pressures (albeit weaker) acting proximate to earlier stages in a particular reproductive process. As a response to these seasonal pressures, the length of time that a process takes may have been extended or delayed so as to coincide with more favorable times earlier in the year without compromising the timing of the end of the process. This can be particularly important in fish because the optimal time for the end of certain important reproductive processes can be immediately preceded by very harsh conditions (e.g., winter or dry periods prior to the spring and rainy seasons, respectively) that are not particularly conducive to the preparatory early stages of these processes. The strategic response to these pressures can involve specific actions such as delayed ovarian development, dormant embryonic stages, increased egg yolk content for extended egg development, and early migration to spawning sites. Many temperate cyprinids (such as *Cyprinus carpio*), for example, grow their eggs to advanced stage of development during the previous season, and then they maintain their ova at a fixed stage of development over the winter ready for final maturation early in the spring.

Since many components of the reproductive cycle have an element of anticipation, they are reliant on temporal control mechanisms that allow prediction of future (optimal) environmental conditions. This phenomenon has important consequences on the subsequent mechanisms that control the timing of reproduction (see Section III,B) and can constrain the degree of flexibility an organism has to respond to environmental fluctuations.

D. Exact versus Flexible Seasonality

While the pattern of variation in solar radiation is constant and predictable, the climatic conditions that arise from these changes can be far from predictable. Subsequently, in most regions of the world, superimposed against an overriding pattern of seasonal change are proximate fluctuations in the timing and amplitude of climatic variations (and thus, ultimate factors). Therefore, there exists a strong evolutionary advantage of synchronizing reproductive processes directly to these fluctuating environmental conditions, i.e., flexible seasonality, and certain processes are indeed synchronized directly to fluctuating environmental changes. Many reproductive processes, however, are timed by mechanisms that tend to ignore these environmental fluctuations and instead synchronize reproduction to an almost exact time of the year, i.e., exact seasonality. The peak spawning times for Southern Bight plaice, Norwegian herring, and Fraser River salmon, for example, showed virtually no change in their mean spawning date during a 70-year period. Nevertheless, all these fish inhabit unpredictable and fluctuating aquatic environments. Since the cost of rigidity can be high in a fluctuating environment, there must exist strong factors that act against flexibility and cause synchronization to an exact time of the year. The reason why certain reproductive processes occur at an exact time of the year, seemingly ignoring environmental fluctuations, is because of the anticipatory nature of these processes. The processes are forced to start in advance of future optimal conditions and since there is no possibility of predicting the actual (and fluctuating) future conditions, they rely on mechanisms that tend be give the organism a sense of time and not the actual environmental state (see Section III,B). These predictive timing mechanisms are disconnected from the actual environmental conditions because the present environmental conditions are usually poor predictors of future conditions. Fluctuations in the environment that are proximate to these earlier anticipatory stages of reproduction are therefore ignored.

There is, however, a cost in having a fixed reproductive season, and there may be considerable benefits to having the flexibility to respond to local fluctuations. If the fish can postpone egg laying during an unusually cold spring, for example, it may gain a significant evolutionary advantage over species that have fixed breeding seasons. The actual timing of short-term and late-occurring reproductive processes (e.g., final oocyte maturation, ovulation, and spawning) is therefore often synchronized to proximate changes in the environment (such as temperature) which can provide better predictors of the optimal environmental conditions for that particular year. In most temperate cyprinid species, for example, spawning will not occur until a certain minimum water temperature is reached. Even for processes that take longer to complete, such as oocyte growth, there can be modifications to the rate at which the process proceeds, particularly at later stages, so that timing of completion is tailored to suit specific local conditions. In many fishes gonadal development proceeds to a certain fixed stage using anticipatory timing mechanisms, and then the fish wait until the actual climatic conditions become optimal before spawning. In many tropical freshwater species the preparatory gonadal growth will, for example, occur prior to the rainy season using anticipatory mechanisms, yet they will release their eggs only when the rains actually arrive using specific local cues that are associated with the flood waters (such as changes in temperature, flow rate, water chemistry, or even abrupt changes in barometric pressure).

Evolutionary pressures therefore act in two directions; one pushing the reproductive process to a specific time or times in the year, the other predominantly modifying the timing of the later stages of the reproductive processes to allow for local fluctuations in the environment. It seems likely that for the vast majority of fish reproductive seasonality is a compromise between rigidity (i.e., synchronized to a particular date) and flexibility (synchronized to a particular change in the environment).

E. Temporal Synchrony with Conspecifics

An additional evolutionary force that may select for certain reproductive processes, particularly spawning, to occur at a specific time of the year is the requirement of temporal synchrony with other conspecifics and the opposite sex. There are strong selective advantages (increased fertilization success, greater genetic recombination, safety in numbers, etc.) for a synchronous breeding season that can, particularly in dispersed species, increase reproductive success. Such temporal synchrony is often associated with migration to specific breeding grounds (i.e., spatial aggregations). These spawning aggregations can be particularly important in many marine species where they gather in large numbers to spawn (up to 100,000+ individuals), sometimes spawning collectively in a single mass frenzy. Some of these aggregations can be spectacular, particularly in the larger solitary species, such as the groupers (*Epinephelus spp.*), larger labrids, and many of the sharks (for mating). For these aggregations to be successful the fish must "agree" on a specific time (and locality). This is probably one of the main causes of seasonal breeding in the tropics (which is surprisingly common), i.e., not to simply coincide with optimal conditions but to synchronize breeding to a single specific time of the year. Furthermore, this probably also accounts for the high degree of reproductive synchrony to lunar cycles in tropical fish since in the tropics there are fewer perceivable seasonal cues (more than 50 species of tropical marine fish have been shown to spawn in synchrony with lunar cycles).

F. Genetic Plasticity

Seasonally dynamic ultimate factors can show a high degree of spatial heterogeneity even at a similar latitude. There is a strong evolutionary pressure for organisms to show a certain degree of genetic plasticity in their reproductive seasonality to permit adaptation to the local environment. Many fishes, therefore, have higher levels of variation in the timing of reproduction between geographically isolated populations than within populations. The mean spawning times of many salmonids, for example, show a high degree of variation in the timing of reproduction between river systems and even between tributaries. Most populations of rainbow trout (*Oncorhynchus mykiss*), for example, spawn in the autumn (September to November), but there are some strains that naturally spawn during the late winter or even in the spring and summer, although these are generally found at the higher latitudes. Furthermore, several species show different reproductive tactics at different latitudes. Several of the temperate cyprinids, for example, show increasingly protracted spawning seasons at lower latitudes. In many of these cases, however, it is not clear whether these differences, e.g., northern vs southern populations, reflect genetic differences or phenotypic plasticity, i.e., responding to local conditions.

III. PHYSIOLOGICAL CONTROL OF REPRODUCTIVE SEASONALITY

Natural selection can be expected to favor those organisms that have responded to seasonally dynamic ultimate factors by developing appropriate adaptive mechanisms that enable them to synchronize their reproduction to an optimal time of the year. Many fish have evolved specific and often intricate mechanisms that enable temporal control over their reproductive physiology, behavior, and, to a lesser extent, embryonic and larval development (see Section III,C). The physiological mechanisms that control the temporal synchrony of the reproductive process can be classified (conceptually) into two main types (Fig. 3): (i) Those that arise mainly from within the organism are based on internal biological clock-type mechanisms which give the organism a sense of time and allow for anticipatory actions, i.e., endogenous control (see Section III,A), and (ii) those that arise solely, or mainly, as a direct response to seasonal environmental signals or cues and give the organism a measure of the current state of abiotic and biotic environmental variations, i.e. exogenous control (see Section III,B).

This classification should be approached with caution, however, since both types of mechanism involve both exogenous and endogenous components within their function (Fig. 3). It is also likely that the temporal control of many reproductive processes involves a combination of both endogenous and exogenous control mechanisms, particularly at different stages of the reproductive cycle. Furthermore, there may be strong interactions between both types of mechanisms. Exactly how these mechanisms operate, particularly with respect to endogenous control, remains to a certain extent both obscure and controversial.

These physiological mechanisms exert control over the reproductive process by specific modulation of the endocrine system at various sites within the brain–pituitary–gonadal axis. Although the temporal pattern of reproduction and its control is variable between species, the general pattern of gamete production and release and its endocrine control is fairly similar in most fishes.

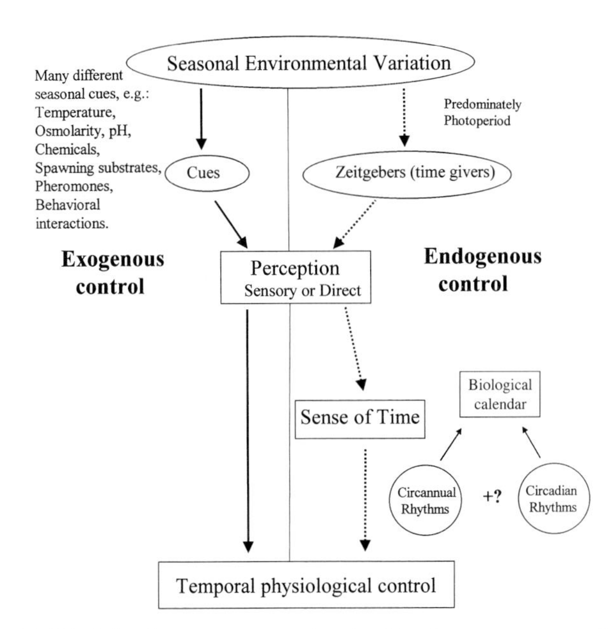

FIGURE 3 Key components involved in the exogenous and endogenous mechanisms that control the timing of reproductive processes.

The reproductive process can be split into two distinct stages: gonadal recrudescence and spawning. The brain, and particularly the hypothalamus, modulates the release of gonadotropins from the pituitary to the peripheral organs mainly by the stimulatory effect of gonadotropin-releasing hormone and in many species by the inhibitory effect of dopamine, although there are several other hormones involved in this neurological modulation. These neurohormones control the release of gonadotropins from the pituitary. Gonadotropin I (GtH I) predominates during oocyte growth (particularly during vitellogenesis) or spermatogenesis (in salmonids and probably most other species, although not for all species investigated, e.g., African catfish) and is involved in both steroid synthesis and uptake of vitellogenin into the developing oocyte. Gonadotropin II (GtH II) is produced later in the gonadal cycle and is involved in gamete final maturation, when it stimulates synthesis of the maturation-inducing steroid(s) which initiates maturation of the germinal vesicle. The main role of gonadotropins is in the stimulation of the gonads and their synthesis of sex steroids, such as 11-ketotestosterone, estradiol, testosterone, 17,20β-dihydroxy-4-pregnen-3-one (17,20βP) or 17,20β,21-trihydroxy-4-pregnen-3-one (20βS). These steroids in turn initiate changes in secondary sexual characteristics, behavior and courtship patterns, in the development and maturation of the gametes and the spawning processes. Steroids feed back to the pituitary–hypothalamus to regulate the release of gonadotropins and the nature of the feedback hormones changes with the maturation of the gonad. In females, plasma estradiol increases during the vitellogenic stage of the oocyte and then declines as the oocytes complete vitellogenesis. In species with synchronous oocyte growth there is then a holding stage during which the activity of the aromatase enzyme declines and the follicle secretes predominantly testosterone. At spawning, a gonadotropin (GtH II) surge is initiated which induces migration of the germinal vesicle to the periphery of the oocyte and stimulates the follicle to produce maturation-inducing steroids (17,20βP or 20βS). This in turn causes the germinal vesicle to break down and the ovulatory process to begin. In males there is a similar sequence, with testosterone and/or 11-ketotestosterone secretion occurring during gonadal recrudescence, followed by secretion of progestogen during the hydration of sperm. Prostaglandins are involved in the final stages of ovulation and spermiation and may also act as pheromones. A full description of these complex endocrine processes is beyond the scope of this article, which concentrates on the temporal mechanisms that modulate these endocrine processes. More detailed accounts of the general endocrinology of fish reproduction can be found elsewhere in this encyclopedia.

Central to the temporal control of all reproductive functions, from gametogenesis to behavior, is the regulatory function of the brain–pituitary–gonadal axis. Most of the temporal control mechanisms probably act by modulation at the brain–pituitary level, although these mechanisms may involve modulation directly at more peripheral sites within the reproductive axis (the exact sites of action and their interaction with the reproductive axis are far from clear).

A. Endogenous Control

1. *Biological Clocks, Calendars, and Photoperiodism*

For the reasons discussed in Section II,C, many stages in the reproductive process are forced to be anticipatory. It is possible to use various external environmental signals (see Section III,B) as predictive cues, provided that they occur at a suitable time of the year, can be perceived by the fish, and are reasonably indicative of future environmental conditions or time. Many environmental cues (with the exception of photoperiod) are poor predictors since they have the tendency to fluctuate (see Section II,D). Anticipatory seasonal reproductive processes, therefore, tend to be disconnected from fluctuating environmental variations and are instead controlled by mechanisms that give the organism a sense of time. The possession of a continuous sense of time (of the year) as a temporal reference point that is unperturbed by local environmental fluctuations is highly advantageous, particularly in fish with their temporally extended and thus highly anticipatory reproductive cycle.

There is increasing evidence that the temporal control of reproduction (particularly at the anticipatory stages) of many fishes is controlled primarily by a biological clock or "calendar"-type mechanism(s). These mechanisms involve the perception of photoperiodic information which is then integrated with various endogenous biological rhythms or biological timekeeping so as to ultimately give the organism a continuous physiological "sense" or measure of the time of year (Fig. 3). There have been many investigations into the effects of photoperiod on the timing of reproduction in fish, probably representing the greatest effort in all fish seasonality research. In common with many other organisms found at higher latitudes, almost every species of temperate fish that has been investigated is photoperiodic to a certain extent. Manipulation of the ambient photoperiod regime frequently results in either a temporal advance or delay in the reproductive process (usually measured by an alteration in the spawning time, although it has also been quantified by hormone measurement or examination of the gonads). The direction and amplitude of the change will depend on the particular photoperiod regime, the fish's reproductive status, and the species under investigation. It was originally hypothesized that these effects could be explained by a passive model whereby the reproductive cycle is simply driven by the environmental photoperiod variations (i.e., exogenous control). More thorough and recent investigations, albeit on a relatively small selection of fish species (primarily the rainbow trout, *O. mykiss*; stickleback, *Gasterosteus aculeatus*, and Indian catfish, *Heteropneustes fossilis*), have found good evidence that these photoperiodic timing mechanisms are driven primarily by endogenous circannual rhythms that are only synchronized by annual changes in the ambient photoperiod, acting as a zeitgebers or "time givers." Evidence for the existence of an endogenous circannual rhythm has come from the discovery that circannual cycles of reproduction persist even under certain constant photoperiodic conditions (i.e., no seasonal change) and that photoperiod manipulations have a different phase shifting ability depending on the phase coincidence between the photoperiod regime and the reproductive cycle or status. Virtually nothing is known about the endocrine basis of these endogenous circannual rhythms either in fish or in other organisms. One possibility is that the endogenous circannual rhythm component could be determined or driven by the actual gonadal cycle itself. Since in many gonadal cycles recrudescence takes place over almost the entire year and during the various phases of gonadal development, the corresponding changing status of steroid feedback to the brain could directly modulate the photoperiodic transduction or response.

All these photoperiodic mechanisms rely on the perception of day length, which has to first be interpreted (i.e., as a long or short day length or a relative change in day length) and then transduced or incorporated into the seasonal control mechanism(s). The perception of photoperiodic information and its interpretation and translation to the brain–hypothalamus probably involve oscillator coincidence models (which are based on endogenous circadian rhythms) or interval timer-type mechanisms. Biological clocks and timekeeping mechanisms (of various sorts and periods) are ubiquitous to all organisms and are implicated in the timing of many seasonal and daily physiological events. However, there is much uncertainty (and controversy) surrounding their exact function, particularly in nonmammalian vertebrates.

Most of the experiments investigating photoperiodism in fish have considered the organism as a "black box" [i.e., manipulating the photoperiod regime (input) and measuring the reproductive response (output)]. They have therefore probably reached a certain limit of understanding in unraveling the mysteries of time measurement. An alternative to this approach has been to examine the neurology and endocrinology involved in the transduction of the photoperiodic and temporal control mechanism(s).

There have been numerous attempts to analyze the role of the pineal organ in the photoperiodic control of reproductive seasonality in fish. The pineal transduces photoperiodic information to the brain directly via neural innervation and release of indoleamines, primarily melatonin, into the circulation. Photoreceptor cells respond to variations in ambient light with a gradual modulation of both neurotransmission to neurons that innervate various brain centers and indoleamine synthesis. Melatonin is produced rhythmically during the dark phase and in some fish species (demonstrated in the pike, *Esox lucius*, and Goldfish, *Carassius auratus*) by an endogenous circadian oscillator that is entrained by photoperiod. Through pinealectomy and/or administration of exogenous indoleamines, the pineal organ has been shown to influence a variety of physiological parameters. These experiments have provided evidence that the pineal has some involvement in the timing of daily and seasonal rhythms in fish, but the effects of pinealectomy or melatonin administration vary with gender, photothermal regimes, reproductive phase, and most particularly with species. In the goldfish, for example, pinealectomy causes a preovulatory surge in gonadotropin and ovulation out of phase with the normal dawn spawning time, whereas in the carp, a closely related species, pinealectomy has no effect on the level or timing of the circadian rhythm of plasma gonadotropin. The pineal is therefore probably only one component in a complex photoperiod-responding system, possibly interacting with circannual rhythms, other photosensory structures such as the retina and extraretinal nonpineal photoreceptors, and other circadian rhythm generators.

2. *Diel and Lunar Periodicity*

This article has concentrated primarily on the temporal control of reproduction with an annual periodicity. There are, however, similar evolutionary advantages in synchronizing spawning to a specific time or times of the day. Spawning often occurs at a specific time of the day, commonly at either dawn or dusk. One of the main advantages of this strategy is to synchronize the spawning process between the sexes and between conspecifics. Similarly, some tropical fish synchronize reproductive processes to lunar cycles (lunar months) to achieve temporal synchrony between conspecifics (see Section II,E and III,B,3). Moreover, many marine species, such as those that inhabit coral reefs or intertidal zones, tend to synchronize spawning with the optimum state of the tidal cycle, mainly to ensure that the tidal currents optimally distribute their eggs. Some species, such as the Californian grunion, *leuresthes tenuis* and Atlantic silverside, *Menidia menidia*, totally rely on tidal rhythmicity because they lay their eggs high in the intertidal zone, thus ensuring protection from aquatic predators. Remarkably, the grunion lays its eggs only on full or new moon phases, thus ensuring maximum development before further inundation by the next high tide.

While little is known about the physiological control of these lunar rhythms, they are probably controlled by some kind of endogenous mechanism. In the case of diel reproductive rhythms, their physiological control is probably based on the same endogenous biological clock-type mechanisms as those described previously (i.e., a circadian biological rhythm synchronized by zeitgebers, such as photoperiod).

B. Exogenous Control

There are many seasonal variations or events in the abiotic and biotic aquatic environment that may be used by an organism to provide a direct measure of the current (or proximate) state of the environment; these are termed proximate factors (Fig. 3). Some of these environmental cues (such as temperature) may be used for predictive control of anticipatory reproductive actions, provided that they are good predictors of either future environmental conditions or time (i.e., stable or predictable) and that they occur at the correct time of the year (see Sections II,C and II,D). In general, however, this exogenous type of control is primarily used to modulate the rate of development at the later stages of gonadal recrudescence or to synchronize the shorter processes associated with gamete maturation and spawning (and also postspawning processes; see Section III,C). This allows the fish to "fine-tune" reproductive processes closer to annual fluctuations in the climate and other seasonal cycles, which can have evolutionary advantages, particularly at the later stages of the reproductive process when ultimate factors are adjacent and exerting their strongest effect (see Section II,D).

With the exception of temperature and pheromonal effects, research efforts have tended to concentrate on the photoperiodic effects on seasonality. While a multitude of environmental factors directly influence both gonadal recrudescence and spawning processes, the mechanisms by which these environmental cues are perceived and then translated into temporal control of the brain–pituitary–gonadal axis are poorly understood. The term exogenous should therefore be used with caution since there obviously exists a strong endogenous element in the perception of the environmental signal and in the translation (and perhaps interpretation) of the cues into the temporal control at the brain–pituitary–gonadal axis. Furthermore, the response to some of these exogenous-type cues is highly dependent on the endocrine status of the individual, which itself may be determined by endogenous circannual and circadian rhythms. In many (or most) fish, reproductive seasonality is probably controlled by a complex of interacting exogenous and endogenous mechanisms and not by mechanisms acting in isolation.

1. *Temperature*

Temperature has a strong influence on the reproductive seasonality of many species of fish. Since temperature can fluctuate in an unpredictable manner, there are limitations on its use as a predictive cue. In some cases, however, it can serve as an effective predictive cue, provided that the environment has either a predictable seasonal cycle of temperature variation, such as in temperate fresh waters in which there are large variations in temperature between summer and winter, or if the organism can somehow ignore the short-term fluctuations in temperature. Some cyprinid species undoubtedly do use temperature as predictive seasonal cue. In some species (such as the shiner, *Notemigonus crysoleucas,* or goldfish) a drop in temperature prior to the winter is necessary to stimulate the onset of the early stages of gametogenesis. Moreover, for most cyprinids a rise in temperature in the spring plays a major role in the development of the later stages of gonadal recrudescence and gamete maturation. Temperature changes are usually only used as a predictive cue in combination with other temporal control mechanisms, especially, photoperiodic control, thereby avoiding costly errors in the timing of reproduction after unusual fluctuations in temperature.

Since temperature is an important ultimate factor, directly effecting the survival of newly released gametes, there are strong evolutionary advantages of being able to respond to short-term fluctuations in temperature at or around spawning time, i.e., flexible seasonality (see Section II,D). This is particularly true for species in which hatching occurs soon after spawning. In many species there is usually a threshold or narrow range of temperatures that must be attained before spawning can occur. When the temperature is too low, the fish will usually maintain its gonads in a semiprepared state ready for when the water warms sufficiently to allow successful spawning. Extreme cold can also stimulate atresia. Temperatures that are too high may also have an inhibitory effect and delay spawning.

Temperature may also modify/modulate reproductive cycles that are determined primarily by other temporal control mechanisms. In temperate regions, low temperatures in the early spring may provide a useful predictor of a later spring; therefore, a long period of cold weather may delay spawning. Salmonids, for example, are normally considered to be almost entirely photoperiodic, although extremely low temperatures delay spawning. This delay in spawning may not be an entirely predictive adaptation (for a late spring) but may simply be caused by the metabolic constraints placed on gonadal recrudescence at low temperatures.

A sudden drop in temperature is also used by some tropical freshwater species as a predictor of forthcoming seasonal floods and can stimulate gonadal development and spawning. The small Amazonian fish, *Paracheirodon innesi*, can be stimulated to undergo final maturation by lowering the water temperature from 25° to 20°C.

Considering the importance of temperature in reproductive seasonality, relatively little is known about the mechanisms by which temperature interacts with the endocrine system to control reproductive seasonality. At the simplest level, temperature could have a direct effect on gonadal development by changing the rate of many different metabolic processes, such as steroid synthesis and metabolism or vitellogenesis. Temperature may also interact with and modulate many different processes at higher levels in the brain–pituitary–gonadal axis. Since temperature has such a widespread effect in a poikilothermic organism, the temporal control of reproduction using temperature may be extremely complex and could involve modulatory processes at many different levels within the organism's reproductive axis. Moreover, many of the endogenous temporal control mechanisms are based on biological clocks or other biological rhythms. These endogenous clock mechanisms are often strongly compensated against changes in temperature so that they function correctly and are independent of any temperature fluctuations.

2. *Physicochemical Environment*

There are many seasonal variations in the physicochemical environment that may be perceived by a fish and can provide useful temporal information in controlling the timing of reproduction, particularly at the later stages of final maturation and spawning. For example, the reproductive cycle of many freshwater fish from lower latitudes is synchronized to the annual pattern of flooding. In these fish, changes in the osmolarity, pH, oxygen, barometric pressure, temperature, flow patterns, and various leached chemicals that occur after a flood can be utilized as cues to induce gametogenesis and spawning. The characoid *Copellea arnoldi*, for example, spawns after a sharp fall in barometric pressure and the cyprinid *Barbus javanicus* can be induced to spawn by simulating the mechanical effect of rainfall by beating the surface of the water with palm leaves. Moreover, it has also been suggested that some fishes (mostly marine species) can detect chemicals that are released during algal blooms in order to predict the future abundance of food, although there has been little research into this area. The spawning time of many fishes is also strongly influenced by the availability of a suitable spawning substrate, which for many of the temperate cyprinids, for example, is determined by seasonal plant growth. In general, the importance of these environmental stimuli as seasonal cues has, in general, been neglected and there is thus very little understanding of their physiological involvement in reproductive seasonality.

Unfavorable water quality, such as high turbidity, low oxygen, high flow rates, and low temperatures, can have a delaying effect on certain reproductive processes, particularly on the timing of spawning when such poor conditions can result in complete reproductive failure. The availability of food has a strong influence on the partitioning of energy into reproductive processes (particularly on gonadal recrudescence) and the level of stress, which may, in some cases, have a modificatory (and complex) effect on reproductive seasonality.

3. *Behavior and Pheromones*

There are strong selective advantages in synchronous spawning (see Section II,E). Many fishes use behavioral (courtship) and/or pheromonal signals to achieve temporally (and spatially) synchronous spawning between sexes and conspecifics (in spawning aggregations). Such behavioral or pheromonal intraspecific signals can be used by either sex to first synchronize the status of each other's reproductive system (a priming effect) and then synchronize the final release of their gametes (a releasing effect). In some species, it seems probable that the male relies almost entirely on behavioral and pheromonal cues from the female for temporal synchrony of sperm maturation, production, and release (and mating behavior) rather than on environmental cues. The complex behavioral and pheromonal interactions that are involved in gamete maturation and spawning, and the endocrinology that controls these processes, have been relatively well researched.

C. Temporal Control of Embryonic and Larval Development

The adaptive mechanisms that control the temporal synchrony of reproduction have so far been discussed from the perspective of the physiological control of the adult gonadal cycle. However, seasonally dynamic selection pressures can also result in mechanisms that control the timing of embryonic and larval development.

For many fish, these developmental stages are short and there is little scope for temporal modification. Moreover, the duration of many of these developmental stages is to a certain extent fixed by overriding genotypic factors which are determined by a particular reproductive strategy of that species. The temporal synchrony of reproduction in most species is therefore primarily determined by the spawning time. The postspawning development of some species, however, can show adaptations for temporal synchrony. Salmonids, for example, often have extended embryonic development, the rate of which can be influenced by temperature, giving a degree of flexibility to respond to proximate conditions. In a particularly cold spring the salmon's eggs and embryos will hatch so as to coincide with the delayed availability of food. For some tropical species of killifish, the ponds in which they live dry up during the dry season and the eggs remain in a static state until the rains arrive and refill the small ponds.

Furthermore, it is important to note that many species of fish (including the majority of elasmobranchs) are ovoviviparous or viviparous. Many of these fish release their gametes seasonally and the gestation period of some species can be considerable (2 years for some elasmobranchs!). Some live-bearing fish may therefore be able to synchronize the timing of gamete release by controlling either the length of gestation (i.e., the rate of internal embryonic development) or the timing of the birth process. However, there have been few investigations on the reproductive seasonality of any of these live-bearing species.

See Also the Following Articles

Circadian Rhythms; Circannual Rhythms; Hormones and Reproductive Behavior, Fish; Pheromones, Fish; Photoperiodism; Seasonal Reproduction, Mammals; Seasonal Reproduction, Birds

Bibliography

Ekstrom, P., and Meissl, H. (1997). The pineal organ of teleost fishes. *Rev. Fish Biol. Fish.* 7, 199–284.

Miller. P. J. (Ed.) (1979). Fish phenology: Anabolic adaptiveness in teleosts. *Symp. Zool. Soc. London* 44.

Munro, A. D., Scott, A. P.; and Lam, L. J. (Eds.) (1990). *Reproductive Seasonality in Teleosts: Environmental Influences*. CRC Press, Boca Raton, FL.

Potts, G. W., and Wootton, R. J. (Eds.) (1984). *Fish Reproduction: Strategies and Tactics*. Academic Press, London.

Seasonal Reproduction, Mammals

Robert L. Goodman
West Virginia University

GLOSSARY

circannual rhythm An endogenously generated rhythm that occurs approximately once a year in the absence of environmental cues.

embryonic diapause Reproductive strategy in which development of the preimplantation embryo is greatly slowed, or arrested, until an environmental cue reactivates it.

leptin Protein produced by adipose that has been postulated to provide information on the metabolic status of mammals to the reproductive system.

melatonin Hormone produced by the pineal gland that provides information on the external photoperiod to the reproductive axis.

photorefractoriness Condition in which a mammal becomes unresponsive to an inhibitory, or stimulatory, photoperiod after prolonged exposure to it.

proximate factor Environmental cue that a mammal uses to time reproductive activity.

ultimate factor Environmental factor that produced the evolutionary pressure causing seasonal breeding.

Seasonal reproduction encompasses any process that results in the birthing period occurring primarily during a specific time of year. This topic thus includes the geophysical factors that influence the timing of reproduction, the energetics of reproduction, the ecological implication of specific reproductive strategies, and the physiological mechanisms that lead to the birth of offspring in a particular season.

I. INTRODUCTION

Ultimately, seasonal reproduction occurs because it favors the survival of the species. For many mammalian species, particularly those living in nonequatorial regions, the young are born in the spring. The waning of the harsh winter climate and the increasing availability of food maximizes the probability of survival of both mother and young. For the mother, lactation is an energy-demanding condition because she provides energy for both herself and her rapidly growing offspring. During lactation, small mammals must more than double their food intake, and large mammals often deplete their fat reserves, to sustain their young. The offspring must be able to maintain their body temperature immediately after birth and have sufficient time for growth and development before the onset of the next winter. Although spring is often the season best suited for both mother and young, environmental conditions sometimes favor a different season. For example, mammals living in arid conditions may reproduce during the rainy season of fall and winter.

The magnitude of the seasonal variation in reproduction varies markedly among mammals (Fig. 1). At one extreme are species, such as some of the seals and ground squirrels, in which births are restricted to a few weeks each year. At the other extreme are humans and some domesticated animals (e.g., cattle and pigs) which show little, if any, seasonal variations in the occurrence of births. The degree of seasonality also varies inversely with the latitude at which the mammal lives. The same animals will have brief, sharp breeding seasons in high latitudes, more

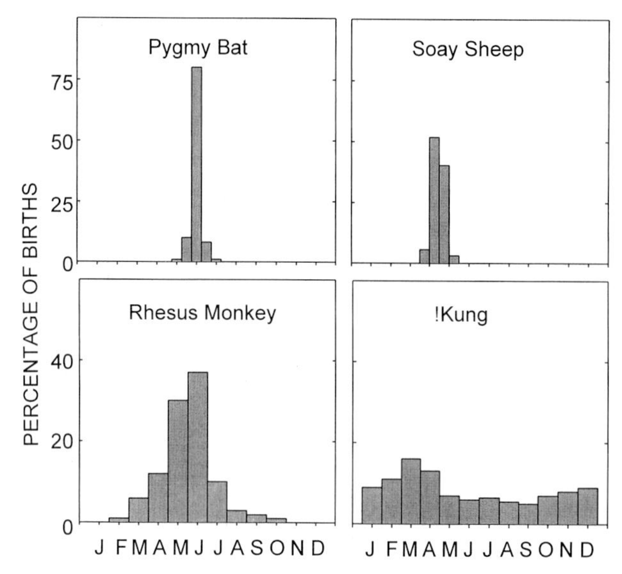

FIGURE 1 Pattern of births in four species selected to illustrate the variability in the duration of seasonal breeding. Data from the tropical Fisher's pygmy bat (*Haplonycteris fischeri*), rhesus monkey (*Macaca mulatta*) living near Puerto Rico, and the desert nomads, !Kung (*Homo sapiens*), are from Bronson and Heideman (1994). Data from the nondomesticated Soay sheep (*Ovis aries*) living on an island off Scotland are redrawn from G. A. Lincoln and R. V. Short, *Recent Prog. Horm. Res.* 36, 1–52, 1980.

prolonged breeding seasons at temperate latitudes, and year-round breeding near the equator. Differences between the sexes in the degree of seasonality are also quite common. A few male mammals invest considerable energy in reproduction [the most extreme example is the brown marsupial mouse (*Antechinus stuartii*) in which all the males die at the end of the breeding season], but for most mammals the male's investment in reproduction ends with insemination. Thus, the consequences to the male of a long breeding season are much less severe than the consequences to the female and, as a general rule, the breeding season of males usually begins before, and ends after, that of females; in some species (e.g., sheep), the females are seasonal breeders but the males are not (Table 1).

While the principles underlying seasonal breeding are fairly straightforward and applicable to most mammals, the reproductive strategies that result in seasonal breeding are diverse and not well understood for the vast majority of mammals. This diversity is illustrated in Table 1, which lists typical reproductive strategies of representative mammals.

II. ENVIRONMENTAL FACTORS CONTROLLING SEASONAL BREEDING

Environmental influences responsible for seasonal breeding can be classified as "ultimate" and "proximate" factors. Ultimate factors produce the evolutionary pressure that caused the species to evolve into a seasonal breeder, whereas proximate factors are the actual environmental cues that the species uses to time reproductive activity each year.

A. Ultimate Factors

Seasonal variations in food availability are undoubtedly the most common ultimate factor responsible for seasonal reproductive patterns. As discussed earlier, adequate food supplies are essential for lactation (particularly for small mammals) and for growth of the infant after weaning. Thus, many mammals time their reproduction so that birth occurs in spring or summer when food is readily available. In other species, however, birth occurs in late winter when food is scarce. The latter strategy maximizes the food available for the growth and development of the offspring after weaning and is usually used by large mammals in which the mother has sufficient energy stored in fat to maintain lactation and the young must lay down similar fat stores before the next winter.

Annual variations in temperature are probably the second most important factor that led to the development of seasonal breeding in nonequatorial regions of the earth. Mammals must expend a considerable amount of energy maintaining a constant body temperature. Because this function is critical for mammalian life, it receives a higher priority than reproduction. This is particularly important for small mammals because the rapid heat loss that occurs as a result of their high surface-to-body ratio requires a high metabolic rate to generate sufficient heat for thermogenesis. This potential problem is further exacerbated by the food scarcity that usually accompanies cold temperatures. Small mammals often use insu-

TABLE 1
Representative Photoperiodic Seasonal Breeders

Species	Common name	Sex[a]	Fertile	Birthing season	Reproductive strategy to time births
Order Monotremata					
Ornithorhynchus anatinus	Duck-billed platypus	F	July–Oct. (S)[b]	Laying: Aug.–Oct.	Limits time of fertility
		M	July–Oct. (S)	Hatching: Aug.–Oct.	
Order Diprotodontia					
Macropus eugenii	Tamar wallaby	F	Jan.–June (S)	Jan.–June	Limits fertility; seasonal embryonic diapause
Setonix brachyurus	Quokka	F	Jan.–June (S)	Jan.–June	Limits time of fertility; lactational diapause
Order Xenartha					
Dasypus novemcinctus	Nine-banded armadillo	F	March–April	March–April	Limits time of fertility; delays implantation
Order Chiroptera					
Myotis lucifugus	Little brown bat	F	Mate: Sept.–April	June–July	Stores sperm in uterus for up to 159 days
			Ovul: April–May		
		M	Sept.–April		
Miniopterus schreibersii	Schreiber's long-fingered bat	F	Aug.–Oct.	June–Aug.	Delays implantation
		M	Aug.–Sept.		
Macrotus californicus	California leaf-nosed bat	F	Aug.–Nov.	May–July	Retards embryonic development
		M	June–Nov.		
Eidolon helvum	Fruit bat	F	April–June	Dec.–May	Embryonic diapause and retards development
		M	April–June		
Order Carnivora					
Vulpes vulpes	Red fox	F	Jan.–April	March–May	Limits time of fertility
		M	Dec.–April		
Ursus americanus	Black bear	F	May–July[c]	Jan.–Feb.	Delays implantation so young are born during hibernation
		M	April–Sept.		
Mustela putorius	Ferret	F	March–Aug.	April–Aug.	Limits time of fertility
		M	March–Sept.		
Mustela vison	Mink	F	late Feb.–April	April–May	Limits time of fertility; delays implantation
		M	Feb.–April		
Martes americana	American marten	F	June–Aug.	March–April	Limits time of fertility; delays implantation
		M	May–Sept.		
Callorhinus ursinus	Northern fur seal	F	June–July	June–July	Mates postpartum; delays implantation so pregnancy lasts 360 days
		M	June–July		
Order Artiodactyla					
Capreolus capreolus	Roe deer	F	July–Aug.[d]	April–July	Limits fertility; usually delays implantation[d]
		M	May–Dec.		
Odocoileus virginianus	White-tailed deer	F	Sept.–Nov.	May–June	Limits time of fertility
		M	Aug.–Feb.		
Ovis aries	Sheep (from high latitudes)	F	Sept.–Jan.	March–June	Limits time of fertility
Order Rodentia					
Mesocricetus auratus	Syrian (golden) hamster	F	March–Oct.	March–Oct.	Limits time of fertility
		M	March–Oct.		

[a] When no M indicated, males are fertile all year.
[b] S, live in Southern Hemisphere; all others live in Northern Hemisphere.
[c] Mates every other year; if little food available skips 1 or 2 years.
[d] Occasionally comes back into heat in October and November and does not delay implantation.

lating behavioral techniques (such as burrows) to help maintain body temperature during winter. However, when they leave these warm environments for extended periods to forage for food, their energy requirements increase proportionately. Consequently, most small mammals are reproductively quiescent during the winter. Cold temperatures are not as stressful for large mammals because of their smaller surface-to-body ratio and the insulating effects of their fat reservoirs. Nevertheless, since their newborn young do not have these ameliorating characteristics, evolutionary pressure to delay birth until spring has probably contributed to the seasonal reproductive patterns of large mammals.

Theoretically, variations in water availability and the number of predators could also serve as ultimate factors. Water may be important in arid environments, but in these conditions seasonal changes in water are so closely associated with changes in food availability that it is impossible to distinguish between the contributions of these two factors. Predator pressure could produce seasonal reproduction by two mechanisms. First, if the predator population shows seasonal fluctuations, it would be adaptive for the prey to bear offspring when predator numbers are low. Alternatively, synchrony of births at one time of year could simply overwhelm the ability of predators to decimate the offspring. Finally, seasonal breeding can be the result of social or behavioral interactions. For example, in many species of seals, the males and females live apart most of the year. Since mating occurs shortly after birth, delivery of the pups must occur during the few weeks when the sexes are together and, consequently, gestation is prolonged so that it lasts a year. The development of seasonal breeding also appears to be linked with the evolution of hibernation in species such as the 13-line ground squirrel (*Spermophilus tridecemlineatus*); in this species changes in ovarian activity are coupled to the occurrence of hibernation.

B. Proximate Factors

While ultimate factors must be inferred from evolutionary and energetic analysis, the proximate factors controlling seasonal breeding can be determined experimentally. Nevertheless, the actual environmental factors timing reproductive function have only been determined for a small percentage of seasonally breeding mammals. Those mammals that have been studied fall into two broad categories: those that use the duration of light (i.e., photoperiod) and those that use food availability to time reproductive function (Fig. 2). The former are usually larger mammals, with a relatively long life span, that use a "K-type" reproductive strategy of producing few offspring at a time and investing considerable energy in their postnatal development. The latter are generally small, short-lived mammals, who use an "r-type" reproductive strategy of producing a large number

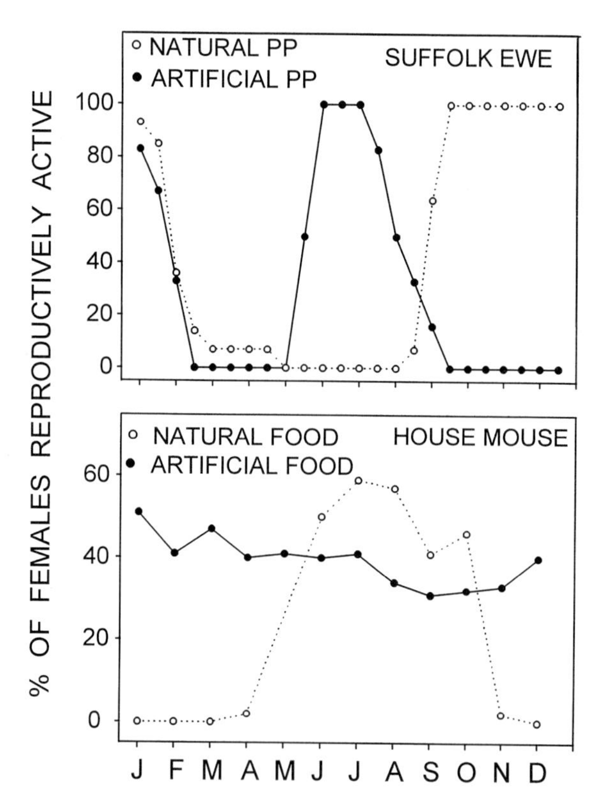

FIGURE 2 Response of a photoperiodic (PP) mammal (top) and a opportunistic mammal (bottom) to artificial manipulation of the environment. Sheep were exposed to natural light (○) or long days (16 hr of light) from November to March (before start of data collection), short days (8 hr of light) from April to July, and then long days through November (●) (redrawn from S. J. Legan and F. J. Karsch, *Biol. Reprod.* 23, 1061–1068, 1980). House mice (*Mus muscularus*) were observed either in the wild (○) with limited food supplies during winter or cohabitating with humans (●) so that food was not limiting (from Bronson and Heideman, 1994).

of offspring at a time and investing almost no energy in them after weaning.

1. *The Planners*

Mammals whose pregnancy lasts for several months must be able to time their reproduction using a geophysical factor that predicts when favorable environmental conditions will occur. Theoretically, any environmental factor that varies in a consistent way with season could be used, but factors such as temperature and rainfall are usually too erratic to provide reliable predictive value. In contrast, the annual variation in the photoperiod is relatively noise free and reproduced exactly from one year to the next. Thus, most mammals living in nonequatorial regions use photoperiod as the proximate cue to time reproduction. There are some exceptions to this general rule. In particular, seasonally breeding mammals with long gestation lengths that live in the tropics are unable to use photoperiod because the length of the day does not vary much in these latitudes. However, there is currently no information on what proximate cues these animals use.

The specific effects of photoperiod on reproductive function depend on the reproductive strategy of a particular species. The most frequently used strategies to control the timing of births are limiting the period of fertility or extending the duration of pregnancy. In species that limit the period of fertility, the males become reproductively active before the females. Thus, almost all females become pregnant shortly after they begin to ovulate and the synchrony among females in the timing of first ovulation results in a synchrony of births. With this reproductive strategy, the breeding season is dependent on the duration of pregnancy. Mammals whose pregnancies last for a few weeks (e.g., hamsters) breed in the spring or summer, those whose pregnancies last 5 or 6 months (e.g., deer and sheep) breed in the fall, and those whose pregnancies last a year (e.g., horses) breed in the spring. Thus, the long photoperiods of spring and summer can be either stimulatory (e.g., in hamsters and horses) or inhibitory (e.g., in deer and sheep) depending on the species.

The most common mechanism for extending pregnancy is embryonic diapause. In species that use this strategy, the embryo develops normally to the blastocyst stage but does not implant. Instead, the embryo remains floating in the uterine lumen and its development slows considerably or even stops. Photoperiod then controls the timing of embryonic reactivation and implantation. Once the embryo reactivates, the course of pregnancy resumes and the embryo develops at its normal rate. Thus, the timing of reactivation and implantation controls the timing of birth. This reproductive strategy is widespread among mammals and is particularly common in marsupials (e.g., wallaby), mustelid carnivores (e.g., mink), and seals. It should be pointed out that these two strategies are not mutually exclusive; some mammals (e.g., roe deer) limit the period of fertility and extend the length of pregnancy using embryonic diapause.

2. *The Opportunists*

Small mammals with life spans of weeks or months have no need to predict the environmental conditions in the future. Instead, many of them have developed an opportunistic reproductive strategy. Because pregnancy lasts for only a few weeks, they can couple their reproductive activity directly to food availability. In these animals, an increase in food supply initiates fertility; pregnancy and birth of offspring follow soon thereafter when there is usually ample food for lactation and growth of the young. Opportunists can easily be distinguished from photoperiodically controlled mammals because they will breed year-round when supplied with sufficient food (Fig. 2). This reproductive strategy is probably particularly common in the tropics where many of these small mammals show changes in reproduction that correlate with the fluctuations in food availability.

Although the dichotomy between the reproductive strategies of large, K-type and small, r-type mammals is a useful generalization, there are numerous exceptions. In particular, the reproductive function of small mammals living in high latitudes with harsh winters is often controlled by photoperiod. The classic example of this is the hamster, which has been used extensively to study the physiological mechanisms underlying seasonal breeding. Similarly, large mammals can sometimes use an opportunistic strategy. This approach is useful in timing births in large mammals with short gestations (e.g., marsupials) but is also sometimes seen in mammals with long gestations (e.g., rhesus monkeys). In the latter, it provides an evolutionary advantage to the adult but is not very useful as a strategy to time births. In addition, lack of food availability can sometimes inhibit reproduction during the breeding season of photoperiodic mammals (e.g., the bear); in this case, the mammal must wait a year to reproduce again.

III. PHYSIOLOGICAL MECHANISMS USED BY PHOTOPERIODIC MAMMALS

The mechanisms underlying seasonal breeding can be divided into two major categories: the photoperiod response system and the reproductive system. The former includes the mechanisms by which the hours of daylight are measured and this information is transduced into an endocrine signal that is used to control reproductive function. Regulation of the reproductive system occurs primarily at hypothalamo–pituitary level, but the critical hormone(s) depends on the reproductive strategy of the species. In mammals that limit the period of fertility photoperiod controls secretion of gonadotropins, whereas in those that use embryonic diapause photoperiod controls prolactin secretion.

A. The Photoperiod Response System

The system for transducing photoperiodic information is the same in all mammals, regardless of the effects of photoperiod on reproduction. This system starts with the retina, which receives the photoperiodic information, and ends with the pineal gland, which transduces this information into an endocrine signal. This system has been found in a wide range of mammals, including marsupials, mink, hamsters, sheep, and ferrets.

1. Anatomy of the Photoperiod Response System

Unlike birds, which have hypothalamic photoreceptors, mammals use retinal photoreceptors to receive the photic information used to time reproduction. This neural information is then transmitted to the pineal gland via a neural pathway (Fig. 3). The retina projects directly to the suprachiasmatic nucleus (SCN), which transmits its output via a multisynaptic pathway to the superior cervical ganglion (SCG). The SCG is the sole innervation of the pineal gland and norepinephrine release from its terminals controls hormone secretion from the pineal.

2. Physiology of the Photoperiod Response System

The first function of this system is to measure the duration of day length or, more specifically, to distinguish between long and short days. In mammals (as in birds) this is accomplished by a circadian rhythm in photosensitivity. The timing of this circadian rhythm is set by the transition from dark to light at dawn and results in a photosensitive window that opens 12 or 13 hr later. If the animal is exposed to light during this photosensitive window, the photoperiod is perceived to be a long day; if there is no light during this window, a short day is perceived. Thus, animals can be "tricked" into perceiving a long day by two brief (15-min long) light pulses 13 hr apart. This endogenous rhythm in photosensitivity resides in the SCN, which controls a number of endogenous circadian rhythms in mammals and has thus been called the "biological clock" of the central nervous system. Neural input from the retina to the SCN sets the rhythm and provides information about the external photoperiod during the period of photosensitivity. The output of the SCN then controls the activity of the pineal gland.

The second function of this system is to transduce this photic information into the endocrine signal that controls reproductive function. This is accomplished by melatonin secretion from the pineal gland. The external photoperiod has two effects on melatonin secretion. First, melatonin release represents an endogenous circadian rhythm that is controlled by the output of the SCN. Thus, photoperiod (acting on the SCN) sets the timing of this rhythm. Second, if a mammal is exposed to light when melatonin is being released, secretion is immediately inhibited. The net effect of these two actions is that melatonin is only secreted at night. Thus, blood melatonin concentrations represent a hormonal analog of the external light–dark cycle (Fig. 3). Classic ablation and re-

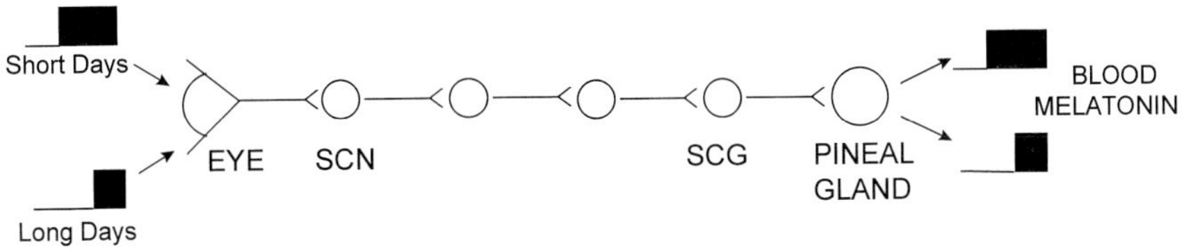

Short Days
EYE SCN SCG PINEAL GLAND
BLOOD MELATONIN
Long Days

FIGURE 3 Photoperiod response system that converts changes in day length in the external environment into a hormonal signal (blood melatonin concentrations). SCN, suprachiasmatic nucleus; SCG, superior cervical ganglion. Shaded blocks depict periods of darkness and elevated melatonin levels.

placement studies have demonstrated that this melatonin pattern controls reproduction in seasonal breeders and that the duration of the nightly melatonin rise is the critical characteristic indicating the length of the day (e.g., if the melatonin rise is longer than 12 hr it is a short day).

The mechanisms by which the duration of melatonin controls reproductive function are not known. Melatonin receptors have been found in a number of neural areas, with the highest concentrations in tissue at the base of the hypothalamus known as the pars tuberalis. Recent work suggests that the pars tuberalis controls seasonal variations in prolactin secretion and may therefore be important for the control of embryonic diapause. However, this area does not appear to be involved in the seasonal changes in gonadotropin secretion that limit fertility to a specific time of year. The sites where melatonin acts to control gonadotropin secretion are species dependent. It appears to act in the SCN of the Siberian hamster, in the anterodorsal region of the medial basal hypothalamus (MBH) of the Syrian hamster, and in the posterior portion of the MBH of the sheep. Since the effects of melatonin on gonadotropin release are species dependent, it is not surprising that it acts in different areas to produce these effects.

B. Changes in the Reproductive Axis Responsible for Seasonal Variations in Fertility

Photoperiod (acting via melatonin) limits fertility by controlling release of gonadotropin-releasing hormone (GnRH) from the hypothalamus. GnRH is secreted episodically and, in those mammals that have been carefully studied, changes in GnRH pulse frequency appear to be critical for seasonal breeding. During the nonbreeding season, GnRH pulse frequency is low and consequently luteinizing hormone (LH) and follicle-stimulating hormone (FSH) secretion is minimal and the gonads are inactive; during the breeding season, GnRH pulse frequency is high, resulting in elevated LH and FSH concentrations and active gonads.

1. *Neuroendocrine Mechanisms*

Seasonal changes in GnRH pulse frequency can be caused by two different neuroendocrine mechanisms. First, in many species there is a seasonal change in response to the negative feedback actions of gonadal steroids (estradiol in females and testosterone in males). During the nonbreeding season the response to this negative feedback is much greater than during the breeding season. Consequently, GnRH secretion is held in check by very low steroid concentrations during the nonbreeding season so that the production of gonadotropic, and gonadal, hormones is insufficient for fertility. During the breeding season, these low steroid levels are not able to inhibit GnRH secretion so that increased GnRH release stimulates gonadotropic, and gonadal, hormone production to levels compatible with fertility. Second, in many other animals seasonal changes in GnRH secretion that are independent of steroid negative feedback appear to be important for seasonal breeding. In these species, marked seasonal changes in GnRH secretion occur in gonadectomized animals; consequently, levels of LH and FSH are often undetectable in the nonbreeding season and elevated in the breeding season. Since these changes occur in the absence of the gonads they cannot be due to changes in steroid negative feedback. The relative importance of these two mechanisms to seasonal control of GnRH varies among mammals, but both mechanisms are evident in those species that have been studied, indicating that they may be functionally, or mechanistically, coupled.

The changes in hypothalamic function responsible for these seasonal variations in GnRH secretion remain unclear, but recent work in the ewe has led to the hypothesis that a set of dopaminergic neurons play a key role in the seasonal changes in estradiol negative feedback (Fig. 4). This model proposes that these neurons inhibit GnRH pulse frequency, are sensitive to estradiol, and are controlled by photoperiod (acting via melatonin) so that they are functional during the nonbreeding season but inactive during the breeding season. Thus, estradiol stimulates these dopaminergic neurons in the nonbreeding season resulting in suppression of GnRH pulse frequency, but it cannot do so during the breeding season when this neural system is not functional. There is now strong evidence for this model in the ewe, and the hypothalamic dopaminergic neurons involved have been identified, but the applicability of this model to other seasonal breeders is unclear.

2. *Photoperiodic Control*

Theoretically, mammals could rely solely on photoperiod to time their periods of fertility. For example, long-day breeders could begin breeding when daylight lasted more than 12 hr and stop breeding when days became shorter than 12 hr. There are, however, no known examples of this simple situation. Instead, a complex interaction between photoperiod and timing mechanisms within the animal occurs. The use of endogenous timing mechanisms introduces considerable flexibility into the timing of breeding and nonbreeding seasons. For example, it is often useful for males to initiate gonadal activity in late winter ahead of photoperiodic changes because it takes about 2 months to produce mature sperm. These mechanisms are also critical for hibernating species, such as the 13-line ground squirrel, which begin gonadal activation before the end of hibernation and thus cannot rely on photoperiodic information for this transition.

Endogenous timing mechanisms fall in to two broad categories: rhythmic and nonrhythmic. Animals that use the former will continue to show fluctuations between fertility and infertility when kept under constant environmental conditions. Since these changes occur with a free-running rhythm that approximates a year they are usually considered endogenous circannual rhythms. Changes in external photoperiod synchronize this rhythm with the seasons, much as the 24-hr light–dark cycle synchronizes endogenous circadian rhythms. In addition, photoperiodic information can be used to set the periodicity of the rhythm to a year and to adjust the

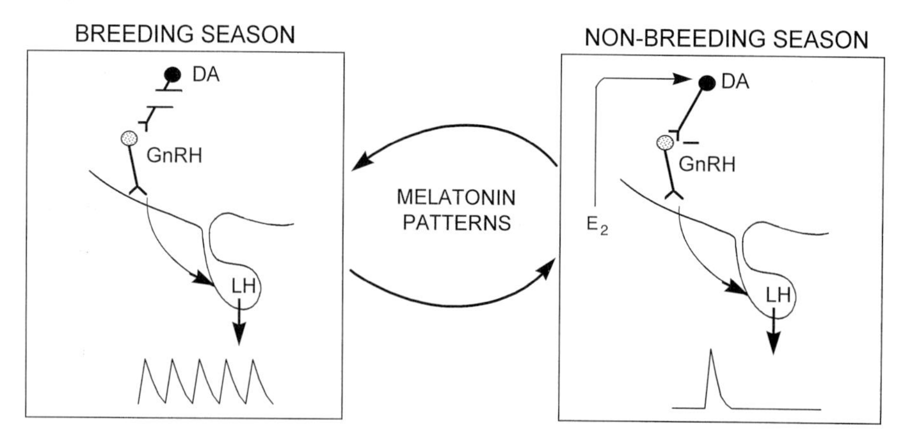

FIGURE 4 Model to account for the seasonal variation in response to estradiol (E_2) negative feedback in the ewe. The pattern of circulating melatonin concentrations controls a set of dopaminergic (DA) neurons so that they are functional in nonbreeding season and inactive during the breeding season. These DA neurons inhibit GnRH pulse frequency and are sensitive to E_2 so that E_2 inhibits GnRH pulse frequency during nonbreeding season but not during the breeding season.

percentage of time that animals spend in a given reproductive state each year. Interestingly, in the few mammals whose circannual rhythms have been studied (e.g., sheep and red deer), thyroid hormones appear to be required for the transition into the nonbreeding season.

Animals that use nonrhythmic timing mechanisms do not show circannual fluctuations when placed in constant photoperiod, but they do undergo one spontaneous change in reproductive activity. For example, if hamsters are placed in short days, testicular activity is inhibited. However, about 4 or 5 months later the testes will recrudesce even though the animals are still exposed to this inhibitory photoperiod. This condition, which is often referred to as photorefractoriness, will then continue indefinitely as long as the hamster is kept in short days. After exposure to long days (which does not alter gonadal activity because long days are stimulatory in hamsters) they lose this photorefractoriness so that testes will involute if they are returned to short days. In the wild, the short days of fall induce testicular regression in hamsters and 5 months later, in February, the testes begin to function because of photorefractoriness. Thus, hamsters initiate testicular activity before they are exposed to long days and are fertile when spring arrives. This type of interaction between photoperiod and nonrhythmic endogenous timers has been found in a number of mammals and is probably widespread.

C. Changes in the Reproductive Axis Responsible for Prolongation of Pregnancy

There are unusual approaches to controlling the interval between insemination and delivery. For example, in some bats the females will mate without ovulation and store the sperm for several months so that newly ovulated ova can be fertilized at the appropriate time of year. In a few cases (often associated with a fall in metabolic rate during hibernation), the rate of fetal development is slowed for the whole duration of pregnancy. This retarded development occurs primarily in bats and may be a response to changes in ambient temperature. However, embryonic diapause, in which the development of the embryo is greatly slowed or completely arrested prior to implantation, is the most common strategy used. Despite its widespread occurrence, the neuroendocrine mechanisms underlying embryonic diapause have only been examined in a few animals. In these mammals, the corpora lutea control reactivation of the embryo. During the period of embryonic quiescence circulating progesterone concentrations remain low; then, in response to a photoperiodic signal, secretion of progesterone and an unidentified peptide hormone from corpora lutea increases rapidly, and these two hormones act in concert to reinitiate normal embryonic development. The photoperiodic information is transmitted to the hypothalamo–pituitary axis using melatonin secretion from the pineal (see Section III,A), but in this case melatonin influences prolactin secretion. In the mink, a long-day melatonin pattern stimulates prolactin secretion, which then acts on corpora lutea to increase production of both progesterone and the second luteal factor. On the other hand, in the wallaby, a daily prolactin surge inhibits progesterone secretion from the corpus luteum and a decrease in photoperiod blocks this prolactin increase, allowing progesterone secretion to increase, which reactivates the embryo. The role of prolactin in other mammals that use embryonic diapause for seasonal breeding remains to be determined.

IV. PHYSIOLOGICAL MECHANISMS USED BY OPPORTUNISTIC BREEDERS: EFFECTS OF NUTRITION

Since opportunists will breed year-round if sufficient food is available, reproductive function must be either actively inhibited by signals associated with insufficient food intake or stimulated by signals that increase with nutritional status. There is considerable interest in this phenomenon because nutritional levels can clearly influence reproductive function in humans. It should be pointed out, however, that it is often difficult to disassociate the inhibitory effects of undernutrition on reproduction from the effects of stress; stressors often inhibit reproductive function [via corticotropin-releasing hormone (CRH) or glucocorticoids] and undernutrition usually activates the hypothalamo–hypophyseal–adrenal axis. Malnutrition probably acts as a stressor to inhibit reproduction in rats, but its suppressive effects on reproduction in sheep and primates are independent of increases in CRH or adrenal steroids. Thus, the role of low nutrition and stress must be evaluated species by species.

A. Neuroendocrine Mechanisms of Nutritional Control of Reproduction

Nutritional control of reproduction occurs via the same basic mechanisms as photoperiodic control. Thus, low levels of food intake inhibit reproductive function primarily by inhibiting GnRH pulse frequency. Moreover, this inhibition can occur either by an increase in response to steroid negative feedback or by steroid-independent mechanisms. The hypothalamic neurotransmitters mediating these inhibitory effects remain to be determined, but there is some evidence that neuropeptide-Y, endogenous opioid peptides, and CRH may be important.

B. Signals of Nutritional Status to the Reproductive Tract

Three types of signals could be used to monitor nutritional status: hormonal signals that regulate metabolism, neural signals from the gastrointestinal tract, and metabolites that are used for energy (Fig. 5).

1. *Hormones Regulating Metabolism*

The metabolic hormones, insulin and glucagon, are potential signals of the nutritional status of animals, but there is no conclusive evidence that either hormone plays an important role in the control of GnRH secretion. Similarly, insulin-like growth factor, growth hormone, and other classical metabolic hormones can also probably be ruled out as important hormonal signals controlling the reproductive axis. Thus, current work in this area focuses on leptin. Leptin is a recently discovered protein produced by fat cells that may provide an index of the energy reservoirs stored in adipocytes. Since gonadal activity is positively correlated with body fat in some circumstances, leptin may be an important regulator of GnRH secretion. While there is some evidence that leptin stimulates reproductive activity, the physiological role of this protein has yet to be established.

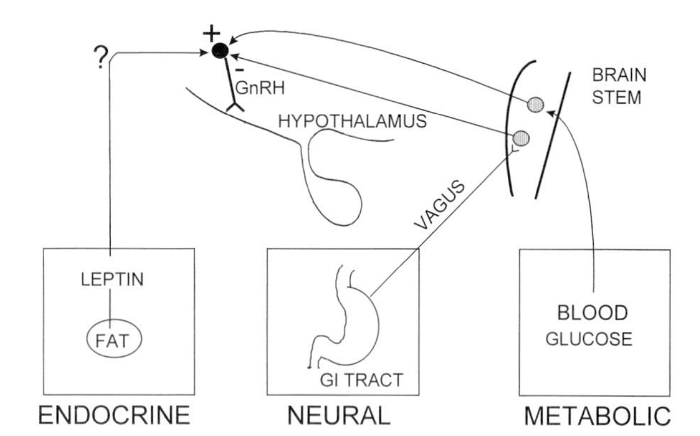

FIGURE 5 Possible signals used by opportunistic breeders to link food intake with GnRH release. Leptin and blood glucose are postulated to increase with a meal and stimulate GnRH secretion, whereas neural signals from the GI tract associated with fasting are postulated to inhibit GnRH release. Both the neural signals and blood glucose are thought to act via the brain stem, but the site of action of leptin is unknown (portions of this figure were redrawn from Maeda and Tsukamura, 1996).

2. *Neural Signals from the Gastrointestinal Tract*

It is well established that sensory input from the gastrointestinal (GI) tract via the vagus nerve provides information important for the control of food intake and this information may also be used to control GnRH secretion. In the rat, transection of the vagal efferents from the upper GI tract produces an immediate increase in episodic LH secretion in the fasted rat. Thus, the vagus actively inhibits GnRH secretion, probably via a pathway that involves the brain stem (Fig. 5). However, the role of this neural pathway in the control of reproductive activity has yet to be determined in other species.

3. *Direct Metabolic Signals*

The simplest mechanism for providing information on the nutritional status of an animal is the circulating concentration of substances that can be used for energy. Since glucose is usually the primary source of energy for the brain in nonruminants, it has received the most attention. In several mammals, postprandial increases in blood glucose levels are required for normal gonadal function. Thus, blockade of glucose utilization with 2-deoxyglucose inhibits episodic LH release. In some ruminants, free fatty acid levels provide information linking gonadotropin secretion to nutrient intake, and in other species both glucose and free fatty acid can be used for this link. Thus, many mammals monitor circulating levels of glucose and/or free fatty acids and use this information to control GnRH secretion. There appear to be important glucoreceptors in the caudal brain stem (area postrema) in rats, sheep, and hamsters, but where free fatty acids act remains unknown.

V. CONCLUSION

Over the course of evolution, mammals have adapted to occupy a multitude of different ecological niches. Given the importance of reproduction to the survival of a species, it is not surprising that a wide range of reproductive strategies have evolved to cope with seasonal fluctuations in the environment. Despite this diversity, a few common physiological mechanisms appear to play a major role in controlling seasonal breeding in mammals: (i) photoperiodic mammals use melatonin to time reproductive function; (ii) changes in GnRH pulse frequency are critical for timing seasonally controlled alterations in fertility; and (iii) increased secretory activity of corpora lutea, in response to changes in prolactin release, reactivates the embryo in species using prolongation of pregnancy to time the season of birth.

Acknowledgment

I could not have written this article without the expertise and advice provided by Dr. Bob Cochrane, Adjunct Professor, Animal and Veterinary Sciences, West Virginia University and the financial support of NIH Grant HD17864.

See Also the Following Articles

Circadian Rhythms; Circannual Rhythms; Cricetidae (Hamsters and Lemmings); Diapause; Microtinae (Voles); Nutritional Factors and Reproduction; Photoperiodism; Seasonal Reproduction, Birds; Seasonal Reproduction, Fish; Sheep and Goats

Bibliography

Bronson, F. H., and Heideman, P. D. (1994). Seasonal regulation of reproduction in mammals. In *The Physiology of Reproduction* (E. Knobil and J. D. Neill, Eds.), Vol. 2, 2nd ed., pp. 541–583. Raven Press, New York.

Goodman, R. L. (1994). Neuroendocrine control of the ovine estrous cycle. In *The Physiology of Reproduction* (E. Knobil and J. D. Neill, Eds.), Vol. 2, 2nd ed., pp. 659–709. Raven Press, New York.

Goodman, R. L., and Karsch, F. J. (1981) A critique of the importance of steroid feedback to seasonal changes in gonadotrophin secretion. *J. Reprod. Fertil. Suppl.* **30**, 1–13.

Hayssen, V., Van Tienhoven, A., and Van Tienhoven, A. (1993). *Asdell's Patterns of Mammalian Reproduction.* Comstock, Ithaca, NY.

Hinds, L. A. (1994). Prolactin, a hormone for all seasons: Endocrine regulation of seasonal breeding in the Macropodidae. In *Oxford Reviews of Reproductive Biology* (H. M. Charlton, Ed.), Vol. 16, pp. 250–301. Oxford Univ. Press, Oxford, UK.

I'Anson, H., Foster, D. L., Foxcroft, G. R., and Booth, P. J. (1991). Nutrition and reproduction. In *Oxford Reviews of Reproductive Biology* (S. R. Milligan, Ed.), Vol. 13, pp. 239–311. Oxford Univ. Press, Oxford, UK.

Karsch, F. J., Bittman, E. L., Foster, D. L., Goodman, R. L., Legan, S. J., and Robinson, J. E. (1984). Neuroendocrine basis of seasonal reproduction. *Recent Prog. Horm. Res.* **40**, 185–225.

Maeda, K., and Tsukamura, H. (1996). Neuroendocrine mechanism mediating fasting-induced suppression of luteinizing hormone secretion in female rats. *Acta Neurobiol. Exp.* **56**, 787–796.

Sundqvist, C., Ellis, L. C., and Bartke, A. (1988). Reproductive endocrinology of the mink (*Mustela vison*). *Endocr. Rev.* **9**, 247–266.

Turek, F. W., and Van Cauter, E. (1994). Rhythms in reproduction. In *The Physiology of Reproduction* (E. Knobil and J. D. Neill, Eds.), Vol. 2, 2nd ed., pp. 487–540. Raven Press, New York.

Seasonal Reproduction, Marine Invertebrates

John S. Pearse
University of California, Santa Cruz

GLOSSARY

endogenous control Control of gametogenesis and other reproductive activities by physiological factors. Generally, these are hormonal, neurohormonal, or neural factors within individuals that are influenced by external, environmental factors.

exogenous control Control of gametogenesis and other reproductive activities by external, environmental factors, which usually act by influencing endogenous factors. In species with seasonal patterns of reproduction, these factors are generally seasonally changing sea temperature, photoperiod, and perhaps salinity and food.

gametogenesis The production of gametes (sperms and eggs) in the gonads; generally synchronized within individuals through endogenous control and among individuals within a population of a species through exogenous control.

photoperiodism Sensitivity to changing day lengths, whereby day lengths provide information to organisms about seasons; used by many organisms to synchronize reproductive and other activities.

proximate factors Factors that directly influence the timing of reproduction (or other activities) divided into endogenous and exogenous components; exogenous factors often provide environmental cues about seasons that organisms can use to synchronize activities such as gametogenesis and spawning.

spawning The release of gametes for subsequent fertilization or of fertilized eggs. Both males and females may release sperms and eggs together so that fertilization occurs in the surrounding seawater (broadcast spawning). Also, only males may spawn gametes, whereas females retain eggs for internal fertilization, either capturing sperms released into the water by males or receiving sperms during copulation. Females later spawn fertilized eggs or release young at later stages of development.

ultimate factors Factors that select for spawning (and other activities) to occur at a particular time or season, and that act on an evolutionary time scale and favor organisms whose reproductive timing maximizes the production of offspring into the succeeding generation.

Most marine organisms, like those on land, undergo seasonal periods of reproduction when gametes are produced and released. The timing of reproduction varies among species both within and among habitats; many species have very precise and predictable reproductive periods. In general, shallow-water, temperate species reproduce in spring and summer, when conditions favor survival of larvae and juveniles. Species in marine environments with less pronounced seasons generally reproduce over more extended periods. Environmental factors regulating the timing of reproductive activities include changes in sea temperature, photoperiod, salinity, and possibly food. Synchronization of reproduction, however, involves both internal (endogenous) and external (exogenous) control mechanisms, neither of which are well understood for most species.

I. INTRODUCTION

Seasonal timing of reproduction is found in most marine invertebrates that have been examined, whether polar, temperate, or tropical; shallow or deep; or nearshore or oceanic. Different species display different patterns of seasonal reproduction but most are usually reproductively active for several months; in extreme cases, however, spawning can last for only a few hours each year. General patterns of seasonal reproduction are well understood, as are the specific patterns of a few well-studied species. However, their selective value and how they are synchronized remain largely unknown.

II. PATTERNS

Seasonal reproduction of marine invertebrates has been studied mainly in shallow coastal waters of the North Atlantic and northeast Pacific Oceans. In the former, most species commence gametogenesis in spring as sea temperatures increase, and spawning occurs in early summer when planktonic food for the larvae and juveniles peaks. This pattern appears to be general for shallow-water temperate regions around the world (South Africa, Australia–New Zealand, Chile, and Japan). However, along the west coast of North America, where temperature changes are minor and upwelling-induced phytoplankton production begins in early spring, many species spawn in late winter to early spring.

Spawning also tends to occur in late winter to early spring at high latitudes (both arctic and antarctic), where development is slow and feeding stages of larvae and juveniles are reached in synchrony with the brief midsummer period of phytoplankton production. At low latitudes in the tropics, where phytoplankton production is more uniform, different species spawn at different times during the year with little overall pattern; some species reproduce throughout the year. Similarly, in deeper waters down to abyssal depths, where conditions tend to not change throughout the year, reproduction tends to occur throughout the year. These overall patterns are very general and exceptions occur for particular species and places; some shallow-water temperate species reproduce throughout the year, and some tropical and deep-sea species have well-defined, discrete spawning seasons.

III. PROXIMATE FACTORS

Discrete spawning seasons usually involve a synchronized series of events whereby all individuals within a population of a species begin gametogenesis at about the same time, gametogenesis proceeds at about the same rate among and within individuals, and spawning is synchronized among individuals to a particular time of the year. Seasonal reproduction is regulated mainly at the level of the initiation of gametogenesis, thereby setting the synchrony for subsequent events. Factors that synchronize gametogenesis can be different from those that synchronize spawning and involve both endogenous and exogenous components.

Hormones and neurohormones are endogenous factors that regulate and synchronize gametogenesis within individuals. These have been little studied among most marine invertebrates, with the notable exception of some polychaete worms and decapod crustaceans, in which neurohormones produced in the brain generally inhibit gametogenesis.

Synchrony among individuals within a population also involves exogenous factors that influence the initiation of gametogenesis. Change in sea temperature is the most common example of such an exogenous factor; the correlation between increasing sea temperature and the initiation of gametogenesis in the spring of many shallow-water, North Atlantic species provides suggestive evidence for the role of temperature. The relationship between changing sea temperature and the initiation of reproductive activities was noted by the British biologist James Orton in the early twentieth century and is often referred to as "Orton's rule." Subsequently, the role of changing temperature in synchronizing reproduction was experimentally shown when gametogenesis in oysters was initiated out of season in midwinter simply by increasing the water temperature. Manipulating the timing of reproduction by changing sea temperature is now standard practice in mariculture. Conversely, some polar barnacles that spawn in midwinter in Britain have been experimentally induced to initiate gametogenesis and spawn in midsummer by decreasing the water temperature. However, the mechanism by which temperature change acts on gametogenesis remains unresolved. Moreover, many species with discrete seasonal breeding periods inhabit regions in which sea temperature changes are small and/or aseasonal, e.g., polar, tropical, and deep seas.

Seasonally changing day lengths also have been shown to synchronize gametogenesis in a wide variety of marine invertebrates. Experimental animals in which reproduction has been manipulated by photoperiod include sea urchins, sea stars, shrimps, polychaetes, squids, tunicates, and cephalochordates. Some appear to respond simply to day length; for example, gametogenesis is initiated in the winter–spring spawning sea urchin *Strongylocentrotus purpuratus* of the eastern North Pacific when day length is less than 12 hr and is inhibited when day length is more than 12 hr. Other species, such as the eastern North Pacific sea star *Pisaster ochraceus,* have an endogenous calendar which is insensitive to day length alone but can be experimentally switched when the phase of the seasonally changing photoperiod is changed (e.g., the longest days are in midwinter and the shortest days are midsummer).

In addition to seasonally changing sea temperatures and photoperiods, other exogenous factors that change seasonally and might be important for synchronizing gametogenesis include salinity (especially in tropical areas experiencing seasonal monsoons) and levels of food (especially phytoplankton in areas of seasonal upwelling or seasonal fluxes of organic material to the deep sea). Although adequate nutrients are necessary for gametogenesis to proceed, and extremes of salinity are detrimental in general, there is no convincing experimental evidence that either salinity or food act directly as exogenous factors synchronizing seasonal gametogenic patterns of marine invertebrates.

IV. ULTIMATE FACTORS

While seasonally changing sea temperatures, photoperiods, or other environmental factors might act as exogenous components of the proximate regulation of reproduction of marine invertebrates, these factors are not necessarily those that select and maintain the seasonal reproductive patterns. Rather, there must be a selective advantage to spawn at particular times during the year for seasonal reproduction to persist; environmental factors selecting favorable times for reproduction are often referred to as "ultimate factors." Because these factors act over evolutionary time, they are challenging to examine experimentally.

Seasonally changing food for planktotrophic larvae is usually considered to select for seasonal reproduction of marine invertebrates, and the correlation between spring–summer spawning and larval food supply in shallow-water, temperate and polar seas supports this notion. The British biologist Dennis Crisp noted this relationship in the mid-twentieth century and proposed that seasonal reproduction in the sea is selected to favor larval survival; this is often referred to as "Crisp's rule."

While adequate larval food is usually treated as the critical selective factor, many species with nonfeeding larvae also have spring–summer spawning periods in temperate regions. In addition, even species with feeding larvae may not depend on seasonal supplies of larval food; larvae of the sea star *P. ochraceus* produced 6 months out of phase (in the fall) through photoperiod manipulation survived as well on ambient food levels of the fall as did those produced during the normal spring period. Other factors that might act on the larvae include seasonally changing salinity (especially in shallow-water areas that experience strongly seasonal rainfall) and seasonally changing currents that transport water and larvae offshore during part of the year (e.g., areas of seasonal upwelling).

Selection could also act on other phases of the life cycle. Larval settlement and early juvenile survival and growth, for example, could be of major importance. In regions with winter storms, open space suitable for settlement could be most available in spring and early summer, and winter–spring spawning would provide larvae for recruitment during that time. Seasonal supplies of food for juveniles, or seasonally changing temperature with higher temperatures favoring juvenile growth, could also be of decisive selective value.

Finally, selection could act on the adults. Food supplies and other conditions that maximize gamete production could be seasonal (in which case proximate and ultimate factors could be the same). Also, selection could act to maximize fertilization by synchronizing spawning among all the individuals in the population (in which case the timing itself would not be of much importance—only that gametogenesis results in animals ready to spawn at the same time). These possibilities have not been critically examined, nor do they appear to be likely in shallow-water temperate habitats in which many species spawn in the late spring and summer, thus placing larvae and juveniles in environments that appear to be most favorable for their survival. However, where food input is strongly seasonal, such as in polar seas and perhaps some deep seas, or where there is little seasonal change, such as in some tropical and deep-sea areas, the ultimate causes of reproductive seasons may rest with selection on the adults. These possibilities need further research.

See Also the Following Articles

Marine Invertebrates, Modes of Reproduction in; Seasonal Reproduction, Mammals; Seasonal Reproduction, Birds; Seasonal Reproduction, Fish

Bibliography

Giese, A. C., and Kanatani, H. (1987). Maturation and spawning. In *Reproduction of Marine Invertebrates. IX. General Aspects: Seeking Unity in Diversity* (A. C. Giese, J. S. Pearse, and V. B. Pearse, Eds.), pp. 251–329. Blackwell, Palo Alto, CA.

Giese, A. C., and Pearse, J. S. (1974). Introduction: General principles. In *Reproduction of Marine Invertebrates. I. Acoelomate and Pseudocoelomate Metazoans* (A. C. Giese and J. S. Pearse, Eds.), pp. 1–49. Academic Press, New York.

Halberg, F., Shankaraiah, K., Giese, A. C., and Halberg, F. (1987). The chronobiology of marine invertebrates: Methods of analysis. In *Reproduction of Marine Invertebrates. IX. General Aspects: Seeking Unity in Diversity* (A. C. Giese, J. S. Pearse, and V. B. Pearse, Eds.), pp. 331–384. Blackwell, Palo Alto, CA.

Palmer, J. D. (1995). *The Biological Rhythms and Clocks of Intertidal Animals*. Oxford Univ. Press, New York.

Pearse, J. S., Eernisse, D. J., Pearse, V. B., and Beauchamp, K. A. (1986). Photoperiodic regulation of gametogenesis in sea stars, with evidence for an annual calendar independent of fixed day length. *Am. Zool.* **26**, 417–431.

Tyler, P. A., and Young, C. M. (1992). Reproduction in marine invertebrates in "stable" environments: The deep sea model. *Invertebr. Reprod. Dev.* **22**, 185–192.

Young, C. M. (1994). A tale of two dogmas: The early history of deep-sea reproductive biology. In *Reproduction, Larval Biology, and Recruitment of Deep-Sea Benthos* (C. M. Young and K. J. Eckelbarger, Eds.), pp. 1–25. Columbia Univ. Press, New York.

Sea Urchins

John S. Pearse

University of California, Santa Cruz

GLOSSARY

aboral ring complex A coelomic ring system around the apical anus with a nerve ring and hemal vessel that connect the five gonoducts; it probably has a role in regulating and synchronizing gametogenesis and spawning.

lecithotrophy Use of maternally derived nutrients (yolk) by embryos and larvae for energy during development.

nutritive phagocytes Somatic cells of the germinal epithelium that accumulate lipids and yolk proteins during gonad growth, decline during gametogenesis, and phagocytize relict gametes after spawning is complete.

ova The haploid female gametes that are spawned by sea urchins. Unlike most animals, sea urchins complete meiosis in the ovary before spawning.

planktotrophy Feeding by larvae on plankton, mainly phytoplankton, to fuel development.

pluteus The typical planktotrophic larva of echinoids, with eight arms supported by calcite rods; also found in ophiuroids.

rudiment A mass of tissue, usually on the left side of the pluteus, that forms most of the body of the juvenile during metamorphosis.

Sea urchins are favored subjects for reproductive studies because they are common and easily collected from most shores, their massive gonads produce millions of small uniform eggs and sperms, fertilization is readily achieved, and the embryos and larvae are easily reared under relatively simple conditions. These studies have provided fundamental insights into the areas of fertilization, early cell cleavage and cell lineages, gene expression during development, and larval ecology.

I. INTRODUCTION

Sea urchins, heart urchins, and sand dollars (phylum Echinodermata, class Echinoidea) have been used as model systems for reproductive and developmental studies for over 150 years. The process of fertilization was first observed with eggs and sperms of sea urchins in 1847; it remains a major research topic today, and these gametes are still used. Early cleavage and regulative development were first described with sea urchin embryos, and today the same systems are fertile areas of cell lineage studies. Sea urchin embryos have been particularly useful for research on gene action and development over the past several decades. The sequences of many regulatory genes are now known and their specific activities are being identified. Moreover, sea urchins have been important subjects for exploring the reproductive ecology and evolution of marine animals.

Of the approximately 900 species of echinoids, relatively few have been studied in detail as model systems. Choice has been determined mainly by their availability to researchers. All are shallow-water species that are easy to collect in large numbers: *Arbacia lixula, Strongylocentrotus droebachiensis, Paracentrotus lividus,* and *Echinocardium cordatum* in Europe, *Arbacia punctulata* and *Lytechinus variegatus* in eastern North America, *Strongylocentrotus purpuratus, Lytechinus pictus,* and *Dendraster excentricus* in western North America, *Hemicentrotus pulcherrimus* in Japan, *Heliocidaris tuberculata* and *H. erythrogramma* in eastern Australia, and *Sterechinus neumayeri* and *Abatus cordatus* in the Antarctic. These species represent a good spread of the taxonomic diversity of the group, yet reproductive events are similar in most of them, so it is reasonable to generalize.

Reproduction is strictly sexual in sea urchins, and although parthenogenesis and polyembryony can be induced in the laboratory, they are seen infrequently in nature. Sexes normally are separate, but sexual dimorphism is slight or absent in most species. The gonads are usually large and dominate the interior of the body; as much as 30% of the wet body weight can consist of gonads. Most species have specific breeding periods, and these periodicities are regulated by changes in sea temperatures and photoperiods. Food, mainly algae and detritus, determines how large the gonads grow and the amount of gametes produced. Broadcast spawning is prevalent, and the embryos and larvae usually are free floating in the sea. Most species produce feeding larvae, which undergo catastrophic metamorphosis when they settle and change into tiny juveniles. The relatively few species without feeding larvae have been subject of productive studies on the evolution of different life history strategies and how macroevolutionary events might be brought about by microevolutionary processes.

II. GONAD ANATOMY AND GAMETOGENESIS

There are five gonads in most sea urchins arranged radially within the voluminous body coelom, each with a gonopore emptying apically near the anus (Fig. 1). They are connected apically at the level of the gonoduct by an aboral ring system that encircles the anus. The ring system contains nerves, hemal vessels, and a separate coelom that connects with each gonad. Each gonad consists of anastomosing tubules enclosed within a layer of the peritoneal epi-

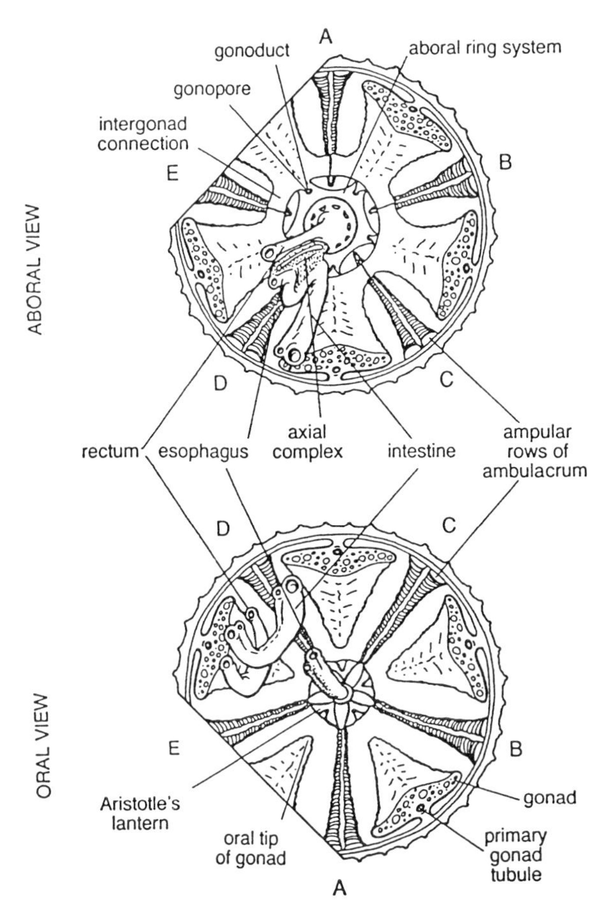

FIGURE 1 Internal anatomy of a generalized sea urchin showing placement of the gonads and associated structures. Figure is drawn as if cut along the equator and opened to show the aboral portion above and the oral portion below. Most of digestive system, which covers the gonads, is not shown. Capital letters indicate five rows of tube feet (reproduced with permission from Pearse and Cameron, 1991).

thelium of the spacious perivisceral coelom. Connective tissue, with muscles, nerves, and a network of hemal spaces, plus a remnant of a second coelom, the genital coelom or sinus, separate the outer peritoneal coelomic epithelium from the inner genital epithelium.

The genital epithelium consists of two cell types: nutritive phagocytes and gametogenic cells. Nutritive phagocytes are somatic cells that are responsible for most of the initial gonad growth. They grow by cell enlargement as they accumulate lipid and protein reserves, including vitellogenin or vitellogenin precursors. At peak development, they fill the gonads. During gametogenesis, they are depleted; presumably their nutrients fuel spermatogenesis in males and are transferred as yolk protein to oocytes in females. When the gonads are ripe and full of gametes, the nutritive phagocytes are small and form a thin lining on the gonad wall. Following spawning, they actively phagocytize relict gametes and gametogenic cells to begin a renewed cycle of growth. There is evidence that steroids, particularly estrogens, regulate the activity of the nutritive phagocytes, and these might be transported in the sea urchin's hemal system, but this has hardly been investigated.

Gametogenesis in sea urchins is typical of that in most animals. In testes, spermatogonia produce spermatocytes which undergo meiosis to form spermatids and finally sperms that accumulate in the lumen (Fig. 2). In ovaries, oogonia produce oocytes which undergo vitellogenesis until full size is reached. Unlike those of most animals, full-grown primary oocytes of echinoids usually undergo meiosis in the ovaries, and ova accumulate in the lumen (Fig. 3). The meiosis-stimulating hormone, 1-methyladenine, increases in the gonads during gametogenesis and may stimulate the production of ova.

The marked gametogenic synchrony seen among the five gonads suggests hormonal control, through either the hemal system or the nervous system. Moreover, synchrony is lost when connection to the aboral ring complex is severed. Changes in seawater temperature and photoperiod also have been shown to affect the timing of gametogenesis in various species, and these factors almost certainly act through hormones. However, the nature of the hormonal control system is unknown.

Sea urchin sperms are unusual in having a pointed, conical nuclear head; other echinoderms have round-headed sperms. In other aspects they are of the typical "primitive" type with a well-formed terminal acrosome, a single toroidal mitochondrion midpiece, and a long flagellar tail. They range in size from 2×0.7 μm in species with small eggs up to 10×2 μm in species with large eggs (*H. erythrogramma*).

Ova of sea urchins typically are around 100 μm in diameter in species with planktotrophic development. Species with lecithotrophic development have much larger ova (up to 2000 μm). The ova, which

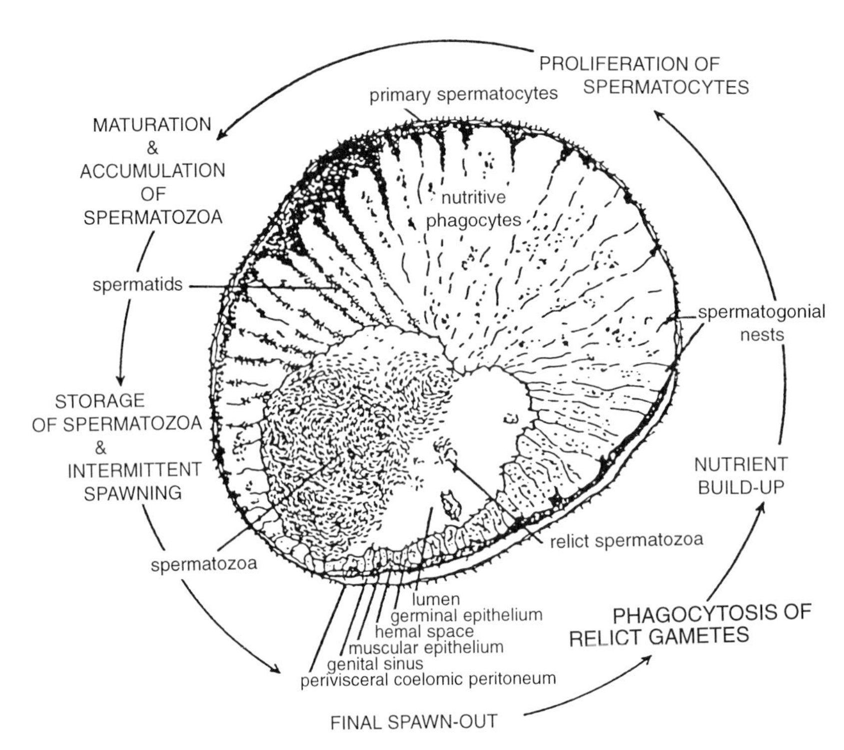

FIGURE 2 Diagram of the changes that take place in a testicular tubule during the spermatogenic cycle of a generalized sea urchin (reproduced with permission from Pearse and Cameron, 1991).

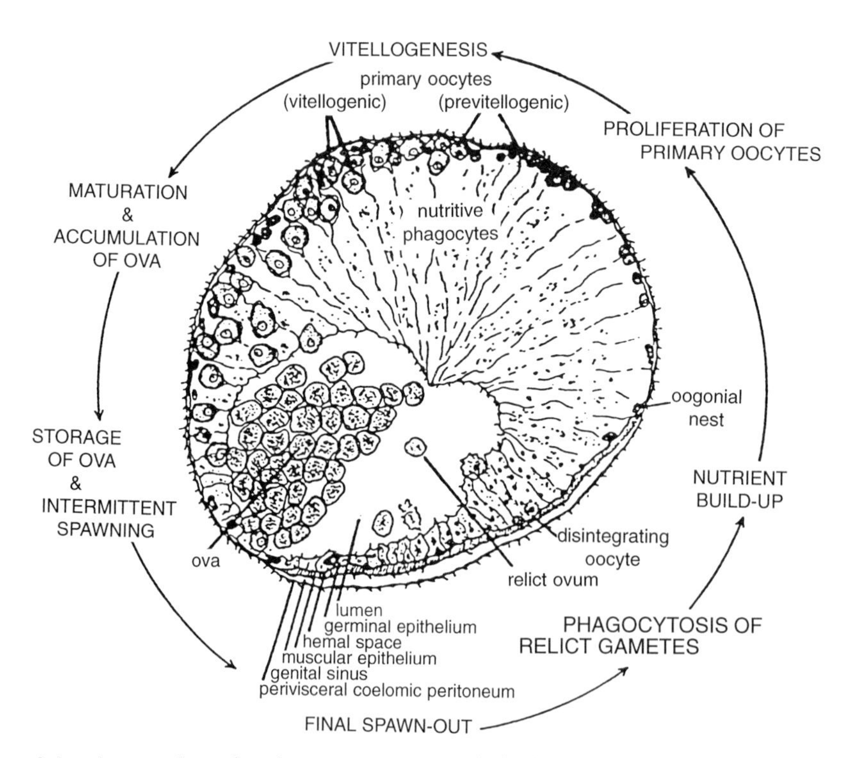

FIGURE 3 Diagram of the changes that take place in an ovarian tubule during the oogenic cycle of a generalized sea urchin (reproduced with permission from Pearse and Cameron, 1991).

are free in the lumen, contain the usual cell organelles, yolk protein platelets, and a layer of cortical granules under the cell surface. Each is covered with a vitelline coat composed of a complex of proteins, over which lies a jelly layer, which contains pigment cells in some species.

III. SPAWNING AND FERTILIZATION

Almost all sea urchins spawn both sperms and eggs freely into ocean waters, where water movement quickly disperses the gametes. Usually gametes are expelled from all five gonopores simultaneously and massive amounts of gametes are released. Factors stimulating and controlling spawning are unknown, but there is evidence that the aboral nerve ring may have a coordinating function. Spawning also can be induced by chemicals released by phytoplankton during spring blooms, thus coordinating seasonal spawning. In the laboratory, spawning is commonly induced by electric shock or by injection of potassium chloride.

Synchronized spawning among individuals has been reported, sometimes among aggregated individuals, but these observations are unusual; more often, isolated individuals are seen spawning by themselves, even when they are within aggregations. This is puzzling because lab and field experiments indicate that high sperm concentrations are necessary to maximize fertilization. The problem of how fertilization is maximized in sea urchins and other broadcast spawning marine invertebrates is a topic of current investigation.

The events during fertilization have been the subject of intensive investigation and will only be summarized here. Sea urchin sperms are attracted to peptides in the egg jelly. When a sperm contacts the jelly layer of an egg, an acrosome reaction ensues in which an acrosomal process forms and facilitates the sperm's movement through the egg jelly and attachment to the vitelline coat. Sperm attachment to the vitelline coat has some species specificity, but hybrids form readily in the laboratory and are known from the field. The sperm passes through the vitelline coat, and immediately upon fusion of the sperm and ovum plasma membranes the cortical granules of the ovum begin to fuse in a wave radiating outward from the point of fusion. Their released contents raise the vitelline layer to form the fertilization envelope, which hardens and prevents polyspermy. After the sperm nucleus enters the ovum, the sperm and egg pronuclei fuse and the resulting nucleus is pushed by microtubules to the center of the zygote.

IV. DEVELOPMENT

Cleavage in sea urchins, as in other echinoderms, is holoblastic and radial. It is highly synchronous through the first four cleavages, resulting in a 16-cell morula. In species with planktotrophic larvae, this stage has 8 equal-sized cells in the animal half (mesomeres) and 4 macromeres and 4 micromeres in the vegetal half, whereas in those with lecithotrophic development, all 16 cells are of similar size. Although sea urchins are usually considered to have indeterminate or regulative development, and normal but smaller larvae are produced when early blastomeres are removed, cell lineage studies show that under usual conditions the fate of descendent cells is highly predictable.

A hollow ball of ciliated cells, the blastula, enzymatically breaks out of the fertilization envelope to continue development while swimming free in the plankton. Gastrulation ensues by invagination, and the blastopore is the future larval anus. Paired syncytia of mesenchyme cells form laterally and initiate spicule formation. The bilaterally symmetrical spicules are formed as closely aligned microcrystals of magnesium-rich calcite and contribute to the characteristic larval shape. Pseudopod-bearing mesenchyme cells guide the archenteron to the oral surface of the ectoderm, to which it fuses then breaks through to form the larval mouth. Sea urchins are classic deuterostomes. Other features during early development include the formation of paired outpockets from the archenteron to form the enterocoels and the concentration of the external cilia into a single band looping around the mouth and extending out onto the developing larval arms.

The planktotrophic larva of echinoids is the pluteus, initially with four arms and when fully developed, eight arms. Each arm contains a calcite spicule and a ciliated band that functions for swimming and feeding. These larvae are feeding machines, using phytoplankton to fuel growth and development. They can remain in the plankton for weeks to months, depending on food and temperature.

Relatively few species (e.g., *H. erythrogramma*) have lecithotrophic development in which the eggs are provisioned with yolk and the larvae go through development without feeding. These larvae have reduced or no spicules and arms and uniform ciliation over the body. They typically settle and metamorphosis within days. Other species have no larvae at all but brood the developing embryos either among the spines or in special brood pouches (e.g., *A. cordatus*), which are released as well-developed juveniles.

During the larval stage, the paired coelomic pouches differentiate asymmetrically with two on the right side and three on the left. The middle of the three on the left combines with the overlying epidermis to form a conspicuous rudiment on the larva's left side. This develops into the oral side of a juvenile urchin, complete with teeth, spines, and tube feet. Metamorphosis is catastrophic and is initiated by contact with a surface filmed with bacteria. Upon metamorphosis, the rudiment opens up, turning much of the larval body inside out, and the juvenile emerges into its new benthic habitat within minutes to hours.

Bibliography

Cameron, R. A., Zeller, R. W., Coffman, J. A., and Davidson, E. H. (1996). The analysis of lineage-specific gene activity during sea urchin development. In *Molecular Zoology: Advances, Strategies, and Protocols* (J. D. Ferraris and S. R. Palumbi, Eds.), pp. 221–243. Wiley-Liss, New York.

Davidson, E. H. (1986). *Gene Activity in Early Development*, 3rd ed. Academic Press, New York.

Emlet, R. B. (1990). World patterns of developmental mode in echinoid echinoderms. In *Advances in Invertebrate Reproduction 5* (M. Hoshi and O. Yamishita, Eds.), pp. 329–335. Elsevier, Amsterdam.

Giudice, G. (1986). *The Sea Urchin Embryo: A Developmental Biological System*. Springer-Verlag, Berlin.

Pearse, J. S., and Cameron, R. A. (1991). Echinodermata: Echinoidea. In *Reproduction of Marine Invertebrates, Vol. VI, Echinoderms and Lophophorates* (A. C. Giese, J. S. Pearse, and V. B. Pearse, Eds.), pp. 513–662. Boxwood Press, Pacific Grove, CA.

Schatt, P., and Färal, J.-P. (1996). Complete direct development of *Abatus cordatus*, a brooding schizasterid (Echinodermata: Echinoidea) from Kerguelen, with description of perigastrulation. *Biol. Bull.* **190**, 24–44.

Wray, G. A. (1996). Parallel evolution of nonfeeding larvae in echinoids. *Syst. Biol.* **45**, 308–322.

Semen

Gail S. Prins

University of Illinois at Chicago

GLOSSARY

accessory sex glands In males, the prostate gland, the seminal vesicles, the bulbourethral glands, the vas deferens, and the epididymis.

semen An admixture of sperm cells and secretions from the male accessory sex glands which are mixed at the time of ejaculation.

seminal plasma The nongamete component of semen composed of secreted fluids, cells, and cellular components from several male accessory sex glands.

-spermia The suffix referring to semen in general. Thus aspermia refers to the absence of semen, oligo or hypospermia refers to low semen volume, and hyperspemia refers to increased semen volume.

-zoospermia The suffix referring to sperm within the semen.

Semen is a body fluid containing mature sperm cells and secretions from the male accessory sex glands. These components are mixed together at the time of emission and ejaculation, when they are released from the body. In mammals, spermatozoa are produced by the testes and transported in minimal fluid through an excurrent duct system to the cauda epididymis, where they are stored until ejaculation. The accessory sex glands that contract during emission and contribute secretions to the ejaculate are the cauda epididymis, vas deferens, ampulla, seminal vesicle, prostate gland, and bulbourethral glands (Cowper's gland). In certain mammalian species, some glands are absent and, in others, additional organs exist such as the coagulating gland. The accessory gland secretions exhibit marked species variability in terms of relative contributions, biochemical composition, and volume of the ejaculate. Along with species variability in sperm production, there are tremendous differences among species in semen volumes and sperm concentrations. Due to the significant variability in semen composition, this article will focus on characteristics of semen in the human and will highlight notable features of semen in other species.

I. SEMINAL PLASMA

A. Physiologic Role and Component Contributions

Seminal plasma is the nongamete portion of semen and is primarily composed of secreted components from the excurrent ducts of the testes and the accessory sex glands. The primary function of this fluid is to serve as a buffered, nutrient transport medium for the sperm cells as they are deposited in the female genital tract. The pH of seminal plasma is slightly alkaline (7.2–7.8) and acts to neutralize the acidic vaginal environment. While individual components have been shown to influence the fertilizing potential of spermatozoa, it is generally accepted that seminal plasma substances are not an absolute requirement for fertilization of the oocyte by the mature spermatozoan. This latter fact has been highlighted in recent years during which normal rates of fertilization have been achieved *in vitro* when epididymal sperm, never exposed to seminal plasma, are used for fertilization. Nevertheless, it is also accepted that specific components of the seminal plasma may enhance the *in vivo* fertilizing capacity of the sperm. Thus, fructose is an energy substrate for sperm, prostaglandins aid in smooth muscle contractions of the female genital tract and assist in sperm transport, whereas decapacitation factors and other protease inhibitors coat the sperm surface and are believed to prevent premature capacitation and activation of enzymes necessary for sperm penetration into the oocyte. Zinc and IgA may act as bacteriostatic factors, whereas antiagglutination proteins derived from the prostate gland prevent sperm from autoagglutination.

At the time of emission/ejaculation, the accessory sex glands do not contract simultaneously but rather in an organ-specific sequence. In man, as in most species, a small initial fraction originating from the glands of Littre's and the bulbourethral gland is released prior to the ejaculate proper and this fluid lubricates the urethra. Next, a sperm-rich fraction is expelled containing components from the epididymis and vas deferens (primarily sperm) along with prostatic secretions. In humans, this fraction is 0.3–0.5 ml in volume or 25% of the ejaculate. The last and largest fraction of the ejaculate originates in the seminal vesicles and varies from 1.0 to 2.5 ml or 75% of seminal volume in man. Exceptions to this sequence include the bull, in which simultaneous contractions of the accessory sex organs occur, and the dog, which has no ampulla or seminal vesicles and thus the postsperm fraction is contributed by the prostate. The contributions of the separate accessory sex glands to the ejaculate differ in their biochemical composition and certain components can be used as markers for those individual glands (Table 1). In humans, assays for these markers can be exploited to identify specific clinical conditions such as congenital absence of the vas and seminal vesicles or prostatic duct obstruction due to prostatitis.

B. Coagulation/Liquefaction

Soon after ejaculation, the semen from most species coagulates and forms a gelatinous mass which

TABLE 1
Biochemical Components of Individual Accessory Sex Glands in Man

Accessory sex gland	*Biochemical marker*
Epididymis	Glycerolphosphorylcholine
	L-Carnitine
	Inositol
Seminal vesicles	Fructose
	Prostaglandins
	Coagulating proteins
Prostate	Prostate-specific antigen
	Acid phosphatase
	Spermine
	Citric acid
	Zinc
	Proteases
Bulbourethral glands	IgA

restricts free movement of the spermatozoa. This process is due to a fibrinogen-like coagulating substrate contributed by the seminal vesicles which is acted upon by a unique clotting enzyme similar to, yet distinct from thrombin. When examined by electron microscopy, the coagulated semen consists of a dense network of long fibers 0.15 μm in length with enmeshed spermatozoa within the narrow spaces. In the bull and dog, the ejaculate remains in the liquid form and does not undergo coagulation. In some rodents, a coagulating gland contributes a secretion (transglutaminase) which catalyzes ε-γ-glutamyl cross-linkages between seminal vesicle proteins to produce a hardened copulatory plug in the vaginal canal which in turn acts to prevent the backflow of deposited sperm. The coagulated semen in man and most other species liquefies after a variable period of time due to proteolytic digestion by prostatic enzymes, including a chymotrypsin-like enzyme termed seminin, urokinase, and distinct plasminogen activators. The liquefaction process allows for sperm to swim out of the coagulum and into the uterus. It is believed that the process of coagulation/liquefaction allows for appropriate exposure of the sperm cells to the seminal plasma constituents, which may in turn affect their fertilizing potential.

C. Biochemical Composition

Mammalian seminal plasma contains a large array of biochemical components and there is considerable variation between species in terms of their presence or absence as well as their concentrations. Some of the components of semen are present in serum and are believed to be exudates from the circulation, whereas others are produced solely by the accessory sex glands and/or the testes and are therefore unique to the seminal plasma. The following discussion will highlight the biochemical composition of human seminal plasma.

1. *Carbohydrates*

Carbohydrates are present in both free and bound forms in the seminal plasma. Fructose is the predominant free carbohydrate, but smaller levels of glucose, sorbitol, and inositol are also present. Pyruvate is present in seminal plasma but its content rapidly decreases as it is metabolized by spermatozoa. Conversely, the concentration of lactate increases with time following ejaculation since it is the end product of fructolysis. Fructose, produced by the seminal vesicles, is considered the principal glycolytic energy source for ejaculated spermatozoa, which can rapidly metabolize this sugar. In man, the average fructose concentration is 2 or 3 mg/ml semen with a range between 0.5 and 5.0 mg/ml. A colorimetric assay based on the resorcinol reaction is used for qualitative fructose measurement, whereas an indol-based reaction is used to quantitate fructose concentration. Fructose measurement in the semen can be a useful test in the clinical laboratory for monitoring seminal vesicle contributions to the ejaculate. In cases of azoospermia, low seminal fructose levels may be indicative of congenital dysgenesis of the seminal vesicle and vas deferens. Since the seminal vesicles contribute the majority of the ejaculate volume, hypospermia (low semen volume) may also be indicative of low seminal vesicle contribution due, perhaps, to infections. Low fructose levels in semen may contribute to low motility due to a deficiency of energy substrate.

Many carbohydrates in semen are in the bound form of glycopeptides and glycoproteins. In man, the average concentration of bound seminal carbohydrates is 0.7 mg/ml. Following acid hydrolysis, the main constituents are galactose, sialic acid, hexosamine, mannose, and fucose. There is minimal glycogen present in human semen. Most of the bound carbohydrates originate in the seminal vesicles and bulbourethral glands.

2. *Lipids*

Human semen contains a high lipid content relative to that seen in other species and most major classes of lipids are present. Phospholipids, contributed primarily by the prostate gland, are the most prevalent lipids in human seminal plasma and include sphingomyelin, phosphatidyl choline, phosphatidyl ethanolamine, and phosphatidyl inositol. Cholesterol is the most prevalent steroid in human seminal plasma and it contributes approximately 25% of the total lipid content.

Palmitic acid is the most prevalent saturated free fatty acid in seminal plasma and since it is the most metabolically active lipid, it is thought to be an important energy source for spermatozoa. Human semen also contains high levels of polyunsaturated fatty acids, in particular linoleic acid, which is required for prostaglandins biosynthesis.

Human seminal plasma contains the highest concentration of prostaglandins of all body fluids. The seminal vesicle is the source of seminal prostaglandins and produces all known forms of this unsaturated fatty acid, including PGA, PGB, PGE, and PGF. Total prostaglandins concentrations are estimated at 500–600 μg/ml in freshly ejaculated semen. It is believed that seminal plasma prostaglandins serve to induce contractions of the female genital tract smooth muscle and thereby aid in sperm transport through the cervix and uterus.

3. *Proteins*

The total protein content in human seminal plasma is between 35 and 50 mg/ml. Many of the proteins are identical to those found in blood plasma and these include albumin, globulins, transferrin, glycoproteins, and immunoglobulins. Other proteins present in seminal plasma are not found in blood and are produced within the male genital tract. Secretory IgA is present in seminal plasma due primarily to contributions from the bulbourethral glands, whereas IgG and small amounts of IgM in the seminal plasma are exudates from blood. The albumins and globulins serve the function of maintaining osmotic equilibrium, which is important for the survival of sperm cells. The normal osmotic pressure of human seminal plasma is slightly hyperosmotic at ~350 mOsm initially after ejaculation; however, this increases to as high as 550–600 mOsm over time. A fibrinogen-like glycoprotein secreted by the seminal vesicle is involved in semen coagulation following ejaculation. Several ion-binding proteins are present in seminal plasma, including calcium-binding proteins, zinc-binding proteins, and lactoferrin, which binds to iron.

There are many enzymes present in seminal plasma and several of them have very high concentrations when compared with other body fluids, including acid phosphatase, prostate-specific antigen, kallikreins, γ-glutamyl transpeptidase, acetylcholinesterase, and creatine phoshokinase. The most notable enzymes, perhaps because they have a clearly defined function, are the proteolytic enzymes responsible for liquefaction of the coagulated semen. In addition to seminin, urokinase, aminopeptidases, and several metalloproteinases which all may be involved in seminal proteolytic activity, the seminal plasma also contains the pepsinogen–pepsin system, although its function is unknown. Proteinase inhibitors are also present in seminal plasma that are important in reversibly regulating acrosin activity in sperm cells. Another group of enzymes present within the seminal plasma is the glycosidases, whose highest activities are in epididymal secretions. These include β-glucuronidase, α-mannosidase, β-galactosidase, β-N-acetylglucosaminidase, and β- and α-glucosidase, and their function is to metabolize complex carbohydrates and thus maintain energy substrates for spermatozoa. Seminal plasma contains several phosphatases, most notably acid and alkaline phosphatase, pyrophosphate, and ATPase. Finally, many nucleolytic enzymes are present in semen, including ribonuclease and NADase, which coat the spermatozoa and may potentially play a role in activating chromatin for DNA synthesis. For detailed descriptions of these enzyme activities, see the book by Mann and Lutwak-Mann.

4. *Ions and Electrolytes*

Seminal plasma contains numerous ions and electrolytes which maintain an isotonic solution. Calcium is absolutely required for normal sperm functions and equivalent amounts are found within sperm cells and seminal plasma. The seminal calcium concentration in man exceeds that in blood by threefold. In contrast, the NaCl concentration is far below that found in blood. Zinc concentrations are very high in seminal plasma, and zinc is believed to play a bacteriostatic role. Zinc is of prostatic origin and most is organically complexed to specific zinc-binding proteins. Iron also occurs in semen, loosely bound to lactoferrin. In humans, as in most species, potassium concentrations are higher within spermatozoa than within seminal plasma.

5. *Low-Molecular-Weight Molecules*

There are several low-molecular-weight molecules within seminal plasma that deserve specific mention. Citric acid, secreted by the prostate gland, is present in high concentrations in semen, in which it plays an important role in the tricarboxylic acid cycle. It is also believed to act as a strong chelator for calcium. Polyamines (spermine and its immediate precursor, spermidine) originate from the prostate and are present in very high concentrations in human seminal plasma. Oxidation of spermine by diamine oxidase in seminal plasma produces the odor unique to semen. Over time, spermine phosphate crystals form in semen, an observation first made by van Leeuwenhoek. Glycerylphosphorylcholine has been used as a marker molecule for epididymal contributions to the ejaculate due to its high production and release from that organ, although its physiologic role is unclear. However, it is of interest that the mammalian female genital tract contains a diesterase which cleaves this molecule to free choline. There is a high content of free amino acids in mammalian seminal plasma and several functions have been attributed to them, including action as chelators of toxic metals or as buffering agents, serving as oxidizable substrates for sperm, and stimulation of sperm motility. In humans, the predominant free amino acids are glycine, glutamic acid, arginine, serine, alanine, and aspartic acid. Finally, it should be noted that human seminal plasma contains a considerable amount of ATP and its concentration has been correlated with the fertility status of the male. A specific test for quantitation of ATP is recommended by the World Health Organization (WHO).

6. *Hormones*

There are many hormones present in human semen, including both steroid and protein hormones and factors. Hormones not produced by the testes are considered to be plasma exudates and reach semen through the accessory sex glands on which they act. Attempts to correlate the levels of seminal plasma hormones with the fertility status of men have produced contradictory findings and the value of their seminal concentrations as predictors of infertility is disputed.

While steroid hormones enter semen as exudates of serum, they also enter through the excurrent ducts of the testes. Testosterone is produced by testicular Leydig cells and concentrations in the seminiferous tubules are several-fold higher than that found in the circulation. A large portion of this testosterone is bound to androgen-binding protein, a specific protein of Sertoli cell origin. Testosterone is metabolized by cells within the excurrent ducts, including spermatozoa, and by the accessory sex glands and thus significant amounts of many steroids are present. In addition to testosterone and dihydrotestosterone and their sulfated forms, semen contains androstenedione, dehydroepiandrosterone (DHEA), DHEA sulfate, estrogens (estradiol-17β, estrone, and estrone sulfate), and pregnenolone sulfate. Data have been presented which show that steroids can affect sperm motility and function; thus, there may be a specific physiologic role for some of the steroids present in semen. In general, estrogens and progesterone have been shown to stimulate sperm, whereas testosterone and DHT have been reported to be inhibitory.

Several protein hormones are present in seminal plasma, some at higher concentrations than that found in blood. Prolactin (PRL), which is present at slightly higher concentrations in semen than in serum, has been shown to stimulate sperm motility and increase longevity. Luteinizing hormone (LH), follicle-stimulating hormone (FSH), and human chorionic gonadotropin (hCG) are also found in similar or higher concentrations as serum. LH and FSH concentrations are higher in seminal plasma of normal men compared to vasectomized men, indicating that some of the LH and hCG may arise from the testicular excurrent ducts. In contrast, PRL and FSH are similar in both groups of men, indicating that they arise from the serum. Seminal plasma also contains inhibin from testicular origin and insulin from the serum—both in significant quantities.

II. SPERM

A. Sperm Properties in Normal Human Men

Sperm make up the second component of semen. Since these gametes are discussed extensively in other sections of the encyclopedia, this portion will summarize the properties of sperm observed in the normal human ejaculate (Table 2). The normal human ejaculate has a sperm concentration of 20 to 150×10^6 sperm cells/ml with a total count varying between 40 and 500 million sperm. At least 50% of the sperm cells should exhibit motility in the neat semen to provide a total motile count of between 20 and 250×10^6 sperm per ejaculate.

Motion analysis of sperm can be performed manually or with computer assistance. Manual or subjective determination of sperm kinetics is made by grading the average quality of motion on a 0–4 scale as follows:

0 Immotile
1 Weak movement with no forward progression
2 Weak to moderate forward progression
3 Good forward progression; active tail movement
4 Rapid forward progression; vigorous tail movement

An average grade of 3 or 4 in the neat liquefied semen is considered normal.

Alternatively, computer-assisted semen analysis (CASA) systems can be employed which objectively monitor sperm motion. For greater detail, the reader is encouraged to study these principles as outlined by Katz. In brief, the commercial systems capture the microscopic sperm image using videography or infrared scatter and convert this image into digital information. Using computerized algorithms, the pixels are evaluated for size and gray-scale intensity to differentiate images as sperm. Sperm counts can then be calculated. The position of the central pixel, the centroid, is determined for each sperm and followed on several frames. CASA systems capture a minimum of 15 sequential frames. If the centroid of a given sperm is observed to move a given distance over several frames, it is classified as motile. Trajectories for that cell are captured and algorithmically analyzed for specific motion characteristics. The greatest strength of CASA is its objective and expanded description of motion including vigor and pattern. Vigor descriptions include velocities and beat frequency of the tail. Pattern of motion describes the straightness of the average path (linearity) and the side-to-side motion of the cell (wobble or yaw), frequently referred to as amplitude of lateral head displacement. The average velocity of normal sperm is considered >30 μm/sec; however, the normal ranges for other parameters differ between systems and have yet to be established worldwide.

Morphologic characterization of sperm has significant correlations to male fertility. Spermatozoa can be classified into one of five categories according to the WHO guidelines: normal, head abnormality, neck/midpiece abnormality, tail abnormality, or im-

TABLE 2
Values for Normal Semen Analysis in Man

Volume	1–5 ml
Sperm concentration	>20 million–ml
Total sperm count	>40 million
Motility	>50%
Motile sperm density	>8 million/ml
Velocity	>30 μm/sec
Morphology	>30% normal forms
Viability	>75%
Liquefaction	10–30 min
Viscosity	3–5 cm threads
pH	7.2–8.0
White blood cells	<1 million/ml

mature. The WHO now considers the normal range for human sperm to be between 0 and 70% abnormal forms. Any sample with <30% normal forms is considered teratozoospermic.

The normal sperm morphology consists of an oval-shaped head 4 or 5 μm in length and 2 or 3 μm in width. The length-to-width ratio should be between 1.5 and 1.75. A well-defined acrosome resides on the anterior head and should comprise 40–70% of the head area. The postacrosomal region (nuclear region) should be rounded. The tail is approximately 45 μm in length and should be straight or loosely bent. The neck and midpiece are short and should contain no cytoplasmic remnant. Head abnormalities include size variations (macrocephalic, microcephalic, and pinhead), shape variations (tapered, pyriform, and amorphous), vacuoles, acrosome deficiency, acrosome absence, and bicephalic. Midpiece defects include distended or irregular shape, bent midpiece, thin midpiece (no mitochondria), bent tail (90% angle to head at the midpiece), and absent tail (free head). Tail defects include multiple, coiled, hairpin, broken (>90% angle), and short tails, or tails with terminal droplet. Immature spermatozoa present with cytoplasmic droplets attached to the midpiece area (or attached to tail).

Teratozoospermia is usually indicative of a testicular disorder due to one of many causes: genetic, hormonal, vascular, stress, infections, radiation, drugs, and toxic exposure. Some causes may be transient, such as stress, infections, and drug therapy, whereas many conditions are more permanent and frequently nontreatable. It has been reported that a prevalence of tapered heads may be indicative of varicocele; therefore, the clinician should be made aware of this. Amorphous heads are seen in patients with allergic reactions. While cytoplasmic droplets may be indicative of a epididymal disorder, they are most frequently associated with recent (<24 hr) or frequent ejaculations.

An alternate morphology grading system, termed strict criteria for sperm morphology, is also in use

TABLE 3
Semen Volume and Sperm Counts in Various Species

	Ejaculate volume		*Sperm concentration*	
Species	*Range (ml)*	*Average (ml)*	*Range (×10⁶/ml)*	*Average (×10⁶/ml)*
Bat	0.03–0.08	0.05	5,000–8,000	6000
Boar	150–500	250	25–300	100
Buffalo	0.5–4.5	2.5	200–800	600
Bull	2–10	4.0	300–2,000	1000
Camel	4–12	8.0	100–700	400
Cat	0.02–0.12	0.03	100–2,600	1700
Deer	2–20	4.0	100–1,300	200
Dog	2–15	9.0	60–300	300
Fox	0.2–4	1.5	30–250	70
Goat	0.2–2.5	1.0	1,000–5,000	3000
Goose	0.4–1.3	1.0	400–1,500	1000
Guinea Pig	0.4–0.8	0.6	5–17	10
Man	2–6	3.0	50–350	80
Pigeon	0.002–0.04	0.016	800–3,800	2000
Rabbit	0.4–6.0	1.0	50–350	150
Ram	0.7–2.0	1.0	2,000–5,000	3000
Rooster	0.2–1.5	0.8	50–6,000	3500
Stallion	30–300	70	30–800	120
Turkey	0.2–0.8	0.3	4,500–18,500	7000

clinically. The defects considered are the same as those of the WHO classification. The difference is that very strict criteria must be met for a sperm to be classified as normal (a platonic form). The head must be exactly within defined limits for size and shape and a micrometer eyepiece is recommended to ensure accurate measurement. Any borderline form is considered abnormal. According to these criteria, >14% normal forms is considered a normal, fertile specimen. The rest are broken into two groups: (i) 5–14% normal forms, G pattern, is either fertile or subfertile with good prognosis, and (ii) 0–4% normal forms, P pattern, is subfertile with poor prognosis.

B. Species Variations

There are considerable species variations in terms of sperm counts, motion characteristics, and morphology within the semen. Table 3 summarizes some of the species differences with regard to sperm concentrations found within the ejaculate. The reader is referred to the articles on spermatogenesis for different species to learn more about different sperm characteristics. Alternatively, more information can be attained in the book by Mann and Lutwak-Mann.

See Also the Following Articles

Epididymis; Prostate Gland; Seminal Vesicles; Spermatogenesis, Overview; Spermatozoa; Sperm Transport; Testis, Overview

Bibliography

Katz, D., and Davis, R. (1987). Automatic analysis of human sperm motion. *J. Androl.* 8, 170–181.

Keel, B., and Webster, B. (1990). *Handbook of the Laboratory Diagnosis and Treatment of Infertility*. CRC Press, Boca Raton, FL.

Mann, T., and Lutwak-Mann, C. (1981). *Male Reproductive Function and Semen*. Springer-Verlag, New York.

World Health Organization (1992). *WHO Laboratory Manual for the Examination of Human Semen and Sperm–Cervical Mucus Interaction* 3rd ed. Cambridge Univ. Press, Cambridge, UK.

Zaneveld, L. J. D., and Chatterton, R. T. (1982). *Biochemistry of Mammalian Reproduction*. Wiley, New York.

Seminal Vesicles

Lawrence S. Ross

University of Illinois at Chicago College of Medicine

GLOSSARY

ejaculation The physiologic process by which the components of the semen are mixed and propulsed through the penile urethra.

fructose A sugar produced exclusively by the seminal vesicle which is the major source of glycolytic energy for spermatozoa.

prostaglandins Acidic lipids based on the 20-carbon skeleton of prostanoic acid.

semen The male reproductive fluid composed of secretions of the prostate, peri-urethral glands (Cowper and Littre), seminal vesicles, and ampulla of the vas deferens, combined with spermatozoa derived from the testis and transported through the epididymis and vas deferens.

Mammalian reproduction requires the transport of the male gamete, sperm, to the female's egg. The fusion of these two haploid cells results in a diploid embryo capable of developing into a complete new member of the species. The sperm is delivered to the female reproductive tract in a supportive and nutrient-containing fluid, the semen. The seminal vesicles are an integral part of the organ system responsible for semen production.

I. DEVELOPMENTAL ANATOMY

The male reproductive system is derived embryologically, from two distinct anatomic sites. The testis develops in the gonadal ridge, a coalescence of mesenchymal tissue in the dorsal abdominal wall adjacent to the developing kidney. The primordial germ cells migrate into the developing testis from the endoderm of the yolk sac. The efferent ducts and proximal epididymis form from the mesonephric ducts. The prostate, seminal vesicles, and vas deferens form as outgrowths of the urogenital sinus in the inferior ventral aspect of the embryo, completing their development in the fourth month in the human. The vas migrates to meet and fuse with the epididymis and the testis begins its descent to the genital tubercle during the sixth through ninth months (Figs. 1 and 2).

II. SECRETORY FUNCTION

The human seminal vesicles are 7.5-cm long, lobulated membranous pouches that produce 45–80% of the ejaculate (2.0–2.5 ml). The secretions of the vesicles are alkaline and contain reducing sugars, prostaglandins, potassium, phosphorylcholine, protein, lactoferrin, and low-molecular-weight proteinase inhibitors.

Fructose is the primary sugar produced in the seminal vesicles and its concentration in the seminal plasma is high (3.5–28 m*M*). Although the vesicles are stimulated by androgen, there is little correlation between serum testosterone levels and fructose concentration. Although fructose is the primary source of energy for the spermatozoa, there is no relation of sperm motility to fructose concentration or consumption. Fructose is a marker for the seminal vesicle and its absence from the semen suggests congenital absence of that organ (usually with simultaneous absence of the vas deferens) or obstruction of the ejaculatory ducts.

Prostaglandins are found in all the accessory sex glands but have their highest concentration in the seminal vesicles. Four types of prostaglandins are generally identified: A, B, E, and F. They have a variety of pharmacologic properties and act at the local area of production rather than via the circulation as with most hormones. Prostaglandin E_1 is found in highest concentration in human seminal plasma followed by E_2. One of the major effects of prostaglandin is smooth muscle contraction and its possible role in reproduction may be to stimulate contraction of the testicular capsule, seminiferous tubules, epididymis, vas deferens, and seminal vesicles. Thus, prostaglandins may facilitate sperm transport. In the female genital tract, seminal fluid prostaglandins may aid passive sperm movement by stimulating cervical and uterine body contractions.

The role of the remaining compounds secreted by the seminal vesicles is not well understood. The semen of many species coagulates on ejaculation, trapping the sperm in a proteinaceous matrix. In primates and man this coagulum dissolves in 5–20 min, releasing the sperm. It is believed by some that the coagulation substrate is derived from the seminal vesicles. The proteins and proteinase inhibitors secreted by the vesicles are likely involved in this process.

III. PHYSIOLOGIC ROLE IN REPRODUCTION

Ejaculation is the process by which the components comprising the semen are mixed and propulsed through the penile urethra to be deposited in the female reproductive tract. The ejaculate consists of the spermatozoa and the seminal plasma. The latter is the combined secretion of all the accessory sex glands. The contents of the seminal plasma are not ejaculated simultaneously. First, the periurethral gland (Cowper and Littre) secretions are emitted, followed, in order, by secretions from the prostate,

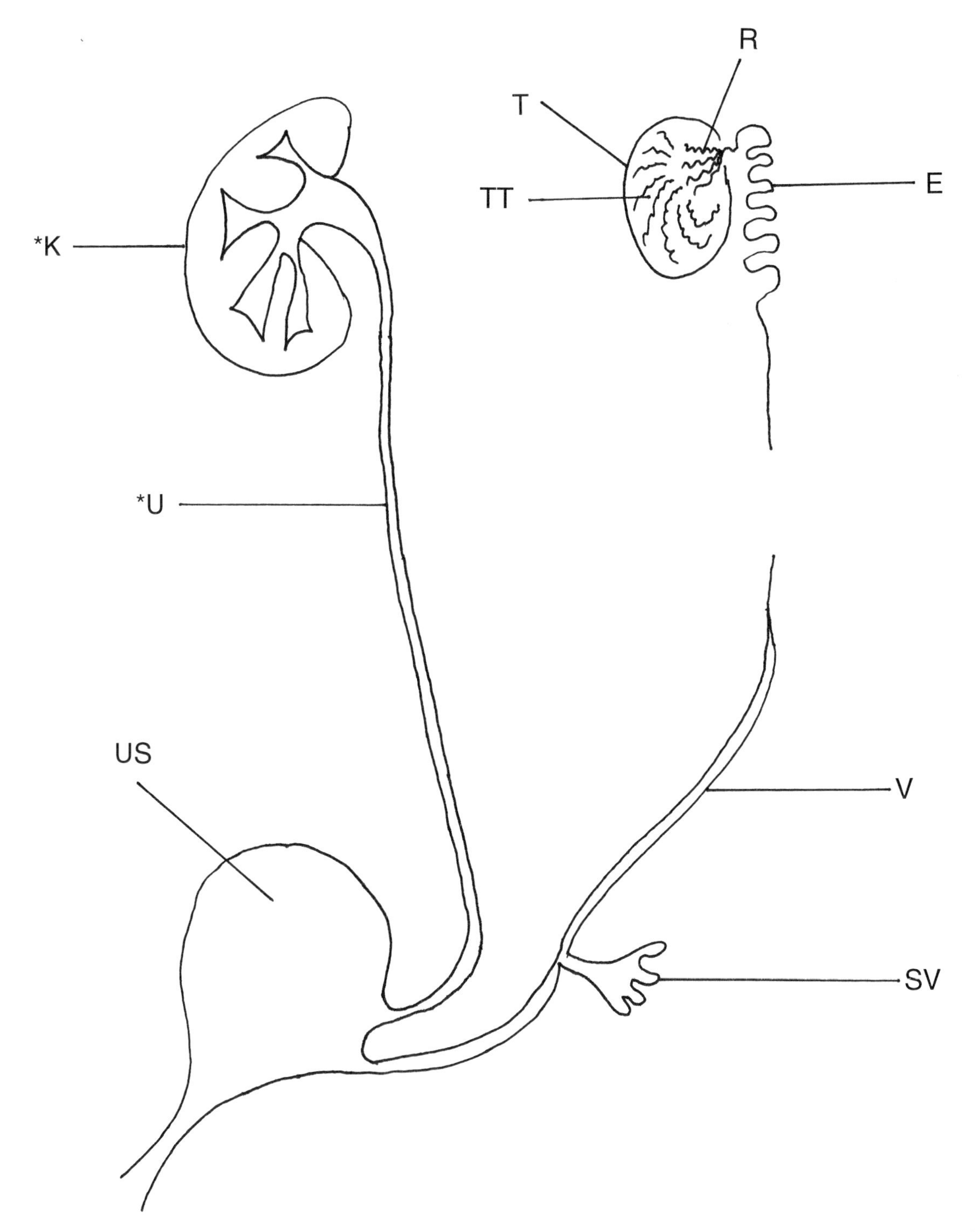

FIGURE 1 Male reproductive tissues at 4 months of embryonic development. T, testis; TT, testicular tubules; E, epididymis (mesonephric origin); V, vas deferens; SV, seminal vesicle; US, urogenital sinus; K, kidney* (metanephric origin); U, ureter*; R, rete testis. *Kidney derives from metanephric tissue; ureter migrates from urogenital sinus (both are shown at a stage later than 4 months for purpose of illustration of relationships to the developing reproductive apparatus).

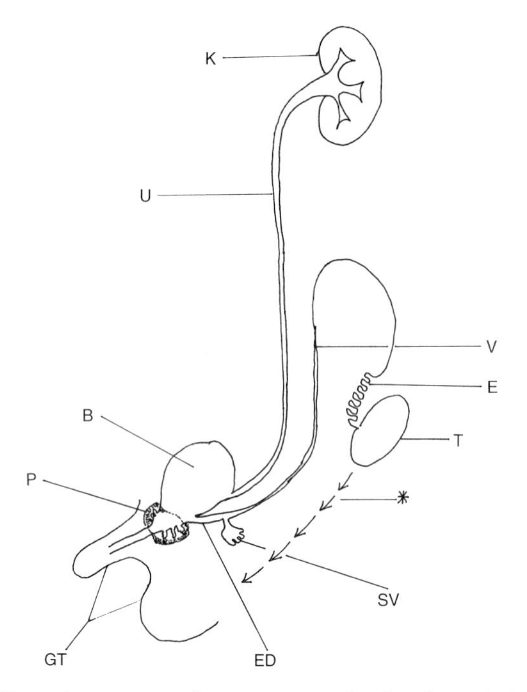

FIGURE 2 Descent of testes to genital tubercle. K, kidney; U, ureter; V, vas; E, epididymis; T, testis; B, bladder; P, prostate; ED, ejaculatory duct; GT, genital tubercle (forms penis and scrotum); SV, seminal vesicle. *Arrows indicate descent of testis.

sperm containing fluid from the epididymis, vasa deferentia, and ampulla of the vas, and, finally, the largest fluid component from the seminal vesicles. In the human and some higher primates the semen coagulates on ejaculation due to a substrate from the seminal vesicles and later lyses to release the sperm in the female genital tract. Lysis is thought to be due to seminal proteinases derived from the prostate.

The seminal vesicles play a pivotal role in the production of semen and in normal reproduction. In humans the seminal vesicles can become infected (usually associated with prostate infection) and the altered state of the semen may cause infertility. Congenital absence of the seminal vesicles is seen in man and is almost always accompanied by the absence of the vasa deferentia. Such patients are azoospermic (seminal fluid does not contain sperm) and cannot reproduce by sexual intercourse. New assisted reproductive techniques that allow *in vitro* combination of sperm and egg aid such men in reproduction.

See Also the Following Articles

Ejaculation; Male Reproductive System, Overview; Semen

Bibliography

Amelar, R. D., Dubin, and Walsh, P. C. (1977). *Male Infertility*. Saunders, Philadelphia.

Hafez, E. S. E. (Ed.) (1976). *Human Semen and Fertility Regulation in Men*. Mosby, St. Louis, MO.

Lipshultz, L. I., and Howards, S. S. (Eds.) (1997). *Infertility in the Male*, 3rd ed. Mosby-Year Book, St. Louis, MO.

Walsh, P. C., Retek, A. B., Vaughn, D., Jr., and Wein, A. J. (Eds.) (1992). *Campbell's Urology*, 7th ed. Saunders, Philadelphia.

Sertoli Cells, Function

Michael D. Griswold
Washington State University

Lonnie D. Russell
Southern Illinois University

GLOSSARY

desmosome–gap junction Sites of adhesion and likely intercellular communication between Sertoli cells and germ cells.

hemidesmosome A type of adhering junction that binds the basal aspect of the Sertoli cell to the underlying connective tissue.

iron shuttle A proposed model for delivery of ferric ions to germ cells in the testis involving transferrin receptors, movement of iron through the cell, secretion of ferric ions associated with a newly synthesized testicular transferrin to germ cells, and incorporation of iron into ferritin in the developing germ cells.

Sertoli–germ cell Ratio is fixed for a species and appears to result from the fixed capability of Sertoli cells to support germ cells.

Sertoli–Sertoli ectoplasmic specialization A specialized region formed by the endoplasmic reticulum and actin filaments that lines the Sertoli cell barrier.

Sertoli–Sertoli tubulobulbar complex An invagination of one Sertoli cell into another at the level of the Sertoli cell barrier that takes the form of a tube connected to a bulb; probably functions in junctional turnover.

Sertoli–spermatid tubulobulbar complex An invagination of the late spermatid head plasma membrane into the apical Sertoli cells that contains both tubular and bulbous portions; probably functions in anchoring the spermatid, elimination of its cytoplasm, and elimination of junctional links that bind the two cells.

tight junction Fusions of the plasma membrane of adjacent Sertoli cells that circumscribe the cell and restrict paracellular movement of materials between Sertoli cells.

transferrin A major secreted glycoprotein of the Sertoli cells that is involved in the delivery of ferric ions to germ cells while navigating around the tight junctional complexes.

Research on the descriptive morphology and endocrine physiology has given rise to an important and fundamental principle, i.e., successful spermatogenesis in higher vertebrates requires the functions of Sertoli cells. The determination of exactly which properties or functions of Sertoli cells are critical in spermatogenesis has been a major goal of research in spermatogenesis. Sertoli cells can influence spermatogenesis by their direct involvement in testis formation and by regulating the immediate environment of the developing germ cells. Sertoli cells are arranged such that they can maintain physical contact with germ cells, they can cordon off germ cell populations into environmentally distinct compartments of the semniferous epithelium, and they can regulate the biochemical surroundings of the germ cells. It is important to note that the regulation of the germ cell environment is likely a joint responsibility of the Sertoli cells and the resident germ cells.

I. EVIDENCE FOR THE REQUIREMENT FOR SERTOLI CELLS IN SPERMATOGENESIS AND TESTIS FORMATION

The relationship between the germinal cells and the Sertoli cells is obligatory in testis development as well as spermatogenesis. Sertoli cells are important in the initial formation of the embryonic testis because they sequester the germ cells (gonocytes) inside of newly formed seminiferous tubules. This process of testis formation requires the expression of specific genes on the Y chromosome. Once the gonocytes are sequestered by the Sertoli cells they enter into a mitotic period. This is in contrast to gonocytes such as those found in the embryonic ovary that are not sequestered by Sertoli cells. Gonocytes in the ovary enter into meiotic prophase shortly after organ formation. A period characterized by proliferation of Sertoli cells and germ cells follows testis formation. Puberty generally involves the cessation of mitosis of Sertoli cells, the formation of tight junctions between adjacent Sertoli cells, and the progress of germ cells through meiosis and differentiation into spermatozoa.

Sertoli cells must provide critical factors necessary for the successful progression and differentiation of germ cells. These critical factors may be physical support, junctional complexes or barriers, or biochemical stimulation in the form of growth factors or nutrients. There is abundant experimental evidence that the persistence of the tight junctional complexes and two-compartment system in the testis is required for spermatogenesis. It is likely that Sertoli cells are required for spermatogenesis because they make specific products that are necessary for germ cell survival and those products combine to form a unique and essential environment in the adluminal compartment. Some of the functions of Sertoli cells may not be absolute requirements for spermatogenesis but rather may influence the efficiency of the process.

There are three basic tenets that underscore the obligatory role of Sertoli cells during spermatogenesis. First, the condition in which testes contain germ cells but no Sertoli cells has never been reported and the development of germ cells in culture independent of coculture with Sertoli cells is very limited. Mammalian spermatogenesis cannot be easily demonstrated in culture, but when claims for success have been published there is always a requirement for Sertoli cells. In the amphibian, germinal cell development will occur in culture and some elements of amphibian germinal cell development are independent of the presence of Sertoli cells. However, the advancement of amphibian spermatocytes through meiosis and some aspects of the maturation of spermatids require the presence of Sertoli cells in the coculture. A similar reliance on Sertoli cells in the mammalian testis can be postulated. The second basic tenet is that the function and efficiency of Sertoli cells appear to be limiting to germ cell numbers. While the maximum number of germ cells supported by Sertoli cells varies between species, it is constant within a species. Support for the notion that Sertoli cells are the limiting factor in spermatogenesis comes from several experiments in which the size of the testis and the spermatogenic output were manipulated by changing the number of Sertoli cells. The third basic tenet is that the endocrine requirement for spermatogenesis in higher vertebrates is a result of the action of reproductive hormones, such as follicle-stimulating hormone and testosterone, on Sertoli cells and not on germ cells.

II. STRUCTURAL INTERACTIONS OF SERTOLI CELLS WITH SERTOLI AND GERM CELLS

The Sertoli cell is a prime example of a cell with regional structural specializations that accomplish a variety of tasks. Sertoli cells simultaneously associate with a minimum of four different germ cell types, with the basal lamina, and with other Sertoli cells. They do so as germ cells progress and differentiate during the spermatogenic cycle. It is obvious that their regional specializations receive orders from a command and control center to do different tasks at the same time. What follows is a listing of the structural relationships of the Sertoli cell, their dynamic or static features, and their proposed function. A generalized diagram showing the sites of specific structural interactions is provided to orient the reader (Fig. 1).

A. The Hemidesmosome

The attachment of Sertoli cells to the basal lamina via hemidesmosome-like structures serves to anchor the Sertoli cell to the underlying connective tissue (Fig. 2A). The connective tissue was initially produced cooperatively by the Sertoli cell and the peritubular myoid cell. Numerous intermediate filaments (IFs) are seen in the vicinity of the hemidesmosome-dense plaques (Fig. 2, arrowheads). The anchorage

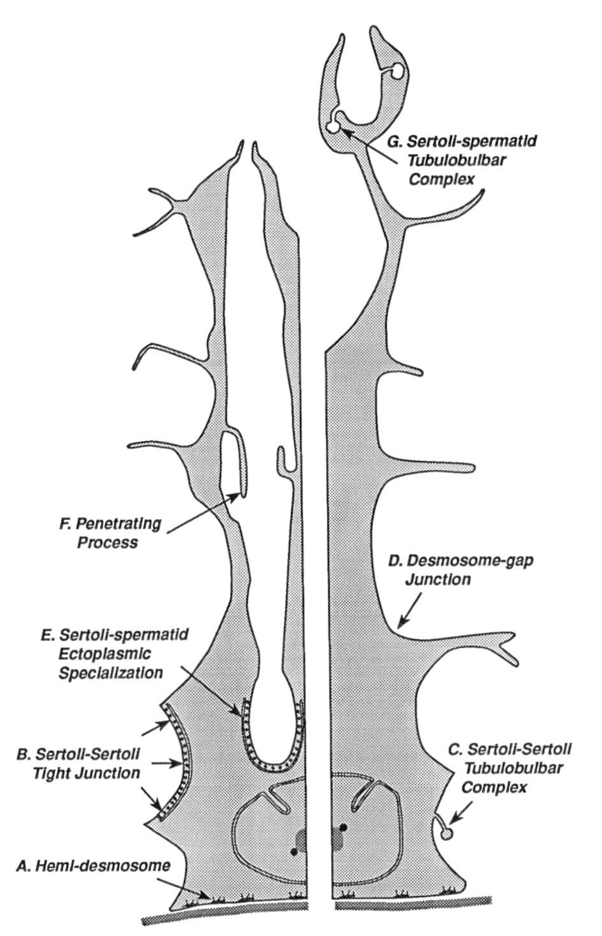

FIGURE 1 Summary diagram of Sertoli and Sertoli–germ cell junctions or specialized relationships. The Sertoli cell is divided into two halves: One half represents the configuration of the cell during the time elongate spermatids are embedded within crypts of the cell (left) and the other half represents the configuration of the cell approaching the time of sperm release as spermatids reside near the tubular lumen (right). The letters indicate the sites at which each relationship occurs and the letters match the designations in Fig. 2.

of numerous Sertoli cells to the basal lamina occurs developmentally and ensures that the Sertoli cell remains a scaffold for germ cell anchorage, germ cell migration, and spatial organization of germ cells.

B. Sertoli–Sertoli Junction

Sertoli cells extensively associate with one another by tight junctions to compartmentalize the seminiferous epithelium, thus forming a Sertoli cell barrier (Fig. 2B). The structural appearance of the Sertoli–Sertoli junction is unusual in that it is associated with an organized region of bundled actin and more deeply positioned endoplasmic reticulum (Figs. 2B-1 and 2B-2). It is significant that the tight junctions isolate the majority of germ cells for two reasons: First, autoantigens develop on the surface of germ cells that may elicit an immune response (orchitis) if ready access of immune cells to germ cells was permitted. Formation of a barrier means that germ cells must at some point cross that barrier to reside in the immunoprivileged environment. The Sertoli cell is most likely the active cell in facilitating germ cell movement across tight junctions. Second, by forming a barrier the Sertoli cells ensure that most products that reach them must do so through the Sertoli cell, in which they may be modified or completely prevented from reaching germ cells. For this reason, toxic chemicals often affect the Sertoli cell rather than more advanced germ cells.

Sertoli cells also form gap junctions with one another at the level of the Sertoli cell barrier. Such junctions are more numerous during development but are found in adulthood. They are presumed to participate in communication between Sertoli cells and the coordination of a Sertoli cell cycle.

C. Sertoli–Sertoli Tubulobulbar Complex

Sertoli cells form an unusual type of relationship with each other at the level of the Sertoli cell barrier. Because this structure possesses a tubular and a bulbous component, it is called a tubulobulbar complex (Fig. 2C). Sertoli–Sertoli tubulobulbar complexes arise and are degraded cyclically. They contain tight and gap junctional elements in their membranes and the cyclic degradation of these structures is assumed to be the means whereby Sertoli–Sertoli junctions are turned over. They may be similar to annular gap junctions in other tissues.

D. Desmosome–Gap Junction

Sertoli cells form junctions with germ cells which have the characteristics of both desmosomes and gap

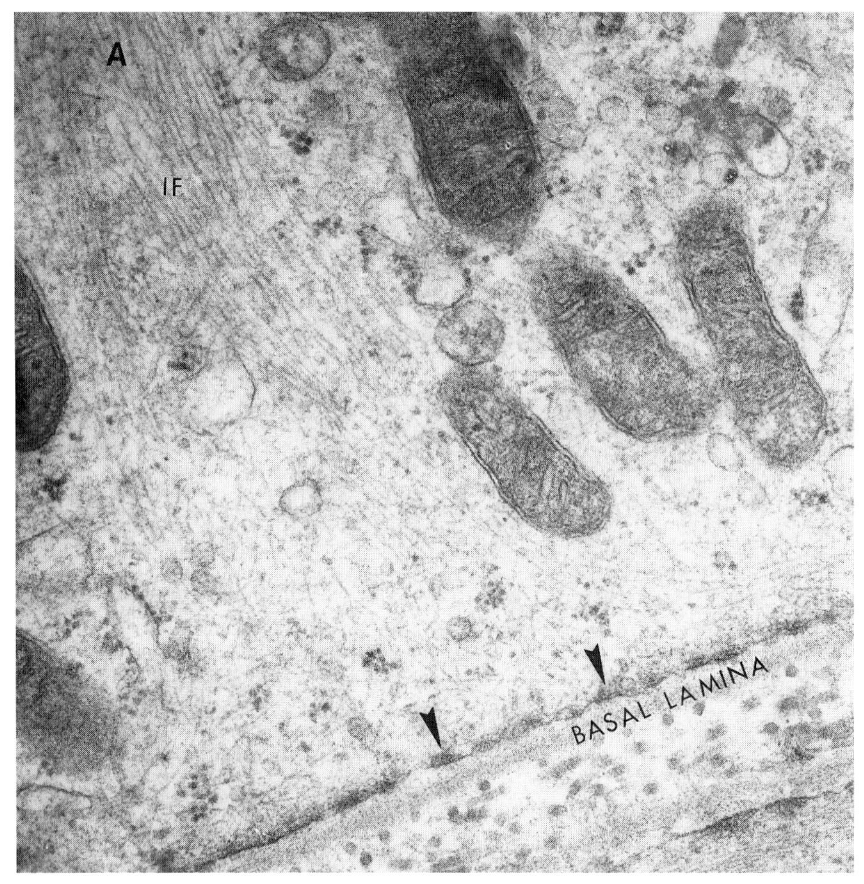

FIGURE 2 (Parts A–G found on pages 374–378) Junctions and specialized relationships of Sertoli cells and between Sertoli cells and germ cells. The letters match those of Fig. 1. (A) Hemidesmosome. The hemidesmosome (arrowheads) is characterized as dense plasma membrane plaques along the basal Sertoli cell in its relationship with the basal lamina. Intermediate filaments are commonly seen impinging on these junctions. (B) Sertoli–Sertoli tight junctions. Both thin-section (B-1) and freeze-fracture (B-2) views are shown. The thin-section view shows multiple Sertoli cells facing each other. There are paired translucencies (arrows) where adjacent Sertoli cell membranes converge and lines (arrowheads) where the plane of section interfaces with the membrane en face. Ectoplasmic specialization (es) in indicated. The membrane translucencies are junctional elements. A freeze-fracture image shows junctions within the plane of the membrane. The number of rows of junctions forming the Sertoli–Sertoli barrier sometimes exceeds 100.

junctions (desmosome–gap junctions; Fig. 2D). The adhering desmosome appears to function in cell-to-cell attachment and movement of cells. The junctions are formed early and persist until the spermatids begin elongation. The numbers and appearance of such junctions vary considerably as germ cells mature, giving rise to the notion that these junctions are dynamic entities. The gap junctions are presumed to facilitate communication between Sertoli cells and germ cells.

E. Ectoplasmic Specialization

Sertoli cells form a unique complex with elongating spermatids called the ectoplasmic specialization, a structure that orients them, holds them, and transports germ cells (Fig. 2E). The ectoplasmic specialization resembles the actin filament–endoplasmic reticulum complex seen at the Sertoli cell barrier, although it is formed only by the Sertoli cell. It first forms as round spermatids begin their elongation

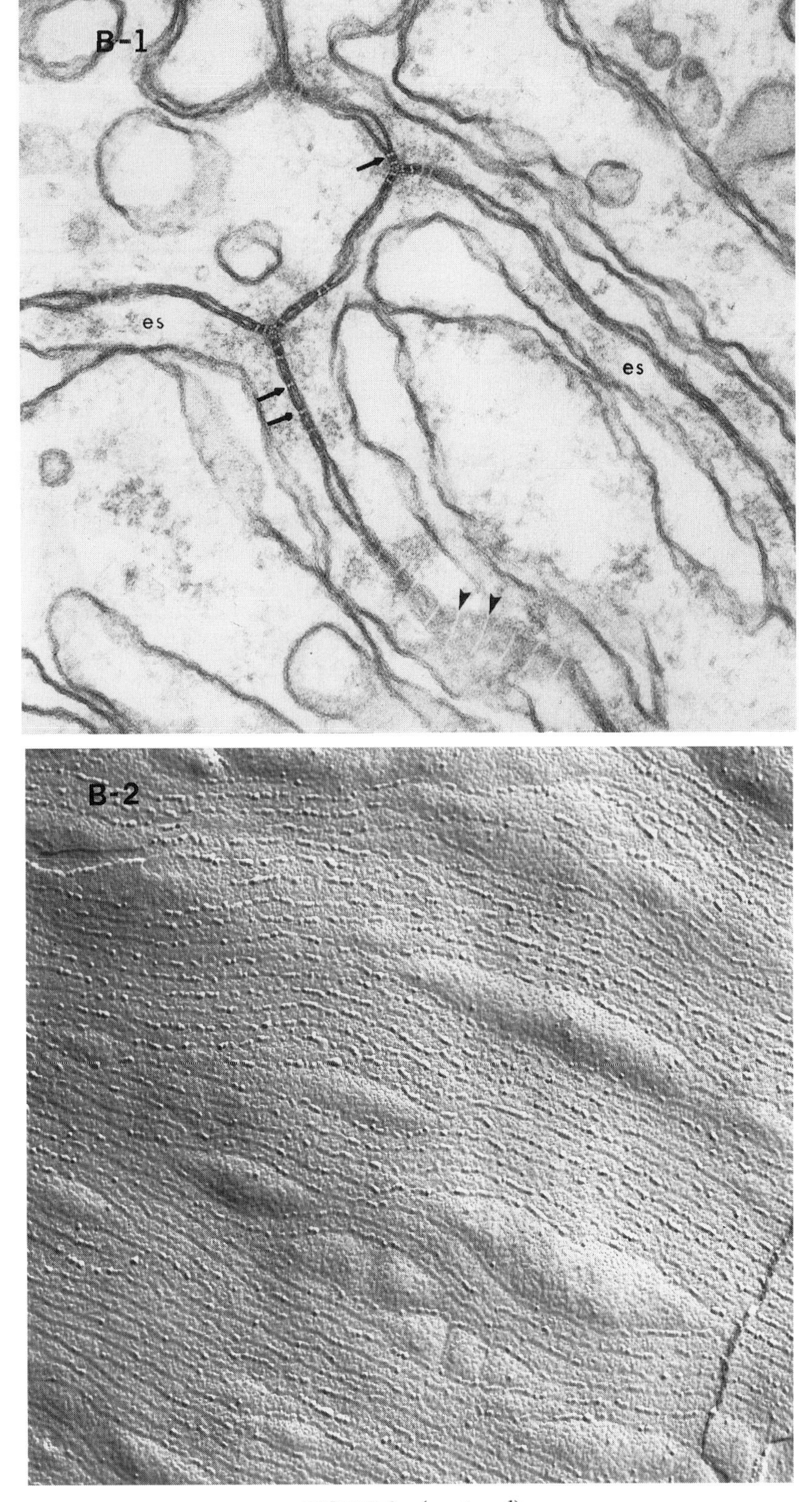

FIGURE 2 (continued)

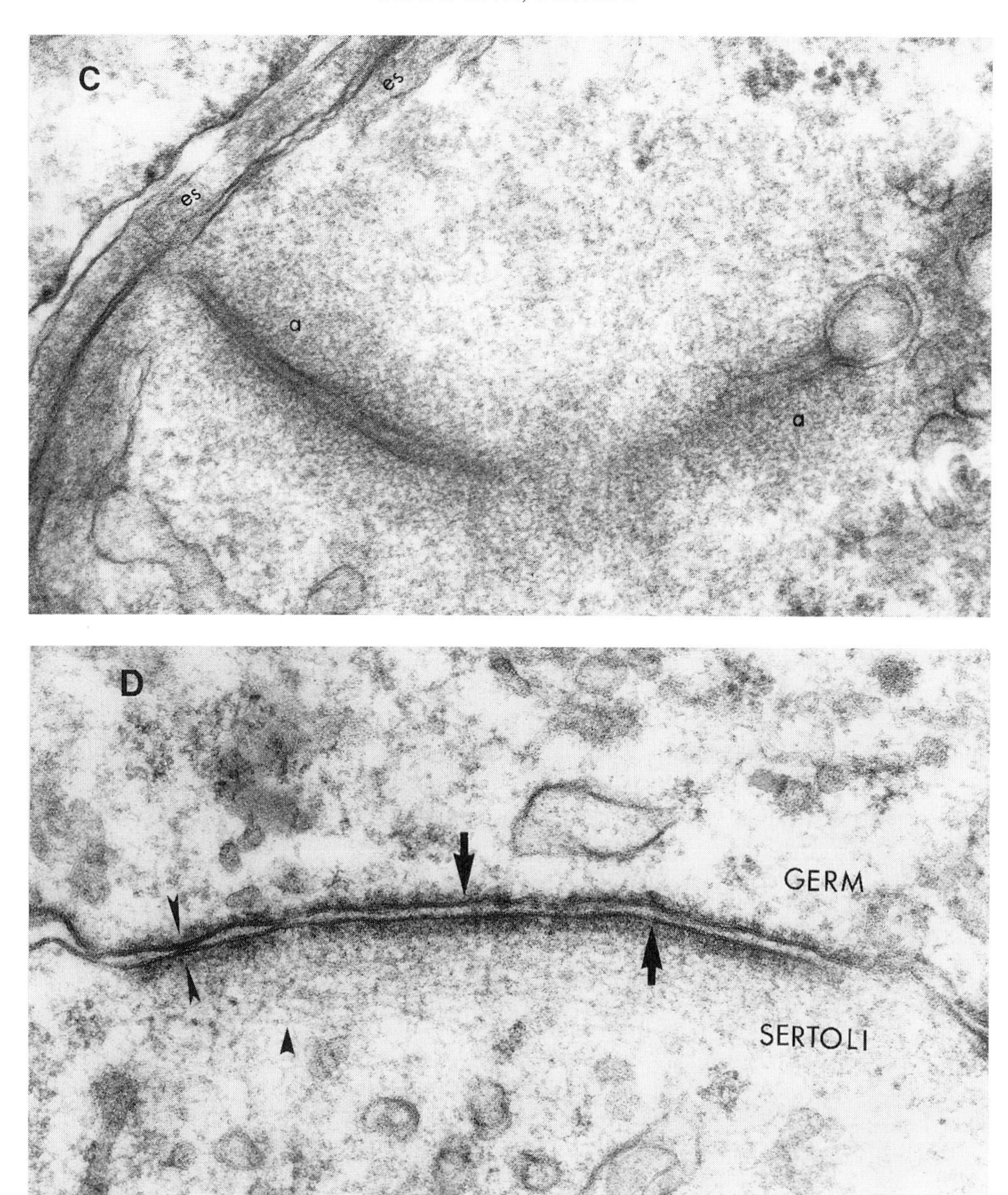

FIGURE 2 (*continued*) (C) Sertoli–Sertoli tubulobulbar complexes. Evaginations of one Sertoli cell into another taking the form of tubes with bulbous endings develop at the Sertoli–Sertoli junctions and the associated ectoplasmic specialization (es). Actin filaments (a) surround the tubular portion of the complex. (D) Desmosome–gap junctions. The desmosome–gap junction is a frequent relationship between Sertoli cells and germ cells. It is characterized by a uniform spacing between cells except where the membranes converge at the small gap junction (opposing arrowheads). Intermediate filaments (arrowhead) are seen in close association with subsurface densities (arrows) on the Sertoli side of the complex. (E) Ectoplasmic specialization. These elongating spermatids, with their nuclei (n), acrosomes (a), and plasma membrane (p), face a surface specialization of the Sertoli cell called the ectoplasmic specialization. The components of the ectoplasmic specialization are the large saccules of endoplasmic reticulum (er) and the actin filaments (arrowhead) that lie near the Sertoli plasma membrane (arrow) (reproduced with permission from Russell, 1993b). (F) Penetrating processes. Several penetrating processes (p) are seen indenting the cytoplasm of elongating spermatids along the region of the flagellum (F) (reproduced with permission from Russell, 1993a).

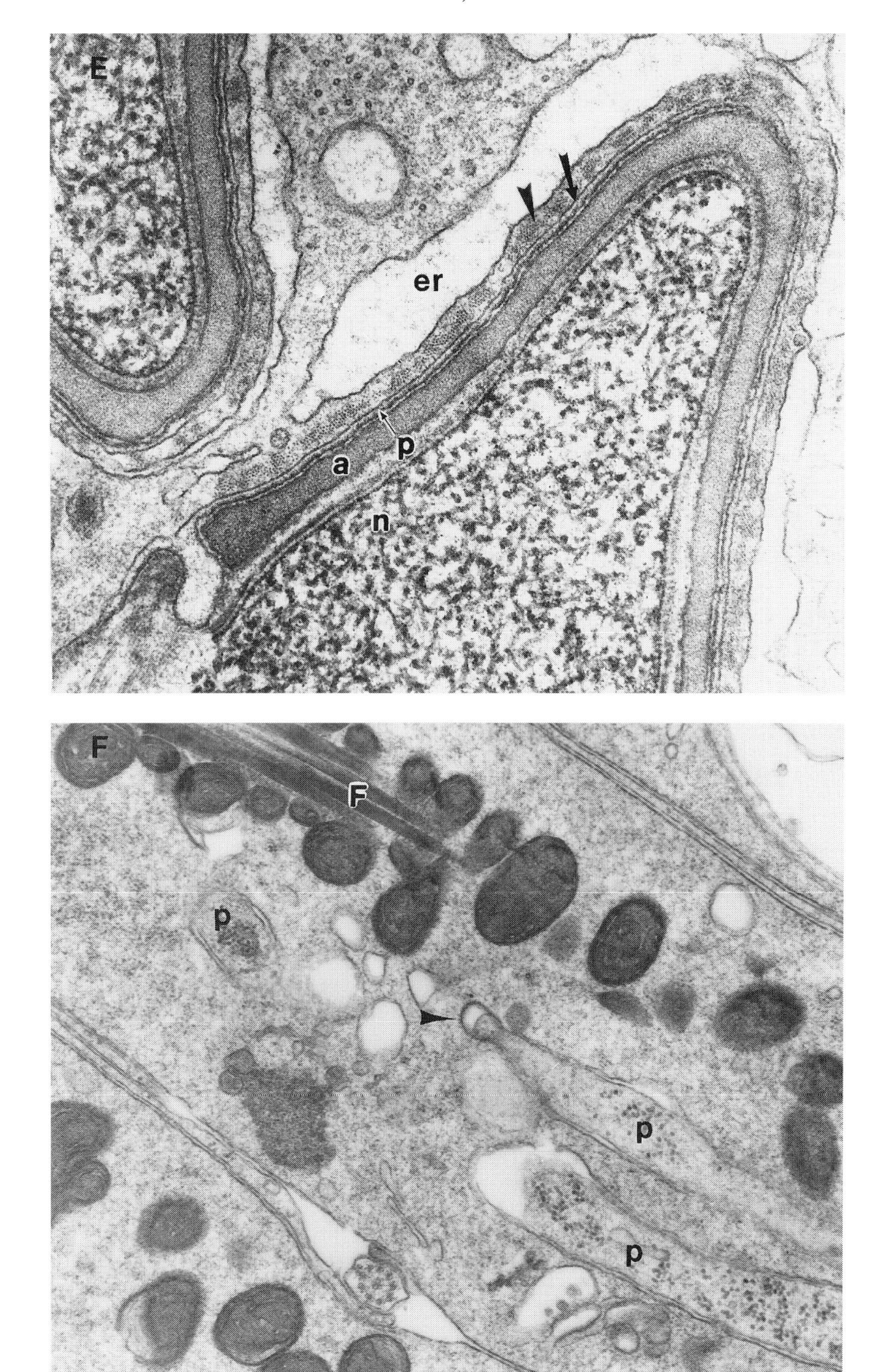

FIGURE 2 (*continued*)

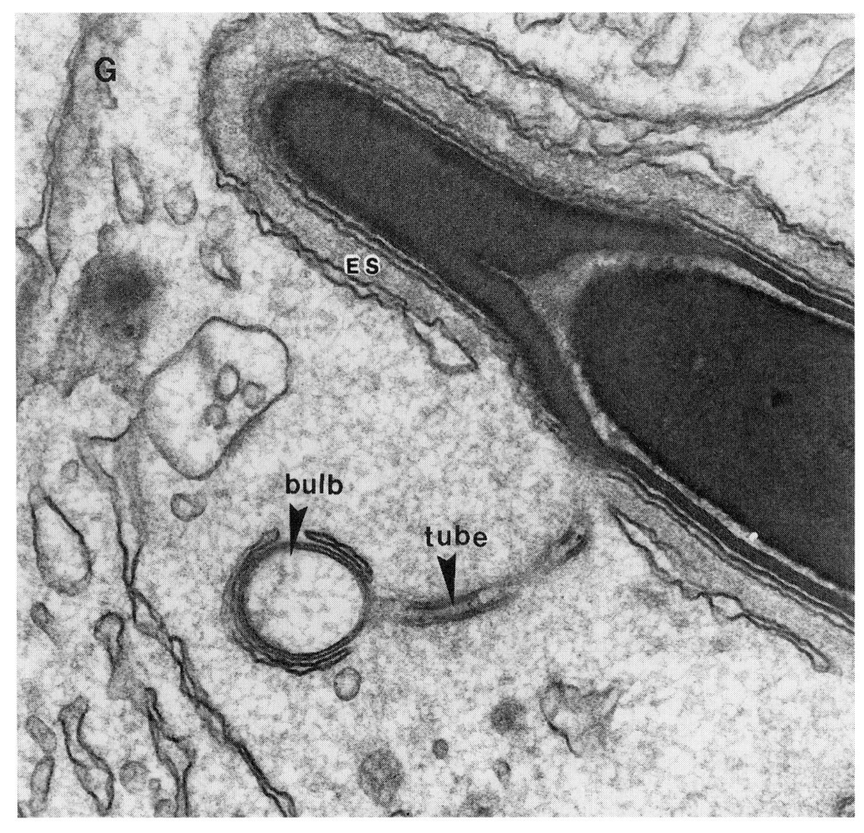

FIGURE 2 (*continued*) (G) Sertoli–spermatid tubulobulbar complexes. Near the time of spermiation, mammalian spermatids form an evagination into the Sertoli cell which has both a tubular and bulbous component. These structures form at sites where ectoplasmic specializations (es) are not present and extend into the Sertoli cell cytoplasm.

phase of development concurrently with the movement of the nucleus of the round spermatid to the cell surface. The acrosomal region of the spermatid is overlain by the ectoplasmic specialization, giving the appearance that all spermatid acrosomes face the base of the Sertoli cells. Once formed there is a tight bond between the two cells and in some micrographs one can see fine filamentous links between the two cells. Also, polarity is imparted to the germ cell after formation of the ectoplasmic specialization since the flagellum exits at the opposite pole of the cell from the acrosome and extends into the tubular lumen.

The round spermatids which lie in cup-shaped recesses of the Sertoli cell shortly after the formation of ectoplasmic specializations find themselves in the crypts of the Sertoli cell. They are drawn within recesses of the Sertoli cell because of attachment of microtubules to the ectoplasmic specializations and movement of the entire complex along microtubule tracts. In most mammalian species the germ cells remain within the deep crypts of the Sertoli cell until near the time of sperm release when they are moved to much shallower crypts near the lumen as the result of microtubules and microtubule motors.

F. Penetrating Process

Apical processes of Sertoli cells extensively indent the cytoplasm of elongating and elongated spermatids to form what are called penetrating processes (Fig. 2F). The function of penetrating processes is unknown but, at a minimum, they substantially increase the surface area of contact of the Sertoli cell with the spermatid to facilitate metabolic exchange. They may be of two types of penetrating processes; some may be phagocytosed by the spermatid and some may simply be withdrawn.

G. Sertoli–Spermatid Tubulobulbar Complex

Sertoli cells form tubulobulbar complexes with spermatids (Fig. 2G). Prior to sperm release the outpocketing of the plasma membrane of the spermatid into the Sertoli cell forms both tubular and bulbous components. Several generations of tubulobulbar complexes form and are phagocytosed by the Sertoli cell. This provides a means to anchor the spermatid as it extends into the lumen prior to sperm release. It appears that cell-to-cell links of the ectoplasmic specialization are internalized by this process, thereby freeing the spermatid from the hold of the Sertoli cell ectoplasmic specialization and triggering spermiation.

III. THE SERTOLI CELL CYTOSKELETON

The asymmetrical and highly columnar shape of the Sertoli cell along with the numerous configurational changes associated with germ cells are a function of an elaborate cytoskeleton. The Sertoli cell is equipped with an extensive array of intermediate filaments that circumscribe the nucleus and travel to sites of cell junctions. The function of intermediate filaments may be to protect the Sertoli cell against mechanical stresses. Foci of actin are associated Sertoli–Sertoli junctions, Sertoli spermatid ectoplasmic specializations, and tubulobulbar complexes. Microtubules extend from a supranuclear position to the apical Sertoli cell and serve as tracts for movement of substances, including the elongate spermatids. Microtubules also likely serve to maintain the tall columnar shape of the Sertoli cell.

IV. BIOCHEMICAL FUNCTIONS OF SERTOLI CELLS

Sertoli cells dedicate a reasonably large percentage of their protein synthesis to the synthesis and secretion of glycoproteins. The glycoproteins secreted by the Sertoli cells can be placed in several categories based on their known biochemical properties (Fig. 3). The first category includes the transport or bio-

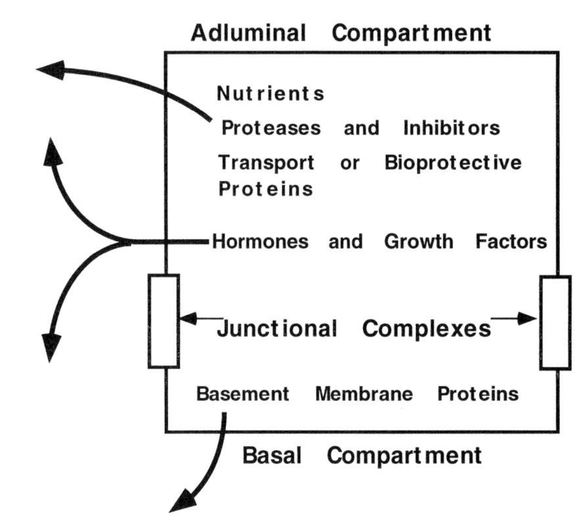

FIGURE 3 The major types of secretions from Sertoli cells. The basal and adluminal compartments of the Sertoli cells are labeled and the major classes of glycoproteins or other secretions are listed. Secretion may occur specifically in one compartment or in both. The type, amount, and direction of secretion may be influenced by stage of testicular development, endocrine status of animal, and stage of cycle of seminiferous epithelium.

protective proteins that are secreted in relative high abundance and include metal ion transport proteins such as transferrin and ceruloplasmin. The second category of secreted proteins includes proteases and protease inhibitors, which are allegedly important in tissue remodeling processes which occur during spermiation and movement of preleptotene spermatocytes into the adluminal compartment. The third category of Sertoli cell secretions includes the glycoproteins which form the basement membrane between the Sertoli cells and the peritubular cells . Finally, the Sertoli cells secrete a class of regulatory glycoproteins that can be made in very low abundance and still carry out their biochemical roles. These glycoproteins function as growth factors or paracrine factors and include products such as Müllerian-inhibiting substance and inhibin. In addition, Sertoli cells may secrete bioactive peptides such as prodynorphin and nutrients or metabolic intermediates.

The requirement for the glycoproteins in the process of spermatogenesis is inferred from the known properties of the proteins. However, *in vivo* evidence for the role of these proteins in spermatogenesis is lacking in most cases. One exception is the putative role of transferrin, which is an iron transport protein also made in the liver and the brain. Sertoli cells make transferrin as part of a proposed iron shuttle system which effectively transports iron around the tight junction complexes to the developing germ cells. The proposed model requires basal transferrin receptors, movement of iron through the cell, secretion of ferric ions associated with a newly synthesized testicular transferrin to germ cells, and incorporation of iron into ferritin in the developing germ cells. Most aspects of this model have been experimentally verified *in vivo*.

Sertoli and germ cells also interact via the c-kit receptor and ligand. Two mouse mutations that result in sterility in the testis are the dominant white spotting (*W/W*) and steel (*Sl*) loci. These loci encode the c-kit receptor and c-kit ligand or stem cell factor, respectively. The gonocytes express the receptor and the ligand is expressed by cells along the migratory pathway, and both are required for successful germ cell migration . However, after testis formation the c-kit ligand is expressed by Sertoli cells and the c-kit receptor is expressed by differentiating germ cells up to the pachytene stages. The biological result of this interaction is currently the subject of investigation.

See Also the Following Articles

Male Reproductive System, Overview; Sertoli Cells, Overview; Sertoli Cells, Regulation; Testis, Overview

Bibliography

Griswold, M. D. (1993). Protein secretion by Sertoli cells: General considerations. In *The Sertoli Cell* (L. D. Russell and M. D. Griswold, Eds.), pp. 195–200. Cache River Press, Clearwater, FL.

Griswold, M. D. (1995). Interactions between germ cells and Sertoli cells in the testis. *Biol. Reprod.* **52**, 211–216.

Russell, L. D. (1993a). Form, dimensions and structure of the Sertoli cell. In *The Sertoli Cell* (L. D. Russell and M. D. Griswold, Eds.), pp. 1–37. Cache River Press, Clearwater, FL.

Russell, L. D. (1993b). Morphological and functional evidence for Sertoli–germ cell relationships. In *The Sertoli Cell* (L. D. Russell and M. D. Griswold, Eds.), pp. 365–390. Cache River Press, Clearwater, FL.

Russell, L. D. (1993c). Role in spermiation. In *The Sertoli Cell* (L. D. Russell and M. D. Griswold, Eds.), pp. 269–303. Cache River Press, Clearwater, FL.

Russell, L. D., and Peterson, R. N. (1985). Sertoli cell junctions: Morphological and functional correlates. *Int. Rev. Cytol.* **94**, 177–211.

Skinner, M. K. (1993). Secretion of growth factors and other regulatory factors. In *The Sertoli Cell* (L. D. Russell and M. D. Griswold, Eds.), pp. 237–248. Cache River Press, Clearwater, FL.

Sylvester, S. R. (1993). Secretion of transport and binding proteins. In *The Sertoli Cell* (L. D. Russell and M. D. Griswold, Eds.), pp. 201–216. Cache River Press, Clearwater, FL.

Vogl, A. W., Pfeiffer, D. C., Redenbach, D. M., and Grove, B. D. (1993). Sertoli cell cytoskeleton. In *The Sertoli Cell* (L. D. Russell and M. D. Griswold, Eds.), pp. 39–86. Cache River Press, Clearwater, FL.

Sertoli Cells, Overview

Lonnie D. Russell
Southern Illinois University

GLOSSARY

adluminal compartment Region of the seminiferous epithelium segregated from intercellular access to blood-borne products by basally positioned Sertoli cell junctions.

basal compartment Region at the base of the seminiferous epithelium which contains spermatogonia and young spermatocytes, all of which have ready access to blood-borne substances.

seminiferous epithelium The cellular composition of the seminiferous tubule composed of germ cells and Sertoli cells.

Sertoli cell or ***blood–testis barrier*** A barrier to the penetration of substances into the seminiferous tubule formed near the base of the seminiferous by tight junctions of adjacent Sertoli cells.

Sertoli cell-only testis A condition in which Sertoli cells are the only cellular element in the seminiferous tubule.

synonyms Also called the sustentacular cell, nurse cell, supporting cell, or ramifying cell.

Sertoli cells are the somatic (nongametogenic) cells within the seminiferous tubule that are closely associated with germ cells and are thought to function by protecting, nourishing, and supporting the development of these cells.

I. HISTORICAL PERSPECTIVE AND COMPARATIVE STUDIES

Sertoli cells make up about 11–40% of the volume of the seminiferous epithelium depending on the species. Sertoli cells are named after their discoverer, Enrico Sertoli, who, in 1865, at the age of 23 provided an elegant description of them. Sertoli's drawings of the cell were amazingly accurate and he insightfully noted that the this cell must be involved in the production of sperm. Although over 5000 research articles have been published about Sertoli cells, scientists have only recently begun to determine how Sertoli cells support spermatogenesis. There is considerable work to be done since many of the details of how this occurs are lacking. There are no examples in mammals in which prolonged germ cell development occurs in the absence of Sertoli cells.

Somatic cells associated with germ cells are present in a variety of species. Some form a simple capsule in which germ cells develop, and others are extremely complex. Sertoli-like cells have been found in animal phylogeny as primitive as Porifera (sponges).

II. LOCATION AND STRUCTURE OF SERTOLI CELLS

Mammalian Sertoli cells are atypical of many other cell types in several respects. They are extremely large cells with a highly unusual shape—one that is continually changing in coordination with the changes in germ cells as they evolve to produce sperm. Their nuclei are irregularly shaped and they show regional specializations that indicate they have differential functions associated with the various

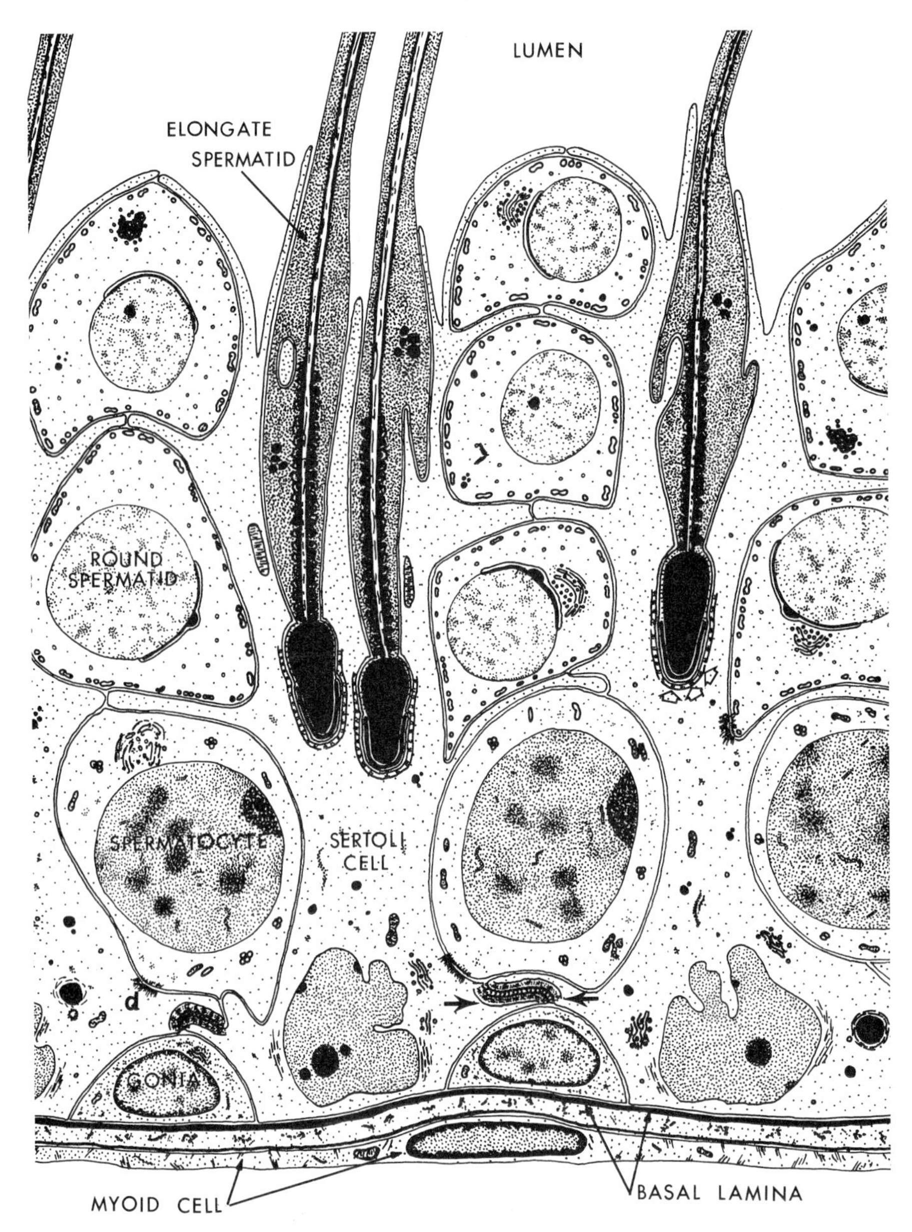

FIGURE 1 Diagrammatic representation of Sertoli cells as they reside among germ cells of the seminiferous tubule. The base of the Sertoli cell resides on the basal lamina along with spermatogonia (gonia). On the lateral surface of the Sertoli cell, a region of tight junctions is noted (arrowheads) that link all Sertoli cells within a specific tubule. The region below the tight junctions and at the periphery of the tubule is known as the basal compartment, and the region above and close to the lumen is known as the adluminal compartment. Sertoli cells are irregularly shaped and, to some degree, their shape is governed by the rounded or elongated nature of the germinal cells. The apical processes of the Sertoli cells extend to the lumen and form recesses for elongated spermatids. Junctions between Sertoli, cells such as desmosome-like junctions (d) and ectoplasmic specializations (open arrows), are indicated. The Sertoli cell nucleus is highly irregular (reproduced with permission from Russell *et al.*, 1990).

germ cell types with which they are simultaneously associated.

The bases of the Sertoli cells reside on a basal lamina. Being generally columnar in form, they extend to reach the lumen of the seminiferous tubule (Fig. 1). Their lateral surfaces are molded to fit the various germ cell types with which they are associated, giving the external aspect of the cell a highly irregular appearance (Fig. 2). The elongate spermatids indent the apical surface and reside in crypts of the apical surface.

The volume of the mammalian Sertoli cell, ranging from 2000 to 8000 μm3, is large compared to that of most other cells of the body. There are about 20 million to 4 billion Sertoli cells per testis depending on the mammalian species being studied.

One particular region of the Sertoli cell makes contact with other Sertoli cells that are adjacent to it. This contact occurs near the base of the tubule in the region traditionally called the "blood–testis" barrier, but because only Sertoli cells form the barrier, this region is more aptly referred to as the Sertoli cell barrier. The full implications of a Sertoli cell barrier are not known. The barrier appears to be immunoprotective against autoantigens that progressively develop on the surface of germ cells as they mature. The barrier ensures that most nutrients and stimulatory/inhibitory factors reaching germ cells do so through the Sertoli cell or are modified in their passage to germ cells. Spermatogonia and very young spermatocytes that are accessible to blood-borne substances are said to reside in the basal compartment. Most spermatocytes are protected by the Sertoli cell tight junctions and reside in the adluminal compartment. Some spermatocytes in their transit from basal to adluminal compartments are said to reside in a transitory compartment called the intermediate compartment.

Generally, Sertoli cells are characterized ultrastructurally by mitochondria with tubular cristae and an abundance of smooth endoplasmic reticulum as well as lipid. They have as much total smooth endo-

FIGURE 2 Two views of a single reconstructed rat Sertoli cell. The reconstructed cell was approximately 90 μm tall and extended from the base to the lumen of the seminiferous tubule. The surface is highly irregular in contour, conforming to the shapes of adjacent Sertoli cells. Ten apical indentations are filled with 10 elongate spermatids (reproduced with permission from Wong and Russell, 1983, p. 143 © 1983 John Wiley & Sons, Inc.).

plasmic reticulum as Leydig cells on a "per cell" basis.

Sertoli cells form numerous types of junctions with germ cells. These are primarily adhering- and communicating-type junctions. Some unusual cell-to-cell relationships between Sertoli cells and germ cells are found only in the testis.

III. HORMONAL REGULATION OF SERTOLI CELL FUNCTION

The adult (mature) Sertoli cell is one which possesses receptors for follicle-stimulating hormone (FSH) on its surface and receptors for testosterone within its nucleus. Testosterone is produced in Leydig cells in response to secretion of and binding of pituitary luteinizing hormone (LH) to the surface of Leydig cells. There is considerable contradictory information about whether or not germinal cells contain receptors for FSH and testosterone. Most investigators believe that hormone action supporting germ cell development is mediated solely by Sertoli cells. It is generally agreed that testosterone is absolutely necessary for spermatogenesis, but the role of FSH remains in question. It is known that FSH has the potential to maintain the viability of the same germ cell types as does testosterone, suggesting that one hormone functions as a backup for the other. After depletion of FSH and LH, the size of the Sertoli cell decreases dramatically as the apoptotic death of germ cells increases.

IV. DEVELOPMENT OF SERTOLI CELLS

The cell population giving rise to mammalian Sertoli cells is first noted in the mesonephros and later as part of the gonadal ridge. Here, they play a pivotal role in the differentiation of the forming gonad into a testis and in suppressing the development of the female reproductive duct system. Cords containing primitive germ cells are formed by the aggregation of Sertoli cells. The cords are surrounded by a basal lamina which is produced by the fetal Sertoli cells. In rodents that begin their gonadal maturation within weeks of birth, the Sertoli cells proliferate in the late prenatal period and postnatally likely through the stimulation of cell FSH receptors . For most species the number of Sertoli cells reaches a peak prior to full gonadal maturation and does not change markedly throughout adult life. The homolog of the Sertoli cell in the female is the granulosa cell.

Postnatal Sertoli cells are characterized by intermediate filaments of the cytokeratin variety such as might be found in epithelial cells. By adulthood, filaments have changed to the vimentin type, which is more characteristic of the mesenchymal-derived cells. The period that Sertoli cells stay immature morphologically is highly species dependent. At first, neonatal Sertoli cells are irregularly cuboidal. With the proliferation of germ cells at the beginning of gonadal maturation the volume and height of the Sertoli cells progressively increase. Coordination of Sertoli cell activity in perinatal development likely takes place due to the presence of Sertoli–Sertoli gap junctions.

Sertoli cells organize the epithelium during gonadal maturation. The large, rounded germinal cells which reside within the cords move peripherally to the basal lamina to coreside with Sertoli cells on this extracellular matrix. After the division of spermatogonia into spermatocytes, adjacent Sertoli cells form extensive rows of tight junctions, establishing a Sertoli cell barrier. Concomitantly, a tubular lumen forms which is an indication that luminally secreted products, unable to flow basally by impeding tight junction, cause an increase in fluid pressure that forces fluid to travel down the duct system of the testis. Cytologically, the Sertoli cell develops an indented nucleus, a distinctive nucleolus, abundant smooth endoplasmic reticulum, distinctive mitochondria, and specialized junctions with germinal cells (Fig. 3). The human Sertoli cell does not fully mature until puberty (11–14 years).

Hormonal responsiveness to FSH in the rodent decreases during gonadal maturation and receptors to androgen progressively increase. Sertoli cells are the main source of estradiol in the developing male. The role of estradiol is not know but it may be involved in maintenance of fluid resorption in the efferent ductules. Inhibin, which is secreted by the Sertoli cell postnatally and which suppresses FSH

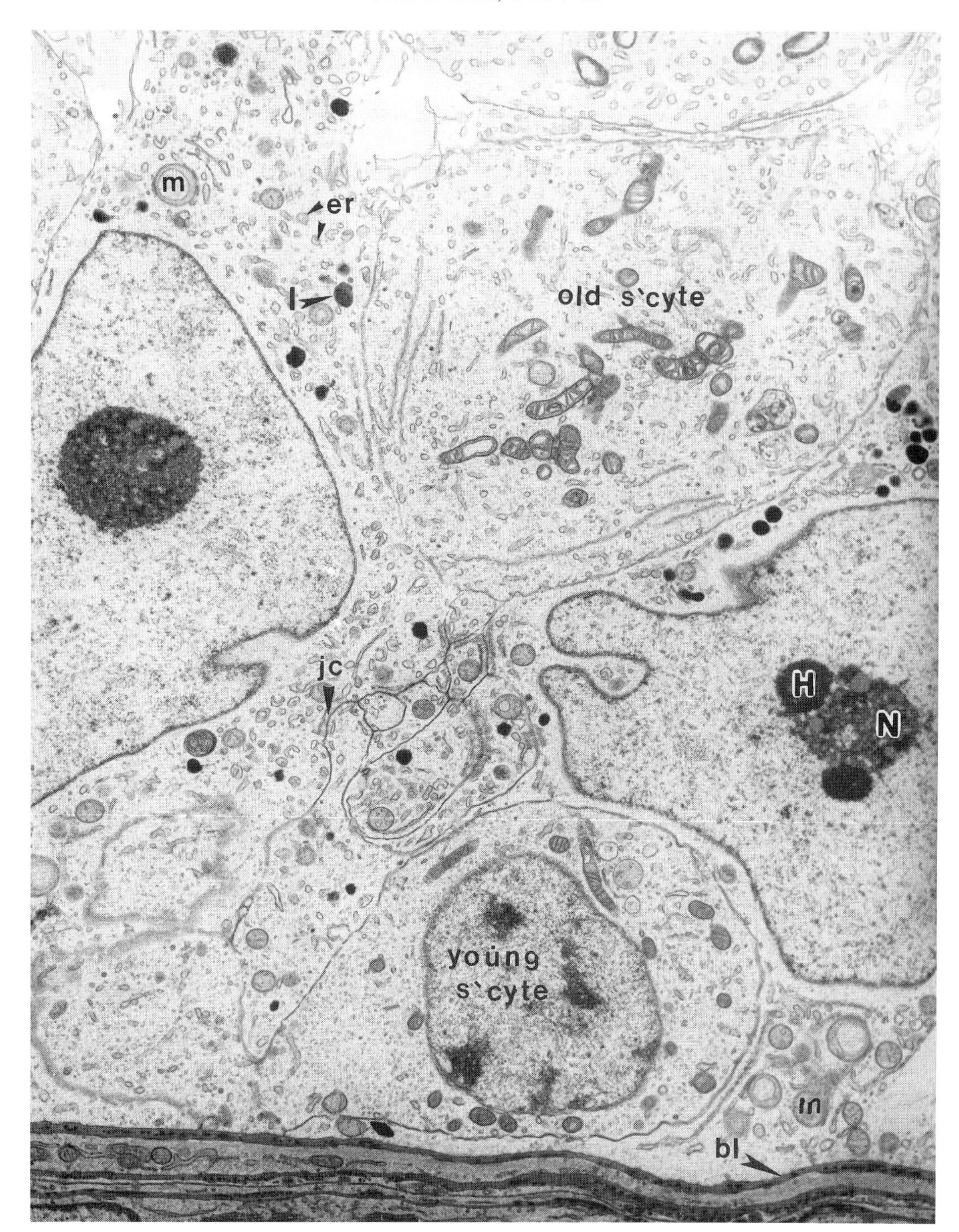

FIGURE 3 The subcellular Sertoli cell is shown in this electron micrograph of the base of the seminiferous epithelium of a mouse seminiferous tubule. In the center, two adjacent Sertoli cells with their irregular-shaped nuclei form typical junctional complexes (jc) demarcating the basal compartment, here containing a young spermatocyte (young s'cyte), and an aduluminal compartment containing a more mature spermatocyte (old s'cyte). Indicated are the Sertoli nucleoli (N) and characteristic heterochromatin (H) that flanks the nucleoli, mitochondria (m), smooth endoplasmic reticulum (er), and lysosomes (l). The micrograph shows only a small portion of the cell as it rests on the basal lamina (bl); the apical aspect extends a considerable distance from the top of the micrograph to reach the lumen of the seminiferous epithelium.

secretion, gradually declines with gonadal maturation. The Sertoli cell begins postnatally to secrete a variety of proteins and growth factors that characterize adult Sertoli cell function.

V. TECHNIQUES TO STUDY SERTOLI CELLS

Sertoli cells are difficult to study since they are among numerous germinal cells and are attached to them. They are enclosed by boundary tissue surrounding the seminiferous tubules and the tubules are, in turn, surrounded by interstitial tissue. Sertoli apical secretions may be studied *in vivo* by micropuncture techniques. It should be noted that Sertoli cells are known to secrete protein-containing fluid basally as well as into the seminiferous tubule lumen.

Culture of Sertoli cells has been the principal method to study isolated Sertoli cell functions. Sertoli cells are often isolated by separation of seminiferous tubules from the interstitial tissue, followed by enzymatic digestions and then elimination of contaminating germ cells by hypotonic shock. Sertoli cells flatten in culture but retain, to some degree, their natural form if they are cultured on an artificial basal lamina-type material. Sertoli cells are most readily isolated from rats between 15 and 20 days of age. Isolation and culture of Sertoli cells from older animals is rarely performed. Most of our knowledge about the biochemical responses of Sertoli cells is derived from cultures of cells taken from immature animals.

Sertoli cells may be studied in seasonal breeding mammals. Here, the testis periodically regresses and regression is followed by recrudescence just prior to the breeding season. The cyclic activity of Sertoli cells is likely in response to cyclically changing hormonal stimuli. Sertoli cells can support the development of transplanted spermatogonia germ cells from another animal of the same species or from another species.

VI. SPECIALIZED FUNCTIONS OF SERTOLI CELLS

The specialized functions of the Sertoli cell make it one of the most complex cells in the body. The Sertoli cell is studied extensively for its ability to reveal fundamental cell processes. The Sertoli cell organizes the seminiferous epithelium to present a coherent layering and/or positioning of germ cell types. By way of junctions and regionally organized underlying cytoskeletal elements, the Sertoli cell attaches to and holds germ cells in position as they move upward in the epithelium or within crypts of the Sertoli cell. The Sertoli cell cycles in coordination with the cyclic events of spermatogenesis.

The Sertoli cell secretes numerous proteins and factors that regulate spermatogenesis. It produces stem cell factor, which serves as a ligand for cell surface receptors on spermatogonia (*c-kit*) that appears necessary for maturation of spermatogonia. Additionally, the Sertoli cell produces proteases, transport and binding proteins. One particular protein produced by the Sertoli cell, androgen-binding protein, binds androgen and is delivered to the epididymis via the excurrent duct system to possibly affect epididymal structure and function. The Sertoli cell cyclically mediates the effects of stimulation by FSH and testosterone and metabolizes steroids.

VII. PATHOLOGY OF SERTOLI CELLS

It is not unexpected that Sertoli cell dysfunction leads to impaired spermatogenesis. Signs of Sertoli cell injury are vacuolation between inter-Sertoli junctions, detachment of Sertoli cells, and retention of late spermatids (failed sperm release). Nonspecific degeneration of germ cells or multinucleation of spermatids may be a consequence of Sertoli cell injury. Although Sertoli cells may be the primary site of injury from a toxicant, they rarely are lost from the seminiferous epithelium. Testes with Sertoli cells but without germ cells are called Sertoli cell-only testes.

See Also the Following Articles

Male Reproductive System, Humans; Sertoli Cells, Function; Sertoli Cells, Regulation; Spermatogenesis, Overview; Testis, Overview

Bibliography

Griswold, M. D. (1995). Interactions between germ cells and Sertoli cells in the testis. *Biol. Reprod.* **52**, 211–216.

Russell, L. D., and Griswold, M. D. (Eds.) *The Sertoli Cell.* Cache River Press, Clearwater, FL.

Russell, L. D., Ettlin, R. A., Sinha Hikim, A. P., and Clegg, E. D. (1990). *Histological and Histopathological Evaluation of the Testis.* Cache River Press, Clearwater, FL.

Sharpe, R. M. (1994). Regulation of spermatogenesis. In *The Physiology of Reproduction* (E. Knobil and J. D. Neill, Eds.), 2nd ed., pp. 1363–1434. Raven Press, New York.

Wong, V., and Russell, L. D. (1983). Three-dimensional reconstruction of a rat stage V Sertoli cell: Methods, basic configuration, and dimensions. *Am. J. Anat.* **167**, 143–161.

Sex Chromosomes

Baccio Baccetti and Giulia Collodel

Institute of General Biology of the University and C.N.R. Center for the Study of Germinal Cells

GLOSSARY

arm Portion of the chromosome from the centromere to the extremities

cluster Group of genes structurally or functionally related.

derived sex chromosome systems Systems differing from the basic XXo–XYo or ZZo–ZW systems are known. In some, the Y chromosome has been lost altogether (XX–XO systems) or the X and Y chromosomes have become completely nonhomologous; that is, unable to pair and undergo crossing over during meiosis. In others, called systems with multiple sex chromosomes, either the X or the Y chromosome is represented by two or more chromosomes.

gene A DNA segment responsible for production of a polypeptide chain.

heterochromatin (heterochromatic) Highly condensed chromatin regions which are not genetically expressed.

homolog, homologous Chromosome which contain the same genetic loci; a diploid cell has two copies of each homolog, one from each parent.

inactivation An aspect of the human X chromosome able to inactivate one of the two X's in females which turns off several thousand genes.

locus Position of a particular gene on the chromosome.

metacentric Chromosome in which the centromere divides in two similar parts along the entire length, forming two similar arms.

pseudoautosomal regions Regions present in the distal part of both Y arms; able to be transferred to X chromosome by meiotic recombination.

recombination The transfer of polynucleotidic sequences between different DNA molecules.

satellite DNA with a large number of repeated sequences.

X-linked characteristics A large majority of the genes are in haploid status in the male and, as a consequence, mutations in the genes of X-linked diseases are present and active in hemizygous males.

I. INTRODUCTION

The sex chromosomes are an exception to the rule that all chromosomes of diploid organisms are present in pairs of morphologically similar homologs. Early microscopic analysis showed that one of the chromosomes in males of some insect species does not have a homolog. This unpaired chromosome was called the X chromosome, and it was present in all somatic cells of the males but in only half the sperm cells. The biological significance of these observations became clear when females of the same species were shown to have two X chromosomes. In other species in which the females have two X chromosomes, a morphologically different chromosome is found in males. This is referred to as the Y chromosome, and it pairs with the X chromosome during meiosis in males because the X and Y share a small region of homology. The difference in the chromosomal constitution of males and females is a chromosomal mechanism for determining sex at the time of fertilization.

The X and Y chromosomes are called the sex chromosomes to distinguish them from other pairs of chromosomes, which are called autosomes. Although the sex chromosomes control the developmental switch that determines the earliest stages of female or male development, the developmental process itself requires many genes scattered throughout the chromosome complement, including genes on the autosomes. The X and Y chromosomes also contain many genes with functions unrelated to sexual differentiation.

II. THE CHROMOSOMAL SYSTEM OF SEX DETERMINATION

Almost all higher animal species have separate sexes and genetic mechanisms which determine them. On the contrary, in several more primitive large groups, such as ctenophores, flatworms, ectoprocts, oligochaetes, leeches, euthyneurous molluscs, and tunicates, hermaphroditism is the rule and bisexuality the exception: Only a few species of flatworms and a few tunicates are clearly bisexual. Benazzi suspects them of having reverted to bisexuality. More probable, in animals hermaphroditism is the primitive condition and bisexuality the derived one. Mollusca include both bisexual and hermaphroditic species, whereas the Arthropoda and Chordata are generally bisexual. There are some exceptions: There are a few primitive crustaceans and a very few insects that are hermaphroditic. Among the chordates hermaphrodites can be found in the subphylum Urochordata and in a small number of fish among vertebrates.

With regard to sex chromosomes, unconvincing claims of vestiges of such sex chromosomes have been proposed in the rare hermaphroditic species which are closely allied to bisexual forms (recently acquired hermaphroditism), but sexual chromosomes have never been proposed for trematodes, gastropods, and oligochaetes (primitive hermaphroditism). Claims by early cytologists were certainly mistaken, either in observation or in interpretation. In some molluscs and fishes the sexual differentiation is consecutive because the individual passes through male and female phases in the life cycle.

We will treat separately the simple system of genetic sex determination, having a single pair of chromosomes involved, the multiple system, having more than a single pair of sex chromosomes, and the haplodiploidy system.

A. The Single Pair of Sex Chromosomes

The two sex chromosomes can be homologous, in length and shape, in one sex (X chromosomes) and heterologous in the other (X and Y chromosomes). Thus, X is present in both sexes, and Y is present only in one. At the meiotic prophase the sex chromosomes associate. In some animal species or groups of species, Y is absent because it was lost during evolution through fusion with the X or with an autosome. Therefore, the diploid number of chromosomes in one sex is uneven, and we distinguish between the XY:XX type of mechanism and XO:XX, where O indicates the absence of Y. The XX sex is said to be "homogametic" because it produces only one type of gametes; the XY or XO sex is called "heterogametic" because it produced two kinds of gametes in equal number. In most groups of bisexual animals, the male sex is heterogametic because it produces two chromosomally different kinds of spermatozoa and one kind of egg. Otherwise, in a few important groups (Trichoptera and Lepidoptera among insects, and birds plus a few fishes, amphibians, and reptiles among vertebrates) the female is heterogametic (XY:

two kinds of eggs) and the male homogametic (XX: one kind of spermatozoa). In this latter condition, some authors speak of ZZ and ZW chromosomes.

Apart from some groups of marine invertebrates in which sex seems to be rather labile and genetic factors seem to play no role in its determination (e.g., the echiuroid worm *Bonellia viridis* and the isopod crustacean *Jone thoracica*), in the higher animals the genetic sex-determining systems are mechanisms switching the pattern of development onto one or other of two alternative pathways. Nevertheless, some exceptions exist, in addition to the primitive hermaphroditism, that are compatible with the presence of sex chromosomes. Goldschmidt distinguishes between the already mentioned hermaphroditism (functional coexistence in the same individual of male and female gametes in the absence of sex chromosomes), gynandromorphism (mosaicism for sex chromosomes), and intersexuality (phenotypes intermediate between maleness and femaleness). While phenotypic intersexuality occurs in the presence of abnormal karyotypes or under the influence of hormonal disorders, gynandromorphism is compatible with the presence of XX and XY cell lineages present in the same individual and is described in Arthropoda (insects, spiders, and ticks) and vertebrates.

It is now recognized that the effect of an XY:XX or XO:XX sex chromosomes mechanism is to produce equal numbers of the two sexes during fertilization. Also, in humans a primary sex ratio close to 1.0 has been demonstrated. Regarding the switch mechanism active in XY systems, two basically different genetic types have been proposed: the "genic balance" system, typical of dipterous insects, and the "dominant Y" system, typical of vertebrates.

The genic balance system was first demonstrated by Bridges (1925) in *Drosophila*. It consists in a sexual phenotype depending on the balance between female-determining genes in the X chromosome and male-determining genes in the autosomes. The Y chromosome, in fact, plays no effects in the determination of the visible sex phenotype but simply interferes with spermatogenesis and the characters of the spermatozoon: XO individuals of *Drosophila* have a normal male phenotype, and XX individuals with supplementary Y chromosomes (one or more) are phenotypically normal and fertile females. In the gypsy moth *Porthetria dispar*, in which, as in all Lepidoptera, the female sex is heterogametic, the X is a male-determining chromosome, whereas the female-determining factors lie in the Y and, perhaps, in autosomes. Obviously the genic balance system is diffused in all organisms in which the male is normally XO: nematodes and, among Arthropoda, the orders Odonata, Hemiptera, Coleoptera, and orthopteroid orders in the class Insecta, as well as the class Arachnida. The same should be true for the mammalian monotremes because they have XO males. The "dominant Y" system has been described in mammals, with male heterogamety, and in the urodele Axolotl, with female heterogamety. Most of the research has been carried out in humans, in which the Y appeared as a powerfully male-determining chromosome. In fact, XO individuals are not sterile males (as occurs in *Drosophila*) but rather sterile females with ovarian dysgenesis (Turner's syndrome) and XYY human individuals are virtually normal males but may be mentally subnormal and aggressive, whereas XXY, XXXY, XXYY, and XXXXY individuals are intersexual (Klinefelter's syndrome) instead of being normal females as occurs in *Drosophila*. All these phenotypes provide evidence of the male-determining properties of Y in humans. The X must have some role in the development of ovaries. The XO individuals comprise 0.03% of all female births; moreover, 4% of spontaneous abortions are XO; the X seems commonly metroclinous. The frequency of XYY and XXY is 0.1–0.2% of male births, whereas XXX females comprise about 0.1% of all female births. The total frequency of human aneuploidism concerning the sex chromosomes is therefore close to 0.3% of all births. Obviously, XO or XXY aneuploidies have been demonstrated in other mammals (cattle, mice, and marsupials).

We have seen that the XY:XX sex chromosomes mechanism is the basic one, in the two modes of male or female heterogamety, and that in both modes a basic system of XO instead of XY can occur. In addition to this, newly formed sex chromosomes can occur in groups belonging to the ancestral heterogametic situation. This new sex chromosomes formation has been described in fishes, crustaceans, and several insects groups.

B. The Multiple Systems of Sex Chromosomes

The systems of the sex chromosomes mechanism in which more than one kind of X or Y is present may be formally classified as X_1X_2Y, $X_1X_2X_3Y$, X_1X_2O, XY_1Y_2, etc., according to the number of nonhomologous, or only partly homologous, X's and Y's in the heterogametic sex. The distinction between X_1X_2Y and XY_1Y_2 can be made with the parallel investigation of the karyotype in the homogametic sex or in related species or genera. This comparison is also useful in understanding the origin of the multiple sex chromosomes mechanism. Also, the type of meiotic association between the heterochromosomes in the heterogametic sex, and particularly according to whether chiasmata are formed between X's and Y's, has been considered, and in many cases it is uncertain whether chiasmata are formed. In all such cases, however, the meiotic mechanisms ensure that all the X's go to one pole and Y's to the other. It can be asked what is the adaptative significance of such elaborate multiple systems of sex chromosomes. White (1977) suggests that they lead to fixation of a great deal of heterozygosity and much genetic recombination in the homogametic sex. Three main types of origin have been distinguished, all involving translocations between sex chromosomes and autosomes.

1. *The Centric Fusions of Neo-X or Neo-Y with Autosomes*

White (1953) and many others have described this phenomenon in Orthoptera, in which neo-X and neo-Y can be formed after fusion with autosomes, giving origin to large metacentric chromosomes. At meiosis, in males a trivalent is formed grouping the three. X_1X_2Y or $X_1X_2X_3Y$ as a result of fusion with autosomes have been described in the beetle families Cicindelicindelidae and Tenebrionidae. The tenebrionid *Blaps polychresta* possesses 12 kinds of X and 6 kinds of Y chromosome. In the isopod crustacean *Jaera*, X autosome fusion has produced XY_1Y_2. Many kinds of centric fusion of various Y's with X or with autosomes are described in marsupials (*Potorous* and *Pratemnodon*) and in rodents (*Sorex* and *Gerbillus*). XY_1Y_2 of this kind are also described in bats. In the reptile *Bungarus*, having female heterogamety, X_1X_2Y has been detected, whereas in *Anolis*, having male heterogamety, the formula X_1X_2Y is found in the males. In both cases, the homogametic sex has $X_1X_1X_2X_2$.

2. *Mutual Translocations between Metacentric Chromosomes*

The classical example is in the insect superfamily Mantoidea (order Orthoptera), in which the basic XO mechanism is replaced in three independent groups by X_1X_2Y as a result of the acquisition of a neo-Y (originally an autosome) and a neo-X by a mutual translocation between the metacentric X and a metacentric autosome. Similar mechanism have been detected in the closely related insect orders Siphonaptera and Mecoptera as well as in the tick *Amblyomma*. In these cases of multiple sex chromosome mechanism, a multivalent is formed at meiosis, in which chiasmata are formed in pairing homologous sex chromosome segments. Nevertheless, many cases are known in which in sex chromosomes arisen by translocation chiasmata are absent. This occurs in crickets, mole crickets (*Gryllotalpa*), in which XO, XY, and X_1X_2Y species are known, many species of *Drosophila* having X_1X_2Y or XY_1Y_2, and other fruit flies.

3. *Dissociation*

This mechanism of multiple sex chromosomes consists of the formation of supplementary heterochromosomes by broken X or Y. It is the most simple way to obtain neo-X or neo-Y. In general, this origin of multiple sex chromosomes is not possible in groups in which the meiotic association of sex chromosomes in the heterogametic sex presents chiasmata in the pairing segment of X and Y confined to one arm. The classic example of dissociation occurs in numerous species of Heteroptera, characterized by failure of synapsis of the holocentric X and Y chromosomes. The multiple sex chromosomes mechanisms in Heteroptera range from X_1X_2Y and X_1X_2O up to $X_1X_2X_3X_4X_5Y$ or to $X_1X_2X_3X_4O$. Other samples of this kind of origin by dissociation of achiasmatic X or Y occur in other insect orders basically having XY male—Dermaptera and Coleoptera, as well as in the females of Lepidoptera. Particularly interesting is the case of Arachnida, in which Y is totally absent and X is represented by several separate elements (X_1, X_2, X_3, etc.), probably originating from the ancestral X. It is important that all the X's at the meiosis pass regularly to the same pole and are all acrocentric. The X*n* arrangement occurs in some members of the insect order Plecoptera, in which the basic type is XO and up to three X's may occur as well as a neo-Y. In nematodes, up to $X_1X_2X_3X_4X_5O$ occurs in *Ascaris* and neo-Y only in two species.

C. The Haplodiploidy

In a few groups of animals, the males are haploid because they develop from unfertilized eggs by parthenogenesis, and the females are diploid because they develop from fertilized eggs. Consequently, females have two parents; males are impaternate. Haplodiploidy is therefore a method of sex determination in which the frequency of males is determined by the frequency with which unfertilized eggs are laid. This form of haploid males producing parthenogenesis is called arrhenotoky, in contrast to the females producing parthenogenesis, named thelytoky. While thelytoky has arisen repeatedly in almost all the phyla of the animal kingdom and can be induced experimentally, arrhenotoky is known to have arisen only eight times in the history of Metazoa: six times in Insecta, one time in Arachnida, and one time in Rotifera. In insects haplodiploidy occurs in almost all the Hymenoptera (including the well-known honeybee), and in many members of the superfamily Coccoidea of the order Homoptera, especially the Aleyrodidea and the Coccidae. In the Diaspidae of the superfamily Coccoidea, haploidy occurs by the elimination of the paternal set of chromosomes from a fertilized egg. In Thysanoptera almost all the species are haplodiploid except for those species that reproduce by thelytoky. Independently the coleopterans *Micromalthus debilis* and *Xyleborus* acquired haplodiploidy. In the several members of the subclass Acari (ticks and mites) of the class Arachnida, as well as in several species of the acoelomate phylum Rotifera, haplodiploid sex determination is present. The haploid males of haplodiploid species obviously produce haploid spermatozoa by an abortive spermatogenesis through an incomplete first meiotic division in which bivalents are not formed and a nonnucleated cytoplasmic bud is eliminated. The spermatocyte nucleus never enters in a true metaphase, and the second meiotic division is a simple mitosis so that each haploid spermatocyte produces two haploid spermatozoa. However, in bees the second meiotic division is also unequal, and each primary spermatocyte produces a single spermatozoon.

Most of the previously mentioned cases of haplodiploidy have developed in animal groups having a basic XO mechanism of sex chromosomes, and we argue that the haploid individuals, possessing only one X, are males, whereas the diploid ones, possessing two X's, are females. In fact, the situation is more complex because in the hymenopteran *Habrobracon*, Whiting (1943) has demonstrated that femaleness seems to depend on heterozygosity for several complementary X factors that can be designated X_1, X_2, X_3, . . . X_9. The same is true for the honeybee: In this species, as in *Habrobracon*, the femaleness depends on heterozygosity in X's factors, whereas the exceptional diploid males obtained by inbreeding are all homozygous for them.

III. THE GENES OF THE SEX CHROMOSOMES

A. The Y Chromosome

1. *Mammals*

Beginning in the 1960s, the presence in the Y chromosome of a gene producing a factor necessary for the differentiation of the embryonal gonad to testis was demonstrated: It was called testis determining factor (TDF), and the responsible gene, when translocated to the X chromosome, was demonstrated to cause the production of XX males. Subsequently, the antigen of histocompatibility (H-Y) was discovered and supposed to be produced by a structural gene in the Y chromosome, possibly corresponding to that of TDF. H-Y was found to be produced by a locus at the extremity of the short arm of the X chromosome and by an autosomic locus, whereas the Y-linked locus was supposedly separate from TDF and had only a regulatory function.

The Y is the only chromosome present exclusively in the male karyotype, and in the humans it is one of the smallest chromosomes. If we omit a small

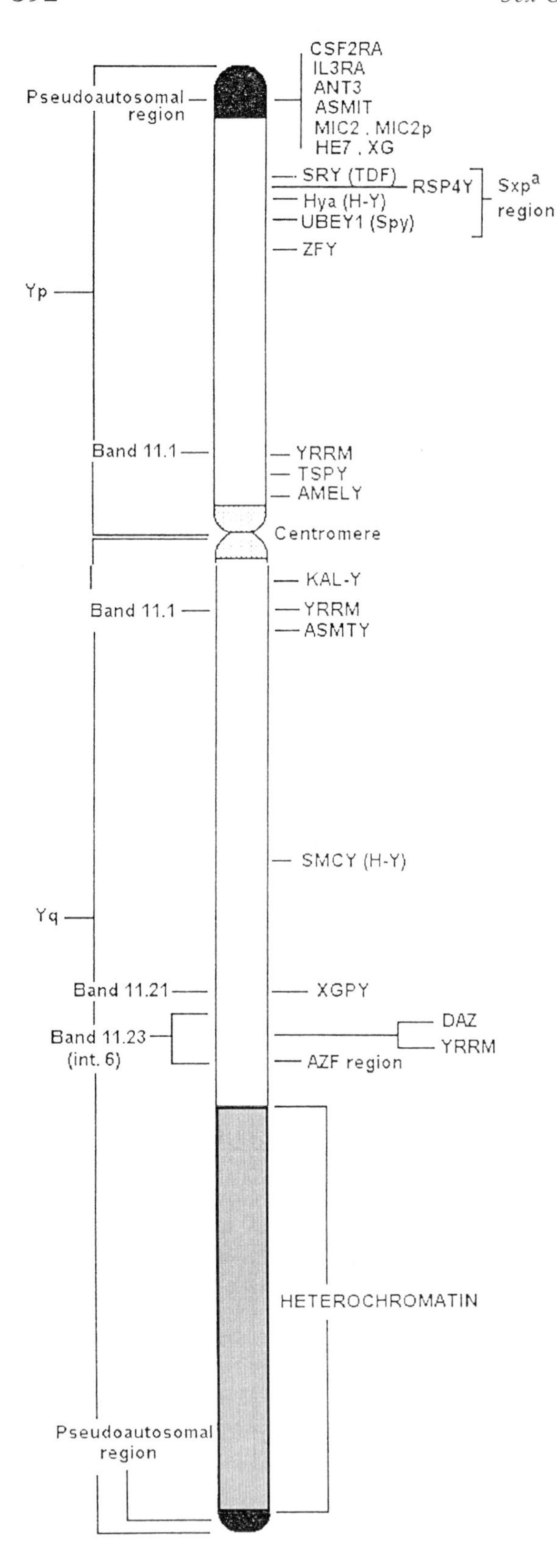

region in Y homologous with the X chromosome, enabling recombination with X during male meiosis, the content of protein-forming genes is minimal and does not recombine in male meiosis. Thus, the majority of the Y chromosome is clonally transmitted through paternal lineages and it essentially contains genes concerned with spermatogenesis.

The human Y chromosome has two arms of different lengths: the minute short arm (Yp) and a long arm (Yq) of variable length according to the individuals (Fig. 1). The distal parts of both arms are defined as "pseudoautosomal" regions and are able to be transferred to X chromosome by meiotic recombination. In the Sxr mouse, a portion of the short arm containing genes involved in spermatogenesis has duplicated, and this portion is transposed and positioned on the tip of the long arm. Thus, it has a 50% chance of being transferred to the X chromosome at meiotic recombination with the pseudoautosomal Yq part, with frequent deletion and loss of spermatogenesis.

Examining the long arm from the tip and proceeding in the direction of the centromere, we find in Yq a heterochromatic region of variable length (causing the variable size of the long arm), almost devoid of genes and dispensable with regard to male fertility. More proximally to the centriole, the euchromatic region (30 Mb long), which includes the proximal Yq, the centromere, and Yp, begins. We can map it by examining the natural deletions occurring in the Y chromosome. All the deletions of this category of genes are *de novo* mutations and are not inherited because they interfere with spermatogenesis. At first, we find close to the heterochromatic region in position 11.23 the band of the Azoospermia Factor (AZF). AZF gene was discovered by Tiepolo and

FIGURE 1 Map of the human Y chromosome. CSF2RA, colony-stimulating factor receptor α chain gene; XG, blood groups locus; SRY(TDF), sex determining region of Y; Hya, locus necessary for the expression of the male-specific H-Y antigen; YRRM, Y located RNA recognition motif family gene; TSPY, a gene family necessary for spermatogenesis; AMELY, amelogenin locus; ASMTY, hydroxynidale-*O*-methyltransferase locus; SMCY, locus for the control of the expression of HY; MIC2, antigen for cell adhesion.

Zuffardi by the microscopic detection of deletions in the distal Yq, close to the heterochromatic segment. The absence of the AZF gene is correlated to azoospermia. In the same region (Yq band 11.23: interval 6) the single-copy gene, Deleted in Azoospermia (DAZ), which encodes an RNA-binding protein, has been described. In interval 6 of Yq, numerous microdeletions have been found associated with azoospermia or severe oligozoospermia. These findings are very important in order to explain many sperm defects implicated in human male sterility. Also, an entire family (at least 15 members) of genes encoding histones involved in RNA trafficking has been localized in the same area [named "Y located RNA Recognition Motif genes" (YRRM) by Ma *et al.* (1992)]. They are essential for the maintenance of normal levels of fertility in primates, including man. It is possible that this localization is incorrect: Reijo et al. (1995) fail to find YRRM genes in the band 11.23 and report a localization, even on the long arm, close to centriole and also in the short arm at position 11.1.

A new pseudogene that is very close to Yq 11.21 has been found—XGPY, originated from the XG gene of the XG blood groups. In the middle of the euchromatic region of Yq are located the locus HY and the SMCY gene, which controls the expression of HY and is present in mouse, man, and marsupials.

Other well-known genes are located in the pericentromeric area: a locus conditioning Kallmann's syndrome (responsible for a hypothalamic deficiency causing hypogonadotropic hypogonadism); a locus ASMTY for the hydroxyindole-*O*-methyltransferase (the last enzyme of the metabolic route to the pineal hormone melatonin); and a locus for the amelogenin, in the short arm, close to the centromere. More distally in Yp, another gene family, TSPY, has been found. It is also needed for spermatogenesis and is present in, and specific for, germ cells in many primates, including man, pigs, and cows.

In the region of the short arm close to the pseudoautosomal segment, the sex determining region of the Y (SRY), identified in 1990 and supposedly corresponding to TDF, is present. SRY is responsible for the transformation of the undifferentiated embryonal gonad in testis and has been found in men, mice, and pigs. In the same apical area other important spermatogenetic genes are present: Hya, needed for the expression of the male-specific H-Y antigen; and UBEY1, needed for Spy, a factor for the proliferation of differentiating A spermatogonia. These three genes constitute the Sxpa region, which can be translocated on the X chromosome, causing a complete arrest of meiotic metaphase. The gene ZFY (Zinc Finger) is located in the same area. This gene encodes a protein (zinc finger) specifically binding the nucleic acids; thus, it regulates gene expression. Most of these sequences are also present in great ape genomic DNAs and are therefore highly conserved, as is the entire euchromatic region. Nevertheless, other sequences are, in humans, more polymorphic, and it has been reported that Y chromosomes from different ethnic backgrounds belong to different chromosomeal groups.

This gene is also present in marsupials and in Monotremata and consequently has been conserved throughout the entire mammalian evolution. Sing (1995) suggests that a DNA minisatellite (Bkm: banded krait minor) provides binding sites for recombination proteins, bringing a coordinate decondensation of the Y sex chromosomes in the heterogametic sex of snakes (female) and mammals (male), thus switching for activation of Y genes.

2. *Insects*

The investigation on the karyological events occurring in Drosophila spermatogenesis has a long tradition. Its origin reaches back to early days of Drosophila genetics, when it was recognized that Y chromosome plays an important role in this developmental process.

The *Drosophila melanogaster* Y chromosome is a submetacentric element which accounts for about 12% of the male genome. It is 12–18% longer than the X chromosome at metaphase and predominantly heterochromatic: Four highly repeated simple sequence satellite DNAs have been identified which together account for about 70% of its length. Moreover, it has been suggested that these satellite DNAs are organized in heterochromatic blocks containing a long and homogeneous array of short sequences tandemly repeated. These blocks have different cytochemical features when stained with Hoechst 33258, quinacrine, and N-banding techniques.

The Y chromosome in *D. melanogaster* is required for male fertility: Males lacking a Y (XO males) are phenotypically normal but sterile. The Y chromosome function occurs only in the male germline since the transplantation of XY cells into XO males results in normal fertility which requires six regions, four on the long arm designated kl-1, kl-2, kl-3, and kl-5, and two on the short arm, ks-1 and ks-2. kl-4 has not been confirmed. Three of these regions (kl-5, kl-3, and ks-1) form lampbrush-type loops in primary spermatocytes. Two additional loci have been identified: bobbed (bb) and Suppressor of Stellate [Su(Ste)]. The first, in the short arm, allelic to the bb locus in the X heterochromatin, encodes clusters of ribosomal RNA genes; the second is required for correct splicing of the gene product of the Ste locus on the X chromosome. Chromosomal deficiency of each of these six regions results in male sterility. In particular, the absence of either the kl-5 or the kl-3 regions damages the structure of the primary spermatocyte nuclei and inhibits the presence of the outer dynein arms associated with the A microtubules of the nine peripheral axoneme doublets. Recently, Gepner and Hays demonstrated that one of the seven different dynein heavy chain genes Dhc-yh3, located in Y chromosome region h3, is contained within kl-5, confirming that Y chromosome is needed for male fertility because it contains conventional genes that function during spermatogenesis. Moreover, the deficiency for the kl-2 region leads to the formation of crystals in the primary spermatocytes and to the abnormal distribution of chromosomes and mitochondria at meiosis; the deficiencies for kl-1 and ks-2 occurring late in spermatogenesis may be secondary consequences of earlier abnormalities, and the absence of the ks-2 region causes alterations of the developing axoneme and the nebenkern. If we assume a uniform distribution of DNA along the Y chromosome (40,000 kb), we can calculate that kl-5, kl-1, and ks-1, which span 8, 3, and 8% of the chromosome length, contain 3200, 1200, and 3200 kb DNA, respectively. We can suppose that the other loci (kl-3, kl-2, and ks-2) have a comparable extension. These genes are greater than the euchromatic genes of *D. melanogaster*.

An important function carried out by the Y chromosome is the meiotic pairing which is attributable to pairing sites or collochores. In *D. miranda* the male karyotype is characterized by a pair of heteromorphic sex chromosomes that are still evolving from a pair of initially homologous autosomes. This characteristic is due to the translocation of one of the autosomes to the chromosome Y, resulting in a neo-Y chromosome and a monosome (neo-X), designated X_2. Lcp_{1-4}, the larval cuticle protein genes, are located on the X_2 and Y chromosomes in this species; in all others they are autosomally inherited. In *D. miranda* all four loci are embedded within a dense cluster of transposable elements. The X_2 Lcp_{1-4} loci are expressed, whereas the Y chromosomal Lcp_3 locus shows only reduced activity and the Lcp_1, Lcp_2, and Lcp_4 are completely inactive. It is possible that the first step in Y chromosome silencing is caused by trapping and accumulation of transposons.

The Y chromosome of *D. hydei* offers some advantages for study due to the presence of lampbrush loops formed by some of the fertility genes with the aid of translocation chromosomes. The linear sequence of the loops and their approximate position within the Y chromosome have been established by Hess (1967). In early spermatocytes, the Y chromosome begins to form five lampbrush loops whose remnants are present during spermatocyte development. Each loop has been designated a name according to its morphology in phase contrast. The nooses, clubs, and threads are always in the neighborhood of nucleolus, and the tubular ribbons and the pseudonucleolus are distal compared to nucleolus. It could also be demonstrated that all loop-forming loci must be present and active to allow normal spermatogenesis and that duplications of one or more loops cannot functionally compensate for the absence of others. There are genetic loci related to lampbrush loop forming sites and others not forming such loops. Each loop is associated with only one complementation group, and the number of fertility genes does not exceed 16. No dominant mutations and no mutations not associated with male fertility were found. The distribution of the genetic loci in the Y chromosome of *D. hydei* resembles closely that of *D. melanogaster*: All the loci are in fact contained in relatively restricted chromosome regions. Mutations of the loci A–C, which include the threads and pseudonucleolus loop pairs, are particularly interesting and most of them are connected with a modified loop morphology or result in an entire inactivation. The genetic studies of the Y chromosome of *D. hydei* revealed another kind of information that cannot be easily obtained with other genetic systems. Many experiments report inactivation of lampbrush loops in X-Y translocations and it can be assumed that the inactivation is a gradual but regular process in X-Y translocations and it depends in part on the particular structure of the translocation chromosomes.

In 1970 the isolation of a Y chromosome of *D. hydei* that is associated with a wm Co chromosome marker derived from the X chromosome wm1 was described. These exchanges are frequently accompanied by mutation of Y chromosome. In 1992 Trapitz *et al.* concluded that at least five different families of repetitive DNA specifically transcribed on the lampbrush loops, nooses and threads are organized as extended clusters of several hundred kb, essentially free from interspersed nonrepetitive sequences. Moreover, the gene Dhmst 101, a member of a small gene family, is specifically expressed in adult testis tissue. The pattern of expression and the particular evaluation of biophysical considerations on the protein sequence data suggest that the Dhmst 101 gene product may have some importance for the structural integrity of the sperm tail. Many observations suggest a regulatory function of the Y chromosome in *D. hydei*. In effect, the binding of specific protein by particular lampbrush loops might have at least three different functions: They may permit a compartimentalization of nuclear protein, provide a sink within nucleus, or have a catalytic function leading to a modification of the bound proteins. For these reasons, blocking of any of these potential functions can occur in Y chromosomal deletions or male sterile mutations which cause an inactivation of lampbrush loop.

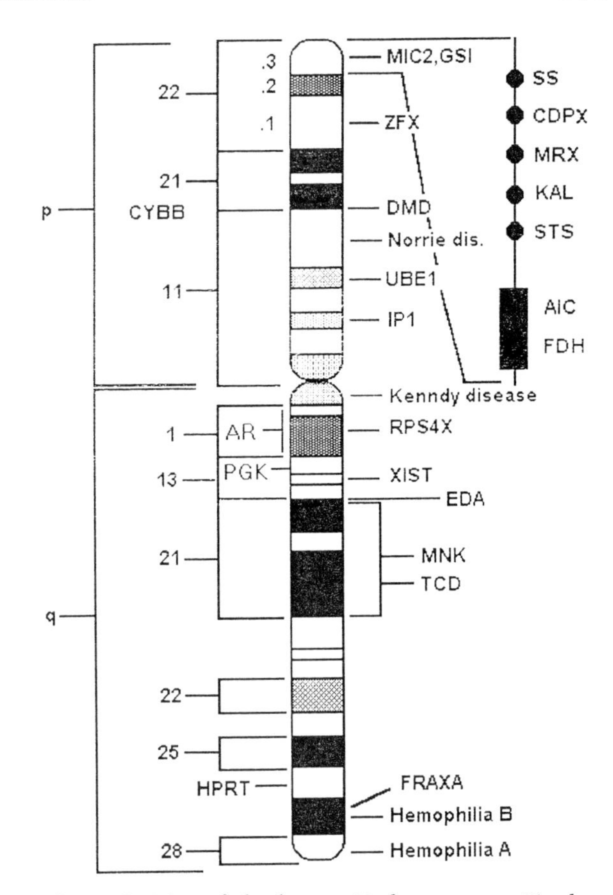

FIGURE 2 Map of the human X chromosome. SS, short stature locus; CDPX, X-linked recessive chondrodysplasia punctata gene; MRX, mental retardation gene; AIC, Aicardi syndrome locus; FDH, focal dermal hypoplasia (Goltz syndrome) gene; KAL, X- linked Kallmann syndrome gene; STS, steroid sulfatase locus; MIC2, antigen for cell adhesion; GSI, glutamine synthetase gene; CYBB, chronic granulomatous disease gene; UBE1, ubiquitin-activating enzyme E1 locus; AR, androgen receptor gene; PGK, phosphoglycerate kinase gene; RPS4, ribosomal protein S4 locus; XIST, X inactive specific transcripts gene; EDA, ectodermal dysplasia locus; HPRT, hypoxanthine–guanine phosphoribosyltransferase locus; FRAXA, fragile X mental retardation syndrome locus; DMD, Duchenne muscular dystrophy gene; IP1, incontinentia pigmenti locus; MNK, Menkes disease gene; ZF, zinc finger gene; TCD, tapeto choroidal dystrophy locus.

B. The X Chromosome

1. *Mammals*

The X chromosome (Fig. 2) is the most extensively studied of mammalian chromosomes, and highly interesting results have been obtained both in the field of X-linked diseases and in the phenomenon of X chromosome inactivation. The two characteristics are correlated to the different quantity of X in males and females.

i. X-Linked Diseases With the exception of genes located in the pseudoautosomal regions and a few others actively present in X and Y chromosomes, a large majority of the genes are in haploid status in the male and, as a consequence, recessive mutations in the genes of X-linked diseases are present and active in hemizygous males and heterozygous and asymptomatic in carrier females. In 1967, Ohno postulated that genes which are X-linked in one mammalian species should be X-linked in all the others. Almost all genes tested as X-linked in humans were also present in mouse X chromosome and five regions showed conserved gene content in the human and mouse X's. Moreover, a novel sequence family (DXF34) located close to the X chromosome centromere in both human and mouse suggested that a block of pericentromeric material is conserved both in man and in mouse. In other instances, in mouse the sequences for steroid sulfatase, Kallmann syndrome (Kal), MIC2 (antigen for cells adhesion), and GSI (gluthamine synthetase) genes that in humans are located in Xp22.3 were found to be absent. It is very interesting that all genes tested and located on the short arm of the human X are autosomal in marsupials and monotremes. This part has an autosomal origin and has been added to the X chromosome in eutherian mammals where the fusion point has been identified by comparative mapping analysis. Moreover, a further X chromosome linkage group was found conserved in placental mammals (cattle, sheep, and goat) by use of painting probes and enabled a better understanding of mammalian sex evolution.

In humans there are a large number of X-linked diseases, whose study has provided diagnostic tools for genetic counseling in affected families. For this reason, the X chromosome was the first to have a genetic map based on restriction fragment length polymorphisms (RFLPs). Mapping based on the early RFLP array was not optimal and was changed by the use of microsatellites which can be rapidly tested by PCR. Three of them have been isolated and characterized in the proximal long arm. However, the pseudoautosomal region is the richest in highly polymorphic minisatellite markers.

The size of the human X is 150 Mb. By the mapping approach the first genes to be isolated were that of chronic granulomatous disease and that of Duchenne muscular dystrophy, both X-linked diseases. The latter is the largest gene known in any organism. Recently, two other X-linked diseases (the fragile X mental retardation syndrome and the spinobulbar muscular atrophy or Kennedy disease) have permitted discovery of a new mutation mechanism: the expansion of trinucleotide repeats. The first disease is caused by an expansion of a CGG repeat in a 5′ exon of the gene FMR-1, and the second one is a more moderate expansion of a CAG repeat in the NH2-terminal code region of the androgen receptor gene.

ii. X Chromosome Inactivation A fascinating aspect of the human X chromosome is the inactivation of one of the two X's in females, a phenomenon which can involve more than 100 Mb of DNA, turning off several thousand of genes. It occurs in three steps: initiation early in embryogenesis in a site (a *cis*-acting locus) called the X inactivation center (XIC), propagation, and stabilization which is very well correlated with DNA methylation (at the cytosine of a CpG dinucleotide). It occurs at CpG-rich regions (CpG islands) near the 5′ end of many genes. This locus has been mapped in man and in mouse is located between the ectodermal dysplasia locus (EDA) and phosphoglycerate kinase gene (PGK). Brown *et al.* (1991) isolated a nonprotein coding gene, XIST, which is expressed only from inactive X chromosomes and is likely to be important in inactivation. Its expression is obligatory for the X inactivation process. Its product is a polyadenylate RNA. The expression of the XIST gene precedes X chromosome inactivation and XIST RNA is found to be bound to the X chromosome at interphase, but only at the *cis* locus. Its binding to the DNA of inactive X chromosome seems to be intrinsically implemented in the process of X inactivation. De Groot and Hochberg suggest that silencing of the X chromosome genes can occur without methylation on the silenced genes and that at least in certain circumstances imprinting itself can occur without differential methylation. It was surprising that several genes along the X escape X inactivation; this characteristic, which was expected for pseudoautosomal region genes [MIC2 and steroid sulfatase locus (STS)], is true also for at least three genes (zinc finger, ZFX in Xp22.1, UBE1 in Xp11.23, and ribosomal protein S4, RPS4X in Xq13.1) that are inactive in the mouse. Another three genes that escape X chromosome inactivation are clustered in Xp11.21–p11.22, suggesting that there are regional control signals as well as gene-specific elements that determine the X inactivation status of genes on the X human proximal short arm.

iii. Mapping and Recombination The development of yeast artificial chromosome (YAC) as a cloning vector has promoted rapid progress in mapping in recent years and YAC libraries have been used to build the X physical map. Also, somatic cell hybrids facilitate the mapping of X; with this method the hypoxanthine–guanine phosphoribosyltransferase (HPRT) locus has been mapped in Xq26. Recently, with this procedure the ubiquitin-activating enzyme E1 (UBE1) locus has also been mapped in Xp11.2.

Increased recombination has been found within the dystrophy gene (DMD) and in regions flanking the fragile site (FRAXA) in Xq27.3. We can observe genes in a number of different positions along the X which cause similar "nonspecific" mental retardation. At least 16 X-linked disease genes have been cloned on the basis of prior knowledge of the definitive protein; in most of the remainder the biochemical defect was unknown and it was thus necessary to use mapping strategies to identify the corresponding genes. Genes for 7 X-linked diseases have been isolated by positional cloning. Linkage studies in affected families are being actively pursued for more than 50 diseases. For about 10 diseases, rare affected females have been found with balanced/X autosome translocations. These patients, in which the normal X is inactive and the translocated X is active, have been important to explain EDA and Menkes disease, which causes abnormalities in copper metabolism. Several other females with different *de novo* X/autosome translocations have been important in the knowledge of a rare disorder, called incontinentia pigmenti. In most cases the lack of function for genes in the deleted region results in a contiguous gene damage. This was first observed in the case of the BB deletion encompassing part of DMD and genes for chronic granulomatous disease associated with cytocrome b deficiency, for McLeod syndrome, a weakened or absent antigenicity in the Kell blood group system, and for retinitis pigmentosa-3, a bilateral progressive retinal dystrophy. A very difficult mapping problem is posed by Rett syndrome, a disabling neurological disease that is only observed in girls, which seems to be an X-linked dominant trait with lethality in males.

In each generation, the number of mutations decreases and some mutations become extinct after a few generations. Clearly, this is not true for mutations that do not affect the reproductive fitness: glucose-6-phosphate dehydrogenase deficiency, color blindness, or some cases of mild hemophilia A or B. Deletion frequency is changeable in the various diseases (ichthyosis, 80–90%; Duchenne muscular dystrophy, 60–70%; ornithine transcarbamylase and HPRT, 5–15%). Point mutations have been established for hemophilia A and B.

iv. The Case of the Distal Short Arm The distal short arm of the X chromosome is a particular region with very characteristic features. It has homology with both the short and the long arm of the Y chromosome, contains genes escaping X inactivation, is not conserved in the mouse, and shows a very high frequency of chromosomal rearrangements. In this region we can find five types of abnormalities: interstitial deletions, terminal deletions, X/Y translocations, X/autosomal translocations, and Xp duplications. Interstitial deletions involve generally the STS and result in isolated X-linked ichthyosis, a severe keratinization disorder. The terminal deletions obviously have terminal breakpoints and the size of the deleted regions is variable; the X/Y translocations can be divided in two groups—Xp/Yp and Xp/Yq translocations—with the first involving the testing-determining factor (TDF) and the second causing variable phenotypes for the different breakpoints on the X chromosome. X/autosomal translocations have been found in females with X-linked dominant disorders. A cluster of sulfatase genes on human Xp22.3 was created though a duplication event which probably occurred in an ancestral pseudoautosomal region and supports the idea that this region has undergone multiple changes during recent mammalian evolution. There have been several reports of female patients with terminal deletions and translocations involving the Xp22 region. Many of them were affected by two Mendelian disorders: Aicardi syndrome, a neuropathological disorder in Xp22.2, and Goltz syndrome or focal dermal hypoplasia in Xp22.3. The recently identified gene for X-linked kal (hypogonadotropic hypogonadism), which has a homolog on the Y chromosome Yq11.2, is located on Xp22.3. In 1994, a new voltage-gated chloride channel (the CIC family), which is encoded by a gene located in this area, was identified. The genetic locus for handedness is located in a X-Y homologous region of the sex chromosomes. Also, granulocyte-macrophage colony-stimulating factor receptor α chain gene (CSF2RA) has been located in this area where Kremer *et al.* (1993) suggest the presence of a cytokine receptor gene cluster. The gene arrestin located to Xcen–Xq21 probably joins the ranks of other X-located retinal genes—the red and green opsins, the genes for choroideremia [tapeto choroidal dystrophy (TCD)], and the Norrie disease, a neurodevelopmental disorder.

v. ***The Area Xq28*** The chromosomal area Xq28 contains a large number of genes responsible for human hereditary diseases. In this area the gene for X-linked myotubular myopathy has been localized.

vi. ***Monosomy and Aneuploidy*** Sex chromosomal monosomy with total loss of an X or Y is frequently observed in malignant gliomas but their changes are not necessarily part of the neoplastic process. A possible relationship between the presence of inactive X chromosomes, forming X chromatin, and the number of autosomes is suggested by the data on karyotyped tumors of the testis. Structural abnormalities of the X chromosome in non-Hodgkin's lymphoma have been suggested at the level of the bands p22 or q28.

Several studies on aneuploidy have shown a significant increase in the loss of chromosomes in both males and females with age and also a significant increase in micronucleus formation has been observed in lymphocytes with age. Hando et al. (1994) found that the X chromosome is present in a high percentage of the micronuclei scored and showed a significant increase with age in the number of them. Recently, the greater susceptibility of X chromosomes to malsegregation compared with autosomes has been confirmed.

2. *Insects*

In Drosophila, males have a single and females have two X chromosomes in their somatic cells, but the products of most X-linked genes are present in equivalent amount in the two sexes. In 1929 Stern noted this equalization, and in 1932 Muller termed it "dosage compensation." It is achieved by an enhancement of the transcriptional rate in males relative to females and it depends on the activity of at least four autosomal genes: male-specific lethal-1 (msl-1), male-specific lethal-2 (msl-2), maleless (mle), and maleless on the third (mle-3 or msl-3). All these genes are called msl genes. Mutations of each of these genes are lethal to males because they result in insufficiency in the level of X-linked gene products; the same mutations do not have any effect on the viability of the females. A master regulatory gene, Sex lethal (Sxl), can enhance the male-specific transcription of the X-linked genes with its product. Sxl is an RNA-binding protein that regulates sexual differentiation by influencing alternative splicing of its own pre-mRNA and that of at least one downstream sex determination gene, the transformer (tra). In addition, Sxl negatively regulates the male mode of dosage compensation in females. The mle gene encodes a polypeptide that contains several short motifs characteristic of a superfamily of DNA and RNA helicases, and it is associated with hundreds of sites along the X chromosome in males but not in females. Similar results have been obtained with the msl-3 gene product. The msl-1 gene encodes a polypeptide probably involved in transcription and chromatin modeling, and it is associated with many sites along the length of the X chromosome in males. The msl-2 gene is a ring finger protein and it has been suggested that msl-2 RNA is the primary target of Sxl regulation. Hilfiker *et al.* (1994) found that these four gene products contribute to the formation of a multisubunit complex. Some evidence can suggest the existence of a second dosage compensation pathway that is independent of the four known msl genes. Kelley *et al.* (1995) strongly favor the hypothesis that the MSL proteins act to increase transcription of most X-linked genes. The most compelling evidence is that the MSL proteins colocalize on the polytene X chromosome rather than the autosomes.

Two potent indicators of X chromosome dose are

sisterless-a (sis-a) and sisterless-b (sis-b). Genetic analysis has shown that a diplo-X dose of these genes activates their regulatory target, the feminizing switch gene Sxl, whereas a haplo-X dose leaves Sxl inactive. Sis-b encodes a transcriptional activator of the bHLH family that dimerized with several other HLH proteins required for the proper assessment of X dose. Erickson and Cline (1993) report that sis-a encodes a bZIP homologous protein that functions in all somatic nuclei to activate Sxl transcription. In contrast with other elements of the sex determination signal, the functioning of this transcription factor in somatic cells may be specific to the X chromosome counting.

The sex chromosomes, especially the X chromosome, are often considered to be of special importance in determining the fertility of hybrids. The evolution of different clusters of genes has been studied to understand some X mechanisms and also the exchanges in the diverse species.

In *D. hydei* transcriptionally active chromosomal materials in spermatogonial nuclei are present: the X chromosome and the autosomes. In early spermatocytes they are found close to the nucleolus. Since the X chromosome carries a nucleolus organizer region, which is supposedly active in spermatocytes, this region is often less condensed than others.

A cluster of genes corresponding to the early ecdysone-stimulated puff 2B of the *D. melanogaster* X chromosome has been localized using *in situ* hybridization in eight Drosophila species. Genes ecs, dor, and swi from this cluster have been mapped in *D. funebris*, *D. virilis*, *D. hydei*, *D. repleta*, *D. mercatorum*, and *D. paranaensis* to the telomeric region of the X chromosome, in *D. kanekoi* to the distal region, and in *D. pseudoobscura* to the proximal region of the X chromosome. It is assumed that organization of this cluster in these species is conserved. The X-linked gene Hmr in *D. melanogaster*, when mutated, rescues otherwise inviable interspecific hybrids from crosses between *D. melanogaster* and any of its three most closely related species—*D. simulans*, *D. mauritiana*, and *D. sechellia*. Three distinct mRNAs are transcribed from this locus, two of which are abundantly expressed throughout life. A third transcript, which is larger but rarer, appears to be disrupted by at least one of the two known mutations of Hmr. The gene encodes a mithocondrial ADP/ATP translocator protein, which plays an essential role in maintaining metabolic energy.

In 1995 Nuzhdin showed that several sites of the base of the X chromosome had higher frequencies of transposable elements compared to distal regions. These data provide evidence for the operation of a force opposing transpositional increase in copy number.

Nine single-copy regions located on the X chromosome have been mapped by *in situ* hybridization in six species of the obscura group of Drosophila. Eight of the regions include known genes from *D. melanogaster* (Pgd, zeste, white, cut, vermilion, RNA polymerase II 215, forked, and suppressor of forked) and the ninth region (lambda DsubF6) has not yet been characterized. Location of these markers has been of great importance to understand the divergency among different species.

See Also the Following Articles

Drosophila; Kallmann's Syndrome; Klinefelter's Syndrome; Sex Determination, Genetic

Bibliography

Atkin, N. B., and Baker, M. C. (1992). X-chromatin, sex chromosomes, and ploidy in 37 germ cell tumors of the testis. *Cancer Genet. Cytogenet.* **59**, 54–56.

Bairati, A., and Baccetti, B. (1966). Observations on the ultrastructure of male germinal cells in the XLCYS mutant of *Drosophila melanogaster* Meis. *Drosophila Information Service* **41**, 152.

Ballabio, A., and Andria, G. (1992). Deletions and translocations involving the distal short arm of the human X chromosome: Review and hypotheses. *Hum. Mol. Genet.* **1**, 221–227.

Ballabio, A., and Willard, H. F. (1992). Mammalian X chromosome inactivation and the XIST gene. *Curr. Opin. Genet. Dev.* **2**, 439–442.

Bell, L. R., Maine, E. M., Schedl, P., and Cline, T. W. (1988). Sex-lethal, a *Drosophila* sex determination switch gene, exhibits sex-specific RNA splicing and sequence similarity to RNA binding proteins. *Cell* **55**, 1037–1046.

Belote, J. M., and Lucchesi, J. C. (1980). Male specific lethal mutations of *Drosophila melanogaster*. *Genetics* **96**, 165–186.

Benazzi, M. (1947). *Problemi Biologici della Sessualità*. Cappelli, Bologna..

Bournier, A. (1956). Contribution à l'étude de la parthénogenèse des Thysanoptères et de sa cytologie. *Arch. Zool. Exp. Gén.* **93**, 219–317.

Bridges, C. B. (1916). Non-disjunction as proof of the chromosome theory of heredity. *Genetics* **1**, 1–52, 107–163.

Bridges, C. B. (1925). Haploidy in *Drosophila melanogaster*. *Proc. Natl. Acad. Sci. USA* **11**, 706–710.

Bridges, C. B. (1932). The genetics of sex in *Drosophila*. In *Sex and Internal Secretions*: *3*, pp. 55–93. Baillère, Tindall & Cox, London.

Brown, C. J., Ballabio, A., Rupert, J. L., Lafreniera, R. C., Grompe, M., Tonlorenzi, R., *et al*. (1991). A gene from the region of the human X inactivation centre is expressed exclusively from the inactive X chromosome. *Nature* **349**, 38–44.

Burgoyne, P. S., Mahadevaiah, S. K., Sutcliffe, M. J., and Palmer, S. J. (1992). Fertility in mice requires X-Y pairing and a Y-chromosomal "spermiogenesis" gene mapping to the long arm. *Cell* **71**, 391–398.

Chandley, A. C., and Cooke, H. J. (1994). Human male fertility. Y-linked genes and spermatogenesis. *Hum. Mol. Genet.* **3**, 1449–1452.

Cline, T. W. (1984). Autoregolatory functioning of a *Drosophila* gene product that establishes and maintains the sexually determined state. *Genetics* **107**, 231–277.

De Groot, N., and Hochberg, A. (1997). Genome imprinting. In *Genetics of Human Male Fertility* (C. Barrat, C. De Jonge, D. Mortimer, and J. Parinaud, Eds.), pp. 308–332. EDK.

Dobzhansky, Th. (1935). *Drosophila miranda*, a new species. *Genetics* **20**, 377–391.

Erickson, J. W., and Cline, T. W. (1993). A bZIP protein, sisterless-a, collaborates with bHLH transcription factors early in *Drosophila* development to determine sex. *Genes Dev.* **7**, 1688–1702.

Francke, U., Ochs, H. D., de Martinville, B., Giacalone, J., Lindgren, V., Distache, C., Pagon, R. A., Hofker, M. H., van Ommen, G. J. B., Pearson, P. L., and Wedgwood, R. J. (1985). Minor Xp21 chromosome deletion in a male associated with expression of Duchenne Muscular dystrophy, chronic granulomatous disease, retinitis pigmentosa, and McLeod syndrome. *Am. J. Hum. Genet.* **37**, 250.

Gatti, M., and Pimpinelli, S. (1983). Cytological and genetic analysis of the Y chromosome of *Drosophila melanogaster*. I. Organization of the fertility factors. *Chromosoma* **88**, 349–373.

Gepner, J., and Hays, T. S. (1993). A fertility region on the Y chromosome of *Drosophila melanogaster* encodes a dynein microtubule motor. *Proc. Natl. Acad. Sci. USA* **90**, 11132–11136.

Goldschmidt, R. (1931). *Die sexuellen Zwischenstufen*. Springer, Berlin .

Gorman, M., Franke, A., and Baker, B. S. (1995). Molecular characterization of the male-specific lethal 3 gene and investigations of the regulation of dosage compensation in *Drosophila*. *Development* **121**, 463–475.

Goyns, M. H., Hammond, D. W., Harrison, C. J., Menasce, L. P., Ross, F. M., and Hancock, B. W. (1993). Structural abnormalities of the X chromosome in non-Hodgkin's lymphoma. *Leukemia* **7**, 848–852.

Hackstein, J. H. P. (1991). Spermatogenesis in *Drosophila*. A genetic approach to cellular and subcellular differentiation. *Eur. J. Cell Biol.* **56**, 151–169.

Hando, J. C., Nath, J., and Tucker, J. D. (1994). Sex chromosomes, micronuclei and aging in women. *Chromosoma* **103**, 186–192.

Hecht, B. K., Turc-Carel, C., Chatel, M., Paquis, P., Gioanni, J., Attias, R., Gaudray, P., and Hecht, F. (1995). Cytogenetics of malignant gliomas. II. The sex chromosomes with reference to X isodisomy and the role of numerical X/Y changes. *Cancer Genet. Cytogenet.* **84**, 9–14.

Henning, W. (1985). Y chromosome function and spermatogenesis in *Drosophila hydei*. *Adv. Genet.* **23**, 179–234.

Hess, O. (1967). Complementation of genetic activity in translocated fragments of the Y chromosome in *Drosophila hydei*. *Genetics* **56**, 283–295.

Hilfiker, A., Yang, Y., Hayes, D. H., Beard, C. A., Manning, J. E., and Lucchesi, J. C. (1994). Dosage compensation in *Drosophila*: the X chromosomal binding of MSL-1 and MLE is dependent on Sxl activity. *EMBO J.* **13**, 3542–3550.

Hutter, P., and Karch, F. (1994). Molecular analysis of a candidate gene for the reproductive isolation between sibling species of *Drosophila*. *Experentia* **50**, 749–762.

La Spada, A. R., Wilson, E. M., Lubahn, D. B., Harding, A. E., and Fischbeck, K. H. (1991). Androgen receptor gene mutations in X-linked spinal and bulbar muscular atrophy. *Nature* **352**, 77–79.

Laval, S. H., and Boyd, Y. (1993). Novel sequences conserved on the human and mouse X chromosomes. *Genomics* **15**, 483–491.

Littlefield, J. W. (1964). Selection of hybrids from mating of fibroblasts in vitro and their presumed recombinants. *Science* **145**, 709–710.

Livak, K. J. (1990). Detailed structure of the *Drosophila melanogaster* Stellate genes and their transcripts. *Genetics* **124**, 303–316.

Kelley, R. L., Solovyeva, I., Lyman, L. M., Richman, R., Solovyev, V., and Kuroda, M. I. (1995). Expression of msl-2 causes assembly of dosage compensation regulators on the X chromosomes and female lethality in *Drosophila. Cell* **81**, 867–877.

Kremer, E., Baker, E., D'Andrea, R. J., Slim, R., Phillips, H., Moretti, P. A., Lopez, A. F., Petit, C., Vadas, M. A., Sutherland, G. R., *et al.* (1993). A cytokine receptor gene cluster in the X-Y pseudoautosomal region? *Blood* **82**, 22–28.

Ma, K., Sharkey, A., Kirsch, S., Vogt, P., Keil, R., Hargreave, T. B., McBeath, S., and Chandley, A. C. (1992). Towards the molecular localisation of the AZF locus: Mapping of microdeletions in azoospermic men within 14 subintervals of interval 6 of the human Y chromosome. *Hum. Mol. Genet.* **1**, 29–33.

Mandel, J. L., Monaco, A. P., Nelson, D. L., Schlessinger, D., and Willard, H. (1992). Genome analysis and the human X chromosome. *Science* **258**, 103–109.

McKusick, V. A. (1970). Human genetics. *Annu. Rev. Genet.* **4**, 1–46.

Meyer, G. F. (1968). Spermiogenese in normalen und Y-defizienten Mannchen von *Drosophila melanogaster* und *D. hydei. Z. Zellforsch. Mikrosk. Anat.* **84**, 141–175.

Muller, H. J. (1932). Some genetic aspects of sex. *Am. Nat.* **66**, 118–138.

Neesen, J., Bunemann, H., and Heinlein, U. A. (1994). The *Drosophila hydei* gene Dhmst 101(1) encodes a testis-specific, repetitive, axoneme-associated protein with differential abundance in Y chromosomal deletion mutant flies. *Dev. Biol.* **162**, 414–425.

Nuzhdin, S. V. (1995). The distribution of transposable elements on X chromosomes from a natural population of *Drosophila simulans. Genet. Res.* **66**, 159–166.

Ohno, S. (1967). *Sex Chromosomes and Sex Linked Genes. Monographs on Endocrinology 1.* Springer-Verlag, Berlin/New York.

Reijo, R., Lee, T. Y., Salo, P., Alagappan, R., Brown, L., Rosenberger, M., Rozen, S., Jaffe, T., Straus, D., Hovatta, O., de la Chapelle, A., Silber, S., and Page, D. C. (1995). Diverse spermatogenic defects in humans caused by Y chromosome deletions encompassing a novel RNA-binding protein gene. *Nat. Genet.* **10**, 383–393.

Ritossa, F. (1976). The bobbed locus. In *The Genetics and Biology of Drosophila* (M. Ashburner and E. Novitski, Eds.), Vol. 1b, pp. 801–846. Academic Press, London.

Sharma, T. (1961). A study on the chromosomes of two Lycosid spiders. *Proc. Zool. Soc. Calcutta* **14**, 33–38.

Singh, L. (1995). Biological significance of minisatellites. *Electrophoresis* **16**, 1586–1595.

Sosnowski, B. A., Belote, J. M., and McKeown, M. (1989). Sex-specific alternative splicing of RNA from the transformer gene results from sequence-dependent splice site blockage. *Cell* **58**, 449–459.

Stern, C. (1929). Untersuchungen uber aberrationen des Y chromosoms von *Drosophila melanogaster. Z. Indukt. Abstamm.-Vererbungsl* **51**, 254.

Suomalainen, E. (1969). On the sex chromosome trivalent in some Lepidoptera females. *Chromosoma* **28**, 298–308.

Tiepolo, L., and Zuffardi, O. (1976). Localization of factors controlling spermatogenesis in the non-fluorescent portion of the human Y chromosome long arm. *Hum. Genet.* **34**, 119–124.

Trapitz, P., Glatzer, K. H., and Bunemann, H. (1992). Towards a physical map of the fertility genes on the heterochromatic Y chromosome of *Drosophila hydei*: Families of repetitive sequences transcribed on the lampbrush loops Nooses and Threads are organized in extended clusters of several hundred kilobases. *Mol. Gen. Genet.* **235**, 221–234.

White, M. J. D. (1953). Multiple sex-chromosome mechanisms in the grasshopper genus *Paratylotropidia. Am. Nat.* **87**, 237–244.

White, M. J. D. (1973). *Animal Cytologic and Evolution.* Cambridge Univ. Press, Cambridge, UK.

White, M. J. D. (1977). *Modes of Speciation* (C. I. Davern, Ed.). Freeman, San Francisco.

Whiting, P. W. (1943). Multiple alleles in complementary sex determination of *Habrobracon. Genetics* **28**, 365–382.

Sex Determination, Environmental

Carlos Augusto Strüssmann
Tokyo University of Fisheries

Reynaldo Patiño
Texas Tech University

GLOSSARY

apoptosis Programmed cell death.

autosome Any chromosome other than sex chromosome.

cryptorchidism Abnormal condition in mammals in which the testes do not descend from the abdomen into a scrotum.

gonochorism Condition in which the male and female sex are expressed in separate individuals.

hermaphroditism Condition in which the male and female sex are expressed in the same individual.

heteromorphic/homomorphic sex chromosomes Homologous sex chromosomes which are morphologically or cytogenetically distinguishable (heteromorphic) or not (homomorphic); the latter are demonstrable through progeny tests of sex-reversed individuals.

homeothermy Having more or less constant body temperature; warm-blooded.

plesiomorphic character Original or ancestral feature.

poikilothermy Having body temperature conforming to surrounding medium; cold-blooded.

sex commitment Ontogenetic loss of sexual plasticity in individuals with environmentally determined or modulated mechanisms of sex determination.

sex determination Genetic or environmental prescription of sex in gonochoristic individuals; the former occurs at fertilization in individuals with genotypic sex determination and the latter during embryonic or larval development in individuals with environmental sex determination.

sex differentiation Developmental expression of phenotypic sex; primary sex differentiation pertains to the gonads, whereas secondary sex differentiation refers to all other tissues.

The mechanisms of sex determination in fishes are typically labile to environmental conditions. Biotic and abiotic environmental factors have been shown or suggested to induce or modulate the functional development of either sex. A distinction must be made between gonochoristic and hermaphroditic fishes. In gonochoristic fishes, the period of sexual plasticity occurs early in development but their sex is stable thereafter. Hermaphroditic fishes include those species in which male and female gametes are produced simultaneously in the same adult individual (simultaneous hermaphrodites) and those in which only one sex can be functionally expressed at any given stage of their life cycle (consecutive hermaphrodites). This functional sexual plasticity is genetically prescribed. However, the precise genetic, molecular, and cellular events involved in sex determination are unclear. Heteromorphic or homomorphic sex chromosomes have been shown in many gonochoristic fishes, but their absence is presumed in hermaphrodites and perhaps also in certain gonochorists in which sex determination may be purely environmentally determined. A role for autosomal sex genes has been suggested even in some gonochorists with known sex chromosomes. Environmental conditions can modulate not only the mechanisms of sex determination but also the process of primary (gonadal) sex differentiation. However, in the absence of sufficient mechanistic information about these events, it is conceptually difficult to distinguish between the process of sex determination and that of primary sex differentiation in animals when the environment is the major sex determinant. This article provides a brief synopsis of the effects of the environment on sex determination and primary sex differentiation in gonochoristic teleosts. The focus is on temperature because this appears to be the most prominent and widespread environmental factor affecting the establishment and development of sex in fishes and in other poikilothermic taxa.

I. EFFECTS OF ENVIRONMENTAL CONDITIONS ON SEX DETERMINATION AND DIFFERENTIATION

As in phylogenetically higher vertebrates, the sex of an individual in most gonochoristic fishes seems to be genetically programmed at fertilization. The process of sex differentiation, however, can be influenced by exogenous environmental factors. Functional sex reversal can be attained by experimental hormonal manipulation and there are recent indications that endocrine disruptors found in the environment also have the potential to influence gonadal sex differentiation. More important, examples of sex determination as a function of environmental factors, including temperature, salinity, pH, social environment, light, water quality, and food abundance, have been recognized in a number of species. When the effects of the environment on sex differentiation are marked and become the major determinant of gonadal sex in natural habitats, the species is said to possess environmental sex determination (ESD). Temperature-dependent sex determination (TSD) seems to be the prevalent form of ESD in most animals with weak or no mechanism of genotypic sex determination (GSD), particularly among reptiles but perhaps also in fishes. In fishes, numerous cases of thermolabile sex determination have been reported, especially among the atherinids, but field studies assessing the ecological significance of this mode of sex determination have been conducted only with the Atlantic silverside, *Menidia menidia* (see Section I,C). However, mounting empirical evidence from laboratory studies indicates that temperature can affect sex determination in many other species, sometimes despite the presence of sex-controlling genes. Except for their relevance within an ecological context, no distinction is made in this article between species in which the thermolability of sex has been established in the natural environment (TSD, as for *M. menidia*) and those in which the thermolability of sex determination has currently been established only by laboratory experimentation. A practical distinction between both situations is probably difficult to achieve since the boundary between "laboratory" and "natural" thermal conditions is unclear.

A. Patterns of Thermolabile Sex Determination

Accurate determination of sex ratios over the entire range of thermal tolerance has been performed only for a few species (e.g., the atherinids *M. menidia*, *Odontesthes bonariensis*, and *Patagonina hatcheri*; Fig. 1). In atherinids, gonadal sex seems to be particularly sensitive to thermal conditions, with male determination occurring at higher temperatures and female determination at lower temperatures. Many other species also show some degree of thermolabile sex determination at conditions ranging from naturally experienced temperatures to near the extremes of their thermal tolerance (Table 1). In addition, high temperatures seem to favor the formation of males in medaka (*Oryzias latipes*), loach (*Misgurnus aguillicaudatus*), and goldfish (*Carassius auratus*) but of females in sockeye salmon (*Oncorhynchus nerka*). In barfin flounder (*Verasper moseri*), the production of females is favored by low temperatures. It is noteworthy that even species with fairly strong genotypic sex determination (e.g., *Oreochromis* sp., *Ictalurus* sp., *Oryzias* sp.) and possibly even heteromorphic sex chromosomes (e.g., as in *Oncorhynchus nerka* and some amphibians) are amenable to some degree of thermolabile sex determination. Also, closely related species or populations may show opposite patterns of thermolabile sex determination. For example, different species of *Oreochromis* seem to have the opposite response to the same temperature change (Table 1).

The data available on the expression of thermolabile sex determination and GSD in gonochoristic teleosts seem to fit into four basic patterns, as summarized in Fig. 2. Thermolabile sex determination may not be operational at all in some species or strains (Fig. 2, I). In contrast, in other fishes such as some atherinids, low and high temperatures have reciprocal effects on sex determination leading from highly female- to highly male-biased sex ratios at the

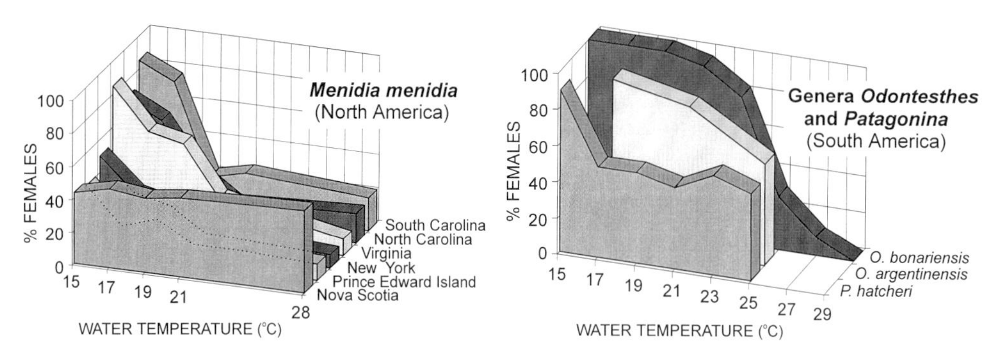

FIGURE 1 Sex ratio response to temperature in atherinid fishes. Populations are arranged in decreasing order of latitude. Typically, northern populations (Nova Scotia) of *M. menidia* have marked GSD, southern populations (South Carolina) have marked TSD, and populations from intermediate latitudes have a mixture of GSD and TSD. In *O. bonariensis*, increasing temperature (15–29°C) induces from 100 to 0% females. In *P. hatcheri*, gonadal sex is under strict genetic control (i.e., sex ratios fluctuate around 1 : 1) at temperatures between 17 and 25°C but becomes highly female biased at 15°C and below. Therefore, data for both North and South American atherinids suggest an increase in the strength of GSD over TSD with increasing latitude (adapted from D. O. Conover and S. W. Heins, Adaptive variation in environmental and genetic sex determination in a fish, *Nature* **326**, 496–498, 1987 (*M. menidia*) © 1987 Macmillan Magazines Ltd.; C. A. Strüssmann, J. C. Calsina Cota, G. Phonlor, H. Higuchi, and F. Takashima, Temperature effects on sex differentiation of two South American atherinids, *Odontesthes argentinensis* and *Patagonina hatcheri*, *Environ. Biol. Fish* **47**, 143–154, 1996; C. A. Strüssmann, T. Saito, M. Usui, H. Yamada, and F. Takashima, Thermal thresholds and critical period of thermolabile sex determination in two atherinid fishes, *Odontesthes bonariensis* and *Patagonina hatcheri*, *J. Exp. Zool.* **278**, 167–177, 1997).

TABLE 1
Response of Sex Ratios to Temperature in Fishes Other Than Atherinids

Species	*Responsive families*[a]	*Type of experimental population tested*	*Experimental temperatures (°C)*[b] *Low*	*High*	*Ratio of % females at high vs low temperatures*[c]
More males at high temperatures and/or more females at low temperatures					
Carassius carassius grandoculis	4/6	Bisexual	19–21	28–31	0.42–0.71
		All female	19–21	28–31	0.47
Gnathopogon caurulescens	1/1	All female	20–30	25–32	0.75
Oreochromis niloticus	24/25	Bisexual	27–29	35–36	0.04–0.67
		All female	27–28	32–37	0.08–0.89
Paralichthys olivaceus	2/2	Bisexual	20	25	0.20
		All female	20	25	0.60
Poecilopsis lucida	1/2	Bisexual	24	30	0.13
More males at low temperatures and/or more females at high temperatures					
Gasterosteus aculeatus	1/1	Bisexual	16–20	22–26	1.43
Ictalurus punctatus	1/1	Bisexual	20	34	1.13
Oreochromis aureus	1/7	All male	26	32	4.17
O. mossambicus	2/2	Bisexual	20	32	1.93
		All female	19	31	9.01

[a] Families with a significant effect of temperature on sex ratio out of the total number of families tested.
[b] Results for intermediate temperatures (when available) are not shown.
[c] Ratios larger than 1 indicate more females at high than low temperature, whereas ratios lower than 1 indicate relatively more males at the higher temperatures. Only the results of responsive families (generally 1.1 ≤ ratio ≤ 0.9) are shown.

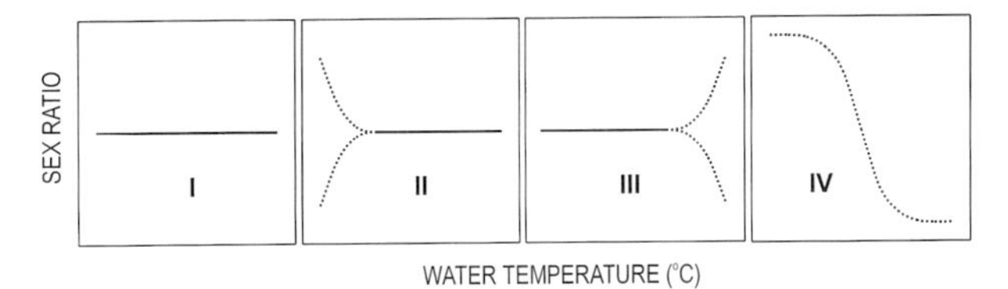

FIGURE 2 Schematic representation of the currently recognized patterns of sex determination in relation to water temperature in fishes. Solid and dotted lines indicate presumable areas of genotypic and thermolabile sex determination, respectively. Patterns I and IV represent species with typically genotypic or thermolabile sex determination, respectively, throughout the whole range of thermal tolerance. In species with both modes of sex determination, thermolabile sex determination is observed at the lower range of temperatures in some species (II) and at the higher range in others (III). Likewise, temperature can induce the formation of females in some species and of males in others. The models are not drawn in proportion to the results of any species in particular.

extremes of their thermal tolerance (Fig. 2, IV). It appears, however, that the effects of temperature in most species are unidirectional and limited to either the low (Fig. 2, II) or the high (Fig. 2, III) range of their thermal tolerance. In these two latter patterns the sex that is produced in excess of the expected distribution varies with the species.

In fishes, the pattern of formation of females at both high and low temperatures and males at intermediate ones observed in some reptiles has not been reported. Also, although considerable shifts in sex ratios may be observed within a 2–6°C range (especially notable when experimental putative monosex broods are used), fishes seem to lack the abrupt thermal thresholds (1 or 2°C) for the all-male and all-female production that are common in reptiles ("pivotal temperatures"). Thus, in most instances of thermolabile sex determination that have been reported in fishes there is a gradual shift in sex ratios with increasing or decreasing temperature. Two lines of evidence indicate that, unlike reptiles, a clear-cut classification of fish species as showing either TSD or GSD may not be tenable. First, there seems to be large individual (as shown by the differential responses of broods from different parental crosses; Table 1) and population (as shown between populations of *M. menidia* living in northern or moderate latitudes; Fig. 1) variations in the thermolability of sex. In other words, whether GSD is operational or not is itself genetically determined at the individual or population levels. Second, it is becoming increasingly evident that GSD in many fishes can be overridden by temperature near the extremes of their thermal tolerance. Therefore, there appears to be an interplay between GSD and TSD in fishes in a manner that has not been described for reptiles. The mechanisms of these apparent differences between fishes and reptiles are unknown.

B. Developmental Aspects of Thermolabile Sex Determination

At a basic level, there are similarities between the effects of temperature on sex determination in fishes compared to TSD in reptiles (and some amphibians). A common feature, for example, is that thermolabile sex determination seems to lead to functional sexuality in all cases. Except in rare cases, thermally sex-determined individuals do not differ sexually from those produced at "neutral" temperatures or from those in which sex determination follows a strict genotypic control. Intersexes have not been reported despite the wide range of thermal regimes tested, including various constant temperatures as well as shifts between low and high temperatures at different times during gonadal formation. This all-or-none determination of sex by temperature in fishes is remarkable if one considers that hormonal manipulation of sex, alone or in conjunction with thermal manipulation, often leads to the abnormal development of hermaphroditism. Another similarity between fishes and reptiles with thermolabile sex determination is that the thermolabile period is restricted to an early ontogenetic stage which, at least partly if not wholly, coincides with the periods of morphological sex differentiation of the gonads and of hormonally inducible sex reversal. Thus, the thermolabile period of sex determination is normally restricted to a few days or weeks during the embryonic, larval, or juvenile stages and appears to be more dependent on body size than on age (Fig. 3). Observations of thermolabile sex determination in *O. nerka* suggest that a short thermal shift during the critical early stage may be sufficient to disrupt putative mechanisms of GSD in

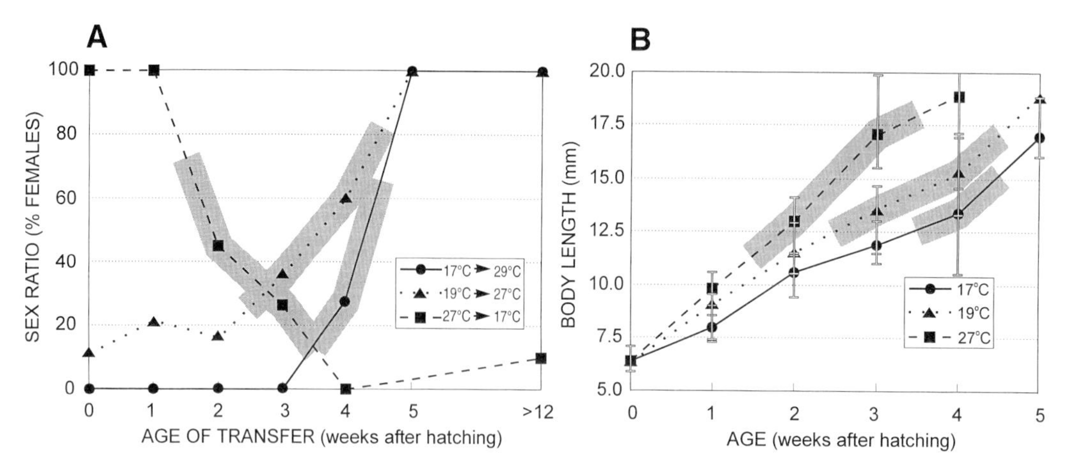

FIGURE 3 Estimation of age and size for thermolabile sex determination of *O. bonariensis* reared at 17, 19, and 27°C. (A) Thermolabile periods (shaded areas) were estimated from the changes in sex ratios of groups transferred at progressively older age (and larger size) between male-producing (27°C) and female-producing (17 and 19°C) temperatures. Individuals transferred before the critical period of sex determination have the gonadal sex favored by the temperature after transfer, whereas those transferred after sex determination have gonads of the sex favored by the initial temperature. (B) The size for sex determination was estimated from age data presented in A. Vertical bars indicate range of body size. It can be seen that the critical periods for sex determination at different temperatures overlap more on the size scale than on the age scale (from C. A. Strüssmann, T. Saito, M. Usui, H. Yamada, and F. Takashima, Thermal thresholds and critical period of thermolabile sex determination in two atherinid fishes, *Odontesthes bonariensis* and *Patagonina hatcheri*, *J. Exp. Zool.* **278**, 167–177, 1997).

this species. In all known cases among fishes as well as reptiles, manipulation of temperature past the (thermolabile) period of sex commitment is unable to reverse the direction of gonadal sex differentiation.

C. Ecological and Evolutionary Aspects of Thermolabile Sex Determination

Thermolabile sex determination is conceivably adaptive for some species, especially for those in which its expression is accentuated. However, the only clearly established example of TSD (shown in natural habitat) in fishes are populations of the Atlantic silverside, *M. menidia*, inhabiting low to moderate latitudes on the eastern coast of North America. The length of the growing season, which varies inversely with latitude, has been suggested as a key factor in the evolution of TSD in this annual species. Because of the different lengths of time available for the growth of fish born early vs late in the season, and because size directly affects the reproductive fitness of females more than males, natural selection for the formation of females at colder temperatures (early breeding season) may allow for the differential, with the larger size of females during the next breeding season optimizing their fecundity. While the case for the adaptive nature of TSD in *M. menidia* seems strong, it is not possible to make generalizations about the origin or the adaptive value of thermolabile sex determination mechanisms for the other species in which it has been demonstrated. Thermolabile mechanisms of sex determination seem to be widespread in some groups of fishes (e.g., atherinids), perhaps reflecting its plesiomorphic origin within these groups, but have also been reported in species from relatively unrelated taxa. These include species from varied habitats (from the tropics to temperate areas) and life histories (from egg layers to livebearers, from short- to long-lived species, etc.). Also, thermolabile sex determination has been expected in species with life history and geographic distribution similar to those of *M. menidia*, but in the studies conducted so far this prediction could not be verified.

Therefore, it is difficult to find a common adaptive value for thermolabile sex determination in fishes. It seems more likely that thermolabile sex determination does not play a significant role in demographic phenomena for most species in which it has been found.

Many fishes are highly prolific and sexually functional over extended periods of their lives and broodstock from different year-classes commonly interbreed. Under these conditions, selective pressure toward balanced sex ratios in each individual generation may be reduced, allowing for the conservation of thermolabile sex-determining mechanisms even if they had only weak or no adaptive value. Moreover, as already discussed, individuals of the same species may present GSD in addition to thermolabile sex determination. Therefore, it is possible that species differences in the thermal sensitivity of sex determination in fishes, and perhaps in other poikilothermic vertebrates, simply reflect differential degrees of evolution of GSD and thus in their capacity to buffer against environmental influences. Furthermore, if the thermolability of sex proves to be a generalized phenomenon among poikilotherms, including those with heteromorphic sex chromosomes, it would be possible to postulate that homeothermy itself is the primary factor that guarantees the stable expression of GSD in warm-blooded vertebrates.

II. EFFECTS OF ELEVATED TEMPERATURE ON GERM CELL AND GONADAL DEVELOPMENT

Environmental temperature can affect not only the direction of sex determination but also the rate at which gonadal sex differentiation occurs. Positive correlations between temperature and rate of differentiation have been observed in a number of species. However, prolonged exposure to elevated temperatures can affect the normal differentiation of the gonads. Namely, whether or not primary gonadal sex is affected by temperature in any given species, there is evidence suggesting that high temperatures can cause the disappearance of germ cells from the gonads possibly leading to the permanent impairment of gonadal function. Interestingly, the abnormal gonads appear to be structurally intact except for the absence of germ cells. Heat-induced germ cell loss has been observed when larvae or juveniles were reared for prolonged periods at 29 and 27 or 28°C for *Odontesthes bonariensis* and *Patagonina hatcheri*, respectively (Fig. 4), at 34–36°C for *Micropterus salmoides*, and at 36 or 37°C for *Oreochromis niloticus*. A common observation among these four examples is that germ cell loss was observed in individuals of both sexes. In mammals, the disappearance of germ cells as a result of high temperatures has been reported only in clinical cases of cryptorchidism (undescended testis; the temperature of the normal scrotal testis is normally 1–4°C lower than core body temperature), after prolonged fever, and after experimental elevations of testicular temperature in scrotal males. Heat-induced inhibition of spermatogenesis has been reported also for the toad, *Bufo melanostictus*. Thus, the observations with fishes indicate that the deleterious effect of high temperatures on germ cells is not restricted to males but occurs also in females. In fishes, the heat sensitivity of germ cells seems to increase after gonadal sex differentiation and also seems to be associated with particular stages of germ cell development. For instance, the ovaries of *M. salmoides* and *P. hatcheri* subjected to sublethal, high thermal conditions after a period of gonadal development at lower temperatures lacked oogonia and young meiotic oocytes (up to pachytene stage) but still retained oocytes at the perinucleolar stage, including newly formed cortical alveoli oocytes (Fig. 4). Moreover, temperatures of 34–36°C caused partial gonadal atrophy and inhibited gametogenesis in male juveniles of *M. salmoides*, suggesting that this thermal regime, in addition to causing degeneration of primary spermatogonia, interfered with the division of primary spermatogonia into secondary spermatogonia and/or the entry into meiosis. The timing and the mechanism of germ cell loss under high temperatures in fishes remain unknown. One hypothesis for germ cell degeneration is heat-induced apoptosis, as suggested by results of experiments with cryptorchid and temperature-manipulated male rodents.

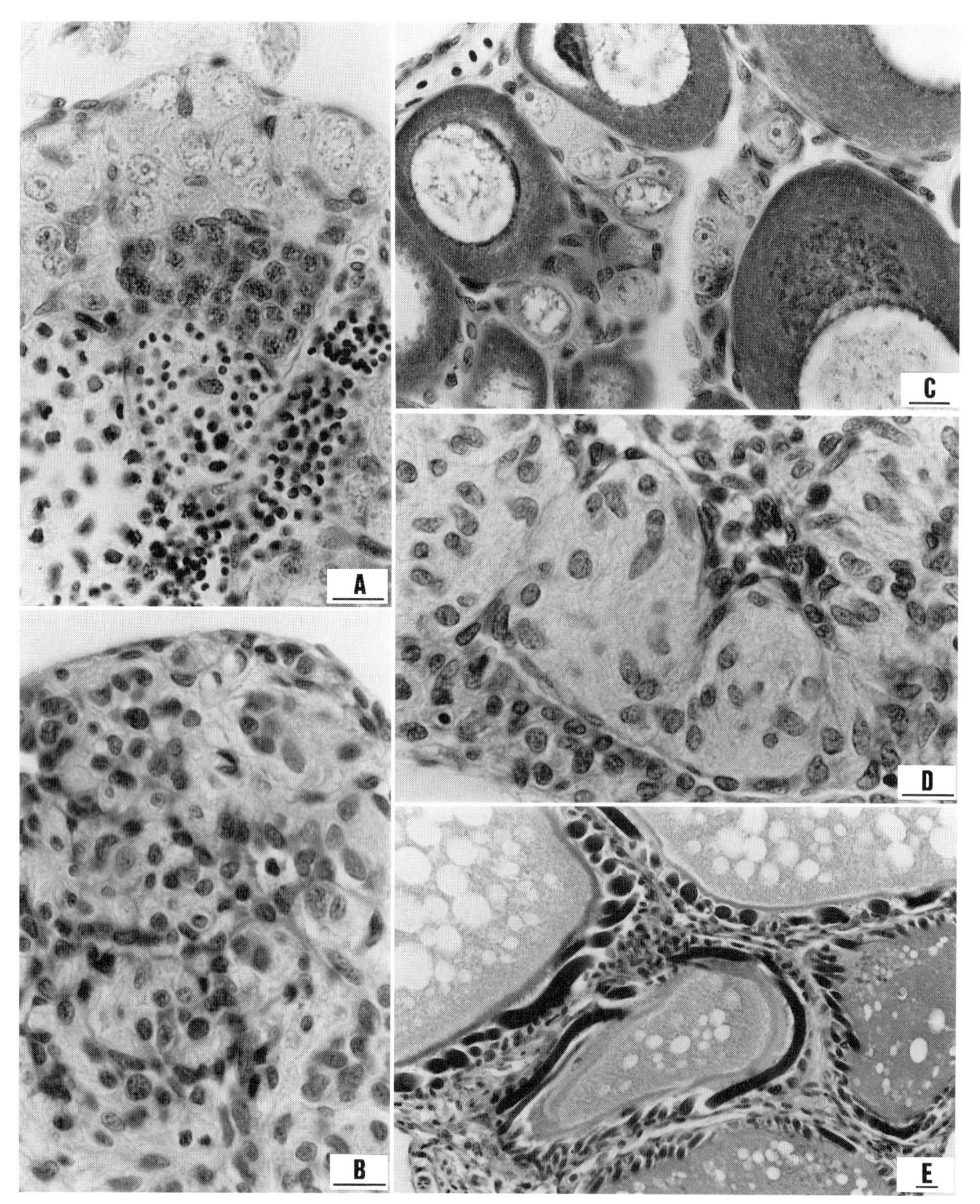

FIGURE 4 Histological appearance of normal and germ cell-deficient gonads of juvenile atherinid fish. (A–D) Details of the germinal epithelium in normal (A) and germ cell-deficient (B) testes and ovarian lamellae in normal (C) and germ cell-deficient (D) ovaries of *O. bonariensis*. (E) Ovarian lamellae of *P. hatcheri* containing only cortical alveoli oocytes. Normal and germ cell-deficient *O. bonariensis* were reared at 17–27 and 29°C, respectively, whereas germ cell-deficient *P. hatcheri* were reared at 27–28.5°C. Scale bars = 10 μm (adapted from C. A. Strüssmann, T. Saito, and F. Takashima, Heat-induced germ cell deficiency in the teleosts *Odontesthes bonariensis* and *Patagonina hatcheri*, *Comp. Biochem. Physiol. A* **119**, 637–644, 1998).

Bibliography

Adkins-Regan, E. (1987). Hormones and sexual differentiation. In *Hormones and Reproduction in Fishes, Amphibians, and Reptiles* (D. O. Norris and R. E. Jones, Eds.), pp. 1–29. Plenum, New York.

Biswas, N. M., Chatterjee, S., Patra, P. B., and Boral, M. C. (1976). Testicular histology in toad (*Bufo melanostictus*) following prolonged heat exposure. *Endokrinologie* **68**, 143–149.

Bull, J. J. (1983). *Evolution of Sex Determining Mechanisms*. Cummings, London.

Bull, J. J. (1985). Sex determining mechanisms: An evolutionary perspective. *Experientia* **41**, 1285–1296.

Bull, J. J., and Bulmer, M. G. (1989). Longevity enhances selection of environmental sex determination. *Heredity* **63**, 315–320.

Chan, S. T. H., and Wai-Sum, O. (1981). Environmental and nongenetic mechanisms in sex determination. In *Mechanisms of Sex Differentiation in Animals and Man* (C. R. Austin and R. G. Edwards, Eds.), pp. 55–111. Academic Press, New York.

Chan, S. T. H., and Yeung, W. S. B. (1983). Sex control and sex reversal in fish under natural conditions. In *Fish Physiology and Reproduction: Behavior and Fertility Control* (W. S. Hoar, D. J. Randall, and E. M. Donaldson, Eds.), Vol. 9B, pp. 171–222. Academic Press, New York.

Charnov, E. L., and Bull, J. J. (1977). When is sex environmentally determined? *Nature* **266**, 828–830.

Conover, D. O. (1984). Adaptive significance of temperature-dependent sex determination in a fish. *Am. Nat.* **123**, 297–313.

Francis, R. C. (1992). Sexual lability in teleosts: Developmental factors. *Q. Rev. Biol.* **67**, 1–18.

Korpelainen, H. (1990). Sex ratios and conditions required for environmental sex determination in animals. *Biol. Rev.* **65**, 147–184.

Patiño, R. (1997). Manipulations of the reproductive system of fishes by means of exogenous chemicals. *Prog. Fish-Cult.* **59**, 118–128.

Shikone, T., Billig, H., and Hsueh, A. J. W. (1994). Experimentally induced cryptorchidism increases apoptosis in rat testis. *Biol. Reprod.* **51**, 865–872.

Strüssmann, C. A., and Patiño, R. (1995). Temperature manipulation of sex differentiation in fish. In *Proceedings of the Fifth International Symposium on the Reproductive Physiology of Fish* (F. Goetz and P. Thomas, Eds.), pp. 153–157. FishSymp95, Austin, TX.

Sex Determination, Genetic

Józefa Styrna
Jagiellonian University

GLOSSARY

autosome Any other chromosome than a sex chromosome.

exon A section of the coding sequence of a gene; together the exons of the one gene constitute the mRNA and are translated into protein.

intersex An individual which has sexual characteristics intermediate between the male and the female.

sex chromosome Chromosome whose presence or absence is correlated with the sex of the bearer; it plays a role in sex determination.

Sex determination is the ensemble of genetic events that gives rise to the choice of male- or female-specific gonad differentiation.

I. INTRODUCTION

The majority of animals have separate sexes (male and female); however, there are some species, which are hermaphroditic, combining both male and female traits within the same individual at the same time or at different times in the life cycle. Hermaphroditism is common in the invertebrates, but there are also a few hermaphroditic vertebrates and virtually all of them are fish. Sex can be determined by various genotypic mechanisms but the two most common chromosomal sex-determining systems found among animal groups are the XX:XY system, in which females are chromosomally XX and males are XY, and the ZZ:ZW system, in which the females have heteromorphic ZW chromosomes, with males being chromosomally ZZ. Mammals and the fruit fly *Drosophila* have the XX:XY system; birds and some insects such as butterflies have the ZZ:ZW system.

The adult sexual phenotype results from a cascade of genetically determined pathways. At some point early in the cascade a primary switch is triggered, resulting in the production of either a male or a female. This switch may be genetically determined by the X:A ratio or factors on the Y chromosome, or it may be triggered by environmental cues such as temperature (in some alligators).

Through the use of combined genetic and molecular approaches many steps in the pathway have been elucidated in model organisms: the fruit fly, the nematode, and some mammals. However, although the genetics of it is relatively well documented, much of the biochemistry is unknown.

II. THE SEX-DETERMINATION PATHWAY IN *DROSOPHILA* AND *CAENORHABDITIS*

Two sexes are present in *Drosophila* and nematodes. In nematodes, XX animals are normally hermaphrodite and XO animals are male. Hermaphrodites are essentially somatic females which can first produce sperm and then oocytes. In *Drosophila,* XX flies are female and XY flies are male; however, the Y chromosome of *Drosophila* is not sex determining. In both species, somatic sex determination is based on the numerical ratio of X chromosomes to autosomes (X:A ratio). The X:A ratio also governs two other related processes: dosage compensation and germline sex determination.

A. Somatic Sex Determination

The X:A ratio in *Drosophila* dictates the activity state of the master regulator gene *Sex-lethal (Sxl).* This gene controls the sex-specific function of a hierarchically organized group of genes and establishes whether a fly will become male or female. A normal 2X *Drosophila* diploid (XXAA) has an X:A ratio of 1.0 and is phenotypically female. An XY diploid (XYAA) has an X:A ratio of 0.5 and is male; an X0AA diploid is also male (although sterile). Triploids with three X chromosomes (XXXAAA) are females. Those with one X (XYYAAA) are males, and those with two X's (XXYAAA) are intersexes. *Sxl* is transcribed in both sexes; in females, however, functional sex-specific activity is achieved through productive RNA splicing. In males, the transcripts are inactive because of the presence of a translational stop codon in an additional exon, which is spliced out in females. Through a positive autoregulatory feedback loop, *Sxl* catalyzes the productive splicing of its own transcripts. Initiation of the autoregulatory loop occurs in females by a system that recognizes the X:A ratio. These are the dose-sensitive X:A numerator genes *sisterless-a*, *sisterless-b*, and *runt* and the denominator gene *deadpan*. The products of these genes, together with the maternal *daughterless* product, generate a functionally active *Sxl* gene product. Genetic data indicate that *Sxl* controls *tra* and *tra-2* (*transformer*). One or both of these genes are involved in the next steps characterized by a series of positive regulatory interactions mediated by RNA splicing which influence a terminal regulator gene *doublesex* (*dsx*). Mutations in *dxs* result in the intersex phenotype rather than complete sex reversal, suggesting final position of this gene in the cascade.

The sexual phenotype in the nematode *Caenorhabdits elegans* is also controlled by a regulated hierarchy of sex-determining genes. The primary sex-determining signal is also the X:A ratio. In response to a low X:A ratio the master regulator gene *xol-1* negatively regulates the sdc genes (*sdc-1*, *sdc-2,* and *sdc-3*), sex determination, and dosage compensation. Based on mutation studies, the sdc genes are predicted to negatively regulate the next gene in the sex-determination cascade, *her-1*. A high level of *her-1* product is found in males, whereas little or none can be detected in hermaphrodites. Genetic analysis shows that *her-1* negatively regulates the activity of *tra-2*, the next gene in the somatic sex-determination pathway. This probably allows female gene products to inhibit the activity of *tra-1*, permitting male-specific development.

B. Dosage Compensation

In both the fruit fly and nematode, about 20% of all the known genes lie on the X chromosome. The resulting large difference in gene dosage between XX and XY or X0 must be compensated to produce a viable organism.

Dosage compensation is controlled in both species by the X:A ratio but the molecular mechanisms are different. In *Drosophila*, dosage compensation acts by a twofold increase of sex-linked gene transcription in male flies relative to the basal level for autosomal gene. This involves at least four autosomal male-specific lethal (msl) genes under the control of the master regulator gene *Sxl*. The dosage compensation mechanism operates independently of the one governing somatic sex determination. *Sxl* controls expression of X-specific transcripts in XX females by activating the transformer gene, and it probably negatively regulates one or more of the msl genes, thereby preventing hypertranscription. In XY flies the *Sxl* gene product is no-functional; consequently, the *msls* gene is switched on to produce hypertranscription of genes on the single X chromosome. The nematode has a different strategy for controlling the levels of X-specific transcripts through downregulation of expression in XX hermaphrodities. The dosage compensation dpy genes halve the expression of each set of X-linked genes in XX animals. How they do it is unknown.

C. Germline Sex Determination

Germline sex determination refers to the developmental choice of the germ cell to enter oogenesis or spermatogenesis. Whereas in higher vertebrates the germline sex is determined by inductive somatic signals, in both *Drosophila* and nematodes additional cell-autonomous genetic signals are also required. Many of the steps necessary for germline sex determination are well characterized in *C. elegans* but relatively few in *Drosophila*. The same genes that control somatic sex in *C. elegans* also control germline sex, but there are some important differences in their functions and interactions. This does not appear to be the case in *Drosophila*, in which none of the genes involved in germ cell determination following *Sxl* activity have been clearly identified, although both inductive signals from the stroma and cell-autonomous components are required. Figure 1 shows an organization of sex-determining gene networks in *Drosophila* and *C. elegans*.

III. MAMMALIAN SEX DETERMINATION MEANS TESTIS DETERMINATION

Mammalian sex determination is well understood, but many of the molecular details remain elusive. The choice of forming either a testis or an ovary by the bipotential, indifferent gonad depends on the presence or absence of a signal encoded by the Y chromosome. Individuals with a normal Y chromosome develop as males, irrespective of the presence of supernumerary X chromosomes. Thus, 47,XXY individuals (Klinefelter's syndrome) have an essentially male phenotype, whereas 45,XO subjects or individuals with X chromosome polysomies such as 47,XXX have a female phenotype. The gonad forms within the first 2 months of human gestation. The formation of the testis is associated with two events. Germ cells and Sertoli cells become enclosed in testicular cords, shortly followed by the differentiation of Leydig cells. Subsequent differentiation of the male genitalia is the result of products secreted by Sertoli and Leydig cells. Leydig cells produce testosterone, which influences the development of both the internal and external genitalia. Therefore, sex determination in mammals can be equated with testis determination. Molecular genetic analysis has identified the locus on the Y chromosome that drives testis forma-

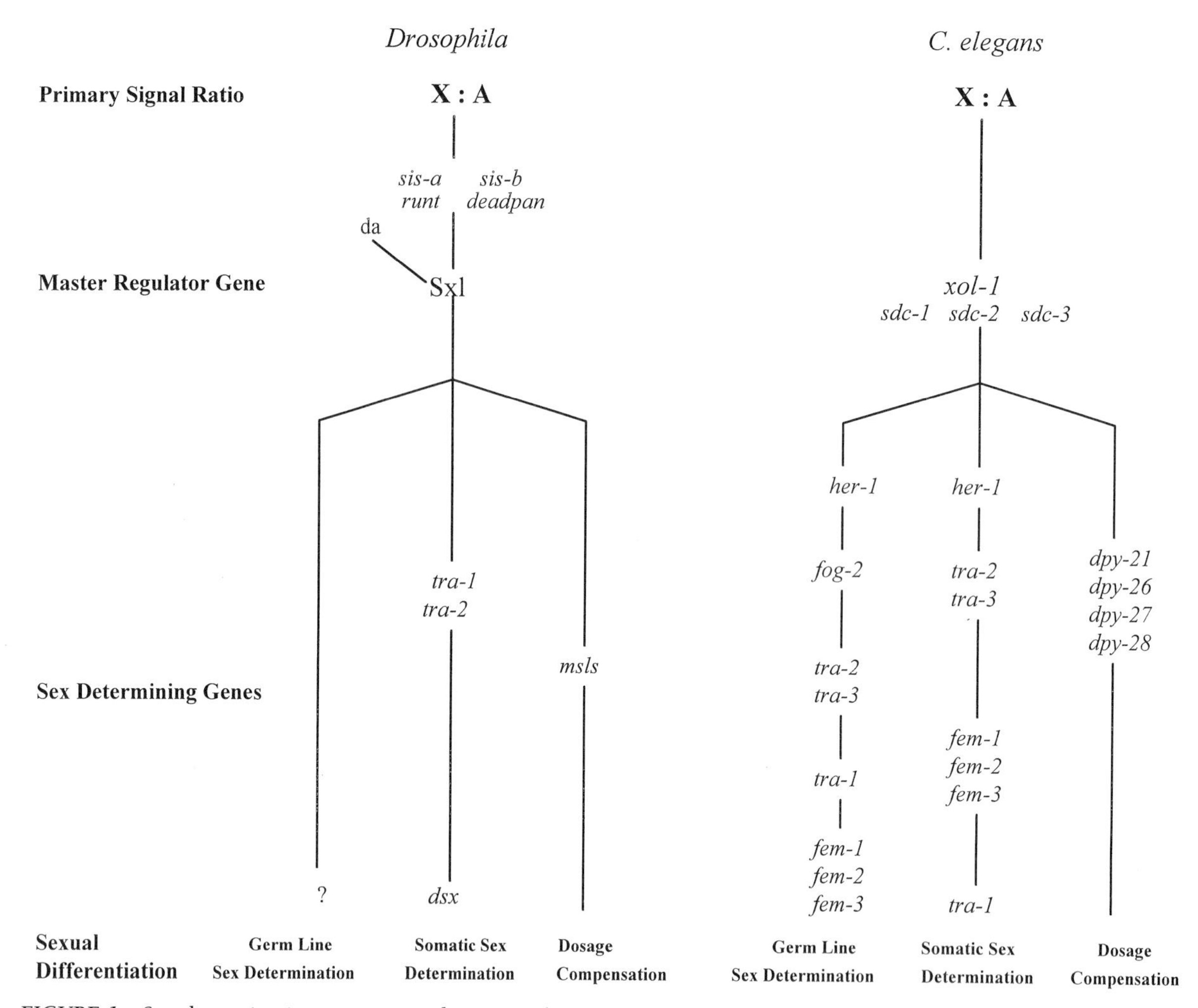

FIGURE 1 Sex-determination gene network in *D. melanogaster* and *C. elegans* (modified from McElreavey *et al.*, 1993).

tion. This hypothetical gene has been called the testis-determining factor on the Y chromosome (*TDF* in humans, *Tdy* in mice). Now it is known to be the same gene as the *SRY/Sry* gene. The SRY/Sry protein seems to be a transcription factor. This protein has a polypeptide sequence called an HMG (high-mobility group) domain which is found in several other transcription factors. Inspection of DNA sequences followed by *in vitro* binding assays have identified high-affinity SRY-binding sites in the promoter regions of a number of genes, including P450 aromatase, AMH, anti-Müllerian hormone, which induces regression of the Müllerian ducts during male development, and *fra-1*, a component of the transcription factor AP-1. *Sry* expression patterns in developing mouse embryos are consistent with a role in testis determination. In mice, gonads developing from the genital ridge show sexually dimorphic differentiation at 12.5 days postcoitum (dpc). In males, *Sry* expression appears in the genital ridge preceding gonad differentiation at 10.5 dpc, peaks at 11.5 dpc, and declines over the following 24 hr. In adult male mice, *Sry* expression is detected exclusively in the testes. Other genes on the Y chromosome are necessary for normal testicular functions such as spermatogenesis, but these genes are not sex determining. Germ cell sex determination in mammals depends only on whether a germ cell is exposed to a testicular environment and therefore is not directly dependent on a signal from the Y chromosome. Nor is dosage compensation linked to sex determination in mammals. The levels of X-linked transcripts are equalized by the independent process of inactivation of one of the two X chromosomes during female development. Reactivation occurs in XX germ cells just prior to birth.

A. Evidence for Other Mammalian Sex-Determining Genes

Genetic crosses between a number of mouse strains have helped to identify a number of non-Y-linked loci which may be involved in sex determination. A normal Y chromosome may fail to induce testis formation if it is introduced into a different genetic background. Several autosomal loci have been identified which interfere with normal sex determination resulting in partial or complete XY sex reversal—that is, a portion of the offspring are either XY females or hermaphrodites—when the Y chromosome from different *Mus domesticus* strains is introduced into the genome of the C57BL/6J inbred strain. This suggests that the *Tdy* allele linked with the Y chromosome in these strains is different, and that C57BL carries an autosomal allele *Tda-1* which is incompatible with *Tdy*. Other interpretations of the data are possible. It has been suggested that there is an incompatibility between C57BL/6J and *M. domesticus Tdy* alleles, which may produce a mismatch of developmental timing. C57BL is a fast-developing strain, and the *M. domesticus* testis-determining gene may not act early enough to divert gonadal development away from the default ovarian pathway.

Two another genes, *WT1* and *SF1*, have been identified in mammals to be necessary for formation of the gonad. Mutations in these genes show that activity of both of them is necessary for normal gonad development.

The study of nephroblastoma or Wilm's tumor (WT), which is a pediatric cancer of stem cells of the developing kidney, suggests the existence of other autosomal genes involved in sex determination. Some patients with Wilm's tumor have malformation of the gonad, resulting in either ambiguous genitalia or complete sex reversal without genital ambiguities. Cytogenetic analysis indicates that, at least in some of the cases, WT is associated with a loss of gene function at chromosome 11q13. One of several genes from this region (*WT1*) has been isolated. The *WT1* gene has a relatively restricted distribution of expression in normal tissues. It is expressed at high levels in the glomerular epithelium, renal vesicle, and condensed mesenchyme of the developing kidney and also in the genital ridge, Sertoli cells, and the ovary. The role of *WT1* in normal gonad formation is unclear.

In the mouse, the gene (designated *Wt1*) is expressed at 9 dpc in the undifferentiated genital ridge of both males and females, and expression in mature gonads occurs in somatic cells of both the testis and the ovary. A null mutation of *Wt1* was introduced into mice by gene targeting and the homozygotes failed to develop both kidneys and gonads, demonstrating the essential role of this gene in development of these organs.

In the adult mouse, one product of the *Ftzf1* gene, the orphan nuclear receptor SF1, is expressed in all primary steroidogenic tissue, where it acts as a key regulator of enzymes involved in steroid production, including the sex hormones. Surprisingly, studies in developing mice show *SF1* expression in the male and female urogenital ridge at the earliest stage of gonad organogenesis. Expression continues until overt gonadal developmental divergence, at which time transcripts disappear in female gonads but persist in the developing testes of males. *SF1* transcripts reappear in the ovary late in fetal life, after gonad development is complete. Targeted disruption of the *Ftzf1* locus, which encodes SF1, results in homozygous null mice which lack adrenals and gonads, with abnormal gonadal development occurring at the stage in which sex differences become manifest.

The study of campomelic dysplasia (CD), a congenital bone and cartilage malformation syndrome with XY sex-reversal occurring in some karyotypic males resulting from mutation of the *SOX9* gene, suggests the role of this gene in testis determination. *In situ* hybridization experiments on a 7-week human embryo show *SOX9* expression in developing bone and in an 18-week human male fetus in the area of the rete testis and seminiferous tubules.

The hierarchy of all known loci involved in mammalian sex determination and molecular interactions of them are not clear. In addition, the problem is complicated by discovery that two species of the rodent vole, species *Ellobius*, do not contain an *SRY* homolog, which seems to be the master gene in sex determination in mammals. This suggests the existence of other mammalian components of sex determination in these animals. The study of this exceptional species may provide unique insights into the mammalian sex-determination pathway.

See Also the Following Articles

Caenorhabditis Elegans; Drosophila; Hermaphroditism; Sex Chromosomes; Sex Ratios; SRY Gene

Bibliography

Griffiths, A. J. F., Miller, J. H., Suzuki, D. T., Lewontin, R. C., and Gelbart, W. M. (Eds.) (1996). *An Introduction to Genetic Analysis*. Freeman, New York.

Hodgkin, J. (1990). Sex determination compared in *Drosophila* and *Caenorhabditis*. *Nature* **344**, 721–728.

McElreavey, K., Vilain, E., Cotinot, C., Payen,. E., and Fellous, M. (1993). Control of sex determination in animals. *Eur. J. Biochem*. **218**, 769–783.

McLaren, A. (1991). Development of the mammalian gonad: The fate of the supporting cell lineage. *BioEssays* **13**, 151–156.

Nagai, K. (1996). Molecular basis governing primary sex in mammals. *Jpn. J. Hum. Genet*. **41**, 363–379.

Reed, K. C., and Graves, J. A. M. (Ed.) (1993). *Sex Chromosomes and Sex-Determining Genes*. Harwood Academic, Reading, UK.

Schafer, A. J., and Goodfellow, P. N. (1996). Sex determination in humans. *BioEssays* **18**, 955–963.

Short, R. V., and Balaban (Eds.) (1994). *The Differences between the Sexes*. Cambridge Univ. Press, Cambridge, UK.

Wood, W. B., Streit, A., and Li, W. (1997). Dosage compensation: X-repress yourself. *Curr. Biol*. **7**, 227–230.

Sex Differentiation in Amphibians, Reptiles, and Birds, Hormonal Regulation

Tyrone B. Hayes
University of California, Berkeley

GLOSSARY

environmental sex determination A mechanism of sex determination in which sex is determined by environmental factors such as temperature. This mechanism is present in some fish, some lizards, some turtles, and all crocodilians.

genetic sex determination Mechanism of sex determination present in the majority of vertebrates in which sex is determined by the genetic constitution of individuals.

gonochorism Condition in which female and male gonads (ovaries and testes, respectively) are present in separate individuals, as opposed to hermaphroditism, in which testes and ovaries reside within a single individual.

primary sex differentiation (gonadal differentiation) Development of testes or ovaries from the undifferentiated (bipotential) gonads in gonochoristic animals.

sex determination Process that determines the direction of primary sex differentiation (whether the gonads will develop into ovaries or testes).

Encyclopedia of Reproduction VOLUME 4

I. GONOCHORISM

Gonochoristic species are those species in which males and females are separate individuals (nonhermaphrodites). The fundamental difference between males and females in gonochoristic species is the presence of testes or ovaries: All other differences between the sexes are considered secondary sex characters and result from exposure to hormones from the differentiated gonads.

II. GONADAL DIFFERENTIATION AND SEX DETERMINATION

Testes and ovaries differentiate from a bipotential gonad that is mesodermally derived. In histological cross section, the undifferentiated (bipotential) gonads contain both a medulla and a cortex in amphibians, reptiles, and birds. The primordial germ cells (which are endoderm derived) migrate into the bipotential gonads. In males, germ cells eventually reside in the lumen of the testicular tubules along with Sertoli (nurse) cells, which provide nutrients for the developing germ cells. The entire tubule is separated from the rest of the gonad, which consists of connective tissue and the steroid-producing interstitial cells of Leydig, by tight junctions. The entire testicular structure is thought to develop from the medulla of the bipotential gonad, and the cortex of the bipotential gonad regresses in males. All primordial germ cells may not complete migration to the medulla before testicular differentiation, however. The fate of the germ cells that do not complete migration to the medulla and remain in the cortex during testicular differentiation is unclear; either they degenerate as the cortex regresses or they migrate into the medulla prior to completion of the cortical regression.

Ovarian development is the opposite of testicular development in that the ovary develops from the cortex of the bipotential gonad and the medulla regresses, often leaving a hole (ovarian vesicles) in the center of the developing ovary (Fig. 1). As the ovary develops, the germ cells become associated with follicles which form in the cortex. The ovarian follicles then become surrounded by connective tissue and sex steroid-producing thecal cells. As in testicular development, the fate of germ cells that migrated into the regressing medulla is unclear.

Sex determination describes the mechanism that directs whether bipotential gonads differentiate into testes or ovaries. In many cases, sex is determined genetically, with little or no environmental influence. In other cases, environmental factors may determine the direction of differentiation. For example, temperature or humidity or some other factor may determine whether the bipotential gonads develop into ovaries or testes. Even in the case of environmental sex determination, however, genetic mechanisms may control gonadal differentiation, and only the direction of differentiation (the switch) is controlled environmentally.

Gonadal differentiation may occur independent of germ cells in nonmammalian vertebrates. In other words, the type of primordial germ cells (male or female) does not determine the fate of the gonad: Oocytes are not predetermined cells that migrate into the gonad and signal it to develop into an ovary, for example. The gonad determines whether the primordial germ cells differentiate into sperm or oocytes. Experiments in which primordial germ cells were extirpated early in embryonic development in salamanders showed that both testes and ovaries (although sterile) were able to develop in the absence of germ cells. These findings suggest that the germ cells are not involved in gonadal differentiation. In addition, sex-reversing frogs or salamanders by steroid treatment can result in fertile animals (genetic females with testes and sperm and genetic males with ovaries and oocytes). Although, in the latter case, it is also possible that steroids affect the primordial germ cells directly rather than "sex reversing" the gonads which then determined the fate of the germ cells, this still provides support for the hypothesis that the gonads determine whether the germ cells differentiate into sperm or oocytes. Endogenous sex steroids from the gonads likely determine the direction of differentiation of the germ cells. Contrary to the previously mentioned data, however, studies in frogs in which the primordial germ cells were destroyed by exposure to UV light suggest that ovaries cannot differentiate properly without germ cells and may eventually revert to testicular development.

What proximal mechanism determines the direc-

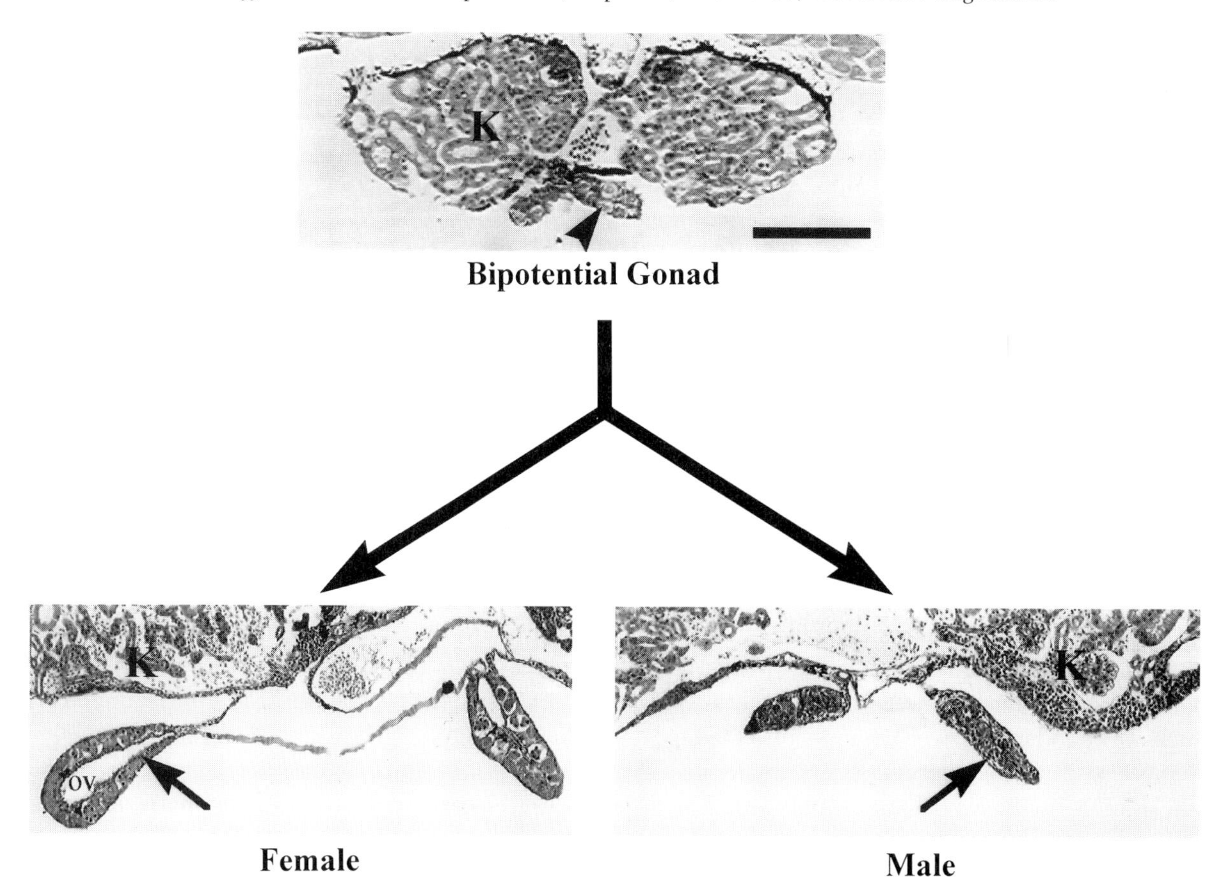

FIGURE 1 Transverse histological cross sections through bipotential gonads (top) ovaries (bottom left) and testes (bottom right) from the African clawed frog (*Xenopus laevis*). Bipotential gonads were dissected from early larvae and differentiated gonads were dissected from newly metamorphosed animals. Ovarian differentiation is characterized by the regression of the medullary portion of the bipotential gonad (resulting in a hole, or ovarian vesicle, in the center) and testicular differentiation is characterized by the regression of the cortical portion. All sections were cut from paraffin-embedded tissues at 10 μg and stained with hematoxylin/eosin. Arrows show gonads in each panel; Bar =250 μm; K, kidney; OV, ovarian vesicle.

tion in which gonads differentiate? It has been proposed that the cortex of the bipotential gonad secretes compounds that inhibit medullary development and/or promote growth and development of the cortex itself in females and/or that the medulla in males secrete compounds that inhibit cortical growth and/or induce its own growth and development (Fig. 2). To date, no compounds that serve these roles have been identified, but steroid hormones are likely candidates. The potential role of steroids in normal gonadal differentiation is suggested by the effects of exogenous steroids on gonadal differentiation. Although gonadal differentiation does not appear to be affected by exogenous steroids in mammals and effects on birds may be minimal, other vertebrates, including amphibians and reptiles, are sex reversed when exposed to exogenous steroids.

In fact, the induction of sex reversal by exposure to steroids in several vertebrates is testimony to the capacity of all individuals to develop either gonad type and clarifies the difference between sex determination and sex (gonadal) differentiation: Although each individual has genes that control sex determination and dictate whether the bipotential gonads develop into testes or ovaries, each individual has the full capacity to develop either gonad type which can be realized by exposure to exogenous hormones. Thus, exogenous hormones can override the sex-

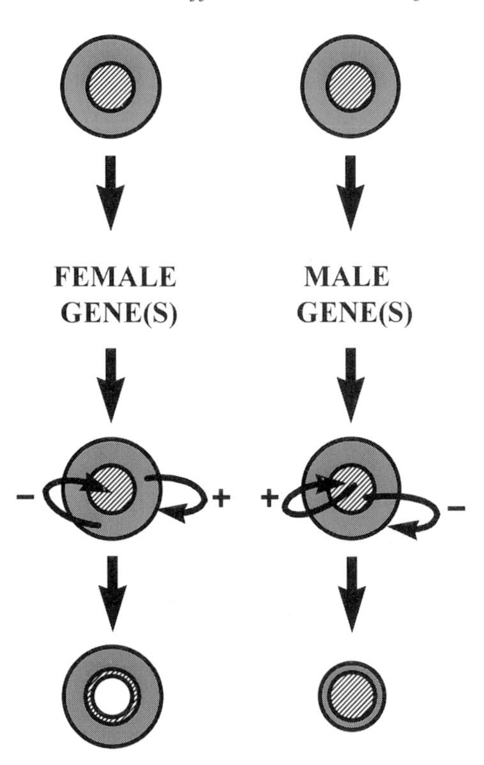

FIGURE 2 Potential mechanism controlling gonadal differentiation (proposed by Witschi, 1947). (Left) Activation of female-producing genes results in the secretion of putative compounds that inhibit (cause regression) of the medulla and/or promote development of the cortex of the bipotential gonad. (Right) Male-determining genes induce the secretion of similar putative compounds that cause regression of the cortex and/or promote development and growth of the medulla which forms the testes.

determining genes and stimulate an alternate path for gonadal differentiation. For example, application of estrogen to tadpoles of the African clawed frog *Xenopus laevis* produces 100% females. These animals mature and are capable of reproducing as adults. When mated with normal males, 50% of the estrogen-treated animals produce a normal sex ratio (50:50). The remaining estrogen-treated animals produce only male offspring, however. The females that produce the normal sex ratio are estrogen-treated genetic females and produce the expected 50:50 sex ratio. The latter type, however, are normal males (ZZ) that developed ovaries as a result of exposure to exogenous estrogen and when bred with normal males (ZZ) produce only male offspring. This experiment shows not only that *X. laevis* has sex-determining genes (despite the absence of morphologically distinguishable sex chromosomes) but also that genetic males have the capacity to develop ovaries (even though they do not normally do so). Thus, the genes control only the switch and all capacity to develop the female gonad is present in the male (Fig. 3).

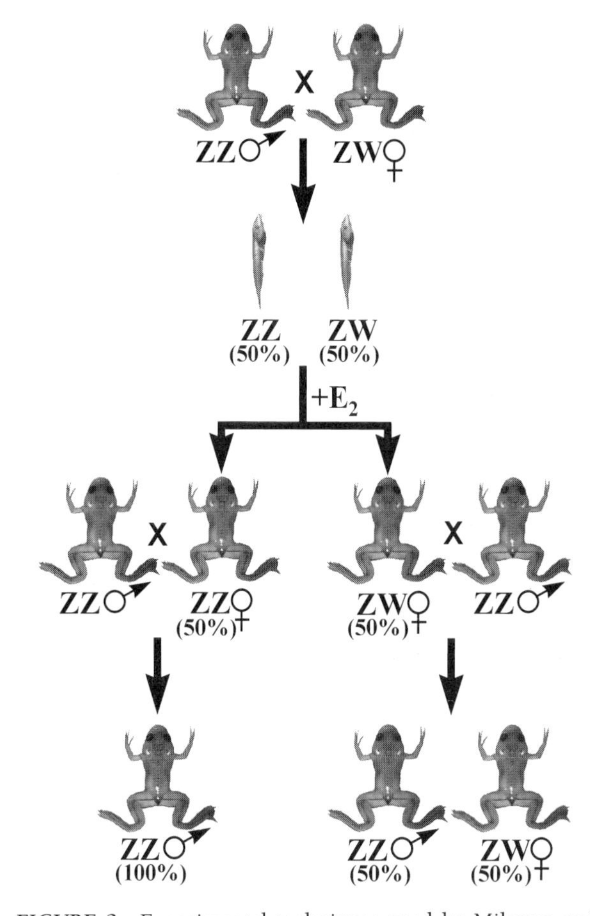

FIGURE 3 Experimental techniques used by Mikamo and Witschi (1964) to determine the genetic sex-determining system of the African clawed frog (*Xenopus laevis*). Normal males and females were bred to produce larvae. The resulting larvae were treated with estrogen and produced 100% females. Fifty percent of the treated larvae were genetic females and the remaining 50% were genetic males, sex-reversed by the hormone treatment such that they developed ovaries. When bred back to normal males, the estrogen-treated females produced a normal sex ratio of 50% male/50% female offspring. The sex-reversed males, however, produced only male offspring when bred back to normal males, indicating that males are homogametic (ZZ) in this species.

III. EFFECTS OF EXOGENOUS STEROIDS ON GONADAL DIFFERENTIATION

In amphibians, only frogs and salamanders have been examined (no studies have addressed effects on caecilians). The data concerning steroidal effects on the gonads of frog larvae are complicated by a great deal of variation with the steroid used as well as variation in the response to steroids between species. For example, a variety of steroids can have the same effect: Testosterone, cortisone, progesterone, pregnenolone, dehydriandrosterone, deoxycorticosterone acetate, and even estrogens can all result in the production of 100% males, depending on the species studied. On the other hand, a single steroid may also have a variety of effects. Estrogens, for example, can result in 100% females, 100% males, have no effects at all, or produce males or females in a single species depending on the dose used. The data for salamanders appear simpler in that androgens typically produce 100% males and estrogens produce 100% females. One possible explanation for this apparent simplicity in salamanders is that fewer studies have been conducted in this order: More thorough studies with a wider variety of steroids may reveal equally variable effects in salamanders.

In reptiles, steroids have more predictable effects than those shown in amphibians: Estrogens produce females and nonaromatizeable androgens produce males. Interestingly, exogenous steroids can override the effects of temperature on sex determination in species in which sex is determined by the temperature at which the eggs are incubated. For example, treatment with estrogens will produce 100% females in turtles (including *Chelydra serpentina, Trachemys sripta,* and *Emys orbicularis*), even when the embryos are incubated at temperatures that normally produces 100% males. Treatment with nonaromatizeable androgens produces 100% males but only at temperatures that normally produce both males and females (i.e., not at temperatures that produce all females). To date, the effects of steroids on gonadal differentiation have been shown only in reptiles with environmental sex determination: Steroidal effects on the gonads of reptiles that do not have environmental sex determination such as snakes (all of which have genetic sex determination and sex chromosomes) and turtles and lizards with genetic sex determination have not been thoroughly examined.

In birds, the effects of steroids on gonadal differentiation are less clear. Although exogenous steroid treatment affects secondary sex characteristics such as the development of the hackle, comb, and wattles in chickens, exogenous steroids do not appear to affect primary sex differentiation in birds as they do in reptiles and amphibians. Some studies showed partial transformation of testes into ovaries with estrogen treatment, but this sex reversal was incomplete and transient, whereas androgens had no effect on developing ovaries. There is some indication that exogenous steroids may be involved in normal sex differentiation in chickens, however: In normal female chickens, the right gonad and oviduct regress during embryonic development. As a result, adult female chickens have a single ovary and oviduct on the left side. When female chickens were treated with an inhibitor of aromatase (the enzyme that produces estrogens) during embryogenesis, the right gonad was retained and both gonads developed into testes. The testes produced in the treated animals was smaller than the testes in normal males, however, and only 50% of the treated animals continued to develop as males. Thus, unlike reptiles and amphibians treated with sex steroids, the sex reversal was not complete. It may be that amniote vertebrates with strong genetic components for sex determination (mammals and birds) are "protected" from influences by exogenous steroids. However, because reptiles with genetic sex determination have not been adequately examined, this idea needs to be tested further.

IV. THE ROLE OF ENDOGENOUS STEROID HORMONES

Based in part on the evidence from exogenous steroid treatments, it has been hypothesized that steroids may be the inducers or inhibitors that determine whether the bipotential gonads develop into testes or ovaries. Steroid hormones from the medulla may inhibit the growth of the cortex and promote medullary development and growth in males, and steroids from the cortex may likewise promote growth of the cortex and inhibit medullary growth and development. The role of endogenous steroids in gonadal differentiation is unclear, however. One problem is that the differentiated gonads (testes and ovaries) are responsible for synthesizing and secreting sex steroids. Thus, endogenous steroid hormones cannot determine sex because differentiated gonads are required to produce sex steroids. In other words, if male and female embryos produce different sex steroids, then they are already different (sex determination has already occurred) even though gonadal differentiation may not be evident morphologically or histologically. In addition, steroidogenic enzymes are not sex specific (e.g., males can and do make estrogens), so it is not likely that the sex-determining genes code for the steroidogenic enzymes. However, sex-determining genes may regulate steroidogenic enzymes that produce the steroid hormones controlling gonadal differentiation. In the latter case, the steroidogenic enzyme genes are involved in sex differentiation (as opposed to sex determination) and steroids are the agents directly responsible for differentiation (inducers and inhibitors of cortical and medullary development).

Examining steroidogenesis in embryos prior to gonadal differentiation is one way to address the role of endogenous steroid hormones in gonadal differentiation. Measurements of endogenous steroids or steroidogenic enzymes prior to histological/morphological differentiation of the gonads can be difficult, however, especially in animals that lack morphologically distinguishable sex chromosomes (as is the case in many reptiles and amphibians). For example, how can one determine if differential expression of enzymes and steroid synthesis precludes gonadal differentiation when in order to conduct the comparison one has to know which animals are male and which are female (genetically) before they differentiate?

Although steroidogenic enzyme activity has been measured in frog and salamander larvae, these measurements have been made typically after gonadal differentiation. Reptiles with environmental sex determination provide a unique opportunity, however, because one can measure steroid production in animals at male- or female-determining temperatures. Studies on turtles with environmental sex determination showed that the gonads have the capacity to produce sex steroids before and during stages in which the embryos undergo sexual differentiation and suggest that embryonic gonads at male- and female-determining temperatures differ in steroidogenic enzyme activities. Also, gonads incubated at male- and female-producing temperatures differ in the ability to respond to gonadotropins as measured by differences in steroidogenesis. Furthermore, measurements of whole body steroid levels *in vivo* from animals incubated at male- and female-producing temperatures showed that animals incubated at male-producing temperatures have higher testosterone and estradiol levels during the temperature-sensitive period compared to animals incubated at female-producing temperatures. Similar studies in alligators also showed that aromatase activity increases in alligator embryos incubated at female-producing temperatures when compared to animals incubated at male-producing temperatures; however, in the case of alligators, this difference was observed only after the period of temperature sensitivity.

In addition to the previously discussed studies on endogenous steroidogenesis, other experimental evidence using steroidogenic enzyme blockers suggests a role for endogenous steroids in gonadal differentiation. For example, treating turtle embryos incubated at female-producing temperatures with an aromatase inhibitor produces 100% males. Similarly, chemical inhibition of the reductase enzyme (which produces nonaromatizeable androgens) results in all female embryos at male-producing temperatures. A study in a parthenogenetic (all female) species of lizard, *Cnemidophorus uniparens*, also showed that treatment with an aromatase inhibitor produced males. Also, a single study in a frog (*Rana catesbeiana*) showed that treatment with an aromatase inhibitor produced all males and similar data from chickens has already been discussed. Thus, it appears that aromatase and estrogen production may be important in ovarian development in nonmammalian vertebrates in general, although the natural role of androgen-producing enzymes has not been as well studied.

V. SUMMARY

Exogenous steroids affect gonadal differentiation in at least two classes of nonmammalian vertebrates (reptiles and amphibians) and in fish (not discussed). In addition, early steroidogenesis in reptiles and the effects of treatment with inhibitors of steroidogenic enzymes in reptiles and amphibians further suggest a role for endogenous steroids in normal gonadal differentiation. Although not affected by exogenous steroid treatment, the impairment of ovarian development and production of all males by treatment with aromatase inhibitors in chickens suggests a role for endogenous steroids in gonadal differentiation in birds.

Exogenous estrogens seem to be more potent and effective at sex-reversing gonads in reptiles and amphibians, and endogenous estrogens may serve an important role in normal gonadal differentiation in these classes and in birds. More thorough studies are needed to further address the role of androgens. Furthermore, most studies addressing the effects of steroids have focused on reptiles with environmental sex determination. A single study examined the effects of estrogens in a turtle that appears to have genetic sex determination (*Trionix spiniferous*). The only other such study focused on a lizard that is a triploid parthenogenic species resulting from natural interspecific hybridization. More studies in reptiles with genetic sex determination are required to determine whether the role of steroids in gonadal differentiation is ubiquitous among nonmammalian vertebrates or important only in animals with environmental sex determination.

Although different sex determination mechanisms exist (genetic and environmental), the mechanism of sex differentiation may be similar across nonmammalian vertebrates. Androgens (or the absence of estrogens) may be important in testicular development and estrogens may be important for ovarian development. In species with genetic sex determination, sex-determining genes may enhance or inhibit the expression and/or activity of the steroidogenic enzymes. In the case of environmental sex determination, the same steroidogenic enzymes may be important, and temperature may regulate the expression/activity of these enzymes directly or regulate sex-determining genes that in turn regulate the steroidogenic enzymes (Fig. 4).

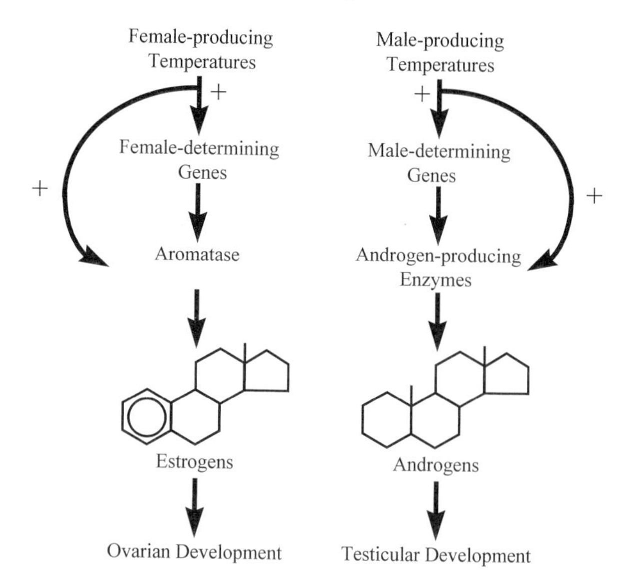

FIGURE 4 Mechanism for the potential role of steroid hormones in gonadal differentiation in species with genetic sex determination and species with environmental sex determination. In species with genetic sex determination, female sex-determining genes may induce expression or enhance the activity of estrogen-producing (aromatase) enzymes, which in turn produce estrogens which may organize the development of the bipotential gonads into ovaries. Alternatively, male-determining genes may induce or enhance the activity of androgen-producing enzymes in males, which in turn produce androgens which organize the testes. In environmental sex determination, gonadal differentiation may still be organized by the same mechanisms, and environmental factors may regulate expression of the sex-determining genes or may induce or enhance the steroidogenic enzymes directly.

See Also the Following Articles

Critical Developmental Periods; Sex Determination, Genetic; Sex Differentiation, Environmental

Bibliography

Bull, J. J. (1993). *Evolution of Sex Determining Mechanisms*. Benjamin/Cummings, London.

Bull, J. J., Gutke, W. H. N., and Crews, D. (1988). Sex reversal by estradiol in three reptilian orders. *Gen. Comp. Endocrinol.* 70, 425–428.

Burns, R. K. (1961). Role of hormone in the differentiation of sex. In *Sex and Internal Secretions* (W. C. Young, Ed.), Vol. 1, 3rd ed. William & Wilkins, Baltimore.
Chardard, D., Desvages, G., Pieau, C., and Dournon, C. (1995). Aromatase activity in larval gonads of *Pleurodeles watl* (Urodele Amphibia) during normal sex differentiation and during sex reversal by thermal treatment effect. *Gen. Comp. Endocrinol.* **99**, 100–107.
Crews, D. (1996). Temperature-dependent sex determination: The interplay of steroid hormones and temperature. *Zool. Sci.* **13**, 1–13.
Elbrecht, A., and Smith, R. G. (1992). Aromatase enzyme activity and sex determination in chickens. *Science* **255**, 467–469.
Hayes, T. B., and Licht, P. (1995). Factors influencing testosterone metabolism by anuran larvae. *J. Exp. Zool.* **271**(2), 112–119.
Mikamo, K., and Witschi, E. (1964). Masculinization and breeding of the WW *Xenopus*. *Experientia* **20**, 622–623.
Pieau, C., Girondot, M., Desvages, G., Dorrizi, M., Richard-Mercier, N., and Zaborski, P. (1994). Environmental control of gonadal differentiation. In *The Differences between the Sexes* (R. V. Short and E. Balaban, Eds.). Cambridge Univ. Press, Cambridge, UK.
Shirane, T. (1984). Regulation of gonadal differentiation in frogs derived from UV-irradiated eggs. *Zool. Sci.* **1**, 281–289.
Wibbels, T., and Crews, D. (1994). Putative aromatase inhibitor induces male sex determination in a female unisexual lizard and in a turtle with temperature-dependent sex determination. *J. Endocrinol.* **141**, 295–299.
Witschi, E. (1947). The inductor theory of sex differentiation. *J. Fac. Sci. Hokkaido Univ. Ser. VI. Zool.* **13**, 428–439.

Sex Differentiation, Psychological

Nancy G. Forger
University of Massachusetts

GLOSSARY

gonads The primary sex organs; ovaries (female) and testes (male).

hypothalamus A small structure at the base of the brain that sits just above the pituitary gland; consists of many different clusters of cells which control functions such as sexual behavior, secretion of hormones, and the regulation of feeding, eating, and emotion.

magnetic resonance imaging A noninvasive procedure using radio waves and a strong magnetic field for two- and three-dimensional imaging of living tissue, including the brain.

motoneuron A specialized nerve cell in the spinal cord or brain stem that directly connects to muscle.

nucleus As used in this article, an identifiable cluster of neural cell bodies in the central nervous system.

sexual dimorphism A sex difference in form or structure.

Sexual differentiation is the process by which the two sexes become different. The psychology of an individual (i.e., his or her mental and behavioral characteristics) is determined by the structure and function of the nervous system. Thus, psychological sexual differentiation can be understood as the process by which the nervous systems of males and females become different. One common misconception is that psychological sex differences are somehow separable from sex differences in neuroanatomy. Although these two phenomena are often studied by different groups of scientists (psychologists and neuroscientists) using very different tools (behavioral and performance measures versus measures of neural structure), any lasting change in behavior logically must be reflected by changes in the brain. At this point in our understanding it is not often clear exactly how observed changes in brain structure affect behavior or, conversely, how sex differences in psychological measures are represented in the brain. Nonetheless, it is evident that at some level these must be two sides of the same coin. With that in mind, the process of sexual differentiation of the body and brain anatomy is presented, followed by a discussion emphasizing sex differences in psychological traits of humans.

I. SEXUAL DIFFERENTIATION OF THE BODY

The process of sexual differentiation is essentially similar in all mammals, including humans. The genetic sex of a mammalian embryo is determined at the moment of fertilization; all embryos inherit an X sex chromosome from the mother and either an X or a Y chromosome from the father. If a sperm carrying a Y chromosome fertilizes the ovum, then the embryo is a genetic male (XY), and if an X chromosome is inherited from the father, the embryo is a genetic female (XX). Nonetheless, as far as anyone can tell, early development proceeds identically for XX and XY embryos. In humans, the sexes are not distinguishable (except by a chromosome test) through the first 6 or 7 weeks of gestation. At that time both sexes possess paired, undifferentiated gonads in the abdomen that can become either testes or ovaries. Similarly, the genitalia of both sexes are identical early on, and prior to sexual differentiation all embryos possess the precursors of both the male and female internal sex organs. For this reason, the mammalian embryo is often described as bipotential; that is, the embryo has the potential to develop along either male or female lines.

In genetic males, a gene on the Y chromosome directs the undifferentiated gonads to develop into testes; in the absence of this gene the gonads become ovaries. The remaining steps in the process of sexual differentiation are driven not primarily by the chromosomes but by hormones produced by the gonads. The most important products of the fetal testes are androgenic steroids such as testosterone. Testosterone masculinizes the internal sexual organs and its metabolite causes the external genitalia to form a scrotum and penis. The fetal ovaries secrete very little hormone prenatally, and in the absence of testosterone normal female development ensues.

An important observation is that testosterone can in many ways override genetic sex in the process of sexual differentiation. That is, if a genetic female (XX) is exposed to testosterone at a critical time in development, male-like genitalia and the development of male internal structures will result. Conversely, removing the testes of an XY embryo will result in normal female development.

II. SEXUAL DIFFERENTIATION OF THE BRAIN: PRINCIPLES DERIVED FROM ANIMAL WORK

A. Scientists Predict That Neural Sex Differences Must Exist

The crucial insight for understanding psychological sexual differentiation came about 40 years ago, when investigators first noticed that the very same hormones that were responsible for differentiating somatic tissues concurrently had permanent effects on behavior. For example, when a normal adult female guinea pig is primed with the appropriate hormones (estrogens and progestins) she will exhibit a posture indicative of sexual receptivity, called lordosis. Female guinea pigs that have been exposed to testosterone prenatally, however, are much less likely than normal females to exhibit lordosis and are more likely to exhibit male-typical sexual behaviors (e.g., mounting of receptive females). Since it was recognized that all behavior is generated by the nervous system, these permanent behavioral effects of early testosterone exposure led investigators to propose that gonadal steroid hormones must have permanent effects on the developing brain. This conviction was strengthened in the late 1960s and the 1970s, when some neurons in the brain were found to make steroid hormone receptors (specialized proteins for binding and responding to a given class of steroid hormones). Androgen and estrogen receptors were found to be especially abundant in the hypothalamus in many animals, including humans. Indeed, a large number of neural sex differences have been described in the past 20 years, and quite a few of these are in the hypothalamus. Some sexual dimorphisms first identified in nonhuman species have also been found in the human central nervous system.

B. The Sexually Dimorphic Nucleus of the Preoptic Area

One of the first areas of the nervous system reported to be sexually dimorphic is found in the hypothalamus of rats and is called the sexually dimorphic nucleus of the preoptic area (SDN-POA). Roger Gorski and colleagues discovered that this nucleus is five or six times larger in volume in adult male rats than in females. The sex difference is not affected by adult hormone treatments but is determined by the hormones present around the time of birth. If female rat pups are treated with testosterone just after birth the size of the SDN-POA is permanently masculinized and, conversely, castration of newborn males results in a significant reduction of SDN-POA volume. Interestingly, it is the local conversion of testosterone to estrogens within the brain (a process known as "aromatization") that is important for sexual differentiation of the SDN-POA and for many other sex differences in the brains of rodents. The sex difference apparently results because estrogens (locally produced from testosterone) prevent cell death in the SDN-POA of males, whereas many SDN-POA neurons die in females. A sexually dimorphic region of the preoptic area of the hypothalamus has also been described in other species, including gerbils, guinea pigs, ferrets, and humans. Although the hypothalamus in general and the preoptic area in particular are crucial brain areas for the expression of sexual behavior, the exact role of the rat SDN-POA is not known. Male rats will no longer copulate if the entire preoptic area is destroyed, but lesions restricted to the SDN-POA have only minor effects on male sexual behavior. Thus, even in laboratory rats the functional meaning of neural sex differences can be difficult to ascertain.

C. The Spinal Nucleus of the Bulbocavernosus

A second well-studied example of sexual differentiation of the nervous system is found in the spinal cord. A cluster of motoneurons, known as the spinal nucleus of the bulbocavernosus (SNB), resides in the lower lumbar spinal cord of rats. These motoneurons innervate the bulbocavernosus muscle, which attaches to the base of the penis and is involved in producing erections and ejaculations. The bulbocavernosus muscle is present in adults male rats but absent in females. Similarly, males have many more motoneurons in the SNB than do females. As with the SDN-POA and the genitalia, the SNB neuromuscular system becomes sexually dimorphic as a result of perinatal hormone exposure. Before birth, male and female rats both have bulbocavernosus muscles, and both sexes have similar numbers of motoneurons in the SNB. Around the time of birth, however, the muscles and the SNB motoneurons begin to die in females. This cell death can be prevented with testosterone: Females treated with testosterone around the time of birth permanently retain the bulbocavernosus muscle and an increased number of SNB motoneurons.

Although the sex difference in the SNB has been most intensively studied in laboratory rats, findings in the rat probably apply much more generally because a similar sex difference in the number of motoneurons innervating the bulbocavernosus muscle has also been described in mice, gerbils, dogs, hyenas, monkeys, and humans.

D. Some Brain Structures Are Larger in Females Than in Males

Although the SDN-POA and the SNB are both larger in males than in females, there are also regions of the nervous system that show the opposite pattern.

For example, a cell group in the preoptic area of the hypothalamus known as the anteroventral periventricular nucleus (AVPv) is larger and has more cells in female rats than in males. In this case, evidence indicates that testosterone increases cell death during prenatal development, resulting in a reduced AVPv in males. Thus, hormonal regulation of neuronal cell death appears to be a common mechanism whereby gonadal hormones sculpt the sexually dimorphic brain and, for reasons that are not yet understood, a given hormone may have opposite effects on cell death in different neural areas.

E. Are There Neural Sex Differences That Do Not Involve Hormones?

Although gonadal hormones have been shown to be responsible for almost all neural sex differences that have been described in nonhuman animals, there is recent evidence for primary genetic control of some aspects of sexual differentiation that may not involve hormones. By definition, only genetic males have a Y chromosome. Therefore, in principle, any gene on the Y chromosome that contributes to neural development could contribute to neural sex differences. There is now evidence that some Y chromosome-specific genes are expressed in the developing brain; how, or if, these genes contribute to sexual dimorphisms is not yet known.

F. Experience Can Alter Sexual Differentiation of the Nervous System

As the foregoing discussion illustrates, differences in gonadal steroid hormone levels between males and females is a crucial factor for sexual differentiation. However, social and environmental factors also have a role in shaping sex differences in the nervous system. For example, Ingeborg Ward and colleagues have shown that if a rat is stressed during pregnancy (e.g., by restraining her under bright lights for a couple of hours each day), her male offspring will show less robust male sexual behavior in adulthood than the offspring of control mothers. Interestingly, the sizes of the SDN-POA and SNB are somewhat feminized in male offspring of stressed mothers. Other behaviors, such as play, that are normally sexually dimorphic in rats are also affected by prenatal stress.

Some neural sex differences depend on the environment in which an animal is reared. For example, rats reared in so-called "enriched" conditions (complex environments with many cagemates and toys) exhibit sex differences in the brain that are not apparent in animals raised in the more standard, solitary, laboratory cages. Since there is a sex difference in the way that rats play when housed in groups, it is possible that the brains of males and females become different, in part, because males and females have different social experiences. Of course, this raises the question of what makes the two sexes play differently in the first place! Even in the laboratory rat, the effect of experience on the brain, and of the brain on experience, illustrates the futility of categorizing influences as purely "nature" or "nurture."

III. SEXUAL DIFFERENTIATION OF THE HUMAN BRAIN

A. What Is the Evidence for Neural Sex Differences in Humans?

In comparison to the very rich literature on sex differences in the rodent nervous system, our knowledge of sex differences in the human brain is quite sparse. In part, the paucity of proven sex differences in the human brain may stem from the difficulty of executing well-designed studies of the human brain with appropriately large sample sizes, matched control groups, and other considerations that are routine in animal work. Alternatively, it may be that sex differences in the human nervous system simply are not as numerous or as pronounced as they are in rodents.

One indication that neurological sexual differentiation does exist in human brains, and is an important factor in human psychology, is the marked sex difference seen in the occurrence of several neurological and psychiatric diseases (Table 1). For example, more than twice as many males as females suffer from dyslexia and schizophrenia, whereas women make up over 90% of all cases of anorexia nervosa. Other disorders, including bulimia, anxiety disorder,

TABLE 1
Neurological and Psychiatric Diseases That Exhibit a Sex Difference in Incidence[a]

Disease	*Ratio (female : male)*
Anorexia nervosa	93 : 7
Bulimia	75 : 25
Anxiety disorder	67 : 33
Depression	63 : 37
Multiple sclerosis	58 : 42
Severe mental retardation	38 : 62
Autism	29 : 71
Stuttering	29 : 71
Schizophrenia	27 : 73
Dyslexia	23 : 77
Sleep apnea	18 : 82
Gilles de la Tourette	10 : 90

[a] Adapted with permission from D. Swaab and M. Hofman, *Trends Neurosci.* **18**, 264, 1995, (see Swaab and Hofman, 1995, for supporting references).

depression, mental retardation, autism, and sleep apnea, are also significantly more prevalent in one sex than in the other. Finding structural differences on which these sex differences are based represents an important challenge that has thus far received relatively little attention.

Many sex differences in the brain are likely to be fairly subtle, consisting of differences in neural connectivity (i.e., the types and numbers of neural connections between brain areas) or differences in the expression of signaling molecules such as neurotransmitters and neuropeptides. These characteristics would not be visible from routine inspection of postmortem tissue, and very few such sex differences have been established in humans. On the other hand, gross differences in structural features are much more amenable to study; a number of structural sex differences that have been reported in the human central nervous system are listed in Table 2. There are several caveats to keep in mind when examining this table. First, many of the listed findings are not firmly established. In some cases there have been multiple studies of a single brain area with conflicting outcomes. For example, a large number of investigators have examined the corpus callosum (the major fiber tract connecting the left and right cerebral hemispheres) in men and women. About half of these studies report a sex difference in the size or shape of the corpus callosum, whereas the other half do not. The reasons for the conflicting outcomes are not clear. For other brain regions listed in Table 2, the initial report of a sex difference simply has not been followed up, so the finding is neither confirmed nor refuted. Second, it is important to keep in mind that it is not clear whether differences in gonadal hormones cause the reported neural differences. Sexually dimorphic peaks of testosterone secretion in humans are found during a prenatal period extending from about Week 10 to Week 20 of gestation, a perinatal period extending through the first several years of life, and throughout the postpubertal period. Any or all of these hormone peaks might contribute to neural sex differences, but a role for hormones has not been definitively established for any proposed sexual dimorphism listed in Table 2. Finally, it is not known how any of the observed sex differences in neural structure affect human behavior.

B. Sex Differences in the Hypothalamus of Humans

One finding that appears solid, in that it has been replicated by three independent groups of scientists, is that there is a sex difference in a region of the hypothalamus of humans known as INAH-3 (the third interstitial nucleus of the anterior hypothalamus). This small brain region is two or three times larger in volume in men than in women. Since the sex difference has been found in postmortem brains of adults, it is not known whether prenatal hormones, social experiences, or both may shape the sex difference, and the function of this nucleus remains completely unknown.

A group of scientists in the Netherlands, led by Dick Swaab, has reported on another region in the hypothalamus that may be sexually dimorphic in humans. This nucleus was named the "SDN-POA of humans," in reference to the proposed homology of this region with the SDN-POA of rats. The existence of a sex difference in the human SDN-POA is somewhat controversial, however, because two other groups have failed to confirm the initial report. One

TABLE 2
Structural Sex Differences Reported in the Human Brain and Spinal Cord

Neural structure	*Nature of the difference*	*Reference*[a]
Whole brain	Larger in men	Many reports over hundreds of years
Regions of the hypothalamus		
SDN-POA	Larger in men	Swaab and Fliers, *Science* **228**, 1112, 1985
INAH-3	Larger in men	Allen *et al.*, *J. Neurosci.* **9**, 497, 1989
BNST	Subarea is larger in men	Allen and Gorski, *J. Comp. Neurol.* **302**, 697, 1990
Structures connecting the left and right brain		
Corpus callosum	Posterior portion is larger and more bulbous in women	de Lacoste-Utamsing and Holloway, *Science* **216**, 1413, 1982
Anterior commissure	Larger in women	Allen and Gorski, *J. Comp. Neurol.* **312**, 97, 1991
Massa intermedia of thalamus	More often present, and larger, in women	Morel, *Acta Anat.* **4**, 203, 1948; Allen and Gorski, *J. Comp. Neurol.* **312**, 97, 1991
Language-related areas		
Planum temporale	Size of left and right sides more symmetrical in women	Wada *et al.*, *Arch. Neurol.* **32**, 239, 1975
Dorsolateral prefrontal cortex and superior temporal gyrus	Proportionally larger in women	Schlaepfer *et al.*, *Psychiatry Res. Neuroimaging* **61**, 129, 1995
Suprachiasmatic nucleus	More elongated in women	Swaab *et al.*, *Brain Res.* **342**, 37, 1985
Spinal Cord		
Onuf's nucleus	More motoneurons in men	Forger and Breedlove, *Proc. Natl. Acad. Sci. USA* **83**, 7527, 1986

[a] Where multiple references exist, only the earliest report is cited.

possible factor contributing to the discrepant findings may be the different ages of the subjects used in each study. Swaab and colleagues report that a sex difference in the human SDN-POA first becomes apparent between 4 years of age and puberty; the magnitude of the difference then decreases between 50 and 80 years of age, as men lose cells from this region. Whether or not the SDN-POA is ultimately confirmed as sexually dimorphic, this work raises several important points. First, sex differences may not be static, and in a long-lived species such as humans a given neural area may exhibit sexual dimorphism during some periods of life and not at other times. A second, related consideration is that sex differences in the human brain may occur relatively late. In the SDN-POA a sex difference seen in adults was not apparent in children. This leaves open the possibility for different socialization experiences of girls and boys to contribute to adult neural sex differences. On the other hand, it is certainly possible that a sex difference appearing at puberty was preprogrammed by gonadal steroids earlier in development. Such delays in the appearance of sex differences that are nonetheless due to perinatal hormone exposure have been seen in rodents. It will be quite difficult to assess the validity of these alternate explanations.

IV. SEX DIFFERENCES IN HUMAN BEHAVIOR AND PERFORMANCE

A. Identifying Reliable Sex Differences

A great many claims have been made (in labs and over backyard fences) about differences in behavior and performance between men and women. The

TABLE 3
Effect Sizes for Sex Differences in Selected Human Behaviors

Behavior	*Approximate effect size*[a]
Aggression	Moderate
Childhood play (level of activity, degree of rough and tumble play, selection of toys and playmates)	Moderate to large
Tasks on which males typically outperform females	
Visuospatial	
Three-dimensional visual rotation	Large
Two-dimensional visual rotation	Small
Spatial perception	Small to moderate
Spatial visualization	Negligible
Quantitative	
Overall ability	Small to moderate
Problem solving	Small to moderate
Tasks on which females typically outperform males	
Verbal	
Overall ability	Neglible to small
Verbal/association fluency	Moderate
Speed production	Small
Perceptual speed	Moderate

[a] Magnitude of effect size was defined according to J. Cohen (1977): Effect sizes of approximately 0.8 standard deviations or greater were considered large, effect sizes of approximately 0.5 were considered moderate, effect sizes of approximately 0.2 were considered small, and effect sizes of less than 0.19 were considered negligible. (Adapted from Collaer and Hines, 1995, with permission from the American Psychological Association; see Collaer and Hines, 1995), for supporting references).

number of behavioral sex differences that have actually been confirmed in carefully designed studies, however, is relatively small. Several of the most reliably seen differences are listed in Table 3. Sex differences have repeatedly been reported for various measures of aggression, play behavior, and for certain specialized cognitive abilities. Males are typically more aggressive, exhibit more active play, and excel at visuospatial skills, particularly those requiring three-dimensional mental rotation. Females typically excel at certain verbal skills, especially verbal fluency and perceptual speed.

It is important to note that most behaviors show considerable overlap for males and females and that the range of differences within each sex is usually larger than the average difference between the sexes. With a large enough sample size even very tiny, functionally meaningless differences may be statistically significant. For this reason it is useful to categorize sex differences by "effect size," as in Table 3. Effect size is a measure of how large the difference between two groups is relative to the variability within groups. When the mean difference between two groups is about half the size of the standard deviation of scores within groups, the effect is said to be "moderate." By way of example, the difference in height between 14- and 18-year-old girls fits this definition of a moderate effect size (Cohen, 1977).

B. Are Human Sex Differences in Behavior and Performance Related to Hormones?

The idea that human sex differences in aggression, play, and spatial learning may be influenced by gonadal hormones has come from studies on nonhuman animals. In at least some laboratory animals each of these behaviors is sexually dimorphic, and the sex difference can be reversed with perinatal hormone treatments.

The study of sex differences in play behavior presents an interesting example of how animal studies may relate to human psychology and will be discussed in some detail here. Many species of mammals display a sex difference in juvenile play. For example, male rats display more active play and engage in play fighting more often than females. This so-called "rough and tumble" play characteristic of males has been shown to be due to pre- or perinatal exposure to testosterone. When female rats are treated with an androgen during the late prenatal or early postnatal period, they show significantly more play fighting as juveniles. Conversely, males that are castrated during the first week of life are feminized with respect to later play behavior, and similar hormone manipulations later in development have no effect on play. One brain area that is altered by early androgens, and that appears to be responsible for sexually dimorphic play in rats, is the amygdala.

Analogous findings have been observed in several nonhuman primates, although in this case the critical period for masculinizing play behavior is exclusively prenatal. Steroid hormones readily cross the placenta; therefore, by injecting a pregnant female monkey with testosterone, her offspring can be prenatally androgenized. Such androgenized female rhesus monkeys show much more rough and tumble play as juveniles than do untreated females. Although prenatal androgen treatments can also masculinize the genitalia, the effect of hormones on play can be separated from effects on the genitalia by taking advantage of the fact that the critical developmental period for formation of the genitalia occurs somewhat earlier than the critical period for masculinizing play behavior. By carefully timing prenatal testosterone exposure, Robert Goy and colleagues were able to produce prenatally androgenized monkeys with normal female genitalia but masculinized juvenile play.

Studies of human play behavior also consistently have found sex differences, with boys displaying a higher level of activity, more rough and tumble play, a preference for male versus female playmates, and a preference for transportation toys and construction toys over dolls, kitchen supplies, and crayons. Similar sex differences are seen across a wide range of cultures. Clearly, children are exposed to many socialization forces which might shape their play behavior and toy choices. Are there also effects of early hormones on human play patterns, as in animals?

Because we cannot treat pregnant women with testosterone simply to test the effect of prenatal hormones on the play behavior of their offspring, we must rely on "experiments of nature" to test this idea. One interesting patient population in this regard is females with congenital adrenal hyperplasia (CAH). Because of an enzymatic defect, CAH girls produce high levels of androgens from their adrenal glands, beginning *in utero*. CAH girls often are born with ambiguous genitalia; that is, an enlarged clitoris and labia that have partially fused to form a pseudoscrotum. After surgical intervention to correct the genitalia, and adrenal hormone replacement which stops the increased androgen production, these individuals usually are raised as girls. In several early studies, CAH girls or their parents reported an increase in childhood tomboyism in CAH versus unaffected girls. One weakness of these studies, however, is that data were collected from interviews rather than from direct observation of play behavior; parents might expect, and therefore report, more masculine behavior from their CAH daughters due to their knowledge of the child's prenatal hormone exposure. A recent study by Sheri Berenbaum and Melissa Hines addressed this concern by videotaping CAH girls and their unaffected male and female relatives during unsupervised play sessions. They found that the play behavior of the CAH girls was significantly masculinized, in that CAH girls spent the majority of their time playing with toys usually preferred by boys. Thus, the play behavior of CAH girls does seem to be different from that of their unaffected female relatives. Of course, it is entirely possible that parental expectations, or the girls' own knowledge of their condition, shaped their play behavior long before the videotaping. As discussed previously, hormone-related sex differences are altered by environmental factors even in rodents, with modification of sex differences observed as a result of different types of maternal care, handling, and overall environmental stimulation. Psychosocial factors in humans are likely to play an even larger role in accentuating or minimizing hormonal or genetic influences. Nonetheless, whether the sex difference in play is due to hormones, to socialization, or to the interaction of both, the brain is still responsible for generating play behavior, and lasting sex differences in play must be a reflection of sex differences some place in the brain. The complexity of psychological sexual differentiation can be appreciated when one realizes that the sexually dimorphic play of boys and girls will lead to sexually dimorphic childhood experiences, which may then have further consequences for neurological and psychological differentiation of the sexes.

C. Imaging the Living Brain

An exciting future direction for the study of human psychology is made possible by the development of brain imaging techniques such as magnetic resonance imaging (MRI). It is now possible to view the living brain with fairly precise anatomical detail and even to identify brain regions activated during the performance of particular tasks. For example, using functional MRI (fMRI) to examine brain activation in men and women, B. A. Shaywitz and colleagues recently reported evidence for a sex difference in the way we use our brains to perform specific language tasks. When making a judgment requiring processing of written language (in this case, deciding whether two nonsense words rhyme), neural activity in the brains of men was lateralized predominantly to a language area in the left cerebral hemisphere, whereas both sides of the brain were used more or less equally by women performing the same task.

In another study, by Ruben and Raquel Gur and colleagues, sex differences in metabolic activity in the human brain were noted. At rest, women had relatively higher activity than men in the cingulate region of the brain and lower relative activity in temporal–limbic regions and the cerebellum. Although the functional consequences of these differences for cognitive or emotional processing are quite obscure, it is nonetheless intriguing that such sex-specific patterns in brain metabolic activity exist.

The great promise of brain imaging techniques which do not rely on postmortem material is that developmental studies of brain structure or function may be feasible. By repeatedly imaging the brains of the same individuals throughout their life span, we may be able to sort out which neural sex differences are inborn, which arise at puberty, and which are predictive of later behavioral outcomes.

V. SEXUAL ORIENTATION AND THE BRAIN

Arguably the largest psychological sex differences in humans are seen in core gender identity, gender role, and sexual orientation. Most, but not all, men view themselves as male, assume a male role in society, and are sexually attracted to women. Most, but not all, women firmly believe themselves to be female, assume a feminine gender role, and are sexually attracted to men. Can brain areas be identified that contribute to any of these sex differences?

Several neuroanatomical correlates of one of these traits—sexual orientation—have recently been reported. As discussed previously, the size of INAH-3 in the general population is sexually dimorphic, with men having a significantly larger nucleus than women. In 1991 Simon LeVay confirmed the sex difference in INAH-3 that was first reported by Laura Allen and colleagues and also reported that the size of INAH-3 was significantly smaller in the brains of homosexual men than in the brains of (presumed) heterosexual men. In fact, the average size of INAH-3 in the gay men did not differ from that in women. This study was not the first to describe a sexual orientation dimorphism in the human brain, but it has garnered the most attention, possibly because of the well-established role of the hypothalamus in sexual behavior and perhaps also because the outcome fit the widely held prejudice that the brains of homosexual men would somehow be "like women." However, the fallacy of this way of thinking is illustrated by the other sexual orientation dimorphisms that have been reported. The first study to report a sexual orientation dimorphism in the human brain found that the size of the suprachiasmatic nucleus is not sexually dimorphic, but that this nucleus is larger in homosexual men than in heterosexual men or women. In another report, the anterior commissure, which is larger in women than in men, was found to be larger still in homosexual men. Thus, although each of these studies awaits replication, the evidence to date suggests that, on average, the brains of homosexual men are different from those of either heterosexual men or women.

It is important to keep in mind that for each of the three papers reporting a sexual orientation dimorphism in the brain, the variability within groups was large compared to the differences between groups, and it is impossible to state the sexual orientation of any individual on the basis of a single brain measurement. In addition, because all three studies used postmortem tissues of adults, it is not known whether life as a homo- or heterosexual influenced the observed brain differences or whether the neuroanatomical differences contributed to sexual orientation. In fact, this kind of question will be impossible to answer, unless some of these structures can be imaged in children, who are then followed to see whether brain anatomy can predict adult sexual orientation.

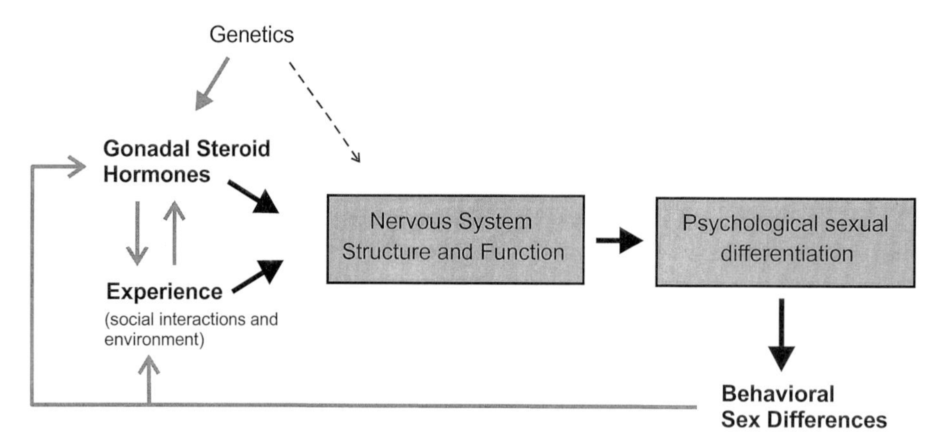

FIGURE 1 Model of the process of psychological sexual differentiation.

VI. A PROPOSED MODEL OF PSYCHOLOGICAL SEXUAL DIFFERENTIATION

A simplified model of psychological sexual differentiation, which incorporates some of the factors discussed in this article, is shown in Fig. 1. At the center of this figure is "nervous system structure and function" since, as we have seen, the psychology of an individual is determined by the nervous system. Direct effects on the nervous system are shown in black and more indirect inputs are depicted in gray. The evidence that sex differences in gonadal hormone levels can cause sex differences in neural structure is very well documented, especially in nonhuman animals. Genetic factors (specifically, genes on the Y chromosome) may also have direct effects on sexual differentiation of the nervous system, although the evidence for this is currently sketchy and circumstantial. Experience (social or environmental factors) is a third input that determines sexually dimorphic neural structure and, hence, sexually differentiated psychology and behavioral output. An individual's behavior can then feed back to impact his or her own nervous system by affecting subsequent experiences and by influencing the endocrine state. In this way, nature and nurture can be seen as interactive and inseparable processes that operate to shape psychological sexual differentiation throughout life.

See Also the Following Articles

Homosexuality; Hypothalamic-Hypophysial Complex; Sex Determination, Genetic

Bibliography

Berenbaum, S. A., and Hines, M. (1992). Early androgens are related to childhood sex-typed toy preferences. *Psych. Sci.* **3**, 203–206.

Breedlove, S. M. (1994). Sexual differentiation of the human nervous system. *Annu. Rev. Neurosci.* **45**, 389–418.

Cohen, J. (1977). *Statistical Power Analysis for the Behavioral Sciences.* Academic Press, New York.

Collaer, M. L., and Hines, M. (1995). Human behavioral sex differences: A role for gonadal hormones during early development? *Psych. Bull.* **118**, 55–107.

Gorman, M. R. (1994). Male homosexual desire: Neurological investigations and scientific bias. *Perspect. Biol. Med.* **38**, 61–81.

Kimura, D. (1992, September), Sex differences in the brain. *Sci. Am.*, 119–125.

Le Vay, S. (1993). *The Sexual Brain.* MIT Press, Cambridge.

Meaney, M. J. (1988). The sexual differentiation of social play. *Trends Neurosci.* **11**, 54–58.

Shaywitz, B. A, Shaywitz, S. E., Pugh, K. R., Constable, R. T., Skudlarski, P., Fulbright, R. K., Bronen, R. A., Fletcher, J. M., Shankweller, D. P., Katz, L., and Gore, J. C. (1995). Sex difference in the functional organization of the brain for language. *Nature* **373**, 607–609.

Sex Hormone-Binding Globulin

see SHBG

Sex Ratios

Ronald J. Ericsson
Gametrics Limited

Scott A. Ericsson
Sul Ross State University

GLOSSARY

cytometer A device for counting and measuring cells.

human serum albumin A sterile preparation of a protein obtained from the blood of healthy donors.

mutagen A chemical or physical agent that induces genetic mutations.

sex chromosomes Chromosomes that are associated with the determination of sex.

skewness The deviation of the curve for a frequency distribution from the symmetrical curve for such a distribution.

Tyrodes Tyrode's solution, a Physiological medium for cells.

The sex ratio is the proportion of males to females and is expressed as the number of males per 100 females or as the percentage of males (e.g., 106/100). Sex ratios can be further defined as primary, secondary, and tertiary sex ratios. The primary sex ratio expresses the proportion of the sexes at the time of fertilization. The secondary sex ratio is the proportion of males to females at the time of birth, although this term has been used to describe the proportion of the sexes during intrauterine development. Lastly, the tertiary sex ratio describes the proportion of the sexes when they reach procreative age. In theory, since the testes produce an equal number of X- and Y-bearing sperm, there should be an exactly equal number of males and females born. In fact, this is not true because there is are variety of factors influencing the sex ratios of both nonhuman mammals and humans. Manipulation of the sex ratio involves an attempted increase in the proportion of either sex.

I. NONHUMAN MAMMALIAN AND BIRD SEX RATIOS

The secondary sex ratios, transformed into percentage of males, will be used to present the nonhuman mammalian and bird sex ratios. Table 1 summarizes the sex ratios of various mammals and birds. Mammalian sex ratios can be modified by a variety of factors such as litter size, maternal age, nutrition, timing of insemination, stress, date of birth, and the maternal parity of the previous breeding season (Table 2). Additional factors include climatic variation, maternal dominance rank, high population density, 6-methoxybenzoxazolinone implantation, age of maturation, excision of accessory sex glands, intra-

TABLE 1
Secondary Sex Ratios for Selected Mammals and Birds

Species	*Sample size*	*Sex ratio (% males)*
Dogs	6,878	52.4
Mice	2,903	52.6
Canaries	200	43.5
Pigs	16,233	48.8
Rats	1,862	46.2
Cattle	1,642,713	52.6
Guinea pigs	7,989	51.7
Horses	135,826	49.1
Sheep	50,685	49.5
Chickens	20,037	48.6

uterine position of the mother among littermates, and exposure to specific wavelengths of light.

The mechanisms involved in varying nonhuman mammalian secondary sex ratios are not known. Possible explanations include differences in the proportion or motility of X- and Y-bearing spermatozoa produced, dissimilarities between the implantation and survival of male and female embryos, and disruptions to the processes involved during sex determination.

II. HUMAN SEX RATIOS

The sex ratio at birth, secondary sex ratios, will again be presented as the percentage of males to illustrate the human sex ratio. A variety of factors have been identified as causing the human sex ratio to fluctuate. There are differences among ethnic groups, as shown by a higher Caucasian (.514) than Negroid (.507) sex ratio, where the races coexist. In addition, there is some evidence that Asian ethnic groups, such as Korean (.5348), have some of the highest sex ratios. The sex ratio varies slightly with the season of the year, with the highest U.S. sex ratio being observed in June (.5140) and the lowest in February (.5116). Birth order affects the sex ratio, with the sex ratio being highest for first-born children (.5157) and declining to .5097 for the fifth and any other subsequent children. The sex ratio appears to decline with maternal age, paternal age, or an increase in the age of both parents. The ratio of sons to daughters is higher for couples with higher coital rates. Likewise, the sex ratio is higher during and after wars (an increase of 0.13 for first-born Caucasian children in the United States in 1946), which might be related to an increase in coital frequency during time of war. There have been conflicting reports pertaining to the effect of the timing of fertilization on sex ratios, with some evidence suggesting that fertilization resulting from natural insemination occurring 3 or more days prior to ovulation results in an increase in the sex ratio (.58) compared to natural insemination at or after ovulation (.46). There is a higher percentage (.593) of boys produced from births resulting from artificial insemination when ovulation is not hormonally induced. When births are due to hormonally induced ovulations, there is a decrease in the sex ratio to .463. There is a difference between dizygotic twins, with there being more same-sex twins (.533) than different-sex twins (.514). Blood group influences the sex ratio, with AB mothers having a greater percentage of boys (.56) than the other groups (.51).

Additional variables that have been suggested to modify the sex ratio include socioeconomic status, prostatic cancer, male infertility drugs, maternal smoking, the nematocide DBCP, non-Hodgkin's lymphoma, toxemia of pregnancy, multiple sclerosis, nutrition and diet, hepatitis and measles, ectopic pregnancies and placenta previa, contraceptive pills, pollution, prior induced abortion, handedness, time trend, and urban vs rural births.

Variations in the human secondary sex ratio suggest that the conditions of the male and female tract could be altered to favor the fertility potential of either the X- or Y-bearing spermatozoa. Differences in implantation, embryonic development, and sexual differentiation are other possible explanations to variations in sex ratios.

III. MANIPULATION OF SEX RATIOS

Considerable folklore has arisen regarding preconception methods to manipulate the human secondary

TABLE 2
Variations in Secondary Sex Ratios for Selected Mammals

Species	Sample size	Sex ratio (% males)
White-tailed deer (captive)		
Litter size		
1	63	66.7
2	213	45.5
3	12	41.7
Maternal age (years)—litter size 1		
1	40	30.0
2.5–6.5	63	66.7
Nutrition (mature animals all litter sizes)		
High	220	43.2
Low	68	72.1
Timing of insemination (hr after onset of estrus)		
13–24	28	14.3
25–36	31	38.7
37–48	40	62.5
49–96	26	80.8
Sheep		
Litter size		
1	88,658	49.2
2	317,787	48.7
3	61,514	48.8
Maternal age (years)		
1	62,185	48.3
2	89,286	48.7
3	101,129	49.0
4	215,359	49.5
Number of lambs weaned previous year		
0	98,759	48.7
1	146,767	49.3
2	205,809	49.1
3	16,624	48.5
Mean lamb weight previous season (kg)		
50	36,526	48.0
46–50	127,027	48.7
41–45	181,451	49.2
36–40	100,423	49.3
36	22,532	49.4
Habitat quality		
Good	137	38.6
Poor	135	59.2
Date of birth (days after January 1)		
<91	6,656	47.5
91–100	35,050	48.7
101–110	122,033	48.7
111–120	169,501	49.0
121–130	99,337	49.2
131–140	28,206	49.8
>140	7,267	49.4
Barbari goats		
Diet (energy)		
High	172	46.5
Medium	159	54.1
Low	84	67.9
Thoroughbred horses		
Litter size		
1	2,783	49.5
2	62	35.5
Cattle		
Litter size		
All births	1,642,713	52.6
Twin births	28,437	51.8
Date of birth (zebu)		
January	124	48.3
February	137	39.4
March	177	41.2
April	249	51.8
May	244	49.6
June	206	47.6
July	367	48.2
August	657	46.3
September	619	51.9
October	340	50.9
November	239	56.5
December	114	60.5
Timing of insemination		
Early estrus	312	42.9
Middle estrus	125	53.6
Late estrus	121	63.6
Barbary macaque		
Maternal parity previous breeding season		
All animals	378	49.3
Nonproducing animals	26	27.0
Mice (laboratory)		
Diet		
Control	119	50.4
Low-fat	41	24.4
Hamsters (laboratory)		
Diet (*ad libitum*)		
100%	305	49.6
75%	549	40.7
Rats (laboratory)		
Maternal stress		
Physical constraint		
Unstressed	142	50.7
Stressed	106	35.8
Temperature		
5°C	448	44.0
22°C	464	51.7

sex ratio, including suggestions that the ingestion of sweet foods will increase the chances of having a female, whereas the ingestion of sour foods will produce males. Even less demanding methods include wearing boots to bed if a boy is desired or hanging the pants of the male on the right side of the bed if a son is wanted and on the left for a daughter. None of these prescriptions are of value and are harmless. The ancient Greeks believed that sperm from the left testicle produced girls, whereas sperm from the right testicle produced boys. Thus, French noblemen who wanted a male heir were advised, even into the late 1700s, to have their left testicle removed. There is no truth to this theory, nor to the theory that the sex of the child is dependent on which ovary produced the ovum. What is scientifically established is that sex determination is contained within the X-bearing and Y-bearing chromosomes which reside in mammalian sperm; the ovum contains only the X chromosome. Fertilization of an ovum by an X-bearing sperm produces an XX (female) individual, whereas union by a Y-bearing sperm results in an XY (male) individual. The genetic sex of a sperm which penetrates an ovum seems to be randomly determined. In all mammalian species studied to date, there have been roughly equal numbers of X- and Y-bearing sperm.

The ability to alter sex ratios would have important implications for both animal and human reproduction. Historical hypotheses and experimentation in the field of sperm separation focused on differences in the weight or density of sperm, the size of the X- and Y-bearing chromosome, or on the purported differences in haploid expression of the two sex chromosomes. In the first instance, the difference depends on the quantity of the material carried by the sperm and in the second on changes in the sperm induced by the contents of the material. None of the previous methods dating back three decades, using a variety of mammalian species for experimentation, resulted in a significant alteration of the secondary sex ratio.

Two methods that focus on either the difference in DNA content between the X- and Y-bearing sperm or on the greater swimming velocity of the Y-bearing sperm have been established to alter the secondary sex ratio of some mammalian species. These methods will be discussed separately.

A. Albumin Separation of Sperm

This method of isolating human sperm for the purpose of influencing the sex ratio is based on the concept that Y-bearing sperm swim downward with greater velocity when placed on top of a vertical column that contains a viscous medium (Fig. 1). Research with this concept started in the early 1970s and has progressed to a level of extensive clinical use. There is scientific debate as to the mechanism which produces an altered sex ratio that routinely results in 75–80% boys being born from sperm isolated via the albumin columns (Fig. 2). One group has DNA data that support the concept that the Y-

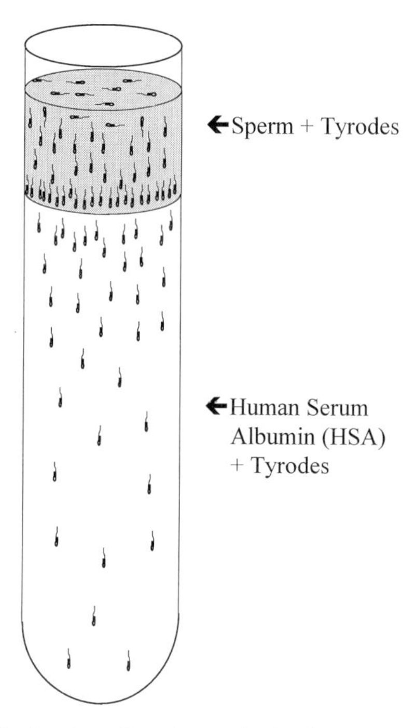

FIGURE 1 In the albumin method of sperm separation, sperm are placed on top of a glass column containing liquid albumin (step 1). Sperm recovered from the albumin layer in step 1 are resuspended over two increasingly higher concentrations of albumin. After the two-step protocol lasting 1 hr 45 min, the sperm remaining above the albumin layer are discarded along with most of the albumin. Sperm then recovered from the residual bottom layer of albumin have been shown to be more than 80% Y-bearing sperm both by DNA tests and by subsequent artificial insemination.

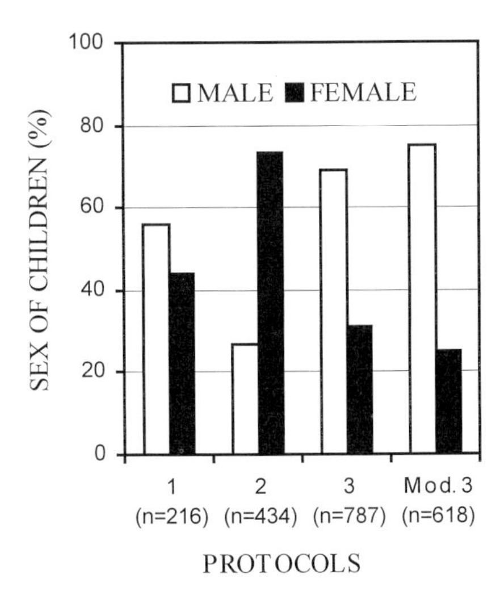

FIGURE 2 The percentage of children born male or female after sperm have been isolated with various albumin separation protocols for the purpose of skewing the sex ratio. Protocol 1, infertile couples (sex of children provided as a comparison to other protocols); protocol 2, female sex selection; protocol 3, male sex selection; modified protocol 3, male sex selection (modification to enhance the chances of male offspring); *n*, the number of children born for each protocol.

bearing human sperm has greater velocity (compared to X-bearing sperm) when subjected to high concentrations of human serum albumin (HSA) and allowed to swim downward with gravity in vertical columns. The enrichment of Y-bearing sperm is documented by both indirect DNA analysis of the fraction of sperm used for insemination and by the direct evidence of the sex of the offspring. Another group accepts that the direct evidence supports that a statistically significant skewness of the sex ratios in favor of males occurs when human sperm are isolated by the albumin columns, but their indirect DNA evidence based on enrichment of Y-bearing sperm can not be confirmed. Both groups do accept that there is good evidence that the human X-bearing sperm is significantly larger and longer than the Y-bearing sperm. The Y-bearing sperm is also known to have a slightly greater swimming velocity when suspended in a medium with a viscosity similar to water. It is therefore possible to postulate that the mechanism for the enrichment of human Y-bearing sperm that occurs in the final fraction of the albumin column involves the two physical parameters of the sperm that differentiate the X- and Y-bearing sperm: namely, the smaller surface area and greater swimming velocity of the Y-bearing sperm. When motile X- and Y-bearing sperm are placed in the viscous HSA medium and allowed to swim with gravity, the differences of size and swimming speed are magnified. This difference can be calculated mathematically; if the recovery of total motile sperm is ≧22% of the initial total motile sperm, there will not be an enrichment of Y-bearing sperm. In fact, the enrichment of Y-bearing sperm goes up as the yield of total motile sperm declines.

The methodology used to significantly increase the expected percentage of human females born from 48.5% to around 73% is more complex than the procedure used to increase the percentage of males and also less well understood (Fig. 2). The motile sperm for female selection are subjected to the same albumin sperm isolation technology with one major change. The recovery of total motile sperm is greater than the 22% mentioned previously and therefore no skewness of X- from Y-bearing human sperm can occur. What does occur in the final fraction of isolated sperm is a population of progressively motile sperm with normal morphology and that are highly fertile. For reasons that remain unexplained, a combination of these isolated sperm inseminated into a woman that has been induced to ovulate with clomiphene citrate (a drug used routinely to hormonally induce ovulation in women that do not ovulate or ovulate irregularly) will produce children that are predominately female. The use of clomiphene citrate does increase the incidence of twins up to 8 or 9%. Twin births have approximately the same percentage females as do the children from single births, and in more than 90% of women with twins at least one of the twins was a female. One popular theory has been that timing of insemination of sperm is responsible for the alteration of the secondary sex ratio. This theory is not supported by results from the sperm isolated on albumin columns for either male or female selection in that insemination is done during the same time of the menstrual cycle regardless of the sex of child desired.

The hormonal induction of ovulation alone statistically changes the percentage of female children born from slightly less than 50% to the mid-50s%.

While this change is not of clinical significance, it is of important biological significance in that it is now known that a hormonal preparation used to bring about ovulation can also alter the normal secondary sex ratio. The second piece of important information is that when induced ovulation is incorporated with artificial insemination, using sperm isolated via the albumin column, the percentage of female children born increases 25% from the expected percentage. Why this increase occurs remains unknown. What is known is that when induced ovulation or insemination with isolated sperm is used separately, this increase in the percentage of females born does not occur. Several hypotheses have been proposed, but none have been subjected to experimental tests. It is apparent that the sperm used for insemination are not a random population of sperm in that these sperm have been preselected for certain important parameters that enhance fertility. Also, the induction of ovulation with a pharmaceutical medication does alter the hormonal environment of the female reproductive system. One clear indication of this influence is that more follicles on the ovary ovulate as measured by the increased rate of twinning. This field of sex selection is still capable of springing biological surprises, and the fact that the combination of isolated sperm in conjunction with induced ovulation brings about a significant skewness toward females is one of the surprises.

B. Flow Cytometry for DNA Analysis of X- and Y-Bearing Sperm

The flow cytometer, also known as a cell sorter, is an instrument used by cell biologists to sort mammalian cells into various fractions based on certain cell parameters. In the case of sorting sperm based on whether or not a sperm has either an X- or Y-bearing chromosome, the amount of DNA contained in an individual sperm is what is measured. It is the difference in the amount of DNA between X- and Y-bearing sperm that allows for the flow cytometer to sort sperm into two fractions. The amount of DNA in the X chromosome is 2.8–4.2% greater than in the Y chromosome, dependent on species. The greater the difference in the amount of DNA between X- and Y-bearing chromosomes the greater the ability to accurately sort the X- and Y-bearing sperm. This technology is now standardized within research facilities to the degree that skewing the sex ratio to 85–90% for either males or females is effective in cattle, pigs, sheep, and rabbits. The effectiveness in humans is on average 75–80% for skewness of the sex ratio in either the X or the Y direction. The lower efficacy rate for humans is due to the smaller difference between the amount of DNA in the X- and Y-bearing chromosomes (2.8%), which is less than the difference in most domestic mammals.

Sperm from any species, including birds, can be analyzed by flow cytometry. A bimodal distribution representing X- and Y-bearing sperm occurs with mammalian species, whereas with birds only a unimodal peak occurs because all sperm contain the same sex chromosome. Successful sorting of sperm depends on the amount of DNA difference, morphology of the sperm, orientation of the sperm during flow cytometric analysis, and the number and viability of the sperm sample. In order to distinguish spermatozoal DNA, the sample is incubated with the *bis*-benzimidazole dye Hoechst 33342, which is a flurochrome that stains the DNA. The sperm are then oriented in the flow cytometric chamber that is equipped with an excitation laser beam that activates the fluorescence of the dye and allows measurement of differences in the quantity of DNA between the X- and Y-bearing sperm. The sperm with less or more DNA are isolated into two fractions. These sperm fractions supposedly contain almost pure populations of X- or Y-bearing sperm. The sex of the offspring resulting from insemination of various mammalian species with flow cytometer sorted sperm has shown that there is enrichment for either X- or Y-bearing sperm using this procedure.

The technology of isolating sperm with the flow cytometer has been available, at different levels of practical use, for more than 10 years. The practical uses of separated sperm are only now beginning to be established in commercial agriculture and in clinical medicine. Other advances in reproductive technology have diminished the need for large numbers of viable sperm for artificial insemination. Nonetheless, sperm isolated via flow cytometry remains a slow process and requires highly trained personnel in the use of expensive equipment. The more problematic area for this technology on a widespread use is the concern that the dye used to stain the chromosomes for identification of differences in DNA amounts may be mutagenic (damage the chromosomes). There have been reports that this dye (Hoechst 33342) has caused chromosomal damage when it was used to stain cells. However, all of the more than 400 resultant offspring from four species, including five generations of rabbits, have been observed to be normal by anatomical criteria. Further developments are needed to eliminate the concern over potential mutagenic damage and to simplify the use of this instrumentation prior to widespread use of this technology for X- and Y-bearing sperm separation. On a practical basis to ensure reasonable rates of conception, the numbers of sperm recovered must be increased significantly and the time required to collect the sperm significantly shortened.

See Also the Following Articles

SEX CHROMOSOMES; SEX DETERMINATION, GENETIC

Bibliography

Beernink, F. J., Dmowski, W. P., and Ericsson, R. J. (1993). Sex preselection through albumin separation of sperm. *Fertil. Steril.* **59**, 382–386.

Bennett, N. G. (1983). Sex selection of children: An overview. In *Sex Selection of Children* (N. G. Bennett, Ed.). Academic Press, New York.

Clutton-Brock, T. H., and Iason, G. R. (1986). Sex ratio variation in mammals. *Q. Rev. Biol.* **61**(3), 339–374.

Ericsson, S. A., and Ericsson, R. J. (1992). Couples with exclusively female offspring have an increased probability of a male child after using male sex preselection. *Hum. Reprod.* 7(3), 372–373.

Glass, R. H., and Ericsson, R. J. (1982). *Getting Pregnant in the 1980s.* Univ. of California Press, Berkeley.

James, W. H. (1987a). The human sex ratio. Part 1: Review of the literature. *Hum. Biol.* 59(5), 721–752.

James, W. H. (1987b). The human sex ratio. Part 2. A hypothesis and a program of research. *Hum. Biol.* 59(6), 873–900.

James, W. H. (1996). Evidence that mammalian sex ratios at birth are partially controlled by parental hormone levels at the time of conception. *J. Theor. Biol.* **180**, 271–286.

Johnson, L. A. (1994). Isolation of X- and Y-bearing sperm. In *Oxford Reviews of Reproductive Biology* (H. M. Charlton, Ed.). Oxford Univ. Press, Oxford, UK.

Sex Skin

Fred B. Bercovitch
Caribbean Primate Research Center

GLOSSARY

consort relationship A temporary heterosexual bond characterized by male mate guarding of a sexually receptive female which is usually found in multimale mating systems.

mating system The social structure resulting from sexual selection yielding variations in heterosexual bonding patterns.

ovarian steroids Hormones produced by the ovaries that regulate the appearance of the sex skin.

secondary sex characteristic A trait that is expressed only in one sex that is not directly associated with the act of reproduction but functions in a reproductive context.

sexual selection An evolutionary process linked to the expression of secondary sex characteristics and consisting of nonrandom mating due to competition for mates and choice of mates.

Sex skin is a secondary sex characteristic consisting of a specialized region of skin that alters in appearance concurrent with fluctuations in sex steroid hormones, specifically estradiol and progesterone. The sex skin most often occurs in the anogenital region and the area surrounding the ischial callosities of female Old World monkeys and apes. The physiological regulation of sex skin activity has been well established, but the evolutionary basis for the origin of this trait is subject to debate. Most primate species possessing large, protuberant sex skin swellings reside in multimale mating systems, but the adaptive significance of this linkage is unclear. The two foremost issues remaining to be resolved in studies of the sex skin are (i) the degree to which changes in the appearance of the sex skin reflect underlying hormonal states that can be monitored and detected by males, and (ii) why some species possess a pronounced ornamental advertisement of proximity to ovulation, whereas others lack obvious visual cues indicating the periovulatory period.

I. DESCRIPTION OF THE SEX SKIN

The sex skin seems to be an extension of the naked vulval and clitidoral region in primates. Hypertrophy of this skin area in response to elevated estradiol during the follicular phase of the ovarian cycle is common. Concurrent with increases in size, the red or pink color often intensifies. Some primates, e.g., gibbons (*Hylobates lar*), develop slight enlargements of the labia minora during the periovulatory period, whereas other primates, e.g., chimpanzees (*Pan troglodytes*), produce bulbous swellings of the sex skin that are the size of a basketball. Sex skin swellings are confined to adolescent rhesus macaques (*Macaca mulatta*), whereas they occur throughout life in savanna baboons (*Papio cynocephalus*). Gelada monkeys (*Theropithecus gelada*) not only have a sex skin in the anogenital region but also the sex skin of their chest produces an edematous beaded necklace in response to ovarian steroids. The sex skin of sooty mangabeys (*Cercocebus torquatus*) increases in size after conception as well as during the periovulatory phase.

A single description of the sex skin is impossible. Primate species vary in the extent to which changes in coloration and enlargement of the sex skin coincide with the periovulatory period. The appearance of the sex skin also varies across individuals and changes with age. In general, younger females tend to have more fully formed, rounded, and larger sex skin swellings than older females. Although the sex skin of primates is marked by variation in structure, size, and color, it functions in all species as a visible advertisement of reproductive state. The origin of sex skin swelling remains contested, but sex skin swelling is linked to circulating concentration of ovarian steroid.

II. PHYLOGENY AND DISTRIBUTION OF THE SEX SKIN

Conspicuous swelling of a sex skin that is visible from a distance characterizes about 15% of extant primate species. Only Old World monkeys and apes possess sex skins that billow in size during the follicular phase of the cycle. Many prosimian and New World monkey species experience vulval pinkening and swelling during the periovulatory period, but none of them have a distinct sex skin. These species are characterized by a well-developed olfactory communication system, and, among primates, an inverse association tends to occur between strength of visual and olfactory signaling channels. While some species of African colobines, e.g., red colobus (*Colobus badius*) and olive colobus (*C. verus*), exhibit cyclic changes in sex skin activity, only one species of Asiatic colobines, i.e., the pig-tailed langur (*Simias concolor*), possesses a sex skin. Phylogenetic analysis suggests that pronounced swelling of the sex skin has arisen at least three or four times in the history of primate evolution.

Species residing in multimale mating systems are significantly more likely to possess sex skins that undergo large, cyclical fluctuations in size than are primate species residing in monogamous or unimale mating systems. Not all species characterized by multimale mating systems have females with pronounced sexual swellings, but all primate species characterized by monogamous mating systems lack pronounced sex skin swellings.

Protuberant sex skin swellings also predominate among terrestrial primates, which also tend to be larger than arboreal primates. However, the tiny, arboreal talapoin monkey, *Cercopithecus talapoin*, which weighs about 1 kg, possesses a sex skin that ballons in size during the follicular phase of the cycle. Most primates reproduce on a seasonal basis, but the extent of reproductive seasonality does not closely correspond with degree of sex skin swellings. Monkeys with a fairly confined birth period, e.g., Barbary macaques (*Macaca sylvanus*), have enormous sex skin swellings during the mating season, whereas those that give birth throughout the year, e.g., savanna baboons, have large sex skin swellings in all months of the year. Finally, sex skin swellings are found in desert-dwelling species, such as hamadryas baboons (*P. hamadryas*), as well as in rain forest-dwelling species, such as gray-cheeked mangabeys (*Lophocebus albigena*).

To summarize, sex skin swellings have emerged during the course of primate evolution at least three or four times, and their origin is not associated with differences in body size, degree of terrestriality, extent of mating seasonality, or type of habitat niche. Possession of a sex skin that balloons in size during the follicular phase of the cycle is most likely to have emerged due to social factors. Pronounced sex skin swellings are significantly more likely to occur among primates living in multimale mating systems than in other types of breeding arrangements.

III. ENDOCRINE REGULATION OF THE SEX SKIN

The sex skin is replete with estrogen and progesterone receptors. The progressive increase in size of sex skin swelling is driven by rising estrogen concentrations, whereas deturgescence of sex skin swelling is instigated by elevations in progesterone concentrations. Progesterone induces estrogen withdrawal from the sex skin target tissues as well as suppresses estrogen receptor levels. Administration of estrogen to ovariectomized females produces swelling of the sex skin, whereas progesterone injections to fully swollen females result in rapid deturgescence of the sex skin swelling. The precise size of sex skin swelling is not a reliable indicator of circulating ovarian steroid levels, but the maximum sex skin swelling size is achieved concurrent with peak estradiol levels. The development of sex skin turgescence and deturgescence during the menstrual cycle follows a pattern that mirrors circulating steroid levels (Fig. 1).

Ovulation is most probable 1 or 2 days prior to onset of deflation of the sex skin. The relatively tight linkage between onset of detumescence and ovulation is due to the hormonal mechanisms responsible for expansion and contraction of the sex skin. A sustained interval between ovulation and deflation of sex skin swelling is unlikely because ovulation follows achievement of peak estrogen levels. The preovulatory plummeting of estrogen levels, combined with rapid discharge of progesterone, has a fairly swift effect on decreasing the swelling of the sex skin.

Among savanna baboons, attainment of maximum sex skin size requires about 20 days, whereas return to the nonswollen condition only takes about 5 days. Maximum sex skin swelling size in this species usually persists for 5 days or less, with the greatest volume of sex skin swelling achieved on the final 2 days prior to deturgescence. Size of sex skin swellings can be estimated using detailed observations under

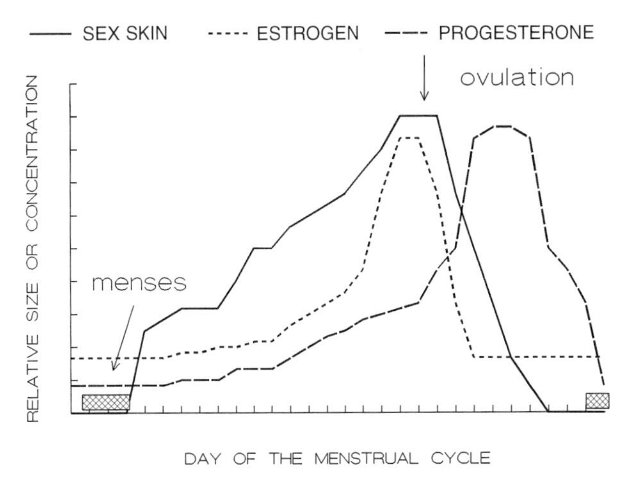

FIGURE 1 A schematic representation of the relationship between circulating ovarian steroid concentrations and the size of sex skin swelling during the menstrual cycle. The steroid concentrations are not drawn to the same scale, but they indicate the relative changes throughout the menstrual cycle.

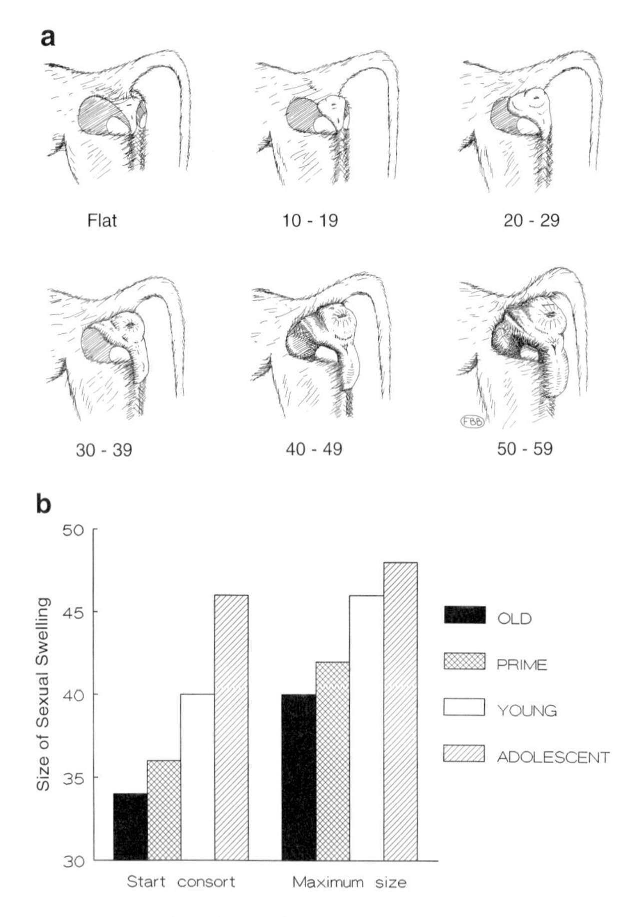

FIGURE 2 The relationship between sex skin size, onset of mating activity, and female age in savanna baboons. (a) A quantitative scale for recording changes in sex skin size. Estimates are made in increments of one unit, with the maximum size dependent on female age. (b) The interaction among size of sex skin, mating activity, and maximum age-related sex skin size. About 5 days separates onset of consortship activity from attainment of peak sex skin swelling size (from F. B. Bercovitch, Reproductive tactics in adult female and adult male olive baboons, PhD thesis, University of California at Los Angeles, 1985, and unpublished data. Drawings by the author).

field conditions based on a quantitative scale (Fig. 2a). A zenith in sexual activity occurs during the period of maximum swelling of the sex skin, and some evidence indicates that copulatory activity among primates is dependent on surpassing a minimum estrogen level and/or a minimum age-related sex skin size (Fig. 2b). For example, the initial day of consortship activity between adult savanna baboons occurs when females are within four points of their own modal maximum size in 80% of cycles, and, in 67% of cases, this is the first day, or the day after, they obtain this size.

The protrusion of the sex skin arises from the accumulation of intercellular fluids, but the resultant size of swelling is not correlated with the amount of water retained by this secondary sex characteristic. Increases in estrogen concentrations are associated with elevations in hyaluronic acid levels which foster edema of the sex skin. A plexus of thin-walled blood vessels lies superficial to the epithelium of the sex skin, which is formed by a thick epidermis resembling the volar surfaces of hands and feet. Incisions of the sex skin produce a gellatinous ooze, with little bleeding, and wounds to the sex skin heal quite rapidly. Water retention during sex skin expansion is alleviated during the initial detumescent phase, when urinary excretion increases in volume. Growth of the sex skin adds 10–20% to body weight, but feeding time accompanying the follicular phase growth of the sex skin substantially declines. Mechanisms regulating energy balance during this period of reduced food intake alongside weight gain due to sex skin swelling remain to be examined.

To summarize, estrogen fosters growth of sex skin swelling, whereas progesterone counteracts that growth. Swelling of the sex skin is an outcome of water retention due to estrogen-facilitated increases in hyaluronic acid levels. The size of sex skin swelling is coordinate with changes in ovarian steroid levels and provides a reasonable indicator of proximity to ovulation, but sex skin size is not a valid quantitative index of circulating steroid levels. Ovulation is most probable on the final 2 days of sexual swelling, when the sex skin achieves its greatest dimensions. Reproductive activity is closely correlated with size of sex skin swelling.

IV. SEXUAL BEHAVIOR AND THE SEX SKIN

Although some primates mate throughout the ovarian cycle, sexual activity peaks during the periovulatory phase of the cycle. Among species with sex skin swellings, male aggression over access to females and male sexual activity with females rise concomitant with increases in sex skin swelling. A number of primate species that are characterized by sex skin swellings (e.g., savanna baboons, gelada monkeys, and chimpanzees) also exhibit multiple menstrual cycles preceding conception; the causes of this phenomenon are not well understood.

Neither the duration of sex skin swelling within a cycle nor the maximum swelling size achieved differ between conception and nonconception cycles in savanna baboons. Similarly, analysis of fecal steroids from savanna baboons living in Tanzania detected no significant differences between conception and nonconception cycles in levels of excreted steroids during the final 7 days of the follicular phase. The number of male partners and rates of ejaculation are also indistinguishable between conception and nonconception cycles among savanna baboons in Kenya. Finally, consort relationships are established for a period of 5–7 days on both conception and nonconception cycles.

Two differences have been reported between conception and nonconception cycles. First, the initial postlactational cycle of savanna baboons tends to be longer or shorter than subsequent cycles and is characterized by a maximum sex skin swelling size that is significantly smaller than that achieved on the conception cycle. Recrudescence of sex skin swelling after weaning results from reactivation of ovarian steroid production, and the initial cycle is often anovulatory. Consort relationships are frequently absent during the initial postlactational cycle. Second, some field investigators have reported that dominant males are more likely to establish consort relationships on conception cycles than on nonconception cycles, although others have failed to find this association. The cause of this discrepancy among studies is unclear, but dominant males could not be detecting a conception cycle in advance because conception is a consequence of mating activity. Mate guarding by a dominant male could reduce the frequency of challenges by subdominant males in a social situation characterized by a stable dominance hierarchy. These circumstances could lessen the degree of stress suffered by the female and increase her chances of conception. The role of stress among wild nonhuman primates as a potential factor affecting the probability of conception has not been addressed.

Swelling of the sex skin provides cues to males about the likelihood of ovulation. In a series of classic experiments, Craig Bielert evaluated the relative importance of sight, sound, and smell as cues arousing sexual interest in male savanna baboons. His initial work pinpointed the visual stimulus as the primary cause of male arousal, and he followed this up with two clever experiments aimed at determining how size and color of sex skin swellings might affect male sexual arousal. Although neither trait varies in tandem with specific endocrine concentrations, both traits change in appearance contemporaneous with systemic changes in ovarian steroid levels.

In his first experiment, Bielert found that male sexual arousal was greater when males were exposed to females with supernormal-sized swellings compared to those with standard-sized swellings. Similar findings emerged from a field study of savanna baboons. Fred Bercovitch reported that males were significantly more likely to alter consort partners from one with a smaller sized sex skin swelling to one with a larger sized sex skin swelling than to shift in the opposite direction. The prime targets of male sexual activity in the Sulawesi crested black macaque (*Macaca nigra*) and the bonobo or pygmy chimpanzee (*Pan paniscus*) are females at or near their maximum sex skin swelling size.

In his second experiment, Bielert attached a thermoplastic model of a swollen sex skin to estradiol-primed, ovariectomized females to gauge male response to color. The artificial sex skin was displayed to males in eight colors: white, red, orange, yellow, green, blue, purple, and black. Males were sexually aroused by all colors, but the red model evoked the highest level of responsiveness. Female behavior toward males was independent of color of their attached sex skin, thereby eliminating female behavioral cues as potential confounding factors explaining the results. Based on both the experimental work and the field studies, one can conclude that size and color of sex skin function as cues to males that modify their sexual and aggressive behavior.

To summarize, maximum sex skin swelling size corresponds to attainment of peak estrogen levels and is closely linked with ovulation. Sex skin swellings provide strong visual cues in both size and color dimensions that stimulate male reproductive behavior as well as male competitive behavior for access to females. The best example of a relationship between sexual behavior and the sex skin comes from numerous studies of savanna baboons in which male aggressive competition over access to females peaks on the final 3 days of sex skin swelling, a period closely matching the length of time that sex skin swellings are at their maximum.

V. THE ADAPTIVE SIGNIFICANCE OF THE SEX SKIN

Regular enlargements of the sex skin have been convincingly described for only four species residing in unimale mating systems: gelada monkeys, hamadryas baboons, mandrills (*Mandrillus sphinx*), and pig-tailed langurs. The remaining primate species with pronounced sex skin swellings live in multimale mating systems. Of the four species characterized by unimale mating systems, three of them (i.e., geladas, hamadryas, and mandrills) reside in unimale systems that are embedded in larger multimale social systems. Retention of sex skin swelling in these unimale species is likely to be a consequence of the maintenance of an ancestral trait.

The absence of a sex skin in Cercopithecines residing in multimale mating systems occurs in only a few species. The vervet monkey, *Cercopithecus aethiops*, lacks a sex skin, but it is an exception to the general guenon monkey lifestyle of living in unimale groups in an arboreal environment. The hanuman langur, *Presbytis entellus*, is one of the few Asiatic colobines found in multimale mating systems, but it is also frequently found in unimale mating systems. The most mysterious group is the sinica subgroup of macaques. These four species (*M. sinica, M. radiata, M. assemensis,* and *M. thibetana*) fail to display edema or color changes during the follicular phase of the cycle, whereas the other 15 macaque species regularly display some degree of color change or swelling during the periovulatory period. Reasons for the disappearance of cyclical sex skin changes in this subgroup are uncertain, but Alan Dixson has noted that an inverse relationship in macaques exists between the degree of sex skin development and the complexity of vaginal, cervical, and penile morphology.

Unimale species exhibiting sex skin swelling tend to reside within larger multimale social systems, whereas multimale species lacking sex skin swelling are either recently established as multimale groups or, in the macaques, have apparently lost the sex skin development in association with development of alternative morphological traits. These patterns, combined with the tight association between sex skin development and residency in a multimale mating system, suggest that the adaptive significance of sex skin swellings is related to the role of sexual selection in the formation of multimale mating systems. However, the precise evolutionary mechanism yielding a pronounced morphological display broadcasting proximity to ovulation in multimale groups is controversial. One camp emphasizes female choice combined with male competition, whereas the alternative school stresses male choice coupled with female competition. Subdivisions also exist within the first camp regarding which male trait is favored by female choice.

Tim Clutton-Brock and Paul Harvey, followed by R. Haven Wiley and Joe Poston, have championed the idea that sex skin swellings have evolved as a mechanism of female choice which creates conditions inciting intense male competition. Females then strive to mate with the victorious male, which is likely to be the dominant or "best" male in the group. Alan Dixson, pursuing a suggestion made in 1925 by Pocock, suggests that large sex skin swellings create a morphological winnow favoring large penises. On average, large penises will belong to bigger males. Smaller males might be able to mate with fully swollen females, but their diminutive penises would not be able to deposit sperm high enough in the female reproductive tract to enable them to successfully impregnate the female. Paul Harvey has speculated that enlarged sex skin swellings are a form of female choice that create a channel to intensify sperm competition. The heightened sperm competition establishes a sieve that enables females to conceive with "good" sperm produced by "high-quality" males. Sarah Hrdy has reasoned that the conspicuous sex skin swellings obscure the precise timing of ovulation, which then functions to attract, and permit, mating with multiple males. Sexual relationships with many males reduces the risk of male infanticide due to paternal uncertainty.

Bill Hamilton arrived at the same position as Hrdy, but based his conclusion on an opposite perspective.

He surmised that the bulbous sex skin swellings are adaptive because they pinpoint, rather than obscure, the timing of ovulation. By advertising proximity to ovulation, females enhance their chances of mating with a single male, which would increase paternal certainty and reduce the risks of infanticide by increasing the extent to which the male mate provides paternal care. Fred Bercovitch, and later Mark Pagel, developed the idea that prominent sex skin swellings indicate fecundability and female "quality." Body ornaments functioning as secondary sex characteristics in one sex are often a consequence of mate choice patterns by the opposite sex. Advertising proximity to ovulation serves as a beacon to male choice, which can be a significant factor in enhancing female reproductive success. Male affiliative relationships with the female's offspring that emerge based on consort relationships could increase infant survivorship prospects.

To summarize, the evolution of a sex skin in primates is linked to the emergence of multimale mating systems. Sexual selection can account for the origin of this conspicuous secondary sexual characteristic among females, but the mode of operation of sexual selection is unclear. The adaptive significance of a sex skin is connected to patterns of mate choice in multimale species of primates, but the precise connection between producing a large sex skin swelling and enhanced reproductive output is unknown. The delicate interplay between male and female choice seems to be a prime factor responsible for the appearance of the sex skin in primates.

See Also the Following Article

PRIMATES, NONHUMAN

Bibliography

Aidara, D., Badawi, M., Tahire-Zagret, C., and Robyn, C. (1981). Changes in concentrations of serum prolactin, FSH, oestradiol and progesterone and of the sex skin during the menstrual cycle in the mandabey monkey (*Cercocebus atys lunulatus*). *J. Reprod. Fertil.* **62**, 475–480.

Bielert, C. (1982). Experimental examinations of baboon (*Papio ursinus*) sex stimuli. In *Primate Communication* (C. T. Snowdon, C. H. Brown, and M. R. Petersen, Eds.), pp. 373–395. Cambridge Univ. Press, Cambridge, UK.

Bielert, C., and Andersson, C. M. (1985). Baboon sexual swellings and male response: A possible operational mammalian supernormal stimulus and response interaction. *Int. J. Primatol.* **6**, 375–391.

Bielert, C., Girolami, L., and Jowell, S. (1989). An experimental examination of the colour component in visually mediated sexual arousal of the male chacma baboon (*Papio ursinus*). *J. Zool.* **219**, 569–579.

Bullock, D. W., Paris, C. A., and Goy, R. W. (1972). Sexual behaviour, swelling of the sex skin and plasma progesterone in the pigtail macaque. *J. Reprod. Fertil.* **31**, 225–236.

Darwin, C. (1876). Sexual selection in relation to monkeys. *Nature* **15**, 18–19.

Dixson, A. F. (1983). Observations on the evolution and behavioral significance of "sexual skin" in female primates. *Adv. Stud. Behav.* **13**, 63–106.

Graham, C. E. (1981). Menstrual cycle of the great apes. In *Reproductive Biology of the Great Apes* (C. E. Graham, Ed.), pp. 1–43. Academic Press, New York.

Hrdy, S. B., and Whitten, P. L. (1987). Patterning of sexual activity. In *Primate Societies* (B. B. Smuts, D. L. Cheney, R. M. Seyfarth, R. W. Wrangham, and T. T. Struhsaker, Eds.), pp. 370–384. Univ. of Chicago Press, Chicago.

Shaikh, A. A., Celaya, C. L., Gomez, I., and Shaikh, S. A. (1982). Temporal relationship of hormonal peaks to ovulation and sex skin deturgescence in the baboon. *Primates* **23**, 444–452.

Sillen-Tullberg, B., and Moller, A. P. (1993). The relationship between concealed ovulation and mating systems in anthropoid primates: A phylogenetic analysis. *Am. Nat.* **141**, 1–25.

Thomson, J. A., Hess, D. L., Dahl, K. D., Iliff-Sizemore, S. A., Stouffer, R. L., and Wolf, D. P. (1992). The Sulawesi crested black macaque (*Macaca nigra*) menstrual cycle: Changes in perineal tumescence and serum estradiol, progesterone, follicle-stimulating hormone, and luteinizing hormone levels. *Biol. Reprod.* **46**, 879–884.

Sexual Attractants

Lee C. Drickamer
Southern Illinois University at Carbondale

GLOSSARY

chemosignal Any chemical substance that produces physiological or behavioral changes in animals. Though this term was often used during the 1970s and 1980s, it has generally been superseded by the use of the term pheromone.

pheromone A species-specific chemical cue released by animals that influences the physiology and behavior of conspecifics. The original criteria for labeling a chemical substance a pheromone included (i) species specificity, (ii) having a well-defined behavioral or endocrinological response function, and (iii) the uniqueness of the compound or a small set of compounds producing the response. Some would also include a fourth criterion—a significant genetic component. Since not all these criteria are generally met by the substances that have been explored in mammalian systems, a broader term has often been used—chemosignal.

Sexual attractants are chemical stimuli deposited via urine, feces, or marking behavior using scent glands. These substances function for species, sex, and, in some instances, individual identification, for informing potential mates regarding reproductive condition, and possibly for mate selection.

I. GENERAL COMMENTS

Chemical stimuli (herein termed pheromones) have a variety of functions in relation to social behavior in many animals. Primer pheromones are those that induce physiological changes in a recipient organism and act over time, often eventually resulting in behavioral changes. Releasing or signaling pheromones are those that result in a relatively rapid change in physiology and behavior in the recipient organism. Sexual attractants are releasing pheromones that provide information to potential mates concerning species, sex, age, and reproductive status.

There is a long history of study of chemically mediated sociospatial behavior in both invertebrates and vertebrates. Among vertebrates, work has now been carried out on pheromones for more than a century. Mammalian pheromone are involved in spatial relations as territory markers, for orientation, for maternal–young relations, with respect to modulation of sexual development and puberty, and with regard to mating and reproduction. In this entry, three aspects of mammalian pheromones that act as sexual attractants are reviewed: (i) sources of the attractant chemicals, (ii) the chemical nature of these attractants, and (ii) functional aspects of these pheromones. It should be noted that for many organisms, including mammals, the same pheromone can have different meanings (effects) in different circumstances. Thus, for example, scents contained in fecal material may act as a repellent for much of the year but could serve as an attractant during the mating season. In the same manner, substances deposited on the substrate or on objects in the environment, such as tree branches, often serve to mark a territory. These same marks may act as orientation cues for the territory holder, its progeny, or its neighbors. Also, at particular times of the year, these same marks may be sexual attractants. Knowing the context for both deposition of the mark and the conditions surrounding its reception by other conspecifics can be important for understanding the interpretation of the message.

A number of mammalian species use both the olfactory epithelium and the vomeronasal organ for reception of odor cues. In general, the more volatile cues are received by the olfactory epithelium and nonvolatile cues are received via the vomeronasal organ. Pheromones that serve as sex attractants can be received via either pathway. The Flehmen behavior exhibited by a wide variety of mammals, ranging from felids to ungulates, involves a curling upward of the upper lip to expose the entrances of the passages that lead to the vomeronasal organ. Flehmen is commonly associated with reproduction, particularly with regard to males detecting estrous females. They often do this by placing their nasal region directly in a urine stream from a female and engaging in the Flehmen behavior pattern.

II. SOURCES OF ATTRACTANTS

There are at least six major sources of pheromones in mammals: skin glands, feces, saliva, respiratory breathing, vaginal secretions, and urine. Products from all these sources can serve as sexual attractants.

A. Skin Glands

The skin, composed of the epidermis and underlying dermis, is the largest organ of the mammalian body. There are two major categories of glands located in the skin. Sebaceous glands are most frequently, but not always, associated with hair follicles. Often, several sebaceous glands open into the same hair follicle. The secretory products, known collectively as sebum, vary widely in chemical composition, but all have a high lipid content. Rats of sebum secretion are generally greater in adult males than for females or young.

The second type of skin gland, tubular glands, consists of two subtypes, apocrine and eccrine glands; together they are often referred to as subodiferous or sweat glands. Their secretion is always watery rather than lipid in nature. Ducts from tubular glands open directly to the skin surface. Mammary glands are apocrine tubular glands. Though apocrine sweat glands are concentrated in axillary, perineal, and pubic areas in humans, they are found over much of the skin surface in many mammals. The characteristic foamy sweat seen on horses is from apocrine glands. Eccrine glands occur on the footpads of many species, for example, in felids and canids.

The next four categories of potential sources of attractants all involve glands derived from epithelial tissue. In each case, the pheromonal products also may involve products from internal organs. In addition to these sources, there are glands associated with the eyes and eyelids that appear to produce and release pheromones. The possible role of these glands as sex attractants remains to be tested.

B. Feces

Substances released in feces that act as pheromones involve both products secreted from glandular complexes in the perianal region and rectum and substances that are metabolites from other organs of the body that are being excreted. Processes that occur in the liver that result in bile production are implicated as a major source of the constituents of fecal pheromones. In some mammals, extensive release of gas, in the form of flatulence, occurs via the rectum and anus.

C. Saliva

Substances that appear to be involved in courtship activities among mammals are found from diverse groups, ranging from pigs to voles. Steroid metabolism, involving androgens, has been reported in males of a number of mammalian species. Hedgehogs deposit a foamy spittle substance on their spines during courtship. Many mammals groom their fur with the aid of saliva. Odors from the saliva contribute to the general body odor of the animal. Considerable work remains to be done on the social significance of saliva and the products that it contains.

D. Respiratory Breathing

Many mammals carry out extensive nose-to-nose sniffing during courtship and mating activities. There are three major sources of odors contained in expired air. One consists of products from bacterial microflora in the oral cavity and pharynx. A second source involves volatile substances that pass across the alveolar membranes in the lungs. Lastly, the sinuses and nasal passages, which are highly vascularized and extensive in terms of surface area, contribute to odors in respired air.

E. Vaginal Secretions

Glandular substances are released from the vaginal opening of numerous mammals. For mammals ranging from hamsters to cows, there are characteristic vaginal secretions associated with estrus cycles. The lining of the cervix contains numerous secretory cells. In addition, the female homologs of the male sex accessory glands release their products into the vagina.

F. Urine

Two major sources of potential pheromone cues contribute to the communications value of urine. Many compounds are filtered through the kidneys and released in the urine. Metabolites of steroid hormones are prominent features of the urine of many mammals. Many of these are sufficiently volatile to convey messages about age, sex, and reproductive condition. During the distinct mating seasons that occur in mammals outside the tropics, urine is a primary cue both for locating members of the opposite sex and for inducing mating activity and related internal physiological changes associated with reproduction. In male mammals, a number of sex accessory glands contribute their secretions to the urine. For males of many species, some urine remains in the prepuce after urination, where bacterial action results in highly odoriferous mixtures.

III. CHEMICAL NATURE OF ATTRACTANTS

Relative to the number of mammalian pheromones that are known from the glandular sources just outlined, very few have been approached from the perspective of determining their chemical composition. A rather lengthy sequence of steps is involved in this process: (i) an observed behavioral or physiological effect that appears to be pheromone mediated; (ii) finding the source to collect material that contains the pheromone; (ii) characterizing and pheromone with regard to whether it is protein, lipid, etc., coupled with fractionation and bioassays to ascertain which fractions contain pheromonal activity; (iv) separation of the active fraction(s) to identify its chemical constituents via procedures such as high-performance liquid chromatography with additional bioassays; and (v) synthesis and testing of the compounds or mixtures of compounds that are believed to constitute the active pheromone.

Using this procedure, or portions of it, several mammalian pheromones have been examined in some detail. Only a few of these may be classified as sexual attractants. One of the best known of these, copulin, is a volatile fatty acid, released from the vaginas of female rhesus macaques, and which attracts males. A number of pheromones affect the reproductive biology of house mice. Some of these influence the timing of puberty and the onset and regularity of estrus cycles. Other house mouse pheromones are involved in identification by age, sex, and reproductive condition and thus can be considered as sex attractants. The adrenal-mediated urinary pheromone that delays the onset of puberty in young female house mice was explored by comparing the urinary profiles of adrenalectomized and intact grouped female mice using high-resolution gas-phase chromatograms (Novtony *et al.*, 1986). The urinary pheromone from adult intact males that accelerates puberty in female house mice has been characterized as androgen dependent, heat labile, nondialyzable, precipitable in ammonium sulfate, and being in the protein fraction of the urine (Vandenbergh *et al.*, 1976).

There are several good examples of sexual attractant pheromones in larger mammals in which the chemical composition has been at least partly identified. Androstenol and andrestenone are both found in the saliva of male pigs. A lactone from the tarsal gland of the black-tailed deer may act as a sexual attractant. It is also possible that the product of the castor glands, called castoreum, in beaver may affect mate finding and/or mate selection. Castoreum contains, among its mixture of compounds, castoramine, which gives the castoreum its pungent odor. Many of these mammalian chemosignals, including those from glands as well as those in urine or feces, are apparently complex mixtures. This makes it rather difficult to obtain complete chemical profiles on these substances. Determining all the active substances may be quite difficult and time-consuming.

IV. FUNCTIONAL ASPECTS

Pheromones function in a wide variety of contexts in mammals. With particular regard to the use of pheromones as sexual attractants, the functional aspects of these substances may be divided into three phases. These phases are not mutually exclusive. First, locating a potential mate may involve the use of pheromones. Deposits of substances from glands, or in urine or feces, can contain information for a receiver concerning the species, age, sex, and reproductive condition of the animal that left the signal. Each of these levels of cue information is, in effect a filter. As females and males seek each other for mating, both may in fact use these types of cues.

Second, pheromones may play several roles with regard to courtship. In some species, variations in particular chemical cues that are released by a sender may communicate, for example, something about its social status or nutritional state. Such qualities may be of significance to potential mates of the opposite sex. The actual process of courtship, including its initiation and maintaining a sequence of events that results in mating, may be, in part, dependent on pheromonal cues that are released by one or both members of the courting pair. Also, the actual copulatory act may involve the release of particular pheromones. Further study is needed to elaborate these various functional roles for pheromones influencing courtship in mammals.

Third, commencing with copulation, pheromones may, in some species, be involved in maintaining some form of bond between the recently mated animals. These are not strictly sexual attractants, but they are very much a continuation of the use of chemical communication in terms of successful reproduction. In the minority of mammals that are monogamous, maintenance of the bond could be important for successful rearing of progeny. The majority of mammals are polygamous. However, that need not mean that pheromones are of little importance. Certainly, there are mother–infant relations that depend on pheromones. It is also quite likely that males of some species can recognize their own progeny using, in part, odor cues from pheromones. In these species as well, some increment of reproductive success must relate to the role of pheromonal cues.

See Also the Following Articles

Mating Behaviors; Olfaction and Reproduction

Bibliography

Albone, E. S. (1984). *Mammalian Semiochemistry*. Wiley, New York.

Bronson, F. H. (1970). Pheromonal influences on mammalian reproduction. In *Reproduction and Sexual Behavior* (M. Diamond, Ed.), pp. 341–361. Indiana Univ. Press, Bloomington.

Bronson, F. H. (1971). Rodent pheromones. *Biol. Reprod.* 4, 344–357.

Bronson, F. H. (1979). The reproductive biology of the house mouse. *Q. Rev. Biol.* 54, 265–299.

Brown, R. E., and Macdonald, D. W. (Eds.) (1985). *Social Odours in Mammals*, 2 Vols. Oxford Univ. Press, Oxford, UK.

Doty, R. L. (1976). *Mammalian Olfaction, Reproductive Processes, and Behavior*. Academic Press, New York.

Drickamer, L. C. (1986). Puberty-influencing chemosignals in house mice: Ecological and evolutionary considerations. In *Chemical Signals in Vertebrates, IV* (D. Duvall, D. Müller-Schwarze, and R. M. Silverstein, Eds.), pp. 441–455. Plenum, New York.

Drickamer, L. C. (1989). Pheromones: Behavioral and biochemical aspects. In *Advances in Comparative and Environmental Physiology, Vol. 3* (J. Balthazart, Ed.), pp. 269–348. Springer-Verlag, Berlin.

Drickamer, L. C. (1992). Chemosignals and reproduction in adult female house mice. In *Chemical Signals in Vertebrates VI* (D. Müller-Schwarze and R. L. Doty, Eds.), pp. 245–251. Plenum, New York.

Ferkin, M. H. (1990). Kin recognition and social behavior in microtine rodents. In *Social Systems and Vole Cycles* (R. H. Tamarin, R. S. Ostfeld, S. R. Pugh, and G. Bujalska, Eds.), pp. 11–24. Birkauser Verlag, Zurich.

Johnston, R. E. (1975). Sexual excitation function of hamster vaginal secretion. *Anim. Learning Behav.* **3**, 161–166.

Johnston, R. E. (1990). Chemical communication in golden hamsters: From behavior to molecules and neural pathways. In *Contemporary Issues in Comparative Psychology* (D. A. Dewsbury, Ed.), pp. 381–409. Sinauer, Sunderland, MA.

Michael, R. P., Bonsall, R. W., and Zumpe, D. (1976). Evidence for chemical communication in primates. *Vitamins Horm.* **34**, 137–186.

Novotny, M., Jemiolo, B., Harvey, S., Wiesler, D., and Marchlewska-Koj, A. (1986). Adrenal-mediated endogenous metabolites inhibit puberty in female mice. *Science* **231**, 722–725.

Shorey, H. H. (1976). *Animal Communication by Pheromones.* Academic Press, New York.

Steiner, J. E., and Ganchrow, J. R. (Eds.) (1982). *Determination of Behaviour by Chemical Stimuli.* IRL Press, London.

Stoddardt, D. M. (1990). *The Scented Ape.* Cambridge Univ. Press, New York.

Vandenbergh, J. G. (Ed.) (1983). *Pheromones and Reproduction in Mammals.* Academic Press, New York.

Vandenbergh, J. G., and Coppola, D. M. (1986). The physiology and ecology of puberty modulation by primer pheromones. *Adv. Stud. Behav.* **16**, 71–108.

Vandenbergh, J. G., Finlayson, J. S., Dobrogosz, W. J., Dills, S., and Kost, T. A. (1976). Chromatographic separation of puberty accelerating pheromone from male mouse urine. *Biol. Reprod.* **15**, 260–265.

Sexual Behavior

see Mating Behaviors

Sexual Dysfunction

Steven M. Petak

Texas Institute for Reproductive Medicine and Endocrinology

GLOSSARY

anorgasmia Inability to achieve orgasm.

dyspareunia Pain with intercourse in women.

impotence Inability to achieve or maintain an erection sufficient for vaginal penetration.

libido Sexual desire which includes sexual thoughts, fantasies, and satisfaction.

potency Ability to obtain and maintain an erection and ejaculate.

vaginismus Spasm of the levator animuscle which makes vaginal penetration difficult or impossible.

Sexual dysfunction encompasses disorders of libido and orgasm in both men and women. It also includes erectile dysfunction in men and dyspareunia and vaginismus in women.

I. INTRODUCTION

Sexual dysfunction may result from medical disorders, psychological problems, or as a result of medical therapy with drugs, surgery, or radiation. It is important to recognize that what constitutes a "dysfunction" may vary depending on the cultural, racial, ethnic, and sexual preference background of the individual.

The initial obstacle to overcome is recognition that there might be a problem. Most patients will present to the professional's office with nonspecific complaints and will usually not volunteer information about problems with sexual function. Such patients may present psychological problems, depression, anxiety disorders, or psychosomatic complaints. Most health professionals will not ask directed questions about sexual function and therefore most patients with such problems will be unrecognized. It is critical for health professionals to question patients felt to be at risk about sexual function as the situation requires.

II. STAGES OF SEXUAL RESPONSE

There are four stages of sexual response. The first stage is desire or libido, which drives the individual to initiate a sexual interaction. Sexual desire most likely has some genetic basis, with psychological factors having a major impact. Previous bad or good experiences set the foundation for what expectations are present and may strongly affect the impact on desire for further experiences. In this manner, a previous medical condition that resulted in sexual dysfunction may continue to have an impact on libido and arousal long after the medical condition is gone. In addition to psychological factors, medical disorders, such as hypogonadism, and side effects of medications may play a major role in affecting libido. The second stage is arousal, in which psychological and physical stimulation leads to vasodilatation and engorgement of the breast and genital regions. In the male, an erection is produced. In the female, the engorgement of the genital region produces vaginal lubrication and swelling of the distal third of the vagina with elevation of the uterus. Arousal may be inhibited by inadequate foreplay, distractions, problems with communication or self image, and neurological problems. Therapy by having the couple touch and use nonverbal means to communicate without proceeding to intercourse at the initial encounters may be helpful to decrease performance anxiety, which may underlie this problem.

The third stage of orgasm results at the peak of excitement and is manifested by a series of rhythmic involuntary contractions at 0.8-sec intervals resulting in ejaculation in the male and bulbocavernosus, ischiocavernosus, and uterine muscle contractions in the female. Anorgasmia may be primary, in which it has always been the case, or secondary, where the inability to achieve orgasm may depend on the partner or is situational. Inadequate arousal is often present as the basis for anorgasmia. Therapy is generally directed to increasing arousal through at-home exercises. Support groups may also be helpful in this setting.

The fourth stage is resolution during which engorgement and muscle tension return to normal levels over a few seconds although complete resolution may take up to an hour. Sexual dysfunction may occur with interruption of any of these stages of sexual response by either organic or psychological factors.

III. PSYCHOLOGICAL ISSUES IN MEN AND WOMEN

Innate sexual responsiveness is distributed throughout the population like many other biological characteristics but is affected by a large psychological component. In both the young and elderly, sexual dysfunction may be caused by prior unresolved problems from divorce, affairs, loss of children, or abuse.

Sexual dissatisfaction may be related to lack of communication, lack of a nurturing environment, inappropriate expectations, inability to use ancillary stimulation such as fantasy, ineffective stimulation, physical discomfort, or fear of pregnancy or infection. Careful attention to the medical history is needed to determine if any of these factors are present and to offer appropriate solutions such as adequate birth control to allay fears of pregnancy.

Childhood sexual abuse, dysfunctional family upbringing, and inadequate sex education have been implicated as important factors predisposing to disorders of libido and orgasm in adult men and women. Consequences of childhood sexual abuse include traumatic sexualization in which there are negative feelings about sex and problems with sexual identity; stigmatization where the person feels guilty and responsible for the act and may exhibit self-destructive behaviors, drug abuse, and suicidal gestures or acts; feelings of betrayal where there is no longer any trust between the person and people in helping roles, resulting in acting out behaviors and unfulfilling relationships; and powerlessness with acting out, identification with the aggressor, anxiety, sleep problems, constipation, and eating disorders.

Recognition of psychological problems as the cause or as a result of medical conditions is important and should prompt referral to an appropriate therapist. It must be emphasized that, where possible, therapy of sexual dysfunction should be couple oriented. Many couples with organic and nonorganic sexual dysfunction will have favorable response to counseling in addition to whatever medical therapy is considered. Behavior modification is generally used initially to reduce anxiety and allow further evaluation and therapy to proceed if necessary.

IV. MEDICATIONS AND SEXUAL DYSFUNCTION

More than two-thirds of patients visiting the physician leave with one or more prescriptions. There is also an increasing use of over-the-counter medications, herbal remedies, illicit drugs, alcohol, and tobacco. Many of these substances have the potential for adversely affecting sexual function. These effects must be recognized by health care professionals and patients as a major cause of sexual dysfunction since such recognition may spare a patient a costly and intensive evaluation for other less likely disorders and produce an easy and cost-effective method of correcting the problem by changing the medication regimen when other therapies in this setting may be ineffective. It is also important that the patient not change his or her own medication regimen because of concern over possible sexual dysfunction. All such changes should be discussed with and changed if indicated under the direct supervision of a physician.

Although many medications and classes of medications are recognized as producing sexual dysfunction, there is still a need for further research on the potential for some drugs to affect sexual function and on drug interactions resulting in sexual dysfunction. The most common drug classes affecting sexual function will be reviewed briefly but this does not constitute an exhaustive list. The interested reader is encouraged to further research the areas in the references provided in addition to other sources.

A. Antihypertensive Medications

The antihypertensives most commonly associated with erectile dysfunction in the male include the diuretics, centrally acting antiadrenergic agents such as clonidine and methyldopa, and guanethidine. Beta blockers are often associated with decreased libido. Hydralazine and prazosin may rarely be associated with priapism in the male. The angiotensin converting enzyme inhibitors (ACE inhibitors) are generally not associated with sexual dysfunction.

B. Psychotropic Medications

Almost any psychotropic medication has the potential of producing sexual dysfunction. It is very important to be aware of these potential side effects since sexual dysfunction may affect the self-esteem and recovery of the patient. The tricyclic antidepressants may produce erectile dysfunction in the male and may produce decreased libido and delayed orgasm or anorgasmia in either sex. The serotonin uptake inhibitors (SSRIs) are common causes of delayed orgasm and in fact may be helpful therapeutically in males with premature ejaculation. Some SSRIs may also cause a decreased libido. Bupropion does not appear to affect sexual function and may be an alternative to the SSRIs in patients manifesting problems with sexual function.

The antipsychotics usually produce erectile dysfunction, delayed ejaculation, and delayed or absent orgasm. About half of patients with schizophrenia have sexual dysfunction with untreated patients noting mainly decreased libido. Therapy appears to improve libido but is often associated with increased erectile and orgasmic dysfunction from medications.

Antianxiety medications such as the benzodiazepenes may adversely affect orgasm. MAO inhibitors may decrease libido and cause erectile dysfunction. Lithium frequently causes erectile dysfunction.

C. Miscellaneous Medications

Acetazolamide may be associated with erectile dysfunction. Antiseizure medications, such as carbamazepine, dilantin, primidone, and phenobarbital, may also cause erectile dysfunction.

Over-the-counter medications may also produce sexual dysfunction. The over-the-counter antiulcer drug cimetidine has some estrogenic activity and may cause erectile dysfunction and inhibit libido in addition to having the potential for gynecomastia in males. The antifungal drug ketoconazole may produce decreased libido and erectile dysfunction. The over-the-counter drug for hyperlipidemia, niacin, may produce decreased libido.

Herbal remedies may also contain substances which affect sexual function. It is very important for the health care professional to question patients on the use of alternative remedies.

D. Substance Abuse: Alcohol, Tobacco, and Drug Use

More than 17 million Americans abuse alcohol. Alcohol abuse is associated with various forms of sexual dysfunction including loss of libido, impotence, and both delayed and premature ejaculation. Chronic alcohol use may result in hypogonadism, gynecomastia, nutritional deficiencies, and sedation.

Tobacco use may cause vasoconstriction resulting in impotence. This is yet another reason for advising patients to stop smoking if they are smoking and never to start if they do not smoke.

Chronic use of drugs such as cocaine and tranquilizers may also cause decreased libido, impaired erections, and delayed or inhibited ejaculation. Amphetamines may result in delayed or absent ejaculation. Marijuana may decrease libido.

An increasing number of people are abusing androgenic steroids for athletic performance enhancement. It is important to understand that the increased circulating levels of androgen will suppress pituitary stimulation of the testes, resulting in a reduced testicular volume and diminished or absent sperm in the ejaculate.

V. MALE SEXUAL DYSFUNCTION: SPECIFIC DISORDERS

A. Impotence

Impotence is the inability to achieve or maintain an erection sufficient for vaginal penetration. Impotence may be organic or psychogenic. A history, physical exam, and selected testing often enable a clinician to come to a diagnosis. Of importance is the need to discuss possible changes in sexual function after surgical procedures such as prostate surgery.

Unfortunately, most of the time patients with organic causes of impotence have an overlying psychological component which must be addressed in addition to the underlying medical disorder.

1. *Psychogenic Impotence*

Symptoms usually begin with a change in lifestyle, change in partner, or increase in stress. The symptoms often depend on the specific partner or situation. Patients or their spouses with angina or a cardiac pacemaker, for example, may have sexual dysfunction and anxiety unless adequate counseling is given.

In psychogenic impotence, nocturnal penile tumescence (NPT) is almost always well preserved. Normal men generally have two or three 20-min erections during REM sleep. Devices for measuring these nocturnal erections include a snap gauge or an electromechanical device designed for this purpose. EEG monitoring may also be indicated since problems affecting REM sleep may prevent nocturnal erections.

In some couples, preexisting vaginismus or dyspareunia in the female partner may be a cause of nonorganic erectile dysfunction in the male. If psychogenic impotence is present or superimposed on organic impotence, it is important to initiate appropriate counseling and, if possible, couple-based therapy.

2. *Organic Impotence*

Impotence affects about 7% of men under age 55, 25% of men by age 70, and more than 75% of men over the age of 80. HIV-infected men, both heterosexual and homosexual, have a higher risk of ejaculatory difficulties, although the reasons remain unclear. Often impotence related to medical problems is gradual. Erectile function is usually affected under all conditions and NPT studies are nearly always abnormal although the vascular disorder, pelvic steal syndrome, may produce a normal result. The history, physical exam, and selected testing will often uncover a specific etiology. A fasting blood glucose, cholesterol, and triglyceride levels should be obtained as part of any preliminary evaluation. Testosterone and prolactin levels should also be done routinely early in the evaluation.

i. Vascular Disease Since an erection is produced by increased arterial inflow and decreased venous outflow from the corpora, any defect in the arterial inflow or venous outflow may affect erectile function. The arterial inflow to the genitals is via the paired pudendal arteries consisting of the dorsal arteries, bulbar arteries, and cavernosal arteries. The cavernosal arteries are responsible for blood flow into the corporal bodies.

Diminishment of blood flow with resulting erectile dysfunction may occur due to hypertensive vascular disease or atherosclerotic plaques of the larger pelvic blood vessels. Renal disease and diabetes are often associated with increased risks of atherosclerosis. Blockage of the iliac vessels may produce a pelvic steal syndrome and may result in significant loss of erectile function predominately in the male-superior position during which more blood flow is needed for the leg muscles thereby reducing penile blood flow.

The corpora are drained by the emissary veins which drain into the deep dorsal vein and then into the pelvic venous plexus. In rare cases, there may be a venous leak which results in excess outflow of blood from the corpora and resulting erectile dysfunction.

It is currently believed that the progressive increase in impotence with age may be related to damage to the corpora from hypoxia. The penis in the normal flaccid state is relatively hypoxic and, over time, scarring with erectile dysfunction may result. Nitric oxide appears to be the critical mediator of relaxation of the corpus cavernosum and new therapies aimed at increasing the local concentrations of nitric oxide are currently being studied as potential therapies for age-related impotence. Priapism, secondary to hypoxia and resulting penile fibrosis, may result in both a painful erection and impotence.

Testing for vascular disease can be complex. Penile blood pressure and blood flow studies using Doppler can be performed initially with other testing, such as cavernosal infusion studies, angiography, or venography, at the discretion of the specialist. Surgical procedures to modify blood flow to the penis are currently considered experimental.

ii. Neurological Disorders Sensory nerves from the penis are carried through the pudendal nerve. Arterial blood flow to the penis is increased through stimulation of parasympathetic fibers from sacral nerves S2, S2, and S4. There are also α-adrenergic fibers present which contract the smooth muscle of the penile corpora and result in detumescence. Men with spinal cord injuries above the level of S2 may have reflexogenic erections from stimulation of the bladder or rectum. Electroejaculation by use of a transrectal electrical probe, generally under anesthesia if sensation is present below the waist, can be used for semen collection from such spinal cord injured men for purposes of fertility. Some men suffering spinal cord injuries below the level of S2–S4 may still achieve psychogenic erections.

Other neurological conditions such as Parkinson's disease and multiple sclerosis are also associated with erectile dysfunction and other forms of sexual dysfunction.

iii. Endocrine Disease Diabetic men have lower levels of libido, arousal, and sexual satisfaction than nondiabetic men. Diabetes mellitus is the most common endocrine disease resulting in impotence. In addition to increased atherosclerotic damage to blood vessels, diabetic sensory neuropathy decreases penile sensation and decreases feedback through the sacral roots, and diabetic autonomic neuropathy prevents proper venous constriction and results in detumescence. Also as a result of autonomic neurpathy, retrograde ejaculation may occur in diabetic males which can result in infertility but is otherwise without consequences. Diabetes in males may initially present as impotence and a fasting glucose should be obtained as a screening test. Very tight glucose control may help limit vascular and neurological damage. Vacuum devices, penile injections with alprostadil, urethral alprostadil, penile implants, and, rarely, vascular surgery may be needed to allow sexual function in diabetic males with advanced vascular or neurogenic impotence. Diabetics frequently have other associated disorders such as depression as a result of chronic disease, obesity with poor self-esteem, and medications which might have a negative impact on sexual function.

Hypogonadism may result from primary testicular disease or secondarily from pituitary disease. In some cases, temporal lobe epilepsy, diagnosed by EEG, may result in hypogonadism. Although males do not have the biologic equivalent of the female menopause, some aging men may have decreased gonadal function resulting in symptoms of fatigue, muscle weakness, and sexual dysfunction. Testing generally includes total testosterone levels, free testosterone levels, follicle-stimulating hormone, luteinizing hormone (LH), and prolactin levels to identify and characterize the nature of the problem. It is of interest that erectile function in adults may be preserved even in the setting of significant acquired hypogonadism above a certain threshold amount of testosterone. Testosterone therapy, by injections, transdermal, or scrotal patches, generally results in complete restoration of sexual function.

Hyperprolactinemia, either from a pituitary tumor or from the effects of certain medications, may cause pituitary dysfunction with hypogonadotropic hypogonadism and sexual dysfunction. Patients with elevated prolactin levels usually present with decreased libido but may also develop galactorrhea and occasionally impotence. Prolactin acts to decrease gonadotropin-releasing hormone secretion and also impairs the testicular response to LH. Prolactin may also have other effects on libido and sexual function independent of the effects on the pituitary–testicular axis. Of interest, this sexual dysfunction is not fully correctable by testosterone therapy until the prolactin levels have been brought down to the near-normal range. Therefore, normalization of prolactin levels is required before assessment and treatment of any remaining testosterone deficiency.

Other endocrine disorders which may also result in sexual dysfunction include hyperthyroidism, hypothyroidism, acromegaly, Cushing's syndrome, and Addison's disease. A TSH level should be done to assess for thyroid disease and any clinical evidence of adrenal or pituitary disease should prompt evaluation by a specialist.

iv. Some Surgical and Radiotherapy Procedures Any procedure resulting in potential damage to the pelvic nerves can produce erectile dysfunction. Such procedures include prostatectomy, cystectomy, proctocolectomy, retroperitoneal lymphadenectomy, sympathectomy, renal transplantation, vascular procedures in the pelvis, and pelvic irradiation. Damage to the testes or castration may result in hypogonadism, which may result in erectile dysfunction if testosterone decreases below a certain threshold.

v. Anatomic Defects Impotence may result from mechanical defects in the penis. Patients with Peyronie's disease, phimosis, some forms of hypospadias, microphallus, or ambiguous genitalia may not have normal erectile function or the ability to achieve penetration.

vi. Infections Some men with infections of the prostate, seminal vesicles, or urinary tract have significant pain with intercourse and may present with decreased libido and other forms of sexual dysfunction.

3. *Treatment*

Therapy must be directed at the underlying cause if possible. Counseling is of prime importance in psychogenic impotence and when a superimposed psychological component has developed in a case of organic impotence. Counseling should be directed at the couple if at all possible as outlined in a previous section.

External vacuum devices are commonly used and

often are effective with minimal side effects. These devices are placed around the penis and create a partial vacuum which draws blood into the penis. While still under the effects of the vacuum, a tension ring is slid onto the base of the penis to retain the erection once the device has been removed. The erection can be maintained for about 30 min in this manner.

Certain substances injected directly into the corpora cavernosa have been noted to create an erection. In the past papaverine and phentolamine were used for this purpose but were not FDA approved for this purpose. Currently, prostaglandin E_1 is FDA approved for therapy in this manner. Prostaglandin E_1 (alprostadil) acts on the smooth muscle in the corpora cavernosa to produce an erection. Until recently, the only way to deliver alprostadil into the corpora was by direct penile injection. In 1996, alprostadil became available as a noninjectable urethral suppository which is placed by a small applicator into the tip of the urethra. The alprostadil is absorbed through the wall of the urethra resulting in relaxation of the smooth muscle in the penis and increased arterial inflow producing an erection. The erections produced generally last from 30 to 60 min.

Hormone replacement therapy with testosterone should be used in patients with hypogonadism only after an appropriate evaluation. Testosterone therapy may be given by intermittent intramuscular injection, in the form of testosterone enanthate or cypionate, or by daily transdermal or scrotal patches. Orally administered alkylated androgens are not recommended because of poor androgen effects, adverse effects on blood lipids, and potential liver toxicity. Periodic follow-up and monitoring of testosterone therapy is required to ensure proper dosing and avoid adverse effects. Testosterone is contraindicated in patients with prostate cancer, male breast cancer, or androgen-dependent sleep apnea.

Yohimbine, an alkaloid related to reserpine and a principal component of the bark of the West African Corynanthe yohimbe tree, has been used for mild cases of erectile dysfunction. Yohimbine acts as a blocker of presynaptic α-2-adrenergic receptors in the autonomic nervous system resulting in increased parasympathetic and decreased sympathetic tone. This action may increase blood inflow to the penis and diminish outflow resulting in an improved erection. The response rate has been estimated at 20–30% in mild cases of erectile dysfunction. Yohimbine also has central nervous system activity and is the source for many of the side effects. The drug may be most useful in cases of suspected psychogenic impotence. Yohimbine is usually given at a dose of three 5.4-mg tablets daily and may produce transient side effects of nausea, dizziness, headaches, and anxiety. It may take 3 or 4 weeks for a response.

Penile implants may be placed as a last resort if less invasive therapies have failed. Semirigid rods or inflatable implants may be used. Vascular procedures to improve blood flow to the penis are currently considered experimental.

As of 1997, oral medications increasing local penile nitric oxide levels are being tested for therapy of impotence. This form of therapy has the potential of revolutionizing the therapy of certain forms of sexual dysfunction. This therapy also holds promise for use in women with sexual dysfunction.

B. Premature Ejaculation

Premature ejaculation is the most common male sexual dysfunction. This disorder is generally characterized by perceived control over ejaculation, latency from penetration to ejaculation, concern over the short time to ejaculation, and dissatisfaction regarding control of ejaculation. Cultural issues may also be of importance since in Scandinavian countries, withdrawal is used as a form of contraception and often results in ejaculatory defects. Counseling and behavior modification techniques are the prime means of therapy. Use of SSRIs or other medications may help to delay ejaculation but may produce other forms of sexual dysfunction.

VI. FEMALE SEXUAL DYSFUNCTION: SPECIFIC DISORDERS

A. Dyspareunia and Vaginismus

Dyspareunia describes pain occurring during intercourse. Failure of vaginal lubrication is a common underlying problem and may result from menopause or other conditions causing inadequate estrogen. Other causes of dyspareunia include endometriosis, pelvic infections, fibroids, and pelvic masses.

Vulvar vestibulitis is a disorder of unknown etiology which results in chronic inflammation of the area between the clitoral hood and the perineum and is a cause of dyspareunia. Women suffering from vulvar vestibulitis generally have situationally dependent symptoms characterized by genital pain, difficulties with lubrication, inhibited libido, and negative feelings about sex. The topical use of 5% lidocaine jelly may be helpful in alleviating symptoms. There is no generally agreed on therapy and spontaneous remission within 6 months is common. Surgery is sometimes recommended as a last resort.

Vaginismus refers to a spasm of the levator animuscle which makes vaginal penetration during intercourse, and sometimes gynecological examination, difficult or impossible and associated with pain. Primary vaginismus refers to symptoms present from the first attempt at intercourse. Often, counseling, pelvic muscle contraction exercises, and graduated vaginal dilation done by the patient may improve the symptoms.

The symptoms of dyspareunia and vaginismus may also be associated with chronic interstitial cystitis. Evaluation by a urologist familiar with this disorder may be indicated. These symptoms may also occur in women with seizure disorders.

B. Pregnancy and Postpartum

Sexual activity decreases during the first and third trimesters though this is thought to be on a psychological basis. Postpartum, sexual activity usually decreases due to physical changes such as healing of a c-section or episiotomy, hormone-related mood disturbances, stress, and self-image problems.

C. Menopause

In addition to self-image problems, women going through menopause generally have symptoms of estrogen deficiency resulting in vasomotor symptoms, sleep problems, fatigue, mood disturbances, and vaginal dryness which contribute to decreased libido, dyspareunia, and decreased arousal. The vaginal dryness and thinning of the vaginal epithelium may result in further pain from abrasion. The spouse may also have a negative response to his partner going through menopause or he may be developing erectile dysfunction himself, resulting in further problems with sexual dysfunction. Hormone replacement therapy is often effective at restoring vaginal lubrication and improving sexual function.

D. Cancer and Cancer Therapy

Among survivors of gynecologic cancers, about half have sexual dysfunction. In addition to the psychological stresses and changes in body image that occur in a women with breast, ovarian, uterine, cervical, bladder, or colorectal cancer, there are potential problems associated with therapy that may affect sexual function. Surgical procedures in the pelvis may damage the autonomic nerves, the blood supply to the genitals, support structures of the genitals, or the genitals directly. Radiation therapy may result in vaginal atrophy which can often be treated with dilators. Radiation may also damage the pelvic autonomic nerves or the ovaries leading to hypoestrogenism resulting in sexual dysfunction. Chemotherapy may cause psychological stress by producing hair loss, gastrointestinal disturbances, and loss of ovarian activity. The partner may also have problems adjusting and may not initiate sexual activity for fear of producing injury to the patient. Counseling of the couple is especially important in the setting of cancer and cancer therapy.

E. Problems in the Male Partner

Sexual dysfunction in the female may be caused by sexual dysfunction or medical problems in her male partner. Erectile dysfunction in the male may result in feelings of rejection or inadequacy in the female partner and inhibit libido and sexual arousal. Relationship problems may then develop with mutual hostility and psychological withdrawal.

See Also the Following Articles

Erection; Impotence; Orgasm

Bibliography

ACOG (1995). ACOG technical bulletin: Sexual dysfunction. *Int. J. Gynecol. Obstet.* **51**, 265–277.

Faller, K. C. (1993). *Child Sexual Abuse: Intervention and Treatment Issues.* U.S. Department of Health and Human Services, Administration for Children and Families, National Center on Child Abuse and Neglect, Washington, DC.

Feldman, H. A., Goldstein, I., Hatzichristou, D. G., Krane, R. J., and McKinley, J. B. (1994). Impotence and its medical and psychosocial correlates: Results of the Massachusetts Male Aging Study. *J. Urol.* **151**, 54–61.

Finger, W. W., Lund, M., and Slagle, M. A. (1997). Medications that may contribute to sexual disorders. *J. Family Practice* **44**, 33–43.

Linet, O. I., and Ogrinc, F. G. (1996). Efficacy and safety of intracavernosal alprostadil in men with erectile dysfunction. The Alprostadil Study Group. *N. Engl. J. Med.* **334**(14), 873–877. [Comment, *N. Engl. J. Med.*, 913–914.

National Institutes of Health (1992). Impotence, NIH Consensus Statement 10(4). NIH, Bethesda, MD.

Petak, S., Baskin, H. J., Bergman, D. A., Dickey, R. A., and Nankin, H. R. (1996). AACE clinical practice guidelines for the evaluation and treatment of hypogonadism in adult male patients. *Endocr. Pract.* **2**, 440–453.

Rajfer, J., Aronson, W. J., Bush, P. A., Dorey, F. J., and Ignarro, L. J. (1992). Nitric oxide as a mediator of relaxation of the corpus cavernosum in response to nonadrenergic, noncholinergic neurotransmission. *N. Engl. J. Med.* **326**, 90–94.

Rosen, C. R., and Leibium, S. R. (1995). Treatment of sexual disorders in the 1990s: An integrated approach. *J. Consult. Clin. Psych.* *63*(6), 877–890.

Sexual Imprinting

David B. Miller
University of Connecticut

GLOSSARY

consolidation A process by which experiential influences cause a reorganization in neuronal and cognitive mechanisms resulting in a change in an organism's perception and/or behavior.

cross-fostering The exchange of offspring between different parents, usually during early development and between different species.

plasticity A property of organisms indicating their malleability or susceptibility to experiential influences, especially during (though not limited to) early development. In its extreme form, plasticity may result in developmental outcomes beyond the range typically occurring in nature.

sensitive period A time-delimited phase during development in which an organism is particularly sensitive to experiential influences; also termed critical period, optimal period, and phase specificity.

species typical That which falls within the normative range of perceptual capabilities and motor activities of members of a particular species in the natural environment.

Sexual imprinting is a process by which experience prior to sexual maturity influences mating preferences on or after sexual maturity.

Encyclopedia of Reproduction VOLUME 4

I. TYPES OF IMPRINTING

In its most general sense, imprinting refers to the development of preferences. Konrad Lorenz, who championed the concept, was primarily concerned with the development of social attachments of two types: (i) filial imprinting, which is the formation of attachments of young organisms to parental objects, and (ii) sexual imprinting, which is the formation of mating preferences by older organisms. Common to both forms of imprinting are the assumptions that (i) species in which imprinting occurs do not instinctively identify or recognize objects of social attachments (which would normally be conspecifics), and (ii) early experience provides the mechanisms underlying the formation of social attachments. In the case of filial imprinting, early experience would have to occur perinatally (around the time of birth or hatching), whereas for sexual imprinting, early experience might involve the perinatal period but also might extend into the juvenile period prior to sexual maturation.

The concept of imprinting has been overextended to include the development of almost any kind of preference, including such nonsocial preferences as diet (food imprinting), location (nest site imprinting), odors (olfactory imprinting), and parasitic preferences (host imprinting).

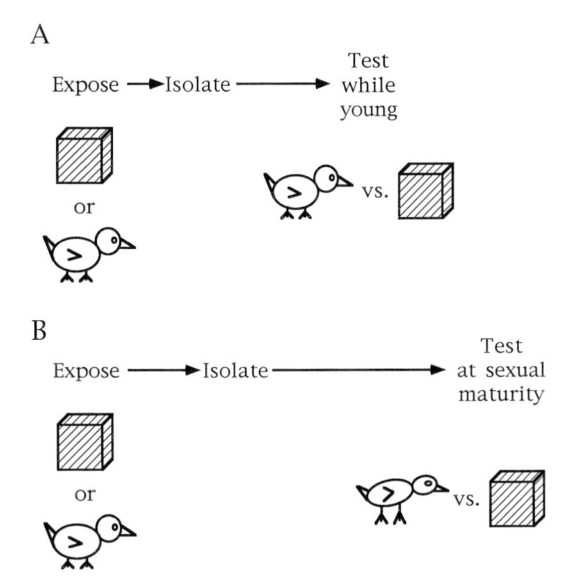

FIGURE 1 Methodological procedures for two types of imprinting. (A) In filial imprinting, an organism is exposed to a stimulus early in life, such as a cube or a conspecific model. Following a period of subsequent isolation from that stimulus, the organism is tested for its preference (measured as approach and/or following) for the imprinted stimulus versus a novel stimulus. (B) In sexual imprinting, an organism is exposed to a stimulus early in life (though the exposure may continue into adolescence). Following a period of subsequent isolation from that stimulus, the organism is tested for its sexual preference for the imprinted stimulus (or one similar to it, such as an adult if the original imprinted stimulus was a young broodmate) versus a novel stimulus.

II. METHODOLOGICAL DEFINITIONS OF IMPRINTING

The study of imprinting, both filial and sexual, has become very popular in laboratories at least in part due to the robustness of the phenomenon. Young organisms exhibit a high degree of plasticity, particularly with regard to the formation of social attachments, and it is rather easy to direct or redirect an organism's existing preferences toward an object to which it has been exposed at an early age.

Most imprinting experiments follow the same basic paradigm of initially exposing an organism to a stimulus object during an alleged sensitive period, then isolating the organism from the stimulus object, and finally testing the organism by presenting it with the object to which it had initially been exposed (or one similar to it) versus one to which it had not been exposed. Figure 1 portrays this paradigm and how it differs in experiments on filial imprinting versus sexual imprinting. The main differences are the length of the period of exposure (which is often somewhat longer in sexual imprinting experiments) and the age at which preferences are assessed (which, in sexual imprinting, is at or beyond the point of sexual maturity). Accordingly, the response measured in sexual imprinting experiments is of a sexual nature (e.g., courtship and copulation), whereas in filial imprinting experiments, the response is usually one of mere proximity (e.g., the organism approaching and/or following the imprinted object).

III. METHODOLOGICAL AND CONCEPTUAL ISSUES IN THE STUDY OF SEXUAL IMPRINTING

The high degree of plasticity exhibited by most species in early development poses some interesting problems in both the conduction of sexual imprinting experiments and the interpretation of their outcomes.

A. Cross-Fostering

Sexual imprinting experiments sometimes use a cross-fostering paradigm, which is a variation of the procedure portrayed in Fig. 1B. An example of cross-fostering in birds would entail swapping the eggs of one species with those of another. The incubating birds of both species now become foster parents such that they hatch and care for the offspring of the alien species in their respective nests. The offspring, in turn, are exposed exclusively to their foster parent species and, if successfully imprinted, will prefer to mate with members of their foster parent species rather than with members of their own species upon reaching sexual maturity. (Because hybridization in nature is rare, such matings will not result in the production of offspring.)

Cross-fostering experiments are performed because of the assumption that the mechanisms underlying the formation of a sexual preference for a foster parent species are the same as those underlying the development of normal sexual preferences for one's own species. However, filial imprinting studies reveal that there are different neuronal mechanisms associated with imprinting preferences formed toward species-typical versus species-atypical objects. Specifically, a lesion to nucleus IMHV (the intermediate and medial part of the hyperstriatum ventrale, the anterior boundary of which is located in front of the midpoint of a line drawn between the anterior and posterior poles of the forebrain hemispheres) disrupts the formation of a chick's (*Gallus gallus*) filial response toward a rotating box (a species-atypical object) but does not interfere with its preference for a stuffed jungle fowl hen (a species-typical object). Although comparable data have not yet been collected on brain mechanisms underlying sexual imprinting, evidence for different brain mechanisms underlying the development of normal, species-typical mating preferences versus abnormal, species-atypical mating preferences would suggest that cross-fostering studies may only superficially enhance our understanding of the development of normal sexual preferences, thereby mirroring the known situation for filial imprinting.

From a biological standpoint, another problem concerning the use of cross-fostering in most (if not all) sexual imprinting studies pertains to the use of species that, in nature, do not cross-foster. The zebra finch (*Taeniopygia guttata*), which is among the most popular species in sexual imprinting research, is not a nest parasite (i.e., species that lay eggs in the nests of other species such that the offspring are hatched and reared by a foster-parent species), nor is the species with which it is often cross-fostered in the laboratory, the Bengalese finch (*Lonchura striata*). Indeed, in the wild, these species occupy different continents; even if they were nest parasites, they would not encounter one another in nature. Although sexual imprinting can be induced in cross-fostered laboratory populations of zebra finches, the extent to which sexual imprinting plays a role in species identification in nature in this species remains unclear. From a biological standpoint, the most interesting studies on sexual imprinting (or species identification in general) are yet to be conducted. Such studies would involve species in which the young have little or no contact with conspecifics, such as Australian mallee fowl (*Leipoa ocellata*), yet when sexually mature, they mate with conspecifics. Perhaps nonlinear (subtle and nonobvious) experiential factors affect the development of conspecific mating preferences in those species in the same manner as they do with regard to certain forms of filial species identification in other precocial species (e.g., wood ducks, *Aix sponsa*; mallard ducks, *Anas platyrhynchos*).

B. Sensitive Periods and Reversibility of Preferences

Sexual imprinting, like its ontogenetic antecedent filial imprinting, is supposed to occur during a sensitive period in the life cycle of the individual. However, the literature regarding the sensitive period, both with regard to imprinting as well as nonimprinting-like phenomena, is fraught with demonstrations indicating the importance of experiential factors operating beyond alleged sensitive periods. Such extrasensitive period experiences can be as important as or, in some cases, more important than those occurring within sensitive periods.

In the case of sexual imprinting, there is evidence indicating the operation of a two-stage process consisting of a period of acquiring knowledge about conspecifics prior to fledging followed by an independent consolidation period after fledging, which most investigators would consider beyond the sensitive period. Klaus Immelmann, who pioneered the experimental analysis of sexual imprinting in zebra finches, and Jaap Kruijt independently demonstrated the reversal of sexual imprinting preferences as a function of courtship experiences that occur well beyond the sensitive period.

A general problem with sensitive period research is that even slight alterations in methodology can render different results pertaining to the existence of sensitive periods and/or the importance of extrasensitive period experience or lack thereof. Another problem is a methodological confound concerning the sensitive period-to-test interval. When different groups are assessed for different sensitive periods (e.g., Week 1 vs Week 2) but are tested at the same age (e.g., Week 10), the sensitive period-to-test intervals are different. Curing that problem by equalizing the sensitive period-to-test intervals only induces a new confound, namely, testing the animals at different ages. There is no way around this problem, and research in other areas indicates that this is not a trivial issue.

C. Sex Differences

Most studies on sexual imprinting involve assessing male preferences toward females. This is perhaps due to the relative ease of quantifying conspicuous male courtship displays (i.e., postures, dances, and songs) directed toward females whose sexual behaviors are often more subtle than those of males. However, in most species females make the ultimate choice with whom to mate after being courted by several males, and investigators gradually realized the importance of assessing whether sexual imprinting plays a role in the development of female mating preferences. Early investigations seemed to indicate that females do not sexually imprint. However, later studies indicated that sexual imprinting does occur in females but that the expression of their preferences is more greatly affected by laboratory methodology than that of males.

Zebra finch males, for example, exhibit sexual imprinting in brief (e.g., 20–30 min) simultaneous choice tests involving two live or even stuffed models of different species (e.g., conspecific versus foster-parent species). Sexual imprinting by females, however, is affected by such factors as test duration, number of males present during the test, the intensity with which they are courted by stimulus males, and the number of times that females are presented with stimulus males. The latter effect has been taken as evidence that sexual imprinting is a two-stage process in females as well as in males, but that the consolidation process might be slower in females. Although more experiments are required to eliminate other interpretations of male and female differences with regard to consolidation, it is clear that sexual imprinting is greatly affected by methodological procedures. Moreover, sex differences in imprintability may be a function of females relying on more cues in mate choice than those presented during a typical laboratory imprinting test, including extramale cues such as quality of territory.

D. Predispositions and Social Experience

A number of investigators studying sexual imprinting in zebra finches (cross-fostered with Bengalese finches) report what appear to be predispositions for conspecifics, which is not altogether surprising given that conspecific predispositions occur in other species in the context of filial imprinting. For example, zebra finch males reared by both zebra finches and Bengalese finches subsequently prefer to mate with conspecifics. Research has shown that such predispositions are probably the result of biased social experience. Adult zebra finch fathers have more social interactions with the young zebra finch males than do adult Bengalese finch males. Increasing Bengalese finch interactions by replacing paternal zebra finch males with nonpaternal zebra finches results in the development of greater sexual preferences for Bengalese finches.

Although such demonstrations are compelling, they primarily illustrate the plasticity of developmental processes by redirecting species-typical preferences toward species-atypical stimuli given that, in nature, zebra finch offspring are reared only by zebra finch parents. These particular experiments, however, also demonstrate the dramatic effect of social experience on behavioral development, an effect that has been shown to be of significant importance in other developmental (nonimprinting) situations in many species.

IV. A REFORMULATION OF SEXUAL IMPRINTING

Konrad Lorenz believed that, especially in birds, conspecifics are not instinctively recognized. He posited the concept of sexual imprinting as a special learning mechanism by which an animal comes to identify a conspecific in the context of mate choice. In other words, he was primarily concerned with species recognition. An alternative role of sexual imprinting has been posited by Patrick Bateson—namely, that of individual recognition. Specifically, Bateson has suggested that sexual imprinting might serve as the means by which individuals learn the identity of close relatives early in life such that they avoid mating with them after reaching sexual maturity—in other words, an incest-avoidance mechanism.

According to Bateson, an animal will choose a mate that is "optimally discrepant" from a rearing partner. The point of optimal discrepancy is that which is neither too similar nor too dissimilar from a familiar object. If the stimulus is too similar to a rearing partner, it might be a close relative and should be avoided as a mating partner; if too dissimilar, it might be a heterospecific, in which case mating would be unsuccessful in terms of producing offspring.

Data have been collected in several species that attest to the validity of Bateson's model. Even Konrad Lorenz expressed his enthusiasm for this reformulation of his original concept as one that involves the early learning of individual characters rather than species characteristics.

V. APPLICATIONS

Although non-conspecific mate choice in nature is rare, mistakes do happen. For example, if a wood duckling happens to be reared by a mallard hen because it was separated from its own mother and broodmates (e.g., by predation), it will form a strong attachment to mallards and exhibit sexual preferences for its foster-parent species upon reaching sexual maturity.

As discussed previously, the laboratory study of sexual (and filial) imprinting represents a powerful demonstration of behavioral plasticity by showing how species-typical preferences can be redirected toward species-atypical objects. With this in mind, conservationists often take painstaking measures to make certain that endangered species reared in zoos do not sexually imprint on human caretakers. When released in the wild as older juveniles or adults, it is important that such human-reared individuals seek conspecifics as mates. Thus, human caretakers often wear elaborate disguises as camouflage and feed young chicks with puppet gloves that mimic the head of adults of the chick's own species. Such procedures have proven effective in ensuring that species-typical sexual preferences are not redirected toward humans.

See Also the Following Articles

Critical Developmental Periods; Mating Behaviors, Birds; Mating Choice, Overview; Sexual Selection

Bibliography

Bateson, P. (1983). Optimal outbreeding. In *Mate Choice* (P. Bateson, Ed.), pp. 257–277. Cambridge Univ. Press, Cambridge, UK.

Immelmann, K. (1972). Sexual and other long-term aspects of imprinting in birds and other species. In *Advances in the Study of Behavior Vol. 4* (D. S. Lehrman, R. A. Hinde, and E. Shaw, Eds.), pp. 147–174. Academic Press, New York.

Immelmann, K., Pröve, R., Lassek, R., and Bischof, H.-J. (1991). Influence of adult courtship experience on the development of sexual preferences in zebra finch males. *Anim. Behav.* **42**, 83–89.

Kruijt, J. P., and Meeuwissen, G. B. (1991). Sexual preferences of male zebra finches: Effects of early and adult experience. *Anim. Behav.* **42**, 91–102.

Lorenz, K. (1937). Companions as factors in the bird's environment. *Auk* **54**, 245–273.

Miller, D. B. (1980). Beyond sexual imprinting. In *Acta XVII Congressus Internationalis Ornithologici*, Vol. 2, pp. 842–850. Deutschen Ornithologen-Gesellschaft, Berlin.

Oetting, S., and Bischof, J.-H. (1996). Sexual imprinting in female zebra finches: Changes in preferences as an effect of adult experience. *Behaviour* **133**, 387–397.

Oetting, S., Pröve, E., and Bischof, H.-J. (1995). Sexual imprinting as a two-stage process: Mechanisms of information storage and stabilization. *Anim. Behav.* **50**, 393–403.

ten Cate, C. (1985). On sex differences in sexual imprinting. *Anim. Behav.* **33**, 1310–1317.

ten Cate, C. (1988). Sexual selection: The evolution of conspicuous characteristics in birds by means of imprinting. *Evolution* **42**, 1355–1358.

ten Cate, C. (1989). Behavioral development: Toward understanding processes. In *Perspectives in Ethology Vol. 8: Whither Ethology?* (P. P. G. Bateson and P. H. Klopfer, Eds.), pp. 243–269. Plenum, New York.

ten Cate, C. (1994). Perceptual mechanisms in imprinting and song learning. In *Causal Mechanisms of Behavioural Development* (J. A. Hogan and J. J. Bolhuis, Eds.), pp. 116–146. Cambridge Univ. Press, Cambridge, UK.

Zann, R. A. (1996). *The Zebra Finch. A Synthesis of Field and Laboratory Studies.* Oxford Univ. Press, Oxford, UK.

Sexual Selection

Anders Pape Møller

Université Pierre et Marie Curie

I. Introduction
II. Expression of Secondary Sexual Characters
III. Functional Explanations for Female Choice
IV. Sperm Competition and Sexual Selection
V. Speciation and Sexual Selection

GLOSSARY

cryptic female choice Postcopulatory mechanism of female choice based on choice of particular sperm, ejaculate or genitalic phenotypes.

differential access Access of females in good condition to attractive males.

differential parental investment Differential parental investment by females in the offspring of attractive mates.

female choice Choice of mates by individuals of the choosy sex.

male–male competition Competition among individuals of the chosen sex for access to individuals of the choosy sex.

sperm competition Competition among gametes of more than a single individual of the chosen sex for fertilization of a clutch of eggs.

Sexual selection is the process by which individuals compete for mating success and access, and ultimately fertilization of the eggs of individuals of the choosy sex (usually females). Two mechanisms result in sexual selection: intrasexual competition among individuals of the same sex for mating access and fertilization of individuals of the choosy sex and mate choice by individuals of the choosy sex for access to attractive individuals of the chosen sex (usually males). The relative reproductive rates of males and females determine whether males or females become the sex that is competing most intensely for mates since individuals of the sex that have reproductive activities of short duration will compete more intensely for access to mates. Sexual selection has given rise to the evolution and maintenance of extravagant characters that are disfavored by natural selection due to their exaggeration: Pheromones, calls and vocalizations, colors, and exaggerated feather ornaments have evolved due to female choice, whereas various kinds of weapons and armament, such as horns in beetles and several mammals, spurs in birds, and antlers in deer, and relatively large male body size have evolved as a consequence of male–male competition.

I. INTRODUCTION

Sexual selection is an important evolutionary factor arising from female choice and male–male competition. Females generally benefit from their mate preferences either in terms of direct fitness benefits obtained from resources under the control of males or indirect fitness benefits from the genes of the sire. Females often pay a cost for their choice of mate relative to individuals that mate indiscriminately. These costs include the use of time and energy searching for mates, the risk of predation during mate search, the risk of not being mated at all, and the reduction in breeding success due to delayed reproduction. Female mate choice will only be evolutionarily stable if females experience a net benefit from their mate choice, with the benefits exceeding the costs. Large costs of mate choice thus require large benefits.

Numerous studies have investigated the premise that secondary sexual characters currently are under sexual selection, and this appears to be the case under both unmanipulated and experimental conditions in a wide range of organisms, including plants and animals, and among the latter invertebrates and vertebrates including humans. Both female choice and male–male competition result in directional selection for increased expression of male traits, giving rise to a sexual selection advantage for males with the most extreme expressions of their secondary sexual characters.

The relative importance of male–male competition and female choice is not readily disentangled but has been studied in a number of cases. There is considerable variation among species in the relative importance of the two processes of sexual selection, apparently causes by ecological conditions. If males compete intensely and create a clear rank of male qualities, females will only pay a small cost for their choice of the highest ranking male.

Many studies have shown that both male secondary sexual characters and female mate preferences have an additive genetic basis, and that there is a considerable amount of genetic variation present. Some studies have also demonstrated a genetic correlation between female mate preference and male trait.

II. EXPRESSION OF SECONDARY SEXUAL CHARACTERS

Secondary sexual characters have an additive genetic basis. The optimum level of expression for a particular individual depends on the level of expression among other individuals. The handicap mechanism predicts that reliable signaling can be achieved if the secondary sexual character is costly and an increment of signaling is more costly for low- than for high-quality individuals. Since the cost function increases more steeply for low- than for high-quality signaling males, only high-quality individuals can afford to develop the most extreme expressions of the secondary sexual character.

Numerous studies have demonstrated that the expression of secondary sexual characters depends on the condition of individual males and current environmental conditions. For example, males with more resources, in better condition, and with fewer parasites generally produce larger secondary sexual characters than the average male in the population. These observations are consistent with the handicap mechanism of reliable signaling.

The evolution of secondary sexual characters should be particularly rapid under intense sexual selection as in species with a lekking mating system. Attractive arbitrary secondary sexual traits or condition-dependent traits should rapidly pass a phase of exaggeration followed by a period of maintenance of relatively costly traits by weak female mate preferences. Since males in such species do not provide any parental care, the relatively low cost of sexual displays should allow more extreme expression of male sexual traits. Comparative analyses have demonstrated that extravagant secondary sexual characters are more common and have more extreme expression, and that multiple sexual signals occur more frequently in species with a high intensity of sexual selection.

All models of sexual selection assume that males incur costs of production and maintenance of secondary sexual characters. These costs can be energy costs of production of a vocalization, a display, or a morphological character; predation costs caused by phonotactic predators, visually searching predators, or predators cueing in on displaying males; parasitism costs caused by parasites using male sexual signals as cues in their host-searching behavior; or immunological costs caused by the tradeoff between investment in secondary sexual display and immune function. Ultimately, investment in secondary sexual characters results in differential male mortality, in particular among males in poor condition with small sexual signals.

III. FUNCTIONAL EXPLANATIONS FOR FEMALE CHOICE

Females may benefit in terms of sexual selection from direct or indirect fitness benefits. Direct fitness benefits are advantageous in the current generation and thus do not need a genetic basis. They include features such as male territories, courtship food, and male parental care, and also the absence of contagious parasites in males. A number of such examples have been described, but equally many examples indicate that females pay a cost for their mate choice in terms of reduced parental care by the most attractive males. Females thus have to make a differential parental investment in reproduction in order to compensate for the lack of male investment. Choice of attractive partners can only be evolutionary stable in this condition if females benefit indirectly in terms of genetic advantages acquired by their offspring.

Indirect fitness benefits acquired by females include attractiveness of sons and viability of sons and daughters. The Fisherian mechanism of mate choice suggests that females will benefit from their mate choice because choosy females will produce attractive sons and choosy daughters if both the female mate preference and the male trait have a genetic basis. The exaggeration of the attractive arbitrary male trait and the female preference will be self-reinforcing in an ever-increasing runaway fashion because the male trait and the female preference will become genetically linked. This process will come to an end when there is no more genetic variation in the male trait or the female preference or when the expression of the male trait is balanced by oppositely directed natural selection. There are no empirical examples of this mechanism, although it will invariably be in action whenever both male trait and female mate preference have an additive genetic basis.

The viability mechanism of female choice suggests that females obtain viability genes for their offspring by choosing the most extravagantly ornamented male. The viability effect may be due to general viability genes or, for example, specific genes for parasite resistance. If only males with viability genes are able to produce an extravagant trait, because these genes confer an advantage in terms of male condition or resistance to parasites, this will result in females indirectly choosing genetically based viability for their offspring by choosing the most ornamented male. This mechanism relies on the presence of additive genetic variation in viability which may be maintained by host–parasite coevolution in the case of resistance genes and by spatially heterogeneous environments and migration in the case of general viability genes. Several studies have demonstrated considerable effects of female mate choice on viability effects for their offspring.

IV. SPERM COMPETITION AND SEXUAL SELECTION

Sexual selection was originally stated by Charles Darwin to arise from the nonrandom differential mating success of males, but male success may also change after acquisition of a mate due to fertilization, differential parental investment, abortion, and infanticide. Sperm competition, which arises from competition among males for access to and fertilization of the eggs of a single female, is a ubiquitous and important component of sexual selection. Sperm competition has been shown to occur to varying degrees in almost all species investigated with the exception of a few species in which males are in complete control of the fertilization process (e.g., fig wasps with single foundresses, pipefishes, and seahorses). The frequency of offspring being fertilized by males other than the resident male exceeds 70% in some bird species and 20% in some human populations. Sperm competition reinforces precopulation sexual selection by increasing the skew in male success in favor of attractive individuals, and species with apparently slight opportunities for sexual selection such as monogamous birds often are exposed to as intense sexual selection as in the most extreme cases of polygyny and lekking due to the effects of sperm competition.

The mechanisms of sperm competition include adaptive changes in features of the ejaculate, such as the number of sperm and their quality and the number of ejaculates; the frequency, timing, and duration of extra-pair copulations relative to the timing of pair copulation; sperm displacement by later copulating males (as in odonates); female differential retention and use of sperms; and sperm loss from the female reproductive tract. Sperm competition potentially allows the females to control the sire of their offspring by the mechanisms discussed previously, and females may thereby secondarily adjust their mate choice at the timing of fertilization.

V. SPECIATION AND SEXUAL SELECTION

Secondary sexual characters diverge much more than characters not subject to sexual selection, as demonstrated by closely related species differing considerably in secondary sexual characters, but usually only to a small degree for other characters. Both the Fisher process and the handicap mechanism may give rise to rapid character divergence, and such divergent secondary sexual characters may eventually act as premating and postmating species isolation mechanisms. Once interbreeding between two divergent populations is reduced, this may result in rapid speciation. Sexual selection is the direct cause of speciation as suggested by some of the extreme radiations (*Drosophila* species in Hawaii and cichlids in large African lakes). Comparative analyses have demonstrated that avian taxa containing species with extravagant male colors or feather ornaments are more speciose than their sister taxa. The mechanism generating this increase in speciation is a greater degree of divergence, as determined by the number of subspecies in ornamented taxa, and a greater degree of speciation in taxa with more intense sexual selection, as determined from the skew of mating success among males. A high rate of speciation also implies a high rate of extinction if the frequency of sexual ornamentation is constant over time. The costs of secondary sexual characters and the resulting increase in male mortality rate give rise to greater probability of extinction due to demographic stochasticity. Species with more extreme sexual selection have suffered from greater extinctions as shown by the paleontological record and introductions of birds to oceanic islands.

See Also the Following Article

Mating Choice, Overview

Bibliography

Andersson, M. (1994). *Sexual Selection*. Princeton Univ. Press, Princeton, NJ.

Birkhead, T. R., and Møller, A. P. (Eds.) (1998). *Sperm Competition and Sexual Selection*. Academic Press, London.

Eberhard, W. G. (1996). *Female Control*. Princeton Univ. Press, Princeton, NJ.

Møller, A. P. (1994). *Sexual Selection and the Barn Swallow*. Oxford Univ. Press, Oxford, UK.

Sexually Transmitted Diseases

Paul R. Summers

University of Utah School of Medicine

GLOSSARY

antepartum A term referring to the time interval prior to the birth process.

colonization The persistence of microbes at a body site without the development of local tissue invasion or symptomatic disease.

intrapartum A term referring to the time interval during the birth process.

mucosal immunity The host's nonspecific as well as directed (from prior exposure) immune response to microbes invading at the skin barrier, involving various types of white blood cells and biochemicals.

pathogenic The ability of a microbe to cause symptomatic disease in the host.

sexually transmitted disease carrier A host who is colonized with a sexually transmitted disease-associated microbe in the genital area and is infectious to others, but has no symptomatic disease.

skin microtrauma Superficial skin damage due to environmental factors.

stratum corneum The outermost layer of the skin, considered to be a key element of the skin barrier, and consisting of nonmetabolically active squamous cells that are slowly sloughed as the underlying skin cells mature.

vulvodynia A syndrome of unexplained vulvar pain that generally precludes intercourse.

Without exception, the spread of disease during sexual contact requires a sequence of events that involve the infectious source as well as the new host. Exposure to a critical number of microbes is an essential requirement before disease transmission can occur. To invade through the protective layer of the stratum corneum, these microbes first must be able to adhere to the exposed epithelium. Adherence requires the presence of specific binding sites that are generally native protein sequences within the skin. Skin microtrauma during sexual relations may compromise the skin barrier by increasing the number of exposed binding sites. The integrity of the local skin barrier and mucosal immunity act as the chief defenses against the sexual transmission of infection. Failure of this defense system in the genital area allows sexual transmission of disease.

I. WHAT IS A SEXUALLY TRANSMITTED DISEASE

A wide variety of microbes are transmitted by sexual contact. The majority of these microbes can colonize the genital area but do not cause disease. A disorder attributed to one of the limited number of microbes that can produce disease is termed a sexually transmitted disease (STD). To be considered a STD, the origin of the infectious microbes as well as the site of microbial invasion in the new host must be the reproductive organs.

Contrary to reasonable expectations, STD transmission usually does not result in symptomatic disease. Thus, many more individuals are STD carriers than symptomatic cases. This frequent delay or failure to develop symptoms is problematic for STD prevention and control. Most infected individuals are not aware of their STD, do not seek treatment, and may spread the infection unknowingly.

Factors that influence development of symptomatic disease are only partially understood. Herpes simplex infection is likely to remain unrecognized because of decreased visibility of the lesions if the vagina or cervix is the primary site of infection. Trichomonas infection is typically symptomatic in the female but asymptomatic in the male. This may be due to a smaller number of the estrogen-dependent trichomonas binding sites in the male genital epithelium. In contrast, there is no explanation why chancroid is generally less severe for the female and is diagnosed 10 times more frequently in the male.

II. WHO IS AT RISK

The physical contact required for reproduction provides an opportunity for transfer of microbes. Any life form that replicates sexually is then theoretically at risk for STDs. Table 1 gives some examples of nonhuman STDs.

Consistent with the observation that humans suffer from many more diseases than lower life forms, the largest majority of recognized STDs afflict man. The small number of STDs that have been recognized in nonprimates are generally viral. Some pathogenic plant viruses infect pollen, and infection can then be spread from plant to plant by pollinating insects. Of the few known STDs in lower mammals, canine herpesvirus infection most accurately mimics the manifestations of its counterpart in humans. The canine virus is biochemically similar to human herpes simplex virus, but dogs are its only host. Bacteria that are significant sexually transmitted pathogens in humans, such as *Neisseria gonorrhea* and *Treponema pallidum*, have no pathologic significance for dogs or cats. Thus, it is difficult to find animal models to study human STDs.

TABLE 1
Some Diseases in the Plant and Animal Kingdoms That Can Be Sexually Transmitted

Various plant viruses if pollen becomes infected
Murine leukemia
Feline leukemia
Bovine and canine herpes virus infection
Simian leukemia and immunodeficiency virus

Populations that are at higher risk for STDs have been identified. Human STD risk factors include youth, low socioeconomic status, drug abuse, multiple sexual partners, or a partner with risk factors. In popular literature, STDs have been inaccurately characterized as a problem limited to night clubs and prostitutes; these social factors only contribute a relatively small number to the large group at risk. Careful investigation may reveal evidence of microbes that cause STDs in 30% of the general population during the reproductive years. However, the range varies widely among different groups, from under 5% to more than 90%. Many in the high-risk groups have markers for multiple STDs.

III. CONSEQUENCES OF STDs

Infertility as a result of STDs is a problem unique to humans, specifically women. STD-associated infertility is generally due to Fallopian tube occlusion from untreated chlamydia or gonorrhea. A chlamydia antibody screen may be part of the routine female infertility evaluation.

When symptomatic, the majority of STDs have local manifestations. Generally, the genital lesions are annoying but not life-threatening. Some are only transient and may resolve spontaneously after a few months (warts and molluscum contagiosum), but others may be recurrent (herpes) or persistent (trichomonas, lymphogranuloma venereum, and granuloma inguinale). Table 2 lists STDs that may have systemic manifestations, often associated with significant disability. These diseases are occasionally life-threatening.

Immunocompromise from chemotherapeutic agents, immunosuppressant drugs, or acquired immunodeficiency syndrome results in more dramatic manifestations of STDs. Genital lesions are larger, more destructive, and are often resistant to therapy.

TABLE 2
Sexually Transmitted Diseases That Have a Systemic Effect

Gonorrhea: Fever, septic arthritis, Fallopian tube infection (salpingitis)
Syphilis: Various severe neurologic problems
Chlamydia infection: Severe destructive arthritis (Reiter's syndrome)
AIDS: Severe immunocompromise
Hepatitis B: Chronic hepatitis, liver failure, liver cancer

Systemic symptoms progress more rapidly and are more severe.

Antepartum or intrapartum spread to the neonate occurs with some STDs. The manifestations of disease are often more severe in the neonate than in the mother. This is partially due to the immaturity of the fetal immune system. Several of these diseases disrupt organogenesis as well and can lead to significant congenital anomalies and mental retardation. Table 3 lists some STDs that have an adverse fetal effect during pregnancy.

IV. IMPORTANT VIRAL STDs IN HUMANS

A. Acquired Immunodeficiency Syndrome

Acquired immunodeficiency syndrome (AIDS) is due to human immunodeficiency virus (HIV), an RNA virus that chiefly infects and eventually destroys lymphocytes in the immune system. Initial infection with this virus may cause a brief illness characterized by rash and malaise, but thereafter the infection typically remains asymptomatic for 5 or more years. During this interval, immunocompromise gradually develops as the number of lymphocytes declines. With a decreased ability to control infection, death may result from various microbes that are easily controlled in normal health. Many antiviral agents are available to prolong life for the AIDS patient and to prevent transmission to the newborn baby during childbirth. Vaccine research is hopeful.

TABLE 3
Sexually Transmitted Diseases That Have an Adverse Fetal Effect during Pregnancy

Gonorrhea: Corneal scarring with blindness
Chlamydia: Pneumonia, conjunctivitis in the newborn period
Syphilis: Multiple skin and skeletal anomalies, mental retardation
Herpes: Serious infection in the newborn period, mental retardation, neurologic problems
Hepatitis B: Chronic hepatitis, liver cancer
AIDS: Childhood AIDS
Genital warts: Laryngeal warts with respiratory distress

B. Herpes Simplex Virus

Herpes simplex virus (HSV) has two serotypes. Type 1 results in oral "fever blisters" and recurrent pharyngitis in childhood. Serotype 2 causes small painful recurrent genital blisters that ulcerate and persist for 5–10 days. Recurrent episodes are most frequent in the first 1 or 2 years after exposure and are less severe than the primary episode. Antiviral agents (acyclovir, valacyclovir, and famciclovir) hasten healing but do not prevent recurrent disease. Disseminated HSV infection in immunocompromised patients is life-threatening. With a relatively immature immune system, newborns may develop serious systemic infection from HSV exposure during childbirth. Antiviral agents during the last weeks of pregnancy may help reduce the risk of neonatal infection if the mother is known to have had HSV infection in the past.

C. Human Papillomavirus

Human papillomavirus(HPV) has a prevalence in the genital tract of 40–90% in some high-risk populations. Only a small portion (possibly 1%) of those who are infected will manifest symptoms, although asymptomatic carriers are considered to be infectious to others. Genital warts, cervical precancer (dysplasia), and squamous cell cancer of the cervix are the common manifestations. Areas of skin involved in reactive or reparative changes are more likely to develop these HPV lesions. Vaginal cancer and approximately one-third of vulvar squamous cancers are attributed to HPV. Penile squamous cancer is also due to HPV. Surgical or destructive therapy (cautery) removes the HPV-associated lesions but does not eliminate the virus from the genital tract. There is a high recurrence rate after therapy. Significant immunocompromise allows cervical cancer to progress more rapidly and increases the likelihood of extensive genital warts that are difficult to control.

D. Hepatitis B

Hepatitis B due to the hepatitis B virus (HBV) is a bloodborne infection that can be spread sexually and during childbirth. Ten percent of infected adults and the majority of infected newborns become chronic HBV carriers. Death from fulminant hepatitis occurs rarely (<3% of symptomatic cases), but chronic hepatitis B results in a significant lifetime risk for liver failure or hepatocellular cancer. An effective vaccine is available for preventive use in high-risk populations.

E. Molluscum Contagiosum

Molluscum contagiosumis a minor genital skin disorder caused by a pox virus. Characteristic lesions are 1 or 2 mm in diameter and are raised with a central indentation that contains a core of cellular debris that can be extruded. Lesions may resolve spontaneously after several months, but cryotherapy with liquid nitrogen is usually employed to hasten resolution. AIDS patients may have extensive molluscum contagiosum lesions that are only made worse by therapeutic attempts.

V. IMPORTANT BACTERIAL STDs IN HUMANS

A. Syphilis

Syphilis is caused by a spirochete, *Treponema pallidum*, that is highly sensitive to low doses of penicillin, although a slow microbial replication rate mandates a 1- to 3-week course of therapy. The longer course of therapy is utilized for syphilis of ≧1 year duration. The initial genital lesion is a 1- or 2-cm painless shallow ulcer that heals spontaneously after 7–10 days. Persistent infection is manifested by the rash of secondary syphilis 6–8 weeks after the primary lesion. If untreated, tertiary syphilis with a multitude of life-threatening manifestations may develop after a few years. Aortic or cerebral aneurysm or neurologic damage that results in loss of balance and dementia are characteristic of end-stage syphilis. Untreated syphilis is generally transmitted to the fetus during pregnancy. Significant birth defects, including tooth and skeletal deformity, and mental retardation occur often. Unfortunately, congenital syphilis occasionally remains unrecognized and untreated until childhood. Screening and treatment for syphilis during pregnancy and the newborn period are mandatory.

B. Gonorrhea

Gonorrhea is a result of infection with the gonococcus (GC), *Neisseria gonorrhea*. Fragile vaginal epithelium prior to puberty allows development of GC vaginitis, but infection after puberty is limited to the cuboidal or columnar epithelium of the urethra and endocervix. Untreated GC urethritis in the male leads to urethral stricture. Endocervical GC may ascend to the endometrium during menses and then may spread further to the Fallopian tube (salpingitis). Increased risk for tubal pregnancy due to tubal mucosal damage, infertility from tubal occlusion, or pelvic abscess are common complications of salpingitis. Disseminated gonococcal infection (DGI) occurs if GC enters the bloodstream. Septic arthritis, high fever, and malaise are characteristic of DGI. Pregnancy increases the risk of DGI, which is typically due to an arginine-, hypoxanthine-, uracil-dependent auxotype of GC. Many antibiotics are effective against GC, although penicillin-resistant strains of GC are rising in prevalence. Single-dose oral therapy with azithromycin or a broad spectrum cephalosporin is currently preferred.

C. Bacterial Vaginosis

Bacterial vaginosis (BV) is a syndrome characterized by overgrowth of anaerobic bacteria that are normally present in the vagina. The patient complains of vaginal discharge and odor. BV is currently attributed to some yet unidentified sexually transmitted microbe that is believed to inhibit hydrogen peroxide-producing lactobacilli in the vagina. This unknown pathogen apparently does not persist in the vagina since suppression of the anaerobic bacteria with the antibiotics, metronidazole or clindamycin, allows lactobacilli to recolonize the vagina with resolution of BV symptoms. With this syndrome, the anaerobic bacteria proliferate to equal the quantity found in an anaerobic abscess. Thus, BV should be treated prior to childbirth or elective pelvic surgery. During pregnancy, BV may contribute to preterm labor. The presumed pathogen that causes BV apparently does not persist in the male, so he generally remains asymptomatic and does not require therapy.

D. Chlamydia trachomatis

Chlamydia trachomatis (CT) typically causes urethritis in the male and asymptomatic endocervicitis in the female. CT may persist in the genital tract for an indefinite time if untreated. Infertility and an increased risk for tubal pregnancy result from chlamydia salpingitis, which appears not to be due to direct microbial invasion but may be immune mediated. All strains of chlamydia are highly sensitive to erythromycin and tetracycline. Current preferred therapy is a single oral dose of azithroymcin. Intrapartum chlamydia exposure may cause conjunctivitis and pneumonia in the newborn. Lymphogranuloma venereum (LGV) is a chronic ulcerative inflammatory condition of the vulva, typically associated with unilateral enlargement of the groin lymph nodes. LGV is due to an L1, L2, or L3 serotype of CT. LGV may mimic the clinical appearance of invasive cancer and therapy includes surgical excision as well as antibiotics.

E. Chancroid

Chancroid typically appears as a deep painful necrotic 1- or 2-cm diameter genital ulcer with drainage. Lymph nodes in the groin are usually enlarged and tender. Fever and malaise are characteristic. Men are diagnosed with this disorder more frequently than women. The pathogen, *Hemophilus ducreyi*, is resistant to penicillin, but it is sensitive to a variety of other antibiotics. Preferred therapy is oral azithromycin or intramuscular ceftriaxone.

F. Granuloma Inguinale

Granuloma inguinale (GI) is a chronic ulcerative condition of the groin and vulva due to *Calymmatobacter granulomatis*. The clinical presentation is similar to invasive cancer, and the diagnosis must be established by biopsy. Therapy includes surgical excision of necrotic tissue areas and oral tetracycline.

VI. OTHER COMMON STDs

A. Trichomonas Vaginalis

Trichomonas vaginalis causes annoying vaginal discharge, but the male is typically asymptomatic. The infected cervix may bleed after sexual contact or when a Pap smear is obtained. Metronidazole is effective therapy.

B. Scabies

Scabies, due to *Sarcoptes scabiei*, is characterized by small red patches that itch in the web space between the fingers, on the genitalia, and around the belt line at the stomach. Microscopic examination of material scraped from a lesion may reveal the scabies mite. A single application of an insecticide in a cream form is usually effective.

C. Pubic Lice

Pubic lice (pediculosis pubis) cause itching in the area of the genital hair. Careful inspection reveals the *Phthirus pubis* mites as well as egg sacks attached to hair shafts. An insecticide in a shampoo form is helpful.

VII. WHAT IS NOT A STD

Many microbes are transmitted by intimate contact, yet all do not qualify as STDs. Our current understanding of the pathophysiology of vulvovaginal candidiasis is best characterized by limited truth but extensive folklore. Thus, the popular belief that candida infection is a STD has little scientific basis. Sexual partners typically are colonized with the same DNA serotype of *Candida albicans* but symptomatic candidia infections are most likely associated with local factors other than intercourse. Similarly, genital group B streptococcal colonization is sexually shared, but it is asymptomatic and it is not generally viewed as a STD.

Vulvodynia, chronic vulvar pain often made worse during intercourse, is not characteristic of any STD. The cause for this unfortunate pain often remains obscure but may be related to vulvar irritant dermatitis or some other dermatopathology. The exacerbation of pain by sexual intercourse often raises the false concern that there may be a STD.

VIII. ALTERNATE MEANS OF STD TRANSMISSION

It is possible for some STDs to be spread by nonsexual contact. Syphilis, HIV, and HBV can be transmitted to laboratory and hospital personnel by blood contact. Blood banks routinely screen donor blood for these diseases. Improperly sterilized surgical instruments may also transmit these disorders. Hand contact with HSV and HPV lesions may spread these infections to other body sites as well as to other individuals. Most STD pathogens are not hardy enough to persist on inanimate objects. Although the theoretical possibility exists at least for HPV and molluscum contagiosum, it is unlikely that a STD will be contracted from contaminated door knobs or toilet seats.

IX. STD TREATMENT VERSUS ERADICATION

Pharmaceutical research is providing better agents for STD treatment. New antibiotics that are effective as a single oral dose, such as azithromycin, substantially improve compliance with therapy. There is renewed interest in vaccines for STD prevention. As research reveals the mechanisms of STD pathophysiology, drugs that inhibit adhesion or improve the immune defenses may become available.

Unfortunately, STDs cannot be eliminated by antimicrobial protocols alone. The more optimal management of STDs includes lifestyle changes that would eliminate exposure. Social planning that could eradicate STDs is unrealistic because of the necessary restrictions of free agency, and community education programs are expensive. STD clinics are often poorly funded, work with high volumes of patients, and use primitive equipment. In this setting, adequate STD therapy remains challenging, and STD prevention often appears impossible. Thus, STDs remain a social problem with no easy solution. The advocacy of abstinence or condom use to limit microbial exposure is the only social planning that has achieved general approval. Even with our expanded technology, social considerations suggest that STDs are likely to persist as a significant health threat.

See Also the Following Article

HIV (AIDS)

Bibliography

Alexander, N. J., Gabelnick, H. L., and Spieler, J. M. (1990). *Heterosexual Transmission of AIDS*. Wiley-Liss, New York.

Aral, S. O., Holmes, K. K., Padian, N. S., and Cates, W. (1996). Overview: Individual and population approaches to the epidemiology and prevention of sexually transmitted diseases and human immunodeficiency virus infection. *J. Infect. Dis.* 174(Suppl. 2), S127–S133.

Catania, J. A., Binson, D., Dolcinc, M. M., Stull, R., Choi, K. H., Pollack, L. M., *et al.* (1995). Risk factors for HIV and other sexually transmitted diseases and prevention practices among US heterosexual adults: Changes from 1990 to 1992. *Am. J. Public Health* 85(11), 1492–1499.

Drugs for sexually transmitted diseases (1995). *Med. Lett. Drugs Ther.* 37, 117–122.

Greene, C. E. (1984). *Clinical Microbiology and Infectious Diseases of the Dog and Cat.* Saunders, Philadelphia.
Holmes, K. K., Mardh, P. A., Sparling, P. F., and Wiesner, P. S. 1990). *Sexually Transmitted Diseases*, 2nd ed. McGraw-Hill, New York.
Johnson, R. B. (1996). Azithromycin. *Curr. Problems Dermatol.* **24**, 184–193.
Woods, G. L. (1995). Update on laboratory diagnosis of sexually transmitted diseases. *Clin. Lab. Med.* **15**(13), 665–684.

SHBG (Sex Hormone-Binding Globulin)

William Rosner
St. Luke's/Roosevelt Hospital Center, Columbia University, College of Physicians and Surgeons

GLOSSARY

binding site The specific area of a molecule to which other molecules can attach.

dimer A molecule that is produced by binding together two molecules of the same chemical structure.

glycoprotein A protein with carbohydrate sugars covalently attached.

hirsutism A condition in human females in which there is excessive hair growth, as on the face or chest.

signal transduction A series of ordered biochemical reactions by which a signal is initiated at the cell surface and transmitted to elicit specific cellular responses.

Sex hormone-binding globulin is a plasma protein that binds a number of steroidal estrogens and androgens with high affinity. Although its most generally accepted appellation is sex hormone-binding globulin, it frequently is called testosterone–estradiol-binding globulin and steroid-binding protein.

I. INTRODUCTION

The initial description of sex hormone-binding globulin (SHBG) followed the discovery of an activity in plasma that bound estradiol or testosterone. It was shown subsequently that both the activities resided in the same molecule, a homodimeric glycoprotein synthesized in the liver. In the early 1970s a similar activity was demonstrated in testis and epididymis. The protein possessing this activity, named androgen-binding protein (ABP), is synthesized and secreted by testicular Sertoli cells and is a product of the same gene as SHBG. Although the amino acid sequences of SHBG and ABP are identical, they are differentially glycosylated and appear to serve different functions. ABP remains largely within the testis, wherein, after its secretion by the Sertoli cell, it is taken up by receptor-mediated endocytosis by both epididymal and germ cells. This article concentrates on SHBG.

II. STRUCTURE OF SHBG

SHBG is a glycoprotein whose amino acid sequence has been determined directly and deduced from its cDNA. The SHBG monomer is a single-chain polypeptide that contains 373 amino acids, a single tyrosine, two disulfide bonds, and three carbohydrate side chains. SHBG forms homodimers and it is probable, but not proven, that it is this state of the protein that binds steroids, 1 mol of steroid/mol dimer. The molecular infrastructure that enables steroid binding to the dimer, but not the monomer, has yet to be elucidated, although there is some indirect evidence that binding takes place at the dimeric interface. Unlike the steroid receptors that have well-demarcated domains, e.g., steroid binding, DNA binding, dimerization, and transactivation, SHBG, although having regions important in steroid binding, receptor binding, and dimerization, is somewhat more loosely organized. Although there are regions of SHBG (amino acids 134–150) that are more important than others in allowing binding of steroids, it appears that SHBG needs most of its sequence to allow the proper conformation for steroid binding to occur. This same general region is involved in dimerization. Mutating amino acids 138–148 impairs but does not abrogate dimerization. Although this suggests that steroid binding is involved in dimerization, the situation is not so simple. SHBG is a dimer in the absence of steroid, but the presence of steroids abrogates the impairment of dimerization consequent to mutations of amino acids 138–148. Ca^{2+} protects SHBG from heat denaturation, indicating that removal of this divalent cation destabilizes dimer formation of mutants with substitutions at residues 140–148.

SHBG is microheterogeneous due to differentially glycosylated species. Protomers of human SHBG with molecular weights of 52 and 48 kDa collapse to a single species of 42 kDa after complete deglycosylation. Removal of the sialic acid moieties from SHBG leaves its ability to bind steroids and antibodies intact but prevents its binding to endometrial membranes. Indeed, SHBG containing no oligosaccharides binds steroids, dimerizes, and binds normally to both monoclonal and polyclonal antibodies.

SHBG binds to receptors on cell membranes. At least part of the receptor-binding domain of SHBG is contained in the decapeptide encompassing amino acids 48–57. This decapeptide and its immediate surroundings are the most highly conserved sequences of SHBG between species as well as between SHBG and a variety of proteins (protein S, laminin, merosin, agrin, and the *Drosophila crumbs* and *slit* proteins), none of which bind steroids but all of which bind to other macromolecules as part of their function. It should be noted that there are no homologies between SHBG and either the androgen (AR) or estrogen (ER) receptors. This notwithstanding, the principal ligands for the AR (dihydrotestosterone and testosterone) and the principal ligand for the ER (estradiol) are the major ligands for SHBG.

III. FUNCTIONS OF SHBG

A. Regulation of the Concentration of Free Steroids in Plasma

Only four steroids in the circulation bind to SHBG with high affinity. Their binding affinities, relative to testosterone, are as follows: dihydrotestosterone, 3.6; 5α-androstane-3α,17β-diol, 1.6; and estradiol, 0.62. Although, in any given tissue, the rate of steroid hormone uptake may be determined by either the free or the total hormone concentration, there is general agreement that steroids must pass through the pool of free extracellular hormone to enter cells. Indeed, in both health and disease, the free hormone concentration generally correlates better than the total with the biological effects of steroids.

Because the plasma concentration of SHBG determines, in large part, the concentration of free hormone, factors that influence this concentration play a role in regulating free testosterone, dihydrotestosterone, and estradiol in plasma. SHBG in plasma is measured either by its steroid-binding capacity or by one or another type of immunoassay. Although there is no universal standard, the concentration range of SHBG in young men is 10–45 n*M* and that in menstruating women 25–90 n*M*. This difference between the sexes is due to differences in circulating estrogens. Substantial increases in estrogen concentrations, as in pregnancy or oral administration of estrogens, are associated with marked increases in SHBG (as much as 8- to 12-fold). The mechanism whereby estrogens affect these changes remains problematic because studies *in vitro* show either no or only modest effects of pharmacological doses of estrogens on SHBG secretion by hepatocytes. Thyroid hormones also cause large increases in plasma SHBG; these increases are probably mediated by a direct effect of thyroid hormones on the synthesis and secretion of SHBG by hepatocytes. Although estrogens and thyroid hormones cause the most impressive increases in SHBG, there are a number of other factors that also do so. As men age, their SHBG rises. This increase, combined with the fall in total plasma testosterone that occurs with age, leads to an exacerbated fall in free testosterone and, taken together with clinical changes in aging, has contributed to a renewed interest in hormonal replacement in the aging male. Just as there are only a small number of factors that result in increases in SHBG, there are a limited number of conditions in which SHBG levels are lowered. Perhaps the most impressive of these is obesity, wherein the lowest concentrations of plasma SHBG are seen. This decrease in SHBG is accompanied by a reduction not only in total testosterone but also in free testosterone; the latter is an effect of obesity on plasma testosterone independent of the effects of obesity on SHBG. Androgens themselves decrease plasma SHBG *in vivo*. As for estrogens, the mechanism is unclear because studies *in vitro* do not show convincing and reproducible decreases in SHBG production and secretion by hepatocytes consequent to exposure to androgens. *In vitro*, two of the most potent inhibitors of SHBG secretion by hepatic cells are insulin and IGF-1. The *in vivo* correlate of the effects of insulin appears in individuals with insulin resistance, e.g., obesity, type II diabetes mellitus, and polycystic ovarian syndrome; they have high plasma insulin and low plasma SHBG. The *in vivo* correlates of the effects of IGF-1 (whose plasma concentration is increased by growth hormone) include depressed SHBG in patients with acromegaly and a fall in SHBG consequent to treatment with growth hormone.

Plasma concentrations of SHBG have been evaluated in numerous human diseases. The most widely studied have been disorders resulting in clinical signs of androgen excess (hirsutism, acne, and alopecia) in women. The conceptual role of SHBG in hyperandrogenic syndromes in women is based on the free hormone hypothesis. That is, modest degrees of androgenization in women are thought to be dependent on the concentration of plasma free testosterone. This hypothesis is buttressed by observances that the clinical state of these patients correlates better with plasma free testosterone than with plasma total testosterone. Coupled with this idea, it is important to recall that in women, unlike in men, there is no homeostatic mechanism that is sensitive to the plasma concentration of free testosterone. An increase in free androgen, to the extent seen in hirsute women, does not downregulate ovarian and adrenal steroid secretion. Additionally, there are a number of events that exacerbate the problem. Some of the syndromes, e.g., polycystic ovarian syndrome, associated with increased androgen concentrations in plasma are also associated with obesity and/or increased plasma insulin that further depress SHBG and elevate the free testosterone concentration. The elevated free testosterone, in turn, depresses SHBG even more. Total plasma testosterone is often normal when there is an increased production of testosterone in these patients. Thus, clinical signs of androgen excess exist because of an elevated free testosterone in the face of a normal total testosterone, and it is free testosterone which is the more sensitive laboratory marker of this group of disorders.

In addition to syndromes in which the primary clinical manifestation relates to the classic, secondary sex characteristic-generating effects of androgens, plasma SHBG has been examined in a number of disorders wherein androgens or estrogens are thought to play an important pathogenic role. In diseases of the prostate, prostate cancer and benign prostatic hypertrophy, SHBG may be high, low, or normal. Similarly, in breast cancer, no definitive alterations in SHBG have been found. A number of investigators, but not all, have found significant correlations between plasma cholesterol and SHBG. The clinical significance of these observations remains obscure. Low plasma SHBG is a powerful independent risk factor/predictor for the development of diabetes mellitus. This risk is independent of body weight, body mass index, and waist-to-hip ratio. Once again, the clinical significance of these observations is not apparent, but it is difficult to entertain reasonable hypotheses that encompass a role for SHBG in the pathogenesis diabetes mellitus. Estrogens are an important regulator of bone density in women and a number of investigations have examined bone density/osteoporosis in relation to SHBG. The driving hypothesis has been that a high SHBG would result in decreased free estrogens and hence decreased bone density. Indeed, there is a strong negative correlation between SHBG and bone density. Most interestingly, the relationship between SHBG and bone density is an independent predictor of osteoporosis, unrelated to steroid concentrations. It has been considered that these observations may be related to the SHBG receptor system that exists in bone.

B. Mediation of Steroid Signal Transduction at the Plasma Membrane

In addition to its role in regulating the plasma concentration of free androgens and estrogens, SHBG is a central component of a steroid signal transduction system that resides in the cell membrane. The standard model of steroid hormone action posits that intracellular receptors mediate their hormonal effects. These receptors have nascent transcriptional activity that steroids unmask, thus allowing the transcription of specific genes. SHBG provides an additional mechanism for steroid hormones to transmit information. It permits certain steroids to generate a second messenger, cAMP, through the intermediacy of an SHBG receptor, R_{SHBG}, on cell membranes.

There are two binding sites on SHBG, one for steroids and the other for the membrane receptor. Upon binding a steroid, SHBG undergoes a conformational change, resulting in masking of the receptor-binding site and inhibiting its interaction with membranes. However, unliganded SHBG maintains the ability to bind steroids after it binds to its receptor. Moreover, exposure of the SHBG–receptor complex to steroids that bind to SHBG results in the dissociation of the complex, but the rate of this dissociation is very slow ($t_{1/2} > 30$ hr). Thus, there is more than adequate time for events other than dissociation to be engendered. Indeed, if the steroid that binds to the preformed SHBG–R_{SHBG} is suitable, then intracellular cAMP accumulation ensues within minutes. It should be emphasized that steroids exert control on this system at two distinct loci. At the first locus, occupation of SHBG's steroid binding site inhibits binding of SHBG to its receptor. This inhibition is an exclusive function of the avidity of the steroid for SHBG; it is independent of the steroid's biological activity. Given this circumstance, it is apparent that the receptor-active moiety of SHBG is the unliganded form. At the second locus, certain steroids that bind to SHBG, after it binds to its receptor, induce the accumulation of intracellular cAMP. There are only three steroids that are effective agonists, i.e., cause SHBG–R_{SHBG} to activate adenylyl cyclase in this system: dihydrotestosterone, estradiol, and 5α-androstane-3α,17β-diol. Any steroid that binds to SHBG is an antagonist of these agonists. Agonists and antagonists alike prevent the binding of SHBG to R_{SHBG}.

Studies addressing events that occur as a consequence of the activation of adenylyl cyclase and the subsequent increase in cAMP are just beginning to appear. The estradiol-engendered increase in the growth of breast cancer cells *in vitro* is inhibited by the SHBG–R_{SHBG} system, whereas the same system stimulates the growth of prostate cancer cells *in vitro*. Both of these studies illustrate the role of SHBG–R_{SHBG} in providing a mechanism for steroids to act in addition to the mechanism provided by classic intracellular steroid receptors. Recently, it has been demonstrated that there is cross talk between the SHBG–R_{SHBG} system and the intracellular androgen receptor. In serum-free organ culture of human prostates, dihydrotestosterone, as expected, caused an increase in prostate-specific antigen secretion. This event was blocked by antiandrogens. In the absence of androgens, estradiol added to prostate tissue, whose R_{SHBG} was occupied by SHBG, reproduced the results seen with dihydrotestosterone. Neither estradiol alone nor SHBG alone duplicated these effects. Antiestrogens did not block the estradiol–SHBG-induced increase in prostate-specific antigen, but the increase was blocked both by antiandrogens and by a steroid antagonist of SHBG–R_{SHBG}, i.e., one that prevents binding of estradiol to SHBG. Furthermore, an inhibitor of protein kinase A (a necessary mediator of the effects of cAMP) prevented the estradiol–SHBG-induced increase in prostate-specific antigen but not that which followed dihydrotestosterone. These results indicate that SHBG–R_{SHBG} is part of a signaling system that amalgamates steroid-initiated intracellular events with steroid-dependent occurrences generated at the cell membrane. Whether the presence of classic steroid receptors is necessary for all effects of SHBG–R_{SHBG} remains to be determined.

See Also the Following Articles

ANDROGENS; ESTROGENS, OVERVIEW; TESTOSTERONE BIOSYNTHESIS AND ACTIONS

Bibliography

Gray, A., Berlin, J. A., McKinlay, J. B., and Longcope, C. (1991). An examination of research design effects on the association of testosterone and male aging: Results of a meta-analysis. *J. Clin. Epidemiol.* **44**, 671–684.

Hammond, G. L. (1990). Molecular properties of corticosteroid binding globulin and the sex-steroid binding proteins. *Endocr. Rev.* **11**, 65–79.

Joseph, D. R. (1994). Structure, function, and regulation of androgen-binding protein/sex hormone-binding globulin. *Vitamins Horm.* **49**, 197–280.

Loukovaara, M., Carson, M., and Adlercreutz, H. (1995). Regulation of production and secretion of sex hormone-binding globulin in HepG2 cell cultures by hormones and growth factors. *J. Clin. Endocrinol. Metab.* **80**, 160–164.

Mendel, C. M. (1992). The free hormone hypothesis. Distinction from the free hormone transport hypothesis. *J. Androl.* **13**, 107–116.

Nakhla, A. M., Khan, M. S., Romas, N. A., and Rosner, W. (1994). Estradiol causes the rapid accumulation of cAMP in human prostate. *Proc. Natl. Acad. Sci. USA* **91**, 5402–5405.

Parker, M. G. (1993). Steroid and related receptors. *Curr. Opin. Cell Biol.* **5**, 499–504.

Petra, P. H. (1991). The plasma sex steroid binding protein (SBP or SHBG). A critical review of recent developments on the structure, molecular biology and function. *J. Steroid Biochem. Mol. Biol.* **40**, 735–753.

Rosner, W. (1990). The functions of corticosteroid-binding globulin and sex hormone-binding globulin: Recent advances. *Endocr. Rev.* **11**, 80–91.

Rosner, W. (1991). Plasma steroid-binding proteins. In *Endocrinology and Metabolism Clinics of North America. Steroid Hormones: Synthesis, Metabolism and Action in Health and Disease* (J. F. Strauss, Ed.), pp. 697–720.

Rosner, W. (1996). Sex steroid transport: Binding proteins. In *Reproductive Endocrinology, Surgery, and Technology* (E. Y. Adashi, J. A. Rock, and Z. Rosenwaks, Eds.), pp. 605–626. Raven Press, New York.

Slemenda, C., Longcope, C., Peacock, M., Hui, S., and Johnston, C. C. (1996). Sex steroids, bone mass, and bone loss. A prospective study of pre-, peri-, and postmenopausal women. *J. Clin. Invest.* **97**, 14–21.

Strel'chyonok, O. A., and Avvakumov, G. V. (1990). Specific steroid-binding glycoproteins of human blood plasma: Novel data on their structure and function. *J. Steroid Biochem.* **35**, 519–534.

Westphal, U. (1986). *Steroid–Protein Interactions II.* Springer-Verlag, New York.

Sheehan's Syndrome

Peter J. Snyder
University of Pennsylvania

GLOSSARY

diabetes insipidus The clinical syndrome of polyuria and polydipsia as a consequence of the inability to concentrate urine due to either deficient secretion of vasopressin or the inability of the renal collecting ducts to respond to it.

hypoadrenalism Deficient secretion of adrenal hormone.

hypopituitarism Subnormal secretion of the anterior pituitary hormones and of the target gland hormones they stimulate.

secondary hypoadrenalism Subnormal secretion of cortisol by the adrenal glands as a consequence of subnormal secretion of adrenocorticotropin by the pituitary gland.

secondary hypogonadism Subnormal secretion of the hormones secreted by the gonads, e.g., estradiol by the ovaries as a consequence of subnormal secretion of follicle-stimulating hormone and luteinizing hormone by the pituitary gland.

secondary hypothyroidism Subnormal secretion of thyroxine by the thyroid gland as a consequence of subnormal secretion of thyrotropin by the pituitary gland.

sella tunica The bony cavity in which the pituitary gland normally sits.

In the late 1930s Sheehan reviewed the case histories of approximately 100 women who developed symptoms and signs of hypopituitarism following childbirth and concluded that hypotension due to hemorrhage during delivery causes necrosis of the pituitary and eventual hypopituitarism. Although Sheehan and others subsequently provided more details about the phenomenon and its etiology, our understanding of this disease remains essentially that first elucidated by Sheehan.

I. THE CLINICAL SYNDROME

The first cases described by Sheehan represented relatively severe examples of the disease and are summarized by the following composite in one of Sheehan's first papers on the subject (Sheehan, 1937, 1939):

> During the puerperium there is complete absence of lactation and sometimes hypoglycaemia. After this, the uterus becomes superinvoluted and the external genitalia atrophic. Menstruation does not return, and libido is absent. There is a gradual loss of axillary and pubic hair. The patient is apathetic. . . . After 10, 20, or 30 years the patient may become more typically myxoedematous, or may develop mental changes with anorexia and some loss of weight. . . . Finally, the patient goes into coma and dies. . . . Postmortem, the anterior pituitary is represented chiefly by the large scar of the original postpartum necrosis, the suprarenal cortex is atrophic, the thyroid usually shows fibrous atrophy, the ovaries and uterus are shrunken.

Subsequently, Sheehan and others described a wide range of hypopituitarism in this syndrome. The most severe cases exhibit hypopituitarism even more profound and sudden in onset than that described previously, especially with regard to hypoadrenalism; these may be accompanied by severe lassitude, hypotension, and death within the first month postpartum. Much less severe cases have also been recognized. Hypothyroidism, hypoadrenalism, and hypogonadism may all be partial, and some patients may have subnormal secretion of some pituitary hormones and normal secretion of others. Partial hypogonadism is most dramatically illustrated by reports of occasional menstruation and even pregnancy. Another feature of partial hypogonadism is a delayed onset of amenorrhea; menses may resume for several cycles before stopping. Diabetes insipidus has rarely been reported in association with Sheehan's syndrome.

II. ETIOLOGY

Sheehan's syndrome occurs when hypotension, usually due to hemorrhage, occurs during childbirth. Sheehan (1939) noted that

> The disease always dates from a delivery at which the patient has been in collapse, usually as a result of severe hemorrhage. In general, the worse the patient's condition at the time of delivery, the severer are the subsequent symptoms. The collapse is due to some obstetrical complication, such as retained placenta, postpartum haemorrhage, placenta praevia. . . . [T]he obstetric history is absolutely characteristic when it can be obtained, and is essential for a definite diagnosis during the patient's life.

Postulation of the causal relationship between hypotension and pituitary necrosis is supported by the observation in a single patient that hypotension due to severe gastric hemorrhage in late pregnancy also led to hypopituitarism. Because hypotension at other times does not usually lead to pituitary necrosis, it is assumed that the hyperplastic pituitary of late pregnancy is unusually susceptible to damage by hypotension, although pathologic examination of the pituitary vasculature of patients who died during the first few hours after postpartum hemorrhage and hypotension revealed no abnormalities.

III. PATHOLOGY

Pathologic examination of the anterior pituitary of patients who died after postpartum hemorrhage reveals necrosis that is typical of that seen in any tissue. The histologic pattern depends on the time since the inciting event, and the extent depends on the severity of the insult. Atrophy and scarring may also occur in the posterior lobe, but it is not seen until years after the insult, is less severe, and is not accompanied by damage to the stalk.

IV. DIAGNOSIS

Diagnosis of Sheehan's syndrome is made first by suspecting the diagnosis because of symptoms and signs of hypopituitarism in a patient who has a history of postpartum hemorrhage and then by confirming hypopituitarism by appropriate laboratory tests.

A. Symptoms

The symptoms of Sheehan's syndrome are similar to those of hypopituitarism of any etiology. One exception is the relationship of the symptoms to postpartum hemorrhage, including a typical delay in onset of symptoms, as described previously. Another exception is the inability to nurse following the inciting event. Typical symptoms of hypogonadism are amenorrhea, decreased libido, decreased vaginal lubrication, and decreased sexual hair. Typical symptoms of hypoadrenalism are lassitude, anorexia, and weight loss. Typical symptoms of hypothyroidism are weakness and cold intolerance.

B. Physical Findings

The physical findings of Sheehan's syndrome are similar to those of hypopituitarism of any etiology. Typical findings of hypogonadism are decreased sexual hair and, after many years, fine, wrinkled skin. A typical finding of severe hypoadrenalism is postural hypotension. Typical findings of hypothyroidism are a dull facial expression, dry skin, and delayed relaxation phase of the deep tendon reflexes.

C. Laboratory Test Results

Laboratory test results in Sheehan's syndrome are typical of those of hypopituitarism of any etiology. Secondary hypogonadism is diagnosed by finding a subnormal serum estradiol concentration associated with follicle-stimulating hormone and luteinizing hormone concentrations that are not elevated. Secondary hypoadrenalism is diagnosed by finding an early morning serum cortisol concentration that is subnormal and an adrenocorticotropin hormone

(ACTH) that is not elevated. If the early morning cortisol is not subnormal, a test of ACTH reserve, such as a metyrapone or insulin tolerance test, is necessary to confirm decreased ACTH secretion. Secondary hypothyroidism is diagnosed by finding a subnormal serum T4 associated with a serum thyroid-stimulating hormone (TSH) that is not elevated.

D. Imaging

Several studies have reported that the size of the pituitary, as judged by computed tomography, is smaller in patients who have Sheehan's syndrome than in age-matched controls, but the size of the sella turcica is usually normal (10–12). As a consequence, the appearance is that of an empty or partially empty sella. Findings by magnetic resonance imaging would presumably be similar, but they have not yet been reported.

E. Differential Diagnosis

Many other conditions cause hypopituitarism, but most are mass lesions, such as pituitary adenomas and craniopharyngiomas, and therefore are readily distinguishable by imaging from that caused by Sheehan's syndrome. Lymphocytic hypophysitis is a cause of hypopituitarism that might be confused with Sheehan's syndrome because of its association with pregnancy, but it is also usually associated with an enlarged pituitary, especially early in its course.

V. TREATMENT

Treatment of hypopituitarism due to Sheehan's syndrome is similar to treatment of hypopituitarism due to other causes. Hypoadrenalism should be treated with hydrocortisone, either as a single dose on arising in the morning or two-thirds of the dose on arising and one-third at noon. For a woman who weighs about 120 pounds, a dose of 20 mg a day is probably sufficient; for 150 pounds, 25 mg; and for 180 pounds, 30 mg. Unfortunately, measurement of serum or urine cortisol is of no value in assessing the adequacy of the dose, nor, of course, is plasma ACTH since the cortisol deficiency is due to ACTH deficiency. Mineralocorticoid treatment is not necessary because aldosterone secretion is not subnormal to a clinically important degree.

Hypothyroidism should be treated with L-thyroxine, and the dose should be monitored by the serum T_4 concentration. Serum TSH cannot be used to monitor the dose because the hypothyroidism is the result of TSH deficiency. L-Thyroxine treatment should not be initiated, however, until adrenal function has been evaluated and, if subnormal, treated.

Hypogonadism should be treated with estradiol transdermally and progestin orally, in cyclic fashion. The dose of estradiol should be monitored by the serum estradiol concentration and the doses of both by restoration of the patient's menses to their premorbid characteristics. If a woman should desire to become pregnant again, ovulation can be induced by gonadotropins in a fashion similar to that for secondary hypogonadism of any etiology.

See Also the Following Article

HYPOPITUITARISM

Bibliography

Aguilo, F., Jr., Vega, L. A., Haddock, L., and Rodriguez, O., Jr. (1969). Diabetes insipidus syndrome in hypopituitarism of pregnancy. Case report and a critical review of the literature. *Acta Endocrinol. (Copenhagen)* **60**(Suppl. 137), 1.

Bakiri, F., Bendib, S. E., Maoui, R., Bendib, A., and Benmiloud, M. (1991). The sella turcica in Sheehan's syndrome: Computerized tomographic study in 54 patients. *J. Endocrinol. Invest.* **14**, 193–196.

Fleckman, A. M., Schubart, U. K., Danziger, A., and Fleischer, N. (1983). Empty sella of normal size in Sheehan's syndrome. *Am. J. Med.* **75**, 585–591.

Koplin, R. S., Rosen, R., Brener, J. L., and Gordon, G. G. (1972). Pregnancy with Sheehan's syndrome after replacement therapy. *N. Y. State J. Med.* **72**, 1157–1159.

Martin, J. E., MacDonald, P. C., and Kaplan, N. M. (1970). Successful pregnancy in a patient with Sheehan's syndrome. *N. Engl. J. Med.* **282**, 425–427.

Sheehan, H. L. (1937). Post-partum necrosis of the anterior pituitary. *J. Pathol. Bacteriol.* **14**, 189–214.

Sheehan, H. L. (1939). Simmonds's disease due to post-partum necrosis of the anterior pituitary. *Q. J. Med.* **32**, 277–309.

Sheehan, H. L., and Stanfield, J. P. (1961). The pathogenesis of post-partum necrosis of the anterior lobe of the pituitary gland. *Acta Endocrinol.* 37, 479–510.

Sheehan, H. L., and Whitehead, R. (1963). The neurohypophysis in post-partum hypopituitarism. *J. Pathol. Bacteriol.* **85**, 145–169

Sherif, I. H., Vanderley, C. M., Beshyah, S., and Bosairi, S. (1989). Sella size and contents in Sheehan's syndrome. *Clin. Endocrinol.* **30**, 613–618.

Taylor, D. S. (1972). Massive gastric haemorrhage in late pregnancy followed by hypopituitarism. *Obstet. Gynaecol.* 79, 476–478.

Tulandi, T., Yusuf, N., and Posner, B. I. (1987). Diabetes insipidus: A postpartum complication. *Obstet. Gynecol.* 70, 492–495.

Sheep and Goats

Duane H. Keisler

University of Missouri at Columbia

GLOSSARY

buck A male goat.
doe A female goat.
ewe A female sheep.
kid A goat that is less than 1 year of age.
lamb A sheep that is less than 1 year of age.
ram A male sheep.

From an agricultural perspective, sheep and goats are among the most versatile livestock species, and certainly ruminant species, in adaptability to diverse production environments and purveyors of products throughout the world. Furthermore, among the 400–800 breeds of sheep and more than 200 breeds of goats, several breeds exist, such as Booroola Merino sheep and Ma T'ou goats, which are functionally litter-bearing ruminants. For these reasons and more, the world's population of sheep has established itself equally in developed and developing countries, whereas goats reside predominantly (by a factor of 10) in developing countries. Despite the reproductive potential among breeds of sheep and goats, reproductive performance is far from optimum. In this article, unique aspects of the reproductive processes in sheep and goats are reviewed.

I. BIRTH TO PUBERTY

At the time of birth of the ewe or doe, all major reproductive structures within the infant females are anatomically identifiable and structurally complete (Fig. 1). In particular, within the ovaries approximately 100,000 eggs exist, which defines the female's limited population of gametes. In reality, during the lifetime of a highly productive female, $<0.05\%$ of the animal's eggs will develop naturally into live offspring. The ability of the female to stand to be mated by a male (i.e., exhibit estrus), ovulate, and conceive begins at puberty. Prior to puberty, hormonal thera-

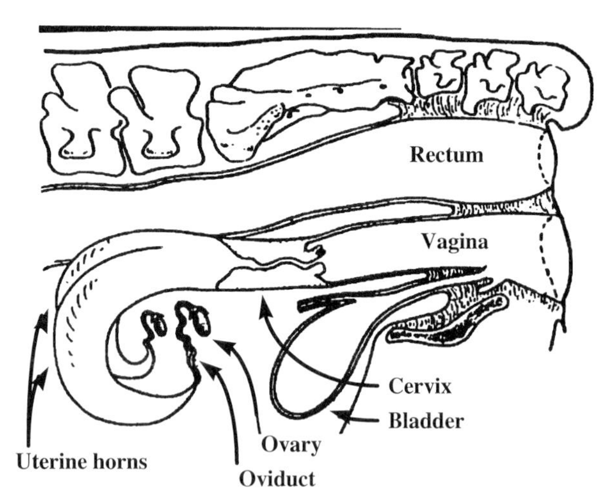

FIGURE 1 Anatomy of the reproductive tract of the ewe and doe.

pies have been used to induce young females to exhibit estrus and/or ovulate with varying levels of efficacy.

In contrast to the female, following the birth of the male lamb or kid, major anatomical changes have yet to be completed (Fig. 2). Constriction of the inguinal canal is incomplete; thus, the testis of the immature male can move freely between the body cavity and the scrotum. Eventual location of the testis in the scrotum is critical to promote spermatogenesis in which the process proceeds at temperatures 3–50C lower than core body temperature. Damage to the scrotum reduces dissipation of heat from the testis and thus is detrimental to the process of spermatogenesis. Developmentally, the growth curve of the testis is similar to that of the bull. Seminiferous tubules (in which sperm are produced and matured) occupy 50% of testicular volume at birth and that proportion increases to 80% at puberty. Movement of the penis within the sheath is restricted by adhesions that break down as the males mature. The urethral process, which is unique to the ram, is freed first from its adhesions followed by the glans penis. At the time of mating, an S-shaped curve (sigmoid flexure) within the fibroelastic penis of the ram and buck straightens to allow the penis to be extended. After mating, the retractor penis muscles contract, withdrawing the penis into the sheath and reshaping the S-shaped sigmoid flexure.

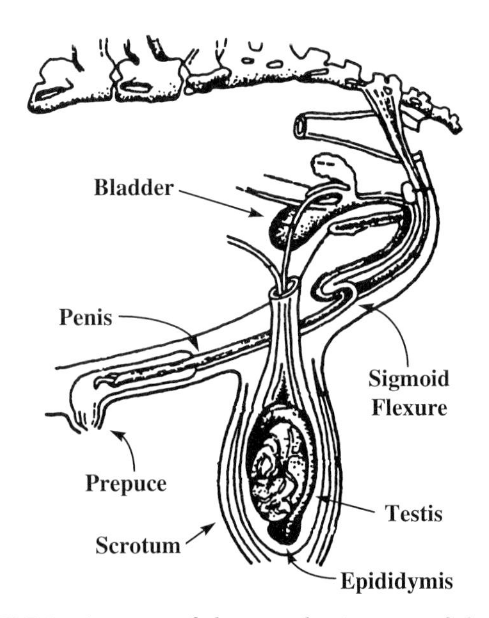

FIGURE 2 Anatomy of the reproductive tract of the ram and buck.

II. PUBERTY

Puberty is defined as the age when the animal is first able to reproduce. The age of puberty in both the male and female sheep and goat is dictated by genetic and environmental influences, which direct physiological mechanisms initiating reproductive activity. Four variables that play a major role in this process are breed, age, weight, and season. In general, with all conditions being optimum, age at puberty occurs earlier in British breeds of sheep, such as Southdown, Shropshire, and Dorset, than in sheep with Spanish ancestry, such as the Merino, Rambouillet, Columbia, and Corriedale. Research has also revealed that ewe lambs sired by rams selected for large testis size reached puberty earlier than ewe lambs from rams selected for small testis size.

Across all breeds of sheep and goats, age at puberty can range from 5 to 12 months, with most sheep and goats reaching puberty at 6–8 months of age, providing other conditions such as photoperiod and nutritional status are favorable. Once the animal has satisfied the minimum age requirement for puberty, the growing animal must also satisfy some minimum weight or body fat content to initiate the reproductive

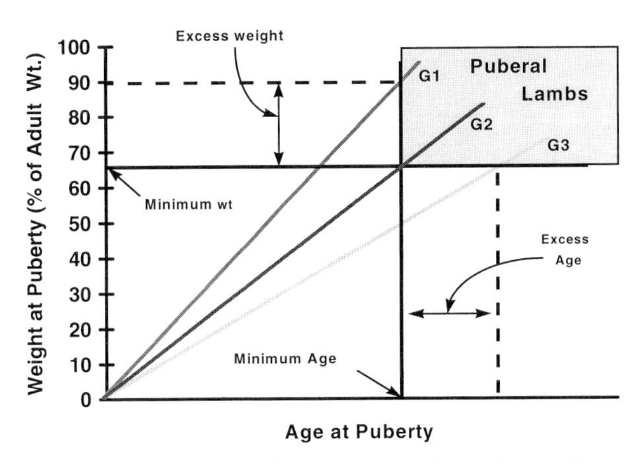

FIGURE 3 Relationship between age and weight at puberty in sheep.

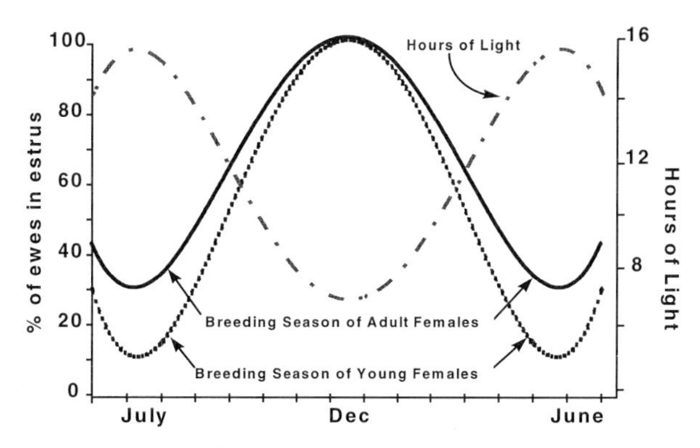

FIGURE 4 Breeding activity in sheep and goats relative to length of day and month of the year. The solid line symbolizes ewes with longer breeding seasons than those represented by the dashed line.

processes. Figure 3 illustrates the relationship between age and weight of offspring: (i) growing at a rapid rate (G_1) exceeding the minimum weight prior to satisfying the minimum age requirement; (ii) growing at an optimum rate (G_2), satisfying both age and weight requirements concurrently; and (iii) growing at a slow rate (G_3), exceeding the minimum age requirement prior to satisfying the minimum weight requirement. The shaded portion of Fig. 3 represents when each animal reaches puberty and the dashed lines indicate the consequence of excessive or insufficient growth rates. In general, the minimum weight requirements of the growing animal are satisfied when it has reached or exceeded two-thirds of its predicted adult body weight. Lambs and kids will not reach puberty unless both the minimum age and weight requirements are satisfied. Once these two minimum provisions are met, seasonal influences, specifically photoperiod, function to permit or inhibit onset of puberty.

Sheep and goats are "short-day" breeders and respond to the decline in length of day to the winter solstice with an increase in reproductive activity (Fig. 4). Conversely, as hours of light per day increase to the summer solstice, breeding activity decreases. The point of critical attention is that following the birth of offspring, those offspring must first be exposed to an increase in hours of light per day followed by a decrease prior to initiating reproductive activity. It is during the latter short-day photoperiodic environment (i.e., "the breeding season") when they must also have satisfied their minimum age and weight requirements. Typically, offspring born prior to late spring ("early born" offspring) reach puberty during the ensuing fall/winter (Fig. 5). In contrast, offspring born after late spring ("late-born" offspring) will not be of sufficient age, weight, or have had sufficient exposure to increasing and then decreasing length of day to reach puberty during the first fall/winter. Consequently, the late-born offspring reach puberty at a subsequent breeding season at an older age and greater body weight than early born offspring. Age at puberty also occurs earliest in offspring born and raised as singles and latest in offspring born as multiples and raised as multiples.

In the male, the point in time that defines the age at puberty is rather nondistinct because the process

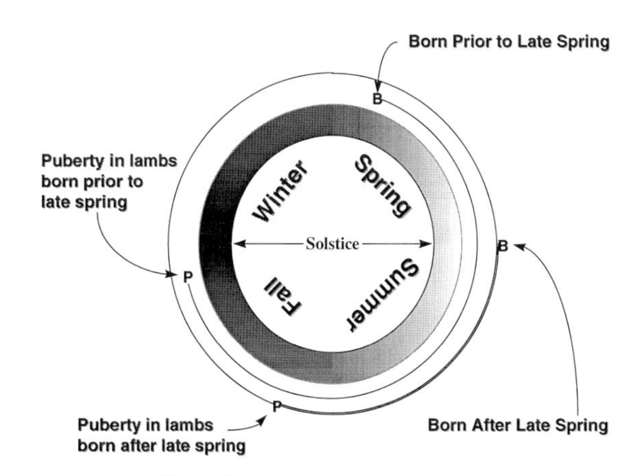

FIGURE 5 Effect of season of birth on age at puberty.

of sperm production begins gradually. In contrast, the event in the life of the developing female that defines when the female transitions from sexual immaturity to maturity is first ovulation. Typically, first ovulations are not preceded by estrous behavior (i.e., they are "silent ovulations") and they result in the formation of corpora lutea that are subnormal in life span and/or secretion of progesterone when compared to reproductively active adult contemporaries. As many as three successive silent ovulations have been reported to occur prior to young females exhibiting estrus accompanied with ovulation, which are the two events that functionally define puberty. Consequently, from a physiological perspective, young females may have begun ovarian cyclic activity as many as 30 days prior to puberty. The significance of this observation is that it serves as an example of the ability to disassociate behavioral events (i.e., estrus) from physiological events (i.e., ovulation).

The physiological mechanisms regulating the onset of puberty have yet to be fully defined. Considerable scientific evidence exists to support the gonadostat hypothesis as an explanation of mechanisms governing developmental and seasonal changes in reproductive activity. The caveats of the gonadostat hypothesis are that developmental (prepubertal) causes of reproductive inactivity as well as seasonal causes of reproductive inactivity are associated with hypothalamic mechanisms that are exquisitely sensitive to inhibition by minute concentrations of steroids (especially estradiol) secreted by the animal's gonads. Consequently, as the animal satisfies age, weight, and/or photoperiodic constraints to the reproductive processes, hypothalamic sensitivity to the inhibitory effects of gonadal steroids decreases, permitting an increase in activity of hypothalamic mechanisms. More specifically, hypothalamic secretion of gonadotropin-releasing hormone (GnRH) increases in pulsatile activity, the consequence of which is an increase in pulsatile secretion, from the pituitary, of the gonadotropins luteinizing hormone (LH) and follicle-stimulating hormone. As secretion of gonadotropins increases, gonadal activity (sperm production or ovarian follicular development) and steroid output increase. In the female, the brain ultimately perceives the increase in blood concentrations of estradiol by permitting the animal to express behavioral estrus and induce a rapid release of gonadotropins from the pituitary, thus initiating the process of ovulation.

While the gonadostat hypothesis provides one explanation of mechanisms governing reproductive activity in sheep and goats, certainly multiple systems exist within the animal to monitor its genetic predisposition to reproduce as well as to monitor the influence of season of the year, age of the animal, body weight/composition, stress, and nutritional status. Collectively, these variables impinge on hypothalamic mechanisms that govern secretion of GnRH. The ultimate response of the hypothalamus may be thought of as a consensus response that is arrived at after the hypothalamus collectively integrates all incoming stimulatory and inhibitory signals. When the consensus of signals is negative (or not positive), the reproductive process is quiescent. Alternatively, when the consensus of signals is positive (or not negative), the animal initiates the process of reproduction.

Offspring that reach puberty at an early age will be more reproductively efficient throughout their lifetime than will offspring that reach puberty at an older age. Consequently, substantial research has provided evidence to support the argument that in sheep (and likely goats), females should be selected, managed, and bred to give birth by 1 year of age. The actual age at which females are first bred, however, rests with the owner's ability to intelligently assess both the risks and the benefits of mating the young animals. The benefits are reduced preproduction costs, shortened generation intervals (thus more rapid genetic gains), and increased total lifetime productivity. The risks are slowed growth rate of the mother, lower initial conception rates (young females in estrus make little or no effort to approach the male), and potential birthing problems as a result of females giving birth to multiple offspring. Key to deciding when to breed young females is planning and management. For example, (i) young females could be bred early in the breeding season to decrease the incidence of multiple offspring (Table 1), potentially reducing birthing problems; (ii) they should be bred as a group separate from the adults to increase conception rates; (iii) they should be bred to breeds of males that are structurally small in size to reduce

TABLE 1
Estrus, Ovulation, and Ovulation Rate in Ewes in Idaho and Texas[a]

	Ewes in estrus (%)		Ewes ovulating (%)		Ovulation rate
Month	Idaho	Texas	Idaho	Texas	Idaho
January	100	100	100	100	1.89
February	100	100	100	94	1.57
March	89	40	94	52	1.50
April	26	38	32	32	1.37
May	2	31	2	31	1.00
June	7	44	7	75	1.00
July	6	94	6	94	1.00
August	12	86	41	100	1.75
September	88	94	100	94	1.72
October	100	94	94	100	1.80
November	100	97	100	91	1.86
December	100	100	100	100	1.88

[a] Adapted from Hulet *et al.* (1974a).

birthing problems; and (iv) young females should be adequately fed so they to continue to grow and produce offspring. Failure to adequately feed females that give birth at 1 year of age is the most common oversight among owners.

III. ADULTHOOD

The major environmental determinant affecting reproductive activity in adults is season, with most sheep and goats exhibiting greatest reproductive activity in the fall. Table 1 illustrates this relationship by depicting the percentage of ewes in estrus and ovulating throughout the year in Idaho and Texas. As ewes and does experience the transition from summer to fall, they experience a decrease in both day length and temperature. The principal seasonal stimulus dictating reproductive activity is day length. Sheep and goats are short-day breeders and consequently reproductive activity increases as length of day decreases (Fig. 4). An animal's ability to perceive photoperiodic changes appears to be mediated through changes in blood concentrations of the hormone melatonin, which is secreted by the pineal gland located in the brain. Melatonin has been called the "hormone of darkness" because when animals are exposed to darkness, blood concentrations of melatonin increase and ultimately signal the sheep or goat to start reproductive activity. A second point illustrated in Table 1 is that (i) a portion of ewes will cycle throughout the year and (ii) a greater proportion of ewes will cycle throughout the year the closer the animals are to the equator (i.e., Idaho vs Texas). Also illustrated in Table 1 is that the proportion of ewes actually ovulating is often greater than the proportion of ewes in estrus, especially as ewes transition into or out of the breeding season. The reason for this disparity between estrus and ovulation may be attributed to either a failure to detect estrus or a failure of the female to exhibit estrus at the time of ovulation (i.e., silent ovulation). The last point to be emphasized in Table 1 is the ovulation rate. It can be interpreted from these data that females bred in the summer months (e.g., June) will produce fewer offspring (i.e., an ovulation rate of 1 implies one egg is released from the ovary) than will females bred in the fall or winter months (ovulation rate = 1.7 or 1.8). Careful attention to this information needs to be considered when planning to bred females "out of season" (i.e., during the summer months).

Treating sheep and goats with melatonin or reducing exposure to long day lengths can be used to induce reproductive activity in males and females. Implementing light-reduction techniques is usually achieved by denying animals exposure to the light of morning and the evening through the use of "light-tight" enclosures. This approach also minimizes an animal's exposure to the adverse effects of the midday high ambient temperatures that accompany long days.

Despite the potential advantages of using breeding schemes that rely on phototherapy or melatonin treatment, two inherent disadvantages of these techniques are the need to treat animals on a daily basis and the long interval from initiating treatments to expression of reproductive activity (i.e., 4–6 weeks).

When seasons change, so does temperature. Temperature does not appear to play a major role in dictating reproductive activity of sheep and goats, but it does have a major effect on fertility and embryo survival. Table 2 illustrates that when ewes and rams were housed in a hot climate vs a cool climate, the hot weather resulted in a lower percentage of eggs fertilized, greater embryonic death, and consequently fewer mothers giving birth. Any condition that limits an animal's ability to cool itself (e.g., high humidity) will adversely affect its reproductive performance.

Two other factors that affect breeding activity of ewes and does are breed and age. Some breeds of sheep, such as Rambouillet, Merino, and Dorset, have longer breeding seasons than other breeds, such as the Southdown and Cheviot. Similarly, breeds of goats, such as the dairy breeds, have longer breeding seasons than Angora. Fortunately, from a genetic perspective, long breeding seasons tend to be genetically dominant; therefore, it is possible to increase the length of the breeding season through genetic selection practices. Within a breed, as an animal's age increases after puberty, length of breeding season also increases (Fig. 4). Young females starting cyclic activity for the first time generally start cycling about 3 weeks after the adults and stop cycling about 3 weeks prior to the adults. As age increases, reproductive efficiency also increases as evaluated in terms of offspring survival (which increases due in part to a learned mothering ability) and number of offspring produced (which is due to an increase in ovulation rate as an animal's age increases).

TABLE 2
Effect of Temperature on Lambing Rates[a]

	Housing for rams and ewes	
	Outside (~85°F)	*Air-conditioned room* (~45°F)
% eggs fertilized	26	64
% embryonic death	49	22
% ewes lambing	13	50

[a] From Dutt and Simpson (1957).

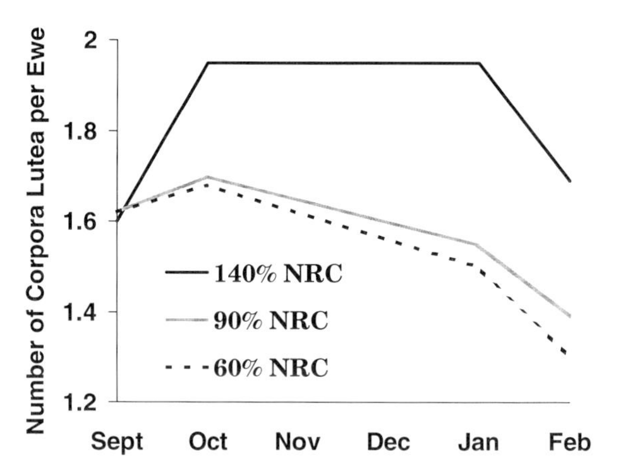

FIGURE 6 Effect of nutrient intake and month of breeding season on ovulation rate. NRC, National Research Counsel (from Hulet *et al.*, 1974a).

In order for animals to produce, those animals must be appropriately fed and maintained in good body condition. Poorly nourished animals cannot meet their needs for maintenance and be expected to produce offspring. Figure 6 illustrates the relationship between nutrient intake and ovulation rate of ewes with respect to the breeding season. Clearly, ewes with the greatest nutrient intake responded most rapidly to the onset of the breeding season and continued to respond with an increase in ovulation rate until the termination of the breeding season. In contrast, ewes receiving less than National Research Council (NRC) recommendations exhibited only a modest increase in ovulation rate at the initiation of the breeding season and failed to maintain the slight increase; most likely due to other energetic demands.

TABLE 3
Effect of Ewe Condition on Ovulation Response to Flushing[a]

Ewe condition	*Treatment*	*Rate of body weight gain (kg/day)*	*No of ewes with single ovulation*	*No. of ewes with double ovulations*
Thin	Flushed	.11	12	8
Thin	Unflushed	.01	20	0
Fat	Flushed	.10	10	10
Fat	Unflushed	.02	8	12

[a] From Clark (1934).

The rapid increase in ovulation rate of ewes and does receiving greater than NRC recommendations has long been recognized and is referred to as flushing. Flushing is usually initiated 2 or 3 weeks prior to the breeding season and is accomplished by supplementing animals with high-energy feed such as 1 or 2 kg of corn/animal/day. The flushing technique is most effective during the periods of the year when ovulation rates are marginal and especially when animals are transitioning into or out of the breeding season. Flushing is also most effective in ewes and does that have an urgent energy demand (Table 3). Clark reported in 1934 that although both fat and thin ewes gained body weight at an equivalent rate in response to flushing, only the thin ewes responded with an increase in ovulation rate. It should also be noted that the increase in ovulation rate among the thin ewes in response to flushing did not exceed the ovulation rate of the fat ewes. Consequently, flushing is a short-lived technique that permits the energy-deficient animal to express its genetic potential. Neither flushing nor feeding the animal in excess of its needs permits the animal to supersede its genetically defined limitations. In contrast, limiting nutrient availability will not only reduce ovulation rates but also shorten the length of the breeding season, limit the animal's milking ability, and compromise other systems related to production. Simply put, when animals fail to be fed, they fail to reproduce.

The final "catch-all" category of factors that influence reproductive performance is stress. The stresses on an animal can be due to disease, parasites, confinement, lactation, suckling, and other factors both known and unknown. Stressful situations typically reduce production performance. For example, Fig. 7 illustrates that ewes which suckle lambs take longer to return to estrus than ewes from which lambs had been weaned at an early age. Selective removal of specific stressful stimuli can produce responses that are ineffective, permissive, or helpful to reproductive performance. The ability to predict the outcome of selectively removing a stressful stimulus relies on the owner's ability to collectively assess the total balance that exists between the animals' inherent drive to reproduce and the demands that keep them from reproducing. For example, although early weaning of offspring may appear to provide a simple solution to accelerate rebreeding of females, the technique is generally ineffective when used as the sole approach among females birthing in the spring or summer months but relatively effective among females birthing in the fall. The difference in response can be attributed to the powerful inhibitory effects of long day lengths. In contrast, by creating an environment in which several inhibitory stressors are

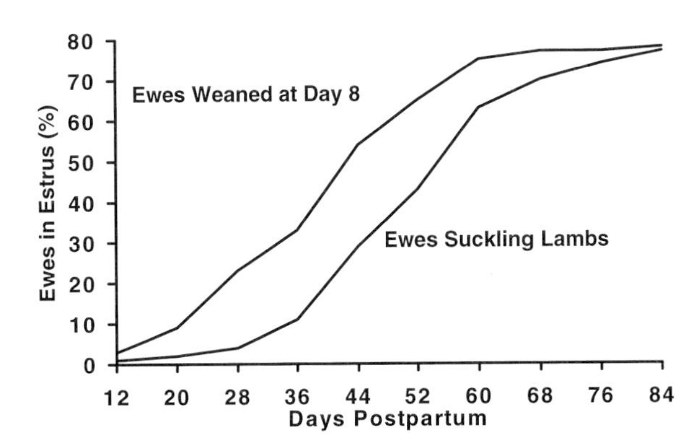

FIGURE 7 Proportion of suckled vs early weaned (Day 8) fall lambing ewes returning to estrus relative to the interval postpartum (from Thibault *et al.*, 1966).

negated simultaneously, the effect may be sufficient to initiate cyclic activity in females even in the presence of other single inhibitory influences such as long day lengths. In reality, it is likely that a far greater proportion of females initiate cyclic activity during the nonbreeding season in response to combined management approaches than is realized. As stated earlier, females initiating cyclic activity after a period of anestrus will typically exhibit a silent first ovulation and thus the owner consequently fails to realize the impact of combined management approaches, therefore relaxing efforts to rebreed the females. This fleeting lack of attention permits the inhibitory influences to again dominate the animals' drive to reproduce. Under the influence of positive environmental conditions, behavioral estrus will generally follow a silent ovulation within 3 weeks. During transitional periods in reproduction activity (e.g., from anestrus to estrus or from prepuberty to puberty) a hastening of the onset of breeding activity can be achieved in a greater portion of females with techniques such as short-term removal of the young and/or flushing. Reports also exist citing the positive effects of transportation stress on inducing females to ovulate. Using transportation stress to evoke a positive response is likely one of the most difficult strategies due to the delicate perception of transportation as a stimulatory vs inhibitory stimulus.

A powerful management tool proven to induce breeding activity in ewes and does is the introduction of a novel male into a herd of females (i.e., "male introduction"). In general, to evoke a male introduction response, females need to be isolated from the sight, sound, and smell of males for approximately 30 days prior to introduction. In reality, the literature varies greatly regarding the completeness and duration of the isolation. Typically, the male introduction technique often involves use of sterile males to hasten onset of the breeding season. Once females are cycling, fertile males are introduced. However, when inducing anestrous or prepuberal females to cycle using this technique, it is highly recommended that the planned herd sires be used to effect the response and not sterile males because the technique, when effective, is often very short-lived and therefore every opportunity should be given to encourage a fertile mating because the likelihood of females recycling following a male-induced ovulation during anestrus is low. In other words, if the objective is to establish pregnancy in females at a time when it is most difficult to breed them, do not use infertile males!

In summary, the factors which influence reproductive performance are many and far-ranging. Recognizing and reducing adverse factors is critical to maximizing animal reproductive efficiency. This task is incumbent upon the animal managers. There is no substitute for good management.

IV. Estrus and Estrous Cycles

A. Detection of Estrus

1. *In the Absence of a Male*

Indications of a ewe or doe in estrus are quite variable. The females may be less eager for food and may isolate themselves from the herd. Attempts by a ewe or doe in estrus to mount or be mounted by other ewes or does are rarely observed. Swelling of the vulva and a mucus flow may be observed but are rarely distinctive indicators of estrus. Very little value should be placed on efforts to identify ewes or does in estrus in the absence of a ram or buck.

2. *In the Presence of a Male*

A ewe or doe in estrus will often seek out the male. In other situations, particularly with young ewes and does, the females may show no signs of estrus until approached and teased by the male. When teased, the female in estrus will stand and refuse to move in response to the male's urging. The female may turn her head toward the male when being teased or rub her head against the male's side. Ewes and does in heat will stand solidly for mounting by a male. Females entering into or exiting estrus will exhibit behaviors similar to females in estrus but will refuse to stand for mounting by a male.

B. Tools to Aid in the Detection of Estrus

To aid in the detection of estrus in ewes and does, a recommended management practice is to affix to the males a commercially available marking harness or apply a grease paint to the brisket of the male. When the male mounts the female, some of the marking pigment will be transferred to the rump of the female. Light-colored markers are used first and exchanged at 2-week intervals thereafter (i.e., less than one estrous cycle length) with successively darker colored markers so as to mask the previous color. This system enables owners to know, with some accuracy, if the male is mating with females (i.e., indicates libido), what females were bred when, and if a problem exists with either female or male fertility. The marking system provides the most readily available evidence of problems with ram and buck fertility. For example, if all females were marked with the first color and all or most females were marked with the second color, this is an immediate indication of a male fertility problem. Another benefit of this marking system is that it provides owners the information needed to predict birthing dates.

During the breeding season, the observation that a female will stand to be mated by a male does not always imply that the ewe is cycling. For example, a proportion of the females that conceive at a previous estrus will subsequently stand to be mated, ovulate, and actually conceive while maintaining a previous pregnancy (i.e., superfetation). Additionally, ewes and does do not need to be in estrus to be mounted by a male. Nonestrus-related marks are more common when females and males are housed or fed in confined areas.

One of the most difficult diagnoses of reproductive importance is the failure to detect estrus. Consequently, if a female is not detected in estrus, careful consideration should be given to the following questions: Is the male capable of detecting estrus, and if so, has a mechanism been enlisted, such as frequent detection of estrus or use of a marking system, which will ensure the detection of estrus?

Finally, it is noteworthy to recognize that sheep and goats will interbreed and conception can occur, but pregnancy is not known to progress beyond the first third of gestation.

C. Duration of Estrus

In a study of the duration of estrus among ewes in Missouri, McKenzie and Terrill observed 1235 occurrences of estrus in 344 ewes. The frequency distribution for the duration of estrus is shown in Fig. 8. These investigators reported that the duration of estrus ranged from 3 to 73 hr in length, with an average duration of estrus of 29.3 hr. Seventy-three percent of the occurrences of estrus were 21–39 hr in length, whereas 92% were 15–45 hr in length. Estrus activity <21 hr in length was observed 18% of the time. Four percent of ewes were observed in estrus 3–14 hr and 4% in estrus 46–75 hr. Typically, the duration of estrus will be shortest among young females exhibiting estrus during the traditional nonbreeding season or females exhibiting their first or last estrus of the breeding season. The use of a marking device affixed to the males has the potential, when used properly, to be the single most important reproductive management tool in the flock to aid in the detection of estrus and infertile or sexually inactive males and in the planning and management of the females for birthing.

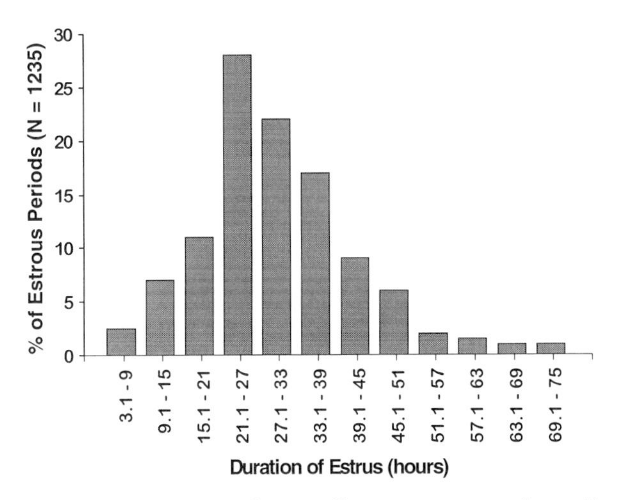

FIGURE 8 Duration of estrus (from McKenzie and Terrill, 1937).

D. Length of the Estrous Cycle

In a study of the length of the estrous cycle of ewes in Missouri, McKenzie and Terrill observed 1038 estrous cycles by 299 ewes. Frequency distribution for the length of the estrous cycles is shown in Fig. 9. These investigators reported observing estrous cycle lengths that varied from 3.5 to 63.5 days; aver-

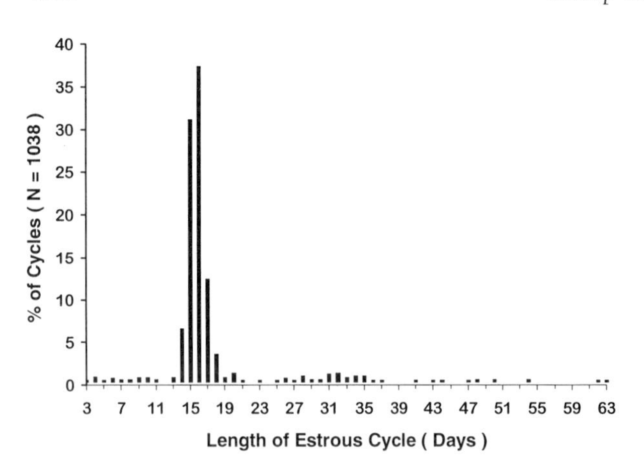

FIGURE 9 Length of estrous cycle (from McKenzie and Terrill, 1937).

age estrous cycle length was 16.7 days and 90% of all cycles ranged from 14 to 19 days in length. Two percent of the cycle lengths were <13 days in length. In other studies, investigators have reported that short estrous cycles (i.e., <14 days) occurred most frequently in young females beginning to cycle for the first time and adults beginning to cycle following periods of anestrus. Estrous cycle lengths >19 days among ewes could be attributed to a failure to detect estrus or a failure of ewes to exhibit estrus (i.e., silent ovulations). Some variation in the length of estrous cycles has been attributed to differences among breed of sheep, with Rambouillets averaging 20–24 day cycles compared to 16- to 19-day cycles for Hampshire and Columbia.

Among goats, length of the estrous cycle is reported to vary greatly, with an average cycle length lasting 19–21 days. Purebred pigmy goats have been reported to have an average cycle length of 24 days, whereas Sicilian breeds were reported to have 8-day cycles.

V. OVERVIEW OF THE ENDOCRINOLOGY OF THE ESTROUS CYCLE, PREGNANCY, AND PARTURITION

Several specific time commitments made by the animal to the reproductive processes are outlined in Table 4. Approximately 9 hr (range, 4–16 hr) after onset of estrus, the pituitary releases a large amount of LH, which results in ovulation approximately 21–26 hr later. Approximately 3 days following ovulation, serum concentrations of progesterone begin to increase beyond detectable limits and parallel the

TABLE 4
Important Time Relationships of Reproductive Functions

Event	*Time required or time occurring*
First estrus	Minimum of 6 months of age, weighing two-thirds of adult body weight
Length of estrous cycle	Average, 17 days; range, 14–19 days (90%)
Length of time in estrus	Average, 30 hr; Range, 3–73 hr
Time of ovulation	Approximately 28 hr after start of estrus; Range, 12–40 hr after the start of estrus
Length of time the ovum is fertilizable after ovulation	12–24 hr
Time to breed	If checking estrus once per day, breed at first detection. If checking estrus twice per day and observe: AM, standing estrus: breed in PM PM, standing estrus: breed AM next day
Time taken by sperm to reach the ovum within the oviduct	Minutes
Viable life of sperm in the reproductive tract	10–12 hr
Time taken by fertilized ovum to reach the uterus	Approximately 4 days
Length of pregnancy	Average, 148 days
Time required for uterine involution	Approximately 21 days

structural growth of the corpus luteum (CL). Concentrations of progesterone, however, cannot be used as an indicator of size or number of CLs. In ewes that are either not bred or fail to establish pregnancy, the uterus begins to secrete prostaglandin F2α (PGF2α) in an episodic manner around Day 11 (Day 0 = estrus), which increases over the next 2 or 3 days in ewes with 17-day estrous cycles and much earlier in ewes that exhibit short estrous cycle lengths. The increase in PGF2α initiates the demise of the CL, and as the CL regresses, serum concentrations of progesterone decline, permitting serum concentrations of gonadotropins to increase. As gonadotropins increase, follicular development again increases, resulting in the concomitant increase in serum concentrations of estradiol. Subsequently, the sequence of events repeats itself whereby estradiol stimulates a surge of LH, which causes ovulation and leads to the formation of the CL.

Should the female be bred and conceive, the process referred to as maternal recognition of pregnancy rescues the CL from regression, thus providing the uterus a constant exposure to progesterone, which serves to subdue uterine myometrial activity. As early as Day 12 following mating of the ewe, the conceptus secretes a number of proteins, most notably interferon-τ (IFN-τ). IFN-τ, when given to nonbred ewes, will extend the life of the CL, presumably by inhibiting uterine secretion of PGF2α. The precise timing of the appearance of the conceptus signaling protein, relative to its ability to extend luteal life span, is critical and likely plays a major role in influencing early embryonic mortality. Asynchronous embryos, embryos delayed in their development, or embryos that fail to secrete sufficient quantities of IFN-τ will likely not be recognized as being present by the mother and will be lost when the CL regresses. Consequently, ewes and does that return to estrus may be judged mistakenly as having failed to conceive.

By 21 days postmating of the ewe, the conceptus ceases secretion of IFN-τ as pregnancy is established. Subsequently, the number of pregnancy-specific proteins increases, including pregnancy-specific protein B (PSPB), pregnancy-associated glycoproteins, ovine placental lactogen, and uterine milk proteins. At least one of these proteins has been exploited to serve as the basis of an early pregnancy test for ewes (i.e., PSPB). Although the precise roles, if any, of these proteins are not known, they may support pregnancy by maintaining CL function and thus secretion of progesterone, aid in fetal development, or enhance uterine–placental function.

As pregnancy progresses, serum concentrations of progesterone, placental lactogen, and PSPB increase to term. Approximately 60 days after mating of the ewe, placental production of progesterone is sufficiently adequate such that a luteal source of progesterone is no longer needed to maintain pregnancy. As the female approaches parturition, the fetal–pituitary–adrenal axis begins to play a critical role in directing the sequence of events leading to parturition. Early in the last trimester (100–130 days), the fetal adrenal possesses the ability to bind adrenocorticotropin hormone (ACTH) but responds poorly to a challenge with ACTH. Subsequently, as the number of ACTH binding sites increase (due to the ability of ACTH to induce expression of its own receptors in the adrenal gland) adrenal responsiveness to ACTH also increases.

Also beginning at approximately Day 100 of pregnancy, the ability of the placenta to synthesize prostaglandins increases markedly. The predominant prostaglandin secreted is PGE2, which is also capable of stimulating secretion of both ACTH, possibly from the placenta, and cortisol from the fetal adrenals. The stimulus for the increase in PGE2 is not known, but it is known that if fetuses are hypophysectomized around Day 75 of gestation, the placenta's ability to synthesize PGE2 does not increase, implying that a pituitary factor is involved in stimulating placental secretion of PGE2. Consequently, as a result of fetal hypophysectomy or disconnection, neither PGE2 nor cortisol increase prior to parturition and thus delivery is delayed.

Approximately 15 days prior to parturition, fetal concentrations of ACTH increase, resulting in an increase in fetal serum concentrations of cortisol. A marked increase in cortisol occurs 3 or 4 days prior to delivery. Coincident with the increase in cortisol, maternal concentrations of progesterone in serum decline, which is paramount to the delivery of the fetus. Treatment of periparturient ewes with 150 mg of progesterone per day blocks cervical dilation and 200 mg/day blocks uterine activity, thus delaying parturition. The decline in progesterone increases the progesterone precursor, pregnenolone, which then serves as a substrate for synthesis of estrogen, resulting in a marked change in the estrogen:progesterone ratio. As estrogen increases, so does the number of uterine receptors for oxytocin. As the result of increasing concentrations of oxytocin, increased tissue responsiveness to oxytocin (as the result of increased oxytocin receptor numbers), or both, peripheral concentrations of the prostaglandins F, E, and 6-keto F1α increase. This increase in prostaglandins is believed to enhance cell to cell communication in the uterus, resulting in enhanced myometrial activity and progressive labor. Cervical ripening proceeds under the influence of PGEs and delivery is imminent.

Following parturition, the female assumes an anestrous condition, uterine involution occurs, and the sequence of events and circumstances described previously repeats itself.

VI. REPRODUCTIVE PROCESSES IN THE RAM AND BUCK

In a natural mating system, variables that affect the reproductive performance of the male have the potential to profoundly affect production performance of the flock/herd. In the ram and buck, as in the ewe and doe, the hypothalamus serves as a focal point to integrate the stimulatory and inhibitory cues that direct the male to initiate, maintain, or cease reproduction.

Developmentally, the growing male and female share considerable endocrine similarities and likely mechanisms leading to the attainment of puberty. The gonadostat hypothesis, described earlier, maintains that in the immature male the hypothalamus is exquisitely sensitive to the negative feedback effects of testosterone. Consequently, early in the life of the male, minute quantities of testosterone have a major inhibitory effect on hypothalamic–pituitary function, resulting in minimal output of gonadotropins from the pituitary. As the male matures, the hypothalamus becomes less sensitive to the negative feedback effects of testosterone, permitting an increased release of LH from the pituitary. Unlike the female, in which the attainment of puberty is rather distinct, the process is more gradual in the male and highly subject to the criteria used to define sexual maturity. Historically, sperm concentration and/or morphology have been used.

Once the ram reaches sexual maturity, the endocrine profile characterized by a pulsatile secretion of LH, followed immediately by a pulse of testosterone, is established and remains unaltered by few internal effectors. An increase in mean blood concentration of testosterone occurs up to approximately 2 years of age. Most of the deviations to the reproductive endocrine profile of the male are in response to effectors such as season, nutrition, and stress-related influences. As in the ewe, long day lengths, poor nutrition and health, or management-related stressors diminish the secretion of LH and are believed to be due in part to an increase in the sensitivity of the hypothalamus to the negative feedback effects of testosterone. As LH decreases, so does secretion of testosterone, consequently resulting in reduced libido and sperm output and a clinically detectable decrease in testicular size.

At sexual maturity (i.e., approximately 1 year of age), testicular size will vary among goats and sheep with breed, as illustrated in Table 5 among breeds of sheep. Testicular size is an important variable to assess in selecting potential herd sires because it reflects potential breeding capacity or mating potential of the male. It has been estimated that in an adult

TABLE 5
Yearling Ram Scrotal Circumference (SC)

Breed	*Live wt (kg)*	*SC + standard deviation (cm)*
Southdown	78.6	32.2 ± 2.3
Montadale	93.2	32.7 ± 2.2
Polled dorset	100.0	37.3 ± 1.9
Shropshire	100.5	33.2 ± 2.6
Rambouillet	108.6	34.7 ± 3.1
Corriedale	112.7	33.5 ± 2.4
Hampshire	113.1	35.6 ± 2.5
Columbia	122.7	35.1 ± 3.1
Suffolk	129.1	37.1 ± 2.3

[a] From Braun *et al.* (1980).

TABLE 6
Quantity and Quality of Semen from Rams and Bucks

	Average and Range	
	Ram	Buck
Volume (ml)	1 (0.8–1.2)	0.8 (0.5–1.0)
Concentration (10^9/ml)	2.5 (1–6)	2.4 (2–5)
Motile sperm (%)	75 (60–80)	80 (70–90)
Normal sperm (%)	90 (80–95)	90 (75–95)

[a] From Evans and Maxwell (1987).

ram or buck, approximately 20 million sperm are produced per gram of testicular tissue per day. Table 6 describes normal quantity and quality characteristics of semen from rams and bucks.

A common reproductive disorder among rams and bucks, known as epididymitis, is due to invasion and inflammation of the epididymis by the Brucella ovis organism, which can render the male permanently sterile. The disease may affect one or both testes and can be diagnosed by clinical tests. Palpation of the testis in the area of the epididymis should be done as a routine management practice to check for lesions and infections (as palpated by "hard" tissue).

Rams and bucks can be rendered sterile, for the purpose of making a "teaser," by performing procedures known as a vasectomy or an epididymectomy. During a vasectomy the vas deferens are cut, and during an epididymectomy the epididymides are cut. The latter procedure is relatively simple to perform through the most distal portion of the scrotum, whereas the vasectomy should be performed by skilled veterinarians familiar with the anatomy of the male.

VII. CONCLUSION

Sheep and goats are versatile animals that represent significant global resources of forage-produced sources of meat animal protein. It is likely that in no other meat animal industry is there the potential to increase reproductive performance among classes of livestock as exists in the sheep and goat industries.

See Also the Following Articles

Breeding Strategies for Domestic Animals; Cattle (Bovidae); Global Zones and Reproduction; GnRH (Gonadotropin-Releasing Hormone); Photoperiodism

Bibliography

Bindon, B. M., and Piper, L. R. (1986). The reproductive biology of prolific sheep breeds. *Oxford Rev. Reprod. Biol.* 8, 414–451.

Braun, W. F., Thompson, J. M., and Ross, C. V. (1980). *Society for Theriogenology Sheep and Goat Manual* (R. S. Ott and M. A. Memon, Eds.), pp. 44–50.

Byatt, J. C., Warren, W. C., Eppard, P. J., Staten, N. R., Krivi, G. G., and Collier, R. J. (1992). Ruminant placental lactogens: Structure and biology. *J. Anim. Sci.* 70, 2911–2923.

Clark, R. T. (1934). Studies on the physiology of reproduction in the sheep. *Anat. Rec.* 60, 125–133.

Downing, J. A., and Scaramuzzi, R. J. (1991). Nutrient effects on ovulation rate, ovarian function and secretion of gonadotropic and metabolic hormones in sheep. *J. Reprod. Fertil. Suppl.* 43, 209–227.

Dutt, R. H., and Simpson, E. C. (1957). Environmental temperature and fertility of southdown rams early in the breeding season. *J. Anim. Sci.* 16, 136–143.

Evans, G., and Maxwell, W. M. C. (1987). *Salmon's Artificial Insemination of Sheep and Goats.* Butterworths, London.

Foster, D. L. (1988). Puberty in the female sheep. In *The Physiology of Reproduction* (E. Knobil and J. D. Neill, Eds.), pp. 1739–1762. Raven Press, New York.

Goodman, R. L. (1988). Neuroendocrine control of the ovine estrous cycle. In *The Physiology of Reproduction* (E. Knobil and J. D. Neill, Eds.), pp. 1929–1970. Raven Press, New York.

Haresign, W. (Ed.) (1983). *Sheep Production.* Butterworths, London.

Haresign, W., McLeod, B. J., Webster, G. M., and Worthy, K. (1985). Endocrine basis of seasonal anoestrus in sheep. In *Endocrine Causes of Seasonal and Lactational Anestrus in Farm Animals* (F. Ellendorff and F. Elsaesser, Eds.), pp. 6–18. Nijhoff, Dordrecht..

Hulet, C. V., Price, D. A., and Foote, W. C. (1974a). Effects of month of breeding and feed level on ovulation and lambing rates of Panama ewes. *J. Anim. Sci.* 39, 73–78.

Hulet, C. V., Shelton, M., Gallagher, J. R., and Price, V. A. (1974b). Effects of origin and environment on reproductive phenomena in Rambouillet ewes. I. Breeding season and ovulation. *J. Anim. Sci.* 38, 1210–1217.

Karsch, F. J., Bittman, E. L., Foster, D. L., Goodman, R. L., Legan, S. J., and Robinson, J. E. (1984). Neuroendocrine basis of seasonal reproduction. *Recent Prog. Horm. Res.* **40**, 185–232.

Lamberson, W. R., and Thomas, D. L. (1982). Effects of season and breed of sire on incidence of estrus and ovulation rate in sheep. *J. Anim. Sci.* **54**, 533–553.

Liggins, G. C., Fairclough, R. J., Grieves, S. A., Kendall, J. Z., and Knox, B. S. (1973). The mechanism of initiation of parturition in the ewe. *Recent Prog. Horm. Res.* **29**, 111–150.

Lindsay, D. R., and Pearce, D. T. (Eds.) (1984). *Reproduction in Sheep.* Australian Academy of Science, Canberra.

Marai, F. M., and Owen, J. B. (Eds.) (1987). *New Techniques in Sheep Production.* Butterworths, London.

Martin, G. B., Oldham, C. M., Cognie, Y., and Pearce, D. T. (1986). Physiological responses of anovulatory ewes to introduction of rams—A review. *Livestock Prod. Sci.* **15**, 219–247.

McKenzie, F. F., and Terrill, C. E. (1937). Estrus, ovulation and related phenomena in the ewe. *Univ. Montana Agric. Exp. Sta. Bull.* **264**, 1–86.

Oldham, C. M., Martin, G. B., and Knight, T. W. (1978/1979). Stimulation of seasonally anovular merino ewes by rams: I. Time from introduction of the rams to the preovulatory LH surge and ovulation. *Anim. Reprod. Sci.* **1**, 283–290.

Oldham, C. M., Martin, G. B., and Purvis, I. W. (Eds.) (1990). *Reproductive Physiology of Merino Sheep. Concepts and Consequences.* School of Agriculture (Animal Science), Univ. of Western Australia, Perth.

Thibault, C., Courot, M., Martinet, L., Mauleon, P., Du Mesnil Du Buisson, F., Ortavant, R., Pelletier. J., and Signoret, J. L. (1966). Regulation of breeding season and estrous cycles by light and external stimuli in some mammals. *J. Anim. Sci. Suppl.* **25**, 119–139.

Thorburn, G. D., Hollingworth, S. A., and Hooper, S. B. (1991). The trigger for parturition in sheep: Fetal hypothalamus or placenta? *J. Dev. Physiol.* **15**, 71–79.

Sipuncula

Mary E. Rice
Smithsonian Marine Station

GLOSSARY

lecithotrophic larva Nonfeeding larva that derives its nutrition from yolk formed in the oocyte during oogenesis.

pelagosphera A larval form unique to the Sipuncula which succeeds the trochophore and in which the prototroch is reduced or lost and replaced by the metatroch as the primary locomotory organ. Pelagosphera larvae frequently possess a terminal organ which may serve for temporary attachment to the substratum.

planktotrophic larva Feeding larva that derives its nutrition from external sources in the seawater.

trochophore A larval form, often oval in shape, with prominent equatorial band of cilia (prototroch) and an anterior tuft of sensory cilia; characteristic of developmental cycles in which the cleavage pattern of eggs is spiral.

In the Sipuncula, a phylum of unsegmented marine worms, individual male and female worms spawn their gametes into the seawater where fertilization occurs. Gametes are formed in minute gonads (ovary or testis) from which they are released into the body cavity, where they undergo the remainder of their development until taken into the nephridial organs. Here, they are stored until the occurrence of spawning, an event usually synchronized by an unknown mechanism within populations. Cleavage is spiral, typically resulting in a trochophore larva. Four developmental patterns are recognized: (i) direct development from the embryo to the juvenile with no swimming larval forms; (ii) development with a single swimming larval stage, the trochophore, which transforms into the juvenile stage; (iii) development with two larval stages, trochophore and lecithotrophic (nonfeeding) pelagosphera; and (iv) development with two larval stages, trochophore and planktotrophic (feeding) pelagosphera. In the latter category, the pelagosphera is known to exist in the oceanic plankton for several months before undergoing metamorphosis to the juvenile, thus providing a means of species dispersal and genetic exchange over wide distances.

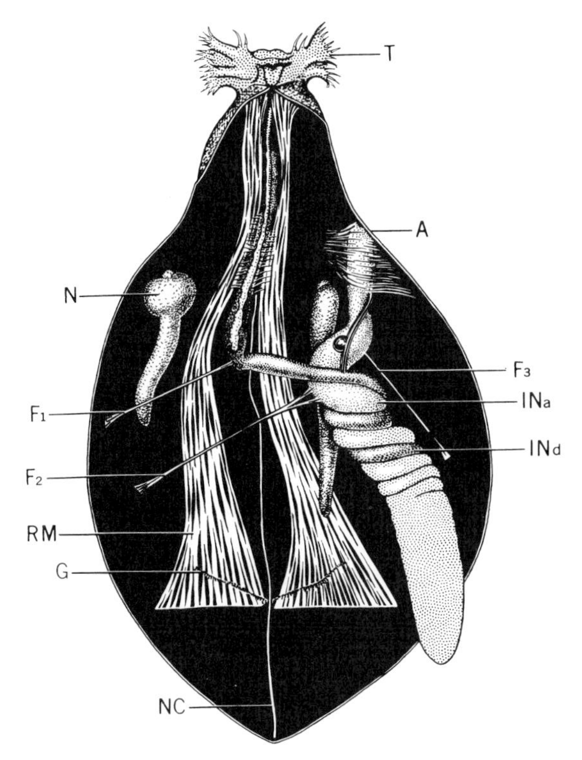

FIGURE 1 General anatomy of an adult sipunculan. Dissected specimen of *Themiste lissum*. A, anus; F_1, F_2, F_3, fixing muscles of the digestive tract; G, gonad; N, nephridium; NC, nerve cord; RM, retractor muscle; T, tentacles; INd, descending intestine; INa, ascending intestine (reproduced with permission from Rice, 1975; redrawn from Fisher, 1952).

I. INTRODUCTION

The phylum Sipuncula is a small group of unsegmented coelomate marine worms, numbering approximately 150 species. Inhabiting all of the world's oceans from tropical to polar seas and from intertidal to abyssal depths, sipunculans are found in habitats of sand and gravel, under rocks, in crevices, in discarded snail shells, in the roots of sea grasses, and among the byssal threads of molluscs. In the tropics, they may occur in association with coral reefs, in which they form borings in coral rubble and calcareous rocks. Sipunculans are distinguished from the closely allied group of annelid worms by an absence of segmentation in the adult as well as in developmental stages.

The body of a sipunculan is divided into two regions (Fig. 1). There is a thickened posterior "trunk," which encompasses a spacious body cavity (coelom) containing most of the internal organs, and a more narrow anterior "introvert," which terminates in a mouth commonly surrounded by tentacles. The introvert may be withdrawn into the trunk by the contraction of prominent retractor muscles which extend from the anterior introvert through the coelomic cavity to attach to the body wall in the mid- to posterior trunk region. An elongated esophagus stretches the length of the introvert from the mouth to a spiraled intestine that is recurved, opening to the exterior through a dorsal introvert usually in the anterior trunk. A ventral unpaired, median nerve cord extends posteriorly along the length of the body from an anterior dorsal supraesophageal ganglion or "brain." The nephridia, which function both in excretion and in the storage of gametes, are usually paired and open to the coelom through a ciliated funnel or "nephrostome" and to the exterior through a "nephridiopore."

II. MODES OF REPRODUCTION

Sexes are typically separate in sipunculans, with eggs and sperm being released by epidemic spawning into the seawater where fertilization takes place.

Commonly, males and females are found in equal numbers; however, in one known parthenogenic species, *Themiste lageniformis*, females comprise over 90% of the population. In this species, eggs may develop in the absence of sperm. In two other species, asexual reproduction may occur by transverse fission or budding. In both instances, the species are known to also reproduce sexually.

III. HERMAPHRODITISM

Separate sexes are the rule in sipunculans; however, hermaphroditism is known in one species, *Nephasoma* (= *Golfingia*) *minuta*. Although developing oocytes and spermatocytes occur simultaneously in the coelom, spermatocytes apparently require a shorter time to mature. Thus, hermaphroditism in this species is considered to be protandric, i.e., individuals function first as males and later as females.

IV. ANATOMY OF THE REPRODUCTIVE SYSTEM

The gonad is a minute strand of tissue at the base of the pair of ventral retractor muscles, extending from the lateral edge of one muscle under the ventral nerve cord to the lateral edge of the other muscle (Fig. 1). Suspended by a peritoneal mesentery from the body wall and enclosed by a peritoneal sheath, the gonadal strand is often composed of digitations within which gametocytes are found in progressive stages of differentiation from proximal to distal regions of the finger-like projections. Gametes are released from the gonad at an early stage into the coelomic cavity where the remainder of their growth and differentiation takes place as freely floating cells in the coelomic fluid along with other cells such as the hemerythrocytes and amebocytes (known as coelomocytes). At the completion of their differentiation, the oocytes are accumulated into the nephridium before spawning into the seawater.

In the ovary, developing stages progress from the more proximal oogonia resembling peritoneal cells through the various stages of the first meiotic prophase. At the distal border of the ovary, oocytes in the diplotene stage are released, either singly or in clumps, into the coelomic cavity. In the testis spermatogonia differentiate into early spermatocytes before their release as clumps of cells into the coelomic cavity.

During coelomic oogenesis oocytes increase in volume and yolk content. In some species they change shape from spheroid to ovoid, and in all species they develop a characteristic thickened, lamellate envelope that is penetrated by pores (Figs. 2a–2d). After release from the ovary, oocytes of some species are covered by a thin layer of scattered follicle cells which are lost at later stages. Another variable character is the development of jelly coverings during later oocyte development.

Spermatocytes break off from the testis in clumps and, as floating cell clusters in the coelomic cavity, undergo meiosis and differentiation into spermatids. In some species spermatid clusters break up into free spermatozoa while still within the coelom, whereas in others free spermatozoa occur after or immediately prior to their uptake by the nephridia. The sipunculan spermatozoan is the so-called "primitive type," characterized by rounded head, mitochondrial midpiece, and elongate tail or flagellum.

Prior to spawning, the fully differentiated gametes are sorted from immature stages and coelomocytes and are taken into the nephridium through its coelomic opening or nephrostome. They are stored in the nephridium for a short time before being expelled through the nephridial pore (nephridiopore) into the surrounding seawater where fertilization takes place. The oocyte, either on exposure to seawater or while within the nephridium, undergoes dissolution of the germinal vesicle (enlarged nucleus) and formation of the first meiotic metaphase.

V. ENDOCRINE CONTROL

The mechanism by which fully differentiated gametes are selected from other cells in the coelomic fluid and accumulated into the nephridium is not known, although endocrine control has been suggested as a possible factor. In some species, nephridial uptake of coelomic oocytes appears to be depen-

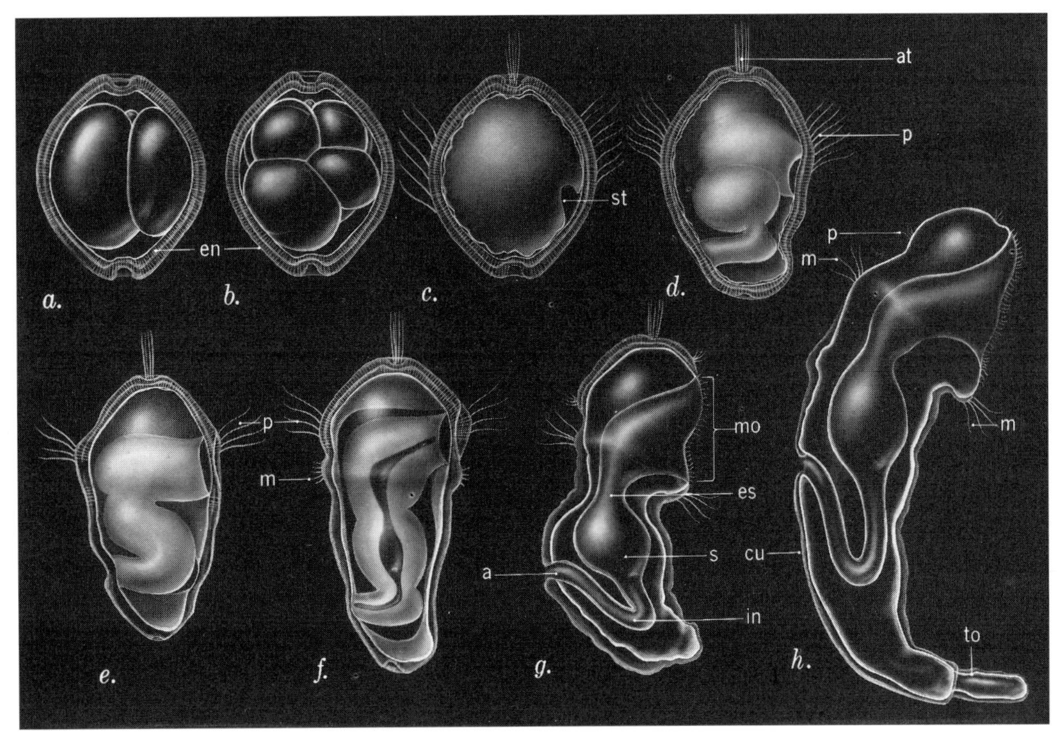

FIGURE 2 Developmental stages of *Phascolosoma perlucens*, from cleavage to trochophore and young pelagosphera larva. (a) two-cell cleavage stage; (b) eight-cell cleavage stage; (c) early trochophore; (d) late trochophore; (e, f) trochophoral metamorphosis; (g) planktotrophic pelagosphera larva immediately after trochophoral metamorphosis (2.5–3 days of age); (h) pelagosphera larva (1 week old). a, anus; at, apical tuft; cu, cuticle; en, egg envelope; es, esophagus; in, intestine; m, metatroch; mo, mouth; p, prototroch; s, stomach; st, stomodeum (position of future mouth); to, terminal organ (reproduced with permission from Rice, 1975).

dent on the maturation of the oocyte, as indicated by dissolution of the germinal vesicle. More evidence is needed for an elucidation of the physiological regulation of this process and for an understanding of the relation of gamete maturation, including germinal vesicle breakdown, to nephridial selectivity.

VI. FERTILIZATION

Fertilization consists of penetration of the egg envelope by the sperm and its entrance into the egg cytoplasm, formation in the egg of the second meiotic metaphase and consequent polar bodies, formation of female and male pronuclei, and, finally, union of the two pronuclei to form the zygote. Sperm penetration in sipunculan eggs is effected by the formation of a hole in the egg envelope. These sperm entry holes may persist for several days during early development of the embryo.

VII. MODES OF DEVELOPMENT

Cleavage in sipunculans is holoblastic, unequal, and, as in other protostomes, spiral (Figs. 2a and 2b). Development may be classified into four developmental patterns (Fig. 3). In the first, the embryo, which may be enclosed in jelly coats, develops directly into a crawling vermiform stage, gradually transforming into a juvenile. In the remaining patterns, the embryo develops into a ciliated, swimming trochophore stage, which may transform directly into a crawling vermiform stage or may metamorphose into a second larval stage, the pelagosphera. In some species the pelagosphera is lecithotrophic or non-

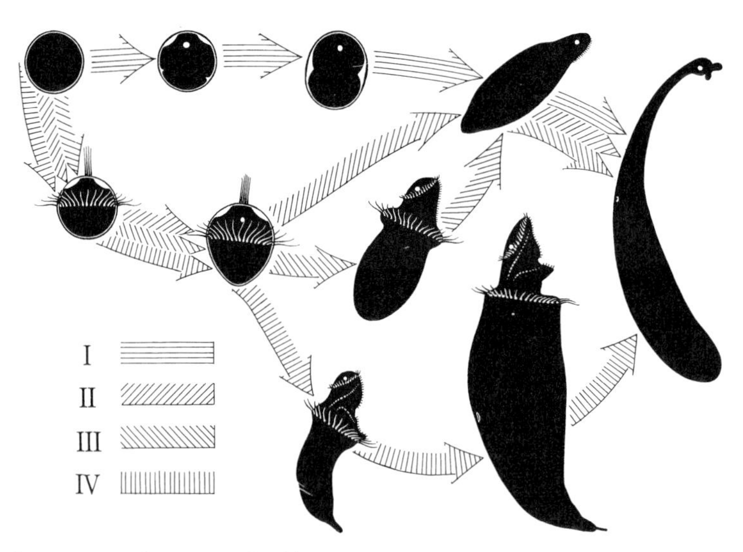

FIGURE 3 Developmental patterns in the Sipuncula. (I) Direct development with no swimming stages. (II) Swimming lecithotrophic trochophore larva which transforms into a vermiform stage. (III) Swimming lecithotrophic trochophore metamorphoses into a second larval stage, the lecithotrophic pelagosphera, which then transforms into the vermiform stage. (IV) Swimming lecithotrophic trochophore metamorphoses into second larval stage, the planktotrophic pelagosphera. After a prolonged existence in the plankton, this feeding larval undergoes a second metamorphosis into the juvenile worm (reproduced with permission from Rice, 1975).

feeding, remaining in the plankton a relatively short time (a few days to 2 weeks) before undergoing transformation into a nonswimming vermiform stage and eventually into a juvenile worm. In other species the pelagosphera is planktotrophic, or feeding, and may persist as a swimming larval stage in the oceanic plankton for several months before undergoing a second metamorphosis into the juvenile worm.

VIII. LARVAE AND METAMORPHOSIS

The trochophore larva of sipunculans is essentially typical of that for other protostomes. It is characterized by apical tuft, apical plate, and equatorial band of relatively large prototrochal cilia (Figs. 2c and 2d). Enclosed by a thickened egg envelope, it is nonfeeding or lecithotrophic. The pelagosphera is defined as a larval type of the phylum Sipuncula that succeeds the trochophore stage and in which there is a loss or reduction of prototrochal cilia, the appearance of a strongly ciliated band of metatrochal cilia which functions as a locomotory organ, and usually a terminal organ by which the larva may attach temporarily to the substratum (Figs. 2e–2h and 4). Whereas the lecithotrophic larva is nonfeeding with a poorly developed gut, the planktotrophic pelagosphera larva has a complete gut with mouth opening onto a ventral ciliated head region, a recurved intestine, and dorsal anus opening in the midtrunk region. Two unique larval organs associated with the head region and presumably serving some function in feeding are the buccal organ, a muscular structure that is protrusible through the mouth, and the lip glands that open through a pore on a lip-like extension posterior to the mouth. At metamorphosis the buccal organ and lip gland are lost, as are the metatrochal band of cilia and the terminal organ (Fig. 4). At the same time, the body is elongated and the mouth, bordered by ciliated tentacular outgrowths, moves from a ventral to terminal position at the end of an elongated, extensible introvert. The resulting juvenile, having lost its swimming capacity, assumes a benthic existence, burrowing into sand or locating in some other cryptic habitat. In laboratory experiments, the process of metamorphosis is enhanced by placing larvae in dishes with sediment in seawater

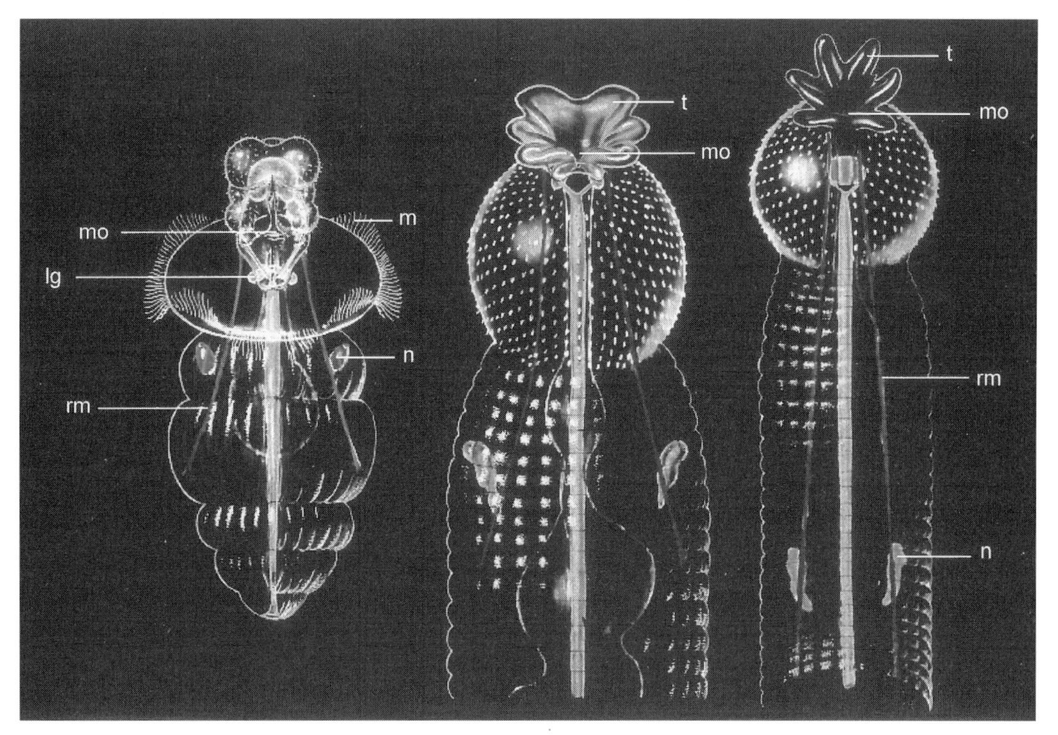

FIGURE 4 Metamorphosis of an oceanic planktotrophic pelagosphera larva of unknown species (family Sipunculidae). Ventral views. (Left) Pelagosphera larva collected from the oceanic plankton. (Middle) First day of metamorphosis, anterior head region. (Right) One week after the beginning of metamorphosis. Note that early in metamorphosis the metatrochal ciliary band is lost, tentacles are formed, and the larval lip glands are reduced. The body is elongated and the mouth, encircled by tentacles, moves from a ventral to terminal position. lg, larval lip gland; m, metatroch (locomotory ciliary band); mo, mouth; n, nephridium; rm, retractor muscle; t, tentacle.

that has been exposed to adult sipunculans. The planktotrophic pelagosphera is of particular interest to zoogeographers because of its lengthy existence in the plankton, its common occurrence in oceanic currents, and thus its potential for dispersal of species over wide distances.

See Also the Following Articles

HERMAPHRODITISM; MARINE INVERTEBRATE LARVAE; MARINE INVERTEBRATES, MODES OF REPRODUCTION IN

Bibliography

Cutler, E. B. (1994). *The Sipuncula: Their Systematics, Biology, and Evolution.* Comstock, Ithaca, NY.

Fisher, W. K. (1952). The sipunculid worms of California and Baja, California. *Proceedings of U.S. National Museum* **102**(3306), 371–450.

Pilger, J. F. (1987). Reproductive biology and development of *Themiste lageniformis*, a parthenogenic sipunculan. *Bull. Mar. Sci.* **41**(1), 59–67.

Rice, M. E. (1975). Sipuncula. In *Reproduction of Marine Invertebrates* (A. C. Giese and J. S. Pearse, Eds.), Vol. 2, pp. 67–127. Academic Press, New York.

Rice, M. E. (1978). Morphological and behavioral changes at metamorphosis in the Sipuncula. In *Settlement and Metamorphosis of Marine Invertebrate Larvae* (F. S. Chia and M. E. Rice, Eds.), pp. 83–102. Elsevier, New York.

Rice, M. E. (1986). Factors influencing larval metamorphosis in *Golfingia misakiana* Sipuncula. *Bull. Mar. Sci.* **39**(2), 362–375.

Scheltema, R. S., and Rice, M. E. (1990). Occurrence of teleplanic pelagosphera larvae of sipunculans in tropical regions of the Pacific and Indian Oceans. *Bull. Mar. Sci.* **47**(1), 159–181.

Social Insects, Overview

Wolf Engels and Klaus Hartfelder

University of Tübingen

GLOSSARY

after swarm Any swarm containing one or a few virgin queens and issued by a honeybee colony after the prime swarm.

alate Winged sexuals, especially in termites.

arrhenotoky Development of males from unfertilized eggs.

caste Polymorphic female adults, queen and worker, in hymenopterans and specialized preimaginal stages of either sex in termites, principally differing in their reproductive capacity.

cerumen Constructional material in stingless bee nests; a mixture of wax and resin.

drone congregation area Special type of lek in honeybees where drones and virgin queens meet for mating.

eusocial Describing an advanced level of social organization with pronounced caste differences and overlapping worker generations.

fitness In genetic terms, the efficacy of genome transfer into the next generation.

foundress queen Mated queen founding a new nest without the help of workers.

gyne Virgin queen in social insects.

intranidal thermoregulation Capacity of social insects to maintain homeostatic conditions within the nest, particularly a widely constant temperature around the brood, by heating and cooling behavior.

lek Male aggregation of numerous drones, waiting for and often attracting a receptive gyne; usually also the mating area (used in a slightly different sense for vertebrates and other nonsocial insects).

mating sign Mucus plug inserted by a copulating male into the female, thus preventing further mating; only in honeybees, multiple matings occur which require removal of the previous mating sign by the next copulating drone.

nymph Last larval instars of hemimetabolous insects with nonfunctional wings.

panoistic-type ovariole An ovariole containing egg follicles without nurse cells.

patriline Progeny of a particular male.

physogastric Describing a hyperfertile state in queens of social insects with much enlarged abdomen.

polydomous colony System of a maternal nest and numerous filial nests in its vicinity with intercolonial worker traffic and exchange of resources.

prime swarm First swarm issued in springtime by honeybee colony, containing the old queen.

primer effect Long-lasting physiological effects of the queen pheromone on the workers.

queen pheromone Multicomponent glandular secretion of queens which, besides the function of female sex pheromone, acts as a dominance signal within the colony.

releaser effect Immediate influence of the queen pheromone on worker behavior.

replacement reproductive In case of queen loss, a queen newly reared from immature instars may become the dominant female.

reproductive dominance Suppression of worker reproduction by the dominant queen, usually by pheromonal control.

reverse molt In termites, advanced larval instars can undergo regressive molting by which the worker capacity of a colony is maintained if no production of sexuals is required.

royal jelly Glandular secretion of worker bees in the honeybee colony, mainly produced by nurse bees in their hyphopharyngeal glands; used for nutrition of all young larvae and exclusively for queen larvae.

single locus sex determination Hymenopteran type of polyallelics single locus gene, which in heterozygous constellation results in a diploid female, in homozygous constellation in a diploid male, and in hemizygous constellation in a haploid male.

spermatheca Receptaculum in the female reproductive tract in which spermatozoa are stored for long periods of time.

supersedure Elimination of an old queen by the workers and replacement by a young gyne.

thelytoky Development of females from unfertilized eggs.

vitellogenin Yolk precursor protein made by the fat body, transported via hemolymph to the ovary, and taken up by receptor-mediated pinocytosis into vitellogenic oocytes.

Social insects are found in two orders of pterygotes: in the hemimetabolous Isoptera or termites, which are all social, and in the holometabolous Hymenoptera, of which all ants are social and many species of bees and wasps have attained differing levels of social organization. Hypotheses on why socioevolution occurred in these two orders will be briefly discussed. Social insects are extremely successful in most terrestrial ecosystems, particularly in the tropics. The ecological impact of ants and termites in recycling of organic materials, and of social bees in pollinating the majority of flowering angiosperm plants, is well recognized. Their importance is reflected, for instance, by the outstanding contribution of social insects to the total biomass of invertebrates in the Amazonian lowland rain forests.

I. SOCIAL INSECTS, OVERVIEW

The characteristic of social life is division of labor within the community, resulting in the evolution of caste syndromes. Castes are principally defined by the tasks to be performed in colonial life, and a specialization into reproductive and nonreproductive members of a colony can be regarded as an early step in socioevolution. The maintenance of within-colony differences in individual participation in the reproductive output of the colony (a reproductive dominance hierarchy) hides basic conflicts over reproduction. Queens, as the dominators, evolved in all groups of social insects. In addition, kings evolved in those termites which have a permanent royal couple. Originally the dominance hierarchy may have been established by aggressive interactions. However, in advanced levels of social organization, dominance is mainly dependent on queen pheromones. By this more subtle influence, the workers, and especially their contribution to reproduction, are effectively controlled.

Reproduction in social insects occurs at two levels: the individual and the colonial. Egg laying is the main task and, at the same time, the privilege, of the queen. Colonies may have one (termed monogynous) or two or more (termed polygynous) queens. Both single mating and multiple mating of queens occur, and the latter results in several patrilines of workers within a colony. Workers do not mate and either produce no eggs at all or only a small number. New colonies may be formed by colony fission or by nest founding by a lone queen. The former is well-known as swarming and involves workers from the parent colony. The construction of filial nests or the establishment of large polydomous colonies, which usually are polygynous and may eventually divide into single nests, are also widespread mechanisms of colony multiplication. Nest initiation by queens which live solitarily until their first worker offspring emerge is called independent foundation. However, young, newly mated queens may also join an existing colony.

Social insects are all organized in family groups, which are termed eusocial if they consist of individuals of overlapping generations: mothers (queens) and their progeny. Workers and males have a shorter life span than queens. In the so-called primitively eusocial species, caste differences may or may not be obvious. Often, allometric differences in body proportions can be recognized in the larger queens, which received more food during the larval stages than the smaller workers. The annual colonies are founded in springtime by queens which possess all the morphological and behavioral requirements to forage and to build a nest. The population of workers in primitively eusocial colonies usually remains small. There is no communication about food sources among nestmates and in most cases no storage of nutrients.

In the highly eusocial species distinct morphological caste characteristics always exist in addition to behavioral and physiological differences. Caste is determined during preimaginal stages, and nutritional factors trigger the switch toward worker or queen development. Both quantity and quality of the larval food affect the caste-specific developmental pathways during critical periods of late larval instars. In some ants a caste predisposition of the eggs has been found.

For the colony, queen rearing requires greater investment than worker production.

Queens in highly eusocial colonies can be regarded as egg-laying machines with the special feature of extreme fertility. In turn, they are unable to survive without workers. In many species, queens are capable of laying thousands of eggs per day, the total mass of which then exceeds their body weight. Consequently, the colonies are perennial and mostly populous, often comprising thousands, tens of thousands, or even millions of individuals. The workers forage, construct, and defend the nest and nurse the brood. They can communicate to each other the location of food sources or suitable new nesting sites. Food stores enable the colony to buffer the effects of temporarily unfavorable conditions. All these advantages lead to the remarkable evolutionary success, especially of the highly eusocial insects, which have become dominant in many terrestrial ecosystems and which are capable of altering the physical environment to an extent comparable only with human activities.

The main aspects of social insect reproduction will be treated in Section II, in which the honeybee is used as the model species. All the other social bees, wasps, ants, and termites, even though much richer in species number and diversity, are given a brief treatment.

II. BEES

Reproduction in the honeybee (*Apis mellifera*) colony, egg laying by the queen and by workers under queenless conditions as well as swarming, was known to beekeepers from ancient times and is depicted in hieroglyphic inscriptions on Egyptian monuments as well as in Aristotle's works. However, it was only a few decades ago that William Hamilton formulated the hypothesis of kin selection, which has contributed substantially to our understanding of the frequent evolution of sociality in hymenopterans. The haplo–diploid system of sex determination, with haploid males and diploid females, results in an especially high relatedness between workers. In the case of single-mated queens, these workers share all the paternal genes and are thus more closely related to their sisters than to their own potential progeny. The difference is great: 75 versus 50% genome identity on average. Thus, the helper function of workers in rearing their mother's offspring results in a higher transfer of their genome into the next generation than that achievable by investment in their own progeny. The kin selection hypothesis explains genetically the evolution of an infertile worker caste and the apparently harmonious cooperation of all the worker members of a colony. This explains the extraordinary evolutionary success of social insects and does not require any exception to the process of Darwinian natural selection. Ultimately, there is thus no altruism operating in the community. The colony is functioning as a superorganism. In many ecological niches, competition between social and nonsocial insects always reveals a huge superiority of the former over the latter.

From this background of genetic preadaptations and resulting competitive advantages, we can understand why evolution of social behavior was repeatedly initiated only within one insect order, the hymenopterans, and especially in many groups of bees and wasps. Social bees are found in several families, though mainly in the Halictidae and Apidae, altogether comprising about 1000 social species. Sweat bees (subfamily Halictinae) contain many primitively eusocial species. Of the Apidae, all the bumblebees (subfamily Bombinae) are also primitively eusocial, with the exception of the secondarily parasitic species. All the stingless bees (Meliponinae) and honeybees (Apinae) have evolved a highly eusocial organization of their colonies. The daily production of large batches of eggs by queens in highly eusocial species requires a correspondingly highly efficient reproductive system. This results from caste-specific morphogenesis in preimaginal development and also from an adaptive optimization of many physiological processes in the adult queen.

Such features have been studied mainly in the honeybee, *A. mellifera*. Today 10 more species of the genus *Apis* are recognized, all occurring in Asia. Except for the structure of the honeycombs and nesting sites, no essential differences in reproduction among the species exist. Morphological caste differences related to reproduction concern especially the size of the ovary and of the spermatheca. In the adult queen about 160–180 ovarioles per ovary are found, but there are only about 3–10 in the adult worker. How is this enormous difference in number achieved, resulting in a more than a 50-fold increased ovarian capacity in functional ovarioles of the queen over her daughter workers? In fourth-instar larvae the numbers of ovariole anlagen are the same in future queens and workers. However, within the long-lasting fifth instar, over 95% of the ovariole primordia undergo degeneration by programmed cell death in worker larvae, and only a few ovarioles survive in the worker ovary. A high juvenile hormone titer in queen larvae has been shown to protect the ovarioles and to hinder cell death and subsequent resorption. Corresponding information on spermathecal morphogenesis is lacking, but the difference in queen versus worker spermathecal diameter is striking. It is a sphere of 1.5 mm diameter in the queen and a rudimentary structure of 2–300 μm in workers, which always remain uninseminated.

The reproductive physiology of the adult queen includes two important differences in comparison to that of the worker: the rate and amount of vitellogenin synthesis and the production of royal pheromones. In honeybees, as in most insects, vitellogenin as the yolk precursor protein is synthesized by the fat body and transported to the ovary via the hemolymph. Vitellogenin uptake into growing oocytes occurs selectively by receptor-mediated pinocytosis. The vitellogenic phase of oogenesis lasts about 3 days. At daily egg-laying rates of 2000 eggs, a normal rate in strong honeybee colonies during springtime, a total of 6000 vitellogenic follicles must be present in the two ovaries of a queen at any time. Therefore, it is not surprising that the vitellogenin fraction of her hemolymph makes up over 50% of all soluble proteins. This requires that vitellogenin synthesis is maintained permanently at extremely high levels. Surprisingly, workers in queenright colonies of honeybees synthesize vitellogenin as well, even though they do not lay eggs. In workers, vitellogenin is found in the hemolymph, especially in nurse bees feeding the brood, but is not incorporated into oocytes. As expected, their rate of vitellogenin production is very low and is at a so-called subfertile status. Only under queenless conditions do some workers become fertile, make more vitellogenin, and produce a small number of eggs. Therefore, the hemolymph vitellogenin titer of a female bee can be regarded as the primary parameter indicating her actual fertility.

The queen as the dominant female is permanently challenged by the workers, her subordinated daughters. This conflict concerns male production and arises because the normally infertile workers are still capable of producing their own progeny by laying unfertilized eggs which develop into drones. However, under queenright conditions the ovaries of most of the workers remain inactive and without vitellogenic follicles. Very few workers lay any eggs, nearly all of which are removed by other workers. This policing effectively prevents successful worker reproduction. The assumed selective background to policing is the closer kinship of a worker to her mother's sons than between a worker and the drone son of an average worker.

The reproductive conflict between the two female castes is suppressed by the gentle but inexorable royal force, the queen's pheromones. In addition, worker fertility is inhibited by brood pheromones. Reproductive dominance is effected by queen pheromonal secretions from three sources: her mandibular glands, her abdominal tergite pocket glands, and her tarsal glands. These all contribute to the queen pheromone, which is a multicomponent mixture of volatiles produced in distinct patterns, varying with age and physiological state of a queen. The inhibition of worker physiology, such as suppression of juvenile hormone synthesis by the corpora allata, is a primer effect of this pheromone, which has releaser functions as well. Within the honeybee colony, an important releaser mechanism ensures the maintenance of the monogynous situation. The queen controls the construction of royal cells; enlarged brood cells usually built at the lower margin of brood combs. Queen cells are regularly constructed prior to swarming and also in the case of queen loss. This emergency queen rearing is only possible if eggs or young worker larvae are present in the brood nest. Some of the worker-sized brood cells on the comb are then enlarged to royal cells and, by changing the diet into pure royal jelly, the female larvae in these cells will become queens. Emergency queen rearing is initiated immediately after a colony realizes that the queen is no longer present. It has been shown that the worker behavior to build queen cells is normally inhibited by the major component of the queen's mandibular gland secretion, the so-called queen substance *(E)*-9-oxo-2-decenoic acid (9-ODA). Only if the colony is empty of queen-laid eggs and young larvae, such as typically occurs before a prime swarm is issued, are queen cells constructed even under queenright conditions.

The queen oviposits normally into large cells. Before laying an egg, the queen inspects the brood cell which has to be prepared properly by the workers. With her forelegs she measures the inner diameter of the upper part of a brood cell. Into the larger cells on drone combs the queen will lay unfertilized eggs, and into the narrower worker cells fertilized eggs are oviposited. What about the large queen cells? Their rim has the same small diameter as a worker brood cell and, consequently, fertilized eggs are likewise laid into the large queen cells, developing into female larvae. The queen is capable of controlling the fertilization process egg by egg. When an egg passes through the common oviduct, a droplet of spermathecal fluid containing a few spermatozoa is pressed by a single contraction of Bresslau's semen pump onto the anterior tip of the egg chorion, where the micropyles allow the sperms to enter the egg just before it is laid. When ovipositing into a drone brood cell, the queen does not operate the semen pump, and the egg will pass unfertilized into the vagina. It should be mentioned that developmental activation of the honeybee egg is not triggered by fertilization but by the compression which occurs during oviposition. By deciding which egg is fertilized, hymenopteran females can precisely control the sex ratio of their progeny. Males originate from unfertilized eggs by arrhenotokous parthenogenesis. This is true for the "normal haploid drones" which are in a hemizygous state for all single copy genes, including the sex determination locus. "Anomalous diploid drones" result from homozygosity at the single but highly polymorphic sex locus. An estimated 15 alleles were found at this locus in local honeybee populations. If by chance, or by inbreeding, the sex locus in the diploid genome of a fertilized egg consists of two identical alleles, a male is produced. In all cases of heterozygosity at the sex locus, females are produced. The molecular mechanisms involved are unknown. Since fertilized eggs are destined to become workers, diploid drones develop in worker-sized comb brood cells. Colony performance is weakened if instead of worker bees many diploid males are produced which contribute neither to reproduction nor to labor. The possible production of diploid males is an apparent constraint of the hymenopteran type of haploid–diploid sex determination based on a single locus with multiple allelic variants. However, under normal conditions, in the honeybee colony diploid drones are eliminated as first-instar larvae by the nursing workers and do not reach the imaginal stage, thus preventing wasted investment in their rearing. It is possible to obtain adult diploid males by experimental rearing of larvae.

Throughout all five larval instars, future queens are fed *ad libitum* and exclusively on royal jelly, a glandular secretion produced by nurse workers, mainly in their hypopharyngeal glands. The trophogenic stimuli emanating from this royal diet trigger reactions of the endocrine system. The prime response is an elevated rate of juvenile hormone production by the corpora allata during the last larval instar followed by an enhanced synthesis of ecdysteroids by the prothoracic gland. This in turn shifts caste development during metamorphosis to the pathway of queen differentiation within the facultatively polymorphic options of the female caste polymorphism. In contrast to this greatly enriched nutritional program, worker and drone larvae in comb brood cells receive jelly only during the first 3 days after hatching, corresponding to larval stages 1–3. Thereafter, during stages 4 and 5, they are fed on a less "expensive" mixture consisting mainly of honey and pollen. This progressive feeding of brood is unique and found only in honeybees and some bumblebees.

The hexagonal brood cells of honeybees are arranged horizontally in two opposite layers, forming a vertical comb made of beeswax. It is repeatedly reused and may also serve to store honey and pollen. The embryonic stage lasts 3 days, but the duration of the larval stages and especially of the pupal stage differs between sexes and castes. The shortest time is required for queen development, which takes only 13 days from hatching of the egg to emergence of the imago from a royal cell. The virgin queen is also called a gyne. In workers the preimaginal period is about 17 or 18 days long, and in drones it is 20 or 21 days. The exact duration depends on the position of the egg within the brood nest. Development is faster in its center, where there is a constant temperature of 35°C. At the periphery, the temperatures vary between 32 and 34°C. Honeybees are capable of exact intranidal thermoregulation, especially in those parts of their nest which are actually used for reproduction. Heat is generated by the workers' shivering of the thoracic flight muscles. Cooling at high ambient temperatures is achieved by fanning with the wings and by evaporation of water droplets which are then also deposited on the brood combs. Because queen cells and especially swarm cells are located in the lower and cooler region of the brood nest, a warming pheromone produced by queen pupae attracts workers and elicits their shivering reaction to warm the precious future princess.

In springtime the beginning of the swarming season is initiated by the construction of drone combs, which then account for about 15% of all brood cells. Some weeks later in strong colonies the workers will build swarm cells and rear a number of queens. These worker preparations for colony reproduction depend on several factors, of which overcrowding, the lack of empty brood cells, the absence of eggs and young larvae, and particularly a reduced level of queen pheromone circulating within the colony are important. Finally, the swarm issues a few days before the first young queen will emerge. The old queen is not fed by the workers for about 1 week prior to swarming. Therefore, she cannot lay eggs but loses weight and is thus able to fly. She leaves the nest with the prime swarm. This swarm comprises about 60% of the colony population including workers of all age classes and several drones, but mostly young bees which engorge on honey prior to leaving the nest. They thus provide the swarm with sufficient energy in case of temporary shortage. This prime swarm often settles on a tree near the original nest site, and a cluster of bees forms. The swarm cluster is stabilized by the queen's pheromones. However, the prime swarm may also fly long distances and thus be able to colonize new areas, or at least lead to a mixing up of the local bee population. The clustered swarm can stay in its bivouac for hours or even several days. Even before the swarm leaves the colony, and more intensively from the bivouac, scout bees search for suitable nesting sites. They look for cavities, which are then inspected.

Dry hollows not too near to the ground and with a volume of 30 liters or more are preferred. On returning, successful scouts dance on the surface of the swarm cluster using the same waggle dance as normally used for communicating profitable food sources. The dance intensity indicates the quality of the detected nesting site, and the worker community of the swarm has to be convinced. When its decision is made, the swarm with the queen will become airborne and fly directly to the new home. There, the workers immediately begin to build new combs, and the queen will start to oviposit even before the cells are fully drawn out. Foraging commences at the same time. Three weeks later the first young worker bees will emerge, and it takes another 3 weeks until they become foragers. Altogether, the swarm has a break of about 2 months until new field bees are available. Presumably, this is the reason that swarming bees are rather unlikely to sting. Stinging leads to worker death and diminishes the bee population of a swarm. Swarming implies a high risk of survival for both the parental and the filial colonies. A worker population sufficient to collect enough pollen and nectar for periods of dearth or cold winters has to be reestablished in time. Only under tropical conditions is swarming not strictly seasonal, and numerous small swarms may be produced in series within 1 year.

In temperate zones, the queen heading the prime swarm was born the year before or even earlier. If her fertility and pheromone production is suboptimal, the colony will replace her after some time by supersedure. She is stung by the workers after having laid fertilized eggs into swarm cells, and one of her daughters emerging later from these royal cells will become the new queen and undertake mating flights from her natal nest. The parent colony can produce additional swarms some time after the departure of the prime swarm and when young gynes have emerged. The first-born princess will try to secure her accession to the throne by killing her sister queens, even while they are still sitting in their royal cells. She pipes, and the yet to emerge queens answer by quaking. This sound communication enables the first emerging gyne to locate rival princesses which, however, are usually protected by the workers. After swarms then depart from the nest, each containing several young sister queens which fight until only one is left. The successor will undertake a nuptial flight from the new nesting site. If she gets lost, the colony is hopelessly queenless and, because no royal and brood pheromones are acting, after some time a few workers will become false queens and start to lay eggs. Since all the workers are unmated, their eggs are unfertilized and develop into drones. Such worker-derived males contribute to the drone population even outside the regular swarming season, when occasional emergency queen rearing may still occur in other colonies. Consequently, worker-produced drones and emergency-reared queens can mate and at least ensure the survival of the colony providing the gyne. The drone-producing queenless colonies can survive for only a limited period and are unable to hibernate.

There is a remarkable exception to colony requeening and the lack of worker reproduction known in only 1 of the about 50 geographic subspecies of honeybees, the South African *Apis mellifera capensis*. In this Cape bee, laying workers can produce diploid eggs. However, they are not fertilized but rather originate from the fusion of one of the normally abortive polar nuclei, products of the meiotic divisions of the oocyte nucleus, with the female pronucleus. Since these postmeiotic genomes differ from the maternal genome because of recombination by crossing over and chromosomal sorting, mostly females and not diploid drones arise from this unique type of worker reproduction, which is a very special case of thelytokous parthenogenesis. Those female larvae which are then fed on royal jelly will become replacement queens.

For a long time the mating biology of the honeybee was regarded as a miracle because copulations were never observed. Gregor Mendel was an experienced beekeeper, and he started his genetic experiments with honeybees. He failed, however, since he was unable to get controlled matings between black German queens and yellow Italian drones. Honeybee sexuals never copulate in the nest. Rather, they fly far away to drone congregation areas. The name indicates that many drones assemble there on warm days, flying around 10 or more meters above the ground. These areas are reused by the drones year after year, which means that physical properties of the routes and the sites must guide the bees. Virgin queens also approach these congregation areas, but workers never do so. The flying drones recognize the queen's silhouette against the blue sky with the dorsofrontal region of their compound eyes, in which the ommatidia have a larger diameter. The males are at the same time attracted by the female sex pheromone, and it has been shown that 9-ODA is one of the volatile compounds releasing this behavior. Within a short time, the queen is mated by about 20 drones in sequence. The first successful male places a mating sign in the queen's vagina which consists of yellowish mucus from his accessory glands. This sign enhances the attraction of the chasing drones, of which a cloud of 20–50 follows the queen. The next male to copulate with the queen has first to remove the plug inserted by the previous drone before inseminating and putting in his own mating sign. During this process the drone becomes paralyzed, falls to the ground, and dies. This consecutive mating is interpreted to increase the fitness of those males who located the rapidly flying gyne and joined the drone cloud. In most hymenopterans, however, such mating plugs serve to inhibit further copulations, thus securing paternity to the first male. Drone abundancy at a particular congregation area is increased by the presence of males which appear to attract others with their mandibular gland pheromones, thus acting as an aggregation lure. The drone congregation area is a special type of lek. Since drones usually fly further to a congregation site than queens, it is very unlikely that brothers and sisters, whose mating could result in diploid drones, will meet. The reproductive strategy of honeybees can be considered an insurance against inbreeding. The mated queen will return to her colony. The workers then help to remove the last mating sign. Most of the sperm of the multiple copulations is ejected from the queen's reproductive tract, but a portion from probably all the drones' ejaculates is retained. The spermatozoa migrate into the spermatheca where they are stored. The queen's longevity may extend over 5 years, and the mixed spermathecal content of about 7 million spermatozoa enables her to fertilize 200,000 or more eggs per year.

Hence, multiple mating results in many patrilines present in her worker progeny at the same time. This seems to be contradictory to the hypothesis that kin selection has favored social behavior since the worker population of a given colony is a mix of a few full sisters and many half sisters. In fact, the moderate intracolonial genetic variability provided by about 10 effective fathers represented in a queen's offspring at any one time increases the fitness of the colony by providing worker bees with individual preferences for the various tasks. The behavior of the individual worker is only in part dependent on age-related polyethism. The intracolonial system of division of labor is regulated by the colony's needs and by the individual worker's juvenile hormone titer. In addition, the decision of a worker to undertake a specific task is controlled by its individual preference thresholds which differ according to its genetic constitution. Thus, the superorganism honeybee colony always contains workers willing to engage in its manifold requirements, especially those related to reproduction.

A. Stingless Bees

The world's approximately 500 species of stingless bees are of pantropic distribution. They all form highly eusocial colonies comprising 500 to over 50,000 individuals. Since their reproductive biology differs from that of honeybees in several aspects, they are considered to have an independent origin of highly eusocial organization. Most stingless bees build horizontal combs of vertical brood cells which are mass provisioned by the workers with all the food for the nourishment of a larva. An egg is positioned on this food by the queen, and the cell is then immediately sealed by a worker. After emergence of the young bee, the brood cell is dismantled. A striking phenomenon is the persistence of worker fertility in queenright colonies. In addition, in many species the workers are capable of producing two types of eggs. One type of worker eggs, the so-called reproductive eggs, remain unfertilized, are oviposited into brood cells, and develop into drones. In many species of stingless bees a considerable portion of the drones originates from such worker reproduction. The other type, the alimentary eggs, differ in size and shape, are laid at unusual places such as the upper rim of a brood cell, and are eaten by the queen before she oviposits into the cell. Alimentary eggs may also be eaten by other workers. This oophagy is considered a relict from simpler stages of social evolution wherein reproduction of subordinate females is hindered by the dominant female who eats the eggs of the subordinates. Queens are reared throughout the year, and most of the gynes are killed by the workers soon after emergence. Swarming is preceded by founding of a filial nest nearby into which a virgin queen moves together with many workers. Drones leave their natal colony when about 3 weeks old and are then "vagabonds."

Males from many different colonies are attracted to nests containing a receptive virgin gyne and form drone aggregations in the vicinity, another type of lek. These aggregations persist for some time until the gyne undertakes a short nuptial flight. She is chased by the waiting males and returns to her nest after a single copulation. Recent molecular analysis of patriline composition of a queen's progeny evidenced that multiple mating, however, of low frequency may also occur. Therefore, the queen mates near her place of birth, but the probability of inbreeding is reduced by the vagabond behavior of the drones. This represents another type of reproductive strategy promoting panmixis which is different from that of the honeybee. A few days later the mated queen becomes physogastric by expanding her ovarioles and she starts to lay eggs. At this stage the new queenright colony is still materially supported by the maternal nest from which the workers carry cerumen and food. The weight of the fully physogastric and fertile queen is three or four times that of a virgin gyne. The heavy queen mother will never again leave the nest; she is unable to fly and may later be replaced by supersedure. Swarms are headed exclusively by virgin queens.

B. Bumblebees

Caste differences are less pronounced in bumblebees, of which some hundred species of the supergenus *Bombus* are known. They all form primitively eusocial colonies. Large and previously mated queens independently found their nests. In temperate climates, the foundress queens will have overwintered. In the reproductive cycle, first small and later larger workers are reared, followed by new daughter queens and males, before the nest collapses at the end of the season. Only in the tropics do long-lasting colonies exist. As the queen ages, some of the workers usually begin to lay eggs, thus participating in reproduction on a regular basis. There are many workerless species which have evolved a social parasitic mode of reproduction. Their mated queens intrude into alien colonies, stay there, and lay eggs. Their offspring are reared by the heterospecific workers, and only reproductive females and males are produced.

C. Halictine Bees

The so-called sweat bees, a subfamily of the Halictidae, are widely distributed and comprise several thousand solitary and social species, the latter showing different levels of socioevolution. In the social species, the colony cycle is annual, similar to that of bumblebees. Foundress queens may produce one or a few generations of workers, the first of which is not inseminated. The last females are produced simultaneously with males and will mate and function as queens in the next colonial cycle. In the case of the primitively eusocial species, only the larger queens are able to copulate.

III. WASPS

The social wasps of the family Vespidae belong to three subfamilies—Polistinae, Stenogastrinae, and Vespinae—of which only the latter exhibit distinct caste differences. Foundress queens independently initiate a colony, which later may comprise thousands of workers before new daughter queens and males are produced. In polistine wasps the nest-founding female is often joined by a few other females which are likewise mated. Social rank is in most cases established by physical dominance, and so a quasi-monogynous status of the colony is attained. In these colonies all the members recognize each other. Suppression does not occur via pheromones, but instead the dominant female often has to eat the eggs of her nestmates in order to exclude their progeny. Consequently, loss of the functional queen has no fatal consequences for the reproductive output of such a colony because a formerly subordinate female can easily assume her position. An exception to this occurs in some species of honey wasps of the genus *Polybia* which occur in the Neotropics and have populous and perennial colonies as well as dependent nest foundation by swarming. In these species, caste differences are more pronounced, and permanent brood rearing is ensured by storage of sometimes large amounts of honey.

IV. ANTS

All of the approximately 10,000 species of ants of the world live in colonies with a eusocial level of organization, and there is considerable variation in reproductive biology. Since worker ants are frequently found embedded in Pleistocene amber, we can conclude that in the family Formicidae colonial life was attained at least in Tertiary times. Colonies can be monogynous, but the larger colonies are polygynous and often also polydomus. Sex and caste differences can be extreme. All the workers are wingless morphs. In many species, the worker population can consist of two or more morphological distinct subcastes, of which only one takes care of the brood. Winged young queens and males undertake mass nuptial flights which often occur on just one day of a year, resulting in a mating strategy promoting panmixis. Mated queens return to the ground and shed their wings. They may independently found a new nest or join an existing colony. In some species, new queens regularly join a nest of closely related species, and the intruding queen replaces the queen of the host nest. Within some months or even years, her progeny will make up an increasing portion of the workers until her offspring have completely replaced the former inhabitants. From this remarkable type of colony reproduction, several types of social parasitism have arisen, ranging from slave-making by robbing pupae from heterospecific nests to a complete loss of the worker caste, as is also observed in parasitic bumblebees. Caste development depends on differential feeding of the female larvae, but in several species a hormonal predisposition of the eggs or other factors such as genetical bias have also been identified. There are some ant species in which queens are unknown, and worker reproduction is regulated in the small colonies by dominance ranking, just as in polistine wasps.

V. TERMITES

The approximately 3000 species of Isoptera of the world are all eusocial. However, there are striking differences between the so-called primitive and higher termites in modes of caste differentiation and reproduction. Termites are hemimetabolous and differ from social hymenopterans because they have a classical type of sex determination with diploid genomes and male heterogamety. This allowed the evolution of male and female castes which, in essence, are all preadult stages. The apparent consequence is that there was no selection for a female-biased sex ratio and for the evolution of female castes which differ in their reproductive capacities. Simply stated, all of the adults are sexuals and active in reproduction.

All termites feed on living or dead plant material and thus require intestinal symbionts in order to digest cellulose and lignin components. These bacterial or protozoan symbionts have to be replaced after each molt, and reinfestation is done by the uptake of feces. The necessary close cohabitation of larvae and adults resulted in social life. Workers are either female or male larvae of later developmental stages. Worker subcastes, small and large types, and special final larval forms such as soldiers and nasutes are also known. The latter are terminally differentiated final larval instars.

The developmental pathway produces adult sexual alates only by larval instars first molting to nymphs, which exhibit the first signs of wing pads. In order to prevent an overproduction of sexuals, reverse molts may occur in the more primitive termites, returning nymphal stages to larval stages, which are the functional workers. In the Kalotermitidae the queen inhibits the formation of sexuals in a fascinating manner. Her feces contain a juvenile hormone-like factor. The preimaginal stages eat the royal feces and thereby increase their hemolymph hormone titer, which in turn results in reverse molting. If the queen is lost, some nymphs immediately undergo a progressive molt and become replacement reproductives.

The sexuals usually perform mass nuptial flights, resulting in the mixing of the local population. Mating, however, occurs later. After landing on the ground, the alates shed their wings, and tandem pairs, with the queen in front and the male marching behind, search for a suitable site to dig a hole into the soil. This is the first step of founding a new colony in which several tandems may participate. Once the nest is established, copulations of the royal pair begin. As soon as the first generation of workers has been reared, the royal pair concentrates on mating and egg laying. This is accompanied by differentiation of numerous panoistic ovarioles and, especially in the higher termites, the queen becomes extremely physogastric. Huge numbers of eggs may be produced which are immediately removed by nursing workers that queue at the abdominal tip of their giant mother. Daily rates of egg laying may range in the tens of thousands, resulting in extremely populous perennial colonies. In some species, there may be 10 million or more workers per colony, all produced by a single royal pair. The ecological impact of such immense populations in recycling of plant biomass, especially of wood, which in many tropical ecosystems is nearly exclusively the task of termites, is unparalleled and indicates the evolutionary success of such complex social and endosymbiontic organization.

Bibliography

Crozier, R. H., and Pamilo, P. (1996). *Evolution of Social Insect Colonies.* Oxford Univ. Press, Oxford, UK.

Engels, W. (Ed.) (1990). *Social Insects. An Evolutionary Approach to Castes and Reproduction.* Springer, Berlin.

Hartfelder, K., and Engels, W. (1998). Social insect polymorphism—Hormonal regulation of plasticity in development and reproduction in the honey bee. *Curr. Topics Dev. Biol.* **40**, 45–77.

Moritz, R. F. A., and Southwick, E. E. (1992). *Bees as Superorganisms. An Evolutionary Reality.* Springer, Berlin.

Seeley, T. D. (1995). *The Wisdom of the Hive: The Social Physiology of Honey Bee Colonies.* Harvard Univ. Press, Cambridge, MA.

Watson, J. A. L., Okot-Kother, B. M., and Noirot, C. (1983). *Caste Differentiation in Social Insects.* Pergamon, Oxford, UK.

Song in Arthropods

Glenn K. Morris
University of Toronto

GLOSSARY

calling song A song broadcast by a lone male, unevoked by receiver-based stimuli, to attract and guide the approach of a female at long range.

courtship song A song given at close range after pairing, influencing mate choice, grading in some species from calling song.

duty cycle The time spent signaling as a fraction of the total time available.

Fourier transform Waves as a sum of sine and cosine functions arrayed incrementally in a spectrum showing the energy at each sound frequency.

near field The region near a sound source where particle velocity and pressure are out of phase.

phonatome All the sound produced by one cycle of the generating apparatus.

phonotaxis Oriented locomotion to a sound source.

pressure difference ear Detecting pressure differences across a tympanum of sound conducted both internally and externally.

prolonged pulse A discrete wave train; the product of a high-Q mechanism in which the radiating structures undergo minimal damping and scraper-tooth contact matches radiator resonance.

pulse An uninterrupted wave train; amplitude modulated from and back to ambient sound.

Q The quality factor of a resonator; it is a measure of tuning sharpness expressed as the relative width of the resonance curve 3 dB below its maximum.

rivalry/territorial song A song exchanged between conspecific males in an aggressive context and often incorporated in calling song; it keeps a singing location free from encroachment by male rivals.

short-circuiting Sound pressure cancellation at the margins of a generating diaphragm.

stridulation Carrier frequency multiplication by exoskeletal friction; the act of contact rubbing of arthropod cuticle to produce sound.

transient pulse A discrete wave train; the product of a low-Q generation mechanism in which scraper-tooth contact rate is far below radiator resonance.

tremulation Oscillations of the body that impart vibrations to substrate via appendages kept in surface contact (nonpercussive).

tymbal Ribbed cuticular region of the exoskeleton generating sounds by sudden distortion.

vibration Mechanical wave disturbance produced in solid substrate.

Arthropod songs are mechanical wave disturbances in air, water, soil, or plants produced most commonly by males; they are signals sent to conspecifics of either sex. They may encode information about the sender's location or species and convey his motivation and capacities, either to females as a potential mate or to males as a rival.

I. SINGING TAXA

Four arthropod "acoustic" taxa are known for calling, courtship, and rivalry songs in air: cicadas, crickets, katydids, and grasshoppers. Their singing occurs in the context of competitive mate attraction. Calling songs are broadcast by males to lure distant females.

In some species the female localizes and moves silently to the singer; in others, she sings in answer and a call exchange ends in pairing. Once male and female come together, courtship singing may occur in advance of mating. Aggression between males can be mediated by the same calling song that attracts a female or by a distinctive rivalry song.

It is impossible to fly without making sound and the flight tones of some Diptera have become airborne signals. Mosquito females use such songs to call their males. However, many more Diptera make courting songs with special wing movements, perceived in the near field as air particle displacement (e.g., Tephritidae and Drosophilidae). Also, near-field signals are part of the dancing of honeybees.

Water boatmen (Corixidae), beetles (Hydrophilidae), and caddis fly larvae (Hydropsychidae) sing in fresh water. In the ocean, pistol shrimp (Alpheidae) make rivalry songs with chelipeds, whereas fiddler and ghost crabs (Ocypodidae) use both airborne sound and vibrations in courtship and agonistic behavior.

Insect vibration singers include stone flies (Plecoptera), lacewings (Neuroptera), pentatomid bugs, and many small Homoptera: leaf-, tree-, and planthoppers. The substrate songs of these insects are as complexly patterned as any made by acoustic Orthoptera or cicadas. Vibrations are important for spiders in the capture of prey, but vibrations also function widely in spider courtship and calling.

II. SONG STRUCTURE

A. Time Domain: Pulse Patterns

The sound pulses in songs vary from transient (Fig. 1A) to prolonged (Fig. 1B). They repeat in trains of varied pulse number and duration. Pulse rates, train rates, amplitude envelopes, changing

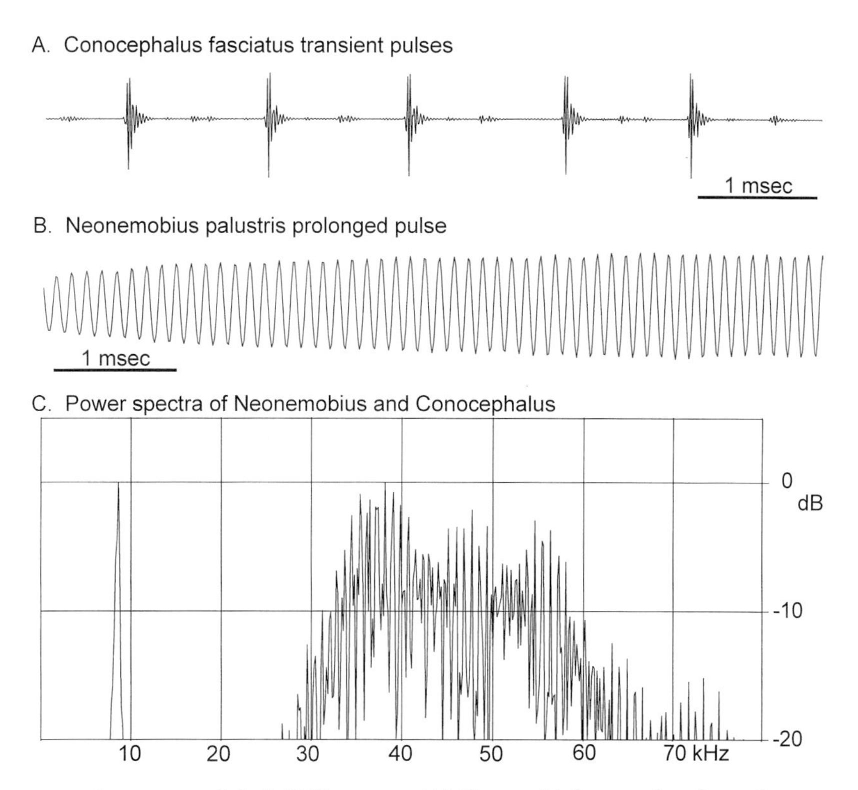

FIGURE 1 Two extreme pulse types and their FFT spectra. (A) Five rapid-decay pulses from the nonresonant generation of a katydid. (B) Portion of a resonant-generated cricket pulse. (C) Power spectrum showing a single narrow cricket peak near 8 kHz and the ultrasonic band of *Conocephalus*.

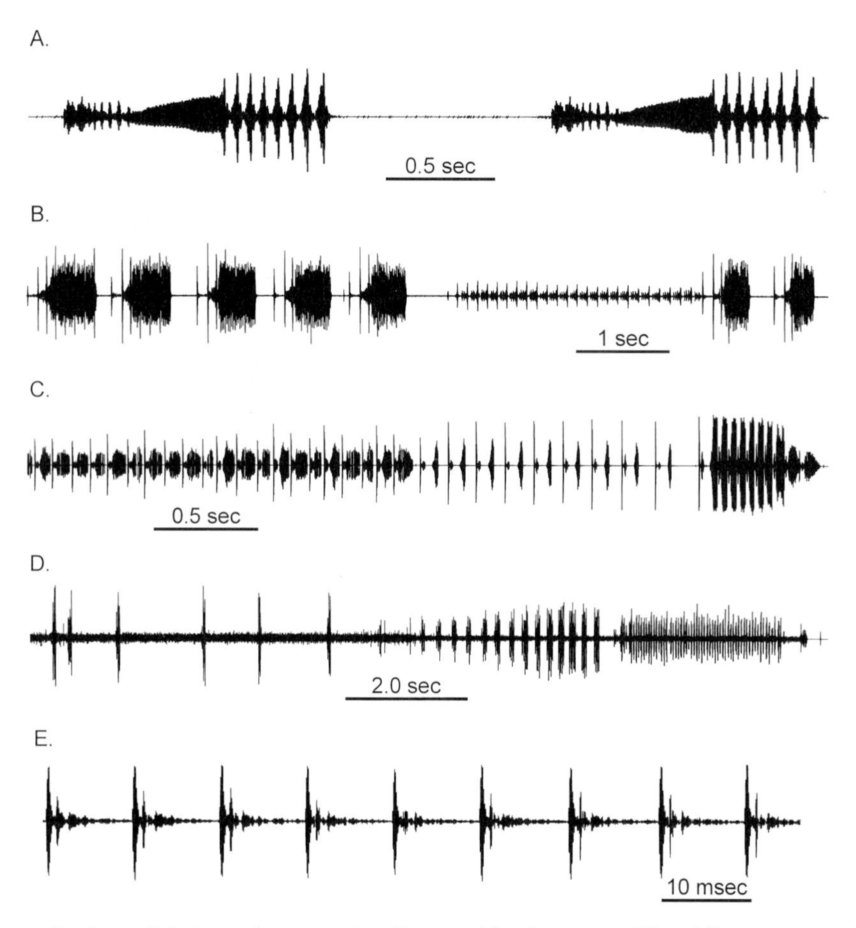

FIGURE 2 Diverse amplitude modulation pulse patterning illustrated by the songs of five different insects. (A) Two vibrational songs of the membracid *Enchenopa binotata* (courtesy Randy Hunt). (B) Part of the complex calling song of the gomphocerine grasshopper *Syrbula fuscovittata*. (C) Three parts of the four-part call of the katydid *Amblycorypha uhleri*. (D) Underwater call of a water boatman (courtesy A. Jansson). (E) Tymbal-generated pulses of a cicada.

rates, and amplitudes occur to form multiple modes or nested groupings. The result for most arthropods is a species-typical amplitude modulation (AM) pattern (Fig. 2).

B. Carrier Frequency

Power spectra of transient and prolonged pulses are contrasted in Fig. 1C. Fourier analysis of five katydid pulses gives a band spectrum (30–60 kHz), whereas the pulse of a nemobiine cricket makes one narrow peak (8.4 kHz).

Transient pulses and band spectra are the rule in grasshopper songs (Fig. 2B), and the same is true of most katydids (Fig. 2C). Grasshoppers show an ultrasonic maximum (20–40 kHz) with a narrower audio peak. Prolonged high-Q pulses, with a modest downward frequency modulation, occur in all cricket calling songs. Also, cricket calling song carriers worldwide lie below 10 kHz.

As a group, katydids have the greatest range in carrier. Many species make high-Q pulses in the high audio and ultrasonic range. The lowest airborne calling song carrier known belongs to a pseudophylline from Malaysia, *Tympanophyllum*, at 600 Hz; the highest occurs in a katydid from the west coast of Colombia—*Arachnoscelis* at 135 kHz.

Frequencies of vibration songs and frequencies used in near-field airborne sound detection are in the very low audio (<1 kHz). Also, the plant substrates through which vibratory signals travel act as unpredictable filters, making it impossible to signal with spectral consistency at the receiver. Carriers underwater are in the low audio: Frequencies used by corixids are between 1 and 12 kHz.

C. Intensity, Range, and Energetics

Cicadas are the loudest acoustic insects: Some Australian species have calling sound levels near 110 dB at 20 cm (re 20 μPa). Levels 10 cm dorsal to katydids are commonly 90–100 dB. Most grasshoppers are quieter: At 10 cm one *Chorthippus* species is only 62 dB.

A portable nerve preparation of the living ear has permitted field assessment of the detection range of some katydids. Under ideal noise conditions *Mygalopsis* hears its species call at distances up to 25 m, much beyond its typical intermale spacing.

The sustained production of these intense songs is energetically costly and because of their small size most arthropods are quite inefficient at converting metabolic energy to acoustic power. Even the loudest insect achieves no more than 10% efficiency.

III. SENDING: GENERATOR FUNCTION

The ready evolution of sound generators, independently, at diverse body locations is a property of arthropod exoskeleton. Incidental sounds, made originally during movement as surfaces touched or were bent by the pull of muscles, became ritualized into sound signals. Cuticular surfaces were elaborated into files, scrapers, and tymbals.

A structure radiating sound that is moved directly by inserted muscles is limited in the frequency of its oscillations to muscle contraction rates. However, suitable intervening cuticular mechanisms can greatly multiply the sound frequencies achieved.

A. Direct Muscle Generation

1. *Flight Tones and Generation for Near-Field Reception*

The whine of a midge has a frequency corresponding to its flight muscle contraction rate. Both sexes in many species are attracted by these tones to form a localized breeding swarm. During courtship *Drosophila* extend one wing in the direction of the female and move it to make a pulse series. The frequency is low, e.g., 166 Hz, and the male fly is very close by—5 mm or about 1/400 the wavelength. The female's arista is stimulated by particle displacement not pressure.

The cricket *Philacris* uses low-frequency air displacement as part of its close-range courtship and aggression. Males pivot their forewings abruptly forward over their head; air vortices are cast off from the wing tips at the end of each flick and move toward the receiver.

2. *Drumming*

Percussive song generation occurs in crabs, ants, termites, book lice, beetles, bugs, and grasshoppers. Spiders drum on webs, retreats, or green or dry leaves using their opisthosoma, legs, or palps. Body surfaces may be specialized for striking, e.g., knobbed hairs on the opisthosoma underside in the wolf spider *Hygrolycosa*.

Stone flies are accomplished drummers. They make vertical strokes with their abdomen, contacting leaf mats and wood debris in litter or live plant tissue with vesicles, lobes, knobs, and hammers. The drumroll of a male evokes a female's drumming response and in most species the male then searches for the waiting female.

Striking with body parts can be directed against another appendage. The movable finger of a pistol shrimp's cheliped bears a cylindrical plunger that fits a cavity on the propodite. As it pops suddenly into the hole, the plunger makes a sound and displaces a stream of water. Also, males of *Hecatesia*, a whistling moth (Noctuidae), strike together castanets on the costal wing margins.

3. *Tremulation*

Because body quivering (tremulation) occurs without percussion, it makes substrate waves quietly. This may be adaptive when courting since airborne sounds can attract the attention of male competitors. Stone flies (e.g., *Sialis*) tremulate as well as drum, and tremulation occurs in many katydids, e.g., *Ephippiger* and *Conocephalus*. The brevity of airborne songs by some neotropical katydids may reduce their vulnerability to eavesdropping bats, but it leaves them transmitting less information: They compensate with elaborate tremulation signals. A few species, *Copiphora* and *Choeroparnops*, engage in calling tremulation.

Male lacewings produce substrate-borne bending waves by abdomen tremulation and are responded to in kind by conspecific females. Tropical wandering male spiders of *Cupiennius* tremulate their opisthosoma to produce vibrations in plants (~76 Hz) attractive to their females.

B. Frequency-Multiplied Generation

1. *Stridulation*

Stridulation requires a file, scraper, and surfaces radiating sound. Grasshoppers use a dual generator: The inner face of each hind femur is lined with pegs. The peg row is drawn against a raised vein of the undeployed forewing which then radiates sound. Corixids stridulate in water by sliding a field of pegs on the prothoracic femur across a flange on the maxillary plate. This oscillates the respiratory air bubble, which acts as both a resonator and a filter, radiating those frequencies that match its natural resonance.

Katydids stridulate when teeth below a transverse vein (file) on the left wing are engaged by the upturned margin (scraper) of the right wing (Fig. 3). Co-opted flight muscles slide the scraper to and fro and wing cells (e.g., mirror) act as radiators. Cricket stridulation is also by forewings but with right above left. This mechanism makes either prolonged or transient pulses depending on the speed of the scraper.

The katydid *M. sphagnorum* makes both pulse types (Fig. 4). Without pause, it alternates every quarter second between two song patterns (Fig. 4A). An audio-dominated phonatome (Fig. 4B), made over the distal half of the file, has only transient pulses. File teeth are met during wing closure at a rate near 2 kHz and each pulse dies away before the next tooth contact. This gives a broad-band frequency spectrum (15–40 kHz; Fig. 4F).

A second phonatome (Fig. 4C) is dominated by five to eight well-spaced, prolonged ultrasonic pulses, generated over the basal half of the file. Within these ultrasonic pulses, the scraper contacts file teeth at a very high rate (35,000 teeth/sec). Radiator movement does not fall away between tooth contacts and the wave train is kept at a uniform high amplitude over 10–12 waves (Fig. 4D), giving rise to a spectrum dominated by a peak at 35 kHz.

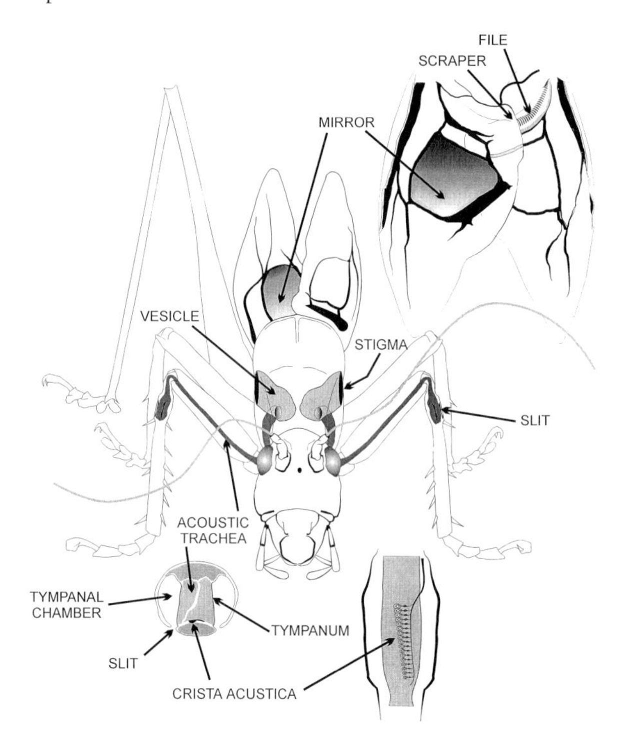

FIGURE 3 Katydid sound signal system. Generator: mirror, scraper, and file of the forewings; Upper inset shows these from beneath. Ear: stigma, acoustic trachea, and tympanal slit; lower left inset, transverse leg section with two slits, two chambers, and two tympana backed by the acoustic trachea; lower right inset, longitudinal view of the crista acustica.

2. *Deformation*

Among oedipodine grasshoppers, crepitation is a common mechanism of sound generation in flight. Membranes associated with thickened anal veins of the hind wing buckle between two positions in antiphase to the wingstroke.

Cicada tymbals are located dorsolaterally on the first abdominal segment. A stiff convex cuticular membrane is coupled to a large abdominal air space that serves as a Helmholtz resonator. The membrane bears a line of ribs which pop inward sequentially on the pull of a large tymbal muscle. Resilin in the membrane and in a dorsal pad acts as the muscle antagonist. The ribs undergo deformations succes-

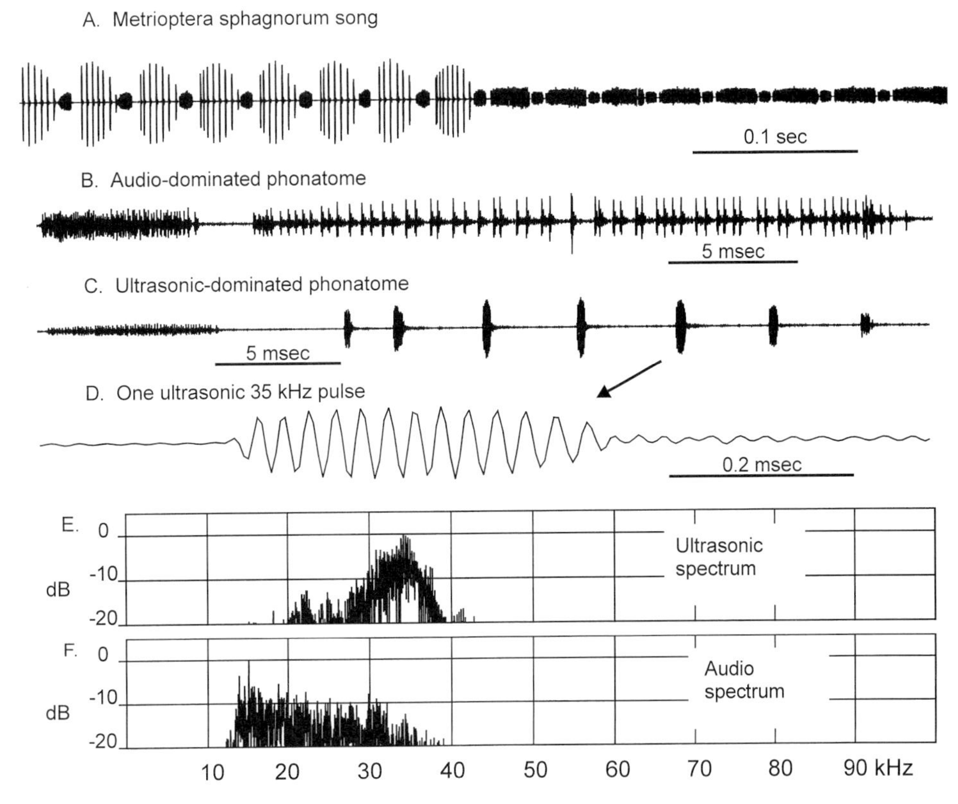

FIGURE 4 Calling song of *Metrioptera sphagnorum*. This katydid uses both resonant and nonresonant generation. A, One complete song; B, nonresonant phonatome; C, phonatome with resonant pulses made on closure; D, high resolution of a single sinsusoid 35-kHz pulse; E and F, distinctive spectra associated with each phonatome.

sively and generate a frequency matching the resonance of the cavity. The nearby eardrums are the major avenue of sound radiation. Though characteristic of cicadas, tymbals are also the basis of sounds in many small Homoptera, and some Lepidoptera (arctiid and noctuid moths) buckle metathoracic episterna.

3. *Burrows and Baffles*

Some arthropods modify their environs to enhance singing power. A tree cricket chews a hole in a leaf and perches with its forewings against the hole to stridulate. The leaf acts as a baffle and reduces short-circuiting at the wing edges. Mole crickets construct a burrow in which horn-shaped entries act as a tuned acoustic transformer. The air of the horns loads the relatively small wings and enables them to generate long wavelength frequencies more efficiently.

IV. RECEIVING: EARS AND LOCALIZATION

Thinned surface cuticle, backed by a tracheal sac, evolved into a tympanum, undamped by body fluids and therefore sensitive to airborne sound. Proprioceptors of the exoskeleton made a transition to chordotonal sensilla and now transduce tympanal movements while tracheae function as waveguides and resonators.

A. Near-Field Receptors

In midges and mosquitoes, Johnston's organ consists of a flagellum ringed with fibrillae atop a basal plate. The plate is recessed in a doughnut-shaped pedicel containing >20,000 radially arranged chordotonal sensilla. This organ serves all insects in flight feedback but readily acts as the near-field directional ear of many Diptera. It is sharply tuned in mosquitoes to the female flight frequency, which it detects as near-field air particle displacement. Localization comes naturally to this near-field receptor because it responds to both the magnitude and the direction of air particle displacement.

The waggle dance of honeybees tells other workers the distance and direction to a food source. Also, in the dark of the hive the movements of a dancer create a near-field sound. Follower bees place their Johnston's organs into a zone to each side of the dancer where acoustical short-circuiting is maximal and air displacements are most intense.

B. Tympanate Hearing

Tympanate ears function in the far field and occur in seven insect orders. Their form varies widely.

1. *Orthopteran Tympanate Ears*

A katydid has four eardrums, two in each foreleg just below the knee (Fig. 3). Each may be recessed within a chamber, opening to the outside by a slit. An acoustic trachea collects sound via a permanently open stigma and horn-shaped vesicle in the thorax and conducts it to the rear of the tympana along a tubular trachea in the foreleg. Within the tibia, a linear array of chordotonal sensilla (crista acustica) attaches to the trachea. Each sensillum is slightly smaller distally and has a different best frequency.

The ears of acridids are on the first abdominal segment. Chordotonal sensilla form a four-part acoustic ganglion attaching by small sclerites to eardrum regions of different thickness. In response to particular sound frequencies these regions develop characteristic patterns of activity which stimulate certain ganglion parts but not others. The differential vibration of the ganglion tissue becomes the basis of frequency discrimination.

Crickets are so small relative to the wavelength of their song that they cannot utilize body diffraction for binaural localization. The distance between a field cricket's ears is about one-seventh its song wavelength. Therefore, a far ear, though the cricket's body intervenes, will not be less stimulated than a near ear. However, a cricket obtains right–left differences in ear activity as it turns relative to the sound source. Phase differences arise at the tympanum between external and internal sound waves because path-length differences change as the cricket turns. A special membrane within a cross-body trachea in *Gryllus bimaculatus* acts to shift the phase and optimize the dynamic range of the ear for directional information. Localization based on phase explains why cricket calling songs universally employ one dominant high-Q carrier frequency.

2. *Song Detection by Parasites*

Female tachinid flies use the songs of crickets and katydids to find and parasitize the singer. They listen with ears located on the frontal aspect of the prothorax. A right and left tympanum are backed by a single air-filled chamber. An unpaired median sclerite within this chamber connects the two tympana across the midline, allowing this ear to integrate extremely small interaural time and intensity differences for localization. Sarcophagid flies parasitize cicadas by eavesdropping and have comparable ear structures.

3. *Simpler Tympanate Ears*

The tympanate ears of moths and mantids function mostly in detection of predators and have only a few sensilla. The right and left tympana of a praying mantis face each other within deep grooves in the ventral thoracic midline. Mantid ears are unable to discriminate frequencies and cannot localize sounds, but they are ultrasonic sensitive and their detection of bat echolocation cries allows evasive flight.

C. Adaptive Degradation

Song structure degrades with distance and some arthropods use this information. Intervening vegetation attenuates higher frequencies more than lower. Therefore, the ultrasonic portion of *M. sphagnorum*'s song is more attenuated with distance than its audio. The insect can hear both song parts at 2 m but only the audio-dominated part at 10 m. Trespass of a singing neighbor is indicated by song degradation in crickets and katydids. In *G. bimaculatus*, lower spectral components are normally lost with distance due to ground reflection; this and the blurring of song elements from reverberation give cues for singer proximity. In field experiments, males of *Tettigonia cantans* withdrew from song models whose band spectrum favored low frequencies but sang more and approached speakers broadcasting songs with higher frequency components indicating a closer male.

D. Localization by Binaural Differences in Intensity

An acoustic trachea of the sort depicted in Fig. 3 will act as an acoustic transformer, imposing a gain upon song carrier frequencies. The sound conducted internally will dominate ear response and for lower ultrasonics (20–50 kHz) localization can be based on bilateral intensity differences arising from diffraction about the thorax.

Directional ear response in some pseudophylline katydids may explain their extremely high carriers (e.g., *Myopophyllum speciosum*, 81 kHz). Throughout this group the stigma is tiny, the slits are ovals, and the tympanal chambers are enlarged. Therefore, external sound input should dominate over that conducted internally by the trachea. The extremely short wavelength of an 80-kHz song will permit detection of interaural intensity differences arising from diffraction about the forelegs and tympanal slits.

E. Localization and Recognition

Song localization and song recognition are done differently in crickets and grasshoppers. In grasshoppers the acoustic information is processed in parallel channels, whereas crickets (and probably katydids) process localization and recognition information serially. In crickets, on each side, the transduced signal first passes a recognition filter and then converges in a "comparator," which makes a side-to-side assessment for localization. In grasshoppers, on each side the incoming signal is separated. One part goes directly to a bilateral comparator while the other goes to a recognition filter.

V. INTERSPECIFIC INTERFERENCE

Noise problems arise among related species that use the same spectrum to make sustained songs. One chirping species of tropical katydid, *Neoconocephalus spiza*, alters the circadian rhythm of its singing wherever it is confronted by the output of three sympatric congeners. In Florida, a bivoltine conehead, *N. triops*, has a different calling song in summer and winter generations. This song change is an adaptive response to interference since in the summer *N. triops* contends with another conehead species, *N. retusus*, having the same song.

The tuning mismatch of a sagine katydid in Australia makes it vulnerable to the noise of other species singing nearby. These are heard more loudly than the calls of its own male. It deals with this by shutting off its acoustic trachea and making its ear less sensitive to the interfering frequencies.

VI. SPECIES DISCRIMINATION

Arthropod songs contain information about the sender's species and many playback experiments have established that females use this information and respond only to the calling song of their conspecific male. There are many instances of species highly similar in morphology, e.g., some *Chorthippus* grasshoppers, but distinguished by very different calling songs. Song specificity may serve some species in premating reproductive isolation.

A *Conocephalus* katydid recognizes the song of another species. The same band spectrum (Fig. 1C) is shared by *C. nigropleurum* and its sympatric congener, *C. brevipennis*. Therefore, *C. nigropleurum* females perceive their own and *C. brevipennis* song equally. However, females who are highly attracted to conspecific song, and so indiscriminate that they will approach random noise, will not go to the distinctive AM pattern of *C. brevipennis* but they will respond to it when overlain sufficiently by rerecording to obscure the species-typical AM pattern.

Comparative studies of the calling songs of *Laupala swordtail* crickets suggest song differences that coevolved as adjustments between interacting species. Thirty-seven species live in rain forests of the Hawaiian archipelago, with up to 4 species coexisting on any given island. The calls of these flightless crickets differ in pulse rate and females prefer the rates of their own species. The pulse periods of species sharing the same singing space differ more than is expected on the basis of chance, indicating song coevolution. Their probable radiation and divergence, tracked by molecular phylogeny, is consistent with this.

VII. RIVALRY

A. Channel Rivalry

Males of the same species calling simultaneously in the same channel provide each other's noise and they are also rivals for mates. They may be selected to time their signal output for best transmission to females. In katydid species whose calling song consists of chirps, females prefer to approach the leading of two call-overlapped males. The males show near synchrony in their calls as they compete to satisfy this female requirement.

B. Rival Exclusion

Calling song may serve simultaneously to attract a mate and to mediate interactions with rival conspecifics. It can evoke aggressive approach in other males. Some *Orchelimum* katydids attack each other in grappling fights and this attack can be evoked to a speaker broadcasting calling song. Male field crickets are influenced in their aggression by perceiving the calling song of their opponent: Deafened males can improve their position in a dominance hierarchy.

In some species of grasshoppers, crickets, katydids, and cicadas, experiments have revealed calling song parts effective only with males. *Teleogryllus oceanicus* has a chirp and trill calling song: Females are attracted by the chirp, whereas males approach the trill. In the 17-year cicada, *Magicicada cassini*, males produce ticks, which function to synchronize their singing, and then a buzz that evokes female phonotaxis.

Distinctive aggressive songs occur in many acoustic Orthoptera and can affect the outcome of fights between males. In *Ligurotettix*, a desert grasshopper, males defend mating territories and engage in overt fighting. A distinctive "shuck" component is added to the calling song as the males enter into alternating song exchanges. The winner of this singing contest is predicted by a combined measure of shuck call number and mean length.

Males of a katydid relative, the haglid *Cyphoderris monstrosa*, come together and engage in singing duels which often escalate to violent fights on the trunks of large trees. The duty cycle is a true indication of a male's quality as an opponent and males use this information in their rivalry.

Carrier is implicated in conveying information about a rival's ability. Males of the katydid *Tettigonia cantans* fight and carrier frequency is significantly lower in males that successfully defend their song perches.

VIII. MATE DISCRIMINATION

Females use the songs of conspecific males to discriminate between them as potential mates. Song components shown by experiment to serve as cues for female choice include intensity, duty cycle, pulse rate, phonatome number, and carrier frequency.

A. Power Preference

The energy males put into singing influences female preference. The more energetic a signal, the more costly and the more reliable it is as an indicator of mate quality. Grasshoppers with only one stridulatory leg convey this lack in their song as silent gaps and female grasshoppers prefer songs without gaps. When other species-specific song parameters are not limiting, song power, the rate of providing sound energy, often predicts female preference.

Females of wolf spiders *Hygrolycosa* prefer higher and more intense (higher power) drumming rates. *Gryllus integer* males show heritable differences in duty cycle; females prefer to approach higher duty cycle singers. Female *G. bimaculatus* show preferential phonotaxis to higher pulse rates for songs of equal intensity. Larger size in this species is predicted by a higher song pulse rate. Individual males of the katydid *Scudderia curvicauda* produce a characteristic number of phonatomes per call. This encodes reliable information about male size: More phonatomes predicts a larger male. Females prefer to respond acoustically to calls with more phonatomes.

The song of the grasshopper *Chorthippus dorsatus* has two distinctive AM patterns—an interrupted, and hence lower power mode of distinct pulses followed by a rapidly stroked higher power mode with no down time. In experiments, females prefer longer durations of the high-power part than occur naturally, an example of directional sexual selection on males to maximize this song component.

B. Pattern Preference Independent of Intensity

Some song features attract females independent of power. Some of these are qualitative and species specific, absolutely required for any attraction and not therefore the object of sexual selection. For example, neither mode in the song of *C. dorsatus* is effective on its own—both must be present and in the proper sequence.

Approaching the higher power of two otherwise equivalent songs corresponds to choosing the nearer of two males under field conditions. However, will a female pass by a near singer of the same species to reach a preferred more distant one?

Females of *G. bimaculatus* walking on a locomotion compensator prefer to approach songs incorporating maximally attractive time domain values even when the sound pressure level of these songs is 5–20 dB lower than the competing stimulus. In some localities *Ephippiger ephippiger* katydid males produce calls with only a single phonatome. Females from these demes prefer to approach short-calling males. Since it has fewer phonatomes, the signal is lower in power. Therefore, females are discriminating in favor of a lower power call typical of their local males. They prefer to approach a single over a multiphonatome call even where the longer call exceeds the shorter in intensity by 10 dB.

C. Carrier Frequency Preference

Tree crickets feed females during courtship, and females that could preferentially approach a larger male, predicting his size from a lower frequency calling song, might benefit nutritionally. *Oecanthus nigricornis* females prefer lower carriers when comparing songs simultaneously, and lower carriers have been shown to correlate with larger males.

See Also the Following Article

SONGBIRDS AND SINGING

Bibliography

Aiken, R. B. (1985). Sound production by aquatic insects. *Biol. Rev.* **60**, 163–211.

Bailey, W. J. (1991). *Acoustic Behaviour of Insects: An Evolutionary Perspective.* Chapman & Hall, London.

Bailey, W. J., and Rentz, D. C. F. (1990). *The Tettigoniidae: Biology, Systematics and Evolution.* Crawford, Bathurst.

Barth, F. G. (1982). Spiders and vibratory signals: Sensory reception and behavioral significance. In *Spider Communication* (P. N. Witt and J. S. Rovner, Eds.), pp. 67–122. Princeton Univ. Press, Princeton, NJ.

Bennet-Clark, H. C. (1984). Insect hearing: Acoustics and transduction. In *Insect Communication* (T. Lewis, Ed.), pp. 49–82. Academic Press, New York.

Dunham, D. W., and Gilchrist, S. L. (1988). Behavior. In *Biology of the Land Crabs* (W. W. Burggren and B. R. McMahon, Eds.), pp. 97–138. Cambridge Univ. Press, Cambridge, UK.

Ewing, A. W. (1989). *Arthropod Bioacoustics: Neurobiology and Behaviour.* Comstock/Cornell Univ. Press, Ithaca, NY.

Greenfield, M. D. (1997). Acoustic communication in Orthoptera. In *The Bionomics of Grasshoppers, Katydids and Their Kin* (S. K. Gangwere *et al.*, Eds.). CAB International, New York.

Gwynne, D. T., and Morris, G. K. (1983). *Orthopteran Mating Systems.* Westview, Boulder, CO.

Huber, F., Moore, T. E., and Loher, W. (1989). *Cricket Behavior and Neurobiology.* Comstock/Cornell Univ. Press, Ithaca, NY.

Kalmring, K., and Elsner, N. (Eds.) *Acoustic and Vibrational Communication in Insects.* Verlag-P. Parey, Berlin.

Otte, D. (1992). Evolution of cricket songs. *J. Orthopt. Res.* **1**, 25–49.

Uetz, G. W., and Stratton, G. E. (1982). Acoustic communication and reproductive isolation in spiders. In *Spider Communication* (P. N. Witt and J. S. Rovner, Eds.), pp. 123–159. Princeton Univ. Press, Princeton, NJ.

von Helversen, O., and von Helversen, D. (1994). Forces driving coevolution of song and song recognition in grasshoppers. In *Neural Basis of Behavioural Adaptations* (K. Schildberger and N. Elsner, Eds.), pp. 253–284. Verlag, New York.

Songbirds and Singing

Eliot A. Brenowitz
University of Washington

GLOSSARY

age-limited learning Also known as sensitive period learning. An ontogenetic pattern in which song learning is limited to the first year of life and new songs are not developed in adulthood.

crystallization The third stage of song production in the sensorimotor phase of song learning, when birds produce a well-structured, stereotyped version of the conspecific song model to which they were exposed during the earlier memory acquisition phase.

open-ended learning An ontogenetic pattern in which new song patterns continue to be developed beyond the first year of life into adulthood.

oscine A suborder of birds in the order Passeriformes, also referred to as "songbirds."

plastic song The second stage of song production in the sensorimotor phase of song learning. Plastic song is louder and better structured than subsong but still variable in form.

sensorimotor The second major phase of song learning which involves ongoing comparison between a sensory model of song acquired during an earlier memorization phase and auditory feedback from a bird's own production of song.

subsong The initial stage of song production in the sensorimotor phase of song learning. Subsong is quiet, poorly structured, and very variable in form.

template A sensory model of song acquired during the initial memory phase of song learning as a result of listening to songs produced by adult conspecific birds.

Songs are long, complex vocal signals that are used in the advertisement of a breeding territory and/or the attraction of a mate. Songbirds must learn the song of their species. Song behavior is regulated by a discrete network of interconnected brain nuclei. Gonadal steroid hormones have pronounced effects on the development and function of these brain nuclei as well as on song behavior.

I. SONGBIRD SYSTEMATICS

There are about 9000 species of living birds. About 5300 species belong to the taxonomic order Passeriformes, which consists of the oscine suborder (the songbirds) and the suboscine suborder, which includes birds such as flycatchers, antbirds, and cotingas. When we speak of "songbirds," we are referring to the oscine birds, such as finches, sparrows, and blackbirds, within this large order. The Passeriformes order is considered to be a single evolutionary lineage derived from a common ancestor (i.e., monophyletic), and the oscines and suboscines are each viewed as monophyletic lineages within the Passeriformes. There are approximately 4000 species of songbirds, essentially all of which use learned vocalizations for communication.

II. GENERAL CHARACTERISTICS OF SONG BEHAVIOR

A. Song Structure

Songs have well-defined acoustic structures that are unique to each species. The structure of a song can be visualized using a sound spectrograph, which plots the sound frequencies against time (Fig. 1). Song consists of different structural components. The

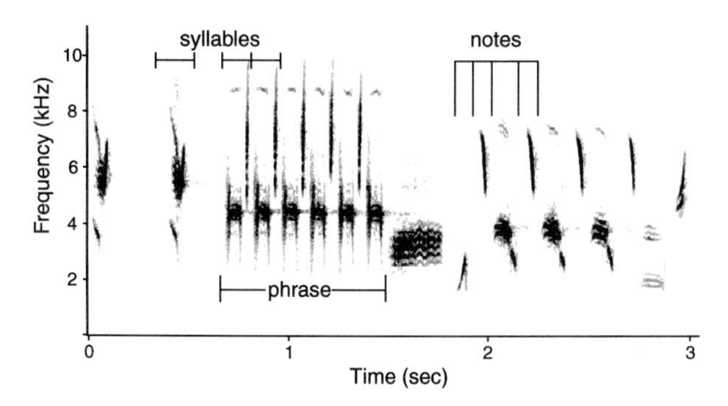

FIGURE 1 Sound spectrogram of a single song produced by a male song sparrow (*Melospiza melodia*) showing a frequency vs time representation. The simplest individual sounds in the song are notes. A sequence of one or more notes that occurs together in a regular manner in song is a syllable. A series of one or more syllables that is repeated in song is a phrase. The entire combination of phrases is a song type that is sung repeatedly. An individual male song sparrow has a repertoire of several different song types.

simplest individual sounds are called song "notes." A sequence of one or more notes that occurs together in a regular manner in song is called a song "syllable." A series of one or more syllables that is repeated in song is referred to as a song "phrase." A specific combination of phrases that occurs repeatedly is a song "type."

There is great diversity among songbird taxa in the structural complexity of song behavior. In some species, such as the white-throated sparrow (*Zonotrichia albicollis*), a male produces only one simple type of song. At the other extreme, a male brown thrasher (*Toxostoma rufum*) has a repertoire of several thousand different types of songs. Most bird species have song repertoire sizes that fall between these extremes.

B. Sex Differences in Song Production

Songbird species show extensive diversity in sex patterns of song behavior. At one extreme are species in which only males sing, such as the zebra finch (*Taeniopygia guttata*). At the other extreme are several tropical species in which females and males contribute equally to elaborate song duets. Between these extremes are species in which females sing, but less commonly and with less complexity than males. In most species, song is a sexually dimorphic behavior; males sing more often and with greater complexity.

C. Behavioral Functions of Song

Song is important in the reproductive behavior of birds and serves two main functions in this context. In many species, song is used to declare a territory from which other birds are aggressively excluded. Breeding occurs in the territory. Muting birds decreases their ability to deter intrusions by other birds. Both males and females may use song for territorial advertisement. Song may also be used by males to attract females to mate with them as well as to stimulate the females' reproductive behavior and physiology.

Each of these functions of song may be influenced by the number of song types that a bird produces (i.e., the repertoire size). Large repertoires are more effective at deterring intrusions on a bird's territory than are small repertoires. Males with large repertoires may mate earlier in the breeding season and/or have more mates than conspecific males with smaller repertoires. Large repertoires may also be more effective at stimulating female reproductive physiology and behavior. Female canaries (*Serinus canaria*), for example, build nests faster and lay more eggs in response to playback of a large song repertoire than they do to a small repertoire. These effects of large song repertoires on females can increase a male's reproductive success.

Song may serve other functions in addition to territorial advertisement and mate attraction. There is often considerable individual variation in song structure, which can enable recognition of individual birds by vocal cues alone. Such individual recognition may be important for recognition of territorial neighbors, mates, kin, and members of a social group. Song may also play a role in maintaining dominance hierarchies in social species such as the brown-headed cowbird (*Molothrus ater*).

The rate at which birds sing may vary seasonally. In temperate zone bird species, song used in either territorial or mate-attraction contexts is produced at higher rates during the breeding season and at lower rates or not at all outside the breeding season. In bird species residing in the tropics, song is often used for territorial defense throughout the year.

III. GENERAL CHARACTERISTICS OF SONG DEVELOPMENT

A. Stages of Song Learning

One of the most interesting features of song is that it is a learned behavior. Song development is characterized by distinct stages. During an initial "memory acquisition" phase, young birds acquire a sensory model of song by listening to adult conspecifics sing. Juvenile birds begin to translate this sensory memory to a motor pattern of song production at some later time, in what is referred to as the "sensorimotor" phase of song development. The sensorimotor phase consists of three stages. Initially birds produce "subsong," which is quiet, poorly structured, and quite variable in form. With continued vocal practice, young birds improve their performance of song and progress to "plastic song," which is louder and better structured but still variable in form. Eventually, song becomes "crystallized" in structure as the birds produce a stereotyped version of the sensory model to which they were exposed earlier. The specific age at which these different phases of song development occur varies considerably between species. Also, some species develop songs by improvisation in addition to copying songs from adult conspecifics.

There is much species diversity in ontogenetic patterns of song development. At one extreme are species referred to as "age-limited" or "sensitive period" learners, in which song learning is limited to the first year of life and new songs are not developed in adulthood. Examples of age-limited learners include the zebra finch and white-crowned sparrow (*Z. leucophrys*). At the other extreme are "open-ended" learning species in which new song patterns can be developed beyond the first year of life. Examples include the canary and European starling (*Sturnus vulgaris*). The development of new songs by adults of open-ended species usually occurs in a restricted seasonal manner. Other species have ontogenetic patterns of song development that fall between the extremes of age-limited and open-ended learning.

B. Auditory Feedback Is Required for Song Learning

In most species, juvenile birds must hear the song of conspecific adults if they are to develop normal song behavior. If a young bird is isolated from hearing the songs of conspecific adults, it will not develop normal song of its species. Young birds must also be able to hear themselves sing in order to develop a crystallized version of the conspecific song learned during the earlier memorization phase. If juvenile birds are deafened after the memorization phase, but before song crystallization, they will not develop normal song. Translating the sensory memory of song to a motor program involves an ongoing comparison between a "template" of song stored during the memorization phase and auditory feedback from a bird's own production of song, hence the description of this phase as being "sensorimotor."

Auditory feedback may also be necessary for the maintenance of crystallized song in adults. Previously crystallized song structure can degrade within weeks to months after an adult bird is deafened. There is variation between species in the continued reliance on auditory feedback.

IV. NEURAL CONTROL OF SONG BEHAVIOR

A. Pathways for Song Learning and Production

Song behavior in songbirds is controlled by a discrete network of interconnected nuclei in the forebrain. Two pathways that are involved in song learning and production are illustrated in Fig. 2. The motor pathway controls song production and is also presumed to participate in the sensorimotor phase of song learning. This circuit consists of projections from the neostriatal nucleus HVc (also known as the high or higher vocal center) to the robust nucleus of the archistriatum (RA) in the telencephalon. RA projects both to the dorsomedial part of the intercollicular nucleus (DM) in the midbrain (not shown in Fig. 2) and to the tracheosyringeal part of the hypoglossal motor nucleus in the brain stem (nXIIts). Motor neurons in nXIIts send their axons to the muscles of the sound-producing organ, the syrinx.

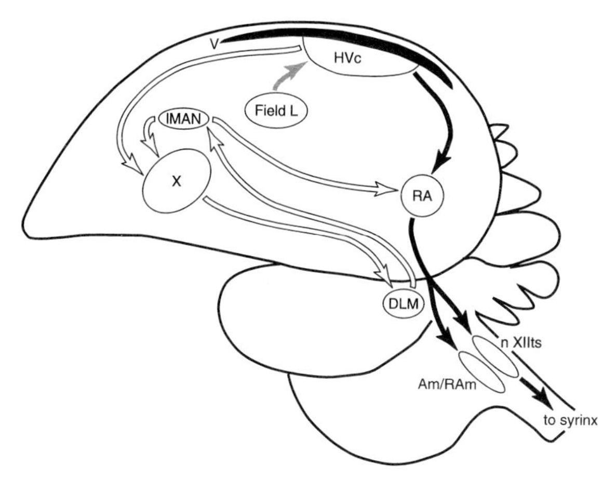

FIGURE 2 A schematic sagittal drawing of the songbird brain showing projections of major nuclei in the song system. The descending motor pathway (solid arrows) controls the production of song and consists of projections from HVc in the neostriatum to RA in the archistriatum and thence to the vocal nucleus nXIIts, the respiratory nucleus RAm, and the laryngeal nucleus Am in the medulla. Motor neurons in nXIIts innervate the muscles of the syrinx, the sound-producing organ. The open arrows indicate the anterior forebrain pathway that is essential for song learning. It indirectly connects HVc to RA via area X in the parolfactory lobe, DLM in the thalamus, and lMAN in the neostriatum. LMAN also projects to area X. Field L is an auditory region in the neostriatum that projects to HVc (gray arrow). Am, nucleus ambiguus; DLM, medial portion of the dorsolateral nucleus of the thalamus; lMAN, lateral portion of the magnocellular nucleus of the anterior neostriatum; RA, robust nucleus of the archistriatum; RAm, nucleus retroambigualis; V, ventricle; X, area X; nXIIts, tracheosyringeal part of the hypoglossal nucleus.

The projection from RA onto the motor neurons in nXIIts is topographically organized according to the distribution of muscle groups within the syrinx. Neuronal activity in HVc and RA is synchronized with the production of sound by the syrinx. If nuclei in this motor pathway are inactivated, a bird is unable to produce song.

RA also projects to nucleus retroambigualis (RAm) and nucleus ambiguus (Am) in the medulla (Fig. 2). RAm contains respiratory-related neurons that fire in phase with expiration. Am contains neurons that innervate the larynx. This pattern of descending projections from RA may be critical for coordination of syringeal, respiratory, and laryngeal muscle activity during song production. Birds only produce song during expiration, and the larynx may filter the sound frequencies produced by the syrinx.

A second, anterior forebrain pathway is thought to be necessary for song learning and recognition (Fig. 2). This pathway consists of projections from HVc to area X, then to nucleus DLM in the thalamus, from DLM to the lateral portion of the magnocellular nucleus of the anterior neostriatum (lMAN), and finally to RA. lMAN neurons that project to RA also send collaterals to area X, thus providing the potential for feedback within this pathway. The projections within this pathway are topographically organized. Inactivation of lMAN, DLM, or area X in adults seems to have no effect on previously crystallized song, whereas the same lesions in juveniles prevent the development of normal song. Nuclei in this pathway may continue to play an important role in adults of species that develop new songs beyond the first year. Lesions of lMAN made in adult male canaries in mid-September, when song is seasonally plastic in structure, lead to a progressive decline in structural complexity.

There are pronounced sex differences in the nuclear size of HVc, RA, and area X in species in which males sing more than females. The extent of morphological sex differences in the song system of a given species is closely related to the degree to which the sexes differ in behavioral song complexity.

B. Auditory Responses in the Song Control Circuits

Auditory information is conveyed from field L in the neostriatum, which is the analog of the auditory cortex, to HVc and then to other nuclei in the song control circuits. In anesthetized birds, neurons in HVc and nuclei in both the descending motor and anterior forebrain pathways show auditory responses to song stimuli, with the most pronounced selectivity to a bird's own song, and weaker responses to other conspecific songs, heterospecific songs, and artificial stimuli. Within the anterior forebrain pathway there is a progressive increase in song selectivity as one proceeds from area X to DLM to lMAN. The responses of neurons in both song control circuits to auditory stimuli appears to be suppressed in awake, nonsinging birds. It has been suggested that the auditory responses of these neurons may be gated by a bird's own song production, through this hypothesis remains untested.

Neurons in HVc, area X, and 1MAN of anesthetized birds initially respond unselectively in young birds prior to the start of sensorimotor song learning. They become more selective as the birds proceeds through the sensorimotor stages to song crystallization.

C. A Circuit for Song Perception and Memory

Subdivisions of the telencephalic auditory region field L also project to the caudomedial neostriatum (NCM) and the caudomedial hyperstriatum ventrale (CMHV), regions of the forebrain not traditionally associated with the song control system. These two regions may play a role in song perception and memory. There is strong induction of the immediate early gene *ZENK* (named *ngfi-a* in rats and *egr-1* in mice) in these regions in response to presentation of conspecific song but only weak induction in response to heterospecific song or artificial sound stimuli. Neurons in NCM show electrophysiological responses to playback of conspecific songs and other complex auditory stimuli. Both the *ZENK* induction and auditory responses in NCM decrease with repeated presentation of the same conspecific song. This decrease in responsiveness is reminiscent of habituation and implies that neurons in NCM form a persistent memory of the stimulus song. This type of memory could be important in the context of recognizing the songs of territorial neighbors and mates.

V. HORMONE EFFECTS

A. Effects on Song Behavior

Gonadal steroids and their metabolites have pronounced effects on the song system of oscine birds, which is consistent with the close relationship between song behavior and reproduction. Proper timing and amount of exposure to estrogenic and androgenic hormones are necessary for normal song learning. Estrogenic and androgenic hormones may play opposing roles in song learning. Circulating estrogen concentrations are high during the memory acquisition phase of song learning in swamp sparrows (*Melospiza georgiana*) and may facilitate the plasticity of the song circuits that underlies this stage. Testosterone, on the other hand, may terminate this plasticity. Young male birds that are castrated or treated with antiandrogenic agents can acquire sensory memories of song and progress to plastic song but will not crystallize song unless they are treated with testosterone. Chronic exposure of juvenile male zebra finches to testosterone severely impairs song learning.

Adult song behavior is also influenced by gonadal steroids. Castration of adult males reduces the rate at which they sing, and treatment with exogenous steroids can reinstate song. It appears that both androgenic and estrogenic hormones must be present for full song activation. In temperate zone birds, seasonal changes in circulating steroid levels are closely correlated with changes in the rate of song production. In some seasonally breeding species, such as the white-crowned sparrow, song continues to be produced at a lower rate after the breeding season, when circulating steroid concentrations are low. In tropical songbirds, testosterone levels tend to be low throughout the year and song behavior appears to be largely independent of this hormone. These observations suggest that while circulating steroids may modulate the level of song behavior, these hormones are not required for song in all species.

B. Effects on the Song Control Circuits

The brain nuclei that control song are extremely sensitive to gonadal steroids and their metabolites. Intracellular receptors for androgenic hormones are present in HVc, RA, lMAN, DM, and nXIIts. Estrogen receptors are also found in HVc and DM in several species. Treatment of young female zebra finches with estradiol masculinizes their song control circuits and enables them to sing as adults, which females are normally incapable of doing. The enzyme aromatase, which converts testosterone to estradiol, is widely distributed in the forebrain of both juvenile males and females. In fact, the brain seems to be the primary source of estradiol in young birds.

Estradiol obtained from aromatization of testosterone may not be the only factor that controls masculinization of the song system, however. Treatment of newly hatched male zebra finches with antiestrogenic agents or inhibitors of estrogen synthesis does not feminize their song nuclei. Furthermore, females that develop functional testicular tissue but not ovaries after embryonic exposure to inhibitors of estrogen synthesis have a feminine song circuit as adults. These and related results imply that normal development of the song system may also be influenced by nonhormonal genetic factors. This unresolved issue is currently a topic of much activity.

Gonadal steroids also have pronounced effects on adult song control circuits. Seasonal changes in circulating testosterone levels induce plasticity in various morphological attributes of the song nuclei, including the size of the song nuclei and individual neurons, the spacing of neurons, the survival of "old" neurons and the incorporation of newly generated neurons in HVc, and various synaptic attributes. These changes in the structure of the song nuclei are functionally related to seasonal changes in different aspects of song behavior, including song rate, stereotypy, and the development of new song patterns.

See Also the Following Article

Avian Reproduction, Overview

Bibliography

Arnold, A. P., Wade, J., Grisham, W., Jacobs, E. C., and Campagnoni, A. T. (1996). Sexual differentiation of the brain in songbirds. *Dev. Neurosci.* **18**, 124–136.

Brenowitz, E. A., and Kroodsma, D. E. (1996). The neuroethology of birdsong. In *Ecology and Evolution of Acoustic Communication in Birds* (D. E. Kroodsma and E. H. Miller, Eds.), pp. 285–304. Comstock, Ithaca, NY.

Brenowitz, E. A., Margoliash, D., and Nordeen, K. W. (Eds.) (1997a). *J. Neurobiol.* **33**(5). [Special issue]

Brenowitz, E. A., Margoliash, D., and Nordeen, K. W. (1997b). An introduction to birdsong and the avian song system. *J. Neurobiol.* **33**, 495–500.

Catchpole, C. K., and Slater, P. J. B. (1995). *Bird Song: Biological Themes and Variations.* Cambridge Univ. Press, Cambridge, UK.

Konishi, M. (1989). Birdsong for neurobiologists. *Neuron* **3**, 541–549.

Kroodsma, D. E., and Miller, E. H. (Eds.) (1996). *Ecology and Evolution of Acoustic Communication in Birds.* Comstock, Ithaca, NY.

Nottebohm, F. (1989, February). From bird song to neurogenesis. *Sci. Am.*, 74–79.

Nottebohm, F. (1991). Reassessing the mechanisms and origins of vocal learning in birds. *Trends Neurosci.* **14**, 206–211.

Schlinger, B. A. (1997). Sex steroids and their actions on the bird song system. *J. Neurobiol.*, in press.

Spawning, Marine Invertebrates

John H. Himmelman
Laval University

GLOSSARY

capsule laying Depositing capsules containing gametes, zygotes, or embryos onto bottom structures.

epidemic spawning Contagious spawning in response to detection of released gametes or larvae from conspecifics or interspecifics. Often chemical substances (pheromones) are involved.

gamete spawning Release of ova (or preova stages) and sperm into the water.

larval spawning Release of larvae, or embryos which develop into larvae, into the water.

proximate spawning cue The environmental factor triggering release of gametes or larvae.

ripe Possessing gonads in condition to spawn. In many groups the ovaries may contain primary or secondary oocytes rather than ova, and meiosis resumes during or after spawning.

spawning Release of gametes, embryos, or larvae into the water or deposition of capsules containing gametes, zygotes, or embryos onto bottom structures.

ultimate spawning cue An environmental condition critical to reproductive success which has led to selection of the proximate spawning cue.

The term spawning is used in various ways. In the strictest sense it refers to the release of ova or preova stages (primary or secondary oocytes) and sperm into the water where external fertilization occurs. This is referred to as gamete spawning (and also as free or broadcast spawning) and is considered a primitive reproductive strategy. A second type of spawning is larval spawning, the release of embryos or larvae into the water. In some instances encapsulated embryos are released into the water and the larvae hatch shortly thereafter (e.g., *Littorina littorea*). Finally, spawning is at times used to refer to capsule laying, the deposition by females of benthic capsules containing gametes, zygotes, or embryos. In some cases the capsule may be just a mucus mass. Larval spawning and capsule laying usually involve internal fertilization and are considered advanced reproductive strategies.

I. INTRODUCTION

This article focuses on gamete and larval spawning. In benthic invertebrates, these are ecologically similar processes representing the important transition to the pelagic phase of the life cycle. Although capsule laying does not involve this transition, in numerous species a pelagic larval stage later emerges from the benthic capsules. This step is thus similar to gamete and larval spawning. For marine invertebrates, spawning is a critical event because the production of the reproductive stages requires high energetic investment. Reproductive events are synchronous in most invertebrates, especially in species with a pelagic larval phase. Such synchrony indicates control by environmental factors because the alternative hypothesis that synchronization is achieved purely by endogenous processes is improbable. Synchrony may be achieved at a number of points during the reproductive cycle and the processes most susceptible to exogenous control are accumulation of reserves, initiation of gametogenesis and its passage to different stages (particularly vitellogenesis), morphological or behavioral changes in anticipation of spawning, and spawning. Stimulation of spawning by environmental factors is particularly likely for species that spawn synchronously in a given location but at variable times between locations and from year to year. In cases in which spawning activity has been documented at frequent intervals, spawning events, whether annual or in successive episodes over several months, have usually been shown to be abrupt (within days and sometimes within hours). External control of spawning is further indicated by the variable delay found in numerous species between the time when individuals become ripe (in spawning condition) and when they actually spawn, and because spawning is delayed when animals are held in the laboratory and separated from natural environmental changes.

Spawning is most likely a response to an environmental event and this response has probably evolved through active selection. Thus, individuals possessing alleles permitting detection of, and appropriate responses to, an environmental signal indicative of a favorable reproductive period should benefit from increased fitness.

Two major strategies have likely been involved in the evolution of restricted spawning periods. The first is to coordinate liberation of gametes, embryos, or larvae by an environmental cue that is a good predictor of conditions favoring their survival. These stages are usually more sensitive to adverse environmental conditions than adults. Potential factors critical to survival are physical and chemical qualities of the water, larval food availability, and the risk of mortality due to either predators (including benthic filter feeders) or hydrographic conditions. The importance of predators should be determined by predator density and feeding rates and also by the duration of the pelagic phase. Hydrographic factors determine whether conditions are favorable for development and growth and also whether pelagic larvae will eventually reach suitable habitats for settlement. The second spawning strategy is to use an environmental signal to ensure synchronous gamete release for optimal fertilization success and enhanced outbreeding. Numerous studies demonstrate that high concentrations of gametes are required for fertilization success, that concentrations decrease rapidly with distance from the source, and that the longevity of gametes is in the scale of minutes to hours. Thus, gametes are likely to be wasted unless a sufficiently high density of individuals release their gametes in unison. In the case of the first strategy, the environmental factor triggering gamete or larval spawning is referred to as the proximate spawning cue and may be different from the ultimate spawning cue, the condition conferring increased survival of the pelagic stages. The proximate factor is a predictor of conditions which ultimately determine survival. In the case of the second strategy, gamete release is triggered by an environmental event and the distinctiveness or abruptness of this event is the factor which confers increased reproductive success. Thus, there is no distinction between proximate and ultimate spawning factors. These two strategies are not exclusive because the cue adapted by gamete spawning species to optimize fertilization may also confer increased larval survival.

II. DOCUMENTING SPAWNING EVENTS

The most common technique used to determine when gamete spawning occurs is by following changes in gonadal size and histological condition over time. Gonadal development varies with body dimensions: Juveniles have virtually no gonad, maturing individuals show a sharp increase in gonadal mass with increasing size, and for fully mature individuals the slope of the relationship of gonadal mass to body dimensions is less than that for maturing individuals (Fig. 1A). Reproductive events are best examined by studying fully mature individuals. Temporal changes in gonadal size are usually evaluated using gonadal indices, such as percentage gonadal mass to total body mass. However, such indices may not completely eliminate the effect of size (Fig. 1B) and in this case it is preferable to correct the index to eliminate allometric effects on gonadal size to increase the ability to detect changes in gonadal size over time (e.g., correcting the index for the difference in slopes in the relationships of gonadal mass and total mass to body dimension; Fig. 1C). Most studies illustrate gonadal cycles by plotting the gonadal index over time. An alternative method is to plot the gonadal mass of a standard-sized adult, and this

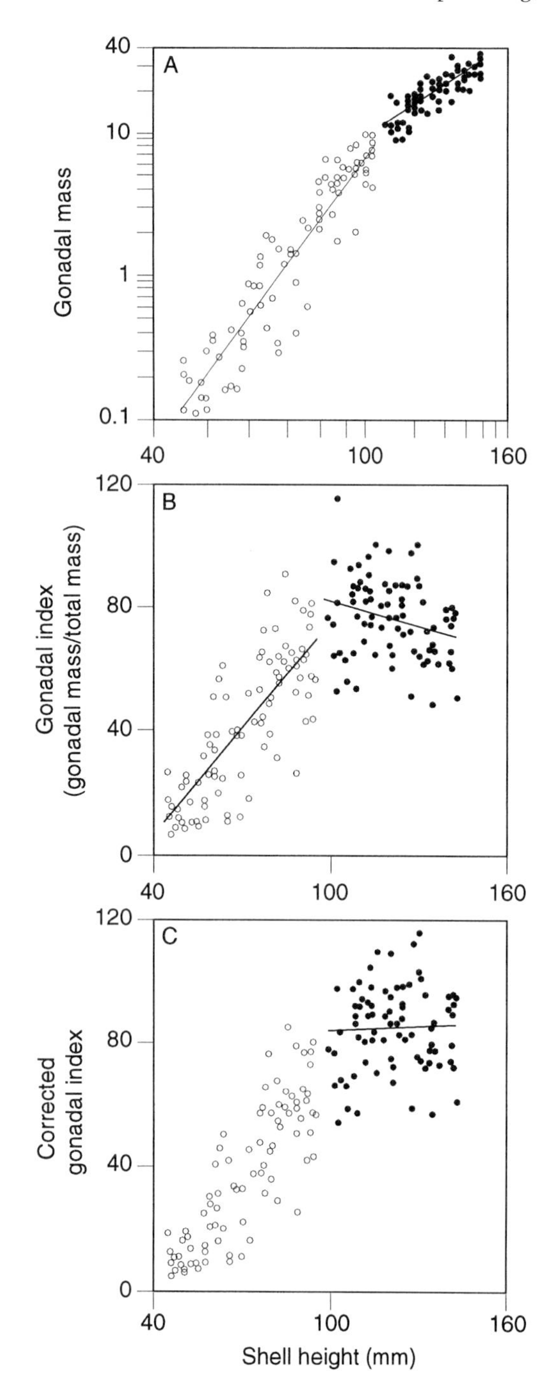

FIGURE 1 Relation of (A) log gonadal to log shell height, (B) gonadal index (gonadal mass as a percentage of total mass) to shell height, and (C) corrected gonadal index to shell height for maturing (○) and fully mature (●) giant scallops *Placopecten magellanicus* (the corrected index adjusts for the difference in slopes for the relationships of gonadal mass and total mass to shell length) (adapted from Bonardelli and Himmelman, 1995).

method has the advantage of illustrating the mass of gametes liberated during individual or sequential spawning events. An important rule in studying changes in gonadal size over time is to consistently collect animals at the same place because slight changes in microhabitat, particularly depth changes, can markedly affect gonadal size.

Gamete release events can often be identified from drops in mean gonadal size over a short period. Lysis and resorption of gametes can only cause a slow decrease in gonadal size. However, if spawning occurs slowly or less synchronously, or if only a few gametes are released (as in some brooding species), histological evidence of loss of mature gametes or of the appearance of spent individuals will be required to confirm spawning activity. Histological analysis is also the main tool for determining when spawning has occurred for species in which the gonad is intertwined with other tissues or too small to be weighed.

Gamete and larval spawning can also be documented by direct observations of individuals spawning in the field. However, such observations are often fortuitous because spawning events are often short-lived and observation time is usually limited (a few hours per day in the subtidal zone). Furthermore, when only a few individuals are observed spawning, it is unclear whether these individuals are incidental spawners or part of a major spawning event. Nevertheless, major spawning events are clearly indicated when large numbers of individuals are observed spawning. One common method used to document larval spawning is to record release by females maintained in outside aquaria supplied with water pumped in from the natural environment. Another approach used to document both gamete and larval spawning is to record the appearance of gametes, embryos, or larvae in the water. In many species gamete spawning can be anticipated by quantifying the proportion of individuals which will spawn in response to artificial stimuli (e.g., KCl, serotonin, H_2O_2, and electric or thermal shock) because release only occurs when gametogenesis is advanced.

III. SPAWNING CUES

Spawning is fascinating because it involves a substantial and abrupt energetic release and because the timing of this event may largely determine reproductive success. However, environmental cues which induce spawning have only been convincingly demonstrated for a small number of species. A diversity of spawning mechanisms is to be expected given the variable spawning times of different species in any given area and because factors limiting reproductive success likely vary in different climatic regions. Moreover, different mechanisms are also indicated because spawning cues have evolved independently in different phylogenetic groups.

A general hypothesis is that proximate spawning cues should readily be distinguishable from the background of complex environmental variations and should confer increased reproductive success because of enhanced fertilization success, outbreeding, or survival of larvae and juveniles. The number of intermediate steps between detection of an environmental cue and release of gametes or larvae will depend on the control pathways in different species. In some species a cue may directly provoke release of gametes or larvae by causing contraction of the body wall or gonadal musculature. However, the spawning mechanism in most species involves more complex neuroendocrine processes which bring a delay (of variable duration depending on the species) between detection of the environmental cue and spawning. The most indirect spawning cues are likely environmental cycles that entrain internal reproductive rhythms which eventually culminate in spawning.

Spawning in most temperate and boreal regions occurs at a particular time of the year and thus coincides with particular climatic conditions. Many studies report reproductive cycles and speculate about spawning cues based on environmental factors associated with spawning. Other studies, often aimed at obtaining gametes for embryological studies or larval culture, report on factors which provoke ripe animals to release gametes in the laboratory (e.g., thermal or electrical shock and chemicals). The identification of spawning cues is a major challenge in that it requires three types of evidence: (i) that an environmental change coincides with spawning in the field or precedes it at a fixed interval, (ii) that the same change can provoke animals to spawn when other conditions are maintained constant, and (iii) that other factors can be eliminated as possible spawning cues. The latter point is important because numerous oceanographic factors vary in parallel.

A. Phytoplankton

In temperate and high-latitude seas, phytoplankton abundance varies in a seasonal pattern, with a low abundance during the winter, a bloom in the spring, and then generally medium to low levels during the summer and autumn, except when occasional secondary blooms occur. The larvae of many invertebrates feed on phytoplankton so that it is essential to coordinate spawning with a period of phytoplankton availability. Although an association between the production of pelagic larvae and phytoplankton blooms has been recognized for many years, and was discussed extensively in the mid-twentieth century by Gunner Thorson in reviews of invertebrate reproduction, the hypothesis that phytoplankton induces spawning was not examined until recently. A number of invertebrates on the Pacific coast of Canada were found to spawn at the onset of the spring diatom bloom (Fig. 2) and laboratory studies on three species, the green sea urchin *Strongylocentrotus droebachiensis* and the chitons, *Tonicella lineata* and *T. insignis*, showed that exposure to natural phytoplankton at bloom concentrations stimulated spawning. A later study showed that the much delayed spawning of *S. droebachiensis* in the St. Lawrence Estuary also coincides with the first increases in phytoplankton. The bloom is delayed until spring freshwater outflow decreases, and thus the larvae are not subjected to low salinities. Phytoplankton is also a spawning cue for some larval spawners because larval abundance in many species has been correlated with blooms, and laboratory studies demonstrate that phytoplankton stimulates larval release by the barnacle *Semibalanus balanoides* and the snow crab *Chionoecetes opilio*. The use of phytoplankton as a spawning cue may not be limited to marine species because it is suggested as a spawning signal for the freshwater zebra mussel *Dreissena polymorpha*. Furthermore, it may also apply to some summer and autumn spawners, for which gamete release or the appearance of larvae has been associated with pulses of phytoplankton.

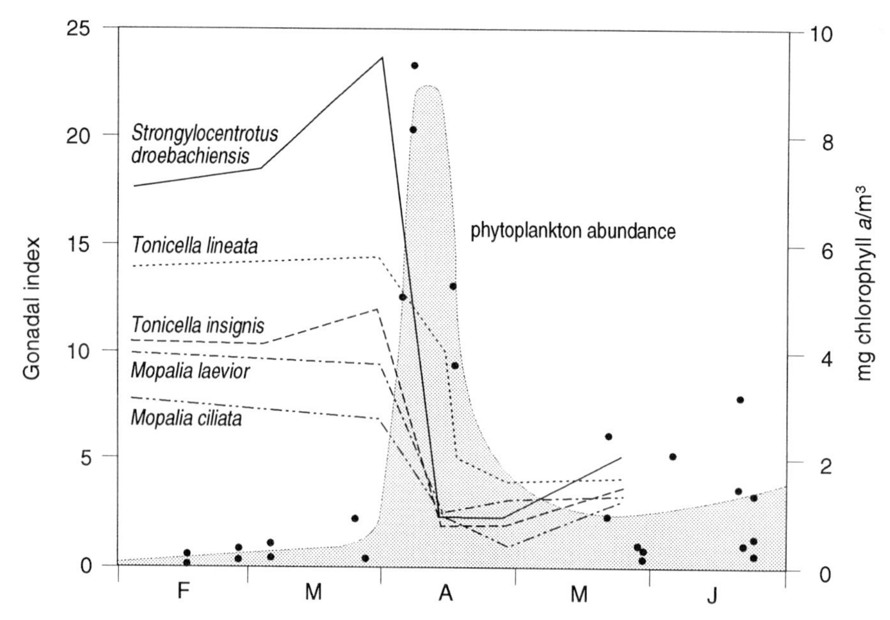

FIGURE 2 Relation between spawning (as indicated by the drop in the gonadal index) and the onset of the spring phytoplankton bloom (increase in chlorophyll *a*) for five benthic invertebrates in coastal British Columbia (adapted from Himmelman, 1981).

Various mechanisms have evolved in the selection of phytoplankton as a spawning cue. For the green sea urchin and several bivalves, spawning is triggered by metabolites released by various species of phytoplankton. In contrast, spawning by the barnacle requires contact with phytoplankton cells. Epidemic spawning may follow as nauplii of conspecifics also stimulate further larval release. Finally, release of larvae by the snow crab is triggered by detection of degrading rather than living phytoplankton, which makes sense because this species occurs mainly below the euphotic zone (at >80 m in depth). Its release of larvae coincides with the arrival on the bottom of particles sedimenting from the spring bloom at the surface (Fig. 3).

In species with planktotrophic larvae, synchronization of the larval phase with phytoplankton blooms might be achieved through the use of climatic factors, such as temperature and photoperiod, but it is ensured when spawning is directly coupled to phytoplankton blooms. Many species spawn at the onset of the bloom, which means that food should be abundant throughout the larval phase. Although phytodetritus signals larval release by snow crabs, the larvae do not feed on this material but rather on microzooplankton. Microzooplankton populations peak in surface waters at about the time required for algal cells to sediment to the bottom. Thus, the snow crab larvae, which rapidly move to the surface once released, should have abundant food.

Interestingly, ensuring larval food availability is not the only factor that has led to the use of phytoplankton as a spawning cue; species with lecithotrophic larvae, which are sustained by egg reserves, also spawn in response to phytoplankton increases. For example, trochophore larvae of chitons (e.g., *Tonicella* spp.), after about a week in the water metamorphose directly into benthic juveniles, thus bypassing the planktotrophic veliger stage. Thus, other factors associated with phytoplankton blooms must explain the selection of phytoplankton as a spawning cue in these species. Several possibilities are suggested by considering environmental conditions prevailing during the spring bloom. The major factors leading to development of the bloom are the vernal increase in sunlight and the stratification of the water column so that phytoplankton cells are retained in the euphotic layer. Larvae originating from spawning in response to phytoplankton likely benefit from favorable temperatures for growth because in most

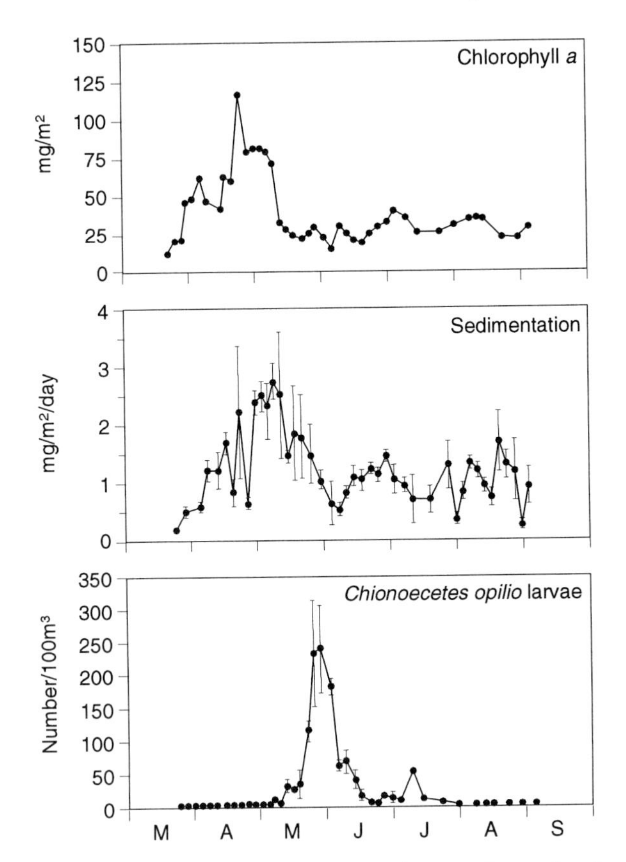

FIGURE 3 Relation of abundance of zoea I larvae by the snow crab *Chionecetes opilio* to the spring phytoplankton increase (chlorophyll *a* concentration at 0–40 m) and to sedimentation of organic particles from the bloom (concentration of chlorophyll *a* and pheopigments collected in sediment traps on the bottom) (adapted from Starr *et al.*, 1994, © Oxford University Press).

areas the stratification is caused by warming at the surface. In fact, because temperature increases are often short-lived, the spring bloom is probably a better predictor of upcoming favorable temperatures than temperature alone. Larvae produced during a bloom are not likely to be transported to deep water, where conditions may be adverse to growth, survival, and settlement. The bloom should be a clear signal for maximizing fertilization success because in most cold-water regions the vernal diatom increase is the most sudden environmental change during the year (water clarity often falls from the annual maximum to the annual minimum within a few days). Finally, predation losses may be minimized by spawning at the onset of the spring bloom because (i) foraging by visual predators is limited by turbidity, (ii) the abundance of zooplankton predators should be at the annual minimum, and (iii) predators are likely satiated because of the high larval abundance resulting from massive and oftentimes multispecies spawnings.

B. Chemical Cues and Gametes

The chemical nature of seawater varies, reflecting its origin, the organisms it contains, their abundance, and the levels of primary and secondary productivity. Many chemicals have been reported to provoke gamete release in ripe invertebrates. A number of these resemble neurotransmitters or hormones and possibly intervene in neuroendocrine pathways leading to spawning (acetylcholine, prostaglandins, serotonin, H_2O_2, which stimulates production of prostaglandins in some molluscs, and dopamine); however, others likely play no role in natural spawning (KCl and pulp mill effluent). Highly effective spawning inducers have been reported in gamete suspensions for an impressive number of marine invertebrates. Such inducers undoubtedly enhance spawning synchrony, fertilization success, and outbreeding and presumably in a species specific fashion.

As mentioned earlier, the spawning trigger for the green sea urchin and a number of bivalves is a metabolite secreted by phytoplankton cells during blooms. Purification studies, using extracts of the diatom *Phaeodactylum tricornutum*, have shown that the inducer of spawning for the green sea urchin is a phenol with a molecular weight between 1.2 and 3.5 kDa. It is likely bound to macromolecules within phytoplankton cells. When released into seawater, this substance readily forms complexes and precipitates, suggesting its soluble phase is short-lived and accurately indicates the presence of phytoplankton. For the green sea urchin, an unknown metabolite in sperm enhances the response of females to the plankton inducer. Thus, phytoplankton probably induces spawning in the most receptive males, and their gametes, together with phytoplankton, stimulate subsequent epidemic spawning.

In some species, chemical signals emitted by ripe individuals have been indicated to stimulate aggregation just prior to spawning. In polychaetes pheromones have been described that invoke the nuptial dance and spawning by swarming heteronereids. The nuptial dance enhances fertilization success by bringing males and females into close proximity.

C. Temperature

Midlatitude regions are characterized by pronounced seasonal temperature cycles and thermal variations are caused by factors such as tidal forcing and stochastic events (unusual weather patterns). Temperature could be an appropriate spawning cue because the timing of the vernal warming and autumnal cooling varies from year to year and likely influences rates of larval growth and survival. Temperature is the factor most often discussed as a potential gamete spawning cue in the literature. This emphasis is in part because temperature is the most easily measured environmental parameter and in part because workers have noted that temperature changes can provoke ripe animals to release gametes in the laboratory. However, the temperature changes used to provoke gamete release are usually much greater than would occur during spawning in the field. The emphasis on temperature is also attributable to Orton's hypothesis (published in 1920) that breeding is stimulated by a critical temperature which is a physiological constant for each species. Orton was vague in his statements; for example, he often used the general term "breeding" and did not distinguish between gametogenetic events and spawning. Although in some species the onset of gametogenesis and spawning is associated with a particular temperature, few species breed continually for as long as temperature is above or below a given level as he inferred. However, Orton's hypothesis is attractive in that it explains why breeding is delayed with increasing latitude.

In view of the emphasis on temperature as a spawning cue, the number of reports providing convincing evidence that a critical temperature triggers spawning is surprisingly low. Many studies are weak because of the lack of data on other environmental factors. Shifts in temperature are often indicative of changes in water masses and associated changes in water chemistry and biota. Unfortunately, little attention has been given to alternative hypotheses about the role of other factors. For numerous species, experimental studies have demonstrated that gonadal maturation is stimulated by temperature increases, but few studies show that gamete or larval release occurs when a particular temperature is attained. A number of studies provide evidence that spawnings coincide with abrupt temperature changes. This is best demonstrated by one study of the giant scallop *Placopecten magellanicus* in which many environmental factors were recorded during spawnings in several locations over 8 years. The temperature changes during spawnings were brought about by downwelling of warm surface water onto the scallop beds following periods of upwelling. Such events could be a clear signal for synchronous gamete release and would place the larvae in warm surface waters favoring enhanced survival (because of a shorter pelagic phase resulting from more rapid development).

D. Coordination of Spawning by Environmental Rhythms

Reproductive cycles involve shifts in energy allocation between somatic and gonadal growth coordinated by neuroendocrine processes. In species displaying interindividual synchrony in reproductive events, such internal production cycles are likely under some control by environmental cues. The annual aspect of reproductive cycles in many species is most likely explained by internal rhythms entrained by seasonal cycles in photoperiod, temperature, and food availability. Whereas spawning in species showing a variable delay between attaining ripeness and spawning is likely triggered by an additional external cue, release of gametes or larvae in species that spawn at a highly predictable time is likely the end result of an endogenous reproductive rhythm. Such rhythms are probably entrained by environmental cycles of shorter periodicity, such as lunar, tidal, and diel cycles. The endogenous nature of such spawning processes can often be demonstrated by observing animals maintained under constant conditions in the laboratory; for a time the reproductive rhythm continues and leads to spawning at the same time as in the field.

Predictable spawnings are often at monthly intervals during the general breeding period and coincide with a particular phase of the moon. Other species have bimonthly spawnings during spring or neap tides. Several instances of predictable spawnings are impressive. For example, the palolo worm swarms and spawns on only a few nights in the first third-quarter moon after early October and the crinoid *Comanthus japonicus* spawns between 2:30 and 4:00 PM. on 1 day in the year during the first- or last-quarter moon in early to mid-October. Also, on coral reefs in Australia, dramatic multispecies spawnings occur at particular tidal phases. Predictable spawning rhythms are most common in tropical and subtropical regions but also exist in temperate species. For some species, the mechanisms determining internal rhythms must vary because the phase of the moon at the time of spawning, or the frequency of spawning (monthly or bimonthly), can vary in different geographical regions. Numerous studies have manipulated environmental variables to identify environmental factors entraining internal spawning rhythms, and among the factors shown to be important are oscillations in sunlight or moonlight, tidal pressure, periods of immersion, noise from waves, and temperature. Manipulation of such factors can reset internal clocks and advance or delay synchronous spawning events. Although most studies suggest that tidal spawnings are controlled by endogenous rhythms, they may also be triggered more directly by a concurrent environmental change because tidal forces can generate changes in many factors, including temperature, salinity, and turbidity. Also, spring tides often cause intrusions of colder nutrient-rich water which stimulate primary production. Such changes are most pronounced in regions in which upwelling is likely.

Numerous hypotheses have been proposed to explain the selective advantage of predictable spawnings. For gamete spawners, predictable release determined by environmental cycles may have evolved as a mechanism to increase fertilization success or outbreeding. This is particularly probable in regions where environmental variations are generally not adverse to larval survival. This may explain why lunar spawnings are most frequent in tropical regions. Nevertheless, if fertilization success is the major determinant of reproductive success, controlling spawning by internal rhythms may also optimize fertilization success in temperate species in highly unpredictable environments. For example, an internal rhythm phased with the lunar cycle could lead to synchronous spawning in a full moon period even if moonlight were obstructed by clouds.

Fertilization success is not the only factor that may have led to the evolution of predictable spawnings; many larval spawners also show lunar or tidal spawning rhythms. A possible explanation, as already mentioned for species which spawn in direct response to environmental changes, is increased larval survival because predators are satiated. Several other hypotheses focus on the advantage of spawning at spring tides which could be a strategy to (i) decrease exposure to low salinities, (ii) increase larval transport, thus preventing stranding of larvae in shallow areas or reduce exposure to coastal planktivorous fish, and (iii) decrease mortality of larvae due to visual predators (because of increased turbidity during spring tides). Also, spawning at neap tides could be a means of retaining larvae near adult habitats in which conditions should favor long-term survival, but this could only apply to species with a short larval life because tidal currents during subsequent spring tides would cancel the larval retention caused by the neap tides. One possibility which has received little attention is that synchronized larval spawning may merely result from previous synchronization, for example, from synchronization to ensure copulation. Larvae would be released after a fixed interval.

E. Light

Photoperiod is the most predictable seasonal signal and experimental studies for a number of shallow water marine invertebrates show that the day length cycle is the major factor determining when gonadal development occurs. Photoperiod may control the annual aspect of the reproductive cycle of predictable spawners; however, other factors, such as lunar, tidal, and diel cycles, appear to determine the precise timing of spawning. Photoperiod is an unlikely spawning cue in species that spawn annually in a particular season but at variable times from year to year. This is logical because in temperate and polar regions, in which seasonal climatic changes are often advanced or delayed in any given year, spawning at a fixed date might result in complete reproductive failure.

On the other hand, correlative and experimental studies show that dark–light cycles determine the precise timing of gamete or larval release for a number of sponges, hydroids, ectoprocts, crustaceans, and ascidians. Exposure to light after a period of darkness, or the inverse, causes gamete release after a given delay. For some organisms, usually rapid growing colonial species, such mechanisms cause spawnings after sunrise or sunset throughout a general reproductive period. Diel mechanisms are also part of the spawning process in many tidal or lunar spawners. Diel spawning mechanisms may have been selected to (i) maximize fertilization success by increasing spawning synchrony, (ii) limit predation on eggs, embryos, or larvae by visual predators (spawning at night or in the absence of moonlight), and (iii) reduce exposure of spawning adults, gametes, embryos, or larvae to periods of peak predator activity.

IV. PROBLEM AREAS

Our understanding of environmental cues for release of gametes or larvae is poor. A vast literature documents synchronous spawnings for many species and suggests environmental cues; however, spawning cues are rarely convincingly demonstrated. A first problem is the inadequacy of correlative studies. Many studies can only indicate the approximate time of spawning because sampling was done at monthly intervals, often with the main objective of documenting annual reproductive or biochemical cycles. It is a major challenge to record both spawning events and a wide range of environmental factors at an adequate frequency to identify particular environmental factors associated with the onset of spawning. For a reasonable identification of the factors associated with spawning, such observations must be made for a number of spawning events. Sampling of animals to determine gonadal size and condition, or to ascertain whether embryos or larvae have been released, should be done at 2- to 4-day intervals over the spawning period and sampling of environmental variables should be done at hourly (or minimally daily) intervals. It is advantageous to study several species at a time, given that the same environmental data serve for all species. Also, this may provide evidence of simultaneous spawning in different species, which should mean that spawning is triggered by the same environmental cue or by interspecies pheromones. Another constraint to our understanding of spawning is the limited number of laboratory studies testing whether potential spawning cues can stimulate spawning when other conditions are controlled. Certain factors may be prerequisites for spawning so that the spawning inducer cannot cause spawning unless the other factor has attained a given level. For example, the triggering of spawning in some species may be inhibited if temperatures are unseasonably cold or warm or salinities too low.

Many studies of gamete spawning focus on species with abrupt and copious spawnings. A question which has been ignored is how fertilization success is achieved in rare species or species with unsynchronized or continuous spawning. Does fertilization in rare species occur only in even rarer instances in which two individuals are found together, and if so, what mechanisms ensure spawning synchrony? Continuous reproduction as indicated for many tropical species would seem to be maladaptive. Possibly, reproductive synchrony occurs in microhabitats, controlled by the specific conditions of the microhabitat or by pheromones. Culture studies on the Caribbean scallop *Euvola ziczac* have provided evidence that gonadal growth and spawning are controlled by conditions in the immediate vicinity of the animals rather than by large-scale climatic factors. In this mollusc, gonadal growth and spawning are highly synchronous in cages on the bottom and in cages held in suspension off the bottom at the same depth, but markedly asynchronous between the two habitats.

The literature on reproductive biology testifies to a striking gap between research on neuroendocrine pathways leading to spawning and research on ecological aspects of spawning. Thus, the final steps in the spawning process have been described in detail for species in a number of phylogenetic groups but fewer studies have focused on the link between the environmental spawning cues and these internal pathways.

We have a poor understanding of ultimate spawning factors. For many gamete spawners, there has likely been strong selection for a well-defined signal to ensure fertilization success because reproductive success is immediately reduced if fertilization is not achieved. This is particularly probable for sessile or slow-moving benthic invertebrates which cannot enhance fertilization success by aggregating prior to spawning. Furthermore, spawning cues should also ensure that conditions are favorable for development, growth, and survival of larvae and juveniles. The importance of the spawning cue as a predictor of favorable conditions for the larval phase is less critical in tropical regions where conditions are probably never really adverse to larval survival. Finally, a factor which may be involved in the selection of spawning cues is how well the signal reflects favorable conditions for larval settlement and for growth and survival of juveniles in the benthic habitat.

See Also the Following Articles

Marine Invertebrate Larvae; Marine Invertebrates, Modes of Reproduction in

Bibliography

Babcock, R., Mundy, C., Keesing, J., and Oliver, J. (1992). Predictable and unpredictable spawning events: *In-situ* behavioral data from free-spawning coral reef invertebrates. *Invertebr. Reprod. Dev.* **22**, 213–227

Bonardelli, J., and Himmelman, J. H. (1995). Examination of assumptions critical to body component indices: Application to the giant scallop *Placopecten magellanicus*. *Can. J. Fish. Aquat. Sci.* **52**, 2457–2469.

Bonardelli, J., Himmelman, J. H., and Drinkwater, K. (1996). Relation of spawning of the giant scallop, *Placopecten magellanicus*, to temperature fluctuations during downwelling events. *Mar. Biol.* **124**, 637–649.

Giese, A. C., and Kanatani, H. (1987). Maturation and spawning. In *Reproduction of Marine Invertebrates, Vol. 9, General Aspects: Seeking Unity in Diversity* (A. C. Giese and J. S. Pearse, Eds.), pp. 251–313. Blackwell. Oxford, UK.

Himmelman, J. H. (1981). Synchronization of spawning in marine invertebrates by phytoplankton. In *Advances in Invertebrate Reproduction* (W. H. Clark and T. S. Adams, Eds.), Vol. II, pp. 3–19. Elsevier/North-Holland, New York.

Morgan, S. G. (1995). The timing of larval release. In *Marine Invertebrate Larvae* (L. MacEdward, Ed.), pp. 157–191. CRC Press, Boca Raton, FL.

Olive, P. J. W. (1992). The adaptive significance of seasonal reproduction in marine invertebrates: The importance of distinguishing between models. *Invertebr. Reprod. Dev.* **22**, 165–174.

Pearse, J. S. (1990). Lunar reproductive rhythms in marine invertebrates: Maximizing fertilization? In *Advances in Invertebrate Reproduction 5* (M. Hoshi and O. Yamashita, Eds.). Elsevier, Amsterdam.

Starr, M., Himmelman, J. H., and Therriault, J.-C. (1990). Marine invertebrate spawning induced by phytoplankton. *Science* **247**, 1071–1074.

Starr, M., Therriault, J. C., Conan, G., Comeau, M., and Robichaud, G. (1994). Larval release in a sub-euphotic zone invertebrate triggered by sinking phytoplankton particles. *J. Plankton Res.* **16**, 1137–1147.

Thorson, G. (1946). Reproduction and larval development of Danish marine bottom invertebrates. *Meddr Kommn Havunders Kbh. Ser. Plankton* **4**, 1–553.

Sperm Activation, Arthropods

Julian Shepherd
Binghamton University

GLOSSARY

acrosome A vacuolar organelle present at the anterior end of most sperm which is involved with sperm–egg fusion.

parasperm A secondary or accessory type of sperm present in some organisms, often deficient or lacking in DNA.

seminal fluid The fluid which surrounds the sperm in the ejaculate.

seminal vesicle A storage organ for mature sperm pending ejaculation.

spermatheca The most widely used term for the sperm storage organ in female arthropods.

spermatophore The capsule in which sperm are transferred to the female in many arthropods.

vas deferens (plural, ***vasa deferentia***) The duct leading from the testis to the seminal vesicle in arthropods.

The term sperm activation is generally used to refer to the induction of motility in spermatozoa (= sperm) at any point after their maturation in the testes until their final fusion with the egg or death. However, it is also used to refer to other changes in the condition of sperm, including changes in morphology, fertilizing potential, or level of activity. Since the arthropods comprise millions of species which have enormously varied lifestyles, they show quite varied patterns of sperm transfer and activation. In general, sperm are quiescent until they are transferred to the female, when they become vigorously active. After their migration to the sperm storage organ(s) in the female, they usually become quiescent until egg deposition begins, when they become active again as they move to fertilize the eggs. These patterns of activity have been described in only a small number of arthropod species, and experimental evidence defining the mechanisms of activation has been elaborated in only a fraction of these species. Therefore, while generalizations can be attempted, any application of these generalizations to other species should be made with caution. Since arthropods include highly divergent phylogenetic groups, the discussion will be separated along the lines of the major classes of arthropods.

I. INSECTS

Because of the abundance, economic importance, and ease of laboratory maintenance of insects, far more is known about their modes of sperm activation than about those of other classes of arthropods.

A. Activity in the Male Reproductive Tract

After their maturation in bundles in the testicular follicles, the sperm exit the testes and descend the vasa deferentia to the storage organs, generally called the seminal vesicles (one example of a male reproductive system is shown in Fig. 1). There is evidence of sperm being weakly active in the testes (in the grasshopper *Chortophaga*, the silkworm *Bombyx mori*, and the fruit flies *Drosophila* spp.) as well as in the vasa of the biting midge *Culicoides melleus*, but it appears that generally sperm remain unactivated before ejaculation. This inactivity would presumably serve to conserve metabolic energy and sperm viability, even though most male insects have lifetimes of only a few weeks.

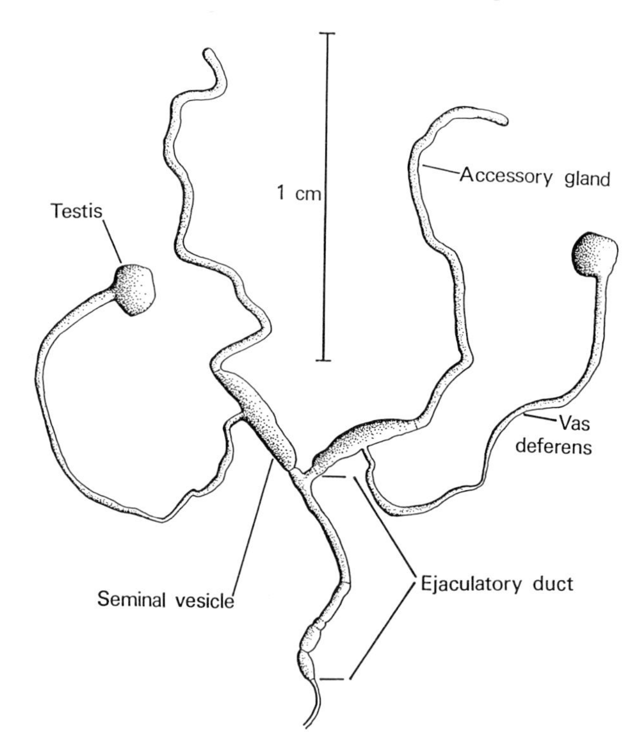

FIGURE 1 Drawing of the reproductive system of a male moth, typical of many insects (reprinted from J. Shepherd, *J. Insect Physiol.*, **20**, p. 2111. Copyright 1974, with permission from Elsevier Science).

B. Activation upon Ejaculation

In almost all species of insects that have been examined, the sperm become vigorously active when ejaculated. Some insects ejaculate sperm into external or internal spermatophores, whereas others ejaculate unencapsulated semen. Whatever the mechanism of transfer, the sperm are diluted during ejaculation by secretions produced by various accessory glands of the male reproductive tract. These male secretions, rather than female secretions, appear to be the principal agents responsible for activation. Evidence for this assertion comes from experiments in many species showing that sperm can generally be activated when dissected from the male tract and diluted by male accessory gland secretions. In fact, sperm dissected from male seminal vesicles of most orders of insects (except Lepidoptera) can usually be activated by dilution in simple saline solutions, though they are rarely viable for long in such diluents. Sperm might be activated in this manner by dilution of inhibitors, increased oxygen tension, or some other mechanism, but this has never been carefully investigated. K. Davey has suggested that the sperm of *Rhodnius prolixus* may be activated by a lowering of pH from neutrality to about 5.5 since the sperm are much more active at the lower pH and because the seminal vesicles have a neutral pH, whereas the ejaculatory bulb measures about pH 5.5. In several other species of insects, however, there are no such significant changes in pH over the course of the activation process. An exception to the general rule of activation during or soon after ejaculation occurs in *Locusta migratoria*: The sperm do not become motile until the elaborate spermatophore of this species everts fully and the sperm reach the vicinity of the spermatheca.

Activation of sperm in Lepidoptera also occurs during ejaculation but is complicated by two phenomena. First, all Lepidoptera produce two types of sperm: a conventionally nucleated sperm (called eupyrene sperm or eusperm) and an anucleate sperm (called apyrene sperm or parasperm). Second, the eusperm are still joined together by a proteinaceous matrix in their original bundles of 256 sperm when they are ejaculated. Consequently, the parasperm are activated as usual immediately by the secretions of the male accessory glands, but the eusperm must dissociate from their bundles before they become active, a process which generally takes several hours. Omura identified the source of the activating substance for the parasperm as the secretion of the most proximal part (the "glandula prostatica") of the ejaculatory duct (secretory in the Lepidoptera, unlike most other insects). Shepherd identified this substance in the wild silkmoth *Antheraea pernyi* as a small heat-stable polypeptide (molecular weight about 3 kDa) which is necessary only to initiate the motility of the parasperm. The ejaculatory ducts of a variety of other Lepidoptera (*Papilio polyxenes, Manduca sexta, Galleria mellonella, Argyresthia thuiella, and Bombyx mori)* also produce proteinaceous activators essential for parasperm activation, but in the cases of *M. sexta* and *B. mori,* the activator is a larger heat-sensitive protein (molecular weight about 25 kDa in *M. sexta*). Osanai and coworkers found that serine proteases will activate *B. mori* parasperm and they have purified a serine endopeptidase (molecular weight 2629 kDa) from the glandula prostatica which they call "initiatorin." However, they have not reported that the purified initiatorin will activate parasperm. In nonconformity with the previous scheme, Thibout has reported that the parasperm of *Acrolepiopsis assectella* cannot be activated by ejaculatory duct secretions but can be activated by a marked lowering of osmotic pressure (to about 100 mOsm) in the absence of any specific secretions. However, it is unknown whether such a decrease in osmotic pressure occurs *in vivo* during ejaculation or in the spermatophore.

The mechanism of activation of eusperm in Lepidoptera is only partially understood. The bundles of eusperm must first be dissociated in the spermatophore; Osanai and coworkers have shown that proteases generally can dissociate the eusperm bundles of *B. mori*, and they ascribe this action in the spermatophore to initiatorin. After the eusperm bundles dissociate, the eusperm do become active in the spermatophore of most species of Lepidoptera, but whether they become active simply as a result of liberation from their bundles or whether they also require specific factors is unknown. Eusperm motility in the spermatophore generally appears rather feeble, which has led to the suggestion that the principal function of the parasperm is to help transport the eusperm in their long migration to the spermatheca. Several other hypotheses for the function of the parasperm have been advanced, but none, including the transport hypothesis, has solid experimental support.

Sperm in some other insects are ejaculated in bundles (some Orthoptera, some Hymenoptera, and many Homoptera) and in many cases are motile in these bundles. Scale insect (Homoptera) sperm, which have microtubular arrays in quite unusual patterns, are usually conjoined in bundles of 10–256 cells surrounded by a sheath. These sperm move coordinately in their bundles, breaking apart as they approach the ovaries, in which fertilization occurs in these insects. In one species, a specialized somatic cell has been reported to assist in this migration.

C. Sustained Motility after Activation

While the factors causing initial activation are generally unknown in insects other than Lepidoptera, studies in a variety of insects have helped elucidate factors necessary for continued sperm activity after activation has taken place. Rao and Davis have shown that sustained motility in bedbug sperm necessitates a brief exposure to the secretions of the male accessory glands, but that motility will then continue and normal aggregation will occur in a solution of 70 m*M* sodium citrate as long as the solution is well oxygenated. Landa has shown that in cockchafer (a beetle) spermatophores, sperm are also activated by dilution with accessory gland secretions during ejaculation. However, while cockchafer sperm will initially become motile under aerobic conditions, subsequent "differentiation" of the spermatophore and sustained vigorous motility of the sperm will occur only under anaerobic conditions. Davey has shown that sustained motility of sperm in the spermatophore of the bug *R. prolixus* is indifferent to the presence of oxygen. It seems likely that generally sperm transferred in spermatophores, especially in the case of internal spermatophores, would be subject to low oxygen concentrations. Presumably in these cases, exogenous substrates for anaerobic metabolism are needed.

The most careful study of nutrients in ejaculated seminal fluids has been carried out in the laboratory of Aigaki, Kasuga, and Osanai. Through amino acid analysis and radiotracer studies, they conclude that the serine endopeptidase initiatorin (whose other functions were elaborated previously) and an exopeptidase found in the spermatophore fluid of *B. mori* cleave proteins in the fluid, producing free arginine. This arginine is then metabolized to glutamate, which, when coupled with pyruvate derived from glycogen by glycolysis, produces alanine and α-ketoglutarate, the latter being oxidized to succinate under anaerobic conditions. In *Drosophila nigromelanica*, high concentrations of glutamate in male accessory gland secretions might similarly be used as metabolic substrates for sperm motility. Another study by Blum and coworkers suggested that fructose might be a source of energy for sperm in the seminal fluid of honeybees.

D. Activity in the Spermatheca

After ejaculation, the sperm migrate to the sperm storage organs in the female, the spermatheca(e), or seminal receptacle(s). Motility may or may not be essential for this migration or sustained during it, but usually sperm activity subsides in the storage organs (although a low level of activity is seen in many species: *Locusta migratoria*, the boll weevil *Anthonomus grandis*, the tsetse fly *Glossina morsitans*, and at least three species of parasitic wasps). Presumably this inactivity preserves sperm viability and conserves energy; this would seem to be particularly important in long-lived female insects, especially the queens of social insects which can be fertile in some cases for more than 10 years without replenishment of their sperm supplies. Spermathecae in many insects have glands attached; these presumably supply nutrients to maintain sperm viability, but their secretions have been little investigated.

E. Activation Prior to Fertilization

In insects, eggs are fertilized in the common oviduct (often called the vestibule or vagina) just prior to their release from the female. It has been assumed that active sperm are essential for fertilization, and observations of sperm during the oviposition process generally show active sperm in the spermathecal duct or the oviduct. How they are activated is generally unknown, however. Oviposition is certainly neurally controlled, and so it is likely that part of that neural program would induce timely contractions of the spermatheca propelling the sperm toward the oviduct, where the changed milieu of the sperm may induce activation. Lensky and Schindler found that dilution of spermathecal sperm of the honeybee with a secretion from a gland attached to the base of the spermatheca would induce vigorous motility. These sperm, like sperm from the male tract, could also be activated by dilute solutions of varied salt concentration and pH, even by pure water. Presumably a pulse of this secretion is released as sperm are expelled. Koeniger found that removal of the gland did not affect sperm viability but did result in the queen laying only drone (uninseminated) eggs. However, the spermathecal glands in most of those species of insects which have them are at the distal end of the spermatheca, where they would not be in a position to play a role in sperm activation.

There are evidently some morphological changes which occur prior to fertilization. In Lepidoptera, the eusperm shed a substantial extracellular coat either during or after their migration to the spermatheca. Presumably this change is essential for fertilization because sperm in the fertilization canal lack this coat. In the cockroach *Periplaneta americana*, Hughes and Davey found that the sperm both increased their tail-beat frequency and underwent a compaction of their heads in the spermatheca. In the housefly *M. domestica*, Leopold and Degrugillier have shown that a female accessory gland secretion causes a morphological change in the sperm acrosome as they are about to fertilize the eggs. By analogy with the sperm of other organisms, changes in the morphological or physiological state of sperm in the female tract, especially just before fertilization, are probably much more widespread among insects than have been documented.

II. ARACHNIDS

Study of the internal events of reproduction in arachnids has been largely concentrated on ticks because of their large size and economic importance. Spermatogenesis in ticks produces an atypical type of large sperm with a large volume of cytoplasm and no flagellum. In the male tract, these sperm are nonmotile and immature; hence, they have been called spermatids or prospermia. After their transfer to the female's uterus or seminal receptacle, the anterior end of the prospermium releases a cap called the operculum (Fig. 2). J. Shepherd and coworkers identified a secretion from the accessory glands of the male *Ornithodoros moubata* and *Dermacentor variabilis* which contains a protein (molecular weight about 12.5 kDa) necessary for this rupture. The prospermia then begin to elongate, turning completely inside out and exposing cellular processes on their exterior which appear to be responsible for their mysterious gliding motility. Although the sperm are competent to become motile at this point, they apparently need some stimulus induced by the female taking a blood meal in order to begin movement out of the spermatophore capsule. In the oviducts, a further small eversion of the posterior end of the sperm exposes the nucleus on the surface of this projection. How this structure fertilizes the egg is unknown.

Because anactinotrichid mites have sperm similar

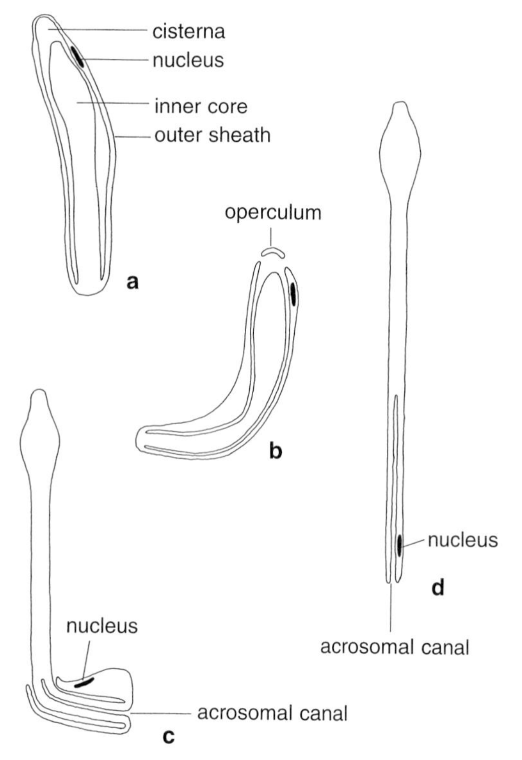

FIGURE 2 Diagram of the metamorphosis of tick sperm after transfer to the female in a spermatophore. a, the prospermium just after transfer; b and c, stages in the elongation of the sperm; d, the mature sperm prior to migration out of the spermatophore (reprinted from J. Shepherd, S. Levine, and J. D. Hall, *Int. J. Invertebr. Reprod.* 4, p. 312. Copyright 1982, with permission from Balaban Publishers.)

in morphology to those of ticks, probably the same phenomena occur. However, there have been no studies of these sperm beyond the extensive ultrastructural studies of Alberti. Actinotrichid mites have quite different, though nonflagellate, sperm—one report noted a change in morphology occurring in the seminal receptacle of the spider mite *Tetranychus urticae*.

III. CRUSTACEANS AND HORSESHOE CRABS

Since most crustacean sperm are aflagellate and nonmotile (except in barnacles, ostracods, and a few others), the only significant changes that could be construed as activation occur when the sperm approach the vicinity of the egg. These changes, as described by Hinsch in the spider crab, constitute an acrosomal reaction. These are comparable to the classical acrosomal reactions observed in other marine invertebrates, such as that described by Tilney in the horseshoe crab (which is an arthropod but not a crustacean). A few studies indicate that barnacle sperm, which are flagellated, are activated earlier, during ejaculation into the pallial cavity of the female. Whether this is induced by a secretion of the male's penial glands or by the female's oviducal glands is unclear.

IV. MYRIAPODS

Several studies have examined the novel ultrastructure and interesting behavior associated with sperm transfer in the centipedes, millipedes, and other myriapods. The only reference, by Baccetti and coworkers, to a maturational change in the sperm of these organisms describes an elongation from a barrel shape in the male tract to a long, flat, spiral ribbon in the female tract of the millipede *Polyxenus* sp. Millipede sperm are aflagellate and apparently nonmotile.

See Also the Following Articles

Crustacea; Insect Reproduction, Overview; Male Reproductive System, Insects; Myriapoda; *Rhodinus prolixus*; Sperm Transport, Arthropods

Bibliography

Adiyodi, R. G. (1985). Reproduction and its control. In *The Biology of Crustacea* (D. E. Bliss and L. H. Mantel, Eds.), Vol. 9, pp. 147–215. Academic Press, New York.

Aigaki, T., Kasuga, H., Nagaoka, S., and Osanai, M. (1994). Purification and partial amino acid sequence of initiatorin, a prostatic endopeptidase of the silkworm, *Bombyx mori*. *Insect Biochem. Mol. Biol.* 24, 969–975.

Davey, K. G. (1965). *Reproduction in the Insects*. Oliver & Boyd, Edinburgh, UK.

Degrugillier, M. E. (1985). *In vitro* release of housefly, *Musca domestica* L. (Diptera: Muscidae), acrosomal material after treatments with secretion of female accessory gland and micropyle cap substance. *Int. J. Insect Morphol. Embryol.* **14**, 381–391.

Engelmann, F. (1970). *The Physiology of Insect Reproduction.* Pergamon, New York.

Gillott, C. (1988). Arthropoda—Insecta. In *Reproductive Biology of Invertebrates. Vol III. Accessory Sex Glands.* (K. G. and R. G. Adiyodi, Eds.), pp. 319–471. Wiley, New York.

Gillott, C. (1995). Insect male mating systems. In *Insect Reproduction* (S. L. Leather and J. Hardie, Eds.), pp. 33–55. CRC Press, Boca Raton, FL.

Hinsch, G. W. (1990). Arthropoda—Crustacea. In *Reproductive Biology of Invertebrates. Vol. IV, Part B. Fertilization, Development, and Parental Care* (K. G. Adiyodi and R. G. Adiyodi, Eds.), pp. 121–155. Wiley, New York.

Kasuga, H., Aigaki, T., and Osanai, M. (1987). System for supply of free arginine in the spermatophore of *Bombyx mori*. Arginine-liberating activities of contents of male reproductive glands. *Insect. Biochem.* **17**, 317–322.

Leopold, R. A. (1976). The role of male accessory glands in insect reproduction. *Annu. Rev. Entomol.* **21**, 199–221.

Mann, T. (1984). *Spermatophores.* Springer-Verlag, Berlin.

Reger, J. F., and Fitzgerald, M. E. C. (1983). Arthropoda—Myriapoda. In *Reproductive Biology of Invertebrates. Vol. II. Spermatogenesis and Sperm Function* (K. G. Adiyodi and R. G. Adiyodi, Eds.), pp. 451–475. Wiley, New York.

Sonenshine, D. E. (1991). *The Biology of Ticks. Vol. 1.* Academic Press, New York.

Thibout, E. (1981). Evolution and role of apyrene sperm cells of lepidopterans: Their activation and denaturation in the leek moth, *Acrolepiopsis assectella* (Hyponomeutoidea). In *Advances in Invertebrate Reproduction* (W. H. Clark, Jr., and T. S. Adams, Eds.), pp. 231–242. Elsevier/North Holland, New York.

Spermatogenesis, Overview

Rex A. Hess

University of Illinois at Urbana

GLOSSARY

acrosomal system A Golgi-derived organelle that forms over the nucleus consisting of a membrane-bound vesicle with dense acrosomal granules that eventually fuse; consists of enzymes necessary for the acrosomal reaction at fertilization.

clonal unit The synchronous group of developing germ cells formed by incomplete cytokinesis during spermatogonial division and held together by intercellular bridges until spermiation.

cycle A complete sequential progression of the cellular associations (or stages) that occur over time is called the cycle of the seminiferous epithelium. The stages follow one another in development over time through an entire cycle, returning to the original stage and repeating this cycle approximately 4.5 times until spermatogonia eventually become spermatozoa and are released.

cytoplasmic lobe A cytoplasmic protrusion of the late step 19 spermatid in stage VII (rat), containing abundant RNA, mitochondria, lipid droplets, and other unused cellular remnants that are eventually phagocytized by the Sertoli cell.

meiosis A specialized process by which one germ cell produces four haploid spermatids after undergoing two meiotic cellular divisions. A long prophase permits the duplication of chromosomes and genetic recombination before these largest of germ cells rapidly divide, producing secondary spermatocytes after meiosis I and small step 1 spermatids after meiosis II.

residual body A large spherical body containing the cytoplasmic remnants of sperm formation which is formed by detachment of the cytoplasmic lobe during sperm release into the lumen. Residual bodies are phagocytized by Sertoli cells in subsequent stages.

seminiferous epithelium Consists of two cell types, a somatic cell, the Sertoli cell, and male germ cells at various steps in development.

Sertoli cell barrier Once called the "blood–testis-barrier," this tight occluding junction is formed between adjacent Sertoli cells separating basal and adluminal compartments. The barrier separates most germ cells from blood-borne substances and lymph, thus requiring the Sertoli cell to sustain germ cell development.

spermiation A complex process by which spermatozoa are released into the seminiferous tubule lumen after detaching from the Sertoli cell junctional complex.

spermiogenesis Cellular differentiation of the spermatids from a small, nondescript round cell to the spermatozoon that has a highly condensed elongate nucleus, unique acrosomic system derived from the Golgi, and a complex flagellum that is motile.

stages A stage (numbered with Roman numerals) is represented by a defined association of spermatogonia, spermatocytes, and spermatids in a cross section of seminiferous epithelium, at a specific phase in time during spermatogenesis. The acrosomal system of the spermatids is commonly used to identify specific stages in the cycle of the seminiferous epithelium.

stem cell Quiescent self-renewing spermatogonia that, with proper stimulation, proliferate in order to renew the germinal epithelium.

steps A unique morphologically identifiable change in the differentiation of a spermatid, based on the acrosomic system formation, sperm head shape, and nuclear condensation. These changes divide spermiogenesis into sequential steps that are numbered with Arabic numbers (e.g., step 1 spermatid).

wave A series of sequential stages in physical space along the length of a seminiferous tubule, formed by the synchronous development of clonal units of germ cells.

Spermatogenesis is the biological process of gradual transformation of germ cells into spermatozoa over an extended period of time within the boundaries of the seminiferous tubules of the testis. This process involves cellular proliferation by repeated mitotic divisions, duplication of chromosomes, genetic recombination through cross-over, reduction-division by meiotic division to produce haploid spermatids, and terminal differentiation of the spermatids into spermatozoa. Thus, spermatogenesis can be divided into three phases: proliferation, reduction-division (or meiosis), and differentiation. These phases are also associated with specific germ cell types, i.e., spermatogonia, spermatocytes, and spermatids, respectively.

I. THE SEMINIFEROUS TUBULE

Spermatogenesis occurs within the extensive seminiferous tubular structures of the testis. Seminiferous tubules are lined by the seminiferous epithelium and contain a fluid-filled lumen, into which fully formed spermatozoa are released. The seminiferous epithelium consists of two basic cell types, somatic and germinal cells. The germ cells (Fig. 1) are found at different levels from the base of the tubule to the lumen and are surrounded by cytoplasm of the somatic cell, the Sertoli cell (Fig. 2). The Sertoli cell cytoplasm extends the entire height of the epithelium because the cell serves to nurture the germ cells through their cycles of development. As the germ cells divide and develop into different types of cells, they move from the basement membrane region through tight junctional complexes of adjacent Sertoli cells until they reside in the adluminal compartment. The Sertoli–Sertoli cell junctions form the blood–testis barrier, which helps to protect the developing germ cells from potentially harmful bloodborne chemicals. The germ cells develop as a syncytium or clonal unit connected to one another by intercellular bridges after cell division (Fig. 3). This unique process of incomplete division ensures synchronous development and permits rapid communication between the cells. Synchrony of germ cell development results in large areas of the seminiferous tubule containing vast numbers of cells at the same level of development, the specific identification of which scientists refer to as stages.

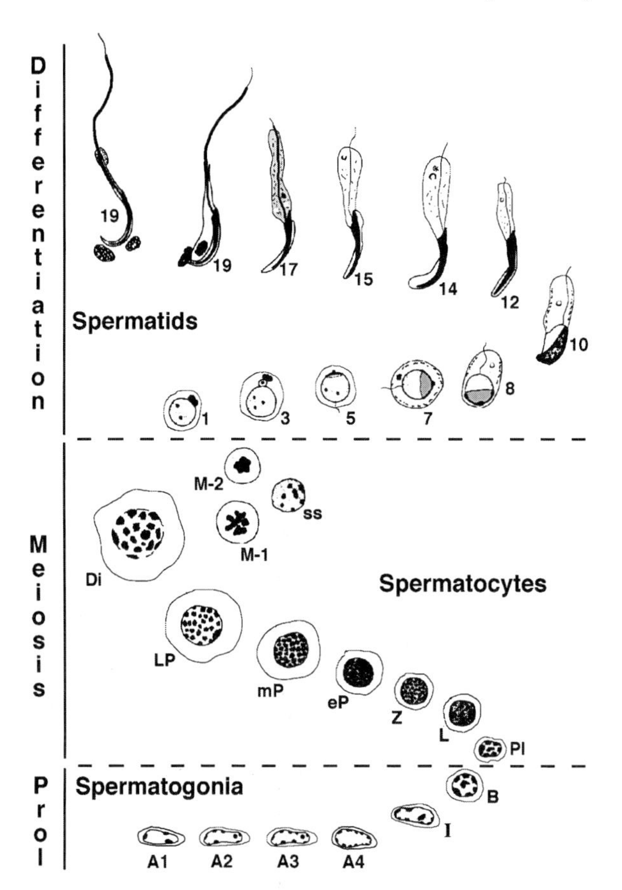

FIGURE 1 Germ cell development in rat spermatogenesis. The proliferation phase (Prol) includes repeated spermatogonial division from type A spermatogonia (A1–A4) to intermediate (I) and B-type cells. Meiosis is an extended phase that begins after B-type spermatogonia divide to produce preleptotene spermatocytes (Pl). Meiotic prophase begins with small leptotene spermatocytes (L). The cells enlarge as prophase continues through zygotene (Z), early, mid-, and late pachytene (eP, mP, LP) spermatocytes. Diplotene cells undergo the first meiotic division (M-1) producing secondary spermatocytes (ss). After the second meiotic division (M-2), haploid cells called spermatids begin the differentiation phase by forming round spermatid steps (1–7). Round spermatids are slowly transformed into elongated cells (steps 8–19) and finally into spermatozoa that are released.

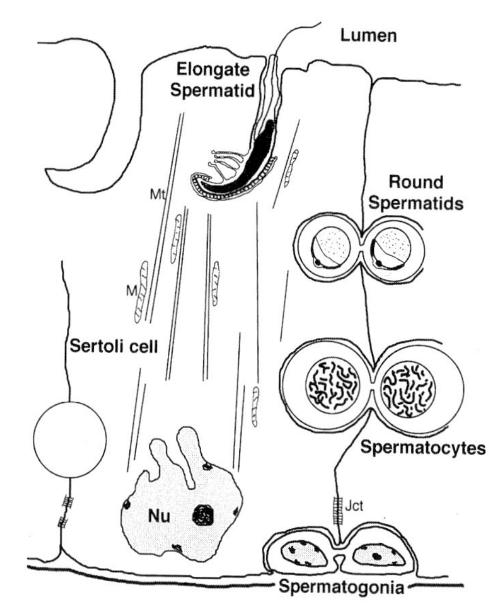

FIGURE 2 The seminiferous epithelium consists of somatic cells, the Sertoli cells, whose cytoplasm surrounds the developing germ cells. Sertoli–Sertoli cell junctions (Jct) separate spermatogonia from the adluminal compartment where spermatocytes and spermatids develop. Microtubules (Mt) are parallel in the Sertoli cell cytoplasm and help to transport germ cells within the epithelium. M, mitochondria; Nu, nucleus of the Sertoli cell.

II. PHASES OF SPERMATOGENESIS

A. Proliferation

Spermatogonia, which constitute the first phase, are the most immature cells and are located along the base of the seminiferous epithelium. They proliferate by mitotic division and multiply repeatedly to continually replenish the germinal epithelium. Spermatogonia are capable of self-renewal and thus also produce stem cells that remain along the base as well as committed cells that are on a one-way tract leading

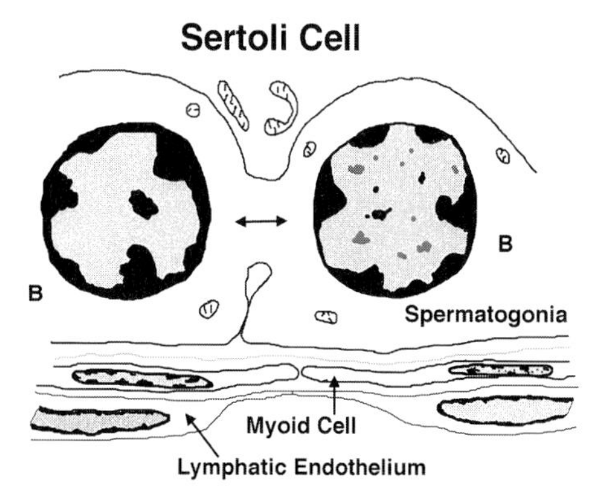

FIGURE 3 Two B-type spermatogonia are connected through the intercellular bridge (double arrow). Sertoli cell cytoplasm helps to hold the bridge in place. Myoid cells have a common basal lamina with spermatogonia and the Sertoli cell.

to spermatozoa. In most species, the B spermatogonia is the last to divide by mitosis. Its division produces the first cell of the second phase, the preleptotene spermatocyte, which migrates upwards away from the base of the seminiferous tubule and crosses through the Sertoli–Sertoli junction.

B. Meiosis

Reduction-division by meiosis involves numerous types of spermatocytes that range in size from cells smaller than a red blood cell (preleptotene) to very large cells (pachytene) that occupy portions of every cross section of seminiferous tubules. Reduction-division is a biological mechanism by which a single germ cell can increase its DNA content, then divide twice to produce four individual germ cells containing a single strand of each chromosome or half the number of chromosomes normally found in cells of the body. The process of meiosis is extended over a long period of time; therefore, spermatocytes are found in every stage of spermatogenesis, and in some stages two different types of spermatocytes are observed. During meiosis, the changes that take place in the chromosomes are easily recognized (Figs. 1 and 7).

DNA synthesis occurs in preleptotene spermatocytes. Prophase of the first meiotic division may last for nearly 3 weeks, during which time the chromosomes first unravel as thin unpaired filaments (leptotene). Homologous chromosomes become paired in the zygotene cell, forming the synaptonemal complex. Pachytene spermatocytes begin as small cells but their nuclei enlarge greatly as the chromosomes become shorter and thicken. Genetic recombination occurs through cross-over between paired chromosomes. Pachytene cells also exhibit an increase in RNA and protein synthesis in preparation for the next phase. Diplotene spermatocytes separate the synaptonemal complexes and the chromosomes are spread apart in the nucleus. In diakinesis the nuclear envelope disappears and chromosomes condense. Both meiotic divisions occurs rapidly, thus limiting these cells to one stage (Fig. 7). Small secondary spermatocytes (2N) are produced by meiosis I which then rapidly divide again by meiosis II, with unique metaphase formations by the chromatin. Meiosis II produces very small haploid (1N) cells called round spermatids that enter the next phase called differentiation.

C. Differentiation

The haploid germ cells undergo a prolonged phase of terminal differentiation known as spermiogenesis. The cells undergo dramatic changes, including the following three major modifications: (i) The nucleus elongates and chromatin condenses into a very dark staining structure having unique shapes that are species specific (Fig. 4); (ii) the Golgi apparatus produces a lysosomal-like granule that elaborates over the nucleus to form the future acrosome (Fig. 5). The acrosomic system contains the hydrolytic enzymes required for sperm–egg interaction and fertilization; and (iii) the cell forms a long tail lined with mitochondria in the proximal region and it loses excess cytoplasm, which is discarded first as the cytoplasmic lobe that eventually is phagocytized by the Sertoli cell as the residual body. Recognizable changes in the differentiation of a spermatid are called "steps" of spermiogenesis. In the rat, the first step is the small round step 1 spermatid produced by meiosis II. Step 1 occurs in the first stage of the cycle. In all species, the late elongate spermatids, steps 15–19 in the rat, overlap with the younger round spermatids. Thus, in some stages two generations of spermatids are present in the same tubule cross section (Figs. 5 and 7).

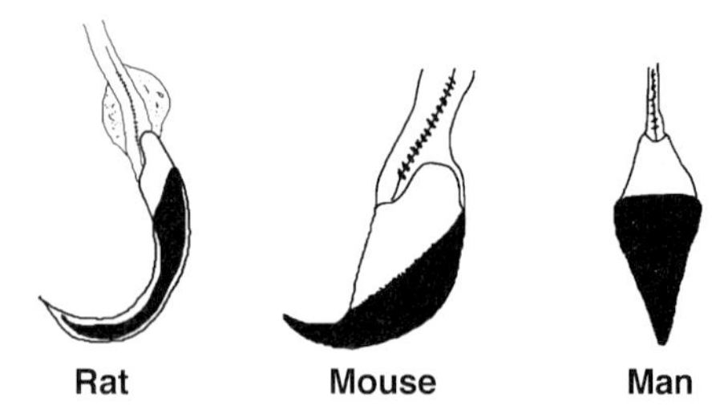

FIGURE 4 Heads of newly released sperm from three species illustrating the variation achieved through differentiation of the haploid spermatid. The black areas represent portions of the nucleus covered by the acrosome.

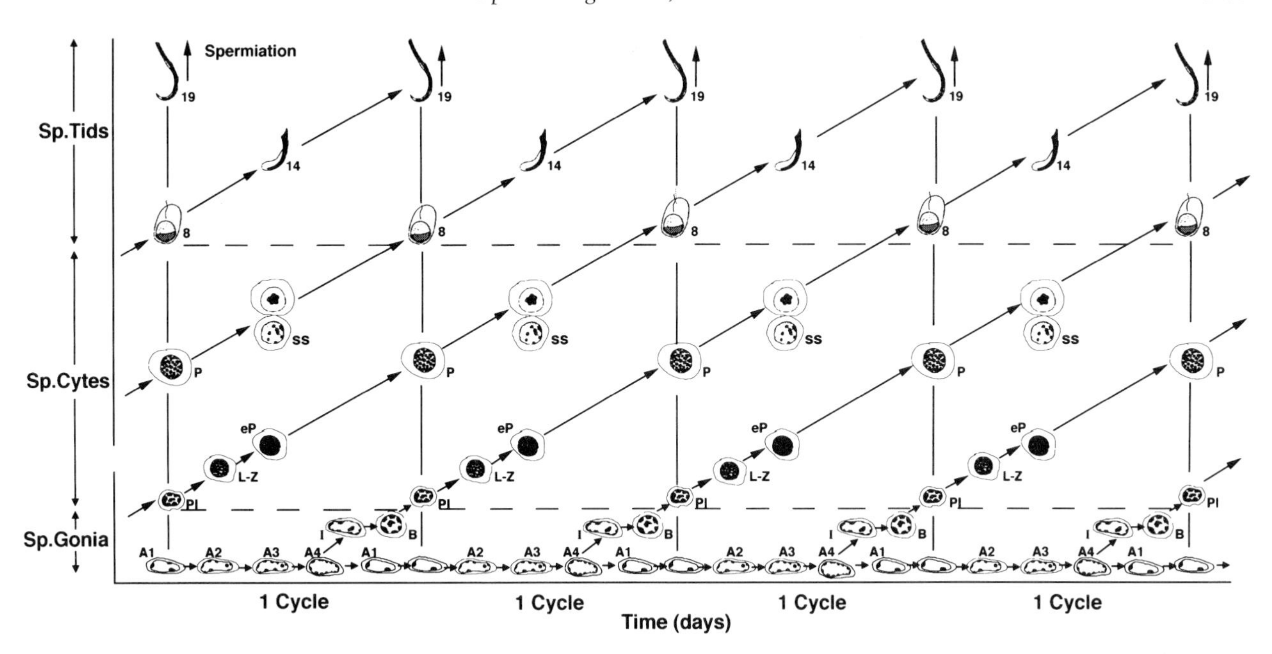

FIGURE 5 Repetitions of the cycle of the seminiferous epithelium are represented in a temporal manner. Each cycle shows the different types of cells and their progeny that would be found in a particular stage of the cycle. The phases of spermatogenesis are represented by the three cell types, spermatogonia (Sp.Gonia), spermatocytes (Sp.Cytes), and spermatids (Sp.Tids). Type A spermatogonia along the first row are self-renewing, but A1–A4 are committed cells in the spermatogenic lineage. Types I (intermediate) and B appear distinctive and are found in greater numbers than the type A spermatogonia. The small preleptotene spermatocyte begins the extended period of meiosis, with modifications producing leptotene (L), zygotene (Z), early pachytene (eP), and pachytene (P) spermatocytes. Meiotic division I produces the secondary spermatocyte (ss). Meiotic division II results in the haploid spermatids, of which three are shown: steps 8, 14, and 19. Step 19 spermatids are released into the lumen through the process of spermiation.

III. STAGES OF THE CYCLE

The synchronized process of spermatogenesis allows germ cells to advance (or change) within the seminiferous epithelium. In a general sense, the more mature cells are found away from the basement membrane and in specific associations with the younger cells that will divide and mature in time. This process of epithelial evolution in a synchronized manner over time produces a cycle because there is a beginning, the entrance of spermatogonia into type A mitosis, and an end, the release of new sperm. Spermatogenesis can be split into repeated cycles of the seminiferous epithelium which are defined by the specific cellular associations established at specific points in time. Over a set period of time, these cellular associations repeat themselves, thus establishing the cycle (Fig. 5). When a cellular association exhibits distinguishing morphological features, it is identified as a different stage of the cycle. Stages are recognized by examining cross sections of seminiferous tubules histologically, with a particular focus on the acrosomic system associated with the spermatids. The acrosomic system is stained using the periodic acid-Schiff's reaction (PAS). The pink PAS stain recognizes the Golgi and acrosomic granule. As the granule flattens against the nuclear envelop the stain picks up the acrosomal vesicle that extends over the nucleus as a cap until finally it forms a very thin layer over the condensed nucleus of the mature sperm (Fig. 6).

The repetitive nature of the cycle is shown in Fig. 5. Although the arrows suggest that the cells move laterally in time, they actually only move upward in the seminiferous epithelium. Over approximately 4.5 cycles the A spermatogonia becomes a spermatozoon that is released, after having gone through six mitoses, two meiotic divisions, and more than 2 weeks of differentiation.

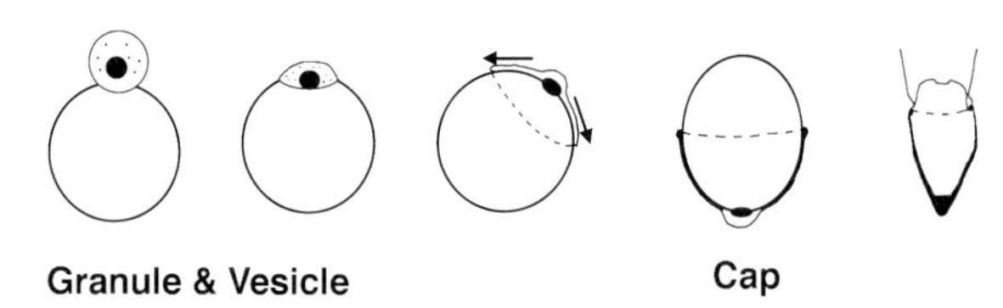

FIGURE 6 The acrosomic system consists of the Golgi apparatus, which produces the acrosomic vesicle, and granules. The granules are small at first, but fuse to form a single large granule that becomes flattened against the nuclear envelop. The vesicle also flattens and spreads across the nucleus (arrows) until a cap is formed that covers nearly one-half of the nucleus. In the mature sperm, the acrosome is tightly bound to the nuclear envelope as a thin covering over a major portion of the sperm head.

Recognition of the stages of the cycle is best performed by comparing histological sections to a "staging map" (Fig. 7). In the map, cells progress from left to right, then move up one row and again progress from left to right. In time, the cells are simply changing into the next cell type through cell division or differentiation, and the cells then move through the epithelium toward the lumen. Because the definition of stages is arbitrary, the length of time that the cells remain in a particular stage is variable and ranges from 0.3 to 2.7 days. Thus, the length of time occupied by a stage will determine the frequency in which that stage is found in seminiferous tubule cross sections of the testis (Fig. 8).

IV. THE WAVE

Cells in the stages do not move laterally along the length of the seminiferous tubule. However, there is an unusual ordering of the stages so that the segments of the tubule contain stages in consecutive order. Although there are short reversals of this segmental order, called modulations, the sequential order of the stages and their repetition along the length of the tubules constitutes the "wave" of spermatogenesis in the seminiferous epithelium. That is, stage I is followed by stage II, which is followed by stage III, etc. through stage XIV, which is followed by stage I. The stages are found in ascending order from the rete testis to the center of the seminiferous tubule, where a reversal site is typically found (Fig. 8). The wave is produced by synchronous development of

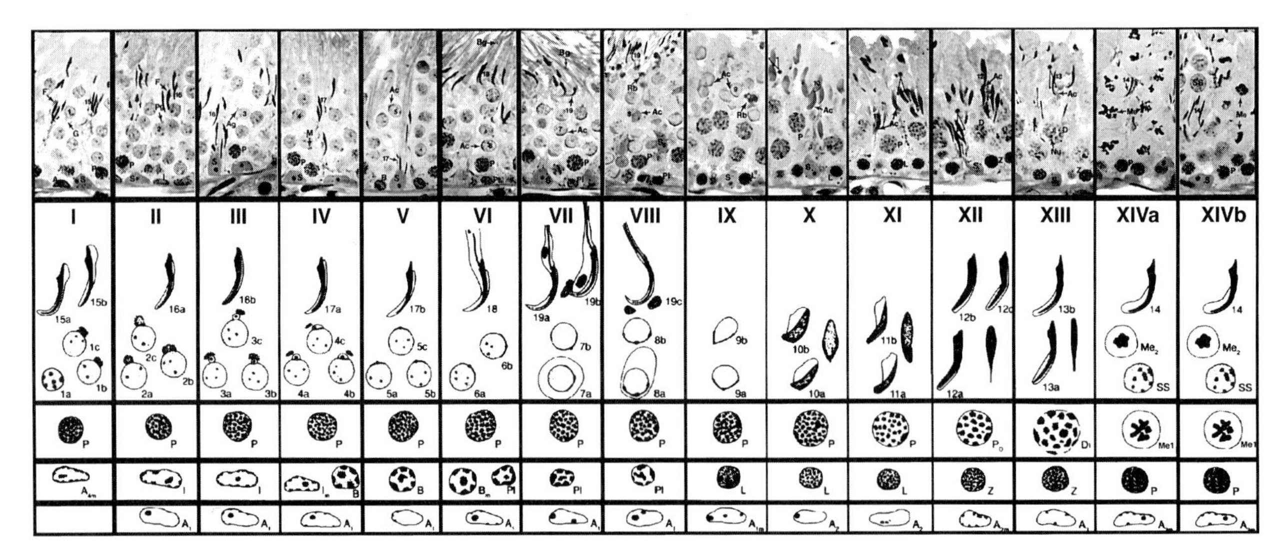

FIGURE 7 A staging map of rat spermatogenesis with actual photos of individual stages (top). The staging map contains illustrations that emphasize the nucleus of all cell types in the cycle of the seminiferous epithelium. Steps of spermiogenesis are split into intermediate steps to demonstrate variations in the morphology within a single stage. Spermatogonia (A1–4, I, B); spermatocytes (Pl, preleptotene; L, leptotene; Z, zygotene; P, pachytene; P_D, diplotene; Di, diakinesis; Me1, meiosis I; Me2, meiosis II; ss, secondary spermatocyte); spermatids (1–19). S, Sertoli cell; F, acrosomal flag; G, Golgi; M, acrosomal margin; Ac, acrosomal system; Bg, basophilic granule; Rb, residual body; Nu, nucleus.

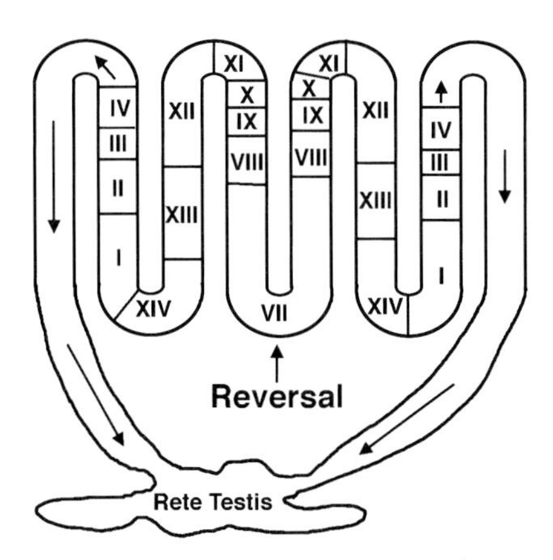

FIGURE 8 The wave of spermatogenesis in the seminiferous epithelium is illustrated with the sequential order of stages, increasing from the reversal site toward the rete testis (arrows).

clonal units of germ cells through a mechanism of biological signaling that is unknown.

Bibliography

Clermont, Y. (1972). Kinetics of spermatogenesis in mammals seminiferous epithelium cycle and spermatogonial renewal. *Physiol. Rev.* **52**, 198–236.

Clermont, Y., and Leblond, C. (1953). Renewal of spermatogonia in the rat. *Am. J. Anat.* **93**, 475–501.

Clermont, Y., and Leblond, C. P. (1959). Differentiation and renewal of spermatogonia in the monkey, *Macacus rhesus*. *Am. J. Anat.* **104**, 237–273.

Courot, M., Hochereau-de Reviers, M., and Ortavant, R. (1970). Spermatogenesis. *Testis* **1**, 339–432.

Dym, M., and Fawcett, D. W. (1970). The blood–testis barrier in the rat and the physiological compartmentation of the seminiferous epithelium. *Biol. Reprod.* **3**, 308–326.

Dym, M., and Fawcett, D. W. (1971). Further observations on the numbers of spermatogonia, spermatocytes, and spermatids connected by intercellular bridges in the mammalian testis. *Biol. Reprod.* **4**, 195–215.

Hess, R. (1990). Quantitative and qualitative characteristics of the stages and transitions in the cycle of the rat seminiferous epithelium: Light microscopic observations of perfusion-fixed and plastic-embedded testes. *Biol. Reprod.* **43**, 525–542.

Leblond, C., and Clermont, Y. (1952a). Spermiogenesis of rat, mouse, hamster and guinea pig as revealed by the "periodic acid-sulfurous acid" technique. *Am. J. Anat.* **90**, 167–215.

Leblond, C., and Clermont, Y. (1952b). Definition of the stages of the cycle of the seminiferous epithelium in the rat. *Ann. N. Y. Acad. Sci.* **55**, 548–573.

Oakberg, E. F. (1971). Spermatogonial stem-cell renewal in the mouse. *Anat. Rec.* **169**, 515–531.

Perey, B., Clermont, Y., and Leblond, C. (1961). The wave of the seminiferous epithelium in the rat. *Am. J. Anat.* **108**, 47–77.

Roosen-Runge, E. C. (1962) The process of spermatogenesis in mammals. *Biol. Rev.* **37**, 343–377.

Russell, L., Ettlin, R., Sinha Hikim, A., and Clegg, E. (1990). *Histological and Histopathological Evaluation of the Testis.* Cache River Press, Clearwater, FL.

Setchell, B. P. (1978). *The Mammalian Testis.* Cornell Univ. Press, Ithaca, NY.

Sharpe, R. (1994). Regulation of spermatogenesis. In *The Physiology of Reproduction* (E. Knobil and J. D. Neill, Eds.), Vol. 2, pp. 1363–1434. Raven Press, New York.

Spermatogenesis, Disorders of

Thorsten Diemer and Claude Desjardins
University of Illinois at Chicago College of Medicine

GLOSSARY

chromosomal aberrations Structural anomalies in either autosomes or sex chromosomes.

Intracytoplasmic sperm injection (ICSI) An assisted reproductive technology involving the intracytoplasmatic injection of an ovum with a spermatozoon.

microdeletions Error in meiosis, detected by recombinant DNA techniques, resulting in the loss of a gene or group of genes.

occupational toxins Chemical substances, present in the workplace, that act on one or more types of cells within the testis to interfere with spermatogenesis or steroidogenesis or both.

orchitis Inflammation of the testis.

spermatogenesis Maturation of the germ cell line beginning with the differentiation of spermatogonia and ending with the release of elongated spermatids from the seminiferous epithelium.

spermatogenic arrest A disorder in germ cell development in which spermatogenesis is halted at one or more steps prior to the release of elongated spermatids from the seminiferous epithelium.

testicular damage Palpable reduction in testicular mass occasioned by a disturbance in spermatogenesis or steroidogenesis or both.

testicular torsion Spontaneous rotation of the spermatic cord causing occlusion of arterial blood supply with ischemia and severe pain.

testosterone Primary sex hormone produced by Leydig cells present in the testis.

varicocele Distension of the spermatic veins or pampiniform plexus; frequently left sided but affecting spermatogenesis in both testes and infertility.

Disorders in spermatogenesis are caused by genetic or somatic factors. Genetic factors manifest themselves at fertilization and are evident early in fetal life with the onset of gonadal development. Affected patients produce few, if any, sperm as adults and spermatogenesis is arrested prior to the onset of spermatogonial meiotic activity. Somatic factors affect spermatogenesis during pre- and postpubertal life due to some secondary disorder leading to the production of reduced or limited numbers of spermatozoa and infertility. In this article, the genetic basis of disorders in spermatogenesis is discussed and we consider somatic factors affecting the proliferation of germ cells due to endocrine failure, exposure to infectious agents, or toxins found in the workplace. This article is limited to a consideration of disorders seen in human spermatogenesis except where supplemental information is required from animal models.

I. DEVELOPMENTAL DISORDERS OF SPERMATOGENESIS

A. Numerical Aberrations in Chromosomes

1. Klinefelter's Syndrome (47,XXY)

Klinefelter's syndrome is the most frequent chromosomal defect causing hypogonadism and infertility in men. Patients with Klinefelter's syndrome have an accessory X chromosome caused by meiotic nondisjunction in parental germ cells; about two-thirds of meiotic nondisjunctions are of maternal origin and about one-third are paternal. The typical karyotype is 47,XXY, but chromosomal mosaics with 46,XY/47,XXY and complements with multiple X chromosomes such as 48,XXXY are known.

Among patients with the typical form of Klinefelter's syndrome, testicular volume is 80–90% below normal and ejaculates contain few or no spermatozoa. Biopsies confirm that seminiferous tubules are fibrotic and hyalinized with infrequent spermatogonia. Spermatogonia typically fail to differentiate beyond the primary spermatocyte stage of spermatogenesis. Spermatozoa have been isolated in exceptional cases and processed for fertilization using ICSI.

Klinefelter's patients with a mosaic karyotype of 46,XY/47,XXY are known to induce pregnancies. Spermatogonia containing either a 46,XY or 47,XXY complement of chromosomes undergo meiosis to form spermatozoa with a normal (23,X or 23,Y) or abnormal (24,XY) karyotype. Recent detection of such abnormal sperm karyotypes, using recombinant DNA technology, underscores the genetic and ethical dilemmas of using spermatozoa from Klinefelter's patients in assisted reproductive technologies.

2. *XYY Syndrome (47,XYY)*

Paternal nondisjunction during meiosis can give rise to a double Y chromosome or 47,XYY. Spermatogenesis is normal among most XYY men since the extra Y chromosome is eliminated during meiosis, but the production of spermatozoa with disomic and hyperhaploid karyotypes, such as 24,XY or 24,YY, is significantly greater than that observed in individuals with the normal complement of 46 chromosomes.

3. *XX Male Syndrome (46,XX)*

Partial translocation of genes on the Y chromosome during paternal meiosis, particularly the SRY gene on Yp (Fig. 1), results in phenotypic males with a complement of female chromosomes (46,XX). Approximately 2% of the men seen at infertility clinics are reported with this disorder. The translocated SRY gene (or testis determining factor) is found on the X chromosome following meiotic error. SRY encodes for a nucleotide-binding protein regulating transcription. Testicular volume is reduced, the seminiferous epithelium is devoid of elongated spermatids, and most ejaculates are azoospermic. Patients have fully differentiated secondary sexual features of males, but gynecomastia is common, with body

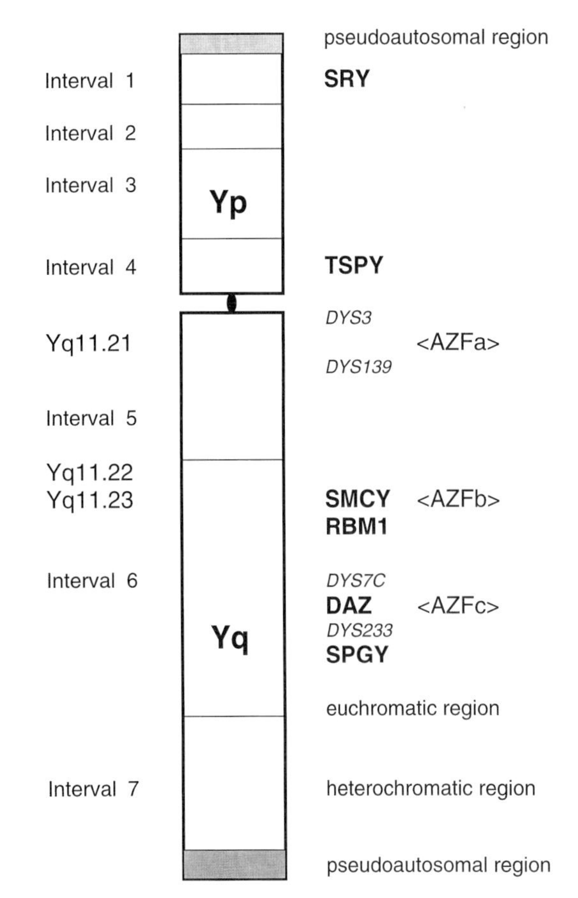

FIGURE 1 Diagrammatic representation of the Y chromosome showing interval numbers based on banding patterns of the Y chromosome (left) and genes involved in spermatogenesis (right). Interval dimensions of the Y chromosome are not drawn to scale to facilitate illustration. Genes known to affect spermatogenesis are designated in bold type; other loci, purported to control spermatogenesis, are listed in brackets, and genes used as markers for specific regions of the Y chromosome are shown in small italicized type. Specific regions enclosed in brackets may be similar to those genes shown in bold type in regions where they appear adjacent to each other. New information on genes mapping to the Y chromosomes can be obtained at the website *http://www.nhgri.nih.gov./Data.*, maintained by the National Human Genome Research Institute.

height and weight below normal and typical of females.

4. *Mixed Gonadal Dysgenesis (Mosaic 45,X0/46,XY)*

Mixed gonadal dysgenesis occurs in patients with a mosaic arrangement of chromosomes such as 45,X0/46,XY. Gonadal and phenotypic sexual features are typically female, but abdominal testicles are present and may occur unilaterally. In exceptional cases males have been described with intraabdominal testes, striking reductions in spermatogenesis, and azoospermia with primary infertility. The endocrine support for spermatogenesis is insufficient and testosterone replacement therapy is required for the completion of secondary sexual differentiation.

B. Structural Aberrations in Chromosomes

1. *Autosomes*

Spermatogenic disorders cannot be predicted on the basis of microscopic anomalies in chromosomal appearance since equivalent chromosomal translocations may be expressed differently in individuals. Chromosomal analyses of infertile men indicate that autosomal translocations occur about 10 times more frequently than in men with normal fertility. Pericentric inversions on chromosome 1, 3, 5, 6, 9, or 10 are typical among infertile men. Hundreds of genes, either on autosomes or sex chromosomes, are thought to be involved in one or more aspects of male reproduction, suggesting that disorders involving a single gene will be difficult to identify since countless possibilities exist.

Unbalanced chromosomal translocations are associated with spermatogenic arrest and defects in several other tissues due to the loss of significant genetic information. Such individuals, however, are rare among infertile men seen at infertility clinics.

i. Robertsonian Translocations Spermatogenesis may be reduced as a result of translocations between two acrocentric chromosomes when centromeres fuse. The impact of such translocations on spermatogenesis is unpredictable since defects in spermatogenesis range from mild to severe losses in germ cell development.

ii. Peri- and Paracentric Inversions Peri and paracentric inversions on chromosomes are associated with reduced spermatogenesis, but the degree of impairment is inconsistent, ranging from severe defects to no clinical change.

iii. Campomelic Dysplasia Campomelic dysplasia is caused by the deletion of a single gene on chromosome 17q24–q25 in individuals with a 46,XY karyotype. Affected patients typically differentiate as females despite a male complement of chromosomes. Spermatogenic arrest is complete in all such individuals even when testicular anlagen are evident.

iv. Autosomal Translocations Over 30 chromosomal translocations are associated with disorders in spermatogenesis and more can be expected to be discovered (Table 1). Certain translocations are limited to a small number of patients or in rare cases involve a single individual. Nearly all detected translocations involve autosomes. Despite exact knowledge of the gene loci involved in a translocation, neither the mechanism nor the cellular lesion has been established for any autosomal translocation affecting spermatogenesis. Understanding the contribution of the affected genes to spermatogenesis can be expected to offer insights about the cellular mechanisms directing the development of germ cells. Alternatively, certain translocations may operate indirectly and exert a secondary influence on spermatogenesis. Chromosomal translocations of this sort may be transferred to subsequent generations when fertilization relies on assisted reproductive protocols.

v. Noonan's Syndrome Noonan's syndrome is a congenital disorder in which half of all cases are due to autosomal-dominant inheritance, whereas the remainder are unexplained and occur spontaneously. The incidence is estimated at 1:1000 to 1:5000 live births. Hallmarks of the disorder include short stature, stenosis of the pulmonary valve, chest deformities, and a facial phenotype with posteriorly rotated ears. About 60% of affected males are bilaterally cryp-

TABLE 1
Summary of Translocations Occurring on or between Somatic (Autosomal) and Sex Chromosomes Reported to Impair Spermatogenesis and Fertility

Translocations on somatic (autosomal) chromosomes	
T (1;9) (q32.1; q34.3)	T (9;12) (q22; q22)
T (1;16) (q22; p12)	T (9;13) (q22; q32)
T (1;16) (q23; p13)	T (9;15) (q22; q15)
T (1;18) (p31.2; q23)	T (9;17) (q11.3; q11.3)
T (1;19) (p13; q13), 1qh+	T (9;17) (q11; q11)
	T (9;20) (q34; q11)
T (2;4) (q23; q27)	
T (2;8) (q23; p21)	T (11;15) (q25-qter; pter)
T (3;4) (p25; q21.3)	T (13;14)
T (3;5) (q29; q14) (p12; q34)	
T (3;14) (q27; q11)	T (14;21) (q13; p13)
T (3;20) (q11; p13)	T (14;22) (p11; q11.1)
T (4;6) (q26; q27)	T (17;21) (p13; q11)
T (4; 13) (q11; q12.3)	T (19;22) (p13.1; q11.1)
Translocations between somatic and sex chromosomes	
T (13q;15q)/ 46, XY	T (Y;11) (q11.2; q24)
T (Y;1) (q12; p34.3)	
T (Y;3) (q12; p21)	
Translocations between sex chromosomes	
T (X;Y)	T (X;Y) (p22.3; q11)

torchid and infertile due to spermatogenic arrest caused by elevated testicular temperature. The gene defect has been localized to chromosome 12q22-qter between D12S84 and D12S366 for the autosomal form of the disorder. Additional gene loci are proposed, including NF1 (neurofibromatosis gene) on chromosome 17 and locus 22q12. No specific disorder in spermatogenesis has been recognized from the set of gene defects seen in Noonan's syndrome since interindividual variation in germ cell differentiation is common.

2. *Sex Chromosomes*

i. X Chromosome Aberrations Major deletions in the X chromosome are incompatible with the development of a male fetus and cause severe defects in multiple organ systems similar to the fragile X syndrome. Functional deletions of the X chromosome have modest consequences, including mild to severe disorders in spermatogenesis.

Xp22-pter Microdeletion (Xp22 Contiguous Gene Syndrome) The Xp22 contiguous gene syndrome is associated with multiple defects, including mental retardation and Kallmann's syndrome due to the loss of the KALIG-1 gene. The impact of Kallmann's syndrome on spermatogenesis is variable and operates secondarily via the failure of hypothalamic neurons to secrete gonadotropin-releasing hormone (GnRH) and ultimately required for the secretion of pituitary gonadotropins supporting the development of germ cells.

Translocations of X Chromosome Genes to Autosomes Severe failure in spermatogenesis is evident when X chromosome genes are translocated to autosomes. Affected patients are either oligozoospermic or azoospermic. The mechanism responsible for spermatogenic arrest is unknown but germ cells do not appear to complete meiosis as evidenced by the absence of pachytene spermatocytes within the seminiferous epithelium.

ii. Y Chromosome Aberrations **Deletion of the SRY Gene** Complete deletion of the SRY gene or loss of selected portions of the Yp interval (Fig. 1) interferes with the regression of the Müllerian ducts and the subsequent differentiation of the testis. Patients develop as phenotypic females with gonadal dysgenesis and a 46,XY karyotype like that of normal healthy men.

The SRY gene, located in the upper part of Yp (Fig. 1), appears to function as a single locus directing male sexual differentiation. The gene encodes a DNA-binding protein transcribed in testicular germ cells. Patients with a 46,XY complement of chromosomes and a normal SRY region may also lack testes, suggesting that genes from other chromosomes are required for male sexual differentiation, including normal spermatogenesis.

A gene family important for spermatogenesis, TSPY, is located within Yp and is selectively transcribed in spermatogonia and spermatocytes (Fig. 1). TSPY appears to function in spermatid differentiation but this gene family may control other aspects of spermatogenesis. The impact of mutations in TSPY is unknown and will require more systematic investigation of patients with spermatogenic disorders.

Deletions of Yq11 Diagnosis of major deletions in chromosomes has relied on karyotyping, whereas microdeletions can be detected only by molecular techniques. Major deletions in the euchromatic part of the Y chromosome at Yq11 (Fig. 1) are known to interfere with spermatogenesis as revealed by severe oligozoospermia or azoospermia. Deletions in Yq indicate that a group of related genes in the region around Yq11.21–23 are required for normal spermatogenesis.

Microdeletions in the Yq11 (Azoospermia Factor) Region Microdeletions of the Y chromosome account for up to 30% of all cases of testicular azoospermia, but the reason for reduced spermatogenesis in the remaining 70% of azoospermic men is still unknown. The possibility that infertility may be inherited in the future generations must be considered since ICSI and other modern reproductive technologies make it possible to inherit infertility occasioned by microdeletions on Yq11 linked to infertility. Genes in the azoospermia factor (AZF) region of the Y chromosome (Fig. 1) are critical to spermatogenesis since microdeletions cause infertility without other apparent defects in the male phenotype. Three to five AZF loci have been identified within intervals 5 and 6 (Fig. 1), suggesting a family of genes directs germ cell proliferation.

Microdeletions in Yq11.21–Yq11.23 mapping to the AZF region are known to cause reduced spermatogenesis ranging from the total absence of all classes of germ cells to reduced and incomplete germ cell maturation with the production of few mature spermatozoa. Microdeletions in the AZF region are absent from the paternal genome and appear to occur as *de novo* mutations causing infertility. Additional studies will be required to establish the precise number and location of genes responsible for AZF and their potential interaction within and among gene families located in intervals 5 and 6 (Fig. 1). The potential for discovery of important new information about the genetic basis for infertility is great since multiple genes within intervals 5 and 6 have been suspected to encode RNA-binding proteins required for germ cell development. Microdeletions have been discovered at the start of Yq11.21 (interval 5) distal of DYS3 associated with severe disorders in spermatogenesis typified by the Sertoli cell-only syndrome. The region identified as AZFa (Fig. 1) appears to contain other undetected genes crucial for spermatogenesis.

The RBM gene family (RBM1 and RBM2) is located at several sites on the Y chromosome due to multiple repeats and maps to AZFb region (Fig. 1). A single gene microdeletion in RBM1 produces a profound disturbance in spermatogenesis resulting in the loss of all classes of germ cells. RBM1 encodes for an RNA-binding protein expressed in fetal germ cells and in spermatogonia and spermatocytes of the adult testis. RBM2 is polymorhic and absent from a group of 20 fertile Japanese men suggesting that this gene may not be essential for spermatogenesis.

RBM1 appears to be highly conserved as evidenced by its presence on the Y chromosome in mammals studied so far. In contrast, other Y chromosome genes, such as DAZ and SPGY, mapping to the AZF region (Fig. 1) are typical of subhuman primates but absent in other mammals and birds. Ancestors of the DAZ gene appear to have migrated to the Y chromosome of primates after amplification from certain somatic chromosomes.

The DAZ gene (deleted in azoospermia), located within interval 6 between DYS7C and DYS233 in the AZFc region (Fig. 1), was discovered among a cohort of azoospermic or severe oligozoospermic men. Deletion of the DAZ gene is associated with azoospermia in all patients studied so far. The gene product of DAZ, an RNA-binding protein, was found to be similar to other RNA-binding proteins in humans. Testicular biopsies reveal that a Sertoli cell-only syndrome (SCO) is common but spermatogenesis may proceed to condensed spermatids in some instances.

The SMCY gene is located proximal to RBM1 and DAZ in the AZFb region (Fig. 1). Deletions of SMCY cause severe defects in spermatogenesis similar to those observed with deletions in RBM1 or DAZ. The precise molecular function of SMCY in spermatogenesis is unknown, but it is known to encode the H-Y antigen, a protein regulating gene expression in several tissues including the testis.

The SPGY (spermatogenesis gene on the Y) gene, located near DAZ in the AZFc region (Fig. 1), is required for spermatogenesis. Deletions of the SPGY gene cause defects in spermatogenesis comparable to those discovered for DAZ gene deletion. SPGY contains repeated 72-base pair exons and shares remarkable sequence homology with the DAZ gene but appears to function as an independent gene.

Translocations Involving the Y Chromosome Reciprocal translocations between the Y and other chromosomes or between the Y and the X chromosomes cause multiple disorders in spermatogenesis (Table 1). The defects in spermatogenesis are not characterized so cellular lesions remain unknown.

C. Spermatogenic Disorders of Unknown Chromosomal Location

1. *Kartagener's Syndrome and Axoneme Defects*

Infertility, bronchiectasis, and situs inversus are the hallmarks of Kartagener's syndrome, an autosomal-dominant defect of unknown chromosomal origin. Ejaculates from patients with Kartagener's syndrome contain normally appearing spermatozoa that are immotile. Spermatozoa lack dynein, the protein connecting microtubules within the axoneme. Cilia of the respiratory tract also lack dynein and are immotile, causing recurrent infections and bronchiectasis.

A second axoneme defect of unknown genetic origin involves the central doublet of microtubules in the sperm's cilia. This defect, known as the 9+0 syndrome, is always associated with the production of immotile spermatozoa and infertility.

2. *Globozoospermia*

Globozoospermia refers to a special condition in which acrosomal caps are absent from spermatozoa. Spermatozoa fail to bind to the zona pellucida to initiate fertilization. Globozoospermia is due to an undefined failure of the Golgi apparatus to direct acrosomal formation. Normal ejaculates may contain a small fraction of spermatozoa without an acrosome but the consistent absence of this organelle from spermatozoa typifies globozoospermia.

II. ENDOCRINE DISORDERS

The differentiation of the male phenotype including the seminiferous epithelium requires testosterone and luteinizing hormone (LH) from the pituitary gland. Male sexual differentiation, including defects such as micropenis and hypospadias, occurs when androgen is insufficient or absent during organogenesis. Patients with a 46,XY karyotype are genetic but not phenotypic males as evidenced by the failure of secondary sexual tissues to differentiate according to a male pattern. Without testes seminiferous tubules fail to differentiate and spermatogenesis is not induced. Defects in spermatogenesis can occur when GnRH is lacking or when either follicle-stimulating hormone (FSH) or LH is insufficient resulting in the absence of testosterone. Endocrine disorders associated with a failure in spermatogenesis are outlined here since they account for disorders in spermatogenic failure seen during puberty or in adults.

A. Primary Endocrine Disorders

1. *Kallmann's Syndrome*

Kallmann's syndrome is caused by the absence or a mutation in the KALIG-1 (KAL-1) gene located on the X chromosome, a defect occurring in 1:7000–8000 births. KALIG-1, a large gene that may be transcribed incompletely, is critical to spermatogenesis since either the loss of the gene or a defect in gene function drastically limits the migration of GnRH neurons to the hypothalamus during the fetal brain development. Differentiation of the male phenotype is normal but infantile since pituitary gonadotropic hormones required for the development of the seminiferous epithelium during puberty are insufficient or absent. Spermatogenesis is absent or reduced due to the almost total lack of FSH and testosterone stimulation. Patients are usually azoospermic but spermatogenesis can be initiated with hormone replacement involving either GnRH or gonadotropins.

2. *Prader–Labhardt–Willi Syndrome*

The Prader–Labhardt–Willi syndrome is due to a deletion or uniparental disomy of chromosome 15q11–13, a defect of paternal origin occurring in 1:15,000 births. An unidentified error in the imprinting center (IC) on chromosome 15q11–13 impairs the regulation and expression of genes in the vicinity of paternal alleles. Multiple genes are unexpressed in patients with the Prader–Labhardt–Willi syndrome, including ZNF127, SNRPN, IPW, PAR-1, and PAR-5. Hypogonadotropic hypogonadism with decreased levels of FSH, LH, and testosterone is evident after failure in one or more of these genes. The primary disorder has been localized to the hypothalamus since replacement of GnRH evokes FSH and LH release with the subsequent induction of spermatogenesis. Short stature, mental retardation, obesity, and cryptorchidism are typical of patients with this syndrome. Cryptorchidism is frequent and testicular biopsies reveal that the seminiferous epithelium is atrophic and disorganized. Leydig cells are undifferentiated and spermatogenesis is arrested at the level of late spermatids. The seminiferous epithelium can be stimulated to produce spermatozoa with appropriate hormonal therapy.

3. *Congenital Adrenal Hypoplasia*

Congenital adrenal hypoplasia occurs in 1:12,500 births, with nearly 70% of patients expressing a mutation in the DAX-1 gene on the X chromosome, Xp21. Congenital adrenal hypoplasia coincides with hypogonadotropic hypogonadism since steroidogenesis is impaired in both the testis and the adrenal gland. Hormone therapy involving glucocorticoids, mineralocorticoids, and testosterone is required to replace hormones made by the adrenal cortex and Leydig cells. Spermatogenesis should respond to hormone replacement but this has not been tested systematically.

4. *Laurence–Moon Syndrome and Bardet–Biedel Syndrome*

Both syndromes are rare and complex with multiple defects, including mental retardation, obesity, neurological problems, and retinal dystrophy. Hypogonadism may occur but is inconsistent and its origin is unknown. Both steroidogenesis and spermatogenesis can be impaired due to the failure in pituitary gonadotropin secretion. Testicular biopsies from patients with the Laurence–Moon and the Bardet–Biedel syndromes contain elongated spermatids, but the number of germ cells that complete spermatogenesis is frequently below normal.

5. *Congenital Hypopituitarism*

Mutations in the transcription factor, PIT-1, cause congenital hypopituitarism with profound structural and functional alterations of the pituitary gland. Aplasia is typical and fusion defects between the anterior and posterior lobes are common. The range of defects is broad, indicating that multiple genes may have a role in explaining this rare disorder.

6. *Isolated Deficiency of FSH and LH*

Isolated FSH deficiency is a rare condition in which LH and testosterone levels remain in the normal range. FSH can be stimulated by repeated, pulsatile treatment with GnRH, implying the defect is probably localized to hypothalamic GnRH neurons. Neither the mechanism nor the genetic origin of the disorder are known. Male differentiation is complete but defects in spermatogenesis can range from complete arrest to a moderate reduction in the number of ejaculated spermatozoa. When spermatogenesis is arrested only Sertoli cells may remain. In other forms of the disorder, spermatogenic arrest may occur at the level of spermatids without the formation of spermatozoa.

Isolated LH deficiency, known as the Pasqualini or fertile eunuch syndrome, is due to insufficient secretion of LH. Neither the genetic nor the molecular basis for the selective insufficient stimulation of LH release in the pituitary gland are known. The syndrome is associated with hypogonadism, subnormal levels of LH and testosterone, Leydig cell hypoplasia, and normal concentrations of FSH. Spermatogenesis is reduced, but all classes of germ cells are present, including spermatozoa in the ejaculate. Secretion of LH can be restored to normal levels by substitution of GnRH, suggesting the disorder is of hypothalamic origin.

B. Genetic Disorders of Testosterone Synthesis and Androgen Action

1. *Defects in 17,20 Desmolase and 17β-Hydroxysteroid Dehydrogenase*

The cytochrome P 450 enzymes, 17,20 desmolase and 17β-hydroxysteroid dehydrogenase, are required for the biosynthesis of testosterone. Autosomal recessive mutations result in a range of enzymatic defects involving the complete lack of testosterone to varying degrees of testosterone insufficiency. Disturbances in male sexual differentiation are common and include pseudohermaphroditism in which patients manifest a complete female phenotype. Wide deviations in secondary sexual differentiation are possible and related to the extent of the enzyme defect and the level of testosterone insufficiency. Testes remain in the intraabdominal or inguinal position and germ cell development fails to proceed in the absence of testosterone or at elevated intraabdominal temperature. Testicles should be removed or surgically relocated to the scrotum to reduce the risk of testicular cancer. Spermatogenesis may be induced with testosterone substitution therapy in the fraction of patients with mild to modest loss of enzyme activity.

2. *Disorders of Androgen Action at Target Organs*

Androgens fail to exert their effect on target organs when the intracellular androgen receptor is absent or impaired. Mutations in the androgen receptor have been localized to the X chromosome, Xq11–12, and are estimated to occur at 1:50,000 births. Testicular feminization occurs when the androgen receptor is absent or defective. Patients with a normal male karyotype of 46,XY may be phenotypic females with inguinal testicles containing only Sertoli cells. Spermatogenesis is also absent in other forms of testicular feminization with moderate resistance to testosterone. In the Reifenstein's syndrome androgen resistance is moderate and spermatogenesis is impaired.

In the undervirilized fertile male syndrome, the mildest form of testosterone insensitivity, spermatogenesis may be reduced but fertility is often unaffected. Patients are not well characterized due to normal appearance and phenotype.

C. Secondary Endocrine Disorders

Tumors, traumas, or systemic diseases may affect the discharge of GnRH from hypothalamic neurons. Interference with GnRH secretion alters the discharge of FSH and LH from the anterior pituitary and ultimately interferes with spermatogenesis.

III. TESTICULAR DISORDERS

A. Orchitis and Epididymoorchitis

1. *Viral-Induced Orchitis*

Several viruses. including the mumps, varicella zoster, herpes, and cytomegaly, may localize within the testis to induce primary orchitis. Orchitis can also occur after exposure to the group of viruses causing hemorrhagic fevers, including Coxackie, Marburg, Dengue, lymphocytic choriomeningitis, arbo, and other viruses, but only as a secondary consequence of systemic infection.

Mumps is the most important virus that acts directly on the testis since orchitis develops in 15–35% of infected males after puberty. Spermatogenesis may be permanently arrested after recovery from the infection due to tubular sclerosis and testicular atrophy. Milder forms of the disease most commonly lead to complete restoration of spermatogenesis with normal numbers of spermatozoa in ejaculates.

Severe disorders in spermatogenesis are evident in HIV-infected patients at the late stages of the disease. Proviral DNA is localized in all classes of germ cells in patients with HIV infection. Spermatogenesis is arrested at the spermatocyte or spermatid stage of germ cell maturation during the AIDS stage of the illness or in testes sampled after death from AIDS.

2. *Bacterial Orchitis and Epididymoorchitis*

Bacterial orchitis and epididymoorchitis are induced by certain enterobacteria, chlamydia, mycoplasms, gonococci, and other bacteria. The epididymis is the primary site of infection but the testis is severely affected if antibiotic treatment is delayed. Persistent epididymoorchitis may lead to fibrosis and hyalinization of seminiferous tubules and eventually cause spermatogenic arrest. Bacterial infections may have a long-lasting impact on spermatogenesis depending on the severity of testicular damage. Bacterial infections of hematogeneous origin such as tuberculosis induce necrosis and calcification of the testicular tissue. Damage to the seminiferous epithelium is not permanent but chronic infectious disease is a common cause of spermatogenic arrest.

B. Varicocele

Varicocele is associated with a tortuous and enlarged venous network of the pampiniform plexus of the spermatic cord usually occurring on the left side. The cause of the disorder is unclear. The severity and extent of venous enlargement differ among individuals as indicated by the array of grades used to estimate the clinical magnitudes of the disorder. Testicular volume and alterations in spermatogenesis leading to the oligoasthenoteratozoospermia syndrome are evident in some patients. Spermatogenic arrest occurs during spermiogenesis and can be reversed in up to 70% of patients by ligation of the spermatic veins at the inguinal level.

C. Testicular Torsion

Complete torsion of the spermatic cord interrupts blood flow to the testis, causing testicular damage primarily in spermatogenesis. Untreated torsion, for 6 hr or more, causes permanent damage to the seminiferous epithelium, and the function of the contralateral testis is compromised as well. Impairment in spermatogenesis depends on the severity and duration of ischemia.

IV. DRUG-INDUCED DISORDERS

Numerous drugs may interfere with spermatogenesis by acting directly on the testis or via the hypothalamus–pituitary gland–testis axis. Drugs that impair spermatogenesis by affecting the endocrine support of spermatogenesis should be avoided at all ages due to the risk of permanent hypogonadism and impotency.

A. Gonadotoxic Drugs

Ketokonazole, an antifungal drug interfering with P450 enzymes of steroidogenesis, has a dose-dependent toxicity on spermatogenesis. Severe oligozoospermia and disturbed spermatogenesis have been observed following chronic treatment with the following drugs: valproic acid, an anticonvulscent; sulfosalazine, used to treat ulcerative colitis; spironolactone, a competitive inhibitor of aldosterone; and allopurinol, a hypoxanthin oxidase blocker.

Among antimicrobial agents, nitrofurane, minocycline, and related drugs affect spermatogenesis by interfering with the structure and motility of spermatozoa. Both radio- and chemotherapeutic protocols used to treat tumors affect spermatogenesis directly. Prolonged exposure to both radiotherapy and chemotherapy can destroy stem cell spermatogonia and result in permanent loss of germ cells from the seminiferous epithelium. Cisplatin, carboplatin, procarbazin, and triethylenemelamine affect spermatogenesis depending on the cumulative dose. Spermatogenic damage may be reversible when spermatogenesis is halted at the level of spermatogonia and spermatocytes. Colchicine also affects spermatogenesis but this effect is reversible after the therapy, which is usually short term.

B. Drugs Affecting Sperm Function

Spontaneous acrosome loss and a reduction in mannose-binding receptors on the sperm surface have been reported among patients using calcium channel blockers for cardiac disorders. Both effects interfere with the fertilizing ability of spermatozoa. Calcium channel blockers act directly on the sperm cell membrane. Cimetidine, reserpine, and phenothiazine all inhibit gonadotropin secretion leading to Leydig cell atrophy and reduction in testosterone production. Cyclosporine A acts via hypothalamic GnRH neurons and via a direct effect on Leydig cells. Drug effects are reversible and testosterone secretion returns to normal levels once therapy is terminated.

C. Drugs Affecting the Sexual Hormones

Abusive use of androgens for muscular hypertrophy suppresses the secretion of pituitary gonadotropins causing testicular atrophy and spermatogenic arrest. The inhibitory effects on spermatogenesis may be irreversible after long-term application of external androgens including Leydig cell insufficiency. Similar effects are evident after use of estrogens and androgen receptor antagonists.

V. ENVIRONMENTAL TOXINS

Spermatogenesis fails following exposure to heavy metals such as lead, manganese, cadmium, and mercury. Heavy metals can act directly on the vasculature of the testis causing severe defects in spermatogenesis. Certain chemical agents, such as 2-methoxyethanol and 2-ethoxyethanol, selectively affect spermatogenesis at the level of pachytene spermatocytes, resulting in a dose-dependent apoptotic arrest of spermatocytes. Pesticides including dibromochloropropane, dichlorodiphenyl trichloroethane, and other organophosphates reduce spermatogenesis quantitatively and exposed workers are frequently infertile. Testicular damage can be permanent among workers who have been exposed for years or more. Other chemicals suspected of interfering with spermatogenesis include organic solvents such as benzene, ethylene glycol, carbon disulfides, and ethylene dibromide.

See Also the Following Articles

Adrenal Hyperplasia, Congenital Virilizing; Infertility; Male Reproductive Disorders; Prader–Willi Syndrome; Spermatogenesis, Overview; *SRY* Gene

Bibliography

Chandley, A. C. (1994). Chromosomes. In *Male Infertility* (T. J. Hargreave, Ed.), 2nd ed. Springer-Verlag, New York.

Delbridge, M. L., Harry, J. L., Toder, R., Waugh O'Neill, R. J., Ma, K., Chandley, A. C., and Marshall Graves, J. A. (1997). A human candidate spermatogenesis gene, RBM 1, is conserved and amplified on the marsupial Y chromosome. *Nature Genet.* **15**, 131–136.

Elliott, D. J., and Cooke, H. J. (1997). The molecular genetics of male infertility. *Bioassays* **19**, 801–809.

Engel, W., Murphy, D., and Schmid, M. (1996). Are there genetic risks associated with microassisted reproduction? *Hum. Reprod.* **11**, 2359–2370.

Guichaoua, M. R., Speed, R. M., Luciani, J. M., Delafontaine, D., and Chandley A. C. (1992). Infertility in human males with autosomal translocations. *Cytogenet. Cell Genet.* **60**, 96–101.

Jaffe, T., and Oates, R. D. (1994). Genetic abnormalities and reproductive failure. *Urol. Clin. North Am.* **21**, 389–408.

Jamieson, C. R., van der Burgt, I., Brady, A. F., van Reen, M., Elsawi, M. M., Hol. F., Jefferey, S., Patton, M. A., and Mariman, E. (1994). Mapping a gene for Noonan syndrome to the long arm of chromosome 12. *Nature Genet.* **8**, 357–360.

Lipshultz, L. I., and Howards, S. S. (1997). *Infertility in the Male*, 3rd ed. Mosby-Year Book, St. Louis.

Liu, H. C., Tsai, T. C., Chang P. Y., and Shib, B. F. (1994). Varicella orchitis: Report of two cases and review of the literature. *Pediatr. Infect. Dis. J.* **13**, 748–750.

Ma, K., Inglis, J. D., Sharkey, A., Bickmore, W. A., Hill, R. E., Prosser, E. J., Speed, R. M., Thomson, E. J., Jobling, M., Taylor, K., Wolfe, J., Cooke, H. J., Hargreave, T. B., and Chandley, A. C. (1993). A Y chromosome gene family with RNA binding protein homology: Candidates for the azoospermia factor AZF controlling human spermatogenesis. *Cell* **75**, 1287–1295.

Manson, A. L. (1990). Mumps orchitis. *Urology* **36**, 355–358.

Maroulis G. B., Parlow, A. F., and Marshall, J. R. (1977). Isolated follicle-stimulated hormone deficiency in man. *Fertil. Steril.* **28**, 818–822.

Muciaccia, B., Uccini, S., Filippini, A., Ziparo, E., Paraire, F., Baroni, C. D., and Stefanini, M. (1998). Presence and cellular distribution of HIV in the testes of seropositive subjects: An evaluation by in situ PCR hybridisation. *FASEB J.* **12**, 151–163.

Nakahori, Y., Kuroki, Y., Komaki, R., Kondoh, N., Namiki, M., Iwamoto, T., Toda, T., and Kobayashi, K. (1996). The Y chromosome region essential for spermatogenesis. *Horm. Res.* **46**(Suppl.), 20–23.

Nieschlag, E., and Behre, H. (1997). *Andrology: Male Reproductive Health and Dysfunction*. Springer-Verlag, Berlin.

Nuovo, G. J., Becker, J., Simsir, A., Margiotta, M., Khalife, G., and Shevchuk, M. (1994). HIV-1 nucleid acids localize to the spermatogonia and their progeny. A study by polymerase chain reaction in situ hybridization. *Am. J. Pathol.* **144**, 1142–1148.

Page, D. C., Mosher, R., Simpson, E. M., Fisher, E. M. C., Mardon, G., Pollack, J., McGillivray, B., de la Chapelle, A., and Brown, L. G. (1987). The sex determining region of the human Y chromosome encodes a finger protein. *Cell* **51**, 1091–1104.

Reijo, R., Lee, T. Y., Salo, P., Alagappan, R., Brown, L. G., Rosenberg, M., Rozen, S., Jaffe, T., Straus, D., Hovatta, O., de la Chapelle, A., Silber, S., and Page, D. C. (1995). Diverse spermatogenic defects in humans caused by Y chromosome deletions encompassing a novel RNA-binding protein gene. *Nature Genet.* **10**, 383–393.

Saitoh, S., Buiting, K., Rogan, P. K., Buxton, J. L., Driscoll, D. J., Arnemann, J., Konig, R., Malcolm, S., Hortshemke, B., and Nicholls, R. D. (1996). Minimal definition of the imprinting center and fixation of chromosome 15q11–q13 epigenotype by imprinting mutations. *Proc. Natl. Acad. Sci. USA* **23**, 7811–7815.

Schill, W. B. (1991). Some disturbances of acrosomal development and function in human spermatozoa. *Hum. Reprod.* **6**, 969–978.

Sikka, S. C. (1997). Gonadotoxicity. In *Male Infertility and Sexual Dysfunction* (W. J. G. Hellstrom, Ed.). Springer-Verlag, Berlin

Vogt, P. H. (1997). Human Y-chromosome deletions in Yq11 and male infertility. *Adv. Exp. Med. Biol.* **424**, 17–30.

Whitaker, M. D., Scheithauer, B. W., Kovacs, K. T., Randall, R. V., Campbell, R. J., and Okazaki, H. (1987). The pituitary gland in the Laurence–Moon syndrome. *Mayo Clinic Proc.* **62**, 216–222.

Yanase, T., Takayanagi, R., Oba, K., Nishi, Y., Ohe, K., and Nawata, H. (1996). New mutations of DAX-1 genes in two Japanese patients with X-linked congenital adrenal hypoplasia and hypogonadotropic hypogonadism. *J. Clin. Endocrinol. Metab.* **81**, 530–535.

Spermatogenesis, Hormonal Control of

Barry R. Zirkin

Johns Hopkins University

GLOSSARY

gonadotropic hormones Follicle-stimulating hormone (FSH) and luteinizing hormone (LH) from the anterior pituitary.

gonadotropin-releasing hormone Hypothalamic decapeptide that stimulates the anterior pituitary to produce FSH and LH.

initiation of spermatogenesis The first complete round of spermatogenesis during the peripubertal period.

interstitial fluid Fluid surrounding Leydig cells in the interstitial compartment of the testis.

maintenance of spermatogenesis Prevention of the loss of germ cells once spermatogenesis has been fully established.

restoration of spermatogenesis Restoration of germ cells to the testis after the loss of germ cells resulting from experimental or natural perturbation of established spermatogenesis.

seminiferous tubule fluid Fluid surrounding germ cells within the seminiferous tubular compartment of the testis.

Spermatogenesis is a complex, cyclic process that involves a number of different cell types interacting over time and space. Spermatogenesis begins with spermatogonial proliferation which, in the rat, comprises a series of six mitotic divisions that ultimately produce progeny that enter meiosis as preleptotene spermatocytes. The two meiotic divisions produce haploid spermatids, and these differentiate during spermiogenesis to become mature spermatids. Any given cross section of seminiferous tubule will contain one or two generations of spermatogonia, one or two generations of spermatocytes, and one or two generations of spermatids. The various generations of germ cells form stages, defined as specific sets of cellular associations that follow each other over time in any given area of seminiferous tubule. Different regions of a given seminiferous tubule will contain different stages of the cycle, the net result of which is continuous and asynchronous production of sperm. Given this complexity, it is not surprising that our understanding of the consequences of hormone action on particular germ cells and cellular events is far from complete.

I. INTRODUCTION

The production of spermatozoa and testosterone, the primary functions of the adult mammalian testis, depends on stimulation of the testes by the gonadotropic hormones, follicle-stimulating hormone (FSH) and luteinizing hormone (LH). Both hormones are produced by the pituitary gland in response to gonadotropin-releasing hormone (GnRH) from the hypothalamus. In response to LH, testosterone is produced by the Leydig cells. It has been clear for decades that testosterone is a necessary prerequisite for the initiation of spermatogenesis in the peripubertal mammal, the maintenance of established spermatogenesis in the adult, and the restoration of spermatogenesis (and fertility) in adults induced experimentally to become oligospermic or azoospermic. It also is well established that FSH is involved in the peripubertal initiation of spermatogenesis. Whether or not FSH is also involved in regulating spermatogenesis in the adult mammal continues to be debated.

Although the effects of testosterone and FSH administration or withdrawal on spermatogenesis have been studied for years, the mechanisms by which they regulate spermatogenesis remain uncertain. Androgen and FSH receptors are localized to Sertoli cells, making it likely that the effects of both testosterone and FSH are mediated via these cells. Over the past 10 years, it has become well established that the testosterone concentration normally present within the testis is very high relative to the concentration that should be required to saturate androgen receptors, and that although intratesticular testosterone concentration can be reduced substantially with no effect on spermatogenesis, the required testosterone concentration is also very high. There is no good explanation for this observation. Recent studies have shown that increased germ cell apoptosis occurs when testosterone is withdrawn from the mammalian testis, pointing to testosterone as a cell survival factor. The molecular mechanisms by which testosterone acts in this way only recently have begun to be considered. FSH also may act to prevent cell death, though this is less certain.

This article focuses on the roles of testosterone and FSH in mammalian spermatogenesis. Discussion will center on the relationship between intratesticular testosterone concentration and spermatogenesis, controversy regarding the requirement for FSH, and the possible mechanisms by which testosterone and FSH act to regulate spermatogenesis.

II. EXPERIMENTAL APPROACHES

Studies of the effects of testosterone and/or FSH on spermatogenesis largely have relied on *in vivo* models in which LH (and therefore testosterone) and/or FSH have been reduced or suppressed. Withdrawal or suppression of gonadotropins can be accomplished by hypophysectomy, active immunization against GnRH or LH, GnRH agonist or antagonist administration, or the administration of exogenous androgens. Each of these approaches has experimental advantages and disadvantages. With hypophysectomy, both LH and FSH are withdrawn completely; but other pituitary hormones are withdrawn as well. Active immunization against GnRH results in undetectable LH and FSH, and active immunization against LH results in undetectable LH without effect on FSH. In both cases, there is reduced intratesticular testosterone. GnRH antagonist or agonist treatment also results in undetectable gonadotropins. With GnRH-withdrawn animals, spermatogenesis is reversibly impaired; cessation of treatment or the administration of testosterone can largely restore spermatogenesis. This approach would seem ideal for studying the effects of testosterone and FSH separately or together. However, there are some reports that testosterone administration to GnRH-immunized rats may result in increased FSH by unknown mechanisms, thus potentially confounding studies of the role of testosterone alone in maintaining or restoring spermatogenesis.

As previously indicated, testosterone is produced by Leydig cells in response to LH. Testosterone and its aromatized product, estradiol, feed back negatively at the level of the hypothalamus and pituitary to suppress LH, thereby transiently suppressing Leydig cell testosterone production. The administration of testosterone to rats or humans similarly can suppress LH and thereby suppress endogenous testosterone production by Leydig cells. Because FSH is relatively unaffected, this approach can be used to examine testosterone effects in the presence of FSH.

III. ARE BOTH TESTOSTERONE AND FSH REQUIRED FOR SPERMATOGENESIS?

A. Testosterone

In the rat, the concentration of testosterone in blood serum is approximately 2 ng/ml. Because Leydig cells are located in the interstitial compartment of the testis, it is not surprising that the concentrations of testosterone in the interstitial fluid (IF; 70 ng/ml) and seminiferous tubule fluid (STF; 50 ng/ml) of untreated rats are far greater than those in serum. Similar relationships between serum and intratesticular testosterone concentrations occur in the human. Interestingly, when testosterone is administered to rats in circumstances in which its local production by Leydig cells is suppressed, testosterone concentration does not equilibrate throughout the body, as might be expected; rather, a concentration gradient is established that is similar to the gradient that occurs normally, with testosterone concentration in IF > STF > serum. Thus, even when there is no local testosterone production, testosterone becomes concentrated within the testis. Whether or not administered testosterone also becomes concentrated in the human testis is not known.

The average testosterone concentration in rat STF (50 ng/ml; $1.7 \times 10-7$ M) is considerably higher than the K_D of testicular androgen receptors ($3 \times 10-9$ M). This suggests that the testosterone concentration in the seminiferous tubules may be considerably in excess of that required to maintain established spermatogenesis in the adult rat. This was found experimentally to be the case. Zirkin and colleagues examined the quantitative relationship between intratesticular testosterone concentration and sperm number per testis in adult rats that received testosterone of increasing doses via testosterone-filled Silastic capsules of increasing size. The administration of testosterone in this way suppressed Leydig cell testosterone production so that the only source of testosterone was from the capsules. As shown in Fig. 1, quantitatively complete spermatogenesis was maintained at a STF testosterone concentration of only 20 ng/ml, which represented a reduction of 60% from controls. Further reduction of testosterone concentration below 20 ng/ml resulted in graded reductions in sperm production, indicative of a dose–response relationship between STF testosterone concentration and sperm produced by the testis. Similarly, in adult rats rendered azoospermic with a

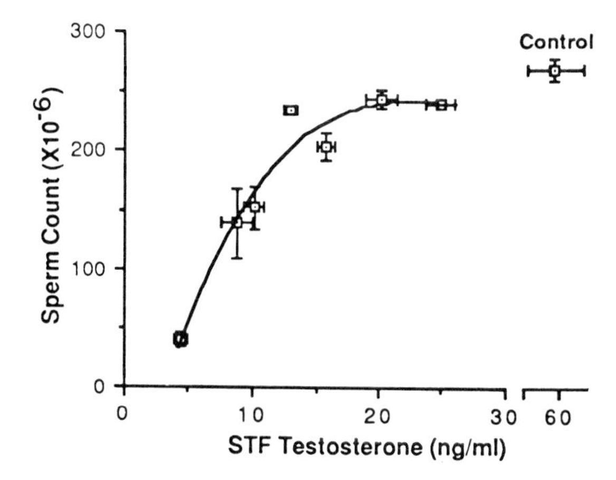

FIGURE 1 Relationship between the mean number of advanced spermatids per testis and the mean testosterone concentration in seminiferous tubule fluid (STF) 2 months after implantation of testosterone-filled plastic capsules of increasing sizes into intact adult rats (reproduced with permission from B. R. Zirkin *et al.*, *Endocrinology* **124**, 3043–3049, 1989 © The Endocrine Society).

contraceptive dose of LH-suppressive testosterone, the intratesticular testosterone concentration found to be required for the restoration of quantitatively complete spermatogenesis also was 20 ng/ml. These observations indicate that spermatogenesis can be quantitatively maintained or restored at testosterone concentrations far lower than those normally present within the testis but still an order of magnitude greater than that present in serum.

These observations pose the following questions: Why is so much testosterone required to maintain or restore spermatogenesis? How does testosterone regulate spermatogenesis?

B. FSH

Though the involvement of FSH in spermatogenesis has been studied extensively, considerable controversy remains about the circumstances in which it is required and what it does. The uncertainty stems in part from apparent species differences in the effects of FSH on spermatogenesis and in part from the time during the life cycle during which FSH is either administered or withdrawn. Numerous studies have shown that FSH is required (in addition to testosterone) for quantitatively normal spermatogenesis in adult nonhuman primates and humans and for testicular recrudescence in seasonally breeding rodents. There is also agreement that FSH is integrally involved in initiating spermatogenesis in immature rats. The receptor for FSH resides with the Sertoli cell. Given the effect of FSH on spermatogenesis initiation, it is not surprising that FSH elicits dramatic cAMP production by Sertoli cells from prepubertal rat testes. Adult rat Sertoli cells also have FSH receptors, but there is a dramatic decrease in the response of adult Sertoli cells to FSH as rats approach adulthood. The altered sensitivity to FSH has been attributed in part to age-related induction of phosphodiesterases, increased activity of G_i proteins, or induction of protein kinase inhibitors. Whatever the mechanism, it is apparent that the response of Sertoli cells to FSH changes from the prepubertal to the adult period.

Controversy about the involvement of FSH in maintaining or restoring adult spermatogenesis has come largely from studies of the adult rat. The administration of testosterone to rats at the time of hypophysectomy, or after hypophysectomy-induced testicular regression, fails to maintain or restore spermatogenesis quantitatively, suggesting that pituitary factors in addition to LH may participate in the regulation of spermatogenesis. A number of studies of this kind have concluded that it is the absence of FSH, rather than the absence of other pituitary factors, that explains the inability of testosterone to sustain or restore spermatogenesis after hypophysectomy. Such a conclusion is consistent with studies showing synergy between FSH and testosterone effects in the adult. For example, the administration of recombinant FSH, by itself or together with testosterone, to hypophysectomized adult rats has been shown in some studies to prevent germ cell loss and in others to restore spermatogenesis (qualitatively) to germ cell-depleted testes. The conclusion from such studies, taken together, is that FSH can affect adult spermatogenesis.

Does FSH affect adult spermatogenesis under non-experimental conditions? Dym and colleagues reported that the administration of FSH antiserum to adult rats did not affect testis weight or germ cell number, concluding that FSH has little or no effect on spermatogenesis. Consistent with this, Awoniyi and colleagues reported that the administration of testosterone to adult rats at the time of their active immunization against GnRH maintained spermatogenesis quantitatively and, similarly, that testosterone alone, when administered to rats made azoospermic by active immunization against GnRH, restored spermatogenesis quantitatively (Fig. 2), in both cases in the absence of detectable FSH. If FSH remained suppressed throughout these immunization experiments, as Awoniyi and colleagues reported, these studies provide conclusive evidence that FSH is not required for the maintenance or restoration of spermatogenesis in adult rats. However, others have reported that at least some FSH is restored by testosterone treatment of GnRH-immunized rats. The issue has yet to be resolved.

The arguments for and against a role for FSH in the adult rat might be reconciled by the thesis that under conditions in which intratesticular testosterone levels are high, FSH may not be required for either the maintenance of spermatogenesis or its res-

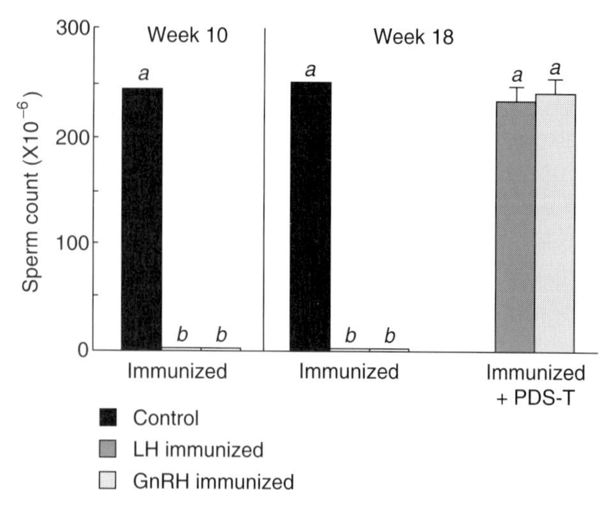

FIGURE 2 Mean number of advanced spermatids per testis in control rats and in LH- and GnRH-immunized rats at Weeks 10 and 18 after immunization and in LH- and GnRH-immunized rats that received testosterone-filled Silastic capsules (PDS-T) from Weeks 10 to 18 postimmunization. Different superscripts above bars indicate significant differences (reproduced with permission from C. Awoniyi *et al.*, *Endocrinology* **124**, 1217–1223, 1989 © The Endocrine Society).

toration. In normal circumstances, total intratesticular testosterone concentration in fact is very high, typically more than twice the concentration found to be required to quantitatively maintain or restore spermatogenesis. In summary, available data suggest (i) that FSH may not be required for spermatogenesis in the normal adult rat because intratesticular testosterone concentrations normally are very high; but (ii) that FSH can have significant effects on spermatogenesis when intratesticular testosterone concentrations fall. However, Griswold and colleagues reported that immunization of adult rats against the FSH receptor resulted in reduced sperm number, presumably despite normal high levels of testosterone. Moreover, in a setting in which there are defects in the genes encoding FSH and its receptor, spermatogenesis and fertility are possible. Thus, although the controversy regarding the role of FSH continues, it is becoming apparent that complete spermatogenesis can occur, and fertility can be realized, in the absence of FSH.

IV. WHY IS SO MUCH TESTOSTERONE REQUIRED FOR SPERMATOGENESIS?

The administration to rats of contraceptive doses of testosterone results in a reduction in germ cell numbers despite intratesticular testosterone concentrations up to three times higher than the testosterone concentration in serum. Why do such high testosterone concentrations fail to maintain spermatogenesis? In serum, free testosterone represents only a small fraction of total testosterone. Roberts and colleagues hypothesized that although total intratesticular testosterone concentration is high following the administration of contraceptive doses of testosterone to rats, the proportion that is free to bind to androgen receptors must be far lower. Indeed, they found that the concentration of androgen-binding protein (ABP) in the STF was about the same as the intratesticular testosterone concentration that resulted from contraceptive testosterone implants (30–40 n*M*); moreover, they found that ABP binding of testosterone suppressed testosterone-dependent gene transcription in a transformed Sertoli cell line cotransfected with the androgen receptor and an androgen-dependent reporter construct (MMTV-CAT) (Fig. 3). This might explain why a contraceptive dose of testosterone fails to sustain spermatogenesis, but this leaves unanswered the question of why so much testosterone (70 n*M* or 20 ng/ml) relative to androgen receptor concentration is minimally required to maintain or restore spermatogenesis.

In fact, does testosterone function via the androgen receptor? Testosterone concentration in the STF of the rat testis is about an order of magnitude greater than the concentration of 5α-dihydrotestosterone (DHT), the major regulatory androgen for most androgen-dependent organs of the male reproductive tract. If intratesticular androgens support spermatogenesis via the androgen receptor, DHT, with its greater affinity for and slower rate of dissociation from the androgen receptor, should more effectively maintain spermatogenesis than testosterone. In fact, Chen and colleagues demonstrated that in experimental circumstances in which endogenous testosterone production was suppressed, the minimal aver-

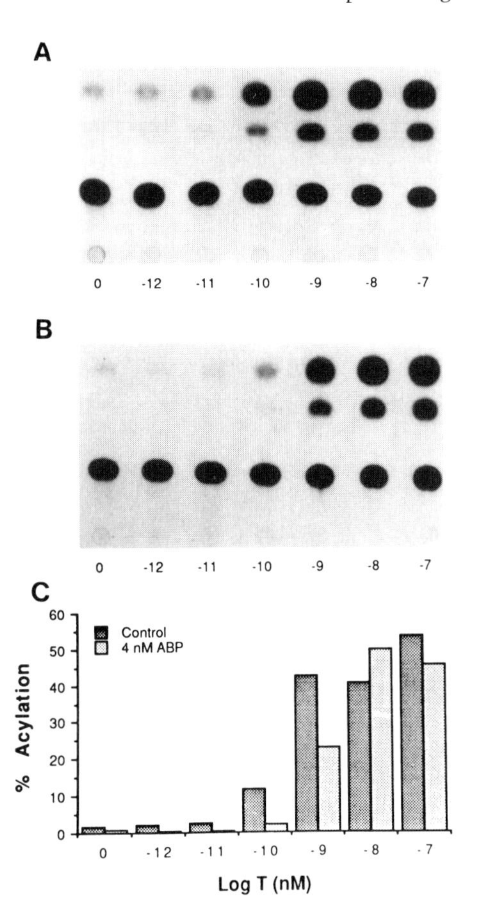

FIGURE 3 Effect of recombinant ABP on androgen-dependent transcription in MSC-1 cells. Cells were treated with increasing concentrations of testosterone (48 hr) in the absence (A) or presence (B) of 4 n*M* ABP. (C) Percentage acetylated chloramphenicol was determined by scintillation counting of the appropriate spots from thin layer plates (reproduced with permission from K. R. Roberts and B. R. Zirkin, *Endocr. J.* 1, 41–47, 1993 © The Endocrine Society).

age STF concentration of DHT required to quantitatively maintain spermatogenesis (7 or 8 ng/ml) was only one-third the required testosterone concentration. These results indicate that DHT more effectively maintains spermatogenesis than testosterone, which is consistent with the greater (at least three- or fourfold) affinity of DHT than testosterone for the androgen receptor.

V. HOW DO TESTOSTERONE AND FSH REGULATE SPERMATOGENESIS?

As previously indicated, it is clear that testosterone is required for appropriate numbers of germ cells to be maintained in the adult testis or to be restored to germ cell-depleted testes. The mechanism by which testosterone acts, however, is far from clear. It is well established that acute testosterone withdrawal depletes the testes of particular germ cells at particular stages of the cycle of the seminiferous epithelium; rapid effects on pachytene spermatocytes and spermatids at stages 7 and 8 typically are seen. Significantly, the conversion of spermatids of stage 7 to stage 8 is inhibited by testosterone withdrawal and restored when testosterone levels are restored. Why are cells lost and why do particular cells fail to proceed in development? In part, testosterone withdrawal results in changes in the adhesion of spermatids to the Sertoli cells with which they are associated, probably via effects on the Sertoli cell cytoskeletal components actin and vinculin filaments and thus on the junction between the spermatids and Sertoli cells. Loss of spermatid adhesion, in turn, precludes further maturation of these cells.

In vitro studies support a synergistic role for FSH in testosterone-mediated spermatid binding, with FSH influencing the distribution of Sertoli cell structural proteins. Additionally, Sharpe and colleagues have shown that androgen depletion markedly reduces the level of synthesis of particular proteins by tubules at defined stages of the cycle. The identity and function of these proteins, which apparently are androgen regulated, are not known. Finally, testosterone is known to stimulate the secretion of some well-known Sertoli cell proteins, including ABP, and to inhibit the production of others (e.g., plasminogen activator). Unfortunately, it is not known whether and how these androgen-dependent proteins function in spermatogenesis regulation.

With respect to FSH, Handelsman and colleagues have shown that injection of gonadotropin-deficient weanling mice with recombinant human FSH (rhFSH) results in increased numbers of spermatogonia and spermatocytes but not elongated spermatids.

In primates, FSH apparently is required for the production of appropriate numbers of spermatogonia, without which quantitatively complete spermatogenesis cannot be achieved. These and related studies suggest that either FSH or testosterone may stimulate initial spermatogenic development, but that only testosterone is able to complete spermatogenesis. This conclusion is consistent with studies of the rat demonstrating that rhFSH results in the maintenance or restoration of all germ cell types prior to elongated spermatids but does not affect the elongated spermatids. Taken together, these studies indicate that FSH is able to maintain or restore spermatogenesis by increasing numbers of spermatogonia and promoting the maturation of the spermatids that result from meiosis through the round spermatid stage but not beyond; apparently, testosterone is required for the completion of spermatogenesis.

It has been appreciated for some time that germ cell losses occur normally during spermatogenesis, but only recently have these losses been attributed to apoptosis. A number of recent studies have shown that the experimental withdrawal of testosterone from the testis leads to increased germ cell apoptosis, suggesting that testosterone may serve as a cell survival factor, protecting germ cells from apoptotic death. In many cells, the interaction of Fas, a transmembrane receptor protein, with its ligand Fas ligand (FasL) triggers the death of cells expressing Fas. Boekelheide and colleagues recently reported that FasL is localized to Sertoli cells and Fas to germ cells in the rat testis, that the expression of FasL and Fas are upregulated following exposure of rats to xenobiotics, and that this leads to germ cell apoptosis. It is possible that this system will prove to be one mechanism by which testosterone can mitigate germ cell apoptosis via its site of action, the Sertoli cell.

Another, perhaps related, possibility is that testosterone will prove to affect the Bcl-2 family of proteins. Bcl-2 is a protooncogene product which, when elevated in cells *in vivo* or *in vitro*, exerts an antiapoptotic effect. For example, overexpression of Bcl-2 has been shown to protect prostatic cancer cells from apoptosis *in vitro* and to protect against apoptosis induced by androgen depletion *in vivo*. It is well established that there is a family of Bcl-2 proteins, some of which inhibit while others promote apoptosis, and that interactions between these groups of proteins antagonize their opposing functions and thereby modulate the sensitivity of cells to apoptosis. For many cells, changes in the ratios of the Bcl-2 family of proteins promote cell survival or cell death. To date, studies have shown that Bcl-2 family members are present in mammalian testes. However, little is known about the molecular consequences of testosterone depletion/repletion with respect to either the Fas/FasL or the Bcl-2 systems.

There is evidence that FSH may also be involved in suppressing cell death. Russell and colleagues have shown that rhFSH partially prevents degeneration of germ cells in the testes of hypophysectomized rats. McLachlan and colleagues similarly have reported that rhFSH partially restores spermatogenesis in rats by supporting the survival of spermatogonia and promoting subsequent maturation steps through round spermatids. A recent study reported that immunoneutralization of FSH in both immature and adult rats causes increased apoptosis, with spermatogonia and pachytene particularly susceptible. These studies all support a role for FSH in cell survival.

Understanding the roles of testosterone and FSH in cell death/cell survival should help us understand some cases of male infertility, including at least some cases of idiopathic male infertility, and perhaps to introduce new treatments. It also should help us understand the mechanisms by which male hormonal contraceptives function.

See Also the Following Articles

Androgens; Apoptosis (Cell Death); FSH (Follicle-Stimulating Hormone); GnRH (Gonadotropin-Releasing Hormone); LH (Luteinizing Hormone); Testosterone Biosynthesis and Actions

Bibliography

McLachlan, R. I., Wreford, N. G., O'Donnell, L., deKretser, D. M., and Robertson, D. M. (1996). The endocrine regulation of spermatogenesis: Independent roles for testosterone and FSH. *J. Endocrinol.* **148**, 1–9.

Roberts, K. R., and Zirkin, B. R. (1991). Androgen regulation of spermatogenesis in the rat. *Ann. N. Y. Acad. Sci.* 637, 90–106.

Sharpe, R. M. (1994). Regulation of spermatogenesis. In *The Physiology of Reproduction* (E. Knobil and J. D. Neill, Eds.), 2nd ed., pp. 1363–1434. Raven Press, New York.

Zirkin, B. R. (1993). Regulation of spermatogenesis in the adult mammal: Gonadotropins and androgens. In *Cell and Molecular Biology of the Testis* (C. Desjardins and L. L. Ewing, Eds.), pp. 166–188. Oxford Univ, Press, Oxford, UK.

Spermatogenesis, in Nonmammals

Gloria Vincz Callard and Ian P. Callard

Boston University

GLOSSARY

autocrine Chemical control of a cell by a secretion that the same cell produces.

cytokinesis Division of the cytoplasm during cell division.

gonadotropin Peptide hormone of the pituitary regulating gonadal function.

hypophysectomy Surgical removal of the pituitary gland.

isogenetic clone Cells of identical genetic composition, derived from a single precursor cell.

Leydig cell The steroidogenic cell of the intertubular (interstitial) compartment of the testis.

paracrine Chemical control of one cell by the secretion of a nearby cell, not requiring transport of the message via a blood vessel.

rete testis Remnant of the embryonic kidney incorporated into the testis during development.

Sertoli cell The somatic cell of the testicular spermatogenic compartment, which is intimately associated with male germ cells.

syncytial Referring to a multinucleate mass of cytoplasm resulting from the fusion of cells.

tight junctions Areas where the surfaces of adjacent cells are so intimately associated that passage of fluids through the space between the cells is prevented.

I. GENERAL CONSIDERATIONS

The study of spermatogenesis, defined as the production and development of male gametes, has a long history. By 300 BC, as revealed in the writings of Aristotle, the testis was recognized as essential for male fertility; however, it was not until 1679 that von Leeuwenhoek first used the microscope to describe the presence of "animalcules" in semen. Promulgation of the cell theory led von Kolliker in 1841 to describe the derivation of sperm from less mature cells of animal testes. This prompted investigations in various species ranging from fish to mammals. From these early comparative studies, spermatogenesis was recognized as an orderly process of differentiation, and the basic concepts of meiosis and its significance for sexual reproduction were formulated. The monograph by Roosen-Runge is a comprehensive account of spermatogenesis throughout the animal kingdom and provides a valuable and still timely synthesis of the subject. Despite taxonomic differences in the superficial organization of the testis, or in the timing of different steps in the spermatogenic sequence, there are striking similarities at the cellular and molecular levels (Tables 1, 2). Thus, the funda-

TABLE 1
Conserved Characteristics of Spermatogenesis in Vertebrates

Basic cytological changes in germ cells during proliferation, differentiation, and meiosis
Segregation of germ cells into synchronously developing isogenetic clones to form a primary germinal unit (spermatocyst) and connectedness of members of the clone by intercellular bridges
Strict temporal and spatial relationships between succeeding germ cell generations
Intimate anatomical and functional relationship between germ cells and somatic (Sertoli) cells
Steroid-rich environment, separated from the general circulation at specific stages

Note. Source: Callard (1991).

mental mechanism which ensures survival of the species in all vertebrates by directing formation of gametes in mature male animals has had a long evolutionary history. This article is intended to emphasize common themes, identify evolutionary trends, and highlight the utility of less well-studied species as animal models for obtaining valuable insights of general relevance.

II. GERM CELL STAGES

Strictly speaking, spermatogenesis is the life history of a single cell, recognizable as the progression of: primordial germ cell → gonocyte → spermatogonium → primary spermatocyte → secondary spermatocyte → spermatid → spermatozoon.

TABLE 2
Organizational Variations in the Testes of Vertebrates

"Open" versus "closed" spermatocysts
The presence or absence or absence of a secondary germinal compartment (tubule or lobule)
The presence or absence of definitive Leydig cells at some stage of spermatogenesis
The permanent, multipotent Sertoli cells versus renewable, unipotent Sertoli cells

Note. Source: Callard (1991).

A. Prespermatogenic Stages

Germ cells are first identifiable in the yolk sac or en route to the hindgut of the developing embryo. These cells are capable of mitosis and ameboid movement as they migrate to the gonadal primordium. Once within the gonadal primordium, they are termed gonocytes, but at this stage they do not necessarily have sex-specific characteristics. In some nonmammals, there is an influence of the somatic environment in dictating the final location of male germ cells within the testis and the expression of other male- versus female-typical characteristics because hormones, temperature, and certain social factors affect sex ratios or sex reversals. Following arrival in the testis, gonocytes transform into spermatogonia and enter into spermatogenesis proper. The period to commitment may be measured in years or in days after birth. In the dogfish shark *Scyliorhinus*, the beginning stages of spermatogenesis are already visible before birth, and proliferation and degeneration continue for at least 2 years, but progression to meiotic stages does not occur until sexual maturity. In seasonal breeders, although spermatogenesis is limited to part of the year, it is continuous during the breeding season. Mechanisms of spermatogenic recrudescence and decline, other than variations in gonadotropin secretion, are not understood.

B. Mitotic Stages

The spermatogenic process begins with a stage during which primitive spermatogonia undergo differentiation, correlated with a series of mitoses. The number of spermatogonial divisions before meiosis begins is a species-specific characteristic and ranges from 4 to 14 in vertebrates. The sequence is readily visualized in elasmobranchs, in which there are anatomically distinct units of spermatogenesis (spermatocysts) and a linear progression of successive stages across the testis. For example, the dogfish *Squalus acanthias* has 13 spermatogonial divisions, whereas estimates of spermatogonial divisions in rats and mice indicate there are 6.

C. Meiotic Stages

The mitotic stage of spermatogenesis is followed by meiosis, the central event of spermatogenesis during which the diploid (2n) set of chromosomes is reduced to the haploid (1n) number characteristic of mature gametes. Meiosis begins with transformation of spermatogonia to primary spermatocytes, then secondary spermatocytes, and finally round spermatids which represent the haploid germ cell population.

D. Spermiogenesis

After meiosis the round spermatids enter a metamorphic stage during which they are transformed into elongated, flagellated spermatozoa. This process occurs in the haploid state and is accompanied by dramatic morphological and biochemical changes. By convention, the term spermatozoon is reserved for the mature gamete after it has been released from the germinal epithelium; therefore, haploid cells undergoing spermiogenesis are designated as early (round) or late (elongated and maturation phase) spermatids.

III. ANATOMICAL AND FUNCTIONAL UNITS OF SPERMATOGENESIS

Spermatogenesis can be subdivided into anatomical and functional units, which comprise both germ cells and associated somatic (Sertoli) cells. Because the boundaries of these subdivisions or compartments are less distinct in some groups than in others, and because their organization is particularly complex in conventionally studied laboratory animals, cross-species homologies are not always evident. The designations originally proposed by Roosen-Runge have been modified to include primary, secondary, and tertiary spermatogenic units: respectively, germ cell clones plus associated Sertoli cells (spermatoblasts and spermatocysts), lobules or tubules, and testes. These divisions, as applied to different vertebrate groups, are summarized in Table 3.

A. The Primary Spermatogenic Unit

1. Germ Cell Clones

All germ cells but those in the very earliest stages of stem cell division (stem cells) progress through

TABLE 3

Comparative Features of Testicular Organization in Vertebrates

- Germ cell clone—component cells derived from the same spermatogonium (isogenetic), linked by intercellular bridges and synchronized through all developmental stages; cell number per clone stage- and species-specific (all vertebrates)
- Sertoli cell—somatic cell intimately associated with germ cells from the earliest stages.
  - Proliferates in adult animals, developmentally synchronized with a single germ cell generation (unipotent), degenerates or is otherwise lost after spermiation (temporary) (fishes and urodelean amphibians)
  - Nonproliferative in adults, simultaneously associated with more than one germ cell generation (multipotent), continuously recycling (anuran amphibians and amniotes)
- Primary germinal unit (spermatocyst)—germ cell clone plus associated Sertoli cells, each Sertoli–germ cell cohort comprising a spermatoblast; ratio of germinal to somatic elements stage and species specific
  - Closed, spherical unit bounded in whole or in part by a basal lamina
    - Present in all developmental stages prior to spermiation (fish and urodelean amphibians)
    - Limited to early (premeiotic and meiotic) stages (anuran amphibians)
  - Irregular, sometimes flattened unit, with limited or no basal lamina; open to the environment
    - Present only during spermiogenesis (anuran amphibians)
    - Present in all developmental stages (amniotes)
- Secondary germinal compartment—delimited by a complex, cellular boundary wall and comprising blind-ended sacs (lobules) or open-ended tubules, both continuous with the intratesticular collecting duct system, and surrounded by Leydig cells
  - None present at any stage—no definitive Leydig cells—spermatocysts embedded in connective tissue stroma and joined to collecting ducts via a short stalk patent only at spermiation (elasmobranchs)
  - Limited to mid- and late developmental stages, with stem cells and spermatogonial-stage spermatocysts embedded directly in connective tissue stroma; continuous formation adjacent to germinal region and regression after spermiation; interstitial Leydig cells limited to lobular–tubular regions (some teleosts and urodele amphibians)
  - Permanent components of adult testis, containing germinal elements and spermatocysts in all developmental stages; interstitial Leydig cells (some teleosts, anuran amphibians, and amniotes)
- Tests—medially fused (cyclostomes) or paired, composite organs (all other vertebrates) made up of multiple primary and/or secondary spermatogenic units and encompassed by a connective tissue capsule

Note. Source: Callard (1991).

spermatogenesis as part of a synchronously developing isogenetic clone of cells. The members of each clone are interconnected by cytoplasmic bridges which result from incomplete cytokinesis during succeeding mitotic and meiotic divisions. Not until spermiation do the syncytially connected spermatids separate from each other and become, at least for a short period of time, the most autonomous of metazoan cells: free-swimming spermatozoa.

In elasmobranchs clone formation begins with a single spermatogonium and terminates with the production of 32,000 spermatozoa in the spotted dogfish (*Scyliorhinus*) and 16,000 in the ray (*Torpedo*), the final clone size reflecting differences in the number of spermatogonial mitoses. Why synchronization is valuable during germ cell development is not known.

2. *Sertoli Cells*

In all vertebrates and many invertebrates, developing germ cell clones are intimately associated with Sertoli cells. The term is applied here to the somatic cell of the germinal compartment, regardless of species or stage of development. Several important characteristics distinguish Sertoli cells of anamniotes from those of the amniote testis. In sexually mature fishes and urodelean amphibians, Sertoli cells proliferate, remain associated, and develop synchronously with a given germ cell clone throughout spermatogenesis; following spermiation, they degenerate. By contrast, Sertoli cells of reptiles, birds, and mammals are permanent elements of the seminiferous epithelium, nurturing generation after generation of developing germ cells. The Sertoli cell of anuran amphibians seems to represent an intermediate condition between the anamniote and amniote Sertoli cell in that the apical portion is pinched off at spermiation but the basal portion remains intact and regenerates. It is generally accepted that Sertoli cells not only provide a structural support and essential nutrients but also mediate between the soma and germinal tissue in all respects.

3. *Spermatocysts and Spermatoblasts*

The syncytial germ cell clone together with its cohort of Sertoli cells form the primary spermatogenic unit in all vertebrates. In fishes and amphibians, this unit comprises an anatomically discrete, closed, spherical compartment termed a spermatogenic cyst (spermatocyst), and all cells contained within it are synchronized in development. Thus, cytological changes occurring in Sertoli cells during the spermatogenic progression are readily visualized. In contrast to anamniote species with readily discerned spermatocysts, and hence a "cystic" mode of spermatogenesis, the homolog of the amniote testis is not closed or spherical but has a more flattened form with an extremely irregular three-dimensional outline and appears to lie in "open communion" with other clones and with the tubular lumen. To understand the key position of the spermatocyst and its amniote homologs in the organization of the vertebrate testis, it is useful to consider the genesis of the association between germ cells and Sertoli cells in elasmobranchs. Cyst formation in *Squalus* begins with a single stem cell of each type. Initially, mitoses are synchronized, maintaining a germ cell/somatic cell ratio of 1:1 through nine divisions; however, Sertoli cells cease dividing prior to the final four spermatogonial mitoses. Upon entry into meiosis, therefore, the germ cell/Sertoli cell ratio is 16:1 and, at this stage, a unit comprising a single Sertoli cell and its 16-member germ cell cohort becomes readily apparent. Following two meiotic divisions and spermiogenesis, there are 64 germ cells per somatic cell. Actual counts are very close to theoretical when tested in a second dogfish species, *Scyliorhinus*, indicating little or no cell death or, alternatively, degeneration of an entire spermatocyst in an all-or-none fashion. Because a single Sertoli cell and its cohort of developing germ cells define a functional subdivision of the spermatocyst, in keeping with von Ebner, who first recognized this organizational unit in the mammalian germinal epithelium, here it is designated a "spermatoblast." The number of spermatoblasts per cyst varies among elasmobranch species. For example, there are 500 in *Scyliorhinus* and 250 in *Torpedo*, each comprising a single Sertoli cell and 64 spermatozoa at maturation. Also, although germ cells within a single spermatoblast are strictly synchronized, they are slightly out of phase with those in adjacent spermatoblasts.

Essentially the same developmental pattern is seen in other groups with a cystic mode of spermatogenesis. In fish and urodelean amphibians the spermatocyst remains a closed sphere until immediately before spermiation. In anurans, however, spermatocysts lose their spherical form and open to the environment somewhat in advance of sperm release. Indeed, the integrity of the cyst boundary is lost in the spermatocyte stage in certain anuran species. As with certain features of the anuran Sertoli cell, this situation can be viewed as intermediate between the closed cysts of fishes and the amniote condition in which the cysts are "open" throughout spermatogenesis. The formation and integrity of the spermatocyst/spermatoblast, and the microenvironments defined by these organizational units, are dependent on three factors: Sertoli cells and the close apposition of germ cells, tight junctions between adjacent Sertoli cells, and the presence of an acellular basal lamina encompassing the group. The relative importance of each factor varies with phylogenetic position and stage of development. In fishes and amphibians, tight junctions are seen only in spermatocysts containing advanced spermatogonia or spermatocytes and are absent in earlier stages. In amniotes in which germ cell clones in different stages share the same set of Sertoli cells, tight junctions form between contiguous Sertoli cells so that the basally located spermatogonial clones are outside the blood–testis barrier, whereas the more centrally placed spermatocyte and spermatid clones are inside the barrier. The most variable mechanism by which the cyst boundary is defined is the formation of a basal lamina. In elasmobranchs the entire cyst is bounded by an acellular basal lamina; however, in this group there are no definite tubules or lobules, and cysts are embedded directly in a connective tissue matrix. In groups in which the cysts are contained in a secondary compartment (e.g., teleosts and amphibians), the basal lamina is present only where the cyst abuts the wall of the lobule.

B. The Secondary Spermatogenic Unit: Lobules and Tubules

Organization of primary germinal units into a secondary unit or compartment does not occur in all vertebrates (Table 3). In elasmobranchs, spermatocysts are embedded directly in the connective tissue matrix of the testis. Each is attached to a short duct, part of a system of intratesticular collecting tubules which become patent only toward the end of spermatogenesis when the mature sperm exit the cyst and the testis via the efferent ducts. In teleosts and more advanced vertebrates, however, primary germinal units (i.e., spermatocysts) are contained within a secondary compartment and this, in turn, is contiguous with the lumen of the collecting duct or rete testis; therefore, sperm must first enter the lumen of the secondary compartment before entering the excurrent ducts. Where this secondary compartment opens into the collecting duct at both ends, it is termed a tubule (mammals, birds, and reptiles); however, in species in which spermatocysts are contained in blind-ended sacs and have only one route of exit, it is preferable to call them lobules (teleosts and amphibians), a term that emphasizes organizational homology with the amniote tubule. In all vertebrates, the tubule/lobule is demarcated by a limiting membrane or lamina propria which is, in part, cellular. The seminiferous tubules of birds and reptiles closely resemble those of mammals in that they form early in development, remain permanent elements of the adult testis, and have a stratified germinal epithelium. This contrasts with the situation in teleosts and amphibians, which have closed cysts through all or some of spermatogenesis and lobules/tubules which regress and regrow in conjunction with seasonal cycles. In teleosts, two main patterns are seen, exemplified by Fundulus and related species and Mugil and most other teleosts. In the former, the pattern is reminiscent of the dogfish, except that in the dogfish the cysts are outside the duct system; primitive spermatogonia are embedded directly in the testicular stroma. The second and most common teleost pattern has an anastomosing and branching network of lobules with stem cells located adjacent to the basement membrane along their length rather than in a defined germinal zone. This second pattern more closely resembles mammals than elasmobranchs.

Lobular development in urodelean amphibians follows one or the other of these teleost patterns. In the newt (*Taricha*) extensively branched lobules are sac-like in form, whereas in the salamander (*Necturus*) elongate lobules radiate from a germinal zone. In contrast to teleosts, however, germinal tissue and immature lobules in *Necturus* are proximate to the collecting ducts, whereas maturing cysts complete their development at the distal ends of the lobules, with the demarcation between immature and mature regions being quite distinct. Thus, when spermatozoa are released, they must traverse the immature lobular regions in order to reach the excurrent ducts. The most conspicuous difference between urodelean and anuran amphibians is the relatively early rupture of the cyst in the latter. In *Rana*, sperm bundles embedded in Sertoli cells line the open cysts, which at this stage constitute the wall of the seminiferous lobule. At the periphery of the lobule, primary spermatogonia surrounded by Sertoli cells are occasionally seen in preparation for the next reproductive period. In both fishes and amphibians, therefore, elements of the mammalian condition can be seen in their lobular organization.

C. The Tertiary Spermatogenic Unit: Testes

The term testis is derived from the Latin meaning "to witness" or "to testify" and is said to originate from a practice in which athletes were verified as males by palpation of their testes. With the exception of cyclostomes, which have a single, medial testis, the testes of most vertebrates are compact, paired organs that are spherical, ovoid, or elongate in shape, although in certain teleosts and amphibians the testis is composed of separate lobes, varying in number and connected by strands of germinal tissue. Testes are generally located in the abdominal cavity but may be quite far anterior (sharks) or seasonally or permanently located in scrotal sacs outside the abdomen (mammals). Testes are enveloped in a connective tissue capsule (tunica albuginea) which may be pigmented in some poikilotherms and is connected to a dorsal mesentery by which the testes are suspended from the body wall. A connective tissue framework penetrating and subdividing the body of the testis is especially obvious in large animals.

IV. KINETICS OF SPERMATOGENESIS: THE SPERMATOGENIC WAVE AND CYCLE OF THE SEMINIFEROUS EPITHELIUM

In all vertebrates, there is a strict temporal and spatial relationship among different germ cell stages; however, species variations are most evident in the timing and anatomical organization of cells in the spermatogenic sequence. These characteristics have been reviewed for vertebrates and for some invertebrates and are remarkably similar. The number of spermatogonial mitoses varies from 13 in *Squalus* to 6 in the rat, whereas the length of a complete cycle is not markedly different in the guppy (36 days) when compared to mouse (34.5 days), rat (48 days), hamster (35 days), bull (49 days), rabbit (43 days), and man (64 days). Although intervals between steps are invariant in continuous breeding mammals such as rat, mouse, and man, elevated temperatures accelerate spermatogenesis in poikilotherms. For example, in the medaka *Oryzias*, the minimum time from preleptotene spermatocytes to early spermatids is reduced from 12 to 5 days at 15 versus 25°C, respectively. Moreover, a period of spermatogenic arrest is a normal part of the testicular cycle in seasonal breeders. In the salamander *Necturus*, germ cells in the premeiotic stage of development proliferate in the fall of the year and remain in a state of arrested development until the following breeding season when spermatogenesis is completed.

As a manifestation of the temporal coordination of developing germ cells, they are arranged in consecutive order in space. This linear sequence is termed the spermatogenic wave. Owing to a less complex testicular organization, the spermatogenic wave of anamniotes is more readily visualized than that in mammals and often leads to a distinct "zonation" of the testis.

Teleosts in which germinal tissue is restricted to a defined region at one end of each lobule have a linear wave resembling that in *Squalus*, but those in which germinal tissue is located at intervals along the length of the lobules have both a spatial and a temporal progression of stages. The latter occurs as one stage succeeds another at a given point on the epithelium. The spermatogenic wave is especially complex in the urodelean amphibian *Necturus* and, like *Squalus*, the spermatogenic wave in the *Necturus* testis results in a spatial separation of germ cells (and associated endocrine elements) in different stages of development and has been used to study stage-related biochemical change.

V. STEM CELLS AND SPERMATOGONIAL RENEWAL

Throughout the life of a mature animal a constant supply of new germ cells is required. Initial stages of cyst formation in anamniotes result in spermatogonial "clusters" and ultimately spherical clones. Stem cell renewal has not been well studied in lower species, despite their having a clearly defined and sometimes large germinal zone. In the killifish, *Fundulus heteroclitus*, stem cells are confined to blind ends of the lobules where they are seen as single cells or clusters of cells and not synchronized in mitoses. Entry of proliferating spermatogonia into spermatogenesis is signaled by a division in which cytokinesis is incomplete, leading to formation of a spermatogonial pair.

VI. REGULATION OF SPERMATOGENESIS

Available information indicates that spermatogenesis is dependent on (i) circulating hormones, primarily gonadotropins; (ii) paracrine and autocrine regulatory factors, including steroids which are considered hormones in other contexts; and (iii) direct cell–cell interactions. It is generally agreed that Sertoli and Leydig cells have a central role in the control of spermatogenesis, either as sources of regulatory molecules or as targets of their action. In addition, the possibility that germ cells have a role in signaling and synchronizing development of somatic and germinal elements is supported by several lines of evidence.

A. Role of Leydig Cells

In contrast to established dogma derived from mammalian studies, there are numerous lower vertebrates in which Leydig cells, because of anatomic position or state of development, are unlikely to play a major role in maintaining germ cell development. For example, in *Squalus* typical vertebrate Leydig cells are absent. Rather, there are Leydig-like cells which resemble Leydig cell precursors in fetal human testis, and these remain small, sparse, and undifferentiated through all spermatogenic stages . Likewise, in *Necturus*, hypertrophy and differentiation of Leydig cells is keyed to spermiation. Thus, Leydig cells interspersed among the seminiferous lobules are indistinguishable from fibroblasts except at the electron microscopic level. After spermiation and tubular degeneration, Leydig cells hypertrophy. Even within a single lobular cross section, Leydig cells adjacent to sites of sperm release show beginning development, whereas those next to areas containing immature germ cells remain undifferentiated. Since the spermatogenic wave and, hence, tubular regression passes longitudinally from posterior to anterior in *Necturus* testis, there is a corresponding gradient of Leydig cell development and steroidogenic activity. This process culminates in a subcapsular glandular tissue comprising Leydig cells entirely and exceptionally rich in steroidogenic enzymes. In addition, Leydig cells frequently form a distinct glandular tissue separated from the germinal region in mesorchium in the bullhead catfish and in a subcapsular band in whiptail lizards. Taken together, these and other comparative studies suggest that Leydig cells evolved primarily for the secretion of steroids into the peripheral circulation and may have secondarily usurped the role of Sertoli cells in supporting spermatogenesis. Steroid receptors resembling those of mammals in all important characteristics have been identified in the testis of several lower vertebrates, suggesting that steroid regulation of gene expression has been widely conserved as a mechanism of control in vertebrate testis. In *Necturus*, estrogen and androgen receptors have been localized in glandular tissue composed almost entirely of Leydig cells. Moreover, increased androgen receptor levels correspond to Leydig cell differentiation following tubular regression, suggesting a role in steroid secretion, as is true for mammals, but not in germinal development.

B. Role of Sertoli Cells

In fishes, amphibians, reptiles, and birds, the appearance and disappearance of 3β-hydroxysteroid dehydrogenase, lipid, and agranular reticulum has been interpreted as evidence of a secretory cycle that waxes and wanes with seasonal spermatogenic recrudescence and decline. Although peak Sertoli activity appears to correspond to high intratesticular androgen levels, often it is temporally separated from the seasonal development of Leydig cells, maximal circulating androgen levels, and sex behavior. In *Squalus*, Sertoli cells are the primary steroidogenic element in the testis at all times of the year. At this phyletic level, Leydig cells are undifferentiated regardless of germ cell stage, but Sertoli cells have typical steroidogenic organelles which increase in abundance as germ cell maturation progresses so that by late spermatid stages the Sertoli cell is virtually a sac of agranular membranes. A nonreceptor steroid-binding protein thought to be the counterpart of mammalian testicular androgen binding protein (ABP) has been characterized in the testis of *Squalus*, *Necturus*, and recently in the salmon trout. Although it is concentrated in regions with mature spermatids and highest androgen biosynthetic activity in *Squalus* testis, it is highest in immature lobular regions in *Necturus* testis. The idea that steroids may somehow support sperm storage or maturation is reinforced by the observation that, in *Squalus*, cytoplasts (anuclear, membrane-bound cytoplasmic remnants) representing the pinched-off apical tips of Sertoli cells are found in great abundance in semen and may account for high levels of steroidogenic activity and endogenous steroids in this tissue.

VII. COMPARATIVE CONTRIBUTIONS

A cursory review of spermatogenesis reveals that state-of-the-art research utilizing the mammalian testis has been hampered primarily by the multiple cellular elements involved and their complex interrelationships. While information from species outside the range of conventional laboratory mammals has expanded our knowledge base sufficiently to permit recognition of unifying principles, these species have been relatively little used to address problems of contemporary interest. As documented here, there is reason to believe that the basic process of spermatogenesis has been highly conserved through vertebrate evolution and has origins traceable to invertebrates. Thus, where there are obvious technical advantages, there is a sound scientific basis for a wider choice of animal models in reproduction research. The simple linear progression of spermatogenesis in fishes and certain amphibians is ideal for studying biochemical and morphological changes stage by stage.

Acknowledgment

This work supported by Grant HD 16715 from the National Institutes of Health and Grant DCB88-09560 from the National Science Foundation to G.V.C.

See Also the Following Article

Spermatogenesis, Overview

Bibliography

Callard, G. (1988). Reproductive physiology: (B) The male. In *Physiology of Elasmobranch Fishes* (T. J. Shuttleworth, Ed.), pp. 292–317. Springer-Verlag, New York.

Callard, G. (1991). Spermatogenesis. In *Vertebrate Endocrinology: Fundamentals and Biomedical Implications*, Vol. 4, Part A. Academic Press, San Diego.

Callard, I. P., Callard, G. V., Lance, V., Bolaffi, J. L., and Rosset, J. S. (1978). Testicular regulation in non-mammalian vertebrates. *Biol. Reprod.* **18**, 16–43.

Roosen-Runge, E. C. (1977). *The Process of Spermatogenesis in Animals*. Cambridge Univ. Press, London.

Spermatogenetic Cycle in Fish

Takeshi Miura
Hokkaido University

GLOSSARY

androgens Male gonadal steroid hormones. Androgens possess the ability to induce male characteristics and maintain male sex accessory glands and ducts. Examples of androgens are testosterone, 11-ketotestosterone, 5α-dihydrotestosterone, and androstenedione.

gonadotropins Pituitary glycoprotein hormones that control the production of sex steroids and gametogenesis. There are two kinds of pituitary gonadotropins (GTHs); follicle-stimulating hormone and luteinizing hormone in higher vertebrates and GTH I and GTH II in some teleosts.

growth factors These proteins stimulate cell division. Examples of growth factors relating to fish spermatogenesis are insulin-like growth factor-I, activin, and fibroblast growth factor.

Leydig cells One kind of testicular interstitial cell. A major function of Leydig cells is to secrete androgens in response to pituitary GTH stimulation.

meiosis The division process by which germ cell chromosomes are reduced from a diploid state to a haploid state during gametogenesis. By this process, a single diploid cell gives rise to four haploid cells.

progestogens Progesterone derivatives produced by the teleost testis which include $17\alpha,20\beta$-dihydroxy-4-pregnen-3-one, 17α-hydroxyprogesterone, and $17,20\beta,21$-trihydroxy-4-pregnen-3-one.

Sertoli cells The only nongerminal elements within the seminiferous epithelium of testes. Sertoli cells have been referred to as nurse cells, sustentacular cells, and supporting cells for germ cell development.

Spermatogenesis is a complex developmental process. It begins with the mitotic proliferation of spermatogonia; it then proceeds through two meiotic reduction divisions and finishes with spermiogenesis during which the haploid spermatid develops into a sperm. Although it is generally accepted that the principal hormones controlling vertebrate spermatogenesis are pituitary gonadotropins (GTHs) and androgens, the specific roles played by individual hormones have not yet been clarified. Progress in this field has been hampered by the complex organization of the testis of higher vertebrates, in which seminiferous tubules contain several successive generations of germ cells. On the other hand, the germ cells within each testicular cyst (spermatocyst in elasmobranchs) formed by Sertoli cells in fish develop synchronously. This system has proven to be advantageous for studying spermatogenesis. This article provides a description of the characteristics of the spermatogenetic cycle and its control mechanism in fish.

I. SPERMATOGENESIS IN FISH

Morphologically and physiologically, the spermatogenetic cycle can be divided into the following stages: proliferation of spermatogonia, two meiotic divisions, spermiogenesis, spermiation, and sperm maturation.

A. Proliferation of Spermatogonia

Spermatogenesis starts with the mitotic proliferation of type A spermatogonia, which are spermatogenetic stem cells. A type A spermatogonium is a relatively large cell (approximately 10 μm in diameter in various species) and has a clear large homogeneous nucleus containing one or two nucleoli (Fig. 1A). Type A spermatogonia occur independently, with each cell almost completely surrounded by Sertoli cells. In some species, early type B spermatogonia can be distinguished from type A spermatogonia. Although early type B spermatogonia are morphologically similar to type A spermatogonia, they tend to form a cyst of two or four germ cells surrounded by Sertoli cells. It is unclear whether early type B spermatogonia represent a renewal of the stem cells or progress further into spermatogenesis. In the Japanese eel, *Anguilla japonica*, type A and early type B spermatogonia are undeveloped and resting spermatogonia.

These types of spermatogonia proliferate rapidly by mitosis and, as a result, appear in the seminiferous lobules or tubules. The morphology of late type B spermatogonia differs from that of their earlier undeveloped spermatogonial counterparts by the fact that

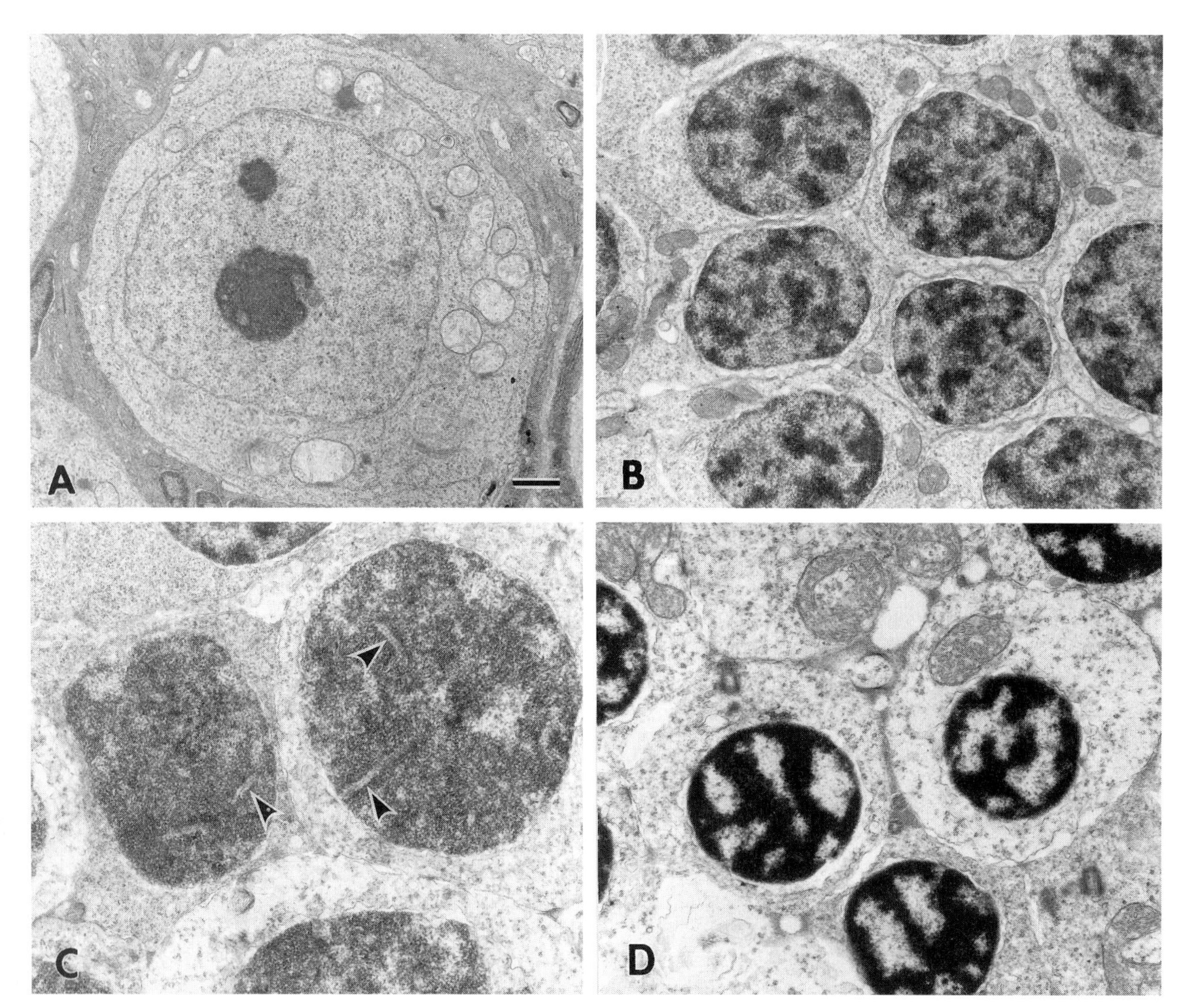

FIGURE 1 Electron micrographs of spermatogenetic cells in Japanese eel. (A) A type A spermatogonium with clear homogeneous nuclei containing one or two nucleoli; (B) late type B spermatogonia with dense and heterogeneous nucleus; (C) spermatocyte with synaptonemal complexes (arrowheads); and (D) spermatid with round and heterogeneous nuclei (scale bar = 1 μm).

their nucleus is denser and more heterogeneous and their mitochondria are smaller (Fig. 1B). After the proliferation of spermatogonia, the germ cells enter meiosis.

In general, the mitotic divisions of spermatogonial stem cells preceding meiosis are species specific. In teleosts, a spermatogonial stem cell of medaka (*Oryzias latipes*) will yield spermatocytes following 9 or 10 mitotic divisions, and those of the guppy (*Poecilia reticulata*) after 14, the zebrafish (*Brachyodanio danio*) after 5 or 6, and the Japanese eel after 10 divisions. However, it is not clear whether the number of mitotic divisions is an inherent property of the type A spermatogonial stem cell, environmentally controlled, or both.

B. Meiosis in Fish Spermatogenesis

Following mitotic proliferation, type B spermatogonia differentiate into primary spermatocytes. These cells enter the first meiotic prophase and then proceed with the first meiotic division to produce secondary spermatocytes. These, in turn, undergo a second meiotic division to produce haploid spermatids, cells with only a single set of chromosomes.

In primary spermatocytes, the leptotene stages are distinguished from the final spermatogonia by their larger and more homogeneous nuclei. In some species, however, it is difficult to distinguish the leptotene spermatocytes from late type B spermatogonia due to their morphological similarities. During the zygotene stage of prophase, spermatocytes can be identified by locating the synaptonemal complex in their nuclei using an electron microscope (Fig. 1C). Because the time between the first and second meiotic divisions is very short, the secondary spermatocyte is difficult to observe. After two meiotic divisions, the germ cells develop into spermatids having small, round, and heterogeneous nuclei (Fig. 1D).

C. Spermiogenesis

During spermiogenesis, the round spermatids transform into spermatozoa. This process is characterized by remarkable morphological changes associated with the formation of the sperm head and its condensed nucleus, with the midpiece, and with the flagellum. Spermatozoan structure varies considerably in complexity among teleost species (Fig. 2). For example, carp (*Cyprinus carpio*), sculpin (*Cottus hangiongensis*), and tilapia (*Oreochromis niloticus*) spermatozoa have spherical heads with the flagellum attached (inserted) to one side. Salmonid spermatozoa have a slightly elongated and cylinder-like head. On the other hand, the spermatozoa of guppy have an elongated head and an extremely well-developed midpiece. Furthermore, eel spermatozoa possess a crescent-shaped nucleus, their flagellum has a 9+0 axonemal structure (generally, the axonemal structure of the flagellum is 9+2), and a single large and spherical mitochondrion with developed tubular cristae is attached to the caput end at one side of the

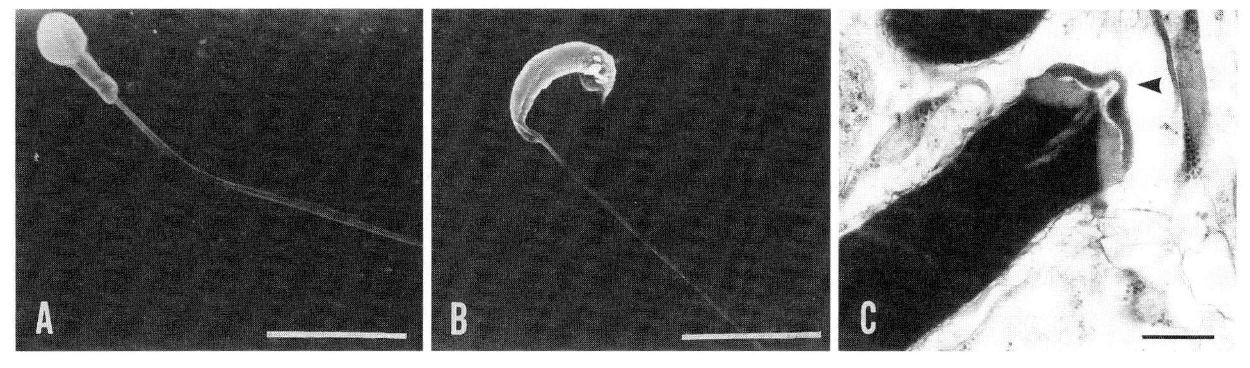

FIGURE 2 Electron micrograph of a spermatozoon of various teleosts. (A) Sculpin (scale bar $=5\ \mu$m); (B) Japanese eel (scale bar $= 5\ \mu$m); (C) sturgeon (scale bar $= 1\ \mu$m). Arrowhead indicates acrosome [courtesy of Drs. G. Quinitio (sculpin) and M. A. Bagher (sturgeon)].

head. An acrosome is absent in the spermatozoa of most teleosts, but it is found in acipenserid fish, lamprey, and shark.

D. Spermiation

In mammals, "spermiation" indicates that embedded bundles of spermatozoa are released from the enveloping Sertoli cell and are swept into the efferent ducts. In most teleosts (except in Poeciliidae), however, spermatozoa are not associated with Sertoli cells. By comparison, spermiation of teleosts indicates the release of spermatozoa from the seminal cysts into the lobular lumen or efferent duct. From a fisheries science point of view, however, "sperm release" or "ejaculation," which occurs after milt hydration and sperm migration down the sperm duct, is more readily observed than spermiation. Therefore, the term spermiation is often used interchangeably with these other terms in fish.

E. Sperm Maturation

Although the spermatozoa in the testis have already completed spermiogenesis, in some species they are still incapable of fertilizing eggs. In salmonids, the sperm in the testis and sperm duct are immotile. If sperm from the sperm duct are diluted with fresh water, they become motile, whereas the testicular sperm, if diluted with fresh water, remain immotile. Thus, spermatozoa acquire their ability to become motile during their passage through the sperm duct.

The acquisition of the motile ability of sperm is different from the initiation of motility. The development from nonfunctional gametes to mature spermatozoa fully capable of vigorous motility and fertilization is referred to as "sperm maturation." Sperm maturation involves physiological, not morphological, changes. In salmonids, sperm maturation (acquisition of sperm motility) is induced by the high pH of the seminal plasma (approximately pH 8.0) in the sperm duct and involves the elevation of intrasperm cAMP levels. Maturation of eel spermatozoa is also induced by the high pH of the seminal plasma and/or HCO_3^-.

II. SPERMATOGENETIC PATTERNS IN FISH

Fish inhabit an enormous range of aquatic habitats. Moreover, their breeding habits are often strongly influenced by environmental factors such as temperature and salinity. Therefore, it is not surprising that fish display a variety of spermatogenetic patterns. However, it appears that these patterns are largely dependent on the length of the spawning period and that they can be classified into three main types which closely correspond to those described by Billard.

In the first type, fish have short spawning seasons lasting several weeks a year. Most anadromous, catadromous, and several cold-water species (i.e., salmonids, sturgeon, flounders, and pike) belong to this type. Their testes have two germ cell groups: a spermatogonial stem cell (type A spermatogonia) group and a major stage group. The germ cells of the major group show almost synchronous development in the testes. The spermatogonial stem cell group forms the next spermatogenetic cycle in the following year. Some species of salmonids only spawn once in a lifetime and die, thus undergoing a single spermatogenetic cycle. These fishes nevertheless possess a spermatogonial stem cell group in their testes.

In the second type, fish have a relatively long spawning season lasting for several months a year. Sea bream, carp, and white perch, which all live in temperate zones, belong to this type. Although these fish have an annual spermatogenetic cycle, the spermatogenesis is not synchronous in the testes. During the spawning season, their testes contain spermatogonia, spermatocytes, spermatids, and numerous spermatozoa.

The third type includes fish that spawn throughout the year, for example, tilapia, Asian catfish, and guppy, which live in subtropical or tropical zones. Spermatogenesis is continuous throughout the year, and all stages of germ cells, from type A spermatogonia to spermatozoa, can be observed in the testis at any time.

III. ENDOCRINE CONTROL OF THE SPERMATOGENETIC CYCLE IN FISH

Fish spermatogenesis is endocrinologically controlled in the same way as it is in other vertebrates. It is well established that in vertebrates, including fish, GTHs are the primary hormones regulating spermatogenesis. However, it appears that GTHs do not act directly but rather work through the gonadal biosynthesis of steroid hormones, which in turn mediate various stages of spermatogenesis. As previously mentioned, fishes are excellent model animals for the study of spermatogenesis since all germ cells within each cyst enveloped by Sertoli cells develop synchronously. In salmonid fish, germ cell development is almost synchronous throughout the whole testis. Taking advantage of this feature, the testicular steroid biosynthetic pathway has been examined in a cell-free system of some salmonid fishes. Using ^{14}C-labeled precursor steroids, it was shown that the testis synthesizes 11-ketotestosterone during spermatogenesis. 11-Ketotestosterone was first identified by Idler and co-workers in the 1960s as a major androgenic steroid in the male sockeye salmon (*Oncorhynchus nerka*). In various teleost fishes, this steroid has since been shown to be synthesized in the testis following GTH stimulation and high levels were detected in the serum during spermatogenesis. A distinct shift in the steroidogenic pathway from production of 11-ketotestosterone to $17\alpha,20\beta$-dihydroxy-4-pregnen-3-one ($17\alpha,20\beta$-DP) occurs in salmonid testis during the spawning season. $17\alpha,20\beta$-DP has also been identified as the maturation-inducing hormone of salmonid oocytes. In Atlantic croaker, spotted seatrout, and several other species $17\alpha,20\beta$-DP, or the related progestogen $17\alpha,20\beta,21$-trihydroxy-4-pregnen-3-one (20β-S), the maturation-inducing hormone can be found at high levels in the serum during the spawning season. It is speculated that these steroids are related to the regulation of spermatogenesis. Therefore, a description of the relationship between the regulation of spermatogenesis and the endocrine system, including the relevant steroid hormones, follows.

Spermatogenesis is regulated at several critical stages of the spermatogenetic cycle. In teleosts, the following critical regulatory stages have been chronicled: initiation of spermatogenesis (the beginning of the proliferation of spermatogonia), entry of spermatogonia into meiosis, induction of spermiation or sperm release, and induction of sperm maturation.

A. Initiation of Spermatogenesis

As already mentioned, there is a relationship between 11-ketotestosterone production and spermatogonial proliferation since high serum levels of the steroid are detected during this period of the spermatogenetic cycle. It has been reported that 11-ketotestosterone induces spermatogenesis directly in the Japanese eel, *Anguilla japonica*. Under aquaculture conditions, the male Japanese eel has immature testes containing only type A and early type B spermatogonia, which are premitotic. A single injection of exogenous human chorionic GTH (hCG) can induce all the stages of eel spermatogenesis *in vivo*. This injection also causes an increase in serum levels of 11-ketotestosterone. A testicular organ culture system has been developed using eel testes which have only undeveloped spermatogonia. Organ cultures provide a simplified experimental system in which the direct effects of various factors upon the testes can be investigated. All stages of spermatogenesis, from spermatogonial proliferation to spermiogenesis, were induced when 11-ketotestosterone was added to this culture system. 11-Ketotestosterone injection also induces spermatogenesis in goldfish, further implicating 11-ketotestosterone as one of the factors involved in the initiation of teleost spermatogenesis.

However, it is believed that the inducing action of 11-ketotestosterone is mediated by other factors produced by Sertoli cells. It is possible that some of these factors are growth factors such as activin B, insulin-like growth factor-I (IGF-I), and fibroblast growth factor (FGF). Activin B is a dimeric growth factor which belongs to the transforming growth factor-β family and is composed of two activin β_B subunits. In the Japanese eel, activin B is present in the testis at the initiation of spermatogenesis after hCG stimulation, with its expression site restricted to Sertoli cells. Both transcription and translation of eel activin B were induced by 11-ketotestosterone stimulation *in vitro*. Furthermore, activin B induced proliferation of spermatogonia.

IGFs are known to be mediators of growth hormone action in fishes. In the rainbow trout testis, IGF-I is expressed in spermatogonia and/or Sertoli cells and binds to a type 1 IGF receptor. Furthermore, IGF-I stimulates DNA synthesis in spermatogonia.

FGF plays an important role in the control of the proliferation and/or differentiation in various cell types. In the immature male white-spotted char (*Salvelinus leucomaenis*), for example, FGF is expressed in the spermatogonia. Following gonadotropin administration, the proliferation of spermatogonia commences. At the same time, FGF expression becomes undetectable in spermatogonia but appears in Sertoli cells. This and other growth factors may act on the early stages of spermatogenesis. It is quite possible that activin B and IGF-I induce spermatogonial proliferation *in situ* since these two growth factors can induce the proliferation of spermatogonia in *in vitro* experimental systems. Further investigation is needed to fully understand the relationship between growth factors and spermatogenesis in fish.

B. Entry of Spermatogonia into Meiosis

In mammalian spermatogenesis, the onset of meiosis is precisely timed and it is speculated that this is regulated by meiosis-activating substances and meiosis-preventing substances. Recently, in the Japanese eel, it was shown that there is a regulatory stage in the fourth or fifth mitotic division on proliferation of spermatogonia prior to the cells entering meiosis. To cross this regulatory stage, some unknown factors other than 11-ketotestosterone may be required.

C. Induction of Spermiation or Sperm Release

During the breeding season, the levels of numerous hormones show remarkable changes in male teleosts which are initiated by an increase in GTH secretion. GTH secretion induces an increase in the production of the testicular steroids 11-ketotestosterone, testosterone, and 17α,20β-DP or 20β-S. In several teleosts, two pituitary GTHs have been characterized: GTH I resembles the tetrapod follicle-stimulating hormone, whereas GTH II resembles tetrapod luteinizing hormone. It has been speculated that GTH II stimulates testicular steroid production in salmonids because of the secretion patterns of GTH I and GTH II and the expression patterns of their receptors. The relationships between GTH and/or testicular steroids and spermiation have been extensively investigated. GTH injection has a positive effect on spermiation and stimulates an increase in plasma levels of the previously mentioned sex steroid hormones in some teleosts. 11-Ketotestosterone injections are effective in inducing spermiation in goldfish and some salmonids, whereas 17α,20β-DP injections are effective in Japanese eel and several salmonid species. Thus, it is certain that GTH and these steroids are related to spermiation, but the action of each hormone on each stage, i.e., milt hydration, sperm migration to the sperm duct, the increase in milt volume, etc., is still unclear.

D. Induction of Sperm Maturation

Since spermiation and sperm maturation occur simultaneously in fish, it is difficult to analyze the hormonal control of sperm maturation. Therefore, few studies exist on the relationship between the endocrine system and sperm maturation in teleost (especially the acquisition of sperm motility). Salmonid fish such as masu salmon, *Oncorhynchus masou*, offer a good system for studying the endocrine mechanisms involved in the acquisition of sperm motility. As previously mentioned, masu salmon spermatozoa taken from the testis are immotile but acquire motility during their passage through the sperm duct. Early in the breeding season, however, sperm from the sperm duct do not gain motility even when they are diluted with fresh water. This indicates that there is a distinct time period during which sperm become motile. Plasma levels of several sex steroids (11-ketotestosterone, testosterone, and 17α,20β-DP) increase during the spawning season. While 17α,20β-

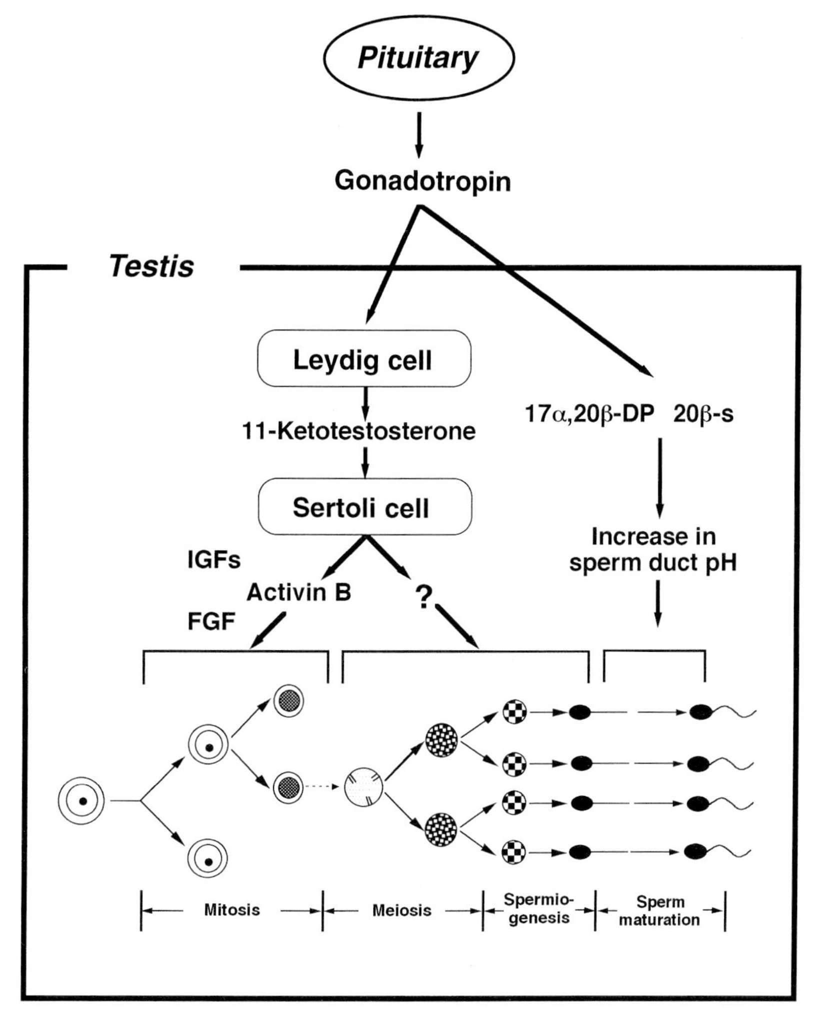

FIGURE 3 Model illustrating hormonal regulation of spermatogenesis in fish.

DP injections induce the acquisition of sperm motility, neither 11-ketotestosterone nor testosterone are effective. In masu salmon, $17\alpha,20\beta$-DP does not act directly on the sperm, but its action is mediated through an increase in the seminal plasma pH, and this in turn increases cAMP content in the sperm allowing the acquisition of sperm motility. It is thought that the Japanese eel possesses similar mechanisms for the acquisition of sperm motility.

Recently, the presence of a 20β-S progestogen binding site was reported on the Atlantic croaker sperm membrane. This raises the possibility of the presence of another mechanism by which the steroid acts directly on the sperm to induce acquisition of sperm motility.

Figure 3 illustrates a generalized spermatogenetic cycle and its predicted regulatory mechanisms in fish. This cycle and its control mechanisms cannot be applied to all fish species. Since fish constitute the largest group of living vertebrates (~24,600 species of which about 23,700 are teleosts), their breeding styles and spermatogenesis tend to be diverse and species specific. The spermatogenetic cycles of fewer than 100 fish species have been investigated. The investigation of these species has led to the discovery of several interesting aspects of the spermatogenetic cycle. It is highly possible that future investigations of fish spermatogenesis will lead to a better understanding of the general aspects of spermatogenesis.

See Also the Following Articles

MALE REPRODUCTIVE SYSTEM, FISH; MODES OF REPRODUCTION, FISH; SERTOLI CELLS, OVERVIEW; SPERMATOGENESIS, OVERVIEW; SPERMIOGENESIS

Bibliography

Billard, R. (1986). Spermatogenesis and spermatology of some teleost fish species. *Reprod. Nutr. Develop.* **26**, 877–920.
Billard, R., Fostier, A., Weil, C., and Breton, B. (1982). Endocrine control of spermatogenesis in teleost fish. *Can. J. Fish. Aquat. Sci.* **39**, 65–79.
Grier, H. J. (1981). Cellular organization of the testis and spermatogenesis in fishes. *Amer. Zool.* **21**, 345–357.
Loir, M., and Le Gac, F. (1994). Insulin-like growth factors—I and II binding and action on DNA synthesis in rainbow trout spermatogonia and spermatocytes. *Biol. Reprod.* **51**, 1154–1163.
Miura, T., Yamauchi, K., Takahashi, H., and Nagahama, Y. (1991). Hormonal induction of all stages of spermatogenesis *in vitro* in the male Japanese eel (*Anguilla japonica*). *Proc. Natl. Acad. Sci. U.S.A.* **88**, 5774–5778.
Miura, T., Yamauchi, K., Takahashi, H., and Nagahama, Y. (1992). The role of hormones in the acquisition of sperm motility in salmonid fish. *J. Exp. Zool.* **261**, 359–363.
Nagahama, Y. (1994). Endocrine regulation of gametogenesis in fish. *Int. J. Dev. Biol.* **38**, 217–229.
Nelson, S. S. (1994). *Fishes of the world,* 3rd Ed. John Wiley & Sons, New York.

Spermatophores in the Arthropods

Heather C. Proctor
Griffith University

GLOSSARY

accessory fluids The liquid part of the ejaculate accompanying and surrounding the spermatozoa (or earlier developmental stages of sperm cells).

direct sperm transfer The male places his ejaculate on or in the female's sperm-receiving structure.

indirect sperm transfer The male places a spermatophore on a substrate (e.g., soil, leaf, or pebble) and the female takes up the sperm by moving her genital opening over it. Transfer may be paired (male courts a given female) or dissociated (no courtship and often no contact at all between the sexes).

sex pheromone A chemical produced by one sex that attracts or arrests the movement of the opposite sex. Pheromones may act at a distance or they may require direct contact.

sperm capsule The rigid, sclerotized covering that often encases a sperm packet.

sperm droplet A naked, unpackaged droplet of spermatozoa and accessory fluids.

sperm packet Spermatozoa and accessory fluids packaged in a flexible membranous covering.

Spermatophore is a general term referring to a male ejaculate that is packaged into a self-contained unit of sperm cells and accessory fluids. This contrasts with the unpackaged, liquid ejaculate common to mammals and most other vertebrates and to the freely spawned sperm of many marine invertebrates. In the strict definition, this ejaculate is encased in a flexible or rigid coat and is often set on a stalk; however, in this article, naked sperm droplets will also be considered spermatophores. Spermatophores serve to protect or aid a male's ejaculate before, during, or after transfer to the female. Spermatophore size and morphology are relatively constant within a species but vary between them, making these structures taxonomically useful for differentiating between closely related species.

I. TAXONOMIC DISTRIBUTION OF SPERMATOPHORES

A. Nonarthropods

In many species of marine invertebrates, males release sperm into the water as spermatophores and females collect them to internally fertilize their eggs (e.g., Pogonophora, Phoronida, and many families of Polychaeta). In an even larger number of invertebrate taxa, males transfer spermatophores directly to the genital openings of females or place them on the female's body (e.g., all Cephalopoda and Onychophora and many species in the Platyhelminthes, Nematoda, Gastropoda, Oligochaeta, and Hirudinea). There are even some aquatic vertebrates that produce spermatophores. Male salamanders and newts (Urodela) deposit spermatophores that females pick up, and males of the cyprinodont fish *Horaichthyes* attach spermatophores near the female's genital opening. Although older literature describes the production of spermatophores as an adaptation to life on dry land, it is clear that sperm packaging has evolved many times in aquatic situations in which desiccation of sperm was not a selective factor.

B. Arthropods

In the phylum Arthopoda, the ancestral mode of sperm transfer probably involved free-spawning of eggs and sperm. Horseshoe crabs (Xiphosura) and sea spiders (Pycnogonida), both considered to be primitive groups of marine arthropods, have external fertilization and unpackaged spermatozoa. However, the vast majority of aquatic and terrestrial arthropods have internal fertilization and transfer sperm in the form of spermatophores (Table 1). This is true for many Crustacea, including crabs, lobsters, and crayfish (Decapoda), as well as less familiar groups such as Copepoda, Isopoda, and Ostracoda. In the Arachnida, transfer of sperm via spermatophores is the rule for scorpions, pseudoscorpions, whipscorpions (Uropygi), tailless whipscorpions (Amblypygi), and microwhipscorpions (Schizomida). With some exceptions, all Acari (mites, including ticks) have spermatophores. Only the harvestmen (Opiliones), sun spiders (Solifugae), ricinuleids (Ricinuleida), and true spiders (Araneae) primarily transfer liquid ejaculates, but even here there is an exception; males of the spider family Telemidae transfer spermatophores rather than free sperm. In the Myriapoda, spermatophores are produced by male centipedes (Chilopoda), symphylans (Symphyla), and pauropods (Pauropoda). Most millipedes (Diplopoda) directly transfer liquid ejaculates with the exception of the family Polyxenidae, in which males deposit sperm droplets on a silken substrate and the females later take up the sperm. In the Hexapoda, almost all apterygote (ancestrally wingless) taxa produce spermatophores and transfer them indirectly. In the pterygote hexapods (winged insects) there are no species with indirect sperm transfer; however, many orders include species that transfer spermatophores directly. Pterygote spermatophores may be extruded by the male prior to copulation and then transferred to the female's genital opening (e.g., katydids, crickets, and grasshoppers: Orthoptera) or may be retained inside the male's genital tract and transferred to the female's tract during copulation (e.g., moths and butterflies: Lepidoptera). The Coleoptera (beetles) includes species that internally transfer spermatophores and those that transfer liquid ejaculates. Spermatophores are seldom produced in two orders of insects, the Diptera (true flies) and the Hymenoptera (ants, wasps, bees, and sawflies), although there are some species in each order that do transfer spermatophores. Males of some other insect orders produce sperm balls, unencapsulated agglomerations of spermatozoa that are transferred directly along with a liquid ejaculate (e.g., some Thysanoptera, Psocoptera, and Pthiraptera); however, such structures are not considered to be true spermatophores.

II. SPERMATOPHORE MORPHOLOGY AND TRANSFER BEHAVIOR

Because there is such a huge diversity of spermatophores in the arthropods, only selected examples of spermatophore structure and function are discussed.

TABLE 1
Types of Spermatophores and Sperm Transfer Modes in the Phylum Arthropoda

Taxon	*Common name*	*Spermatophore morphology*	*Transfer mode*	*Signaling structure*	*Transfer mechanism*
Subphylum Crustacea					
Copepoda	Copepods	Flask-shaped sperm capsule; unstalked, but sometomes with large adhesive plate	PDE	None	Osmotic expulsion of sperm
Ostracoda	Seed shrimps	Balloon-shaped sperm "bladder" connected to long tail	PDI?	None	None
Isopoda	Sowbugs, pillbugs, slaters, sea lice	Extracellular matrix surrounding sperm bundle; unstalked; many spermatophores per ejaculate	PDI	None	None
Cumacea	Cumaceans	Tube-shaped sperm packet; unstalked	PDE	None	None?
Mysidacea	Opossum shrimp	Tube-shaped sperm packet; unstalked	PDE?	None	None?
Euphausiacea	Krill	Unstalked packet?	PDE	None	None?
Decapoda	Crabs, lobsters, crayfish, prawns, shrimp	Small unstalked packets; many stalked packets with common base; tubular multilayered capsules	PI, PDI, PDE	None	Female may release substance that dissolves spermatophore coating
Subphylum Chelicerata					
Scorpiones	Scorpions	Two rigid hemispermatophores glued together medially; stalked	PI	None	Mechanical extrusion of sperm
Uropygi	Whipscorpions	Two sperm packets mounted on single stalk	PI	None	None
Amblypygi	Tailless whipscorpions	Two sperm packets or masses mounted on single stalk	PI	None	None
Schizomida	Microwhipscorpions	Two sperm packets mounted on single stalk	PI	None	None
Araneae[a]	Spiders	Unstalked?	PDI	None	None?
Pseudoscorpions	False scorpions, book scorpions	Sperm droplets or capsules; simple or elaborate stalk	Dis, PI	Pheromones, threads	Osmotic and mechanical extrusion of sperm
Acari	Mites, ticks	Sperm droplets, packets, or capsules; simple, complex, or no stalk	Dis, PI, PDE	Pheromones, threads, secretion tracks	Gaseous expulsion of sperm
Class Myriapoda					
Diplopoda	Millipedes	Sperm droplets on threads	Dis	Pheromones, threads	None
Pauropoda	Pauropods	Sperm droplets on threads	Dis	Pheromones?	None?
Chilopoda	Centipedes	Sperm droplets or capsules supported by threads	PI, PDE	Pheromones?, threads	None?
Symphlya	Symphylans	Sperm droplets; simple stalk	Dis	Pheromones?	None
Class Hexapoda					
Collembola	Springtails	Sperm droplets; simple stalks or no stalks	Dis, PI, PDI	Pheromones	Nuptial gift
Diplura	Diplurans	Sperm packets; simple stalks	Dis	Pheromones?	None
Thysanura	Silverfish	Flask-shaped sperm capsule; unstalked	PI	Threads	Osmotic or mechanical expulsion of sperm?
Microcoryphia	Bristletails	Sperm droplets on threads; stalked droplets on substrate; droplets transferred directly	PI, PDI	Threads	None
Pterygote Insecta[b]	Winged insects	Sperm packet or capsule; unstalked	PDE, PDI	None	Osmotic explusion of sperm?, nuptial gift

Note. Question marks indicate subjects that are poorly described in the literature. Dis, dissociated; PI, paired-indirect; PDE, paired, direct, spermatophore transferred external to male reproductive tract; PDI, paired, direct, spermatophore transferred internally within the male reproductive tract.

[a] Only a single family, Telemidae.

[b] See text for orders of winged insects that transfer spermatophores.

A. Subphylum Crustacea

1. *Subclass Copepoda*

Spermatophores have been described from four orders of copepods: Calanoida, Harpacticoida, Cyclopoida, and Siphonostomatoida. Depending on the species, the male extrudes a flask-shaped spermatophore from his genital opening either before or after having grasped a female. While holding onto the female with his modified antennae and/or thoracic appendages, he uses his thoracic legs to move the spermatophore to the female's venter and glues the open neck of the flask to the female's gonopore. In some calanoids, the male produces an articulated plate that fits the ventrolateral surface of the female's genital segment like a shield. The sperm-bearing flask of this spermatophore complex hangs freely from the plate, clasped by the neck at the point of articulation between anterior and posterior sections. The neck of the flask extends between the plate and the female's venter, terminating in her genital opening. In both calanoids and cyclopoids, the distal section of the sperm-containing flask is filled with hygroscopic Q-bodies (Q = quellen; i.e., "swelling"). The Q-bodies absorb water and swell, thereby pumping the spermatozoa into the female's genital opening (Fig. 1a). The male may release the female immediately after affixing the spermatophore (Calanoida) or remain in contact with the female for minutes to hours (Harpacticoida and Siphonostomatoida). After the flask has emptied itself, the female pries the empty spermatophore from her venter. The loosening of the spermatophore adhesive may be hastened by solvents secreted by pit pores on the female's ventrolateral surface.

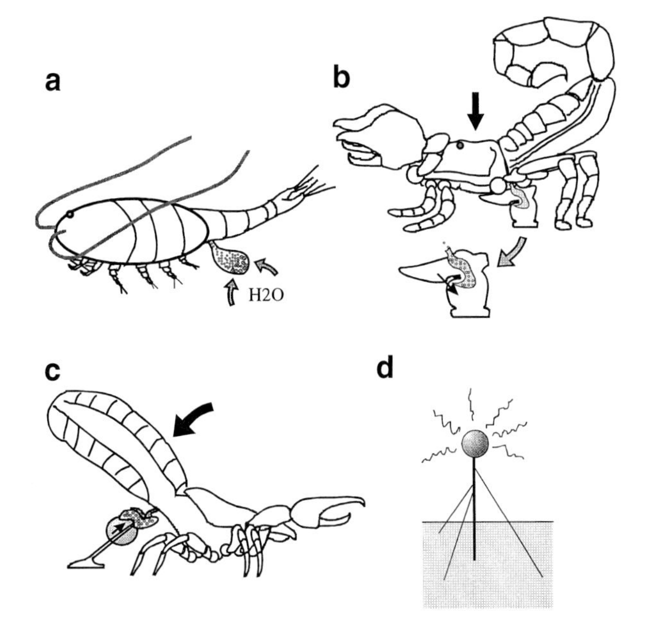

FIGURE 1 Transfer mechanisms and signaling structures in arthropod spermatophores. (a) Hygroscopic expansion of Q-bodies forces sperm out of a copepod spermatophore; (b) weight of female scorpion forces lever arm down to squeeze sperm out of spermatophore (male partner omitted); (c) female pseudoscorpion presses sperm packet into droplet of hyposmotic fluid, which enters the sperm packet to cause expulsion of sperm (male partner omitted); (d) spermatophore of dissociated mite showing signal threads and pheromones.

2. *Class Malacostraca*

Within the Malacostraca, spermatophores have been described in the orders Euphausiacea, Decapoda, Isopoda, Cumacea, and Mysida. Although they are widespread within this class, spermatophores are best described within the shrimps, crabs, lobsters, and crayfish (Decapoda). Decapods are particularly interesting because fertilization may be internal or external. The assembly of spermatophores in this group typically occurs within the male's genital tract. Assembled spermatophores are stored either in the vas deferens or in a special spermatophore ampulla until the time of sperm transfer. In some brachyuran crabs, the spermatophores are extruded through the male gonopores at the bases of the fifth walking legs and are introduced directly into the female's genital ducts. However, in other brachyurans and in lobsters, crayfish, and penaeid shrimp, the spermatophores are deposited externally in a groove or pouch near the female's genital opening, and fertilization occurs externally. In hermit crabs (infraorder Anomura), males and females of most species partially emerge from their shells, appress their venters, and simultaneously release spermatophores and eggs. However, in one species (*Pagurus prideaux*) the male attaches spermatophores to the internal wall of the appropriated snail shell occupied by the female.

Three general morphologies of decapod spermatophores have been described. Brachyuran crabs produce small spherical or ellipsoid sperm masses each surrounded by a thick matrix of material. Hermit crabs produce stalked spermatophores in clusters of several spermatophores attached to common gelatinous bases. The number of spermatophores per cluster may vary intraspecifically, e.g., 2–14 in *Pagurus novizealandiae*. Multilayered tubular spermatophores are produced by most crayfish, lobsters, and shrimp. The mechanism of rupture for spermatophores deposited externally on female decapods is unknown but it is possible that secretions produced by the female prior to egg release cause swelling of the inner layers of the spermatophore, leading to rupture of the spermatophore walls and liberation of spermatozoa.

B. Class Chelicerata, Subclass Arachnida

1. *Order Scorpiones*

Scorpion spermatophores are complex, stalked structures that are deposited on a substrate by the male during a mating dance with the female. The spermatophore is composed of two preformed hemispermatophores. A single pair of hemispermatophores is stored in the male's paraxial organs at any time. During spermatophore deposition, the hemispermatophores slide out of their respective paraxial organs, are glued together medially in the genital atrium, and are attached to the substrate. The sperm mass is contained in a central chamber that is not exposed to desiccation. There are two general morphologies of scorpion spermatophores: flagellar (in the Buthidae) and lamellate (in all other families). After depositing the spermatophore, the male encourages the female to walk over it by pulling on her pedipalps with his own. There are two mechanisms of sperm removal that correspond to spermatophore shape. In species with lamellate spermatophores, when the male first pulls the female over the spermatophore, hooks on the dorsal surface of the spermatophore engage with flap covering the female's genital opening. When she subsequently moves backwards, the flap is pulled open. The male then presses down on the female or the female presses down without the male's help. The weight of the female's body depresses the lamella of the spermatophore, causing the spermatophore to buckle and the sperm mass to be expressed into the female's genital chamber (Fig. 1b). In species with flagellate spermatophores, after depositing the ejaculate the male walks backwards while the flagellar extensions are still attached to his genital opening. This causes the spermatophore to bend, allowing the female to more readily engage hooks on the spermatophore with her genital opercula. This pulls the female's opercula open and allows the sperm-conducting tube of the spermatophore to enter her genital opening. After mating it may take a male from 3 days to 2 weeks to regenerate another pair of hemispermatophores.

2. *Order Pseudoscorpiones*

In addition to the paired-indirect mode of sperm transfer shown by scorpions, many species of pseudoscorpions engage in dissociated transfer. In this mode, a male does not court a specific female but rather deposits spermatophores on a substrate, departs, and females later encounter the spermatophores and take up sperm without assistance from the male. Spermatophores produced by pseudoscorpions come in a great variety of shapes, particularly with regard to the apical sperm-bearing structures, although spermatophores of most species have similar thin, solid stalks. Unlike scorpions, pseudoscorpions do not preform all components of their spermatophores but rather shape the stalk from a liquid matrix as it is extruded; however, very complicated apical portions are likely to be preformed in some species.

i. Pairing Species In most pseudoscorpions that pair during sperm transfer (superfamily Cheliferoidea), the male grasps the female's pedipalps, legs, or body with his pedipalps and leads her about in a courtship "dance" until he is ready to deposit a spermatophore containing hundreds of sperm cells. In the family Cheliferidae, the male attracts a female to his spermatophore by everting two sacs on his abdomen that are coated with attractive pheromones. When the female walks over the spermatophore, he grasps her and presses her down on it, and uses his forelegs to help push the sperm masses into her genital aperture. In the other cheliferoid families (Chernetidae, Atemnidae, and Withiidae) the male pulls the female over the spermatophore but does not push sperm in with his forelegs. In pairing pseudoscorpions there are several mechanisms that aid in sperm transfer. In the Cheliferidae and Chernetidae, there is a large droplet of fluid on the stalk of the spermatophore that is apparently hypotonic to the accessory fluids inside the sperm packet. When the female presses (or is pressed) down on the spermatophore, this brings the sperm packet and the stalk droplet into contact (Fig. 1c). In the Chernetidae, this results in the extrusion of a tube from the sperm packet that directs the expanding fluids and sperm cells into the female's genital opening. In cheliferids, males simply stuff the mass in. In the Atemnidae and Withiidae, the sperm stalk lacks this hyposmotic droplet. Rather, much as in scorpions, the male presses the female down on lever-like extensions of the sperm capsule, which results in deformation of the capsule and expression of sperm into the female's genital opening. Most cheliferoids deposit only one spermatophore per courtship bout, with deposition and uptake taking 10–30 min, but in some species (e.g., *Epactiochernes tumidus*) several are produced per bout.

ii. Nonpairing Species Spermatophores of nonpairing pseudoscorpions are morphologically much simpler than those of pairing species. The stalk is a thin, unelaborated filament without a fluid droplet, and the sperm is contained in a simple spherical sperm packet rather than in the often convoluted or lever-bearing capsules of the Cheliferoidea. Some of these species do have the tip of the stalk modified into a cup or crown that cradles the sperm droplet/packet. The number of sperm cells/spermatophore is strikingly low, ranging from 8 to 100. With very few exceptions, males do not require physical contact with a female before beginning a bout of spermatophore deposition. In some species, the male guards a small territory in which he maintains a "garden" of spermatophores, pruning them as they age and replacing them with fresh ones (e.g., *Chthonius tetrachelatus*). In most species, males react aggressively to unfamiliar or old spermatophores and tear them up or crush them when encountered, usually depositing their own beside the ruined ejaculate. There have also been observations of females destroying spermatophores by breaking the stalks with their mouthparts. Females are probably attracted to spermatophores by volatile pheromones associated with the ejaculates. In one species (*Serianus carolinensis*), the female is funneled to the exact site of the spermatophore by silken strands deposited by the male. To take up sperm, the female simply walks over a spermatophore until her genital opening is positioned above it, lowers herself, then walks away leaving the emptied spermatophore behind. It is hypothesized that accessory fluids in the sperm packet are hygroscopic, and that the humid environment of the female's genital atrium causes the packet to swell and burst.

3. *Subclass Acari*

The two groups that comprise most of the Acari, the superorders Parasitiformes and Acariformes, are readily distinguished by their modes of sperm transfer.

i. Superorder Parasitiformes As far as is known, all species in the Parasitiformes (including the ticks, Order Ixodida) show paired, direct transfer in which the male inserts all or part of an unstalked spermatophore into the sperm-receiving structure of the female. In most taxa, the female receives sperm in the "true" genital opening; that is, the opening through which eggs are laid. In the Dermanyssina and Parasitina of the Order Mesostigmata, however, the female's genital opening is displaced laterally from the normal site to the axillar region of a leg or even to the leg itself. Spermatophores of ticks have been particularly well studied. Although they appear externally simple, tick spermatophores have a complex internal structure and mechanism. When a male finds a female, he climbs beneath her and inserts his chelicerae into her genital opening, where he detects chemicals that indicate the female's species and readiness for mating. If these factors are satisfactory, the male begins to extrude a large spermatophore from his genital opening. First a droplet of mucopolysaccharides is produced, then a droplet of proteins is extruded inside this. A mass of sperm cells and accessory fluids is ejected into the protein droplet, followed by adlerocysts (yeast-like symbionts of the tick). All these layers together are termed the ectospermatophore. Finally, the male injects a multicomponent structure called the endospermatophore that caps off the ejaculate. Producing the spermatophore is a relatively rapid procedure, lasting about 30 sec. The male grasps the completed spermatophore in his chelicerae and applies the neck to the female's genital opening. At this point, through an unknown reaction, CO_2 is generated between the two layers of the ectospermatophore wall. The pressure of the expanding gas causes the endospermatophore to extrude its components into the female's reproductive tract and forces the contents of the ectospermatophore (sperm cells, fluids, and adlerocysts) through the neck of ectospermatophore. This explosive mechanism is not known to occur in spermatophores of other parasitiform taxa.

ii. Superorder Acariformes Mites in this superorder show a much greater diversity of sperm transfer modes than do those in the Parasitiformes. The ancestral behavior is probably paired-indirect transfer of stalked spermatophores. From this state there have been many independent evolutions of direct transfer of spermatophores, insemination of liquid ejaculates, and dissociated transfer of spermatophores. In the beetle mites (Order Oribatida), sperm transfer in sexual species is almost invariably dissociated via unadorned stalked spermatophores with spherical sperm packets or droplets. Although the sperm packets appear superficially simple, SEM has shown that they are internally complex. The stalks are generally unelaborated, but in one species of *Pergalumna*, the male deposits several lines of signaling structures that radiate from the central spermatophore. This arrangement likely increases the probability of females contacting the ejaculate, either through pheromones or by intercepting her as she walks nearby. Similar signaling structures in the forms of threads, trails of secretions, or spermless (but presumably fragrant) stalks are produced by males of many species in the Order Prostigmata that show dissociated transfer, particularly in terrestrial members of the cohort Parasitengona (Fig. 1d). Dissociated species of aquatic Parasitengona (water mites) do not produce such obvious signaling devices but do deposit large "fields" of spermatophores that probably produce a strong pheromonal aura to attract females swimming or crawling past. Spermatophores of species that live in dry environments often have sperm packets that contain hygroscopic fluids that absorb water vapor from the atmosphere and prevent desiccation. As in pseudoscorpions, destruction or deposition of new spermatophores on top of those of other males appears to be a common behavior in dissociated species of Parasitengona and other Prostigmata. All members of the Order Astigmata transfer liquid ejaculates via copulation.

C. Hexapoda

1. *Apterygote Hexapoda*

i. Order Collembola With one exception, (*Sphaeridea pumilis*), all springtails transfer sperm indirectly, most with simple stalked spermatophores with spherical sperm droplets and some with unstalked sperm droplets deposited on the substrate. There is a slight elaboration of the stalk base in the aquatic species *Podura aquatica* that allows the spermatophore to float on the water's surface. Pairing and courtship occurs in *P. aquatica* and in a number of other species, but dissociation appears to be more common. Male *Dicyrtomina minuta* seem to show a mixed strategy. When population density is low, a male *D. minuta* lucky enough to encounter a female "fences her in" with a circle of spermatophores; when densities are high, the male ignores individual females and deposits fields of ejaculates. Males of dissociated species do not produce obvious signaling structures (e.g., threads), although spermatophores and other male secretions appear to have pheromonal properties. Like those of some terrestrial mites, spermatophores of some dissociated springtails are able to maintain hydration by taking up water vapor. In some pairing taxa, the male directs the female over his spermatophore by enticing her forward with a blob of sperm held in his crooked antenna; when the female has her genital opening over the top of the spermatophore, the male stops and allows her to feed on the sperm he holds. In *S. pumilis*, the only springtail known to transfer sperm directly, the male uses his hind legs to push a sperm droplet into the female's genital opening.

ii. Orders Thysanura and Microcoryphia Silverfish (Thysanura) and most bristletails (Microcoryphia) show paired, indirect transfer of sperm. Both groups also make use of guiding threads that indicate to the female where the spermatophore lies (silverfish) or that actually bear the sperm droplets (bristletails). A courting male silverfish walks forward with his female partner following behind until he encounters a vertical surface. He then deposits a stalkless, flask-shaped spermatophore on the substrate, surrounding it with threads produced from glands near the tip of his abdomen, and raises his abdomen to attach threads to the vertical surface that he then runs diagonally to the substrate, thus creating a "lean-to" under which the female walks. When the cerci on the female's upraised abdomen encounter the main signal thread that descends from the wall, she halts and lowers her abdomen, thereby bringing her ovipositor into contact with the spermatophore. Although the opening of the spermatophore is initially far from the genital aperture, within three minutes the neck of the flask elongates until sperm can be released into the female's genital opening. It is thought that pumping movements of the female's abdomen help bring this about, but it is not clear whether the mechanism is purely mechanical or osmotic, as in some pseudoscorpions.

Most male bristletails also make use of a major guiding thread, but rather than depositing an encapsulated spermatophore on the substrate, the male attaches the thread to the ground and then deposits several sperm droplets on the thread as it is being extruded, the resulting product looking like beads on a string. He then pushes his female partner around until he can bring the sperm-bearing thread in contact with the base of her ovipositor. In some genera of bristletails, males deposit stalked spermatophores on the ground rather than on a thread, and in a few species the male extrudes a droplet of sperm directly onto the female's ovipositor.

2. *Pterygote Hexapoda (Winged Insects)*

All winged insects transfer sperm directly by apposition of the primary genitalic openings, with the exception of the Odonata (dragonflies and damselflies), in which males transfer sperm to secondary genitalia prior to copulating with the female. Most orders of pterygotes include at least a few species that produce spermatophores (see Section I). Typically, the spermatophores are small and transferred internally (within the male's copulatory appendage); however, many katydids (Orthoptera: Ensifera) externally transfer enormous ejaculates, up to 30% of the male's own body weight. The majority of such an ejaculate is composed of a sticky proteinaceous mass (the spermatophylax) that the female eats. The spermatophylax is both a nutritious "nuptial gift" from the male to the female and has the additional role of occupying the female long enough to allow the contents of the sperm ampulla to be evacuated into the female's genital tract. Evacuation may be aided by the expansion of a hygroscopic "pressure body" in the ampulla. After the female consumes the spermatophylax, she reaches back to remove the emptied capsule of the ampulla and eats it. There is a great deal of evidence for the existence of nutritional substances, anaphrodisiacs, and oviposition-inducing chemicals in the spermatophores of many groups of winged insects.

III. MORPHOLOGICAL AND BEHAVIORAL TRENDS

Although our knowledge is restricted to relatively few taxa, several patterns emerge from the diversity of spermatophore morphology and sperm transfer behavior in the arthropods. These include the repeated evolution of: physical signaling structures associated with spermatophores for both dissociated taxa (Myriapoda, Pseudoscorpiones, and Acari) and paired taxa with indirect transfer (Thysanura and Microcoryphia); spermatophore-associated pheromones (Myriapoda, Pseudoscorpiones, and Acari); osmotic mechanisms for maintaining hydration (Collembola and Acari); osmotic, mechanical, or gaseous mechanisms for expelling sperm from sperm capsules (Copepoda, Scorpiones, Pseudoscorpiones, and Ixodida); spermatophores as nuptial gifts (Collembola and Orthoptera); external sperm competition (Pseudoscorpiones and Acari); and dissociated species producing many, small spermatophores and paired ones producing few, large ones (Collembola, Pseudoscorpiones, and Acari). Much research is needed to understand how these morphological and behavioral adaptations help to ensure sperm transfer in terrestrial and aquatic environments that are often inimical to unprotected sperm.

See Also the Following Article

CHELICERATE ARTHROPODS

Bibliography

Blades, P. I., and Youngbluth, M. J. (1988). Morphological, physiological, and behavioral aspects of mating in calanoid copepods. In *Proceedings of the 3rd International Conference on Copepoda* (G. A. Boxshall and H. K. Schminke, Eds.), pp. 39–51. Kluwer, Dordecht.

Feldman-Muhsam, B. (1986). Observations on the mating behaviour of ticks. In *Morphology, Physiology, and Behavioral Biology of Ticks* (J. R. Sauer and J. A. Hair, Eds.), pp. 217–232. Horwood, Chichester, UK.

Fernandez, N. A., Alberti, G., and Kümmel, G. (1991). Spermatophores and spermatozoa of oribatid mites (Acari: Oribatida). Part I: Fine structure and histiochemistry. *Acarologia* **32**, 261–286.

Mann, T. (1984). *Spermatophores: Development, Structure, Biochemical Attributes, and Role in the Transfer of Spermatozoa.* Springer-Verlag, New York.

Proctor, H. C. (1992). Mating and spermatophore morphology of water mites (Acari: Parasitengona). *Zool. J. Linnean Soc.* **106**, 341–384.

Proctor, H. C. (1998). Indirect sperm transfer in arthropods: Behavioral and evolutionary trends. *Annu. Rev. Entomol.* **43**, 153–174.

Schaller, F. (1971). Indirect sperm transfer by soil arthropods. *Annu. Rev. Entomol.* **16**, 407–446.

Weygoldt, P. (1969). *The Biology of Pseudoscorpions.* Harvard Univ. Press, Cambridge, MA.

Witte, H. (1991). Indirect sperm transfer in prostigmatic mites from a phylogenetic viewpoint. In *The Acari: Reproduction, Development and Life-History Strategies* (R. Schuster and P. W. Murphy, Eds.), pp. 107–176. Chapman & Hall, New York.

Spermatozoa

Clarke F. Millette
University of South Carolina School of Medicine

GLOSSARY

acrosome A membrane-limited vesicle covering the anterior surfaces of the sperm nucleus, which contains hydrolytic enzymes required for penetration of the protective investments of the ovum.

axoneme The central portion of the sperm flagellum containing a microtubular array with associated proteins; directly responsible for sperm motility.

flagellum The motile apparatus of the spermatozoon with associated accessory structures.

membrane polarity Topographical compartmentalization of lipids and proteins on the surface of cells.

motility Locomotive ability conferred by whip-like motions of the sperm flagellum.

spermatozoon The male gamete; produced by spermatogenesis in the seminiferous epithelium of the testis.

Spermatozoa are the male gametes. They contain one-half of the normal somatic DNA complement and are highly specialized in terms of their function and structure. Unlike other mammalian cells, they are destined to leave the body in order to complete their physiological role of uniting with the egg to form the zygote. To fulfill this role, spermatozoa have evolved a uniquely polarized structure designed to facilitate passage through both the male and the female reproductive tracts. Simultaneously, the genetic material carried by the sperm must be protected from adverse influences. Therefore, both the sperm head containing the DNA and the sperm tail responsible for cell motility exhibit specialized morphological features not present elsewhere. Moreover, the topographical positioning of many if not most sperm proteins and lipids also reflects an extraordinary high degree of compartmentalization.

I. INTRODUCTION

Spermatozoa are the male gametes, haploid cells with one-half of the usual genetic DNA complement. Sperm are destined for fusion with the egg at fertilization to form the zygote. In mammals, sperm are formed in the testis during the process of spermatogenesis. Still maturing sperm are then released into the excurrent duct system of the testis and pass into the epididymis. During transit through the relatively lengthy epididymal duct, spermatozoa undergo substantial biochemical and morphological alterations which finally render them capable of fertilization. Additional changes occur to spermatozoa during ejaculation when they come into contact with secretions of the male accessory reproductive glands and also as they traverse the female reproductive tract before contacting the ovum. It appears, however, that these later effects modulate sperm–egg interactions but are not absolutely necessary for effective fertilization.

Sperm exhibit a variety of features which are either unique or which are found in few other cell types, including (i) a haploid genome, (ii) a highly condensed and transcriptionally inactive nuclear DNA content, (iii) specialized nuclear proteins involved in protection of this genetic material, (iv) an acrosomal package of hydrolytic enzymes located on the anterior aspect of the sperm head, (v) an axoneme equipped for motility with accessory structural features not seen in flagella of other eukaryotic or prokaryotic cells, and (vi) a plasma membrane exhibiting extreme polarity in the arrangement of its protein and lipid constituents. It should also be noted that the function of spermatozoa is unique. In mammals, they are the only cell type designed to leave the organism in order to justify their existence. Many of their specialized physical and biochemical traits have evolved to ensure successful passage through both the male and the female reproductive tracts, to protect the DNA in the sperm head during this passage, and to assist in the penetration of the egg's protective investments and final fusion with the egg plasma membrane.

This article will concentrate on sperm structure as it relates to sperm function. Spermatozoa of mammals will be emphasized, but the reader is cautioned that nonmammalian spermatozoa, including plant spermatozoa, are of great importance and interest. These gametes exhibit a tremendous diversity of structural features, which have also evolved to subserve the particular requirements of each species.

II. HISTORY OF SPERMATOLOGY

Spermatozoa were one of the first animal cell types ever examined using a microscope. Sperm are not only intimately involved in reproduction, a subject that has long fascinated both scientists and philosophers, but also they are easy to obtain. As a result, many of the earliest advances in microscopic science were made using the male gamete as the experimental subject of interest.

Mammalian spermatozoa were first observed by Anton van Leeuwenhoek in 1679. His observations were conducted on a variety of species but were generally unrecognized for over 100 years. Other early microscopists subsequently examined spermatozoa. The work Ledermuller published in 1758 was particularly thorough, although he misjudged the relative lengths of sperm tails in relation to the dimensions of the sperm heads. Little further progress in the morphological assessment of sperm structure was made until the middle of the nineteenth century. With the development of improved equipment and an increased general pace of overall scientific activity in Europe, investigators began to describe the detailed structural characteristics of male gametes.

Prevost and Dumas in 1821 correctly reported that guinea pig sperm were larger than mouse spermatozoa. In 1837, Dujardin and also Wagner first used achromatic lenses for the study of spermatozoa. Independently, these workers accurately described the falciform shape of the heads of rodent gametes and published the first descriptions of both the sperm flagellar midpiece and the cytoplasmic droplet. Dujardin also identified an apical substructure on the guinea pig sperm head. This region is now recognized as the acrosome. Additional observations by other investigators soon followed. Valentin, Gerber, and Pouchet described bear, guinea pig, and human sperm, respectively, between 1839 and 1847. Pouchet's work postdated the seminal discovery by Kolliker in 1841 that spermatozoa arise by cell division and subsequent differentiation in the testis. In 1847, it was established that spermatozoa were cells and not independent organisms.

Further progress was made in 1865 by Schweiggar-Seidel, who stained vertebrate sperm with ammonia carmine and aniline red. This study was one of the first to apply a synthetic dye for the examination of biological material. Coupled with the use of a newly developed water immersion lens, this author described the sperm flagellar midpiece for the first time. In 1887, Jensen pursued these observations and by chemically treating sperm discovered that the midpiece exhibited spirally organized cross-striations, now known to be individual mitochondria. Jensen also noted for the first time that the sperm tail was not structurally homogeneous but was composed of a number of separate filaments.

Finally, in 1909 Retzius published a distinctive series of light microscopic studies on spermatozoa. His observations were so thorough and accurate that important new advances in our understanding of sperm structure did not occur until the invention of the transmission electron microscope about 40 years later. Ultrastructural descriptions of the fine details of spermatozoan cellular organization have since proceeded at a great pace due in large part to the application of advanced histochemical and electron microscopic technology. Prominent researchers who have produced important contributions to our knowledge base are extremely numerous and by the 1970s most of the intricate details of sperm structure had been described. The work of Leblond, Clermont, Hancock, Anderson, Afzelius, Holstein, Colwin and Colwin, Hadek, Bedford, Yanagimachi, Phillips, and Fawcett stands out particularly, although this list is far from complete. Today, substantial new insights into sperm structure are being described continually by observers applying ever more powerful microscopic procedures.

III. STRUCTURAL CHARACTERISTICS OF SPERMATOZOA

A. Overall Description

A typical mammalian spermatozoon has three major subcellular regions: the sperm head, the midpiece, and the principal piece. The midpiece and the principal piece together constitute the tail of the sperm (Fig. 1). Exact details of cellular morphology, for example, the size and shape of the head or the length of the tail, vary greatly even between closely related species. Furthermore, within an individual one often finds significant numbers of abnormally shaped cells, including sperm exhibiting double nuclei or several tails. Therefore, only the basic traits of general sperm structure will be outlined here.

The overall length of mammalian spermatozoa varies considerably. Human spermatozoa have a total length of approximately 60–70 μm, whereas mouse and rat sperm have a length of about 130 and 190 μm, respectively. Spermatozoa of some amphibians may reach a length of several millimeters and male gametes of the hemipteran insect *Notonecta glauca* have a length of more than 1 cm.

Reliable estimates of the volume of individual mammalian spermatozoa are few, but investigators have inferred from the overall cellular dimensions that bull sperm occupy about 30 cubic μm. The dry weight of bull sperm is about 16.5×10^{-12} g with approximately half of the cellular mass being contributed by the sperm tail. For comparison, the ram spermatozoon has a dry weight of 7.2×10^{-12} g. These values are probably typical for most mammalian spermatozoa. It is also of note that in contrast to most somatic cells, in which 80–90% of the cell is water, mammalian sperm are about 50% dry matter. The relative lack of cytoplasm and the uniquely

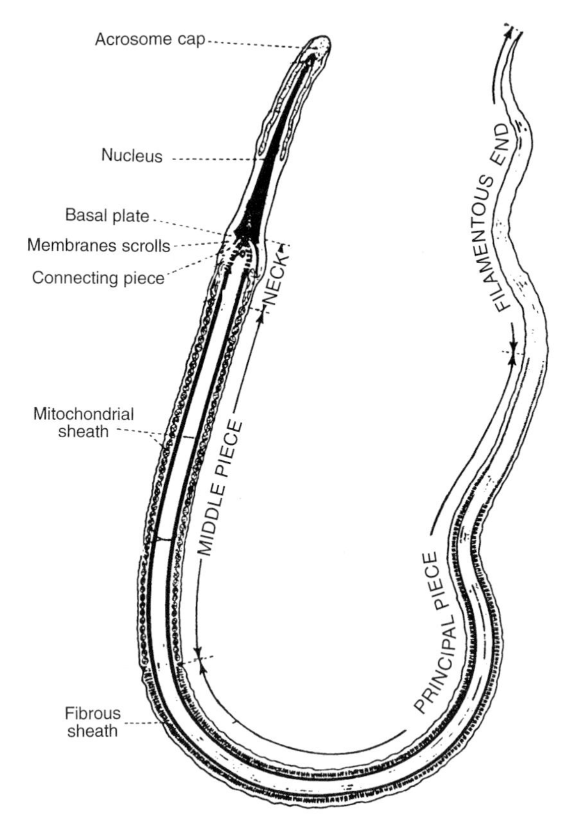

FIGURE 1 A typical mammalian spermatozoon. Important components of the sperm head include the nucleus and the acrosome. Major subdivisions of the sperm tail include the neck, middle piece, principal, and end piece. Sperm from different mammalian species exhibit modifications of this basic structural pattern. Nonmammalian animal or plant spermatozoa are similar in concept but, depending on the species, could exhibit a strikingly different morphology. Note the highly polarized organization of the spermatozoon.

condensed state of chromatin in the sperm nucleus account for this difference.

B. Morphology of the Sperm Head

The mammalian sperm head includes the nucleus and acrosome, which occupy most of the volume, as well as cytoskeletal elements and scant cytoplasm. The acrosome lies anteriorly and usually caps the nucleus with a deep indentation. Cytoskeletal materials lie in the space separating the inner aspect of the acrosome and the outer surface of the nuclear membrane. Some cytoskeletal elements are found immediately subjacent to the plasma membrane. In most species the sperm head is spatulate, being flattened in the plane of the anterior–posterior axis. The overall shape of the head in mammals varies from falciform (rodents) to ovate (human, guinea pig, and rabbit), but in lower animals or plants there is much more diversity. Typical lengths of the mammalian sperm head range from about 4.6 μm in man to 8.3 μm for *Mus musculus* and 12.1 μm for *Rattus norvegicus*, respectively. Again, nonmammalian sperm heads are often quite different, with very long lengths being found in marine animals, for example.

1. *The Sperm Nucleus and Associated Structures*

The sperm nucleus is unlike that usually encountered in somatic cells. Its volume is less and its chromatin is much more highly condensed. Not only does the nucleus contain only one-half of the usual DNA complement but also this DNA is associated with specialized proteins, termed protamines, which are strongly basic in overall amino acid composition. Protamines replace the normal somatic cell histones during the final stages of spermiogenesis prior to sperm release from the testicular epithelium. A strong association between the basic protamines and the acidic nucleic acids is aided by extensive covalent disulfide linkages. Although a number of detailed models have been proposed for the three-dimensional relationships between protamines and sperm DNA, the precise arrangement of these molecular elements remains obscure.

The nuclear envelope is also unusual. Somatic cell nuclei usually have numerous pores scattered over most of their surface. In sperm, however, pores in the nuclear membrane are absent from all of the surface covering the genetic material. Only in the most posterior aspect of the nucleus does one find pores. Here, immediately caudal to the posterior ring of the sperm nucleus, the nuclear envelope is formed into scrolled sheets. Termed the "redundant nuclear envelope," this membrane is covered by numerous pores often organized in an hexagonal array. Any physiological function for these pores and for the redundant nuclear envelope itself remains a matter of conjecture. The inner and outer leaflets of the sperm nuclear envelope are also unique in that they are closely apposed over most of the nuclear surface, separated by only 7–10 μm. In the posterior region containing pores, however, the two envelope leaflets are separated by a normal distance of 40–60 μm.

The nuclear lamina of the sperm head is a protein network that lines the inner surface of the nuclear envelope. It probably provides structural support for the membrane and also allows anchoring of the chromatin to its inner surface. Proteins comprising the nuclear lamina include lamins, which share extensive homology with proteins found in intermediate filaments of somatic cells. Differences in the precise nature and functions of the lamins found in mammalian sperm of varying species remain in most instances to be determined.

Cytoskeletal elements found in sperm heads are located in a variety of places and in a variety of identifiable structures. The perinuclear theca, for example, is composed of the subacrosomal and postacrosomal cytoskeletons. Formed from a number of proteins ranging in molecular weight from 8 to over 80 kDa, this theca seems similar to the nuclear matrix of somatic cells. Note, though, that in sperm the matrix material is external to the nuclear envelope. Exact definition of the role of the perinuclear theca is not available. Since this structure maintains its original dimensions after biochemical isolation, it is often presumed that it is an important determinant of the overall nuclear shape.

Other cytoskeletal proteins are located in the perforatorium, or subacrosomal cytoskeleton, of the sperm head. This structure is found in a narrow region between the inner acrosomal membrane and the outer leaflet of the nuclear membrane. Prominent in rodent spermatozoa, the perforatorium is not a major structural component of nonfalciform sperm, including those of man. The function of the perforatorium is unknown. Although named after similarly appearing entities in sperm of toads and birds that have a mechanical role in penetration of the egg at fertilization, the perforatorium of rodent sperm seems inert. Molecular analysis of the perforatorium in rodent spermatozoa shows that cysteine-rich proteins are present.

2. *The Sperm Acrosome*

The acrosome begins as a vesicular extension of the Golgi apparatus of developing spermatids during the later phases of spermatogenesis. It contains hydrolytic enzymes required for passage of sperm through the protective investment, or zona pellucida, surrounding the ovum. Other enzymes found in the acrosome may aid fusion of the sperm and egg plasma membranes and perhaps assist in cortical granule exocytosis by the egg to help prevent polyspermy. The acrosome is a membrane-limited vesicle which forms an anterior cap covering one-half to two-thirds of the sperm nucleus. The sperm plasma membrane is closely placed atop the outer acrosomal membrane, whereas the inner acrosomal membrane lies almost immediately on the outer leaflet of the nuclear envelope.

The size and general shape of the acrosome is highly species dependent, with that of the guinea pig being particularly prominent. Usually an equatorial segment of the acrosome forms a band covering the middle zone of the nucleus in spatulate sperm heads. Falciform sperm exhibit an equatorial segment which includes almost all the lateral surfaces of the nucleus. In contrast, some discoid sperm such as found in the wooly opossum show little or no identifiable equatorial segment of the acrosome. Apical protrusions of the acrosome which extend beyond the front limits of the nucleus are found in guinea pigs, chinchillas, and squirrels but are not large in human, primate, bull, boar, rabbit, or bat spermatozoa. Forces regulating the shape of acrosome in divergent species are not well understood.

Often the contents of the acrosome are not distributed uniformly. Crystalline arrays of proteins are sometimes apparent, as are cobblestone-like patterns found inside the vesicular membranes. The acrosomal membranes themselves also often exhibit such nonhomogeneity, with the equatorial segment usually involved. Species whose sperm show such features include the human, rat, mouse, guinea pig, rabbit, degu, and rhesus monkey.

Molecular identification of acrosomal contents remains an active area of study. Previously considered to be a modified lysosome, the acrosome is now known to contain a complex array of proteins and enzymes that participate in events immediately prior to final formation of the zygote. Acrosomal proteins include many if not most of the normal hydrolytic substances enclosed inside somatic cell lysosomes.

Note, however, that the sperm acrosomes do not function in phagocytosis or autophagocytosis as do their somatic counterparts. Instead, acrosomal enzymes are destined to function extracellularly.

Some acrosomal constituents are unique in that they are not found in lysosomes. The proteins acrosin and hyaluronidase fall into this class of proteins, and they are two of the most important molecules found inside the acrosome. Acrosin is a trypsin-like serine proteinase that is unlike other such proteinases with regard to most aspects of its molecular structure. Functional, but not structural, similarity is maintained between acrosins isolated from sperm of different mammals. Molecular weights are usually in the range of 25–60 kDa. Functional acrosins are processed from larger precursor forms, termed proacrosins, during the mid- to late stages of spermatogenesis. Acrosomal hyaluronidase is distinct in structure from the normal lysosomal form of this protein and like acrosin seems to represent a spermatogenic-specific isozyme. Hyaluronidase is a very abundant acrosomal component located mainly in the anterior acrosome and/or tightly bound to the interior aspect of the inner acrosomal membrane. Hyaluronidase bound to the inner acrosomal membrane is exposed to the egg zona pellucida or egg plasma membrane following rupture and fusion of the sperm plasma membrane with the outer acrosomal membrane occurring just prior to fertilization. Molecular weights for hyaluronidases found in mammalian sperm are in the range of 60–70 kDa.

The list of additional acrosomal proteins is long. The following enzymes have been detected in a variety of mammals: β-*N*-acetylglucosaminidase, acid phosphatase, arylamidase, arylsulfatase A, aspartylamidase, calpain II, a cathepsin D-like protease, collagenase-like proteins, various esterases, β-glucuronidase, neuraminidases, and phospholipases, especially phospholipase A. In most instances, the physiological roles of these molecules are unknown. It is likely that additional acrosomal components remain to be discovered.

C. The Sperm Flagellum

The function of the flagellum is to provide the locomotive force for the spermatozoon. This force is required to enable sperm to reach the ovum and also for the actual process of egg penetration at fertilization. Mammalian spermatozoa have a long flagellum which can usually be subdivided morphologically into four sections. Proceeding from anterior to posterior aspect, these segments are the connecting piece, the middle piece, the principal piece, and the end piece (Fig. 1). Within the flagellum itself, the major structural features are the axoneme, the mitochondrial sheath, the outer dense fibers, and the fibrous sheath. Sperm axonemes closely resemble those of other flagellate organisms with microtubules comprising a "9+2" array occupying the central portion of the flagellum throughout its length. Slightly posterior to the base of the sperm nucleus, a spirally organized sheath of mitochondria wraps around the axoneme. This region is defined as the middle piece of the flagellum. Replacing the mitochondrial wrapping, the fibrous sheath surrounds the axoneme in the principal piece section of the sperm tail. When this fibrous layer ends, the axoneme quickly tapers to its end as the filamentous end piece. The outer dense fibers originate from the area of the connecting piece, immediately at the base of the nucleus in the sperm head, and cover the axoneme in both the middle piece and portions of the principal piece. The outer dense fibers, therefore, lie between the axonemal microtubules and the mitochondrial and fibrous sheathes. Exterior to all these structures lies the continuous sperm plasma membrane. Representing most of the length of the sperm, mammalian flagella vary greatly in length. Human sperm flagella are about 60 mm long, but Chinese hamster sperm have tails which may exceed 250 mm. Tails of nonmammalian spermatozoa exhibit much variability in length as well as structure. They often contain additional structural elements or modifications which facilitate unique needs.

Abnormal sperm flagella are commonly seen, especially in humans. Although not necessarily an indication of a pathological condition, in infertile men up to 30% of their ejaculated sperm may have abnormal flagella. Even fertile men, however, often exhibit malformed tails or misconstructed central axonemes. Some genetic diseases are indeed based on disruptions of flagellar structure. Kartagener's syndrome, for example, is a consequence of an autosomal recessive condition yielding cilia and flagella that are mostly or completely immotile. A lack of dynein arms and radial spokes is often noted in flagella from patients with this disease. Kartagener's syndrome causes male sterility and chronic respiratory problems. Other malformations of sperm flagellar structure have been described both in humans and in experimental animal models, particularly the mouse. Both axonemal elements and periaxonemal structures may be absent or defective.

1. *The Connecting Piece*

Like the entire flagellum, the connecting piece region may be further subdivided. Its main components consist of the capitulum and the segmented columns. The capitulum is a fibrous structure that assumes the deeply indented shape of the implantation fossa, which itself consists of the posterior section of the nuclear envelope and a dense accumulation of material organized into the basal plate. Filaments of unknown identity connect the basal plate of the sperm head to the capitulum of the sperm tail. This area is usually affected by experimental treatments or genetic conditions resulting in separation of sperm heads from tails.

Two major and five minor segmented columns extend posteriorly from the capitulum. Initially, these columns are only a few millimeters long, but then the major columns each form two columns. These segmented columns, and the original minor segmented columns, finally fuse to yield the nine outer dense fibers which extend down the flagellum. Segmented columns show cross-striations with strict periodicity, but little is known regarding their molecular structure. During early development of the sperm tail, centrioles serve as nucleation sites, or organizing centers, for the axoneme and the segmented columns of the connecting piece.

2. *The Axoneme*

As already stated, the mammalian sperm axoneme shares its basic structure with the axoneme of most cilia and flagella in both plants and animals. Two central microtubules are encircled by nine doublets of microtubules. The doublets, in turn, have one complete microtubule exhibiting 13 protofilaments and one incomplete or C-shaped microtubule which is closed by its apposition to the wall of the complete tubule. Two arms extend from the complete tubule toward the incomplete tubule of the adjacent ring doublet. These arms are arranged in a clockwise manner as viewed from the base of the sperm tail. Radial spokes extend from the central pair of microtubules to the outer ring of doublets and form a central helix in the axoneme. Circumferential links connect the outer doublets to each other.

Much has been learned about the molecular composition and function of these various axonemal moieties. The microtubules themselves are composed of α- and β-tubulin subunits that have molecular weights of about 56 and 54.5 kDa, respectively. Other proteins which associate with microtubules of the sperm flagellum include a variety of motor proteins. Kinesin, for instance, is a large multimolecular array consisting of at least two subunits. Kinesin attaches to the surface of microtubules and serves to abet motion. This molecule is a member of a large family of microtubular-associated motor proteins now being identified in cells of many types and in many species. It is likely that more sperm motor proteins, perhaps some unique to the sperm tail, will soon be identified.

Dynein and nexin are two other important axonemal proteins. Dynein is a protein with ATPase activity that is found in the arms extending from the microtubular doublets. This protein of about 500 kDa is thought to be involved in the generation of force needed for the sliding of axonemal filaments resulting in locomotion. Nexin is a protein originally identified in sea urchin spermatozoa. Probably also necessary for mammalian sperm flagella, it has a molecular weight of about 170 kDa and forms the structural links between the outer microtubular doublets. A number of additional proteins have been discovered in mammalian sperm axonemes, but their precise functions remain clouded.

3. *The Mitochondrial Sheath*

In the flagellar middle piece mitochondria wrap helically around the outer dense fibers. The number and organization of the helices vary between species, as does the overall length of the entire middle piece. Mitochondria are usually situated end-to-end into parallel helices, with two being present in mouse and three parallel arrays found in bull. Each mammalian species examined seems to have something unique regarding either the exact number or placement of flagellar mitochondria. The functional importance, if any, of these differences is unknown. The most obvious functional importance of the flagellar middle piece relates directly to motility. The mitochondria here produce the ATP required for the generation of force by sliding of the axonemal microtubules.

4. *The Outer Dense Fibers*

Extending well down the length of the flagellum, the outer dense fibers are found mainly in mammalian sperm, although nonmammalian cells exhibit structures with some similarity. As is the case with most other flagellar components, the precise sizes and shapes of the outer dense fibers vary significantly between species. Often they are teardrop in form. In most species, particular outer dense fibers are distinguished by a greater diameter than others within the same flagellum. Human and bat sperm have outer dense fibers which terminate in the proximal half of the tail. In rodents, however, these fibers extend farther in the posterior direction. Probably the most important trait is that in both humans and rats the outer dense fibers extend to about 60% of the total length of the principal piece. Presumably, this consistency relates to the physiological function of the outer dense fibers. They are thought not to play any active role in axonemal or flagellar movement. Protein constituents of the outer dense fibers, for example, show little or no molecular homology to known contractile proteins. Cross-linked by extensive disulfide bridges, these fibers may instead serve to stiffen or provide elastic recoil properties. Biophysical measurements and time-lapse cinematography studies have confirmed that there is a direct correlation between the size of outer dense fibers and the radius of curvature in sperm tails of different mammals. Human sperm outer dense fibers are flexible, whereas those of species such as the rat seem relatively inflexible.

5. *The Fibrous Sheath*

The fibrous sheath occupies the entire length of the flagellar principal piece. This entity is composed of cytoskeletal-like proteins but forms a unit that is unique apparently to mammals and some birds. The fibrous sheath lies immediately subjacent to the sperm plasma membrane. It is not attached to this membrane as are some cytoskeletal elements in somatic cells. Suborganization of the fibrous sheath is obvious, with two longitudinal columns connected by circumferentially located ribs. The columns and ribs together form a tapering cylinder defining the general shape of the sperm flagellum. Individually, the longitudinal columns are about 20 nm in diameter and they run alongside and peripheral to particular outer dense fibers. In sperm of certain species, such as the opossum, these columns are quite large and result in an elliptical cross section for the principal piece. In contrast, longitudinal columns in guinea pig sperm are narrow. Most species exhibit columns with a morphology between these extremes. The circumferential ribs connecting the longitudinal columns are complicated in structure, composed of multiple filaments and flaring where they join the columns. Again, the exact form of these ribs differs between species.

Fibrous sheaths may be isolated in a relatively intact form. This has greatly facilitated analysis of their biochemical composition. The sheaths are highly resistant to solubilization by acid but may be separated into component elements by reducing agents since constituent proteins are linked by disulfide bridges. A number of proteins have been identified using biochemical and immunological approaches. Ranging in size from small to large (11–>80 kDa), the physiological roles of these proteins are unclear. As a unit, the fibrous sheath is thought to regulate the spatial plane in which the flagellum beats. Since the fibrous sheath is physically attached to some of the outer dense fibers its involvement in microtubular sliding must be minimal.

D. The Sperm Plasma Membrane

Covering the entire surface of the spermatozoon, from the most anterior aspects of the head to the flagellar end piece, a continuous plasma membrane modulates cellular interactions with the surrounding environment. This plasma membrane is highly specialized and in many ways unlike that found in most somatic cells. The sperm surface is organized into a series of restricted regional domains. These domains contain a unique panoply of lipids and proteins which together function to regulate particular aspects of sperm physiology and function. Each region of the sperm surface is so specialized. The membrane covering the middle piece of the flagellum contains proteins presumably vital for mitochondrial production of ATP and necessary for the modulation of microtubule sliding. Discrete domains on the sperm head contain particular proteins and, probably, lipids important in the cell–cell recognition and membrane fusion events occurring at fertilization. Significant alterations in sperm plasma membrane composition also occur during transit down the epididymis and up the female reproductive tract. These changes undoubtedly assist in ensuring successful fertilization, although many of the exact physiological details are not well understood.

Due to the extraordinary biochemical characteristics of the sperm surface, it has received much experimental attention. Many data have accumulated during the past 15 years which support the contention that considerations of this plasma membrane will provide important insights not only into unique factors related to fertilization but also into general biological questions concerning the mechanisms by which cells relate to and react with their immediate surroundings. Here, only a brief survey of the current knowledge about the biochemical composition of sperm surfaces is presented.

1. *Membrane Lipids*

The sperm plasma membrane is unlike somatic cell membranes with respect to many aspects of its lipid composition. Somatic cells have almost all their membrane lipids free to diffuse in the plane of the membrane. In striking contrast, many of the lipids in sperm plasma membranes cannot freely diffuse. Typically, 30–50% of the sperm surface lipids are somehow anchored in position. This is a dramatic difference and one which must relate importantly to sperm function. Probably, individual sperm proteins with restricted topographical localizations on the sperm surface require particular local lipid environments. It must also be remembered that fertilization involves fusion of the sperm plasma membrane with the outer acrosomal membrane of the sperm head as well as eventual fusion of the sperm and egg plasma membranes. These fusion events potentially also require the participation of particular lipid microenvironments.

Investigators have conducted detailed studies of the lipid compositions of sperm plasma membranes. Spermatozoa generally have extraordinarily high levels of ether-linked lipids. Highly unsaturated fatty acyl groups are also detected in unusually high concentrations compared to somatic cells. Some glycolipids are seemingly important for sperm function, although they are much less abundant than are phospholipids or sterols. Of the glycolipids, sulfatoxygalactosylacylalkylglycerol (SGG) is found in the highest amounts. SGG is localized to the sperm head and tail and may function during sperm–egg interactions. Finally, morphological techniques such as freeze-fracture coupled with filipin treatment demonstrate clearly that sperm plasma membrane lipids are organized into large domains. Guinea pig sperm heads, for example, show a quilt-like pattern of lipid-containing domains.

Important changes in overall membrane lipid composition occur as sperm transit the epididymis. These alterations must involve direct importation of lipids from the epididymal lumen since sperm themselves cannot synthesize lipids because their DNA is transcriptionally inactive. Furthermore, changes in sperm membrane lipid composition occur during passage through the female reproductive tract. Relative concentrations of cholesterol to phospholipids, for example, change significantly during this period. Such changes might relate to capacitation of the spermatozoon, in readiness for fertilization. More study is required, however, to elucidate the timing, extent, and physiological relevance of these changes in membrane lipids.

2. *Membrane Proteins*

Of all cell types, spermatozoa probably exhibit the highest degree of polarity, or mosaicism, with respect to their plasma membrane proteins. Virtually all proteins that have been characterized show some localization to the membrane overlying the sperm head, middle piece, or principal piece. Moreover, significant reorganizations of these localizations may occur during sperm development, transit through the epididymis, passage up the female tract, or during the final stages of capacitation and fertilization. Alterations in surface protein mobility, or the ability to diffuse in the plane of the membrane, are common. This again contrasts sperm with most somatic cells.

Space does not permit adequate evaluation of all the individual proteins that have been identified. Most workers have concentrated on the study of proteins localized to the sperm head in the hope that information regarding the control of fertilization will be obtained. A number of interesting molecules have been characterized. Members of the fertilin superfamily, for example, are clearly important during sperm–egg interactions. Surface proteins that undergo dramatic shifts in topographical positioning include proteins termed PH-20 in the guinea pig, M-42 in mouse sperm, and ESA-152 in male gametes of the ram. Many other examples could be cited. Note, however, that few detailed biochemical studies of human sperm plasma membranes are available.

Sperm plasma membrane proteins involved as direct mediators of fertilization are of great interest. Such sperm surface molecules are probably important during both the initial binding of sperm to the protective zona pellucida enveloping the ovum and the final fusion of sperm and egg plasma membranes. A large body of somewhat controversial data has arisen concerning the identity of the molecule(s) responsible for binding to the zona pellucida. A large glycosylated protein of the zona pellucida, termed ZP-3, is the egg investment constituent to which mammalian sperm attach prior to fertilization. Accordingly, researchers have been attempting to identify the sperm surface molecule which binds to ZP-3. The following sperm surface molecules are some of the many which have been described in this regard: β-1,4-galactosyltransferase, fucosyltransferase, α-D-mannosidase, sp56, p95, and a trypsin-inhibitor-insensitive site. These proteins have been described in a variety of species, usually rodents, and with differing degrees of detail. Galactosyltransferase is by far the best characterized potential candidate for binding to the zona pellucida, but important questions persist as to the necessity for this sperm membrane protein during interaction with the egg. Likewise, significant questions have been raised regarding the involvement of each of the other proposed sperm–egg-binding molecules based on biochemical or genetic concerns. It is most likely that (i) additional sperm plasma membrane proteins modulating interaction with the zona pellucida remain to be identified, and (ii) no single sperm surface molecule is all-important in this event. Rather, multiple constituents of the sperm plasma membrane probably interact to ensure successful formation of the zygote.

Binding of spermatozoa to the zona pellucida triggers a series of events which lead to fertilization. Most prominent among these happenings is the acrosome reaction. The anterior region of the sperm plasma membrane overlying the head fuses with the outer acrosomal membrane, thereby releasing acrosomal enzymes to the extracellular milieu. The acrosomal reaction is effected by changes in intracellular pH and a variety of membrane-mediated signal transduction events. Plasma membrane proteins regulating the acrosomal reaction are now under intense study. Protein tyrosine kinases have been described, including protein p95 found in mouse sperm. G proteins have been well described on the surface of mammalian spermatozoa. In particular, Gi proteins are important for release of acrosomal enzymes. In mouse sperm, for example, a 41-kDa protein is a substrate for pertussis toxin-mediated ADP ribosylation. Furthermore, Giα protein subunits have been detected using immunofluorescence microscopy on intact mouse, guinea pig, and human sperm. These subunits are located on the plasma membrane overlying the acrosome, another example of the polarized compartmentalization of proteins and lipids on the sperm surface. In contrast to the G_i protein , however, other forms of G proteins have not be identified in mature spermatozoa. Some of these constituents have been detected on developing spermatogenic cells and perhaps play important physiological roles during sperm maturation, if not at fertilization.

Exocytosis of acrosomal contents subsequent to binding to the zona pellucida is also triggered by a strong influx of calcium ions . Sperm plasma membrane ion channels regulate this process. Very recently, low-voltage gated T-type Ca^{2+} channels have been identified and isolated from mammalian spermatozoa. It appears that only the T-type channels are found on sperm membranes. This is in stark contrast to somatic cells in which additional high-voltage stimulated Ca^{2+} channels coexist with the T-type proteins. Mammalian spermatozoa, in fact, may prove to be an excellent model system for the detailed biochemical, biophysical, and physiological behavior of T-type activity. Other systems, such as skeletal muscle, which also exhibit high levels of T-type channels present technical difficulties not encountered with spermatozoa due to the presence of multiple channel types. Thus, analysis of ion channels on sperm plasma membranes promises to yield important new data contributing to the general knowledge of cellular activity. The significance of such data with regard to the control of fertilization is clear.

Lastly, there are virtually no data available regarding mammalian sperm plasma membrane molecules directly involved in the actual fusion of sperm and egg membranes immediately prior to egg penetration by the sperm nucleus. Following the acrosome reaction, the sperm plasma membrane remaining intact on the equatorial region of the sperm head joins with the surface membrane of the egg. This allows direct passage of the sperm nucleus into the egg cytoplasm. A putative sperm plasma membrane protein recognizing egg membranes, as opposed to the zona pellucida, has been described in sea urchins. Mammalian homologs of this protein have not been reported. This represents an important gap in our understanding of sperm function. Clearly, much remains to be learned about the molecular composition of the sperm surface.

See Also the Following Articles

Andrology: Origins and Scope; Spermatogenesis, Overview; Spermiogenesis

Bibliography

Bellve, A. R., and O'Brien, D. A. (1983). The mammalian spermatozoon: Structure and temporal assembly. In *Mechanisms and Control of Animal Fertilization* (J. F. Hartman, Ed.), pp. 56–137. Academic Press, New York.

Eddy, E. M. (1988). The spermatozoon. In *The Physiology of Reproduction* (E. Knobil and J. D. Neill, Eds.), Vol. 1, pp. 27–68. Raven Press, New York.

Fawcett, D. W. (1970). A comparative view of sperm ultrastructure. *Biol. Reprod. Suppl.* **2**, 90–127.

Fawcett, D. W. (1975). The mammalian spermatozoon. *Dev. Biol.* **44**, 394–436.

Gibbons, I. R. (1981). Cilia and flagella of eukaryotes. *J. Cell Biol.* **91**, 107s–124s.

Grudzinskas, J. G., and Yovich, J. L. (Eds.) (1995). *Gametes: The Spermatozoon*. Cambridge Univ. Press, Cambridge, UK.

Wassarman, P. M. (1995). Towards molecular mechanisms for gamete adhesion and fusion during mammalian fertilization. *Curr. Opin. Cell Biol.* 7, 658–664.

Sperm Capacitation

J. Michael Bedford
Cornell University Medical College

Nicholas L. Cross
Oklahoma State University

GLOSSARY

acrosome A membrane-bound sac containing hydrolytic enzymes that is arrayed over the front half of the sperm head.

ampulla The wide upper segment of the Fallopian tube where fertilization occurs.

epididymis The duct in which spermatozoa finally mature and are stored after leaving the testis.

estrogen Steroid hormone(s) that influences female secondary sex characteristics.

eutheria Group comprising the true placental mammals.

in vitro A situation maintained outside the body (literally "in glass").

ionophore A molecule that transports specific ions across cell membranes.

isthmus The narrow lower segment of the Fallopian tube.

ligand A molecule that binds to a specific site on a protein or other macromolecule.

pH A numerical measure from 1 to 10 used to describe relative acidity or alkalinity.

progesterone The steroid hormone primarily responsible for the maintenance of pregnancy.

seminal plasma The fluid portion of the ejaculate contributed by the male accessory sex glands.

signal transduction The biochemical pathway by which proteins send signals from the cell surface to the nucleus.

theria Animal group comprising marsupial and eutherian/placental mammals.

zona pellucida The egg coat which in most mammals comprises three glycoproteins.

Spermatozoa ejaculated by fertile mammals are not able to fertilize immediately. In order to penetrate an egg, mammalian spermatozoa must first undergo a physiological change (capacitation). This is brought about by the milieu of the female tract during their transport to the site of fertilization in the ampulla of the Fallopian tube and can be achieved by their incubation in appropriate conditions *in vitro*. Capacitation involves specific changes in the spermatozoon which allow it to undergo the acrosome reaction and to develop hyperactivated motility in appropriate conditions.

I. INTRODUCTION

Compared to other animals, the events leading to conception in mammals are unduly complex in several puzzling respects. In particular, the maturation of both spermatozoa and eggs and their interactions leading to fertilization involve various novel features, prominent among which is the phenomenon of sperm capacitation. Historically, fertilization was first studied primarily in invertebrates and subsequently in some amphibia, in large part because of the ease of collecting and manipulating their gametes. In mammals, not only was it more difficult initially to obtain eggs but also spermatozoa did not fertilize when placed with eggs in the Fallopian tube or *in vitro*. This failure to fertilize was explained in 1951 by C. R. Austin and independently by M. C. Chang in experiments using the rat and rabbit. By varying the time of tubal insemination in relation to ovulation, and also by using sperm populations collected from the female some hours after mating, they showed that mammalian spermatozoa need to spend some hours in the female reproductive tract before they can fertilize eggs. Such results clearly implied that spermatozoa undergo a physiological change while in the female tract, conferring upon them the capacity to fertilize, for which Austin coined the term "capacitation." In the decades since then, capacitation has proved to be a requisite for fertilization in all mammals studied in this respect, both for mature epididymal and for ejaculated spermatozoa. Moreover, recognition of this phenomenon has been a key in analysis of the mechanisms involved in mammalian fertilization and in development of techniques for *in vitro* fertilization in animals and man.

II. SPERM CAPACITATION IN THE FEMALE TRACT

The rate of capacitation varies in different animals. For example, when spermatozoa are surgically inseminated directly into the Fallopian tube containing eggs, fertilization occurs after about 2 hr in the sheep and ferret, by 4 or 5 hr in the hamster, and not for about 10 hr in the rabbit. Different regions of the female reproductive tract vary in their ability to capacitate spermatozoa. In the rabbit, hamster, and rat at least, spermatozoa that experience first the uterine and then the tubal environment develop the ability to fertilize sooner than spermatozoa exposed to the tubal environment alone. This suggests a synergism between capacitation factors expressed, respectively, in the uterus and Fallopian tubes.

The capacitation potential of the female tract is sensitive to the hormonal status of the female. In rabbits, the capacitating ability of the uterus is low in the estrogen-deficient female and is absent during a state of progesterone domination. On the other hand, the capacitating ability of the Fallopian tubes is not abolished by progesterone. Such evidence of hormonal effects and of synergy between compartments of the tract implies that capacitation *in vivo* is not merely a matter of keeping spermatozoa alive—the female tract appears to produce factors that regulate capacitation. No such factors have been identified as yet, but it has been shown that they are not species specific and nor are they obligatory since capacitation can occur *in vitro* in the absence of any material from the female tract.

It is generally believed that onset of the capacitated state not only confers the ability to fertilize but also limits the sperm's functional life thereafter. However, it is not clear precisely where in the female tract spermatozoa switch to a fully capacitated state. Based on studies in the hamster, it is likely that the final switch to a fully capacitated state may occur *in vivo* only as spermatozoa are released from the isthmus and ascend to the ampulla. In accord with this, among the very few spermatozoa which ever reach the site of fertilization in the ampulla of the Fallopian tube, essentially all appear to be functionally capacitated.

III. CAPACITATION *IN VITRO*

The development of techniques for capacitation *in vitro* has allowed dissection of different steps in the fertilization process and has provided a setting for analysis of the cellular change(s) that capacitation entails. The first demonstration of capacitation *in vitro* was the report of Yanagimachi and Chang in 1963 of fertilization by hamster epididymal spermatozoa incubated at 37°C for several hours with eggs in Tyrode's solution or Medium 199. Capacitation of human spermatozoa *in vitro* was demonstrated clearly a few years later, and reports of capacitation *in vitro* have been published for many mammals since then. The methods are largely empirical and use various calcium-containing culture media such as Tyrode's or Krebs–Ringer solutions controlled for pH generally with bicarbonate/CO_2, together with protein sources such as homologous serum, bovine serum albumin, or fetal calf serum. Additional factors have proven helpful in specific situations. For example, *in vitro* capacitation of bovine spermatozoa is optimized by including heparin in the medium, capacitation of the spermatozoa of some monkey species is aided by the presence of caffeine or cyclic 3′,5′-adenosine monophosphate (cyclic AMP), and pig spermatozoa were first capacitated most effectively *in vitro* when incubated at a high concentration. However, experience dictates that no one medium is optimal or even appropriate for all species; therefore, precise tailoring of conditions is important in that respect.

As in the female tract, there is also species variation in the time required for capacitation *in vitro*. Judged by the time of egg penetration, some cat epididymal spermatozoa may become functionally capacitated *in vitro* in less than 30 min, some human spermatozoa in less than 1 hr, and mouse spermatozoa in no more than 2 hr. At the other extreme, rabbit spermatozoa prove to require rigorous *in vitro* conditions and in many successful studies their capacitation finally required about 12 hr. However, such records of fertilization timing really reflect only the interval required for capacitation of a vanguard population. Indeed, a significant proportion of an apparently mature population may not even be susceptible to capacitation *in vitro*. In the human particularly, acrosomal responsiveness reflective of the capacitated state seems to develop, at best, in no more than about 30% of the population in fertile ejaculates. There is no clue as to the basis for this. It has seemed a likely possibility that such heterogeneity might reflect individual differences in aging or maturity of spermatozoa within one sample. However, in addressing this question, the differential rate at which individual hamster spermatozoa became able to undergo the spontaneous acrosome reaction during capacitation *in vitro* was maintained when they were first aged for up to 3 weeks in the cauda epididymidis. Thus, the heterogeneity does not seem to simply reflect a relative maturity.

IV. THE CELLULAR NATURE OF CAPACITATION

The front half of the sperm head is invested by the acrosome (a membrane-bound "cap" that contains hydrolytic enzymes). Before fertilization can occur, the sperm head plasma membrane undergoes point fusions with the underlying outer membrane of the acrosome, excepting that over the nonreactive posterior region that constitutes the equatorial segment of the acrosome. This fusion step is accompanied by a mobilization and then progressive loss of soluble enzymatic components of the acrosome and is followed eventually by loss of the carapace of fused membranes which remains at the surface of the egg coat. This sequence—the acrosome reaction—is required for sperm penetration of the zona pellucida. The acrosome reaction not only exposes the inner acrosome membrane and creates a sharper head profile but also brings an associated functional change in the persistent segment of plasma membrane over the stable equatorial segment that does not take part in the acrosome reaction.

After the discovery of capacitation, there was at first a sense that the acrosome reaction per se was the cellular basis of capacitation. However, subsequent studies have pointed to capacitation as representing a set of biochemical changes in the spermatozoon, with the ability to undergo the acrosome reaction being one consequence. Among capacitated populations, some of the spermatozoa may react spontaneously *in vitro*, and many become responsive to inducers of the reaction such as the zona pellucida of the egg, progesterone, and calcium/proton exchanging ionophores.

Another primary consequence of capacitation is a switch to "hyperactivation"—a thrashing pattern of motility which is illustrated in Fig. 1. The advent of hyperactivated motility as a function of capacitation may parallel the development of acrosomal responsiveness. However, the onset of hyperactivation is by no means necessarily linked to the occurrence of the acrosome reaction *in vitro*, and free hyperactivated spermatozoa that have yet to react have been observed in the ampulla of the Fallopian tube. The underlying function of hyperactivation remains unclear. Among several ideas, it has been suggested that hyperactivation could function to "shake" spermatozoa loose from their tenacious hold on the epithelium of the lower tubal isthmus. It seems most likely, however, that the hyperactivated pattern functions to provide the particular type of motility required for penetration of the unusually resilient zona pellucida that characterizes the egg in eutherian mammals.

Research into capacitation has focused on spermatozoa of several species and it has employed different techniques, but it has been difficult to elucidate the precise nature of the changes in the spermatozoon

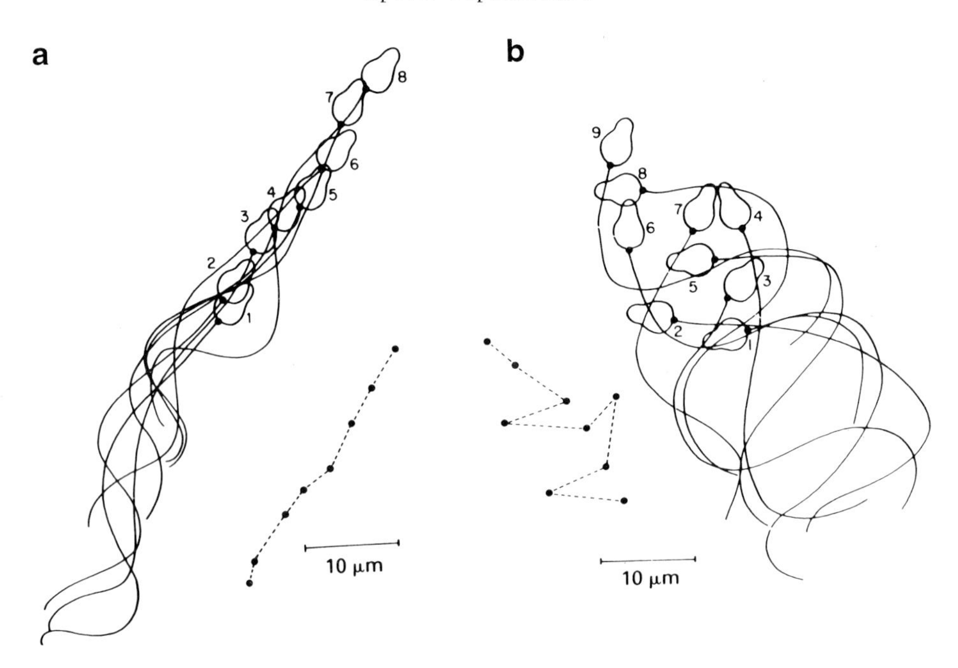

FIGURE 1 Movement characteristics of human spermatozoa capacitated at 37°C in medium BWW: (a) before and (b) on hyperactivation *in vitro*. The numbered positions in a were taken 0.032 sec apart and in b were 0.049 sec apart. The points connected by dotted lines represent sequential positions of the sperm head (reproduced with permission from P. Morales, J. W. Overstreet, and D. F. Katz, *J. Reprod. Fertil.* **83**, 119, 1988).

that constitute capacitation. However, there is considerable evidence that modifications of the protein and lipid components of the sperm plasma membrane are important elements in this functional transformation of the spermatozoon. According to species, certain macromolecules acquired by or modified on the sperm surface in passing through the epididymis may be changed during capacitation—in accord with proposals discussed below as to the significance of capacitation. It has been suggested that membrane-associated proteins are either lost or modified during capacitation. Second, and of major interest currently, it appears that the cholesterol/phospholipid ratio of the membrane must decrease for spermatozoa to become capacitated. Cholesterol efflux begins soon after separation of spermatozoa from seminal plasma and seems to be driven not by internal events but by mass action and the action of lipophilic molecules in the environment, such as albumin, lipoproteins, and lipid transfer proteins, which tend to elute membrane cholesterol as an early step in the capacitation process. Prevention of this decrease in membrane cholesterol makes spermatozoa unable to fertilize or to respond to inducers of the acrosome reaction. Conversely, acceleration of cholesterol loss increases the proportion of spermatozoa that become acrosomally responsive.

Several molecular consequences appear to follow cholesterol efflux and to be modulated by it. First, loss of cholesterol permits a rise in the sperm intracellular pH, the alkalinization then making it acrosomally responsive. Alkalinization of the internal milieu during capacitation has been best documented in bovine sperm, during which the pH within the sperm head shifts from 6.70 to 6.92. That may lead in turn to an elevation of intracellular bicarbonate, which has been implicated in the increasing concentration of Ca^{2+} and of the activity of adenylyl cyclase in sperm. The product of adenylyl cyclase activity (cyclic AMP) may increase the rate of capacitation. In the case of human spermatozoa, it has been suggested that this cholesterol loss reduces the thickness of the membrane bilayer, as it can in other systems, helping to bring to expression sperm head receptors for mannose ligands at the zona surface and apparently nonnuclear progesterone receptors. Such exposure of receptors at the sperm surface for mannosyl residues may help the spermatozoon to bind to the surface of the zona pellucida and, coincidentally, lead to intramembrane rearrangements that are conducive to fusion.

Another important event noted in mouse and bovine spermatozoa incubated *in vitro* is an increase in sperm membrane potential from about −30 to about −60 mV. This shift is due in part to an increased permeability of the membrane to K^+. The hyperpolarization, which is also required for acrosomal responsiveness, may cause Ca^{2+} channels to become voltage sensitive such that the depolarization brought by an agonist permits an influx of Ca^{2+}. Elevated intracellular Ca^{2+} levels are essential for onset of the acrosome reaction.

Finally, recent work has demonstrated that a set of sperm proteins is phosphorylated on membrane tyrosine residues during capacitation. (Tyrosine phosphorylation often signals a change of protein function.) Interestingly, much of the tyrosine phosphorylation is dependent on cyclic AMP, suggesting a link between cyclic AMP-dependent protein kinase A and protein tyrosine kinase(s)/phosphatases. Reactive oxygen species (hydrogen peroxide and superoxide radical) also stimulate tyrosine phosphorylation. Identifying the phosphorylated proteins and their roles in sperm function are topics of great current interest.

It should be appreciated in the case of several (e.g., rabbit and man), though not all species (e.g., dog and cat) that mixing capacitated spermatozoa with seminal plasma produces what seems to be a reversible inhibition of the capacitated state. This effect appears to be due not only to reinsertion of seminal cholesterol into sperm membranes but also probably in some cases to adsorption by spermatozoa of certain molecules that are present in secretions of the epididymis and/or male accessory sex glands and are normally removed during capacitation.

V. SIGNIFICANCE OF THE NEED FOR CAPACITATION

Among invertebrate species used for fertilization research, spermatozoa are able to penetrate the egg immediately on their release from the male. Capacitation is not required in several submammalian vertebrates—notwithstanding the fact that the spermatozoa of some must traverse the female tract to fertilize internally. Thus, the limited comparative evidence seems consistent with the possibility that the need for capacitation has arisen *de novo* in therian mammals. However, because the information about most other animal groups in this regard is sparse at best, it has been difficult to understand the underlying adaptive or functional significance of the species-specific pattern of change(s) in the sperm membrane that constitutes capacitation. Now, a variety of observations suggest that an important clue to the functional significance of capacitation may lie in the prior sperm membrane changes imposed in preparation for regulated sperm storage in the epididymis.

In mammals, sperm membrane change in the epididymis involves complex integral molecular modification of its constituent lipids and sterols, as well as its integral proteins and in some cases their glycosylated moieties, and this is a key element in the process of sperm maturation. It is true that a sperm surface binding of secreted proteins has also been demonstrated in the epididymis in a few selected reptiles and birds. However, therian mammals express an additional function in the lower (cauda) region of the epididymis—a regulated maintenance of the viability of mature spermatozoa (sperm storage) that is acutely dependent on both testicular androgens and the lower temperature of the scrotum. As a corollary of this feature, it has been shown for at least three mammals that locally secreted macromolecules unrelated to fertilizing ability bind to the plasma membrane of spermatozoa entering that storage region. Since spermatozoa are demonstrably fragile once they mature, such surface components acquired in the distal region of the epididymis could represent membrane-stabilizing elements required to promote viability of spermatozoa for at least the period of their storage or until they are ejaculated.

On the other hand, it appears unlikely that a sperm plasma membrane stabilized for storage could also participate immediately in the fusion events required for fertilization. Hence, at least in regard to the acrosome reaction, the need for capacitation may well be a consequence of the stable state imposed on spermatozoa as a function of their regulated storage in the epididymis—representing a necessary destabilization of the sperm cell membrane. In accord with that idea, it has been observed that (hamster) epididymal spermatozoa are capacitated more rapidly where the function of the cauda is suppressed. Moreover, fertile (pig) spermatozoa from the upper epididymis fertilize sooner when introduced into the ampulla of the Fallopian tube than do those from the cauda, the storage region of the epididymis.

See Also the Following Articles

ACROSOME REACTION; EPIDIDYMIS; IN VITRO FERTILIZATION SPERMATOGENESIS, OVERVIEW; SPERM TRANSPORT

Bibliography

Bedford, J. M. (1997). Capacitation and the acrosome reaction in human spermatozoa. In *Infertility in the Male* (L. I. Lipschultz and S. S. Howards, Eds.), 3rd ed. Mosby, St. Louis.

Benoff, S. (1993). The role of cholesterol during capacitation of human spermatozoa. *Hum. Reprod.* **8**, 2001–2006.

Drobnis, E. Z. (1993). Capacitation and acrosome reaction. In *Reproductive Toxicology and Infertility* (A. R. Scialli and M. J. Zinaman, Eds.), pp. 77–132. McGraw-Hill, New York.

Florman, H. M., and Babcock, D. F. (1991). Progress toward understanding the molecular basis of capacitation. In *Elements of Mammalian Fertilization* (P.M. Wassarman, Ed.), Vol. 1, pp. 105–132. CRC Press, Boca Raton, FL.

Langlais, J., and Roberts, K. D. (1985). A molecular model of sperm capacitation and the acrosome reaction of mammalian spermatozoa. *Gam. Res.* **12**, 183–224.

Oliphant, G., Reynolds, A., and Thomas, T. (1985). Sperm surface components involved in the control of the acrosome reaction. *Am. J. Anat.* **174**, 269–283.

Rogers, B. J. (1978). Mammalian sperm capacitation and fertilization *in vitro:* A critique of methodology. *Gam. Res.* **1**, 165–223.

Yanagimachi, R. (1994). Mammalian fertilization. In *The Physiology of Reproduction* (E. Knobil and J. D. Neill, Eds.), 2nd ed., pp. 189–317. Raven Press, New York.

Spermiogenesis

Richard Oko and Yves Clermont
Queen's University and McGill University

GLOSSARY

acrosome A membrane-bound organelle closely applied to the surface of the nucleus of spermatids and spermatozoa. Formed by the Golgi apparatus of early spermatids, it is rich in hydrolytic enzymes and plays a major role in fertilization.

annulus A ring-like structure composed of electron-dense material associated with the plasma membrane. It surrounds the axoneme just below the centrioles in elongating spermatids and later it migrates distally along the sperm tail and forms the junction of the middle piece and principal piece.

axoneme Axial component of the tail of spermatozoa composed, as in cilia, of α and β tubulin organized in the form of nine microtubular doublets and two central microtubules. It is the contractile component of the tail responsible for the anterograde movement of spermatozoa.

chromatoid body A small chromophilic mass present in the

cytoplasm of spermatocytes and early spermatids. It is composed of electron-dense material and small vesicles. In round spermatids it migrates toward the centrioles and contributes material to the developing annulus. It disappears from the cytoplasm in elongated spermatids.

fibrous sheath A complex cytoskeletal structure present along the principal piece of the tail of spermatozoa. It is composed of longitudinal columns that run along microtubular doublets 3 and 8 and numerous arched ribs that bridge the longitudinal columns.

manchette A curtain-like, conical, and rigid structure composed of microtubules which appears in elongating spermatids. Inserted at the equator of the nucleus, it surrounds the forming tail and disappears during the late steps of spermiogenesis.

middle piece The segment of the tail of spermatozoa located between the neck and the principal piece. It is composed of the axoneme and associated outer dense fibers and is characterized by the presence of numerous, elongated, transversely arranged mitochondria side by side to form a mitochondrial sheath.

neck (synonym, ***connecting piece)*** A short segment of the tail that bridges the nucleus and the middle piece. Solidly attached to the nucleus within an implantation fossa, the neck is composed of electron-dense material in the form of a striated collar seen around cavities that in the spermatozoa of mice and rats contain remnants of the proximal and distal centrioles. The neck is fully formed late in spermiogenesis.

outer dense fibers Major cytoskeletal elements of the tail of elongated spermatids and spermatozoa. In the middle piece, there are nine coarse outer dense fibers (ODFs), one per axonemal doublet, each having a characteristic size and shape. In the principal piece, only seven ODFs are present and they decrease in caliber in a proximal–distal direction. In this segment, ODFs 3 and 8 are replaced by the longitudinal columns of the fibrous sheath.

perinuclear theca A layer of proteinaceous material present around the nucleus except over an area surrounding the implantation fossa in which the neck piece of the tail is inserted. The perinuclear theca shows two distinct regions: the subacrosomal region, known as the perforatorium in rats and mice, and a postacrosomal sheath also designated as the postacrosomal dense lamina. The perinuclear theca serves to bind the acrosome and the plasma membrane to the nucleus.

principal piece The segment of the tail of the late spermatid and spermatozoa distal to the middle piece and characterized by the presence of the axoneme associated with outer dense fibers and the fibrous sheath.

residual body The mass of cytoplasm containing residual organelles (i.e., mitochondria, vesicles, ribosomes, and lipid droplets) which detaches from the late spermatid just before it is released from the seminiferous epithelium. Residual bodies are eliminated from the seminiferous epithelium by Sertoli cell phagocytosis.

Sertoli cell A large stellate and columnar somatic cell present in the seminiferous epithelium of mammals. Solidly attached to the basement membrane of seminiferous tubules, it is clearly related structurally and functionally to all germinal cells, in particular the elongated spermatids.

spermatid The haploid germinal cell arising from meiotic divisions of spermatocytes which differentiates, within the seminiferous epithelium, into a spermatozoon.

spermiogenesis The third and last phase of spermatogenesis during which the newly formed spermatids metamorphose into spermatozoa.

Spermiogenesis is the phase of spermatogenesis during which the spermatid differentiates into a spermatozoon. It is the last one of the three main phases of spermatogenesis: The first phase involves the spermatogonia, which proliferate, renew, or differentiate to produce spermatocytes; in the second phase primary and secondary spermatocytes undergo meiotic divisions, yeilding haploid cells, the spermatids; and the third phase is spermiogenesis. During this cellular metamorphosis the major structural changes observed in the spermatid are formation of the acrosome, which derives from the Golgi apparatus, nuclear elongation and chromatin condensation, formation of the axoneme triggered by the centriolar complex; and assembly of sperm-specific cytoskeletal elements and the mitochondrial sheath around the axoneme. These changes will be briefly reviewed using human spermiogenesis as a model.

I. FORMATION OF THE ACROSOME

Soon after the formation of the spermatid, several small secretory-like granules, rich in glycoproteins (mainly hydrolytic enzymes), form on the *trans* face of the Golgi stacks (Fig. 1, step 2). These proacrosomic granules coalesce to give a single larger acrosomic vesicle which associates with the nuclear mem-

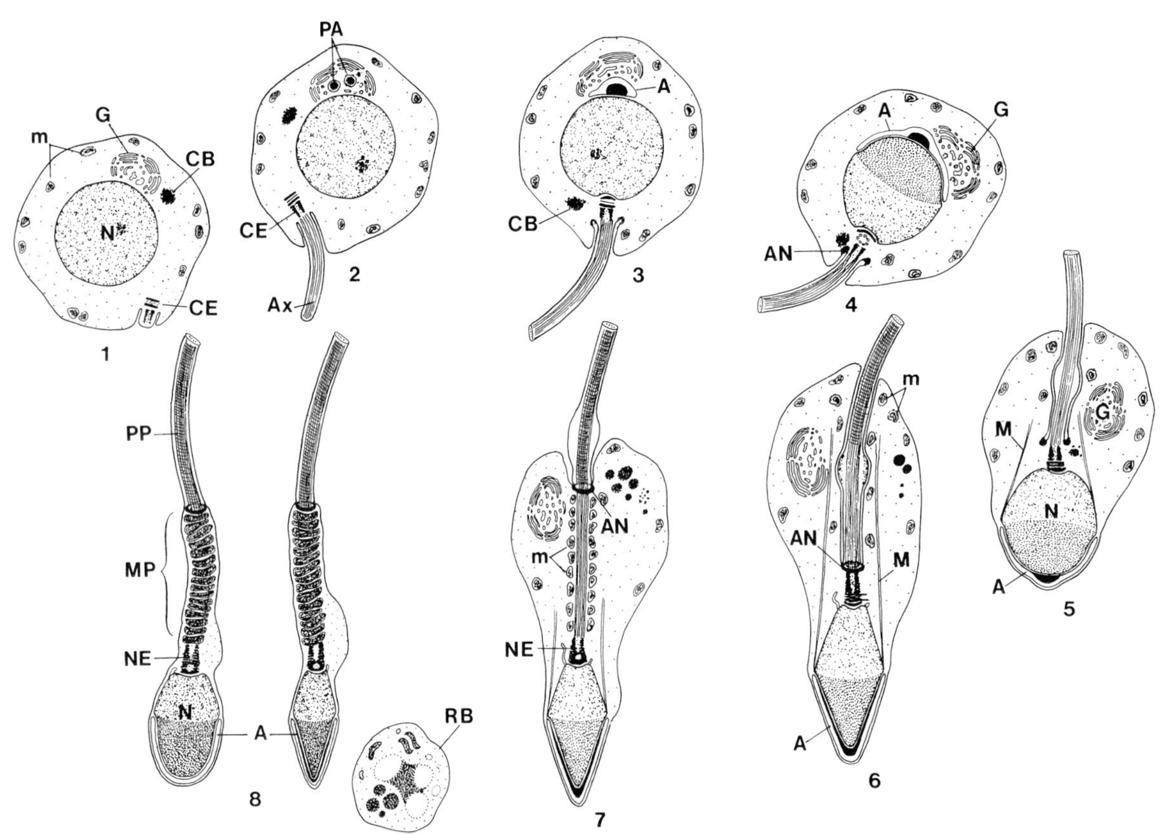

FIGURE 1 Spermiogenesis in man. (1) Newly formed haploid spermatid resulting from the second meiotic division of a secondary spermatocyte. In addition to a centrally located spherical nucleus (N), the cell contains a pair of centrioles (CE), a juxtanuclear Golgi apparatus (G), a chromatoid body (CB), and peripherally located mitochondria (m). The cytoplasm also contains cisternae of endoplasmic reticulum and ribosomes (not shown). (2) Secretory-like proacrosomic granules (PA), rich in glycoproteins, appear in the center or medulla of the Golgi apparatus. The centrioles (CE) trigger the formation of microtubules, which compose the contractile axoneme (Ax) of the tail and migrate toward the nucleus. In so doing, they pull with them the plasma membrane, which forms a double-wall invagination surrounding the axoneme. (3) In the Golgi apparatus, the proacrosomic granules fuse to form a single, membrane-delimited, acrosomic vesicle. This vesicle attaches to the nuclear envelope to become the acrosome (A). The chromatoid body (CB) migrates toward the centrioles and becomes associated with the nucleus at the pole opposite to the one occupied by the acrosome. At this step the axoneme has completed its growth and reached its full length of approximately 50 μm. (4) The Golgi apparatus (G) continues to contribute glycoproteins to the acrosome by means of numerous small carrier vesicles. As a result the head cap portion of the acrosome (A) grows and covers half of the nuclear surface. The chromatoid body contributes some of its material to the plasma membrane which is in proximity of the centrioles to form the annulus (AN). (5) The spermatid starts to elongate and the bulk of the cytoplasm migrates toward the caudal pole of the cell and surrounds the forming tail. At the apical pole of the cell, the plasma membrane approximates the acrosome (A). The Golgi apparatus detaches from the acrosome and stops contributing glycoproteins to it. A manchette or caudal tube (M), made up of microtubules which surround the proximal part of the forming tail, develops and is inserted at the equator of the nucleus (N). (6) The spermatid continues to elongate and most of the cytoplasm forms a lobule along the developing tail. Additional cytoskeletal elements, i.e., the outer dense fibers and the fibrous sheath (not shown), are deposited along the axoneme. The manchette elongates and the mitochondria (m) migrate from the periphery of the cytoplasm toward the forming tail. The annulus (AN) is still seen next to the centrioles now forming the neck region of the future tail. The chromatin initiates its condensation within the nucleus, which takes an elongated piriform shape. Its pointed extremity is covered by the condensing acrosome (A). (7) As the nucleus continues to condense, the tail completes its formation. Following the migration of the annulus (AN) away from the neck (NE) along the proximal portion of the tail for a distance of approximately 6 or 7 μm, the mitochondria (m) align side by side along the portion of the axoneme now uncovered by the plasma membrane thus demarcating the middle

brane and tightly adheres to it to become an acrosome. This early acrosome shows an electron-dense core and a less dense cortex when examined with the electron microscope (Fig. 1, step 3). As the Golgi apparatus continues to deliver glycoproteins to the acrosome, via numerous small carrier vesicles, the lighter portion of the acrosomic system expands at the surface of the nucleus and takes the shape of a cap which covers approximately half of the nucleus (Fig. 1, step 4). The tight adhesion of the acrosomal membrane to the nuclear envelope is due to the presence of a proteinaceous substance that later condenses to form part of a rigid capsule or perinuclear theca that covers the nucleus of spermatozoa.

With the displacement of most of the cytoplasm toward the caudal pole of the spermatid, the Golgi apparatus separates from the acrosome and stops contributing glycoproteins to it (Fig. 1, step 5). The Golgi apparatus later regresses and disappears from the cytoplasm. At later steps of spermiogenesis and as the nucleus takes its species-specific shape, the acrosome also condenses and conforms to the changing form of the apex of the nucleus (Fig. 1, steps 6–8). The acrosome plays a major role during the fertilization of the oocyte.

II. NUCLEAR ELONGATION AND CHROMATIN CONDENSATION

During spermiogenesis, as the nucleus elongates and takes its species-specific shape (i.e., piriform in man, flattened or paddle shape in ram, bull, and dog, and falciform in many rodents), the nuclear chromatin condenses. Two main sequential stages of chromatin reorganization are observed. In the first stage there is a remodeling of chromatin from its normal nucleosomal form to a thread-like filamentous form. In the second stage, the chromatin filaments thicken, become coarse, and aggregate into compact masses that coalesce to form a dense chromatin mass (Fig. 2A). Eventually, this chromatin mass condenses even more to give a homogeneous chromatin in which chromosomal filaments can no longer be resolved (Fig. 2B). The transformation of the nucleosomal to the filamentous form of the chromatin has been correlated with a modification or loss of H1t histone, whereas the progressive condensation of the chromatin filaments into the compact chromatin appears to be influenced by the presence of transition nucleoproteins TP1 and TP2. Once the chromatin condensation has occurred and the definitive shape of the nucleus has been acquired, the transition proteins TP1 and TP2 are replaced late in spermiogenesis by arginine- and cysteine-rich protamines. While the species-specific shapes of the nuclei of spermatozoa are obviously genetically determined, it has been suggested that external pressures exerted by the cytoskeletal elements of the Sertoli cells, by the microtubules of the manchette, or by the perinuclear theca may be involved in nuclear morphogenesis.

III. FORMATION OF THE TAIL

During the early steps of spermiogenesis, the pair of centrioles migrates toward the nucleus and eventually solidly binds to it (Fig. 1). Concurrently, one of two centrioles triggers the formation of the axoneme, which rapidly reaches its full length. As is true for

piece of the tail. During this step of spermiogenesis the manchette regresses. Toward the end of this step and as the spermatid prepares to be released from the seminiferous epithelium, the cytoplasmic lobule detaches from the cell to yield a residual body (RB) that will be phagocytosed by Sertoli cells. Only a small vestige of cytoplasm, delimited by the plasma membrane, remains at the surface of the late spermatid or spermatozoon. (8) Two diagrams of the mature spermatid or spermatozoon showing the cell face view on the left and side view on the right. The condensed nucleus, which is slightly flattened, shows a paddle shape on the left and a piriform shape on the right. The apical extremity of the nucleus is covered by the acrosome (A). The tail, attached to the base of the nucleus, is composed of three segments: the neck (NE), the middle piece (MP) containing closely aligned condensed and elongated mitochondria along the outer dense fibers which are associated with the axoneme, and the principal piece (PP) made up of the axoneme and associated cytoskeletal elements. The whole cell is covered by a thin layer of cytoplasm and the plasma membrane.

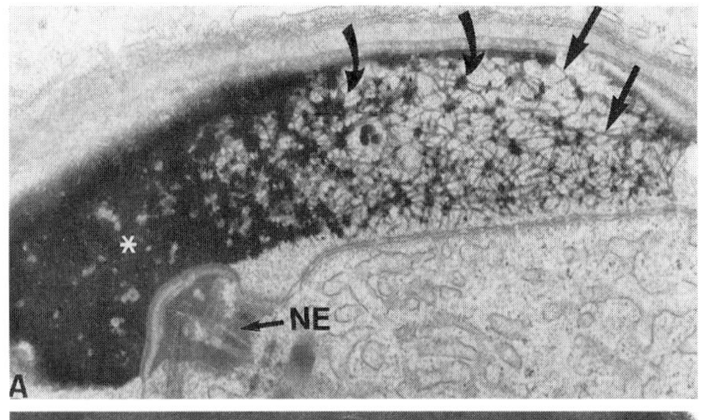

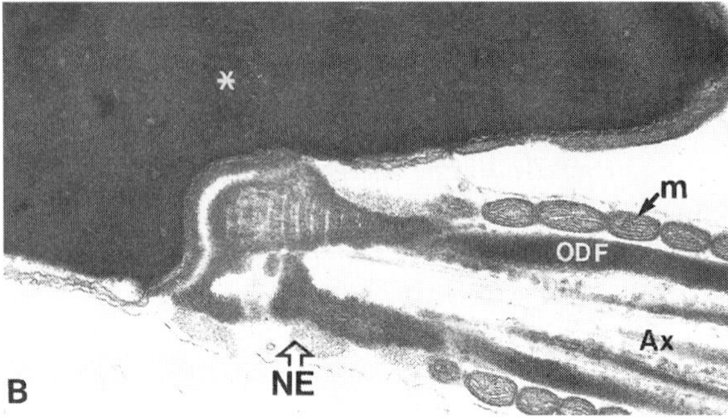

FIGURE 2 (A) Electron micrograph showing the caudal extremity of the nucleus of an elongated rat spermatid. Progressive condensation of the chromatin is seen. Coarse chromatin filaments are visible (straight arrows). These filaments aggregate into clusters of various sizes (curved arrows). Such clusters coalesce to form a continuous dense chromatin (*) which still shows less dense cavities inside. The forming neck of the tail is indicated (NE). (B) The caudal extremity of the nucleus of a mature spermatid or spermatozoon showing the homogenous dense chromatin in which no internal substructure can be resolved (*). The fully formed neck region of the tail (NE) is indicated. The proximal portion of the middle piece of the tail is also visible in which the mitochondria (m), outer dense fibers (ODF), and axoneme (Ax) are seen in a longitudinal section of the tail. Magnification, ×50,000 (photographed by Dr. M. Lalli).

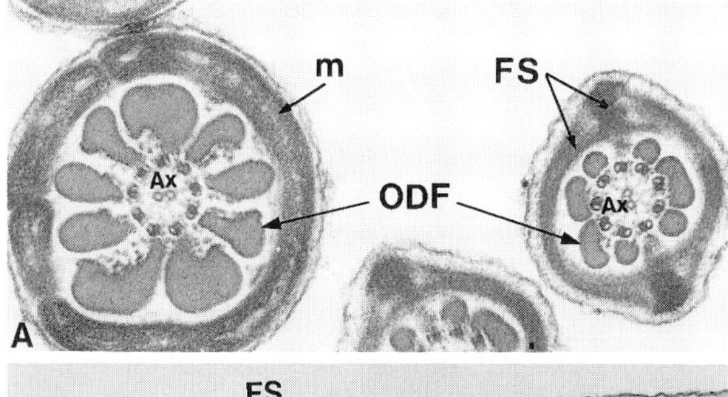

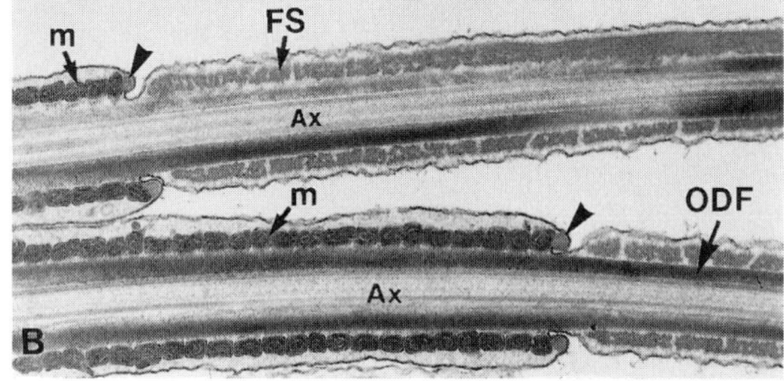

FIGURE 3 (A) Electron micrograph showing cross sections of the tail of a rat spermatozoon at the level of the middle piece (left) and the principal piece (right). In both, the central axoneme made up of microtubules are seen (Ax). In the middle piece (left) the outer doublets are associated with nine coarse outer dense fibers (ODF). These are surrounded by condensed mitochondria (m), seen here cut longitudinally. In the principal piece, with the exception of two ODFs (Nos. 3 and 8) no longer present, the ODFs are distinctly smaller than those in the middle piece and are surrounded by the arches of the fibrous sheath (FS), seen here cut longitudinally. Magnification, ×60,000. (B) Electron micrograph showing longitudinal sections of tails of rat spermatozoa. These sections show the junctions of the middle piece (left) and the principal piece (right) with cross sections of the annulus (arrowheads) at the interface. In the middle piece region the condensed mitochondria (m) are seen transversally cut, whereas the cross sections of the arches of the fibrous sheath (FS) in the principal piece are visible. Longitudinal sections of outer dense fibers (ODF) are also indicated. Magnification, ×30,000.

the acrosome nuclear docking, the association of the centriolar apparatus with the nuclear envelope represents a unique case of nuclear and cytoplasmic organelle association. As the centrioles move toward the nucleus, they pull with them the plasma membrane, which then forms a double-wall sleeve around the growing axoneme (Fig. 1). In the elongated spermatid, electron-dense material is deposited around the centrioles and forms a striated collar that characterizes the neck portion of the tail (Figs. 1 and 2B).

The tail of spermatozoa cannot be equated to a cilium or flagellum. While both cilia and the spermatozoon's tail contain a central contractile axoneme composed of microtubules, the tail of spermatozoa contains in addition major and sperm-specific cytoskeletal elements, i.e., the outer dense fibers (ODFs) and the fibrous sheath (FS) (Figs. 3A and 3B). These cytoskeletal components are deposited along the axoneme soon after it terminates its growth. Their assembly continues contemporaneously and terminates toward the end of spermiogenesis. Although the exact role of these cytoskeletal structures is not yet elucidated, they may contribute to the characteristic helical movement of the tail of free spermatozoa.

The evolution of mitochondria during mammalian spermiogenesis is interesting. Located peripherally close to the plasma membrane in early spermatids, these organelles migrate toward the forming tail as the manchette appears in elongating spermatids (Fig. 1). Following the sliding of the annulus (which derives in part from the chromatoid body) along the complex made by the axoneme and associated cytoskeletal structures, the mitochondria line up in an orderly manner along the ODFs, between the neck and the annulus. Later during spermiogenesis, the mitochondria condense, form a tubular–crescentic shape, and dispose themselves side by side and tip to tip in a spiral manner. Such a mitochondrial covering demarcates the middle piece of the tail, the length of which is species specific (e.g., 7 μm in man and 80 μm in the rat) (Figs. 1 and 3). These closely packed mitochondria constitute the respiratory organ of the spermatozoon.

IV. OTHER CHARACTERISTICS OF SPERMIOGENESIS

A. Intercellular Bridges

All the spermatids originating from the same spermatogonial stem cell are connected to each other by open intercellular bridges. This is the consequence of incomplete cytokinesis of the spermatogonia following their mitoses and of spermatocytes at the end of their respective meiotic divisions. Such cytoplasmic bridges remain open throughout spermiogenesis and disappear at the very end of the differentiation of the spermatid with the elimination of the residual cytoplasm or the residual body. Such bridges may contribute to synchronize the evolution of the interconnected spermatogonia, spermatocytes, and spermatids.

B. Duration

Spermiogenesis is a rigidly timed process and every step in the differentiation of the spermatid has a fixed duration. The duration of spermiogenesis is species specific (e.g., 24 days in man and 20 days in the rat). This feature explains why large numbers of spermatids are seen at the same steps of their evolution within the seminiferous epithelium. The duration of spermiogenesis or of its individual steps have been found to be invariable and not extended or shortened by extracellular factors such as temperature and hormonal growth factors.

C. Relations with Sertoli Cells

Spermatids have close structural and functional relations with Sertoli cells. For example, elongated spermatids are deeply inserted within invaginations seen in the supranuclear cytoplasm of the Sertoli cells. The Sertoli cell's functions, which are maintained by steroid hormones secreted by the Leydig cells, are essential for the maintenance of spermiogenesis. The dependence of spermatids on Sertoli cells and their stage-specific interactions may explain the failure so far to maintain spermiogenesis *in vitro*.

V. TRANSCRIPTIONAL AND TRANSLATIONAL REGULATION DURING SPERMIOGENESIS

Most of the integral proteins present in the specialized components of spermatozoa are synthesized during spermiogenesis, implying a process of haploid-regulated transcription and translation. During mammalian spermatogenesis two notable bursts of RNA synthesis occur, one in midpachytene spermatocytes and the other at the earlier steps of spermiogenesis in round spermatids. A major purpose of these two bursts of RNA activity is to contribute to the synthesis of proteins incorporated in the components of the differentiating spermatids (e.g., glycoproteins added to the acrosome and cytoskeletal proteins added to the sperm tail). Since gene transcription terminates at the beginning of the nuclear elongation in midspermiogenesis, the proteins required for the formation of cytoskeletal components of the tail (ODF and FS), for nuclear condensation (transition proteins and protamines), and for adhesion of mitochondria to the ODF must be translated from stored mRNAs.

Several mechanisms to regulate translation of these proteins have been proposed. For example, in the case of the protamines, it has been shown in transgenic mice that *cis* elements of the 3′ untranslated region of protamine-1 mRNA act to delay the translation of fusion construct transcripts during spermiogenesis. In this context several sequence-dependent germ cell-specific proteins that bind to the 3′ untranslated region of protamine and transition protein mRNAs may be candidates for translation repressors.

Regarding the transcriptional control of spermiogenesis, it has been shown that transcriptional activator cyclic AMP-responsive element modulator (CREM) may be responsible for the activation of a wide range of haploid germ cell-specific genes. The genes known to be inactivated in CREM mutant mice encode the protamines, calspermin, the major 27-kDa outer dense fiber protein, and mitochondrial capsule selenoprotein. Most of these genes have been shown to contain cyclic AMP-responsive elements on their promoters and are essential for the structuring of the spermatid.

VI. CONCLUSION

Considering the multiplicity and complexity of the cellular events taking place during spermiogenesis, one would expect a multitude of elaborate regulatory processes at transcriptional, translational, and even posttranslational levels (e.g., during multiple steps in the deposition of ODF and FS along the axoneme). These regulatory mechanisms, of the type described previously, are just beginning to be explored and clarified by genetic and molecular approaches, and these will be the subject of future studies on spermiogenesis.

Acknowledgments

This work was supported by grants from NSERC and MRC of Canada.

See Also the Following Articles

Acrosome Reaction; Sertoli Cells, Overview; Spermatogenesis, Overview; Spermatozoa

Bibliography

Barth, A. D., and Oko, R. (1989). Normal bovine spermatogenesis and sperm maturation. In *Abnormal Morphology of Bovine Spermatozoa*, pp. 19–88. Iowa State Univ. Press, Ames.

Blendy, J. A., Kaestner, K. H., Weinbauer, G. F., Nieschlag, E., and Schütz, G. (1996). Severe impairment of spermatogenesis in mice lacking the CREM gene. *Nature* **380**, 162–165.

Braun, R. E., Peschon, J. J., Behringer, P. P., Brinster, R. L., and Palmiter, R. D. (1989). Protamine 3′ untranslated sequences regulate temporal translational control of growth hormone in spermatids of transgenic mice. *Genes Dev.* **3**, 793–802.

Burfeind, P., and Hoyer-Fender, S. (1991). Sequence and developmental expression of a mRNA encoding a putative protein of rat sperm outer dense fibers. *Dev. Biol.* **148**, 195–204.

Cataldo, L., Baig, K., Oko, R., Mastrangelo, M.-A., and Kleene, K. C. (1996). Developmental expression, intracellular localization, and selenium content of the cysteine-rich protein associated with the mitochondrial capsules of mouse sperm. *Mol. Reprod. Dev.* **45**, 320–331.

Clermont, Y. (1972). Kinectics of spermatogenesis in mammals: Seminiferous epithelium cycle and spermatogonial renewal. *Physiol. Rev.* **52**, 198–236.

Clermont, Y., Oko, R., and Hermo, L. (1993). Biology of mammalian spermiogenesis. In *Cell and Molecular Biology of the Testis* (C. Desjardins and L. L. Ewing, Eds.), pp. 332–376. Oxford Univ. Press, Oxford, UK.

Cole, A., Meistrich, M. L., and Trostle-Weige, P. K. (1988). Nuclear and manchette development of normal and *azh/azh* mutant mice. *Biol. Reprod.* **38**, 385–401.

de Kretser, D. M., and Kerr, J. B. (1994). The cytology of the testis. In *The Physiology of Reproduction* (E. Knobil and J. D. Neill, Eds.), pp. 1177–1290. Raven Press, New York.

Fajardo, M., Butner, K., Lee, K., and Braun, R. E. (1994). Germ cell-specific proteins interact with the 3′ untranslated region of Prm-1 and Prm-2 mRNA. *Dev. Biol.* **166**, 643–653.

Fawcett, D. W. (1975). The mammalian spermatozoon. *Dev. Biol.* **44**, 394–436.

Hecht, N. B. (1989). Mammalian protamines and their expression. In *Histones and Other Basic Nuclear Proteins* (L. S.

Hnilica, G. S. Stein, and J. Stein, Eds.), pp. 347–373. CRC Press, Boca Raton, FL.

Kistler, M. K., Sassone-Corsi, P., and Kistler, W. S. (1994). Identification of a functional cyclic adenosine 3′, 5′-monophosphate response element in the 5′-flanking region of the gene for transition protein 1 (TP1), a basic chromosomal protein of mammalian spermatids. *Biol. Reprod.* **51**, 1322–1329.

Kleene, K., and Flynn, J. F. (1987). Characterization of a cDNA clone encoding a basic protein, TP2, involved in chromatin condensation during spermiogenesis in the mouse. *J. Biol. Chem.* **262**, 17272–17277.

Kleene, K. C., Borzorgzadeh, A., Flynn, J. F., Yelick, P. C., and Hecht, N. B. (1988). Nucleotide sequence of a cDNA clone encoding mouse transition protein 1. *Biochem. Biophys. Acta* **950**, 215–220.

Kleene, K. C., Smith, J., Bozorgzadeh, A., Harris, M., Hahn, L., Karimpour, I., and Gerstel, J. (1990). Sequence and developmental expression of the mRNA encoding the seleno-protein of the sperm mitochondrial capsule in the mouse. *Dev. Biol.* **137**, 395–402.

Kwon, Y. K., and Hecht, N. B. (1993). Binding of a phosphoprotein to the 3′ untranslated region of the mouse protamine 2 mRNA temporally represses its translation. *Mol. Cell. Biol.* **13**, 6547–6557.

Meistrich, M. L., Trostle-Weige, P. K., and VanBeek, M. (1994). Separation of specific stages of spermatids from vitamin A-synchronized rat testes for assessment of nucleoprotein changes during spermiogenesis. *Biol. Reprod.* **51**, 334–344.

Monesi, V. (1965). Synthetic activity during spermiogenesis in the mouse. RNA and protein synthesis. *Exp. Cell Res.* **225**, 46–55.

Morales, C., and Clermont, Y. (1993). Structural changes of the Sertoli cell during the cycle of the seminiferous epithelium. In *The Sertoli Cell* (L. D. Russell and M. D. Griswold, Eds.), pp. 305–329. Cache River Press, Clearwater, FL.

Morales, C. R., Oko, R., and Clermont, Y. (1994). Molecular cloning and developmental expression of an mRNA encoding the 27 kDa outer dense fiber protein of rat spermatozoa. *Mol. Reprod. Dev.* **37**, 229–240.

Nantel, F., Monaco, L., Foulkes, N. S., Masquilier, D., LeMeur, M., Henriksän, K., Kierich, A., Parvinen, M., and Sassone-Corsi, P. (1996). Spermiogenesis deficiency and germ-cell apoptosis in CREM-mutant mice. *Nature* **380**, 159–162.

Oko, R. (1995). Developmental expression and possible role of perinuclear theca proteins in mammalian spermatozoa. *Reprod. Fertil. Dev.* **7**, 777–797.

Oko, R., and Clermont, Y. (1990). Mammalian spermatozoa: Structure and assembly of the tail. In *Controls of Sperm Motility: Biological and Clinical Aspects* (C. Gagnon, Ed.), pp. 3–27. CRC Press, Boca Raton, FL.

Oko, R., and Maravei, D. (1995). Distribution and possible role of perinuclear theca proteins during bovine spermiogenesis. *Microsc. Res. Technol.* **32**, 520–532.

Oko, R., and Morales, C. R. (1996). Molecular and cellular biology of novel cytoskeletal proteins in spermatozoa. In *Cellular and Molecular Regulation of Testicular Cells* (C. Desjardins, Ed.), Serono Symposia USA, pp. 135–165. Springer-Verlag, New York.

Oko, R., Jando, V., Wagner, C. L., Kistler, W. S., and Hermo, L. S. (1996). Chromatin reorganization in rat spermatids during the disappearance of testis-specific histone, H1t, and the appearance of transition proteins TP1 and TP2. *Biol. Reprod.* **54**, 1141–1157.

Otani, H., Janaka, O., Kasai, K.-I., and Yoshioka, T. (1988). Development of mitochondrial helical sheath in the middle piece of the mouse spermatid tail: Regular dispositions and synchronized changes. *Anat. Rec.* **222**, 26–33.

Russell, L. D., and Griswold, M. D. (Eds.) (1993). *The Sertoli Cell.* Cache River Press, Clearwater, FL.

Sharpe, R. M. (1994). Regulation of spermatogenesis. In *The Physiology of Reproduction* (E. Knobil and J. D. Neill, Eds.), 2nd ed., pp. 1363–1434. Raven Press, New York.

Sun, Z., Sassone-Corsi, P., and Means, A. R. (1995). Calspermin gene transcription is regulated by two cyclic AMP response elements contained in an alternative promoter in the calmodulin kinase IV gene. *Mol. Cell. Biol.* **15**, 561–571.

van der Hoorn, F. A., and Tarnasky, H. A. (1992). Factors involved in the regulation of the PT7 promoter in a male germ cell-derived in vitro transcription system. *Proc. Natl. Acad. Sci. USA* **89**, 703–707.

van der Hoorn, F. A., Tarnasky, H. A., and Nordeen, S. K. (1990). A new rat gene RT7 is specifically expressed during spermatogenesis. *Dev. Biol.* **142**, 147–154.

Vogl, A. W., Pfeiffer, D. C., Redenback, D. M., and Grove, B. D. (1993). Sertoli cell cytoskeleton. In *The Sertoli Cell* (L. D. Russell and M. D. Griswold, Eds.), pp. 39–86. Cache River Press, Clearwater, FL.

Sperm Transport

Mary A. Scott and James W. Overstreet
University of California, Davis

GLOSSARY

acrosome The caplike, membrane-bound structure that surrounds the anterior head of the sperm. The acrosome contains hydrolyzing enzymes that are released during the acrosome reaction thereby facilitating sperm–oocyte interaction.

ampulla The thin-walled expanded region of the oviduct extending from the isthmus to the infundibulum. The ampulla is the site of fertilization for most mammals.

capacitation The physiological process by which a sperm becomes capable of fertilizing the oocyte.

caudal isthmus The caudal region of the thick-walled, narrow isthmus of the oviduct, which is located between the oviductal ampulla and the uterus. In many species, the caudal isthmus is a site of sperm storage prior to ovulation.

estrus The period of the estrous cycle during which a female is receptive to mating.

external os The vaginal opening of the cervix.

hyperactivation Sperm motility that is characterized by a highly vigorous whip-like flagellar motion with an intermittent nonprogressive trajectory. The development of hyperactivation is associated with sperm capacitation and allows sperm to gain release from reservoirs and to penetrate the outer vestments of the oocyte.

luteal phase The stage of the estrous cycle during which a corpus luteum is actively secreting progesterone.

pericoital Taking place at or around the time of coitus or copulation.

periovulatory Occurring around the time of ovulation.

rheological Pertaining to the flow characteristics of fluids.

seminal plasma The fluid component of semen. Seminal plasma includes secretions of the testes, the epididymides, and accessory glands of the male reproductive tract.

uterotubal junction The anatomical region at which the uterine tube, or oviduct, joins the uterus.

Successful sperm transport within the female reproductive tract results in an adequate number of fertilization-competent sperm reaching the site of fertilization within the functional life span of the ovulated egg(s). Sperm transport comprises not only the migration of sperm from the site of insemination to the site of fertilization but also the induction of functional changes in sperm cell physiology that are required for fertilization to proceed. Both male and female components are integral to the success of this fundamental process of mammalian reproductive biology.

I. INTRODUCTION

When viewing a drop of semen on a microscope slide, it is easy to be impressed by the vigor and trajectory of sperm motility, and the observer may be tempted to conclude that this form of locomotion is the principal mode of sperm transit through the confines of the female reproductive tract. One might also surmise that sperm arrival at the site of fertilization in the oviduct is purely a random event that is guaranteed by the superabundance of sperm in the ejaculate. Although the dynamics of this process have not been fully elucidated in any species, it is clear that the process of sperm transport is not a random one but rather involves a concert of interactions between sperm and the female tract. Sperm motility is important during certain phases of transit, but also sperm are distributed mechanically by contractile activity of the female musculature. Sperm motility is modulated by a changing luminal environment including variations in spatial constraints, epithelial surface characteristics, and the rheological characteristics of luminal fluids. The events of sperm transport proceed with a variable timetable that is driven by the duration of time between insemination and ovulation. Accordingly, many features of this process may be closely regulated by the female reproductive tract.

The animal models for which sperm transport has been characterized most completely include laboratory animals (rabbit, mouse, and hamster) and several food animal species (cow, pig, and sheep).

II. REPRODUCTIVE BEHAVIOR AND SPERM TRANSPORT BIOLOGY

Mammalian sperm exhibit a remarkable variation in size and morphology. These physical characteristics are likely to reflect significant species differences in sperm transport biology. Other factors that are likely to influence the species-specific mechanisms of sperm transport include the duration of estrus, or receptivity to mating, the timing of ovulation with respect to estrus and mating, the anatomical site of insemination, and the life span of sperm in the female tract.

For many domestic animal species, mating will occur only during a strict period of estrus, which may be relatively short (e.g., the cow, 18 hr) or which may last for days (e.g., the mare, 4–7 days). The timing of ovulation with respect to insemination will influence the duration of sperm residence within the female tract and may reflect differences in the functional life span of sperm after insemination or the time required for sperm to attain the capacity for fertilization (capacitation). Species variation in these characteristics will be associated with corresponding species differences in mating strategies.

The anatomical site of semen deposition defines the anatomical barriers that sperm must cross to reach the oviducts and the site of fertilization. By restricting sperm access to the upper tract, these barriers effectively create a gradient in sperm numbers along the length of the female tract. For species with uterine deposition of semen (e.g., pig, horse, and dog), the initial barrier to the oviducts is the uterotubal junction (UTJ), and a significant reduction in sperm numbers occurs cranial to this barrier. For species with vaginal deposition of semen (e.g., ruminants, primates, and rabbit), the cervix is the initial anatomical barrier, and the UTJ serves to further restrict sperm access to the oviducts. The cervical canal may be filled with mucus (ruminants and primates), which is secreted locally by the epithelium. The biophysical characteristics of cervical mucus are affected by the endocrine status of the female and may serve to block sperm passage (luteal phase, progesterone dominance), thereby functioning as an absolute barrier. Under estrogen dominance (such as in the periovulatory period), mucin production and hydration increase, and the mucus matrix can be penetrated by sperm.

The site of semen deposition also appears to relate to corresponding species differences in the number and concentration of sperm in the ejaculate, as well as the physical and biochemical characteristics of semen. For example, when the site of insemination is the vagina, semen characteristically has a low volume and a high sperm concentration and may have little accessory fluid (ruminants) or may develop a coagulum (primates). Following insemination into the cranial vagina, this concentrated semen bathes the external os of the cervix, providing sperm access to the cervical canal and direct contact with cervical mucus (ruminants and primates). In contrast, the semen of other species reaches the uterus almost immediately after ejaculation either because the uterus is the site of insemination (pig, horse, and dog) or because female contractions result in uptake of whole semen across the cervix (rodents). For many of these species, the ejaculate has a large volume with a correspondingly lower sperm concentration. Visceral contractions of the female tract distribute the inseminate throughout the uterine lumen.

III. PERICOITAL EVENTS

Sperm transport is initiated at the time of insemination. In all species examined, sperm can be recovered from the upper oviducts within minutes of mating or artificial insemination. Careful study of these rapidly transported sperm has demonstrated that they are moribund, have disrupted membranes, or are nonviable, indicating that these sperm do not contribute to the fertilizing population in the oviduct. Because the rate of transport is much faster than sperm swimming speeds, and because immobilized or dead sperm are similarly transported, this phenomenon has been attributed to muscular contractility of the female tract and concomitant modulations of intraluminal pressures. Rapid sperm transport may be universal in mammals. This universality would imply that rapid transport has a specific physiological role in mammalian reproductive biology. One hypothesis is that the interaction between these vanguard sperm and the female tract provides a message to the upper tract that influences subsequent events of sperm transport and fertilization.

Sperm are deposited in large numbers and are suspended in fluids (seminal plasma) that are secretory products of the male tubular tract and accessory glands. These fluids have direct effects on sperm and on the female tract. Ejaculated sperm become coated with seminal proteins that prevent premature capacitation and enhance sperm survival in the female tract. Constituents of seminal plasma, such as prostaglandins and other biochemicals, can stimulate smooth muscle activity of the female reproductive tract. For some species (e.g., pig, dog, and llama), mating and the ejaculatory process are prolonged, resulting in a continual mechanical stimulation of the female tract. The fractional delivery of seminal components over the prolonged coitus may also have differential stimulatory effects that influence sperm distribution.

IV. SPERM MIGRATION INTO THE OVIDUCTS

A prolonged phase of sperm migration follows the rapid transport phase. During this stage, competent sperm become distributed along the female tract and may accumulate in specific regions which function as sperm reservoirs. In ruminants and primates, sperm migrate actively from the vagina into the cervical mucus. Because semen and mucus do not mix, the cervical mucus prevents the passage of any appreciable amount of seminal plasma into the uterus and upper tract. As sperm swim through cervical mucus, they must pass through spaces that are similar in size to the sperm head and, as a consequence, there is a very close association between the sperm membranes and the structural elements of the mucus (Fig. 1). The sperm tail pushes against the mucus structure to propel the sperm forward (Fig. 2). Periovulatory cervical mucus supports sperm viability, and motile sperm have been recovered from human cervical mu-

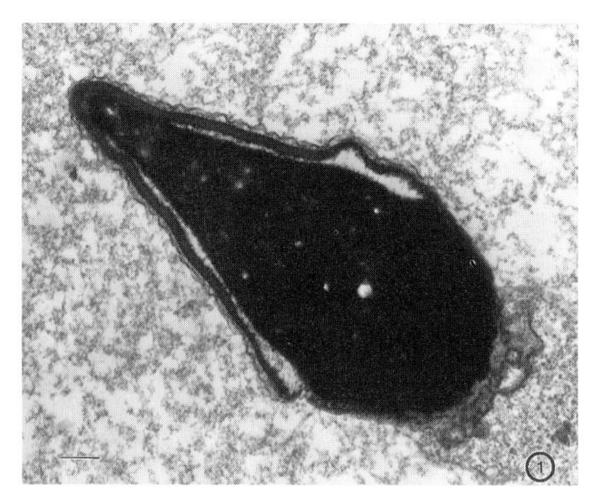

FIGURE 1 Transmission electron micrograph of a human sperm in human cervical mucus. The membranes of the sperm head are closely associated with the structural elements of mucus. Bar = 0.2 μm.

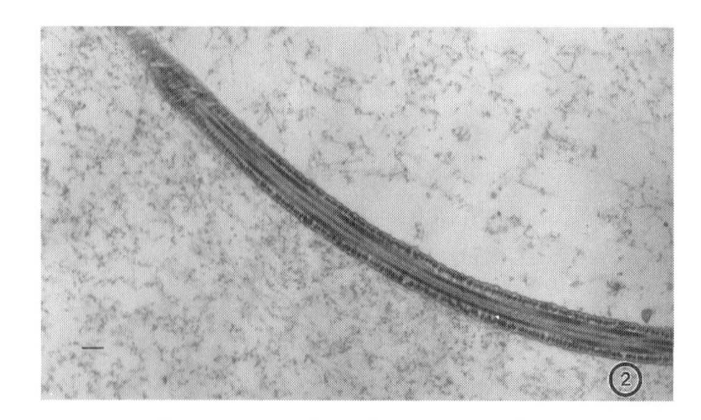

FIGURE 2 The sperm tail pushes against the mucus microstructure with each beat. On the convex side of the beat the microstructure is compressed and on the other side the mucus is stretched. Bar = 0.2 μm.

cus up to 120 hr after insemination. Prolonged conservation of viable sperm in the cervix suggests that this region acts as a sperm reservoir in ruminants and primates. The cervix is an efficient barrier to large numbers of sperm, and only a small percentage of the sperm ejaculated into the vagina continue to migrate into the upper tract. In species with vaginal insemination, the cervix also functions as a biological filter by restricting the passage of weak or morphologically abnormal sperm.

Sperm transport through the uterus occurs mechanically by contractile activity of the tubular tract and by active sperm motility. Sperm accumulation occurs again at the next anatomical barrier, the UTJ. Sperm motility is considered a requirement for sperm passage through this complex portal to the oviduct. When whole semen passes directly into the uterus, the initial barrier encountered is the UTJ, and this region must serve as an important regulator of sperm entry into the oviduct for those species.

The time required for a population of viable and functionally competent sperm to reach the oviducts varies by species. Fertilizing sperm reach the oviduct within 1 or 2 hr for the sow, by 4 hr after insemination for the mare, and by approximately 6–8 hours for sheep and cows.

V. PREOVULATORY SPERM STORAGE

When insemination occurs prior to ovulation, sperm accumulate in reservoirs where they reside until an unknown signal stimulates them to continue their ascent to the site of fertilization. The existence of sperm reservoirs in ruminants, pigs, rabbits, and hamsters has been well characterized. For ruminants, the cervix with its cervical mucus accumulates and maintains a population of viable sperm that provides a continued source of sperm for the upper tract. The persistence of motile sperm in the primate cervix suggests that it may function similarly. In the mare, some sperm develop a close association with the epithelium on the uterine side of the UTJ that is suggestive of sequestration there (Fig. 3). It is unknown if these sperm can subsequently detach, enter the oviduct, and participate in fertilization. However, in the sow, a well-studied animal model, the UTJ has been identified as a preovulatory sperm reservoir.

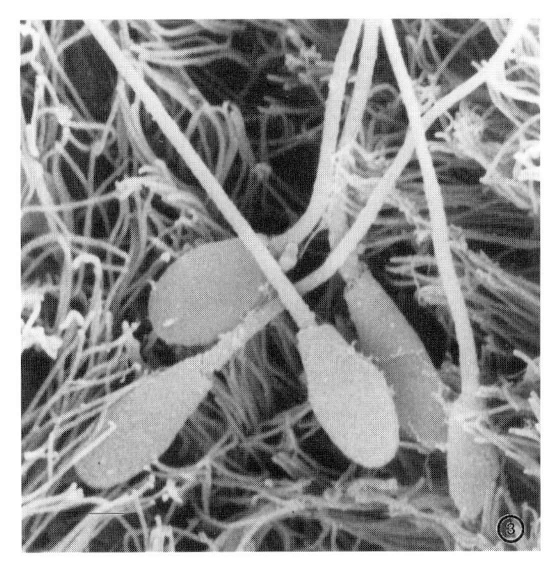

FIGURE 3 Scanning electron micrograph of equine sperm at the uterine side of the mare's uterotubal junction. The sperm heads are closely associated with the ciliated epithelium. Bar = 2.0 μm.

In ruminants, pigs, rabbits, and hamsters, functional sperm that cross the UTJ accumulate in the caudal isthmus of the oviduct by developing a specific association with the epithelium of this region. This intimate contact appears to be important for maintaining sperm viability during storage (experimental evidence in the hamster) and has been visualized in the cow and pig caudal isthmus using scanning electron microscopy. In the rabbit, motility is suppressed in sperm recovered from the caudal isthmus in native fluid, and this is a proposed mechanism for their retention in this region.

VI. PERIOVULATORY EVENTS

A. Changes in Sperm Physiology

A sperm must be fully mature and functionally competent to interact with the oocyte and achieve fertilization. At ejaculation, normal sperm have the potential to fertilize but are not yet fully mature. Fertilization competence is gained after a period of residence within the female tract and is induced or facilitated in competent sperm as they interact with the luminal fluids and epithelial surfaces during transit. Accordingly, these maturational changes, collectively known as capacitation, are an integral component of the sperm transport process. As a result of capacitation, a sperm can acrosome react (for most species this occurs when the sperm contacts the zona pellucida of the oocyte), the sperm can penetrate the zona pellucida, and the sperm can fuse with the oolemma. Associated with capacitation are changes in motility, called hyperactivation, a highly vigorous flagellar motion that generates thrusting forces great enough to enable a sperm to gain release from sperm reservoirs and to penetrate the vestments of the oocyte.

In ruminants and primates, the shearing force generated as sperm swim through the interstices of the cervical mucus removes surface proteins and initiates capacitation in these sperm (Fig. 1). Experimental evidence in the hamster model suggests that capacitation may be completed in the caudal isthmus, and that duration of sperm residence in this region is regulated by the interval between insemination and ovulation.

B. Periovulatory Redistribution of Sperm in the Oviduct

There is a temporal relationship between the resumption of sperm ascent to the site of fertilization in the ampulla and the occurrence of ovulation. The coordination of these events is the subject of continued study. In the sow, the anatomical arrangement of the vascular supply to the caudal isthmus is such that the blood within reflects the changing hormone production from the ipsilateral ovary. Thus, the transition in hormone production that occurs as a follicle approaches ovulation could signal impending ovulation to sperm stored in the caudal isthmus.

There is *in vitro* evidence of specific factors in human follicular fluid that are chemotactic for sperm. Because it is likely that follicular fluid enters the oviduct at ovulation, these factors may also coordinate sperm–oocyte interaction.

VII. SPERM SELECTION

Millions to billions of sperm may be inseminated, but relatively few will gain access to the oviducts. There are several routes for the elimination of excess sperm. Sperm are rapidly eliminated from the lower tract by retrograde efflux through the cervix and vagina. In addition, a physiological inflammatory response occurs in response to insemination and phagocytosis of sperm by leukocytes occurs. Excess oviductal sperm are eliminated via passage into the peritoneal cavity. The heterogeneity of seminal sperm, reduction in sperm numbers from the site of insemination to the site of fertilization, and uniform morphologic quality of oviductal sperm are evidence of sperm selection during transport. The most obvious benefit of selection is the increased likelihood of successful fertilization, followed by successful pregnancy, and perpetuation of the species.

VIII. CONCLUDING COMMENTS

Sperm transport is an aspect of *in vivo* sperm function that is fundamental to fertilization success and that has both male and female components. The complexity of this dynamic process, and its fundamental association with the final stages of sperm maturation and function *in vivo*, suggests that abnormal sperm transport might occur as a result of a defect in the male or female component of the process. A failure of normal sperm transport can be a primary cause of reproductive failure.

Acknowledgment

The authors thank Ashley Yudin for providing the transmission electron micrographs of human sperm in human cervical mucus.

See Also the Following Articles

Spermatozoa; Sperm Capacitation

Bibliography

Drobnis, E. Z., and Overstreet, J. W. (1992). Natural history of mammalian spermatozoa in the female reproductive tract. In *Oxford Reviews of Reproductive Biology* (S. R. Milligan, Ed.), Vol. 14, pp. 1–45. Oxford Univ. Press, Oxford.

Harper, M. J. K. (1994). Gamete and zygote transport. In *The Physiology of Reproduction* (E. Knobil and J. D. Neill, Eds.), 2nd ed., pp. 123–187. Raven Press, New York.

Hawk, H. W. (1983). Sperm survival and transport in the female reproductive tract. *J. Dairy Sci.* 66, 2645–2660.

Hunter, R. H. F. (1975). Transport, migration, and survival of spermatozoa in the female genital tract: Species with intra-uterine deposition of semen. In *The Biology of Spermatozoa: Transport, Survival, and Fertilizing Ability: Proceedings of the INSERM International Symposium, Nouzilly, November 4–7, 1973 Basel* (E. S. E. Hafez and C. G. Thibault, Eds.), pp. 145–155. Karger, New York.

Katz, D. F., Drobnis, E. Z., and Overstreet, J. W. (1989). Factors regulating mammalian sperm migration through the female reproductive tract and oocyte vestments. *Gamete Res.* 22, 443–469.

Overstreet, J. W. (1983). Transport of gametes in the reproductive tract of the female mammal. In *Mechanism and Control of Animal Fertilization* (J. F. Hartmann, Ed.), pp. 499–543. Academic Press, New York.

Sperm Transport, Arthropods

Julian Shepherd
Binghamton University

GLOSSARY

copulatory bursa In female insects, a pouch or an expansion of the median oviduct into which sperm are deposited by the male insect during copulation.

palps The appendages near the mouth used to transfer sperm in mites and spiders.

seminal fluid The fluid which surrounds the sperm in the ejaculate.

seminal vesicle A storage organ for mature sperm pending ejaculation.

spermatheca The most widely used term for the sperm storage organ in female arthropods.

spermatophore The capsule in which sperm are transferred to the female in many arthropods.

vas deferens (plural, ***vasa deferentia***) The duct leading from the testis to the seminal vesicle in arthropods.

The journey of the sperm (= spermatozoa) from their site of production in the testes to their ultimate union with the egg comprises the subject of sperm transport. Since the most conspicuous feature of sperm is their motility, it would seem that this would be the major way they are transported. However, muscular action of both the male and the female ducts play at least a major role, if not a dominant role, in transporting the sperm in arthropods. Generally, within the male, it is clear that muscular action by the male ducts plays by far the dominant role in transport. Within the female, however, both sperm motility and muscular action by the female ducts play important roles in sperm transport. Teasing out the contribution of each has been difficult, but it is interesting not only for its physiological understanding but also for the evolutionary insights it may give in understanding male/female conflict and male/male sperm competition. Since arthropods comprise strikingly different taxonomic groups and millions of species with a multiplicity of lifestyles, the discussion of sperm transport will be divided along major taxonomic lines but will attempt to generalize from the regrettably very few species in which these phenomena have been investigated.

I. INSECTS

A. Emergence from the Testes

Sperm in insect testes mature in cysts consisting of bundles of 2^n cells, where $n = 4$–11 or even more depending on the order of insect. The cyst cells encapsulating these bundles are shed as the sperm leave the testes, usually liberating the individual sperm. Just how this departure is regulated is not understood. It would seem likely that there is generally a block at the junction of the testis and vas deferens, but even the extensive histological literature is rarely clear on this point. In most cases this block may simply be a constriction in the tubules, but in the Lepidoptera, there is a distinct cellular layer ("basilar membrane") through which the sperm penetrate. In his seminal works on the reproductive system of the commercial silkworm *Bombyx mori*, Omura presented histological evidence showing that as the bundles pass through the basilar membrane, they contort and rotate 180°, which he concluded indicated motility. Others have suggested that the process is passive, mediated by microvilli on the testicular side of the membrane which capture the cysts. Often, the cyst cells become inserted into the membrane, suggesting a role for these in the extrusion process. Riemann, Thorson, Giebultowicz, and coworkers have found a striking circadian rhythm in the emergence of sperm from the testes in Mediterranean flour moths and gypsy moths and have shown further the surprising fact that this rhythm will continue in isolated reproductive tracts. This results in pulses of sperm appearing in the uppermost vasa and then passing down the tract to the seminal vesicles. The physiological effector of these pulses is unknown, although agents which block action potentials in muscle by blocking sodium channels did not affect these rhythms. The pulses are correlated with a rhythm of secretory activity in the upper vasa deferentia which probably provide important chemical components. This could be related to the major rearrangement of the extracellular coats of the eusperm which occurs at this point.

B. Progress within the Male Tract

However the sperm may emerge from the testes, there is almost uniform agreement that sperm are quiescent in the vasa deferentia and seminal vesicles before ejaculation, implying that muscular action in the walls of these ducts must be responsible for transport of the sperm. However, this needs to be substantiated experimentally.

C. Transfer to the Female

Male insects may transfer sperm as a spermatophore, i.e., in encapsulated form, or as free semen. Generally, Orthoptera, many Hemiptera, Neuroptera, Trichoptera, Lepidoptera, many Coleoptera, some Hymenoptera, a few Diptera, and some other smaller orders use spermatophores. This leaves predominantly the Hymenoptera, Diptera, and some Coleoptera and Hemiptera which transfer free semen. Typically (but not universally) the latter group transfer sperm relatively quickly (within seconds or minutes), whereas those that make spermatophores copulate for longer periods (up to many hours). One small group, the bedbugs and some of their close relatives, have the peculiar (though not unique among invertebrates) technique of hemocoelic insemination, i.e., sperm are introduced by the male into the hemolymph by puncture of the body wall, often through a specialized receiving structure. The sperm then migrate to and penetrate into the female reproductive tract.

Ejaculation from the male duct is uniformly due to the strong muscular contractions which occur during the process of sperm transfer to spermatophores or directly into the female. Characteristically, there are heavily muscularized ejaculatory bulbs for rapid sperm transfer as occurs in Hymenoptera and Diptera, or muscular linings to ejaculatory ducts, seminal vesicles, and accessory glands which form semen or spermatophores in those insects with less rapid sperm transfer. Some insects (e.g., some Coleoptera) stay coupled for long periods of time but transfer pulses of sperm several or many times during coupling. Sperm motility almost certainly plays little or no role in this transfer process.

D. Transport in the Female

Sperm are typically deposited by the male in an expanded portion at the base of the female reproductive tract, generally called the copulatory bursa (= pouch) or vagina (see Fig. 1 for a diagram of one type of female reproductive system). Major exceptions are the male apterygote insects (springtails, silverfish, etc.), which deposit spermatophores on the substrate for pick up by females, and many male Orthoptera, which deposit spermatophores on the outside of the female genitalia. Many of this latter type of spermatophore have special devices for pushing the sperm into the female (e.g., in crickets), and in some cases they deliver the sperm all the way to the sperm storage organ, the spermatheca (e.g., in *Locusta migratoria*).

However, it is perhaps remarkable that in almost all insects, the sperm must move, often a considerable distance, from the site of deposition to the sperm storage organs (sometimes called the seminal receptacle, but more often, and subsequently in this article, the spermatheca). It is also striking that only some (usually a minority) of the sperm succeed in reaching the spermatheca, with the rest being left in the bursa or oviduct and usually rapidly voided or degraded. An exception to this generalization is probably very small insects (e.g., the biting midge *Culicoides melleus*), which have only small numbers of sperm.

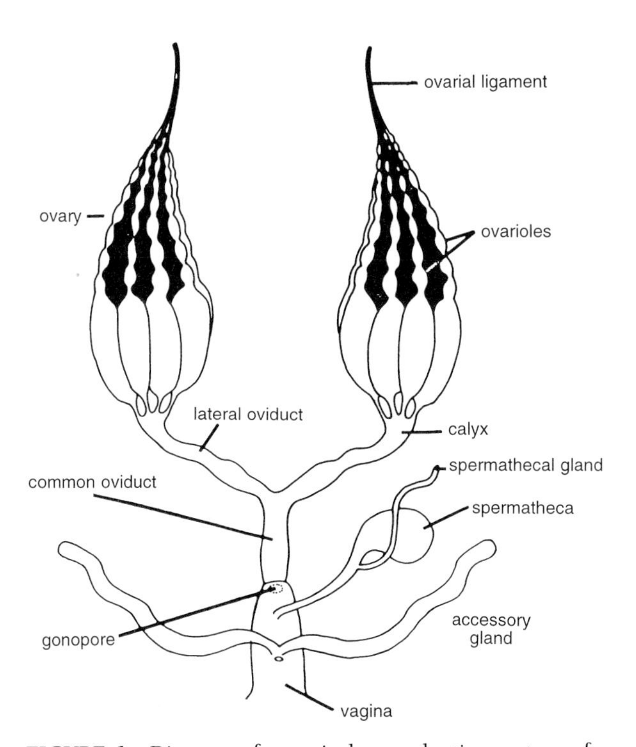

FIGURE 1 Diagram of a typical reproductive system of a female insect. Reprinted with permission from Gillott (1995) which was based on Snodgrass *The Principles of Insect Morphology*, p. 553, figure 284, Cornell University Press. Ithaca, New York. © Cornell University Press.

There have been numerous studies on sperm movement in the female. These include many observational studies, which permit some insights, and only a few using experimental techniques, which allow deeper discrimination of mechanisms. All too frequently, observations or partial experiments form the basis for assertions which are not really substantiated by the evidence. A discussion of sperm movement in the female can be conveniently divided into (i) departure from the site of deposition and (ii) spermathecal filling, though these are much the same process in a number of species.

1. *Departure from the Site of Deposition*

Since the sperm are almost always vigorously active at the site of deposition in the female, it might be assumed that their motility plays an important role in transport. This assumption is reinforced by the observation that most or all the seminal fluid is left behind by the sperm. There are several experimental studies in different insects which support the notion that sperm motility is important at least at the first stage of sperm movement from the bursa. Huignard has shown that if females of the seed beetle *Acanthoscelides obtectus* are injected with 2% parathion (which paralyzes muscles but not sperm), the sperm still move posteriorly in the spermatophore to their point of emergence but cannot progress further. Several studies in female Lepidoptera using parathion (*Acrolepiopsis assectella*) or anaerobiosis (*Pieris brassicae* and *Manduca sexta*), both of which paralyze muscles but do not affect sperm motility, show much the same result: Sperm move posteriorly and exit the spermatophore but do not leave the bursa.

On the other hand, there is both observational and experimental evidence that muscular action by the female tract is important in this process in other insects. Davey has shown through ablation experiments in the bug *Rhodnius prolixus* that the secretions of the opaque accessory glands of the male induce muscular contractions in the female tract, which carry the sperm, even if they are dead, to the spermathecae. Vigorous muscular contractions have been reported in the female tract, including the bursal areas, in other species of insects. In several Lepidoptera, these bursal contractions are reputed to squeeze out the sperm, but, as noted previously, the only experimental tests indicate that sperm motility is the principal force. Furthermore, muscular pressure would force out the copious seminal secretions along with the sperm, and this does not happen in most cases.

It seems reasonable to conclude that in many species of insects, sperm take the initial plunge through their own action. It is possible that females allow this as a means of discriminating more viable from less viable sperm.

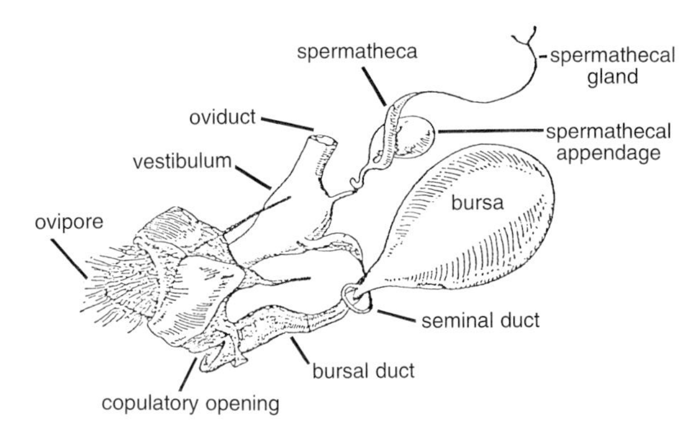

FIGURE 2 Diagram of the reproductive system of a female moth, with the ovaries and lateral oviducts removed. The copulatory opening and the egg-laying opening (ovipore) are separate. Reproduced from H. Eidmann (1929), *Zeitschrift fuer Angewandte Entomologie* **15**, p. 26. © Blackwell Wissenschafts-Verlag, Berlin.

2. *Filling of the Storage Organs*

Numerous reports from most of the major orders of insects indicate that the oviducts, spermathecal ducts, and other ducts involved in the transit of sperm from the site of deposition show rhythmic, sometimes spasmodic, contractions. Many reports also cite observations of sperm moving as discrete masses through these ducts, as if they are being transported en masse by muscular action. The relatively sparse experimental evidence supports this notion: In the studies on *A. obtectus, R. prolixus,* and various Lepidoptera cited previously, sperm movement through the ducts is blocked by muscle paralysis. In intact animals of two of these species, *R. prolixus* and *M. sexta*, dead sperm or inanimate objects have been shown to be transported to the spermatheca(e). In the Lepidoptera, as can be seen from Fig. 2, this migration is particularly complex as sperm are deposited into a bursa whose opening is separate from that of the oviduct. Thus, the sperm must first migrate through the seminal duct to reach the oviduct, then cross it and make their way up the spermathecal ducts. However, the scanty evidence available indicates that it is principally muscular action that carries the sperm over even this long trek. Some Lepidoptera have a lateral pouch, called the bulla seminalis, off the seminal duct which might act as a pump, but most Lepidoptera do not have this organ; in those that do, eggs tend to accumulate in it toward the end of oviposition, suggesting a resorptive role for this organ. In many Diptera, including many *Drosophila* (but not *D. melanogaster*), a swelling of the vagina after mating, called the "insemination reaction," may represent some kind of assistance to the muscular action.

Many workers, using a variety of insects, have suggested that the sperm are chemotactically drawn to the spermatheca. However, in a one-dimensional system, such as the spermathecal ducts or other narrow ducts, there seems little need for chemotactic guidance of sperm. Also, in no insect species has there been demonstrated a directed orientation of sperm to a chemical stimulus. It does seem likely, however, that in some cases the spermathecal secretions do stimulate sperm motility, and this may help the sperm in the stages of spermathecal filling. Observations of enhanced sperm motility in the vicinity of the spermatheca have been reported in the locusts *Locusta migratoria* and *Schistocerca gregaria* and in several Lepidoptera. Frequently, there are glands associated with the spermathecae, though whether these play a role in sperm motility, sperm maintenance, sperm activation, or some other function is largely conjectural. The only experimental support comes from the experiments of Villavaso on the boll weevil *Anthonomus grandis* (Coleoptera). He found that removal of the spermathecal gland reduced sperm motility and prevented spermathecal filling in young females, though removal does not have this same effect in females older than 3 days. He suggested that the gland produces substances which move into the spermatheca in the early stages of female life, and that these substances then act to facilitate sperm movement to the spermatheca.

A novel mechanism for spermathecal filling has been suggested by the careful work of Linley and Simmons with the biting midge *C. melleus*. Their measurements of the dimensions of the spermathecal ducts and the volumes of sperm and spermathecal lumen led them to propose that a withdrawal of fluid from the spermatheca sucks the sperm passively up into the spermatheca.

In Lepidoptera, in which two kinds of sperm are produced and transferred to the female, the spermatheca usually has two lobes. Several reports indicate that most or all the eusperm, the fertilizing sperm, end up in the main part of the spermatheca which is attached to the gland (Fig. 2). The parasperm, which lack a nucleus, end up largely in the lateral appendage, sometimes called the "lagena," where they die and in some cases degenerate completely. How this is accomplished is unknown.

E. Fertilization of the Eggs

Fertilization of eggs in insects occurs as each egg passes down the oviduct on its way to being laid. Since egg passage often occurs fairly rapidly, particularly when eggs are laid in batches, it is obvious that a repeated, neural program must coordinate the release, transport, and deposition of each egg. It seems logical to assume that release of sperm from the spermatheca is somehow coordinated as a part of this process. In the few cases in which this process has been viewed internally while it is happening, such rhythmic contractions of the spermatheca have been observed. In the cricket *Teleogryllus commodus*, Sugawara reported that the spermathecal duct shows a continuous myogenic peristalsis but produces a twitch when stimulated during the passage of an egg. He postulates that the myogenic contractions propel the sperm down the duct and the twitches propel the sperm into the genital chamber (part of the oviduct) where the eggs are fertilized.

In *D. melanogaster*, in which the use of genetic mutants so often enhances understanding, a mutant which lacks a lipolytic esterase ("esterase 6"), in the ejaculated seminal fluid, also shows a slower release of sperm from the ventral receptacle (the sperm storage organs in *D. melanogaster* comprise this ventral receptacle plus two spermathecae). It has been suggested that lipid release enhances lipid availability in wild-type flies and that this would enhance sperm motility, thus facilitating release from the receptacle. However, it has not been ascertained that this enzyme enhances the amounts of lipid which the sperm can use or that sperm motility plays a role in sperm release.

In the Lepidoptera, the lumen of the spermathecal duct is bifurcated into a larger, elastic lumen and a very narrow, heavily sclerotized channel, which has been called the "fertilization canal." Most observers agree that the former is used in spermathecal filling and the latter during fertilization (one report of the moth *Plodia interpunctella* makes the opposite case). Just how sperm passage in these ducts is controlled is unknown.

Actual penetration of the eggs has rarely been observed, presumably because of observational difficulties created by the presence of a hard eggshell. Sperm in the ducts and around the eggs during fertilization are generally motile, and it is widely assumed that sperm motility is essential for penetration of the eggs. However, the scanty evidence suggests that muscular action is largely responsible for passage down the spermathecal duct.

F. Sperm Competition in the Female Tract

The consequences for paternity of multiple matings in insects have been a subject of intense interest to evolutionary biologists, especially since Parker's detailed review of the subject in 1970. Numerous experiments have investigated the paternity of offspring after various patterns of multiple mating, but very few have examined the mechanisms within the female tract. In most cases, the second male to mate usually sires most of the future offspring ("second male precedence"), at least for the initial period after the second mating. The simple explanation of "last in, first out" of the spermatheca seems to explain many such results. However, in at least some cases, it is clear that sperm of a previous mating must be evicted from the sperm storage organs in order for second male precedence to be effected. The most dramatic manifestation of such an action has been reported in the Odonata, in which Waage, Michiels, and others have reported mechanisms by which the penis simultaneously or sequentially removes the sperm of previous matings and deposits a new batch. Apparently, penes do not do this in other insects, but other mechanisms may accomplish the same goal. In *D. melanogaster*, Harshman and Prout have used various genetic varieties to show that it is the seminal fluid, not the sperm, which is important in displacing the sperm of a previous mating. Clearly, more work on mechanisms of sperm displacement in insects would be of interest to biologists.

II. Arachnids

Although some aspects of arachnid reproduction, such as morphology and behavior, have received considerable attention, very little is known about their reproductive physiology. Sperm transfer from male to female in various species of arachnids has been well described, but sperm transport has received significant attention only in the ticks since this is the only group within the Arachnida which combines large size and economic importance.

A. Ticks

As in insects, mature sperm are released from the testes and pass down to a pair of seminal vesicles prior to mating. During mating, a spermatophore is extruded by the male from his genital aperture and transferred to the female's genital aperture in just a few minutes. Within a fraction of a minute, the spermatophore then explodes down the female tract, carrying the sperm to the seminal receptacle or to the vagina. Since the sperm are immature and nonmotile until this point, this transfer is accomplished passively. Within the spermatophore inside the female, the sperm undergo a radical transformation over several hours and then become motile and leave the spermatophore, traveling up the oviducts. Some workers believe this migration is accomplished by the motility of the sperm, whereas others believe it results from female muscular action; no definitive experiments have been performed. It is known that this migration occurs only if the female has had a good blood meal, but the nature of the stimulus to either the sperm or the female's muscles is unknown. Most of the sperm collect in expansions of the oviduct called the ampullae, but a few can be found in the ovaries. It is unknown whether the eggs are fertilized in the oviducts or ovaries or how they are fertilized.

B. Other Arachnids

Since ticks are just one superfamily of the large and diverse order of mites (Acari), more variety in mechanisms of sperm transfer is seen across the whole order. Some male mites transfer spermatophores while mating like ticks, but others leave spermatophores on the substrate for females to pick up. Others have a process of copulation and internal insemination like most insects, though many, like spiders, use their palps to accomplish this. Due mostly to the small size of mites, there has been almost no work done on the internal phenomena of transport or fertilization.

Spiders are another large and ecologically important group of arachnids, but knowledge about their internal transport phenomena is almost completely unknown. Males transfer sperm by ejaculating sperm onto a small web constructed for this purpose, then they pick up the sperm with their palps and introduce their palps into the female's genital opening. The male's palps inflate through blood pressure during this process, and a special structure, the embolus, picks up the sperm. However, even the mechanism of this latter well-known action is unelucidated; it has been suggested to occur by capillarity or by suction, maybe involving displacement of fluid.

III. CRUSTACEANS

With the exception of those in barnacles, ostracods, and a few other groups, the sperm of most crustaceans are aflagellate and apparently nonmotile when mature. They accumulate in the vasa deferentia of the male, where they are encapsulated in spermatophores bathed in a seminal fluid. Frequently using their abdominal appendages, males transfer these spermatophores to the external abdominal appendages of the female or into a seminal receptacle (or spermatheca), which may be separate from or attached to the female reproductive tract. Wherever they are deposited, the sperm generally must enter the female tract to fertilize the eggs, although in some cases fertilization is external. The exact site of fertilization is unknown in most Crustacea: It may occur in the ovaries, in the oviducts as the eggs pass the spermathecal opening, or even externally after the egg has passed to the exterior. The process is different in barnacles (Cirripedia): The male mates with the female by inserting a long penis into her pallial cavity and then releasing sperm which make their way into the female tract.

See Also the Following Articles

Crustacea; Female Reproductive System, Insects; Insect Reproduction, Overview; Male Reproductive System, Insects; Sperm Activation, Arthropods

Bibliography

Davey, K. G. (1958). The migration of spermatozoa in the female of *Rhodnius prolixus*. *J. Exp. Biol.* **35**, 694–701.

Giebultowicz, J., Riemann, J. G., Raina, A. K., and Ridgway, R. L. (1989). Circadian system controlling release of sperm in the insect testis. *Science (Washington, DC)* **245**, 1098–1100.

Gillot, C. (1995). *Entymology*, 2nd ed. Plenum Publishing, New York.

Gillott, C. (1995a). Insect male mating systems. In *Insect Reproduction* (S. L. Leather and J. Hardie, Eds.), pp. 33–55. CRC Press, Boca Raton, FL.

Harshman, L. G., and Prout, T. (1994). Sperm displacement without sperm transfer in *Drosophila melanogaster*. *Evolution* **48**, 758–766.

Hinsch, G. W. (1990). Arthropoda—Crustacea. In *Reproductive Biology of Invertebrates. Vol. IV, Part B. Fertilization, Development, and Parental Care* (K. G. Adiyodi and R. G. Adiyodi, Eds.), pp. 121–155. Wiley, New York.

Hinton, H. E. (1964). Sperm transfer in insects and the evolution of haemocoelic insemination. In *Insect Reproduction*, pp. 95–107. Second Symposium of the Royal Entomological Society of London.

Huignard, J. (1971). Utilisation d'une substance paralysante, le parathion, pour l'étude de la migration des spermatozoides vers la spermathèque chez *Acanthoscelides obtectus* Say (Coleoptère Bruchidae). *C. R. Acad. Sci. Paris Sér. D* **273**, 2557–2559.

Linley, J. R., and Simmons, K. R. (1981). Sperm motility and spermathecal filling in lower Diptera. *Int. J. Invertebr. Reprod.* **4**, 137–146.

Pochon-Masson, J. (1994). Gametogenesis and fertilization. In *Traité de Zoologie. Tome VII. Crustacés. Fasc. 1*, pp. 727–805. Masson, Paris.

Rao, H. V., and Davis, N. T. (1969). Sperm activation and migration in bed bugs. *J. Insect Physiol.* **15**, 1815–1832.

Reger, J. F., and Fitzgerald, M. E. C. (1983). Arthropoda—Myriapoda. In *Reproductive Biology of Invertebrates. Vol. II. Spermatogenesis and Sperm Function* (K. G. Adiyodi and R. G. Adiyodi, Eds.), pp. 451–475. Wiley, New York.

Sonenshine, D. E. (1991). *The Biology of Ticks. Vol. 1*. Academic Press, New York.

Sugawara, T. (1993). Oviposition behaviour of the cricket *Teleogryllus commodus*: Mechanosensory cells in the genital chamber and their role in the switch-over of steps. *J. Insect Physiol.* **39**, 335–346.

Tschudi-Rein, K., and Benz, G. (1990). Mechanisms of sperm transfer in female *Pieris brassicae* (Lepidoptera: Pieridae). *Ann. Entomol. Soc. Am.* **83**, 1158–1164.

Waage, J. K. (1979). Dual function of the damselfly penis: Sperm removal and transfer. *Science (Washington, DC)* **203**, 916–918.

Spiders

see Chelicerate Arthropods

SRY Gene

Grace Lee and Mert Bahtiyar
Yale University School of Medicine

GLOSSARY

HMG domain High-mobility group domain is a DNA-binding motif that is associated with several eukaryotic regulatory proteins, including SRY and SF-1. Attachment of the HMG domain-containing protein to DNA causes conformational changes and regulates RNA transcription.

pseudoautosomal region The region on the distal end of the short arm of Y chromosome which pairs with X chromosome during meiosis and where crossing over takes place.

SOX *SRY* HMG b*ox*-related genes. A family of HMG box-containing genes whose predicted peptide sequence within the box shows more than 60% similarity to *SRY*. They are found on autosomal chromosomes.

SRY *S*ex-*r*elated gene on *Y* chromosome in human.

Sry *S*ex-*r*elated gene on *Y* chromosome in mouse.

TDF Testis-determining factor in human.

Tdy Testis-determining factor in mouse.

Sex-determining region on Y chromosome is the gene that initiates the genetic cascade leading to testis development in mammals. Sex-determining genes on the Y chromosome induce testicular development. Genetic modulators of sexual differentiation may also be found on autosomes. The human testis produces hormones necessary for male phenotype development. Sertoli cells secrete Müllerian-inhibiting factor, whereas Leydig cells secrete testosterone which induces Wolfian duct development. The presence of the ovary is not necessary for Müllerian duct differentiation.

I. MOLECULAR CHARACTERISTICS OF THE *SRY* GENE

The *SRY* gene (*s*ex-*r*elated gene on *Y* chromosome) is located at the short arm of the Y chromosome (Yp, localized to pY53.3). The SRY gene product contains a sequence of 80 amino acids which has a high homology to high-mobility group (HMG) proteins. This group of proteins is important in the regulation of RNA transcription. Sequence characterization of *SRY* in human and mouse differs. Human *SRY* binds to the AACAATG sequence, whereas murine *Sry* binds to the CATTGTT sequence on DNA (Fig. 1). Human *SRY* gene is a single exon of 850 bp and it is a major transcription initiation site at 91 bp upstream from the first methionine. Mouse *Sry* gene also en-

CONSERVED HMG-like Box

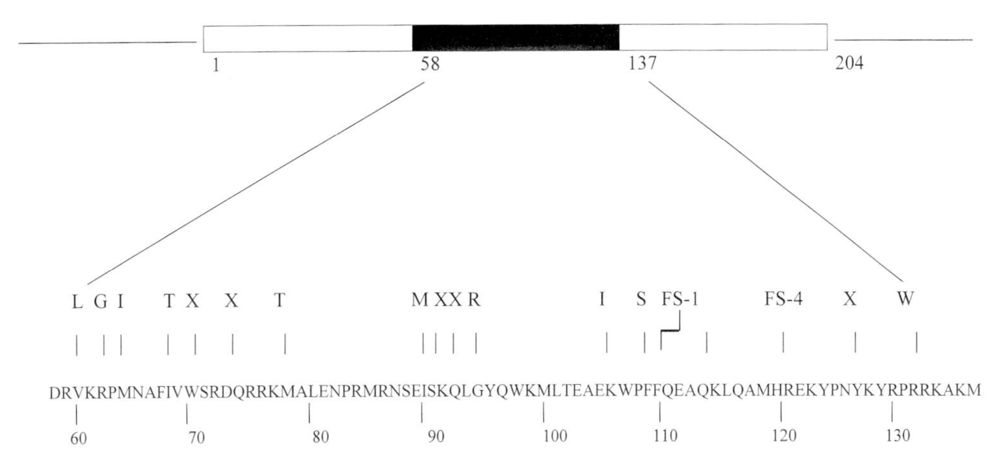

FIGURE 1 Reproduced with permission from McElreavey *et al.* (1995). The genetic basis of murine and human sex determination; a review *Heredity*, 75, 599–811.

codes for a single exon product. *SRY* gene is conserved among mammalian species. HMG regions have 71% similarity between human and mouse *SRY*. Outside this region there is very low homology (Fig. 2). While mouse *Sry* transcripts are linear in the fetus, they are circular in adult testicular tissue. *SRY* transcripts can also be found in adult human testicular tissue but so far no circular transcripts have been described. *Sry* is expressed for a short time after fertilization. *Sry* expression starts at Postcoital Day 10.5, it reaches its peak at Day 11.5, and by Day 12.5 it decreases to nondetectable levels again. *SRY* gene contains GC-rich regions which are important in gene inactivation in the adult life. This region methylates after embryonic life and inactivates the gene.

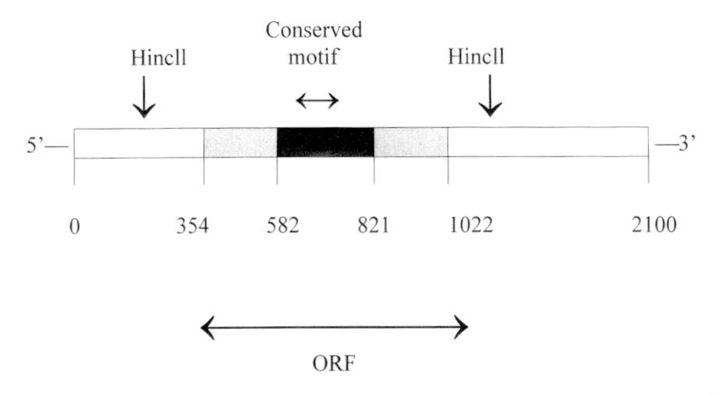

FIGURE 2 Reproduced with permission from Sinclair *et al.* and *Nature*.

II. HISTORICAL DEVELOPMENT OF THE *SRY* GENE

In early 1950s the relation between sex determination and the presence of an intact testicular tissue in rabbits was shown. When male rabbit embryos were castrated before gonadal differentiation, all embryos had female genitalia. It has also been shown that the presence of the ovaries was not necessary for the sex differentiation. Demonstration of XO genotype with female characteristics and XY, XXY, and XXXY genotypes with male characteristics indicated that the Y chromosome is necessary for sex determination.

It was originally believed that the small Y chromosome did not play a role in sex determination. Since the late 1950s it has been accepted that Y chromosome is male determining and it acts as the primary signal for male development. It has been suggested that human hermaphroditism and XX Klinefelter's syndrome are generated by a rare and illegitimate X-Y chromosomal interchange during paternal meiosis when the X and Y chromosomes pair, enabling the Y-specific sequence to be transferred to the short arm of the paternal X chromosome. This finding led to the hypothesis that a gene termed testis-determining factor (TDF) resides in the short arm (Yp) of the Y chromosome. From banding and DNA sequencing

TABLE 1
SRY-Related HMG Box Genes[a]

Gene	*Data*
Mouse	
Sox1, Sox2, Sox3	Located on the X chromosome between Hprt and Dmd
	Most closely related to *Sry*
	Has a role in neural development (*Sox3*)
	Xq26–27
	Deleted in patients with mental retardation and hemophilia B
Sox4	HMG1B similar structure
	474 amino acids
	Chromosome 6p distal to MHC region
	Transcriptional activator in lymphocytes
Sox6	High homology with *Sox5*
	Expressed highly in adult mouse testis
	AACAAT binding ability
	CNS development role
xSox7	*Xenopus laevis* ovary
	ORF codes for 362 amino acids
	ORF has an HMG motif
	HMG region shows 90% similarity with mouse *Sox7*
	Capable of binding to AACAAT
xSox12	*Xenopus laevis*
	amino acids
	Contains leucin zipper motif
	Glutamin-rich segments
Sox17	Encodes a protein for 419 amino acid
	Contains a single HMG box
	Subject to alternative splicing
Sox18	378 amino acids
	Heart, lung, and skeletal muscle in adult mouse
Human	
SOX5	Subject to alternative splicing
	Pseudogene located on 8qw21.1
	Gene maps to 12p12.1
	High similarity between *SOX5* and *Sox5*
SOX11	Maps to 1p25
	Related to *SOX4*
	Important in the development of CNS
SOX19	Member of Sox subfamily B
	Subfamily B contains *Sox1*, *Sox2*, and *Sox3*
	Earliest molecular marker of the CNS in vertebrates
SOX20	Cross-hybridize to a *SOX9* cDNA probe
	Amino acid sequence has some similarity to *SOX12* and *Sox16* HMG domains
	Maps to 17p13

[a] They are not necessarily involved in sexual determination and differentiation.

studies, investigators constructed deletion maps from analysis of the genomes of sex-reversed XX males and XY females, a meiotic map of the pseudoautosomal region which is shared by the X and Y chromosomes, and a long-range restriction map linking the first two maps. It has been shown that the existing piece of DNA is Yp specific, with sequences present on paternal X chromosome. Recently, it has been shown that TDF is a short segment of the Y chromosome adjacent to the pseudoautosomal boundary. Sinclair and coworkers in 1990 showed that the breakpoints in the XX males were clustered around the region ≈35-kb proximal to the pseudoautosomal boundary. Sequencing this region revealed an open reading fragment, which was called the *SRY* gene. It was shown that this region was necessary for sex determination.

Zinc finger protein gene on the Y chromosome (*ZFY*) was an early candidate for TDF. Later data showed that this gene was not the initiative signal for the testicular development. Today *SRY* gene is the leading candidate for the TDF. It has been shown in knockout mouse models that male mice lacking this gene develop as females, and when this gene was injected into female embryo they developed as males.

III. SEX-RELATED GENES

Contrary to what was believed in 1950s the *SRY* gene is not the only gene responsible for sex differentiation. Other genes are also involved in testicular development. *SRY* is believed to initiate a very complex cascade of events and leads to sex differentiation. One of the candidates for this sex differentiation-related genes is Wilm's tumor gene (*WT1*). Mutation in *WT1* causes nephropathy associated with sex reversal, which is called Denys–Drash syndrome. Xp21–22 (*DSS*) cytogenetic abnormality associated with XY sex reversal implies the existence of a dose-sensitive gene involved in sex determination. In some autosomal chromosomal abnormalities, namely, monosomy *9p* and *10q*, 46,XY sex reversal has been described. Autosomal chromosomes which have high levels of similarity to *SRY* in the HMG region were also identified. They are called *SRY* box (*SOX*) genes. Their role is not fully understood.

TABLE 2
Other Genes Shown to Play a Role in Sex Determination and Differentiation

Gene
WT1
DSS
9p
10q

SOX genes share a high sequence identity with the HMG box present in the testis-determining factor gene *SRY*. A large family of genes sharing a high similarity with the *SRY* HMG box and named *Sox* (*Sry*-related HMG box) in mouse and *SOX* in human has been identified from various organisms (Tables 1 and 2).

IV. APPLICATIONS

The *SRY* gene is used not only to explain the sex reversal cases in the clinic but also as a diagnostic tool for preimplantation genetic diagnosis. Isolated genetic material from biopsied blastomers was amplified with polymerase chain reaction (PCR) for the presence of *SRY* gene. The presence of the gene was accepted as an indication of male gender. Some investigators used single primer sets to amplify the sequence. Others used nested PCR with two different sets of primers in order to increase the sensitivity specificity of this procedure. Two-step PCR greatly reduces the generation of nonspecific PCR products which may characteristically result when a large number of amplification cycles are carried out.

V. FUTURE PROSPECTIVE

Mammalian sex differentiation is not yet fully understood. Explanation of the whole cascade of events in sexual development will lead to a wide range of applications from *in vitro* fertilization and embryo transfer to genetic counseling.

See Also the Following Articles

Sex Chromosomes; Testis, Overview

Bibliography

Affara, N. A., Ferguson-Smith, M. A., Tolmie, J., *et al.* (1986). Variable transfer of Y specific sequences in XX males. *Nucleic Acids Res.* **14**, 5375–5387.

Bennett, C. P., Docherty, Z., Robb, S. A., Ramani, P., Hawkins, J. R., and Grant, D. (1993). Deletion 9p and sex reversal. *J. Med. Genet.* **30**, 518–520.

Ford, C. E., Jones, K. W., Polani, P. E., de Almeida, J. C., and Briggs, J. H. (1959). A sex-chromosome anomaly in a case of gonadal dysgenesis (Turner's syndrome). *Lancet* **1**, 711–713.

Han, Y., Yoo, O., and Lee, K. (1993). Sex determination in single mouse blastomeres by polymerase chain reaction. *J. Assisted Reprod. Genet.* **10**, 151–156.

Jacobs, P. A., and Ross, A. (1966). Structural abnormalities of the Y chromosome in man. *Nature* **210**, 352–354.

Jacobs, P. A., and Strong, J. A. (1959). A case of human intersexuality having a possible XXy sex determination mechanism. *Nature* **183**, 302–303.

Jost, A., Vigier, B., Prepin, J., and Perchellet, J. P. (1973). Studies on sex differentiation in mammals. *Recent Prog. Horm. Res.* **29**, 1–41.

Koopman, P., Munsterberg, A., Capel, B., Vivian, N., and Lovell-Badge, R. (1990). Expression of a candidate sex-determining gene during mouse testis differentiation. *Nature* **348**, 450–452.

Koopman, P., Gubbay, J., Vivian, N., Goodfellow, P., and Lovel-Badge, R. (1991). Male development of chromosomally female mice transgenic for Sry. *Nature* **351**, 117–121.

Kunieda, T., Xian, M., Kobayashi, E., Imamichi, T., Moriwaki, K., and Toyoda, Y. (1992). Sexing of mouse preimplantation embryos by detection of Y chromosome-specific sequences using polymerase chain reaction. *Biol. Reprod.* **46**, 692–697.

Ogota, T., Hawkins, J. R., Taylor, A., Matsuo, N., Hata, J., and Goodfellow, P. N. (1992). Sex reversal in a child with a 46 X,Yp+ karyotype: Support for the existence of a gene(s) located in distal Xp, involved in testis formation. *J. Med. Genet.* **29**, 226–230.

Page, D. C., Mosher, R., Simpson, E. M., *et al.* (1987). The sex-determining region of human Y chromosome encodes a finger protein. *Cell* **51**, 1091–1104.

Palmer, M. S., Sinclair, A. H., Berta, P., *et al.* (1989). Genetic evidence that ZFY is not the testis determining factor. *Nature* **342**, 937–939.

Sinclair, A. H., Berta, P., Palmer, M. S., *et al.* (1990). A gene for the human sex-determining region encodes a protein with homology to a conserved DNA-binding motif. *Nature* **346**, 240–244.

Welshons, W. J., and Russell, L. B. (1959). The Y-chromosome as the bearer of male determining factors in the mouse. *Proc. Natl. Acad. Sci.* **45**, 560–566.

Werner, M. H., Huth, J. R., Gronenborn, A. M., and Clore, G. M. (1995). Molecular basis of human 46,XY sex reversal from the three-dimensional solution structure of human SRYT-DN complex. *Cell* **81**, 705–714.

Wilkie, A. O. M., Campell, F. M., Daubeney, P., *et al.* (1993). Complete and partial XY sex reversal associated with terminal deletion of 10q—Report of 2 cases and literature review. *Am. J. Med. Genet.* **46**, 597–600.

Sterility

Bradley Shawn Hurst
University of Colorado Health Sciences Center

GLOSSARY

infertility Inability to conceive after an adequate exposure to unprotected intercourse; usually considered at least 12 months.

sterility Inability to conceive.

sterility, absolute Inability to conceive in any circumstance; no treatment is available to enable conception.

sterility, reversible (or ***relative sterility***) Inability to conceive without intervention, but treatment is available which may allow pregnancy to occur.

sterilization Surgical procedure performed to render an individual sterile.

An individual who is unable to conceive is sterile. Sterility may be voluntary or involuntary and may be congenital or acquired. Voluntary sterility can be accomplished by occluding the Fallopian tubes in the female or the vas deferens in the male to permanently prevent an undesired pregnancy. Involuntary sterility, or infertility, generally implies that an individual or couple is unable to conceive despite a desire for pregnancy.

I. INTRODUCTION

Our understanding of sterility has changed considerably in recent years. Conception is now possible with modern therapy for individuals once considered to be sterile. Reports of menopausal women who deliver following oocyte donation are the most extreme examples of the changing definition of sterility. Sterility now may be considered either absolute or reversible. Pregnancy is impossible despite medical or surgical intervention with absolute sterility. Reversible sterility, on the other hand, may be treated. Fortunately, most cases of undesired sterility can be overcome if resources are available.

Sterility is caused by any condition which disrupts the reproductive system. Clearly, an individual is sterile if no gametes are produced. Sterility is the consequence of interrupted gamete transport by surgical sterilization, congenital absence of the vas, or occlusion of the Fallopian tubes. Sterility may occur if gametes are not capable of fertilization, even if gamete production and transport are not hindered. Finally, since implantation must occur for a pregnancy to be established, uterine abnormalities may result in sterility.

II. FEMALE STERILITY

Female fertility requires the incredible synchrony of the menstrual cycle. The cervix must change to allow sperm transport into the upper genital tract before ovulation. The endometrium must cyclically proliferate and mature to allow implantation at the proper time for the embryo. Functional Fallopian tubes must be able to capture the oocyte at ovulation, transport sperm from the uterus toward the distal tube, and transport the oocyte and early embryo toward and eventually into the uterus. At the same time, the tube provides a nurturing environment for early embryonic development. This process is orchestrated by the ovarian hormonal changes that occur during development of the follicle, ovulation, and formation of the corpus luteum. All these hormonal changes occur as a result of an intricate feedback loop between the ovary, the hypothalamus, and the pituitary gland. These hormonal changes are necessary to achieve ovulation and oocyte maturation, which render the oocyte capable of fertilization. Female infertility or sterility may occur from a disruption of any of these interrelated events (Fig. 1).

A. Oocyte Depletion

Sterility results when the supply of gametes are depleted. Age is the most common cause of oocyte depletion. It is estimated that 40% of women are sterile at the age of 40, despite the fact that many continue to ovulate. By the age of 45, approximately 85% of women are sterile. However, oocyte depletion

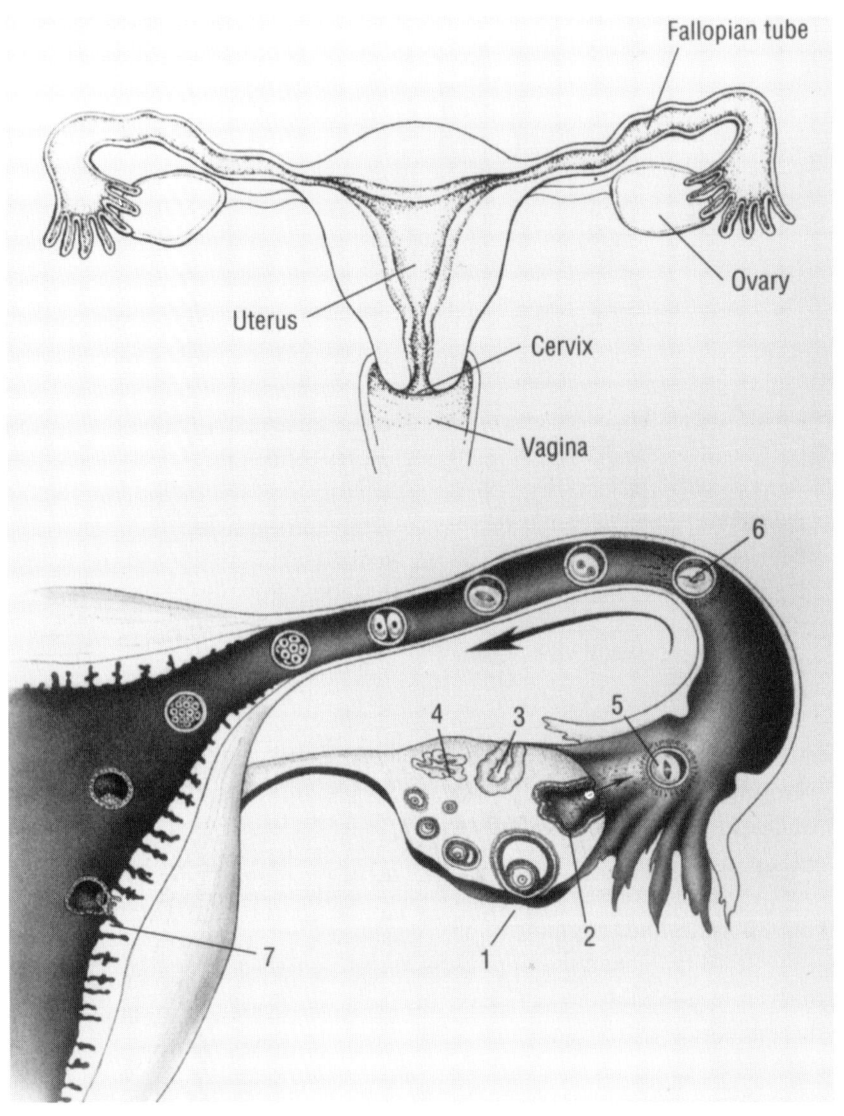

FIGURE 1 The female reproductive system. The dominant follicle is formed (1), followed by ovulation (2). After ovulation, the corpus luteum produces estrogen and progesterone to support the endometrium (3). If pregnancy does not occur, the corpus luteum degenerates (4). The ovulated oocyte is picked up by the Fallopian tube and matures to the stage of meiosis (5). Fertilization occurs in the Fallopian tube, and the embryo is transported to the uterus (6). Implantation requires synchrony between the embryo and the endometrium (7). Reproduced from The American College of Obstetricians and Gynecologists. Preconceptional Care. (Patient Education Pamphlet No. AB012). Washington, DC, © ACOG 1995.

may occur at any age. An accelerated loss of oocytes may occur *in utero*, during childhood, or during the reproductive years. Gonadal agenesis or dysplasia is a condition in which the ovaries fail to form *in utero*. Infants are phenotypically normal females. These girls fail to develop secondary sexual characteristics since ovarian hormonal function cannot occur. A congenital depletion of oocytes may occur as a result of genetic anomalies such as Turner's syndrome (45,X karyotype), galactosemia, or may be unexplained. Gamete depletion during childhood is rare but can occur with autoimmune diseases. Chromosomal anomalies such as Turner's mosaicism (46,XX; 45,X) accelerate ovarian failure. Loss of ovarian function during childhood may occur due to iatrogenic causes. Pelvic irradiation with exposure of the ovaries to doses over 800 rad will cause ovarian failure. Some chemotherapeutic agents, especially the alkylating drugs, may cause ovarian failure depending on the dosage, duration of therapy, and gonadal susceptibility during treatment. Surgical removal of the ovaries, or the loss of ovarian function by extensive damage to the ovarian vasculature during surgery, results in sterility. Loss of ovarian function during childhood is not symptomatic until the time of expected puberty, when sexual maturation fails to occur. After childhood, premature ovarian failure may occur at any time during the reproductive years. Although the cause may be unknown, chromosomal, autoimmune, or iatrogenic causes may be involved.

Individuals with ovarian dysgenesis and those with ovarian failure were considered to be absolutely sterile until recent years. However, with oocyte donation, pregnancy may be achieved if the uterus can function normally in response to exogenous hormonal manipulation. Estrogen and progesterone uterine priming and supplementation in the first trimester are essential to establish and maintain an oocyte donation pregnancy.

B. Anovulation

Ovulation is mandatory for fertility, and conditions which prevent ovulation result in sterility. Ovulation requires complex interactions between the ovary, hypothalamus, and pituitary. Anovulation occurs in conditions which interfere with normal feedback. Hypothalamic anovulation may be due to conditions such as anorexia, severe physical or mental stress, or starvation. Since these conditions usually are not permanent, infertility is more common than sterility. However, congenital absence of the hypothalamic gonadotropin-releasing hormone (GnRH) neurons, as occurs in Kallman's syndrome, is permanent and sterility is expected. Pituitary causes of anovulation may be due to anatomic abnormalities. Tumors that distort the pituitary or its portal circulation may lead to anovulation. Hyperprolactinemia can prevent ovulation by disrupting the normal hypothalamic–pituitary feedback system. The entire hypothalamic–pituitary–ovarian axis may be disrupted in polycystic ovarian syndrome. This condition is associated with a hyperandrogenic state. Affected individuals are typically estrogenized, unlike those with hypothalamic or pituitary anovulation. Women with polycystic ovarian syndrome may ovulate sporadically; therefore, sterility is not always the result.

Ovulation induction is almost always possible. Therefore, sterility in anovulatory women is considered reversible. Pulsatile GnRH can be administered to those with hypothalamic amenorrhea. Follicle-stimulating hormone can be given to those with pituitary dysfunction if the underlying abnormalities cannot be corrected. Those with estrogenized amenorrhea, such as polycystic ovarian syndrome, will generally ovulate with clomiphene citrate. However, gonadotropin stimulation may be necessary in some.

C. Tubal Sterility

Tubal damage or pelvic adhesions may cause sterility by preventing oocyte pickup or gamete or embryo transport. Tubal occlusion is the major cause of involuntary female sterility. Ironically, tubal occlusion is the most common form of elective female sterilization. Tubal sterility is caused by an inflammatory process that damages the cilia needed for gamete and embryo transport. In severe cases, the inflammatory process results in closure of the Fallopian tube (Fig. 2). Infections known to cause tubal infections include chlamydia and gonorrhea, although salpingitis usually is caused by a polymicrobial infection. Pelvic adhesions caused by endometriosis, salpingitis, and

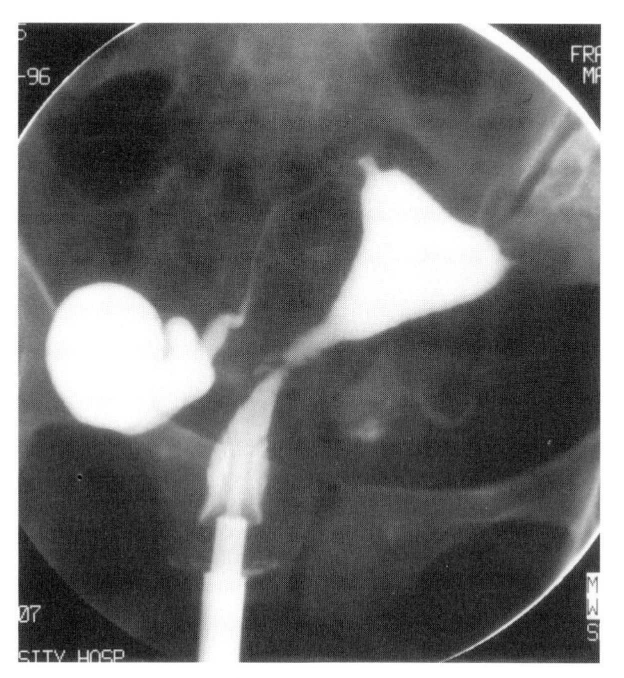

FIGURE 2 Hysterosalpingogram showing a large hydrosalpinx, with incomplete filling of the contralateral tube.

surgery cause sterility even if tubal patency is confirmed.

Tubal occlusion is the most commonly performed method of sterilization. Postpartum tubal sterilization is generally performed by removing a segment of the Fallopian tube through a small skin incision at the umbilicus. Laparoscopic tubal sterilization is usually performed at other times, and the tubes are occluded by cautery, clips, bands, or by removing a tubal segment. Tubal sterilization is sometimes performed as an open-abdominal procedure and can be performed vaginally by colpotomy. New techniques of tubal sterilization are being developed. Transcervical tubal occlusion by plugs, wires, chemicals, laser, or cautery have been tried. Unfortunately, the failure rate of most of these new approaches has been unacceptably high, and these techniques remain investigational. On the other hand, traditional sterilization techniques are quite effective, with long-term success rates of 96–99%. Of the available techniques, unipolar cautery and postpartum partial salpingectomy have the lowest failure rates (<1%). The highest failure rate occurs with surgical clips (3 or 4%). Young women who are sterilized by bipolar cautery or surgical clips can expect a 5% long-term failure rate. Fortunately, complications are uncommon. Ectopic pregnancy occurs in <1% of women following tubal sterilization, although the risk is as high as 3% following sterilization by bipolar cautery. Chronic pelvic pain was once thought to be a complication of sterilization procedures, but well-designed studies have not shown this to increase after tubal ligation.

Tubal infertility of any cause is treatable by surgical correction, *in vitro* fertilization, or both. The overall success of surgery depends on the extent of tubal damage and the skill of the surgeon. Success can range from <10% for women with a large hydrosalpinx and extensive adhesions to over 80% for some types of sterilization reversals. *In vitro* fertilization (IVF) success is 20–25%, although the outcome with this technique is steadily improving. The success of IVF may be compromised when a hydrosalpinx is present, and some physicians now recommend surgery to remove or correct the hydrosalpinx prior to attempting IVF.

D. Uterine Sterility

Proliferation and maturation of the endometrium is necessary for normal implantation and subsequent development of pregnancy. Sterility or infertility is expected if these physiologic changes cannot occur. Since the endometrium is controlled by ovarian or extrinsic estrogen and progesterone, abnormalities in these reproductive hormones may reduce the endometrial receptivity to an implanting embryo. For example, chronic progestin exposure, as occurs with progestational contraception, reduces endometrial receptivity. Receptivity is restored when the progestational exposure is removed if cyclic ovulation returns. Fertility requires an intact vascular supply to the endometrium. Conditions which interfere with the endometrial blood flow, such as uterine myomas, occasionally cause sterility. Physical or mechanical destruction of the endometrium by endometrial ablation, Asherman's syndrome, or tuberculosis prevents pregnancy, possibly irreversibly. Furthermore, cell adhesion molecules such as integrins and cadhedrins appear to be required for implantation. The absence of these may reduce fertility or cause sterility in extreme situations.

Uterine sterility may be treated only if a functional endometrium can be restored. Sterility is absolute when there is congenital absence of the uterus or when the uterus has been surgically removed. Sterility is absolute when the endometrium has been irreversibly destroyed by surgery. Fortunately, however, in most cases of Asherman's syndrome, uterine myomas or other uterine defects can be surgically repaired. Integrins, when absent, may be restored in many circumstances by hormonal manipulation.

E. Cervical/Vaginal Sterility

Abnormalities of the cervix and vagina are rare causes of sterility. Cervical or vaginal agenesis is an extreme example of cervical sterility. Loss of the endocervical mucus-producing glands can occur with surgery such as an extensive cervical cone biopsy. However, sterility would be a very rare complication. Cervical stenosis resulting from cervical cryotherapy, cervical cancer, or radiation therapy may rarely prevent pregnancy. Infections that cause intense endocervical inflammation, conditions which alter the pH or consistency of the cervical mucus, or antisperm antibodies may reduce fertility but are not usually the cause of sterility.

Most forms of cervical infertility can be treated. Cervical agenesis is the exception. Several deaths have occurred from attempts to surgically create a cervical canal. These deaths usually are attributed to the lack of cervical mucus and local immunity, which provide a defense against ascending infections. Almost all other causes of cervical sterility can be bypassed by intrauterine insemination, gamete intrafallopian transfer, or and zygote intrafallopian transfer if surgery is not feasible or successful.

III. MALE STERILITY

Male fertility requires an intact hypothalamic–pituitary–testicular axis. Germ cells must be present and must be able to undergo maturation from spermatogonia to motile sperm. An intact sperm transport and storage system is required (Fig. 3). Erection and ejaculation must be sufficient to allow the deposition of viable sperm to the vagina and the cervix.

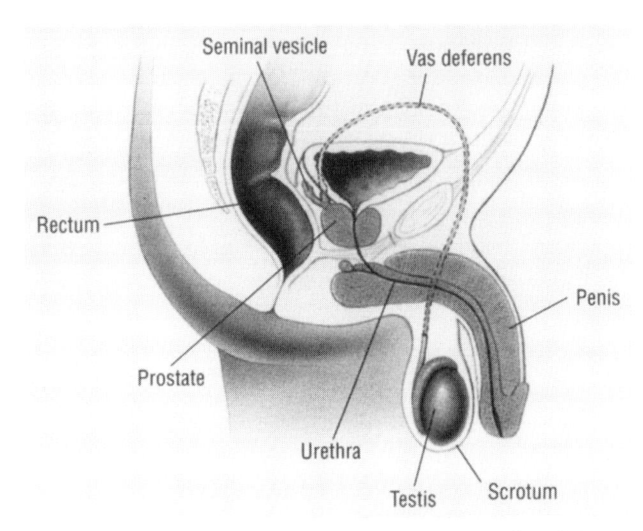

FIGURE 3 The male reproductive tract. Reproduced from The American College of Obstetricians and Gynecologists. Preconceptional Care. (Patient Education Pamphlet No. AB012). Washington, DC, © ACOG, 1995.

Male sterility occurs if these events are disrupted or if inadequate numbers of motile sperm are produced.

A. Testicular Failure/Gonadal Dysgenesis

Spermatogenesis requires the production and recognition of a high local testicular concentration of testosterone and dihydrotestosterone, produced by Leydig cells under the stimulation of pituitary luteinizing hormone. Sertoli cells provide support to spermatogonia and maturing sperm cells. Sperm continue to mature as they pass from the seminiferous tubules to the epididymis and are stored in the vas deferens.

The absence of Sertoli, Leydig, or germ cells results in gonadal failure. Gonadal failure can be congenital or acquired. If testosterone and dihydrotestosterone are not produced by the testes or are not perceived by the target cell receptors, masculinization of the fetus will not occur. Regression of the Müllerian system is dependent on production of Müllerian-inhibiting factor by intact testes. Failure of these embryonic events to occur results in feminization, or male pseudohermaphroditism. Sterility is absolute since functional testicular function cannot be restored. Testicular failure may occur after birth due to chemotherapy, especially with alkylating agents, testicular radiation therapy, or autoimmune disease. Surgical castration also results in absolute sterility.

While sterility is absolute when sperm are lost from iatrogenic causes, future fertility can be preserved if sperm are cryopreserved prior to chemotherapy, radiation therapy, or castration. When feasible, semen should be frozen and stored prior to any treatment that may cause sterility.

B. Maturation Arrest

Causes of maturation arrest are not well understood. At one time, maturation arrest was always associated with absolute sterility. However, successful pregnancies have been established with intracytoplasmic sperm injection (ICSI) into mature oocytes. Round sperm cell secondary spermatocyte injection has been successful in some ICSI cases. Since ICSI is a relatively new procedure, it is not known if individuals born from these procedures will have normal fertility. However, there is evidence that some azospermia gene defects may be passed to offspring by ICSI.

C. Sperm Transport

Obstruction of sperm transport results in sterility. Blockage may be congenital, such as in cases of vasal agenesis, or intended in male sterilization. Although azoospermia is the result, spermatogenesis is expected to be complete. For those who desire fertility, sperm can be surgically aspirated from the epididymis and used for ICSI in most cases. Thus, although these azospermic men used to suffer absolute sterility, the sterility is now considered reversible. Care should be taken for men who have vasal agenesis since a high percentage may be carriers for cystic fibrosis. Genetic counseling and screening should be offered prior to epididymal aspiration for those with congenital absence of the vas deferens.

Voluntary male sterilization is a highly effective form of contraception with a low failure rate and low rate of complications. Vasectomy is effective in over 99% of cases, although unprotected intercourse should be avoided until azoospermia is confirmed by semen analysis. Surgical complications are rarely severe and include infection and hematoma. Once a concern, there is now little evidence that a vasectomy raises the risk of cardiovascular disease or prostate cancer. Less than 5% of men who have a vasectomy request reversal at a later date. Antisperm antibodies may form after vasectomy and may be the cause of continued sterility in some men who fail to conceive after vasectomy reversal. Nevertheless, vasectomy reversal is effective in approximately 60–80% of men, depending on the skill of the surgeon, length of time since vasectomy, type of anastomosis, and fertility of the female partner. Fortunately, sperm recovery for ICSI preserves fertility for those couples unable to conceive following vasectomy reversal.

D. Semen Abnormalities

Occasionally, sperm are present but sterility is caused by low semen concentration, low motility, or severely abnormal morphology. Although norms for semen parameters have been established, sterility may be difficult to determine if sperm are present. Semen parameters vary widely over time, and pregnancy sometimes occurs unexpectedly. In general, semen concentrations above 20 million per milliliter are considered normal, but 10% of those with concentrations below 5 million per milliliter eventually conceive. Pregnancy is unlikely as the semen concentration falls below 1 million per milliliter, although pregnancies sporadically occur. Motility is considered normal when 50% or more of sperm are moving and 25% of sperm movement is rapid. Sperm motility correlates less with fertility than concentration, and sterility is expected only when motility falls to a very low percentage. Morphology, assessed by strict criteria, appears to correlate well with fertilization *in vitro*. Fertility is compromised when more than 95% of sperm are abnormal.

Therapy is now available in most cases when semen parameters are abnormal. Intrauterine insemination is used for subfertility, and IVF with ICSI is used for oligospermia. The rate of birth defects is not greatly increased when pregnancy is achieved by insemination, IVF, or ICSI.

E. Erection and Ejaculatory Sterility

Erection and ejaculatory abnormalities were once considered a cause of sterility. Erection and ejaculation must occur for transmission of sperm to the cervix. Erection can be inhibited by medical illnesses such as diabetes mellitus, medications, and neuropathic and psychogenic causes. Ejaculation can be prevented or abnormal following some reproductive surgical procedures, such as prostate resection, or spinal cord injuries.

Fortunately, most cases of erectile or ejaculatory sterility can now be treated. Erection can be accomplished by penile implants or by injection of prostaglandin E_1. Sperm can be harvested from the bladder for men who have retrograde ejaculation and the sperm can be used for insemination or IVF. If necessary, ICSI can be performed. Men who are unable to spontaneously ejaculate can be stimulated by a rectal probe, in a procedure known as electroejaculation. The harvested sperm can be used for insemination, IVF, or ICSI.

IV. CONCLUSION

Although absolute sterility once was common, advances in the treatment of infertile couples have allowed many previously sterile couples to conceive. With modern therapy, female sterility now is absolute only when the uterus is absent or the endometrium is irreparably damaged. Male sterility is absolute only when no sperm can be found in the testes. *In vitro* fertilization and ICSI have revolutionized infertility therapy. Further advances must be realized to increase the success of these complex procedures.

See Also the Following Articles

Erection; Fallopian Tube; Infertility; Sperm Transport; Turner's Syndrome; Uterine Anomalies

Bibliography

Adashi, E. Y., Rock, J. A., and Rosenwaks, Z. (1995). *Reproductive Endocrinology, Surgery, and Technology*. Lippincott-Raven, Philadelphia.

The American College of Obstetricians and Gynecologists (1995). *Planning for Pregnancy, Birth, and Beyond*, 2nd ed. American College of Obstetricians and Gynecologists, Washington, DC.

ESHRE Capsi Workshop (1996).. Prevalence, diagnosis, treatment and management of infertility. *Hum. Reprod. Excerpts* 4.

Peterson, H. B., Zia, Z., Hughes, J. M., Wilcox, J. S., Tylor, L. R., Trussell, J. and U.S. Collaborative Review of Sterilization Working Group (1996). The risk of pregnancy after tubal sterilization: Findings from the U.S. Collaborative Review of Sterilization. *Am. J. Obstet. Gynecol.* 174, 1161–1170

Speroff, L., Glass, R. H., and Kase, N. G. (1994). *Clinical Gynecologic Endocrinology and Infertility*, 5th ed. Williams & Wilkins, Baltimore.

Steroid Hormones, Overview

Terry R. Brown
Johns Hopkins University

GLOSSARY

cholesterol A primary lipid found in blood and cells that serves as the precursor for biosynthesis of all steroid hormones.

estradiol A member of the C18 family of steroids synthesized and secreted by the ovarian follicular cells and the primary estrogen in the blood of women.

progesterone A member of the C21 family of steroids synthesized and secreted by the ovarian corpus luteum and the primary progestin in the blood of women.

steroid receptor A member of the family of nuclear receptors that binds specifically to a class of steroid hormones and interacts with the cellular genome to regulate gene transcription.

testosterone A member of the C19 family of steroids synthesized and secreted by the testicular Leydig cells and the primary androgen in the blood of men.

The sex steroids function as hormones to control or influence every aspect of reproduction. The steroids are synthesized in the gonads, ovaries, and testes and either promote oogenesis or spermatogenesis locally within the gonads or are secreted into the peripheral circulation where they influence the reproductive tract, sex accessory tissues, the sex phenotype, and the secondary sexual characteristics of males and females. There are three general classes of sex steroids, the androgens, estrogens, and progestins, based on their chemical structures and biological activities. The purpose of this chapter is to review and summarize the historical background and experimental basis for our current understanding of steroid hormones, their chemical structures, classification, biosynthetic pathways, mode of action, and biological effects.

I. HISTORICAL PERSPECTIVE

It has been known from ancient times, dating back as far as the Neolithic age (ca. 7000 BC), that removal of the gonads resulted in the loss of sexual activity and infertility. Accurate accounts of the hormonal effects of castration were available during Aristotle's time (400 BC). In the modern era of scientific discovery, experiments by Berthold in 1849 showed that castration caused the loss of comb size and crowing in roosters, an effect that was reversed by transplantation of the testes. Similarly, Knauer prevented atrophy of the uterus upon removal of the ovaries from guinea pigs by grafting ovarian pieces back to the same animals. In 1850, Leydig described cells containing fatty vacuoles and pigment inclusions occupying the spaces between the seminiferous tubules of the testes. Several investigators subsequently suggested that Leydig cell lipid droplets provided the substrate from which testicular hormones were synthesized and Bouin and Ancel reported that Leydig cells produced the hormonal stimulus for spermatogenesis and maintenance of secondary sexual characteristics. Leydig cell inclusions were revealed to contain cholesterol esters and McGee demonstrated that lipid extracts from bovine testes could stimulate male development when administered to other animals.

Marshall and Jolly showed that estrus could be induced in ovariectomized dogs by injecting extracts of the ovary removed from another dog during estrus. These workers recognized that the ovary produced two different hormones and that the secretion causing estrous was different from that of the corpora lutea. In the 1920s, Allen and Doisy discovered that follicular fluid from the pig ovary caused vaginal cornification in the rat, leading to the isolation and identification of estrone. This was followed by the demonstration by Corner and Allen that extracts from the corpora lutea of pigs would cause progestational proliferation of the rabbit uterus. This observation led to the isolation and identification of progesterone. In 1935, David *et al.* identified testosterone in the lipid extracts from testes and the synthesis of testosterone from cholesterol was soon demonstrated. Further studies to demonstrate that the gonads were the predominant sites for sex steroid biosynthesis and elucidation of the relevant steroid biosynthetic enzyme pathways within the testes and ovaries did not occur, however, until the 1960s. Over the past 30 years, studies of enzyme kinetics and of steroid precursor–product relationships have helped to clarify the enzymatic pathways responsible for conversion of cholesterol to the active steroid hormones. The actions of the various classes of sex steroids are distinguished by their tissue-specific actions mediated via specific intracellular steroid hormone receptors. The genes for the enzymes of sex steroid biosynthesis and steroid hormone receptors have been cloned, leading to a rapid advancement in our understanding of the molecular mechanisms for sex steroid synthesis and function.

II. CHEMICAL STRUCTURES

Steroid hormones are classified on the basis of their chemical structure or on their principal biological functions. In regard to reproduction, we are concerned in this chapter with the sex steroids belonging to one of three major classes: progestins, androgens, and estrogens. Progestins are for the most part represented by compounds belonging to the C21 (pregnane), androgens by the C19 (androstane), and estrogens by the C18 (estrane) series, respectively. Steroids belonging to all three classes are synthesized by the ovaries and testes but the primary secretory end products of the ovaries are progesterone and 17β-estradiol, whereas the testes secrete large amounts of testosterone and small amounts of 17β-estradiol.

The chemical classification system (Fig. 1) relates all steroids to one of several parent compounds, all of which comprise a ring complex made up of three cyclohexane rings (A–C) and a cyclopentane ring (D). To this fully saturated ring complex, referred to as the perhydrocyclopentanophenanthrene nucleus (or more simply, the steroid nucleus), are attached additional components that vary according to steroid class. Cholestane, which comprises the steroid nucleus with methyl groups at the junctions between the A and B rings (at C-10) and the C and D rings (at C-13) and an eight-membered side chain attached at C-17, is the parent compound of cholesterol and other sterols, the biosynthetic precursors of all steroids. Fission of the side chain between C-20 and C-22 leads to the formation of the C21 steroids (pregnane series), the chemical class to which the progestins belong. Further cleavage of the side chain between C-17 and C-20 produces the C19 steroids (androstane series) to which the androgens belong. Finally, removal of the angular methyl group at C-10 leads to formation of C18 steroids (estrane series), the class to which the estrogens belong.

A systematic method of nomenclature is used to identify each steroid according to which of the three series it belongs as well as the nature and location of various modifications and chemical substitutions on the parent structure. Modifications of the parent structure include (i) introduction of double bonds between adjacent carbon atoms, in either the ring structure or the side chain; (ii) hydroxyl (OH) substituents; and (iii) carbonyl groups resulting from oxidation of the hydroxyl substituents. For example, the androgen, testosterone, and the progestin, progesterone, each have double bonds between carbon atoms 4 and 5 in the A ring, whereas the estrogens, estrone and 17β-estradiol, each have an aromatic A ring with three carbon–carbon double bonds. In addition, the estrogenic compounds are distinguished by the presence of $-OH$ groups at C-3 in estrone, at C-3 and C-17 in estradiol, and at C-3, C-

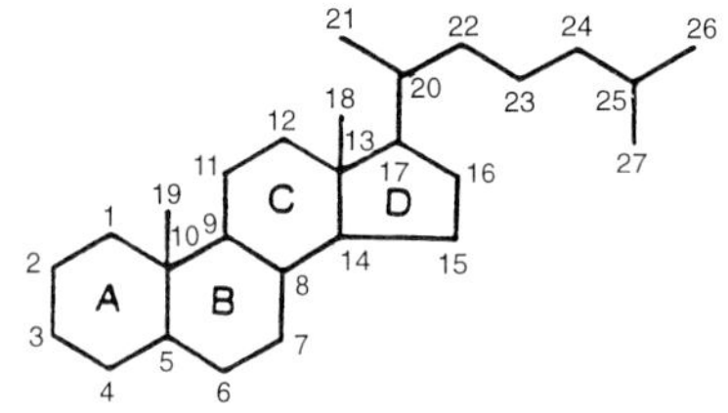

Cholestane
(C-27 "Steroid Nucleus")

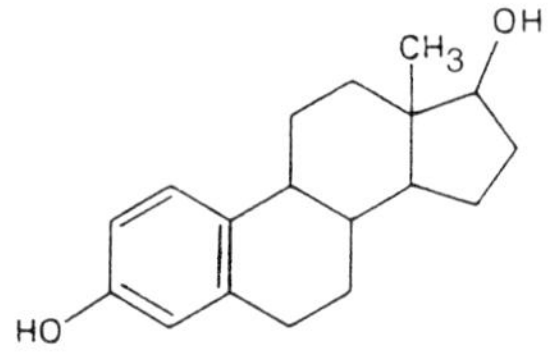

17β-Estradiol
(C-18 Estrogen)

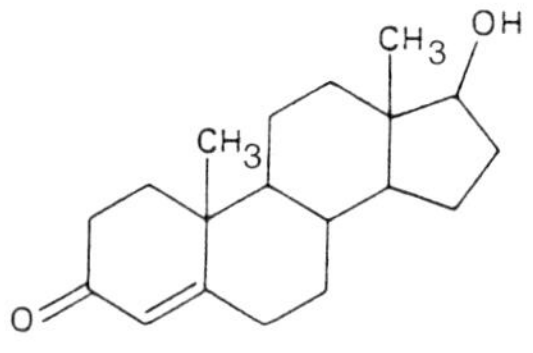

Testosterone
(C-19 Androgen)

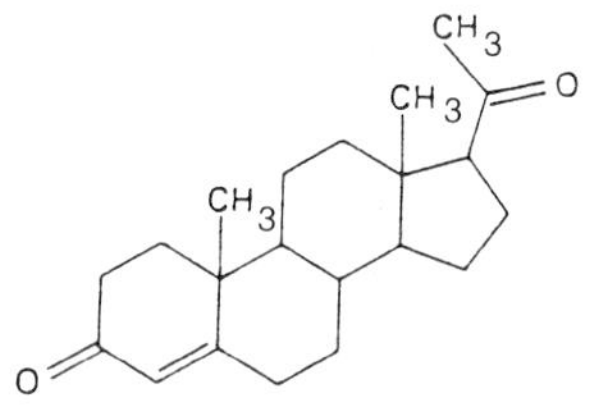

Progesterone
(C-21 Progestin)

FIGURE 1 Chemical structures of cholestane (top) which serves as the 27-carbon atom, four-ring structure that forms the nucleus for all steroids represented by the three classes of sex steroids: 17β-estradiol, the primary C18 estrogen synthesized by the ovarian follicles; testosterone, the primary C19 androgen synthesized by the testis; and progesterone, the primary C21 progestin synthesized by the ovarian corpus luteum.

16, and C-17 in estriol. Testosterone and progesterone each have a carbonyl group at the C-3 position.

Stereoisomers resulting from substituents on the steroid nucleus are distinguished from each other by nomenclature that signifies the side of the plane on the molecule to which the substituent is bound. By convention, substituents on the same side of the plane as the angular methyl group at C-10 are considered as having the β-configuration; those with substituents on the opposite side of the plane have the α-configuration. For many steroids, isomerization at C-3 and C-17 is of particular importance. For example, 17β-estradiol is the most potent naturally occurring estrogen, whereas 17α-estradiol is only weakly estrogenic.

The other position of asymmetry of special interest is that which occurs at C-5 as a result of orientation of the hydrogen atom there, which can occur on either the α or β side of the molecule. This results in *cis–trans* isomerism about the A–B ring junction, the *trans* isomer resulting when the C-5 hydrogen atom is on the α side of the molecule, i.e., *trans* to the orientation of the angular methyl group at C-10, and the *cis* isomer resulting when this hydrogen is in the 5β configuration. The conformation of the steroid nucleus differs markedly, depending on the orientation of the hydrogen at C-5: The A and B rings of 5α-reduced steroids lie in essentially the same plane, whereas those of 5β-reduced steroids are approximately at right angles to each other. For example, the 5α-reduced product of testosterone, 5α-dihydrotestorone, is a potent androgen by virtue of its high binding affinity for the androgen receptor, whereas 5β-dihydrotestosterone is only weakly androgenic.

III. BIOLOGICAL CLASSIFICATION

A. Androgens

Androstenedione, a weak androgen, and testosterone, a potent androgen, are produced by the ovary and the testis. In the testis, the primary end product of steroid biosynthesis and the major secretory product of Leydig cells is testosterone, synthesized from its precursor androstendione. Ovarian follicles utilize androstenedione produced by thecal cells as a precursor for conversion to testosterone by granulosa cells and subsequent aromatization to 17β-estradiol. Alternatively, granulosa cells can aromatize androstenedione to estrone. Small amounts of 17β- estradiol are also synthesized by the testis and peripheral tissues, such as brain, skin, and adipose tissue, and in males and females are able to form estrogens from the precursors, androstenedione or testosterone. 5α-Dihydrotestosterone is another biologically important androgen formed as the 5α-reduced metabolite of testosterone in peripheral tissues, such as the prostate and seminal vesicles.

B. Estrogens

Physiologically, the estrogens, estrone and 17β-estradiol, are the most important of the ovarian follicular steroids. Their trivial names are reflections of their roles in induction of sexual receptivity (estrus) in female mammals, and they play key roles in many other aspects of female reproductive physiology. Estrone was the first steroid to be isolated and identified. 17β-Estradiol is approximately 10 times as potent as estrone in most biological assays and, on a molar basis, is the most active of all steroids produced by the ovary. Several hydroxylated derivatives of these C-18 steroids, all of which possess an aromatic A ring, have been identified in follicular fluids and tissues. These include 2-hydroxyestrone, 2-hydroxyestradiol, and their 2-methylated derivatives, as well as 4-OH, 6-OH, and 16-OH derivatives of estrone and 17β-estradiol.

C. Progestins

Pregnenolone is an important progestin produced by the ovary and testis because of its key position as the precursor of all steroid hormones. However, the most abundant C21 product in the ovary and testis is progesterone, produced as a biosynthetic intermediate by ovarian follicles at all growing stages of development and by testicular Leydig cells and as a secretory end product of the corpus luteum in the peri- and postovulatory periods. Other C21 steroids of ovarian and testicular origin include 17-OH-progesterone, the immediate precursor of androgens which are further aromatized in the ovary, 20α-dihydroprogesterone and its 20β-epimer, and 17α,20α- and 17α,20β-pregnanediols.

IV. BIOSYNTHETIC PATHWAYS

The enzymatic pathways, including their subcellular organization and control mechanisms, leading to the synthesis of steroid hormones are common to each of the steroidogenic tissues, such as the adrenal cortex, placenta, testis, and ovary. This section will provide a summary of the biosynthetic pathways in the testis and ovary, identifying the individual enzymatic steps and processes that are subject to physiological regulation (Fig. 2).

Steroids are produced from cholesterol derived from one of three potential sources: (i) preformed cholesterol taken up from the blood, primarily in the form of circulating lipoproteins; (ii) preformed cholesterol stored within steroidogenic cells as free cholesterol, a constituent of cell membranes, or liberated from cholesterol esters stored within cytoplasmic lipid droplets; and (iii) cholesterol synthesized *de novo* in steroidogenic cells from two-carbon components derived from metabolism of carbohydrate, fat, or protein within the cell. Depending on the vascular supply and the prevailing physiologic conditions, the source of cholesterol may vary be-

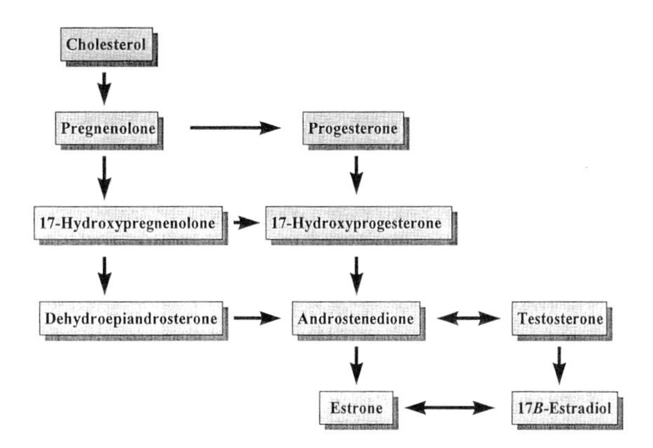

FIGURE 2 Steroidogenic pathways leading to the biosynthesis of testosterone, estrogens (17β-estradiol and estrone), and progesterone via the Δ^5 pathway (left) or the Δ^4 pathway (right).

tween species and between tissues. Cholesterol uptake from blood involves binding of extracellular lipoproteins via their apoprotein component to specific receptors located on the cell membrane, followed by internalization of the lipoprotein–receptor complex and uptake into lysosomes, degradation of lipoproteins by lysosomal esterases, and release of free cholesterol which is available to steroidogenic enzymes. Cholesterol from both low-density lipoproteins and high-density lipoproteins has been implicated, with the former being a more important source in humans. Steroidogenic cells also store substantial amounts of cholesterol in intracellular lipid droplets, primarily as esters of long-chain fatty acids. An equilibrium between the cholesterol fatty acyl esters in these intracellular depots and free cholesterol is maintained by a balance between two enzymes, acyl coenzyme A:cholesterolacyl transferase (commonly called cholesterol ester synthetase) and sterol ester hydrolase (cholesterol esterase), the former favoring storage of excess cholesterol and the latter catalyzing release of stored cholesterol. The activities of both of these enzymes, as well as 3-hydroxy-3-methylglutaryl coenzyme A-reductase, the rate-limiting step in cholesterol biosynthesis, are under hormonal control in addition to being regulated by intracellular levels of cholesterol. The relative contributions of each of these sources of cholesterol to steroidogenesis may differ considerably under different physiological states.

The first step in the steroidogenic enzyme pathway is the cleavage of the C-20–22 bond of cholesterol involving a multienzyme complex to form pregnenolone and the six-carbon fragment, isocaproic aldehyde. The multienzyme complex is composed of three components: cytochrome P450 side chain cleavage ($P450_{scc}$), which is the terminal (electron acceptor) oxygenase; a flavin adenine dinucleotide-containing flavoprotein; and the sulfur-containing heme protein adrenodoxin, which serves to shuttle an electron between the other two components. The reaction utilizes nicotinamide adenine dinucleotide phosphate (NADPH) generated within the mitochondria by oxidation of Krebs cycle intermediates or fatty acids. Three moles each of NADPH and oxygen are utilized per mole of cholesterol undergoing side chain cleavage. The enzyme machinery that catalyzes this reaction is located on the matrix side of the inner mitochondrial membrane which invokes the requirement for cholesterol to be transported to the mitochondrion and across its membrane. Whereas it was originally believed that the $P450_{scc}$ enzyme was the rate-limiting step for steroid biosynthesis, recent studies have suggested that the steroid acute regulatory protein and/or the peripheral benzodiazepine receptor are critical factors in the transport of cholesterol to the inner mitochondrial membrane, and it is their function that is rate limiting and under the control of luteinizing hormone.

Pregnenolone is the key intermediate common to all classes of steroid hormones produced by steroidogenic tissues. It is converted to progesterone by a microsomal enzyme, or enzyme complex, Δ^5-3β-hydroxysteroid dehydrogenase:Δ^{5-4}-isomerase. The placenta and corpus luteum secrete progesterone and this enzyme is responsible for the final steps in that biosynthesis. The enzyme catalyzes two reactions, the dehydrogenation of 3β-equatorial hydroxysteroids and the subsequent isomerization of the Δ^5-3-ketosteroid products, to yield the α,β-unsaturated ketones. Although separation of the dehydrogenase from the isomerase activity has been achieved for a bacterial enzyme system, the two enzyme activities in mammalian steroidogenic tissues have not been separated and appear to function as a single entity. The enzyme utilizes NAD^+ as an electron acceptor, and the reaction is essentially irreversible under physiological conditions. These enzymes also bring about the conversion of 17α-OH-pregnenolone and dehydroepiandrosterone (DHEA) to 17α-OH progesterone and androstenedione, respectively. At least two different isoforms are expressed in human tissues: Type I predominates in the placenta and type II in the adrenal and gonads.

The action of the microsomal cytochrome P450 17α-hydroxylase:C-17,20-lyase enzyme complex is required to convert pregnenolone or progesterone to their respective products, DHEA and androstenedione. These two alternative pathways are referred to as the 5-ene-3β-hydroxy (or Δ^5) pathway and 4-en-3-oxo (or Δ^4) pathway, respectively, although it is uncertain whether the same or separate enzymes are involved. The thecal cells of the tertiary ovarian follicle in the human produce mostly 17β-estradiol prior to ovulation and the thecal cells primarily utilize the Δ^5 pathway to produce androstenedione prior to aromatization, whereas cells of the immediate preovulatory follicle and the corpus luteum that produce larger amounts of progesterone and smaller amounts of 17β-estradiol utilize the Δ^4 pathway. The cytochrome P450 acts as a mixed function oxidase requiring NADPH, oxygen, and cytochrome P450 reductase as an electron donor. The two reactions occur in a concerted fashion in which the 17α-hydroxy intermediate probably remains bound to the enzyme complex with subsequent removal of its side chain by the lyase activity. This enzymatic step is subject to hormonal feedback regulation and is one of the key points at which physiologic control of steroid secretion occurs.

The microsomal enzyme 17β-hydroxysteroid dehydrogenase (17β-HSD) catalyzes the interconversion of the weak androgen, androstendedione, and the potent androgen, testosterone, in the testis. By contrast, 17β-HSD in the placenta and ovary catalyzes the interconversion of the weak estrogen, estrone, and the potent estrogen, 17β-estradiol. Four isozymes of 17β-HSD have been identified in humans: Type I is a soluble form in placenta and ovary that converts estrone to 17β-estradiol using NADPH as a cofactor; type II is a microsomal form in placenta, endometrium, and liver that uses NAD^+ as a cofactor to oxidize testosterone and 17β-estradiol to androstenedione and estrone, respectively; type III is found in testicular microsomes and reduces androstenedione to testosterone using NADPH as a cofactor; and type IV is similar to the type II isoform in activity. Therefore, the type III isoform in testicular Leydig cells is essential for normal testosterone production in the male.

The 4-ene-C19 steroids, androstenedione, testosterone, and 16α-hydroxydehydroepiandrosterone sulfate, are converted to the estrogens, estrone, 17β-estradiol, and estriol, respectively, by an enzyme complex located in the endoplasmic reticulum. The ovary produces primarily 17β-estradiol, the placenta mainly estriol, and adipose tissue mostly estrone. The enzyme complex is referred to as aromatase because of the aromatic structure of the products, and its activity is regulated by follicle-stimulating hormone in ovarian granulosa cells. Aromatase is a cytochrome P450-containing mixed function oxidase that catalyzes a multistep reaction leading to removal of the methyl group at C-10 as formic acid followed by rearrangement of the A ring to the aromatic structure. The reaction requires NADPH as a cofactor for NADPH–cytochrome P450 reductase to transfer reducing equivalents to the enzyme, and 3 moles of oxygen are consumed in the sequence of hydroxylation reactions.

V. MODE OF ACTION

Steroids are lipophilic molecules that enter most cells by diffusion, although in some cases active uptake may be involved (Fig. 3). In target cells that are sensitive to the hormone, the steroid binds to intracellular proteins, termed receptors, that function as transcription factors. The receptors are specific for a given class of steroid and hence cells may contain androgen, estrogen, and/or progesterone receptors. Receptors are synthesized in the cytoplasm and immediately form large macromolecular complexes with other proteins, most notably members of the family of heat shock proteins, that ensure proper protein folding. Most receptors become predominantly localized to the nuclear compartment of the cell, although some receptors may also remain in the cytoplasm. Binding of the steroid to its receptor forms an active complex induced by a conformational change in the receptor structure and dissociation of the heat shock proteins from the macromolecular complex. Steroid receptors are also phosphoproteins and although the overall level of receptor phosphorylation increases following hormone binding, the exact role of this protein modification remains unclear. Within the nuclear compartment, the activated receptor–steroid complexes bind as dimers to DNA regulatory sequences, called steroid response elements, in the 5′ regions of hormonally responsive genes. The receptor functions as a transcription factor through its interactions with other factors bound to the DNA and with various recently discovered cellular coactivator or corepressor proteins that link with, or stabilize, the RNA polymerase II transcriptional initiation complex. Coactivators and corepressors have recently been demonstrated to possess his-

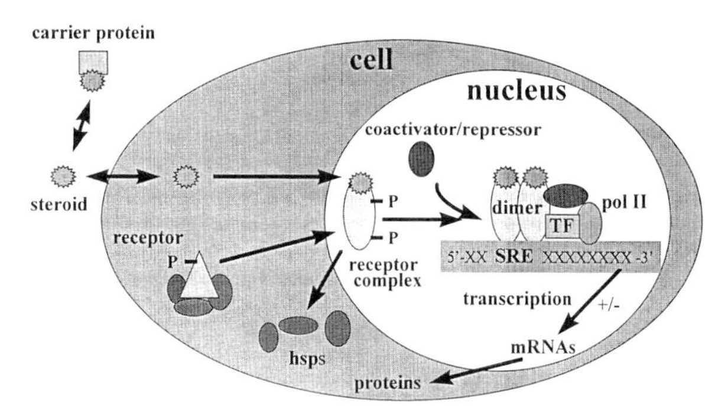

FIGURE 3 Cellular mechanism of steroid hormone action. Steroids circulate in the blood predominantly bound to carrier proteins and the "free" steroid enters cells by passive diffusion. The cytoplasmic steroid receptors are phosphoproteins that form a large macromolecular complex with various proteins, including specific heat shock proteins (hsps). The receptors reside within the nucleus and undergo a conformational alteration upon binding of steroid and the release of hsps. The receptor–steroid complex becomes increasingly phosphorylated and acquires increased avidity for binding to chromosomal DNA. The activated receptor–steroid complexes bind as dimers to specific steroid response elements (SRE) composed of defined nucleotide sequences in regulatory regions of hormonally regulated genes. The chromatin-bound receptor complexes with other nuclear proteins that function as coactivators or corepressors of transcription that may modify the chromatin structure and/or interact with the transcriptional initiation complex composed of various transcription factors (TF) and RNA polymerase II. Transcription of specific genes may be either activated (+) or repressed (−) to regulate the level of mRNAs and their subsequent translation into proteins.

tone acetylase and deacetylase activities, respectively, and thereby function to promote or repress chromatin remodeling and facilitation of transcription. Although receptors are most often regarded as activators of gene transcription that increase the levels of specific mRNAs, receptors may also repress gene transcription. The result of steroid action is the presence or absence of new protein synthesis that alters cell function, growth, or differentiation.

In general, the theory underlying steroid hormone action is supported by the classical concepts and properties of receptors. Hormonally responsive target tissues contain receptors and the biological response within that tissue is correlated with the total number of receptors and their level of occupancy by different concentrations of steroid. Furthermore, the biological response to steroid hormones is a saturable phenomenon that becomes maximal when the finite number of receptors within a cell or tissue are occupied. Only low levels of endogenous hormone are available to cells from the peripheral circulation, and therefore steroids entering target cells are bound with high affinity to receptors. In order to generate biological responses that are specific, different receptors must be available to ensure that the binding of steroid to a receptor is specific for each class of hormone. Just as steroid binding to receptors is specific, the expression of each class of steroid receptor must be specific to a given tissue or cell type. Although cells may obviously express numerous different receptors, the biological response of a given tissue to the general hormonal milieu will be determined by the physiologic levels of available steroid as well as the level and specificity of receptor expression.

The sex steroid receptors for androgens, estrogens, and progesterone are members of the large superfamily of nuclear hormone receptors that include those for glucocorticoids, mineralocorticoids, vitamin A, thyroid hormone, and retinoids. These receptors have been cloned and sequenced and their structural domains have been correlated with their biological functions. The nuclear steroid receptors have a modular structure reflected in their common functional domains for DNA and hormone binding as well as transcriptional activity (Fig. 4).

The amino-terminal transcriptional activation domain (A/B domains) of the nuclear receptors is the most variable region in terms of amino acid sequence and in actual length. The function of the amino-terminal region is the least well-defined but appears to mediate a ligand-independent transactivation function (TAF1) of the receptors when linked to a heterologous DNA-binding domain.

The nuclear hormone receptors have a highly conserved 66- to 68-amino acid DNA-binding domain (C domain) that contains two looped nonhomologous zinc fingers. The zinc atom is coordinated tetrahedrally to cysteine residues. The regions between the two zinc fingers contain amino acids that can discriminate between the DNA sequences of the respective DNA response elements to which these nuclear hormone receptors bind. In fact, studies have demonstrated that a region of 3 amino acids in the

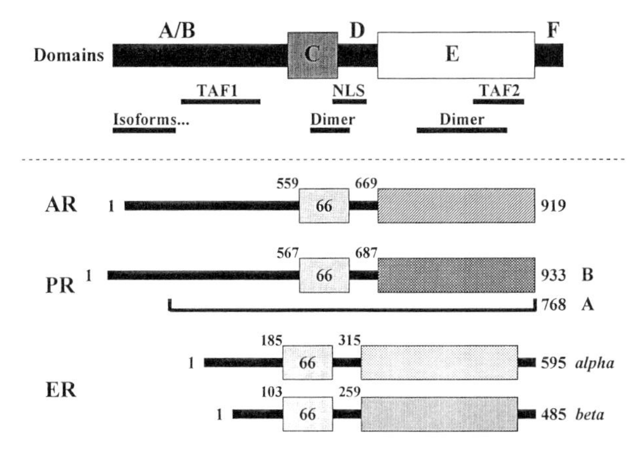

FIGURE 4 The human nuclear steroid receptor family. The receptor amino acid sequence (top) can be divided into functional domains. These domains include the following: (i) The A/B domain encodes a transcriptional activation function (TAF) and varies in length according to specific isoforms; (ii) the C domain encodes the DNA-binding specificity and includes a dimerization function; (iii) the D domain, or hinge region, encodes a nuclear localization signal (NLS) for these nuclear receptors; (iv) the E domain encodes serval functions including specificity of hormone binding, as well as additional hormone-dependent transcriptional activation (TAF2) and dimerization functions; and (v) some of the receptors contain an additional carboxyl-terminal regulatory domain F. The comparative peptide lengths of the various human sex steroid receptors (AR, androgen receptor; PR, progesterone receptor; ER, estrogen receptor) are shown with their homologous functional domains indicated. The human PR occurs in two isoforms: PR-A is a shorter splice variant of the PR-B. The human ER occurs as two isoforms transcribed from two different genes, termed ER-alpha and ER-beta.

so-called P box at the bottom of the first DNA loop is sufficient for this discrimination. Androgen, progesterone, glucocorticoid, and mineralocorticoid receptors form a subfamily of receptors that share amino acid homology in this region and can bind to a common consensus steroid response element in DNA. By contrast, the estrogen receptor contains a different amino acid sequence in the P box and recognizes a separate steroid response element in DNA. The structural interaction of the DNA-binding domain fragment from the glucocorticoid, estrogen, and retinoic acid receptors bound to DNA has been solved by X-ray crystallography and confirms the direct interaction of amino acids with the DNA response element.

The steroid-binding domain (domain E) of the nuclear receptors is approximately 210 amino acids in length at the carboxyl terminus of the receptor and determines the specificity and affinity of agonist and antagonist hormone binding and the dimerization of receptor monomers, and it confers a hormone-dependent transactivation function (TAF2). Steroid binding induces a conformational change which transforms the nuclear receptor from a low-affinity DNA-binding state to one with high affinity, thus enabling transcriptional activation. In fact, the binding of steroid to the receptor can be regarded to relieve repression within the hormone-free receptor which is further demonstrated by the constitutive transcriptional activity of steroid receptors with C-terminal truncations of the steroid-binding domain. The steroid-binding region for each class of receptor is well conserved across species. Recently, the crystal structures of the steroid-binding domains of several members of the nuclear receptor family have been determined.

Between the DNA-binding and steroid-binding domains is a region known as the hinge domain (domain D). This region contains a putative nuclear localization peptide sequence for importation of these relatively large proteins through nuclear pores.

A. Steroid 5α-Reductase and Androgen Receptor

In tissues such as sexual skin, prostate, and seminal vesicles, testosterone serves as a prohormone that is converted to 5α-dihydrotestosterone by the enzyme steroid 5α-reductase. Two genes encode different isoforms of steroid 5α- reductase: The type 1 enzyme is present in skin, whereas the type 2 enzyme is expressed in the prostate. Deficiency of the type 2 isoform leads to abnormal sex differentiation apparent at birth in some subjects with a 46,XY karyotype.

Testosterone and 5α-dihydrotestosterone bind to the same intracellular androgen receptor, although the latter has greater androgenic potency due to its higher binding affinity. The human androgen receptor gene on the X chromosome encodes a protein with a molecular weight of 110 kDa. Local, high concentrations of testosterone are required for normal spermatogenesis within the seminiferous tubules of the testis where the steroid binds to androgen receptors in Sertoli cells and promotes paracrine effects on germ cells. Tissue-specific formation of 5α-dihydrotestosterone and its binding to androgen receptors is required for development of the prostate, seminal vesicles, and external genitalia. Deficiency of androgen action due to mutations in the androgen receptor causes feminization of the external genitalia in 46,XY subjects with androgen insensitivity.

B. Estrogen Receptor

The recent discovery of a second estrogen receptor (ER) gene suggests that estrogenic activity is regulated by the tissue-specific expression of two different ERs, the classical ERα that has been studied for many years and the newer isoform, ERβ. ERα is expressed in uterus, testis, pituitary, ovary, kidney, epididymis, and adrenal, and ERβ is expressed in prostate, ovary, lung, bladder, brain, uterus, and testis. The two ER isoforms are highly homologous, particularly in the DNA and steroid-binding domains, but differ by truncation of the N terminus of the ERβ protein which has a molecular weight of 45 kDa compared to the 66-kDa ERα protein. The binding affinity of 17β-estradiol is four times higher for ERα than for ERβ, and the ligand specificity for a number of other compounds differs between the two isoforms. The presence of two different ERs expands the diversity of potential responses to estrogenic and antiestrogenic compounds and leads to the possibility that ER homodimers and/or heterodimers may bind to estrogen response elements to provoke estrogen-regulated gene expression.

C. Progesterone Receptor

Progesterone receptor (PR) synthesis is induced by estrogens and appears in humans as two molecular forms encoded by a single gene. The 81-kDa PR-A and the 115-kDa PR-B proteins are under the control of distinct promoters, each of which gives rise to a distinct subgroup of PR mRNA species. Although both PR-A and PR-B bind progestins and interact with DNA on progesterone response elements as dimers, there is increasing evidence that they function differently. The two proteins exhibit promoter- and cell-specific differences in their abilities to activate transcription, suggesting that cellular responsiveness to progestins may be modulated by alterations in the ratio of PR-A and PR-B expression and the binding of homodimeric or heterodimeric PRs to DNA. PR expression has been described in tissues known to be progesterone responsive, such as uterus (endometrium and myometrium), ovary (luteinized granulosa cells and corpus luteum), preovulatory granulosa cells, breast, brain (pituitary, ventromedial hypothalamus, and preoptic areas), vascular endothelium, and osteoblasts.

D. Nongenomic Actions

A large body of experimental evidence suggests steroid effects that do not obey the properties of the classical receptor–steroid complex interaction with the cellular genome. These effects are best explained by a model whereby steroid "receptors" on the cell surface generate signals by a nonclassic, nongenomic pathway. These effects have the following characteristics: (i) too rapid (seconds to minutes) to be compatible with changes in mRNA and protein synthesis; (ii) occur in highly specialized cells (e.g., spermatozoa) unable to synthesize mRNAs and proteins or that lack nuclear steroid receptors; (iii) can be stimulated even when steroids are coupled to high-molecular-weight substances that do not cross the cell membrane; (iv) are not blocked by inhibitors of mRNA or protein synthesis; (v) are not blocked by antagonists that bind the classic nuclear receptors; and (vi) are highly specific to a given class of chemically related steroids that exhibit differential potencies at physiologic concentrations. Experimental evidence exists for the nongenomic effects of androgens, estrogens, and progesterone in granulosa cells, endometrial cells, oocytes, spermatozoa, and Sertoli cells. The steroids appear to elicit cellular responses by altering plasma membrane calcium channels, a membrane-associated protein tyrosine kinase, and/or a plasma membrane chloride channel. The apparent localization of these steroid binding sites within the cell membrane in the proximity of sites for peptide hormone binding suggests the possibility for interplay between these hormones in their similar signal transduction pathways.

VI. BIOLOGICAL EFFECTS

A. Androgens

Testosterone is the primary androgen synthesized by Leydig cells and secreted by the testis. Much smaller amounts of other androgens, including androstanediol, androstenedione, dehydroepiandrosterone, and dihydrotestosterone, also leave the testis. The total daily production rate of testosterone by the human testis is 6 or 7 mg. The spermatic vein concentration of testosterone is 40–50 μg/100 ml and is approximately 75 times the peripheral concentration of 600 ng/100 ml. Only 2% or less of the total serum testosterone is not bound to proteins, testosterone–estradiol binding globulin or albumin. Therefore, the serum concentration of free testosterone available to cells is approximately 15 ng/100 ml or <1 n*M*. In the prostate and other sex accessory tissues, testosterone serves as a prohormone that is converted to the more potent androgenic metabolite, dihydrotestosterone. This conversion takes place primarily within certain target tissues, and therefore prostatic tissue concentrations of androgen favor dihydrotestosterone (5 ng/g tissue weight) over testosterone (1 ng/g), whereas testosterone concentrations (600 ng/100 ml) in serum are much greater than those of dihydrotestosterone (50 ng/100 ml). Dihydrotestosterone is 1.5–2.5 times more potent than testosterone in most bioassays (e.g., rat prostate growth) and binds to the androgen receptor with greater affinity. Androgens are responsible for the development and maintenance of the internal and external genitalia, appearance of the secondary sexual characteristics, development of the musculoskeletal system, feedback inhibition of the hypothalamic–pituitary axis, and stimulation of spermatogenesis. In addition, androgens stimulate a variety of other tissues, including the skin, kidneys, and hair growth.

B. Estrogens

The ovary is the major site of synthesis and secretion of estrogen and progesterone and gives rise to cyclical fluctuations in the levels of these hormones in the circulation. Primary follicles play a dual role in secreting both hormones as well as being responsible for the release of the ovum during the normal cycle. Before ovulation, granulosa cells in the follicle biosynthesize and secrete estrogen. Fluid from large follicles in the late follicular phase contains mean estradiol concentrations in the range of 1500–2400 ng/ml, whereas small antral follicles contain <200 ng/ml estradiol. Levels of estradiol and estrone are about 15- and 7-fold higher, respectively, in venous blood from ovaries containing large antral follicles (>8 mm diameter) compared to small antral follicles (<8 mm diameter). The level of estradiol in peripheral blood varies from 6 ng/100 ml during the early follicular phase of the menstrual cycle to 30–60 ng/100 ml in the late follicular phase and 20 ng/100 ml in the midluteal phase. Estradiol binds to testosterone–estradiol binding globulin (TEBG) in serum; the levels of TEBG are higher in women than in men, but estradiol binds less avidly to TEBG than does testosterone. Estrogens are responsible for the appearance of the secondary sexual characteristics, feedback inhibition of the hypothalamic–pituitary axis, stimulation of oogenesis, regulation of the uterine endometrium, and bone and cardiovascular homeostasis.

C. Progesterone

After follicle rupture and release of the ovum, these granulosa cells mature to form the corpus luteum, which is responsible for secretion of progesterone and estrogen in the latter part of the cycle. During the late follicular phase, progesterone concentrations are significantly higher in large antral follicles (1300 ng/ml) than in small ones (250 ng/ml) and the increase in large antral follicles occurs during the transition from midfollicular to late follicular phases. In the human, if fertilization does not occur within 1 or 2 days, the corpus luteum will continue to enlarge for 10–12 days followed by regression of the gland and concomitant cessation of estrogen and progesterone release. If fertilization occurs, the corpus luteum will continue to grow and function for the first 2 or 3 months of pregnancy. After this time it will slowly regress as the placenta assumes the role of hormone biosynthesis and maintenance of pregnancy. Progesterone levels in peripheral blood vary between 50–100 ng/100 ml in the follicular phase and 1000–1500 ng/100 ml in the luteal phase of the menstrual cycle. Progesterone is carried in the blood bound to transcortin (corticosteroid-binding globulin). Progesterone plays a physiological role in the uterus and ovary for release of mature oocytes, for facilitation of implantation and maintenance of pregnancy, and for uterine growth; in the mammary gland for development in preparation for milk secretion; and in the brain for mediation of sexually responsive behavior.

Bibliography

Beato, M., and Sanchez-Pacheo, A. (1996). Interaction of steroid hormone receptors with the transcription initiation complex. *Endocr. Rev.* **17**, 587–609.

Graham, J. D., and Clarke, C. L. (1997). Physiologic action of progesterone in target tissues. *Endocr. Rev.* **18**, 502–519.

Hammond, G. L., and Bocchinfuso, W. P. (1995). Sex hormone-binding globulin/androgen-binding protein: Steroid-binding and dimerization domains. *J. Steroid Biochem. Mol. Biol.* **53**, 1–6.

Horwitz, K. B., Jackson, T. A., Bain, D. L., Richer, J. K., Takimoto, G. S., and Tung, L. (1996). Nuclear receptor coactivators and corepressors. *Mol. Endocrinol.* **10**, 1167–1177.

Katzenellenbogen, B. S. (1996). Estrogen receptors: Bioactivities and interactions with cell signaling pathways. *Biol. Reprod.* **54**, 287–293.

Penning, T. M. (1997). Molecular endocrinology of hydroxysteroid dehydrogenases. *Endocr. Rev.* **18**, 281–305.

Quigley, C. A., DeBellis, A., Marschke, K. B., El-Awady, M. K., Wilson, E. M., and French, F. S. (1995). Androgen receptor defects: Historical, clinical and molecular perspectives. *Endocr. Rev.* **16**, 271–319.

Revelli, A., Massobrio, M., and Tesarik, J. (1998). Nongenomic actions of steroid hormones in reproductive tissues. *Endocr. Rev.* **19**, 3–17.

Simpson, E. R., Mahendroo, M. S., Means, G. D., Kilgore, M. W., Hinshelwood, M. M., Graham-Lorence, S., Amarneh, B., Ito, J., Fisher, C. R., Michael, M. D., Mendelson, C. R., and Bulun, S. E. (1994). Aromatase cytochrome P450, the enzyme responsible for estrogen biosynthesis. *Endocr. Rev.* **15**, 342–355.

Stocco, D. M. (1997). The steroidogenic acute regulatory (StAR) protein two years later. *Endocrine* **6**, 99–109.

Steroidogenesis, Overview

Margaret M. Hinshelwood

University of Texas Southwestern Medical Center

GLOSSARY

adrenal An endocrine gland composed of two parts, the cortex and the medulla. The cortex is derived from embryonic mesoderm and secretes corticosteroid hormones. The medulla has an origin similar to that of the sympathetic nervous system and secretes catecholamine hormones.

androgens Steroid hormones secreted by the testes that control development and maintenance of masculine characteristics. The primary androgen secreted by the testes is testosterone. Testosterone can be converted into other active metabolites, such as 5α-dihydrotestosterone, in the brain, prostate, and other target organs.

corticosteroid Any of a class of steroid hormones secreted by the adrenal cortex. They consist of glucocorticoids, such as cortisol, and mineralocorticoids, such as aldosterone.

estrogens Steroid hormones secreted by the ovaries that control development and maintenance of feminizing characteristics. Estradiol-17β is the primary estrogen secreted by the ovary.

glucocorticoid Any of a class of steroid hormones secreted by the adrenal cortex that affects metabolism of glucose and other organic molecules. Glucocorticoids also have antiinflammatory and immunosuppressive effects. They include cortisol and corticosterone.

gonads A collective term for testes and ovaries.

mineralocorticoid Any of a class of steroid hormones secreted by the adrenal cortex that regulates Na^+ and K^+ balance. Aldosterone is the most potent mineralocorticoid.

ovary Gonad of the female that produces ova and secretes female sex steroids such as estrogen and progesterone.

placenta An organ which serves for the exchange of materials between the fetal and maternal circulations during pregnancy. It develops primarily from embryonic tissue; however, it also has a component derived from the uterus. The fetal portion of the placenta secretes steroids.

steroidogenesis The process by which a lipid, derived from cholesterol, is converted to steroid hormones by such glands as the adrenal, gonads, and placenta.

steroids Compounds with three six-sided carbon rings and one five-sided carbon ring.

testis The gonad of the male that produces male gametes (sperm) and secretes male sex steroids such as testosterone.

Steroid hormones are derived from the precursor cholesterol and regulate numerous developmental and physiological processes in species ranging from plants to primates.

I. INTRODUCTION

In vertebrates, steroid hormones are synthesized in a number of endocrine tissues, most notably the gonads, adrenals, and placenta. In addition to these tissues, there are many extraglandular sites of steroid synthesis where steroids may have very profound effects on local cell function. These extraglandular sites of steroidogenesis include the brain, adipose tissue, skin, bone, and numerous fetal tissues. The type of steroid hormone produced in a tissue is dependent on the steroidogenic enzymes present as well as the precursor molecule(s) available to these enzymes. Understanding the tissue-specific, developmental, and cyclic expression of these various enzymes would allow greater insight into the regulation of steroid hormone synthesis.

Steroid biosynthesis is accomplished by the actions of two major families of enzymes. The first are the hydroxylase enzymes, encoded by genes belonging to the cytochrome P450 superfamily. The P450 enzymes localized in the mitochondria utilize ferredoxin and ferredoxin reductase as electron donors, whereas those localized to the endoplasmic reticulum (microsomal) use the ubiquitous electron donor, the P450 flavoprotein reduced nicotinamide-adenine dinucleotide phosphate (NADPH) reductase. The second family, the steroid dehydrogenase enzymes, belongs to one of two distinct groups, the short-chain alcohol dehydrogenase/reductase family or the aldo-keto reductase superfamily. Both these families of steroid dehydrogenases employ NAD^+/NADH or $NADP^+$/NADPH as cofactors. Together, the P450 and dehydrogenase enzymes are necessary for steroidogenesis in the gonads, placenta, and adrenals (Fig. 1).

II. CHOLESTEROL

Cholesterol, the substrate from which all steroid hormones are derived, comes from two sources in steroidogenic cells: (i) endogenous synthesis from acetyl-CoA and (ii) exogenous sources by receptor-mediated uptake of low-density lipoproteins (LDL) and/or high-density lipoproteins (HDL; major lipoprotein-type species dependent). Intracellular cholesterol levels are tightly regulated and a balance in the concentration of intracellular cholesterol is maintained by two means. First, the 3-hydroxy-3-methylglutaryl coenzyme A (HMG CoA) reductase enzyme is regulated, which is the rate-limiting step in cholesterol biosynthesis. Second, uptake of LDL and HDL is controlled at the level of their receptors. In addition to these mechanisms, any excess cholesterol in the cell is esterfied and stored in lipid droplets, where it functions as a reserve of free cholesterol. Hydrolysis of cholesterol ester to cholesterol is catalyzed by cholesterol esterase, an enzyme that can be activated by luteinizing hormone (LH).

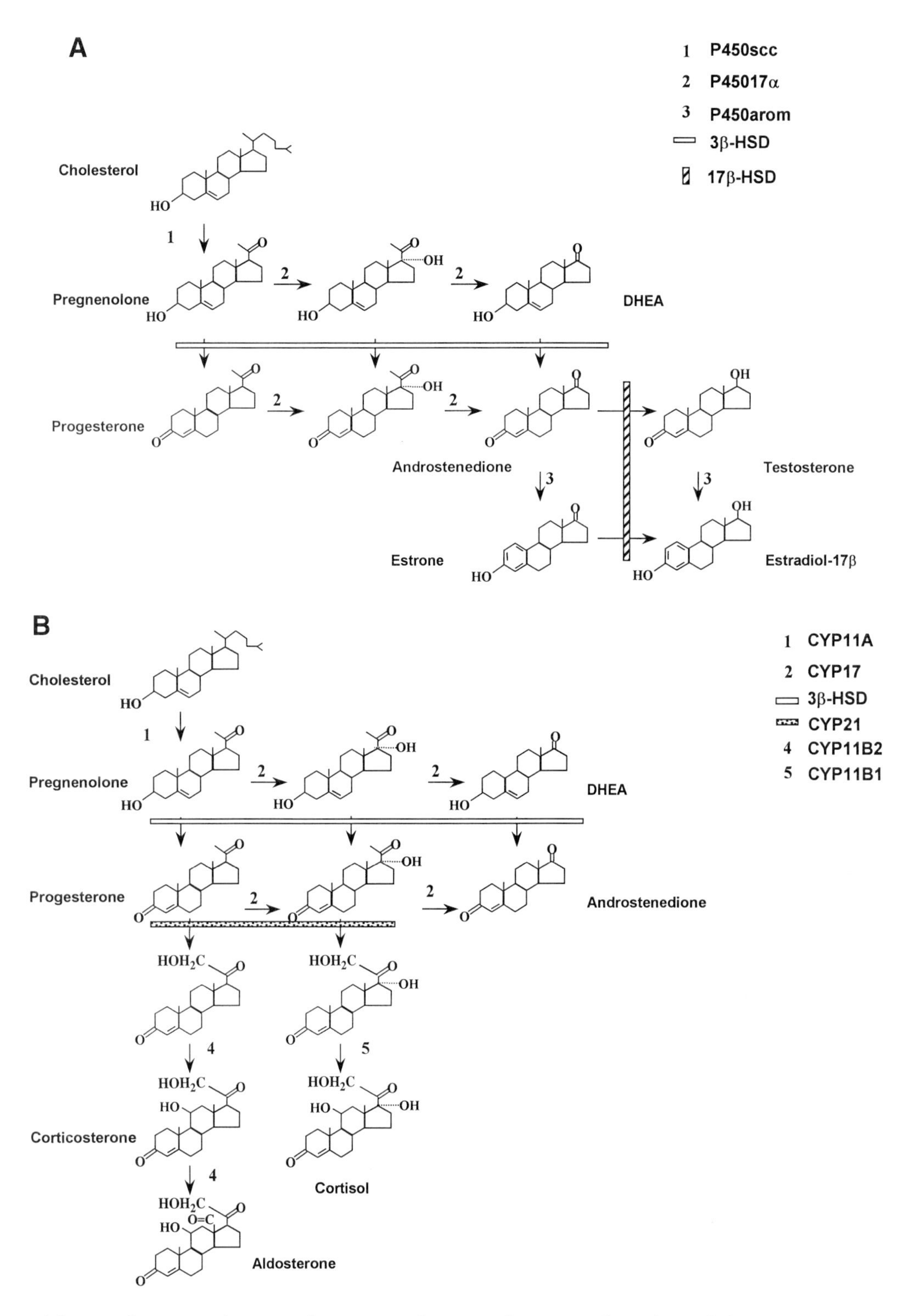

FIGURE 1 (A) Steroidogenic pathway in the ovary and testes. The principal product of the ovary is estradiol-17β and progesterone, whereas the principal product of the testes is testosterone. (B) Steroidogenic pathway in the adrenal gland. The production of steroids is zone specific, such that mineralocorticoids (aldosterone) are produced in the ZG, glucocorticoids are produced in the ZF (cortisol in most species; corticosterone in the rat, no CYP17 present in the adrenal), and in some species C_{19} steroids in the ZR. (C) Steroidogenic pathway in the human placenta. The principal product of the placenta is progesterone. In primates and ungulates, the estrogens, estrone, estradiol-17β, and estriol, are also formed. Substrates for estrogen formation vary with species. In species with $P450_{arom}$ expression in the placenta (primates and ungulates) the sources of C_{19} substrates for aromatization are maternal and fetal (adrenal) for DHEAS and fetal liver for 16α-DHEA(S). In humans and other higher primates, $P450_{17\alpha}$ is not expressed in placenta, whereas in other animals, such as cattle and sheep, placentas expresses P450 $_{17\alpha}$, but only near the end of gestation.

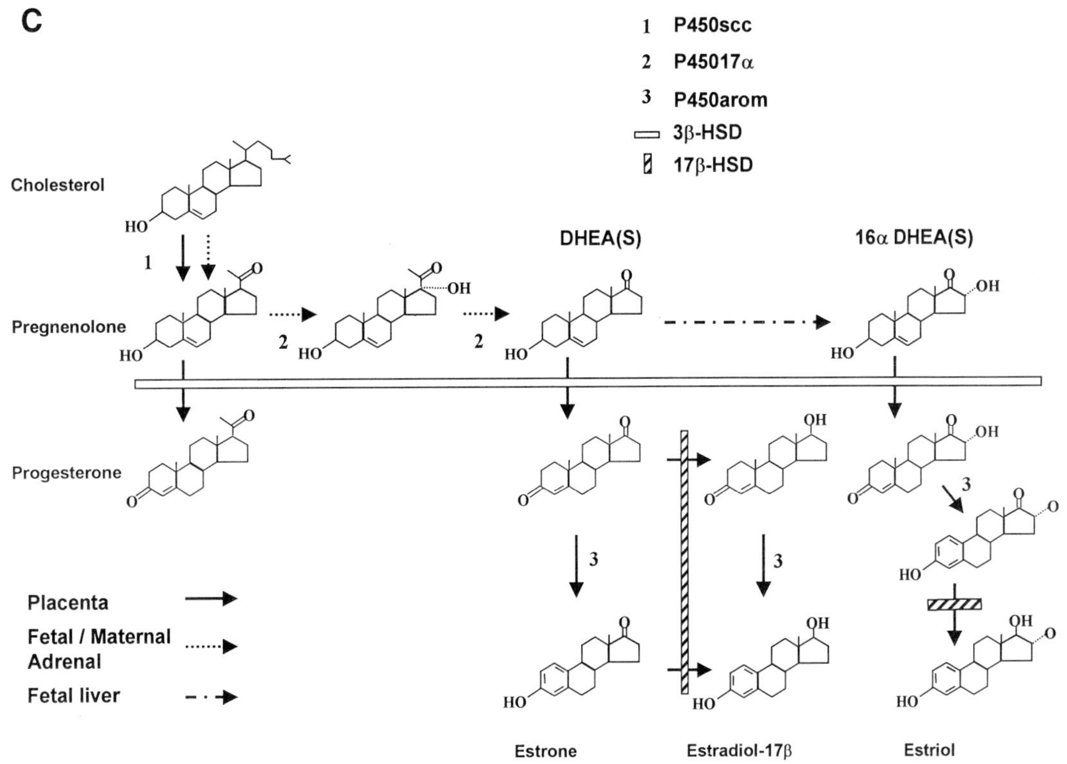

III. P450 FAMILY

A. Cholesterol Side-Chain Cleavage Enzyme (the Product of the CTP11A Gene)

The first step in steroid hormone biosynthesis is the conversion of cholesterol (a C_{27} steroid) to pregnenolone (a C_{21} steroid). The mitochondrial cholesterol side-chain cleavage enzyme ($P450_{scc}$), encoded by the CYP11A gene, catalyzes this rate-limiting reaction. The conversion of cholesterol to steroid hormones is influenced by expression of $P450_{scc}$ and the rate of cholesterol delivery to $P450_{scc}$, an integral protein on the inner mitochondrial membrane. $P450_{scc}$ is localized to the adrenals and gonads, as well as the placenta, brain, fetal gut, skin, and central and peripheral nervous systems. Cholesterol delivery to the inner mitochondrial membrane is thought to be the true rate-limiting step in steroidogenesis. The two proteins implicated in the transport of cholesterol to the inner mitochondrial membrane are steroidogenic acute regulatory protein (StAR) and the mitochondrial peripheral-type benzodiazepine receptor (PBR).

Steroidogenesis may be enhanced in two stages: acute and long term. The acute phase occurs within minutes of tropic hormone stimulation and does not appear to involve gene transcription but rather an enhancement of cholesterol delivery to the inner mitochondrial membrane. This enhanced cholesterol delivery in response to trophic hormones appears to be through translational effects on the StAR protein and posttranslational modifications of PBR to increase binding affinity of cholesterol. In contrast to acute regulation, the long-term (hours or days) regulation of steroidogenesis is mediated by trophic factors, such as LH in the gonads and adrenocorticotropin hormone (ACTH) in the adrenals, through the actions of cAMP on transcription of $P450_{scc}$, ferredoxin, and the cholesterol delivery system. Regulation of this first step in steroidogenesis, however, may be very different in tissues other than adrenals or gonads. For example, $P450_{scc}$ expression does not appear to be regulated via cAMP in all tissues, and in a number of species, StAR is not present in the placenta. Reports of mutations in CYP11A, the gene that encodes $P450_{scc}$, are rare, potentially due to lethality. There were reports of cases of $P450_{scc}$ defects as a cause of congenital lipoid adrenal hyperplasia (lipoid CAH) in humans; however, it was found that the mutations were in the StAR gene. In severe forms, this disease is characterized by impairment of the formation of all adrenocortical and gonadal steroids. These observations suggest a critical role for StAR in the process of steroidogenesis. Genetic males (46,XY) with defects in the StAR protein present with male pseudohermaphroditism. Salt wasting occurs in all cases in both genetic males and females (46,XX).

B. 17α-Hydroxylase/17,20 Lyase (the Product of the CYP17 Gene)

The 17α-hydroxylation of either pregnenolone (or alternatively progesterone) and subsequent cleavage of the C17–20 bond to form dehydroepiandrostenedione (DHEA) or androstenedione (C_{19} steroids) is catalyzed by a single enzyme, 17α-hydroxylase/17,20 lyase ($P450_{17\alpha}$). This enzyme is localized in the endoplasmic reticulum of cells in which it is expressed. $P450_{17\alpha}$ is present in the cortex of the adrenal, most notably the zona fasciculata (ZF) and zona reticularis (ZR), of most species except rats. $P450_{17\alpha}$ is also expressed in the gonads, in Leydig cells of the testes, in theca interna of the ovarian follicle, and, depending on the species, in the corpora lutea (CL). $P450_{17\alpha}$ is also localized in the embryonic central nervous system. Placental expression of $P450_{17\alpha}$ is species specific and stage of development specific. In humans and other higher primates, $P450_{17\alpha}$ is not expressed in placenta. In placentae lacking the ability to synthesize C_{19} steroids, the sources of C_{19} substrates are maternal and fetal (adrenal) for DHEA sulfate (DHEAS) and fetal liver for 16α-DHEA(S). In other animals, such as cattle and sheep, the placenta does express $P450_{17\alpha}$, although not until near the end of gestation. This increased expression of $P450_{17\alpha}$ near the end of gestation results in greater C_{18} steroid production at the expense of progesterone production.

Expression of $P450_{17\alpha}$ is regulated by such trophic factors as ACTH in the adrenals and LH in the gonads via cAMP, whereas activators of the protein kinase C pathway tend to block stimulation by cAMP. In species that express $P450_{17\alpha}$ in their placenta, regulation of expression does not appear to be via cAMP, so it differs greatly from regulation in the adrenal and gonad.

$P450_{17\alpha}$ is one enzyme capable of catalyzing two reactions, namely, 17α-hydroxylation and 17,20-lyase activities, and the ratio of these two activities may vary between tissue types. In the gonads, the 17,20-lyase activity is very high, resulting in an end product of either DHEA or androstenedione. In the adrenal gland, steroid production is limited to the cortical region, and this region may be further subdivided into three zones: the ZG, the ZF, and the ZR. These zones not only differ in histology but also in steroidogenesis; the ZG produces mineralocorticoids, the ZF produces glucocorticoids, and, depending on the species, the ZR synthesizes C_{19} steroids. The 17,20-lyase activity in human adrenals, therefore, is low in the ZF to allow cortisol production, but it is high in the ZR to allow the synthesis of C_{19} steroids. This tissue-specific difference in the 17, 20-lyase activity is not due to multiple species of mRNA for $P450_{17\alpha}$ because there is only one gene encoding one transcript for $P450_{17\alpha}$ in human and rat. This is less clear in other species such as in cattle, in which there might be three CYP17 genes. What may play a much larger role in determining the comparative amount of 17α-hydroxylase activity versus 17,20-lyase activity, however, is the abundance of electron-donating redox partners present in relation to the amount of $P450_{17\alpha}$ protein The ratio of P450 NADPH reductase to $P450_{17\alpha}$ is much greater in the testes than in the adrenal and the ratio of 17α-hydroxylase activity to 17,20-lyase activity can be manipulated *in vitro* to produce the activities seen in either testes or adrenal microsomes by adding varying quantities of purified $P450_{17\alpha}$ and P450 NADPH reductase. Besides P450 NADPH reductase, cytochrome b_5 is also a potential source of electrons for the lyase reaction. Not only may there be a shift in the ratio of 17α-hydroxylase to lyase activities due to the abundance of electron donors present in relation to the amount of $P450_{17\alpha}$ but also specific electron donors may preferentially enhance one reaction versus the other.

Mutations in the CYP17 gene, the gene encoding

$P450_{17\alpha}$, manifest themselves in differing human syndromes, depending on whether 17α-hydroxylase and/or 17,20-lyase activities are affected. In the case of a deficiency in 17α-hydroxylase activity, CAH results, along with hypertension, as a consequence of overproduction of mineralocorticoids. In addition, the impaired production of androgens and estrogens is the cause of primary amenorrhea in genetic females and a lack of pubertal development. In genetic males, male pseudohermaphroditism results. There have been fewer clinical cases of 17,20-lyase deficiency noted, but it appears to be similar to that of 17α-hydroxylase deficiency, only without hypertension.

C. Aromatase (the Product of the CYP19 Gene)

The aromatization of C_{19} steroids to estrogens involves the conversion of the Δ^4-3-one A-ring of the androgens to the corresponding phenolic A-ring characteristic of estrogens (C_{18} steroids). This reaction is catalyzed by aromatase ($P450_{arom}$), an enzyme localized in the endoplasmic reticulum. $P450_{arom}$ is present in the gonads and brain of all vertebrates studied to date. However, in a much smaller group, including humans, nonhuman primates, and ungulates, $P450_{arom}$ expression occurs in the placenta as well. In humans, an even broader tissue distribution of $P450_{arom}$ exists, which includes adipose, bone, skin, and fetal liver. The CYP19 gene, which encodes $P450_{arom}$, differs from many other genes encoding P450 enzymes because there are multiple untranslated first exons that appear in $P450_{arom}$ transcripts in a tissue-specific manner. The presence of these tissue-specific first exons is due to differential splicing, and this differential splicing results from the use of tissue-specific promoters. Transcription in gonads utilizes a promoter just upstream of the start site of translation (promoter II) that is regulated principally by cAMP. In the placenta, transcription is driven by a promoter (promoter I.1) at least 40 kb upstream from the start site of translation and retinoids can regulate transcription. Promoter I.4 is utilized for $P450_{arom}$ expression in the stromal cells of adipose tissue and fetal liver. In adipose, transcription via promoter I.4 can be enhanced by glucocorticoids, class I cytokines including interleukin-6 (IL-6), IL-11, oncostatin-M, and leukemia-inhibitory factor, as well as tumor necrosis factor-α. To a lesser extent, promoters I.3 and II are also employed. In the brain, various promoters may be used, including promoters II, I.4, and a "brain-specific" promoter. The use of different promoters in the brain may be loci specific and may explain why several protein kinase pathways (e.g., PKA, PKC, and PKG) may enhance $P450_{arom}$ expression in differing regions of the brain. These untranslated first exons are all incorporated into a common splice site, just upstream of the start site of translation, so the coding region of the transcripts and consequently the protein are the same, regardless of the tissue site of expression or promoter employed.

The use of these specific promoters explains, in part, the tissue-specific expression of $P450_{arom}$ in normal tissue. Through the use of the reverse-transcription polymerase chain reaction technique, however, examples of changes in the untranslated exons, and hence the promoter utilized, have been demonstrated in cases of human disease. In breast cancer, there is a switch by adipose stromal cells from predominantly promoter I.4 employed for $P450_{arom}$ expression to promoter II. Promoter II was also utilized to drive $P450_{arom}$ transcription in an estrogen-producing hepatocelluar carcinoma. This is in contrast to adult liver, which does not express $P450_{arom}$, and fetal liver, which uses promoter I.4. Normal endometrium does not express $P450_{arom}$, but promoter II drives transcription of $P450_{arom}$ in the case of endometriosis. Thus, local production of estrogen may play a role in the pathology of some disease states.

There have been only a few cases of aromatase deficiency reported. Genetic females present with primary amenorrhea, sexual infantilism, and cystic ovaries. At birth, ambiguous external genitalia are evident. Genetic males display no obvious phenotype until after puberty, when linear growth does not cease. This continued growth is due to failure of epiphyseal fusion and is accompanied by osteopenia and a lack of bone mineralization.

D. 21-Hydroxylase (the Product of the CYP21 Gene)

21-Hydroxylase ($P450_{C21}$) catalyzes the conversion of progesterone and 17-hydroxyprogesterone to 17α,21-dihydroxypregn-4-ene-3,20-dione(11-deoxycortisol) and 21-hydroxypregn-4-ene-3,20-dione(11-deoxycorticosterone) (DOC), respectively. 11-Deoxycortisol is a precursor for aldosterone, and 11-deoxycorticosterone is a precursor for cortisol. $P450_{C21}$ is localized in the endoplasmic reticulum and present almost solely in the ZG and ZF of the adrenal cortex. In humans, $P450_{C21}$ is also present in skin and B lymphocytes in culture, but to a much lesser extent than in adrenals. There are two genes in the CYP21 family: $P450_{C21}$ and its pseudogene $P450_{C21P}$. Transcription of $P450_{C21}$ is enhanced in adrenal cells by ACTH via the PKA pathway.

Deficiencies in $P450_{C21}$ may manifest in several forms, depending on the severity of the mutation. There is a simple virilizing form, without salt wasting or a combination of both virilization and salt wasting. In either case a genetic female presents as a female pseudohermaphrodite. These are both considered classical forms of $P450_{C21}$ deficiency. There is also a nonclassical form, which is an attenuated, late-onset form of CAH with virilization in genetic females. This form is due to an obligate heterozygote of a severe $P450_{C21}$ mutation.

E. CYP11B1 CYP11B2

In some species, there are two closely related isozymes responsible for the synthesis of corticosterone, aldosterone, and cortisol. These two isozymes of 11β-hydroxylase are CYP11B1 (also called 11β-hydroxylase) and CYP11B2 (also called aldosterone synthase). They are encoded by two related (95% identity between human B1 and B2 cDNAs) but distinct genes in the rat, mouse, and human, but it is not apparent if there are separate enzymes for these two activities in cattle and sheep. In those species in which two enzymes have been identified, their localization is mitochondrial and primarily adrenal cortex specific. CYP11B1 is restricted to the ZF, whereas CYP11B2 is restricted to the ZG of the adrenal cortex of mice and rats. In culture, human ZG cells express both CYP11B1 and CYP11B2, but it is not clear if this occurs *in vivo*. In humans, CYP11B1 can 11β-hydroxylate 11-deoxycorticosterone to corticosterone and 11-deoxycortisol to cortisol. It can also 18-hydroxylate 11-deoxycorticosterone to 18-hydroxy-11-deoxycorticosterone. CYP11B2 also has 11β-hydroxylase activity but can 18-hydroxylate and 18-oxidize corticosterone and cortisol to aldosterone and 18-oxo-cortisol, respectively. Control of CYP11B1 expression, and hence cortisol synthesis, is by ACTH, primarily through the PKA pathway. Angiotensin II and potassium levels regulate aldosterone biosynthesis. CYP11B2 expression is enhanced by angiotensin II via actions on phospholipase C.

Deficiencies in 11β-hydroxylase (CYP11B1) result in CAH and hypertension. Genetic females display female pseudohermaphroditism. In the nonclassic form of the deficiency, genetic females are normal at birth but virilize postnatally. Hypertension is not associated with the nonclassical form of the deficiency. For deficiencies in CYP11B2, there is also CAH with salt wasting. Virilization is also common and prominent in females. A third syndrome, called glucocorticoid-suppressible hyperaldosteronism (GSH), is a form of hypertension inherited in an autosomal-dominant manner. Individuals present with high blood pressure but display no impairment of normal growth or sexual development. Aldosterone production can be suppressed by treatment with glucocorticoids and enhanced by treatment with ACTH. All patients with GSH were found to have the same mutation—three CYP11B genes instead of two. The first and third genes were normal for CYP11B2 and CYP11B1, but the middle gene was a chimera with 5′ and 3′ ends identical to CYP11B1 and CYP11B2, respectively. This gene is generated by unequal crossing over, and due to the location of the breakpoint, this chimeric gene is regulated like CYP11B1 (expressed in the ZF and regulated by ACTH) but maintains CYP11B2 activity (aldosterone synthase). This unusually regulated gene is sufficient to cause the disorder of GSH.

IV. STEROID DEHYDROGENASE FAMILIES

A. 3β-Hydroxysteroid Dehydrogenase/$\Delta^{5(R)4}$-Isomerase

The enzyme responsible for the oxidation and isomerization of Δ^5-3β-hydroxy steroid precursors into Δ^4-3-ketosteroids is 3β-hydroxysteroid dehydrogenase (3β-HSD). This reaction is essential for the formation of androgens and estrogens as well as for mineralocorticoid and glucocorticoid biosynthesis. The exact intracellular location of 3β-HSD is a matter of debate. It is, however, membrane bound and associated with both endoplasmic reticulum and mitochondrial fractions. In several species, there are multiple forms of 3β-HSD and these different but closely related proteins are encoded by multiple genes. In the human genome, five 3β-HSD genes have been identified (only two are functional), whereas in the mouse six functional genes for 3β-HSD have been cloned. It is thought that multiple copies of the 3β-HSD genes arose as a result of gene duplication events because both the 3β-HSD genes and pseudogenes are arranged in a cluster and share a high amino acid sequence identity (72–93% for mouse cDNAs). These different forms of 3β-HSD exhibit varying substrate affinities as well as tissue-specific expression. The nomenclature for these different forms of 3β-HSD varies depending on the species since in many cases they were named as they were cloned and not named by tissue location.

In humans, the type I isoform of 3β-HSD is specific for the syncytiotrophoblast, chorion laeve, and villous cytotrophoblasts of the placenta. 3β-HSD I is also in the sebaceous glands in human skin. The type II isoform is present in adrenal cortex, in follicles and CL of the ovary, and in Leydig cells of the testes. Expression of 3β-HSD II is enhanced by activating the PKA pathway (via ACTH in adrenals and LH in gonads) as well as activating the PKC pathway.

In mice, the six isoforms may be divided into two functional groups. The first group operates as dehydrogenase/isomerases (requires NAD^+ as a cofactor) and produces active steroids. This group includes 3β-HSD type I (gonads, adrenal, and neonatal liver), type III (adult liver and kidney), type VI (embryonic cells, uterine tissue during pregnancy, placenta, adult skin, and some in testes), and probably type II (kidney and liver). The second group has 3-ketosteroid reductase activity (requiring NADPH as a cofactor) and includes type IV (kidney and some in testes) and type V (male liver). The type V is capable of converting dihydrotestosterone (DHT) to 5α-androstane-3β,17β-diol, a process that may be very important in inactivating DHT. Type I in the mouse corresponds to type II in the human, and type VI in the mouse is more closely related to type I in the human. In addition to these tissue sites noted previously, there have been several studies noting 3β-HSD activity in the rat brain.

In humans, mutations have been described in 3β-HSD II but not in the 3β-HSD I gene. 3β-HSD deficiency is transmitted as an autosomal recessive disorder and may result in various degrees of salt wasting and CAH. Genetic males exhibit male pseudohermaphroditism, whereas genetic females display either normal sexual development or a mild virilization.

B. 17β-HSD

17β-HSD is responsible for the final step in the production of androgens and estrogens. 17β-HSD catalyzes the interconversion of 17-ketosteroids with their corresponding 17β-hydroxysteroids [androstenedione $\leftrightarrow$ testosterone, or estrone (E_1) $\leftrightarrow$ estradiol-17β (E_2)]. There are a number of 17β-HSD isoforms which exhibit tissue-specific expression and differing substrate specificities. Reactions can be bidirectional if the proper substrate and cofactor are present with the purified enzyme. A specific 17β-HSD isoform in intact cells, however, catalyzes a reaction that is in essence unidirectional [e.g., either reductive ($E_1 \rightarrow E_2$) or oxidative ($E_2 \rightarrow E_1$)]. These isoforms, therefore, are important for either the formation or inactivation of C_{19} and C_{18} steroids.

Five isoforms of 17β-HSD have been characterized thus far in humans. 17β-HSD type 1 is cytosolic and preferentially catalyzes the reduction of E_1 to E_2 in intact cells with NADPH as the cofactor. It is present at low levels in numerous tissues, but it is in greatest abundance in the syncytiotrophoblast of the placenta and granulosa cells of the ovarian follicle. Moreover, it is expressed in some cases of malignancies in epithelial cells of the breast and endometrium. 17βHSD type 2 is microsomal in location and is expressed in a variety of extraglandular tissues. It preferentially catalyzes the oxidization of C_{18}, C_{19}, and C_{21} hydroxysteroids at the C17 position with NAD^+ as the cofactor. This enzyme is principally expressed in the endometrium and to a lesser extent in placenta and liver. 17β-HSD type 3 is also a microsomal enzyme, present primarily in the testes, and it catalyzes the reduction of C_{18} and C_{19} steroids, principally androstenedione to testosterone with NADPH as a cofactor. Type 4 17β-HSD catalyzes the oxidative conversion of E_2 to E_1 in the presence of NAD^+. It is present in the peroxisomes of almost every tissue and might provide one means to inactivate estradiol-17β in peripheral tissues. A 17β-HSD type 5 has been described as a cytosolic enzyme expressed primarily in liver and skeletal muscle. This enzyme catalyzes the C17 reduction of C_{19} and C_{21} steroids in the presence of NADPH.

Mutations have been described in the 17β-HSD type 3 gene which cause male pseudohermaphroditism in genetic males. These males have testes, male internal genitalia (structures derived from the Wolffian duct), no prostate, and female external genitalia.

V. OTHER STEROIDOGENIC ENZYMES

In addition to the previously mentioned enzymes, there are numerous other steroidogenic enzymes that play important roles in various physiological processes through activating or inactivating steroids. 5α-Reductase, which among its other actions converts testosterone to dihydrotestosterone, is important in male reproductive development and function. 5α-Reductase also affects the availability of substrate for aromatization; therefore, it may also play a significant role in female reproduction. The 11β-HSD family of enzymes is important in regulating actions of corticosteroid hormone through the interconversion of cortisol to cortisone. Many of the 3α-HSD enzymes inactivate steroid hormones in the circulation and modulate the occupancy of steroid hormone receptors in cells. Steroid sulfotransferase and sulfatase, which sulfoconjugate or deconjugate steroids, respectively, also play an important role in the availability and clearance of steroids, especially during pregnancy. In addition to the previously mentioned enzymes, there are numerous other enzymes involved in the catabolism of steroids, which help to maintain levels of steroid hormones appropriate for a given physiological function, and that play essential roles in the deactivation and clearance of steroids.

A. Steroidogenic Factor-1

Regulation of steroid hormone biosynthesis occurs at many levels: substrate availability, rate of transcription and translation of steroidogenic enzymes, and the presence or absence of additional steroidogenic enzymes to further activate or metabolize a specific steroid hormone. One factor shown to be extremely important in the transcriptional control of many of the steroidogenic enzymes present in the gonads and adrenals is *steroidogenic factor-1* (SF-1). SF-1 is a member of the orphan nuclear receptor family of transcription factors. SF-1 regulates in a coordinate fashion the expression of steroid hydroxylases as well as 3β-HSD and the StAR protein. Through studies aimed at inactivation of this gene, it was found that SF-1 works at all levels of the reproductive axis, being necessary for the formation of the ventral medial hypothalamus, the adrenal glands, and gonads. Understanding the regulation of SF-1 expression, as well as other genes SF-1 may affect, may provide greater insight into development and control of the endocrine system.

VI. CONCLUSIONS

There are numerous enzymes implicated in the synthesis and metabolism of an active steroid hormone, and there is much to be learned regarding the regulation and function of these enzymes. We know what constitutes an active hormone in many tissues; however, this may not be the same in all tissues. It is clear that some steroids once considered inactive metabolites may have very profound effects in specific tissues; therefore, the isolation and characterization of the enzymes responsible for metabolite formation may provide greater insight into the regulation of these effects. The transcription factor SF-1 is extremely important for the expression of many steroidogenic enzymes as well as for development and function of the gonads and adrenals. Other correlates to SF-1 may exist to regulate both the expression of steroidogenic enzymes and the development and function of other steroidogenic tissues. We need to know how expression of a gene regulated by SF-1 in the gonads and adrenals (i.e., $P450_{scc}$ and $P450_{17\alpha}$

is also regulated in the placenta, which has no SF-1. Answers to these questions regarding regulation of steroidogenic enzymes will greatly advance our understanding of the physiology and pathophysiology of steroid hormone metabolism.

Bibliography

Andersson, S., and Moghrabi, N. (1997). Physiology and molecular genetics of 17β-hydroxysteroid dehydrogenases. *Steroids* **62**, 143–147.

Caron, K. M., Clark, B. J., Ikeda, Y., and Parker, K. L. (1997). Steroidogenic factor 1 acts at all levels of the reproductive axis. *Steroids* **62**, 53–56.

Chung, B.-C., Guo, I.-C., and Chou, S.-J. (1997). Transcriptional regulation of the CYP11A and ferredoxin genes. *Steroids* **62**, 37–47.

Goldstein, J. L., and Brown, M. S. (1984). Progress in understanding the LDL receptor and HMG-CoA reductase, two membrane proteins that regulate plasma cholesterol. *J. Lipid Res.* **25**, 1450–1461.

Mahendroo, M. S., Cala, K. M., Landrum, C. P., and Russell, D. W. (1997). Fetal death in mice lacking 5α-reductase type 1 caused by estrogen excess. *Mol. Endocrinol.* **11**, 917–927.

Mason, J. I., Keeney, D. S., Bird, I. M., Rainey, W. E., Morohashi, K.-I., Leers-Sucheta, S., and Melner, M. H. (1997). The regulation of 3β-hydroxysteroid dehydrogenase expression. *Steroids* **62**, 164–168.

Mellon, S. H., and Zhang (1997). The testes and the adrenal are (transcriptionally) the same. *Steroids* **62**, 46–52.

Miller, W. L., Auchus, R. J., and Geller, D. H. (1997). The regulation of 17,20 lyase activity. *Steroids* **62**, 133–142.

Papadopoulos, V., Amri, H., Boujrad, N., Cascio, C., Culty, M., Garnier, M., Hardwick, M., Li, H., Vidic, B., Brown, A. S., Reversa, J. L., Bernassau, J. M., and Drieu (1997). Peripheral benzodiazepine receptor in cholesterol transport and steroidogenesis. *Steroids* **62**, 21–28.

Penning, T. M., Bennett, M. J., Smith-Hoog, S., Schlegel, B. P., Jez, J. M., and Lewis, M. (1997). Structure and function of 3α-hydroxysteroid dehydrogenase. *Steroids* **62**, 101–111.

Simpson, E. R., Zhao, Y., Agarwal, V. R., Michael, M. D., Bulun, S. E., Hinshelwood, M. M., Graham-Lorence, S., Sun, T., Fisher, C., Qin, K., and Mendelson, C. R. (1997). Aromatase expression in health and disease. *Recent Prog. Horm. Res.* **52**, 185–214.

Stocco, D. M., and Clark, B. J. (1997). The role of the steroidogenic acute regulatory protein in steroidogenesis. *Steroids* **62**, 29–36.

White, P. C., Curnow, K. M., and Pascoe, L. (1994). Disorders of steroid 11β-hydroxylase enzymes. *Endocr. Rev.* **15**, 421–438.

Yamamoto, T., Chapman, B. M., Clemens, J. W., Richards, J. S., and Soares, M. J. (1995). Analysis of cytochrome P450 side-chain cleavage gene promoter activation during trophoblast cell differentiation. *Mol. Cell. Endocrinol.* **113**, 183–194.

Steroid Hormone Receptors

Nancy H. Ing
Texas A&M University

GLOSSARY

aporeceptor A steroid receptor not bound by hormone.

cofactor A molecule that cooperates with others to effect an action.

domain A region of a protein with a defined function.

enhancer Regions of genomic DNA, more distant (up- or downstream) than the "promoter," that upregulate expression of a gene.

gene A contiguous sequence of cellular DNA that encodes an RNA product and the sequences which regulate its synthesis.

gene expression Synthesis of products from a gene, including gene transcription to RNA and mRNA translation to protein.

hormone response element A finite sequence within a gene that confers binding of an activated hormone receptor and subsequently modulates the transcription rate of the gene.

ligand A molecule with a high affinity for a receptor molecule.

messenger RNA A product of transcription from a protein-encoding gene that has been processed to be a template for translation of protein.

promoter Casually defined as the first 100 bases of genetic DNA sequence upstream of the transcription start site; contains orientation-dependent *cis* elements and usually a TATAA box about 20 bases upstream of the start site.

steroid hormone receptor One of the intracellular proteins that binds a specific steroid hormone with high affinity and subsequently carries out the biological effects of that hormone.

transactivation The increase in the rate of gene transcription brought about by the action of a transcription factor.

transcription RNA synthesis by RNA polymerase using genetic DNA as a template

transcription factor A protein that participates in the process of gene transcription.

translation Synthesis of protein from a messenger RNA by ribosomes.

The steroid hormone receptors are a family of intracellular proteins that bind to freely permeable hormones and transduce these signals by altering cellular gene expression to bring about the ultimate actions of hormones on cells. The cloning of sex steroid hormone receptors has led to identification of a large group of related molecules, called the steroid/thyroid hormone receptor superfamily. In addition to playing important roles in normal development and reproductive physiology, these receptors are also of great interest for their role in uncontrolled growth of hormone-responsive tumors. Thus, therapeutic use of the hormones and synthetic antagonists ("antihormones") for the control of steroid hormone receptor functions is a large and growing industry. Current investigations on steroid hormone receptors focus on precise description of the molecules they interact with (cofactors) to activate or inactivate gene transcription and how they interact with other cell signals in normal physiology to explain such phenomena as tissue specificity of steroid effects, delayed responses to steroid hormones, and even hormone-independent activation of hormone receptors.

I. INTRODUCTION

The steroid hormones of mammals direct and regulate many aspects of development and reproduction; for example, maturation of sperm (testosterone), mating behavior (estrogen), pregnancy (progesterone), and parturition (glucocorticoids). These circulating hormones exert their effects by binding specific intracellular proteins, called steroid hormone receptors. Steroid hormone receptors are present in most cells of the body, but their concentrations are much higher in tissues that demonstrate the most pronounced responses to steroid hormones ("target organs"). The hormone-activated receptors alter the pattern of gene expression in the cell, which modifies its complement of proteins and, thereby, its function.

II. THE STEROID HORMONE RECEPTOR SUPERFAMILY

A. Definition

The steroid hormone receptor superfamily is a large group of related proteins. The hallmark of the superfamily is the protein domain which confers the DNA-binding activity, with the distinctive two cysteine-rich zinc fingers. There are other areas of homology as well which dictate other common functions of the receptors: nuclear localization, binding of fat-soluble ligands, dimerization, and activation of transcription. Variations in the domains create many diverse members of the superfamily: true steroid hormone receptors [estrogen receptor (ER), progesterone receptor (PR), androgen receptor (AR), and glucocorticoid receptor (GR)], receptors that bind nonsteroid ligands (retinoids, thyroid hormone, and vitamin D), and related proteins with no identified ligand ("orphan receptors").

B. Subclasses of the Steroid Receptor Superfamily

The true steroid hormone receptors are probably the most evolved of the steroid hormone receptor superfamily. They have the largest and most complex genes and proteins (Fig. 1). It is thought that the

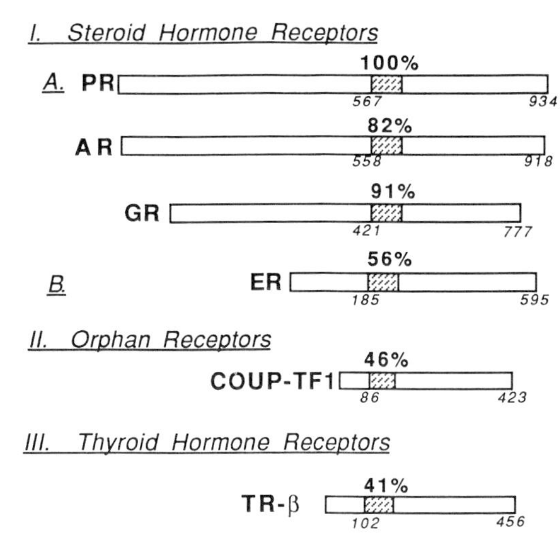

FIGURE 1 The steroid hormone receptor superfamily. The steroid hormone receptor superfamily is a group of intracellular proteins that function as transcription factors. Their hallmark domain is the DNA-binding domain (stippled). Sequence similarities (percentage identical amino acids) to the DNA-binding domain of the progesterone receptor (PR) are indicated above the DNA-binding domains of various receptors. PR is highly related to androgen receptor (AR) and glucocorticoid receptor (GR), whereas estrogen receptor (ER) is a more distant relative. Orphan receptors and thyroid (and retinoid) hormone receptors, typified by chicken ovalbumin upstream promoter–transcription factor-1 (COUP-TF1) and thyroid receptor-β (TR-β), are even more distantly related to PR. Sequence position numbers for the first amino acid in the DNA-binding domain and the last in the entire receptor are indicated below the binding domains.

ancestral steroid hormone receptor was a simple molecule that could help a cell monitor its metabolism by sensing a metabolite and altering gene expression, as many feedback systems do in bacteria. However, with advent of large integrated systems of cells in animals and complex development and reproductive strategies, especially in mammals, it was necessary for subsets of cells to respond to signals given by distant cells. Hormonal systems developed. The relationship between members of the superfamily can be determined by amino acid sequence similarities. Figure 1 shows the percentage of identical amino acids at positions within the DNA-binding domain. Of the steroid hormone receptors, ER diverges from PR, AR, and GR in its amino acid sequence. Orphan receptors and thyroid receptor subclasses of the steroid receptor superfamily are even less similar to the PR, AR, and GR group than they are to each other (not shown). The true steroid hormone receptors will be the focus of this article. Compared to peptide hormones, the signal transduction system of the steroid hormones is extremely simple, revolving around the function of one protein—the steroid hormone receptor.

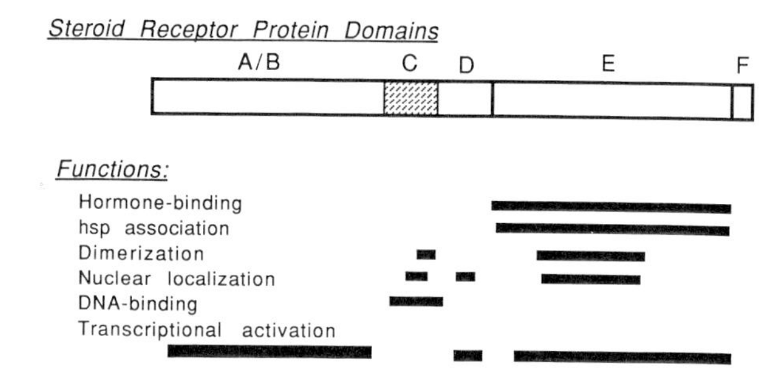

FIGURE 2 Domain structure of steroid hormone receptors. A generic steroid hormone receptor protein is shown, with amino terminus at left and carboxy terminus at right. Positions of domains (A–F) are indicated with their letter codes: A/B is the variable amino terminal, C is the DNA-binding (stippled for reference to Fig. 1), D is the hinge, E is the hormone- (or ligand-) binding, and F is a variable domain of unknown function (Tsai and O'Malley, 1994). The functions of receptor proteins are mapped to the responsible amino acid sequences below. While some functions are limited to one domain (e.g., hormone binding to E), several functions reside in more than one domain (e.g., transcriptional activation in A/B, D, and E).

III. RECEPTOR FUNCTIONS: FROM HORMONE BINDING TO GENE ACTIVATION

A. Subcellular Location and Association of Heat Shock Proteins

Prior to hormone binding, steroid hormone receptors are termed "aporeceptors." GRs exist in the cytoplasm as a large complex with heat shock proteins (hsp90, hsp70, and hsp59). The hsp's are ubiquitous proteins that function as chaperones of protein folding during translation. However, ER is primarily located in the nucleus and has weaker interactions with hsp's, which are also prevalent in the nucleus. The amino acid sequences responsible for interactions with hsp's are located within DNA-binding and hormone-binding domains ("C" and "E" in Fig. 2). Probably, a dynamic equilibrium exists between cytoplasmic and nuclear localization for the aporeceptors that is determined by the cell and the steroid receptor.

B. Hormone Binding

The high-affinity binding of steroid hormones to their receptors is reflected in their nanomolar dissociation constants. This confers sensitivity of responsive tissues to very low levels of circulating hormone (nanomolar to picomolar). The binding specificity of a hormone or a class of hormones to a steroid hormone receptor is also high, resulting in distinct cellular responses to each hormone. The fat-soluble steroids freely permeate the cells to interact at the hormone-binding domain, which includes the largest part of the carboxy-terminal half of the receptor molecules (Fig. 2). The binding of hormone alters the conformation of the receptor protein and dissociates various heat shock proteins from it so that different domains are accessible. Thus, the aporeceptor is transformed to an activated steroid receptor.

C. Phosphorylation

Steroid hormone aporeceptors are phosphoproteins and, upon hormone binding, undergo "hyperphosphorylation" on specific amino acids. Three specific amino acid residues of the PR have been identified as sites phosphorylated in response to progesterone binding. Surprisingly, mutation of individual sites of hyperphosphorylation does not have much effect on the ability of the PR to transactivate genes. Changes in phosphorylation are probably also involved in the conformational change initiated by hormone binding, resulting in receptor activation.

D. Dimerization

The hormone-bound receptors join together to form dimeric molecules. The amino acid sequences responsible for dimerization are located in DNA-binding and hormone-binding domains (Fig. 2).

Within the true steroid hormone receptors, only like molecules dimerize: ER with ER, for example, to form homodimers. In the case of retinoid and thyroid receptors, different partners (heterodimers) are most common. These fully active forms of steroid receptors are found in the nucleus by function of their nuclear localization domains (Fig. 2).

E. DNA Binding

The DNA-binding domains of steroid receptor proteins are centrally located, 60-amino acid sequences (Figs. 1 and 2). Eight conserved cysteine residues form two finger-like structures that each coordinate one zinc ion (Fig. 3). Structural analyses have shown that these fingers orient the receptor dimers so that the intervening amino acid sequences bind DNA. It is intriguing that mutation of only three amino acids at the carboxy terminus of the first zinc finger of ER alters its DNA-binding specificity from the estrogen to the progesterone/glucocorticoid type of hormone response element. Therefore, estrogens bind this mutant receptor, but it transactivates progesterone and glucocorticoid-responsive genes.

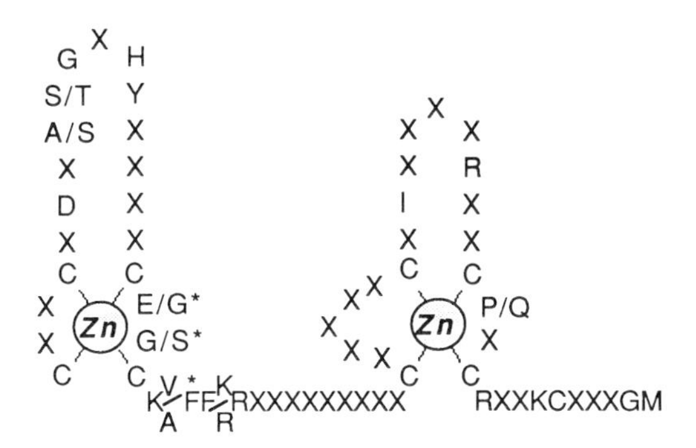

FIGURE 3 Structure of the DNA-binding domain of steroid hormone receptors. The amino acid sequence of the DNA-binding domain, which confers DNA-binding affinity and specificity of the steroid hormone receptors, is shown. The two groups of four conserved cysteines coordinate zinc ions to create the two "zinc fingers." Conserved residues throughout the steroid hormone receptor superfamily are indicated in single-letter amino acid code, whereas less conserved residue positions are indicated by "X." Mutation of only three GR amino acids at the base of the first finger to the residues marked by asterisks alters the receptor's DNA-binding specificity from glucocorticoid response elements to estrogen response elements.

Receptor	Hormone Response Element
PR, GR, AR	TGTACAnnnTGTTCT
ER	AGGTCAnnnTGACCT
COUP-TF1	AGGTCAnAGGTCA
TR, RAR, VDR (N=3, N=4, N=5)	AGGTCA(n)$_N$AGGTCA

FIGURE 4 Steroid hormone response elements. Steroid receptor proteins are listed with their consensus hormone response elements. The latter are short DNA sequences of responsive genes that bind the active receptor dimers. The nucleotide sequences of the upper DNA strand are shown and form a 5′ to 3′ orientation. The two half-sites, each bound by half of the receptor dimer, are indicated by arrowheads. The arrowheads indicate the orientation of the (T/A)G(T/A)ACA half-site of the PR, GR, and AR subfamily or the AGGTCA half-site of the others. When the arrow is pointing right to left, for example, the half-site sequence is on the bottom strand (not shown). Positions indicated by "*n*" (any nucleotide) are primarily important for spacing of the half-sites. Spacing and orientation of half-sites is critical to determining which hormone receptors bind: thyroid (TR), retinoid (RAR), and vitamin D receptor (VDR) response elements differ in spacing ($N = 3$, 4, or 5, respectively), whereas estrogen response elements and thyroid response elements differ in the orientation of the second half-site. G residues that contact receptor proteins directly are indicated by asterisks.

The DNA binding is sequence specific for "hormone response elements" (Fig. 4). These are composed of two "half-sites," each bound by one receptor molecule of the dimer. The half-sites have nucleotide sequences of (5′)(T/A)G(T/A)ACA for PR, GR, and AR or (5′)AGGTCA for ER and all other receptors. The orientation and spacing of the half-sites is critical to the response element. Steroid hormone response elements have antiparallel half-sites with three intervening nucleotides. The other members of the superfamily bind a variety of half-site orientations and spacings, demonstrating much more flexibility of the dimeric protein.

Although the "consensus" hormone response elements (Fig. 4) demonstrate very strong DNA binding and gene activation *in vitro*, they are rarely found in natural genes. Hormone response elements found in natural genes usually consist of one consensus half-site associated with a divergent one. Flanking sequences and divergent nucleotides within half-sites modulate the binding of steroid receptors. Hormone response elements in responsive genes are usually found as multiple arrays, across which receptor dimers bind synergistically. Because of these complexities, hormone response elements must be identified by functional assays for DNA binding and gene activation. Many now believe that the affinity and specificity of steroid receptor binding to genetic elements is as crucial to steroid hormone action as are those properties regarding hormone binding to receptor.

F. Activation of Gene Transcription

Positioned upon a responsive gene, the receptor dimer associates transcription factors with its gene activation domains. The amino acid sequences involved in transactivation are dispersed throughout the entire length of the receptor protein (Fig. 2). The carboxy-terminal portion is hormone dependent, whereas the amino-terminal transactivation domain is constitutive. These interact with general transcription factors TFIIB and TFIID, which are cofactors of RNA polymerase II, to signal assembly of other general transcription factors and RNA polymerase II on the start site of transcription (Fig. 5). However, ste-

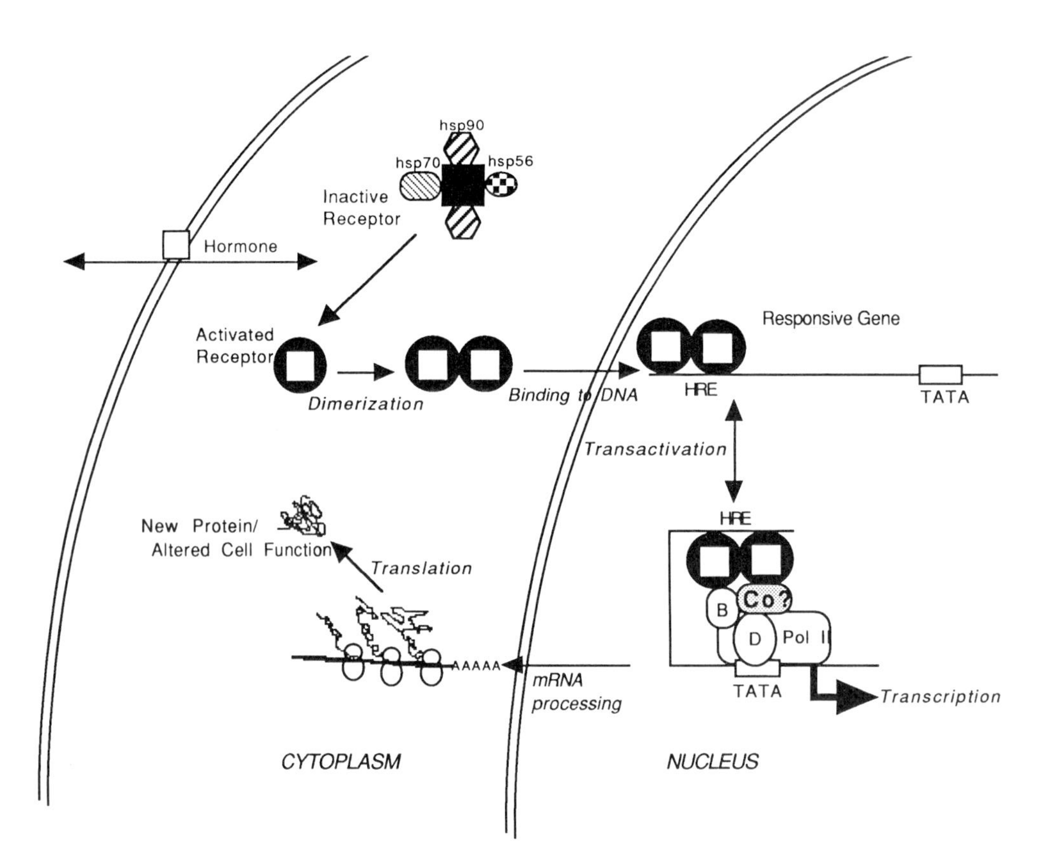

FIGURE 5 A model for how steroid hormone receptors activate transcription. Steps in gene activation by steroid hormones are shown in cell cytoplasm and nucleus. Steroid hormones (□) freely permeate cell and nuclear membranes. They bind aporeceptors (■) that are associated with heat shock proteins 56, 70, and 90. The aporeceptors and dimerization steps may occur in cytoplasm, as for GR (shown), or in the nucleus, as for ER (not shown). Hormone binding induces conformational changes in the receptors (●). These active receptors dimerize and translocate to the nucleus if they are cytoplasmic. They bind hormone response elements (HRE) of responsive genes. The intervening DNA between the HRE and TATA box at the transcription start site loops out to bring the receptors close to the latter. There, the receptors interact with general transcription factors TFIIB (B) and TFIID (D) and cofactors (Co?). This enhances assembly of functional RNA polymerase II (Pol II) complexes on the gene and activates transcription (large arrow). The resultant RNA is processed, exported to the cytoplasm, and translated to proteins that give the cell new functions. Thus, the steroid hormone signal is transduced by steroid hormone receptors.

roid hormone response elements are generally far removed from the start site of transcription. Therefore, the intervening genomic DNA is believed to bend out in a loop to bring the DNA-bound steroid receptors into close association with general transcription factors at the start site. In addition, other cofactors interact with steroid hormone receptors and RNA polymerase II to further modulate the rate of transcription initiation on responsive genes.

IV. GENE REPRESSION BY STEROID HORMONES

Although many steroid hormones activate transcription of specific genes, transcriptional repression is also common. Negative regulatory regions of steroid receptors have been mapped to amino acid sequences within the hormone-binding domain. Glucocorticoids are best known for transcriptional repression, and, via the actions of GR, are antiinflammatory agents. The repressive actions of GR usually occur on composite hormone response elements, which contain one GR half-site adjacent to an enhancer element such as AP-1. The family of transcription factors that bind AP-1 elements include c-*fos* and c-*jun*, which are strong activators of transcription. They interact with hormone-activated GR both on and off the DNA so that the basal, AP-1-driven activity of the gene carrying the composite element is repressed by glucocorticoids. In another example of repression of specific genes, the PR-A represses the activation of some progesterone-responsive genes by PR-B. These repressive effects could result from a shift in the equilibrium between receptor interactions with coactivators toward those with corepressors.

V. STEROID RECEPTOR ACTIVATION IN THE ABSENCE OF HORMONE

Steroid hormone aporeceptors can be activated in the absence of hormone by agents that stimulate intracellular phosphorylation pathways. Some examples are dopamine activation of PR in rat brain to stimulate sexual behavior and epidermal growth factor activation of ER to enhance uterine growth. Specific inhibitors demonstrated that both steroid hormone receptors and plasma membrane receptors (for dopamine and epidermal growth factor, respectively) were required for the effects. Such interactions between cell signaling pathways ("cross-talk") determine the ultimate response of a cell to a natural, complex array of stimuli.

VI. SPECIFICITY OF STEROID HORMONE RESPONSE

A critical question in endocrinology is the following: How are hormonal responses limited to a specific subset of cells and, within them, to a limited number of responsive genes? Also, since all cells of an organism have an identical set of genes, why don't all cells respond alike?

To a large extent, the set of cells that responds to a steroid is specified by expression of steroid hormone receptor genes. The biology of the steroid hormone receptor superfamily is complex, with examples of multiple genes for a receptor (designated α, β, and γ), multiple messenger RNAs from a gene (generated by differential promoter usage and alternative splicing), and multiple protein forms (designated A, B, and C). The genes encoding the true steroid receptor proteins are large, spanning 100–150 kilobases of genomic DNA. Unlike the thyroid receptor subfamily, which has multiple genes for most family members, only one multiple gene has been identified for a true steroid receptor—ER. Both ER-α and ER-β receptors (65 kDa) are high-affinity receptors for estrogens, but their expression is independently regulated. Messenger RNAs of sex steroid receptors are also large, ranging from 6.5 (ER-α and -β) to 11 (AR) kilobases. The PR gene has two promoters that, with other posttranscriptional modifications, produce multiple messenger RNAs, ranging from 2.5 to 11 kilobases in length. These are translated into PR-A, PR-B, and PR-C protein forms (94, 116, and 60 kDa, respectively), which share some but not all amino acids sequences. Most important, the different forms of PR evoke gene- and cell-specific responses to progesterone stimulation.

Most cells express steroid receptor proteins, but those of extremely responsive "target" organs (such as breast and uterus) express very high levels in

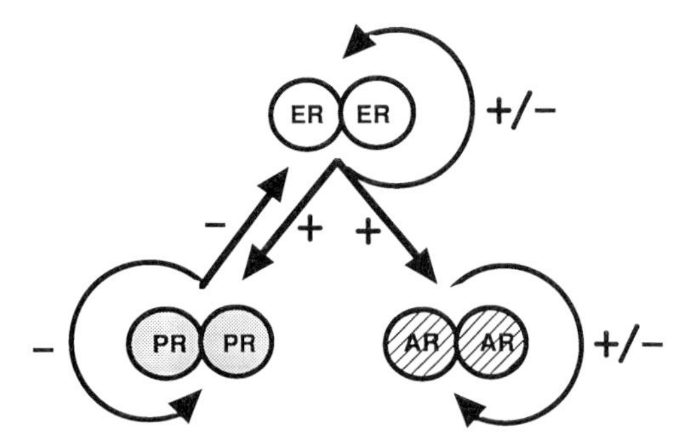

FIGURE 6 Regulation of steroid hormone receptor genes by steroid hormones. The major determinant of how steroid hormones affect a cell is the concentration of steroid hormone receptors. Steroid hormones themselves regulate the levels of steroid hormone receptors. In female animals (left), estrogens upregulate (+) ER and PR concentrations, whereas progesterone downregulates (−) them. In some cases, estrogens may downregulate ER. In males (right), estrogen upregulates AR concentrations. The effect of androgens on AR gene expression depends on the cell type and its state of growth.

comparison to other tissues. Although it is not known how many steroid hormone receptors are required for a response, it is known that cells with more receptors respond more strongly to hormonal challenges. Interestingly, steroid hormone receptor levels are regulated in target cells by several factors, including steroid hormones themselves (Fig. 6). In the female, the interplay of estrogen and progesterone concentrations during the estrous/menstrual cycle alters ER and PR concentrations in target organs. In general, ER and PR concentrations are upregulated by estrogens and downregulated by progesterone, so their levels vary accordingly during the reproductive phases of the female. In the male, AR is also upregulated by local production of estrogen, from aromatization of testosterone, acting on ER in target organs. The positive or negative nature of autologous regulation of ER and AR depends on the cell and its growth state, with autologous upregulation of ER being predominant in target tissues during reproductive phases. Also, while ER concentrations are high in prepubertal animals, disease and aging decrease steroid receptor gene expression.

However, even in cells expressing steroid hormone receptor genes, not all genes that contain hormone responsive elements are activated. Most genes are silenced by their packaging in chromatin structure. Thus, the physiology of the cell regulates chromatin structure, which changes during development and growth to vary accessibility of responsive genes to steroid hormone receptors. The ability of steroid receptors to alter the chromatin structure of responsive genes is a critical area of research. The actions of other transcription factors and cofactors acting on the gene and/or interacting with the steroid receptors are also critical to the response.

VII. UNDER- AND OVEREXPRESSION OF STEROID HORMONE RECEPTOR GENES

Many important lessons can be learned from humans and animals that have genetic mutations of steroid receptor genes that either greatly lower or raise the concentrations of functional hormone receptors. Many natural AR mutants have been identified in man because they cause the androgen insensitivity syndrome, typified by feminization of a genetic male in the presence of normal or elevated testosterone levels. Of other steroid hormone receptors, only one has been identified medically: An ER mutation in a man caused bone effects. The man was abnormally tall (because estrogens are involved in closure of the growth plates of long bones) and had poorly mineralized bones. These natural mutations have helped identify critical physiological functions and domains of the steroid receptors (Fig. 2).

Modern technology has allowed radical manipulations of steroid hormone receptor gene expression in mice. Before gene "knockouts" created mice that lack ER or PR genes, it was debated whether the receptors were critical to life or normal development. Surprisingly, the ER and PR knockout mice live and grow relatively normally. However, their reproductive functions are compromised. Both male and female ER knockout mice are infertile, with poorly developed testes and hemorrhagic, cystic ovaries, respectively. The PR knockout females are also infertile, with oocyte development blocked at the secondary follicle stage. Mating behavior of the knockout mice is also inhibited. Overexpression of ER in transgenic mice also had reproductive effects: The females demonstrated reduced fertility, delayed parturition, and prolonged labor.

VIII. DELAYED EFFECTS OF STEROID HORMONES

Many effects of steroid hormones in vivo require days or weeks—much longer than the 30-min to 2-hr time periods used in assays *in vitro* and expected from molecular events outlined in Fig. 6. Some natural genes require repeated doses of steroid treatment over days to weeks to induce marker genes. In addition, some single-dose effects (such as estrogen priming) appear to influence subsequent steroid responses over the following weeks. This is best explained by a cascade of steroid hormone effects, with gene products directly induced by steroid having effects on expression of a second set of genes, and so on. Upregulation of primary response genes is distinguished from later effects of steroids by its rapid time scale and independence from synthesis of new protein. Note that mitogenic and tissue-remodeling effects of estrogens may cause alterations in cell populations over longer time courses. The ultimate response to a steroid hormone stimulus is, therefore, a combination of both acute and delayed responses from prior steroid exposure of the tissue.

See Also the Following Articles

Androgens; Estrogen Action, Behavior; Testosterone Biosynthesis and Actions

Bibliography

Beato, M., Candau, R., Chavez, S., Mows, C., and Truss, M. (1996). Interaction of steroid hormone receptors with transcription factors involves chromatin remodeling. *J. Steroid Biochem. Mol. Biol.* **56**, 47–59.

Dean, D. M., and Sanders, M. M. (1996). Ten years after: Reclassification of steroid-responsive genes. *Mol. Endocrinol.* **10**, 1489–1495.

Enmark, E., and Gustafsson, J. A. (1996). Orphan nuclear receptors—The first eight years. *Mol. Endocrinol.* **10**, 1293–1307.

Horwitz, K. B., Jackson, T. A., Bain, D. L., Richer, J. K., Takimoto, G. S., and Tung, L. (1996). Nuclear receptor coactivators and corepressors. *Mol. Endocrinol.* **10**, 1167–1177.

Katzenellenbogen, J. A., O'Malley, B. W., and Katzenellenbogen, B. S. (1996). Tripartite steroid hormone receptor pharmacology: Interaction with multiple effector sites as a basis for the cell- and promoter-specific action of these hormones. *Mol. Endocrinol.* **10**, 119–131.

Korach, K. S., Couse, J. F., Curtis, S. W., Washburn, T. F., Lindzey, J., Kimbro, K. S., Eddy, E. M., Migliaccio, S., Snedeker, S. M., Lubahn, D. B., Schomberg, D. W., and Smith, E. P. (1996). Estrogen receptor gene disruption: Molecular characterization and experimental and clinical phenotypes. *Recent Prog. Horm. Res.* **51**, 159–186.

Lydon, J. P., DeMayo, F. J., Conneely, O. M., and O'Malley, B. W. (1996). Reproductive phenotypes of the progesterone receptor null mutant mouse. *J. Steroid Biochem. Mol. Biol.* **56**, 67–77.

O'Malley, B. W., Schrader, W. T., Mani, S., Smith, C., Weigel, N. L., Conneely, O. M., and Clark, J. H. (1995). An alternative ligand-independent pathway for activation of steroid receptors. *Recent Prog. Horm. Res.* **50**, 333–347.

Stress and Reproduction

Thomas H. Welsh, Jr., C. Nann Kemper-Green, and Kimberly N. Livingston
Texas A&M University

GLOSSARY

homeostasis Relatively stable condition of an animal's internal environment that occurs as a result of compensatory physiologic reactions.

hypothalamic–pituitary–adrenal axis The neuroendocrine linkage of the hypothalamus, the anterior pituitary gland, and the adrenal glands.

hypothalamic–pituitary–gonadal axis The neuroendocrine linkage of the hypothalamus, the anterior pituitary gland, and the gonads (ovaries and/or testes).

stress Circumstance in which an individual animal or person is required to make functional, structural, or behavioral adjustments or responses as a coping reaction.

stressful Describing an environment that places unusual demands on an individual.

stressor An environmental factor that contributes to a stressful circumstance or elicits stress responses.

stress physiology The study of physiological, biochemical, and behavioral responses to factors that comprise an animal's physical, chemical, and biological environment.

Survival of an individual animal depends on maintaining homeostasis, and survival of a species depends on reproductive success. Reproductive failure may result from exposure to environmental, physical, or psychological stressors.

I. INTRODUCTION

Stress was originally described as the sum of the biological reactions to physical, mental, or emotional stimuli that disturb an organism's homeostasis and change the balance of hormones in the body. Stressors are factors that threaten the health of the body or adversely affect physiological functions such as reproduction. The well-documented adverse effects of environmental stressors, such as overcrowding, nutritional deprivation, or thermal stress on reproductive functions (e.g., ovarian cyclicity, spermatogenesis, or embryo development), are amply reviewed in many textbooks. This article emphasizes the need for integrative studies that will more fully define the influence of the immune and adrenal systems on reproductive processes. To prevent undesirable influences of stressors on reproductive function or even to prove the existence of a stress and reproduction linkage, one must initially understand the normal regulatory mechanisms that control the adrenal and gonadal axes. The following sections outline the hormonal control of the adrenal and reproductive axes and highlight significant findings relative to intercommunication of these systems while pointing out deficiencies or conflicts in the current knowledge base.

II. REGULATION OF HYPOTHALAMIC–PITUITARY–ADRENAL AXIS

Extrinsic stimuli that affect homeostasis may be viewed as stressors, and the changes in biological function that occur as a result can be considered an individual's response to a stressor or stressful event. Environmental, physiological, or psychological stressors can induce the "fight or flight" and "general adaptation" syndromes. Physiological changes that prepare an animal to respond to a stressor include increases in heart rate, respiration, and blood pressure and repartitioning of oxygen and nutrients to organs that require additional energy to function during a stressful event. The physiological response to stress begins with the higher centers of the brain.

Neurotransmitters such as cholecystokinin, serotonin, and norepinephrine stimulate the synthesis and release of the protein neurohormones corticotropin-releasing hormone (CRH) and vasopressin (VP) from the hypothalamus into the hypothalamic–hypophyseal portal blood system. Both CRH and VP stimulate the release of adrenocorticotropin (ACTH) from the anterior pituitary gland's corticotroph cells. ACTH in turn stimulates the synthesis and release of glucocorticoids (cortisol and corticosterone) from the cortex of the adrenal gland. Elevated plasma concentrations of stress-related hormones (e.g., epinephrine, cortisol, and ACTH) have been classically used as indicators of stress (i.e., increased levels are associated with stress response). An individual animal's perception of a stressful event and the subsequent response can be influenced by many variables, including previous experience, genetic makeup, age, gender, and the animal's physiological state at the time it is exposed to a stressor. Much work has been done to characterize the physiological changes involved in the stress response. In recent years, the impact of these changes on reproductive function has been studied.

A. Hypothalamic Level

A major advance in stress physiology research was the characterization and sequencing of ovine CRH. This peptide has been identified in various regions of the brain and hypothalamus, including the amygdala, stria terminalis, septal nucleus, and preoptic area; however, the primary source has been localized in the parvocellular neurons of the paraventricular nucleus (PVN) of the hypothalamus. The CRH neurons can be divided into two types: the VP-containing neurons and the VP-deficient neurons. The axons that transport CRH from these neurosecretory cells originate in the medial parvocellular division of the PVN and project to the external zone of the infundibulum [median eminence (ME)] in the tuberohypophyseal tract where CRH is stored and released into the extracellular space surrounding the fenestrated capillaries of the hypothalamo–hypophyseal portal system. Vasopressin has been localized mainly in two regions of the hypothalamus: the magnocellular neurons of both the PVN and the supraoptic nucleus (SON). VP has also been colocalized with CRH-containing neurons in the PVN. The neurons that contain VP in the PVN and SON project into the paraventriculohypophyseal and supraopticohypophyseal tracts, respectively, and terminate in the neural lobe of the pituitary gland.

CRH is synthesized as part of a prohormone and then is modified enzymatically to produce the active form of the molecule which is stored within secretory granules. VP is synthesized as a 166-amino acid precursor molecule that codes for VP and its carrier protein neurophysin-II, which is essential for axonal transport. After posttranslational modification and subsequent release from its carrier, VP becomes active as a 9-amino acid peptide. Stress-induced neurotransmitters stimulate the release of CRH and VP from these granules by the classic mechanism of membrane fusion.

B. Pituitary Level

CRH binds to its specific G-protein-associated receptor in the corticotrophs of the anterior pituitary gland and activates the protein kinase A pathway, whereas VP binds to the corticotrophs VP receptor system to activate the protein kinase C pathway. These two neuropeptides stimulate expression of the proopiomelanocortin (POMC) gene. POMC is processed to yield several peptides, including ACTH and β-endorphin (β-END). Several neuropeptides and neurotransmitters (e.g., VP and catecholamines) are capable of stimulating the release of ACTH and β-END individually and may enhance CRH-stimulated secretion of ACTH both *in vitro* and *in vivo*.

Bovine, ovine, porcine, and human ACTH are all 39-amino acid, single-chain polypeptides. The N-terminal amino acids (1–24) are identical across these species and this portion of the sequence contains the full biological activity of the molecule. Consequently, synthetic ACTH(1–24), CRH, and VP are used for biological and immunological studies of the adrenal axis and to investigate the effects of stress-related hormones on male and female reproductive functions.

The CRH receptor was initially identified in the hypothalamus, cerebral cortex, limbic system, cerebellum, spinal cord, and anterior pituitary gland. CRH receptors have since been identified in many other areas, including the peripheral nervous system, heart, lung, liver, gastrointestinal tract, immune system, placenta, and gonads. The biological roles of CRH and its receptors (and binding proteins) in these peripheral areas are not well understood, although CRH may be involved in the autocrine/paracrine regulation of ACTH and β-END release that may modulate reactions in reproductive tissues.

C. Adrenal Gland Level

The adrenal gland is surrounded by a fibrous membrane called the capsule. Inside the capsule are three morphologically distinct zones (zona glomerulosa, zona fasiculata, and zona reticularis) of the adrenal cortex. The main product of the zona glomerulosa is the mineralocorticoid aldosterone, whereas the primary product of the fasiculata and reticularis is the glucocorticoid cortisol. The fasiculata is responsible for secretion of cortisol in response to acute ACTH stimulation. The cells of the reticularis maintain basal secretion of cortisol and respond to chronic ACTH stimulation. The reticularis is also responsible for the adrenal production and secretion of other steroids such as the androgens [dehydroepiadrosterone (DHEA) and testosterone] and estrogens (estradiol and estrone). Cortisol and other steroids (progesterone and DHEA) from the cortex are secreted into the blood, which flows inwardly toward the adrenal medulla and empties into the central vein. Recent evidence suggests that the cellular arrangement of the adrenal cortex arises from a centripetal migration of cells. This implies that cell proliferation occurs mainly at the periphery of the cortex and that cells are subsequently displaced through the fasciculata to the reticularis where they are eventually eliminated. Therefore, the function of each cell changes throughout its lifetime and is influenced by its immediate environment (i.e., internal milieu). These shifts in intraadrenal dynamics are now considered of importance to physical and mental well-being, especially during adolescent and postpartum periods. Hence, adrenal gland development and function may impact reproductive health or competence at several points in one's life span.

The adrenal medulla is surrounded by the adrenal cortex. The medulla is a densely innervated region containing chromaffin cells of two distinct types that secrete the catecholamines epinephrine and norepinephrine. These cells are characterized by the presence of large, dense granules that are the sites for synthesis and storage of the catecholamines. Secretion from these granules involves calcium-dependent exocytosis and is part of the response to emergency situations as described by the fight or flight syndrome. Peripherally produced catecholamines may directly influence pituitary gland secretion of ACTH, gonadotropins, and prolactin during episodes of acute stress response because this gland lies outside the blood–brain barrier.

The main products of the adrenal cortex, the glucocorticoids, are known to have a wide variety of effects throughout the body. Many cell types contain cytoplasmic receptors that bind glucocorticoids. This receptor–ligand complex can then transverse the nuclear membrane and interact with specific regulatory elements to affect gene transcription. The products of these genes can then elicit or reflect the biological effects commonly ascribed to glucocorticoids. Glucocorticoids exhibit negative feedback regulation of the hypothalamic–pituitary–adrenal (HPA) axis at the levels of the hypothalamus, the pituitary, and the adrenal. Specific inhibitory effects of glucocorticoids on reproductive functions [such as decreased gonadotropin-releasing hormone (GnRH) and luteinizing hormone (LH) secretion] are thought to be mediated by the glucocorticoid receptor family which has been localized in critical cell types in tissues of the reproductive system. Glucocorticoid-induced repression of proteins needed for formation of hormone receptors, intracellular signaling molecules, or steroidogenic enzymes may be a component of stress-mediated suppression of reproductive cell types.

A recent concept with implication to human medicine is that developmental experiences of the fetal adrenal gland may impact the health and well-being of the adolescent and mature person. Therefore, the physiologic linkage of the adrenal and gonadal axes may begin as a relationship of embryonic origin but it is also a relationship that persists over the lifetime of the individual. One may project that over the next decade very intriguing interrelationships of the reproductive, adrenal, and immune systems will be identified.

III. REGULATION OF HYPOTHALAMIC–PITUITARY–GONADAL AXES

The anatomy and physiology of each component of the hypothalamic–pituitary–gonadal (HPG) axis are thoroughly reviewed elsewhere in this encyclopedia. Also, other articles extensively describe the GnRH pulse generator and cellular mechanisms regulating production and action of reproductive hormones. In brief, the hypothalamic neurohormone GnRH stimulates the synthesis and release of pituitary gonadotropins, which in turn control steroidogenic and gametogenic functions of the ovary and testis. The interstitial cells of Leydig in the testis produce steroids, especially androgens such as testosterone which possesses anabolic as well as androgenic capability. The Sertoli cells located within the seminiferous tubule of the testis are responsible for production of anti-Müllerian hormone during fetal development. In addition, the Sertoli cell can metabolize steroids (e.g., testosterone to dihydrotestosterone via 5α-reductase and testosterone to estradiol via aromatase) as well as produce inhibin, which interferes with secretion of follicle-stimulating hormone (FSH) but not LH. The gametogenic function of the testis depends on secretion of LH, FSH, and prolactin plus locally produced intratesticular testosterone and growth factors. With respect to the female, pituitary-derived FSH stimulates folliculogenesis and steroid production (especially estrogen production via activation of the aromatase enzyme complex). The process of oocyte maturation and ovulation appears to be greatly affected by LH. The formation and function of the corpus luteum is influenced by LH as well as by various growth factors or cytokines of systemic or local origin. Furthermore, endocrine input from the uterus (prostaglandins) or conceptus (chorionic gonadotropin) signals the cell types of the corpus luteum to produce more or less progesterone depending on pregnancy status. As outlined elsewhere in this encyclopedia, the gonadal hormones modulate synthesis and secretion of GnRH and the gonadotropins in males and females via negative and positive feedback systems. This brief overview of the major endocrine features controlling the HPG axis in males and females highlights the multiple sites whereby exteroceptive stimuli or pathogenic factors may disrupt so-called normal reproductive processes. Many cell types that comprise the HPG axis also express the signaling receptors for products of both the immune and adrenal systems. The presence of the glucocorticoid receptor system in reproductive tissues leads to consideration that stress-induced hypersecretion of cortisol may negatively affect reproductive functions. The following sections briefly introduce potential influences of the stress axis on elements of the HPG axes, whereas a later section addresses immunopeptide modulation of reproductive function.

IV. INTERACTIONS OF THE HPA AND HPG AXES

Environmental, physiological, and managerial stresses have often been implicated as causes of reproductive disorders and decreased fertility in animals and humans. Hypothetically, deleterious effects of stress on reproductive performance are mediated primarily via the HPA axis. Instances of pathologically (adrenal hyperplasia) and pharmacologically (synthetic glucocorticoids) induced suppression of reproductive function support the concept of an adrenal–gonadal relationship. However, a true physiological association of the endocrine functions of the adrenal glands and gonads has only recently been appreciated. The concept of an endocrine relationship between these glands is based on the mesodermal origin of their steroidogenic tissue, possession of similar steroidogenic pathways, and analogous neuroendocrine control mechanisms. This concept has attracted increasing attention as advances in hormone assay technology facilitate descriptive characterization of relationships between the HPA and HPG axes in males and females of several species. The biologic and economic importance of potential deleterious influences of HPA axis products on reproductive processes of people, domestic livestock, and captive animals dictate a thorough evaluation of adrenal–gonadal relationships.

A. Hypothalamic Level

GnRH is a 10-amino acid hypothalamic-derived peptide hormone, also referred to as luteinizing hormone-releasing hormone (LHRH). GnRH neurons develop in the vomernasal and olfactory structures and migrate caudally during embryogenesis. In mammalian species, hypothalamic areas rich in GnRH neurons include the preoptic area, the medial basal hypothalamus, and the diagonal band of Broca. Therefore, stressors may affect the hypothalamic component of the male or female reproductive system by altering GnRH neuron structure (e.g., migration and morphology of neurons) and function (e.g., GnRH gene expression and content of GnRH). GnRH travels via the hypophyseal portal system to stimulate secretion of LH and FSH by pituitary gonadotrophs.

The effects of stress on the HPG axis are influenced by the type and duration of the stimulus as well as the gender, age, and species of the animal. Although all the exact mechanisms whereby stressors disrupt reproductive functions are not fully understood, the main effects seem to involve suppressing the synthesis and release of GnRH from the hypothalamus, which results in decreased synthesis of LH and FSH. The existence of a specific FSH-releasing hormone has been proposed but remains elusive. Limited information is available regarding effects of stress on FSH secretion. In some instances treatment with a synthetic glucocorticoid such as dexamethasone reduced secretion of LH but not FSH. The inability of dexamethasone to inhibit the postcastration rise in FSH suggests that gonadotroph synthesis and secretion of LH in response to GnRH may be more susceptible than is the synthesis of FSH to inhibitory actions of glucocorticoids (or that an FSH-releasing hormone exists that is less susceptible than LHRH to stress-induced levels of glucocorticoids).

The biosynthesis of the GnRH decapeptide depends on the combined actions of neurotransmitters and steroids as well as immunopeptides, growth factors, cytokines, and perhaps metabolites (e.g., glucose, fatty acids, and/or specific amino acids). Hence, environmental stimuli and nutritional stressors may directly and indirectly affect the neuroendocrine control of reproductive function. In particular, the actions of stressors may disrupt the actions or timing of regulatory mechanisms that control maturation and function of the GnRH system (from developmental stages to functionality of the GnRH pulse generator). GnRH stimulates secretion of LH and FSH via plasma membrane receptor-mediated pathways. Therefore, recognition of genetic and molecular mechanisms controlling GnRH receptor gene expression and synthesis/secretion of GnRH is important from biological and toxicological perspectives, particularly as we gain a greater appreciation of the influence of stressors and environmental toxicants on the adrenal, immune, and gonadal axes.

Recent studies suggested that CRH plays a primary role in suppressing reproductive function. Because the hypothalamus is a central controlling region for both reproductive function and the stress response, it is not surprising that a hypothalamic factor or a "hypothalamic-acting" factor could play a pivotal role in the cross talk of the two systems. GnRH neurons have been shown to have synapses with CRH-containing neurons in the preoptic area. Several studies have shown that CRH can inhibit GnRH release, and CRH-induced β-END has also been shown to suppress GnRH and LH secretion. Additionally, CRH and its receptor have been found in both the ovary and the testis. CRH has been shown to inhibit testosterone synthesis in the rat Leydig cell yet stimulate steroidogenic acute regulatory protein and steroid synthesis in the murine Leydig cell.

Depressed synthesis and secretion of GnRH and LH underpin the predominant concept that CRH-driven pathways suppress reproductive function via action at the hypothalamic level. *In vivo* and *in vitro* evidence indicates that CRH may directly and/or indirectly affect secretion of GnRH and LH at the level of the hypothalamus, ME, or pituitary. Evidence derived from castrate nonhuman primate, sheep, and cattle models demonstrates that a POMC-derived peptide such as β-END may mediate CRH- or stress-induced suppression of GnRH and LH secretion. Furthermore, CRH-independent POMC peptides may directly affect GnRH neuronal activity. Production of CRH at the gonadal level may also be associated with alterations in ovarian and testicular steroidogenesis because both inhibition and stimulation have been reported *in vitro*. The demonstration of CRH binding at various target levels in the reproductive system, coupled with the ability of CRH antagonists and antisera to negate CRH action, suggests that a sufficiently elevated concentration of peripherally circulating or locally produced CRH may modulate reproductive axes of males and females. Some species differences may exist and CRH's action on the reproductive system may vary according to the acute or chronic nature of the stressors/stimuli.

B. Pituitary Level

An oft-cited consequence of stress is the disruption of reproductive functions. Increased cortisol in male rats, bulls, and men as well as males of other species can decrease secretion of LH and testosterone. However, acute increases in stress-related hormones (ACTH or cortisol) are associated with increased plasma concentrations of LH and testosterone in the monkey and boar. Although reproductive function in both the male and the female is affected by stress, many specialists suggest that female reproductive function is more susceptible to stress because of the dependency of successful reproduction on carefully timed hormonal secretions. For ovulation to occur, the HPG axis must be functioning such that the necessary signals are released from the hypothalamus (GnRH) and pituitary (LH and FSH) at the proper time to affect ovarian structures.

Environmental or managerial stressors have been implicated or suspected as being causative or contributing factors to menstrual or estrous cycle dysfunction. A stressful event may create a transient depression in fertility which may adversely affect reproductive efficiency of a herd, especially in management schemes which employ artificial insemination or estrus synchronization protocols. Hence, several investigators devised treatments (provision of exogenous adrenal axis hormones) or handling procedures to mimic HPA axis response to stressful events. For example, natural and synthetic glucocorticoids and neurohormones (e.g., ACTH, CRH, and VP) have been administered to experimentally mimic activation of the adrenal axis in an effort to determine temporal responses, if any, of the reproductive axis. Furthermore, handling protocol, such as transportation, electroshock, and nutrient restriction/realimentation studies, have been used to investigate whether perturbations of the stress axis elicit short- or long-term negative consequences on the reproductive system. Less effort has been placed toward identification of the impact of imposition of stressful events on gametogenesis, fertilization, or pregnancy. In most cases, such experimentation successfully demonstrates short-term modulation of the profile of reproductive hormones, leading numerous authors to speculate that, in the long-term, one may expect adverse consequences in terms of fertility or reproductive efficiency.

Several investigators reported that imposition of transportation stressors (e.g., movement, crowding, fumes, and water and feed restriction) results in diminished release of LH in cows and ewes. Specifically, the physical and psychogenic elements involved with animal response to this particular stressor diminished the ability of exogenous estradiol to induce the preovulatory surge of LH in cows or ewes. Also, a similar handling protocol was associated with reduced LH response to exogenous GnRH. In some studies, supplementation with naloxone (opioid antagonist) blocked these inhibitory actions, whereas in other cases naloxone treatment was ineffective in preventing stressor-induced suppression of induced LH release. Therefore, it appears that some stressful events in particular individuals result in opioid-mediated inhibition of GnRH secretion or action; however, one cannot exclude the potential direct involvement of CRH and/or ACTH on GnRH and gonadotropin secretion. The findings that exogenous CRH, ACTH, or dexamethasone transiently reduce LH secretion in castrate and intact females and males suggest a direct action of these stressor-induced factors *in vivo*. Treatment of gonadotrophs *in vitro* demonstrates the *in vivo* potential for direct inhibitory action of ACTH, CRH, or cortisol on LH secretion. A stressor such as electroejaculation of bulls reduces LH peak amplitude and frequency. Because the pituitary gland may be exposed to high levels of catecholamines in response to an acute stressor, the concept is emerging that the initial diminution of LH secretion by stressors may be affected by adrenomedullary derived catecholamines. Evidence is accruing that the longer term suppressive action of various stressors involves suppressed secretion of GnRH, decreased pituitary gland production of fully functional GnRH receptors, as well as reduced synthesis and/or secretion of gonadotropins (perhaps via the combined effects of POMC-derived peptides and adrenal steroids).

In view of species differences and interindividual variation in responsiveness to stressors, it is likely that stressors may affect reproduction via action of more than one messenger and by exerting inhibitory effects at more than one tissue or cell level. Specifically, stressors may exert inhibitory actions at the hypothalamus and/or pituitary gonadotroph level to modulate gonadotropin secretion and support of the gonads. Even a transient diminution of the GnRH–gonadotroph linkage can have adverse effects on reproductive success if the stress response occurs at a critical point in the reproductive cycle or stage of development.

Further evaluation of a coordinated endocrine regulation of gonadal function during and following exposure to stressors should not ignore brain peptides such as β-END, which has been described as the body's endogenous opioid and is secreted in response to stress and pain. It is intriguing that ACTH, melanocyte-stimulating hormone (MSH), and β-END are derived from the same precursor prohormone (POMC), have a partially homologous amino acid sequence, and are secreted concomitantly from the same pituitary cell. Potentiation of ACTH-induced adrenal steroidogenesis has been observed in rats treated with an MSH-derivative and exogenous MSH-stimulated corticosteroidogenesis in neonatal sheep. Potential synergism of MSH and ACTH and simultaneous release of β-END, ACTH, and MSH are particularly intriguing in view of the marked elevation of glucocorticoids and progesterone and the subsequent suppressed secretion of LH and testosterone after exposure of bulls to stressful procedures. A physiological role for pituitary and adrenal-derived β-END also may be established with further evaluation and clarification of adrenal-mediated endocrine influences on gonadal function.

C. Gonadal Level

Delineation of the how the HPA axis influences reproductive processes, such as steroidogenesis and gametogenesis, has attracted considerable attention because exposure to environmental challenges such as thermal stress can adversely affect fertility. As a result of these efforts, negative effects of glucocorticoids on synthesis and secretion of GnRH, LH, and testosterone are now recognized. A single injection of ACTH stimulates testosterone secretion from the testes of boars and rabbits. In contrast, acute treatment with ACTH inhibits testosterone secretion by the bovine, ovine, and primate testis. The suppressive effects of ACTH or dexamethasone on testosterone production are associated with the ability of glucocorticoids to inhibit LH secretion in the bull and ram. However, the stimulatory effect of ACTH on testosterone secretion by the boar testis is not directly mediated by LH since the effect of ACTH occurs independently from any detectable increase in LH secretion *in vivo*. *In vitro* studies with isolated perfused rabbit testes suggest that ACTH can act directly on the testis to increase testosterone secretion.

Perhaps some optimal level of LH maintains Leydig cell structure–function relationships which may facilitate or permit the stimulatory effect of ACTH. Leydig cell production of POMC-derived hormones has been reported for several species, including the rabbit, guinea pig, hamster, and mouse. These peptides, including ACTH and β-END, have been implicated in paracrine as well as autocrine regulation of testicular function. It has been suggested that locally produced testicular β-END has a possible inhibitory action on testosterone production. However, ACTH and β-END appear to have opposing effects on Sertoli cell growth because ACTH is stimulatory and β-END is inhibitory. The localization of POMC-derived peptides in the testes and the ability of these peptides to have opposite functions at the testicular level further suggest a role for ACTH and β-END in the endocrine regulation of the testis development and function.

The glucocorticoid receptor system is present in the Leydig cell, which likely facilitates the inhibitory action of naturally occurring and synthetic glucocorticoids. *In vivo* and *in vitro* treatment with glucocorticoids can lower LH receptor number, cAMP levels, and steroidogenic enzyme synthesis, which results in diminished androgen production. In females, glucocorticoid treatment has been shown to decrease gonadotropin secretion and FSH induction of aromatase activity, which results in decreased circulating estradiol concentrations. The presence of cortisol-metabolizing enzymes in reproductive tissues may be a protective system in that cortisol is metabolized to less potent steroids. This reduces the targeted cell's chances of exposure to antireproductive actions of glucocorticoids.

Many reports highlight the existence of numerous paracrine and autocrine mechanisms within the hypothalamic–pituitary–testis axis. In addition, an extrapituitary role for the neurohormones GnRH and CRH has been suggested. Whether CRH or CRH-like peptides are also responsible for testicular synthesis of ACTH, ACTH-like, or opioid substances remains to be determined.

Intriguing, but conflicting, reports state that CRH directly affects gonadal steroid production. Specifically, both enhancement and inhibition of gonadotropin-stimulated steroid production have been observed in response to CRH treatment of cultured testicular Leydig cells and ovarian granulosa cells. These effects are considered to be due to a CRH receptor-mediated pathway because a CRH antagonist negated CRH action on gonadal steroidogenesis. Further studies are needed to determine if gonadal cell types produce CRH-like peptides and whether locally produced CRH induces production of mediators, such as cytokines or opioids, that actually are responsible for altered gonadal function.

In most cases, experimental elevation of ACTH and cortisol depresses testosterone concentration but does not adversely affect sperm output, concentration, or viability. The number of immature sperm and number of sperm with abnormal head structures may be slightly elevated following prolonged elevation in plasma concentration of cortisol and depressed plasma concentration of testosterone. Stress-susceptible individuals whose sperm production parameters are of borderline adequacy may be at greater risk of experiencing stressor-induced depression of spermatogenesis. The available data provide an endocrine basis for future research into adrenal and stress-mediated influences on spermatogenesis. Similarly, from the female perspective, reports indicate that hypercortisolemia is associated with a reduction in number of follicles recruited by exogenous FSH, suggesting that additional studies should focus on the impact of stress on folliculogenesis, ovulation induction, and development of corpora lutea.

V. IMMUNE SYSTEM ASPECTS OF HPA AND HPG AXES INTERACTION

The immune system is subject to endocrine and neural regulation and, in turn, the immune system can influence neuroendocrine functions. Many cell types of the immune system have receptors for products of the HPA and HPG axes and can produce stress-related factors. Some products of the immune system, including interleukin-1 (IL-1) and IL-6, stimulate secretion of CRH, ACTH, and cortisol. Localization of IL-1β in regions of the hypothalamus similar to the locations of CRH and GnRH neurons suggests that the hypothalamus may be an important site of interaction between the neuroendocrine and immune systems. Receptors for IL-1, IL-2, and IL-6 have been found in the pituitary gland, adrenal gland, and gonads, and these cytokines have been shown to affect secretion of ACTH, gonadotropins, and adrenal and gonadal steroids. Investigators are pursuing the intriguing observation that cytokines are produced by cell types that comprise the HPA and HPG axes.

Ovarian function can be influenced by cytokines. IL-1 and tumor necrosis factor-α (TNF-α) have proliferative effects on developing follicles and inhibitory effects on differentiation and steroid production in those same follicles. At the final stages of follicular development, cytokines may be involved in the ovulatory process because ovulation has been described as an inflammatory event. Furthermore, corpora lutea regression may be regulated in part by an increased production of cytokines derived from infiltrating macrophages. Perturbations in the immune system such as immunosuppression or hyperactivation due to prolonged or acute stressors may modulate the concentration of cytokines influencing ovarian function. Upsetting this balance could possibly disrupt the sequential events of follicular development, ovulation, corpora lutea formation, and subsequent luteal regression leading to reproductive cycle disruption or failure.

Although peripheral CRH is thought to have proinflammatory actions by stimulating cytokine production by immune cells or by modulating the effects of cytokines at sites of inflammation, the best understood influences of the HPA axis on the immune system are those of the glucocorticoids. At plasma concentrations generally associated with stress, glucocorticoids can have immunosuppressive and anti-inflammatory effects. These include interference with cell-mediated immunity, decreased cytokine production, and inhibition of cytokine effects on target tissues. Glucocorticoids decrease the numbers of circulating monocytes and lymphocytes as well as inhibit the production of IL-2, which is an important T-cell growth factor, and inhibit expression of IL-1. Natural killer cell activity has also been reported to decrease after treatment with glucocorticoids. Dexamethasone inhibits production of IL-2 and interferon-γ in cultured lymphocytes and inhibits lipopolysaccharide-stimulated TNF-α production by macrophages. These generally suppressive effects of glucocorticoids on the immune system suggest that glucocorticoids may play an important role in regulation of the interaction of the endocrine and immune systems. Analogous to the circumstance in reproductive tissues, these effects are likely mediated by the interaction of this ligand with the glucocorticoid receptor and subsequent regulation of specific genes controlled by glucocorticoids. A recent consideration is the role of nongenomic actions of stress-induced or pharmacologically administered glucocorticoids on the reproductive, adrenal, and immune systems and the interrelationships of these three physiologically important systems. Innervation of the reproductive system provides a pathway by which exteroceptive and enteroceptive stimuli may directly impact function of reproductive cell types. The recent observations that immune system tissue receives neural input and that cytokines are produced by and affect neural cells provide yet another pathway to integrate the neuroimmunoendocrine functions. This type of physiological networking provides the opportunity for particular inflammatory events to distort reproductive function or fertility.

An example of the linkage of the neural, immune, and endocrine systems is demonstrated by the finding that intracerebroventricular infusion of IL-1α in the ovariectomized monkey stimulates the adrenal axis and inhibits pulsatile secretion of the gonadotropins LH and FSH. Experimental exposure to gram-negative bacteria elicits a marked cytokine response and interferes with LH secretion in cattle and pigs. Consequently, investigators are pursuing the concept that the systemic inflammatory response syndrome may alter secretion of the hypothalamic, pituitary, adrenal, and gonadal hormones in a manner that is conducive to survival of the individual but does not necessarily support fertility.

VI. IMPACT OF BEHAVIORAL AND PSYCHOLOGICAL STRESSORS ON REPRODUCTIVE FUNCTION

Psychological stressors may alter the activity of both the autonomic nervous system and the HPA axis, with consequent adverse effects on the reproductive system. This concept, whose predictability and mechanistic nature have proven difficult to precisely pinpoint, remains popular among clinicians and the laypublic. For instance, observations that stressful events, such as immobilization, surgery, and social disruption, sometimes are associated with diminished body mass, a lower production of testosterone, and decreased copulatory activity in animals fuel the effort to attribute individual cases of human infertility to adverse effects of stress on reproduction. Although mild- to severe emotional stressful events may transiently diminish production of testosterone and disrupt spermatogenesis in men, assignment of a specific psychopathologic circumstance as the cause of infertility is not often attempted or accomplished in many cases. There is concern among animal behaviorists that stressors associated with intensive livestock management procedures may reduce reproductive efficiency. In the human clinical setting and the commercial livestock production setting, mechanisms by which stressors disrupt normal reproductive processes must be identified before appropriate actions can be taken to alleviate the potential adverse effects of behavioral or psychological stressors on reproduction. An additional difficulty in this area is that limited data are available regarding whether psychological factors affect testicular morphology. Limited morphologic responses at the gonadal level prompt most investigations of psychological aspects of stress and reproduction to focus on the neuroendocrine interface. Hence, in addition to consideration of the gonadal actions of ACTH, CRH, β-END, and cortisol, the actions of these hormones are under study at the neuroendocrine level.

The physiological coupling of adrenal and gonadal function may also include the phenomenon of puberty and adrenarche. Young adolescent humans may be subject to a so-called "vicious cycle" of stress-induced adrenal activity that deters gonadal maturation, which may heighten anxiety and the stressfulness of the situation. For instance, gonadal and adrenal hormones have been linked to social stressors and adjustment difficulties experienced by young adolescents. Specifically, these individuals experienced a lower peripheral concentration of gonadal steroids and a higher concentration of adrenal steroids. This inverse relationship is similar to that described previously for males of several species. In these circumstances, clinicians suggest that later gonadal maturation may be a result of stress-mediated suppression of the reproductive axis. Further extension of this concept includes suggestions by zoologists that some species of captive animals have difficulty becoming reproductively competent due in part to the stress of confinement or captivity.

Emotional distress has been anecdotally linked to infertility. This belief has been revived recently as techniques to assist human patients or couples seeking therapy for infertility issues have improved. Recent advances in assisted reproductive technology (i.e., *in vitro* fertilization, intracytoplasmic sperm injection, etc.) provide additional options for infertile couples, but there are possible emotional stressors. Patients/clients who seek fertility assistance often experience significant emotional stress, particularly if reproductive success is elusive. Specialists in assisted reproductive technologies have begun to consider whether stress susceptibility is associated with successful outcome of *in vitro* fertilization procedures. The stressful state experienced by the patients is usually attributed to being infertile rather than to stress causing the infertility. However, as reviewed previously, impaired fertility may be caused by the action of stress-related hormones (catecholamines, opioids, and adrenal steroids) acting at one or more levels of the HPG axis in men and women.

Because assisted reproductive technologies may be viewed as a couple's last option, stress management is becoming a consideration to optimize success in achieving and sustaining a pregnancy. The impetus for this philosophy is the observation that the women undergoing *in vitro* fertilization procedures who experienced a greater increase in blood pressure and heart rate in response to a stress test had fewer fertilized oocytes and transferred embryos relative to the women who experienced a less pronounced cardiovascular response. In addition, having a career occupation outside the home may be associated with lower success rates in *in vitro* fertilization and embryo transfer programs. In such an instance, there is concern that the duress experienced by some infertile individuals further compounds the infertility.

VII. ASPECTS OF STRESS DURING GESTATION

The environment may affect the establishment, progression, and outcome of a pregnancy. Concern that environmental stressors (e.g., high ambient temperature and pyrogens) may adversely affect gamete transport, early embryonic development, and maternal recognition of pregnancy leads livestock producers to avoid imposing managerial stressors (such as transportation) on newly inseminated females until after the time of implantation. With respect to humans, discussions of stressful events during gestation often focus on the health of the mother, pregnancy status, and well-being of the fetus. Cardiovascular and hemodynamic stressors, such as placental hemorrhage or hypertensive disorders, contribute to the risk of preterm delivery. Also, intrauterine infections may be linked with cytokine production which activates premature parturition. Behavioral factors, such as poor nutritional habits and use of tobacco, drugs, or alcohol, also contribute to low fetal birth weight and incidence of preterm delivery. Exposure of pregnant women to psychosocial stressors at 28–30 weeks of gestation has been associated with shortened gestation length and lower body weight of newborns. Although fetal production of CRH, ACTH, and cortisol increases in late gestation and triggers parturition, maternal stress during this time frame is associated with increased maternal plasma levels of ACTH and cortisol. An elevated secretion of ACTH and cortisol by the mother may be amplified by the placental release of CRH, which when coupled with increased fetal HPA axis activity may advance the time of parturition. Consequently, physiologic, psychologic, or pathogenic activation of the HPA axes in the mother and fetus is viewed as contributing to the risk of preterm delivery. Therefore, gestation length as well as fetal development and well-being may be subject to maternal responsiveness to physiological and environmental stressors.

In the modern workplace, occupational specialists are concerned about the potential for noise to induce a stress response in the mother, possibly causing reproductive disturbances for her, with possible direct effects on the fetus. For instance, subjecting pregnant rats to stressors adversely affects the development and sexual behavior of the male offspring by diminishing production of androgens by the developing fetal testis. Because noise stress often occurs in organizations that employ shift workers, the stressors of work and noise in the workplace may interact to alter the individual's circadian rhythm and affect aspects of the ovarian and testicular axes and result in diminished reproductive function or fecundity. Women who work full time under stressful conditions (e.g., lengthy work weeks and high-strain careers) during pregnancy have a higher incidence of preterm delivery. However, the individual's perception of the stressful event or environment and the nature of that individual's social support network throughout gestation may influence the outcome of the pregnancy. Experts concur that additional solid data are needed and strongly advocate further study because workplace stressors (e.g., high-level noise and shift work) represent environmental and psychosocial challenges to reproductive well-being.

The secretion of maternal, placental, and fetal stress-related hormones is relevant to maturation of behavioral and reproductive competence in the offspring. Inappropriate social behaviors, attention deficits, anxieties, and impaired coping abilities of children have been linked to prenatal stressors. Impaired coping ability of adult monkeys and rodents that were prenatally stressed is associated with prolonged elevation of plasma glucocorticoids in response to a stressor. This disturbance in regulation of the HPA axis includes decreased feedback inhibition of CRH. Higher, sustained levels of CRH and cortisol may expose central and peripheral components of the HPG axis of an affected individual to the potential antireproductive actions of these two elements of the HPA axis. Emergent data suggest that prenatal stress affects the adrenal axis development and functionality of the progeny at both prenatal and neonatal periods. The longer range issue that has arisen in the area of maternal stress during pregnancy is whether gestational stress impacts mental and physical health of progeny in the postnatal, adolescent, and adult phases of life.

VIII. SUMMARY

Both acute and prolonged psychosocial stressors can interfere with normal patterns of secretion of reproductive hormones in many species. The primary mediators of stress-related impairment of reproductive hormone secretion appear to be catecholamines, POMC-derived peptides, and adrenal steroids. The prime tissue targets appear to range from the central nervous system to the pituitary gonadotrophs and specific gonadal cells. The entire process of reproduction may be subject to adverse consequences of environmental, thermal, pathogenic, and psychologic stressors. For example, gamete production and transport, mating behavior, fertilization, conceptus development, placental function, pregnancy status, parturition, and health of the offspring all appear somewhat susceptible to the consequences of stress responsiveness of the host.

Over the past two decades physiologists and clinicians have taken advantage of the power of technological advances to acquire the information needed to convincingly demonstrate the distribution of receptors for stress-related hormones in reproductive tissues. The temporal aspects of the secretion of stress and reproductive hormones have been characterized in efforts to establish causal, physiologic relationships of gonadal and adrenal axes. Rodent models (e.g., intact and castrate male and female mice and rats) have been the most thoroughly studied animals with regard to investigating the impact of stress on reproduction. However, interspecies variation reemphasizes the need for species-specific data when trying to establish causal links between two integrative endocrine systems (namely, the HPA and HPG axes).

CRH can interact with GnRH neurons to inhibit gonadotropin secretion by the pituitary gland. Opioids of neural or pituitary origin may also inhibit GnRH neurons. Via these routes, pulsatile gonadotropin secretion is inhibited by stressful events which may lead to impaired steroidogenesis or gametogenesis. Disturbances in follicle growth, maturation, or ovulation constitute the primary cause of stress-related impairment of reproductive dysfunction in humans. Although both plasma concentrations of ACTH and glucocorticoids are elevated in response to stressors, whether either of these hormones directly affect LH secretion or ovulation is unclear. In some instances, it has been noted that exogenous ACTH or CRH reduce GnRH secretion *in vitro* or reduce basal or GnRH-induced secretion of LH by nonrodent male and female mammals in *in vivo* and *in vitro* circumstances. However, even within the same species a direct action of cortisol or ACTH on gonadotroph function during or after periods of stress or mimicking of stress is not consistently demonstrated. However, secretion of gonadotropin or altered follicular or luteal function may occur in concert with or subsequent to mental or physical stressors, such as electroshock, transportation stress, or inflammatory events. Because the elevation in ACTH and cortisol following such events (deemed to be due to induction of CRH and/or VP, for example) is temporally associated with diminution of LH output, we return to the concept that at one or multiple levels of the HPG axis, one or more hormones of the HPA axis (e.g., β-END, cortisol, and catecholamines) negatively influence reproduction. This simplified view, however, is tempered by observations that secretion of LH is increased in response to acute stressors in some species. Clarification of these discrepancies awaits systematic assessment of the impact of each HPA axis factor on cell types and hormones that control each level of the HPG axis in major species aside from rodent models.

We must remain cognizant that each exposure to a stressor does not automatically equate to impairment of reproductive function or fertility. Despite the extensive list of many adverse events, the impact of the stress response on reproductive processes is a function of individual responsiveness. Interindividual variation in stress responsiveness depends on the type, magnitude, and duration of exposure to the stressor, the individual's control and perception of the stressful circumstance, recent health, and immune and reproductive status. In addition, genotype and environmental interactions warrant consideration in future assessment of how stress affects reproduction.

IX. FUTURE CHALLENGE

Although anecdotal accounts abound that stressful events have adverse consequences on libido, ovulation, pregnancy, etc., precise delineation of the cellular mechanisms whereby any of the stress-related hormones impede functionality of any reproductive tissue or cell type remains elusive. Much work has been done to characterize the physiological changes involved in regulating reproductive and stress responses. In addition, whether (and, if so, how) genotypic and psychologic factors control the mechanisms that lead to physiological changes is not well understood. Of particular importance is the need to improve our understanding of the intercommunication of the immune, adrenal, and gonadal axes during periods of eustress as well as distress.

In this age of molecular endocrinology and molecular genetics, it is likely that detailed information regarding the genetic mechanisms whereby particular stress-related hormones inhibit synthesis, secretion, or action of reproductive hormones will be acquired at a more rapid rate. The imminent challenge may not be to acquire new mechanistic information but rather to make more astute use of the current information to ameliorate if not prevent adverse consequences of stress on reproduction.

See Also the Following Articles

Adrenarche; Fetal Adrenals; FSH (Follicle-Stimulating Hormone); GnRH (Gonadotropin-Releasing Hormone); Hypothalamic–Hypophysial Complex; LH (Luteinizing Hormone); Pituitary Gland, Overview

Bibliography

Berga, S. L. (1997). Behaviorally induced reproductive compromise in women and men. *Sem. Reprod. Endocrinol.* **15**, 47–53.

Cameron, J. L. (1997). Stress and behaviorally induced reproductive dysfunction in primates. *Sem. Reprod. Endocrinol.* **15**, 37–45.

Chatterton, R. T. (1990). The role of stress in female reproduction: Animal and human considerations. *Int. J. Fertil.* **35**, 8–13.

Dobson, H., and Smith, R. F. (1995). Stress and reproduction in farm animals. *J. Reprod. Fertil. Suppl.* **49**, 451–461.

Dorn, L. D., and Chrousos, G. P. (1997). The neurobiology of stress: Understanding regulation of affect during female biological transitions. *Sem. Reprod. Endocrinol.* **15**, 19–35.

Facchinetti, F., Matteo, M. L., Artini, G. P., Volpe, A., and Genazzani, A. R. (1997). An increased vulnerability to stress is associated with a poor outcome of in vitro fertilization-embryo transfer treatment. *Fertil. Steril.* **67**, 309–314.

Fraser, D., Ritchie, J. S. D., and Fraser, A. F. (1975). The term "stress" in a veterinary context. *Br. Vet. J.* **131**, 653–662.

Greenfeld, D. A. (1997). Does psychological support and counseling reduce the stress experienced by couples involved in assisted reproductive technology? *J. Assist. Reprod. Genet.* **14**, 186–188.

Lederman, R. P. (1995). Treatment strategies for anxiety, stress, and developmental conflict during reproduction. *Behav. Med.* **21**, 113–122.

Liptrap, R. M. (1993). Stress and reproduction in domestic animals. *Ann. N. Y. Acad. Sci.* **697**, 275–284.

McGrady, A. V. (1984). Effects of psychological stress on male reproduction: A review. *Arch. Androl.* **113**, 1–7.

Moberg, G. P. (1991). How behavioral stress disrupts the endocrine control of reproduction in domestic animals. *J. Dairy Sci.* **74**, 304–311.

Negro-Vilar, A. (1993). Stress and other environmental factors affecting fertility in men and women: Overview. *Environ. Health Perspect.* **101**(Suppl. 2), 59–64.

Nurminen, T. (1995). Female noise exposure, shift work, and reproduction. *J. Occup. Environ. Med.* **37**, 945–950.

Rivier, C., and Rivest, S. (1991). Effect of stress on the activity of the hypothalamic–pituitary–gonadal axis: Peripheral and central mechanisms. *Biol. Reprod.* **45**, 523–532.

Sandman, C. A., Wadhwa, P. D., Chicz-DeMet, A., Dunkel-Schetter, C., and Porto, M. (1997). Maternal stress, HPA activity, and fetal/infant outcome. *Ann. N. Y. Acad. Sci.* **814**, 266–275.

Schenker, J. G., Meirow, D., and Schenker, E. (1992). Stress and human reproduction. *Eur. J. Obstet. Gynecol. Reprod. Biol.* **45**, 1–8.

Susman, E. J., Nottelmann, E. D., Dorn, L. D., Inoff-Germain, G., and Chrousos, G. P. (1988). Physiological and behavioral aspects of stress in adolescence. *Adv. Exp. Med. Biol.* **245**, 341–352.

Weinstock, M. (1997). Does prenatal stress impair coping and regulation of hypothalamic–pituitary–adrenal axis? *Neurosci. Biobehav. Rev.* **21**, 1–10.

Xiao, E., and Ferin, M. (1997). Stress-related disturbances of the menstrual cycle. *Ann. Med.* **29**, 215–219.

Yousef, M. (1985). *Stress Physiology in Livestock*, Vols. 1–3. CRC Press, Boca Raton, FL.

Substance Abuse and Pregnancy

Sara J. Marder and Mark A. Morgan
University of Pennsylvania School of Medicine

GLOSSARY

abruptio placentae Premature detachment of a normally implanted placenta.

congenital anomaly Structural abnormality which may be hereditary or due to some influence occurring during gestation.

meconium Fetal or neonatal greenish-colored intestinal discharge consisting of epithelial cells, mucus, and bile.

neonatal The period immediately following birth through the first 28 days of life; newborn.

perinatal mortality The number of fetal deaths after 28 weeks gestation plus neonatal deaths up to 6 days of age.

placenta previa Implantation of the placenta in the lower segment of the uterus and covering all or part of the internal os of the cervix.

teratogen An agent that causes abnormal development.

Substance abuse in pregnancy includes the use of illicit drugs, the abuse of legal substances, and the misuse of medical prescriptions during gestation. The medical, social, and psychological consequences of this complicated problem are related to the potential effects on maternal health, fetal development, perinatal outcome, and long-term behavioral and mental development.

I. INTRODUCTION

According to the National Institute on Drug Abuse, 5.5% of women use an illicit drug during pregnancy. Drug use can be found across all socioeconomic levels and does not discriminate according to maternal education level, race, or marital status. On the other hand, there are specific maternal characteristics which are associated with drug abuse during pregnancy (Table 1).

The effects of substance abuse in pregnancy are numerous. First, substance abuse is potentially harmful to maternal health and the pregnant woman is as susceptible to these adverse effects as any adult. These consequences may include poor nutrition, HIV infection, hepatitis, paraphernalia complications, and life-threatening withdrawal.

Furthermore, the dramatic physiologic adaptations to pregnancy often alter a drug's pharmacological properties, potentially worsening the negative effects. In addition, the cardiovascular effects of certain drugs, as well as the physical effects of withdrawal, may change the characteristics of uterine blood flow leading to an inadequate supply of oxygen and nutrients to the fetus. Placental abnormalities causing antepartum hemorrhage, such as abruptio placentae and placenta previa, may also be related to the cardiovascular changes induced by substance abuse.

Second, most substances of abuse are transferred across the placenta allowing the potential for an effect on fetal formation and development. As a result, an increased incidence of miscarriage, fetal death, and congenital anomalies can be seen in association with drug use in pregnancy. This passive transfer to the fetus may have a direct effect on the fetal brain resulting in abnormal neurobehavioral development. Transfer of the substance across the placenta may also result in neonatal withdrawal after delivery.

Third, substance abuse in pregnancy is associated with increased rates of premature rupture of the membranes and preterm delivery, whether related to drug abuse per se or a result of other factors associ-

TABLE 1
Risk Factors for Substance Abuse in Pregnancy

Cigarette smoking
Partner or family member who abuses drugs or alcohol
Unstable living situation
Depression
Lack of social support
History of substance abuse
No prenatal care
History of sexually transmitted diseases

ated with a drug-abusing lifestyle. The frequency of low birth weight is also increased in the offspring of drug-abusing women. Several neonatal complications may result from prematurity and low birth weight.

In the following sections, the effects of abusing various drugs in pregnancy will be discussed. Importantly, there are a number of difficulties faced by investigators in establishing a clear cause-and-effect relationship between a specific substance and a specific outcome. The investigators' ability to accurately identify the drug abuser can be problematic and exposure dosage and timing can be difficult, if not impossible, to assess. Polysubstance abuse is common in drug abuse, further clouding the picture. Nutritional status and social and other lifestyle-related factors will also influence the outcomes of interest.

II. SUBSTANCES OF ABUSE

A. Alcohol

Ethyl alcohol, a central nervous system (CNS) depressant, is a frequently used substance in pregnancy, with 16.3% of pregnant women reporting having a drink during pregnancy and 3.5% reporting frequent drinking (*Morbidity Mortality Weekly Rep.*, 1997). Maternal alcohol abuse may be complicated by potentially severe withdrawal symptoms in addition to the effects of chronic alcohol use on the coagulation profile, heart, liver, and other organ systems. Pregnancy-related gastrointestinal changes and alcohol's effect on this organ system theoretically increases the risk of pulmonary aspiration during pregnancy.

Alcohol is rapidly transferred across the placenta and is responsible for the most common cause of teratogenic mental retardation, the fetal alcohol syndrome (FAS). FAS is characterized by growth deficiency in the pre- and postnatal periods, facial abnormalities, and CNS dysfunction. Congenital anomalies, such as cardiac and genitourinary abnormalities, are also found in FAS, which has a wide range of severity. There is no known "safe" amount of alcohol consumption in pregnancy, and the risk of FAS increases with increasing consumption. Approximately 2 or 3% of pregnant woman daily consume alcohol in amounts associated with FAS (Abel and Sokol, 1987). Alcohol effects not associated with FAS include mild to moderate growth deficiency, mild neurobehavioral delay, and an increased incidence of congenital anomalies, such as heart, kidney, and skeletal abnormalities. Nutritional deficiency, genetic predisposition, and the use of other substances seem to also play a role in the teratogenic effects of alcohol.

Alcohol abuse is associated with an increased risk of spontaneous abortion, stillbirth, preterm delivery, and low birth weight. Acute maternal withdrawal may compromise uteroplacental blood flow due to maternal catecholamine release. Neonatal depression and symptoms of neonatal withdrawal, including excessive crying, agitation, irritability, and seizures, may result from third-trimester alcohol consumption.

B. Tobacco

Although nicotine is the chief active substance in tobacco, there are many more harmful chemicals found in cigarette smoke, including carbon monoxide. Nicotine stimulates the release of vasoactive catecholamines and peptides. The primary metabolite of nicotine is cotinine, which has a longer half-life.

Fifteen to 30% of women smoke during pregnancy, making nicotine the most common drug used in pregnancy. Nicotine readily crosses the placenta, but an increase in the rate of congenital anomalies has not been observed.

However, women who smoke have an increased risk for spontaneous abortion. Increases in premature rupture of membranes, preterm delivery, placenta previa, abruptio placentae, and low birth weight are also associated with smoking. Smoking cessation early in pregnancy seems to decrease the risk of low birth weight and abruptio placentae to that observed in a nonsmoker. Cigarette smoking during pregnancy and passive smoking exposure to the neonate may increase the risk of sudden infant death syndrome (SIDS). Other neonatal and childhood effects include asthma and frequent upper respiratory tract infections. Although nicotine and cotinine have been measured in the hair of neonates of mothers exposed to passive smoking, the clinical effects remain unknown.

C. Marijuana

The cannabis (marijuana) plant is the source of cannabinoids, of which Δ^9-tetrahydrocannabinol has the greatest activity. Marijuana, hashish, hash oil, and THC are all forms of cannabinoids. Marijuana is used by 3–20% of women during pregnancy (Gilstrap and Little, 1992).

Marijuana use does not result in a withdrawal syndrome, although psychological dependence may occur. Frequent use may cause pulmonary abnormalities in the adult.

Marijuana does cross the placenta but an increase in congenital anomalies has not been reported. Some features of FAS have been observed in the offspring of marijuana users (Qazi *et al.*, 1985), but this finding may be due to the frequent concomitant alcohol use. Because polysubstance abuse is common, it is difficult to distinguish whether perinatal outcomes, such as preterm delivery, low birth weight, or meconium staining of the amniotic fluid, are related to marijuana or other drugs. Nutritional and socioeconomic factors may also affect these outcomes. Interestingly, long-term development seems to be largely unaffected.

D. Amphetamines

Amphetamines are used illicitly as stimulants. This drug has been shown to cross animal placentas, but teratogenicity has not been reported in humans. Pregnancies complicated by amphetamine use are at an increased risk for preterm delivery and perinatal mortality. The potential for intrauterine growth restriction exists as amphetamines markedly reduce uterine blood flow. Neonatal withdrawal symptoms can occur in the neonates of mothers using amphetamines.

E. Cocaine

Cocaine, or benzoylmethylecgonine, is an alkaloid derived from the *Erythroxylon coca* plant. When cocaine is dissolved in hydrochloride, it becomes a water-soluble white powder. It may be ingested orally, intranasally, intravaginally, intravenously, subcutaneously, or smoked. "Crack" or "rock," the alkaloid base form of cocaine, is highly addictive. Cocaine, a CNS stimulant, prevents the reuptake of norepinephrine, epinephrine, dopamine, and serotonin at the nerve terminal. The effects of cocaine include vasoconstriction, hypertension, cardiac arrhythmias, and seizures. Plasma cholinesterases, which are involved in the detoxification of cocaine, are reduced during pregnancy, potentially increasing cocaine's effects. Fetuses also seem to metabolize cocaine slowly.

Approximately 10% of women use cocaine during pregnancy. Pregnant women are at risk for the complications of cocaine use seen in other adults, such as myocardial infarction, arrhythmia, stroke, and seizures. A decrease in uterine blood flow occurs with cocaine use as well as an increase in uterine tone and contractility. This may be related to an increase in catecholamines or oxytocin concentration. Cocaine use is associated with a higher incidence of sexually transmitted diseases.

Cocaine readily crosses the placenta potentially eliciting vasoconstriction within the fetus, affecting the developing organs. Cocaine use has been associated with an increase in congenital anomalies, particularly genitourinary malformations. Abnormalities of the brain, heart, intestines, and limbs have been less clearly linked to cocaine use.

The woman using cocaine during her pregnancy seems to be at an increased risk for abruptio placentae, possibly related to cocaine-induced vasoconstriction and hypertension. Spontaneous abortions, intrauterine growth restriction, fetal death, meconium stained amniotic fluid, fetal intolerance of labor, premature rupture of the membranes, and preterm delivery occur with increased frequency in cocaine-abusing pregnant women. Precipitous deliveries have been frequently observed with cocaine use. The incidence of SIDS and neurobehavioral abnormalities may be increased in the offspring of women using cocaine during pregnancy.

F. Misused Prescription Medicines: Barbiturates and Benzodiazepines

Barbiturates are used medically for their analgesic, sedative, and anticonvulsant properties. Benzodiazepines are also used as sedatives, anticonvulsants, and tranquilizers. Conclusions regarding problems in pregnancy related to misuse of these prescriptions cannot be reliably extrapolated from investigations on their medical use in pregnancy. The amount and duration of exposure may vary and the use of other substances complicates the results.

Barbiturates are reported to cross the placenta of animal models and benzodiazepines are known to cross the human placenta. Examples of misused prescription medications are phenobarbital and valium. Phenobarbital, a barbiturate, in addition to other anticonvulsants, has been associated with an increased risk of congenital anomalies when used to treat seizure disorders. It has not been resolved if this risk is due to the medication, the disorder, the patient's genetic makeup, or a synergistic effect of this combination. The medical use of valium, a benzodiazepine, is not believed to be a teratogen.

The "floppy infant syndrome," neonatal hypotonia and hypothermia, has been described in neonates of mothers taking valium regularly close to delivery. Neonatal sedation and withdrawal can also occur with the use of phenobarbital. Phenobarbital may also result in a hemorrhagic disorder of the newborn due to its effects on the fetal liver.

G. Opiates (Heroin)

Papaver somniferum is the poppy plant from which opiates are derived. The different drugs within this class may be natural, semisynthetic, or synthetic and differ accordingly in their pharmacologic properties. The effects of sedation and euphoria are achieved with the binding of certain specific CNS receptors.

Heroin also causes respiratory depression and physical dependence. Acute maternal withdrawal from heroin, characterized by severe neurologic, cardiovascular, and gastrointestinal symptoms, has been associated with fetal heart rate abnormalities and fetal demise. Heroin users may also experience serious bacterial and viral infections associated with needle use during intravenous administration.

Heroin rapidly crosses the placenta. It has not been reported to be teratogenic in humans. However, several adverse perinatal outcomes are attributed to heroin use during pregnancy. These pregnancies have a higher frequency of spontaneous abortion, abruptio placentae, premature rupture of the membranes, preterm delivery, meconium stained amniotic fluid, and intrauterine growth restriction. Passively acquired heroin results in neonatal respiratory depression, and frequently neonatal withdrawal occurs. SIDS, mild developmental delay, and behavioral problems occur more commonly in the offspring of women who use heroin.

H. Inhalants

Inhalants include benzene, gasoline, toluene, and xylene. Their use during pregnancy is uncommon compared to other substances. Cerebral atrophy, renal dysfunction, seizures, coma, and death may result from inhalant abuse. Inhalants can cross the placenta and a fetal solvent syndrome has been described, presenting as growth restriction and facial and digit malformations (Gilstrap and Little, 1992). Preterm delivery and intrauterine growth restriction are risks with toluene abuse and respiratory problems may be seen in neonates from inhalant-abusing mothers.

I. Lysergic Acid Diethylamide

Lysergic acid diethylamide (LSD) is a chemically synthesized powerful hallucinogen to which rapid tolerance develops. It is not commonly used in pregnancy. Transfer across the placenta is known to occur in animal models. An increase in the frequency of chromosomal breakage has been reported in somatic cells of LSD users (Gilstrap and Little, 1992), but

LSD does not appear to be a human teratogen. Cessation of LSD use does not result in withdrawal symptoms.

J. Phencyclidine

Phencyclidine (PCP), or Angel Dust, may be ingested intranasally, orally, intravenously, or via smoking. It freely crosses the placenta; however, it has no clear association with a specific pattern of fetal anomalies. PCP use during pregnancy seems to increase the risk of intrauterine growth restriction. Symptoms of neonatal withdrawal have been reported.

K. Pentazocine and Tripelennamine (T's and Blues)

Pentazocine tablets, which are narcotics, and tripelennamine, an antihistamine, are crushed, dissolved in water, and injected intravenously for a narcotic-like effect. Placental transfer does occur; however, little information on teratogenicity is known. Perinatal complications associated with the use of this drug during pregnancy include growth restriction, meconium-stained amniotic fluid, stillbirth, and neonatal withdrawal.

III. CLINICAL EVALUATION AND MANAGEMENT

The first step in caring for pregnant women abusing substances is the identification of the problem. Early identification allows the clinician to counsel the patient and institute a multispecialty approach to her treatment. Although a history of drug abuse should be obtained at the first prenatal visit, patients may have many reasons to keep this information from their caregiver. Numerous behavioral, social, and medical characteristics may alert the clinician toward further investigation of the potential for substance abuse. Patients who are abusing drugs may demonstrate jitteriness, depression, or frequently miss prenatal appointments. There may be a history of physical abuse, prostitution, or substance abuse in a family member or partner. Cigarette smoking, poor nutritional status, sexually transmitted diseases, and viral infections such as HIV or hepatitis should raise the index of suspicion for substance use. A precipitous or unplanned delivery outside the hospital, premature labor, abruptio placenta, unexplained fetal growth restriction or death, characteristic congenital anomalies, neonatal withdrawal, or SIDS are obstetric complications that may be clues toward current or past maternal drug use.

Urine toxicology can be used to confirm and follow the patient who reports drug abuse or to detect drug abuse when suspected. Because of the legal implications of drug testing, consent for testing should be obtained. The utility of urine toxicology is limited by timing of last drug use prior to the urine sample. Most drugs may be detected in the urine for at most 72–96 hr after exposure. Meconium or neonatal hair testing allows detection of cocaine use in the latter half of pregnancy. Relationship building with these patients will assist in allowing frequent urine drug screens.

Counseling should provide information regarding the risks of substance abuse to the mother and the developing fetus. Abstinence is recommended and patients should be given referrals to appropriate support groups and/or rehabilitation programs. Reinforcement of being clean is extremely important.

With the exception of heroin, withdrawal is usually recommended in such a way as to minimize fetal and maternal risks. Due to the fetal risks of heroin withdrawal, a program of methadone maintenance during pregnancy has been advocated. Methadone, a synthetic opiate, is administered, potentially preventing dangers of withdrawal and the need to engage in the risky behaviors associated with drug seeking. Benefits include improved compliance with prenatal visits, the opportunity for social and psychiatric intervention, reduced relapse, and improved nutrition. Prolongation of gestation and an increase in birth weight have been observed compared to heroin-complicated pregnancies (Edelin *et al.*, 1988). There is no increased risk of congenital anomalies with methadone; however, methadone maintenance does not eliminate the risk for neonatal withdrawal, decreased birth weight, or SIDS.

A multidisciplinary approach to the care of these patients is important. Members of the team include an obstetrician, psychologist, specialist in chemical dependency, social worker, nutritionist, and pediatrician. Patients should be counseled regarding nutritional needs during pregnancy and the warning signs for preterm labor and abruptio placentae. Maternal screening for sexually transmitted diseases, HIV, and hepatitis is recommended. Fetal screening with a maternal serum α-fetoprotein value and ultrasound examination for structural anomalies is also recommended. Estimation of gestational age as early as possible is important for establishing appropriate fetal growth. Antenatal fetal heart rate testing is recommended if intrauterine growth restriction or unexplained third-trimester bleeding are detected. Lastly, drug rehabilitation should continue beyond pregnancy.

IV. SUMMARY

Substance abuse during pregnancy is a complicated medical and social issue with numerous implications for the health of the mother and fetus, including congenital anomalies, perinatal morbidity and mortality, and abnormal development. Identification of the problem is critical in achieving an optimal perinatal outcome. Comprehensive treatment of pregnant substance-abusing patients can be obtained through a multidisciplinary approach.

See Also the Following Articles

FETAL ALCOHOL SYNDROME; FETAL ANOMALIES; PLACENTAL NUTRIENT TRANSPORT; PRETERM LABOR AND DELIVERY; SEXUALLY TRANSMITTED DISEASES

Bibliography

Abel, E., and Sokol, R. (1987). Incidence of fetal alcohol syndrome and economic impact of FAS-related anomalies. *Drug Alcohol Depend.* **19**, 51–70.

Clarren, S. K., and Smith, D. W. (1978). The fetal alcohol syndrome. *N. Engl. J. Med.* **298**, 1063.

Edelin, K. C., Gurgonious, L., Golar, K., *et al.* (1988). Methadone maintenance in pregnancy: Consequences to care and outcome. *Obstet. Gynecol.* **71**, 399.

Eriksson, M., Larsen, G., and Zetterson, R. (1981). Amphetamine addiction and pregnancy. II. Pregnancy, delivery, and the neonatal period. *Acta Obstet. Gynaecol. Scand.* **60**, 253.

Forster, R. M., and Albright, G. A. (1997). Anesthetic management of the high-risk mother and fetus. In *Gynecology and Obstetrics* (J. J. Sciarra, Ed.), Vol. 3. Lippincott-Raven, Philadelphia.

Gilstrap, L. C., and Little, B. B. (Eds.) (1992). *Drugs and Pregnancy*. Elsevier, New York.

Lambers, D. S., and Clark, K. E. (1996). The maternal and fetal physiologic effects of nicotine. *Sem. Perinatol.* **20**, 115–126.

Lutiger, B., Graham, K., Einarson, T. R., and Koren, G. (1991). Relationship between gestational cocaine use and pregnancy outcome: A meta-analysis. *Teratology* **44**, 405–414.

Macgregor, S. N. (1997). Substance abuse in pregnancy: Obstetric management. In *Gynecology and Obstetrics* (J. J. Sciarra, Ed.), Vol. 2. Lippincott-Raven, Philadelphia.

Morbidity and Mortality Weekly Report (1997). Alcohol consumption among pregnant and childbearing-aged women—United States, 1991–1995. *Morbidity Mortality Weekly Rep.* **46**, 346.

Qazi, Q. H., Mariano, E., Milman, D. H., Beller, E., *et al.* (1985). Abnormalities in offspring associated with prenatal marijuana exposure. *Dev. Pharmacol. Ther.* **8**, 141.

Walsh, R. A. (1994). Effects of maternal smoking on adverse outcomes: Examination of the criteria of causation. *Hum. Biol.* **66**, 1059–1092.

Suckling Behavior

Edward O. Price
University of California, Davis

GLOSSARY

altricial Offspring born when in a relatively undeveloped stage of maturation in terms of mobility and sensory capacity.

cross-suckling or ***intersuckling*** Sucking by one animal on the body parts of another.

nursing behavior Maternal behaviors that facilitate suckling by young mammals.

precocious Offspring born in a relatively developed stage of maturation in terms of mobility and sensory capacity.

suckling behavior Behavior associated with the extraction of milk from the mammary gland of lactating female mammals.

teat order An established order of teat preferences among the young of a litter of mammals.

weaning Termination of nursing behavior.

Suckling behavior concerns the process by which young mammals extract milk from the mammary gland of lactating females, usually their mother (dam). The term nursing refers to the mother's role in this process. Milk is extracted from the dam by creating a vacuum (negative pressure) around the nipple. In some ungulate mammals (e.g., cattle), milk is extracted both by suction and by squeezing milk out of the teat cistern. The latter is accomplished by compressing the neck of the teat between the tongue and hard palate and using the tongue to squeeze the teat from neck to tip. When the pressure on the teat is released, the teat cistern fills up again.

I. DEVELOPMENT OF SUCKLING BEHAVIOR

Suckling is considered an innate behavior; it is fully developed at the time of birth. This is not to say that it suddenly appears at birth. Human fetuses are known to suck their thumbs while *in utero*. The healthy newborn mammal is highly motivated to suckle and will do so at the first opportunity. If suckling is denied (by feeding the animal through a stomach tube) the motivation to suckle will eventually disappear. Experiments with stomach-fed kittens indicate that they lose their motivation to suckle in about 22 days.

II. STIMULI ELICITING SUCKLING BEHAVIOR

Tactile stimulation of the head (forehead, cheek, lips, and nose) readily elicits nipple-seeking and suckling behavior. Warm surfaces elicit a stronger response than cooler surfaces. Young born in a relatively developed state (precocious) such as common farm animals are attracted to any large prominent object in their immediate environment particularly if that object has motion and utters sounds. In ruminants, once the neonate contacts its dam, it will follow a thermal gradient (cold to warm) with its nose. The inguinal region (groin) of the mother is attractive to the newborn because of its smooth texture, warmth, and the odor of a wax produced by the inguinal gland. Exploration of this area with its

FIGURE 1 A neonatal lamb contacts the inguinal region of its dam shortly after birth. The higher temperature and surface texture (smoothness) of this region and the odor of the inguinal gland at the base of the udder are attractive to the newborn.

head (Fig. 1) results in discovery of the mammary gland (udder) and nipples (teats). Softness (i.e., surface yield) of the udder is also attractive to the young neonate.

The dam can assist its offspring in finding teats by positioning herself to facilitate contact by the neonate and by standing immobile while the offspring explores. Experienced mothers are better at assisting the neonate than first-time mothers. The newborn ungulate may require a few minutes to several hours to attain its first suckling bout. Finding a teat takes longer if the udder is pendulous as in many dairy cows.

Mammalian young born in a relatively helpless state (altricial) are often born in a nest in a den or confined area. The mother positions herself over the nest and young to facilitate suckling. Again, odors and warmth stimulate the young to attain nipple attachment and commence suckling.

Although the suckling reflex is considered innate, the young mammal must learn the characteristics of its mother and the sensory cues signaling the mother's intent to nurse. Precocious young will not only suckle their own mothers but also initially attempt to suckle other adult females. Other females, however, will actively avoid or exhibit aggression toward alien young. Eventually, the young learn to only approach their own mothers for suckling. Nursing bouts in the sow are signaled by the initiation of rhythmical grunting. Piglets soon learn to associate her grunting with a suckling opportunity and will position themselves on nipples well before milk ejection occurs. For altricial young (e.g., puppies and kittens), suckling opportunities prior to eye opening are signaled by a variety of nonvisual cues (e.g., auditory, olfactory, tactile, and thermal). Suckling may be socially facilitated in species bearing multiple offspring. Suckling by one young will encourage others to do likewise.

Newborn ungulates will quickly learn to drink milk or water from a pail to which an artificial nipple or teat has been attached (Fig. 2). They can also be taught to drink directly from a pail by encouraging them to suck on one or two fingers while lowering their head into the pail (Fig. 3). Lactating females of other species may be used as surrogate mothers (Fig. 4).

III. IMPORTANCE OF COLOSTRUM

It is important that the newborn mammal engages in suckling very soon after birth to obtain colostrum ("first" milk). Colostrum contains protective antibodies from the mother and maximum absorption of these antibodies by the neonate is achieved in the first few hours following birth. The greater disease resistance attained by colostrum consumption is achieved even if it is obtained from the female of a different species and fed artificially. There is some evidence in farm animals that artificially fed colos-

FIGURE 2 Young Hereford calves are fed milk in pails equipped with artificial nipples. The calves will continue to suck the nipples for a while after the milk is consumed, thus increasing their secretion of digestive hormones.

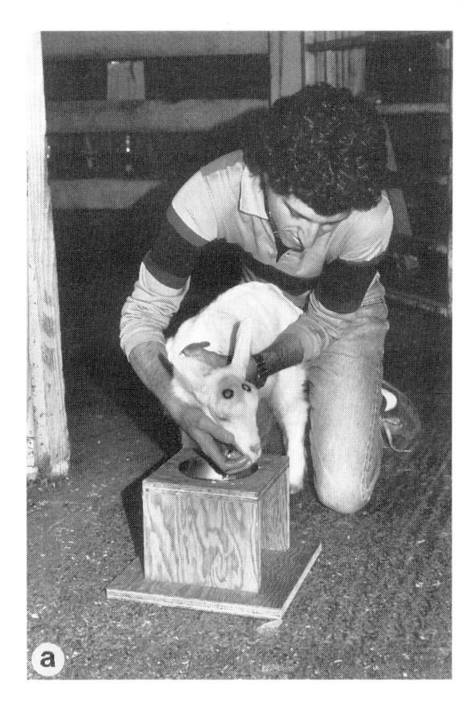

FIGURE 3 (a) A young kid goat is taught to drink from a bowl mounted in a stand. The handler puts his fingers in the kid's mouth, which stimulates sucking. (b) The kid's head is lowered into the bowl while sucking on the handler's fingers. Once the mouth is submersed, the fingers are slowly withdrawn from the kid's mouth. Several repetitions may be necessary before the kid drinks on its own.

trum is of greater benefit if the neonate is allowed to remain with its dam even when it cannot suckle her.

IV. TEAT PREFERENCES

Young mammals may develop teat preferences particularly if they are part of a litter. The "teat order" of young piglets is based on teat preferences established within the first few days following birth (Fig. 5). Preferred teats are defended against littermates and if each piglet stays with its own teat, the efficiency of nursing bouts is maximized. The sow nurses her piglets every 45–90 min and milk ejection normally lasts about 30 sec. Hence, it is important that her piglets be present on a nipple when milk ejection

FIGURE 4 A Jersey dairy cow is used as a surrogate mother for a group of young lambs.

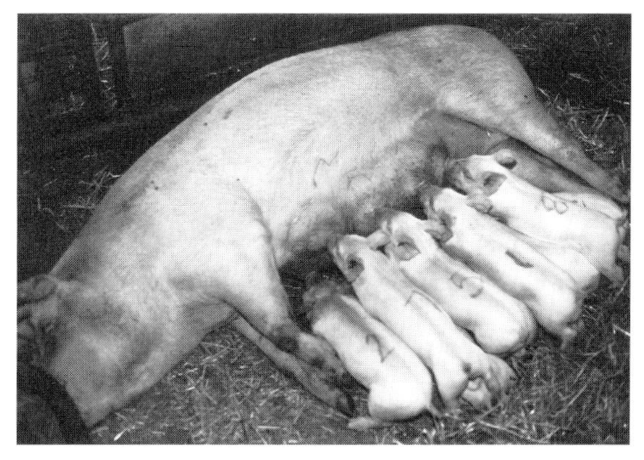

FIGURE 5 A sow nurses her litter. Each piglet has a preferred teat, which is identified by its unique odor.

begins. A feeding opportunity could easily be missed if littermates were constantly fighting over teat positions. Once the teat order has been established, contests over teat rights become relatively infrequent. Piglets identify their preferred teats by position on the sow and by salivary and other body odors left behind from previous suckling bouts. If the nipples of the sow are washed between suckling bouts, the piglets become confused and may fail to suckle for several nursing bouts. Young kittens will also form teat preferences and preferred teats are identified from olfactory cues.

V. SUCKLING FREQUENCY

In precocious young, the duration of suckling bouts and the intervals between bouts vary considerably with species and age of offspring. Newly born young tend to suckle more frequently and for shorter durations than older young. For newly born altricial young, the dam may nurse her offspring many times in a 24-hr period. However, the female rabbit nurses her relatively helpless neonates only once per day for just a few minutes.

Although milk ejection in the sow is very brief, her piglets will suck their preferred teats long before and after milk ejection. The motivation to suckle will result in nonnutritive sucking when the animal's milk is delivered quickly through artificial feeding (hand rearing). Studies with dairy calves have shown that nonnutritive sucking is stimulated by the taste of milk, an effect that lasts about 10 min whether milk is consumed or not. Water has no such effect. Furthermore, it was found that the time taken to ingest a meal is more important than the quantity consumed in determining the duration of nonnutritive sucking. These facts suggest that neither stomach distension nor the oral sensations from ingesting milk immediately inhibit nonnutritive sucking in calves. Sucking, whether nutritive or nonnutritive, facilitates the secretion of digestive hormones and leads to an increase in insulin and cholecystokinin which are involved in the satiety response.

The young of ungulates will frequently butt the mother's udder while sucking and when the milk supply becomes depleted. Such massaging of the udder is believed to facilitate the ejection of residual milk. It may also have a positive effect on milk production.

Young mammals will continue to suckle their dam even after they are eating substantial amounts of solid foods. Eventually, the dam will wean (terminate suckling behavior) her young by actively avoiding suckling attempts and by walking away from her offspring after a few seconds of nursing. In some species of mammals, weaning may not occur until the dam prepares to give birth in the next reproductive cycle.

VI. ATYPICAL SUCKLING BEHAVIOR

When reared communally, nonnutritive sucking may be manifested by sucking on penmate's mouths, ears, prepuce, umbilical region, or other body parts (Fig. 6), a phenomenon sometimes referred to as cross-sucking or intersucking. Intersucking can cause skin inflammation and other health problems.

Offspring of domestic ruminants (cattle, sheep, and goats) that have not been accepted by their mothers will learn to suckle from between her hindlegs while a second offspring is suckling in a more normal lateral position at a right angle to the mother's body (Fig. 7). Young not getting enough milk from their mothers may attempt to steal milk from other lactating females by this same rear orientation when the female's natural young is suckling. Young ungulates

FIGURE 6 Two calves engage in intersucking after being fed milk.

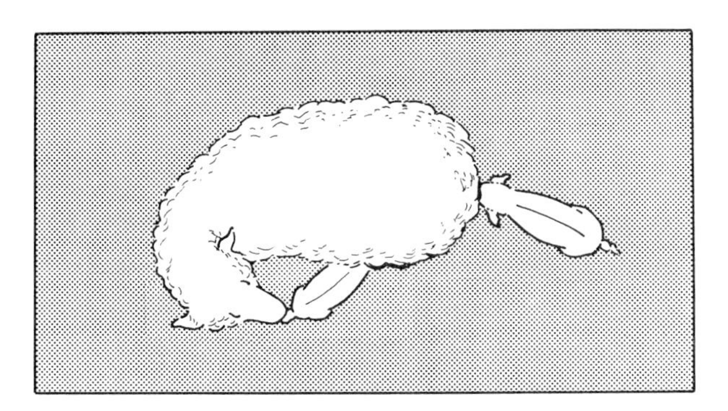

FIGURE 7 Rear-oriented suckling is commonly seen in young ungulates (e.g., sheep) that have not been accepted by their dam or that are stealing milk from another female. The dam identifies the second offspring as her own by olfactory cues but cannot smell or see the young behind her.

FIGURE 8 This young twin lamb with feces on its head was not accepted by its mother and used a rear-oriented approach when suckling. Its body weight at weaning was in the normal range.

with feces on their heads is a likely sign of maternal nonacceptance or milk stealing (Fig. 8).

See Also the Following Articles

LACTOGENESIS; MAMMARY GLAND, OVERVIEW; MILK, COMPOSITION AND SYNTHESIS; MILK EJECTION; PARENTAL BEHAVIOR, MAMMALS

Bibliography

Billing, A. E., and Vince, M. (1987a). Teat-seeking behaviour in newborn lambs. I. Evidence for the influence of maternal skin temperature. *Appl. Anim. Behav. Sci.* **18**, 301–313.

Billing, A. E., and Vince, M. (1987b). Teat-seeking behaviour in newborn lambs. II. Evidence for the influence of the dam's surface textures and degree of surface yield. *Appl. Anim. Behav. Sci.* **18**, 315–325.

de Passille, A. M., and Rushen, J. (1997). Motivational and physiological analysis of the causes and consequences of non-nutritive sucking by calves. *Appl. Anim. Behav. Sci.* **53**, 15–32.

de Passille, A. M., Rushen, J., and Janzen, M. (1997). Some aspects of milk that elicit non-nutritive sucking in the calf. *Appl. Anim. Behav. Sci.* **53**, 167–174.

Fraser, A. F., and Broom, D. M. (1990). *Farm Animal Behaviour and Welfare,* 3rd ed. Bailliere Tindall, London.

Gonyou, H., and Stookey, J. (1987). Maternal and neonatal behavior. In *Food Anim. Practice* **3**, 231–249.

Rushen, J., and de Passille, A. M. (1995). The motivation of non-nutritive sucking in calves, *Bos taurus*. *Anim. Behav.* **49**, 1503–1510.

Wolff, P. H. (1968). Sucking patterns in infant mammals. *Brain Behav. Evol.* **1**, 354–367.

Suprachiasmatic Nucleus

Robert Y. Moore

University of Pittsburgh

GLOSSARY

circadian rhythm A cyclic biological event which has a period of about 24 hr.

entrainment A process that precisely sets the period of a pacemaker, or a rhythm. For circadian rhythms the light–dark cycle typically entrains the pacemaker.

melatonin A hormone produced by the pineal gland that modulates pacemaker function and controls reproduction in seasonal breeders.

pacemaker A set of neurons that establishes a biological rhythm with a predictable period.

pineal gland A neural derivative of the posterior diencephalic roof that is under the control of the suprachiasmatic nucleus to produce a circadian rhythm in melatonin secretion. The pineal gland may be a circadian pacemaker in some animals, e.g., birds.

suprachiasmatic nucleus A hypothalamic nucleus lying dorsal to the optic chiasm and lateral to the third ventricle that acts as a pacemaker for circadian rhythms in mammals.

The suprachiasmatic nucleus (SCN) of the hypothalamus consists of distinctive, paired nuclei lying dorsal to the optic chiasm and lateral to the third ventricle. The SCN is derived from the ventral hypothalamic germinal epithelium in late gestation as a component of the periventricular nuclear group. It is present and well developed in all mammals, including humans. Although the critical role of the SCN in circadian function is a relatively recent finding, the SCN has been known to be important in the regulation of reproduction for more than 50 years.

In the interval surrounding the discovery of the hypothalamic control of the anterior pituitary, approximately 1940–1960, it was found that hypothalamic lesions could cause distinctive types of gonadal syndromes. Lesions destroying the median eminence resulted in anestrus with uterine and ovarian atrophy, reflecting a loss of hypothalamic drive of the anterior pituitary. In contrast, lesions placed more anteriorly in the midline, in the region of the optic chiasm, resulted in persistent vaginal estrus with polyfollicular ovaries. An elegant study by Hillarp first associated this syndrome with the suprachiasmatic nucleus (SCN) and this conclusion was confirmed and extended subsequently by Brown-Grant and Raisman. Although there was substantial speculation about the mechanism of action of such lesions, it was not until the role of the SCN in circadian function was recognized, and we understood that the estrous cycle in polyestrous animals is made up of a series of circadian cycles, that the syndrome was explained. The significance of circadian control was presaged by the classic study of Everett and Sawyer showing that barbiturate blockade of the luteinizing hormone (LH) surge blocks ovulation for exactly 24 hr. Subsequent studies demonstrated that the daily proestrus-like LH surges occurring in ovariectomized rats primed with estrogen are abolished by SCN lesions which do not affect gonadotropin-releasing hormone (GnRH) levels or pulsatile patterns of LH secretion.

These findings clarify an important point: The effects of SCN lesions on reproduction are purely on its circadian regulation, not on pulsatile LH release or on any of the fundamental neuroendocrine mechanisms. From this we derive a series of pertinent questions regarding the SCN and reproduction: What is the evidence that the SCN is a circadian pacemaker? What is the functional organization of the SCN? How is a circadian signal transmitted from the SCN to the hypothalamic GnRH system and integrated with other inputs? and What is the role of the SCN in seasonal reproduction?

I. THE SCN IS A CIRCADIAN PACEMAKER

Four lines of evidence demonstrate the critical role of the SCN in the generation and regulation of circadian rhythms in mammals. First, the SCN is the major site of termination of a visual pathway crucial for the conveyance of photic information from the retina to the circadian system. The retinohypothalamic tract (RHT) arises from a discrete subset of type III retinal ganglion cells and terminates in the core of the SCN. Transection of all central retinal projections beyond the optic chiasm results in a loss of visual discrimination and visual reflexes but no alteration of circadian function. In contrast, transection of the RHT at its entry into the SCN, but sparing the SCN, does not affect visual discrimination or reflexes but abolishes entrainment. Thus, the RHT is a central retinal projection which transmits the entraining effects of photic information to the SCN but is not involved in other visual functions.

The second piece of evidence that the SCN is a circadian pacemaker comes from an analysis of its neuronal firing patterns and metabolism over time. Both *in vivo* and *in vitro,* the SCN shows a high firing rate and high glucose utilization during subjective day and a low firing rate and glucose utilization during subjective night. Since this occurs in the SCN even when isolated from the rest of the brain, it indicates that the SCN is capable of maintaining circadian function independent of other brain structures.

The third line of evidence is a large number of studies showing that ablation of the SCN results in a loss of behavioral, hormonal, and physiological circadian rhythms. Animals subjected to SCN lesions are arrhythmic in the expression of these functions and the loss of rhythmicity is permanent. Furthermore, when the lesions are made at 2 days of age, before the initiation of any circadian function beyond that expressed in the SCN, the animals fail to develop circadian rhythms throughout their lives. These animals also fail to develop either estrous cycles or daily LH surges.

Fourth, transplantation of fetal SCN into the brains of animals rendered arrhythmic by SCN lesions restores circadian rhythmicity in locomotor activity and drinking and, presumably, in sleep–wake behavior. This effect of transplantation is dependent on the presence of SCN in the transplants and the transplanted SCN drives circadian function in the host brain. Taken together, these lines of evidence establish the SCN as the principal circadian pacemaker in the mammalian brain.

II. FUNCTIONAL ORGANIZATION OF THE SCN

The SCN is made up of paired nuclei, one on each side of the hypothalamus. Each SCN has two functional subdivisions, a core which receives direct and indirect photic input and a shell which receives nonphotic input from other brain areas. The core contains neurons that are slightly larger than those in the shell and which produce the peptides, vasoactive intestinal polypeptide and gastrin-releasing peptide. Shell neurons produce vasopressin (Fig. 1). Although core and shell neurons are characterized by a different peptide phenotype, all SCN neurons appear to produce the small molecule neurotransmitter, GABA. Thus, the output of the SCN should be inhibitory on areas innervated. The neurons of the SCN project predominantly to hypothalamus. Within hypothalamus, the densest projections are to the area directly dorsal and caudal to the SCN, the subparaventricular zone. Other projections extend rostrally to the medial preoptic area and anterior hypothalamic area and caudally to the retrochiasmatic area, dorsomedial hypothalamic nucleus, and the ventral tuberal area. Since the projections of the subparaventricular zone essentially overlap those of the SCN, the output of the circadian system appears quite restricted.

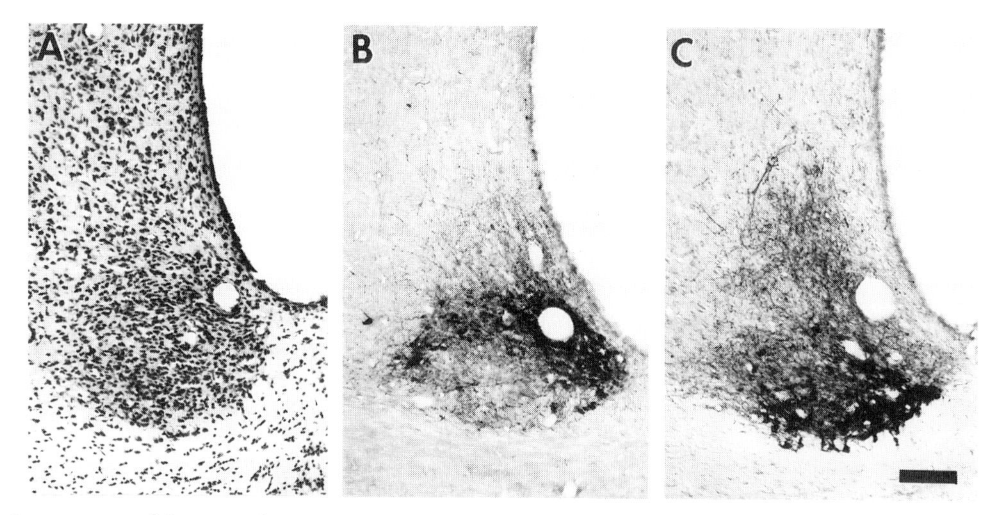

FIGURE 1 Organization of the suprachiasmatic nucleus (SCN). (A) The SCN is a compact cell group dorsal to the optic chiasm and lateral to the third ventricle (Nissl stain). (B) Vasopressin neurons in the SCN shell. (C) Vasoactive intestinal polypeptide neurons in the SCN core. Scale bar =100 μm.

III. COUPLING OF THE SCN TO THE REPRODUCTIVE SYSTEM

The GnRH system in rodents is composed of neurons largely located in the medial preoptic area, in contrast to primates in which the predominant population is in the arcuate nucleus. GnRH neurons project to the median eminence (Fig. 2), and the control of GnRH secretion is both hormonal and neural. GnRH neurons do not have steroid receptors and, hence, must receive hormonal information from neurons that do receive such inputs. Although GnRH neurons have a relatively sparse supply of afferents, neural inputs come from a number of sources, particularly from the preoptic area. Recent work has also shown a direct input from the vasoactive intestinal polypeptide neurons of the SCN to the GnRH neurons. Axons of SCN neurons come into direct synaptic contact with GnRH neurons in the medial and lateral preoptic area. These axons exit the SCN rostrally, traverse the anterior hypothalamic area ventrally, and extend into the medial and ventral preoptic area and the lateral preoptic area, where they come into contact with a large proportion of the total population of GnRH neurons.

How does the circadian system, specifically the SCN, affect GnRH release? The LH surge resulting in ovulation and daily LH surges in ovariectomized rats treated with estrogen occur at the light–dark transition in animals maintained in a light–dark cycle. Shifting of the light–dark cycle shifts the timing of the LH surge, a mechanism mediated by the SCN. Vasoactive intestinal polypeptide inhibits the LH surge. Since that peptide is colocalized with GABA in SCN neurons, and colocalized peptides typically have an action that prolongs or enhances that of the colocalized small molecule transmitter, SCN input on GnRH neurons inhibits GnRH release. This is in accord with the daily profile of firing of SCN neurons. The firing rate of these neurons is lowest at the middle of the dark period. It increases during the late dark period and into the light period to reach a peak at the middle of the light period (Fig. 3). The firing rate then decreases and this rhythm persists on a circadian basis. It is reasonable to assume that the downslope of the firing rate rhythm releases the GnRH neuron from a tonic inhibition by SCN neurons. This would allow excitatory inputs from other areas, including steroid responsive neurons, to produce a phasic firing of GnRH neurons with release of GnRH into the portal circulation to affect pituitary gonadotropes to produce an LH surge. The rhythmic tonic inhibition of the GnRH neurons by the SCN appears critical to the process. If the SCN is ablated, or its activity disrupted by exposure to constant light, the LH surge is lost and animals exhibit the constant

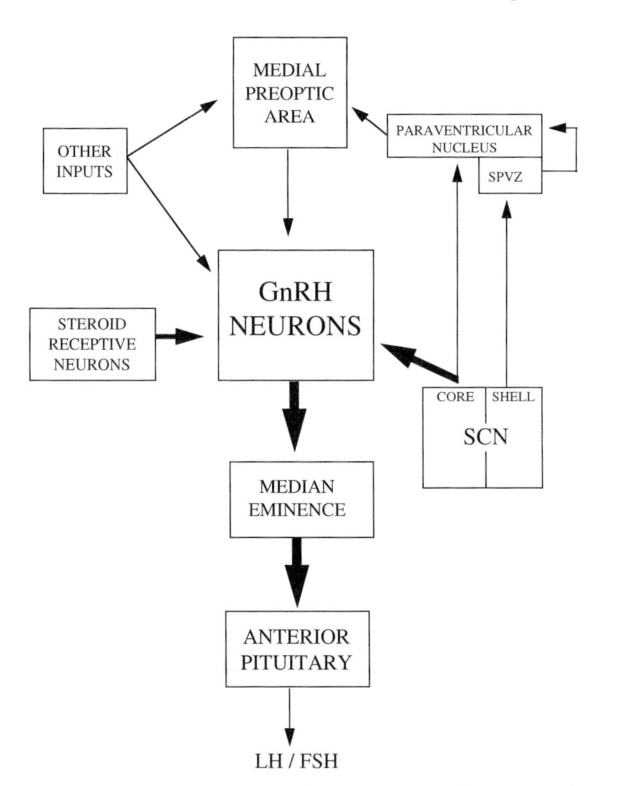

FIGURE 2 Neural control of reproductive function. This diagram shows a scheme of the neural pathways involved in control of the GnRH neurons projecting on the median eminence.

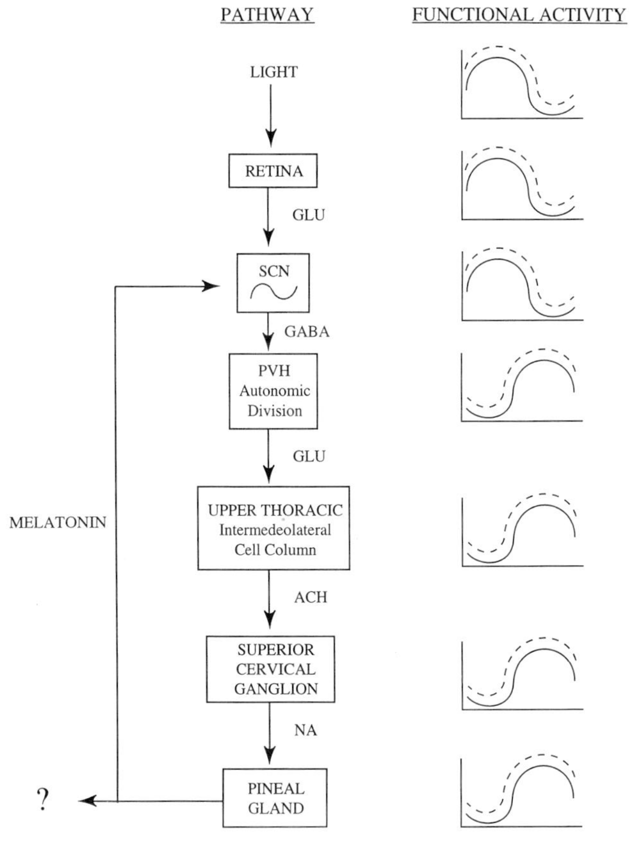

FIGURE 3 Neural control of pineal function. (Left) The pathway controlling melatonin production in the pineal; light entraining the SCN pacemaker through photic activation of the retina, SCN projecting on the autonomic division of the paraventricular nucleus, paraventricular nucleus projecting to the intermediolateral cell column, which projects to the cells in the superior cervical ganglion projecting to the pineal. (Right) Representation of the functional activity of each component of the pathway in short days (solid line) and long days (dashed line). Melatonin feeds back on the SCN. It has other effects which are not completely understood, hence the question mark.

estrus, polyfollicular ovary syndrome. Thus, circadian regulation is a crucial aspect of reproductive function to the female rat. In the male rat, the same circadian pattern of input to the GnRH system occurs. It does not affect reproductive function because of the organization of the male reproductive axis. As noted previously, GnRH neurons in the primate brain are concentrated in the arcuate nucleus and work by Knobil and colleagues and others has shown that they are not under circadian regulation. Photoperiodic regulation of reproduction is another phenomenon requiring SCN input.

IV. THE SCN AND PHOTOPERIODIC CONTROL OF REPRODUCTION

One of the most interesting facets of adaptation in reproductive function is seasonal breeding. Animals may breed preferentially in the spring or fall depending on the length of uterine gestation and the time required for young to become self-sufficient. Typically, animals that breed preferentially in the fall have a long gestation and the young require several months to mature. Sheep are exemplary of this group. In contrast, animals that breed in the spring typically have a short gestation period and the young mature rapidly. A number of rodents follow this pattern. The signal for seasonal reproduction is day length. For spring breeders, lengthening days initiate reproductive function, whereas for fall breeders decreasing day length initiates the reproductive function. How does this work? What are the pathways and mechanisms by which photic information is utilized to control reproduction?

The first insight into these problems came from the work of Russel Reiter and colleagues investigating the role of the pineal in seasonal reproduction in the Syrian hamster. Their studies showed that a short photoperiod resulted in marked testicular regression in the male hamster with loss of spermatogenesis. This effect was totally blocked by pinealectomy, demonstrating that the pineal was essential to the effect of short photoperiod on reproduction. Subsequent studies demonstrated that the SCN was a critical part of the pathways controlling pineal function. The pineal produces melatonin in a circadian pattern with high levels of secretion at night and low levels during the day. In addition, light exposure at night results in a rapid fall in melatonin production and secretion. The circadian rhythm in melatonin production is generated by the SCN. Although the intrinsic period of the rhythm in SCN neuron firing rate is determined by molecular and cellular events, the pooled rhythm of all SCN neurons appears to be environmentally determined. For example, the phase of the SCN firing rate rhythm is established by the entraining effects of the light–dark cycle. Although this has not been definitively shown, it is likely that the length of the peak of the firing rate rhythm is set by day length (Fig. 3). Under constant conditions (e.g., constant darkness), the area-under-the-curve of the peak of the firing rate rhythm will be maximal. As day length increases, the onset of the peak will be delayed and the effect advanced. The SCN projects directly to the autonomic division of the hypothalamic paraventricular nucleus, which we assume has a tonic firing pattern that is inhibited when the SCN is firing at a high rate. Thus, the paraventricular nucleus will show the inverse of the SCN firing pattern, and this pattern will be maintained through the pathway to the pineal (Fig. 3).

During short days, the SCN firing peak will be short and components of the paraventricular-to-pineal circuit will have a long peak, including the period of melatonin production. Prolonged melatonin secretion in the hamster inhibits reproduction. During long days, the SCN firing peak is long, the melatonin peak is short, and reproductive function is maximized in the male hamster, a long-day breeder (Fig. 3). The sequence of events is exactly the same in short-day breeders such as sheep, but melatonin activates reproductive function in these species. Melatonin also feeds back on the SCN as one component of pacemaker regulation.

V. SUMMARY: CIRCADIAN CONTROL OF REPRODUCTION

Reproductive function is under circadian control in those animals in whom the timing of the gonadotrophin system is precisely set in a circadian pattern by the SCN. In this situation, a complex array of neural and endocrine signals is orchestrated by SCN output. These events presumably provide an adaptive advantage by precisely timing mating and the birth of young. Such adaptive mechanisms are used widely in living organisms as divergent as insects (*Drosophila*) and mammals. In higher mammals, particularly primates, the circadian control of reproduction is lost in favor of longer period rhythms. The circadian mechanism of reproductive regulation is also expressed in seasonal breeders. Here, the circadian clock, the SCN, is used both to provide a circadian signal to produce the rhythm in melatonin secretion and to modulate that rhythm by photoperiodic time (day length) measurement. Together, these account for important aspects of adaptation to the solar cycle in the control of reproduction.

See Also the Following Articles

Circadian Rhythms; Cricetidae (Hamsters and Lemmings); GnRH (Gonadotropin-Releasing Hormone; Hypothalamic–Hypophysial Complex; LH (Luteinizing Hormone); Rodentia; Seasonal Reproduction, Mammals

Bibliography

Brown-Grant, K., and Raisman, G. (1977). Abnormalities in reproductive function associated with destruction of the suprachiasmatic nuclei in female rates. *Proc. R. Soc. London* **198**, 129–296.

Klein, D. C., Moore, R. Y., and Reppert, S. M. (Eds.) (1991). *The Suprachiasmatic Nucleus: The Mind's Clock.* Oxford Univ. Press, New York.

Ma, Y. J., Kelly, M. J., and Ronnokliev, O. K. (1990). Pro-gonadotrophin-releasing hormone (pro-GnRh) and GnRH content in the preoptic area and basal hypothalamus of anterior medial preoptic nucleus/suprachiasmatic nucleus lesioned rats. *Endocrinology* **127**, 2654–2664.

Meijer, J. H., and Rietveld, W. J. (1989). The neurophysiology of the suprachiasmatic circadian pacemaker in rodents. *Physiol. Rev.* **69**, 671–702.

Moore, R. Y. (1996a). Neural control of the pineal gland. *Behav. Brain Res.* **73**, 125–130.

Moore, R. Y. (1996b). Entrainment pathways and the functional organization of the circadian system. *Prog. Brain Res.* **111**, 103–119.

Moore, R. Y., and Speh, J. C. (1993). GABA is the principal neurotransmitter in the mammalian circadian system. *Neurosci. Lett.* **150**, 112–116.

Mosko, S. S., and Moore, R. Y. (1979). Neonatal suprachiasmatic nucleus lesions: Effects on the development of circadian rhythms in the rat. *Brain Res.* **164**, 17–38.

van der Beek, E., Weigant, V. M., van der Donk, H., van den Hurk, R., and Buijs, R. M. (1993). Lesions of the suprachiasmatic nucleus indicate the presence of a direct vasoactive intestinal polypeptide-containing projection to gonadotropin-releasing hormone neurons in the female rat. *J. Neuroendocrinol.* **5**, 137–144.

Surfactant

Aron B. Fisher

University of Pennsylvania School of Medicine

GLOSSARY

alveoli The smallest air sacs of the lung where gas exchange occurs.

amphipathic Describing a molecule composed of hydrophobic and hydrophilic domains. In phosphatidylcholine, the acyl chains comprise the hydrophobic domain and the phosphorylcholine is hydrophilic.

hypophase The aqueous layer in the alveolar space between the surfactant monolayer and the epithelial cell membrane.

lamellar bodies The membrane-bound subcellular organelles that are the intracellular storage sites of lung surfactant.

lung lavage The procedure of washing of the lung alveoli by instilling saline or other liquid via the trachea.

phosphatidylcholine A phospholipid composed of glycerophosphate, choline, and two fatty acyl chains.

phospholipase A_2 An enzyme that degrades phospholipids by removing the fatty acid attached to the second carbon position.

type II alveolar epithelial cell The cuboidal cell in the lung alveolus that is responsible for lung surfactant production. Along with the thin, membranous type I alveolar epithelial cell, it forms the cellular lining on the air side of the alveolus.

Lung surfactant is a complex mixture of lipids and proteins that is secreted by the lung epithelium into the alveolar air spaces. It is required to maintain patency of lung alveoli by promoting a low intraalveolar surface tension.

I. LUNG SURFACTANT OVERVIEW

The presence of surfactant facilitates lung gas exchange and is essential for life. However, its existence was only suggested (by von Neergard) in the 1920s and was not confirmed until the studies of Pattle, Clements, Radford, and others in the late 1950s. Why is surfactant necessary? Formerly, the alveoli of the lung were regarded as air sacs lined by lung epithelial cells in direct contact with the air phase. It was assumed that alveolar stability is maintained by the inherent mechanical properties of the alveolar septum. It is now known that the epithelial cell lining of the alveolus is bathed by an extracellular water layer. This water layer is extremely thin so as not to interfere with gas exchange. It is the presence of this water layer and the surface tension that results from the air–water interface that leads to the requirement for lung surfactant.

II. SURFACE TENSION

A. LaPlace Relationship

Surface tension is the result of attractive forces between water molecules. This effect can be conceptualized by considering the forces acting on two hypothetical water molecules: one in the midpart of an aqueous solution and another at the air–water interface. The former molecule would be subjected to attractive forces that are evenly distributed and thus there is no net force for vectorial displacement. In contrast, the water molecule at the interface is subjected to uneven attractive forces with a net force that tends to prevent it from leaving the interface. The summation of these attractive forces on water molecules at the interface is called surface tension. A common example of surface tension at work is a soap bubble in which the surface tension in the wall leads to progressive contraction until it bursts because of the rising internal pressure.

The magnitude of surface tension in an enclosed spherical structure such as a soap bubble can be calculated from the LaPlace equation, which states that the internal pressure is proportional to the surface tension and inversely proportional to the radius of the sphere:

$$P = 2T/r,$$

where T is the surface tension, P is the internal pressure, and r is the radius of the alveolus. Surface tension is usually expressed in units of dynes/centimeter or Newtons/meter.

B. Measurement of Surface Tension

There are many ways to measure surface tension *in vitro*. A popular method has been the Wilhelmy balance, in which a thin plate is placed in a liquid with an air interface and the force (pressure) required to move the plate and just overcome the adhesive forces at the interface is measured. This has been combined with a Longmuir trough that allows contraction and expansion of the surface exposed to air in order to measure the relationship between surface tension and surface area. An example of surface tension measured by this method is shown in Fig. 1. Another method, called the pulsating bubble tech-

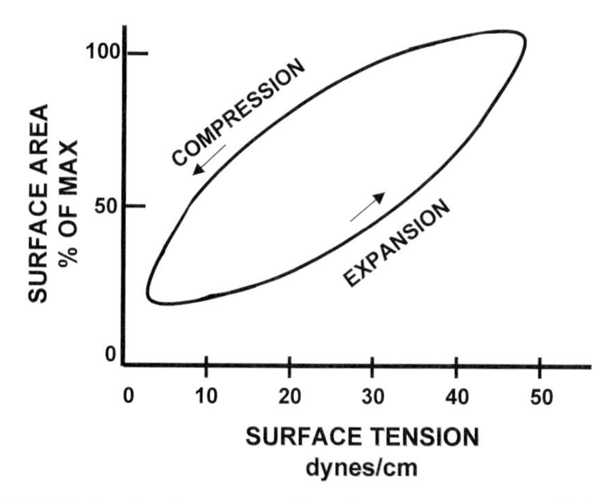

FIGURE 1 *In vitro* assay of surfactant using a Wilhelmy balance and Longmuir trough. The surfactant sample is placed on the surface of the water and surface tension is measured as a function of surface area. The surface area is varied from 100% to approximately 20% of the maximal surface area and the force just necessary to overcome surface tension is continuously measured. The surface tension decreases during the compression phase (as surface area is decreased) and increases during the expansion phase (as surface area is increased).

nique, calculates surface tension from direct measurements of pressure and radius of a bubble observed microscopically. Pure water has a surface tension of 70 dyn/cm at body temperature (37°C).

C. What Is a Surfactant?

A surfactant is an agent that acts at the air–liquid interface to lower the surface tension. Common examples include soap and household detergents. A surface-active substance (surfactant) is generally an amphipathic molecule in which the hydrophilic portion is anchored in the aqueous phase, whereas the hydrophobic portion is excluded and is positioned at the interface. Thus, the presence of a surfactant at the interface dilutes the concentration of water molecules, therefore decreasing the intermolecular attractions and diminishing the surface tension. The degree of surface tension lowering is a function of a particular surfactant, representing its ability to insert into the air–liquid interface. Lung surfactant is particularly effective and can lower the surface tension *in vitro* from the 70 dyn/cm of pure water to less than 5 dyn/cm when maximally compressed as shown in Fig. 1. For comparison, Tween 80, a commonly used laboratory detergent, when added to water will produce a surface tension of approximately 25 dyn/cm.

D. Surface Tension in the Lung

Assuming spherical geometry for a typical lung alveolus, surface tension can be estimated using the LaPlace relationship. At the resting lung volume, the average radius for a lung alveolus is approximately 5×10^{-3} cm and the normal elastic recoil pressure gradient is about 4 cm H_2O. Approximately one-half of the recoil pressure is related to surface tension (the other half is due to interstitial collagen and elastic fibers). Therefore, the calculated surface tension is 5 dyn/cm (remembering that 1 cm H_2O = 980 dyn/cm^2). Experimental study has provided evidence in support of this low value for surface tension, although definitive proof has not yet been obtained in the intact lung. The low value for surface tension in the lung indicates the presence of a surfactant.

III. LUNG SURFACTANT *IN VITRO*

A. Isolation of Lung Surfactant

Surfactant can be isolated from lungs in an intracellular or extracellular (secreted) form. The intracellular form is obtained by isolation of the secretory organelles (lamellar bodies) from homogenates of lung tissue. The extracellular form is isolated by lavage of lungs through the trachea. Both methods require gradient centrifugation to separate the relatively less dense surfactant from tissue contaminants. The intracellular and extracellular forms are essentially equivalent in terms of composition and biophysical properties. Surfactant also has been isolated from amniotic fluid and represents the material secreted by the fetal lung.

B. Composition of Lung Surfactant

Surfactant is a complex mixture of various lipids and proteins. The lipid composition is relatively well defined, whereas the protein composition varies with the source and the possible presence of contaminants. For the intracellular (lamellar body) form, the lipid to protein ratio is approximately 10. The lipid fraction is dominated by phospholipids which account for about 80% of the total lipid pool. The major phospholipid component of surfactant is dipalmitoylphosphatidylcholine (DPPC), which accounts for approximately 45–50% (by weight) of surfactant. DPPC is a phosphatidylcholine in which both acyl chains are palmitate—a 16-carbon, fully saturated fatty acid. Additional phospholipid components include phosphatidylcholine species with one or more unsaturated fatty acids, phosphatidylglycerol, and phosphatidylethanolamine. The presence of phosphatidylglycerol is of special interest since this phospholipid is relatively uncommon in other mammalian tissues. Cholesterol is the predominant neutral lipid and comprises 5–10% of surfactant by weight. Minor quantities of other phospho and neutral lipids are also present. DPPC is primarily responsible for the surfactant (surface tension lowering) properties of the material, whereas the other lipids play accessory roles related to formation of the monolayer or surfactant metabolism.

The protein components comprise approximately 10%, by weight, of intracellular (lamellar body) surfactant. Surfactant obtained by lung lavage generally has a higher protein content, probably reflecting contaminants coisolated during lavage and centrifugation. Three major surfactant-associated proteins have been identified and the molecular sequence for each has been determined. The proteins are called surfactant protein A (SP-A), surfactant protein B (SP-B), and surfactant protein C (SP-C). Other proteins have been identified (including one with the misnomer SP-D) but are not considered as components of the secreted product.

SP-A, the major surfactant-associated protein by weight, is synthesized as a 26-kDa peptide, posttranslationally modified to a mass of approximately 33–36 kDa, and secreted as an octadecamer with molecular mass of approximately 600 kDa. Collagenous and lectin-binding domains in SP-A have been identified and it has been assigned to the collectin family of proteins. SP-A is recovered in the aqueous phase following extraction of lung surfactant with a mixture of organic and aqueous solvents. SP-B is secreted as a dimeric protein with two identical 9-kDa subunits linked through a disulfide bond. The primary translation product is a 42-kDa protein that is processed intracellularly before secretion. The secreted protein is hydrophobic and segregates with the organic fraction following extraction of lung surfactant. SP-C is an even more hydrophobic protein that also segregates with the organic fraction. It is synthesized as a 21-kDa protein and is cleaved posttranslationally to a 3.5-kDa secreted form.

C. Surface Active Properties of Lung Surfactant

Lung surfactant *in vitro* displays strong surfactant properties, i.e., it lowers surface tension of an aqueous solution measured in a Wilhelmy balance to an equilibrium value of approximately 25 dyn/cm. An important feature of lung surfactant is its dynamic properties. Compression of the surface film in a Longmuir trough or by the pulsating bubble method can generate a surface tension of <5 dyn/cm, whereas values of about 50 dyn/cm are obtained on maximal expansion. The monolayer is unstable and returns to the equilibrium value when cycling is interrupted. Continuous recording of surface tension during a compression–expansion cycle indicates hysteresis, i.e., at equivalent surface areas, the surface tension during compression is lower than that demonstrated during the expansion phase. This observation has been explained by the rate of recruitment of surface active molecules from the hypophase into the monolayer. An analogy has been made to the hysteresis normally observed in the pressure vs volume relationships of the lung during the inspiratory/expiratory phases of the respiratory cycle (Fig. 2).

IV. PHYSIOLOGICAL ROLE OF LUNG SURFACTANT

A. Surface Tension and Lung Surfactant

Distensibility of the lung is a reflection of its elastic properties and determines the pressure (effort) that must be generated by the respiratory muscles for lung ventilation. Distensibility is determined by the elasticity (or compliance), which is commonly evaluated by the transpulmonary pressure vs air volume characteristics of the lung. Elasticity in turn has approximately equal contributions from the lung extracellular matrix and the surface tension. Thus, lung inflation requires the generation of a pressure to stretch interstitial collagen and elastic fibers as well as a pressure to overcome the retractive forces of surface tension. Inflation of a lung with saline or other aqueous solution abolishes the effect of surface tension so that only about half of the pressure is required compared to inflation with air (Fig. 2). The presence of lung surfactant, by maintaining a low surface tension, minimizes the effort required for lung distention during air breathing. An increase of surface tension due to surfactant deficiency increases the required distending pressure to achieve a given degree of lung inflation (Fig. 2). Expressed in terms of lung mechanical properties, the presence of lung surfactant helps to maintain normal lung compliance; the corollary is that surfactant deficiency results in decreased lung compliance.

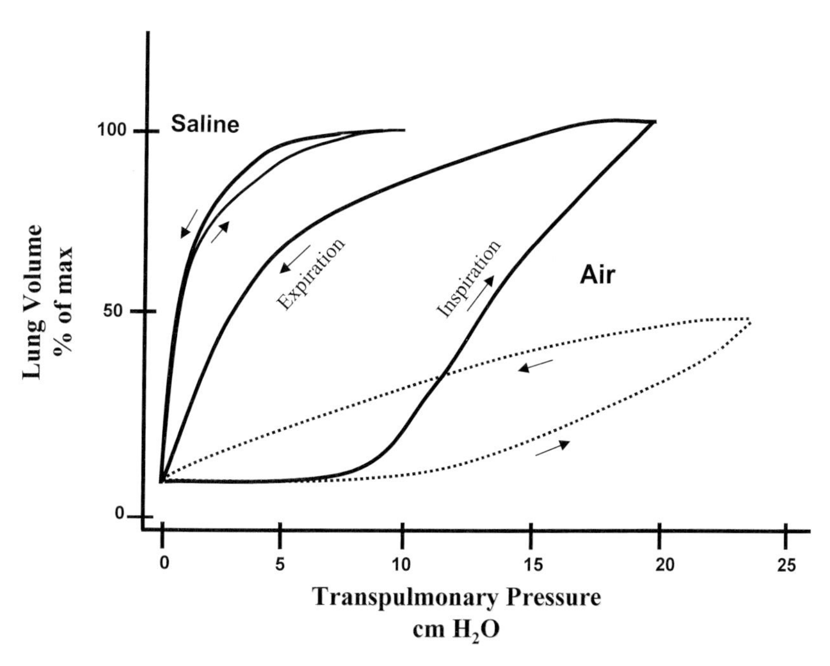

FIGURE 2 Pressure versus volume characteristics of an isolated lung. With air inflation, a significant pressure is required to open the alveoli (opening pressure), which is not seen during the expiratory phase. With saline inflation, there is no opening pressure requirement and the pressure required to inflate the lung is significantly less compared with air inflation. For air inflation, the solid line indicates the pressure versus volume relationship for a normal lung, whereas the dashed line indicates the relationship for a lung that is surfactant deficient. Note the higher pressure requirement for lung inflation with surfactant deficiency indicating decreased lung compliance.

B. Lung Surfactant and Lung Stability

Lung surfactant also functions to maintain lung stability, especially at low lung volumes, and helps to prevent lung collapse (atelectasis). To understand this role, consider the LaPlace relationship applied to a spherical object such as a soap bubble. With constant surface tension, a sphere of smaller radius will have a higher internal pressure than a larger sphere. Translating this concept to the lung alveoli in which many air sacs are in communication, smaller alveoli would empty into larger alveoli because of their higher internal pressure. This effect would create instability in the gas exchange region of the lung. Unequal alveolar size is expected in lungs, particularly during the expiratory phase of the respiratory cycle.

How does surfactant help? The presence of the lung surfactant by lowering the surface tension decreases the alveolar internal pressure and minimizes the potential pressure gradients among alveoli. Furthermore, the *in vitro* behavior of lung surfactant indicates a greater effect on surface tension as the surface area decreases (see Section III,C). Assuming that this *in vitro* behavior applies to the lung, compression of the surface during expiration will lower surface tension to a progressively greater extent as the alveoli decrease in volume. Therefore, the presence of surfactant tends to minimize the differences in pressure among various alveoli, thus promoting stability. For this reason, surfactant has been called the "antiatelectasis" factor.

C. Structure–Function Relationships of Surfactant Components

The ability of lung surfactant to lower surface tension is due predominantly to the presence of DPPC. DPPC is an amphipathic compound with nonpolar fatty acyl and polar glycerophosphorylcholine moieties. It can insert into the air–liquid interface, where it forms a lipid monolayer that lowers surface tension. The saturated fatty acids (palmitate) in the acyl chain allow close packing of adjacent DPPC molecules which accounts for its high surface activity. By contrast, acyl chains with unsaturated fatty acids have larger molecular radii, cannot pack as tightly at an interface, and are less effective as surface tension lowering agents.

While DPPC is an effective surfactant, its phase-transition temperature is 41°C so that it is solid at normal body temperature. Consequently, DPPC in solution spreads very slowly and inserts slowly into the air–liquid interface. The presence of other phospholipids effectively lowers the phase-transition temperature and promotes the spreading and surfactant function of DPPC. The surfactant function of DPPC also is accelerated by the presence of SP-B, which facilitates insertion of the lipid into the interface. This lipid "shuttle" appears to be the major function for SP-B. SP-C may help to generate the monolayer, although its precise function is less well defined. Thus, DPPC, SP-B, SP-C, and associated phospholipids all play a role in promoting surfactant function and maintaining normal lung alveolar surface tension.

SP-A does not have a direct role related to surface tension but may be the regulatory protein for surfactant metabolism. With *in vitro* systems, SP-A inhibits DPPC secretion, stimulates DPPC removal, and inhibits DPPC degradation. In addition, SP-A can serve as an opsonin for some bacteria and may have other functions related to host defense against infection. The physiological relevance of these proposed roles for SP-A is currently under investigation.

V. CELLULAR PROCESSING OF LUNG SURFACTANT

A. Synthesis and Secretion

Surfactant components are synthesized in the endoplasmic reticulum of the lung alveolar epithelial type II cell (granular pneumocyte) and packaged in the lamellar body, the characteristic secretory organelle (Fig. 3). These latter structures in fixed lung sections examined by electron microscopy are approximately 1 to 2 μm in diameter and are filled with arrays that appear as lipid bilayers. Lamellar bodies are released from the cells at a constitutive rate as well as by regulated Ca^{2+}-dependent exocytosis. Exocytosis occurs into the liquid alveolar lining layer, designated as the alveolar hypophase. The physiologic factors that control regulated secretion are not well understood. Experimentally, mechanical stretch of lung cells leads to exocytosis of surfactant, suggesting that hyperventilation can result in surfactant secretion. Tissue alkalosis associated with hyperventilation also can trigger exocytotic activity. Mediators that experimentally increase surfactant secretion include β-adrenergic agonists and activators of protein kinase C. The importance and coordination of these as well as various other potential mediators remains to be determined.

Once secreted, the lamellar arrays unravel in the alveolar hypophase by a Ca^{2+}-dependent mechanism to generate tubular myelin, a structured extracellular repository of surfactant. It is this form that subsequently gives rise to lipids and proteins for generation of the surfactant monolayer. Subfractionation of material obtained by lung lavage yields tubular myelin and also protein-poor lipid vesicles which may represent lipids derecruited from the surfactant monolayer into the hypophase.

B. Surfactant Removal

The type II epithelial cell not only secretes surfactant but also is the cell predominantly responsible for its removal. Surfactant components are taken up by both receptor-mediated and bulk-phase endocytosis. Clathrin-associated SP-A receptors are present on the type II cell membrane and lead to SP-A internalization. The high affinity of SP-A for DPPC-containing vesicles facilitates DPPC uptake through the SP-A receptor mechanism. Lipid vesicles also bind to specific domains on type II cell membranes and are internalized through endocytic mechanisms specialized for cell membrane retrieval. This process may reflect retrieval of lamellar body membrane inserted into plasma membrane during exocytosis. Endocytosis of surfactant is stimulated by agonists that promote secretion, such as β-adrenergic mediators and activators of protein kinase C.

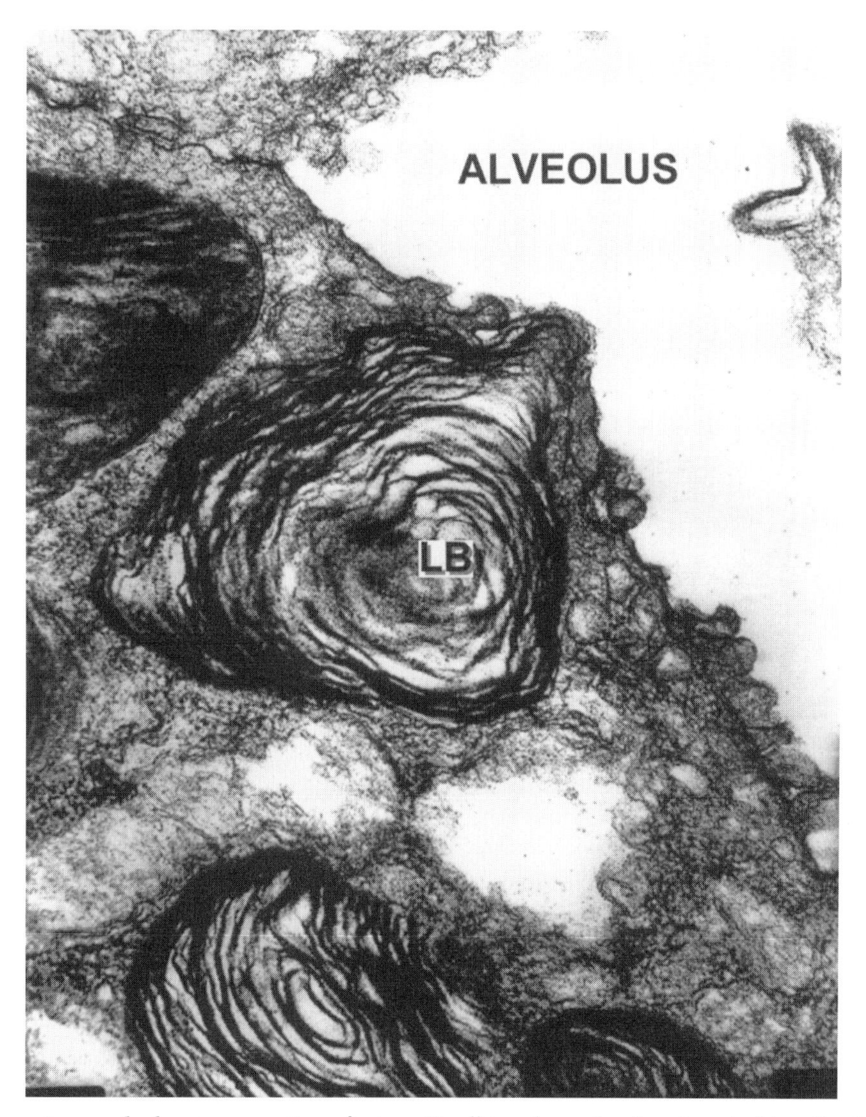

FIGURE 3 Electron micrograph showing a portion of a type II cell in a lung alveolus. A typical lamellar body, the surfactant storage organelle, is indicated by LB. The labeled lamellar body appears to be in the early phase of exocytosis. Several additional lamellar bodies are evident in the field. ALV indicates an alveolar air space. The alveolar hypophase and lung surfactant which normally cover the surface of the cell have been removed during processing.

Internalized surfactant is reprocessed by the type II cell. SP-A is returned to lamellar bodies for resecretion. DPPC is either resecreted through lamellar bodies or degraded intracellularly by phospholipase A_2. Some of the liberated components, especially choline, are reutilized for resynthesis of surfactant or are directed into other cellular metabolic pathways. The signals that target DPPC for resecretion or degradation are not well understood. The rate of resecretion is relatively high where demand for surfactant is high, such as in the neonatal lung, whereas degradation of DPPC is the primary route in the normal adult lung.

C. Lung Surfactant Turnover

The cycle of surfactant synthesis, secretion, removal, and degradation/resynthesis is coordinated in order to regulate the amount of surfactant in the extracellular space. This process is important since either a deficit or an excess of surfactant can lead to altered lung function. Coordination of the surfactant cycle is indicated by the observation that surfactant secretagogues also stimulate exocytosis, although the precise mechanisms for coordination are not known. These signals serve to maintain a rapid rate of surfactant turnover. Using labeled surfactant, the surfactant turnover time has been estimated at 5–10 hr. Thus, approximately 10% of extracellular surfactant is replaced per hour through surfactant secretion and removal. The rapid turnover may be a mechanism to replace "damaged" material but could also represent a mechanism to facilitate the regulation of extracellular surfactant concentration.

VI. THE SURFACTANT SYSTEM DURING GESTATION

The liquid-filled lung of the fetus *in utero* does not require the presence of surfactant since there is no air–liquid interface in the alveolus. Consequently, the lung surfactant system develops and matures relatively late in gestation in preparation for air breathing. The appearance of lamellar bodies in fetal human lung tissue occurs at approximately 24 weeks of gestation and DPPC and surfactant-associated proteins can be detected in the amniotic fluid shortly thereafter. For diagnostic purposes, a specific increase of DPPC can be determined by comparison to a nonsurfactant lipid such as sphingomyelin. The ratio of DPPC to sphingomyelin in amniotic fluid increases dramatically at approximately 32 weeks of gestation, indicating increasing maturation of the surfactant system at that stage of fetal development. A more marked increase in surfactant secretion occurs during parturition in preparation for onset of air breathing by the newborn.

Insufficient lung surfactant associated with prematurity is one of the major etiologic factors in neonatal respiratory distress syndrome (RDS). Adrenocortical steroids accelerate maturation of the lung surfactant system and maternal administration of dexamethasone or analogs prior to the onset of labor can decrease the incidence of RDS in newborns. Artificial surfactants are now available to treat RDS; the most effective have been organic extracts of animal lungs that contain SP-B and SP-C in addition to the surfactant lipids. A genetic form of RDS due to deficiency of surfactant protein B has been recognized and is refractory to surfactant replacement therapy.

See Also the Following Article

Respiratory Distress Syndrome

Bibliography

Goerke, J., and Clements, J. A. (1986). *Alveolar Surface Tension and Lung Surfactant,* Handbook of Physiology. Section 3: The Respiratory System, Vol. 3, pp. 247–261. American Physiological Society.

Herzog, H., Muller, B., and von Wichert, P. (1994). *Lung Surfactant: Basic Research in the Pathogenesis of Lung Disorders,* Progress in Respiration Research, Vol. 27. Karger, Basel.

Hill, B. A. (1988). *The Biology of Surfactant.* Cambridge Univ. Press, Cambridge, UK.

King, R. J., and Clements, J. A. (1985). *Lipid Synthesis and Surfactant Turnover in the Lungs,* Handbook of Physiology. Section 3: The Respiratory System, Vol. 1, pp. 309–336. American Physiological Society.

Long, W. (1993). *Surfactant Replacement Therapy,* Clinics in Perinatology, Vol. 20, No. 4. Saunders, Philadelphia.

Robertson, B., and Taeusch, H. W. (1995). *Lung Biology in Health and Disease,* Surfactant Therapy for Lung Disease, Vol. 84. Dekker, New York.

Rooney, S. A. (1998). *Lung Surfactant: Cellular and Molecular Processing.* Landes Bioscience, TX.

Suidae

see Pigs

Symbiosis

Mary Beth Saffo
Arizona State University West

GLOSSARY

aposymbiotic Lacking symbionts.

commensalism Symbiosis in which one species partner is benefited and the other partner unaffected by the association.

dinoflagellates A group of protists; photosynthetic dinoflagellates are commonly found as photosynthetic symbionts among several groups of marine invertebrate animals, including reef-building corals and some species of sea anemones.

ectosymbiont A symbiont living in close association with another organism, but outside the body of that organism.

endosymbiont An organism which lives inside the body of another organism (the "host").

endosymbiosis Organism of one species living inside the body of another species.

host The larger partner in an endosymbiotic association.

parasitism Antagonistic symbiosis in which one partner is harmed and one partner benefits by the association.

protist A eukaryotic microbe, including protozoa, unicellular algae, and fungus-like microorganisms.

symbiosis Intimate association—"living together"—of two or more species; includes parasitic, commensal, and mutualistic symbioses, as well as many intimate associations is which the relationship between symbiotic partners is not well understood.

vestimentifera A group of gutless, worm-shaped animals which inhabit sulfide-rich hydrothermal vents in the deep sea; their organic carbon is supplied by endosymbiotic, chemosynthetic bacteria.

I. MUTUALISTIC AND OTHER NONPATHOGENIC SYMBIOSES

Mutualistic symbioses are very widespread among animals. For example, at least 13 animal phyla include species which are hosts to chronic endosymbioses—that is, to endosymbionts which inhabit all individuals of a particular species. A ubiquitous infection of this sort suggests that the animal host benefits from the association. Although the benefits of symbiosis to the endosymbionts are sometimes a matter of debate, such chronic endosymbioses are usually considered to be mutualisms. Several chronic endosymbioses, such as reef-building corals (which all harbor intracellular photosynthetic dinoflagellate symbionts), vestimentiferan "worms" and other deep-sea marine invertebrates with chemosynthetic bacteria, and wood-eating termites harboring bacterial and protistan symbionts in their guts, are of considerable ecological importance, and the animal hosts are known to be closely dependent on their symbionts. There are also numerous symbioses, not studied well enough to define unequivocally as mutualisms, in which, nevertheless, the symbionts are thought not to harm the animal hosts. This suite of probably mutualistic and possibly mutualistic or -commensalistic symbioses raises the question: How do nonpathogenic symbionts affect host animal reproduction?

II. ANEMONES, WEEVILS, AND BEETLES: DO NONPATHOGENIC SYMBIONTS ENHANCE REPRODUCTION IN ANIMAL HOSTS?

Despite the considerable body of experimental work on nutrient exchanges in animal–microbial mutualistic symbioses, surprisingly little is known about the specific effects of microbial symbionts on the fecundity of their animal hosts.

In many such symbioses, such as termite–microbial symbioses, and the associations between the Vestimentifera and chemosynthetic bacteria, the animal host is so clearly dependent on microbial symbionts for basic nutritional input that any specific symbiont effects on reproduction are assumed to be a secondary consequence of their key nutritional roles. Furthermore, extreme nutritional dependence of the host on its microbial symbionts precludes most experimental investigations of reproductive effects which compare aposymbiotic and symbiotic animals since aposymbiotic animals do not survive long enough to assess reproduction . The few direct investigations of symbiont effects on animal reproduction have necessarily focused on animals with less extreme nutritional dependence on their hosts or on systems in which artificial diets can substitute for the presence of symbionts.

A. Asexual Reproduction

The intertidal sea anemone, *Anthopleura elegantissima,* receives a substantial (about 40%) part of its organic carbon from the photosynthetic products of its dinoflagellate symbionts and the rest from a largely carnivorus diet. Although the *Anthopleura*–dinoflagellate symbiosis is considered a mutualism, aposymbiotic animals can be grown in the laboratory and are even occasionally found in nature. Aposymbiotic animals were used in recent (1994) investigations by C. Tsuchida and D. Potts, who compared growth rates in symbiotic and aposymbiotic *A. elegantissima* in various environmental conditions. Tsuchida and Potts found that light and symbionts each enhanced growth in individual *A. elegantissima*, but that neither light nor symbionts affected clonal growth—the rate of asexual fission—in these animals. In contrast, external feeding of *A. elegantissima* with brine shrimp increased both anemone body size and asexual fission rate of anemones, whether or not symbionts were present. Thus, in the case of these symbiotic anemones, it is not clear whether algal symbiosis affects asexual reproduction at all.

B. Sexual Reproduction

Though symbiosis has been shown to enhance sexual reproduction in at least two endosymbioses, even in these cases reproductive enhancement is linked closely to nutritional contributions of the symbionts.

In nature, the rice weevil *Sitophilus oryzae* is usually infected with intracellular gram-negative bacteria, which are transmitted to the next host generation via weevil oocytes. The bacteria are thought to provide B vitamins and several amino acids to the weevil hosts, which subsist on a grain diet; the bacteria also affect the morphology, pigmentation, and energetics of their hosts, as well as host development and reproduction. Laboratory experiments by P. Nardon and A. M. Grenier have demonstrated that development time of symbiotic *S. oryzae* (27 days) is much faster than that of its aposymbiotic counterparts (44 days) and that developmental rate is directly correlated with the number of symbiotic bacteria. Elimination of bacteria from *Sitophilus* also reduces fertility by 30%.

Ambrosia beetles such as *Xyleborus ferrugineus* breed in the wood of dead or dying trees, in obligate symbiosis with ectosymbiotic fungi lining the wood "galleries" bored by the beetle host; these "symbionts" are also the primary food source for most ambrosia beetles. Beetles can be reared without fungal symbionts only if supplied with complex artificial diets including ergosterol and 10 amino acids among its several components; in intact symbioses, these compounds are provided by the fungal symbionts. Of these compounds, ergosterol seems an especially key compound for reproduction because the progeny of fungus-free adult females, raised on a sterol-free diet, did not pupate.

Numbers of offspring can also be affected by the species of symbiotic fungus utilized by ambrosia beetles. Feeding of adult female ambrosia beetles on "secondary" ambrosia fungal species only, in the absence of the primary fungal symbiont, can reduce brood production by 50%.

Ambrosia beetles have also been reported to contain bacterial symbionts which are transmitted with beetle oocytes. Rearing of ambrosia beetles on antibiotics reduced female fertility and prevented completion of larval development, suggesting that ambrosia beetles may be dependent not only on ectosymbiotic fungi but also on endosymbiotic bacteria for successful reproduction and development.

III. SEX AND SYMBIOSIS: MATERNALLY TRANSMITTED BACTERIA IN FLIES AND OTHER ARTHROPODS

Evolutionary biologists have long been intrigued by the several animal species in which sex ratios differ strikingly from the 50% male:50% female theoretical "norm." Reproductive incompatibility between populations of the same species has also been the subject of evolutionary investigations into the process of speciation. Unidirectional incompatibility (e.g., females from population A mated with males from population B might yield offspring, but not females from B mated with males from A) has presented an especially puzzling problem. Recent work has yielded the surprising suggestion that, at least in some arthropods, phenomena such as high female:male sex ratios, parthenogenesis (the production of only female offspring), feminization of genetic males and reproductive incompatibility can be "cured" by antibiotics. For several arthropods, symbiotic intracellular bacteria, transmitted in insect ovaries, can kill male eggs, resulting in mostly female or all-female populations. Others can abort crosses of bacteria-infected males with bacteria-free females (but sometimes not crosses of uninfected males with bacteria-infected females), or they abort crosses between males and females harboring different genetic strains of symbiotic bacteria; such phenomena can result in unidirectional (asymmetric) or bidirectional reproductive isolation. In inducing the production of functional females from genetic males, maternally transmitted bacterial symbionts of terrestrial isopods can serve as the chief factors of sex determination in infected isopod populations.

Cases of bacterial-induced reproductive isolation, feminization or parthenogenesis have been described from at least four species of *Drosophila* and from species of mosquitoes, moths, weevils, beetles, parasitic wasps, and terrestrial isopods. In all these cases, bacterial endosymbionts are transmitted maternally, usually cytoplasmically, with the host egg but sometimes by other methods as well. Recent molecular studies suggest that most of these bacterial endosymbionts belong to the genus *Wolbachia*; these symbionts cause reproductive ("cytoplasmic") incompatibility by disrupting incorporation of paternal chromosomes in fertilized eggs of uninfected females, and they induce parthenogenesis by preventing chromosome segregation in unfertilized (male) eggs. In terrestrial isopods, *Wolbachia* infections induce feminization in genetic males, in part by causing inhibition of "male-determining genes" that control development of the androgenic gland.

Despite the potentially important evolutionary impact of such symbioses, especially on our understanding the evolutionary implications of parthenogenesis and processes of speciation, little is known about other effects of these bacterial–insect symbioses; neither the metabolic activities of the bacteria nor their effect on the general biology of the animal hosts (other than the documented effects on sex ratios and mating incompatibility) have been investigated in much detail. From what is known, infection by *Wolbachia* does not seem to be strongly pathogenic in nature. Although they do not always infect 100% of the host population, the bacteria have rarely been documented to harm their host. In short, it is not clear whether these bacterial–arthropod symbioses are mutualistic in all, or even any, cases: indeed, *Wolbachia*–arthropod symbioses have been described by Bourtzis and O'Neill (1998) as "reproductive parasitism." But, whatever their overall effects on host biology, their effects on the population biology of several terrestrial arthropods are striking and instructive.

The wasp *Nasonia vitripennis*, which parasitizes fly pupae, shows wide variety in the sex ratios of its offspring, which can vary from 0 to 90% female. Some of this sex-ratio variability is attributable to the presence of "son-killer" bacteria (Enterobacteriaceae), which kill male eggs early in development. These bacteria are transmitted maternally, into the fly pupa, as the female wasp lays its eggs in the pupa. Bacteria are then ingested by developing wasp larvae in the pupa. Bacteria can be transmitted to new wasp lineages when an uninfected wasp lays eggs in a pupa already containing eggs of a bacteria-infected wasp. In other species of parasitic wasps, parthenogenesis can be induced by hereditarily-transmitted *Wolbachia*. Some species of the parasitic wasp *Trichogamma* which are 100% parthenogenetic can be restored to normal sex ratios through antibiotics.

Cytoplasmically transmitted bacterial infections can also spread through populations through reproductive incompatibility between infected and uninfected animal hosts. The percentage of *Wolbachia*-infected *Drosophila simulans* has grown rapidly in several California populations because uninfected females crossed with *Wolbachia*-infected males produce fewer viable progeny than (for instance) infected–infected crosses; thus, infected females have a reproductive advantage over their uninfected counterparts in polymorphic populations.

As awareness of symbiont-caused reproductive isolation, feminization and parthenogenesis has grown, so have the number of biological examples of these phenomena. Given these growing number of examples, biologists should consider the possibility of bacterial symbiosis whenever they encounter parthenogenetic populations or reproductive isolation between individuals of the same species, especially in arthropods.

IV. CONCLUSION

Lynn Margulis has described symbiosis as a "parasexual phenomenon." Indeed, sex and symbiosis do share the general characteristic of close association or genetic mixing of different genotypes, albeit with the important distinction of sex as an intraspecific combination, and symbiosis as an interspecific combination of genotypes. As demonstrated in *Wolbachia*–insect interactions, hereditarily or maternally transmitted symbioses can combine the effects of sexual and symbiotic "combination" in particularly startling ways. The effects of endosymbiosis on various aspects of animal reproduction clearly warrant continued, and deeper, investigation.

See Also the Following Article

Parasitoids

Bibliography

Beaver, R. A. (1989). Insect–fungus relationships in the bark and ambrosia beetles. In *Insect–Fungus Interactions* (N. Wilding, N. M. Collins, P. M. Hammond, and J. F. Webber, Eds.), pp. 121–143. Academic Press, London.

Bourtzis, K. and O'Neill, S. (1998). *Wolbachia* infections and arthropod reproduction. *BioScience* **48**, 287–293.

Douglas, A. E. (1994). *Symbiotic Interactions*. Oxford Univ. Press, Oxford, UK.

Hickey, D. A. (1993). Molecular symbionts and the evolution of sex. *J. Heredity* **84**, 410–414.

Margulis, L. (1993). *Symbiosis in Cell Evolution*. Freeman, New York.

Nardon, P., and Grenier, A.-M. (1991). Serial endosymbiosis theory and weevil evolution: The role of symbiosis. In *Symbiosis as a Source of Evolutionary Innovation* (L. Margulis and R. Fester, Eds.). MIT Press, Cambridge.

O'Neill, S. L., Hoffmann, A. A., and Werren, J. H. (1997). *Influential Passengers: Inherited Microorganisms and Arthropod Reproduction*. Oxford University Press, Oxford, UK.

Saffo, M. B. (1992a). Coming to terms with a field: Words and concepts in symbiosis. *Symbiosis* **14**, 17–31.

Saffo, M. B. (1992b). Invertebrates in endosymbiotic associations. *Am. Zool.* **32**, 557–565.

Thompson, J. N. (1994). *The Coevolutionary Process*. Univ. of Chicago Press, Chicago.

Turelli, M. (1994). Evolution of incompatibility-inducing microbes and their hosts. *Evolution* **48**, 1500–1513.

Tardigrada

Roberto Bertolani and Lorena Rebecchi
University of Modena

GLOSSARY

autobivalents Oocyte chromosomes similar in shape to bivalents but that originate by a double replication of the same chromosome and not by pairing of homologous chromosomes.

coelomic pouches Mesodermic paired structures positioned laterally along the archenteron.

cryptobiosis Capacity to enter into a latent life state, suspending animation during periods of environmental adversity (e.g., drying, freezing, and lack of oxygen).

hydrobios The organisms which live in water, especially soil organisms active only when water is present.

iteroparous Breeding occurring several times per lifetime.

morphospecies Organisms characterized by the same morphological pattern regardless of whether they are genetically connected.

semelparous Breeding only once per lifetime.

semiterrestrial Describing land organisms belonging to the hydrobios but needing at least a film of water when active.

thelitoky A kind of parthenogenesis in which unfertilized eggs produce females only.

Tardigrada (called "water bears" for their shape and way of moving) are microscopic, worm-like, metameric invertebrates belonging to hydrobios. Mostly known to be moss- and lichen-dwelling animals, they are also present in leaf litter, turf, and aquatic habitats, especially marine and freshwater sediments. Tardigrades, or at least tardigrades that colonize land (semiterrestrial), are capable of cryptobiosis. In this state of life they can survive extreme conditions and can be passively dispersed, colonizing distant territories. Tardigrades have several conservative characters, e.g., a small body size, a fixed number of metamers, a limited and almost constant cell number (even if mitoses also occurs in the somatic tissues of the adult), a sucking buccopharyngeal apparatus, a small genome size, and a relatively uniform chromosome set.

I. INTRODUCTION

Tardigrada are a minor phylum of Metazoa, already present in the Cambrian, and are currently represented by more than 600 species. According to some authors, tardigrades are related to arthropods, annelids, or both, even though others have noted relationships with nematodes. Tardigrades probably originated in the sea and subsequently colonized freshwater sediments and terrestrial habitats. Very rarely, they can be ectoparasites of marine animals. The phylum is subdivided into three classes: Heterotardigrada, composed of both marine (Fig. 1A) and semiterrestrial (Figs. 2A and 2B) species; Mesotardigrada, represented by one species only (found once in thermal hot waters); and Eutardigrada (Figs. 1B and 2C), composed by limnic and semiterrestrial, rarely marine species. The length of tardigrades typically ranges from 150 to 800 μm, even though some

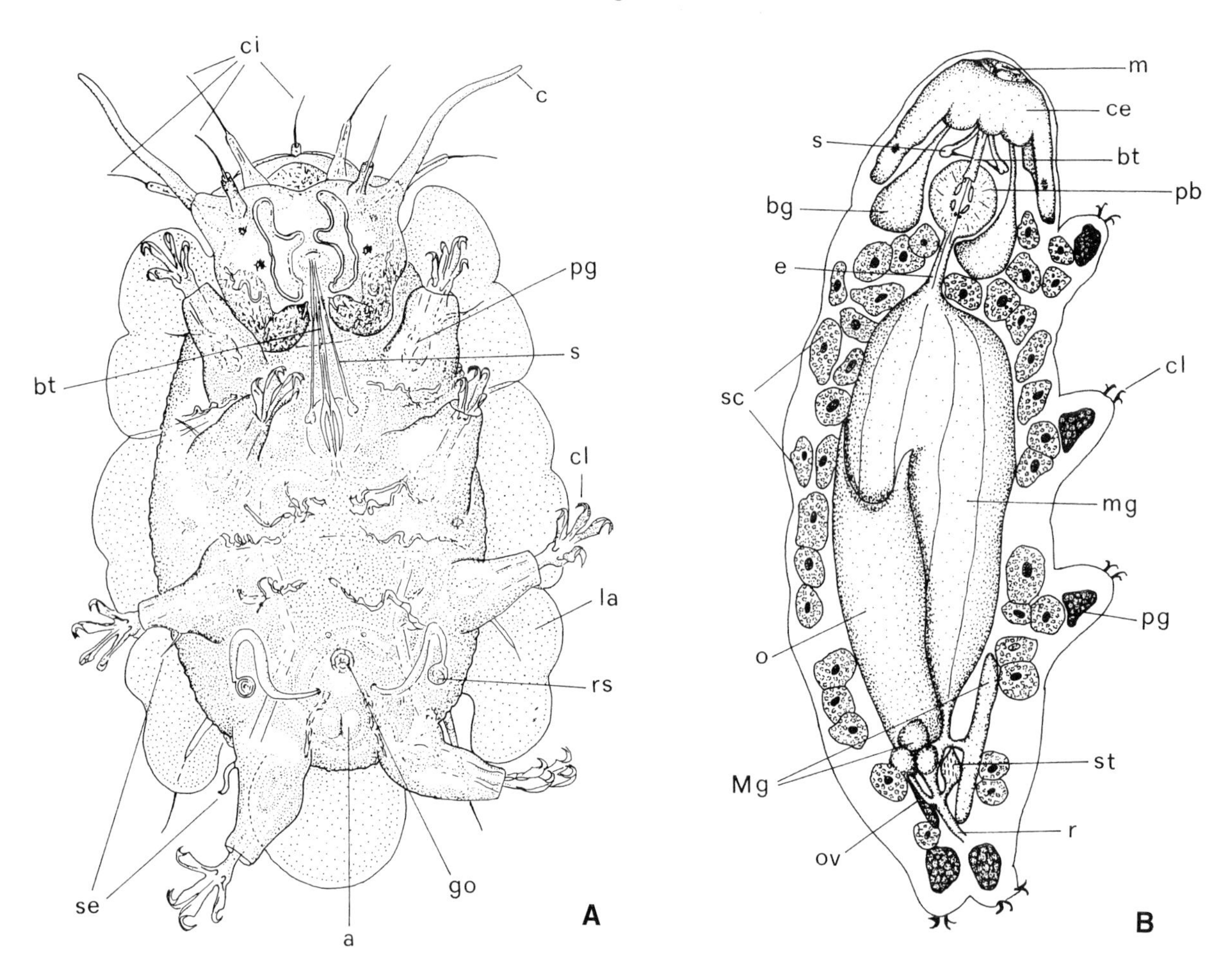

FIGURE 1 A female marine heterotardigrade (A) and a female semiterrestrial eutardigrade (B); a, anus; bg, buccal glands; bt, buccal tube; c, clavae; ce, cerebrum; ci, cirri; cl, claws; e, esophagus; go, gonopore; la, lamina; m, mouth; Mg, Malpighian glands; mg, midgut; o, ovary; ov, oviduct; pb, pharyngeal bulb; pg, pedal glands; r, rectum; rs, seminal receptacles; s, stylets; sc, storage cells; se, sensory organs; st, spermatheca (A, adapted with permission from R. M. Kristensen, On the biology of *Wingstrandarctus corallinus* nov. gen. et spec., with notes on the symbiontic bacteria in the subfamily Florarctinae (Arthrotardigrada), *Vidensk. Meddr. Dansk Naturh. Foren.* **145**, 210–218, 1984; B, modified with permission from R. Bertolani, Guide per il riconoscimento delle specie animali delle acque interne italiane, Quaderni CNR, AQ/1/168, 15, pp. 104, 1982).

specimens can be <100 μm or exceed 1 mm in length. The body is bilaterally symmetrical and metameric. The body shape, generally worm-like, is more variable in marine species than in freshwater and semiterrestrial ones, but the general body organization is basically the same in all tardigrades (Figs. 1 and 2). The animal is limited by a thin chitinous cuticle renewed at each molt and secreted by a thin layer of epidermal cells. In some heterotardigrades the dorsal cuticle is thickened and subdivided into distinct individual plates (Fig. 2A). A cephalic segment is followed by four trunk segments, each bearing a pair of legs (parapods) in the second half. The legs end with claws or more rarely with digits (in some marine species). The claws and the digits are renewed at each molt and secreted by pedal glands. Eye spots and cuticular sensory organs (papillae and, limited to heterotardigrades, filaments and cephalic clavae) may be present. The wide body cavity is not coelomic because it is not lined by an epithelium; it is considered to be a hemocoel. It contains many free cells, named storage cells, floating in a fluid. The

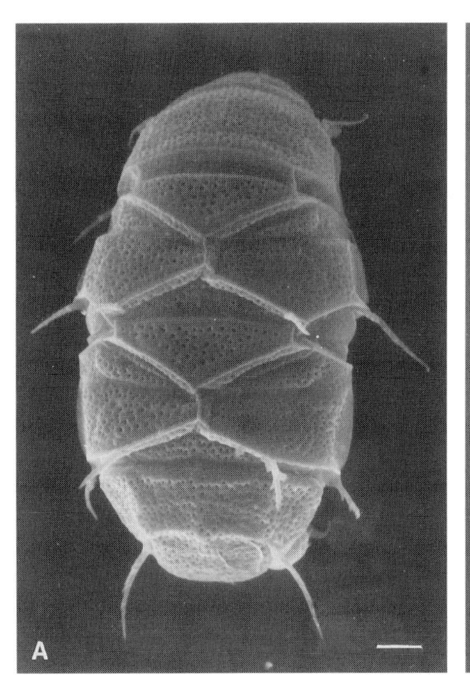

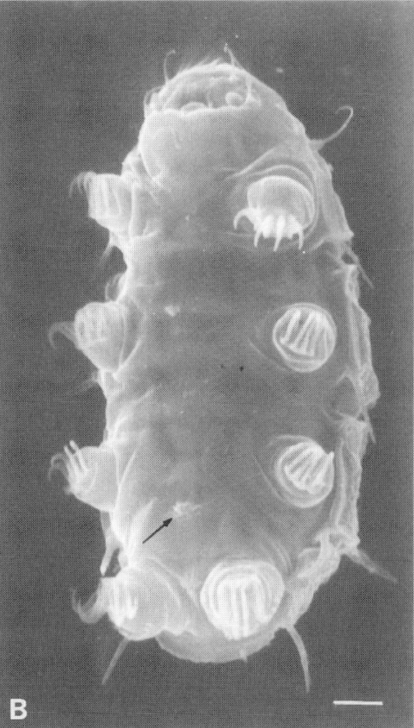

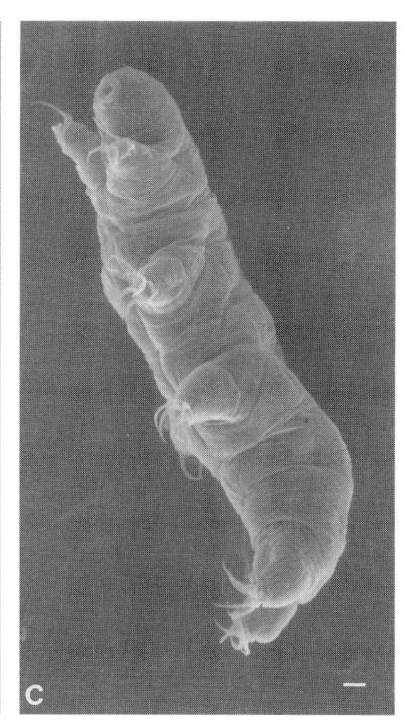

FIGURE 2 Scanning electron micrographs of tardigrades. (A, B) The semiterrestrial heterotardigrade *Echiniscus duboisi*; dorsal view (A), note the plates of the armature; ventral view (B), note the female gonopore (arrow). (C) The freshwater eutardigrade *Pseudobiotus megalonyx* (lateral view). Scale bars = 10 μm.

body cavity is crossed by longitudinal muscles and transverse leg muscles connected with the cuticle but not circular muscles. The central nervous system is composed by a multilobed brain with a ring neuropil and four bilobed ventral ganglia. Tardigrades swallow fluid or microcorpuscolate food of vegetal or animal origin by a buccopharyngeal apparatus, composed of a mouth, a cuticular gullet supported by two protrusible stylets, and a sucking mioepithelial pharyngeal bulb. Gullet and stylets are reconstructed at every molting by two big buccal glands. Through a narrow esophagus, food is brought into a wide midgut, where it is digested. The midgut is followed by a rectum that ends in an anus (heterotardigrades) or in a cloaca (eutardigrades). Three Malpighian glands connected with the hindgut (eutardigrades and mesotardigrades) or two ventromedial organs related to the epidermis (heterotardigrades, at least moss-dwelling) can be present for excretory/ion osmoregulatory functions. Malpighian glands may play a role as accessory glands by either secreting glycoproteins to be added to the seminal fluid or intervening in ejaculation of spermatozoa by increasing fluid pressure. Respiration takes place through the epidermis.

II. MODES OF REPRODUCTION AND SEX

A. Reproduction

Reproduction occurs only through fertilized or unfertilized eggs, and therefore only through gametes. Amphimixis can be due to cross-fertilization or self-fertilization. In marine tardigrades males and females are always present and, as in most other marine animals, parthenogenesis is unknown. These tardigrades are unitary animals. On the contrary, parthenogenesis is most widespread in the less stable nonmarine environments, both in hetero- and eutardigrades, even if amphimixis is maintained in several species. In most cases, the tardigrades are modular and clonal. For all tardigrades heterogony is unknown. Parthenogenesis is always bound to thelitoky. In tardigrades the cytology of parthenogenesis is varied and known almost exclusively for eutardigrades. In most cases parthenogenesis is ameiotic and produces clones. It is often associated with triploidy, sometimes with tetraploidy, and more rarely with diploidy. Meiotic parthenogenesis is known for two diploid species of eutardigrades in which males have never been found. Chiasmata can be observed in one of the species. In both species the diploid chromosome number is restored after meiosis. In a few tri- and tetraploid populations, restoration of the chromosome number is premeiotic due to a supernumerary duplication of the oocyte chromosomes with formation of autobivalents. The absence of chiasmata indicates the same genetic consequence as in ameiotic division. The unique data on heterotardigrades (a semiterrestrial species of *Echiniscus*) indicates an unequal distribution of the chromosomes in the first anaphase.

B. Sex

Tardigrades can be gonochoric (bisexual and unisexual) or hermaphroditic. Sex chromosomes are unknown. In most marine species in which the presence of males and females was ascertained, the sex ratio was close to 1 : 1; only one case of hermaphroditism is known. Even in nonmarine tardigrades, gonochorism is the most frequent sexual condition, even if it is preferentially bound to unisexuality, i.e., to the presence of females only. There are several morphospecies that have both gonochoric amphimictic (diploid; sex ratio 1 : 1) and thelytocous populations (often triploid and rarely tetraploid); the latter are the most widespread. Some cases of sympatry were observed. Hermaphroditism is the less frequent sexual condition in tardigrades. Although sporadic, it was found in several families both in limnic and in semiterrestrial eutardigrades. Although in hermaphroditic tardigrades spermatozoa begin to mature earlier than oocytes, a concurrent presence of mature male and female gametes without limits between them can be observed (simultaneous hermaphroditism). Our recent laboratory data demonstrated that self-fertilization is possible.

C. Evolution of Reproduction and Sex

The absence of parthenogenesis and, practically, also of hermaphroditism in marine tardigrades indicates that both events represent apomorphic conditions that appeared after the transfer from the sea to a less stable environment. Considering that hermaphroditism is simultaneous and that it allows self-fertilization, the advantages of this kind of sexuality and especially of this way of reproduction, such as parthenogenesis, can be explained by the low-density model advanced by Ghiselin. It should be recalled that limnic and semiterrestrial tardigrades are subject to passive dispersal and that in this way they can often become isolated. The capability for self-fertilization or parthenogenesis allows tardigrades to colonize new territories with a single specimen. Parthenogenesis is more widespread than cross-fertilization and self-fertilization, indicating that it is favorably selected. Its permanence across many generations may have been aided by cryptobiosis, which through the capability to survive drying and freezing well allows tardigrades to be subjected to a lower environmental selective pressure.

From a point of view of reproduction and sex, we can observe different situations in tardigrades. Several species are known only as amphimictic. Some of them are only gonochoric, whereas only one presents both gonochoric or hermaphroditic conditions. A larger number of species (or species complexes) can be amphimictic or parthenogenetic. In any case parthenogenesis and hermaphroditism have never been noted simultaneously. Therefore, we can hypothesize that (i) some species never reach hermaphroditism or thelitoky; (ii) many others increase their dispersal ability by being able to colonize a new territory with only one specimen through self-fertilization or parthenogenesis; (iii) when both events take place, parthenogenesis is the most efficient way of dispersal and therefore is favorably selected; and (iv) in the species (or, more properly, species complexes, because the possibility of gene flow has been hypothesized but not demonstrated) characterized by more than one mode of reproduction, cross-fertilization is maintained because it is favored under climax conditions.

III. ANATOMY OF THE REPRODUCTIVE SYSTEMS AND GERM CELL MORPHOLOGY

A. Reproductive Systems

In tardigrades the reproductive system is formed by a single gonad, one or two gonoducts, a gonopore or a cloaca, and possible seminal receptacles and seminal vesicles (Figs. 3, 4 and 5A–5D). The gonad (ovary, testis, or ovotestis) is always a dorsocaudal sack overlying the midgut and anteriorly suspended at the cuticle by ligaments (Figs. 3A, 3B, 4A, 4B, and 10). Principally in females, its size varies according to reproductive activity. In heterotardigrades the gonad is limited by a basal membrane, whereas in eutardigrades a discontinuous epithelium with bundles of contractile-like filaments has been described. The gonad is followed caudally by one (females and hermaphrodites) or two (males) short and narrow gonoducts, which turn ventrally and medially (Fig. 3B). The two deferents join to form a short common duct. Heterotardigrades, both marine and nonmarine, have a cuticular ventral gonopore which opens separately from the anus, whereas eutardigrades have a cloaca. In males of heterotardigrades the gonopore is gener-

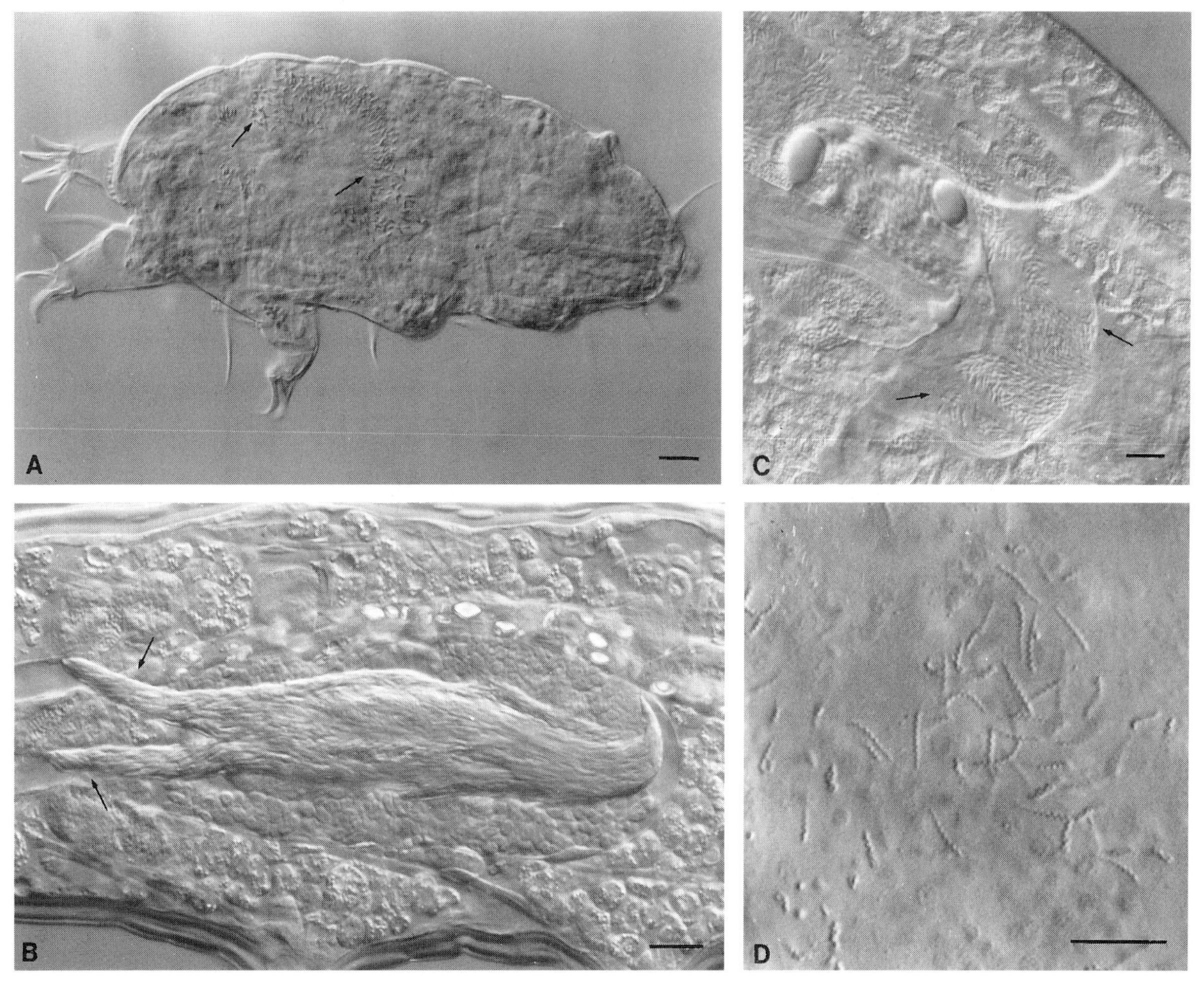

FIGURE 3 Male reproductive system. (A) Testis with spermatozoa (arrows) in the heterotardigrade *Echiniscus duboisi*; orcein, differential interference contrast (DIC). (B) Mature testis and deferents (arrows) in the eutardigrade *Diphascon humicus*; *in vivo*, DIC. (C) Seminal vesicles (arrows) containing spermatozoa in the eutardigrade *Platicrista angustata*; *in vivo*, DIC. (D) Detail of the testis with helical nuclei of spermatozoa in the eutardigrade *Macrobiotus macrocalix*; orcein, DIC. Scale bars = 10 μm.

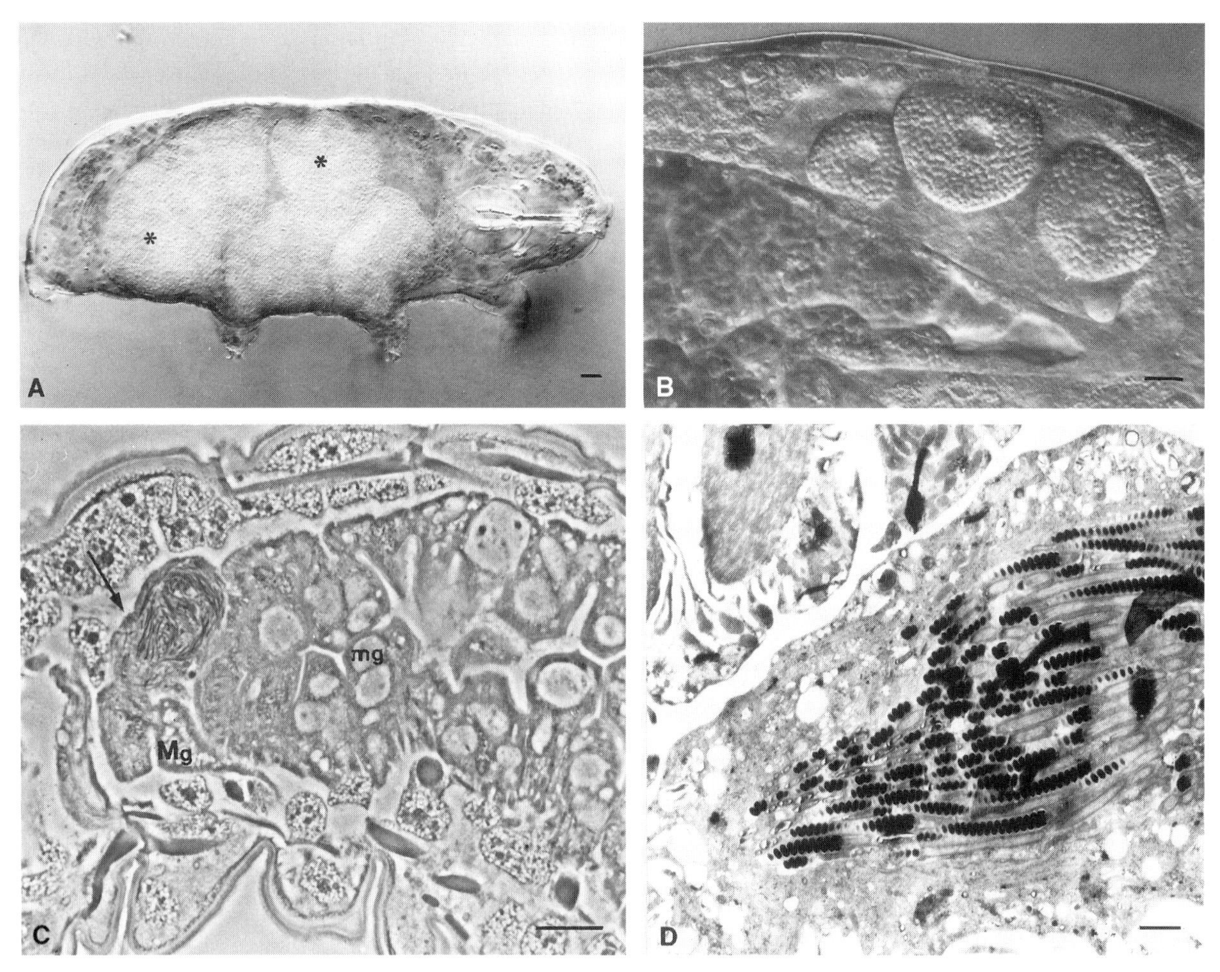

FIGURE 4 Female reproductive system. (A) Large ovary with mature oocytes (asterisks) in the eutardigrade *Amphibolus volubilis*; orcein, DIC. (B) Oocytes in vitellogenesis in the eutardigrade *Macrobiotus macrocalix*; *in vivo*, DIC. (C, D) Longitudinal section of the spermatheca (arrow) located near the Malpighian glands (Mg) and the midgut (mg) in the eutardigrade *Xerobiotus pseudohufelandi*; note the bundles of spermatozoa. (C) Iron hematoxylin, phase contrast; (D) TEM. Scale bars = 10 μm (A–C) and 1 μm (D).

ally tubular and located just anterior to the anus (Figs. 5A and 5B), whereas in females it has the shape of a six-petal rosette and is located a little more anterior to the anus (Figs. 5C and 5D), between the third and the hindlegs. In contrast, in eutardigrades gonoducts join to the rectum, which ends in a ventral cloaca similar in both sexes. In some species males have seminal vesicles. In heterotardigrades they are represented by two caudal lateral bulges of the testis, whereas in eutardigrades they are due to a small swelling on the distal part of each deferent (Fig. 3C). In most marine heterotardigrades females have two cuticular, ventrolateral seminal receptacles with variably long and convoluted genital tubes that open exteriorly via an independent outlet at a distance from the gonopore (Fig. 1A). These organs may be related to the difficulty in finding a partner due to a low population density. A single epithelial and internal spermatheca connected with the rectum is present in a few nonmarine eutardigrades (*Macrobiotus*, *Xerobiotus*, and *Ramazzottius*; Figs. 4C and 4D) and in one marine heterotardigrade (*Batillipes*).

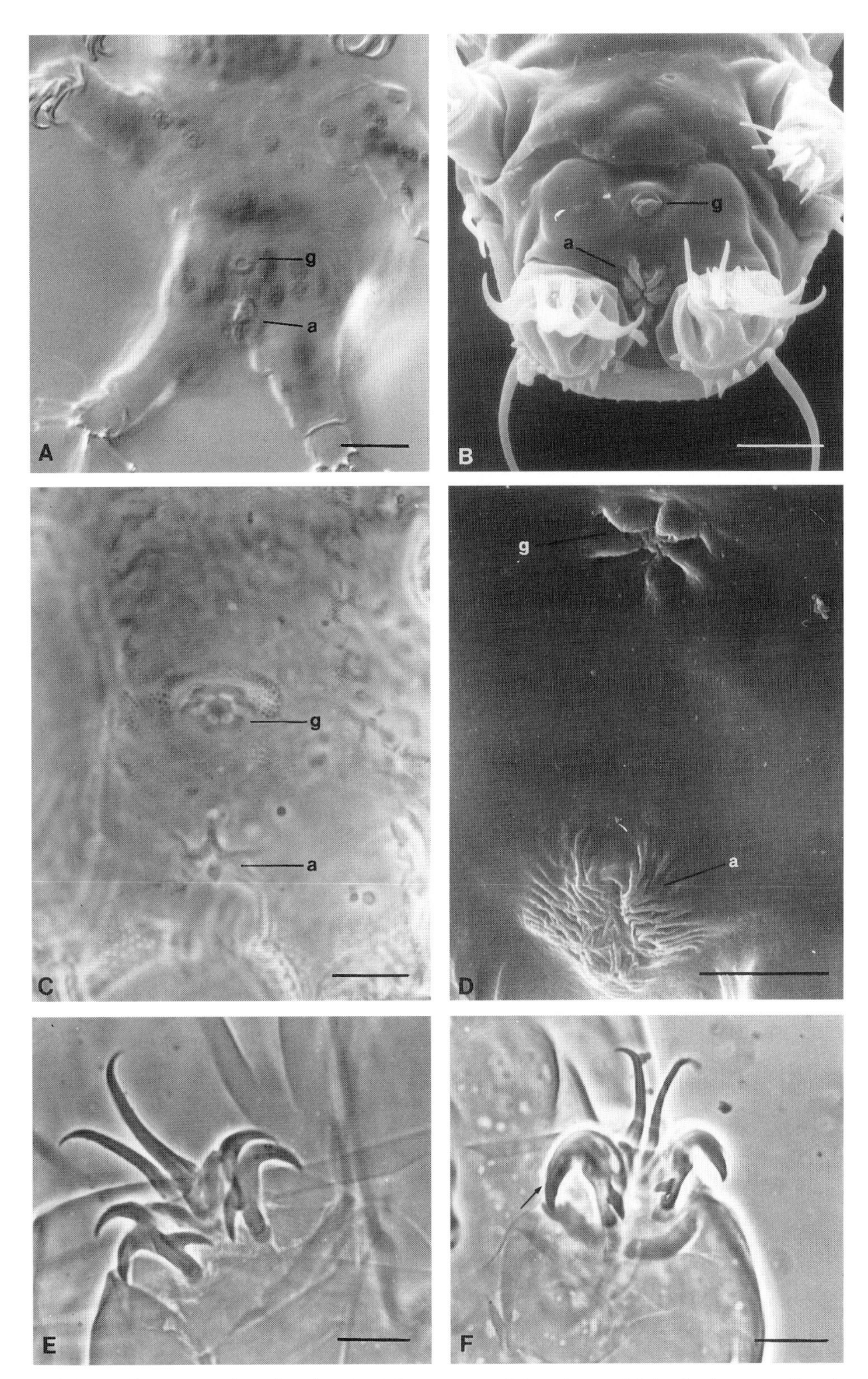

FIGURE 5 Sexual dimorphism in tardigrades. (A, B) Male gonopore (g) and anus (a) in the heterotardigrades. *Antechiniscus parvisetus* (A), DIC; and *Echiniscus duboisi* (B), SEM. (C, D) Female gonopore (g) and anus (a) in the heterotardigrades *Pseudechiniscus juanitae* (C), phase contrast; *Echiniscus* sp. (D), SEM. (E, F) Diplo-claws on the front legs in the eutardigrade *Milnesium tardigradum*. Normal claws of the female (E) and modified claws in the male (F) with the basal branches shaped like a hook (arrow) bearing a small spur; phase contrast. Scale bars = 10 μm.

B. Sexual Dimorphism

Other than the gonopore of the heterotardigrades, secondary sexual dimorphic characters can also exist. In several cases, males are clearly smaller than females. Sometimes there are dwarf males. Moreover, in a few species of eutardigrades the claws on the front legs in males and those on the hindlegs in females are modified (Figs. 5E and 5F). Even cephalic clavae of heterotardigrades, which are much longer and larger in male than in female, represent a secondary sex character.

C. Spermatogenesis and Spermatozoa

Spermatogenesis usually begins after the first or the second molt of the animal. Only in the ectoparasite marine species *Tetrakentron synaptae*, which has two kinds of males, are spermatozoa probably present at hatching in dwarf males. The testis of young males of the eutardigrade *Macrobiotus* is characterized by a wall partly lined by a monolayered cubic epithelium, whose cells are detaching, giving rise to the germinal cell line. In adults, cells exhibiting the various steps of spermatogenesis display a zonal arrangement. Spermatocytes can be recognized by their larger size compared to spermatogonia or by star-shaped metaphases. Each spermatogonial mitosis and the two subsequent meiotic divisions give rise to a multinucleate cell that then changes into a cluster of eight spermatids joined by cytoplasmic bridges. Spermiogenesis follows the typical pattern described for most metazoans. The amount of cytoplasm decreases, the Golgi apparatus gives rise to the acrosome, the flagellum develops from the unique centriole, and the nucleus condenses its chromatin, changing in most eutardigrades from roundish to elongated and helicoidal. In the heterotardigrade *Batillipes* the nucleus develops an anterior snout, which undergoes rotation, probably due to the action of three "Saturn rings" formed in the cytoplasm in close contact with the nucleus, each consisting of two unit membrane and a thin undulate membrane.

Tardigrade spermatozoa are always flagellate, with a 9 + 2 axoneme. The male gametes of heterotardigrades are small (14–30 μm in length; Figs. 6A and 6B) and have a globose or slightly elongated head (comprehensive of an acrosome, sometimes abortive) with an ovoid nucleus containing homogeneous, and sometimes heterogeneous, electron-dense chromatin; the nucleus is surrounded by an appreciable amount of cytoplasm. The head is followed by two elongated mitochondria extending from the main axis of the cell and by a flagellum that tapers out to its termination. This kind of spermatozoon can be attributed to a primitive type, even if specialization exists. On the contrary, spermatozoa of eutardigrades can be attributed to a derivative type and display a remarkable morphological heterogeneity in different genera and sometimes within the same genus (Figs. 6C–6F and 7). They are longer (from 25 to 100 μm in length) than those in heterotardigrades and in many cases filiform. Their elongated head always bears a helicoidal piece that corresponds to the acrosome or, more frequently, to the nucleus. The nucleus, composed of electron-dense and homogeneous chromatin, is often surrounded by a very scanty amount of cytoplasm (Fig. 7B). A neck or midpiece can be present or absent. It is small and cup-shaped with large mitochondria around the unique centriole, as in *Amphibolus* (Figs. 7D and 7E), or elongated with many ovoidal elements with a dense core around a mitochondrial sleeve, as in *Macrobiotus* and *Xerobiotus* (Figs. 7A and 7B). The flagellum always ends with a tuft of filaments (Figs. 6C–6F).

In both hetero- and eutardigrades, the spermatozoa stored in the seminal receptacles or in the spermatheca of the females are immobile and appear to be modified in comparison with testicular spermatozoa (Fig. 7C).

D. Oogenesis and Egg

Oogenesis occurs usually after the second molt. In most marine heterotardigrades maturation of the oocytes is asynchronous, with the most mature oocyte located caudally in the ovary. In all other tardigrades, oocytes develop synchronously and are laid in groups from 2 or 3 eggs to more than 30 eggs. In young females and those with synchronous oocytes, the content of the ovary after oviposition appears undifferentiated, sometimes characterized by mitotic divisions. These divisions lead to clusters of cells connected to each other by cytoplasmic bridges. Only

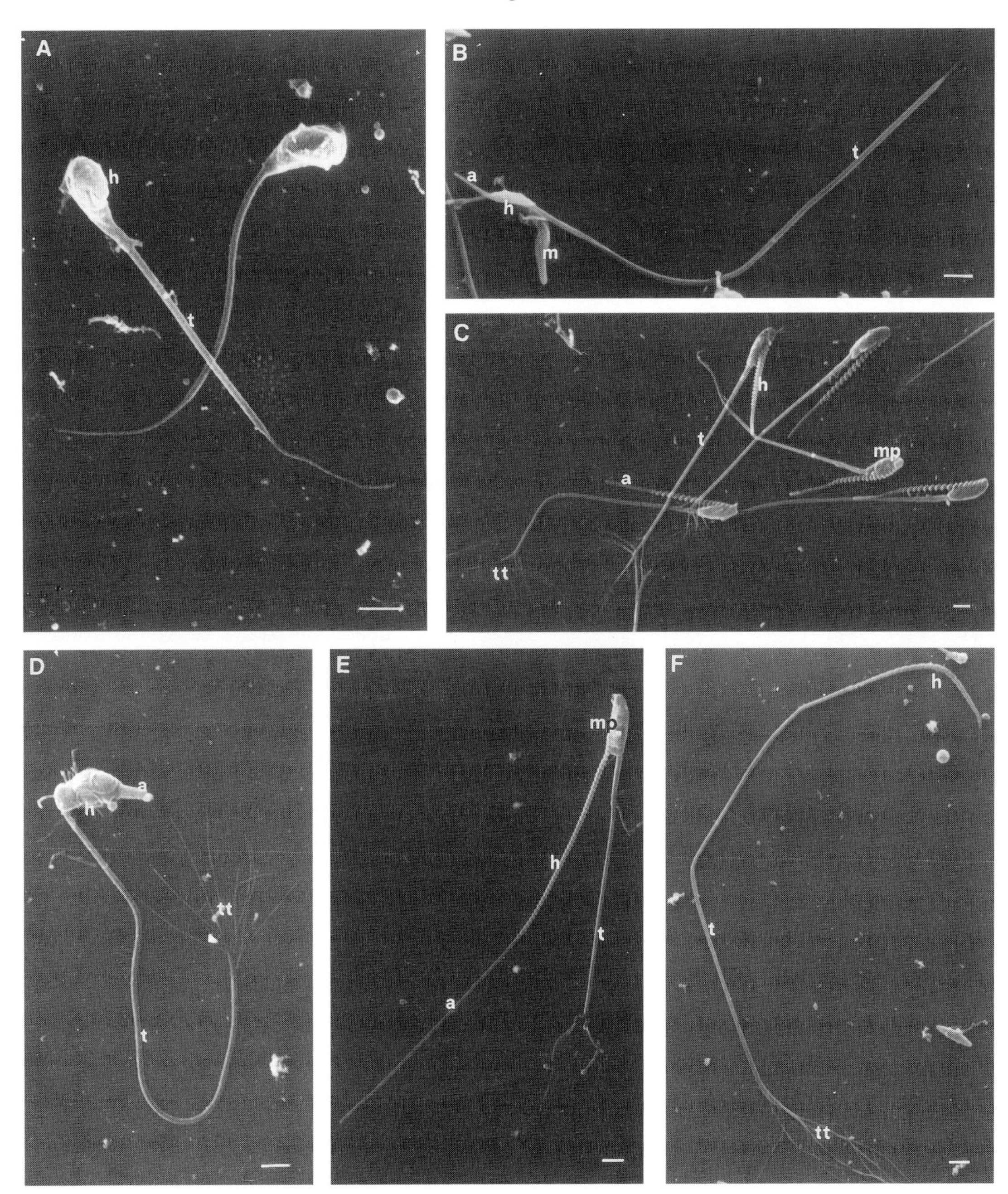

FIGURE 6 Scanning electron micrographs of testicular spermatozoa. (A) *Batillipes pennaki* (marine heterotardigrade). (B) *Pseudechiniscus facettalis* (semiterrestrial heterotardigrade). (C) *Macrobiotus sandrae* (semiterrestrial eutardigrade). (D) *Amphibolus volubilis* (semiterrestrial eutardigrade). (E) *Macrobiotus islandicus* (semiterrestrial eutardigrade). (F) *Isohypsibius monoicus* (limnic eutardigrade) note the filiform shape of the head. a, acrosome; h, head; m, mitochondria; mp, midpiece; t, tail; tt, terminal tuft. Scale bars = 1 μm. (D, reproduced with permission from Rebecchi and Guidi, 1995. Copyright © The Royal Swedish Academy of Sciences.)

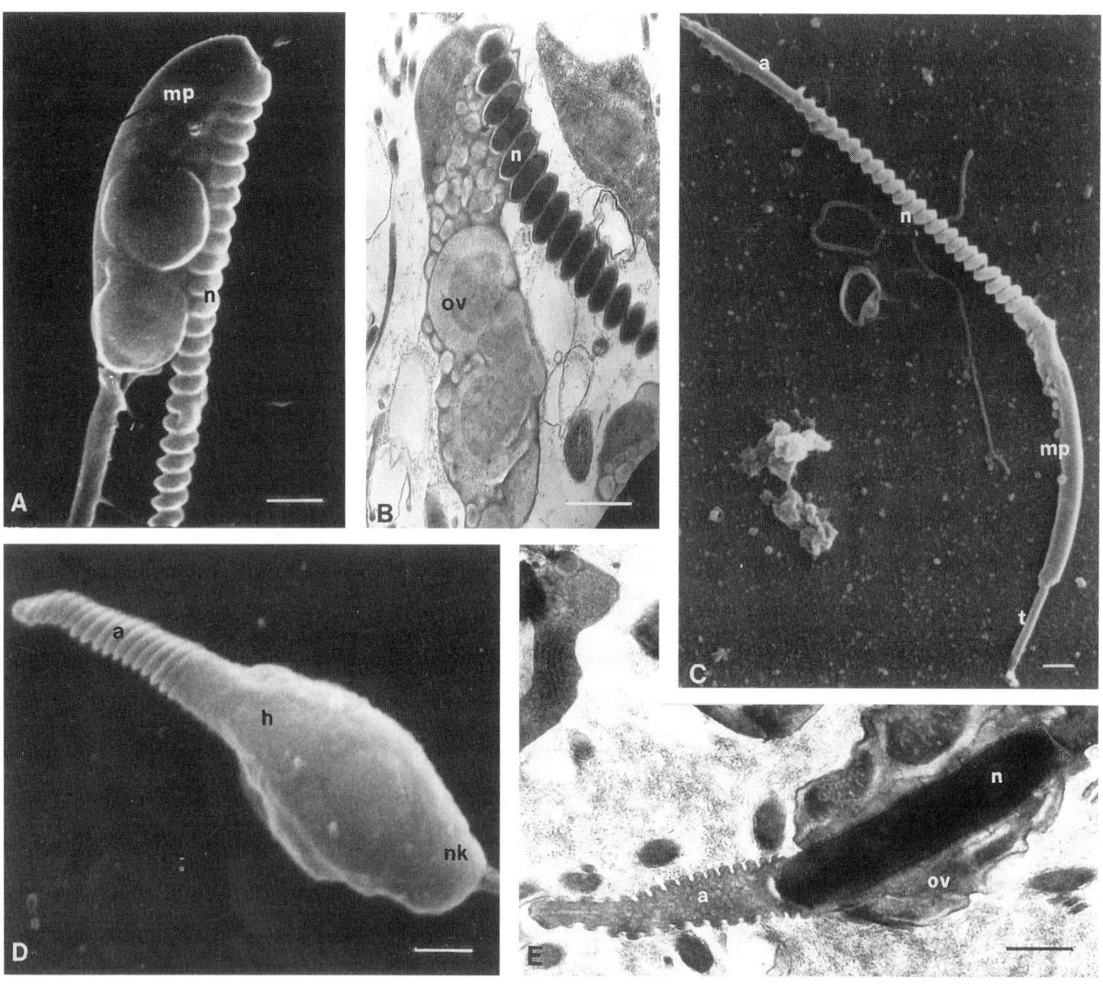

FIGURE 7 Eutardigrade spermatozoa. (A, B) Testicular spermatozoa of *Xerobiotus pseudohufelandi* showing the nutcracker profile with the hinge between the helical and the voluminous midpiece with hemispherical/ovoid elements. (C) Spermathecal spermatozoon of the previous species; note the linear profile, the narrow midpiece, and the tail reduced to a short stub. (D, E) Detail of the testicular spermatozoon head of *Amphibolus volubilis*; note the corkscrew-shaped acrosome and the cylindrical nucleus surrounded by cytoplasmic ovoid elements. a, acrosome; h, head; mp, midpiece; n, nucleus; nk, neck; ov, ovoid elements; t, tail. (A) SEM, (B) TEM, (C), SEM, (D) SEM, (E) TEM. Scale bars = 0.5 μm. (B, reproduced with permission from Rebecchi, 1997. Copyright © John Wiley and Sons, Inc. D, E, reproduced with permission from Rebecchi and Guidi, 1995. Copyright © The Royal Swedish Academy of Sciences.)

one cell per cluster increases its volume and develops into an oocyte. The other cells develop in nurse cells which remain joined to the oocyte by intercellular bridges. The cytoplasm of the young oocytes displays both rough and smooth endoplasmic reticulum, numerous mitochondria, dyctiosomes, and annulated lamellae. During vitellogenesis the germinative vesicle of the oocyte grows considerably, the nucleolus is very large, and the chromosomes are hardly recognizable. At the same stage, the nucleus of the nurse cells is sometimes at prophase and bivalents and chiasmata may be recognized. Oocyte chromosomes are again evident at a diplotene stage only at the end of vitellogenesis. Meiosis (or a corresponding ameiotic oocyte division in many parthenogenetic strains) stops at the first metaphase and starts again only after oviposition. Autosynthesis, micropinocytosis, and nurse cells contribute to the formation of yolk. In mature oocytes the number of mitochondria decreases; the ooplasma displays many yolk bodies

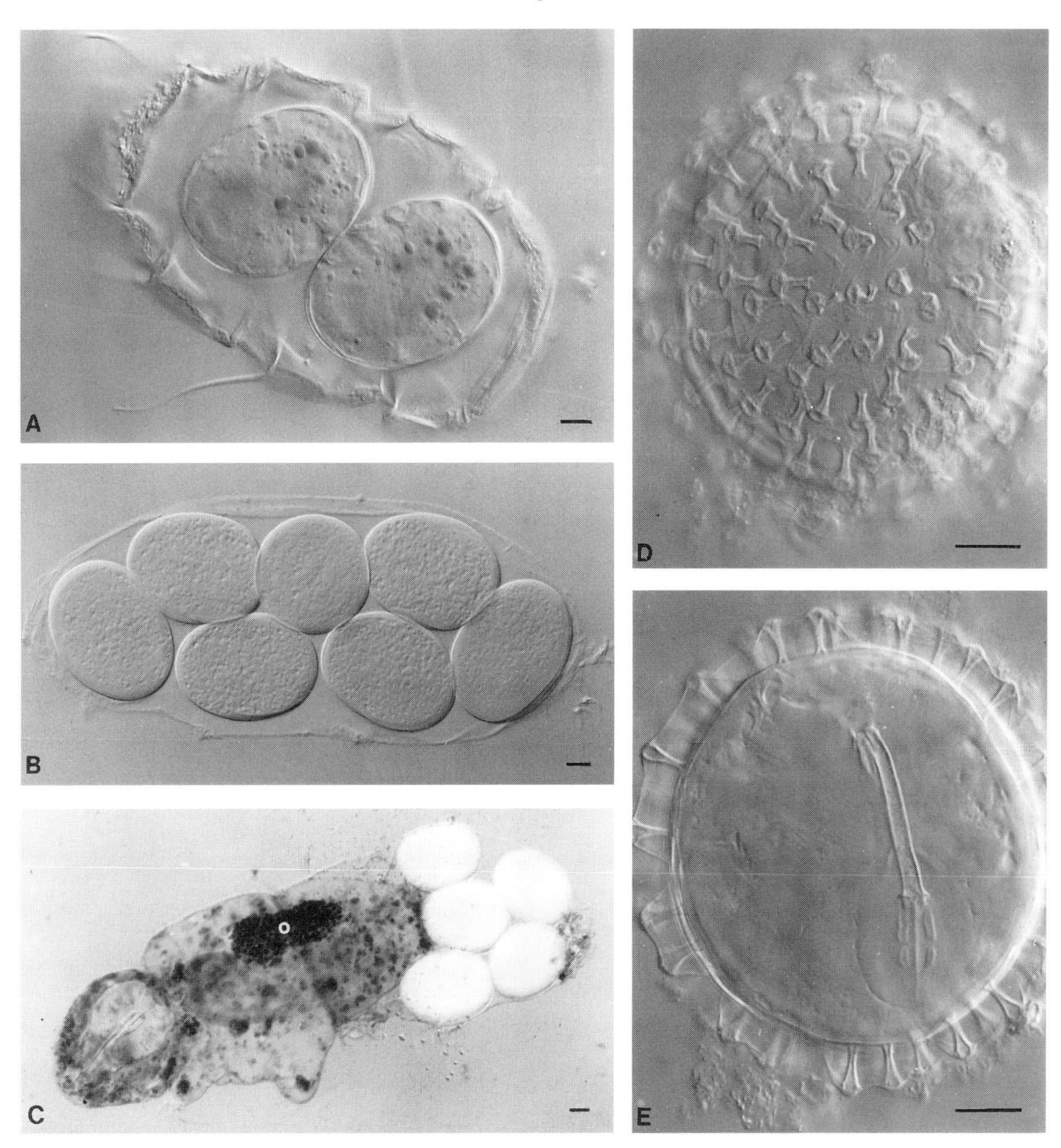

FIGURE 8 (A, B) Smooth eggs laid into the exuvium in the heterotardigrade *Echiniscus* sp. (A) and the eutardigrade *Diphascon scoticum* (B). (C) Egg-filled exuvium attached to the female of the eutardigrade *Pseudobiotus megalonyx*. After oviposition, the content of the ovary (o) looks undifferentiated; orcein, DIC. (D, E) Eggshell with processes in the eutardigrade *Murrayon hastatus*. In E, the egg contains an animal at the end of embryonic development. Note the complete buccopharyngeal apparatus: DIC. Scale bars = 10 μm.

with a granular or spongy texture and rich in glycoproteins with a basic proteinaceous moiety. Acid lipids, which do not seem to be linked to proteins, are also noticeable. A vitelline membrane is present in the mature oocytes; vacuoles and cortical granules are visible near the oolemma.

The eggs of tardigrades are spherical or ovoidal, with a diameter up to 235 μm but usually between 50 and 100 μm (Figs. 8 and 9). Those of marine heterotardigrades are surrounded by a sticky shell and are laid freely. The eggs of the nonmarine heterotardigrades and those of the eutardigrades are sur-

FIGURE 9 Scanning electron micrographs of ornamented eggs in eutardigrades showing the different types of processes. (A) *Macrobiotus persimilis*, (B) *Macrobiotus joannae*, (C) *Macrobiotus richtersi*, and (D) *Ramazzottius oberhaeuseri*. Scale bars = 10 μm.

rounded by a sclerified shell. They have a variously ornamented shell (with spines, cones, laminae, reticulations, and so on; Figs. 8D, 8E, and 9) when laid freely and a smooth shell when laid in the exuvium (Figs. 8A–8C). In a very few cases smooth egg can be laid freely and ornamented eggshells can be occasionally laid in the exuvium. In any case a micropile has never been observed. Ornamented and smooth eggs can be present in both classes. This situation was interpreted as a consequence of two evolutionary steps. With the passage from marine to nonmarine environment the female gamete required more protection, and for this reason a hard and ornamented eggshell developed. Later, with the synchronization between molt and oviposition, the exuvium was used to protect the eggs and subsequently shell ornamentations were lost, and this probably occurred independently in hetero- and eutardigrades.

IV. MATURATION PATTERN AND REPRODUCTIVE PHYSIOLOGY

Most marine female heterotardigrades mature oocytes singularly, whereas nonmarine female heterotardigrades and eutardigrades are iteroparous and

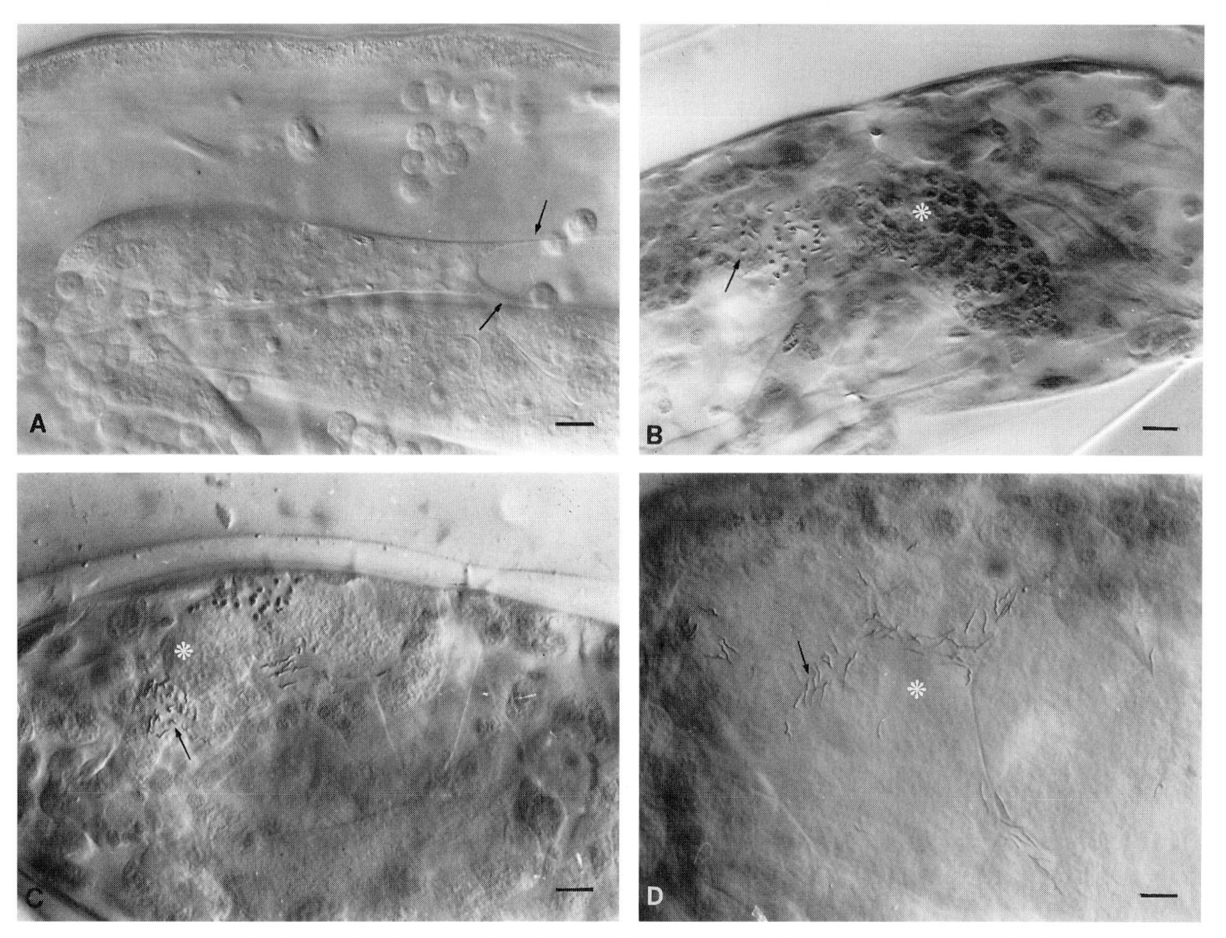

FIGURE 10 Ovotestis in the hermaphroditic eutardigrade *Macrobiotus joannae*. (A) Young ovotestis; note the ligaments (arrows). (B) Gonad containing spermatids, spermatozoa (arrow), and young oocytes (asterisk). (C) Ovotestis containing immature male (arrow) and female (asterisk) germinal elements. (D) Mature oocytes (asterisk) surrounded by mature spermatozoa (arrow). Orcein, DIC. Scale bars = 10 μm.

mature their oocytes in groups, filling all the ovary and using the cavity of their small body alternately to increase the gonad or the midgut. Four maturative stages of the oocyte can be distinguished in the ovarian cycle: previtellogenesis, early and late vitellogenesis (the latter during molting; Fig. 4B), and mature oocyte (Fig. 4A).

In males, three maturation patterns of the testis are found depending on taxa. In some species (marine heterotardigrades and limnic eutardigrades) males are semelparous with a progressive maturation of gametes that leads to a testis full only of spermatozoa (Fig. 3B). Other male eutardigrades have a continuous maturation, which permits them to always have mature and immature male elements in the gonad (Fig. 3D). Lastly, in some species of eutardigrades, males are iteroparous, like females, with cyclical maturation of sperm cells in the testis.

In hermaphrodites the ovotestis begins to mature male germ cells. Subsequently, female elements prevail and fill most of the gonad during the rest of the life, with the exception of one case in which there is an alternate maturation of the two types of gametes. Like females, hermaphrodites are iteroparous and oocyte maturation is synchronous with molting (Fig. 10).

Synchronization of molting and oviposition (also supported by laying of eggs in the exuvium in many species) certainly needs endocrine control. Unfortunately, to date very little is known on this topic.

The only data indicate an increase of neuroendocrine activity in the cells of the middle and lateral areas of the brain that could be related to early/middle oogenesis and spermatogenesis, in the first case in relationship with the molting cycle.

V. FERTILIZATION

Spermatozoon morphology of both marine and semiterrestrial heterotardigrades seems to be related to an external fertilization. This phenomenon was observed in *Batillipes*; moreover, it is deducible in many other marine heterotardigrades by the presence of cuticular seminal receptacles with an independent external opening. On the contrary, the spermatozoon of the eutardigrades seems to be adapted to internal fertilization. In support of this hypothesis are the sporadic presence of an internal epithelial spermatheca and some observation on copulation. In particular, in *Pseudobiotus* we observed males which introduced their sperm cells into the exuvium still attached to the female, from where they rapidly reached the female cloaca and disappeared. In this case there is external insemination, even if protected, and internal fertilization. Therefore, the path followed by tardigrades, or at least eutardigrades, in the evolution of fertilization seems to go from external to internal fertilization.

VI. DEVELOPMENT

A. Embryonic Development

Only oviparous tardigrades are known. All tardigrades possess a homolecithal egg. Cleavage is total, equal, and asynchronous. In particular, the slight twisting of the spindles during the cleavage, observed by some authors, indicates a modified spiral cleavage, even if the clarity of the cleavage pattern is obscured by the asynchrony of early cleavage steps, the equal size of the resulting blastomeres, and the absence of reliable polar bodies that could provide a basis for orientation of the embryo. The cleavage leads to a blastula without blastocoel. Gastrulation takes place by delamination at a very early stage of development. Authors of the past century and of the first half of this century described the origin of the mesoderm in tardigrades by enterocoely with the formation of five paired coelomic pouches connected with the archenteron. Recent studies revealed that mesodermal cells are present before the occurrence of the coelomic pouches and that there are no connections between these pouches and the archenteron. Therefore, these pouches could arise by schizocoely from preexisting mesoderm. The four anterior coelomic pouches desegregate very early, whereas the posterior one merges dorsally into a single anlage that originates the gonad, the only eucoelomic structure in tardigrades.

B. Postembryonic Development

The postembryonic growth takes place by molts, without evident variation in cell number. A difference exists between heterotardigrades and eutardigrades (mesotardigrade development is unknown). Heterotardigrades, both marine and semiterrestrial, are characterized by indirect development. Newborns always lack an anus and gonopore and can differ to various extents from adults in some cuticular structures (at least two claws or two digits less per leg). After the first molt the anus is present, whereas the gonopore is still absent or present as a very small pore. Some authors call these two stages "larval stages." Only after the second molt does the animal assume the shape of the adult. The only exception is *Tetrakentron*, an ectoparasite on holothurians, which has two types of males, one (dwarf male) probably already complete and sexually mature at hatching. Eutardigrades have a direct development. The newborns are similar in shape to the adults, including the presence of a cloaca. Only secondary sex characters, when present, appear in the adult at the life history.

C. Parental Care

Tardigrades limit their parental care to the eggs. It was previously mentioned that many species lay their eggs in the exuvium. Moreover, in a few limnic species belonging to the genera *Isohypsibius* and *Pseudobiotus*, the egg-filled exuvium remains attached to the female until hatching (Fig. 8C). In the marine genus *Echiniscoides* the female transports some eggs on the dorsocaudal part of its body.

Bibliography

Bertolani, R. (1983). Tardigrada. In *Reproductive Biology of Invertebrates Vol. 1: Oogenesis, Oviposition, and Oosorption* (K. G. Adiyodi and R. G. Adiyodi, Eds.), pp. 431–441. Wiley, New York.

Bertolani, R. (1990). Tardigrada. In *Reproductive Biology of Invertebrates Vol. 4, Part B: Fertilization, Development, and Parental Care* (K. G. Adiyodi and R. G. Adiyodi, Eds.), pp. 49–60. Wiley, New York.

Bertolani, R. (1992). Tardigrada. In *Reproductive Biology of Invertebrates Vol. 5: Sexual Differentiation and Behavior* (K. G. Adiyodi and R. G. Adiyodi, Eds.), pp. 255–266. Wiley, New York.

Bertolani, R. (1994). Tardigrada. In *Reproductive Biology of Invertebrates Vol. 6, Part B: Asexual Propagation and Reproductive Strategies* (K. G. Adiyodi and R. G. Adiyodi, Eds.), pp. 25–37. Wiley, New York.

Bertolani, R., Rebecchi, L., and Claxton, S. (1996). Phylogenetic significance of egg shell variation in tardigrades. *Zool. J. Linnean Soc.* **116**, 139–148.

Dewel, R. A., Nelson R. D., and Dewel, W. C. (1993). Tardigrada. In *Microscopic Anatomy of Invertebrates Vol. 12: Onychophora, Chilopoda, and Lesser Protostomata* (F. W. Harrison and M. E. Rice, Eds.), pp. 143–183. Wiley-Liss, New York.

Eibye-Jacobsen, J. (1996–1997). New observations on the embryology of the Tardigrada. *Zool. Anz.* **235**, 201–216.

Ghiselin, M. T. (1969). The evolution of hermaphroditism among animals. *Q. Rev. Biol.* **44**, 189–208.

Nelson D. R. (1982). Developmental biology of the Tardigrada. In *Developmental Biology of Freshwater Invertebrates* (F. W. Harrison and R. R. Cowden, Eds.), pp. 363–398. A. R. Liss, New York.

Pollock, L. W. (1975). Tardigrada. In *Reproduction of Marine Invertebrates Vol. 2: Entoprocts and Lesser Coelomates* (A. C. Giese and J. S. Pearse, Eds.), pp. 43–54. Academic Press, New York.

Rebecchi, L. (1997). Ultrastructural study of spermiogenesis and the testicular and spermathecal spermatozoon of the gonochoristic tardigrade *Xerobiotus pseudohufelandi* (Eutardigrada, Macrobiotidae). *J. Morphol.* **234**, 11–24.

Rebecchi, L., and Bertolani, R. (1994). Maturative pattern of ovary and testis in eutardigrades of freshwater and terrestrial habitats. *Invertebr. Reprod. Dev.* **26**, 107–117.

Rebecchi, L., and Guidi, A. (1995). Spermatozoon ultrastructure of two species of *Amphibolus* (Eutardigrada, Eohypsibiidae). *Acta Zool.* **76**, 171–176.

Rebecchi, L., Guidi, A., and Bertolani, R. (1998). Tardigrada. In *Progress in Male Gamete Biology: Reproductive Biology of the Invertebrates* (B. G. M. Jamieson, Ed.). Oxford & IBH, New Delhi, in press.

Teleosts, Viviparity

John P. Wourms

Clemson University

GLOSSARY

histotrophe A term describing the supplemental nutritive substances, usually fluid, supplied to the embryos of matrotrophic fishes.

lecithotrophy A form of embryonic nutrition occurring in oviparous and viviparous species in which the embryo depends solely on its yolk reserves for nutrients.

matrotrophy A form of embryonic nutrition confined to viviparous species in which the developing embryo receives supplemental supplies of maternal nutrients during gestation.

pericardial amnion A term that refers to the anterior fold of pericardial somatopleure (ectoderm and parietal mesoderm) that encloses the head of the embryos of viviparous fishes.

propagule A collective term that refers to the offspring of sexually reproducing organisms without specifying their stage of development, e.g., egg, zygote, embryo, hatchling, neonate, larva, and juvenile.

trophotaeniae Perianal rosette or ribbon-like embryonic appendages composed of a surface epithelium derived from the intestine and a vascularized connective tissue core that function in nutrient uptake and gas exchange.

Oviparity, literally egg laying, and viviparity, literally live bearing, are successful modes of reproduction that occur among both vertebrates and invertebrates. Oviparity is considered to be the unspecialized, primitive mode of reproduction. Viviparity is a derived, specialized mode of reproduction that has independently evolved from oviparity many times in numerous taxonomically divergent groups. Viviparity, defined in the context of fishes, is a reproductive mode in which eggs are fertilized internally and are retained and complete their embryonic development within the maternal reproductive system. Hatching (i.e., eclosion from an egg envelope, if one is present) precedes or coincides with parturition and the result is a free-living fish. The evolution of viviparity from oviparity involves (i) a shift from external to internal fertilization, (ii) retention of embryos in the female reproductive system, (iii) utilization of the ovary or oviduct as a site of gestation, (iv) structural and functional modification of the embryo and of the female reproductive system, and (v) modification of extant endocrine systems controlling reproduction.

I. EVOLUTION OF TELEOST VIVIPARITY

A. Selective Advantages and Disadvantages

Viviparity evolved from oviparity because the selective advantages of a viviparous mode of reproduction for a specific taxon outweigh its disadvantages and are greater than the advantages of an oviparous mode. The following are advantages of piscine viviparity:

1. Development within the female reproductive tract reduces the risk of egg/embryo predation.
2. Maternal homeostasis of the embryonic environment protects propagules from environmental perturbations.
3. Viviparity allows embryos to develop into precocious neonates, thus enhancing their postpartum survival.
4. Because of enhanced survival of offspring, the total number of surviving offspring is greater than that of low-fecundity congers.
5. In very small fishes, evolution of reproductive modes such as viviparity that enhance offspring survival will be favored to compensate for the intrinsic low fecundity that is the consequence of reduced female body size.
6. Viviparity enhances reproductive success by eliminating competition for external spawning sites because the use of the female reproductive tract as a site of gestation in effect amplifies the number of available, unutilized reproductive niches.
7. Viviparity permits exploitation of the pelagic niche and the adoption of a completely pelagic lifestyle without evolution of pelagic eggs.
8. Viviparity permits colonization of new habitats by small numbers of gravid founder individuals.
9. Matrotrophic viviparity permits efficient allocation of maternal energy to propagules during both oogenesis and gestation.
10. By permitting the maternal transport of embryos from the geographical site of fertilization to the site of parturition, viviparity facilitates a migratory lifestyle with its implied benefits of resource acquisition over a broad geographical range.

On the other hand, the following are the major disadvantages of viviparity:

1. Reduced fecundity
2. Increased energetic cost to the female
3. Increased risk of predation on gravid fish
4. Increased potential of brood loss resulting from extreme physiological stress of gravid fish or maternal death.

B. Incidence of Viviparity in Fishes

Viviparity has evolved in five of the major groups of fishes, all within the Chondrichthyes (cartilaginous fishes) and Osteichthyes (bony fishes). Viviparity is the dominant mode of reproduction among the cartilaginous fishes in which 510–520 among 900–1100 extant species (about 55% of living species) are known to be oviparous. Among the major groups of Osteichthyes, viviparity is widespread but far less prevalent. About 510 of an estimated 25,000–28,000 species (about 2 or 3%) are viviparous. Nearly all the viviparous species occur in 13 families of teleost fishes. *Latimeria*, the living coelacanth, which is a sarcopterygian (flesh-finned) fish and part of the lineage from which the terrestrial vertebrates evolved, is viviparous, as were some of its fossil relatives. Viviparity also evolved among extinct chondrosteans (sturgeon-like fishes). The relative paucity of viviparous species among teleost fishes seems puzzling. There are several possible explanations. For many species, oviparity, the primitive mode of reproduction, is successful, i.e., selectively advantageous. Second, the acquisition of internal fertilization seems to have been a necessary prerequisite to the evolution of viviparity in fishes. Internal fertilization is not at all common in teleost fishes and has evolved independently in several different lineages. Although only about 2 or 3% of all teleost fishes are viviparous, 57% of those species with internal fertilization are viviparous. This ratio is very close to the 55% value for the number of viviparous species in the chondrichthyans, all of which have internal fertilization. Once internal fertilization has evolved in a particular lineage of fishes, there appears to be a high degree of probability that viviparity will also evolve in that lineage. Finally, because viviparity is a form of parental care, some of its advantages such as enhanced survival of offspring can be achieved by other reproductive strategies, e.g., external brooding or egg guarding. In fact, other forms of parental care among teleosts are more widespread than viviparity. There are more teleosts that brood their young, externally or in pouches, e.g., pipefishes and seahorses, than there are viviparous species.

C. Families of Viviparous Fishes

Viviparity has independently evolved an estimated 13–23 times in the following 13 families of teleost fishes:

1. Bythitidae or viviparous brotulas are predominantly marine (Atlantic, Pacific, and Indian Oceans), rarely freshwater; 90/90 species are viviparous. Examples: *Oligopus*, *Microbrotula*, *Ogilbia*, *Lucifuga*, and *Bythites*.
2. Aphyonidae, also known as viviparous brotulas, are marine (Atlantic, Pacific, and Indian Oceans) and mostly found at depths >700 m; 21/21 species are viviparous.
3. Parabrotulidae or false brotula are small (6 cm) deep-water, mesopelagic, marine (Atlantic, Pacific, and Indian Oceans) fishes; 3/3 species are viviparous.
4. Hemiramphidae or halfbeaks are marine (Atlantic, Pacific, and Indian Oceans) or freshwater fishes. There are three Indopacific viviparous genera, viz. *Dermogenys*, *Hemirhamphodon*, and *Nomorhamphus*; 21/85 species are viviparous.
5. Anablepidae include two genera, viz. *Anableps*, the four-eyed fishes, and *Jenynsia*, the jenynsiids. The former occur in fresh and brackish waters of the Atlantic slope of southern Mexico and northern South America and the latter occur in fresh water in southern South America; 8/9 species are viviparous.
6. Poeciliidae or poeciliids are small fresh- or brackish water fishes widely distributed from the eastern United States through South America; 190/293 species are viviparous. Viviparity among teleosts is frequently associated with the Poeciliidae because of their ubiquity as home aquarium fishes. Examples: *Gambusia*, *Heterandria*, *Poecilia*, and *Xiphophorus*.
7. Goodeidae or splitfins are freshwater fishes from the southwestern United States and Mexico; 36/40 species are viviparous. Examples: *Ameca*, *Ataeniobius*, and *Goodea*.
8. Scorpaenidae include the scorpion fishes and rockfishes, they are marine, and they occur in all tropical and temperate seas; 110+/388 species are viviparous. Examples: *Sebastes* and *Helicolenus*.
9. Comephoridae or Baikal oilfishes are pelagic freshwater fishes endemic to Lake Baikal, Siberia; 2/2 species are viviparous. Example: *Comephorus*.
10. Embiotocidae or surfperches are coastal marine, rarely freshwater fishes that occur in the Northern Pacific Ocean, especially off the coast of western North America; 24/24 species are viviparous. Examples: *Cymatogaster* and *Rhacochilus*.
11. Zoarcidae or eelpouts are marine fishes, usu-

ally benthic, that occur from the Arctic to the Antarctic; 3/220 known to be viviparous. Example: *Zoarces viviparus*.

12. Labrisomidae or labrisomids are marine fishes that occur mostly in the tropical waters of the western Atlantic and eastern Pacific Oceans; 21/102 species are viviparous. Examples: *Starksia* and *Xenomedea*.

13. Clinidae or clinids are marine (Atlantic, Pacific, and Indian Oceans) fishes that primarily occur in the temperate waters in both Southern and Northern Hemispheres; 60/73 species are viviparous. Example: *Clinus superciliosus*.

II. ESTABLISHMENT OF NOVEL MATERNAL–EMBRYONIC RELATIONSHIPS

During the evolution of viviparity from oviparity there is a shift in the site of embryonic development from an external abiotic environment to the internal biotic environment of the female reproductive system. The transition to viviparity is marked by the establishment of novel maternal–embryonic relationships that involve modification of the trophic, osmoregulatory and excretory, respiratory, endocrinological, and immunological processes that are common to the development of both oviparous and viviparous fishes. The eggs of oviparous fishes develop in an aquatic environment which is conducive to development. The environment, however, is subject to change in temperature, salinity, pH, ion and oxygen content, and pressure. Oxygen is generally plentiful in water but there may be significant differences between the osmolarity and ionic content of the water and the egg. The developing embryo of oviparous species is metabolically autonomous. Osmo- and ionic regulatory processes and respiration are primarily linked to conditions extrinsic to the egg, whereas trophic, i.e., embryonic energy budget, endocrinological, and immunological processes are primarily associated with intrinsic events of development and growth. In viviparous fishes, all these processes are retained but are modified because they now involve both the embryo and the maternal fish. The female reproductive system, which is the new environment, is regulated by the maternal organism to provide conditions suitable for development. For its part, the developing embryo will, to varying degrees, lose its metabolic autonomy and become increasingly dependent on the maternal organism.

There are two stages in the evolution of viviparity and the establishment of new maternal–embryonic relationships. The first stage involves egg retention in the female reproductive tract and an unspecialized form of lecithotrophic viviparity, i.e., the embryo depends solely on its yolk reserves for its nutrition, as do oviparous embryos. This transition involves substantive changes in the morphology, endocrinology, and physiology of the female reproductive system but is accompanied by little change in the structure and function of the embryo, i.e., the embryo tends toward metabolic autonomy. Egg retention and development to term is brought about by a modification of the extant endocrine mechanisms controlling reproduction in the female. Retention of the developing embryos requires the coopting of the oviduct, follicle, or ovarian lumen as sites of gestation and often involves morphological changes in these structures that accommodate their new function. Still another suite of structural and functional changes may occur because the maternal organism now assumes the function of regulating the environment of the developing embryo, i.e., gas exchange and osmo- and ionic regulation.

A. Ionic and Osmoregulation

Comparative studies of ionic and osmoregulation in elasmobranchs have revealed a shift from embryonic physiological autonomy in oviparous species to varying degrees of maternal dependency in viviparous species. Physiological relationships are less clear in developing teleosts. Eggs of some oviparous fishes, such as *Fundulus*, are well-known for their tolerance of ionic and osmotic fluxes. Early embryos of the guppy, *Poecilia reticulata* (a lecithotrophe) and a surfperch, *Cymatogaster*, and clinid, *Clinus superciliosus* (both matrotrophes) are physiologically dependent and are unable to iono- or osmoregulate. Physiological autonomy and the ability to iono- and osmoregulate are acquired later in gestation.

B. Immunological Relationships

Retention of developing embryos also poses an immunological problem for the maternal organism because the embryos constitutes an allograft. Except in highly inbred matings, the embryo inherits paternal histocompatibility antigens that are foreign to the mother but fails to inherit all the maternal histocompatibility antigens. There is little knowledge of how this challenge is resolved in fishes. In many instances of lecithotrophic viviparity, embryos are enclosed in egg envelopes, an extracellular matrix of maternal origin, throughout gestation. Because these envelopes seem to be impermeable to the passage of large molecules, such as protein antigens and antibodies, the envelopes in effect serve as an immunological barrier. Such, however, is not the case in many matrotrophic fishes, in which there is often direct contact between maternal and embryonic tissues. In addition, several other processes are of immunological consequence. In some fishes, sperm can be stored in the female reproductive tract for long periods (10 months or more). Since fish sperm is antigenic and the sperm constitutes an allograft, why is it not rejected? Finally, evidence for maternal–embryonic passive transfer of immunity has been documented in some poeciliids and embiotocids. One route seems to be the uptake of immunoglobulin M by the egg during oogenesis. Other routes seem to exist in matrotrophic species that endocytose proteins from a histotrophe that is composed of serum proteins, e.g., surfperches (Embiotocidae).

C. Trophic Relationships: Lecithotrophy and Matrotrophy

Lecithotrophic viviparity is the terminal stage in the evolution of viviparity in some teleosts such as the rockfish, *Sebastes* and many poeciliids. Although primitive and unspecialized, it obviously is a successful reproductive style. The selective advantages of increased postpartum survival that accrue to large, well-developed young, even of lecithotrophic fishes, should favor a continued shift from altricial to precocial development and a further increase in propagule size. The production of larger, more mature offspring has been effected in some lecithotrophic species through the evolution of egg gigantism, e.g., the living coelacanth, *Latimeria,* whose eggs are 9 cm in diameter with a dry weight of 185 g. More commonly, there is a transition from lecithotrophy to matrotrophy and a corresponding shift from embryonic nutritional autonomy to dependency that brings about an increase in size and degree of maturity of offspring. This set of changes is the second stage in the evolution of viviparity. In those species in which the embryo shifts to a metabolic dependency on the maternal organism, the size of the young at term is no longer limited by the initial supply of yolk, and the rate of embryonic development is no longer limited by the ability of the embryo to regulate its environment. Rather, the size of the maternal organism and its ability to provide supplemental nutrients and regulate the embryonic environment become the limiting factors in regulating embryonic growth and neonatal size at the cost of reducing the number of young produced.

An increase in embryo size and the shift to embryonic metabolic dependency are causally linked to the evolution of distinctive maternal and embryonic specializations in matrotrophic viviparity. Two critical phases during the development of viviparous fishes require special physiological adaptations of the ovary as well as the developing embryos. The first phase occurs when the demand for additional nutrients has to be met after the embryonic yolk reserves have been absorbed completely. The second phase occurs when the respiratory requirements of the embryo exceed the transport capacity of unspecialized exchange surfaces. Maternal tissues at the site of gestation in matrotrophic species become modified for nutrient transfer and gas exchange to a far greater extent than do those in lecithotrophic species. Similarly, the most elaborate embryonic specializations for metabolic exchange are associated with matrotrophy. A distinction has been made between primary and secondary embryonic exchange surfaces. The former category includes the yolk sac, pericardial sacs and their derivatives, the pericardial chorioamnion, portal blood networks, gill filaments, and the general surface epithelium. Secondary exchange surfaces are found in the vertical fin system of embryonic surfperches, the trophotaeniae of goodeids, ophidioids, and parabrotulids, and the yolk sac placenta and umbilical cord appendiculae of sharks. All these structures seem to incorporate a basic principle of design: the possession of a thin surface epithelium, whose surface area is often amplified and beneath which there is the capillary network of a hypertrophied vascular supply. In many instances, the maternal and embryonic structures in question may serve a dual role both in gas exchange and in nutrient transport. It should be kept in mind that in addition to structural specializations, functional specialization have occurred at the tissue, cellular, and molecular levels.

The trophic relationship, linked as it is to embryonic size, emerges as a dominant theme in the evolution of piscine viviparity. In their primitive state, embryos are lecithotrophic; that is, they depend entirely on their yolk reserves to supply the energy and anabolic metabolites for development and growth. Yolk reserves are metabolized much as they are in oviparous species and there is a net loss in dry weight during gestation. In contrast, matrotrophic embryos receive a steady supply of supplemental maternal nutrients during gestation. Other substances that regulate development, differentiation, or growth also may be transferred. The extent of matrotrophy, in both absolute and relative terms, varies from species to species. Extreme matrotrophes undergo enormous increases in dry weight, e.g., 842,900% in the four-eyed fish, *Anableps*, a teleost with follicular gestation, and $5.8 \times 10^6\%$ in the spadenose shark, *Scolodion laticaudus*, a placental species. Three different patterns of embryonic nutrient transfer occur in matrotrophic fishes: (i) oophagy and adelphophagy, (ii) trophodermy, and (iii) placentotrophy. Oophagy and adelphophagy refer to the ingestion of ova or siblings during gestation. In trophodermy, maternal nutrients are transferred from their epithelial site of origin across intratissue spaces to distally located embryonic epithelial absorptive sites that are not in intimate association with maternal tissues. A further distinction can be made between dermotrophy, i.e., transfer across the embryonic body surface and derivatives of it (such as the classical extraembryonic membranes), and enterotrophy, i.e., transfer across the gut epithelium and derivatives of it (such as trophotaeniae). Placentotrophy is the transfer of maternal nutrients to the embryos via a placenta.

III. SITES OF GESTATION AND MATERNAL–EMBRYONIC METABOLIC EXCHANGE

A. Teleost Ovary as a Site of Gestation

The morphology of the female reproductive system determines the sites of gestation in viviparous fishes. Two sites are utilized: the uterus and the ovary. In viviparous chondrichthyans and the living coelacanth, fertilized eggs are retained and develop to term in the highly modified posterior region of the oviduct that is called the uterus. Oviducts are embryological derivatives of the Müllerian ducts. Gestation in these two lineages is referred to as uterine gestation. Teleost viviparity is anomalous because gestation in these fishes takes place in the ovary and is referred to as ovarian gestation. Thus, the ovary of viviparous teleost fishes differs from that of other vertebrates inasmuch as it is the site of both egg production and gestation. Ovarian gestation is the consequence of the evolutionary modification of the female reproductive system. The currently accepted view is that teleosts lack an oviduct, i.e., a structure derived from the Müllerian duct. In viviparous and some oviparous teleosts, a gonoduct is present, but it is derived from ovarian tissue. The consensus is that oviducts derived from Müllerian ducts, which are present in primitive actinopterygian fishes, were lost during evolution of the teleosts.

Ovarian gestation in teleosts occurs either in the ovarian lumen (intralumenal gestation) or in the ovarian follicle (intrafollicular gestation). Intralumenal gestation is the prevalent mode of development, occurring in 10 of the 13 families in which viviparity is known. In most species with intralumenal gestation, fertilization and embryonic development commence in the ovarian follicle and development is completed in the ovarian lumen. Intrafollicular gestation occurs in clinids, some labrisomids, the poeciliids, and the anablepid, *Anableps*. In these fishes, ova are fertilized within the ovarian follicle, and the developing embryos are retained within it until they are released at the time of parturition. After fertilization, the follicle remains intact and undergoes changes to accommodate the requirements of the developing embryos. In both intralumenal and intrafollicular gestation, specializations of maternal and embryonic tissues have evolved to facilitate physiological exchange.

B. Follicular Gestation

When the follicle is the site of gestation, the follicular epithelium and its associated vasculature are involved in the maintenance of the embryonic environment and maternal–embryonic metabolite exchange. In clinids, the follicular epithelium produces a follicular fluid that contains high concentrations of lipids and amino acids but levels of protein lower than those in maternal plasma. In *Anableps*, an extreme matrotrophe, the follicle wall is greatly modified and forms villi. Among poeciliids the extent of follicle wall modification varies. In lecithotrophes such as *Poecilia*, the follicle cells are not specialized for nutrient transport but may play a role in gas exchange. In some matrotrophic poeciliids, the follicle wall is heavily vascularized and develops elongated villi covered by "secretory" cells. Recently, the follicle wall of *Heterandria formosa*, another matrotrophic poeciliid, has been shown to display several cellular features typical of a metabolically active, transporting epithelium. Ultrastructural observations and experimental data indicate that the follicular epithelium is a selective barrier that permits the passage of some serum macromolecules as well as low-molecular-weight nutrients, e.g., monosaccharides and amino acids from the maternal circulation to the follicular fluid.

C. Intralumenal Gestation

Fishes with intralumenal gestation possess a number of structural and physiological specializations of the ovary and gonoduct that are implicated in nutrient transfer or metabolic exchange. In many instances, evidence for function is circumstantial. There appear to be four major classes of specializations. The modified lumenal epithelium may form a transporting epithelium that regulates the selective passage of proteins, amino acids, and other molecular species from the maternal vascular system to the ovarian fluid, e.g., goodeids. Second, hypertrophied regions of the lumenal epithelium and their associated vascular supply form ovarian folds or flaps that serve as the maternal portion of a branchial placenta, e.g., *Jenynsia*. In the surfperches (Embiotocidae) hypertrophied ovigerous folds form sheet-like structures that enclose the embryos in compartments. These highly vascularized structures appear to function in gas exchange and also may be involved in the synthesis or transport of histotrophe proteins as well as the transport of amino acids from the maternal serum. The ovary of the eelpout, *Zoarces viviparous,* elaborates an unusual modification during gestation. Following ovulation, empty follicles do not regress but are retained throughout gestation. These follicles undergo a remarkable reorganization and hypertrophy to form extensively vascularized, villous projections termed "calyces nutriciae." These structures seem to function in the transfer of low-molecular-weight nutrients from the blood plasma into the histotrophe. The morphology of the calyces nutriciae is strikingly similar to that of the ovigerous bulbs that occur in some viviparous brotulas (Bythitidae), e.g., *Ogilbia*, *Lucifuga*, and *Bythites*. Late-term embryos of *Ogilbia* have been reported to mouth the bulbs in such a manner as to constitute a buccal placenta.

IV. EMBRYONIC SITES OF NUTRIENT UPTAKE AND METABOLIC EXCHANGE

Embryonic uptake of maternally derived nutrients and maternal–embryonic metabolic exchange takes place across two major classes of embryonic epithelial surfaces: the integument and the gut. Integumental or dermotrophic transfer takes place across the epithelium of the general body surface and its derivatives. Transfer sites include (i) general body surface, e.g., early clinid and *Heterandria* embryos; (ii) gill filaments, e.g., sharks and rays; (iii) finfolds and fin epaulets, e.g., some goodeids and surfperches; (iv) yolk sac surfaces and yolk sac placenta, e.g., in sharks, rays, and the living coelacanth; (v) pericardial sac surfaces, e.g., some poeciliids and goodeids; (vi) surfaces of the pericardial amniochorion, e.g., *Heterandria* and other poeciliids; and (vii) the pericardial trophoderm, e.g., the four-eyed fish, *Anableps*. Gut-associated or enterotrophic transfer takes place across the epithelium of the gut and gut derivatives. Transfer sites include (i) the gut, especially the intestine and intestinal villi, e.g., surfperches and *Anableps*; (ii) the branchial portion of the branchial placenta, e.g., *Jenynsia* and some rays; and (iii) trophotaeniae, i.e., perianal extensions of the intestinal epithelium present in goodeids, a parabrotulid, a surfperch, and two brotulas.

A. Trophotaeniae

Trophotaeniae (literally "growth ribbons") are external rosette- or ribbon-like structures that project from the terminal end of the embryonic hindgut into the fluid-filled ovarian lumen. Trophotaeniae have independently evolved in the parabrotulid (*Parabrotula*), the surfperch (*Rhacochilus*), and in two genera of brotulas (*Oligopus* and *Microbrotula*). Their structure and function have been most extensively investigated among goodeid fishes. Trophotaeniae facilitate substantial nutrient transfer to developing embryos, resulting in dry mass increases in excess of 15,000%. Among goodeids, two types of trophotaeniae occur: rosette trophotaeniae that consist of a series of short, lobulated processes and ribbon trophotaeniae that consist of long, slightly flattened ones. Trophotaeniae of both types consist of a simple surface epithelium surrounding a highly vascularized core of connective tissue. Trophotaenial epithelial cells are derived from the embryonic hindgut and exhibit the structural and functional characteristics of intestinal absorptive cells. Cells of ribbon trophotaeniae take up macromolecules, possess an apical endocytotic complex, and are structurally organized in the "open" configuration characteristic of neonatal mammalian intestinal absorptive cells. Rosette cells transport small molecules, lack an apical endocytotic complex, and are structurally organized in the "closed" configuration characteristic of adult mammalian, intestinal absorptive cells. Intestinal cells of both embryonic and adult goodeids endocytose macromolecules and are in the open configuration. Thus, the closed configuration of the rosette cells is an evolutionary innovation for the transport of small molecules.

B. Placental Associations

The juxtaposition of maternal and embryonic sites of metabolic exchange and the selective advantages occurring from more efficient transport processes have led to the evolution of four different classes of placental relationships in viviparous fishes. The term placenta is used in its modern sense; that is, a placenta is an intimate apposition or fusion of the fetal (or embryonic) organs to the maternal (or paternal) tissues for physiological exchange. Four classes of placental relationships have evolved among viviparous fishes: (i) the yolk sac placenta, apposition of a modified yolk sac with the uterine wall, which occurs in some sharks; (ii) follicular placenta, an intimate apposition between the follicular epithelium and embryonic surface in poeciliid and other fishes; (iii) branchial placenta, an association either between villi of the oviductal lumenal epithelium and the branchial region of embryonic rays or between ovarian flaps, i.e., projections of the ovarian lumenal epithelium, and the branchial region of *Jenynsia* embryos; and (iv) trophotaenial placenta, i.e., embryonic trophotaeniae lying in close proximity to the maternal ovarian lumenal epithelium of goodeids and several other fishes.

C. Fate of Exchange Sites

Although absorptive and exchange structures are advantageous during gestation, they present a potential hazard as gestation ends. At parturition, the full-term embryo leaves the isosmotic female environment and enters either a hypoosmotic (freshwater) or hyperosmotic (seawater) environment. The neonate could suddenly experience an increased osmotic stress due to either the influx or the efflux of water and ions across exchange structures. The problem with absorptive and exchange structures suddenly becoming osmotic liabilities is "solved" by reducing the surface : volume relationship of absorptive surfaces or by rendering embryonic exchange surfaces nonfunctional at birth or by discarding them. In embiotocids, blood is shunted away from the capillary web of the hypertrophied vertical finds. In goodeids, the trophotaeniae are autotomized at the site of the nascent anus.

Bibliography

Grove, B. D., and Wourms, J. P. (1994). The follicular placenta of the viviparous fish, *Heterandria formosa*. II. Ultrastructure and development of the follicular epithelium. *J. Morphol.* **220**, 167–184.

Hollenberg, F., and Wourms, J. P. (1994). Comparative studies of the ultrastructure and protein uptake of the absorptive epithelia of ribbon and rosette trophotaeniae of goodeid fishes. *J. Morphol.* **219**, 105–126.

Hollenberg, F., and Wourms, J. P. (1995). Embryonic growth and maternal nutrient sources in goodeid fishes. *J. Exp. Zool.* **271**, 379–394.

Knight, F. M., Lombardi, J., Wourms, J. P., and Burns, J. R. (1985). Follicular placenta and embryonic growth of the viviparous four-eyed fish (*Anableps*). *J. Morphol.* **185**, 131–142.

Korsgaard, B. (1994). Proteins and amino acids in maternal–embryonic trophic relationships in viviparous teleost fishes. *Israel J. Zool.* **40**, 417–429.

Lombardi, J., and Wourms, J. P. (1988). Embryonic growth and trophotaeniae development in goodeid fishes (Teleostei: Atheriniformes). *J. Morphol.* **197**, 193–208.

Meisner, A., and Burns, J. R. (1997). Viviparity in the halfbeak genera *Dermogenys* and *Nomorhamphus* (Teleostei: Hemiramphidae). *J. Morphol.* **234**, 295–317.

Mor, A., and Autalion, R. R. (1990). Transfer of antibody activity from immunized mother to embryo in tilapias. *J. Fish Biol.* **37**, 249–255.

Nagahama, Y. (1983). The functional morphology of teleost gonads. In *Fish Physiology* (W. S. Hoar, D. J. Randall, and E. M. Donaldson, Eds.), Vol. 9A, pp. 223–275. Academic Press, New York.

Schindler, J. E., and Hamlett, W. C. (1993). Maternal–embryonic relationships in viviparous teleosts. *J. Exp. Zool.* **266**, 378–393.

Webb, P. W., and Brett, J. R. (1972). Respiratory adaptations of prenatal young in the ovary of two species of viviparous seaperch, *Rhacochilus vacca* and *Embiotoca lateralis*. *J. Fish Res. Board Can.* **29**, 1525–1542.

Wourms, J. P. (1981). Viviparity: The maternal–fetal relationship in fishes. *Am. Zool.* **21**, 473–515.

Wourms, J. P. (1994). The challenges of piscine viviparity. *Israel J. Zool.* **40**, 551–568.

Wourms, J. P., and Lombardi, J. (1992). Reflections on the evolution of piscine viviparity. *Am. Zool.* **32**, 276–293.

Wourms, J. P., Grove, B. D., and Lombardi, J. (1988). The maternal–embryonic relationship in viviparous fishes. In *Fish Physiology* (W. S. Hoar and D. J. Randall, Eds.), Vol. 11B, pp. 1–134. Academic Press, San Diego.

Temperature, Effects on Testicular Function

Jeffrey B. Kerr

Monash University

GLOSSARY

apoptosis A specific type of cell death with DNA fragmentation, chromatin clumping, cell shrinkage, and condensation and phagocytic disposal of pyknotic debris.

germ cells Cells undergoing proliferation and maturation into spermatozoa.

inguinal canal An extension of the peritoneal cavity containing the vas deferens and neurovascular supply of the testis.

Leydig cells Steroidogenic cells in the testicular intertubular tissue; they produce testosterone.

peritubular tissue A layer of connective tissue and modified smooth muscle cells surrounding the seminiferous tubules.

scrotum A sac of skin, muscle, and connective tissues in which the testis is suspended by the spermatic cord.

Sertoli cells Supporting or sustentacular cells in the seminiferous tubules which direct the process of spermatogenesis.

spermatogenesis The transformation by cell division and maturation of a spermatogonium into spermatozoa.

The temperature within the testis of mammals with a scrotum is several degrees Celsius below that of core body temperature. In these mammals the lower temperature afforded by the intrascrotal environment is essential for normal testicular development during postnatal and pubertal development and is an absolute requirement for quantitatively normal sperm production and fertility in adult life. All of the cellular elements within the testis exhibit abnormalities of morphology and function in response to sustained or frequent but intermittent episodes of elevated temperature. The germ cells are particularly sensitive in this regard and show early degeneration in response to induced rises in temperature, which in chronic conditions may result in infertility and an increase in the risk or incidence of pathological changes including malignant tumors.

I. INTRODUCTION

In most mammals including man, the process of spermatogenesis is dependent on a temperature environment which is cooler than that of deep-body or core temperature. In humans the former is around 33°C compared to 36 or 37°C in the latter. Precisely why the testis requires the lower temperature within a scrotum remains unknown, but the association between nondescent of the testis (and thus elevated extrascrotal temperature) and infertility has been recognized for over 150 years. Initial experiments exploring the relationship between elevated temperature and testicular function began in the late nineteenth century using "experimental cryptorchidism" in dogs in which the testis was surgically relocated into the abdominal cavity. Since then, this technique has been the main approach used in the study of temperature vs testicular function, although water baths, saunas, insulated clothing, and the heat generated from electric light bulbs have all been used to demonstrate that heat is detrimental to testicular function. The scientific interest in the effects of temperature on the testis is of importance since the incidence of maldescended testes in infant boys is about 1% and, if untreated, the affected testis or testes will show impaired spermatogenesis and are more likely to develop tumors compared to normal scrotal testes.

II. NATURAL AND EXPERIMENTAL VARIATION IN TESTICULAR TEMPERATURE

Although primates and numerous domestic, farm, and wild mammals have scrotal testes, others such as aquatic mammals (dolphins, whales, manatees, and dugong) do not. The elephant and hyrax have no scrotum, bats and sloths have testes in inguinal locations, and monotremes (echidna and platypus) have intraabdominal testes. Partially descended testes, at times covered by abdominal skin, are noted in seals, sea lions, walrus, anteaters, and aardvarks. The function of the scrotum is for thermoregulation of the testis since the failure of normal testicular descent in scrotal mammals (or its experimental induction) ensures that testicular function is compromised, with spermatogenic failure and infertility being inevitable if both testes are chronically exposed near to or at peritoneal cavity temperature. Spermatogenesis is disrupted if testicular temperature is raised but the absolute increase in temperature necessary to impair the testis is variable between species. In man, elevation of scrotal temperature by 1°C for 1 year does not affect sperm function but higher temperatures, closer to core temperature, suppress sperm production. In laboratory animals, the difference between intrascrotal and peritoneal temperature may be 2 or 3°C (rabbit and guinea pig) or 4–6°C (rat). Movement of the testis in these species from the lower temperature within the scrotum to inguinal locations can occur in response to lower ambient temperature, with the reverse occurring at higher ambient temperatures.

The scrotum shows the following anatomical features which contribute to the regulation of testicular temperature and thus normal spermatogenesis: (i) Cremaster muscle of the pendulous scrotal component may contract and relax to elevate or lower the scrotal sac, respectively, raising or reducing internal temperature; (ii) the temperature of blood within the testicular artery (arising from the abdominal aorta) is cooled by countercurrent exchange by the pampiniform venous plexus within the spermatic cord; (iii) superficial fascia of scrotal skin is thin and the skin is convoluted by contraction of Dartos muscle, thus dissipating heat; and (iv) sweat glands of the scrotum provide cooling via evaporation of sweat on the skin surface. Changes in thermoregulation of the testis are well-developed in some seasonal breeders (many rodents, insectivores, and bats), in which the testes are scrotal at the commencement of the breeding season but return to the abdomen out of season.

Experimental elevation of scrotal and therefore testicular temperature has long been known to impair testicular function by reducing sperm production. The two most widely used approaches have been surgical cryptorchidism or immersion of the scrotum in water baths of known temperature.

A. Laboratory Animals

The response of the testis to hyperthermia induced by local heating in water baths was reported in the 1920s using guinea pigs, in which scrotal testis temperature is about 35°C. A 10-min exposure of the testis to 42°C caused local histological damage to the seminiferous tubules just deep to the testicular capsule. When the same bath temperature was applied for four 5-min periods separated by intervals allowing return to normal scrotal temperature, the tissue damage was less extensive, indicating that intermittent applied heat is less harmful compared to continuous exposure. Similar multiple heat exposure at 44 or 45°C resulted in greater morphological damage and at 47°C maximum degeneration resulted. Severe degeneration of seminiferous tubules occurred after 15–30 min immersion in water at 46 or 47°C, but several months later some seminiferous tubules showed histological recovery indicating the reversibility of heat-induced damage due to the regenerative capacity of the seminiferous epithelium. Heat applied to the testis of the rat (immersion in water) also produces histological damage, particularly above 42°C (normal scrotal testis temperature is 33°C) and with increasing duration of exposure. Mice appear to be more sensitive to elevated ambient temperatures. Air temperature of 32°C for 10 days caused infertility with testicular damage evident on the 15th day. A temperature of 33°C (normal scrotal temperature 29°C) for 10, 15, or 20 days resulted in up to 80% abnormal spermatozoa and, after a 60-day recovery period, up to 40% of spermatozoa were abnormal. The rabbit testes (normal scrotal temperature 37°C) are less sensitive to elevated temperature compared to mice and rat testes. Sperm motility and concentration are unaffected by up to 2 hr at 43°C ambient temperature, but higher temperatures cause spermatogenic disruption.

B. Domestic and Farm Animals

Exposure of the dog testis (normal scrotal temperature 35°C) to applied heat at 38–40°C resulted in reduced sperm activity in the ejaculate after 5 days and zero activity after 10 days. In cattle, breeding efficiency is reported to wane during the hotter months of the year compared to the spring season. Semen volume, motility, and quality seem to be impaired. Normal scrotal temperature in bulls is about 35°C and exposure to elevated ambient temperatures of variable durations and frequency may result in a decline in seminal parameters. Seasonal changes in sperm production and fertility are well documented in sheep and elevated temperatures induced by scrotal insulation or various regimens of high ambient temperatures can cause temporary impairment of seminal characteristics. The domestic pig also shows a decline in fertility during summer months, and although changes in the endocrine status of the hypothalamic–pituitary–testicular axis may contribute to these changes (this also applies to sheep, in which food intake and photoperiod are influencing factors), experimental elevation of temperature may result in a decline in sperm quality and concentration in the ejaculate.

C. Primates Including Man

Immersion of testes of adult monkeys in water baths up to 44°C for brief periods daily but repeated for several months has been reported to cause testicular atrophy and degeneration of the seminiferous epithelium. In man, hyperthermia induced by confinement to fever cabinets or sauna baths may, depending on the length and frequency of exposure and the temperature, cause a temporary decline in sperm numbers in the ejaculate together with increases in sperm with abnormal motility. Scrotal insulation with extra clothing or athletic supports, if maintained for several months, has been reported to reduce sperm counts in the ejaculate, but data on actual temperature rise have been difficult to obtain. In addition, these earlier studies showed similar ranges of sperm concentrations in the treated vs control (loose-clothing) groups. Recent studies in which temperature was recorded in men wearing insulated athletic supports for 23 hr/day for 52 weeks found an increase of 1°C in scrotal temperature but seminal parameters and *in vitro* sperm penetration tests of hamster oocytes were not affected. Other studies have shown that daily heat exposure by scrotal immersion in hot water can reversibly suppress sperm counts in men. A 30-min exposure to 43–47°C in water for 6–12 consecutive days suppresses spermatogenesis, with fluctuations in sperm count. These effects are reversed following treatment discontinuation.

III. EFFECTS ON SPERMATOGENESIS

The seminiferous epithelium in scrotal testes is sensitive to hyperthermia if heat is applied externally or if the testes are exposed to abdominal temperature in cases of surgically induced or naturally induced cryptorchidism. All cellular types within the seminiferous tubules, including the cellular and extracellular components of the peritubular tissue, are structurally and functionally damaged or altered in response to elevated temperature. In earlier studies, histological degeneration of the germ cells and peritubular fibrosis were consistent findings, the extent of which was species-dependent and variable in relation to the duration of exposure and actual temperature. Prior to the introduction of electron microscopy and biochemical and physiological studies of the specific environment within the seminiferous epithelium, it was thought that the germ cells were vulnerable to increased temperature, whereas the Sertoli cells were relatively resistant since they persisted as many of the germ cells degenerated. Studies during the past two decades have shown that the Sertoli cells are acutely sensitive to temperature elevation, exhibiting ultrastructural and functional abnormalities previously unsuspected. On the other hand, Sertoli cells may remain in the testes of severely damaged heat-treated testes but their morphology and function is markedly altered. Similarly, some spermatogonia persist in such testes as exhibited by adult human congenitally cryptorchid testes, in which the seminiferous tubules contain Sertoli cells and occasional spermatogonia but all other germ cells may be absent. Why these cells survive heat treatment is unknown and, conversely, why most germ cells readily degenerate in response to elevated temperature is also not known.

By far the most widely used method to study spermatogenic disruption in response to heat is the surgical cryptorchid testis model. The morphological and functional assessments of spermatogenesis are more readily comparable between different laboratories using this approach, whereas more variation is apparent in studies in which heat is applied using water immersion. Although a comprehensive ultrastructural and functional analysis of the earliest lesions of spermatogenesis in response to heat is not available, it is known that spermatocytes and round spermatids are particularly sensitive to experimental cryptorchidism, with most of the data obtained from the rat and mouse. Within 12 hr, degeneration of meiotic division primary and secondary spermatocytes and their progeny, the early round spermatids, has been noted. At 24 and 48 hr, germ cell degeneration spreads to include many primary spermatocytes and most spermatids. Elongated spermatids show progressive shape alterations and degeneration at subse-

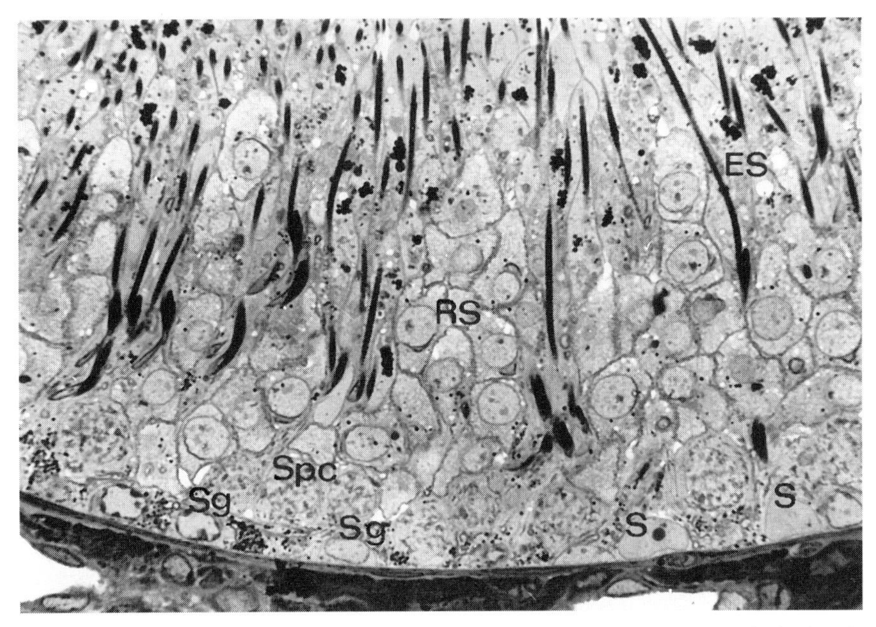

FIGURE 1 Normal rat seminiferous epithelium showing spermatogonia (Sg), Sertoli cells (S), spermatocytes (Spc), and round (RS) and elongated (ES) spermatids.

quent times. Spermatogonia remain largely unaffected but in cryptorchid testes beyond 7 days of abdominal confinement, many of them degenerate (although some may remain for weeks or months). The complex histological response of the seminiferous epithelium to heat can be summarized as follows: (i) The normal seminiferous epithelium shows very little, if any, cellular degeneration (Fig. 1); (ii) spermatocytes and round spermatids degenerate, showing shrinkage, nuclear pyknosis, and condensation into cellular debris (Fig. 2); (iii) based on morphology, histochemical analysis of DNA, and electrophoretic separation of extracted DNA, this degenerative process is cell apoptosis; (iv) many of the spermatids become confluent, presumably a result of their multiple associations via intercellular bridges, resulting in the formation of multinucleated giant bodies (Fig. 3); (v) elongating spermatids show abnormal nuclear shapes, pyknotic condensation, and abnormal retention within the Sertoli cell cytoplasm (Fig. 4); (vi) the epithelium shows numerous vacuoles or empty-appearing spaces as single or clustered structures (Figs. 2 and 4); and (vii) these vacuoles often aggregate focally (Fig. 5) and ultrastructural studies show that they are multiple extracellular dilations of the apposed plasma membranes of Sertoli cell tight junctions.

Together with increased lipid inclusions (Fig. 6), phagocytosed germ cell debris, and numerous other structural alterations to the cytoplasm of the Sertoli cell, the vacuole-type spaces are indicative of dysfunctional Sertoli cells within heat-exposed testis. Fluid production and protein secretion by these Seroli cells, including androgen-binding protein, are impaired in response to heat and probably represent a small fraction of abnormal metabolic, synthetic, and secretory activities of the Sertoli cells. The mechanisms by which the seminiferous epithelium undergoes degeneration and the Sertoli cells become dysfunctional remain unknown. Oxidative stress via increased production of reactive oxygen species, activation or inactivation of heat shock proteins, and abnormal upregulation of insulin-like growth factor receptors and p53, bcl-2, bax, and macrophage colony-stimulating factor have all been implicated in regulating germ cell apoptosis.

Evaluation of the sites of cellular damage in the heat-stressed human testis is more difficult compared to that of animal studies. Daily heat exposures by immersion of the scrotum in heated water for several

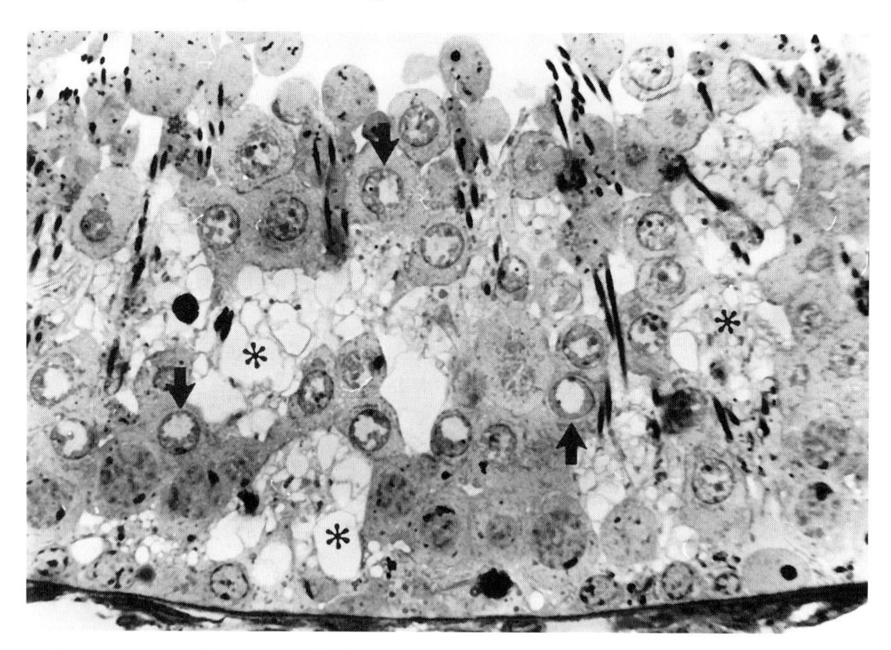

FIGURE 2 Five-day cryptorchid rat testis. Note nuclear chromatin clumping in round spermatids (arrows) and large and small empty vacuoles (asterisks) in the seminiferous epithelium.

days reveal arrest of spermatogenesis observed in testicular biopsy specimens. Athletic supports, or pushing the testes back into the inguinal canal to increase testicular temperature, also show suppression of spermatogenesis in biopsy specimens. Much of the data on the effects of hyperthermia on human spermatogenesis have been gained from histologic and ultrastructural analysis of the undescended testis. The initial studies showed that maturation of the seminiferous epithelium in cryptorchid boys was

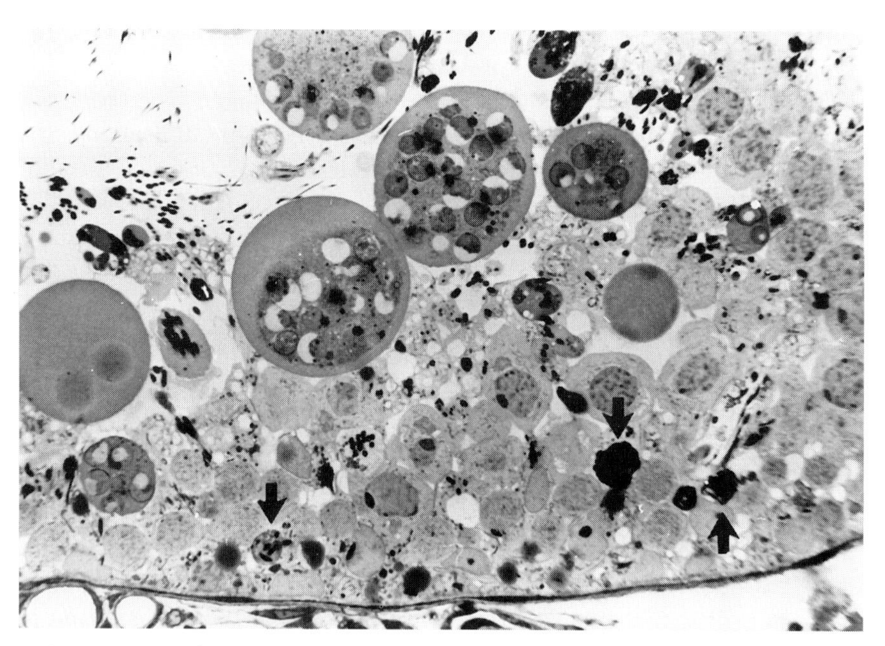

FIGURE 3 Five-day cryptorchid rat testis showing condensed clumps of degenerating cells (arrows) and large circular bodies containing multiple spermatid nuclei.

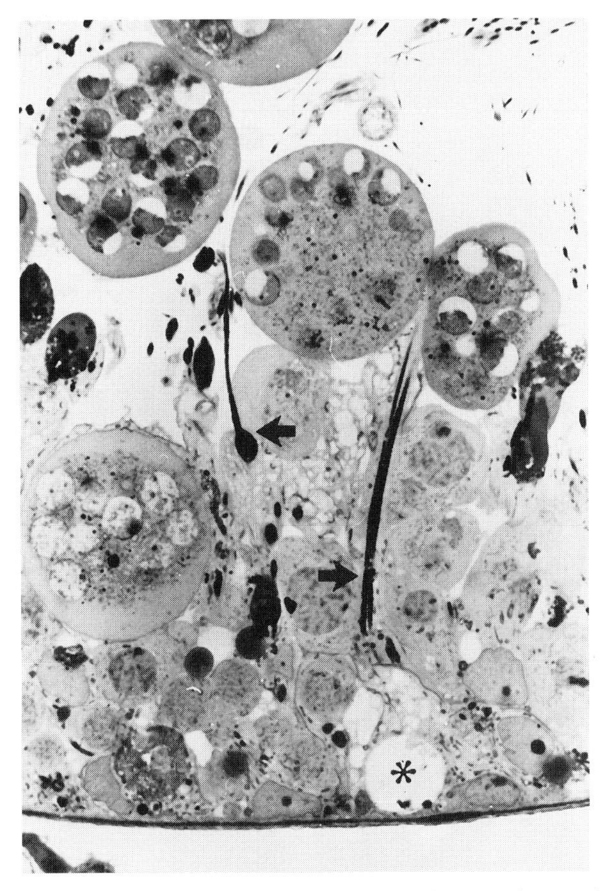

FIGURE 4 Five-day cryptorchid rat testis showing abnormal elongated spermatids (arrows) and a large, circular empty-looking vacuole near the basement membrane (asterisk). The luminal multinucleated bodies contain spermatids separated into nuclei and particulate cytoplasmic matter.

impaired by age 10, and during puberty, these testes exhibited arrested spermatogenesis and peritubular fibrosis. Other studies on infants with cryptorchidism have reported that histologic alteration to the seminiferous cords may occur in the second year of life, with reduction in spermatogonia and increased frequency of degenerating cells. If untreated, the human cryptorchid testis loses almost all germ cells beyond puberty. In adults the typical presentation of the heat-stressed, undescended testis is markedly different from normal and shows shrunken seminiferous tubules with peritubular fibrosis containing Sertoli cells and a few spermatogonia, some of which are enlarged with resemblance to germ cells thought to be precursor types to tumor formation, including carcinoma *in situ* (Figs. 7 and 8).

IV. EFFECTS ON INTERTUBULAR TISSUE

The response of the intertubular tissue to raised temperature has been extensively investigated, mainly in experimental animal studies. Attention has been chiefly directed to the Leydig cells, which do not degenerate but show dysfunctional responses with regard to morphology and secretory capacity. The specific changes in the Leydig cells following hyperthermia remain controversial since in most experiments it is not known if temperature affects Leydig cells directly or if the rapid damage to the seminiferous epithelium results in secondary, local alterations in the intertubular compartment. Thus, in short-term cryptorchid testes or those testes subjected to brief hot-water immersion, the Leydig cells may show hypertrophy, increased or decreased *in vitro* testosterone production in response to human chorionic gonadotropin (hCG) stimulation, and a decline in content of hCG receptors. Regardless of whether the structure and/or functions of Leydig cells are altered by paracrine and/or direct effects of heat, these cells persist in heat-stressed testes of animals including man. They continue to secrete androgens, albeit at reduced or low-normal levels, and serum testosterone levels, secondary androgen-dependent tissues, and libido are generally not markedly affected.

V. REVERSAL OF THE EFFECTS OF ELEVATED TEMPERATURE

In man the cessation of applied heat to the testes, provided the insult is not sufficient to destroy most germ cells, results in restoration of spermatogenesis and return of transient decline in seminal parameters toward normal. Similarly, brief exposure of testes to heat, such as in water immersion of the rat scrotum, is followed by recovery of histological parameters and fertility. Reversal of heat stress is accomplished more readily by testes of sexually immature animals. For example, rats made cryptorchid at 14 days of age and reversed on Day 35 showed complete restoration of testicular function when examined as adults. In adult animals, cryptorchidism for 2 weeks results in poor recovery of spermatogenesis if orchidopexy

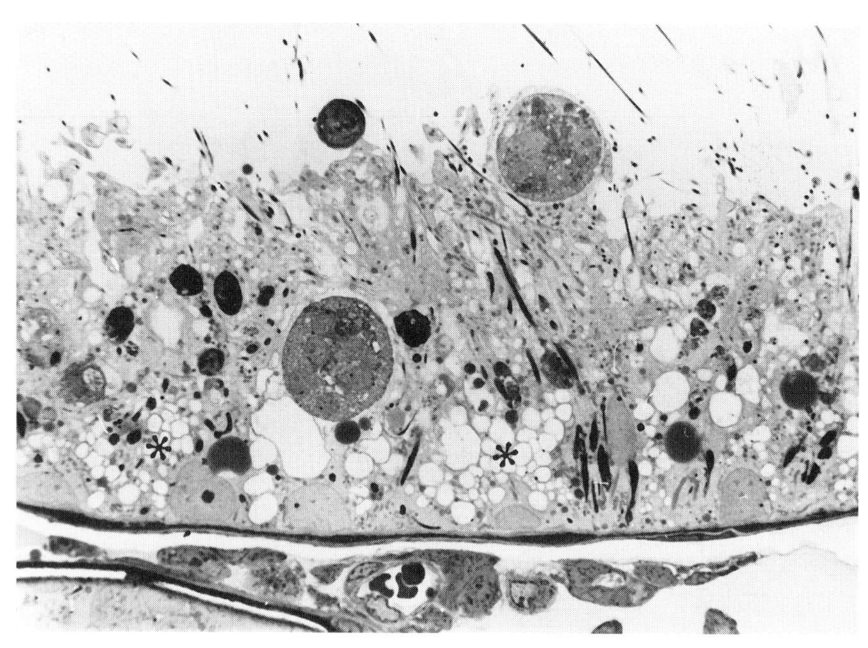

FIGURE 5 Seven-day cryptorchid rat testis showing the loss of most germ cells with the Sertoli cells exhibiting many small aggregated vacuoles (asterisk). Numerous pyknotic bodies and spermatids are retained in the seminiferous epithelium.

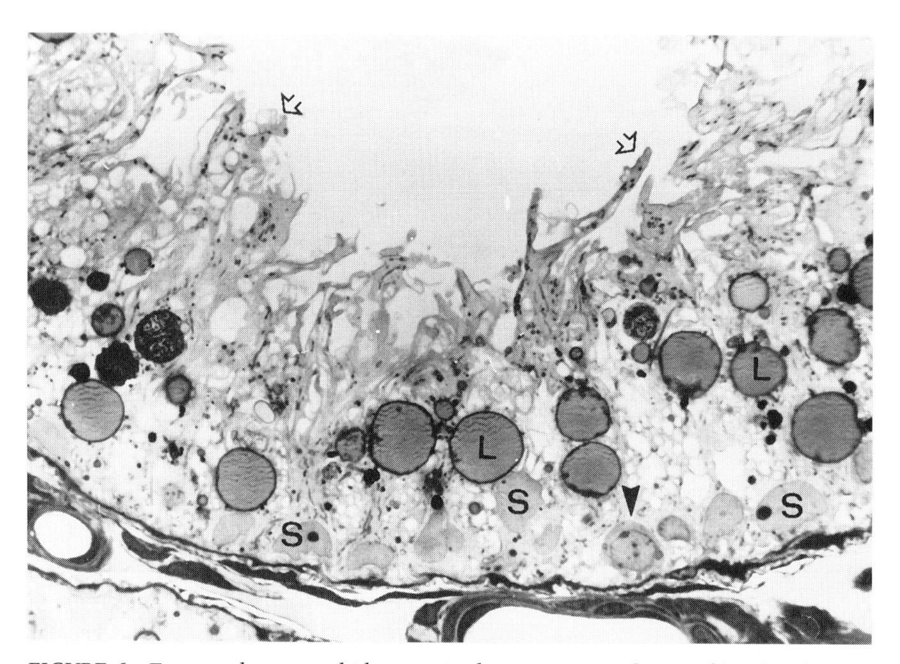

FIGURE 6 Four-week cryptorchid rat testis showing accumulation of lipid inclusions (L) within the cytoplasm of Sertoli cells (S). The disappearance of the spermatocytes and spermatids leaves extensions of Sertoli cell cytoplasm (arrows) facing the tubule lumen. A spermatogonium (arrowhead) survives in the epithelium.

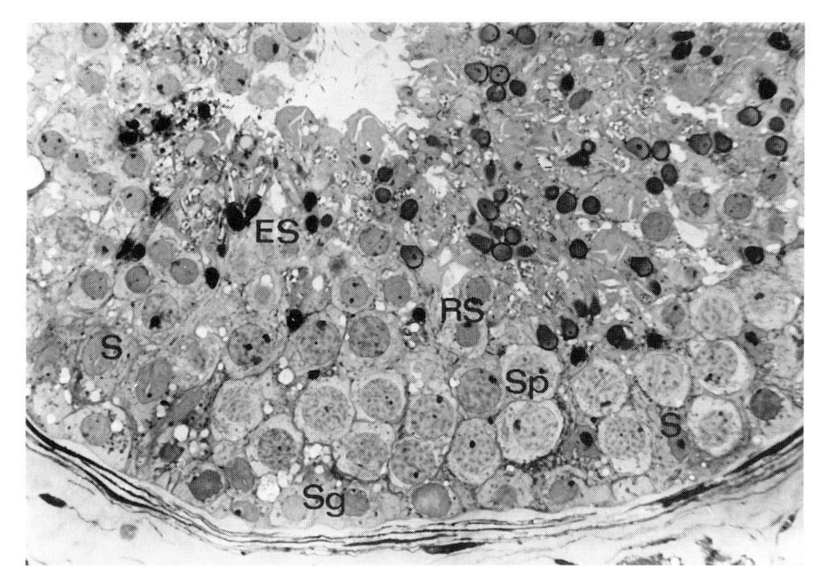

FIGURE 7 Normal human seminiferous epithelium showing Sertoli cell nuclei (S), spermatogonia (Sg), spermatocytes (Sp), and round (RS) and elongating spermatids (ES).

is performed. When cryptorchidism is induced for 10 days or less, partial reversal of testicular function is achieved by orchidopexy. The decision to perform orchidopexy in boys is a controversial matter for the pediatric surgeon, but the available evidence suggests that relocation of the cryptorchid testis to the scrotum is desirable prior to puberty and often is preferred in early childhood, i.e., before the age of 6. In patients with poor semen characteristics and varicocele, increased testicular temperature has been ob-

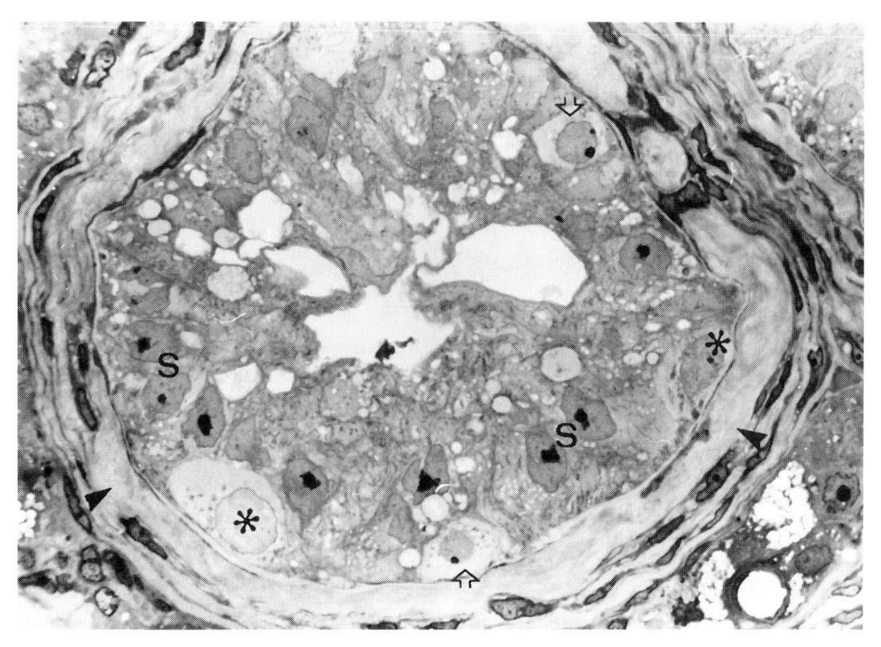

FIGURE 8 Cryptorchid testis from an adult man showing a shrunken seminiferous tubule with peritubular fibrosis (arrowheads). Sertoli cell nuclei (S) are indicated, together with spermatogonia of two types—one appearing normal (arrows) and the other somewhat enlarged (asterisks).

served. The application of cooling to the scrotum has been reported to improve semen quality and to increase fertility.

VI. THE EFFECT OF COLD

The thermoregulatory capacity of the scrotum is obviously an effective mechanism to prevent exposure of the testis to very low temperature, and in nonscrotal species testicular function is protected against extremes of low temperature such as in polar environments, high-altitude, and cold aquatic environments. In rats exposed to subzero ambient temperatures with the testes surgically anchored in the scrotum, testicular function is rapidly impaired due to reduction in androgen production. When the testis is cooled directly by ice application or immersion in subzero water baths, spermatogenesis is impaired and fertilizing capacity is reduced. These effects are thought to occur in response to ischemia in which blood flow to the inner parenchyma of the testis is progressively impeded with lowering of temperature.

See Also the Following Articles

Cryptochordism; Leydig Cells; Sertoli Cells, Function; Spermatogenesis, Overview; Testis, Overview

Bibliography

Abney, T. O., and Keel, B. A. (Eds.) (1989). *The Cryptorchid Testis*. CRC Press, Boca Raton, FL.

Jegou, B., Laws, A. O., and deKretser, D. M. (1984). Changes in testicular function induced by short-term exposure of the rat testis to heat: Further evidence for interaction of germ cells, Sertoli cells and Leydig cells. *Int. J. Androl.* 7, 244–257.

Mieusset, R., and Bujan, L. (1995). Testicular heating and its possible contributions to male infertility: A review. *Int. J. Androl.* **18**, 169–184.

Setchell, B. P. (1978). The scrotum and thermoregulation In *The Mammalian Testis* (B. P. Setchell, Ed.), pp. 90–108 Cornell Univ. Press, Ithaca, NY.

VanDemark, N. L., and Free, M. J. (1970). Temperature effects. In *The Testis* (A. D. Johnson, W. R. Gomes, and N. L. VanDemark, Eds.), Vol. 3, pp. 233–312. Academic Press, New York.

Waites, G. M. H. (1970). Temperature regulation and the testis. In *The Testis* (A. D. Johnson, W. R. Gomes, and N. L. VanDemark, Eds.), pp. 241–279. Academic Press, New York.

Wang, C., *et al.* (1997). Effect of increased scrotal temperature on sperm production in normal men. *Fertil. Steril.* **68**, 334–339.

Yin, Y., *et al.* (1997). Heat stress causes testicular germ cell apoptosis in adult mice. *J. Androl.* **18**, 159–165.

Teratogens

Robert L. Brent and David A. Beckman

Thomas Jefferson University and Alfred I. duPont Hospital for Children

GLOSSARY

developmental toxicity Adverse effects on development manifested as death, malformation, growth retardation, and functional deficit.

malformation Irregular or anomalous formation or structure of parts of an organism.

organogenesis Formation and development of the different organs; the period of greatest sensitivity for the production of major malformations.

pharmacokinetics The absorption, distribution, metabolism, and excretion of a substance and its metabolites.

threshold dose The lowest dose at which the incidence of death, malformation, growth retardation, or functional deficit is statistically greater than that of controls.

The field of teratology includes the study of (i) normal and abnormal embryonic and fetal development, (ii) the causes and mechanisms of action of environmental agents that are responsible for the induction of congenital malformations, and (iii) the epidemiology of congenital malformations. Teratologists utilize this information (i) for counseling families regarding the reproductive risks of particular environmental agents, (ii) for attempting to determine the etiology of a child's malformation, or (iii) to introduce measures that might prevent recurrences in an individual family or on a population basis.

I. ETIOLOGIES OF CONGENITAL MALFORMATIONS

There have been dramatic changes in the explanation of the causes of human birth defects in the twentieth century. Since ancient times, the causes of birth defects were predominantly based on superstition, ignorance, and prejudice rather than on scientific knowledge. The stigma associated with birth defects has primitive beginnings and persists today. In the minds of many, even the most sophisticated, a birth defect is believed to be some form of punishment. At the beginning of this century the predominant cause was believed to be genetic and the remainder were totally unsolvable clinical problems.

At this point in the history of birth defects research, the etiology of congenital malformations can be divided into three broad categories: unknown, genetic, and environmental factors (Table 1). The etiology of the majority of human malformations (approximately 65–75%) is unknown; however, a significant proportion of congenital malformations of unknown etiology is likely to be polygenic; that is, due to two or more genetic loci or at least have an important genetic component. Malformations with an increased recurrent risk, such as cleft lip and palate, anencephaly, spina bifida, certain congenital heart diseases, pyloric stenosis, hypospadias, inguinal hernia, talipes equinovarus, and congenital dislocation of the hip, can fit the category of multifactorial disease as well

TABLE 1
Etiology of Human Malformations and Reproductive Toxicity Observed during the First Year of Life[a]

Suspected cause		*Percentage of total*
Unknown	65–75	
Polygenic		
Multifactorial (gene–environmental interactions)		
Spontaneous errors of development		
Synergistic interactions of teratogens		
Genetic	10–25	
Inherited autosomal dominant, recessive, and sex-linked genetic disease		
New mutations		
Cytogenetic (chromosomal abnormalities)		
Environmental	10	
Maternal conditions: Nutritional perturbations and deficiencies, alcoholism, diabetes, endocrinopathies, maternal phenylketonuria, smoking, starvation		4
Infectious agents: Rubella, toxoplasmosis, syphilis, herpes, cytomegalic inclusion disease, varicella, Venezuelan equine encephalitis, parvovirus B19		3
Mechanical problems (deformations): Amniotic band contrictions, umbilical cord constraint, disparity in uterine size and uterine contents, multiple pregnancy, vascular dysruption		1–2
Chemicals, prescription drugs, ionizing radiation, hyperthermia		<1

[a] From Beckman *et al.* (1997) Brent (1995), Brent and Holmes (1988).

as the category of polygenic inherited disease. The multifactorial/threshold hypothesis involves the modulation of a continuum of genetic characteristics by intrinsic and extrinsic (environmental) factors. Although the modulating factors are not known, they probably include placental blood flow, placental transport, site of implantation, maternal disease states, infections, drugs, chemicals, and spontaneous errors of development.

Spontaneous errors of development may account for some of the malformations that occur without apparent abnormalities of the genome or environmental influence. We postulate that there is some probability for error during embryonic development based on the fact that embryonic development is a complicated process, similar to the concept of spontaneous mutations. It has been estimated that up to 50% of all fertilized ova in the human are lost within the first 3 weeks of development. It is estimated that 15% of all clinically recognizable pregnancies end in spontaneous abortion, whereas 50–60% of the spontaneously aborted fetuses have chromosomal abnormalities. This means that, as a conservative estimate, 1173 clinically recognized pregnancies will result in approximately 173 miscarriages and 30 of the infants will have serious congenital malformations in the remaining 1000 live births. The true incidence of pregnancy loss is much higher, but undocumented pregnancies are not included in this risk estimate. The 3% incidence of seriously malformed offspring represents the background risk for human maldevelopment. While we know little about the mechanisms which result in the *in utero* death of defective embryos, it is perhaps more important to understand the circumstances which permit abnormal embryos to survive to term.

Understanding the pathogenesis for the large group of malformations with unknown etiology will depend on identifying the genes involved in polygenic or plurogenic processes, the interacting genetic and environmental determinants of multifactorial traits, and the statistical risks for error during embryonic development.

The known etiologies of teratogenesis include genetic and environmental agents or factors that affect the embryo during development (e.g., drugs, chemicals, radiation, hyperthermia, infections, abnormal maternal metabolic states, or mechanical factors). Environmental and genetic causes of malformations have different pathologic processes that result in abnormal development. Congenital malformations due to genetic etiology have a spectrum of pathologic processes that are the result of a gene deficiency, a gene abnormality, chromosome deletion, or chromosome excess. The pathologic process of genetic mutations is preconceptional, i.e., determined before conception, because of newly acquired genetic abnormalities present in all or most of the cells of the embryo. New mutations account for only a small proportion of hereditary disease (<3%) (Table 2), but they can contribute to mosaicism when they occur in the preimplantation stages of embryonic development. Although environmental factors may modify the development of the genetically abnormal embryo, the genetic abnormality is usually the predominant contributor to the pathologic process.

II. OVERALL TERATOGENIC RISK

In order to appreciate the difficulty of predicting the effect that an exposure to a drug or therapeutic agent will have on the developing embryo, we will briefly discuss factors which influence this prediction.

The baseline risk of human reproduction is based on epidemiological studies which have determined the incidence of reproductive failure and maldevelopment (Table 2). A much higher proportion of fertilized ova are lost than the estimate of 15%, which is the incidence of spontaneous abortions in known pregnancies, since 50% are lost within the first 3 weeks, before pregnancy has been recognized. Of the liveborn infants, 3% will be recognized as having serious congenital malformations.

III. ENVIRONMENTAL RISK PARAMETERS OR MODIFIERS

Regarding environmental influences or agents which interfere with embryonic development, there are several scientific or embryological principles that have an important impact on the effect of various environmental agents on the developing embryo, including the impact of (i) embryonic stage, (ii) dose or magnitude of the exposure, (iii) threshold phenomena, (iv) pharmacokinetics and metabolism of the agent, (v) placental transport, and (vi) inter- and intraspecies differences. The complexity of evaluating item vi, species differences, is made clear by the fact that more than 30 drug-related disorders are

TABLE 2
Frequency of Reproductive Risks in the Human

Reproductive risk	*Frequency*
Immunologically and clinically diagnosed spontaneous abortions per 10^6 conceptions	350,000
Clinically recognized spontaneous abortions per 10^6 pregnancies	150,000
Genetic diseases per 10^6 births	110,000
Multifactorial or polygenic (genetic–environmental interactions)	90,000
Dominantly inherited disease	10,000
Autosomal and sex-linked genetic disease	1,200
Cytogenetic (chromosomal abnormalities)	5,000
New mutations	3,000
Major congenital malformations per 10^6 births	30,000
Prematurity per 10^6 births	40,000
Fetal growth retardation per 10^6 births	30,000
Stillbirths per 10^6 pregnancies (>20 weeks)	2,000–20,000
Infertility	7% of couples

related to genotype, and while not proven in the human, genetic variations alter drug or chemical teratogenicity in experimental animals.

Finally, maternal disease states may produce deleterious effects on the fetus which are difficult to separate from a possible teratogenic effect of a therapeutic agent. This is an especially relevant consideration for long-standing conditions such as diabetes.

When counseling patients, especially in our litigious climate, three confounding influences are at work. One is the necessity to critically evaluate reported associations because of the anxiety created by unfounded reports and misinformation. The second is the fact that pregnancy is not without risk and congenital malformations occur in the absence of drug or chemical exposures. The third is that there are teratogenic agents and new ones could be introduced. Therefore, we must continually be on guard because each child malformed *in utero* by an environmental agent is a preventable tragedy.

TABLE 3

Developmental Stage Sensitivity to Thalidomide-Induced Limb and Ear Malformations in the Human[a]

Days from conception for induction of defects	*Limb reduction defects*
21–26	Thumb aplasia
22–23	Microtia
23–34	Hip dislocation
24–29	Amelia, upper limbs
24–33	Phocomelia, upper limbs
25–31	Preaxial aplasia, upper limbs
27–31	Amelia, lower limbs
28–33	Preaxial aplasia, lower limbs; phocomelia, lower limbs; femoral hypoplasia; girdle hypoplasia, triphalangeal thumb

[a] Modified from Brent and Holmes (1988).

IV. PRINCIPLES OF ENVIRONMENTALLY INDUCED MALFORMATIONS

A basic tenet of environmentally produced malformations is that teratogens or teratogenic milieu have certain characteristics in common and follow certain basic principles. While these common characteristics lay the foundation for understanding teratogenesis, there are many exceptions to the "rules" which in themselves reinforce the principles which guide teratologists in their research and clinical activities. These principles determine the quantitative and qualitative aspects of environmentally produced malformations and allow teratologists to determine reproductive risk on an individual or population basis.

A. Embryonic Stage

The induction of malformations by environmental agents usually results in a spectrum of malformations which varies somewhat because of variations in stage of exposure and dose. The developmental period at which an exposure occurs will determine which structures are most susceptible to the deleterious effects of the drug or chemical and to what extent the embryo can repair the damage. Furthermore, the period of sensitivity may be narrow or broad, depending on the environmental agent and the malformation in question. Limb defects, produced by thalidomide, have a very short period of susceptibility (Table 3), whereas microcephaly produced by radiation has a long period of susceptibility.

During the first period of embryonic development, from fertilization through the early postimplantation period, the embryo is most sensitive to the toxic effects of drugs and chemicals resulting in embryo lethality. Surviving embryos have malformation rates similar to the controls not because malformations cannot be produced at this stage but because significant cell loss or chromosome abnormalities at theses stages have a high likelihood of killing the embryo. Because of the omnipotentiality of early embryonic cells, surviving embryos have a much greater ability to have normal developmental outcome. Wilson *et al.* demonstrated that the "all or none phenomenon" or marked resistance to teratogens disappears over a period of a few hours in the rat during early organogenesis utilizing ionizing X-irradiation as the experimental teratogen. The term all or none phenomenon has been misinterpreted by some investigators to indicate that malformations cannot be produced at this stage. On the contrary, it is likely that certain

TABLE 4
Estimated Outcome of 100 Pregnancies versus Time from Conception[a]

Time from conception	Percentage survival to term	Percentage death during interval	Last time for induction of selected malformations
Preimplantation			
0–6 days	25	54.55	
Postimplantation			
7–13 days	55	24.66	
14–20 days	73	8.18	
3–5 weeks	79.5	7.56	23 days: cyclopia; sirenomelia 26 days: anencephaly 28 days: meningomyelocele 34 days: transposition of great vessels
6–9 weeks	96	6.52	36 days: cleft lip, limb reduction, defects 6 weeks: diaphragmatic hernia, rectal atresia, ventricular septal defect, syndactyly 9 weeks: cleft palate
10–13 weeks	92	4.42	10 weeks: omphalocele
14–17 weeks	92.26	1.33	12 weeks: hypospadias
18–21 weeks	97.56	0.85	
22–25 weeks	98.39	0.31	
26–29 weeks	98.69	0.30	
30–33 weeks	98.98	0.30	
34–37 weeks	99.26	0.34	
38+ weeks	99.32	0.68	38+ weeks: CNS cell depletion

[a] From Beckman *et al.* (1997).

drugs, chemicals, or other insults during this stage of development can result in an increase in surviving malformed embryos, but the nature of embryonic development at this stage will still reflect the basic characteristic of the all or none phenomenon which is a propensity for embryo lethality rather than surviving malformed embryos.

The period of organogenesis (from Day 18 through about Day 40 of post conception in the human) is the period of greatest sensitivity to teratogenic insults and the period when most gross anatomic malformations can be induced. Most major malformations are produced before the 36th day of post conception in the human. The exceptions are malformations of the genitourinary system, the palate, the brain, or deformations due to problems of constraint, disruption, or destruction. In midgestation, severe growth retardation in the whole embryo or fetus may also result in permanent deleterious effects in many organs or tissues.

The fetal period is characterized by histogenesis involving cell growth, differentiation, and migration. Teratogenic agents may decrease the cell population by producing cell death, inhibiting cell division, or cell differentiation. There is, of course, some overlap in that permanent cell depletion may be produced earlier than the 60th day. Effects such as cell depletion or functional abnormalities, not readily apparent at birth, may give rise to changes in behavior or fertility that may be apparent only later in life. The last gestational day on which selected malformations may be induced in the human is presented in Table 4.

B. Dose or Magnitude of the Exposure

The dose–response relationship is extremely important when comparing effects among different species because usage of mg/kg doses are, at most, rough approximations. Dose equivalence among species can only be accomplished by performing pharmacokinetic studies, metabolic studies, and dose–response investigations in the human and the species being studied.

Several considerations affect the interpretation of dose–response relationships:

1. The concentration of active metabolites may be more pertinent than the dosage of the original chemical, i.e., the metabolite phosphoramide mustard and acrolein may produce maldevelopment resulting from exposure to cyclophosphamide.
2. A chronic exposure at a low dose can contribute to an increased teratogenic risk, i.e., anticonvulsant therapy.
3. Pregnancy alters the distribution and metabolism of drugs.
4. It may be difficult to determine whether a maternal condition contributes to the etiology of malformations associated with the treatment for that condition during pregnancy, i.e., etiological factors which cause epilepsy may also contribute to the maldevelopment associated with exposure to diphenylhydantoin.
5. Fat-soluble substances, such as polychlorinated biphenyls and etretinate, can produce fetal maldevelopment for an extended period after the last ingestion or drug exposure in a woman because they have an unusually long half-life.

Furthermore, the response should be interpreted in a biologically sound manner. One example is that a substance given in large enough amounts to cause maternal toxicity is likely to also have deleterious effects on the embryo such as death, growth retardation, or retarded development. Another example is that because the steroid receptors that are necessary for naturally occurring and synthetic progestin action are absent from nonreproductive tissues in early development, the evidence is against the involvement of progesterone or its synthetic analogs in nongenital teratogenesis.

An especially anxiety-provoking concept is that the interaction of two or more drugs or chemicals may potentiate their developmental effects. Although this is an extremely difficult hypothesis to test in the human, it is an especially important consideration because multichemical or multitherapeutic exposures are common. Fraser warns that the actual existence of a threshold phenomenon when nonteratogenic doses of two teratogens are combined could easily be misinterpreted as potentiation or synergism.

C. Threshold Phenomena

The threshold dose is the dosage below which the incidence of death, malformation, growth retardation, or functional deficit is not statistically greater than that of controls. The threshold level of exposure is usually from less than one to three orders of magnitude below the teratogenic or embryopathic dose for drugs and chemicals that kill or malform half the embryos. An exogenous teratogenic agent therefore has a no-effect dose compared to mutagens or carcinogens, which have a stochastic dose–response curve. Threshold phenomena are compared to stochastic phenomena in Table 5. The severity and incidence

TABLE 5

Comparison of Stochastic Phenomena and Threshold Phenomena in the Etiology of Diseases Produced by Environmental Agents and the Risk of Occurrence[a]

Relationship	*Pathology*	*Site*	*Diseases*	*Risk*	*Definition*
Stochastic phenomena	Damage to a single cell may result in disease	DNA	Cancer, mutation	Exists at all exposures, although at low exposure the risk is below the spontaneous risk	The incidence of disease increases with exposure but the severity and the nature of the disease in the patient remain the same
Threshold phenomena	Multicellular injury	Great variation in etiology affecting many cell and organ processes	Malformation, death, growth retardation, chemical toxicity, etc.	Completely disappears below a certain threshold dose	Both the severity and incidence of disease increase with higher exposures

[a] From Brent (1986).

of malformations produced by every exogenous teratogenic agent that has been appropriately tested have exhibited threshold phenomena during organogenesis.

Because of space constraints we do not discuss the importance of the pharmacokinetics and metabolism of teratogenic agents, placental transport, and species differences.

V. MECHANISMS OF TERATOGENESIS

Based on his review of the literature, Wilson provided a format of theoretical teratogenic mechanisms: mutation; chromosomal aberrations; mitotic interference; altered nucleic acid synthesis and function; lack of precursors, substrates, or coenzymes for biosynthesis; altered energy sources; enzyme inhibition; osmolar imbalance, alterations in fluid pressures, viscosities, and osmotic pressures; and altered membrane characteristics. Even though an agent can produce one or more of these pathologic processes, exposure to such an agent does not guarantee that maldevelopment will occur. Furthermore, it is likely that a drug, chemical, or other agent can have more than one effect on the pregnant woman and the developing conceptus and therefore the nature of the drug or its biochemical or pharmacologic effects will not in themselves predict a human teratogenic effect. In fact, the discovery of human teratogens has come primarily from human epidemiological studies. Animal studies and *in vitro* studies can be very helpful in determining the mechanism of teratogenesis and the pharmacokinetics related to teratogenesis. We have proposed a list of mechanisms (Table 6) which we will use in our discussion of the known teratogenic drugs and therapeutic agents in man. However, even if one understands the pathologic effects of an agent, one cannot predict the teratogenic risk of an exposure without taking into consideration the developmental stage and the magnitude of the exposure and the repairability of the embryo.

TABLE 6
Mechanisms of Action of Environmental Teratogens

Cell death, delayed differentiation, mitotic delay beyond the recuperative capacity of the embryo or fetus (ionizing radiation, chemotherapeutic agents, alcohol)
Biologic and pharmacologic receptor-mediated developmental effects (i.e., etretinate, isotretinoin, retinol, sex steroids, streptomycin, thalidomide)
Metabolic inhibition (i.e., warfarin, anticonvulsants, nutritional deficiencies)
Inhibition of cell migration, differentiation, and cell communication
Physical constraint, vascular disruption or insufficiency, inflammatory lesions, amniotic band syndrome
Interference with histogenesis by processes such as cell depletion, necrosis, calcification, or scarring

VI. IDENTIFICATION OF A HUMAN TERATOGEN

Without evidence of human teratogenicity it is difficult to provide estimates of the hazard that exposures to specific agents present to the human fetus. Uncritical evaluation of single reports suggesting causal associations between suspected agents and human malformations can be misleading, such as the erroneous association of Bendectin with congenital defects.

It should be noted that most human teratogens have been identified by alert physicians or scientists. Epidemiologic studies have been most helpful in understanding the frequency, trends, and incidence of congenital malformations. While animal studies have been most useful in understanding the mechanism of action of known human teratogens, they also lend support to epidemiologic studies by the development of animal models. Although methods of extrapolation are improving, an animal model cannot be extrapolated with certainty to the human condition if one has no information on the teratogenicity in the human.

Several criteria are required to establish that a drug or chemical exposure causes maldevelopment in the human (Table 7): (i) Epidemiologic studies should repeatedly report that exposure to a drug or chemical is associated with an increased incidence of a specific malformation or group of malformations; (ii) for common exposures, secular trend data should support the allegation; (iii) an animal model should be

TABLE 7
Proof of Teratogenesis in the Human

Controlled epidemiologic studies consistently demonstrate an increased incidence of a confined group of congenital malformations in the exposed human population.

Secular trends demonstrate a relationship between the incidence of particular malformations and a reduction or increase in exposure to an environmental agent in human populations. This analysis can only be performed when a high proportion of the population has been exposed and there has been a marked change in exposure.

An animal model mimics the human malformations at clinically comparable exposures
- Without evidence of maternal toxicity
- Without reduction in food and water ingestion
- With careful interpretation of malformations that occur in isolation, such as anophthalmia in the rat, cleft plalate in the mouse, vertebral, limb, and rib malformations in the rabbit, and omphalocele in the ferret. These malformations are epigenetic malformations that can be altered nonspecifically by environmental stresses.

The teratogenic effects increase with dose within the usual range of human exposures.

The mechanisms of teratogenesis are understood and/or the results are biologically plausible based on the principles of developmental biology and teratology.

developed utilizing exposures comparable to therapeutic doses in the human; (iv) the teratogenic effects should increase with dose; and (v) the alleged teratogenic response should be biologically plausible and not contradict proven scientific principles.

VII. TERATOGENIC DRUGS, CHEMICALS, INFECTIOUS DISEASES, THERAPEUTIC AND DIAGNOSTIC AGENTS, AND OTHER ENVIRONMENTAL FACTORS RESPONSIBLE FOR REPRODUCTIVE TOXICITY AND TERATOGENESIS IN HUMANS

Table 8 lists those agents and environmental milieu that have been demonstrated to be associated with an increase in the incidence of congenital malformations. Since over a thousand drugs and chemicals have been reported to produce congenital malformations in animal species, it is important to understand the discrepancy in the number of "proven teratogens" in humans and the reported teratogens in animals. The principles of teratology readily explain this apparent discrepancy. Many animal teratogens are administered at exposures that never will occur in human populations or at exposures that are maternally toxic. Thus, the fact that the dose–response curve for teratogens has a sigmoidal toxicologic relationship resulting in threshold exposures below which no effect will occur readily explains why drugs and chemicals which have been demonstrated to produce malformations in animals have not been reported to do so in humans. These concepts can be converted to definitions as follows:

Human teratogen: An agent or milieu that has been demonstrated to produce permanent alterations in the embryo or fetus following intrauterine exposures that usually occur or are readily attainable in the human. These are the agents or milieu described in Table 8.

Potential human teratogen: An agent or milieu that has not been demonstrated to produce permanent alterations in the embryo or fetus following intrauterine exposures that usually occur or are attainable but can affect the embryo or fetus if the exposure is substantially raised above the usual human exposure. Thousands of chemicals and drugs are included in this category, including aspirin, vitamin A, and cortisone.

Nonhuman Teratogen: An agent or milieu that has no embryotoxic or fetotoxic potential because it is nontoxic at any dose or because it is so toxic to the mother that it kills the mother before or at the same dose that it begins to affect the embryo. Atropine and morphine are drugs that could probably be placed in this category.

The general principles of teratology previously discussed cannot be universally applied. For example, the concept of a sensitive stage of susceptibility may not be appropriate for metabolic diseases, chronic exposure to drugs, or exposure to infectious agents. Some teratogens, e.g., warfarin, may have more than

TABLE 8
Developmental Toxicants: Risks of Congenital Malformations and Abortion in the Human[a]

Developmental toxicant	*Reported effects or associations and estimated risks*	*Comments[b]*
Alcohol	Fetal alcohol syndrome: intrauterine growth retardation, maxillary hypoplasia, reduction in width of palpebral fissures, characteristic but not diagnostic facial features, microcephaly, mental retardation. An increase in spontaneous abortion has been reported but since mothers who abuse alcohol during pregnancy have multiple other risk factors, it is difficult to determine whether this is a direct effect on the embryo. Consumption of 6 oz of alcohol or more per day constitutes a high risk but it is likely that detrimental effects can occur at lower exposures.	Quality of available information: Good to excellent Direct cytotoxic effects of ethanol and indirect effects of alcoholism. While a threshold teratogenic dose is likely, it will vary in individuals because of a multiplicity of factors.
Aminopterin, Methotrexate	Microcephaly, hydrocephaly, cleft palate, meningomyelocele, intrauterine growth retardation, abnormal cranial ossification, reduction in derivatives of first branchial arch, mental retardation, postnatal growth retardation. Aminopterin can induce abortion within its therapeutic range; it is used for this purpose to eliminate ectopic embryos. Risk from therapeutic doses is unknown but appears to be moderate to high.	Quality of available information: Good Anticancer, antimetabolic agents; folic acid antagonists that inhibit dihydrofolate reductase, resulting in cell death.
Androgens	Masculinization of female embryo: clitoromegaly with or without fusion of labia minora. Nongenital malformations are not a reported risk. Androgen exposures which result in masculinization have little potential for inducing abortion. Based on animal studies, behavioral masculinization of the female human will be rare.	Quality of available information: Good Effects are dose and stage dependent; stimulates growth and differentiation of sex steroid receptor-containing tissue.
Angiotensin converting enzyme (ACE) inhibitors	The therapeutic use of ACE inhibitors has neither a teratogenic effect nor an abortigenic effect in the first trimester. Since this group of drugs does not interfere with organogenesis, they can be used in a woman of reproductive age; if the woman becomes pregnant, therapy can be changed during the first trimester without an increase in the risk of teratogenesis. Later in gestation these drugs can result in fetal and neonatal death, oligohydramnios, pulmonary hypoplasia, neonatal anuria, intrauterine growth retardation, and skull hypoplasia. Risk is dependent on dose and length of exposure.	Quality of available information: Good Antihypertensive agents; adverse fetal effects are related to severe fetal hypotension over a long period of time during the second or third trimester.
Caffeine	Caffeine is teratogenic in rodent species with doses of 150 mg/kg. There are no convincing data that moderate or usual exposures (300 mg per day or less) present a measurable risk in the human for any malformation or group of malformations. On the other hand, excessive caffeine consumption (exceeding 300 mg per day) during pregnancy is associated with growth retardation and embryonic loss.	Quality of available information: Fair to good Behavioral effects have been reported and appear to be transient or temporary; more information is needed concerning the population with higher exposures.
Carbamazepine	Minor craniofacial defects (upslanting palpebral fissures, epicanthal folds, short nose with long philtrum), fingernail hypoplasia, and developmental delay. Teratogenic risk is not known but is likely to be significant for minor defects. There are too few data to determine whether carbamazapine presents an increased risk for abortion. Since embryos with multiple malformations are more likely to abort, it would appear that carbamazepine presents little risk because an increase in these types of malformations has not been reported.	Quality of available information: Fair to good Anticonvulsant; little is known concerning mechanism. Epilepsy may itself contribute to an increased risk for fetal anomalies.
Cocaine	Preterm delivery; fetal loss; placental abruption; intrauterine growth retardation; microcephaly; neurobehavioral abnormalities; vascular disruptive phenomena resulting in limb amputation, cerebral infarctions, and certain types of visceral and urinary tract malformations. There are few data to indicate that cocaine increases the risk of first-trimester abortion. The low but increased risk of vascular disruptive phenomena due to vascular compromise of the pregnant uterus would more likely result in midgestation abortion or stillbirth. It is possible that higher doses could result in early abortion. Risk for deleterious effects on fetal outcome is significant; risk for major disruptive effects is low but can occur in the latter portion of the first trimester as well as the second and third trimesters.	Quality of available information: Fair to good Cocaine causes a complex pattern of cardiovascular effects due to its local anesthetic and sympathomimetic activities in the mother. Fetopathology is likely to be due to decreased uterine blood flow and fetal vascular effects. Because of the mechanism of cocaine teratogenicity, a well-defined cocaine syndrome is not likely. Poor nutrition accompanies drug abuse and multiple drug abuse is common.

continues

TABLE 8 Continued

Developmental toxicant	Reported effects or associations and estimated risks	Comments[b]
Chorionic villous sampling (CVS)	Low, but increased risk of orofacial malformations and limb reduction defects of the congenital amputation type as seen in vascular disruption malformations in some series. The risk of abortion following CVS is quite low.	Quality of available information: Fair Excessive bleeding from the chorion is probably related in part to the experience of the operator. Further research is necessary to determine whether CVS is safer for the fetus at certain stages of gestation.
Coumarin derivatives	Nasal hypoplasia; stippling of secondary epiphysis; intrauterine growth retardation; anomalies of eyes, hands, neck; variable central nervous system anatomical defects (absent corpus callosum, hydrocephalus, asymmetrical brain hypoplasia). Risk from exposure is 10–25% during Weeks 8–14 of gestation. There is also an increased risk of pregnancy loss. There is a risk to the mother and fetus from bleeding at the time of labor and delivery.	Quality of available information: Good Anticoagulant; bleeding is an unlikely explanation for effects produced in the first trimester. Central nervous system defects may occur anytime during second and third trimester and may be related to bleeding.
Cyclophosphamide	Growth retardation, ectrodactyly, syndactyly, cardiovascular anomalies, and other minor anomalies. Teratogenic risk appears to be increased but the magnitude of the risk is uncertain. Almost all chemotherapeutic agents have the potential for inducing abortion. This risk is dose related; at the lowest therapeutic doses the risk is small.	Quality of available information: Fair Anticancer, alkylating agent; requires cytochrome P450 monooxydase activation; interacts with DNA, resulting in cell death.
Diethylstilbestrol (DES)	Clear cell adenocarcinoma of the vagina occurred in about 1 in 1000–10,000 females who were exposed *in utero*. Vaginal adenosis occurs in about 75% of females exposed *in utero* before Week 9 of pregnancy. Anomalies of the uterus and cervix may play a role in decreased fertility and an increased incidence of prematurity although the majority of women exposed to DES *in utero* can conceive and deliver normal babies. *In utero* exposure to DES increased the incidence of genitourinary lesions and infertility in males. DES can interfere with zygote survival, but it does not interfere with embryonic survival when given in its usual dosage after implantation. Offspring who were exposed to DES *in utero* have an increased risk for delivering prematurely, but do not appear to be at increased risk for first-trimester abortion.	Quality of available information: Fair to good Synthetic estrogen; stimulates estrogen receptor-containing tissue, may cause misplaced genital tissue which has a greater propensity to develop cancer.
Diphenylhydantoin	Hydantoin syndrome: microcephaly, mental retardation, cleft lip/palate, hypoplastic nails, and distal phalanges; characteristic but not diagnostic facial features. Associations documented only with chronic exposure. Wide variation in reported risk of malformations but appears to be no greater than 10%. The few epidemiological data indicate a small risk of abortion for therapeutic exposures for the treatment of epilepsy. For short-term treatment, i.e., prophylactic therapy for a head injury, there is no appreciable risk.	Quality of available information: Fair to good Anticonvulsant; direct effect on cell membranes, folate, and vitamin K metabolism. Metabolic intermediate (epoxide) has been suggested as the teratogenic agent.
Electromagnetic fields (EMF)	The data pertaining to video display terminals indicates that the electromagnetic field exposures from these units do not present an increased risk for abortion or congenital malformations. The data on power line and appliance exposures are too varied to draw any conclusions, although the risks appear to be small or nonexistent. Human exposures to video display terminals and power lines are quite low and are unlikely to have reproductive effects.	Quality of available information: Fair Pregnant animals exposed to EMF do not exhibit consistent or reproducible reproductive effects. There are still questions about biologic effects for frequencies and waveforms of magnetic fields that have not been adequately studied.
Folic acid, decreased availability	Populations of patients who parented offspring with neural tube defects have significantly decreased the recurrence rate by use of periconceptional folic acid administration. Other studies have indicated that prospective administration of folic acid also reduces the frequency of neural tube defects	Quality of available information: Fair
Infectious agents	The cytotoxic effects and inflammatory responses resulting from fetal infections interfere with organogenesis and/or histogenesis. Fetopathic syndromes are related to the specific tissue localization, pathologic characteristics of the infectious agent, and the duration of the infection in the embryo and fetus. In some instances the infection may be debilitating to the mother and indirectly result in pregnancy loss.	
Cytomegalovirus (CVM)	Fetal cytomegalovirus infection presents an increased risk for abortion but it does not appear that maternal genital infection increases the risk of abortion. Fetal infection occurs in about 20% of maternal infections. Intrauterine growth retardation; risk of brain damage is moderate after fetal infection early in pregnancy; characteristic parenchymal calcification.	Quality of available information: Good to excellent CMV damages organs principally by cellular necrosis.

continues

TABLE 8 Continued

Developmental toxicant	*Reported effects or associations and estimated risks*	*Comments*[b]
Herpes simplex virus	Generalized organ infections, microcephaly, hepatitis, eye defects, vesicular rash. Maternal infection can be transmitted *in utero* or perinatally. Herpes simplex 2 is one of the few infections for which it is agreed that the risk of abortion is increased.	Quality of available information: Good Fetal anomalies associated with herpes simplex virus infection appear to be due to disruption rather than malformation.
Human immunodeficiency virus (HIV)	The overall risk of vertical transmission of HIV is 25–40%. Fetal HIV infection and asymptomatic maternal HIV infection are not associated with adverse effects on fetal growth or development. Symptomatic maternal HIV infection, other sexually transmitted diseases, and opportunistic infections may increase the risk of low birth weight or perinatal morbidity.	Quality of available information: Fair to good Prophylactic treatment with zidovudine does not appear to cause permanent adverse effects in the fetus and is reported to decrease the incidence of transplacental infection of the newborn.
Parvovirus B19	Infection can result in erythema infection in children but in the fetus can result in hydrops fetalis and fetal death; congenital anomalies are likely to be very rare. The risk for stillbirth with hydrops has been clearly substantiated. An increased risk for abortion has been suggested but is more difficult to substantiate.	Quality of available information: Fair to good Fetal infection is not common. Infection of red blood cell precursors causes severe anemia.
Rubella virus	Greater than 80% incidence of embryonic infection with exposure in first 12 weeks, 54% at 13 or 14 weeks, 25% at end of second trimester, and 100% at term. Defects include mental retardation, deafness, cardiovascular malformations, cataracts, glaucoma, and microphthalmia. Diabetes mellitus or rubella panencephalitis may develop later in life. The abortigenic risk of maternal rubella is uncertain.	Quality of available information: Excellent Rubella has an affinity for specific tissues. Damage is caused by mitotic inhibition, cell death, and interference with histogenesis by repair processes, resulting in calcification and scarring.
Syphilis	Defects in 50% of offspring after early exposure to primary or secondary syphilis and 10% after late exposures. Defects include maculopapular rash, hepatosplenomegaly, deformed nails, osteochondritis at joints of extremities, congenital neurosyphilis, abnormal epiphyses, and chorioretinitis. Syphilis can increase the incidence of abortion.	Quality of available information: Good Fetal pathology is associated with maturation of the fetal immune system at about the 20th week.
Toxoplasmosis	Hydrocephaly, microphthalmia, and chorioretinitis. Risk is predominantly associated with pregnancies in which the mother acquires toxoplasmosis. Epidemiological studies do not indicate that toxoplasmosis increases the incidence of early abortion, but congenital toxoplasmosis may be responsible for the stillbirth of severely affected fetuses.	Quality of available information: Good to excellent Toxoplasmosis is unlikely to contribute to the risk of repeated abortion.
Varicella–Zoster	Skin and muscle defects, intrauterine growth retardation, and limb reduction defects. No measurable increased risk of early teratogenic effects. Incidence of maternal varicella during pregnancy is low but risk of severe neonatal infection is high if maternal infection occurs in last week of pregnancy. There does not appear to be an increased risk of first-trimester abortion.	Quality of available information: Fair to good Virus infection of fetal tissues can cause cellular necrosis.
Venezuelan equine encephalitis	Hydroanencephaly; microphthalmia; central nervous system destructive lesions; luxation of hip. There are not enough data to determine whether infection presents an increased risk of abortion.	Quality of available information: Poor to fair Infection can cause cellular necrosis in fetal tissues but fetal infection is rare.
Lead	There is no indication that serum lead levels below 50 μg% result in congenital malformations in exposed embryos and fetuses but the developing central nervous system in the fetus and child may be susceptible to lead toxicity resulting in decreased IQ and behavioral effects. Lead levels above 50 μg% result in anemia and encephalopathy can have serious effects on central nervous system development. Lead levels below 50 μg% do not increase the risk of abortion.	Quality of available information: Good While there are human studies indicating small deficiencies in IQ in patients with lead levels above 10 μg%, there could be other explanations for these IQ differences. Furthermore, pathological findings have not been described in the brain at these levels.
Lithium carbonate	Although animal studies have demonstrated a clear teratogenic risk, the effect in humans is uncertain. Early reports indicated an increased incidence of Ebstein's anomaly, other heart and great vessel defects, but as more studies are reported the strength of this association has diminished. Lithium levels within the therapeutic range (<1.2 mg%) do not increase the risk of abortion.	Quality of available information: Fair to good Antidepressant; mechanism has not been defined.
Maternal conditions		
Diabetes	Caudal hypoplasia or caudal regression syndrome; congenital heart disease; anencephaly. Vascular lesions in long-standing diabetics may produce placental dysfunction and result in fetal growth retardation. Documented significant increased risk of abortion. The risk is greatest in untreated diabetics or patients who are poorly controlled; insulin therapy protects the fetus.	Quality of available information: Good The results of *in vitro* studies suggest that diabetic embryopathy has a multifactorial etiology. Adverse fetal effects have not been demonstrated with gestational diabetes.

continues

TABLE 8 *Continued*

Developmental toxicant	*Reported effects or associations and estimated risks*	*Comments*[b]
Endocrinopathy	If condition is compatible with pregnancy, effects are similar to those following administration of high or low doses of the hormone. Hypo- and hyperthyroidism may increase the risk of abortion. Cushing's disease, pituitary tumors, hypothalamic tumors, and androgen-producing tumors do not appear to increase the risk of abortion but may contribute to infertility.	Quality of available information: Good Receptor-mediated exposures to high levels of hormone.
Maternal Phenylketonuria	Mental retardation; microcephaly; intrauterine growth retardation. Documented significant increased risk of abortion. The risk is greatest in pregnant women who were not treated for their phenylketonuria.	Quality of available information: Good Very high levels of phenylalanine interfere with embryonic cell metabolism.
Nutritional deprivation	Central nervous system anomalies; intrauterine growth retardation; increased morbidity. In some instances of teratogenesis, abnormal nutrition may be the final common mechanism. Severe malnutrition can contribute to pregnancy loss at any stage of pregnancy.	Quality of available information: Fair to good The high mitotic rate of the fetus in general and the central nervous system in particular is very sensitive to severe alterations in nutrient supply.
Mechanical problems	Birth defects such as club feet, limb reduction defects, aplasia cutis, cranial asymmetry, external ear malformations, midline closure defects, cleft palate, and muscle aplasia. Submucosal and intramural myomata, amniotic bands, multiple implantations, bifid uterus, or infantile uterus which contribute to mechanical problems do not result in early abortions.	Quality of available information: Good Physical constraint can result in distortion and a reduction in blood supply and is more frequent in pregnancies with multiple conceptuses, abnormal uterus, amniotic abnormalities, certain placental abnormalities, and oligohydramnios.
Methyl mercury	Minamata disease: cerebral palsy, microcephaly, mental retardation, blindness, cerebellar hypoplasia. Does not appear to decrease fertility unless the mother becomes clinically ill from methyl mercury poisoning. At low exposures the teratogenic effect predominates and there are few human data to indicate the risk of abortion.	Quality of available information: Good Organic mercurials accumulate in lipid tissue causing cell death due to inhibition of cellular enzymes, especially sulfhydryl enzymes. Since most cases result from accidental environmental exposure, risk estimation is usually retrospective.
Methylene blue	Hemolytic anemia and jaundice in neonatal period after exposure late in pregnancy. There may be a small risk for intestinal atresia but this is not yet clear. No indication of increased risk of abortion.	Quality of available information: Poor to fair Used to mark amniotic cavity during amniocentesis.
Misoprostol	Misoprostol is a synthetic prostaglandin analog that has been used by millions of women for illegal abortion. A low incidence of vascular disruptive phenomenon, such as limb reduction defects and Mobius syndrome, has been reported.	Quality of available information: Fair Classical animal teratology studies would not be helpful in discovering these effects because vascular disruptive effects occur after the period of early organogenesis.
Oxazolidine-2,4-diones (trimethadione, paramethadione)	Fetal trimethadione syndrome: V-shaped eye brows, low-set ears with anteriorly folded helix, high-arched palate, irregular teeth, CNS anomalies, severe developmental delay. Wide variation in reported risk. Characteristic facial features are documented only with chronic exposure. The abortifacient potential has not been adequately studied but appears to be minimal.	Quality of available information: Good to excellent Anticonvulsants; affects cell membrane permeability. Actual mechanism of action has not been determined.
D-Penicillamine	Cutis laxa, hyperflexibility of joints. Condition appears to be reversible and the risk is low. There are no human data on the risk of abortion.	Quality of available information: Fair to good Copper chelating agent; produces copper deficiency inhibiting collagen synthesis and maturation.
Polychlorinated biphenyls	Cola-colored babies: pigmentation of gums, nails, and groin; hypoplastic deformed nails; intrauterine growth retardation; abnormal skull calcification. Although abortion can be induced at high exposures, most human exposures from environmental contamination are unlikely to increase the risk of abortion.	Quality of available information: Good Environmental contaminants; polychlorinated biphenyls and commonly occurring contaminants are cytotoxic. Body residues in exposed women can affect pigmentation in offspring for up to 4 years after exposure.
Progestins	Masculinization of female embryo exposed to high doses of some testosterone-derived progestins and may interact with progesterone receptors in the liver and brain later in gestation. The dose of progestins present in modern oral contraceptives presents no masculinization or feminization risks. All progestins present no risk for nongenital malformations. Many synthetic progestins and natural progesterone have been used to treat luteal phase deficiency, embryos implanted via IVF, threatened abortion, or bleeding in pregnancy with variable results. Conversely, synthetic progestins that interfere with progesterone function may cause early pregnancy loss; RU 486 is currently used specifically for this purpose.	Quality of available information: Good Stimulates or interferes with sex steroid receptor-containing tissue.

continues

TABLE 8 Continued

Developmental toxicant	*Reported effects or associations and estimated risks*	*Comments*[b]
Radiation (external irradiation)	Microcephaly; mental retardation; eye anomalies; intrauterine growth retardation; visceral malformations. Teratogenic risk depends on dose and stage of exposure. Exposures from diagnostic procedures present no increased risk of abortion, growth retardation, or malformation. No measurable risk with exposures for 5 rad (5 mGy) or less of X-rays at any stage of pregnancy. In contrast, exposure of the pregnant uterus to therapeutic doses of ionizing radiation significantly increases the risk of aborting the embryo; the fetus is more resistant.	Quality of available information: Good to excellent Diagnostic and therapeutic agents; produce cell death and mitotic delay.
Radioactive isotopes	Tissue- and organ-specific damage is dependent on the radioisotope element and distribution, i.e., ^{131}I administered to a pregnant woman can cause fetal thyroid hypoplasia after Week 8 of development. Radioisotopes used for diagnosis present no risk for inducing abortion because the dose to the embryo and implantation site is too low. There may be unusual circumstances wherein isotopes are introduced into the abdominal cavity in a pregnant woman for the treatment of malignancy. If the resulting dose to the embryo or fetus is substantial, the risk for abortion is increased.	Quality of available information: Good to excellent Higher doses of radioisotopes can produce cell death and mitotic delay. Effect is dependent on dose, distribution, metabolism, and specificity of localization.
Retinoids, systemic (isotretinoin, etretinate)	Increased risk of central nervous system, cardioaortic, ear and clefting defects. Microtia, anotia, thymic aplasia, and other branchial arch, aortic arch abnormalities, and certain congenital heart malformations. Exposed embryos are at greater risk for abortion. This is plausible since many of the malformations, such as neural tube defects, are associated with an increased risk of abortion.	Quality of available information: Fair Used in treatment of chronic dermatoses. Retinoids can cause direct cytotoxicity and alter programmed cell death; they affect many cell types but neural crest cells are particularly sensitive.
Retinoids, topical (tretinoin)	Epidemiological studies, animal studies, and absorption studies in humans do not suggest a teratogenic risk. Regardless of the risks associated with systemically administered retinoids, topical retinoids present little or no risk for intrauterine growth retardation, teratogenesis, or abortion because they are minimally absorbed and only a small percentage of skin is exposed.	Quality of available information: Poor Topical administration of tretinoin in animals in therapeutic doses is not teratogenic, although massive exposures can produce maternal toxicity and reproductive effects. More important, topical administration in humans results in nonmeasurable blood levels.
Smoking and nicotine	Placental lesions; intrauterine growth retardation; increased postnatal morbidity and mortality. While there have been some studies reporting increases in anatomical malformations, most studies do not report an association. There is no syndrome associated with maternal smoking. Maternal or placental complications can result in fetal death. Exposures to nicotine and tobacco smoke are a significant risk for pregnancy loss in the first and second trimester.	Quality of available information: Good to excellent While tobacco smoke contains many components, nicotine can result in vascular spasm vasculitis which has resulted in a higher incidence of placental pathology.
Sonography (ultrasound)	No confirmed detrimental effects resulting from medical sonography. The levels and types of medical sonography that have been used in the past have no measurable risks. The current clinical use of diagnostic ultrasound presents no increased risk of abortion.	Quality of available information: Good to excellent It appears that if the embryonic temperature never exceeds 39°C, there is no measurable risk.
Streptomycin	Streptomycin and a group of ototoxic drugs can affect the eighth nerve and interfere with hearing; it is a relatively low-risk phenomenon. There are not enough data to estimate the abortigenic potential of streptomycin. Because the deleterious effect of streptomycin is limited to the eighth nerve, it is unlikely to affect the incidence of abortion.	Quality of available information: Fair to good Long-duration maternal therapy during pregnancy is associated with hearing deficiency in offspring.
Tetracycline	Bone staining and tooth staining can occur with therapeutic doses. Persistent high doses can cause hypoplastic tooth enamel. No other congenital malformations are at increased risk. The usual therapeutic doses present no increased risk of abortion to the embryo or fetus.	Quality of available information: Good Antibiotic; effects seen only if exposure is late in the first or during second or third trimester since tetracyclines have to interact with calcified tissue.
Thalidomide	Limb reduction defects (preaxial preferential effects, phocomelia); facial hemangioma; esophageal or duodenal atresia; anomalies of external ears, eyes, kidneys, and heart; increased incidence of neonatal and infant mortality. The thalidomide syndrome, while characteristic and recognizable, can be mimicked by some genetic diseases. Although there are fewer data pertaining to its abortigenic potential, there appears to be an increased risk of abortion.	Quality of available information: Good to excellent Sedative–hypnotic agent. The etiology of thalidomide teratogenesis has not been definitively determined.
Thyroid: iodides, radioiodine, antithyroid drugs (propylthiouracil), iodine deficiency	Fetal hypothyroidism or goiter with variable neurologic and aural damage. Maternal hypothyroidism is associated with an increase in infertility and abortion. Maternal intake of 12 mg of iodide per day or more increases the risk of fetal goiter.	Quality of available information: Good Fetopathic effect of endemic iodine deficiency occurs early in development. Fetopathic effect of iodides, antithyroid drugs, and radio iodine involves metabolic block and decreased thyroid hormone synthesis and gland development.

continues

TABLE 8 Continued

Developmental toxicant	*Reported effects or associations and estimated risks*	*Comments*[b]
Toluene	Intrauterine growth retardation; craniofacial anomalies; microcephaly. It is likely that high exposures from abuse or intoxication increase the risk of teratogenesis and abortion. Occupational exposures should present no increase in the teratogenic or abortigenic risk. The magnitude of the increased risk for teratogenesis and abortion in abusers is not known because the exposure in abusers is too variable.	Quality of available information: Poor to fair Neurotoxicity is produced in adults who abuse toluene; a similar effect may occur in the fetus.
Valproic acid	Malformations are primarily neural tube defects and facial dysmorphology. The facial characteristics associated with this drug are not diagnostic. Small head size and developmental delay have been reported with high doses. The risk for spina bifida is about 1% but the risk for facial dysmorphology may be greater. Because therapeutic exposures increase the incidence of neural tube defects, one would expect a slight increase in the incidence of abortion.	Quality of available information: Good Anticonvulsant; little is known about the teratogenic action of valproic acid.
Vitamin A deficiency	While there have been case reports of apparently vitamin A deficient mothers delivering babies with congenital malformations, there are no quality epidemiological studies evaluating the magnitude of the risk of vitamin A deficiency. There are excellent nutritional studies in a number of mammalian species that clearly indicate that vitamin A deficiency is teratogenic.	Quality of the data: Poor The magnitude of this clinical problem has not been elucidated.
Vitamin A excess	The same malformations that have been reported with the retinoids have been reported with very high doses of vitamin A (retinol). Exposures below 10,000 IU present no risk to the fetus. Vitamin A in its recommended dose presents no increased risk for abortion.	Quality of available information: Good High concentrations of retinoic acid are cytotoxic; it may interact with DNA to delay differentiation and/or inhibit protein synthesis.
Vitamin D excess	Large doses given in vitamin D prophylaxis are possibly involved in the etiology of supravalvular aortic stenosis, elfin faces, and mental retardation. There are no data on the abortigenic effect of vitamin D.	Quality of available information: Poor Mechanism is likely to involve a disruption of cell calcium regulation with excessive doses.

[a] From Beckman *et al.* (1997).
[b] Quality of available information is modified from TERIS (1994).

one mechanism of action: metabolic interference in the earlier stages of pregnancy and bleeding in the latter portion of pregnancy. The angiotensin converting enzyme inhibitors actually are teratogenic in the latter part of gestation and relatively innocuous during organogenesis. Some teratogens produce teratogeneis not by interfering with organogenesis or differentiation but by producing physiologic or biochemical effects that are similar to effects in the adult organism, e.g., antithyroid medication or sex steroids.

The clinical manifestation for the human teratogens and teratogenic milieu listed in Table 8 follow important principles that should guide the analysis of teratology studies and the interpretation of epidemiology studies and individual patient exposure to teratogens:

1. Exposure to teratogens follows a toxicological dose–response curve. There is a threshold below which no effect will be observed and as the dose of the teratogen is increased, both the severity and frequency of reproductive effects will increase.
2. The period of exposure during embryonic development is critical in determining which effects will be produced and whether any effects can be produced even by a known teratogen.
3. Most teratogens have a confined group of congenital malformations that result after exposure during the critical period of embryonic development: the teratogenic syndrome.
4. Even the most potent teratogenic agent cannot produce every malformation. (It is very unlikely that an isolated cleft palate or postaxial limb defect would be caused by thalidomide even if a thalidomide exposure occurred during pregnancy.)
5. While a group of malformations may presumptively suggest the possibility of a particular teratogenic exposure, they cannot definitively confirm the causal agent because there are genocopies and pheno-

copies from exposure to other teratogenic agents that result in similar malformations.

6. The presence of certain malformations can eliminate the possibility that a particular teratogenic agent was responsible. (An isolated unilateral congenital amputation in a child with normal head size, normal birth weight, and normal intelligence could not have been caused by ionizing radiation, regardless of the exposure.)

These principles are frequently helpful when analyzing experimental teratology studies, individual patient's malformations, or epidemiological studies.

VIII. SUMMARY

Environmental influences that produce abnormal development have certain characteristics in common and follow certain basic principles. These principles determine quantitative and qualitative aspects of teratogenesis and include (i) embryonic stage, (ii) the dose or magnitude of the exposure, (iii) threshold phenomena, (iv) pharmacokinetics and metabolism of the agent, (v) placental transport, and (vi) species differences. Environmental causes of human malformations account for approximately 10% of human malformations and fewer than 1% of all human malformations are related to prescription drug exposure, chemicals, or radiation. Other environmental factors, such as maternal infections, maternal disease states, and maternal nutritional states, account for a greater proportion of environmentally induced human malformations and these causes may also be preventable. However, malformations caused by drugs and other therapeutic agents are also important because these exposures are preventable as well. As we develop a better understanding of the mechanisms of teratogenesis from all etiologies we may learn how to best predict and test for teratogenicity and eventually reduce the incidence of human congenital malformations. When an environmental agent or milieu is alleged to be responsible for human congenital malformations, the allegation can be evaluated in an organized and scientific manner using data obtained from epidemiological studies, animal studies, and by applying the basic principles of developmental biology and teratology as well as the clinical principles that are part of the dogma of clinical evaluation. It is not unusual that these basic principles can support or refute the allegation on the basis of biological plausibility.

Acknowledgments

This work was supported in part by funds from Nemours Foundation, NIH Grant HD29902-02, Foerderer Foundation, and Harry Bock Charities.

See Also the Following Articles

FAMILY PLANNING; FETAL ANOMALIES; REPRODUCTIVE TOXICOLOGY

Bibliography

Beckman, D. A., Fawcett, L. B., and Brent, R. L. (1997). Developmental toxicity. In *Handbook of Human Toxicology* (E. J. Massaro, Ed.), pp. 1007–1084. CRC Press, New York.

Brent, R. L. (1986). Editorial comment: Definition of a teratogen and the relationship of teratogenicity to carcinogencity. *Teratology* **34**, 359–360.

Brent, R. L. (1995). The application of the principles of toxicology and teratology in evaluating the risks of new drugs for the treatment of drug addiction in women of reproductive age. In *Medications Development for the Treatment of Pregnant Addicts and Their Infants* (C N. Chiang and L. P. Finnegan, Eds.), Research Monograph Series No. 149, pp. 130–184. National Institute on Drug Abuse, Rockville, MD.

Brent, R. L., and Holmes, L. B. (1988). Clinical and basic science lessons from the thalidomide tragedy: What have we learned about the causes of limb defects? *Teratology* **38**, 241–251.

Fraser, F. C. (1976). The multifactorial/threshold concept—Uses and misuses. *Teratology* **14**, 267–280.

Fraser, F. C. (1977). Relationship of animal studies to man. In *Handbook of Teratology* (J. G. Wilson and F. C. Fraser, Eds.), pp. 75–76. Plenum, New York.

Heinonen, O. P., Sloane, D., and Shapiro, S. (1977). *Birth Defects and Drugs in Pregnancy*. Publishing Sciences Group, Littleton, MA.

McKusick, V. A. (1988). *Mendalian Inheritance in Man: Catalogs of Autosomal Dominant, Autosomal Recessive, and X-Linked Phenotypes*, 8th ed. Johns Hopkins Univ. Press, Baltimore.

TERIS (1994). *Teratogenic Effects of Drugs: A Resource for Clinicians* (J. M. Friedman and J. E. Polifka, Eds.). Johns Hopkins Univ. Press, Baltimore.

Warkany, J. (1978). Aminopterin and methotrexate: Folic acid deficiency. *Teratology* 17, 353–358.

Wilson, J. G. (1973). *Environment and Birth Defects*. Academic Press, New York.

Wilson, J. G. (1985). Misinformation about risks of congenital anomalies. In *Prevention of Physical and Mental Congenital Defects. Part C. Basic and Medical Science, Education, and Future Strategies* (M. Maurois, Ed.), pp. 165–169. A. R. Liss, New York.

Wilson, J. G., Brent, R. L., and Jordan, H. C. (1953). Differentiation as a determinant of the reaction of rat embryos to x-irradiation. *Proc. Soc. Exp. Biol. Med.* 82, 67–70.

Termites

see Social Insects

Territorial Behavior, Overview

Judy Stamps
University of California, Davis

GLOSSARY

conspecific A member of the same species.

home range The area used by an animal over the course of its daily activities.

reproductive success The number of an animal's offspring that survive to maturity relative to the number produced by others in the same population.

resource Any environmental factor that enhances reproductive success and that is in short supply relative to the number of potential users.

territory The portion of a home range that is defended against particular categories of conspecifics.

I. INTRODUCTION

Over 50 years ago, Noble defined a territory as "any defended area." This definition captures several of the key elements in any territorial system. First, defense is important: The behavior of the territory owner discourages other individuals from using an area. Second, space is important in that the type and outcome of social interactions depend on the location in which they occur. Finally, Noble's definition implies that animals need not defend every area they use on a regular basis, thus anticipating the later distinction between a home range (the area used by an animal over the course of its daily activities) and territory (the portion of the home range that is defended against other individuals).

Since Noble's time, this definition has been expanded and elaborated to suit particular species or situations. For instance, today most workers focus on space that is defended for long periods relative to an animal's life span to differentiate territory defense from the defense of emphermal resources, such as localized food or basking sites. It is also clear that most territorial animals do not indiscriminately defend space against every member of their species (conspecifics). In many birds a male, female, and their young share a territory while defending this space against nonfamily members, and species such as wolves, ants, and Florida scrub jays maintain group territories, which are jointly defended against neighboring territorial groups and wandering intruders.

To date, territorial behavior has been described in a bewildering array of animals, including sea anemones, molluscs (limpets), insects, spiders, fish, amphibians, reptiles, birds, and mammals. There is also considerable variation among species and taxonomic groups in the age, sex, and timing of territory defense. Territory defense by juveniles occurs in many animals in which individuals are self-sufficient from the time of hatching, as in barnacles, lizards, or salmon. Territory defense by adult males has received the most attention from field workers, but in many mammals adult females defend territories, whereas males have large undefended home ranges, and in many birds both sexes jointly defend the same territory (although in this case, males typically defend against males, whereas females attack females). Variation in the onset, duration, and cessation of defense is also common, with some animals such as tawny owls or ant colonies establishing territories in which they will spend their entire lives, whereas the members of other species acquire and then abandon territories on a seasonal basis.

II. TERRITORIES AND REPRODUCTION

From the perspective of this encyclopedia, the important question is how territorial behavior affects and is affected by reproductive activities. Thus, the remainder of this chapter considers territories that are established prior to the beginning of reproductive activity and that are held during the period when animals pair, mate, or raise offspring. In some cases, the reproductive significance of these territories is obvious because the only events which transpire on the territory are related to reproduction. For instance, for many years students of territorial and mating behavior have been fascinated by leks, which are clusters of small territories held by males, whom females visit to choose a mate. Studies of lekking birds, fish, butterflies, lizards, and deer indicate that the area encompassed by the lek contains no resources of interest to females or relevant to offspring production, and in many lekking species, males leave their territories when feeding or evading predators. Thus, the importance of lek territories for mate attraction and mating is particularly clear.

In other species, the members of one sex (typically but not always the males) defend an area that contains receptive females or resources that are attractive to receptive females. Some male dragonflies, for example, defend patches of floating vegetation upon which females will lay their eggs, and the males thereby gain access to receptive females. Similarly, male cicada-killer wasps mature before the females and then establish territories in areas where females will eventually emerge, with each owner gaining priority mating access to the females that appear within his territory.

Still other species defend small territories centered around a location where offspring are protected and/or fed. Small nesting territories are common in fish-eating birds which travel far from the nest in search of food; examples include colonial seabirds such as gulls or pelicans. In these species, both parents defend a small area around the nest, and these nests are aggregated within a larger colony. Nesting territories are also common in other animals. Indeed, some of the most popular subjects for laboratory studies of territorial behavior are fish that establish and defend small nesting territories. Familiar examples include sticklebacks or Siamese fighting fish, in which males attract females to their nests for egg laying and fertilization, after which the males care for the eggs and protect the newly hatched young from predators.

It is easy to discern the reproductive significance of small territories that only function in mating and offspring production and that are only held for the duration of reproductive activities. However, in many territorial animals, reproduction is only one of a number of activities that occur within the confines of territory boundaries. In vertebrates, adults often establish and defend large "all-purpose" territories, in which the territory owners forage, survive the onslaughts of inclement weather, predators, and disease, and raise their offspring. The members of these species often begin defending territories well in advance of the onset of mating or egg production and continue to defend their territories after reproduction has ceased. Indeed, climate permitting, many species with all-purpose territories occupy and defend them on a year-round basis. To readers from temperate regions of the world, the most familiar examples of all-purpose territories are those of songbirds. For instance, the onset of singing by white-crowned sparrows in early spring marks the establishment of all-purpose territories on which the males, their mates, and their offspring will spend the rest of the season. Other examples of all-purpose territories can be found in lizards and rodents, in which males establish large territories that encompass the home ranges of several females and then defend these against male conspecifics.

III. THEORETICAL STUDIES OF TERRITORIAL BEHAVIOR

Individuals in a territorial species face a number of decisions, including (i) where to settle (i.e., habitat selection), (ii) whether or not to defend a territory, and (iii) size, shape, and location of a territory. For instance, a young male bird might settle in either an oak woodland or a pine forest; join a flock or establish a territory; establish a territory near older, established territory owners or off by himself; or defend a larger or a smaller territory at a given location. In keeping with the general framework of theory in behavioral ecology, workers assume that each of these options affects individual reproductive success, where reproductive success is estimated by the number of offspring produced over a lifetime relative to the number of offspring produced by other individuals in the same population.

Reproductive success, in turn, is important because it determines which individuals will contribute the most genes to the next generation, i.e., it predicts the future direction of evolution for the members of that species. In practice, it is usually difficult, if not impossible, to measure all the effects of a particular behavior on an animal's lifetime reproductive success. Thus, most behavioral ecologists instead measure "components of fitness" (factors that contribute to reproductive success) over periods much shorter than a lifetime. For instance, they might study the relationship between territory size and the number of independent offspring per territory owner measured over a single breeding season.

Conversely, behavioral ecologists assume that the proximate mechanisms responsible for producing the behavior of present-day individuals are an outcome of past selective pressures on the members of that species, selective pressures which in the past favored individuals that exhibited some behavior patterns rather than others. Thus, a preference for oak woodland versus pine forest exhibited by birds today might reflect the fact that in the past, individuals with woodland territories produced more offspring (and hence left more descendants) than equivalent birds who established territories in forests. It is important to note the important distinction between past and current selective pressures because, as a result of recent widespread habitat destruction and other human-induced changes in predators, food supplies, and competing species, today many territorial animals are experiencing different selective pressures than those that shaped the behavior of their ancestors. Hence, although the members of a species of bird might still prefer to settle in woodland rather than forest, today their reproductive success might be lower in the former than in the latter, e.g., if most wooded areas now occur near human habitations that support large populations of fledgling-eating, feral cats.

Modern theoretical approaches to territorial defense are based on economic models of territorial behavior first developed in the early 1970s. These models assume that defense of space is a surrogate for the defense of resources located within that space, where "resource" can be loosely defined as any environmental factor that enhances reproductive success and that is in short supply, relative to the number of potential users. Thus, resources relevant to reproduction might include suitable nesting burrows or cavities, food for females preparing eggs or provisioning dependent young, or space to ensure that young are not injured by neighboring territory owners, as has been reported for several species of gulls.

The question of whether members of the opposite sex can be considered to be a resource is controversial, since recent studies show that females "defended" by one male territory owner often literally go out of their way to acquire and use sperm from other males. As a result, the number of females within a male's territory may be unrelated to the number of offspring actually fathered by that male. Thus, male defense of females (or vice versa) does not fit neatly into the classic theoretical framework for territory defense, which assumes that resources within a territory do not change their own availability as a function of the identity or characteristics of the territory owner.

The economic approach to territoriality assumes that the effects of various behavioral options on reproductive success can be divided into potential costs and benefits of those behaviors. In theory, the effect of a given option on reproductive success is a function of the benefits minus the costs of that option, relative to the benefits and costs of other behavioral options that are available to that same individual. In the classic formulation, the benefits of defense are related to access to resources that, over the period of territory tenure, tend to increase reproductive success. In turn, the costs of defense are related to behavior that is required for territory acquisition and maintenance and that would, over the period of territory tenure, be likely to reduce reproductive success. Thus, when comparing the "payoffs" of defending a territory versus adopting a nonterritorial (floating) strategy for dragonflies, one might measure the number of females inseminated per unit time by territory owners and floaters, the energetic costs of chasing intruders and fighting with neighbors, and whether owners have lower daily survival rates than floaters, e.g., because owners spend more time foraging in risky locations or are more conspicuous to predators as a result of their territorial behavior.

The traditional economic approach focuses on individuals and the costs and benefits to each individual of defending a particular type or size of territory. However, most territory owners are not solitary but instead live near neighbors with which they interact on a regular basis over the period of territory tenure. Recently, workers have extended cost–benefit analysis to consider the implications of settling and living in neighborhoods for animals in territorial species. This new line of research suggests a number of ways that territorial animals may benefit by settling and living near neighboring territory owners at the same time as they benefit by keeping neighbors and intruders out of their own territories. The notion that territorial animals might actually benefit from the presence of conspecific neighbors was originally suggested by observations showing that territorial animals belonging to several different taxa (e.g., barnacles, birds, fish, and lizards) prefer to settle and establish territories next to previously established territory owners. That is, some territorial animals are attracted to conspecifics while settling, forming clusters of territories separated by unoccupied, suitable habitat. In turn, observations of conspecific attraction have caused workers to consider the ways that territorial animals might benefit from the presence of neighboring conspecifics, either while choosing habitats or territories or while living in established territorial neighborhoods.

There are at least two reasons why conspecific attraction should be particularly advantageous for species in which males establish territories, to which they then attract reproductive females. First, recent theoretical and empirical studies of female mate-search behavior suggest that females should prefer to search for prospective mates within clusters of male territories in order to reduce their search costs and increase their chances of obtaining a high-quality mate. Thus, as suggested by David Lack and other early workers, males might be forced to establish breeding territories in aggregations because males with neighbors attract more females (or higher quality females) on a per capita basis than do males living on isolated territories. Second, when females eventually settle on a male's territory, females might prefer males in aggregations because they can eventually copulate with males other than the ones with whom they share a territory. In this case, a female could "eat her cake and have it too," obtaining resources from her home territory but sperm from attractive males on neighboring territories.

IV. IMPLICATIONS OF TERRITORIALITY FOR STUDIES OF REPRODUCTION

Territoriality is important for students of reproductive biology because so many of the favorite experimental subjects for reproductive research are derived from territorial ancestors. For example, rats, mice, and dogs are all domesticated strains of species whose members defend all-purpose group territories in nature. Similarly, domesticated birds, such as canaries or zebra finches, are derived from wild species whose members defend small nesting territories within larger colonies. Thus, regardless of the focus of a particular study, some knowledge of territorial behavior may be useful, if not essential, for interpreting the behavior and physiology of the experimental subjects.

In some cases, an appreciation of the natural territorial behavior of a species might help increase the proportion of individuals who voluntarily engage in reproductive activities under captive conditions. Increasing the frequency of reproduction in captivity is of obvious importance to conservationists whose goal is the captive breeding of endangered species and to biomedical researchers interested in animal models that can provide data relevant to human reproductive biology.

For example, if free-living males establish territories several weeks before attracting and pairing with receptive females, one would not expect normal reproductive events to ensue if males and females were unceremoniously dumped into a novel habitat and then observed for the next 2 hr. More important, if establishing stable social relationships with neighbors is an important precursor to a species' reproductive activity under natural conditions, then reproductive events might proceed more quickly and smoothly under laboratory conditions if prospective breeders were allowed to interact aggressively with conspecific "neighbors" prior to the onset of mating, nest building, or engaging in other reproductive activities. Indeed, pioneers of behavioral biology, Konrad Lorenz and Niko Tinbergen, long ago emphasized the intimate relationship between aggression and reproductive activities in cichlid fish and herring gulls, animals which defend small nesting territories in colonies under natural conditions.

The notion that territorial animals benefit by settling and living near neighbors has additional implications for reproduction in captivity and nature. For instance, the classic view of territoriality argues that reducing interference from potential competitors is a primary goal for territorial animals, in which case isolated male–female pairs should reproduce well in captivity or in patches of suitable habitat in the field. In contrast, if territorial animals prefer to live near conspecific neighbors (albeit at a "polite" distance), then normal reproduction may require signals or cues that are indicative of established neighbors living at an appropriate distance from the territory owners. It is clear that this is true for animals whose territories are small by human standards, e.g., studies have shown that social stimulation enhances reproduction in animals that establish small territories in colonies under natural conditions. However, stimuli from neighboring territory owners may also be important in animals that defend larger territories in nature. In this situation, isolated pairs reluctant to breed in captivity or in the field might be induced to do so by playbacks of attenuated, low-intensity songs or aerosols of faint pheromonal odors, i.e., cues that in nature would be indicative of established neighboring pairs living at an appropriate distance from the pair's territory.

The economic approach to territorial behavior offers additional insights for those studying territorial species in captivity. This literature emphasizes the large array of costs and benefits of territorial behavior, but only a subset of these costs and benefits may be apparent under captive conditions. For instance, territory owners that vigorously attack conspecifics may gain larger or higher quality territories in the field, but a variety of studies have shown that the time, energy, and circulating hormones required to support high levels of aggression can reduce foraging rates, curtail courtship of members of the opposite sex, or interfere with care of the young. In addition, behavior associated with territorial defense (including fighting, chasing, advertisement, or patrolling the territory boundaries) may increase susceptibility to predators or parasites, thus reducing survival rates for individuals who engage in extensive aggressive activities. Under the simplified conditions of most captive studies—food *ad lib*, no predators or parasites, little or no opportunity for mate choice by either sex, and few opportunities for parents to raise young to independence—the potential benefits of territorial behavior may more apparent than the costs. Thus, when interpreting the behavior of species with reproductive territories under artificial conditions, one should not conclude that the most aggressive individuals, or the individuals with the largest territories, would necessarily produce the most offspring under natural conditions.

A final point is the critical importance of spatial context when studying reproduction in territorial animals. Typically, territory owners become strongly attached to familiar areas (in nature, these would be their own territories) and conversely reluctant to venture into novel areas, especially if these are inhabited by potentially aggressive conspecifics. Thus, reproduction in a territorial species requires either that members of the two sexes establish overlapping home ranges prior to beginning reproductive activities or that the members of one sex stay at home while the members of the other sex travel abroad in search of mates. An appreciation of the spatial context of courtship, pairing, and mating under natural conditions can go a long way to ensuring that these events unfold naturally in captivity. For instance, male and female anoles (a type of lizard) typically have overlapping territories during an extended breeding season so that males and females repeatedly encounter one another in an area familiar to both of them. In these animals, normal reproductive cycling in females, courtship, and mating are most likely to occur if the sexes are housed together and allowed to mate in their shared home cage. In contrast, female pine squirrels defend long-term territories in which they remain while in estrus, when they are visited by a number of males from surrounding areas. If one wanted to study the reproductive physiology and behavior of this species in the laboratory, it might be advisable to introduce sexually active males into an estrus female's home cage rather than vice versa. Thus, in these and other territorial animals, captive animals may be more likely to exhibit a normal range of reproductive physiology and behavior if they are kept and studied in conditions which mimic the spatial context in which reproductive events unfold under natural conditions.

See Also the Following Articles

Aggressive Behavior; Altruistic Behavior, Vertebrates

Bibliography

Brown, J. L., and Orians, G. H. (1970). Spacing patterns in mobile animals. *Annu. Rev. Ecol. Syst.* **1**, 239–269.

Davies, N. B., and Houston, A. I. (1984). Territory economics. In *Behavioural Ecology: An Evolutionary Approach* (J. R. Krebs and N. B. Davies, Eds.), 2nd ed., pp. 148–169. Blackwell Scientific, Oxford.

Gibbs, H. L., Weatherhead, P. J., Boag, P. T., White, B. N., Tabak, I. M., and Hoysak, D. J. (1990). Realized reproductive success of polygynous red-winged blackbirds revealed by DNA markers. *Science* **250**, 1394–1397.

Muller, K. A., Stamps, J. A., Krishnan, V. V., and Willits, N. L. (1997). The effect of conspecific attraction and habitat quality on habitat selection in territorial birds (Troglodytes aedon). *Am. Nat.* **150**, 650–661.

Stamps, J. A. (1994). Territorial behavior: Testing the assumptions. *Adv. Stud. Behav.* **23**, 173–232.

Wagner, R. (1997). Hidden Leks: Sexual selection and the clustering of avian territories. In *Avian Reproductive Tactics: Female and Male Perspectives* (P. G. Parker and N. Burley, Eds.), Ornithological Monographs. AOU, Washington, DC.

Testicular Cancer

Gary D. Steinberg
The University of Chicago Hospital

GLOSSARY

cryptorchidism Abnormality of normal testicular descent, with testicle residing outside of the scrotum.

epididymis The portion of spermatic cord arising from the cephalad portion of the testis, where sperm is processed and matured and leads to the vas deferens which leads to the ejaculatory duct.

gonadal–hypothalamic–pituitary axis Feedback loop of the testis, hypothalamus, and pituitary which regulates spermatogenesis and androgen production.

radical inguinal orchiectomy A surgical procedure for removing a cancerous testis; it requires an inguinal incision and delivery of the entire testis and spermatic cord from the scrotum and ligation of the structures at the level of the internal ring of the abdominal cavity.

Raynaud's phenomenon Hypersensitivity to cold temperatures in the distal digits of the hands.

retroperitoneum An abdominal cavity posterior to the cavity of the body which contains the gastrointestinal tract, through which the great vessels, lymphatics, and nerves from the spinal cord course.

seminiferous tubules The sperm-producing organ of the testis.

The diagnosis and treatment of testis cancer is a modern medical success story. While the incidence of testicular cancer continues to rise, the death rate has fallen dramatically due to improvements in medical knowledge and medical care. Surgery in combination with chemotherapy or radiation therapy has led to dramatic long-term survival rates. Investigations to identify the risk factors as well as the molecular genetic alterations that lead to testis cancer are ongoing. In addition, newer treatments to decrease the morbidity as well as lessen the duration of therapy are currently being studied. This article provides an overview of the diagnosis and treatment of germ cell cancers arising in the gonad and does not address extragonadal or mediastinal germ cell tumors

I. EPIDEMIOLOGY AND ETIOLOGY

A. Epidemiology

The estimated number of new cases of testicular cancer in the United States is 7000 per year with a 5-year overall survival rate of 93%. Carcinoma of the testis is the most common malignancy in males 15–35 years of age and accounts for approximately 1 or 2% of all neoplasms in men. Testis cancer is approximately three times more common in Caucasians than African Americans in the United States; the incidence is also low in Asians.

Testis tumors are rare before puberty, although there is a small peak during the first year of life. Seminomas typically occur in the 31- to 42-year-old age group and then after age 50; embryonal cell cancer peaks from ages 26 to 33 years and for choriocarcinoma from ages 24 to 26 years.

B. Etiology

Testis cancer is believed to be due to a number of factors, including testicular atrophy and injury, reduced feedback inhibition of the gonadal–

hypothalamic–pituitary axis, and increased gonadotropin relapse (specifically follicle-stimulating hormone). This is thought to accelerate cellular proliferation of the remaining spermatogonia and increase molecular genetic alterations in the cells leading to malignant transformation from increased environmental mutagenesis.

There are multiple possible causes of testicular injury and atrophy. Cryptorchidism has been consistently associated with testicular cancer, with a relative risk approximately five times greater. It appears that the risk is highest for patients with abdominal cryptorchid testes. It also appears that if a cryptorchid testis is surgically repaired prior to puberty, the risk of testicular cancer approaches that for patients with normally descended testes. Adult patients that present with an absent testis may have either an intraabdominal testis or a "vanishing testis" from an *in utero* torsion of the testicular artery. The risks of surgical exploration for a presumed intraabdominal testis versus dying from testicular cancer intersect at approximately 32–34 years of age. Thus, patients <32–34 years of age should undergo an evaluation for the location or presence of the testis.

Other causes of testicular injury may be environmental. Exogenous estrogen from the ingestion of oral contraceptives, during pregnancy, from the poultry, dairy, and livestock industry, or from pesticides and herbicides may be related to an increased incidence of germ cell tumors. Other chemicals, such as lubricating oils, paints, degreasing agents, exhaust fumes, and petroleum products, may also be chemical carcinogens of the testis. Viruses, especially mumps, may also cause testicular atrophy and lead to testicular cancer.

II. ANATOMY AND PATHOLOGY OF GERM CELL CANCERS

A. Anatomy

The testis is made up of the seminiferous tubules, which combine to open into the rete testis which then empties into the ductuli efferentes of the epididymis. The seminiferous tubules are lined by Sertoli cells and germ cells in different stages of maturation. The tubules are surrounded by a basal lamina. The interstitium contains Leydig cells, blood vessels, lymphatics, and nerves. The surface of the testicle is covered by the tunica albuginea, followed by the tunica vaginalis, the cremaster muscles, and the scrotal skin.

B. Pathology

The majority (94–97%) of testis tumors arise from the germ cells. The germ cell tumors are classified into two major groups: the seminomas (40%) and the nonseminomatous (35%) germ cell tumors (NSGCT)—approximately 15% are mixed seminomas and NSGCT. The differentiation of the two types is made by microscopic examination of the tumors. The biological behavior and malignant potential of testicular cancer depend on the cell type, the presence of vascular or lymphatic invasion, serum tumor marker level, and pathologic stage. In addition, three rare tumor types distinct from seminomas and NSGCT are teratomas, spermatocystic seminomas, and yolk sac tumors. Non-germ cell tumors arise from either the Sertoli, Leydig, granulosa, or primitive gonadal stromal cells or the nonspecific stromal cells, i.e., fibroblasts, angioblasts, nerve sheath cells, muscle cells, or the mesothelium of the tunica vaginalis.

1. Seminoma

Seminoma is the most common testicular tumor. Grossly, the tumor appears grayish white or yellow and may be lobulating or irregularly nodular. Histologically, the tumor is composed of large polygonal cells in broad sheets separated by connective tissue and infiltrated by lymphocytes, macrophages, and plasma cells. At least 15% of seminomas contain trophoblastic giant cells which may produce human chorionic gonadotropin (hCG). Seminoma cells also react with antibodies to placental alkaline phosphatase (PLAP).

2. Spermatocystic Seminoma

This tumor occurs in elderly patients. Histologically, the tumor is composed of a mixed population of small, medium, and large cells which may be multinucleated. The tumors do not stain with PLAP and are associated with a good prognosis.

3. *Embryonal Carcinoma*

Embryonal carcinoma is a tumor composed of cells that correspond to early embryonic cells. These cells are undifferentiated but could differentiate into all three germ cell layers (endoderm, ectoderm, and mesoderm) and extraembryonic derivatives (trophoblast and yolk sac). Grossly, the tumors appear as an irregular, brown-tannish mass with areas of necrosis. Histologically, the cells are anaplastic with irregularly shaped nuclei and indistinct borders. The cells may grow in sheets or form papillary or glandular structures. Rarely are tumors pure embryonal cell carcinoma; rather, they are usually a mix of germ cell types.

4. *Teratoma*

Teratomas are tumors composed of benign somatic tissues. Mature testicular teratomas are typically found in children. Postpubertal teratomas may be composed of the same mature tissues as the benign teratomas in children; however, they may present with metastatic disease. Thus, postpubertal teratomas should be considered mature teratocarcinomas that have lost the embryonal carcinoma stem cells.

5. *Teratocarcinoma*

Teratocarcinomas are composed of embryonal carcinoma cells plus various somatic (endoderm, ectoderm, and mesoderm) and extraembryonic derivatives (trophoblast and yolk sac). Classically noted are embryoid bodies which resemble early embryos. The malignant stem cells of teratocarcinoma may metastasize to distant sites, producing tumors that may have a similar histologic appearance or be composed of embryonal cells or teratoma cells exclusively. The embryonal cells are exquisitely sensitive to chemotherapy; however, the teratoma cells are generally chemoresistant. In addition, teratocarcinomas may contain yolk sac and trophoblast cells which may produce α-fetoprotein (AFP) and hCG in approximately two-thirds of patients. The presence of vascular invasion and the percentage of embryonal cells have prognostic significance for staging and help predict the likelihood of lymphatic metastases.

6. *Yolk Sac Tumor*

Yolk sac carcinoma or endodermal sinus tumor occurs in two forms; a pure yolk sac tumor of infancy or a component of teratocarcinoma. Histologically, the cells are flattened and cuboidal epithelial cells arranged in a complex manner and resemble the yolk sac in extraembryonic parts of the early conceptus. Yolk sac tumors of infancy and childhood are the most common testicular tumor in children under 5 years old. The tumors have a high degree of variability histologically, with multiple patterns described. Patients with yolk sac tumors in infancy and childhood have an excellent prognosis.

7. *Choriocarcinoma*

Choriocarcinoma accounts for only 0.5% of all testicular tumors. The tumor is composed of cytotrophoblastic and syncytiotrophoblastic cells. Typically, patients present with distant metastases and have a poor prognosis. Choriocarcinomatous elements are found in approximately 10% of NSGCT and can produce hCG.

8. *Gonadal Stromal Tumors*

Gonadal stromal tumors account for 3–5% of all testicular tumors. Leydig cell tumors are usually small and are composed of cells that have the features of Leydig cells. Approximately 10% are believed to be malignant, with no reliable histologic features to suggest malignancy. Prognosis depends on the amount of tumor spread.

Sertoli cell tumors can occur in any age group with approximately 10% considered malignant. Granulosa cell tumors are rare benign tumors of the testis and resemble granulosa cell tumors of the ovary.

9. *Carcinoma in Situ of the Testis*

Carcinoma *in situ* (CIS) is a preinvasive stage and precursor of testicular cancer. CIS cells are usually located along a thickened basement membrane of the seminiferous tubules at the site occupied by spermatogonia. Histologically, CIS resembles seminoma, with large irregular nuclei and prominent nucleoli. CIS cells may invade into the interstitial tissue. CIS cells share many structural and immunologic features of early embryonic germ cells and may originate from malignant transformation of fetal gonocytes under the influence of various maternal hormonal agents or environmental factors.

In adults, nearly 100% of patients with testicular tumors have associated CIS. However, the spontaneous course of CIS is unknown, and the cells may lie dormant for many years. Approximately 50% of patients with CIS will develop testicular tumors.

III. PRESENTATION AND DIAGNOSIS

Symptoms may be due to either the local manifestations of testicular cancer or the effects of testicular cancer that has spread to other parts of the body. The most common symptom at the time of diagnosis is painless swelling or enlargement of the testis. Less common is pain in the testis, which may be due to infarction or hemorrhage of the tumor in the testis, or what appears to be acute epididymitis. Some patients report feeling a nodule in the testis or a change in the texture—becoming more firm. Others report a dull ache in the scrotum. Approximately 3% report breast tenderness and 4% have a trauma history. Approximately 3% of patients present with infertility. Symptoms due to disease dissemination include a neck mass, shortness of breath, chest pain, and abdominal or back pain due to extensive retroperitoneal disease.

Careful examination of the scrotum and testes is an important clinical skill. The testis must be palpated in its entirety and separated from the epididymis and spermatic cord. The differential diagnoses include hydrocele, spermatocele, hematocele, varicocele, or epididymal orchitis. Transscrotal ultrasound scanning has a 96% sensitivity and specificity for diagnosing scrotal and testicular masses. An intratesticular mass is felt to be testis cancer until proven otherwise. The signs of disseminated disease include masses in the abdomen or supraclavicular region or signs of pleural effusions or other pulmonary findings. Enlargement or breast tenderness may be due to elevated levels of β-hCG, which is produced by some testis tumors.

A. Tumor Markers

The two most important testis tumor markers are β-hCG and AFP. Other useful serum markers are lactic acid dehydrogenase (LDH) and PLAP. β-hCG is elevated in 40–60% of patients with testicular cancer. It is elevated in all patients with choriocarcinoma, 50% of patients with embryonal cell carcinoma, and 10–25% of patients with pure seminoma. The serum half-life is 24–36 hr. AFP is elevated in 50–70% of patients with testis tumors. Embryonal and yolk sac tumors can have elevated AFP, whereas seminoma and choriocarcinoma will not. The serum half-life of AFP is 5–7 days. If the serum level does not reduce to the normal range within five half-lives, then there most certainly remains residual or metastatic disease. A normal level of AFP or β-hCG does not imply the absence of metastatic disease. LDH is not specific for histologic type but correlates with bulk disease. PLAP is elevated in 30–50% of seminomas and in 100% of patients with advanced seminomas.

B. Staging

After the diagnosis of a testicular mass has been made by history, physical examination, and ultrasonography of the scrotum, testicular tumor markers are obtained and surgery is performed to confirm the diagnosis. At the time of surgery an inguinal incision is made and the entire testis and spermatic cord is removed. The surgical specimen is examined grossly and microscopically. The histologic cell type and the presence or absence of vascular and lymphatic invasion is noted.

Computerized tomography (CT) of the chest, abdomen, and pelvis is obtained to assess for hematogenous or lymphatic metastases. Clinical stages are the following: Stage A or I is tumor confined to the testis, stage B or II is tumor confined to the retroperitoneal lymph nodes (left-sided testis tumors predominantly affect the paraaortic lymph nodes below the left renal hilum and right-sided tumors commonly metastasize to the interaortocaval and precaval lymph nodes), and stage C or III is tumor involving the abdominal viscera or above the diaphragm. The lymphatic drainage of the testis follows the arterial and venous supply of the testis. The left and right testicular arteries originate from the aorta just caudal to the origin of the left and right renal artery. The left testicular vein drains into the left renal vein, and the right testicular vein drains into the vena cava.

Repeat serum tumor markers are obtained at least five half-lives (5–7 days for β-hCG and 25–35 days for AFP) after surgery to assess for the presence of residual biochemical disease in patients with clinical stage I tumors and negative CT scans. Evidence from multiple institutions reveals that 30–40% of clinical stage I NSGCT patients and 15–20% of clinical stage I seminoma patients have microscopic metastatic disease in the retroperitoneal lymph nodes not detected by CT scanning.

Surgical excision, i.e., retroperitoneal lymph node dissection (RPLND), serves two purposes in clinical stage I NSGCT patients. First, since approximately 30–40% of clinical stage I patients have lymphatic metastases, RPLND provides more accurate staging information to further guide therapy. Second, in approximately 50–70% of patients it is curative in that no additional therapy is necessary and disease does not recur. Moreover, if disease does recur, it recurs in the chest and thus follow-up is simplified, requiring only chest X-rays and serum tumor markers. However, complications of RPLND include wound infection, bowel obstruction, and infertility due to loss of antegrade emission of semen with ejaculation. With the introduction of modified dissections and nerve sparing techniques the lumbar sympathetic nerves can routinely be spared, with avoidance of the latter complication.

IV. CLINICAL MANAGEMENT OF TESTICULAR CANCER

A. Seminomatous Germ Cell Tumors

1. *Management of Stage I and II Seminoma*

Traditionally, patients with stage I and II seminoma have been treated with radical inguinal orchiectomy and radiation therapy, with 95–100% long-term survival rate. Currently, however, possible approaches include surveillance for stage I patients and chemotherapy for stage II patients.

Standard postorchiectomy radiation therapy in stage I seminoma consists of 25–35 Gy directed at the paraaortic and ipsilateral pelvic lymph nodes. The low incidence of pelvic lymph node involvement in stage I seminoma has led some centers to irradiate the paraaortic lymph nodes alone. Prophylactic mediastinal irradiation has been demonstrated to be unnecessary and harmful. The in-field disease control rate is close to 100%, with approximately 5% of patients having disease relapse outside of the irradiated field. The cause-specific survival rate in stage I seminoma approaches 100%. Prognostic factors that predispose patients to distant relapse include anaplastic histology and locally extensive tumor with invasion into the tunica albuginea and cord structures; however, the positive predictive value is relatively low.

Several prospective nonrandomized studies have been conducted to determine if surveillance is a reasonable alternative for patients with stage I seminoma. Avoiding retroperitoneal irradiation would decrease the risk of subsequent secondary radiation-induced abdominal malignancies. The relapse rate in these series ranges from 15 to 20%, with the majority of relapses in the retroperitoneum within 12–18 months. However, late relapses were noted. The vast majority of patients are cured with chemotherapy and the death rate remains <1%. Surveillance regimens require frequent office visits and radiological examinations and must be adhered to strictly.

Patients with stage II seminoma are currently treated with radiation therapy or chemotherapy. Most patients with stage II seminoma have local control after radiotherapy; however, failure to achieve complete tumor regression in the retroperitoneum may be due to the possible presence of NSGCT. If the residual mass is small and continues to regress, observation is recommended. Masses 3 cm or larger usually are resected.

The radiation treatment failure rate for stage II seminoma depends on the extent of retroperitoneal disease, with the failure rate as high as 50% for patients with bulky disease. However, most patients with stage II seminoma, especially bulky nodal disease (lymph nodes 5 cm or larger in size), are now treated with chemotherapy with excellent long-term results.

2. *Management of Advanced Seminoma Stage III*

Patients with advanced seminoma, bulky stage II and stage III disease, receive combination chemotherapy without pretreatment radiotherapy. Using cisplatin-based chemotherapy regimens the survival rates are approximately 85%, whereas previously response rates to irradiation alone or chemotherapy following irradiation were only 40–70%. Currently, most patients receive three or four cycles of bleomycin, etoposide, and cisplatin (BEP). Patients failing this regimen are candidates for salvage chemotherapy regimens and or high-dose chemotherapy and autologous bone marrow transplantation or stem cell rescue.

Bulky seminomas often do not completely regress after chemotherapy or radiation therapy; however, only a minority of residual masses contain tumor. Surgical resection of residual seminomas is complicated by the dense fibrotic reaction in these tumors, making dissection extremely difficult without injuring major vascular structures and excess morbidity. In patients with a discrete mass >3 cm in size, surgical excision may be indicated; alternatively, these patients may be followed radiographically and operated on only if the mass increases in size.

B. Nonseminomatous Germ Cell Tumors

1. *Management of Stage I NSGCT*

Patients with a NSGCT, negative CT scan, and normal serum tumor markers can either be surgically staged with a RPLND, followed by a surveillance regimen or treated with two cycles of chemotherapy. Problems with loss of antegrade emission after RPLND have largely been eliminated because of surgical modifications. The anxiety produced by frequent office visits and CT scans can be difficult for patients being observed and patient compliance with the surveillance regimen is mandatory. In addition, late recurrences may be less chemosensitive and/or patients may present with bulky disease which may lead to an adverse outcome. Lastly, there is significant morbidity from chemotherapy and it may only be necessary for 30–40% of patients.

Currently, many urologists and oncologists recommend RPLND in clinical stage I NSGCT, especially in patients with vascular or lymphatic invasion or >50% embryonal carcinoma cells. RPLND provides accurate staging information to guide further therapy, simplifies follow-up, reduces anxiety, may be curative, and is associated with low morbidity and almost no additional loss of fertility potential.

2. *Management of Stage II NSGCT*

The optimal treatment for stage II disease is controversial. Options include full bilateral RPLND with nerve sparing if possible, followed by either two cycles of BEP or careful surveillance or three or four cycles of BEP, which is associated with acute toxicity and potential for late complications. In addition, patients with serum marker elevation only most likely have systemic disease and may be best treated with chemotherapy. Lastly, patients with teratomatous elements and surgically resectable stage II disease may benefit from RPLND followed by chemotherapy. Teratoma is typically chemoinsensitive and residual masses following primary chemotherapy will require postchemotherapy RPLND, which is associated with increased morbidity and a more difficult dissection. Some patients, approximately 20–30% of clinical stage II patients, are found to be pathologic stage I patients after staging RPLND. Thus, RPLND may help avoid chemotherapy in these patients.

3. *Management of Stage II and Good-Risk Stage III NSGCT*

The development of cisplatin-based combination chemotherapy has yielded a probability of cure in 90% or more of good-risk patients. Good-risk patients are defined as having an absence of mediastinal primary tumor; no liver, bone, or brain metastases; serum hCG < 1000 ng/ml; serum AFP < 1000 ng/ml; and serum LDH < 1.5 times upper limit of normal. Approximately two-thirds of patients are in the good-risk category. BEP is the preferred chemotherapy induction regimen and three cycles may be as effective as four cycles. Trials are ongoing to determine if it is possible to decrease the toxicity of the regimen without reducing the efficacy. Long-term complications of BEP include pulmonary fibrosis, renal dysfunction, peripheral neuropathy, hearing loss, and the possibility of second malignancies.

Following initial treatment patients undergo repeat clinical staging with history, physical examination, CT scans of the chest, abdomen, and pelvis, and serum tumor markers. Approximately 70% of patients are complete responders and are observed. However, 10% of these patients will relapse and require salvage chemotherapy, usually with the addition of ifosfamide and carboplatin. Patients that have a partial response to this regimen undergo surgical resection of residual mass(es). If residual cancer is noted the patient receives a third-line chemotherapy regimen. Patients who are initial partial responders to the BEP undergo RPLND. Approximately 20–30% are found to have residual cancer and undergo additional chemotherapy.

Complications of postchemotherapy RPLND include pulmonary, gastrointestinal, and vascular complications and, rarely, death. The surgery is considerably more complicated secondary to the toxicities of the previous chemotherapy. I have had to replace portions of the aorta and vena cava as well as remove kidneys or reconstruct ureters in some cases.

4. *Management of Poor-Risk Patients*

Approximately one-third of patients who present with metastatic NSGCT are "poor-risk" disease patients. Approximately 50% of these patients will either never achieve complete remission or will rapidly relapse from remission. The vast majority of these patients will die of their testis cancer. These patients are currently treated with four cycles of BEP followed by two cycles of cisplatin, etoposide, and cyclophosphamide plus autologous bone marrow transplantation. However, the toxicity of these regimens is significant and investigational strategies and randomized trials are currently ongoing.

V. TOXICITIES OF TREATMENT OF TESTICULAR CANCER

Cisplatin-based chemotherapy has revolutionized the treatment of testicular cancer and is responsible for the excellent long-term cure rates seen in patients with metastatic disease. Short-term acute toxicities are well recognized; however, physicians are beginning to describe and study late effects as well. Short-term myelosuppression occurs in 40–50% of patients; however, toxic deaths from neutropenia should occur only rarely. The short-term gastrointestinal toxicities include severe nausea and vomiting, diarrhea, ileus, and mucositis. Recently, the introduction of newer antiemetic medications such as ondansetron have helped patients greatly. Cisplatin is nephrotoxic to the tubules and collecting ducts in the kidney. However, the incidence of renal complications has been reduced dramatically by improved schedules of administration and intravenous hydration and forced diuresis. Moreover, there is a dose-related persistent decrease in renal function after cisplatin.

Small vessel vascular toxicity is common and is manifest clinically as Raynaud's phenomenon. Approximately 25–45% of patients have a persistent problem; however, this has not led to an increase in large vessel or coronary artery disease. It is unclear if the Raynaud's phenomenon is due to the combination of drugs or secondary to vinblastine or bleomycin alone. Neural toxicity after chemotherapy for testicular cancer may take the form of chronic sensory peripheral neuropathy, autonomic neuropathy, or auditory damage. In addition, short-term paresthesias are reported by up to 80% of patients, but the majority of patients report resolution of symptoms after cessation of therapy. The impairment of auditory function is the most common persistent toxicity after cisplatin-based chemotherapy. Most patients experience tinnitus and sensorineural hearing loss.

Pneumonitis and pulmonary fibrosis are dose related to bleomycin administration. This disorder may be exacerbated by cigarette smoking. Pulmonary symptoms are uncommon after cessation of therapy; however, persistent reductions in lung diffusion capacity seem to be confined to smokers.

With increasing numbers of long-term survivors, there has been associated increasing numbers of second malignancies reported in testicular cancer patients. Testis cancer in the contralateral testis is the most common malignancy in these patients and does not appear to be treatment related. Prior abdominal radiotherapy predisposes to solid tumor development in the radiation portal. Alkylating agents such as etoposide predispose to the formation of acute leukemia. Etoposide-induced leukemias occur sooner after therapy and are frequently associated with cytogenetic abnormalities.

VI. FERTILITY IN TESTIS CANCER PATIENTS

Many studies have demonstrated that 40–80% of patients with testicular cancer have oligo- or azoospermia at the time of diagnosis as well as endocrine and exocrine gonadal dysfunction. Each modality of treatment of testis cancer, chemotherapy, radiotherapy, or surgery can exacerbate the impaired fertility. Both radiotherapy and chemotherapy impair spermatogenesis. Chemotherapy induces an elevation of serum gonadotropins. However, many patients have a return to spermatogenesis by 1–3 years after completion of therapy, and 50% have a return to pretreatment sperm counts within 2–5 years after therapy. The reported incidence of azoospermia more than 2 years after therapy is 0–40% and is related to pretreatment gonadal function, age >30, history of radiotherapy, or therapy lasting longer than 6 months.

VII. MOLECULAR GENETICS OF TESTICULAR CANCER

A measure to assess genetic alteration of a tumor is the allelotype, which measures loss of heterozygosity at multiple polymorphic loci spanning the chromosomes. Studies on germ cell tumors have detected deletions affecting chromosomes 1, 12q, 3p, and 11p. In addition, other studies have noted deleted tumor suppressor genes RB1, DCC, NME, and as well as 3P, 9p and q, 10q, and 17p. Of note, the NME genes may be metastasis suppressor genes. The putative molecular mechanism of testicular cancer has not been elucidated; however, putative genetic models have been proposed. Germ cell tumors exhibit tetraploidy, express p53, have a high frequency of allelic loss of RB, DCC, and NME, and exhibit multiple copies of 12p. Overexpression of genes on chromosome 12p such as CCND2 (D2 cyclin) may reinitiate the cell cycle and rescue a cell otherwise destined to undergo cell death and apoptosis. Downstream events such as Rb or DCC may lead to neoplastic proliferation.

VIII. SUMMARY

The incidence of testicular cancer continues to rise in the United States, and it is the most common solid tumor in young men. Advances in the diagnosis and treatment of testicular cancer are truly remarkable, with survival rates approaching 95%. Research into the etiology, epidemiology, and molecular genetic mechanisms of testicular cancer is ongoing.

See Also the Following Articles

FSH (Follicle-Stimulating Hormone); Male Reproductive Disorders; Testis, Overview; Tumors of the Female Reproductive System; Yolk Sac

Bibliography

Donohue, J. P., Foster, R. S., Rowland, R. G., *et al.* (1990). Nerve-sparing retroperitoneal lymphadenectomy with preservation of ejaculation. *J. Urol.* **144**, 287.

Einhorn, L. H., and Williams, S. D. (1980). The management of disseminated testicular cancer. In *Testicular Tumors: Management and Treatment* (L. H. Einhorn, Ed.), p. 117. Mason , New York.

Einhorn, L. H., Williams, S. D., Loehrer, P., *et al.* (1989). A comparison of four courses of cisplatin, VP-16, and bleomycin in favorable prognosis disseminated germ cell tumors: A Southeastern Cancer Study Group protocol. *J. Clin. Oncol.* **7**, 387.

Foster, R. S., Bennett, R. M., Bihrle, R., *et al.* (1993). A preliminary report: Postoperative fertility assessment in nerve-sparing RPLND patients. *Eur. J. Urol.* **23**, 165.

Klein, E. A. (1993). Tumor markers in testis cancer. *Urol. Clin. North Am.* **20**, 67–73.

Mostofi, F. K., and Sobin, L. H. (1977). *International Histologic Classification of Tumors: XVI. Histologic Typing of Testis Tumours.* World Health Organization, Geneva.

Moul, J. W., McCarthy, W. F., Fernandez, E. B., and Sesterhenn, I. A. (1992). Percentage of embryonal carcinoma and of vascular invasion predicts pathological stage in clinical stage I nonseminomatous testicular cancer. *Cancer Res.* **54**, 362–364.

Murphy, B., Breeden, E. S., and Donohue, J. P. (1993). Surgical salvage of chemorefractory germ cell tumors. *J. Clin. Oncol.* **11**, 324.

Murty, V. V. V. S., Houldsworth, J., Baldwin, S., *et al.* (1992). Allelic deletions in the long arm of chromosome 12 identify sites of candidate tumor suppressor genes in male germ cell tumors. *Proc. Natl. Acad. Sci. USA* **89**, 11006–11011.

Samaniego, F., Rodriguez, E., Houldsworth, J., *et al.* (1990). Cytogenetic and molecular analysis of human male germ cell tumors: Chromosome 12 abnormalities and gene amplification. *Genes Chromosome Cancer* **1**, 289–300.

Sesterhenn, I. A., Weiss, R. B., Mostofi, F. K., *et al.* (1992). Prognosis and other clinical correlates of pathologic review in stage I and II testicular carcinoma: A report from the Testicular Cancer Intergroup Study. *J. Clin. Oncol.* **10**, 69–78.

Testicular Developmental Anomalies

Jayant Radhakrishnan
University of Illinois College of Medicine

GLOSSARY

anorchia The absence of both testicles.

cryptorchidism Testis not present in the scrotum.

ectopic testis Testis that has exited the inguinal canal but is not in the scrotum.

epididymis Oblong structure attached to the testicle which carries sperm from the testis to the vas deferens.

Leydig cells Testosterone-producing testicular cells.

macroorchidism Larger than normal testis.

mesonephric duct A duct of the mesonephros which in the male becomes the vas deferens and in the female is obliterated.

microorchidism Smaller than normal testis.

monorchia The absence of one testis.

Müllerian duct An embryonic duct that develops into the oviducts, uterus, and vagina.

polyorchidism More than two testes.

primordial germ cells The primitive precursor cells of the testis.

retractile testis Testis that moves in and out of the scrotum.

Sertoli cells Müllerian-inhibiting substance-producing testicular cells.

testicular torsion A twisted testis which will lose its blood supply.

undescended testis Testis which is retained in the abdomen or the inguinal canal.

vas deferens The excretory duct of the testis which leads to the urethra.

Developmental anomalies of the testicle occur as a result of abnormal formation, improper migration into the scrotum, or inadequate fixation once it reaches the scrotum.

I. DEVELOPMENT OF THE TESTIS

Gonads develop from three sources: the mesodermal epithelium lining the posterior abdominal wall, mesenchyme underlying this structure, and primordial germ cells. Initially, during the fifth week, mesodermal epithelium on the medial side of the mesonephros thickens and the underlying mesenchyme proliferates to produce the gonadal ridge. Soon primary sex cords grow into the underlying mesenchyme, producing an "indifferent gonad." In embryos with an XX sex chromosomal pattern the gonadal cortex differentiates while the medulla regresses, thus forming an ovary. On the other hand, in embryos with an XY pattern the gonadal medulla differentiates into a testis and the cortex regresses, leaving a few vestigial remnants. The gene for a testis determining factor (TDF) has been localized in the sex-determining region of the Y chromosome (SRY). Under the influence of TDF primary sex cords develop from the rete testis. The sex cords, known as seminiferous cords, then lose their connection with surface epithelium when the tunica albuginea develops. These cords form the seminiferous tubules into which the interstitial cells of Leydig develop. By the eighth week Leydig cells produce testosterone, which induces differentiation of the mesonephric ducts and the external genitalia. Müllerian-inhibiting substance (MIS) is elaborated from the Sertoli cells and induces regression of the Müllerian ducts by approximately Week 16 of gestation.

A. Vestigial Remnants of Genital Ducts

1. *Müllerian Remnants*

The cranial end of the Müllerian duct often persists as the appendix testis. The prostatic utricle, which is homologous to the vagina, is believed by some to be a remnant of the caudal end of the Müllerian duct, whereas others believe that it develops from the vaginal plate.

2. *Mesonephric Remnants*

The blind cranial end of the mesonephric duct may persist as the appendix epididymis, whereas caudal to the efferent ductules mesonephric tubules may persist as a paradidymis.

3. *Descent of the Testis*

By about 28 weeks of gestation the testis descends from the posterior abdominal wall to the deep inguinal ring as a result of elongation of the lumbar region of the embryo. During this phase of descent the distance between the testis and the internal inguinal ring does not decrease. Descent through the inguinal canal and into the scrotum occurs between the seventh month and birth. The mechanism of this phase is not clearly understood but it appears to be controlled by androgens produced by the fetal testis. Passage of the testis through the inguinal canal takes only 2 or 3 days.

II. ABNORMALITIES OF NUMBER

A. The Absence of Testes

True agenesis of one or both testicles is probably extremely uncommon since most patients with unilateral (monorchia) or bilateral (anorchia) absence of the testes are not feminized and do not demonstrates ipsilateral (in monorchia) or bilateral (in anorchia) persistence of Müllerian ducts. The absence of Müllerian ducts is indicative of early elaboration of MIS by testicular Sertoli cells since the Müllerian duct loses its sensitivity to MIS before the 30-mm stage in humans. In such instances the term atrophy or extreme dysgenesis of the testicle is probably more accurate than absence of the testicle.

Unilateral absence (monorchia) may be associated with ipsilateral renal agenesis. The absence of the testis is noted in 40% of patients with a nonpalpable testis. Compensatory contralateral testicular hypertrophy may occur in up to 12% of monorchid patients. In a neonate with a solitary palpable testis, if it is >2 cm in length it is considered to be hypertrophied. Bilateral absence of the testes (anorchia) is much less common: There is 1 anorchid patient for every 30 monorchid patients. Anorchia has also been termed the "vanishing testes syndrome," which has a reported frequency of 1 in 20,000 males. It has been reported in identical twins. These children demonstrate no testosterone response to stimulation by exogenous human chorionic gonadotropin.

B. Polyorchidism or Supernumerary Testes

This uncommon condition may arise due to division of the genital ridge, an aberrant peritoneal band dividing the ridge, or subdivision of a large testicle and bifurcation of its vessels. It occurs more commonly on the left side and typically a single supernumerary testis (triorchidism) is noted. The supernumerary testicle is intrascrotal in half the cases and undescended in the rest. In most instances the two ipsilateral testes share a common epididymis and vas deferens but have separate blood supplies. Complete duplication occurs in <10% of reported cases. Some supernumerary testicles are structurally abnormal and the epididymis or vas deferens may be malformed.

C. Testicular Fusion

Fusion of both testes into a single mass may be confused with monorchia or a testicular tumor. This single mass demonstrates two epididymides and may be associated with fused kidneys, suggesting a deformity of the urogenital ridges.

III. ABNORMALITIES OF SIZE

A. Microorchidism (Hypoplasia)

Congenital unilateral testicular hypoplasia occurs in patients with abnormalities of descent. Bilateral hypoplasia occurs due to functional, structural, or complete hypogonadism. Numerous congenital conditions result in hypogonadism, e.g., male psuedohermaphroditism with hypospadias, Klinefelter's syndrome, Noonan's syndrome, XX males, Kallmann's syndrome, Lawrence–Moon–Biedl syndrome, Prader–Willi syndrome, Del Castillo syndrome, rudimentary testis syndrome, hypogonadotropic hypogonadism, and Reifenstein's syndrome, which is now believed to encompass the previously separate entities of Reifenstein, Gilbert–Dreyfus, Rosewater, and Lubs syndromes.

B. Macroorchidism

A testis is considered to be larger than normal if its size exceeds the 95th percentile for age in the prepubertal child or it is >30 ml in volume after puberty. The condition may be unilateral or bilateral. In patients with unilateral enlargement one must rule out lesions of the epididymis or spermatic cord, adrenal rests, and testicular tumors. Unilateral compensatory hypertrophy occurs in monorchid patients and those with an undescended testis. Bilateral macroorchidism prior to puberty suggests congenital adrenal hyperplasia and precocious puberty. Other causes of bilateral enlargement are juvenile hypothyroidism, isosexual precocity, and fragile X syndrome.

IV. DISORDERS OF LOCATION

Cryptorchidism or "hidden testis" is a term generally used for undescended testes. However, it encompasses not only true undescended testes but also ectopic, retractile, and secondarily ascended testes.

A. Undescended Testis

Undescended testes are noted in approximately 30% of premature infants and 3% of full-term infants. During the first year spontaneous descent often occurs so that by 1 year of age only about 0.8% of boys are still cryptorchid. This incidence does not change in the school-age child nor does it change in adulthood. Bilaterally undescended testes are noted in about 15% of all cases of undescended testes. According to Kleinteich *et al.*, who have the largest cumulative series of cryptorchid patients, the testis is found to be intraabdominal in 12.21%, inguinal in 67.54%, prescrotal in 23.89%, and ectopic (including the superficial inguinal pouch) in 11.52%. The undescended testis is smaller than the contralateral descended testis and this loss of volume is evident within the first year of life. In addition, the number of spermatogonia per tubule and the diameter of the seminiferous tubules is greater in boys under 1 year of age when compared to older cryptorchid children. The previous findings suggest that early intervention may help preserve function. Relative infertility in cryptorchid patients is multifactorial and not due only to thermal injury because of its abnormal location. An inherent defect in germ cell maturation may also be present. In addition, approximately 20% of undescended testes either have an abnormal fusion of the epididymis with the testis or an abnormality in suspension of the epididymis. These defects affect sperm transport.

Cryptorchid testes have at least a 10 times greater risk of developing a tumor than is seen in the general male population. Up to 10% of all testicular cancers originate in patients who had or have a cryptorchid testis. Interestingly, in patients with unilateral cryptorchidism, 20% of testicular tumors occur in the contralateral normally descended testis. Although orchiopexy does not alter this malignant potential, it permits self-examination and early detection.

In addition, 60–90% of patients with an undescended testis have an associated inguinal hernia. A higher incidence of testicular torsion and trauma to the undescended testis has also been reported.

B. Ectopic Testis

An ectopic testis is misdirected after traversing the inguinal canal. The most common site of ectopia is the superficial inguinal pouch of Dennis Browne, but it could also be located in the perineum, the prepenile area, the thigh, and the contralateral scrotum (transverse testicular ectopia). It is unclear whether the testis does not enter the scrotum because of an obstruction at the scrotal inlet (third inguinal ring of McGregor) or due to an abnormal attachment of the gubernaculum. Since these testes do not respond to hormonal treatment, mechanical obstruction to their descent is probably a major factor. Furthermore, upon surgical exploration they are generally of normal size and appearance and minimal dissection and lengthening of the spermatic cord is required to transpose the testicle into the scrotum.

C. Retractile Testis

The retractile testis is a normal testis which has descended into the scrotum but retracts into the inguinal canal due to a strong cremasteric reflex. These testicles are normal and no treatment is required other than annual follow-up until growth and development is complete. In adult life these patients have a normal potential for fertility.

D. Ascended Testis

Although some cases of secondarily ascended testes may in fact be misdiagnosed undescended testes, it is unquestionably true that on occasion a well-developed testicle which was previously in the scrotum will ascend into the inguinal area as the child grows. This may be due to the inability of the spermatic cord to elongate in proportion with somatic growth. Upon exploration these testicles are normal in shape and size.

V. ABNORMALITIES OF FIXATION

Improper attachment of the testicle can result in testicular torsion, which could be intravaginal or supravaginal.

A. Intravaginal Torsion

Normally, the testicle is anchored in the scrotum by a band of tissue which connects it to the epididymis and then connects the epididymis to the vas deferens and the posterolateral wall of the processus vaginalis. Improper fixation at any of these points results in a predisposition to testicular torsion. Such a testicle hangs horizontally in the scrotum and is called a "bell-clapper" testicle. These patients present with excruciating scrotal pain of sudden onset. There may be a past history of similar lesser episodes that resolved spontaneously.

Immediate treatment is required to salvage the torsed testicle. Since the bell-clapper is often bilateral, the contralateral testicle should be prophylactically fixed at the time of surgery. Sympathetic immunologic damage to the contralateral testis has been observed in some laboratory animals and is probably a reperfusion injury following treatment.

B. Supravaginal or Extravaginal Torsion

In these patients, the spermatic cord twists between the external inguinal ring and the insertion of the tunica vaginalis. This form of torsion occurs essentially in the neonate. It can occur prenatally, intranatally, or in the immediate postnatal period. Its etiology is unclear but it could be due to a weak attachment of the tunica vaginalis to the scrotal wall combined with strong cremasteric activity. It is uncommon for these testicles to be salvageable upon exploration, indicating that a large number occur in the prenatal period.

VI. ABNORMALITIES OF THE VAS DEFERENS AND EPIDIDYMIS

A. Absent or Dysplastic Vas Deferens

Less than 1% of the population demonstrates either hypoplasia or the absence of a single vas deferens. The left side is affected four times more often than the right. The caput epididymis is usually present in these patients. Since one-third of patients with renal dysplasia have an absent vas deferens, it is important to perform a renal sonogram in patients with a missing vas deferens.

Bilateral agenesis of the vas deferens is common

in patients with cystic fibrosis. Patients with cystic fibrosis may also have a beaded, atretic vas deferens.

B. Ectopic Vas Deferens

Since the ureter and the vas deferens arise from the same embryologic structure, an ectopic vas deferens may insert into the ureter or into a common channel which enters the bladder along with the ureter. Embryologically, such an anomaly would occur if the ureteric bud had a more cranial origin than normal, thus preventing separation of the vas deferens. Typically, these patients present with epididymitis or recurrent urinary tract infections.

C. Epididymal Anomalies

Up to 32% of patients with maldescended testes demonstrate gross epididymal abnormalities. Four main types of such anomalies have been identified based on continuity of the epididymis with the testis. In 15%, the head and the tail are attached to the testicle, 13% have attachment of the head only, 2% have attachment of the tail only, and 2% have complete separation of the testis and epididymis. These anomalies may help explain the increased incidence of infertility in patients with undescended testes. In addition, 63% of these patients have an unfurled epididymis located distal to the testis in the inguinal canal.

See Also the Following Articles

Cryptorchordism; Epididymis; Primordial Germ Cells; Testis, Overview

Bibliography

Berta, P., Hawkins, J. R., Sinclair, A. H., *et al.* (1990). Genetic evidence equating SRY and the testis-determining factor. *Nature* **348**, 448.

Bloom, D. A., and Semm, C. (1991). Advances in genitourinary laparoscopy. In *Advances in Urology*, Vol. 4. Yearbook Med. Pub., Chicago.

Bloom, D. A., Ayers, J. W., McGuire, E. J. (1988).The role of laparoscopy in the management of nonpalpable testes. *J. Urol* **94**, 465.

Canavese, F., Lalla, R., Linari, A., *et al.* (1993). Surgical treatment of cryptorchidism. *Eur. J. Pediatr.* **152**(Suppl. 2), S43.

Crawford, D. S., and Bastable, J. R. G. (1989). Triorchidism with torsion. *Br. J. Urol.* **63**, 553.

Feldman, S., and Drach, G. W. (1983). Polyorchidism discovered as testicular torsion. *J. Urol.* **130**, 976.

Heath, A. L., Man, D. W. K., and Eckstein, H. B. (1984). Epididymal abnormalities associated with maldescent of the testis. *J. Pediatr. Surg.* **19**, 47.

Hutson, J. M., and Beasley, S. W. (1988). Embryological controversies in testicular descent. *Sem. Urol.* **6**, 68.

Johnson, D. E., Woodhead, D. M., Pohl, D. R., *et al.* (1968). Cryptorchidism and testicular tumorigenesis. *Surgery* **63**, 919.

Kleinteich, B., Hadziselimovic, F., Hesse, V., *et al.* (1979). *Kongenitale Hodendystopen.* Thieme, Stuttgart.

Kogan, S. J., Tennenbaum, S., Gill, B., *et al.* (1990). Efficacy of orchiopexy by patient age one year for cryptorchidism. *J. Urol.* **144**, 508.

Moore, K. L., and Persaud, K. V. N. (1993). The urogenital system. In *The Developing Human*, 5th ed., pp. 265–303. Saunders, Philadelphia.

Nguyen, L. B., Lievano, G., Radhakrishnan, J., Fornell, L. C., and John, E. G. (1997). Testicular blood flow and endothelin and prostacyclin release during and after unilateral testicular torsion. *Surg. Forum* **XLVIII** 773.

Puri, P., and Nixon, H. H. (1977). Bilateral retractile testes—Subsequent effects on fertility. *J. Pediatr. Surg.* **12**, 563.

Radhakrishnan, J., and Donahoe, P. K. (1981). The gubernaculum and testicular descent. In *The Undescended Testis* (E. W. Fonkalsrudand W. Mengel, Eds.), pp. 30–41. Yearbook Med. Pub., Chicago.

Scorer, G. C., and Farrington, G. H. (1971). *Congenital Deformities of the Testes and Epididymis.* Butterworth, London.

Skandalakis, J. E., Gray, S. W., Parrott, T S., and Ricketts, R. R. (1994). The ovary and testis. In *Embryology for Surgeons* (J. E. Skandalakis and S. W. Gray, Eds.), 2nd ed., pp. 736–772. Williams & Wilkins, Philadelphia.

Thompson, M. W., McInnes, R. R., and Willard, H. F. (1991). *Thompson and Thompson Genetics in Medicine*, 5th ed. Saunders, Philadelphia.

Testicular Feminization Syndrome

see Androgen Insensitivity Syndromes

Testis, Overview

Larry Johnson, Tobin A. McGowen, and Genevieve E. Keillor

Texas A&M University

GLOSSARY

acrosome The cap-like bag of hydrolytic enzymes in the head of the spermatozoon necessary for penetrating the corona radiata and zona pellucida of the ova.

annulus The structure that marks the termination of the mitochondrial helix around the middle piece and that is located between the middle and principal pieces of the spermatozoon.

axoneme The core of cilia and flagella composed of nine doublet microtubules encircling a central pair of microtubules that provides their motile properties.

lipofuscin Indigestible compounds within a residual body from a secondary lysosome (fused cellular material with primary lysosome that contains hydrolytic enzymes).

mediastinum An extension of the connective tissue from the capsule (tunica albuginea) into the testicular parenchyma site of rete testis tubules that connects the seminiferous tubules with the efferent ductus that carry spermatozoa from the rete testis to the epididymis.

meiosis The development (exchange of genetic material between mother and father chromosomes) and division of primary spermatocytes to produce secondary spermatocytes that divide to produce haploid spermatids.

pachytene The third of four steps during prophase I of meiosis, during which exchange of genetic material occurs; means "thick threads."

paired homologous chromosomes Joined pairs of chromosomes, one from the father and one from the mother, that facilitate exchange of genetic material (genes) between father and mother chromosomes during prophase of meiosis.

pampiniform plexus A venous plexus that forms a meshwork around the coiled testicular artery; it functions in thermoregulation and countercurrent heat exchange.

preleptotene The period of meiosis during which DNA synthesis occurs in primary spermatocytes, preceding the first step of prophase I of meiosis.

primary spermatocytes The earliest germ cell that does not divide by mitosis but has entered in meiosis I in which chromosomes are duplicated and exchange of genetic materials occurs between homologous chromosomes. They divide at the first meiotic division to produce secondary spermatocytes.

secondary spermatocytes Cells that divide at the second meiotic division quickly without DNA synthesis to produce haploid (n) spermatids.

spermatids The most mature germ cells in the seminiferous tubules. Without division, they undergo differentiation from round cells to those shaped like the spermatozoon of that species.

spermatocytogenesis Mitotic activity of spermatogonia that produces stem cells to carry on the lineage of these cells and that increases the number of later forms of spermatogonia and produces primary spermatocytes.

spermatogenesis The process of germ cell division and differentiation that produces spermatozoa in seminiferous tubules.

spermatogonia Diploid (2n) cells; the most immature form of male germ cells; they divide by mitosis to produce stem cells (that can continue to divide throughout the life of the male) and primary spermatocytes.

spermiation The release of spermatozoa from seminiferous tubules; the counterpart to ovulation in the female.

spermiogenesis The differentiation of spermatids from round cells to streamlined cells shaped like male gametes of that particular species.

tunica albuginea A connective tissue capsule around the outside of the testis delineating the outer limits of the testis.

varicocele A swollen venous cavity returning blood from the testis toward the vena cava.

zygotene The second of four steps during prophase I of meiosis during which the first pairing of the synaptonemal complex occurs.

Paired testes are male gonads that produce the male haploid gametes, spermatozoa. Spermatozoa are produced by seminiferous tubules in the testis by a process called spermatogenesis. Seminiferous tubules are composed of three cell types: Sertoli cells and myoid cells of the limiting boundary tissue (both somatic cells) and germ cells. Germ cells themselves are of three types: spermatogonia, spermatocytes, and spermatids. Spermatogonia located at the base of the seminiferous tubules divide to produce spermatogonial stem cells as well as spermatocytes. Spermatocytes undergo meiosis to produce haploid (half chromosome number) spermatids which differentiate into spermatozoa. Spermatogenesis takes 74 days to complete and produces 6×10^6 spermatozoa per gram parenchyma or 250×10^6 per man each day. In humans, paired testes weigh an average of 47 g, and the paired testicular parenchyma (testis without the connective tissue capsule) weighs about 41 g. The length of the seminiferous tubules is $\approx$600 m, and they occupy 62% of the testicular parenchyma. Seminiferous tubules contain 947×10^6 myoid cells that form the outer limits of the tubules and 977×10^6 Sertoli cells (nurse cells) that facilitate development of germ cells during spermatogenesis. In the interstitium between the tubules, 432×10^6 Leydig cells per man produce testosterone that contributes to male body shape, behavior, accessory sex gland development, and spermatogenesis. Spermatozoa are released into the lumen and are transported out of the testis by the tubules known as the rete testis. The rete testis tubules attach to the efferent ducts that carry spermatozoa to the epididymis for maturation.

I. INTRODUCTION

Two testes located in a cutaneous–fibroelastic scrotum and the penis comprise the external genitalia of the male reproductive system. Because testes develop retroperitoneally and carry part of the peritoneum into the scrotum upon descent, this outpocketing of the peritoneum (called the tunica vaginalis) envelopes the mature testis on the anterior and lateral surfaces. The parietal layer of the tunica vaginalis is juxtaposed to the visceral layer of the tunica albuginea. Both of these layers receive lubricating fluid from the peritoneum. This allows for some movement of the testis in the scrotum to escape damaging pressure to the testis. The tunica albuginea is a thick fibrous capsule enclosing the testicular parenchyma. Aging in humans induces a thickening of the tunica albuginea, and an increased percentage of the testis is occupied by it.

On the posterior side of the human testis, the connective tissue of the capsule extends into the parenchyma to create the mediastinum (Fig. 1a). This is the region where blood vessels and nerves enter the parenchyma and where spermatozoa exit from the rete testis through the efferent ducts. Unlike the rat testis, the human testis is divided into partial lobes by connective tissue septa which are complete near the mediastinum but may be incomplete in the anterior portion of the testis. In fixed testes, the lobes can be separated at the region of the connective tissue septa (Fig. 1a). Sectioning of these lobes near the rete testis yields a high density of tubules cut in cross section (Fig. 1b). The parenchyma of the testis is

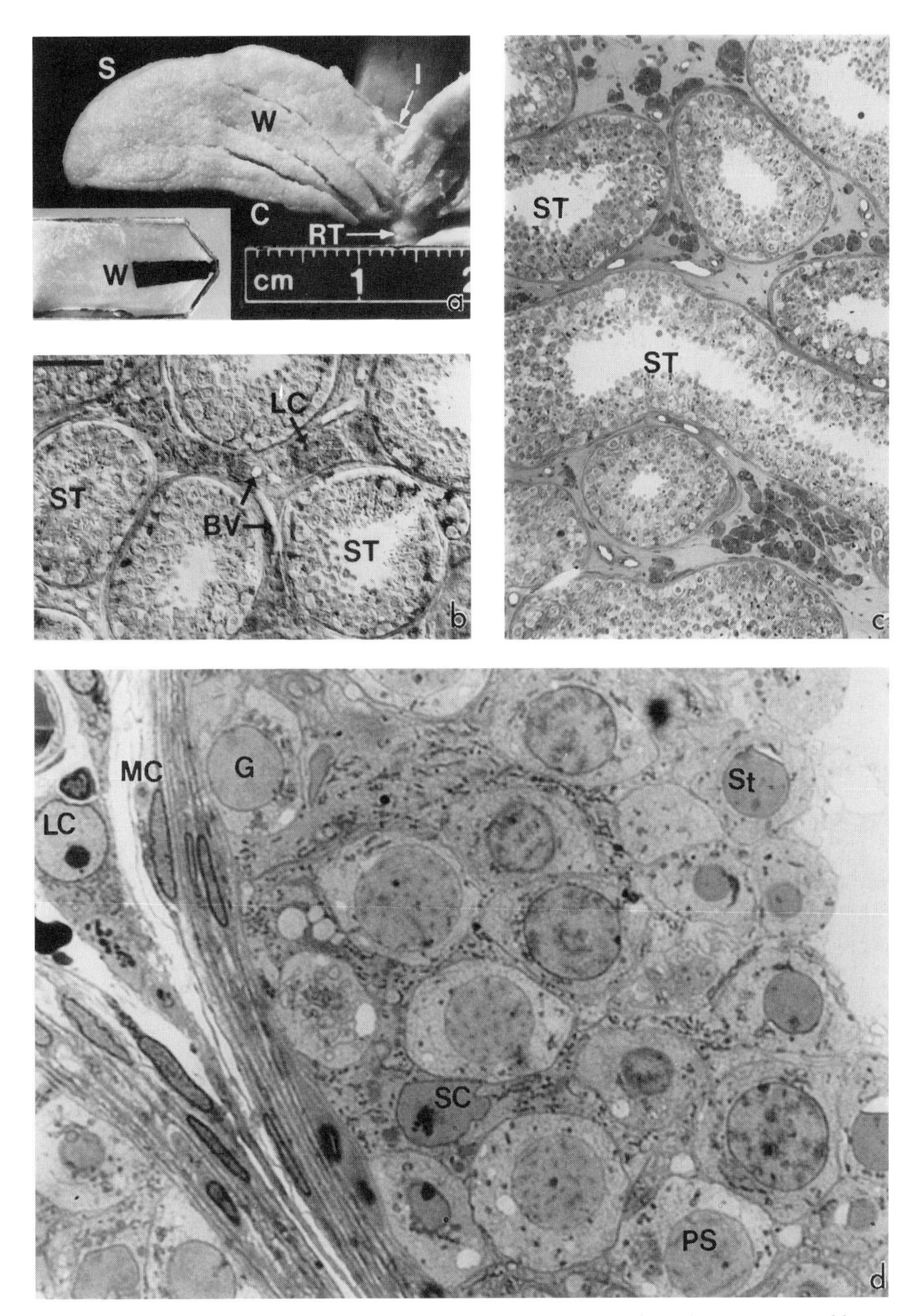

FIGURE 1 Human testicular parenchyma. (a) Groups of tubules are teased apart where they are separated by connective tissue of testicular septa into lobes. The wedges (W), composed of groups of tubules, extend from the rete testes (RT) to the surface (S), where the tunica albuginea is removed. Coils of seminiferous tubules (C) and individual tubules (I) are visible. Once fixed in osmium, the wedge of tubules (W) is oriented and embedded in Epon 812, with regions closest to the rete testis at the tip of the block. (b) Low-magnification (16× objective) photomicrograph of a section of oriented tubules cut perpendicularly to the long axis of the wedge starting near the rete testis and viewed by Nomarski optics. Several cross sections of seminiferous tubules (ST), blood vessels (BV), and Leydig cells are observed. (c) Seminiferous tubules (ST) and interstitium between tubules are observed by bright field microscopy of 0.5-μm Epon sections. (d) Leydig cells (LC) located in the interstitium, myoid cells (MC), Sertoli cells (SC), spermatogonia (G), primary spermatocytes (PS), and spermatids (St) in the tubules are visible (reproduced with permission from Johnson *et al.*, 1980, and Johnson, 1994).

composed of seminiferous tubules (50–90% depending on the species) and the interstitium (Figs. 1b and 1c).

II. SEMINIFEROUS TUBULES

A. General

Spermatogenesis (production of spermatozoa) occurs in seminiferous tubules, which are composed of two somatic cells (Sertoli cells and myoid cells) and germ cells (Fig. 1d). The myoid or peritubular cells mark the outer limits of the tubule (Figs. 1d and 2). Inside the myoid cell boundary, the seminiferous epithelium is composed of Sertoli cells (nurse cells) and germ cells. Germ cells are of three types: spermatogonia, spermatocytes, and spermatids.

B. Myoid Cells

Myoid cells (peritubular cells) ensheathe the seminiferous epithelium and define the boundary tissue on the outer limits of seminiferous tubules (Fig. 2). Myoid cells, like Sertoli cells and Leydig cells, are influenced by testicular growth factors. In many species, myoid cells form a single layer, but there are three to five layers in humans. These cells contain fine cytoplasmic filaments believed to be actin and are thought to be involved in slow, weak, rhythmical contractions of seminiferous tubules. Myoid cells maintain the integrity of the seminiferous tubules, and their contraction may aid in movement of spermatozoa and fluid from the seminiferous tubules. No nerve endings have been found in the boundary tissue of seminiferous tubules; however, blood vessels (i.e., capillaries) can penetrate the myoid cell boundary.

In many instances of human male infertility and aging, the boundary tissue layer of tubules becomes greatly thickened. Age-related thickening of the boundary tissue in humans results from age-related reduced total seminiferous tubular length without loss in the total boundary tissue volume.

C. Sertoli Cells

Sertoli cells are the somatic cellular component of the seminiferous epithelium, and they are critical to germ cell development (Figs. 1d and 2). Sertoli cells form a syncytium around the lumen of seminiferous tubules. The plasma membranes of Sertoli cells and the tight junctions between adjacent cells create the adluminal compartment of the blood–testis barrier. In several species including humans, each Sertoli cell supports 6–12 elongated (Sd) spermatids. In the horse, each Sertoli cell supports an average of 22–28 (depending on the season) germ cells, which are themselves in different steps of germ cell development at the same time. Sertoli cells have nuclei that are homogeneous and composed of euchromatic nucleoplasm. Numerous long, thin mitochondria are oriented along the long axis of the cell that extends from the base of the tubule to the lumen. Many microtubules also parallel the long axis of the cell. Also, actin filaments and intermediate filaments are prevalent, and actin filaments are likely involved in changes of the Sertoli cell shape during migration of germ cells from the base to the lumen of the seminiferous tubule. The Golgi apparatus is large, and smooth endoplasmic reticulum is abundant in the region of the Sertoli cell cytoplasm next to the developing acrosome of a spermatid. Though endoplasmic reticulum is relatively sparse in Sertoli cell cytoplasm, Sertoli cells have secondary lysosomes associated with residual body digestion, but the amount of lipofuscin is remarkably low.

The number of Sertoli cells in an individual is important because it is related to the spermatogonial number, spermatid number, or spermatozoan production rates in rams, bulls, horses, rats, and humans. This relationship exists because Sertoli cells function in structural support and nutrition of germ cells (lactate), spermiation of mature spermatids, movement of young germ cells, phagocytosis of degenerating germ cells and residual bodies left by released spermatozoa, secretion of luminal fluid and proteins, formation of the blood–testis barrier, and cell-to-cell communication. The Sertoli cell secretes lactate (produced from serum glucose as energy for germ cells) and produces mitogenic polypeptides.

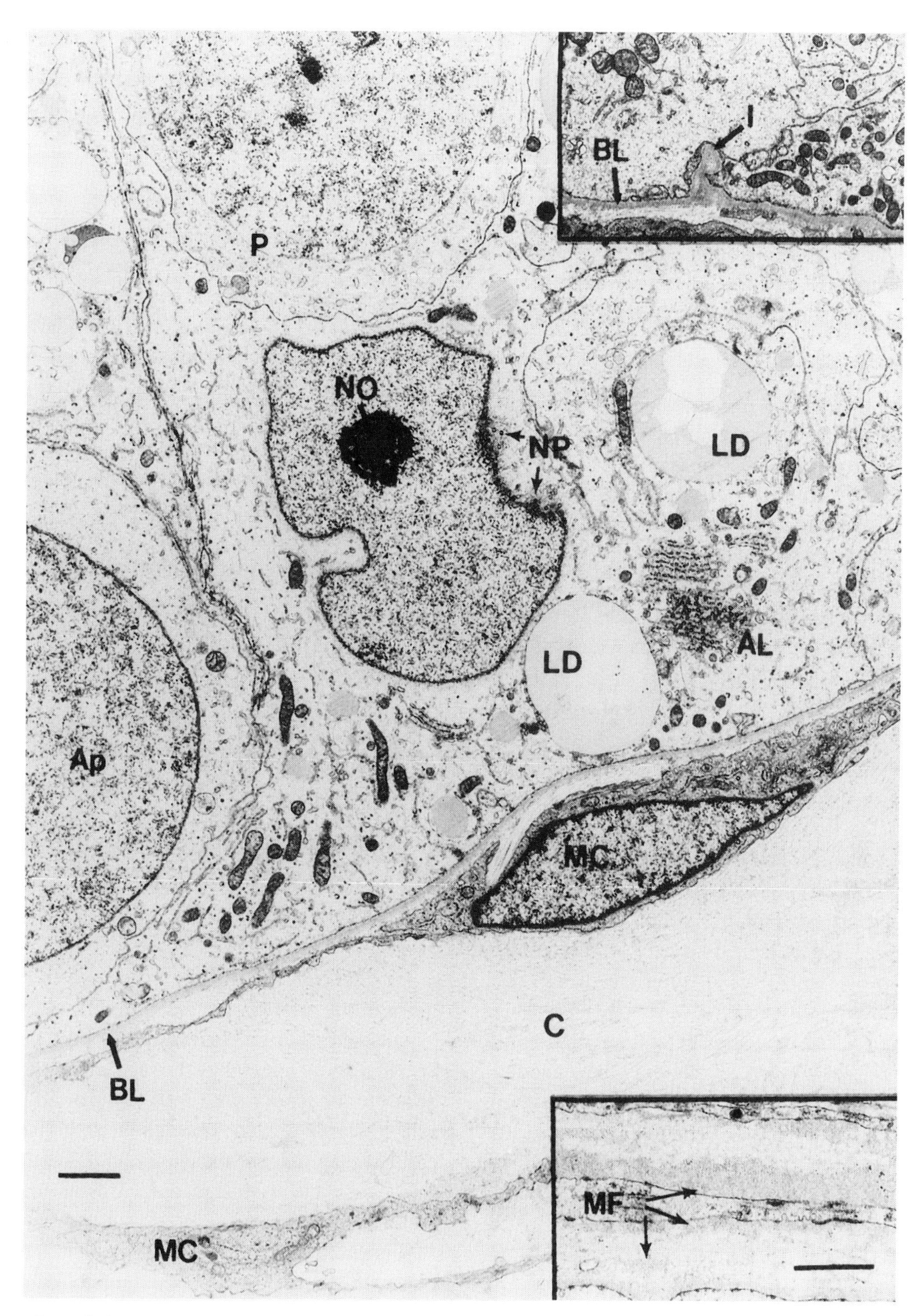

FIGURE 2 Plate of electron micrographs of the basal region of the seminiferous epithelium and boundary tissue in a 59-year-old man. A large region of collagen (C) is located between layers of myoid cells (MC). In the top insert, involution (I) of the basal lamina (BL) in the extracellular space between a Sertoli cell and a spermatogonium is shown. In the bottom insert, microfibrils (MF) are observed between myoid cells. A characteristic profile of a Sertoli cell nucleus with numerous nuclear pores (NP) and the central nucleolus (NO) with the surrounding satellite karyosomes is indicated. Sertoli cell cytoplasmic structures include typical cytological organelles plus large lipid droplets (LD) and annulate lamellae (AL). An A-pale spermatogonium (Ap) and pachytene primary spermatocyte (P) are present in this portion of the seminiferous epithelium. Scale bar = 2 μm in the main micrograph and top insert and 1 μm in the bottom insert (reproduced with permission from Johnson *et al.*, 1988).

Sertoli cells provide structural support and communicate with developing germ cells through intimate contact and gap junctions. The tight junctional complexes between adjacent Sertoli cells constitute the final obstacle of the blood–testis barrier. This barrier restricts the flux of serum components to spermatocytes and spermatids. It also produces the unique environment in which spermatocytes and spermatids develop.

The Sertoli cell junctional complexes divide the seminiferous epithelium into basal and adluminal compartments. Spermatocytogenesis occurs in the basal compartment where preleptotene primary spermatocytes are produced. Newly formed preleptotene primary spermatocytes migrate through the blood–testis barrier into the adluminal compartment where meiosis and spermiogenesis occur. The blood–testis barrier isolates spermatocytes and spermatids from serum-borne components and provides an immunologic isolation for these germ cells. Also, it prevents loss of specific concentrations of androgen-binding protein, inhibin, and enzyme inhibitors from the luminal compartment. Rete testis fluid has been used to evaluate luminal contents of seminiferous tubules. Rete testis fluid differs from serum in that potassium, glutamate, and inositol are higher in concentration; however, concentrations of glucose and protein (especially immunoglobulins) are lower.

Sertoli cells, like other nurse cells in the body, function to provide material to stimulate growth but also must function in removing waste. Degenerating germ cells and residual bodies left behind when spermatids are released (spermiated) are phagocytized by Sertoli cells. The lipid content of Sertoli cells, resulting from phagocytosis of residual bodies, varies with the stage of the cycle of the seminiferous epithelium (stages of the spermatogenic cycle are described under Section V).

D. Germ Cells

1. *Spermatogonia*

In humans, types A light, A dark, and B spermatogonia can be distinguished (Figs. 1d, 2, and 3). It is not clear whether A dark or A light include the stem cells. However, the A spermatogonia divide by mitosis to produce stem cells, other A spermatogonia, and B spermatogonia. Division of B spermatogonia produces primary spermatocytes that enter meiosis.

2. *Spermatocytes*

Primary spermatocytes are the largest cells and have the largest nuclei of all germ cells and are located between the basal-placed spermatogonia and the luminal-placed spermatids in the seminiferous epithelium (Figs. 1d, 2, and 3). Primary spermatocytes (Fig. 3) duplicate chromosomes (preleptolene) and exchange genetic material between paired homologous chromosomes (zygotene, pachytene, and diplotene primary spermatocytes).

Primary spermatocytes, in the first meiotic division, rapidly undergo nuclear membrane dissolution, metaphase (tetrad alignment), anaphase (separation of dyads of sister chromatids), and telophase (complete separation of dyads) to produce secondary spermatocytes. Secondary spermatocytes contain a haploid number of duplicated chromosomes (i.e., XX or YY) and have spherical nuclei with chromatin flakes or aggregates of varying sizes. These cells undergo a short interphase with no DNA synthesis prior to the second meiotic division which produces spermatids with a haploid number of chromosomes (Fig. 3).

3. *Spermatids*

Spermiogenesis is the morphologic differentiation of spermatids into spermatozoa (Figs. 1d and 3). Spermatids, produced by the second meiotic division, differentiate from round cells with round nuclei into cells that have a streamlined head consisting of a condensed nucleus and an enzyme-filled acrosome as well as a tail that is necessary for motility of spermatozoa.

Spermatids (Golgi, cap, acrosomal, and maturation phases) can be distinguished histologically based largely on the development of the acrosome in the different spermatid types. Compared to domestic species, these phases (especially the acrosome phase) are more pronounced in rodents, the animals in which they were first described. However, the early phases of development and the sequencing of events

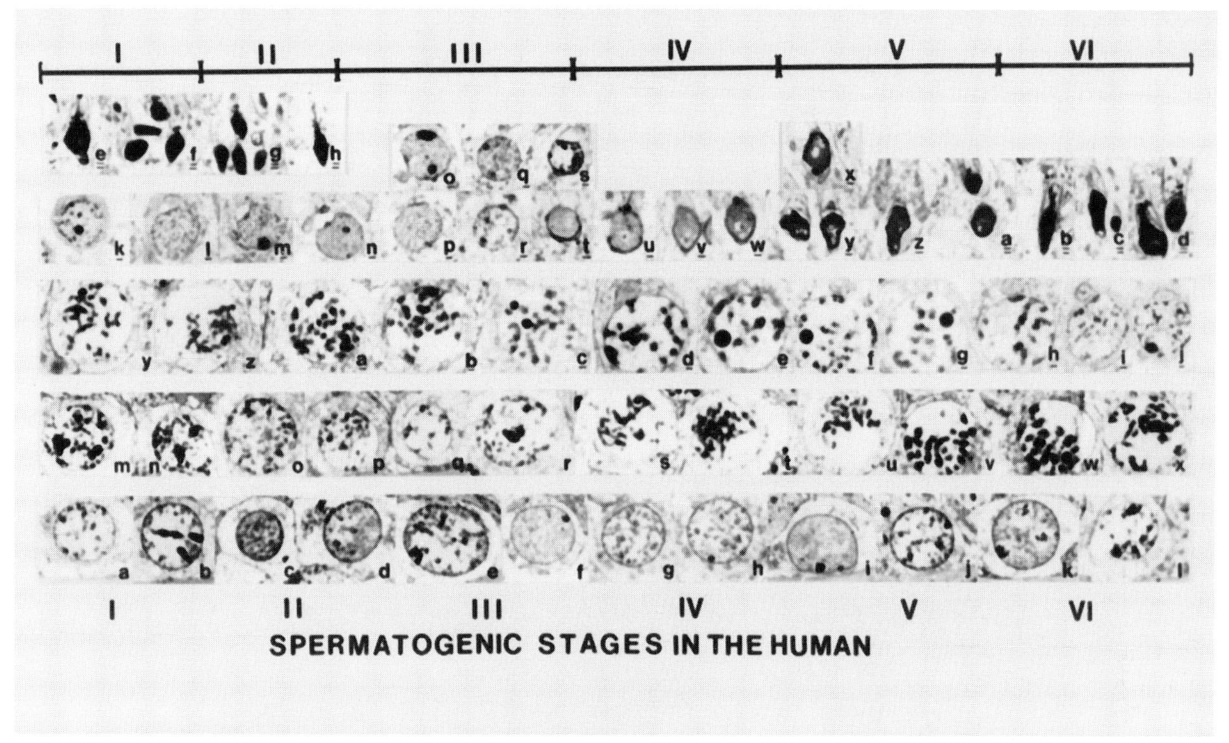

FIGURE 3 Composition of germ cells in human spermatogenesis. These nuclear profiles characterize germ cell development through spermatogenesis as viewed by bright field microscopy of 0.5-μm Epon sections stained with toluidine blue. (a–k) Various nuclear profiles of A dark and A pale spermatogonia characterize some of the spermatogonia found in specific stages. However, all types for a given stage are not shown. (l–o) Type B spermatogonia are first seen in stage IV but also are found in stages I and II. (p, r) Preleptotene primary spermatocytes result from the mitotic division of B spermatogonia. Nuclear size, chromatin distribution, and chromatin clump size distinguish the (s, t) leptotene primary spermatocyte, (u–x) zygotene primary spermatocyte, (y–g) pachytene primary spermatocyte, (h) diplotene primary spermatocyte, and i, j) secondary spermatocytes. (k–n) Sa spermatids have spherical nuclei and acrosomic vesicles. (o–s) Sbl spermatids have spherical nuclei. The acrosomic development ranges from acrosomic vesicle formation to acrosomic cap formation. (t–y) Sb2 spermatids have the manchette and nuclei which are engaged in elongation. (z–d) Sc spermatids and (e, f) Sd1 spermatids have elongating nuclei with condensing chromatin. (g, h) Sd2 spermatids have mitochondria around the middle piece. Scale bar = 10 μm (reproduced with permission from Johnson *et al.*, 1992).

in domestic species and humans are similar to those in rodents.

III. INTERSTITIUM STRUCTURE

A. General

The interstitial tissue between seminiferous tubules is composed of clusters of Leydig cells, mesenchymal cells, macrophages, occasional mast cells, blood vessels, and lymphatics (Figs. 1c and 1d). Nerves are difficult to observe in the interstitium but are more common in the connective tissue near the mediastinum testis than elsewhere in the testicular parenchyma.

B. Blood Vessels and Lymphatics

Branches of the internal spermatic or testicular artery (a branch of the dorsal aorta) penetrate the capsule of the testis to form the centripetal branches. The centripetal branches course through the connective tissue of the testis toward the mediastinum. Major bundles of centripetal branches run in the opposite direction as centrifugal arterioles located in the

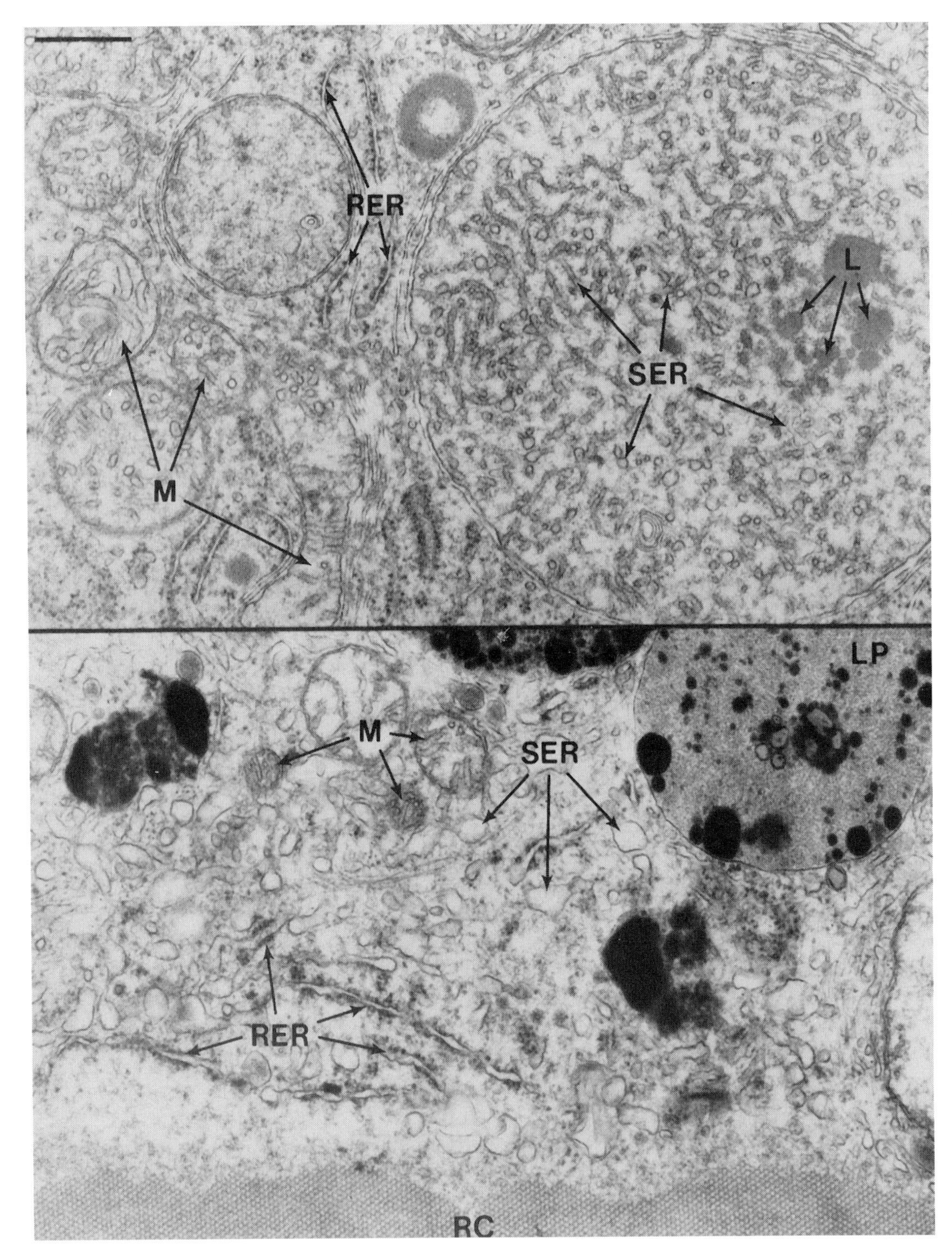

FIGURE 4 Transmission electron micrographs of human Leydig cells. Cytoplasmic components include smooth endoplasmic reticulum (SER), rough endoplasmic reticulum (RER), mitochondria (M), lipofuscin pigment (LP), lipid (L), Reinke crystals (RC), Golgi bodies, and other organelles. Scale bar = 1 μm (reproduced with permission from Johnson *et al.*, 1990).

column (septa) of connective tissue between seminiferous tubules. Ladder-like, peritubular, or intertubular capillaries form arches to partially encircle seminiferous tubules.

Postcapillary venules join collecting venules which join the centrifugal veins that course toward the capsule or the centripetal veins that course toward the mediastinum testis (Fig. 1a). The pampiniform plexus surrounds the testicular artery in the spermatic cord. Blood drains from the right pampiniform plexus into the inferior vena cava via the internal spermatic vein; however, the left pampiniform plexus joins the left renal vein. The draining veins of the left pampiniform plexus are torturous and varicose, and they are a common site for development of a varicocele. The countercurrent heat-exchange system afforded by the pampiniform plexus cools arterial blood entering the testis and allows the core testicular temperature to be a few degrees lower than that of the body cavity. A cooler temperature of the testis is essential for spermatogenesis. Thin-walled lymphatic vessels drain into larger lymphatics in the septula testis. These vessels lead to paraaortic lymph nodes and nodes associated with the renal blood vessels. Usually, Leydig cells are not located directly within the lymphatic space (Fig. 1). However, in some species (e.g., rats), Leydig cells are located directly in the lymphatic space. Leydig cells are closely (but not intimately) associated with blood vessels. Although some of the testosterone exits the testis via rete testis fluid and epididymis (facilitating maintenance of the epididymal epithelial height), most of the testosterone produced by Leydig cells exits the testis via both blood and lymph vessels.

C. Leydig Cells

Leydig cells, in clusters of various sizes, are located among the very loose connective tissue in the testicular interstitium where they constitute the major endocrine compartment of the testis (Fig. 1). Binucleate Leydig cells often are found; however, these constitutively secreting cells have no secretory granules because they do not accumulate or store secretions. Leydig cells have an abundance of smooth endoplasmic reticulum, have many mitochondria with tubular cristae, lipid droplets, and lysosome/peroxisomes, and may accumulate lipofuscin granules (Fig. 4). In humans, Leydig cells have characteristic Reinke crystals of unknown function, but the number and size of these crystals increase with age. Leydig cells are dispersed throughout the testicular parenchyma, bathing adjacent seminiferous tubules with a concentration of testosterone that is 10- to 100-fold higher than that in the peripheral blood. Leydig cells are produced by mitosis and/or differentiation of mesenchymal cells.

D. Other Interstitial Cells

Other interstitial cells are mostly mesenchymal cells (e.g., fibroblasts), but this category includes all cells that cannot be classified as Leydig cells, macrophages, or vascular cells. In humans, the loss of Leydig cells with aging does not result in an augmentation of other interstitial cells, making it unlikely that Leydig cells persist in the aged human testis as dedifferentiated mesenchymal cells.

IV. RETE TESTIS TUBULES

Spermatozoa are produced in seminiferous tubules and flow through tubuli recti (short, straight ducts) and into the main rete testis tubules, which carry them outside the testis on the posterosuperior surface of the testis and into the efferent ducts (Fig. 1a). The rete testis tubules are composed of a highly anastomotic network of channels in the mediastinum region lined by cuboidal epithelium. In the transitional zone between the seminiferous tubules and tubuli recti, only Sertoli cells comprise the tubular epithelium. This abruptly changes to the simple cuboidal epithelium of the tubuli rectus and rete testis tubules. A single cuboidal epithelial cell of the rete testis has a single, very long (longer than the height of the cell), true cilium. Its function is unknown. There is no evidence of a secretory function for these cells. However, the tight junction at the apical border between adjacent cells maintains the blood–excurrent duct barrier sequestering spermatozoan antigens from the immune system.

In humans, loops of seminiferous tubules are directed toward the rete testis at small angles (Fig. 1a). Cross section of the lobe between the adjacent connective tissue septa separating lobes of the testis yields a high percentage of seminiferous tubules cut in cross section (Fig. 1b). Blood vessels and nerves (which are hard to find in the testicular parenchyma) traverse the mediastinum region before diving deeper into the testicular parenchyma.

V. SPERMATOGENESIS

A. General

Spermatogenesis is a lengthy but chronological process (Figs. 3 and 5). A few stem cell spermatogonia lining the base of seminiferous tubules divide by mitosis to maintain their own stem cell numbers and to cyclically (every 16 days, the duration of the spermatogenic cycle in humans) produce primary spermatocytes. These undergo meiosis to produce haploid spermatids, which differentiate into spermatozoa released into the tubular lumen. Hence, spermatogenesis is the division and differentiation process in seminiferous tubules of the testis. Seminiferous tubules constitute the major component of the testis (62% in humans).

For most species, the duration of spermatogenesis can be fairly equally divided into three phases: spermatocytogenesis (mitosis), meiosis, and spermiogenesis (Fig. 5). In spermatocytogenesis, stem cell spermatogonia divide by mitosis to produce other stem cells in order to continue the lineage throughout the adult life of the male and divide cyclically by mitosis to produce committed spermatogonia which immediately proliferate to produce primary spermatocytes. During meiosis, the exchanged genetic material between homologous chromosomes of primary spermatocytes is followed by a reductive division producing haploid spermatids. Spermiogenesis is the differentiation of spherical spermatids into mature spermatids, which are released at the luminal-free surface as spermatozoa.

The spermatogenic cycle is superimposed on the major divisions of spermatogenesis (spermatocytogenesis, meiosis, and spermiogenesis; Fig. 5). The spermatogenic cycle, also known as the cycle of the seminiferous epithelium, is defined as a series of changes in a given region of seminiferous epithelium

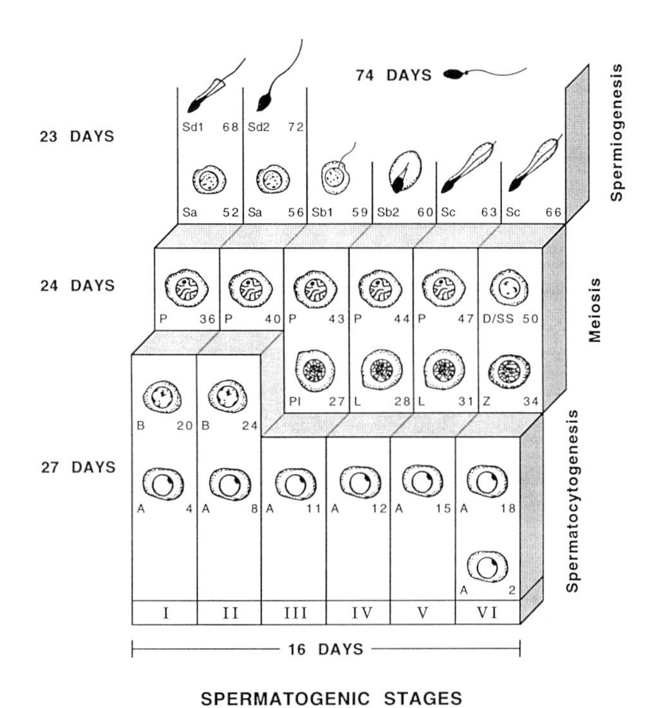

FIGURE 5 Diagram illustrating the timing of germ cell development in the six spermatogenic stages (I–VI) and how the three major developmental divisions (spermatocytogenesis, meiosis, and spermiogenesis) of spermatogenesis are superimposed on the spermatogenic cycle. The letter on the lower left side of each cell type identifies the cell type using the same terminology as that in Fig. 3, in which the actual cells are shown and described. The number on the lower right of each cell indicates the average developmental age of the cell type in that spermatogenic stage. During the 27 days of spermatocytogenesis, A (A) spermatogonia divide to form B (B) spermatogonia, which divide to form preleptotene primary spermatocytes (P1). During the 24 days of meiosis, preleptotene primary spermatocytes differentiate to produce leptotene (L), zygotene (Z), pachytene (P), and then diplotene (D) primary spermatocytes. These divide at the first meiotic division to produce secondary spermatocytes (SS) which divide at the second meiotic division to produce Sa spermatids. During the 23 days of spermiogenesis, Sa spermatids differentiate into Sb1, Sb2, Sc, Sd1, and then Sd2 spermatids. The spermatozoon is spermiated after completion of spermatogenesis in 74 days (reproduced with permission from Johnson *et al.*, 1997).

between two appearances of the same developmental stages. Using spermiation as a reference, the cycle is composed of all the events that occur in a given region of the tubule between two consecutive spermiations. Using the human as a model, the 27 days of spermatocytogenesis, 24 days of meiosis, and 23 days of spermiogenesis compose the 74-day duration of spermatogenesis. These three major divisions of spermatogenesis are combined (overlapped) to create 4.5 generations of germ cell that last 16 days each. Each cycle is divided into six spermatogenic stages of uneven length (Figs. 3 and 5).

The classification of spermatogenic stages is based on observations made by conventional histology (Figs. 3 and 5). Spermatogenic stages represent man-made divisions of naturally occurring and continuously changing cellular associations in the seminiferous tubules. Classification of stages has been based on the morphological changes of germ cells or on the development of acrosomes of spermatids. In addition, classification of stages of the spermatogenic cycle has been based on the presence and location of specific germ cells (i.e., B spermatogonia, one or two generations of primary spermatocytes or spermatids, or secondary spermatocytes) and on changes associated with nuclear elongation, location, and release of spermatids.

To understand the spermatogenic cycle and development of germ cells throughout spermatogenesis, it is useful to compare spermatogenesis with a 4-year college model (Fig. 6). Using graduation as the reference point in the college model, the cycle would be all the events between two consecutive graduations. In both processes, the cycle length (time between two consecutive releases) is dictated by the frequency of cells or classes entering the process

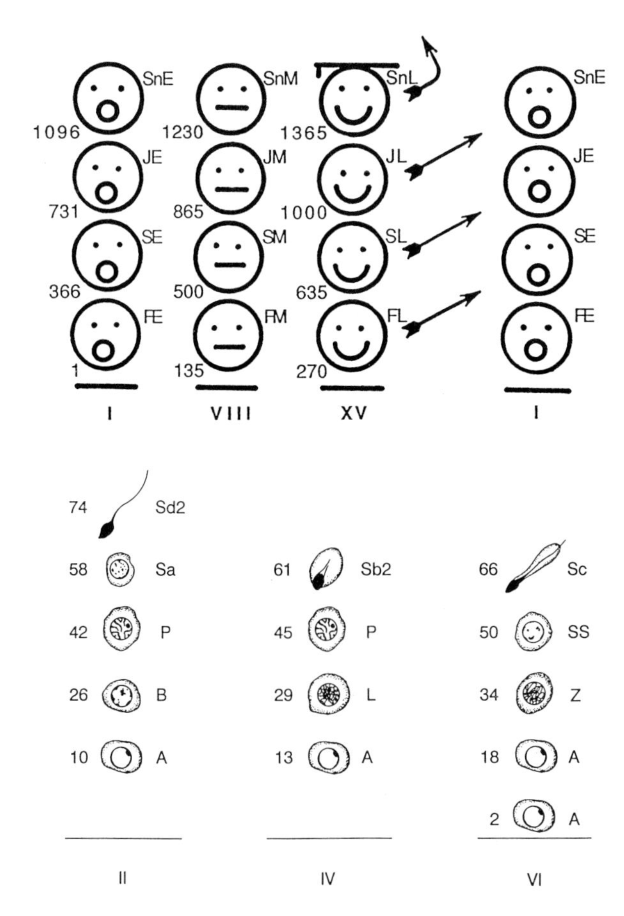

FIGURE 6 Comparison of the kinetics of spermatogenesis in humans with timing of student development in 4 years of college. To facilitate comprehension of the orderly process of spermatogenesis, 4 years of college (top) and spermatogenesis (bottom) have been divided into classes or generations (rows) and stages (columns). In the college model, 1 year of college has been divided into 15 stages (3 shown here), and the placement of the same students in the first stage of the second year is indicated by the arrows. Three of the 8 stages in spermatogenesis in horses are illustrated. Numbers indicate the developmental age (in days) of A or B spermatogonia (A, B), leptotene (L), zygotene (Z), or pachytene (P) primary spermatocytes, secondary spermatocytes (SS), and spermatids of various types (Sa, Sb2, Sc, and Sd2). In the college model, the numbers represent days of completion as freshmen (FE, FM, or FL), sophomores (SE, SM, or SL), juniors (JE, JM, or JL), or seniors (SnE, SnM, or SnL) and progress through the early (FE, SE, JE, or SnE), middle (FM, SM, JM, or SnM), or late (FL, SL, JL, or SnL) period or stage of that year, respectively. College and spermatogenesis are similar in that (i) time between consecutive releases is less than the duration of the entire process, (ii) multiple classes or generations of cells occur simultaneously, (iii) the amount of time between classes or generations is one cycle length, and (iv) the process continues at a defined rate such that students or cells in specific developmental steps are always associated with other students or cells at respective developmental steps. Since the early period of the year occurs at the same time for each class, incoming freshmen are in the same stage of the cycle as are incoming sophomores, juniors, and seniors. Hence, any combination of students or cell types that differ in developmental age by a multiple of the cycle length plus a constant will constitute a stage of the cycle. The cycle refers to the maturation of students within one academic year between two consecutive graduations or maturation of cells between two spermiations (modified with permission from Johnson, 1991).

and is less than the duration of the entire process. Although spermatozoa are released every 16 days in humans from a given region of the seminiferous tubule, the duration of spermatogenesis is 74 days. Although graduation occurs yearly, college takes 4 years to complete. Hence, multiple generations of germ cells (spermatogonia, spermatocytes, and spermatids) or classes (freshmen, sophomores, juniors, and seniors) must occur simultaneously. The amount of time between consecutive generations or classes is one cycle length. Germ cell degeneration in spermatogenesis and dropouts in college reduce the product yield (number of cells spermiated and number of students graduated, respectively).

In both college and spermatogenesis (Fig. 6), once the process starts, it continues at a defined rate such that cells (Golgi-phase spermatids) or students (incoming freshmen) at a given developmental step are almost always associated with other cells (maturation-phase spermatids) or students (incoming sophomores, juniors, or seniors) at respective developmental steps. Indeed, any combination of germ cells or classes that differ in developmental age by a multiple of the cycle length plus a constant will constitute a stage of the cycle. A stage in the cycle is defined as associations of four or five generations of germ cells formed in specific, chronological developmental steps whose developmental age differs by a multiple of the cycle length plus a "constant value." Differences among stages of the cycle result in all the cells in that group (association) being in steps of development that are different from those of other stages. Hence, the value of the constant is different for each stage of the cycle.

Differences between the college model and spermatogenesis illustrate even more dynamics of the process of spermatogenesis (Fig. 6). Cell division magnifies the yield in spermatogenesis; however, no multiplying component occurs in college. Each committed spermatogonium entering spermatogenesis has a potential yield of 64 spermatozoa; however, less than one student graduates for each student entering college. In spermatogenesis, the frequency of product release is greater than that in college model. As a result of college largely beginning only once a year in the fall, graduation for most students occurs only once a year, usually in late spring. In spermatogenesis, the production of newly committed spermatogonia is not synchronized among tubules and not simultaneous along the length of the same tubule. Therefore, somewhere in the testis, newly committed spermatogonia enter the process of spermatogenesis each second. This creates a continual release of spermatozoa into the lumen of seminiferous tubules. This results from the production of about 40 committed spermatogonia entering the 74-day process of spermatogenesis each second in humans. If college started each second instead of once in the fall, graduation would occur each second instead of once in the late spring. Although this continual release seems wasteful, this ensures that at least some male gametes are always available. Even after exhaustive sexual activity, some spermatozoa are available whenever female gametes are available for fertilization.

The length of cycle is the amount of time between consecutive releases of spermatozoa (spermiations) at one region in the seminiferous tubule and is dictated by the frequency of newly committed spermatogonia entering the process of spermatocytogenesis. Therefore, unlike the classification of stages, the cycle length is not a man-made division of the duration of spermatogenesis. The cycle lengths in days for various species are 7.2 for the prairie vole, 8.6 for the boar, 8.7 for the hamster, 8.9 for the mouse, 10.4 for the ram, 13.5 for the bull, 13.6 for the beagle dog, 9.5 for the rhesus monkey, 12.2 for the horse, 12.9 for the rat, and 16 for man.

B. Spermatocytogenesis

Spermatocytogenesis involves spermatogonia, derived from embryonic gonocytes, which are located in the basal compartment of seminiferous tubules in adults (Figs. 1–3). Stem cell spermatogonia divide by mitosis to produce other stem cells (to continue the lineage of stem cells throughout the adult life of males) and produce committed spermatogonia which immediately begin the clonal expansion characteristic of germ cell production.

Spermatogonia (like spermatocytes, spermatids, or residual bodies) are attached to one another by intercellular bridges. These bridges between cells in the same developmental step result from incomplete cytokinesis during cell division. The intercellular bridges may function to facilitate synchronous development or degeneration of similar germ cells. Since the presence of intercellular bridges is among the earliest features of committed spermatogonia (i.e., paired), they may allow separation of committed spermatogonia from single stem cell spermatogonia. Also, these bridges allow differentiation of haploid spermatids (which now have only one sex chromosome) and/or phagocytosis and digestion of residual bodies left behind at spermiation of mature spermatids during spermiogenesis. Since germ cells arise as a clone from a single stem cell during spermatocytogenesis, a defect in this single cell could result in the loss (degeneration) of the entire progeny of the clone. Toxicants or other detrimental influences (that alter mitosis or differentiation of spermatogonia) would be expected to have a major influence on the efficiency and total productivity of spermatogenesis.

C. Meiosis

In meiosis, genetic material is exchanged between paired homologous chromosomes, and haploid spermatids are produced by two meiotic divisions. Meiosis occurs only in germ cells (male and female), and in males these cells are spermatocytes (Figs. 1–3). Following the mitotic division of spermatogonia, preleptotene primary spermatocytes result. These cells are very similar in appearance to spermatogonia, but they begin meiosis by active DNA synthesis. Their rapid incorporation of [^{3}H]thymidine during DNA synthesis facilitates the timing of the kinetics of spermatogenesis when the labeled cells are followed throughout spermatogenesis. Genetic exchange occurs at the synaptonemal complex between paired chromosomes in zygotene and pachytene primary spermatocytes. In females, the two X chromosomes pair and segregate like somatic homologs. In males, the X and Y chromosomes partially synapse, but this partial synapsis is sufficient to keep the X and Y chromosomes paired on the spindle for the first meiotic division. The pairing of sex chromosomes ensures segregation of these homologous chromosomes in meiosis I and is essential to ensure that spermatozoa each contain either one X or one Y chromosome and not both X and Y or neither chromosomes.

RNA synthesis is highest during the midpachytene phase and the end of the meiotic prophase, and mRNA specific for structural proteins increases during the prophase of meiosis. During the first meiotic division, pachytene primary spermatocytes rapidly undergo metaphase (chromosomal alignment), anaphase (separation of chromosomes), and telophase (complete separation of chromosomes) to produce secondary spermatocytes. Histologically, these cells have spherical nuclei with chromatin flakes of varying sizes (Fig. 1–3). These cells are in a brief interphase with no major DNA synthesis prior to the second meiotic division to produce spermatids. Hence, one duplication of chromosomes followed by two divisions (reduction division and equatorial division) produce the haploid genome of spermatids and spermatozoa. Early spermatocytes are active in both RNA and DNA synthesis. Meiotic divisions are especially vulnerable in humans: Young men have losses in production of 35–40%, and aged men (78 ± 2 years) lose >75% of their germ cell production potential. Toxicants or detrimental influences (that alter the cell division process or exchange of genetic material during the long prophase of meiosis) significantly decrease spermatogenic potential.

D. Spermiogenesis

Spermiogenesis is the morphologic differentiation of spermatids into spermatozoa (Figs. 1–3). Spherical cells with spherical nuclei change shape to form a streamlined head with a condensed nucleus, an acrosomal covering, and a tail that is necessary for cellular motility. Based on development of the acrosome, spermatids are classified into Golgi, cap, acrosomal, and maturation phases. In the spermatozoon, the acrosome is a membrane-bound, enzyme-containing bag over the nucleus. The Golgi apparatus of new spermatids gives rise to the acrosome as it does to lysosomes in all cells. The Golgi-phase spermatid has a prominent Golgi apparatus that produces membrane-bound granules on its "mature face." Small vesicles fuse to form the acrosomal vesicle adjacent to the nucleus. Viewed by light microscopy, the acrosome appears to form almost on top of the nucleus.

As spermatid maturation continues, the centrioles migrate to a region near the nuclear envelope where the distal centriole gives rise to the developing axoneme inside the cytoplasm before its growth is projected away from the cell body toward the lumen of the seminiferous tubules (Fig. 3). The "flagellar canal" is a tubular infolding of the spermatid plasma membrane from the membrane's surface to the annulus located just below the attachment of the flagellum to the nucleus. The flagellum extends through the flagellar canal toward or into the lumen of the seminiferous tubule. The flagellar canal provides a mechanism by which new growth of the flagellum can extend from the spermatid cell body to allow subsequent flagellar growth. The flagellar canal also prevents mitochondria from gaining access to the developing tail until it has reached its maximal length. This could prevent the presence of aberrantly placed mitochondria in the principal piece of the spermatozoon.

As the acrosomal cap extends over the nucleus and the nucleus begins its elongation, the spermatid is classified as being in the acrosome phase (Fig. 3). Acrosome-phase spermatids are embedded in deep recesses within the Sertoli cell apex. The manchette, a transient organelle found only in spermatids, is composed of microtubules attached by linking arms and arranged in a sheath which forms around the developing flagellum and extends about half way up the elongating nucleus. Nuclear elongation begins when the manchette appears at the caudal region of the spermatid where the flagellum is attached. The manchette is not normally found in the spermatozoon, but this transient organelle apparently helps direct nuclear elongation and tail development.

The maturation phase of spermatid development is the final phase of spermiogenesis (Fig. 3). The manchette migrates caudally where it may provide a shaft that supports the flagellar canal and freedom of movement for the flagellum during development. Dissolution of the manchette corresponds to the migration of the annulus to its permanent location at the junction of the middle and principal pieces of the spermatozoon and the shortening/disappearance of the flagellar canal. Mitochondria quickly reorient around the flagellum in the middle piece region following annulus migration.

Before spermiation, a large portion of spermatid cytoplasm, composed of organelles not needed by spermatozoa, is left behind. These residual bodies of Regaud are formed by remaining spermatid cytoplasm, which is phagocytized, moved to a basal location, and digested by Sertoli cells which may have stored lipid from previous cellular digestion (Fig. 2). The developing spermatid is attached to its residual body by cytoplasmic stalks. The neck region of spermatids is attached to this excess cytoplasm remaining as a cytoplasmic droplet located on the middle piece of the luminally released spermatozoa.

The spermatozoan nucleus contains the male genetic material (male haploid genome composed of the X or Y chromosome and a haploid number of somatic chromosomes) to be delivered to the egg. It also has an acrosome which contains hydrolytic enzymes necessary for penetration of the egg vestments during fertilization. The tail connects to the head at the neck region, where it is attached to the implantation fossa of the head. The middle piece contains mitochondria for energy production. The principal piece contains the fibrous sheath, and the end piece is at the end of the tail.

A plasma membrane encloses both the head and the tail. At the apical region of the head, the plasma membrane covers the underlying acrosome. The size and shape of the apical ridge vary widely among domestic species. The acrosome contains hydrolytic enzymes necessary for penetration of the layers of the egg. Attached to the base of the head is the neck region of the tail. The proximal centriole remains attached at the implantation fossa in the neck, but the distal centriole gives rise to the axoneme (nine microtubule doublets surrounding a central pair of microtubules) during tail development. The axoneme is surrounded by nine dense fibers which arise from dense bodies attached to the nine doublets. Although biochemical studies have been conducted on the outer dense fibers, their functional role remains uncertain. They do, however, function as passive elastic structures. The annulus marks the end of the middle piece and the beginning of the principal piece. The principal piece is lengthy and contains a fibrous sheath with rib-like structures, which facilitate bending during movement, and dense fibers that extend to different lengths. The fibrous sheath is proteinaceous and is composed of two longitudinal columns bridged by attaching ribs. The axoneme continues through the middle piece, principal piece, and end piece. The end piece is composed strictly of the axoneme or single microtubules and a plasma membrane.

Size and shape of spermatozoa vary among species. Human spermatozoa have a more tapered nucleus in the acrosomal region of the head. In most species, spermatozoa are not considered to be mature when they are released from the seminiferous tubules. Each newly spermiated spermatozoon contains a cytoplasmic droplet on its middle piece which normally is lost during epididymal maturation. Other maturational changes in spermatozoa during their migration in the epididymis include progressive motility, structural stability of chromatin and the tail, and gaining the ability to fertilize an egg (oocyte).

See Also the Following Articles

Gonadogenesis, Male; Male Reproductive System, Overview; Penis; Sertoli Cells, Overview; Spermatogenesis, Overview; Spermiogenesis

Bibliography

Amann, R. P. (1970). Sperm production rates. In *The Testis, Vol. 1* (A. D. Johnson, W. R. Gomes, and N. L. VanDemark, Eds.), pp. 433–482. Academic Press, New York.

Amann, R. P. (1981). A critical review of methods for evaluation of spermatogenesis from seminal characteristics. *J. Androl.* **2**, 37–58.

Berndtson, W. E., Igboeli, G., and Parker, W. G. (1987). The numbers of Sertoli cells in mature Holstein bulls and their relationship to quantitative aspects of spermatogenesis. *Biol. Reprod.* **37**, 60–67.

Clermont, Y. (1963). The cycle of the seminiferous epithelium in man. *Am. J. Anat.* **112**, 35–51.

deKretser, D. M., Kerr, J. B., and Paulsen, C. A. (1975). The peritubular tissue in the normal and pathological human testis. An ultrastructural study. *Biol. Reprod.* **12**, 317–324.

Fawcett, D. W. (1975). Ultrastructure and function of the Sertoli cell. In *Handbook of Physiology* (D. W. Hamilton and R. O. Greep, Eds.), pp. 21–55. American Physiological Society, Washington, DC.

Fawcett, D. W. (1994). Male reproductive system. In *Bloom and Fawcett: A Textbook of Histology* (D. W. Fawcett, Ed.), 12th ed., pp. 796–850. Chapman & Hall, New York.

Garner, D. L., and Hafez, E. S. E. (1980). Spermatozoa. In *Reproduction in Farm Animals* (E. S. E. Hafez, Ed.), 4th ed., pp. 167–188. Lea & Febiger, Philadelphia.

Griswold, M. D. (1995). Interactions between germ cells and Sertoli cells in the testis. *Biol. Reprod.* **52**, 211–216.

Heller, C. G., and Clermont, Y. (1964). Kinetics of the germinal epithelium in man. *Recent Prog. Horm. Res.* **20**, 545–575.

Hochereau-de Reviers, M. T., and Courot, M. (1978). Sertoli cells and development of seminiferous epithelium. *Ann. Biol. Anim. Biochem. Biophys.* **18**, 573–583.

Johnson, L. (1986). Review article: Spermatogenesis and aging in the human. *J. Androl.* **7**, 331–354.

Johnson, L. (1991). Spermatogenesis. In *Reproduction in Domestic Animals* (P. T. Cupps, Ed.), 4th ed., pp. 173–219. Academic Press, New York.

Johnson, L. (1994). A new approach to study the architectural arrangement of spermatogenic stages revealed little evidence of a partial wave along the length of human seminiferous tubules. *J. Androl.* **15**, 435–441.

Johnson, L. (1995). Efficiency of spermatogenesis. *Microsc. Res. Tech.* **32**, 385–422.

Johnson, L., Petty, C. S., and Neaves, W. B. (1980). A comparative study of daily sperm production and testicular composition in humans and rats. *Biol. Reprod.* **22**, 1233–1243.

Johnson, L., Zane, R. S., Petty, C. W., and Neaves, W. B. (1984). Quantification of the human Sertoli cell population: Its distribution, relation to germ cell numbers, and age-related decline. *Biol. Reprod.* **31**, 785–795.

Johnson, L., Petty, C. S., and Neaves, W. B. (1986). Age-related variation in seminiferous tubules in men: A stereologic evaluation. *J. Androl.* **7**, 316–322.

Johnson, L., Abdo, J. G., Petty, C. S., and Neaves, W. B. (1988). Effect of age on the composition of seminiferous tubular boundary tissue and on the volume of each component in humans. *Fertil. Steril.* **49**, 1045–1051.

Johnson, L., Grumbles, J. S., Chastin, S., Goss, H. F., Jr., and Petty, C. S. (1990). Leydig cell cytoplasmic content is related to daily sperm production in men. *J. Androl.* **11**, 155–160.

Johnson, L., Chaturvedi, P. K., and Williams, J. D. (1992). Missing generations of spermatocytes and spermatids in seminiferous epithelium contribute to low efficiency of spermatogenesis in humans. *Biol. Reprod.* 47, 1091–1098.

Johnson, L., Welsh, T. H., Jr., and Wilker, C. E. (1997). Anatomy and physiology of the male reproductive system and potential targets of toxicants. In *Reproductive and Endocrine* Toxicology—Male *Reproductive Toxicology, Vol. 10 in Comprehensive Toxicology* (K. Boekelheide and R. Chapin, Eds.), pp. 5–61. Pergamon, New York.

Russell, L. D. (1984). Morphological and functional evidence for Sertoli–germ cell relationships. In *The Sertoli Cell* (L. D. Russell and M. D. Griswold, Eds.), pp. 365–390. Cache River Press, Clearwater, FL.

Russell, L. D., and Peterson, R. N. (1984). Determination of the elongate spermatid–Sertoli cell ratio in various mammals. *J. Reprod. Fertil.* 70, 635–641.

Setchell, B. P. (1991). Male reproductive organs and semen. In *Reproduction in Domestic Animals* (P. T. Cupps, Ed.), 4th ed., pp. 221–250. Academic Press, New York.

Testosterone Biosynthesis

Douglas M. Stocco

Texas Tech University Health Sciences Center

GLOSSARY

cholesterol A C27 compound which is the precursor of all steroid hormones.

3β-hydroxysteroid dehydrogenase The enzyme which converts pregnenolone to progesterone.

17β-hydroxysteroid dehydrogenase The enzyme which converts androstenedione into testosterone.

Leydig cells The cells found in the interstitial compartment of the testis which synthesize testosterone.

$P450_{scc}$ The enzyme which converts cholesterol to pregnenolone, the first steroid formed. This enzyme is located on the matrix side of the inner mitochondrial membrane.

$P450_{17\alpha}$ An enzyme that exhibits two functions. The first is the 17-hydroxylation of pregnenolone or progesterone and the second is the conversion of C21 steroids to C19 steroids.

StAR Steroidogenic acute regulatory protein, the protein which regulates the transfer of cholesterol from the outer mitochondrial membrane to the inner mitochondrial membrane and the $P450_{scc}$ enzyme.

testosterone A C19 androgenic steroid responsible for maintenance of male secondary sex characteristics and spermatogenesis.

Testosterone is a steroid hormone which is synthesized in the Leydig cells of the male testis. It is often referred to as the male hormone because of its androgenic biological actions which include its essential role in spermatogenesis and the maintenance of secondary sex characteristics in the male. Testosterone is synthesized from C27 cholesterol through the action of several steroidogenic cell enzymes which convert cholesterol first to C21 steroids and eventually to C19 androgens, of which testosterone is the most prevalent. This series of reactions takes place in the Leydig cells, which are housed in the interstitial compartment of the testis. In most species, the Leydig cells comprise $<10\%$ of the total volume of the testis but synthesize virtually all the steroids. Testosterone biosynthesis in the Leydig cell is regulated by the anterior pituitary glycoprotein, luteinizing hormone (LH). This regulation is brought about by the interaction of LH with specific Leydig cell LH receptors, which in turn results in the intracellular accumulation of cyclic AMP, the second messenger which mediates LH action. This article summarizes the key enzymes involved in the biosynthesis of testosterone and the manner in which both the acute and the chronic regulation of testosterone synthesis occurs.

I. INTRODUCTION

The biosynthesis of steroid hormones occurs in specific steroidogenic cells in the adrenal, ovary, testis, and placenta of most species. Smaller amounts of steroids are synthesized in the brain and other peripheral tissues, but the function of such steroids, especially neuroactive steroids, has not yet been well-defined. The biological roles of steroid hormones represent indispensable functions for the homeostasis of the organism, with the adrenal mineralocorticoid, aldosterone, being involved in salt balance and blood pressure control; the adrenal glucocorticoids, cortisol or corticosterone, being involved in carbohydrate metabolism and stress management; and reproductive function being dependent on the ovarian steroids progesterone and estrogen and the testicular steroid testosterone. Indeed, without the proper concentrations and actions of steroid hormones, life itself is not possible. All steroid hormones, regardless of species or tissue, utilize cholesterol as the substrate for their formation. Cholesterol is a C27 multiringed compound which is converted in a series of tissue-specific reactions to one of the C21 or C19 steroid hormones listed previously. The source of cholesterol for steroid hormone biosynthesis can vary from species to species and even organ to organ, but cholesterol comes from one of three main sources, namely, (i) from *de novo* intracellular synthesis from acetate; (ii) from hydrolysis of cholesterol esters which are stored in the cell, usually in lipid droplets; or (iii) from cellular uptake of low-density lipoprotein and/or high-density lipoprotein and utilization of the cholesterol present in these circulating lipoproteins.

Perhaps the most important of the C19 steroids is the androgenic hormone, testosterone. In the human male, this androgen represents the major steroid found in the circulation and more than 95% of the testosterone found in the blood is secreted by the testis. Secretion of testosterone by the testis is normally on the order of 6 or 7 mg per day. The testis is divided into two compartments: the seminiferous tubules, which house the Sertoli cells, peritubular, or myoid cells and the germinal elements, and the interstitial compartment, which consists of blood vessels, lymphatic tissue, macrophages, and Leydig cells. Testosterone is produced almost entirely by the interstitial Leydig cells. Testosterone biosynthesis in the Leydig cell occurs in a pulsatile manner, each pulse being the result of the stimulation of the Leydig cells with the gonadotropin luteinizing hormone (LH), a glycoprotein hormone secreted from specific gonadotrophs located in the anterior pituitary. Interaction of LH with the LH receptor, specific to the Leydig cell in the testis, results in the activation of adenylyl cyclase activity, through the intermediacy of stimulatory G-protein, and the subsequent increase in intracellular cAMP levels. This in turn activates a cascade of events which ultimately result in the synthesis and secretion of testosterone from the Leydig cell within a few minutes following the initial stimulation. The conversion of cholesterol to testosterone occurs via the activity of a series of proteins whose characteristics and enzymatic activities are well-known (Fig. 1). Since there is no convincing evidence that testosterone is stored in cellular pools in the Leydig cell awaiting release upon LH stimulation, its production and secretion appears to be entirely the result of *de novo* synthesis. Thus, in response to LH stimulation and through the intermediary cAMP, the Leydig cell quickly converts cholesterol to testosterone and then secretes this androgen into the blood. Circulating testosterone is taken up by target tissues, in some cases converted to its more potent androgenic form, dihydrotestosterone, binds to the androgen receptor, and exerts its biological function in the nucleus of the cell by interacting with specific sequences in the DNA. Brief descriptions of the enzymes involved in the conversion of cholesterol to testosterone, their location in the Leydig cell, and the acute and chronic regulation of this process will be provided.

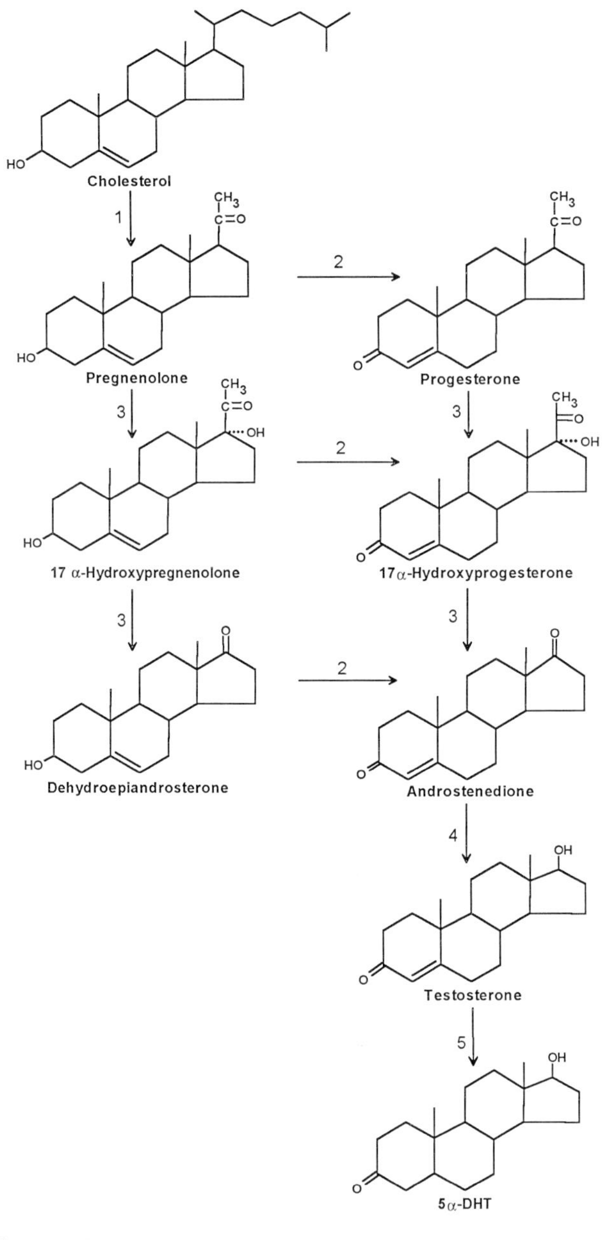

Cholesterol
Pregnenolone
Progesterone
17 α-Hydroxypregnenolone
17α-Hydroxyprogesterone
Dehydroepiandrosterone
Androstenedione
Testosterone
5α-DHT

FIGURE 1 Testosterone biosynthetic pathway. This figure illustrates the steps found in the enzymatic conversion of cholesterol to testosterone. Once delivered to the cholesterol side chain cleavage cytochrome P450 enzyme, cholesterol is quickly converted to pregnenolone. Then, if the 5-ene pathway is employed, pregnenolone is converted to 17α-hydroxypregnenolone by the enzyme cytochrome 17α-hydroxylase/17,20 lyase cytochrome P450 and eventually to dehydroepiandrosterone by the same enzyme. If the 4-ene pathway is used, pregnenolone is converted to progesterone through the action of 3β-hydroxysteroid dehydrogenase and then to 17α-hydroxyprogesterone and androstenedione by cytochrome 17α-hydroxylase/17,20 lyase cytochrome P450. Androstenedione is then converted to testosterone by the action of 17-ketosteroid reductase. Testosterone can also be converted to dihydrotestosterone in target tissues by the enzyme 5α-reductase. In general, cholesterol side chain cleavage cytochrome P450 is found in the mitochondria and the remainder of the enzymes are found in the microsomal compartment of steroidogenic cells. However, there is now good evidence that 3β-hydroxysteroid dehydrogenase can also be found in the mitochondria of some species and some tissues. 1, Cholesterol side chain cleavage cytochrome P450 ($P450_{scc}$); 2, 3β-hydroxysteroid dehydrogenase/Δ^5-Δ^4 isomerase (3β-HSD); 3, 17α-hydroxylase/17,20 lyase cytochrome P450 ($P450_{17\alpha}$); 4, 17β-hydroxysteroid dehydrogenase (17β-HSD); 5, 5α-reductase (5α-RED).

II. BIOSYNTHESIS OF TESTOSTERONE

A. $P450_{scc}$

Cytochrome P450 side chain cleavage enzyme is part of an enzyme system whose function is to cleave a six-carbon unit from cholesterol to yield pregnenolone. This enzyme system consists of the catalytic $P450_{scc}$ enzyme and two additional proteins, a flavoprotein reductase, and an iron sulfur protein, whose function is to transport electrons from reduced pyridine nucleotide to the $P450_{scc}$, a process required for this conversion. It essentially represents a mini electron transport system. During the conversion of cholesterol to pregnenolone, three separate events occur. First, cholesterol is hydroxylated at position C22, followed by a second hydroxylation at position C20 and finally by a scission of the 20–22 bond to yield pregnenolone. All three reactions are performed by the $P450_{scc}$ enzyme. This enzyme system is located on the matrix side of the inner mitochondrial membrane and its action has often been referred to as the rate-limiting step in steroid hormone biosynthesis. While it is undoubtedly the rate-limiting enzymatic step, it is, in fact, not the true rate-limiting step in steroidogenesis. Following the removal of the six-carbon isocaproic acid moiety from cholesterol, the more hydrophilic pregnenolone is free to diffuse out of the mitochondria to the microsomes for further conversion to other steroids by enzymes located in this compartment of the cell. However, it is also possible that pregnenolone can be converted to progesterone within the mitochondria.

B. 3β-Hydroxysteroid Dehydrogenase

The enzyme 3β-hydroxysteroid dehydrogenase (3β-HSD)/Δ^5-Δ^4 isomerase is responsible for the conversion of pregnenolone to progesterone in tissues which utilize the 4-ene pathway and for the conversion of pregnenolone to hydroxypregnenolone if the 5-ene pathway is utilized. This enzyme can also effect the conversion of dehydroepiandrosterone to androstenedione. These reactions constitute the dehydrogenase activities found in this enzyme. Also found in this single protein, which has a molecular weight of approximately 45 kDa, is an isomerase activity which rearranges the double bond found between C5 and C6 to form the double bond found between C4 and C5. While both dehydrogenase and isomerase activities are found in a single protein, it is now clear that there exists a number of isoforms of this enzyme. These isoforms are all products of distinct genes and have been shown to be expressed in a tissue-specific manner. One group of 3β-HSD isoforms (isoforms I–III) appears to function as dehydrogenase/isomerases as indicated previously, whereas a second group (forms IV and V) appears to have ketosteroid reductase activity and is thought to function in the degradation of active steroid hormones. While the intracellular location of the family of 3β-HSD enzymes was once thought to be specifically microsomal, recent evidence has shown that in addition to cytoplasmic 3β-HSD, this enzyme is also localized to the mitochondria of steroidogenic tissue. Thus, in some tissues it might be possible to convert cholesterol to progesterone prior to its exit from the mitochondria.

C. $P450_{17\alpha}$

This enzyme, in a manner similar to 3β-HSD, also contains two separate functions within a single protein. The first of these functions is the hydroxylation of the substrates pregnenolone or progesterone to 17α-hydroxypregnenolone or 17α-hydroxyprogesterone, respectively. The second activity found within this protein is the 17,20 lyase activity which can convert the C21 steroids hydroxypregnenolone and hydroxyprogesterone to their C19 products dehydroepiandrosterone and androstenedione, respectively. It was originally believed that the hydroxylation and lyase activities were performed by separate proteins until the enzyme was cloned and expressed in COS-1 cells and shown to harbor both activities. Finally, it was demonstrated that both the hydroxylation and lyase activities were located in the same active site of the protein. This enzyme is localized to the smooth endoplasmic reticulum of steroidogenic cells.

D. 17β-HSD

The last reaction in the testosterone biosynthetic pathway is the reduction of androstenedione to testosterone by 17β-hydroxysteroid dehydrogenase. This enzyme has 17-ketosteroid reductase activity and has also been referred to as 17-ketosteroid reductase. This is a reversible reaction with the concentration of the substrate and product determining direction. This enzyme also resides in the smooth endoplasmic reticulum, but unlike other steroidogenic enzymes in the testis, its localization is not strictly confined to the Leydig cell. A significant portion of testicular 17β-HSD activity can be found in the seminiferous tubule compartment. Like 3β-HSD, several isoforms of 17β-HSD exist which give rise to proteins of 327, 214, and 310 amino acids. These isoforms are characterized by differences in substrate specificity and tissue distribution. The type 3 isoform, which is 310 amino acids in size, appears to be specific to the testis. Mutations in the type 3 isoform appear to be responsible for most of the defects in testosterone biosynthesis found in humans.

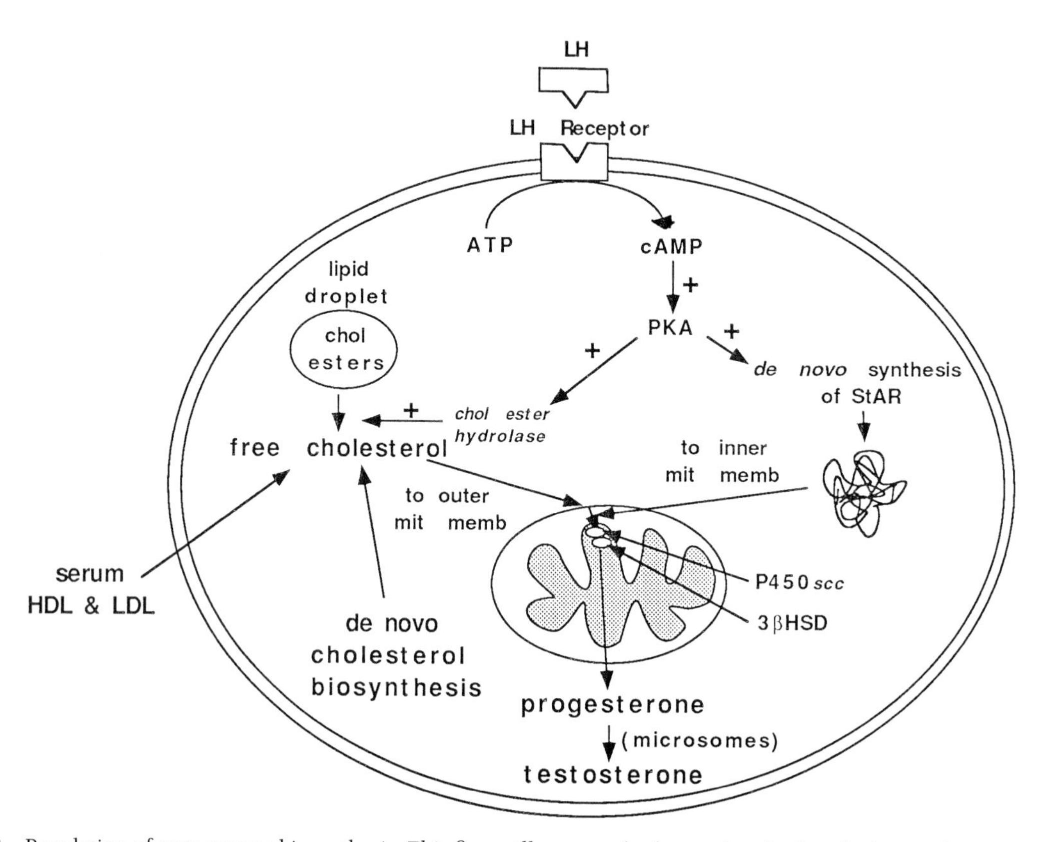

FIGURE 2 Regulation of testosterone biosynthesis. This figure illustrates the factors involved in the biosynthesis and regulation of testosterone by the testicular Leydig cell. The sources of substrate cholesterol are mainly from one or more of three sources depending on the tissue in question. Serum low-density lipoprotein (LDL) and high-density lipoprotein (HDL), *de novo* cholesterol biosynthesis, and the utilization of cholesterol esters stored in lipid droplets all constitute sources of this steroid hormone precursor. LH interaction with its receptor on the cell surface elicits an increase in intracellular cAMP. This in turn activates cholesterol ester hydrolase, which results in the formation of free cholesterol. Additional, but not well understood, events result in the delivery of cellular cholesterol to the outer mitochondrial membrane. Furthermore, cAMP action also results in the *de novo* synthesis of the StAR protein, which regulates the transfer of cholesterol from the outer mitochondrial membrane to the inner mitochondrial membrane and the $P450_{scc}$ enzyme in which it is converted into pregnenolone. In addition, the conversion of pregnenolone to progesterone may also occur in the mitochondria of some species or tissues. Progesterone then freely diffuses out of the mitochondria to the microsomes, where it is further converted to testosterone. In some target tissues, testosterone can be converted to dihydrotestosterone, a biologically more potent form of this steroid.

III. REGULATION OF TESTOSTERONE BIOSYNTHESIS

As in all metabolic pathways, much attention has been directed to the manner in which the biosynthesis of testosterone is regulated. The synthesis of steroid hormones is regulated by trophic hormones which originate in the anterior pituitary [adrenocorticotropin hormone (ACTH), follicle-stimulating hormone (FSH), and LH] and act specifically on steroidogenic cells (adrenal cortex cells, ovarian thecal and granulosa cells, and testicular Leydig cells, respectively) via the action of specific receptors (ACTH receptor, FSH receptor, and LH receptor, respectively). As pointed out earlier, the synthesis of all steroids regardless of tissue begins with the conversion of cholesterol to pregnenolone. This conversion was found to be the rate-limiting enzymatic step as early as 1954. However, with subsequent studies of the steroid pathway in adrenal as well as in gonadal tissues, it became apparent that the activity of the enzyme responsible for the conversion of cholesterol to pregnenolone, $P450_{scc}$, was not the real rate-limiting step. Rather, it became apparent that it was the delivery of the substrate cholesterol to the site of the $P450_{scc}$ which was the pivotal rate-limiting step in the steroidogenic cascade. The manner in which cholesterol is kept from the $P450_{scc}$ is simply a reflection of the inability of this hydrophobic compound to traverse the aqueous space between the inner and outer mitochondrial membranes. Later studies indicated that the *de novo* synthesis of a new protein was required for the transfer of cholesterol to the inner mitochondrial membrane for its subsequent cleavage. While several protein candidates have been put forth as the acute regulator of cholesterol transfer and hence steroidogenesis, only one candidate appears to satisfy all the proposed characteristics of the putative regulatory protein. This protein, named the steroidogenic acute regulatory (StAR) protein, has fulfilled all the criteria deemed essential for the acute regulator. The role of other proteins which have been purported to play a role in cholesterol transfer, such as the steroidogenesis activator polypeptide, sterol carrier protein-2, and the peripheral benzodiazepine receptor and its ligand, the diazepam-binding inhibitor, remain to be elucidated.

In addition to the acute regulation of steroid hormone biosynthesis (Fig. 2), it is well documented that the long-term maintenance of steroid hormone biosynthesis is also dependent on the trophic hormone-induced synthesis of the enzymes located in the steroidogenic pathway. Thus, on the order of tens of hours, transcription and translation of the genes encoding the steroid hydroxylases and dehydrogenases listed previously are essential for the maintenance of maximal steroidogenesis.

Acknowledgments

I thank Dr. Gwynne Little for help with the artwork, and the support of NIH Grant HD17481 is gratefully acknowledged.

See Also the Following Articles

Anterior Pituitary; Leydig Cells; Steroidogenesis

Bibliography

Jefcoate, C. R., McNamara, B. C., Artemenko, I., and Yamazaki, T. (1992). Regulation of cholesterol movement to mitochondrial cytochrome $P450_{scc}$ in steroid hormone synthesis. *J. Steroid Biochem. Mol. Biol.* 43, 751–767.

Miller, W. L. (1988). Molecular biology of steroid hormone synthesis. *Endocr. Rev.* 9, 295–318.

Miller, W. L. (1995). Mitochondrial specificity of the early steps in steroidogenesis. *J. Steroid Biochem. Mol. Biol.* 55, 607–619.

Payne, A. H., and O'Shaughnessy, P. J. (1996). Structure, function and regulation of steroidogenic enzymes in the Leydig cell. In *The Leydig Cell* (A. Payne, M. Hardy, and L. Russell, Eds.), pp. 259–285. Cache River Press, Vienna.

Stocco, D. M. (1997a). A StAR search: Implications in controlling steroidogenesis. *Biol. Reprod.* 56, 328–336.

Stocco, D. M. (1997b). The steroidogenic acute regulatory (StAR) protein two years later: An update. *Endocrine* 6, 99–110.

Stocco, D. M., and Clark, B. J. (1996). Regulation of the acute production of steroids in steroidogenic cells. *Endocr. Rev.* 17, 221–244.

Theca Cells

Katherine F. Roby and Paul F. Terranova
University of Kansas Medical Center

GLOSSARY

cholesterol side chain cleavage A cytochrome P450 enzyme that converts cholesterol into pregnenolone. This enzyme is found in thecal cells and other steroid-producing cells.

17α-hydroxylase:C-17,20-lyase enzyme complex A cytochrome P450 enzyme that converts progesterone to 17α-hydroxyprogesterone and subsequently to androstenedione. This enzyme is located in theca cells and is a major control point for regulation of androgen biosynthesis.

interstitial cells Within the stroma, lying between developing follicles, lie the interstitial glands. These cells are steroidogenic-like cells that can produce progestins and androgens and are sensitive to luteininzing hormone. Upon disassociation of the ovary using enzyme digest, the ovarian cells can be separated into thecal–interstitial cells and granulosa cells using Percoll gradient centrifugation.

theca A loose matrix of spindle-shaped cells surrounding the follicle. Thecal cells can be distinguished as a *theca interna* and *theca externa*. *Theca interna* is an inner glandular, highly vascular layer that produces large amounts of steroids and is located adjacent to the follicular basement membrane. An outer *theca externa* is adjacent to the theca interna and consists largely of a fibrous capsule.

thecal cell hypertrophy Enlargement of theca cells during the process of follicular atresia.

I. ANATOMY/ HISTOLOGY/MORPHOLOGY

A. Development and Morphology

The fully developed theca consists of two major layers, theca interna and externa, is supplied by blood and lymph vessels, and has adrenergic and cholinergic nerve supplies. The theca develops and differentiates as the follicle grows from a primordial to preovulatory state (Fig. 1).

Prior to the initiation of follicular development primordial follicles in the ovarian cortex consist of an ovum surrounded by a single layer of squamous granulosa cells and rest in dormancy. Primordial follicles have no identifiable theca. Those stromal cells nearest the primordial follicles exhibit the same ultrastructural features as all stromal cells. As primordial follicles enter the growing pool, these fibroblast-like stromal cells align to form a concentric layer surrounding the follicle. Although the specific signals initiating development of the theca are poorly understood it is generally believed that these signals originate from the granulosa and/or the oocyte.

As follicles continue to grow the concentric layers of thecal cells increase in number. Differentiation of these theca cells into theca interna and theca externa does not occur until later in follicular development. There is some species variation as to exactly when during the developmental continuum differentiation of the theca occurs. In general theca interna and theca externa differentiate near the time the antrum forms.

1. *Theca Externa*

The theca externa is predominantly a collagenous connective tissue consisting of fibroblasts, epitheli-

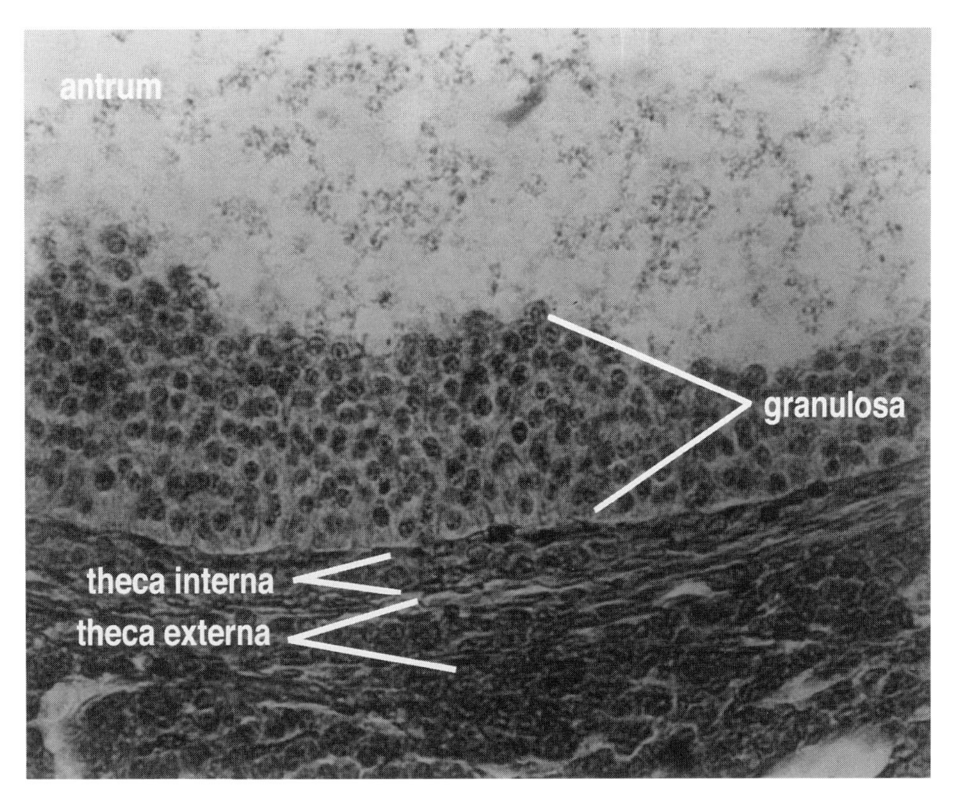

FIGURE 1 The wall of a mature follicle. The healthy granulosa are encircled by the theca interna and the theca externa.

oid cells, and a transitional or intermediate cell type. It is this thecal tissue that provides most of the strength to the follicle wall. The theca externa is relatively well demarcated from the theca interna and appears to merge into the surrounding stroma at its periphery. Immunofluorescence studies have demonstrated the presence of actin and myosin in cells of the theca externa indicating the presence of muscle cells in this layer. Muscle cells in the theca are believed to play a role in contraction of the follicle and expulsion of the oocyte during ovulation.

2. *Theca Interna*

The theca interna is bounded by the granulosal basement membrane and the fibroblast cells of the theca externa. The theca interna is only a few cell layers thick and is characterized by large vacuolated cells in a reticulum of connective tissue fibroblasts. These vacuolated secretory cells have large oval nuclei with prominent nucleoli. Their cytoplasm contains abundant lipid granules, numerous mitochondria, Golgi apparatus, and a few small lysosomal-like bodies. Abundant smooth endoplasmic reticulum is indicative of the steroid biosynthetic properties of these cells.

The capacity of the theca interna cells to respond to luteinizing hormone (LH) and secrete steroid develops as the follicle grows. During the early antral stages of follicle development the theca interna cells begin to express the genes for LH receptors and the enzymes P450 cholesterol side chain cleavage (P450scc) and 3β-hydroxysteroid dehydrogenase (3βHSD). Thus, at this early stage the theca interna cells respond to LH and synthesize progesterone. As the follicle continues to grow and reaches the preovulatory stage, further differentiation of the theca interna leads to expression of 17α-hydroxylase:C17,20-lyase enzymes and androgen secretion. When the follicle reaches the preovulatory stage of development the theca is actively producing androgen.

Development of a technique for separation of ovarian cells utilizing enzyme digestion and Percoll gradient centrifugation has allowed for the separation of theca interstitial cells (TIC) from other ovarian cell types. Subsequent studies utilizing this technique have furthered considerably our understanding of thecal cell biology.

B. Blood Supply

Primordial follicles do not have an independent blood supply; it is not until the follicle enters the growing pool and the theca begins to develop that a blood supply is established. An extensive vascular wreath develops in the theca interna up to but not through the basement membrane. The vasculature of the theca interna is linked to a series of arterioles and venules in the theca externa. The vascular compartment grows along with the rest of the follicle although the pattern of vascular supply is unchanged. Angiogenic activity, stimuli driving the growth of blood vessels toward the follicle, is present in theca of small follicles and probably plays a role in the attraction of blood vessels toward the follicle. Following the LH surge angiogenic activity is found in both the thecal and the granulosal compartments. The presence of angiogenic factors in the granulosa would function to drive blood vessel growth into the developing corpus luteum. Basic fibroblast growth factor (bFGF) and vascular endothelial growth factor (VEGF) are potential mediators of angiogenesis in the ovarian follicle and the corpus luteum.

The blood supply to the follicle provides nutrients to the steroidogenic theca interna cells and indirectly supplies the metabolic needs of the avascular granulosal compartment, including the oocyte. Cholesterol substrate for thecal steroidogenesis is largely provided by lipoproteins delivered into the theca via the rich blood supply. Blood supply to the follicle is increased following the LH surge and stays elevated throughout the ovulatory process. As the time of ovulation nears, the blood vessels become leaky and the thecal tissue edematous. Some studies indicate that this increase in local ovarian blood flow may be mediated by eicosanoids.

A reduction in blood flow to follicles undergoing atresia has been noted in several species. During the process of atresia, thecal vessels appear to accumulate cellular debris. Altered blood flow does not appear to be a primary cause of atresia; however, more studies are warranted to determine the role of changes in blood flow during a timed onset of atresia.

C. Ovarian Nerves

Adrenergic nerves are present in the blood vessels of the theca interna, theca externa, and secondary interstitial cells as well as throughout the stroma of the rat ovary. Theca externa is also innervated by cholinergic fibers. Some of these nerve fibers may terminate on the muscle-like fibroblasts in the theca externa and these nerves may regulate contractility of the theca cells and aide in the mechanism of ovulation. Although there is evidence for muscle-like contractility of the theca many studies have shown that functionally intact nerves are not needed for apparently normal ovulation.

Immunoreactive VIP is present in small numbers of fibers in the ovaries of several species including humans. VIP increased follicular steroidogenesis in whole preovulatory follicles and increased cAMP indicating the possibility of an action on theca.

Nerve growth factor (NGF) mRNA and protein are present in ovarian sympathetic nerves of the juvenile rat. Denervation increases ovarian NGF concentrations; however, mRNA levels are unaffected. Injection of anti-NGF during the first 3 days postnatally blocks development of the ovarian sympathetic innervation, reduces the number of large antral follicles, and increases the number of small preantral follicles at 30 days of age (ratio of small to large increases). Anti-NGF also delayed the time of the first ovulation and results in only 50% successful pregnancies. Expression of NGF in the ovary virtually disappears at ovulation.

Neurotransmitters present in the ovary can alter steroid production by TIC *in vitro*. Norepinephrine and catacholamines increase basal androgen production and norepiniphrine and isoproterenol increase LH-stimulated androgen production by TIC *in vitro*. In contrast, other studies have shown that cellular extracts prepared from superior ovarian nerves inhibit LH-stimulated androgen production. Although the roles of nerves and neurotransmitters in the function of the ovary are still unclear, it does appear from these and other studies that nervous input to growing follicles may play a role in thecal steroidogenesis and thus follicle development.

D. Ovulation

During the process of ovulation, in response to the LH surge, several ultrastructural and functional changes occur in the theca. Fibroblasts in the theca externa and the tunica albuginea change from a resting to an active, proliferating state. The fibroblasts begin to dissociate from each other and take on a looser connective tissue appearance. Degradative events spread up to the surface of the ovary where the cuboidal epithelial cells become vacuolated and necrotic. In the last few minutes prior to rupture, there are changes at the apex of the follicle. The fibroblasts become much more dissociated and the apex balloons out. The surface epithelial cells slough off the ovary where the rupture will occur. The cells in the theca interna and granulosa layers break apart and retract away from the site of rupture. Rupture of the ovarian follicle was proposed to be a proteolytic mechanism as early as 1916 by Schochet. Recently, evidence has accumulated indicating that the proteolytic events of ovulation involve the plasminogen activator (PA), plasminogen inhibitor (PAI), the matrix metalloproteinase (MMP), and tissue inhibitor of MMP (TIMP) systems. Thecal urokinase (u)PA is stimulated by LH, both *in vitro* and *in vivo*. In rats, uPA activity was increased dramatically in preovulatory follicles beginning about 8 hr after hCG injection. Coordinate inhibition of PAI during the time of ovulation has also been demonstrated. Clearly, the PA system is not the only proteinase involved in ovulation. Inactivation of both PA genes [uPA and tissue (t)PA] results in only approximately a 26% decrease in the ovulation rate in mice. Other potential mediators of the tissue degradation during ovulation include the MMP/TIMPS and the kallikrein/kinin systems. The precise enzymatic systems involved in the process of ovulation are currently under intense investigation.

After ovulation and the breakdown of the basement membrane, the blood vessels from the theca interna invade the granulosal compartment and set up an extensive blood supply to the developing corpus luteum. Cells from the theca interna migrate into the developing corpus luteum and become part of the luteal tissue. As an acute response to the LH surge, thecal cells lose the capacity to express P450 17α-hydroxylase (P450$17\alpha$) and thus synthesize progesterone. In humans, as the theca cells luteinize they once again express P450$17\alpha$ and secrete androstenedione.

E. Theca from Atretic Follicles

The vast majority of follicles undergo the degenerative process of atresia before follicle growth is complete and ovulation occurs. Although these atretic follicles will not provide a healthy ovum for fertilization, they can contribute significantly to the steroid production by the ovary. Atresia is characterized by the apoptosis of the granulosa cells and ovum. Unlike the granulosa, the theca maintain viability and in fact thecal hypertrophy is a characteristic of follicular atresia in many species (Fig. 2). Thecal cells from atretic follicles exhibit numerous lipid droplets, smooth endoplasmic reticulum, and mitochondria with tubular cristae, all characteristics of healthy steroidogenic cells. During the early stages of atresia, prior to morphological changes in the granulosa, an increased progesterone production by theca in response to LH has been observed *in vitro*. As atresia proceeds, blood flow to the follicle is reduced, and fewer erythrocytes are observed in theca of late atretic follicles than early atretic follicles. Theca from atretic follicles secrete large amounts of progesterone in comparison to theca from healthy follicles. As the remainder of the follicular components degenerate and are removed from the ovary by cellular scavengers, the theca is maintained. These cells, called secondary interstitial cells (SICs), maintain the expression of LH receptors and the enzymes needed to synthesize androgens. SICs are maintained in the interstitial portion of the ovary.

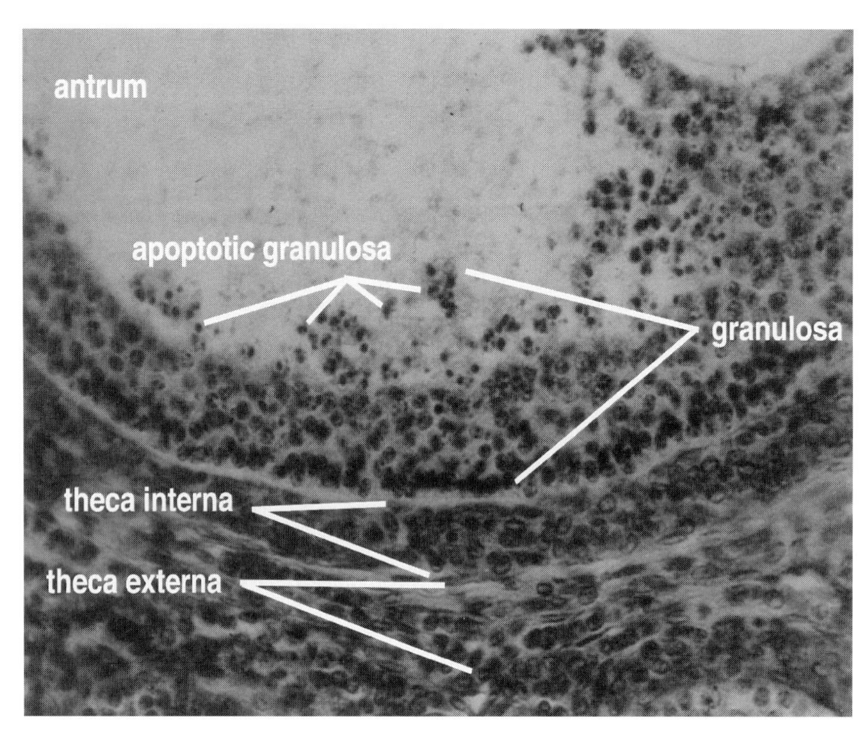

FIGURE 2 The wall of an atretic follicle. Apoptotic granulosa cells are abundant in the granulosa cell layer of this atretic follicle. The theca interna layer is hypertrophied.

II. PHYSIOLOGY: ENDOCRINE REGULATION BY GONADOTROPINS AND ESTROGEN

There is abundant data describing the regulatory function of LH on thecal differentiation and function. However, there is also substantial evidence that LH does not function alone. Prolactin and estradiol, as discussed below, play a role in the action of LH. Clearly, other factors not yet uncovered will be added to this list at a later time. Recent studies have detected progesterone receptor in theca interna of healthy follicles and atretic follicles at all stages of the cycle. The significance of these recent studies awaits further understanding.

A. LH

Early in the development of the theca, LH receptor gene expression is initiated and the cells gain functional LH receptors. The low levels of serum LH present during the follicular phase of the cycle maintain the development and differentiation of the theca. Many studies support the hypothesis that LH is the primary regulator of thecal development. LH mediates both increased theca cell proliferation and differentiation of the steroidogenic capabilities of the theca.

LH binding to thecal membrane receptors initiates a cascade of intracellular events. Stimulation with LH induces thecal expression of each of the enzymes necessary for androgen synthesis: P450scc, 3βHSD, and P45017α. *In vitro* studies have shown that undifferentiated theca express LH receptors and the LH receptors are coupled to cAMP production but not to androgen biosynthesis. Only after approximately 20 hr of continuous exposure to LH do the TICs differentiate into active androgen-producing cells. hCG binding (LH receptors) increases as the follicle continues to develop. The ability of theca to synthesize and secrete proper amounts of androgen is essential for normal follicular development. LH-mediated thecal androgen production is further regulated by multiple autocrine and paracrine factors within and around the developing follicle.

SICs maintain LH receptor expression and respond

to LH with androstenedione synthesis. The steroidogenic function and biosynthetic pathways of SICs and TICs are very similar. However, immunocytochemical studies have indicated the expression of P450scc in the SICs is greater than that seen in theca interna cells. In addition, the expression of this enzyme in the SICs does not appear to be dependent on gonadotropins, as it is in the theca interna. Although levels of enzyme expression appear to be different any actual difference in the enzymatic activity between the two tissues has not been shown.

B. Prolactin

Autoradiographic and membrane-binding studies have demonstrated the presence of prolactin (PRL) receptors on TICs and SICs. Quantitative binding studies indicate TICs express a single class of high-affinity prolactin receptors. The presence of prolactin receptors suggests a role for prolactin in thecal development and/or function. When TICs are treated with PRL *in vitro* there is a rapid and irreversible inhibition of LH-stimulated steroid synthesis. PRL functions in a dose-dependent manner to inhibit androgen biosynthesis. *In vivo* studies indicate a similar function for PRL. When serum levels of PRL are high, such as during lactation, folliculogenesis is inhibited and estrogen biosynthesis is suppressed. Thus, inhibition of thecal androgen synthesis by PRL reduces androgen substrate and subsequently estrogen synthesis is reduced. The opposite condition is also observed. Androgen biosynthesis stimulated by 8-bromo-cyclic AMP and other adenylate cyclase activators was inhibited by PRL treatment. Thus, the inhibitory action of PRL on LH-stimulated thecal steroidogenesis appears to be exerted at a step after adenylate cyclase. Bromoergocryptine-induced hypoprolactinemia caused increased secretion of 5α-androstane-3α,17β-diol, the major androgen secreted by the prepubertal rat ovary in response to hCG stimulation.

C. Estradiol

The presence of estrogen receptors was indicated in early studies when an accumulation of labeled estradiol was observed in the nuclei of both TICs and SICs in rats. Subsequently, both *in vivo* and *in vitro* studies have supported a role for estradiol in thecal function. Estradiol inhibits thecal androgen production. *In vivo* experiments have shown that estradiol inhibits the ability of LH to stimulate ovarian androgen synthesis but not progesterone production. When immature rats were treated with estradiol, hCG-stimulated androgen synthesis was inhibited. This study concluded that estradiol blocked ovarian androgen production by reducing the activity of the P450$17\alpha$ enzyme.

In vitro experiments have demonstrated a direct effect of estradiol on the ovary. Estradiol treatment of TICs caused a rapid and selective inhibition of P450$17\alpha$ enzyme activity. In the same studies there were no changes in the LH receptor, LH-stimulated cAMP production, or progesterone production. Other *in vitro* studies assessing P450$17\alpha$ in ovarian homogenates indicate that estradiol does not directly inhibit the activity of the P450$17\alpha$ enzyme but estradiol may block the transcription or translation of the gene and thus reduce enzyme protein levels.

Still other studies indicate that thecal steroidogenesis is not altered by estradiol but that metabolites such as the catecholestrogens mediate the actions of estradiol. Catecholestrogens inhibited androgen production by porcine theca *in vitro*.

There is some evidence that estrogens may regulate androgen metabolism by a direct action on 5α-reductase. Studies using rat ovarian microsomes demonstrated that estradiol inhibited 5α-reductase activity. Furthermore, the concentration of estradiol required to cause a reduction in enzyme activity was in the range of that measured in the follicular fluid and thus this function of estradiol may have physiological relevance.

The physiological effects of estradiol on inhibition of thecal androgen synthesis probably occur with every cycle. As the preovulatory follicle continues to grow and estradiol levels increase, positive feedback and the LH surge are initiated. With the LH surge, estradiol levels increase even further before dropping. This latter increase in estradiol may act on theca to shut down the P450$17\alpha$/lyase enzymes. Thus, the supply of androgen for aromatization to estrogen is reduced and the follicle differentiates into the corpus luteum, whose major secretory product is progesterone.

TABLE 1
Hormones and Potential Regulators of Thecal Function

Hormone	*Cytokine*	*Growth factor*	*Neurotransmitter*	*Other peptides*
LH	TNF	bFGF	VIP	PA
PRL	IL-1	aFGF	Norepinephrine	PAI
Estradiol	IL-2	VEGF	Catecholamines	MMP
hCG		NGF		TIMP
Insulin		IGF		Inhibin
GH		IGFBP		Activin
Progesterone		KGF		MIS
Catecholestrogens		HGF		Follistatin
		TGF-β		Prorenin
		EGF		Renin
		TGF-α		Angiotensin

III. GROWTH FACTORS: AUTOCRINE AND PARACRINE FACTORS AFFECTING THECAL FUNCTION

LH is the primary factor regulating thecal cell function and differentiation; however, it is also clear that many other factors impinge on the theca to interact with the action of LH. These interacting factors are numerous (Table 1) and although many are produced by other cells types there is ample evidence to indicate many of these factors are produced by the theca (Table 2) and thus function in a autocrine manner on the theca.

A. The Insulin-like Growth Factor System

The insulin-like growth factor (IGF) family of proteins consisting of the IGF-I and -II, type I and II receptors, and several binding proteins (IGFBP) have been localized to ovaries of various species and implicated in ovarian function. The role of the IGF system in ovarian function is obviously very important. This review will focus on the role of IGF in thecal cell function and where pertinent interaction with the granulosal cell compartment will be discussed. Thus, it is beyond the scope of this review to cover the granulosa cell IGF system.

Insulin-like growth factor-I is a protein consisting of a single chain of 70 amino acids with a molecular size of 7.5 kDa. It is similar to proinsulin, can produce effects similar to those of insulin, and is produced by many cells types. Upon binding to its receptors, consisting of two α and two β subunits, IGF-I induces protein tyrosine kinase activity in the β subunits. The major known effects of IGF-I and -II on theca cells in culture are to increase the responsiveness to LH by upregulating LH receptors and amplifying cAMP secretion and subsequently increasing androgen production. Those effects are largely universal across species. However, species specificity becomes an important consideration as to which IGF, IGF receptors, and their binding proteins are expressed by theca. For example, thecal cells of

TABLE 2
Hormones and Growth Factors Produced by Theca

Androgen	PA
Progesterone	MMP
IGF	TIMP
TGF-β	IL-1
TGF-α	Inhibin
EGF	Prorenin
bFGF	
HGF	

sheep express IGF-II mRNA but not IGF-I mRNA, whereas pigs express IGF-I and -II mRNAs.

The genes encoding IGF-I and -II and their two receptors, type I and II, have been detected in theca. In pigs, IGF-I and -II mRNA increase in theca interna as follicles grow and this is paralleled by increases in thecal LH receptor, 3βHSD, P450scc, and P45017α. In rats, IGF-II and type I and II IGF receptor genes are expressed in theca. In sheep, type I and II IGF receptors are present in theca of healthy follicles. Type II receptors were low in theca of atretic follicles.

Growth hormone (GH), a known stimulator of IGFs, stimulates androgen production by theca cells in culture in a time- and dose-dependent manner, and its effects were not due to increases in cAMP and cell number. Interestingly, anti-IGF-I did not modify the GH effect on thecal androgen production, indicating that the effects of GH were not likely to be mediated by IGFs.

1. *IGFBPs*

Rat theca interstitial cells express the genes for IGFBP-2 and -3 which are hypothesized to be antigonadotropic *in vitro*. Since it is well-known that granulosa cells synthesize and secrete IGF-I, and IGF-I synergizes with follicle-stimulating hormone (FSH) in induction of progesterone secretion and aromatase activity within that cellular compartment, it is plausible to hypothesize that thecal IGF-binding proteins regulate the response of the granulosa cells to FSH by regulating the bioavailability of IGFs. Studies in humans have shown that immunoreactive IGFBP-4 is present in high amounts in atretic theca but largely absent in healthy antral and preantral follicles. It was hypothesized for the human that the increase in IGFBP-4 may be causally related to a decrease in IGF-I activity resulting in reduced estradiol production and an increased androgenic environment characteristic of follicular atresia. IGFBP-2, -4, -5, and -6 mRNA have been detected in fresh human theca; IGFBP-1 and -3 mRNA were not detected. However, in cultured human theca, IGFBP-1, -3, and -4 were induced by the culture conditions and LH treatment *in vitro* increased IGFBP-5. Growth hormone, a known stimulator of IGF secretion, together with LH had no effect on IGF or IGFBP expression in thecal tissue. Thus, it appears that there are differences in the expression of IGFs and IGFBPs in cultured and fresh human theca. In contrast to human, IGFBP-4 mRNA expression in porcine theca closely parallels that of the dominant follicle.

B. The Transforming Growth Factor Family

The transforming growth factor family includes five known forms of transforming growth factor-β (TGF-β_{1-5}), the activins, inhibins, Müllerian-inhibiting substance (MIS), and their receptor systems. These factors are well-known for their effects on growth and differentiation of cells. All of the receptors exhibit serine/threonine kinase activity.

1. *TGF-β*

TGF-β1(human and rat) and -β2 (human) mRNAs and proteins have been localized to theca and both TGF-β1 and -β2 are secreted by thecal cells *in vitro*. Although rat theca express both TGF-β1 and TGF-β2, only TGF-β2 is regulated by gonadotropins. Currently, there are three known receptors for TGF-β, types 1–3; all the receptors exhibit serine/threonine kinase activity in response to TGF-β binding.

TGF-β induces thecal secretion of progesterone but at the same time has inhibitory effects on LH-stimulated androgen production. *In vitro* studies indicate the inhibitory effects of TGF-β on androgen production are mediated by a direct noncompetitive inhibition of P45017α enzyme. In contrast to its effects on theca, TGF-β is a stimulator of granulosa cell functions. TGF-β increased LH and TGF-α receptor expression, enhanced expression of the P450scc and the P450 aromatase enzymes, and increased mitosis of granulosa cells. Thus, it has been hypothesized that TGF-β emanating from theca may enhance granulosa cell functions by promoting differentiation and also inducing differentiation of thecal cells. Estradiol inhibits TGF-β synthesis in TICs and thus, as the follicle grows, inhibition of TGF-β becomes advantageous in order to allow thecal cells to produce more androgen precursors for conversion to estradiol. It appears that TGF-β function is advantageous in the small growing follicle and is then turned off when its function would be inhibitory to further development of the follicle.

2. *Inhibin*

There are two forms of inhibin, inhibin A and inhibin B. Inhibin A consist of an α subunit and the βA subunit. Inhibin B consists of an α subunit and the βB subunit. Inhibins are largely produced by granulosa cells although there is some evidence indicating the presence of mRNA in theca. Inhibin α and βA subunits are transiently expressed by TICs and SICs during the luteal phase of the estrous cycle. Inhibin is a potent stimulator of LH-induced androgen production by rat and human theca cells. Inhibin A alone increased androgen production in TICs. In fact, inhibin synergized with forskolin in a time- and dose-dependent manner in increasing androgen production from human ovarian thecal tumor cells; progesterone secretion decreased by 40% in this model. Inhibin also synergizes with insulin, IGF-I, and -II in enhancing androgen production by human theca. In cows, inhibin also enhanced LH-stimulated androstenedione secretion and its action was blocked by activin but not follistatin, an activin-binding protein.

One hypothesis is that inhibin functions, at least in part, as an intraovarian regulator of follicular dominance. *In vivo* inhibin acting on the theca would function to increase the supply of androgen to the granulosa of growing follicles, which in turn would produce more estrogen. As the follicles continue to grow, increasing production of inhibin and estradiol would decrease FSH secretion and thus restrict the growth of other nondominant antral follicles in the ovary. The dominant follicle would be hypothesized to have a supply of other local factors such as IGF-I which would increase sensitivity to FSH, resulting in promotion of dominant follicle growth. Interestingly, immunization of animals against inhibin results in an increase in ovulation rate probably by increasing secretion of FSH.

3. *Activin*

Activin is produced by granulosa cells and although no specific receptors have been shown on theca thus far, activin is believed to exert paracrine action on the theca. In rats, activin A alone has no effect on TIC androgen production but in the presence of LH, activin A increased androsterone production. In contrast, previous reports using human theca and rat stromal cells have observed that activin suppresses the LH-stimulated androgen production. In human thecal cells, activin A inhibited LH/IGF-I-stimulated androgen production in a dose dependent manner and inhibition of steroidogenesis occurred without inhibition of DNA synthesis. Similarly in bovine theca, addition of activin A inhibited LH-stimulated and estradiol-stimulated androstenedione production, and LH-stimulated progesterone production was unaffected by activin A. In human ovarian theca tumor cells, activin A had no significant effect on basal levels of steroid production, enzyme activities, or steroidogenic enzyme mRNA levels; however, in the presence of forskolin, a stimulator of adenylate cyclase, activin A dose dependently inhibited androstenedione production while progesterone secretion was maintained at control levels. The latter was attributed to a reduction in P45017α.

4. *MIS*

MIS, a product of granulosa cells, increases androsterone production and 3βHSD mRNA in TICs. Interestingly, MIS had no synergistic effect on LH-stimulated androsterone.

5. *Follistatin*

Follistatin is an activin-binding protein produced by granulosa that could regulate the bioactivity of inhibin and activin in the TICs. Follistatin alone increased bovine thecal progesterone production *in vitro* and addition of activin A reversed the increase to control levels. Follistatin had no effect on androstenedione production.

C. Transforming Growth Factor-α

EGF and TGF-α are distinct proteins encoded for by different genes; however, because EGF and TGF-α bind the same receptor, they are discussed together.

The expression of TGF-α or EGF by the follicle appears to be species specific. For example, the bovine ovary produces TGF-α but not EGF. With the exception of the hamster, where EGF was localized in TICs and also in granulosa cells of small and medium preantral follicles, EGF and TGF-α are produced by the theca. The expression of TGF-α in rat theca is upregulated by FSH. FSH receptors are present only on granulosa cells and TGF-α is produced only by theca in the rat; the indirect mechanism whereby FSH-upregulated TGF-α expression is unknown; however, this is a good example of granulosal–thecal interaction. In general, localization and regulation studies indicate that the expression of TGF-α and EGF is inversely related to follicle size and that the growth factor is expressed primarily in theca of small follicles with the expression decreasing as the follicle develops.

The EGF receptor has been well characterized and binds both TGF-α and EGF and has intrinsic tyrosine kinase activity. EGF receptors are present on granulosa cells of several species with levels being highest in small follicles. An exception has been observed in the bovine, in which binding has been observed in the theca. Both FSH and estradiol have been reported to increase granulosal EGF binding.

In general, TGF-α and EGF are negative regulators of TICs. TGF-α and EGF each inhibit LH-stimulated TIC androgen production. In the hamster, EGF immunostaining was absent in granulosa cells from atretic follicles but in the atretic follicles EGF immunostaining was present in the theca, suggesting a role for EGF in atresia. In contrast, TGF-α was mitogenic to bovine theca cells *in vitro* and TGF-α staining was most intense in theca of rapidly growing follicles.

D. The Fibroblast Growth Factor Family

The fibroblast growth factor family consists of seven members, three of which are present in the ovary [acidic and basic FGF (aFGF and bFGF) and keratinocyte growth factor (KGF)]. Members of this family bind heparin sulfate proteoglycan on cell surfaces, in basement membranes, and in the extracellular matrix. Several high-affinity membrane receptors for the FGFs have been identified and linked to tyrosine kinases.

bFGF is produced by the bovine CL, in which the expression is upregulated by LH. Recent studies using PCR have shown that bFGF mRNA is detectable in the rat whole ovary but not in granulosa cells, indicating bFGF may be produced by the stromal–interstitial tissue. The greatest amounts of aFGF mRNA are detected in the rat stromal tissue. KGF mRNA has been detected in the stroma but not the granulosa of rat.

The major role for these proteins in the ovary is believed to be in promoting angiogenesis; however, this is an active area of research and other recently found factors such as VEGF may be equally if not more important in ovarian angiogenesis.

bFGF has no effects on basal or LH-stimulated androgen production by rat TICs *in vitro;* however, bFGF can inhibit hCG-stimulated androgen production in the presence of IGF-I. In human theca FGF reduced both forskolin- and cAMP-stimulated androgen production.

E. Hepatocyte Growth Factor

Hepatocyte growth factor (HGF) is produced by both granulosa and theca cells of the rat. HGF does not alter basal levels of androgen but HGF reversibly inhibits LH-dependent androgen production. The reduction in androgen was associated with an increase in progesterone and decreased levels of P45017α.

F. Cytokines

Cytokines are proteins that effect proliferation, differentiation, and the expression of mature phenotypes of various target cells throughout the body. Cytokines are produced by multiple cell types. Immune cells produce an array of cytokines, and because the ovary contains the greatest number of immune cells compared to any other endocrine gland, the ovary and ovarian cells have the potential to encounter all the cytokines at some time. It has become clear that several ovarian cells types also express cytokines. The cytokines that fall into this category include tumor necrosis factor (TNF) and interleukin-1 (IL-1).

1. *Tumor Necrosis Factor*

TNF has been localized in ovarian theca and granulosa cells, the oocyte, and ovarian macrophages. TNF functions in isolated theca to inhibit androgen production. TICs isolated from immature rats exhibited dose-dependent inhibition of LH-stimulated androsterone production in response to TNF. This inhibitory effect was within physiological range of the TNF receptor dissociation constant, was reversible by the removal of TNF, and was neutralized by antiserum to TNF. The effects of TNF were mediated by inhibition of cAMP generation. However, TNF also acts in isolated TICs at sites distal to cAMP.

Interestingly, TNF was shown to induce clustering of TICs from immature rats when cultured *in vitro*; clustering was dependent on the dose of TNF. It was hypothesized that TNF, present in the oocyte, might act as a organizing factor as the granulosa and theca orient around the growing follicle.

When whole preovulatory follicles are cultured under conditions that maintain healthy follicles, TNF stimulates follicular progesterone, androstenedione, and estradiol production. When the culture conditions are altered such that the follicles become atretic during the culture period, TNF still has the ability to stimulate progesterone synthesis but conversion to androstenedione and estradiol does not occur.

2. *Interleukin-1*

The two types of IL-1, IL-1α and IL-1β, exhibit only 26% sequence homology but bind the same receptor and appear to elicit the same biological responses. Both IL-1α and IL-1β have been detected in porcine and human follicular fluid but the IL-1β form is much more abundant and may be the physiologically relevant form. The ovary contains the complete IL-1 system, including IL-1, the IL-1 receptor type I and the IL-1 receptor agonist. IL-1 receptor has been localized to both granulosa and theca interstitial cells in the human ovary. IL-1β mRNA is not found in immature rat ovaries but it was observed in theca 6 hr after hCG; this suggested that the LH surge may induce thecal IL-1β gene expression in theca of dominant follicles. Other studies indicate that hCG does not directly increase IL-1β mRNA, but IL-1β mRNA is increased by IL-1 itself. Some studies show that low doses of estradiol or progesterone stimulate ovarian IL-1 production while at high concentrations the steroids inhibit production.

TICs produce IL-1 as ovulation nears, and IL-1 acts in an autocrine manner to inhibit LH- and IGF-I-stimulated steroidogenesis. IL-1β had no effect on basal or LH-stimulated androgen, similar to the effects of FGF.

Perfusion of rabbit ovaries with IL-1β can induce ovulation. Fertilization and development occurred after IL-1β induced ovulation but the rates were lower than with hCG-induced ovulation.

3. *Interleukin-2*

An ovarian cell type expressing IL-2 has not been identified; however, IL-2 and its receptor, possibly of immune cell origin, have been localized to the follicular fluid. Some studies indicate IL-2 stimulates thecal androgen production.

G. Prorenin

Prorenin, renin, and angiotensin II have been immunohistochemically localized to the theca in human and rat ovaries. Granulosal immunoreactivity was observed only in immediately preovulatory follicles and in granulosa of atretic follicles. Immunoreactivity, although weaker, was also seen in the stroma and the corpus luteum. Evidence for local expression of these peptides has been provided by Northern and *in situ* hybridizations. Northern analysis indicated the ovary expressed message for angiotensinogen and renin in the rat. *In situ* hybridization revealed the presence of renin mRNA in the corpus luteum of the rat. These mRNAs, along with the presence of angiotensin-converting enzyme in the ovary, allow for the local synthesis of angiotensin II.

Expression of the ovarian renin angiotensin system in the ovary is regulated by gonadotropins and angiotensin II has been shown to regulate granulosal and luteal steroidogenesis. Although a direct effect has not been described in the theca, data have shown that prorenin is produced by theca in response to LH. The cytokines TNF, TGF-α, and bFGF inhibited LH-stimulated but not basal prorenin production at a site distal to cAMP, and TGF-β increased LH-stimulated prorenin secretion. Angiotensin II binding sites on theca and were upregulated by LH or cAMP.

IV. BIOCHEMISTRY: SIGNAL TRANSDUCTION OF GONADOTROPINS AND GROWTH FACTORS

Signal transduction mechanisms functional in the TICs in response to the gonadotropins and growth factors described previously are numerous. The data suggest that the most important mechanism is the cAMP/protein kinase A pathway. Other pathways described in the TICs shown to regulate thecal functions include the tyrosine kinase pathway, the serine/threonine kinase pathway, and the protein kinase C (PKC) pathway (Fig. 3).

A. cAMP/Protein Kinase A Pathway

Theca cells express LH/hCG receptors on the plasma membrane and these receptors are a member of the G-protein-coupled family of cell surface receptors. The LH receptor is coupled to adenyl cyclase by Gsα, the stimulatory guanosine triphosphate-binding protein. Once generated in the cell cAMP complexes with protein kinase A (PKA). There is extensive evidence that LH mediates TIC function after binding to the membrane receptor and increasing intracellular cAMP. This is substantiated by experiments demonstrating the ability of cAMP analogs to mimic the effects of LH on thecal steroidogenesis. The genes for protein kinase A isoenzymes, type I and type II, are expressed in TICs. Further studies have indicated that both forms are functional in the theca. Steroidogenesis and expression of the steroidogenic enzymes P45017α, 3βHSD, and P450scc

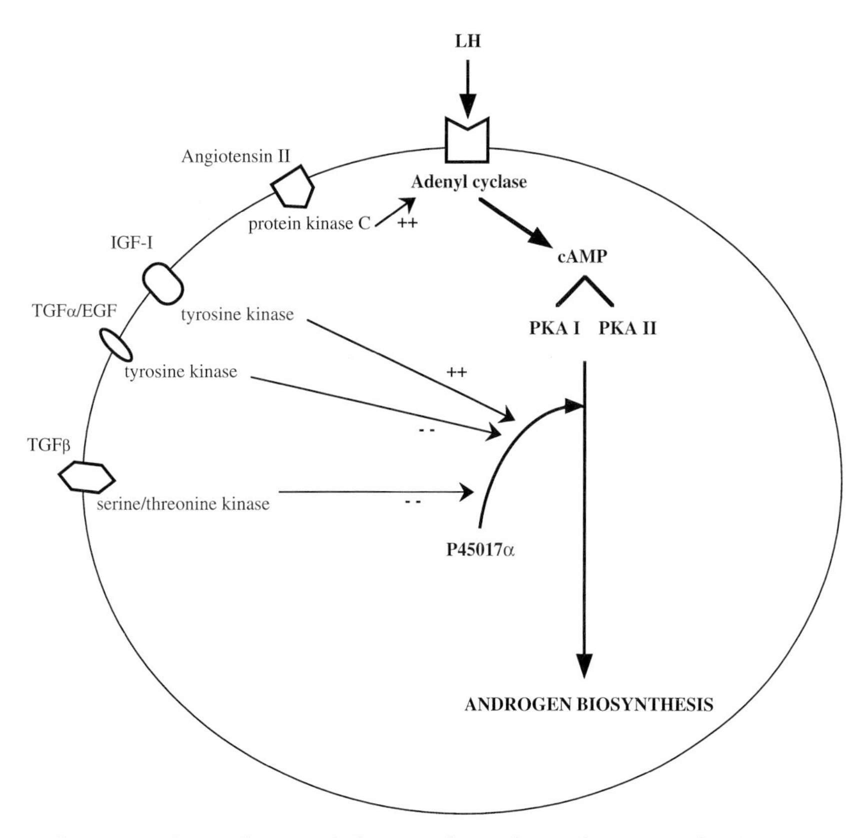

FIGURE 3 Schematic drawing outlining the growth factor and signal transduction mechanisms interacting with LH/cAMP-stimulated thecal steroidogenesis.

are regulated by analogs of cAMP that selectively activate the type I or type II form of protein kinase A, suggesting that both isoforms are involved in LH-mediated thecal function.

B. Tyrosine Kinase Pathway

The insulin/IGF-I and TGF-α/EGF families of proteins bind to cell surface receptors with intrinsic tyrosine kinase activity. As already discussed, both these growth factor families play a role in the development and function of the theca. The fact that IGF-I and EGF use the same signaling mechanism is somewhat puzzling because these growth factors have opposing effects on thecal steroid production. IGF-I increases LH-stimulated androgen production, whereas TGF-α inhibits androgen production.

Activation of the tyrosine kinase pathway alone has no effects on thecal steroidogenesis. However, in the presence of LH, or other activators of the cAMP/PKA pathway, IGF-I causes a synergistic stimulation of androgen production and at the same time progesterone synthesis increases. In contrast to IGF-I, TGF-α, in the presence of LH, causes inhibition of androgen production and decreases the sensitivity of TICs to LH stimulation of progesterone production. Although the mechanisms of these differential effects are unknown, it has been hypothesized that the substrate proteins of the receptor tyrosine kinases may be different or there may be differences in tyrosine phosphatase activity. Differential effects on steroidogenic enzyme gene expression by the two receptor tyrosine kinases have been observed. In the absence of LH, both IGF-I and TGF-α increase the expression of P45017α and P450scc enzyme proteins; however, P450scc remains in the inner mitochondrial membrane and is thus inactive. In addition, the tyrosine kinase pathway alone does not increase the transfer of cholesterol across the mitochiondrial membrane. Only when the cAMP/PKA pathway is concurrently activated does the activity of P450scc increase. Another disparity occurs at the levels of P45017α expression. IGF-I augments LH-stimulated P45017α expression, whereas TGF-α blocks LH-stimulated P45017α expression. It has been hypothesized that tyrosine kinase-mediated signals function to fine-tune the differentiated state of TICs by increasing androgen for subsequent aromatization or functioning to keep androgen production at normal levels. Interestingly, interstitial cells that do not express LH receptor respond to activation of the tyrosine kinase pathway with increased P450scc and P45017α expression and thus it may be these growth factors that provide the primary signals to initiate thecal differentiation.

C. Serine/Threonine Kinase Pathway

The serine/threonine kinase pathway is activated in TICs by TGF-β and activin. When protein kinase A, also a serine/threonine kinase, is activated by LH, thecal steroidogenesis is stimulated. In contrast, in the absence of LH, when the serine/threonine pathway is activated by TGF-β, there is no effect on basal thecal steroidogenesis. When both the cAMP/PKA and the serine/threonine kinase pathways are activated, serine/threonine kinase blocks the stimulatory action of cAMP/PKA in regard to androgen biosynthesis but progesterone synthesis is increased. This and other data indicate that TGF-β serine/threonine kinase has little effect on TIC gene expression but does specifically inhibit P45017α activity. The physiological significance of this action of TGF-β and its serine/threonine kinase activity are unclear. However, it has been hypothesized that TGF-β may function to maintain follicular dominance if TICs from one follicle suppress the androgen produced in a neighboring follicle and thus perpetuate its own dominance.

D. Protein Kinase C Pathway

It is well-known that phorbol esters inhibit LH-stimulated androgen production. In many cell types long-term exposure to phorbol esters depletes intracellular protein kinase C; thus in thecal cells PKC may be important for maintaining LH responsiveness. Angiotensin II, a known stimulator of PKC, increases the sensitivity of TICs to LH. The mechanism by which this occurs is unknown but may involve inhibition of the G_i protein or phosphorylation of adenylate cyclase.

See Also the Following Articles

ACTIVIN AND ACTIVIN RECEPTORS; GROWTH FACTORS; IGF (INSULIN-LIKE GROWTH FACTORS); INHIBINS; LH (LUTEINIZING HORMONE); OVULATION; PROLACTIN, OVERVIEW

Bibliography

Byskov, A. G., and Hoyer, P. E. (1994). Embryology of mammalian gonads and ducts. In *The Physiology of Reproduction* (E. Knobil and J. D. Neill, Eds.), pp. 487–540. Raven Press, New York.

Erickson, G. F., Magoffin, D. A., Dyer, C. A., and Hofeditz, C. (1985). The ovarian androgen producing cells: A review of structure/function relationships. *Endocr. Rev.* **6**, 371–399.

Espey, L. L., and Lipner, H. (1994). Ovulation. In *The Physiology of Reproduction* (E. Knobil and J. D. Neill, Eds.), pp. 725–780. Raven Press, New York.

Findlay, J. K., Xiao, S., Shukovski, L., and Michel, U. (1993). Novel peptides in ovarian physiology: Inhibin, activin, and follistatin. In *The Ovary* (E. Y. Adashi and P. C. K. Leung, Eds.), pp. 413–432. Raven Press, New York.

Magoffin, D. A. (1991). Regulation of differentiated functions in ovarian theca cells. *Sem. Reprod. Endocrinol.* **9**, 321–331.

Magoffin, D. A., and Erickson, G. F. (1994). Control systems of theca-interstitial cells. In *Molecular Biology of the Female Reproductive System* (J. K. Findlay, Ed.), pp. 39–65. Academic Press, Orlando, FL.

Magoffin, D. A., Weitsman, S. R., Hubert-Leslie, D., and Zachow, R. J. (1996). IGF-I regulation of thecal differentiation and function in developing ovarian follicles. In *Frontiers in Endocrinology* (D. LeRoith, Ed.), Vol. 19, The Role on Insulin-like Growth Factors in Ovarian Physiology, pp. 117–132. Ares-Serono Symposia, Rome.

Richards, J. S., Fitzpatrick, S. L., Clemens, J. W., Morris, J. K., Alliston, T., and Sirois, J. (1995). Ovarian cell differentiation: A cascade of multiple hormones, cellular signals, and regulated genes. *Rec. Prog. Horm. Res.* **50**, 223–254.

Vinatier, D., Dufour, Ph., Tordjeman-Rizzi, N., Prolongeau, J. F., Depret-Moser, S., and Monnier, J. C. (1995). Immunological aspects of ovarian function: Role of cytokines. *Eur. J. Obstet. Gynecol. Reprod. Biol.* **63**, 155–168.

Zachow, R. J., and Magoffin, D. A. (1997). Ovarian androgen biosynthesis: Paracrine/autocrine regulation. In *Androgen Excess Disorders in Women* (R. Azziz, D. Dewailly, and J. E. Nester, Eds.). Lippincott-Raven, New York.

Theca Cell Tumors

Richard E. Leach and Nilsa C. Ramirez

Wayne State University School of Medicine

GLOSSARY

ascites Accumulation of a serous fluid in the peritoneal cavity.

endometrium The tissue lining the internal cavity of the uterus whose superficial layer is sloughed at menses.

hyperplasia A pattern of increased cell proliferation resulting in an increased number of cells that may have varying degrees of atypical morphology.

leiomyoma A benign smooth muscle tumor of the uterus.

ovarian stroma The stroma of the ovarian cortex, composed of tightly packed connective tissue cells that surround the follicles and form the theca.

peritoneum The single cell mesothelial layer which covers the entire interior surface of the abdomen and its contents.

sex steroid hormones Estrogen, testosterone, and progesterone which are secreted by the ovary during different intervals of the menstrual cycle.

Encyclopedia of Reproduction VOLUME 4

Theca cell tumors (TCTs) or thecomas are invariably benign and derived from specialized ovarian stromal cells. The latter are present around the periphery of developing follicles within the ovary. The TCTs which originate from these specialized cells can make a full range of sex steroid hormones.

I. CLINICAL CHARACTERISTICS

Theca cell tumors (TCTs) were first described by Loffler and Priesel in 1932 and are included in a classification of ovarian tumors named the sex cord stromal tumors (Table 1). TCTs comprise <1% of ovarian tumors and can be found at any age but have a peak incidence during the sixth decade of life, later than other sex cord stromal tumors. Less than 10% of patients present before age 30 and 80% are postmenopausal, with 60–70% presenting with postmenopausal vaginal bleeding. This is evidence of resumption of estrogen stimulation of endometrial growth. Other manifestations of TCT estrogen production include hyperplasia or adenocarcinoma of the endometrium and may influence uterine leiomyoma and polyp growth. One pure TCT has been documented to cause virilization due to increased androgen production.

These tumors are unilateral 97% of the time and equally distributed between right and left sides. They range in size from 1 to 40 cm and are usually solid, rubbery in consistency, and typically yellow in color when sectioned due to the high fat content (Fig. 1). TCTs are encountered early in their natural course, with nearly 97% diagnosed at the earliest possible stage (stage 1a). That is, the tumor is confined to one ovary, there are no protrusions over the ovarian surface, and there are no ascites.

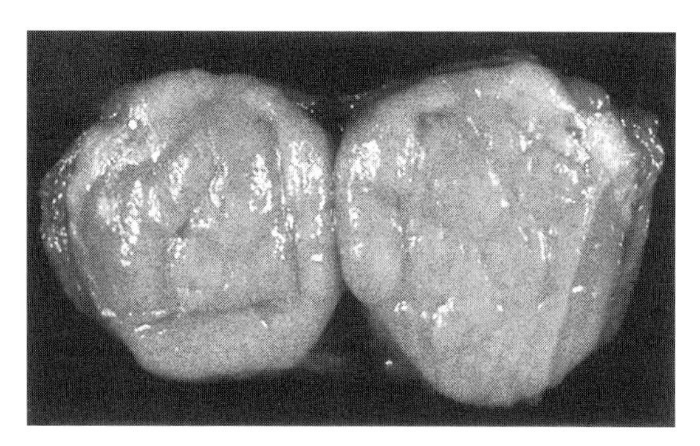

FIGURE 1 Gross view of ovarian theca cell tumor. Sectioned surface reveals yellow, firm, and dense tumor (provided by Dr. Nilsa Ramirez).

TABLE 1
Classification of Sex Cord Stromal Cells

Sex cord stromal tumors
Granulosa cell tumor
Thecoma–fibroma group
Thecoma
Fibroma
Stromal tumors
Sertoli stromal cell tumors
Sertoli cell tumor
Leydig cell tumor
Sertoli–Leydig cell tumor

The use of transvaginal sonography (TVUS) has enhanced our ability to evaluate pelvic masses due to characteristic sonographic findings with different tumor types. Pure TCTs are highly correlated with the presence of diffusely hypoechoic ovarian masses with no posterior echo enhancement. Although TVUS findings are nonspecific, they can provide useful information to detect these rare ovarian tumors preoperatively. Luteinized TCTs are uncommonly associated with sclerosed peritonitis, which is the result of proliferation of submesothelial spindle cells or myofibroblasts below the peritoneum. This results in thickening of the peritoneum which is most apparent in the small intestine and omentum. The sclerosed peritoneum is responsible for the small bowel obstruction and pain, the most common presenting complaints. The pathogenesis of the peritoneal reaction is currently unknown but appears dependent on the presence of luteinized TCTs and mediated by some secretory product.

The treatment of TCTs is unilateral surgical removal in young women of childbearing age when the tumor is confined to one ovary. Endometrial biopsy should be performed due to the association of endometrial adenocarcinoma and hyperplasia with TCTs. Total abdominal hysterectomy and bilateral salpingooopherectomy is recommended for TCTs discovered outside the confines of the involved ovary or in postmenopausal women. Tumor recurrences are rare, supporting the belief that TCTs are typically benign.

II. MICROSCOPIC FEATURES

Pure ovarian TCTs feature mainly round to oval lipid-laden cells, with little or absent nuclear atypia, and absent or rare mitoses. They may be associated with hyaline plaques and varying numbers of spindle-shaped fibroblasts (Fig. 2). Electron microscopy (EM) confirms the presence of smooth endoplasmic reticulum (SER), evidence of steroid synthesis, in oval tumor cells of women with elevated estrogen levels. Fat droplets are found in and around cells (Fig. 3). Reticulum stain often demonstrates reticulum fibers surrounding individual cells, a pattern which is absent in granulosa cell tumors. Often there is an associated stromal cell hyperplasia in the same and/or uninvolved ovary.

TCTs may also have a histological pattern in which large eosinophilic cells resemble luteinized theca cells called luteinized TCTs. In one series of 46 patients with ovarian luteinized TCTs, 50% of patients had signs of estrogen action, 39% had no hormonal activity, and 11% had clinical signs of androgen excess. The higher frequency of androgen activity of luteinized TCTs is contrasted by the rarity of this activity in simple TCTs. They also appear to be more common in a younger age group.

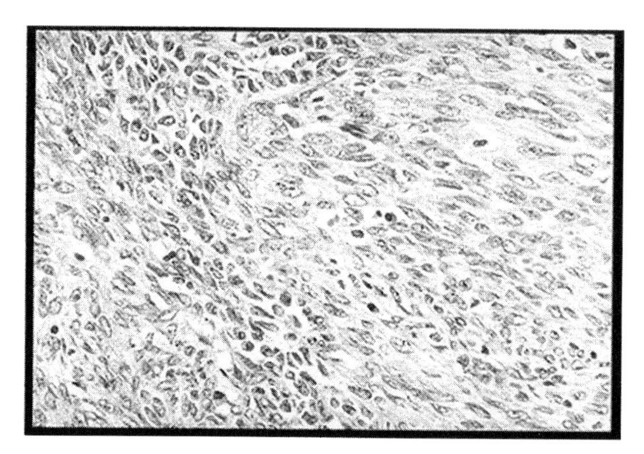

FIGURE 2 Cross section of theca cell tumor with H&E stain. Magnification, ×400 (provided by Dr. Takuja Ichihashi).

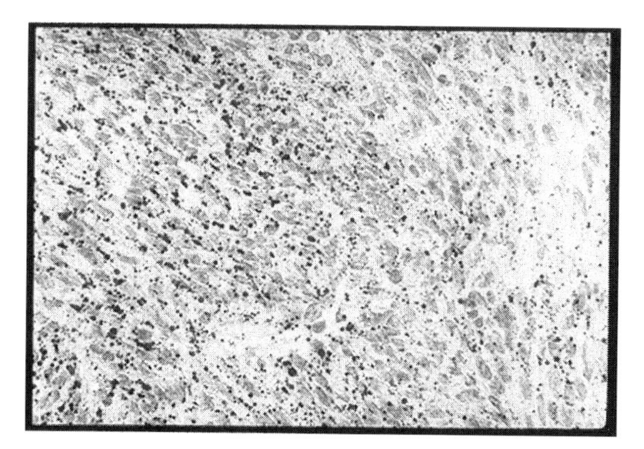

FIGURE 3 Cross section of theca cell tumor with Sudan red fat stain. Dark stained structures represent fat droplets. Magnification, ×400 (provided by Dr. Takuja Ichihashi).

III. HORMONAL ACTIVITY

Despite the clinical evidence for hormone production by TCTs, little is known about the hormonal levels or the specific cell source within the tumor. In one study of 12 patients with TCTs, 7 had significantly elevated urinary estrogen and five did not. This is consistent with the clinical finding of postmenopausal vaginal bleeding, a sign of estrogen action, in 60–70% of TCTs. One strategy to identify the cells within the tumor responsible for sex steroid synthesis is to determine the expression of enzymes required for the various steps of the synthetic process. P450 side chain cleavage enzyme and 17α-hydroxylase, which are required enzymes for sex steroid synthesis, are expressed in TCTs determined by immunolocalization. However, aromatase, which is required to convert androgen to estrogen, is not. This expression pattern of enzymes by TCTs would promote elevated androgen and not estrogen synthesis. 3β-Hydroxysteroid dehydrogenase (3β-HSD) is an enzyme that catalyzes the essential step of the steroid formation through the synthesis of progesterone and androstenedione. Using immunolocalization, 3β-HSD is identified only in tumor cells with abundant cytoplasm and not in spindled tumor cells with small to moderate amounts of cytoplasm. This is consistent with the EM findings of SER in large oval cells. Taken together, it appears that the large eosinophilic oval cells primarily secrete sex steroids. Whether these cells primarily secrete androgen, which is converted to estrogen in other cells within the ovary, remains to be determined.

IV. PATHOGENESIS

Trisomy 21 is a nonrandom chromosomal abnormality that has been reported in benign solid tumors, including leiomyomas, endometrial polyps, and sex cord stromal tumors, which include TCTs. In one study, 9 of 10 thecoma–fibromas exhibited Trisomy 21. Furthermore, the tumors were characteristically benign both histologically and clinically, proving that Trisomy 21 detection is of little prognostic usefulness. However, the high frequency with which it occurs in this and other types of tumors argues for its crucial role in the early mechanism of benign cell proliferation.

Further insight into the pathogenesis of TCTs is demonstrated by transplantation of ovarian grafts under the splenic capsule in rats. This maneuver allows for hepatic conversion of estrogen to its metabolites, thereby releasing the pituitary gland from negative feedback. The lack of estrogen inhibition results in increased synthesis and secretion of follicle-stimulating hormone (FSH) and luteinizing hormone (LH) from the pituitary. Both FSH and LH levels and pulse frequency are exquisitely modulated to control ovarian follicle development. Using the transplantation method, animals with elevated FSH and LH for >90 days developed TCTs, with an increase in tumor proliferation independent of FSH and LH after 240 days. Animals with elevated levels of FSH only, due to incomplete loss of estrogen negative feedback, were tumor free. Furthermore, inhibiting both LH and FSH synthesis with gonadotropin-releasing hormone agonist resulted in tumor-free animals. These experiments implicate elevated LH levels for inducing TCTs in this model system. Similar to their human counterparts, the TCTs in this model were benign.

See Also the Following Articles

Endometrium; Leiomyoma; Theca Cells; Tumors of the Female Reproductive System

Bibliography

Clement, P. B., Young, R. H., Hanna, W., and Scully, R. E. (1994). Sclerosing peritonitis associated with luteinized thecomas of the ovary: A clinicopathological analysis of six cases. *Am. J. Surg. Pathol.* **18**, 1–13.

Jèger, W., Dittrich, R., Recabarren, S., Wildt, L., and Lang, N. (1995). Induction of ovarian tumors by endogenous gonadotropins in rats bearing intrasplenic ovarian grafts. *Tumor Biol.* **16**, 268–280.

Klemi, P. J., and Gronroos, M. (1979). An ultrastructural and clinical study of theca and granulosa cell tumors. *Int. J. Gynaecol. Obstet.* **17**, 219–225.

Russell, P., and Bannatyne, P. (1989). *Thecoma–Fibroma Group of Tumors*. Churchill Livingstone, London. [See Chap. 34]

Sasano, H., Mason, J. I., Sasaki, E., Yajima, A., Kimura, N., Namiki, T., Sasano, N., and Nagura, H. (1990). Immunohistochemical study of 3β-hydroxysteroid dehydrogenase in sex cord-stromal tumors of the ovary. *Int. J. Gynecol. Pathol.* **9**, 352–362.

Schats, R., and Schoemaker, J. (1994). *Ovarian Endocrinopathies. Studies in Profertility Series*, Vol. 2. Parthenon , New York.

Young, R. H., and Scully, R. E. (1994). Sex cord-stromal and steroid cell ovarian tumors. In *Blaunstein's Pathology of the Female Reproductive Tract* (R. J. Kurman, Ed.), 4th ed., pp. 783–847. Springer-Verlag, New York.

Thyroid Hormones, in Subavian Vertebrates

David O. Norris
University of Colorado at Boulder

GLOSSARY

corticotropin A pituitary glycoprotein hormone that stimulates the adrenal cortical tissue or its homolog (interrenal) to secrete glucocorticosteroids (cortisol and/or corticosterone).

corticotropin-releasing hormone A hypothalamic neuropeptide that stimulates release of corticotropin from the pituitary gland and that also may stimulate release of thyrotropin from the pituitary.

goitrogen A chemical that lowers circulating thyroid hormones by inhibiting thyroid gland function.

gonadotropin A pituitary glycoprotein hormone that stimulates the gonads to make gametes and to secrete reproductive steroid hormones (estrogens, androgens, and progestogens).

hypophysectomy The surgical removal or chemical inactivation of the pituitary gland (hypophysis).

hypothalamus The region of the brain controlling pituitary function.

pituitary gland (hypophysis) A gland that secretes hormones which stimulate other endocrine glands to secrete hormones.

radioimmunoassay A technique employing radioisotopes that allows scientists to precisely measure hormone levels in blood or other body fluids.

thyroidectomy The surgical removal or destruction of the thyroid by radioactive iodine or inactivation by chemical means.

thyrotropin A pituitary glycoprotein that stimulates thyroid gland to secrete thyroid hormones.

thyrotropin-releasing hormone A hypothalamic neuropeptide responsible for stimulating thyrotropin release in birds and mammals; possibly active in reptiles and amphibians.

thyroxine The major circulating form of thyroid hormone.

triiodothyronine The more active circulating form of thyroid hormone.

vitellogenesis Estrogen-dependent synthesis of yolk protein precursors by the liver.

All subavian vertebrates have thyroid glands. The jawed subavian vertebrates [including a great array bony fishes (sturgeons, teleosts, lungfishes, etc.), elasmobranchs (sharks, rays, and skates), ratfishes (chimeras), amphibians, and reptiles] all have well-developed thyroid glands throughout development, subadult, and adult life. However, among the jawless lampreys, which are members of the agnathan fishes (*a*, "without"; *gnathos*, "jaw"), the thyroid gland first differentiates when the larval lamprey undergoes metamorphosis to the adult. The thyroid system of lampreys appears to inhibit the transformation of the larva to the adult body form (metamorphosis), which is a marked contrast to its stimulatory roles reported for metamorphosis of fishes and amphibians.

I. THE VERTEBRATE THYROID SYSTEM

The thyroid glands of subavian vertebrates, for the most part, resemble their mammalian counterparts in both structure and function. Typically, the thyroid gland consists of units called follicles. Each follicle consists of a single layer of thyroid cells organized into hollow spheres that are filled with a proteinous fluid called colloid. Thyroid hormones are synthesized in the thyroid follicular cells and are stored in the colloid until released into the blood. They are formed by combining two molecules of the amino acid tyrosine in a unique manner with the inclusion of three or four iodine atoms attached to the finished product. There are two forms in which thyroid hormones occur: Thyroxine (T_4) contains four iodine atoms and triiodothyronine (T_3) has only three iodine atoms but is considered to be the more active form. T_3 enters cells more rapidly and binds more strongly to thyroid hormone receptors in target cells than does T_4. Thyrotropin (TSH), a glycoprotein hormone secreted by the pituitary gland (also called the hypophysis), is responsible for stimulating synthesis, storage, and release of thyroid hormones from the follicles. Most of the circulating T_3 actually is made by peripheral deiodination of T_4 by enzymes located in the liver.

Release of TSH from the pituitary gland in reptiles appears to be controlled by thyrotropin-releasing hormone (TRH), a neuropeptide made in the hypothalamus of the brain, as it is in mammals and birds. Secretion of TRH is affected by seasonal environmental factors often related to controlling reproductive cycles. TSH release among amphibians, however, appears to be under the influence of another hypothalamic neuropeptide, corticotropin-releasing hormone (CRH), which also stimulates release of the polypeptide hormone adrenocorticotropin (ACTH) from the pituitary. In jawed fishes, the regulation of TSH release is unclear, but some studies suggest that CRH may stimulate whereas TRH may inhibit its release.

As in birds and mammals, the hypothalamic–pituitary–thyroid (HPT) axis of all vertebrates is implicated in the regulation of development, metabolism, growth, and activity levels. In subavian vertebrates, thyroid hormones regulate additional processes, such as osmoregulation in teleostean fishes, molting in amphibians and reptiles, and temperature regulation in reptiles. Many actions of thyroid hormones involve synergisms with the actions of other hormones. Furthermore, the activity of the HPT axis is affected by seasonal environmental factors including photoperiod and temperature. Reproductive cycles are also under seasonal influences and can be strongly affected by factors such as nutrition, metabolism, growth, and temperature. Because of the widespread nature of thyroid hormone effects in vertebrates and their actions on factors that may alter reproductive events (e.g., nutrition, growth, and temperature regulation), thyroid hormones can have both important direct effects on reproductive hormone actions and indirect effects through their more general actions. Consequently, it has been difficult to ascribe specific roles for thyroid hormones in reproduction per se.

Assessment of endogenous thyroid activity classically has been accomplished by examining thyroid glands histologically, by determination of the uptake of radioactive iodine by thyroid follicles, or by measuring circulating levels of thyroid hormone. The height of the thyroid follicular epithelium is a direct reflection of how much the thyroid gland is being stimulated by TSH. Thyroid follicles or the region of the animal containing them is removed, preserved, embedded in paraffin, and sectioned very thinly, and the sections are applied to a glass slide for staining and later microscopic examination and measurement. Radioiodide uptake is determined by measuring the radioactivity of the thyroid or thyroid region of the animal following administration of a dose of radioactive iodine. Increased thickness of the thyroid follicle epithelium or increased uptake of radioactive iodine indicates increased stimulation of the thyroid by TSH. Circulating levels of thyroid hormones can be determined from a small blood sample using a technique called radioimmunoassay. This technique allows the investigator to precisely measure the amount of each thyroid hormone and compare levels to different environmental and/or reproductive states. Radioimmunoassay is a more sensitive indicator of thyroid state than either histology or radioactive iodine uptake. Reproductive state typically is assessed by observation of mating or spawning, determination of time for oviposition or birth, measurement of the size of gonads or of oocytes, etc., or radioimmunoassay of blood levels of pituitary and/or gonadal steroids.

Although past studies have linked thyroid hormones to reproductive function in subavian vertebrates, most of the data are correlative, e.g., comparisons of thyroid hormone levels of intact animals with stages of reproductive development (Table 1). Additional studies are needed to verify cause–effect relationships between thyroid state and reproductive functions. Once a cause–effect relationship is estab-

TABLE 1
Selected Experimental Results Relating Thyroid and Reproduction in Subavian Vertebrates

Vertebrate	*Treatment/observation*	*Effect on reproduction*
Agnathan fishes		
Sea lamprey	Elevated thyroid hormones	Occurs during spermiation and at ovulation
Jawed fishes		
Sharks	Thyroidectomy	Prevents annual gonadal recrudescence[a] by blocking vitellogenesis or incorporation of yolk into oocytes
	Stimulated thyroid histology	Coincides with sexual maturation
Teleosts	Elevated thyroid hormones in blood	During gonadal development
	Decreased thyroid hormones in blood	At spawning
	Goitrogen exposure	Inhibits gonadal recrudescence; can cause gonadal atrophy in both sexes
	Presence of goiters in salmon from Great Lakes	Associated with low egg production; low gonadal steroid levels in blood
	Thyroid hormone treatments	No effect or mild enhancement of gonadal maturation
Amphibians	Surgical or chemical thyroidectomy of frogs	Prevents ovulation
	Treatment with thyroxine of thyroidectomized frogs	Permits ovulation
	Treatment of anuran liver with thyroid hormones	Enhances action of estrogen on vitellogenesis
	Thyroid histology	Not correlated with reproductive activities
	Goitrogen treatment of tadpoles of some anurans	Prevents testicular development; may allow sex reversal to females
	Elevated thyroxine in the blood of salamander	Corresponds to period of gonadal growth in males and females
Reptiles	Excess or insufficient thyroid hormones	Impaired ovarian growth; loss of steroid-secreting cells in testes
	Thyroxine treatment	Causes atrophy of steroid-secreting cells in testes
	Decreased thyroid hormones in blood and thyroid gland histology	Correlated with peak reproductive activity and high androgen levels
	Elevated thyroxine levels in blood	Decreased androgen levels in males

[a] Normal regrowth of the gonad during the next breeding season.

lished, it will be necessary to identify the mechanisms through which these effects are manifest. The following sections provide an overview of our limited knowledge in this field.

II. THYROIDS AND REPRODUCTION IN FISHES

Data concerning the relationship of thyroid hormones to reproduction in lampreys are scarce. A reduction in thyroid function occurs at metamorphosis and it is not clear how thyroid hormones might be related to reproduction in these fishes. Thyroid hormone levels are elevated in spawning sea lampreys during spermiation and ovulation, although earlier studies using thyroid histology (a less sensitive measure of thyroid gland activity) suggest low thyroid activity at this time. Detailed studies are needed to understand this relationship. Studies of the jawless hagfishes also would be instructive since hagfishes are closely related to lampreys (although they lack bony vertebrae protecting the dorsal nerve cord as in vertebrates), and studies might provide insight into the evolution of this relationship.

Activity of the thyroid gland of the cartilaginous

fishes (sharks, rays, skates, and ratfishes) is correlated positively with reproduction. Histological evidence of increased thyroid gland activity is associated with sexual maturation in sharks. Surgical thyroidectomy of sharks prevents oocyte growth although the mechanism of this action is unknown. Much more work is needed in this group of fishes before a thorough understanding of the role of thyroid hormones in reproduction will be possible.

In most teleostean fishes (a group of more than 20,000 species comprising the majority of all living fish species), the thyroid follicles are not organized into a discrete gland but rather are scattered mostly among the connective tissue surrounding the ventral aorta anterior to the heart, between the second and fourth aortic arches. Discrete thyroid glands surrounded by a connective tissue covering as in tetrapods have been described for only a few teleosts (e.g., tuna and Bermuda parrot fish). Consequently, thyroidectomy (surgical removal of the thyroid glands) has rarely been employed to examine thyroid relationships to reproduction. Because the thyroid gland selectively accumulates radioactive forms of iodine, the thyroid can be destroyed following administration of a large dose of radioactive iodine. This process, called radiothyroidectomy, rarely has been employed in fishes, in part because of the hazards of working with large amounts of radioactive iodine and because radioactive iodine is also incorporated into the ovary (especially in egg-laying species), which can damage reproductive development. Surgical removal of the pituitary gland (a process called hypophysectomy) has occasionally been employed to render a fish hypothyroid. Hypophysectomy, however, removes not only the source of thyroid gland stimulation (i.e., TSH) but also several other important hormones, including the gonad-stimulating hormones (the gonadotropins, such as follicle-stimulating hormone and luteinizing hormone), prolactin (an important osmoregulatory hormone in freshwater fishes), and ACTH (which controls the stress response). Consequently, most investigations have employed intact or hypophysectomized fishes and have relied on correlations between reproductive activities and circulating hormone levels. The effects of treatments to intact fishes by injections of thyroid hormones or immersion of fishes in solutions of thyroid hormones also have been reported.

Some studies have employed application of artificial chemical inhibitors of thyroid hormone synthesis such as thiourea or propylthiouracil (PTU). Such chemicals often are called goitrogens because of their ability to cause an abnormal enlargement of the thyroid gland known as a goiter. Goitrogens may be employed with or without thyroid hormone replacement therapy. Goiter formation is caused by excessive release of TSH in the absence of thyroid hormones and consequent overstimulation of the thyroid cells. This activation of the hypothalamus and pituitary also occurs following surgical thyroidectomy or radiothyroidectomy. Normally, high levels of thyroid hormones inhibit TSH release and prevent excessive amounts of thyroid hormones from being secreted. This action of thyroid hormones to regulate their own production is termed feedback.

Interpretation of studies using goitrogens is difficult. In some cases, the goitrogen (e.g., thiourea) may produce other toxic effects which are not seen following surgical thyroidectomy. Other goitrogens (e.g., PTU) may affect multiple sites in the HPT axis (e.g., thyroid hormone synthesis in the thyroid and conversion of T_4 to T_3 in the liver). Although newer goitrogens have been developed that have fewer side effects than thiourea and PTU (e.g., methimazole and carbimazole) and liver deiodinase blockers (e.g., ipodate and amiodarone), these have been little employed to study the relationship between thyroid hormones and reproduction.

The possible roles of thyroid hormones in the development and reproduction of teleostean fishes have been reviewed numerous times during the past 20 years, but relatively little new information has appeared except in relation to early development. In bony fishes, thyroid hormones are sequestered by growing oocytes and are important in early development. Furthermore, the thyroid is involved in the regulation of both metabolism and ionoosmotic balance, and circulating levels of thyroid hormone are strongly influenced by temperature. It is important to separate these seasonally related involvements of thyroid hormones from their strictly reproductive roles; however, such separation is difficult to achieve.

Most studies have employed relatively toxic thyroid inhibitors such as thiourea or PTU, observed effects of thyroid hormone injections, or simply measured circulating thyroid hormone levels in relation to ongoing reproductive activities. In general, thyroid hormones enhance gonadal maturation and vitellogenesis (estrogen-dependent production of yolk proteins by the liver and their incorporation into growing oocytes in the ovaries) by working in concert with reproductive hormones. Treatments with goitrogens (chemical thyroidectomy) result in gonadal atrophy of both male and female teleosts and can prevent regrowth of the gonads (recrudescence) prior to a second breeding season. Creation of hypothyroid fish with specific liver deiodinase inhibitors such as ipodate confirms that T_3 enhances the actions of gonadotropins on ovarian growth and estrogen secretion. Indeed, some observations indicate that thyroid hormones are essential for these processes to occur at all in some species. Numerous correlations have been made between seasonal reproductive events and heightened thyroid activity, although it has been difficult to verify a cause–effect relationship.

III. THYROIDS AND REPRODUCTION IN AMPHIBIANS

Paired thyroid glands are typical of amphibians (like other tetrapods), making surgical thyroidectomy feasible. Still, most studies relating thyroid to reproduction in these groups have also relied on correlations between thyroid hormone levels and reproductive events, hormone treatment of intact or hypophysectomized animals, and the effects of goitrogens.

Some studies suggest thyroid hormones do not enhance reproduction in amphibians and may even be antagonistic. Measurements of thyroid histology, thyroid activity, and circulating thyroid hormones either do not correlate with gonadal events or are inversely related. In contrast, a few studies report an enhancement of reproductive events by thyroid hormones. For example, thyroid hormones enhance estrogen-induced vitellogenesis in the frog *Xenopus laevis*, and higher thyroxine levels were observed in the tiger salamander when the gonads were growing. Because of the various methods employed, it is not clear whether these are simply related to species differences or to other factors that were not controlled adequately (e.g., photoperiod, temperature, and nutrition). Although a positive relationship between circulating thyroidal and gonadal hormones is reported for *Bufo japonicus* and *Bufo bufo*, a negative relationship is suggested in other species of the same genus (*Bufo regularis* and *Bufo viridis*).

Thyroid gland activation occurs just prior to metamorphosis but is not necessarily accompanied by increased gonadal development. In various species of frogs and toads, sexual differentiation and gonadal development may begin before, during, or after metamorphosis. There is evidence to suggest that thyroid hormones may be essential for gonadal differentiation in some species.

IV. THYROIDS AND REPRODUCTION IN REPTILES

Although there are relatively few studies that have investigated the roles of thyroid hormones in reptiles, some of these reports show a positive relationship between thyroid activity and reproduction. In lizards, snakes, and turtles, seasonal cycles and thyroid activity (usually based on histological examination of thyroid glands in intact animals) are correlated positively with reproductive events, including spermatogenesis, ovulation, and mating. Surgical thyroidectomy decreases spermatogenesis and androgen secretion in male lizards and causes follicular atresia in females. Furthermore, surgical thyroidectomy causes premature expulsion of eggs in viviparous lizards. Data on crocodilians and the primitive tuatara (*Sphenodon*) are lacking.

In contrast, several studies have found that there are either no correlations between thyroid activity and reproduction or that these events are negatively correlated. Furthermore, there is no evidence for a relationship of thyroid function with pregnancy or birth in viviparous species. Some studies indicate that thyroid hormone levels in the blood are highest during the nonbreeding period and that thyroid hormones are high when gonadal steroids are low and vice versa. Obviously, more species need to be examined employing experimental manipulations and with more precise measurements of thyroid and reproductive activities before a definitive picture of their relationship in reptiles will be achieved.

See Also the Following Articles

AMPHIBIAN REPRODUCTION, OVERVIEW; FISH, MODES OF REPRODUCTION; REPTILIAN REPRODUCTION, OVERVIEW

Bibliography

Blaxter, J. H. S. (1988). Pattern and variety in development. In *Fish Physiology, Vol. XI. The Physiology of Developing Fish, Part A. Eggs and Larvae* (W. S. Hoar and D. J. Randall, Eds.), pp. 1–58. Academic Press, San Diego.

Dodd, J. M. (1983). Reproduction in cartilaginous fishes (Chondrichthyes). In *Fish Physiology: Reproduction* (W. S. Hoar, R. J. Randall, and E. M. Donaldson, Eds.), Vol. 9a, pp. 31–96. Academic Press, New York.

Dodd, J. M., and Sumpter, J. P. (1984). Fishes. In *Marshall's Physiology of Reproduction. Vol. 1. Reproductive Cycles of Vertebrates* (G. E. Lamming, Ed.), pp. 1–126. Churchill Livingston, Edinburgh, UK.

Leatherland, J. F. (1987). Thyroid hormones and reproduction. In *Hormones and Reproduction in Fishes, Amphibians, and Reptiles* (D. O. Norris and R. E. Jones, Eds.), pp. 411–431. Plenum, New York.

Licht, P. (1984). Reptiles. In *Marshall's Physiology of Reproduction. Vol. 1. Reproductive Cycles of Vertebrates* (G. E. Lamming, Ed.), pp. 206–282. Churchill Livingston, Edinburgh, UK.

Lynn, W. G. (1970). The thyroid. In *The Biology of the Reptilia* (C. Gans, Ed.), Vol. 3, pp. 201–234. Academic Press, New York.

Norris, D. O. (1997). *Vertebrate Endocrinology*, 3rd ed. Academic Press, San Diego.

Ticks

see Chelicerate Arthropods

Tocolytic Agents

Martha E. Rode and George A. Macones
University of Pennsylvania Health System

GLOSSARY

preterm delivery Delivery between 20 and 37 completed weeks of gestation.

preterm labor Uterine contractions accompanied by cervical change at <37 completed weeks of gestation.

tocolysis The inhibition of uterine contractions.

Preterm labor is a major problem in the United States. One of the mainstays of the current treatment of preterm labor is the pharmacologic inhibition of preterm labor (tocolysis).

I. INTRODUCTION

A. Incidence and Importance of Preterm Labor and Delivery

Preterm delivery, that occurring between 20 and 37 completed weeks of gestation, is one of the most pressing problems facing obstetricians worldwide. Reports of the incidence of preterm delivery vary somewhat by the population studied and by the methods of investigation and ascertainment employed. Recent data from the United States March of Dimes initiative suggests a risk of preterm delivery of approximately 9.6%. This is based on a large multicenter evaluation of well-dated pregnancies and represents a reasonable estimate of the incidence of preterm delivery in the United States. Based on these data, there are approximately 384,000 preterm deliveries in the United States per year. While there are many possible etiologies for a preterm delivery, "idiopathic premature labor" (i.e., preterm labor that begins without an identifiable etiology such as spontaneous rupture of amniotic membranes) is one of the most common and most vexing, accounting for approximately 50% of preterm deliveries.

Preterm delivery is an important issue from a societal standpoint not only because of its high incidence but also because of the significant associated morbidity and mortality. In fact, preterm delivery is the most common cause of perinatal death in nonanomalous newborns. Furthermore, it has been well documented that both morbidity and mortality increase with decreases in gestational age at delivery. Even at gestational ages at which survival is relatively satisfactory (i.e., >30 weeks), significant morbidity can still develop in infants delivered prematurely. For example, Robertson *et al.* reported that at 30 weeks gestation, the risk of respiratory distress syndrome is approximately 50%, necrotizing enterocolitis 11%, and intraventricular hemorrhage 5%. Thus, preterm delivery is a major cause of morbidity and mortality in the United States.

Because of the frequency of preterm labor as well as the neonatal complications that can arise should a preterm birth occur, a great deal of investigative effort has surrounded the prevention and treatment of idiopathic preterm labor. Among the most common modalities for dealing with idiopathic preterm labor is the use of pharmacological agents designed to arrest preterm labor once it has begun (so-called "tocolysis"). The ultimate goal of tocolysis is the prevention of perinatal deaths and major neonatal morbidities due to prematurity. However, an intermediate goal has become delay of delivery for 24–48 hr—the time necessary to administer and receive maximal benefit from antenatal corticosteroids, which have been documented to improve perinatal outcome. Despite the evaluation of multiple candidates in past decades, there are still many unanswered questions about the relative efficacy and safety of these agents.

B. Methodological Considerations of Studies of Tocolytics

The search for the ideal tocolytic has been complicated by several factors. The first problem is how best to define preterm labor. Classically, preterm labor is defined as documented uterine contractions at 20–37 weeks of gestation with either ruptured or intact membranes, accompanied by documented cervical change. This definition, used in many studies of preterm labor, undoubtedly results in the inclusion of many women who do not have preterm labor (e.g., those with minimal cervical change). In fact, in the trials of β agonists for preterm labor, up to 50% of patients treated with placebo actually delivered at term. This diagnostic misclassification, in effect, lowers the statistical power of the efficacy studies, thereby biasing the results toward the null (i.e., no effect). Clearly, a future challenge in perinatal medicine is to better define preterm labor, perhaps through the use of biochemical markers.

A second important consideration when evaluating the studies of tocolysis is the primary endpoint for these trials. While it may seem apparent that the ultimate goal of tocolysis is to reduce death and morbidity from prematurity, few studies have actually had the statistical power for the evaluation of such uncommon outcomes. Instead, most trials use delivery delay as the primary endpoint. While it is generally believed that even short delivery delays with tocolysis are beneficial (because of steroid use), this hypothesis is largely untested (given that virtually no tocolytic has been studied in combination with steroid use).

II. β MIMETICS

A. Mechanism of Action

The β mimetics, such as ritodrine (the only Food and Drug Administration-approved tocolytic), terbutaline, and isoxuprine, have been the most commonly used parenteral tocolytics in the United States over the past three decades. β Mimetics are structurally related to catecholamines. Importantly, there are two types of β receptors (β-1 and β-2) to which these agents bind. β-1 receptors are located in the small intestine, heart, and adipose tissue. Their stimulation leads to increased cardiac automaticity, positive chronotropic and inotropic effects, and elevated free fatty acids. The β-2 receptors are located in the smooth muscle of the uterus, blood vessels, and bronchioles. Accordingly, their stimulation leads to uterine relaxation, vasodilatation, and bronchodilation. Activation of these receptors leads to an elevation in cyclic AMP (cAMP), mediated through adenylate cyclase. In turn, increased levels of cAMP prevent myosin light-chain kinase (MLCK) activity through both decreased phosphorylation and inhibition of release of stored intracellular calcium. Because MLCK phosphorylation is the key step in the actin–myosin interaction, it is thought that uterine contractions are thus inhibited. Although the tocolytics of this class commonly in use are β-2 selective, they do retain some β-1 activity, which accounts for their side effect profile. An ideal β agonist tocolytic would be completely β-2 selective; however, no such agent exists. Thus, the benefit of uterine relaxation is often accompanied by side effects attributable to β-1 activity.

B. Efficacy

Many randomized, placebo-controlled studies have been performed evaluating the efficacy of the β mimetics as acute tocolytics. Overall, these placebo-controlled studies demonstrate that β mimetics are effective in the delay of delivery for 24–48 hr. This is borne out by a recent meta-analysis in which 16 placebo-controlled trials using β mimetics for the acute treatment of preterm labor were included. The analysis revealed a statistically significant reduction in delivery in <48 hr with β agonists. No advantage was found for treatment with β agonists over placebo for (i) reduction in risk of preterm delivery, (ii) frequency of birth weight <2500 g, (iii) neonatal respiratory distress syndrome, or (iv) perinatal death. While one must question whether these agents should be used in light of these results, it is important to remember that few trials employed concomitant use of steroids. Current thinking asserts that the primary benefit of tocolysis is to allow for the administration of corticosteroids.

C. Safety

Maternal side effects with the treatment of β mimetics are common. Those less life-threatening, but nevertheless bothersome enough to occasionally require discontinuation of therapy, include emesis, fever, headaches, and tremulousness. Cardiovascular complications, such as tachycardia, arrhythmias, and ischemia, are common. Metabolic disorders may also be encountered, such as glucose intolerance, hypokalemia, and sodium retention.

The most frequently reported major complication is pulmonary edema, occurring in as many as 5% of patients. The etiology of pulmonary edema is not clear and is probably due to a number of factors, including decreased colloid oncotic pressure, increased permeability of the pulmonary vasculature, and inappropriate use of iv fluids. The β agonists decrease the release of antidiuretic hormone, resulting in a decrease in renal function that can produce water retention, further increasing the risk of pulmonary edema. It is theorized that patients at increased risk include those with multiple gestations, hypertension, anemia, treatment extending longer than 24 hr, and infection.

Complications in the neonate have mainly been limited to hypoglycemia (secondary to fetal hyperglycemia and hyperinsulinemia) and ileus. However, there are conflicting reports regarding the incidence of periventricular and intraventricular hemorrhage among infants exposed to β adrenergic tocolysis. In a large retrospective study of patients delivering between 25 and 36 weeks, the authors concluded that the use of β mimetic tocolysis was significantly associated with an increase in the incidence of periventricular and intraventricular hemorrhage. Other reports, with fewer subjects, have found no association.

III. MAGNESIUM SULFATE

A. Mechanism of Action

Long familiar to obstetricians secondary to its use for seizure prophylaxis in preeclampsia, magnesium sulfate has gained popularity as a tocolytic due to its ease of administration and relatively low risk of significant side effects. It is now used as the first-line agent of choice for patients with preterm labor at many medical centers. Its mechanism of action still remains unclear, but it is believed to exert its action at two possible sites. Magnesium has been shown to decrease acetylcholine release at motor endplates at the neuromuscular junction and to block nerve transmission by preventing calcium entry. Magnesium also acts as a calcium antagonist, both at the intra- and extracellular levels. Hypocalcemia is produced through suppression of the parathyroid and prevention of renal reabsorption of calcium at the renal tubules. Elevated magnesium levels also compete with cellular calcium binding sites, resulting in a decrease of adenosine triphosphate levels. This renders the cell unable to bind calcium, and therefore unable to activate the actin and myosin complex to initiate uterine contraction.

B. Efficacy

As with other tocolytic agents, there are few randomized, controlled studies of magnesium sulfate, and these in general have enrolled few patients. The first U.S. study of magnesium was a comparison to alcohol, prompted by the observation that preeclamptics being treated with magnesium for seizure prophylaxis often demonstrated a slowing of uterine contractions. This study claimed magnesium sulfate to be the superior tocolytic but only if used early in the treatment of preterm labor (PTL). Tocolysis was achieved in 96% of patients with a cervical dilation of 1 cm or less but only in 25% if dilation was 2–5 cm. The efficacy of magnesium was compared to no tocolytic therapy in the 1990 study by Cox *et al.* No significant difference was found between the two groups in terms of gestational age at delivery, birth weight, neonatal morbidity, or perinatal mortality. This study has been faulted for its limitation of dosages of magnesium administered because it has been shown that magnesium serum levels do not necessarily correlate with tocolytic effect. Studies have also been performed comparing magnesium to the β mimetics; these have generally shown equivalent efficacy in achieving 2 or 3 days of tocolysis with decreased serious maternal side effects. Macones *et al.* examined the available evidence regarding the efficacy of magnesium sulfate for acute tocolysis compared to placebo and β agonist agents in their meta-analysis. In a review of the eight randomized controlled trials of magnesium sulfate for tocolysis which met their criteria for inclusion, there was no significant difference between magnesium and placebo for any of the measured outcomes for delay in delivery (presumably because of the low number of patients enrolled). Comparing magnesium sulfate to ritodrine or β agonists did not demonstrate any differences between the agents in achieving clinically significant tocolysis. A difference did exist between magnesium and the β agonists in the frequency of medication discontinuation secondary to side effects (more common with the β agonists), but the overall rate of major adverse effects was equitable.

C. Safety

Magnesium is excreted by the kidney, and unlike patients being treated for preeclampsia, patients with preterm labor generally have normal renal function. Magnesium toxicity with the treatment of preterm labor is therefore rare; serum levels are usually between 4 and 9 mg/dl. Those side effects most commonly encountered with magnesium include flushing, nausea and vomiting, headache, generalized muscle weakness, diplopia, and shortness of breath. Patellar reflexes disappear with plasma levels of 9–13 mg/dl, and respiratory depression occurs at 14 mg/dl. Toxic levels are treated with 1 g of calcium gluconate iv, with rapid reversal of symptoms.

Pulmonary edema, occurring in approximately 2% of patients, is the most serious associated drug effect with magnesium sulfate. It generally responds well to discontinuing magnesium and treatment with diuretics. Although the etiology is uncertain, increased iv fluids, decreased oncotic pressure, increased pulmonary capillary permeability, and infection have all been assigned possible etiologic roles.

The use of intravenous magnesium sulfate therapy appears to involve little fetal/neonatal risk. Radiographic bone changes have been described in neonates of patients receiving long-term intravenous magnesium infusion. These include radiographic bony abnormalities, rachitic changes of the calvaria and long bones, abnormal fetal bone mineralization, and parietal bone thinning. Although these changes disappear rapidly after birth, long-term therapy with magnesium should not be considered harmless.

Recent studies have also shown a possible benefit of magnesium in reducing the risk of cerebral palsy and intraventricular hemorrhage in premature neonates, separate from its tocolytic benefits. Nelson and Grether, in a case-control study, followed neonates born to women who had received magnesium sulfate for treatment of either preterm labor or preeclampsia. At 3–7 years after birth, the incidence of cerebral palsy was found to be 7.1% in magnesium-exposed children versus 36% in controls. Bottoms reported a decreased incidence of intraventricular hemorrhage and improved survival in premature neonates <1000 g if exposed to magnesium sulfate antepartum.

IV. INDOMETHACIN

A. Mechanism of Action

Prostaglandins are known to be important mediators in myometrial contractility. Term labor has been found to be associated with increased concentrations of arachidonic acid and prostaglandins E_2 and $F_{2\alpha}$. Prostaglandins enhance myometrial gap junction formation and increase intracellular free calcium levels, thereby increasing the activation of MLCK and the frequency of uterine contractions. Prostaglandins are frequently used for induction of abortion and cervical ripening or induction of labor. Therefore, prostaglandin synthetase inhibitors appear to be natural candidates for tocolysis. The focus thus far has been on the nonsteroidal antiinflammatory drug indomethacin.

B. Efficacy

Numerous articles on the efficacy of indomethacin have been published since it was first reported by Zuckerman in 1974. As with other studies on other tocolytics, most of these have lacked adequate sample size, uniform diagnosis of preterm labor, and adequate blinding or control groups. Niebyl and Zuckerman reported randomized, double-blinded, placebo-controlled studies. Both studies suggested that indomethacin is more effective than placebo in delaying delivery for 48 hr. However, there was no documentable difference in terms of neonatal outcome, gestational age at delivery, or birth weight. It is believed that this finding may be attributable to inadequate sample size.

Indomethacin may also be the tocolytic of choice for patients with polyhydramnios and resultant preterm labor. A decrease in fetal urine production mediated by the release of the normal prostaglandin-mediated block of antidiuretic hormone in the fetus, disturbances in autoregulation of renal blood flow, and increased fetal breathing and swallowing are thought to lead to the decrease in amniotic fluid volume experienced by women undergoing treatment with indomethacin at usual doses given for preterm labor. Because these patients are often treated for longer than the usual 2- or 3-day course, close fetal surveillance is necessary.

C. Safety

Treatment with indomethacin is remarkable for the relative infrequency of serious maternal side effects. Gastrointestinal disorders such as heartburn, nausea and vomiting, and peptic ulcer disease can be encountered, although serious complications are uncommon when indocin is used for a short course. Thrombocytopenia and an increased bleeding time have also been documented. Acute renal failure, especially when indomethacin is administered with other nephrotoxic drugs, can also complicate treatment. Hypertensive women have been noted to suffer acute elevations in blood pressures. Therefore, aspirin-induced asthma, kidney or liver disease, coagulation disorders, poorly controlled hypertension, and active peptic ulcer disease are all felt to be maternal contraindications to treatment.

In contrast, neonatal side effects from treatment with indomethacin are a focus of concern. Worrisome side effects have been oligohydramnios, constriction of the ductus arteriosus, and neonatal pulmonary hypertension. As stated previously, oligohydramnios is a consequence of decreased fetal urine output, which usually resolves rapidly with the cessation of treatment. Ductal constriction occurs secondary to the inhibition of prostacyclin and PGE_2 formation, which are responsible for the maintenance of ductal patency. Moise noted in Doppler echocardiography studies that the risk of ductal constriction rises rapidly after 32 weeks of gestation. Ductal constriction can lead to cardiac failure with hydrops or pulmonary hypertension and persistent fetal circulation in the newborn period. It also may lead to decreased ability of the ductus to constrict in response to oxygen at birth, leading to a paradoxically increased risk for patent ductus arteriosus. For these reasons, treatment with indomethacin should be limited to gestations <32 weeks and without evidence of oligohydramnios. Ultrasound should be performed after 48 hr of therapy to rule out oligohydramnios, and ductal flow or evidence of tricuspid regurgitation should be examined after 3 days of therapy. Continuation of therapy beyond 48–72 hr has become rare; if such therapy is deemed necessary, amniotic fluid volume should be checked twice weekly, and Doppler studies should be performed at least weekly. Therapy should be discontinued if oligohydramnios or ductal constriction are encountered.

Recently, concerns have arisen over other possible neonatal complications resulting from use of indomethacin at more premature gestational ages. In a retrospective cohort study, infants antenatally exposed to indomethacin prior to 30 weeks of gestation demonstrated poorer renal function, a higher rate of necrotizing enterocolitis, more cases of intracranial hemorrhage grades II–IV, and patent ductus arteriosus compared to nonindomethacin-exposed infants. Major *et al.*, in their examination of the effect of indomethacin on the incidence of necrotizing enterocolitis among low-birth-weight infants, also obtained concerning results. Neonates born to patients in preterm labor who received indomethacin tocolysis were compared to preterm neonates whose mothers did not receive indomethacin. The incidence of necrotizing enterocolitis in neonates delivered within 24–48 hr of treatment with indomethacin was significantly higher than that of nonindomethacin treated infants. Future prospective trials may determine whether these risks are outweighed by the efficacy of indomethacin as a tocolytic agent. However, neither of these studies are conclusive because of some methodological considerations. Specifically, in both studies, patients with "worse" or refractory (failing conventional therapy) preterm labor are those most likely to be treated with indomethacin. Thus, it is unclear whether the observed increases in necrotizing enterocolitis and intraventricular hemorrhage are due to indomethacin exposure or to some other underlying reason for more serious preterm labor (such as an intraamniotic infection).

V. ORAL TOCOLYTICS FOR MAINTENANCE THERAPY

Although commonly utilized, there is currently no support in the literature for routine use of oral tocolytics (most commonly β mimetics) after parenteral treatment of preterm labor. A recent meta-analysis of the four randomized trials of oral β agonist maintenance therapy revealed no benefit in terms of reducing the incidence of preterm delivery, increasing the interval to delivery, or reducing the incidence of recurrent PTL. Importantly, long-term β agonist use is increasingly associated with maternal glucose intolerance as well as with the more mild side effects listed previously. These findings may also be confounded by the issues of tachyphylaxis, subtherapeutic blood levels when compared to parenteral therapy, and inaccurate diagnosis of preterm labor. For these reasons, physicians must weigh the risks of oral β mimetics against the uncertain benefits.

VI. ANTENATAL GLUCOCORTICOIDS

Since 1972, when Liggins and Howie reported a 50% reduction in hyaline membrane disease/respiratory distress syndrome with antepartum glucocorticoid treatment, many more benefits have been revealed. Such treatment has now been shown to have the additional benefits of a decreased incidence and severity of intraventricular hemorrhage, necrotizing enterocolitis, and perinatal mortality. It is believed this is mediated through enhanced cell maturation and differentiation rather than cell growth. In the lungs, there is enhanced production of surfactant, neonatal lung compliance is increased, and alveolar leakage is decreased. Maturation of the fetal brain, gastrointestinal tract, and skin is also enhanced. A meta-analysis of 12 controlled trials involving over 3000 patients confirmed these findings and noted no strong evidence suggesting adverse effects of corticosteroids.

The current recommendation by the American College of Obstetricians and Gynecologists and the NICHD Consensus Development Conference on Antenatal Steroids (1994) is to increase the use of antenatal steroids for patients at <32 weeks of gestation expected to deliver imminently, regardless of the status of fetal membranes. For patients at 32–34 weeks of gestation, both organizations recommended antenatal steroid treatment for mothers with intact membranes who were felt to be at risk for delivery within the following week.

See Also the Following Article

Preterm Labor and Delivery

Bibliography

American College of Obstetricians and Gynecologists (ACOG) Committee on Obstetric Practice (1994, December). Clinical opinion 147. Antenatal corticosteroid therapy for fetal lung maturation. ACOG.

Boyle, J. (1995). Beta-adrenergic agonists. *Clin. Obstet. Gynecol.* 38, 688–696.

Copper, R. L., Goldenberg, R. L., Creasy, R. K., DuBard, M. B., Davis, R. O., Entman, S. S., Iams, J. D., and Cliver, S. P. (1993). A multicenter study of preterm birth weight and gestational age specific mortality. *Am. Obstet. Gynecol.* **168**, 78–84.

Crowley, P., *et al.* (1995). Antenatal corticosteroid therapy—A meta-analysis of the randomized trials. *Am. J. Obstet. Gynecol.* **173**, 322.

Elliott, J. P. (1985). Magnesium sulfate as a tocolytic agent. *Cont. Obstet. Gynecol.* **25**, 49–61.

Gordon, M. C., and Iams, J. D. (1995). Magnesium sulfate. *Clin. Obstet. Gynecol.* **38**, 706–712.

Gordon, M. C., and Samuels, P. (1995). Indomethacin. *Clin. Obstet. Gynecol.* **38**, 697–705.

Higby, K., Xenakis, E., and Pauerstein, C. (1993). Do tocolytic agents stop preterm labor? A critical and comprehensive review of efficacy and safety. *Am. J. Obstet. Gynecol.* **168**, 1247–1259.

Hill, W. C. (1995). Risks and complications of tocolysis. *Clin. Obstet. Gynecol.* **38**, 725–745.

King, J. F., Grant, A., Keirse, M. J. N. C., and Chalmers, I. (1988). Beta-mimetics in preterm labor: An overview of the randomized controlled trials. *Br. J. Obstet. Gynecol.* **95**, 211–222.

Macones, G. A., Berlin, M., and Berlin, J. (1995). Efficacy of oral beta-agonist maintenance therapy in preterm labor: A meta-analysis. *Obstet. Gynecol.* **85**, 313.

Macones, G. A., Sehdev, H. M., Berlin, M., Morgan, M., and Berlin, J. (1997). The evidence for magnesium sulfate as a tocolytic agent. *Obstet. Gynecol. Surv.* **52**, 652–658.

National Institutes of Health Consensus Development Conference (1994). Effect of corticosteroids for fetal maturation on perinatal outcomes. February 28–March 2. *Am. J. Obstet. Gynecol.* **173**, 246.

Petrie, R. (1981). Tocolysis using magnesium sulfate. *Sem. Perinatal.* **5**, 266–273.

Robertson, P. A., Sniderman, S. H., Laros, R. K., Cowan, R., Heilbron, D., Goldenberg, R. L., Iams, J. D., and Creasy, R. K. (1992). Neonatal morbidity according to gestational age and birthweight from five tertiary care centers in the United States, 1983 through 1986. *Am. J. Obstet. Gynecol.* **166**, 1629–1643.

Toxemia of Pregnancy

see Preeclampsia/Eclampsia

Transgenic Animals

Karen Moore and Jorge A. Piedrahita
Texas A&M University

GLOSSARY

chimera An animal composed of two or more genetically distinct types of cells. These individuals are produced by mixing cells of two embryos or more commonly by mixing embryonic stem cells with host embryo cells.

concatemer A joining of DNA constructs in tandem, usually in a head-to-tail array but may also be head-to-head or tail-to-tail.

construct A piece of genetically engineered DNA (recombinant DNA), containing a promoter, the coding region of a functional gene, and a polyadenylation signal, used for producing transgenic animals.

embryonic stem cells Cell lines derived from the inner cell mass or primordial germ cells of an embryo that are maintained in an undifferentiated, proliferative state in culture and can differentiate into any of the three germ layers when reintroduced into a host embryo.

gene targeting Site-specific genetic alteration of a particular gene; this may involve insertional inactivation, deletions, or small changes within the gene of interest.

genome The genetic makeup of an individual.

germline transmission Incorporation of foreign DNA within the genome so that it is transmitted to the transgenic animals offspring.

heterozygotes Diploid organisms, those with two sets of chromosomes, which have different forms of a gene (alleles) at a particular locus.

homologous recombination A method used for making specific modifications within a gene of interest. It is a repair mechanism initiated by double strand breaks in the DNA and 100% homology between the targeting construct and the host gene of interest.

homozygotes Diploid organisms which have the same form of a gene (allele) at a particular locus.

insertional inactivation Insertion of foreign DNA into a functional gene causing its loss of function or inactivation.

messenger RNA A complementary copy of the DNA transcribed from a gene that serves as a template from which a protein is translated.

positional effect The effect of surrounding chromatin and regulatory elements on the expression pattern of a transgene; this can be positive or negative.

pronucleus (pronuclei) Nuclear membrane-bound structures that contain the maternal or paternal chromosomes in the early fertilized one-cell stage embryo.

retroviral vector An RNA viral vector used for shuttling transgenes into embryonic or somatic cells. After infecting the cell, the virus converts itself to DNA, inserts itself into the host chromosome at a random location, and is then called a provirus.

somatic cell Any cell in the body that is not a germ cell.

Transgenic animals are animals that have been genetically modified, through the introduction of foreign DNA or RNA, so that they overproduce, underproduce, lack the production of, or have modified expression of a particular protein. While most transgenic animal studies involve mice, other transgenic species, such as rabbits, rats, hamsters, sheep, goats, swine, cattle, chicken, and fish, have also been produced. Transgenic animals have been and continue to be generated to improve animal and human health and welfare. This is accomplished through the production of animal models of human diseases and new gene therapies, production of pharmaceuticals and vaccines by the mammary glands, production of xenografts, production of blood substitutes, improvement in disease resistance and improvement in milk composition as well as wool and meat production and composition. Each of these will be addressed in the this article along with an introduction into methods for producing transgenic animals and considerations for improved efficiencies.

I. INTRODUCTION

Transgenic animals were first produced nearly two decades ago. Merging recombinant DNA technologies with embryology has allowed for the isolation, cloning, and modification of genes followed by their transfer into the genome of developing embryos to produce transgenic animals. This technology allows the transfer of genes of interest from one species to another, thus permitting genetic improvements as well as a better understanding of how genes function within an individual. Transgenes can also be expressed in a specific tissue at a given time or at a particular stage of development. The first animals to be generated were transgenic mice, which remain the major species utilized to date due to their ability to produce large litters with a short generation interval, and low cost of maintenance. While much information continues to be gained from transgenic mice, they do have limitations. Often, mice are not suitable transgenic models, for example, in some human genetic disorders or for the production of large quantities of biopharmaceuticals, where larger, more closely related species would be more effective. Moreover, transgenesis offers the potential for great genetic advances in livestock production through improvements in animal health and production traits such as growth, meat, milk, and wool production. As efficiencies of this technology improve, other species of transgenics will become more commonplace and enhance our lives through their contributions to medicine and agriculture.

II. METHODS OF GENE TRANSFER

There are five general methods for production of transgenic animals: pronuclear injection, viral vectors, gene targeting in embryonic stem cells, sperm-mediated gene transfer, and biolistics. Each method will be briefly described.

A. Pronuclear Injection

Pronuclear injection remains the predominant method for generating transgenic mice and is the only method currently available for production of transgenic livestock. Fertilized pronuclear-stage embryos are flushed from the oviducts of the species of interest. Pronuclei are visualized with differential interference contrast (DIC) microscopy while holding the embryo in place with a holding pipet. Cattle and pig embryos, however, have very dense cytoplasm, making pronuclear visualization impossible without prior centrifugation (3–5 min, 15,000g). One or both pronuclei are injected with 1 or 2 picoliters of DNA construct (1 or 2 μg/ml) just until the pronucleus visibly swells (Fig. 1). This is enough volume to deliver from 100 to 1000 copies of the transgene. Pronuclear swelling is also thought to induce chromosomal breakage which improves chances of transgene incorporation. Injected embryos are either surgically transferred to recipients or cultured for a period of time and then transferred to recipients where they are carried to term. Of those transferred, only 10–30% survive to term. This can be due to physical damage to the embryos or to insertional inactivation. The percentage surviving that are transgenic is even lower and varies for different species, averaging $\approx$5% for laboratory animals (mice, rats, and rabbits) and $\approx$1% or less for farm animals (cattle, sheep, goats, and pigs).

The initial transgenic animals produced are called founders. All founders are different due to (i) random integration of the transgene and (ii) variable copy number due to integration of head-to-tail concatemers of the transgene. Mosaicism may also be a problem if transgene incorporation into the genome does not occur until after one or more cell divisions have occurred. Founder animals are bred to create transgenic lines for a particular transgene. The transgene should be transmitted to 50% of the offspring in a normal Mendelian fashion. Efficiencies may be further reduced if founders are mosaics and do not incorporate the transgene into the germline

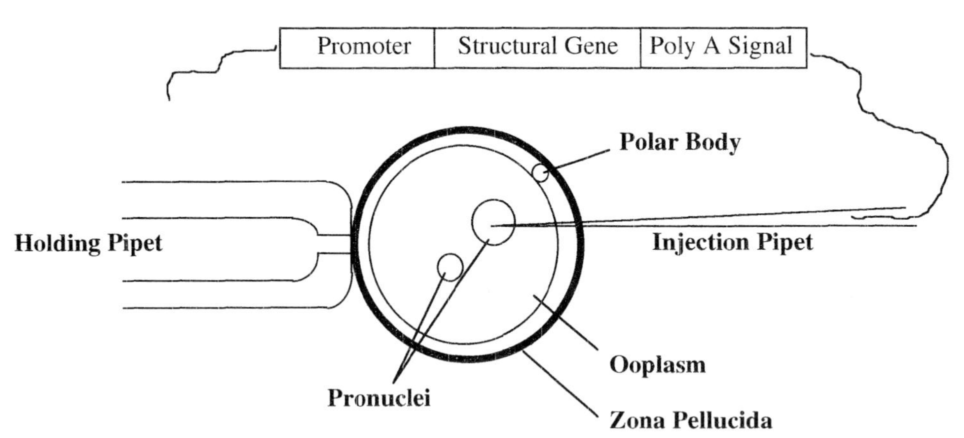

FIGURE 1 Microinjection of pronuclear-stage embryos for production of transgenic animals.

or if the transgene has detrimental effects on fertility. Further breeding of the heterozygous offspring can generate homozygous transgenic lines. Unfortunately, transgenes are only expressed in about half of the transgenic lines tested. While generating transgenic animals by pronuclear injection is very inefficient, 100–200 embryos can be injected in a single day by a researcher, making the procedure tedious yet feasible.

The majority of transgenic animals produced by pronuclear injection exhibit a gain of gene function. That is, a gene not normally expressed in a particular tissue or even within the animal is being expressed. This methodology is also traditionally used to examine effects of overexpression of a particular gene. While this provides valuable information about gene function, some questions cannot be addressed using this type of system. There are many times when loss of gene function is necessary. While pronuclear injection is not the best system for this, partial loss of function experiments can be generated. These animals are generated by one of four methods: (i) dominant negative, (ii) antisense, (iii) ribozymes, or (iv) cell ablation. The dominant negative is the most effective method when trying to remove a multiple subunit protein or receptor. When a protein is a conglomerate of many pieces, it is important that each piece be properly synthesized and folded. If a single subunit is defective, by joining with the others it creates a defective protein that is nonfunctional. Therefore, if the animal overproduces a single defective subunit due to transgenesis it will effectively cause the loss of function of the entire protein. Antisense and ribozyme procedures are used to block a particular messenger RNA (mRNA) such that it cannot be translated into protein. The antisense approach involves production of a complementary copy of the mRNA of interest, which binds the target mRNA making it double stranded and unable to be translated. Ribozymes, on the other hand, are catalytic enzymes that recognize and bind the mRNA of interest and cleave it, effectively destroying the message. Although all three of these methods can effectively reduce protein expression, there is never complete loss of gene function due to the leaky nature of these systems.

Transgenic cell ablation is another technique for destruction of a particular cell lineage. Tissue-specific regulatory elements regulate expression of toxin genes, such as diptheria toxin-A (DT-A) or inducible thymidine kinase (tk), in specific cells, effectively killing them. This approach is useful for studying animal development without a particular cell lineage. Several animal models for human diseases which lack particular cell types have been generated in this manner. More impressive is its use for killing tumor cells while leaving normal tissues unharmed. However, this procedure does not result in complete ablation of particular cell lineages and, if not properly regulated, can destroy other cell types as well.

B. Viral Vectors

Viral vectors are another mode for introducing foreign genes into animals. They are not used as frequently as pronuclear injection due to the complexity of generating replication-deficient vectors and the species specificity of these vectors. Viral vectors are, however, the method of choice for chickens as well as for gene therapy protocols. The following sections summarize the two most prevalent types of viral vectors utilized for producing transgenic animals.

1. *Retroviral Vectors*

Retroviruses are single-stranded RNA viruses that infect dividing cells in a species-specific manner. Once inside the host, the RNA virus is reverse transcribed to DNA which can then be incorporated into the host genome. Retroviruses incorporate one viral copy at a particular insertion site but will normally have multiple insertion sites within a host. The incorporated virus is called a provirus, which is replicated by the host and can bud from the infected cell to infect neighboring cells. The retrovirus is composed of several structural genes, such as *gag*, *pol*, and *env*, which are flanked by long terminal repeat (LTR) sequences which serve as strong promoters for transcription of the structural genes and for incorporating the virus into the chromosome of the host. The structural genes code for a polymerase that converts the RNA to DNA as well as proteins that make up the coat proteins of the viral particle which allow infection. Retroviruses have been genetically engineered to serve as vectors for production of transgenic animals by removing the structural genes *gag*, *pol*, and *env* and inserting RNA for the gene of interest (Fig. 2). It is best to utilize regulatory elements other than the LTR because these are inacti-

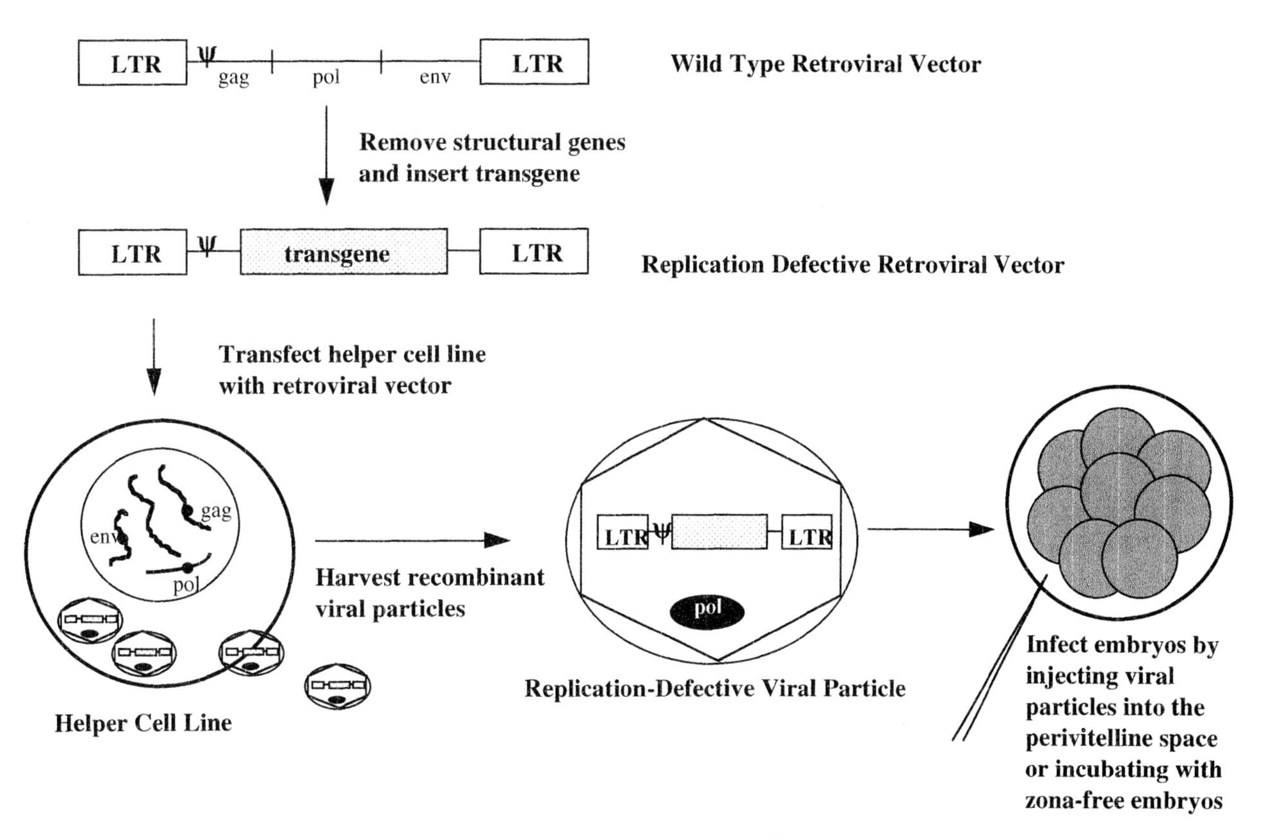

FIGURE 2 Preparation of retroviral vectors for producing transgenic animals. Ω, the packaging signal which allows incorporation of the vector into the viral particle; LTR, long terminal repeats that allow chromosomal integration and act as strong promoters; *gag*, *pol*, and *env*, structural genes coding for coat proteins and a polymerase.

vated by the host over time, resulting in loss of transgene expression.

The retroviral vector is transfected into a helper cell line where it is packaged into infective but replication-defective viral particles. The helper cell line produces the gag, pol, and env proteins but lacks all other viral components. The viral particles can then be harvested for infection of target tissue. Embryos are infected with the retroviral vector particles by microinjection into the perivitelline space or by culturing zona pellucida-free embryos with the viral particles for 16–24 hr. This system is frequently utilized in somatic cell gene therapy protocols as well. Retroviral systems are very efficient, infecting nearly 100% of dividing cells. However, the disadvantages are in the difficulty of making the recombinant vectors. They can only be of limited size (9 kb) and are species specific. Additionally, transgenic animals produced by retroviral vectors are mosaics, which require additional breeding for segregation of the transgene. There are also concerns with possible recombination of the retroviral components, producing a virulent form of the retrovirus. For these reasons, pronuclear injection is still the preferred method for producing transgenic animals.

2. *Adenoviral Vectors*

Adenoviruses are DNA viruses that infect only nondividing cells. These viruses, once within the host cell, do not incorporate into the host chromosome but rather replicate episomally (independent of the chromosomes). They are relatively nonpathogenic, however, since they are not stably integrated into the host genome and can be lost when cells divide. This gives rise to a dilution of transgene expression and mosaicism over time. Though their use is limited in conventional transgenesis, they are receiving increased attention in gene therapy procedures.

C. Gene Targeting in Embryonic Stem Cells

Gene targeting is another powerful means for producing genetically modified animals. This procedure involves genetically modifying embryonic stem (ES) cells through homologous recombination for the production of transgenic animals. This procedure allows for the complete loss of gene function (gene knockout), subtle modification of genes, insertions, deletions, or even translocations, making this technology incredibly powerful in the field of transgenesis. However, currently ES cells are only available for use in the mouse, thus limiting its applications. More than 300 different knockout animal lines have been produced to study effects of loss of gene function. The following sections will briefly describe ES cells and homologous recombination and their applications for production of transgenic mice.

1. *Introduction into ES Cells*

Embryonic stem cells are unique cell lines derived from the inner cell mass of early preimplantation embryos of the mouse. These cells can be maintained in culture indefinitely in an undifferentiated yet proliferative state. This status is maintained by culturing ES cells in the presence of proteins called cytokines, such as leukemia inhibitory factor, or feeder cells which secrete these cytokines. ES cells can be genetically manipulated in culture, selecting for those cells that have incorporated the transgene properly, and then reintroduced into host blastocyst-stage embryos by microinjection (Fig. 3). ES cells are pluripotent because they incorporate into the developing embryo and contribute to all cell types within the body, including the germline. Animals produced by this procedure are chimeras, made up of ES cells and host embryo cells. Germline chimeras provide a vehicle for transmitting the transgene to the offspring, which through inbreeding can bring the transgene to homozygosity.

2. *Homologous Recombination*

Gene targeting by homologous recombination has been utilized to modify the genome of mice for more than a decade. Two phenomena drive this recombination reaction: regions of complete homology and double-strand breaks within the DNA. When gene targeting constructs are introduced into the ES cell, normally through electroporation, calcium phosphate precipitation, or lipofection, they either find their way into the nucleus or are degraded. The DNA making its way to the nucleus is perceived by the cell to be broken and is quickly repaired by either

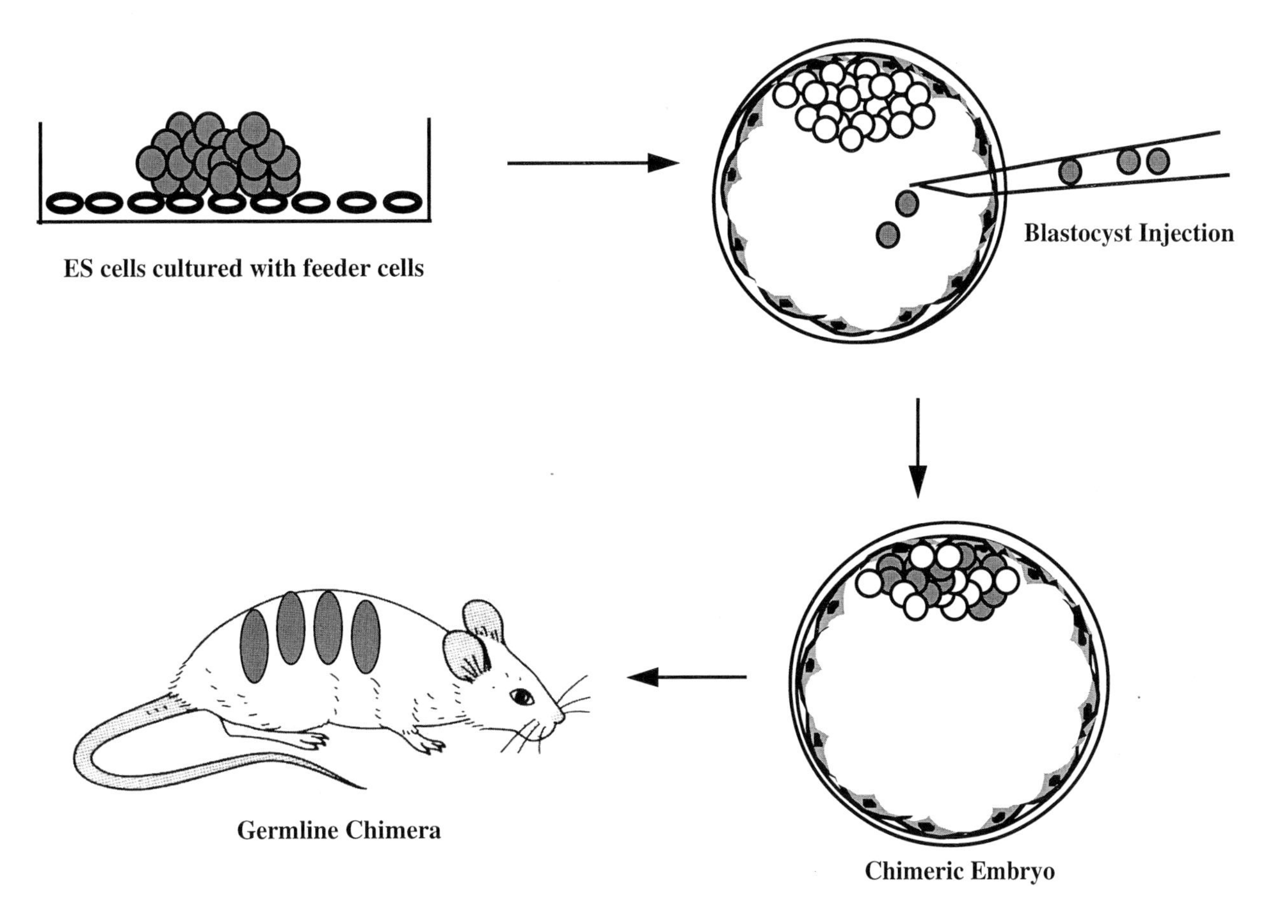

FIGURE 3 Blastocyst injection with embryonic stem (ES) cells to produce germline chimeras.

random insertion into the chromosomal DNA or through homologous recombination at a region homologous to the targeting construct. This method is very inefficient, yielding 1 in 10^6 targeted events, so it is not practical for use on embryos. However, ES cells can be grown in large numbers, making this procedure quite useful.

There are two types of targeting constructs utilized in gene targeting events (Fig. 4). The first is the insertional or O-type targeting vector. This vector is linearized within the region of homology, resulting in a single crossover event and gene duplication. It also incorporates vector DNA into the endogenous gene which may have detrimental effects. The replacement or Ω-type vector is the second and most common type of targeting vector utilized. This vector is linearized outside the regions of homology and requires a double-crossover event to replace the endogenous gene with the homologous targeting vector. Vector DNA is outside the region of homology and is therefore never incorporated into the targeted gene.

Since homologous recombination is very inefficient, targeting constructs also include selectable markers. Table 1 lists the selectable markers currently used in gene targeting experiments. Positive selection markers allow for selection of those ES cells that have incorporated the transgene by either homologous recombination or random insertion. These cells can be expanded and tested by PCR and/or Southern analysis to determine those that have been properly targeted. To enrich for homologous recombination events, negative selection markers are utilized to flank the targeting construct outside the region of homology (Fig. 4B). Previously, it was noticed that vector DNA lying outside the regions of homology were never incorporated into a targeted gene but rather were clipped off and degraded. Only

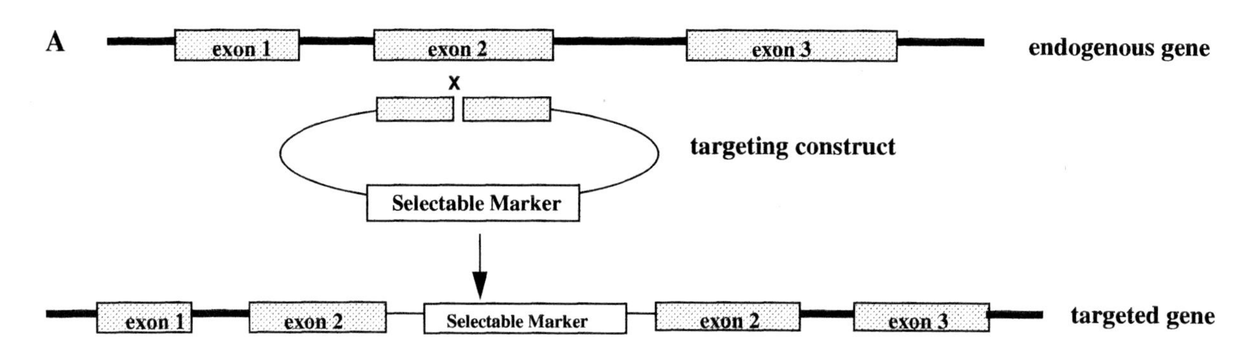

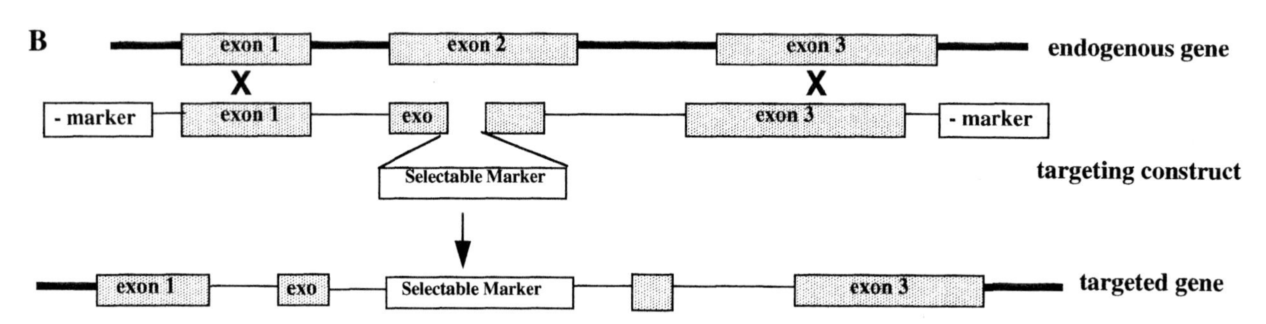

FIGURE 4 Gene targeting by homologous recombination. (A) Insertion or O-type homologous recombination. (B) Replacement or Ω-type homologous recombination.

random integrants incorporated the vector DNA. In this way, the negative selection cassette is removed in the gene targeting event and survives in negative selection medium, whereas the random integrant incorporates the negative marker and is eliminated. This has allowed for improved screening efficiencies and is utilized in most gene targeting experiments.

There are three strategies currently utilized in gene targeting experiments for modifying a particular gene: (i) in and out or hit and run, (ii) tag and exchange, or (iii) plug and socket. The first method, in and out, utilizes the insertional-type vector to introduce a small mutation (i.e., addition of 4 bp) and a selectable marker in the first of two events.

TABLE 1
Selectable Markers Utilized in Gene Targeting

	Selection	
Gene	*Positive*	*Negative*
Neomycin phosphotransferase (Neo)	Yes; G418	No
Thymidine Kinase (tk)	Yes; HAT medium	Yes; Gancyclovir
Diptheria Toxin (DT)	No	Yes
Hygromycin B phosphotransferase (hph)	Yes; Hygromycin B	No
Hypoxanthine phosphoribosyl transferase (hprt)	Yes; HAT medium	Yes; 6-thioguanine
Xanthine/Guanine phosphoribosyl transferase (gpt)	Yes; HAT medium	Yes; 6-thioguanine

After the initial homologous recombination event there is a duplication of the targeted gene, the result of using insertion-type vectors. This duplication drives the second event, which is an intrachromosomal recombination event between the duplicated regions, resulting in the excision of the selectable marker yet leaving the small mutation in the targeted gene.

Another strategy used in homologous recombination is called tag and exchange. This method uses the replacement-type vector to insert a positive–negative selection (PNS) cassette into the target gene in the first recombination event. Tagged colonies are selected using the positive selection marker and used in a second targeting event in which a new targeting construct with a small mutation replaces the PNS cassette. These exchange colonies are selected using negative selection. Theoretically, only those colonies undergoing complete homologous recombination, losing the PNS cassette in the second event, should survive. However, this does not work as well in practice due to reversion to the wild type and the high incidence of mutations in the tk negative selection marker. Research with DT-A and development of new markers may improve this strategy.

The plug and socket is a recent advance in gene targeting. This strategy also uses replacement-type vectors in two events. In the first event a selectable marker and a truncated hypoxanthine phosphoribosyl transferase (hprt) "socket" are inserted into the target gene. The socket colonies are selected using positive selection and used in the "plug" targeting event. The plug contains the other portion of the truncated hprt which serves as a crossover point in the second homologous recombination event to remove the initial selectable marker and any other region selected but produce a functional hprt. This allows for removal of large pieces of DNA (20 kb). One disadvantage to this strategy, however, is that it leaves a functional transcriptional unit within the gene of interest which, depending on the locus, could detrimentally affect expression of neighboring genes.

3. *Site-Specific Recombination*

The most recent advancement in gene targeting is the use of site-specific recombination systems. Two of the systems being utilized are Cre lox-P and frt flipase. This procedure involves a recombinase enzyme (Cre or flipase) which recognizes and binds two short (34-bp) recognition sequences (lox-P or frt sites) inducing a recombination event between them. This results in either an excision or an insertion event depending on whether it is an intramolecular or intermolecular event, respectively (Fig. 5). While kinetics favors the excision event, researchers are currently inducing point mutations in the recognition sequences such that insertion can be favored.

Site-specific recombination offers great potential with ES cell technologies. Targeting constructs can be made that allow for gene inducibility or for inducible knockouts after the introduction of the recombinase enzyme. For instance, an artificial stop codon can be inserted into the beginning of a gene that is flanked by lox-P sites, thus inactivating the gene until the Cre enzyme induces excision of the stop. Several exons coding for the gene of interest can also be flanked by lox-P sites, or "floxed," so that introduction of Cre causes excision, inducing a gene knockout. Another exciting possibility, although still in experimental stages, is production of a marked gene for multiple targeted insertions. Inserting a single lox-P site into a particular gene allows later insertion of any structural gene into that location (Fig. 5B). This could be very useful for production of biopharmaceuticals in milk because a mammary-specific gene can be tagged with a lox-P site. These founder animals can then be superovulated and mammary gene-tagged embryos can be collected and injected with new constructs containing another lox-P site plus the gene of interest along with a transient expression Cre plasmid or the Cre protein. This will induce the insertion of the gene of interest into the mammary gland, and these animals should express the new protein in their milk. While all examples here address the Cre lox-P system, the frt flipase system is analogous. Other recombinase systems are also being tested for use in site-specific recombination.

D. Sperm-Mediated Gene Transfer

Sperm-mediated gene transfer is theoretically a very exciting means for production of transgenic animals. However, it has not been proven effective.

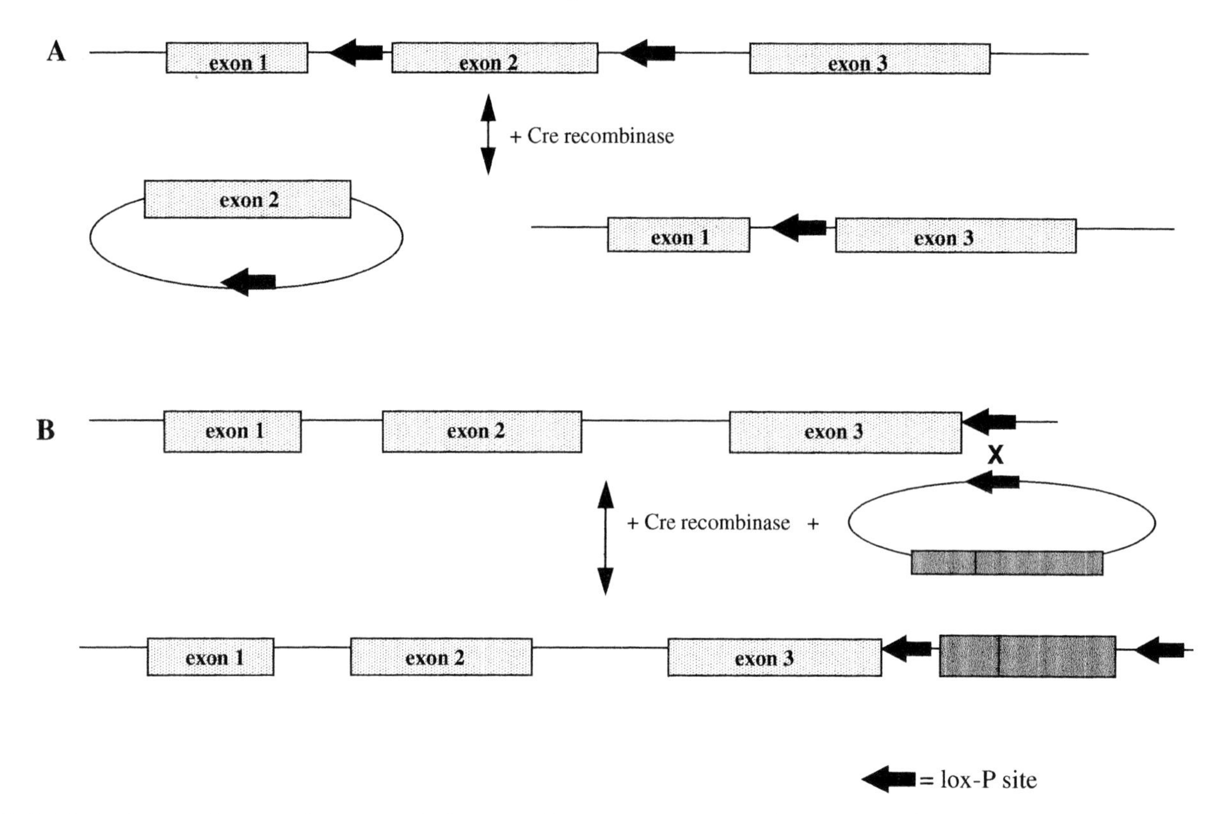

FIGURE 5 Site-specific recombination with Cre recombinase and lox-P sites. (A) Gene knockout or repair using site-specific recombination. (B) Cre-mediated transgene insertion into a lox-P tagged locus.

E. Biolistics

The gene particle gun or biolistics is another option for introducing foreign DNA or RNA into cells. Small gold particles are coated with the transgene and are shot into the target tissue. Some of the transgene will be propelled into the cytoplasm and make its way into the nucleus, where it can be integrated and expressed. It has been utilized mainly in gene therapy and vaccine protocols, but some have used biolistics for introducing transgenes into developing embryos.

III. FACTORS AFFECTING TRANSGENE EXPRESSION

There are several elements one must consider when preparing a construct to be used for generating a transgenic animal. The transgene must insert itself into the genome and be expressed in a predictable fashion. While many facets of regulating transgene expression are still unknown, there are three major elements that a construct must include for proper expression: (i) regulatory element(s), (ii) the structural gene, and (iii) a polyadenylation signal.

A. Regulatory Elements

Considerable research continues to define regulatory elements required for proper insertion and expression of transgenes. Several elements have been discovered that may have a role in appropriate gene regulation. These include promoter regions, enhancer elements, and transcription control elements. The most important regulatory element to include within a construct is the promoter region. With multiple possibilities to choose from, one can select promoters that allow for inducible, tissue-specific, stage-specific, or ubiquitous expression of the transgene. Promoter sizes and complexities vary and are not correlated. Elements within the promoter allow for their unique expression patterns, and failure to include these components can change expression patterns completely. These elements, such as the Goldberg–Hogness or TATA box, CCAAT box, and GC box, are binding sites for transcription factors which direct the start of mRNA transcription. Transcriptional control elements, such as the metal responsive element or the glucocorticoid responsive element, are also located in the 5′ promoter region and are important for binding factors that induce gene expression. Enhancer sequences are additional regulatory elements that have a positive impact on transgene expression. These sequences act over long distances and can be found 5′ to the gene, 3′ to the gene, or within introns of the gene. These elements are orientation independent and may also exhibit tissue or cell-type specificity. Therefore, defining promoter and enhancer elements for a transgene is essential and may result in a 1000-fold difference in transgene expression as well as proper specificity and regulation.

B. Gene of Interest

Most early constructs were prepared from complementary DNA (cDNA) for the gene of interest. cDNA is reverse transcribed from an mRNA of interest, making it a complementary copy of DNA that lacks introns. Expression level for most cDNA constructs was very low in transgenic animals even when the construct was quite efficient in tissue culture. However, when genomic DNA was utilized this problem was overcome. It was determined that many introns contain enhancer elements as well as unknown sequences that affect proper transport and processing of newly transcribed mRNA. Therefore, whenever possible, genomic DNA is utilized in construct preparation. When the gene of interest is too large, however, minigenes are constructed. Minigenes contain cDNA for the gene of interest along with introns that have known enhancer properties for improving transgene expression.

C. Reporters

Reporter genes allow us to determine the level of expression of a transgene, both temporally and spatially. They are included in many constructs or coinjected to improve transgene diagnostics. When a reporter construct is cointroduced it is almost always incorporated with the transgene of interest due to concatemer formation. Reporters may also be utilized as selectable markers in gene targeting. The most common reporters in use are chloramphenicol acetyl transferase, luciferase, β-galactosidase, and green fluorescent protein.

D. Polyadenylation Signal

The most conserved recognition element in DNA is the polyadenylation signal, AATAAA. This 3′ regulatory element is necessary for the proper addition of the poly A tail approximately 10–30 nucleotides downstream. During posttranscriptional modification this element signals cleavage of the mRNA downstream followed by the addition of 50–250 adenine nucleotides, with longer poly A tracts conferring greater mRNA stability. Therefore, it is essential to include a poly A signal when designing constructs, preferably one that produces enhanced mRNA stability such as the SV40 late gene poly A signal.

E. Chromatin Structure

Chromatin structure, or location within the chromosome where the transgene integrates, also has a profound effect on transgene expression. This is referred to as the positional effect, in which the chromatin environment governs transgene expression. In order to achieve expression, transgenes must be integrated into open, actively transcribed regions of a chromosome. Several elements that appear to regulate chromatin structure have now been identified by their hypersensitivity to DNase I. These sites are (i) locus control regions (LCRs) or dominant control regions, (ii) matrix attachment regions (MARs) or scaffold attachment regions, and (iii) specialized chromatin structure (SCS elements). The LCRs are large elements located far upstream of a gene and appear to behave as super enhancers. However, very few of these elements have been identified. Other elements that improve position independence are the MARs. These elements attach to the nuclear matrix, creating DNA loops or domains which are insulated from neighboring regions allowing for gene expression. When used to make transgenic mice, they increase the number of transgenic animals expressing the transgene; however, they do not improve the level of expression. SCS elements have also been shown to confer transgene insulation. They do not seem to have enhancer activity but rather restrict repressive effects from passing through from neighboring regions. Use of one or more of these elements in transgene design should improve position-independent transgene expression by providing an independent chromatin domain. Future discoveries of regulatory elements will improve transgene regulation capabilities and allow for copy number-dependent expression.

IV. FACTORS AFFECTING GENE TARGETING EFFICIENCY

Gene targeting in ES cells, while providing a powerful means for making subtle genetic modifications, is still quite inefficient. Approximately one in a million cells will have a targeted locus. While improved understanding of the mechanisms involved in DNA repair and homologous recombination will continue to benefit these procedures, there are several known strategies for improving targeting efficiency.

A. Length of Homology

Homology between the gene of interest and the targeting construct is essential for effective gene targeting. Most targeting vectors are composed of two arms of homologous DNA and an internal selectable marker. It has been shown that increasing the length of homology of these arms can improve targeting efficiencies. Regions of homology need to be at least 1 kb in length, with increases up to 14 kb giving improved gene targeting efficiencies. However, when yeast artificial chromosomes were used to give regions of homology $\geqq$18 kb, there was no further benefit and in some there was a reduction in efficiency due to rearrangements.

B. Isogenic versus Nonisogenic DNA

In early gene targeting experiments, targeting efficiencies were even lower and more variable than those of today. Targeting constructs were generated from genomic DNA from various strains of mice and used for targeting within and across different strains. Over time it was realized that targeting constructs prepared from DNA of a different strain than that of the ES cell line to be targeted (nonisogenic) yielded very low or, more commonly, no targeting events. However, when genomic DNA from the same strain as ES cells was utilized to produce the construct to be targeted (isogenic), efficiencies were dramatically improved. These dramatic differences are due to the very slight differences between the endogenous and the targeting DNA sequences, which are called polymorphisms. Today, all gene targeting constructs are prepared using isogenic DNA. This factor may need to be addressed in the future as ES cell lines become available for other species in which true inbred lines are not available.

V. TRANSGENIC ANIMALS IN MEDICINE

The production of transgenic animals has the greatest application in human and veterinary medicine. Many advancements have been made toward understanding how genes regulate developmental pathways, normal and abnormal physiology, and their role in disease. Several areas that have been topics of intensive research for enhancing human and animal health are reviewed in the following sections.

A. Animal Models for Human and Animal Diseases

With the advent of gene targeting, subtle modifications can be generated in an animal genome for generation of animal models for human and animal diseases. Genetic disorders modeled in the mouse through gene knockouts or modifications allow for development of new drug and nutritional therapies for treating these diseases. These animals are also being utilized to generate new somatic cell gene therapies that may lead to cures of genetic disorders in humans. A comprehensive list of knockout mouse lines currently available is referenced in the Bibliography. Those interested may also want to review transgenic lists available on the Internet.

B. Biopharmaceuticals and Oral Vaccines

The most active area in transgenic livestock research is for production of pharmaceuticals or vaccines in milk. Through the use of mammary-specific regulatory elements, large quantities of pharmaceuticals or vaccines can be harvested from a lactating animal in a noninvasive manner. Transgenic animals provide an ideal bioreactor by producing mature proteins that are more similar to human proteins than those produced by bacterial and yeast systems and more cost-effective than animal cell culture. These recombinant proteins also provide a safer source of factors, such as insulin or growth hormone, which are traditionally obtained from slaughterhouses and cadavers. More efficient production of these biopharmaceuticals will allow for an affordable source of medications and vaccines worldwide. Transgenic cows, pigs, sheep, goats, rabbits, and mice have been generated that produce blood-clotting factors and anticoagulation factors such as tissue plasminogen activator, factors VIII and IX, fibrinogen, protein C, and α1-antitrypsin, a drug for emphysema and cystic fibrosis patients. While this list is incomplete, it is apparent that the choice of drug or vaccine has infinite possibilities. The only drawback to date is the variable response of transgenes and inefficiencies of harvesting these proteins. Future improvements in transgene regulation and advancements in protein purification will lead to major advances in medicine through better, more cost-effective drugs and vaccines.

C. Xenotransplantation

Inadequate supplies of human organs for transplantation result in over 3000 deaths annually. Transgenic technologies in xenotransplantation may alleviate this problem in the future. Xenotransplants are the transfer of organs or tissues from one animal to another. Discordant transplants between dissimilar species, such as the primate and human or pig and human, result in violent hyperacute rejection within minutes to hours of transfer due to an antibody and complement cascade reaction directed against the foreign tissue or organ, resulting in cell lysis and tissue destruction. However, transgenic animals are being generated that produce human proteins, such as decay accelerating factor (DAF) and complement inhibitors CD59 and CD46, which inhibit this initial rejection response. Organ perfusion studies have shown an increased survival time with this technology over nontransgenic controls. Future improvements in this area, along with traditional immunosuppressive drugs such as cyclosporine and leflunomide, should allow for controlled suppression of organ rejection. Initially, this will provide patients with temporary organ transplants until human organs become available. However, as the technology improves the animal organs may become permanent solutions to this problem.

VI. TRANSGENIC ANIMALS IN AGRICULTURE

Improvements in animal productivity and health are key reasons for producing transgenic livestock. While agriculture has not benefited as greatly from transgenic technology to date, further improvements in transgenic efficiencies will make this a viable means for enhancing livestock genetics. Areas of transgenic technology being heavily pursued include improvements in milk quality, improvements in viral resistance and natural immunity, and improvements in production traits, such as wool and carcass characteristics.

A. Improved Milk Quality

Modification of milk protein composition is possible through transgenic animal technology. Several applications of agricultural benefit may be realized when transgenesis becomes more efficient and cost-effective. Improvements in cheese production may be made through increasing casein expression. This will improve thermal stability, thus improving curd formation. Changes in phosphorylation of casein would also allow for cheeses to be made either softer or harder. Subtle changes in the sequence of the milk genes will allow for improved proteolysis, making cheese maturation more rapid. Another possibility is to "humanize" the milk of farm animals. By replacing animal milk protein genes with human genes, better milk substitutes could be generated for infants. Others are investigating the reduction or deletion of milk components, such as lactose or fat. With an estimated 90% of adults having lactose intolerance, modification of this sugar or its cleavage to glucose and galactose would provide a useful alternative. However, deletion of lactose may not be an option since α-lactalbumin knockout mice did not lactate properly due to viscous milk production from the lack of lactose. Finally, several researchers are working to improve antibacterial proteins in milk. Lysozyme concentration is 3000-fold higher in human milk than in that of cows. This protein is effective against bacteria causing food spoilage and food-borne disease. These antibacterial protein genes may also reduce the incidence of mastitis, a disease which results in large financial losses to the dairy industry every year.

B. Improved Disease Resistance

Another means by which transgenic animals may benefit animal agriculture is through improved disease resistance. Several groups are taking different approaches to enhance the animals immune response, including transfer of genes for natural immunity, antibodies, and viral proteins. Natural immunity to diseases may eventually be transferred from one species to another as genes regulating disease resistance are determined. Several studies have utilized the *Mx-1* influenza viral resistance gene in mice and pig transgenesis with variable results. Proper regulation of these resistance genes will allow further progress. Transfer of preformed antibodies is another approach for improving disease resistance. Transgenic mice, rabbits, sheep, and pigs produce light- and heavy-chain components of antibodies. Mice have been produced that have antibodies toward the bacterial surface antigen phosphorylcholine. However, other species have produced hybrid antibodies that do not recognize the antigen. Researchers have also produced transgenic sheep and mice that produce the *env* gene of the visna virus and chickens which express avian leukosis viral envelope proteins. These transgenic animals produce antibodies to these proteins, which may improve resistance to disease by blocking the entrance of the virus into the cell.

C. Improved Production Traits

Improvements in animal productivity by transgenesis have been limited due to the polygenic nature of most production traits and the lack of knowledge of genes controlling them. Genome mapping projects currently under way should improve our understanding in this area by identifying genes responsible for economically important traits. Recently, however, there have been advances to improve wool production. Researchers produced transgenic sheep that have a 6.2% increase in fleece weight due to an insulin-like growth factor-1 (IGF-1) gene driven by a keratin promoter. This is the first known improvement in a production trait. Unfortunately, transgenic improvements in meat production have been less successful. Several groups have utilized the growth hormone (GH), growth hormone-releasing factor (GRF), IGF-1, and c-*SKI* genes driven by several different promoters to improve overall growth and muscling in transgenic pigs, cattle, and sheep. While several reports indicated improved rate of growth, improved feed conversion, and decreased fat, they were not without consequence. All researchers reported health problems, including joint problems, kidney disease, gastric ulcers, infertility, and muscle weakness or atonia. These problems may be associated with improper regulation of the transgene. Future improvements rely on elucidation of factors controlling positional independence and proper regulation of transgene expression.

VII. FUTURE AVENUES FOR TRANSGENIC ANIMAL TECHNOLOGY

For nearly two decades researchers have been producing transgenic animals using a variety of approaches. There continues to be problems generating these animals, most notably due to inefficiencies in (i) transgene insertion or gene targeting, (ii) regulation of transgene expression, and (iii) generating transgenic livestock at a reasonable expense in both time and capital. Several possibilities for improvements are addressed here.

A. Transgenic Embryonic Stem Cells or Somatic Cells in Domestic Livestock

Recent advances in ES cell technology in livestock species may enhance the ability to produce transgenic livestock. Pluripotent stem cells are being generated from primordial germ cells (EG cells) and from preimplantation embryos (ES cells) in pigs, sheep, and cattle. Recently, it has been shown that even somatic cells from adult sheep can be utilized in nuclear transfer procedures to produce live offspring. Advances in these areas indicate that improvements in production of transgenic livestock are within reach.

1. *Nuclear Transfer*

Nuclear transfer or cloning is a procedure for transferring the nucleus of a blastomere from an embryo or a somatic cell from a fetus or an adult to the cytoplasm of an enucleated unfertilized egg by electrical fusion. This technique allows for reprogramming of the cell's genetic material to generate a new animal with all the characteristics of the nuclear donor. This is a very powerful technique with exciting implications for the field of transgenics. It may be utilized to make multiple copies of a transgenic founder embryo. More important, it has potential for generating transgenic animals from ES or EG cells of livestock. Currently, mouse ES and EG cells are injected into host embryos to produce chimeric offspring which must be bred to generate heterozygotes and homozygotes for the transgene. This is time-consuming and expensive for livestock production due to their long generation interval. Generation of homozygotes is also potentially problematic due to inbreeding depression associated with noninbred species. If successful, nuclear transfer would eliminate two generations of animal breeding. Currently, this avenue is being thoroughly pursued as a means to improve livestock transgenesis.

2. *Tetraploid Embryos*

Another area for improved production of transgenic livestock is through the use of tetraploid embryos. Tetraploid embryos are produced by fusing the cells of a two-cell embryo to produce a one-cell embryo with twice as much genetic material. This can be accomplished chemically or through electrical fusion. These embryos have been used to study events in early embryogenesis and it has been determined that cells from these embryos develop primarily into trophoblast, the placenta-forming cells, while those comprising the fetus die off after a short period of time. They have been utilized as host embryos for mouse ES cell injections, providing a means of generating mice derived solely from ES cells. This may be an excellent alternative for the development of completely ES cell-derived livestock, eliminating the need for breeding lines to homozygosity. Further testing of this method is currently under way for enhanced livestock production.

B. Improved Regulation of Transgene Integration and Expression

Greater understanding of the factors affecting gene expression is paramount for improvements in all species of transgenic animal production. Current investigations in gene mapping and genetic regulation should prove fruitful. Future goals include the complete regulation of transgene expression, allowing for accurate inducibility of tissue- and stage-specific expression. Generation of universal regulatory elements and matrix attachment regions that can be utilized in all constructs may also be realized. Improvements in transgene integration within the genome will also become available as mechanisms for DNA repair and homologous recombination are more fully understood. This will enhance efficiencies in both random and targeted gene insertions. Additionally, improvements in screening for transgene integration at early stages of development will improve efficiencies through reduced recipient costs. Embryo biopsy and PCR screening for transgenes are currently unreliable due to transient transgene expression. Recent work using selectable markers has shown promise by allowing only transgenic cells to survive to the blastocyst stage. Early determination by either method reduces the number of transfers resulting in decreased recipient costs. Finally, improvements in superovulation and *in vitro* techniques, such as oocyte maturation, fertilization, and embryo culture, are being investigated. Further advancements in all these areas will enhance the field of transgenesis within the next decade or two.

See Also the Following Articles

Cloning Mammals by Nuclear Transfer; Immunology of Reproduction; Reproductive Technologies, Overview

Bibliography

Brandon, E. P., Idzerda, R. L., and McKnight, G. S. (1995). Targeting the mouse genome: A compendium of knockouts. *Curr. Biol.* **5**, 625–881.

Campbell, K. H. S., McWhir, J., Ritchie, W. A., and Wilmut, I. (1996). Sheep cloned by nuclear transfer from a cultured cell line. *Nature* **30**, 64–66.

First, N. L., and Haseltine, F. P. (1988). *Transgenic Animals.* Proceedings of the Symposium on Transgenic Technology in Medicine and Agriculture. Butterworth-Heinemann, Stoneham, MA.

Hogan, B., Beddington, R., Costantini, F., and Lacy, E. (1994). *Manipulating the Mouse Embryo: A Laboratory Manual,* 2nd ed. Cold Spring Harbor Laboratory Press, Cold Spring Harbor, NY.

Houdebine, L. M. (1994). Production of pharmaceutical proteins from transgenic animals. *J. Biotechnol.* **34**, 269–287.

Joyner, A. L. (1993). *Gene Targeting: A Practical Approach.* Oxford Univ. Press, New York.

Murphy, D., and Carter, D. A. (1993). *Transgenesis Techniques: Principles and Protocols.* Humana Press, Totowa, NJ.

Pinkert, C. A. (1994). *Transgenic Animal Technology: A Laboratory Handbook.* Academic Press, San Diego.

Pursel, V. G., and Rexroad, C. E. (1993). Status of research with transgenic farm animals. *J. Anim. Sci.* **71**(Suppl. 3), 10–19.

Rosen, J. M., Li, S., Raught, B., and Hadsell, D. (1996). The mammary gland as a bioreactor: Factors regulating the efficient expression of milk protein-based transgenes. *Am. J. Clin. Nutr.* **63**, 627S–632S.

Sambrook, J., Fritsch, E. F., and Maniatis, T. (1989). *Molecular Cloning: A Laboratory Manual.* Cold Spring Harbor Laboratory Press, Cold Spring Harbor, NY.

Wall, R. J. (1996). Transgenic livestock: Progress and prospects for the future. *Theriogenology* **45**, 57–68.

Wasserman, P. M., and DePamphilis, M. L. (1993). *Guide to Techniques in Mouse Development,* Methods in Enzymology, Vol. 225. Academic Press, San Diego.

Wilmut, I., Hooper, M. L., and Simons, J. P. (1991). Genetic manipulation of mammals and its application in reproductive biology. *J. Reprod. Fertil.* **92**, 245–279.

Trophoblast to Human Placenta

Harvey J. Kliman

Yale University School of Medicine

I. Formation of the Placenta
II. Structure and Function of the Placenta
III. Complications of Pregnancy Related to Trophoblasts and the Placenta

GLOSSARY

amnion The inner layer of the external membranes in direct contact with the amnionic fluid.

chorion The outer layer of the external membranes composed of trophoblasts and extracellular matrix in direct contact with the uterus.

chorionic plate The connective tissue that separates the amnionic fluid from the maternal blood on the fetal surface of the placenta.

chorionic villous The final ramification of the fetal circulation within the placenta.

cytotrophoblast A mononuclear cell which is the precursor cell of all other trophoblasts.

decidua The transformed endometrium of pregnancy.

intervillous space The space in between the chorionic villi where the maternal blood circulates within the placenta.

invasive trophoblast The population of trophoblasts that leaves the placenta, infiltrates the endo- and myometrium, and penetrates the maternal spiral arteries, transforming them into low-capacitance blood channels.

junctional trophoblast The specialized trophoblast that keeps the placenta and external membranes attached to the uterus.

spiral arteries The maternal arteries that travel through the myo- and endometrium which deliver blood to the placenta.

syncytiotrophoblast The multinucleated trophoblast that forms the outer layer of the chorionic villi responsible for nutrient exchange and hormone production.

The precursor cells of the human placenta—the trophoblasts—first appear 4 days after fertilization as the outer layer of cells of the blastocyst. These early blastocyst trophoblasts differentiate into all the other cell types found in the human placenta. When fully developed, the placenta serves as the interface between the mother and the developing fetus. The placental trophoblasts are critical for a successful pregnancy because they mediate such critical steps as implantation, pregnancy hormone production, immune protection of the fetus, increase in maternal vascular blood flow into the placenta, and delivery.

I. FORMATION OF THE PLACENTA

As early as 4 days after fertilization, the trophoblasts—the major cell type of the placenta—begin to make human chorionic gonadotropin (hCG), a hormone which ensures that the endometrium will be receptive to the implanting embryo. Over the next few days, these same trophoblasts attach to and invade the uterine lining, beginning the process of pregnancy (Fig. 1). Over the next few weeks the placenta begins to make hormones which control the basic physiology of the mother in such a way that the fetus is supplied with the necessary nutrients and oxygen needed for successful growth. The placenta also protects the fetus from immune attack by the mother, removes waste products from the fetus, induces the mother to bring more blood to the placenta, and, near the time of delivery, produces hormones that mature the fetal organs in preparation for life outside of the uterus.

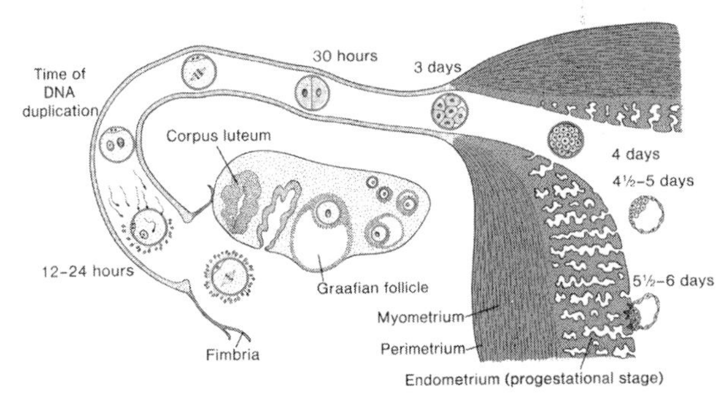

FIGURE 1 From ovulation to implantation. Ovulation occurs around Day 14 of the menstrual cycle, followed by fertilization within 24 hr. The first 3 days of development occur within the Fallopian tube. Upon arrival within the uterus the conceptus has developed into a blastocyst (see Fig. 2) and has already begun to make mRNA for human chorionic gonadotropin, the first hormone signal from the early embryo. By Day 6 after fertilization the blastocyst has initiated implantation into the maternal endometrium (modified with permission from T. W. Sadler, *Langman's Medical Embryology*, 5th ed., Williams & Wilkins, Baltimore, 1985).

A. Early Development

Within a few days of fertilization the embryo develops into a blastocyst, a spherical structure composed on the outside of trophoblasts and on the inside of a group of cells called the inner cell mass (Fig. 2). The inner cell mass will develop into the fetus and, ultimately, the baby. In addition to making hCG, the trophoblasts mediate the implantation process by attaching to, and eventually invading into, the endometrium (Fig. 3).

Implantation is regulated by a complex interplay between trophoblasts and endometrium. On the one hand, trophoblasts have a potent invasive capacity and, if allowed to invade unchecked, would spread throughout the uterus. The endometrium, on the other hand, controls trophoblast invasion by secreting locally acting factors (cytokines and protease inhibitors), which modulate trophoblast invasion. Ultimately, normal implantation and placentation is a balance between regulatory gradients created by both the trophoblasts and endometrium (Fig. 4).

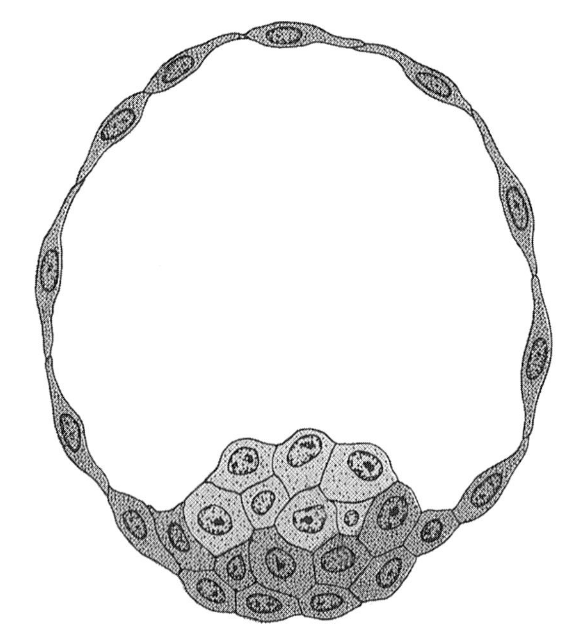

FIGURE 2 Blastocyst. By 4 or 5 days after fertilization the embryo has differentiated into two distinct cell types: inner cell mass (lighter cells), which will develop into the fetus and eventually become the newborn, and trophoblasts (darker cells), which will develop into the placenta and external membranes. Even by this stage the trophoblasts have begun to make their hallmark hormone: human chorionic gonadotropin (the hormone of pregnancy-test fame) (modified with permission from T. W. Sadler, *Langman's Medical Embryology*, 5th ed., Williams & Wilkins, Baltimore, 1985).

B. Formation of the Early Placenta

Once firmly attached to the endometrium the developing conceptus grows and continues to expand into the endometrium. One of the basic paradigms which is established even within the first week of gestation is that the embryonic/fetal cells are always separated from maternal tissues and blood by a layer of cytotrophoblasts (mononuclear trophoblasts) and syncytiotrophoblasts (multinucleated trophoblasts) (Fig. 5). This is critical not only for nutrient exchange but also to protect the developing fetus from maternal immunologic attack.

Within the first 2 weeks of development, the invading front of trophoblasts has penetrated the endometrial blood vessels, forming intertrophoblastic maternal blood-filled sinuses (Fig. 6). The trophoblastic shell continues to grow and develop until 3 weeks after fertilization, when the earliest evidence of fetal circulation can be identified and, with this, the earliest evidence of the chorionic villi formation (Fig. 7).

Despite the fact that the entire conceptus is less than 2 cm in diameter at 4 weeks, the basic structure of the mature placenta has been laid out: A fetal circulation that terminates in capillary loops within chorionic villi which penetrate a maternal blood-filled intervillous space which is supplied by spiral arteries and drained by uterine veins (Fig. 8).

The chorionic villi closest to the maternal blood supply will continue to develop and expand into a mass of chorionic tissue which we identify as the placenta. The chorionic villi farthest away from the maternal blood supply are slowly pushed into the uterine cavity by the expanding amnionic sac which surrounds the embryo. These villi eventually degenerate and form the chorionic layer of the external membranes. At around 20 weeks of gestation the combined amnion–chorion membrane makes con-

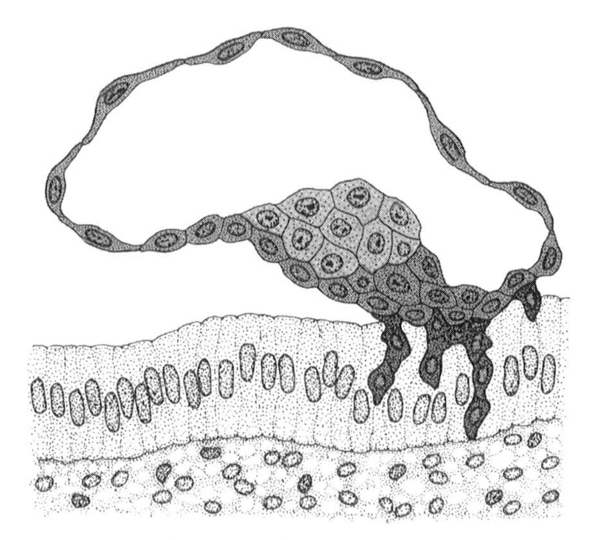

FIGURE 3 Implantation. The most difficult hurdle for the floating blastocyst is to become attached to the uterine lining (endometrium). Like trying to dock a tanker coming into port, the blastocyst is first slowed down by long molecules that extend from the endometrium (mucins), followed by a cascade of molecules that bring the trophoblasts into closer contact with the endometrium. Once intimate contact is made the trophoblasts begin to invade into the endometrium, beginning the process of placentation (modified with permission from T. W. Sadler, *Langman's Medical Embryology*, 5th ed., Williams & Wilkins, Baltimore, 1985).

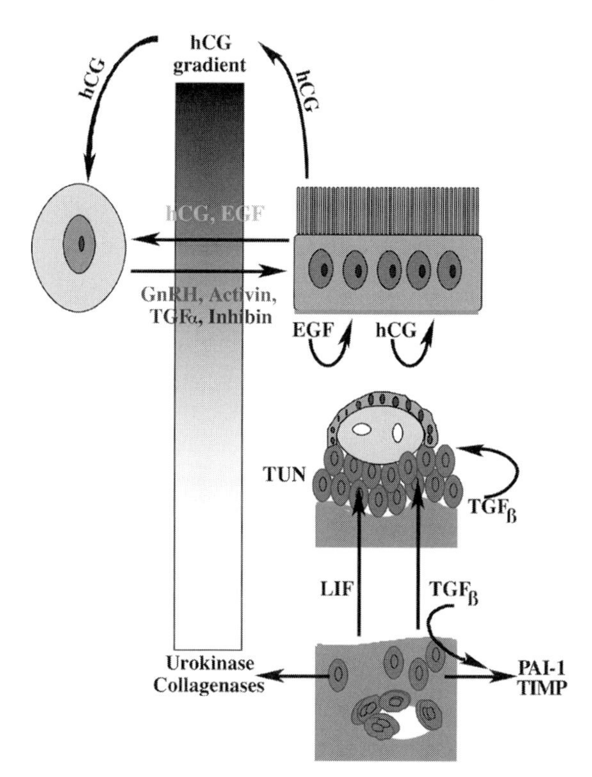

FIGURE 4 Regulation of trophoblast invasion by an hCG gradient. Within the placenta the syncytiotrophoblasts generate high levels of hCG, which shifts cytotrophoblast differentiation toward a noninvasive hormone-secreting villous-type trophoblast. The closer the trophoblasts are to the endometrium, the less hCG is made, allowing the trophoblasts to differentiate into anchoring-type cells which make the placental glue protein trophouteronectin. Trophoblasts that leave the placenta and migrate within the endo- and myometrium are induced to make proteases and protease inhibitors, presumably to facilitate trophoblast invasion into the maternal tissues.

tact with the opposite side of the uterus, where it fuses with the decidualized maternal endometrium, forming the complete external membrane consisting of amnion, chorion, and decidua layers.

II. STRUCTURE AND FUNCTION OF THE PLACENTA

A. Basic Structure

The placenta is the fetus' extension into the mother, where it functions as the interface between

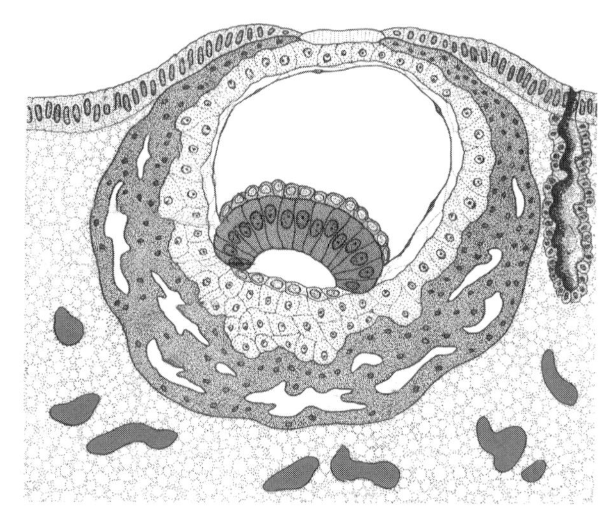

FIGURE 5 Day 9 implantation site. By 9 days the embryo is surrounded by two layers of trophoblasts: the inner mononuclear cytotrophoblasts and the outer multinucleated syncytiotrophoblast layer. This arrangement of embryo, trophoblasts, and maternal tissue remains the paradigm throughout gestation. This trophoblast interface not only serves as the means to extract nutrients from the mother but also protects the embryo and fetus from maternal immunologic attack (modified with permission from T. W. Sadler, *Langman's Medical Embryology*, 5th ed., Williams & Wilkins, Baltimore, 1985).

the two. Like the radiator of a car—which is a heat exchanger—the placenta is a nutrient and waste exchanger. The fetal circulation enters the placenta much like the water of an automobile engine enters the radiator—via the umbilical arteries embedded within the umbilical cord. Once in the placenta, the fetal circulation branches into units called cotyledons, structures similar to inverted trees (Fig. 9). The finest branches of the fetal circulation are made up of capillary loops within the chorionic villi (Fig. 10). Once nutrients have been absorbed and waste products released, the fetal blood ultimately collects into the umbilical vein, where it returns to the fetus via the umbilical cord.

If the fetal circulation is analogous to the circulating water in an engine, the maternal circulation is analogous to the cool air rushing by the fine fins of the radiator. The maternal blood enters the placenta via the spiral arteries of the uterus. At the point where the spiral arteries make contact with the pla-

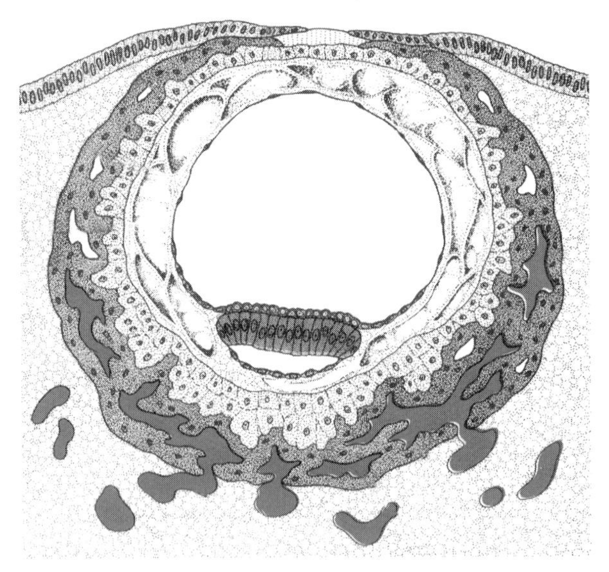

FIGURE 6 Day 12 implantation site. The invading trophoblasts have penetrated the maternal vessels, forming pools of maternal blood which surround the growing trophoblasts (modified with permission from T. W. Sadler, *Langman's Medical Embryology*, 5th ed., Williams & Wilkins, Baltimore, 1985).

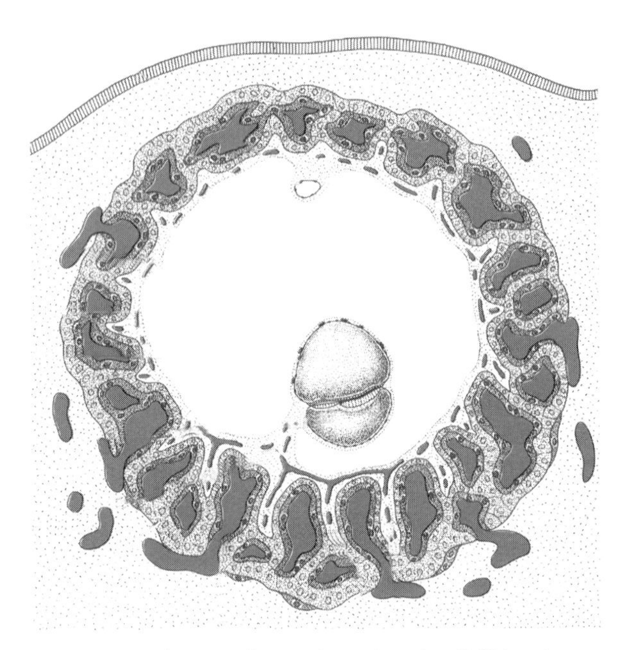

FIGURE 7 Three-weeks implantation site. Still barely perceptible to the naked eye, the embryo has already begun to make an early circulatory system. Again, embryonic tissue and maternal blood are separated by a layer of cytotrophoblasts and syncytiotrophoblasts (modified with permission from T. W. Sadler, *Langman's Medical Embryology*, 5th ed., Williams & Wilkins, Baltimore, 1985).

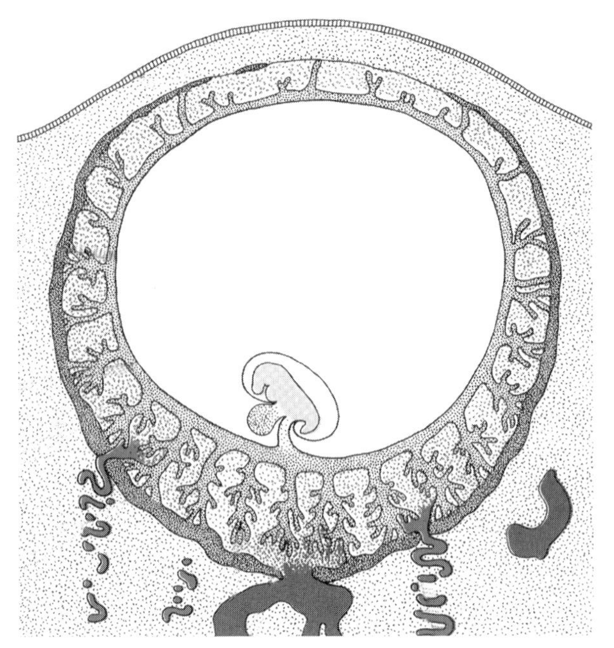

FIGURE 8 Four-weeks implantation site. The basic structure of the placenta has been formed with maternal blood being delivered to the forming placenta via spiral arteries while being drained away via uterine veins. Like the roots of a tree, the developing chorionic villi remain immersed in a space filled with the nutrient-rich maternal blood (modified with permission from T. W. Sadler, *Langman's Medical Embryology*, 5th ed., Williams & Wilkins, Baltimore, 1985).

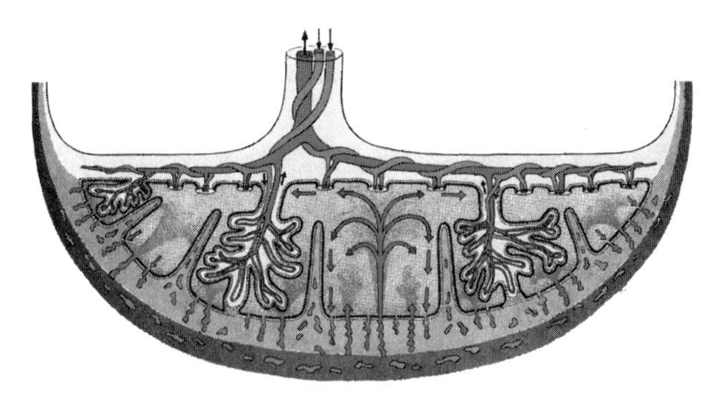

FIGURE 9 Maternal and fetal circulations within the placenta. Maternal blood is fountained into the placenta through the uterine spiral arteries, where it circulates around the chorionic villi—much like ocean water circulating around sea anemone. The fetus pumps blood into the placenta via two umbilical arteries that branch over the fetal surface of the placenta. The fetal arteries then dive into the placental mass, continuously branching until the blood reaches the capillary loops of the chorionic villi—much like a branching tree with leaves (modified with permission from K. L. Moore, 1993).

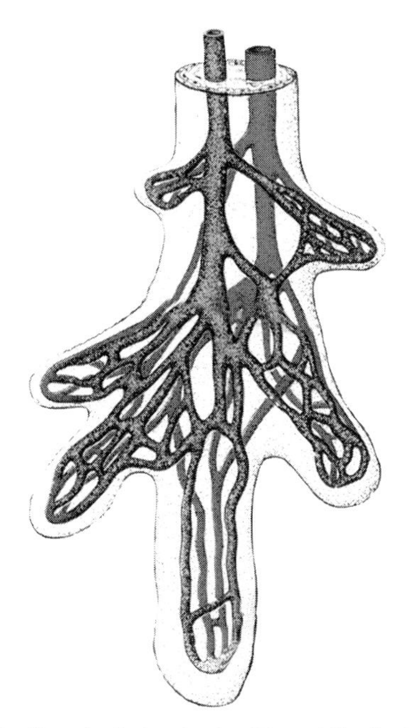

FIGURE 10 Terminal chorionic villous. The fetal circulation branches until it reaches the capillaries of the chorionic villi (Latin for leaf) where exchange of nutrients takes place between the mother and fetus. Like a growing branch, new villous branches bud off of the larger villi to increase the mass and exchange surface area of the placenta (modified with permission from T. W. Sadler, *Langman's Medical Embryology*, 5th ed., Williams & Wilkins, Baltimore, 1985).

centa, they end in open channels, fountaining maternal blood into the intervillous space (Fig. 9). The intervillous blood is returned to the maternal circulation via drain-like uterine veins. In order to support the developing fetus, especially at term, up to 35% of the maternal blood flow courses through the intervillous space.

The finger-like chorionic villi are the main functional units of the placenta (Fig. 10)—mediating nutrient absorption, waste elimination, and generating the bulk of the hormones produced by the placenta during pregnancy. Cross sectioning a chorionic villous reveals the basic components of this part of the placenta (Fig. 11). Toward the end of the first trimester a cross section of a chorionic villous reveals a central mesenchymal core with embedded fetal capillaries surrounded by a layer of cytotrophoblasts and syncytiotrophoblasts (Fig. 11A). At term a chorionic villous cross section reveals the same basic structure with some noteworthy differences (Fig. 11B). A term chorionic villous exhibits increased numbers of fetal capillaries, some of which are very close to the outer edge of the villous to facilitate nutrient exchange. The syncytiotrophoblast layer, a flat, multinucleated cell sheet for most of pregnancy, develops grape-like nucleated clusters within its cytoplasm called syncytial knots near term. Although there are fewer

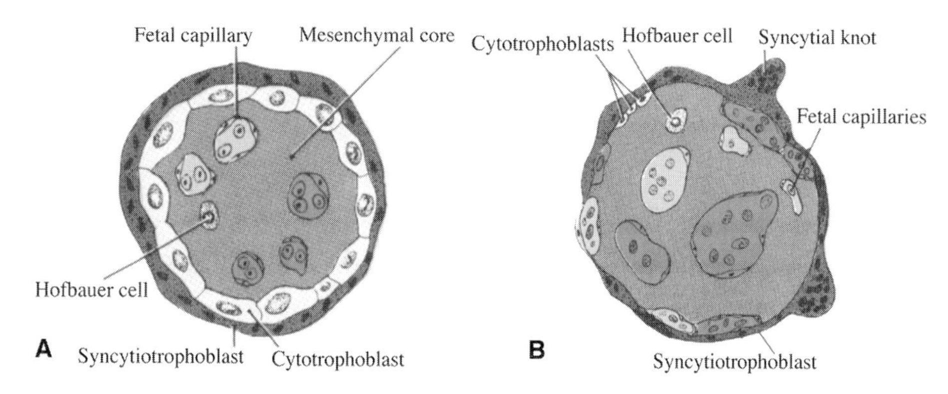

FIGURE 11 Diagrammatic cross sections of first- (A) and third- (B) trimester chorionic villi. The villous core of villi from all gestational ages beyond 5 weeks of gestation contains fetal capillaries embedded in a loose matrix which contains fibroblasts and macrophages (also called Hofbauer cells). In the first trimester a villous cross section (A) reveals two distinct trophoblast layers: the outer syncytiotrophoblast layer, which is in direct contact with maternal blood, and the inner cytotrophoblast layer, the source of new trophoblasts. A cross section of a third-trimester villous still exhibits a distinct syncytiotrophoblast layer, but fewer cytotrophoblasts can be identified, although they persist throughout pregnancy (see also Figs. 13 and 14) (modified with permission from T. W. Sadler, *Langman's Medical Embryology*, 5th ed., Williams & Wilkins, Baltimore, 1985).

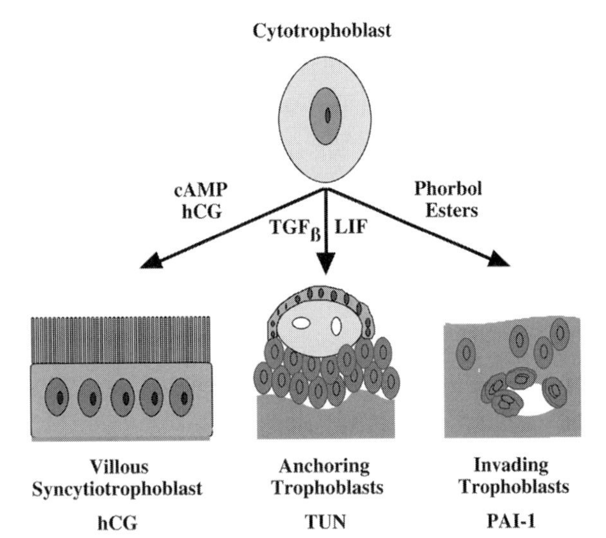

FIGURE 12 Pathways of trophoblast differentiation. Just as the undifferentiated basal layer of the skin gives rise to differentiated keratinocytes, the cytotrophoblast—the stem cell of the placenta—gives rise to the differentiated forms of trophoblasts. (Left) Within the chorionic villi, cytotrophoblasts fuse to form the overlying syncytiotrophoblast. The villous syncytiotrophoblast makes the majority of the placental hormones, the most studied being hCG. Cyclic AMP and its analogs, and recently hCG itself, have been shown to direct cytotrophoblast differentiation toward a hormonally active syncytiotrophoblast phenotype. (Center) At the point where chorionic villi make contact with external extracellular matrix (decidual stromal ECM in the case of intrauterine pregnancies), a population of trophoblasts proliferates from the cytotrophoblast layer to form the second type of trophoblast—the junctional trophoblast. The junctional trophoblasts make a unique fibronectin (TUN) that appears to mediate the attachment of the placenta to the uterus. TGF-β and, recently, LIF have been shown to downregulate hCG synthesis and upregulate TUN secretion. (Right) Finally, a third type of trophoblast (the invasive intermediate trophoblast) differentiates toward an invasive phenotype and leaves the placenta entirely. In addition to making human placental lactogen, these cells also make urokinase-type plasminogen activator (uPA) and type 1 plasminogen activator inhibitor (PAI-1). Phorbol esters have been shown to increase trophoblast invasiveness in *in vitro* model systems and to upregulate PAI-1 in cultured trophoblasts.

cytotrophoblasts visible at term, they are still present, as they have been throughout gestation, functioning as the precursor cells for all the other trophoblast types.

B. Trophoblast Differentiation Pathways

Concomitant with the overall development of placental architecture is the differentiation of three distinct trophoblast types. Depending on their subsequent function *in vivo*, undifferentiated cytotrophoblasts can develop into (i) hormonally active villous syncytiotrophoblasts, (ii) extravillous anchoring trophoblastic cell columns, or (iii) invasive intermediate trophoblasts (Fig. 12). Within the villi of the human placenta—at all gestational ages—there always exists a population of cytotrophoblasts which remain undifferentiated and available for differentiation as necessary.

1. *Villous Syncytiotrophoblast*

Within the chorionic villi, cytotrophoblasts fuse to form the overlying syncytiotrophoblast. The villous syncytiotrophoblast makes the majority of the placental hormones, the most studied being hCG. hCG (Fig. 13) is critical to pregnancy since it rescues the corpus luteum from involution, thus maintaining progesterone secretion by the ovarian granulosa cells. Its usefulness as a diagnostic marker of pregnancy stems from the fact that it may be one of the earliest secreted products of the conceptus. Researchers have

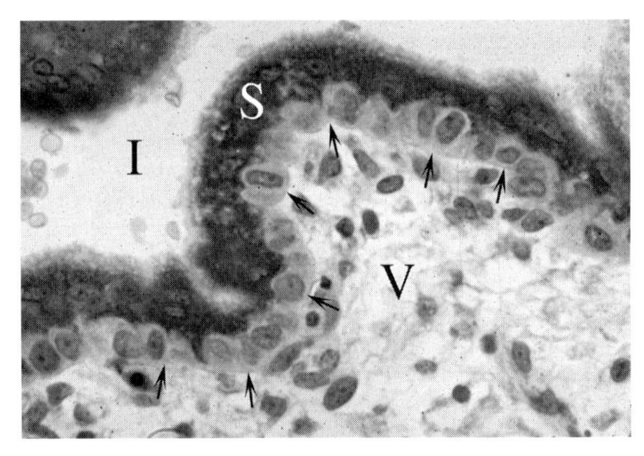

FIGURE 13 Immunohistochemical identification of hCG in first-trimester chorionic villi. The villous core (V) is covered by two layers of trophoblasts. The intensely stained syncytiotrophoblast layer (S) is in direct contact with the intervillous space (I) which contains the maternal blood. Underlying the syncytiotrophoblast layer is a continuous layer of cytotrophoblasts (arrows), the proliferative cell type of the placenta.

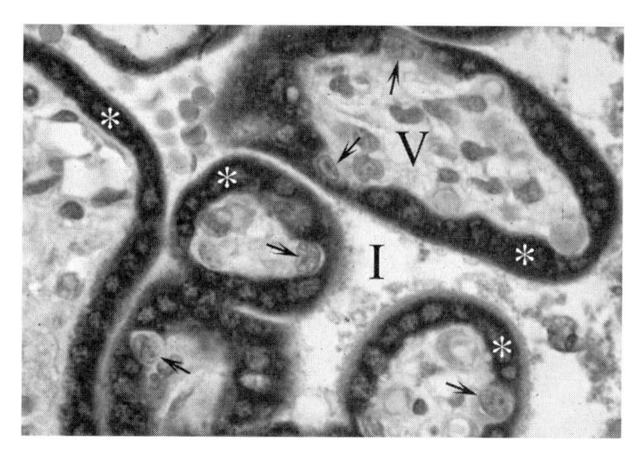

FIGURE 14 Immunohistochemical identification of hPL in third-trimester chorionic villi. The villous cores (V) are surrounded by a continuous layer of syncytiotrophoblasts (asterisks) which are in direct contact with the intervillous space. Underlying the syncytiotrophoblasts are scattered cytotrophoblasts (arrows). Although fewer in number during the third trimester, cytotrophoblasts are still easily identified, especially when the overlying syncytium is stained so as to highlight the less differentiated, hormonally negative cytotrophoblasts.

demonstrated by *in situ* hybridization that β-hCG transcripts are present in human blastocyst trophoblasts prior to implantation. Placental production of hCG peaks during the 10th to the 12th week of gestation and tends to plateau at a lower level for the remainder of pregnancy.

Human placental lactogen (hPL) is a potent glycoprotein made throughout gestation, increasing progressively until the 36th week, where it can be found in the maternal serum at a concentration of 5–15 μg/ml, the highest concentration of any known protein hormone. The major source of hPL appears to be the villous syncytiotrophoblasts, where it is made at a constant level throughout gestation. In addition to the villous syncytiotrophoblast, hPL has been identified in invasive trophoblasts during the first trimester. hPL appears to regulate the lipid and carbohydrate metabolism of the mother (Fig. 14). Placental researchers have also demonstrated by immunohistochemical studies that villous syncytiotrophoblasts contain prolactin, relaxin, and chorionic adrenocorticotropin, an ACTH-like protein. The physiological role of placental ACTH is unclear, but as with other placental hormones, all these hormones may represent a shift from maternal to placental control.

The significance of placental elaboration of progesterone was revealed when it was demonstrated that bilateral oophorectomy between 7 and 10 weeks of gestation had little impact on the conceptus or urinary pregnanediol levels. Recently, researchers have been able to directly demonstrate progesterone secretion by cultured term trophoblasts. The placenta does not have all the necessary enzymes to make estrogens from cholesterol or even progesterone. Human trophoblasts lack 17α-hydroxylase and therefore cannot convert C_{21} steroids to C_{19} steroids, the immediate precursors of estrogen. To bypass this deficit, dehydroisandrosterone sulfate from fetal adrenal is converted to estradiol-17β by trophoblasts. Not surprisingly, trophoblasts contain the necessary enzymes to make this conversion, namely, sulphatase, 3β-hydroxysteroid dehydrogenase/$\Delta^{5\rightarrow 4}$-isomerase, and aromatase.

The placenta also appears to produce a number of hypothalamic–pituitary hormones, including gonadotropin-releasing hormone (GnRH) and corticotropin-releasing hormone (CRH). GnRH was first identified within villous cytotrophoblasts by immunochemical staining of intact placentae. CRH is also made and secreted by cultured trophoblasts and glucocorticoids stimulate this secretion. In addition to the hypothalamic factors GnRH and CRH, pituitary growth hormone is synthesized and secreted by first- and third-trimester cultured trophoblasts. It appears from these studies that the placenta, in addition to replacing much of the women's pituitary function during pregnancy, also replaces critical hypothalamic functions so as to maintain control and feedback loop mechanisms close to the conceptus.

2. *Anchoring Trophoblasts*

At the point where chorionic villi make contact with the uterus, a population of trophoblasts proliferates from the cytotrophoblast layer to form the second type of trophoblast—the junctional trophoblast. Like the pylons of a bridge that attach to a riverbed, these cells form the anchoring cell columns that can be seen at the junction of the placenta and endometrium throughout gestation (Fig. 15). Similar trophoblasts can be seen at the junction of the chorion layer of the external membranes and the decidua (Fig. 16). The junctional trophoblasts make a unique fibronec-

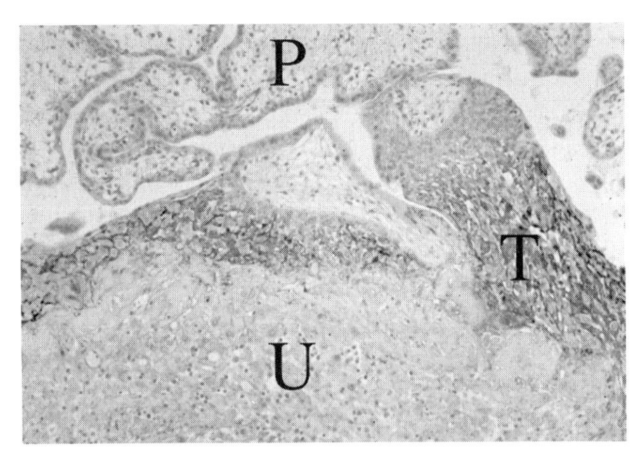

FIGURE 15 Trophouteronectin (pregnancy glue) expression by the junctional trophoblasts (T) that mediate the attachment of the placenta (P) to the uterus (U).

tin—trophouteronectin (TUN)—that appears to mediate the attachment of the placenta to the uterus. Transforming growth factor-β (TGF-β), and, recently, leukemia inhibitory factor, have been shown to downregulate hCG synthesis and upregulate TUN secretion. The premature loss of attachment of the developing conceptus or placenta to the uterus can terminate the gestation. Therefore, the anchoring trophoblast cell columns and the extracellular matrix proteins that promote this attachment are critical to the developing pregnancy.

3. *Invading Trophoblasts*

A third type of trophoblast, the invasive intermediate trophoblast, differentiates toward an invasive phenotype and leaves the placenta entirely. Studies using human specimens 3 or 4 weeks after implantation show that as gestation progresses, these invasive populations of extravillous trophoblasts attach to and interdigitate through the extracellular spaces of the endo- and myometrium. The endpoint for this invasive behavior is penetration of maternal spiral arteries within the uterus. Histologically, trophoblast invasion of maternal blood vessels results in disruption of extracellular matrix components and development of dilated capacitance vessels within the uteroplacental vasculature (Fig. 17). Biologically, trophoblast-mediated vascular remodeling within the placental bed allows for marked distensibility of the uteroplacental vessels, thus accommodating the increased blood flow needed during gestation. Abnormalities in this invasive process have been correlated with early and midtrimester pregnancy loss, preeclampsia and eclampsia, and intrauterine growth retardation.

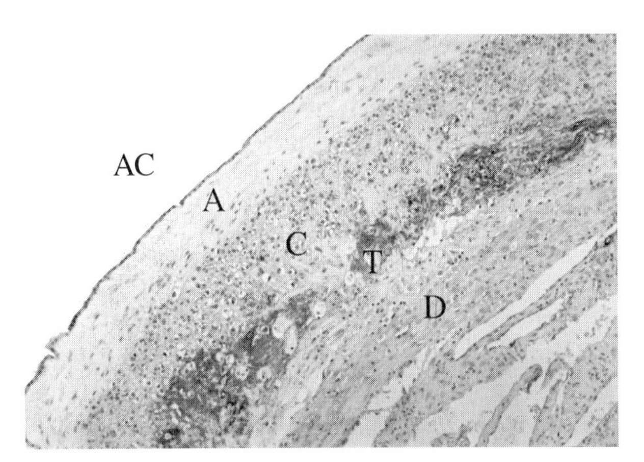

FIGURE 16 Trophouteronectin (pregnancy glue) expression by membrane trophoblasts (T). The external membranes consist of two layers: amnion epithelium and its underlying connective tissue (A) and chorion trophoblasts (C). Where the chorionic trophoblasts make contact with the maternal decidua (D), a layer of trophouteronectin can be identified (T). AC, amnionic cavity.

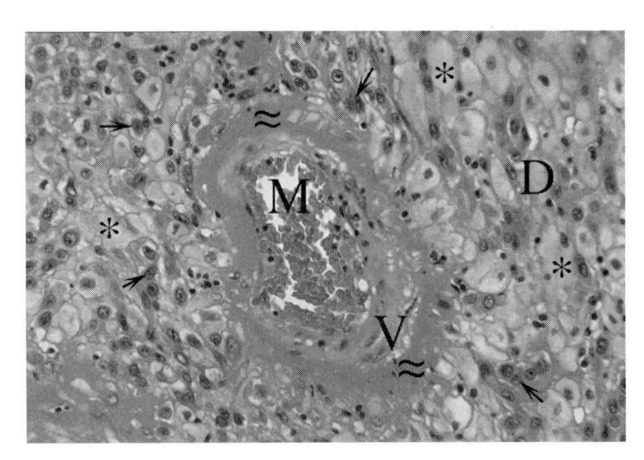

FIGURE 17 Invasive trophoblasts. Uterine spiral artery (V) containing maternal blood (M) from a 4-week pregnancy. The maternal endometrium (D) has become decidualized, meaning that the stromal cells have been transformed into large, pale cells (asterisks). Infiltrating between these decidual cells are the invasive trophoblasts, which have begun to modify the vessel wall ($\approx$).

III. COMPLICATIONS OF PREGNANCY RELATED TO TROPHOBLASTS AND THE PLACENTA

As in any complicated system, problems can arise. It is not possible to discuss all the pathologic states of the placenta, but the three most important complications of pregnancy related to the placenta will be outlined.

A. Diseases of Trophoblast Invasion: Preeclampsia and Gestational Trophoblastic Neoplasia

Preeclampsia, the clinical state prior to full-blown eclampsia (seizures), is one of the "toxemias" of pregnancy. The basic clinical definition is a pregnancy-specific condition of increased blood pressure accompanied by proteinuria, edema, or both. Despite the simplicity of this description of these clinical signs and symptoms, the etiology of the disease has remained elusive. In fact, a seal over the University of Chicago Lying-in-Hospital remains blank awaiting the name of the person(s) who elucidate the cause of this disease. Many phenomena have been investigated, but the recurring theme appears to be an abnormally low blood flow into the placenta. One of the difficulties has been to distinguish between primary cause and secondary effects. Part of this difficulty may be attributable to the fact that the common end result of low uteroplacental blood flow may be caused by many primary defects. Possibly, therefore, preeclampsia/eclampsia is not a disease but a syndrome with many causes. Significantly, one of the most frequent findings in preeclampsia is decreased or absent trophoblast invasion of the maternal spiral arteries.

Decreased or absent trophoblast invasion may be a consequence of primary defects in the invasive trophoblasts or in the environment that the trophoblasts are attempting to invade. Studies have shown that in some cases of preeclampsia there are abnormalities in trophoblast function, including, but not limited to, integrin expression, glycogen metabolism, and decreased galactose α-1,3-galactose expression. In addition, preeclampsia has been associated with trisomy 13, the chromosome that carries the gene for type IV collagen. Placental bed biopsy in a case of preeclampsia in a multiparous woman carrying a trisomy 13 fetus showed lack of trophoblast invasion of maternal spiral arteries. These trophoblasts may have had difficulty invading through the maternal extracellular matrix because of increased type IV collagen production. In addition to primary trophoblast defects, many cases of preeclampsia appear to be related to maternal immunologic reaction against the invading trophoblasts. A common clinical finding in these cases is that the invasive trophoblasts have reached the vicinity of the spiral arteries but have not penetrated them. In addition, the unconverted arteries are often surrounded by lymphocytes, presumably attacking the foreign-appearing invasive trophoblasts. As can be seen from a placental bed biopsy in a typical case of preeclampsia, the invasive trophoblasts have invaded through the endo- and myometrium, but have failed to complete their journey into the spiral arteries (Fig. 18). Failure to convert the maternal spiral arteries into low-resistance

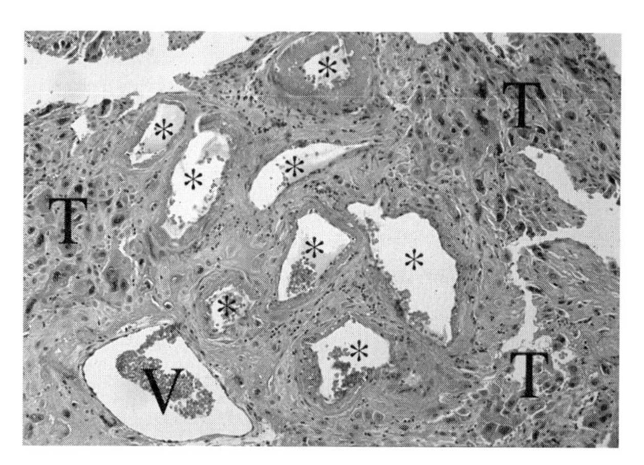

FIGURE 18 Failure of invasive trophoblasts to penetrate the maternal spiral arteries. Normally the invasive trophoblasts (T) infiltrate through the endo- and myometrium, reach the spiral arteries (asterisks), and convert their muscular walls into pliant channels. In cases of preeclampsia, the trophoblasts often do not complete the final arterial penetration, possibly due to the maternal lymphocytes that often surround the spiral arteries. Compensatory maternal hypertension can lead to spiral artery damage or even occlusion. V, maternal uterine vein.

channels can induce the placenta to secrete vasoactive substances, leading to maternal hypertension. If the maternal blood pressure rises significantly, the spiral arteries can be damaged and may even become occluded, leading to placental infarction.

In contrast to the clinical syndrome of decreased trophoblast invasion, gestational trophoblastic disease represents increased and uncontrolled trophoblast invasion. Expanded trophoblast invasion includes an exaggerated placental site with increased numbers of benign intermediate trophoblasts, placental site trophoblastic tumors, invasive moles, and frank choriocarcinoma. Morphologic distinction between these forms of trophoblast proliferation can be difficult, but it appears that the normal mechanisms that stop trophoblast invasion are defective in choriocarcinoma cell lines. Normal cytotrophoblast differentiation can be shifted toward a villous syncytiotrophoblast and away from an invasive trophoblast phenotype by cAMP analogs, whereas this treatment does not affect choriocarcinoma invasiveness, suggesting a primary defect in differentiation signaling in the malignantly invasive trophoblast.

B. Infection

More than a third of all preterm births are associated with labor initiated by acute chorioamnionitis (inflammatory infiltrates in the chorionic plate and chorion and amnion layers of the external membranes). Not only does chorioamnionitis have severe consequences for the fetus through the initiation of preterm delivery but chorioamnionitis also increases the risk for cerebral palsy by a factor of at least four. Elucidation of the role of cytokines in controlling the inflammatory process during pregnancy has facilitated the development of diagnostic tests which can detect the earliest phases of chorioamnionitis as well as point the way toward effective treatment modalities for this disease process.

The Collaborative Perinatal Study (CPS) of the National Institute of Neurological and Communicative Disorders and Stroke followed the course of over 56,000 pregnancies in the United States between 1959 and 1966. In the CPS, more than a third of all preterm births were associated with labor initiated by acute chorioamnionitis. This study also revealed that acute chorioamnionitis was the most frequent cause of stillbirth and neonatal death. Chorioamnionitis does not only have severe consequences for the fetus through the initiation of preterm delivery but also may—through the initiation of the inflammatory cascade in the placenta and decidua—have direct deleterious effects on the fetus. The CPS showed clearly that acute chorioamnionitis was followed by a 20% greater than expected frequency of neurologic abnormalities at 7 years of age.

Infections of the amniotic fluid arise by a variety of routes, including from the abdominal cavity through the Fallopian tube, via the maternal bloodstream through the placenta, or iatrogenically following amniocentesis or funipuncture, but the most common route is an ascending infection through the cervix. It is not surprising, therefore, that the most common organisms cultured from amnionic fluid are commonly found in the vagina. There are clinical reports, however, of a wide variety of organisms found to cause intrauterine infections, including group B streptococci, *Listeria* monocytogenes, *Morganella morganii*, *Ureaplasma urealyticum*, Herpes simplex virus, parvovirus, *Chlamydia* species, *Capnocytophaga*, adeno-associated virus, and human immunodeficiency virus.

Once bacteria enter the amniotic fluid, they quickly multiply. Within 30 hr of the initiation of the infection, maternal neutrophils begin to be chemoattracted from the maternal circulation toward the amniotic cavity to fight off the bacterial infection. The extent to which these maternal neutrophils migrate toward the amniotic fluid has been used to estimate the severity and timing of intraamniotic infections.

The basic paradigm for inflammation-induced cytokine release and leukocytic chemoattraction has been elucidated in a number of organs and is being actively studied in the placenta. Breakdown products of the growing bacteria—especially the lipopolysaccharides (LPS) of the bacterial cell walls (endotoxin)—appear to play an important role in the initiation of this process. LPS initiates the inflammatory response by triggering the release of a cascade of cytokines from a variety of cell types which mediate

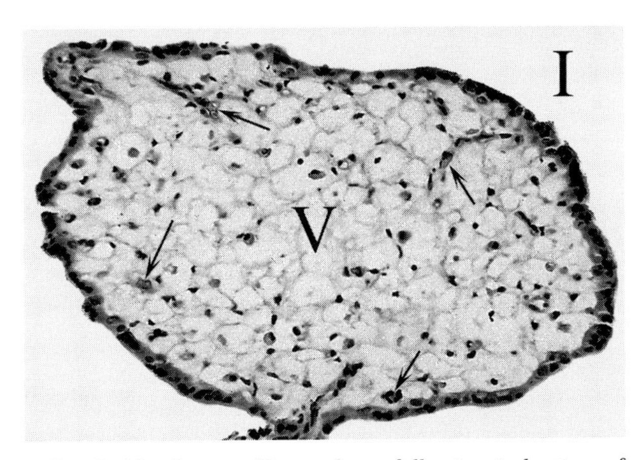

FIGURE 19 Severe villous edema following induction of intrauterine inflammation. Cross section of a markedly edematous chorionic villous 16 hr after the initiation of an intrauterine infection. Note the very pale, fluid-filled villous core (V). The edema fluid has compressed the fetal vessels (arrows) so severely that only one or two erythrocyte cross sections can be seen in each capillary. I, intervillous space.

the physiological process of inflammation. Components of this process include the chemotaxis and activation of inflammatory cells, such as the neutrophil, and the leaking of fluid from vascular spaces into the extracellular space, possibly in response to tumor necrosis factor-α (TNF-α). In experimental models, peak TNF-α production occurs within 1 hr of LPS treatment. TNF-α then induces the prolonged production of IL-8, a neutrophil chemoattractant and activator. In addition to chemoattracting leukocytes to the focus of inflammation, this cytokine cascade—as is succinctly described in Celsus's first-century phrase rubor, tumor, calor, and dolor—also results in swelling due to edema. In the placenta this swelling is manifested by villous edema—fluid accumulation within the mesenchymal cores of chorionic villi of the placenta. It has been hypothesized that villous edema makes the fetus hypoxic by compressing fetal blood vessels inside the villi and by increasing the diffusion barrier for oxygen between the maternal and fetal circulations (Fig. 19). These effects make villous edema the most frequent cause of stillbirth, neonatal death, and neonatal morbidity, especially cerebral palsy, in children born before 28 weeks.

C. Immunologic Rejection

Despite the fact that the placenta and fetus are "foreign" to the mother, most pregnancies show no evidence of "immunologic rejection." When immunologic reactions do occur, they can be against any of the components of the gestation (placenta and fetus). These reactions can occur at all stages of pregnancy and can occur repeatedly, pregnancy after pregnancy.

Examination of hematoxylin and eosin-stained histologic sections of placental, abortion or dilation, and curettage material can often reveal evidence of immune damage. However, more specialized techniques are needed to reveal the specific immune cells involved in these reactions. Immunohistochemistry of formalin-fixed, paraffin-embedded tissues can be used to identify specific cell types involved in immunologic reactions of pregnancy.

In approximately 1 or 2% of all gestations, mononuclear cells can be seen infiltrating into the chorionic villi of the placenta. Until the work of Redline and Patterson, however, the origin of these cells had been controversial, with some arguing for a fetal origin and some for a maternal origin. Since immunochemistry alone could not answer this question they utilized *in situ* hybridization for Y and X markers in male gestations to demonstrate that the lymphocytes present in cases of chronic villitis are maternally derived, thus allowing us to focus on the causes of this apparent maternal immunologic reaction against trophoblast and/or villous antigens. Immunohistochemistry of such cases has shown that the cells within the villous core are T cells.

Occasionally, placentas manifest an intervillous space that is filled with mononuclear cells. When immunohistochemically stained, these cells are revealed to be monocytic/macrophage in origin. This monocytic intervillositis has been associated with IUGR, preeclampsia, recurrent pregnancy loss, and intrauterine fetal demise.

Dizygotic twins offer a unique opportunity to examine maternal immunologic reactions against the placenta and fetus. Examination of dizygotic twins can sometimes reveal discordant chronic villitis between the two placentas, with concomitant discrepancies in fetal growth. Detailed analysis of each of such twins' HLA expression and cytokine expression may help to elucidate the causes of these immunologic reactions.

Although most cases of first-trimester pregnancy loss are the result of genetic defects in the fetus and/or placenta, some patients have recurrent pregnancy loss due to repeated maternal immunologic reactions. These reactions can be directed against villous core antigens, against antigens of the syncytiotrophoblast surface manifested as an intervillositis, or against invasive intermediate trophoblasts. Some have suggested a variety of therapies for these conditions, including treatment with intravenous immunoglobulins, immunization with paternal or allogenic leukocytes, or exposure to semen through vaginal or rectal suppositories. However, the scientific basis of many of these approaches remains controversial and the efficacy of the therapies proposed has been questioned.

See Also the Following Articles

Blastocyst; Decidua; Endometrium; Placenta: Implantation and Development; Placental and Decidual Protein Hormones; Preeclampsia/Eclampsia

Bibliography

Benirschke, K., and Kaufmann, P. (1995). *Pathology of the Human Placenta*, 3rd Ed. Springer-Verlag, New York.

Boyd, J. D., and Hamilton, W. J. (1970). *The Human Placenta*. Heffer, Cambridge, UK.

Feinberg, R., Kliman, H. J., and Lockwood, C. (1991a). Oncofetal fibronectin: A trophoblast "glue" for human implantation? *Am. J. Pathol.* **138**, 537–543.

Feinberg, R., Kliman, H. J., and Cohen, A. W. (1991). Preeclampsia, trisomy 13, and the placental bed. *Obstet. Gynecol.* **78**, 505–508.

Kliman, H. J. (1994a). Trophoblast infiltration. *Reprod. Med. Rev.* **3**, 137–157.

Kliman, H. J. (1994b). Placental hormones. *Infertil. Reprod. Med. Clin. North Am.* **5**, 591–610.

Kliman, H. J., Nestler, J., Sermasi, E., Sanger, J., and Strauss, J. (1986). Purification and in vitro differentiation of cytotrophoblasts from human term placentae. *Endocrinology* **118**, 1567–1582.

Kliman, H. J. (1993). The placenta revealed. *Am. J. Pathol.* **143**, 332–336.

Kliman, H. J., and Feinberg, R. F. (1990). Human trophoblast–extracellular matrix (ECM) interactions in vitro—ECM thickness modulates morphology and proteolytic activity. *Proc. Natl. Acad. Sci. USA* **87**, 3057–3061.

Kurman, R. J., Main, C. S., and Chen, H. C. (1984). Intermediate trophoblast: A distinctive form of trophoblast with specific morphological, biochemical, and functional features. *Placenta* **5**, 349–369.

Moore, K. L. (1993). *The Developing Human*, 5th ed. Saunders, Philadelphia.

Naeye, R. L. (1992). *Disorders of the Placenta, Fetus, and Neonate: Diagnosis and Clinical Significance*. Mosby/Year Book, St. Louis, MO.

Redline, R. W., and Patterson, P. (1993). Villitis of unknown etiology is associated with major infiltration of fetal tissue by maternal inflammatory cells. *Am. J. Pathol.* **143**, 473–479.

Sadler, T. W. (1990). *Langman's Medical Embryology*, 6th ed. Williams & Wilkins, Baltimore.

Tsetse Flies

R. H. Gooding
University of Alberta

GLOSSARY

adenotrophic viviparity A reproductive method in which a chorionated egg is fertilized, embryonated, and hatched within the female's uterus and the resulting larva is nourished by material from the female's reproductive accessory glands until it has completed its development.

choriothete A muscular and glandular invagination of the anterioventral wall of the uterus forming a tongue-shaped structure that anchors the egg and assists hatching.

Hippoboscoidea Four families of higher flies (Order Diptera) that include tsetse flies, louse flies, keds, and bat flies.

oviductal shelf A shelf-like structure on the dorsal wall of the uterus at the junction of the uterus and the common oviduct. It carries the openings of ducts from the spermathecae and the female accessory glands (the so-called "milk glands").

pupariation The process by which a mature third-instar larva of a higher fly forms the puparium, a hard case within which the true pupa and, ultimately, the adult are formed.

spermatophore A structure made from secretions of the male reproductive accessory glands that is used to transfer sperm to the female during mating.

Tsetse flies, and other Hippoboscoidea, are the only insects to reproduce by adenotrophic viviparity. The female gives birth to an offspring that, on average, weighs more than the female did at eclosion. After parturition the larva does not feed; its only activities are to burrow into the soil, defecate, and pupariate. This reproductive strategy is associated with significant modifications to the structure and physiology of female and larval tsetse flies. The process imposes a major limitation on the reproductive capacity of these flies: Only 1 offspring is produced at a time and female fecundity is relatively low. A theoretical maximum of about 20 offspring may be produced by a female. However, even in a well-managed colony females produce an average of only 6–8 offspring; in nature the mean, averaged over all seasons, is about 4 offspring per female.

I. INTRODUCTION

There about 30 species and subspecies of tsetse flies; all are members of the genus *Glossina.* Most tsetse flies are confined to Africa south of the Sahara, but a few species are found also in restricted pockets in the southwestern part of the Arabian peninsula. Each species inhabits a particular environment within rain forests, along edges of rivers or lakes, or in savannah. Both male and female tsetse flies feed exclusively on blood; each species prefers certain species of wild and domestic animals. Most, if not all, will feed on humans. The frequency with which tsetse flies feed varies; on average it is approximately every second day, but the feeding frequency of females is influenced by the stage of the pregnancy cycle. Tsetse flies are best known as the vectors of trypanosomes that cause sleeping sickness in man and a wasting disease, Nagana, in domestic animals in Africa. It is their importance as vectors that has stimulated most of the work on these insects.

II. ANATOMY OF THE REPRODUCTIVE TRACTS

A. Male Reproductive Tract

The male reproductive tract is similar to that of most male insects. Each testis is tubular, coiled to form a superficially spherical structure, and connected to a vas deferens (Fig. 1). The vasa deferentia and the two accessory glands empty into the ejaculatory duct, which, as in other higher flies, loops around the hindgut as it leads to the penis or intromittant organ. The peritoneal or testicular sheath contains a large amount of xanthommatin, a pigment derived from tryptophan, and thus the testes are dark brown or chestnut colored. Two notable features of the male reproductive system are that the vasa deferentia are not expanded to form distinct seminal vesicles and the accessory glands are relatively large. The accessory glands produce material that is transferred during copulation to form a spermatophore; other possible functions for these materials, such as inducing reproductive refractoriness, remain speculative.

B. Female Reproductive Tract

The female reproductive system (Fig. 2) includes the structures normally found in female insects: ovaries, bursa copulatrix, accessory glands, and sperma-

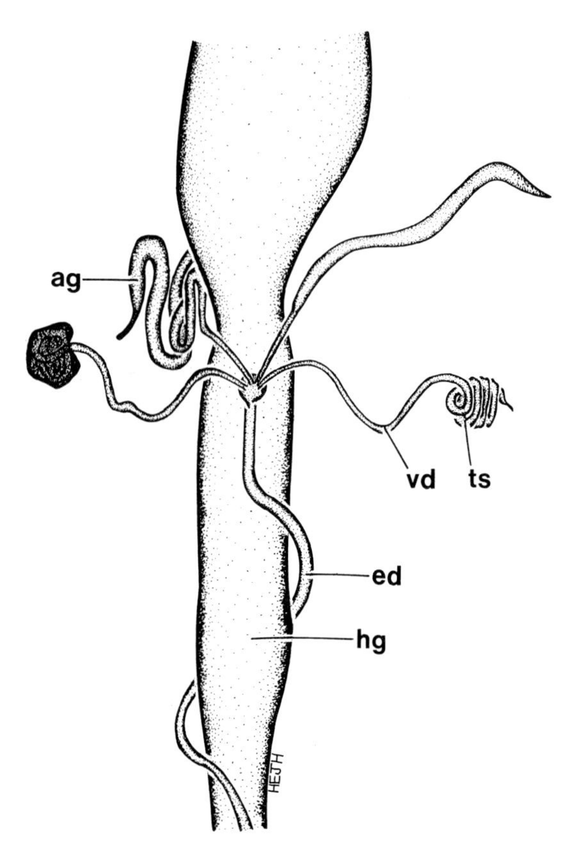

FIGURE 1 Male reproductive system in *G. m. morsitans* (ventral view) (freehand drawing by H. E. J. Hammond). The accessory gland shown on the right has been stretched from its normal configuration (on left) and the testicular sheath has been removed from the testis on the right to illustrate the coiled, tubular nature of the testis. ag, accessory gland; ed, ejaculatory duct; hg, hindgut; ts, testis; vd, vas deferens.

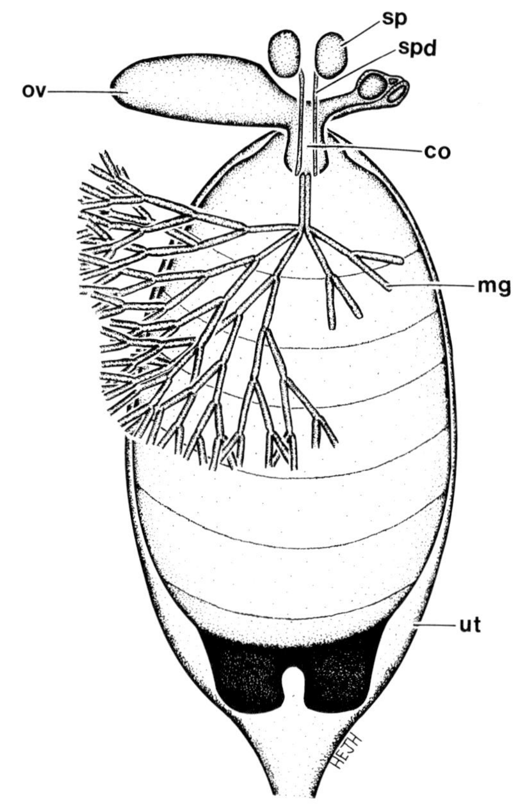

FIGURE 2 Female reproductive system in *G. m. morsitans*, during latter part of the pregnancy cycle (dorsal view) (camera lucida drawing by H. E. J. Hammond). Milk glands are drawn diagrammatically because of their complex ramifications within the fat body. co, common oviduct; mg, milk glands; ov, ovary; sp, spermatheca; spd, spermathecal duct; ut, uterus containing a late third-instar larva.

thecae. Each ovary is composed of two polytrophic ovarioles, each of which has two follicles. Development of the follicles is coordinated in such a way that at all times each is at a different stage of development. At eclosion the most developed follicle occurs, almost invariably, in one ovariole in the right ovary, the second most developed is in the left ovary, and the third most developed and the least developed are the other ovarioles in the right and left ovary, respectively. The sequence of ovulation thus created is maintained throughout the female's life. The mechanism by which an ovariole in the right ovary is first to develop has not been established. The bursa copulatrix and the accessory glands are highly modified to accommodate and nourish the developing larva.

III. GAMETOGENESIS

A. Spermatogenesis

In *G. m. morsitans* at 25°C, all mitotic divisions necessary for development of the testes and for production of a full complement of sperm have been completed within 3 days of ovulation. Premeiotic chromosome replication occurs 2 or 3 days later, i.e., it is complete 4 days prior to parturition. The testes develop within 2 days of pupariation and all spermatogenesis is complete prior to eclosion: Spermatogonia are formed by Day 5, meiosis occurs between Days 7 and 9, spermatids are found beginning on Day 11, and the full complement of mature sperm is formed prior to eclosion.

B. Oogenesis

The ovaries develop during the first quarter of the female's life within the puparium. Within the germarium, a cell undergoes three mitotic divisions to produce a cluster of 8 cells; one cell becomes the oocyte and the remainder are destined to become nurse cells. This cluster of 8 cells becomes surrounded by the follicular epithelium and descends farther into the follicular tube. Each mature follicle contains an ovum and 14 nurse cells, the latter having arisen by an additional mitotic division of the 7 nurse cells found in the young follicle. Vitellogenesis begins before eclosion in some species (e.g., *G. morsitans* and *G. pallidipes*) and after it in others (e.g., *G. austeni*). In most insects juvenile hormone (JH) is responsible for spaces appearing between follicular cells thus allowing direct contact of the hemolymph with developing oocytes during vitellogenesis. Such spaces do not form in tsetse and JH does not appear to be necessary for vitellogenesis, although it is necessary for previtellogenic growth. The bulk of the material for vitellogenesis comes from the follicular epithelium and nurse cells, but some yolk may be synthesized *de novo* within the oocyte. The follicular epithelium produces the chorion.

As the most mature oocyte acquires yolk and grows, so do the others, but only one egg is mature at any time. At 26 or 27°C it takes a female *G. m. morsitans* approximately 4 weeks for an egg to mature, and the sequence of development provides one mature egg every 9 or 10 days. Ovulation tears the pedicel portion of the follicular tube so that after ovulation the ovariole is attached to the ovary at the anterior end only; presumably it is the ovarian sheath that ensures that subsequent ova are guided to the common oviduct.

Oogenesis can be disrupted by the presence of antibiotics or coccidiostats (e.g., sulfaquinoxaline) in the flies' diet. The first detectable effect of these compounds is degeneration of bacteroid symbionts in the midgut mycetome; about 1 week later the chromatin in the polytene nurse cells begins to fragment but the microbial symbionts (*Wolbachia*) in the nurse cells and oocytes are not affected. Southern (1980) proposed that symbiotic bacteroids in the midgut produce most of the folic acid that serves as a precursor for purines and thymine. The deficiency of these bases results in inadequate DNA replication, death of the nurse cells and oocytes, and finally sterility, if a sufficiently high concentration of antibiotic or coccidiostat is administered.

IV. MATING AND INSEMINATION

Tsetse flies have well-developed vision which is used for host location by most, if not all, species. Males of several of the savannah-inhabiting species (*morsitans* group) often follow large, slow-moving objects (both animals and motor vehicles) without making any attempt to feed on the animals and will rest near these when they cease moving. Such "following swarms" are believed to be part of the mechanism by which males find virgin females when the latter come to feed. Males will approach, follow, and attempt to contact flies that approach large animals and mating pairs of some, but not all, species of savannah-inhabiting tsetse may be found near potential host animals. The mechanisms by which males and females of the forest-dwelling (*fusca* group) and riverine (*palpalis* group) tsetse encounter each other are less well understood.

Regardless of the mechanisms employed to bring the sexes together, males recognize potential mates by detecting specific hydrocarbons on the female's cuticle. These hydrocarbons are sex pheromones that induce copulatory attempts by males. In addition to the sex pheromones, sound production and other female behavior may have a role in identifying the species of the female. Nonetheless, males of the *palpalis* and *morsitans* groups may mate with heterospecific females, and in some cases viable offspring are produced.

Immediately after establishing contact with a female, the male mounts her and copulation begins. Copulation has been studied only in the laboratory where, if the female is receptive, it typically consists of three phases: a period of several minutes in which the male's tarsi and tibia contact various parts of the female, a relatively long quiescent period, and a brief period during which the male resumes some behaviors displayed early in copulation and in some species the male changes the alignment of the flies' bodies or repeatedly flexes his abdomen. The male's activities during the first phase of copulation may have an appeasement role and may influence postcopulatory activities of the female, including ovulation or refractory behavior. During the second phase, secretions from the male's accessory glands form a spermatophore; sperm transfer occurs during the relatively short final phase of copulation. The duration of copulation varies among species, strains, and individuals; typically it lasts an hour or more.

In the laboratory, very young males will copulate with females, but for most species the males must be several days old before they are able to inseminate females. In the field, females are rarely found *in copula* with very young males. Usually females are receptive at the time of their first meal, or shortly thereafter. However, *G. pallidipes* females are not receptive until 7–9 days of age, and males of this species must be about 12 days old before insemination is ensured. The question of whether sexual appetitiveness is influenced by the male's nutritional state is unresolved.

Following mating, sperm are transferred from the spermatophore to the spermathecae by a mechanism which is yet to be elucidated. A single mating provides sufficient sperm to last a female throughout her reproductive life, which may last several months. Under laboratory conditions, females may copulate more than once and sperm from more than one mating may be used to produce offspring.

As female *G. m. morsitans* age, their mating receptivity declines: This occurs slowly among virgins and rapidly among mated females. The decline in receptivity of mated females is due to the act of mating (possibly mediated through tactile stimulation during the early phases of mating) and by material from the male's accessory glands (transferred during the final phase of copulation). Both the nature of the chemicals involved in inducing refractoriness and their sites of action are unknown. Assessing the decline in receptivity of mated females may be complicated by transfer of antiaphrodisiacs (abstinon) from males to females during mating (Carlson and Schlein, 1991). These compounds, alkenes whose action may be enhanced by alkanes, may function by masking the effects of the sex pheromone that is produced by females.

V. OVULATION, FERTILIZATION, AND EMBRYONIC DEVELOPMENT

At ovulation, the ovum is passed to an expanded portion of the bursa copulatrix, usually referred to as the uterus in tsetse. Remnants of the nurse cells and follicular epithelium are passed later into the uterus. Tsetse flies vary in whether virgins ovulate: In some species (e.g., *G. p. palpalis, G. m. morsitans,* and *G. austeni*) virgins retain eggs in the ovaries but in other species (e.g., *G. m. centralis* and *G. pallidipes*) most virgins ovulate. Ovulation is induced in females in the former group by tactile stimulation that occurs during the early stages of mating rather than by chemicals transmitted from the males. In most species in which ovulation is stimulated by mating, a week may elapse between copulation and ovulation; this raises the question of how these neurally transmitted stimuli can induce ovulation. The simplest explanation is that the stimuli initiate a process that takes up to a week to complete; candidates are egg maturation and synthesis and release of an ovulation hormone. During the days following mating, there is a hemolymph-borne factor that helps induce ovulation. This hormone may come from median neurosecretory cells in the brain and be released into the hemolymph via the corpus cardiacum. For *G. f. fuscipes* it has been proposed that the first ovulation is regulated as follows (Robert *et al.*, 1984): Mechanical stimuli received during copulation result in activation of neurosecretory cells in the brain. Neurotransmitters are carried via nerve XVII to, and released from, two types of neuromuscular junctions in the ovary and oviduct. cAMP appears to be the second messenger at the receptors.

What controls the second and subsequent ovulations? The anterior end of a developing larva pushes the oviductal shelf into a position that blocks the opening of the common oviduct; thus, premature ovulation is physically impossible. Ovulation occurs about 90 min after each parturition, and it is possible that mechanical stimulation and realignment of the common oviduct and the uterus provide the cue and make it physically possible for ovulation to occur. The median neurosecretory cells in the brain may also be involved in regulating the second and subsequent ovulations.

The posterior end of the egg enters the uterus first; thus, the micropyle end of the egg comes to lie near the oviductal shelf. A valve, consisting of columnar cells surrounded by circular and longitudinal muscles, is located at the junction of each spermatheca and its duct. The valve opens at ovulation and closes at gastrulation, i.e., after 2.5 days, thus permitting sperm to leave the spermathecae over a relatively long time period. The mechanism by which sperm are moved, or are induced to move, from the spermathecae to the uterus has not been determined, but muscles in the spermathecal ducts may play a role.

Embryological studies of tsetse have been hampered by several factors: Embryonation is internal, only one embryonating egg is obtained by dissecting each female, and the timing of the first ovulation is variable and not easily determined. Studies of tsetse embryology are, therefore, best carried out on eggs ovulated after the first, or a subsequent, full-term parturition. The above notwithstanding, the average times at which major embryonic events occur during the first pregnancy cycle in *G. m. morsitans* have been estimated: syncytial blastoderm, $\approx$1 day; late blastoderm, $\approx$1.75 days; gastrulation begins on approx Day 2.5; early organogenesis (with midgut enclosing yolk), $\approx$3.5 days; and late organogenesis, $\approx$4.25 days. Embryonic development takes almost as long as larval development.

VI. HATCHING, POSTEMBRYONIC DEVELOPMENT, AND PARTURITION

Hatching and molting are physically complex processes in tsetse because the egg and larva are in a confined space. The first-instar larva splits the chorion using a chitinized "egg tooth" which appears to block the larva's mouth. The larva cannot crawl free of the chorion. The chorion is pulled off the larva by muscular contractions of the choriothete, to which it is attached by a mucoprotein, and by muscles that extend from the choriothete to the abdominal wall of the female. The choriothete may function in a similar manner (although no mucoprotein is involved) to remove the integument when the first instar molts. The chorion and the first instar's integument are stored between the egg and/or larva and the ventral wall of the uterus. The second molt is incomplete: The integument of the second instar is thin and flexible, and although it splits as the third instar grows, the integument of the second instar in not completely shed. At parturition the third-instar larva emerges posterior end first and the chorion and integument of the first-instar larva are then expelled.

The uterine wall is richly tracheated and some of the oxygen for young larvae may be provided by diffusion from the uterus. However, by the late third instar, the highly sclerotized and melanized polypneustic lobes, bearing the posterior spiracles, are well developed and gas is exchanged directly with the atmosphere, through the open genital pore of the female.

During its development, the larva is nourished by secretions from the female accessory glands, or milk glands, that empty onto the oviductal shelf (often referred to as a papilla because of its appearance in sagital section). Much of the food obtained by the female, during the first half of a pregnancy cycle, is converted to lipids and stored in the abdominal fat body, whereas some of the food is converted to protein and stored in the abdominal integument (Solowiej and Davey, 1987). These lipids and proteins are mobilized for inclusion in milk that is produced after the egg hatches. Tsetse milk is composed almost entirely of lipids and proteins, in nearly equal concentrations. Proteins are synthesized by the milk glands from amino acids derived mainly from the large meal ingested on the sixth day of the pregnancy cycle. A net synthesis and storage of milk is indicated by an increase in the diameter of the milk gland tubules during early stages of larval development. A net loss of material from the glands occurs during the rapid growth of the third-instar larva. A peptide, or protein, found in the midbrain, but not in other endocrine organs such as the corpora cardiaca, stimulates amino acid and protein synthesis by the milk glands (Pimley, 1983). Elevated levels of this putative hormone in the brain during the first half of the pregnancy cycle, and low levels when the milk glands are active, suggest that this peptide is responsible for controlling the synthetic activities of the milk glands.

Larvae digest milk in the midgut, which has no opening to the hindgut. Although larvae must produce considerable quantities of nitrogenous wastes during development, they do not excrete wastes while in the uterus; the anus is closed and the hindgut serves as a reservoir to store nitrogenous wastes. Most intraspecific variation in the size of larvae is due to variation in the quantity of food consumed by their mothers. Failure of females to obtain a sufficiently large amount of blood results in aborting eggs or immature larvae or production of undersized larvae. At least three meals are required to produce a mature larva, with the most critical one being a large meal on Day 6 of the 9-day pregnancy cycle. Because the abdominal integument has a finite ability to stretch, females feed to a constant abdominal volume (i.e., as the larva grows the volume of the meal decreases). There is also a thinning of the abdominal integument late in the pregnancy cycle which may aid in accommodating the rapid and large amount of growth of the third-instar larva.

Third-instar larvae must be nourished by their mothers until very near the time of parturition to ensure that they have the maximum resources for metamorphosis and early adult life. Therefore, the obligatory feeding phase is relatively long compared to the facultative feeding phase. The commitment to metamorphosis, which occurs at the end of the obligatory feeding phase, occurs about 6 hr prior to parturition. The relatively short facultative feeding phase results in a narrow weight range for viable tsetse larvae and adults compared to the size range in other flies. Zdárek and Denlinger (1993) point out that this restriction is no doubt imposed by adenotrophic viviparity: Small tsetse do not have the option of producing fewer offspring, as small females do in most insect species; small tsetse females simply must produce smaller offspring and these have reduced fitness.

About 5 or 6 hr prior to parturition the larva shifts posteriorly in the uterus, the female becomes more active and searches for a larviposition site, oxygen consumption rises and then declines, and during the 2 hr preceding parturition the pressure in the hemocoel rises in a pulsating manner. Ecdysteroids may have a role in parturition: Hemolymph ecdysteroid levels are high at parturition, injection of ecdysteroids induces abortion early in the pregnancy cycle, and ecdysone increases uterine contractions but 20-hydroxyecdysone decreases them. However, the mechanisms by which ecdysteroids influence parturition and their differential effects on the physiology of the female and the larva are not yet known.

As the time for parturition approaches, females select a suitable larviposition site. Particularly among the savannah species there is a clumping of puparia at particular sites. Parturition requires only a few seconds and usually occurs in late afternoon; the timing is entrained by light received by both the female and the larva, i.e., they both contribute to the timing of parturition. The larva takes an active part in parturition, using strong peristaltic contractions to back out of the genital tract.

Parturition in *G. f. fuscipes* is influenced by external and internal factors which affect the brain (Robert *et al.*, 1984). Pregnancy is maintained by secretion of a "parturition-inhibiting" hormone, probably from the corpus cardiacum or perisympathetic organs. The rapid growth of the third-instar larva stimulates stretch receptors, probably located in the uterus. These receptors provide the stimulus to turn off secretion of the parturition-inhibiting hormone and to turn on release of a parturition hormone. The latter is also synthesized by neurosecretory cells of the brain and is synthesized by, or stored in, the anterior part of the fused thoracic ganglia. The hormone is transmitted by nerve XVII to, and released from, neurohemal areas near the uterus; cAMP appears to be the second messenger in the target. Nerve XVII also terminates in two types of neuromuscular junctions in the vaginal sphincter and one type in muscles that run from the uterus to the abdominal wall. A cephalic inhibition of parturition is accomplished by contraction of the vaginal sphincter until the female has selected a suitable site in which to deposit her larva.

VII. PUPARIATION, METAMORPHOSIS, AND ECLOSION

After parturition there is a brief wandering period during which the larva responds mainly to humidity and tactile stimuli as it locates a suitable place to burrow into the substrate. The larva moves in an anterior direction, but does so on its dorsal surface. The duration of the wandering phase is influenced by environmental factors, but usually ends abruptly and is promptly followed by contraction of the anterior part of the body. If, at this time, the larva is not disturbed its ability to move again is lost quickly. The hindgut is evacuated and its contents cover the surface of the larva. The functions (larviposition pheromone and protection from predators or pathogens) of this material remain the subject of speculation and investigation. Within a few minutes of finding a suitable pupariation site, muscle contractions convert the larva into the barrel shape that is typical of the puparium (Fig. 3). Sclerotization of the integumentary proteins which has, until now, been inhibited by the central nervous system, is complete within 2 hr and the puparium is a black, brittle structure.

FIGURE 3 Photo of a recently deposited larva and a fully formed puparium of *G. m. morsitans* (photo by R. Mandryk).

Prior to parturition the critical amount of ecdysone necessary for molting has already been received by the epidermal cells. This hormone probably comes from the larva's prothoracic gland, but the female may also contribute ecdysone to the larva because there is a pulse of ecdysone in the female's hemolymph at parturition. Formation of the true pupa (i.e., separation of pupal integument from the integument of the third-instar larva) occurs about 3 days after pupariation; the pupal to adult molt occurs 3 days later. The development of the adult takes another 3–4.5 weeks under optimal conditions and may take more that twice as long at lower temperatures. Females develop slightly faster than males. The pattern and sequence of development of the tsetse adult is similar to that in other flies, but the process takes much longer among the Hippoboscoidea than it does in other flies of similar size. Prolonged development may result from the need of these flies to save space by packing tissues close together and this may extend the time required for movements of tissues within the puparium (Zdárek and Denlinger, 1993).

As with other higher flies, tsetse use the ptilinum, a sac-like structure within the head, to force the cap off the puparium and to force obstacles out of the way as the newly emerged adult makes its way to the soil surface. Tsetse usually emerge from the puparium during the afternoon, the stimulus being rising temperatures within the range 18–32°C. After reaching the surface, the tsetse expands its body to nearly double its length by swallowing air and expands its wings by pumping hemolymph into them. Sclerotization of proteins in the integument begins after expansion of the body. Because of limited reserves, the development of the integument and flight muscles is incomplete at eclosion and the process continues with material acquired from the first meals taken by the adult.

VIII. CONCLUSION

Evolution of adenotrophic viviparity has been accompanied by many structural and physiological modifications, primarily in females and larvae but also in the pupae and adults as they develop within the puparium. One of the most striking results of this reproductive strategy is low female fecundity. This is, however, adequately compensated for by the protection and nurturing that the larva receives and tsetse flies are well adapted to their environment. The unusual method used by tsetse to reproduce has created many challenges for those who study these interesting and economically important insects, and many details of tsetse biology remain to be elucidated.

Bibliography

Buxton, P. A. (1955). *The Natural History of Tsetse Flies,* London School of Hygiene and Tropical Medicine, Memoir No. 10. Lewis, London.

Carlson, D. A., and Schlein, Y. (1991). Unusual polymethyl alkenes in tsetse flies acting as abstinon in *Glossina morsitans. J. Chem. Ecol.* **17**, 267–284.

Ejezie, G. C. (1983). Hormones and reproduktion [sic] in tsetse flies. *Z. Angewandte Zool.* **70**, 1–12.

Glasgow, J. P. (1963). The distribution and abundance of tsetse. Pergamon. Oxford, UK.

Gooding, R. H. (1990). Postmating barriers to gene flow among species and subspecies of tsetse flies (Diptera: Glossinidae). *Can. J. Zool.* **68**, 1727–1734.

Gooding, R. H., Feldmann, U., and Robinson, A. S. (1997). Care and maintenance of tsetse colonies. In *Molecular Biology of Insect Disease Vectors: A Methods Manual* (J. M. Crampton, C. B. Beard, and C. Louis, Eds.), pp. 41–55. Chapman & Hall, Cambridge, UK..

Pimley, R. W. (1983). Neuroendocrine stimulation of uterine gland protein synthesis in the tsetse fly, *Glossina morsitans. Physiol. Entomol.* **8**, 429–437.

Robert, A., Grillot, J. P., Guilleminot, J., and Raabe, M. (1984). Experimental and ultrastructural study of the control of ovulation and parturition in the tsetse fly *Glossina fuscipes* (Diptera). *J. Insect Physiol.* **30**, 671–684.

Solowiej, S., and Davey, K. G. (1987). Variation in thickness and protein content of the cuticle of the female of *Glossina austeni. Arch. Insect Biochem. Physiol.* **4**, 287–296.

Southern, D. I. (1980). Chromosome diversity in tsetse flies. In *Insect Cytogenetics. Symposia of the Royal Entomological Society of London: Number 10* (R. L. Blackman, G. M. Hewitt, and M. Ashburner, Eds.), pp. 225–243. Blackwell Scientific, London.

Tobe, S. S. (1974, November/December). How tsetse flies reproduce. *Insect World Digest,* 9–17.

Tobe, S. S., and Langley, P. A. (1978). Reproductive physiology of *Glossina. Annu. Rev. Entomol.* **23**, 283–307.

Wall, R., and Langley, P. A. (1993). The mating behaviour of tsetse flies (*Glossina*): A review. *Physiol. Entomol.* **18**, 211–218.

Zdárek, J., and Denlinger, D. L. (1993). Metamorphosis behaviour and regulation in tsetse flies (*Glossina* spp.) (Diptera: Glossinidae): A review. *Bull. Entomol. Res.* **83**, 447–461.

Tubal Surgery

Alan DeCherney and Mikio A. Nihira

University of California, Los Angeles

GLOSSARY

assisted reproductive technologies Procedures for the treatment of infertility that involve retrieval of oocytes from the ovary.

fecundability The probability of achieving a pregnancy within one menstrual cycle. It is approximately 20% in the general population.

fecundity The probability of achieving a live birth within one menstrual cycle.

in vitro fertilization A process in which oocytes are harvested from the ovary, fertilized in the laboratory, and the resulting embryos are transferred into the uterus.

microsurgery Surgery performed under magnification.

The structure of the Fallopian tubes may be disrupted by congenital errors in development, infections, endometriosis, or intentional ligation for the purpose of sterilization. Approximately 25–40% of female infertility results from Fallopian tubal pathology The first reported reconstructive tubal surgery for the treatment of infertility is attributed to Schrüder in 1884, and the first live birth was attributed to a successful procedure in 1909. Despite these successes, it was recognized that the potential complications of tubal surgery, notably, ectopic pregnancy, questioned the therapeutic value of "plastic tubal procedures" for the treatment of infertility. In 1937, J. P. Greenhill reviewed the available literature on reconstructive tubal surgery and polled 107 gynecologic surgeons. He concluded that tubal surgery for the treatment of infertility should be performed only on a small, select group of women who met very specific criteria. Given the availability of *in vitro* fertilization, this strategy should be applied today.

I. INDICATIONS AND CLASSIFICATION OF TUBAL SURGERY

Before the development of assisted reproductive technologies, reconstructive tubal surgery was the only alternative for patients with tubal infertility. During the 1960s, 1970s, and 1980s, the following procedures were engineered to address specific structural lesions of the Fallopian tubes:

1. Implantation: Indicated for bilateral interstitial/cornual occlusion. After the obstructed area is excised, either the isthmic or ampullary region of the Fallopian tube is implanted directly into the coruna of the uterus (Fig. 1).
2. Anastomosis: Indicated after excision of a segmental occlusion or partial salpingectomy/ligation for sterilization. There are five possible types: intramural–isthmic, intramural–ampullary, isthmic–isthmic, isthmic–ampullary, and ampullary–ampullary anastomosis. A number of studies published in the 1980s demonstrated greatly diminished pregnancy rates if the residual ampulla is <4 cm (Fig. 2).

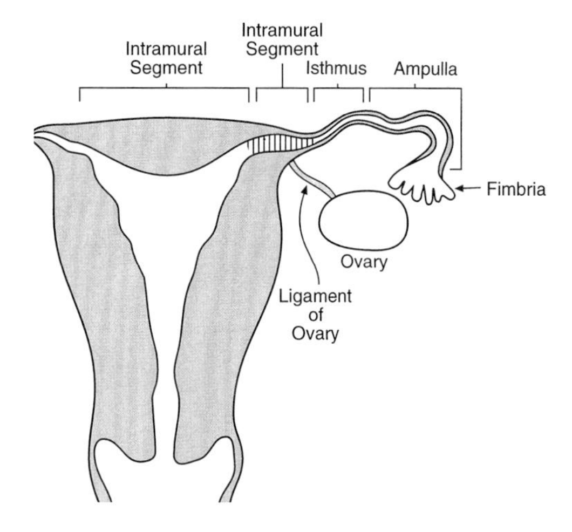

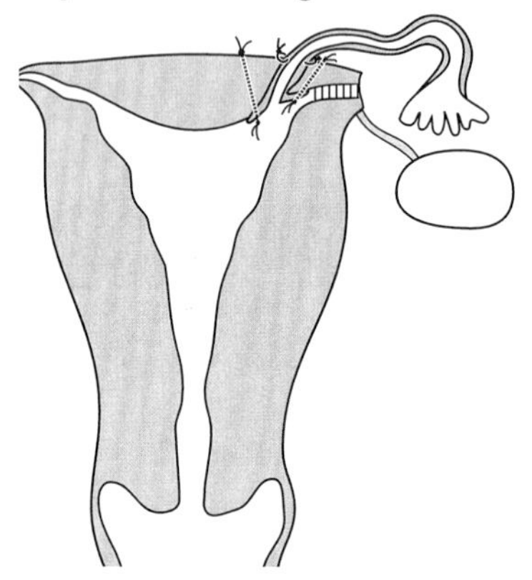

FIGURE 1 Implantation procedure.

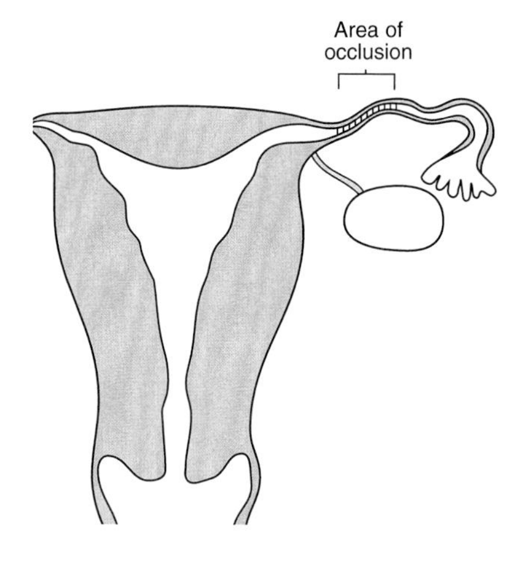

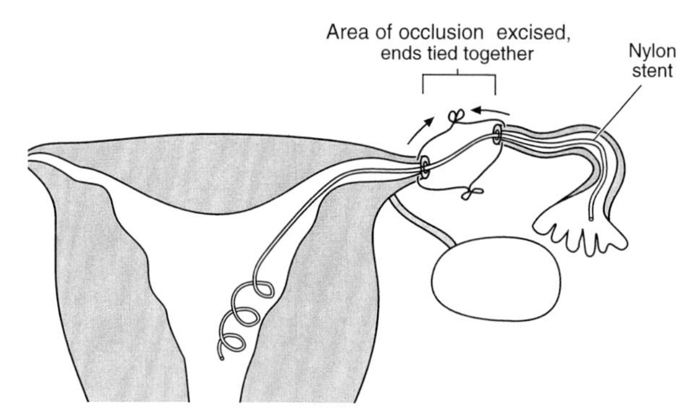

FIGURE 2 Anastomosis procedure.

3. Salpingoneostomy: Indicated for distal tubal obstruction in which no identifiable fimbriae or tubal opening is available. Commonly it is required in the reconstruction of a hydrosalpinx that has resulted from salpingitis. After the tube is freed of all adhesions, an ostium is created at the distal end (Fig. 3).

4. Fimbrioplasty: Often performed with salpingoneostomy to free agglutinated fimbriae. This permits the fimbriae access to the ovary.

5. Salpingolysis: Indicated to remove peritubal or periovarian adhesions that distort adnexal anatomy and prevent fimbrial–ovarian juxtaposition.

6. Combinations of previous procedures: May be indicated for patients with multiple sites of structural pathology.

7. Tuboscopy: Transcervical passage of a catheter to evaluate and possibly to recanalize an obstructed Fallopian tube. Dilation is performed either with a balloon catheter or with direct probing. This procedure is distinguished from falloposcopy, which is laproscopic, retrograde passage of a catheter for evaluation of the ampullary mucosa.

II. OPEN MICROSURGERY OR OPERATIVE LAPROSCOPY?

Oelsner, *et al.* demonstrated that laproscopic salpingolysis offers similar results to open microsurgery with a subsequent average intrauterine pregnancy rate of 56.7%. However, laproscopic tubal anastomosis is uniformly associated with poorer results than open microsurgery. Effectiveness of laproscopic salpingostomy is debatable. In some hands, it may have equal efficacy to open procedures. Overall, the conclusion for the utility of operative laproscopy, like

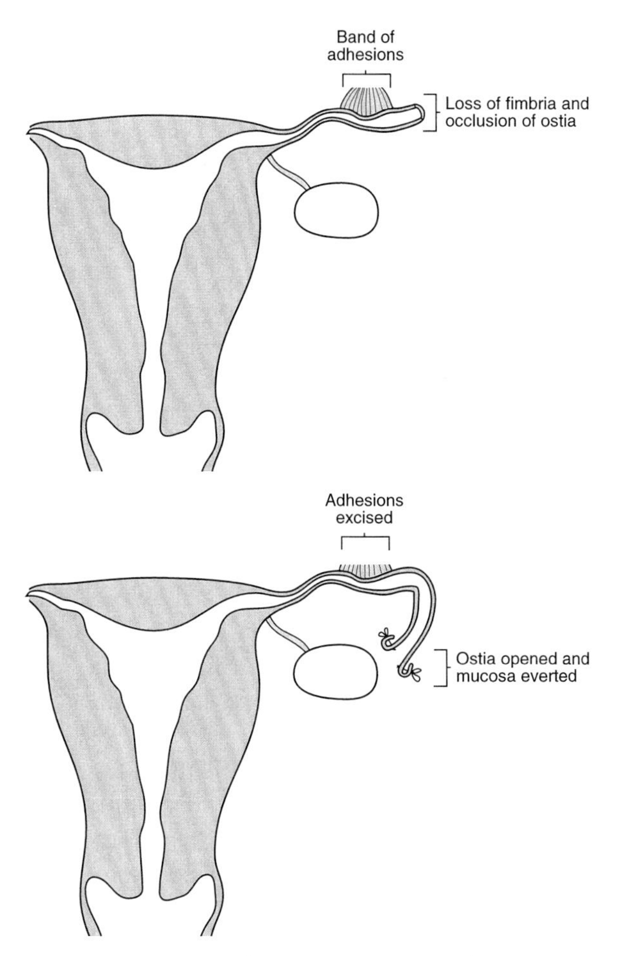

FIGURE 3 Salpingoneostomy procedure.

that of open microsurgery, is that the probability of a successful procedure is dependent on the severity of preexisting adnexal damage both external and internal.

III. QUANTIFICATION OF TUBAL PATHOLOGY

The following conditions are associated with reduced intrauterine pregnancy (IUP) rates and increased ectopic pregnancy rates after tubal surgery: extensive adnexal adhesions, even if filmy; fixed, dense adhesions; fibrotic thickening of the tubal wall; and severe endotubal mucosal adhesions giving an alveolar or "honeycomb" appearance by HSG or after opening the tube surgically. Boer-Meisel *et al.* prospectively analyzed a group of 108 infertile women who were treated for hydrosalpinx. Based on the presence or absence of the previous factors they were subsequently able to divide their patients into three groups—good, intermediate, and poor prognosis—with a probability of term pregnancy of 59, 16, and 3%, respectively. Furthermore, an examination of their results reveals a directly inverse relationship between factors predicting term pregnancy and the probability of developing an ectopic pregnancy.

IV. FECUNDITY

The basic measure of a successful intervention for the treatment of infertility is an improvement in fecundity. Fecundity is the probability of achieving a live birth within one menstrual cycle. This calculation permits comparison of the per cycle rate of fertility for the general population to that of a therapeutic intervention. Expected fecundity among normal reproductive-aged couples is approximately 20%. The overall 2-year cumulative intrauterine pregnancy rate is 25% after reconstructive tubal surgery. Given this rate of intrauterine pregnancy, the calculated fecundity is <1.0%. Two retrospective studies have evaluated the results of reconstructive tubal surgery in unselected populations with procedures performed in nonspecialized hospitals in the United Kingdom and Denmark. Their results provided a realistic estimate of the outcome that most patients may expect from tubal surgery. Mosgaard *et al.* reported on 236 patients who received tubal surgery. After a 2-year follow-up, 58 (25%) had one or more deliveries, 142 (60%) had no clinical pregnancies, and 37 (16%) had ectopic pregnancies. Notably, tubal anastomosis was associated with a 47% rate of subsequent ectopic pregnancy. Watson *et al.*, in a study of 82 patients, identified a lower rate of delivery (12%). Best results were associated with salpingolysis in the presence of patent tubes with a delivery rate of 21%. The worst results were associated with repair of bilateral distal tubal obstruction with a delivery rate of 5%. The calculated fecundity based on the results of these two studies is 2.5 and 0.3%, respectively (maximum estimates based on 24 months of follow-up).

V. REVERSAL OF STERILIZATION

Tubal patency after reversal of tubal sterilization approaches 90%. Pregnancy rates, however, are much lower. Henderson *et al.* completed a follow-up study of 102 patients who received reversal of sterilization. They observed an overall pregnancy rate of 68%, with a term delivery rate, spontaneous abortion rate, and ectopic pregnancy rate of 52, 11, and 5%, respectively. The calculated fecundity is approximately 2%.

VI. CONSEQUENCES OF RESTORING TUBAL PATENCY: ECTOPIC PREGNANCY

Marchbanks *et al.* performed a retrospective, case-controlled study of 274 patients with ectopic pregnancies compared with 548 controls matched for approximate date of conception, age, and race. Previous tubal surgery (tuboplasty, salpingectomy, and sterilization) was associated with a relative risk of 4.5 [95% confidence interval (CI), 1.5–13.9]. Notably, a history of pelvic inflammatory disease (PID) was associated with a relative risk of 3.3 (CI, 1.6–6.6). In a survey of the available literature on incidence of ectopic pregnancy following tubal surgery, Lavy *et al.* noted a reported incidence of ectopic pregnancy following reconstructive tubal surgery between 3 and 20%. Most important, they pointed out that "Ectopic pregnancy is the shady companion of tubal surgery." Reconstructive tubal procedures cannot restore normal tubal function. Even with the use of magnification, surgeons cannot hope to repair mucosal deciliation or a dysfunctional tubal muscular contraction.

VII. COMPARISON TO *IN VITRO* FERTILIZATION

In vitro fertilization (IVF) is expensive. It requires specialized technology and often involves repeated cycles of ovulation induction. The risks include ovarian hyperstimulation and complications with ovum recovery, gamete/zygote replacement, and multiple gestation. However, IVF is associated with a success rate of 20–30% per cycle in women with tubal disease. The estimated fecundity of IVF based on the 1993–1994 report from the Society for Assisted Reproductive Technology is 15.9%. The ectopic pregnancy rate is 1.4%.

VIII. POTENTIAL FOR MULTIPLE PREGNANCIES

Proponents of reconstructive tubal surgery argue that successful tubal surgery provides the opportunity for potential repeated natural conceptions. It is therefore less costly for multiple conceptions and has a decreased risk of multiple gestation compared with IVF. Upon examination of the available follow-up data, it is clear that multiple pregnancies following reconstructive surgeries are uncommon. For example, in the study by Watson *et al.*, they observed 9 deliveries and only a single case of 2 deliveries after an average of 40 months of follow-up. Similarly, in the study by Mosgaard *et al.*, out of a total of 58 deliveries, seven women delivered twice and one three times after an average follow-up period of 57 months.

IX. CONCLUSION

Reconstructive tubal surgery is very successful at restoring tubal patency but not for improving fecundity. It should be reserved only for reversal of sterilization and laproscopic salpingolysis in cases in which tubal fibrosis, mucosal damage, and intratubal adhesions are minimal. IVF should be planned if no intrauterine pregnancy occurs after 2 years. Close follow-up for the identification of ectopic pregnancies is a crucial requirement. When compared to IVF, reconstructive tubal surgery offers patients a diminished chance of a successful live birth and an increased risk of ectopic pregnancy, and it requires a relatively long time interval to evaluate the success of the procedure. As a result, this interval may reduce the success of ovulation induction for IVF.

See Also the Following Articles

Ectopic Pregnancy; Fallopian Tube; *In Vitro* Fertilization

Bibliography

Boer-Meisel, M. E., te Velde, E. R., Habbema, J. D. F., and Dardaun, J. W. P. F. (1986). Predicting the pregnancy outcome in patients treated for hydrosalpinx: A prospective study. *Fertil. Steril.* **45**, 23–29.

Donnez, J., and Casanas-Roux, F. (1986). Prognostic factors of fimbrial microsurgery. *Fertil. Steril.* **46**, 200–204.

Gomel, V. (1975). Laproscopic tubal surgery in infertility. *Obstet. Gynecol.* **46**, 47–48.

Greenhill, J. P. (1937). Evaluation of salpingostomy and tubal implantation for the treatment of sterility. *Am. J. Obstet. Gynecol.* **33**, 39–51.

Henderson, S. R. (1984). The reversibility of female sterilization with the use of microsurgery: A report on 102 patients with more than one year of follow-up. *Am. J. Obstet. Gynecol.* **149**, 57–61.

Holst, N., Maltau, J. M., Forsdahl, F., and Hansen, L. J. (1991). Handling of tubal infertility after introduction of in vitro fertilization: Changes and consequences. *Fertil. Steril.* **55**, 140–143.

Hull, M. G., and Fleming, C. F. (1995). Tubal surgery versus assisted reproduction: Assessing their role in infertility therapy. *Curr. Opin. Obstet. Gynecol.* **7**, 160–167.

Lavy, G., Diamond, M. P., and DeCherney, A H. (1986). Pregnancy following tubocornual anastomosis. *Fertil. Steril.* **46**, 21–25.

Lavy, G., Diamond, M. P., and DeCherney, A. H. (1987). Ectopic pregnancy: Its relationship to tubal reconstructive surgery. *Fertil. Steril.* **47**, 543–556.

Lilford, R. J., and Dalton, M. E. (1987a). Effectiveness of treatment for infertility. *Br. Med. J.* **295**, 155–156.

Lilford, R. J., and Dalton, M. E. (1987b). Effectiveness of treatment for infertility. *Br. Med. J.* **295**, 609.

Mage, G., Pouly, J., de JoliniÅre, J. B., Chabrand, S., Riouallon, A., and Bruhat, M. (1986). A preoperative classification to predict the intrauterine and ectopic pregnancy rates after distal tubal microsurgery. *Fertil. Steril.* **46**, 807–810.

Marchbanks, P. A., Annegers, J. F., Coulam, C. B., Strathy, J. H., and Kurland, L. T. (1988). Risk factors for ectopic pregnancy. *J. Am. Med. Assoc.* **259**, 1823–1827.

McEwen, D. C. (1966). Reconstructive tubal surgery. *Fertil. Steril.* **17**, 39–48.

Mosgaard, B., Hertz, J., Steenstrup, R. B., Sørensen, S. S., Lindhard, A., and Andersen, A. N. (1996). Surgical management of tubal infertility, a regional study. *Acta Obstet. Gynecol. Scand.* **75**, 469–474.

Oelsner, G., Sivan, E., Goldenberg, M., Carp, H., Admon, D., and Mashiach, S. (1994). Should lysis of adhesion be performed when in-vitro fertilization and embryo transfer are available? *Hum. Reprod.* **9**, 2339–2341.

Penzias, A. S., and DeCherney, A. H. (1996). Is there ever a role for tubal surgery? *Am. J. Obstet. Gynecol.*, 1218–1221.

Reich, H., McGlynn, F., Parente, C., Sekel, L., and Levie, M. (1993). Laproscopic tubal anastomosis. *J. Am. Assoc. Gynecol. Laprosc.* **1**, 16–19.

Siegler, A. M. (1967). Surgical treatments for tubo-pertoneal causes of infertility since 1967. *Fertil. Steril.* **28**, 1019–1032.

Society for Assisted Reproductive Technology, American Society for Reproductive Medicine (1995). Assisted reproductive technology in the United States and Canada: 1993 results generated from the American Society for Reproductive Medicine/Society for Assisted Reproductive Technology Registry. *Fertil. Steril.* **64**, 13–21.

Speroff, L., Glass, R. H., and Kase, N. G. (1994). *Clinical Gynecologic Endocrinology and Infertility*. Williams & Wilkins, Philadelphia.

Thompson, J. D., and Rock, J. A. (1992). *Te Linde's Operative Gynecology 7th edition*. Lippincott, Philadelphia.

Trimbos-Kemper, T. C. M. (1990). Reversal of sterilization in women over 40 years of age: A multicenter survey in the Netherlands. *Fertil. Steril.* **53**, 575–577.

Watson, A. J., Gupta, J. K., O'Donovan, P., Dalton, M. E., and Lilford, R. J. (1990). The results of tubal surgery in the treatment of infertility in two non-specialist hospitals. *Br. J. Obstet. Gynaecol.* **97**, 561–568.

Tumors of the Female Reproductive System

Th. Agorastos and B. C. Tarlatzis
Aristotle University of Thessaloniki

GLOSSARY

benign Not malignant, i.e., describing a tumor that is confined to its original site and that does not invade surrounding tissue or spread from this site to other distant sites.

invasive Describing a tumor that spreads from the primary site into surrounding tissue.

malignant Harmful; describing a tumor that is capable of invading surrounding tissue and/or spreading from its original site to other distant sites.

neoplasm Literally, new formation; an abnormal growth of cells, i.e., a mass of tissue exhibiting the general characteristics of uncontrolled proliferation of cells, an accelerated rate of growth, an abnormal structural organization, and a lack of coordination with the surrounding tissue.

neoplastic Relating to or producing neoplasms.

polyp An abnormal, typically benign mass of tissue projecting outward from a mucous membrane.

The genital organs of the female, contrary to those of the male, are relatively often altered through benign or malignant neoplastic transformations of their tissues. Thus, the knowledge about these forms of diseases of the female genitalia is of particular importance because of the significance of these organs for the reproductive potency of young women, and because the majority—with the exception of tubal or ovarian neoplasias—represent early detectable or even preventable diseases.

I. TUMORS OF THE VULVA

Embryologically the vulva is derived from the ectoderm, but it also contains neuroectodermal and mesenchymal components. Histologically the vulvar alterations often simulate those of the vagina and cervix; on the other hand, most of the dermatological disorders have also been described on the vulva.

A. Epithelial Disorders

According to the last nomenclature of the International Society for the Study of Vulvar Diseases, the nonneoplastic disorders of the vulvar epithelium [lichen sclerosus and atrophicus, squamous cell hyperplasia (former hyperplastic dystrophy), and other dermatoses (lichen planus, contact dermatitis, dermatitis medicamentosa, atopic dermatitis, psoriasis, vulvar vestibulitis, Crohn's disease, Bechet's syndrome, mites and lice, hidradenitis suppurativa, and others)] are distinguished from vulvar intraepithelial neoplasia (VIN), which, with its subclassification of VIN 1, 2, and 3, represents the different grades in quantity and quality of the dysplastic (neoplastic) cellular abnormalities in the vulvar epithelium. Pigmented lesions of the vulva with cellular dysplasia, formerly described as Bowen's disease, Bowen's dysplasia, Bowenoid papulosis, and erythroplacia of Queyrat, are now diagnosed as VIN (1–3). On the contrary, Paget's disease of the vulva represents a distinct, mostly intraepithelial neoplastic lesion with an often occurring underlying malignancy.

The incidence of VIN has markedly increased during the past few decades, especially in mid-aged women, and its presence is often concomitant with similar epithelial abnormalities of the anogenital region, mostly with cervical intraepithelial neoplasia (CIN). The putative causal role of the human papillomavirus (HPV) infection has been suggested by many investigators after HPV DNA was demonstrated in most cases of VIN. Spontaneous regression or persistence of VIN is the most frequent finding, especially in young women, whereas progression to invasive disease is observed mainly in older or immunosuppressed women.

B. Vulvar Cancer

Vulvar cancer represents about 4% of all female genital tract malignancies and, in approximately 90% of the cases, is of the squamous cell type.

Unlike VIN, the overall incidence of vulvar invasive squamous carcinoma has not increased significantly in the past few decades. However, there is evidence that vulvar squamous carcinoma can be separated in two groups, one observed in relatively young women (mean age 55 years), of warty or basaloid type, and often coexisting with intraepithelial or invasive neoplasias of the cervix, vagina, and anus, and the other referring to older women (mean age 75 years), of squamous cell type, without concomitant lesions in the other parts of the anogenital area but often associated with lichen sclerosus or hyperplastic dystrophy. The age distribution, the concomitant epithelial disorders in the region, and the marked difference in positive detection of HPV DNA between the two groups (about 75% for the first group and very rarely for the second) further support the notion that HPV infection probably plays a causative role in the carcinogenesis process only in the first group of women.

Pruritus is the leading but not pathognomonic symptom of vulvar cancer. Discharge, pain, and ulceration may also be present. The diagnosis is always based on histological examination of the biopsies taken from the suspicious areas.

Vulvar cancer spreads by direct extension to vagina, urethra, perineum, and anus; by lymphatic route to regional lymph nodes; and by the hematogenous route to liver, lungs, and bones. According to the classification of the International Federation of Gynecology and Obstetrics (FIGO) from 1990, the clinical stages (St) of the vulvar cancer are defined as follows: St 0 = carcinoma *in situ*; St I = tumor <2 cm confined to vulva and/or perineum without nodal metastases; St II = tumor >2 cm confined to vulva and/or perineum without nodal metastases; St III = tumor of any size with (i) spread to the lower urethra and/or the vagina, the perineum, and the anus and/or (ii) unilateral regional lymph node metastases; St IV = tumor of any size with (i) spread to the upper urethra, bladder mucosa, rectal mucosa, pelvic bone, and/or bilateral regional node metastases and/or (ii) other distant metastases including pelvic lymph nodes.

Radical local excision of the lesion extended with inguinal lymphadenectomy is the treatment of choice in most St I–III tumors. However, radiation therapy, with or without additional chemotherapy, is likely to have an increasingly important role in the management of patients with vulvar cancer.

Vulvar cancer, when treated appropriately, has a relatively good prognosis with a 5-year survival in groin nodes-negative cases of 77–100% and overall 5-year survival in operable cases of about 70%.

C. Malignant Melanoma

Melanomas of the vulva account for approximately 1–7% of all melanomas in women and for 9 or 10% of all malignant tumors of the vulva, with increasing incidence. Two-thirds of the cases occur in women older than 50 years; however, there are also cases with melanoma of the vulva documented in girls younger than 20 years. Because of the worse prognosis of these aggressive tumors, any changing dark pigmented vulvar lesion should be widely excised for histologic verification. The so-called ABCD rule by Friedman (asymmetry, border irregularity, color variation, and diameter > 6 mm) may be helpful for differentiating a benign nevus from a melanotic lesion suspect for malignancy.

Radical vulvectomy with bilateral groin dissection has been the most advocated mode of treatment in cases of malignant melanoma, with pelvic lymphadenectomy in patients with clitoral melanoma. However, some authors could not find a significant relation between the type of surgery and the survival of the patients. Postoperative radiation may decrease the incidence of local recurrences, but chemo- or immunotherapy have been proved of limited value in improving the survival of these patients.

The overall prognosis of malignant melanoma of the vulva is poor, and the 5-year survival rate varies between 20 and 70%. According to some authors, the prognosis of vulvar melanoma is worse than that of nongenital skin melanomas.

D. Other Vulvar Malignancies

In addition to squamous cell carcinoma and melanoma of the vulva, there are also other rare malignant tumors documented at the vulva either of epithelial (adenocarcinomas of Bartholini, Skene, or sweat gland origin as well as Paget's disease) or mesenchymal origin (sarcomas, malignant rhabdoid tumor, malignant fibrous histiocytoma, aggressive angiomyxoma, schwannoma, and others). Finally, endodermal sinus tumor, primary malignant lymphoma, merkel cell tumor, and metastatic tumors have also been rarely identified at the vulva.

II. TUMORS OF THE VAGINA

Despite the long persistent debates about the embryologic origin of the vagina, the contribution of both the Müllerian ducts and the urogenital sinus, for the formation of this organ is generally accepted and explains the complex interrelationship of normal tissues and their neoplastic alterations derived from these different germ cell layers.

A. Benign Tumors

Squamous inclusion cysts, mesonephric cysts, Müllerian cysts, and bartholin gland cysts form the first group of benign vaginal tumors; there are also other benign tumors or tumor-like lesions in the vagina, including those of epithelial origin (most common) (squamous papillomas, condylomata accuminata, adenosis, and endometriosis), of mesenchymal origin (mesodermal stromal polyp, leiomyoma, rhabdomyoma, neurofibroma, and hemangioma), and, very rarely, of mixed or unknown origin (benign mixed tumor, mature cystic teratoma, adenomatoid tumor, Brenner tumor, and others).

These lesions are often asymptomatic or, when enlarged, may cause dyspareunia and should be surgically excised. Marsupialization is the treatment of choice for the Bartholini gland cysts, whereas laser vaporization is often used for the treatment of papillomas and condylomata accuminata.

B. Vaginal Cancer

Vaginal intraepithelial neoplasia (VAIN) is very often associated with CIN and, hence, it is more accepted as an extension of the cervical lesion rather than a primary, autonomous premalignancy of the vaginal epithelium. Therefore, it is not surprising that most of the VAIN cases occur in the upper part of the vagina, especially after treatment of CIN or following hysterectomy for CIN or invasive cancer.

Invasive cancer of the vagina accounts for 1 or 2% of cancers of the female genital organs. Histologically, more than 80% of vaginal carcinomas are of squamous cell origin (followed by adenocarcinoma, sarcoma, and malignant melanomas) and most of them are considered to be metastatic from cervical or vulvar cancers.

Bleeding is the most common symptom of vaginal cancer, often accompanied by discharge, whereas pain and palpable mass are seldom reported. The majority of carcinomas are localized in the posterior upper part of the vagina and spread primarily by local extension and secondary through the lymphatic channels to the pelvic nodes (cancers of the upper vagina) or to the inguinal nodes (cancers of the lower vagina). According to the latest FIGO staging, primary carcinoma of the vagina is of stage I if it is limited to the vaginal wall, stage II if it has extended to the paravaginal tissues but not to the pelvic wall, stage III if it has extended to the pelvic wall, and stage IV if it has extended to the bladder or rectum or beyond the true pelvis.

Laser vaporization, electrocautery, local excision, partial vaginectomy, and irradiation are the different treatment possibilities for VAIN, taking into account the extent and localization of the lesion as well as the age and gynecologic history of the patient. In some cases of early invasion (microcarcinoma) of the vagina, the previously mentioned conservative treatment modalities can also be used.

With the exception of stage I disease involving the upper vagina in which surgical treatment can be used, radiotherapy is the treatment of choice for invasive vaginal cancer, with combined use of teletherapy and intracavitary brachytherapy. On the other hand, chemotherapy and immunotherapy do not appear to have any positive impact on the prognosis of the disease. The overall 5-year survival rate for invasive vaginal cancer is only 42% (69% for stage I disease).

C. Sarcoma Botryoides

Sarcoma botryoides, a distinct type of rhabdomyosarcoma, is the most common primary sarcoma of the lower female genital tract, and it occurs classically in the vagina of infants as grape-like masses. It is a highly malignant tumor. Radical surgery showed poor survival results; recently, more conservative surgery in combination with chemotherapy and radiotherapy has been used with markedly better results.

III. TUMORS OF THE CERVIX

The cervix, the most inferior part of the uterus, has its origin in the fused Müllerian ducts, which form the uterovaginal canal. After migration of squamous cells from the urogenital sinus to the lower part of the uterovaginal canal, the original columnar epithelium of the Müllerian ducts is replaced by squamous epithelium. At the vaginal portion of the cervix, the border between the stratified squamous epithelium of the exocervix and the columnar epithelium of the endocervix form the so-called original squamocolumnar junction.

A. Benign Tumors of the Cervix

Benign proliferations of the epithelium and stroma of the cervix are often observed in women of reproductive age but also during menopause. Polyps, mainly derived from the endocervical mucosa, as well as fibrous, vascular, or mesodermal stromal polyps are, together with Nabothian cysts, the most common benign tumors. However, the definite diagnosis of the nature of polyps can only be made by histological examination; therefore, because of the possibility of existing malignancy in whole or in part of the epithelium or stroma of the polyp, every polyp oid protrusion of the cervix should be excised and evaluated microscopically.

Other polyp-like lesions of the cervix, frequently seen in pregnant patients, are mesodermal stromal polyps (pseudosarcoma botryoides), decidual pseudopolyps, and placental site trophoblastic nodules; benign tumors of primarily mesenchymatic origin, such as leiomyomas, adenofibromas, adenomyomas, fibroadenomas, hemangiomas, and lymphangiomas, can also rarely be found in the cervix.

B. Precancerous Lesions of the Cervix

During the past few decades the older and widely used World Health Organization (WHO) definition of two types of precancerous lesions of the cervical epithelium (i.e., dysplasia and carcinoma *in situ*) was replaced by the concept of CIN proposed by Richart, introducing the theory of the continuum in the neoplastic transformation of the cervical epithelial cells. According to this classification, intraepithelial lesions are categorized into three grades: CIN 1 corresponds to mild dysplasia, CIN 2 corresponds to moderate dysplasia, and CIN 3 corresponds to severe dysplasia and carcinoma *in situ*. Recently, and after recognition of the crucial role of HPV infection in carcinogenesis, it has been suggested that the terminology should be changed to better reflect the biologic procedure than the morphologic features. Thus, Richart proposed replacing the CIN 1–3 classification with the terms low-grade CIN (corresponding to lesions formerly classified as condyloma and CIN 1) and high-grade CIN (corresponding to lesions formerly classified as CIN 2 and CIN 3). On the other hand, the three-tier CIN terminology is incorporated into the two-tier Bethesda system of cytological diagnosis, with the terms low-grade squamous intraepithelial lesion and high-grade squamous intraepithelial lesion.

According to evidence from recent epidemiological, clinical, and experimental studies, cervical HPV lesions and CIN are one and the same disease; the risk for progression obviously increases with the severity of the dysplasia, but it also seems to be closely linked with the type of HPV involved and is particularly high in lesions induced by high-risk ("oncogenic") HPV types, mainly HPV-16. However, progression does not occur in every case, and even high-grade lesions may, rarely, spontaneously regress.

Cytology, colposcopy, histology, and HPV testing are well-established diagnostic methods for evaluating the risk of prospective malignancy (i.e., the risk of progressing into an invasive cancer) of a cervical intraepithelial lesion. In general, the clinical management of intraepithelial cervical lesions includes a conservative follow-up for the low-grade cases and an ablative or excisional treatment (with CO_2 laser, cold knife, or loop electrosurgical procedures) for the high-grade cases. Nevertheless, special emphasis should be placed on identifying among the women with low-grade lesions those at risk for having high-grade or even invasive lesions.

Dysplasias of the glandular epithelium of the endocervix are defined as cervical glandular intraepithelial neoplasia (CGIN). In cases of adenocarcinoma *in situ* of the cervix, a high rate of HPV DNA has been detected, often with involvement of the HPV type 18. It may be that both the squamous and glandular lesions develop from HPV-infected subcolumnar reserve cells through a process of bidirectional differentiation.

C. Cervical Cancer

During the past 50 years, enough evidence has been provided indicating that cervical cancer can be considered a preventable disease. However, despite the decrease in its incidence in developed countries as a result of the introduction of cytology screening programs, worldwide cervical cancer is the second most prominent cancer in women, after cancer of the breast.

Cancer of the cervix has a higher incidence in middle-aged women. Unfortunately, in the early stages of the disease there are no symptoms which could lead the woman to the physician. Vaginal bleeding, more often in the form of postcoital bleeding, as well as malodorous vaginal discharge are the only symptoms, but they mostly occur in women with advanced disease.

Pathological cytologic findings in cervical smears or suspicious macroscopic and colposcopic appearance of the cervix during gynecological examination lead to the diagnosis of invasive disease only on the basis of the histologic evaluation of the preformed small size or cone biopsies. About 60–80% of all cervical carcinomas are of squamous cell type, followed by adenocarcinomas, adenosquamous, glassy cell, mucoepidermoid, small cell carcinomas, mesenchymal, and mixed and secondary tumors.

According to the last FIGO staging system, cervical cancer can be divided into four stages: Stage I includes all tumors confined strictly to the cervix (Ia, lesions with depth of invasion <5 mm and horizontal spread <7 mm; Ib, lesions of greater diameter than those of Ia), stage II includes tumors that extend beyond the cervix to the upper vagina (IIa) and/or to the parametria (IIb) but not to the lower vagina or to the pelvic wall, stage III includes tumors that extend to the lower vagina or to the pelvic wall or cause hydronephrosis, and stage IV tumors extend to the bladder or rectal mucosa or beyond the true pelvis.

Cervical carcinoma spreads principally through local invasion of the surrounding tissues and secondarily through the lymphatic vessels to the pelvic and paraaortic lymph nodes; a hematogenous dissemination occurs less frequently. Obstruction of one or both ureters as well as formation of vesicovaginal and/or rectovaginal fistulas may occur in relatively advanced stages of the disease.

The basic therapeutic modalities for primary treatment of cervical cancer include surgery, radiotherapy, chemotherapy, and a combination of these. Especially in the early stages, each patient must be evaluated and treated individually. If distant spread of malignant cells seems very unlikely (as in some cases of stage Ia), simple but complete excision of the lesion, e.g., via conization, can represent a sufficient treatment. However, surgery involving radical hysterectomy and pelvic lymphadenectomy is the typical treatment for stages Ib and IIa. Radiation therapy in the form of tele-60Co-therapy and intracavitary brachytherapy can be used in all stages of the disease but more frequently is used in the advanced stages

(>IIa). Chemotherapy, before or after surgery and in combination with radiotherapy, can also be used as neoadjuvant or adjuvant treatment, respectively.

Lymph node involvement represents one of the most crucial parameters affecting the prognosis of the disease. Thus, survival rates can be decreased even in early stages when lymph node metastases are present. In general, the overall 5-year survival rate for treated stage I cases is about 90–95%, for stage II cases 50–70%, for stage III cases 30%, and for stage IV cases <20%.

Although adenocarcinomas of the cervix (of mucinous, endometrioid, clear cell, serous, mesonephric, or minimal deviation type) comprise a heterogeneous group of malignant neoplasms, their spread mode and clinical behavior are generally similar to those of squamous carcinoma and, consequently, they are also treated similarly. Earlier reports showing lower 5-year survival rates for patients with treated adenocarcinomas than those with squamous carcinomas of the cervix have not been confirmed by several recent population-based studies.

IV. TUMORS OF THE UTERINE CORPUS

Embryologically, the two compartments of the uterine corpus, the endometrium (i.e., the internal columnar mucosa) and the myometrium (i.e., the surrounding muscular layers), have their origin in the two fused Müllerian ducts. During the reproductive period of a woman's life, the endometrium undergoes cyclic changes as a result of the cyclic effect of estrogens and progesterone. In the postmenopausal period, because of the hormonal deprivation, no more proliferative or secretory activity can be achieved by the atrophic endometrium.

A. Benign Tumors of the Endometrium

Endometrial polyps are a relatively common finding during hysteroscopic examination of the uterine cavity or histologic evaluation of curettage material. These polyps mostly originate from the basal layer of the endometrium and contain stroma and vessels, covered by the superficial endometrial layers. There is no evidence to suggest that polyps have a greater probability for developing carcinoma than the adjacent endometrium. Atypical polypoid adenomyoma, Brenner's tumor, endometrial stromal nodule, papillary serous tumor, as well as mature and immature teratomas are also endometrial tumors with benign behavior that very rarely occur within the uterus.

B. Benign Tumors of the Myometrium

Leiomyomas are the most frequent uterine neoplasms, noted in about one in four women older than 30 years of age. They are hormone (estrogen)-dependent tumors and often decrease in size after the menopause. Typical leiomyomas, histologically characterized as smooth muscle tumors, their specific subtypes (mitotically active, cellular, hemorrhagic, atypical, and epithelioid), as well as other types of leiomyomas and related entities (myxoid, vascular and lipoleiomyoma, diffuse, intravenous and disseminated peritoneal leiomyomatosis, and benign metastasizing leiomyoma) show a benign clinical behavior and must be histologically distinguished from leiomyosarcoma, metastasizing leiomyosarcoma, or endometrial stromal sarcoma.

Most leiomyomas are asymptomatic and represent incidental findings during gynecological examination, ultrasonographic exploration of the uterus, or histological evaluation of hysterectomy specimens. However, abnormal uterine bleeding, pelvic or abdominal pain, disturbances of miction, and a sensation of pressure are common symptoms, especially if large myomas are present. Torsion of pedunculated leiomyomas, infarction, secondary infection, calcification, lipoid or cystic degeneration, and necrosis are possible complications mostly noted in large leiomyomas. In some instances, infertility as well as complications of pregnancy are attributed to the presence of leiomyomas.

Surgical enucleation of one or more leiomyomas (myomectomy) via laparotomy, laparoscopy, or hysteroscopy in premenopausal symptomatic or asymptomatic but subfertile women is frequently recommended. However, sometimes large or multiple leiomyomas can only be removed by hysterectomy. Recently, it has been advocated to treat myomas before surgery with gonadotropin-releasing hormone agonists to decrease the uterine and tumor size and to reduce the risk of intraoperative hemorrhagia.

C. Endometrial Hyperplasias

Endometrial hyperplasias are defined as abnormally increased proliferation of the endometrial glands as a response to unopposed endogenous or exogenous estrogenic stimulation. According to the last WHO classification, they have been divided into two categories, based on the presence or absence of cytological atypia, and further classified as simple or complex according to the extent of architectural abnormalities of the endometrium.

Previously, endometrial hyperplasias have routinely been treated either by gestagen substitution, in order to oppose the estrogenic effect, or by hysterectomy because of the suspected probability of progression to carcinoma. Nevertheless, in clinical practice the vast majority of hyperplasias without atypia treated with cyclic progestins (mainly medroxyprogesteronacetate) regress; on the contrary, atypical hyperplasias of the endometrium can be considered as true premalignant lesions and, therefore, especially in peri- and postmenopausal women, should be mainly treated by surgical procedures.

D. Endometrial Carcinoma

Endometrial carcinoma is the most common malignant tumor of the female genital tract and the fifth leading cancer in women, occurring with increasing incidence in developed countries during recent decades.

The most typical epidemiological characteristic of women with endometrial carcinoma, regardless of menopausal status, is obesity. Obesity, often associated with diabetes mellitus, hyperinsulinemia, and dyslipidemia, leads to an increased endogenous production of estrogen through increased conversion of androgens to estrogens in the adipose tissue and by suppression of the sex hormone-binding globuline capacity. The prolonged and continuous effect of estrogens without the cyclic countereffect of progestins in patients with anovulatory cycles or in the postmenopause increases the risk of endometrial carcinoma.

More than three-fourths of all endometrial carcinoma cases are of endometrioid histologic type, whereas serous, clear cell, mucinous, and squamous carcinomas, as well as mixed types and undifferentiated carcinomas, form the rest of the malignant tumors of the endometrium.

The disease is relatively uncommon in premenopausal women (mean age 59 years). Irregular or postmenopausal uterine bleeding is frequently the most common or sole symptom in patients with invasive endometrial cancer. Diagnosis is only made by histological verification of endometrial tissue removed either by biopsy through the hysteroscope or by curettage after dilatation of the cervical canal. Transvaginal ultrasonography to evaluate the thickness and status of the endometrium can contribute to the detection of asymptomatic women with early stages of endometrial cancer.

The former FIGO clinical staging for endometrial carcinoma was replaced in 1988 by a surgical staging system. Stage I involves tumors limited to the uterine corpus (Ia without myometrial invasion, Ib invasion to less than one-half of the myometrium, Ic invasion to more than one-half of the myometrium); stage II involves tumors extending to the cervical glandular epithelium (IIa) or stroma (IIb); stage III includes tumors extending to the uterine serosa and/or adnexa (IIIa), to vagina (IIIb), and to pelvic and/or paraaortic lymphnodes (IIIc); and stage IV refers to tumors extending to the bladder and/or bowel mucosa (IVa) and/or with distant metastases (IVb). In this revision of the FIGO staging system of uterine carcinoma, it is recommended that tumors should also be graded using both architectural and nuclear criteria. The architectural grade of the tumor, determined by the proportion of solid cell masses compared to well-defined endometrial glands, should be increased by one grade if a marked nuclear atypia is present.

Endometrial carcinoma spreads primary by direct extension to the adjacent tissues and by transtubal passage of malignant cells. After tumor spread to the cervical stroma, lymphatic and hematogenous dissemination frequently occurs.

There are well-demonstrated prognostic factors in endometrial carcinoma, first recognized after using the surgical–pathological staging and grading of the surgical specimen. Thus, the size and volume of the tumor, the cervical involvement, the depth of invasion, the cell type and grading, the DNA ploidy, the nodal status, and the extrauterine spread are factors which have been shown to correlate independently with the outcome of the patients; therefore, their assessment is essential for individualization of the treatment in order to minimize the treatment-related morbidity and mortality for low-risk cases and for identifying the high-risk cases, which are likely to benefit from adjuvant therapy.

Abdominal hysterectomy with bilateral salpingooophorectomy has long been the typical treatment of patients with stage I endometrial carcinoma, which accounts for more than 80% of primary diagnosed endometrial cancers. However, pelvic or even pelvic and paraaortic lymphadenectomy is indicated in patients with unfavorable prognostic factors. Surgery plus postoperative irradiation or radiotherapy alone (combined external-beam radiation plus intracavitary radiation) can be performed in patients with advanced stages of the disease. Cytotoxic chemotherapy in cases with endometrial cancer is of only palliative value and the treatment results are not very encouraging. In addition, the combination of progestogens with cytotoxic chemotherapy does not seem to improve the response and survival rates. Similarly, no specific benefit from adjuvant treatment with progesterone or from progesterone therapy in advanced or recurrent endometrial cancer has been shown.

The FIGO clinical staging correlates directly with the 5-year survival rates: 72% in stage I, 56% in stage II, 31% in stage III, and 10% in stage IV. Age, race, histologic subtype, grade, and hormone-receptor status are factors that also influence the survival rates.

E. Uterine Sarcomas

Uterine sarcomas account for approximately 3 or 4% of uterine malignancies, with leiomyosarcoma arising from the myometrium and the stromal sarcomas or the mixed mesodermal tumors arising from the endometrium. They can also be divided into pure sarcomas, with only mesodermal malignant elements (e.g., leiomyosarcoma), and mixed sarcomas, with mesodermal and epithelial malignant elements (e.g., carcinosarcoma), and subdivided into homologous and heterologous, depending on whether the malignant mesodermal elements are normally present in the uterus.

Irregular enlargement of the uterus, tumor protrusion through the cervix, and abnormal uterine bleeding are symptoms which may be present in a proportion of patients with uterine sarcoma. Although the number of mitoses per 10 high-power fields after Taylor and Norris seems to be the most reliable predictor of the biologic behavior of these tumors, mitosis counts are not always reproducible.

Staging of the uterine sarcomas is identical to the FIGO staging system of endometrial carcinoma. Also, the spread patterns of uterine sarcomas are similar to those of endometrial carcinoma, with a propensity for early hematogenous and lymphatic dissemination.

Total abdominal hysterectomy with bilateral salpingooophorectomy is the only treatment with any proven curative value for the frankly uterine sarcomas. Aggressive adjuvant treatment with cytotoxic agents and irradiation has often been used and it appears to improve tumor control and reduce pelvic recurrences but without influencing the final outcome of the patients. The 5-year survival rate for stage I tumors reaches only 50%, whereas for the more advanced stages it does not exceed 20%.

V. TUMORS OF THE FALLOPIAN TUBES

Embryologically, the Fallopian tubes are first formed during the cephalocaudal differentiation of the Müllerian ducts. They are characterized by a serosal membrane, a layer of smooth muscle, and a mucosal columnar epithelium, composed of ciliated, secretory, and intercalary cell types, that may undergo metaplastic changes.

A. Benign Tumors of the Fallopian Tubes

Adenomatoid tumors (benign mesotheliomas) represent the most common type of this rare group of benign tumors. Inclusion cysts of the tubal serosa as well as paratubal and paraovarian cysts and tumors, epithelial polyps or papillomas, leiomyomas, adenomyomas, teratomas, and other benign mesenchymal and mixed tumors of the Fallopian tubes are also very rarely found, mostly as incidental findings at laparotomy.

B. Cancer of the Fallopian Tubes

Carcinoma *in situ*, invasive adenocarcinoma, sarcomas, and mixed tumors of the Fallopian tubes are uncommon (0.2–0.5% of primary female genital malignancies). More often, the tubes are involved in metastatic processes by local extension of uterine or ovarian malignant neoplasms.

The disease is mostly unilateral and located near the ampullary part of the tube, which is usually swollen with fimbrial occlusion. There are no typical symptoms; abdominal pain, serosanguineous vaginal discharge, abnormal uterine bleeding, and a pelvic mass may be noted in patients with tubal cancer. Diagnosis is very difficult before surgery because of the lack of specific symptoms, objective signs, or imaging findings. Tubal cancer is classified as stage I when the tumor is confined to the tube(s), stage II when the tumor extends beyond the tube(s) but is confined to the pelvis, stage III when the tumor extends beyond the pelvis but is confined to the abdomen, and stage IV when the tumor extends beyond the abdominal cavity.

There is no established method of treatment for tubal cancer and the current mode of therapy with total hysterectomy, bilateral salpingooophorectomy, and adjuvant chemotherapy or radiotherapy fails to achieve good results. Thus, the overall 5-year survival rate is about 35–40%, but the survival rate for stage I is about 50–60%, for stage II 40%, and for stages III and IV 0–10%.

VI. TUMORS OF THE OVARIES

The clinical or ultrasonographic manifestation of an ovarian enlargement is a very common finding, especially during the reproductive age of the woman. The differential diagnosis of these tumors or tumor-like lesions is quite difficult but at the same time very important for the appropriate treatment of the patient, if such a treatment is really necessary.

A. Nonneoplastic Tumors and Tumor-Like Lesions of the Ovary

Solitary cysts of follicular origin, follicle and corpus luteum cysts (mostly mediated through disturbances in gonadotropin secretion), large solitary follicle cyst of pregnancy and puerperium, multiple cysts of follicular origin secondary to ovarian stimulation by endogenous human chorionic gonadotropin (hCG) (hyperreactio luteinalis) or by exogenous gonadotropin and/or clomiphene citrate administration (ovarian hyperstimulation syndrome), and multiple atretic cystic follicles (such as those in polycystic ovary syndrome and 17-hydroxylase-deficiency syndrome) are the most common nonneoplastic tumors or tumor-like lesions of the follicular elements of the ovary.

Stromal hyperplasia and stromal hyperthecosis, HAIR-AN (hyperandrogenism, insulin resistance, and acanthosis nigricans) syndrome, massive ovarian edema, ovarian fibromatosis, pregnancy luteoma with or without virilization, and hilar or stromal Leydig cell hyperplasia are the most common diseases accompanied by tumor-like lesions of the ovarian stromal elements. Some of these ovarian tumors or tumor-like enlargements are discovered incidentally and mostly regress spontaneously, whereas others represent only a part of a much or less severe endocrinopathy, which requires appropriate treatment; rarely, complications of such ovarian tumors (torsion, rupture with intraabdominal hemorrhage, hemoconcentration, oliguria, and thromboembolic phenomena) may lead to life-threatening conditions for the patients.

B. Neoplastic Tumors of the Ovary

The neoplastic tumors of the ovary can be divided into (i) those derived from the surface epithelium or stroma; (ii) those containing granulosa, theca, Sertoli or Leydig cells or fibroblasts of gonadal stromal origin (sex cord–stromal, steroid cell, and other tumors); (iii) those derived from the primitive germ cells of the embryonic gonad (germ cell tumors); (iv) those nonspecific to the ovary tumors, including mesenchymal tumors and malignant lymphomas; and (v) metastatic tumors.

The most frequently found ovarian tumors derive from the surface epithelium or stroma (about 60% of all ovarian neoplasms and 80–90% of primary ovarian malignancies). Germ cell tumors [the vast majority are benign teratomas (dermoid cysts)] account for about 20% of all ovarian neoplasms, sex cord–stromal tumors (mainly benign fibromas and granulosa cell tumors with malignant potential) account for about 10%, and nonspecific mesenchymal and metastatic tumors account for the remaining 10%.

Surface epithelial–stromal tumors of the ovary are classified as serous, mucinous, endometrioid, clear cell, transitional (Brenner) tumors; mixed epithelial with benign, borderline, or malignant clinical behavior; Müllerian mesenchymal, nonteratomatous, undifferentiated, and unclassified epithelial tumors.

Whereas in benign tumors the epithelium is made up of a single layer of normal-appearing cells, epithelial proliferation with papillary formation and pseudostratification, nuclear atypia with increased mitotic activity and absence of stromal invasion (with some exceptions) are the characteristics of tumors with borderline malignancy (tumors with low malignant potential or atypically proliferating tumors). In malignant tumors, stromal invasion is the main diagnostic criterion.

C. Ovarian Cancer

Ovarian cancer is the most common malignancy of the female genital tract to result in death in most Western countries. More than 80% of epithelial ovarian cancers occur in postmenopausal women since the mean age of patients with benign tumors is 45 years, with borderline tumors at 50 years and with malignant tumors at about 60 years. There is a risk for ovarian carcinoma in women with certain family histories, such as site-specific familial ovarian cancer, breast/ovarian familial cancer syndrome, or Lynch II syndrome (familial colon, endometrial, ovarian, breast, and other gastrointestinal and genitourinary cancers). Similarly, the risk for ovarian cancer seems to be increased by infertility and milk consumption, or use of talc, and decreased by pregnancy and oral contraceptives. On the other hand, a causal relationship between the use of drugs for ovarian stimulation and ovarian carcinoma has not been documented and remains controversial.

The diagnosis of ovarian cancer can only be made during or after surgery. Unfortunately, there are no symptoms or signs indicating a malignant ovarian disease, especially in its early stages. Thus, when symptoms do develop, ovarian cancer has mostly reached advanced stages (III or IV). Transvaginal ultrasonography as well as measurements of the tumor marker CA 125 have been used unsuccessfully to screen for ovarian cancer in asymptomatic postmenopausal women. However, serum CA 125 levels were shown to be useful in distinguishing between benign and malignant ovarian tumors, particularly in the postmenopause.

The main spread mode of ovarian cancer is by exfoliation of cells into the abdominal cavity; lymphatic and hematogenous dissemination occurs less frequently. Surgical staging is of crucial importance because subsequent treatment will be determined by the stage of the disease. According to the FIGO surgical staging, in stage I the growth is limited to the ovaries; in stage II the growth involves one or both ovaries with pelvic extension; in stage III the tumor involves one or both ovaries with abdominal peritoneal surface metastases, positive retroperitoneal lymph nodes, and/or superficial liver metastases; and in stage IV parenchymal liver metastases and/or other distant metastases are also present.

In cases of borderline ovarian tumors, total surgical resection of the primary tumor with preservation of the uterus and the contralateral ovary is the principal mode of treatment. The same surgical procedure can also be performed in selected cases of women with well-differentiated stage Ia tumors who desire to preserve fertility. In all other cases of ovarian carcinoma, cytoreductive surgical treatment with removal of as much disease as possible is the main principle of the primary therapy of the patients. Chemotherapy as monotherapy or mostly as combined therapy is also the standard treatment for metastatic ovarian cancer. The addition of abdominal irradiation after chemotherapy is of no apparent benefit; however, the use of whole-abdominal radiation as an alternative to combination chemotherapy in selected patients with metastatic ovarian cancer has showed relatively acceptable results.

The prognosis of ovarian cancer is poor, mostly because the majority of cases are diagnosed in advanced stages. Thus, although the survival rates for stages I and II are more than 80%, the 5-year survival rates for most patients, who are primarily treated in stages III or IV, are 20–40% and about 5%, respectively. The age of the patient, the grade of the tumor, the residual disease after the primary surgical treatment, and the second-look status are factors also affecting the prognosis.

VII. GESTATIONAL TROPHOBLASTIC TUMORS

Gestational trophoblastic tumors can be classified according to WHO as hydatidiform mole (categorized as either complete or partial), invasive mole, gestational choriocarcinoma, and placental-site trophoblastic tumor (PSTT).

The overall incidence of complete hydatidiform mole is about 1 per 1000 pregnancies, whereas the partial mole was found to be even less frequent. Two or three percent of patients with hydatidiform mole develop choriocarcinoma, whereas placental-site trophoblastic tumors are extremely rare. The highest incidence of mole pregnancies has been reported from Africa and Southeast Asia, particularly from Japan.

With respect to cytogenetic and biochemical studies, complete hydatidiform mole originates from the fertilization of an ovum by a single (>90% of moles) or two (4–8% of moles) spermatozoa, in which the maternal genetic material has been lost or inactivated. On the contrary, in partial hydatidiform moles the maternal genetic material is retained. Thus, the conceptus in complete hydatidiform moles is paternally derived and results in XO or YO (90%) and XY or XX (10%) configurations of the sex chromosomes, depending on whether one or two spermatozoa, respectively, have penetrated the oocyte. On the other hand, partial moles are triploid, with two paternal and one maternal set of chromosomes, e.g., 69,XXX or -XXY or XYY.

In invasive moles, a penetration of the myometrium by trophoblastic villous structures of a complete or partial hydatidiform mole is observed. Choriocarcinomas as well as placental-site trophoblastic tumors mostly derive from complete moles or even from a normal conception. Pathologically complete moles lack identifiable embryonic or fetal tissues and are characterized by hydropically dilated trophoblastic villi with hyperplastic syncytio- and cytotrophoblast, edematous mesenchyme, and the absence of blood vessels. However, twin pregnancies with a complete mole and a normal fetus have also been reported. In partial hydatidiform moles, the trophoblast contains some normal and some abnormal villi—hence, embryonic or fetal tissues can be identified; hydatidiform swelling and surface (syncytiotrophoblast) hyperplasia are mostly focal and, if the fetus is viable, there are blood vessels in the villi containing nucleated fetal erythrocytes. Invasive forms of hydatidiform mole occasionally progress to choriocarcinoma or metastasize to the vulva, vagina, cervix, and lungs.

Choriocarcinoma is characterized by rapidly dividing cytotrophoblasts with extensive areas of hemorrhage and necrosis and by the absence of chorionic villi, stroma, and vascularization. It metastasizes to the lungs, pelvic organs, and brain. In placental-site trophoblastic tumors, the placental-site trophoblast separates from the villous trophoblast and infiltrates the surrounding tissues.

Vaginal bleeding during the first trimester of pregnancy, occasionally characterized by transcervical passage of molar tissue vesicles, is the most common symptom in complete molar pregnancies, followed by uterine enlargement, hyperemesis gravidarum, toxemia, hyperthyroidism, and theca lutein ovarian cysts. Partial moles frequently present with the symptoms of a missed abortion, but the final diagnosis can be made only after histologic evaluation of the surgically removed intrauterine material. Ultrasonography and measurements of serum levels of hCG are the main diagnostic criteria used for discriminating between normal and hydatidiform molar pregnancies.

Choriocarcinoma occurs much more frequently (about 1500×) after abortion of a molar pregnancy than after a term delivery, mostly within the first year. Vaginal bleeding or discharge, abdominal pain, and an extrauterine mass are frequent signs; however, in about 30% of the cases the first symptom may originate from a distant metastatic tumor localization. Similarly, in women with PSTT, vaginal bleeding is the typical symptom, but amenorrhea also occurs after term delivery, abortion, or mole pregnancy. Together with ultrasonography, hCG measurements and histologic examination of curettage material, radiography of the chest, nuclear magnetic resonance, or computer tomography are often used to detect pelvic or distant metastases of malignant trophoblastic tumors.

Suction evacuation of the uterine cavity is the first main step in the therapeutic management of a trophoblastic disease, followed by monitoring the urine or serum hCG levels. Chemotherapy is only performed when an invasive mole or choriocarcinoma is suspected. Hysterectomy is less effective than chemotherapy and also undesirable, especially in young nulli- or primiparous patients.

The age of the patient, the type of the antecedent pregnancy (mole, abortion, or term), the interval between the end of the antecedent pregnancy and the beginning of chemotherapy, the hCG levels, the ABO group, the largest tumor size, the site and number of metastases, and prior chemotherapy are the main parameters taken into account for categorizing the cases as low, medium, and high risk. Methotrexate plus folic acid is usually successful for the treatment of low-risk cases, with eradication of the tumor in more than 75% of patients. In medium- and high-risk cases, a combination chemotherapy is always needed and achieves complete remission of the disease in 50–74% of patients. In cases with chemotherapy-resistant disease, surgical resection of localized tumors can sometimes lead to sustained response. Placental-site trophoblast tumors, if localized to the uterus, should be treated by hysterectomy. Nevertheless, if metastases occur the prognosis is very poor.

See Also the Following Articles

Breast Cancer; Cervical Cancer; Leiomyoma; Ovarian Cancer

Bibliography

Baak J. P. A. (Ed.) (1991). *A Manual of Quantitative Pathology in Cancer Diagnosis and Prognosis*. Springer-Verlag, Heidelberg.

Beral, V. (1987). The epidemiology of ovarian cancer. In *Ovarian Cancer: The Way Ahead* (F. Sharp and W. P. Soutter, Eds.), pp. 21–31. Royal College of Obstetricians and Gynaecologists, London.

Berek, J. and Hacker, N. F. (Eds.) (1994). *Practical Gynecologic Oncology*, 2nd ed. Williams & Wilkins, Baltimore.

Burghardt, E. (Ed.) (1993). *Surgical Gynecologic Oncology*. Thieme-Verlag, Stuttgart.

Franco, E., and Monsonego, J. (Eds.) (1997). *New Developments in Cervical Cancer Screening and Prevention*. Blackwell, Oxford, UK.

Friedman, R. J., Rigel, D. S., and Kopf, A. W. (1985). Early detection of malignant melanoma: The role of physician examination and self-examination of the skin. *Cancer J. Clin.* **35**, 130–151.

Harlap, S. (1993). The epidemiology of ovarian cancer. In *Cancer of the Ovary* (M. Markman and W. J. Hoskins, Eds.), pp. 79–93. Raven Press, New York.

Kurman, R. J. (Ed.) (1994). *Blaustein's Pathology of the Female Genital Tract*, 4th ed. Springer-Verlag, New York.

Richart, R. M. (1968). Natural history of cervical intraepithelial neoplasia. *Clin. Obstet. Gynecol.* **10**, 748–784.

Richart, R. M. (1990). A modified terminology for cervical intraepithelial neoplasia. *Obstet. Gynecol.* **75**, 131–133.

Shepherd, J. and Monaghan, J. (Eds.) (1990). *Gynecologic Oncology*. Blackwell, Oxford, UK.

Tarlatzis, B. C., Grimbizis, G., Bontis, J., and Mantalenakis, S. (1995). Ovarian stimulation and ovarian tumours: A critical reappraisal. *Hum. Reprod. Update* **1**, 284–301.

Taylor, H. B., and Norris, H. J. (1996). Mesenchymal tumors of the uterus. IV. Diagnosis and prognosis of leiomyosarcoma. *Arch. Pathol.* **82**, 40–44.

World Health Organization (WHO) (1993). Gestational trophoblastic disease, Tech. Rep. Ser. 692. WHO, Geneva.

Zur Hausen, H. (1977). Human papillomaviruses and their possible role in squamous cell carcinomas. *Curr. Topics Microbiol. Immunol.* **78**, 1–30.

Tunicata (Urochordata)

Andrew Todd Newberry
University of California, Santa Cruz

GLOSSARY

genet Life cycle; genetic individual; all the solitary or multiple ramets mitotically derived from a zygote.

hermaphrodite A module, ramet, or genet containing both ovary and testis.

module The multicellular functional unit of a clone or colony, replicated by asexual reproduction.

myoplasm The peripheral region of an ascidian egg, containing many organelles.

oviparous Spawning unfertilized eggs.

ovotestis A gonad comprising both ovary and testis.

ovoviviparous Retaining zygotes, brooding young without nourishing them via special membrane.

ramet Body; anatomical individual, sometimes comprising multiple modules.

tadpole The larva of an ascidian, usually but not always tailed.

zooid The module in an ascidian colony.

The wholly marine phylum Urochordata (tunicates) comprises four classes; members of the class Ascidiacea, or ascidians ("sea squirts"), are by far the most familiar. The phylum is famous for its apparently chordate traits. Ascidian larvae (tadpoles) have a dorsal nerve cord and notochord, both lost at metamorphosis, and adults have chordate-like gill slits and a pharyngeal endostyle strikingly similar to that of cephalochordates and even similar to the endostyle of the ammocoete larvae of lampreys. Ascidians are esteemed, as well, among a smaller circle of cognoscenti for their odd ways: a heart that periodically reverses the direction of its blood flow, cellulose in the tunic, high hemal vanadium concentrations in some species, and other curiosities.

I. THE PHYLUM

Urochordata is a big phylum: By one recent authoritative estimate it contains approximately 3000 benthic and pelagic species, about half of them described. Tunicates can exert a powerful influence on the communities to which they belong. Seasonally, pelagic salps achieve enormous abundance, excluding much zooplankton and claiming a significant fraction of the phytoplankton as food, whereas benthic ascidians spread vigorously over rocky substrates and eat much of the plankton at the bottom of the water column.

In addition to E. G. Conklin's classic studies of ooplasmic segregation at fertilization and of cell lineage patterns in early development, research using ascidians has shed light on many biological phenomena of wide interest: simple immune responses, symbiotic interactions, the tumor-suppressing potential of some ascidian metabolites, and patterns of marine biogeographical spread. The phylum has not received the general attention that it merits perhaps because its members are anatomically perplexing, mostly not pretty externally even if lovely inside, soft-bodied and thus subject to gross distortions when clumsily handled, taxonomically a challenge even to the group's few experts, and, medical promise aside, without much popular impact on human affairs (though a few are delicious).

Two urochordate classes are pelagic: two dozen species of small to large salps, doliolids, and pyrosomes (class Thaliacea) and approximately 80 species of mostly tiny larvaceans (class Appendicularia). These low diversities are typical of holopelagic taxa. Two classes are benthic: a recently established class comprising a dozen abyssal sorberaceans (class Sorberacea) and the tremendous array of ascidians—approximately 1400 species known—widespread from low intertidal habitats to the abyss.

II. NON-ASCIDIANS

Pyrosomid thaliaceans, entirely colonial, are extraordinary creatures of the open sea. In each zooid (i.e., each modular unit of the colony) a budding region near the heart generates more zooids, which align themselves in the sheath of a condom-like, jet-propelled colony. As do ascidians, each pyrosomid zooid generates its own weak ciliary current to draw outside water into its pharynx, in which the water is filtered for food and gas exchange. The water then leaves the zooid through a cloacal aperture that opens into the lumen of the colony. The combined effluent water flow of thousands of these tiny modules is expelled through an orifice to the rear of the motile colony, thus driving it forward. These fast-swimming and intensely luminescent colonies may be just a few centimeters long or, in some tropical species, can grow to several meters long—in one species (*Pyrosoma spinosum*) to an astonishing 10 m.

A pyrosomid zooid produces one big, yolky, internally fertilized egg at a time or even just one total. The zygote remains in its mother zooid, where it develops into a colony of four feeding zooids plus a modified zooid transformed into a structure that adjusts the new colony's common cloacal opening, its rocket nozzle. At this stage the little colony leaves its mother, assumes a swimming life, and begins to generate its own multitude of zooids by prolific budding.

The barrel-like salps and doliolids, ringed by hoops of strong propulsive muscles, are colonial (modules connected) at some time in their life cycle and clonal (modules dispersed) after that. The cycle of colony formation and cloning is extremely complex. The oozooid (the module, directly derived from the embryo, that initiates a colony) generates strands of blastozooids (budded modules). In doliolids, but not in salps, these zooids are polymorphic. The modular strands break up into clones of jet-swimming bodies that are each a few millimeters or centimeters long.

Doliolids and salps generally produce just one large egg per clonal module. Doliolids release a heavily encased tadpole that becomes the founder of a new colony and clone. Salps placentally nurture embryos in the cloacal cavity; the young fatally rupture their mothers as they escape to the open sea to found new colonies and clones.

While the thaliacean life cycle reduces or omits its larval stage, larvaceans, as the class's vernacular name implies, seem to use it as their basic bauplan. However, despite their name, larvaceans have no larval stage. Also, their tailed adults differ profoundly both in their development and in their anatomy from the ascidian tadpole with which they sometimes are confused. The resemblances are superficial, secondary, and coincidental; the ascidian larva and the larvacean adult evidently are independent inventions of urodele body plans among the tunicates.

Larvaceans have no cellulose-laden tunic. Instead, they construct and inflate around themselves—and then repeatedly discard only to rebuild—a remarkable mucopolysaccharide "house" complete with windows and an escape hatch. This structure dwarfs its resident: A larvacean with a body about 7 mm long may build a house a few centimeters across, and one big abyssal larvacean with a body 75 mm long crafts a delicate house that spans fully a meter. A larvacean's house not only provides it with a measure of protection but also filters food particles through intricate openings and channels. The larvacean inside then captures its food with a mucous net that it attaches to the inner walls of its house and periodically swallows. The little animal abandons its house—escapes through the hatch—when disturbed or when its steady filtering of its surroundings has fouled its house beyond use, occasions that in some species arise several times daily.

Larvaceans do not bud. Their eggs are few and tiny, but even so, once they mature, the eggs rupture their mother's body during their release. They are fertilized in the open sea, and the embryos grow quickly and directly to solitary adulthood.

Sorberaceans are so inaccessible in their abyssal habitats that nothing is known of their reproductive biology beyond what can be inferred from collected specimens. These inferences do not set them markedly apart from ascidians.

Because ascidians are by far the most diverse, the best known, and the most accessible tunicates to land-based investigators, they are the focus in the rest of this article.

III. ASCIDIANS

A. Ascidian Orders

There are three ascidian orders based on two broad traits: placement of the gonads and structure of the pharyngeal wall. The latter can be simple (aplousobranch), finely pleated (phlebobranch), or deeply folded (stolidobranch). Although the orders get their names from pharyngeal traits, only their gonadal traits need concern us. The aplousobranchs and phlebobranchs are both enterogonid orders: Their gonads are associated with the gut loop. All aplousobranchs are colonial, except for the famous but systematically puzzling *Ciona*, which despite Kott's persuasive case is still widely deemed a phlebobranch. An aplousobranch zooid's single but often extensive ovotestis typically lies in the abdominal gut loop or in a postabdominal extension of the zooid that receives gonads and the heart from the abdomen. In phlebobranchs and stolidobranchs, compressed into a design with a relatively immense thorax and no abdomen at all, most of the gut loop lies beside the pharynx. In phlebobranchs, solitary or colonial, the gonad lies on the gut itself, which attaches to the body wall; again, it is the body's single (albeit elaborate) sexual organ. The stolidobranchs, also solitary or colonial, are pleurogonid: Gonads lie in the thoracic body wall, independent of the gut loop. Stolidobranch gonads are usually multiple, and they often show a bilateral symmetry or an arrangement evidently derived from that.

B. Appearance

Solitary ascidians range in size from a few millimeters to several centimeters across. Most are somewhat potato shaped and broadly cemented to hard substrates, but some rise on stalks of tunic and others extend hairs or tendrils to bury themselves in loose sediments. Colonies may be only a few millimeters across or cover many square centimeters. According to species, colonies adopt a great array of forms: erect "bouquets" of zooids, dendritic encrustations of zooids in a thin basal tunic, or thick sheets or lobes of deeply embedded zooids. The zooids of ascidian colonies are mostly quite small compared to solitary forms, and some are truly minute (<1 mm long). With the support of a substantial colonial tunic, many aplousobranch zooids become elongate, delicate entities. Phlebobranch and stolidobranch colonies contain zooids that look like miniature solitary forms.

The enveloping tunic may be thin and membranous, thick and firm, softly yielding, or very tough, almost resembling wood. With such a range of appearance, it is no wonder that ascidians are often mistaken for sponges, algal holdfasts, patches of mucus or detritus, and even rocks. The saccular body within the tunic is revealed, like a soft kernel, only by extracting it from its husk after a specimen is relaxed and fixed, and the beauty of the pharynx, the gonads, and other ascidian structures becomes clear only by patient dissection and by staining the body's delicate and virtually transparent tissues.

IV. ASEXUAL REPRODUCTION

Asexual reproduction is a mitotic growth phenomenon unrelated to meiotic–syngamic sexual reproduction except that both processes proliferate bodies. The "offspring" of asexual reproduction are repeated modules—multicellular functional units that are genetically identical to one another. They may be monomorphic or polymorphic and they may be dispersed as a clone or linked into a colony. One may speak in a vernacular way of "asexual generations," but in fact, of course, all the modules generated in asexual reproduction come from the same initial zygote and so are multiple representatives of just one genet— that is, of just one life cycle. Salps and doliolids eventually clone themselves into many entirely separate bodies after a transient colonial phase. Pyrosomids and ascidians replicate themselves into the zooids of colonies that stay more or less intact. Larvaceans do not replicate themselves asexually at all.

From the variety of modes by which ascidians replicate asexually, Nakauchi discerned just two functional themes. On the one hand, zooids in vigorously growing colonies make new zooids in a bewildering variety of ways: strobilation of the abdomen or postabdomen or both, herniation of vascular processes extended onto the substrate, accumulation of cell masses in the tunic's blood vascular system, projection of the thoracic wall into the surrounding basal tunic, double herniation of a narrow thoracoabdominal "waist" to make one daughter zooid's new thorax and another's new abdomen, and variations on these variations.

On the other hand, sometimes budding produces resistant bodies analogous to those by which members of many aquatic taxa survive hard times—the gemmules of sponges, the statoblasts of bryozoans, and the hibernacula of kamptozoans. Again, one species or another uses a wide variety of tissues and body regions to form these types of buds: stolon tips, seemingly aborted abdominal strobilae, and postabdominal fragments. No matter the device or tactic, the result is to accumulate and store a mass of trophocytes in an epithelial envelope until, with improved conditions, the cells can organize themselves into a new, functional zooid that initiates a new colony.

V. SEX

Ascidians are hermaphrodites; when the sexual cycles of allegedly gonochoric (dioecious) species of certain aplousobranch genera have been followed long enough, genets have proved to be sequential hermaphrodites. In most ascidian species the adult solitary body or mature colony taken as a whole has recognizable testes and ovaries simultaneously even if these are not both full of ripe gametes. While various mechanisms (sequential hermaphroditism and blocks within the egg's chorion) reduce self-fertilization in some species, it appears to be widespread.

A. Gonadogenesis and Gametogenesis

Ascidian gonadogenesis has been described only occasionally, mostly long ago. Ascidians raise old puzzles about how cells, especially cells in modular organisms, become germinal at all and how the sex of gonads is determined in a hermaphrodite. In solitary and colonial ascidians alike, the gametes appear to arise from masses of "undifferentiated" amoebocytes in the blood. However, behind their morphological appearances, what these cells may really be remains unknown. No precociously segregated germline has yet been demonstrated in ascidians, despite the group's mosaic development; yet subtlety, more than absence, may account for this. Cell lineage studies have not aided as much as might be expected, but perhaps the cell markers have been inadequate to the task.

The typical ascidian gonad is an ovotestis. In the few enterogonids whose gonadal development has been studied, a single tissue mass develops into the gonad's male and female parts. In pleurogonids, the cell masses that will become gonads, sometimes a great many, establish themselves separately from the start.

Usually the ovarian and testicular elements of ascidian ovotestes are closely juxtaposed, though structurally distinct. Despite this anatomical separation, the gametogenic processes inside them may present a certain confusion. For example, in the stolidobranch *Stolonica* one regularly finds small oocytes with the tailed sperm inside testes and spermatocytes with the advanced oocytes inside closed ovaries. Occasionally, however, testes and ovaries are widely separated. Thus, the widespread stolidobranch genus *Distomus* is male on the left side and female on the right side (though tiny, aborted gonads of the "wrong" sex occur rarely on either side), and in the New Caledonian aplousobranch *Citorclinum* abdominal ovaries lie in a different body region from the postabdominal testes. Of course, the sexes are largely or wholly separated temporally in sequential hermaphroditism.

It is hard to imagine a determinative role for sex chromosomes in hermaphrodites. The sex of particular gonads or their parts in hermaphrodites seems to reflect a richer, more open interplay between genotype and environment. Both the interior, cellular–physiological environment of the developing gonads and gametes and the exterior, ecological environment of the entire animal seem to figure more prominently than they do in more rigidly dioecious designs. The reason for this remains largely obscure.

Unless one has previously dealt with modular hermaphrodites, germinalization and sex determination in ascidians present unfamiliar but enticing circumstances. Colonial ascidians are attractive experimental animals, and these two developmental processes are central to understanding the biology of sex itself. However, apparently no investigation to date has gone beyond merely pattern descriptions to clarify the processes of ascidian gonadogenesis or gametogenesis.

B. Hormonal Control of Reproduction

The control of gonadal and gametic maturation in ascidians has been much more assiduously attended to than the dynamics of germinalization or sex determination. A prominent dorsal ganglion and gland together form the adult ascidian's dorsal "neural complex," and a "dorsal strand" of nonaxonic neural cells extends from the neural complex into the viscera. The phylogenetic affinities between urochordates and vertebrates suggest there may be parallels between the tunicate neural complex and the vertebrate brain; this has led to comparing the neural gland to the vertebrate pituitary gland. No wonder, then, that ascidians have been probed for endocrine hormones like those of humans.

Despite much study and debate, only ambiguous evidence ties the neural gland to the production of gonadotropic compounds. Most of the (somewhat contradictory) embryological studies conclude that the neural gland is homologous only with the posterior lobe, the pars nervosa, of the vertebrate pituitary gland, and the weight of physiological evidence suggests that the ascidian neural gland is an exocrine organ, not an endocrine one. Despite many studies, virtually every aspect of the neural gland's behavior and function remains deeply puzzling. The dorsal ganglion shows a likely, if problematic, clue to neurosecretory activity: electron-dense secretory granules resembling ones found in endocrine cells, some with an immunoreactivity that may imply a gonadotropic function.

Recently, attention has shifted to the "dorsal strand," especially to a nerve plexus associated with it. Immunofluorescence reveals a gonadotropin-releasing hormone-like hormone in this plexus. The hormone is present both in enterogonid ascidians (*Ciona* and *Chelyosoma*), in which the dorsal cord and its plexus extend right into the gonads around the gut, and in a pleurogonid (*Dendrodoa*), in which the gonads, lying in the lateral body wall, are not closely associated with the dorsal cord or its plexus. The hormone seems to be released into blood sinuses, which would give it wide circulation.

Goodbody and Mackie have reviewed the history of investigations that have led to our current hard-won, if still only partial, understanding of the hormonal control of ascidian sexual reproduction.

C. Maturation of Gametes

The ascidian sperm has two especially notable features: (i) an enormous, greatly elongated nucleus in the head, with a single big mitochondrion beside it, and (ii) the absence of a middle body. Seminiferous tubules pack the testis. Sperm toward the center of a tubule generally mature before those near the tubule's wall.

The ascidian egg tends to be smaller (100–200 μm) in solitary, oviparous species than in colonial, ovoviviparous ones (in which some eggs approach 700 μm). A mature ascidian's ovary, like its testis, usually contains sex cells in all stages of maturation. A more or less yolky endoplasm lies within a peripheral "myoplasm" that contains mitochondria, pigment granules, an endoplasmic reticulum, and ribosomes, all "tethered in the cortex by a meshwork of cytoskeletal filaments, which is linked to the plasma membrane" (Jeffery and Swalla, 1997). Between the egg's plasma membrane and its surrounding chorion lies a loose array of little "test cells" of unknown origin. The chorion is produced by a two-layered jacket of follicle cells whose source—whether they are transformed germinal cells or invasive somatic cells—also remains hotly debated. The inner follicle cells accompany the released egg; along with the chorion and the test cells, they constitute the egg's persistent vitelline envelope. The collapsed outer follicle layer remains in the ovary after ovulation.

In those few species in which meiosis has been followed, the egg pauses (at sea in oviparous species and in the oviduct in ovoviviparous species) at metaphase II, awaiting fertilization. Upon entry of a sperm, the egg completes its meiosis and, like the sperm, forms a pronucleus, The two pronuclei then fuse. Since the sperm does not carry its mitochondrion beyond the chorion of the egg it penetrates, in the zygote these organelles and their products are entirely maternal.

The follicle cells of the unfertilized egg not only attract sperm but also activate them at fertilization. In oviparous species the perivitelline space expands and the inner follicle layer cells vacuolate after fertilization to increase the egg's buoyancy. In ovoviviparous species, which brood their young in a confined chamber, zygotes tend to remain compact.

D. Seasonality

In temperate and polar waters ascidians typically are summer breeders, although there are plenty of exceptions among common species that have been surveyed. In the tropics, year-round breeding is the rule. Solitary ascidians appear to release gametes daily through their breeding season, and in many species adults probably breed for several years. In colonies, individual zooids seem mostly to deteriorate after their gonads are played out and their brooded larvae released. However, such zooids may then recover and resume colony-building asexual reproduction, whereas other zooids replace decrepit ones in sex so that the colony breeds for far longer than any of its constituent zooids.

E. Release of Eggs or Larvae

Some solitary ascidians (e.g., *Corella willmeriana* and various species of *Molgula*) brood their young, but most are oviparous, releasing their eggs to be fertilized at sea. Development of their zygotes to hatched larvae is very fast, taking scarcely a day, after which the tadpole may spend hours or a few days in the plankton before settling. Colonial ascidians are ovoviviparous, brooding their offspring in the cloacal cavity or in an auxiliary brood pouch and releasing them as swimming tadpoles. These tadpoles emerge a few at a time and then swim only briefly—in some species for merely minutes—before settling.

Light levels often control the release of gametes in oviparous species. A light shock after prolonged darkness initiates synchronous spawning after delays of minutes or hours. As a result, some species are morning spawners and others are late afternoon spawners, even though the spawning process may have been triggered by the same dawn. In some brooding species larval release also seems to favor one or another time of day. However, careful comparative and experimental studies of the influence of light on larval release remain to be performed; our knowledge of this matter, as of so much else about ascidians, is still limited.

VI. EARLIEST DEVELOPMENT

Bonner emphasized that sex, for whatever reasons, employs single cells—eggs and sperm—while the plants and animals and many of the algae that engage in sex are, for other but equally compelling reasons, multicellular. One consequence of unicellular sex by multicellular organisms is that the filial life cycle, to achieve a renewed, orderly multicellularity, must "recover" from its sexual, zygotic start through embryonic development. Thus, the life cycle's initial development is part and parcel of sexual reproduction.

Ascidians undergo a bilateral cleavage that cuts up the zygote's strikingly regionalized cytoplasm. Early development is mosaic, but certainly the blastomeres interact with one another to considerably soften the developmental rigidity such a term implies. Further mitosis leads to a ball of such large blastomeres, with such a small blastocoel, that it is effectively a stereoblastula. Despite its bulk, the blastula gastrulates partly by invagination as well as by mesodermal involution and ectodermal epiboly. Neurulation proceeds in a chordate-like fashion, establishing the spinal complex of the tadpole larva (thoracic brain and caudal nerve cord). Gastrulation and neurulation reveal the embryonic axis that was laid down long before, during ooplasmic segregation in the zygote. Tunicates are deuterostomes; the blastopore disappears but a tiny posterior embryonic neuropore for a while marks its former vicinity, and in any case the eventual mouth lies far away. Enterocoely, so often linked with deuterostomy as a developmental trait of phylogenetic significance, is really a moot point in tunicates. What uncertain candidates there are for coelomic structures (epicardia, pericardium, and molgulid renal sac) are probably mere analogs of other bauplans' coeloms and do not appear until larval metamorphosis, if at all.

Differentiation of the rest of the tadpole larva follows and produces a stout trunk, a prominent tail with a dorsal nerve cord, a notochord of vacuolated cells beneath it, and strong propulsive muscles on either side of the notochord. The test cells, which lay beneath the chorion of the unfertilized egg, now come into play, forming the larva's outermost tunic, especially the prominent tail fin. It may seem odd that these cells, sequestered at the chorion so early in gametogenesis, have been carried along solely for this task. However, early development has proceeded entirely inside the vitelline envelope; the test cells could not be better placed to do their vital but peripheral work.

VII. LARVAE

The ascidian tadpole does not feed; it does not even pass a respiratory current over its developing gills. Both its apertures are sealed by its outermost envelope of tunic, which is shed at metamorphosis to open them. As it approaches the crisis of settling, the tadpole responds to light with an ocellus, to gravity with an otolith, to chemical stimuli by unknown means, and in some species apparently to acceleration as it is swept along the sea bottom. These sensitivities combine to direct it not only to the bottom generally but also particularly to vertical surfaces, the undersides of rocks and overhangs, or other physically appropriate sites for subsequent development. Settlement involves behavioral changes (reduced swimming effort and avoidance of light) and sometimes the release of a glue that may help the larva adhere to rocks in high-energy habitats. Instructively, a few molgulid ascidians that inhabit flat subtidal sediments have tailless larvae that settle to the bottom as other larvae might, not with the gymnastics that tailed tadpoles employ. Paradoxically, still another molgulid, one that inhabits habitually violent surf surge channels, also releases tailless larvae; these are extremely sticky and adhere to the first surfaces they meet.

When an attractive substrate is found, the larva attaches to it firmly with its anterior adhesive papillae. Its tail collapses rapidly and dramatically as contractile elements in it crumple the neural cord, notochord, musculature, and epidermis. The larval brain gives way to a dorsal neural complex of very different structure. The pharynx multiplies and transforms its gills. The cellophane-like outermost larval tunic sloughs off, and a renewed tunic, usually irrigated by vascular elements, encloses the settled body. Throughout this process, the attached larval trunk slowly rotates in place so that its oral and cloacal pores, now open, point away from the substrate. Within a week the ascidian is a robust little member of the benthos and on its way to its own sexual maturity.

See Also the Following Articles

Asexual Reproduction; Marine Invertebrate Larvae; Marine Invertebrates, Modes of Reproduction in

Bibliography

Berrill, N. J. (1975). Chordata: Tunicata. In *Reproduction of Marine Invertebrates* (A. C. Giese and J. S. Pearse, Eds.). Academic Press, New York.

Bonner, J. T. (1958). *The Evolution of Development*. Cambridge Univ. Press, Cambridge, UK.

Bonner, J. T. (1965). *Size and Cycle*. Princeton Univ. Press, Princeton, NJ.

Brien, P. (1948). Embranchement des tuniciers. In *Traité de Zoologie* (P. Grassé, Ed.), Vol. 11, pp. 553–930. Masson & Cie, Paris.

Brusca, R. C., and Brusca, G. J. (1990). *Invertebrates*. Sinauer, Sunderland, MA.

Conklin, E. G. (1905). The organization and cell-lineage of the ascidian egg. *J. Acad. Natl. Sci. Philadelphia Ser.* 2 8, 1–119.

Goodbody, I. (1974). The physiology of ascidians. *Adv. Mar. Biol.* **12**, 1–149.
Jeffery, W. R., and Swalla, B. J. (1997). Tunicates. In *Embryology: Constructing the Organism* (S. F. Gilbert and A. M. Raunio, Eds.). Sinauer, Sunderland, MA.
Kott, P. (1990). The Australian Ascidicea Part 2, Aplousobranchia (1). *Mem. Queensland Mus.* **29**, 1–266.
Mackie, G. (1995). On the "visceral nervous system" of *Ciona*. *J. Mar. Biol. Assoc. UK* *75*, 141–151.
Millar, R. H. (1971). The biology of ascidians. *Adv. Mar. Biol.* 9, 1–100.
Monniot, C., Monniot, F., and Laboute, P. (1991). *Coral Reef Ascidians of New Caledonia.* ORSTOM, Paris.
Pearse, V., Pearse, J., Buchsbaum, M., and Buchsbaum, R. (1987). *Living Invertebrates.* Blackwell/Boxwood, Boston/Pacific Grove, CA.
Satoh, N. (1994). *Developmental Biology of Ascidians.* Cambridge Univ. Press, Cambridge, UK.

Turner's Syndrome

David H. Barad
Montefiore Medical Center

GLOSSARY

amniocentesis The sampling of amniotic fluid to determine the karyotype of a pregnancy.
anaphase lag The loss of a chromosome from one or both daughter cells during cell division.
aneuploidy A total number of chromosomes that is different from the normal 46,XX or 46,XY.
aneurysm A swelling of a blood vessel.
chorionic villus biopsy A sampling of pregnancy tissue early in pregnancy to determine the karyotype.
cubitus valgus A term used to describe the exaggerated angle found at the elbow of women with Turner's phenotype when the arms are relaxed at the sides and the palms face forward.
dysmorphic Having an abnormal form.
gonadotropins Follicle-stimulating hormone and luteinizing hormone, which are produced by the pituitary and act on the gonads.
growth hormone A hormone produced by the pituitary gland that is responsible for regulating growth.
haploid Having half the chromosomal material of a normal somatic cell.
karyotype The chromosomal complement of a cell.
mosaicism Having two or more cell lines of different chromosomal identity.
nondisjunction The failure of homologous chromosomes or sister chromatids to separate during anaphase.
pterigium colli Folds of skin that stretch from the neck to the shoulders.

Turner's syndrome results from the loss of part or all of one X chromosome. The characteristic features of Turner's syndrome are short stature, lack of secondary sexual development, and structural anomalies.

I. INTRODUCTION

In 1762, Morgagni reported findings at autopsy of a woman with an immature uterus and "corpuscles of white" appearance in place of ovaries. Other reports of similar abnormal ovarian development appeared throughout the eighteenth century. In the early twentieth century, reports began to appear associating this type of abnormal ovarian development with short stature and anomalies that we now associate with Turner's syndrome. Turner's syndrome is also known as Ullrich–Turner syndrome, in recognition of Dr. Otto Ullrich, who published reports discussing hypothesis about this constellation of symptoms as early as 1930. In 1938, Dr. Henry Turner described a syndrome of short stature, sexual infantilism, and skeletal anomalies. Turner thought that the lack of sexual development in the cases he described was secondary to pituitary failure; later it was determined that patients with Turner's syndrome had elevated gonadotropins consistent with ovarian failure. Dr. Turner's report triggered other investigations of this syndrome, resulting in many publications over the next few years. In recognition of the impact of his 1938 report, the syndrome continues to bear Turner's name.

II. EPIDEMIOLOGY

The incidence of Turner's syndrome is about 1 of 2500 live births. There is no evidence for a relationship between parental age or birth order on the incidence of Turner's syndrome. Some reports have noted an association of Turner's syndrome with increased paternal age; however, Turner's syndrome has also been reported in the offspring of artificially inseminated pregnancies, in which the sperm donors were presumably young and healthy.

Corothers's group in England reported a trend for a higher incidence of identical twinning in Turner's sibships then for the general population. Identical twining occurs when cells of a newly formed embryo break apart into two individuals. "Identical" twins have also been reported in which one twin has Turner's syndrome and the other has a normal female karyotype.

III. PATHOGENESIS

A. X Chromosome Abnormalities

Turner's syndrome develops because of the presence of a single X chromosome. Just how alone the X chromosome is may determine how the syndrome is expressed. Some women with Turner's syndrome will have incomplete pieces of the missing X chromosome. Women who are missing just a small part of the X chromosome will have few of the characteristics that we normally associate with Turner's syndrome. Studies of individuals with Turner's syndrome and its variants have been important in the development of our understanding of the molecular basis for normal sexual differentiation.

The typical karyotype arising from the loss of one X chromosome is 45,X. The X chromosome may be lost during meiosis or during early embryonic mitotic divisions of either oogenesis or spermatogenesis. The 45,X karyotypes comprise 50–60% of most reported cohorts with the other chromosomal abnormalities being present in varying frequencies. Turner's syndrome may also be caused by structural anomalies of the X chromosome such as 46,Xi (Xq) or from mosaicism. Mosaicism is a condition in which the 45,X cell line may be mixed with other normal cell lines such as 45,X/46,XX and 45,X/46,XY.

B. Meiosis and Aneuploidy

Meiosis is the process of cell division that leads to the formation of haploid gametes. During meiosis, gamete-producing cells undergo two divisions. In the first meiotic division, the paired homologous chromosomes line up in the center of the cell and are separated, with each member of the homologous pair going to a different daughter cell. In the second division, each chromosome divides at its centromere, with a single chromatid going to each daughter cell.

Aneuploid conditions such as Turner's syndrome can arise from either nondisjunction or anaphase lag. Nondisjunction may occur during anaphase of meiotic or mitotic division. When nondisjunction occurs during the first meiosis one daughter cell receives a double dose of the chromosome and the other cell gets none. Anaphase lag is the loss of a chromosome. With anaphase lag, either one or both daughter cells may lose the chromosome. Anaphase lag is thought to occur because of failure of the chromosome to position itself properly on the equatorial plate during metaphase. Nondisjunction and anaphase lag may occur in either oogenesis or spermatogenesis. Thus, the single X chromosome in the Turner's karyotype may be of either maternal or paternal origin. Complimentary 45,Y karyotypes must also occur, but they appear to be lethal.

C. Mitosis and Mosaicism

After the zygote has formed it continues to divide by the process of mitosis. In normal mitotic division, each daughter cell is a genetic replicate of the cell from which it arose. Abnormal mitotic division can give rise to two or more cell lines in the developing embryo. This condition, known as mosaicism, arises from errors in mitosis after the zygote has been formed.

Mosaicism is the presence of two or more cell lines with different chromosomal composition within an individual. Multiple cell lines may continue into adulthood or some of the cell lines may die out. In the case of mosaic Turner's syndrome, the presence of a 46,XX cell line allows development of height closer to normal and some gonadal function. Ovarian volume is greater in women with mosaic Turner's syndrome than in those with 45,X or X chromosome structural anomalies. Women with mosaic Turner's are also more likely to have undergone spontaneous menses and breast development.

IV. DIAGNOSIS

A. Prenatal Screening

Prenatal ultrasound may detect Turner's syndrome by finding a thickening of the nuchal fold, cystic hygroma, horseshoe kidney, and left-sided cardiac abnormalities. Positive multiple-marker screening test (maternal serum α-fetoprotein, conjugated estriol, and human chorionic gonadotropin) is associated with a higher incidence of Turner's syndrome. However, these screening tests are much more strongly associated with the presence of cystic hygroma, only one of the characteristics of fetal Turner's syndrome, than with the presence of Turner's syndrome itself.

Clinicians can accomplish prenatal diagnosis of Turner's syndrome using cytogenetic studies of fetal cells removed by chorionic villus sampling (CVS) or amniocentesis. Physicians perform CVS by passing a small tube through the cervix and sampling a few milligrams of fetal tissue. Amniocentesis involves aspiration of amniotic fluid and cells by a needle passed through the abdominal wall and uterus into the pregnancy sac. Obstetricians perform amniocentesis several weeks later in pregnancy than CVS. A recent evaluation of all cases of Turner's syndrome listed in the *Danish Cytogenetic Registry* found a higher incidence of Turner's syndrome diagnosed at CVS than at amniocentesis. Both CVS and amniocentesis revealed higher incidence of Turner's syndrome than at live birth. Some have used this data to question the accuracy of prenatal diagnosis of Turner's syndrome. However, this apparent decreased incidence of Turner's syndrome as pregnancy progresses only emphasizes the tendency of Turner's syndrome pregnancies to abort spontaneously early in gestation.

B. Diagnosis in Childhood

Diagnosis at birth or during childhood is based on the presence of characteristic dysmorphic features of the syndrome. Diagnosis must be confirmed by karyotype. Karotyping must be done with care to discover possible mosaic and structural X chromosome anomalies.

C. Diagnosis in Late Childhood or Young Adulthood

The characteristic presentation after childhood will be primary amenorrhea or premature menopause associated with short stature. Diagnosis must be confirmed by chromosomal studies. Individuals with this presentation are less likely to have the features usually characteristic of Turner's syndrome and are more likely to be mosaic or have X chromosome structural anomalies.

V. CLINICAL FEATURES

A. Structural Abnormalities

Turner's syndrome patients have a characteristic development of their bodies which is different from

TABLE 1
Clinical Features of Turner's Syndrome

Clinical feature	%
Short stature	98
Ovarian failure	95
Low hairline	80
Low-set, rotated ears	80
Broad chest, widely spaced nipples	75
Small jaw	70
Inner canthal folds	70
Arms turned out at elbows	70
Soft upturned nails	70
Webbed neck	50
Pigmented nevi	50
Hearing loss	50
Shortened fourth fingers	50
Narrow, high arched palate	40

the general population (Table 1). They are short and do not possess features associated with normal sexual maturation. Half of Turner's syndrome girls develop small brown moles called nevi. Lynne-Georgia Tesch, president of the Turner's Syndrome Society, writes, "With a 45,X karyotype, I am 4' 9" tall (?), have webbing of the neck, low ears, broad chest, moles, and slightly turned out arms."

As infants, Turner's syndrome girls have swelling of their hands and feet as well as loose folds at the nape of the neck. An abnormality of the lymphatic system during intrauterine development appears to cause these neonatal features. Older children and adults have a short, broad neck. Almost half of girls affected with Turner's syndrome will have a weblike ridge of skin between the neck and shoulders called a pterygium. Eighty percent have puffiness of the backs of the hands and feet. The pterygium and puffy hands and feet are due to the alteration in the lymphatic circulation.

Turner's syndrome is associated with changes in the development of the skeleton. Turner's syndrome individuals have distinctive facial features, including a small jaw, epicanthal folds, and rotated low-set ears. Their chest is broad with widely spaced nipples. Seventy percent of Turner's syndrome individuals have arms that are slightly turned out at the elbows (cubitus valgus). Many have a small lower jaw (micrognathia), a narrow, high-arched palate, and short fourth fingers secondary to a shortened fourth metacarpal bone.

B. Short Stature

Short stature is present in almost all women with Turner's syndrome. Children with the syndrome are usually small at birth. Weight and length at birth are affected equally. The childhood pattern of growth continues longer than that for the normal population. This leads to a progressive decline in growth velocity through childhood and almost no pubertal growth spurt. In a given population the average adult height of women with Turner's syndrome is about 20 cm less than that of normal adult women. The result of this is that the average height of an adult woman with Turner's syndrome in the United States is 143 cm (56 in.). Parental height is a major factor determining how tall a girl affected by Turner's syndrome will be when she is an adult. Turner's syndrome patients with tall parents tend to be taller themselves.

The final height of Turner's syndrome patients is proportional to the average height of normal females in the country of origin. For example, the mean height of normal Japanese women is about 10 cm less than that of normal Danish women. Likewise, the mean height of Japanese women with Turner's syndrome is about 10 cm less than their Danish counterparts. This observation is important in evaluating the results of therapy for short stature among Turner's syndrome cohorts from different countries.

Recently a 170-kb locus was mapped within the pseudoautosomal region of the human X chromosome which is associated with adult height determination. This locus was deleted in patients with short stature and the deletion did not occur in relatives who were of normal height. Other studies have shown, by the use of fluorescence *in* situ hybridization (FISH) techniques, that Turner's syndrome patients with Y-derived chromosomal fragments had milder phenotypic abnormalities than those with X-derived marker chromosomes.

Since short stature is one of the universal symptoms for patients with Turner's syndrome, many investigators have attempted to treat girls with Turner's so that a greater adult height could be achieved. Growth hormone, oxandrolone (an anabolic steroid), and estrogens have been used alone and in various combinations for this purpose.

C. Gonadal Function

Unless Turner's syndrome girls are given sex hormone replacement, they never undergo changes of the body that are normally associated with puberty. They will have no breast development and will not undergo menstruation. With rare exception, Turner's syndrome individuals cannot become pregnant unless they undergo *in vitro* fertilization and ovum or embryo donation.

1. *Endocrine*

Women with Turner's syndrome demonstrate a spectrum of gonadal function, ranging from the onset of spontaneous puberty and the potential for fertility to complete gonadal failure. There is evidence of the loss of gonadal function among women with Turner's syndrome in the elevated levels of the plasma gonadotropins. These hormones, luteinizing hormone and follicle-stimulating hormone, may be detected in infants with Turner's syndrome as early as 5 days after birth.

2. *Amenorrhea and Premature Menopause*

Most women with Turner syndrome have ovarian failure, though 10% may experience up to a few years of menstrual cycles. Most of these women will be mosaics or have X chromosome partial deletions. In Turner's syndrome patients with normal ovarian function the gonads are smaller than usual and there are fewer primordial and primary follicles. Even if the Turner's syndrome patient experiences menarche, she is most likely destined to have premature menopause. A rare case of endometrial cancer has been reported in a woman with 45,X Turner's syndrome, who had never taken hormone replacement, in whom graafian follicles were confirmed at the time of hysterectomy and oophorectomy.

3. *Gonadal Development and Morphology*

In most women with Turner's syndrome the gonads appear as a small fibrous white band of tissue 2 or 3 cm in length and 0.5 cm in diameter. These fibrous bands are known as streak "gonads" because they look very much like white streaks. Streak gonads are characterized by interlacing waves of dense fibrous stroma. There are no follicles or oocytes. Ovarian rete tubules may be found at the medial border of the streak gonad.

Ovaries of 45,X embryos appear normal at 14–18 weeks of gestation with normal differentiation of primordial germ cells and otherwise normal morphology. The streak gonads represent late fetal neonatal degeneration after formation of the medullary cord. The loss of oocytes begins after formation of the primordial germ cells and is completed between the prenatal period and the first few years of life. Carr and associates have postulated that 45,X germ cells may be unable to organize primordial follicles. The consequence of this failure is the accelerated loss of primordial germ cells. A critical region has been isolated to the long arm of the X chromosome, between q13 and q26, which appears to be responsible for ovarian dysfunction and infertility.

4. *Gonadal Tumors*

The presence of a cell line with a Y chromosome places the Turner's syndrome individual at risk of developing a gonadal neoplasm. A woman with a 45,X/46,XY karyotype has a 10–20% chance of developing a dysgerminoma or gonadoblastoma. The entire Y chromosome or a portion of it is present in 90% of Turner's syndrome patients who develop gonadal tumors. Recent studies have implicated genes close to the centromere of the Y chromosome as a cause of gonadoblastoma. FISH techniques have recently been described to accurately diagnose the presence of Y chromosomal markers. Gonadectomy should be performed early in puberty for women with positive Y chromosome markers.

D. Psychosocial Development

There is no connection between Turner's syndrome and mental retardation. Math or spatial problems may be difficult but verbal learning comes easily. Turner's syndrome individuals are at risk of social, behavioral, and educational problems. Young women with Turner's syndrome speak of feelings of isolation from their peers. Some women express feelings of depression and/or sadness. For adult women these feelings are often tied to disappointment about their infertility. Women with Turner's syndrome tend to exhibit more conservative sexual attitudes and a more negative body image and are less likely to have been sexually active when compared to unaffected women. Psychologists in Seattle found that sexual satisfaction among women with Turner's syndrome was associated with a higher self-reported health status. For children, self-esteem appears to be related to height attainment. Children who resume a more normal growth rate after treatment with growth hormone report thinking of themselves as more intelligent, more attractive, and having more friends and greater popularity. Child psychologists recommend that intellectual, learning, motor skills, and social maturity of girls with Turner's syndrome should be assessed before enrollment in kindergarten. If cognitive strengths and weaknesses are identified, early preventative and intervention strategies, if needed, can be implemented in a timely fashion.

VI. ASSOCIATED DISORDERS

A. Cardiovascular

A significant cardiac anomaly occurs in up to 16% of 45,X Turner's syndrome individuals. The most common anomalies are coarctation of the aorta and ventricular septal defects. Turner's syndrome patients can also have mitral valve prolapse, cardiomesoversion, and dissecting aneurysms of the aorta. Turner's syndrome patients often have idiopathic hypertension (Table 2).

B. Urologic

Urologic anomalies occur in 40–60% of 45,X patients. Most common are horseshoe kidney, ureteral duplication, and the absence of one kidney.

Women with Turner's syndrome are now able to establish pregnancies by egg donation. Because pregnancy represents an added stress, it is important to assess the cardiovascular and renal systems before pregnancy is attempted.

TABLE 2
Associated Disorders of Turner's Syndrome

Associated disorder	*%*
Kidney anomalies	60
Cardiac anomalies	50
Hypothyroidism	20–30
Impaired glucose tolerance	25–60
Diabetes mellitus	5
Ulcerative colitis	?

C. Autoimmunity

1. Thyroid Disease

Turner's syndrome is associated with autoimmune conditions. The incidence of thyroid antibodies was significantly higher in both patients with Turner's syndrome and their mothers than in controls. Turner's syndrome patients are more likely to develop autoimmune thyroid disease in childhood. Clinical hypothyroidism will be present in 20–30%. Women with 46,Xi (Xq) are at greater risk of developing Hashimoto's thyroiditis or Grave's disease.

2. Diabetes Mellitus

Impaired glucose tolerance occurs in 25–60% of patients.

3. Other

Inflammatory bowel disease, such as Crohn's disease or ulcerative colitis, is more common in individuals with 46,Xi (Xq). Some authors have reported a greater incidence of rheumatoid arthritis.

VII. MANAGEMENT

A. Treatment of Short Stature

1. Growth Hormone

Secretion of growth hormone is normal in girls with Turner's syndrome. Provocative testing of growth hormone secretion is not indicated, unless growth velocity falls below that expected for girls with Turner's syndrome. Henry Turner was the first to attempt to treat short stature with growth hormone. He used extract of bovine pituitary hormone but was unable to demonstrate an increase in the height of his three subjects. In 1996, the Food and Drug Administration approved use of recombinant DNA human growth hormone for treatment of patients with Turner's syndrome. Most studies have shown that there is a significant increase in growth rate following administration of exogenous growth hormone and an increase in adult height. Growth hormone therapy is begun between the age of 2 and 5, when the growth velocity usually drops below the fifth percentile of the normal female growth curve. The usual initial dose of growth hormone is 0.5 mg per kilogram body weight per day. The maximum height increase attained after growth hormone use has been less than about 3.5 in. (9 cm), which is greater than the height predicted before treatment. The average gain in adult height is about 2 in. In some randomized trials, investigators have been unable to detect a significant increase in adult height after treatment with growth hormone. It appears that girls with Turner's syndrome must start therapy early in childhood to achieve a maximal effect. They experience less improvement of growth if they start therapy after the age of 14 years.

Growth hormone may accelerate the growth of pigmented nevi in Turner's syndrome girls. Extreme growth hormone excess, as occurs with the pituitary disorder acromegaly, can result in cardiac hypertrophy that can progress to cardiac failure. Girls with Turner's syndrome take growth hormone in smaller, more physiologic doses without significant cardiac effects. Growth hormone treatment in girls with Turner's syndrome has no effect on their carbohydrate or lipid metabolism.

2. *Anabolic steroids*

Oxandrolone alone or in combination with growth hormone or estrogens will increase adult height. The usual dose of oxandrolone is 0.625 mg per kilogram per day.

3. *Estrogen*

Estrogen is used at puberty to induce normal development of secondary sexual characteristics. If girls with Turner's syndrome start estrogen replacement too early in childhood, their adult height will be less. However, recent studies have demonstrated no difference in adult height between girls who were started on low-dose conjugated estrogen therapy in early puberty and those who waited until their midteens to start estrogen.

B. Treatment of Sexual Infantilism

Turner's syndrome patients may experience significant psychological benefit by starting estrogen and allowing puberty to develop at the same age as their peers. Girls with Turner's syndrome should start estrogen therapy between the ages of 12 and 14. A low daily dose of estrogen should be used in the first 2 years to help promote skeletal growth. Usual doses are 0.3 mg of conjugated estrogen (Premarin), 0.5 mg of micronized estradiol (Estrace), or 0.3 mg of esterified estrogens (Estratab). After 2 years, the dose is increased to 1.25 mg conjugated estrogens, with 10 mg per day of medroxyprogesterone acetate being added for 16–25 days of the cycle to induce menses. This allows for normal progression of female sexual development. Some investigators are considering adding an ultra-low dose of estrogen (25–50 ng/kg/day) together with supplemental growth hormone from age 5 to 11 years.

C. Fertility

Spontaneous pregnancy has been reported in both 45,X and mosaic Turner's patients. These pregnancies are at increased risk of spontaneous abortion (26%), stillbirth (6%), and congenital anomalies (29%). Among the women with 45,X, 35% of the observed congenital anomalies in the offspring were sex chromosome abnormalities. This suggests that Turner's syndrome patients who become pregnant spontaneously are themselves at risk of having aneuploid eggs.

In vitro fertilization (IVF) and embryo transfer techniques using donated eggs have enabled women with Turner's syndrome to become pregnant and give birth. IVF clinics use donated eggs, which they fertilize with sperm from the patient's husband. The uterus must be prepared with exogenous estrogen and progesterone before transfer of the embryos. Clinics usually transfer two or three at a time. Clinics performing donor embryo procedures report pregnancy rates per transfer of between 35 and 55%. The risk of birth abnormalities in these pregnancies is dependent on the characteristics of the donors. In general, risk is less than or equal to that of the general population.

Maternal abnormalities can also place the pregnancy at risk. In cases of preexisting coarctation of the aorta, the incidence of maternal mortality is between 5 and 8%, with an 11% risk of fetal loss. Turner's patients who become pregnant are also at increased risk of dissecting aortic aneurysm, bacterial endocarditis, and intracranial aneurysms. They may have worsening of hypertension and carbohydrate intolerance. Since Turner's syndrome patients are usually short, they are at an increased risk of needing a cesarean section for delivery.

Some IVF clinics have recently reported successful techniques for freezing oocytes. Frozen oocytes may be thawed later and fertilized to achieve a pregnancy. It is possible that as laboratories advance this technique, oocytes may be harvested from girls with mosaic Turner's syndrome in early puberty so that they can use them later in their lives.

VIII. TURNER'S SYNDROME SOCIETY

In 1987, a small group of women in Minneapolis founded the Turner's Syndrome Society to help people affected with Turner's syndrome contact and help each other and their families to better understand their condition. Lynn-Georgia Tesch, President of the Turner's Syndrome Society, writes,

> The real problem with having Turner's syndrome is this sense of being different. We grow differently, look slightly different than normal women, and the problems we have with our bodies and heath are beyond those of the rest of the population. When normal friends tell me they have never had a blood test, it is hard to believe, and when they talk about their periods or pregnancies, I might as well be in another world. It is not the specific physical effects of this condition that create the problems, but how it sets one apart from the rest of society.

The Turner's Syndrome Society has created a community of women and girls affected with Turner's syndrome. Today, instead of having to battle with ignorance, intolerance, and shame, the Turner's Syndrome Society has given women affected by this condition a resource for support and sharing information.

The Turner's Syndrome Society can be contacted at Turner's Syndrome Society of the United States, 1313 Southeast 5th street, Suite 327, Minneapolis, MN 55414 USA or by e-mail at *http://www.turner-syndrome.org*.

See Also the Following Articles

AMNIOCENTESIS; FETAL MONITORING AND TESTING; PRENATAL GENETIC SCREENING; ULTRASOUND

Bibliography

Albertsson-Wiklund, K., and Ranke, M. B. (1997). *Turner's Syndrome in a Life Span Perspective: Research and Clinical Aspects*. Elsevier, Oxford, UK.

Blum, K., and Kambich, M. P. (1997). Maternal genetic disease and pregnancy. *Clin. Perinatol.* **24**, 451–465.

Pavlidis, K., McCauley, E., and Sybert, V. P. (1995). Psychosocial and sexual functioning in women with Turner's syndrome. *Clin. Genet.* **47**, 85–89.

Ranke, M. B. (1994). Growth in Turner's syndrome. *Acta Paediatr.* **83**, 343–344.

Rovet, J., and Holland, J. (1993). Psychological aspects of the Canadian randomized controlled trial of human growth hormone and low-dose ethinyl oestradiol in children with Turner's syndrome. The Canadian Hormone Advisory Group. *Horm. Res.* **39**(Suppl 2), 60–64.

Saenger, P. (1996). Turner's syndrome. *N. Engl. J. Med.* **335**, 1749–1754.

Schwartz, S., Depinet, T., Leana-Cox, J., *et al.* (1997). Sex chromosome markers: Characterization using fluorescence in situ hybridization and review of the literature. *Am. J. Med. Genet.* **71**, 1–7.

Tesch, L. G. (1989). Turner syndrome: A personal perspective. *Adolescent Pediatr. Gynecol.* **2**, 186–188.

Twinning

Kurt Benirschke
University of California, San Diego

GLOSSARY

amnion The innermost layer of the placental membranes.

anastomoses Blood vessel connections in the twin placenta surface.

chorion One layer of the placental surface that carries the blood vessels.

dizygotic twins Two egg-derived ("fraternal") twins.

monozygotic twins Single egg-derived ("identical") twins.

placenta Afterbirth.

I. INCIDENCE AND TYPES OF TWINS

The occurrence of multiple gestations varies greatly among various races and for several environmental reasons. This variation in the incidence of multiple gestation is largely due to differences in the occurrence of dizygotic (DZ) or fraternal twins; that of monozygotic (MZ) or identical twins is nearly similar throughout the world. About 1.05–1.35% of pregnancies result in twins, with the highest incidence of multiple pregnancies occurring in the Western Yoruba tribe of Nigeria, with twinning as high as 5.3%; 91% of these are DZ twins. In contrast, in Japanese (and presumably in other Orientals) the overall twinning rate is only 0.61 per 100 births, and most Japanese twins are monozygotic. The incidence of MZ twinning in Japanese and Yoruba, on the other hand, is nearly similar, representative of "identical" twinning throughout the world.

MZ twins have the same sex, female or male, whereas fraternal twins have an expected sex distribution of 25% female, female: 50% female, male: 25% male, male. This allows one to roughly infer the distribution of MZ:DZ twins in a neonatal population by employing Weinberg's calculation:

$$\text{MZ twins} = \text{All twins} - \frac{\text{Unlike sex twins}}{2\,pq}$$

where p is the frequency of male births, and q is the frequency of female births in a population.

Furthermore, estimates made from birth statistics suggest that the incidence of triplets and higher multiple births in a population may be calculated by Hellin's rule as follows: If twins occur with $1/N$, then triplets occur with $(1/N)^2$ and quadruplets with $(1/N)^3$. Higher multiples follow the same rule, unless fertility enhancement agents [follicle-stimulating hormone (FSH) and clomiphene] were used. In that case the frequency of fraternal twins increases. The highest number of multiples recorded in human pregnancies is nine fetuses.

Because animal models play an important role in research, it is necessary to recognize that, in animals, many aspects of multiple gestation obey quite different rules. Only one mammalian taxon, the *Dasypodidae* (two species of armadillo), regularly has MZ multiple offspring (quadruplets or higher numbers). Most other mammals have singletons, DZ twins, or higher multiple births (litters). We know only from the occurrence of the occasional conjoined twins in other species that MZ twinning also occurs occasionally. Also, the armadillo quadruplets, being totally identical genetically, can be considered as having been cloned, not unlike the cloned sheep in England.

II. THE CAUSES OF TWINNING

Human identical (MZ) twins occur sporadically. There is no concrete evidence of heritability, and environmental circumstances are also not known to induce this type of embryonic duplication in human pregnancies. Most induced multiples, on the other hand, are polyovular and thus of dizygotic or multizygotic origin.

Fraternal (DZ) twins occur more frequently, representing approximately two-thirds of twins in the United States. Their incidence is higher with advancing maternal age, with the administration of FSH and clomiphene. In addition, in some families this form of twinning is apparently inherited. The latter finding suggests an element of general heritability as the cause of double ovulation. The high frequency of DZ twinning in the Yoruba tribe (and somewhat less so in other black populations) and the larger size of the pituitary glands in blacks led investigators to study the hormone levels of these populations. Their findings suggest that the familial and racial incidence of DZ twinning is related to spontaneously variable (inherited) rates of FSH and luteinizing hormone secretion. Other investigators have confirmed these results with different populations and by employing different methods of analysis. It thus appears that the ability of women to produce DZ multiple offspring is related to the structure and function of their pituitary gland. It is not known whether there exists a larger number of gonadotropin-releasing hormone-producing cells in twin-prone individuals, whether they produce more hormone, or whether the primary alteration resides in the pituitary cells. The higher rate of fraternal twins in women with advancing age is also due to their elevated gonadotropin levels.

III. IDENTIFICATION OF TWIN TYPE

Because twins often have a discordant development, zygosity diagnosis is important. Moreover, most parents are anxious to know whether their twins are identical or fraternal. Such a discrimination is best begun at the time of delivery since placentation is one important aspect, and this diagnostic tool becomes unavailable soon after birth. Because of the higher incidence of congenital anomalies in twins and their frequent discordance (i.e., one normal and the other abnormal), especially among MZ twins, and because of the gross discordance in identical twins affected by the twin transfusion syndrome (TTS), careful attention must be exercised when making the zygosity diagnosis. When considering only normal twins, it is obvious that those with different sex are DZ or fraternal twins, whereas all MZ twins must have the same sex. Also, all monochorionic placentas (in humans) belong to MZ twins because no unlike-sex twins associate with such a placenta. Thus, for most newborns one proceeds first with efforts to identify zygosity of a set of twin zygosity with an examination of the placenta, as did Cameron in a British population of 688 sets of twin neonates. He set aside the monochorionic twins as being MZ and the unlike-sex twins as being DZ. There then remained approximately 45% of twins in whom genotyping was necessary for the identification of their zygosity. Cameron did this by ascertaining genetically determined placental enzyme markers. Others have used blood groups for the differential diagnosis, and other characters may also be usefully examined, especially the conformation of ear lobes. Currently, restriction fragment length polymorphism of DNA fragments from small samples of tissue have been effectively used.

IV. THE PLACENTATION OF TWINS

There are principally two types of twin placentas—dichorionic and monochorionic—with several minor subdivisions (Fig. 1). The following scheme is an appropriate guide for their frequency distribution as ascertained in the group of 668 twins studied by Cameron in England:

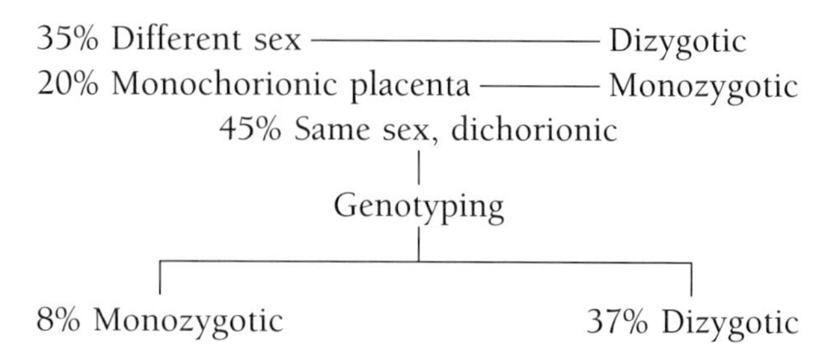

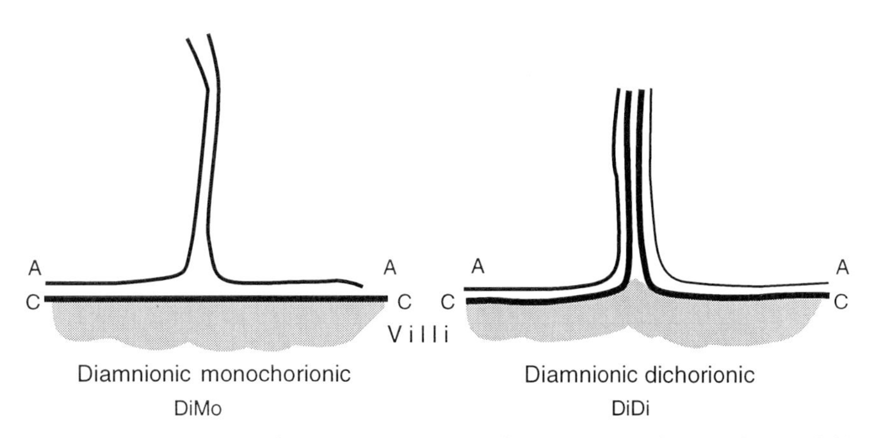

FIGURE 1 Schematics of the arrangement of a diamnionic, monochorionic twin placenta (DiMo; left) and of diamnionic, dichorionic twins (DiDi; right).

Dichorionic placentas may be fused or separated organs, and the place of fusion may be minimal, depending on where the primary placental implantations have occurred. In my experience, they are as often separate as fused. Conversely, nearly all monochorionic twin placentas are fused. Importantly, dichorionic twin placentas have membranous partitions (the so-called dividing membranes) that are thick and generally opaque. They are composed of four layers of membranes—two amnions and two chorions. Sonographic identification of the nature of these dividing membrane is generally possible and this is helpful in the subsequent management of twin pregnancies.

Monochorionic twin placentas are always from identical (monozygotic) twins and they may be monoamnionic or diamnionic. Monoamnionic, monochorionic (MoMo) twins are contained within a single amnionic sac. This type of twin placentation is the least common (0.5%) and entangling of the two umbilical cords is a common cause of their high prenatal fetal death (~40%). Much more commonly, monochorionic twin placentas are diamnionic (DiMo). They have two amnionic sacs but possess one fused placental mass, the chorion. Their "dividing membranes" are thin and translucent, being composed of only two layers of an avascular amnion. Virtually all monochorionic twin placentas have blood vessel communications between the two fetal circulations. Also, because of these anastomoses between fetal circulations of monochorionic twins, their blood is mingled. One prospective study found that 29% of twins were monochorionic, 28% were DZ because of unlike sex, 25% were of like sex but had different genotypes (i.e., being DZ), and 18% were of like sex and monozygotic because of identical genotypes (a few were undecided).

The finding of different types of twin placentas among MZ twins has been explained by the difference in the variable timing of the MZ twinning event. It is hypothesized that this accidental splitting of an embryonic mass can occur only during the first 14 days of development. Later in the growth of the embryo its organization has advanced too much for a separation to take place. If it were to occur, only conjoined twins might develop because an already axiated embryo cannot be divided any longer. Furthermore, it is assumed that a cavity (such as the amnionic sac or the chorion) cannot split spontaneously once it has formed. As such, when one analyzes the frequencies of dichorionic and monochorionic twin placentas and correlates the results with the known development of embryonic cavities, an approximate concordance of these events results. That is, approximately one-third of MZ twins have a dichorionic twin placenta because splitting has occurred early and before the chorionic cavity had developed, and two-thirds have monochorionic twin placentas because of later embryonic splitting. These assumptions also explain why MoMo twins are so uncommon—because it is a late event and because it becomes quite difficult to split an embryo at so late a stage.

The following findings were made in a survey of our own local experience from 1970 to 1985.

MoMo	*DiMo*	*DiDi (separate)*	*DiDi (fused)*
11	221	175	183
2%	37%	30%	31%

Note. Triplets 29; Quadruplets 2; Quintuplets 1.

V. ANASTOMOSES

Without a doubt, blood vessel anastomoses between the fetal circulations are the most important aspects of the monozygotic twin placentation. They strongly influence intrauterine development because of the mingling of blood, because they not only cause TTS but also are responsible for the development of acardiac twins, a form of deformity that occurs only in MZ twins. There may be artery-to-artery (AA) anastomoses (the most common), arteriovenous (AV) communications, and uncommon vein-to-vein (VV) connections in the placental surface of monochorionic twin placentas (Fig. 2). AV anastomoses always occur only through the sharing of one placental cotyledon (one lobule). Blood enters such a cotyledon from one twin and is then drained to the other twin. This commonly leads to TTS in which one twin (the donor) constantly loses some blood into the other twin (the recipient). It makes the recipient plethoric and is the reason for severe growth retardation of the donor. The imbalance caused by this AV shunt usually leads to very early birth, unless it is surgically corrected.

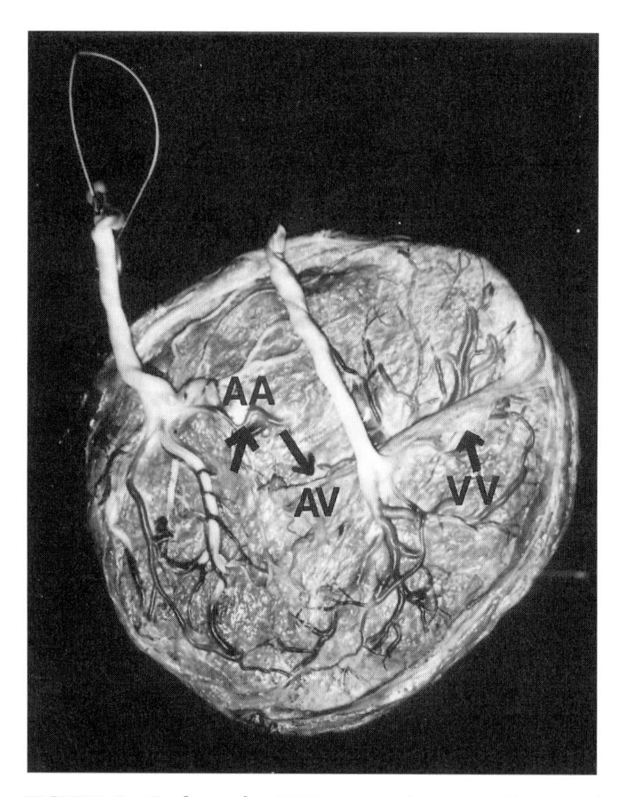

FIGURE 2 Surface of a DiMo twin placenta with several blood vessel connections. AA, artery-to-artery; AV, artery-to-vein (through cotyledon); VV, vein-to-vein anastomosis.

VI. ACARDIAC TWINS

Acardiac fetuses represent the most severe congenital anomaly. The deformity cannot be survived and, strangely, it occurs only in one of identical twins. That is but one reason why the gemellologist rarely refers to an identical twin but prefers the designation "monozygotic" instead. These anomalous fetuses develop because of their AA and VV anastomoses that carry blood between the two fetal circulations. Most possess no heart and their growth depends on being perfused, through placental anastomoses, by the cardiac activity of the normal monozygotic cotwin. It follows that the placentas of acardiac twins invariably have AA and VV anastomoses to allow for this twin reversed arterial perfusion phenomenon to take place. Indeed, it is this reversal of circulation that is responsible for the degeneration of the acardiac twin's heart early in the embryonic development.

VII. PROGNOSIS WITH TYPE OF PLACENTATION

Fetal and neonatal survival of twins are strongly influenced by the type of placentation they possess. MoMo twins have the highest perinatal mortality, whereas the lowest is experienced in twins with separated DiDi placentas. MoMo placentas frequently lead to an entangling of the twins' umbilical cords, with a fetal death rate of ~40%. The entanglement is also the cause of cord knotting, and obstruction of venous return with placental surface venous thrombosis may also occur in MoMo twins for the same reason.

DiMo twin placentation follows MoMo in impacting fetal development adversely. Approximately 25% of DiMo twins suffer perinatal death. The most important untoward sequel of anastomoses in DiMo twin placentas is the occurrence of TTS. It develops in monochorionic twins when an "unbalanced" partition of blood is present through which one twin transfuses the other, usually through an AV anastomosis in the placenta. The donor, losing blood (fluid), constantly stops urinating and becomes "stuck" in its diminutive amnionic sac. This is easily identified sonographically. The recipient becomes hypervolemic, hypertensive, and urinates excessively with hydramnios (excess amnionic fluid) developing. Hydramnios is then the cause of premature delivery because of abdominal distension of the pregnant woman. It can now be promptly averted by laser obliteration of the connecting placental vessels while the twins are still *in utero*. In the fully developed transfusion syndrome, the donor's placental villous tissue is pale, whereas that of the recipient is dark because the color of the villous tissue is wholly dependent on the fetal hemoglobin content.

Another serious sequel of major anastomoses in the monochorionic twin placenta is central nervous system destruction of twins, often occurring long before birth. This becomes especially important when, for whatever reason, one twin dies before birth. The dead twin then changes into a compressed "fetus papyraceus" and can be found in the placenta at birth. When fetal death of one twin occurs *in utero*, the blood pressure can fall very rapidly because much blood may be transferred quickly from the survivor through the large surface anastomoses in the placenta. This results in acute hypotension in the ultimately surviving twin and brain necrosis may ensue in that twin. This has been witnessed sonographically before birth. The live twin may survive, only to exhibit cerebral palsy in later life. To further understanding of this important feature, Lander and colleagues showed that within 6 min of the death of one twin (at 27 weeks) remarkable changes occurred in the velocity waveforms of the circulation in the survivor, an indication of markedly altered blood flow. This aspect of monochorionic twin placentation has special medicolegal implications and represents a challenge for the future management of multiple gestations. Also because of this, more attention is being paid to the placenta than ever before.

See Also the Following Article

Placenta: Implantation and Development

Bibliography

Benirschke, K., and Kim, C. K. (1973). Multiple pregnancy. *N. Engl. J. Med.* **288**, 1276–1284, 1329–1336.

Cameron, A. H. (1968). The Birmingham twin survey. *Proc. R. Soc. Med.* **61**, 229–234.

Lander, M., Oosterhof, H., and Aarnoudse, J. G. (1993). Death of one twin followed by extremely variable flow velocity waveforms in the surviving fetus. *Gynecol. Obstet. Invest.* **36**, 127–128.

Twinning, in Animals

see Freemartin

Ullrich Syndrome

James Aiman
Medical College of Wisconsin

GLOSSARY

gonadal dysgenesis General term for all patients with any form of aberrant gonadal development.

mixed gonadal dysgenesis Subjects with one streak gonad and the other a testis. These individuals have a 45,X/46,XY karyotype.

pure gonadal dysgenesis Women with streak gonads but normal stature and few or none of the phenotypic features of Ullrich's syndrome.

Xp, Xq The short arm of the X chromosome and the long arm of the X chromosome, respectively.

Ullrich (Ullrich's) syndrome is one form of gonadal dysgenesis in which affected subjects are phenotypic females with primary amenorrhea and increased serum concentrations of follicle-stimulating hormone and luteinizing hormone. The gonads are fibrous streaks with few or no germ cells. Puberty is delayed until hormone replacement is begun because the streak gonads are incapable of secreting androgens or estrogens. The karyotype is most often 45,X, but a variety of chromosomal abnormalities may be found. Those with this syndrome also have short stature, webbed neck, shield chest and wide-spaced nipples, a short fourth metacarpal, and other physical characteristics.

I. BACKGROUND

A. Historical

Gonadal aplasia was recognized by the late nineteenth century and was the subject of a publication in 1922 by R. I. Rossle, who classified the disorder under "sexagen dwarfism." In 1930, O. Ullrich described an 8-year-old girl with short stature, lymphedema, webbed neck, and cubitus valgus. He also recognized that a mouse model described by Bonnevie represented an analogous animal condition. Ullrich's 8-year-old girl was restudied at the age of 66 years. She had a 45,X karyotype, primary amenorrhea, failure of development of secondary sexual characteristics, and an adult height of 144.5 cm. In 1938, H. H. Turner described seven females, ages 15–23 years, with short stature, sexual infantilism, primary amenorrhea, cubitus valgus, and webbed neck. The common syndrome described by these investigators is known variously as Ullrich's syndrome, Turner's syndrome, Ullrich–Turner syndrome, and Bonnevie–Ullrich syndrome. In 1954, women with this disorder were found to be X chromatin negative. M. M. Grumbach and coworkers introduced the term gonadal dysgenesis in 1955 to include all women with anomalous development of the gonads. C. E. Ford reported the first karyotype of a woman with Ullrich's syndrome (it was 45,X). Studies reported in the 1960s and 1970s documented that a variety of chromosomal anomalies were present in women with Ullrich's syndrome. The appearance of the gonads was also described during this time, including bilateral streak gonads, unilateral streak and contralateral testis, and, rarely, bilateral testes. R. P. Singh and D. H. Carr in 1966 demonstrated that germ cell migration to the gonadal ridge occurs normally in 45,X stillborns. The disorder therefore is one of accelerated germ cell destruction within the gonads. In the 1980s and 1990s advances in our understanding of this syndrome have been principally in the realms of molecular genetics and molecular biology.

B. Karyotype

Streak gonads and gonadal dysgenesis are associated with a large number of sex chromosomal abnormalities. With current techniques such as fluorescence *in situ* hybridization and polymerase chain reaction (PCR), it is preferable to characterize these disorders on the basis of genetic makeup. As such, the eponyms should be abandoned or strictly defined according to the appearance of the gonads (bilateral or unilateral streaks), and then according to the karyotype.

A 45,X karyotype is present in about 1% of conceptions but only 1:1500 to 1:2500 births, with the remainder being lethal and resulting in spontaneous pregnancy loss before 28 weeks' gestation. Pure gonadal dysgenesis occurs in approximately 1:4000 to 1:10,000 live births. Approximately 53% of affected fetuses have a 45,X karyotype. The frequency of various karyotypes is summarized in Table 1. Using Southern blot analysis with several Y chromosome probes and PCR of specific sequences, Coto *et al.* identified at least one Y sequence in 11 of 18 subjects with Ullrich's syndrome. Other case reports demonstrating similar findings suggest that Y-chromosomal material may be present in at least 40% of subjects with gonadal dysgenesis. From a review of 651 cases of Y-chromosomal aneuploidy, Hsu concluded that the presence of the entire short arm of the Y chromosome (Yp) or a region of it that included the sex-determining region resulted in a male phenotype. Conversely, lack of Yp or the presence of a 45,X cell line (even with Yp, Yq, both, or the entire Y chromosome in another cell line) resulted in a female phenotype.

TABLE 1
Sex Chromosome Constitution of Patients with Gonadal Dysgenesis (n = 588)[a]

Karyotype	*%*	*Range of percentages*
Nonmosaic		
45,X	53.0	21.2–61.2
Xi(Xq)	6.1	5.5–73
X del(Xp)	1.6	0.8–2.6
X del(Xq)	1.1	1.6–2.4
Mosaic		
X/XX	9.9	8.2–14.7
X/Xi(Xq)	8.5	4.7–12.9
X/Xr(X)	3.9	1.7–5.5
X/XXX	1.7	0.8–3.5
X/X + mar	4.2	0.0–20.0
Complex mosaic	4.9	3.4–8.2
X/XY	3.4	2.6–5.9
Other	1.7	0.7–3.3

[a] From Plouffe and McDonough (1969, p. 1369).

II. PATHOPHYSIOLOGY

A. Germ Cell Migration

Germ cell migration is normal in 45,X fetuses. To survive, the oocytes must be completely enveloped by granulosa cells within the developing ovary. This envelopment and subsequent oocyte survival appears to require the presence of two normal X chromosomes. Although inactivation of one X chromosome occurs by Day 50 of embryonic life in somatic cells, such inactivation does not occur in oocytes since there is full expression of both glucose-6-phosphate-dehydrogenase alleles. The absence of a part or whole X chromosome may relate to a dosage effect for critical genes on the X chromosome or to destruction of meiotic cells that carry chromosomes that are unpaired or incompletely paired during the pachytene stage of meiosis I. By whatever mechanism, phenotypic females with X-chromosomal aneuploidy (45,X, 45,X/46,XX, etc.) have streak gonads.

B. Ribosomal Proteins

A role for ribosomal proteins has been suggested to explain the phenotypic manifestations of Ullrich's syndrome. The concept is that two homologous genes, *RPS4X* and *RPS4Y*, respectively located on Xq and the Y chromosome, encode for ribosomal protein S4. In humans, *RPS4X* escapes inactivation and the genes from both X chromosomes are normally active and constantly expressed. Individuals with X-chromosome aneuploidy should have a relative deficiency of these proteins, which may account for the Turner phenotype. However, there are reports of individuals with 46,Xdel(Xp) who display most of the phenotypic features of Ullrich's syndrome. Since the long arm of both X chromosomes is present in these subjects, both *RPS4X* alleles are also present. Also against the concept of deficient ribosomal proteins in this syndrome is a recent report of increased *RPS4X* mRNA levels in four subjects with 46,X,i(Xq) who had the Ullrich–Turner phenotype. (Thus, the molecular basis for the phenotypic abnormalities in Ullrich's syndrome is not established.

C. Etiology

The abnormal karyotypes in subjects with gonadal dysgenesis may occur as a result of one of several abnormalities. The frequency of each cause is usually not known because affected subjects appear as case reports or in series with small numbers. Meiotic nondisjunction is a ready explanation for gonadal dysgenesis but probably accounts for few instances. Pairing errors between the telomeric portions of the X and Y chromosomes during meiosis I or abnormal meiotic pairing between two X chromosomes is another mechanism to explain the loss of genetic information necessary for normal gonadal development. Loss of short or long arm material from the X chromosomes is associated with streak gonads and suggests that ovarian determinants are present on both arms of the X chromosome.

Mitotic errors early in embryogenesis can result in mosaic forms of gonadal dysgenesis. During an early mitotic division, if the Y chromatid in one of the blastomeres does not migrate to the daughter cell, an embryo that was initially a normal 46,XY develops a 45,X cell line and the karyotype becomes 45,X/46,XY. Autosomal aneuploidy such as trisomy 13, 18, or 21 have been associated with gonadal dysgenesis. Translocation between an X chromosome and various autosomes has also been associated with gonadal dysgenesis.

Metabolic diseases, such as steroid 17α-hydroxylase deficiency, galactosemia, cystinosis, and Prader–Willi syndrome, are associated with gonadal agenesis. Environmental factors, such as radiation, chemotherapy, and profound gonadal infection (e.g., mumps oophoritis), have resulted in the disappearance of germ cells and postnatal hypergonadotropic hypogonadism. The syndrome of embryonic testicular regression results in gonadal aplasia. This is a poorly understood syndrome occurring in 46,XY males. The later in embryonic life the testes regress, the more virilized the external genitalia at birth. Very early testicular regression, before secretion of Müllerian-inhibiting hormone, would result in a phenotypic female with a uterus and a 46,XY karyotype. This is the set of findings of 46,XY gonadal dysgenesis.

III. DIAGNOSIS

A. Clinical

Common clinical features are listed in Table 2. Other features may be present but are reported to occur in fewer than 30% of subjects. Gonadal dysgenesis with X-chromosome aneuploidy is the likely diagnosis in any phenotypic female shorter than 155 cm (61 in.) with primary amenorrhea and increased serum concentrations of follicle-stimulating hormone (FSH). Because the gonads consist only of fibrous streaks, estrogen production does not occur and breasts consequently fail to develop. An important clinical finding in all subjects with gonadal dysgenesis is the presence of a uterus and cervix. This and the sexual infantilism eliminate complete testicular feminization and Müllerian agenesis as possible causes of primary amenorrhea. The high concentration of FSH after the age of puberty eliminates hypothalamic and pituitary causes of primary amenorrhea and sexual infantilism. All women with hypergonadotropic primary amenorrhea should probably have a karyotype of peripheral lymphocytes because the correlation between phenotype and genotype is imprecise. In the absence of genetic material from the

TABLE 2
Frequency (%) of Common Anomalies Reported in Subjects with Gonadal Dysgenesis

	Karyotype		
Phenotypic feature	*Xdel*	*Xp⁻*	*Xq⁻*
Short stature (<155 cm)	99	90	40
Growth failure	99	20	20
Infertility	99	?	?
Gonadal failure	90	65	9
Otitis media	75	?	?
Facial and ear anomalies with micrognathia	60	10	1
Cardiovascular anomalies	55	2	1
Cubitus valgus	50	25	15
Hearing deficit	50	5	5
Renal and renovascular anomalies	40	10	10
Carbohydrate intolerance	40	10	10
Short neck	40	40	20
Low hairline (posterior)	40	40	20
Short fourth metacarpal	40	30	10
Hashimoto's thyroiditis	35	50	50
High arched palate	35	10	10
Genu valgum	35	?	?
Multiple pigmented nevi	35	30	20

[a] From Plouffe and McDonough (1969, p. 1374).

Y chromosome it is not necessary to biopsy or remove the streak gonads.

B. Gonadal Tumors

Approximately 30% of women with gonadal dysgenesis and a Y chromosome or fragment thereof will develop a gonadal tumor by the third decade of life. For this reason, the gonads of such women should be removed by the late teens. This can be done laparoscopically. There are approximately 10 women with gonadal dysgenesis and no evidence of a Y chromosome by karyotype who have developed a gonadal tumor. Most of these women presented with findings of excess androgen or estrogen secretion. As the use of Y-chromosome probes becomes clinically more available, more women with gonadal dysgenesis will likely be identified as having genes from the Y chromosome.

Gonadoblastomas are the most common tumors that develop in gonads of women with gonadal dysgenesis, but these are exceedingly rare in women with the phenotypic features of Ullrich's syndrome. Calcification is usually present in these tumors. An abdominal X ray or ultrasound of the pelvis may be useful in women with gonadal dysgenesis who have normal stature and few features of Ullrich's syndrome. A small percentage of women with a gonadoblastoma will develop a dysgerminoma. The gonadoblastoma and dysgerminoma may be bilateral.

C. Growth Failure

The mean adult height in untreated women with Ullrich's syndrome is 143 cm. The growth failure seems to be the result of a combination of factors: underlying skeletal dysplasia, growth hormone (GH) secretory abnormalities, and estrogen deficiency. Though these women do not have classic GH deficiency, they do develop a functional deficit in GH secretion with increasing age. By the age of 8 years, GH levels are below normal. Patients with Ullrich's syndrome may also have a partial resistance to the action of GH since higher doses of GH appear to be necessary to stimulate a growth response.

IV. MANAGEMENT

A. Estrogen and Progesterone

Estrogen plus progestin therapy should be started at the average age of pubertal onset. This can be in the form of a low-dose oral contraceptive or menopausal hormone replacement therapy. Inclusion of a progestin is essential because endometrial adenocarcinoma has occurred in women with Ullrich's syndrome treated with estrogen alone. Steroid hormone therapy will also help promote linear growth and is important for minimizing the development of osteoporosis.

B. Growth Hormone

Treatment with GH significantly increases growth velocity and adult height. Achieved height in one study of 30 women with Ullrich's syndrome who have completed growth hormone therapy was 151.9 ± 4.9 cm. This is significantly taller than the average of 143 cm in untreated adult women with Ullrich's syndrome. GH therapy should be instituted in those whose height falls below the 5th percentile. It should be started at approximately 4–6 years of age and continued until the bone age exceeds 14 years and the growth rate is <2.5 cm/year. An anabolic steroid such as oxandrolone has been reported to stimulate growth rates in conjunction with GH. However, anabolic steroids should not be used routinely because of side effects. Estrogen therapy has no role in the prepubertal therapy of growth failure and should not be started until the age of puberty.

C. Fertility

About 50 women with Ullrich's syndrome have been reported who have ovulated and conceived. Presumably, there was a 46,XX cell line within the gonads. However, women with gonadal dysgenesis should be counseled that spontaneous ovulation and conception are extremely unlikely because there are few to no oocytes in the gonads. However, pregnancy with oocyte donation, *in vitro* fertilization (IVF), and embryo transfer (ET) timed to estrogen/progesterone stimulation of the endometrium offers a probability of pregnancy equal to that of other women undergoing IVF–ET. Women with a risk to develop a gonadoblastoma and dysgerminoma should have both gonads removed. However, the uterus should not be removed because these women can have IVF with donated oocytes and embryo transfer.

V. OTHER DISORDERS

Women with Ullrich's syndrome more commonly have certain medical problems than women with a normal karyotype and normal reproductive function. Otitis media should be treated and hearing loss should be monitored. Strabismus, nystagmus, and cataracts are also more common in women with gonadal dysgenesis. Frequent ophthalmic exams and early intervention should be performed as indicated. Patients with structural cardiac defects should receive prophylactic antibiotics for any procedure at risk for bacteremia. Periodic cardiac assessment should be done because of the risk of coarctation and aortic dissection. Hashimoto's thyroiditis, diabetes mellitus, and other autoimmune disorders occur more frequently in women with gonadal dysgenesis. Thyroid function and glucose tolerance should be monitored periodically.

Bibliography

Coto, E., Toral, J. F., Menendez, M. J., Hernando, I., Plasencia, A., and Lopez-Larrea, C. (1995). PCR-based study of the presence of Y-chromosome sequences in patients with Ullrich–Turner syndrome. *Am. J. Med. Genet.* 57, 393–396.

Geerkens, C., Just, W., Held, K. R., and Vogel, W. (1996). Ullrich–Turner syndrome is not caused by haploinsufficiency of RPS4X. *Hum. Genet.* 97, 39–44.

Haeusler, G., Frisch, H., Schmitt, K., Blumel, P., Plochl, E., Zachmann, M., and Waldhor, T. (1995). Treatment of patients with Ullrich–Turner syndrome with conventional doses of growth hormone and the combination with testosterone or oxandrolone: Effect on growth, IGF-I and IGFBP-3 concentrations. *Eur. J. Pediatr.* **154**, 437–444.

Hsu, L. Y. (1994). Phenotype/karyotype correlations of Y chromosome aneuploidy with emphasis on structural aberrations in postnatally diagnosed cases. *Am. J. Med. Genet.* 53, 108–140.

Kohn, G., Yarkoni, S., and Cohen, M. M. (1980). Two conceptions in a 45,X woman. *Am. J. Med. Genet.* **5**, 339–343.

Neely, E. K., and Rosenfeld, R. G. (1996). Medical management of gonadal dysgenesis. In *Reproductive Endocrinology, Surgery, and Technology* (E. Y. Adashi, J. A. Rock, and Z. Rosenwaks, Eds.), pp. 1411–1427. Lippincott-Raven, Philadelphia.

Pierga, J. Y., Giacchetti, S., Vilain, E., Extra, J. M., Brice, P., Espie, M., Maragi, J. A., Fellous, M., and Marty, M. (1994). Dysgerminoma in a pure 45,X Turner syndrome: Report of a case and review of the literature. *Gynecol. Oncol.* **55**, 459–464.

Plouffe, L., Jr., and McDonough, P. G. (1996). Ovarian agenesis and dysgenesis. In *Reproductive Endocrinology, Surgery, and Technology* (E. Y. Adashi, J. A. Rock, and Z. Rosenwaks, Eds.), pp. 1265–1285. Lippincott-Raven, Philadelphia.

Saenger, P. (1993). Clinical Review 48: The current status of diagnosis and therapeutic intervention in Turner's syndrome. *J. Clin. Endocrinol. Metab.* 77, 297–301.

Wiedmann, H. R., and Glatzl, J. (1991). Follow-up of Ullrich's original patient with "Ullrich–Turner" syndrome. *Am. J. Med. Genet.* **41**, 134–136.

Ultradian Hormone Rhythms

Johannes D. Veldhuis
University of Virginia Health Sciences Center

GLOSSARY

episodic pulses Abrupt increases in neurohormone concentrations occurring at less predictable intervals; thus, they are not strictly periodic or regularly recurring.

feedback and ***feedforward*** Respectively, a glandular product's acting negatively (to inhibit) output of a neuroendocrine gland or ensemble and a hormone agonist's acting positively (trophic or stimulatory effect) on a target gland.

neuroendocrine axis or ***network*** Represents the overall system of two or more neuroendocrine glands or ensembles, which are related via feedback and feedforward in the generation, maintenance, and modulation of ultradian, circadian, and episodic hormone release (e.g., the hypothalamus, pituitary gland, and gonad constitute principal nodes in the neuroendocrine reproductive axis at glandular levels).

ulradian rhythm A short-term recurring variation in hormone concentrations (or behavioral measures, etc.) at intervals of <24 hr (e.g., circhoral, nearly hourly; sesquihoral, approximately every 1.5 hr). Distinguished from circadian rhythms, which represent recurring variations with an approximately 24-hr periodicity.

The neuroendocrine axes in the male and female have as their hallmark episodic and rhythmic communication via neurally and glandularly derived signaling molecules (neurotransmitters and hormones). The signaling molecules (neurohormones) serve to direct secretory activity, maintain homeostatic hormone output, and generate, where appropriate, either abrupt or gradual variations in hormone release patterns; for example, episodic luteinizing hormone (LH) pulses wherein LH is secreted at varying amplitudes and frequencies in females across the menstrual cycle and in a more sustained mode during the monthly preovulatory LH surge. Ultradian rhythms are defined by relatively regularly recurring variations in neurohormone concentrations with cycle lengths (periods) of <24 hr, e.g., circhoral or sesquihoral rhythms in which a hormone concentration increases and then decreases approximately every hour or 90 min. Many neuroendocrine rhythms are not explicitly periodic or regularly recurring but rather more nearly episodic. Episodic variations or pulses (abrupt and unpredictably timed increases followed by decreases) in reproductive hormone concentrations are cardinal signaling features within both the male and the female reproductive axes. Indeed, all measurable output within the reproductive axis exhibits such pulsatility or rhythmicity.

I. INTRODUCTION

Rhythmicity of reproductive hormones is both ultradian (recurring on a time scale of <24 hr) and, in many cases, circadian (exhibiting approximately a 24-hr periodicity, more particularly with temperature compensation, phase entrainment by environmental cues, and a free-running cyclicity that approximates 24 hr). Notably, the expression of both ultradian and circadian rhythms by neuroendocrine hormones within the male and female reproductive axes raises important mechanistic questions, such as whether and how such short-term and long-term rhythms might be linked within the organism. In addition, physiological regulation of episodic and rhythmic neurohormone secretion within the reproductive axis is strongly evident throughout the normal (monthly) human menstrual cycle and throughout the life span of the healthy male, with quantifiable changes during healthy aging and at puberty. In contrast, numerous stressors and various diseases disrupt ultradian (and circadian) rhythms pathologically.

The essential mechanistic basis for short-term neuroendocrine rhythms is believed to originate in a neuronal pulse-generator system and time-delayed network or feedback/feedforward control linkages within the axis, e.g., whereby a particular hormone directs responses in a remote target gland, which in turn after a time delay produces a second chemical signal that feeds back to readjust secretion by the parent gland. In the case of the reproductive axis, at least three primary signaling units or nodes are linked within a network or axis, namely, the hypothalamus, pituitary gland, and gonad. The individual and overall outputs of these three functional nodes are complex, interactive, and nonlinear and serve as the essential basis for normal physiology as well as provide markers of disruptive influences of diseases.

In examining the reproductive axis, several time scales of rhythmicity become evident (Fig. 1). At the shortest time scale, namely, over seconds to minutes, neurons exhibit recurrent firing episodes that are associated with the release of regulatory molecules, such as in the case of hypothalamic gonadotropin-releasing hormone (GnRH)-secreting neurons and pulses of GnRH (Fig. 1). GnRH pulses have been inferred both electrophysiologically and by direct hormone measurements of effluent blood sampled near the hypothalamus and/or pituitary gland. The relatively brief GnRH-release episodes with duration on the order of minutes are followed by more prolonged pulses of GnRH-driven (feedforward) gonadotropin [luteinizing hormone (LH)] secretion, lasting several to many minutes, as inferred by monitoring the hormone content of pituitary and/or jugular blood in the rat, sheep, horse, monkey, and human. A longer time scale of pulsatility applies to the peripheral blood concentrations of the two gonadotropins, LH and follicle-stimulating hormone (FSH), due to the slow metabolic clearance of these glycoproteins from the circulation (half-lives range from approximately 20 to 150 min in the case of LH to several or many hours in the case of FSH). The release of LH and FSH into the bloodstream in turn feeds forward on the testis or ovary to stimulate follicle growth, spermatogenesis, and the secretion

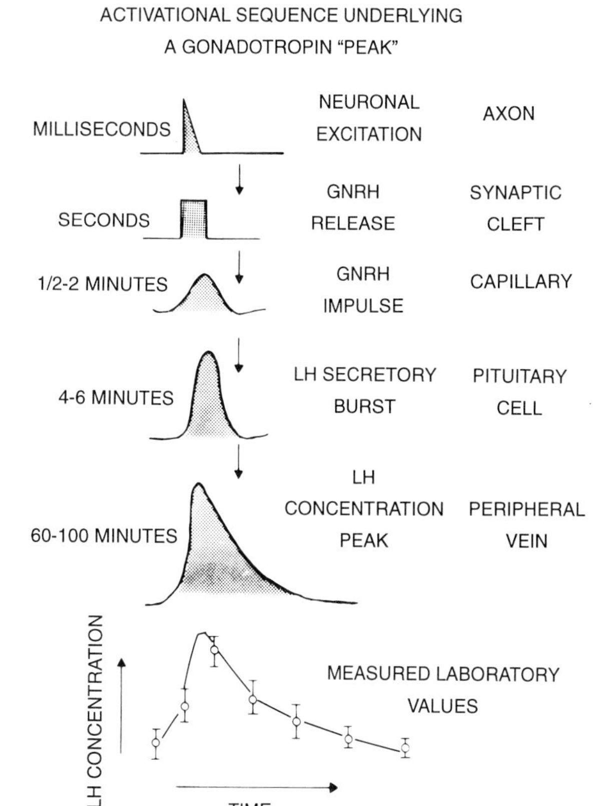

FIGURE 1 Schematic representation of range of time scales involved in a portion of a neuroendocrine axis, namely, that of the hypothalamo–pituitary (GnRH–LH) unit. Rapid activation of hypothalamic neurons occurs over seconds or minutes, with subsequent release of a pulse of the decapeptide-releasing factor, gonadotropin-releasing hormone (GnRH), into portal blood. In turn, GnRH activates pituitary LH synthesis and secretion with a resultant LH secretory burst lasting several minutes, which evokes a rise in blood LH concentrations. Circulating LH levels are sustained for many minutes by the relatively long half-life of this hormone, thereby resulting in a prolonged serum LH concentration peak.

of steroid and protein regulatory molecules such as the sex steroids, testosterone, estradiol, and progesterone in the male and female, and inhibin as well as other regulatory proteins (e.g., transferrin by the Sertoli cell, Müllerian-inhibiting substance by the granulosa cell, etc.). The secreted gonadal hormones themselves feed back via the circulation on both the hypothalamus and the pituitary gland (in the case of certain sex steroids) and/or on the pituitary gland alone (probably relevant in the case of some glycoprotein hormones, such as inhibin B).

The complexity of time scales illustrates the concept of a network or feedback/feedforward signal control system, which will be discussed further in Section VII. Most important, pulsatile neurohormone output by the functionally linked multiglandular reproductive axis exhibits strong physiological regulation in health and provides numerous examples of pathological disruption in disease.

II. MECHANISMS OF ULTRADIAN RHYTHM GENERATION

As intimated in the Introduction, neural signals in the form of predominantly GnRH pulses (but not necessarily limited to GnRH) are produced by an ensemble of hypothalamic neurons, the so-called GnRH pulse generator system. Experimental evidence indicates that the preponderance of ultradian rhythmicity of (pulsatile) LH and FSH secretion in various experimental animals and the human originates from the intermittent or episodic output of hypothalamic GnRH, which is transported axonally from the somata of neurons in the mediobasal hypothalamus to the median eminence for release into small portal blood vessels that carry the peptide signal to the anterior pituitary gland. GnRH delivered to the anterior pituitary gland drives pulsatile gonadotrope cell output of LH (and FSH) via strong feedforward actions that are dose dependent (Fig. 2). Thus, humans or mice born without normal GnRH neuronal migration into hypothalamic sites and/or with failed GnRH neuronal activation at puberty exhibit essentially undetectable or low and pulseless (nonrhythmic) blood concentrations of LH and FSH, fail to progress through puberty, and have immature reproductive capacity. The restoration of reproductive potential in such individuals (patients with Kallmann's syndrome or "hpg" mice) can be accomplished simply by long-term infusion of GnRH in pulses administered every 90–120 min (humans) or by engrafting GnRH neurons into the third ventricle (mice); such procedures reconstitute pulsatile LH (and FSH) output and result in appropriate maturation of gonadal function. In brief, a clarion example of neuronally driven ultradian rhythmicity is imparted by the GnRH–LH/FSH (hypothalamo–pituitary) unit, which is capable of issuing episodic neuroendocrine signals endowed by the neural (GnRH) component.

Whereas GnRH is linked positively to (by feedforward actions on) responsive gonadotrope cells, GnRH may also autoregulate its own release via (negative) feedback actions on GnRH neurons. A smaller additional contribution of possible "basal" (nonpulsatile) LH and FSH release by the isolated pituitary gland *in vitro* in perifusion systems has also been suggested, but the role, if any, *in vivo* of this non-GnRH-dependent, nearly continuous, irregular, low-amplitude, and high-frequency LH/FSH release in reproductive function remains unknown. Basal LH/FSH secretion may arise from autocrine and paracrine interactions within and/or constitutive gonadotropin release by the pituitary gland since it is defined primarily *in vitro* in the absence of known hypothalamic input.

Although endowed by the hypothalamic GnRH pulse-generator neurons, pulsatile LH (and FSH) release is strongly modulated by feedback linkages, which modulate pulse amplitude and frequency of gonadotropin secretion.

III. MECHANISTIC LINKAGES BETWEEN ULTRADIAN AND CIRCADIAN RHYTHMS

By way of longer term nonultradian rhythms, 24-hr (circadian) variations are well recognized in various endocrine axes, especially the ACTH–cortisol axis. However, in experimental animals, as well as to a lesser degree in the human, there are also strong 24-hr variations in LH release patterns. In the rat, the

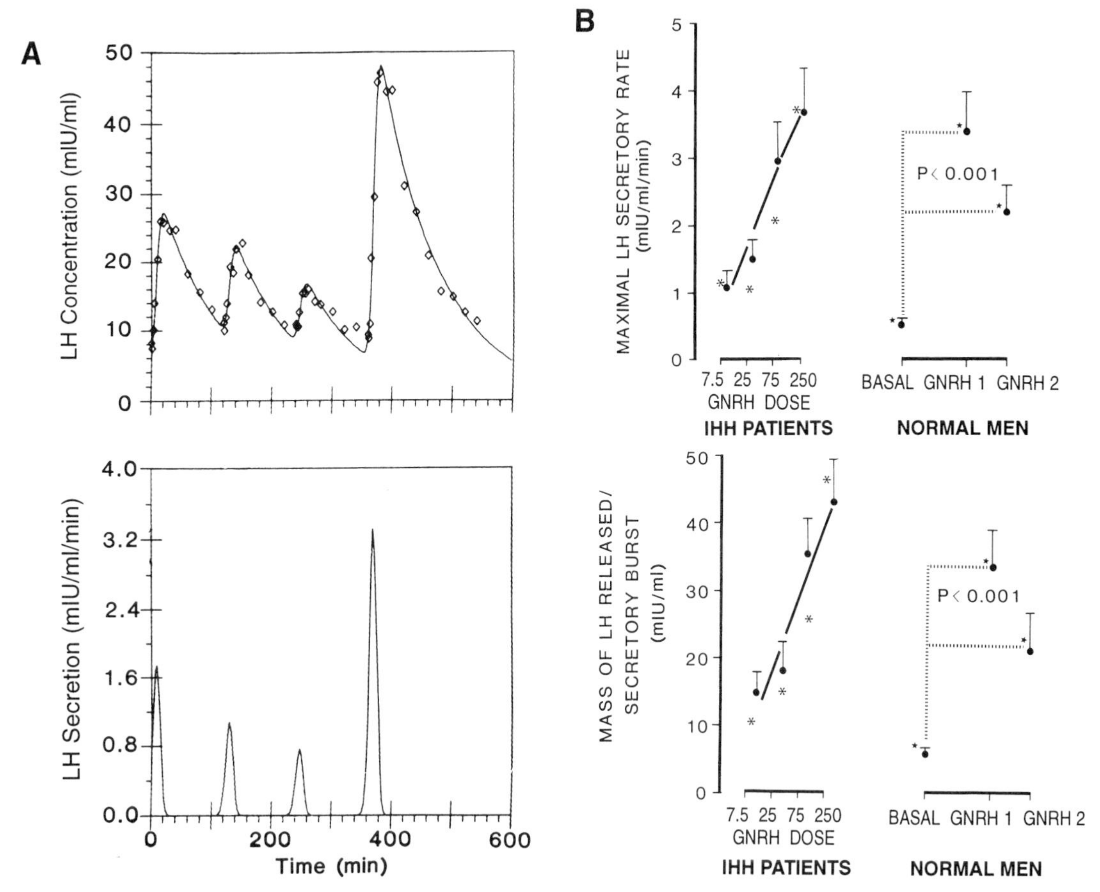

FIGURE 2 Responses of serum LH concentrations and calculated LH secretion rates of randomly ordered doses of exogenous GnRH injected every 2 hr in a man with inborn GnRH deficiency, namely, idiopathic hypogonadotropic hypogonadism. (A) Deconvolution analysis was used to calculate the apparent rate of LH secretion from the serum LH concentration pulse profile, assuming a burst model of hormone release and a subject-specific hormone half-life. The data values in the upper panel are serum LH concentrations connected by a continuous curve predicted by deconvolution analysis; the inferred LH secretion rates are shown below. (B) Relationship between injected GnRH dose (ng/kg by iv bolus) and the calculated maximal LH secretory rate (burst amplitude) and mass of LH secreted per burst for a group of human subjects with isolated GnRH deficiency. The figure (right-hand side) also gives data for normal men given two consecutive 10-μg pulses of GnRH iv 2 hr apart (adapted with permission from J. D. Veldhuis, L. O'Dea, and M. L. Johnson, The nature of the gonadotropin-releasing hormone stimulus-luteinizing hormone secretory response of human gonadotrophs in vivo, *J. Clin. Endocrinol. Metab.* 68, 661–670, 1989).

afternoon proestrous surge in ultradian LH release is clearly coupled to circadian phase. This preovulatory ultradian–circadian relationship is not known to occur in the human. Thus, the question arises how the short-term variations described previously (e.g., episodic gonadotropin pulses driven by GnRH from the hypothalamus) are linked to the longer term 24-hr rhythmicity. This is illustrated in Fig. 3 for the more well-described 24-hr (adrenal) cortisol rhythm. The circadian variations in blood cortisol concentrations can be accounted for largely (>85%) by strong 24-hr changes in the mass of cortisol secreted per individual release episode (pulse) and to a lesser extent by 24-hr variations in cortisol pulse frequency. For the LH release profile in the human, variations in LH secretory burst mass also dominate the 24-hr rhythm in serum LH concentrations. Diurnal variations in LH pulse frequency, albeit small, are evident in healthy young men and in women in the late follicular and midluteal phases of the normal men-

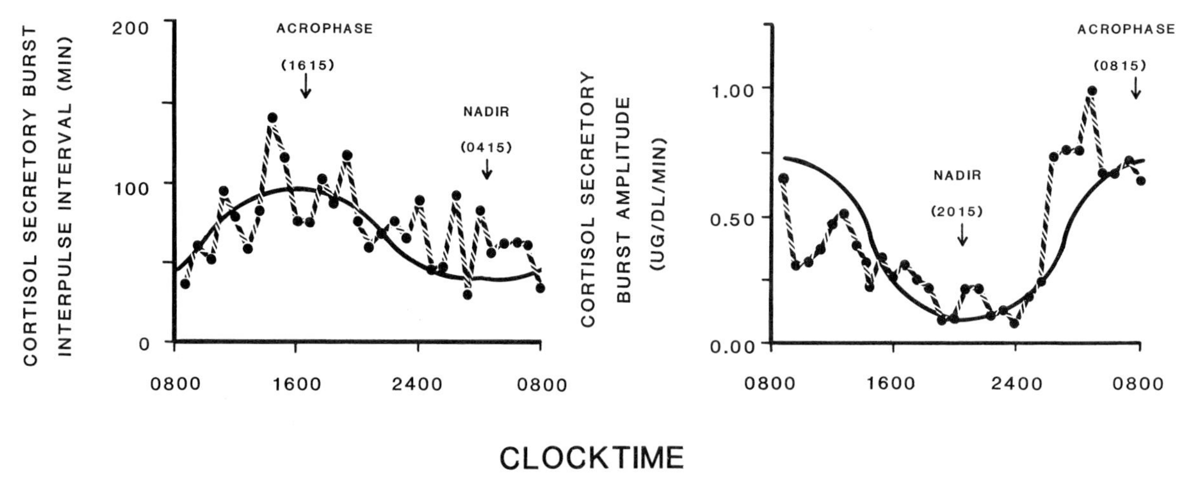

FIGURE 3 Twenty-four-hour rhythmicity in calculated cortisol secretory burst amplitude (right) or intersecretory burst interval (left). Both measures are plotted as functions of 24 hr (clock times) in a group of eight healthy men. Volunteers underwent blood sampling at 10-min intervals for 24 hr, and the sera were analyzed by RIA for adrenal cortisol concentrations. The resultant 24-hr serum cortisol concentration profiles were deconvolved assuming a simple model of burst-like hormone release. For the group of eight men as a whole, median values are shown for the circadian variations in cortisol secretory burst amplitude and interpulse intervals (adapted with permission from J. D. Veldhuis, A. Iranmanesh, G. Lizarralde, and M. L. Johnson, Amplitude modulation of a burst-like mode of cortisol secretion subserves the circadian glucocorticoid rhythm in man, *Am. J. Physiol.* **257**, E6–E14, 1989.)

strual cycle. In particular, there is nighttime slowing of the apparent GnRH pulse generator/LH pulse frequency in both men and women. This slowing is of unknown mechanistic origin but does not appear to result from increased opioid or dopaminergic inhibition of GnRH/LH pulsatility at night. In brief, for the LH/sex steroid and the ACTH/cortisol axes, modulation of ultradian secretory burst amplitude and frequency may jointly control the 24-hr variations in serum hormone concentrations.

Less is known about the coupling between ultradian and circadian rhythms for the sex steroid hormones. Data in healthy men sampled every 10 min for 24–36 hr reveal strong 24-hr variations in serum total testosterone concentrations. Although not well studied mechanistically, testosterone's circadian rhythmicity appears to reflect 24-hr variations in testosterone pulse amplitude and frequency and possibly variable basal (interpulse) testosterone release. On the other hand, circadian sex steroid variations are not so well studied in the human female for either estradiol or progesterone and remain essentially unstudied for inhibin B. Since sex steroid hormones are synthesized and diffuse out of the steroidogenic cell in response to FSH/LH stimuli, a likely model of ultradian steroid hormone secretion would encompass combined basal (interpeak) and pulsatile sex steroid release. Twenty-four-hour variations in serum sex steroid concentrations could then in principle arise from circadian control of pulse frequency and amplitude as well as basal secretion rates.

IV. PHYSIOLOGICAL COUPLING AMONG ULTRADIAN RHYTHMS

As illustrated in Fig. 4, when serum LH and testosterone concentrations are measured in blood samples collected frequently (e.g., every 15 min), strong ultradian variations are evident in healthy young men and in LH, estradiol, and progesterone concentrations in healthy young women studied in the midluteal phase of the menstrual cycle. By several methods of analysis, the ultradian pituitary and gonadal sex steroid secretory activities can be shown to be significantly (nonrandomly) coupled.

At least three independent techniques have been used to demonstrate physiological coupling between

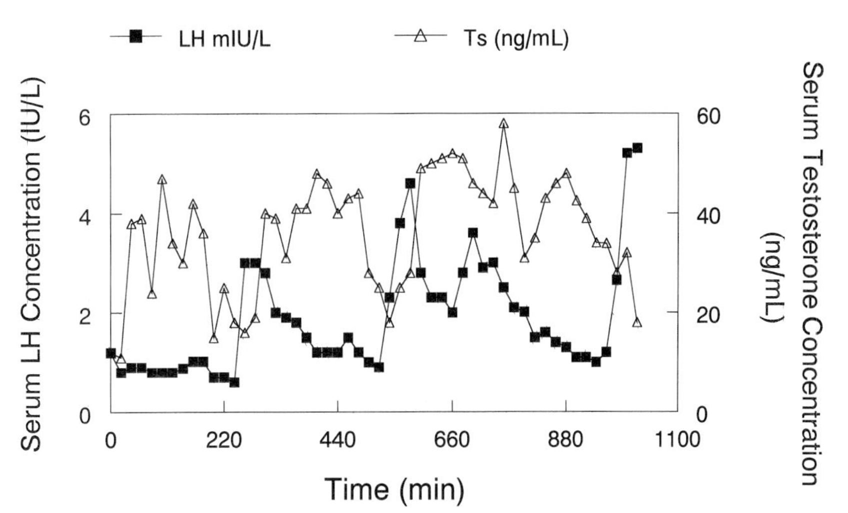

FIGURE 4 Plot of simultaneously measured serum LH and testosterone (Ts) concentrations in a middle-aged man with varicocele who underwent spermatic vein blood sampling at 15-min intervals for approximately 18 hr. Note the tendency of ultradian LH and testosterone fluctuations to occur in parallel (see text for quantitative analysis) (J. Veldhuis and de Foresta, unpublished results.)

two, or among three or more, hormones within the reproductive (or other) axis. For example, the expected probability of chance coincidence between individual release episodes (pulses) in paired hormone series can be evaluated under a null hypothesis of purely random relationships. The question can be posed: "Do LH and testosterone copulsate to an extent greater than that expected by random pulse associations alone (the latter assuming independent pulse activity of the two hormones)?" Similarly, one can ask whether LH, estradiol, and progesterone pulses are nonrandomly linked in time (e.g., double or even triple copulsate) to an extent greater than expected by chance association under the null hypothesis that these rhythms are entirely independent. This statistical approach to physiological coupling has been called discrete pulse coincidence probability testing, and an appropriate mathematics has been formulated and confirmed by computer simulations. In the case of the foregoing examples, LH and testosterone copulsate nonrandomly to a highly significant degree, as do LH and estradiol, LH and progesterone, estradiol and progesterone, and all three hormones (a triple copulsatility phenomenon). This provides evidence for nonrandom associations among the gonadotropin and gonadal sex steroid pulse events per se, which would be expected on the basis of a network (nodally connected) structure of the reproductive axis.

One can also compare hormone concentrations (rather than pulses) in two series over 24 hr or less by way of cross-correlation analysis. This analysis does not identify discrete peaks within the data but simply relates by linear correlation coefficients the extent of similar or directionally opposing changes in the concentrations of paired hormone series, e.g., LH and testosterone. By cross-correlation analysis, serum LH concentration increases tend to precede serum testosterone concentration increases by 20–120 min, and this correlation is significantly nonrandom and positive (feedforward). The correlation is observed consistently in young men, but it is significantly diminished in older individuals. Similarly, cross-correlation shows that serum LH and estradiol, LH and progesterone, and estradiol and progesterone concentrations are strongly positively coupled (feedforward relationship) at lags of 15–75 min in luteal-phase women.

A third approach to evaluating physiological coupling is to compare the relative orderliness of hormone release in the two (paired) series. This approach has been referred to as cross-approximate entropy (cross-ApEn). For LH and testosterone time series in men, cross-ApEn is significantly nonrandom, thus showing statistically important pattern repetition in the two hormone series. Of considerable interest, this synchrony between the regularity of LH and testosterone release patterns is substantially degraded in healthy older men. The synchrony between LH and FSH release also becomes progressively reduced in both men and women with aging. Based on mathematical considerations, these changes imply altered network performance of the human reproductive axis. The mechanisms underlying such alterations and their pathophysiological implications are not yet known.

Collectively, the three independent perspectives in evaluating physiological coupling between ultradian endocrine rhythms (i.e., discrete coincidence of peaks, variably lagged cross-correlation of hormone concentrations, and cross-ApEn of pattern reproducibility independently of lag) establish a high degree of network coupling, which is presumptively enacted via feedforward and feedback relationships within the GnRH–LH/FSH–gonadal axis. In this context, various sampling studies in experimental animals have shown essentially a one-to-one coupling between electrophysiological discharges in the mediobasal hypothalamus (putatively linked to the firing of the GnRH pulse generator), episodic GnRH peptide release, and matching LH pulses (in the rat, sheep, rhesus monkey, and goat).

V. PHYSIOLOGICAL REGULATION IN HEALTH

As shown in Fig. 5, ultradian rhythms in biologically active LH release are highly modulated in both amplitude and frequency throughout the normal human female menstrual cycle. Measuring LH by way of a bioassay (namely, via *in vitro* Leydig cell secretion of testosterone in response to serum containing unknown or standard amounts of LH) discloses marked dynamic variations in blood LH concentrations, bioactive LH pulse frequency, bioactive LH pulse amplitude, mass, duration, and interpulse basal bioactive LH secretion rates across the normal human menstrual cycle. Such physiological regulation is widely supported by numerous in studies by RIA, IRMA, chemiluminescence, or bioassay of LH (and FSH) throughout the mammalian species. In contrast to varying pulse features per se, the apparent half-life of LH and its daily (total) secretion rate are relatively constant throughout the human menstrual cycle, except during the preovulatory LH surge. Moreover, evident disruption of this organized pulsatile (ultradian) output of the GnRH–LH/hypothalamo–pituitary–gonadal unit is the hallmark of numerous diseases in the human as well as in experimental animals.

As expected in a feedback-control network or axis, loss or reduction of gonadal negative feedback (e.g., following surgical removal of the gonads, spontaneous menopause-associated ovarian failure in women, and/or administration of nonsteroidal pharmacological agents to block sex steroid hormone action at the receptor level) will tend to increase the irregularity, accelerate the frequency, and amplify the absolute amplitude of circhoral LH and, to a lesser extent, FSH pulsatility (the latter seems to occur in combination with an increase in basal hormone release). Reinfusion of gonadal sex steroid hormones will restore LH pulses toward a normal (gonad-intact) frequency and amplitude while also reinstating a baseline degree of orderliness or serial regularity of LH release profiles, e.g., as quantified by an approximate entropy statistic. In the dramatic physiological state of puberty, marked (30-fold) augmentation of LH secretory pulse amplitude with modest acceleration (30–50%) of LH burst frequency underscore the dynamic range of physiological ultradian rhythmicity.

VI. PATHOLOGICAL DISRUPTION IN DISEASE

Several clinical conditions can result in disruption of normal LH (and hence, by inference, GnRH) pulsatility, despite preserved pituitary responsiveness to the injection of small doses of synthetic GnRH (indicating intact pituitary gonadotrope cell responsiveness, when GnRH is made available). These diseases in women have been referred to as "hypothalamic amenorrhea" in an effort to suggest (reversible) loss of organized (pulsatile) output of the hypothalamic GnRH pulse-generator system. For example, as illustrated in Fig. 6, in strenuously exercising

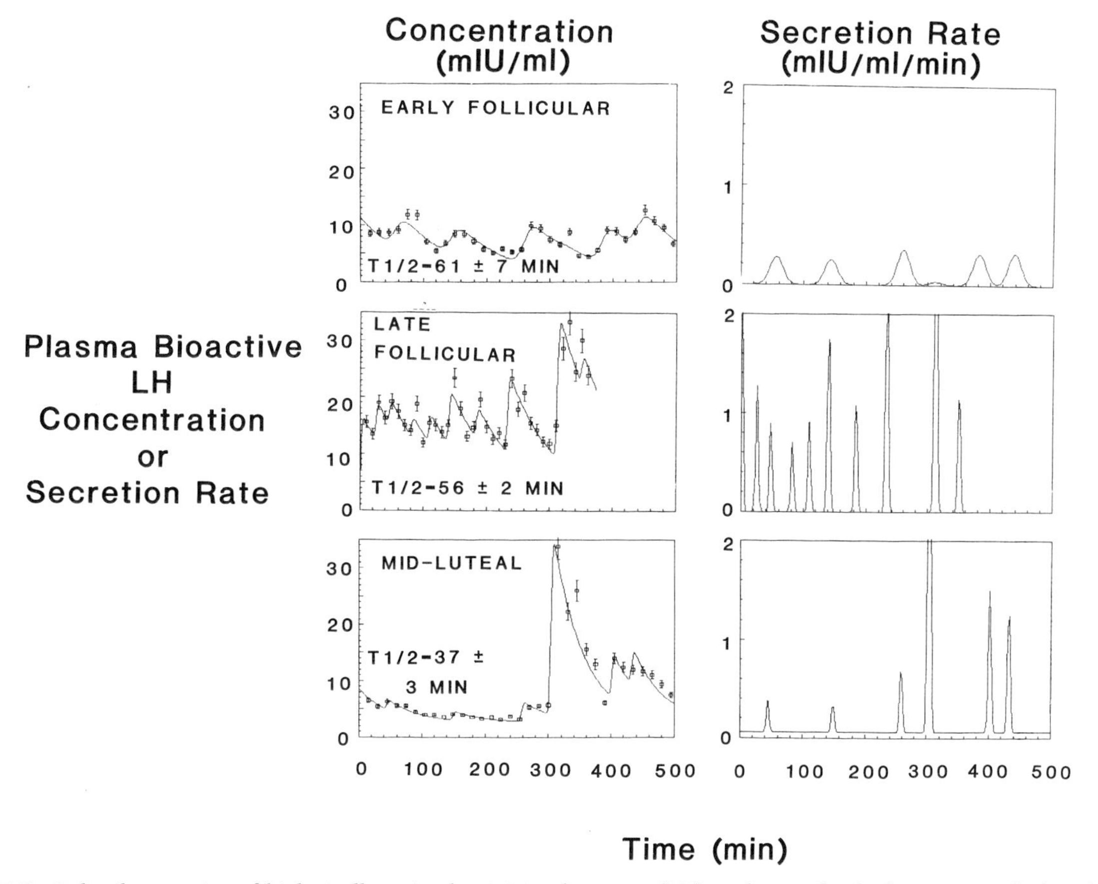

FIGURE 5 Pulsatile secretion of biologically active luteinizing hormone (LH) in three individual women studied at different times within the normal menstrual cycle, namely, in the early follicular, late follicular, and midluteal phases. (Left) The measured plasma bioactive LH concentrations were assessed *in vitro* in a rat Leydig cell bioassay. The fitted curves through the data are predicted by deconvolution analysis, assuming a predominantly burst-like model of hormone release. Corresponding half-life estimates and their confidence intervals are given. (Right) The corresponding calculated bioactive LH secretion rates plotted over the same time interval. In these studies, young women underwent blood sampling at 10- or 15-min intervals for 6–8 hr (adapted with permission from Veldhuis *et al.*, 1984).

women, e.g., long-distance runners, experiencing cessation of menstrual cycles (amenorrhea), the pulsatile output of LH over 24 hr is severely disrupted in approximately 70% of individuals, with significantly fewer LH pulses, many of which are restricted to the nighttime. This "hypopulsatile" GnRH/LH release pattern in amenorrheic women athletes is similar to that observed in healthy young girls and boys in the early stages of puberty, when nighttime activation of ultradian LH release first becomes evident. Pathophysiological mechanisms underlying this particular clinically inferred disruption of pulsatile (neuronal) GnRH drive to the gonadotrope cell population are not known but might involve endogenous opiates (which are potent inhibitors of the GnRH neuronal system), dopamine, and/or withdrawal of excitatory amino acid input. Further investigations are required in the human to determine the specific neurotransmitter pathways involved in the pathophysiology of various reversible GnRH pulse-generator disturbances. On the other hand, structural diseases of the hypothalamus, such as a craniopharyngioma (Rathke's pouch remnant), parasellar meningiomas, traumatic contusion of the hypothalamus, metastatic cancer, subarachnoid cysts, and ependymomas, can damage the hypothalamic GnRH-neuronal ensemble,

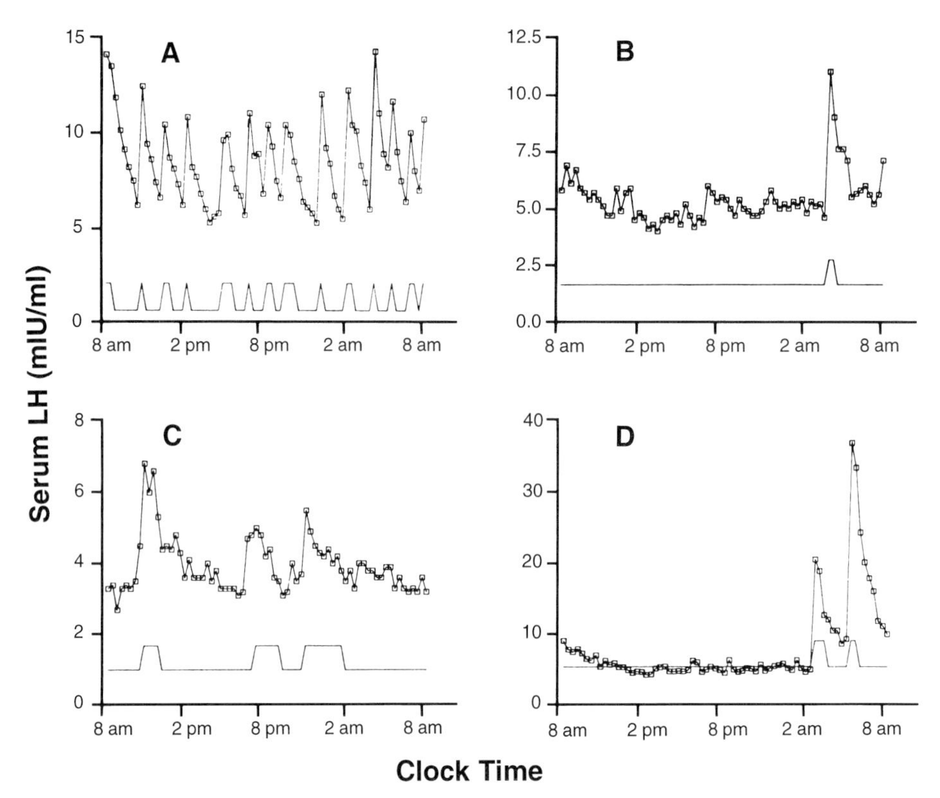

FIGURE 6 Twenty-four-hour serum LH concentration profiles determined by RIA in four women long-distance runners (A–D) with amenorrhea. Blood was sampled at 20-min intervals for 24 hr in each of the four volunteers. Below each measured serum LH concentration profile is the computerized indication of significant hormone pulsatility. Note that in three of the four women, LH pulsatility was impoverished and largely restricted to the hours of sleep (adapted with permission from J. D. Veldhuis, W. S. Evans, L. M. Demers, M. O. Thorner, D. Wakat, and A. D. Rogol, Altered neuroendocrine regulation of gonadotropin secretion in women distance runners, *J. Clin. Endocrinol. Metab.* **61**, 557–563, 1985).

resulting in hypothalamic amenorrhea. Likewise, neoplasms metastatic to and primary tumors within the pituitary gland, infarction ("stroke") of the pituitary gland, and hypothalamo–pituitary stalk injuries can also produce hypogonadotropism, i.e., deficiency of LH and FSH, in men and women.

Diseases or stressors that disrupt only the hypothalamic input of GnRH to the pituitary gland and leave the pituitary gland otherwise intact and functional can be treated with pulses of GnRH injected every 60–120 min to reestablish ultradian gonadotropin pulsatility. This "GnRH pump" therapy provides an appropriate signal to the pituitary gland and gonads for maturation of gametes and the production of sex steroid hormones with resultant restoration of adult reproductive potential. Indeed, even the prepubertal primate pituitary–gonadal axis can be activated fully by pulsatile GnRH infusions administered circhorally.

VII. NETWORK AND FEEDBACK CONCEPTS WITHIN THE REPRODUCTIVE AXIS

As adumbrated in the Introduction, the hypothalamus, anterior pituitary gland, and gonad in both the male and the female constitute primary regulatory nodes within a highly organized control system or feedback/feedforward axis. This notion identifies a network concept for the reproductive axis. In this construction, episodic (GnRH) output of the hypothalamus directs the pituitary gland by feedforward actions, and the secreted gonadotropins (LH and

FSH) in turn supervise gonadal activation by way of further feedforward (agonistic) effects. Relevant signals released by the gonads (e.g., sex steroid hormones and glycoproteins such as inhibin) exert time-delayed feedback actions on the hypothalamus (sex steroid hormones) and pituitary gland (both sex steroid hormones and gonadal glycoproteins). Given intermittent (feedforward) input in the form of hypothalamic GnRH to drive ultradian gonadotropin and gonadal–steroid pulse episodes and relevant feedback activities, this feedforward/feedback network sustains episodic pulsatile patterns of the blood concentrations of LH (and FSH) and sex steroids. (Less is known about the regularity of bioactive inhibin release over 24 hr and how inhibin in the systemic circulation participates in feedback control, in contrast to likely intrapituitary actions of inhibin, follistatin, and the activin peptides.) As illustrated in Fig. 7, the network concept of the (male) reproductive axis allows one to formalize possible quantitative alterations in uni- or multinodal function, as well as feedback or feedforward coupling, and to evaluate the predicted pathophysiological changes. For example, a systematic network structure will allow one to postulate and explore several possible specific neuroendocrine mechanisms (e.g., shifting of an *in vivo* dose–response curve) by which the cross-correlation and cross-ApEn (physiological bihormonal synchrony) between LH and testosterone might be progressively disrupted in healthy aging men or explain the importance of proposed GnRH autofeedback (ultrashort loop inhibition by secreted GnRH of GnRH neuronal activity), as has been suggested in the rodent but is not clearly evident in the monkey or human.

VIII. SUMMARY

The reproductive axes in men and women are typified by physiological ultradian rhythms in GnRH, LH, FSH, and gonadal sex steroid (and possibly in-

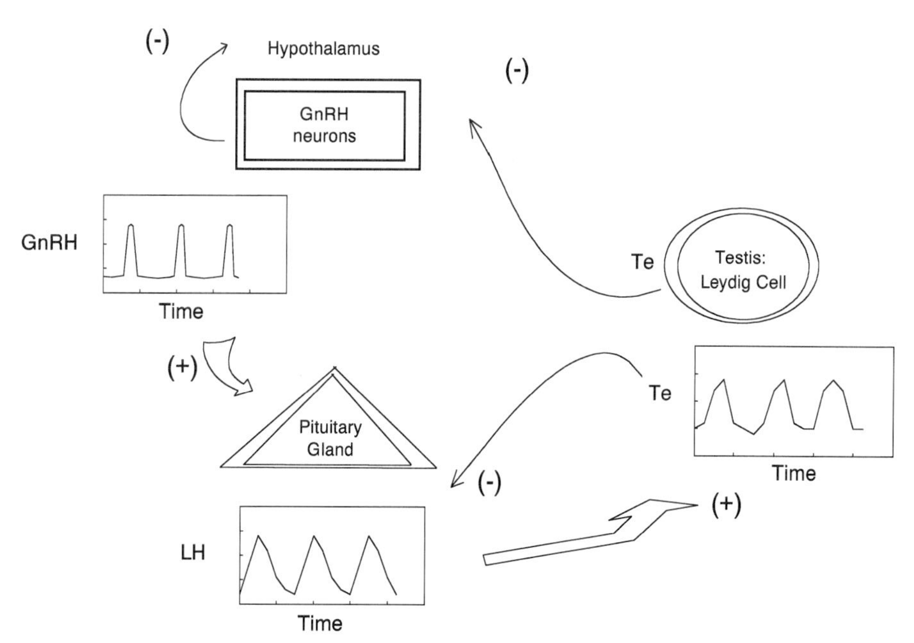

FIGURE 7 Schematized feedback and feedforward model of the GnRH–LH–Leydig cell (testosterone) axis for the male reproductive system. The arrows denote agonistic (feedforward) or inhibitory (feedback) connections among the three primary control nodes within the male reproductive axis, namely, the hypothalamic GnRH neuronal ensemble, the pituitary LH-secreting gonadotrope population, and the testicular testosterone (Te)-secreting Leydig cells. Specific dose–response curves interface the feedback and feedforward signals with the target endocrine gland, as discussed in the text. GnRH autofeedback on the hypothalamus, although suggested in the rodent, is not clearly, if at all, defined in the monkey or the human.

hibin) secretion. These short-term variations in blood hormone concentrations recur over minutes to hours and are linked to longer term trends over 24 hr by selective variations in the frequency and/or amplitude of the circhoral pulses. In addition, various diseases disrupt these short-term reproductive rhythms. Ultradian hormone release is clearly important physiologically since in pathological states restitution of pulsatile GnRH input will restore full function of the entire reproductive axis. Recent concepts define the reproductive axis as a network of feedback/feedforward control nodes, wherein more subtle derangements are manifested by alterations in network function.

Acknowledgments

We thank Patsy Craig for her skillful preparation of the manuscript and Paula P. Azimi for the data analysis, management, and figures. This work was supported in part by NIH Grant MO1 RR00847 (to the General Clinical Research Center of the University of Virginia Health Sciences Center), Research Career Development Award 1-KO4-HD-00634, the Baxter Healthcare Corporation (Round Lake, IL), the University of Virginia Pratt Foundation and Academic Enhancement Program, the National Science Foundation Center for Biological Timing (Grant DIR89-20162), and the NIH NICHD P-30 Center for Reproduction Research (Grant HD-28934).

See Also the Following Articles

Circadian Rhythms; Circannual Rhythms; FSH (Follicle-Stimulating Hormone); GnRH (Gonadotropin-Releasing Hormone); LH (Luteinizing Hormone); Menstrual Cycle

BIBLIOGRAPHY

Booth, R. A. J., Weltman, J. Y., Yankov, V. I., Murray, J., Davison, T. S., Rogol, A. D., Asplin, C. M., Johnson, M. L., Veldhuis, J. D., and Evans, W. S. (1996). Mode of pulsatile FSH secretion in gonadal-hormone sufficient and deficient women. *J. Clin. Endocrinol. Metab.* **81**, 3208–3214.

Clarke, I. J., and Cummins, J. T. (1987). Pulsatility of reproductive hormones: Physiological basis and clinical implications. *Bailliere's Clin. Endocrinol. Metab.* **1**, 1–21.

Conn, P. M., and Crowley, W. F. (1991). Gonadotropin-releasing hormone and its analogs. *N. Engl. J. Med.* **324**, 93–103.

Evans, W. S., Christiansen, E., Urban, R. J., Rogol, A. D., Johnson, M. L., and Veldhuis, J. D. (1992). Contemporary aspects of discrete peak detection algorithms: II. The paradigm of the luteinizing hormone pulse signal in women. *Endo. Rev.* **13**, 81–104.

Keenan, D. M., and Veldhuis, J. D. (1998). A biomathematical construct embodying time-delayed feedback and feedforward in the human male hypothalamic–pituitary–Leydig axis: Pulsatile and circadian elements. *Am. J. Physiol.*, in press.

Knobil, E. (1980). The neuroendocrine control of the menstrual cycle. *Recent Prog. Horm. Res.* **36**, 53–88.

Pincus, S. M., Mulligan, T., Iranmanesh, A., Gheorghiu, S., Godschalk, M., and Veldhuis, J. D. (1996). Older males secrete luteinizing hormone and testosterone more irregularly, and jointly more asynchronously, than younger males: Dual novel facets. *Proc. Natl. Acad. Sci. USA* **93**, 14100–14105.

Rossmanith, W. G., Laughlin, G. A., Mortola, J. F., Johnson, M. L., Veldhuis, J. D., and Yen, S. S. (1990). Pulsatile co-secretion of estradiol and progesterone by the midluteal phase corpus luteum. *J. Clin. Endocrinol. Metab.* **70**, 990–995.

Sollenberger, M. L., Carlson, E. C., Johnson, M. L., Veldhuis, J. D., and Evans, W. S. (1990). Specific physiological regulation of LH secretory events throughout the human menstrual cycle: New insights into the pulsatile mode of gonadotropin release. *J. Neuroendocrinol.* **2**(6), 845–852.

Spratt, D. I., O'dea, L. L., Schoenfeld, D., Butler, J., Rao, P. N., and Crowley, W. F. (1988). Neuroendocrine–gonadal axis in men: Frequent sampling of LH, FSH, and testosterone. *Am. J. Physiol.* **254**, E658–E666.

Urban, R. J., Evans, W. S., Rogol, A. D., Kaiser, D. L., Johnson, M. L., and Veldhuis, J. D. (1988). Contemporary aspects of discrete peak detection algorithms: I. The paradigm of the luteinizing hormone pulse signal in men. *Endocr. Rev.* **9**, 3–37.

Veldhuis, J. D., Beitins, I. Z., Johnson, M. L., Serabian, M. A., and Dufau, M. L. (1984). Biologically active luteinizing hormone is secreted in episodic pulsations that vary in relation to stage of the menstrual cycle. *J. Clin. Endocrinol. Metab.* **58**, 1050–1058.

Veldhuis, J. D., King, J. C., Urban, R. J., Rogol, A. D., Evans, W. S., Kolp, L. A., and Johnson, M. L. (1987). Operating characteristics of the male hypothalamo–pituitary–gonadal axis: Pulsatile release of testosterone and follicle-stimulating hormone and their temporal coupling with luteinizing hormone. *J. Clin. Endocrinol. Metab.* **65**, 929–941.

Veldhuis, J. D., Iranmanesh, A., Johnson, M. L., and Lizarralde, G. (1990). Twenty-four hour rhythms in plasma concentrations of adenohypophyseal hormones are generated by distinct amplitude and/or frequency modulation of underlying pituitary secretory bursts. *J. Clin. Endocrinol. Metab.* **71**, 1616–1623.

Wu, F. C. W., Butler, G. E., Kelnar, C. J. H., Huhtaniemi, I., and Veldhuis, J. D. (1996). Patterns of pulsatile luteinizing hormone secretion from childhood to adulthood in the human male: A study using deconvolution analysis and an ultrasensitive immunofluorometric assay. *J. Clin. Endocrinol. Metab.* **81**, 1798–1805.

Ultrasound

Edith Diament Gurewitsch and Frank A. Chervenak

The New York Hospital–Cornell Medical Center

GLOSSARY

amniocentesis Withdrawal of amniotic fluid through a needle inserted into the uterus through the mother's abdomen.

arthrogryposis A congenital defect of the limbs characterized by contractures, flexion, and extension.

biparietal diameter A measurement of the fetal head taken between the parietal bones of the skull.

cardiotocography A simultaneous graphic representation of fetal heart rate and uterine contraction patterns obtained from Doppler and pressure monitors applied to the maternal abdomen.

chorionic villus sampling A technique for biopsy of placental tissue in early pregnancy, usually for karyotype analysis.

diastole The portion of the cardiac cycle when the muscles of the ventricles relax.

echogenicity The property of or degree to which tissue reflects sound waves.

ectopic pregnancy A conception that implants outside the uterine cavity.

endometrium The lining of the uterus.

follicle A self-contained sphere of cells surrounding a single oocyte within the ovary.

follicular phase The period of time from the onset of menses until ovulation, during which a dominant follicle with a maturing ovum is biologically recruited for ovulation.

human chorionic gonadotropin The "pregnancy hormone" produced by the developing placenta, which sustains early pregnancy.

hydronephrosis Dilation of the collecting system of the kidneys resulting from obstruction to the flow of urine.

hysterosalpingogography A technique for producing an X-ray image of the uterine cavity and the oviducts following injection of a radiopaque dye.

myoma (myomata) A benign tumor of smooth muscle tissue of the uterine wall.

nomogram A graphic representation of serial measurements representing a range of normal measurements, usually generated from data from large population studies.

nonstress test A cardiotocographic test to determine fetal well-being.

oligohydramnios A condition of low amniotic fluid.

papillary Having a small nipple- or finger-like shape.

piezoelectric Pertaining to electric current generated by pressure on certain crystals, e.g., quartz.

polyp Any mass of tissue that projects outward or upward from the normal surface which may sometimes be premalignant.

synechiae Any adhesion or band of scar data tissue; in gynecologic cases these usually traverse the uterine cavity and can interfere with fertility.

systole The portion of the cardiac cycle when the muscles of the ventricles contract.

I. PHYSICS OF ULTRASONOGRAPHY

Ultrasound is the most commonly used imaging modality in assessment of the female reproductive tract. As illustrated in Fig. 1, pulsed sound waves are produced when piezoelectric elements in the ultrasound transducer are electrically stimulated. The sound waves penetrate the tissues under study, and their reflection is detected by the transducer and converted to a gray-scale image representing the echogenicities of the intervening structures. If the interface under study is moving, the reflected echo frequency will increase or decrease relative to the incident sound wave by an amount proportional to the velocity along the direction of the beam. This shift is known as the Doppler shift, and this principle can be utilized to assess blood flow within vessels in various anatomical locations.

A. Safety

At diagnostic frequencies, there are no known examples of tissue damage resulting from ultrasound exposure. According to the American Institute of Ultrasound in Medicine's *Official Statement on Clinical Safety*, current data indicate that the benefits to patients of the prudent use of diagnostic ultrasound (including use during human pregnancy) outweigh the risks, if any, that may be present." For this reason, it has been widely applied in obstetrics to evaluate the fetus, placenta, and amniotic fluid.

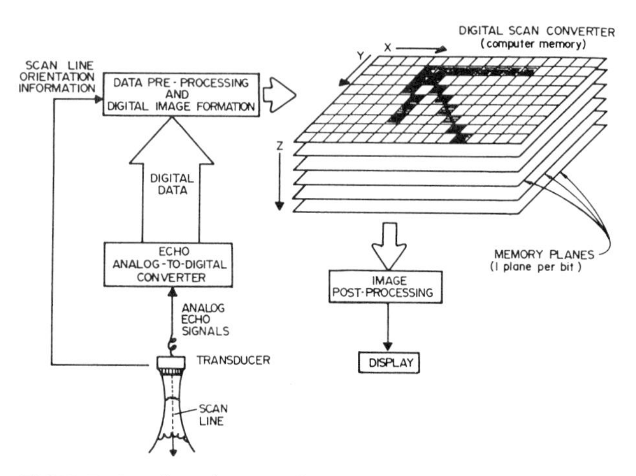

FIGURE 1 The ultrasound transducer both emits sound waves and detects their reflection. These signals are then digitized and converted to a gray-scale image on a video monitor [reprinted with permission from R. R. Price and A. C. Fleischer, Sonographic instrumentation, In *The Principles and Practice of Ultrasonography in Obstetrics and Gynecology* (A. C. Fleischer *et al.*, Eds.), p. 32].

II. OBSTETRIC APPLICATIONS

Specific applications of ultrasound to obstetrics include the determination of gestational age, fetal growth assessment, detection of structural anomalies, fetal biophysical evaluation, and use as an adjunct to invasive diagnostic and therapeutic procedures. With the relatively recent addition of transvaginal and Doppler capability, ultrasound examinations have expanded to include study of the very early pregnancy, dynamic changes in the cervix throughout pregnancy, and flow characteristics in fetal, umbilical, and uterine vessels.

A. Detection of Fetal Anomalies

Before the advent of ultrasound for prenatal diagnosis, the discovery of an anomalous fetus occurred unexpectedly at birth. Today, most fetal anomalies can be detected prenatally by ultrasound, and the number is increasing rapidly with new advances in ultrasound technology. An anatomical survey of discrete regions, including the head, neck, spine, thorax, heart, abdomen, and extremities, is best performed in the mid-second trimester, when the fetus is large enough to have most structures well imaged but has less calcification of bones creating less artifact and other acoustical "noise." This timing also allows for further diagnostic and therapeutic intervention prior to fetal viability, which increases the management options for prospective parents and their physicians. The spectra of diagnosable anomalies include structural defects, such as central nervous system malformations (Figs. 2A–2C) and cardiac malformations; functional aberrations, such as cardiac arrythmias or arthrogryposis; and a combination of structural and functional abnormalities, such as bladder outlet obstruction (Fig. 3) causing hydronephrosis and oligohydramnios (Fig. 4). Correct and early diagnosis lays

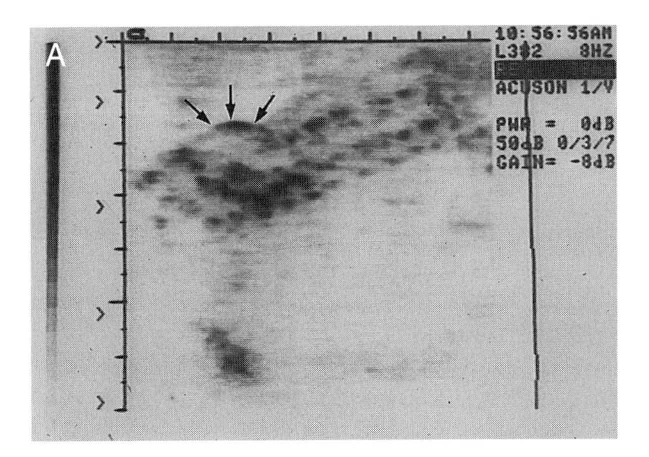

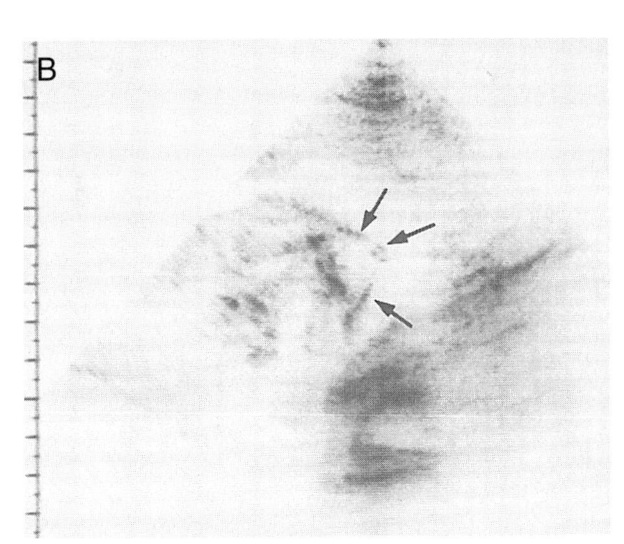

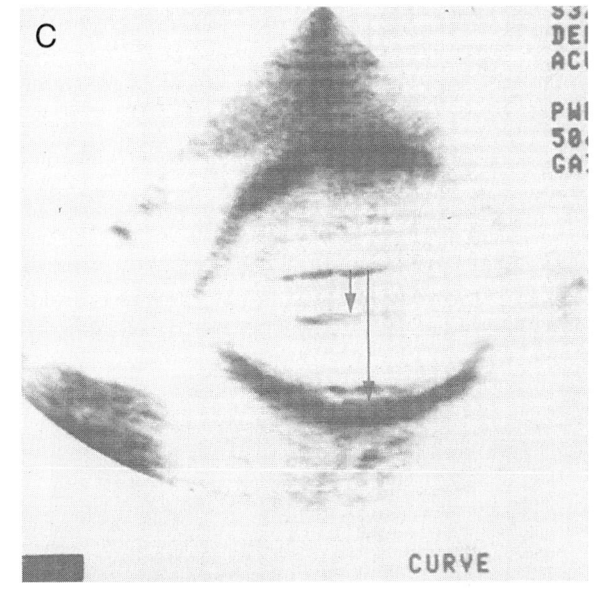

FIGURE 2 (A) Sagittal view of the spine demonstrating a defect in the lumbosacral area, with herniation of the spinal cord (arrows). This malformation is known as spina bifida. (B) A transverse view of the spine in a fetus with spina bifida, again showing herniation of a sac derived from the meninges, which cover the spinal cord (arrows). The defect results from the failure of the neural tube, an embryological precursor of the vertebral column and spinal cord, to fuse during the first few weeks after conception. (C) A transverse view of the fetal head demonstrating a dilated cerebral ventricle. This malformation is known as ventriculomegaly, which when severe can cause the head to become abnormally large, a condition termed hydrocephalus. This defect is most often caused by obstruction of normal flow of cerebrospinal fluid.

the foundation for appropriate patient counseling regarding management options.

B. Pregnancy Dating

Accurate determination of gestational age is perhaps the most important assessment made in standard obstetric practice. Knowledge of the precise pregnancy dating is critical to both normal pregnancy management and the evaluation of suspected deviations from normal. Traditional dating of pregnancy by the last normal menstrual period can yield inaccuracies owing to such clinical factors as variable lengths of the follicular phase of the menstrual cycle, abnormal patterns of bleeding, and recent use of oral contraceptives. Estimation of gestational age by

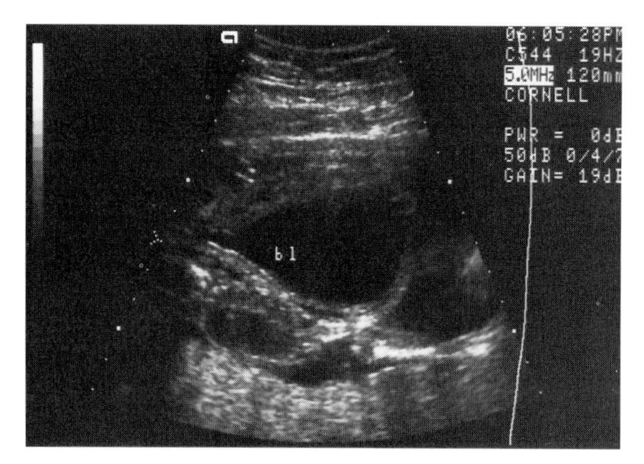

FIGURE 3 A coronal view of the fetal abdomen and pelvis demonstrating a dilated bladder resulting from outlet obstruction at the level of the urethra.

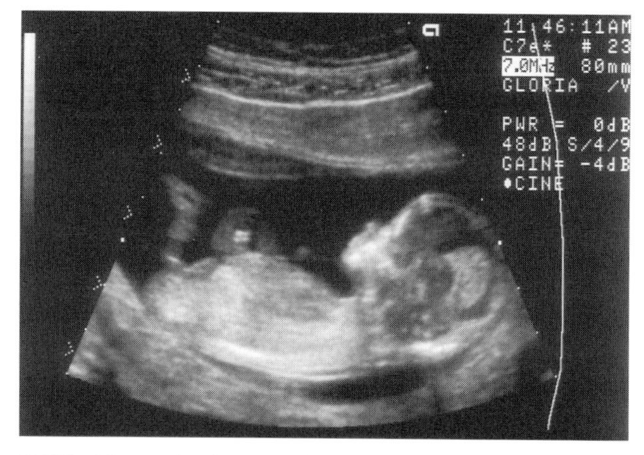

FIGURE 5 Early ultrasound image of a fetus demonstrating the view from which a measurement from the top of the fetal head (crown) to the rump can be made in order to determine gestational age.

ultrasound is based on known relationships between the age of the fetus and its size. Therefore, a number of physical measurements, such as the crown–rump length (Fig. 5) in early pregnancy and the biparietal diameter or femur length (Fig. 6) after the first trimester, can be compared to nomograms to correlate with gestational age. Because biological variability in fetal size within a population supervenes to a greater extent as pregnancy progresses, the accuracy of gestational age assessment by ultrasound diminishes over time. Whereas dating performed at 6–12 weeks is accurate to within 3 or 4 days, the margin of error in third-trimester dating increases to >3 weeks.

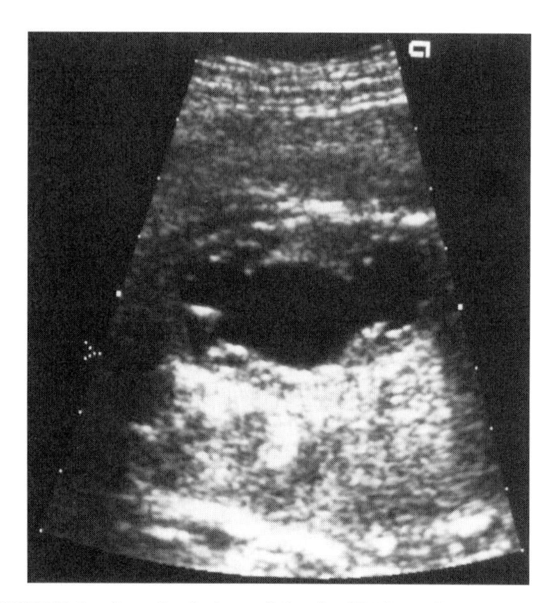

FIGURE 4 A sagittal view of the fetal kidney demonstrating severe dilatation of the collecting system. This condition, known as hydronephrosis, is caused by the backward pressure generated from an obstruction to urine flow. Note the lack of amniotic fluid surrounding the fetus. A significant source of amniotic fluid is fetal urine passed *in utero*.

C. Fetal Growth and Biophysical Assessment

Determination of the fetal abdominal circumference combined with at least one other biometric

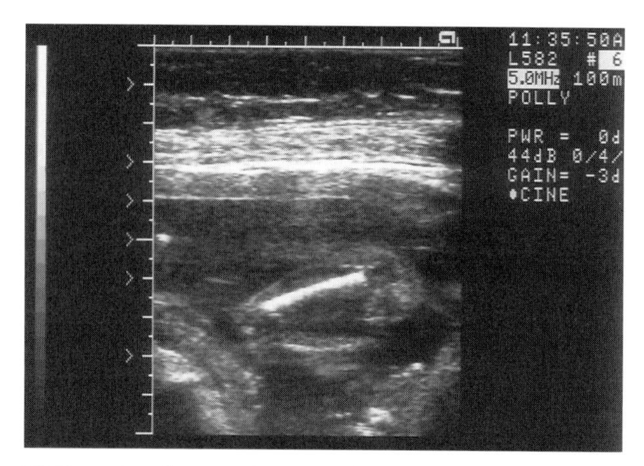

FIGURE 6 Ultrasound image of the femur in a second-trimester fetus. The length of the femur correlates with gestational age.

parameter mentioned previously allows estimation of the fetal weight. Serial ultrasound assessments can detect the abnormally large and abnormally small fetus, both of which are at greater risk for poor pregnancy outcome.

Biophysical evaluation of the fetus includes assessment of amniotic fluid, fetal breathing, fetal tone, and fetal movements. These, combined with nonstress testing by external cardiotocography, comprise a biophysical profile score that is reflective of fetal well-being. This application of ultrasound is usually employed in the ongoing evaluation of high-risk pregnancies, most commonly those in which maternal disorders may potentially affect the fetus adversely. Doppler evaluation of umbilical artery blood flow velocity can provide a measure of uteroplacental resistance by determining the ratio of blood flow velocity during systole to the blood flow velocity during diastole. This measurement is used adjunctively in the overall evaluation of the fetus at risk for growth impairment resulting from poor placental blood flow.

D. Invasive Diagnostic and Therapeutic Applications

Special applications of ultrasound in obstetrics include its adjunctive role in the performance of invasive diagnostic and therapeutic procedures. Prenatal diagnostic techniques, such as amniocentesis, chorionic villus sampling, placental biopsy, fetal tissue biopsy, and fetal blood sampling, employ ultrasound guidance of invasive instruments to their appropriate targets. Therapeutic interventions, such as intrauterine blood transfusion or shunt placement, are also aided by ultrasound guidance.

Among the newest advances in obstetrical sonography is the employment of computer technology to serial images taken in different planes to generate three-dimensional images. The precise role of three-dimensional sonography in fetal evaluation is under investigation.

III. GYNECOLOGIC APPLICATIONS

Ultrasound has a number of applications in the gynecologic population, including evaluation of normal and pathological anatomy of the genital tract, diagnosis of ectopic pregnancy, screening for malignancies, and as an adjunct to assisted reproduction technology.

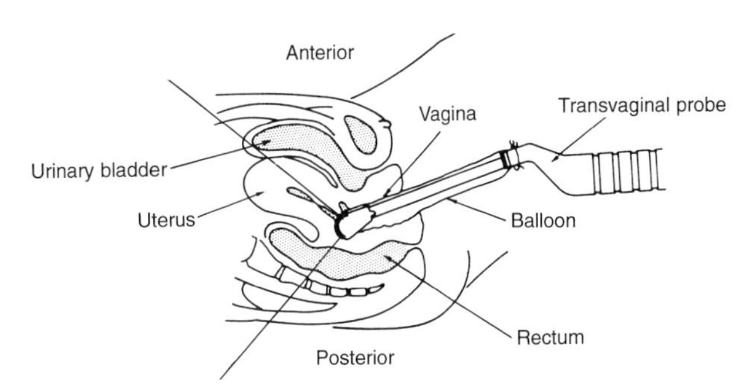

FIGURE 7 Diagram of the approximate field of view obtained with the transvaginal probe [reprinted with permission from R. R. Price and A. C. Fleischer, Sonographic instrumentation, In *The Principles and Practice of Ultrasonography in Obstetrics and Gynecology* (A. C. Fleischer *et al.*, Eds.), p. 35].

Evaluation of the female reproductive tract by ultrasound can be accomplished by transabdominal scanning. However, the use of transvaginal probes (Fig. 7) allows for the greater proximity of the transducer to the structures under study so that higher frequency probes can be used to achieve greater resolution without the expense of depth attenuation of the ultrasound beam. Evaluation of the uterine corpus can detect myomata. Defects within the uterine cavity, such as septa, synechiae, and polyps, can now be detected with use of sonohysterography, which employs the instillation of normal saline and avoids the more risky oil-based contrast media of traditional hysterosalpingography. Evaluation of the adnexae can detect masses such as ovarian cysts or ectopic pregnancy.

A. Evaluation of Ectopic Pregnancy

Ultrasound can serve a critical role in the evaluation of a suspected ectopic pregnancy. Clinical indices such as vaginal bleeding in the first trimester, lateralized pelvic pain, and inappropriate increases in serum human chorionic gonadotropin or progesterone can raise the suspicion for ectopic pregnancy. Transvaginal sonography should document an intrauterine gestation by five menstrual weeks or at a serum human chorionic gonadotropin concentration of 1500 IU/ml. The inability to visualize an intrauterine gestational sac in these circumstances is highly suggestive of ectopic pregnancy, as is the detection of an adnexal mass. The detection of a fetal heartbeat in the adnexae is, of course, definitive.

B. Screening for Pelvic Malignancy

Ultrasound can be used as a screening modality for malignancies of the ovary and endometrium. Suspicious characteristics of ovarian masses include complex cystic and solid appearance and the detection of papillary projections. Doppler evaluation of ovarian vessels can be correlated with the degree of suspicion for malignancy. Low resistive index correlates fairly well with malignancy. The measurement of endometrial thickness, particularly in postmenopausal women, has been shown to correlate with cancer of the endometrium.

C. Applications in Assisted Reproduction Technology

Ultrasound is used to follow recruitment and development of ovarian follicles in ovulation induction protocols of assisted reproduction. Measurement of endometrial response can also be used to assess the preparedness of the endometrial lining for implantation. In *in vitro* fertilization, transvaginal aspiration of mature ova from ovarian follicles is performed under ultrasound guidance.

Ultrasound has extensive application in the field of obstetrics and gynecology. Advances in technology have vastly increased the scope of diagnostic and therapeutic capabilities, providing safer and less invasive options to patients and their providers.

See Also the Following Articles

Amniocentesis; Fetal Diagnosis, Invasive; Fetal Monitoring and Testing; Fetal Surgery

Bibliography

Callen, P. (Ed.) (1994). *Ultrasound in Obstetrics and Gynecology*. 3rd ed. Saunders, Philadelphia.

Chervenak, F. A., Isaacson, G., and Campbell, S. (Eds.) (1994). *Ultrasound in Obstetrics and Gynecology*. Little, Brown, Philadelphia.

Manning, F. A. (1995). The anomalous fetus. In *Fetal Medicine. Principles and Practice* (F. A. Manning, Ed.). Appleton & Lange, Norwalk, CT.

McGahan, J. P., and Porto, M. (Eds.) (1994). *Diagnostic Ultrasound*. Lippincott, Philadelphia.

Nyberg, D. A., Mahoney, B. S., and Pretorius (Eds.) (1990). *Diagnostic Ultrasound of Fetal Anomalies. Text and Atlas*. Year Book Med. Pub., Chicago.

Romero, R., Pilu, G., Jeanty, P., *et al.* (1988). *Prenatal Diagnosis of Congenital Anomalies*. Appleton & Lange, Norwalk, CT.

Sabbagha, R. (Ed.) (1994). *Diagnostic Ultrasound Applied to Obstetrics and Gynecology*. Lippincott, Philadelphia.

Umbilical Cord

Harvey J. Kliman
Yale University School of Medicine

GLOSSARY

allantois Primitive excretory duct.

cord prolapse Passage of the umbilical cord through the cervix prior to delivery of the infant.

funisitis Inflammatory cell infiltrate in the umbilical vessels walls and Wharton's jelly.

insertion Point at which the umbilical cord attaches to the placenta.

Meckel's diverticulum Persistent outpouching of bowel contents through the abdominal wall at the umbilicus.

omphalomesenteric duct remnant Remains of the yolk sac stalk within the proximal portion of the umbilical cord.

vasa previa Umbilical cord vessels, usually in a case of a velamentous insertion, which are overlying the internal cervical os.

velamentous Insertion of the umbilical cord into the external membranes.

Wharton's jelly Proteoglycan-rich matrix in which the umbilical vessels are embedded.

yolk sac Outpouching of the endoderm which serves as the site of initial blood cell formation.

The umbilical cord is the lifeline between the fetus and placenta. It is formed by the fifth week of development and it functions throughout pregnancy to protect the vessels that travel between the fetus and the placenta. Compromise of the fetal blood flow through the umbilical cord vessels can have serious deleterious effects on the health of the fetus and newborn.

I. FORMATION AND STRUCTURE OF THE UMBILICAL CORD

By the end of the third week of development the embryo is attached to placenta via a connecting stalk (Fig. 1). At approximately 25 days the yolk sac forms and by 28 days, at the level of the anterior wall of the embryo, the yolk sac is pinched down to a vitelline duct, which is surrounded by a primitive umbilical ring (Fig. 2A). By the end of the fifth week the primitive umbilical ring contains (i) a connecting stalk within which passes the allantois (primitive excretory duct), two umbilical arteries, and one vein;

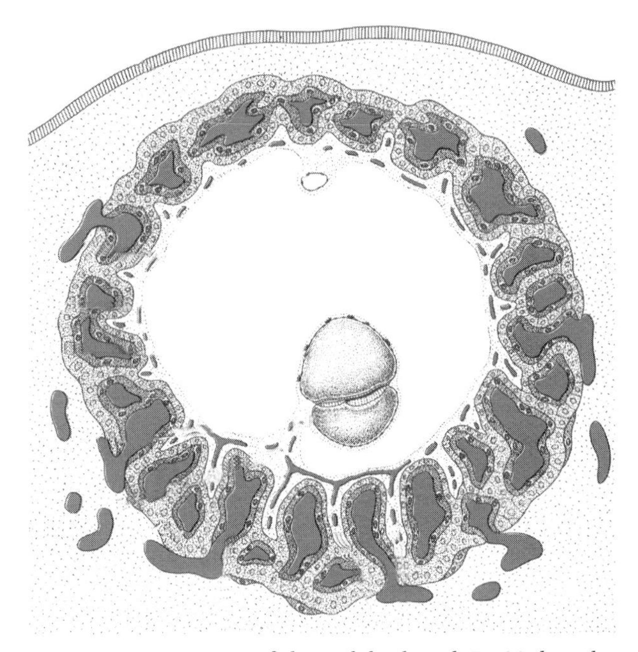

FIGURE 1 Beginning of the umbilical cord. By 21 days the embryo has begun to separate from the developing placenta by a connecting stalk. Within this stalk are the beginnings of the early circulatory system (modified with permission from T. W. Sadler, *Langman's Medical Embryology*, 5th ed., Williams & Wilkins, 1985).

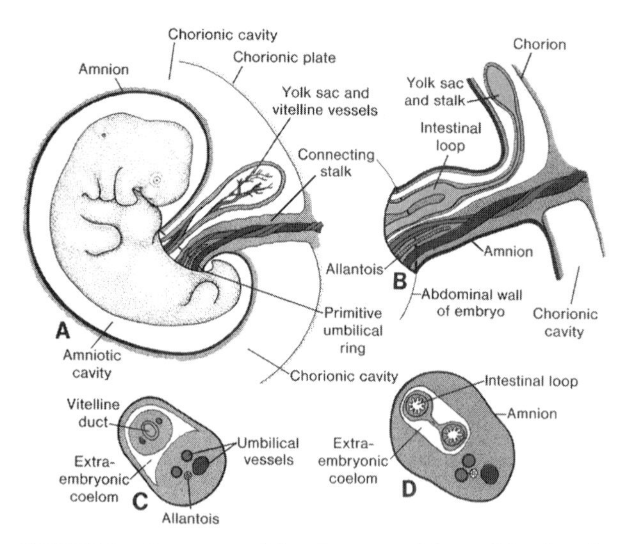

FIGURE 2 Contents and development of the umbilical cord. (A, C) At 5 weeks of development the embryo is connected to the placenta by a stalk which contains the umbilical vessels and allantois. Adjacent to this stalk is the yolk sac stalk which consists of the vitelline duct (yolk sac duct) and the vitelline vessels. These structures all pass through the primitive umbilical ring. (B, D) By 10 weeks of development the yolk sac duct has been replaced by loops of bowel within the umbilical cord. These will normally regress back into the peritoneal cavity by the end of the third month (reproduced with permission from T. W. Sadler, *Langman's Medical Embryology*, 5th ed., Williams & Wilkins, 1985).

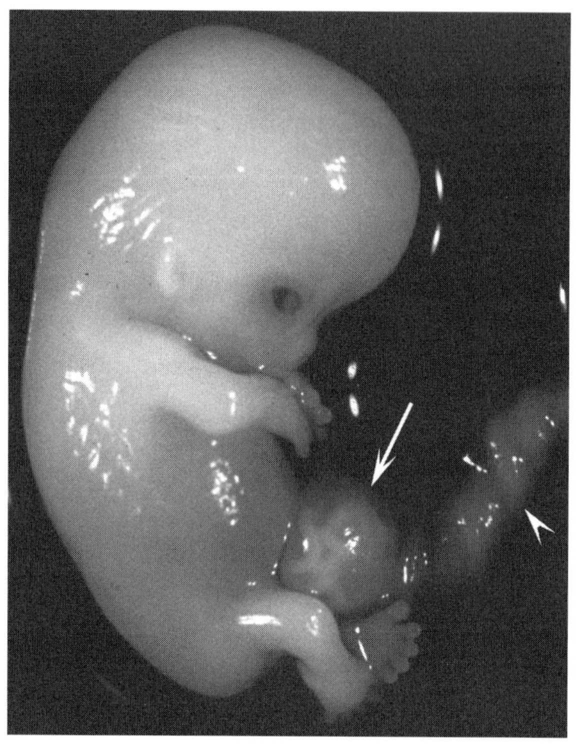

FIGURE 3 Fetus at ~53 days postovulation (21.5-mm crown–rump length) showing distinct intestinal herniation into proximal umbilical cord (arrow). Note twisting of umbilical cord (arrowhead).

(ii) the vitelline duct (yolk sac stalk); and (iii) a canal which connects the intra- and extraembryonic coelomic cavities (Fig. 2C). By the 10th week the gastrointestinal tract has developed and protrudes through the umbilical ring to form a physiologically normal herniation into the umbilical cord (Figs. 2B, 2D, and 3). Normally, these loops of bowel retract by the end of the third month. Occasionally, residual portions of the vitelline and allantoic ducts, and their associated vessels, can still be seen even in term umbilical cords, especially if the fetal end of the cord is examined.The umbilical cord normally contains two umbilical arteries and one umbilical vein. These are embedded within a loose, proteoglycan-rich matrix known as Wharton's jelly (Fig. 4). This jelly has physical properties much like a polyurethane pillow, which is resistant to twisting and compression. This property serves to protect the critical vascular lifeline between the placenta and fetus (Fig. 5).

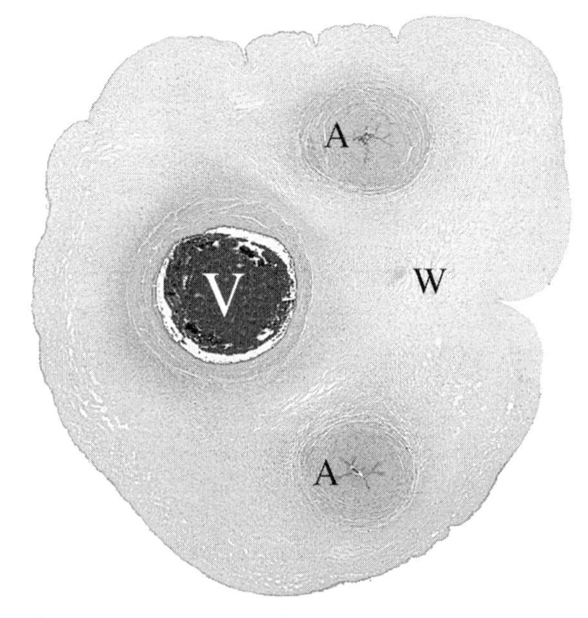

FIGURE 4 Cross section of normal umbilical cord. Embedded within a spongy, proteoglycan-rich matrix know as Wharton's jelly (W) are normally two arteries (A) and one vein (V).

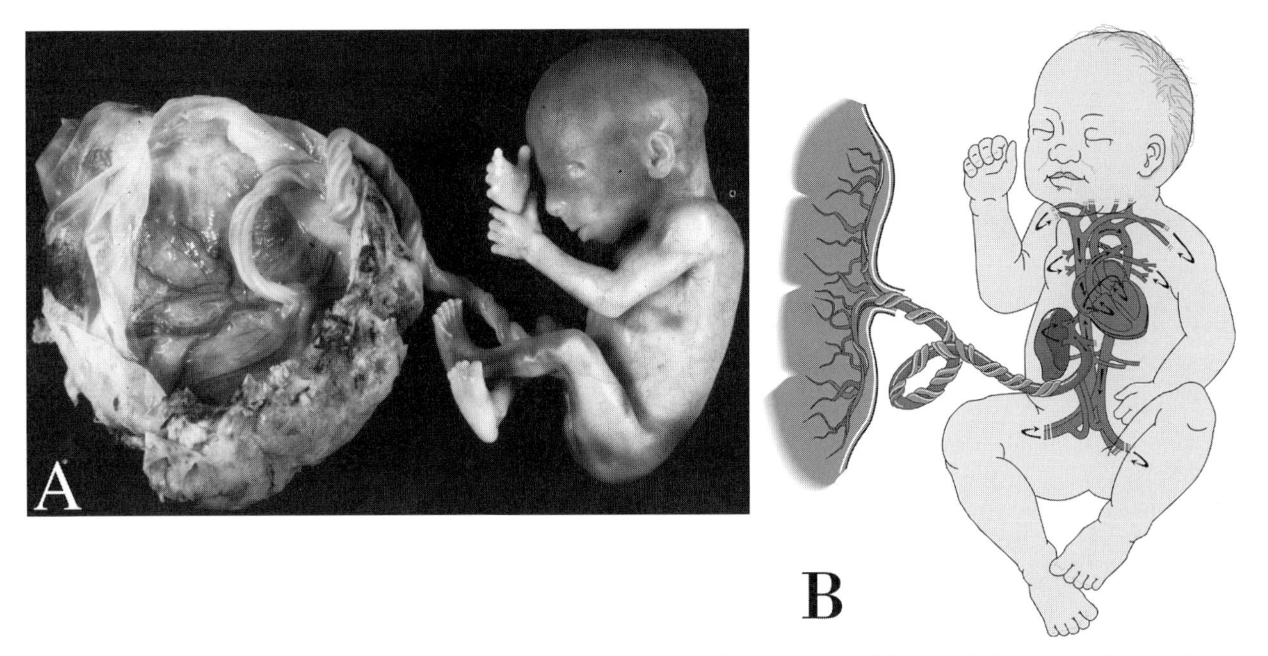

FIGURE 5 The umbilical cord protects the fetal vessels that connect the placenta and fetus. (A) Fetus and placenta from a 17-week gestation. (B) Diagram of the circulation within the fetus, umbilical cord, and placenta.

II. ABNORMAL UMBILICAL CORD DEVELOPMENT

Approximately 1% of all umbilical cords contain only one artery rather than the normal two. Although many infants born with a single umbilical artery have no obvious anomalies, single umbilical artery has been associated with cardiovascular anomalies in 15–20% of such cases. While these anomalies could be the result of genetic factors alone, environmental factors may also play a part. For example, Naeye has shown an association between a single umbilical artery and maternal smoking during pregnancy.

As stated previously, loops of bowel can be found within the proximal portion of the cord up until the end of the third month (Figs 2. and 3). When this regression does not take place and herniation of peritoneal contents persists to term, a condition known as Meckel's diverticulum exists. Occasionally, only a small portion of the vitelline duct may persist to term, leading to a vitelline cyst or fistula, which may need to be surgically removed after birth.

III. PATHOLOGIC PROCESSES AFFECTING THE UMBILICAL CORD

As with any organ or tissue, the umbilical cord can be subjected to both intrinsic and extrinsic pathological processes. Intrinsic processes include inflammation, knots, and torsion, whereas extrinsic damage can occur iatrogenically following invasive, diagnostic procedures.

The most common pathological finding in the umbilical cord is funisitis (Latin for "cord inflammation"). Funisitis is the result of neutrophils being chemotactically activated to migrate out of the fetal circulation toward the bacterially infected amnionic fluid (Fig. 6). Since the ability of neutrophils to respond to chemokines and endotoxin is dependent on cellular maturation, it is not surprising to note that funisitis is only seen commonly after 20 weeks of gestation.

Less commonly, but with potentially devastating consequences, the umbilical cord can become knotted (Fig. 7). If the knot is loose, fetal circulation is maintained. However, if the knot is tightened, for example, at the time of fetal descent through the birth

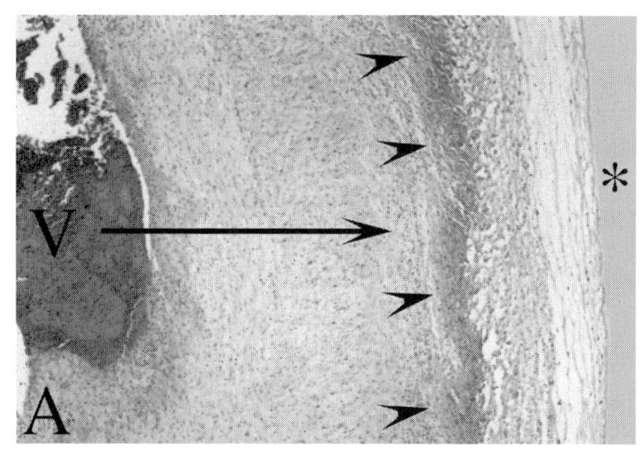

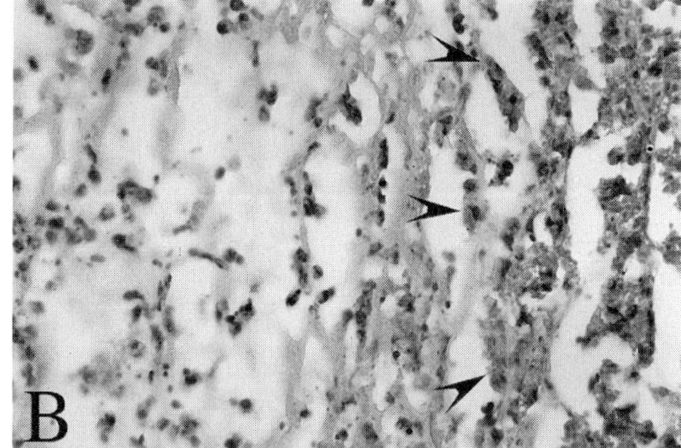

FIGURE 6 Fetal neutrophil migration through the umbilical cord (funisitis). (A) In the presence of bacterial growth within the amnionic fluid (*), fetal neutrophils leave the umbilical vessels (V) and migrate toward the amnionic cavity (arrow). In this case of severe funisitis, a wave of neutrophils and neutrophil breakdown products can be seen (arrowheads). (B) Higher magnification of the edge of the neutrophil wave (arrowheads).

canal, the tightening knot can occlude the circulation between the placenta and fetus, resulting in an intrauterine demise. The Wharton's jelly surrounding the fetal vessels is capable of withstanding significant torsional and compressional forces, as shown in Fig. 8. Occasionally, however, Wharton's jelly does not develop in all portions of the cord. When this occurs, the fetal vessels are no longer protected from torsional forces and they can become occluded if twisted sufficiently (Fig. 9), again leading to an intrauterine demise.

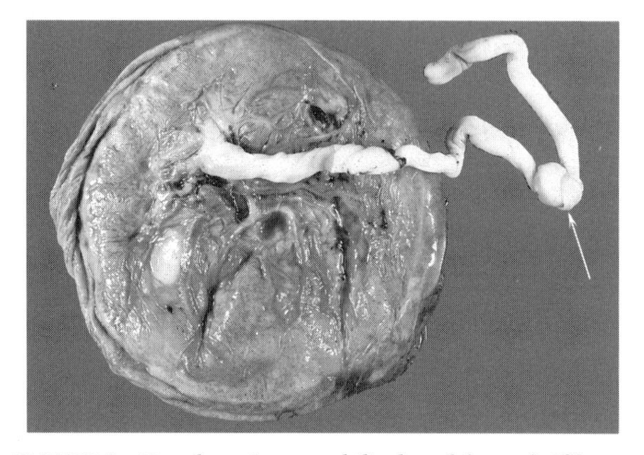

FIGURE 7 True knot in an umbilical cord (arrow). If loose, a true knot will not lead to fetal compromise. However, if the knot tightens, for example, at the time of delivery, fetal blood flow through the umbilical cord vessels can become occluded, leading to fetal demise.

IV. UMBILICAL CORD LENGTH AND TWISTING

Analogous to how a phone cord becomes longer with increased use, umbilical cord length is dependent on fetal movements—the more movement, the longer the cord. The converse is also true—less intrauterine movement leads to shorter umbilical cords (as attested to by animal experiments in which induced fetal muscle paralysis led to shortened umbilical cord length). Normally, the human umbilical cord reaches a length of 60–70 cm at term. Although the length of the umbilical cord has no intrinsic effect on fetal blood flow, a longer cord is more susceptible to knotting, entanglement around the fetus (especially the neck), and even prolapse out of the uterus during delivery (Fig. 10), any of which can lead to intrauterine fetal demise.

An intriguing association between umbilical cord length and mental and motor development has been suggested by Naeye. As part of the Collaborative Perinatal Study, Naeye correlated 35,799 umbilical cord lengths with clinical, demographic, and social data. He found that decreased cord length was correlated with decreased IQ and a greater frequency of motor abnormalities. Very long cords, on the other hand, were associated with abnormal behavior control and hyperactive behavior.

Intrauterine movement, in addition to controlling

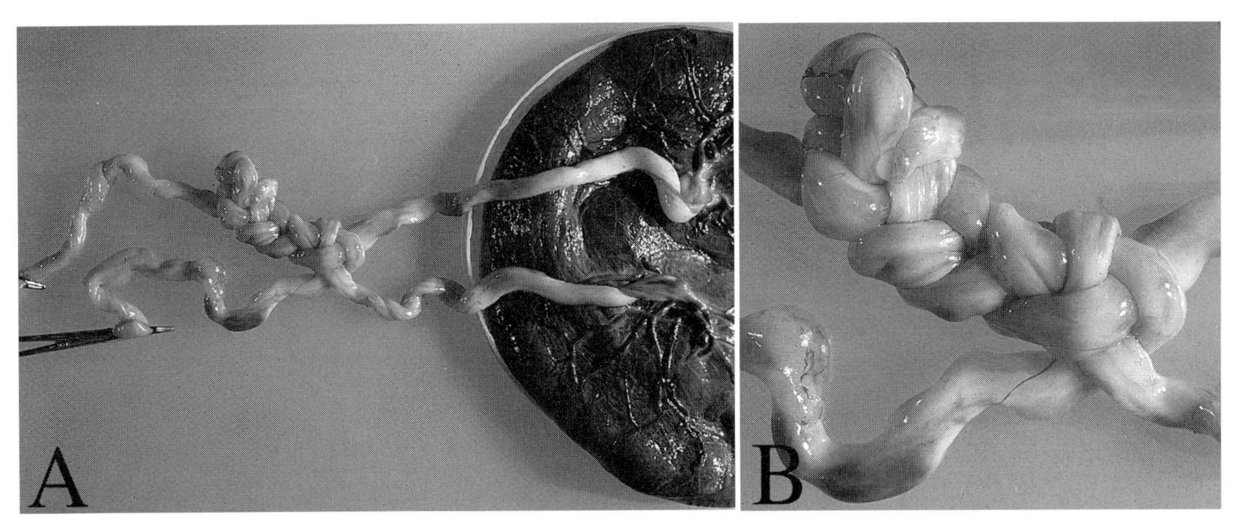

FIGURE 8 Umbilical cord braiding in a monochorionic–monoamnionic twin placenta at 34 weeks gestation. (A) This braid was diagnosed by ultrasound at approximately 32 weeks. The fetuses were monitored continuously and when they showed signs of stress were delivered successfully by emergency cesarean section. (B) Closer examination of the braided umbilical cord shows that despite the marked compression of the Wharton's jelly, the fetal vessels were still protected from complete occlusion.

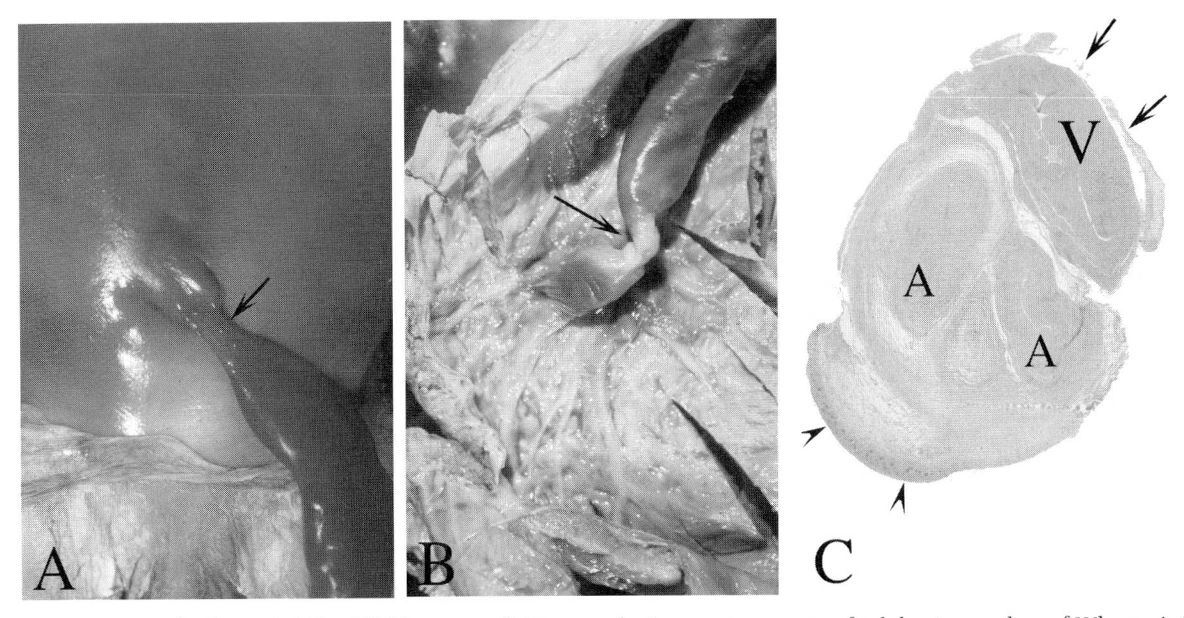

FIGURE 9 Loss of Wharton's jelly. (A) The cause of this second-trimester intrauterine fetal demise was loss of Wharton's jelly near the fetal insertion (arrow). (B) Although loss of Wharton's jelly is most often seen near the fetal insertion, occasionally loss and subsequent torsion of the fetal vessels can occur near the placental insertion point (arrow). (C) Cross section of umbilical cord at fetal insertion with marked loss of Wharton's jelly. The umbilical arteries (A) and vein (V) have little protective matrix beyond their vascular walls (arrows), making these vessels, especially the vein, susceptible to compression. Note the fetal epidermal vessels at one edge of the tissue section (arrowheads).

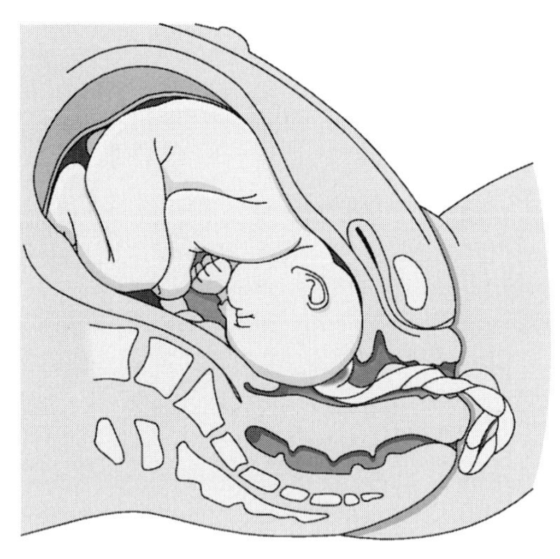

FIGURE 10 Umbilical cord prolapse. During delivery the umbilical cord, especially if excessively long, may deliver prior to the fetus. Folding and compression of the umbilical cord can lead to fetal stress and in some case fetal demise.

umbilical cord length, also appears to control cord twisting. Cord twisting can be seen as early as the sixth week and is well established by the ninth week of development. One might imagine that the umbilical cord twist—either counterclockwise (left) or clockwise (right)—might be random, but left twisting outnumbers right by a ratio of approximately 7 : 1 (in other words, ~85% are left and 15% are right twisted). Since this ratio is similar to the ratio for right to left handedness (approximately 15% of the population is left-handed), some authors have suggested that handedness may be the determining factor for umbilical cord twisting. This has proven not to be true, however. What is clear, nevertheless, is that the degree of twisting does relate to intrauterine movement and, as with short umbilical cords, cords with little twisting are associated more frequently with compromised fetuses. Finally, hypertwisting can lead to intrauterine fetal demise by compressing the fetal vessels beyond the capacity of the Wharton's jelly to protect them (Fig. 11).

V. UMBILICAL CORD INSERTION

The umbilical cord normally inserts near the center of the placenta (Fig. 7). However, in approximately 7% of single births the insertion point occurs at the very edge of the placenta (marginal insertion) and in about 1% of cases, the umbilical cord does not insert into the placenta at all, but the fetal vessels ramify through the external membranes before entering the placenta (velamentous insertion). When the umbilical cord inserts into the chorionic plate of the placenta (Fig. 12), the fetal vessels are stabilized and thus protected from torsional and shear forces. On the other hand, insertion into the membranes exposes the fetal vessels to the potential for rupture due to shearing forces (Fig. 13) or if the vessels pass near the internal cervical os (vasa previa) by rupture due to an ascending inflammation prior to the time of delivery (Fig. 14).

VI. DIAGNOSTIC UTILITY OF THE UMBILICAL CORD

Increasingly, noninvasive procedures are being utilized to assess fetal well-being *in utero*. Assessment of fetal blood flow through the umbilical cord has

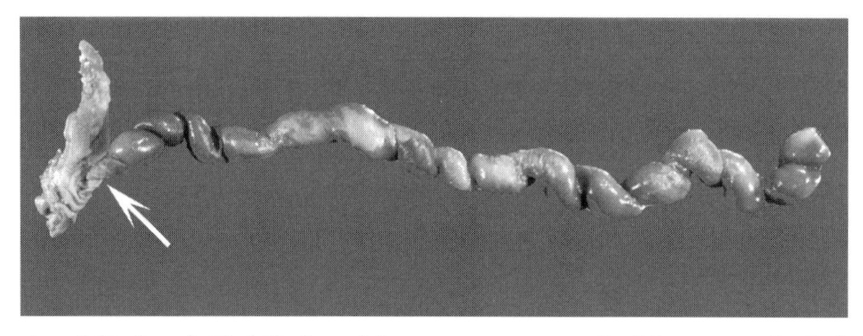

FIGURE 11 Hypertwisted umbilical cord. Umbilical cord from an intrauterine fetal demise in which the cord has been markedly twisted. Note the decreased Wharton's jelly at the fetal insertion point (arrow).

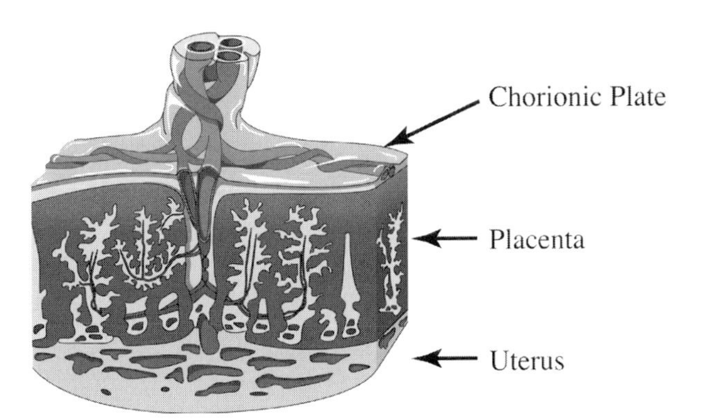

FIGURE 12 Insertion of umbilical cord into chorionic plate. Normally the umbilical cord inserts near the center of the chorionic plate, which stabilizes the fetal vessels as they leave the umbilical cord. Like the roots of a tree, the fetal vessels branch over the surface of the chorionic plate and then dive into the placental parenchyma.

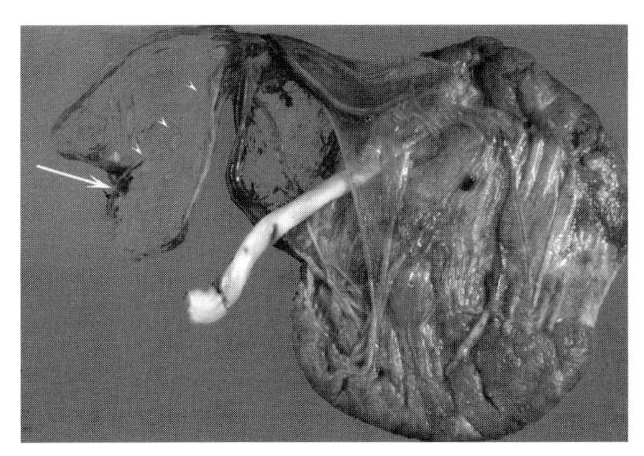

FIGURE 13 Rupture of a fetal vessel within the external membranes. The umbilical cord of this placenta inserted at the placental margin. The fetal vessels emanating from the insertion point did not traverse into the placenta, as is the usual case, but instead traveled through portions of the external membranes (arrowheads). This velamentous vessel (arrow), overlying the cervical os (vasa previa), was inadvertently ruptured at the time of delivery. Although the fetus lost a significant amount of blood, it survived and did well due to the rapid delivery by the obstetrician following the vascular rupture.

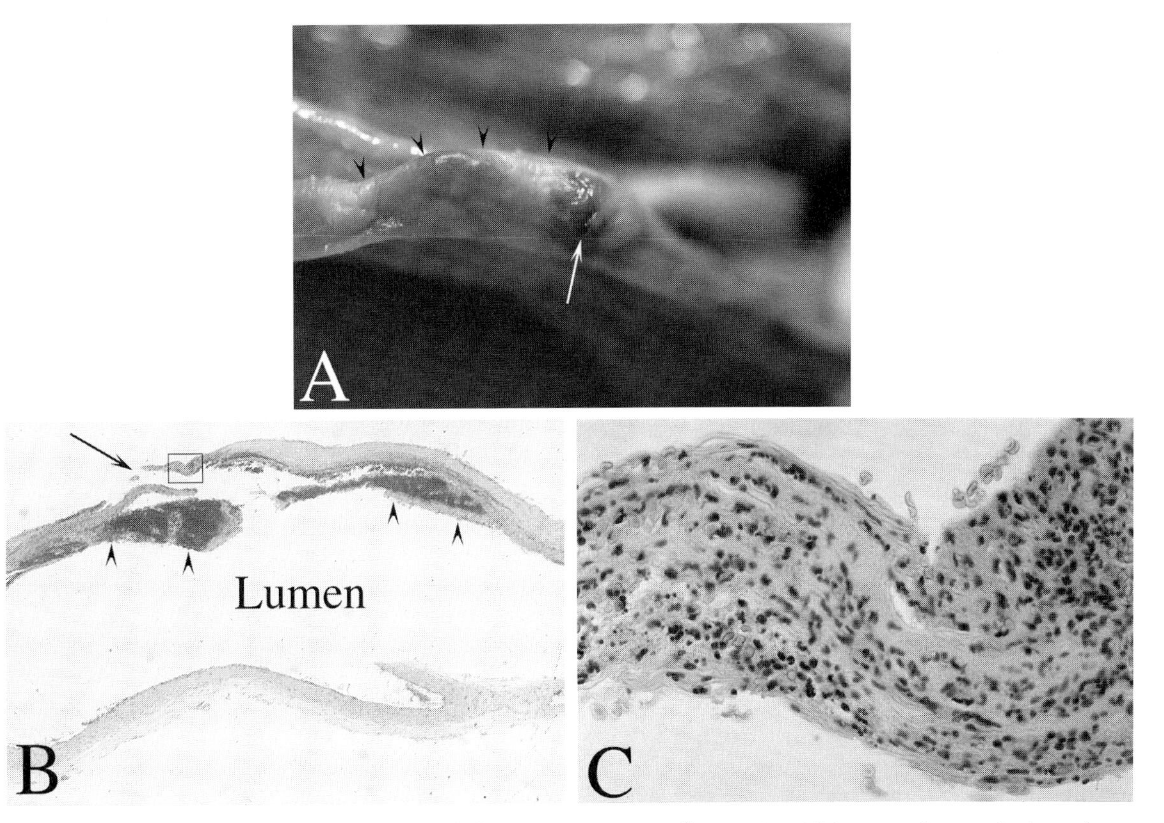

FIGURE 14 Rupture of a velamentous fetal vessel due to necrotizing inflammation. This term placenta had a velamentous insertion of the umbilical cord. As with the placenta shown in Fig. 13, one of the fetal vessels passed over the internal cervical os. In this case an ascending infection developed several days prior to delivery which eventually eroded the vessel wall until it ruptured. (A) Gross exam of the rupture site (arrow) of the vasa previa vein (arrowheads). (B) Microscopic exam of a longitudinal section through the rupture site (arrow). Note fibrin clot (arrowheads) attempting to stop the hemorrhage near the site of rupture. Inset showing acute inflammatory infiltrate is demonstrated at higher magnification in C.

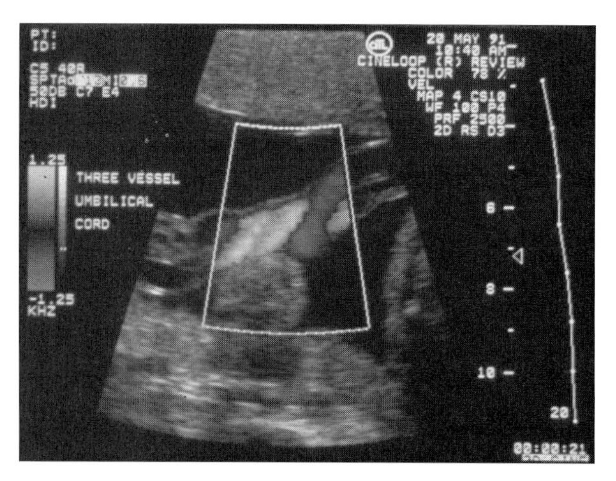

FIGURE 15 Doppler ultrasound of the umbilical cord. By measuring the shift in frequency of the reflected ultrasound, blood flow in the umbilical vessels can be visualized. In this example the two umbilical arteries and one vein can be easily seen within the marked off region in the center of the ultrasound image.

proven to be one such measure. Utilizing ultrasound and Doppler flow measuring techniques, not only can the umbilical cord be visualized (Fig. 15) but also the flow of fetal blood through these vessels can be assessed. By measuring the amount of forward blood flow through the umbilical artery during both fetal systole and diastole, an overall measure of fetal health can be obtained. In general, the more forward blood flow from the fetus to the placenta through the umbilical artery, the healthier the fetus (Fig. 16).

Certain clinical situations, however, necessitate a more invasive approach. At these times the fetus's survival may be dependent on directly evaluating or giving the fetus blood. In these cases, direct puncture of the umbilical cord vessels (cordocentesis) may become necessary. This more invasive approach to fetal therapy is used only in the most serious cases since these procedures carry the risk of rupture of the umbilical vessels, which can lead to thrombosis, hemorrhage, or even vascular tamponade (Fig. 17).

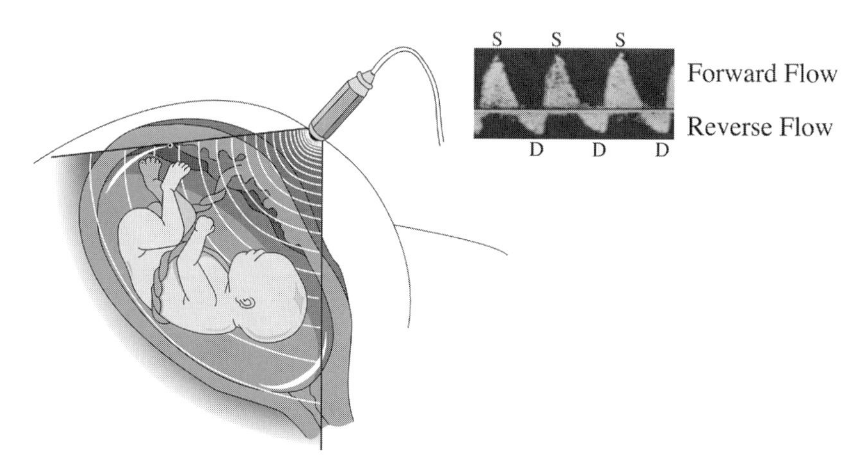

FIGURE 16 Doppler flow ultrasound of the umbilical cord can also be used to quantitatively assess umbilical artery blood flow. By directing the Doppler ultrasound measurement in the path of umbilical artery blood flow, a measurement of both systolic (S) and diastolic (D) flow through the umbilical artery can be made. Normally, a high forward flow signal is seen during systole, followed by a lesser, but still forward flowing, diastolic pulse. In cases of severe fetal compromise, reverse flow may be seen during diastole (as shown in the inset).

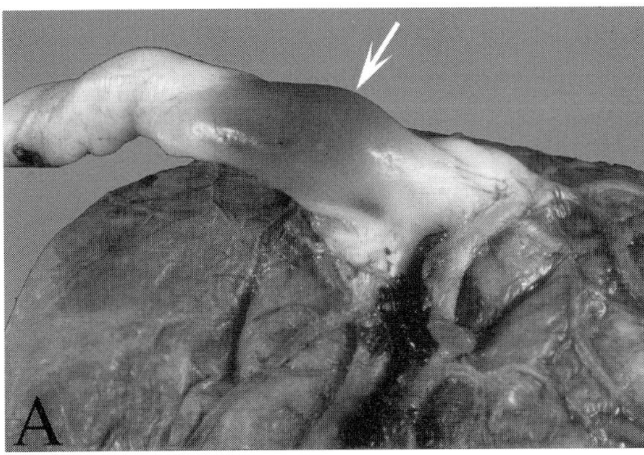

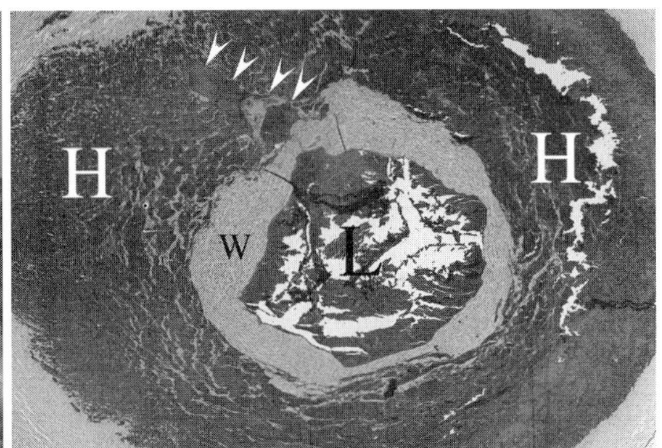

FIGURE 17 Therapeutic cordocentesis occasionally leads to umbilical vessel hemorrhage. (A) Site of umbilical cord puncture (arrow) during an intrauterine fetal transfusion to treat severe fetal anemia. Note bulging and discoloration of cord at site of puncture. (B) Cross section of umbilical cord at point of puncture which demonstrates the needle tract (arrowheads) through the artery wall (W). Rupture of the vessel resulted in perivascular hemorrhage (H) with tamponade. L, artery lumen. The fetus experienced acute distress and was delivered by emergency cesarean section, but unfortunately died within an hour of birth.

See Also the Following Articles

ALLANTOIS; FETUS, OVERVIEW; PLACENTA: IMPLANTATION AND DEVELOPMENT; YOLK SAC

Bibliography

Benirschke, K., and Kaufmann, P. (1995). *Pathology of the Human Placenta*, 3rd ed. Springer-Verlag, New York.

Boyd, J. D., and Hamilton, W. J. (1970). *The Human Placenta*. Heffer & Sons, Cambridge, UK.

Moore, K. L. (1993). *The Developing Human*, 5th ed. Saunders, Philadelphia.

Naeye, R. L. (1992). *Disorders of the Placenta, Fetus, and Neonate: Diagnosis and Clinical Significance*. Mosby/Year Book, St. Louis.

Sadler, T. W. (1990). *Langman's Medical Embryology*, 6th ed. Williams & Wilkins, Baltimore.

Uterine Anomalies

Bradley S. Hurst
University of Colorado Health Sciences Center

John A. Rock
Emory University School of Medicine

GLOSSARY

agenesis Failure of an organ to develop.

dysgenesis Abnormal formation of an organ.

hematocolpus Blood-filled vaginal cavity.

hematometria Blood-filled uterine cavity.

hemiuterus A functional uterus formed by a single Müllerian duct.

hemivagina Incomplete formation of a vagina in the case of cervical and vaginal duplication.

hypoplasia Incomplete formation of an organ.

menarche The initial episode of menstrual or uterine bleeding.

metroplasty Surgical repair of the uterus.

Encyclopedia of Reproduction VOLUME 4

Uterine developmental anomalies have a profound impact on reproduction. Some of these anomalies preclude pregnancy, and others may cause recurrent pregnancy losses, premature labor, or poor pregnancy outcomes. Occasionally, some of these conditions will produce frank obstetric emergencies. Although the exact incidence is unknown, uterine anomalies may occur in as many as 2 or 3% of all women. Knowledge of embryologic development is necessary to fully comprehend these disorders. The classification system used to describe uterine anomalies, their reproductive implications, and the possible role for surgical correction will be reviewed.

I. EMBRYOLOGY

The female reproductive organs are composed of the external genitalia, gonads, and an internal duct system between the two. The Müllerian (paramesonephric) duct system is responsible for development of the Fallopian tubes, uterus, cervix, and a portion of the vagina. These Müllerian ducts first appear approximately 37 days after fertilization as invaginations of the dorsal coelomic epithelium (Fig. 1). The origin of the invaginations ultimately forms the fimbriated ends of the Fallopian tubes. The Müllerian ducts form as a solid bud at the point of origin, and each of these buds elongates. A lumen appears in the cranial part and gradually extends to the caudal growing tips of the ducts. The Müllerian ducts continue to grow in a medial and caudal direction until they eventually meet in the midline and become fused together. The septum between the Müllerian ducts eventually is resorbed, leaving a single uterovaginal canal. The caudal segments of the fused Müllerian ducts form the uterus, cervix, and upper vagina. Uterine anomalies may occur as a result of failure of the Müllerian duct system to form, failure to descend, failure of the ducts to fuse, or failure of the septum between the ducts to resorb.

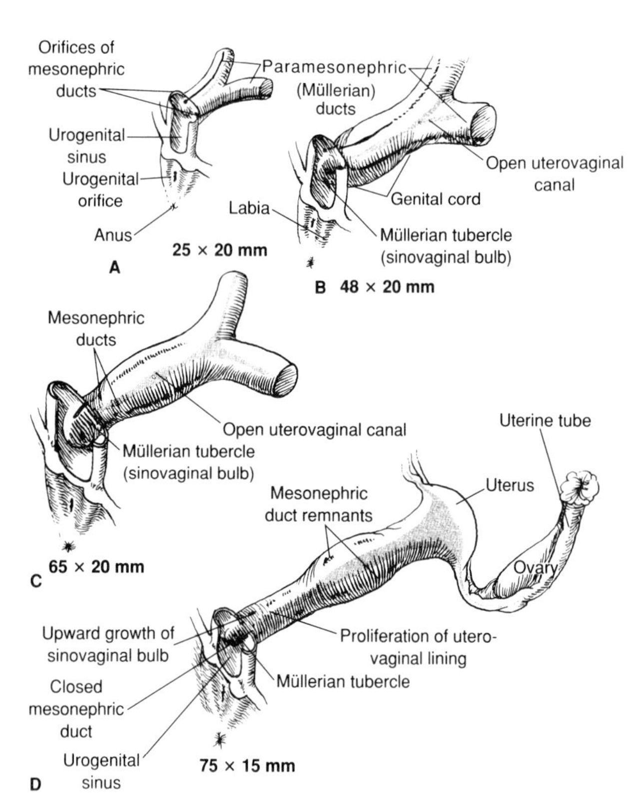

FIGURE 1 Embryologic development of the Müllerian duct system. See text for details [reproduced with permission from J. A. Rock and J. D. Thompson (Eds.), *TeLinde's Operative Gynecology*, 8th ed., Lippincott-Raven, Philadelphia, 1996].

II. CLASSIFICATION OF UTERINE ANOMALIES

Various schemes have been used to classify uterine anomalies, but the 1988 American Fertility Society classification system is recognized as the standard (Fig. 2). This classification system separates types of Müllerian anomalies into the following categories: (i) hypoplasia/agenesis, (ii) unicornuate, (iii) didelphus, (iv) bicornuate, (v) septate, (vi) arcuate, and (vii) diethylstilbestrol (DES). This classification system allows the user to indicate the uterine malformation type and, if present, describe corresponding anomalies of the vagina, cervix, and Fallopian tubes.

A. Hypoplasia/Agenesis

1. *Uterine and Vaginal Hypoplasia/Agenesis*

Women with hypoplasia or agenesis of the Müllerian duct system may present with vaginal, cervical, fundal, tubal, or combined defects. Hypoplasia or agenesis of the Müllerian system preclude natural reproduction.

Abnormalities in the development of the uterus

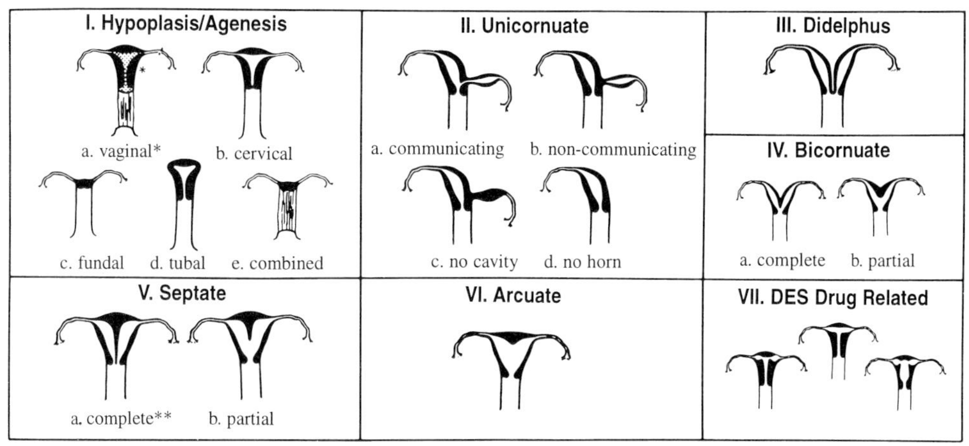

*Uterus may be normal or take a variety of abnormal forms.
*May have two distinct cervices.

FIGURE 2 The American Fertility Society classification of uterine anomalies. Reproduced with permission from American Society for Reproductive Medicine (*Fertility and Sterility*, 1988, 49, 952).

and vagina occur if the Müllerian ducts fail to develop at any time between their origin at 5 weeks of embryonic age and their fusion with the urogenital sinus at 8 weeks. Development of the Müllerian ducts is necessary at this crucial time to promote proliferation of the sinovaginal bulbs from the urogenital sinus. When these events do not occur, the uterus and vagina fail to develop. Congenital absence of a uterus and vagina is the most common example of this anomaly and is known as the Mayer–Rokitansky–Kuster–Hauser syndrome. The estimated incidence of Müllerian agenesis is 1 in 10,000 female births.

A woman with Müllerian agenesis usually has small rudimentary uterine bulbs with rudimentary Fallopian tubes. Normal ovarian function as well as ovulation is expected. The phenotypic sex is clearly female since affected individuals experience normal breast and sexual hair development, and external genitalia are normal. The karyotype is normal female 46,XX. Since almost all individuals with Müllerian agenesis have a normal female karyotype, the cause of these abnormalities is not clear. Leading theories include exposure to a terotogenic agent that results in arrest of Müllerian duct development. Gene mutation is another suggested etiology. Other theories include the inappropriate production of Müllerian-inhibiting factor in the female embryonic gonad, regional absence or deficiency of estrogen receptors in the lower Müllerian duct, or mesenchymal defects.

Women with Müllerian agenesis have rudimentary uterine bulbs that vary in size. These are not usually palpable. However, these rudimentary uterine bulbs occasionally contain a cavity lined by endometrial tissue. In rare circumstances, a large hematometria (blood-filled uterine bulb) will develop due to cyclic accumulation of trapped blood. Only in this unusual circumstance is surgical removal of uterine bulb necessary.

Other congenital anomalies are frequently associated with Müllerian agenesis. Approximately one-third of affected individuals will be found to have urinary anomalies including renal agenesis, a unilateral or bilateral pelvic kidney, horseshoe kidney, hydronephrosis, ureteral duplication, or other anomalies. Approximately 10–15% have skeletal anomalies, most involving the spine, but the limbs and digits can be involved.

In most circumstances, an individual with this condition will seek medical attention between the ages of 14 and 16 for primary amenorrhea. The diagnosis of Müllerian agenesis is made by a careful history and physical examination. Normal pubertal milestones, with the exception of menstruation, are expected. If pubic and axillary hair development is complete, there should be no confusion between Müllerian agenesis and androgen insensitivity syndrome. However, if hair distribution is not mature, a karyotype may be necessary to differentiate these two conditions. Individuals with incomplete androgen insensitivity syndrome may present with amenorrhea, Müllerian agenesis, and incomplete development of sexual hair. A 46,XX karyotype confirms Müllerian agenesis and excludes the diagnosis of androgen insensitivity syndrome.

If a pelvic mass is discovered on physical examination, an ultrasound can help identify a hematometria, hematocolpos, endometrial or ovarian cyst, and a pelvic kidney. If ultrasound findings are not definitive, a magnetic resonance imaging (MRI) scan performed and interpreted by an experienced radiologist should help. An intravenous pylogram (IVP) should be performed in all cases of Müllerian agenesis. The IVP identifies associated urinary tract anomalies and provides an opportunity to evaluate spinal anomalies.

Women with Müllerian agenesis may choose between vaginal dilatation or surgical creation of a vagina before initiating intercourse. With either of these choices, psychological preparation and emotional maturity are necessary to understand her important role with the success of these approaches. Generally, vaginal dilatation or surgery are started between the ages of 17 and 20 years old.

Vaginal dilatation has been used for approximately 60 years, since the first description by Frank in 1938. The Ingram technique of progressive dilatation is the procedure of choice for patients who wish to avoid surgery and are willing to devote the time and effort needed to dilate a satisfactory vagina. This method uses a set of 19 progressively longer and wider dilators for use with a bicycle seat. A woman begins with the smallest dilator and places this dilator against the vaginal dimple. The dilator may be held in place with a light girdle and regular clothing worn over this. She is instructed to sit leaning slightly forward on a racing-type bicycle seat attached to a stool with the dilator for at least 2 hr per day at interval of 15–30 min. She can be expected to graduate to the next size dilator approximately every month. Continued dilatation is recommended if intercourse is infrequent. Overall, success rates with this method approach 80%. If dilatation is unsuccessful, surgery is indicated.

Although several surgical techniques to create a new vaginal space have been described, the Abbe–Wharton–McIndoe procedure is chosen in most cases of vaginal agenesis. This technique involves dissection of an adequate space between the rectum and the bladder. An inlay split-thickness skin graft is placed into the neovaginal space. Following surgery, continuous and prolonged dilatation is necessary. Otherwise, severe contraction of the vagina will result in surgical failure.

Results with this operation have improved over the years. Recent reports of satisfaction range from 80 to 100%. Serious complications occasionally occur, including postoperative urethravaginal, vessicovaginal, or rectovaginal fistula formation. Postoperative infection or intraoperative or postoperative hemorrhage can occur. Failure of the graft to take is an occasional complication. However, these complications have become unusual with surgical advances including the use of a soft foam rubber form to hold the graft in place and distend the vagina.

2. *Cervical Agenesis*

Isolated congenital absence of the uterine cervix is rare. Fewer than 30 cases have been described in the literature. Cervical anomalies include agenesis caused by inadequate length or width in the development of the cervix. Affected individuals may have small endocervical buds within the endocervical stroma or a complete absence of cervical or endocervical tissue. The vagina is usually, but not always, normal.

A woman with cervical agenesis and a normally functioning uterus and ovaries has normal sexual development until the time of expected menarche. Typically, she presents to her physician with severe lower abdominal pain and amenorrhea due to obstructed uterine blood and formation of a hematometrium, or blood-filled uterine cavity. Endometriosis due to retrotubal menstruation is common.

Pelvic bimanal or rectal examination can distinguish a congenitally absent cervix from an imperforate hymen or a low or midtransverse vaginal septum. A mass is consistent with a hematocolpos, or blood-filled vagina, and is expected with vaginal obstruction. This distinction can be a difficult one to establish with a high transverse vaginal septum or agenesis of the upper vagina. A hematometrium may be present in any of these situations.

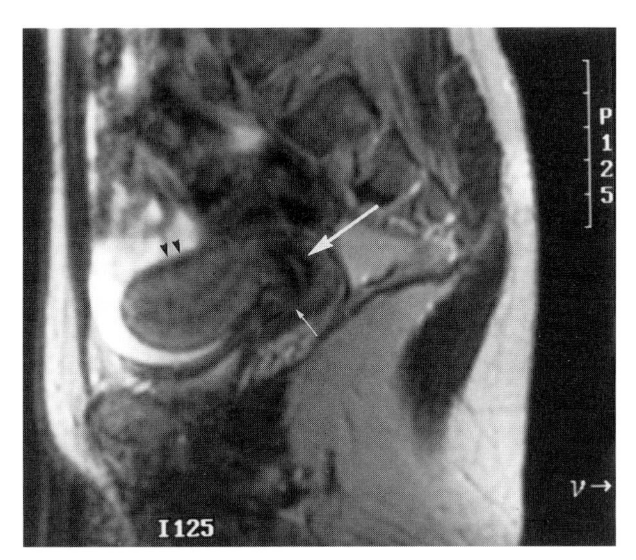

FIGURE 3 MRI scan showing a young woman with cervical agenesis. The uterus (double arrowhead) is normal, as is the upper cervix (large arrow), but the lower cervical canal (small arrow) is absent.

Ultrasound examination is recommended as the initial screening study. Cervical atresia can be identified with visualization of an atretic, uncanalized cervix. However, MRI of the cervix, uterus, and upper vagina is generally needed to distinguish cervical atresia from vaginal agenesis when ultrasonography is inconclusive (Fig. 3). Adequate diagnosis is essential to ensure proper counseling and therapy.

Surgery is usually necessary during adolescence with cervical agenesis because of severe cyclic pain. Unfortunately, procedures to anastomose the endometrial cavity to the vaginal apex have been met with an almost universally poor outcome. Because of an extremely poor prognosis with reconstructive surgery coupled with a high rate of complications, most experts recommend hysterectomy for definitive therapy.

3. *Fundal Hypoplasia and Agenesis*

Congenital absence or hypoplasia of the uterus is associated with the absence of the upper vagina due to failure of the Müllerian ducts to develop and descend. In the absence of estrogen exposure, the uterus occasionally may be identified as severely hypoplastic. A "hypoplastic" uterus of this type will generally develop normally in response to estrogen and progesterone. True isolated fundal hypoplasia or agenesis is very rare.

Diagnostic evaluation of suspected fundal agenesis or hypoplasia is similar to that for Müllerian agenesis. Normal pubertal landmarks are expected, with the exception of menstrual bleeding. Sonography or MRI imaging should define pelvic anatomy. An IVP and axial radiographic studies are necessary to evaluate urinary and skeletal anomalies.

B. Unicornuate Uterus

Asymmetric uterine anomalies are caused by incomplete or absent descent and fusion of the Müllerian ducts. The unicornuate uterus arises from normal development of one Müllerian duct and inadequate development of the contralateral duct. Several subtypes may be seen with this class of anomalies. The rudimentary horn may communicate with the unicornuate uterus or it may be obstructed. The incompletely formed uterine bulb may also be solid or absent.

The unicornuate uterus comprises 5–10% of all uterovaginal anomalies. Only 25% of rudimentary horns communicate with the functional hemiuterus. Ipsilateral renal anomalies are very common. Approximately two-thirds of these women have ipsilateral renal agenesis and 15% have an ipsilateral pelvic kidney.

The unicornuate uterus usually is diagnosed after menarche. These adolescents experience normal growth and secondary sexual development, and onset of menarche is normal. Those with an obstructed hemiuterus develop dysmenorrhea soon after menarche. Their dysmenorrhea is typically unilateral and increases in severity with subsequent menstrual periods. A pelvic mass may be present, and endometriosis caused by retrograde menstruation is likely when there is an obstructed uterine horn. On the other hand, patients with a unicornuate uterus with no endometrium in the hypoplastic uterine bulb are asymptomatic. Therefore, the diagnosis of a unicornuate uterus may be overlooked.

A high index of suspicion is necessary to make the diagnosis of a unicornuate uterus. Uterine anomalies should be considered in patients with an early onset of severe dysmenorrhea, repeated miscarriage, preterm labor, malpresentation, or fetal growth restriction. The diagnosis of a unicornuate uterus is readily established in the nonpregnant state. A hysterosalpingogram (HSG) shows a solitary, fusiform, deviated uterine cavity which terminates into a single Fallopian tube. Therefore, the HSG is considered the diagnostic test of choice. Sonography typically shows an asymmetric uterus but may incorrectly diagnose a rudimentary horn as a myoma or degenerating myoma. MRI may be used to visualize the single uterine horn. Failure of the endometrial cavity to widen toward the fundus is typically seen. Laparoscopy can confirm the diagnosis when there is still a question following a radiologic examination. Most women with a unicornuate uterus do not require surgery. However, if a partially functional uterine horn is identified, surgery is advised to avoid the risk of pregnancy in the rudimentary horn and reduce the risk of endometriosis. Pregnancy in the rudimentary uterine horn is a particularly dangerous condition. Pregnancy may occur as a result of transperitoneal migration of sperm or through a fistula site between the normal hemiuterus and the rudimentary horn. Rapid expansion of the rudimentary horn occurs, usually resulting in rupture before 24 weeks.

Women with a unicornuate uterus are usually fertile, although endometriosis due to retrotubal menstruation from an active rudimentary horn may impair fertility. Pregnancy wastage is common with a unicornuate uterus, and premature labor occurs with an overall incidence of 20%. Malpresentations and intrauterine growth restriction are increased. Despite these complications, 80% of those with a unicornuate uterus will deliver a surviving infant with modern high-risk obstetric management.

C. Didelphus

Uterus didelphus, or duplication of the uterus and cervix, results in failure of fusion of the Müllerian ducts. Uterus didelphus has an estimated incidence of 1 in 2000 women. This condition is associated with a vaginal septum in 75% of cases and occasionally with an obstructing hemivagina. Renal agenesis occurs in 10–20%. However, 100% of women with uterus didelphus and an obstructed hemivagina have ipsilateral renal agenesis. Although rare autosomal dominant defects have been identified in some familial aggregates of Müllerian fusion defects, most consider incomplete Müllerian fusion to be a polygenic or multifactoral problem.

In general, there are no unusual symptoms related to menses in women with a didelphyc uterus, although some may not tolerate tampon use if a longitudinal vaginal septum is present. Vaginal dyspareunia may occur in those with a vaginal septum. Often, the diagnosis of a didelphyc uterus is not considered until a first pelvic examination is performed. Typical findings include identification of a double cervix, irregular uterine contour, and possibly a vaginal septum. However, a woman with a complete septate uterus can also have a double cervix and a longitudinal vaginal septum. Because of the reproductive implications associated with these anomalies, pelvic examination alone should not be relied on to differentiate a didelphyc uterus from other uterine anomalies.

The hysterosalpingogram does not differentiate a uterus didelphus from a complete septum with cervical duplication since two separate uterine cavities, each arising from its own cervix, are present for each. Proper diagnosis requires determination of the uterine fundal contour. An ultrasound may show two widely spaced uterine horns and two distinct cervices (Fig. 4). However, no set of criteria has

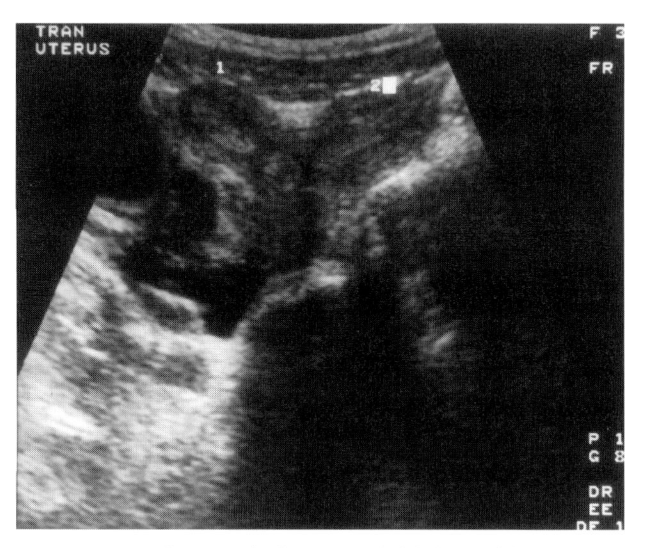

FIGURE 4 Ultrasound of a uterus didelphus. Although two cervices are not seen, two separate uterine horns (1 and 2) are clearly identified.

proven to be completely accurate. Alternately, MRI appears to be a highly accurate tool to evaluate the uterine fundus. When the diagnosis remains uncertain after a noninvasive evaluation, laparoscopy may be necessary.

Infertility is not associated with a didelphyc uterus. However, uterus didelphus is associated with a higher than normal incidence of first- and second-trimester pregnancy losses. Because of these complications, pregnant women with a didelphyc uterus should be considered a high-risk group. Overall, a live birth rate of 50–75% has been reported. Surgery is not recommended since there is no improvement in outcome following surgery.

D. Bicornuate Uterus

A bicornuate uterus is characterized by fusion of the uterine horns caudal to the normal site, resulting in a "rabbit ear" external appearance. A vaginal septum is seen in <5% of women with a bicornuate uterus, and urinary tract anomalies are found in approximately 10%. A bicornuate uterus may be suspected when two palpable horns are noted on pelvic examination, although examination can be confused with uterine fibroids. A bicornuate uterus may be present with a history of recurrent pregnancy losses, preterm labor, or a history of fetal malpresentations.

Hysterosalpingogram demonstrates separation of the uterine horns when a bicornuate uterus is present. However, the HSG does not identify the external uterine shape, and the image of a bicornuate and subseptate uterus may be exactly the same (Fig. 5).

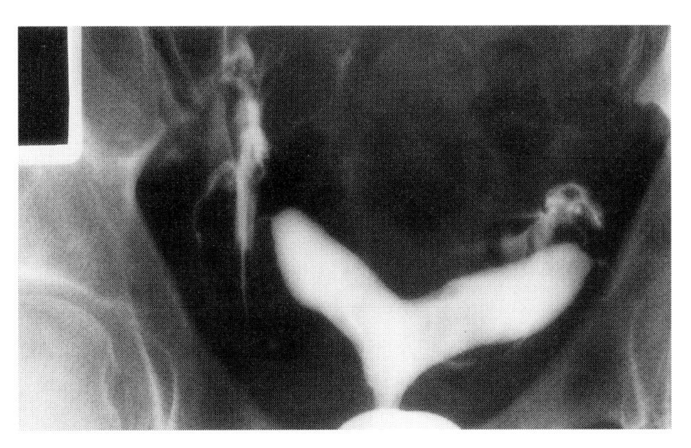

FIGURE 5 HSG of a bicornuate or septate uterus. Definitive diagnosis cannot be made at the time of HSG alone.

In fact, a septate uterus is often incorrectly diagnosed as a bicornuate uterus by HSG.

Visualization of the uterine fundus is necessary to differentiate the bicornuate from the septate uterus. Ultrasound may be helpful, but uterine retroversion or fibroids may lead to an incorrect diagnosis. MRI demonstrates a clear separation of the endometrial cavities in the fundal region, whereas only a low-intensity septum separates the two cavities in the septate uterus. When the diagnosis is uncertain, surgery is still the gold standard used to differentiate bicornuate from septate uterus. Two distinct horns are seen with a bicornuate uterus, and a septate uterus will have a flat fundus.

Primary infertility should rarely, if ever, be an indication for uterine unification surgery with a bicornuate uterus. On the other hand, women with recurrent pregnancy losses or major obstetric complaints may benefit from surgery. Surgery is performed by dividing the uterine fundus transversely, followed by a multiple-layer closure in an anterior–posterior direction. This technique is known as a Straussman metroplasty. Cesarean section is recommended following a Straussman metroplasty because of the risk of uterine rupture during labor at the site of the repair.

The bicornuate uterus does not cause infertility. However, pregnancy complications are common. Overall, a 30% first-trimester pregnancy loss rate is expected, and pregnancy loss may also occur in the second trimester. Preterm labor is common. Pregnancies that progress beyond 26 weeks often require tocolysis, approximately 25% have malpresentation, and as many as 50% require cesarean section. With intensive obstetrical management, a live birth rate with an uncorrected bicornuate uterus is 50–85%. The prognosis for pregnancy is greatly improved following metroplasty. Pregnancy rates range from 75 to 94% following metroplasty, and those who conceive achieve a term delivery rate of 75–85%.

E. Septate Uterus

The septate uterus is the most common congenital uterine anomaly. A septum results from failure of midline resorption of the fused Müllerian ducts and may be complete or partial. A complete septum, present in <10% of those with a septate uterus, divides the entire uterine cavity. Duplication of the cervix and vagina is occasionally seen. A partial septum divides only a portion of the uterine cavity and is far more common. A single fundus is seen with a septate uterus, differentiating the septate from the bicornuate or didelphyc uterus.

In most circumstances, a nonpregnant woman with a septate uterus is completely asymptomatic. On occasion, a septate uterus is found when a longitudinal vaginal septum or cervical duplication is identified during a pelvic examination. However, since most have a normal exam and are asymptomatic, the true incidence of a septate uterus is difficult to determine. Women with a history of recurrent pregnancy wastage, preterm labor, growth restriction, or malpresentation may have a uterine anomaly, including a septate uterus.

The HSG plays an important role in the diagnosis of a septum, but the interpretation may be misleading. Often, the radiologic diagnosis of a bicornuate uterus is incorrectly made. Sonography may allow the differentiation between these two conditions (Fig. 6). Newer techniques such as fluid contrast ultrasound may be a more sensitive test. When these techniques are inconclusive, MRI is helpful. However, if the diagnosis is still in doubt following MRI, laparoscopy demonstrates a flat fundus with a septate uterus as opposed to the characteristic fundal abnormalities seen with other defects of midline fusion.

Treatment of the septate uterus was first reported

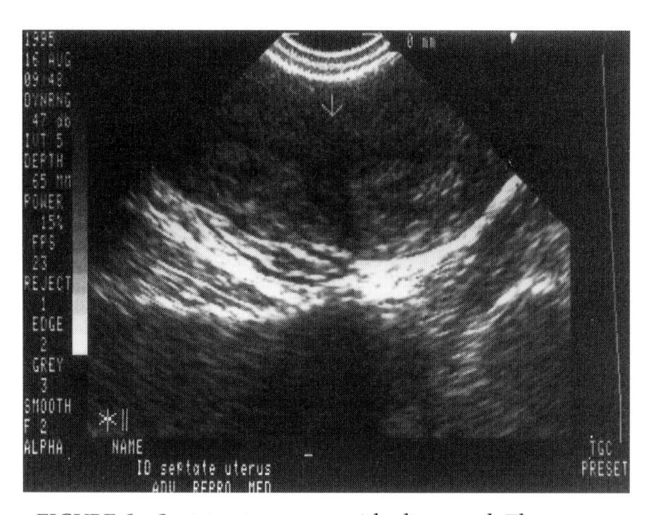

FIGURE 6 Septate uterus seen with ultrasound. The septum (arrow) separates the two uterine cavities.

in 1882 when Schroeder blindly divided a uterine septum with scissors in a woman with a history of pregnancy wastage. The patient conceived within 1 month and delivered at term. Despite this early success, abdominal metroplasty was the first widely accepted surgical approach for correction of the septate uterus. However, abdominal metroplasty has now been replaced by the hysteroscopic approach for the septate uterus.

A modern metroplasty is performed by hysteroscopic resection of the septum with scissors, laser, or a modified resectoscope. With each cut, the septum retracts and the cavity opens. Minimal bleeding is encountered. The resection continues until the fundus has been reached and both ostia are seen. Modifications in the procedure are made in instances of a complete septum and a double cervix since division of the cervix or the internal os may increase the risk of cervical incompetence. In these cases, the septum is perforated above the level of the internal os and resected to the fundus. Healing occurs completely by 2 months.

The incidence of infertility is estimated to be 8% with a septate uterus, similar to the incidence of infertility seen in the general population. On the other hand, pregnancy outcome is poor with a uterine septum. Pregnancy wastage rates of more than 67% have been published. Others have estimated first- and second-trimester loss rates of 40–50%. Live birth rates <30% seem extreme but have been described in some studies. Fortunately, results following hysteroscopic resection of a septate uterus are gratifying. Successful pregnancy rates of 75% or more are expected following surgery.

F. Arcuate Uterus

The arcuate uterus is a mild variant of a septate uterus since the arcuate uterus has a unified fundus. The arcuate uterus is generally an incidental finding at the time of hysteroscopy or HSG and is identified by midline curve indentation of the uterine fundus. Most authorities believe that the arcuate uterus is not the cause of impaired reproduction. The arcuate uterus should be considered a normal uterine variant with no need for therapy.

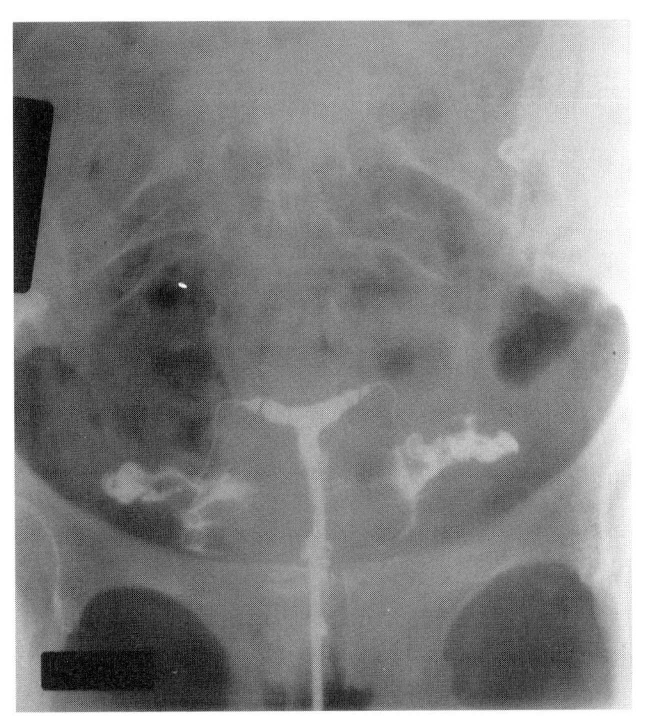

FIGURE 7 T-shaped uterus, typical of DES exposure.

G. DES Related

Exposure of the female fetus to DES can cause significant anomalous development of the uterus. The T-shaped uterus is the most common variant seen (Fig. 7). It is associated with an increased rate of spontaneous pregnancy wastage, preterm deliveries, and ectopic pregnancies. The hysterosalpingogram is the diagnostic test most commonly advised to evaluate a DES-related uterine anomaly. Often, characteristic anomalies are seen despite a negative history of DES exposure.

Women with DES exposure must be monitored closely for evidence of premature dilatation and effacement of the cervix in pregnancy. Cervical cerclage may be helpful in some patients. Surgery has been attempted in an effort to incise constriction rings and septations. However, information about surgical outcome is limited. Therefore, surgery is not advised for correction of DES-associated uterine anomalies to enhance pregnancy outcome.

V. CONCLUSION

Uterine and vaginal anomalies have a profound impact on reproduction. Women with reproductive tract hypoplasia and agenesis are sterile, and those with septate, bicornuate, unicornuate, or didelphyc uterine anomalies may be at higher risk for spontaneous abortion, preterm labor, and malpresentation. A conclusive evaluation of suspected malformations of the uterus is essential to provide accurate information about reproductive consequences and provide optimal care.

See Also the Following Articles

Cervix; Endometriosis; Fallopian Tube; Sterility; Vagina

Bibliography

American Society for Reproductive Medicine (1988). American Fertility Society classification of Müllerian anomalies. *Fertil. Steril.* 49, 952.

Golan, A., Langer, R., Bukovsky, I., and Caspi, E. (1989). Congenital anomalies of the mullerian system. *Fertil. Steril.* 51, 747–755.

Hurst, B. S., and Schlaff, W. D. (1993). Congenital anomalies of the uterus. In *Congenital Malformations of the Female Reproductive Tract and Their Treatment* (B. S. Verkauf, Ed.). Appleton & Lange, East Norwalk, CT.

Jones, H. W., Jr., and Rock, J. A. (1983). *Reparative and Constructive Surgery of the Female Generative Tract.* Williams & Wilkins, Baltimore.

Ludmir, J., Samuels, P., Brooks, S., and Mennuti, N. T. (1990). Pregnancy outcome of patients with uncorrected uterine anomalies managed in a high risk obstetric setting. *Obstet. Gynecol.* 75, 906–910.

Rock, J. A. (1996). Surgery for anomalies of the mullerian ducts. In *TeLinde's Operative Gynecology* (J. A. Rock and J. D. Thompson, Eds.), 8th ed. Lippincott-Raven, Philadelphia.

Rock, J. A., and Schlaff, W. D. (1985). The obstetric consequences of uterovaginal anomalies. *Fertil. Steril.* 43, 681–692.

Uterine Contraction

Venu Jain, George R. Saade, and Robert E. Garfield
The University of Texas Medical Branch, Galveston

GLOSSARY

action potential The rapid and transient change in membrane potential which occurs in nerve, muscle, and other excitable tissue in response to excitation.

calcium pump A protein structure which utilizes metabolic energy from adenosine triphosphate to actively transport calcium against a chemical gradient across a biological membrane.

gap junction An intercellular junction consisting of nonselective ion channels which produce electrical and metabolic coupling between adjacent cells.

ion channel A protein structure which when open allows passive flux of ions (specific or nonspecific) across a biological membrane.

myofilaments The contractile filaments of muscle cells: Thick filaments contain myosin, whereas the thin filaments contain actin.

myometrium The muscular wall of the uterus.

parturition The process of delivery of the products of conception including the fetus and the placenta.

smooth muscle An involuntary muscle which is characterized by the absence of striations due to random arrangement of its contractile proteins, slow and prolonged contractions, and spontaneous or agonist-induced contractile activity.

Uterine contraction is necessary for expulsion of the products of conception at the term of pregnancy. The preponderance of smooth muscle in the uterine wall enables the uterus to perform this function. However, unlike other smooth muscle organs such as the gastrointestinal tract, blood vessels, and airways, the uterus exhibits significant contractile activity only for a brief period of time, i.e., during labor, and must maintain quiescence during the rest of gestation.

I. INTRODUCTION

The uterus is an important component of the female reproductive tract. It is the site of implantation of the embryo. The fetus continues to develop, grow, and differentiate in the favorable environment of the uterus, drawing its nutrition from the rich maternal blood supply until the time when it is able to survive in the external environment. At this stage of gestation, the uterus undergoes the process of parturition, i.e., the delivery of the products of conception. Therefore, contractility of the uterus plays a key role in enabling it to serve a dual function. The uterus is a relatively quiescent organ during pregnancy when it serves the function of harboring the developing fetus. In contrast, at the time of labor, it is converted to a very active and reactive state, characterized by forceful, rhythmic, and synchronous contractility. This change in the uterine contractility is thought to be the result of complex interaction between an array of systems and events.

II. ANATOMICAL BASIS OF UTERINE CONTRACTION

The uterine wall is composed of three layers. The innermost layer, the endometrium, lines the lumen of the organ and consists of columnar epithelium and underlying stromal tissue. The middle layer, the myometrium, which consists predominantly of smooth muscle cells, also contains blood and lymph vessels, nerves, immune cells, and connective tissue. The outer layer, the serosa, is a thin layer which covers most of the uterus and is composed of mesothelial cells.

A. Structure of the Myometrium

The myometrium of most species consists of two distinct layers of smooth muscle with a vascular zone in between. The outer longitudinal and the inner circular layers have muscle fibers arranged parallel to or concentrically around the long axis of the uterus, respectively. The spindle-shaped smooth muscle cells are embedded in a connective tissue matrix and arranged into bundles of 10–50 partially overlapping cells. The muscle bundles in the outer longitudinal layer are interconnected, forming a network over the surface of the uterus, whereas those of the inner circular layer are more diffusely arranged.

B. The Uterine Smooth Muscle

The individual smooth muscle cells are the physiological units of uterine contraction. They are small, spindle-shaped cells, 50–800 mm long and 2–10 mm wide. Advancing gestation is accompanied with hypertrophy as well as hyperplasia of the uterine smooth muscle. The increase in size can be up to three- to fivefold by the end of gestation. Structurally, uterine smooth muscle is very similar to other types of visceral smooth muscle. Each cell is bound by a plasma membrane and contains an elongated central nucleus and various cytoplasmic organelles (Fig. 1). The ultrastructure of smooth muscle cells is similar to that of other cell types with the exception of two noteworthy modifications: the contractile apparatus and the sarcoplasmic reticulum.

1. *The Contractile Apparatus*

Contraction in the smooth muscle cells occurs by the interaction of the myofilaments: the thick filaments and the thin filaments. Thick filaments are

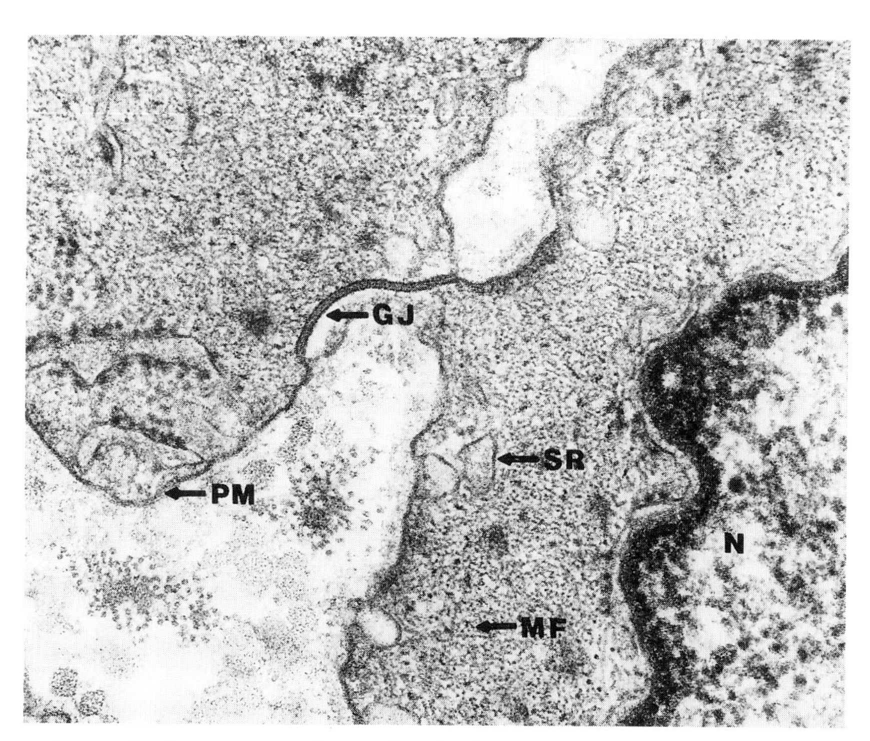

FIGURE 1 Electron micrograph of uterine smooth muscle. The figure shows the apposing plasma membranes (PM) of two myometrial cells forming a gap junction (GJ). Enclosed by the plasma membrane is the nucleus (N) and the cytoplasmic organelles including sarcoplasmic reticulum (SR) and myofilaments (MF) (magnification, ×66,000).

composed of myosin, whereas the thin filaments are composed predominantly of actin but also contain other proteins such as tropomyosin, caldesmone, and filamin. The arrangement of myofilaments is random; hence, development of force in smooth muscle is slow as opposed to striated (skeletal or cardiac) muscle cells in which force is generated rapidly in regularly arranged bundles of myofilaments. The thick filaments form bridges with thin filaments, which in turn are attached to dense bodies, forming longitudinal or oblique fibrils in the smooth muscle cell. The dense bodies contain α-actinin, a protein which binds vinculin, which in turn binds actin in thin filaments. At the ends and sides of the muscle cells, the thin filaments are inserted into dense cytoplasmic patches formed by association of dense bodies with the plasma membrane. The cytoplasm also contains intermediate filaments which are composed of desmin. These are believed to form a cytoskeleton which supports the contractile apparatus.

2. *Sarcoplasmic Reticulum*

The sarcoplasmic reticulum is a network of tubules and sacs in the cytoplasm of the smooth muscle cells. It serves as a storage site for calcium, which plays a key role in muscle contraction. In contrast to striated muscle, however, the sarcoplasmic reticulum of smooth muscle is not very well developed.

C. Cell-to-Cell Contacts

Neighboring cells within muscle bundles come in close apposition in certain specialized regions of their plasma membranes forming cell-to-cell contacts. Except for these specialized regions, smooth muscle cells within the muscle bundles are separated from each other by a space $\approx$ 50–100 nm. These contacts, which are functionally important in the control of smooth muscle contractility, have been termed gap junctions (Fig. 1).

Gap junctions are modifications of the apposing plasma membranes of the adjacent cells which serve to couple them electrically and metabolically. They are present ubiquitously in all types of tissues. Under electron microscope, they appear in regions of close apposition between cells as zones of paired parallel membranes of unusually smooth outline separated by a narrow space of about 2 or 3 nm (the gap). Each gap junction is composed of a few thousand channels and each channel is formed by two symmetrically aligned hemichannels, i.e., connexons, one in each of the two apposing cell membranes. A connexon is formed by six connexin proteins, the major connexin in the uterus being connexin 43. The channel formed by the connexons allows diffusion of molecules less than $\approx$1 kDa. Thus, inorganic ions and small molecules can readily pass from one cell to the other.

III. MECHANISM OF UTERINE CONTRACTION

The uterine smooth muscle, like other types of visceral smooth muscle, is spontaneously active. The uterus consists of billions of smooth muscle cells. During pregnancy, the contractile activity of these cells is poorly coordinated, resulting in ineffective uterine contractions and, consequently, relative uterine quiescence. However, at the time of labor, enhancement of cell-to-cell coupling results in formation of a functional syncitium. As a result, the uterine contractions are well synchronized and effective in expulsion of the products of conception. The sequence of events that result in contraction of the uterus is described in detail.

A. Spontaneous Electrical Activity of Uterine Smooth Muscle

The contractile activity in the uterus is a direct consequence of the electrical activity in the uterine smooth muscle cells. During periods of quiescence, the plasma membrane of the smooth muscle cell maintains a potential difference across it of -40 to -70 mV, inside negative (resting membrane potential). This drop in potential across the membrane is the result of a differential permeability of the plasma membrane to ions, especially potassium (K^+), sodium (Na^+), and calcium (Ca^{2+}). K^+ is present in higher concentrations intracellularly, whereas Na^+ and Ca^{2+} are present in higher concentrations in the extracellular space. The membrane is normally more permeable to K^+ which moves down the electrochemical gradient, i.e., from the intracellular space to the extracellular space, thereby creating a negative potential inside the cell.

The uterine smooth muscle displays spontaneous electrical activity characterized by cyclic depolarization and repolarization of the plasma membrane and termed action potentials. The electrical activity in the longitudinal muscle is in form of intermittent bursts of spike action potentials. The circular muscle displays single plateau-type action potentials during pregnancy, changing to action potentials similar to those in the longitudinal muscle at the time of labor. The action potentials result from voltage- and time-dependent changes in the membrane permeability to the different ions (Fig. 2). Depolarization of the membrane is mainly due to an increased permeability to Ca^{2+} and, to a lesser extent, to Na^+. Both ions have higher concentrations in the extracellular space and hence move intracellularly, making the membrane potential more positive. The membrane repolarizes by increasing the permeability to K^+. The resultant outward movement of K^+ consists of a voltage-dependent (fast) and a Ca^{2+}-activated (slow) component. It may be noted that the movement of only a few ions across the plasma membrane is sufficient to change the membrane potential without significant changes in the intracellular/extracellular ion concentrations.

B. Myometrial Pacemakers

Although contractile activity is an inherent property of all uterine smooth muscle cells, some of them are specialized for pacesetting the uterine contractions. These are known as the pacemaker cells. Pacemaker regions are 2–4 mm in size and contain more than one of these specialized cells. However, unlike the pacemaker of the heart, the sinuauricular node, which is a well-defined region, the pacemakers in the uterus are not represented by a discreet anatomical location. Any myometrial cell is capable of assuming the role of a pacemaker; hence, pacemaker regions can shift from one site to another. In animal studies, some authors have suggested that pacemaker potentials originate at a specific location in the uterus. The majority of investigators, however, have failed to identify a constant, localized pacemaker in the uterus.

The pacemaker cells have a lower resting membrane potential rendering them more excitable. Their

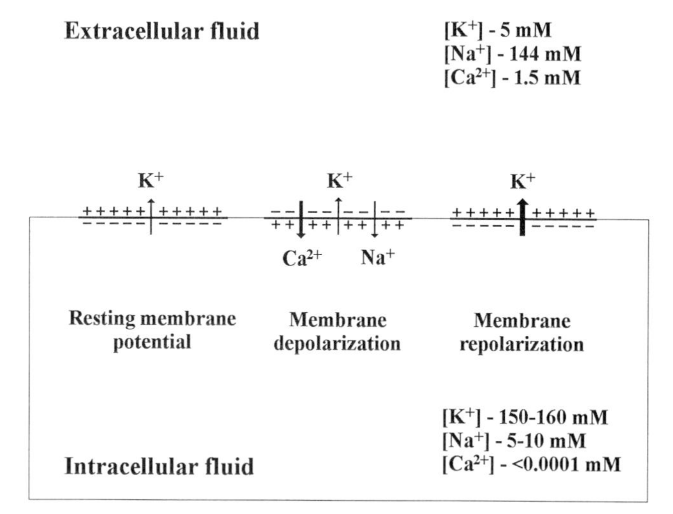

FIGURE 2 Schematic illustration of the mechanism of generation of action potential in smooth muscle cell membrane. The approximate ionic concentrations of the cations important in this process are illustrated in the extracellular and intracellular fluid. Resting membrane potential is maintained by higher permeability to K^+. Opening of Ca^{2+} and, to a lesser extent, Na^+ channels causes membrane depolarization. A further increase in open probability of K^+ channels causes membrane repolarization. Note: Small fluxes of ions across the membrane are sufficient to cause the changes in membrane potential; no significant change in the ionic composition of intracellular/extracellular fluids occurs.

electrical activity is characterized by two types of subthreshold oscillations in their membrane potential, i.e., pacesetter potentials and prepotentials. Pacesetter potentials consist of a burst of spike discharges 3–20 mV in amplitude and 10–60 sec in duration. Prepotentials, which have an amplitude of 2–7 mV and a duration of 0.8–1.2 sec, trigger an action potential within a spike burst. Thus, the frequencies of the pacesetter potentials and the prepotentials determine the burst discharge frequency and the spike frequency, respectively.

C. Propagation of Electrical Activity and the Role of Gap Junctions

Electrical activity propagates from the active regions of the tissue to the surrounding resting tissue by local circuit current flow. Although the individual smooth muscle cell is the unit for excitation of the myometrium, a bundle of smooth muscle cells is the unit of propagation of the electrical stimulus. This is supported by the fact that the length constant for the spatial decay of electronic potentials along the muscle bundles is much larger than the length of a muscle cell. Gap junctions, which couple the neighboring smooth muscle cells electrically by allowing free movement of current-carrying ions between the cells, are the sites of intercellular propagation of action potentials. In addition to ions, the large pore size of the gap junction channels (connexons) also allows small molecules to be exchanged between cells, thereby coupling them metabolically.

The gap junction channels exhibit rapid transformations between open and closed states. The functionality of the gap junctions, i.e., propagation of action potentials, is dependent on the open probability of these channels. Increases in intracellular cAMP, Ca^{2+}, or pH cause a decrease in open probability and hence cause uncoupling of the cells. Hence, the number of gap junctions and their permeability determines the efficiency of electrical and metabolic coupling of cells in the myometrium and, in effect, the speed of conduction of action potentials.

Conduction velocity of action potentials in the myometrium varies in different species and at different stages of gestation in the same species. Progression of gestation is associated with an increase in the conduction velocity. Also, the values for conduction velocity and the length constant are higher along the long axis of the smooth muscle bundle, reflecting a lower resistance due to fewer cells and cell junctions per unit length compared to the tangential directions.

Thus, the electrical activity spreads from cell to cell, along the muscle bundles which are in turn linked with other bundles. There is also electrical coupling between the longitudinal and the circular muscle layers. Hence, gap junctions are of singular importance in cell-to-cell communication in the uterus, enabling it to function as a syncitium, with the efficiency of this syncitium being dependent on the number of gap junctions present in the myometrium.

D. Excitation–Contraction Coupling

An increase in intracellular Ca^{2+} forms the underlying basis for excitation–contraction coupling in all types of contractile cells, including the smooth muscle cells. In the resting stage, the intracellular Ca^{2+} in the smooth muscle cell is below 10^{-7} M. Contraction is associated with an increase in intracellular Ca^{2+} above 10^{-7} M. The main source of this increase in intracellular Ca^{2+} is the extracellular space which has a Ca^{2+} concentration of $\approx 10^{-3}$ M. This huge chemical gradient is maintained by a low permeability of the plasma membrane to Ca^{2+} as well as efficient Ca^{2+} sequestration mechanisms in the cell.

Contraction of the smooth muscle is preceded by an increase in free intracellular Ca^{2+} levels. Ca^{2+} combines with a cytoplasmic Ca^{2+}-binding protein, calmodulin. Ca^{2+}–calmodulin forms an active complex with myosin light-chain kinase (MLCK) and this complex phosphorylates the myosin light chain. The myosin Mg^{2+}-ATPase can now hydrolyze ATP. This causes cycling of cross-bridges between myosin and actin, leading to sliding of the thin filaments over the thick filaments and contraction of the smooth muscle cell. Calmodulin is probably the main Ca^{2+} receptor of the contractile proteins, but there is evidence that Ca^{2+} can also directly activate the MLCK. In addition, MLCK can be phosphorylated by various protein kinases like cyclic adenosine monophosphate (cAMP)- and cyclic guanosine monophosphate (cGMP)-dependent protein kinases, Ca^{2+}-calmodulin-dependent protein kinase II, and protein kinase C. Phosphorylated MLCK requires a higher concentration of Ca^{2+}–calmodulin for its activation. Caldesmon and calponin, intrinsic proteins of thin filament, also have an inhibitory effect on myosin Mg^{2+}-ATPase.

Increase in intracellular Ca^{2+} can occur by two mechanisms, i.e., electromechanical coupling and pharmacomechanical coupling (Fig. 3). In electromechanical coupling, membrane depolarization associated with action potentials causes influx of Ca^{2+} through voltage-gated L-type Ca^{2+} channels. This is the main mechanism operative in the myometrium. However, unique to the smooth muscle is the ability to respond to chemical stimuli with contraction without a change in its membrane potential. This phenomenon is termed pharmacomechanical coupling. This mechanism involves an increase in intracellular Ca^{2+} by influx of calcium through receptor-operated Ca^{2+} channels or release of Ca^{2+} from the intracellular stores (mainly sarcoplasmic reticulum) mediated by a receptor-operated increase in inositol triphosphate (IP_3). An increase in the sensitivity of the contractile apparatus to calcium is another form of pharmacomechanical coupling.

During the repolarization phase of action potential, the membrane conductance to calcium is decreased, thereby restricting the Ca^{2+} influx. Now the calcium sequestration mechanisms of the cell become

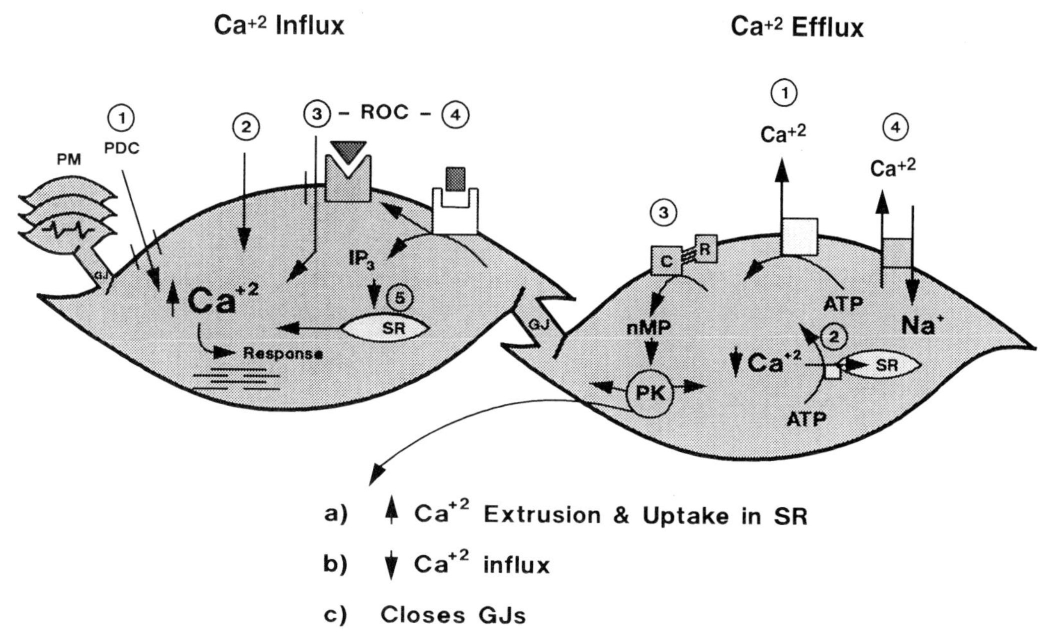

FIGURE 3 Diagram illustrating the relationship between gap junctions (GJ) and calcium regulatory systems in the uterine smooth muscle cells. The influx and efflux pathways are shown in two separate cells coupled together via gap junctions and to a pacemaker (PM). Although both pathways exist within every myometrial cell, they are illustrated separately to emphasize the importance of gap junctions. Pathways for Ca^{2+} influx [1, potential-dependent channels (PDC); 2, passive influx; 3, receptor-operated channels; and 4, receptor-operated IP_3 generation and release from sarcoplasmic reticulum (SR)] cause an increase in intracellular Ca^{2+} and contraction, whereas pathways for Ca^{2+} efflux [1, sarcolemmal Ca^{2+} pump; 2, sarcoplasmic reticulum Ca^{2+} pump; 3, receptor (R = C)-mediated cyclic nucleotide (nMP, cAMP, or cGMP) generation and protein kinase (PK) interaction; and 4, sarcolemmal Na^+/Ca^{2+} exchanger] inhibit the contraction as well as close the gap junctions. The most important mechanism involved in contraction is opening of PDCs during the passage of an action potential. Gap junctions are involved in regulating both contraction and inhibition. The agents that affect contractility may modulate any of the pathways but generally their actions are superimposed on the ability of action potentials to propagate between cells [reprinted with permission from R. E. Garfield and C. Yallampalli, Structure and function of uterine muscle, In *Cambridge Reviews in Human Reproduction* (T. Chard and J. G. Grudzinskas, Eds.), p. 77. Cambridge Univ. Press, Cambridge, UK, 1994].

the determinants of the intracellular Ca^{2+} levels. Intracellular Ca^{2+} is lowered to the resting level by (i) extrusion of Ca^{2+} from the cell by the sarcolemmal Ca^{2+} pump and the Na^+/Ca^{2+} exchanger (which utilizes the inward gradient of Na^+ to remove intracellular Ca^{2+}) and (ii) reuptake of calcium into the sarcoplasmic reticulum (SR) by the SR Ca^{2+} pump. Decrease in intracellular calcium leads to dephosphorylation of the myosin light chain by the myosin light-chain phosphatase, thereby causing relaxation.

A single action potential can generate a twitch contraction. Repetitive discharge of action potentials causes more pronounced contraction because of an increase in the intracellular calcium as well as summation of the twitch contractions. At an action potential discharge rate $\approx$1 cycle per second, a fused tetanic type of contraction is produced. Synchronized stimulation of large areas of the myometrium can increase the force of contraction without an increase in the spike frequency.

E. Transmission of Force

The force generated during uterine contraction is a result of contraction of the individual smooth muscle cells. The tension in the myofilaments is transmitted to the wall of the cell through the dense bodies which form dense bands on plasma membrane. This leads to shortening of the cell. Cell-to-cell transmission of force occurs through cell-to-cell interaction. Force is also transmitted to the connective tissue stroma. Thus, the synchronized contraction of the multitude of uterine smooth muscle cells (estimated to be billions) produces a decrease in uterine lumen. It has been suggested that spread of electrical activity is from the cervical end prior to labor and from the ovarian end at the time of labor. If such a directionality exists, it may be useful in the expulsion of the products of conception out of the uterus during labor while helping to maintain them *in utero* prior to labor.

IV. MODULATION OF UTERINE CONTRACTILITY

The change in the uterine function from one of harboring and nurturing the growing fetus to that of contracting vigorously to expel the products of conception involves a major shift in its contractile activity. During pregnancy, the uterine smooth muscle is relatively quiescent, displaying weak, localized, and poorly coordinated contractions. In contrast, during parturition, the contractions are forceful, sustained, regular, and well synchronized. A variety of intrinsic as well as extrinsic factors are involved in this significant change that occurs over a relatively short period of time. For sake of discussion, these different factors, though acting in concert, will be classified into three groups: myogenic, neurogenic, and hormonal.

A. Myogenic Factors

These are the intrinsic properties of the uterine smooth muscle which include its excitability, ability to contract spontaneously, and ability to propagate the electrical activity. Some of the important changes that occur toward the end of pregnancy and which are believed to be causal in initiating the process of labor are an increase in the number of gap junctions, decreased production of nitric oxide, and stretch of the myometrium. In the following sections, we will elaborate on each of these.

1. Gap Junctions

The magnitude of uterine contractions is dependent on the total number of simultaneously and synchronously active smooth muscle cells. Propagation of the pacemaker electrical activity from cell to cell in the myometrium is essential for achieving the synchronicity of contraction of the smooth muscle cells. As previously discussed, gap junctions are important for cell-to-cell coupling, enabling the propagation of the action potentials across the myometrium.

Studies in various species have shown that gap junctions are either absent or present in very low numbers throughout most of the pregnancy but increase tremendously at the time of labor. Thus, a low level of cell-to-cell communication during pregnancy keeps the uterus quiescent, whereas increased efficiency of communication due to an increase in gap junctions enhances uterine contractility.

Coupling of myometrial cells by gap junctions can be regulated at various levels: (i) rate of synthesis and assembly of gap junction protein (connexin), (ii) permeability of gap junction channels (connexons), and (iii) rate of degradation of gap junctions (Fig. 4). Change in gap junction permeability can be achieved rapidly, over a period of seconds, whereas the change in the number by increased synthesis/assembly or decreased degradation takes longer (12–24 hr). These are important mechanisms by which some of the neurohumoral factors modulate uterine contractility.

2. *Nitric Oxide*

First discovered in the studies on vasculature as an endothelium-derived relaxing factor and a potent relaxant of the vascular smooth muscle, it has since been identified as an important regulator of uterine contractility. Nitric oxide is a small, electroneutral lipophilic molecule which can rapidly diffuse through the aqueous media and plasma membranes of biological systems. It has a short half-life of a few seconds and a high affinity for the heme iron. It is synthesized from L-arginine by the nitric oxide synthase (NOS). The constitutive isoforms of NOS, which are found in endothelial cells (eNOS) and neurons (bNOS), are dependent on Ca^{2+} for their activity. The inducible isoform (iNOS), which is more ubiquitous, is Ca^{2+} independent because of its inherently tight binding to calmodulin and can synthesize nitric oxide at very high rates. The iNOS is present in a variety of cells in the myometrium, including the uterine smooth muscle and the macrophages.

Nitric oxide binds to the heme iron in the soluble guanylate cyclase, activates it, and increases the pro-

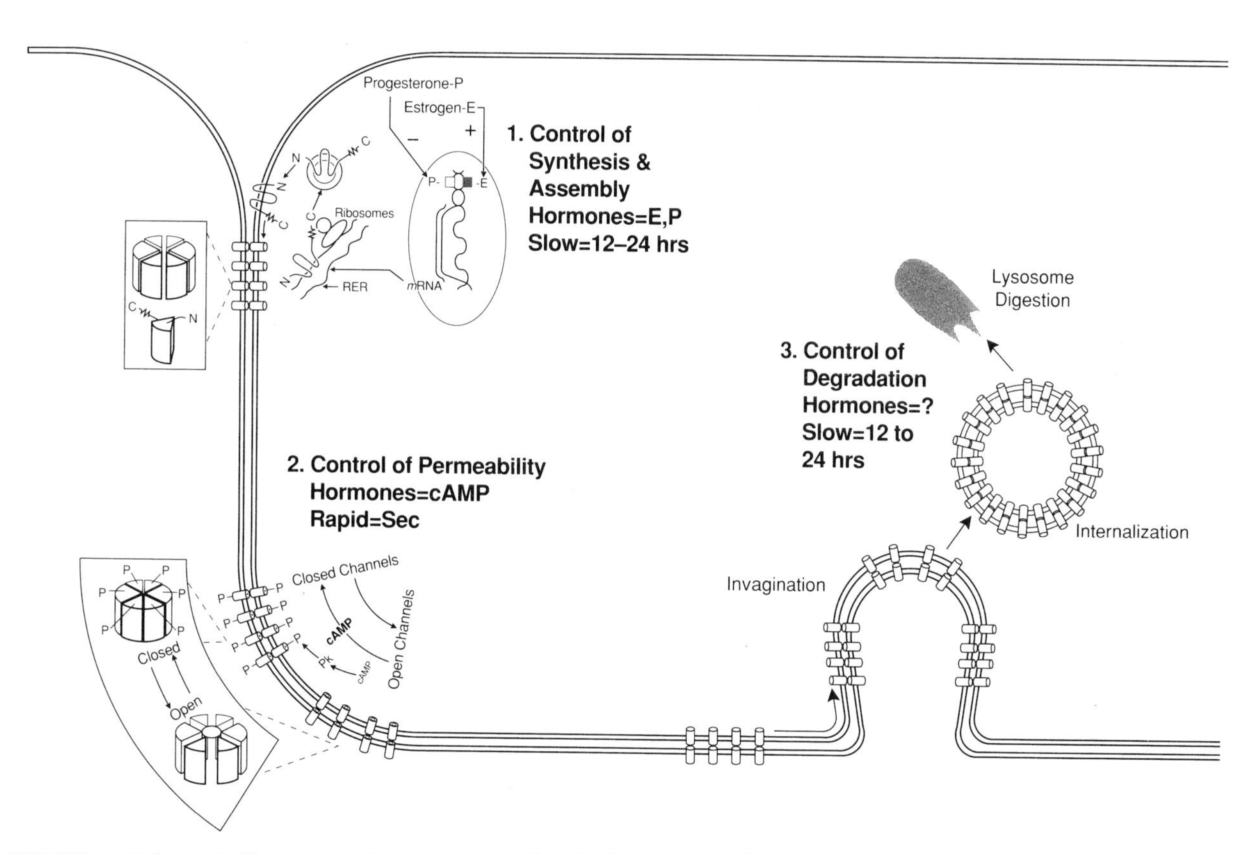

FIGURE 4 Schematic illustration of two myometrial cells showing sites for the control of the synthesis, permeability, and degradation of the junctions. Estrogen (E), progesterone (P), and possibly the prostaglandins regulate site 1. An increase in cAMP closes the junctions (site 2). Degradation of the junctions involves invagination, internalization, and digestion (site 3) [reprinted with permission from E. Garfield and C. Yallampalli, Structure and function of uterine muscle, In *Cambridge Reviews in Human Reproduction* (T. Chard and J. G. Grudzinskas, Eds.), p. 73. Cambridge Univ. Press, Cambridge, UK, 1994].

duction of cGMP. Increase in cGMP activates cGMP-dependent protein kinase (PKG), which inhibits the Ca^{2+} influx from the extracellular space and Ca^{2+} release from the intracellular stores. It also activates the Ca^{2+} pump thereby increasing Ca^{2+} extrusion across the plasma membrane and Ca^{2+} reuptake by the sarcoplasmic reticulum. PKG can also activate Ca^{2+}-activated K^+ channels (resulting in membrane hyperpolarization) and decrease the sensitivity of the contractile apparatus to Ca^{2+}. In addition to activation of guanylate cyclase, nitric oxide can inhibit the smooth muscle by other pathways such as inhibition of voltage-dependent Ca^{2+} channels or direct activation of Ca^{2+}-activated K^+ channels. The end result of all these actions is potent relaxation of the uterine smooth muscle by nitric oxide.

During pregnancy, the myometrium is very sensitive to the relaxant effect of nitric oxide. However, this sensitivity is substantially decreased at the time of labor. In addition, nitric oxide production in the rat myometrium has been shown to be increased during pregnancy compared to the nonpregnant state and decreased during parturition. Hence, the nitric oxide–cGMP system is upregulated in the uterus during pregnancy, thereby maintaining the uterine quiescence, and is downregulated at the time of labor.

3. *Mechanical Stretch*

Progression of gestation is associated with an increase in the intrauterine volume, which is quite substantial in the later stages of pregnancy. Although this is compensated for by myometrial hypertrophy and hyperplasia, it is inevitably associated with mechanical stretching of the uterine smooth muscle. A stretch-sensitive channel has been described in smooth muscles. Stretch causes depolarization of the smooth muscle, increases frequency of action potentials, and enhances contractility. Hence, this could be another mechanism by which uterine contractility is enhanced at the term of gestation.

B. Neurogenic Factors

In most species, the uterus is innervated by postganglionic adrenergic and cholinergic nerve fibers. In addition, peptidergic (vasoactive intestinal peptide, substance P, neuropeptide Y, gastrin-releasing peptide, calcitonin gene-related peptide, and galanin) nerves have been shown to be present in the uterus. However, during pregnancy there is a physiological denervation of the uterus with a disappearance of almost all nerves.

Norepinephrine, the neurotransmittor released by the adrenergic nerves, acts on the β-adrenergic receptor of the uterine smooth muscle. This results in stimulation of the adenylate cyclase, causing an increase in the cAMP levels, which in turn activates the cAMP-dependent protein kinase in a manner similar to cGMP and causes relaxation of the uterine smooth muscle. Additionally, cAMP may also decrease the open probability of gap junction channels. In contrast, activation of the α-adrenergic receptor causes influx of calcium through receptor-operated Ca^{2+} channels as well as IP_3-mediated calcium release from the sarcoplasmic reticulum, thereby causing contraction of the myometrium. In pregnant rats, norepinephrine causes relaxation of longitudinal muscle and contraction of circular muscle except at term when it is inhibitory to both. Cholinergic neurotransmission is excitatory to the uterine smooth muscle. Vasoactive intestinal peptide and calcitonin gene-related peptide inhibit myometrial contractility. Substance P, gastrin-releasing peptide, and galanin stimulate contractility. Neuropeptide Y inhibits neurally evoked contractions. However, it is likely that the effect of these neurotransmittors on uterine contractility is minimal, especially at term when the nerve content of the uterus is sparse. In addition to effecting uterine contractility, these nerves could be important in regulating the uterine and ovarian blood flow.

C. Hormonal Factors

Hormones are very important in modulating uterine contractility by their direct effects on the uterine smooth muscle as well as through indirect effects on the other neurohumoral factors. Important agents and their possible influences on the myometrium are discussed in the following sections.

1. *Steroid Hormones*

Female steroid hormones, namely, estrogens and progesterone, play a key role in regulating uterine contractility. Progesterone suppresses the formation of gap junctions thereby decreasing the coupling between the cells. It may also decrease uterine excitability by hyperpolarizing the smooth muscle plasma membrane. Estrogens, on the other hand, induce the formation of gap junctions, depolarize the membrane, increase prostaglandin production, and enhance the expression of oxytocin and relaxin receptors in the smooth muscle. High progesterone levels during pregnancy are believed to maintain the quiescence of the uterus, whereas a decrease in progesterone and an increase in estrogens at term favor increased contractility.

2. *Prostaglandins*

Prostaglandins markedly enhance uterine contractility by causing membrane depolarization, influx of calcium, release of calcium from intracellular stores, and inhibition of calcium extrusion. They also enhance the reactivity of the uterine smooth muscle to other contractile agents such as oxytocin, increase oxytocin release, inhibit progesterone synthesis, and modify steroid receptors. Prostaglandins may also affect the number of gap junctions in the myometrium. Prostaglandin production by the uterine wall and the products of conception is increased at term in response to estrogens and during parturition due to mechanical factors such as stretch. Prostaglandins are believed to be important in increasing the force of uterine contractions during parturition.

3. *Oxytocin*

Oxytocin is a potent stimulant of the myometrium. It stimulates uterine contractions by increasing the calcium influx through the receptor-operated Ca^{2+} channels, by IP_3-mediated release of Ca^{2+} from sarcoplasmic reticulum, and by inhibiting Ca^{2+} extrusion and reuptake by the Ca^{2+} pump. Oxytocin also increases the sensitivity of the contractile apparatus to Ca^{2+}. In addition, oxytocin stimulates release of arachidonic acid and thus formation of prostaglandins. As pregnancy advances, sensitivity of the uterine smooth muscle to oxytocin is increased due to upregulation of oxytocin receptors. Also, oxytocin release is increased during parturition. Oxytocin is believed to be more important for progression rather than onset of labor.

4. *Endothelins*

Endothelins stimulate uterine contractility by membrane depolarization and influx of Ca^{2+} through voltage-dependent Ca^{2+} channels. Endothelins are synthesized by the placenta and are present in the amniotic fluid in high concentrations. They may have a role in enhancing uterine contractility at term.

5. *Relaxin*

Relaxin inhibits uterine contractility. Its levels are elevated during pregnancy and decrease at the time of labor. It may be important for maintaining uterine quiescence during pregnancy.

V. SUMMARY

The instability of membrane potential of the uterine smooth muscle forms the basis for the spontaneity of contractions in the uterine smooth muscle. A prerequisite for the effectiveness of these contractions is the ability of the myometrium to propagate its electrical activity, wherein lies the importance of gap junctions. The absence of gap junctions, as well as inhibitory influence of nitric oxide, is important in maintaining the uterine quiescence during pregnancy. Parturition is marked by a favorable estrogen:progesterone ratio and an upregulation of gap junctions, rendering the myometrium a functional syncitium. Superimposed upon this enhanced activity and reactivity of the uterus is an array of uterine stimulants such as prostaglandins and oxytocin which, in the absence of uterine relaxants, result in forceful, well-sustained, rhythmic, and regular uterine contractions, bringing new life into the world and marking the end of another cycle of biological perpetuation.

See Also the Following Articles

Labor and Delivery, Human; Oxytocics; Oxytocin; Uterus

Bibliography

Barany, M. (1996). *Biochemistry of Smooth Muscle Contraction*. Academic Press, San Diego, CA.

Carsten, M. E., and Miller, J. D. (1990). *Uterine Function: Molecular and Cellular Aspects*. Plenum, New York.
Csapo, A. I. (1981). Force of labor. In *Principals and Practice of Obstetrics and Perinatology* (L. Iffy and H. A. Kamientzky, Eds.), pp. 761–799 Wiley, New York.
Garfield, R. E. (1984). Myometrial ultrastructure and uterine contractility. In *Uterine Contractility* (S. Bottari, J. P. Thomas, A. Vokaer, and R. Vokaer, Eds.), pp. 81–109. Mason, New York.
Garfield, R. E. (1994). Role of cell-to-cell coupling in control of myometrial contractility and labor. In *Control of Uterine Contractility*(R. E. Garfield and T. N. Tabb, Eds.), pp. 40–81. CRC Press, Boca Raton, FL.
Garfield, R. E., and Hertzgerg, E. L. (1990). Cell-to-cell coupling in myometrium: Emil Bozler's predictions. In *Frontiers in Smooth Muscle Research* (N. Sperelakis and J. D. Wood, Eds.), p. 673. Wiley-Liss, New York.
Garfield, R. E., and Somlyo, A. P. (1985). Structure of smooth muscle. In *Calcium and Contractility: Smooth Muscle* (A. K. Grover and E. E. Daniel, Eds.), pp. 1–36. Humana Press, Clifton, NJ.
Garfield, R. E., Blennerhasset, M. G., and Miller, S. M. (1988). Control of myometrial contractility: Role and regulation of gap junctions. *Oxford Rev. Reprod. Biol.* **10**, 436–490.
Hille, B. (1992). *Ion Channels in Excitable Membranes*, 2nd ed. Sinauer, Sunderland, MA.
Kao, C. Y. (1989). Electrophysiological properties of uterine smooth muscle. In *Biology of Uterus*(R. M. Wynn, and W. P. Jollie, Eds.), 2nd ed., pp. 403–454. Plenum, New York.

Uterus, Human

David A. Grainger
University of Kansas School of Medicine

GLOSSARY

allograft A graft to a different individual of the same species (the human fetus would represent a semiallograft because half of the antigens would be paternally derived).

cytokines A group of substances produced by macrophages, monocytes, and activated T cells that possess diverse functions, including immunomodulation, growth promotion, and inflammatory response. Examples include interleukins-1–6 and interferon-γ.

endometrium Inner lining of the uterus, consisting of glands and stroma, which serves as the site for embryonic implantation and undergoes cyclic changes as directed by ovarian sex steroid hormone secretion.

major histocompatability complex Six major antigens that are markers of "self" derived from a group of genes located on the short arm of chromosome 6 (also know as human leukocyte antigens).

mesonephric (Wollfian) ducts Pair of ducts draining the mesonephroi (or "midkidney") into the urogenital sinus, eventually becoming the epididymis, vas deferens, and ejaculatory duct in the male.

myometrium Smooth muscular wall of the uterus which provides the propulsive forces necessary for parturition.

paramesonephric (Müllerian) ducts Paired ducts developing lateral and parallel to the mesonephric system, eventually becoming the paired Fallopian tubes, crossing medially to become the fused uterovaginal primordium.

The function of the human uterus is twofold. First, it serves as the home for the developing embryo and provides nutrition during gestation. Second, the uterus, because of the large component of smooth muscle, provides the means for parturition. Developmental abnormalities during embryogenesis can impede both the establishment of pregnancy and subsequent successful gestation. The uterus arises from fusion of the paired paramesonephric ducts. The blood supply to the uterus—the uterine artery—is a branch of the anterior division of the internal iliac artery (also referred to as the hypogastric artery). The venous drainage of the uterus is into the vena cava on the right and the renal vein on the left. Structurally, the uterus is divided into three layers: the endometrium, the myometrium, and the serosa. The inner lining, or endometrium, responds to the ovarian hormonal fluctuations and serves as the site for embryo implantation. The myometrium consists of smooth muscle and provides the expulsion forces required for parturition. Smooth muscle contractility is regulated primarily by gradients in calcium concentrations between the extracellular and intracellular spaces. Steroid hormones, particularly progesterone, may influence uterine contractility by uncoupling the excitation–contraction process. Immune mechanisms active in the uterus modulate the implantation of a semiallograft, the embryo. Failure of immune modulation may lead to implantation failure and subsequent abortion. All layers of the uterus are susceptible to common benign conditions, including polyps of the endometrium and smooth muscle tumors of the myometrium.

I. INTRODUCTION AND HISTORICAL PERSPECTIVE

Prior to anatomic dissection, the uterus was described as an "independent animal" (from the Ebers Papyrus, 1550 BC), an "animal within an animal" (Aretaeus, second century AD), and as an organ requiring feeding by male semen to prevent it from going wild (Hippocrates). Following the animalistic concept came the "seven cells doctrine," which partitioned the uterus into seven compartments. Male fetuses were thought to originate in the three cells on the right and female fetuses from the three cells on the left. Hermaphrodites were consigned to the middle cell The first accurate description of the human uterus came from Soranus (first and second century AD). His descriptions—clearly the work of extensive cadaver dissection—are detailed and anatomically correct. Despite his advances in gynecologic descriptions, he retained some interesting misconceptions. He believed (as did Aristotle) that the cotyledons were nipple-like projections used for intrauterine suckling by the developing fetus.

Galen, while not having any particular interest in the female reproductive tract, did have two distinctive observations. First, he believed that the tubes contained lumina through which the ova were conducted to the uterine cavity. Second, he believed that the vessels supplying menstrual blood opened into the crypts of the cotyledons. He erroneously believed that the uterus was multilocular.

Following Galen's death, there was a marked decline in gynecologic scientific investigation. It was not until the establishment of universities in the 1300s that there was a resurgence in scientific curiosity. During the Renaissance work was carried out by such investigators as Mondino dei Luzzi of the University of Bologna and Leonardo da Vinci with his detailed and, for the most part, anatomically correct drawings. Vesalius represents the peak of Renaissance anatomy, and his publication, *De Humani Corporis Fabrica,* changed anatomic science forever. The term "uterus" first appears in the writings of Vesalius. By the end of the sixteenth century, the morphologic characteristics of the human uterus had been accurately described.

II. EMBRYOLOGY OF THE HUMAN UTERUS

The developing human embryo has two sets of genital or sex ducts, regardless of sexual genotype. The mesonephric (Wollfian) ducts play an important role in the development of the male reproductive system, whereas the paramesonephric (Müllerian) ducts have a leading role in female reproductive development. During the fifth and sixth weeks of gestation, both sets of ducts are present (undifferentiated or indifferent stage).

The proximal mesonephric ducts, originally draining urine from the mesonephric kidneys, become the epididymis under the influence of testosterone produced from the testes. The remainder of the duct becomes the vas deferens and ejaculatory duct. In the absence of testosterone (the presence of ovaries), the mesonephric ducts almost completely disappear. Classic studies of the active role of male-determining factors were performed by Alfred Jost and published in the early 1970s.

The paramesonephric ducts develop lateral to the mesonephric ducts and the developing gonad. These ducts result from longitudinal invaginations of the mesothelium just lateral to the mesonephros. As each paramesonephric duct grows caudally, the adjacent mesonephric duct serves as a guide. The funnel-

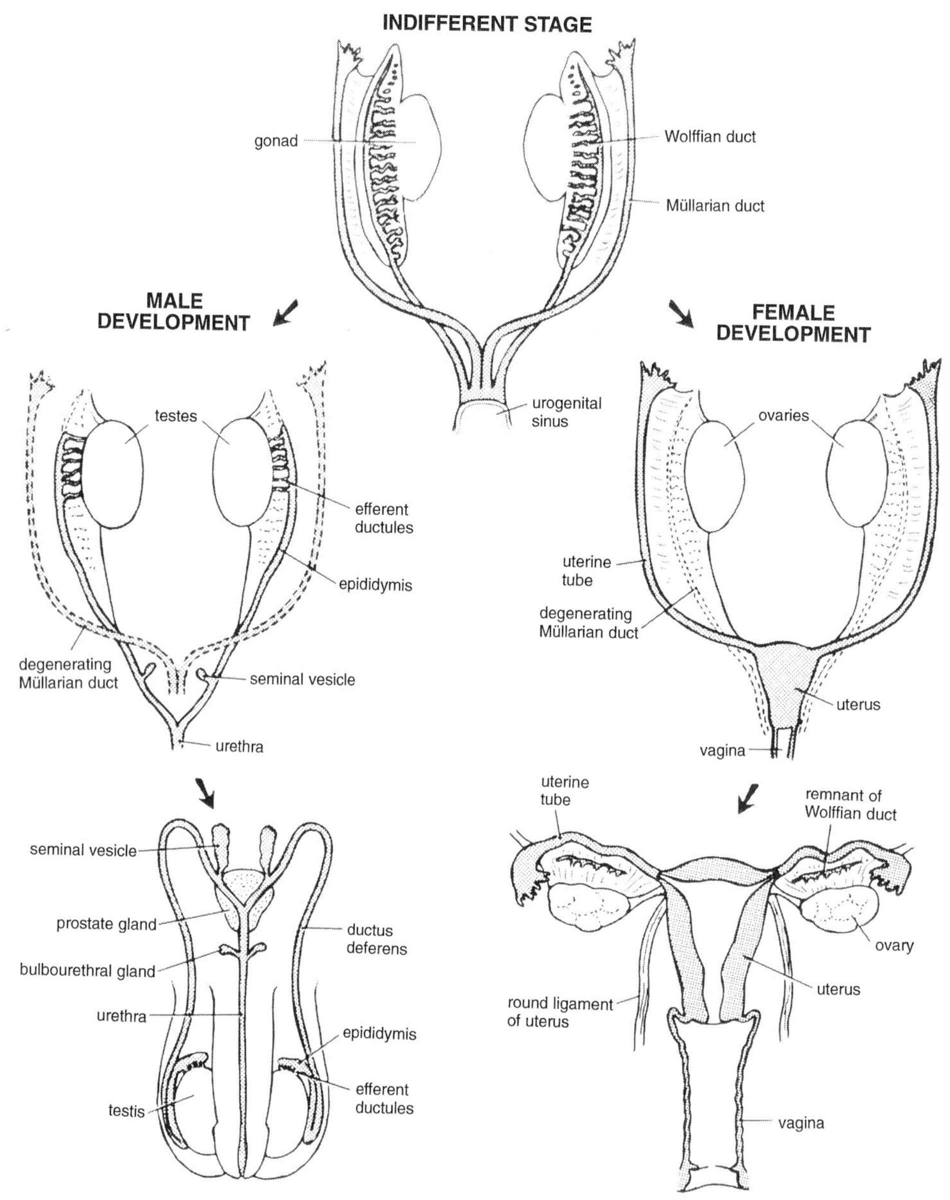

FIGURE 1 Embryonic development of the female and male internal reproductive tracts. Note the parallel and lateral development of the paramesonephric ducts (Müllerian) to the mesonephric ducts (Wolffian). Under the influence of testosterone and Müllerian-inhibiting factor, the Müllerian ducts degenerate and the Wolffian system predominates. In the female, the Müllerian system forms the paired Fallopian tubes, and the medial fusion results in a single uterus (reproduced with permission from Spence and Mason, 1987).

shaped cranial ends form openings into the developing peritoneal cavity and become the eventual fimbriated portion of the Fallopian tubes. The paramesonephric ducts parallel the mesonephric ducts as they pass caudally into the pelvis, eventually crossing ventrally to form a Y-shaped uterovaginal primordium. This tubular structure projects into the dorsal wall of the urogenital sinus, forming the Müllerian (or sinusal) tubercle (Fig. 1).

The caudal fusion of the paramesonephric ducts gives rise to the uterus, cervix, and upper vagina. The endometrial stroma and myometrium are derived from the adjacent splanchnic mesenchyme. The midline fusion of the parmesonephric ducts also brings together two folds of peritoneum, resulting in the formation of the broad ligaments and anterior and posterior cul-de-sacs. Failure of central fusion or resorption of the fused medial portions of these ducts may result in a wide range of uterine anomalies, including uterine didelphys, bicornuate uterus, and septate uterus. Failure of caudal development of the mesonephric ducts is accompanied by the absence of the adjacent paramesonephric duct; hence, unilateral renal agenesis is a frequent accompaniment of a uterus unicornis.

III. ANATOMY/PHYSIOLOGY OF THE UTERUS

The uterine corpus lies between the folds of peritoneum created by the midline fusion of the paremesonephric ducts. These folds, called the broad ligament, enclose laterally the spaces for the vascular supply to the uterus. Anteriorly, the reflection of this peritoneum over the bladder creates the anterior cul-de-sac. Posteriorly, the reflection extends onto the rectum, creating the posterior cul-de-sac. The ligamentous support of the human uterus includes the uterosacral ligaments, the round ligaments, the pubocervical ligaments, the cardinal ligaments, and, to a lesser degree, the uteroovarian ligaments.

The uterus varies in size, shape, location, and structure depending on age and pregnancy. Normally, the uterus is pear shaped, with the narrow end directed downward and backward to form an angle of approximately 90° with the vagina. Figure 2 is a diagrammatic representation of the human uterus, demonstrating the inner mucosal lining (the endometrium), the muscular wall (myometrium), and the outer peritoneal covering (the serosa or perimetrium). The nonpregnant uterus is about 8 cm long, 4 cm wide, and 2 cm thick and weighs about 90 g. It is divided into four parts as shown in Fig. 3: the fundus, the body, the isthmus, and the cervix. The fundus is the rounded part of the uterus that lies above the junction of the Fallopian tubes and the uterus. The body is the main part of the uterus to which the broad ligaments attach. The isthmus lies between the body and the cervix, is about 1 cm long, and expands into the lower uterine segment during gestation. The cervix, anchored by the cardinal ligaments, is the least freely movable part of the uterus and represents the portion of the uterus that projects into the vagina.

The major blood supply of the uterus is the uterine artery, a branch of the anterior division of the internal iliac (hypogastric) artery (Fig. 4). Collateral blood supply to the uterus arise from numerous anastomoses along the lateral border of the uterus with the ovarian arteries (arising from the aorta). The uterine artery perforates into the myometrium, giving rise to a system of smaller arteries (arcuate, radial, and basal, which are nonhormone dependent) and eventually the coiled terminal spiral arterioles which are hormone responsive (Fig. 5). The venous drainage of the uterus follows the arterial supply, with the

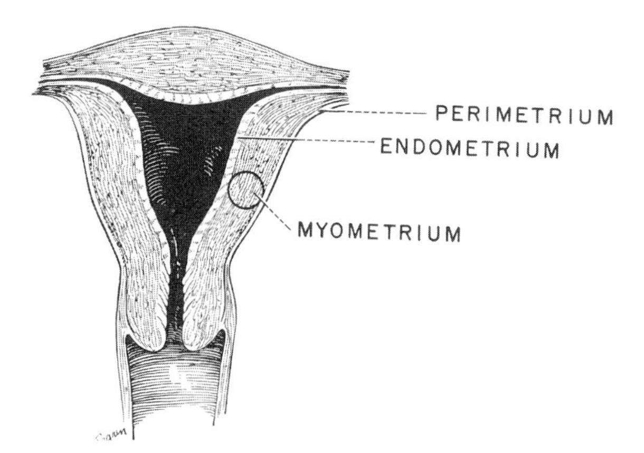

FIGURE 2 Basic structure of the human uterus showing the endometrium, myometrium, and perimetrium or serosa.

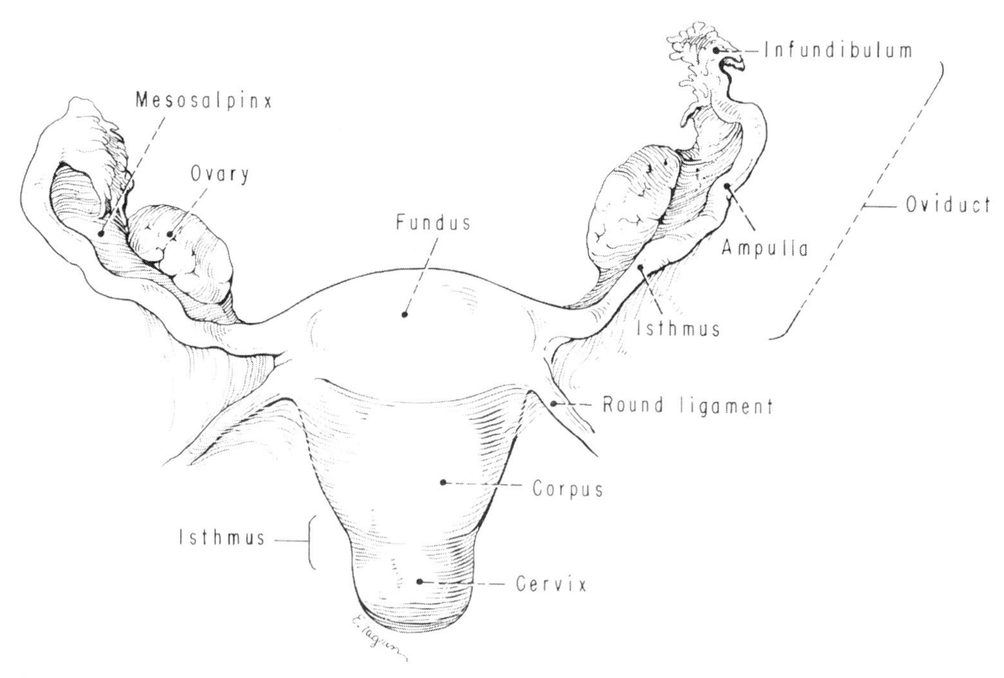

FIGURE 3 The human uterus and adnexa (Fallopian tubes and ovaries) as viewed from the front. The four parts of the uterus (fundus, corpus or body, isthmus, and cervix) are seen (reproduced with permission from Y. Kistner).

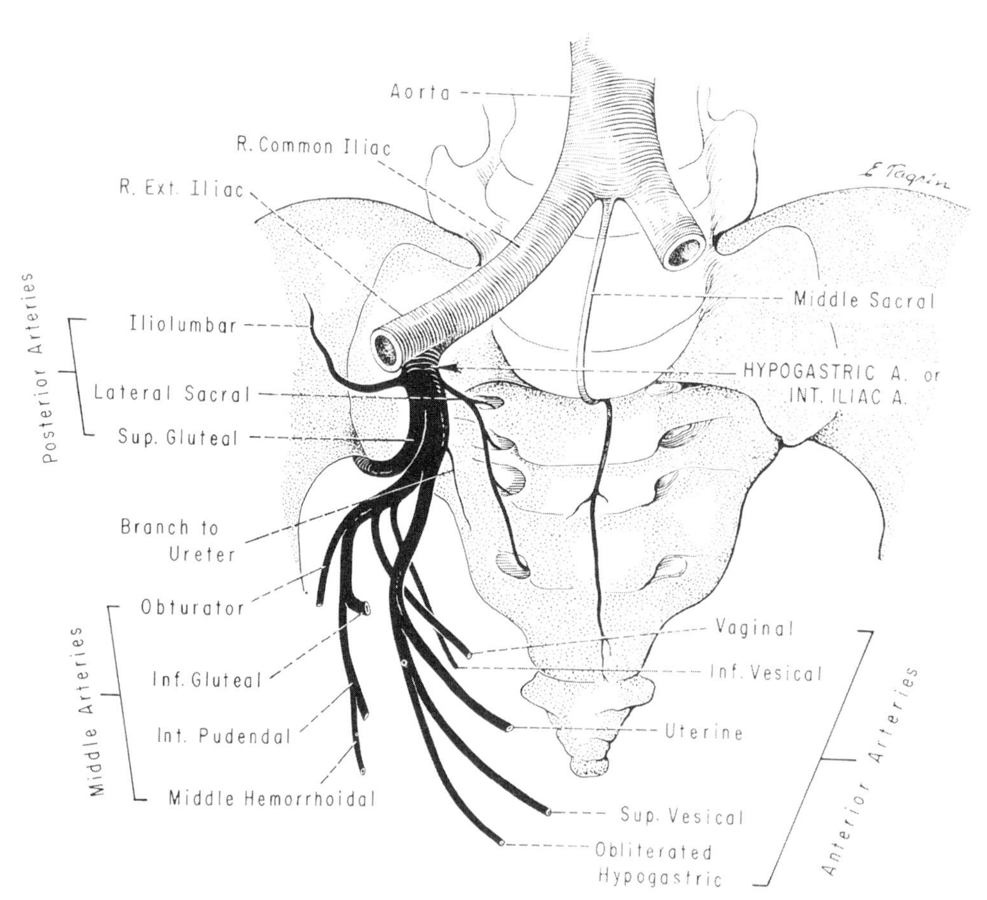

FIGURE 4 Branches of the internal iliac or hypogastric artery. The uterine veins parallel the uterine artery in the pelvis but empty into the vena cava on the right and the renal vein on the left (reproduced with permission from Y. Kistner).

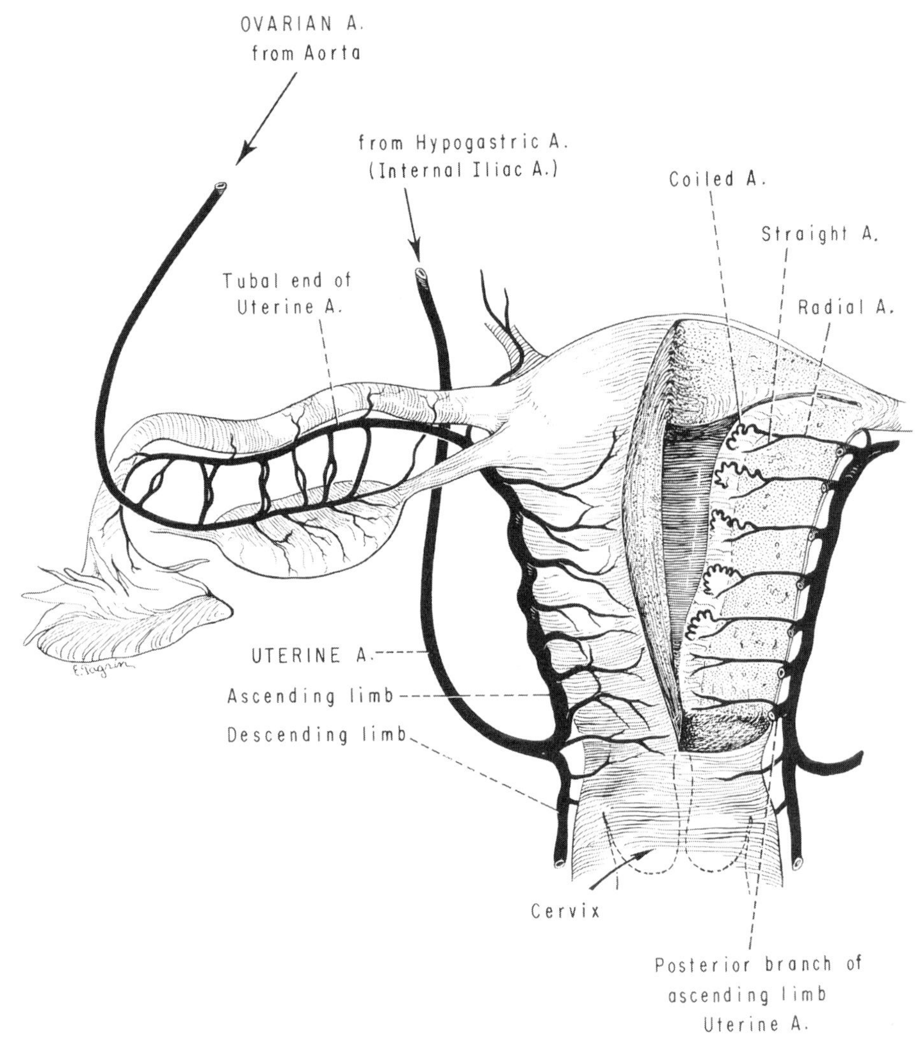

FIGURE 5 Blood supply to the uterus showing the radial arteries, straight arteries, and coiled arteries and arterioles (reproduced with permission from Y. Kistner).

right uterine vein emptying into the vena cava and the left uterine vein emptying into the left renal vein.

The uterus, as with other pelvic organs, is innervated by both sympathetic and parasympathetic fibers. The major nerve bundle supplying the uterus is the presacral plexus. These sympathetic fibers arise from the spinal cord (L1–L4) and divide at the pelvic brim into a left and right bundle. They pass deep into the pelvis and innervate the uterus through the broad and uterosacral ligaments. Parasympathetic fibers arising from S2–S4 also enter the uterus via the pelvic plexus through the broad ligament. Sensory fibers from the uterus are carried along the sympathetic distribution and enter the spinal cord at the L2–L4 level. Uterine discomfort is often perceived in the lower abdominal or hypogastric region. Sensory fibers from the cervix are registered through the sacral sympathetic chain (S2–S4), thus cervical pain is frequently referred to the lumbosacral region.

The morphological changes in the endometrium as a result of ovarian cycling are well documented and consistent. These characteristics include changes in the glands (mitotic activity, nuclear pseudostratification, subnuclear glycogen vacuoles, and secretory products), the stroma (mitoses, edema, and collapse), and the vasculature (decidual reaction and blood lakes). Proliferation of the endometrium occurs under the influence of increasing serum concentrations of estrogen in the 10–12 days following menstruation. After ovulation, the corpus luteum secretes increasing amounts of progesterone, inducing differentiation of the endometrium (secretory changes). The glandular production of proteins, glycoproteins, and proteolgycans is maximal 7 days postovulation. This bathing of the endometrium with secretory products coincides with the arrival and sebsequent implantation of the human embryo. In the absence of implantation (and the lack of β-chorionic gonadotropin production by the trophoblast), the corpus luteum undergoes demise. Decreasing serum concentrations of estrogen and progesterone result in menstruation.

The endometrium not only responds to ovarian hormone secretion but also actively secretes many substances that may modulate growth and differentiation of the epithelium. These substance include lipids (prostaglandins, throboxanes, and leukotrienes), cytokines (interleukins, interferons, and colony-stimulating factor), and peptides (prolactin, relaxin, renin, insulin-like growth factors, transforming growth factors, and many more). Many of these products are also expressed in myometrial cells. This complex autocrine/paracrine and probably endocrine pattern of hormone production is regulated by (and may in turn regulate) sex steroid production by the ovary. The ultrastructure of the endometrium and myometrium is demonstrated in Fig. 6.

The myometrium is composed of smooth muscle cells imbedded in connective tissue in an arrangement that facilitates the transmission of contractile forces generated by each muscle cell. The smooth muscle cells communicate with each other by means of gap junctions, whose functions are regulated by estrogen, progesterone, and prostaglandins. Additionally, the physical stretch of the uterus during pregnancy affects the formation of gap junctions. The disordered array of smooth muscle cells within the myometrium necessitates communication via gap junctions. Indeed, the actin–myosin bundles within smooth muscle cells are randomly aligned, allowing exertion of a pulling force in any direction. These contractions are highly dependent on calcium gradients between the extracellular (high) and intracellular (low) spaces. A 1000-fold difference in calcium concentrations exists between these spaces. This gradient is tightly controlled via a calcium pump–ATPase system (efflux) and calcium channels (influx). The changes in intracellular calcium induced by physical stimuli or contractile agents (such as acetylcholine or norepinephrine) are detected by calmodulin. Activation of calmodulin by calcium binding induces activation of other phophotases and kinases (myosin light-chain kinase) important in smooth muscle contractility. There is evidence that steroid hormones, particularly progesterone, can induce a relative quiescence and uncoupling of the excitation–contraction process.

IV. IMMUNOLOGY OF THE UTERUS

The primary function of the immune system is to protect self from nonself. Considering the pivotal role of the uterus—providing a home for the developing fetus—it is not surprising that there exists an elaborate mechanism to protect the implanting semiallograft. The antigenic determinants of the fetus are derived from both the maternal (self) and the paternal (nonself) genetic contributions. These antigenic differences between mother and fetus have been hypothesized to be critical for successful implantation. Likewise, antigenic similarities between mother and fetus may lead to rejection of the embryo and subsequent miscarriage. The antigenic dissimilarities are thought to be responsible for the elicitation of "blocking antibodies." These antibodies are thought to prevent a cell-mediated immune response against the fetus that could result in abortion. These antibodies may "hide" the fetoplacental antigens from the maternal lymphocytes. Alternatively, these induced blocking antibodies may bind to and inhibit maternal lymphocyte function. The detection of these blocking antibodies is reliant on indirect immunologic testing, primarily the mixed lymphocyte reaction. It is important to consider, however, that these initial immunologic interactions occur at the interface between the embryo and the endometrium.

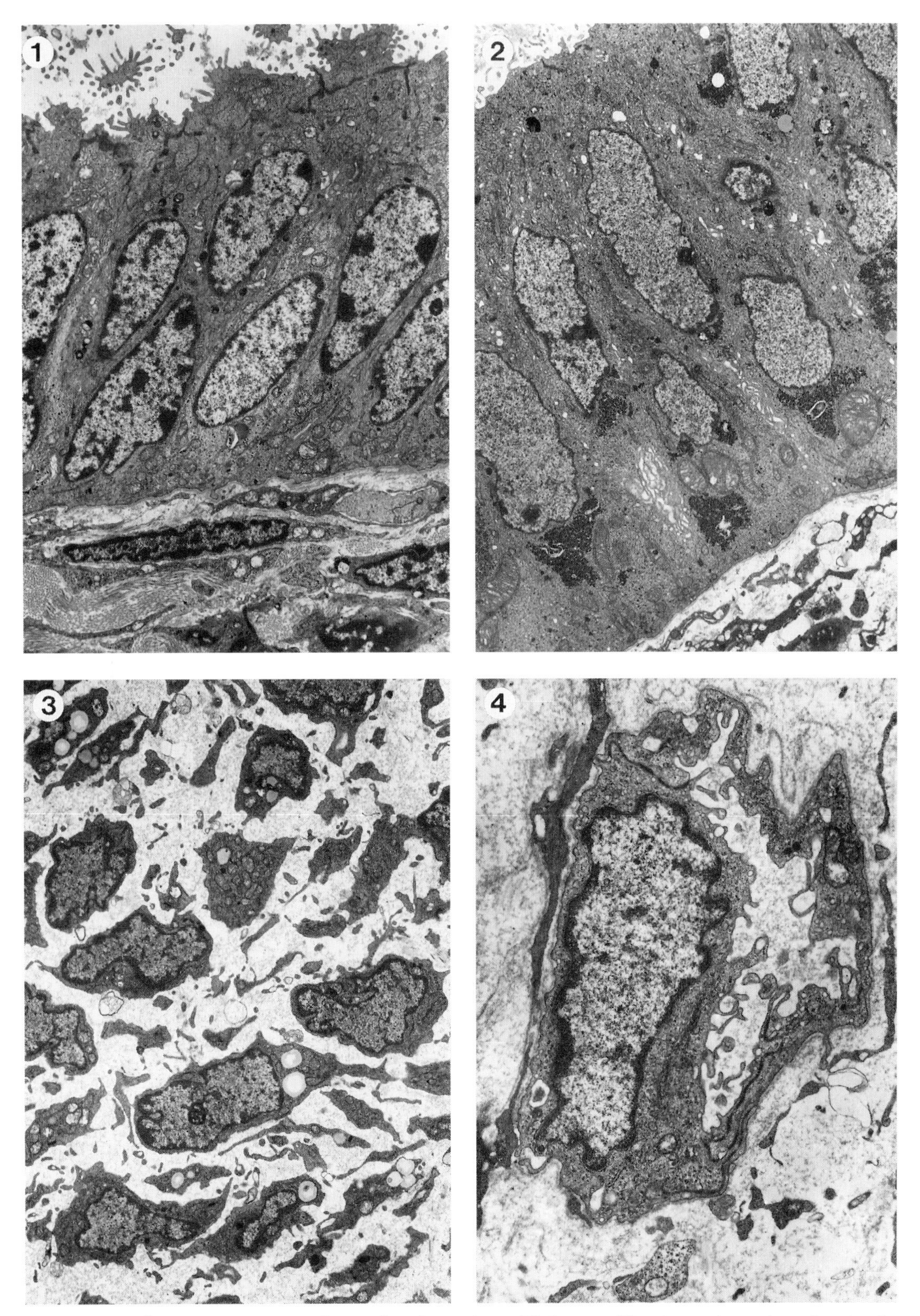

FIGURE 6 Scanning electron microscopy of normal endometrium. (1) Late proliferative endometrium showing large, oval-shaped nuclei with prominent multiple nucleoli (×4838). (2) Early secretory endometrium showing characteristic subnuclear patches of glycogen and giant mitochondria (×4838). (3) Late proliferative stroma showing moderately dense connective tissue with large nuclear/cytoplasmic ratio (×4838). (4) Early secretory capillary showing luminal surface blebbing and vacuole formation (×11,340) (photos courtesy of Wesley Medical Center Electron Microscopy Laboratory, Wichita, KS).

See Also the Following Article

UTERUS, NONHUMAN

Bibliography

Clark, A., and Gall, S. A. (1997). Clinical uses of intravenous immunoglobulin in pregnancy. *Am. J. Obstet. Genecol.* **176**, 241–253.

Giudice, L. C., and Polan, M. L. (1995, May). Growth factors in the endometrium. In *Seminars in Reproductive Endocrinology* (L. Speroff, Ed.).Thieme, New York.

Jost, A., Vigier, B., Prepin, J., and Perchellet, J. P. (1973). Studies on sex differentiation in mammals. *Recent Prog. Horm. Res.* **29**, 1.

Moore, K. L. (1992). *Clinically Oriented Anatomy*, 3rd ed. Williams & Wilkins, Baltimore.

Spence, A. P., and Mason, E. B. (1987). *Human Anatomy and Physiology*, 3rd ed., pp. 809–812. Benjamin/Cummings, Menlo Park, CA.

Wynn, R. M. (1989). The human endometrium: Cyclic and gestational changes. In *Biology of the Uterus* (R. M. Wynn and W. P. Jollie, Eds.), 2nd ed., pp. 289–331. Plenum, New York.

Young, R. C., and Hession, R. O. (1997). Paracrine and intracellular signaling mechanisms of calcium waves in cultured human uterine myocytes. *Obstet. Gynecol.* **90**, 928–932.

Uterus, Nonhuman

Frank F. Bartol
Auburn University

GLOSSARY

conceptus Products of conception, including the embryo and extraembryonic membranes, that become the fetal component of the placenta.

corpus luteum A transient ovarian structure responsible for production of the steroid hormone progesterone.

cytodifferentiation The process through which a cell or tissue acquires its characteristic biochemical phenotype.

embryotrophic Conditions supportive of and/or stimulatory to growth and development of the embryo or conceptus.

morphogenesis The process through which a tissue or organ acquires its characteristic form or structure.

placenta A transient organ of pregnancy in eutherian (placental) mammals that forms as a consequence of the interaction of conceptus and maternal uterine tissues for the purpose of physiological exchange and nutrient transport necessary to support development and growth of the fetus.

The uterus is part of the female internal tubular genital tract which, in mammals, consists of paired oviducts (uterine tubes), the uterus, cervix, and vagina. A developmental specialization of the oviduct, the uterus is a completely mesodermal organ. In adult mammals the uterus consists of a gestational portion, or uterus, and the uterine cervix. Uterus and cervix are continuous with one another and can be distinguished grossly in most species owing to the greater density of cervical connective tissues. Absolute delineation between uterus and cervix requires histological evaluation of these tissues at their junction. Histologically, the uterine wall consists of three tissue layers: (i) endometrium, a serous-type mucosa; (ii) myometrium, which consists of an inner (adluminal) layer of smooth muscle oriented circularly and an outer layer of smooth muscle oriented longitudinally; and (iii) perimetrium, the serous peritoneal coat of the uterus. Functionally, the uterus is an essential reproductive organ. In many mammalian species the uterus is required to support normal ovarian cyclicity. In viviparous species the uterus serves as a conduit for transport and maturation of spermatozoa, provides a unique embryotrophic environment required to support conceptus development and fetal growth throughout gestation, and generates much of the expulsive force required to deliver the fetus and placenta at parturition.

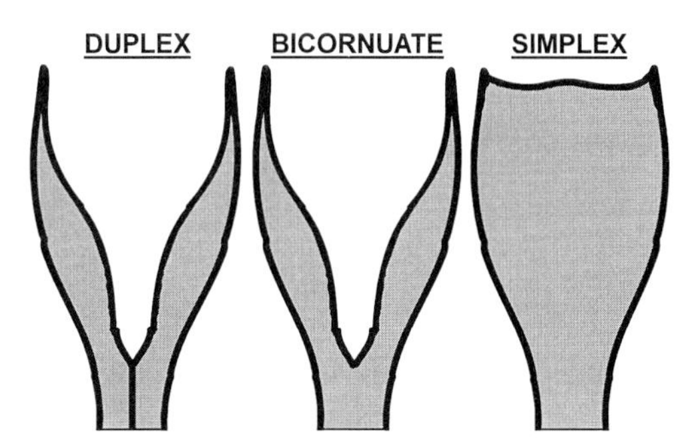

FIGURE 1 Diagrams of the three principle forms of uteri found in mammals (duplex, bicornuate, and simplex). Categories are defined by the degree of paramesonephric duct fusion that occurred during uterine organogenesis. Fusion can be incomplete, partial, or complete. Incomplete fusion produces a duplex uterus in which a septum partitions the two uterine horns or cornua and there is no common uterine lumen. Partial fusion produces a bicornuate uterus in which the cavities of the two ducts merge, forming a common uterine lumen or corpus. Complete fusion produces a simplex uterus, in which there are no uterine cornua.

I. ORGANOGENESIS

A. Embryonic Origins

Timing of uterine development varies considerably among species. However, in all mammals, two internal embryonic genital duct systems arise from the mesonephros (intermediate mesoderm) prior to gonadal differentiation in both males and females. These are the mesonephric (Wolffian) ducts, which give rise to internal genitals in normal males, and the paramesonephric (Müllerian) ducts, which are paired mesodermal tubes that give rise to the infundibula, oviducts, uterus, cervix, and anterior vagina in normal females.

The paramesonephric ducts arise from invaginations of coelomic epithelium on the lateral aspects of the embryonic urogenital ridges. They are found lateral to and closely associated with the mesonephric ducts. In normal females, lacking testes, mesonephric ducts ultimately regress. However, though mechanisms remain unclear, each mesonephric duct affects caudal growth of the paramesonephric duct adjacent to it. Arrest of early mesonephric growth is accompanied by developmental failure in the adjacent paramesonephric duct at the same level. This can produce an abnormal condition known as uterus unicornis (single uterine horn). In normal circumstances the paramesonephric ducts grow caudally, meet, and begin to fuse. The degree of fusion that occurs is species specific and defines gross morphological characteristics of adult uteri, which are diverse.

Paramesonephric duct fusion can be incomplete, partial, or complete. Incomplete fusion involves formation of a septum between the two paramesonephric ducts. In this situation the uterine horns or cornua that result remain separate but are fused longitudinally over a portion of their length. Partial fusion occurs when the cavities of the two ducts merge to form a common lumen called the uterine body or corpus. In this situation uterine horns are joined and intercornual migration of embryos is possible. Complete fusion occurs when the cavities of the two ducts fuse over their entire length. In this situation there are no uterine horns. Types of adult uteri that result from these events are described below and illustrated in Fig. 1.

B. Histogenesis

Organizational events leading to differentiation of paramesonephric duct tissues into endometrial and myometrial tissue layers begin during fetal life but are completed postnatally in most if not all species. Broadly, ordered events characteristic of uterine histogenesis can be characterized to include (i) stratification and spatial organization of subepithelial mesenchymal cells to form endometrial stroma, (ii) organization and differentiation of a prospective myometrium, and (iii) coordinated development of species-specific endometrial histoarchitecture. Genesis and proliferation of coiled tubular uterine glands, characteristic of the endometrium in all mammals, occurs during fetal life in the human uterus but is a postnatal event in most other species studied, including the rat, mouse, sheep, and pig. Development of the inner circular myometrial layer, derived from cells of the intermediate layer of ductal mesenchyme, precedes that of the outer longitudinal myometrium, which is derived from subperimetrial mesenchyme.

C. Organizational Mechanisms

Mechanisms regulating organization and differentiation of uterine tissues are not well defined. Indeed, more may be currently known about what is not required for uterine organogenesis.

1. *Gonad Independence*

It has been known for almost 50 years, since the classic experiments of Alfred Jost showing that female genital phenotype developed in either male or female rabbits castrated during fetal life prior to gonadal differentiation, that prenatal uterine development is an ovary-independent process. Furthermore, ovariectomy at birth does not affect patterns of uterine growth or endometrial histogenesis for a defined period of time postnatally in several species, including mice, rats, sheep, and pigs. Thus, prenatal and early postnatal uterine organizational events appear to be gonad independent.

2. *Molecular and Cellular Mechanisms*

Recent gene targeting or "knockout" (KO) experiments produced mice lacking functional receptors for estrogen (ERKO) or progesterone (PRKO). Both ERKO and PRKO females are born with complete genital tracts. Thus, at least individually, embryonic ER, PR, and, by inference, their cognate ligands are not necessary for uterine organogenesis in this species. Similarly, targeted disruption of the receptor for 1α,25-dihydroxy vitamin D_3, a member of the steroid–thyroid hormone receptor superfamily, did not affect murine uterine organogenesis. Collectively, such observations have been interpreted to suggest that uterine organogenesis during fetal and early neonatal periods is developmentally programmed and steroid receptor independent. In contrast to ERKO mice, homozygous mutant mice lacking insulin-like growth factor-I (IGF-I) are dwarfs. These animals have proportionately hypoplastic, infantile uteri in which myometrial hypoplasia is pronounced. However, uteri in ERKO mice are also hypoplastic. Thus, while not required for organogenesis, functional ER and IGF-I appear to be necessary for normal uterine growth and myometrial development.

Experiments in which the consequences of stromal and epithelial recombination on morphogenesis and cytodifferentiation of uterine and uterovaginal tissues have been studied provide strong support for the idea that mechanisms regulating uterine wall development involve reciprocal paracrine communication between developing epithelium and underlying stroma. Uterine stroma can direct and specify the developmental program of overlying epithelium and ensure establishment of uterine-specific epithelial phenotype. Reciprocally, uterine epithelium is necessary to ensure stromal organization and myometrial growth.

Paracrine effectors of uterine wall development are far from being completely defined. However, in the mouse, stromal–epithelial interactions were recently shown to induce uterine epithelial expression of homeobox-containing genes. Epithelial expression of *Msx1* was dependent on the presence of stromal mesenchyme and its expression of *Wnt-5a*. Similarly, the *Hoxa-11* gene was recently shown to be necessary for differentiation of uterine stromal and glandular epithelial cells during pregnancy. These and related genes, the expression of which appears to be dependent on appropriately timed interactions between developing epithelium and stroma, are among the first to be implicated in maintenance of uterine developmental responsiveness and may prove to be essential for developmental success. Pleiotropic effectors of uterine organogenesis such as stromally derived hepatocyte growth factor, recently shown to effect the organization of uterine epithelium into glandular structures *in vitro*, are being identified rapidly. Much work remains to determine how uterine morphogenesis and cytodifferentiation are orchestrated at cellular and molecular levels and how failure or disruption of these mechanisms may affect the function of adult uterine tissues.

II. CLASSIFICATIONS

The range of uterine shapes, sizes, and histoarchitectural characteristics that can be found among vertebrate species is almost overwhelming. Still, categorization is possible and even necessary to facilitate study and functional comparisons. Consequently, uteri have been classified based on gross morphology, type of placentation supported (deciduate vs adeciduate), and endometrial histology.

A. Gross Morphology

Three major types of mammalian uteri have been defined to include duplex, bicornuate, and simplex forms (Fig. 1). Classifications are based on degree of paramesonephric duct fusion as reflected in gross uterine morphology. The duplex uterus has two separate uterine horns (cornua) and no uterine body (corpus). Cornua can be joined externally at their cervical ends but open independently into two cervical canals. These can join within the cervix but usually open separately into the vagina. The bicornuate uterus consists of two tubes that are joined externally from their cervical ends for a portion of their length and are fused internally to form a common uterine body. Here, the uterine lumen opens into the vagina through a single cervical canal. The simplex uterus consists of a single corpus, reflecting complete ductal fusion. Rudiments of cornual lumina may be found internally in some simplex uteri. The simplex uterus also opens into the vagina through a single cervical canal. Duplex and bicornuate uteri can be further categorized based on uterine horn length (long, medium, or short) and relative size of the corpus (small or large). A great diversity of uterine shapes and sizes can be found across mammalian genera. Table 1 lists

TABLE 1
Types of Uteri in Major Groups of Eutherian Mammals—General Distribution[a]

Long cornua		*Medium cornua*		*Short cornua*	*No cornua*
No corpus (duplex)	*Small corpus (bicornuate)*	*No corpus (duplex)*	*Small corpus (bicornuate)*	*Large corpus (bicornuate)*	*Corpus only (simplex)*
Dermoptera (flying lemur) Megachiroptera (fruit bats) Lagomorpha (rabbit) Rodentia (beaver, mice, rat, pocket gopher, vole, guinea pig) Tubulidentata (aardvark)	Artiodactyla (pig) Carnivora (dogs, cats, seals) Hyracoidea (hyrax) Insectivora (moles, hedgehogs) Megachiroptera (fruit bats) Rodentia (Jerboa)	Artiodactyla (gnu) Microchiroptera (bats) Rodentia (North American porcupine)	Artiodactyla (sheep, cow) Cetacea (dolphins, whales) Insectivora (tree shrews) Megachiroptera (bats) Microchiroptera (bats) Pholidota (pangolins) Proboscidea (elephants) Prosimii (lemurs) Sirenia (dugongs, manatees)	Perissodactyla (horses, rhinos)	Anthropoidea (human, apes, monkeys) Edentates (sloths, armadillos) Microchiroptera (bats)

[a] Organized on the scheme proposed by, and compiled from the work of H. W. Mossman (1987). Uterine type categories are based on relative length (long, medium, and short) or the presence of uterine horns (cornua). Duplex uteri have no body or corpus. Simplex uteri have no cornua. Common name(s) of representative species is given in parentheses.

types of uteri and major groups of eutherian mammals in which they are found.

B. Placentation

Uteri are also classified according to whether the type of placentation supported is deciduate or adeciduate (also called nondeciduate). Decidua, or deciduomata, form as a result of the proliferation, hypertrophy, and differentiation of endometrial stromal cells; infiltration of the endometrium by a variety of bone marrow-derived cells; localized endometrial edema; and remodeling of the extracellular matrix. Decidual transformation of the endometrium (decidualization) is steroid dependent, requiring both estrogen and progesterone. Decidualization occurs naturally at specific endometrial sites in response to an implanting embryo but can also be induced artificially by various nonspecific stimuli administered at the appropriate stage of the reproductive cycle. Decidua serve as the site of conceptus attachment, endometrial invasion (nidation), and placental development. Decidual cells are endocrinologically active and decidualization can prolong the life span of ovarian corpora lutea, as seen in the rat. Artificial induction of decidualization in ovariectomized animals given an appropriate regimen of steroid hormones is an important experimental system for studies of conceptus–endometrial interactions and implantation in species that display the "decidual cell reaction."

In animals with true deciduate uteri a substantial amount of endometrial tissue is lost when the placenta is shed at parturition and regenerated during the postpartum period. Similarly, only species with true deciduate uteri, particularly humans and closely related menstruating primates, display cyclical loss, regeneration, and remodeling of endometrial tissue. True deciduate uteri are found in several mammalian groups, including Anthropoidea, Rodentia, Lagomorpha, Hyracoidea, Dermoptera, Chiroptera, and in some Insectivora, Edentata, and Tarsiidae. Dogs (Carnivora) have been described to have atypical decidua since they shed a small amount of glandular endometrium at parturition. However, they do not produce true decidual cells. Adeciduate uteri are found in Artiodactyla, Perissodactyla, Cetacea, Tubulidentata, Sirenia, and Pholidota.

C. Endometrial Histology

In all viviparous mammals the endometrium is lined with a simple cuboidal or columnar epithelium and contains simple coiled, tubular glands that can be branched. Generally, three zones of endometrial stroma are defined based on patterns of stromal cell distribution. Classically, these are the (i) stratum compactum, in which stromal cells are distributed densely in the adluminal endometrium between the necks of glands; (ii) stratum spongiosum, characterized by more loosely arranged stromal cells in the deeper mucosa between the glands; and (iii) stratum basale, a more densely arranged zone of stromal cells adjacent to the myometrium. In deciduate species, two endometrial compartments are typically defined to include the superficial lamina functionalis and the deep lamina basalis. Quadrapartite zonation is described for menstruating primates such as the human and rhesus monkey. In this case, the functionalis and basalis are subdivided histologically and functionally into functionalis zone I (luminal epithelium and highly vascular subadjacent stroma) and zone II (straight upper portions of endometrial glands) and basalis zone III (coiled segments of glands) and zone IV (basal regions of glands and stroma adjacent to myometrium). In these species, tissues in functionalis zones I and II are eliminated cyclicly and regenerate from the basalis, which serves as the germinal tissue compartment. In most other mammals, which are adeciduate and do not shed uterine tissues cyclicly or at parturition, remodeling of the endometrium occurs without loss of endometrial integrity.

Tremendous natural diversity is found in endometrial histoarchitecture. Nevertheless, endometria are described and categorized based on regional differences in thickness between mesometrial and antimesometrial sides (i.e., adjacent to and opposite from the site of attachment of the mesometrium or supportive ligament of the uterus), relative number, length and density of endometrial glands, and on whether aglandular ridges or caruncles are present. As a rule, endometrial thickness and gland distribution are uniform in species with simplex uteri, whereas regional differences in these characteristics can be pronounced in some species with duplex and bicornuate uteri. However, uniform endometrial thickness and gland density are also characteristic of several species with duplex or bicornuate uteri. More familiar species in this category include pigs, horses, camels, dogs, cats, ferrets, and lemurs. Asymmetry in endometrial thickness and gland distribution is seen classically in the uteri of rodents, rabbits, shrews, and vespertillionid bats. Localized aglandular caruncles or aglandular endometrial ridges are found in uteri of many common domestic ungulates, including sheep, cattle, and goats, as well as in cervids (deer and antelope), tupaiids (tree shrews), and giraffes. Photomicrographs illustrating histoarchitecture of the uterine wall are shown for the pig and sheep in Fig. 2.

III. VASCULARIZATION AND INNERVATION

In female mammals, internal genital organs are suspended in the pelvis by mesenteries. The uterus is supported by the mesometrium, a caudal anatomic division of the broad ligament. The line along which the mesometrium attaches to the uterus is referred to as the mesometrial border, in contrast to the opposite or antimesometrial border. Orientation and attachment of the mesometrium are uterine type specific. Mesometrial smooth muscle is continuous with the longitudinal muscle layer of the myometrium.

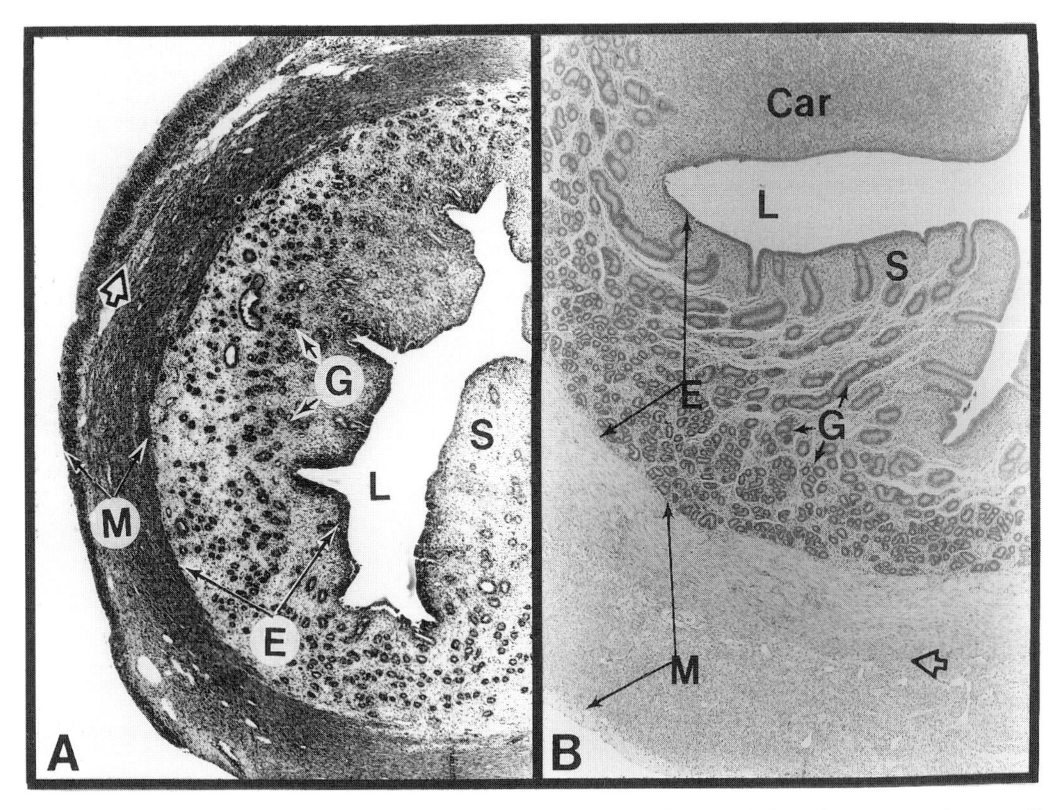

FIGURE 2 Photomicrographs depicting the uterine wall of the pig (*Sus scrofa*; A) and sheep (*Ovis aries*; B). As in all mammals, the uterine wall in these species consists of perimetrium (outer peritoneal coat), myometrium (M), and endometrium (E). The interface of inner circular (adluminal) and outer longitudinal myometrial smooth muscle layers is indicated by an open arrow. In the pig (A), endometrial folds are seen as regular invaginations along the uterine lumen (L), and uterine glands (G) are distributed uniformly throughout the endometrial stroma (S). In the sheep (B), intensely glandular intercaruncular endometrial areas separate numerous raised aglandular caruncles (Car). These endometrial structures serve as specialized sites for placental attachment in sheep.

Serosal layers of the mesometrium are continuous with the visceral peritoneum of the uterus or perimetrium. Blood vessels, lymphatics, and nerves are conveyed to the uterus through the mesometrium and join the uterus along the mesometrial border in the parametrium.

A. Vascularization

Urogenital organs are vascularized by caudal branches of the descending aorta, including the paired ovarian and external and internal iliac arteries. Venous and lymphatic drainage generally parallel arterial supply. Primary branches of the ovarian arteries supply the ovaries, whereas uterine branches of the ovarian arteries deliver blood to the uterine corpus and cranial portions of uterus or uterine horns directly. Uterine arteries arise as branches of the internal or, as in the horse, external iliac arteries. From these vessels a series of anastomosing branches arise to supply the uterine body and cranial portions of the uterus or uterine horns. Most, if not all, mammals have functional anastomoses between cranial branches of the uterine artery and branches of the ovarian artery. Uterine arteries may also branch to supply the ovaries directly. The urogenital artery, derived from the internal iliac, supplies blood to the caudal uterus as well as to the cervix and vagina. Collectively, these vessels ramify within the uterine wall to serve myometrial and endometrial tissues through a highly organized and dynamic microvascular system, the architecture and specific functions of which can vary dramatically among species.

In some species, including the guinea pig, rat, hamster, Mongolian gerbil, and rabbit, as well as the cow, sheep, goat, pig, and horse, the uterus produces a luteolysin required for regression of the ovarian corpus luteum (luteolysis) and thereby maintains ovarian cyclicity. In the guinea pig, rat, hamster, gerbil, cow, sheep, goat, and pig, local uteroovarian relationships exist whereby luteal life span in each ovary is regulated by the uterine horn adjacent (ipsilateral) to it. In such circumstances, local uteroovarian venoarterial vascular anatomy is functionally critical. Studies of both laboratory and farm animals in which such mechanisms have been described indicate that the uterine luteolysin, prostaglandin $F_{2\alpha}$ ($PGF_{2\alpha}$), is most likely transported to the ipsilateral ovary through a venoarterial pathway involving countercurrent exchange of $PGF_{2\alpha}$ between the uteroovarian vein and its uterine branch, and the ovarian artery. The latter is intimately apposed to and coiled tortuously about the uteroovarian venous surface. The integrity of this vascular arrangement, involving extensive contact between the walls of the veins carrying $PGF_{2\alpha}$ from the uterus and the wall of the artery supplying the ipsilateral ovary, must be maintained to ensure normal ovarian cyclicity in these species.

B. Innervation

The uterus is well supplied by autonomic nerves representing adrenergic, cholinergic, and peptidergic systems. Innervation is parasympathetic via pelvic nerves and postganglionic sympathetic from the caudal mesenteric ganglion via hypogastric nerves. Preganglionic parasympathetic fibers project from the intermediate gray matter of sacral spinal segments through the ventral branches of sacral spinal nerves and leave to become the pelvic nerve(s). These join with the hypogastric postganglionic sympathetic axons to form the pelvic plexus and distribute to viscera of the pelvic cavity where they synapse on ganglia in the walls of the viscera, including the uterus. Short postganglionic parasympathetic axons innervate both smooth muscle and glands. Parasympathetic preganglionic neurons in the spinal cord that project axons to the uterus are cholinergic and nitric oxide (NO) containing. Sympathetic nerve fibers directed to the uterus originate in lumbar spinal segments and course, via lumbar spinal nerves and rami communicans, through ganglia of the sympathetic chain and lumbar splanchnic nerves to the caudal mesenteric ganglion for synapse. Postganglionic sympathetic axons destined to provide adrenergic innervation to the pelvic viscera constitute the hypogastric nerves. These proceed caudally to join the pelvic nerve, where they form the pelvic plexus (plexus of Frankenhèuser). Generally, postganglionic autonomic nerve fibers project to myometrial and vascular smooth muscle as well as to vascular and nonvascular cells of the submucosal endometrium.

Layer-specific variations in myometrial innervation have been described for some species in which circular smooth muscle is primarily endowed with cholinergic innervation, whereas longitudinal smooth muscle receives adrenergic innervation. Sensory nerves, originating primarily from neurons in lumbar and sacral dorsal root ganglia, can be both peptidergic and NO containing.

IV. DIVERSITY, EVOLUTION, AND FUNCTION

Among higher vertebrate classes, including Reptilia (lizards, snakes, and turtles), Monotremata (egg-laying mammals), Marsupialia (pouched mammals), and Eutheria (placental mammals), both oviparous (egg-bearing) and viviparous (live-bearing) species are found. Aves (birds) is the only class of higher vertebrates in which only oviparous species exist. In true oviparous species, including birds, many reptiles, and monotremes, the uterus is not a gestational organ. Instead, this segment of the oviduct or uterine tube serves primarily as a site for accumulation of albumen in the amniote egg, formation of a calcareous shell, and transport of the egg to the vagina for oviposition. In this context, the term "shell gland" rather than "uterus" is the more correct term. However, the term uterus has come to be applied here as well as in reference to analogous tubular organs in most vertebrate species, and even some invertebrates such as *Glossina* (Tsetse fly), that display adenotrophic viviparity. This is a variation of vertebrate ovoviviparity in which insect larva hatch in the uterus and are nourished to a state of advanced maturity by specialized glands in the uterine wall. In all viviparous species, successful embryo development requires some type of intimate contact between the conceptus and the uterine mucosa.

A. Evolutionary Considerations

The uterus of viviparous eutherian mammals may have evolved from the oviduct of ancient therapsid reptilians through a process parallel to that presumed to have produced viviparous reptiles. In modern reptiles (lizards and snakes), in which extraembryonic (placental) membranes are developmentally and structurally homologous to those of mammals, the most common form of viviparity is ovoviviparity. Here, specialized areas of the uterine mucosa make intimate contact with the surface of a thin shell membrane to establish an interface for exchange of gases between conceptus and maternal systems. As with oviparity, eggs in ovoviviparous species contain sufficient nutrients in the yolk to support embryonic development and do not depend on the uterus as an organ of nutrition. In monotremes, in which eggs are small and yolk content is reduced by comparison with reptilian eggs of similar size, as well as in marsupials, nutrients lacking *in utero* are provided postnatally through the mammary gland—a mammalian adaptation. In some marsupials, however, extraembryonic membranes are mildly invasive of the uterine mucosa and are presumed to take up embryotrophic nutrients of maternal (uterine) origin at the conceptus–maternal interface. This pattern has been interpreted to suggest that eutherian mammals evolved through a monotreme–marsupial–placental transition in which gestational nutrition of the developing embryo shifted gradually to the uterus. The exact course of evolutionary events that led to development of the gestational uterus of eutherian mammals remains speculative.

B. Functional Considerations

Functions of the uterus in eutherian mammals include transport, storage (some species) and maturation of spermatozoa, spacing (in litter-bearing or polytocous species), recognition and reception of embryos, provision of an embryotrophic environment for conceptus development, and expulsion of the fetus and placenta at parturition. Additionally, as indicated previously, the uterus can play an important role in regulation of ovarian function and maintenance of normal reproductive cyclicity. These processes involve systematic biochemical, physiological, and morphological changes in both endometrial and myometrial tissues that are programmed, to a large extent, by the ovarian hormones estrogen and progesterone. Consequently, patterns of expression and cellular distribution of receptors for these steroids in uterine tissues are critical determinants of uterine function.

1. *Uterine Environment and Sperm Transport*

Uterine response to endocrine cues characteristic of specific periods of the reproductive cycle dictate the physical and biochemical condition of uterine tissues as well as that of the uterine luminal environment. Around the time of mating these conditions typically facilitate the transport of sperm through the uterine lumen to the oviducts (uterine tubes), where fertilization occurs. Although spermatozoa are motile, transport of sperm through the uterine lumen depends primarily on organized myometrial contractions, which may be effected by neuroendocrine substances released at or around the time of coitus and/or delivered to the uterine lumen in semen. Uterine response to seminal plasma may be especially important for species such as the mouse, guinea pig, ferret, and pig, in which the uterus serves as the natural site of semen deposition. Quality of the uterine luminal environment is important. Once deposited, sperm are suspended in uterine fluid, which serves as the medium for their transport. Constituents of uterine fluid affect patterns of sperm motility and maturation, including changes in the molecular characteristics of plasma membranes required for capacitation and hyperactivation of these gametes. Uterine storage of sperm, for hours or days, occurs to some extent in many vertebrate species, most commonly in specialized epithelial "crypts" or glands found at uterovaginal or uterotubal junctions. In some species, however, the uterus serves as an important site for long-term storage of ejaculated spermatozoa. In vespertillionid bats, for example, sperm deposited at mating are stored in the uterine lumen for months before ovulation and fertilization occur. This storage mechanism is undefined. More commonly, spermatozoa become trapped in endometrial glands and mucosal folds and are phagocytosed by leukocytes drawn into the uterine lumen in response to semen deposition. In this regard, the endometrium acts as a barrier to sperm, reducing the number of male gametes that reach the oviducts and thereby contributing to prevention of polyspermy (fertilization by more than one spermatozoan).

2. *Conceptus–Maternal Interactions*

To survive and be born, embryos of viviparous animals must establish a functionally cooperative relationship with their mother through the uterine wall. Placentation, the process through which this relationship is established, begins when the conceptus enters the uterus and becomes apposed to the endometrial surface where it will attach and develop. The capacity of uterine tissues to support this process, and to provide a functionally efficient maternal interface for exchange of gasses with and delivery of nutrients to the developing conceptus(es), can determine patterns of fetal growth, litter size (in polytocous species), and postnatal performance of offspring.

i. Uterine Immune System The uterus is not an immunologically privileged site. Bone marrow-derived cells, including macrophages, cytolytic/suppressor and helper T lymphocytes, large granular lymphocytes of the natural killer subtype, and antibody-producing B lymphocytes, populate uterine wall tissues, especially the endometrium. The number, relative proportion, distribution in specific uterine compartments, and state of activation of these lymphoid cells is relatively species specific and affected dramatically by both maternal endocrine status and pregnancy (the presence of a conceptus). Uterine macrophages and lymphocytes are likely to serve as an important line of defense against infectious microorganisms that gain access to the uterine environment. Additionally, the uterus is an inductive site for immune responses that can be elicited by sperm- and conceptus-associated antigens. The extent to which such antisperm and/or anticonceptus uterine immune responses affect the overall success of pregnancy remains under intense investigation. However, maternal immune recognition of the conceptus via the uterus may have significant embryotrophic potential through growth-promoting effects of cytokines encountered by the conceptus at its interface with the endometrium. Cytokines described originally as products of immune cells, such as granulocyte-macrophage colony-stimulating factor, leukemia inhibitory factor , various interleukins, colony-stimulating factor-1, and transforming growth factor-β, are also products of the endometrium, which may be the primary source of such soluble factors in the uterine environment. These and other uterine-derived lymphokines, cytokines, growth factors and proteins of endometrial origin are thought to affect behaviors of both conceptus and uterine immune cells as part of a dynamic paracrine communication network that provides fine regulatory control of conceptus development, endometrial invasiveness, and potential cytolytic activity of endometrial lymphocytes toward the conceptus. Mechanistic details of this system are being defined for many species.

ii. Uterine Receptivity Mammalian embryos enter the uterus at the morula or early blastocyst stage. In species with bicornuate uteri, intercornual migration of embryos is common and serves to effect balance in embryo numbers between uterine horns. Mechanisms regulating movement of embryos within the uterine lumen and their apposition to receptive endometrial attachment sites characteristic of the species are not defined precisely. However, it is clear that such mechanisms exist. Exotic species with bicornuate uteri, such as certain African antelopes and vespertillionid bats, are potentially useful models for study of such phenomena. These animals display partial or complete obligatory attachment of embryos in the right uterine horn, although eggs are released from both (antelope) or only the left (bat) ovary.

First contact between maternal uterine and embryonic cells occurs at the apical border of endometrial epithelium. Uterine receptivity to conceptus signals at this interface is necessary for establishment and maintenance of pregnancy. Receptivity is programmed by ovarian estrogen and progesterone and is thought to develop through their action on endometrial epithelial and stromal cells and their resultant interactions. Endometrial response to these steroids not only defines but also limits the period during which embryos can establish an effective gestational relationship with the uterine wall. Therefore, developmental synchrony between embryos and the uterine environment is required for reproductive success. Much attention has been focused on the identification of structural and biochemical indices of endometrial receptivity as potential biologic markers of uterine capacity to support embryo development. Species variations are significant. However, transformation of the mammalian uterus from a nonreceptive to a receptive state generally involves changes in endometrial synthesis, secretion, distribution, and orientation of proteins, glycoproteins, and glycosaminoglycans that, collectively, contribute to alterations in the quality of the apical plasma membrane and glycocalyx (glycoprotein coat) of uterine epithelial cells and the uterine luminal environment. These changes are reflected by alterations in epithelial fine structure and surface contour, a generalized decrease in epithelial surface negativity and thickness of the glycocalyx, secretion of proteinases that may limit or regulate conceptus invasiveness, and expression of molecules on apical and basolateral endometrial cell surfaces that mediate cell–cell interactions and facilitate adhesion.

Removal of the uterus (hysterectomy) does not affect patterns of ovarian cyclicity in many higher primates (e.g., human, rhesus monkey, and cynomologous monkeys), carnivores (e.g., dog, ferret, and ground squirrel), or marsupials (e.g., opossum). In such species, life span of the corpus luteum is not uterine dependent. However, hysterectomy extends luteal life span in many species, including pseudopregnant rats, mice, hamsters, and rabbits, as well as in the normal cyclic guinea pig, cow, sheep, goat, pig, and horse. In these species, as indicated previously, the uterine endometrium is responsible for luteolysis and maintenance of ovarian cyclicity. Regardless of whether luteal life span is uterine dependent or uterine independent, establishment and maintenance of pregnancy in all species in which the duration of uterine gestation exceeds that of a normal ovarian cycle (e.g., primates, rodents, and domestic ungulates) requires that cyclic regression of the corpus luteum be prevented. This ensures a continued source of progesterone and maintenance of an embryotrophic uterine environment. Rescue of the corpus luteum requires that the endometrium be receptive to species-specific conceptus signals that can be physical and chemical in nature. Regardless of the type of signal, uterine response to the conceptus and its products sets into motion those luteotrophic and/or antiluteolytic events that ensure a continuous source of progesterone for pregnancy maintenance. The capacity of uterine tissues to recognize and integrate maternal and conceptus signals as necessary for gestational success is a critical determinant of overall reproductive efficiency.

See Also the Following Articles

Cervix; Corpus Luteum; Endometrium; Uterus, Human; Vagina

Bibliography

Baker, J., Hardy, M. P., Zhou, J., Bondy, C., Lupu, F., Bellve, A. R., and Efstratiadis, A. (1996). Effects of an Igf1 gene null mutation on mouse reproduction. *Mol. Endocrinol.* **10**, 903–918.

Bartol, F. F. (1993). Early uterine development in pigs. *J. Reprod. Fertil. Suppl.* **48**, 99–116.

Gendron, R. L., Paradis, H., Hsieh-Li, H. M., Lee, D. W., Potter, S. S., and Markoff, E. (1997). Abnormal uterine stromal and glandular function associated with maternal reproductive defects in Hoxa-11 null mice. *Biol. Reprod.* **56**, 1097–1105.

Ginther, O. J. (1976). Comparative anatomy of uteroovarian vasculature. In *Veterinary Scope. XX*, No. 1. Upjohn, Kalamazoo, MI.

Glasser, S. R., Mulholland, J., Mani, S. K., Julian, J., Munir, M. I., Lampelo, S., and Soares, M. J. (1991). Blastocyst–endometrial relationships: Reciprocal interactions between uterine epithelial and stromal cells and blastocysts. *Troph. Res.* **5**, 229–280.

Hansen, P. J. (1995). Interactions between the immune system and the ruminant conceptus. *J. Reprod. Fertil. Suppl.* **49**, 69–82.

Jost, A. (1953). Problems of fetal endocrinology: The gonadal and hypophyseal hormones. *Recent Prog. Horm. Res.* **8**, 379–418.

Korach, K. S., Couse, J. F., Curtis, S. W., Washburn, T. F., Lindzey, J., Kimbro, K. S., Eddy, E. M., Migliaccio, S., Snedeker, S. M., Lubahn, D. B., Schomberg, D. W., and Smith, E. P. (1996). Estrogen receptor gene disruption: Molecular characterization and experimental and clinical phenotypes. *Recent Prog. Horm. Res.* **51**, 159–188.

Lydon, J. P., DeMayo, F. J., Conneely, O. M., and O'Malley, B. W. (1996). Reproductive phenotypes of the progesterone receptor null mutant mouse. *J. Steroid Biochem.* **56**, 67–77.

Mossman, H. G. (1987). *Vertebrate Fetal Membranes.* Rutgers Univ. Press, New Brunswick, NJ.

Murphy, C. R., and Shaw, T. J. (1994). Plasma membrane transformation: A common response of uterine epithelial cells during the peri-implantation period. *Cell Biol. Int. Rep.* **18**, 1115–1128.

Pavlova, A., Boutin, E., Cunha, G. R., and Sassoon, D. A. (1994). Msx1 (Hox-7.1) in the adult mouse uterus: Cellular interactions underlying regulation of expression. *Development* **120**, 335–345.

Robertson, S. A., Seamark, R. F., Guilbert, L. J., and Wegmann, T. G. (1994). The role of cytokines in gestation. *Crit. Rev. Immunol.* **14**, 239–292.

Sugawara, J., Fukaya, T., Murakami, T., Yoshida, H., and Yajima, A. (1997). Hepatocyte growth factor stimulates proliferation, migration, and lumen formation of human endometrial epithelial cells in vitro. *Biol. Reprod.* **57**, 936–942.

Taggart, D. A. (1994). A comparison of sperm and embryo transport in the female reproductive tract of marsupial and eutherian mammals. *Reprod. Fertil. Dev.* **6**, 451–472.

Whitelaw, P. F., and Croy, B. A. (1996). Granulated lymphocytes of pregnancy. *Placenta* **17**, 533–543.

World Association of Veterinary Anatomists, International Committee on Veterinary Gross Anatomical Nomenclature (1983). *Nomina Anatomica Veterinaria*, 3rd ed. International Committee on Veterinary Gross Anatomical Nomenclature, Ithaca, NY.

Wynn, R. M., and Jollie, W. P. (Eds.) (1989). *Biology of the Uterus*, 2nd ed. Plenum, New York.

Yoshizawa, T., Handa, Y., Uematsu, Y., Takeda, S., Sekine, K., Yoshihara, Y., Kawakami, T., Arioka, K., Sato, H., Uchiyama, Y., Masushige, S., Fukamizu, A., Matsumoto, T., and Kato, S. (1997). Mice lacking the vitamin D receptor exhibit impaired bone formation, uterine hypoplasia, and growth retardation after weaning. *Nature Genetics* **16**, 391–396.

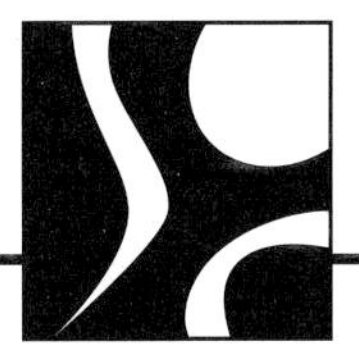

Vagina

Raymond E. Papka
Northeastern Ohio Universities College of Medicine

Sonya J. Williams
Yale University

GLOSSARY

desquamate The shedding of the most superficial layers of cells from the lining epithelium of the vagina. These cells can be collected and examined for signs of pathology of the vagina.

fornix An arched space between the uterine cervix (as it projects into the vagina) and the vaginal wall.

hymen A thin fold of tissue forming a membrane that closes the vaginal orifice in the virginal female.

innervation The nerve supply to the vagina.

Müllerian ducts A pair of ducts existing during development of the genitourinary system that ultimately fuse and, in part, give rise to the vagina.

Pap smear Samplings of cells that are placed on a microscope slide and include uterine epithelial as well as vaginal epithelial cells. The cells can be stained and examined for evidence of pathology (including precancerous conditions).

perineum The space between the external genitalia and the anus.

prolapse The slipping down of the vagina due to loss of supporting structures; the vagina may be exposed to the exterior at the vestibule.

vaginitis Inflammation or infections of the vagina; may be caused by a fungus, bacterium, or virus.

vestibule The space between the labia minora at the entrance to the vagina.

vulva The female external genitalia.

The female reproductive system or reproductive organs of the internal group consist of the ovaries, the uterine tubes, the uterus, uterine cervix, and the vagina; these organs are located within the pelvis.

I. INTRODUCTION

The vagina (meaning sheath) is a thin-walled tube connecting the vestibule externally with the uterine cervix internally. The vagina is lined by stratified squamous epithelium resting on a dense connective tissue that contains a vascular venous plexus. The venous plexus acts as an erectile tissue. Intermixed with the connective tissue are slips of smooth muscle. The vagina is capable of dilation and constriction, as a result of the action of these supporting muscles and erectile tissues, which subserves its function as the female organ of copulation. The squamous epithelium lining the mucous membranes of the vulva and vagina is subject to alteration by estrogenic substances.

II. GROSS ANATOMY

The vagina is 8–10 cm in length and lies in large part within the female pelvis. It passes upward and backward at an angle of 45° from the perineum to

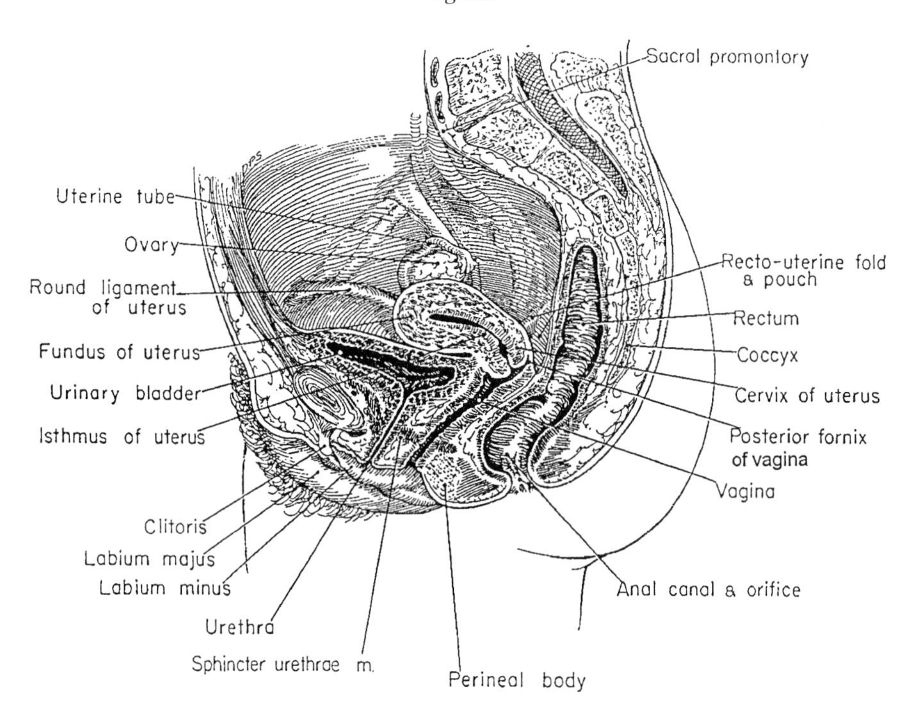

FIGURE 1 A semischematic median section of the female pelvis (reproduced with permission from R. T. Woodburne, *Essentials of Human Anatomy*, 7th ed., Oxford Univ. Press, New York, 1983).

connect the vestibule externally with the internal genital canal, the uterine cervix (Fig. 1). The spaces (or recesses) between the projection of the uterine cervix into the vaginal vault are the fornices (singular = fornix) (Figs. 1, 2B, and 3C). There are two lateral fornices, an anterior fornix and a posterior fornix. The posterior fornix is deeper than the other fornices and abuts the peritoneum in the pelvis, which dips inferiorly forming the rectouterine pouch (Fig. 1). The angle between the uterus and the vagina is approximately 90°. The vagina lies posterior to the urethra and urinary bladder but anterior to the rectum and anal canal.

III. EMBRYOLOGY

In the sixth week of development (human), the gender-indifferent embryo has two pairs of genital ducts, the Wolffian (mesonephric) and Müllerian (paramesonephric). Both sets of ducts do not complete development; if the embryo is to be female, the Müllerian ducts develop. The Müllerian ducts are paired initially, but fuse to form the uterovaginal canal (Fig. 2); the uterovaginal canal extends caudally to an opening called the urogenital sinus (Fig. 3). Eventually (in humans, by the end of the third month of development) the fusion septum of the two Müllerian ducts recedes and a single uterovaginal canal results (Fig. 3). Endoderm at the urogenital sinus proliferates forming an enlargement of tissue called the sinovaginal bulb (Figs. 2A, 3A, and 3B). The sinovaginal bulbs extend from the urogenital sinus to the uterovaginal canal. The sinovaginal bulb (plate) expands, eventually canalizes, and represents the primordium of much of the vagina. The upper one-third of the vagina is derived from the uterovaginal canal (Müllerian duct origin), whereas the lower two-thirds are derived from the endoderm of the sinovaginal bulb (urogenital sinus origin). The hymen is a membrane covering the orifice of the vagina; it marks the junctional area between sinovaginal plate and urogenital sinus (Fig. 3). The hymen usually ruptures in the perinatal period but remains as

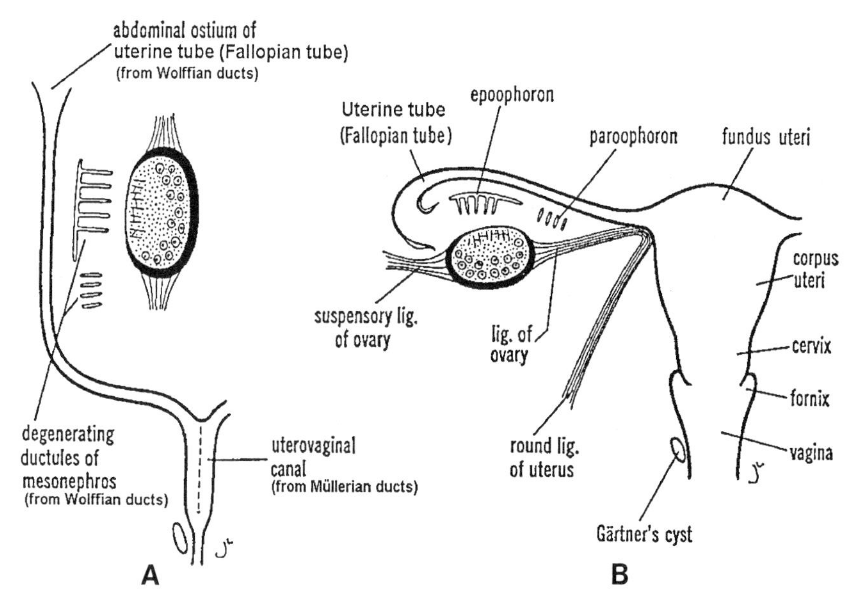

FIGURE 2 (A) Diagram of the genital ducts in the female in the fourth month of development. Note the epoophoron, paroophoron, and Gärtner's cyst, as remnants of the mesonephric system. (B) The genital ducts after descent of the ovary. Note the suspensory ligament of the ovary, the ligament of the ovary proper, and the round ligament of the uterus (modified after Starck) (reproduced with permission from Langman, 1963).

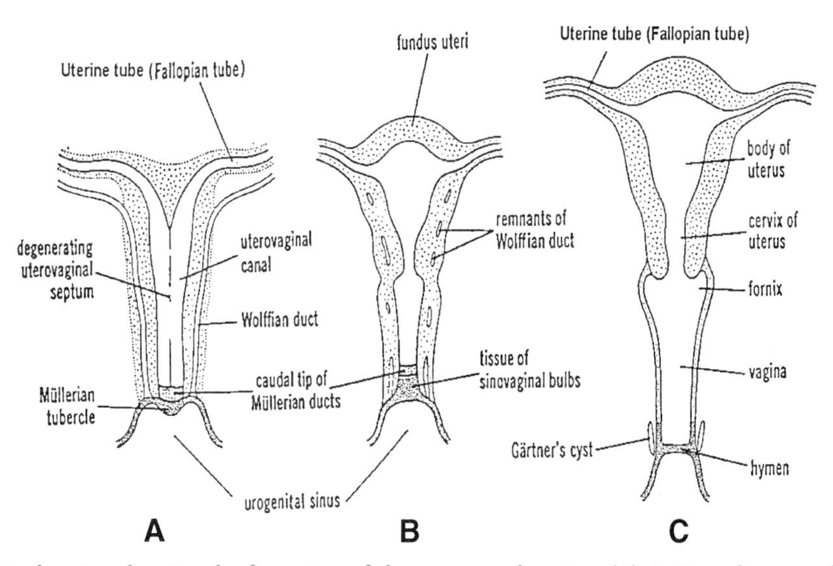

FIGURE 3 Schematic drawing showing the formation of the uterus and vagina. (A) At 9 weeks; note the fusion of the two Müllerian ducts. (B) At 11 weeks; note the solid tissue bar between the lumen of the uterovaginal canal and that of the urogenital sinus. (C) Newborn; the vagina is formed by the uterovaginal canal and the sinovaginal bulbs (reproduced with permission from Langman, 1963).

a thin fold of tissue just within the entrance to the vagina. The fibromuscular wall of the vagina develops from the splanchnic mesenchyme that surrounds the epithelium of the developing vagina.

At birth the stratified squamous epithelium lining the vagina is from 6 to 10 cells thick. The vaginal epithelium remains thin during childhood but thickens at puberty due to the stimulation of estrogen. Thereafter, the epithelial thickness varies somewhat with the menstrual cycle and with pregnancy and atrophies after menopause to its thin prepubertal state.

IV. HISTOLOGY

Three layers can be seen when a cross section is taken through the vaginal wall. These layers consist of the tunica mucosa or mucosa, tunica muscularis or muscularis, and the tunica adventitia or adventitia (Fig. 4).

A. Tunica Mucosa

The mucosa consists of the lining epithelium and the underlying lamina propria (connective tissue). The epithelium is a nonkeratinized stratified squamous epithelium that is 150–200 μm thick. In humans, the epithelium undergoes some changes with the hormonal fluctuations of the menstrual cycle, but these changes are not as marked as in some species. During the follicular phase (rising estrogen levels) the epithelium is about 45 cells thick, but during the luteal phase (lowered estrogen but upregulated progesterone) the epithelium is about 30 cells thick (Fig. 5). Glycogen content of the epithelial cells is significant and varies with the cycle. The epithelial cells are glycogen rich at midcycle (under the stimulus of estrogen) (Fig. 5), and this diminishes in amount later in the cycle. Epithelial cells are joined by desmosomes (mechanical, adherent junctions) and gap junctions (communication junctions), primarily in the basal layers. There are tight junctions

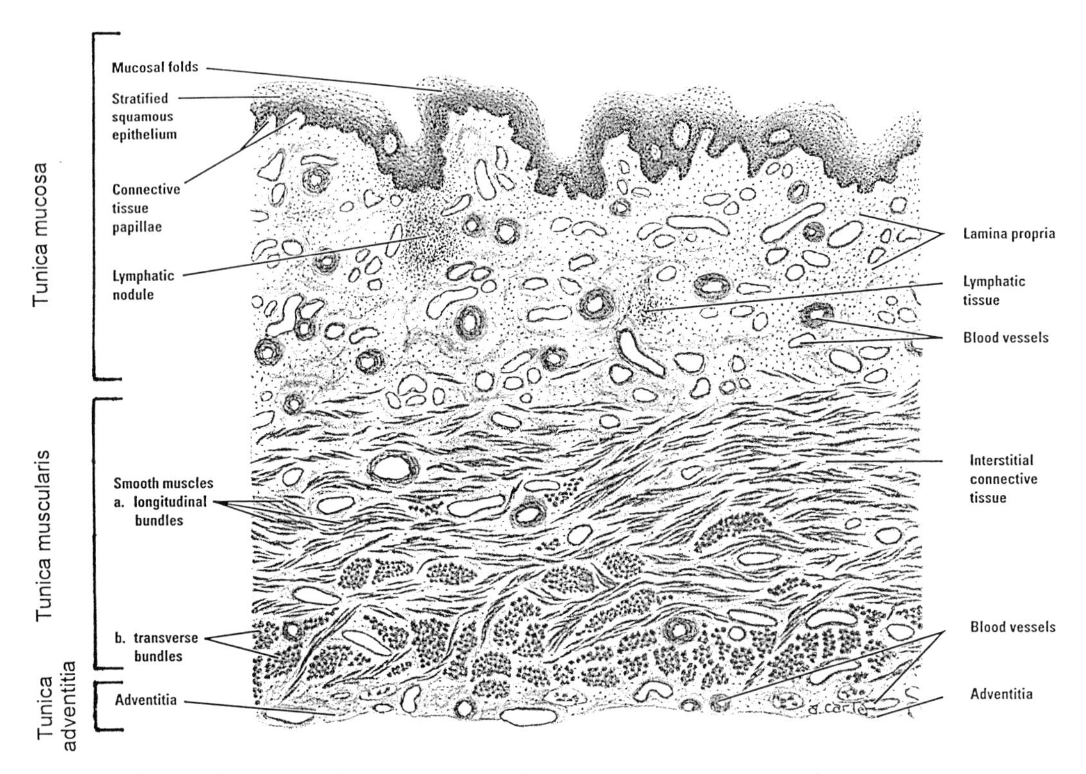

FIGURE 4 Vagina (longitudinal section). Stain: hematoxylin-eosin; low magnification (reproduced with permission from V. P. Eroschenko, *diFiore's Atlas of Histology with Functional Correlations*, Williams & Wilkins, Baltimore, 1996).

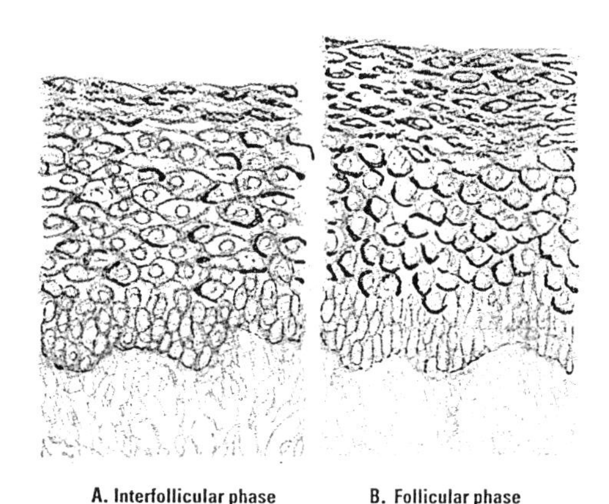

FIGURE 5 Glycogen in human vaginal epithelium. Stain: Mancini's iodine technique; medium magnification (reproduced with permission from V. P. Eroschenko, *diFiore's Atlas of Histology with Functional Correlations*, Williams & Wilkins, Baltimore, 1996).

at the superficial layers, which appear to form a permeability barrier to large water-soluble molecules. There are intercellular spaces among the epithelial cells through which blood-derived mononuclear leukocytes (mainly small lymphocytes) move. These cells are involved in immune surveillance of the vaginal epithelium.

The lamina propria lies subjacent to the epithelium. It is a dense connective tissue layer supported by elastic fibers crossing through the connective tissue to the underlying muscle. The lamina propria becomes less dense as it approaches the muscle. Within the lamina propria is a network of large thin-walled veins (Fig. 4), giving it the appearance of erectile tissue. The lamina propria lacks glands and thus the vagina derives much of its lubrication from the glands of the uterine cervix. However, there are suggestions that the rich vascularization of the vagina serves as a source of a fluid exudate that seeps through the squamous epithelium during sexual stimulation. Great elasticity of the vagina is related to the large number of elastic fibers in the connective tissues of its wall and its extensive venous plexus.

B. Tunica Muscularis

This layer (Fig. 4) consists of loosely arranged, spiraling bundles of smooth muscle cells. Though spirally arranged, the smooth muscle comprises largely inner circular and outer longitudinal layers. There are also skeletal muscle fibers (bulbospongeosus muscle) at the vaginal opening that forms a sphincter.

C. Tunica Adventitia

This is a thin layer (Fig. 4) of connective tissue that is rich in elastic fibers. It merges with the connective tissues of the urethra, rectum, and anal canal. Nerve bundles course through the adventitia to reach the muscle layers and mucosa.

V. CYTOLOGY

It is useful to discuss briefly the vaginal epithelium and the changes it expresses under the influence of the hormones of functioning ovaries and under the influence of exogenously administered hormones.

Cellular differentiation takes place in the vaginal epithelium. For example, the superficial epithelial cells continuously differentiate and are shed during the menstrual cycle with the greatest desquamation at the latter part of the luteal phase and during menstruation. Vaginal smears can be taken and studied as an indicator (based on the cytology of cells) of the functional state of the ovaries (in regard to the secretion of hormones). Thus, examination of cells collected from vaginal smears gives information of clinical importance (Pap smear; Papanicaolau and Traut, 1943). The morphology of the most superficial layers of the epithelium and the different types of cells found in the smears are studied (for the presence of leukocytes and indications of early cancer).

During midcycle or premenstruation, the desquamation of superficial cells is more pronounced shortly before the onset of menstruation. During the first trimester of pregnancy, the surface appears relatively smooth. Postmenopause is characterized by fragmentation of superficial cell groups. Finally, as aging progresses, the vaginal smear reveals continued exfoliation of single superficial cells. The size of superficial vaginal cells varies with age; large cells are found in young women (17 years) and smaller ones in older women (47 years). The normal vaginal mucosa after menopause consists of a thin squamous epithelium with basal cells and no glycogen present.

VI. HORMONAL INFLUENCE

During the functional reproductive years with an active sex life the vaginal epithelium is under normal estrogen stimulation and is thickened and slightly cornified but not keratinized. During the proliferative phase (follicular stage) of the menstrual cycle, when estrogen levels are rising, the superficial epithelial cells become large and flattened, the nuclei become pyknotic, the cytoplasm shows increased acidophilia, and glycogen accumulates in the cytoplasm. At midcycle, estrogen stimulation maximizes epithelial thickness and glycogen accumulation (Fig. 5B). The glycogen of desquamated cells is fermented to lactic acid by bacteria resulting in a decrease in pH of the vaginal fluid at midcycle (this helps resist infections). During the secretory phase (luteal) of the cycle estrogen levels decrease, the epithelial thickness decreases due to sloughing of surface cells, and desquamated cells appear irregular, elongated with folded and curled edges. Glycogen is less abundant within the cells; thus, the pH of vaginal fluid increases with increased chances for infections.

Because of its influence on the epithelium, administration of exogenous estrogen is useful as part of the therapy for certain vaginal infections. Hyperestrogenism, caused by excessive stimulation with estrogen, results in an increased thickness, cornification and exfoliation of epithelial cells, and very acidic pH of the vaginal fluid. Hypoestrogenism is the reverse condition; the epithelium atrophies and vagina fluid pH increases.

VII. INNERVATION

The vagina receives innervation from the autonomic nervous system and the pudendal nerve. Parasympathetic supply to the vagina is via pelvic splanchnic nerves (from sacral spinal cord levels S2–S4 in the human) and the uterovaginal plexuses. Abdominal sympathetic trunks continue caudally into the pelvis by passing over the ala of the sacrum and converging into the ganglion impar near vertebral level S5. Two or three sacral splanchnic nerves are given off which enter the inferior hypogastric plexus. Mixed branches from the inferior hypogastric plexus distribute with branches of the internal iliac artery to the vagina. The pudendal nerve provides general visceral efferent fibers that pass to the uterovaginal plexus.

Many fibers in the uterovaginal plexuses are autonomic efferent and supply the vaginal smooth muscle and vascular smooth muscle. Most of these nerves travel through and in the lamina propria (Fig. 6). Some intraepithelial nerve fibers are apparent (Fig. 6) and appear as free nerve endings among the epithelial cells (Hilliges *et al.*, 1995). Some fibers penetrate nearly to the surface of the epithelium, but most end among the basal two-thirds of the epithelium; these are considered to by nociceptive and are located mainly near the introitus of the vagina.

The nerve fibers in the lamina propria and tunica muscularis are much more dense than in the epithelium and are more numerous in the anterior wall of the vagina than posterior wall (Hilliges *et al.*, 1995). Apparently there is also a lower threshold to pain stimulation in the anterior wall compared to the posterior wall. Nerve fibers in the vagina utilize a variety of classical (catecholamine and acetylcholine) and nonclassical neurotransmitters (neuropeptides and nitric oxide) (Fig. 6). The neuropeptides include neuropeptide Y (NPY), calcitonin gene-related peptide (CGRP), substance P (SP), and vasoactive intestinal polypeptide (VIP) (Fig. 6). All of these substances are found in nerves close to arteries and capillaries in the connective tissue papilla evaginating into the epithelium (Fig. 4). The neuropeptides and nitric oxide influence capillary permeability. VIP has a high content in the vagina, has a reduced effect postmenopausally, increases the rate of vaginal exudation (vaginal fluid production), and increases blood flow during sexual arousal. CGRP and SP are usually associated with sensory nerves and are vasodilators. NPY potentiates the vasoconstrictor activity of noradrenaline. Most of these neuropeptides and nitric oxide are likely to have roles in controlling blood flow and capillary permeability in the vagina (Hoyle *et al.*, 1996).

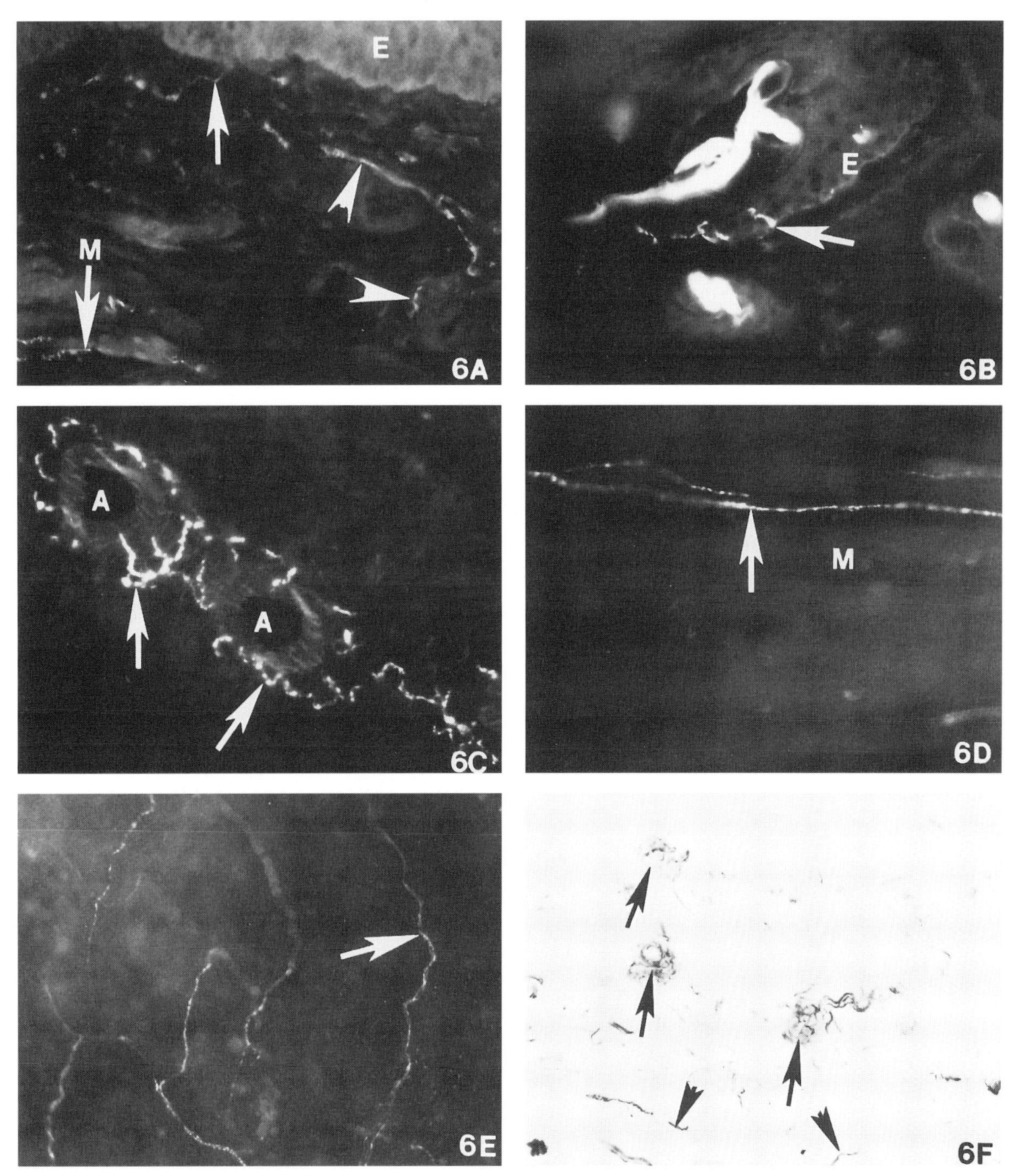

FIGURE 6 (A) Immunofluorescence micrograph showing substance P (SP)-containing sensory nerve fibers adjacent to smooth muscle (M) of the rat vagina, adjacent to arteries (arrowheads), and coursing near the vaginal epithelium (E) (arrow). SP acts as a vasodilator of arteries and stimulates contraction of nonvascular smooth muscle. Cryostat section. (B) Immunofluorescence micrograph showing calcitonin gene-related peptide (CGRP)-immunoreactive sensory nerve fibers (arrow) penetrating the vaginal epithelium (E). Cryostat section. (C) Immunofluorescence micrograph showing tyrosine hydroxylase-immunoreactive (noradrenergic vasoconstrictor) sympathetic postganglionic nerve fibers (arrows) adjacent to arteries (A) in the vaginal wall. Cryostat section. (D) Immunofluorescence micrograph showing neuropeptide Y (NPY)-containing postganglionic autonomic nerve fibers (arrow) adjacent to smooth muscle (M) of the vaginal wall. NPY has relaxing effects on the smooth muscle. Whole mount preparation of part of the wall of the rat vagina.(E) Immunofluorescence micrograph showing vasoactive intestinal polypeptide (VIP)-containing postganglionic nerve fibers in the wall of the rat vagina. VIP is a relaxor of vascular and nonvascular smooth muscle. VIP release is thought to play a role in sexual arousal. Whole mount preparation of part of the wall of the rat vagina. (F) Brightfield micrograph showing NADPH-*d*-positive nerve fibers adjacent to smooth muscle (arrowheads) of the rat vagina as well as adjacent to arteries (arrows). NADPH-*d* is a cofactor in the synthesis of neuronal messenger nitric oxide. Nitric oxide is a potent relaxor of smooth muscle and a vasodilator. Cryostat section.

The lower vagina also receives contributions from the pudendal nerves; these nerves presumably carry somatic pain from the lower vagina.

VIII. FUNCTIONAL ANATOMY

The female equivalent of erection is vaginal lubricity, the vaginal "sweating" resulting from vascular engorgement of the vaginal bulb subsequent to sexual arousal. The female equivalent of ejaculation is orgasm that involves rhythmic contractions of the orgasmic platform and contractions from the uterine fundus inferiorly. These effects are believed to be mediated by the sympathetic nervous system.

IX. PATHOLOGY

Failure of fusion of the Müllerian ducts developmentally results in duplication of the vagina. Lack of the development of the sinovaginal bulb or canalization of the bulb results in the absence of a vagina. Faulty development of a urorectal septum results in a connection of the rectum and vagina (rectovaginal fistula).

Damage (e.g., during childbirth) to the supporting structures (transverse cervical, pubocervical, and sacrocervical ligaments) of the uterine cervix, and thus the vagina, within the pelvic cavity can lead to prolapse of the vagina (exposure to the exterior). Perforation of the posterior fornix can occur during an artificial abortion by the instrumentation being passed through the vaginal wall instead of into the external os of the uterine cervix.

Diseases of the vagina are usually infections that result in vaginitis. Common agents include viruses (herpes simplex virus and human papilloma virus) and fungi (*Candida albicans*). The normal flora of the vagina includes the bacteria *Lactobacillus acidophilus*, which is responsible for production of lactic acid from glycogen and maintenance of the proper pH of the vagina; this helps protect the vagina from most bacteria. Other organisms (e.g., viruses and the bacterium gonoccoccus) can infect the vagina producing vaginitis (infections). These often are transmitted through unprotected sex.

Bibliography

Cormack, D. H. (1987). *Ham's Histology*, 9th ed. Lippincott, Philadelphia.

Fawcett, D. W (1994). *A Textbook of Histology* (Bloom and D. W. Fawcett, Eds.), 12th ed. Chapman & Hall, New York.

Hebel, R., and Stromberg, M. W. (1986). *Anatomy and Embryology of the Laboratory Rat,* pp. 85–86. BioMed Verlag, Germany. (ISBN 3-9801234-0-5)

Hilliges, M., Falconer, C., Ekman-Ordeberg, G., and Johansson, O. (1995). Innervation of the human vaginal mucosa as revealed by PGP 9.5 immunohistochemistry. *Acta Anat.* **153**, 119–126.

Hoyle, C. H., Stones, R. W., Robson, T., Whitley, K., and Burnstock, G. (1996). Innervation of vasculature and microvasculature of the human vagina by NOS and neuropeptide-containing nerves. *J. Anat.* **188**, 633–644.

Knobil, E., and Neill, J. D. (1994). *The Physiology of Reproduction*, 2nd ed., Vol. 2, pp. 1261–1262.

Langman, J. (1963). *Medical Embryology. Human* Development—*Normal and Abnormal.* Williams & Wilkins, Baltimore.

Marshall, F. H. A. (1922). *The Physiology of Reproduction.* Longmans, Green, London.

Netter F. H. (1992). *Reproductive System,* Vol. 2, pp. 139–151. Acme, New York.

Papanicaolau, G., and Traut, H. (1943). *Diagnosis of Uterine Cancer by the Vaginal Smear.* Commonwealth Fund, London.

Papka, R. E. (1995). *Anatomy, Oklahoma Notes.* Springer-Verlag, New York.

Reiffenstuhl, G. (1975). *Atlas of Vaginal Surgery—Surgical Anatomy and Technique.* Saunders, Philadelphia.

Varicocele

Terry T. Turner
University of Virginia School of Medicine

GLOSSARY

countercurrent heat exchange The process of temperature regulation that occurs within the pampiniform plexus. Cooler venous blood acts as a heat sink for the warmer arterial blood arriving from the abdomen, thus lowering the temperature of the arterial blood as it reaches the testis.

infertility The condition of a couple or partner in a couple who have attempted to achieve a pregnancy for at least 1 year without success.

Ivanissevitch procedure Spermatic vein ligation and division through an inguinal incision—a "low ligation." At this level the spermatic vessels are still retained within the spermatic cord.

Palomo procedure Spermatic vein ligation and division through a suprainguinal incision—a "high ligation." At this level the spermatic vessels are found in the retroperitoneum just under the external oblique fascia.

pampiniform plexus Spermatic venous plexus immediately superior to the testis. The plexus invests the spermatic artery and is important for the countercurrent heat exchange mechanism.

retrograde blood flow The flow of blood in the opposite direction from that expected; in varicocele, the flow of blood in the spermatic vein toward the testis rather than away from it.

reflux Movement of content of a duct back toward its origin; occurs during retrograde blood flow.

Tanner stage State of puberty based on pubic hair growth and development of genitalia in boys or breast development in girls.

varicosity Abnormal and permanent dilatation of a vein.

Varicocele refers to an abnormal dilatation of the spermatic vein, conventionally noted at the level of the pampiniform plexus. This lesion is detected clinically as a palpable and sometimes visible mass of engorged vessels above or sometimes lateral to the testis itself. The varicosity of these vessels is easily demonstrated by retrograde venography or, less invasively, ultrasonography. There is a widespread interest in this lesion because it is thought to be a cause of male infertility subject to surgical correction.

I. HISTORICAL PERSPECTIVE

The medical literature contains descriptions of varicocele as far back as the first century AD, but interest in the lesion because of male infertility is much more recent. In the latter half of the nineteenth century the major reasons for varicocele repair were either voluminous deformity of the scrotum or pain caused by grossly distended vessels. Varicocele repair was by excision of the dilated plexus, literally a varicocelectomy. During the late nineteenth century occasional reports appeared which noted that the varicocele operation had improved reproductive function of previously infertile patients; nevertheless, over a half a century elapsed before Tulloch drew significant attention to the association of varicocele with subfertility and ushered in the modern era of varicocele repair as a treatment for testicular dysfunction.

II. THE VARICOCELE REPAIR

A. Adults

The typical surgical repair of the varicocele in an infertile patient is no longer technically a varicocelec-

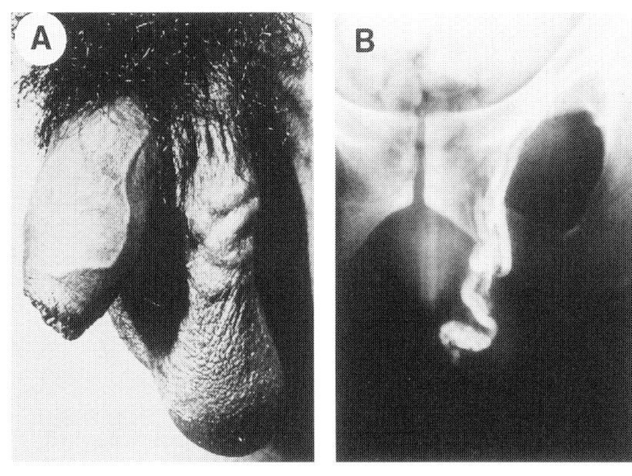

FIGURE 1 The appearance of varicocele. (A) Left varicocele in an adult male. The varicose pampiniform plexus can be seen underlying the scrotal skin above the testis (reproduced with permission from D. C. Saypol *et al.*, In *Infertility in the Male* (L. I. Lipshultz and S. S. Howards, Eds.), p. 300, Churchill Livingstone, New York, 1983). (B) Radiographic demonstration of left varicocele. Contrast medium was injected into the spermatic vein in a retrograde fashion via the left renal vein. The left varicosity extends into the scrotum, but contrast medium does not reach the testis.

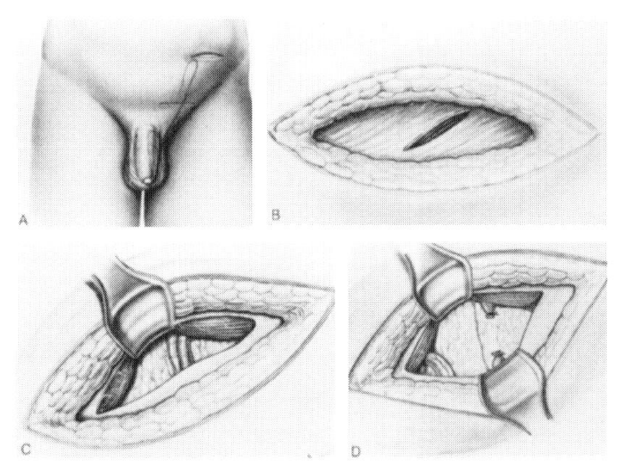

FIGURE 2 Illustration of a surgical method to perform a retroperitoneal, "high-ligation" varicocele repair. (A) The incision for the retroperitoneal approach is above the location of an inguinal incision (- - -). (B) Exposure of the external oblique faschia to reach the retroperitoneal spermatic vessels. (C) Exposure of the spermatic vein under the external oblique muscle. (D) Removal of a short segment of the spermatic vein between two ligatures of the vessel (reproduced with permission from Pryor and Howards, 1987).

tomy, that is, the engorged vessels of the pampiniform plexus are not surgically removed (Fig. 1). Rather, in surgical treatments, the spermatic vein is typically exposed through either a retroperitoneal approach (the Palomo procedure) or an inguinal approach (the Ivanissevitch procedure). These procedures have been modified over time, but the retroperitoneal approach provides exposure for the so-called "high ligation," and the inguinal approach provides exposure for a "low ligation" of the spermatic vein. While Fig. 2 illustrates the retroperitoneal approach, the location of an inguinal incision is also shown (Fig. 2a). Using the retroperitoneal approach the spermatic vein is visualized under the external oblique fascia (Figs. 2b and 2c) and is ligated twice, and a short segment is removed between the two ligatures (Fig. 2d). This eliminates the direct venous communication between the testis and the left renal vein and typically leads to detumescence of the scrotal vessels over time. The varicocele repair operation is simple but, like all surgery, it has its risks and difficulties. In particular, the surgeon must be aware that the spermatic vein is often branched in the region of the operation and care must be taken to ligate and divide all branches of the vessel.

Varicocele repair can also be achieved with interventional radiologic techniques rather than surgery. Such techniques occlude the spermatic vein with sclerosing agents, wire coils, or detachable balloons. Efficacy of the radiologic techniques in experienced hands appears similar to that of the surgical techniques, but there appears to be an increased failure rate with the radiologic techniques when performed by less experienced hands. Additionally, some have concerns about the radiation exposure required.

B. Adolescents

The operation to repair a varicocele is typically performed in adults, but whether adolescent varicoceles also should be repaired has recently become an issue. While varicoceles in adolescents have historically been untreated unless they were the cause of pain or disfigurement, the current debate is driven by modern concerns for the adolescent's future fertility. The reasoning has been that if a varicocele is associated with infertility in the adult, it might be expected to affect testicular development during puberty as well. If a varicocele is detected and repaired early enough, then future infertility might be avoided.

The problem with this logic is that a minority of men with varicocele are infertile and currently there is no method to predict which adolescents with varicocele will be infertile as adults. When this uncertainty of benefit is considered along with the risk of surgical complications or of recurrence of the varicosity, most urologists still choose to leave adolescent varicoceles untreated unless there is a significant retarded growth of the ipsilateral testis. A significant disparity in testicular size documented on serial examination is indicative of a significant pathological lesion and is accepted as an indicator for surgical intervention. Such differences have to be arrived at with the use of orchidometry or volume determinations with ultrasound since simple palpation can lead to inaccurate estimation. What constitutes a "significant disparity" differs among experts, but a 10–20% difference between ipsilateral and contralateral testes is generally accepted as significant.

In all cases, parents or guardians must be presented with the biological and psychological issues surrounding adolescent varicocele so that joint decisions about treatment are made on a fully informed basis.

III. VARICOCELE INCIDENCE AND ETIOLOGY

It has been estimated that between 20,000 and 40,000 varicocele operations take place every year. Despite this fact and despite the several decades of attention being paid to the association between varicocele and infertility, many questions remain regarding how varicocele develops, whether varicocele actually causes testicular dysfunction, and, if so, how the varicocele causes that testicular dysfunction.

A. Development of the Varicocele

Results of physical exams have shows that 1% or less of prepubertal males have a varicocele. This incidence increases with puberty to a point that approximately 15% of the adult male population has a clinically evident varicocele. Surveys of the adolescent population generally report the same frequency of varicocele as in the adult population, so the appearance of the lesion must be quite rapid at the initiation of puberty. For this reason, the appearance of varicocele is likely more related to Tanner scale (pubertal development) than to age.

The estimate that 15% of the adult male population has a varicocele is derived from the grand mean of six different studies (range of estimates, 4–23%). A recent study combining physical exam with both gray-scale and color Doppler sonography resulted in a similar conclusion. Interestingly, 90–95% of these varicoceles occur on the left side only. Abnormalities of venous valves, degenerative changes in the venous walls, and increased intravenous hydrostatic pressure in the scrotal vessels due to man's upright posture have been offered as general reasons for the development of varicocele. The tendency for the varicosity to form on the left side is believed to be due to the longer length of the spermatic vein on the left than the right, the insertion of the left spermatic vein into the left renal vein (the right inserts into the inferior vena cava), or, relatedly, the obstruction of the left renal vein by the superior mesenteric artery. All of these causes are hypothetical and are usually accompanied by little evidence. Hypotheses about causation cannot be tested directly in humans, but it is clear from studies in a variety of experimental animals that partial obstruction of the left renal vein can cause the development of a left spermatic varicosity. Such obstruction is proposed to happen in humans when the left renal vein is compressed against the aorta by the superior mesenteric artery, the so-called "nutcracker" phenomenon. Renal vein compression occurs when the origin of the superior mesenteric artery at the aorta occurs too close to the upper margin of the left renal vein as the vein passes in front of the aorta. In such cases, the left renal vein can be compressed in the tight angle formed between the aorta and the superior mesenteric artery, and this compression can lead to partial obstruction of the vein and to an increase in intravenous pressure lateral to the obstruction. Since the left spermatic vein inserts to the left renal vein, also laterally to the obstruction, any pressure increases in the left renal vein can be transmitted directly to the left spermatic vein.

Such pressure increases eventually caused the development of the left spermatic varicosity in experimental animals, and this is consistent with the hypothesis that such occurs in humans. This is only one scenario. In reality there is probably more than one cause for the development of varicocele, and the prominent cause in one individual might not be the prominent cause in another.

IV. CLINICAL EFFECT OF VARICOCELE: FACT OR FICTION?

While approximately 15% of human males have a varicocele, it is generally accepted that only 5% of men suffer from infertility. Clearly, the presence of a varicocele does not predict infertility. On the other hand, the incidence of varicocele is increased in the infertile population. A survey of nine different studies indicates the presence of varicocele approximately 31% of infertile males, which is slightly more than twice that in the general population. Such evidence indicates, but clearly does not prove, a relationship between varicocele and infertility. Also, the effect of varicocele on fertility appears to be progressive. That is, the longer the patient has had the varicocele, the greater the apparent testicular dysfunction, again suggesting, but not proving, a relationship between varicocele and infertility.

The common reports of improvement in semen parameters or increases in fertility after varicocele repair have been taken as evidence that varicocele is directly related to testicular dysfunction, but the data supporting this conclusion are not straightforward. While the majority of reports do show an improvement of fertility or ejaculate quality after varicocele repair, some reports also show no improvement in the same parameters. It is important to recognize that individual investigations vary regarding the repair procedure used, the specifics of the populations treated, the parameters evaluated, and the standards used to claim an "improvement." More important, the vast majority of studies contain no internal control group with which to make data comparisons. These factors have made it difficult to reach any conclusion based on information in the literature. Because of this unresolved question about the effect of varicocele on infertility, a World Health Organization Task Force undertook a large, multicenter study of the association between varicocele and infertility. Their report contained data from 9034 males presenting for infertility at 34 centers in 24 countries. The task force concluded that varicocele is clearly associated with impairment of testicular function and infertility. Finally, two controlled, randomized trials are now available in the literature. Both studies found significant improvement in semen parameters after varicocele repair, whereas similar patients given no surgery showed no improvement. Madgar *et al.* also evaluated fertility in their patients and found a significant benefit with varicocele repair. Thus, current data indicate there is an association between varicocele and testicular dysfunction, and there is a benefit to varicocele repair. The mechanism underlying these associations has been the focus of considerable investigation.

V. PATHOPHYSIOLOGY OF VARICOCELE

Many hypotheses of the mechanism relating varicocele and testicular dysfunction have been examined. By far, most investigations have occurred in human subjects in which experimentation is limited by ethical and medical concerns. In a growing number of areas, however, animal models have been used to allow investigation in greater detail. Most clinical investigations have examined hypotheses that fall generally under four headings: (i) insufficiency along the hypothalmo–hypophyseal–gonadal axis, (ii) retrograde blood flow down the spermatic vein allowing reflux of renal or adrenal metabolites to the testis, (iii) elevation of testicular temperature leading to decreased spermatogenesis, or (iv) reduced blood flow to the testis leading to oxygen deficiency, carbon dioxide excess, or the accumulation of testicular metabolites.

A. Insufficiency along the Hypothalmo–Hypophyseal–Gonadal Axis

Plasma testosterone concentrations in varicocele patients usually fall within normal ranges, but it has been reported that *in vitro* testosterone synthesis by biopsy specimens from testes with varicocele is significantly less than that by specimens from controls. Also, the general conclusion of endocrine investigations from several labs is that an endocrinopathy exists in at least some men with varicocele. Whether this endocrinopathy is a result of the varicocele is unknown. Data from laboratory animals with experimental left varicocele imposed by a partial obstruction of the left renal vein have shown that intratesticular testosterone concentrations do decline significantly bilaterally within weeks of the surgery to cause the varicocele. Follow-up experiments examining pituitary gonadotroph, responsivity to gonadotropin-releasing hormone, and Leydig cell responsivity to luteinizing hormone found no differences between control animals and those with experimental varicoceles, however. It has been postulated that the decrease in intratesticular testosterone concentrations in the experimental model is due to a washout phenomenon subsequent to an increased testicular blood flow. This might be a factor in the human as well and could portend lower intratesticular testosterone in the human despite a normal testosterone concentration in peripheral plasma.

B. Retrograde Blood Flow Down the Spermatic Vein: Reflux of Adrenal or Renal Metabolites to the Testis

Retrograde flow in the spermatic vein is presumed *a priori* to be a part of the condition of varicocele. Venographic evidence has reinforced this impression as has the filling of the varicocele during the Valsalva maneuver or when the patient stands upright from a horizontal position. This presumed retrograde flow has been used to explain how renal or adrenal metabolites supposedly arrive at the testis, or both testes for that matter, and cause changes in various aspects of testicular physiology. Often, the following factors are not taken into account: that retrograde flow is not the only explanation for scrotal filling during Valsalva or after rising, that proper perfusion technique is important in studies of blood flow using venography, that common venographic techniques block normal venous outflow, and that perfusion pressures (never reported in retrograde venography studies) can artifactually cause retrograde filling of vasculature with contrast dye. Furthermore, radiocontrast dyes have a much higher specific gravity than blood and can gravitate downward against the direction of positive left spermatic vein blood flow.

In any case, studies of the incidence of retrograde flow in normally fertile patients, those with varicocele and normal semen, and those with varicocele and abnormal semen have concluded that the incidence of reflux in these groups is not different. Furthermore, specific studies of retrograde blood flow in the rat model of varicocele, a model which shows decreases in spermatogenesis and other testicular alterations after establishment of the varicocele, concluded that reflux of blood down the spermatic vein does not occur in that model and that reflux is not necessary for the testis to respond to the presence of the varicocele.

While it is possible for retrograde blood flow to occur in the human, it cannot be used to claim unique presentation of toxic metabolites to the testis. First, the incidence of reflux is similar in varicocele patients and normal patients as mentioned previously. Second, radiocontrast dyes injected in a retrograde fashion into the left spermatic vein have not been shown to reach the testis itself. It is likely that any refluxing blood eventually encounters a collateral vessel which will carry the blood away from the testis (Fig. 3). Finally, the search for adrenal or renal metabolites in spermatic vein or varicocele patients has never produced consistent evidence that such metabolites reach significant concentrations in spermatic vein blood, let alone testicular blood. Reflux is unlikely to be a significant link in the direct mechanism of testicular dysfunction and is not specifically associated with varicocele.

C. Elevation of Testicular Temperature Leading to Decreased Spermatogenesis

Zorgniotti and MacLeod reported a decrease in sperm concentrations associated with increases in testicular surface temperature in men with varicocele. This has been reported in other studies as well, but often as an ipsilateral effect only. Direct, intratesticular thermistor probes in experimental animals have documented a bilateral increase in testicular

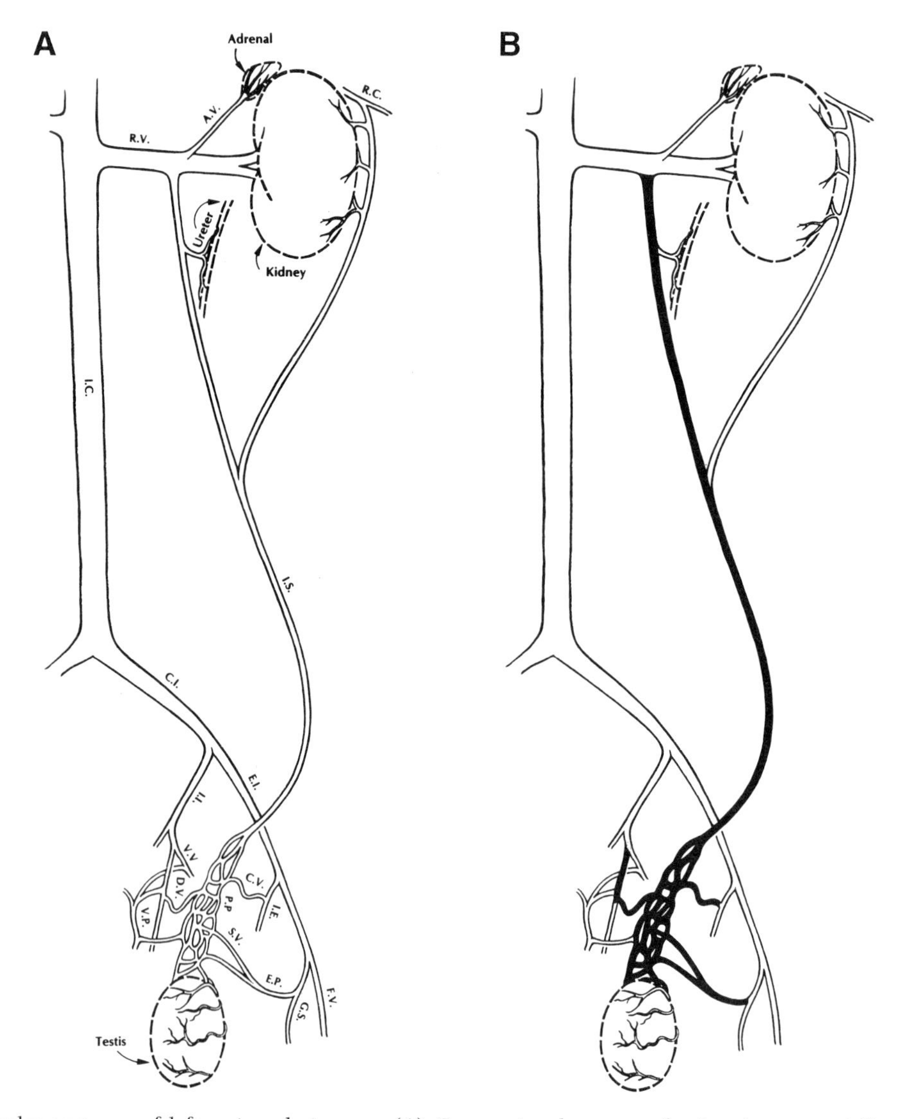

FIGURE 3 Vascular anatomy of left varicocele in man. (A) Conventional pattern of veins that potentially serve as routes of effluent blood flow from the left testis. I.C., inferior vena cava; R.V., renal vein; C.I., common iliac vein; I.E., inferior epigastric vein; V.P., vesicular plexus; P.P., pampiniform plexus; E.I., external iliac vein; I.I., internal iliac vein; F.V., femoral vein; C.V., cremasteric vein; D.V., deferential vein; V.V., vesicular vein; E.P., external pudendal vein; S.V., scrotal vein; R.C., renal capsular vein; A.V., adrenal vein; G.S., great saphenous vein; I.S., internal spermatic vein. (B) Vessels consistently noted in previous studies as being major effluent routes in men with varicocele are shown in black. Vessel sizes are not to scale. Reprinted with permission from the American Society for Reproductive Medicine (*Fertility and Sterility*, 1994, **62**, 869–875).

temperature after surgery to impose a unilateral varicocele. Later use of such probes intraoperatively in human patients with varicocele has confirmed that varicocele increases testicular temperature, often bilaterally. The increases in testicular temperature are moderate but sufficient to bring temperatures within ranges known to cause a reduction in spermatogenesis.

D. Reduction in Testicular Blood Flow Causing a Carbon Dioxide and Oxygen Imbalance or the Accumulation of Testicular Metabolites

It has sometimes been presumed that blood flow up the varicose left spermatic vein is inefficient and that testicular blood flow, at least in the ipsilateral testis, is subsequently reduced from normal. Testicular hypoxia has not been detected in either human patients with varicocele or in the experimental models, however. Rather, testicular blood flow has been shown to be significantly increased bilaterally in the varicocele model, and recent work by Ross *et al.* indicates the same possibility in humans with varicocele. While this finding is contrary to the common presumption of reduced testicular blood flow in the human with varicocele, it is important to realize that only an increase in testicular blood flow is physiologically consistent with the increase in testicular temperature found in both humans with varicocele and in animals with experimental varicocele. This is because increased blood flow through the pampiniform plexus decreases the efficiency of the countercurrent heat exchange mechanism and causes an increase in testicular temperature. Reduced blood flow would cool the testis, not heat it.

VI. PHYSIOLOGY OF VARICOCELE REPAIR: A THEORY

Figure 3a shows that the internal spermatic vein is not the only route for blood to leave the testis. In several studies in the literature there appear to be the following four commonly noted collateral effluents, all arising at the level of the pampiniform plexus: (i) the cremasteric vein that contributes to the inferior epigastric vein, which contributes to the external iliac; (ii) the scrotal veins that contribute to the external pudendal, which contributes to the great saphenous, which then contributes to the external iliac; (iii) the deferential vein that contributes to the vesicular vein, which contributes to the internal iliac; and (iv) occasional collaterals to the vesicular plexus, which also eventually contributes to the internal iliac via other vessels. Left varicocele causes a dilatation of the left spermatic vein and the pampiniform plexus and, variably, the collateral vessels in the routes mentioned (Fig. 3b). The collateral vessels consistently noted in a variety of studies to be enlarged in patients with varicocele, in order of their general (though not universally accepted) prominence, are (i) the cremasteric vein, (ii) the external pudendal vein, and (iii) the deferential vein (Fig. 3b). Appreciation of this anatomy helps explain, first, why ligation of the spermatic vein does not cause testicular necrosis: Collateral vessels continue to drain the arterial input. Second, it illustrates how any refluxing blood can exit the system without ever reaching the testis. The blood exits through the collateral veins. Third, it suggests how higher than normal testicular blood flow might occur despite increased resistance to flow up the internal spermatic vein. If the collateral vessels are expanded under varicocele, there is possibly more capacity to carry blood from the testis than was present when the spermatic vein was the major effluent and the collaterals were relatively tiny. Local mitogenic factors or vasoactive substances which might increase in-flow on the arterial side of the varicocele testis have not been found, however.

If there is an increased capacity to carry blood away from the testis under varicocele, then ligation of the spermatic vein should reduce that capacity and return blood flow to within normal ranges. This might explain why varicocele repair in the varicocele model returns testicular blood flow and spermatogenesis to normal and why varicocele repair in the human patient increases semen quality and fertility potential. A significant weakness in this proposition is that it does not explain bilateral testicular responses to unilateral varicocele or varicocele repair. Results from the experimental model do not generally support a hypothesis of humoral factors affecting the contralateral testis.

VII. CONCLUSIONS

Varicocele is a lesion that occurs in a high proportion of infertile men, but not all men with varicocele are infertile. It seems probable that varicocele alone is not sufficient to render an otherwise reproductively healthy male infertile; however, varicocele can occur in tandem with some other condition. That other condition might not in itself render the male infertile, but it might put him at the margin of infertility with regard to sperm concentration, sperm motility, or sperm morphology. The effect of varicocele, whether affecting only the ipsilateral testis or affecting bilateral testes, could add a further decrement in overall testis function that, when summed with the underlying defect, could induce infertility. Thus, in some males varicocele repair will allow a return to fertility but in males in which the primary problem is not the varicocele, varicocele repair will not improve fertility. One of the most important advances, both clinically and scientifically, would be improvement in the prediction of which males will benefit from varicocele repair and which will not. This increase in predictive capacity will not occur without improved understanding of the mechanisms which produce the testicular response to varicocele.

Direct research on the human testis with varicocele is understandably limited. Unfortunately, animal models cannot fully answer all questions about human varicocele; thus, it is likely that information about some aspects of this lesion will remain, at best, approximations. Nevertheless, important questions remain. Does a unilateral varicocele cause bilateral testicular dysfunction in humans or is the response typically only unilateral? How does unilateral varicocele cause a bilateral testicular effect? Is varicocele alone enough to cause the loss of fertility in some males or is varicocele only a problem when partnered with another pathology? What are the cellular mechanisms by which varicocele exerts its effects on the testis, i.e., which specific cell types in the testis are affected and what is the molecular basis for this effect? These are only examples of the many unresolved questions about varicocele, and research at both the physiological and cell biological levels must continue before we fully understand this enigmatic lesion.

See Also the Following Articles

MALE REPRODUCTIVE DISORDERS; PAMPINIFORM PLEXUS

Bibliography

Goldstein, M., and Eid, J. F. (1989). Elevation of intratesticular and scrotal skin-surface temperature in men with varicocele. *J. Urol.* **142**, 743–745.

Hurt, G. S., Howards, S. S., and Turner, T. T. (1988). Repair of experimental varicocele in the rat: Long term effects on testicular blood flow and temperature and cauda epididymal sperm concentration and motility. *J. Androl.* 7, 271–276.

Laven, J. S. E., te Velde, E. R., Haans, L. C. F., Wensing, C. J. C., Mali, W. P. T. M., and Eimern, J. M. (1992). Effects of varicocele treatment in adolescents: A randomized study. *Fertil. Steril.* **58**, 756–762.

Madgar, I., Karasik, R., Weissenberg, R., Goldwasser, B., and Lunenfeld, B. (1995). Controlled trial of high spermatic vein ligation for varicocele in infertile men. *Fertil. Steril.* **63**, 120–124.

Mali, W. P. T. M., Arndt, J. W., Coolsaet, B. L. R. A., Kremer, J., and Oei, H. Y. (1984). Haemodynamic aspects of left-sided varicocele and its association with so-called right-sided varicocele. *Int. J. Androl.* 7, 297–308.

Meacham, R. B., Townsend, R. R., Rademacher, D., and Drose, J. (1994). Incidence of varicoceles in the general population when evaluated by physical exam, gray scale sonography, and color doppler sonography. *J. Urol.* **151**, 1535–1538.

Meacham, R. B., Lipschultz, L. I., and Howards, S. S. (1997). Male infertility. In *Adult and Pediatric Urology* (J. Y. Gillenwater, J. T. Grayback, S. S. Howards, and J. W. Duckett, Eds.), 2nd ed. Mosby, Baltimore.

Nagler, H. M., Luntz, R. K., and Martinis, F. G. (1997). Varicocele. In *Infertility in the Male* (L. I. Lipshultz and S. S. Howards, Eds.), 3rd ed. Mosby, Baltimore.

Pryor, J. L., and Howards, S. S. (1987). Varicocele. *Urol. Clin. North Am.* **14**, 499–513.

Ross, J. A., Watson, N. E., and Jarow, J. P. (1994). The effect of varicoceles upon testicular blood flow in man. *J. Urol.* **44**, 535–539.

Saypol, D. C. (1981). Varicocele. *J. Androl.* **2**, 16–71.

Saypol, D. C., Howards, S. S., Turner, T. T., and Miller, E. D. (1981). Influence of surgically induced varicocele on testicular blood flow, temperature, and histology in adult rats and dogs. *J. Clin. Invest.* **68**, 39–45.

Tullock, W. S. (1952). A consideration of sterility factors in the light of subsequent factors. *Trans. Edinburgh Obstet. Soc.* **59**, 29–33.

World Health Organization (1992). The influence of varicocele on parameters of fertility in a large group of men presenting to infertility clinics. *Fertil. Steril.* **57**, 1289–1293.

Zorgniotti, A. W., and MacLeod, J. (1973). Studies in temperature, human semen quality, and varicocele. *Fertil. Steril.* **24**, 854–863.

Vas Deferens

see Male Reproductive System

Vasectomy

Sherman J. Silber
Infertility Center of St. Louis, Saint Luke's Hospital

GLOSSARY

epididymis A coiled, 20-ft long, fragile, microscopic tubule that carries sperm from the testicle into the vas deferens.

epididymitis Inflammation and dilation of the epididymis.

microscopic vasectomy reversal A procedure in which the tiny inner canal of the vas deferens—which is roughly the size of a pinpoint—is microsurgically reconnected to restore fertility. Then the relatively thick muscular wall is sutured to ensure proper muscular contraction forming the sperm into the ejaculate. About 98% of patients develop normal sperm count after reversal.

sperm granuloma A lump at the vasectomy site caused by sperm leakage.

testicles The glands responsible for the production of sperm and the male hormone testosterone, which is needed for the development of male sexual characteristics and behavior.

vas deferens A sperm duct about one-eighth of an inch in diameter. The inner canal of the duct is about 1/70 to 1/100 in. diameter and carries the sperm out of the testicle into the ejaculate.

vasectomy The cutting and sealing of the male sperm duct (vas deferens) for the purpose of sterilization.

I. WHAT IS VASECTOMY?

Vasectomy is the cutting and sealing of the male sperm duct (vas deferens) for the purpose of sterilization. Vasectomy is the most popular method of birth control in the world today. It is a simple operation that can be completed in a few minutes in the doctor's office under local anesthesia with just a handful of surgical instruments. It can just as easily be performed in a tent or a hut in developing world countries and therefore is the bulwark of massive population-control programs in countries such as China, India, Bangladesh, and those in Southeast Asia. It involves a simple severing of the vas deferens, which carries the sperm from the testicle and epididymis into the ejaculatory duct. Because this duct can easily be palpated just underneath the scrotal skin, the operation requires only a tiny eighth-inch incision and can be accomplished simply in a matter of minutes. It can even be performed with just sharpened hemostats with no incision. The only drawback to vasectomy has been its irreversibility.

There are about 1 million sterilizations performed in the United States every year, and almost half of those are vasectomies. Most men who undergo vasectomy are over 30 years of age, have two or three children, have already used other methods of birth control with their partner, and have decided they want a quick, simple, safe, permanent end to their worry about having more children. In many cases their wives have suffered from ill effects from birth control pills or have had painful cramping periods or infections from an IUD. Some already have unwanted children.

There is no effect on the man's sex drive or sexual ability. The patient who has had a vasectomy notices no difference in his orgasm or the amount of fluid in his ejaculation. It simply is all fluid and no sperm. He notices no physical change in any aspect of sex, except that he is unable to impregnate. Ninety-five percent of the fluid of the ejaculate issues from the prostate gland and seminal vesicles. At the time of ejaculation, normally, sperm are pumped from the epididymis, through the vas deferens, and up to the area of the ejaculatory duct at the base of the penis. There is very little fluid carrying the sperm to the ejacultory duct. Most of what constitutes the ejaculatory fluid meets the sperm at the ejaculatory duct and literally carries the sperm out, propelled with force from the penis. When the vas deferens is interrupted by vasectomy, none of this changes except that sperm never arrive at the ejaculatory duct. The ejaculation otherwise functions as though nothing were different.

The testicles produce testosterone, which continues to provide the men with their sex drive and secondary sex characteristics such as a beard, low voice, increased musculature, higher red blood cell count, increased facial oiliness compared to women, and, even to some extent, aggressiveness. Hormone levels in the blood are unaffected by vasectomy because the hormones are released directly into the bloodstream.

II. REVERSIBILITY OF VASECTOMY

Vasectomy has been considered irreversible in the past because of the difficulty of reuniting the tiny, delicate, inner canal, which is about 1/70 in. in diameter, or smaller than a period on this page. However, with microsurgical techniques, vasectomy can now usually be reversed. Unfortunately, the reversal operation is very intricate because after vasectomy the tiny amount of fluid that carries sperm from the testicles begins to accumulate in the vas deferens. The pressure from this accumulation is not felt at all because the canal is very small. However, pressure does build up, and sperm that otherwise would have been ejaculated instead stagnate and die of old age (the longevity of the sperm in the vas deferens is about 2 weeks to a month). The sperm then gradually decompose. Fluid and sperm accumulate in the epididymis, the tiny, coiled, delicate 20-ft-long tubule that carries sperm out of the testicle into the vas deferens.

Although the amount of fluid from the testicle is tiny compared to the volume of the ejaculate, the pressure that builds up eventually results in microscopic ruptures or "blowouts" in this delicate epididymal duct, causing blockage in that region. Thus, in time sperm can no longer even get to the vas deferens. With extremely refined microsurgical techniques, the damage in this area can be bypassed and the success rate for reversal of vasectomy can still be good. However, the technique is difficult. Thus, vasectomy, for practical purposes, still has to be considered a potentially permanent procedure unless it is performed in such a way as to prevent pressure buildup. Thus, vasectomy should still not be encouraged for those who do not wish permanent sterilization.

III. TECHNIQUE OF VASECTOMY

Vasectomy is a very simple procedure. It is performed in the doctor's office using local anesthesia only. It is much less painful than a simple dental extraction. The surgery requires only 10 min, although somewhat more time may be needed for vari-

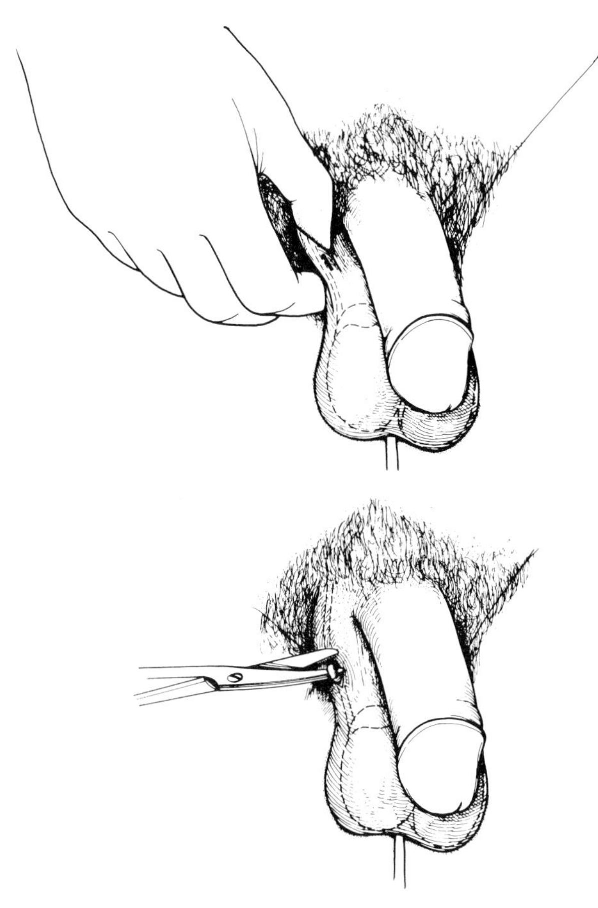

FIGURE 1 (A) Vasectomy incision (feeling vas under skin). (B) Freeing up vas through tiny incision.

ous preparations, such as washing and shaving the scrotum. The incisions are in the scrotum (not on the penis) and generally are only 1/8 in. long. They will hardly be visible postoperatively. The patient will generally feel pain for about 3 sec on each side when the local anesthetic is first injected.

The surgeon will feel the vas between his or her thumb and forefinger and then inject local anesthesia over the area of scrotal skin under which the vas is lying. The vas can be mobilized painlessly by the surgeon's fingers to any area of scrotal skin where he or she wishes to make the incision. Once the local anesthesia has been injected and the area is numb, the surgeon then clamps the vas deferens to the scrotal skin. This secures it just underneath the skin in that particular region where the anesthesia was injected.

A $\frac{1}{8}$-in. incision is then made just over the vas. The tiny incision goes through multiple layers of connective tissue that have all been squeezed between the vas and the scrotal skin by the clamp. The surgeon knows he or she has reached the surface of the vas deferens when these layers of "shiny" connective tissue are no longer seen and then he or she sees the more "dull" surface of the vas deferens. The physician then grasps the vas deferens through this incision with another tiny clamp, which allows him or her to pull it right out of the tiny incision (Fig. 1). The surgeon then either divides the vas with a scissors or cuts a piece of it out. The reason for removing a piece of vas is to ensure a large enough space between the two cut ends so they are less likely to heal together spontaneously and risk sperm getting through to the other side. The gap created by removing a large area of vas at the time of vasectomy can always be bridged during a reversal operation. Other considerations are more important regarding the issue of making vasectomy easily reversible, mainly preventing pressure buildup. Of course, a large area of vas need not be removed, but often it is; if so, reversibility should not be a problem.

If the vas were simply cut and nothing further done, sperm would continue to leak out of the end of the vas coming from the testicle and form an inflammatory cluster of scar tissue known as a sperm granuloma.

A better way to prevent this spontaneous "recanalization" is to seal the ends of the vas properly. In the late 1960s and early 1970s, when vasectomy was beginning to surge in popularity, the two ends of the vas were sealed simply by tying surgical thread around each end. This probably is the least effective way of sealing the vas. The problem is that when the surgical thread, or suture, is tied, the stump on the other side of the tie loses its blood supply, dies, withers away, and then sperm begin once again to leak out of the stump. This problem was not at first appreciated by surgeons because this method has been used successfully for a hundred years to occlude blood vessels. However, from the mid-1970s to the mid-1980s, another method of sealing the vas, cautery, became more popular because it was much more reliable.

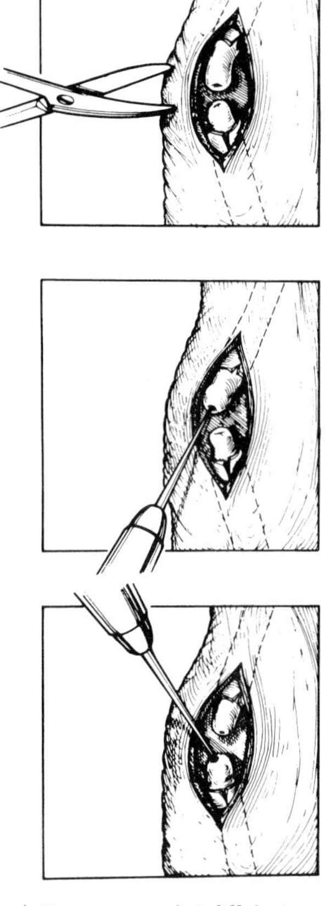

FIGURE 2 (Top) Cutting vas. (Middle) Cauterizing vas side toward ejaculatory duct only (open-ended). (Bottom) Cauterizing testicular end of vas and vas side (close-ended).

Cautery simply means burning. When the vas is cut, a little wire (or needle electrode) is slipped into the opening down a distance of about one-half inch into the vas canal. A button is pressed and the wire heats up, thus burning the delicate mucosal lining but leaving the outer musculature of the vas unburned. The inside of the vas becomes sealed solidly and does not allow sperm to leak.

The only problem with cautery is that it is "too good." Since its widespread adoption, there has been a much greater (and earlier) increase of pressure on the testicular side of the vas deferens after vasectomy. This has resulted in an earlier occurrence of blowouts in the more delicate epididymal duct near the testicle. Thus, much more difficult microsurgery must be performed to reverse this type of vasectomy. This greater pressure buildup from cautery also causes more postoperative pain. Nonetheless, cautery is very appealing because it so efficiently seals the vas that the chance for spontaneous recanalization (and thus failure of the sterilization procedure) is extraordinarily low. Another approach is to seal only the "receiving" end, leaving the testicular end of the vas open so that sperm can continue to leak and thus prevent pressure buildup (Fig. 2). This is the so-called "open-ended" approach.

Despite the fact that vasectomy is an extraordinarily simple operation, it is possible for some disastrous complications to occur if performed incorrectly. Several patients actually lost one of their testicles because the spermatic blood supply was accidentally tied off. This is a rare occurrence and is likely to happen only if the doctor is having difficulty locating the vas. This is such a rare complication because the vas has a characteristic "feel" to it that is unmistakably like a copper wire underneath the scrotal skin. The blood vessels to the testicle are always soft and pliable. That is why it is so easy to isolate the vas through a tiny little incision and not to have to visualize any of the other important structures in the scrotum when doing a vasectomy.

IV. DISAPPEARANCE TIME OF SPERM (AND RISK OF REAPPEARANCE)

The first ejaculate after vasectomy will normally have a sperm count of about 35% of what it was before the vasectomy. The next ejaculate after that will have a sperm count of about 35% of the previous ejaculate. After about 10–12 ejaculations there should be close to zero sperm in the ejaculate. For example, if one started with an average count of about 60 million per cubic centimeter, the first ejaculate after vasectomy would have a sperm count of about 21 million cm^2. The second ejaculate after vasectomy would have a sperm count of about 7 million cm^2. The third ejaculate after vasectomy would have a count of about 2.5 million sperm cm^2. The fourth ejaculate would have about 900,000 sperm cm^2. By the 10th to 12th ejaculate, it would be approaching zero. If spontaneous recanalization has not occurred by the 12th ejaculate, the vast ma-

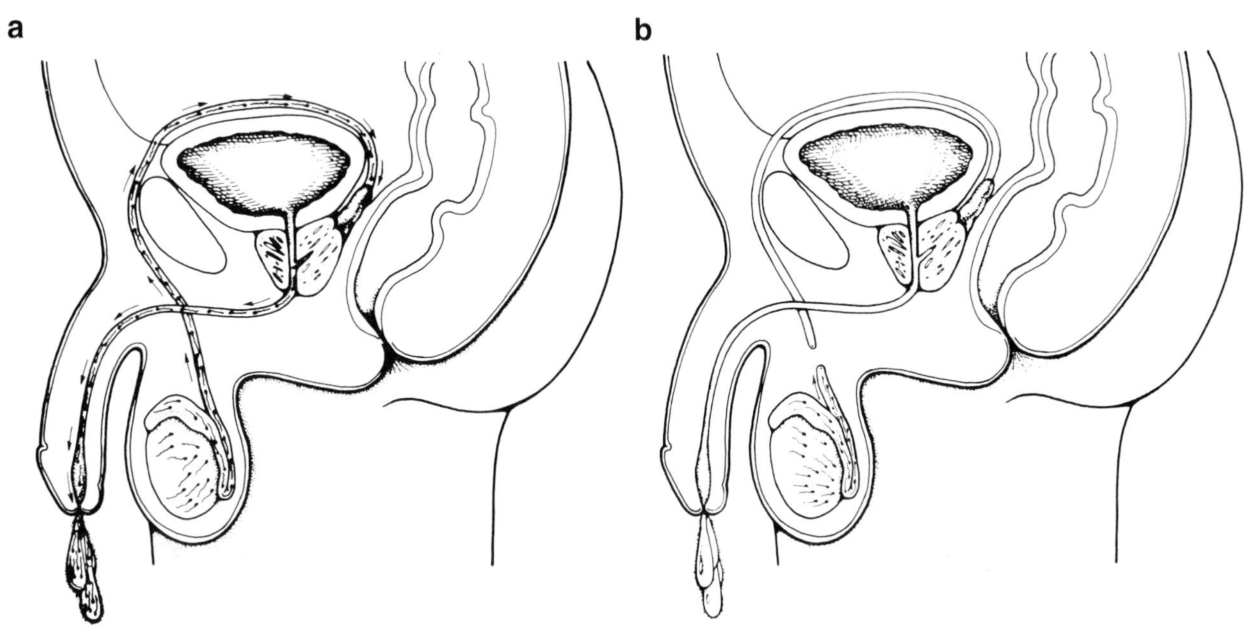

FIGURE 3 (Top) Before vasectomy. (Bottom) After vasectomy.

jority of patients no longer have any sperm whatsoever in their ejaculate (Fig. 3).

V. POSTOPERATIVE COMPLICATIONS

A. Scrotal Swelling

The most terrifying-looking complication after vasectomy is a swollen scrotum, or "hematoma." Because the vasectomy is performed through a tiny incision, in most cases there will be virtually no sign of any surgery having been performed. However, if one is unfortunate enough to have a hematoma, the scrotum will swell up to the size of a grapefruit or possibly even a football. Despite the grim appearance, it is eventually harmless and will heal with no residual problems, but very slowly.

The scrotal tissue is unique compared to that of any other part of the body. It is loose and incredibly expandable. In most areas of the body, the rigidity of the tissue itself stops the bleeding and minimizes the swelling. In the scrotum the tiniest little "bleeder" that would be inconsequential in many other parts of the body keeps on bleeding, and the scrotum keeps on expanding as it fills with as much as a pint of blood.

Despite the painful image that a scrotal hematoma presents, it is truly not a dangerous complication and, if left alone, it eventually will resolve completely over the next 3 months, and the scrotum and testicles will look and feel normal.

What can the doctor do to minimize this risk? He or she should be a good technical surgeon and should make sure to stop all bleeders. The doctor should not put the patient on pain killers such as aspirin that act as anticoagulants. He or she should leave the tiny incision in the scrotum open because if the little incision is closed with a stitch, there is no escape route for blood to drain if bleeding should develop.

B. Sperm Antibodies

There has been a great deal of publicity about the so-called "autoimmune" consequences of vasectomy.

Autoimmune diseases are common. They occur when an immune response to something foreign creates a peculiar type of chronic allergic reaction, and the antibodies so produced actually damage one's own tissue.

Occasionally, the immune system may produce adverse rather than beneficial effects. If the patient has a kidney transplant, the immune system may recognize the kidney as a foreign enemy and attack it. One might wonder whether after vasectomy a man's sperm (which are slightly different genetically from all the other cells in his body) would not be thought of as foreign invaders and stimulate an antibody response. Could this autoimmune response after vasectomy be harmful, or are we protected from it just as the fetus is protected against the mother's immune system?

The body has a remarkable and as yet totally inscrutable mechanism for recognizing that a new baby living within the mother's womb is not to be rejected or attacked by antibodies. The same protection appears to be conferred on sperm so that usually they do not stimulate an antibody response in the woman with each episode of intercourse.

How sure can we be that the rare case of illness occurring after vasectomy might not in truth be caused by it? The only way to find the answer is to have large-scale population studies involving hundreds of thousands of people, some of whom have been vasectomized and some of whom have not, to see whether there is any difference in their health. There is no cause for alarm if a percentage of men get arthritis or heart attacks after their vasectomy if the same percentage of men get arthritis or heart attacks after they start using a condom. The fact that a man may have had a vasectomy prior to developing a health problem does not mean that the vasectomy caused it. Rather, it simply means that a certain percentage of men in any population at any time, whether or not they have had a vasectomy, are going to develop heart disease, arthritis, or other diseases.

About one-half to two-thirds of men will develop sperm antibodies following vasectomy. Widespread publicity concerning this finding created almost hysterical fears that vasectomized men would not only develop autoimmune disease but also have no possibility for restoration of fertility. Both of those notions have been disproven. We can return fertility to the great majority of men who have had a vasectomy, and epidemiological studies have proven that vasectomy results in no autoimmune disease such as arthritis. What about the monkey scare? In 1978 and in 1980, studies suggested that vasectomized monkeys on a high-fat diet developed arteriosclerosis (and presumably heart disease) at a much greater rate than monkeys that were fed the same high-fat diet but that had not been vasectomized. A new wave of hysteria developed not because physicians seriously believed that vasectomy could cause heart disease, but rather because such a study in monkeys created a legal liability risk if patients were not so informed. It was theorized that sperm antibodies created by vasectomy could lodge in the inner wall of the blood vessels and cause arteriosclerosis. What was needed was a large, properly controlled scientific epidemiological study of thousands of patients so it could be determined whether vasectomy should be abandoned as a method of birth control or whether it could be performed safely without risk of endangering health.

Five separate, large-scale epidemiological studies were undertaken. The results showed that vasectomized men had absolutely no increased risk of heart disease, arteriosclerosis, strokes, arthritis, or any other major health disorder. Why did those monkeys get sick? One possibility is that monkeys may be different from humans. The other possibility is that these monkey studies were done with very small numbers, and since the "control" monkeys were not randomly alternated with the vasectomized monkeys to ensure that all other factors were the same, it could very well be that the vasectomized group was at greater risk for other reasons, such as diet or even lifestyle in the cage.

C. Failure of Vasectomy

The major risk of vasectomy is, of course, that it will fail. If there is even a small number of sperm in the ejaculate, there is a risk of pregnancy, and so the operation has failed unless it reduces the sperm count absolutely and unquestionably to zero. Urologists are literally terrified of this complication. It is the largest single cause of lawsuits against urologists, particularly if a defective child is born as a result. This is the so-called "wrongful birth" lawsuit.

The major cause of failure is that sperm manage to leak out of the testicular end of the vas and grind their way through the small amount of scar tissue to get to the other side. When it occurs, this "spontaneous recanalization" almost always happens within 6 months after vasectomy. Different vasectomy techniques have different rates of recanalization. For example, if the doctor simply cuts the vas and ties it off with a suture, 3% of the patients will still have sperm postoperatvely because of recanalization. If the doctor takes out a large segment of vas; adding space between the two cut ends will reduce the risk of recanalization to <0.5%. If the doctor seals the inside opening of the vas by burning it with an electrocautery needle rather than "tying it off," this is much more secure and leads to a recanalization rate of <1/1000. Therefore, the recommended approach to having permanent sterilization with no spontaneous recanalization is to seal the inside of the vas with cautery, with no need to remove a large section.

VI. OPEN-ENDED VERSUS CONVENTIONAL VASECTOMY

There are two reasons to perform an open-ended vasectomy. One is to reduce the occasional risk of long-term aching scrotal pain caused by the inevitable pressure buildup that occurs after regular vasectomy. The other is to increase the ease of subsequent reversibility.

The continual production of fluid from the testicle, with no place to escape, causes a gradual pressure buildup within the vas and the microscopic epididymal tubule. Eventually, there are blowouts and perforations in this delicate epididymal tubule, and this relieves some of the pressure. It is this dilatation and "blowing up" of the epididymis that results in the mild but aggravating pain that occasionally bothers a small percentage of men who have had a vasectomy.

Epididymitis (i.e., inflammation and dilation of the epididymis) occurs in virtually every patient who undergoes vasectomy but usually produces no symptoms. Pain from epididymitis occurs in only about 8% of patients. In a very small percentage of these men the discomfort can be extremely irritating and aggravating.

Vasectomy produces no change in hormone or sperm production of the testicle. The testicle continues to produce fluid and sperm that are transferred into the epididymis and from there into the vas deferens, which is now completely blocked and unable to let any of the fluid out. The question should not be why a small number of patients have pain after vasectomy, but rather why everybody does not have constant, chronic pain after vasectomy.

The reason is that with complete blockage, the tubules of the epididymis dilate quickly, and the muscle wall thins out. As it becomes thinned out and more able to expand, there is less pain from pressure within it. The initial pain is masked by the pain that the patient feels because of the surgery, and within a week the epididymis has dilated so much that it no longer hurts. However, for 8% of the patients it does hurt.

What exactly is a sperm granuloma? It is a lump at the vasectomy site caused by sperm leakage. More than 15 years ago, sperm granuloma occurred in 33% of men after vasectomy. It represents continual sperm leakage at the vasectomy site, thus preventing pressure buildup in the epididymis. If leakage at the vasectomy site were ever to stop, the sperm granuloma would disappear within weeks. The presence of this lump at the vasectomy site does not represent a permanent glob of scar tissue but rather a dynamic pressure-releasing valve. It was present in literally millions of men who had vasectomies in the 1960s and 1970s when the vasectomy was performed with suture only. Over the past 20–40 years these granulomas have caused no harm.

Why did sperm granulomas occur more frequently in vasectomies performed with older techniques than they do now? The common technique in the past was to divide the vas and put a tie of surgical string, or suture, around each end. This tie was meant to seal the two ends of the vas to prevent sperm from leaking out of the testicular side. In case sperm did manage to leak out of the testicular side, the tie on the other side would prevent the sperm from getting into the end of the vas going toward the ejaculate. However, this method of sealing the vas proved to be very poor. Once the vas is tied off, there is no longer a blood supply going to the stub of vas on the other side of the tie. Thus, the little stub on the other side of the tie withers away and falls off, leaving behind a raw end of vas that is able to leak sperm freely. The modern method of sealing the vas with cautery burns the lining of the canal so that it fills up solidly with scar, forming a permanent block. Using this technique, sperm granulomas almost never occur.

The presence of sperm leakage at the vasectomy site does not increase the likelihood of having sperm antibodies. The incidence of sperm antibodies in patients with or without sperm granuloma is about the same. This is because once one has a vasectomy, one physically must have sperm leakage somewhere. The leakage is going to occur either at the vasectomy site, forming the sperm granuloma, or from blowouts in the epididymis. Therefore, no matter how one does the vasectomy, the risk of formation of sperm antibodies is the same.

In summary, a sperm granuloma represents a continual tiny leakage of sperm from the vasectomy site that protects against pressure buildup that would automatically occur after vasectomy. The presence of a sperm granuloma would result in a decreased risk of pain from pressure buildup and would make subsequent reversal of vasectomy much easier to perform.

VII. REVERSAL OF VASECTOMY

For reversal of vasectomy, obstruction is either at the site of the vas reconnection (vasovasostomy) or in the epididymis from blowouts that have been caused by the pressure buildup. If the patient has obstruction at both sites, then reconnection of the vas deferens alone would not restore fertility.

The length of vas removed at the time of vasectomy and the area of vas that was cut should have no effect on success rate. If a large segment of vas has been removed, the gap can be bridged by making the incision larger and freeing up a large enough segment of vas. The success rate should not be any lower in cases in which large portions of the vas have been removed.

In patients undergoing vasectomy reversal with no damage in the epididymis, more than 98% have adequate postoperative sperm counts, and 88% are able to eventually impregnate their wives. This statistic is not significantly different from a normal population of couples trying to get pregnant. The results are certainly not as good when there is epididymal damage. Of patients who have had previous failures of vasectomy reversal who come to our clinic for reoperation, 81% have impregnated their wives. Many of these patients impregnate their wives with a moderately low sperm count.

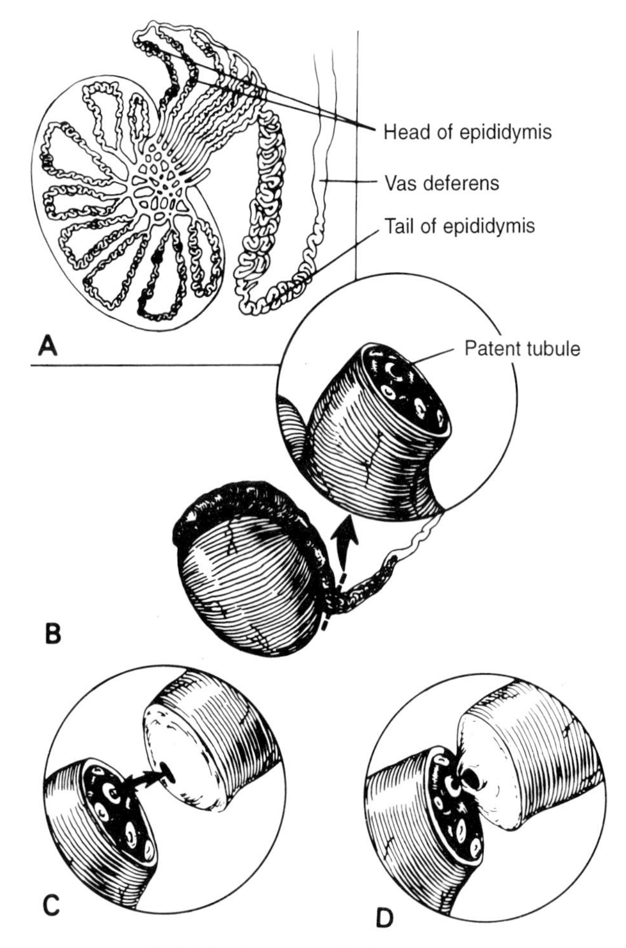

FIGURE 4 Reversal of vasectomy.

To reestablish fertility after vasectomy, the tiny, delicate duct must be microsurgically reconnected with extreme accuracy. If there are epididymal blowouts causing blockage in the ductwork closer to the testicle, the operation becomes 10 times more difficult (Fig. 4). The delicate wall of the epididymis is a thin, filmy membrane $\frac{1}{1000}$ in. thick. The diameter is $\frac{1}{300}$ in., or roughly one-third the size of a pinpoint. If there is blockage in the epididymis caused by pressure-induced blowouts, the vas deferens must be microsurgically reconnected to this much more delicate epididymis, bypassing the blowouts. This means stitching together a $\frac{1}{1000}$-in.-thick tubule. With this sort of very fine microsurgery, most vasectomies can be reversed, despite the secondary epididymal obstruction.

See Also the Following Articles

Contraceptive Methods and Devices, Male; Epididymis; Family Planning; Female Sterilization

Bibliography

Silber, S. J. (1978). Vasectomy and vasectomy reversal. *Fertil. Steril. Mod. Trends* **29**, 125–140.
Silber, S. J. (1979a). Epididymal extravasation following vasectomy as a cause for failure of vasectomy reversal. *Fertil. Steril.* **31**, 3098–3015.
Silber, S. J. (1979b). Open-ended vasectomy, sperm granuloma, and postvasectomy orchialgia. *Fertil. Steril.* **32**, 546–550.
Silber, S. J. (1980). *How to Get Pregnant*. Scribner, New York.
Silber, S. J. (1987). *How Not to Get Pregnant*. Warner, New York.
Silber, S. J. (1989a). Pregnancy after vasovasostomy for vasectomy reversal: A study of factors affecting long-term return of fertility in 282 patients followed for 10 years. *Hum. Reprod.* **4**, 318–322.
Silber, S. J. (1989b). Results of microsurgical vasoepididymostomy: Role of epididymis in sperm maturation. *Hum. Reprod.* **4**, 298–303.
Silber, S. J. (1991). *How to Get Pregnant with the New Technology*. Warner, New York.

Vitellogenins and Vitellogenesis

Gary J. LaFleur, Jr.
Brown University

The increase in size of egg-laying (oviparous) animals has long been a simple indicator of reproductive health. On a molecular scale, most of this size increase is due to an increase of yolk content in growing oocytes. Thus, the formation of yolk or vitellogenesis represents one of those physiological processes that was likely to have been studied since the earliest days of biological observation, even before recorded time. As early as 300 BC it was realized that the yolk of an egg was not part of the embryo proper but rather acted as nourishment for the developing embryo. Aristotle (as quoted in Peck, 1990) offered the following insight while considering the differences in viviparous and oviparous transfer of nutrition to embryos:

> The nourishment for the young of viviparous animals, what we call milk, is formed in breasts, a different part of the body altogether; but for birds Nature provides this inside their eggs.... It is not the white of the egg that is the milk, but the yolk, because it is the yolk that is the nourishment for the chicks. (p. 287)

It was not until 21 centuries later that the mechanisms responsible for yolk formation were begun to be truly understood. In recent years vitellogenesis has received attention both from researchers interested in yolk formation per se and from researchers who have chosen vitellogenesis as a model system to study other biological processes, such as protein evolution, hormonal regulation of transcription and translation, and receptor-mediated endocytosis.

I. HISTORY, TERMINOLOGY, AND THE SCOPE OF THIS ARTICLE

The yolk or vitellus of an egg consists of material that has been stored within the growing oocyte before fertilization to be utilized by the nascent embryo after fertilization as a source of nutrition supporting the processes of early development. Thus, the term "yolk" could theoretically include a large mixture of materials, such as proteins, vitamins, sugars, and lipids, that are all necessary for embryonic development and deposited in the oocyte in storage form (Fig. 1). However, it is the protein component of yolk that has received by far the most investigative attention, and it is yolk protein and its precursors that will be the primary subject of this article. In most vertebrates, such as the African clawed frog, *Xenopus laevis*, more than 90% of the yolk protein is derived from the large yolk precursor protein, vitellogenin (Vtg). Thus, in the following description of yolk formation, I will concentrate primarily on the role that the yolk protein precursor, Vtg, plays in vitellogenesis, as it occurs in vertebrates. For more in-depth treatments of this subject, the reader is referred to reviews by Wallace (1985), Mommsen and Walsh (1988), Byrne *et al.* (1989), and White (1991).

The term "vitellogenin" (L. yolk origin) was coined long after the process of vitellogenesis had been established as a model for study. Pan *et al.* first used vitellogenin to designate the serum form of a yolk protein precursor isolated from the *Cecropia* moth; vitellogenin was thereafter adopted to represent all types of yolk protein precursors in all animals. Now, however, the term vitellogenin is being reserved for yolk protein precursors that have been shown by sequence similarity to belong to an ancient gene family occurring over a vast range of metazoans from nematodes to insects and vertebrates. All members of the Vtg gene family code for proteins that follow a common theme of molecular and cellular characteristics: They are synthesized by a central organ (heterosynthetic origin) under hormone induction, transported by the blood, taken up by the oocyte through receptor-mediated endocytosis, and stored in a membrane-bound compartment (Fig. 4). The most noted yolk proteins that do not adhere to this "Vtg profile" are the yolk protein precursors of *Drosophila melanogaster* that were shown to be more

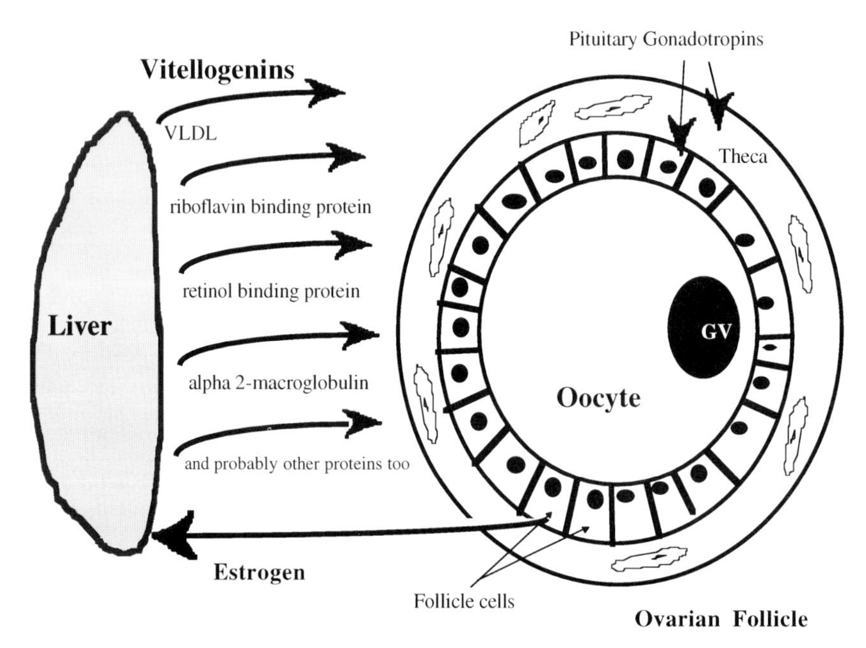

FIGURE 1 Yolk formation or vitellogenesis involves communication between several different cell types, including secretory cells of the hypothalamus and pituitary, germ and somatic cells of the ovarian follicle, and the hepatocytes of the liver. Though a diverse set of proteins are included in yolk, vitellogenins appear to be the major yolk protein precursor in most vertebrates. GV, germinal vesicle.

similar to triacylglycerol lipases than to vertebrate Vtgs (for review see Bownes, 1992). Though these alternative yolk precursor proteins may still be referred to as vitellogenins by some authors, this terminology is becoming less favored. Another type of exception to the Vtg profile also exists: These are the mammalian serum lipoproteins, von Willebrand factor, and apolipoprotein B-100, which are members of the Vtg gene family but do not function as yolk precursors.

II. VITELLOGENIN PRIMARY STRUCTURE

Vitellogenins are large lipophosphoglycoproteins (~200 kDa) that belong to an ancient gene family. The three major yolk proteins that historically have been described as being derived from Vtg can be easily picked out as domains along the primary structure of the cDNA translations. At the N terminus is the domain representing lipovitellin I, at the C terminus lies the smaller but related domain representing the yolk protein lipovitellin II, and these are bisected by a polyserine domain representing the yolk protein phosvitin (Fig. 2). The lability that Vtg displays during laboratory isolations led to the original hypothesis that these three yolk proteins existed as separate entities in the serum of laying hens. However, once protease inhibitors and more gentle isolation techniques were utilized, the native full-length Vtg polypeptides were successfully isolated. Thereafter, it was found that vertebrate Vtgs are cleaved only once they are taken up into the oocyte by receptor-mediated endocytosis.

Once Vtg cDNAs began to be isolated, researchers realized that most animals possessed a whole family of related genes that were expressed as a suite of Vtgs. So far, several vertebrate species have been shown to possess multiple Vtg mRNAs: four Vtgs were reported from *X. laevis*, three from chicken, two from *Fundulus heteroclitus*, and at least two from *Anolis pulchellus*. Other vertebrates whose single Vtg sequences have been reported include the silver lamprey *Ichthyomyzon unicuspis*, the white sturgeon *Acipenser transmontanus*, the rainbow trout *Oncorhynchus mykiss*, and the tilapia *Oreochromis aureus*. The two Vtg sequences that have been completed from *F. heteroclitus* offer the first opportunity in vertebrates to examine the divergence of Vtgs, along their full length, that has occurred within one species. While *F. heteroclitus* Vtgs are about 29–32% identical to the Vtgs of other vertebrates, the two *F. heteroclitus* Vtgs share 45% identity with each other. A comparison of the codon usage and clustering in

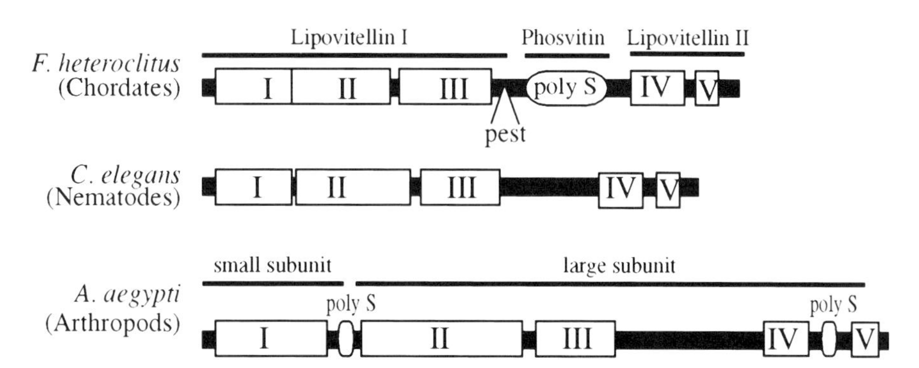

FIGURE 2 Vitellogenins from selected organisms representing three distant phyla and three basic Vtg organization schemes. The five basic subdomains of sequence identity that were described by Chen *et al.* are shown here graphically. Chordate Vtgs can be divided into three primary domains according to their major yolk protein products, lipovitellin I, phosvitin, and lipovitellin II, though some vertebrates such as *Fundulus heteroclitus* derive a more complex suite of yolk proteins. Here the position of a single PEST site is denoted that may affect the processing of YP120 (equivalent to lipovitellin II) in *F. heteroclitus*, though this site does not appear to be conserved among other reported Vtgs. The *Caenorhabditis elegans* Vtgs contain no polyserine domains, but the five conserved subdomains are apparent. The *Aedes aegypti* Vtg contains two polyserine domains and the largest Vtg polypeptide sequence so far reported. The positions of the large and small subunits are shown above the Vtg (modified from Chen *et al.*, 1997).

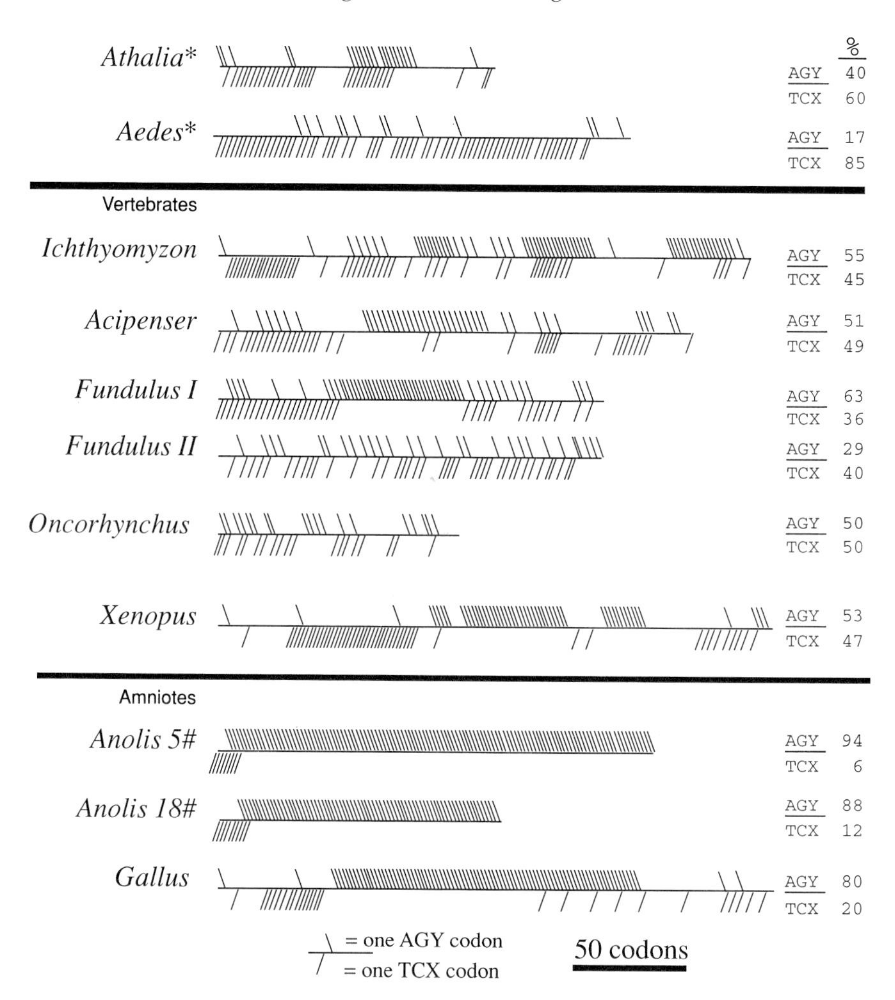

FIGURE 3 The polyserine domain of selected Vtgs, showing the positions of only the serines according to which codon (AGY or TCX) is used. This graphic scheme is used to emphasize the variability in length, codon clustering, and codon usage. Domain boundaries were chosen according to the alignment convention used in LaFleur *et al.* (1995), except for those marked by asterisks, which were chosen according to only the N-terminal polyserine runs of Chen *et al.* (1997), and those marked by #, which were chosen according to only partial sequence database entries. Percentage codon usage tabulations represent number of AGY or TCX codons divided by total number of serine codons. Vtg accession numbers: sawfly, *Athalia rosae*: AB007850; mosquito, *Aedes aegypti*: U02548; silver lamprey, *Ichthyomyzon unicuspis*: M88749; white sturgeon, *Acipenser transmontanus*: U00455; mummichog, *Fundulus heteroclitus* I: U07055, *F. heteroclitus* II: U70826; rainbow trout, *Oncorhynchus mykiss*: X92804, the African clawed frog, *Xenopus laevis*: A2:M18061; lizard, *Anolis pulchellus* 5:U46857, *A. pulchellus* 18:U46856 ; chicken, *Gallus domesticus*: II X13607.

the polyserine domains of these sister Vtgs shows a higher degree of divergence in this domain than is found in the flanking lipovitellin domains (Fig. 3). The disparity in sequence identity between these sister polyserine domains is consistent with the hypothesis that the polyserine domains of Vtgs have evolved independently and at a faster rate than the lipovitellin domains. Even though several cDNAs have been reported from various vertebrates, it appears that more data will be required before orthologous and paralogous relationships can be determined. For instance, it remains unclear whether *F. heteroclitus* Vtg I and Vtg II are more similar to the A subgroup or B subgroup of *X. laevis* Vtgs.

One of the more surprising discoveries with regard to Vtg structure that was recently reported is the occurrence of polyserine domains in the Vtgs of several arthropod species, including the mosquito *Aedes aegypti* (Fig. 3) and the sawfly *Athalia rosae*, with vestigial polyserine domains occurring in the boll weevil *Anthonomus grandis* and the silkmoth *Bombyx mori*. Although the five subdomains that can be aligned between nematodes, arthropods, and vertebrates are convincing evidence that these Vtgs are related, Chen *et al.* argue that the polyserine domains of vertebrates and arthropods are not likely to be of common ancestry. In Fig. 3, I have graphically displayed the codon usage and clustering for serines in the polyserine domain of several vertebrates along with those of two insects. The clusters of AGY or TCX codons suggest that amplification of trinucleotide repeats has played a part in the variability that characterizes this domain. Whereas the insect serine repeats occur in the space intervening subdomains I and II or IV and V and are coded for by a majority of TCX serine codons, vertebrate serine repeats lie between subdomains III and IV and are typically coded for by a majority of AGY codons, though the polyserine domains from *F. heteroclitus* Vtg II and the rainbow trout, *O. mykiss*, are exceptions to this trend (Fig. 3).

III. THE ROLE OF VITELLOGENINS IN VITELLOGENESIS

The role that Vtgs play in the process of vitellogenesis can be divided into three primary steps that are summarized in Fig. 4: (i) estrogen-induced hepatic synthesis, (ii) oocyte-specific uptake by the oocyte vitellogenesis receptor (OVR), and (iii) yolk protein processing and storage. Each of these processes has received varying degrees of examination in only a handful of vertebrate model organisms. Thus, a complete characterization in any one organism is not yet available. The reader should be aware that though not all aspects of vitellogenesis will be conserved throughout vertebrates, common themes can be discerned, and these will be described.

A. Estrogen Induces the Hepatic Synthesis of Vitellogenin

One of the key concepts to be revealed by early studies of vitellogenesis was that the majority of yolk protein contained in the oocyte was not synthesized by the oocyte itself, nor by any other cell of the ovary, but that it was derived heterosynthetically and transported to the ovary, where it was specifically

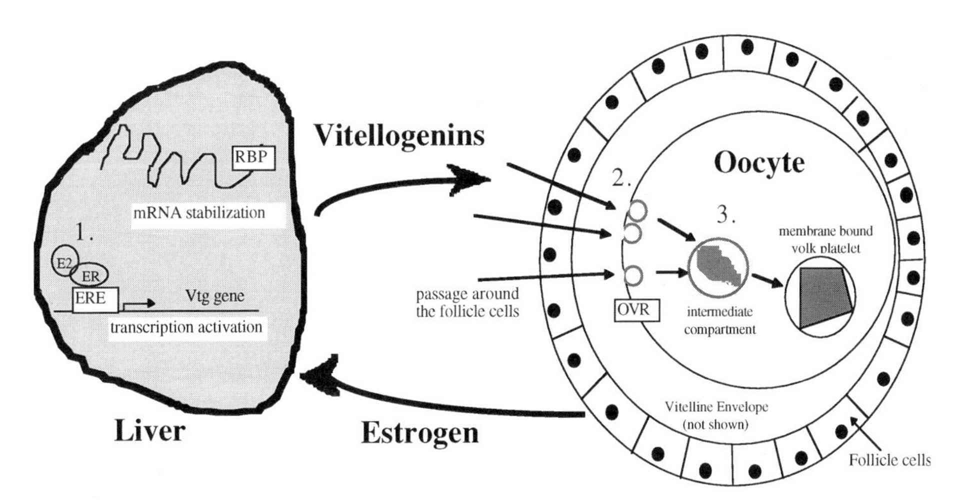

FIGURE 4 A schematic depiction of the three progressive steps that vitellogenins are subject to through vitellogenesis. RBP, RNA-binding protein; E2, estradiol-17β; ER, estrogen receptor; ERE, estrogen response element; OVR, oocyte vitellogenesis receptor. 1, estrogen-induced hepatic synthesis; 2, receptor-mediated uptake via the OVR; 3, processing by cathepsin D and storage in vesicles.

taken up by oocytes. In the insects, this organ was shown to be the fat body, in the nematode *Caenorhabditis elegans* it was the intestine, and in the vertebrates the source of yolk was shown to be the liver. A second key concept was that the normally female-specific synthesis of Vtg could be induced in males by treatment with the female hormone, estradiol-17β. Thereafter, the synthesis of Vtg became a popular model system for the study of steroid-induced protein synthesis.

Eventually, the molecular mechanism of estrogen induction was shown to rely on a pair of 13-base pair (bp) palindromic elements about 325 bp upstream of exon 1 in the chicken Vtg II gene and the four *X. laevis* Vtg genes: A1, A2, B1, and B2. The DNA site supporting interaction with the estradiol-bound estrogen receptor was designated as the estrogen response element (ERE), a *cis*-acting element lying upstream of the Vtg promoter region. The consensus sequence N-G-G-T-C-A-N-N-N-T-G-A-C-C-N showed similarity to the glucocorticoid response element, suggesting that these response elements belong to a common family of *cis*-acting elements. It was later found that proper function of the two EREs in the *X. laevis* Vtg B1 gene is potentiated by a nucleosome-mediated static loop in the chromatin which leads to proper spacing of EREs and their proximal promoter elements. Other proteins whose hepatic synthesis is induced by estrogen, presumably through the action of similar upstream ERE elements, include the estrogen receptor itself, apo-VLDL-II, transferrin, riboflavin-binding protein, α2-macroglobulin, and, in teleosts, a set of vitelline envelope protein precursors (choriogenins).

Besides affecting transcription activity through interaction with upstream DNA elements, estrogen has also been found to augment translation by stabilizing Vtg mRNA. This effect is believed to be maintained in *X. laevis* by a protein designated the RNA-binding protein, whose synthesis is also induced by estrogen. RNA-binding protein was found to increase the half-life of Vtg mRNAs by binding a 27-nucleotide segment of the 3′ untranslated region. Eventually, it was found that proteins similar to this RNA-binding protein occur in all tissue types and these are now suspected to fulfill a broad role in mRNA metabolism.

B. Oocyte-Specific, Receptor-Mediated Endocytosis Sequesters Vtg from the Bloodstream via the Oocyte Vitellogenesis Receptor

Once Vtg is secreted by hepatocytes, it travels in the bloodstream as a dimer and gains entry to the ovary through the vascularized thecal layer of the ovarian follicle. Next, a morphological change takes place whereby follicle cells loosen their connective junctions with adjacent follicle cells, allowing free movement of extracellular proteins from the thecal layer through the interstitial spaces between follicle cells to the surface of the oocyte. Studies using *in vitro* incubated oocytes of *X. laevis* showed that under certain incubation conditions Vtg was taken up 20–50 times more efficiently than other serum proteins. Additionally, it was found that Vtg receptors could recognize exogenous Vtgs (originating from different species). When in tact ovarian follicles (containing oocytes surrounded by follicle cells and theca) were incubated *in vitro*, Vtg uptake could be stimulated by the addition of gonadotropin I. However, when trout oocytes were denuded of both thecal cell layers and follicle cell layers, the gonadotropin induction was lost, thereby suggesting that the action of gonadotropins on oocyte uptake of Vtg is mediated by the somatic cells of the ovarian follicle.

A recent revolution in understanding Vtg uptake has emerged from studies in the lab of W. Schneider where it was found that a strain of chicken deficient in yolk formation (the restricted ovulator strain) possesses a single nucleotide substitution in the coding sequence of the its Vtg receptor. The 95-kDa receptor, now called the OVR, was later found to not only be responsible for Vtg uptake but also to act as the receptor for very low-density lipoprotein (VLDL) as well as α2-macroglobulin, all three of these proteins being estrogen-induced hepatically derived components of chicken yolk protein. The exact mechanism by which one receptor handles the daunting traffic of several incoming ligands has not yet been elucidated; however, recent reports suggest that a second protein, the "receptor-associated protein" (39 kDa), may play a regulatory role in OVR-binding processes.

C. After Uptake, Yolk Proteins Are Processed by Cathepsin D

In the presence of protease inhibitors, native, unprocessed vertebrate Vtgs can be isolated intact from serum; however, full-length Vtg has never been convincingly isolated from ovarian follicles, leading to the belief that vertebrate Vtgs are proteolytically processed as part of their natural uptake and storage in the oocyte. In chicken and *X. laevis*, isolation of yolk proteins typically yields three major yolk species: lipovitellin I, phosvitin, and lipovitellin II (Fig. 2). Over the past 10 years, studies in chickens and *X. laevis* suggested that the protease responsible for this limited and specific event of yolk protein processing is the aspartyl protease cathepsin D. These studies showed that cathepsin D was present in a compartment intermediate between endocytotic vesicles and mature yolk platelets, and that incubations of native Vtg with the isolated enzyme resulted in cleavage products that were virtually identical to native yolk proteins.

In some teleosts such as *F. heteroclitus*, yolk proteins do not strictly adhere to the scenario found in chicken and *X. laevis*. Rather than a distinct set of three yolk proteins, a suite of several yolk proteins is typically found, suggesting a higher degree of proteolytic processing. Also characteristic, but not unique, to the yolk of *F. heteroclitus* is that the yolk proteins are not condensed into a crystalline lattice as exemplified by the yolk granules of chickens or the yolk platelets of *X. laevis*. A third characteristic of *F. heteroclitus* yolk is that besides the initial proteolytic cleavage of Vtg, a second round of proteolysis occurs during maturation. Most notable is that the largest yolk protein YP120 (120 kDa, equivalent to lipovitellin I) can be observed to diminish as maturation proceeds, whereas levels of several of the smaller yolk proteins appear to increase. By performing N-terminal sequencing of isolated yolk proteins, and comparing these sequences to the cDNA translations of two separate *F. heteroclitus* Vtgs, we were able to estimate where precursor–product cleavages take place. These analyses uncovered a PEST sequence within the predicted YP120 C terminus that may offer clues as to why YP120 is subject to further proteolytic processing during maturation (Fig. 2). PEST domains have been defined by Rogers *et al.* as regions of polypeptides that are rich in Pro, Glu, Ser, and Thr. Most proteins containing PEST domains are rapidly degraded in eukaryotic cells; however, these are usually cytosolic proteins, and whether PEST domains can influence the degradation of proteins in a vesicle compartment remains unclear.

IV. WHAT IS THE FUNCTION OF YOLK PROTEINS?

Although as long ago as 300 BC, Aristotle hypothesized that yolk was present in eggs to act as nutrition for developing embryos, the process of yolk utilization by vertebrate embryos has not received vigorous investigation.

General hypotheses for the function of yolk are that it is used as a source of free amino acids for protein synthesis or as a carrier protein for ions, lipids, sugars, and perhaps vitamins. Of these two possibilities, functioning as a carrier protein of specific compounds would provide the more specific set of evolutionary constraints that could be responsible for maintaining even the modest amount of conservation found in Vtgs from insects and nematodes to vertebrates. Recent advances in understanding how arthropods and lower vertebrate larvae metabolize yolk proteins may lead to a reexamination of this question in the near future.

There have been several hypotheses for alternative uses of Vtgs and yolk proteins. One suggestion is that the degradation of yolk proteins in the oocytes of marine teleosts may be involved in a process called oocyte hydration, in which fish that spawn pelagic eggs, such as the Atlantic croaker *Micropogonias undulatus*, decrease the total density of their oocytes by increasing their intracellular volume, providing buoyancy against the saline conditions of the ocean. Greeley *et al.* suggested that some but not all of the hydration force in *F. heteroclitus* eggs may be due to the degradation of yolk proteins and release of free amino acids. Others have proposed that hydration is regulated by ionic fluxes, but neither proposal has been shown to be the dominant effector. If yolk proteins are used in this role in marine teleosts, it would represent an interesting example of a departure from the traditionally expected role of yolk as a nutritive source.

Another function of yolk has been proposed by Terasaki *et al.*, who have observed the Ca^{2+}-dependent resealing of large tears in oocyte plasma membranes by an immediate coalescence of yolk vesicles. This observation may imply an ancient function for yolk vesicles in regulating cytoplasmic damage control.

A third hypothesis that actually involves nutrition, but evokes an alternative delivery system, is that proposed by Kishida and Specker concerning the feeding of tilapia young on the mucus of their mother. Because Vtgs have been measured in the surface mucus of brooding females, and it is common for young to nip at the mothers skin, it was proposed that young fish may gain nutrition by ingesting yolk precursors orally, but this too requires further examination.

A final scenario to consider in reference to the conventional uses of yolk is the evolutionary shift of some animals away from yolk usage, as must have occurred in the rise of viviparous organisms from oviparous ancestors. One component of such a shift in reproductive strategy would be an alteration in hormone regulatory events. Accordingly, Callard *et al.* proposed that the regulatory activity of progesterone in viviparous sharks is twofold: (i) to induce accommodation of the reproductive tract for developing embryos and (ii) to inhibit the processes of vitellogenesis.

Besides the exceptions to the rule presented by the egg-laying monotremes, platypus and echidna, it is generally considered that yolk proteins are nonexistent in mammals. Yolk inclusions similar to those observed in the oocytes of lower vertebrates admittedly cannot be detected in mammalian oocytes. However, after fertilization, the nascent human embryo survives up to 9 days before establishing a utero-placental connection with the mother that could provide blood-borne nutrition. It seems feasible that a small amount of maternal nutritive stores might be transferred to the growing oocytes before ovulation, and that these could be utilized by the nascent embryo before placentation, but currently no Vtgs have been identified in therian mammals. Two human proteins, apo B-100, a component of VLDL and LDL, and von Willebrand factor, have been identified as members of the Vtg protein family. These Vtg-related polypeptides may represent proteins whose functions have shifted over evolutionary time; as the demand for yolk proteins diminished, various domains may have been preserved within a class of new proteins with different functions. On the other hand, if these proteins were able to gain access from the mother to the embryo through the placenta, the opportunity may exist that they would be utilized in their ancestral Vtg-like roles as a source of nutrition for the developing fetus.

Acknowledgments

The author thanks Professor R. A. Wallace for his kindness and generosity with ideas concerning germ cells and livers, and Professor G. M. Wessel for his gracious contribution of advice during the preparation of the manuscript.

See Also the Following Article

Egg, Avian

Bibliography

Baker, M. E. (1988a). Is vitellogenin an ancestor of apolipoprotein B-100 of human low-density lipoprotein and human lipoprotein lipase? *Biochem. J.* **255**, 1057–1060.

Baker, M. E. (1988b). Invertebrate vitellogenin is homologous to human von Willebrand factor. *Biochem. J.* **256**, 1059–1063.

Bownes, M. (1992). Why is there sequence similarity between insect yolk proteins and vertebrate lipases? *J. Lipid Res.* **33**, 777–790.

Brock, M. L., and Shapiro, D. J. (1983). Estrogen stabilizes vitellogenin mRNA against cytoplasmic degradation. *Cell* **34**, 207–214.

Byrne, B. M., Gruber, M., and AB, G. (1989). The evolution of egg yolk proteins. *Prog. Biophys. Mol. Biol.* **53**, 33–69.

Chen, J.-S., Sappington, T. W., and Raikhel, A. S. (1997). Extensive sequence conservation among insect, nematode, and vertebrate vitellogenins reveals ancient common ancestry. *J. Mol. Evol.* **44**, 440–451.

Callard, I. P., Riley, D., and Perez, L. (1990). Vertebrate vitellogenesis: Molecular model for multihormonal control of gene regulation. *Prog. Clin. Biol. Res.* **342**, 343–348.

Dodson, R. E., and Shapiro, D. J. (1994). An estrogen-inducible protein binds specifically to a sequence in the 3′ untranslated region of estrogen-stabilized vitellogenin mRNA. *Mol. Cell. Biol.* **14**, 3130–3138.

Dodson, R. E., Acena, M. R., and Shaprio, D. J. (1995). Tissue distribution, hormone regulation and evidence for a human homologue of the estrogen-inducible *Xenopus laevis* vitellogenin mRNA binding protein. *J. Steroid Biochem. Mol. Biol.* **52**, 505–515.

Fawcett, D. W. (1986). *Histology*, 11th ed., pp. 886–896. Saunders, Philadelphia.

Greeley, M. S., Jr., Calder, D. R., and Wallace, R. A. (1986). Changes in teleost yolk proteins during oocyte maturation: Correlation of yolk proteolysis with oocyte hydration. *Comp. Biochem. Physiol.* **84**, 1–9.

Handley, H. H., and Bradley, J. T. (1987). Vitellogenin degradation in the cricket embryo. *J. Ala. Acad. Sci.* **58**, 77.

Heisberger, T., Hermann, M., Jacobsen, L., Novak, S., Hodits, R. A., Bujo, H., Meilinger, M., Httinger, M., Schnieider, W. J., and Nimpf, J. (1995). The chicken oocyte receptor for yolk precursors as a model for studying the action of receptor-associated protein and lactoferrin. *J. Biol. Chem.* **270**, 18219–18226.

Heming, T. A., and Buddington, R. K. (1988). Yolk absorption in embryonic and larval fishes. In *Fish Physiology Vol XIA. The Physiology of Developing Fish* (W. S. Hoar and D. J. Randall, Eds.), pp 407–446. Academic Press, New York.

Hyllner, S. J., Oppen-Berntsen, D. O., Helvik, J. V., Walther, B. T., and Haux, C. (1991). Oestradiol-17 beta induces the major vitelline envelope proteins in both sexes in teleosts. *J. Endocrinol.* **131**, 229–236.

Jacobsen, L., Hermann, M., Vieira, P. M., Schneider, W. J., and Nimpf, J. (1995). The chicken oocyte receptor for lipoprotein deposition recognizes alpha 2-macroglobulin. *J. Biol. Chem.* **24**, 6468–6475.

Kishida, M., and Specker, J. L. (1994). Vitellogenin in the surface mucus of tilapia (*Oreochromis mossambicus*): Possibility for uptake by the free-swimming embryos. *J. Exp. Zool.* **268**, 259–268.

LaFleur, G. J., Jr. (1996). Estrogen-induced hepatic contributions to ovarian follicle development in *Fundulus heteroclitus*: Vitellogenins and choriogenins. PhD dissertation, University of Florida, Gainesville.

LaFleur, G. J., Jr., and Thomas, P. (1991). Evidence for a role of Na+, K+-ATPase in the hydration of Atlantic croaker and spotted seatrout oocytes during final maturation. *J. Exp. Zool.* **258**, 126–136.

LaFleur, G. J., Jr., Byrne, M. B., Kanungo, J., Nelson, L. D., Greenberg, R. M., and Wallace, R. A. (1995a). Fundulus heteroclitus vitellogenin: The deduced primary structure of a piscine precursor to noncrystalline, liquid phase yolk protein. *J. Mol. Evol.* **41**, 505–521.

LaFleur, G. J., Jr., Byrne, B. M., Haux, C., Greenberg, R. M., and Wallace, R. A. (1995b). Liver-derived cDNAs: Vitellogenins and vitelline envelope protein precursors (choriogenins). In *Proceedings of the Fifth International Symposium on the Reproductive Physiology of Fish* (F. W. Goetz and P. Thomas, Eds.). FishSymp 95, Austin, TX.

Mommsen, T. P., and Walsh, P. J. (1988). Vitellogenesis and oocyte assembly. In *Fish Physiology Vol XIA. The Physiology of Developing Fish* (W. S. Hoar and D. J. Randall, Eds.), pp. 348–406. Academic Press, New York.

Murata, K., Sugiyama, H., Yasumasu, S., Iuchi, I., Yasumasu, I., and Yamagami, K. (1997). Cloning of cDNA and estrogen-induced hepatic gene expression for choriogenin H, a precursor protein of the fish egg envelope (chorion). *Proc. Natl. Acad. Sci. USA* **94**, 2050–2055.

Pan, M. L., Bell, W. J., and Telfer, W. H. (1969). Vitellogenic blood protein synthesis by insect fat body. *Science* **165**, 393–394.

Peck, A. L. (1990). *Arisotle Generation of Animals, English Translation*. Harvard Univ. Press, Cambridge, MA.

Retzek, H., Steyrer, E., Sanders, E. J., Nimpf, J., and Schneider, W. J. (1992). Molecular cloning and functional characterization of chicken cathepsin D, a key enzyme for yolk formation. *DNA Cell Biol.* **11**, 661–672.

Rogers, S., Wells, R., and Rechsteiner, M. (1986). Amino acid sequences common to rapidly degraded proteins: The PEST hypothesis. *Science* **234**, 364–368.

Schild, C., Claret, F. X., Wahli, W., and Wolffe, A. P. (1993). A nucleosome-dependent static loop potentiates estrogen-regulated transcription from the *Xenopus* vitellogenin B1 promoter *in vitro*. *EMBO J.* **12**, 423–433.

Selman, G. G., and Pawsey, G. J. (1965). The utilization of yolk platelets by tissue of *Xenopus* embryos studied by a safranin staining method. *J. Embryol. Exp. Morphol.* **14**, 191–212.

Shibata, N., Yoshikuni, M., and Nagahama, Y. (1993). Vitellogenin incorporation into oocytes of rainbow trout, *Oncorhynchus mykiss, in vitro*: Effect of hormones on denuded oocytes. *Dev. Growth Differ.* **35**, 115–121.

Speith, J., Nettleton, M., Zucker-Aprison, E., Lea, K., and Blumenthal, T. (1991). Vitellogenin motifs conserved in nematodes and vertebrates. *J. Mol. Evol.* **32**, 429–438.

Stifani, S., Barber, D. L., Nimpf, J., Schneider, W. J. (1993). A single chicken oocyte plasma membrane protein mediates uptake of very low density lipoprotein and vitellogenin. *Proc. Natl. Acad. Sci. USA* **87**, 1955–1959.

Tata, J. R., and Smith, D. F. (1979). Vitellogenesis: A versatile model for hormonal regulation of gene expression. *Recent Prog. Horm. Res.* **35**, 47–95.

Terasaki, M., Miyake, K., and McNeil, P. L. (1997). Large plasma membrane disruptions are rapidly resealed by

Ca2+-dependent vesicle–vesicle fusion events. *J. Cell Biol.* **139**, 63–74.
Wahli, W. (1988). Evolution and expression of vitellogenin genes. *Trends Genet.* **4**, 227–232.
Wallace, R. A. (1985). Vitellogenesis and oocyte growth in nonmammalian vertebrates. In *Developmental Biology. Vol. 1. Oogenesis* (L. W. Browder, Ed.), pp. 127–177. Plenum Press, New York.
Wallace, R. A., and Jared, D. W. (1968). Estrogen induces lipophosphoprotein in serum of male *Xenopus laevis*. *Science* **160**, 91–92.
White, H. B. (1991). Maternal diet, maternal proteins and egg quality. In *Egg Incubation: Its Effect on Embryonic Development in Birds and Reptiles* (D. C. Deeming and M. W. J. Ferguson, Eds.). Cambridge Univ. Press, Cambridge, UK.
Yamashita, O., and Indrasith, L. S. (1988). Metabolic fates of yolk proteins during embryogenesis in arthropods. *Dev. Growth Differ.* **30**, 337–367.
Yoshizaki, N., and Yonezawa, S. (1994). Cathepsin D activity in the vitellogenesis of *Xenopus laevis*. *Dev. Growth Differ.* **36**, 299–306.

Viviparity and Oviparity: Evolution and Reproductive Strategies

Daniel G. Blackburn
Trinity College

GLOSSARY

chorioallantois The vascularized respiratory membrane of the amniote egg which contributes to placental formation in viviparous squamates, eutherian mammals, and the marsupial bandicoot.

extraembryonic membranes Tissues that lie external to the embryo and that function in embryo protection, physiological exchange, and placentation; among them are the yolk sac, amnion, chorion, and chorioallantois.

lecithotrophy A developmental pattern in which the yolk of the ovum provides nutrients for embryonic development.

matrotrophy A developmental pattern in which the mother provides nutrients during gestation by a means other than the yolk of the ovum (e.g., oviductal secretions and placental tissues).

oviparity A reproductive mode in which females lay unfertilized or developing eggs that complete their development and hatch in the external environment; also known as "egg-laying" reproduction.

oviparous egg retention A form of oviparity in which the fertilized eggs are retained and begin to develop inside the maternal reproductive tract, are oviposited, and complete their development in the external environment; sometimes termed "extended oviparity."

ovoviviparity An archaic term, now seldom used, that was applied in the past to a wide variety of reproductive patterns including some that are mutually exclusive; these patterns include lecithotrophic viviparity, oviparous egg retention, and pseudoviviparity.

ovuliparity Reproduction by the laying of unfertilized eggs; a form of oviparity.

parition Expulsion by a female of the reproductive product, be it an egg or a neonate; this term subsumes oviposition and parturition.

placenta An organ formed through apposition of embryonic and parental tissues that functions in physiological exchange; mainly found in viviparous species.

pseudoviviparity An unusual form of oviparity in which the eggs are fertilized externally and are brooded in some parental structure such as the stomach, vocal sacs, skin pouches, branchial chambers, or gastrovascular cavity; found in a few anurans, teleosts, and invertebrates.

viviparity A reproductive mode in which females retain developing eggs inside their reproductive tracts and give birth to their offspring; also known as "live-bearing" reproduction.

yolk sac Extraembryonic tissues that surround and digest the yolk and that contribute to placental organs in certain viviparous fishes, most therian mammals, and all viviparous reptiles.

zygoparity Reproduction by the laying of fertilized eggs; zygoparity grades into oviparous egg retention.

Viviparity is a reproductive pattern in which females retain developing eggs inside their reproductive tracts or body cavity and give birth to offspring capable of a free-living existence. Oviparity, in contrast, is a pattern in which females deposit eggs that develop and hatch in the external environment. These patterns can be viewed as "reproductive strategies," patterns that have advantages as well as disadvantages that affect their evolution. An advantage of viviparity, for example, is that embryos are protected and physiologically maintained by the pregnant female. In many viviparous species, the mother provides nutrients to the embryo during gestation, a pattern known as "matrotrophy." Viviparity has originated over 160 times among animals and is found among bony fishes, cartilaginous fishes, amphibians, mammals, and squamate reptiles, as well as in several invertebrate groups. Viviparity and matrotrophy are phenomena of considerable biological interest. They have been studied from the standpoints of morphology, physiology, endocrinology, ecology, and evolution.

I. REPRODUCTIVE MODES OF ANIMALS

A. Oviparity and Viviparity

In the great majority of animal species, females reproduce by laying eggs that develop and hatch in the external environment. However, in some species, females retain fertilized eggs inside their bodies and give birth to their young. These reproductive patterns are known respectively as "oviparity" (egg-laying reproduction) and "viviparity" (live-bearing reproduction). Although viviparity is sometimes viewed as a mammalian phenomenon, many other groups of vertebrates also contain live-bearing species, including sharks, batoids, teleosts, caecilians, anurans (frogs and toads), salamanders, snakes, and lizards (Table 1). In some major taxa oviparity is predominant (frogs and teleosts) or universal (birds and turtles), whereas in others (chondrichthyans and mammals) viviparity is widespread. Several major invertebrate groups contain both oviparous and viviparous species, including insects, arachnids, crustaceans, tunicates, onychophorans, scorpions, sea cucumbers, and molluscs.

TABLE 1
Phylogenetic Distribution of Viviparity among Major Groups of Vertebrates

Taxon	*Frequency (%)*	*Examples of viviparous species*
Agnathans	0	—
Chondrichthyans	65	Mackerel sharks, frilled sharks, whale sharks, butterfly rays, torpedo rays
Osteichthyans	2.8	Coelacanth, surf perch, poeciliids, half-beaks, four-eyed fish
Amphibians	3	*Salamandra atra*, East African frogs (*Nectophrynoides*), various caecilians
Mammals	99	Marsupials, eutherians
Chelonians	0	—
Squamate reptiles	20	Lizards and snakes of numerous families
Crocodilians	0	—
Birds	0	—

Note. Nonviviparous species are classified as oviparous.

The existence of viviparity in so many unrelated animal groups raises a multitude of functional and evolutionary questions. What advantages and disadvantages accrue to viviparous and oviparous reproduction? What selective pressures have led some lin-

TABLE 2
Phylogenetic Distribution of Reproductive Modes and Embryonic Nutritional Patterns among Vertebrates[a]

	Oviparity	*Viviparity*
Lecithotrophy	Crocodilians, chelonians, birds; most squamates and anamniotes	Many squamates, very few amphibians; some chondrichthyans and osteichthyans
Matrotrophy[b]	Monotreme mammals	A few squamates; some caecilians, frogs, and salamanders; some chondrichthyans and osteichthyans; therian mammals

[a] Lecithotrophy and matrotrophy represent extremes of a continuum. Except for therian mammals, viviparous species lie between these extremes and are classified by the predominant source of nutrients for development.
[b] "Matrotrophy" refers to substantial provision of extravitelline nutrients to the embryo prior to birth of hatching. Provision of nutrients after birth or hatching is universal among birds and mammals.

eages to evolve viviparity and what factors have constrained its development in others? By what sequence of evolutionary steps has viviparity evolved from oviparity? How are oxygen and nutrients for development provided in viviparity and oviparity, and how can we account for the diversity in evolutionary terms? Such questions have intrigued reproductive and evolutionary biologists for the greater part of this century. Only in recent decades have we begun to answer such questions with confidence.

B. Embryonic Nutritional Patterns

Not only can we distinguish between species according to whether they lay eggs or give birth to their offspring but also we can characterize them according to how nutrients are provided to the embryos. In most animals, nutrients for development are contained within the yolk of the egg. This embryonic nutritional pattern is termed "lecithotrophy": The word *lecithos* is Greek for "yolk." This pattern contrasts with "matrotrophy," in which the female provides nutrients for embryonic development by means of a placenta ("placentotrophy") or by some analogous arrangement. Lecithotrophy and matrotrophy are extremes of a continuum. In many viviparous species, for example, nutrients for early development are derived from the yolk and later are supplemented by placental sources.

Reproductive mode and embryonic nutritional pattern can be used in combination to characterize animal species (Table 2). Most animals exhibit lecithotrophic oviparity; upon deposition by the female the egg is nutritionally autonomous. Excluding mammals, most viviparous vertebrates are also lecithotrophic; although the egg is retained to term in the female reproductive tract, nutrients are still supplied via the ovulated yolk. In animals with matrotrophic viviparity, nutrients are supplied by the mother during the course of gestation via a placenta or products of the reproductive tract (e.g., uterine or ovarian secretions and sibling yolks). Matrotrophic oviparity is rare but occurs in the monotreme mammals; the eggs absorb large quantities of oviductal secretions before being deposited. The confusing and archaic word "ovoviviparity" has been applied in the past to both oviparous and viviparous lecithotrophy but is now avoided by most reproductive biologists in favor of the bipartite classification system shown in Table 2.

C. Other Reproductive Distinctions

Just as fundamental as the distinction between oviparity and viviparity is a distinction based on the site of fertilization. In external fertilization, the male fertilizes the eggs after the female lays them. In contrast, internal fertilization occurs when males introduce sperm into the female reproductive tract. Accordingly, we can distinguish the oviposition of unfertilized eggs (ovuliparity) from the laying of fertilized eggs (zygoparity). In sexually reproducing animals internal fertilization is a prerequisite for viviparity.

In some oviparous species, females retain their fertilized eggs for some period before laying them; thus, the embryos can be in an advanced stage of development at oviposition. This pattern is termed "oviparous egg retention." Although in theory this pattern grades into viviparity, oviparous amniotic species with prolonged egg retention are rare. We reserve the term viviparity for species in which the embryos are retained to term by the female and in which hatching (if it occurs) precedes or accompanies emergence of the offspring from the mother. In contrast, in oviparous egg retention, the female typically deposits an intact egg that continues to develop in the external environment before it finally hatches.

In some oviparous invertebrates (e.g., certain molluscs and anemones) externally fertilized eggs are taken into the body (i.e., gastrovascular cavity and branchial chambers) of one of the parents. This pattern is best viewed as a type of egg tending. It is analogous to a pattern called "pseudoviviparity" in anamniote vertebrates and can be referred to by this same term.

II. OVIPARITY AND VIVIPARITY COMPARED

A. Site of Development

In oviparous species, egg development occurs in the external environment after oviposition. Fish eggs can be buoyant (most marine osteichthyans), nonbuoyant (freshwater fishes), attached to marine plants via their egg casings (various sharks and skates), buried in gravel (salmonids), or deposited in nests (lampreys and sticklebacks). Eggs of amniotes are terrestrial and can be deposited in protected environments, buried, or laid in nests. Amphibian eggs usually develop in breeding ponds or streams, although in some oviparous anurans, eggs are deposited on terrestrial plants, laid in bubble nests, or carried by one of the parents.

In all viviparous animals, eggs develop inside the body of the maternal parent. However, the site of development varies. In most live-bearing vertebrates embryonic development occurs in the oviduct or its mammalian homolog, the uterus. This situation characterizes the viviparous chondrichthyans, coelacanths, amphibians, squamates, and therian mammals. Thus, an organ that primitively provided a site of fertilization and deposited the egg encasements (e.g., eggshell or jelly coat) has been recruited evolutionarily for accommodation of the egg during development. Teleosts lack a homolog to the oviduct and gestation commonly occurs in the lumen of the hollow ovary. Moreover, in some teleost species (e.g., viviparous poeciliids and labriosomids), gestation occurs within the ovarian follicle; fertilization and hatching therefore precede ovulation. In others (embiotocids and goodeids), early development begins in the ovarian follicle and the embryo is then ovulated into the ovarian lumen where it completes its development. In the absence of a true oviduct, teleost viviparity seemingly evolved through recruitment of the only available reproductive structure capable of developing features to maintain the embryos—the ovary.

B. Extent and Duration of Development Inside the Female

In oviparous animals, most or all of embryonic development typically occurs in the external environment; thus, the eggs at oviposition are entirely undeveloped (ovuliparous species) or laid in an early stage of development. In birds, for example, females lay their eggs within a day or two of ovulation, at the preprimitive streak stage. Female turtles and crocodilians deposit eggs at the gastrula stage, and monotremes do so at the 18-somite stage. As a rule, the duration of gestation in the maternal reproductive tract is much shorter in oviparous species than in related viviparous forms.

Nevertheless, viviparous species vary widely in length of gestation and degree of development of the neonates. In eutherian mammals, for example, gestation lengths range from 20 days [shrews (*Sorex*)] to 660 days [the African elephant (*Loxodonta africana*)] and birth weight ranges from 7 to 91,000 g. The newborns range from very altricial (most rodents and bats) to precocial (ungulates). Gestation length bears an approximate relationship to birth weight in these matrotrophic species. Marsupials give birth to very tiny, altricial neonates (5 mg–0.5 g) after a pregnancy of 12–38 days, depending on the species. Viviparous squamates universally give rise to precocial offspring; pregnancy commonly is 2–4 months, but gestation periods of more than a year are well documented in some lizards. Chondrichthyans also give birth to well-developed neonates and gestation lengths range from 2 or 3 months (some rays) to about 4 years (the soupfin shark *Galeorhinus*). In teleosts and amphibians, offspring commonly are born at a larval stage after a period of pregnancy that is relatively short by chondrichthyan standards. A notable exception is the Eurasian urodele *Salamandra atra*, in which neonates undergo metamorphosis in the oviduct, where they develop for a period of 2–5 years. Among viviparous invertebrates, offspring can be released at larval, pupal, or more advanced developmental stages.

C. Viviparity and Oviparity as Reproductive Strategies

Traditionally, viviparity has been assumed to be inherently superior to reproduction by laying eggs. This assumption reflects a bias toward therian mammals and fails to consider the biological significance of reproductive diversity. A more sophisticated approach considers oviparity and viviparity as reproductive "strategies" that entail both advantages and disadvantages, each of which may differ in their applicability to particular species.

Clearly, oviparity can entail disadvantages not found in viviparity. Oviposited eggs may be subject to a variety of predators as well as microbial attack, flooding, dehydration, ultraviolet light, and temperature extremes. A viviparous female can protect her developing eggs from many sources of mortality, both biotic and abiotic, while buffering her eggs from environmental fluctuations and maintaining them physiologically.

However, viviparity can entail significant disadvantages as well. During pregnancy, the ability of a female to locomote and feed may be affected, with consequent effects on maternal and offspring survival. Space constraints may limit litter size in viviparous forms more than in oviparous relatives because wet mass and volume of a neonate exceeds that of the ovulated yolk. Furthermore, pregnancy may prevent a female from starting her next litter of offspring, with corresponding effects on reproductive output.

Therefore, from the standpoint of life history theory we would expect viviparity to evolve only when the benefits of stages that are evolutionarily intermediate outweigh the costs. Furthermore, viviparity is but one way in which developing eggs can be protected and should not be thought of as the best way to achieve that end in all circumstances. Birds are notable by having alternative ways of achieving the benefits that accrue to viviparity. Avian eggs are cared for and thermoregulated by one or both parents; protected from dehydration and microbial attack by the albumen, eggshell, and shell membranes; and can be protected from predators by parental care, camouflaged shells, and oviposition into constructed nests that are located in arboreal or inaccessible environments. The Antarctic emperor penguin, *Aptenodytes forsteri*, offers a case in point. Paternal care is critical in this species; following oviposition the male incubates the egg for 2 months while the female is away feeding. Egg retention could be severely disadvantageous to the female by requiring her to take on the entire burden of care for the developing egg, with corresponding effects on her own survivorship and reproductive output. The highly specialized nature of avian oviparity may help explain why oviparous egg retention and viviparity have not evolved among birds.

III. EVOLUTION OF VIVIPARITY

A. Reconstructing the Evolution of Viviparity

Questions about how and why viviparity has evolved traditionally have been approached by examining habitats of extant live-bearing species. Unfortunately, simply to consider the ecological conditions under which living viviparous species are found does not allow us to infer the circumstances or selective pressures under which that pattern originated. After all, we need look no further than mammals to see that viviparous vertebrates have radiated into virtually every conceivable habitat.

However, if we could identify and define each of

the independent origins of viviparity that have occurred among animals, we might be able to reconstruct the environmental circumstances in which this pattern evolved each time it originated. From such information we could assess hypothetical selective pressures and constraints. We could also test scenarios for the evolution of viviparity against information on actual lineages. The tools of cladistic analysis have allowed exactly this sort of analysis. Outgroup analysis indicates that oviparity is ancestral for each of the major vertebrate groups (including the traditional vertebrate classes). By superimposing data on reproductive modes over established phylogenies, we have been able to define the independent origins of viviparity. We can also reconstruct stages in the evolutionary transition to viviparity and infer habitats in which this pattern has originated.

B. Origins of Vertebrate Viviparity

Phylogenetic analyses have revealed that viviparity has evolved convergently on at least 142 separate occasions in vertebrate history (Table 3), far more frequently than previously suspected. At least 26 additional origins of viviparity are identifiable among invertebrates. The majority (<100) of these independent origins have occurred among lizards and snakes. In squamates, viviparity often has evolved at subfamilial and subgeneric levels and, in some cases, at the subspecific level. One of the latter origins (represented by the lizard *Lacerta vivipara*) is thought to have occurred during the Pleistocene. At the other extreme, the earliest origin of viviparity among reptiles is represented by the Mesozoic ichthyosaurs; pregnant females with developing fetuses have been found as fossils that date back 180 million years.

The other vertebrate origins of the live-bearing mode are distributed as follows. Viviparity has evolved at least 5 times in living amphibians—once among urodeles, twice among anurans, and at least twice among caecilians. In the osteichthyan fishes, viviparity has evolved at least 15 times. One of these origins is represented by the coelacanth, *Latimeria*, whose viviparity was discovered by dissection of a pregnant female at the American Museum of Natural History. Other origins are represented by poeciliids (e.g., swordtail guppies) and anablepid fishes, as well as goodeids, surf perches, half-beaks, and scorpaenids. Among chondrichthyans, viviparity appears to have evolved on 20 separate occasions. Independent origins of viviparity are represented by the mackerel sharks, carpet sharks and their allies, frilled sharks, angel sharks, torpedo rays, butterfly rays, and a Paleozoic holocephalan.

TABLE 3

Evolutionary Origins of Viviparity and Matrotrophy among the Vertebrates

	Minimum number of origins[a]	
Taxon	*Viviparity*	*Matrotrophy*
Osteichthyans[b]	15	12
Chondrichthyans	20[c]	5
Lissamphibia	5	4
Mammals	1	1
Ichthyosaurs	1	
Squamate reptiles	>100	3
Total	>142	25

Note. All data represent conservative estimates based on phylogenetic analyses. The analysis is based on the assumption that viviparity evolves irreversibly from oviparity.

[a] From Blackburn (1992; unpublished data).

[b] Represented by the clades Teleostei and Actinistia and some extinct basal chondrostei.

[c] Chondrichthyans can be interpreted as exhibiting 11 origins if it is assumed that skates (Rajidae) have reverted to oviparity.

The origins of viviparity that are now recognized (Table 3) are based on minimum estimates; actual numbers are probably higher. Mammals offer a case in point. Marsupial and placental mammals are conservatively interpreted as derivatives of a single Mesozoic origin of viviparity. However, viviparity may well have originated independently in the two groups, which differ in their placental membranes, reproductive physiology, early development, and the uterine sites at which the embryos undergo gestation.

C. Why Has Viviparity Evolved?

Because viviparity in squamate reptiles has evolved so frequently and at such low taxonomic levels, in many cases we can reconstruct the environmental circumstances in which it originated. A recurring correlate of its origin among squamates is cold climate. Squamate viviparity frequently has evolved at high altitudes and high latitudes. Several lines of evidence—circumstantial, descriptive, and experimental—support the hypothesis that viviparity is adaptive under such conditions due to its thermoregulatory benefits. Because development rate is affected by temperature, eggs laid in cold environments may develop too slowly to hatch before the onset of winter or may succumb to low temperatures. As ectotherms, viviparous squamates can act as mobile incubation chambers, regulating their eggs at optimal temperatures for development. Perhaps thermoregulatory benefits also help account for viviparity in the high-altitude *S. atra*. However, cold climate cannot explain the origins of viviparity in tropical and subtropical caecilians and squamates, where its evolution presumably conferred other benefits. In therian mammals, the origin of viviparity may be associated with the pattern of nutrient provision, if it was preceded evolutionarily by altriciality and matrotrophy.

The selective pressures that led to viviparity in fishes are uncertain. Given that predation can take an enormous toll on aquatic eggs, perhaps retention of the eggs inside the female enhances egg survival. Why does such a minuscule proportion of amphibians and osteichthyans give birth to their young? The answer partly lies in the fact that most species in these groups have external fertilization. For viviparity to evolve requires that the eggs be fertilized inside the female reproductive tract. Internal fertilization is probably more than a prerequisite; it may also predispose lineages toward the evolution of live-bearing habits. Evidence for this supposition derives from the fact that most of the lineages of bony fishes that have evolved internal fertilization also include viviparous representatives. In addition, chondrichthyans (in which internal fertilization is universal) exhibit a high number of origins of viviparity and are second only to mammals in the percentage of viviparous species (Tables 1 and 2).

D. How Has Viviparity Evolved?

Viviparity is widely viewed as a result of a progressive, gradualistic, and unidirectional increase in the proportion of egg development that takes place inside the female. Accordingly, over evolutionary time oviparous females would retain their eggs for progressively longer periods, laying them in more advanced stages of development. The culmination of this evolutionary trend would be viviparity, in which embryos are retained to term inside the female. Two main lines of evidence have been invoked to support this scenario in squamates: (i) the existence of oviparous species that lay eggs in early, middle, and advanced stages of development, and (ii) the existence of viviparous species that retain shelled eggs to term that hatch at the time of birth. As applied to particular taxa, other features have been added to this scenario, notably that both matrotrophy and simple placentae that function in gas exchange only evolve subsequent to viviparity.

However, consideration of the transition to viviparity in actual lineages has challenged aspects of the traditional scenario. Evidence from mammals indicates that matrotrophy can evolve prior to viviparity. Studies of squamates suggest that functional placentation as well as incipient matrotrophy can evolve simultaneously with viviparity. Thus, the chronological sequence by which viviparity has evolved differs between taxa (Fig. 1).

In addition, contrary to common belief, the eggs of oviparous squamates do not exhibit a full contin-

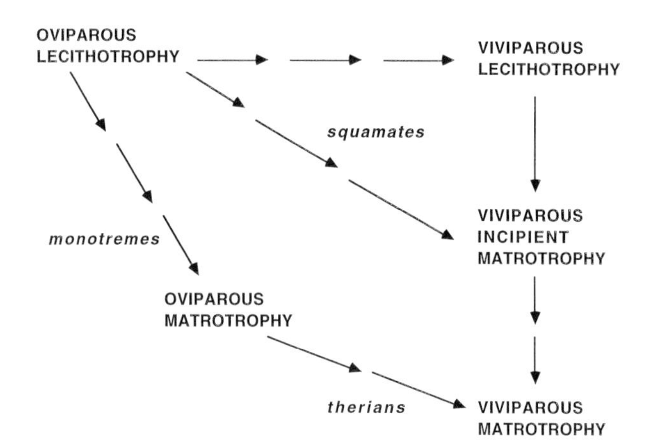

FIGURE 1 Multiple scenarios for reproductive evolution. Traditionally, matrotrophy was thought to have originated after viviparity. However, recent analysis suggests that in mammals matrotrophy evolved prior to viviparity and in squamates incipient matrotrophy commonly has evolved simultaneously with viviparity.

uum of developmental stages at the time of oviposition. Instead, most squamates either deposit their eggs at the limb bud stage of development (typical oviparity) or retain them to term (viviparity). On this basis, viviparity and placentation in squamates have been suggested to evolve in accord with a punctuated equilibrium model, in which intermediate evolutionary stages either never occur (aplacental viviparity) or are evolutionarily unstable and transient (i.e., oviparous egg retention). This model has been proposed recently and its implications and validity are under investigation. Application of the punctuated equilibrium model to anamniotes seems less likely; a broad spectrum of possible evolutionary stages is represented among living species. Due to differences in the tertiary egg membranes and the fact that aquatic offspring can be born at a larval stage of development, perhaps oviductal gestation of teleost and amphibian eggs does not involve the same physiological limitations that affect squamates.

IV. EVOLUTIONARY ASPECTS OF EMBRYO MAINTENANCE IN VIVIPAROUS SPECIES

A. Evolution of Matrotrophy

The origins of matrotrophy can be analyzed by using the same approach used for the evolution of viviparity. Phylogenetic analysis reveals that matrotrophy has evolved on 25 or more occasions among the vertebrates (Table 3). Most of these origins have occurred among the bony fishes, and only 4 have occurred among the amniotes, with mammals representing 1 such origin.

Given the multiple evolutionary origins of matrotrophy, this nutritional pattern presumably can have significant advantages over lecithotrophy. Although we do not fully understand the selective pressures that have led to its evolution, some plausible advantages of this pattern can be identified. In lecithotrophy, females have invested their nutrients into reproduction by the time of fertilization. In contrast, matrotrophy may permit females to control the provision of nutrients during gestation according to environmental circumstances and nutrient availability. In addition, when circumstances warrant, in some matrotrophic species females can terminate reproduction early in gestation (via spontaneous abortion or embryo resorption) with relatively little energetic loss. Termination of reproduction in lecithotrophic species can result in loss of the energy invested into the yolk. Furthermore, matrotrophy minimizes the weight of the conceptus during much of gestation since its mass will not approach that of the newborn until late in development. Consequently, females may be able to circumvent a major cost of viviparity—the detrimental effects of pregnancy on maternal mobility.

B. Evolution of Specializations for Gas Exchange

In viviparous vertebrates with lecithotrophy, the immediate need of an embryo retained within the maternal reproductive tract is respiratory gas exchange. Structures that function in gas exchange under oviparous conditions commonly have been recruited to perform analogous functions in embryos of live-bearing forms. One example of such a structure is the chorioallantois, a vascularized extraembryonic membrane that is responsible for gas exchange in the eggs of all oviparous reptiles, birds, and mammals. This same structure forms an intimate placental relationship with the uterine tissues in each of the many lineages of viviparous squamates as well as in the eutherian mammals. Other respiratory structures shared by oviparous and viviparous embryos include the yolk sac of squamates, gills of fishes and caecilians, the "belly sac" (pericardial somatopleure) of cyprinodont fishes, and possibly the skin of larval amphibians. In various teleosts with ovarian gestation, the ovarian epithelium is vascularized and hypertrophied to provide for gas exchange with the embryos.

C. Evolution of Specializations for Nutrient Provision

Convergent evolution not only characterizes viviparity, matrotrophy, and specializations for embryonic gas exchange but also the specific structures by which nutrients are provided to developing embryos.

For example, in three unrelated genera of lizards, the chorioallantois has been recruited evolutionarily for placentotrophic nutrient provision. Chorioallantoic placentae also provide nutrients for development in eutherians and some marsupials, whereas a yolk placenta provides nutrients in most eutherians and all marsupials.

Like the chorioallantois, embryonic structures that function in gas exchange often have taken on a role in matrotrophy. Among these structures are the yolk sacs of certain sharks and poeciliids (which form yolk sac placentae), the belly sac of cyprinodonts (which also contributes to placentae), and gills of typhlonectid caecilians. Multiple lineages of teleost fishes have evolved trophotaeniae, outgrowths of the embryonic hindgut that absorb maternal secretions. Several viviparous lineages (*S. atra*, lamniform sharks, and multiple clades of osteichthyans) have evolved a pattern in which developing fetuses feed on siblings or ovulated yolks. Some lineages, including the viviparous caecilians and the frog *Nectophrynoides occidentalis*, have evolved a form of matrotrophy in which embryos ingest maternal tissues or secretions. These are but some of the adaptations for matrotrophy that have evolved among vertebrates. Diverse invertebrate specializations also abound.

V. OVIPAROUS ANALOGS TO VIVIPARITY AND MATROTROPHY

A. Parental Care, Egg Brooding, and Pseudoviviparity

As noted previously, viviparity is but one way for a parent to protect eggs from the vicissitudes of the environment. Parental care of eggs has evolved many times in oviparous animals. It is found in birds, monotreme mammals; crocodilians; various lizards, snakes, and salamanders; numerous lineages of frogs; and some teleost fishes. Care of eggs can range from simple proximity of the mother (as in crocodilians) to thermoregulatory incubation (birds, monotremes, pythons, and some scincid lizards). Care of eggs is sometimes labeled "brooding" and "egg tending."

In some anamniote vertebrates, the association of the eggs and parent is physiologically intimate. Among teleosts, for example, externally fertilized eggs can be brooded in the mouth (e.g., certain cichlids and catfish) or carried in an integumentary pouch (male seahorses and pipefish). In anurans, eggs can be brooded in the stomach, vocal sacs, or depressions in the dorsal integument, or they can be carried on the back or legs by one of the parents. These patterns are types of egg brooding and it would be inappropriate to refer to them as viviparity. However, the term pseudoviviparity can be applied to such situations in recognition of their superficial similarity to cases in which the eggs develop to term in the female reproductive tract. Pseudoviviparity also can be applied to molluscs, in which externally fertilized eggs are taken into the branchial chamber of one of the parents, and cnidarians, which take hatched larvae into their gastrovascular cavities.

B. Lactation and Other Means of Nutrient Provision

Just as viviparity has its analogs among egg-laying species, matrotrophic nutrient provision has equivalents among oviparous forms. In a broad sense, matrotrophy could be applied to all cases in which females provide extravitelline nutrients to their offspring. Therefore, it can be useful to distinguish matrotrophy that precedes parition (oviposition or parturition) from that which follows it. Perhaps the most striking case of preparitive matrotrophy to have evolved among oviparous vertebrates is that of monotremes, in which the eggs absorb large quantities of nutritious oviductal fluids prior to their oviposition.

The most elaborate manifestation of postparitive matrotrophy is lactation, a phenomenon whose extraordinary nature tends to be taken for granted because it is so familiar. From the time of birth or hatching until weaning, a mammalian mother provides for the nutritional needs of her offspring by her own glandular secretions. The lactating female converts her own dietary nutrients and fat reserves into a secretion that is rich in lipids, carbohydrates, protein, and antimicrobial agents and that can offer immunological protection to the suckling young. Lactation reaches a pinnacle of complexity in marsupials; not only does the milk provide most of the nutrients for development but also female kangaroos simultaneously can produce milks of markedly different composition to sucklings of different ages. Lactation in marsupials performs some of the same functions as does placentation in eutherian mammals, which commonly give birth to more advanced offspring.

Analogs to lactation have rarely evolved. They include the glandular secretions of the crop with which pigeons feed their young and the integumentary mucous which some teleosts secrete for ingestion by their hatchlings. Pipefish may offer another example because evidence indicates that the male parent provides nutrients to the eggs while carrying them in his brood pouch. This pattern and analogous cases in pseudoviviparous frogs can be called "patrotrophy." Parental feeding of the young by birds and mammals serves the same general function.

See Also the Following Article

Placenta and Placenta Analogs in Reptiles and Amphibians

Bibliography

Blackburn, D. G. (1992). Convergent evolution of viviparity, matrotrophy, and specializations for fetal nutrition in reptiles and other vertebrates. *Am. Zool.* **32**, 313–321.

Blackburn, D. G. (1995). Saltationist and punctuated equilibrium models for the evolution of viviparity and placentation. *J. Theor. Biol.* **174**, 199–216.

Blackburn, D. G., and Evans, H. E. (1986). Why are there no viviparous birds? *Am. Nat.* **128**, 165–190.

Hogarth, P. J. (1976). *Viviparity,* Institute of Biology's Studies in Biology No. 75. Camelot Press, Southampton, UK.

Renfree, M. B. (1983). Marsupial reproduction: The choice between placentation and lactation. In *Oxford Review of Reproductive Biology,* Vol. 5, pp. 1–29. Clarendon, Oxford, UK.

Shine, R. (1985). The evolution of viviparity in reptiles: An ecological analysis. In *Biology of the Reptilia* (C. Gans and F. Billet, Eds.), Vol. 15, pp. 605–694. Wiley, New York.

Wourms, J. P., Grove, B. D., and Lombardi, J. (1988). The maternal embryonic relationship in viviparous fishes. In *Fish Physiology* (W. S. Hoar and D. J. Randall, Eds.), Vol. XI, pp. 1–119. Academic Press, New York.

Voles

see Cricetidae

Vomeronasal Organ

Michael Meredith
Florida State University

GLOSSARY

accessory olfactory system The vomeronasal organ, its associated glands, blood vessels, capsule, etc., and its central neural connections to the accessory olfactory bulb with onward connections (in mammals) to parts of the corticomedial amygdala and bed nucleus of the stria terminalis.

chemosignal A chemical released externally by one individual that can convey information to another individual (compare with *pheromone*).

lipocalin A class of proteins which bind lipophilic molecules, frequently carrying lipids in an aqueous medium. In the vomeronasal system, lipocalins are present in some stimulus sources, in nasal mucus, and in vomeronasal mucus, and thus they may help to hold, deliver, and/or remove stimuli from vomeronasal receptors.

nervus terminalis A ganglionated cranial nerve of unknown function that connects the nasal epithelium and anterior forebrain in all vertebrate groups and contains gonadotropin-releasing hormone-immunoreactive (and other) neurons and fibers in most groups. Also called *terminal nerve* and, sometimes, *cranial nerve zero*.

pheromone A chemical released externally by one individual of a species that communicates information to another member of the species to their mutual benefit (compare with *chemosignal*).

priming pheromone A pheromone that produces a longer latency, often hormonally mediated, response in the recipient.

signaling pheromone A pheromone that produces an immediate, or short, latency, behavioral, or physiological response in the recipient.

vomeronasal basal cell A cell located at the base of the vomeronasal sensory epithelium, next to the basal lamina, that is a persistent stem cell capable of dividing to produce new sensory neurons (and possible supporting cells).

vomeronasal organ (VNO) The sensory end organ of the accessory olfactory system. Also called Jacobson's organ (more usually used for nonmammalian species). Chemosensory neurons within the VNO have axons that extend to the accessory olfactory bulb of the brain and carry information about chemical stimuli entering the organ.

The vomeronasal organ is present in most terrestrial vertebrates with the exception of birds. It is implicated in reproduction as a sensory organ for chemical signals from conspecific animals of both sexes.

I. INTRODUCTION

Chemical signals that may elicit hormonal or behavioral responses are often called pheromones. Definitions of the word "pheromone" vary, but the implication of the word is generally that the chemical signals are part of a communication system that has evolved because it confers advantage on sender, receiver, or both. We might expect a communication system that facilitated reproduction to be advantageous to both, although there are probably opportunities for pheromonal deception there, as for other signals. Defined as a chemical signal that is mutually beneficial to sender and receiver, it is clear that some pheromones operate via the vomeronasal system but that some operate via the main olfactory system. Equally, there are also examples of the vomeronasal organ (VNO) used for detecting nonpheromone chemicals in some species, particularly in snakes, which use the organ for trailing prey. The vomeronasal organ is the sensory end organ for the accessory olfactory system. Its receptor neurons project to the accessory olfactory bulb, which in turn projects to the corticomedial amygdala and bed nucleus of the stria terminalis. Both of these brain structures have been implicated in reproductive behavior and both are sexually dimorphic. The central vomeronasal projection areas connect to the medal preoptic area and hypothalamus, which are involved with reproductive behavior and physiology. However, as indicated previously, it would be a mistake to consider the vomeronasal organ to be exclusively concerned with reproduction or to consider chemosensory communication related to reproduction to be exclusively the province of the vomeronasal organ.

II. VOMERONASAL ORGAN STRUCTURE AND STIMULUS ACCESS

A. The Peripheral Organ

In mammals, the paired vomeronasal organs consist of elongated tubes, within a bony or cartilaginous capsule, lying along the base of the nasal septum and opening at the anterior end only via a narrow duct. The lumen of the organ is partially lined with vomeronasal sensory epithelium. The bipolar receptor neurons bear microvilli on their apical surface (unlike olfactory receptor neurons which bear cilia). Receptor molecules are presumably embedded in the microvillar membrane. Supporting cells, which may also be secretory, surround the receptor neurons. Both types of cells span the depth of the pseudostratified epithelium, with the cell bodies of receptor cells forming several layers below a single superficial row of nuclei indicating the cell body locations of the supporting cells. Basal cells are less numerous than in the olfactory epithelium but are similar in that they are persistent stem cells retaining the ability to divide and generate new neurons throughout life. Receptor cells die if their axons are cut but new neurons are formed and can grow axons back to the brain. However, axon bundles from a small remnant left after experimental removal of most of the organ frequently form large peripheral neuromata, with few if any axons reaching the brain. The sensory epithelium in mammals mainly lines the medial wall of the lumen; the lateral wall is lined with a ciliated epithelium which may contain trigeminal sensory fibers from the nasopalatine nerve and possibly nervus terminalis fibers. Segovia and Guillamon report that the vomeronasal organ in rats is larger in males than females (not adjusted for body size) and contains more receptor neurons.

B. Stimulus Access

In rodents the vomeronasal duct opens onto the floor of the nasal cavity a few millimeters into the nostril. Stimuli reach the organ in the mucus stream, which flows along the floor of the nasal cavity past the duct. Mucus glands opening via ducts near the nostril, especially the lateral nasal glands, contain proteins of the lipocalin family that can bind potential olfactory and vomeronasal stimuli. Nonvolatile stimuli entering the nostril after contact with the stimulus source may dissolve in mucus with the aid of the binding proteins. Volatile stimuli can dissolve in the mucus, again possibly with the aid of binding proteins, or (unlikely) might enter the organ in vapor form. Lipocalin-binding proteins are also present in stimulus sources. Major urinary proteins in the mouse are lipocalins that bind small molecules with possible pheromonal functions (Pelosi, 1995). Aphrodisin, a lipocalin in hamster vaginal fluid, was originally identified as the aphrodisiac pheromone which acts through the vomeronasal organ to facilitate male mating behavior. Later work showed a small lipid was also required for full activity. Lipocalins are also present in the vomeronasal glands and delivered directly to the organ. Binding proteins may prolong release of stimulus molecules from the source, carry stimuli to the receptors, assist in the stimulus transduction, or remove stimuli from receptor molecules after stimulation: or a sequence of proteins may do all of these things.

Stimuli are drawn into the organ in hamsters and other rodents when pressure is reduced by vasoconstriction and the collapse of cavernous vascular tissue within the vomeronasal capsule. Mucus and dissolved stimuli are drawn in as the lumen expands. Stimuli are removed when pressure is reexerted on the organ as the constriction relaxes or by active vasodilation. This system constitutes a vascular pump controlled by autonomic fibers in the nasopalatine nerve (Cr. N. V), which enters the posterior end of the capsule. Mucus from vomeronasal glands in the nasal septum enters the lumen at the posterior end of the organ and may also help to flush out stimuli. The pump appears to operate in response to novel situations (Meredith, 1998). Pump movements have a period of 2 or 3 sec and are capable of delivering stimuli to the sensory epithelium much faster than by diffusion through the narrow duct (although diffusion in the absence of pump action may still occur). The pump rate is still slow relative to nasal respiratory rate, but vomeronasal receptor neurons may be more sensitive to tonic stimulation.

In carnivores, ungulates, and Old World primates, the vomeronasal ducts open into the nasopalatine canal (also known as the nasoincisor canal) which joins the nose and mouth. Stimuli can enter the organ from the mouth in these species, possibly following licking of a stimulus source. Most of these species also have large vomeronasal blood vessels and may have a functional pump. Several species (e.g., ferret) have an incomplete capsule at the posterior end which might allow alterations in intranasal pressure to operate the pump, especially if the nostrils were closed. Rodents have a patent nasopalatine canal but it opens into the nasal cavity posterior to the vomeronasal duct and is probably not a route for vomeronasal stimulation: Stimuli picked up by licking could be transferred to the nostril, however. In reptiles, the vomeronasal organs open into the roof of the mouth and stimuli are delivered by the tongue, especially in snakes and some lizards, in which stimuli are picked up by tongue flicking (Fig. 1).

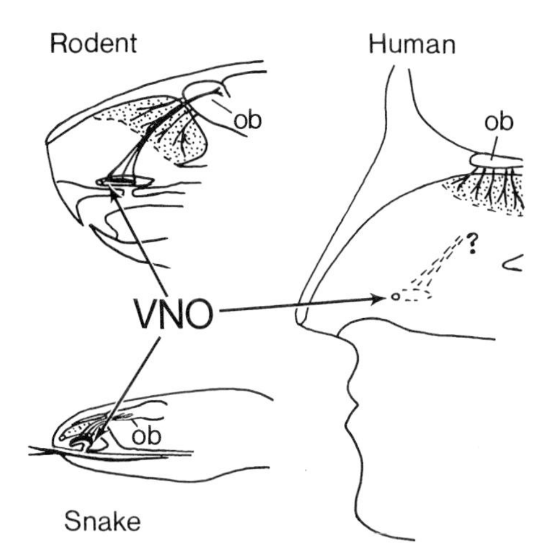

FIGURE 1 The location of the vomeronasal organ (VNO) at the base of the nasal septum in rodent, snake, and human. The vomeronasal nerves project to the accessory olfactory bulb, dorsocaudal to the main olfactory bulb (ob). In humans, the existence of a neural connection with the brain is uncertain (?) and no accessory bulb is identifiable. The stippled areas show the extent of olfactory epithelium.

C. Neural Projections

Vomeronasal sensory axons are unmyelinated and pass in a few large bundles under the mucosa of the nasal septum through the cribriform plate into the cranial cavity. The nerves then pass between the olfactory bulbs (except in guinea pig, in which they pass around the lateral side of the bulbs) to the accessory olfactory bulbs (AOBs), ending in glomeruli. The AOB is divided into anterior/dorsal and posterior/ventral sectors in several species of rodents and one opossum. The two sectors receive input from two different types of sensory neurons, largely segregated to different layers of the vomeronasal epithelium. AOB relay cells receive input through several glomeruli but only within one or the other sector of the AOB. It is not clear whether there is any selective projection of the two AOB sectors among known AOB targets. The central projections of the AOB are separate from, and more limited than, those of the main olfactory bulb. They extend to the ipsilateral corticomedial amygdala (the medial and posteromedial cortical nuclei in rodents), nucleus of the accessory olfactory tract, part of the bed nucleus of the stria terminalis (BNST), and a small dendritic zone of the superoptic nucleus. There are intraamygdaloid connections between the vomeronasal division of the corticomedial amygdala and the adjacent but nonoverlapping main olfactory division (anterior and lateral cortical nuclei in rodents). In some cases, olfactory input can sustain a chemosensory-dependent behavior after removal of the vomeronasal organ. The amygdala is the first site at which information from the two systems converges.

The amygdaloid regions receiving vomeronasal input project to the medial BNST and to medial preoptic area. These regions are implicated in the control of chemosensory investigation and reproductive behavior and these same areas contain both androgen and estrogen receptors. Some androgen receptor containing neurons in the medial amygdala are activated by mating in male hamsters.

III. RECEPTOR FUNCTION/ TRANSDUCTION

A. Receptor Molecules: Two Classes of Receptor Neurons

Recent work from several laboratories has identified two families of genes coding for putative vomeronasal receptor molecules, expressed in different populations of mammalian vomeronasal neurons. These are seven-transmembrane domain receptors that vary one from another within a putative ligand binding region; within the plane of the membrane for the first family discovered by Dulac and Axel. Members of the other family, discovered independently by three labs, code for proteins predicted to have a large extracellular N-terminal region, with sequence similarity to metabotropic glutamate receptors and parathyroid calcium-sensing proteins. This N-terminal region may bear the ligand binding site. The first discovered receptors have been called V1R receptors and the new type are called V2R receptors. Both types have sequences consistent with G protein activation regions in intracellular domains. The two types are expressed in two populations of cells that also express different G proteins and project to separate regions of the AOB (in rodents). The V1R receptors are expressed in receptor cells with cell bodies closer to the surface of the receptor epithelium. This population of cells also expresses Giα2 and its axons project to the anterior part of the AOB. The V2R receptors are expressed in cells with cell bodies deeper in the epithelium (with one exception), which also express Goα, and which have axons projecting to the posterior AOB. The conjunction of two putative receptor molecules and two different G proteins suggests that V1R and V2R receptors may respond to different types of stimuli. Both populations of cells have dendrites that reach the epithelial surface and they are generated independently. They do not appear to be different developmental stages of the same population of cells. Vomeronasal receptor neurons do turn over but it is not clear whether V1R and V2R receptor neurons differ in this characteristic. There is initial indication that some V2R receptors may be expressed differentially in males and females. The receptor function of V1Rs and V2Rs has not yet been demonstrated.

B. Transduction

A number of studies have addressed aspects of the transduction mechanism in mammals. Isolated vomeronasal receptor cells from mouse have been patch clamped to record their electrophysiological properties and responses to potential chemical stimuli, but any differences in the properties of V1R and V2R receptors have not yet been reported. Like olfactory receptor cells, vomeronasal receptor cells have high input resistances and are easily depolarized, suggesting that the opening of a few channels would be sufficient to fire either cell type. However, vomeronasal receptor cells respond with a continuous train of spikes to depolarization through the recording electrode, unlike main olfactory receptors which fire one or a few spikes and then become silent. The difference may be due to differential expression of various channels in these two types and may be reflected by a difference in the nature of response to chemical stimulation (Liman, 1996). Response to potential stimulus chemicals *in vitro* varies. At least some laboratories report a hyperpolarization and/or an outward current bias in mouse VNO cells in response to substances found in mouse urine. Other laboratories report excitatory response, including an increased impulse frequency in response to urinary components delivered to rat vomeronasal neurons patch clamped in slices. A depolarization has also been reported in isolated human neurons obtained by aspiration from the putative vomeronasal organ (see Section VIII).

In the main olfactory epithelium, receptor neurons show changes in their second messenger cascades in response to odors that can certainly involve an increase of cAMP and can probably involve an increase in IP_3. Olfactory neurons have a specific stimulatory G protein (G_{olf}), a specific adenylyl cyclase, and cyclic nucleotide gated (CNG) channels, providing for excitatory response to many odorants. The vomeronasal receptor cells lack the G_{olf}, the specific adenyl cyclase, and all but the β (noncyclic nucleotide sensitive) subunit of the CNG channels. There is evidence for IP_3 involvement but, with one exception, responses to isolated substances proven to be natural vomeronasal stimulants have yet to be recorded. In hamsters, aphrodisin isolated from female vaginal fluid has been demonstrated to facilitate mounting of females by males via activation of the vomeronasal system. Kroner *et al.* found that it modulates IP_3 but not cAMP levels in hamster vomeronasal membranes. Unspecified urinary or semen components are reported to increase IP_3 in a dose-dependent and GTP-dependent way in membrane fractions from pig vomeronasal organ. Intracellular dialysis of IP_3 also can increase conductance of rat vomeronasal receptor cells in slices; and a firing rate increase elicited by urine in these cells can be blocked by ruthenium red, an IP_3 channel inhibitor, or by phospholipase C inhibitors. In isolated mouse vomeronasal receptors, behaviorally active compounds from mouse urine can produce a decrease in cAMP.

In lower vertebrates IP_3 has also been implicated in the transduction process. In garter snakes, ES20, a protein extracted from earthworm prey and shown to be attractive via vomeronasal activation, increases IP_3 levels and decreases cAMP. In turtles, indirect evidence suggests that IP_3 participates in vomeronasal responses to general odorants, although the mechanisms for cyclic nucleotide gated depolarization of vomeronasal receptor cells also seem to be available. There is also potential for novel transduction mechanisms, including modulation of a sodium–potassium pump current, nitric oxide-mediated activation of the β subunit of the CNG channel, and second messenger cascades involving Gq proteins.

IV. AFFERENT SENSORY CODING

The nature of the afferent signal encoding important vomeronasal sensory stimuli is not clear. In the main olfactory system receptor neurons that express the same receptor proteins project to a pair of glomeruli in each main olfactory bulb where they synapse with relay neurons and interneurons. However, odors appear to activate many more bulbar neurons than would be connected to just four glomeruli, implying that single odors activate several types of receptor neurons, projecting to different glomeruli (Buck, 1996). The situation in the accessory olfactory system is unknown. There are two classes of vomeronasal receptor neurons in rodents and some other species that project to anterior and posterior subdivisions of the accessory olfactory bulb, but the relative specificity of function of the two classes is unknown. If hyperpolarization of receptor neurons in response to stimuli turns out to be a normal mode of response, afferent signals may involve a decrease in firing rate. The few reports of electrophysiological recordings from the intact epithelium, nerve, or accessory olfactory bulb do not suggest such a response. Responses of the intact system do not appear to be highly selective. However, responses from intact systems have not been recorded with well-characterized, isolated, biologically relevant stimuli. It remains possible that the natural response in vertebrates is highly sensitive and highly selective, as are pheromone responses in insects; but this has yet to be demonstrated. An indication of selectivity comes from experiments with equivalent chemosignals from related species. Female voles are dependent on vomeronasal input of male stimuli to trigger uterine growth preparatory to breeding. Tubbiola and Wysocki found that male vole urine triggers uterine growth and activates vomeronasal relay neurons in the accessory olfactory bulb. Male mouse urine does not trigger uterine growth and does not activate the AOB. Fiber *et al.* previously showed a different result in male Syrian hamsters, which respond more strongly to vaginal fluid from female Syrian hamsters than they do to vaginal fluid from Djungarian hamsters. Both of these stimuli activate neurons in the AOB. The finding suggests selectivity in vomeronasal receptor activation or in the network response in the AOB. Discrimination at the AOB level is also implicated in the Bruce effect. Female mice abort their pregnancy if exposed to male chemosignals (urine) from a male of a different strain than the impregnating male. Extensive investigation of this phenomenon by Keverne and by Kaba suggests that the impregnating male's chemical signature is learned by the female during some hours following mating. Successive experiments blocking circuits in the accessory olfactory bulb, amygdala, and hippocampus indicated that the learning takes place in the accessory bulb. Keverne suggests that major urinary proteins, which act as carrier proteins for small chemosignal molecules, may provide information on strain (or individual) identity, whereas the small molecule provides information on gender, reproductive status, etc.

V. RESPONSES DEPENDENT ON VOMERONASAL INPUT

A. Reproductive Physiology and Behavior

The vomeronasal system has been implicated in many behavioral/physiological responses to chemical signals in several species and both sexes. These responses may be of the type attributed to priming pheromones, where there is some delayed physiological effect usually mediated by a hormonal change, or they may be of the type attributed to signaling pheromones, where a rapid behavioral change can be seen (Wysocki and Meredith, 1987; Wysocki and Lepri, 1991). However, not all putative pheromone communication is mediated by the vomeronasal system, nor is all vomeronasal chemosensory function necessarily related to pheromone communication. Communication between mother and young appears to be mediated by olfactory and not vomeronasal signaling in at least two examples. Levy *et al.* showed that vomeronasal lesions did not prevent odor-dependent induction of maternal behavior in sheep, and Hudson and Distel found that vomeronasal organ removal did not disrupt odor-dependent nipple finding by young rabbits. Recently, Dorries *et al.* showed that the induction of the mating posture in female pigs by salivary signals from the male, a classical sex pheromone effect, was unaffected by blockage of the vomeronasal duct. Priming-type responses that are dependent on an intact vomeronasal system in females are well documented. These include modulation of estrus in mice by urine signals from both males and females, modulation of estrus in rats, the acceleration of puberty in female mice and voles, and the block to pregnancy in mice produced by "strange" males (Bruce effect): All are mediated by signals in male urine and all are prevented by lesions of the vomeronasal system. A similar dependence on intact vomeronasal organs is seen in the induction of estrus in prairie voles (but not in meadow voles), in Monodelphis opossums, and in the induction of ovulation in light-induced, or estrogenized, persistent-estrus rats.

A signaling function is seen in lactating female mice, whose aggression toward intruders is reduced or eliminated by vomeronasal lesions, regardless of experience. An additional indication of a signaling function in females comes from Rajendran and Moss's tests of repetitively mated female rats, where vomeronasal lesions reduce the "lordosis quotient" displayed at the end of six repeated mating tests.

Among males, there are clear examples of signaling functions mediated by the vomeronasal organ but also examples of hormonal modulation by vomeronasal input. Golden hamsters, mice, and prairie voles are particularly dependent on chemosensory input for normal reproductive behavior. Vomeronasal lesions can produce deficits in mating and courtship behaviors (and a reduction in aggression), especially in inexperienced males. Sexual behavior and aggression are also reduced by vomeronasal organ removal in the male mouse lemur, a prosimian primate, although Aujard found in this species that there was a general reduction in activity. None of these behavioral changes is a consequence of changed androgen levels. Androgen/testosterone levels are not affected by vomeronasal lesions in the hamster or mouse lemur; testosterone injections do not restore ultrasonic courtship vocalizations in mice with vomeronasal lesions. Vomeronasal function in nonmammalian vertebrates has been less studied but in snakes it clearly involves both reproductive and nonreproductive functions. For example, Halpern and colleagues have shown that both courting behavior and prey trailing are impaired by vomeronasal occlusion in garter snakes.

B. Gonadotropin-Releasing Hormone and Behavior

Many species and both sexes show increased serum luteinizing hormone (LH) in response to contact with the opposite sex. In male mice and hamsters, there is also an increase in testosterone (presumably triggered by the LH increase). Exposure to female chemical signals (urine or vaginal fluid) produces a similar hormonal response that is eliminated by vomeronasal lesions. It is important that the response to female chemosignals appears whether or not males have sexual experience because other evidence shows that the LH response can be conditioned. Vomeronasal lesions eliminate the hormonal response to female chemosignals regardless of experience but do not eliminate the hormonal response to the females themselves in experienced mice and hamsters, indicating a multisensory activation of the hormonal response. Pfeiffer and Johnston found that lesions of the main olfactory system sufficient to produce behavioral anosmia, by intranasal infusion of zinc sulfate, did not affect male hamsters' hormonal response to female chemosignals, reinforcing the view that vomeronasal input is the more salient. Combined lesions of vomeronasal and olfactory systems, which eliminate mating in male mice and hamsters regardless of experience, eliminated the hormonal response to females in inexperienced but not experienced hamsters. Chemosensory input must be important for more than the hormonal response since mating behavior is lost after these lesions but the hormonal response is not.

Gonadotropin-releasing hormone (GnRH) release presumably precedes the vomeronasal-dependent hormonal responses, and in female rats there is a correlation, in some circumstances, between male stimulation, behavior, and activation of GnRH neurons in the brain (see Section VI). GnRH also facilitates mating behavior in a number of circumstances, and Meredith found that GnRH substantially restores mating behavior deficits in male hamsters with vomeronasal lesions. It is not clear whether GnRH is a mediator of the vomeronasal facilitation of behavior, i.e., so that exogenous GnRH restores the original pathway. GnRH could be an alternative facilitator which bypasses the sensory pathway. It seems unlikely that the facilitation of behavior by GnRH is a result of LH secretion. It seems more likely that GnRH has a direct effect on the brain. GnRH analogs with no LH releasing activity can facilitate behavior in female rats and can restore behavior in male hamsters, although not in an identical fashion to that of GnRH. Possibly, the LH secretion is a "spill over" effect of GnRH release into the brain, at least with regard to correlation of LH secretion and behavior.

VI. CENTRAL NERVOUS SYSTEM RESPONSES AND LESIONS

A. Fos Studies

There have been few reports of central nervous system (CNS) electrophysiological response to vomeronasal sensory input. The advent of c-*fos* gene expression as a measure of neuronal response has produced numerous studies of brain activation during mating. Some of these also examined the effect of chemosensory input. Imunocytochemistry with antibodies against Fos protein stains the nuclei of all neurons that have been activated sufficiently strongly to express the immediate early gene, c-*fos*, and generate its product, the DNA-binding protein, Fos. Although the precise function of c-*fos* is not known, the most likely cause for an increase in Fos protein expression in neurons of the mature, undamaged brain appears to be depolarization and the entry of calcium (Morgan and Curran, 1991). The pattern of Fos expression forms a map of brain areas activated at a high level during particular behaviors if animals are prepared for Fos staining 60–180 min after some significant behavior such as mating or investigation of chemosignals.

1. *Mating*

Mating animals of both sexes show intense Fos expression in central vomeronasal pathways, including the AOB, medial amygdala (Me), and BNST, as well as in the medial preoptic area (MPOA), subparafasicular nucleus, and, in females, the ventromedial hypothalamus and premammillary nucleus. Beyond the AOB, much of this activation is not dependent on vomeronasal input and is not chemosensory at all. Chemosensory input is essential for mating in male hamsters, however, and Wood and Newman show that this input is effective only when integrated with intracerebral steroid hormone action in the same hemisphere. The important mating-related, but nonchemosensory contributors to Fos expression in Me and elsewhere, include genital somatosensory input and possibly a separate input related to ejaculation. These inputs appear to activate a more lateral part of the posterior dorsal Me as well as other discrete areas in BNST. Coolen *et al.* suggest that both these inputs may be relayed via the subparafasicular nucleus. Androgen receptor-containing neurons in Me, BNST, and MPOA are activated by mating in male hamster but it is not clear if any of these are driven by chemosensory inputs specifically.

2. *Chemosensory Input*

A chemosensory contribution to Me activation can be demonstrated in animals exposed to pheromone-containing stimuli in the absence of mating. An explicitly vomeronasal contribution can be seen in male hamsters after a unilateral vomeronasal organ lesion. This lesion does not alter behavior but produces a significant decrease in Fos expression on the lesion side. Fos activation by purely chemosensory stimuli does appear to extend to the bed nucleus of the stria terminalis and to the medial preoptic area in animals exposed to reproductively relevant chemosignals. These signals include bedding from estrus females, in the case of rats, and female hamster vaginal fluid (HVF), in the case of hamsters. Experience may alter the pattern of activation as it alters behavior. Inexperienced, intact male hamsters show low levels of activation in the medial preoptic area after they have investigated HVF (barely above the level in unexposed controls), and activation in inexperienced males with vomeronasal lesions is zero. Sexually experienced, intact males show a high level of activation after equivalent exposure to HVF, and vomeronasal lesions made after experience do not reduce it. Experience appears to sensitize the medial preoptic area to chemosensory input and to reroute olfactory input so that it can substitute for vomeronasal input in driving the MPOA. This change reflects the behavioral change. Chemosensory input is essential for mating even for experienced males, but vomeronasal lesions made after experience do not affect mating: The olfactory input is sufficient to maintain the behavior, whereas vomeronasal lesions made before experience can impair mating behavior. Other laboratories find variable degrees of impairment after vomeronasal lesions, from zero to severe, but all agree that olfactory input is never necessary in male hamsters, although it may be sufficient to sustain mating.

In one area, the posteromedial (pm) BNST, Fos activation may reflect the control of chemosensory investigation rather than chemosensory input per se. Activation there persists after olfactory peduncle lesions in rats and after either vomeronasal organ lesions or olfactory lesions in male hamsters.

3. *Interaction with Steroid Hormones*

In both rats and hamsters, testosterone modulates the degree of Fos activation in animals exposed to chemosensory stimuli. Four weeks after gonadectomy, male (or female) rats showed activation throughout the central vomeronasal pathway if injected with testosterone and exposed to estrous female bedding. They showed no significant activation on exposure to estrous bedding if injected with oil vehicle. In hamsters, testosterone was necessary for Fos activation in MPOA of males but was insufficient to permit activation of MPOA in females, when both were exposed to HVF. On the other hand, testosterone was not necessary for Fos activation in the male hamster Me or BNST, regions that receive explicit chemosensory input. It is not entirely clear whether these results reflect some steroid modulation of chemosensory pathways or of c-*fos* gene expression itself or whether they are related to the way the animals delivered stimuli to their chemosensory receptors via sniffing or vomeronasal pump activation. Circulating steroids increased the animals' interest in the chemical stimuli in both of these experiments so part of the steroid-dependent response could be due to changes in the chemosensory investigation rather than chemosensory stimulation.

4. Interaction with GnRH

In repetitively mated female rats, vomeronasal lesions reduce Fos expression in GnRH neurons as well as preventing the increase in lordosis quotient which intact females show in these tests (see Section V). In reflex ovulator species, such as rabbits and ferrets, Fos expression in GnRH neurons is increased by mating (or by vaginocervical stimulation). Spontaneous ovulators, such as mice and rats, can also show increased Fos expression after mating. Whether there is any chemosensory component to this response (when triggered by a male) is not clear. Activation of GnRH neurons, either by chemosensory input or mating, has not been found in male mice, hamsters, or ferrets, despite the fact that LH/testosterone responses in males exposed to females are dependent on an intact vomeronasal system in mice and hamsters.

VII. NERVUS TERMINALIS

The nervus terminalis (NT) or terminal nerve is embedded in the vomeronasal nerve in most mammals. It is generally involved when peripheral lesions are made to assess vomeronasal function, so it should be mentioned here. The NT nerve is present in all vertebrate groups and connects the mucosa of the nasal cavity with the anterior midline of the brain (lamina terminalis). In elasmobranchs, where it was first described, it carries a peripheral ganglion and there is an equivalent group of nerve cells (ganglion terminale) embedded in the forebrain in mammals. Many, but not all, of the ganglion cells are GnRH immunoreactive (ir), and there are GnRH-ir (and other) fiber connections both peripherally and more centrally. The GnRH cells appear to have been left behind during the migration of GnRH cells from the olfactory placode into the brain. Their persistence in all vertebrate groups argues for some adult function. The penetration of some NT fibers into the nasal mucosa suggests a possible chemosensory function but this has yet to be demonstrated. Preliminary evidence suggests that electrical stimulation of the peripheral NT in elasmobranch fish results in an increase in GnRH concentration in the cerebrospinal fluid (Moeller and Meredith, unpublished). In male hamsters, Wirsig-Weichmann found that lesions of NT that did not involve the vomeronasal system resulted in small deficits in mating behavior. Similar lesions did not prevent the normal (presumably GnRH/LH-dependent) testosterone response on stimulation with female chemosignals. Humans have a nervus terminalis which innervates the vomeronasal organ during development and may continue to innervate the residual vomeronasal organ in human adults. Marine mammals that lack any vomeronasal or olfactory structures as adults retain the NT.

VIII. HUMAN VOMERONASAL ORGAN

Human adults do not have a vomeronasal organ easily recognizable as similar to that of other mammals. There is no bony or cartilaginous capsule and no associated large blood vessels. However, there is a "pit" opening and a tubular structure in the nasal septal mucosa in adult humans that can be seen in almost all subjects on routine nasal endoscopy. Almost certainly it represents the simplified remnant of the human vomeronasal organ. Recent reports claim that this structure is functional for detection and discrimination of a specialized set of compounds, but there are difficulties with this interpretation.

A. Development

The vomeronasal organ is clearly present during fetal development in humans and, as Boehm and Gasser show, contains bipolar cells expressing neuron-specific enolase (NSE). As in other mammals, the vomeronasal epithelium appears to be the source of GnRH neurons that migrate into the forebrain. At around 4 or 5 months gestation the organ is sometimes described as degenerating and, in Kajer and Fisher-Hansen's recent study, could not be found in 17- to 19-week-old fetuses. Other studies find the organ present at 30 weeks (Smith *et al.* 1997), although Boehm and Gasser failed to find characteristic NSE stained bipolar cells at this age. One study reports bipolar cells in the adult human vomeronasal organ that stain with the neural marker, PGP 9.5. The vomeronasal organ is reported as absent in Old

World primates, but researchers were probably not looking for a simplified structure similar to that in adult humans, so descriptions of primate development are not available for comparison with human development. The situation in New World primates is also not helpful in understanding the human condition: A normal mammalian vomeronasal organ develops and is functional in the adult. Clearly, more studies of vomeronasal organ development in humans are needed.

B. Adult Connectivity and Function?

Although there seems to be a persistent vomeronasal organ in adult humans, its function is unclear. The single report of bipolar cells in adults did not find axons leaving the epithelium and no vomeronasal nerve has been described. There are nerve bundles below the vomeronasal mucosa but these could be autonomic or NT bundles. There is also no identifiable accessory olfactory bulb in the adult human brain. The accumulated anatomical evidence does not encourage a belief in human adult vomeronasal function. However, Monti-Bloch *et al.* have recorded electrophysiological responses to chemicals from the vomeronasal pit in awake humans. These recordings are similar to the electroolfactogram that can be recorded from the surface of the olfactory epithelium in response to odors. The chemicals stimulating the vomeronasal organ (termed "vomeropherins") did not necessarily stimulate an EOG from the olfactory epithelium, nor did odors necessarily produce a response from the pit. Recent abstracts report an extension of these studies to isolated cells aspirated from the vomeronasal pit of volunteers and a depolarizing response to vomeropherins. The vomeropherins used in these experiments are certain steroids similar to those isolated from human skin. These substances have also been incorporated into commercial perfumes and claimed to be pheromones, although there is no scientific evidence for a pheromonal function. Despite this, the physiological experiments are intriguing. They have no obvious fatal flaws and should be repeated by other laboratories. Sexually dimorphic responses and autonomic responses to some chemicals have been reported, although subjects are apparently unable to detect the stimulation at a conscious level. A failure of conscious perception is not inconsistent with known central vomeronasal projections, which do not have any direct pathway to neocortex. However, such projections have not been demonstrated in adult humans, so it is unclear whether any kind of central neural representation of vomeronasal stimulation could be expected. Recent reports that intranasal vomeropherin stimulation slightly perturbed LH hormone pulsatility provide less convincing evidence of vomeronasal function because stimulation was not confined to the vomeronasal organ.

See Also the Following Articles

Bruce Effect; GnRH (Gonadotropin-Releasing Hormone); Lordosis; LH (Luteinizing Hormone)

Bibliography

Aujard, F. (1997) Effect of vomeronasal organ removal on male socio-sexual responses to female in a prosimian primate (*Microcebus murinus*). *Physiol. Behav.* **62**, 1003–1008.

Bargmann, C. I. (1997). Olfactory receptors, vomeronasal receptors, and the organization of olfactory information. *Cell* **90**, 585–587.

Boehm, N., and Gasser, B. (1993). Sensory receptor-like cells in the human foetal vomeronasal organ. *Neuroreport* **4**, 867–870.

Buck, L. B. (1996). Information coding in the olfactory system. *Annu. Rev. Neurosci.* **19**, 517–544.

Coolen, L. M., Peters, H. J. P. W., and Veening, J. G. (1997). Fos immunoreactivity in the brain following consummatory elements of sexual behavior: A sex comparison. *Brain Res.* **738**, 67–82.

Halpern, M. (1987). The organization and function of the vomeronasal system. *Annu. Rev. Neurosci.* **10**, 325–362.

Keverne, E. B. (1998). Vomeronasal/accessory olfactory system and pheromonal recognition. *Chem. Senses*, **23**.

Kroner, C., Breer, H., Singer, A. G., and O'Connell, R. J. (1997). Pheromone induced second messenger signaling in the hamster vomeronasal organ. *Neuroreport* **7**, 2989–2992.

Levy, F., Locatelli, A., Piketty, V., Tillet, Y., and Poindron, P. (1995). Involvement of the main but not the accessory olfactory system in maternal behavior of primiparous ewes. *Physiol. Behav.* **57**, 97–104.

Liman, E. R. (1996). Pheromone transduction in the vomeronasal organ. *Curr. Opin. Neurobiol.* **6**, 487–493.

Meredith, M. (1998a). Vomeronasal function. *Chem. Senses*, **23**.

Meredith, M. (1998b). Olfactory, vomeronasal, hormonal convergence in the brain, co-operation or coincidence. Olfaction and Taste XII. *Ann. N. Y. Acad. Sci.*, in press.

Monti-Bloch, L., Jennings-White, C., Dolberg, D. S., and Berliner, D. L. (1994). The human vomeronasal system. *Psychoneuroendocrinology* **19**, 673–686.

Morgan, J. I., and Curran, T. (1991). Stimulus transcription coupling in the nervous system: Involvement of the inducible proto-oncogenes fos and jun. *Annu. Rev. Neurosci.* **14**, 421–451.

Paredes, R. G., Lopez, M. E., and Baum, M. J. (1998). Testosterone augments neuronal Fos responses to estrous odors throughout the vomeronasal projection pathway of gonadectomized male and female rats. *Horm. Behav.* **33**, 48–57.

Pelosi, P. (1995). Perireceptor events in olfaction. *J. Neurobiol.* **30**, 3–19.

Segovia, S., and Guillamon, A. (1993). Sexual dimorphism in the vomeronasal organ and sex differences in reproductive behaviors. *Brain Res. Rev.* **18**, 51–74.

Smith, T. D., Siegel, M. I., Mooney, M. P., Burdi, A. R., Burrows, A. M., and Todhunter, J. S. (1997). Prenatal growth of the human vomeronasal organ. *Anat. Rec.* **248**, 447–455.

Tubbiola, M. L., and Wysocki, C. J. (1997). FOS immunoreactivity after exposure to conspecific or heterospecific urine: Where are the chemosensory cues sorted? *Physiol. Behav.* **62**, 867–870.

Wysocki, C. J., and Lepri, J. J. (1991). Consequences of removing the vomeronasal organ. *J. Steroid Biochem. Mol. Biol.* **39**(4B), 661–669.

Wysocki, C. J., and Meredith, M. (1987). Function of the vomeronasal system. In *Neurobiology of Taste and Smell* (T. E. Finger and W. L. Silver, Eds.), pp. 125–150. Wiley Interscience, New York.

Whales and Porpoises

Daniel P. Costa and Daniel E. Crocker

University of California, Santa Cruz

GLOSSARY

Mysticeti The baleen whales, characterized by their feeding apparatus—a series of transverse plates of comb-like baleen which are used to strain plankton. Other distinctive characteristics include a pair of nasal openings and a symmetrical skull, lacking a melon.

Odontoceti The toothed whales. Either teeth are numerous, conical, and uniform (homodont dentition) or there is only a single tooth. Other distinctive characteristics include a single nasal opening and an asymmetrical skull.

The Cetacea are an order of aquatic mammals that includes the suborder Mysticeti, or baleen whales, and the Odontoceti, or toothed whales. Cetaceans give birth in the water to a single precocial offspring. Cetacean mating systems range from the loosely solitary or gregarious mysticetes to the extremely social odontocetes. Mysticete whales are highly migratory, typically breeding in the warm tropical to subtropical waters in winter and then moving to the cold temperate and polar water in summer. Reverse sexual dimorphism (females may be larger than males) is common in mysticetes. Cetacean social systems appear to have evolved in response to the types of prey consumed and habitats occupied. The solitary or gregarious mysticete whales feed on small prey that are consumed in great quantities, whereas odontocetes tend to feed on larger individual prey that often requires a number of individuals working collectively as a highly coordinated unit or pack. The reproductive pattern of odontocetes is intimately related to their social systems and stable social units.

I. CETACEAN BIOLOGY

Mysticete whales are exclusively marine and feed by using baleen to sieve their food out of the water. There are three families of baleen whales. The Balaenoptera, or rorquals, include the blue, fin, sei, Bryde's, humpback, and minke whales. These are fast-swimming whales that feed by gulping large prey-filled volumes of water. These large gulps of prey-filled water are then sieved through their baleen plates. The Balaenidae include the right and bowhead whales, which are slow-moving whales that feed by skimming prey-containing water through their very long baleen plates. The family Eschrichtiidae is composed of one living species, the California gray whale. Mysticete whales are highly migratory, typically breeding in the warm tropical to subtropical waters in winter and then moving to the cold temperate and polar water in summer. It is believed that most if not all of the food requirements of these whales are met during the brief summer feeding season. This includes the energy required to give birth and suckle the calf until the mother returns to the summer feeding grounds with the calf the following season. During the nonfeeding periods whales obtain energy from fat reserves laid down in their blubber during the summer.

The odontocete or toothed whales include the sperm whale, the smaller killer and pilot whales, as well as dolphins and porpoises. This group is composed of 10 families. Members of this suborder range from the sperm whale, the largest odontocete, to the harbor porpoise, the smallest. Toothed whales inhabit a broad range of habitats, including fresh water and marine. The definitive characteristic of this group is their ability, like bats, to use highly sophisticated biosonar to find their prey and to sense their environment, in addition to the highly evolved social system which equals that found in primates and social carnivores. The highly evolved social system of odontocetes appears to be associated with environmental factors such as susceptibility to predation and availability of prey.

A. Mysticete Breeding Systems

Mysticete cetaceans comprise the largest animals currently alive. For their body size they have a very short period of maternal dependence with a relatively short gestation and lactation period. Baleen whales are considered to be gregarious rather than social. This is thought to be due to their need to feed low on the food chain, consuming large quantities of very small schooling prey. Since they engulf large portions of the school they do not have a need for a highly integrated cooperative feeding, which is typical of toothed whales. Although baleen whales have been seen to cooperatively feed (humpback whales cooperatively feed using bubble nets), these aggregations are not stable through time and thus do not require the development of a strong social unit or bond. Because the young do not learn or need to become integrated into a strong social unit, there is no need for a prolonged dependency period. Baleen whales are generally believed to tend toward a monogamous mating system (Fig. 1).

Although little is known about the breeding behavior of large cetaceans, it is thought that males go to where the females are located and set up territories. Courtship behaviors have been reported for many species. Complex vocalization may also play a role in courtship. The well-known song of the humpback whale is virtually the same between males in one breeding area and can change during or between breeding seasons. Their songs may function much like male song birds competing to have the best song. Females may choose mates on the basis of the quality or complexity of their call. It is also possible that the song may be used in male–male competition or to act as a spacing signal. Other mysticete whales are known to produce sounds that travel great distances in the ocean. Blue, fin, minke, bowhead, and right whales all produce sounds that may serve an important role in finding or competing for mates.

FIGURE 1 Gray whale mother and recently born calf in one of the breeding lagoons of Baja California. Gray whales are a mysticete cetacean; notice the two openings to the blowhole.

FIGURE 2 Pacific bottlenose dolphins are an example of an odontocete cetacean. Male bottlenose dolphins often travel together in pairs.

B. Odontocete Breeding Systems

The breeding systems of odontocete cetaceans are tightly correlated with the highly integrated social system of this group. Typically toothed whales exist in highly structured social units that enable them to feed on schooling prey or to reduce the risk of predation. Because most toothed whales are small and exposed in the three-dimensional world of the open ocean, they are quite susceptible to predation. The highly structured social groups of schooling dolphins can significantly reduce the risk of predation, especially for the young. Such a highly structured social unit requires a long development and learning period. In contrast to baleen whales, the social aggregations of toothed whales are extremely stable, lasting years and in many cases an entire lifetime. Because toothed whales feed during lactation there is no premium on large body size in females and thus there is a more typical mammalian trait of males having a tendency to be larger than females (Fig. 2).

The reproductive biology of odontocetes is best understood for the bottlenose dolphin. In this species females form bands that are based along matrilineal lines. These groups are usually made up of mothers and daughters and their offspring. It is not unusual to see multiple generations swimming together in a group. While females tend to stay with their mothers, males leave the bands by the time they reach puberty. Upon leaving the bands, males pair up with other males. These pairs are often formed with males from the same band and often last their entire lifetime. Breeding takes place as males patrol the area looking for estrous females. Once located, the males escort the females, apparently taking turns copulating. Many species of dolphins appear to exhibit sperm competition and have extremely large testis for their body mass.

Because sound travels well in water, acoustic cues are important in coordinating the social system of toothed whales. Many dolphins have signature whistles that are unique among different individuals but can be mimicked by other members of their band. Signature whistles are thought to be important in coordinating and maintaining these associations. Interestingly, the signature whistle develops after birth and is a learned behavior. Although common in birds, among mammals such a trait is unique to humans and dolphins.

II. PARENTAL INVESTMENT

The cost of reproduction for females can be broken down into the energetic requirements for gestation (the cost of producing a fetus) and for lactation (the cost of nursing young until weaning). Although no direct measurements of the cost of gestation are available for marine mammals, investigations of terrestrial eutherians suggest that the cost of producing a fetus is insignificant relative to the costs associated with lactation. Fetal mass at birth relative to maternal mass can be used as a relative index of the cost of gestation. Among cetaceans, there is no difference in birth mass relative to maternal mass. Conversely, lactation strategies and maternal investment vary markedly between mysticetes and odontocetes.

A. Mysticetes

Mysticetes have shorter lactation durations for their body size than do other marine mammals with the exception of phocid seals. Like phocid seals, the shorter lactation duration is compensated for by an increased growth rate. Fasting during lactation is a unique component of the life history pattern of whales. With the exception of bears and pinnipeds, no other mammal is capable of producing milk without feeding. By undertaking this energetic challenge, mysticetes are able to temporally and spatially separate feeding from breeding. Separation of lactation from feeding allows mysticete whales to feed in the seasonally productive polar regions of the world's oceans but to retain the advantage of breeding in the calm, warm tropical regions of the world. Migrating to warmer water for parturition reduces the thermal demands on the newborn calf and provides additional thermoregulatory and energetic advantages for the mother.

The ability of a whale to fast while providing milk to her offspring is related to the size of her energy and nutrient reserves and the rate at which she utilizes them. When food resources are far from the breeding grounds, as may occur for large mysticete whales, the optimal solution is to maximize the amount of energy and nutrients provided to the young and to minimize the amount expended on the mother. One mechanism for reducing the metabolic overhead is to attain large body size. This results from the relationship between maintenance metabolism, which scales as $mass^{0.75}$, and fat stores which scales as $mass^{1.0}$. These relationships demonstrate that as body size increases, energy reserves increase proportionately faster with mass than does maintenance metabolism. This implies that larger females have a greater ability to provision their young from stored body reserves than small females. Large mysticetes whales are therefore able to maintain longer lactation intervals without eating since their metabolic rate is so low relative to their energy stores simply due to their extremely large body mass.

In the case of mysticetes, parental care is provided solely by the mother. Calves are weaned early, allowing for a short intercalf interval. For example, blue and humpback whales have a gestation period of 11 months and wean their calves at 7 and 11 months, respectively. The reproductive cycle has a minimum duration that is controlled by gestation time, which is about 1 year in all Mysticeti. Because conceptions are seasonal, the reproductive interval in mysticetes is usually a multiple of 1 year.

B. Odontocetes

Odontocetes exhibit longer lactation durations and slower growth rates compared to other marine and terrestrial mammals. Toothed whales have prolonged gestation and calf-dependency periods compared to marine and terrestrial mammals. The age at weaning is thus quite old, which leads to a longer intercalf interval. For example, the gestation period of the sperm whale is 15 or 16 months and the age at weaning varies from 2 to 13 years. In bottlenose dolphins the gestation period is 12 months and weaning occurs between 18 and 20 months. In pilot whales gestation lasts up to 15 months and the age at weaning is quite protracted, ranging from 4 or 5 years in one species to 13–15 years in another. The long calf dependency period is further supported by the observation of reproductively senescent pilot whale females that are still lactating. Observations have been made of milk in the stomachs of a 15.5-year-old pilot whale and a 7.5-year-old female and a 13-year-old male sperm whale. There is evidence that some cetacean species, such as harbor porpoises, may alter the age of sexual maturity in response to changes in prey availability or population density.

C. Milk Composition and Suckling

A consistent adaptation among all marine mammals is that they produce a high-fat and therefore energy-rich milk. A high-fat, energy-rich milk allows the mother to transfer high levels of energy in a very short period of time. Reported mean values for the lipid content of baleen whales range from 24 to 53%. This is an important adaptation in the aquatic environment in which suckling while swimming is difficult at best. Because the young lack proper lips, gripping the nipple is difficult, and it is believed that cetaceans may squirt the milk into the mouth of the young, probably using contractions of the cutaneous muscles or contractions of the myoepithelial cells surrounding the aureoli.

III. REPRODUCTIVE PHYSIOLOGY

Cetacean reproductive anatomy mirrors their terrestrial origins because they are thought to have evolved from a terrestrial ancestor common to modern ungulates. It should not be surprising that cetaceans share a placental structure similar to that of modern ungulates. There are differences from the control of estrous cycles in typical terrestrial mammals. The time period between birth and estrus is associated with active follicular growth in the ovaries. There is, on average, one large follicle remaining in

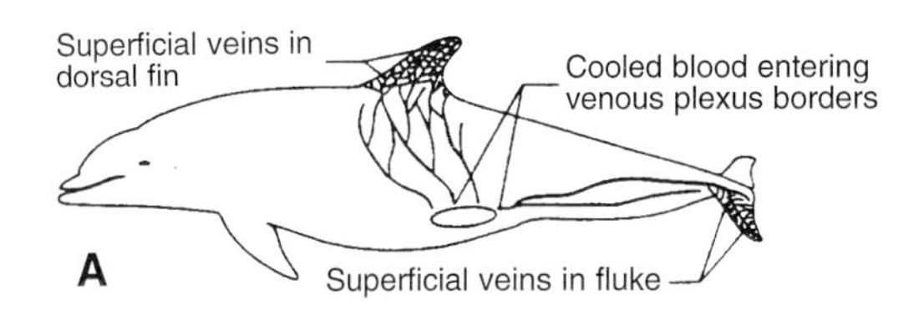

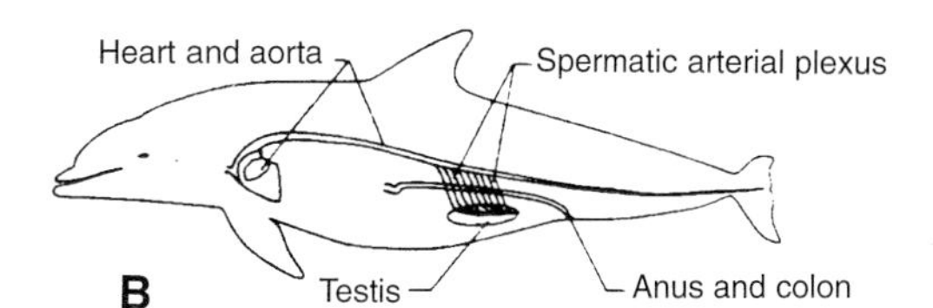

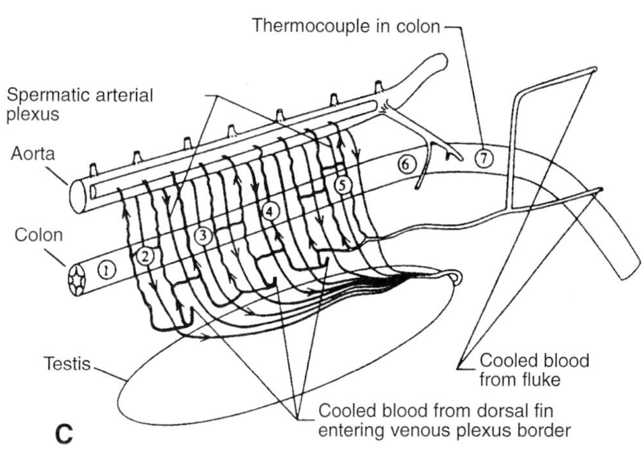

FIGURE 3 Schematic topography of the countercurrent heat exchanger at the dolphin testis. (A) Blood in the superficial veins of the dorsal fin and the tail flukes is cooled by exposure to ambient water. Blood from these extremities is drained by relatively thick-walled, large veins that remain superficial (i.e., just beneath the blubber layer and superficial to the vertebral muscles) as they coalesce and course toward the abdominal cavity. These veins enter the posterior abdominal cavity, near the pelvic vestiges, and feed directly into the lateral and posterior margins of the lumbocaudal venous plexus. Thus, relatively cool blood can be introduced into the deep posterior abdominal cavity near the testes. (B) The spermatic arterial plexus is a unique arrangement of arteries extending ventrolaterally from the lumbar aorta. The vessels are organized into a single layer and are oriented approximately parallel to each other. At the distal margin of the plexus, the arteries coalesce to form a cone-shaped structure, from which a single testicular artery continues posteriorly to enter the testis. (C) Oblique lateral view of the left half of the countercurrent heat exchanger. The juxtaposition of the lumbocaudal venous plexus to the spermatic arterial plexus suggests that dolphins use countercurrent heat exchange to regulate the temperature of arterial blood flow to the testis. Arrowheads indicate the direction of blood flow. Cut ends at the lateral border of the lumbocaudal venous plexus show where the superficial veins from the dorsal fin enter the abdominal cavity (see A). Note that the countercurrent heat exchanger flanks a region of colon in the posterior abdominal cavity. The numbers refer to locations of temperature sensors used to measure blood cooling. (Reproduced with permission from D. A. Pabst, S. A. Rommel, W. A. McLellan, T. M. Williams, and T. K. Knowles, *J. Exp. Biol.* 198, 221–226, 1995. © The Company of Biologists Ltd.).

one ovary when estrus behavior occurs. This coincides with peak estrogen secretion. Progesterone levels elevate at the presumed time of ovulation. Typically, a single ovulation occurs and twinning is highly unusual. Ovulation results in the formation of a corpus luteum, with a resultant rise in progesterone levels, potentially signaling a mother to initiate weaning.

In males, spermatogenesis appears to follow the typical mammalian pattern. Testicular mass and spermatogenesis increase prior to and peak during the breeding season. The testes then regress. These seasonal alterations in male endocrine physiology are most likely important in stimulating the reproductive behaviors such as territory defense found in the males of some species. However, because testes are internal in cetaceans, one might expect some novel adaptation since spermatogenesis in most mammals only occurs at temperatures significantly below core body temperature. In most mammals temperature reduction is achieved by placing the testes externally. Recent work has shown that at least bottlenose dolphins have a novel circulatory system that cools the internal testes (Fig. 3). In this species the vessels that drain the dorsal fin go deep into the body and form a retie around the testes prior to joining the central venous pool. This novel circulatory structure keeps the testes a few degrees below core body temperature. Like primates, cetacean males lack an os penis or baculum.

See Also the Following Article

SEALS (PINNIPEDIA)

Bibliography

Bryden, M. M., and Harrison, R. J. (1986). Gonads and reproduction. In *Research on Dolphins* (M. M. Bryden and R. J. Harrison, Eds.), pp. 149–162. Clarendon, Oxford, UK.

Leatherwood, S., and Reeves, R. R. (1989). *The Bottlenose Dolphin.* Academic Press, San Diego.

Perrin, W. F., Brownell, R. L., Jr., and DeMaster, D. P. (Eds.) (1984). *Reproduction in Whales, Dolphins and Porpoises*, Reports of the International Whaling Commission, Special Issue 6. International Whaling Commission. Cambridge, UK.

Whitten Effect

John G. Vandenbergh

North Carolina State University

GLOSSARY

estrus A period of sexual receptivity associated with ovulation in spontaneously ovulating female mammals, often associated with characteristic vaginal cellular changes.

gonadotropin-releasing hormone A hormone produced in the hypothalamus of the brain that activates the release of gonadotropins from the pituitary.

menstrual synchrony The occurrence of menstruation coordinated in time among a group of female primates, including humans, following synchronous ovulation.

pheromone A substance or blend of substances produced by one organism, transmitted through the environment, inducing a behavioral, developmental, or physiological change in the recipient of the same species.

puberty acceleration The induction of early puberty in females of a number of mammalian species following exposure to adult male pheromones.

vomeronasal organ A bilaterally symmetrical pair of tubes opening to the oral and/or nasal cavities, containing specialized olfactory receptor epithelium, on either side of the nasal septum; a part of the accessory olfactory system that sends afferent signals to the accessory olfactory bulb and onto parts of the brain involved in reproduction.

The ovaries of many placental mammals cycle rhythmically as a result of neural and endocrine feedback systems. These rhythms are not fixed but show some adaptability to environmental signals received and processed by the female. For example, severe physical or psychological trauma can induce arrhythmia. Social stimuli from other individuals can also influence the regularity of cycles and in some cases make females cycle in synchrony. One such well-studied effect is the synchronization of ovulation in house mice (*Mus musculus*) named the "Whitten effect" for its discoverer, Wesley K. Whitten.

I. INTRODUCTION

Whitten was breeding laboratory mice for his research at the Australian National University when he noted that females housed in a dense group showed prolonged periods of anestrus, some as long as 40 days (Whitten, 1959). Isolated females cycled with a 4- or 5-day periodicity. He then found that by placing a male confined to a wire basket into the group, the formerly anestrous females exhibited estrus 3 or 4 days later. Normally about 20% of females with a regular cycle of 5 days would be expected to conceive each night. However, when the females were individually paired with a male, over 50% conceived on the third night after pairing, thus revealing a synchronous estrus among the females.

Whitten's discovery and the almost simultaneous report that pregnancy in mice can be blocked by olfactory stimuli from strange males stimulated a number of studies on the role of olfaction in reproduction . The mechanisms involved in mediating the Whitten effect, its ecological implications, and its generalizability to other species have been explored by Whitten and others, and will be briefly reviewed here.

II. MECHANISMS

A. Identification of Stimulus

Whitten's studies showed that the male signal responsible for inducing the female's ovarian cycles involved pheromonal regulation. His original findings have been upheld, but apparently describe only a part of the stimulus coming from the male since physical contact between the sexes also contributes to the synchronizing effect. The male produces an olfactory stimulus only when gonadally intact. Castrated males lose the ability to induce estrus in females about 1 week after castration and can have the ability restored by injections of testosterone. These findings suggest that the source of the pheromone is either metabolite of testosterone itself or, more likely, the production of an androgen-dependent substance in a gland or organ of the male mouse. The olfactory signal is present in the urine of an intact male mouse, a very complex fluid containing several hundred substances. Two volatile constituents of urine from an intact male have been shown to enhance estrous synchrony in mice but only do so when added to the urine of a castrate male. This finding suggests that the stimulatory substance is a complex including at least the two substances identified as 2-(*sec*-butyl)-4,5-dihydrothiazole and dehydro-*exo*-brevicomin, plus an unknown component in the urine that is not androgen dependent (Jemiolo *et al.*, 1986). This work is preliminary and awaits confirmation.

B. Reception

It is likely that the active ingredient has its effect on the female after being received by the vomeronasal organ (VNO), a part of the accessory olfactory system. Reception by the VNO is a possibility because a number of pheromones operate via the VNO to regulate reproduction. However, there is no direct experimental evidence implicating the VNO. The reason for this gap in our knowledge is that destruction of the olfactory bulb containing the accessory olfactory bulb (which receives its signals from the VNO) causes the disappearance of cycles among female mice. Thus, there is no way to confirm that the accessory olfactory system specifically underlies the Whitten effect. It is thought, but with only modest experimental support, that the olfactory signal is processed by the brain, triggering neuroendocrine changes in the hypothalamus, which induces the release of gonadotropin-releasing hormone, which in turn induces the anterior pituitary to release gonadotropins. This cascade of events results in increasing levels of gonadotropins available to the ovary, which in turn induces the secretion of ovarian steroids such as estrogen and progesterone. These steroids then feed back on the brain and pituitary to initiate the next ovarian cycle.

III. OVARIAN RESPONSE

A. Synchronization

The notion that ovarian cycles can be synchronized stimulated a series of studies on rats. It was soon discovered that female rats held in a cage "downwind" from another female rat soon began to cycle in synchrony. Some female rats are more "effective" than others in inducing this change and are called "drivers." As a result of a series of studies, McClintock (1983) suggested that the synchrony in these rats is the result of airborne signals and that there are two signals involved. One phase advances ovulation, and one delays ovulation. A balance among these two signals results in the induction of estrus at a synchronous time. In the rat, the relevant chemical signals influencing ovarian cyclicity seem to be produced by the female, not the male. These chemical signals are likely to be airborne and received by the main olfactory system.

The possibility that human females' menstrual cycles can be synchronized has not escaped scientific attention. In pioneering work, Martha McClintock showed that the cycles of women living in a dormitory became more synchronous during the course of a school year (McClintock, 1971). The synchrony seemed to be most intense among women who were in close physical contact rather than among women who were close social friends. Many explanations are possible for this result, but one likely explanation is that there is a pheromone or mixture of chemical signals which can initiate the endocrine changes resulting in synchronous ovulation. Additional work since that time suggests that an alcohol-soluble component of axial perspiration may be involved. In one experiment, women who were exposed to an alcohol extract of axial cotton pads worn by cycling women tended to show some degree of menstrual synchrony. Some caution should be applied in interpreting these data since there are problems with small sample size, statistical concerns, and other difficulties associated with research on human subjects.

B. Puberty Acceleration

Another set of studies that grew out of Whitten's initial report are those of Vandenbergh and others on the role that males play in accelerating the onset of puberty (Vandenbergh, 1994). A component of male mouse urine, probably a small peptide, induces early onset of puberty in juvenile females in several rodents, swine, and nonhuman primates. This pheromonal effect, termed the "Vandenbergh effect," is described in greater detail elsewhere in this encyclopedia.

IV. EXTENSION TO OTHER SPECIES

The Whitten effect, or something closely resembling it, has been observed in the Norway rat, both the golden and the Dungerian hamster, the prairie vole, goat, cow, sheep, and perhaps human. In all these species, the critical signal from the environment seems to be olfactory. The females synchronize their ovarian cycles in response to a pheromonal signal. In most cases, there is also an additive effect of tactile or other social stimuli.

The finding that the ovarian cycle could be manipulated through modification of the social environment has resulted in a number of practical improvements to animal husbandry. First, in laboratory mice, breeding can be synchronized so that 50–70% of the females conceive on a single night (Scharmann and Wolff, 1980). This is a significant advantage when timed pregnancies are important for a research project. The technique is basically to use Whitten's procedure: Female mice are grouped six or more to a cage for 10 days or more and then paired with a male. A high proportion of the females mate on the third night. Second, among domestic farm animals, an adult male has been used to initiate earlier onset of puberty in sows and to synchronize the date of conception in both sheep and cattle. In cattle, and to some extent in sheep, the synchronization is often done via administering hormones or their analogs rather than social manipulation. The ovarian cycles of cows can be induced by a specific regimen of injections of progesterone or its analogs. It is interesting to note that the early basic scientific studies of Whitten, Bruce, Vandenbergh, and others on estrus chronicity in rodents now have practical implications for animal husbandry that were unanticipated.

V. ECOLOGICAL AND EVOLUTIONARY IMPLICATIONS

The ecological and evolutionary implications of synchronizing estrous cycles have not been explored from an experimental point of view. However, theoretical studies have suggested that female reproductive synchrony is likely to reduce male desertion because the probability of the male finding another mate in heat somewhere else would be reduced if all females were in estrus at the same time. This may also favor high-ranking males in mate competition because the effort to mate would have to be concentrated in a shorter period of time, and so it would be relatively easier to reduce subordinant males' access to the females in the group. Lions are a particularly interesting example of this theoretical phenomenon. The females in a pride come into heat in a synchronous wave, mate frequently over a short period of time, and then deliver pups at about the same time. This provides the female with the advantage described previously and permits communal nursing, a behavior shown by this species.

VI. CONCLUSIONS

The original work of Whitten, showing that the ovarian cycle of a mouse could be altered by environmental stimuli, enhanced the awareness of reproductive physiologists and those dealing with animal husbandry about the important role that the social environment plays in regulating reproduction. The reproductive success of many animals also depends on factors more subtle than food availability and housing conditions. Social stimuli are now known to be important contributors to regulation of reproductive events in all mammals, including humans.

See Also the Following Articles

Bruce Effect; Estrus; GnRH (Gonadotropin-Releasing Hormone); Mating Behavior; Pheromones; Sexual Attractants; Vomeronasal Gland

Bibliography

Brown, R. E. (1985). The rodents I: Effects of odours on reproductive physiology. In *Social Odours in Mammals, Vol. 1* (R. E. Brown and D. W. Macdonald, Eds.), pp. 245–344. Clarendon, Oxford, UK.

Bruce, H. M. (1959). An exteroceptive block to pregnancy in the mouse. *Nature (London)* **184**, 105.

Izard, M. K. (1983). Pheromones and reproduction in domestic animals. In *Pheromones and Reproduction in Mammals* (J. Vandenbergh, Ed.), pp. 253–286. Academic Press, New York.

Jemiolo, B., Harvey, S., and Novotny, M. (1986). Promotion of the Whitten effect in female mice by synthetic analogs of male urinary constituents. *Proc. Natl. Acad. Sci. USA* **83**, 4576–4579.

McClintock, M. K. (1971). Menstrual synchrony and suppression. *Nature* **229**, 244–245.

McClintock, M. K. (1983). Pheromonal regulation of the ovarian cycle: Enhancement, suppression and synchrony. In *Pheromones and Reproduction in Mammals* (J. Vandenbergh, Ed.), pp. 113–150. Academic Press, New York.

Scharmann, W., and Wolff, D. (1980). Production of timed pregnant mice by utilization of the Whitten effect and a single cage system. *Lab. Anim. Sci.* **30**, 206–208.
Vandenbergh, J. G. (1967). Effect of the presence of a male on the sexual maturation of female mice. *Endocrinology* **81**, 345–349.
Vandenbergh, J. G. (1994). Pheromones and mammalian reproduction. In *Physiology of Reproduction* (E. Knobil and J. D. Neill, Eds.), Vol. 2, pp. 343–362. Raven Press, New York.
Whitten, W. K. (1959). Occurrence of anoestrus in mice caged in groups. *J. Endocrinol.* **18**, 102–107.

Wolffian Ducts

Terry W. Hensle and Harry Fisch
Columbia College of Physicians & Surgeons

GLOSSARY

cryptorchidism A condition in a phenotypically male individual in which one or both gonads are not fully descended into their normal position in the scrotum.
epididymis A tortuous ductular structure which receives sperm from the testis. The epididymis is divided into a head, body, and tail.
hydrocele A benign fluid-filled collection within the parietal and visceral layers of the tunica albuginea of the testis.
Leydig cell An eponym for the primary androgen-secreting cell of the testis.
mesonephric duct The embryological ductal system which ultimately differentiates into the tubular structures of the male reproductive tract.
prostate A gland in the male reproductive system which is anatomically found at the base of the bladder and adds its secretions to the ejaculate.
seminal vesicle A gland in the male reproductive system situated posterior to the prostate gland. The seminal vesicle contributes the majority of the ejaculate volume and empties into the ejaculatory duct.
Sertoli cell An eponym for the primary germinal cell in the testis with which spermatic maturation occurs.
testosterone The primary male sex hormone produced primarily in the testis but also in the adrenal cortex.
ureter A urinary tubular structure which acts as a conduit for urine between the kidney and the bladder.
vas deferens A muscular tube which serves as a conduit for the sperm from the epididymis to the ejaculatory duct.
Wolffian duct An eponym for the mesonephric duct.

Wolffian ducts, otherwise known as the mesonephric ducts, are the main embryological elements that ultimately give rise to the glands and tubular structures of the male reproductive system. Specifically, the Wolffian ducts in the developing male fetus eventually differentiate into the epididymis, vas deferens, ejaculatory ducts, and seminal vesicles.

I. EMBRYOLOGY OF THE WOLFFIAN DUCTS

A. Origins of the Primitive Genitourinary Ductal Systems

The Wolffian ducts originate as mesenchymal tissue around the cervical somites, or neck region, of the embryo. The Wolffian ducts reach their ultimate destination in the hind region of the fetus by a relative elongate growth of the rest of the embryo. The

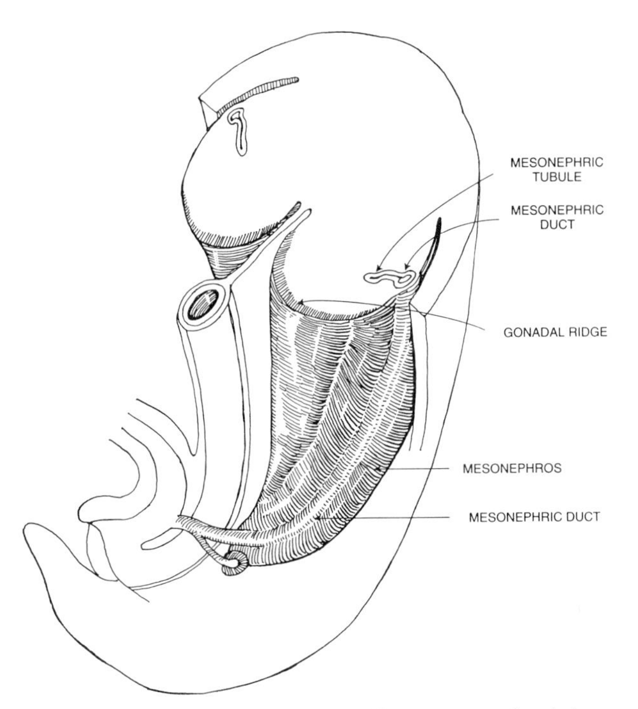

FIGURE 1 Relationship of the Wolffian (mesonephric) duct and ureteral bud at 5 weeks of gestation [reproduced with permission from T. W. Hensle and E. K. Seaman, Embryology of the male reproductive tract, In *Atlas of Surgical Management of Male Infertility* (A. J. Thomas and H. M. Nagler, Eds.), Igaku-Shoin, New York, 1995].

Wolffian ducts are situated next to the mesoderm of the embryo and interposed between the rudimentary coelom (body cavity) and the somites. Aspects of the Wolffian system will ultimately give rise to the ductal systems of the urinary tract as well as the genital tract. The caudal aspect of the Wolffian ducts develops an outpouching known as the ureteral bud which separates from the mesonephric duct at approximately 7 weeks gestation. Further ureteral bud differentiation into the mesonephric tubules occurs and the surrounding mesodermal tissue differentiates into the primitive kidney (or metanephros) by approximately 37 days gestation (Fig. 1). Once fully differentiated, the tissue of the ureteral bud infiltrates into the metanephros to become the collecting system of the urinary tract. Concomitantly, the Wolffian duct also undergoes development to become the genital ductal system in the male.

B. Influences of Gender-Determining Factors on Development of the Genital Ductal System

Upon fertilization, the presence or absence of a Y chromosome determines chromosomal gender. Although the function of the Y chromosome is poorly elucidated, it is clear that the Y chromosome does harbor the genes involved in testicular development. Encoded on the Y chromosome is the gene for testicle-determining factor which stimulates the differentiation of the primitive gonad into the testis. Once testicular differentiation is under way, the secretory function of testicular cells begins and subsequently influences the differentiation of the Wolffian and Müllerian (or paramesonephric) systems. At approximately the 7th week of gestation, the Sertoli cells of the testis produce Müllerian-inhibiting substance which leads to the regression of the Müllerian ducts, leaving only the appendix testis and the prostatic utricle as remnants in the male. Meanwhile, the Leydig cells of the testis produce testosterone, which stimulates further development of the Wolffian duct as the major ductal system of the male reproductive system as well as the virilization of the male external genitalia.

C. Role of the Wolffian Ducts in the Development of the Male Reproductive System

The caudal aspect of the Wolffian ducts become continuous with the urogenital sinus by 28 days of gestation, thus leading to an upper and lower division of the urogenital sinus (Fig. 2 and Fig. 3). The urogenital sinus undergoes differentiation between Weeks 9 and 12 of gestation. Relative to the insertion of the Wolffian duct, the upper portion of the urogenital sinus becomes the bladder base and neck, whereas the lower aspect leads to the prostate and external genitalia.

The development of the caudal aspect of the Wolffian duct leads to the ejaculatory ducts and seminal vesicle. Lateral outgrowths arise from the caudal mesonephric ducts at about 10 weeks of gestation and ultimately differentiate into the seminal vesicle. While the seminal vesicle develops from buds in the

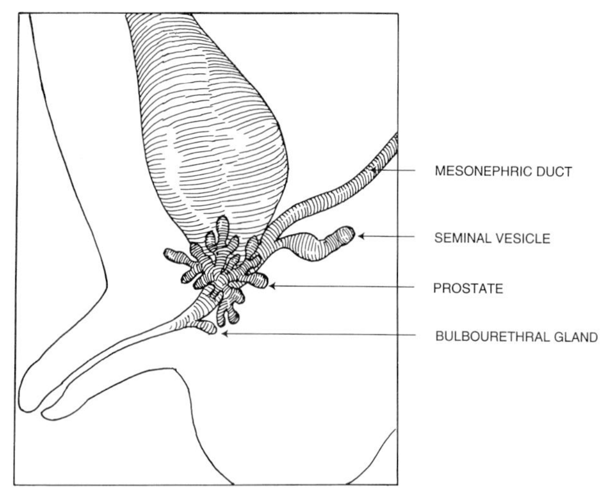

FIGURE 2 Differentiation of the lower Wolffian duct segments [reproduced with permission from T. W. Hensle and E. K. Seaman, Embryology of the male reproductive tract, In *Atlas of Surgical Management of Male Infertility* (A. J. Thomas and H. M. Nagler, Eds.), Igaku-Shoin, New York, 1995].

distal Wolffian duct, the prostate and bulbourethral glands develop simultaneously from buds arising from the adjacent urethra. Once development of the seminal vesicle is established, all remaining Wolffian duct tissue is thereafter referred to as the ejaculatory duct (Fig. 2). The secretory products of the seminal vesicles, prostate, and bulbourethral glands combine together with the spermatozoa to produce the male ejaculate.

Wolffian duct development at its cranial end is also essential in the proper development of the male genital system. Cranial mesonephric tubules come into contact with the germinal epithelium, which ultimately leads to the testes. These mesonephric tubules differentiate to become the efferent tubules, carrying sperm away from the testis during ejaculation. The area of the Wolffian duct adjacent to the mesonephric tubules likewise undergoes a differentiation leading to a series of tubular structures called the epididymis. The remaining central portion of the mesonephric duct then becomes invested in a thick smooth muscular layer and ultimately becomes known as the vas deferens (Fig. 3).

II. CLINICAL RELEVANCE OF WOLFFIAN DUCT DEVELOPMENT

Wolffian duct development is relevant to both the male reproductive system and the urinary tract in both genders. Since the ureteral bud separates itself from the mesonephric duct at 7 weeks of gestation, the timing with which developmental abnormalities occur is critical. Errors in development occurring after Week 7 of gestation will usually only affect the genital tract. Conversely, abnormalities in differentiation that occur prior to Week 7 of gestation may result in abnormalities of both the genital and urinary systems. Both the urinary and genital systems should be investigated when abnormalities of Wolffian duct development become evident.

A. Abnormalities of the Vas Deferens

Abnormalities of the vas deferens are the most common congenital problem identified as resulting in male factor infertility. While testicular function may be normal, mechanical difficulties arise with the transport of sperm in the ejaculate. Abnormalities seen may range from complete absence of the vas deferens to hypoplasia to the presence of diverticulae. Physical examination of the vas deferens is very important in the evaluation of the infertile patient. Upon palpation, the vas deferens has a distinctive firm tubular character in the scrotum. Physical examination will often reveal abnormalities such as the absence or hypoplasia when evaluating the palpable scrotal portion of the vas deferens. A normal scrotal examination, however, does not rule out an abnormality of the pelvic vas deferens since this portion is not felt on physical examination. In such cases, the pelvic vas deferens should be evaluated with transrectal ultrasound or other imaging studies.

B. Abnormalities of the Epididymis

Abnormalities of the epididymis include the absence of hypoplasia, failure of attachment to the testicle, and the presence of cysts. The clinical relevance is the consequent difficulties with infertility. Epididy-

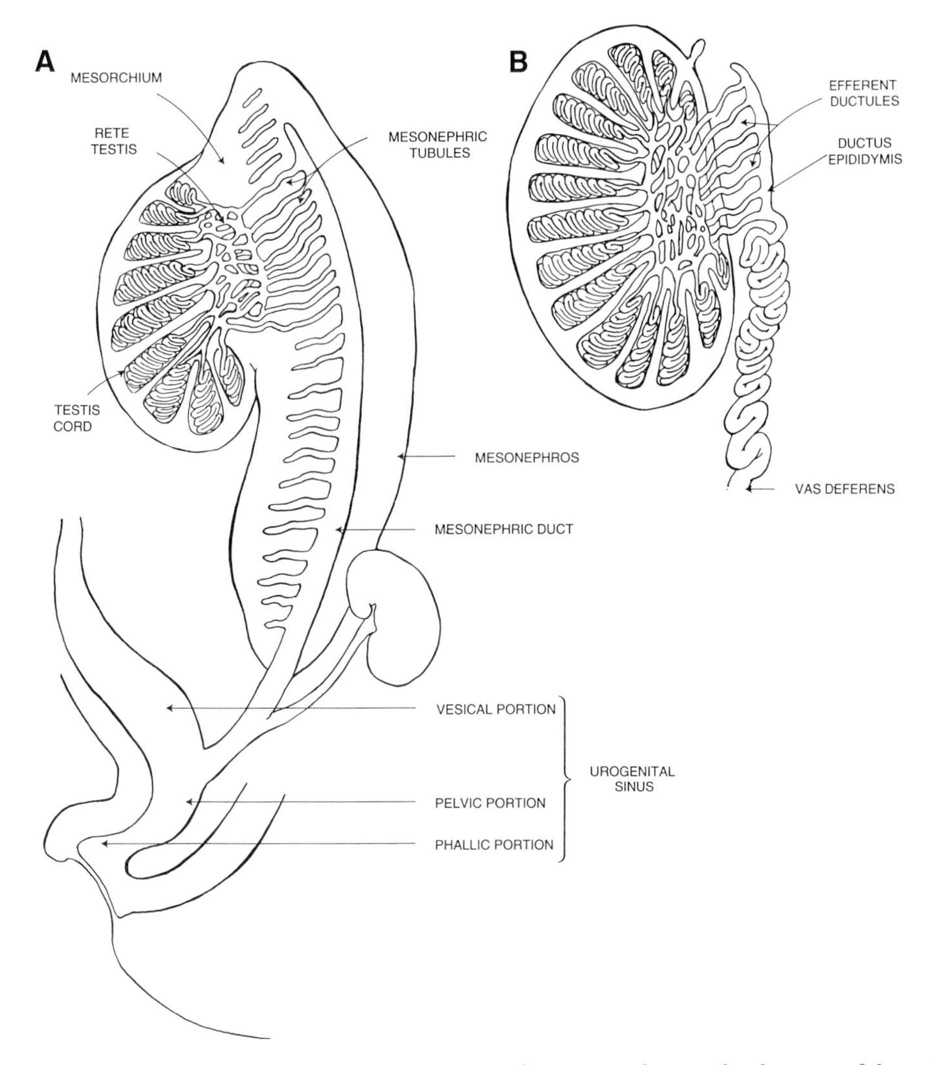

FIGURE 3 Cross-sectional view of the Wolffian duct inserting into the urogenital sinus; development of the male genital ductal system [reproduced with permission from T. W. Hensle and E. K. Seaman, Embryology of the male reproductive tract, In *Atlas of Surgical Management of Male Infertility* (A. J. Thomas and H. M. Nagler, Eds.), Igaku-Shoin, New York, 1995].

mal abnormalities are frequently found in conjunction with other Wolffian duct developmental anomalies, such as those of the vas deferens or the seminal vesicle. Unlike abnormalities of the scrotal vas deferens, however, abnormalities of the epididymis are notoriously difficult to diagnose on physical examination given the normal variation in epididymal anatomy between individuals. Epididymal anomalies are also frequently found in the pediatric population as an incidental finding during scrotal surgery for unrelated problems such as cryptorchidism or hydroceles.

C. Abnormalities of the Seminal Vesicle

The seminal vesicles contribute to the ejaculate by adding nutritive fluid to the sperm. Abnormalities of the seminal vesicle range from the absence or hypoplasia to the presence of cysts. Currently, the clinical evaluation of the seminal vesicles is performed almost exclusively using transrectal ultrasound. Like epididymal anomalies, anomalies of the seminal vesicles frequently occur concurrently with other Wolffian duct anomalies, with these cases resulting in infertility. Isolated abnormalities of the seminal vesicles may not, however, uniformly result in infertility since the seminal vesicles are anatomically extrinsic to the path of sperm traveling between the testis and the urethra.

See Also the Following Article

MALE REPRODUCTIVE SYSTEM, HUMANS

Bibliography

Hensle, T. W., and Seaman, E. K. (1995). Embryology of the male reproductive tract. In *Atlas of Surgical Management of Male Infertility*. (A. J. Thomas, Jr. and H. M. Nagler, eds.) pp. 1–8. Igaku-Shoin, New York and Tokyo.

Moore, K. L., and Persaud, T. V. N. (1993). The urogenital system. In *The Developing Human*. Saunders, Philadelphia.

Vohra, S., and Morgentaler, A. (1997). Congenital anomalies of the vas deferens, epididymis, and seminal vesicles. *Urology* 49(3), 313–321.

Yolk Proteins, Invertebrates

Gerard R. Wyatt
Queen's University

GLOSSARY

ecdysteroid The molting hormone of insects and other arthropods such as Crustacea, comprising several closely related hydroxylated C_{27} steroids, produced in the prothoracic gland; the major active form is 20-hydroxyecdysone.

fat body A mesodermal tissue in the form of lobes dispersed in the body cavity of insects that is both a storage organ for nutrient reserves and the major source for synthesis of hemolymph proteins, to some degree analogous with the liver of vertebrates.

follicle The ovarian cell layer of epithelium that immediately surrounds the oocyte, serving to regulate transport between the blood and the oocyte and to secrete certain components of the egg, such as (in insects) the chorion or eggshell.

hemolymph The circulating fluid or blood of insects which circulates freely in the body cavity or hemocoel and serves to transport nutrients, proteins, and hormones.

juvenile hormone In insects and Crustacea, a lipoidal hormone, comprising several closely related sesquiterpenoids, that serves two distinct functions: (i) to maintain juvenile character in the larval molts and (ii) to stimulate and coordinate reproductive maturation and function in the adult.

vitellin The principal yolk protein of invertebrates, derived from hemolymph vitellogenin and deposited in yolk granules or platelets within the eggs.

vitellogenin The major yolk precursor protein, which is synthesized in the fat body of insects and other tissues of other invertebrates, transported in the hemolymph, and taken up through the follicle into the growing oocytes.

In the eggs of most invertebrates, with the exception of viviparous species and endoparasites, the yolk provides the principal nutrient reserve for growth of the embryo and immediate posthatching development and activity.

The yolk contains dense quasi-crystalline protein granules, or yolk platelets, as well as lipid droplets and glycogen. The major component of the protein granules is vitellin (Vn), a macromolecular glycolipoprotein derived by slight modification from vitellogenin (Vg), which is synthesized extraovarially and circulates in the coelomic fluid or hemolymph before being taken up into the oocytes during their vitellogenic phase of development. The Vgs constitute a superfamily, those of insects and other invertebrates being homologous with Vgs of oviparous vertebrates (amphibia, reptiles, and birds). One insect group, the higher Diptera (flies), however, utilizes a distinct family of lower molecular weight yolk proteins related to lipases. Yolk also includes other proteins, which may have special importance in certain species. In insects, Vg is produced in the fat body (in insects), usually under the control of juvenile hormone, but controlled by ecdysone in the Diptera. In other invertebrates, Vg is produced in various nonovarian tissues, under diverse endocrine control. Uptake into the ovary requires passage through the follicular epithelium which is made possible in insects by juvenile hormone-induced shrinkage of the follicular cells, followed by receptor-mediated endocytosis into the oocyte itself. In early embryogenesis, the yolk proteins are digested by several enzymes, including a cathepsin-like cysteine protease released by activation of a proenzyme. Some vitellin often persists within the gut of the newly hatched larva to support early larval life.

I. PROTEIN COMPONENTS OF YOLK

A. Insects

Most insect eggs contain abundant yolk, and in various species the major yolk protein, Vn, has been reported to account for 40–90% of the total egg protein, usually in the upper part of this range. Vn is generally antigenically identical with its hemolymph precursor, Vg, from which it may be modified by proteolytic cleavage and loss of some lipid. Vg was discovered by W. H. Telfer as a female-specific antigen in silkworm hemolymph which was taken up selectively by the growing oocytes, and the name vitellogenin, applied to this protein in 1969, was subsequently adopted for the yolk precursor proteins of vertebrates. In insects, Vg is generally produced only in the female fat body. Vn can be solubilized from yolk granules by extraction with high-salt buffers, and both Vn and Vg precipitate at low ionic strength. They are phosphoglycolipoproteins of high molecular mass (400–600 kDa) and upon analysis by denaturing electrophoresis (sodium dodecyl sulfate–polyacrylamide gel), usually yield two or more fragments which fall into large (120–200 kDa) and small (40–60 kDa) classes. Although referred to by different authors as subunits, apoproteins, or heavy chain and light chain, these are not primary subunits like those from which many complex proteins are assembled. The large and small components have been shown by amino acid labeling experiments and by peptide and gene sequencing in several species to be products of proteolytic cleavage from larger precursors or preproteins, encoded within one gene. In many insects (Lepidoptera, the boll weevil, and the mosquito *Aedes aegypti*) only one large and one small fragment are observed; in some (locusts and cockroaches) there is more than one fragment in each size class, indicating multiple cleavages or products from different genes; and in a few (the honeybee and carpenter ants) only a single component is seen, derived from the uncleaved translation product. The native molecules are usually dimers of the translation product. The amino acid composition is generally quite similar to that of average proteins, but Vn of the mosquito, *A. aegypti*, contains unusually high proportions of serine (much of it phosphorylated) and tyrosine. The carbohydrate moiety is of the high-mannose type, containing a smaller proportion of *N*-acetylglucosamine, bound to asparagine residues. The bound lipid, analyzed for locusts and silk moths, is principally diacylglycerol and phospholipids, with a small proportion of cholesterol and mere traces of triacylglycerol. Vn of several species carried less lipid, especially diacylglycerol, than Vg, suggesting that some lipid was released during deposition of the protein in the egg. In some insects, the Vn carries significant amounts of covalently bound phosphate and sulfate.

Genes or cDNAs have been cloned for several insect Vgs, and complete coding sequences, permitting deduction of amino acid sequences, are available for four species: *Anthonomus grandis* (boll weevil), *Bombyx mori* (silkworm), *Lymantria dispar* (gypsy moth; Hiremath and Lehtoma, 1997), and *A. aegypti* (yellow fever mosquito; Romans *et al.*, 1995). The migratory locust has two Vg genes, *A. aegypti* has five, and several other species appear to have single genes. Gene lengths range from 7 to 12 kb, include two to six introns, and encode mRNAs in the range 5.4–6.5 kb. Amino acid sequence comparison indicates homology among the Vgs of insects and other invertebrate and vertebrate animals, although sequence divergence has been relatively rapid: thus, mosquito Vg has about 50% amino acid similarity with boll weevil and silkworm Vgs and much lower, but still significant, similarity with the Vgs of the nematode, *Caenorhabditis elegans*, and the amphibian, *Xenopus laevis*. The Vg family also shows homology with human apolipoprotein B-100, indicating that both belong to a lipoprotein superfamily. In *A. aegypti*, *B. mori*, and *A. grandis*, the sequence corresponding to the small cleavage fragment lies at the N terminus and the consensus sequence Arg-X-Arg-Arg for cleavage by a processing endoprotease is found in the expected positions. In *L. dispar*, on the other hand, the cleavage site is located toward the C terminus. Although the insect Vgs, in contrast to vertebrate Vgs, do not give rise to the phosphorylated serine-rich protein, phosvitin, they do include one or more serine-rich regions or polyserine tracts, the role of which is not clear.

Surprisingly, one group of insects, the higher Diptera (suborder Cyclorrhapha), which includes the fruit fly, *Drosophila melanogaster*, and other fruit flies as well as houseflies and blowflies (but not mosquitoes), utilizes yolk proteins of a different type. The yolk proteins of this group, when denatured, release several (one to six) polypeptides of 42–54 kDa, with associated carbohydrate and lipid, which are not products of processing from larger precursors but are encoded by separate genes. In the native proteins, they form associations of 200–300 kDa. They comprise a family which shows homology with mammalian triacylglycerol lipases and, to distinguish them from the typical Vgs, they are usually referred to simply as yolk proteins or yolk polypeptides (YP). Recent sequence analyses by H. H. Hagedorn *et al.* have found significant similarity between the cyclorraphan YPs and a conserved region within the Vgs, suggesting common origin from an ancient antecedent. It is not apparent, however, why the higher Diptera, an evolutionarily advanced insect group, should have evolved a unique, truncated type of yolk protein. In *D. melanogaster*, there are three yolk protein genes, designated *YP1*, *YP2*, and *YP3*, which share about 50% sequence identity. They are located on the X chromosome: *YP1* and *YP2* are closely linked and divergently transcribed, the intervening 1225 bp of DNA containing promoter and regulatory elements for both genes, whereas *YP3* is removed by about 1000 kb. In the Mediterranean fruit fly, *Ceratitis capitata*, two apparent yolk protein components of 46 and 49 kDa are encoded by four genes, in two divergent pairs (Rina and Savakis, 1991). These share extensive sequence identity with one another and with the YPs of *D. melanogaster*; suggesting the conservation of structural and functional domains. Synthesis of YPs has been demonstrated both in fat body and in ovarian follicles.

Insect yolk also contains other proteins, the nature and amounts of which vary greatly among species. Several Lepidoptera possess yolk proteins of about 30 kDa, called microvitellogenins, which originate in the fat body and represent a family with related sequences. While the microvitellogenins are usually minor yolk components, in the silkworm, *B. mori*, a set encoded by five related genes, referred to as 30K proteins, are the principal proteins in late larval hemolymph and make up about 30% of the yolk protein. The proteins belonging to this family are female-specific in some species but not in others.

A number of other proteins of extraovarian (usually fat body) origin are found in eggs in relatively small amounts. These include lipophorin (the major hemolymph lipid transport protein), cyanoproteins that contribute blue-green color to the eggs of the tobacco hornworm and the brown bean bug, hemoglobins or degradation products that confer pink coloration on the eggs of certain midges and blood-sucking bugs, a 21-kDa female-specific protein in *Locusta,* and a vitellogenic carboxypeptidase in mosquitoes. In the cochineal insect (*Dactylopius confusus*, Homoptera) the usual vitellin is apparently replaced by a unique yolk protein with 45- to 56-kDa subunits that form polymers of 270 kDa upwards, seen in the electron microscope as helical ribbons (Ziegler *et al.*, 1996). These variations illustrate the diversity that insects exhibit in their biochemistry, just as in their morphology and lifestyles.

The ovarian follicular cell layer also makes a significant protein contribution to the egg. In *B. mori*, a trimer of 72-kDa glycosylated polypeptides of follicular origin, designated egg-specific protein (ESP) contributes to the yolk granules, making up about 25% of the protein of newly laid eggs. Yolk proteins of follicular origin, not necessarily homologous with the ESP of *B. mori* and usually present in smaller amounts, have been identified in several other Lepidoptera.

B. Other Invertebrates

Other groups of invertebrates utilize yolk proteins that are similar in their biochemical properties to those of insects, and probably belong to the Vg superfamily, although this relationship has been established by sequence in only a few instances. Among Crustacea, vitellins from the ovary and Vgs from the blood have been described as glycolipoproteins with native M_r 260–540 kDa, yielding two or more unequal "subunits," often in the size range 75–190 kDa, upon denaturation. These fragments are derived by proteolytic cleavage from precursors of >200 kDa which is initiated, usually with a single cut, in the tissue of synthesis and carried further after uptake into the oocyte so that vitellin yields more and smaller fragments than Vg. In other invertebrate groups, in contrast to the insects, the principal neutral lipid component is chiefly triacylglycerol, not diacylglycerol. Among Crustacea, there is evidence for synthesis of Vg in several tissues, including the hepatopancreas and the ovary itself.

In the soft tick, *Ornithodorus moubata*, two Vgs, 300 and 600 kDa, probably monomer and dimer, are found which yield six fragments of 100–215 kDa; the vitellin, containing 12.4% carbohydrate and 7.6% lipid, yields smaller fragments. The hard tick, *Dermacentor variabilis*, has similar characteristics. Tick yolk proteins are colored reddish-brown by heme derived from hemoglobin in the ingested blood. The main site of synthesis is the fat body, which in ticks is a faint tissue attached to the tracheae.

In annelids, the yolk proteins are glycolipoproteins of high molecular mass. The polychaete, *Nereis*, has in its coelomic fluid a Vg of 530 kDa which yields a single 175-kDa polypeptide that is cleaved further in the ovary. In nereids, Vg is synthesized and secreted into the coelomic fluid by the eleocytes, a class of coelomocytes. In the leech, *Theromyzon*, while 60% of the yolk protein is a vitellin similar to that of polychaetes, 30% is a hemerythrin—a pink 14-kDa non-heme iron-binding protein which may have a role in sequestering excess iron from the blood diet.

A sea urchin, *Strongylocentrotus purpuratus*, produces a 195-kDa Vg which is synthesized, surprisingly, in the intestine and gonads of both sexes; Vg synthesis in these tissues is confirmed by the presence of specifically hybridizing 5.1-kb mRNA. The Vg is converted in the ovary to a 180-kDa vitellin.

The free-living soil nematode, *C. elegans*, has received special attention as a model for molecular genetic analysis of development. In this species, two macromolecular yolk complexes, estimated as M_r 450 and 250 kDa with about 14% lipids, are composed of polypeptides of 83–188 kDa, of which the smaller are products of proteolytic cleavage. Of six identified genes, three have been completely sequenced and show clear homology with the Vg superfamily and closer sequence conservation with vertebrate than insect Vgs. A Vg gene from a distantly related nematode of the same family shows sequence homology but a surprising degree of divergence including differently located introns (Winter *et al.*, 1996). The comparisons indicate an ancient gene undergoing rapid evolutionary change, as would be expected of proteins whose chief role is to provide nutrition for embryos, but with conservation of motifs that may be needed for functions such as lipid binding and specific uptake into the oocyte.

II. FUNCTIONAL ASPECTS OF YOLK PROTEINS

A. Production and Regulation by Hormones

The yolk proteins are principally produced heterosynthetically—that is, in tissues other than the ovary which can provide the massive synthesis that is needed during peak vitellogenesis—and they are transported to the ovary in the coelomic fluid or blood. In insects and ticks, the fat body is generally the exclusive site of synthesis of Vg, and in some insects the inner fat body, closer to the ovary, is found to be more active in this function than the peripheral fat body. The yolk proteins of the higher Diptera can be made both in the fat body and in the ovarian follicular cells. In the nematode, *C. elegans*, Vg is made in the intestine; in nereid worms, in coelomocytes; and in Crustacea, both in the hepatopancreas and within the ovary.

The cellular and molecular processes in yolk protein production and its regulation have been studied in detail in certain insects, particularly cockroaches, locusts, mosquitoes, and fruit flies. Vg is generally synthesized only in adult female fat body, though in a few species (the bug, *Rhodnius*) Vg is found in male hemolymph, and in several, synthesis can be induced in larval or male tissue by high hormonal doses. In preparation for a vitellogenic cycle, the fat body cells (trophocytes), which serve at other times to store reserves of fat, glycogen, and protein, must tool up for protein synthesis. This involves depletion of the nutrient reserves, replication of DNA (producing polyploidy), and massive proliferation of ribosomes and endoplasmic reticulum. At the appropriate endocrine signal, transcription of the Vg genes commences and Vg mRNA accumulates, often to become the most abundant mRNA in the cells: In vitellogenic locusts, Vg mRNA of 6.3 kb has been recorded as 50% of the total fat body mRNA. The intracellular steps in Vg production have been analyzed in detail in the yellow fever mosquito (*A. aegypti*) by A. S. Raikhel and co-workers. Translation leads to release of the precursor polypeptide (pre-pro-Vg; 224 kDa) into the cisternae of the endoplasmic reticulum. Glycosylation occurs cotranslationally and is followed, within the cisternae and Golgi complex, by phosphorylation, sulfation, and cleavage to give the two Vg subunits of 200 and 66 kDa. The mature Vg is secreted into the hemolymph.

As an energy-demanding process that is required only during the vitellogenic phase of oogenesis, the synthesis of yolk proteins is subject to hormonal regulation. In most insects, this is a function of the juvenile hormone (JH), a terpenoid hormone which regulates the onset of metamorphosis in larval insects and in the adult stage is the principal coordinator of various aspects of reproduction. Only in certain insects such as silk moths, which complete their oogenesis during morphogenesis of the adult and are ready for mating and oviposition immediately after emergence, is vitellogenesis independent of JH. In the fat body, JH is required in two steps: first, to stimulate the preparative phase, in which the cells are readied for active protein synthesis and secretion, and second, in most insects, to activate transcription of the Vg genes themselves. In the order Diptera, while the preparative phase remains a function of JH, the transcriptional control of the yolk protein genes (both the Vgs of the lower Diptera and the distinct YPs of the higher Diptera) depends principally on the steroidal molting hormone, ecdysone. The system is orchestrated by the brain, which utilizes a variety of neuropeptides, differing in different insect groups, both to modulate the production of JH and ecdysone and for direct influences upon fat body function. In *Locusta*, there is also evidence for control of translation so that Vg mRNA is conserved from one vitellogenic cycle to the next, and neuropeptides may be involved in its regulation.

Molecular genetic studies in *Drosophila* by P. Wensink, M. Bownes, and others have gone some way toward revealing the basis for the sex-, tissue-, and hormone-specific regulation of transcription of the three yolk protein genes. In the DNA flanking these genes, fat body enhancer elements have been identified that include binding sites for several transcription factors as well as sites for the protein product (DSX) of *doublesex*, the terminal gene of the sex-determination pathway. DSX exists in female- and male-specific forms, DSX^F and DSX^M, both of which bind to the yp genes, but differences in the proteins cause DSX^F to activate and DSX^M to repress transcription. In the fat body, yp gene expression is induced by both ecdysone and JH, and ecdysone response elements have been identified. High doses of ecdysone can bring about expression in males and are apparently able to overcome the repressive effect of DSX^M. Expression of the yp genes in the ovarian follicle cells is differently regulated, apparently keyed to differentiation of the cells, independently of the sex determination pathway or influence of ecdysone, and here the observed stimulation by JH may be a result of furthering the program of cellular differentiation.

In the Crustacea, the regulation of Vg synthesis appears to involve both neuropeptides and the JH homolog, methyl farnesoate. In ticks, some reports that vitellogenesis is regulated by JH have not been confirmed, and there appears to be a two-step controlling system, the initiating factor being a neuropeptide from the synganglion, followed by action of a posterior factor, possibly ecdysteroid. Little is known about the regulation of yolk protein production in other invertebrate groups.

B. Uptake into the Oocyte

For deposition in the growing oocytes, Vgs and YPs are taken up from the hemolymph through the ovarian wall and the follicular epithelium. In insects, JH is required to bring both the follicle cells and the oocyte to the stage of differentiation required for yolk protein uptake, whether JH or ecdysone (as in the Diptera) is used for the control of synthesis in the fat body. Vg or YP uptake involves several steps that have been studied in some detail (the cell biology especially in silk moths and mosquitoes and the hormonal control especially in *Rhodnius*), but the process appears to be essentially similar in different insect groups. The follicular exterior membrane, the basal lamina, is permeable to molecules up to M_r 500,000. Passage through the follicular epithelium takes place via intercellular channels that are opened at the beginning of vitellogenesis by shrinkage of the cells, creating a condition known as patency. The development of patency is induced by JH, which, according to studies by K. G. Davey, acts in this process directly at the cell membrane without need for gene transcription or new protein synthesis, in contrast to its mode of action in cellular differentiation and Vg synthesis.

Entry into the oocyte itself proceeds by receptor-mediated endocytosis, which is selective principally for Vg and YP but also favors some other hemolymph proteins, such as lipophorin, over nonspecific proteins. Vg receptors have been purified from oocyte membranes of several insect species as proteins of about 200 kDa, and sequence comparison from cloned cDNAs shows that they belong to the VLDL receptor family which also includes chicken and *Xenopus* Vg receptors. The YP receptor of *Drosophila* and the Vg receptor of *Aedes* share 42% amino acid identity, which is consistent with there being an ancient relationship between these two classes of yolk proteins (Sappington *et al.*, 1996). The Vg-receptor complexes are enclosed in clathrin-coated pits, which are internalized into the oocyte as coated vesicles and then converted to endosomes, along with release of the receptor which returns to the cell surface for recycling. The Vg-containing endosomes coalesce into yolk bodies, in which condensation and crystallization of the Vg takes place, producing yolk granules.

C. Utilization of Yolk Proteins

The degradation and utilization of the stored yolk proteins as a source of amino acids for the developing organism may involve several distinct processes. In insects, the blastoderm is formed by migration of cleavage nuclei to produce a peripheral layer which surrounds a central yolk mass. Some nuclei that remain within the yolk, and others that return to it, form vitellophages or yolk cells, which have been shown in cockroaches to take up vitellin and are believed to digest it for use by the embryo. Later, as the embryonic organs are formed, the midgut surrounds the remaining yolk and digestion is completed within the gut of the late embryo and newly hatched larva.

The different classes of yolk proteins may be consumed with different timing, which suggests some specificity in their functions. In the silkworm, the ESP of follicular origin disappears first, followed by vitellin and then the 30k protein so that about one-third of the vitellin and more than half of the 30k protein are still present at hatching. Several types of protease have been implicated in yolk protein digestion. In the early developing eggs of the silkworm, a cathepsin-type cysteine protease of broad substrate specificity, which was contributed to the oocyte by the follicle cells and stored as an inactive proenzyme, becomes activated proteolytically to initiate yolk protein digestion. Similar enzymes have been described in *Drosophila*, tick, and brine shrimp eggs. Acidification of the membrane-bound yolk granules may also contribute to activation of these enzymes. In the mosquito, *Aedes*, a serine carboxypeptidase, called vitellogenic carboxypeptidase, produced in the adult female fat body during vitellogenesis and taken up into the oocyte as a proenzyme, becomes activated early in embryogenesis, possibly serving for proteolytic activation of other enzymes. In *Drosophila*, a serine protease found with low activity in yolk granules in the oocyte shows increased activity and a shift to the soluble fraction during embryonic development. In the silkworm during embryonic differentiation, two substrate-selective trypsin-type serine proteases are produced as a result of transcriptional activation in the embryonic tissues: from Day 4 an enzyme with specificity for ESP and from Day 8, in the embryonic midgut, one favoring vitellin. These complete the digestion of these proteins.

Although the role of yolk as a nutrient reserve for early development seems self-evident, the importance of specific yolk proteins has been questioned. In the silkworm, *B. mori*, O. Yamashita has shown that when ovaries were transplanted into male larvae and allowed to complete metamorphosis, eggs formed which, after parthenogenetic activation, developed to apparently normal larvae. This showed that Vg, which is not produced in males, is not essential. However, *B. mori* is exceptional in the high proportion of other proteins (30k and ESP) in the yolk, which could apparently replace vitellin. The experiment is consistent with the evolutionary divergence of Vg sequences and the replacement of vitellin by other proteins in certain species and groups in showing that the functions of nutrient reserve proteins do not impose extensive constraints upon their structure.

See Also the Following Articles

DROSOPHILA; ECDYSIOTROPINS; ECDYSTEROIDS; JUVENILE HORMONE; RHODNIUS PROLIXUS

Bibliography

Bowness, M. (1994). The regulation of the yolk protein genes, a family of sex differentiation genes in *Drosophila melanogaster*. *BioEssays* **16**, 745–752.

Chinzei, Y., and Taylor, DeM. (1994). Hormonal regulation of vitellogenin biosynthesis in ticks. *Adv. Dis. Vector Res.* **10**, 1–22.

Engelmann, F. (1979). Insect vitellogenin: Identification, biosynthesis and role in vitellogenesis. *Adv. Insect Physiol.* **14**, 49–108.

Gruber, M., Byrne, B. M., and Ab, G. (1989). The evolution of egg yolk proteins. *Prog. Biophys. Mol. Biol.* **53**, 33–69.

Hagedorn, H. H., Maddison, D. R. and Tu, Z. (1998). The evolution of vitellogenins, yolk proteins and related molecules. *Adv. Insect Physiol.* **27**, 335–384.

Hiremath, S., and Lehtoma, K. (1997). Complete nucleotide sequence of the vitellogenin mRNA from the gypsy moth: novel arrangement of the subunit encoding regions. *Insect Biochem. Mol. Biol.* **27**, 27–35.

Izumi, S., Yano, K., Yamamoto, Y., and Takahashi, S. Y. (1994). Yolk proteins from insect eggs: Structure, biosynthesis and programmed degradation during embryogenesis. *J. Insect Physiol.* **40**, 735–746.

Kunkel, G. G., and Nordin, J. H. (1985). Yolk proteins. In *Comprehensive Insect Physiology, Biochemistry and Pharmacology* (G. A. Kerkut and L. I. Gilbert, Eds.), Vol. 1, pp. 84–111. Pergamon, Oxford, UK..

Postlethwait, J. H., and Giorgi, F. (1985). Vitellogenesis in insects. In *Developmental Biology, a Comprehensive Treatise* (L. W. Browder, Ed.), Vol. 1, pp. 85–126. Plenum, New York.

Raikhel, A. S., and Dhadialla, T. S. (1992). Accumulation of yolk proteins in insect oocytes. *Annu. Rev. Entomol.* **37**, 217–251.

Rina, M., and Savakis, C. (1991): A cluster of vitellogenin genes in the Mediterranean fruit fly, *Ceratitis capitata*: sequence and structural conservation in dipteran yolk proteins and their genes. *Genetics* **127**, 769–780.

Romans, P., Tu, Z., Ke, Z., and Hagedorn, H. H. (1995). Analysis of a vitellogenin gene of the mosquito, *Aedes aegypti* and comparisons to vitellogenins from other organisms. *Insect Biochem. Mol. Biol.* **25**, 939–958.

Sappington, T. W., Kokoza, B. A., Cho, W. L., and Raikhel, A. S. (1996). Molecular characterization of the mosquito vitellogenin receptor reveals unexpected high homology to the *Drosophila* yolk protein receptor. *Proc. Natl. Acad. Sci. USA* **93**, 8934–8939.

Valle, D. (1993). Vitellogenesis in insects and other groups—A review. *Mem. Inst. Oswaldo Cruz (Rio de Janeiro)* **88**, 1–26.

Wahli, W. (1988). Evolution and expression of vitellogenin genes. *Trends Genet.* **4**, 227–232.

Winter, C. E., Penha, C., and Blumenthal, T. (1996). Comparison of a vitellogenin gene between two distantly related rhabditid nematode species. *Mol. Biol. Evol.* **13**, 674–784.

Wyatt, G. R., and Davey, K. G. (1996). Cellular and molecular actions of JH. II. Roles of juvenile hormone in adult insects. *Adv. Insect Physiol.* **26**, 1–155.

Ziegler, R., *et al.* (1996): A new type of highly polymerized yolk protein from the cochineal insect *Dactylopius confusus*. *Insect Biochem. Mol. Biol.* **31**, 273–287.

Yolk Sac

Robert W. McGaughey
Arizona State University

GLOSSARY

blastocoelic cavity The fluid-filled cavity of the mammalian blastocyst.

cavitation The developmental process that forms spaces among cells of a tissue and by which these developing spaces coalesce to form a cavity.

chorioallantoic circulation The major fetal–maternal circulation system that employs the umbilical vein and artery for exchange of blood from the fetal circulatory system to the placental region of the chorionic villi for exchange of gases, nutrients, and wastes with the maternal blood.

coelomic cavity A body cavity that is completely surrounded by mesodermal tissue.

ectoderm The embryonic tissue type from which outer skin (epidermis) derivatives and most nervous tissues differentiate; in mammals, the dorsal cells of the embryonic disk from which most fetal tissues differentiate following gastrulation.

embryonic disk The inner cell mass after its cells have differentiated to form two layers: one of dorsal embryonic ectoderm and another of ventral embryonic endoderm cells.

endocytosis A process by which extracellular materials are taken inside a cell by including the material within deeply invaginating regions of plasma membrane that subsequently bud off internally to form membrane-bound, intracytoplasmic vesicles.

endoderm The embryonic tissue type from which digestive and respiratory epithelial tissues differentiate.

endothelial The cell type that forms the lining of blood vessels.

hematopoiesis The differentiation process that forms blood cells.

inner cell mass The group of cells inside the trophectodermal layer of the mammalian blastocyst stage embryo that contain the stem cells for fetal tissues.

mesenchyme An embryonic tissue type; usually composed of mesodermal cells that migrate among the cells of other embryonic tissues, often epithelia.

mesenteries Supporting structures for the visceral organs; derived from lateral plate mesoderm that lines the coelomic cavities.

trophectoderm The outer epithelial cell layer of the mammalian blastocyst; these cells differentiate into placental chorion tissues.

The yolk sac is derived from extraembryonic endoderm cells that grow out ventrally from the developing early embryo to form a membranous lining of the yolk sac cavity. Unlike the yolk sac of nonplacental vertebrates, such as birds, reptiles, and prototherian mammals, the yolk sac cavity in eutherian mammals contains fluid instead of yolk.

I. INTRODUCTION

The following description of the yolk sac provides discussion of the general processes shared by most mammalian embryos. The entire developmental sequence for the yolk sac, including its subsequent disappearance, varies in structural detail and timing for different major taxa of mammals and even within the eutherian group. Much of the description provided here is based on work with human embryos. The bibliography provides the reader with information on the yolk sac in different mammals for comparison. Although specific reference to work relating to the yolk sac of reptiles and birds is not included, the interested reader shall find from a study of those phylogenetically related groups that the yolk sac of mammals bears striking resemblance both structurally and functionally to that of the close evolutionary relatives of mammals.

II. EMBRYONIC ORIGIN OF THE YOLK SAC

The first indication for the formation of the mammalian yolk sac is differentiation of a layer of inner cell mass into primitive endoderm (Fig. 1A). These endoderm cells arise and proliferate during the expanded blastocyst stage at which the embryo is composed of inner mass cells and surrounding, outer trophectoderm cells that encapsulate the fluid-filled blastocoelic cavity. Differentiation of endoderm cells from the inner cell mass results in the formation of a double-layered or bilaminar embryonic disk composed of upper (i.e., dorsal) embryonic ectoderm adjacent to the ventral endoderm layer underneath (Fig. 1). The newly formed endodermal layer proliferates outward and ventrally, away from the embryonic disk, to form a complete layer around the inside of the trophectodermal epithelium. The completed endodermal layer constitutes the yolk sac membrane and encloses the yolk sac cavity (Fig. 2A). The early, primitive yolk sac forms while the blastocyst is undergoing implantation into or against (depending on the species) the uterine endometrium. Coincidental with formation of the primary yolk sac, the amniotic membrane and its enclosed amniotic cavity form on the dorsal surface of the embryonic disk.

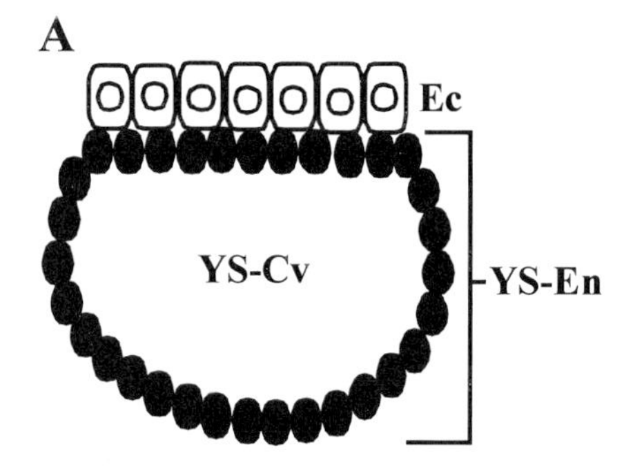

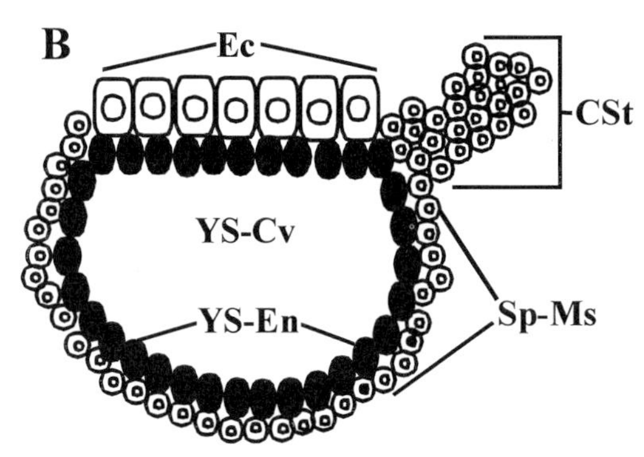

FIGURE 2 Diagrammatic representations of the essential components in yolk sac formation in mammals. In A, embryonic endoderm is shown to proliferate ventrally to form a complete endoderm-lined primary yolk sac (YS-En) that encloses the fluid-filled yolk sac cavity (YS-Cv). In B, extraembryonic splanchnic mesoderm (Sp-Ms) is shown to proliferate around the outside of the yolk sac endoderm layer (YS-En). Also shown in B is the location at which extraembryonic splanchnic and somatic mesoderm layers join to form the connecting stalk (CSt).

FIGURE 1 Diagrammatic representation of the inner cell mass (A) that is located inside the mammalian blastocyst and that is completely surrounded by the trophectodermal epithelium (not shown). The inner cell mass differentiates to form the embryonic ectoderm (Ec) and the embryonic endoderm (En) that proliferates on the ventral side of the embryonic ectoderm (B).

After the endodermal lining cells of the yolk sac proliferate around the inside of the trophectodermal layer, other cells differentiate and migrate from the region of embryonic ectoderm of the embryonic disk to lie between the endodermal yolk sac membrane and the trophectoderm cells. These intermediate cells constitute extraembryonic mesoderm that subsequently undergoes cavitation to form small coalescing fluid-filled spaces between the yolk sac and the trophectoderm. When the cavitation process is complete, a layer of extraembryonic mesoderm cells, called extraembryonic splanchnic mesoderm or visceral mesoderm, lines the outside of the yolk sac endodermal membrane (Fig. 2). Outside that mesodermal lining (toward the trophectoderm) is the extraembryonic coelomic cavity separating the yolk sac from the trophectoderm layer and its immediately internal extraembryonic somatic or parietal mesoderm layer. The ventrally located primary yolk sac and the dorsal amnion expand to fully surround the embryonic disk.

III. STRUCTURAL CHANGES OF THE YOLK SAC

After the primitive or primary yolk sac has formed, its lower or ventral portion narrows and pinches off, becoming isolated from the dorsal yolk sac and finally completely separates to lie in the extraembryonic coelom. The remaining dorsal yolk sac develops a nearly spherical architecture and is called the secondary yolk sac (Fig. 3). While this change takes place in the yolk sac, the embryonic disk begins gastrulation and the embryonic anterior–posterior axis develops. As gastrulation progresses, the roof of the secondary yolk sac constricts, is surrounded by gastrulating embryonic tissues, and develops into the fetal midgut. At this stage the secondary yolk sac is connected to the midgut by means of a narrow yolk stalk that gradually becomes positioned posteriorly relative to the embryonic disk. The yolk stalk associates with the extraembryonic mesoderm where the splanchnic and somatic layers meet to form the connecting or body stalk. Posterior to the yolk stalk, a region of the embryonic hindgut proliferates to form the allantois, an outpocketing into the mesodermal connecting stalk. The connecting stalk, containing the allantois and the yolk sac stalk, narrows to form the umbilical cord (Fig. 4). Near the middle of gestation, the yolk stalk narrows further and subsequently disappears in many mammalian species.

IV. FUNCTIONAL SIGNIFICANCE OF THE YOLK SAC

The mammalian yolk sac provides nutrition for the early embryo, it provides stem cells that differentiate into blood cells and blood vessel tissues, and it is the location at which germ cells are first observed in mammals. The primary yolk sac initially forms in close association with the inside surface of the trophectodermal epithelium and is thought to facilitate uptake of nutrients through that epithelium. Evidence suggests that the extraembryonic endoderm cells of the yolk sac membrane function in endocytosis of macromolecules, including proteins, and in degrading those macromolecules to precursors, such as amino acids, for assimilation by the developing embryo. The yolk sac has been shown to

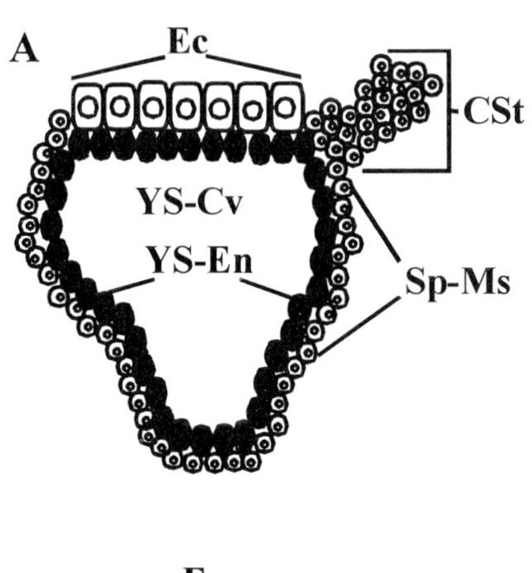

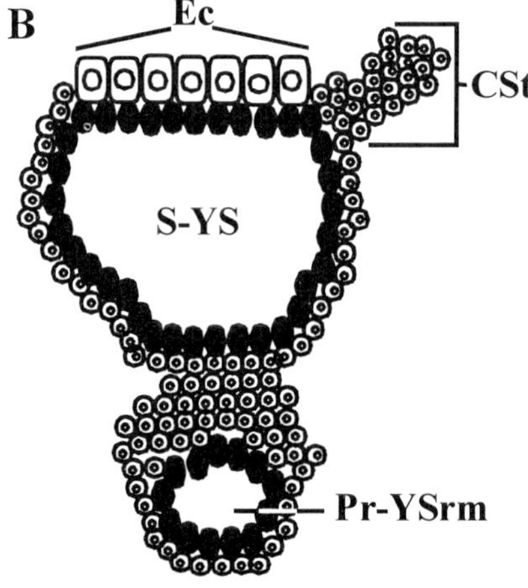

FIGURE 3 Diagrammatic representation of the formation of the secondary yolk sac from the primary yolk sac. In A, the ventral portion of the primary yolk sac has begun to narrow, and in B the ventral portion of the primary yolk sac has separated to form a remnant (Pr-YSrm) embedded in extraembryonic splanchnic mesoderm, leaving the dorsal portion of the yolk sac to constitute the secondary yolk sac (S-YS). During these stages, the embryonic ectoderm cells (Ec) have begun gastrulation.

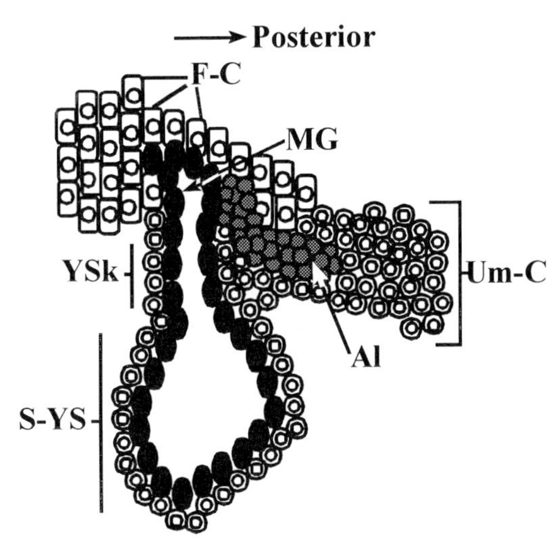

FIGURE 4 Diagrammatic representation of the posterior portion of a developing mammalian embryo, showing fetal cells (F-C), the ventrally protruding yolk stalk (YSk), the secondary yolk sac (S-YS) that is continuous dorsally with the developing embryonic midgut (MG), and the developing allantois (Al) that forms as an outpocketing from the embryonic hindgut. The allantois is contained within the connecting stalk extraembryonic mesoderm that has become the umbilical cord (Um-C). Subsequently, the yolk stalk becomes encapsulated within the umbilical cord and the secondary yolk sac becomes closed off and disappears in many mammals.

selectively take up certain harmful exogenous substances that are then transported to the developing embryo resulting in abnormal development (i.e., teratogenesis).

Mesenchyme tissues differentiate between the splanchnic layer of mesoderm and the yolk sac extraembryonic endoderm. Within that mesenchyme the first stem cells for blood cell formation are observed along with stem cells that differentiate to form blood vascular endothelial tissues. Although still somewhat controversial, it is likely that the yolk sac wall is the first location for blood cell and blood vessel stem cell formation. It is thought that later in development, these blood-forming stem cells migrate into the developing embryo and become established in the liver where they undergo the process of hematopoiesis. These stem cells differentiate to form the yolk sac blood vascular network and the vitelline vessels during the stage at which the heart and umbilical blood vascular systems become established.

In mammals the first unambiguous observation of primordial germ cells (i.e., stem cells of the gametogenic germ line) occurs after the yolk sac has formed. These early germ cells are identified by their specific staining characteristics and by their larger size compared to associated somatic cells. The region in which these primordial germ cells are first observed is in the dorsal, posterior wall of the yolk sac membrane, near the allantoic outpocketing of the hindgut. At a later stage of embryogenesis the germ cells migrate dorsally along developing mesenteries to the fetal gonads. Some evidence suggests that the developmental establishment of the primordial germ cell lineage occurs during cleavage although detailed evidence for an origin of germ cells in mammals before their first appearance in the yolk sac membrane remains obscure.

V. TIMING FOR STAGES OF YOLK SAC DEVELOPMENT

As pointed out in the Introduction, details of structure and timing vary somewhat among different mammals. However, the stages of early human development are relatively well defined. In human blastocysts endodermal cells differentiate on the ventral surface of the embryonic disk on the eighth day following fertilization and the primary yolk sac is completed by Day 10. By Day 13, the ventral portion of the primary yolk sac has begun to collapse and by Days 15 or 16 after fertilization, vascularization of the secondary yolk sac begins. In Day 18.5 human embryos, the stalk of the yolk sac is in contact with the extraembryonic mesoderm of the developing umbilical cord and subsequently becomes incorporated into it. The distal secondary yolk sac and yolk stalk persist into later fetal stages; however, with the development of the chorioallantoic circulation the yolk sac circulation decreases in importance. Although usually no longer observable in preterm fetal tissues, the yolk sac occasionally persists to the time of birth.

See Also the Following Articles

Allantois; Blastocyst; Embryogenesis, Mammalian

Bibliography

Balkan, W., Phillips, L. S., Goldstein, S., and Sadler, T. W. (1989). Role of the mouse visceral yolk sac in nutrition: Inhibition by a somatomedin inhibitor. *J. Exp. Zool.* **249**, 36–40.

Båvik, C., Ward, S. J., and Chambon, P. (1996). Developmental abnormalities in cultured mouse embryos deprived of retinoic acid by inhibition of yolk-sac retinol binding protein synthesis. *Proc. Natl. Acad. Sci. USA* **93**, 3110–3114.

Beck, F., Moffat, D. B., and Davies, D. P. (1985). *Human Embryology,* pp. 76–109. Blackwell Scientific, Oxford, UK.

Carlson, B. M. (1996). *Patten's Foundations of Embryology,* pp. 267–290. McGraw-Hill, New York.

FitzGerald, M. J. T. (1978). *Human Embryology. A Regional Approach,* pp. 33–56. Harper & Row, New York.

Freeman, S. J. (1990). Functions of extraembryonic membranes. In *Postimplantation Mammalian Embryos. A Practical Approach* (A. J. Copp and D. L. Cockroft, Eds.), pp. 249–265. Oxford Univ. Press, Oxford, UK.

Hunter, R. H. F. (1995). *Sex Determination, Differentiation and Intersexuality in Placental Mammals,* pp. 70–79. Cambridge Univ. Press, Cambridge, UK.

Kalthoff, K. (1996). *Analysis of Biological Development,* pp. 318–320. McGraw-Hill, New York.

Moore, K. L. (1973). *The Developing Human. Clinically Oriented Embryology,* pp. 30–107. Saunders, Philadelphia.

Mossman, H. W. (1987). *Vertebrate Fetal Membranes.* Rutgers Univ. Press, New Brunswick, NJ.

Palacios, R., and Imhof, B. A. (1993). At day 8–8.5 of mouse development the yolk sac, not the embryo proper, has lymphoid precursor potential *in vivo* and *in vitro. Proc. Natl. Acad. Sci. USA* **90**, 6581–6585.

Vögler, H. (1987). *Human Blastogenesis. Formation of the Extraembryonic Cavities,* pp. 51–101. Karger, Basel.

Zygotic Genomic Activation

Ginger E. Exley and Carol M. Warner

Northeastern University

GLOSSARY

chromatin A complex of DNA and basic proteins containing subunits called nucleosomes.

genomic imprinting Gene expression based on the maternal or paternal origin of the gene.

histone The major basic protein in chromatin that is found in all eukaryotic cells except sperm.

pronucleus A membrane-bound structure in the fertilized one-cell embryo containing the haploid genome inherited from either the sperm or the egg.

protamine A basic protein that replaces histones in the chromatin of sperm.

transcription Synthesis of RNA, encoded by DNA, by RNA polymerase.

translation Synthesis of proteins encoded by messenger RNA.

trophectoderm The outer cell layer of the mammalian blastocyst that gives rise to extraembryonic membranes.

zygote A fertilized one-cell embryo.

Zygotic genomic activation (ZGA) is the point during preimplantation embryonic development at which the embryonic genome, inherited from transcriptionally silent gametes, becomes transcriptionally active and begins to control development of the embryo. Prior to the onset of ZGA, the development of the embryo is controlled by maternally inherited proteins and RNAs. ZGA has been most extensively studied in the mouse, which requires an early contribution from the embryonic genome for preimplantation development.

I. GENE EXPRESSION IN GAMETES

A. Gene Expression in the Oocyte

The process of oogenesis has evolved to create a molecular environment in which developmental events within a single cell can be regulated both temporally and spatially. During the development of the oocyte, meiosis is arrested (at different stages in different organisms) and transcription is suppressed. The oocyte must undergo the final phases of cellular differentiation in the absence of new gene expression. When the oocyte is fertilized, all genes are transcriptionally silent and must be activated in the absence of external contributions from the mother. Even embryos from mammals, which are normally in contact with the mother until birth, can develop normally in chemically defined media during the preimplantation period. Thus, a newly formed embryo must be able to initiate the developmental program autonomously using components synthesized and stored by the oocyte.

One aspect of oogenesis that has been evolutionarily conserved in both vertebrates and invertebrates is the increase in size of the maturing oocyte, accompanied by the stockpiling of RNAs and proteins that are stored in inactive form. Stored mRNAs are often nonadenylated, bound to proteins in translationally inactive complexes. In *Xenopus* and the mouse, consensus sequences have been identified in the 3′ untranslated regions of stored mRNAs that bind the masking proteins that prevent translation of the mRNA. Unmasking of the mRNAs is often accompanied by cytoplasmic polyadenylation to activate translation of the message. This process is developmentally regulated such that proteins are synthesized at discrete time points when needed. Some of these proteins are utilized for the last stages of oocyte maturation following transcriptional arrest. Other stored molecules are used to direct the first stages of development of the zygote. On the other hand, transcripts whose translation products would be inappropriate in the zygote can be inactivated via deadenylation.

Although the basic mechanism of packing the egg with regulatory and informational molecules is highly conserved, the extent of the maternal contribution varies considerably among different organisms. This is reflected by the difference in the number of embryonic cell divisions that can be sustained by the maternal contribution alone, when embryonic transcription is blocked. In general, mammalian embryos require embryonic expression products after only a few cell cycles. Nonplacental organisms that produce eggs with a large store of yolk often do not require embryonic gene expression until 10–14 cell divisions have been completed. In embryos that cleave asymmetrically, it is thought that zygotic genomic activation (ZGA) occurs at different time points in different cell lineages.

B. Gene Expression in the Sperm

During spermatogenesis, the immature sperm cells are transcriptionally active. The haploid spermatids generated by meiosis continue to actively transcribe some genes as they go through a prolonged differentiation phase, called spermiogenesis, during which the cells are transformed into mature sperm. Transcription originating from a haploid genome is unique to the sperm. During this time, some transcripts are stored as inert ribonucleoprotein particles for translation at a later time. Messages can be stored as long as a month before translation. Thus, both the sperm and the egg rely on posttranscriptional control mechanisms to direct the final stages of germ cell differentiation.

At the end of spermiogenesis, the chromatin is condensed, then packed even more tightly by replacing histones with protamines. Protamines are thought to bind DNA along the minor groove, neutralizing its charge and allowing DNA strands to interact with each other through Van der Waal's forces. This has the net effect of restructuring the chromatin so that it is very densely packed and completely suppresses transcription. The mature sperm has lost most of its cytoplasm and, with the exception of the mitochondrial genome, is transcriptionally and translationally dormant. However, it has recently been found that mature sperm do contain low levels of some mRNAs. It is presumed that these mRNAs are residual products of transcription that occurred during the differentiation process.

II. GENE EXPRESSION IN THE ZYGOTE

A. Fertilization

In both vertebrates and invertebrates, fertilization of the egg by the sperm triggers the resumption of meiosis by the maternal genome, and in mammals a second polar body is extruded. The male genome begins to replace protamines with histones immediately after fertilization. The haploid maternal and paternal genomes then form separate pronuclei. Each haploid genome undergoes DNA synthesis within the confines of its own distinct pronuclear environment.

B. ZGA Is Biphasic

The onset of zygotic gene expression has been assessed using a variety of techniques. Treatment with α-amanitin, an inhibitor of transcription, has been used to determine the point at which embryonic development is blocked in the absence of embryonic transcription. Comparison of the two-dimensional electrophoretic patterns of proteins translated at each cleavage stage relative to the proteins translated in the presence of α-amanitin has been used to determine the point at which a unique set of proteins is synthesized in untreated embryos. The specific stage at which protein expression shifts dramatically from maternal to embryonic origin generally corresponds to the stage at which development is blocked by α-amanitin treatment. However, this method pinpoints only the time at which embryonic proteins are required and does not necessarily reflect the time at which the corresponding genes are transcribed. As in gametogenesis, posttranscriptional regulation may temporally disconnect transcription from translation.

At the level of transcription, mRNA hybridization techniques and cDNA subtractive hybridization techniques have been used to detect embryonic transcripts of specific genes. Recently, more sensitive assays such as reverse transcription-polymerase chain reaction, reporter gene assays, and detection of labels incorporated into new transcripts *in situ* have been used to detect embryonic transcription at the earliest stages. Taken together, these studies have shown that a large burst of embryonic gene expression initiates at a stage of development that is species specific (Table 1). However, use of sensitive techniques has shown that mammalian embryos are transcriptionally competent as early as the one-cell stage. For example, the major burst of ZGA occurs at the two-cell stage in the mouse, but labeling of mRNA synthesized *in situ* has shown that transcription begins during S phase of the one-cell embryo and that by the end of the first cell cycle, the level of embryonic mRNA accumulated is approximately 20% of the level found in late two-cell embryos. Hence, a minor burst of ZGA occurs prior to the first cell division. Recently, the expression of specific genes has been confirmed in one-cell embryos, such as the *ZFY* and *SRY* genes in human pronuclear-stage embryos.

TABLE 1
Variation of the Onset of ZGA among Different Species

Species	*Global ZGA*	*Minor ZGA*
Human	4-cell	1-cell
Mouse	2-cell	1-cell
Rat	2-cell	ND[a]
Rabbit	8- to 16-cell	1-cell
Sheep	8-cell	ND
Cow	8-cell	2-cell
Pig	4-cell	ND[a]
Xenopus	~4000 cell (midblastula)	ND
Drosophila	~6000 cell (syncytial blastoderm)	64 nucleus
C. elegans	~90 cell	~30 cell

[a] ND, not determined.

C. ZGA in Mammals Is Independent of Cleavage but Linked to DNA Replication

Treatment of one-cell mouse embryos with aphidicolin, an inhibitor of DNA polymerase, during S phase of the first cell cycle does not prevent the expression of several proteins (the transcription requiring complex) considered to be indicators of the major burst of ZGA that takes place at the two-cell stage. These proteins are expressed at the appropriate time chronologically, even though mitosis has not occurred. A time-dependent mechanism for the activation of transcription (zygotic clock) has been proposed. According to this model, ZGA may be dependent on posttranslational modifications of transcription factors. However, this model is controversial because some researchers have found that adding aphidicolin during mid-S phase of the first cell cycle did not prevent ZGA, but addition prior to initiation of S phase significantly inhibited expression of the marker proteins (~90%). These results indicate that embryonic transcription is linked to DNA replication.

It has been shown that expression of injected reporter genes requires strong promoters or enhancers after the two-cell stage of preimplantation development. In agreement with these results, the expression of some endogenous genes that are activated at the two-cell stage drops at the four-cell stage. It appears that global activation of the genome is followed by the establishment of a transcriptionally repressive state that requires gene-specific activation. If the second round of DNA replication is inhibited with aphidicolin, this state of repression is not established. Both the onset of zygotic gene expression and the restriction of expression that follows may depend on changes in chromatin structure that either permit or restrict access to transcription factors.

D. Regulation via Chromatin Remodeling

It is likely that chromatin remodeling plays an important role in the regulation of gene expression in the early embryo. The fertilized zygote inherits two haploid genomes that have been transcriptionally silenced during passage through the gametes. In the mouse, embryonic transcription begins in the male pronucleus which appears to exist in a more transcriptionally permissive state than the female pronucleus. It has been hypothesized that the male pronucleus is more transcriptionally active during this stage because replacement of protamines with histones leaves the DNA more accessible to transcription complexes for a short time following fertilization.

Proteins generated from maternal mRNAs and maternal ribosomes are thought to activate expression of the zygotic genome. However, an alternate model has recently been proposed in which activation of the embryonic genome is postulated to be dependent on the titration of maternal histones. As the embryo replicates its DNA during the first few cell cycles, the new DNA is packed with maternally stored histones. As the stockpile of histones is depleted, newly synthesized DNA is no longer bound by histones and becomes accessible to transcription factors. This model would explain why not all genes are activated during the first cell cycle, even though transcription factors and RNA polymerase II are present. It would also explain why the embryonic genomes of organisms that produce larger eggs, which presumably contain more histones, require more cell cycles before activation.

As previously described, the zygotic genome is activated in a more gradual manner than once thought, and in mammals a repressive environment then develops that suppresses transcription of genes that lack strong promoters or enhancers. This repression is thought to be mediated by changes in chromatin structure. Recent advances suggest that specific chromatin modification could also play a role in the initial activation of the zygotic genome and subsequent repression.

Recent evidence has shown that chromatin structure is dynamic, and that certain modifications of histones result in changes in the efficiency of transcription. Of primary interest is histone acetylation, which has been shown to facilitate the binding of certain transcription factors and to correlate with increased transcriptional activity. Each core histone has a domain that interacts with other histones and a tail that interacts with DNA and regulatory proteins. Acetylation does not disrupt the basic structure of the nucleosome but reduces the affinity of the tail for DNA, "loosening" the wrapping of the DNA around the histone octamer. Acetylation probably enhances transcription by allowing access to transcription factors. In fact, it has recently been shown that several coactivators bound by sequence-specific transcription factors are in fact histone acetylases. Completing the picture is the discovery that other transcriptional regulators are histone deacetylases. If, as proposed in one model, maintenance of transcriptional repression through the action of histone deacetylases is the default state, then transcription would require the continued presence of the coactivating acetylases. This model could explain the establishment of the transcriptionally repressive state that develops in the early embryo, during which the requirement for enhancers and strong promoters is established. In addition, histone acetylation has been linked to cellular differentiation.

Another aspect of transcriptional regulation by chromatin structure involves histone H1, which binds to the linker region between nucleosomes. H1 has been shown in some cases to act as a transcriptional repressor. In *Drosophila*, H1 does not bind DNA until the tenth cleavage, the time point at which ZGA also occurs. In other organisms, different subtypes of H1 are utilized during early embryonic development. In the mouse, H1 is not detected until the four-cell stage, the time at which a transcriptionally repressive state develops. Hence, tightening of chromatin architecture by the addition of H1 is developmentally regulated and probably serves to restrict access to transcription factors.

III. IMPRINTING

During passage through oogenesis and spermiogenesis, a specific gene can acquire an imprint, mean-

TABLE 2
Imprinted Genes Identified in Mammals

Gene	Product	Species	Expression	Embryonic tissues with imprint
H19	Nontranslated RNA	Human, mouse	Maternal	Blastocyst, most fetal tissues, placenta
Igf2	Insulin-like growth factor II	Human, mouse, rat	Paternal[a]	Most fetal tissues and placenta
Igf2r	Insulin-like growth factor II receptor	Mouse[b]	Maternal[a]	Fetal tissues, tissue specific
Ins1/Ins2	Insulin-1 and -2, growth factors	Human, mouse	Paternal[a]	Yolk sac, tissue specific
IPW	Nontranslated RNA	Human	Paternal	Fetal tissues and placenta
Mas	Angiotensin receptor	Mouse[b]	Paternal	Fetal tissues, developmentally regulated
Mash-2	Trophoblast-specific transcription factor	Human, mouse	Maternal	Early extraembryonic tissues, transient
p57^{KIP2}	Cyclin-dependent kinase inhibitor	Human, mouse	Maternal	Fetal tissues, tissue specific
Peg-1/Mest	Epoxide hydrolase	Mouse	Paternal	Many fetal tissues, especially mesodermal
Peg-3	Zinc finger protein	Mouse	Paternal	Many fetal tissues
RT1.A, Pa	MHC class Ia antigens	Rat	Paternal	Placenta
Snrpn	RNA splicing factor	Human, mouse	Paternal	Most fetal tissues
U2afbp-rs	RNA splicing factor	Mouse	Paternal	Preimplantation embryo, all fetal tissues
Wt1	Transcription factor	Human	Maternal	Placenta, fetal brain (polymorphic)
Xist	Nontranslated RNA	Human, mouse	Paternal	Preimplantation embryo, placenta
Znf-127	Zinc finger transcription factor	Human, mouse	Paternal	Fetal tissues, tissue specific

[a] Monoallelic expression is preceded by a period of biallelic expression. Other genes may be expressed biallelically in certain tissues.
[b] These genes are not imprinted in the human.

ing that only one allele will be expressed after fertilization, depending on whether inheritance was maternal or paternal. Imprinting is primarily restricted to mammals, and at least 16 imprinted genes have been identified (Table 2). Differential expression of imprinted alleles may be tissue specific and/or developmentally regulated. Expression of several imprinted genes has been detected in preimplantation embryos (e.g., *U2afbp-rs*), and many are expressed at later stages of gestation.

Imprinted genes appear to play important roles during embryonic development. The first evidence for the importance of imprinting during development came from pronuclear transplantation studies in which the male pronucleus of a zygote was replaced with a female pronucleus, and in the reciprocal experiment the female pronucleus was replaced with a male pronucleus. These two types of embryos, both of which have a complete set of diploid chromosomes, die at midgestation. Clearly, genetic contribution from both the mother and the father is required for normal development, and these contributions are not equivalent (i.e., they are imprinted). The most striking characteristic of embryos reconstructed with two maternal pronuclei is the failure of the extraembryonic tissues to develop. In contrast, the extraembryonic tissues of embryos reconstructed with two male pronuclei develop normally, but development of the embryo proper is rudimentary. It is not surprising, then, that imprinted genes expressed only as the paternal allele appear to play a role in the growth and invasiveness of the trophectoderm during implantation. For example, the paternally expressed growth factor *Igf2* is highly expressed in the human placenta. In addition, some human birth defects and diseases have been linked to chromosomal abnormalities that only exert their deleterious effects when inherited through one of the sexes.

The mechanism by which genes are imprinted is not well characterized but may involve methylation of the unexpressed allele. Allele-specific methylation patterns have been observed for all imprinted genes identified to date. These patterns are maintained in daughter cells following mitosis by the action of DNA methyltransferase, an enzyme that recognizes hemimethylated DNA during DNA replication and restores methylation to the newly synthesized strand. In mice that lack methyltransferase, methylation patterns are lost and expression patterns of imprinted genes are altered.

In one experiment, the methylation patterns on an inherited, imprinted transgene were fully established in the mature egg and did not change during embryogenesis, but methylation patterns established in the sperm were found to change at the end of the preimplantation period. It has also been shown that additional parental-specific methylation sites are gradually added to the *Igf2* and *Igf2r* genes after fertilization has occurred. This evidence supports a model in which the early embryo plays an active role in fine-tuning parental imprints on transcription. This idea is corroborated by the fact that in some cases, the parental-specific restriction on expression is established after the initiation of transcription. For example, *Igf2* is expressed biallelically in the preimplantation embryo but only from the paternal allele in the postimplantation embryo.

During the first few cell divisions, the genome of the mouse embryo undergoes global demethylation. The DNA in cells that will contribute to the embryo proper is then remethylated at the time of implantation. If parental imprints are established by methylation, it is not known how these imprints are maintained in the face of embryonic demethylation and remethylation. In any case, some methylation patterns on imprinted genes appear to be preserved throughout development, either by protection from demethylation or by reestablishment of the earlier pattern.

The imprint that is established during gametogenesis must be reversible in the next generation. For example, if an allele of an imprinted gene that is paternally expressed is passed from father to daughter, the daughter will express that allele. However, her progeny will not express the same allele when they inherit it from her because it is now maternally inherited. Therefore, imprints must be removed and reestablished in the gonads of the parents. It remains a mystery how a reversible imprint can maintain its integrity throughout the dramatic changes that take place in the early embryo, such as chromatin remodeling, genomewide activation of embryonic transcription, and global demethylation at a later stage.

See Also the Following Articles

Embryogenesis, Mammalian; Transgenic Animals

Bibliography

Gold, J. D., and Pedersen, R. A. (1994). Mechanisms of genomic imprinting in mammals. *Curr. Topics Dev. Biol.* **29**, 227–280.

Goshen, R., Ben-Rafael, Z., Gonik, B., Lustig, O., Tannos, V., de-Groot, N., and Hochberg, A. A. (1994). The role of genomic imprinting in implantation. *Fertil. Steril.* **62**, 903–910.

Miller, D., Tang, P.-Z., Skinner, C., and Lilford, R. (1994). Differential RNA fingerprinting as a tool in the analysis of spermatazoal gene expression. *Hum. Reprod.* **9**, 864–869.

Nothias, J.-Y., Majumder, S., Kaneko, K. J., and DePamphilis, M. L. (1995). Regulation of gene expression at the beginning of mammalian development. *J. Biol. Chem.* **270**, 22077–22080.

Schultz, R. M., and Worrad, D. M. (1995). Role of chromatin structure in zygotic gene activation in the mammalian embryo. *Sem. Cell Biol.* **6**, 201–208.

Stebbins-Boaz, B., and Richter, J. D. (1997). Translational control during early development. *Crit. Rev. Eukaryotic Gene Expression* **7**, 73–94.

Wade, P. A., Pruss, D., and Wolffe, A. P. (1997). Histone acetylation: Chromatin in action. *TIBS* **22**, 128–132.

Ward, W. S., and Zalensky, A. O. (1996). The unique, complex organization of the transcriptionally silent sperm chromatin. *Crit. Rev. Eukaryotic Gene Expression* **6**, 139–147.

Wieschaus, E. (1996). Embryonic transcription and the control of developmental pathways. *Genetics* **142**, 5–10.

Contributors

Vicki Abrams-Motz
Boston University, Boston, Massachusetts
Hormonal Control of the Reproductive Tract, Subavian Vertebrates

Terry S. Adams
U. S. Department of Agriculture
Fargo, North Dakota
Oostatins, Folliculostatins, and Antigonadotropins, Insects

Eli Y. Adashi
University of Utah, Salt Lake City
Granulosa Cells
Infertility
Menopause

N. Scott Adzick
University of Pennsylvania
Philadelphia, Pennsylvania
Fetal Surgery

Th. Agorastos
Aristotle University
Thessaloniki, Greece
Tumors of the Female Reproductive System

Katsumi Aida
University of Tokyo
Female Reproductive System, Fish

James Aiman
Medical College of Wisconsin, Milwaukee
Androgen Insensitivity Syndromes
Ullrich Syndrome

R. Michael Akers
Virginia Polytechnic Institute, Blacksburg, Virginia
Lactogenesis

Benson T. Akingbemi
The Population Council
New York, NY
Leydig Cells

Eugene D. Albrecht
University of Maryland School of Medicine, Baltimore
Placental Steroidogenesis in Primate Pregnancy
Protein Hormones of Primate Pregnancy

Rupert P. Amann
Biopore, Inc.
State College, Pennsylvania
Cryopreservation of Sperm

Lloyd L. Anderson
Iowa State University, Ames, Iowa
Pregnancy in Other Mammals

O. Roger Anderson
Columbia University, New York, NY
Protozoa

Russell V. Anthony
Colorado State University, Fort Collins
Relaxin, Mammalian

Stephen Arch
Reed College, Portland, Oregon
Aplysia

Brian J. Arey
Wyeth-Ayerst Research, Radnor, Pennsylvania
Neuropeptides

Claude Arnaud
University of California, San Francisco
Osteoporosis

Michael L. Arnold
University of Georgia, Athens
Hybridization

Cheryl S. Asa
St. Louis Zoological Park, St. Louis, Missouri
Dogs

Tzazil Ayala
University of Pennsylvania
Philadelphia, Pennsylvania
Amniocentesis

Gabriella Chieffi Baccari
University of Naples, Italy
Corpora Lutea of Nonmammalian Species

Baccio Baccetti
Center for the Study of Germinal Cells, Siena, Italy
Sex Chromosomes

Nilima K. Badwaik
Cornell University Medical College, New York, NY
Discoidal Placenta

Janice Bahr
University of Illinois, Urbana-Champaign
Ovary, Overview

Mert Bahtiyar
Yale University, New Haven, Connecticut
SRY Gene

Philip L. Ballard
University of Pennsylvania
Philadelphia, Pennsylvania
Fetal Lung Development

Amy Banulis
University of Colorado Health Sciences Center, Denver
Lactational Amenorrhea

David Barad
Albert Einstein College of Medicine, Bronx, New York
Turner's Syndrome

Marylynn Barkley
University of California, Davis
Chorionic Gonadotropins, Nonhuman Mammals

Joseph D. Barnes
Vanderbilt University, Nashville, Tennessee
Parental Behavior, Arthropods

Frank F. Bartol
Auburn University, Auburn, Alabama
Uterus, Nonhuman

Michael J. Baum
Boston University, Boston, Massachusetts
Mating Behaviors, Mammals

Fuller W. Bazer
Texas A&M University, College Station, Texas
Allantochorion (Chorioallantois)
Allantoic Fluid
Allantois
Lymphokines
Pregnancy, Maintenance of

Nancy Beckage
University of California, Riverside
Parasitoids

David A. Beckman
Alfred I. Dupont Institute, Wilmington, Delaware
Teratogens

J. Michael Bedford
Cornell University Medical College, New York, NY
Sperm Capacitation

Ricardo Beduschi
University of Michigan, Ann Arbor
Prostate-Specific Antigen

Andrew N. Beld
Vanderbilt University, Nashville, Tennessee
Parental Behavior, Arthropods

Kurt Benirschke
University of California, San Diego
Placenta: Implantation and Development
Twinning

Fred B. Bercovitch
Caribbean Primate Research Center, Puerto Rico
Sex Skin

Sarah Berga
University of Pittsburgh, Pittsburgh, Pennsylvania
Amenorrhea
Menstrual Cycle

Alan S. Berkeley
New York University
Diapause
In Vitro Fertilization

Karen J. Berkley
Florida State University, Tallahassee, Florida
Autonomic Nervous System and Reproduction
Pelvic Nerve

Peter Berthold
Max Planck Society
Radolfzell, Germany
Migration, Birds

Roberto Bertolani
University of Modena, Italy
Tardigrada

Shalender Bhasin
Charles R. Drew University, Los Angeles, California
Androgens, Effects in Mammals

R. John Bicknell
Babraham Institute, Cambridge, England
Magnocellular System

Nadine Binart
INSERM, Paris, France
Prolactin, Overview

Stephen Birken
Columbia University College of Physicians and Surgeons, New York, NY
Chorionic Gonadotropin, Human

Timothy R. Birkhead
University of Sheffield, England
Male Reproductive System, Birds

Daniel G. Blackburn
Trinity College, Hartford, Connecticut
Placenta and Placental Analogs in Reptiles and Amphibians
Viviparity and Oviparity: Evolution and Reproductive Strategies

Charles A. Blake
University of South Carolina, Columbia
Gonadotropin Secretion, Control of

Jorge D. Blanco
University of Texas Medical School, Houston
Puerperal Infections

George R. Bousfield
Wichita State University, Wichita, Kansas
LH (Luteinizing Hormone)

Robert A. Brace
University of California, San Diego
Amniotic Fluid

William J. Bremner
University of Washington, Seattle
Contraceptive Methods and Devices, Male

Peter Brennan
Cambridge University, England
Bruce Effect

Eliot Brenowitz
University of Washington, Seattle
Songbirds and Singing

Robert L. Brent
Alfred I. Dupont Institute, Wilmington, Delaware
Teratogens

Franklin Bronson
University of Texas, Austin
Global Zones and Reproduction
Rodentia

C. H. Brown
University of Edinburgh, Scotland, UK
Endogenous Opioids

Janine Brown
National Zoological Park, Front Royal, Virginia
Cats
Elephants

Terry R. Brown
Johns Hopkins University, Baltimore, Maryland
Steroid Hormones, Overview

Wendy C. Brown
Washington State University, Pullman
Lymphokines

Peter J. Bruns
Cornell University, Ithaca, New York
Conjugation in Ciliates

Erdal Budak
University of Pennsylvania
Philadelphia, Pennsylvania
Fetal Membranes
Hemochorial Placentation

Serdar E. Bulun
University of Texas Southwestern Medical Center, Dallas
Estrogen Action, Breast

John D. Buntin
University of Wisconsin, Milwaukee
Parental Behavior, Birds

Carol Burdsal
Tulane University, New Orleans, Louisiana
Embryogenesis, Mammalian

Joanna Burger
Rutgers University
Piscataway, New Jersey
Nesting, Birds

Thomas Burris
R. W. Johnson Pharmaceutical Research Institute
Raritan, New Jersey
Progestins

William J. Butler
Medical University of South Carolina, Charleston
Hirsutism

Maria Byrne
University of Sydney, New South Wales, Australia
Echinodermata

Anne Byskov
University Hospital, Copenhagen, Denmark
Oocyte, Overview

William H. Cade
Brock University, St. Catherines, Ontario, Canada
Mating Behaviors, Insects

Gloria V. Callard
Boston University, Boston, Massachusetts
Spermatogenesis, in Nonmammals

Ian Callard
Boston University, Boston, Massachusetts
Estrogen Effects and Receptors, Subavian Species
Hormonal Control of the Reproductive Tract, Subavian Vertebrates
Progesterone Effects and Receptors, Subavian Species
Spermatogenesis, in Nonmammals

Robert E. Canfield
Columbia University College of Physicians and Surgeons, New York, NY
Chorionic Gonadotropin, Human

Jacob A. Canick
Brown University
Providence, Rhode Island
Alpha-Fetoprotein and Triple Screening

Cynthia Carey
University of Colorado, Boulder
Energetics of Reproduction

Steve N. Caritis
University of Pittsburgh
Pittsburgh, Pennsylvania
Preterm Labor and Delivery

R. W. Caron
University of Edinburgh, Scotland, UK
Endogenous Opioids

Bruce R. Carr
University of Texas Southwestern Medical Center, Dallas
Female Reproductive System, Humans
Fetal-Placental Unit

Douglas T. Carrell
University of Utah, Salt Lake City
Artificial Insemination, in Humans
Cryopreservation of Embryos

Daniel D. Carson
M. D. Anderson Cancer Center, Houston, Texas
Implantation

Carol Sue Carter
University of Maryland, College Park
Oxytocin

Deborah Cartmill
University of Utah, Salt Lake City
Artificial Insemination, in Humans

Vivienne Cass
Corporate Pyschology Consultants
South Perth, Australia
Homosexuality

Kevin J. Catt
National Institutes of Health, Bethesda, Maryland
Receptors for Hormones, Overview

Robert E. Chapin
National Institute of Environmental Health Sciences
Research Triangle Park, North Carolina
Reproductive Toxicology

Gin-Den Chen
University of Minnesota, Minneapolis
Follicular Atresia
Graafian Follicle

Frank A. Chervenak
Cornell University Medical College, New York, NY
Ultrasound

Giovanni Chieffi
University of Naples, Italy
Corpora Lutea of Nonmammalian Species

Gwen Childs
University of Texas Medical Branch, Galveston
Gonadotropes

George J. Christ
Albert Einstein College of Medicine, Bronx, New York
Erection

Jennifer Claman
University of California, Los Angeles
Postdate (Postterm) Pregnancy

Mertice M. Clark
McMaster University, Hamilton, Ontario, Canada
Intrauterine Position Phenomenon

Eloise D. Clawson
Brigham and Women's Hospital
Boston, Massachusetts
Breastfeeding

Jeffrey W. Clemens
Duquesne University, Pittsburgh, Pennsylvania
Androgens, Subavian Species

Yves Clermont
McGill University, Montreal, Quebec, Canada
Spermiogenesis

Robert J. Collier
Monsanto Company, St. Louis, Missouri
Lactation, Nonhuman

Giulia Collodel
Center for the Study of Germinal Cells, Siena, Italy
Sex Chromosomes

Patrick W. Concannon
Cornell University, Ithaca, New York
Pregnancy in Dogs and Cats

Alan J. Conley
University of California, Davis
Aromatization

P. Michael Conn
Oregon Regional Primate Research Center
Beaverton, Oregon
GnRH (Gonadotropin-Releasing Hormone)

Martin Connaughton
Washington College, Chestertown, Maryland
Female Reproductive System, Fish

Elizabeth Connell
Emory University, Atlanta, Georgia
Ovarian Function in the Perimenopause

Trevor G. Cooper
Münster University, Germany
Epididymis

Elizabeth M. Coscia
University of California, Berkeley
Hyenas

Daniel P. Costa
University of California, Santa Cruz
Seals
Whales and Porpoises

Raymond Counis
Université Pierre et Marie Curie, Paris
Gonadotropin Biosynthesis

Christos Coutifaris
University of Pennsylvania
Philadelphia, Pennsylvania
Ectopic Pregnancy

Elysse M. Craddock
State University of New York, Purchase, NY
Egg Coverings, Insects

David Crews
University of Texas, Austin
Reptilian Reproduction, Overview

Daniel E. Crocker
University of California, Santa Cruz
Seals
Whales and Porpoises

D. W. T. Crompton
University of Glasgow, Scotland, UK
Acanthocephala

Nicholas L. Cross
Oklahoma State University, Stillwater
Sperm Capacitation

Horacio B. Croxatto
Unidad de Reproducción y Desarrollo, Santiago, Chile
Oocyte and Embryo Transport

Diana O. Cua
Stanford University, Stanford, California
Breast Disorders

Timothy A. Cudd
Texas A&M University, College Station, Texas
Fetus, Overview

Noemi Custodia
Boston University, Boston, Massachusetts
Estrogen Effects and Receptors, Subavian Species
Progesterone Effects and Receptors, Subavian Species

Robert A. Dailey
West Virginia University, Morgantown
Female Reproductive System, Nonhuman Mammals

Ivan Damjanov
University of Kansas School of Medicine, Kansas City
Male Reproductive Disorders

Joseph Dancis
New York University
Apgar Score

Vibeke Dantzer
Royal Veterinary and Agricultural University, Copenhagen, Denmark
Endotheliochorial Placentation
Epitheliochorial Placentation

Kenneth G. Davey
York University, North York, Ontario, Canada
Hemocoelic Insemination
Insect Reproduction, Overview
Rhodnius prolixus

Alistair Dawson
Institute of Terrestrial Ecology, Cambridgeshire, UK
Seasonal Reproduction, Birds

Alan H. DeCherney
University of California, Los Angeles
Tubal Surgery

Nava Dekel
Weizmann Institute of Science, Rehovot, Israel
Meiotic Cell Cycle, Oocytes

Johanne DeLisle
Natural Resources Canada, Ottawa, Ontario
Pheromones, Insects

Francesco DeMayo
Baylor College of Medicine, Houston, Texas
Progesterone Actions on Reproductive Tract

David L. Denlinger
Ohio State University, Columbus
Diapause

Claude Desjardins
University of Illinois College of Medicine, Chicago
Spermatogenesis, Disorders of

A. Courtney DeVries
University of Maryland, College Park
Oxytocin

Michael P. Diamond
Wayne State University, Detroit, Michigan
Fallopian Tube

Thorsten Diemer
University of Illinois College of Medicine, Chicago
Spermatogenesis, Disorders of

Gregory A. Dissen
Oregon Regional Primate Research Center Beaverton, Oregon
Ovarian Innervation

Steven G. Docimo
Johns Hopkins University, Baltimore, Maryland
Hypospadias

Mark K. Dodson
University of Utah, Salt Lake City
Ovarian Cancer

Peter J. Donovan
Thomas Jefferson University Philadelphia, Pennsylvania
Primordial Germ Cells

Christine M. Drea
University of California, Berkeley
Hyenas

Lee C. Drickamer
Northern Arizona University, Flagstaff, Arizona
Intrauterine Position Phenomenon
Sexual Attractants

Deborah A. Driscoll
University of Pennsylvania Philadelphia, Pennsylvania
Prenatal Genetic Screening

Stuart Dryer
University of Houston, Texas
Pineal Gland, Melatonin Biosynthesis and Secretion

Charles A. Ducsay
Loma Linda University, Loma Linda, California
Hypoxia, Effect on Reproduction

Donald Dudley
University of Utah, Salt Lake City
Erythroblastosis Fetalis

Alfred M. Dufty, Jr.
Boise State University, Boise, Idaho
Brood Parasitism in Birds

Kent D. Dunlap
University of Texas, Austin
Hormones and Reproductive Behaviors, Fish

Victoria L. Dunning
Columbia University, New York, NY
Family Planning

Barbara S. Durrant
Zoological Society of San Diego, California
Captive Breeding of Wildlife

Philip J. Dziuk
University of Illinois, Urbana-Champaign
Fertility and Fecundity

Francis J. P. Ebling
Cambridge University, England, UK
Puberty, in Nonprimates

Sherrill E. Echternkamp
U. S. Agricultural Research Service
Clay Center, Nebraska
Freemartin

David Edwards
Emory University, Atlanta, Georgia
Aggressive Behavior

Scott V. Edwards
University of Washington, Seattle
Birds, Diversity of

Diaa M. El-Mowafi
Wayne State University, Detroit, Michigan
Fallopian Tube

Wolf Engels
University of Tübingen, Germany
Social Insects, Overview

Adviye Ergul
University of Georgia, Athens
Hormone Receptors, Overview

Ronald J. Ericsson
Gametrics Limited, Alzada, Montana
Sex Ratios

Scott A. Ericsson
Sul Ross State University, Alpine, Texas
Sex Ratios

Mary S. Erskine
Boston University, Boston, Massachusetts
Pseudopregnancy

Lawrence L. Espey
Trinity University, San Antonio, Texas
Ovulation

M. Sean Esplin
University of Utah, Salt Lake City
Circumcision

Laura Esserman
University of California, San Francisco
Breast Cancer

Ann Etgen
Albert Einstein College of Medicine, Bronx, New York
Progesterone Actions on Behavior

Anthony C. Evans, Jr.
University of Utah, Salt Lake City
Choriocarcinoma

Claude Everaerts
Centre National de la Recherche Scientifique, Paris
Pheromones, Insects

John Ewer
York University, North York, Ontario, Canada
Drosophila

Ginger Exley
Northeastern University, Boston, Massachusetts
Zygotic Genomic Activation

Gabrielle U. Falk
Texas A&M University, College Station, Texas
Male Reproductive System, Nonhuman Mammals

Harold M. Farrell, Jr.
U. S. Department of Agriculture
Wyndmoor, Pennsylvania
Milk, Composition and Synthesis

Christopher Faulkes
Zoological Society of London, England, UK
Naked Mole Rats

Daphne Fautin
University of Kansas, Lawrence
Cnidaria

Asgerally T. Fazleabas
University of Illinois College of Medicine, Chicago
Growth Factors

Paul E. Fell
Connecticut College, New London, Connecticut
Porifera

Michel Ferin
Columbia University College of Physicians and Surgeons, New York, NY
Ovarian Cycle, Mammals

Michael Fields
University of Florida, Gainesville
Parturition, Nonhuman Mammals

Phillip A. Fields
University of South Alabama, Mobile, Alabama
Corpus Luteum Peptides

George Fink
Medical Research Council
Edinburgh, Scotland, UK
Neuroendocrine Systems

Patricia D. Finn
University of Washington, Seattle
Estrus

Harry Fisch
Columbia University College of Physicians and Surgeons, New York, NY
Wolffian Ducts

Aron B. Fisher
University of Pennsylvania
Philadelphia, Pennsylvania
Surfactant

Rafael A. Fissore
University of Massachusetts, Amherst
Gametes, Overview

Jodi A. Flaws
University of Maryland School of Medicine, Baltimore
Reproductive Senescence, Nonhuman

William L. Flowers
North Carolina State University, Raleigh
Artificial Insemination, in Animals

Roy Fogwell
Michigan State University, East Lansing
Cattle

Peter P. Fong
Gettysburg College, Gettysburg, Pennsylvania
Rhythms, Lunar and Tidal

Stephen P. Ford
Iowa State University, Ames
Cotyledonary Placenta

Nancy G. Forger
University of Massachusetts, Amherst
Sex Differentiation, Psychological

Iraj Forouzan
University of Pennsylvania
Philadelphia, Pennsylvania
Fetal Monitoring and Testing

Isabel A. Forsyth
Babraham Institute, Cambridge, England, UK
Mammary Gland, Overview

Douglas L. Foster
University of Michigan, Ann Arbor
Puberty, in Nonprimates

George Foxcroft
University of Alberta, Edmonton, Alberta, Canada
Breeding Strategies for Domestic Animals

David A. Freeman
University of California, Berkeley
Cricetidae (Hamsters and Lemmings)

Ellen Freeman
University of Pennsylvania
Philadelphia, Pennsylvania
PMS (Premenstrual Syndrome)

Michael Freemark
Duke University, Durham, North Carolina
Human Placental Lactogen (Human Chorionic Somatomammotropin)

Andrew J. Friedman
Boston Regional Center for Reproductive Medicine
Stoneham, Massachusetts
Hypogonadism

Anna-Riitta Fuchs
Cornell University Medical College, New York, NY
Parturition, Nonhuman Mammals

Vivian L. Fuh
Merck Research Laboratories, Rahway, New Jersey
Androgen Inhibitors/Antiandrogens

Victor Y. Fujimoto
University of Washington, Seattle
Eunuchoidism

Peter Funch
University of Aarhus, Denmark
Cycliophora

Cindee R. Funk
Baylor College of Medicine, Houston, Texas
Progesterone Actions on Reproductive Tract

Harold Gainer
National Institutes of Health, Bethesda, Maryland
Neurosecretion

Bennett G. Galef, Jr.
McMaster University, Hamilton, Ontario
Intrauterine Position Phenomenon

Robin Gandley
University of Pittsburgh, Pittsburgh, Pennsylvania
Cardiovascular Adaptation to Pregnancy

William F. Ganong
University of California, San Francisco
Neurotransmitters

Robert E. Garfield
University of Texas Medical Branch, Galveston
Uterine Contraction

Ren-Shan Ge
The Population Council, New York, NY
Leydig Cells

Rodney D. Geisert
Oklahoma State University, Stillwater
Pigs

Michael T. Ghiselin
California Academy of Sciences, San Francisco, CA
Onychophora

Geula Gibori
University of Illinois College of Medicine, Chicago
Deciduoma

Glenys D. Gibson
Acadia University, Wolfville, Nova Scotia, Canada
Poecilogony

Georgia Giannoukos
Boston University, Boston, Massachusetts
Hormonal Control of the Reproductive Tract, Subavian Vertebrates

Cedric Gillott
University of Saskatchewan
Saskatoon, Saskatchewan, Canada
Insect Accessory Glands
Locusts
Male Reproductive System, Insects

Christopher Gilfillan
Prince Henry's Institute of Medical Research
Victoria, Australia
Follistatin

Daniel H. Gist
University of Cincinnati, Ohio
Male Reproductive System, Reptiles

Linda C. Giudice
Stanford University, Stanford, California
Decidua
Endometrium
Leiomyoma

Stephen E. Glickman
University of California, Berkeley
Hyenas

Vincent Goffin
INSERM, Paris, France
Prolactin, Overview

Bruce D. Goldman
University of Connecticut, Storrs
Cricetidae (Hamsters and Lemmings)

Irwin Goldstein
Boston University, Boston, Massachusetts
Impotence

Frank Gonzalez
State University of New York at Buffalo
Adrenarche

R. H. Gooding
University of Alberta, Edmonton, Alberta, Canada
Tsetse Flies

Arnold L. Goodman
Jones Institute for Reproductive Medicine
Norfolk, Virginia
Reflex (Induced) Ovulation

Robert L. Goodman
West Virginia University, Morgantown
Seasonal Reproduction, Mammals

Pal Gööz
Semmelweis University, Budapest, Hungary
Prolactin Secretion, Regulation of

Aubrey Gorbman
University of New Hampshire, Durham
Agnatha

Roger G. Gosden
University of Leeds, England, Uk
Oogenesis, in Mammals

Patricia Adair Gowaty
University of Georgia, Athens
Mate Choice, Overview

David A. Grainger
University of Kansas School of Medicine, Kansas City
Uterus, Human

E. Gordon Grau
University of Hawaii, Kaneohe
Prolactin, in Nonmammals

John T. Grayhack
Northwestern University Medical School
Chicago, Illinois
Benign Prostatic Hyperplasia (BPH)

Gilbert Greenwald
University of Kansas School of Medicine, Kansas City
Luteotropic Hormones

Mervyn Griffiths
CSIRO, Division of Wildlife and Ecology
Canberra, Australia
Monotremes

Michael Griswold
Washington State University, Pullman
Regulation of Sertoli Cells
Sertoli Cells, Function

Yan Gu
University of Illinois College of Medicine, Chicago
Deciduoma

Edith D. Gurewitsch
Cornell University Medical College, New York, NY
Ultrasound

Brett B. Gutsche
University of Pennsylvania
Philadelphia, Pennsylvania
Obstetric Anesthesia

Susan Guttentag
University of Pennsylvania
Philadelphia, Pennsylvania
Fetal Lung Development

Henry H. Hagedorn
University of Arizona, Tucson
Ecdysiotropins
Ecdysteroids

Lawrence S. Hakim
University of Miami, Florida
Impotence

Béla Halász
Semmelweis University, Budapest, Hungary
Pituitary Gland, Overview

Lisa Halvorson
Harvard Medical School, Boston, Massachusetts
Kallmann's Syndrome

William C. Hamlett
Indiana University School of Medicine, Indianapolis
Placenta and Placental Analogs in Elasmobranchs

Geoffrey L. Hammond
University of Western Ontario
London, Ontario, Canada
CBG (Corticosteroid-Binding Globulin)

James M. Hammond
Pennsylvania State University
Hershey Medical Center, Hershey, Pennsylvania
IGF (Insulin-Like Growth Factors)

Victor Han
University of Western Ontario
London, Ontario, Canada
Intrauterine Growth Restriction and Mechanisms of Fetal Growth

Stuart Handwerger
University of Cincinnati
Cincinnati, Ohio
Placental and Decidual Protein Hormones, Human

Wilfrid Hanke
University of Karlsruhe, Germany
Interrenal Gland, Stress Response and Reproduction

John D. Harder
Ohio State University, Columbus
Opossums

Jim Hardie
Imperial College of Science, Technology and Medicine
Silwood Park, UK
Aphids

Cheryl F. Harding
Hunter College, City University of New York
Androgens, Effects in Birds

Matthew P. Hardy
The Population Council, New York, NY
Leydig Cells

Carla D. Harris
Brigham and Women's Hospital
Boston, Massachusetts
Breastfeeding

Klaus Hartfelder
University of Tübingen, Germany
Social Insects, Overview

Harry Hatasaka
University of Utah, Salt Lake City
Menarche

William W. Hay, Jr.
University of Colorado School of Medicine, Denver
Fetal Growth and Development
Pregnancy, Metabolic Changes in

Tyrone B. Hayes
University of California, Berkeley
Sex Differentiation in Amphibians, Reptiles, and Birds, Hormonal Regulation

Linda J. Heffner
Harvard Medical School, Boston, Massachusetts
Cesarean Delivery

Karen Hendricks-Muñoz
New York University
Apgar Score

Terry W. Hensle
Columbia University College of Physicians and Surgeons, New York, NY
Wolffian Ducts

David L. Hepner
University of Pennsylvania
Philadelphia, Pennsylvania
Obstetric Anesthesia

Rex A. Hess
University of Illinois, Urbana-Champaign
Spermatogenesis, Overview

Richard A. Hiipakka
University of Chicago, Illinois
Dihydrotestosterone

John H. Himmelman
Laval University, Sainte-Foy, Quebec, Canada
Spawning, Marine Invertebrates

Margaret M. Hinshelwood
University of Texas Southwestern Medical Center, Dallas
Steroidogenesis, Overview

Anne N. Hirshfield
University of Maryland School of Medicine, Baltimore
Reproductive Senescence, Nonhuman

Penelope J. Hitchcock
National Institutes of Health, Bethesda, Maryland
HIV (AIDS)

Keith Hodges
German Primate Centre, Göttingen, Germany
Elephants

Alan N. Hodgson
Rhodes University, Grahamstown, South Africa
Paraspermatozoa

Harald H. J. Hoekstra
University Hospital, Groningen, The Netherlands
Castration, Effects in Humans (Male)

Richard L. Hoffman
Virginia Museum of Natural History, Martinsville, VA
Myriapoda

Glen E. Hofmann
Bethesda Hospital, Cincinnati, Ohio
Hormones of Pregnancy

Katalin M. Horváth
Semmelweis University, Budapest, Hungary
Prolactin Secretion, Regulation of

Erwin Huebner
University of Manitoba, Winnipeg, Manitoba, Canada
Female Reproductive System, Insects

Joan S. Hunt
University of Kansas Medical Center, Kansas City
Immunology of Reproduction

R. H. F. Hunter
Royal Veterinary and Agricultural University
Copenhagen, Denmark
Intersexuality in Mammals
Polyspermy

William W. Hurd
Indiana University School of Medicine, Indianapolis
Hypopituitarism

Bradley Hurst
University of Colorado Health Sciences Center, Denver
Sterility
Uterine Anomalies

Reinhold J. Hutz
University of Wisconsin, Milwaukee
Guinea Pig, Female

Luisa Iela
University of Naples, Italy
Female Reproductive System, Amphibians

Nancy H. Ing
Texas A&M University, College Station, Texas
Steroid Hormone Receptors

Juan C. Irwin
Stanford University, Stanford, California
Decidua

C. Todd Jackson
Vanderbilt University, Nashville, Tennessee
Parental Behavior, Arthropods

G. Marc Jackson
University of Pennsylvania
Philadelphia, Pennsylvania
Erythroblastosis Fetalis
Morning Sickness and Hyperemesis Gravidarum

Leslie M. Jackson
Ohio State University, Columbus
Opossums

Robert B. Jaffe
University of California, San Francisco
Fetal Adrenals

Venu Jain
University of Texas Medical Branch, Galveston
Uterine Contraction

A. Janovick
Oregon Regional Primate Research Center
Beaverton, Oregon
GnRH (Gonadotropin-Releasing Hormone)

Lynn Janulis
Northwestern University Medical School
Chicago, Illinois
Prostate Gland

Stefanie Jeffrey
Stanford University, Stanford, California
Breast Disorders

J. L. Jennes
University of Connecticut, Storrs
GnRH (Gonadotropin-Releasing Hormone)

Alan L. Johnson
University of Notre Dame
Notre Dame, Indiana
Ovarian Cycles and Follicle Development in Birds

Donald C. Johnson
University of Kansas Medical Center, Kansas City
Hypophysectomy

Jason L. Johnson
University of Utah, Salt Lake City
Ovarian Cancer

Larry Johnson
Texas A&M University, College Station, Texas
Male Reproductive System, Nonhuman Mammals
Testis, Overview

Patricia Johnson
Cornell University, Ithaca, New York
Female Reproductive System, Birds

Peter M. Johnson
University of Liverpool School of Medicine, England, UK
Immunology of Reproduction

Richard F. Johnston
University of Kansas, Lawrence
Pigeons

Gerald H. Jordan
Eastern Virginia Medical School, Norfolk
Penis

Jennifer Juengal
Colorado State University, Fort Collins
Corpus Luteum (CL)

Michael P. Kambysellis
New York University
Egg Coverings, Insects

George W. Kaplan
University of California, San Diego
Cryptorchidism

W. Reuben Kaufman
University of Alberta, Edmonton, Alberta, Canada
Chelicerate Arthropods

Genevieve E. Keillor
Texas A&M University, College Station, Texas
Testis, Overview

Duane Keisler
University of Missouri, Columbia
Sheep and Goats

Paul A. Kelly
INSERM, Paris, France
Prolactin, Overview

Nann Kemper-Green
Texas A&M University, College Station, Texas
Stress and Reproduction

Jeffrey B. Kerr
Monash University, Melbourne, Victoria, Australia
Temperature, Effects on Testicular Function

Izhar A. Khan
University of Texas Marine Science Institute
Port Aransas, Texas
Ovarian Cycle, Teleost Fish

Signe M. Kilen
Northwestern University, Evanston, Illinois
Estrous Cycle

J. Julie Kim
University of Illinois College of Medicine, Chicago
Growth Factors

Gary M. King
Darling Marine Center, University of Maine, Walpole
Hemichordata

John D. Kirby
University of Arkansas, Fayetteville
Internal Fertilization, Birds and Mammals

Takeo Kishimoto
Tokyo Institute of Technology
Oocyte Maturation and Spawning in Starfish

Thomas Klein
Thomas Jefferson University
Philadelphia, Pennsylvania
Puberty, in Humans

Harvey Kliman
Yale University, New Haven, Connecticut
Trophoblast to Human Placenta
Umbilical Cord

Marc J. Klowden
Westmead Hospital, Sydney, Australia
Ovulation and Oviposition, Insects

Thomas Koob
Shriner's Hospital for Children, Tampa, Florida
Elasmobranch Reproduction
Relaxin, Nonmammalian

Brian J. Koos
University of California, Los Angeles
Postdate (Postterm) Pregnancy

Gregory S. Kopf
University of Pennsylvania
Philadelphia, Pennsylvania
Acrosome Reaction

Kenneth S. Korach
National Institute of Environmental Health Sciences
Research Triangle Park, North Carolina
Estrogen Action on the Female Reproductive Tract

James M. Kozlowski
Northwestern University Medical School
Chicago, Illinois
Prostate Cancer

W. Krause
University of Marburg, Germany
Orchitis

Lewis Krey
New York University
Critical Period, Estrous Cycle
In Vitro Fertilization

Reinhardt Møbjerg Kristensen
University of Copenhagen, Denmark
Cycliophora

Jacek Z. Kubiak
Institut Jacques Monod, Paris, France
Meiosis

Ursula Kuhnle
Universitätskinderklinik, München, Germany
Clitoris

Gary J. LaFleur, Jr.
Brown University
Providence, Rhode Island
Vitellogenins and Vitellogenesis

Valentine Lance
Center for Reproduction of Endangered Species
San Diego, California
Female Reproductive System, Reptiles
Reptilian Reproductive Cycles

Matthew Landau
Richard Stockton College
Pomona, New Jersey
Crustacea

Bill Lasley
University of California, Davis
Primates, Nonhuman

Hans Laufer
University of Connecticut, Storrs
Crustacea

Richard Leach
Wayne State University, Detroit, Michigan
Theca Cell Tumors

Chung Lee
Northwestern University Medical School
Chicago, Illinois
Prostate Gland

Grace Lee
Yale University, New Haven, Connecticut
SRY Gene

Mary Min-chin Lee
Harvard Medical School, Boston, Massachusetts
Genitalia
Gonadogenesis, Male

Peter A. Lee
University of Pittsburgh, Pittsburgh, Pennsylvania
Adrenal Hyperplasia, Congenital Virilizing

Theresa M. Lee
University of Michigan, Ann Arbor
Microtinae (Voles)

Florence Le Gac
Institut National de la Recherche Agronomique
Rennes, France
Male Reproductive System, Fish

Sandra Legan
University of Kentucky, Lexington
Castration, Effects in Nonhumans (Female)

Richard S. Legro
Pennsylvania State University
Hershey Medical Center
Hershey, Pennsylvania
Polycystic Ovary Syndrome

Christian Lemburg
University of Göttingen, Germany
Priapulida

Janet Leonard
University of California, Santa Cruz
Mating Behaviors, Invertebrates Other Than Insects

Charles A. Lessman
University of Memphis, Tennessee
Oogenesis, in Nonmammalian Vertebrates

Jon E. Levine
Northwestern University, Evanston, Illinois
GnRH Pulse Generator

Shutsung Liao
University of Chicago, Illinois
Dihydrotestosterone

Jonathan Lindzey
National Institute of Environmental Health Sciences
Research Triangle Park, North Carolina
Estrogen Action on the Female Reproductive Tract

Daniel I. H. Linzer
Northwestern University, Evanston, Illinois
Placental Lactogens

Kimberly N. Livingston
Texas A&M University, College Station, Texas
Stress and Reproduction

Leslie Lobel
Columbia University College of Physicians and Surgeons, New York, NY
Chorionic Gonadotropin, Human

Rogerio A. Lobo
Columbia University College of Physicians and Surgeons, New York, NY
Estrogen Replacement Therapy

Maurice Loir
Institut National de la Recherche Agronomique
Rennes, France
Male Reproductive System, Fish

Lawrence D. Longo
Loma Linda University, Loma Linda, California
Placental Gas Exchange

Francisco J. López
Ligand Pharmaceuticals, San Diego, California
Neuropeptides

Barry G. Loughton
York University, North York, Ontario, Canada
Corpus Cardiacum, Insects

Jack Ludmir
Harvard Medical School, Boston, Massachusetts
Infections in Pregnancy

George A. Macones
University of Pennsylvania
Philadelphia, Pennsylvania
Tocolytic Agents

Everett F. Magann
University of Mississippi Medical Center, Jackson
Preeclampsia/Eclampsia

Lucia Magliulo-Cepriano
State University of New York, Farmingdale
Pituitary Gland, in Fish

Jennifer M. Malvey
Colorado State University, Fort Collins
Radioimmunoassay

Robert Marcus
Stanford University, Stanford, California
Estrogen Action, Bone

Sara Marder
University of Pennsylvania
Philadelphia, Pennsylvania
Substance Abuse and Pregnancy

Lukas Margaritis
University of Athens, Greece
Egg Coverings, Insects

Graeme B. Martin
University of Western Australia
Nedlands, Australia
Castration, Effects in Nonhumans (Male)

James N. Martin
University of Mississippi Medical Center, Jackson
Preeclampsia/Eclampsia

Vicki J. Martin
University of Notre Dame
Notre Dame, Indiana
Hydra

Sharmin Maswood
Texas Woman's University, Denton, Texas
Estrogen Action, Behavior

Ranjiv Mathews
Johns Hopkins University, Baltimore, Maryland
Hypospadias

George I. Matsumoto
Monterey Bay Aquarium Research Institute
Moss Landing, California
Ctenophora

Irene M. McAleer
Children's Hospital & Health Center
San Diego, California
Cryptorchidism

John A. McCracken
Worcester Foundation for Biomedical Research
Shrewsbury, Massachusetts
Local Control Systems in Reproduction
Luteolysis

Donald McDonnell
Duke University, Durham, North Carolina
Antiestrogens

Robert W. McGaughey
Arizona State University, Tempe
Yolk Sac

Tobin A. McGowen
Texas A&M University, College Station, Texas
Testis, Overview

Damhnait McHugh
Harvard University, Cambridge, Massachusetts
Annelida

Eric McIntush
Colorado State University, Fort Collins
Corpus Luteum (CL)

Kevin E. McKenna
Northwestern University Medical School
Chicago, Illinois
Ejaculation
Orgasm

Margaret K. McLaughlin
University of Pittsburgh, Pittsburgh, Pennsylvania
Cardiovascular Adaptation to Pregnancy

F. M. Anne McNabb
Virginia Polytechnic Institute, Blacksburg, Virginia
Altricial and Precocial Development in Birds

Jeremy N. McNeil
Laval University, Sainte-Foy, Quebec, Canada
Pheromones, Insects

Kevin T. McVary
Northwestern University Medical School
Chicago, Illinois
Benign Prostatic Hyperplasia (BPH)

Ramkrishna Mehendale
University of Illinois College of Medicine, Chicago
Oxytocics

Robert Meisel
Purdue University, West Lafayette, Indiana
Copulation, Mammals

Michael Meredith
Florida State University, Tallahassee, Florida
Vomeronasal Organ

M. Cristina Meriggiola
University of Bologna, Italy
Contraceptive Methods and Devices, Male

Samuel Mesiano
University of California, San Francisco
Fetal Adrenals

Dietrich L. Meyer
University of Göttingen, Göttingen, Germany
Olfaction and Reproduction

David B. Miller
University of Connecticut, Storrs
Sexual Imprinting

Josephine B. Miller
University of Illinois College of Medicine, Chicago
Rabbits

Clarke F. Millette
University of South Carolina, Columbia
Spermatozoa

Shahab Minassian
Allegheny University of the Health Sciences
Philadelphia, Pennsylvania
Prader-Willi Syndrome

Takeshi Miura
Hokkaido University, Japan
Spermatogenetic Cycle in Fish

Akiyasu Mizukami
University of Utah, Salt Lake City
Cryopreservation of Embryos

Charles V. Mobbs
Mount Sinai School of Medicine, New York, NY
Reproductive Senescence, Human

Kamran S. Moghissi
Wayne State University, Detroit, Michigan
Cervix

Anders P. Möller
Université Pierre et Marie Curie, Paris, France
Sexual Selection

Karen Moore
Texas A&M University, College Station, Texas
Transgenic Animals

Lorna Grindlay Moore
University of Colorado Health Sciences Center, Denver
Altitude, Effects on Humans

Robert Y. Moore
University of Pittsburgh, Pittsburgh, Pennsylvania
Suprachiasmatic Nucleus

Mark A. Morgan
University of Pennsylvania
Philadelphia, Pennsylvania
Pelvimetry
Substance Abuse and Pregnancy

Glenn K. Morris
University of Toronto, Ontario, Canada
Song in Arthropods

Joseph F. Mortola
Cook County Hospital, Chicago, Illinois
Postpartum Depression
Pseudocyesis

György Nagy
Semmelweis University, Budapest, Hungary
Prolactin Secretion, Regulation of

Jon P. Nash
University of Sheffield, England, UK
Seasonal Reproduction, Fish

Peter W. Nathanielsz
Cornell University, Ithaca, New York
Labor and Delivery, Human

Andrés F. Negro-Vilar
Ligand Pharmaceuticals, San Diego, California
Neuropeptides

Ajay Nehra
The Mayo Clinic, Rochester, Minnesota
Impotence

Linda R. Nelson
University of Illinois College of Medicine, Chicago
Menstruation

Randy Nelson
Johns Hopkins University, Baltimore, Maryland
Photoperiodism, Vertebrates

John Edwin Nestler
Medical College of Virginia, Richmond, Virginia
DHEA (Dehydroepiandrosterone)

Terry M. Nett
Colorado State University, Fort Collins
Radioimmunoassay

Birger Neuhaus
Institute for Systematic Zoology, Berlin, Germany
Kinorhyncha

Peggy Neville
University of Colorado Health Sciences Center, Denver
Lactation, Human

Andrew Todd Newberry
University of California, Santa Cruz
Tunicata (Urochordata)

Camran Nezhat
Stanford University, Stanford, California
Endometriosis

Ceana Nezhat
Stanford University, Stanford, California
Endometriosis

Farr Nezhat
Stanford University, Stanford, California
Endometriosis

Cheryl Niemuller
Stouffville, Ontario, Canada
Elephants

Mikio Nihira
University of California, Los Angeles
Tubal Surgery

Fred Nijhout
Duke University, Durham, North Carolina
Metamorphosis, Insects

Gordon Niswender
Colorado State University, Fort Collins
Corpus Luteum (CL)

Marcelo F. Noguera
University of Pennsylvania
Philadelphia, Pennsylvania
Morning Sickness and Hyperemesis Gravidarum

Claire Norris
University of California, Berkeley
Hormonal Contraception

David O. Norris
University of Colorado, Boulder
Thyroid Hormones, in Subavian Vertebrates

Michael Numan
Boston College, Boston, Massachusetts
Parental Behavior, Mammals

Jack O'Brien
University of South Alabama, Mobile, Alabama
Parasites and Reproduction

John O'Connor
Columbia University College of Physicians and Surgeons, New York, NY
Chorionic Gonadotropin, Human

Joseph E. Oesterling
Ann Arbor, Michigan
Prostate-Specific Antigen

Sergio Ojeda
Oregon Regional Primate Research Center
Beaverton, Oregon
Ovarian Innervation

Richard Oko
Queen's University, Kingston, Ontario, Canada
Spermiogenesis

William C. Okulicz
University of Massachussetts Medical School
Worcester, Massachusetts
Immunocytochemistry

Rush Oliver
University of Minnesota, Minneapolis
Follicular Atresia
Graafian Follicle

Christopher J. Ormandy
INSERM, Paris, France
Prolactin, Overview

B. Hannah Ortiz
Stanford University, Stanford, California
Cervical Cancer

Troy L. Ott
University of Idaho, Moscow, Idaho
Interferons
Pregnancy in Farm Animals

James W. Overstreet
University of California, Davis
Sperm Transport

David Owens
Texas A&M University, College Station, Texas
Migration, Reptiles

Laurence Packer
York University, North York, Ontario, Canada
Altruism in Insect Reproduction

Robert B. Page
Pennsylvania State University
Hershey Medical Center, Hershey, Pennsylvania
Anterior Pituitary
Hypothalamic-Hypophysial Complex (Pituitary Portal System)

Nancy Pahle
Medical College of Virginia, Richmond
DHEA (Dehydroepiandrosterone)

Marina Paolucci
University of Naples, Italy
Estrogen Effects and Receptors, Subavian Species
Progesterone Effects and Receptors, Subavian Species

Raymond Papka
Northeastern Ohio Universities College of Medicine
Rootstown, Ohio
Vagina

Jeffrey A. Parrott
Washington State University, Pullman
Gonadogenesis, Female

Samuel Parry
University of Pennsylvania
Philadelphia, Pennsylvania
Pelvimetry

Reynaldo Patin~o
Texas Tech University, Lubbock
Sex Determination, Environmental

Jonathan J. H. Pearce
Mount Sinai Hospital, Toronto, Ontario, Canada
Germ Layers

John S. Pearse
University of California, Santa Cruz
Orthonectida
Rhombozoa
Seasonal Reproduction, Marine Invertebrates
Sea Urchins

Vicki Pearse
University of California, Santa Cruz
Placozoa

Jan Pechenik
Tufts University, Medford, Massachusetts
Marine Invertebrates, Modes of Reproduction in

Gerald J. Pepe
Eastern Virginia Medical School, Norfolk, Virginia
Placental Steroidogenesis in Primate Pregnancy
Protein Hormones of Primate Pregnancy

Richard D. Peppler
University of Tennessee College of Medicine, Memphis
Armadillo

Sally D. Perreault
U.S. Environmental Protection Agency
Research Triangle Park, North Carolina
Internal Fertilization, Birds and Mammals

Steven Petak
Texas Institute for Reproductive Medicine, Houston
Sexual Dysfunction

Matthew Peterson
University of Utah, Salt Lake City
Cryopreservation of Embryos

Donald W. Pfaff
Rockefeller University, New York, NY
Lordosis

Helen M. Picton
University of Leeds, England, UK
Oogenesis, in Mammals

Jorge A. Piedrahita
Texas A&M University, College Station, Texas
Gene Transfer, Sperm-Mediated
Transgenic Animals

Riccardo Pierantoni
University of Naples, Italy
Male Reproductive System, Amphibians

Dave Pilgrim
University of Alberta, Edmonton, Alberta, Canada
Caenorhabditis elegans

Tony M. Plant
University of Pittsburgh, Pittsburgh, Pennsylvania
Puberty, in Nonhuman Primates

Ed Plotka
Marshfield Medical Research Foundation
Marshfield, Wisconsin
Deer

Leo Plouffe, Jr.
Eli Lilly & Company
Indianapolis, Indiana
Puberty, Precocious

V. Polanski
Jagellonian University, Krakow, Poland
Meiosis

Gary A. Polis
Vanderbilt University, Nashville, Tennessee
Parental Behavior, Arthropods

Alberta M. Polzonetti-Magni
University of Camerino, Italy
Amphibian Ovarian Cycles

Tom E. Porter
University of Maryland, College Park
Lactotrophs

Malcolm Potts
University of California, Berkeley
Contraceptive Methods and Devices, Female
Hormonal Contraception

Randall Prather
University of Missouri, Columbia
Cloning Mammals by Nuclear Transfer

Brian J. Prendergast
University of California, Berkeley
Circannual Rhythms

Edward Price
University of California, Davis
Suckling Behavior

Gail Prins
University of Illinois College of Medicine, Chicago
Semen

Heather C. Proctor
Griffith University, Queensland, Australia
Spermatophores, Arthropods

Jeffrey Pudney
Harvard Medical School, Boston, Massachusetts
Leydig and Sertoli Cells, Nonmammalian

J. David Puett
University of Georgia, Athens
Gonadotropin Receptors
Hormone Receptors, Overview

Theresa M. Quinn
University of Pennsylvania
Philadelphia, Pennsylvania
Fetal Surgery

Jay Radhakrishnan
University of Illinois College of Medicine, Chicago
Testicular Developmental Anomalies

Alexander S. Raikhel
Michigan State University, East Lansing
Aedes aegypti

Jeffrey L. Ram
Wayne State University, Detroit, Michigan
Oviposition in Molluscs

Nilsa Ramirez
Wayne State University, Detroit, Michigan
Theca Cell Tumors

Jagdeece J. Ramsoondar
Texas A&M University, College Station, Texas
Gene Transfer, Sperm-Mediated

Rakesh K. Rastogi
University of Naples, Italy
Female Reproductive System, Amphibians
Olfaction and Reproduction

John J. Rasweiler, IV
Cornell University Medical College, New York, NY
Discoidal Placenta

Robert W. Rebar
University of Cincinnati, Cincinnati, Ohio
Female Reproductive Disorders, Overview

Lorena Rebecchi
University of Modena, Italy
Tardigrada

John F. Redman
University of Arkansas for Medical Sciences
Little Rock, Arkansas
Male Reproductive System, Humans

E. Albert Reece
Temple University, Philadelphia, Pennsylvania
Fetal Anomalies

Carmen L. Regan
University of Pennsylvania
Philadelphia, Pennsylvania
Pregnancy in Humans, Overview

Leo E. Reichert, Jr.
Albany Medical College, Albany, NY
FSH (Follicle-Stimulating Hormone)

Rudolf Reinboth
Institute for Zoology, University of Mainz, Germany
Fish, Modes of Reproduction

Marilyn Renfree
University of Melbourne, Parkville, Victoria, Australia
Marsupials

Mary E. Rice
Smithsonian Marine Station, Fort Pierce, Florida
Sipuncula

John W. Riggs
University of Texas Medical School, Houston
Puerperal Infections

James A. Rillema
Wayne State University, Detroit, Michigan
Prolactin, Actions of

Bernard Robaire
McGill University, Montreal, Quebec, Canada
Androgens

James A. Roberts
Stanford University, Stanford, California
Cervical Cancer

David M. Robertson
Prince Henry's Institute of Medical Research
Victoria, Australia
Follistatin

Sarah A. Robertson
University of Adelaide, South Australia
Cytokines

James M. Robl
University of Massachusetts, Amherst
Gametes, Overview

Katherine F. Roby
University of Kansas Medical Center, Kansas City
Theca Cells

John Rock
Emory University School of Medicine
Atlanta, Georgia
Uterine Anomalies

Martha E. Rode
University of Pennsylvania
Philadelphia, Pennsylvania
Tocolytic Agents

Nancy C. Rose
University of Pennsylvania
Philadelphia, Pennsylvania
Amniocentesis
Fetal Alcohol Syndrome

Allan Rosenfield
Columbia University, New York, NY
Family Planning

Janet F. Roser
University of California, Davis
Equine Chorionic Gonadotropin (ECG)

William Rosner
St. Luke's/Roosevelt Hospital Center
New York, NY
SHBG (Sex Hormone-Binding Globulin)

Lawrence S. Ross
University of Illinois College of Medicine, Chicago
Seminal Vesicles

John Russell
University of Edinburgh, Scotland, UK
Endogenous Opioids

Lonnie Russell
Southern Illinois University, Carbondale
Sertoli Cells, Function
Sertoli Cells, Overview

Irma H. Russo
Fox Chase Cancer Center
Philadelphia, Pennsylvania
Mammary Gland Development

Jose Russo
Fox Chase Cancer Center
Philadelphia, Pennsylvania
Mammary Gland Development

Travis J. Ryan
University of Missouri, Columbia
Migration, Amphibians

George R. Saade
University of Texas Medical Branch, Galveston
Uterine Contraction

Stephen H. Safe
Texas A&M University, College Station, Texas
Environmental Estrogens

Mary Beth Saffo
Arizona State University West, Phoenix, Arizona
Symbiosis

M. Ram Sairam
Clinical Research Institute of Montreal
Montreal, Quebec, Canada
Gonadotropins, Overview

Colin J. Saldanha
University of California, Los Angeles
Estrogen, Effects in Birds

Saber Saleuddin
York University, North York, Ontario, Canada
Dorsal Bodies in Mollusca
Mollusca

Thomas W. Sappingto
Michigan State University, East Lansing
Aedes aegypti

Gerald Schatten
Oregon Regional Primate Research Center
Beaverton, Oregon
Fertilization

Paul F. Schelhammer
Eastern Virginia Medical School, Norfolk, Virginia
Penis

Isaac Schiff
Harvard Medical School, Boston, Massachusetts
Menstrual Disorders

William D. Schlaff
University of Colorado Health Sciences Center, Denver
Lactational Amenorrhea

Barney Schlinger
University of California, Los Angeles
Estrogen, Effects in Birds

Kearston Schmidt
National Institutes of Health, Bethesda, Maryland
HIV (AIDS)

Andreas Schmidt-Rhaesa
University of South Florida, Tampa, Florida
Nematomorpha
Priapulida

Martin P. Schreibman
Brooklyn College, New York
Pituitary Gland, in Fish

Richard Schultz
University of Pennsylvania
Philadelphia, Pennsylvania
Blastocyst

Ralph H. Schwall
Genentech Inc., San Francisco, California
Activin and Activin Receptors
Inhibins

Marlene Schwanzel-Fukuda
State University of New York at Brooklyn
Nervus Terminalis

Neena B. Schwartz
Northwestern University, Evanston, Illinois
Estrous Cycle

Mary A. Scott
University of California, Davis
Sperm Transport

Shimon Segal
Harvard Medical School, Boston, Massachusetts
Menstrual Disorders

Harish M. Sehdev
University of Medicine and Dentistry of New Jersey
Piscataway, New Jersey
Puerperium

George E. Seidel, Jr.
Colorado State University, Fort Collins
Embryo Transfer

Kyle Selcer
Duquesne University, Pittsburgh, Pennsylvania
Androgens, Subavian Species

Michael Selmanoff
University of Maryland School of Medicine, Baltimore
Prolactin Inhibitory Factors

Raymond D. Semlitsch
University of Missouri, Columbia
Migration, Amphibians

Brian P. Setchell
University of Adelaide, South Australia
Blood-Testis Barrier
Pampiniform Plexus

Fady I. Sharara
University of Maryland School of Medicine, Baltimore
Klinefelter's Syndrome

Dan C. Sharp
University of Florida, Gainesville
Horses

Howard T. Sharp
University of Utah, Salt Lake City
Hysterectomy

Peter J. Sharp
Roslin Institute, Edinburgh, Scotland, UK
Chickens, Control of Reproduction in

Geoffrey Shaw
University of Melbourne, Parkville, Victoria, Australia
Marsupials

Julian Shepherd
State University of New York at Binghamton
Sperm Activation, Arthropods
Sperm Transport, Arthropods

Celeste Sheppard
University of Maryland School of Medicine, Baltimore
Fetal Diagnosis, Invasive

O. D. Sherwood
University of Illinois, Urbana-Champaign
Corpus Luteum Peptides

Susan Shideler
University of California, Davis
Primates, Nonhuman

George L. Shinn
Truman State University, Kirksville, Missouri
Chaetognatha

Colin P. Sibley
University of Manchester, England, UK
Placental Nutrient Transport

Sherman J. Silber
St. Luke's Hospital, St. Louis, Missouri
Vasectomy

Theresa Siler-Khodr
University of Texas Health Science Center, San Antonio
Fetal Hormones

Ann-Judith Silverman
Columbia University College of Physicians and Surgeons, New York, NY
Median Eminence

William J. Silvia
University of Kentucky, Lexington
Eicosanoids

Rebecca A. Simmons
University of Pennsylvania
Philadelphia, Pennsylvania
Respiratory Distress Syndrome

Evan R. Simpson
University of Texas Southwestern Medical Center, Dallas
Estrogen Action, Breast

Michael K. Skinner
Washington State University, Pullman
Gonadogenesis, Female

Collin B. Smikle
University of California, San Francisco
Adrenal Androgens

Carolyn L. Smith
Baylor College of Medicine, Houston, Texas
Estrogens, Overview

Gordon C. S. Smith
Cornell University, Ithaca, New York
Labor and Delivery, Human

Samuel Smith
Harbor Hospital Center
Baltimore, Maryland
Gynecomastia

Peter J. Snyder
University of Pennsylvania
Philadelphia, Pennsylvania
Sheehan's Syndrome

Thomas E. Snyder
Bowman Gray School of Medicine
Winston-Salem, North Carolina
Pelvic Inflammatory Disease (PID)

Richard Soderstrom
University of Washington School of Medicine, Seattle
Female Sterilization

Marla B. Sokolowski
York University, North York, Ontario, Canada
Drosophila

Steven J. Sondheimer
University of Pennsylvania
Philadelphia, Pennsylvania
Abortion

Lisa A. Sorbera
Boston University, Boston, Massachusetts
Hormonal Control of the Reproductive Tract, Subavian Vertebrates

Peter Sorensen
University of Minnesota, Minneapolis
Pheromones, Fish

Michael R. Soules
University of Washington, Seattle
Eunuchoidism

Stacia A. Sower
University of New Hampshire, Durham
Agnatha

Thomas E. Spencer
Texas A&M University, College Station, Texas
Pregnancy, Maternal Recognition of

Genevieve E. Spoede
Texas A&M University, College Station, Texas
Male Reproductive System, Nonhuman Mammals

Irving M. Spitz
Shaare Zedek Medical Center
Jerusalem, Israel
Antiprogestins

Norman Stacey
University of Alberta, Edmonton, Alberta, Canada
Pheromones, Fish

David M. Stamilio
University of Pennsylvania
Philadelphia, Pennsylvania
Fetal Alcohol Syndrome

Judy Stamps
University of California, Davis
Territorial Behavior, Overview

Barbara Stay
University of Iowa, Iowa City
Diploptera punctata

Gary D. Steinberg
University of Chicago
Testicular Cancer

Wolfgang Sterrer
Bermuda Natural History Museum
Flatts, Bermuda
Gnathostomulida

Jeffrey S. Stevenson
Kansas State University, Manhattan
Lactational Anestrus

Colin L. Stewart
Frederick Cancer Research and Development Center
Frederick, Maryland
Leukemia Inhibitory Factor

Douglas Stocco
Texas Tech University, Lubbock
Testosterone Biosynthesis

Dale Stokes
Hopkins Marine Station, Stanford University
Pacific Grove, California
Cephalochordata

Elizabeth Stoner
Merck Research Laboratories
Rahway, New Jersey
Androgen Inhibitors/Antiandrogens

Richard L. Stouffer
Oregon Regional Primate Research Center
Beaverton, Oregon
Corpus Luteum of Pregnancy

Jerome F. Strauss, III
University of Pennsylvania
Philadelphia, Pennsylvania
Fetal Membranes
Hemochorial Placentation

Steven A. Stricker
University of New Mexico, Albuquerque
Brachiopoda

Maria Strömstedt
University Hospital, Copenhagen, Denmark
Oocyte, Overview

Carlos A. Strüssman
Tokyo University of Fisheries
Sex Determination, Environmental

Józefa Styrna
Jagiellonian University, Krakow, Poland
Sex Determination, Genetic

Paul Summers
University of Utah, Salt Lake City
Sexually-Transmitted Diseases

William F. Swanson
National Zoological Park, Front Royal, Virginia
Cats

Seiji Tanaka
National Institute of Sericultural and Entomological Science, Ibaraki, Japan
Diapause

Basil C. Tarlatzis
Aristotle University, Thessaloniki, Greece
Tumors of the Female Reproductive System

Christopher C. Taylor
University of Kansas Medical Center, Kansas City
Apoptosis (Cell Death)

Salli Tazuke
Stanford University, Stanford, California
Leiomyoma

Jose Teixeira
Harvard Medical School, Boston, Massachusetts
Gonadogenesis, Male

Paul F. Terranova
University of Kansas Medical Center, Kansas City
Apoptosis (Cell Death)
Theca Cells

William W. Thatcher
University of Florida, Gainesville
Ruminants

Peter Thomas
University of Texas Marine Science Institute
Port Aransas, Texas
Ovarian Cycle, Teleost Fish

Robert C. Thommes
Laboratory of Developmental Endocrinology
Sarasota, Florida
Avian Reproductive System, Developmental Endocrinology

Christopher W. Thompson
Washington State Department of Fish and Wildlife
Mill Creek, Washington
Molt and Nuptial Color

Janice E. Thornton
Oberlin College, Oberlin, Ohio
Estrus

Stephen S. Tobe
University of Toronto, Ontario, Canada
Allatostatins
Corpus Allatum

Béla E. Tóth
Semmelweis University, Budapest, Hungary
Prolactin Secretion, Regulation of

Amanda L. Trewin
University of Wisconsin, Milwaukee
Guinea Pig, Female

Philip Troen
University of Pittsburgh, Pittsburgh, Pennsylvania
Andrology: Origins and Scope

Alan O. Trounson
Monash University School of Medicine
Clayton, Victoria, Australia
Reproductive Technologies, Overview

Wenbin Tuo
Washington State University, Pullman
Lymphokines

James M. Turbeville
University of Arkansas, Fayetteville
Nemertea

Fred W. Turek
Northwestern University, Evanston, Illinois
Circadian Rhythms
Pineal Gland, Regulatory Function

Terry T. Turner
University of Virginia, Charlottesville
Varicocele

Seth Tyler
University of Maine, Orono
Platyhelminthes

Lawrence C. Udoff
University of Utah, Salt Lake City
Infertility
Menopause

Lynda Uphouse
Texas Woman's University, Denton, Texas
Estrogen Action, Behavior

Michael J. VandeHaar
Michigan State University, East Lansing
Nutritional Factors and Lactation

John G. Vandenbergh
North Carolina State University, Raleigh
Intrauterine Position Phenomenon
Pheromones, Mammals
Puberty Acceleration
Whitten Effect

Mels F. van Driel
University Hospital, Groningen, The Netherlands
Castration, Effects in Humans (Male)

Bradley Van Voorhis
University of Iowa, Iowa City
Follicular Development
Follicular Steroidogenesis

Luis Velasquez
Unidad de Reproducción y Desarrollo
Santiago, Chile
Oocyte and Embryo Transport

Johannes D. Veldhuis
University of Virginia, Charlottesville
Ultradian Hormone Rhythms

John Verstegen
University of Liege, Belgium
Pregnancy in Dogs and Cats

Manuel Villalon
Unidad de Reproducción y Desarrollo
Santiago, Chile
Oocyte and Embryo Transport

Carol M. Vleck
Iowa State University, Ames
Egg, Avian

Frederick S. vom Saal
University of Missouri, Columbia
Intrauterine Position Phenomenon

Robert C. Vrijenhoek
Rutgers University, New Brunswick, New Jersey
Parthenogenesis and Natural Clones

George N. Wade
University of Massachusetts, Amherst
Energy Balance, Effects on Reproduction

Marvalee H. Wake
University of California, Berkeley
Amphibian Reproduction, Overview

Jonathan B. Wakerley
University of Bristol, England, UK
Milk-Ejection

Hope Wallace
University of California, San Francisco
Breast Cancer

Robert Lee Wallace
Ripon College, Ripon, Wisconsin
Rotifera

Karen W. Walters
University of California, Davis
Aromatization

Carol Warner
Northeastern University,, Boston, Massachusetts
Zygotic Genomic Activation

Kerstin Wasson
Humboldt State University, Arcata, California
Asexual Reproduction
Hermaphroditism
Kamptozoa (Entoprocta)

Gregory M. Weber
North Carolina State University, Raleigh
Prolactin, in Nonmammals

Wolfgang Weidner
University of Giessen, Germany
Orchitis

Carl P. Weiner
University of Maryland School of Medicine, Baltimore
Fetal Diagnosis, Invasive

Thomas H. Welsh
Texas A&M University, College Station, Texas
Stress and Reproduction

David Wildt
National Zoological Park
Front Royal, Virginia
Cats

R. Haven Wiley
University of North Carolina, Chapel Hill
Altruistic Behavior, Vertebrates

Pax H. B. Willemse
University Hospital, Groningen, The Netherlands
Castration, Effects in Humans (Male)

Gary L. Williams
Texas A&M University, College Station, Texas
Nutritional Factors and Reproduction

Sonya J. Williams
Yale University, New Haven, Connecticut
Vagina

Tony D. Williams
Simon Fraser University
Burnaby, British Columbia, Canada
Avian Reproduction, Overview

Kenneth Wilson
University of Stirling, UK
Migration, Insects

Laird Wilson, Jr.
University of Illinois College of Medicine, Chicago
Oxytocics

Selma F. Witchel
University of Pittsburgh, Pittsburgh, Pennsylvania
Adrenal Hyperplasia, Congenital Virilizing

Arnon Wiznitzer
Ben Gurion University
Beer-Sheva, Israel
Fetal Anomalies

Douglas A. Woelkers
University of Pittsburgh, Pittsburgh, Pennsylvania
Preterm Labor and Delivery

Diana S. Wolfe
University of California, Berkeley
Contraceptive Methods and Devices, Female

Kenneth H. H. Wong
University of Utah, Salt Lake City
Granulosa Cells

James E. Woods
Laboratory of Developmental Endocrinology
Sarasota, Florida
Avian Reproductive System, Developmental Endocrinology

Robert Woollacott
Harvard University, Cambridge, Massachusetts
Bryozoa (Ectoprocta)

John P. Wourms
Clemson University, Clemson, South Carolina
Teleosts, Viviparity

Denis J. Wright
Imperial College of Science, Technology and Medicine, Silwood Park, UK
Nematodes and Related Phyla

G. R. Wyatt
Queen's University
Kingston, Ontario, Canada
Juvenile Hormone
Yolk Proteins, Invertebrates

Humphrey H. C. Yao
University of Illinois, Urbana-Champaign
Ovary, Overview

John Yeh
University of Minnesota, Minneapolis
Follicular Atresia
Graafian Follicle

Shao-Yao Ying
University of Southern California
Los Angeles, California
Ovarian Hormones, Overview

Koji Yoshinaga
National Institutes of Health
Bethesda, Maryland
Estrogen Secretion, Regulation of

Craig M. Young
Harbor Branch Oceanographic Institution
Fort Pierce, Florida
Marine Invertebrate Larvae

Graham Young
University of Otago, Dunedin, New Zealand
Migration, Fish

Howard A. Zacur
Johns Hopkins University, Baltimore, Maryland
Galactorrhea
Hyperprolactinemia

Harold H. Zakon
University of Texas, Austin
Hormones and Reproductive Behaviors, Fish

Mary B. Zelinski-Wooten
Oregon Regional Primate Research Center
Beaverton, Oregon
Chorionic Gonadotropins, Nonhuman Mammals

Anthony J. Zeleznik
University of Pittsburgh, Pittsburgh, Pennsylvania
Luteinization

Zhong Zhang
University of Southern California
Los Angeles, California
Ovarian Hormones, Overview

Russel Zimmer
University of Southern California
Los Angeles, California
Phoronida

Hans H. Zingg
McGill University, Montreal, Quebec, Canada
Neurohypophysial Hormones

Barry R. Zirkin
Johns Hopkins University, Baltimore, Maryland
Spermatogenesis, Hormonal Control of

Irving Zucker
University of California, Berkeley
Circannual Rhythms

Glossary of Key Terms

A

abortifacient an agent producing an abortion.

abortion expulsion from the uterus of an embryo or fetus prior to the stage of viability.

abruptio placenta the premature separation of a normally implanted placenta from the uterine wall.

acanthor the stage in an acanthocephalan life history that results from embryogenesis in the egg shells within the female worm, adapted for the infection of the intermediate host.

accessory fertilization cells in chaetognaths, paired somatic cells in the ovary to which eggs are attached and which sperm must pass through in order to fertilize the eggs.

accessory fluid in arthropods, the liquid part of the ejaculate accompanying and surrounding the spermatozoa (or earlier developmental stages of sperm cells).

accessory olfactory bulb the first stage of processing of the pheromonal information from the vomeronasal nerves; it has a similar structure to the main olfactory bulb, but it projects subcortically.

accessory olfactory system the vomeronasal organ, its associated glands, blood vessels, capsule, etc., and its central neural connections to the accessory olfactory bulb with onward connections (in mammals) to parts of the corticomedial amygdala and bed nucleus of the stria terminalis.

accessory sex glands in males, a collective term for the prostate gland, the seminal vesicles, the bulbourethral glands, the vas deferens, and the epididymis.

acclimatization physiological adjustment to a new environment, occurring over a period of hours to weeks. See also ADAPTATION.

acidemia an increased level of hydrogen ions in the blood.

acidosis an increased level of hydrogen ions in body tissue.

acromegaly a condition caused by excessive secretion of growth hormone in adults, leading to hyperplasia of the nose, jaws, fingers, and toes.

acrosomal system see ACROSOME.

acrosome a caplike, membrane-bound structure that overlies the anterior head of mammalian sperm and that plays a major role in fertilization. Formed by the Golgi apparatus of early spermatids, the acrosome contains hydrolytic enzymes that are necessary for the acrosome reaction, thereby facilitating sperm-oocyte interaction.

acrosome reaction an exocytotic event that is an absolute pre-requisite to successful fertilization and that occurs following sperm binding to the zona pellucida of the egg.

ACTH adrenocorticotropic hormone, the anterior pituitary hormone that promotes adrenocortical growth and controls corticosteroid biosynthesis and secretion.

actinotroch the characteristic larva of the phylum Phoronida.

action potential the rapid and transient change in membrane potential which occurs in nerve, muscle, and other excitable tissue in response to excitation.

activation a condition in which target tissues exhibit a response to a hormone only when the hormone is present (e.g., circulating plasma levels are elevated) and the response disappears when the hormone is absent or at basal levels.

activational or **activating effects** temporary effects of sex hormones on brain structures and thus on brain functions and behaviors, typically exhibited during adulthood.

activation function one of two specific regions within the estrogen receptor that contributes to the ability of the receptor to activate target gene transcription. One activation function is located in the amino-terminus of the estrogen receptor. The other overlaps the ligand binding domain in the carboxy-terminus.

activin a protein hormone produced widely with effects on development, the pituitary, gonads, and other organs.

adaptation a characteristic of structure, function, or behavior that enables an organism to live and reproduce in a given environment, an alteration occurring over years to generations. See also ACCLIMATIZATION.

adelphophagy a form of embryonic nutrition in which offspring feed on extraembryonic yolk.

adenohypophysis the anterior lobe of the vertebrate pituitary, composed of the the pars tuberalis, the pars distalis, and the pars intermedia and consisting of a mixed population of cells that secrete a cell-specific complement of hormones, such as LH (luteinizing hormone), growth hormone, and prolactin.

adenotrophic viviparity a reproductive method in invertebrates in which a chorionated egg is fertilized, embryonated and hatched within the female's uterus and the resulting larva is nourished by material from the female's reproductive accessory glands until it has completed its development.

adjuvant any substance that when mixed with an antigen, enhances antigenicity and produces an enhanced immune response.

adjuvant therapy treatment that is given to patients to prevent or delay the spread or recurrence of cancer. Adjuvant therapy is given to patients who do not have any indication that their cancer has spread.

adluminal compartment a region of the seminiferous epithelium segregated from intercellular access to blood-borne products by basally positioned Sertoli cell junctions.

adolescence the period of time from the appearance of secondary sex attributes until the completion of physical maturation.

adrenal (gland) one of a pair of flattened endocrine glands located above each kidney and composed of two parts, the cortex and the medulla. The cortex is derived from embryonic mesoderm. The medulla has an origin similar to that of the sympathetic nervous system.

adrenal androgens steroid hormones secreted primarily by the adrenal gland. They include dehydroepiandrostenedione (DHEA) dehydroepiandrostenedione sulfate (DHEAS), 11 hydroxy-androstenedione, and androstenedione.

adrenal cortex the outermost portion of the adrenal constituting 90% of the gland and composed of the zona glomerulosa, the zona fasciculata, and the zona reticularis; the adrenal cortex secretes important corticosteroid hormones.

adrenal medulla the central compartment of each adrenal gland composed of neuroendocrine chromaffin cells that secrete catecholamine hormones.

adrenarche a condition of intensified secretion by the adrenal cortex, especially androgens, occurring in humans at about eight years of age in both sexes. Often used synonymously with PUBARCHE.

adrenocorticotropic hormone see ACTH.

aedeagus the extensible portion of the intromittent organ or penis of most insects.

afferent moving toward a certain region, specifically referring to sensory nerve fibers that project from peripheral structures to the spinal cord or brain.

afferent pathway a neural pathway carrying impulses (usually derived from the nipples) towards the oxytocin-releasing cells of the hypothalamus.

affinity a measure of the ability of a moiety, or ligand, to bind reversibly to a receptor, and an indication of potency or effectiveness.

AFI see AMNIOTIC FLUID INDEX.

after swarm any bee swarm containing one or few virgin queens, issued by a honey bee colony after the prime swarm.

AGA appropriate for gestational age, a term used in characterizing or describing fetal devlopment.

agametes in some invertebrates, cells or groups of cells within the hypertrophied host cell (plasmodium), developing into ciliated, vermiform adults that swim free of the host.

age-limited learning in songbirds, an ontogenetic pattern in which song learning is limited to the first year of life and new songs are not developed in adulthood. Also, SENSITIVE-PERIOD LEARNING.

agenesis the failure of an organ to develop.

agonist a chemical compound that evokes the same or very similar actions as another compound in a target cell.

agonistic behavior a collective term for behaviors shown when animals are in conflict, usually applied to conflict within the same species. Common forms include threatening gestures and vocalizations, physical attack, and defensive reactions such as counterattack, submission, and escape.

AI see ARTIFICIAL INSEMINATION.

air cell the gas-filled space that forms at the blunt end of the avian egg between the inner and outer shell membrane as water evaporates from the egg.

alate a term for winged sexuals, especially in termites.

albumen egg white, a mixture of more than 40 proteins secreted by the oviduct.

albumen gland a female sex gland concerned with the secretion of perivitelline fluid.

allantois or **allantoic membrane** an extraembryonic membrane that arises as a highly vascular, sac-like outgrowth of the embryonic hindgut; in humans its vessels develop into the vessels of the umbilical cord.

allatostatin in invertebrates, a neuropeptide that inhibits the synthesis of juvenile hormone by the corpus allatum; it is also found in nerve cells other than those innervating the corpora allata and appears to function as a neuromodulator and myomodulator.

allatotropin a neuropeptide that stimulates the synthesis of juvenile hormone by the corpus allatum.

allodynia pain due to a stimulus that does not normally provoke pain.

allograft a graft of tissue obtained from a donor genetically different from, though of the same species as, the recipient individual. (The human fetus would represent a **semiallograft**, as half of the antigens would be paternally derived.)

allometry the study or measurement of body size and its consequences. Thus, **allometric**.

alopecia the loss of hair, baldness.

altricial describing or relating to an immature state at birth that requires a period of dependency on adult care; specifically, referring to a pattern of development in birds, in which the young are blind and helpless when hatched and remain in the nest for some time, during which they are completely dependent upon their parents for warmth, protection, and food. See also PRECOCIAL.

alveolar relating to or involving the alveoli.

alveolus *plural,* **alveoli** a small sac-like structure that serves as the basic functional unit of many tissues; e.g., one of a large number of air cells in the lung through which the gas exchange of respiration takes place; or, the milk-producing unit of the mammary gland, composed of a hollow sphere of secretory epithelial cells, surrounded by a network of myoepithelial cells and blood vessels, and drained by a terminal duct.

Alzheimer's disease (AD) a severe, progressive disorder involving major neuronal degeneration in specific regions of the brain, identified by the presence of distinctive plaques and tangles; characterized by loss of memory, diminished cognitive function, behavioral abnormalities, and eventual death. The main risk factor for the disease is age; early-onset AD typically occurs before age 50 and late-onset AD after age 60. (Described by Alois *Alzheimer,* German neurologist.)

ambiguous genitalia a form of physical development resulting from virilization of external genitalia in females or inadequate virilization of genitalia in males. In the case of the virilized female infant, the clitoris is enlarged with varying degrees of labial fusion which may resemble a scrotum.

ambisexuality the normal occurrence of male and female function in one and the same individual, either as simultaneous or as sequential hermaphroditism.

amenorrhea the absence or cessation of menstrual periods for an extended period of time, either normal (as during pregnancy) or abnormal.

amictic not able to be fertilized; specifically, referring to a phase in the life cycle of monogonont rotifers in which reproduction takes place by parthenogenesis.

amniocentesis a technique performed to obtain fetal cells and amniotic fluid for prenatal testing, e.g., for genetic study, involving the passage of a needle through the mother's abdomen into the amniotic sac.

amnion the innermost membrane that encloses the developing fetus prior to birth, composed of a layer of epithelial cells resting on a basement membrane which overlies a dense collagen-rich matrix and fibroblasts. The human amnion is adherent to the chorion.

amniote one of the Amniota, the higher terrestrial vertebrates (i.e., mammals, birds, reptiles) in which the embryo is enclosed in a fluid-filled membrane (amnion). The amnion is considered a major character in the evolution of vertebrates, contributing to their radiation into the land environment.

amniotic relating to or involving the amnion.

amniotic fluid the liquid that surrounds the embryo/fetus prior to birth.

amniotic fluid index (AFI) a measurement of the largest amniotic fluid pocket in each of the four abdominal quadrants. The AFI normally ranges between 8 and 26 cm. An AFI < 8 represents reduced amniotic fluid volume; values < 5 cm are associated with significantly increased perinatal morbidity.

amniotic vertebrate see AMNIOTE.

amniotomy artificial rupture of the fetal membranes.

amphiblastula a hollow larva with anterior flagellated cells and posterior non-flagellated cells.

amphimixis sexual reproduction; the mixing of the genes from two distinct individuals, involving the recombinational effects of meiotic reduction and fusion of gametes.

amphoteric referring to or describing a type of monogonont female capable of producing both AMICTIC and MICTIC eggs.

ampulla a general anatomical term to designate a flasklike dilation of a tubular structure; specifically, the thin-walled expanded region of the oviduct extending from the isthmus to the infundibulum. The ampulla is the site of fertilization for most mammals.

anabolic steroids testosterone derivatives that are claimed to promote nitrogen retention in the body and thereby enhance the growth of muscle and bone tissue.

anabolism the net synthesis and growth of tissue.

anadromy a type of life cycle in which fish migrate from freshwater to seawater and then return to freshwater to spawn.

analgesia the absence of pain in response to a stimulus that would normally be painful; lack of sensitivity to pain.

analgesic a substance or agent capable of alleviating pain without loss of consciousness.

analog a chemical compound that is chemically similar to another compound.

anamorphosis a developmental pattern in invertebrates, in which body segmentation is completed after escape of the embryo from the egg capsule, by successive addition of somites and appendages from a growth zone at the posterior end of the body.

ancestrula the founding member of a colony, derived from metamorphosis of a larva.

ANDI aberrations of normal development and involution; a benign (non-cancerous) departure from the normal tissue structure, related to normal growth or development patterns.

androgen one of a class of steroid hormones that have the ability to induce masculine characteristics and maintain male sex accessory glands and ducts, but which have physiological and pharmacological effects in both sexes. Testosterone is the primary androgen produced in the testes in men.

androgenic describing a substance that is able to produce masculine characteristics or effects such as, in humans, increased facial and body hair, male body habitus, deepening of voice, temporal recession of hair, and phallic enlargement.

androgenic gland a ductless gland, now known only in certain crustaceans, that is found in association with the sperm duct, producing androgenic hormones.

androgen inhibitor a substance that decreases androgen production or expression.

androgen insensitivity syndrome a genetic disorder in which the tissues lack androgen receptors and do not respond to testosterone. It is also called TESTICULAR FEMINIZATION SYNDROME because the affected persons have cryptorchid abdominal testes, but otherwise appear female. Genetically they are male (46, XY).

androgen receptor intranuclear proteins that, when activated by binding to androgens, will bind to nuclear DNA and regulate specific gene expression to effect the physiological actions of androgens.

androstenedione an androgenic steroid secreted by the ovaries and adrenals that is involved in the synthesis of testosterone.

anesthesia a lack of sensation, usually produced by loss of consciousness, induced for the performance of surgery.

anestrus an infertile state in the female during which sexual receptivity (estrus) is not observed. See also ANOVULATION.

aneuploidy a total number of chromosomes that is different from the normal 46, XX or 46, XY.

aneurysm the swelling of a blood vessel.

angiogenesis the formation of new blood vessels, or neovascularization, which begins with capillary proliferation and culminates in the formation of a new microcirculatory bed, composed of arterioles, capillaries and venules.

anisogamy reproduction in which the two haploid gametes that unite in fertilization are of very unequal size (egg and sperm). See also ISOGAMY.

anlage or **anlagen** an embryonic precursor; the earliest discernible development of an organ or part.

annulus a ring-like structure composed of electron dense material associated with the plasma membrane. It surrounds the axoneme just below the centrioles in elongating spermatids and later it migrates distally along the sperm tail and forms the junction of the middle piece and principal piece of the spermatozoon.

anogenital distance (AGD) the length of the perineal tissue, which becomes the scrotum in males, separating the anus and the genital tubercle (penis in males and clitoris in females). AGD is longer in males than in females.

anorchia the absence of both testicles.

anorgasmia the inability to achieve orgasm.

anovulation lack of ovulation in a woman with an ovary or ovaries; it may occur as amenorrhea, oligomenorrhea, or eumenorrhea.

anovulatory uterine bleeding the shedding of endometrial tissue that occurs on a sporadic basis in the absence of ovulatory cycles.

antagonist a chemical compound that counteracts or blocks the actions of another chemical.

antepartum a term referring to the time interval prior to the birth process.

anterior lobe or **anterior pituitary** the adenohypophysis, a glandular region of the pituitary producing hormones that stimulate various target organs throughout the body; e.g., FSH(follicle-stimulating hormone).

antiandrogen a substance capable of antagonizing and reducing the biological effects of androgenic hormones.

antibody a protein produced in response to the introduction of an antigen and having the ability to combine specifically with the antigen that stimulated its production.

antigen a high molecular weight substance, usually protein or protein-carbohydrate complex, that enters an animal's body and is there perceived as foreign, stimulating the production of specific antibodies.

antigenicity the capacity to act as an antigen.

antigenic site the portion of a molecule that stimulates production of an antibody or to which an antibody binds.

antigen-presenting cells (APCs) cells that take up, process, and present antigens in a form that can initiate protective or destructive immune responses.

anti-Müllerian hormone see MÜLLERIAN-INHIBITING SUBSTANCE.

antiprogestin a substance that inhibits the synthesis of progesterone, its transport or stability in the blood, or that reduces its uptake by or effects on target organs..

antisense oligonucleotide a small piece of artificially generated DNA that prevents the production of a particular protein by binding to the mRNA sequence that codes for the protein.

antrum the fluid-filled compartment of the mature follicle.

Apgar score a scoring system applied immediately after birth to assess a newborn's condition. The parameters used are: heart rate, respiratory effort, muscle tone, reflex irritability, and color. (Developed by Virginia *Apgar,* U.S. anesthesiologist.)

apocrine describing a type of secretion in which the apical portion of a cell is shed, e.g. the secretion of fat into milk by cells of the mammary gland.

apomixis asexual reproduction without chromosome reduction or fusion of gametes.

apoptosis programmed cell death, a naturally occurring phenomenon in developing or diseased tissues, in which scattered cells undergo a genetically regulated process of spontaneous death characterized by highly specific phenomena such as DNA fragmentation, chromatin clumping, and cell shrinkage and condensation.

aporeceptor a steroid receptor not bound by a hormone.

apyrene sperm a type of sperm unique to Lepidoptera in which there is no nucleus, and the sperm do not fertilize the oocytes.

arachidonic acid the most common substrate for eiconsanoid biosynthesis. It is a 20-carbon polyunsaturated fatty acid with double bonds at positions 5, 8, 11, and 14.
architomy a form of asexual reproduction in which simple subdivision of the worm is followed by regeneration of the missing parts.
area postrema a neural structure located in the floor of the fourth ventricle in the caudal hindbrain; situated so that it is able to sample the composition of both cerebrospinal fluid and blood.
areola the pigmented area surrounding the mother's nipple which overlies the milk sinuses of the breast.
aromatase a cytochrome P-450 steroidogenic enzyme that converts androgens to estrogens.
aromatizable describing androgens, such as testosterone, that can be metabolized (aromatized) to estrogens.
aromatization a process mediated by the enzyme cytochrome P-450 that converts the androgens testosterone and androstenedione to the estrogens 17β-estradiol and estrone, respectively. Some testosterone effects such as those on the bone and in sexual differentiation of the brain require aromatization of testosterone to estradiol.
arrhenotokous a process of reproduction in invertebrates, in which an unfertilized egg gives rise to a male haploid individual and a fertilized egg to a female diploid individual.
arrhenotoky the development of males from unfertilized eggs.
arthrogryposis a congenital defect of the limbs characterized by contractures, flexion, and extension.
artificial insemination (AI) the placement of spermatozoa into the female reproductive tract by any method other than sexual intercourse. In domestic animals, it is a process by which spermatozoa are collected, diluted with semen extenders for either short or long-term preservation and then manually inseminated into the reproductive tract of sexually receptive females.
asexual reproduction reproduction without the union of gametes having taken place.
aspartate an acidic amino acid with excitatory or neurotransmitter activity, released from presynaptic nerve terminals within the brain.
asphyxia a severely decreased level of oxygen in tissue.
assisted reproduction another term for ASSISTED REPRODUCTIVE TECHNOLOGY.
assisted reproductive technology (ART) a collective term for various procedures developed to enhance reproduction in infertile, subfertile, or geographically distant individuals.
associated reproductive pattern a pattern of reproduction in which the secretion of gonadal sex steroid hormones, the production of gametes, and the display of sexual behavior occur synchronously. Species exhibiting this pattern of reproduction tend to be found in predictable environments with prolonged conditions beneficial to breeding.
asthenozoospermia poor motility of sperm within the ejaculate.
asymptomatic infection the presence of a virus, bacteria, or protozoa in the mother and/or fetus, without causing symptoms or manifestations of clinical disease.
atelectasis the collapse of airspaces.
atresia the degenerative process that occurs in ovarian follicles that fail to ovulate or fail to reach maturity. The basement membrane surrounding the oocyte breaks down, granulosa and macrophages invade, and the yolk is eventually resorbed. Atresia is mediated by apoptosis (programmed cell death) and is initiated within the granulosa cell layer. Also, FOLLICULAR ATRESIA.
atretic follicle see CORPUS ATRETICUM.
atrial gland an elaboration of the secretory epithelium that secretes several peptides related to the egg-laying hormone.
attrition a term for the depletion of oogonia and primary oocytes during fetal development.
autocrine relating to or acting upon itself; e.g., an autocrine product affects the same cell that produces it.
autocrine secretion the release of a chemical messenger by a cell, into the extracellular space, that has as its target the secretory cell itself.
autoecious of a parasite, remaining on one host throughout the year; not having an alternation between primary and secondary hosts.
autogamous capable of or involving autogamy.
autogamy sexual reproduction by fusion of gametes from a single parent organism, resulting in self-fertilization.
automixis asexual reproduction with chromosomal reduction but without fusion of gametes.
autosomal dominant inheritance a disorder expressed when only one abnormal autosomal allele is present and the corresponding gene on the homologous chromosome is normal.
autosomal recessive inheritance a disorder apparent only when both alleles at a particular autosomal genetic locus are mutant.
autosome any chromosome other than a sex chromosome.
axon the major process of a neuron along which action potentials as well as secretory materials are propagated.
axonal transport the mechanism by which proteins made in the nerve cell body can be delivered to the axon and nerve terminals.
axoneme the central portion of the sperm flagellum, containing a microtubular array with associated proteins; directly responsible for sperm motility.
azoospermia a lack of spermatozoa in the semen; the failure to form spermatozoa.

B

baby blues an informal term for a self-limited episode of unexplained sadness experienced by as many as 60% of women in the post-partum period, usually beginning three days to a week after delivery.
balanitis inflammation of the glans penis.
barrier method any technique of contraception that is based on providing a physical or chemical barrier to the passage of sperm and thus preventing fertilization.

basal compartment a region at the base of the seminiferous epithelium that contains spermatogonia and young spermatocytes, all of which have ready access to blood-borne substances.

basal (basement) membrane a noncellular structure that isolates and contains the follicle.

B cells lymphocytes that produce antibodies following encounter with foreign substances (antigens).

Bcl-2 a family of proteins in mammals that is homologous to the Ced family of proteins in *C. elegans*. The bcl-2 family may be apoptotic and anti-apoptotic.

benign not malignant; specifically, describing a tumor that is confined to its original site, and that does not invade surrounding tissue or spread from this site to other distant sites.

benign prostatic hyperplasia (BPH) involutional changes occurring in specific regions of the prostate, thought to occur early in adult life and intensifying in late middle age, resulting in a marked increase in cell number; when seen in conjunction with certain symptoms, noted as clinical BPH.

benthic relating to or found in the benthos, the region at the bottom of an ocean or at the soil-water interface of an ocean.

bihormonal containing two hormones; specifically, in the case of gonadotropes, a cell that contains both LH and FSH.

binary fission asexual reproduction yielding two daughter cells by a nearly equal division of the parent cell.

binding site the specific area of an enzyme molecule to which other molecules can attach in a chemical reaction.

biological clock a term implying an underlying physiological mechanism that times a measurable rhythm or other type of biological timekeeping, as in animal reproductive cycles.

biome a region of the earth having a distinct form of vegetation; e.g., a tropical forest.

biophysical profile (BPP) an ultrasonographic evaluation of fetal movements, amniotic fluid volume, and heart rate activity, based on a combined score for various factors.

biopsy the removal of a sample of tissue which is microscopically examined; e.g., for cancer cells.

biparietal diameter a measurement of the fetal head taken between the parietal bones of the skull.

bisexuality the fact or condition of experiencing sexual desire and/or romantic attraction for both opposite-sex and same-sex individuals.

bladder neck the smooth muscle at the base of the bladder. When contracted, it blocks the flow of urine into the urethra or prevents the flow of semen into the bladder.

blastocoel or **blastocoelic cavity** the fluid-filled cavity that occupies the center of the blastula and that is a result of fluid transport by the trophectoderm cells.

blastocoelic relating to or involving the blastocoel.

blastocyst the stage of embryonic development following cleavage of a fertilized ovum and preceding implantation. It consists of an outer layer of trophectoderm, which is destined to become the placenta, a fluid-filled cavity or blastocoel, and an inner cell mass which will give rise to the fetus.

blastocyst hatching the process in which the blastocyst emerges from the zona pellucida prior to implantation.

blastomere a cell in the early preimplantation embryo.

blastula stage the stage of an embryo folowing cleavage and preceding gastrulation.

blood-testis barrier see SERTOLI CELL BARRIER.

body plumage all the feathers on a bird except for the flight feathers of their wing (primaries and secondaries, collectively called remiges) and tail (rectrices).

BPH see BENIGN PROSTATIC HYPERPLASIA.

Braxton-Hicks contractions low amplitude uterine contractions in human pregnancy that occur throughout pregnancy and give way to labor contractions at delivery.

breast duct a mainframe tubular structure of the glandular component of the breast tissue.

breech describing a birth process in which the feet or buttocks emerge as the presenting fetal part rather than the head.

broodiness or **broody behavior** in birds, female maternal behavior, including incubating eggs and caring for young.

brood parasitism a technique utilized by some species of birds and insects in which the parasite lays its eggs among the clutch of the host species, which then cares for the young of the parasite as if it were its own.

brood patch or pouch in birds, a thickened, edematous, and highly vascularized patch of featherless skin that facilitates the transfer of heat from the incubating or brooding parent to the eggs or nestlings.

Bruce effect the termination of pregnancy prior to implantation when a female is exposed to the odors of an unfamiliar, reproductively active male. (First identified by Hilda *Bruce*.) Also, PREGNANCY BLOCKAGE.

Buck's fascia the fascial support structure of the deep structures of the penis.

budding an asexual form of reproduction that generates genetically identical individuals.

bursa copulatrix in insects and certain other invertebrates, the portion of the female tract that receives the male genitalia, analogous to the vagina of mammals. Also, GENITAL CHAMBER.

bursting activity a brief episode of high frequency firing characteristically displayed by oxytocin neurons prior to milk ejection.

C

Caenorhabditis elegans a nematode that has been extensively studied, leading to the identification of a family of genes that direct apoptosis (programmed cell death) and that have mammalian homologues.

calcification an abnormal formation or deposit of calcium deposits in body tissue; e.g., the breast.

Call-Exner bodies the fluid-filled spaces between folliclular granulosa cells that coalesce to form the antrum.

calling song a type of song broadcast by a lone male songbird, unevoked by receiver-based stimuli, to attract and guide the approach of a female at long range.

cannibalism the eating of members of the same species; e.g., an adult bird killing and eating eggs or young birds that are not yet fully grown.

capacitation a series of chemical processes usually occurring within the female reproductive tract, and a precursor to the acrosome reaction, resulting in an influx of calcium ions into the sperm and increased membrane fluidity.

cardiotocography a simultaneous graphic representation of fetal heart rate and uterine contraction patterns obtained from monitors applied to the maternal abdomen.

cardiovascular pertaining to or involving the heart and blood vessels.

caruncle a convex connective tissue (stromal) structure of the endometrium that is covered by a layer of luminal epithelium; characteristic of the uterus of ruminants and developing into the maternal half of the placentome.

CASA computer-assisted semen analysis, the automated analysis of sperm swimming patterns, used to help identify treatment effects on sperm. Poor CASA values may correlate with reduced fertility.

casein a white, amorphous, colloidal aggregate, soluble in acids, that is found in milk and that is composed of several proteins together with calcium and phosphorus.

CASH cortical androgen stimulating hormone, a human pituitary factor proposed as the primary regulator of adrenal androgens due to its ability to stimulate adrenal androgen secretion *in vitro*.

caste any set of individuals that are specialized in behavior, as in social insects.

castration the surgical removal of the testes.

catabolism net breakdown and loss of tissue.

caudal isthmus the caudal region of the thick-walled, narrow isthmus of the oviduct, which is located between the oviductal ampulla and the uterus. In many species, the caudal isthmus is a site of sperm storage prior to ovulation.

cavitation the developmental process that forms spaces among cells of a tissue and by which these developing spaces coalesce to form a cavity.

cell lineage the map or description of the lineal descent of the cells in the adult from the single cell of the zygote. The somatic cell lineage of *C. elegans* is largely invariant from individual to individual.

central zone (CZ) the region of the prostate thought to be derived from the mesonephric duct, accounting for 20% of prostate volume.

centripetal circulation the adrenal's unique blood supply, characterized by a continuous intraglandular capillary and venous network extending from the cortex to the centrally located medulla and ultimately coalescing into a large central vein.

centrosome a cell's microtubule organizing center that organizes the poles for mitotic and meiotic spindles, establishes the axis for cell divisions, and serves as the cell's internal compass endowing it with directionality for locomotion. During oogenesis, the maternal centrosome is destroyed and the sperm contributes the precursor centrosome (sperm centrosome). After the recruitment of maternal gene products and post-translational modifications, the sperm centrosome is transformed into the zygote centrosome, capable of directing microtubule assembly and reproducing at each cell cycle.

cephalopelvic disproportion a disparity during labor between the dimensions of the fetal head and those of the maternal pelvis.

cervical relating to or affecting the cervix; e.g., cervical cancer.

cervical conization the removal of a cone-shaped wedge of cervix, involving both ectocervix and endocervix.

cervical effacement the process in which the uterine cervix changes from a 3–4 cm fibrous cylinder to a lumen in continuity with the lower pole of the uterus.

cervical softening or **ripening** a process that transforms the cervix from an inextensible to a highly compliant tissue in order to allow dilation for the passage of the term fetus.

cervicitis inflammation of the cervix.

cervix a necklike part that forms the lower and narrower end of the uterus, separating the vagina from the uterine cavity. It is the site of numerous glands secreting cervical mucus.

CG see HUMAN CHORIONIC GONADOTROPIN.

chemical castration the induction of gonadal atrophy by prolonged treatment with female sex hormones (estradiol), anti-androgen or GnRH (LHRH) analogs.

chemical messengers substances produced by endocrine or non-endocrine cells that bind to a receptor on or in a target cell to elicit control or function of the cell.

chemokine a cytokine with chemotactic properties; e.g. activity in the regulation of target cell motility.

chemosignal a chemical released externally by one individual that can induce physiological or behavioral change in another individual; pheromone.

chemotactic relating to or able to produce chemotaxis.

chemotaxis the directional migration of cells, especially in response to concentration gradients of a chemical.

chemotherapy a treatment for cancer that involves the use of certain chemicals to kill rapidly proliferating cells.

childbed fever an older term for PUERPERAL FEVER.

chimera or **chimaera** an animal composed of two or more genetically distinct types of cells. These individuals are produced by mixing cells of two embryos or more commonly by mixing embryonic stem cells with host embryo cells.

Chlamydia trachomatis a microorganism that lives as an intracellular parasite in humans and other species and that is identified as the causative organism in many cases of STDs (sexually transmitted diseases).

cholesterol a primary lipid found in blood and cells that is the precursor for the biosynthesis of all steroid hormones.

chordate one of the Chordata; an animal having at some stage of its development a dorsal nerve cord, a notochord, and pharyngeal gill slits; e.g., mammals, birds, reptiles, amphibians, fish, and certain marine invertebrates.

chordee an abmormal bending of the penis during erection. Significant bends may affect the ability to have normal intercourse.

chorioallantois the placental tissue layer formed following the fusion of two fetal membranes, the nonvascular chorion and the vascular allantois.

chorion in placental mammals, the outermost fetal membrane, composed of trophoblast and extraembryonic mesoderm. In insects, the shell surrounding the insect egg.

chorionic relating to or involving the chorion.

chorion frondosum the part of the chorion provided with different types of projections that create a substantial surface area increase.

chorionic gonadotropin see HUMAN CHORIONIC GONADOTROPIN.

chorionic villus *plural,* **villi** the final ramification of the fetal circulation within the placenta. The primate placenta is organized as a tree-like tubular structure. The branches, or villi, become progressively smaller with each branching, ending in terminal villi where gas and nutrient exchange occurs between maternal and fetal blood.

chorionic villus sampling (CVS) a procedure performed in the late first trimester (10–12 weeks) to obtain fragments of chorionic villi from the placenta for prenatal testing.

chorion laeve the smooth part of chorion.

choriothete in some insects, a muscular and glandular invagination of the anterioventral wall of the uterus, forming a tongue-shaped structure that anchors the egg and assists hatching.

choriovitelline placenta the placental type found in most marsupial taxa, formed from two embryonic membranes, the chorion and the vitelline membrane, or yolk sac.

chromatin a complex of DNA and basic proteins containing subunits called nucleosomes.

chromatoid body a small chromophilic mass present in the cytoplasm of spermatocytes and early spermatids. It is composed of electron dense material and small vesicles. In round spermatids it migrates toward the centrioles and contributes material to the developing annulus. It disappears from the cytoplasm in elongated spermatids.

chromosome one of the discrete rod-like bodies (normally numbering 46 in humans) found in the cell nucleus that is the bearer of genetic information.

chronobiotic capable of inducing change in circadian rhythms or other biological rhythms.

circadian clock an internal timing device that regulates the expression of circadian rhythms. In mammals, a master circadian clock is located in the hypothalamic suprachiasmatic nucleus (SCN).

circadian oscillator a self-sustaining biological clock that acts to drive circadian rhythms. It continues to function in environments devoid of external time cues, but can be entrained by external cues known as ZEITGEBERS.

circadian rhythm(s) a cycle of biochemical, physiological or behavioral events that reoccur with a period of about 24 hours (one day) when under constant conditions.

circalunar rhythm(s) a series of repeated responses that occur over a period of time of about 34 days (approximating one lunar cycle).

circannual rhythm(s) a cycle of biochemical, physiological or behavioral events that reoccur with a period of about one year in the absence of environmental cues.

circumcision the removal of the foreskin, or prepuce, that covers the tip, or glans, of the penis.

circumflex vein the veins that travel circumferentially around the corpora cavernosa, joining the emissary veins to the deep dorsal vein of the penis.

CL see CORPUS LUTEUM.

clade see MONOPHYLETIC GROUP.

cleavage a specialized series of mitotic cell divisions that occur following fertilization.

clinical pelvimetry measurement of the diameters of the pelvis obtained by pelvic examination.

clitellum the glandular girdle involved in cocoon formation in oligochaetes and leeches.

cliteromegaly abnormal enlargement of the clitoris.

clitoris a small, sensitive erectile organ in the female corresponding to the male penis and located at the ventral end of the vulva.

cloaca in many vertebrates, the common body cavity into which the genital, urinary, and intestinal canals discharge; in certain invertebrates, the respiratory, excretory, and reproductive duct chamber.

cloacate having a cloaca; involving or transported by the cloaca.

clonal animal a clone; i.e., an animal in which the genetic entity is composed of multiple "bodies" that are physiologically separated.

clone a group of animals that are not physically connected and are able to function independently, but that are genetically identical units derived asexually from a single progenitor. More generally and in popular use, a genetic copy of another organism; a genetically identical individual.

cloning a reproductive technique that involves the insertion of nuclei from an embryo, embryo cell line, fetal cell line, or adult somatic cell line into the cytoplasm of a mature egg that has had its own chromatin removed. These embryos may develop to term in some animal species.

clutch a number of eggs accumulated in a nest prior to incubation, which are incubated together, and then usually hatch within a few days of each other.

coactivators adaptor proteins that act as a bridge between nuclear receptors and general transcription factors to enhance transcription.

COCs combined oral contraceptives; contraceptive pills that combine an artificial progestin and an artificial estrogen.

coeloblastula a simple blastula-like larva.

coelom in some animals, a fluid-containing body cavity that is completely lined by a mesodermally derived epithelium.

coitus sexual union between a female and male involving insertion of the penis into the vagina.

colonial relating to or involving a colony or coloniality.

colonial animal an animal that is part of a colony; one of many physiologically connected individuals.

coloniality the behavior of birds (or other animals) whereby many individuals breed in close proximity, interact with each other, and usually join in group defense against predators.

colonization the persistence of microbes at a body site without the development of local tissue invasion or symptomatic disease.

colony a multiple group of organisms that are physically connected to one another; (e.g. most hydroids and reef-forming corals).

colostrum the fluid in the breast at the end of the pregnancy and during the first few days following the birth; the first milk. It is viscous yellow, high in protein, low in fat and has high concentrations of protective immunoglobulins.

colposcopy the process of evaluating the cervix with magnification to identify abnormal cells.

colpotomy an incision made at the top of the vagina between the cervix of the uterus and the rectosigmoid colon.

commensal in close association with another; relating to or living in a condition of commensalism.

commensalism an ongoing association of two organisms of different species, in which one species partner is benefited by the association and the other is not affected by it.

common oviduct in insects, an ectodermal invagination that connects the genital opening with the lateral oviducts.

common penile artery the continuation of the deep internal pudendal artery after the departure of the perineal artery and the labial scrotal artery.

compaction a process that occurs during embryogenesis in placental mammals, in which cells flatten upon one another and display increased cell-to-cell adhesion.

competent larva a term for a larva that is physiologically and morphologically capable of undergoing metamorphosis.

competition the situation that exists when two or more individuals use the same environmental resource that is in short supply, such as breeding sites, mates, or food.

complex life cycle a life cycle with at least two distinct post-embryonic morphologies, each of which is adapted for a fundamentally different way of life or niche (e.g., aquatic larvae and terrestrial adults).

conception the fertilization of an oocyte by a spermatozoon, resulting in pregnancy.

conceptus the products of conception; the sum of derivatives from a fertilized ovum at any stage of development from fertilization until birth, including extraembryonic membranes as well as the embryo or fetus.

confined placental mosaicism (CPM) a dichotomy between the chromosomal constitution of the placental tissues, both cytotrophoblast and chorionic connective tissue, and the embryonic/fetal tissues.

congenital present at, and usually before, birth.

congenital adrenal hyperplasia a genetic disorder associated with an enzymatic defect such that, beginning before birth, the adrenal glands of affected individuals produce excessively high levels of androgens.

congenital anomaly a structural abnormality that may be hereditary, or that may be due to some influence occurring during gestation or in the process of birth.

consort relationship in some species, e.g., nonhuman primates, a temporary heterosexual bond characterized by male mate guarding of a sexually receptive female, which is usually found in multimale mating systems.

conspecific an animal that is a member of the same species as another animal.

constant reproductive pattern a pattern of reproduction in which the secretion of gonadal sex steroid hormones, the production of gametes, and the display of sexual behavior do not occur synchronously. Species exhibiting this pattern of reproduction tend to be found in unpredictable environments with brief periods suitable for breeding.

contraception the fact or process of preventing pregnancy, whether by the use of certain physical agents or devices, or by certain procedures.

contraction a painful tightening of the muscular outer wall of the uterus that occurs at regular intervals during labor; as childbirth approaches, these contractions generally increase in duration and intensity, at decreasing intervals. Contractions serve to decrease the size of the uterus and move the fetus through the birth canal.

contraction stress test (CST) a type of test in which an external monitor records fetal heart rate responses to spontaneous or induced contractions.

contractures long-lasting, low-amplitude epochs of myometrial activity that occur throughout pregnancy.

cooperative breeding a social system displayed by some species in which members of a stable social group assist in rearing young that are not their own offspring.

copulatory bursa in female insects, a pouch or an expansion of the median oviduct into which sperm are deposited by the male insect during copulation.

cordocentesis a technique for obtaining a sample of fetal blood from the umbilical vein for fetal diagnosis or therapy.

cord prolapse the passage of the umbilical cord through the cervix prior to delivery of the infant.

corepressors adaptor proteins that act as a bridge between nuclear receptors and general transcription factors to inhibit transcription.

corona radiata the innermost layer of granulosa cells surrounding an oocyte, in contact with the zona pellucida.

corpora the plural form of *corpus*. See CORPUS entries below.

corpus albicans the white fibrous scar in an ovary, produced by involution of the corpus luteum.

corpus allatum *plural,* **corpora allata** an endocrine gland, usually paired but sometimes fused, located behind the insect brain and linked to it by nerve tracts; it is the primary site of production of juvenile hormone.

corpus atreticum *plural,* **corpora atretica** a degenerating structure developing from ovarian follicles that do not undergo ovulation. Also, ATRETIC FOLLICLE.

corpus bursa a component of the female reproductive tract where the spermatophore is deposited.

corpus cardiacum *plural,* **corpora cardiaca** one of a pair of glands that are the source of some peptide hormones in insects, consisting of the secretory terminals of neurosecretory cells in the brain together with intrinsic neurosecretory cells.

corpus cavernosum *plural,* **corpora cavernosa** one of the specialized paired vascular tissues that are the primary modulators of erectile capacity of the penis.

corpus luteum (CL) *plural,* **corpora lutea** a yellow endocrine structure formed on the ovary following ovulation, in the site of a ruptured follicle; a major source of the steroid hormone progesterone, which is primarily responsible for the maintenance of pregnancy. (From Latin; literally, yellow body.)

corpus spongiosum a spongy body that contains the urethral bridge from the prostate to the tip of the glans penis.

cortical granules organelles located near the surface of the egg. The contents of the cortical granules are released during fertilization and, in what is known as the **cortical reaction,** modify the extracellular matrix of the egg (zona pellucida, vitelline layer, fertilization coat) to make it impenetrable by other sperm, thus preventing polyspermy.

corticotroph an anterior pituitary cell type that produces ACTH (adrenocorticotropic hormone).

corticotropin see ACTH.

corticotropin-releasing hormone (CRH) a hypothalamic neuropeptide that stimulates the release of ACTH (adrenocorticotropic hormone) from the pituitary gland.

cortisol the major glucocorticoid produced by the adrenal cortex in humans.

costs of mating or reproduction the negative consequences of mating in terms of the energy and time expenditure involved, and the increased chance of being located by predators and parasites.

cotyledon in domestic ruminants, fetal placental structures which form on areas of the chorioallantois overlying maternal caruncles and which ultimately develop into the maternal half of a placentome during pregnancy.

countercurrent heat exchange a process of temperature regulation that occurs within the pampiniform plexus. Cooler venous blood acts as a heat-sink for the warmer arterial blood arriving from the abdomen, thus lowering the temperature of the arterial blood as it reaches the testis.

countercurrent transfer a close anatomical apposition of arteries and veins which allows substances secreted into the venous return of an organ to pass directly into the adjacent arterial supply, and thus feed back to the same or adjacent organ at a higher concentration than would reach it via the systemic circulation.

courtship behavior behaviors exchanged between a male and a female of the same species prior to mating.

courtship song in songbirds, a type of song given at close range after pairing, influencing mate choice.

cradle position a common position for breastfeeding in which the baby is cradled in the mother's arms and brought to the level of her breast for feeding.

crop milk a nutritive substance consisting of epithelial cells sloughed from the crop sac wall of incubating and brooding pigeons and doves, regurgitated to the nestlings at hatching.

crop sac an outpocketing of the esophagus that is used as a seed storage organ in many birds and as a source of crop milk in pigeons and doves.

cross-fostering the exchange of offspring between different parents, usually during early development and between different species.

cryopreservation preservation by freezing; a controlled procedure for the long-term storage of cells or groups of cells (e.g., sperm, embryo) at extremely low temperatures.

cryoprotectant a chemical (e.g., glycerol) used to protect cells during cryopreservation.

cryptic choice a mechanism of mate choice involving between-sex interactions that are difficult to observe by investigators, such as post-mating mechanisms that allow a female to accept or reject the use of sperm from one male or another.

cryptobiosis the capacity of an organism to enter in a latent life state suspending animation during periods of environmental adversity (e.g., drying, freezing, lack of oxygen).

cryptomenorrhea see HYPOMENORRHEA.

cryptorchidism an abnormal condition in males in which the testes are not fully descended from the abdominal cavity into their normal position in the scrotum.

cubitus valgus a term used to describe the exaggerated angle found at the elbow of women with Turner's phenotype when the arms are relaxed at the side and the palms face forward.

cumulus cells or **investment** see CUMULUS OOPHORUS.

cumulus/oocyte complex see CUMULUS OOPHORUS.

cumulus oophorus the several layers of granulosa cells surrounding an oocyte and supporting it in the antrum of a Graafian follicle.

CVS see CHORIONIC VILLUS SAMPLING.

cyanosis bluish discoloration of the skin resulting from an inadequate amount of oxygen in the blood.

cyclical parthenogenesis an alternation between parthenogenesis and sexual reproduction.

cyphopods sclerotized receptacular structures at the external end of the oviducts of millipeds; they grasp the gonopods during mating and contain seminal receptacles for temporary sperm storage; anatomically they may be derived from basal elements of an otherwise disappeared pair of legs.

cystacanth the end product of acanthocephalan development in an intermediate host, adapted for transmission to a definitive host.

cystid in colonial animals, living and nonliving parts of the body wall of an individual colony member.

cystitis an inflammation or infection of the urinary bladder.

cystotomy an incision into the wall of the bladder.

cytochrome P450 a superfamily of heme-containing enzymes that catalyze the oxidation of a large variety of organic substrates. Three of the enzymes required for estrogen biosynthesis (cholesterol side chain cleavage enzyme, 17α-hydroxylase and aromatase) are members of this superfamily.

cytodifferentiation the process through which a cell or tissue acquires its characteristic biochemical phenotype.

cytogamy a self-mating process in paired cells that yields whole genome homozygotes.

cytogenetics the study of chromosomal structure.

cytokine one of a group of mitogenic compounds synthesized by cells of the immune system as well as in certain other tissues, among them endometrium and placenta. Cytokines have numerous actions and serve as signals between different cells, for such functions as immunomodulation, growth promotion, and inflammatory response.

cytokine receptor a membrane-bound protein or glycoprotein that binds a specific cytokine and triggers intracellular signal transduction.

cytokinesis division of the cytoplasm during cell division.

cytometer a device for counting and measuring cells.

cytoplasm the substance of a cell, exclusive of the nucleus, containing various organelles.

cytoplasmic maturation the process that accompanies oocyte nuclear maturation to prepare the cytoplasm for fertilization and embryo development.

cytoskeleton the cytoplasmic architectural system of fibers responsible for shape and motion. It is composed of three major networks: microfilaments (7 nm diameter), intermediate filaments (10 nm), and microtubules (25 nm), along with a myriad of accessory proteins including motor proteins, stabilizing proteins, and cargo docking proteins.

cytostatic factor (CSF) activity responsible for the arrest at the second meiotic metaphase in amphibian and mammalian oocytes.

cytotoxic harmful to cells.

cytotrophoblast cells derived from the outer wall of the blastocyst which act as a stem cell population that undergoes rapid division and fusion to form the syncytiotrophoblast layer of the placenta.

D

dartos fascia the fascial support structure of the penis, intimately associated with the skin and vascularity to the skin.

decidua the transformed endometrium of pregnancy; a mass of cells produced by the uterine stroma and also containing many blood cells derived from the bone marrow. The decidua is formed after implantation and surrounds and supports the early growth and development of the embryo. In mammals where the trophoblast invades the uterus, the decidua restricts trophoblast growth.

decidual relating to the decidua or decidualization.

decidual cell a lipid- and glycogen-rich secretory cell, formed within the endometrium from stromal cells in response to implantation of the conceptus and the presence of progesterone.

decidualization the process of transformation of the stromal cells of the endometrium into the morphologically and functionally distinct decidual tissue, involving the differentiation of endometrial cells, and infiltration by large numbers of lymphoid cells. It can be induced by either the implanting embryo in pregnancy or by artificial stimuli to the uterus in pseudopregnant animals.

decidual prolactin a protein hormone synthesized and released by human endometrial stromal cells during the luteal phase of the menstrual cycle and during pregnancy; identical in structure to pituitary prolactin.

deciduoma *plural,* **deciduomata** an intrauterine mass containing decidual cells; specifically, the decidual tissue artificially induced in pseudopregnant animals in the absence of embryo. This is comparable to the decidua of pregnant animals in its formation, regression, and secretory capacity.

deep dorsal vein the major component of the intermediate venous drainage of the penis, beginning subcoronally as the retrocoronal plexus extending proximally to join the preprostatic plexus.

deferred sexual maturation in birds, the delaying of the first breeding attempt for one or more years, and up to 12 years, after hatching.

definitive host the host in which a parasite having an indirect life-history pattern attains maturity and participates in sexual reproduction.

definitive zone a term for the narrow cortical zone that surrounds the fetal zone of the fetal adrenal gland.

degrowth in some invertebrates, the ability to shrink or reduce in size when nutrients are scarce.

delayed implantation a period of time when the fertilized egg lies in a resting state in the uterine cavity without attachment. By delaying implantation, certain mammals regulate the duration of pregnancy, thus ensuring the young are born at the most optimal time and conditions for their survival.

delayed pseudopregnancy a phenomenon in which the initiation of pseudopregnancy by cervical stimulation is delayed until the following ovulation when luteinization can occur.

depolarization a reversal of the electric potential present across the cell membrane.

desmosome-gap junctions sites of adhesion and likely intercellular communication between Sertoli cells and germ cells.

desquamation the shedding of the most superficial layers of cells from the lining epithelium of the vagina. These cells can be collected and examined for signs of pathology of the vagina. Also, EXFOLIATION.

deuterostome one of the Deuterostomia, those higher animals in which the mouth forms at the opposite end of the body from the blastopore, which is the first opening to the embryonic digestive tract.
development all of the processes involved in the differentiation of specific tissues plus growth in body mass.
developmental migration the movement of animals from one foraging area to another as animal size and feeding preferences change. This is often a one-way migration.
developmental toxicity a general term for adverse effects on development, manifested as death, malformation, growth retardation and functional deficit.
DHEAS dehydroepiandrosterone sulfate, the primary circulating adrenal androgen.
diabetes insipidus antidiuretic hormone insufficiency resulting in polyuria and polydipsia. This may be a result of hypopituitarism or secondary to a kidney defect and can be mimicked by excessive fluid intake related to psychiatric causes.
diabetogenic tending to cause diabetes mellitus, including insulin resistance and glucose intolerance.
diabetogenic hormones chemical messengers that antagonize the actions of insulin on peripheral tissues and indirectly promote an increase in pancreatic insulin secretion.
diagonal conjugate the distance between the promontory of the sacrum and the lower margin of the pubic symphysis, which can be measured clinically.
diandry the existence of two types of males which originate in two different ways.
diapause a dormant period of arrested growth and development, with greatly slowed metabolism; specifically, such a stage in insect development or reproduction that is used to circumvent inimical seasons. See also EMBRYONIC DIAPAUSE.
diaphragm a form of barrier contraceptive consisting of a flexible rubber dome-shaped device with a rubber-covered metal or plastic rim.
diestrus in veterinary species, the period of ovarian cycles dominated by corpus luteum secretion of progesterone.
differentiation the fact of becoming different; the process of acquiring a character or function different from that of the original type; e.g., the specialization of a cell for a specific function.
dihydrotestosterone the 5α-reduced metabolite of testosterone that is synthesized in many androgen target tissues and is the naturally produced steroid that binds most avidly to the androgen receptor.
dilatation and curettage (D and C) a technique involving the mechanical stretching of the cervical os, allowing access to the uterine cavity for scraping of the wall to remove the contents or lining of the uterus.
dimer a molecule that is produced by binding together two molecules of the same chemical structure, either two identical molecules (homodimers) or two different molecules (heterodimers).
dimerization the coupling of two molecules (monomers) to form dimers.
dimorphic or **dimorphous** having two different forms; exhibiting dimorphism.
dimorphism the fact of occurring in two distinct forms; e.g., the physical or biochemical differences between males and females, or between young and mature, in a given species.
dioecious describing species in which there are individuals of two separate sexes, male and female.
diploblastic consisting of two germ layers, an ectoderm and an endoderm, which give rise to two adult epithelia, the epidermis and gastrodermis, respectively.
diploid of a cell or organism, having two copies of each chromosome; physiologically one comes from the mother and the other from the father.
diplosegments the body unit in Diplopoda, composed of two adjoining embryonic somites fused with the dislocation of legs and ganglia into the posterior subunit of each.
diplotene a stage in the first meiotic prophase in which DNA is duplicated, after which the oocyte enters a resting phase
direct development a type of life cycle that includes neither a distinct larval form nor a dramatic metamorphosis. In direct development, the embryo develops directly into a juvenile by a series of gradual changes.
discoidal placenta a type of chorioallantoic placenta in which the surface that is amplified (usually in the form of villi or as a labyrinth) to promote physiological exchange between the maternal and fetal circulations occupies a disc-shaped area.
disposition a term describing the orientation of the fetal axis and extremities, in addition to presentation and position.
dissociated reproduction a pattern of reproduction in which the secretion of gonadal sex steroid hormones, the production of gametes, and the display of sexual behavior do not occur synchronously. Species exhibiting this pattern of reproduction tend to be found in predictable environments with restricted breeding opportunities.
dissogony the ability to exhibit two periods of reproductive activity (larval and adult) separated by a period of gonadal regression.
dizygotic twins "fraternal" twins; i.e, twins that develop from two separate eggs.
DNA deoxyribonucleic acid, the substance that constitutes the genetic material of cellular organisms.
DNA-DNA hybridization a molecular technique for assessing the genetic divergence between two complex genomes belonging to two species.
DNA methylation a process which has been implicated in the control of genomic imprinting.
DNA repair process in which erroneous sequences of DNA are removed and substituted by correct ones; it involves the cutting and reannealing of DNA strands.
domain a region of a protein with a defined function.
dominant follicle of the cohort of follicles that begin final maturation, a follicle that is destined to ovulate.
donor a female from whom embryos are recovered, or a male provider of spermatozoa used for insemination.

Down (or **Down's**) **syndrome** a chromosomal abnormality producing severe mental retardation, caused by the presence of an extra copy of the number 21 chromosome in each cell of the affected individual. (First described by J. L. H. *Down,* English physician.)

ductal carcinomal in situ (DCIS) cancer cells that develop from the lining of the milk duct but are confined to the ducts of the breast. DCIS is considered to be a precursor to invasive cancer.

ductus arteriosus a vascular connection in the fetus joining the left pulmonary artery directly to the descending aorta.

ductus deferens another name for VAS DEFERENS, part of the male reproductive tract.

ductus venosus a vascular connection in the fetus passing throught the liver and joining the umbilical vein with the inferior vena cava.

dysgenesis the abnormal formation of an organ.

dyspareunia in women, pain with intercourse.

dystocia difficult labor and delivery, generally caused by an abnormality of the mother or fetus, or by fetal-pelvic disproportion.

E

ecdysiostatin a hormone that inhibits the production of ecdysteroids by endocrine glands.

ecdysiotropin a hormone that stimulates the production of ecdysteroids by endocrine glands.

ecdysone a steroid hormone produced by an endocrine gland in insects and other arthropods such as Crustacea as a precursor to the active form, 20-hydroxyecdysone, which controls molting.

ecdysteroid a generic term for a family of steroid hormones related to ecdysone, that induce molting and that in some insects are involved in reproduction.

eCG see EQUINE CHORIONIC GONADOTROPIN.

ECM see EXTRACELLULAR MATRIX.

ectadenia accessory glands of male insects derived from ectoderm.

ectoderm the embryonic tissue type from which outer skin (epidermis) derives and most nervous tissues differentiate; in mammals, the dorsal cells of the embryonic disk from which most fetal tissues differentiate following gastrulation.

ectolecithal a condition of eggs of platyhelminths in which the yolk cells are separate from the oocyte but packaged with it in the egg.

ectopic pregnancy a pregnancy established at any site outside the uterine cavity, most commonly within the Fallopian tube (oviduct).

ectopic testis testis that has exited the inguinal canal but is not in the scrotum.

ectosymbiont a symbiont living in close association with another organism, but outside the body of that organism.

ectotherm an animal that has a variable body temperature, usually close to the ambient temperature (e.g., all invertebrates and vertebrates except for birds and mammals).

ectothermic having body temperature that varies according to the surrounding environmental temperature.

edema swelling of tissues related to increased water content in cells and interstitial spaces.

efferent moving away from a region or site.

efferent fibers nerve fibers that convey information by means of action potentials (or axonal transport) from the central nervous system to the periphery.

egg a female gamete involved in sexual reproduction; ovum. In certain animals, e.g., birds, an external shelled body expelled from the female's body, consisting of this reproductive element and protective and nutritive materials..

egg cell the female germ cell arrested in metaphase of the second meiotic division.

egg-laying hormone (ELH) in certain animals, e.g., gastropods, peptides of approximately 4,500 daltons in size, found in clusters of neurons in the nervous system. Upon secretion into the circulation, these peptides activate egg laying and related egg-release behavior.

egg predator a symbiont that lives within and feeds upon the egg mass of its host; many nemertean worms have adapted this manner of existence in crustacean hosts.

egg resorption in parasitism, a process in which eggs are recycled and are broken down during periods when parasitoids are unable to oviposit and fail to find hosts to parasitize.

eicosanoids a generic term for a series of 20 carbon unsaturated fatty acids. An example is eicosatetraenoic acid which is the chemical name for arachidonic acid, the precursor for prostaglandins, thromboxanes, leukotrienes and epoxides containing two double bonds in the aliphatic side chains.

ejaculation the physiologic process by which the components of the semen are mixed and emitted from the penis.

elasmobranch a cartilaginous fish; a member of the class of fish which includes the sharks, skates, and rays, the most ancient jawed vertebrates, originating about 400 million years before present.

electric organ discharge the electrical signal produced by modified neural and muscle structures in some fish. In some species, the discharge is sexually dimorphic and is used in gender recognition.

electrocyte one of the cells of the electric organ of electric fish that generate the electric organ discharge.

embryo a fertilized egg or ovum; for vertebrates the term usually describes the developing organism in the period from when the long axis of the body appears until all major organs are represented (which in humans is typically from the second week after fertilization until the seventh or eighth week.)

embryonic relating to, describing, or involving the embryo.

embryonic diapause a reproductive process in which development of the preimplantation embryo is greatly slowed, or arrested, until an environmental cue reactivates it.

embryonic disk the inner cell mass after its cells have differentiated to form two layers; one of dorsal embryonic ectoderm and another of ventral embryonic endoderm cells.

embryonic germ (EG) cells see EMBRYONIC STEM CELLS.

embryonic stem (ES) cells cell lines derived from inner cell mass or primordial germ cells of an embryo that are maintained in an undifferentiated, proliferative state in culture and that can differentiate into a variety of cell types when reintroduced into a host embryo. In laboratory mice, they are primarily used to derive lines of mice carrying mutations in a gene of interest.

embryo transfer a general term encompassing superovulation, recovery of ova from donors, and transfer of embryos to the reproductive tract of recipients.

embryo transport the passage of the embryo from the site of fertilization, which is the ampullary segment of the oviduct, to the site of implantation in the uterus.

embryotropic relating to or describing conditions supportive of or stimulatory to growth and development of the embryo.

emigration the movement of individuals away from a certain site; specifically, movement away from a breeding site once reproduction has concluded.

endocervix the membrane lining the canal of the cervix; the region of the opening of the cervix into the uterine cavity.

endocrine relating to or involving the secretion of a substance, usually a hormone, directly into the bloodstream via which it is transported to a distant target organ or tissue to initiate a cellular response, e.g., the secretion of luteinizing hormone (LH) from the anterior pituitary into the bloodstream via which it acts on the ovaries or testes.

endocrine action or **communication** the activation of a cell or tissue by a chemical messenger produced in cells of another tissue and then secreted into the blood.

endocrine gland a ductless gland whose secretion enters small blood vessels and is then conveyed by the blood stream to another site of action, where it influences body processes.

endocrine secretion the release of a chemical messenger by a cell into the bloodstream to act on a distant site.

endocrine system a system through which chemical substances called hormones regulate their secretion by feedback. Endocrine systems are equipped with mechanisms for monitoring the magnitude of the biological effects controlling their secretory rate.

endocrinology the discipline in which hormones (chemical messengers) are studied.

endocytosis a process by which extracellular materials are taken inside a cell by including the material within deeply invaginating regions of plasma membrane that subsequently bud off internally to form membrane-bound vesicles.

endoderm the inner of three layers of cells during early embryonic development; the embryonic tissue type from which digestive and respiratory epithelial tissues differentiate.

endoduplication the duplication of the entire chromosomal set without cell division prior to meiosis.

endogenous from within; internal; occurring or controlled from within the organism.

endogenous control the control of gametogenesis and other reproductive activities by physiological factors. Generally, these are hormonal, neurohormonal, or neural factors within individuals that are influenced by external, environmental factors.

endogenous factors the array of internal factors affecting reproduction, including regulatory mechanisms provided by the central nervous system via hypothalamic secretions and by hormones secreted by the pituitary and the gonads.

endogenous rhythms biological processes ultimately controlled from within the organism by some kind of physiological timekeeping mechanism (biological clock).

endogenous selection natural selection that acts against genotypes regardless of the environment.

endometritis a clinically apparent, postpartum infection that involves the decidua and myometrium, and can also involve the parametrium.

endometrium the inner lining of the uterus, consisting of luminal and glandular epithelium and the underlying stroma, which serves as the site for embryonic implantation and undergoes cyclic changes as directed by ovarian sex steroid hormone secretion. In the absence of pregnancy it is shed and then regenerated each menstrual cycle.

endonuclease an enzyme that cleaves DNA into approximately 185-200 base segments at internucleosomal sites.

endosalpinx the inner layer of mucous membrane lining the oviductal lumen, composed of epithelium, basal membrane and subepithelial connective tissue, blood and lymphatic vessels and nerve fibers.

endoscopy an operative procedure in which a fiberoptic camera is used to visualize any potential body cavity.

endosymbiont an organism that lives inside the body of another organism.

endosymbiosis a condition in which an organism of one species lives inside the body of another species; usually used of a relationship that is described as mutually beneficial.

endothelial cell one of the layer of epithelial cells that line the cavities of the heart and the linings of the blood and lymph vessels.

endothelium the inner cell surface of the heart, blood vessels, and lymph vessels.

endotherm an animal (e.g., mammals, birds) that has a high and relatively constant body temperature produced by internal heat production and relatively unaffected by changes in the ambient temperature.

endothermic having body temperature determined by heat energy generated internally.

energy balance the difference in total nutrients ingested as food input minus the amount of nutrients excreted as the result of various physiological functions (i.e., urination, digestion, milk synthesis, respiration).

engorgement a temporary swelling of the breasts due to vascular dilation and the production of early milk.

enhancer a region of genomic DNA, more distant (upstream or downstream) than the promoter sequence, that up-regulate expression of a gene.

enterocoely in animals, a method of coelom formation in which the epithelial lining of the coelom develops as an elaboration of the embryonic digestive tract; this involves folding of the epithelium comprising the embryonic gut.

entolecithal referring to a condition of eggs of platyhelminths in which the yolk is packaged within the oocyte.

entrainment a process by which an internal circadian clock is entrained, or synchronized, to the period of an external stimulus that usually has a period of about 24 hours. In nature, the light-dark cycle is the primary external environmental signal that entrains the circadian clock to the 24-hour period of the day that is due to the rotation of the earth on its axis.

environmental sex determination a mechanism of sex determination in which sex is determined by environmental factors such as temperature. This mechanism is present in some fish, some lizards, some turtles, and all crocodilians.

epiblast the outer layer of the blastoderm, which will later segregate into ectoderm and mesoderm.

epidemic spawning contagious spawning in response to detection of released gametes or larvae from conspecifics or interspecifics. Often chemical substances (pheromones) are involved in the process.

epidemiology the study of the distribution and determinants of health-related events in specified populations, and the application of this study to control health problems.

epidermis the outer epithelium of the body wall.

epididymal relating to or involving the epididymis.

epididymal fluid or **plasma** the fluid in which epididymal sperm are bathed and obtained by centrifugation of luminal contents to remove spermatozoa.

epididymal sperm mature or immature sperm contained within the epididymal canal.

epididymis a long, coiled, fragile, microscopic tubule attached to the testicle that carries sperm from the testis into the vas deferens; responsible for the sustenance, protection, transport, maturation, and storage of spermatozoa.

epididymoorchitis inflammation of the testis and epididymis, representing usually a complication of a sexually transmitted disease, such as gonorrhea or syphilis.

epidural space a space between the dural membrane and vertebral column, containing nerve roots that are the sites of action for local anesthetics; e.g., during childbirth.

epigamic selection mechanisms of sexual selection depending on between-sex interactions in which one sex behaviorally or physiologically discriminates among individuals of the opposite sex to affect mating outcomes.

episomal integration "foreign" DNA that enters the cell but is not integrated into the host genome. The extrachromosomal DNA may replicate independently.

epitheliochorial placenta a placental type characterized by six tissue layers separating maternal and fetal blood (fetal vascular endothelium, stroma and chorionic epithelium, maternal uterine epithelium, stroma and vascular endothelium).

epithelium the tissue cells composing the skin; also, the cells lining all the passages of the hollow organs of the respiratory, digestive, reproductive, and urinary systems.

epitoky the reproductive phenomenon seen in some polychaetes, in which a benthic adult is morphologically transformed to produce a pelagic worm that swims to the surface of the water to spawn gametes.

epitope the portion or structural element of an antigen that an antibody recognizes (binds).

equine chorionic gonadotropin (eCG) a gonadotropin-like hormone secreted by embryonic cells in the mare, similar in structure and function to human chorionic gonadotropin; involved in the maintenance of early pregnancy. Formerly known as PREGNANT MARE SERUM GONADOTROPIN (PMSG).

ER see ESTROGEN RECEPTOR.

erection the complex series of neurovascular events resulting in arterial and corporal smooth muscle relaxation, and thus increased blood flow to the penis, causing it to become rigid.

erectile dysfuntion see IMPOTENCE.

ERKO estrogen receptor knockout mouse, a mouse in which the gene coding for the estrogen receptor-α has been altered.

ERT see ESTROGEN REPLACEMENT THERAPY.

erythroblastosis fetalis a fetal condition in which maternal antibodies cross the placenta and destroy fetal red blood cells, leading to hydrops fetalis.

erythropoiesis the development of erythrocytes from multipotential stem cells.

estradiol a steroid hormone produced at high levels by cells in the ovary during the first half of the menstrual cycle (after puberty). During pregnancy in humans the placenta also produces estradiol. Estradiol is one of the estrogen family of hormones; estradiol-17β is the most important estrogen in premenopausal women.

estriol the most abundant estrogen in pregnancy; its measurement can be used as an index of fetal/placental functions.

estrogen one of a large class of female sex steroid hormones synthesized and secreted primarily by the ovary, also present in low levels in males. Estrogens are responsible for female secondary sexual development and play a crucial role in the reproductive cycle.

estrogenic acting as or having the effects of an estrogen.

estrogen receptor (ER) a cellular structure possessing a specific affinity for the binding of estrogens.

estrogen replacement therapy (ERT) a technique for relieving various symptoms associated with estrogen deficiency; e.g., as experienced following menopause.

estrogen response element a palindromic nucleotide sequence that is specifically recognized and bound by the DNA binding domain of estrogen receptors.

estromedin a factor that mediates the effect of estrogen.

estromimetic a substance that mimics the effect of estrogen.

estrone a relatively potent form of estrogen found in human pregnancy urine.

estrous cycle the period between estrus of one ovarian cycle and estrus of the subsequent cycle for animals exhibiting distinct periods of estrus behavior (female sexual receptivity). The length of the estrous cycle is a characteristic of a given species; e.g., estrus occurs every 17 to 24 days in domestic cattle.

estrous synchronization in social groups of some female mammals, a correspondence in time of the estrous cycle in various individuals within the group. Also, WHITTEN EFFECT.

estrus the period of sexual receptivity in the female; the portion of the estrous cycle characterized by high levels of follicular estrogen production, during which time the female will accept the male for mating.

eunuchoidism a condition like that of a eunuch (castrated male), associated with a deficiency of the testes or testicular function and characterized by excessive long bone growth and a poor progression of primary and secondary sexual characteristics.

euprolactinemic having a serum or plasma prolactin concentration within the normal range.

eupyrene sperm the nucleate sperm of Lepidoptera, used to fertilize the oocytes.

eusocial describing a social strategy observed in certain colonial insects such as bees and termites, involving overlapping generations, cooperative care of young, and a reproductive division of labor.

euspermatozoa sperm that fertilize the egg and therefore contribute to the zygote genome.

eutherian one of the Eutheria, all those mammals that have a placenta during fetal development; i.e., mammals excluding the monotremes and marsupials.

evolutionary arms race a term for an evolutionary process whereby two sets of elaborate adaptations evolve as counters to each other, in a moving stalemate.

excurrent canals the ducts of the male reproductive tract, derived embryologically from the Wolffian duct, through which sperm are transported from the testis.

exfoliation see DESQUAMATION.

exocervix the portion of the cervix lined with squamous epithelial cells.

exocrine the secretion of a substance into a glandular duct which opens directly onto an epithelial surface; i.e., not into the bloodstream; e.g., the secretion of proteins by uterine glands into the uterine lumen.

exocytosis the process by which chemical substances within membrane-bounded vesicles in the cell can be secreted into the extracellular space. This involves fusion of the spherical vesicle membrane with the planar plasma membrane.

exogenous from without; external; occurring or controlled from outside the organism.

exogenous control the control of gametogenesis and other reproductive activities by external, environmental factors, which usually act by influencing endogenous factors. In species with seasonal patterns of reproduction, these factors are generally seasonally changing temperature, photoperiod, and perhaps availability of food.

exogenous DNA "foreign" DNA sequences that are not homologous to the host genome.

exogenous factors environmental variables such as humidity, light cycles, temperature and nutritional state that are sensed and neurally integrated to mediate the endogenous regime of an animal.

exogenous rhythms biological processes that arise solely, or mainly, as a direct response to transient environmental signals or conditions.

exogenous selection natural selection for or against genotypes, depending upon environment-genotype interactions.

exon the sequences of the primary RNA transcript (or the DNA that encodes them) that are spliced together to form a messenger RNA molecule. In the primary transcript neighboring exons are separated by introns.

ex situ outside the native habitat or site.

extender a medium used to prolong the interval over which sperm can be stored with maintenance of fertilizing potential, and also used to dilute semen to a number of sperm per milliliter appropriate for storage or artificial insemination.

external fertilization fertilization occuring outside the body of the female, although possibly still in contact with or sheltered by the female's body.

external os the vaginal opening of the cervix.

external pipping the first break of the eggshell by the hatchling when it begins to emerge from the egg.

extracellular matrix (ECM) a matrix secreted from cells that serves mainly as attachment for cells and surrounds primary body cavities.

extraembryonic membranes specific tissues lying outside of the body of the embryo proper (such as the yolk sac, amnion, and chorioallantois) which can function in embryo protection, physiological exchange, and placental formation.

extragonadal reserve the sperm stored in the epididymis, amounting to several days' worth of testicular production, available for ejaculation.

eyestalk in crustacea, a movable stalk (or peduncle) that supports the compound eyes; eyestalks not only aid the animal in sensing, but also house complex endocrine tissues.

F

facultative capable of living under conditions other than those which are typical; adaptable to changes in the environment.

facultative brood parasitism a reproductive strategy in which females sometimes lay eggs in other birds' nests but also raise a clutch of their own; or, some females lay eggs in other birds' nests and other females raise their own young.

Fallopian or **fallopian tube** either of a pair of tubes that extend from the uterine cavity to the ovary. They serve as a conduit for sperm, eggs, and the fertilized egg. (First described by Gabriele *Fallopio,* Italian anatomist.)

farrowing the birth process in swine.

fascia strong connective tissue lying deep beneath the skin and forming a protective covering for muscles and body organs.

fat body a mesodermal tissue in the form of lobes dispersed in the body cavity of insects, both a storage organ for nutrient reserves and the major source for synthesis of hemolymph proteins, to some degree analogous with the vetebrate liver.

fecundability the fact or probability of achieving a pregnancy within one menstrual cycle.

fecundity the fact or probability of achieving a live birth within one menstrual cycle. More generally, the production of sperm or eggs and of viable embryos.

feedback a reciprocal process in which a chemical agent produces a change or activity, which in turn regulates the further action of that agent.

feedforward the anticipatory effect that one intermediate element of a control system, e.g., the endocrine system, exerts on another intermediary farther along the pathway.

felid one of the Felidae (cat family); e.g., domestic cats, tigers.

female choice the selection of a male by a female based on different levels of criteria such as the male belonging to the same species, being sexually mature and competent, and that he is the most suitable male available compared to others.

fertility the capability to reproduce; the ability to produce offspring. This can be expressed or measured as the number of live births produced by a group per unit of time.

fertilization a union of male and female gametes; the process that culminates in the union of a single sperm nucleus with the egg nucleus within the activated egg cytoplasm.

fertilization cone the microfilament-containing eruption on the egg's surface that assembles around the entering, successful sperm. The fertilization cone is responsible for the physical incorporation of the sperm into the egg cytoplasm in most nonmammalian systems.

fetal occurring in, involving, or affecting the fetus.

fetal adrenal an organ located at the top of the kidney, composed in the primate of a medulla and a cortex made up of a fetal zone which produces steroid precursors for estrogen biosynthesis and a definitive "adult-type" zone which produces cortisol important for fetal maturation.

fetal alcohol effects (FAE) birth defects that may be attributed to maternal alcohol use but that do not manifest as the complete fetal alcohol syndrome.

fetal alcohol syndrome (FAS) a series of birth defects attributed to excessive maternal alcohol use during pregnancy, including prenatal and postnatal growth restriction, central nervous system abnormalities and midface hypoplasia.

fetal anencephaly a congenital condition occurring in the fetus, in which there is a developmental absence of portions of the cranial structures including the pituitary gland.

fetal echocardiography a noninvasive technique used to identify structural cardiac defects and disturbances of cardiac rhythm.

fetal growth restriction see INTRAUTERINE GROWTH RESTRICTION.

fetal hydrops an accumulation of fluid in the entire body of the fetus.

fetal Leydig cell a Leydig cell that differentiates from Leydig stem cells during embryogenesis, containing numerous lipid droplets and low aromatase activity; it produces testosterone needed for masculinization during fetal and neonatal life.

fetal membrane one of the extraembryonic membranes that usually participate in protection of the conceptus and/or the formation of placental structures that are sites of physiological exchange between conceptus and mother; these include the allantois, amnion, chorion, and yolk sac.

fetal-pelvic disproportion a condition in which the relative size of the fetus exceeds maternal pelvic diameters, often precluding vaginal delivery and associated with increased risks of labor dystocia and birth trauma.

fetal-pelvic index a method that combines maternal X-ray pelvimetry with fetal sonography to predict fetal-pelvic disproportion.

fetal-placental unit see FETO-PLACENTAL UNIT.

fetal position the direction that a given fetal landmark is facing in the maternal pelvis.

fetal presentation the relationship of the long axis of the fetus to that of the mother, also called **lie**. The presenting part of the fetus is that which enters the maternal pelvis first.

fetal surgery an operation performed on a fetus while the fetus remains inside the uterus.

fetal zone the unique large central compartment of the fetal adrenal cortex which secretes androgens and which regresses following birth

fetomaternal hemorrhage the transfer across the placenta of fetal red blood cells into the maternal circulation.

feto-placental unit the functional relationship between the fetal adrenal cortices and the placenta such that androgens produced by the adrenals are utilized as essential substrates for placental estrogen production.

α-fetoprotein a glycoprotein synthesized by the embryonic yolk sac and fetal liver during human pregnancy. It is used clinically in second trimester prenatal screening of fetal open neural tube defects and Down syndrome through measurement of its levels in maternal serum.

fetoscopy an operative technique for fetal surgery using tiny uterine puncture holes through which are placed instruments and a telescope attached to a fiberoptic camera.

fetus the stage of prenatal development from major organ formation to birth.

FGM fat globule membrane, the membrane that surrounds fat droplets of milk as they approach the apex of the plasma membrane of a milk-secreting cell.

fibroadenoma a growth of breast tissue that is discrete, well-circumscribed, movable, and firm or rubbery in consistency, of uncertain cause but associated with an imbalance of progesterone.

fibrous sheath a complex cytoskeletal structure present along the principal piece of the tail of spermatozoa. It is composed of longitudinal columns that run along microtubular doublets 3 and 8, and numerous arched ribs that bridge the longitudinal columns.

fimbria a complex expansion of the mucous membrane of the oviduct outside the lumen at the ovarian end.

fimbriectomy the surgical removal of the end of the Fallopian tube.

final oocyte maturation (FOM) the final stage of oocyte development prior to ovulation, characterized by germinal vesicle breakdown, the completion of the first meiotic division, and hydration, resulting in a mature oocyte capable of being fertilized. This process is triggered by a maturational surge of gonadotropins.

FISH fluorescent in-situ hybridization, the hybridization of a specific DNA probe to a standard metaphase preparation on a microscope slide, used for the detection of abnormal chromosomal numbers. Site-specific labeling of only that area of chromosome complementary to the probe occurs.

fission the division of a somatic individual into two or more fragments that become independent new individuals, sometimes after a period of regeneration.

fistula a connection between a closed organ or space in the body and the outside.

fitness in genetic terms, the differential viability and fecundity of genotypes, leading to differences in offspring numbers.

flaccidity the normal state of the penis in which the smooth muscle is contracted, and blood flow and intracavernous pressure are maintained at low levels.

flagellum *plural,* **flagella** a long, tail-like part of certain cells; specifically, the motile apparatus of the spermatozoon, with its associated accessory structures.

flehmen a characteristic behavioral posture (including flaring of the nostrils and curling of the upper lip) carried out by many male mammals to detect chemical signals from estrous females.

fluorescent in-situ hybridization see FISH.

follicle a spherical structure in the ovary consisting of an oocyte encircled by a single or multiple granulosa cells, which, in turn, are surrounded by androgen-secreting theca and stroma cells; the cells of the follicle secrete androgens, estrogens, and progestins. As the follicle matures, it develops a fluid-filled cavity and just before ovulation it is termed a Graafian follicle. Also, OVARIAN FOLLICLE.

follicle-stimulating hormone (FSH) a peptide hormone produced by the anterior pituitary gland which stimulates the growth and development of follicles within the ovary; also responsible for the initiation of spermatogenesis during puberty, and largely responsible for Sertoli cell function in the testes.

follicular relating to or involving a follicle or the follicular phase.

follicular antrum a fluid-filled cavity of the growing ovarian follicle.

follicular atresia see ATRESIA.

follicular epithelium a monolayer of somatic cells surrounding the oocyte and its nurse cells. One of its major functions is to secrete the chorion around the oocyte.

follicular phase the ovarian events that occur during the first half of the menstrual cycle, namely, the growth and development of a primary follicle into a preovulatory or Graafian follicle, leading to ovulation. This estrogen-dominated phase of the cycle corresponds to the proliferative phase of the uterus.

follicular waves cohorts of ovarian follicles that emerge every eight to nine days, from which one follicle becomes dominant and eventually is capable of ovulation.

folliculogenesis the maturation of the follicle; the process by which a primordial follicle grows and develops into a specialized follicle with the potential to ovulate its egg or to die by atresia.

follistatin a single-chain glycoprotein that acts as a binding protein for activin and neutralizes its diverse functions.

food chain a term for the feeding pattern of various groups of organisms in a given habitat; e.g., plants are consumed by plant-eating mammals, which in turn are eaten by other, carnivorous mammals.

foramen ovale an opening between the vena cava and the left atrium in the fetus that permits blood to bypass the right ventricle and proceed directly to the left heart.

fornix an arched space between the uterine cervix (as it projects into the vagina) and the vaginal wall.

fos a family of genes that produce the Fos proteins, which serve as transcription factors that activate or inhibit the expression of other genes.

fossorial having the behavior pattern of digging or burrowing; living in holes or burrows.

founder a wild-born individual assumed to possess maximum genetic diversity.

foundress in social insects, a mated queen founding a new nest without the help of workers.

freemartin an intersex or hermaphroditic genotypic female born as co-twin to a male.

free PSA the proportion of total serum PSA that is not complexed with serine protease inhibitors present in the serum.

free-running a term used to describe circadian rhythms persisting in the absence of any 24-hour synchronizing information from the external environment. The expression of free-running circadian rhythms demonstrates the endogenous nature of the circadian clock.

free testosterone the fraction of testosterone in blood that is not bound to binding proteins and is presumably biologically active at the tissue level.

FSH see FOLLICLE-STIMULATING HORMONE.

functional luteolysis the decline of progesterone secretion by the corpus luteum, the first sign of impending luteolysis.

fundus the thick contractile portion at the top of the uterus.

G

galactopoiesis the maintenance of milk secretion.

galactorrhea inappropriate secretion of milk from the nipples of the breast, unrelated to pregnancy or to postpartum time periods.

gamete a specialized product from the gonads (testis or ovary) of an organism, designed to transfer genetic material while participating in fertilization.

gamete intrafallopian transfer see GIFT.

gametogamy sexual reproduction by fusion of gametes to produce a zygote.

gametogenesis the production of female and male sex cells, or gametes.

gamma globulin a fraction of blood plasma rich in antibodies.

gamontogamy sexual reproduction by the pairing of parent cells and exchange of gametes, usually resulting in cross fertilization.

gap junctions specialized regions in adjacent membranes of neighboring cells, through which small water-soluble molecules, including ions, pass. They have a defined structure and likely function in intercellular communication.

gastrodermis in hydra, the inner epithelium of the body wall that surrounds the gastrovascular cavity.

gastrula the early embryonic stage that follows the blastula.

gastrulation the process that generates the three primary germ layers of the embryo: the ectoderm, the endoderm, and the mesoderm.

gender the fact of being male or female; one's sex. In current usage, the term *gender* is often distinguished from sex by referring to the cultural expression of biological sex in terms of masculine or feminine behavior, as opposed to anatomical and physiological features.

gender identity the feeling that one is male or female.

gene a hereditary unit; a contiguous sequence of cellular DNA that encodes an RNA product and the sequences which regulate its synthesis.

gene expression the synthesis of products from a gene, including gene transcription to RNA and mRNA translation to protein.

gene flow the transfer of DNA that results from matings between individuals belonging to different populations.

general anesthesia pain relief over the entire body induced by drugs that produce unconsciousness.

generation in birds, a term for all the feathers grown during a single molt.

genet a genetic individual; the entire mitotic product of one zygote.

gene targeting a technique used for the site-specific genetic alteration of a particular gene; this may involve insertional inactivation, deletions, or small changes within the gene of interest.

genetic relating to or involving genes.

genetic recombination a sum of processes resulting in the appearance of new combinations of genes in the progeny.

genetics the scientific study of genes.

genetic sex determination a mechanism of sex determination present in the majority of vertebrates, in which sex is determined by the genetic constitution of individuals.

gene transfer any of various methods used for carrying new or altered genes into cells.

genital chamber see BURSA COPULATRIX.

genitalia or **genitals** a collective term for all reproductive structures, especially the external organs.

genital ridge the embryonic structure that is the precursor of the gonad of the adult animal. Before the time of sexual differentiation the genital ridge of male and female embryos are indistinguishable and may be referred to as the indifferent gonad. Upon sexual differentiation the testis is easily distinguishable by the formation of testis cords.

genital tubercle the embryonic primordium of the phallus. Seen in both males and females, it enlarges in the male to form the penis. In the female significant growth does not occur and it forms the clitoris.

genome the genetic make-up (the set of all genes and gene signals) of cells and viruses. It represents the total number of chromosomes which is the diploid number in somatic cells and the haploid number in gametes.

genomic relating to or involving the genome.

genomic exclusion in certain invertebrates, mating between normal cells and sterile cells; this induces genetic events that yield germinal nuclei with whole genome homozygotes and parental somatic nuclei.

genomic imprinting a phenomenon whereby a specific DNA segment is differentially modified during gametogenesis and therefore functions differently depending on the parental origin of the chromosome.

genotype the entire genetic makeup of an individual.

germ cell in general, a gamete (sex cell), or any cell capable of producing a gamete. See also PRIMORDIAL GERM CELL.

germinal disc a 2–3 mm region of the surface of the avian oocyte that contains the female pronucleus, cellular organelles, and the only continuous oolemma with microvilli.

germinal epithelium a layer or bed of germ cells in the ovary that divide mitotically and give rise to primary oocytes.

germinal vesicle the diploid nucleus of the oocyte.

germinal vesicle breakdown (GVBD) the breakdown of germinal vesicle (nuclear membrane) after it migrates to the periphery on the resumption of meiosis during final oocyte maturation.

germ layer a morphologically distinguishable sheet of cells present in the early embryo that will form one of the body's three principal organ systems.

germline the cells from which gametes are derived.

germline transmission in transgenic animals, the incorporation of foreign DNA within the genome, so that it is transmitted to the offspring.

germogen the germinal area of the ovary.

gestation the period of embryonic and fetal development in the uterus, from fertilization to birth.

gestational age the age of a fetus, usually expressed in weeks as measured from the time of fertilization, or, more typically, from the first day of the mother's last menstrual period which actually precedes conception by about two weeks.

gestational choriocarcinoma a malignancy that arises from trophoblastic tissue of term pregnancies, ectopic gestations, spontaneous/induced abortions, or molar elements and contains malignant cytotrophoblasts and syncytiotrophoblasts

gestational trophoblastic disease a spectrum of disease characterized by abnormal trophoblastic proliferation and including hydatidiform moles, invasive mole, and choriocarcinoma.

GIFT gamete intrafallopian transfer, a surgical procedure in which egg(s) and sperm are laparoscopically placed in the Fallopian tube.

glucocorticoid any of a class of steroid hormones secreted by the adrenal cortex that affect metabolism of glucose and other organic molecules. They also have a wide range of anti-stress, anti-inflammatory, and immunosuppressive effects. Glucocorticoids include cortisol and corticosterone.

gluconeogenesis the production of glucose from non-glucose precursors such as amino acids and lactate.

glucoprivation a reduced availability of glucose for intracellular oxidation and ATP (adenosine triphosphate) production.

glutamate an acidic amino acid with excitatory or neurotransmitter activity, released from presynaptic nerve terminals within the brain.

glycoprotein a protein with carbohydrate sugars covalently attached.

glycoprotein hormone one of a family of four related hormones containing protein and sugar components; human chorionic gonadotropin, luteinizing hormone, follicle-stimulating hormone, thyroid-stimulating hormone.

GnRH gonadotropin-releasing hormone, a peptide hormone composed of 10 amino acids that is produced by neurons in the hypothalamus of the brain; it regulates the synthesis and secretion of the hormones LH (luteinizing hormone) and FSH (follicle-stimulating hormone) by the anterior pituitary gland and has other central nervous system actions.

GnRHa gonadotropin-releasing hormone agonist, a drug that lowers estrogen to menopausal levels for the purpose of controlling endometriosis.

GnRH pulse generator a group of pulsatile, interconnected neurons located in the mediobasal hypothalamus of the brain that synthesize and release gonadotropin-releasing hormone (GnRH) into the pituitary portal circulation to stimulate pituitary gonadotropin synthesis and secretion.

Golgi apparatus or **complex** a complex cytoplasmic organelle consisting of a series of layered fluid-containing sacs and associated small vesicles; involved in the delivery of cellular products to the cell surface or to an intercellular destination. (After Camillo *Golgi,* Italian neurologist.)

gonad a gamete-producing gland; an ovary or testis.

gonadal activity a term for the growth and development of external and internal sexual characters as well as levels of hormones and events of reproduction.

gonadal dysgenesis a general term for any form of aberrant gonadal development.

gonadal-hypothalamic-pituitary axis the feedback loop of the testis, hypothalamus, and pituitary which regulates spermatogenesis and androgen production.

gonadal involution the degeneration of testes and ovaries to a state containing primarily primordial germ cells, leaving the animal infertile with very low concentrations of gonadal steroids.

gonadarche the onset of clinical evidence of increased gonadal sex steroid secretion or (in males) increased testicular size.

gonadectomy the removal of the gonads of either sex; the term *castration* is usually used for males, and *ovariectomy* for females.

gonadostat the hypothalamic-pituitary mechanism that regulates gonadotropin secretion; modulated by various hormonal and neural influences.

gonadostat hypothesis the theory that the increase in GnRH secretion which underlies puberty results from a decrease in the ability of circulating sex steroids to inhibit GnRH release.

gonadotrope or **gonadotroph** a cell type of the anterior pituitary gland that produces and secretes gonadotropin hormones.

gonadotropic stimulating the gonads (sex organs).

gonadotropic hormone see GONADOTROPIN.

gonadotropin (GTH) generally, any hormone that stimulates the gonads (ovary or testis); specifically, either of two such substances secreted by the the anterior pituitary, luteinizing hormone (LH) and follicle-stimulating hormone (FSH), that differentially affect the various cell types in the ovary and testes. The hormones GTH-I and GTH-II correspond to FSH and LH, respectively, in some fishes.

gonadotropin-releasing hormone see GnRH.

gonads a collective term for the primary reproductive organs; the ovaries (female) and testes (male).

gonad-stimulating substance (GSS) in starfish, a primary hormone released from radial nerves that causes spawning and oocyte maturation. GSS acts directly on gonadal follicle cells to produce a secondary hormone, a maturation-inducing substance which is identified as 1-methyladenine.

gonochoric or **gonochoristic** possessing either female or male gonads, not both; single-sexed.

gonochorism a condition in which female and male gonads (ovaries and testes, respectively) are present in separate individuals; as opposed to hermaphroditism, in which testes and ovaries reside within a single individual.

gonopods the sperm-transfer organs of most diplopods, composed of appendages of the seventh body segment modified during the final one or several immature stadia; highly specific in structure and mandatory for identification.

gonopore in some invertebrates, a ciliated, discrete invagination of the epidermis where the eggs and sperm are released.

gonostylus a specialized structure on the ninth abdominal segment of male insects, used to grasp the female during copulation.

gonotrophic cycle the series of events involved in the synchronous maturation of a single batch of eggs.

G-protein any of various guanine nucleotide-binding proteins that associate with hormonal receptors. G (guanine) proteins are involved in many aspects of cell regulation, and are of major importance in signal transduction at the plasma membrane level.

Graafian or **graafian follicle** a multilayered ovarian structure containing an oocyte and a fluid-filled antrum. A Graafian follicle responds to the preovulatory surge of pituitary gonadotrophic hormones by a series of biochemical, endocrine, and structural changes culminating in shedding of the oocyte (ovulation). (From R. R. de *Graaf,* Dutch anatomist.)

granulosa cell a female somatic cell type of epithelial origin in the ovarian follicle that surrounds the oocyte and provides the required microenvironment for its maturation.

growth factor generally, any compound that facilitates or that is required for cell division or whole animal growth; specifically, certain polypeptides that bind to target cell membrane receptors and promote cell division and various other cellular functions.

growth hormone a hormone produced by the pituitary gland that is responsible for regulating growth.

growth hormone-variant (GH-V) a protein hormone synthesized and released by the human placenta that has biologic actions similar to pituitary growth hormone.

GSS see GONAD-STIMULATING SUBSTANCE.

GTH see GONADOTROPIN.

gubernaculum a mesodermal derivative located at the lower end of the testis and epididymis.

GVBD see GERMINAL VESICLE BREAKDOWN.

gynandromorph the individual mosaic for male and female tissues.

gyne a virgin queen in social insects.

gynecomastia the enlargement of glandular breast tissue in males.

gynogenesis sperm-dependent parthenogenesis; sperm are used to activate embryogenesis but fusion of egg and sperm nuclei does not occur.

H

half-life the time required for a substance to reach half of its concentration in a medium or system.

half-time the time required for half of any starting amount of a substance to disappear.

Hamilton's Rule a theory that describes the evolution of altruistic social behavior as a means of promoting the survival of close relatives, thus increasing the fitness of the group even if it is at the expense of individual fitness.

haplodiploidy the sex determining system by which unfertilized eggs develop into males and fertilized eggs usually develop into females.

haploid having one copy of each chromosome, rather than two (diploid).

hCG see HUMAN CHORIONIC GONADOTROPIN.

hectocotylus the structurally modified arms of certain male cephalopods involved in spermatophore transfer to females.

heifer a bovine female that has not produced a calf.

hematocolpus an accumulation of blood in the vaginal cavity.

hematometria an accumulation of blood in the uterine cavity.

hematopoiesis the differentiation process that forms and develops blood cells.

hemiclone a haploid clonal genome that is transmitted without recombination by hybridogenetic females.

hemi-desmosome an adhering junction that binds the basal aspect of the Sertoli cell to the underlying connective tissue.

Hemimetabola insects with incomplete metamorphosis. These insects molt to the adult stage directly from the larval stage.

hemiuterus a functional uterus formed by a single Mllerian duct.

hemivagina incomplete formation of a vagina in the case of cervical and vaginal duplication.

hemochorial placenta a placental type characterized by three tissue layers separating maternal and fetal blood (fetal vascular endothelium, stroma, and fetal chorionic epithelium).

hemocoel the body cavity of insects within which the hemolymph, or blood, circulates freely.

hemolymph the circulating fluid or blood of insects, which is pumped by a dorsal tubular heart and circulates freely in the body cavity, or hemocoel; it serves to transport nutrients, immunity factor, and hormones but not oxygen; several types of hemocytes, but no erythrocytes, are present.

hemorrhagic producing hemorrhage.

hemostatic preventing hemorrhage.

hemotroph or **hemotrophe** nutrients provided by the maternal blood across the interhemal barrier of the placenta.

hepatopancreas the site of production of many crustacean digestive enzymes; it is also the likely source of extraovarian yolk lipoproteins in at least some crustaceans.

hermaphrodite an animal possessing gonads of both male and female type, often at the same time.

hermaphroditic describing or referring to individuals that are both male and female, either simultaneously or in sequence.

hermaphroditism the fact or condition of having both the male and female reproductive organs expressed in the same individual, either simultaneously or in sequence.

hernia a bulging of an organ or tissue through a defect in its surrounding wall; e.g., in the groin (inguinal hernia) or the abdominal wall (ventral hernia).

herp or **herptile** an artificial taxonomic group composed of reptiles and amphibians.

heterochromatin highly condensed chromatin regions that are not genetically expressed.

heterochrony the fact of exhibiting changes in the timing of appearance of features during development (or in the rate of development of an organism); e.g., the precocious formation of adult structures in a larval stage.

heterodimer a molecule composed of two dissimilar subunits.

heteroecious describing a parasitic life cycle with a seasonal alternation between primary and secondary hosts.

heterogony the alternation of generations between asexual and sexual phases.

heteromorphic describing homologous sex chromosomes that are morphologically or cytogenetically distinguishable.

heterosexuality the fact of experiencing sexual and/or romantic attraction to individuals of the opposite sex to oneself.

heterozygotes a diploid organism, with two sets of chromosomes, which have different forms of a gene (alleles) at a particular locus.

hGH human growth hormone. See GROWTH HORMONE.

hGH gene cluster the region on the long arm of chromosome 17 that encodes the genes for pituitary and placental growth hormones and human placental lactogen (hPL).

hibernaculum the winter dwelling place of an animal during hibernation.

hierarchy a term for the arrangement of follicles by size on the avian ovary with the largest follicle destined to ovulate first, the second largest next, and so on.

higher vertebrates a collective term for mammals and birds.

hirsutism excessive hair growth; specifically, a condition in human females of excessive hair on the face or chest.

histone the major basic protein in chromatin that is found in all eukaryotic cells except sperm.

histotroph or **histotrophe** often called uterine milk; the sum of supplementary nutritive substances derived directly from maternal tissues other than blood, including glandular secretions and cell fragments.

HMG domain high mobility group domain, a DNA-binding motif that is associated with several eukaryotic regulatory proteins, including SRY and SF-1. Attachment of the HMG-domain containing protein to DNA causes conformational changes and regulates RNA transcription.

holocrine secretion a type of secretion in which the whole cell is shed while filled with its component substances, e.g. the secretion of sebaceous glands.

holocyclic describing life-cycles in which parthenogenetic and sexual modes of reproduction alternate.

Holometabola those insects with complete metamorphosis. They molt from a larval to a pupal stage before molting to an adult.

homeorhesis orchestrated changes in metabolism of a species to support a specific physiological process.

homeostasis internal balance; a relatively stable condition of an animal's internal environment that occurs as a result of compensatory physiologic reactions.

homeothermic having a relatively constant body temperature; warm-blooded.

homeothermy the fact of having a regulated, constant body temperature relatively independent of changes in environmental temperature; i.e. using heat produced by metabolism.

homodimer a molecule composed of two identical subunits.

homologous corresponding to or in common with another; e.g., of anatomical strucures in different species, having a common evolutionary origin.

homologous chromosomes chromosomes carrying the same genetic loci. Diploid organisms contain pairs of homologous chromosomes, one deriving from each parent.

homologous insemination the use of the semen of the spouse or partner for artificial insemination.

homologous recombination a method used for making specific modifications within a gene of interest. It is a repair mechanism initiated by double strand breaks in the DNA and 100% homology between the targeting construct and the host gene of interest.

homology the fact of being homologous; having a common or corresponding origin, structure, organization, etc.

homomorphic describing homologous sex chromosomes that are not morphologically or cytogenetically distinguishable.

homononymous the condition of an animal body being composed of a series of similar structural units without modification into regional functional units (e.g., as in earthworms).

homophobia a general term for various forms of prejudice and discrimination against homosexuals.

homosexuality the fact of experiencing sexual and/or romantic attraction to individuals of the same sex as oneself.

homozygotes diploid organisms that have the same form of a gene (allele) at a particular locus.

hormone any of a number of molecules with diverse chemical structures that regulate a variety of cellular functions. In the classical sense, the term was used to denote a compound that is biosynthesized and secreted by one cell type and that travels through the bloodstream to act on another cell type at a distant site. It is now recognized that a hormone can act on neighboring cells within the same tissue, i.e., paracrine function, or even on or within the same cell in which it is biosynthesized, i.e., autocrine function. Cell responsiveness to a hormone depends upon the presence of a specific receptor, and the response can be either stimulatory or inhibitory to a biochemical process.

hormone agonist an analog that binds to receptors and elicits the same biological response as the natural hormone.

hormone antagonist a molecule that binds to hormone receptors without eliciting the normal cellular response.

hormone replacement therapy (HRT) the use of estrogens and/or other hormones in treatment of the consequences of a hormone deficiency state such as menopause.

hormone response element a finite sequence within a gene that confers binding of an activated hormone receptor and that subsequently modulates the transcription rate of the gene.

hormone therapy a treatment for cancer that involves the removal, blocking, or addition of hormones.

host the larger partner in a living association of two organisms of different species; e.g., parasitism.

hot flash or **flush** a brief, subjective sensation of intense heat mainly of the upper body; associated in the human female with a decline in circulating estrogen levels during or after menopause.

hourglass model a model of photoperiodic time measurement depending upon the accumulation (or depletion) of some physiological product during the light or dark portion of the light-dark cycle.

HPA axis see HYPOTHALAMIC-PITUITARY-ADRENAL AXIS.

HPG axis see HYPOTHALAMIC-PITUITARY-GONADAL AXIS.

hPL see HUMAN PLACENTAL LACTOGEN.

HPV see HUMAN PAPILLOMA VIRUS.

HRT see HORMONE REPLACEMENT THERAPY.

HSD hydroxysteroid dehydrogenase; the 3β-HSD enzyme converts pregnenolone to progesterone, and 17β-HSD converts androstenedione into testosterone.

human chorionic gonadotropin (hCG) a glycoprotein subunit hormone that is synthesized by the trophoblast layer of the placenta during pregnancy.

human papilloma virus (HPV) a double-stranded DNA virus of approximately 8000 base pairs, which forms part of a family of papillomaviruses that include common skin or genital wart viruses as well as the virus thought to produce cervical cancer.

human placental lactogen (hPL) a single-chain polypeptide secreted by the trophoblast that exhibits growth hormone and lactogenic activity; important in fetal growth regulation.

humoral circulating or found in the blood or another fluid of the body (these fluids were once called *humors*).

hyaline membrane disease see RESPIRATORY DISTRESS SYNDROME.

hybrid anything of heterogeneous origin or composition; i.e., an offspring of two genetically different parents, or even different species.

hybrid zone a region in which two populations of individuals that are distinguishable on the basis of one or more heritable characters overlap spatially and temporally, mate and form viable (and at least partially fertile) offspring.

hydatidiform mole abnormal placental tissue exhibiting trophoblastic hyperplasia and hydropic chorionic villi.

hydrobios those organisms that live in water oxygen, including soil organisms active only when at least a layer of water is present.

hydrocele a benign fluid-filled collection within the parietal and visceral layers of the tunica albuginea of the testis.

hydronephrosis dilation of the collecting system of the kidneys resulting from obstruction to the flow of urine.

hydropic involving an abnormal collection of fluid.

hydrops (fetalis) fetal distress manifested by fluid collections around the heart and lungs, in the abdomen, or in the skin; it can be caused by maternal antibodies cross-reacting with the fetus or related to fetal heart failure.

hymen a membranous fold of tissue that partly occludes the external opening of the vagina in a virginal female.

hyperactivation sperm motility that is characterized by a highly vigorous whip-like flagellar motion with an intermittent non-progressive trajectory. The development of hyperactivation is associated with sperm capacitation, and allows sperm to penetrate the outer vestments of the oocyte.

hyperalgesia an increased response to a stimulus that is normally painful.

hyperandrogenism a clinical state characterized by an excessive production or expression of androgens.

hyperemesis gravidarum a syndrome of nausea and vomiting of great intensity requiring prompt correction of fluid and electrolyte imbalance.

hypergonadotropic hypogonadism a disease state in which hypogonadism occurs due to primary gonadal failure.

hypermenorrhea increased menstrual flow and/or duration. Also, MENORRHAGIA.

hyperphagia abnormally increased food intake.

hyperplasia a pattern of increased cell proliferation resulting in an increased number of cells of varying degrees of atypical morphology.

hyperprolactinemia elevation of the serum or plasma prolactin concentration above the normal range.

hypertrophy an increase in cell size.

hypoadrenalism inadequate function of the adrenal glands.

hypoalgesia a decreased response to a stimulus that is normally painful.

hypodermic insemination in certain invertebrates, a process in which sperm are introduced into the body of a female not through a specialized reproductive opening but by forcing their way through the body wall, either mechanically or chemically. In some species they are met by specialized cellular adaptations of the female.

hypodermis the external epithelium, which surrounds the animal and secretes the cuticle. The hypodermis is composed of a sheet of cells, many of which fuse during development to form multinucleate syncytia.

hypogonadal affected by or involving a deficiency in the secretions of the ovary or testis.

hypogonadism a reproductive condition characterized by impaired or inadequate secretory function of the ovary or testis. Primary hypogonadal states lead to lack of secondary sexual development. Secondary hypogonadal states are due to conditions acquired after puberty has occurred.

hypogonadotropic hypogonadism a disease state in which hypogonadism occurs due to a lack of gonadotropins, associated with hypothalamic or pituitary lesions that decrease gonadotropin production.

hypogonadotropism a hormonal state characterized by low circulating concentrations of the pituitary gonadotropins.

hypomenorrhea decreased menstrual flow and/or duration. Also, CRYPTOMENORRHEA.

hypophysectomy the surgical removal or chemical inactivation of the pituitary gland (hypophysis).

hypophysial or **hypophyseal** relating to, involving, or affecting the pituitary gland.

hypophysiotropic relating to or involving hormones or other biochemical agents that stimulate or inhibit the secretion of pituitary hormones.

hypophysis another name for the PITUITARY GLAND.

hypopituitarism inadequate secretion of the anterior pituitary hormones and of the target gland hormones they stimulate.

hypoplasia the incomplete formation of an organ.

hypothalamic relating to or involving the hypothalamus.

hypothalamic amenorrhea the absence of menstrual bleeding based on failure of the hypothalamus to produce appropriate amounts of GnRH (gonadotropin-releasing hormone).

hypothalamic-hypophysial axis the neuronal and vascular associations between the hypothalamus at the base of the brain and the hypophysis (pituitary gland).

hypothalamic-pituitary-adrenal axis (HPA) a collective term for the neuroendocrine linkage of the hypothalamus, anterior pituitary gland, and adrenal glands.

hypothalamic-pituitary-gonadal axis (HPG) a collective term for the neuroendocrine linkage of the hypothalamus, anterior pituitary gland, and gonads (ovaries or testes).

hypothalamic-pituitary unit the ventral portion of the brain with the anatomically attached endocrine gland (pituitary) that controls the reproductive system.

hypothalamus a small region at the base of the brain, joined to and controlling the pituitary gland and regulating sexual behavior and other functions. It consists of numerous, discrete brain nuclei that serve to maintain the body's internal balance by regulating vital phenomena such as temperature, heart rate, sleep, appetite, and blood pressure. Also, the hypothalamus coordinates the endocrine and nervous systems, by producing the hormones of the neurohypophysis and the regulating factors that control the functions of the adenohypophysis.

hypoxemia a decreased level of oxygen in the blood.

hypoxia a decreased level of oxygen in the body tissues; also, low oxygen content in the ambient air; e.g., at high altitude.

hysterectomy the surgical removal of the uterus.

hysterosalpingography a technique for producing an X-ray image of the uterine cavity and the oviducts following injection of a radiopaque dye.

hysteroscopy endoscopy of the uterine cavity.

hysterotomy a uterine incision.

I

iatrogenic describing an unintended adverse condition induced during medical treatment; e.g., the introduction of an infection.

ICC see IMMUNOCYTOCHEMISTRY.

ICM see INNER CELL MASS.

ICSI intracytoplasmic sperm injection, an assisted reproductive technology used in men with severe male factor infertility to aid fertilization. It involves the use of a single spermatozoon that is microinjected into the oocyte's cytoplasm using a micromanipulator.

IG see IMMUNOGLOBULIN.

IGFBPs insulin-like growth factor binding proteins, a family of six or more soluble proteins secreted in IGF target tissues that bind with the IGFs with high affinity and regulate their bioavailability.

IGF receptors a family of cell membrane proteins with high affinity binding sites for the IGFs (insulin-like growth factors) and signal transducing capability, including the Type I (IGF-I) receptor and the Type II (IGF-II, mannose-6-phosphate) receptor.

IGFs see INSULIN-LIKE GROWTH FACTORS.

imaginal disc tissues in holometabolous insect larvae that will give rise to adult appendages and other structures. Imaginal discs can exist as slightly thickened epidermal placodes or as relatively complex invaginated pouches of epidermal cells.

immature Leydig cell a cell that differentiates directly from the progenitor Leydig cell, containing high levels of testosterone-metabolizing enzyme activities.

immigration the movement of individuals toward a certain site; specifically, movement toward a breeding site in advance of reproductive activity.

immune deviation the process by which the immune system is biased towards a TH1 or TH2 cytokine pattern, or a protective or destructive response.

immune privileged sites tissues and organs that prevent the entrance of immune cells. These include the ovary, testis, uterus, and placenta.

immune system the cells (white blood cells and other progenitor cells) and antibodies that serve to protect an organism from undesirable foreign cells, viruses, and other molecules.

immunization the process in which an organism becomes immune; e.g., by means of injection of a specific antigen foreign to the organism to promote the formation of antibodies.

immunocytochemistry (ICC) a method in which certain known antibodies to specific proteins of interest are used in combination with specific tags so as to identify the site of production or action of the protein in question; e.g., a hormone, growth factor, or other regulatory product.

immunoglobulin (Ig) one of a distinct class of proteins found in plasma and other body fluids that exhibit antibody activity and bind with other molecules with a high degree of specificity; these are divided into five classses (IgM, IgG, IgA, IgD, IgE) on the basis of structure and biological activity.

immunohistochemistry see IMMUNOCYTOCHEMISTRY.
immunologic response the reaction of an organism to injection of an antigen.
immunoreactivity the ability of an antigen (usually a protein or peptide) to react with a specific antibody (usually an immunoglobulin); this interaction is usually visualized by tagging the antibody with an enzyme or dye.
implantation the process by which the embryo establishes an intimate connection with the maternal tissues, involving the penetration of the embryo through the endometrial epithelium and its invasion into the underlying stroma.
impotence the consistent or recurring inability to attain and maintain a penile erection sufficient to permit satisfactory sexual intercourse. Also, ERECTILE DYSFUNCTION.
inbred strain a strain of laboratory mouse that has been inbred for at least twenty generations of sibling matings and is thus genetically homozygous.
incidence rate the number of people who develop a certain disease or condition in a given amount of time.
inclusive fitness the concept that an animal's fitness (the ability of genetic material to perpetuate itself in the course of evolution), depends not only on its own individual reproductive success, but also that of its close relatives who share many genes in common with the individual; postulated as the basis for altruistic behavior.
incorporation cone the structure that forms as a result of new microfilament and myosin II assembly, at the cortex overlying the position of the incorporated sperm nucleus. It is the result of the sperm nucleus incorporation.
incubation the behavior pattern of sitting on eggs to provide the correct environment for embryo development; also, the provision of a comparable artificial environment for eggs to hatch.
indeterminate layer a species that lays additional eggs if some eggs are removed from the nest before the clutch is complete.
indicator trait a characteristic in the chosen sex that signals underlying genetic or phenotypic quality of significance to the choosing sex.
indirect development a life cycle that includes a larval stage and metamorphosis.
indirect sperm transfer in invertebrates, a process in which sperm (or spermatophores) do not move directly from the male's body into the body of the female but come in contact with a third structure, usually either the substrate (e.g. soil, leaf, pebble) or an appendage of the male's body; the rule in arachnids.
induced ovulation the ovulation of follicles in response to a coitally induced surge of LH (luteinizing hormone). See also OVULATION INDUCTION.
infertility a lack of fertility; the failure to conceive; in humans, clinically defined as inability to conceive after an adequate exposure to unprotected intercourse, usually considered at least twelve months.
inflammation a protective response of tissues to disease or damage; a complex sequence of metabolic changes leading to vasodilatation, hyperemia, exudation, edema, proteolysis, and eventual tissue remodeling in response to microbial invasion, radiation, friction, chemical irritation, or other factors such as acute stimulation of target tissues by glycoprotein hormones.
inflatable penile prosthesis an artificial device placed into the corpora to achieve girth and rigidity for the treatment of impotence (erectile dysfunction).
infochemical a chemical substance involved in the transfer of information between two individuals (of the same or different species) that results in a behavioral and/or physiological response in the receiver that is adaptive for one or the other (or both) of the individuals.
infundibular in anatomy, acting as or shaped like a funnel.
infundibular process a lobular region that is the major portion of the neurohypophysis (posterior pituitary gland). Also, NEURAL LOBE.
infundibular stem or **stalk** the narrowed, intermediate region region adjoining the median eminence and connecting the infundibular process to the floor of the hypothalamus, via the median eminence.
infundibulum a funnel-shaped structure; e.g., a region at the tip of the oviduct that picks up the egg when released by the ovary. See also INFUNDIBULAR PROCESS/STEM, MEDIAN EMINENCE.
infusoriform larvae in mesozoans, free-swimming, non-feeding larvae that develop from zygotes within the axial cell of the rhombogen and escape the host excretory organ, presumably to infect new hosts.
infusorigen in mesozoans, a stage within the axial cell of the rhombogen with an outer layer of oocytes surrounding an axial cell containing non-flagellate sperms; sometimes considered to be a hermaphroditic gonad of the rhombogen.
inguinal canal an extension of the peritoneal cavity containing the vas deferens and neurovascular supply of the testis; it is the conduit enabling the testes to descend from their initial abdominal location into the scrotal sac.
inhibin a glycoprotein that is produced primarily in the ovary; it feeds back to the pituitary to selectively inhibit the release of FSH (follicle-stimulating hormone).
inhibitor any agent that restrains or retards physiological, chemical, or enzymatic action.
inner bud in Cycliophora, a cluster of cells with the ability of regenerating new feeding structures and new stages .
inner cell mass (ICM) a group of cells that are present on the inside of the blastocyst and that will give rise to the future embryo.
innervation the nerve supply to a certain area of the body.
insemination any process by which spermatozoa are placed into the female reproductive tract, either naturally or artificially.
insertion a term for the point at which the umbilical cord attaches to the placenta.

insertional inactivation the insertion of foreign DNA into a functional gene, causing its loss of function, or inactivation.

in situ in site; within the native habitat or locale; specifically, referring to cancers that have not grown beyond their original site.

***in situ* hybridization** a method for localizing a specific gene or DNA sequence within a chromosome based on the binding of a complementary, radioactively labeled segment of RNA or DNA to it.

***in situ* ligand binding** a procedure for localization of receptor binding sites by placing a drop of labeled ligand onto a histological section of tissue and then microscopically determining the distribution of the label.

instar in arthropods, any subadult life stage betwen molts. For example, a first instar would be the stage between hatching or birth and the next molt.

insulin-like growth factor binding proteins see IGFBPs.

insulin-like growth factors (IGFs) peptide growth factors (IGF-I, IGF-II) with homology to insulin and widespread expression and action in reproductive tissues; produced primarily in the liver, but also throughout other tissues and organs, acting to mediate the local effects of growth hormone.

insulin resistance a subnormal biologic response to a given amount of insulin.

integrins transmembrane proteins that are receptors for extracellular linkers at the cell surface.

interleukins a group of protein factors involved in signaling between cells of the immune system.

intermediate host the host in which a parasite having an indirect life-history pattern undergoes development but would not normally attain sexual maturity.

internal fertilization fertilization occuring within the body of the female.

internal pipping the breaking into the air cell several hours to days before hatching by the embryo, which then begins to breathe convectively from the air cell.

internal reproductive organs a collective term for: in females, the ovaries, oviducts, uterus, cervix and vagina; in males, the testis, vas deferens, epididymis, seminal vesicles, prostate, and some accessory sex glands.

interrenal gland another term for the adrenal cortex, so-called in nonmammalian vertebrates because of the distribution of steroid and catecholamine producing cells in these groups of vertebrates. Both cell types are more or less randomly mixed and not associated as a cortex and a medulla.

intersex an individual having both female and male sexual characteristics.

intersexuality gender ambiguity; the intermingling in one individual, to various degrees, of both female and male sexual characteristics; may result from a disturbance in sex determination during embryonic development.

intersexual selection a process in which variation in reproduction of members of one sex occurs because of behavioral, physiological, or morphological interactions with members of the opposite sex.

interspecific involving two different species.

interstice a narrow space or gap in a tissue or structure.

interstitial relating to or occurring in an interstice; i.e., in a narrow space or gap in a tissue or structure.

interstitial cells cells of mesenchymal origin that fill the spaces between follicles and receive innervation from the peripheral nervous system. Some of these have steroidogenic properties.

interstitial fluid the fluid surrounding Leydig cells in the interstitial compartment of the testis.

interstitial membrane prominent basal lamina that forms between the maternal endothelium and trophoblast of the interhemal barrier in many endotheliochorial placentae.

interstitial tissue loose connective tissue that lies between loops of the seminiferous tubule.

intervillous space a space surrounding the chorionic villi in some hemochorial placentae through which maternal blood circulates.

intracavernosal artery the extension of the common penile artery, traversing through the substance of the corpus cavernosum.

intracrine describing a system in which a substance produced by a cell acts on a receptor or membrane within the cell without being secreted by the cell.

intracytoplasmic sperm injection see ICSI.

intramembranous exchange any direct exchange of water and/or solutes between amniotic fluid and fetal blood.

intranidal thermoregulation the capacity of social insects to maintain homeostatic conditions within the nest, particularly a widely constant temperature around the brood, by heating and cooling behavior.

intraovarian factors a term for steroid hormones and growth factors synthesized and released within the ovary to act as paracrine or autocrine regulators of follicle development.

intrapartum a term referring to the time interval during the birth process.

intrasexual selection a process in which variation in reproduction of members of one sex and species occurs because of behavioral, physiological, or morphological interactions among individuals of that sex.

intraspecific within a single species.

intraspecific variability marked inter-individual or between-female variation in reproductive traits (egg size, clutch size, timing of laying) typical of all or most wild birds, often repeatable within individual females.

intrasyncytial lamina an interrupted extracellular layer in the interhemal barrier of many hemochorial placentae that is derived initially from the basal lamina of maternal endothelial cells and engulfed by processes of the syncytiotrophoblast.

intrauterine located within or occurring within the uterus.

intrauterine device (IUD) an object inserted into the uterus for the purpose of contraception.

intrauterine growth restriction or **retardation (IUGR)** a condition in which the fetus is less than the optimal size for survival; usually defined as a pregnancy in which the fetal size is less than the tenth percentile of the expected estimated size for gestational age; a severe IUGR is defined when the pregnancy is associated with fetal size of less than the third percentile). Also, FETAL GROWTH RESTRICTION.

intrauterine insemination (IUI) a process that utilizes special laboratory techniques (such as washing, swimup or gradients) to separate from a semen sample a population of sperm that may have a higher fertilization potential. These sperm are resuspended in a small volume of media and are then introduced into the uterus by the placement of a small catheter through the cervix.

intrauterine position (IUP) in certain litter-bearing animals, the position of a fetus, when there is more than one fetus, within the uterus relative to adjacent male and female fetuses.

introgressive hybridization the transfer of DNA between genetically distinguishable populations or groups of populations through repeated matings of hybrid offspring with parental individuals.

intromittent organ a term for the male copulatory organ; e.g., in reptiles.

intron a sequence in the transcriptional unit of a gene that is excised from the primary transcript during messengrer RNA maturation within the nucleus.

intussusception the growth of an organ or part by means of the reception of new matter from an external source.

in utero occurring or located within the uterus.

invasive describing a species in which the trophoblast invades the uterine wall to establish direct contact with maternal circulation and initiate placentation; e.g., humans, rodents. Also, describing a tumor that spreads from the primary site into surrounding tissue.

in vitro literally "in glass;" outside the body of an organism, or, more specifically, in a laboratory medium or setting as opposed to a living body.

***in vitro* fertilization (IVF)** an assisted reproductive technology in which oocytes are retrieved from the ovaries, usually after pharmacological stimulation to produce multiple mature oocytes. The oocytes are fertilized in the laboratory and allowed to grow and develop in culture. Usually, a few of the highest quality embryos are then chosen for placement into the uterus with the goal that one of these will successfully implant.

***in vitro* maturation (IVM)** in reproductive technology, the completion of cellular events in the oocyte that allows for fertilization and full developmental competence.

in vivo in life; occurring within a living organism.

involution the coordinated regression of the secretory component of the mammary gland at the cessation of milk secretion and removal. When complete, only the ductal structure of the mammary gland is apparent.

ion channel a protein structure that when open allows the passive flux of ions (specific or non-specific) across a biological membrane upon appropriate cell activation.

ionophore a molecule that transports specific ions across cell membranes.

irruption a sudden rapid increase in local population density, well above the typical numbers found during peaks in population cycles; population explosion.

ischemia insufficiency of blood flow to a given organ or tissue.

isoenzymes enzymes catalyzing the same reaction but encoded by different genes.

isogamy reproduction in which the two haploid gametes that unite in fertilization are of of equal size (isogametes). See also ANISOGAMY.

isogenetic describing cells of identical genetic composition, derived from a single precursor cell.

isthmus or **isthmic segment** the narrow lower segment of the Fallopian tube attached to the uterus.

iteroparous describing a reproductive pattern in which breeding occurs several times per lifetime.

IUGR see INTRAUTERINE GROWTH RESTRICTION.

IVF see IN VITRO FERTILIZATION.

J

Jacobson's organ see VOMERONASAL ORGAN.

JH see JUVENILE HORMONE.

junctional trophoblast the specialized trophoblast that keeps the placenta and external membranes attached to the uterus.

juvenile in indirect development, a stage of the life cycle following settlement and resembling the adult, yet not reproductively mature. In direct development, any pre-reproductive stage resembling the adult.

juvenile hormone (JH) a major insect hormone that is used in maintaining juvenile characteristics during larval development prior to metamorphosis, and by many insects in the adult stage as a gonadotropic hormone and a regulator of vitellogenin synthesis.

juvenile pause a period of slow growth without sexual maturation, between age 2 years and the onset of puberty, accompanied by low gonadotropin levels. Also, PREPUBERTAL HIATUS.

juxtacrine describing a system in which a substance produced by a cell is bound to the plasma membrane or extracellular matrix of the cell and acts on an adjacent cell by direct contact, e.g. growth hormone.

K

karyotype the chromosomal constitution of a cell; the representation of a complete set of chromosomes prepared from a photomicrograph of chromosomes within a dividing cell. A normal human karyotype consists of 46 chromosomes: 22 pairs of autosomes and a pair of sex chromosomes.

11-ketotestosterone a potent androgen of teleost fishes that promotes spermatogenesis. Once thought to be a male-specific androgen, it has also been found in substantial quantities in females of certain species.

kin selection changes in the spread of alleles in a population as a result of their influence on the survival or reproduction of genealogical relatives.

Kleihauer-Betke test a measure of the percentage of fetal blood in maternal blood, used to calculate the needed dose of Rh-immune globulin.

Klinefelter syndrome a developmental disorder characterized by gynecomastia, tall extremities, small atrophic testicles, elevated gonadotropins, and usually azoospermia. The underlying defect is the presence of an extra X chromosome (47, XXY) in the male karyotype. (From Harry *Klinefelter*, U.S. physician.)

knockout mouse a laboratory mouse created from embryonic stem cells in which a specific gene has been deleted or "knocked out" by special gene-targeting procedures.

L

labor forceful uterine contractions resulting in progressive cervical dilatation, fetal descent, and delivery of the fetus; also, the period of time during which this takes place.

labrum the anteriormost edge (or separate sclerite) of an arthropod head, functionally a kind of pre-oral upper lip.

α-lactalbumin specific milk whey protein, induced by prolactin, and part of the enzyme lactose synthetase necessary for the production of lactose (the primary milk carbohydrate).

lactation the production and secretion of milk by the mammary glands of the breast.

lactational anestrus the absence of ovulatory estrous cycles during lactation.

lactogen a factor that stimulates milk production; specifically, one of the lactogenic hormones.

lactogenesis the onset of secretory activity in the mammary gland; stage I lactogenesis is the first appearance of milk-specific products during pregnancy; stage II lactogenesis (the more common use of the term) is the initiation of milk secretion at parturition.

lactogenic hormones a family of peptide hormones, such as pituitary prolactin and placental lactogens, which not only stimulate lactogenesis, but also act on the brain, in conjunction with steroid hormones (estradiol and progesterone), to promote maternal responsiveness.

lactose the carbohydrate found in milk, which is composed of a molecule of glucose and galactose. It is the primary osmotic determinant of milk.

lactotrope or **lactotroph** a cell within the anterior pituitary gland that synthesizes and secretes prolactin.

laparoscope a slender fiberoptic telescope, placed through a small incision, to view or operate within the abdominal and pelvic cavities.

laparoscopically-assisted vaginal hysterectomy the use of a laparoscope and endoscopic instrumentation inserted through small ports in the abdominal wall to assist in the vaginal removal of the uterus.

laparoscopy a surgical procedure for the visual evaluation of the abdominal and pelvic cavities, involving the insertion of a fiberoptic telescope into the abdominal cavity through a small umbilical incision.

laparotomy a large incision through the abdominal wall, made to operate on or view the abdominal contents.

large granular lymphocytes phenotypically unique lymphocytes that have roles in cytokine secretion and immunoregulation in the decidua.

larva *plural,* **larvae** in various invertebrates, an independent, often free-living (free-swimming in marine forms) stage of development that metamorphoses to adult form and habitat.

larval relating to or occurring as larvae.

larval ecology the study of factors influencing the distribution and abundance of marine larvae and of processes occurring during the larval stage that influence the distribution and abundance of juveniles and adults.

larval spawning in marine invertebrates, the release of larvae, or embryos that develop into larvae, into the water.

lateral line organ a sensory organ in many fish that detects low frequency vibrations in the water. It is sometimes used in communication during courtship.

lateral oviduct in insects, branches from the ovary connecting to the common oviduct.

lecithotrophic relating to or describing a form of development in embryos and larvae, in which the sole source of nutrition is maternally-derived yolk reserves.

lecithotrophic viviparity a live-bearing reproductive pattern in which the ovulated eggs contain all of the essential organic nutrients for embryonic development but development to term takes place in the uterus.

lecithotrophy a developmental pattern occurring in oviparous and viviparous species, in which the embryo depends solely on its yolk reserves for nutrients; in organisms with a larval stage, a mode of development in which the larva relies on maternal yolk stores for nutrition.

leiomyoma see MYOMA.

lek a communal area in which various males of a species (e.g., birds) assemble during the mating period to attract females; also, the aggregation of males waiting in such an area.

Leopold's manuevers a series of four specific maneuvers performed during abdominal examination to determine the fetal position.

leptin a hormonal product of the obese (*ob*) gene and expressed by adipocytes; believed to be involved in regulating food intake, metabolism, and reproductive function.

leptotene the first stage of prophase I of meiosis.

Leydig cells the primary androgen-secreting cells of the testis; round or ovoid cells found in clusters in the interstitial compartment of the testis between the seminiferous tubules; they produce the preponderance of testosterone in the male in response to gonadotropin stimulation. (Discovered by Franz von *Leydig,* German anatomist.)

Leydig stem cell a mesenchymal cell from which the Leydig cell lineage originates, and which does not yet express specific proteins that are characteristic of adult Leydig cells.

LGA large for gestational age, a descriptive term for newborns whose weight is above the 10th percentile for gestational age.

LH see LUTEINIZING HORMONE.

LHRH see LUTEINIZING HORMONE RELEASING HORMONE.

LH surge a large release of LH (luteinizing hormone) that is triggered by exponentially rising estradiol concentrations in the circulation and that causes follicular rupture and release of the oocyte.

libido sexual energy, which includes sexual thoughts, fantasies and satisfaction.

ligand a naturally occurring or synthetic molecule that binds to a receptor and either activates the receptor for binding and responding to an activating molecule (agonist function) or prevents the receptor from binding and responding to an activating molecule (antagonist function).

limbic forming or occupying a limbus; i.e., a border.

limbic system a collective term for an array of interconnected neural structures concerned primarily with smell, emotion, memory, and neuroendocrine control; so termed from the sense that it forms a border between the old brain (paleoencephalon) and new brain (neoencephalon).

linkage analysis a method used to trace the inheritance of a disease-causing gene within a family based on the coinheritance of two or more genes because they are in close proximity on the same chromosome.

lipocalin a class of proteins which bind lipophilic molecules, frequently carrying lipids in an aqueous medium. In the vomeronasal system, lipocalins are present in some stimulus sources and thus may help to hold, deliver or remove stimuli from vomeronasal receptors.

lipogenesis the formation of fatty acids.

lipolysis the breakdown of triglycerides to release fatty acids.

lipoma a usually benign growth of breast tissue, typically composed of mature fat cells.

lipomastia excessive adipose (fat) tissue in the breast, commonly seen in obese men.

lipophilic literally, fat-loving; describing a substance that is highly soluble in fats.

lipoprivation a reduced availability of fatty acids.

lipoxygenase an enzyme that catalyzes the dioxygenation of an unsaturated fatty acid.

litter a number of young produced at a single birth.

LMP last menstrual period, a date calculation used in defining gestational age.

lobular carcinoma in situ (LCIS) abnormal cells that develop from the lining of the lobules in the breast. LCIS is not considered to be a precursor of cancer, but it is a marker of high risk.

lobule a small lobe; specifically, such a structure of the breast.

lobulo-alveolar units collections of ductules and differentiated apocrine glands that secrete milk.

local anesthetic a drug that blocks the neural impulses, therefore blocking nerve conduction, and preventing pain in a particular area of the body.

local therapy treatment that is used to remove or kill cancer cells in the specific region; e.g., surgery or radiation.

lochia vaginal discharge in the initial postpartum period resulting from the sloughing of decidual tissue.

lophophore a circumoral ring of ciliated tentacles arising from the mesosome in brachiopods, bryozoans and phoronids; sometimes used for the entire mesosome in these groups.

lordosis the distinctive component of reproductive behavior in female mammals produced in response to copulatory stimulation (e.g., mounting), consisting of immobilization, concave flexion of back muscles, extension of the limbs, lateral deviation of the tail, and elevation of the head and rump.

lorica a number of cuticularized plates in priapulid larvae covering the abdomen and probably having a protective function.

low density lipoprotein a complex macromolecule composed of apolipoprotein and lipid subunits, including cholesterol, a substrate for steroidogenesis.

lower vertebrates a collective term for reptiles, amphibians, and fish (cartilaginous and bony); i.e, vertebrates other than mammals and birds.

luminal epithelium the single layer of cells lining the cavity or lumen of the uterus. This tissue is central to regulating the occurrence of blastocyst implantation stimulating the underlying stromal cells to decidualize.

lumpectomy the surgical removal of a breast lump and also a margin of healthy breast tissue.

luteal related to the function of the CORPUS LUTEUM, a progesterone-releasing body formed on the ovary after ovulation.

luteal phase the phase of the female reproductive cycle following ovulation, when the corpus luteum produces and actively secretes progesterone.

luteal-placental shift the interval in pregnancy when the essential activities (i.e., progesterone production) of the corpus luteum are assumed by the placenta.

luteal regression the demise of a corpus luteum, indicated morphologically by apoptosis and biochemically by a decline in progesterone and increase in DNA laddering.

luteinization the physical and metabolic transformation of a mature ovarian follicle into a corpus luteum, principally characterized by a marked increase in progesterone synthesis in response to an ovulatory surge in luteinizing hormone.

luteinizing hormone (LH) a gonadotropic hormone produced by the anterior pituitary gland which, together with FSH (follicle-stimulating hormone), stimulates estrogen secretion and triggers ovulation of the oocytes from mature follicles in the ovary; it also stimulates the Leydig cells of the testis to produce testosterone.

luteinizing hormone releasing hormone (LHRH) another term for GnRH (gonadotropin-releasing hormone).

luteolysis the regression of the corpus luteum (CL), occurring in two phases: functional luteolysis involves drastic reduction in progesterone synthesis, followed by structural luteolysis which entails morphological destruction of the CL. In many, but not all, mammalian species uterine production of prostaglandin $F_{2\alpha}$ triggers luteolysis.

luteolytic causing regression or destruction (lysis) of the corpus luteum.

luteolytic hormone a hormone that causes decreased secretion of progesterone from the corpus luteum and luteal cell death.

luteotropic or **luteotrophic** having the ability to stimulate the corpus luteum, including cellular growth and hormone synthesis.

luteotropic factors factors intrinsic to the corpus luteum, acting as local luteotropins by either paracrine or autocrine pathways.

luteotropic hormone any of a variety of blood-borne substances acting directly on the corpus luteum (CL) to maintain its function and sustain the secretion of progesterone; e.g., luteinizing hormone (LH).

luteotropic signal a substance that supports corpus luteum function and/or inhibits mechanisms of luteolysis.

luteotropin or **luteotrophin** see LUTEOTROPIC HORMONE.

lymph node or **gland** one of the glands found throughout the body which defend it from bacteria or other foreign invaders. A lymph node dissection is usually done in the axilla (or the underarm area) during a lumpectomy or mastectomy to determine the extent of the spread of cancer outside of the breast area.

lymphatic relating to or involving the lymphatic system.

lymphatic system a system of ducts and nodules that help filter out invaders and transport fluid and protein in the body.

M

macrogynecomastia a dome-shaped breast enlargement greater than 5 cm in diameter that resembles the middle and late stages of normal female development.

macronucleus the somatic nucleus, usually highly polyploid, although the amount of polyploidy and the size of the chromosomes is species-specific.

macroorchidism a condition of larger than normal testis size.

macrophage a distinctive type of large cell that can ingest extracellular particles; e.g., microorganisms and other antigens, by engulfing them.

macrosomia the condition of having greater than normal fetal size or weight; commonly defined as a weight of 4,500 g or more after 38 weeks of gestation.

macrosomic relating to or characterized by macrosomia; i.e., greater than normal fetal size.

magnetic resonance imaging (MRI) a noninvasive procedure using radio waves and a strong magnetic field for two- and three-dimensional imaging of living tissue, including the brain and other body parts; e.g., a tumor site.

magnocellular neurosecretory system neurons whose large cell bodies (hence *magno-*) are located primarily in the paraventricular and supraoptic nuclei of the hypothalamus. These cells make oxytocin and vasopressin and send axons to the posterior pituitary.

magnum a region of the avian oviduct where the preponderance of egg white proteins (albumen) are deposited on the nascent oocyte.

maintenance of pregnancy a series of interactions between the embryo, uterus, and/or ovarian CL(corpus lutem) that prevent regression of the CL, or luteolysis.

maintenance of spermatogenesis the process of preventing the loss of germ cells once spermatogenesis has been fully established.

major histocompatibility complex (MHC) a large cluster of genes located on the short arm of chromosome 6 that code for many activities of the immune system, including the determination of tissue compatibility.

major urinary proteins (MUPs) small pheromone binding proteins that are present in large quantities in rodent urine and that are thought to convey the genetic individuality of the pheromonal signal.

male-male competition competition among individuals of the chosen sex for access to individuals of the choosy sex.

male pseudohermaphroditism a condition in 46XY males who have defective virilization of the external (and possibly internal) genitalia and/or the presence of Müllerian duct structures.

malformation irregular or anomalous formation or structure of parts of an organism.

malignant harmful; describing a tumor that is capable of invading surrounding tissue and/or spreading from its original site to other distant sites.

mammal one of the Mammalia, a large class of vertebrate animals (including humans) characterized by various common anatomical features and specifically by the nourishment of young with milk from the mother's mammary glands.

mammalian relating to or designating a mammal.

mammary relating to or involving the mammary glands.

mammary duct a hollow tube composed of a double layer of non-secreting epithelial cells, which directs mammary secretions from the alveolar lumen to the teat or nipple.

mammary explant a small portion of intact mammary tissue (~3 mg) removed from the mammary gland and often used in tissue culture to determine effects of hormones and other substances on milk component biosynthesis or secretory cell differentiation.

mammary gland a distinctive milk-secreting structure of adult female mammals, also present in rudimentary form in males and immature females; in the human female, one of a pair of glands overlying the pectoralis major muscle on either side of the chest, composed of 15 to 25 lobes arranged radially around the nipple and separated by connective and adipose tissue, with each lobe having a secretory duct opening on the nipple.

mammogenesis the growth and differentiation of the mammary gland under the influence of the hormones of puberty and pregnancy.

mammogram a low dose X-ray used to identify abnormalities in the breast.

manchette a curtain-like, conical, rigid structure composed of microtubules, which apears in elongating spermatids. Inserted at the equator of the nucleus, it surrounds the forming tail and disappears during the late steps of spermiogenesis.

mandibular organ (MO) a small endocrine gland in crustacea associated with the mandibular tendon, the source of methyl farnesoate, which is a major hormone in crustacean reproduction.

mantle in certain shelled invetebrates, a bilayered region of body wall that secretes the calcified shell.

marsupial one of the Marsupialia, those mammals that carry and nourish their developing young in a pouch, including koalas, kangaroos, wombats, and oppossums.

masculinize to cause an individual to develop the secondary sexual characteristics of a mature male, as by the administration of androgens.

mastalgia a general term for any recurring or persistent sensation of pain in the breast that is not associated with an immediate trauma.

mastectomy the surgical removal of the breast.

mastitis inflammation of the breast, usually due to bacterial infection.

mate attraction the use of specialized acoustical, visual, or chemical signals (pheromones) to attract mates from a distance, sometimes over several hundred meters or more.

mate guarding aggressive behavior by a male who has just mated toward other nearby males to prevent or discourage their mating with the same female.

maternal relating to, involving, or derived from the female parent.

maternal constraint non-genomic limitation of fetal growth by maternal size (and thus uterine size) and related physiological factors such as uterine blood flow and uterine capacity to support placental growth and function.

maternal effect a mechanism by which a female affects the characteristics of her offspring that does not involve the transmission of nuclear genes. Typically this is achieved through the transmission of chemicals to the egg.

maternal-placental-fetal unit a collective term for mother, placenta, and fetal communication and interrelated production of hormones during pregnancy.

maternal recognition (of pregnancy) maternal physiological changes, induced by signals emanating from the conceptus, which result in maintenance of corpus luteum function and facilitate establishment and maintenance of pregnancy.

mating the sexual coupling of male and female individuals for the purpose of reproduction, as well as other behaviors that promote or are associated with this sexual union.

mating sign in social insects, a mucus plug inserted by a copulating male into the female, thus preventing further mating; in honeybees, multiple matings require the removal of the previous mating sign by the next copulating drone.

mating system the typical pattern of reproductive behavior and parental care in a given species, or the social structure resulting from this.

matrotrophic relating to or characterized by matrotrophy.

matrotrophy or **matrotrophic viviparity** a viviparous developmental pattern in which the mother provides nutrients during gestation by a means other than the yolk of the ovum (e.g., oviductal secretions, sibling yolks, placental tissues).

maturation-inducing hormone another term for MATURATION-INDUCING SUBSTANCE.

maturation-inducing substance or **steroid (MIS)** a steroid hormone that is produced in the follicular tissue under the influence of a gonadotropin, responsible for the control of the final stages of oocyte maturation in various animals; e.g., teleost fish.

maturation-promoting factor (MPF) a complex of proteins originally identified as a key regulator of final oocyte maturation, composed of the regulatory cyclin B and the catalytic $p34^{cdc2}$ kinase; also subsequently found to be a general inducer of the M (mitotic) phase of the cell cycle common to all eukaryotic cells.

mature follicle an ovarian follicle that has acquired an adequate concentration of gonadotropin receptors to undergo an ovulatory response when stimulated by a surge in endogenous LH (and/or FSH), or by an adequate amount of exogenous chorionic gonadotropin such as hCG.

mature milk milk that is usually present after approximately ten days following birth. Volume and content of the mature milk adapts to the growth needs of the infant.

MCR metabolic clearance rate, the rate at which a given chemical product is removed from the body or converted to another substance.

meconium a dark-green fetal or neonatal material consisting mainly of gastrointestinal secretions, cellular debris, bile acids, mucus, and water that appears in the fetal ileum by 16 weeks of gestation. The passage of meconium into amniotic fluid occurs in up to 22% of live births and is more commonly seen in post-date pregnancies.

meconium aspiration a syndrome in the neonate characterized by severe respiratory distress resulting from the aspiration of meconium. This produces a mechanical obstruction of airways as well as a diffuse chemical pneumonitis.

medial preoptic area a forebrain region critical for the expression of copulatory behavior in male mammals and certain maternal behaviors in female mammals.

median eminence the extension of the neurohypophysis, or posterior pituitary gland, that actually connects with (and is sometimes described as part of) the hypothalamus. It contains capillaries that receive nerve fiber output from the cells in the hypothalamus, thus serving to convey hypothalamic regulating factors such as GnRH to the endocrine cells of the pituitary.

medusa the pelagic (free-swimming) phase in the typical cnidarian life cycle; these animals reproduce primarily or exclusively by sexual reproduction. (From the Greek mythological figure *Medusa*, whose snaky locks were thought to resemble these animals.)

megalopa in certain crustaceans, a larval form that appears after the zoea, which swims with its abdominal appendages, and is restricted to the malacostracans since they are the only group with true pleoplods.

meiosis a specialized process of nuclear division of germ cells, in which the DNA content is first duplicated and then the chromosome number is halved; an essential step in the formation of gametes. By this process, a single diploid cell gives rise to four haploid cells.

meiotic relating to or involving meiosis.

meiotic maturation the final step of oogenesis that includes condensation of chromatin, nuclear envelope breakdown, and separation of homologous chromosomes; and results in the formation of female gametes capable of being fertilized by sperm.

melatonin the primary hormone product of the pineal gland, released almost exclusively into the circulation during the night in both day-active (diurnal) and night-active (nocturnal) species; it has an important role in the regulation of daily and seasonal biological rhythms; e.g., seasonal reproductive patterns.

menarche the onset of menses; the initial episode of menstrual or uterine bleeding.

menopausal having to do with, involving, or occurring at menopause.

menopause the permanent termination of the menstrual cycle in the human female, in the contemporary era occurring at a mean age of about 51 years. Menopause is clinically defined by the World Health Organization as the final menstrual period following which no menses occur for 12 consecutive months.

menopause transition the point of time encompassing the beginning signs of ovarian failure such as shortened or irregular menstrual cycles and fluctuations in hormone levels, up to the onset of menopause.

menorrhagia see HYPERMENORRHEA.

menotropins a term for luteinizing hormone (LH) and follicle-stimulating hormone (FSH) obtained from the urine of menopausal women.

menses the flow of tissues, blood, and mucus that occurs during menstruation.

menstrual cycle the reproductive cycle of the human female, characterized by monthly shedding of the endometrium (the lining of the uterus) resulting in a period of vaginal bleeding. (From Latin *mensa,* month.)

menstrual synchrony the occurrence of menstruation coordinated in time among a group of female primates, including humans, following synchronous ovulation.

menstruation the periodic shedding of endometrial tissue that occurs when an ovulatory cycle has not resulted in a pregnancy.

meroistic ovary a type of ovary found in many insects, in which nutritive germ cells supply the developing oocyte with RNA and organelles through cytoplasmic bridges.

mesadenia accessory glands of male insects derived from mesoderm.

mesenchymal cell a spindle-shaped cell, and principal constituent of mesenchyme, the first tissue to appear in the embryo; it has the ability to differentiate into all connective tissue cell types, including Leydig cells in the testis.

mesenchyme an embryonic tissue type; usually composed of mesodermal cells that migrate among the cells of other embryonic tissues, often epithelia.

mesenteron a capacious non-digestive anterior portion of the midgut of some insects, in which undigested food is stored and from which food is released gradually to the intestine for digestion.

mesentery a tissue consisting of flattened longitudinal muscle cells that connects the urogenital system to the body wall.

mesoderm the middle of the three primary layers of the embryo, which forms the skeletal, muscle and vascular systems.

mesoglea the central body layer of a cnidarian, which varies from being an acellular adhesive holding together the two cellular layers (ectoderm and endoderm) in Hydrozoa, to a substantial layer containing many cells in Anthozoa, to being the thick, mostly acellular central layer of Scyphozoa that gives "jellyfish" their name.

mesometrial relating to the mesometrium or located in or near the mesometrium.

mesometrium the lower portion of the broad ligament of the uterus, composed of layers of peritoneal cells that enclose the uterus and attach it to the lateral wall of the pelvis.

mesonephric duct another name for the WOLFFIAN DUCT, an embryological duct draining the mesonephros.

mesonephros the second of three sets of excretory organs that develop sequentially during human embryogenesis. The mesonephros functions as an interim kidney until the permanent kidneys evolve, then partially regress. The glomeruli and tubules of the mesonephros connect to the mesonephric (Wolffian) duct.

mesosalpinx the external layer of mesothelium and loose connective tissue covering the oviduct.

messenger RNA (ribonucleic acid) a strand of nucleotide bases that code for particular proteins made by the cells. It is produced in the nucleus by the transcription of the genetic code on DNA molecules. Messenger RNA (mRNA) travels to the cytoplasm where it is decoded by ribosomes and transfer RNA (tRNA).

metabolic relating to, involved in, or affected by a process of metabolism.

metabolic overhead the proportion of a lactating female's energy expenditure that is used to meet her own metabolic needs as opposed to milk production.

metabolic rate the sum total of all metabolic or biochemical processes going on in an animal per unit time. The lowest metabolic rate necessary to keep an animal alive is termed the basal or standard metabolic rate; it is usually estimated by measuring oxygen consumption, heat production, or carbon dioxide production. Energy expenditures for reproduction, activity, thermoregulation (in birds and mammals), and growth elevate the metabolic rate above basal levels.

metabolism a broad term for the array of complex physical and chemical processes involved in the maintenance of life by an organism.

metacentric describing a chromosome in which the centromere divides in two similar parts along the entire length, forming two similar arms.

metamorphic relating to or involved in a process of metamorphosis.

metamorphosis in many animals, a life cycle in which there is a significant difference in form between the adult life stage and preadult stages; also, any such process of change in form; e.g., the transition from a larval to a juvenile state in certain marine invertebrates.

metanephridia a common excretory structure in invertebrates, characterized by a ciliated funnel opening to the visceral coelom.

metastasis the spread of cancer beyond the primary site. The extent of the spread is measured by the number of lymph nodes that have cancerous cells in them.

metestrus a term classically used to designate the period of subsiding follicular function or rest following estrus in female mammals.

methyl farnesoate (MF) in crustaceans, a hormone produced by the mandibular organ (MO) that is very similar to the juvenile hormone (JH) produced by insects, and that is associated with reproduction in both males and females.

methyl palmoxirate a compound that inhibits fatty acid transport into mitochondria, inhibits fatty acid oxidation, and induces lipoprivation.

metorrhagia bleeding between menstrual periods.

metritis inflammation of the uterus.

metroplasty surgical repair of the uterus.

MHC see MAJOR HISTOCOMPATIBILITY COMPLEX.

micelle a spherical arrangement formed by a group of lipid molecules in an aqueous environment.

microcotyledon a placental exchange unit created by interdigitation of trophoblastic microvilli and apposing endometrial epithelium.

microdeletion an error in meiosis, detected by recombinant DNA techniques, resulting in the loss of a gene or group of genes.

microfilament a thin intracellular fiber composed primarily of actin, involved in many aspects of cell strucure and activity.

microorchidism smaller than normal testis.

micropyle or **micropylar apparatus** in insects, the localized structure on the egg surface that permits sperm to traverse the chorion to fertilize the oocyte.

microsurgery surgery performed under magnification.

microtubules long filamentous skeletal structures in cells that help to maintain the shape of the cell and that also are responsible for various other important cellular processes; e.g., moving DNA during the cell division cycle.

mictic involving sexual reproduction (mixis); specifically, referring to a phase in the life cycle of monogonont rotifers in which reproduction is sexual, resulting in a diapausing, diploid embryo.

midpelvis one of two clinically important pelvic planes, which is bounded by the sacrum posteriorly, the inferior margins of the ischial spines laterally, and the lower margin of the pubic symphysis anteriorly.

mifepristone (RU-486) the first synthetic antiprogestin steroid produced; competes with progesterone for binding to the progesterone receptor and blocks the action of progesterone.

migration in general, any sustained movement of an animal to a relatively distant location, not in response to an immediate transitory stimulus such as a potential threat; specifically, a regular movement of a group of animals of a given species from one site to another distant site (e.g., a reproductive site), occurring on a regular basis and associated with seasonal changes in light, temperature, precipitation, etc.

migration drive an innate behavior pattern that causes animals to begin migration.

migration syndrome a suite of interacting physiological, morphological, life-history and behavioral adaptations that have evolved to maximize the benefits of the migratory lifestyle.

migratory disposition a state of readiness for migration, which involves many morphological, physiological, and behavioral adaptations.

migratory drive see MIGRATION DRIVE.

migratory restlessness the characteristic restless or incessant activity of certain migrant species (e.g., birds) while in captivity, an expression of migration drive.

milk ejection the contraction of myoepithelial cells surrounding the alveoli and ducts that brings about expulsion of milk.

milk-ejection reflex a reflex initiated by the suckling of the infant at the breast, causing the release of the hormone oxytocin and the ejection of milk from the collecting ductules in the breast.

milk transfer the process of by which milk is moved from the mammary gland to the mouth of the offspring.

mineralocorticoid any of a class of steroid hormones secreted by the adrenal cortex that regulate Na^+ and K^+ balance.

MIS see MATURATION-INDUCING SUBSTANCE; MÜLLERIAN-INHIBITING SUBSTANCE.

missense mutation a single base DNA mutation changing the codon for one amino acid to another.

mitochondria *singular,* **mitochondrion** the self-replicating organelles found in the cytoplasm of all eukaryotic cells that direct ATP production and have a genome separate from the nucleus.

mitogen any factor that induces mitosis or cell replication; e.g., a protein derived from plants that is used in the laboratory to stimulate cells to divide.

mitogenic relating to or stimulating cell division.

mixed development a developmental mode of marine invertebrates that includes a brooded or encapsulated embryonic stage as well as a free-swimming larval stage.

mixis sexual reproduction.

modular animal a genet composed of multiple modules; a collective term for all clonal and colonial animals.

module the fundamental functional unit that is repeated in sequential iterative fashion in the development of a colony or clone of organisms.

molecular clock a condition in which the probability of nucleotide changes in a sequence of DNA is approximately constant per unit time. If this condition holds, and if the number of changes is known by calibration with the fossil record, then the number of DNA differences between species can be used to estimate the time of divergence of those species.

monandrous describing species in which females generally mate only once during their lifetime.

monestrous or monoestrous having a single fertile cycle during a given mating season.

monestrum the ovulatory pattern in which there is only one ovulation per breeding period.

monoclonal antibodies (MA) identical antibody molecules produced from a single immune cell; antibodies produced by one cell recognize only a single antigenic site.

monocyte a large white blood cell active in destroying and devouring bacteria or foreign matter that has entered the body.

monogamous having the behavior pattern of monagamy; mating with only one of the opposite sex.

monogamy a mating pattern in which one male mates with one female for one or more reproductive cycles; in some species this behavior also involves cohabitation, the cooperative rearing of young, and the exclusion of strange males and females from the territory.

monohormonal containing only one hormone; specifically, in the case of gonadotropes, a cell containing only LH or FSH.

monokine a type of polypeptide produced by the monocyte/macrophage lineages that mediates signaling in a number of cellular systems; however, some traditional monokines are from other cell lineages.

monophyletic derived from one ancestral taxon; sharing a single common ancestor and including all the descendants of this ancestor.

monophyletic group any group of organisms including all the descendants, living and extinct, from a common ancestral species. Recent trends in taxonomy designate monophyletic groups of organisms, rather than the traditional Linnaean categorical ranks of genus, family, order, etc. This trend recognizes that different groups of the same categorical rank may nonetheless differ drastically in age and time of origin, and some traditional taxonomic groups do not include all the descendants from a particular common ancestor (i.e., are not monophyletic). Also, CLADE.

monorchism or **monorchia** the condition of having one testis.

monotocous producing one offspring at a time.

monotreme one of the Monotremata, primitive egg-laying mammals including the echidna and the platypus.

monozygotic twins "identical" twins; i.e, twins that develop from one single egg.

morning sickness nausea and vomiting during the first half of pregnancy, typically commencing between the first and second missed period.

morph the form or shape of an organism; e.g., the specific form of a given individual or the characteristic form of a species.

morphogen a substance existing in a concentration gradient along the body axis that specifies position along the axis.

morphogene a gene that directly or indirectly controls the development of body form.

morphogenesis the process through which an organism, or a body part, acquires its characteristic form or structure.

morphological relating to or based on body shape or form.

morphology the form or shape of an organism; also, the scientific study of the form of organisms, especially external form.

morphospecies organisms characterized by the same morphological pattern, regardless of whether they are genetically connected.

morphotype a morphologically distinct body form present within populations whose polymorphic features undergo a generational, non-genetic change in phenotype.

morula *plural,* **morulae** the stage of embryonic development between the eight-cell and blastocyst stages, when cells are too numerous to count easily.

mosaic an individual with cells of different genotypes.

mosaicism the fact of having two or more cell lines of different genetic identity.

motility locomotive ability conferred by whip-like motions of the sperm flagellum.

motoneuron a specialized nerve cell in the spinal cord or brainstem that directly connects to muscle.

mounting behavior a component of masculine sexual behavior that brings the male's body into contact with the female in an orientation that facilitates penile intromission into the female's vagina.

MPF see MATURATION-PROMOTING FACTOR.

MRI see MAGNETIC RESONANCE IMAGING.

mRNA see MESSENGER RNA.

MSAFP maternal serum alpha-fetoprotein; used as a screening test for fetal neural tube defects. It can also be used in tests to identify pregnancies at risk for aneuploidy.

mucosal immunity the host's nonspecific as well as directed (from prior exposure) immune response at the mucosal level to microbes invading at the skin barrier, involving various types of white blood cells and biochemicals.

Müllerian anomaly an abnormality of the developing female reproductive organs (Müllerian ducts).

Müllerian or **müllerian duct** one of the pair of embryonic ducts that develop into the upper vagina, cervix, uterus, and oviducts (Fallopian tubes) in females and degenerate into a vestigial structure (appendix testis) in males. (From Johannes Peter *Müller,* German physiologist.) Also, PARAMESONEPHRIC DUCT.

Müllerian-inhibiting factor (MIF) see MÜLLERIAN-INHIBITING SUBSTANCE.

Müllerian-inhibiting substance (MIS) a glycoprotein hormone of the TGF (transforming growth factor)-β multigene superfamily that is produced by the fetal Sertoli cells and that causes regression of the Müllerian ducts in the male fetus. Also, ANTI-MÜLLERIAN HORMONE.

Müllerian structures a collective term for the internal organs that develop from the Müllerian ducts of the fetus. See MÜLLERIAN DUCT.

multiple fission asexual reproduction by successive or simultaneous divisions of the parent cell resulting in numerous genetically identical daughter cells.

murid an animal of the family Muridae, the Old World mice and rats.

murine relating to or describing the murids.

musth a physiological and psychological condition in the male elephant characterized by elevated androgens, increased aggression, and mating behavior.

mutagen a chemical or physical agent that induces genetic mutations.

mutualism an ongoing association of two organisms of different species, in which both benefit by the association.

MYA millions of years ago.

myoepithelial cell any of various epithelial cells that facilitate secretion of sweat, saliva, or milk.

myofilaments the contractile filaments of muscle cells; thick filaments contain myosin while thin filaments contain actin.

myoma *plural,* **myomata** a benign tumor of the smooth muscle tissue of the uterine wall. Also, LEIOMYOMA.

myomectomy the surgical removal of leiomyoma without the removal of the entire uterus.

myometrial relating to or involving the myometrium, the muscular wall of the uterus.

myometrium the muscular outer wall of the uterus, consisting of an inner, circular layer of smooth muscle and an outer, longitudinal layer of smooth muscle.

myosalpinx the muscle layer of the oviduct.

myotropin a chemical agent that causes muscles to contract.

N

natural hybridization successful (i.e., producing viable and at least partially fertile offspring) matings in nature between individuals from two populations, or groups of populations, which are distinguishable on the basis of one or more heritable characters.

natural selection the processes of differential survival and reproduction depending on non-random interactions between individuals and their environments, such that environmental factors (pressures) differentially favor the survival and reproduction of some individuals rather than others on the basis of variation among individuals in heritable phenotypic traits.

neck a short segment of the spermatic tail bridging the nucleus and the middle piece. Solidly attached to the nucleus within an implantation fossa, the neck is composed of electron dense material in the form of a striated collar seen around cavities that in the spermatozoa of mice and rats contain remnants of the proximal and distal centrioles. The neck is fully formed late in spermiogenesis.

negative feedback a process in which secretory products released by a target organ in response to a stimulating hormone prevent or limit further release of that hormone.

nematocyst one of three types of cnidae, and the only one to occur in members of all four classes.

nematocyte a stinging cell of cnidarians.

neoblasts quiescent undifferentiated cells capable of differentiating into other cell types, including gametes.

neonatal relating to or occurring in the period immediately following birth through the first 28 days of life.

neonate a newborn.

neoplasm literally, new formation; an abnormal growth of cells; i.e., a mass of tissue exhibiting the general characteristics of uncontrolled proliferation of cells, an accelerated rate of growth, an abnormal structural organization, and a lack of coordination with the surrounding tissue.

neoplastic relating to or producing neoplasms.

nervus terminalis a ganglionated cranial nerve of unknown function that connects the nasal epithelium and anterior forebrain in all vertebrate groups, and that contains GnRH-immunoreactive (and other) neurons and fibers in most groups.

neural relating to a nerve or the nervous system.

neural-hemal see NEUROHEMAL.

neural hormone see NEUROHORMONE.

neural lobe another term for the INFUNDIBULAR PROCESS.

neural plasticity long-term changes in peripheral or central innervation density, synaptic efficacy or phenotypic action of neurons (e.g., neurotransmitter and neuromodulator characteristics) that can be induced by pathophysiology and other means.

neural tube defect an abnormal opening in the brain or spine.

neuroendocrine relating to the interaction of the nervous system and endocrine system; specifically, relating to the release of a hormone into the circulation from a nerve terminal.

neuroendocrine axis a network of two or more neuroendocrine glands or ensembles, which are related via feedback and feedforward in the generation, maintenance, and modulation of hormone release; (e.g., the hypothalamus, pituitary gland, and gonads constitute principal nodes in the neuroendocrine reproductive axis at glandular levels.

neuroendocrine reflex a physiological mechanism by which an external or internal stimulus activates afferent and then central neurons to alter the secretory rate and blood levels of a specific hormone; i.e., a neural stimulus triggers a stereotypical endocrine response.

neuroendocrine system the overall complex of nervous and hormonal pathways that act in concert to regulate many body processes and activities; e.g., reproduction, growth and differentiation of tissues, water balance, and various behaviors.

neuroendocrinology the scientific study of the interactions between the nervous system and endocrine system.

neurohemal relating to or describing the interrelated functioning of the nervous and endocrine systems.

neurohemal junction a specialized junction between nerve terminals and a plexus of capillaries which facilitates the transmission of neurohormones from the nervous system into the bloodstream or, conversely, the transfer of hormones from the bloodstream into the nervous system.

neurohemal organ a specialized site of release for neurohormones; e.g., the median eminence.

neurohormone a chemical messenger that is released from nerve terminals into the bloodstream, which then transports the substance over some distance to its target cell. Neurohormones are often the same chemical substance as neurotransmitters.

neurohypophysial relating to, occurring in, or involving the neurohypophysis.

neurohypophysial hormones hormones synthesized in the brain and stored in the neurohypohysis; they control water balance and smooth muscle contraction (e.g., in the uterus).

neurohypophysis the posterior neural lobe of the hypophysis (pituitary gland). Hormones produced in the hypothalamus (e.g., oxytocin, vasopressin) are transported down axonal fibers to terminals in the neurohypophysis, where they are stored until released.

neuromodulator a substance secreted by a neuron that does not change the membrane potential of other neurons but alters their electrical activity.

neuron a nerve cell.

neuronal relating to nerve cells.

neuropeptides small molecules in neurons made from amino acids that are connected by peptide bonds and are secreted as intercellular messengers; i.e., as hormones, neurotransmitters, etc.

neuropeptide Y an appetite-stimulating agent linking nutrient status to hypothalamic regulation of feeding behavior and function of the hypothalmic-hypophyseal axis; an intermediary of leptin activity.

neurophysin the carrier protein for oxytocin and vasopressin that increases their half-life in blood.

neurosecretion the process by which a neuron is able to release soluble messengers from its nerve endings; specifically, the process by which specific neuroendocrine cells release hormones directly into the blood stream. Neurons of the hypothalamo-neurohypophysial system have served as an important model system for the development of the concept of neurosecretion.

neurosecretory relating to or participating in neurosecretion.

neurosecretory cell a neuroendocrine cell that has characteristics of both endocrine cells and neurons, located in or associated with the central nervous sytem, containing peptides or amines that are released into general circulation.

neurosecretory hormone a hormone that is produced by specialized neuronal cells in the hypothalamus.

neurosecretory vesicles or **granules** large intracellular vesicles, 100–350 nm in diameter, usually with electron-dense cores containing peptides and/or proteins that are released as secretory material.

neurotransmission a process in which a substance produced by a nerve cell is secreted so as to influence the activity of a neighboring nerve cell or other cell.

neurotransmitter a chemical compound released by nerve cells that either excites or inhibits a target cell. The target cell could be another nerve, a glandular cell, or a muscle fiber.

nipple the pigmented, protruding coneshaped structure at the center of each mammary gland, with a series of tiny openings through which milk passes.

nipple discharge the spontaneous escape of fluid from the nipple.

nipple secretion fluid within the ducts of the breast collected by nipple aspiration or breast massage.

node a common ancestor, depicted as the intersection of two branches in a phylogenetic tree.

non-disjunction the failure of two like chromosomes to separate during cell division, such that one cell receives both chromosome copies and the other cell receives neither.

non-invasive describing a species in which the trophoblast does not invade the uterine wall to establish direct contact with maternal circulation, but remains essentially in the uterine lumen, and placentation involves only superficial contact with the maternal tissue; e.g., ruminants.

nonsense mutation a single-base DNA mutation that changes the codon for an amino acid to a stop codon.

non-steroid hormones a class of widely distributed, water-soluble polypeptides that are either growth factors or growth factors-related and that often regulate cellular proliferation or differentiation through paracrine and/or autocrine actions.

normospermia a condition in which most ejaculated sperm are structurally normal.

notochord a supportive, longitudinal rod dorsal to the nerve cord in chordates; only present in embryos of vertebrates.

NSAID non-steroidal anti-inflammatory drug; e.g., aspirin, indomethacin, naproxen and ibuprofen.

nullipara a woman who has not had a child.

nulliparity no occurrence of child birth.
null mutation the targeted disruption (knockout) of a single gene in mice.
nuptial plumage in many birds, especially males, an array of brilliant or showy feathers assumed at the onset of breeding, associated with increases in pituitary or gonadal hormones.
nurse cell in invertebrates, a cell that contributes nutrients and possibly organelles to growing oocytes, embryos, or developing gemmules.
nurse egg a nondeveloping egg that is ingested by developing larvae.
nursing in mammals, the act of feeding offspring by enabling them to suck milk from the female nipples.
nursing behavior any of various maternal behaviors that facilitate suckling by young mammals.
nymph the last larval instars of hemimetabolous insects with still non-functional wings.

O

obligate brood parasitism in birds, a reproductive strategy in which all females of a species lay their eggs in other species' nests.
obstetric relating to or involved with obstetrics.
obstetrics the medical management of pregnancy, labor, and delivery, and postpartum care.
occupational toxin a harmful chemical substance, present in the workplace, that acts on one or more types of cells within the testis to interfere with spermatogenesis or steroidogenesis or both.
olfactory relating to or involved with olfaction; i.e., the sense of smell.
olfactory bulb the brain structure where most primary olfactory nerve fibers are first relayed.
olfactory pit either of two structures formed during early development when the olfactory placodes sink below the surface. The olfactory pit then forms two recesses, one medial and one lateral. The lateral recess of the olfactory pit gives rise to the olfactory nerves, and the medial recess gives rise to the vomeronasal and terminal nerves.
olfactory placode a thickening of the ectoderm on the lateral sides of the head of vertebrate embryos that gives rise to the nasal cavity.
olfactory system a collective term for the olfactory receptor cells in the olfactory epithelium of the nose, their axons that form the olfactory nerves, and the main olfactory bulb.
oligohydramnios a disorder of extremely reduced amniotic fluid volume, usually defined as an index of 5 cm or less.
oligomenorrhea a condition in which menstrual cycles occur at infrequent or irregular intervals; specifically, intervals greater that 35 days apart.
oligospermia or **oligozoospermia** a low sperm count, usually defined as fewer than 20 million spermatozoa per milliliter of ejaculate.
oncogene a gene that influences cell proliferation and that can cause or contribute to tumor formation.
oncogenic causing or contributing to tumor formation.
ontogeny or **ontogenesis** the development and changes of an individual organism during its particular lifetime; specifically, such development at the embryonic/fetal stage.
oocyte a female germ cell prior to fertilization. A developing egg is termed a **primary oocyte** after being derived from the oogonia and entering meiotic division. Following an extended resting period, meiosis is then resumed at ovulation time; this completes the first meiotic division, resulting in a large haploid **secondary oocyte** and a tiny haploid first polar body. Only a very small percentage of the oocytes that enter meiosis will ovulate; all others will disappear by atresia or apoptosis.
oocyte maturation the transformation of an oocyte into a fertilizable egg, occuring upon cessation of the first prophase arrest and progression through metaphase I to the second metaphase of meiosis.
oocyte maturation inhibitor (OMI) a factor possibly generated by the granulosa cells, that has been sugggested as the agent maintaining oocytes in their prolonged resting period at the prophase stage of the first meiotic division.
oocyte transport the passage of the oocyte from the ovarian surface into and through the oviduct, with further passage into uterus and expulsion via the vagina.
oogenesis the process of oocyte growth and development, or, more generally, the entire series of developmental events and cellular transformations, beginning with the formation of primordial germ cells and then progressing to oogonia and oocytes, up to the production of a highly specialized gamete in the adult that is capable of transmitting genetic information to the embryo.
oogenesis-flight syndrome an explanation for the temporal and physiological relationship between migration and reproductive development in female insects.
oogonia *singular,* **oogonium** stem cells of the ovary, derived from primordial germ cells, which increase in number by mitosis and then enter the meiotic cell cycle as primary oocytes.
oolemma the limiting plasma membrane of oocytes.
oostegites platelike structures on the ventral thoracic surface of some crustaceans, used in the incubation of developing eggs or young.
ootype the egg-packaging region of the oviduct in neodermatan platyhelminths; the site where the eggshell develops and encloses oocytes and yolk cells.
open neural tube defects (ONTDs) a term for congenital abnormalities caused by incomplete closure of the neural tube (spinal cord) during embryogenesis, the two most common forms being anencephaly (failure of the cephalic end of the neural tube to close) and open spina bifida (failure of inner portions of the neural tube to close).

operculum the specialized structure in many insect eggs that facilitates the escape of the larva or nymph from the eggshell at the time of hatching. This region of the egg surface may also function in respiration.

opiate a drug having actions like those of morphine, one of the many similar natural alkaloids present in extracts of the opium poppy, *Papaver somniferum,* having a range of actions through opioid receptors.

opioid a peptide produced by neurons, and other cell types, having actions like morphine (which is an opiate), belonging to one of three groups: enkephalins, endorphins, dynorphins.

opisthogoneate describing a condition in certain invertebrates in which the reproductive systems debouch to the exterior at the posterior end of the body.

opisthosoma the posterior body segment of chelicerates, corresponding to an abdomen. In scorpions, the opisthosoma comprises two regions, a proximal mesosoma and a distal metasoma.

opportunistic breeding the characteristic in some species of having reproductive behavior that varies according to the immediate, variable conditions at hand; e.g., food availability or the effect of predators, as opposed to being strictly regulated by more general factors of season, period of daylight, temperature, etc.

orchidectomy or **orchiectomy** the surgical removal of one or both testes.

orchiopexy a surgical procedure that frees the undescended testis from the structures preventing its reaching a dependent scrotal position.

orchitis an inflammation of the testis.

organification the process of original organ formation that occurs during the embryonic period.

organization see ORGANIZATIONAL EFFECTS.

organizational or **organizing effects** long-term or permanent changes in structure and function, typically due to exposure to elevated levels of a hormone during discrete critical periods of development.

organogenesis the formation and development of the different organs, or of a particular organ.

osmotic shock cellular damage resulting from the rapid flow of extracellular fluid into the cell.

osteoblast the primary bone forming cell, derived from stem cells in marrow stroma. Osteoblasts control bone remodeling by elaborating cytokines that suppress or recruit osteoclast production and maturation.

osteoclast multinucleated giant cells of bone reabsorption, derived from macrophage/monocyte precursors. Osteoclasts respond to cytokines secreted by osteoblasts in response to hormonal or other signals.

osteoporosis an abnormal loss of density and weight in bone, associated with the postmenopausal stage in women and typically characterized by loss of stature, various deformities, and fractures from insignificant or incidental injury.

osteoporotic relating to or affected by osteoporosis (loss of bone tissue).

osteoporotic fracture a fracture occurring as a result of minimal trauma, such as a fall from standing height.

outer dense fibers (ODF) major cytoskeletal elements of the tail of elongated spermatids and spermatozoa. In the middle piece, there are nine outer dense fibers, one per axonemal doublet, each having a characteristic size and shape. In the principal piece, only seven such fibers are present and they decrease in caliber in a proximal-distal direction.

ova the plural form of OVUM (egg).

ovalbumin egg-white protein synthesized and secreted in large amounts upon estrogenic stimulation of tubular gland epithelial cells in the avian oviduct.

ovarian relating to, involving, or affecting the ovary.

ovarian cycle the cyclical sequence of correlated events within the ovary, leading to the release of a mature egg (oocyte) that is capable of being fertilized.

ovarian follicle the oocyte and its surrounding granulosa and theca layers. See FOLLICLE.

ovarian hyperstimulation syndrome (OHSS) a syndrome of massive enlargement of the ovaries, a potential complication of ovulation induction.

ovariectomy the surgical removal of the ovaries.

ovaries see OVARY.

ovariole in insects, one of the individual tubular structures of the ovary within which eggs are developed in a linear series.

ovary the female gonad; one of a pair of internal genital organs in which oocytes or eggs mature and are then released on a cyclic basis each month during the reproductive years; also the site of secretion for female sex steroids such as estrogen and progesterone.

oviduct literally, egg duct; a tubal structure that is the passageway for the egg (ovum) from the ovary to the uterus and that is the site of internal fertilization. In humans also called the FALLOPIAN TUBE.

oviductal shelf in certain invertebrates, a shelf-like structure on the dorsal wall of the uterus at the junction of the uterus and the common oviduct. It carries the openings of ducts from the spermathecae and the female accessory glands.

oviductal transport the stage of oocyte or embryo transport that takes place in the oviduct.

oviparity egg-laying as reproduction; the reproductive mode in which females lay unfertilized or developing eggs that complete their development and hatch in the external environment. In certain species the fertilized eggs may be retained in the oviduct for some time before laying.

oviparous relating to or describing a mode of reproduction in which the female lays eggs that develop and hatch outside the body; e.g., birds.

oviparous egg retention a form of oviparity in which the fertilized eggs are retained and begin to develop inside the maternal reproductive tract, are oviposited (laid), and complete their development in the external environment.

oviparous species animals producing eggs that develop and hatch outside the body.

oviposition the fact or process of egg-laying; the passage of an egg from the female reproductive tract to the external environment; also, the selective placement (laying) of eggs by the female at a particular locality.

ovotestis a gonad of certain invertabrates (e.g., the *C. elegans* hermaphrodite) that supports the production of both eggs and sperm.

ovoviviparity a broad term for various reproductive patterns in which fertilized eggs are retained within the maternal body and develop there, until the young hatch and are born; more specific (and now preferred) terms for this pattern include LECITHOTROPHIC VIVIPARITY, OVIPAROUS EGG RETENTION, and PSEUDOVIVIPARITY.

ovoviviparous relating to or describing a mode of reproduction in which the female lays eggs that develop and hatch within the body, with the subsequent release of live offspring.

ovoviviparous species animals producing eggs that develop within the maternal body to be born as live young.

ovulation the release of the egg from the ovary; the final stage of oogenesis involving the separation of a mature oocyte from the surrounding follicular cells, rupture of the follicular wall, and the expulsion of the oocyte into the oviduct for (possible) fertilization.

ovulation induction any of various medical or surgical techniques used to cause ovulation.

ovulatory relating to or involved in ovulation; i.e., the release of a mature egg from the ovary.

ovulatory process see OVULATION.

ovuliparity reproduction by the laying of unfertilized eggs; a form of oviparity.

ovum *plural,* **ova** a Latin term meaning egg; in some contexts limited specifically to an unfertilized egg.

oxygen consumption the amount of oxygen used by cells to meet metabolic needs.

oxygen transport system the lungs, heart, and blood vessels that conduct oxygen from the atmosphere to the mitochondria of cells.

oxyhemoglobin hemoglobin chemically bound to oxygen.

oxytocin a nine amino acid peptide hormone that is secreted by the neural lobe of the pituitary in all mammalian species and also by the corpus luteum in some mammals; oxytocin stimulates milk ejection and uterine contractions.

P

pachytene the third of four stages during the prophase I portion of meiosis; literally, "thick threads."

paedogenesis a process in certain invertebrates in which reproductive organs undergo an accelerated development in relation to the rest of the body.

pair bond(ing) the persistent social relationship between a male and a female whereby the sexual and parental aspects of reproduction are enabled.

paired homologous chromosomes joined pairs of chromosome, one from the father and one from the mother, that facilitate exchange of genetic material (genes) between father and mother chromosomes during prophase of meiosis.

palindromic sequence a nucleic acid sequence that is identical to its complementary strand when each is read in the 5′ to 3′ direction (e.g., TGGCCA).

palp an appendage near the mouth used to transfer sperm in mites and spiders.

pampiniform plexus a highly convoluted arrangement of testicular artery and corresponding vein in immediate proximity to the gonad, enabling a countercurrent transfer of heat and hormones across the walls of the two blood vessels.

Pandora larva an asexual formed larva in Cycliophora involved in the colonization of the host mouthparts.

panhypopituitarism inadequate or absent secretion of all anterior pituitary hormones.

panoistic ovariole an ovariole with a developing chain of oocytes, each encased by follicle cells.

papillary having a small nipple-like or finger-like shape.

Pap smear or **test** a collection of uterine and vaginal epithelial cells that can be stained and examined for evidence of pathology, including precancerous conditions. (From George *Papanicolaou,* U.S. physiologist.)

paracrine relating to or acting on sites that are near to itself; e.g., paracrine cells secrete hormones that affect target cells in the immediate vicinity, whereas endocrine cells secrete hormones into the bloodstream that affect distant cells.

paracrine action or **communication** the chemical control of one cell by the secretion of a nearby cell, sometimes of a different type, not involving transport of the chemical message via a blood vessel.

paracrine secretion the release by a cell of a chemical messenger that has an adjacent cell as its target. Contrasted with AUTOCRINE SECRETION, in which the target is the cell itself.

paramesonephric duct see MÜLLERIAN DUCT.

parametrium tissues adjacent to the uterus and cervix which include the broad ligament, bladder pillars, uterosacral ligaments, and cardinal ligaments.

paraphimosis retention of the preputial ring proximal to the coronal sulcus, causing swelling of the foreskin and glans penis.

parasite in a relationship of parasitism, the organism that benefits from the association.

parasitic being a parasite; living as a parasite in a host-parasite relationship.

parasitic castrator a parasite that in addition to acquiring its nutritional requirements from the host actively inhibits the reproductive development of the host usually by either destroying or inhibiting maturation of the gonads.

parasitism a close living relationship of two organisms of different species, in which one partner (the parasite) benefits, and the other partner (the host) does not benefit (and usually is harmed) by this association.

parasperm a secondary or accessory type of sperm present in some organisms, often deficient or lacking in DNA.

parasympathetic nerve fibers a division of the autonomic nervous system composed of those nerve fibers innervating internal organs whose preganglionic nerve cell bodies are located in the medulla or in the sacral region and that synapse with postganglionic neurons in ganglia located adjacent to or in the walls of the target organ.

paratenic host a host which may be essential for the transport of a parasite between intermediate and definitive hosts but which is not thought to be necessary for development.

paratomy a form of asexual reproduction in which complete individuals are produced prior to subdivision of the worm.

paraurethral glands prostatic glands nearest or surrounding the prostatic urethra which are generally thought to develop into BPH (benign prostatic hyperplasia).

parenchyma a collective term for the glandular tissue of the mammary gland.

parenchymella a usually solid flagellated larva.

parition the expulsion by a female of a reproductive product, be it an egg or a neonate; a collective term for both oviposition (egg laying) and parturition.

pars a division or part.

pars distalis the region of the adenohypophysis, or anterior pituitary, lying within the sella turcica (a transverse depression) and below the diaphragm sella, separated from the infundibular process by the hypophysial cleft.

pars intermedia the region of the adenohypophysis lying within the sella turcica, applied to the lower infundibular stem and the infundibular process and separated from the pars distalis by the infundibular cleft.

pars nervosa see INFUNDIBULAR PROCESS.

pars tuberalis the tubular region of the adenohypophysis that lies above the sella turcica and that is applied to the median eminence.

parthenogenesis a process of reproduction in females without fertilization by males; found in some lizards and fish and common in various invertebrates; e.g., arthropods and some groups of mollusks.

parthenogenetic or **parthogenic** relating to or characterized by parthenogenesis; reproducing without the fertilization of eggs. Parthenogenetic species are generally polyploid and there are no known males.

partial hypopituitarism inadequate or absent secretion of one or more (but not all) anterior pituitary hormones.

parturition the act or process of giving birth, involving the expulsion of the products of conception, including the fetus and the placenta.

passerine any bird of the largest avian order, Passeriformes, approximately 5,200 species of generally small-bodied birds with a foot structure evolved for perching.

paternal relating to or derived from the male parent.

patency the fact or process of development of spaces between the follicular epithelial cells surrounding the oocyte.

pathogenic of a microbe, having the ability to cause symptomatic disease in the host.

patriline the progeny of a particular male.

PCR see POLYMERASE CHAIN REACTION.

pedicle a posteriorly positioned anchoring organ that typically attaches to hard substrata; e.g., in deer, a bony protuberance on top of the skull from which the antlers grow annually.

pedipalps sensory and protective appendages associated with the mouthparts of chelicerates.

pelvic ganglion or plexus an aggregation of nerve cell bodies located adjacent to the cervix in females or the prostate in males that receives preganglionic fibers from the pelvic and hypogastric nerves and gives rise to postganglionic fibers supplying many of the internal pelvic organs.

pelvic inflammatory disease see PID.

pelvimetry the measurement of the diameters of the pelvis.

peptide hormone a subgroup of hydrophilic, polypeptide hormones composed of a sequence of fewer than 50 amino acids.

pericoital taking place at around the time of copulation.

perinatal relating to or occurring in the period around the time of childbirth; usually described as the 28th week of gestation up to 6 days following birth.

perineal relating to or located in the region of the perineum.

perineum the area between the thighs from the coccyx to the pubis, below the pelvic diaphragm; or, the region extending from the external genitalia to the anus.

periodomorphosis a phenomenon occurring in some families of Diplopoda, in which males alternate reproductive and non-reproductive forms after becoming sexually mature.

perioocytic space a space between the oocyte and follicle cell surfaces that is created at the beginning of vitellogenesis to allow hemolymph proteins to come in contact with the oocyte plasma membrane.

periovulatory occurring around the time of ovulation.

peritoneal relating to or involving the peritoneum.

peritoneum a strong colorless membrane that covers the abdominal organs and the walls of the abdomen and pelvis.

peritubular tissue a layer of connective tissue and modified smooth muscle cells surrounding the seminiferous tubules.

perivitelline space the fluid-filled space in an ovulated egg (secondary oocyte) between the vitelline (plasma) membrane and zona pellucida, containing the first polar body. Upon activation of the oocyte by a fertilizing spermatozoon, contraction of the vitellus significantly increases the volume of the perivitelline space and its accumulated fluid.

Peyronie's disease abnormal penile curvature due to fibrotic plaque in the tunical lining of the corpora, typically as a result of injury.

PGC see PRIMORDIAL GERM CELL.

pH potential (of) Hydrogen, a numerical measure from 1–14 used to describe the relative acidity or alkalinity of an aqueous solution.

pharmacokinetics the absorption, distribution, metabolism, and excretion of a substance and its metabolites.

pharmacological effect a term for a condition in which a certain hormone activates a behavior pattern, although it is not the hormone responsible for stimulating the behavior under normal circumstances.

phase-response curve (PRC) in circadian biology, a graphic representation of the different effects that a periodic environmental signal (e.g., light) has on biological rhythms. For example, light pulses typically will evoke a phase delay if applied early in nighttime, a phase advance if applied late in nighttime, and little or no effect if applied during daytime.

phenotype an observable physical or other (e.g., biochemical) characteristic of an individual as determined by the interaction of its genetic makeup with its environment.

pheromonal involving or describing a pheromone; i.e., a chemical used to transmit information between animals.

pheromone a chemical substance released externally by one individual of a species and perceived by another individual of the same species, so as to produce a behavioral, developmental, or physiological change in the other individual, usually so as to benefit one or both individuals.

pheromonostasis the temporary or permenant inhibition of pheromone production following mating.

philopatry recurring use of the same reproductive sites during an individual's lifetime.

phimosis scarring or stenosis of the preputial ring with resultant inability to retract the fully differentiated foreskin back to expose the glans penis.

phonotaxis locomotion oriented to a sound source.

photoperiod the light phase of a light-dark cycle, or the entire cycle; specifically, the period of time within a 24-hour day in which an organism is exposed to a significant amount of daylight, varying in higher latitudes with the season of the year; i.e., longer photoperiods occur during summer and shorter photoperiods during winter. Changes in photoperiod trigger certain physiological changes to prepare organisms for a change in season.

photoperiodic relating to or caused by a change in photoperiod (light-dark cycle).

photoperiodism sensitivity to changing day lengths; the ability of organisms (both animals and plants) to determine and respond to the seasonal changes in the amount of daylight that occur in all non-equatorial regions; used by many organisms to synchronize reproduction and other activities.

photopic involving substantial amounts of light; e.g., daylight.

photorefractoriness a condition in which an animal becomes temporarily unresponsive to a photoperiodic condition after prolonged exposure to it; also, a photoperiod-dependent mechanism by which birds will spontaneously regress their gonads and terminate their breeding attempt, thus ensuring that young are not hatched and reared at an inappropriate time late in the breeding season.

photorefractory relating to or caused by photorefractoriness (lack of response to a light-dark cycle).

photoresponsive see PHOTOSENSITIVE.

photosensitive relating to or displaying photosensitivity; having patterns of reproductive behavior that are affected by changes in the amount of available daylight.

photosensitivity a physiological state in which the reproductive system responds to a stimulatory environmental cue (usually the increasing photoperiod during spring) with growth and development.

phylogenetic or **phylogenic** relating to or indicating phylogeny; i.e., the evolutionary history of a group.

phylogenetic tree a treelike diagram displaying the evolutionary relationships in a group of organisms, with ancestral groups shown closer to the "trunk" of the tree and their derived descendants at the end of branches.

phylogeny the history of ancestor-descendant relationships for a species or other defined group of organisms; genealogy or historical descent.

physogastric describing a perfertile state in queens of social insects, with a much enlarged abdomen.

phytoplankton a type of plant plankton, such as algae, that is the basic food source in many marine ecosystems.

PID pelvic inflammatory disease, an infection originating from cervicovaginal flora that ascends along mucosal surfaces to the uterus and Fallopian tubes.

pineal gland an endocrine gland lying outside of, but attached to, the brain. The pineal gland secretes melatonin under the rhythmic control of the circadian clock.

piracy a behavior pattern in which an animal steals resources, such as food or nest material, from another.

pituitary relating to or involving the pituitary gland or the hormones produced by it; also, the gland itself. See PITUITARY GLAND.

pituitary adenoma a benign neoplasm of the anterior pituitary, usually made up of functioning secretory cells of one of the types normally found in the pituitary.

pituitary apoplexy a hemorrhagic infarction of the anterior pituitary, usually related to infarction of a pre-existing pituitary adenoma. It usually presents with severe headaches and visual disturbances and may culminate in coma.

pituitary crisis acute systemic decompensation occurring in a patient with untreated panhypopituitarism that may be fatal if not rapidly diagnosed and treated.

pituitary gland a gland composed of a heterogeneous population of cells lying in a bony cavity at the base of the skull just below the brain; attached by a stalk to the hypothalamus, from which it receives neural and hormonal input and with which it shares its blood supply. The pituitary gland is often referred to as the "master gland" because of its role in many key physiological functions within the body. It is divided into two main regions, the anterior lobe (adenohypophysis) and posterior lobe (neurohypophysis). The anterior pituitary secretes several hormones, three of which, luteinizing hormone (LH), follicle-stimulating hormone (FSH), and prolactin, are central to normal reproductive function. Also, HYPOPHYSIS.

pituitary-testicular axis the control of testicular function by means of pituitary hormones and the feedback regulation of the pituitary gland by products of the testis.

placenta a transient organ of pregnancy in most mammals; a structure formed by fusion of the chorion with the wall of the uterus that serves to attach the embryo to the uterine wall and to exchange nutrients, wastes, and gases between the maternal blood and the embryonic blood, thus supporting the development and growth of the fetus.

placenta accreta the erosion of placental tissue into the endometrial lining of the uterus.

placental relating to the placenta or to a placental mammal.

placental analog an invertebrate counterpart to the mammalian placenta.

placental labyrinth a system of tubular channels found in many chorioallantoic placentae, through which the maternal blood flows.

placental lactogen (PL) see HUMAN PLACENTAL LACTOGEN.

placental mammal a mammal whose unborn young are nourished within the maternal uterus by means of a placenta; i.e., all mammals except marsupials and monotremes.

placental viviparity the bearing of young nourished within the maternal uterus by transfer of nutrients via a placenta.

placenta previa the implantation of the placenta in the lower segment of the uterus, covering all or part of the internal os of the cervix.

placentome a placental structure consisting of maternal (caruncular) and fetal (cotyledonary) tissue that functions to bring maternal and placental blood vessels in close proximity for transfer of nutrients and wastes and for production of steroid and protein hormones. Between 50 and 100 placentomes develop on the placenta of domestic ruminants.

placentotrophy a type of maternal nourishment in which the nutrients for development are supplied by placental organs.

placode a platelike epithelial thickening marking in the embryo the anlage (precursor) of an organ or part.

plankton a collective term for the wide variety of plant and animal organisms, often microscopic in size, that float or drift freely in water because they lack the ability to control their own movement; found worldwide in marine and aquatic environments and an important food source for many larger organisms.

planktotrophic relating to or characterized by plankotrophy; feeding on phytoplankton and other particulates in seawater.

planktotrophic larva a feeding larva that obtains at least part of its nutritional needs from plankton and other exogenous sources (e.g., bacteria, detritus). Planktotrophic larvae generally hatch from small, transparent eggs.

planktotrophy the fact or process of feeding on plankton (and other external matter) in the marine or aquatic environment.

planula *plural,* **planulae** the typical larva of cnidarians which, in most species, is planktonic.

plasma membrane the phospholipid bilayer membrane that surrounds the cytoplasm of all eukaryotic cells. It contains numerous proteins that span the lipid bilayer, including a panoply of receptors, ion channels, and transporters.

plasticity a property of organisms indicating their malleability or susceptibility to experiential influences, especially during their early development. In its extreme form, plasticity may result in developmental outcomes beyond the range that typically occurs in nature. See also NEURAL PLASTICITY.

pleiomorphic having multiple structural forms within one life cycle.

plesiomorphic describing an original or ancestral feature.

pluteus the typical planktotrophic larva of echinoids, with 8 arms supported by calcite rods; also found in ophiuroids.

PMDD premenstrual dysphoric disorder, the diagnostic term for PMS (premenstrual syndrome).

PMS see PREMENSTRUAL SYNDROME.

podomere any one of the several structural units composing the leg of an arthropod.

poecilogony variability of reproductive pattern within a single species.

poikilothermic or **poikilothermous** relating to or describing animals having a body temperature that varies with the surrounding environment, as in the "cold-blooded" reptiles and amphibians.

poikilothermy the fact of having a body temperature conforming to the surrounding medium; being cold-blooded.

polar body one of the two small cells produced and discarded during the first and second meiotic divisions that yield the haploid ovum.

polyamine a compound that facilitates the formation of proteins and RNA in cells. A variety of growth-promoting factors increase the synthesis of these substances.

polyandrous relating to or describing a species in which the females mate several times with two or more males during the same reproductive period.

polyandry a mating system in which one female mates with two or more males.

polyclonal antibodies (PA) antibodies produced by multiple immune cells; each immune cell produces antibodies that recognize different antigenic sites on the same molecule.

polycystic ovarian disease (PCOD) a heterogeneous group of disorders that may result in oligomenorrhea, anovulation, obesity, bilateral ovarian enlargement, and hirsutism.

polycystic ovaries (PCO) enlarged ovaries due to increased central stroma; the subcapsular area is strewn with multiple small (2–8mm) follicles ("necklace of pearls").

polycystic ovary syndrome (PCOS) unexplained hyperandrogenic chronic anovulation.

polydomous colony in social insects, a system of a maternal nest and numerous filial nests in its vicinity with intercolonial worker traffic and exchange of resources.

polyembryony an asexual reproductive process whereby multiple individuals are produced from a single ovum.

polyestrus having two or more fertile cycles during a given mating season.

polygamous relating to or characterized by polygamy; mating with several members of the opposite sex.

polygamy a mating system in which a single member of one sex mates with a number of individuals of the opposite sex.

polygynous relating to or describing a species in which the males mate several times with two or more females during the same reproductive period.

polygny a mating system in which one male mates with two or more females.

polyhydramnios a disorder of pathologically large amounts of amniotic fluid.

polymerase chain reaction (PCR) an enzymatic process that allows the amplification and identification of specific DNA or RNA sequences; a widely used technique for the copying of a sequence of DNA.

polymicrobial infection an infection characterized by the growth of multiple pathogenic organisms.

polymorphic relating to or characterized by polymorphism; having two or more forms; e.g, in the same species or the same individual.

polymorphism the fact or condition of having different forms; e.g., different body forms within the same individual during different stages of the life cycle, or different body forms in different individuals within the same species.

polyorchidism the condition of having more than two testes.

polyp in certain marine invertebrates, an adult form with the shape of an elongated cylinder; the aboral end (the foot), is attached to the substrate, while the opposite free (oral) end, bears a mouth and tentacles. Also, in medical terminology, an abnormal, typically benign mass of tissue projecting outward from a mucous membrane; it may be pre-malignant.

polyphenism the ability of individuals with identical genotypes to develop into one or more discrete alternative phenotypes, usually in response to an environmental factor.

polypide in some marine invertebrates, the basic feeding unit of an individual colony member, made up of the lophophore, digestive tract, and associated musculature and ganglion.

polyspermy the (abnormal) fertilization of the oocyte by more than one sperm.

polytocous producing more than one offspring at a time.

polytrophic ovariole a type of ovariole having nurse cells associated with each oocyte and each unit enclosed in a follicular epithelium. A chain of successive stages make up the ovariole.

POPs progesterone-only pills; contraceptive pills that contain only a progestin.

portal system a vascular arrangement by which a substance secreted into the venous return of one organ or tissue is transported directly in the venous blood to an adjacent organ or tissue on which it then acts; e.g., the transport of GnRH from the hypothalamus to the anterior pituitary.

portal vessel a vessel that transports blood from one capillary bed to a second bed before joining the systemic circulation.

positive feedback a process in which secretory products released by a target organ in response to a stimulating hormone enhance or increase the further release of that hormone.

posterior lobe or **posterior pituitary** the NEUROHYPOPHYSIS, the posterior region of the pituitary gland, or, more specifically, the INFUNDIBULAR PROCESS of the neurohypophysis.

posthitis an inflammation of the prepuce, or foreskin, alone.

postmenopausal following menopause; relating to or occurring in the period of female life after the natural age-related cessation of menstruation.

postmenstrual relating to or occurring in the days following menses and before ovulation, in clinical assessments usually about days 6-12 (day 1 is the first day of menses).

postnatal after birth.

postnuptial molt a process in which many birds replace their showy plumage associated with breeding with more conservatively-colored feathers. Timing of the postnuptial molt is important so that feather replacement occurs before fall migration and/or the onset of adverse weather.

postovulatory after ovulation; relating to or describing the events that follow expulsion of the oocyte from the follicle.

postovulatory follicle another term for CORPUS LUTEUM.

postpartum after parturition or birth.

postpartum depression or **blues** a term for a transient state of irritation, anxiety, restlessness, and tearfulness that occurs in the first few days after birth.

postpartum estrus a condition in which a female ovulates and becomes sexually receptive immediately after parturition.

postprandial describing the period after a meal.

posttranslational processing mechanisms enzymatic steps that serve to modify new synthesized proteins so as to influence their functions. Peptides are made from precursor proteins that are proteolytically cleaved, amidated, sulfated, etc. in the cell to produce a biologically active secreted peptide.

potency the ability of a male to obtain and maintain an erection and ejaculate.

pouch an abdominal fold of skin around the mammary glands on a marsupial.

precocial describing or relating to a relatively mature state at birth that requires minimal dependency on adult care; specifically, referring to a pattern of development in birds, in which the young can leave the nest shortly after hatching and are capable of feeding and thermoregulating largely independently, although some parental care, such as brooding or guarding, may still occur. See also ALTRICIAL.

precocious advanced in early development; precocial.

predation the action of a predator; the killing and eating of an animal by another animal., usually of a different species.

preeclampsia an abnormal condition of pregnancy characterized by high blood pressure, edema, and excessive protein in the urine; progression from this condition to eclampsia poses significant risk of fetal or maternal death.

preembryo the pluripotent dividing cells of the fertilized ovum prior to the beginning of cellular differentiation.

pregnancy the condition of having a developing fetus or embryo in the body.

pregnancy blockage another term for the BRUCE EFFECT.

pregnancy-specific (associated) protein a glycoprotein hormone produced by the trophoblast/chorion in cattle and assayable in the maternal circulation as a method of pregnancy diagnosis.

pregnant mare serum gonadotropin (PMSG) see EQUINE CHORIONIC GONADOTROPIN.

preimplantation embryo a collective term for the fertilized egg and developing embryo up to the blastocyst stage.

preleptotene the period of meiosis during which DNA synthesis occurs in primary spermatocytes, preceding the first stage of prophase I of meiosis.

premature earlier than normal or expected in a pregnancy.

prematurity early birth; usually clinically defined as a birth prior to 37 weeks of gestation.

premenstrual referring to the days preceding the menstrual flow, usually about one week in clinical assessments, but symptomatic days may range from several days to two weeks.

premenstrual syndrome (PMS) a syndrome associated with the period immediately before menstruation, of uncertain cause; generally involving an array of symptoms including especially nervous tension, irritability, mood swings, and depression, as well as fatigue, headache, breast discomfort, food cravings, appetite changes, and swelling or bloating of the abdomen or extremities.

prenatal before birth; during pregnancy.

preoptic area the area anterior to the hypothalamus.

preovulatory before ovulation; relating to or describing the events that precede ovulation.

preovulatory follicle see CORPUS ATRETICUM.

prepartum before parturition or birth.

preprohormone the entire polypeptide encoded by a mRNA for a peptide hormone(s) before processing by proteases to remove signal sequences and other nonfunctional regions.

prepubertal before puberty.

prepubertal hiatus see JUVENILE PAUSE.

prepuberty the period before puberty; especially the period immediately before the onset of puberty.

prepuce the foreskin; the free fold of skin that covers the glans penis; excised during circumcision.

preputial ring the ring-like opening formed by the prepuce at the tip of the penis.

preputial space the space between the glans penis and the prepuce.

preterm before term; less than the normal or expected term of pregnancy.

preterm delivery a delivery between 20 and 37 completed weeks of gestation.

preterm labor uterine contractions accompanied by cervical change, at less than 37 completed weeks of gestation.

previtellogenic follicle a follicle in the slow-growth stage of development.

primary afferent (fiber) a nerve fiber that conveys information about events occurring in peripheral tissues by means of action potentials (or axonal transport) traveling from the tissue to the central nervous system.

primary antibody the antibody used to measure the concentration of ligand in biological samples.

primary capillary plexus the capillary bed of the neurohypophysis that extends throughout the median eminence, the infundibular stem, and infundibular process (neural lobe).

primary host in parasitism, the host organism on which the sexual part of the life-cycle occurs.

primary infertility an existing condition of infertility with no prior history of conception.

primary oocyte see OOCYTE.

primary sex differentiation the development of testes or ovaries from undifferentiated gonads in gonochoric (single-sexed) animals.

primary sexual characteristics a term for those characterisitcs that are essential for reproduction; i.e., the gonads and gametes.

primary spermatocyte the earliest germ cell that does not divide by mitosis, but that has entered in meiosis I where chromosomes are duplicated and the exchange of genetic materials occurs between homologous chromosomes. Primary spermatocytes divide at the first meiotic division to produce secondary spermatocytes.

primer effect a relatively long-lasting physiological response to a pheromone.

prime swarm the first swarm issued in springtime by a honey bee colony, containing the old queen.

priming pheromone a chemical signal produced by one or more individuals resulting in a developmental or physiological change, usually over the long term, in the recipient animal(s) of the same species.

primiparous describing a female who has delivered one offspring.

primitive spermatozoon a special type of spermatozoon with a round body and a long cilium, thought to be phylogenetically plesiomorphic and connected to free spawning.

primitive streak a region in the mammalian embryo from which prospective mesoderm and endoderm cells migrate and differentiate during gastrulation.

primordial relating to or occurring in the very earliest stages of development.

primordial follicle a dormant ovarian follicle consisting of a single oocyte that is surrounded by a single or partial layer of squamous "pregranulosa cells" and that will initiate follicular development and ovulate or degenerate. The pool of primordial follicles contains all the available oocytes that a female will ever have.

primordial germ cell (PGC) an undifferentiated, mitotically dividing stem cell found in both sexes, the earliest identifiable precursor to the female ovum or male spermatozoon.

primordium the anlage, or earliest precursor.

prisoner's game or **dilemma** a well-known model in game theory for repeated social interactions, in which individuals can play either of two strategies, "cooperate" (accept benefits and return them) or "defect" (accept but not return benefits); used to analyze patterns of altruistic behavior. (Based on the idea of a prisoner cooperating with his captor.)

PRL see PROLACTIN.

proceptive behavior various species-specific behaviors by the female which serve to initiate or maintain sexual contact with the male.

proctodeum the most external of the three chambers comprising the avian cloaca.

proctotomy an incision into the wall of the rectum.

proestrus the transition period between anestrus or diestrus (the luteal phase) and estrus.

progestagen a progesterone-like compound used to regulate reproductive activity.

progestational having the effect of progesterone; referring to the ability of an agent to stimulate alterations in the uterus necessary for implantation and development of a fertilized ovum.

progesterone a hormone secreted by the corpus luteum (CL), placenta, and in minute amounts by the adrenal cortex. It prepares the uterus for the reception and development of the fertilized ovum by transforming the endometrium from the proliferative to the secretory stage and is also essential in the maintenance of an optimal intrauterine environment for sustaining pregnancy.

progesterone response element (PRE) a specific DNA sequence that is recognized and bound by the progesterone receptor.

progestin a progestional agent; a substance, either natural or artificial, that is able to induce the modifications in the uterus necessary for the implantation and development of a fertilized ovum. Progesterone is the primary progestin in mammals.

progestogen a progesterone derivative produced by the testis in vertebrates.

progestomedin a factor that mediates the effects of progesterone.

progoneate describing a condition in which the reproductive systems open to the exterior near the anterior end of the body; e.g., in Diplopoda, through or behind the coxae of the second pair of legs.

progyny a form of sex differentiation in which all developing animals are at first female.

prohormone a precursor of a hormone.

prolactin (PRL) a peptide hormone produced by the anterior pituitary gland that stimulates and sustains lactation (milk production) in mammals; it also appears to have a variety of other roles throughout vertebrates; e.g., maintaining corpus luteum function after mating in the rat and nesting behavior in birds.

prolapse the slipping down of the vagina or uterus due to loss of supporting structures; the vagina may be exposed to the exterior at the vestibule.

proliferative phase the descriptive term for the alterations in the endometrium that occur during the preovulatory or follicular phase of the menstrual cycle, under the influence of increasing circulating estradiol secreted by the growing dominant ovarian follicle.

promiscuity a breeding pattern of multiple males mating with the same female, such that the female will likely be impregnated by more than one male.

promoter a region of DNA in which RNA polymerase binds before initiating the transcription of DNA into RNA. Most factors that regulate gene transcription bind at or near the promoter sequence and thereby affect the initiation of transcription.

pronucleus the haploid nucleus of a male (spermatozoon) or a female (egg) germ cell present in the fertilized egg before the male and female pronuclei fuse.

propagule a collective term that refers to the offspring of sexually reproducing organisms without specifying their stage of development.

prophase I the first phase of meiosis in which genetic recombination and crossing over occurs; it consists of leptotene, pachytene, zygotene, and diplotene stages.

prosoma the anterior body segment of chelicerates, roughly corresponding to a combined head and thorax (cephalothorax).

prostaglandin $F_{2\alpha}$ the luteolytic hormone that causes functional and structural demise of the corpus luteum.

prostaglandins a family of twenty carbon compounds, derivatives of arachidonic acid formed from polyunsaturated free fatty acids. They are widely distributed in cells and have a role in various autocrine, paracrine and sometimes endocrine activities, including the initiation and progression of labor.

prostate (gland) a complex muscular and glandular organ in the male reproductive system that surrounds the urethra at its juncture with the bladder neck and that adds its secretions to the ejaculate.

prostate cancer cancer of the prostate gland, the most frequently detected malignancy in men and the second leading cause of cancer death in the male population.

prostate-specific (membrane) antigen see PSA.

prostatic involving or affecting the prostate.

prostatic ductal system a functional unit of the prostate. All glandular structures within a ductal system share a single ductal orifice which opens into the urethra.

prostatic urethra the portion of the urethra that penetrates the prostate.

protamines a highly basic protein that replaces somatic histones in the chromatin of maturing spermatids and facilitates packaging of the sperm nucleus in a compact and transcriptionally inert state. In mammals, protamines are stabilized by disulfide bonds.

protandry the fact of possessing both male and female gonads, with the male gonads occurring first and the female gonads appearing later.

protegulum the first shell secreted after larval settlement.

protein hormones hormones made from amino acids that usually but not always have sugar molecules attached. Protein hormones act by binding to receptors on the cell surface, or within the cell binding to receptors in the membranes of intracellular organelles including the nucleus.

protein kinase C an intracellular serine-threonine kinase that catalyzes phosphorylation of intracellular proteins and alters their activities.

protein phosphorylation the covalent attachment of phosphate groups to serine, threonine, and tyrosine groups of proteins by specific kinases. Protein phosphorylation often leads to activation or inhibition of enzymatic and other regulatory functions, and to interactions between signal transduction proteins.

prothoracic gland an endocrine gland in the prothorax of insects that produces ecdysteroids.

protist a eukaryotic microbe, including protozoa, unicellular algae, and fungus-like microorganisms.

protogyny the fact of possessing both male and female gonads, with the female gonads occurring first and the male gonads appearing later.

proto-oncogene a cellular gene normally involved in the regulation of cellular functions. Most of these are involved in the regulation of cellular growth and differentiation. A proto-oncogene has the potential to become oncogenic (cancer-causing) by mutation, deletion, or overexpression.

proximate factor a cue that directly influences the timing of reproduction or other activities, divided into endogenous and exogenous components; exogenous factors often provide environmental cues about seasons that organisms can use to synchronize activities such as gametogenesis and spawning.

proximate spawning cue the environmental factor triggering the release of gametes or larvae.

PSA prostate-specific antigen, a transmembrane glycoprotein produced by prostate epithelial cells, recognized as the best tumor marker available for prostate cancer.

pseudoautosomal region the region on the distal end of the short arm of the Y chromosome which pairs with the X chromosome during meiosis and where crossing-over takes place.

pseudocoelom in certain invertebrates, a space between body wall and alimentary canal that is surrounded by tissues of mixed lineage.

pseudocyesis the false idea that one is pregnant, a syndrome involving signs associated with pregnancy in the absence of such a pregnancy; e.g., menstrual cessation or irregularity, adominal enlargement, secretion of milk, and often a sense of fetal movement. See also PSEUDOPREGNANCY.

pseudogamy sperm-dependent parthenogenesis in plants; pollen is required to activate seed development, but the seed nucleus is produced clonally.

pseudogene a nonfunctional DNA sequence that is very similar to the sequence of a known gene. Some pseudogenes probably arise from gene duplications that become nonfunctional because of loss of regulatory elements or accumulation of mutations.

pseudomaturation oocyte maturation during atresia.

pseudopregnancy false pregnancy; a condition that occurs in several species (e.g., rats, rabbits) in which ovarian follicular development ceases and functional activity of the corpus luteum is maintained. Pseudopregnancy is usually induced by an infertile mating; it typically is of shorter duration than pregnancy. See also PSEUDOCYESIS.

pseudospecies a group of related organisms produced strictly by asexual reproduction but otherwise resembling a (biological) species.

pseudoviviparity an unusual form of oviparity in which the eggs are fertilized externally and are brooded in some parental structure such as the stomach, vocal sacs, skin pouches, branchial chambers, or gastrovascular cavity; found in a few anurans, teleosts, and invertebrates.

pterygium colli folds of skin that stretch from the neck to the shoulders, associated with Turner's syndrome.

pubarche the initiation of growth of pubic hair.

pubertal relating to or occurring at puberty.

pubertal growth spurt a steroid-dependent process of accelerated bone growth during adolescence.

puberty the period of life in which the ability to reproduce is attained, culminating in the production of mature gametes and reproductive behavior; occurring in humans in adolescence and at a comparable period in other species.

puberty acceleration the acceleration of puberty in females of many mammalian species through direct contact with the adult male, or by exposure to a male urinary pheromone. Also, VANDENBERGH EFFECT.

puberty inhibition the delay of puberty among female mammals housed in groups due to direct physical interactions or to a female urinary pheromone.

pubic symphysis the juncture between the two pubic bones in the pelvis which in some mammals is necessarily modified in order to facilitate delivery of relatively large fetuses.

pudendum or **pudenda** a collective term for the external genital organs, especially those of the female.

puerpera a woman who has just given birth.

puerperal relating to or occurring in the time immediately or shortly after birth (**puerperium**).

pulsatile neurosecretion rhythmic releases of neurohormone pulses from neurovascular junctions in the median eminence, into the hypothalamic-hypophysial portal vessels.

pulse generator the cellular and molecular mechanisms that produce and sustain pulsatile neurosecretion; these mechanisms are currently unknown.

pulse pressure the difference between systolic and diastolic blood pressure in a vessel.

pupariation the process by which a mature third instar larva of a higher fly forms the **puparium**, a hard case within which the true pupa and, ultimately, the adult are formed.

pure gonadal dysgenesis a condition in women with streak gonads but normal stature and few or none of the phenotypic features of Ullrich's syndrome.

pyelonephritis an inflammation of the kidney, particularly one due to local bacterial infection.

pyknosis a shrunken nucleus with condensed darkened chromatin, usually characteristic of apoptosis.

pyriform cell a unique type of granulosa cell that forms an intercellular bridge with the oocyte. These occur only in the previtellogenic follicle of snakes and lizards and degenerate once vitellogenesis begins.

Q

queen in a colony of social insects (e.g., bees, ants), a mature, fertile female that lays eggs.

queen pheromone a multicomponent glandular secretion of queens that, besides functioning as female sex pheromone, acts as a dominance signal within the colony.

R

radial symmetry symmetry in which the body parts are arranged radially around a central oral-aboral axis.

radiation therapy a local cancer treatment consisting of short X-ray treatments to the affected area, used either to shrink a tumor before surgery or to kill any cancer cells that remain after surgery.

radical inguinal orchiectomy a surgical procedure for removing a cancerous testis; it requires an inguinal incision and involves delivery of the entire testis and spermatic cord from the scrotum and ligation of the structures at the level of the internal ring of the abdominal cavity.

radical prostatectomy surgical removal of the prostate; the most used definitive treatment for prostate cancer.

radioimmunoassay a technique employing radioisotopes that allows scientists to precisely measure hormone levels in blood or other body fluids.

ramet a somatic individual; a coherent "body" able to function independently; in purely clonal animals, a single module is the ramet, while in colonial animals, the colony is the ramet.

raphe a ventral ridge on the penis denoting the edge of fusion of the two urethral folds.

rapid yolk development the final stage of oocyte maturation during which large amounts of yolk material are sequestered from the blood and deposited in the developing follicle.

rare-male mating advantage a condition in which males of rarer phenotypes mate more often than males with more common phenotypes, occuring when preference is determined by frequency dependence of male phenotypes.

Rathke's pouch the embryonic fold in the oral ectoderm that develops to form the adenohypophysis (anterior pituitary gland) during embryogenesis. (First described by Martin Heinrich *Rathke,* German antomist.)

ratite any one of about 60 species of extant birds belonging to the clade Paleognathae, one of two primary branches in the avian tree, distributed primarily in Africa, Australia and South and Central America. Most ratite birds, such as ostriches, emus, or kiwis are flightless, but some, such as New World tinamous, have limited capabilities of flight.

Raynaud's phenomenon hypersensitivity to cold temperatures in the distal digits of the hands. (Named for Maurice *Raynaud,* French physician.)

reactive oxygen species (ROS) molecules produced by the transfer of an electron to molecular oxygen. ROS are associated with apoptosis by initiating DNA strand breaks. 30

receptive behavior species-specific behavior by the female, usually involving immobility and orientation of the perineal region towards the male conspecific in a posture that facilitates mounting and penile intromission into the female's vagina.

receptor a protein, located within the cell or localized on the cell surface, that recognizes and binds to specific chemical messengers (e.g., hormones, neurotransmitters). The binding of the messenger to its specific receptor leads to a biological response within the cell.

receptor downregulation a reduction in the number of functioning receptors on a cell, reducing the maximal effect of a chemical messenger.

receptor-mediated endocytosis a specific cellular mechanism for internalizing macromolecules, in which a membrane-bound receptor binds a ligand, collects with other receptors in a specially coated pit, and is then internalized in a coated vesicle.

reciprocal altruism the return of benefits to the original actor by the beneficiary of an altruistic act.

recombinant DNA technology the procedures used to insert (clone) DNA segments (genes) of interest into DNA molecules (i.e., cloning vectors) that are able to replicate when placed within the environment of a cell that permits replication and expression of the cloned gene.

recombinant protein a product engineered for large-scale production by manipulating genes.

recombination the transfer of polynucleotidic sequences between different DNA molecules.

recrudescence the reinitiation of gonadal development in seasonally breeding reptiles.

5α-reductase an enzyme that converts testosterone to DHT(dihydrotestosterone). Two isoforms of the enzyme have been described. 5α-reduction of testosterone is necessary for mediating some of its effects on the prostate and skin.

5α-reductase inhibitor an androgen inhibitor that blocks the enzymatic conversion of testosterone to DHT (dihydrotestosterone) in target tissues, decreasing the physiological effect of DHT without inhibiting the effect of testosterone at the androgen receptor.

reflux the movement of the content of a duct back toward its origin.

refractory period a period of time following a response to a stimulus, when the response cannot be duplicated; e.g., the several weeks or months following the breeding season during which the gonads fail to respond to environmental stimuli, or a period during which pheromone production is inhibited following mating.

regional anesthesia the blockading of nerves that provides pain relief or prevents pain in a particular area of the body while maintaining consciousness.

relatedness the proportion of genes shared among individuals that are identical by descent (rather than simply identical in state).

relaxin a peptide hormone with a structure similar to insulin that is produced by the corpus luteum and other portions of the reproductive tract during pregnancy in mammals; it stimulates softening of the cervix and preparation of the birth canal for parturition.

releaser effect a rapid behavioral response to a pheromone.

releasing/release-inhibiting hormone a product of the parvicellular neurosecretory system that stimulates secretion from the anterior pituitary or inhibits such release, respectively.

remodeling a continuous process of breakdown and renewal in bone; the final common pathway for adult bone loss or gain.

reproduction the ability of a species to produce offspring,or the fact or activity of producing offspring.

reproductive axis the integrated system of reproductive organs and control systems.

reproductive competition the various behavior patterns such as territoriality and aggression that males use to compete for access to females and for the opportunity to mate and to prevent other males from mating.

reproductive cycle a recurring, predictable series of changes in physiological activity that characterize the process of reproduction in a given species; can be short-term or over the lifetime of the organism.

reproductive diapause an arrested development of the reproductive system induced by environmental factors.

reproductive dominance in social insects, the suppression of worker reproduction by the dominant queen, usually by pheromonal control.

reproductive effort the proportion of the total energy budget of an animal allocated to reproductive processes.

reproductive health a general term for health issues related to human sexuality and reproduction; e.g., family planning, sexually transmitted diseases, pregnancy, childbirth, prenatal care, reproductive tract infections, reproductive cancers, infertility, menopause, and sexual dysfunction.

reproductive mode the combination of features such as egg size, number, and oviposition; developmental biology; hatching or birth; and parental care that characterize species or groups of species.

reproductive skew the partitioning of direct reproduction among individuals in groups. Where all individuals have equal chances of reproduction, skew is said to be low, whereas in groups in which reproduction in restricted to one or a small number of individuals, reproductive skew is high.

reproductive strategy a unique assembly of co-adapted reproductive traits characteristic of an individual from a common gene pool, which enables the organism to optimize its reproductive success in a particular environment.

reproductive success the number of an animal's offspring that survive to maturity, relative to the number produced by others in the same population.

reproductive suppression a failure or delay in the ability of an animal to reproduce, resulting from a disruption of normal behavioral and physiological processes brought about by environmental or social cues.

residual body a large spherical body containing the cytoplasmic remnants of sperm formation, which is formed by detachment of the cytoplasmic lobe during sperm release into the lumen. Residual bodies are are eliminated from the seminiferous epithelium by Sertoli cell phagocytosis in subsequent stages.

resource a general term for any environmental factor that enhances reproductive success and that is in short supply, relative to the number of potential users.

respiratory distress syndrome (RDS) the condition of a newborn infant in which the lungs are imperfectly expanded. Also, HYALINE MEMBRANE DISEASE.

resting egg a resistant (encysted), thick-walled, diploid embryo with highly sculptured outer surfaces, usually produced by sexual reproduction in monogonont rotifers; these eggs always hatch into amictic females.

restoration of spermatogenesis the process of restoring germ cells to the testis after the loss of germ cells resulting from experimental or natural perturbation of established spermatogenesis.

retained placenta a condition in which the natural expulsion of the placenta is delayed for 48 hours or longer after birth.

rete a network of blood vessels, nerve fibers, or other strands of interlacing tissue in an organ.

rete testis a network of sperm-transporting ducts that are contained within the mediastinum of the testis and connect the collecting ducts of the testes with the epididymal tubule; a remnant of the embryonic kidney incorporated into the testis during development.

retinohypothalamic tract a direct projection from the retina that terminates in the suprachiasmatic nucleus of the hypothalamus.

retractile testis testis that moves in and out of the scrotum.

retrocerebral complex those structures immediately posterior to the insect brain, including the corpus cardiacum, corpus allatum, and hypocerebral ganglion.

retrograde blood flow a flow of blood in the opposite direction from that expected; e.g., the flow of blood in the spermatic vein toward the testis rather than away from it.

retroperitoneum an abdominal cavity posterior to the cavity of the body which contains the gastrointestinal tract.

retroviral vector a RNA viral vector used for shuttling transgenes into embryonic or somatic cells. After infecting the cell, the virus converts itself to DNA, inserts itself into the host chromosome at a random location, and is then called a provirus.

retrovirus an RNA virus carrying an enzyme that makes a DNA copy of the viral genetic information.

reverse transcriptase polymerase chain reaction see RT-PCR.

rheological relating to the flow of blood and other fluids.

ribosome the site where the message on mRNA is read and decoded to produce proteins.

rigidity the entrapment of blood within the penis under high-pressure, low-flow conditions that are sufficient to permit coitus.

ring gland an endocrine ring-like structure, found in larvae and pupae of higher Diptera, such as Drosophila, which lies behind the brain and around the aorta; it contains three different types of endocrine cells corresponding to the corpora allata, corpora cardiaca, and the prothoracic glands.

ripe possessing gonads in condition to spawn.

rivalry song see TERRITORIAL SONG.

RNA ribonucleic acid; it transfers the DNA code to the cytoplasm to direct protein synthesis.

role-reversal the opposite of the usual condition in mate selection in which females are choosy about the male they mate with and males compete for mates; in role-reversal the males are the choosy sex.

rough endoplasmic reticulum a cellular organelle arranged in sacs or vacuoles that may be flattened or dilated. The outside surface is covered with ribosomes which help to organize the synthesis of proteins. Newly synthesized proteins are inserted through a groove in the ribosome and stored in the rough endoplasmic reticulum.

royal jelly a glandular secretion of worker bees in the honey bee colony, mainly produced by nurse bees in their hypopharyngeal glands, used for nutrition of all young larvae, and exclusively for queen larvae.

RT-PCR reverse transcriptase-polymerase chain reaction, a molecular technique in which repeated cycles of DNA synthesis are carried out to produce a large number of a specific DNA sequence.

RU-486 see MIFEPRISTONE.

ruminant one of the Ruminantia, a suborder of hoofed, even-toed, usually horned mammals (e.g., domestic cattle), characteristically having a stomach divided into four compartments and chewing a cud consisting of regurgitated, partially digested food.

runaway selection a selective process in which preferences for traits in choosers coevolve with trait expression in a reinforcing process of positive feedbacks.

rut the period of sexual excitability of a male mammal, corresponding to the estrus period in a female.

S

salpingectomy the surgical removal of the Fallopian tube.

salpingo-oophorectomy surgical removal of the Fallopian tube and ovary.

salpingostomy a surgical incision through the full thickness of the Fallopian tube wall; usually performed to extract an unruptured tubal pregnancy or to surgically repair (open) a distally occluded Fallopian tube.

satellite DNA DNA having a large number of repeated sequences.

sauropsid relating to or resembling modern reptiles and birds.

scalids regularly arranged spines at the head of juvenile and adult Kinorhyncha.

SCN see SUPRACHIASMATIC NUCLEUS.

scotopic involving minimal amounts of light.

scrotum a sac of skin, muscle and connective tissues in which the testis is suspended by the spermatic cord.

seasonal breeding the characteristic in some species of having reproductive behavior that varies according to season, as by not reproducing in winter.

seasonality the fact of varying according to the season of the year; specifically, the characteristic in some species of having reproductive behavior that varies according to season, as by not reproducing in winter.

seasonal rhythms behavioral and physiological rhythms that vary throughout the year in such a way that specific functions occur only during certain seasons of the year (e.g. reproduction, hibernation, migration).

SEC secretory epithelial cell, one of the system of mammary cells involved in the secretion of milk.

secondary hypoadrenalism subnormal secretion of cortisol by the adrenal glands.

secondary hypogonadism subnormal secretion of the hormones secreted by the gonads, e.g. estradiol by the ovaries.

secondary hypothyroidism subnormal secretion of thyroxine by the thyroid gland.

secondary infertility an existing condition of infertility but with at least one prior conception having been documented.

secondary oocyte see OOCYTE.

secondary sex(ual) characteristic or character a trait (or a behavior) that is expressed only in one sex, and that is not directly associated with the act of reproduction, but that functions in a reproductive context.

secondary spermatocyte a cell that divides at the second meiotic division quickly without DNA synthesis to produce haploid (n) spermatids.

secretion the process by which the hormone product is sent into the extracellular space. It usually involves the movement of storage granules to the plasma membrane, the joining of the membrane to that of the storage granule, and the release of the material inside the granule. The product may move into the blood vessels (endocrine secretion). Or, it may move in the extracellular space to a neighboring cell (paracrine secretion).

secretory relating to or involved in secretion.

secretory granule a specific intracellular particle surrounded by a bilayer membrane containing secretory peptides.
secretory phase the alterations in the endometrium that occur during the post-ovulatory or luteal phase of the menstrual cycle; i.e., the transformation of proliferative endometrium into secretory endometrium.
selective reproductive toxicant a compound that causes reproductive toxicity at doses that do not detectably affect other systems in the body.
semelparity the fact of being semelparous; having only a single breeding season in a lifetime; e.g., certain invertebrates.
semelparous describing a reproductive pattern in which there is only a single spawning season per lifetime of an individual; usually the adults die after breeding.
semen the male reproductive fluid comprised of secretions of the prostate, peri-urethral glands (Cowper, Littre), seminal vesicles, and ampulla of the vas deferens, combined with spermatozoa derived from the testis and emitted from the urethral opening of the penis during ejaculation.
semialtricial describing a reproductive pattern, e.g., in birds, that is partially altricial; i.e, the young are capable of leaving the nest soon after hatching (as are precocial birds) but are still dependent upon their parents for food for several days or weeks (as are altricial birds).
seminal relating to semen or its ejaculation.
seminal fluid the fluid surrounding the sperm in the ejaculate.
seminal plasma the nongamete component of semen, composed of secreted fluids, cells, and cellular components from several male accessory sex glands.
seminal vesicle a gland in the male reproductive system situated posterior to the prostate gland. The seminal vesicle contributes the majority of the ejaculate volume and empties into the ejaculatory duct. In invertebrates, it is a site of storage for mature sperm prior to mating.
seminiferous epithelium the cellular composition of the seminiferous tubule, composed of male germ cells and Sertoli cells.
seminiferous tubule the sperm-producing organ of the testis.
seminiferous tubule fluid (STF) the fluid surrounding germ cells within the seminiferous tubular compartment.
semisocial describing a condition in certain insects in which a reproductive division of labor occurs among individuals of the same generation.
semiterrestrial describing land organisms belonging to the hydrobios but needing at least a film of water when active.
senescence the process of growing old; aging.
sensitive period a time-delimited phase during development in which an organism is particularly sensitive to experiential influences.
sensitive-period learning see AGE-LIMITED LEARNING.
sensitization a term for the process by which a nonpregnant and nonlactating female rodent can be induced to display maternal behavior as a result of exposing her to pups over a period of days.
sensorimotor in songbirds, the second major phase of song learning; it involves ongoing comparison between a sensory model of song acquired in an earlier memorization phase and auditory feedback from the bird's own production of song.
septation in fetal lung development, subdivision of the alveoli that increases alveolar number and surface area. Primary septa are produced by dichotomous branching of saccules. Secondary septa are formed when one of the two capillary layers of a primary septum rise up into the alveolar space.
sequentially hermaphroditic having temporally separate female and male phases; i.e., both types of gametes are produced by the same body, but not at the same time.
serine proteinase inhibitors (serpins) an extensive superfamily of protein substrates for several classes of proteinases, including serine proteinases and metalloproteinases. Many act as proteinase inhibitors but some function primarily as plasma transport proteins for hormones or cytokines.
SERM selective estrogen receptor modulator, one of a class of drugs that can selectively act as agonists or antagonists of estrogen receptors (ERs) in various tissues; e.g., raloxifene is an ER agonist in bone tissue and an antagonist in breast or endometrial tissue.
serosa a thin, slippery coating that covers all organs within the abdominal cavity.
serotonin a biogenic hormone (5-hydroxytryptamine) that in humans acts as a vasoconstrictor and neurotransmitter and also has various other physiological roles; present in high concentrations in the intestines, brain, and blood platelets. It is also found in a range of other animals (and in plants); e.g., it causes reinitiation of oocyte maturation and spawning in many bivalves.
Sertoli cell a large stellate and columnar somatic cell present in the seminiferous epithelium of mammals. Solidly attached to the basement membrane of seminiferous tubules, it is clearly associated structurally and functionally with male germ cell development, the elongated spermatids in particular. (From Enrico *Sertoli,* Italian histologist.)
Sertoli cell barrier a barrier to the penetration of blood-borne substances and lymph into the seminiferous tubule, formed near the base of the seminiferous tubule by tight junctions of adjacent Sertoli cells. Also, BLOOD-TESTIS BARRIER.
Sertoli cell-only testis a condition in which Sertoli cells are the only cellular element in the seminiferous tubule.
Sertoli-germ cell ratio a ratio that is fixed for a species and appears to result from the fixed capability of Sertoli cells to support germ cells.
Sertoli-Sertoli ectoplasmic specialization a specialized region formed by the endoplasmic reticulum and actin filaments that lines the Sertoli cell barrier.
Sertoli-Sertoli tubulobulbar complex an invagination of one Sertoli cell into another at the level of the Sertoli cell barrier that takes the form of a tube connected to a bulb; probably functions in junctional turnover.

Sertoli-spermatid tubulobulbar complex an invagination of the late spermatid head plasma membrane into the apical Sertoli cells having both tubular and bulbous portions; probably functions in anchoring the spermatid and elimination of its cytoplasm and the junctional links binding the two cells.

severe oligospermia a condition of less than 3 million per milliliter of spermatozoa in the ejaculate.

sex the reproductive identity of an individual; usually a classification of male or female based on various factors (the type of gametes produced, external and internal anatomy, chromosomal makeup).

sex allocation the allocation of available resources (especially time and energy) to male versus female function.

sex chromosome a chromosome whose presence or absence is correlated with the sex of the bearer.

sex commitment the ontogenetic loss of sexual plasticity in individuals with environmentally determined or modulated mechanisms of sex determination.

sex determination the genetic or environmental prescription of sex in gonochoristic individuals; the former occurs at fertilization in individuals with genotypic sex determination and the latter during embryonic or larval development in individuals with environmental sex determination.

sex differentiation the distinction between male and female; primary sex differentiation pertains to whether the gonads develop into ovaries or testes and secondary sex differentiation refers to all other tissues.

sex hormone binding globulin (SHBG) a glycoprotein found in the blood that binds some androgens and estrogens with relatively high affinity. This protein contributes to the regulation of the amount of free estrogen present in plasma.

sex pheromone a chemical produced by one sex that attracts or arrests the movement of the opposite sex. Pheromones may act at a distance or they may require direct contact.

sex-specific behavior a behavior shown exclusively by members of one sex and largely dictated by physical factors, e.g. suckling in female mammals.

sex steroid (hormone) a collective term for steroidal hormone products of the gonads.

sex-typic behavior a behavior shown primarily by one sex but which may also be shown by the other.

sexual characteristic a discernable anatomical, physiological, or behavioral characteristic that differs between the female and male sexes.

sexual conflict a situation in which the members of a mating pair have conflicting or divergent interests due to differing selective pressures; usually involves a male wanting to mate with a non-receptive female.

sexual determination see SEX DETERMINATION.

sexual dialectics conflict between the sexes, distinguished from other modes of between-sex conflict because it is an inevitable consequence of the benefits of mate choice. Because some mates (usually males) will be rejected, they will be under selection to manipulate or control the other sex's reproductive decisions to their own advantage, while the more choosy sex, usually females, will be under selection to resist manipulation or control by others.

sexual differentiation see SEX DIFFERENTIATION.

sexual dimorphism a sexual difference in form or structure, by extension applied to any observable distinction between males and females, including differences in behavior.

sexually antagonistic coevolution a process of selection and counter-selection between the sexes in which pairs or potential pairs have opposing survival or reproductive interests; occurs via sexual conflict.

sexually transmitted disease (STD) a disease process (e.g., syphilis, genital herpes, AIDS) that is transferred from one individual to another by means of sexual intercourse or other sexual contact.

sexual orientation a pattern in an individual of persistent and consistent sexual and romantic attraction, either to the opposite sex, to one's own sex, or to both.

sexual reproduction reproduction involving union of gametes derived from two genomes.

sexual segment a portion of the kidney found only in male lizards and snakes that hypertrophies in response to testosterone during the spermatogenetic cycle. It is believed to contribute lipid secretions to the semen.

sexual selection an evolutionary process linked to the expression of secondary sex characteristics and consisting of non-random mating due to competition among members of the same sex and species for mates and choice of mates.

SGA small for gestational age, describing newborns whose weight is below the 10th percentile for their gestational age.

Sheehan's syndrome a condition of panhypopituitarism secondary to pituitary necrosis as a result of hypovolemic shock, usually related to postpartum hemorrhage. Symptoms include amenorrhea, failure of puerperal lactation, hypothyroidism, and adrenal insufficiency. (From Harold *Sheehan*, British pathologist.)

shell conductance a physical property of the avian eggshell, determined by shell thickness and the number and geometry of pores in the egg shell that, along with the gradient in gas pressures, determines the rate at which gases moves by diffusion between the egg and environment.

signaling (cascade) a series of biochemical reactions by which extracellular signals exert their action(s) on specific intracellular target(s). This process involves specific recognition of the signal by a specific receptor, transduction into the cell (via different mechanisms) and the activation or induction of cellular and/or nuclear proteins.

signaling pheromone a pheromone that produces an immediate, or short latency, behavioral or physiological response in the recipient.

signal transduction a series of ordered biochemical reactions by which a signal initiated at the cell surface is transmitted inside the cell to elicit specific cellular responses; e.g., to regulate gene expression.

simultaneously hermaphroditic simultaneously female and male; i.e., both types of gametes are produced by the same body at roughly the same time.

sister chromatids identical copies of a chromosome, derived by DNA replication.

small for gestational age see SGA.

social facilitation in avian behavior, a situation in which the presence of many birds is so stimulating that there are more behaviors (or more intense behaviors) than might occur otherwise. This presumably leads to earlier egg-laying, and in general, early nesting pairs are more successful than later-nesting pairs. Social facilitation occurs in courtship, feeding, and preening.

solitary describing a behavior pattern in which a male-female pair of birds nests alone in the habitat, separated from other members of its species; territories may abut, but nests are far apart.

soma the whole body of an organism, excluding the reproductive cells.

somatic cell any body cell that is not a germ cell.

somatostatin the hypothalamic growth hormone inhibiting hormone that inhibits GH secretion from the anterior pituitary gland.

somatotrope or **somatotroph** a cell of the adenohypophysis (anterior lobe of the pituitary gland) that produces and secretes growth hormone.

somite an embryonic structural unit in various animals.

sonic muscle a muscle found in some teleosts that contracts at very high frequency and causes the swim bladder to vibrate. The vibrations produce a sound that is used as a courtship call.

sonohysterography a procedure in which the uterine cavity is filled with a saline solution and evaluated by transvaginal ultrasound.

spadix the male reproductive organ in molluscs, used for sexual arousal in females.

spawn a large number of eggs released into water for hatching by fishes, amphibians, molluscs, and other such animals.

spawning a broad term for the release of a large mass of reproductive material into the water by certain animals; e.g., both males and females may release sperms and eggs together so that fertilization occurs in the surrounding water (broadcast spawning). Or, only males may spawn gametes, while females retain eggs for internal fertilization, either capturing sperm released into the water by males or receiving sperm during copulation. Females later spawn fertilized eggs, or release young at later stages of development.

species the fundamental level of classification for organisms; the category that provides, along with the next higher level of genus, the scientific name of an organism. Various criteria are used to describe a group of organisms as a species, the most typical being common ancestry and the ability to reproduce freely within the group, but not outside it.

species-typical describing phenomena that fall within the normal range of perceptual capabilities and motor activities of members of a particular species in the natural environment.

sperm the mature germ cells of the male that serve to fertilize the egg of the female, i.e., the spermatozoa; or, the entire male ejaculate; i.e., the semen. (Used as both a singular and plural form.)

spermalege a specialized structure found on the abdominal wall only in cimicoid bugs through which the male inserts the spermatozoa.

spermatheca the sperm storage organ in female insects; a sac or receptacle in a female or hermaphrodite animal in which sperm cells received from a male or another individual are stored until required to fertilize the eggs.

spermatic cord the structure formed by the pampiniform plexus and the coils of the internal spermatic artery between the inguinal canal and the testis.

spermatid the haploid germinal cell arising from meiotic divisions of spermatocytes, which differentiates, within the seminiferous epithelium, into a spermatozoon.

spermatocyst a unit of spermatogenesis present in anamniote testis.

spermatocytes differentiated spermatogonia that have entered the first meiotic division of the spermatogenic cycle and moved into the adluminal compartment of the tubule.

spermatocytogenesis the process by which spermatids elongate, develop acrosomes and flagella, and undergo volume reductions to become morphologically mature spermatozoa; this occurs within Sertoli cell vacuoles.

spermatogenesis sperm production; a complex process in which primitive germinal stem cells (spermatogonia) develop into highly specialized cells called spermatozoa. This process takes place in the seminiferous tubules of the testis.

spermatogenic arrest a disorder in germ cell development in which spermatogenesis is halted at one or more steps prior to the release of elongated spermatids from the seminiferous epithelium.

spermatogonia *singular,* **spermatogonium** the most immature form of male germ cells; they divide by mitosis to produce stem cells (that can continue to divide throughout the life of the male) and primary spermatocytes.

spermatophoral gland an epidermal gland in which sperm are fashioned into spermatophores.

spermatophore a coated or encapsulized structure produced by the males of many animals that encloses sperm and accessory gland secretions to be transferred (directly or indirectly) to the female for mating.

spermatozeugmata a bundle of sperm not surrounded by any defined sheath or covering, which is transferred to a mating partner.

spermatozoa *singular,* **spermatozoon** the male gametes; the aggregation of motile cells produced in the testis, capable of rising in the female genital tract, recognizing the egg, penetrating the zona pellucida, and then fusing with the ovum. A normal human ejaculate may contain 40 to 300 million of spermatozoa.

spermatozoon the singular form of SPERMATOZOA.

sperm capacitation the changes undergone by mammalian sperm within the female tract that make the sperm capable of fertilization. Capacitation is a prerequisite for the acrosome reaction, a phenomenon (not unique to mammals) in which the acrosomal cap and its lytic enzymes are shed and the sperm becomes able to fuse with the egg membrane.

sperm capsule the rigid, sclerotized covering that often encases a sperm packet.

sperm competition the competition within a single female's reproductive tract between sperm of different mates for the fertilization of ova, often resulting in the last male to mate fertilizing most of the eggs.

sperm droplet a naked, unpackaged droplet of spermatozoa and accessory fluids.

sperm duct in Lepidoptera, a structure connecting the bursa copulatrix of the copulatory opening (which is separate from the ovipositioning opening) with the common oviduct so that sperm can pass to the spermatheca.

sperm granuloma a circumferential wall of phagocytic monocytes and neutrophils surrounding a core of sperm that have escaped their normal tubular environment. When this occurs in the excretory ducts, it may block sperm movement through those ducts, and reduce fertility, even though spermatogenesis is normal.

-spermia a suffix referring to sperm or semen.

spermiation the release of spermatozoa from the seminiferous tubules; the counterpart to ovulation in the female.

spermicide a substance or agent that acts to kill sperm; i.e., for the purpose of contraception.

sperm incorporation the physical incorporation of the sperm into the egg cytoplasm.

spermiogenesis the third and last phase of spermatogenesis, during which the newly formed spermatids metamorphose into spermatozoa.

spermiophagy the phagocytic ingestion of spermatozoa.

sperm maturation the process occurring in the epididymis during which immature spermatozoa produced in the testis gain the potential to fertilize eggs.

sperm migration the movement of sperm from the point of insemination to the site of fertilization.

sperm packet spermatozoa and accessory fluids packaged in a flexible membranous covering.

sperm wash an artificial insemination technique involving the removal of seminal plasma by centrifugation of semen diluted with culture medium followed by resuspension of the sperm in culture medium.

squamocolumnar junction a portion of the cervix where the squamous cells meet the glandulae cells; this is the site at which most cervical cancers begin.

SRY gene sex related gene (or sex-determining region) on the Y chromosome, a Y-linked gene (i.e., a gene carried on the Y chromosome) central to the process of sex determination in mammals; e.g., testicular development.

stage a method used to describe the extent or spread of cancer. Categorization by stage is determined by the tumor size and whether it has spread to lymph nodes or other distant sites around the body.

StAR steroidogenic acute regulatory protein, the protein that makes cholesterol available to the cholesterol side chain complex and regulates testosterone biosynthesis.

STD see SEXUALLY TRANSMITTED DISEASE.

STD carrier a host who is colonized with an STD-associated microbe in the genital area and is infectious to others, but has no symptomatic disease.

stem cell a cell that is capable of extensive proliferation, creating more stem cells (self-renewal), as well as differentiation into cellular progeny.

sterile not able to reproduce; unable to produce offspring.

sterility a failure or inability to produce offsping; in humans **absolute sterility** is described as the inability to conceive under any circumstances (i.e., no treatment is available to enable conception), while **reversible (relative) sterility** is the inability to conceive without intervention, but with the possibility that treatment is available which may allow pregnancy to occur.

sterilization a surgical procedure performed to render an individual sterile.

steroid a lipid compound generally derived from cholesterol. See STEROID HORMONE.

steroid gradient a progressive increase in corticosteroid concentration from the outer cortex to the inner cortex due to steroid hormone production from concentric adrenocortical layers and inward secretion into the adrenal's centripetal vasculature.

steroid hormone a member of a class of lipid compounds that have a basic structure of three six-sided carbon rings and one five-sided carbon ring, referred to as the the cyclopentano-perhydro-phenanthrene nucleus. The steroid hormones are derived from cholesterol in the gonads, adrenal glands, and placenta; they include the progestins (e.g., progesterone), the estrogens, and the androgens (e.g., testosterone). Steroid hormones have various crucial effects on reproduction.

steroid hormone receptor see STEROID RECEPTOR.

steroid hydroxylase the sub-class of cytochrome P450 enzymes involved in the synthesis and metabolism of steroid hormones.

steroidogenesis the biochemical synthesis of steroid hormones from cholesterol.

steroidogenic relating to or involved in the synthesis of steroid hormones.

steroid receptor one of the intracellular proteins that binds a specific steroid hormone with high affinity and subsequently carries out the biological effects of that hormone.

stomatogastric describing the portion of the nervous system of arthropods that pertains to the esophagus and gut, particularly the anterior portion of the gut.

stratum corneum the outermost layer of the skin, considered to be a key element of the skin barrier, and consisting of non-metabolically active squamous cells that are slowly sloughed as the underlying skin cells mature.

stress any circumstance in which an individual animal or person is required to make functional, structural or behavioral adjustments or responses as a coping reaction; this circumstance may be immediate, e.g., the sudden appearance of a predator to a prey species, or ongoing, e.g., the overcrowding of a species in a given area.

stressful describing an environment that places demands on an individual.

stressor an environmental factor that contributes to a stressful circumstance or elicits stress responses.

stress physiology the study of physiological, biochemical and behavioral responses to factors that comprise an animal's physical, chemical, and biological environment.

stroma or **stromal cells** the supporting tissue or matrix of an organ or structure; e.g., in the uterus, the central cell mass separating the luminal epithelium and myometrium.

structural luteolysis a later involution of the corpus luteum in the ovary to form a small white connective tissue scar, the corpus albicans.

subadult an animal more advanced than the juvenile stage but that has not yet fully attained its adult stage. The extent to which subadults are physiologically and behaviorally capable of breeding differs dramatically among species.

subitaneous egg a thin-walled, rapidly-developing egg produced in rotifers by an amictic female that hatches into a young, diploid female that is either amictic or mictic.

subsong in songbirds, the initial stage of song production in the sensorimotor phase of song learning. Subsong is quiet, poorly structured, and highly variable in form.

substrate a lower surface on which something is located or takes place; e.g., the surface on which a bird builds its nest. Also, a substance that when acted on by an enzyme is converted to a product with different chemical properties.

subunit a polypeptide chain linked to other polypeptide chains to form a protein.

sucking or **suckling** the act of milk removal from the breast in humans and other mammals.

suckling behavior a behavior associated with the extraction of milk from the mammary gland of lactating female mammals.

superovulation the induction of multiple follicular growth with fertility drugs which are usually purified preparations of FSH (follicle-stimulating hormone).

supersedure in social insects, the elimination of an old queen by the workers and replacement by a younger successor.

suppressor T lymphocytes that dampen or shut down the immune response to an infection or other foreign body.

suprachiasmatic nucleus (SCN) a structure of the hypothalamus that is identified as the source of an important monitor of circadian rhythms in mammals and birds.

surfactant a wetting agent; a substance, such as detergent, that has the property of reducing surface tension. Pulmonary surfactant is a complex mixture of lipid and protein that is responsible for reducing surface tension in pulmonary alveoli, which is essential for early pulmonary function.

surge mode a brief (hours long) episode of accelerated gonadotropin secretion that produces a transient, monophasic pulse or "surge" in blood hormone levels many times above the tonic levels that ordinarily prevail.

symbiont an organism living in a relationship of symbiosis.

symbiosis the intimate association (living together) of organisms of different species; this may be a relationship in which one partner benefits while the other is harmed (parasitism), both benefit (mutualism), or one benefits while the other is relatively unaffected (commensalism), as well as other associations in which the relationship between the partners is not clearly understood.

symbiotic living together; characterized by symbiosis.

symmetric IUGR IUGR(intrauterine growth restriction) involving an infant who is proportionately small with both weight and length below the normal standards for gestational age, and organs proportionately reduced in size.

synapse a specialized junction between two nerve cells where signaling occurs by way of the release of neurotransmitters.

synaptic relating to or involving a synapse.

synclone a clone derived from isolated mating pairs.

syncytial describing a multinucleate mass of cytoplasm resulting from the fusion of cells.

syncytiotrophoblast the multinucleated trophoblast cells that form the outer layer of the chorionic villi and function as the site of hormone synthesis and of exchange of chemicals and nutrients between the maternal and fetal blood.

syncytium a mass of cytoplasm containing multiple nuclei, not divided by membranes into separate cells; a "super cell" formed by the fusion of uninucleate cells.

synechiae *singular,* **synechia** an adhesion of parts, especially an unnatural adhesion.

syngamy the fusion of gametes (ovum and sperm) initiating fertilization.

syrinx the avian voice box; a structure unique to birds, situated ventral to the point where the trachea splits into two bronchi and capable of producing a wide variety of sounds by air passing through a series of complex membranes.

systemic affecting or involving the entire organism; i.e, the whole system.

systemic therapy therapy that is given to a patient to treat the whole body.

systole the portion of the cardiac cycle in which the muscles of the ventricles contract.

T

Tanner stage or **staging** an assessment of the state of puberty based on pubic hair growth and on development of genitalia in boys or breast development in girls.

target an organ or cell stimulated by a chemical messenger.

T cells multifunctional lymphocytes that can be subdivided into helper (Th) cells identified by the presence of CD4 on their surface and cytotoxic T lymphocytes (CTL) identified by CD8 on their surface.

teleost one of the Teleosti, the dominant bony fishes; i.e., a fish other than the cartilaginous fishes, lobe-finned fishes, and sturgeons.

teloblastic describing a postembryonic developmental pattern in certain invertebrates in which body units are produced by an embryonic growth zone located at the rear of the body.

telotrophic describing a type of ovary in which trophocytes in a common chamber provide nutrients to developing oocytes via a trophic cord.

temperature-dependent sex determination mechanisms of sex determination found in many reptiles, in which the temperature of the incubating egg during the mid-trimester of embryogenesis establishes the gonadal sex of the individual.

template in songbirds, a sensory model of song acquired during the initial memory phase of song learning as a result of listening to songs produced by adult conspecific birds.

terato- a prefix meaning malformed or deformed.

teratocarcinoma a malignant form of teratoma.

teratogen an agent or factor that causes physical defects in the developing embryo/fetus.

teratogenic causing abnormal development.

teratoma a tumor composed of many differentiated cell types, none of which is native to the site of the tumor; typically found in the testis or ovary.

teratospermia a condition in which most ejaculated sperm are structurally abnormal.

terminal endbud a specialized transient multilayered epithelial structure at the growing ends of mammary ducts.

terminal nerve a cranial nerve that may mediate information on pheromones.

territorial behavior the behavior pattern of identifying a certain area and defending this area against intrusions by others of the same species.

territorial song in songbirds, a type of song exchanged between conspecific males in an aggressive context and often incorporated in calling song; it keeps a singing location free from encroachment by male rivals. Also, RIVALRY SONG.

territory a more or less defined space within an animal's total home range, within which certain individuals are aggressive toward, and usually dominant over, other members of the same species; a territory is typically used for courtship, mating, and care of young.

testes the male gonads; the plural form of TESTIS.

testicle another term for TESTIS, the male reproductive gland.

testicond having the testes normally retained within the abdominal cavity, a characteristic of various mammals.

testicular relating to, involving, or affecting the testes (male reproductive glands).

testicular feminization syndrome another (earlier) term for ANDROGEN INSENSITIVITY SYNDROME.

testicular torsion a condition of a twisted testis that will lose its blood supply, associated with a spontaneous rotation of the spermatic cord and involving ischemia and severe pain.

testis one of the two male reproductive glands, located in the the scrotum; responsible for the production of male gametes (sperm) and the secretion of the male hormone testosterone, which is needed for the development of male sexual characteristics and behavior.

testosterone the principal male sex hormone; a steroid hormone produced at high levels by the Leydig cells of the testis, both during fetal life (when it causes masculinization of the fetal brain and reproductive system) and after puberty (when it causes additional changes associated with puberty in boys). Much lower amounts of testosterone are also produced in the adrenal cortex, and, in females, by the ovaries. Testosterone is one of a family of hormones called androgens.

TGF see TRANSFORMING GROWTH FACTOR.

Th1/Th2 cells a classification of cytokines into one of two groups, usually mutually antagonistic, based on their association with or capacity to favor immune deviation towards cell-mediated and inflammatory immunity (Th1) or toward antibody-mediated humoral immunity (Th2). T cells that produce both Th1 and Th2 cytokines simultaneously are described as Th0 cells.

theca or **thecal cell** a female somatic cell type that surrounds and provides structural integrity for the ovarian follicle, functionally divided into theca interna and theca externa. Theca cells produce androgens and receive innervation from the peripheral nervous system. Theca-granulosa cell interactions are essential for ovarian follicular development.

thecal cell hypertrophy enlargement of theca cells during the process of follicular atresia.

thelarche the onset of female breast development.

thelectomy surgical removal of the nipples.

thelitoky or **thelytoky** a type of parthenogenesis in which unfertilized eggs produce females only.

theria a classification of animals including all those that give birth to live offspring; i.e., the eutherian/placental mammals and the marsupials, but not the monotremes..

theriogenology the diagnosis, treatment, and prevention of reproductive diseases in animals.

thermoregulation the regulation of a relatively constant body temperature by metabolic activities that will either produce or dissipate heat, depending on the environmental temperatures to which the animal is exposed.

threshold dose the highest dosage level of a toxic substance at which the incidence of harm or damage is not statistically greater than that of a neutral control susbtance.

thyroid (gland) an endocrine gland at the base of the neck that plays an important role in metabolic regulation.

thyroidectomy surgical removal or destruction of the thyroid gland, or inactivation by chemical menass.

thyroid hormone any hormone secreted by the thyroid gland and functioning in metabolism; specifically, one of two substances (thyroxine, triiodothyronine) crucial to the development, growth, and functioning of various tissues, e.g., of the reproductive system.

thyroid-stimulating hormone (TSH) a pituitary glycoprotein that stimulates the thyroid gland to secrete thyroid hormones (thyroxine, triiodothyronine) and also functions as a general trophic factor for the thyroid gland itself. Also, THYROTROPIN.

thyrotrope or **thyrotroph** a cell type of the adenohypophysis of the pituitary gland that produces and secretes thyroid-stimulating hormone (TSH).

thyrotropin see THYROID-STIMULATING HORMONE.

thyrotropin-releasing hormone (TRH) a peptide hormone synthesized in the brain that stimulates secretion of thyroid-stimulating hormone (thyrotropin) from cells in the pituitary; also involved in neurotransmission.

thyroxine the major circulating form of thyroid hormone; it is converted to a biologically active form, triidothyronine, in peripheral tissues.

TIDA neuron tuberoinfundibular dopaminergic neuron, neurons found in the arcuate nucleus of the hypothalamus that release dopamine into the pituitary portal system, thereby inhibiting prolactin secretion from the anterior pituitary.

tight junction a term for a cellular area where the surfaces of adjacent cells are so closely associated that passage of fluids through the space between cells is prevented; specifically, the fusions of the plasma membrane of adjacent Sertoli cells that circumscribe the cell and restrict the paracellular movement of materials between Sertoli cells.

titer the concentration of a given substance in a solution; e.g., the concentration of a hormone in the circulatory system or of antibody molecules in serum.

tocolysis the inhibition of uterine contractions.

tornaria a ciliated, motile, free-living planktonic larval stage characteristic of echinoderms and enteropneusts.

total birth rate the number of births in a given year per 1,000 persons.

total fertility rate the average number of children that would be born alive to a woman during her lifetime if she were to pass through child-bearing years conforming to age-specific fertility rates of a given year.

toxicant a poison, especially one of man-made origin.

toxin a poisonous or harmful agent; a poison, especially one of natural origin.

transactivation an increase in the rate of gene transcription brought about by the action of a transcription factor.

transcription the process of synthesizing mRNA(messenger ribonucleic acid) from a DNA template by RNA polymerase.

transcription factor a protein factor that binds to a specific DNAsequence and regulates the transcription of the adjacent gene from DNA to mRNA.

transcytosis a process by which a substance in the blood is bound and transported across endothelial cells to target cells or tissues, e.g. endothelial transport of gonadotropins in the gonads.

transferrin a major secreted glycoprotein of the Sertoli cells that is involved in the delivery of ferric ions to germ cells while navigating around the tight junctions.

transformation zone the portion of the cervix where the squamocolumnar junction began and is now located; the change in position results from squamous metaplasia.

transforming growth factor (TGF) one of a diverse family of polypeptides that generate the phenotype of transformed cells from certain normal cells. One of these in particular, **TGF-β**, has an array of functions in many normal cellular processes, e.g., the proliferation of epithelial cells, and it is known to be a major inhibitor of cell proliferation.

transgenic animal an individual whose genetic composition (genome) has been artificially modified by the addition or deletion of a specific gene sequence. The transgenic animal called the founder transmits the genetic mutation to its progeny in subsequent generations.

transitional milk breast milk produced within the first few days following birth as the continuum between colostrum and mature breast milk.

translation the synthesis of proteins from messenger RNA by ribosomes.

translocation an exchange of chromosomal material between two or more chromosomes.

transmembranous across membranes; specifically, describing the exchange of water and solutes across the amnniotic and chorionic membranes between amniotic fluid and maternal blood within the wall of the uterus.

transport protein a protein that catalyzes the movement of a molecule across the plasma membrane.

transrectal ultrasound (TRUS) an imaging technique that allows visualization of the prostate zones and precise tissue sampling.

transsexual an individual who identifies strongly with the opposite sex or has a feeling of having been born the wrong sex. Transsexuals often undergo surgery for a sex change.

TRH see THYROTROPIN-RELEASING HORMONE.

trinucleotide repeat a short, repetitive sequence of DNA composed of three base pairs that occur throughout the genome but that can change in size on transmission from parent to offspring.

triple screen(ing) a blood assay to measure molecules associated with α-fetoprotein (AFP), unconjugated estradiol, and hCG (human chorionic gonadotropin), which has shown to be useful for diagnosis of likely Down pregnancies as well as for other problem pregnancies.

trochal larva a larva with cilia arranged in distinct bands.

trochophore a larval form, often oval in shape, with prominent equatorial band of cilia (prototroch) and an anterior tuft of sensory cilia; characteristic of developmental cycles in which the cleavage pattern of eggs is spiral.

trophectoderm the outermost layer of epithelial cells of the mammalian blastocyst; the trophectoderm will give rise to the trophoblast cells that are the precursors of the placenta..

trophectodermal (TE) cells see TROPHECTODERM.

trophic relating to or involved with food or feeding.

trophic level one aspect of a food chain; e.g., large predatory mammals such as the wolf and wolverine can be described as being at the same trophic level.

trophoblast the outer layer of cells of the blastocyst, the major cell type that will differentiate into all the other cell types that form the placenta. In humans the placental trophoblasts are crucial to a successful pregnancy by mediating such critical steps as implantation, pregnancy hormone production, the immune protection of the fetus, and delivery.

trophocytes sister cells of an oocyte which are specialized for nourishing the growing oocyte and which are connected to it by intercellular bridges.

TSH see THYROID-STIMULATING HORMONE.

tuberhypophyseal dopamine neurons dopamine-releasing neurons whose cell bodies are located within the arcuate and periventricular nuclei of the hypothalamus and whose terminals are located in the posterior and intermediate lobes of the pituitary.

tuberinfundibular dopamine neurons neurons whose cell bodies are located within the arcuate and periventricular nuclei of the hypothalamus and whose terminal areas are within the median eminence. Dopamine released by these neurons reaches the anterior pituitary through the pituitary-portal vessels.

tubo-ovarian abscess (TOA) inflammation and abscess formation involving the ovary and the Fallopian tube.

tuboplasty surgery on a tubular body; e.g., the Fallopian tube, most commonly to restore tubal patency or reduce scarring to promote fertility.

tubulins the 55-kDa proteins that assemble into microtubules. The majority of a microtubule is composed of a heterodimer of α-tubulins and β-tubulins; γ-tubulins are a recently discovered rare, but ubiquitous, tubulin serving as the nucleating site for each microtubule.

tumescence the fact or condition of being enlarged or swollen; specifically, the property of the erectile bodies of the penis in which the vascular spaces dilate causing engorgement.

tumor a growth that arises from normal tissue, but that is abnormal in its growth rate and structure and that can interfere with the tissues surrounding it.

tumor grade the level of aggressiveness of a tumor.

tumor marker a specific and easily accessible protein that can be used for the evaluation and management of a certain malignancy.

tumor necrosis factor (TNF) one of a family of membrane-anchored cytokines produced by cells of the immune system that can orchestrate host defenses against disease, through a corresponding group of specific cell surface receptors (**TNF receptors**). Members of the TNF family have potent effects in both cell death (apoptosis) and cell growth.

tunica albuginea a dense, whitish fibrous sheath covering a part or organ; e.g., a connective tissue capsule surrounding the outside of the testis, or a similar covering of the corpora cavernosa of the penis.

tunica propria the extracellular basal lamina that surrounds ovarioles.

tunicate one of a subphylum of solitary or colonial chordates with sac-like bodies in which the notochord appears early in development but disappears in the adult form.

TURP transurethral prostatectomy, a surgical procedure on the prostate gland to alleviate obstructive tissue causing adverse symptoms or conditions in men with enlarged prostates.

tychoparthenogenesis occasional or accidental parthenogenetic development in unfertilized eggs.

U

ultimate factor a factor that selects for breeding (and other activities) to occur at a particular time or season, acting on an evolutionary time scale and favoring organisms whose reproductive timing maximizes the production of offspring into the succeeding generation.

ultimate spawning cue an environmental condition critical to reproductive success that has led to selection of the proximate spawning cue. See also PROXIMATE SPAWNING CUE.

ultradian rhythm a short-term recurring variation in hormone concentrations (or behavioral measures, etc.) at intervals of less than 24 hours; i.e., a pattern that recurs but at less than circadian (daily) intervals.

ultrasound or **ultrasonography** the use of high frequency sound waves that are converted from sound into electrical energy to form an image; specifically, such a procedure in a noninvasive technique used to date a pregnancy, diagnose fetal structural anomalies, identify multiple gestation, etc.

umbilical cord the stalk connecting the fetus to the placenta and giving passage to two arteries and a vein that are involved in fetal circulation.

umbilicus the point at which the umbilical cord attaches to the fetal abdomen, or the umbilical cord itself.

undescended testis a testis that is retained in the abdomen or the inguinal canal.

uniparental disomy the inheritance of two copies of a chromosome from one parent.

unitary animal an animal that does not undergo asexual reproduction; with a single ramet per genet.

univalent describing a single chromosome present during the first meiotic division (when homologous chromosomes are arranged in bivalents).

urachus a duct that transports fetal urine (i.e., the water and nutrients cleared by the fetal kidney into the bladder) to the allantoic sac.

ureter a urinary tubular structure that acts as a conduit for urine from the kidney to the bladder.

urethra a small urinary structure that drains urine from the bladder.

urethral relating to or involving the urethra.

urethral or **urinary meatus** the external tip of the urethra.

urogenital ridge a primordial gonadal structure formed by the enlargement of the coelomic epithelium in the early embryo and from which the gonad and internal reproductive tract ducts are formed.

urogenital sinus an elongated sac formed by the division of the cloaca in early embryonic development, giving rise to the urethra in females and part of the urethra in males. Also, in other animals, a common chamber and external opening for both the urinary and reproductive tracts.

urogenital system a collective term for the urinary and reproductive organs and their associated structures, especially in an organism in which the ducts of these systems are fused and release their products through the same pore.

uterine relating to or affecting the uterus; e.g., uterine cancer.

uterine atony an ongoing inability of the uterus to maintain its muscular tone in the postpartum state, usually resulting in hemorrhage.

uteroglobin in some animals, the major protein in the uterine fluid at the time of implantation.

uterotubal junction the anatomical region at which the uterine tube, or oviduct, joins the uterus.

uterus the hollow, muscular nutritive and protective organ in which the fertilized ovum implants and then develops into an embryo/fetus; eventually the uterus provides the means for birth.

V

vagina the female copulatory organ and birth canal.

vaginal relating to or affecting the vagina.

vaginal cycle the cyclic changes in cell type that are seen in the vagina and that have been correlated to cyclic changes in the ovary.

vaginal smear see PAP SMEAR.

vaginitis inflammation or infections of the vagina; may be caused by a fungus, bacterium, or virus.

Vandenbergh effect see PUBERTY ACCELERATION.

variable penetrance variable expression of a heritable trait between individuals despite identical genotype.

varicocele a palpable distension of the spermatic veins or the pampiniform plexus, frequently left-sided but affecting spermatogenesis in both testes and fertility.

varicosity the abnormal, chronic swelling and twisting of a vein.

vasa previa the blood vessels of the umbilical cord, usually presented in front of the fetal head during labor.

vascular relating to or involving the blood vessels.

vascular resistance a measure of how difficult it is for blood to flow between any two points at a given blood pressure. Total peripheral vascular resistance is the sum of the resistance to flow offered by all the blood vessels of the system.

vascular smooth muscle the cells in the vascular wall that constrict and relax to control vessel diameter.

vasculogenesis the new or first development of blood vessels from undifferentiated mesodermal cells.

vas deferens *plural,* **vasa deferentia** a thin white muscular tube containing the duct for sperm transport from the testes to the urethra; also, a similar structure in other animals; e.g., arthropods.

vasectomy the cutting and sealing of the male sperm duct (vas deferens) for the purpose of sterilization.

vasoactive having the property of affecting vascular tone.

vasoactive-intestinal polypeptide see VIP.

vasoperitoneal tissue peritoneal tissue that serves both as the lining of a coelom and the lining of certain blood vessels.

vasotocin the most ubiquitous of the nonmammalian neurohypophysial hormones; causes contraction of smooth muscle of the oviduct.

VBAC vaginal birth after cesarean, a successful trial of labor after prior cesarean delivery.

vector navigation a form of navigation in which migratory birds fly according to a vector that consists of genetically determined migration directions and programs governing the duration of migration; the only form of bird navigation that has yet been explained.

vegetative nucleus the main nucleus of the vitellogenic oocyte encompassing, but not including, the germinal nucleus that contains the chromosomes.

velamentous describing the insertion of the umbilical cord into the external membranes.

veno-occlusive mechanism the closure of the emissary veins due to their compression against the tunica albuginea by the relaxed corporal smooth muscle cells.

ventral tegmental area an area of the ventral midbrain which is an important mediator of progesterone facilitation of lordosis behavior in female hamsters and possibly other species.

ventromedial nucleus a region of the hypothalamus (basal forebrain) that contains high levels of estradiol-induced progesterone receptors and that mediates many of the effects of progesterone on reproductive behaviors in female animals.

vermiform worm-like; having a shape suggestive of a worm.

vertex the top, used to refer to the fetal head when presenting in labor.

vertical transmission the transmission of an infectious agent from mother to fetus either through the placental circulation or through the birth canal at the time of birth.

verumontanum the portion of the prostatic urethra into which the ejaculatory ducts empty.

very low-density lipoprotein (VLDL) one of the two main yolk precursors in birds (the other is vitellogenin) and the main source of yolk lipids; synthesized in the liver and transported to the developing follicle in the blood.

vesicoureteral reflux a reversed flow of urine from the bladder towards the kidney; due to abnormalities in the ureteral insertion in the bladder.

vestibule the space between the labia minora at the entrance to the vagina.

VIP vasoactive-intestinal polypeptide, a 28-amino peptide. It is present in neurons of the basal hypothalamus with projections to the median eminence. VIP acts as a neurotransmitter and, in birds, is the releasing hormone for prolactin.

virilization in females, clinical features indicative of excessive androgen secretion; if prenatal, ambiguous genitalia including clitoromegaly results; if postnatal, clitoromegaly, premature sexual hair, or hirsutism may occur.

vitellarium or **vitelline gland** a large yolk gland situated between the ovary and the oviduct in certain invertebrates; together the ovary and the vitellarium make up the germovitellarium.

vitellin (Vn) the principal protein found in the yolk of eggs, derived from vitellogenin and deposited in yolk granules or platelets within the eggs.

vitelline envelope or **membrane** the innermost layer of many invertebrate egg coverings, directly in contact with the plasma membrane of the oocyte.

vitellogenesis yolk formation; a complex process in various nonmammalian vertebrates, e.g., birds, fishes, in which yolk precursor protein (vitellogenin) is synthesized in the liver in response to estrogen and transported via the bloodstream to the developing oocyte for the formation of (vitellin); also, a similar process in invertebrates.

vitellogenic relating to or involved in vitellogenesis.

vitellogenin the major yolk precursor protein, synthesized in response to estrogen induction in the liver in vertebrates and synthesized in the fat body of insects and other invertebrates, then transported in the bloodstream (or hemolymph) and taken up into the growing oocytes within the ovary.

vitellus the egg proper in an unfertilized condition, enclosed by its plasma membrane (vitelline membrane) and encompassed by the zona pellucida.

vitrification in cryobiology, a process of freezing into a glassy, vitreous state; this technique depends on the exposure of cells to high concentrations of cryoprotectants.

viviparity the birth of live young; the mode of reproductiion in which embryos receive nutrition and develop within the female body and are born as viable offspring; seen in mammals and some reptiles and fishes; also, many chelicerate arthropods give birth to live offspring, e.g., scorpions. Generally contrasted with OVIPARITY in which the female lays eggs that hatch in the external environment.

viviparous relating to or describing a reproductive mode in which the female nurtures embryos internally and gives birth to live young; e.g., the placental mammals and some reptiles and fishes, such as most sharks.

viviparous species animals producing embryos that develop within the maternal body and then are born as live young.

VNO see VOMERONASAL ORGAN.

vomeronasal basal cell a cell at the base of the vomeronasal sensory epithelium, next to the basal lamina, that is a persistent stem cell capable of dividing to produce new sensory neurons (and possible supporting cells).

vomeronasal organ (VNO) the sensory end organ of the accessory olfactory system; chemosensory neurons within the vomeronasal have axons that extend to the accessory olfactory bulb of the brain and carry information about chemical stimuli entering the organ. The vomeronasal organ is present in terrestrial vertebrates (other than birds) and is involved in reproductive behavior and physiology as a sensory organ for chemical signals (pheromones) that elicit hormonal or behavioral responses in others of the same species. Also, JACOBSON'S ORGAN (usually for nonmammalian species).

vulva the external genitalia of the female, including the labia majora and labia minora, the vaginal opening, and a number of glands located in the region. Also, in invertebrates, a specialized opening in the ventral hypodermis of the hermaphrodite that serves as a passageway for male sperm during copulation and eggs following fertilization.

vulvar relating to or affecting the vulva.

vulvodynia a syndrome of unexplained vulvar pain that generally precludes intercourse.

W

weaning the termination of nursing behavior; specifically, the change of diet for domestic animals from milk to solid feed.

wedge resection a surgical procedure that removes a wedge-sized portion of the ovary in women with PCOS (polycystic ovary syndrome) to cause spontaneous ovulation.

wether a castrated male sheep.

Wharton's jelly a proteoglycan-rich matrix in which the umbilical vessels are embedded. (From Thomas *Wharton*, English anatomist.)

Whitten effect see ESTROUS SYNCHRONIZATION.

Wolffian duct one of a pair of embryonic ducts draining the mesonephros; the Wolffian ducts regress in females but are the precursor of the internal accessory reproductive organs in the male. (Described by Kaspar Friederich *Wolff*, German embryologist.)

Wolffian structures the internal organs that develop from the Wolffian ducts in the fetus; i.e., in males, the epididymis, vas deferens, and seminal vesicles.

X

X/A ratio the ratio of the number of X chromosomes to sets of autosomal (A) chromosomes.

X chromosome a designation for the sex chromosome that is paired in the sex producing gametes with chromosomes of the same type, and that occurs with another sex chromosome (designated as the Y chromosome) in the sex that produces gametes with sex chromosomes of two different types; thus female mammals are XX and males are XY.

xenoestrogen an anthropogenic compound that binds to the ER (estrogen receptor) and elicits ER agonist or antagonist responses.

X-linkage the fact of being X-linked; i.e., present on the X chromosome and not on the Y chromosome.

X-linked sex-linked; relating to or involving the transmission of a gene that is located on the X chromosome; most X-linked diseases and conditions, e.g., hemophilia, are manifested mainly or exclusively in males (who are XY) while females (XX) inherit the condition but its effect is masked by the second X chromosome.

X-linked disorder a disorder due to an abnormal gene on the X-chromosome and therefore invariably expressed in males.

Y

Y chromosome a designation for the sex chromosome that is paired with another sex chromosome (designated as the X chromosome) in the sex that produces gametes with sex chromosomes of two different types, and is absent in the sex producing gametes with chromosomes of one type; thus male mammals are XY and females are XX.

yolk a mass of material, rich in fats and proteins, contained within the ovum and serving to supply nutrients to the developing embryo; the amount and distribution of yolk varies considerably with the type of animal; in the avian egg it is a large spherical body, while humans and most other mammals have minimal yolk reserves because the embryo receives its nutrition directly from the mother via the placenta. (From the Old English word for yellow.)

yolk protein proteins stored in the eggs of oviparous and ovoviviparous vertebrates, and utilized during embryonic development.

yolk sac extraembryonic tissues that surround and digest the yolk and that contribute to placental organs in most mammals, certain viviparous fishes, and all viviparous reptiles.

Y-organ an endocrine gland in the thorax of crustacea that produces ecdysteroids.

Z

zeitgeber literally, time giver; an external cue that brings an animal's internal periodic controls (e.g., circadian rhythms) into synchronization with the periodic changes of the external environment. The most important zeitgeber for most organisms is the light-dark cycle (day-night).

ZIFT zygote intrafallopian transfer, the surgical placement of a fertilized egg into the Fallopian tube.

zinc fingers highly conserved amino acid sequences containing four cysteines that complex with zinc to form a fingerlike process; present in the central DNA binding domain of nuclear receptors where they mediate the specific, high affinity binding of nuclear receptors to target genes.

zona pellucida a relatively thick acellular glycoprotein coat that surrounds the oocyte and young embryo, composed in most mammals of three major glycoproteins. It plays a key role in sperm recognition and binding, induction of the acrosome reaction, and the prevention of polyspermy.

zooid an individual organism within a larger unit, especially an individual member of a colony. Feeding zooids are known as **autozooids** and zooids modified for other functions are kenozooids.

-zoospermia a suffix referring to sperm within the semen.

zygoparity reproduction by the laying of fertilized eggs; zygoparity grades into oviparous egg retention.

-zygosity a suffix referring to the characteristics of a zygote.

zygote the product resulting from the fusion of a male gamete (spermatozoon) and female gamete (oocyte); i.e., a single-cell fertilized egg with a male and a female pronucleus.

zygotene the second of four stages during prophase I of meiosis, during which the first pairing of the synaptonemal complex occurs.

zygote intrafallopian transfer see ZIFT.

zygote nucleus the diploid nucleus formed during the first cell cycle by the fusion between the male and female pronuclei. Many organisms, including mammals, do not have a zygote nucleus, and instead the separate, but adjacent, male and female pronuclei undergo synchronous nuclear breakdown at the time of first mitosis.

Subject Index

Volume numbers are boldface, separated from the first page reference with a colon. Subsequent references to the same volume are separated by commas.

A

B

C

D

E

F

G

H

J

K

L

M

N

O

Q

R

S

T

U

V

W

X

Y

Z

ISBN 0-12-227024-X